AF212500

H. Hayakawa N. Koshizuka (Eds.)

Advances in Superconductivity IV

Proceedings of the 4th International Symposium on Superconductivity (ISS '91), October 14-17, 1991, Tokyo

ISTEC

With 1315 Figures

Springer-Verlag Tokyo Berlin Heidelberg New York
London Paris Hong Kong Barcelona

Prof. Dr. Hisao Hayakawa
School of Engineering
Nagoya University
Chikusa-ku, Nagoya
464 Japan

Dr. Naoki Koshizuka
Deputy Director
Superconductivity Research Laboratory
ISTEC
Koto-ku, Tokyo
135 Japan

ISBN-13:978-4-431-68197-7 e-ISBN-13:978-4-431-68195-3
DOI: 10.1007/978-4-431-68195-3

© Springer-Verlag Tokyo 1992
Softcover reprint of the hardcover 1st edition 1992

This work is subject to copyright. All rights are reserved, whether the whole or part of the material is concerned, specifically the rights of translation, reprinting, reuse of illustrations, recitation, broadcasting, reproduction on microfilms or in other ways, and storage in data banks.
The use of registered names, trademarks, etc. in this publication does not imply, even in the absence of a specific statement, that such names are exempt from the relevant protective laws and regulations and therefore free for general use.

Foreword

Five years have passed since the breakthrough in the critical temperature for superconductors. During this period, many superconducting materials have been discovered and developed, and our knowledge of the physical and other properties of oxide superconductors has deepened through extensive and intensive research. This knowledge has advanced superconductivity science and technology from the initial questioning stage to a more developed but still uncertain second stage where research activity in superconductivity now overlaps with fields of application.

Generally speaking, science resonates with technology. Science not only complements but also competes with or stimulates technology. New scientific knowledge has triggered the second technological research stage. Much progress has been made in the development of practical devices, encouraging the application of superconductors in areas such as human levitation, a high speed levitated bearing, large current transforming leads, and high frequency devices. This technological progress has increased our understanding of the science involved, such as flux pinning and dynamics, and anomalous long-range superconducting interactions.

At this important stage, international cooperation and collaborative projects can effectively sustain aggressive research and development in order to advance superconductivity to the next stages. The ISS Symposium is expected to serve as a venue for increasing our knowledge of superconductivity and for exchanging visions for future research and applications, through the presentation and discussion of the latest research results. These proceedings also aim to summarize annual progress in high-T_c superconductivity in all fields.

Shoji Tanaka
Vice President, ISTEC
Chairperson, Steering Committee

Preface

Advances in Superconductivity IV is published as the proceedings of the 4th International Symposium on Superconductivity (ISS '91) held October 14-17, 1991 in Tokyo, under the cosponsorship of the International Superconductivity Technology Center (ISTEC) and the Greater-Kanto Industrial Advancement Center. ISS has been held annually since 1988 as a forum for discussion of the most recent progress in superconducting science and technology, and to offer opportunities to promote international cooperation among scientists, engineers, business administrators, and entrepreneurs. ISS '91 saw the presentation of 263 papers, both oral and poster contributions, including 110 from industry, 76 from academia, 34 from public institutes, and 43 from the Superconductivity Research Laboratory of ISTEC.

The volume begins with the organization of ISS '91 and a list of the committees. Two Special Plenary Lectures given at the opening of the Symposium by A.M. Campbell (IRC in Superconductivity) and V.O. Heinen (NASA) as well as four Plenary Lectures are included at the beginning of the volume in the order in which they were presented at the Symposium. The other contributions are arranged in the following sections: (a) physics, (b) chemistry, (c) flux pinning, (d) wires, tapes, and bulk, (e) preparation and properties of thin films, (f) device application, and (g) system application. Invited papers, which are distinguished by their length of six pages from the four-page contributed papers, are arranged at the beginning of sections and subsections.

Much progress was reported in all research fields. In particular, remarkable progress has been made in the field of superconductor applications, including integrated SQUIDs, passive microwave devices, and superconducting bearings, which are almost at the stage of practical utilization.

The editors are indebted to a number of people for their valuable assistance at all stages of preparation for the Symposium. The International Advisory Committee helped us in selecting invited speakers. The Program Committee selected the invited papers, arranged the program, and refereed the contributed papers in cooperation with many of the scientists who participated in the Symposium. We would like to acknowledge all the people who helped to make the Symposium a success. Finally, we express our thanks to the authors for the cooperation in carefully preparing their manuscripts for Advances in Supercoductivity IV.

Hisao Hayakawa
Naoki Koshizuka

Organization of ISS '91

Sponsored by:
International Superconductivity Technology Center

Cosponsored by:
Greater-Kanto Industrial Advancement Center

Supported by:
Ministry of International Trade and Industry Agency of Industrial Science and Technology
Kanto Bureau of International Trade and Industry
Science and Technology Agency
Ministry of Education
Ministry of Transport
Ministry of Posts and Telecommunications
New Energy and Industrial Technology Development Organization
The Japan Key Technology Center

Approved by:
The Ceramic Society of Japan
The Chemical Society of Japan
The Cryogenic Society of Japan
The Federation of Electric Power Companies
The Institute of Electrical Engineers of Japan
The Institute of Electronics Information and Communication Engineers of Japan
The Iron and Steel Institute of Japan
Japan Chemical Industry Association
The Japan Electrical Manufacturers' Association
Japan Electronic Industry Development Association
Japan Fine Ceramics Association
The Japan Institute of Metals
The Japan Iron and Steel Federation
Japan Mining Industry Association
The Japan Society of Applied Physics
The Japanese Electric Wire & Cable Makers' Association
The Magnetic Society of Japan
Physical Society of Japan
Research and Development Association for Future Electron Devices

Organizing Committee

Chairman:	
G. HIRAIWA	President, ISTEC
Members:	
J. AOI	Vice-President, ISTEC
K. FUEKI	Professor, Science University of Tokyo
S. ISHIHARA	Director-General, AIST, MITI
H. KASHIWAGI	Director-General, Electrotechnical Laboratory
Y. MUTO	Professor Emeritus, Tohoku University
S. NAKAJIMA	Professor, Tokai University
S. NASU	Chairman, Greater-Kanto Industrial Advancement Center
S. SAITO	Chairman, The Society of Non-Traditional Technology
T. SEKIZAWA	Vice-President, ISTEC
T. SUGANO	Dean of Faculty of Engineering, The University of Tokyo
S. TANAKA	Vice-President, ISTEC
S. YAMAMURA	Professor Emeritus, The University of Tokyo

International Advisory Committee

A. BARONE	University of Naples
B. BATLOGG	AT&T Bell Laboratories
M.R. BEASLEY	Stanford University
P. CHAUDHARI	IBM T.J. Watson Research Center
C.W. CHU	University of Houston
J. CLARKE	University of California, Berkeley
R.C. DYNES	University of California-San Diego
J.E. EVETTS	University of Cambridge
D.K. FINNEMORE	Ames Laboratory
K. FOSSHEIM	Norwegian Institute of Technology
K. FUEKI	Science University of Tokyo
W. KLOSE	Karlsruhe Nuclear Research Center
P. KOMAREK	Karlsruhe Nuclear Research Center
D.C. LARBALESTIER	University of Wisconsin
W.Y. LIANG	University of Cambridge
M.B. MAPLE	University of California-San Diego
K.A. MÜLLER	IBM, Zürich
Y. MUTO	Tohoku University
S. NAKAJIMA	Tokai University
B. RAVEAU	University of Caen
J.M. ROWELL	Conductus Inc.
A.W. SLEIGHT	Oregon State University
M. SUENAGA	Brookhaven National Laboratory

Steering Committee

Chairman:	
S. Tanaka	Vice-President, ISTEC
Members:	
T. Fujiwara	The University of Tokyo
H. Hayakawa	Nagoya University
T. Hirayama	Tokyo Electric Power Company
N. Imamura	Tosoh Corporation
E. Masada	The University of Tokyo
T. Nakada	Ishikawajima-Harima Heavy Industries Co., Ltd.
H. Osanai	Fujikura Ltd.
M. Ozeki	Railway Technical Research Institute
J. Seto	Sony Corporation
Y. Tanaka	Greater-Kanto Industrial Advancement Center
K. Uchinokura	The University of Tokyo
M. Yuge	Mitsubishi Materials Corporation

Program and Editing Committee

Chairman:	
H. Hayakawa	Professor, Nagoya University
Members:	
K. Ando	Toshiba Corporation
Y. Bando	Kyoto University
J. Fujie	Railway Technical Research Institute
A. Fujimori	The University of Tokyo
H. Igarashi	NEC Corporation
I. Iguchi	University of Tsukuba
Y. Iye	ISSP, The University of Tokyo
T. Kawai	Osaka University
N. Kobayashi	Tohoku University
S. Kosaka	Electrotechnical Laboratory
N. Koshizuka	SRL, ISTEC
S. Matsuda	Hitachi Ltd.
T. Matsushita	Kyushu Institute of Technology
M. Nakao	Sanyo Electric Co., Ltd.
K. Sato	Mitsubishi Electric Corporation
K. Sato	Sumitomo Electric Industries, Ltd.
K. Sawano	Nippon Steel Corporation
Y. Shiohara	SRL, ISTEC
Y. Tanaka	The Furukawa Electric Co., Ltd.
K. Togano	National Research Institute for Metals
K. Uyeda	Central Research Institute of Electric Power Industry
A. Yamaji	Tokyo Institute of Technology
K. Yamaji	Electrotechnical Laboratory
T. Yamamoto	Tokyo Electric Power Company

General Affairs Committee

Chairman:	
T. Fujiwara	Professor, The University of Tokyo
Members:	
T. Hatogai	The University of Tokyo
M. Matsumoto	The University of Tokyo
A. Maeda	The University of Tokyo
Y. Niwano	Toshiba Corporation
H. Osanai	Fujikura Ltd.
S. Saotome	Ishikawajima-Harima Heavy Industries Co., Ltd.
K. Sugimoto	Greater-Kanto Industrial Advancement Center
N. Takahashi	Tokyo Gas Co., Ltd.
N. Toyokura	Fujitsu, Ltd.
M. Ueda	Tokyo Electric Power Company

Table of Contents

1 Plenary Lectures

2 Physics

2.1 Theoretical Studies

2.2 Electron Spectroscopy

2.3 Transport and Thermal Properties

2.4 Magnetic Resonance and Mössbauer Studies

2.5 Optical Properties

2.6 Magnetic Properties

3 Chemistry

3.1 Synthesis

3.2 Substitution and Doping Effects

3.3 Structures and Properties

3.4 Crystal Growth

4 Flux Pinning

4.1 Flux Pinning Properties

5.1 Bulk — (2) BSCCO

5.2 Wires and Tapes — (1) YBCO

5.2 Wires and Tapes — (2) BSCCO

5.2 Wires and Tapes — (3) TBCCO

5.3 Characteristics

6 Preparation and Properties of Thin Films

6.1 YBCO and Related Films

6.2 BSCCO and Related Films

6.3 Other Superconducting Films

7 Device Application

7.1 Multilayer Structures and Junctions

8 System Application

8.1 Application for Electric Power System

8.2 Application of Electric Railway and Other Industry

1 Plenary Lectures

The Pinning and Motion of Flux Vortices

A.M. CAMPBELL

IRC in Superconductivity, Cambridge, UK

ABSTRACT
The differences between conventional and high T_c superconductors are discussed, with particular reference to the behaviour of vortices and the critical current density. It is shown that the description as a hysteretic pinned system, Josephson Junction array, or spin glass, are equivalent, and lead to the same experimental predictions. The main limitations to J_c are granularity at grain boundaries and the irreversibility line. The first causes a dramatic decrease in J_c at all temperatures. The cause is unclear, but it may be soluble in layered compounds by mechanical deformation. The second causes J_c to vanish at a combination of field and temperature which can be much lower than the critical value.The most likely explanation is decoupling between the planes.

KEY WORDS:- Superconductors, Pinning, Spin Glass, Granularity, Irreversibility Line.

Introduction

The initial reaction to the discovery of high Tc superconductors was that they were completely different from conventional superconductors. However it has become more clear that the difference is mainly one of degree than of character. In particular the vortex lattice and magnetisation curves are very similar to those in conventional superconductors. This is a result of the very general applicability of the Ginzburg-Landau theory, which is independent of the mechanism causing superconductivity. However apart from conventional flux pinning there are two very important other effects which were absent, or negligible in conventional materials, but which are the main barrier to the application of high Tc materials. The first is the granular nature of these materials which is due to the fact that grain boundaries act as weak links, or Josephson Junctions. This dramatically reduces the critical current density in a magnetic field. The second is the irreversibility line which separates two regimes of the superconductor. Above the line the material is completely reversible and useless for practical purposes, since there is no pinning of flux lines. Below the line pinning is effective and the material can be used like a conventional superconductor.

Simple Ideal Pinning

It is customary to start with a theoretical account of a subject. As in many subjects the initial simple theory of a topic has proved helpful, but further more complex theories are mainly of academic interest. We can calculate the ideal pinning strength on a vortex by considering a cavity down its core. If the vortex is moved off this cylindrical cavity we have to drive the core normal in the superconductor, which requires the condensation energy multiplied by the volume. This must be done over a distance equal to the coherence length, and putting in the numbers leads to a maximum critical current comparable with the depairing current Hc/λ which is about 10^8 A/cm^2. From this we conclude that for high critical currents we should look for materials with large critical fields and short penetration depths. However it is also clear that these are far higher current densities than we need, and the limitation in high T_c materials is not low pinning forces. It is remarkable that thin films of YBCO have been made with critical current densities approaching this value and it is not profitable to look for further large improvements in J_c in epitaxial films. It is worth pointing out that these are regarded as very 'good' films, although the high J_c implies that they are full of inhomogeneities and therefore rather bad for physical measurements. However the low critical currents in so-called 'bad' films are probably caused by poor stoichiometry and low condensation energy so they are therefore even worse.

It is extremely difficult to identify the cause of pinning in a superconductor unless the pinning centres have been deliberately added, and even then the increased J_c may be caused by a change in the reversible parameters, rather than the pinning centre density. Recently discovered screw dislocations have been suggested as the cause of the high current densities (1), but the density is rather too low and recent pictures of similar dislocation densities in single crystals with much lower current densities cast doubt on this mechanism (2). Oxygen vacancies have been suggested (3) but J_c does not seem to be as closely correlated with oxygen content as this mechanism would require. Intrinsic pinning by the atomic planes can only work in one direction. It is likely that all mechanisms contribute and that we should be looking at a model with a continuously varying background potential due to all sorts of inhomogeneities, rather than individual pinning centres.

General Pinning Theory.
The calculation of pinning by individual pinning centres is the easiest part of pinning theory, and for practical purposes probably all that is needed. However the next step, which is to calculate the critical current density in a random array of pinning centres, brings us to a very general area of Physics in which superconductors can provide a link between previously unrelated topics. Systems with random potentials have been studied in a number of areas, and are becoming of great theoretical interest. It is recognised that there are many common features, but it is probably in superconductors that the widest range of properties have been calculated.

The properties of high T_c materials have been explained in terms of spin glasses, critical state model, and Josephson Junction arrays. In spite of the apparent differences between these systems it is remarkable that no experimental test could distinguish between them. It now appears that this is because these different models are in fact different ways of formulating the same physical ideas. This means that we can now bring together the two large areas of hysteretic systems which have previously been quite separate. The first are pinned systems in which a driving force is resisted by a pinning force up to a certain value. The second are spin glasses which have received much more theoretical treatment. We can show that a typical force balance equation leads to a spin glass type of Hamiltonian and so the systems are identical. The Hamiltonian is the most powerful formulation above the glass temperature, but the force balance is certainly easier to follow below it. This may be why concepts like frustration and logarithmic creep were first recognised and explained in force balance systems. Such systems are characterised by force-displacement curves like figure 2, velocity force diagrams like the dashed line in figure 4, and logarithmic creep.

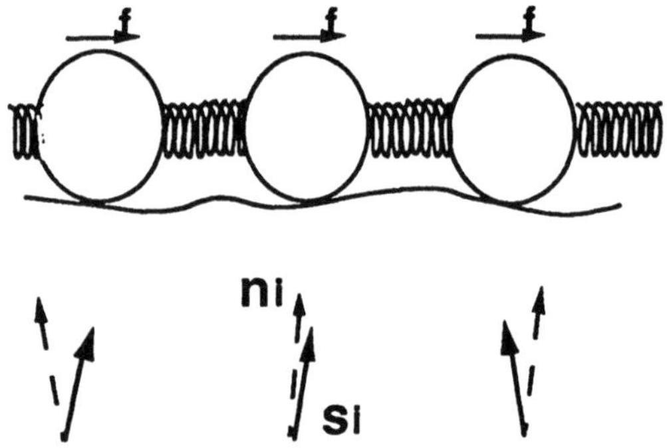

Figure1
Flux lines represented by balls on random potentials. Below is the equivalent spin system.

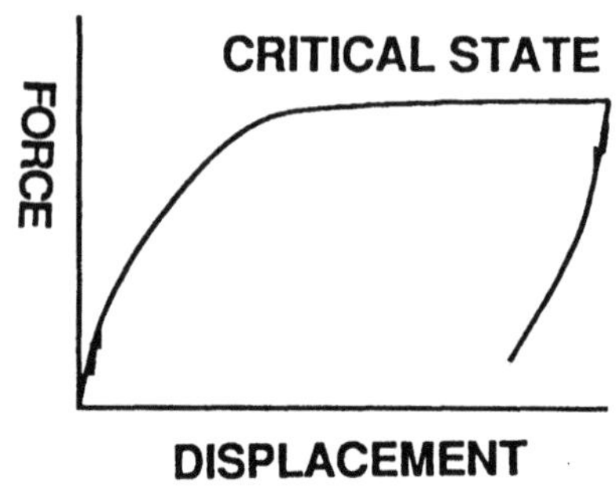

Figure2
The force displacement curve for a typical hysteretic system

The simplest model system is represented in figure 1 where the elasticity of the vortex lattice is represented by springs. Consider a line of vortices which start at the same spacing but which are placed on a weak random potential so that after they have relaxed to equilibrium local deviations from the equilibrium vortex spacing are small. (This does not preclude large collective displacements and the destruction of long range order in the process of relaxation). If we now apply an external force f to each, and include thermal oscillations, so that the displacement of vortex i is x_i the equation of equilibrium is :-

$$C(x_i-x_{i-1}) - C(x_{i+1}-x_i) + d_i x_i + f = 0$$

Here C is the stiffness of the vortex lattice, and d_i the curvature of the random potential under the ith vortex. This force balance equation can be derived by minimising a Hamiltonian of the form:-

$$H = {}^1/_2\, \Sigma C(x_i-x_{i-1})^2 + C(x_{i+1}-x_i)^2 + d_i xi^2 + 2fa_i.$$

This can be put in the same form as the Hamiltonian of an anisotropic XY spin glass.(25) The physical picture is to put a spin at each vortex position as in Figure.1. The energy of this system contains an elastic term dependent on the difference in angle of neighbouring spins. This exchange interaction corresponds to the elasticity of the vortex lattice. We then add an energy dependent on the angle between the i th. spin and the local easy magnetisation direction **n**. We take the anisotropy D to be constant and **n** is random. Finally there is an external field term proportional to the difference in angle between the i th spin and the external field H_o. The energy terms are proportional to the differences in angle squared which, if small, can be written as the cosine of the angle. This gives an identical Hamiltonian to the pinned system.

$$H = \Sigma J_{ij} s_i . s_j + D(n_i . s_i)^2 + H_0 . s_i$$

It is not clear whether this link between pinning systems and spin glasses will lead to a better quantitative understanding. Complex spin glasses are just as insoluble as random pinning systems. However it greatly extends the number of systems which can be regarded as spin-glass like, since it is already well known that the pinning of flux lines is similar to pinning of dislocations, domain walls in ferromagnets, charge density waves, and sand piles.

Arrays of Josephson Junctions probably provide the best practical system to study these effects and they have already been treated as both spin glasses (4) and pinning arrays. First we can show that the equation of motion of the vortex lattice is the same as the RSJ model of a single Josephson Junction (5). Suppose a pinning centre is in a vortex lattice so that its energy varies as $\varepsilon \mathrm{Cos}\, kx$ where x is the distance moved by the lattice (Figure 3).
We apply a force f to the lattice, and assume a mass m and a viscosity η. The equation of motion is:-

$$m\ddot{x} + \eta\dot{x} + k\varepsilon \,\mathrm{Sin}\, kx = f$$

Now the velocity is proportional to the voltage, the force to the current and ε to the critical current so with a suitable scaling factor this can be written

$$C\dot{V} + V/R + I_c \,\mathrm{Sin}\phi = I$$

This is just the circuit equation for the RSJ model of a single Josephson junction. It follows that the magnetic behaviour of an array of junctions is exactly the same as an array of pinning centres in a vortex lattice with the critical current determining the pinning force and the activation energy for flux creep $i_c\phi_o$.

Figure 3
A Single Pinning Centre in a Vortex Lattice

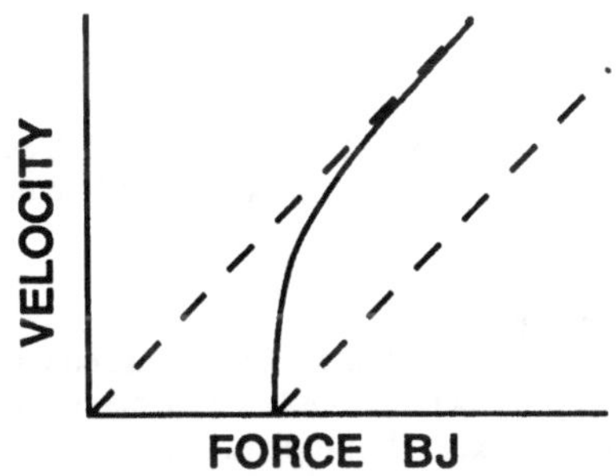

Figure 4
Voltage-Current, or Velocity-Driving Force
The continuous line is that for a single centre.
The line through the origin is with no pinning and the other dashed line that for random pins.

We therefore regard the granular superconductor as a composite of strong and weak pinning centres, where the weak pinning centres are Josephson Junctions. This allows us to apply the critical state and flux creep theory to both sets of pinning centres. Figure 5. illustrates the two vortex arrays in a very simple geometry.

One practical consequence of this is that we can draw some conclusions about the VI characteristics of junctions. The Josephson equations and the equation of motion of a single pinning centre predict a VI characteristic like the continuous line in Figure 4. However in practice what is observed is the dashed line, and according to the theory of Yamafuji and Irie (6), this is predicted for an array of pinning centres. It follows that if this is observed in a weak link, as often happens in high T_c materials, it means that several Josephson vortices are pinned along the width of the junction. This will occur if the junction is wide in the direction perpendicular to the current, and the faces are rough.

Some of the common concepts which have different names in the three descriptions are in Table I.

Superconductor	Josephson array	Spin Glass
Lattice Rigidity	Phase Interference	Frustration
Pinning Force	Critical Current	Anisotropy Energy
Reversible Vortex Movement	Dynamic Inductance	AC Susceptibility
Activation Energy	Josephson Coupling Energy	Activation Energy
Critical State	Phase Coherence	FC and ZFC Susceptibility
Flux Creep	Flux Creep	Log decay of M

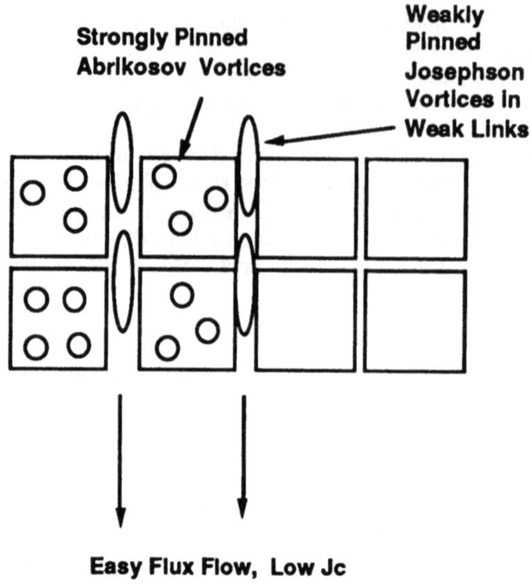

Figure 5.

Other vortex states have been suggested such as vortex entanglement(7) and a vortex glass state(8). Experimental proof of these appears to depend on details in the V-I characteristics so they are unlikely to be of much practical importance.

Granularity

Granular superconductors have been studied for many years, but usually in a different regime from that encountered in oxide superconductors. Studies were mainly on small grains of Type I superconductor embedded in a normal, or insulating matrix. The properties can be averaged over many grains and the material behaves magnetically like a homogeneous material. In oxide superconductors the grains are much larger than the the penetration depth and the material must be treated as a mixture of two materials, the weak links at the grain boundaries and the individual grains. The two materials are demonstrated clearly in AC susceptibility measurements in which the derivative of the flux entering is plotted as a function of AC amplitude. This shows how far the flux has penetrated, and it is found that there is a very sharp discontinuity separating penetration between the grains and penetration into the grains. This sharp discontinuity is perhaps the clearest indication of granularity. It is not seen in cast A15 compounds, nor in Chevrell phases which can show many of the other magnetic characteristics of granular materials. Granularity in Oxide superconductors was first recognised in the BaPbBiO system, (28), but seems to be a general feature of all oxides and the effect is a catastrophic fall in current density at low applied fields.

The reasons for this type of granularity are not clear. The most obvious cause is the short coherence length, and this must be a major factor. It makes the superconductivity sensitive to much smaller scale disruption than is the case with conventional materials. However it is difficult to make the numbers fit. The critical current is reduced from the depairing current by a factor of 10^5 which requires a very thick barrier. Ginzburg Landau calculations assuming a continuous variation of properties across the barrier show that we need severe damage over several coherence length to explain the low current densities observed. Similarly a large variation of T_c with shear strain is needed to explain the results on a dislocation model.(9) Better agreement can be obtained by modelling the system as a sharp SIS junction but it is not clear that this is a very realistic model of a real grain boundary (10). A factor which must be taken into account is the depression of the order parameter at a free surface,(11) but this will not be more than a factor of ten and most calculations give a smaller reduction in value.

The first direct experimental work on high Tc materials was by Dimos et. al.(12) who showed in a series of experiments on YBCO bicrystals that the critical current dropped rapidly as the grain boundary angle increased. Although this indicated that it was purely the change in angle which caused the weak link, since their boundaries appeared very clean, a curious feature of their results was that the depression did not depend on the orientation of the crystal axes to the grain boundary although the material is very anisotropic. This suggests chemical imbalance rather than crystalline misalignment as the cause.

These results prompted attempts to align grains in bulk materials, but this has not worked in YBCO. There have been reports of grain boundaries in YBCO which do not act as weak links (13), but they seem rare and the problem remains an intractable one. It must be significant that no energy gap can be seen on the surface of single crystals by vacuum tunnelling, and it has not been possible to make tunnel barriers on them, which suggests that surfaces and grain boundaries deteriorate in some way. On the other hand very thin films have carried high current densities (14) and the Bismuthates, which have had tunnel barriers made to them, are granular.

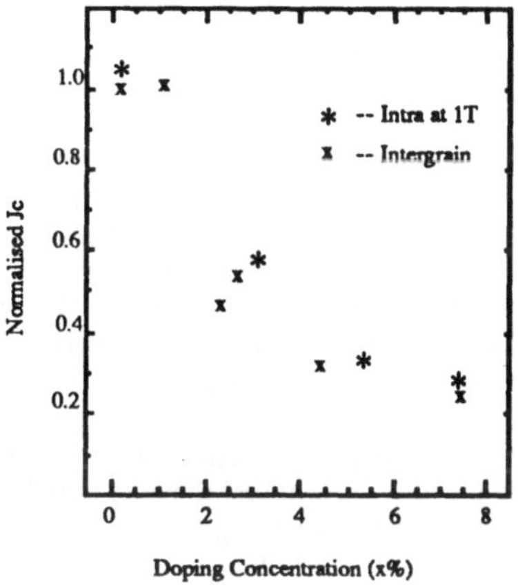

Figure 6.
A Comparison of Inter and Intra grain Critical Currents in Co Doped YBCO.

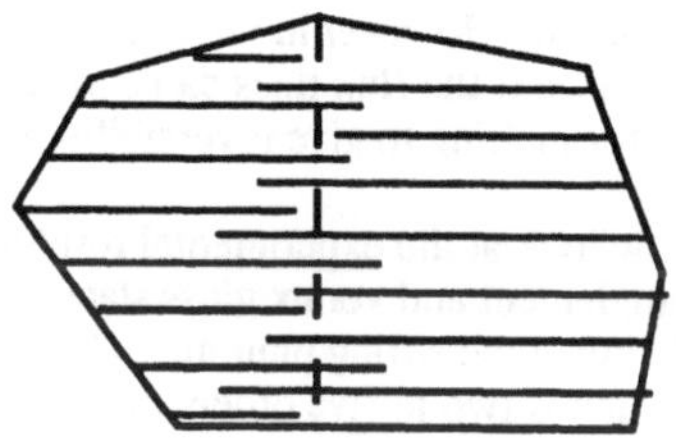

Figure 7.
An Illustration of how plastic flow could smooth grain boundaries

It is important to look at different materials to try to get some clues. Figure 6 shows the effect of Cobalt doping. It is found that both inter and intra grain critical current densities vary in a similar way with doping. This is presumably because if nothing else is changed both are determined by the same microscopic parameters. This correlation is common, but not always present. In the septernary compound $Tl_{0.5}Pb_{0.5}Sr_2Ca_{1-y}Y_yCu_2O_7$, changing y reduces the number of holes taking it from a more metallic type of material to a semiconducting one. Although T_c and the intragrain J_c go through a maximum, the intergrain J_c decreases continuously with doping. (15) (The effect on the irreversibility line is considered again below.) In both these cases there is no correlation between J_c and changes in the coherence length as measured from B_{c2}.

One of the very strange things about the granularity is that it seems less of a problem in BSCO. Present processing methods use cold plastic work to obtain optimum properties so that it may be that we should look at the mechanical properties, rather than the electrical ones. If grains are already in reasonable alignment plastic work could wipe out the grain boundary by the interleaving of layers on a very fine scale, as shown in Figure 7.

The Irreversibility Line

One of the most important new features of high Tc materials is the irreversibility line. It is found that above certain fields and temperatures the pinning vanishes, rendering the material useless for practical applications. The graph of this field against temperature is known as the irreversibility line.

In sintered materials there are two such lines, one for the intergrain currents and one for the intragrain currents. It was the first of these that was described by Bednorz and Müller but in fact this should be expected in granular materials. It was already well known that when the Josephson coupling energy $i_c\phi_o$ was less than kT thermal activation reduced the critical current density to zero, and this is what happens to the Josephson links in the sintered material.

However a similar effect was observed at much higher fields in single crystals (16) and this has been the source of much speculation, as well as setting rather severe limits to the practical applications of BSCO. In principle the effect should have been expected since it is inevitable that pinning energies become less than kT before T_c is reached. However it occurs very close to B_{c2} in conventional materials and was not looked for until recently. In high T_c materials the irreversibility line can be a very long way from the critical temperature, for reasons which are not yet clear.

Three possible explanations have been suggested. The first is that the pinning centres have a low activation energy. This could be caused by the the very small coherence length leading to very small vortex cores. The energy of a void a coherence length in diameter will be quite low, even although the pinning force is high.

The second possibility is that the superconducting layers become decoupled so that we have a two dimensional system instead of a three dimensional system. (17) The vortices become pancakes instead of continuous lines and we would expect the activation energy of a disc of atomic thickness to be much less than that of a section of a line. Therefore at the two dimensional crossover thermal activation will increase by a large factor.

The final hypothesis is that the lattice melts so that the shear modulus goes to zero. (27) For this to lead to a reduction in pinning requires rather special circumstances. The conventional view has always been that a reduction in shear modulus increases J_c by removing the frustration caused by the rigidity of the flux line lattice. The first suggestion of this effect, which led to the peak effect, was by Pippard (18) but numerous later models have also predicted an increase in J_c as c_{66} decreases. The melting hypothesis therefore seems unlikely on theoretical grounds, but there are ways in which it can be sustained. The first requirement is for point pinning centres. Planar barriers will pin a liquid without problems as can be seen at the Hoover dam. In a two dimensional system point pins could not pin a liquid. However in three dimensions every flux line is pinned somewhere so they cannot flow. Flow can only take place if the flux lines can curt each other, and experiments in longitudinal fields show that this is not very easy. So for melting to give reversibility needs a rather special pinning array and very easy flux cutting.

Let us now look at the experimental results. The question to be addressed is not whether concepts like flux melting, flux entanglement and vortex glass states occur in an ideal Type II superconductor, but whether they occur in high Tc materials given the strong pinning. (In conventional materials weak pinning means something between 1 and 100 A/cm^2 which is two to five orders of magnitude below that of high T_c single crystals and films.).

Comparisons of experimental results can only be approximate since different techniques give a different line. This is quite easy to explain since the effect is due to thermal activation and therefore a band rather than a line. However this imprecision is not sufficient to alter the conclusions.

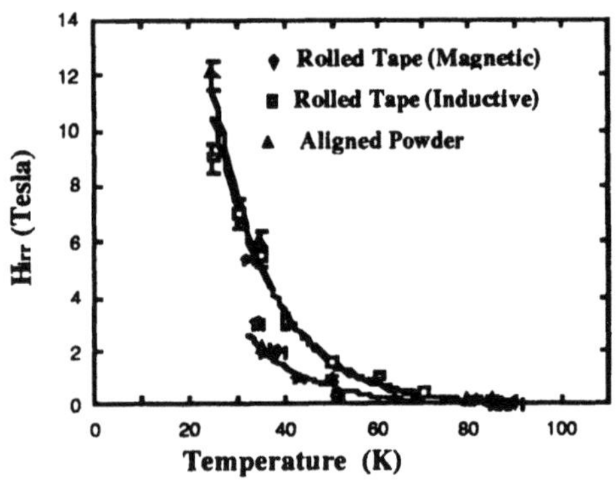

Figure 8.
Reversibility Line for two silver clad tapes and aligned powders.

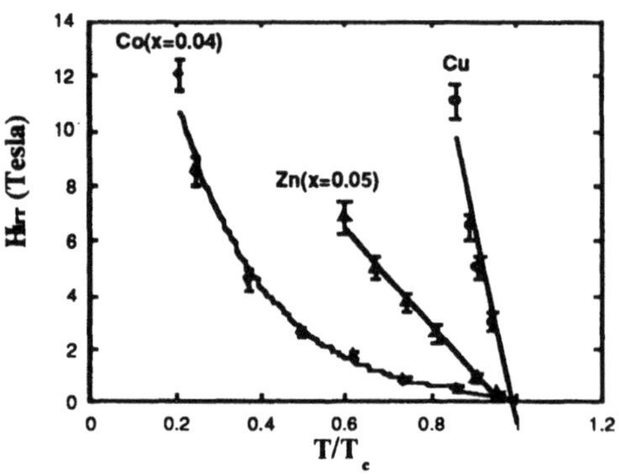

Figure 9.
The effects of doping on the irreversibility line.

Most experimental results show that the irreversibility line is relatively insensitive to pinning. There is certainly some effect, but this may be in part due to changes in the reversible parameters. However given the wide range of pinning strengths measured, a much wider range of reversibility lines would be expected if the effect was due to low energy pins. Another indication is that doping with Cobalt lowers the reversibility line although measurements of B_{c2} indicate a decrease with doping, so that the coherence length must be increased by the doping . These results suggest that the effect is essentially intrinsic to the material and not determined by pinning. However two recent papers show significant changes in the line after neutron irradiation of BSCO (19) and Sn ion irradiation of YBCO.(20)

Nearly all the evidence is consistent with decoupling of the planes. Most striking is the difference between YBCO and BSCO. BSCO has much looser coupling between planes and is much more two dimensional . Figure. 8. shows the line for two silver clad tapes and for powder aligned in wax. Although different criteria are used it is clear that the processing to produce high Jc has not changed the reversibility line significantly. Experiments with dopants confirm the connection between the reversibility line and low dimensionality. Figure 9 shows results on YBCO doped with Cobalt and with Zinc, plotted as a function of T/Tc (21). It can be seen that the Cobalt has a far greater effect than the Zinc. In resistive measurements Cobalt greatly increases the fluctuation effects on the resistivity which are normally associated with a two dimensional structure, and the same effect is seen in specific heat measurements. We can attribute all these effects to the hypothesis that Co decouples planes but Zn does not. In YBCO the very thin films made by Schauer et. al. (14) have a resistive transition.transition in an applied field like that of BSCO so it appears to be characteristic of 2D systems. The results of Kim et. al. (22) are also interpreted in terms of decoupling of planes, and several experiments on multilayers give similar results.(23)

Other dopants and concentration changes can affect the irreversibility line An interesting system is the septernary Tl based compound described above. (15) As y is changed from zero to 0.4 the hole density is decreased and the T_c goes through a maximum when the carrier density is zero. The overdoped material is more semiconducting than metallic. Table I shows the critical temperature and the width of the resistive transition at 7T for three values of y.

It can be seen that the more we get away from the metallic end of the phase diagram the wider the resistive transition and it is found that the other features of oxide superconductors, low dimensionality, granularity etc. also become more marked. We do not know why this is so but it mirrors in a limited way the more extreme transition from real shiny metals to oxides.

y	0	0.2	0.4
T_c	67	110	90
ΔT_c	12	45	65

TableII

More indirect evidence comes from a number of results showing that the resistive transition is not altered if the the current is applied parallel to the field so that there is no Lorentz force. This is what would be expected if flux cutting was extremely easy and a decoupled set of discs is an extreme example of this.

Finally on a more optimistic note recent results from ETL have shown that if the field is applied parallel to the planes, and the current is also in the planes, but perpendicular to the field, large current densities can be carried well above the generally accepted irreversibility line of BSCO. This is understandable if the planes are independent superconductors, but inconsistent with a phase transition in the vortex lattice which would depend on the direction of the magnetic field but not of the test current. Most inductive measurements cannot measure this critical current sense they are determined by the lower of two critical currents, one of which is perpendicular to the planes, but recently it has shown that the reversibility line as measured by the inductive transition is changed if the direction of the AC field is changed. (24) This is in agreement with the the transport measurements and the result is of considerable practical importance since it shows the irreversibility line is not an absolute barrier to applications..

These last two experiments are very hard to explain by flux melting, and we have seen that there are strong theoretical grounds for doubting it. Perhaps the strongest evidence in favour is that calculations using the Lindemann criterion predict a temperature fairly close to that observed. Other measurements of details in V-I characteristics can also be explained by pinning models of varying complexity. A recent observation by Alloul (26) is that the width of the NMR resonance becomes very small above the reversibility line, indicating that flux is moving around rapidly even without a current. This is what would be expected from lattice melting, but is not inconsistent with decoupled planes since very great mechanical stability is required to ensure that the relative movement between the sample and the external field is less than a vortex spacing.

It cannot be said that any experiment so far carried out distinguishes unambiguously between the models but the case for plane decoupling appears strongest. Whatever the mechanism there is clearly important work to be done in this area to improve performance.

Conclusions and Future Prospects

Most of the superconducting properties of high T_c superconductors are much better than those of conventional ones. The large energy gap means that both analogue and digital systems should work ten times faster than with Niobium technology. The upper critical field is ten times higher and we have already seen thin films which carry 10^6 A/cm^2 in fields of 5T at 77K, with useful current densities at much higher fields.. The Lorentz force per unit volume is many times larger than can be achieved in Nb_3Sn at 4.2K and indicates the possible power density of a motor built with this material. The rate of progress has been much faster than that of conventional materials but the problems that remain in realising this potential are not easily solved. The problems are the reversibility and granularity. Solving latter is probably also the key to making good SIS Josephson Junctions Different materials have widely differing irreversibilty lines and degrees of granularity and if we could find a material with properties between those of BSCO and YBCO we would have a really useful product. There are signs a that some Tl based materials will fill this position and there are many materials with lower T_c which could be equally important. If half the effort had been put into maximising J_c that has been put into increasing T_c we would be a lot further forward by now.

References

1)G.Gerber, D.Anselmetti, J.G.Bednorz, J.Mannhart, and D.G. Schlom, Nature,**350**, 279, (1991).
2) A.V.Narlikar, to be published.
3) C.J. van der Beek and P.H.Kes, Phys. Rev. **B43**, 13032 (1991)
4) T.C. Halsey, Phys. Rev. Lett., **55**, 1018, (1985).
5) A.M.Campbell, Jap. J. Appl. Phys,**26**, Supplement 26-3, 2053, (1987).
6) K.Yamafuji,,and F.Irie, Phys. Lett. A. **25**, 387, (1967).
7) M.C. Marchetti and D.Nelson, Phys.Rev. B,**42**,9938,(1990).
8) M.P.A. Fisher, Phys.Rev.Lett., **62**, 1415,(1989).
9) A.M. Campbell, Supercon. Sci. and Tech., **2**, 287, (1989)
10) B.Dwer, Physica C, **168**, 109, (1990).
11) G. Deutscher and K.A. Müller, Phys. Rev. Lett., **59**,1745.,(1987)
12) D.Dimos, P. Chaudhari, J. Mannhart and F.K. LeGoues, Phys. Rev. Lett, **61**, 219, (1988).
13) S.E. Babcock, X.Y.Cai, D.L.Kaiser, and D.C. Larbalestier, Nature, **347**, 167, (1990).
14) W. Schauer, X.X.Xi, V.Windte, O.Meyer, G.Linker, Q. Li and J. Geerk, Cryogenics,**30**, 561, (1990).
15) J.D.Johnson,A.R.Perry, A.M.Campbell, R.S.Liu, and D.N.Zheng, Proc. 6th Int. Workshop on Critical Currents, Cambridge 1991, to be published in Super. Sci. and Tech.
16) Y.Yeshrun, A.P. Malozemoff, T.K. Worthington, R.F.Yandrofski,, L.Krusin-Elbaum, F.H.Holtberg, T.R.Dinger and G.V.Chandrasekhar, Cryogenics, **29**, 258,(1989).
17) L.J. deJongh, Sol. State. Comm., **70**, 955, (1989)
18) A.B.Pippard, Phil. Mag., **19**, 217, (1969).
19) H.W.Weber, H.P. Wiesinger, W.Kritscha, and F.M.Sauerzopf, Proc. 6th Int. Workshop on Critical Currents, Cambridge 1991, to be published in Super. Sci. and Tech.
20) L.Civale, A.D.Marwick, T.K.Worthington, M.A.Kirk, J.R.Thompson, L. Krusin-Elbaum, Y.Sun, J.R.Clem, and F.Holtzberg, Phys. Rev. Lett., **67**, 648, (1991).
21)D.N. Zheng, J.D. Johnson and A.M. Campbell, Proc. 6th Int. Workshop on Critical Currents, Cambridge 1991, to be published in Super. Sci. and Tech.
22) D.H.Kim, K.E.Gray, R.T.Kampwirth, J.C.Smith, D.S.Richieson, T.J.Marks, J.H.Kang, J.Talvaccio, and M.Eddy, Physica C, **177**, 431, 1991.
23) P.Przyslupski, G.Jakob, C.Stolzel, C.Tome-Rosa, A.Walkenhorst, M.Schmitt, and H.Adrian, Proc. 6th Int. Workshop on Critical Currents, Cambridge 1991, to be published in Super. Sci. and Tech.
24)F. de la Cruz and C Duran, Proc. 6th Int. Workshop on Critical Currents, Cambridge 1991, to be published in Super. Sci. and Tech.
25) "Spin Glasses and Other Frustrated Systems', D. Chowdhury, World Scientific,Singapore 1986, p111.
26)H.Alloul, to be published.
27)P.L.Gammel, L.F.Schneemeyer, J.V. Waszczak and D.J.Bishop, Phys.Rev. Lett., **61**, 1666, (1988).
28) Y.Enomoto, M.Suzuki, T.Murakami, T.Inuki, and T.Inamura, Jap. J. Appl. Phys. **20**, L661,(1981).

N.A.S.A. Space Applications of High-Temperature Superconductors

VERNON O. HEINEN[1], MARTIN M. SOKOLOSKI[2], PAUL R. ARON[1], KUL B. BHASIN[1], EDWIN G. WINTUCKY[1], and DENIS J. CONNOLLY[1]

[1] National Aeronautics and Space Administration, Lewis Research Center, Cleveland, OH 44135, USA
[2] Washington, DC 20546, USA

ABSTRACT

The application of superconducting technology in space has been limited by the requirement of cooling to near liquid helium temperatures. The only means of attaining these temperatures has been with cryogenic fluids which severely limits mission lifetime. The development of materials with superconducting transition temperatures (T_c) above 77 K has made superconducting technology more attractive and feasible for employment in aerospace systems. In this paper, potential applications of high-temperature superconducting technology in cryocoolers and remote sensing, communications and power systems will be discussed.

KEY WORDS: superconductivity, space applications, cryocoolers, communication satellites, space power.

INTRODUCTION

There has been a great deal of interest in low temperature technology for space applications from the beginning of the space age. The initial interest was chiefly in using cryogens for rocket fuel and life support systems. The interest shifted later to scientific instruments. An example of such an application is Gravity Probe B (GPB) which is being developed to test certain tenets of Einstein's general theory of relativity. The experiment is designed to measure the gyroscopic precession which should result from the influence of the mass of the Earth on space-time. GPB makes extensive use of superconducting technology. It will have superconducting lead sleeves to attenuate the Earth's magnetic field. Rotating superconducting niobium-coated quartz spheres will act as gyroscopes, and SQUID magnetometers will measure their magnetic moments. There are many other examples of past and planned space missions which require low temperatures. One of the common elements of these missions is the use of liquid helium (up to 4000 liters) to reach the low temperatures. Use of helium limits the lifetimes of the missions to less than 5 years.

This limited mission lifetime has been one of the key factors which has limited the applications of superconducting technology. Until the discovery of high temperature superconducting (HTS) materials, liquid helium was necessary to reach the temperatures at which materials were in their superconducting states. High temperature superconducting (HTS) materials with transition temperatures above liquid nitrogen temperature significantly reduce the cooling required to use these materials in space. A satellite using liquid nitrogen or hydrogen as a cryogen would have a much longer lifetime than a satellite which had the same volume of liquid helium. Mechanical coolers capable of cooling to 80 K are much further along in development than coolers capable of cooling to < 10 K. Passive cooling can also be used to reach the 70 to 80 K range for a very low heat loads. Because of the dramatic increase in transition temperature, superconducting technology may now be feasible for use in space for communication, power, and propulsion systems. Before HTS materials will be used in space, their long term stability must be demonstrated, especially in the space environment. Also, the added weight, power requirements and complexity of the necessary cooling must be included when estimating the benefits of using superconducting technology. Many applications have been suggested and investigated to determine the benefits of incorporating superconducting technology in space systems and are described below.

APPLICATIONS OF HTS MATERIALS

Cryocoolers

Space cryocoolers typically operate at or below 77 K and provide a natural cryogenic environment for HTS applications, with no weight/energy penalty for added refrigeration. Applications are being developed that potentially can significantly enhance cryocooler performance, efficiency, reliability and lifetime, such as HTS electrical leads for thermal isolation in space borne liquid helium dewars, damping of vibrations induced in the focal plane by Stirling cycle mechanical cryocoolers and passive magnetic bearings for turboexpander-type cryocoolers.

Some of the future NASA mission such as the Earth Observing System (EOS) will employ cooling of infrared and other types of sensors in order to obtain the required sensitivity and energy resolution. A large number of electrical leads, typically manganin wire, are needed to connect the sensors situated on the cooled focal plane at <10 K to instrumentation at a much higher temperature located outside the dewar. Another mission, the Advanced X-ray Astrophysics Facility (AXAF) will have immersed in liquid helium an adiabatic demagnetization refrigerator requiring nearly 3 A of current for operation. Although the sensor leads usually carry only microamps or less of current, they nevertheless comprise a significant parasitic heat load. For example, in the case of the Spectroscopy of the Atmosphere by Far Infrared Emission (SAFIRE) mission, the sensor leads constitute 33% of the instrument and 17% of the total heat load.

A high degree of thermal isolation can be achieved if the sensor leads are replaced by HTS copper oxide ceramics, which have thermal conductivities of less than 1/4 that of manganin. Using $YBa_2Cu_3O_{7-\delta}$ for just the 158 sensor leads in the SAFIRE dewar would reduce the total heat load by at least 8% and extend the mission life by more than 5%. Much larger extensions of lifetime have been estimated for other missions. Another important benefit of using HTS electrical leads, in addition to thermal isolation, is an estimated 10 to 100 times reduction in signal-to-noise ratio.

Stirling cycle mechanical cryocoolers are a leading candidate to provide the detector cooling required for the many far-infrared (>10 μm) experiments planned for the EOS missions of the late 1990s and early 21st century. In order to achieve precision imaging and high optical resolution, vibrations of the detectors induced by the cryocooler operation, as well as spacecraft induced resonances, must be damped. The generally accepted goal is 0.01 to 0.02 pounds of force at the focal plane. One approach that is showing some success is back-to-back cooler operation supplemented by active electronic suppression of the fundamental cryocooler frequency (~ 40 Hz) and higher harmonics.[1]

This approach, however, is unlikely to satisfy all the vibration damping requirements and it may be that an additional, and preferably passive, damping mechanism will still be required. $YBa_2Cu_3O_{7-\delta}$ materials, which have an inherent magnetic damping capability due to dissipation of energy by flux motion,[2] could potentially provide this mechanism. Magnetic Technologies, Inc. laboratory experiments using non-optimum materials and magnetic field geometry have demonstrated that presently available $YBa_2Cu_3O_{7-\delta}$ materials can provide at least an order of magnitude greater damping at 77 K than with no superconducting material.

Advanced turboexpander-type cryocoolers based on the Brayton thermodynamic cycle are being developed by NASA for future long duration missions. Important advantages offered by the turboexpander type, which use non-contact gas bearings, over the Stirling cycle mechanical cryocooler are potentially vibration free performance and greater reliability and lifetime. A disadvantage of the gas bearings is the need to run warm, which causes a large heat leak between the warm bearing zone and the cold expansion turbine, resulting in inefficient operation at low temperatures. Therefore, the need exists to improve the efficiency. A promising approach to eliminating the heat leak is to replace the rotary gas bearings with HTS passive magnetic (Meissner) bearings. A study by Creare, Inc. of a 1W, 10 K cryocooler, where the gas bearings for the coldest turboexpander were replaced by Meissner bearings fabricated from currently available HTS materials, showed that a 40% decrease in input power was possible.

With $YBa_2Cu_3O_{7-\delta}$ materials prepared by the melt quench process, levitation forces are now adequate for cryocooler applications although some improvements in stiffness may be required. [3] A rotational speed of 450,000 revolutions per minute (surface speed of 150 m/s) has been demonstrated by Creare, Inc. which is more than adequate for cryocooler applications (90 m/s).

Remote Sensing Applications

Many space-based sensing systems are cooled for improved sensitivity and decreased noise. This is another area of application where including HTS materials would have no cooling penalty when compared against nonsuperconducting systems. Three devices which would provide great benefit to near earth and deep space missions are bolometers, mixers and signal processors.

In missions to the outer planets lasting ten years or longer, passive radiative cooling is presently the only satisfactory option for cooling the focal plane. This means the detector temperature is 70 K or higher. Studies of planetary atmospheres at wavelengths greater than 20 μm will require the use of thermal detectors.[4] A promising candidate for this application transition-edge bolometer that uses the resistive transition of a HTS element as the temperature sensing element. Such bolometers have been demonstrated to be superior to pyroelectric detectors. Another type of bolometer is the kinetic inductance bolometer which potentially has much lower noise and higher sensitivity. The operation of this bolometer depends on the temperature dependence of the magnetic penetration depth just below T_c.

Submillimeter spectroscopy is an important objective of future missions for astrophysics and earth atmospheric studies. A principle component for these studies is mixer. Thus far the most sensitive mixers in the millimeter to near submillimeter range have been Nb based superconductor-insulator-superconductor (SIS) tunnel junctions. HTS tunnel junctions could possible extend this frequency range well into the THz range. However, SIS tunnel junctions fabricated from HTS materials have not yet been demonstrated. An alternative approach is the superconductor-normal-superconductor (SNS) microbridge. Progress in HTS SNS is proceeding rapidly.

Locating a superconducting signal processor on a focal plane has many advantages over a semiconducting processor in terms of speed, power consumption and thermal loading. The higher speed would allow fewer data lines to the focal plane which means lower thermal loading. The development of such a processor is further in the future than development of HTS bolometers or mixers.

Space Communications Applications

In the past, the need for heavy, short life and complicated cryogenic systems near 20 K prohibited the use of superconducting electronics circuits in space communications systems in spite of their outstanding advantages when compared to room temperature semiconductor electronics circuits. Since this discovery, NASA has been very active in identifying the applications of HTS materials in space communications systems. The approach has been twofold: first to identify subsystems and communications scenarios where superconducting devices and circuits will have the largest impact and second to design, fabricate and characterize generic superconducting devices and components and compare their performance with conventional circuits to quantitatively determine their advantages.

Under the first approach, in-house application analysis[5] and studies conducted by TRW, Ball Aerospace and Varian were used to identify key applications and are listed below.

1. Ultra Low Noise Hybrid Superconductor/Semiconductor Receivers
2. Phased Array Antennas from 1 GHz to 50 GHz
3. Digital Signal Processors
4. Traveling Wave Tube Amplifiers

These studies included the impact of cryocoolers in their comparison which has often been ignored. One of the key finding is that superconducting circuits can offer outstanding advantages in reducing space craft's power, size and weight in high performance space communications systems. It can also provide low loss, low noise, high data rate subsystems. However, only in a communication scenario where such high performance is desired, can one justify the use of cryocoolers.

Under the second approach, NASA has designed, fabricated and characterized microstrip and coplanar resonator circuits from X-band to Ka band frequencies,[6],[7] X-band filters, phase shifter switches,[8] X-band superconducting resonator stabilized oscillators, Ka-band passive microstrip antenna [6] and hybrid superconductor/semiconductor low noise amplifiers using $YBa_2Cu_3O_{7-\delta}$ and thallium thin films on lanthanum aluminate substrate. Superconducting resonator circuits have shown higher "Q" values than gold circuits which translates into low loss passive microwave circuits.[9] Superconducting filters have shown low insertion loss and higher "Q"s. [7] Early evidence of superior performance of high T_c superconducting passive microwave components at liquid nitrogen temperatures has led Naval Research Laboratories (NRL) to evaluate their performance in several space experiments and expedite technology development. [10] NASA has participated in NRL's High Tc Superconducting Space Experiment (HTSSE) I. NASA is also planning to participate in HTSSE II expected to be launched in 1996. For this experiment NASA Lewis Research Center and the Jet Propulsion Laboratory will be jointly developing X-band hybrid HTS/semiconductor receiver. Also under planning is Ka-band superconducting receiver antenna experiment as a part of a link between shuttle and Advanced Communications Technology Satellite.

For space communication applications, the discovery of high Tc superconductors has led to the development of high performance microwave components operating at 77 K. It is possible that these components may find use in real space communication systems demanding extraordinary performance. However, only the discovery of superconductors operating at 150 K can make their use possible in most of the space communications systems.

Space Power Applications

When high current density superconductors appeared in the early 60's their potential for application to space power technologies was carefully explored since it seemed certain that a zero resistance conductor could lower the launch weight by reducing the masses of both the conductors and the power system. Further, it could allow the wide use of magnets which would no longer need large and heavy power supplies and/or massive amounts of iron. The price to be paid however was the need to maintain the conductor below its transition temperature. For the conductors then available this meant refrigeration to well below 20 K (typically 4.2 K). This constraint introduced weight, complexity and reliability issues that effectively foreclosed the use of superconducting technology in space power systems. As was found with terrestrial power applications, only space power systems, with capacities much greater than ones that have been of interest, could justify the use of superconductors.

Now, in the 90's, this issue is being reexamined as the result of two developments. First, we are seriously considering space missions, such as Moon colonization and Mars exploration, that will require large capacity (>100kW) electric power systems and second, the discovery of HTS materials.

The most straight forward application of SC is direct power transmission. In this case we look at the wiring harness which carries power and control signals within the power management circuitry and between it and its electrical sources and loads. In ref [11] it was shown that, at a current density of $2X10^4$ amps/cm^2 a superconducting wire can reduce the launch weight by $1.3X10^{-3}$ gm/cm-W. Since large space systems will require many kilometers of wire carrying many kilowatts, substantial savings are possible if it all could be operated below the SC transition temperature. In general, this is not possible, but it might be practical if the power source must be located a long distance from its load. Not only would a minimum weight penalty be incurred, but the system designer would have additional flexibility in choosing the system operating voltage. A recent design study [12] has shown that it is possible to use directional radiators to cool such a line to less than 77 K in orbits as close as 1000 km from the earth and to even lower temperatures in deep space. The weight of such a system, which uses radiation shields and commercially available HTS materials is still substantially less than system with copper transmission lines and it requires no active cooling system.

About 30% of a transformer's weight is conductor, therefore it is another element of a power distribution system where superconductors might be usefully employed. Because the useful current density in superconducting materials is at least 100X greater than that of copper, the use of superconductors could eliminate more than 99% of this mass. In the case of capacitors only a small fraction of a capacitor's mass is conductor and therefore little is to be gained by the direct substitution of superconducting material. Of course, transformers and capacitors are elements of AC systems and the largest impact of superconducting technology on power systems might come from their elimination as the use of superconductivity could make DC systems more interesting. [11]

This long term storage of electrical energy as a magnetic field is only feasible if Sc inductors are used since the I^2R losses of normal conductors would quickly dissipate the energy. Early examination [13] of its potential for use as a space "battery" showed that it has some important advantages. Such a storage system has virtually no losses, allowing the recovery of 95% of the input, and would tolerate a wide range of charge/discharge rates. Further, it would have an unlimited cycle life. The specific energy, which is limited by the forces that the field exerts on the conductor, has been estimated at near 50 W-hrs/kg, which is competitive with NiH_2 batteries.

Presently most space systems depend on photovoltaic arrays for primary electrical power and with increased power demands the arrays can reach large sizes. Also, in many orbits there is sufficient plasma density such that arcing can be a problem at >150 volts. Therefore power from these large arrays may have to be carried to their loads at high currents. The weights of these conductors, presently 3-5% [14] of the array mass and the 1-2% of the array lost to I^2R can be substantially reduced by the use of superconductors. Although the percentages are small, in large or multiple systems the absolute weight and money savings can be non-trivial. It remains to be seen if the system complexity introduced by the need to thermally shield the superconducting components will discourage its use in this application.

Many other space applications of superconducting technology in addition to the ones discussed above have also been identified. These include active bearings for rocket engines, magnets for magnetoplasmadynamic thrusters,

CONCLUSIONS

The discovery of materials which are superconducting above liquid nitrogen temperature has made superconducting technology more attractive and feasible for use in aerospace systems. For some future missions where cooling is necessary to fulfill the mission objectives, the HTS materials meet all the technical requirements and are likely to be used after the material stability issues have been settled. For other missions, the added complexity of cooling must be justified by the improved performance offered by superconducting technology. If materials are developed with higher transition temperatures, cooling will be much less of an issue, and the use of superconducting technology should become much more widespread in space-based systems.

REFERENCES

1. Ross RG, Johnson DL, Sugimura RS (1991) Proceedings of the 6th International Cryocooler Conference, DTRC91/002: 27

2. Moon FC, Weng KC, Chang PZ (1989) J. Appl. Phys. **66**: 5643

3. Carlson L (1991) Superconductor Industry **4**: 22

4. Castles S, Breon S, Viens M, Allen C, Patschke R, Steward K, Mosely H, Brasunas J, Hilliard L, Richards C, Flom Y (1988) Presented at the First World Congress on Superconductivity, Houston, Texas

5. Heinen VO, Bhasin KB, Long KJ (1990) Presented at Second World Congress on Superconductivity, Houston, Texas and NASA TM 103629

6. Richard MA, Bhasin KB, Gilbert C, Metzler S, Claspy PC (1991) Presented at Antenna and Measurement Techniques Association Meeting, Boulder, Colorado and NASA TM 105183

7. Chew W, Riley AL, Roscoe DL, Hunt BD, Foote MC, Cooley TW, Bajuk LJ (1991) IEEE Trans on Microwave Theory and Tech. **39**: 1455-1461

8. Bhasin KB, Warner JD, Chorey CM, Romanofsky RR, Heinen VO (1990) Proceedings of the SPIE **1292**: 77-82

9. Chorey CM, Kong KS, Bhasin KB, Warner JD, Itoh T (1991) IEEE Trans on Microwave Theory and Techn. **39**: 1480-1487

10. Webb DC, Nisenoff M (1991) Microwave Journal **34**: 85-90

11. Aron PR, Myers IT (1988) Proceedings of the 23rd IECEC and NASA TM 100901

12. Shimko MA, Crowley CJ, Wallis PN (1991) Presented at the 5th Conference on Superconductivity and Applications, Buffalo, New York

13. Fickett FR (1979) Cryogenics **19**: 691-701

14. Meulenberg Jr. A, Private communication

Microstructural Issues Concerning the Critical Current Properties of High Temperature Superconductors

DAVID C. LARBALESTIER

Applied Superconductivity Center and Department of Materials Science and Engineering
University of Wisconsin, Madison, WI 53706, USA

ABSTRACT

Principal tasks in the development of a reliable HTS magnet technology are the control or elimination of weak links, the increase of the flux pinning strength so that high critical current density is possible without significant flux creep and the engineering of these attributes into an appropriate conductor form. These developments require microstructural understanding and control at both the nanometer and the much larger, macrostructural scale. The state of knowledge of HTS macro- and micro-structures is discussed, partly from the perspective of Nb-Ti, since its optimization offers useful lessons for HTS conductor development.

KEY WORDS: Microstructures, critical currents, Nb-Ti, BSCCO, YBCO

INTRODUCTION

Gaining control of the microstructure of the Niobium-based low temperature superconductors (LTS) was crucial to maximizing the performance of superconducting magnets. At one extreme are very high field 13T Nb-Ti alloy magnets [1] and hybrid Nb-Ti/Nb_3Sn 20T magnets [2]: another extreme is represented by very high J_c, fine filament accelerator magnets which have provided the stimulus for so much superconductor wire development. High temperature superconductors (HTS) have a J_c and H_{c2} potential that in principle permits magnets whose limits are only determined by structural considerations. This potential is subject to the condition that high J_c values can be developed in long lengths of conductor which can be wound into magnets while retaining these properties. In this paper some of the features which mark the community's understanding of Nb-Ti alloys and HTS materials are compared, with a view towards suggesting some of the areas of microstructural study that may yield rapid advance for HTS materials.

Optimization of Niobium-Titanium Based LTS Materials

Many of the key advances took place in the early 1980s [3-9]. An important step was recognizing that the J_c of many wires was not determined by their microstructure, but by their macrostructure [3,9]. Macrostructural variations of composition and cross-section produce comparatively long range variations of the critical current which are quite independent of the local critical current density produced by flux pinning interactions. The natural length scales of the macrostructure are typically 1-10μm while those for flux pinning are the coherence length (ξ) and the fluxon spacing (a_0) and these are both orders of magnitude smaller than the macrostructural scale. For Nb47wt%Ti ξ is ~5nm. a_0 is ~$1.07(\phi_0/B)^{1/2}$, thus being ~50nm at 1T and ~15nm at 10T. Consequently, significant macro-variations of structure render the influence of the nanometer-scale microstructure hard to determine. Until these macro-scale variations were identified, controlled and reduced to insignificant levels, many of the fundamental microstructure-J_c properties were not deduced. Some of the macro problems were chemical in origin (for example, the "coring" associated with ingot solidification) and some were mechanical. For example, bcc alloys do not deform homogeneously to very large strains and filament-matrix chemical reactions produce intermetallics that embed in the filament and further aggravate the non-uniformity of deformation [9]. All of these effects can be revealed by standard metallographic observation but such observation is always time- and people-intensive and thus tends to be minimized in favor of quicker tests. A critical current measurement is the standard evaluation for superconducting wires. However, the word critical implies a precision which is frequently not present: in a standard resistive transition, the critical current is quite sensitive to the criterion chosen and no single value completely describes wire performance. A key step was the recognition that the shape of the V/I curve in the resistive transition could be an excellent diagnostic [3,9,10] of whether the wire was in the extrinsic limit (where macrostructural irregularities control I_c) or in the microstructural limit (where the I_c is determined by the elementary fluxoid-pinning center interactions). In Nb-base materials the resistive transitions are well described by a power law $V \propto I^n$, where n is the resistive transition index. The shape of the n-H curve is particularly interesting (Fig. 1). In the intrinsic limit, the plot is steeply increasing with decreasing H; in the extrinsic limit, the plot forms a plateau at intermediate fields[9]. Such plots demonstrated that practical conductors of Nb-Ti always reach maximum J_c as the conductor crosses from the intrinsic to the extrinsic limit. The variable that controls this transition is the degree of deformation applied to the wire. Eventually the filaments always deform non-uniformly and J_c then becomes extrinsically limited. This is almost certainly the case for HTS conductors too.

Knowing where the intrinsic to extrinsic transition occurs makes it possible to devise experiments that explicitly study the microstructural dependence of the flux pinning force. In this way it has been possible to show that the J_c is proportional to the volume fraction of α-Ti precipitate [11,12] and to address the fluxon summation relationships that exist in this system [8,13]. Figure 2 shows a micrograph of an optimized, high J_c wire. In transverse section, the flux pinning precipitates of α-Ti appear as very high aspect ratio ribbons some 100-200nm wide, 1-5nm

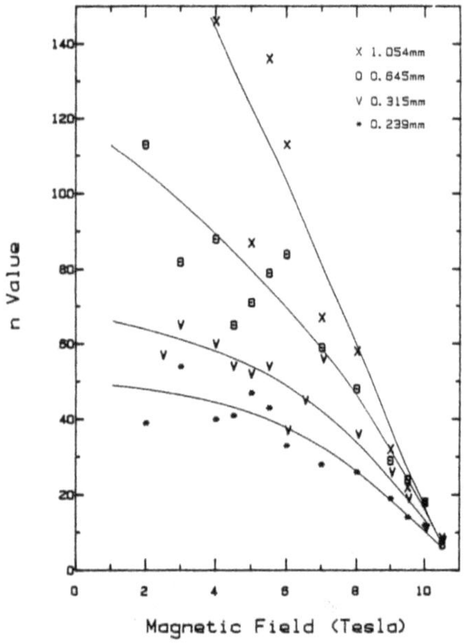

Fig. 1 Resistive transition index (n) vs. magnetic field (H) for a Nb-Ti multi-filamentary composite. The sizes are wire diameters. As the wire is drawn to smaller size filament diameters become irregular and n declines.

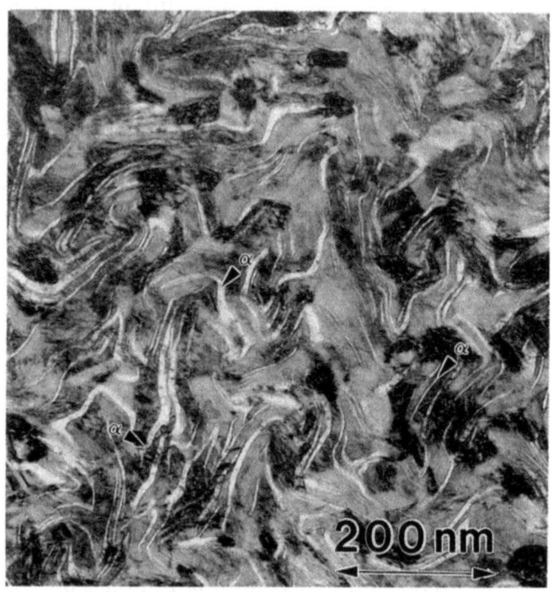

Fig. 2 Transverse section micrograph of portion of one filament of an optimized Nb47wt% multifilament composite having ~25vol% of α-Ti ribbons (white ribbons) in the superconducting β-Nb-Ti matrix. (Courtesy P. J. Lee)

thick and some 5-10 nm apart [5]. They are generally several µm long and constitute 15-20vol.% of the cross-section. Being normal, they exert a strong elementary pinning force (f_p) and, being very dense, they exert a large bulk pinning force (F_p). One of the big advantages of this system is that the volume fraction of α-Ti can be varied from 0-25%. The size and spacing of the precipitates can also be controlled to be less than, comparable to or more than ξ. Although this makes the system very complex, it is possible to make a quantitative description of the precipitate microstructure and to relate this to the measured bulk pinning forces [7,8]. As recent work shows [12,13], this is providing a strong conceptual basis for designing yet more optimized microstructures, including those that are produced artificially which escape some of the natural limitations of the Nb-Ti system [14]. Some of the lessons learned from Nb-Ti development are undoubtedly useful for HTS systems too.

High Temperature Superconductors

However complex the situation in Nb-Ti, it is far more complex in HTS. In order to make the present discussion even moderately tractable, we discuss here only the YBCO-123 and BSCCO-2212 and -2223 materials, confining our discussion to a few features of the granularity and flux pinning in the compounds. The immediate issue is the superconducting granularity [15] which is associated with grain boundaries or other internal interfaces. This means that two (or occasionally more) levels of currents exist within the sample and that various forms of percolation and multiple-connectivity of the material exist. This immediately produces uncertainty about the significance of measured J_c values. These effects are very strong in polycrystalline 123 compounds and can also be seen in good 123 single crystals, when explicitly searched for [16]. At first it was proposed that this granularity at internal interfaces and grain boundaries was a natural consequence of the short coherence length of HTS [17]. However, the phenomenon is much more prevalent in 123 compounds (ξ_{ab} ~2-3nm) than in the even shorter ξ BSCCO-2212 and -2223 materials. Granularity is not seen in full density Chevrel phase materials possessing short ξ values of order 2-3nm [18], while being found in BaPbBiO having ξ of order 7nm [19]. Thus microstructural descriptions of the granular behavior are still needed. Our group has been working hard on this problem with YBCO-123 bicrystals, first making electromagnetic measurements on single boundaries and then examining the same boundaries by TEM [20-22]. However, we are not yet able to formulate an explicit microstructural description of the difference between a strongly-coupled flux pinning boundary and a weakly-coupled Josephson junction grain boundary. Figure 3 illustrates the problem. This shows a high resolution electron micrograph in a section of grain boundary bicrystal which acts as a Josephson Junction; deviations from the expected cation positions only become significant within about 0.5nm of the grain boundary, a distance which is significantly less than ξ_{ab}(1.5-3nm). In fact, the most common feature of all studies of well-made HTS is that their grain boundaries are not perturbed in any immediately obvious way. Of course, electron microscopy is not very sensitive to oxygen on the atomic scale and this is currently one of the major limitations of all (HR)(S)TEM-(EDXS) studies of HTS materials. It is tempting and entirely plausible to believe that the local oxygen content of the boundary controls the properties, as seen in both bulk-scale and thin film studies [21,23,24]. However, explicit microstructural proof is very desirable. A novel alternative is provided by the Position Sensitive Atom Probe (POSAP). In this instrument a sharp tip is dissected one atom at a time, thus in principle permitting a three-dimensional reconstruction of the tip layer by layer both with respect to position (to about 1nm) and chemical species [25]. In its original version the instrument is rather slow, since the atoms are desorbed by pulsing the tip electric field, an inherently slow process. My Madison colleague, T.F. Kelly, is developing a thermally pulsed instrument [26], which will be many orders of magnitude faster which we hope to apply to HTS studies.

Fig. 3 HRTEM image of a 26° [001] tilt boundary. (courtesy Na Zhang ref. 21)

Although granularity is evidently much less severe in 2212- and 2223-BSCCO than in 123, there is evidence of granularity in BSCCO when it is explicitly searched for [27]. One intriguing question is whether the connections across BSCCO grain boundaries are inherently different than those across 123 grain boundaries. No systematic experiments have explicitly answered this question yet. At least two features of BSCCO are different; the micaceous cleavage character of BSCCO encourages the formation of high aspect ratio grains that can be idealized as a "brick wall" microstructure. The more two-dimensional nature of BSCCO also changes the electromagnetic properties of the structure, which can be regarded as a c-axis Josephson- coupled stack of a-b planes [28,29]. At present a variety of views are possible concerning the grain boundary connections: they may be as poorly connected as 123 grains or, on the contrary, they may be well-connected. A recent model [30] emphasizes the brick wall, Josephson-coupled picture in which current flows preferentially in the c direction across the generally only slightly mis-aligned (001) twist boundaries that are the most common grain boundaries in the 2212 and 2223 compositions. The long range perfection of the grain stacking is crucial to such a model. At the high magnification of the TEM (Fig. 4), such a grain morphology is generally seen [31]. However, at the scale of 10-1000μm, the grain structures generally exhibit much less perfection (Fig. 5). Undissolved or precipitated second phase particles, sausaging irregularities from the silver sheath and a variety of other extrinsic factors conspire to disrupt the long range alignment of the grain structure [31]. Understanding what factors control the J_c of BSCCO tapes is crucial to the magnetic field applications of HTS and the time is thus appropriate for developing a detailed description of the macro- and micro-structure of high J_c BSCCO tapes. This work should include detailed analyses of the resistive transitions, data useful both for the critical current distribution models already cited (Fig. 1) [9,31], as well as for more basic models of the fluxon dynamics of HTS [33]. Prior work on Nb-Ti is particularly relevant to this development phase [3,9].

The tantalizing issue facing the further development of HTS materials in bulk form is well exemplified by comparing the best J_c values possessed by the 123 and the 2223 compounds at 77K. 123 compounds are still weak-linked and thus are limited by granularity. The highest J_c values obtained across high angle grain boundaries of YBCO reach about 10^3 A/cm^2, a value obtained within the grains too [20]. In such bicrystals the flux pinning is dominated by oxygen vacancies and is not inherently strong [16]. Much stronger pinning has been developed by second phase insulating particles of Y_2BaCuO_5 [34] yielding ~10^5 A/cm^2 (though this view of 211 pinning has also been challenged [40]), by controlled cation doping and phase decomposition [35] yielding ~2 x 10^5 A/cm^2 and by neutron irradiation yielding up to 10^6 A/cm^2 [36,37]. In none of these strong

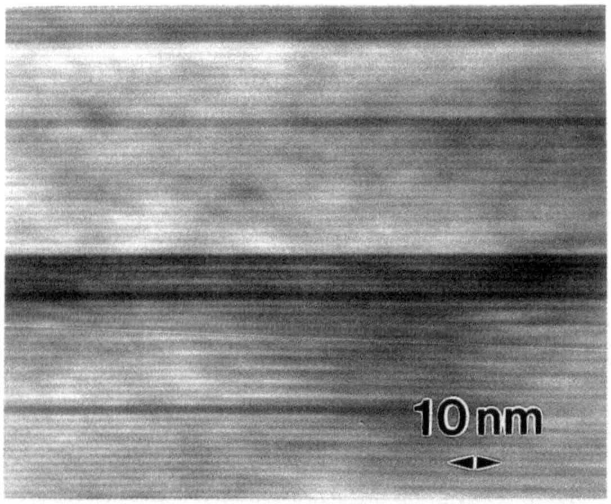

Fig. 4 Sideview of parallel grain boundaries in a silver sheathed BSCCO tape (courtesy Y. Feng)

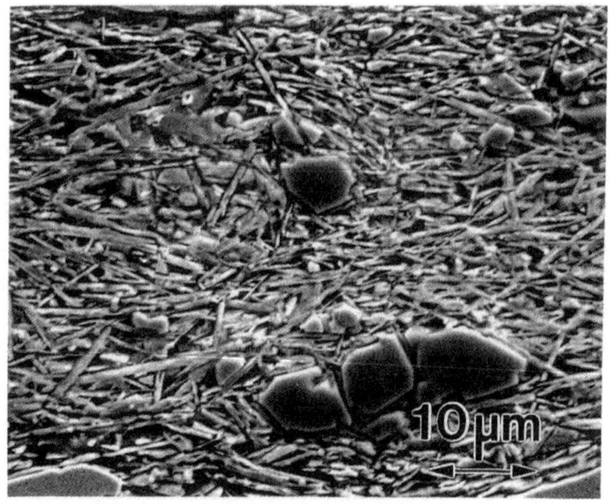

Fig. 5 Scanning electron micrograph of long range grain structure in a silver sheathed BSCCO tape. (courtesy Y. High)

pinning cases, however, is J_c measured crossing high angle grain boundaries. By contrast, BSCCO-2223 has reached J_c values of 2-5x10^4 A/cm^2 when made into a polycrystalline oxide-powder-in-tube (OPIT) tape [32,38,39]. However, the J_c of such tapes suffers a relatively much more rapid fall-off with field than does 123. This fall-off is generally ascribed to the weak flux pinning of the BSCCO compounds and the very easy flux creep which is the consequence. Whether it will be possible to strengthen the pinning of the very anisotropic BSCCO materials is not clear. In the pancake vortex model [29], the poor interplanar coupling along the c-direction effectively subdivides the fluxon into a set of pancakes lying within each ab plane which are only weakly coupled to each other. In such a model the weak flux pinning is inherent and the gains to be obtained from strongly pinning each pancake vortex are not expected to be large.

Systematic experiments to control the pinning in BSCCO have lagged those in 123. In part this is due to the much greater phase complexity of the BSCCO system, though one study has shown that the flux pinning can be significantly enhanced by cation vacancies [41]. Whatever the theoretical predictions, it is impossible to under-emphasize how important it is to do experiments in which controlled changes in the microstructure are made, because *a priori* flux pinning calculations are seldom accurate to better than factors of 5 or so. BSCCO J_c values now have a sufficient magnitude that further increases of this order would revolutionize their applications potential. Confronted with the type of misoriented grain structure seen in Fig. 5, it is not hard to imagine that a better aligned microstructure would have a much higher J_c. Although this sample had a J_c(0 T, 77 K) of only 10^4 A/cm^2, the published micrographs for samples with 5 times higher J_c are not greatly different [39]. Systematic experiments which explicitly distinguish between intrinsic and extrinsic properties and which systematically vary the flux pinning and granular microstructure are, in this author's opinion, the key to raising the J_c of HTS materials to values having real practical importance.

ACKNOWLEDGEMENTS

I am grateful to P.J. Lee, E.E. Hellstrom, S.E. Babcock and many other colleagues for valuable discussions. Our studies have been supported by the Dept. of Energy- High Energy Physics (Nb-Ti work) and NSF, EPRI and DARPA (HTS work).

REFERENCES

1. Canavan ER, Larbalestier DC, Van Sciver SW (1988) IEEE Trans. on Magnetics MAG-24: 1082
2. 20T magnets have been constructed at KfK Karlsruhe, Oxford Instrument Co. and Intermagnetics General Corp.
3. Larbalestier DC (1985) IEEE Trans. on Magnetics MAG21: 257
4. Larbalestier DC, West AW (1984) Acta Metallurgica 32: 1871
5. Lee PJ, Larbalestier DC (1987) Acta Metallurgica 35: 2523
6. Li C, Larbalestier DC (1987) Cryogenics 27 (4): 171
7. Meingast C, Lee PJ, Larbalestier DC (1989) J. Applied Physics 66(12): 5962
8. Meingast C, Larbalestier DC (1989) J. Applied Physics 66(12): 5971
9. Warnes WH, Larbalestier DC (1986) Cryogenics 26: 643
10. Ekin J (1987) Cryogenics 27: 603
11. Lee PJ, McKinnell J, Larbalestier DC (1989) IEEE Trans. on Magnetics MAG-25: 1918
12. Stejic G, Cooley LD, Joynt R, Larbalestier DC, Takacs S (1991) Proc. of Cambridge Flux Pinning Conference July 1991 Supercond. Sci. Technol. (to be published)
13. Cooley LD, Stejic G, Larbalestier DC, Takacs, S (1991) Proc. of Cambridge Flux Pinning Conference July 1991 Supercond. Sci. Technol. (to be published)
14. Motowidlo LR, Kanithi HC, Zeitlin BA (1990) Adv. in Cryogenic Engineering 36A: 311
15. Küpfer H, Apfelstadt I, et al (1988) Cryogenics 28: 650
16. Daeumling M, Seuntjens J, Larbalestier DC (1990) Nature 346: 332
17. Deutscher G, Muller KA (1987) Phys. Rev. Lett. 59: 1160
18. Lelay L, Willis TC, Larbalestier DC (1991) Applied Phys. Lett. (submitted)
19. Cooley L, Daeumling M, Willis TC, Larbalestier DC (1989) IEEE Trans. on Magnetics 25(2): 2314
20. Babcock SE, Cai XY, Kaiser DL, Larbalestier DC (1990) Nature 347: 167
21. Larbalestier DC, Babcock SE, Cai XY, Field MB, Gao Y, Heinig NF, Kaiser DL, Merkle K, Williams LK, Zhang N (1991) Proc. of M^2S-HTSC July 1991 Physica C (to be published)
22. Babcock SE, Zhang N, et al (1991) J. Adv. Science (to be published)
23. Chaudhari P Physica C (to be published)
24. Lathrop DK, Russek SE, Moeckly BH, Chamberlain D, Pesenson L, Buhrman RA, Shin DH, Silcox J (1991) IEEE Trans. on Magnetics 27(2): 3203
25. Cerezo A, Godfrey TJ, Smith GDW (1988) Rev. Sci. Instrum. 59: 862
26. Kelly TF, Mancini DC, McCarthy JJ, Zreiba NA (1991) Surface Science 246: 396
27. Matsushita T et al (1991) Superconductor Science and Technology (to be published)
28. Kim DH, Gray KE, et al (1991) Physica C 177: 431-437
29. Clem J (1991) Phys. Rev. B 43 7837
30. Bulaevskii LN, Clem JR, Glazman LI, Malozemoff AP (1991) Phys. Rev. Lett. (submitted)
31. Feng Y, Hautanen KE, High YE, Larbalestier DC, Ray II R, Hellstrom EE, Babcock SE (1991) Physica C (submitted)
32. Heine K, Tenbrink J, Krauth H, Wilhelm M (1992) Adv. in Cryogenic Engineering (to be published)
33. Malozemoff AP (1991) Physica C (to be published)
34. Murakami M (1991) Physica C (to be published)
35. Tallon JL, Pooke DM et al (1991) Appl. Phys. Lett. 59 1239
36. Weber H (1991) Physica C (to be published)
37. Van Dover RB, et al (1989) Nature 342: 55-59
38. Yamada Y, Obst B, Flukiger R (1991) Supercond. Sci. Technol. 4: 165-171
39. Sato K, et al (1991) IEEE Trans. on Magnetics 27: 1231
40. Jin S, Tiefel TH, Kammlott GW (1991) Appl. Phys. Lett. 59(5): 540
41. Nomura S, Chiang YM Appl. Phys. Lett. 58: 768

Flux Pinning in Bulk Oriented-Grained YBa_2Cu_3Ox/Ag Composites: Effects of Ag and Y_2BaCuO_5 Inclusions

K. SALAMA, D.F. LEE, and X. CHAUD

Texas Center for Superconductivity at University of Houston, Houston, TX 77004, USA

ABSTRACT

Transport critical current density (J_{ct}) and its anisotropy in the presence of an applied magnetic field have been investigated in $YBa_2Cu_3O_x$/Ag composites with and without Y_2BaCuO_5 additions. J_{ct} is found to exhibit two characteristic peaks; a sharp J_{ct} peak centered on $B \perp c$ and a smaller and blunter peak centered on $B \parallel c$. Ag and Y_2BaCuO_5 inclusions have no apparent effect on the anisotropy, but they do influence J_{ct}. Small Y_2BaCuO_5 inclusions enhance J_{ct} while Ag has the opposite effect.

KEY WORDS: $YBa_2Cu_3O_x$/Ag composites, Y_2BaCuO_5 addition, melt-textured, transport anisotropy

INTRODUCTION

Most practical applications of high temperature superconductors (HTS) in their bulk form will utilize melt-textured materials due to their large current carrying capability. Since most HTS are anisotropic in nature, melt-textured HTS inevitably exhibit this anisotropy in their properties. Recently, transport critical current density (J_{ct}) of bulk melt-textured $YBa_2Cu_3O_x$ (123) [1] and thin films [2] have been shown to exhibit anisotropy in the presence of a magnetic field (B). Two peaks in J_{ct} were detected when the angle θ between the direction of applied magnetic field and c-axis is varied. One narrow J_{ct} peak is located at $B \perp c$, and is related to intrinsic pinning by the anisotropic crystal structure [3]. Another less commonly observed peak can be found at $B \parallel c$ and may be associated with suitably oriented extrinsic pinning centers. Even though the anisotropic behavior is qualitatively similar, J_{ct} of melt-textured 123 is two orders of magnitude lower than that of thin films. This suggests that some background pinning mechanisms other than those accounted for by the anisotropy are responsible for the difference in J_{ct}.

There has been numerous efforts to introduce different types of defects into the superconducting matrix to act as pinning sites for critical current density enhancement. These range from irradiation damage [4] to $BaSnO_3$ [5] addition and achieve various degree of success. As the J_{ct} is increased to a point where operation in high magnetic field is feasible, the mechanical integrity of the HTS becomes increasingly important due to the large Lorentz force the HTS will experience. It had be shown that ductile inclusions such as silver in the 123 matrix can improve the mechanical properties of the HTS [6]. By melt-texturing pre-sintered 123/Ag composites, we have manufactured oriented-grained 123/Ag composites with large J_{ct} and enhanced mechanical properties [7]. However, the anisotropic properties of these oriented-grained composites are as yet unknown. In this paper, we present the J_{ct} anisotropic behavior of these composites. In addition, the effects of Ag and Y_2BaCuO_5 (211) inclusions on J_{ct} and its anisotropy are examined.

EXPERIMENTAL

Specimens used in the present investigation were manufactured using the liquid-phase processing method developed by Salama et al. [8]. Stoichiometric 123 powder was prepared by solid-state reaction from Y_2O_3 (99.99%), $BaCO_3$ (99.9%) and CuO (99.9%). The calcined 123 powder was ball milled and ground with appropriate amount of Ag powder (99.9%) of 2 - 5 μm size and compacted into 123/Ag bars. In addition, 10 wt.% 211 powder of 2 - 5 μm size obtained through solid-state reaction were added to some composite powders to form 123/Ag/211 composites. The composite bars were sintered at 925°C for 24 hours followed by furnace cooling. These sintered composites have typical relative density of 90 to 92%.

The sintered composites were melt-processed in air as described in details elsewhere [9]. The specimens were rapidly heated to 1020 - 1075°C and held for 10 - 20 minutes. The temperature was then rapidly lowered to 970 - 1000°C where slow cooling commenced. The specimens were cooled at a rate of 1°C/hr to 900°C, and subsequently brought to room temperature at 60°C/hr. The melting temperature and temperature at which slow cooling started depend on Ag content.

Specimens used for J_{ct} measurements were single oriented grains cut from these composites, and exhibit typical stacked-plate morphology of melt-textured 123. These J_{ct} specimens were cut and polished with the aid of an optical microscope equipped with polarized light such that the a-b planes were parallel to the length of the specimen to within

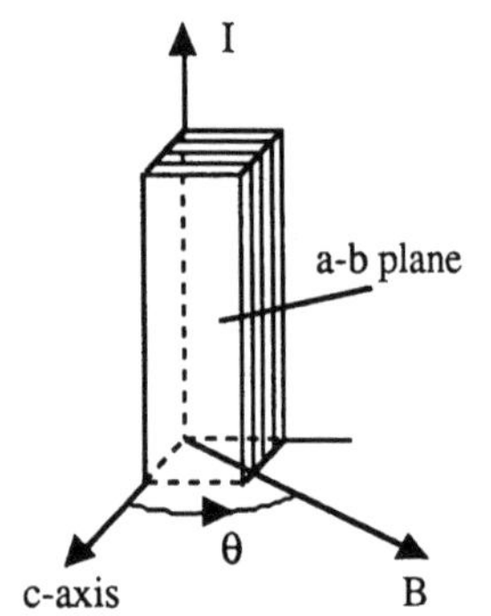

Fig. 1 Specimen - current - magnetic field configuration. Transport current flows in the a-b planes and is always perpendicular to applied field.

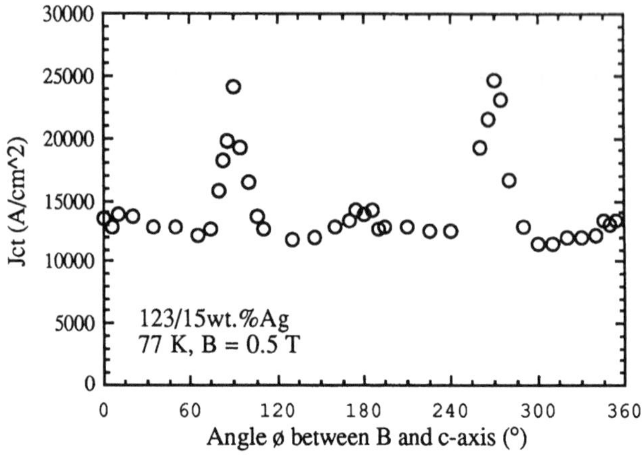

Fig. 2 Angular Dependence of J_{ct} in 123/15 wt.%Ag composite at 77K and 0.5 Tesla.

5° as shown in Fig. 1. These specimens were 5 to 7 mm long, and their cross section areas ranged from 0.2 x 0.2 to 0.3 x 0.5 mm^2. Such small areas were used in order to reduce contact heating that developed as a result of the large current needed to drive the specimens to their normal state. Silver leads were attached to the J_{ct} specimens using silver paste in a four probe configuration. The specimens together with the current leads were annealed in flowing oxygen at 500°C for 48 hours to allow oxygen uptake and reduction in contact resistance. The J_{ct} measurements were carried out at 77K by immersing the specimens in liquid nitrogen between the coils of an electromagnet. J_{ct} of the specimens in the presence of an applied field were determined using continuous d.c. current and a electric field criterion of 1 μV/cm.

RESULTS AND DISCUSSION

J_{ct} Anisotropy

Figure 2 shows a typical behavior of J_{ct} as a function of the angle θ between the applied magnetic field and the c-axis for a 123/15wt.%Ag specimen at 77K and B = 1.5 Tesla. It is found that a symmetric and strong angular dependence of J_{ct} appears with two characteristic peaks; an intense narrow peak centered on B ⊥ c, and a smaller and blunter peak at B || c. All of the specimens investigated show the same symmetric behavior throughout a complete revolution of θ = 360° and therefore in subsequent figures, J_{ct} variations of half a revolution will be shown for clarity. Variations in J_{ct} value of up to 10% may occur between different specimens with identical Ag addition. However, the anisotropic behavior of these specimens is qualitatively similar. The difference in J_{ct} values indicate the influence of sample quality, and in turn the importance of material processing.

The variations of J_{ct} with θ for a 123/25wt.%Ag composite at 77K under different applied fields is shown in Fig. 3. It is found that for B ⊥ c, an intense peak with a value roughly 3 to 4 times larger than the minimum J_{ct} appears. This narrow peak typically has a full width half max of 5° - 10°. Due to the rapid change in J_{ct}, it is possible that the value detected at B ⊥ c is less than the true J_{ct} because of limitation in angular resolution of 0.5°. The second characteristic peak is located at B || c. This peak is weaker than the B ⊥ c peak, and has a full width half max of 30° - 40°. In addition to the J_{ct} anisotropy, it can be seen from Fig. 3 that the J_{ct} at a given angle decreases with increasing magnetic field. This is expected because of the increase in the Lorentz force. Although J_{ct} decreases with B, the ratios between the peaks and the minimum J_{ct} remain constant.

The B ⊥ c peak has previously be discussed by Tachiki and Takahashi [3] in terms of intrinsic pinning by weakly superconducting layers between the Cu-O planes. The maximum pinning is in effect when B is parallel to the a-b planes and the flux lines are pinned along their entire length. If the direction of B deviates from this orientation, the flux lines will only intersect the weakly superconducting layers at discrete points, and the pinning effect will be drastically reduced. The narrow J_{ct} peak at B ⊥ c and the sharp drop in J_{ct} tend to support this intrinsic pinning model. Recent finding that the B ⊥ c peak broadens when the degree of a-b alignment is reduced [1] gives additional support to this model.

When B is parallel to the c-axis, the resultant Lorentz force lies in the a-b plane, and the intrinsic model cannot account for pinning in this configuration. Suitably oriented extrinsic defects such as stacking faults and twin boundaries [10]

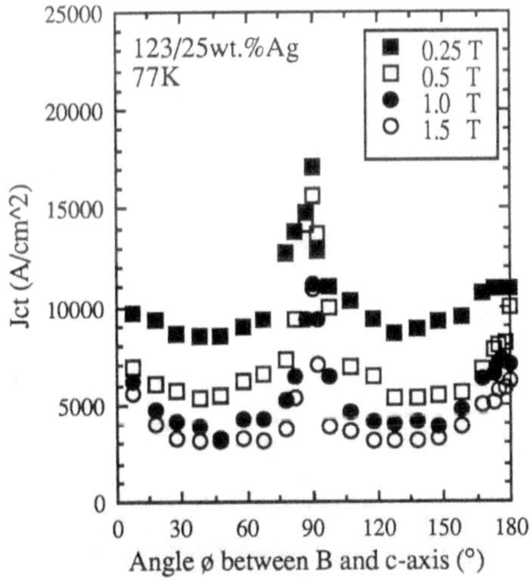

Fig. 3 Variation in J_{ct} as a function of θ in 123/25 wt.%Ag composite at 77K and different applied fields.

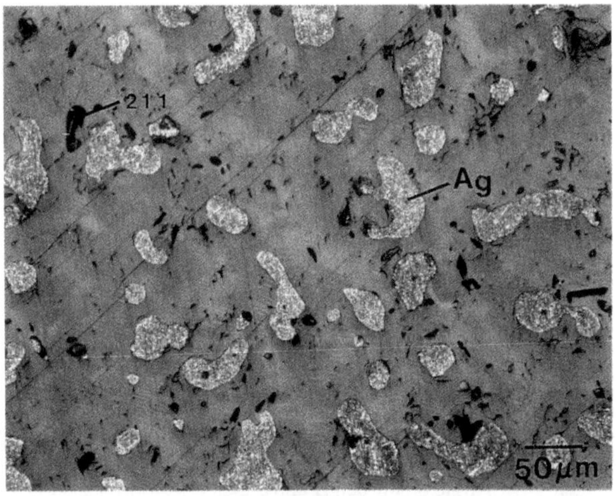

Fig.4 Microstructure of a melt-textured 123/25 wt.%Ag composite. The white particles are Ag inclusions and the dark ones are 211.

seem more likely to be responsible for the pinning. Recent evidence that the intensity of B ll c peak varies with twin spacing [1] suggests that the (110) type twin boundaries may be involved in this pinning process. Ideally, (110) type boundaries are parallel to the c-axis. In actuality, however, these twin boundaries are found to deviate from the c-axis, and the tilt angle is distributed in some region around the c-axis [3]. If this were the case, it will lead to a broad B ll c peak as we have found in our specimens. Even though the angular dependence of J_{ct} are similar in thin films and bulk melt-textured 123, there is still a 2 orders of magnitude difference in their J_{ct} values. This suggests that the pinning mechanisms involved in J_{ct} anisotropy are different from those responsible for the J_{ct} values.

Effects of Ag Inclusions

An optical micrograph of a 123/25wt.%Ag composite is shown in Fig. 4. The Ag inclusions range from 20 - 50 μm in size, and are randomly distributed throughout the 123 matrix. The coarsening of Ag particles is typical for 123/Ag composites processed at elevated temperatures. In addition, 3 vol.% 211 particles of 5 - 15 μm are randomly embedded within the 123 matrix due to incomplete peritectic recombination.

Angular dependence of J_{ct} is shown in Fig. 5 for specimens with 0, 15 and 25 wt.%Ag additions at 77K and 1.5 Tesla. The anisotropic behavior of J_{ct} in these specimens are similar. The B ll c peaks for the 0 and 15%Ag specimens are weak compare with the peak for 25 wt.%Ag. However, there is no evidence that this is due to the presence of Ag. In fact, a substantial B ll c peak was detected in 0%Ag specimens when the twin boundary spacing is reduced [1]. When the individual J_{ct} were normalized with respect to their zero applied field J_{ct} values, the normalized J_{ct} of specimens with different Ag content coincide showing that Ag inclusions have no effect on J_{ct} anisotropy. This is logical since the normal phase inclusions are randomly distributed and do not favor pinning in any particular field and crystallographic orientation.

It can be seen from Fig. 5 that the J_{ct} value at any given angle decreases with increasing Ag content. The size of Ag particles (20-50μm) and also the 211 (5-15μm) is much larger than the coherence length, and these inclusions should be ineffective as flux pinning centers. If the normal-phase inclusions do not participate in the pinning process, their influence will be in the reduction of the effective cross-section area of the superconductor. This will lead to a decrease in apparent J_{ct} as the Ag content is increased. Even though these large particles do not enhance J_{ct}, they are big enough to bridge the 10-15 μm thick 123 grains and aid in decreasing the amount of cracks along the a-b planes. This should improve the current carrying capability in the off-ab direction and the mechanical integrity of the HTS.

Effect of Y_2BaCuO_5 Inclusions

Since large normal-phase particles do not appear to contribute toward flux pinning, inclusions of micron-size were considered. 123 is notorious in its reactivity with most other materials. Therefore, 211 phase is a logical choice as additive to the stoichiometric mix since it has high decomposition temperature and it is one of the phases present in the partial melting stage. Fig. 6 shows the fracture surface of a 123/15wt.%Ag/10wt.%211 composite. The distribution and size of the Ag inclusions are similar to that of the 123/Ag composites. There are, however, two distinct types of

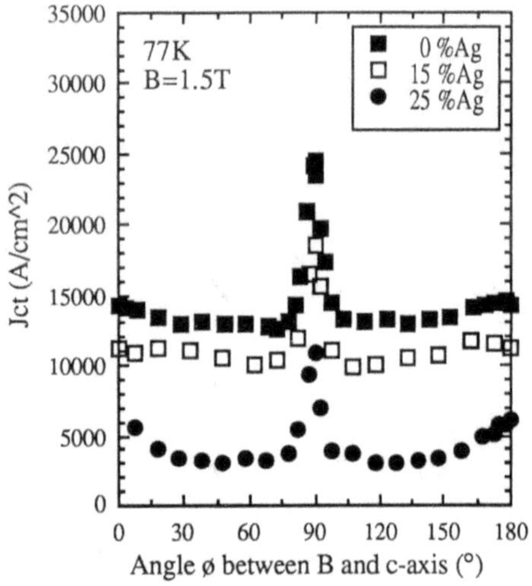

Fig. 5 Angular dependence of J_{ct} in 0, 15 and 25 wt.% 123 composites at 77K and 1.5 Tesla.

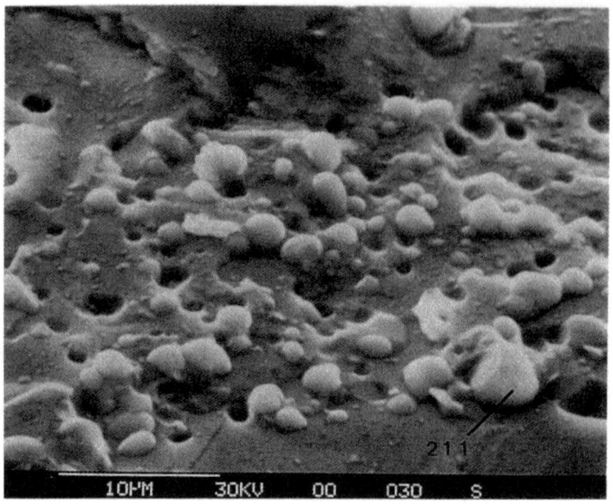

Fig. 6 SEM micrograph of the fracture surface of a 123/15 wt.%Ag composite. The 211 addition appears as spherical particles after melt-texturing.

211 inclusions. There is approximately 3 - 4 vol.% of rod shape 5 - 15 μm 211 due to incomplete peritectic recombination. Another 10 vol.% of spherical 211 of 0.5 - 2 μm size is believed to be the unreacted 211 addition.

Figures 7 and 8 show the angular dependence of J_{ct} at 0.5 and 1.5 Tesla for 123/15wt.%Ag and 123/25wt.Ag, both with and without 211 addition. From these figures, it is clear that the specimens with 211 addition has higher J_{ct} throughout the whole range of θ. As to the effect of micron-size 211 on J_{ct} anisotropy, Fig. 9 shows J_{ct} of 123/25wt.%Ag composites normalized with respect to individual minimum J_{ct} value at 77K and 0.5 Tesla. This behavior is typical for all specimens investigated at fields up to 1.5 Tesla. It is clear that the normalized J_{ct} values of composites with and without 211 addition have identical anisotropic behavior, and randomly distributed micron-size 211 inclusions have no effect on J_{ct} anisotropy.

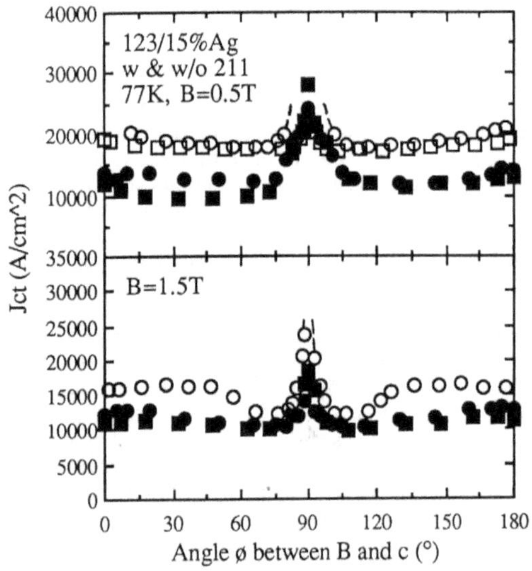

Fig. 7 Angular dependence of J_{ct} of 123/15 wt.%Ag and 123/15 wt.%Ag/10 wt.%211 composites at 0.5 and 1.5 Tesla (Closed symbol: no 211 addition. Open symbol: with 211 addition).

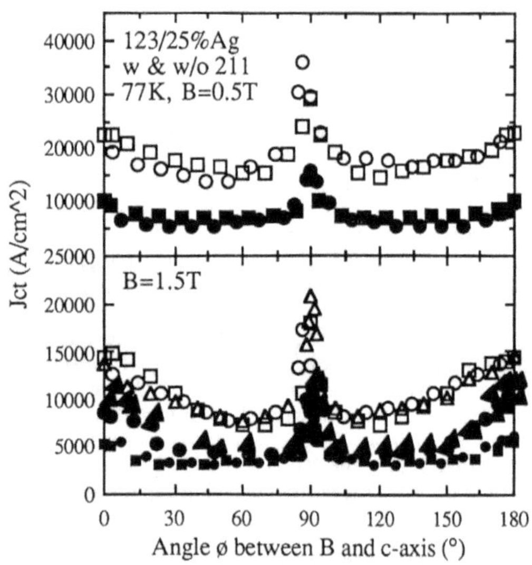

Fig. 8 Angular dependence of J_{ct} of 123/25 wt.%Ag and 123/25 wt.%Ag/10 wt.%211 composites at 0.5 and 1.5 Tesla (Closed symbol: no 211 addition. Open symbol: with 211 addition).

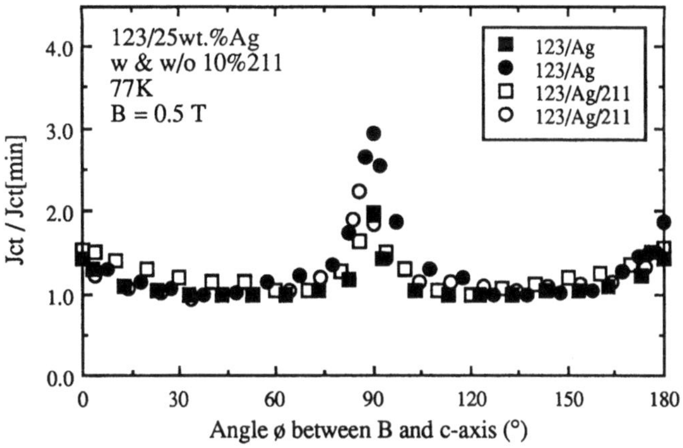

Fig. 9 Angular dependence of normalized J_{ct} for 123/25 wt.%Ag composites at 77K and 0.5 Tesla.

Although micron-size 211 inclusions do not influence J_{ct} anisotropy, they have dramatic effect on J_{ct} at low fields (B = 0.5T). The enhancement in J_{ct} appears to diminish at higher fields (B = 1.5T). The improvement in J_{ct} by small 211 inclusions observed in this study may be due to pinning by 211, or by the cleaner 123 grains with lesser amount of cracks. Murakami et al.[11] have suggested that micron-size 211 inclusions are effective pinning centers, and J_{cm} obtained from magnetization data increases with 211 content. On the other hand, Jin et al. [12] found no enhancement in J_{cm} by varying the 211 content. There is as yet no agreement on whether 211 enhances J_c, and further investigations are needed to establish the effectiveness of small 211 inclusions as pinning centers.

CONCLUSIONS

Strong angular dependence of J_{ct} under the influence of an applied magnetic field is found in oriented-grained 123/Ag composites. Two J_{ct} peaks are detected when the angle between applied field and crystallographic direction is varied. A sharp J_{ct} peak is found to center on $B \perp c$ and is approximately 3 to 4 times larger than the minimum J_{ct} value. Another shallower and blunter peak is found to locate at $B \parallel c$. Randomly distributed normal-phase inclusions such as Ag and 211 have no effect on the J_{ct} anisotropy. However, large Ag inclusions are ineffective as pinning centers and reduce the J_{ct}, whereas the addition of micron-size 211 enhances J_{ct}. An appropriate combination of these additions should result in HTS composites with large J_{ct} and enhanced mechanical properties.

ACKNOWLEDGEMENT

This work is supported by the Texas Center for Superconductivity at the University of Houston under Prime Grant MDA 972-88-G-0002 from the Defense Advanced Research Projects Agency and the State of Texas.

REFERENCES

1. Selvamanickam V, Forster K, Salama K (1991) Physica C 178: 147-157
2. Roas B, Schultz L, Saemann-Ischenko (1990) Phys. Rev. Lett. 64: 479-482
3. Tachiki M, Takahashi S (1989) Solid State Comm. 72: 1083-1086
4. Van Dover RB, Gyorgy EM, White AE, Schneemeyer LF, Felder RJ, Waszczak JV (1990) Appl. Phys. Lett. 56: 2681-2683
5. McGinn P, Chen W, Zhu N, Tan L, Varanasi C, Sengupta S (1991) Appl. Phys. Lett. 59: 120-122
6. Singh JP, Leu HJ, Poeppel RB, Van Voorhees E, Goudey GT, Winsley K, Shi D (1989) J. Appl. Phys. 66: 3154-3159
7. Lee DF, Salama K (1990) Jpn. J. Appl. Phys. 29: L2017-L2019
8. Salama K, Selvamanickam V, Gao L, Sun K (1989) Appl. Phys. Lett. 54: 2352-2354
9. Lee DF, Chaud X, Salama K (1991) Physica C 180: in print
10. Gyorgy EM, Van Dover RB, Schneemeyer LF, White AE, O'Bryan HM, Felder RJ, Waszczak JV, Rhodes WW (1990) Appl. Phys. Lett. 56: 2465-2467
11. Murakami M, Gotoh S, Fujimoto H, Yamaguchi K, Koshizuka N, Tanaka S (1991) Supercond. Sci. Technol. 4: S43-S50
12. Jin S, Kammlott GW, Tiefel TH, Kodas TT, Ward TL, Kroeger DM (1991) Physica C 180: in print

Superconducting Properties and Microstructure of $YBa_2Cu_3O_{7-\delta}$ / $PrBa_2Cu_3O_{7-\delta}$ Superlattices

DOUGLAS H. LOWNDES[1], DAVID P. NORTON[1], SHEN ZHU[2], and X.-Y. ZHENG[3]

[1] Solid State Division, Oak Ridge National Laboratory, Oak Ridge, TN 37831-6056, USA
[2] Department of Physics and Astronomy, The University of Tennessee, Knoxville, TN, USA
[3] Health and Safety Research Division, Oak Ridge National Laboratory, Oak Ridge, TN 37831-6056, USA

ABSTRACT

Epitaxial $YBa_2Cu_3O_{7-\delta}$/$PrBa_2Cu_3O_{7-\delta}$ (YBCO/PBCO) superlattices are tools for systematic, fundamental studies of high-temperature superconductivity. The variation of T_c in YBCO/PBCO superlattices can be understood as arising from changes in the interlayer phase coupling between YBCO layers that are highly two-dimensional when they are very thin (~1–2 *c*-axis unit cells) and completely isolated from each other. Single-cell-thick YBCO layers, containing isolated pairs of CuO_2 planes, are found to be superconducting at $T_c \sim 20$ K, in a PBCO matrix. The resistance in the superconducting transition region scales with temperature as expected for the (flux flow) resistance produced by thermally generated 2D vortices, or for a 2D array of superconducting weak links. Relative to both thin-film and single-crystal HTSc specimens, the thin superconducting layers in YBCO/PBCO superlattices exhibit a greatly expanded temperature range over which characteristic 2D dissipation can be observed, as a consequence of the enhanced anisotropy and reduced dimensionality of the YBCO layers. Scanning tunneling microscope studies reveal that YBCO films and YBCO/PBCO superlattices grow unit cell-by-unit cell by a terraced-island growth mechanism. On miscut, near-(001) substrates the terraces are epitaxially aligned with the substrate crystal lattice and spiral growth structures (screw dislocation-mediated growth) are not seen. These observations explain the steps or "kinks" that are seen in cross-section Z-contrast TEM images of YBCO/PBCO superlattices. The kinks may correspond physically to regions where the supercurrent must tunnel along the *c*-axis, and thus may be weak-link barriers.

KEY WORDS: superlattice, superconductivity, $YBa_2Cu_3O_{7-\delta}$, $PrBa_2Cu_3O_{7-\delta}$, two-dimensional, growth mechanism

INTRODUCTION

For $YBa_2Cu_3O_{7-\delta}$ (YBCO), the zero-temperature *c*-axis superconducting coherence length, $\xi_c(0) \sim 2$–4 Å, [1–3] is comparable to the ~3.9 Å spacing of the two CuO_2 planes within a unit cell and somewhat smaller than the $c \sim 11.7$ Å separation of successive CuO_2 bilayers in adjacent cells. This comparison indicates we should expect strong coupling of the two CuO_2 planes within the YBCO unit cell, as well as significant coupling of CuO_2 bilayers in adjacent *c*-axis unit cells. However, YBCO's effective anisotropy can be greatly increased by artificially separating thin YBCO layers by varying thicknesses of insulating materials, in epitaxial superlattice structures. Consequently, the *c*-axis coupling can be systematically and quasi-continuously varied, and the effect on high-temperature superconductivity studied, including determining the limiting superconducting behavior of isolated, single-cell-thick YBCO layers.

Recent work in several laboratories has demonstrated that high quality high-temperature superconductor (HTSc) superlattices can be fabricated by several different methods [4–12]. Superlattice research using 123-family materials has focused on epitaxial superconductor/insulator structures in which variable-thickness *c*-axis-perpendicular ($c_\perp$) layers of superconducting YBCO alternate with insulating $PrBa_2Cu_3O_{7-\delta}$ (PBCO) [4,6,7,10–12] or its alloys [9]. For example, we have used these YBCO/PBCO structures to show that for a fixed YBCO layer thickness (M = 1, 2, 3,... unit cells), the zero-resistance transition temperature (denoted here by T_c) initially decreases rapidly as the PBCO "barrier" layer thickness is increased, but for a large PBCO layer thickness T_c saturates at a non-zero value that is characteristic of the (fixed) YBCO layer thickness [4,10,11]. In particular, single-cell-thick YBCO layers, containing isolated pairs of CuO_2 planes, were found to be superconducting at $T_c \sim 20$ K in a PBCO matrix [4]. However, the existence of the CuO_2 bilayer structure made it unclear whether the superconductivity of isolated single-cell YBCO layers should be considered two- or three-dimensional (2D or 3D).

Rasolt, Edis, and Tesanovic recently suggested [13] that these YBCO/PBCO superlattices provide a 3D system in which the interlayer (*c*-axis) coupling can be weakened all the way to zero, as indicated by the saturation of T_c values that occurs for large PBCO thicknesses [4]. Thus, a crossover to 2D behavior *must* occur, accompanied by characteristic 2D dissipation [13]. Minnhagen and Olsson also find that the resistive transition data for superlattices containing the thinnest, best-isolated YBCO layers (one or two cells thick) are well explained by the 2D Ginzburg-

Landau Coulomb Gas model [14]. Finally, Ariosa and Beck have suggested [15] that each superconducting CuO_2 bilayer in a superlattice structure should be modeled as a 2D array of ultrasmall Josephson junctions. Thus, several recent papers suggest that reduced dimensionality plays an important role in the resistive behavior of these structures.

It is the purpose of this paper to provide a concise summary of analyses and microstructural studies that we and others have carried out recently, in order to test several models that have been proposed for the highly systematic behavior of T_c and of the resistive transitions in YBCO/PBCO superlattices, as functions of the layer thicknesses and separations. Particular emphasis is placed here on exploiting the variable anisotropy and resulting "tunable dimensionality" of superlattice structures, in order to reveal the physical mechanisms responsible for their behavior.

SUPERLATTICE RESISTIVE TRANSITIONS

Figure 1 displays the normalized resistance $R_n(T)$ = R(T)/R(100 K) for 1×*N* and 2×*N* YBCO/PBCO superlattices [4]. The 1x1, 1x2, and 1x4 superlattices have the same average composition as the 2x2, 2x4, and 2x8 structures, respectively, but have very different T_c values and resistive transition shapes. This establishes that the superconducting behavior is that of periodically layered structures, and not simply random alloys of some average composition. Figure 1 also reveals the importance of having YBCO in adjacent *c*-axis unit cells in order to raise T_c. The other striking feature of Fig. 1 is the broadening of the resistive transition that is so prominent for the thinnest YBCO layers.

Figure 2 (derived from data similar to that in Fig. 1) shows that for a fixed YBCO layer thickness, increasing the PBCO thickness produces an initial rapid decrease of T_c, *followed by saturation of* T_c at a nonzero value for large PBCO thicknesses, for all of the *M*×*N* structures studied [4]. The saturation values are $T_c \sim$ 19, 54, 71, 80, or 87 K for structures with well-isolated 1-, 2-, 3-, 4-, or 8-cell-thick YBCO layers, respectively. Thus, although intercell coupling increases T_c, this work clearly shows that *interlayer coupling is not necessary for superconductivity to occur* and that *single-cell-thick YBCO layers are superconducting* in a PBCO matrix [4].

In contrast to the behavior shown in Figs. 1 and 2, Kanai et al. [8] have used $Bi_2Sr_2(Ca_{1-x}Y_x)_1Cu_2O_8$-based superlattices, in which superconducting (x=0.15) and semiconducting (x=0.5) layers alternate, to show that there is almost no change in the midpoint transition temperature ($T_{c(mid)} \sim$ 65 K) with decreasing superlattice period. Taken together with our results for YBCO/PBCO superlattices, these results emphasize again that high-T_c superconductivity is an intrinsic property of CuO_2 bilayers, but that 3D interactions are present and significant for 123-family materials.

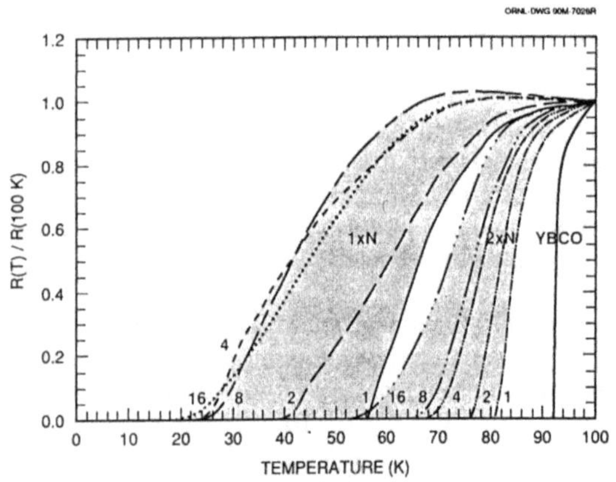

Fig. 1. Normalized resistance vs temperature for 1 × *N* and 2 × *N* superlattices.

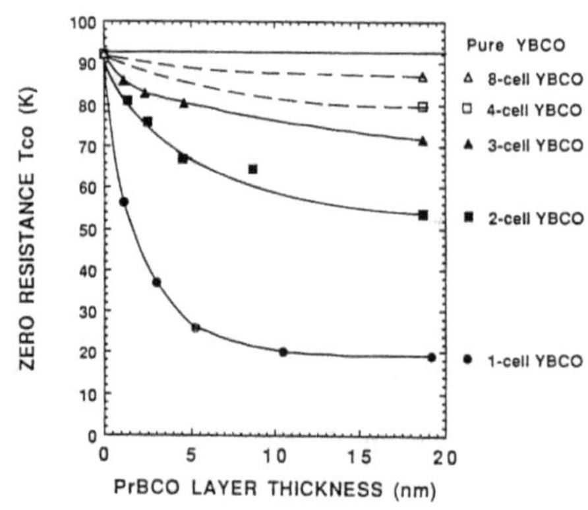

Fig. 2. T_c vs PBCO layer thickness for superlattices with 1- through 8-cell-thick YBCO layers.

EVIDENCE OF CHARACTERISTICALLY 2D DISSIPATION

YBCO/PBCO superlattices provide a 3D system in which the interlayer coupling can be weakened all the way to zero, as evidenced by the saturation of T_c values with increasing PBCO thickness (Fig. 2). Consequently, as Rasolt et al. [13] and Minnhagen and Olsson [14] have pointed out, a crossover to 2D resistive behavior *must* occur for sufficiently thin YBCO layers, as the PBCO thickness is increased. When crossover occurs, characteristic 2D dissipation processes should be observed, specifically resistance in the 2D superconducting state due to free vortices that are produced by thermal unbinding of vortex-antivortex pairs [16–19]. A low temperature Kosterlitz-Thouless (KT) transition also may occur, if a 3D superconducting transition caused by residual interplane coupling does not intervene first [14] (for intermediate PBCO thicknesses). We emphasize that this is most interesting behavior for crystalline epitaxial specimens, since most previous observations of characteristically 2D dissipation, in non-HTSc materials, have required using granular or amorphous thin films in the dirty limit [20–23].

We define T_{KT} as the *KT* transition temperature, above which thermal unbinding of vortex-antivortex pairs produces free vortices [16,17]. One of the characteristic "signatures" of a 2D conductor is that in the temperature range well below the mean-field (bulk, or BCS) superconducting transition temperature, T_{cB}, but above T_{KT}, the thermally generated vortices produce a nonzero resistivity with temperature dependence [18,19]

$$R(T)/R_N \sim \xi_+(T)^{-2} = \xi_{ab}(T_c)^{-2} \exp[-2b' (\tau_c/\tau)^{1/2}]. \tag{1}$$

In Eq. (1) R_N is the normal-state resistance, $\xi_+(T)$ is the phase correlation length for the superconducting order parameter [18,19], $\xi_{ab}(T)$ is the Ginzburg-Landau correlation length within the *a-b* plane, $\tau_c = (T_{cB} - T_{KT})/T_{KT}$, $\tau = (T - T_{KT})/T_{KT}$, and b' is a constant of order unity. From Eq. (1) it follows that log R(T) should scale as $\tau^{-1/2}$.

Figure 3 shows the result of plotting log $R_n(T)$ vs $\tau^{-1/2}$, where $R_n(T) = R(T)/R(100\ K)$, for the M(YBCO) × 16 (PBCO) superlattice structures [4,10]. These are the structures with the thickest PBCO layers, for which the saturation of T_c suggests that a crossover to 2D resistive behavior is required. Very good agreement with the expected scaling behavior is observed, and from these plots we extract least-squares best-fit values for the *KT* transition temperature of T_{KT} = 14, 44, 70, and 86.5 K, for the structures with 1-, 2-, 3-, and 8-cell-thick YBCO layers, respectively.

Moreover, examination of the resistive behavior of a wide range of other superlattices, for which the barrier layers are not so thick, shows that 2D fluctuations are quite generally important in these structures. For 2- and 3-cell-thick YBCO layers, good straight-line behavior is obtained in plots of log $R_n(T)$ vs $\tau^{-1/2}$, until the PBCO barriers become

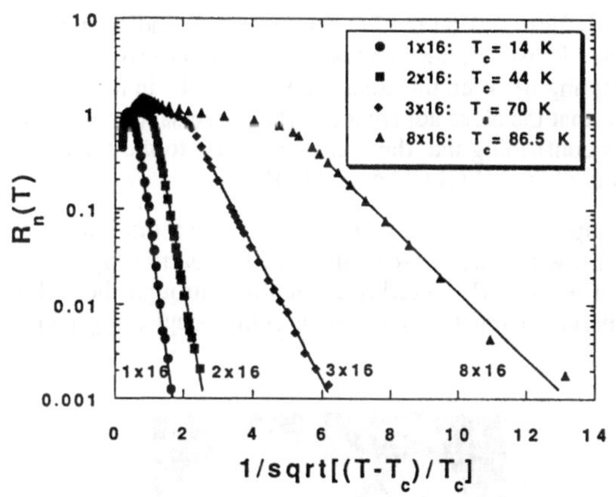

Figure 3. Log $R_n(T)$ vs $\tau^{-1/2}$ for $M \times 16$ superlattices.

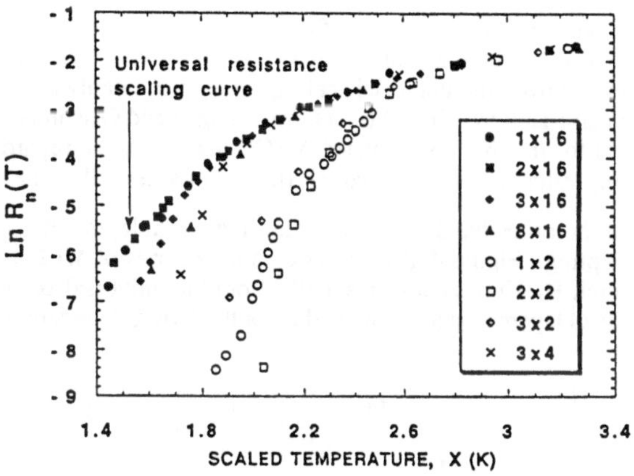

Fig. 4. Logarithm of the normalized resistance vs X(T) for selected YBCO/PBCO superlattices. The X(T) values are calculated from Eq. (3), using fitted values of T_{co} and T_{KT}.

very thin ($N \sim 2$) [4,10]. The best-fit T_{KT} values rise again rapidly toward $T_{cB} \sim 92$ K for a thick YBCO film, either as the YBCO superconducting layers are made thicker or as the PBCO barriers are made thinner, corresponding to a reduced temperature range for 2D behavior, as *c*-axis interlayer coupling is introduced [10].

PURE 2D BEHAVIOR AND 2D–TO–3D CROSSOVER DUE TO INTERLAYER COUPLING

At this point we wish to show the deviations from pure 2D behavior that occur as interlayer coupling is restored between initially isolated YBCO layers (anisotropy reduced). Minnhagen and Olsson have shown how the onset of 3D behavior can be displayed by using the 2D Coulomb gas scaling concept and the 2D Ginzburg-Landau Coulomb Gas (GLCG) model to analyze the superlattice data [14]. The key point, due to Minnhagen [23,24] is that the GLCG model describes the vortex fluctuations in a 2D conductor in terms of a temperature scaling variable, X(T). Consequently, the normalized resistance $R_n(T)$ for all 2D superconductors that are correctly described by the GLCG model should fall on a single, "universal" curve when plotted versus the scaling variable *X* [14,24]. The form of this scaling curve (see Fig. 6) has been well-established from data for type-II superconductor films and also has been related to the GLCG model through Monte Carlo simulations [14].

The scaling variable X(T) is defined by

$$X(T) = [\, T / n^{2D}(T) \,] / [\, T_{KT} / n^{2D}(T_{KT}) \,] \,, \qquad (2)$$

where n^{2D} is the areal density of supercurrent carriers for the 2D superconductor (not renormalized with respect to vortex fluctuations) [14] and T_{KT} is the Kosterlitz-Thouless transition temperature. Minnhagen and Olsson assume a standard Ginzburg-Landau (GL) description for the temperature-dependence of n^{2D} in the absence of vortex fluctuations, so that $n^{2D}(T) \propto T_{co} - T$, where T_{co} is the GL temperature ($\sim T_{cB}$). Consequently, the scaling variable X(T) becomes

$$X(T) = [\, T / (T_{co} - T) \,] / [\, T_{KT} / (T_{co} - T_{KT}) \,] \,, \qquad (3)$$

where T_{co} and T_{KT} are used as fitting parameters in the $R_n(T)$ data analysis.

According to Minnhagen and Olsson, T_{co} should be regarded as the temperature at which the superconducting transition would have occurred, if the superconductor was well described by the GL model and there were no vortex fluctuations [14]. T_{KT} is the temperature at which the *KT* phase transition (the onset of vortex-antivortex unbinding) would take place [20, 23]. However, they also point out that layered materials in general (and superlattices in particular) will usually have a 3D phase transition that is caused by residual interplane coupling at a temperature $T_c > T_{KT}$, and that below T_c there is true long-range (3D) order. In summary, T_{co} can be thought of as parameterizing the temperature-dependence of n^{2D}, T_c is the true (experimental) 3D critical temperature, and T_{KT} is the temperature at which the *KT* transition would have occurred (for zero interplane coupling) if 3D ordering had not intervened. Thus, as Minnhagen and Olsson have emphasized, [14] the universal resistance scaling curve (Fig. 4) does not reflect any critical phase transition properties of the vortex fluctuations. It describes only non-critical properties, outside of the *KT* transition region. Nevertheless, T_{KT} and T_{co} can be estimated by fitting the $R_n(T)$ data in this region. The $R_n(T)$ data then can be plotted versus the scaling variable X, calculated now from the fitted values of T_{co} and T_{KT}, using Eq. (3).

As shown in Fig. 4, the $R_n(T)$ data for the 1×16 and 2×16 superlattices fall on the 2D scaling curve, indicating that isolated 1-and 2-cell-thick YBCO layers behave as 2D conductors. However, the data for the 3×16 and 8×16 structures deviate from the universal scaling curve, presumably due to coupling between the adjacent YBCO cells in the thicker but still-isolated YBCO layers. Minnhagen and Olsson point out that the close adherence to the universal curve by the data for the thinnest, isolated YBCO layers, can be regarded as confirming that the resistance close to the transition temperature is due to 2D vortex fluctuations, as well as the validity of the 2D GLCG model [14].

On the other hand, we can exploit the variable anisotropy of superlattices to deliberately strengthen the interlayer coupling. Figure 4 also shows data for a series of M×2 structures, with only two-cell-thick PBCO barrier layers, for which $R_n(T)$ deviates drastically from the universal scaling curve. Clearly, interlayer coupling through these thin PBCO layers is significant, and all of the YBCO layers in the superlattice must be regarded as being coupled together.

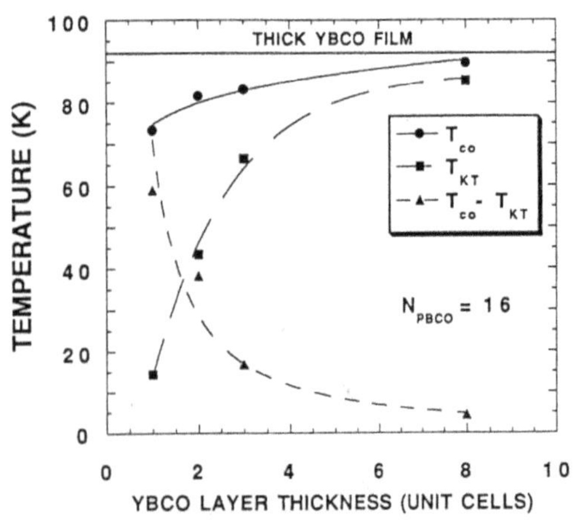

Fig. 5. Transition temperatures and widths vs YBCO layer thickness, for superlattices with well-isolated YBCO layers (N = 16).

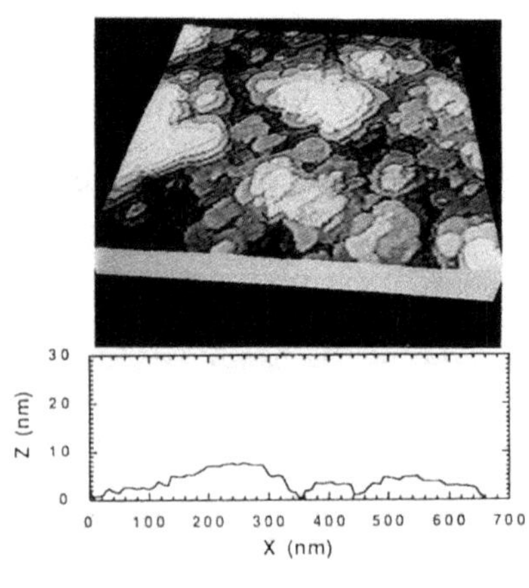

Fig. 6. [Top] STM image of a ~200 nm thick, c⊥ YBCO epitaxial film grown on (001) $SrTiO_3$ at T_{heater} ~730°C. The dimensions of the region shown are 670 × 670 nm^2. [Bottom] Line-scan surface-height profile of the film.

Figure 5 summarizes the effect of 2D excitations in broadening the superconducting transitions of the thinnest YBCO layers. Here we plot the values of T_{KT} (representing the onset temperature for 2D dissipation) and T_{co} (the 3D Ginzburg-Landau temperature) for superlattices with isolated YBCO layers (N = 16 unit cells) of variable thickness (M = 1 to 8). T_{KT} rises rapidly toward a thick YBCO film's $T_c \sim 92$ K, and the transition broadening, $\Delta T_c = T_c - T_{KT}$, which is so prominent for the thinnest YBCO layers and is due to the 2D excitations, nearly vanishes for a YBCO layer only 8-unit-cells thick. This figure illustrates YBCO's transition from a highly anisotropic 2D conductor to a less anisotropic 3D conductor, over a thickness range of only a few unit cells.

MICROSTRUCTURE AND GROWTH MECHANISM

We recently used scanning tunneling microscopy [25] (STM) and cross-sectional Z-contrast scanning transmission electron microscopy [26] (Z-STEM) to study both near-surface and sub-surface microstructures in "123" films and YBCO/PBCO superlattices. From these studies we were able to determine their growth mechanism.

Our STM images show that $c_\perp$ YBCO films deposited on accurately cut (001) substrates grow by an island growth mechanism. The islands consist of stacks of terraces with step heights that are multiples of the c-axis lattice parameter [25]. Figure 6 shows a STM image of the surface of a $c_\perp$ YBCO film grown at a heater temperature of 730°C on (001) $SrTiO_3$. Well-defined islands are present, each island consisting of a stack of terraces. Variations of this terraced-island morphology are common to all of the $c_\perp$ films that we have studied on a variety of (001) substrates, including $LaAlO_3$, $KTaO_3$, MgO and Y_2O_3-stabilized ZrO_2.

However, we find that the surface microstructure and growth mechanism of YBCO films is extremely sensitive to the substrate miscut angle [27]. In particular, the spiral growth structures [25, 28, 29] that are so prominent (see Fig. 6) in YBCO films grown on accurately-cut (001) substrates, can be completely eliminated by growing YBCO on a substrate that is miscut by only 2 – 3 degrees away from (001). Figure 7 shows the striking effect that substrate miscut angle, θ_m, has in modifying the granularity and growth mechanism of a nominally 16-cell (188 Å) thick YBCO film grown at 720°C on a $LaAlO_3$ substrate with $\theta_m = 2.7°$. The film consists of overlapping tilted platelets; a line scan profile of the surface shows that each platelet is exactly one c-axis unit cell (~ 11.7 Å) thick. The exposed terrace widths of successive platelets are ~30 nm. The tilt angle of the one-cell-thick platelets, determined directly from Fig. 7, is 2.5° (±0.6°), showing that the platelets are epitaxially aligned (within measurement error) with the underlaying $LaAlO_3$ crystal lattice.

There is no sign in Fig 7 of the spiral growth structures that are commonly seen in YBCO films grown on accurately-cut (001) substrates [28, 29]. Instead, another type of terraced, unit cell-by-unit cell growth occurs. The rapid growth direction for YBCO films is perpendicular to the c-axis, i.e., material is most easily added at the rapid-growth a-c and b-c faces of the unit cell. Fig. 7 shows that on a substrate with sufficiently large miscut angle, YBCO apparently takes advantage of the high density of surface steps to nucleate and align the rapid-growth a-c and b-c crystal faces; the platelets then rapidly grow out atop one another in the a-b directions. For $\theta_m = 2.7°$, the average spacing between

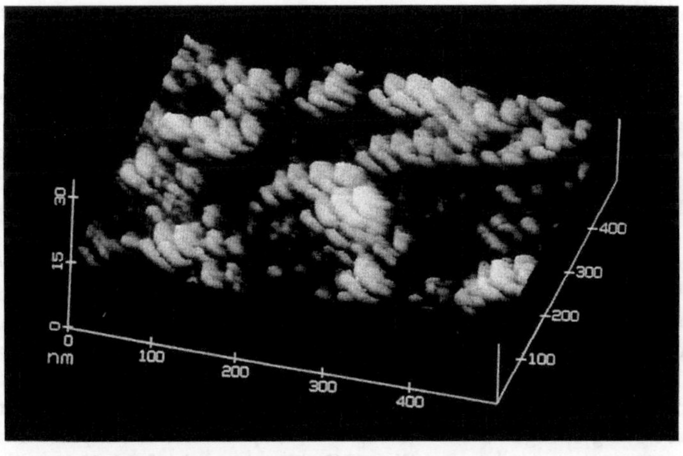

Fig. 7. STM image of surface microstructure of anominally 16-cell-thick YBCO film grown at 720°C on a miscut $LaAlO_3$ substrate with $\theta_m = 2.7°$.

successive 0.39 nm tall $SrTiO_3$ unit cells is ~8.3 nm, corresponding to an average minimum horizontal separation of ~25 nm between successive single-cell-thick YBCO layers. This minimum separation is in good agreement with the actual ~28 nm widths of the exposed terraces at the early stage of growth shown in Fig. 7. A significantly miscut surface simultaneously exposes a large number of nucleation and rapid-growth sites, from which simultaneous growth of many terraces can occur, closely analogous to the lateral growth that can occur from a stack of upward-spiraling terraces. Thus, both the spiral-mediated and the miscut growth mechanisms result in grains that are stacks of terraces one unit cell in height, and both are rapid-growth mechanisms, in that numerous terrace-edge sites are continuously made available for growth.

The terraced structure in these images clearly shows that YBCO islands grow unit cell-by-unit cell. Terashima et al. [30] also concluded that YBCO films grow unit cell-by-unit cell, based on their observations of RHEED oscillations during YBCO growth on (001) $SrTiO_3$ by reactive evaporation. They used precise x-ray measurements of film thickness to show that each RHEED oscillation corresponded to a layer thickness increment of one c-axis lattice constant (~1.17 nm). However, they interpreted the RHEED oscillations as being due to nucleation of 2D islands and to their growth in flat layers in a cyclic fashion. Norton et al. [25] have pointed out that the strictly layer-by-layer description proposed by Terashima et al. is not completely consistent with the granular, variable-height surface that STM reveals for ~200 nm-thick films. The resolution of this apparent difference probably is that Terashima et al. reported RHEED oscillations only during the initial stages of epitaxy, and at a low substrate temperature (~ 680°C). Terashima et al. also have pointed out that unit cell-by-unit cell growth implies that for YBCO the minimum structural unit that can satisfy the requirements of stoichiometry and charge neutrality is the full unit cell [30].

The individual terraces on a particular island/grain are flat on nearly the atomic scale. However, the terraced-island growth mechanism leads naturally to a gradual "roughening" of the growing surface (gentle undulations of surface height or film thickness), as one moves from the center to the edge of an island/grain that has coalesced with an adjacent island/grain. The height variation is a function both of film thickness and substrate temperature. For ~200 nm-thick films, the surface "roughness" is ~10 nm for films grown at a heater temperature of 730°C, and ~30 nm at 780°C [25].

CONSEQUENCES OF MICROSTRUCTURE: WEAK-LINK BEHAVIOR?

The surface microstructures observed by STM for $c_\perp$ YBCO films may have important implications for the transport properties of YBCO/PBCO superlattices, because both the terraced steps and larger-scale undulations on the growing surface should be translated into the individual YBCO layers in a superlattice structure. If we imagine a PBCO layer with terraced steps on it at the time when the laser ablation targets are exchanged in order to grow the next YBCO layer, then it is clear that the individual YBCO layers in superlattices will not be flat but will have steps or "kinks" in them wherever they cross the ends of PBCO terraces: The crystalline continuity of the superlattice is unaffected, but the chemical contents of a particular one-unit-cell crystalline layer will change from Y to Pr. We have observed these steps in the YBCO layers of superlattices, using cross-section Z-STEM [10, 11, 26]. The average lateral distance, L_s, between the steps (or kinks) kinks in cross-section Z-STEM images was $L_s \sim 20$ nm, consistent with the average separation of 15 ± 5 nm between successive terrace steps, observed via STM for films grown at a heater temperature of 730°C.

The STM images also indicate that the individual YBCO layers in superlattices should not be expected to be flat, but will undulate up and down by ~10–30 nm (depending on growth temperature) over lateral distances corresponding to a grain/island diameter (~100–500 nm, also depending on growth temperature); the undulation amplitude is largest for the YBCO layers closest to the film surface. These undulations also can be seen clearly in cross-section Z-STEM images (e.g., Fig. 1 of Ref. 26).

The c-axis superconducting coherence length, $\xi_c \sim 2$–4 Å [1–3], is much smaller than $c \sim 11.7$ Å. Consequently, the existence of "kinks" in the chemical continuity of thin YBCO layers means that the supercurrent must tunnel at least one cell along the c-direction for every distance L_s that it travels laterally. For single-cell-high kinks in an M-cell-thick YBCO layer, this should have no serious effect on the critical current density, J_c, or transition temperature, T_c, unless $M = 1$ or 2, because the average "overlap" of the two adjacent pieces of the YBCO layer at the kink will be M-1 cells. However, for a 1×N superlattice, the average overlap of the YBCO layer at the kink is zero, and its continuity then depends entirely on there being enough "chance" areas of overlap in different in-plane directions to produce a continuous percolative path, or else on c-axis tunneling [26]. Consequently, for 1×N structures, one expects clear weak-link behavior, including a significant reduction of J_c (which we observed [26]) and a depression of T_c (as described above), due to the disruption of the phase of the superconducting order parameter at these weak-link kinks.

Based on these STM observations of surface microstructures of YBCO films, together with Z-STEM images of YBCO/PBCO superlattices, it appears that any theory of superlattice transport properties must include the effects of the "kinks" in the YBCO layers. A very thin (~ single-cell) YBCO layer probably should be regarded as an array of superconducting regions interconnected by weak-link barriers at the "kinks". Lobb, Abraham, and Tinkham have considered in detail how a vortex-unbinding phase transition would affect various properties of a square lattice of superconducting weak links [31]. They found that the free vortex density is the dominant temperature-dependent factor in the expression for the weak-link array's resistance. Their resulting expressions for the temperature-dependent phase correlation length and resistance are very similar to Eq. (1), except for an additional temperature-dependent factor inside the exponential. This additional factor comes from the temperature-dependent coupling energy, $E_J(T)$, between adjacent superconducting islands [31], which causes the resistance to be expressed in terms of a dimensionless temperature parameter, $k_BT/E_J(T)$. However, Lobb et al. also point out that if $T_c << T_{cB}$ then $E_J(T) \sim E_J(0) \sim$ constant, so that the expressions for $R(T)$ become entirely equivalent for fitting purposes [31].

Thus, the characteristically 2D behavior that we observe in YBCO/PBCO superlattices appears to be entirely consistent both with STM and Z-STEM identification of actual defects in the layers, on the one hand, and with the expected behavior of superconducting layers that are composed of an array of Josephson junctions, on the other. We note that Ariosa and Beck recently suggested that each conducting CuO_2 bilayer in a YBCO/PBCO superlattice should be modeled as a 2D array of ultrasmall in-plane Josephson-coupled junctions [15]. Their model is able to fit the observed variation of T_c with YBCO and PBCO layer thicknesses. However, they did not suggest any particular defect that would correspond physically to the Josephson barrier, nor did they consider the important effect of the barriers on the shape of the resistive transition.

ACKNOWLEDGMENT

This research was sponsored by the Division of Materials Sciences, and by the Superconducting Technology Program for Electric Energy Systems, Advanced Utility Concepts Division, Conservation and Renewable Energy, U.S. Department of Energy under contract DE-AC05-84OR21400 with Martin Marietta Energy Systems, Inc.

REFERENCES

1. A. G. Aronov, S. Hikami, and A. I. Larkin, (1989) *Phys. Rev. Lett.* **62**: 965
2. U. Welp, W. K. Kwok, G. W. Crabtree, K. G. Vandervoort, and J. Z. Liu, (1989) *Phys. Rev. Lett.* **62**: 1908
3. Y. Matsuda, T. Hirai, S. Komiyama, T. Terashima, Y. Bando, K. Iijima, K. Yamamoto, and K. Hirata, (1989) *Phys. Rev. B***40**: 5176
4. D. H. Lowndes, D. P. Norton, and J. D. Budai, (1990) *Phys. Rev. Lett.* **65**: 1160
5. J.-M. Triscone, M. G. Karkut, L. Antognazza, O. Brunner, and Ø. Fischer, (1989) *Phys. Rev. Lett.* **63**: 1016
6. J.-M. Triscone, Ø. Fischer, O. Brunner, L. Antognazza, A. D. Kent, and M. G. Karkut, (1990) *Phys. Rev. Lett.* **64**: 804
7. Q. Li, X. X. Xi, X. D. Wu, A. Inam, S. Vadlamannati, W. L. McLean, T. Venkatesan, R. Ramesh, D. M. Hwang, J. A. Martinez, and L. Nazar, (1990) *Phys. Rev. Lett.* **64**: 3086
8. M. Kanai, T. Kawai, and S. Kawai, (1990) *Appl. Phys. Lett.* **57**: 198
9. D. P. Norton, D. H. Lowndes, S. J. Pennycook, and J. D. Budai, (1991) *Phys. Rev. Lett.* **67**: 1358
10. D. H. Lowndes and D. P. Norton, "Kosterlitz-Thouless-Like Behavior Over Extended Ranges of Temperature and Layer Thickness in Crystalline $YBa_2Cu_3O_{7-\delta}/PrBa_2Cu_3O_{7-\delta}$ Superlattices," 1991 (in press), In: *University of Miami Workshop on Electronic Structure and Mechanisms for High-Temperature Superconductivity*, Plenum Publishing Co.
11. D. H. Lowndes, D. P. Norton, J. D. Budai, D. K. Christen, C. E. Klabunde, R. J. Warmack, and S. J. Pennycook, (1990) In: R. Singh, J. Narayan, and D. T. Shaw, (ed) in *Progress in High-Temperature Superconducting Transistors and Other Devices*, (1990) *SPIE Proceedings* **1394**: 150
12. Ø. Fischer, J. M. Triscone, L. Antognazza, O. Brunner, A. D. Kent, L. Mieville, and M. G. Karkut, (1990) *J. Less Common Metals* **164–165**: 257
13. M. Rasolt, T. Edis, and Z. Tesanovic, (1991) *Phys. Rev. Lett.* **66**: 2927
14. P. Minnhagen and P. Olsson, (1991) *Physical Review Letters* (to be published)
15. D. Ariosa and H. Beck, (1991) *Phys. Rev. B* **43**: 344
16. J. M. Kosterlitz and D. J. Thouless, *J. Phys. C* **6**, 1181 (1972); J. M. Kosterlitz and D. J. Thouless, (1978) *Prog. Low Temp. Phys.* **7**: 373
17. J. M. Kosterlitz, (1974) *J. Phys. C* **7**: 1046.
18. B. I. Halperin and D. R. Nelson, (1979) *J. Low Temp. Phys.* **36**: 599
19. V. Ambegaokar, B. I. Halperin, D. R. Nelson, and E. D. Siggia, (1980) *Phys. Rev. B* **21**: 1806
20. M. R. Beasley, J. E. Mooij, and T. P. Orlando, (1979) *Phys. Rev. Lett.* **42**: 1165
21. J. E. Mooij, (1983) In: B. Deaver and J. Ruvalds (ed) in *Advances in Superconductivity*, Plenum Press, New York, p. 433

22. J. E. Mooij, (1984) In: A. M. Goldman and S. A. Wolf (ed) in *Percolation, Localization, and Superconductivity*, Plenum Press, New York, p. 325
23. P. Minnhagen, (1987) *Rev. Mod. Phys.* **59**: 1001
24. P. Minnhagen, (1981) *Phys. Rev. B* **24**: 6758
25. D. P. Norton, D. H. Lowndes, X. -Y. Zheng, Shen Zhu, and R. J. Warmack, (1 November 1991) *Phys. Rev. B: Rapid Commun.* **44**
26. S. J. Pennycook, M. F. Chisholm, D. E. Jesson, D. P. Norton, D. H. Lowndes, R. Feenstra, and H. R. Kerchner, (1991) *Phys. Rev. Lett.* **67**: 765.
27. D. H. Lowndes, X. -Y. Zheng, Shen Zhu, J. D. Budai, and D. P Norton (in preparation)
28. M. Hawley, I. D. Raistrick, J. G. Beery, and R. J. Houlton, (1991) *Science* **251**: 1587
29. C. Gerber, D. Anselmetti, J. D. Bednorz, J. Mannhart, and D. G. Scholm, (1991) *Nautre* **350**: 279
30. T. Terashima, Y. Bando, et al., (1990) *Phys. Rev. Lett.* **65**: 2684 .
31. C. J. Lobb, D. W. Abraham, and M. Tinkham, (1983) *Phys. Rev. B* **27**: 150.

Superconducting Magnetic Energy Storage (SMES) Application in Japan

EISUKE MASADA

Department of Electrical Engineering, The University of Tokyo, Bunkyo-ku, Tokyo, 113 Japan

ABSTRACT

The applicability of the superconducting magnetic energy storage in Japan is reviewed mainly on the basis of the study carried out in ISTEC. Three types of SMES, the small, medium and large scale, are defined for the study. For each of them, the requirements for the technological developments and their difficulties are evaluated. The benefits obtained by their introduction are compared with alternative equipments existed. The planning of its development is discussed for the practical application in the electric power system.

KEY WORDS: superconducting magnetic energy storage, power system

INTRODUCTION

For the society depending strongly on electricity and electronic equipments as in Japan, to secure high reliability of electrical power supply to it, as well as to utilize the primary energy as efficient as possible against the variable demand, the energy storage facilities are required essentially in the modern power system. As the solution to this problem, the battery station and the fuel cell plant have been developed by electric power industries.

The superconducting magnetic energy storage (SMES) has been also much interested for such purpose. Experimental studies have been carried out with model superconducting magnets in many universities, research institutes, utility companies and manufacturers. Table 1 summarizes recent activities of utilities in this field. Its practical applicability to the power system has been also evaluated by various organizations, as well as applications at the demand end such as steel industries and railways.

Table 1. Model study of SMES in electric utilities

	SMES				converter	research project
	type	capacity	voltage/current	material		
Tohoku Electric	solenoid	1MJ	25V / 1000A	NbTi	GTO current fed	enthalpy protection scheme
Chubu Electric	"	1MJ	50V / 1000A	NbTi	GTO current fed	technical feasibility
Kansai Electric	troid (6 coils)	2.5MJ	400V / 350A	NbTi NbsSn	voltage fed + chopper	"
Kyushu Electric	solenoid	0.1MJ	230V / 328A	NbTi	GTO current fed	stability test in actual system
Elec. Power Development Co.	"	10KJ	100V / 52.3A	NbTi	GTO current fed	test of basic characteristics
CRIEPI	"	25KJ	500V / 100A	NbTi	GTO current fed	controllability

With the rapid development of the high temperature (HTc) superconductors, International Superconductivity Technology Center (ISTEC) has carried out a feasibility study on electric power apparatus including superconducting magnetic energy storage for three years (1988 - 1990). The project is supported by Agency of Resources and Energy, Ministry of International Trade and Industry (MITI). The applicability of SMES is considered as the most important subject in the study. In the followings, the result of this study related to the applicability of SMES is summarized with the author's own view on that in industries.

OVERVIEW OF ISTEC-SMES STUDY

The outline of the feasibility study of SMES is shown in Fig. 1 as a flow graph between subjects studied. At first existing technologies and the state of developments related to SMES are reviewed and evaluated. Then the applicability of SMES to the power system is analyzed. Its benefits are assessed with conceptional design of SMES and the computer simulation of its effects. As the goal of the study, it is required to conclude with a proposal of the development program of SMES for the practical application in the power system in the next stage.

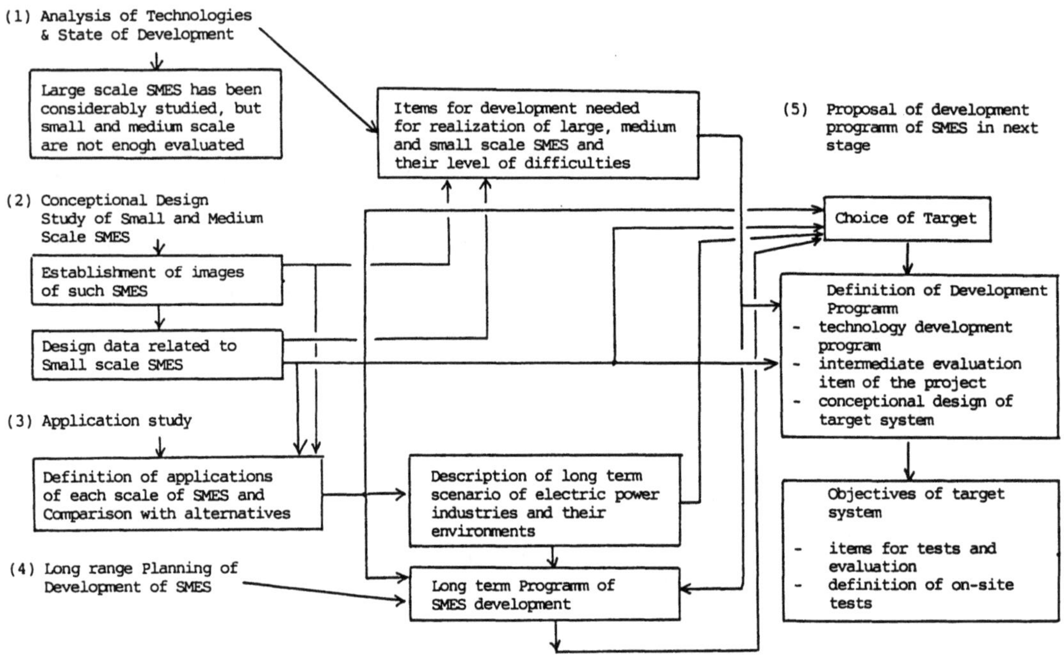

Fig. 1. Outline of the SMES study in ISTEC

The study is carried out by a committee, which consists of specialists from academic institutions, utility companies and manufacturers.

The difference of our study from precedent ones is that it stresses the applicability of the small and medium scale SMES. The definition of the scale is considered as follows on the basis of its energy storage capacity.

large scale SMES	1000 MWh class
medium scale SMES	10 MWh class
small scale SMES	0.1 MWh class

After our analysis of technologies needed for the realization of SMES, it is concluded that there exist so many research and development items to be completed for the medium and large scale SMES, that the development program should be started with the small scale SMES and extended to enlarge the capacity step by step. The feasibility study and conceptional design of such SMES has not been done so much. Then the study is decided to focus on the detailed analysis in this area.

ANALYSIS OF APPLICABILITY OF SMES IN POWER SYSTEM

The scale of electric power system is steadily growing along with the increase in power demands, while the influences of lords on the power quality at the demand end becomes serious. Effective use of energy is still more sought from social and environmental view points. These have given rise to versatile diff--icult problems in the power system. The superconducting technology is expec-ted to provide a break through to solve them. Especially SMES is considered to offer a whole new function other than storage of electrical energy, which allows both real and reactive power to be freely regulated.

General functions and benefits of SMES to the power system are:
As for storage (KWh),
(1) high energy storage efficiency
(2) fast energy charge/discharge capability
(3) independent controllability of real and reactive power during charge and discharge process
As for power quality controller (KW),
(1) load levelling
(2) compensation for load fluctuation
(3) improvement of power system stability
(4) frequency regulation
(5) countermeasures against instantaneous voltage drop
(6) countermeasures against blackout
(7) storage of natural energy
Another feature of SMES is that these functions could be achieved not only by the large scale SMES but also by the distributed arrangement of small and medium scale SMES.

The applicability of each type and arrangement of SMES on these functions are compared with competitive technologies as Table 2. The results show versatile possibilities of applications of small and medium scale SMES in distributed arrangement.

Table 2. General applicability of SMES in power system

Name	Type of installation		Usable capacity			Application								Competitive technology	Remarks
	Arrange-ment	Location	Stored energy (Mwh)	Scale	Output (MW)	Response time level	Load levelling	Compen-sation for load fluctuation	Improve-ment of stability	Frequency regulation	Counter-measure against instanta-neous voltage drop	Counter-measure against blackout	Storage of natural energy		
SMES for large-scale electric power storage	Concent-ric	In the trunk transmission system	5000 class	Large	1000 class	Hour	◎			○		○		Pumped hydro power generator, Compressed air energy storage	Location on rock bed
Distributed configuration type, multi-purpose SMES	Separate	In substation	10 class	Medium	10 to 100 class	Hour ~ Min. ~ Sec.	◎	◎	△	○	△	○		Secondary Cell, Com-pressed air energy storage	Superspeed Maglev transport load, etc.
		At load end	0.1 class	Small	10 to 20 class			◎			△	○	△	Flywheel, Secondary battery	Steel plant load, etc.
SMES for power system stabilization	Separate	In power plant	0.1 class	Small	100 to 500 class.	Sec.			◎		△			System damping resistor, Static var compensator (SVC), Series capacitor compensation, Power system stabilizer (PSS), Early valve actuation	

Note) ◎:Primary purpose of use ○: Effective △: Effective conditionally (location, time period, etc.)

APPLICABILITY STUDY OF SMALL SCALE SMES

From the preliminary study of the application of SMES shown in Table 2, it is clear that the small scale SMES with high power capacity for the distributed arrangement can contribute to solve the problems which the power system is facing. It is also useful to verify the benefits of the medium scale SMES. These points are analyzed in details with the conceptional design of several types of small scale SMES [1] and the computer simulation.

Power System Stabilization

In order to evaluate the benefit of SMES and compare it with ones obtained by existing means, the model system of Fig. 2 is chosen as the objective of the analysis. For the control system of both SMES and existing equipments, the typical configuration and control parameters are assumed on the basis of expe-rimental studies and practical experiences. The difference in the installation capacity required of SMES and existing equipments, which gives the equivalent

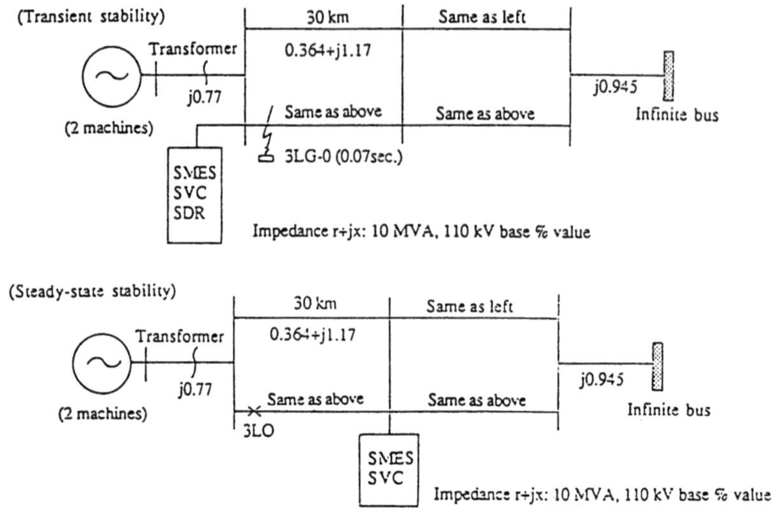

Fig. 2. Model system for system stabilization

effect of stability improvement in the simulation, is considered as the measure to evaluate the benefit of SMES. The results are shown in Table 3 on the index of the stabilty defined in the neighbouring figure to the table.

Table 3. Results of simulation study on system stability

(Transient stability)

Case	Measures for stability	Results of study	
		Capacity (MVA)	Δ δ (degree)
1	No measures	-	31.4
2	SMES	20	22.3
3	Super high-response excitation (with PSS)	-	18.9
4	SVC	50	22.1
5	SDR	120	22.8

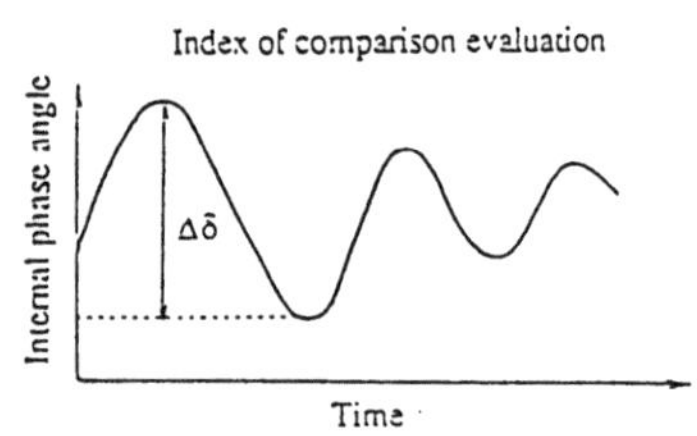

(Steady-state stability)

Case	Measures for stability	Results of study	
		Capacity (MVA)	Is
6	No measures	-	0.056
7	SMES	10	0.030
8	Super high-response excitation (with PSS)	-	0.033
9	SVC	30	0.031

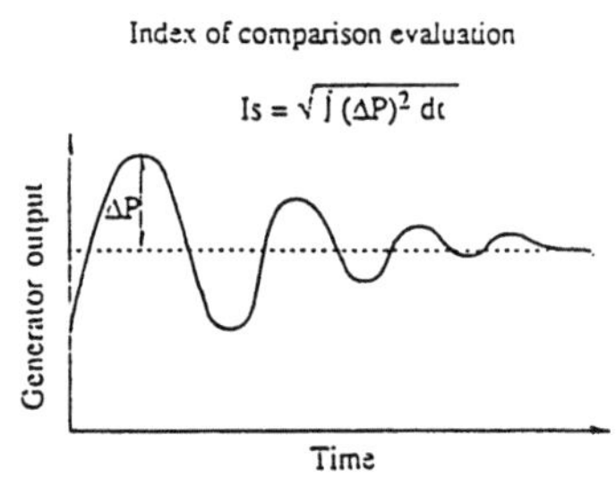

As shown in the table, in order to realize with the static VAR compensator (SVC) the same effect on the transient and steady state stability improvement attained by SMES, a capacity of 2.5 to 3 times that of SMES is needed. In the case of the system damping resistor (SDR), nearly 6 times greater capacity is required. The benefit of SMES is also clearly shown in comparison with the case without such measures. It is equivalent to the case that the generator is equipped with the super high response excitation combined with the power system stabilizer (PSS).

Load Fluctuation Compensation

The benefit of SMES is also studied for the compensation of voltage fluctuation and line power flow fluctuation caused by loads in the iron and steel making process, which are connected to the end of the secondary transmission system, for the comparison with the existing measures as the precedent section. The case of the weak power system is assumed to evaluate the suppression of the frequency fluctuation against the same disturbances. The synchronous phase modifier, SVC and flyweel. The simulated results are evaluated on the voltage and frequency fluctuation of the system agaist 20MW load fluctuation. The qualitative evaluation of the controlled response of these fluctuations is also done from the observation of the simulated transients.

From the result of simulation, the benefit of SMES is clearly verified both for the voltage and frequency fluctuation caused by load variation. Both SMES and other equipments existed are required to install the same capacity to suppress the stady state voltage fluctuation. SMES and SVC are confirmed to be able to respond to disturbance faster than the synchronous phase modifier. SMES and flywheel can only compensate the active power flow fluctuation in the system. SMES can suppress the frequency fluctuation enough even for the weak power system.

Marginal Cost for Multi-purpose SMES

As shown in Table 2, the application of SMES to the power system is versatile. However, the analysis of the required characteristic of SMES for each case shows that it would be enough for the evaluation of the small scale SMES with the distributed arrangement, if we study the cases stated above only, since other applications are included into them on its behaviour. Then, using the obtained results, the marginal cost of such multi-purpose SMES is proposed and evaluated. This proposal is such that; assuming that each function expected from the introduction of SMES can be realized by the alternative existing technologies, their capacity and cost required for such realization are calculated and summed up to determine the marginal cost with the weighting parameters, which are supposed in consideration with the realizability of functions for each purpose of applications based on the technical aspects peculiar to SMES and its operation. This marginal cost depends on the existing technology and considered to be easily understandable for users of SMES.

As the result of the trial computation, the marginal cost comes up to 300,000 ¥/kW or there about, if the multi-purpose application of the small scale SMES with the distributed arrangement is supposed.

APPLICATIONS IN INDUSTRIES

SMES is also applicable to the suppression of the loads fluctuation in the system of customers by themselves. It can make the disturbances in the power system in the customer's plant small enough to secure its own operation. The location of the plant is not effected with the restriction of the power supply system with such installation. The cost needed for the power quality may give some additional benefits for the introduction of SMES to the industrial plants. Of course the scale of the plant should be large enough to install the expensive SMES by itself.

Two possible applications are the railway and steel industry. The high speed guided transport in future should require much power comparing with railways today. For example the maglev transport connecting Tokyo and Osaka in the 21st centry is considered to cause 100MW power fluctuation to the connected network at the worst condition. It restricts the connecting points of the system to the power grid and needs the long connecting line between them. Its costs and the improvement of the operational characteristics of the system might rationalize the introduction of SMES into substations. |2| Another possibility is in smoothing the power variation in mills of the steel plant. Introduction of highly qualified production process and addition of flexibility to it causes frequent load variation in the plant and needs higher quality of power supply. SMES can negotiate the load fluctuation very fast as shown in Fig. 3 and contributes to suppress disturbances to the process. Both cases in future would use the superconducting technology in itself, then it is not so difficult to use SMES in itself. |3|

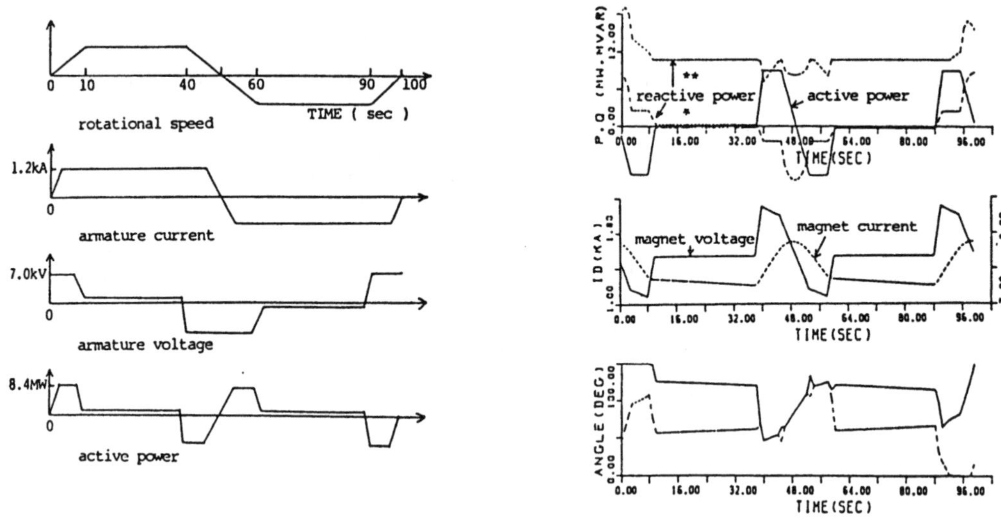

(a) Operational pattern of a large reversible DC mill drives

(b) A simulation of compensation against the load variation modelled in (a)
* reactive power in load
** total reactive power

Fig.3 simulation of load compensation for steel mill with SMES |4|

CONCLUSION

As the conclusion of the study in ISTEC, such scenario is proposed; the development of SMES for the power system is necessary and it should start from the small scale system step by step. The first stage for the establishment of the small scale SMES applicable to the power system would take about a decade, since the system requires well developed enough to be compatible with its matuality. Based on this study, the study on the components technology for the practical developments has started from the fiscal year 1991.

References

1. ISTEC Report on Feasibility Study on Superconducting Electric Power Apparatus in 1990, March 1991.
2. Masada E, et. al., Proc IPEC Tokyo Conf., 612-617, 1991.
3. ISTEC Report on Feasibility Study in Application Fields on Superconductivity
4. Masada E, et. al., PESC'88 Record, 342-350, 1988.

2 Physics

2.1 Theoretical Studies

Present Status of the Theories on High T_c Superconductors

S. MAEKAWA

Department of Applied Physics, Nagoya University, Nagoya, 464-01 Japan

ABSTRACT

Present status of the theories on high T_c superconductors is reviewed with particular emphasis on the electronic structure. It is shown that the reconstruction of the electronic structure occurs near the charge transfer gap upon doping of carriers and a new state emerges in the gap. We propose that this state provides the stage of high T_c superconductivity.

1. INTRODUCTION

The discovery of high T_c superconducting cuprates has posed many important problems to the theory of solid state physics. The parent compounds are antiferromagnetic insulators. There exists the small charge transfer gap between copper and oxygen ions which is dependent on the electrostatic Madelung potential. The antiferromagnetism is caused by the strong electron correlation on copper ions.

The crystal structure of the cuprates is unique. There exists CuO_2 plane as a structural unit. It is known that the CuO_2 plane carries the superconductivity. The superconductivity is induced by doped carriers in the plane. Let us consider La_2CuO_4 as an example. When La^{3+} ions are replaced with Sr^{2+} ions, hole-carriers are introduced into the CuO_2 plane. In this case, the doped holes are almost free from the potential of the substituted ions since the ions are not in the CuO_2 plane. We note that although holes are introduced into oxygen sites in the plane, the magnetism on copper sites is sensitively dependent on the doping. Therefore, the electronic structure of the CuO_2 plane with carriers is derived by the subtle balance between the electron correlation and the crystal structure.

Here, the present status of the theories on high T_c superconductors is reviewed with particular emphasis on the electronic structure. The electronic structure of the doped CuO_2 plane is examined in the cluster model approach. It is shown that the so-called gap state emerging in the charge transfer gap upon doping of carriers is caused by the reconstruction of the electronic state near the gap. We propose that the gap state provides the stage of the superconductivity.

2. ELECTRONIC STRUCTURE

The model Hamiltonian suitable for describing the electronic structure of the CuO_2 plane is the three-band Hubbard Hamiltonian,

$$H = -T_{pd}\sum_{mls}(d^\dagger_{ms}p_{ls} + h.c.) - T_{pp}\sum_{ll's}p^\dagger_{ls}p_{l's}$$
$$+U_d\sum_m n^d_{m\uparrow}n^d_{m\downarrow} + U_p\sum_l n^p_{l\uparrow}n^p_{l\downarrow} + \varepsilon_d\sum_{ms}n^d_{ms} + \varepsilon_p\sum_{ls}n^p_{ls},$$

where the operators $d^\dagger_{is}$ and $p^\dagger_{is}$ create holes in Cu3$d(x^2-y^2)$ and O2p_σ orbitals with spin s at the i site, respectively, and $n^d_{ms}=d^\dagger_{ms}d_{ms}$ and $n^p_{ls}=p^\dagger_{ls}p_{ls}$. The m, l and l' summations are over the orbitals at the neighboring sites. T_{pd} and T_{pp} are the hopping parameters between the neighboring $2p_\sigma$ and $3d(x^2-y^2)$ orbitals and between the neighboring $2p_\sigma$ orbitals, respectively. U_d and U_p denote the intra-orbital Coulomb repulsive interactions on copper and oxygen ions, respectively. ε_p and ε_d are the energy-levels of O2p_σ and Cu3$d(x^2-y^2)$ orbitals, respectively, and the charge transfer (CT) energy Δ is defined as $\Delta \equiv \varepsilon_p - \varepsilon_d$. Among these parameters, T_{pd}, T_{pp} and Δ are assumed to depend on materials: T_{pd} and T_{pp} have the bond length dependence of $d^{-4}_{\mathrm{Cu-O}}$ and $d^{-3}_{\mathrm{O-O}}$, respectively, with $d_{\mathrm{Cu-O}}$ and $d_{\mathrm{O-O}}$ being the bond lengths between copper and oxygen and between oxygen ions, and Δ is determined from the difference in Madelung site potentials for a hole between copper and in-plane oxygen [1]. For example, the values of parameters for Nd_2CuO_4 are taken to be $T_{pd}=0.95$ eV, $T_{pp}=0.40$ eV, $\Delta=2.39$ eV, U_d=8.5 eV and U_p=4.1 eV.

Let us examine a Cu_4O_{13} cluster with the free boundary condition, where four CuO_4 clusters are linked linearly by sharing corner oxygens [2]. The clusters with 3 , 4 and 5 holes corespond to 25 % electoron-doped system, insulator and 25 % hole-doped system, respectively. Note that these doped systems are close to those with maximum T_c.

Figure 1 shows the one-particle excitation spectra in the Cu_4O_{13} cluster, where one-hole additional and one-hole removal spectra correspond to photoemission (PES) and inverse photoemission (BIS) spectra, respectively [3]. In the figure, the vertical dotted lines show the position of the Fermi level. The ground state in the insulating state is mainly composed of configuration having one hole at each copper site so that we symbolize the state d^9 (Fig. 1(a)). There exist two final states represented by d^8 and $d^9\underline{L}$ ($\underline{L}$: O2p hole). The former appears at ~ -11 eV. The latter makes two large peaks located at ~ -0.5 eV and ~ -3 eV. Here, the state at ~ -0.5 eV is the split off state from O2p main band at ~ -3 eV, and is mainly composed of the linear combination of the so-called Zhang-Rice (ZR) singlet state [4] . Therefore, we call the split off state the ZR singlet band. This band is caused by the exchange interaction between Cu3$d(x^2-y^2)$ hole and O2p_σ hole. The final state of BIS at ~ 0.5 eV is mainly composed of the linear combination of d^{10} configuration so that we call the state the upper Hubbard (UH) band. The Figures 1(b) and 1(c) show the one-particle excitation spectra in the 25% hole-doped and 25% electron-doped systems, respectively. We note that the Fermi level moves to the ZR singlet band and the UH band upon doping of hole and electron carriers, respectively. As a result, new states denoted by $d^9\underline{L} \rightarrow d^9$ and $d^{10} \rightarrow d^9$ appear in BIS and PES, respectively. The shift of E_F is not a simple rigid band shift, but the reconstruction of the spectral weights occurs in both the ZR singlet and UH bands [3,5-7] .

In Fig. 2, white circles and white squares show the O1s threshold energies of core-level electron-energy-loss spectroscopy (EELS) for insulators and hole-doped samples, respec-

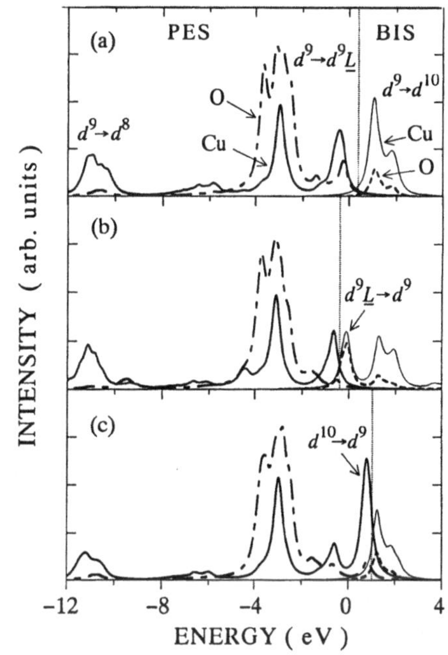

Fig. 1. PES and BIS spectra calculated in the Cu_4O_{13} cluster. (a) insulator, (b) 25% hole-doped system, and (c) 25% electron-doped system. The solid and dot-dashed lines denote $Cu3d(x^2-y^2)$ and $O(P)2p_\sigma$ PES, and the thin-solid and dashed lines denote $Cu3d(x^2-y^2)$ and $O(P)2p_\sigma$ BIS spectra. The spectra of copper and oxygen in the middle of the cluster were calculated in order to avoid the boundary effect. The spectra with the Lorentzian convolution with the half-width γ=0.2 eV are shown. The vertical dotted lines indicate the Fermi level defined as the center of the first ionization and first affinity states.

tively [8] . Also shown by solid circles and solid squares are the theoretical values obtained in the Cu_4O_{13} cluster for insulators and 25% hole-doped systems, respectively [3] . Note that the theoretical values give the peak positions, whereas the threshold values in experiments give the energies of the raising edges at half height. The threshold energy gives the transition of electrons from O1s level to unoccupied O2p orbitals. In the insulators, the transition from O1s level to $Cu3d(x^2-y^2)$ orbital is seen since the latter orbital is hybridized by $O2p_\sigma$ orbitals. We note that the threshold energy depends on materials as the CT gap does. This is because of the difference of the Madelung potentials between oxygen and cupper sites in the CuO_2 plane. When holes are introduced into the insulators, the gap states appear so that the transition of electron from O1s level to the states becomes possible, resulting in the shift of the threshold energy. The similar shift of the energy upon hole-doping has been observed in the soft x-ray absorption experiments (XAS) [9] . Also shown in Fig. 2 are the threshold energy of EELS in insulating and electron-doped Nd compounds. In the compounds, there is little shift of the energy upon electron-doping. This is consistent with theory.

Optical conductivity also changes dramatically upon carrier-doping. Figure 3(a), (b) and (c) show the conductivity calculated in the cluster Cu_4O_{13} in the cases of insulating, 25% hole-doped and 25% electron-doped systems, respectively [2] . When holes are introduced into the insulator, much of the spectral weight of the CT peak is transferred

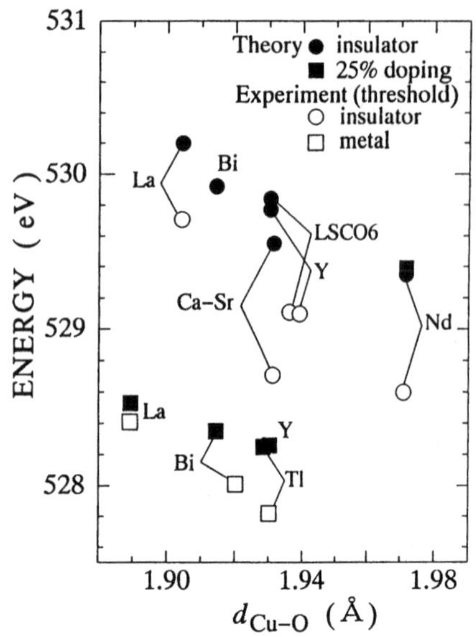

Fig. 2. O1s threshold energy for various undoped and carrier-doped compounds [3]. The white circles and white squares denote the experimental results in undoped and carrier-doped samples, respectively [8]. The solid circles and solid squares denote the theoretical results calculated in the Cu_4O_{13} cluster in undoped and 25% carrier-doped systems, respectively. La: $La_{2-x}Sr_xCuO_4$, Bi: $Bi_2Sr_2(Ca\text{-}Y)Cu_2O_8$, Ca-Sr: $(Ca\text{-}Sr)CuO_2$, Y: $YBa_2Cu_3O_{6+\delta}$, LSCO6: $La_2SrCu_2O_{6.2}$, Tl: $Tl_2Ba_2CaCu_2O_8$, Nd: $Nd_{2-x}Ce_xCuO_4$.

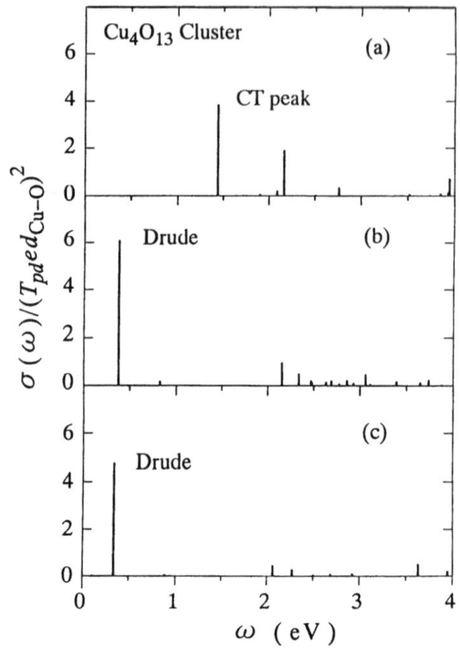

Fig. 3. Optical conductivity calculated in the Cu_4O_{13} cluster . (a) insulator, (b) 25% hole-doped system, and (c) 25% electron-doped system. In (a), a peak at ω=1.4 eV gives the CT gap. In (b) and (c), peaks at ω=0.4 eV correspond to the Drude ones [2].

to the peak at ω=0.4 eV, which corresponds to the Drude one. Consulting the one-particle spectral density in Fig. 1, we may understand that such transfer of the spectral weight of the optical conductivity implies that holes localized on copper sites in the undoped state contribute to the formation to the Drude peak together with doped-holes in the CuO_2 plane. The effects of the electron-doping on the conductivity are very similar to those of the hole-doping. The similarity of the conductivity between electron- and hole-doped systems has been observed experimentally [10].

We have found that the electron- and hole-dopings bring about the formation of the gap states in the low and high energy sides in the CT gap, respectively. This theoretical result is in accord with EELS and XAS experiments. On the other hand, it has been shown in PES experiments that the gap state emerges near the center of the CT gap in both electron- and hole-doped systems [11,12]. At the present stage, it is not clear why the position of the state in the latter experiments is different from the theoretical results.

3. DISCUSSION

Search for the mechanism of high T_c superconductivity in the cuprates has attracted much attention not only because T_c is high but because the cuprates are strongly correlated electron systems so that they contain many essential problems in solid state physics. There exists a variety of the high T_c superconductors. Therefore, it is possible to study the condition of the superconductivity through the examination of the characteristics of materials.

We have shown that the novel electronic properties are brought by the gap state which is given by reconstructing the electronic state near the CT gap upon doping of carriers. The state is well defined in the moderately doped systems and will be sweared out in the over-doped systems. Therefore, we may consider that the gap state causes the superconductivity. However, the mutual interaction between the carriers in the gap state which will bring about the superconductivity is still a controversial question. The gap state is based on the electrostatic Madelung potential and the antiferromagnetic interaction as well as the unique crystal structure. Therefore, in addition to the lattice distortion, spin and charge distortions around the carriers are being discussed to be candidates of the origin of the interaction. Although the conclusion has not been reached yet, much study is now in progress[13,14].

ACKNOWLEDGEMENTS

I am grateful to T. Tohyama for his collaboration and discussion for preparing this paper. This work has been supported by a Grant-in-Aid for Scientific Research on Priority-Areas from the Ministry of Education, Science and Culture of Japan.

REFERENCES

[1] Y. Ohta, T. Tohyama, and S. Maekawa, Phys. Rev. Lett. **66** (1991) 1228.
[2] T. Tohyama and S. Maekawa, J. Phys. Soc. Jpn. **60** 53 (1991).
[3] T. Tohyama and S. Maekawa, to be published in Physica C (1991).
[4] F. C. Zhang and T. M. Rice, Phys. Rev. B **37** (1988) 3759.
[5] W. Stephan and P. Horsch, *Dynamics of Magnetic Fiuctuations in High Temperature Superconductors*, ed. G. Reiter, P. Horsch and G. Psaltakis (Plenum, New York, 1990).
[6] H. Eskes, M. B. J. Meinders and G. A. Sawatzky, Phys. Rev. Lett. **67** (1991) 1035.
[7] E. Dagotto, A. Moreo, F, Ortolani, J. Riera and D. J. Scalapino, Phys. Rev. Lett. **67** (1991) 1918.
[8] M. Alexander *et al.*, Phys. Rev. B **43** (1991) 333.
[9] C. T. Chen *et al.*, Phys. Rev. Lett. **66** (1991) 104.
[10] S. Uchida, T. Ido, H. Takagi, T. Arima, Y. Tokura and S. Tajima, Phys. Rev. B **43** (1991) 7942.
[11] H. Namatame *et al.*, Phys. Rev. B **41** (1990) 7205.
[12] J. W. Allen *et al.*, Phys. Rev. Lett. **64** (1990) 595.
[13] See, for example, *Storong Correlation and Superconductivity*, ed. H. Fukuyama, S. Maekawa and A. P. Malozemoff (Springer Series in Solid State Sciences **89** (1989))
[14] See, for example, *Physics of High Temperature Superconductors*, ed. S. Maekawa and M. Sato (Springer Series in Solid State Sciences, to be published)

Superconducting Phase Diagram of the Two-Band Model Investigated by the Exact Diagonalization Method

K. YAMAJI, K. HARIGAYA, and T. YANAGISAWA

Electrotechnical Laboratory, Tsukuba, Ibaraki, 305 Japan

ABSTRACT

We have studied the finite-size two-band model with the interband pair-transfer interaction by calculating the superconducting (SC) correlation function P_s etc. in the ground state by means of the Lanczos diagonalization method. When the second band approaches the Fermi level lying in the first band, the increase of P_s clearly indicates occurrence of SC. The diagram of the SC phase is obtained in a parameter space. Although calculation is limited up to 7 sites and 6 electrons, the SC phase is wide enough to justify serious application of the model to the cuprates. Assuming that SC occurs by the present mechanism in a conducting channel made of the O $2p_\sigma$ orbitals, we qualitatively explain the observed systematics on T_c with the Madelung potential differences and with the doped hole density.

KEY WORDS: two-band model, interband pair-transfer interaction, diagonalization method, superconducting pair-correlation function, phase diagram

INTRODUCTION

The two-band model [1,2] is known to promote superconductivity (SC) through the interband pair-transfer interaction arising from the off-diagonal element, called exchange-like integral, of the Coulomb interaction [2]. This effect is asserted to be greatly magnified by the interband ladder process [3] and to bring about the high T_c in cuprate superconductors [4], in which there are two possible sets of two-band situations; firstly two bands mainly consisting of the in-plane O $2p_\sigma$ and $2p_\pi$ orbitals, respectively, [5,6] and secondly two bands mainly consisting of Cu $d_{x^2-y^2}$ and d_{z^2} orbitals, respectively [7]. In the first case the coupling constant K of the interband pair-transfer interaction is $K \cong 0.88$ eV and in the second case $K \cong 1.09$ eV [8]. The cuprates, however, have considerable magnitudes of on-site Coulomb energies for both O 2p and Cu 3d orbitals, which should work against SC promoted by the present mechanism.

This study has two aims: first to demonstrate in an unambiguous way that the two-band mechanism of SC works and secondly to find up to what value of on-site Coulomb energies the SC is sustained. We have investigated the ground state properties of the model by the Lanczos diagonalization method [9]. The obtained SC correlation function clearly indicates occurrence of SC. Although with increase of the system size the SC region is gradually expanded, the available size is too restricted for extrapolation to infinity. However, the results already reveal that SC occurs in a reasonably wide parameter range, justifying serious application of the model to the cuprates. The interband on-site Coulomb energy denoted below by U' was found to enhance SC in accord with the result of the interband ladder process, although quantitative comparison is not possible. Assuming that SC in the cuprates occurs by the present mechanism in the conducting network webbed by the O $2p_\sigma$ orbitals in the CuO_2 plane, we qualitatively explain the systematics known on T_c in relation with the Madelung potential differences ΔV_M [10] and ΔV_A [11] and with the doped hole density [12].

MODEL

Our model Hamiltonian is defined by

$$H = \sum_{l,\sigma} \left[t(a_{l\sigma}^\dagger a_{l+1\sigma} + \mathrm{H.c.}) + t'(c_{l\sigma}^\dagger c_{l+1\sigma} + \mathrm{H.c.}) + E_2 c_{l\sigma}^\dagger c_{l\sigma} \right] + K \sum_l (a_{l\uparrow}^\dagger a_{l\downarrow}^\dagger c_{l\downarrow} c_{l\uparrow} + \mathrm{H.c.})$$
$$+ \sum_l (U_1 a_{l\uparrow}^\dagger a_{l\uparrow} a_{l\downarrow}^\dagger a_{l\downarrow} + U_2 c_{l\uparrow}^\dagger c_{l\uparrow} c_{l\downarrow}^\dagger c_{l\downarrow} + U' \sum_{\sigma,\sigma'} a_{l\sigma}^\dagger a_{l\sigma} c_{l\sigma'}^\dagger c_{l\sigma'}), \quad (1)$$

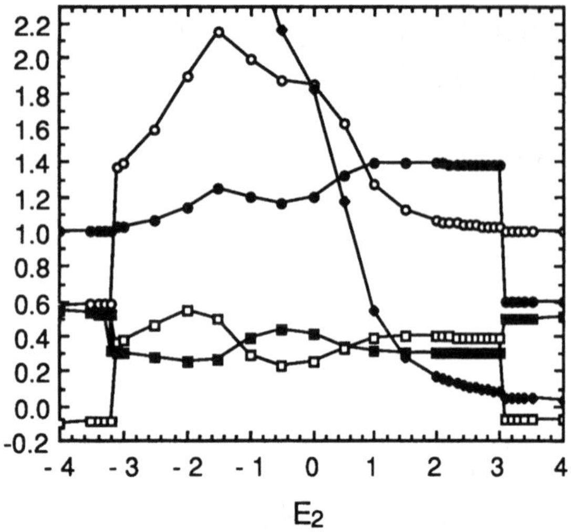

Fig. 1. E_2-dependence of P_s (•), P_{s2} (o), n_2 (♦), S_{2kF} (■) and D_{2kF} (□) in the case of N=5, N_{el}=6, K=1, U≡U_1=U_2=U'=0.2, t=1 and t'=0.1

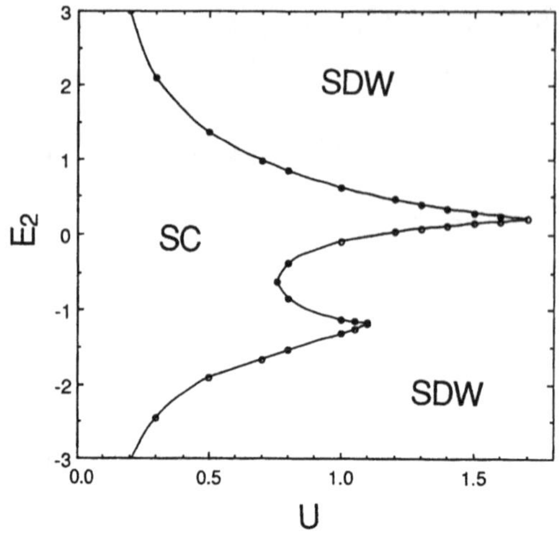

Fig. 2. Upper (•) and lower (o) bounds of the SC region are plotted in the E_2 vs U plane in the case of N=5, N_{el}=6, K=1, t=1 and t'=0.1

where $a_{l\sigma}$ and $c_{l\sigma}$ are the annihilation operators of the orbital-1 and -2 at site l with spin σ which form the tight-binding Band-1 and -2, respectively. t and t' are the transfer energies. The center of Band-2 is offset by E_2 from that of Band-1. K is the coupling constant of the interband pair-transfer interaction. U_i (i=1, 2) is the on-site Coulomb energy for Band-i and U' is that between Band-1 and -2. Our model consists of N sites making a ring. When U_1=U_2=U'=K=0, the two bands are given with $k=2\pi\nu/N$ (ν=0, 1, 2, ..., N-1) by

$$\varepsilon_{1k} = 2t\cos k \quad \text{and} \quad \varepsilon_{2k} = E_2 + 2t'\cos k. \qquad (2)$$

CORRELATION FUNCTIONS AND PHASE DIAGRAM

By using the Lanczos method [9] we have obtained the ground-state energy and wave function |g> of our model Hamiltonian with N sites up to 7 (14 orbitals) and N_{el} electrons up to 6. We have calculated the s-wave superconducting pair correlation function defined by

$$P_s = \langle g|(\Delta^\dagger\Delta+\Delta\Delta^\dagger)|g\rangle, \quad \Delta = (1/\sqrt{N})\sum_l a_{l\downarrow}a_{l\uparrow} = (1/\sqrt{N})\sum_k a_{-k\downarrow}a_{k\uparrow} \qquad (3)$$

for Band-1. We have also calculated the similar pair correlation function P_{s2} for Band-2, $2k_F$ SDW, $2k_F$ CDW and $4k_F$ CDW correlation functions (S_{2kF}, D_{2kF}, D_{4kF}) and the occupancy (n_2) of Band-2. In this report we restrict ourselves to the case of U_1=U_2=U'≡U. In Fig. 1 the results are plotted against E_2 for the case of N=5, N_{el}=6, K=1, U=0.2, t=1 and t'=0.1. For E_2>3.0, P_s≅0.5, P_{s2}≅1 and n_2<<1. As Band-2 approaches the Fermi energy ε_F lying in Band-1 with decrease of E_2, P_s increases jump-wise to ~1.4 around E_2=3.0. The increase of P_s clearly indicates that the ground state becomes of SC nature. This interpretation is supported also by the value P_s≅1.5 in the case of the single-band negative-U Hubbard model with N=5 and N_{el}=6. In the SC state D_{2kF} also increases, suggesting the coexistence of $2k_F$ CDW, while S_{2kF} decreases, suggesting that the SDW correlation is replaced by the SC correlation. With further decrease of E_2, n_2 gradually increases indicating crossing of Band-2 with the Fermi energy ε_F. When E_2 reaches ~-3.2, the SC features of P_s and P_{s2} vanish with Band-1 being empty.

Identifying the E_2 region featured by increased P_s and P_{s2} as SC, we have plotted the SC phase in the E_2 vs U plane in Fig. 2. For U≥0.8 the SC region narrows; SC disappears as the Band-2 lowest levels go away below the Band-1 levels partially occupied before, while for U<0.8 the SC region is much wider since the lower Band-1 levels work here to drive SC. Thus the wavy lower bound of the SC region is an artifact due to the finite size of the model. At U=1.7, the SC region is around E_2=0.2, which is appreciably lower that the value of $E_2 \cong 0.8$ at which the levels of both bands coincide in the case of U=K=0. This difference is due to the Coulomb interaction. The SC region is observed to expand in the U variable with increase of the system size defined by N and N_{el}. Since the size is quite restricted due to the necessary huge memory size, we cannot extrapolate the results for the infinite size. However, the SC region up to U=1.7 in Fig. 2 is already remarkably wide.

The overall U-dependence of the upper bound E_{2up} is roughly given by $E_{2up} = (K^2/U-U)/2+E_2^{(0)}$ with $E_2^{(0)}$ being the E_2 value where both levels coincide in the non-interacting limit. This relationship is especially well obeyed in the half-filled cases of $N=N_{el}=2$ and 4. This means that SC appears when the effective U is negative i.e. $U-K^2/[2(E_2-\varepsilon_F)+U]<0$ is satisfied; the second term is the attractive interaction originating in virtual excitation of two Band-1 electrons to Band-2 noticed by Kondo [2]. $n_2<<1$ near $E_2=E_{2up}$ means that the Band-2 levels are above ε_F. Clearly SC persists when the Band-2 levels submerge below ε_F. One of the present authors claimed that the effect of K is remarkably magnified by the interband ladder process [3,4]. The SC region in E_2 is expanded with increase of U' in accord with the above claim. However, we cannot reach to clear judgement due to restriction to the system size.

RELEVANCE TO CUPRATE SUPERCONDUCTORS

We apply the above model to the CuO_2 plane, assuming that the network consisting of the O $2p_\sigma$ orbitals provides the essential part of the conducting system and that the O $2p_\pi$ orbital plays the role of Band-2 in an electron-hole reversed picture. In this situation since $K \cong 0.88$ eV the upper bound $U \cong 1.7$ for the SC phase in Fig. 2 corresponds to 1.5 eV. Further, the O $2p_z$ orbital is considered to join in driving SC; its level must lie as close as to the O $2p_\sigma{}^*$ band as the O $2p_\pi$ orbital, since the three O 2p levels are identical from the viewpoint of the ionic model and only the p_σ level is shifted upwards due to the covalency with the Cu d_{x2-y2} orbital. Since the effect of the second auxiliary bands are additive, the upper bound of U for the SC phase must be expanded up to 3 eV. This value is comparable with $U_p=4$ eV which is suggested for the on-site Coulomb energy for the O $2p_\sigma$ orbital in ref. [13]. Thus, the present two-band model with the O conducting network deserves serious consideration as the essential mechanism driving high-T_c SC. By the way, the possibility for this mechanism to work with the Cu d_{x2-y2} and Cu d_{z2} orbitals is not hopeful in view of the large on-site Coulomb energy $U_d=10.5$ eV [13] for the Cu d_{x2-y2} orbital.

INTERPRETATION OF SYSTEMATICS OF T_c

It is not easy to prove the present mechanism for high-T_c SC nor to determine the magnitude of the coupling constant K of the interband pair-transfer interaction by means other than calculation. However, on the basis of our proposed scheme three important observations stated below on the systematics on T_c can be qualitatively understood, which encourages further investigations based on our scheme.

T_c vs ΔV_M: Torrance and Metzger [10] observed that the maximum value of T_c of cuprates increases with decrease of ΔV_M, which is defined as the electron Madelung potential at the Cu site minus that at the in-plane O site for the parent materials before doping. We interpret this tendency as coming from decrease of the separation between the O $2p_\sigma$ level and the O $2p_\pi$ and $2p_z$ levels, which plays the role of E_2 in the present two-band model; its decrease is considered to bring about higher T_c. The dominant contribution to the separation among the O 2p levels occurs since a doped hole located in the $2p_\sigma$ orbital can make virtual transition to the Cu d_{x2-y2} orbital on the both sides of the O due to the transfer energy t_{pd}, while such transfer energies vanish for the $2p_\pi$ and $2p_z$ orbitals due to symmetry. Therefore, we get $E_2 \cong 2t_{pd}{}^2/(\varepsilon_d+U_d-\varepsilon_p{}^{(0)})$, where ε_d and $\varepsilon_p{}^{(0)}$ are the hole level for the Cu 3d and O 2p orbital, respectively, and U_d is the on-site Coulomb energy at the Cu site; they satisfy $\varepsilon_d < \varepsilon_p{}^{(0)} < \varepsilon_d+U_d$. Since the decrease of ΔV_M decreases $\varepsilon_p{}^{(0)}-\varepsilon_d$, it entails the decrease of E_2 as postulated above.

T_c vs ΔV_A: Ohta et al. [11] noticed that T_c of cuprates increases with increase of ΔV_A defined by the electron Madelung potential at the in-plane O site minus that at the apical O site. From the viewpoint of our scheme, this tendency of decreasing T_c with decrease of ΔV_A is understood as due to the decrease of the effective coupling constant K_{eff} of the interband pair-transfer interaction with decrease of ΔV_A. In the cuprates the actual Bloch function for the doped hole is considered to extend from the in-plane O $2p_\sigma$ orbitals not only to the Cu d_{x2-y2} but also to the apical O $2p_z$ orbitals due to overlaps with them. Since the pair-transfer interaction with the O $2p_\pi$ and $2p_z$ orbitals works only at the in-plane O site, K_{eff} should decrease with increased extention of the Bloch function outside the conducting network consisting of the O $2p_\sigma$ orbitals. The decrease of K_{eff} is prevented with increased difference of the levels between the in-plane p_σ and apical p_z, which is predominantly controlled by ΔV_A, so that T_c should increase with increase of ΔV_A with tendency of saturation as observed. Incidentally, the extention of the Bloch function is prevented also by the separation of the apical O for the CuO_2 plane.

T_c vs Doped Hole Density: It is almost established that SC in the hole-type cuprates occurs only when the doped hole density in the CuO_2 plane lies in a narrow range from 0.05 to about 0.3 per Cu [12]. This is confirmed on $La_{2-x}Sr_xCuO_4$, $YBa_2Cu_3O_{7-\delta}$, $Tl_2Ba_2CuO_{6+\delta}$ etc., although the upper and lower bounds slightly depend on the materials. It is natural for T_c to increase with increase of the doped hole density. However, it is an intriguing feature that SC disappears for highly doped systems with a rapid fall of T_c with increase of doped hole density. It is too rapid to understand as due to increase of the band separation E_2 for any reasonable model which allows the change of E_2 as a function of doped hole density. However, we can understand this rapid fall by ascribing it to decrease of K_{eff}. We speculate that when the doping level is low, the doped holes are almost confined to the conducting channel made of O $2p_\sigma$ orbitals, probably due to the finite value of ε_p-ε_d, with only a small part of doped holes spilt to the Cu sites; as the doped hole density increases, the Bloch function for the doped hole is considered to extend more and more to the Cu $d_{x^2-y^2}$ orbitals, since the separation ε_p-ε_d decreases due to the intersite Coulomb energy V between the O $2p_\sigma$ and Cu $d_{x^2-y^2}$ orbitals. In the process K_{eff} should decrease, finally becoming insufficient to sustain SC. This interpretation is consistent with the decrease of the Hall coefficient R_H on the doped hole density; R_H is hole-like when the hole-type cuprates are superconducting while in the overdoped systems it finally becomes electron-like as the band calculations predict. Thus, for the electronic structure to be superconducting our scheme requires a situation intermediate between highly-correlated and conventional-band schemes, since the Cu site should be almost singly occupied by a hole while the O $2p_\sigma$ network forms a band-like conducting channel. At the beginning of the present work we presumed that the antibonding band made of O $2p_\sigma$ and Cu $d_{x^2-y^2}$ provides the conduction band in which SC occurs but the above reasoning forced us to take the above-mentioned picture. It is a test to our scheme whether such an electronic structure is theoretically obtained or not when we include the Cu $d_{x^2-y^2}$ orbital into our model on the next step.

SUMMARY

We have studied the nature of the ground state of the finite-size two-band model with the interband pair-transfer interaction by calculating the superconducting(SC), SDW and CDW correlation functions obtained by using the Lanczos diagonalization method. When the second band whose relative position is defined by E_2 approaches or crosses the Fermi level lying in the first band, the SC correlation function jump-wise increases, clearly indicating the occurrence of SC. The diagram of the SC phase thus recognized was obtained in the space of E_2 versus on-site Coulomb energy U. It is close to that expected from the mean-field theory and also has a feature expected from the interband ladder process. The SC region tends to expand with increase of the system size but the calculation is limited up to 7 sites and 6 electrons. It was found to be already wide enough in the region of on-site Coulomb energies to justify serious application of the model to the oxides. Assuming that SC occurs by the present mechanism in a conducting channel made of the O $2p_\sigma$ orbitals in the CuO_2 plane, we argued that we can qualitatively understand the observed systematics on T_c with the Madelung potential differences ΔV_M and ΔV_A and with the doped hole density.

REFERENCES

1 Suhl H, Matthias BT, Walker LR (1959) Phys Rev Lett 3: 552-554
2 Kondo J (1963) Progr Theor Phys 29 : 1-9
3 Yamaji K (1987) Solid State Commun 64: 1157-1160
4 Yamaji K (1991) Physica C 177: 303-309
5 Guo Y, Langlois JM, Goddard WAIII (1988) Science 239: 896-899
6 Asai Y (1989) J Phys Soc Jpn 58: 3264-3269
7 Bianconi A et al (1990) In: Bednorz JG, Müller KA (eds) Earlier and Recent Aspects of Superconductivity, Springer Series in Solid-State Sciences, Vol. 90, pp. 407-421
8 Yamaji K (1990) J Phys Soc Jpn 59: 677-691
9 Yanagisawa T (1991) J Phys Soc Jpn 60: No. 10 in press
10 Torrance JB, Metzger RM (1989) Phys Rev Lett 63: 1515-1518
11 Ohta Y, Tohyama T, Maekawa S (1991) Phys Rev B 43: 2968-2982
12 Uchida S (1991) Physica C in press (Proc Int'l Conf M^2S-HTSC III, Kanazawa)
13 Hybertson MS et al (1990) Phys Rev B 41: 11068-11072

A Microscopic Theory of Tunneling Effect of High T_c Superconductors

YUKIO TANAKA

Department of Physics, Faculty of Science, Niigata University, Niigata, 950-21 Japan

ABSTRACT

A microscopic model of tunneling effect of High T_C superconductors(HTSC) - Insulator - Normal metal is presented. In this model, tunneling into the c axis of High T_C superconductor is treated by taking into account both the periodic distribution of the pair potential and the surface degradation effect. The obtained tunneling conductance is influenced by the spatial dependence of the pair potential near the interface. The finite phase increment ϕ of the pair potential over one unit cell produces drastical change of the tunneling conductance.

Key Words: tunneling, proximity effect, Andreev reflection, quasiparticle

INTRODUCTION

It is important to study the quasiparticle tunneling of high T_C superconductors(HTSC) for the investigation of the superconducting mechanisms of HTSC and for the development of HTSC devices. Recently the fabrication techniques for HTSC tunnel junctions have been improved and the tunnel junctions with an artificial barrier have been fabricated utilizing an all-in-situ technique[1]. The YBCO/I/Pb junctions with an artificial barrier, and those with a native barrier which was formed by an ex-situ growth technique have been developed by Iguchi[1,2]. The tunneling conductance shows different results between these two junctions especially on the behavior around the zero bias. The Andreev reflection in the c axis tunneling process is considered to produce an anomalous contribution to the tunnel conductance near the zero bias[3]. So it is interesting to investigate the tunneling process into the c axis based on the microscopic basis.

One of the important feature of HTSC is the short coherence length. Especially the coherence length along the c axis is nearly 3 Å. It is considered that the superconducting proximity effect becomes significant due to the short coherence length and the the pair potential is drastically reduced near the interface. Due to the very short coherence length of HTSC, the tunneling process detects the degradation of the HTSC films at the interface. The second point is that HTSC have stacking structure of superconducting and insulating layers. The superconducting paring takes place only in the superconducting layers(CuO_2 layers) and the inter layer coupling between adjacent superconducting layers is very weak. So the pair potential changes periodically along the c axis. In such a case, it is expected that the Andreev reflection which originates from the multi-layered structure forms energy bands[4].

To investigate the tunneling process into the c axis, both the periodic distribution of the pair potential and the depletion of the pair potential near the interface must be considered. In the model calculation given by Tachiki et al[5]., they consider an ideal situation where the degradation of surface is neglected, and only the periodic distribution of the pair potential is taken into account. So in this paper, we will make a theory of tunneling effect in HTSC taking into account both the periodic distribution of the pair potential and the depletion of the pair potential near the interface.

MODEL AND METHOD

To calculate the tunneling current into the c axis, we have used the Normal metal-Insulator-HTSC-Insulator-Normal metal system (NM-I-HTSC-I-NM) as shown in Fig.1. The unit cell of HTSC consists of the superconducting region(S) and normal(insulating) region(N) with the thickness d_S and d_N respectively. It is very difficult to obtain the wave function of infinite HTSC, where both the periodic distribution of the electron ion potential and the pair potential appear. So the number of the unit cell along the c axis, M, is chosen to be finite. However M is chosen to be sufficiently large so that the finite size effect does not appear. The Hamiltonian in this system is given as

$$H = \begin{pmatrix} -\frac{\hbar^2}{2m}\frac{d^2}{dx^2} + U(x) - \mu_{\|} & \Delta(x) \\ \Delta^*(x) & -(-\frac{\hbar^2}{2m}\frac{d^2}{dx^2} + U(x) - \mu_{\|}) \end{pmatrix}, \tag{1}$$

in the matrix form. In eq.(1), $U(x)$ is the electron-ion potential and $\mu_{\|}$ is expressed as follows,

$$\mu_{\|} = \mu - \frac{\hbar^2(k_y^2 + k_z^2)}{2m}, \tag{2}$$

with use of the chemical potential μ. The spatial dependence of $U(x)$ along c axis is shown in Fig.1, which is expressed as

$$U(x) = U_0\theta(na - x)\theta(x - na + d_S) + U'[\delta(x) + \delta(x - Ma - d_S)], (1 \leq n \leq M),\ a = d_S + d_N. \tag{3}$$

In the above, U' and U_0 represents the strength of the delta function barrier of the insulator and the difference of the chemical potential between the S and N region of HTSC respectively. The pair potential $\Delta(x)$ is assumed as

$$\Delta(x) = \theta(x - na)\theta(na + d_S - x)\theta(n - n_0)\theta(M - n - n_0)\exp(in\phi)$$
$$\times \{\Delta_1 + (\Delta_0 - \Delta_1)[\tanh(a(n - n_0)/\xi) + \tanh(a(M - n - n_0)/\xi) - \tanh(a(M - n_0)/\xi)]\}. \tag{4}$$

In the above, ξ is the coherence length of the pair potential, and n_0 is the number of the unit cell counted from the interface of HTSC, where $\Delta(x)$ vanishes in the S region. In eq.(4), the quantity ϕ expresses the difference of phase between the adjacent superconducting layers. Since in HTSC, the interlayer coupling is very weak, the case of finite ϕ may be easily attained by some external perturbation.

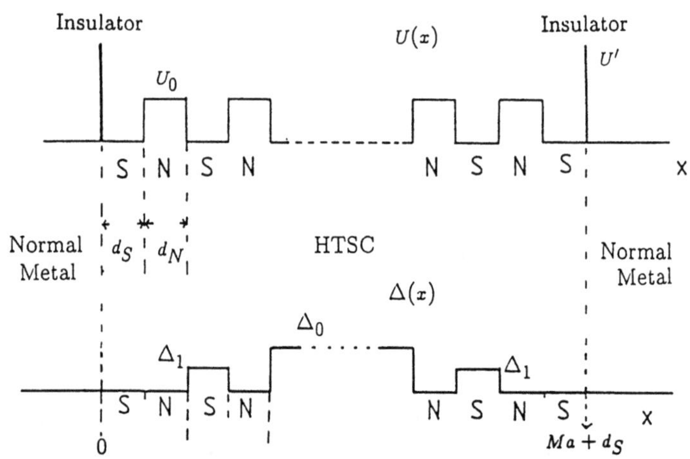

Fig.1
A model of the Normal metal-Insulator-HTSC-Insulator-Normal metal junction for $n_0 = 1$, $\Delta_1 = 0.5\Delta_0$, and $\xi = 0.6d_S$

There are four kinds of eigenfunctions corresponding to the different kind of the asymptotic behavior. The asymptotic form of the one of the eigen functions of eq.(1) are given as

$$\Psi(x) = \begin{cases} \exp(ik_+x)\begin{pmatrix}1\\0\end{pmatrix} + a_1(E,k_\|)\exp(ik_-x)\begin{pmatrix}0\\1\end{pmatrix} + b_1(E,k_\|)\exp(-ik_+x)\begin{pmatrix}1\\0\end{pmatrix} & x \ll -a \\ c_1(E,k_\|)\exp(ik_+x)\begin{pmatrix}1\\0\end{pmatrix} + d_1(E,k_\|)\exp(-ik_-x)\begin{pmatrix}0\\1\end{pmatrix} & x \gg a \end{cases} \quad (5)$$

where the quantities $a_1(E,k_\|)$ and $b_1(E,k_\|)$ represent the coefficient of the Andreev reflection, and the normal reflection, respectively. In the above, k_+ and k_- are given as

$$k_+ = \sqrt{[2m(\mu_\| + E)}, \quad k_- = \sqrt{[2m(\mu_\| - E)}. \quad (6)$$

The tunneling conductance are expressed as

$$(\frac{dI}{dV})R = -\frac{e}{2\pi^2\hbar}\int_{-\infty}^{+\infty}\int_0^{k_F} dEdk_\|\{1 + \frac{k_-}{k_+} \mid a_1(E,k_\|)\mid^2 + \mid b_1(E,k_\|)\mid^2\}\frac{\partial f(E-eV)}{\partial V}, \quad (7)$$

where R is the resistance in the normal state[6].

TUNNELING CONDUCTANCE

With use of eq.(7), the tunneling conductance in the NM-I-HTSC-I-NM system are calculated. Since HTSC behaves like semiconductor along the c axis, E_F lies in the energy gap of the energy band along k_x direction in Fig.1. To meet this fact, we have chosen $U = 1.0eV$, $E_F = 0.75eV$ and $d_S = d_N$=5.0 Å, for example. The quantity ξ in eq.(4) is chosen as 3Å, which roughly corresponds to the coherence length of YBCO along c axis and the pair potential Δ_0 is chosen as $20meV$, which is the energy gap of YBCO observed by tunneling effect into ab plane.

In Fig.2, the obtained tunneling conductance are plotted in the case of $\phi = 0.0$ as a function of energy. From (a) to (c), (dI/dV) becomes maximum at $\mid E \mid = E_C$. Since there is normal region in the HTSC and U' is not so large, (dI/dV) does not vanish for $\mid E \mid < E_C$. In the case of (a), where the depletion of the pair potential does not occur at the interface, E_C is $1.3\Delta_0$. In the case of (b), where $\Delta_1 = 0.0$ and $n_0 = 0$ are satisfied, E_C becomes $1.2\Delta_0$, which is smaller than that in the case of (a). In the case of (c), where $\Delta_0 = 0.0$ and $n_0 = 9$, and E_C becomes $0.4\Delta_0$. As seen from Fig.2, E_C decreases with the increase of n_0 and the spurious energy gap which appears in (dI/dV) becomes smaller. It can be concluded that the degradation effect of the surface significantly influences the tunneling conductance.

Finally we investigate the effect of ϕ, which is the phase difference of the pair potential $\Delta(x)$ between the adjacent S layers. We have chosen $\phi = \pi$ in Fig.3, and other parameters are the same as in Fig.2. The case of (b) takes minimum value and the case of (c) becomes larger than that of (a) except for $-1.1 > (E/\Delta_0) > -1.3$. As seen in (a) to (c), clear gap structure does not appear, and (dI/dV) becomes maximum at the zero voltage regardless of n_0 and Δ_1. In this case, E_F lies in the energy band of quasiparticle due to the Andreev reflection, and superconducting state becomes gapless on the light of our previous theory of superconducting superlattice[4]. So energy gap structure in (dI/dV) can not be observed.

CONCLUSION

In this paper, tunneling conductance is calculated in a NM-I-HTSC-I-NM model. The obtained tunneling conductance is influenced by the spatial dependence of the pair potential near the interface. When the pair potential vanishes in close vicinity of the interface, the observed energy gap becomes smaller than Δ_0, which is the pair potential in the S region of bulk HTSC. Due to the degradation effect of the surface and the periodic distribution of the pair potential near the interface, the observed energy gap of the experiments of c axis tunneling does not always become the pair potential in the S region far from the interface. Further the finite phase increment ϕ of the pair potential over one unit cell produces drastical change of the tunneling conductance. As seen in Fig.3, the tunneling conductance has a peak structure near zero voltage. Although the case of $\phi = \pi$ is unstable as compared to that of $\phi = 0.0$, it may be possible to attain the $\phi = \pi$ state by some perturbation, due to the very weak coupling between adjacent S layers. The appearance of $\phi = \pi$ state can become one of the candidates to explain the zero bias anomaly observed by Iguchi et.al[3]. Anyway it should be remarked that the tunneling conductance into the c axis are sensitive to the spatial dependence of the pair potential. At this stage, our model is very simple. So we are planning to extend our theory to more realistic material parameters region.

ACKNOWLEDGMENT

The author would like to thank to Prof.I.Iguchi, Dr.S.Kurihara, Prof.H.Fukuyama and Prof.M.Tsukada for valuable discussions. This work was supported by the Grand - in Aid from the Ministry of Education, Science and Culture of Japan.

REFERENCES

1. I.Iguchi,M.Furuyama,T.Kusumori,K.Shirai,S.Tomura,M.Nasu and H.Ohtake, (1990) Jpn.J.Appl.Phys. 29: L614
2. M.Gurvitch, J.M.Valles Jr., A.M.Cucolo, R.C.Dynes, J.P.Garno, L.F.Schneemeyer and J.V.Waszcak (1989) Phys.Rev.Lett. 63: 1008
3. I.Iguchi, and Z. Wen (1991) Physica C 178: 1
4. Y.Tanaka and M.Tsukada (1991) Solid state Commun. 77: 129
5. M.Tachiki, S.Takahashi, F.Steglich and H.Adrian (1990) Z.Phys. B 80: 161
6. G.E.Blonder, M.Tinkham, and T.M.Klapwijk (1982) Phys.Rev.B 25: 4515

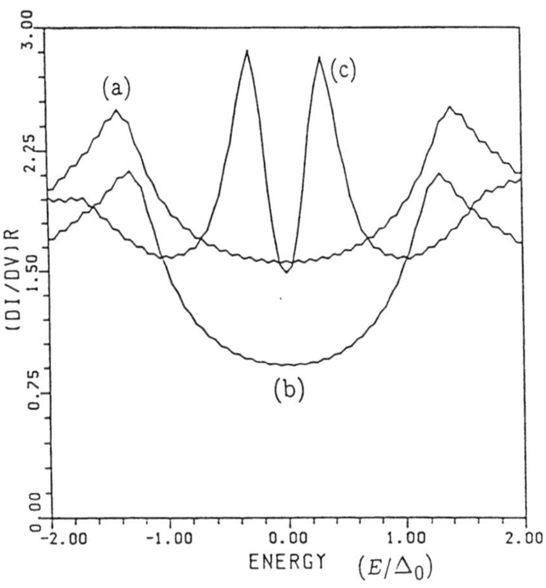

Fig.2
Typical normalized conductance for $\phi = 0.0$
(a)$\Delta_1 = \Delta_0$, $n_0 = 0$
(b)$\Delta_1 = 0.0$, $n_0 = 0$
(c)$\Delta_1 = 0.0$, $n_0 = 9$.

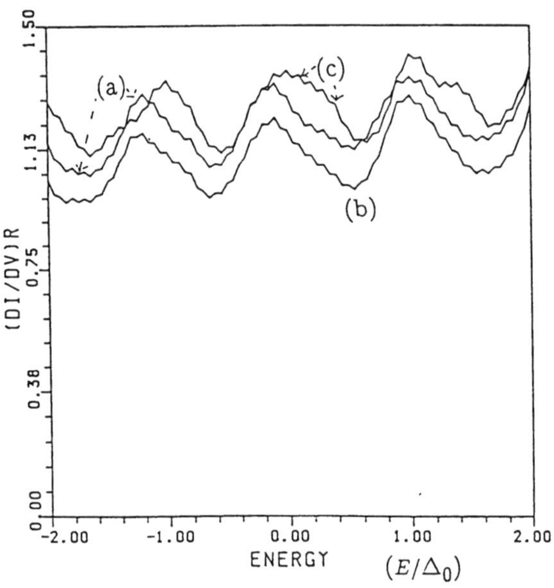

Fig.3
Typical normalized conductance for $\phi = \pi$
(a)$\Delta_1 = \Delta_0$, $n_0 = 0$
(b)$\Delta_1 = 0.0$, $n_0 = 0$
(c)$\Delta_1 = 0.0$, $n_0 = 9$

Superconductivity in Pseudodegenerate Overlapping Bands Systems: Copper Oxides and $C_{60}K_x$

YOSHIHIRO ASAI

Electrotechnical Laboratory, Tsukuba, Ibaraki, 305 Japan

ABSTRACT

The clear enhancements of the superconducting pair-field susceptibilities of the dp+π model at low temperature have been obtained in one and two dimensional lattices. This model includes the second band coming from in-plane oxygen sites which submerges below or cuts the Fermi level in addition to the dp (three band Hubbard) Hamiltonian. There is a pair transfer interaction between the two bands at the oxygen sites. We have presumed that on site Coulomb repulsion at copper site is very large. Here, we will review our results and give some justifications of this model as a model of the high-T_c copper oxides. We will mention about an analogy of the superconducting mechanisms in the copper oxides and the doped C_{60}.

KEY WORDS: Quantum monte carlo simulation, Lattice fermion model, two band model, Superconducting pair-field susceptibility.

INTRODUCTION

The mechanism of high-T_c in the cooper oxides has been studied with various theoretical models. Hubbard model and the dp model (three band Hubbard model) are the models which have been received much attention. These models well explain magnetic properties of the copper oxides. Numerical calculations of these models, which do not include artificial approximations, have been giving negative conclusions to the possible superconductivity in these models in one and two dimensions. t-J model with very large J has been believed to involve a d-wave like superconducting phase, but it must be required that unusual large J must be explained from more fundamental models. Hubbard and the dp models fail to be mapped to the t-J model with large J. Something additional may be required to explain s-wave superconductivity in high-T_c copper oxides.

At the very early stage of the high-T_c studies, there are some arguments that some states other than the states described by Hubbard model and the dp model exist. Some author showed that holes are doped into an $O2p_{\pi x,y}$ state. Some others showed that holes are doped into a $d_{3z^2-r^2}$ state. The possibilities that holes are doped exclusively into the states with symmetries other than the $d_{x^2-y^2}$ symmetry have been believed to be excluded by an old study of anisotropies of the Knight shifts. It should be noted that this experiment does not exclude the possibility that the relatively small amount of holes are doped into the $O2p_{\pi x,y}$ state nor there is a low lying excitation where holes are excited into the $O2p_{\pi x,y}$ state. It has been pointed out that systematic change of anisotropies of the shift accompanying the change of carrier concentration must be measured to decide the symmetry of the doped holes.[1] In the light of negative results on numerical study of the Hubbard-like models, it is interesting to study the role of the second bands. In this context, we review our results on the dp+π model. We will also give some justifications of this model as a model of the high-Tc copper oxides.

ELECTRONIC STRUCTURE OF THE COPPER OXIDES

First, we will discuss the electronic structure of the copper oxides to justify our model. In our dp+π model, it is essential that the second band coming from in-plane $O2p_{\pi x,y}$ orbitals are close to the Fermi level. (not necessarily holes are actually doped into the $O2p_{\pi x,y}$ state) Even when all the holes are doped into the $O2p_{\sigma x,y}$ state, virtual excitation taking place at the oxygen sites between the two bands enhances the superconducting pairing.

It may be worthwhile to remind that lots of chemical systematics of the high-T_c copper oxides in the metallic (superconducting) phase can be well explained in terms of Madelung potential. For example, we had succeeded in explaining the experimental phase boundary between insulating and superconducting $Y(La_xBa_{1-x})_2Cu_3O_y$ and $(Y_xCa_{1-x})Ba_2Cu_3O_y$ in terms of Madelung potential.[2] This suggests that strong ionicity is preserved even in the metallic state of the copper oxides. In this context, we had carried out a ground state electronic structure calculation of a copper oxide in terms of ion-cluster model. In our results, the $O2p_{\pi x,y}$ hole state is more stable than the $O2p_{\sigma x,y}$ state by 0.44eV.[3] If we impose a charge consistency condition in our ion-cluster model, the $O2p_{\pi x,y}$ hole coexists with the $O2p_{\sigma x,y}$ hole. If we remember our approximation used (ion-cluster model) , it may not be justified to believe the results in too details. But it is also very hard to think that the $O2p_{\pi x,y}$ state are far from the Fermi level, because there remains strong ionicity in the metallic state of the copper oxides. Recently, systematic change of the anisotropies of the Knight shift accompanying the change of the oxygen content of YBCO has been measured by Yoshinari, Yasuoka and Ueda. They concluded that holes are doped into the in-plane $O2p_{\pi x,y}$ states.[1] A photoinduced absorption measurement indicates that there are two states coming from in-plane oxygen sites separated by 0.7eV in insulating YBCO.[4] Experimentally, the possibility that the holes are doped into the $O2p_{\pi x,y}$ state has been getting strong. These are strong experimental evidences which stand for our model.

dp+π MODEL WITH A PAIR TRANSFER INTERACTION

We have emphasized that the essential feature of the electronic structure of the copper oxides is that the nearly degenerated hole states coming from in-plane oxygen sites are close to the Fermi level. With this regard, it is important to include low lying interband transition in the model of the copper oxides. The dp+π model described below is the model which represents this two band feature:

$$H = -t_{pd} \sum_{<i,j>\sigma} (d^{\dagger}_{i\sigma} p_{j\sigma} + h.c.) + \varepsilon_p \sum_j n^p_j + \varepsilon_d \sum_i n^d_i + U_d \sum_i n^d_{i\uparrow} n^d_{i\downarrow} - t_{\pi\pi} \sum_{<j,l>\sigma} (\pi^{\dagger}_{j\sigma} \pi_{l\sigma} + h.c.) + \varepsilon_\pi \sum_j n^\pi_j$$
$$+ K \sum_j (\pi^{\dagger}_{j\uparrow} \pi^{\dagger}_{j\downarrow} p_{j\downarrow} p_{j\uparrow} + h.c.). \qquad (1)$$

We use the same notations given in ref. [5,6]. We call the K term of the third line of eq.(1) a pair transfer interaction. This represents interband pair hopping. This term is essential in the superconductivity with this model, which will be mentioned later again. The existence of this term should be checked by experiments. But there are two realistic origins of this interaction. One is the electronic origin. In this case the coupling constant K takes the same value with the direct exchange integral between the two atomic orbitals on an oxygen site.[7] Another origin is the nonadiabatic electron-phonon interaction.[8] The nonadiabatic coupling constant $\tilde{\lambda}$ can be expressed by this formula:

$$i\tilde{\lambda}^{ij}_{l} = i\sqrt{\omega_l/2}\, \frac{\langle \varphi_{il\sigma} | \frac{\partial h}{\partial Q} | \varphi_{jl\sigma} \rangle}{\varepsilon^j_l - \varepsilon^i_l} \qquad (|\varepsilon^j_l - \varepsilon^i_l| > \omega) \quad , \qquad (2)$$

where we have used the same notation in ref.[8]. The coupling constant $\tilde{\lambda}$ can be very large if there are nearly degenerated hole states near the Fermi level. We expect that there is the pair transfer interaction with a substantial magnitude of the coupling constant.

QUANTUM MONTE CARLO CALCULATIONS OF THE dp+π MODEL

We have been studying the dp+π model in one and two dimensional lattices in terms of quantum monte carlo methods.[5,6] We will discuss the results obtained in one dimensional lattice here, because the various parameter regions have been studied in this case. We have obtained similar results also in two dimensional lattice, which will be published elsewhere. The temperature dependencies of the s-wave pair-field susceptibilities have been calculated. If there is an s-wave superconducting transition at finite temperature in the bulk three dimensional system, we expect that there is a divergence of the susceptibility at this temperature. Since we study the model in the low spatial dimensions, we may only expect the divergent behavior at T=0. We should be also careful to artificial divergent behaviors due to finite size effects. We will use the criterion for the superconducting transition (at T=0) by which we expect enhancements of the susceptibilities get large as we decrease temperature.

(a) O2pσ-O2pσ

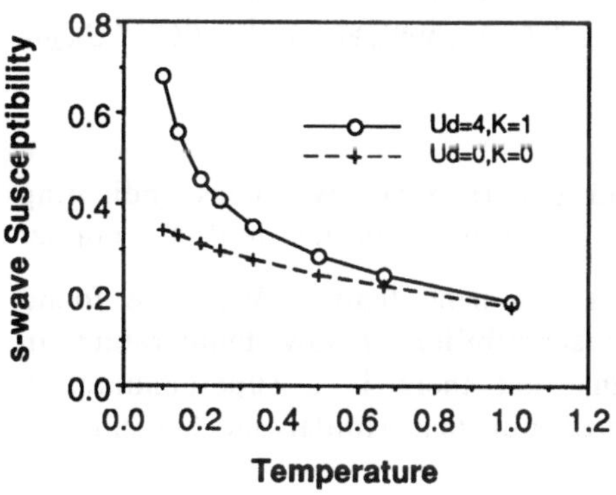

(b) O2pπ-O2pπ

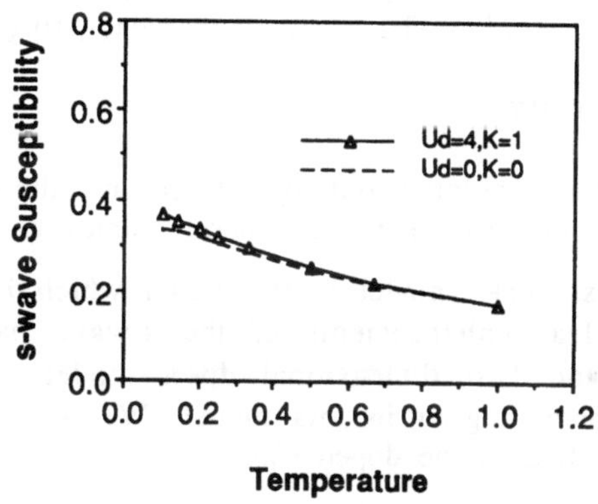

(c) Cu-Cu

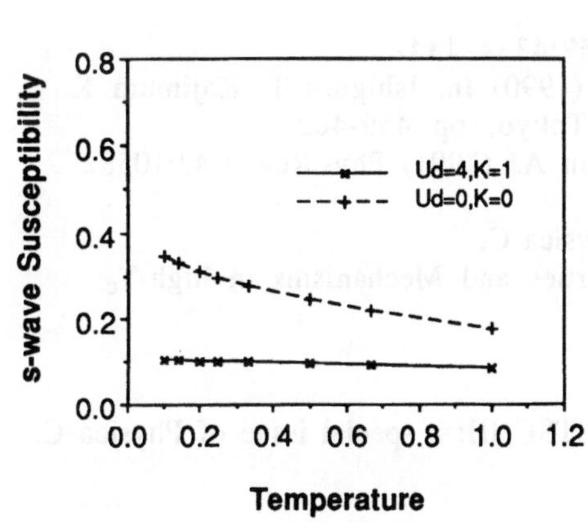

Fig.1 Temperature dependencies of the s-wave superconducting susceptibilities (a) over O2pσ sites (b) over O2pπ sites (c) over Cu sites of the one dimensional dp+π model

We say that there is an enhancement of the susceptibility when the susceptibility of a model is larger than that of the model with no fermion interactions. We have imposed a periodic boundary condition to the finite chain. We show the results obtained with a parameter set $t_{pd}=t_{\pi\pi}=1$, $U_d=4$, $K=1$, $\varepsilon_p-\varepsilon_d=0$, $\varepsilon_p-\varepsilon_d=2$ in the Figs. 1. In this case, holes are not actually doped into the $O2p_\pi$ state. The broken lines show the results obtained in the non-interacting case, where we set U_d and K, 0. The real lines show the results obtained with the parameter set. The lines are drawn only for the guide of eye. Fig. 1(a), 1(b) and 1(c) show the on-site susceptibilities over $O2p_\sigma$ sites, over $O2p_\pi$ sites and over Cu sites, respectively. We find the enhancement at $O2p_\sigma$ sites and it gets large as we decrease temperature, which means that there may be the superconducting transition at T=0 and/or there may be the transition at a finite temperature in the higher dimensions. There is a wide parameter region for the superconductivity. However we could not find the clear enhancement when we use $\Delta=\varepsilon_p-\varepsilon_d=3.5$ down to the temperature T=0.2t. (We have found only very small enhancements within this temperature range.) [5,6] There seems to be some mechanisms for suppressing the superconducting order in this large Δ case. This comes from reduction of double occupancy at O site, which is a consequence of the local singlet formation. So we expect if we add intersite Coulomb repulsions between Cu and O sites and between the two oxygen orbitals [9] to our model, the superconducting parameter region may be enlarged. Our model may be the model of high-T_c in the copper oxides.
There is another nearly degenerate overlapping band system: the doped C_{60}. We expect that short-range Coulomb interactions at each C_{60} site are very small, due to the large size of C_{60}. The nonadiabatic electron-phonon coupling constant may be very large in this system.[8] The similar superconducting mechanism can be applied to the doped C_{60}.

SUMMARY

We have pointed out that there are theoretical and experimental evidences indicating that there are nearly degenerate states close to the Fermi level in the high-T_c copper oxides. dp+π model is the model which includes this two band feature. We have found the clear enhancements of the s-wave pair field susceptibility at low temperature in one and two dimensional dp+π models. This means that there is a supercondtucting parameter region in this model. We have also pointed out that similar mechanism can be applied to the doped C_{60}.

REFERENCES

[1] Yoshinari Y, Yasuoka H, Ueda Y, Proc. M^2S-HTSC III: a special issue of Physica C and a private communication.
[2] Kondo J, Asai Y, Nagai S (1988) J.Phys.Soc.Jpn. **57**:4334-4342
[3] Asai Y (1989) J.Phys.Soc.Jpn. 58:3264-3269 and (1990) In: Ishiguro T, Kajimura K (eds) Advances in Superconductivity. Springer, Tokyo, pp 459-462
[4] Leng JM, Ginder JM, Farneth JM, Shah SI, Epstein AJ (1991) Phys.Rev. B43:10582-10588
[5] Asai Y, Proc. M2S-HTSC III: a special issue of Physica C.
[6] Asai Y, Proc. Int. Workshop on Electronic Properties and Mechanisms in high-T_c Superconductors (Elsevier)
[7] Kondo J (1963) Progr.Theor.Phys. 29:1-9
[8] Asai Y, submitted
[9] Yamaji K, Yanagisawa T, Harigaya K, Proc. M2S-HTSC III: a special issue of Physica C.

Role of Ion-Oscillations and Electronic Excitation for the Pairing Interaction in the Oxide Superconductors

RYOZO AOKI[1], KAZUO SAKAI[2], TETSURO NAKAMURA[3], HIROSHI KAWAJI[3], and MITSURU ITOH[3]

[1] Research Center for Superconducting Materials and Electronics, [2] Department of Electrical Engineering, Osaka University, Suita, Osaka, 565 Japan
[3] Research Laboratory of Engineering Materials, Tokyo Institute of Technology, Yokohama, 227 Japan

§ ABSTRACT

A common expression of critical temperature Tc for most of the p-type oxide supercoductor was presented as $Tc=\Theta_D \exp[-1/g]$ and $g=c \cdot \omega_0$, in terms of the Debye temperature $\Theta_D=9x10^2K$ for $[CuO_2]_n$ net-plane and mid infrared absorption edge frequency ω_0. Dual character of this Tc expression regarding the phonon(Θ_D)and electronic excitation(ω_0) was investigated, and a phonon assisted carrier conduction was proposed. Recently reported long range proximity was discused in terms of this conduction mechanism.

KEY WORDS: Oxide superconducting Tc,Debye temperature,IR optical absorption, Phonon assisted conduction,Long range proximity effect

§ INTRODUCTION

The high T_c superconducting mechanism in the cuprous oxides is much interested, however, not yet been clarified. Several evidences suggesting phonon contribution to the pairing mechanism have been discussed[1], and a theoretical calculation also predicts capability of the phonon mediation but not in main role [2].

§ The pairing parameter,g

Accordingly in terms of the following BCS expression

$$T_c= \Theta_D \exp [-1/g] ,$$

we attempted to elucidate the pairing parameter g with use of the critical temperature T_c and the Debye temperature Θ_D, which was estimated at $\Theta_D= 9x10^2$ K for the essential $[CuO_2]_n$ plane in the oxides in terms of a harmonic oscillator model[3]. As the result, the parameter g was evaluated in the range of 0.27 to 0.47 for the oxides as seen in Fig.1. On the other hand, appearance of the high T_c superconductivity in the oxide is accompanied with an occurrence of optical absorption in mid-infrared region(MIR), and the correlation of the T_c with this MIR absorption edge frequency ω_0 was indicated by Tajima et.al.[4] and Arima et.al[5]. In order to investigate the role of this characteristic frequency ω_0 to the new pairing mechanism, we tried to plot the estimated g value against to the ω_0 value for each oxide, and Fig.1 shows a well linear relation between them for most of the p-type oxides. In consequence, a common expression for T_c of the oxides was obtained[6] as

$$T_c= \Theta_D \exp [-1/g] , \quad \text{and} \quad g=c \cdot \omega_0 \qquad (1)$$

with $c=4.4x\ 10^{-5}$ cm.

§ Electronic excitation

The MIR absorption edge energy $h\omega_0$ has been discussed as corresponding to an electronic excitation regarding the mid gap state[7]. The direct relation $g=c\cdot\omega_0$ suggests that the Cooper pairing will be mediated essentially by the exciton $\hbar\omega_0$, and then the T_c expression in eq.1 has an interesting characteristic of both relating to the ion-oscillation ($k_B\Theta_D=\hbar\omega_D$) and the electronic excitation($\hbar\omega_0$). It has been discussed about this relation being not trivial but an essential property[6]. The electronic state in the $[CuO_2]_n$ plane, which is responsible to the superconductivity, has an ambiguous feature of the Fermi degenerated band structure and also the semi-localized state with strong charge transfer(CT) coupled by optical phonon modes.

§ Carrier transport model

Electronic conduction mechanism in a typical CT compound of halogen bridged mixed valence metal complex(HMMC) has been investigated in terms of a polaronic model including ionic valence fluctuation propagation assisted by counter ion oscillation mode[8]. From these view points the carrier transfer in the $[CuO_2]_n$ plane has been discussed[9] as due to the Cu ion charge fluctuation coupled with oxygen ion oscillation, which propagate along the Cu-O chains giving rise to the carrier transport as depicted in Fig.2. In that case, the conduction band width ΔE is limited by the phonon assisted carrier hopping period $\tau\sim\omega_D^{-1}$ as $\Delta E\cdot\tau\sim h$. Hence the validity of the prefactor $k_B\Theta_D$ in eq.1 in replace of the exciton energy $\hbar\omega_0$ can be explained[9]. This phonon assisted carrier conduction model is considered as noteworthy, however, more direct evidence is required to confirm that really it exists in the present high T_c oxide system. Recently, an interesting phenomenon has been observed in the oxides of superconducting(S)/ normal conducting(N) system as called the long range proximity effect, and it may support this phonon assisted conduction model as follows.

§ The long range proximity effect

At the contact boundary of superconductor(S) and normal conductor(N), proximity effect takes place by mutual propagation of the electronic state through the boundary, and the range of the superconducting electronic state at $T<T_c$ into the N is characterized with a coherence length $\xi_N=\sqrt{\hbar D/2\pi k_BT}$ where $D=v_Fl/3$ for the dirty limit. Accordingly, when S/N/S multilayer contact system is composed and the N layer thickness d_N is thinner than this ξ_N, superconducting current can flow through the composite system. This proximity effect has been investigated also for the oxides with different substances such as YBCO(S)/PrBCO(N)[10], YBCO(S)/PrBCO(N)[10], YBCO(S)/LCMnO(N)[11], HoBCO(S)/LBCO(N)[12], and BSCCO(S)/BSCO(N) [13,14].

In case of those oxide system, ξ_N has been estimated as extremely small as $10^{-1}\sim10^{1}$ Å as due to the small carrier concentration ($p\sim10^{21}$) and short mean free path ($l\leq10^{1}$ Å). Nevertheless, several investigators have reported the fact that even in the case of $d_N=10^2\sim10^4$ Å($\gg\xi_N$) the superconducting state was really recognized through over the S/N/S oxide system by the Josephson effect. If this long range proximity phenomenon was confirmed as due to non-trivial cause but based on some intrinsic property, it will provide a fundamental new scope to the oxide superconductivity.

§ Phonon assisted conduction model

For consideration of this problem, the following two facts which have been observed must be taken into account.

i)The normal conductivity layer N has an enhanced conductivity character even at temperatures of $T_c<T$ by the contact with S, in comparison with the case of N alone. For instance, Kasai et al.[11] reported the fact that $La_{0.7}Ca_{0.3}MnO_z$ shows semiconducting character and its resistivity attains to $2x10^5$ Ω · cm at 77K. However, when it was sandwitched by YBCO films the resistivity over the S/N/S layers showed metallic temperature dependence and it was lowered to $1x10^3$ Ω · cm near 100K.

ii)The long range proximity effect does not takes place in the system composed of different kind oxides e.g. YBCO/Y_2O_3, Al_2O_3[11], or MgO[15]/YBCO but only with the same type structure oxides of YBCO/LCMnO/YBCO or HoBCO/LBCO /HoBCO.

Here we discuss about the S/N/S conduction mechnism taking these facts into consideration. Superconducting electronic state is composed of wide spread wave packet with usually long range coherence length ξ. In contrast, the single electron wave function in the normal conducting state has short wave length of $\lambda_F=10^{-1}\sim10^0$ Å. Hence, the proximity effect between normal conducting metals can hardly be observed. Accordingly, the conductivity enhancement phenomena above T_c cannot be understood only in terms of the electronic state, since any modification of the electronic wave function at the boundary is damped out only by the range of λ_F, and we must consider more long range interaction process through the boundary. The fact (ii) suggests some contribution of the phonon propagation through the system. The S/N/S each layer has common $[CuO_2]_n$ or $[MnO_2]_n$ net-plane structure, and the LO phonon mode, which may contribute to the carrier conduction along the O-Cu-O chains due to the charge fluctuation of the Cu ion as depicted in Fig.2. In case of the same structure S/N/S system, this phonon mode can propagate through the layer contact boundary, while for the different structure system it cannot propagate at the boundary due to lack of the matching phonon mode and electron tunneling process is only permitted for the case of ultra thin($d_N\leq 10^2$Å) d_N layer[15].

Another remarkable property of the oxide superconductor, the critical current density is easily weakened at the site of any grain boundary dislocation structures, may be also interpreted this phonon assisted conduction mechanism. As the next step, the high T_c superconducting pairing mechanism, which is mediated by electronic exiton $\hbar\omega_0$, must be considered by taking this phonon-assisted conduction into account and it will be reported in a forthcoming paper.

In summary,

i) A common T_c expression for most of the p-type oxide superconductors is presented.

ii) Its dual character depending both of phonons and electronic excitation was investigated, and a specific conduction model of phonon assisted carrier transport in the $[CuO_2]_n$ net-plane is presented.

iii)Recently reported new phenonmenon of the long range proximity effect may support this conduction mechnism.

References

[1]Timusk T, Porter C D, Tanner D B (1991)Phys.Rev.Lett.66 663
[2]Anderson O K(1991) Reports in M^2S-HTSC III S4-b
[3]Nakamura T, Liang R, Kawaji H, Itoh M, Aoki R, Sakai K(1991) Physica C180 No1/3 (in print)
[4]Tajima S, Nakahashi T, Uchida S, Tanaka S(1988)Physica C156 90
[5]Arima T, Nobumasa H, Shimizu K, Kawai T(1989)Jpn.J.Appl.Phys.28 L913
[6]Aoki R, Sakai K, Murakami H, Nakamura T, Kawaji H, Itoh M(1991) Proc. of M^2S-HTSC III(in print on Physica C)
[7]Fujimori A(1989)Springer Series in Solid-State Science 89 300
[8]Aoki R, Hamaue Y, Kida S, Yamashita M, Takemura T, Furuta Y, Kawamori K (1982) Mol.Cryst.Liq.Cryst.81 301
[9]Aoki R, Sakai K, Nakamura T, Kawaji H, Itoh M(1991) Proc.of Internat. Workshop on Elec. Prop. and Mechanism in High T_c SC.(Tsukuba)(Elsevier Pub)
[10]Rogers CT,Inam A,Eedge MS,Dutta B,Wu XD,and Venkatesan T(1989) Applied Phys.Lett. 55 2032.
[11]Kasai M, Ohno T, Kanke Y, Kozono Y, Hanazono M, Sugita Y(1990)Jpn. J.Appl.Phys.29 L2219
[12]Tarutani Y, Fukazawa T, Kabasawa U, Tsukamoto A, Hiratani M, Takagi K (1991)Appl.Phys.Lett.58
[13]Setsune K, Mizuno K, Higashino H, Wasa K(1990) Proc of SPIE vol 1394 79
[14]Fujiwara S, Yuasa R, Shikichi K, Matsuta Y, Nakao M, Suzuki S(1990)Jpn. J.Appl.Phys.29 L2307
[15]Tanaka S,Nakanishi H,Matsuura T,Higashi K,Itozaki H,and Yazu S(1991) IEEE trans MAG 27 1607
[16]Aoki R, Sakai K, Nakamura T, Kawaji H, Itoh M [submitled to Physica C]

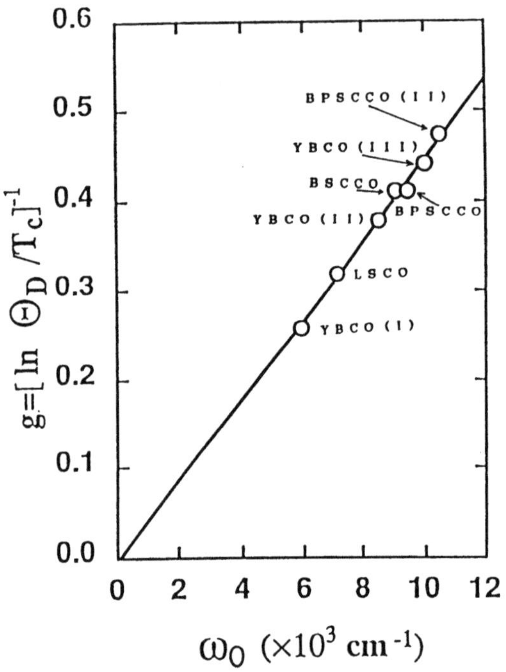

Fig 1. Linear relation-ship between the superconducting pairing parameter g and the optical absorption(MIR) frequency ω_0

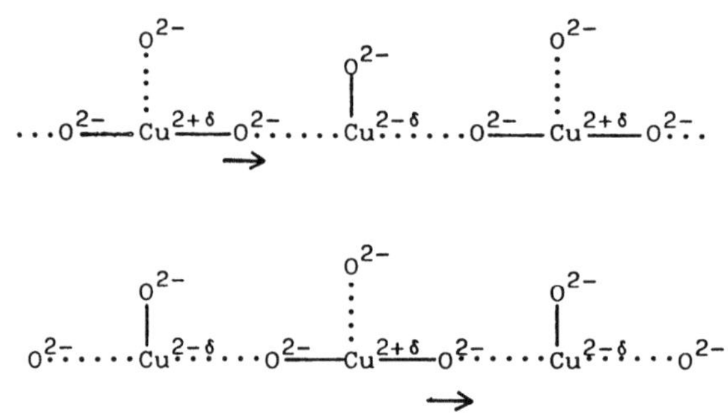

Fig 2. Schematic diagram of the Cu ion charge fluctuation propagation accompanied with oxygen ion oscillation

2.2 Electron Spectroscopy

Experimental Observation of Superconducting Gap Anisotropy in the a-b Plane and Its Theoretical Implications

Z.-X. SHEN

Stanford Electronics Laboratories, Stanford University, Stanford, CA 94305, USA

Abstract:

In this short paper, we give a brief account of our recent experimental observation of superconducting gap anisotropy in the a-b plane and its theoretical implications.

Key Words: Superconductivity, Photoemission

One of the fundamental properties of the conventional superconductors is the existence of a superconducting gap. Within the framework of the traditional BCS theory, the Fermi sea of the quasiparticles is unstable against the attractive interactions mediated by the phonons. As a consequence, the quasiparticles form Cooper pairs and a minimum gap energy Δ is required to create excitations.

The information about the superconducting gap is very important to the understanding of the nature of the superconducting state. Tunnelling and infra-red (IR) absorption are traditional tools to probe the superconducting gap. However, for the high-temperature superconductors, these two techniques are to some extend limited by practical difficulties. The tunnelling data are hindered by the difficulties in making good tunnelling junctions and electrical contacts. The IR data are complicated by the large anisotropies in the dielectric constants of the superconductors. Recently, high-resolution photoelectron spectroscopy has been successfully demonstrated to be able to detect the superconducting gap. [1-5] The reasons for this success are the improved energy resolution of the spectrometer and the stable high-quality sample surface found in $Bi_2Sr_2CaCu_2O_{8+\delta}$.

An unique advantage of PES is that the experiment can be performed in the angle-resolved mode so that one can measure the information of the superconducting gap at different part of the k-space [2-5]. In this paper, we discuss the implications of our recent experimental observation that the superconducting gap in the a-b plane is anisotropic. [6]

Fig.1 shows typical photoemission spectra recorded along Γ-X direction from $Bi_2Sr_2CaCu_2O_8$ at three different temperatures. The spectra showed a clear change as the sample went through the superconducting transition. This change of the spectra is due to the opening of the superconducting gap with the spectral weight in the superconducting gap region been depleted and a pile-up peak appeared below the gap. The position of the leading edge in the spectra taken at the superconducting state reflects the size of the superconducting gap.

Fig.2 reproduces the experimental data we obtained from the $Bi_2Sr_2CaCu_2O_8$ system at the superconducting state. [6] We can see that the leading edges for the spectra taken along Γ-X(Y) direction are shifted to lower binding energy by about 10 meV as compared with the leading edges of the spectra taken along Γ-M direction. This observation suggests that the superconducting gap along Γ-M direction is larger than the gap along Γ-X. A simple estimation based on a fitting of the experimental data indicates that the gap along Γ-M is about 1.5 to 2 times of the gap size along Γ-X.

We believe that the observed superconducting gap anisotropy in the a-b plane is important to the understanding of the superconducting mechanism in these materials. As we have discussed in our earlier paper, [6] there has been no single, conclusive experiment that has shown an in-plane anisotropic gap. However, taking experimental data as whole from tunnelling, infra-red absorption, Raman spectroscopy, and penetration depth measurements, we believe that there is substantial evidence that an anisotropic gap is a general bulk feature of the cuprate superconductors.

It is presently unclear what the origin of the gap anisotropy is. However, several possible explanations of our experimental data have been discussed. First of all, we have to consider the possible complications due to the Bi-O bands along Γ-M direction. As a guide, we depict the Fermi surface of the compound as calculated by A.J. Freeman et al. [7] According to this calculation, the Fermi surface which surrounds the M point is due to the Bi-O bands. Even though it is presently controversial whether there is an independent Bi-O Fermi surface near M, we have experimental evidence supporting that there is at least more Bi-O character in the bands along Γ-M. [4] A recent paper by V.H. Crespi and M.L. Cohen suggested a strong anisotropic electron-phonon coupling with Cu-O or Bi-O planes resulting a strong gap anisotropy in the a-b plane. [8] A possible problem with this interpretation is that one would expect, in contrary to our data, that the gap associated with the Bi-O plane is smaller than the gap associated with the Cu-O plane.

The second possible explanation of the data can be related with the spin-correlations of the carriers in the superconductor. T. Schneider and M.P. Sïrensen have made numerical calculations based on a simple one band tight binding fitting to the Fermi surface and a pairing model of on-site and next nearest neighbor interaction. [9] Their numerical result suggested that the superconducting gap near the M point is about two times the gap along Γ-X, consistent with our data (although the k space locations probed in the experiment and the theory are not exactly the same). This model does have some attractive aspects. Although the authors did not explicitly state what was the origin of the pairing interaction, the mathematical form of the interaction and the numerical values used suggest that it is dominated by the next-nearest neighbor spin-spin interaction of the carriers. [10] This pairing interaction naturally explains why the coherence length in these superconductors is so short, and why the antiferromagnetic correlation persists into the superconducting state. This model also has the virtue of providing a consistent picture between photoemission, tunnelling, and transmission I.R. data. Although this model has some quite nice features as discussed above, there are several problems that have to be addressed.

(1) It is a one band fit to the Fermi surface, the possible complication from the Bi-O bands has not been considered. (2) Their model of the pairing interaction may not be realistic. In particular, the on-site interaction in this material is more likely to be repulsive than attractive, as assumed. (3) Their numerical result was a combined effect of the pairing interaction model and the specific topology of the Fermi surface. It is not clear whether a combination of the same Fermi surface but a different pairing interaction will give qualitatively the same results. In other words, their result may not be unique.

Besides above two examples, there are other possible explanations to our data, including the scenario of having a van Hove singularity near the M point. However, the above discussion shows that the observation of the superconducting gap anisotropy in the a-b plane puts a strong constrain on the microscopic theory of superconductivity.

It should also be noted that the observation of the superconducting gap anisotropy in the a-b plane does not necessary imply a d-wave pairing. As a matter of fact, the second example discussed above has an extended s-wave pairing behavior.

In Summary, we discuss our experimental observation of the superconducting gap anisotropy in the a-b plane from our angle-resolved photoemission data and its theoretical implications. This result demonstrates the unique capability of photoemission as a tool to study the high-temperature superconductors.

I would like to thank D.S. Dessau, B.O. Wells, W.E. Spicer and A.J. Arko, D.B. Mitzi, L. Lambardo, and A. Kapitulnik for very fruitful collaborations. This work is supported by US NSF grant # DMR-8913478.

Reference:

1. Imer et al., Phys. Rev. Lett. 62, 336 (1989)
2. Olson et al., Science 245, 731 (1989); C.G. Olson et al., Sol. St. Comm. 76, 411 (1990)
3. Manzke et al., Europhys. Lett. 9, 477 (1989)
4. B.O. Wells, Z.-X. Shen, D.S. Dessau, W.E. Spicer, C.G. Olson, R.S. List, A.J. Arko, D.B. Mitzi, and A. Kapitulnik, Phys. Rev. Lett. 65, 3056 (1990)
5. D.S. Dessau, B.O. Wells, Z.-X. Shen, W.E. Spicer, A.J. Arko, R.S. List, D.B. Mitzi, and A. Kapitulnik, Phys. Rev. Lett. 66, 2160 (1991)
6. B.O. Wells, Z.-X. Shen, D.S. Dessau, W.E. Spicer, D.B. Mitzi, L. Lombardo, A. Kapitulnik, and A.J. Arko, Phys. Rev. Lett., submitted
7. A.J. Freeman et al., Physica C153-155, 1225 (1988)
8. V.H. Crespi and M.L. Cohen, preprint
9. T. Schneider and M.P. Sörensen, Z. Phys. B - Condensed Matter 80, 331-337 (1990)
10. If one uses effective one band Hubbard model, then this interaction is among the nearest neighbor.

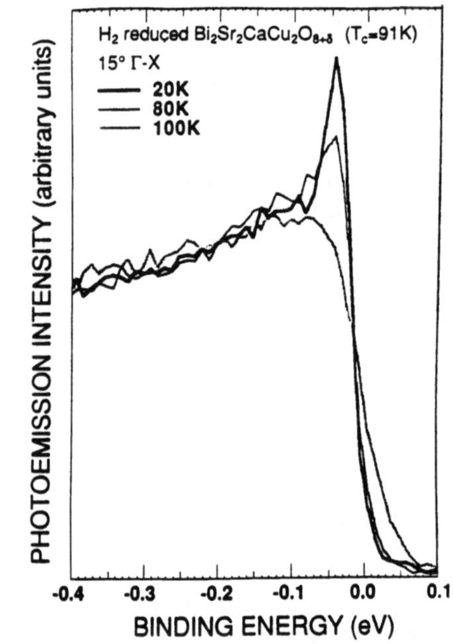

Fig.1 High-resolution angle-resolved photoemission data recorded along Γ-X direction at different temperatures.

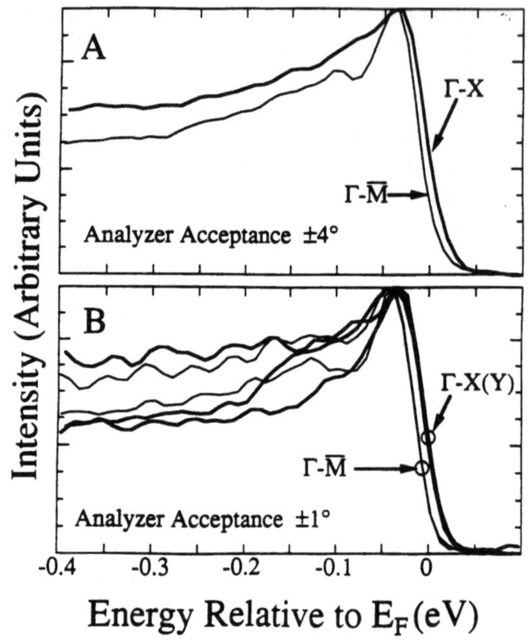

Fig.2 Comparison of photoemission data from the superconducting state along Γ-M and Γ-X directions. The two panels show spectra recorded with different angular resolutions.

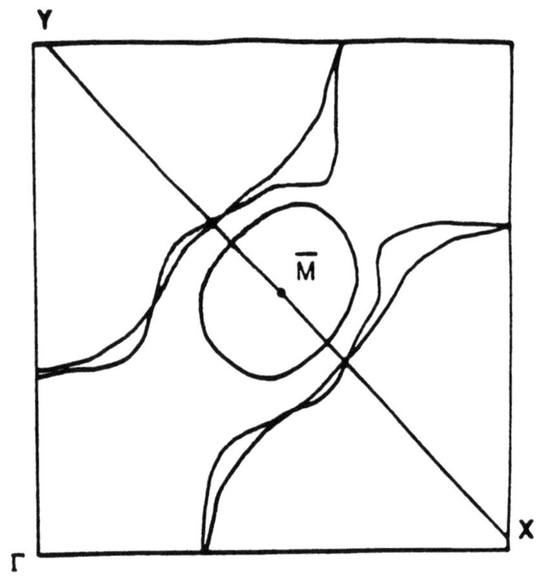

Fig.3 Illustration of the calculated Fermi surface from reference 7.

Electron Energy-Loss Spectroscopy of $Pb_2Sr_2Eu_{(1-x)}Ca_xCu_3O_{(8+z)}$ (x=0.0, 0.1, 0.3, 0.5) Superconductors

P.P. VAISHNAVA[1], J. KELLY[1], and C.L. FOILES[2]

[1] GMI Engineering & Management Institute, Flint, MI 48504, USA
[2] Physics Department, Michigan State University, East Lansing, MI 48824, USA

ABSTRACT

Several studies of high T_c materials have reported a pre-edge structure in the electron energy-loss spectral data for the oxygen edge. The systematic of this pre-edge structure in $Pb_2Sr_2Eu_{(1-x)}Ca_xCu_3O_{8+z}$ (x= 0.0, 0.1, 0.3 and 0.5) samples have been studied to gain knowledge about the effect of hole concentration on the transition temperature of these superconductors. By using the spot mode in our analytical electron microscope, the energy dispersive X-ray (EDX) and Electron Energy-Loss Spectroscopy (EELS) data were obtained for identifiable and well defined small regions of the sample. EDX spectra were used to determine the stoichiometry of the samples. A detailed electron energy-loss spectra for the oxygen edge revealed a pre-edge structure in x =0.1, 0.3 and 0.5 samples.The nature and the intensity of the peak is found to depend upon the amount of Ca substituted in the compounds.

KEY WORDS: PbSrEuCaCuO superconductors, Electron Energy-Loss Spectroscopy

INTRODUCTION

A family of high Tc superconductors with the formula $Pb_2rSr_2ACu_3O_{8+z}$ (where A is a lanthanide or a mixture of Ln + Sr or Ca) has been discovered by Cava et al (1,2) and independently by Subramanian et al (3). This new family has attracted attention from various scientists because of its very unique properties. These are: a) when A is a lanthanide and z = 0, the compound is non- superconducting. It can be made superconducting by doping with extra oxygen or substituting with an alkali earth such as Ca in the lattice, b) this family displays the same kind of substitutional chemistry as exhibited by $LnBa_2Cu_3O_{7-y}$ superconductors i.e. A could be any one of the rare earth elements or a mixture of Ln + Sr or Ca, c) In the new family the average valency for Cu is less than 2+(4). This property is unique for the new series as the previously known 1:2:3 superconductors always have valency greater than 2+, d) In many respect the new family is structurally similar to the 1:2:3 superconductors. Both families have a pair of CuO_2 layers (commonly called as CuO_2 planes) separated by an oxygen free "A" layer. Both the families have an oxygen deficient copper layer which absorbs and/or releases oxygen atoms. The oxygen deficient copper layer can be divided into blocks such as PbO-CuO_z-PbO and BaO-CuO_y-BaO in two families, respectively. These blocks, which are also known as the charge reservoirs or CuO_z/CuO_y chains, perform the most important function in the superconducting properties of the material. They supply or receive holes to or from the CuO_2 planes. One structural difference between the two families is the presence of a SrO layers which separates the CuO_2 planes from the charge reservoir blocks. Thus, the distance between the charge reservoir in the new family is much greater than the 1:2:3 superconductors. Furthermore, since the CuO_z chain layers are sandwiched between two PbO layers in the new family, it produces approximately 5% increase in the c-axis interplanar spacing.

Several studies have been performed to investigate the nature of the states close to the Fermi level and the changes in the states produced by the stoichiometric variations. It is generally believed that these states are located on the CuO_2 planes. It has been found that if the hole states at the Fermi level are of primarily O-2p character, a pre-edge peak appears in the O-1s excitation for the material. Such a peak has been observed in several high Tc compounds. Nucker et al (5), using electron energy-loss spectroscopy observed a pre-edge peak in $YBa_2Cu_3O_{6+y}$ (y = 0.2, 0.5, 0.7, 0.8, 1.0) and a systematic gain of the pre-edge peak as y increased beyond y = 0.5. Vaishnava et al (6) observed a pre-edge peak in the Fe doped $YBa_2Cu_{3-x}Fe_xO_{7-y}$ superconductors and found a concentration dependence of the intensity and the nature of the peak with the transition temperature of the material.

In this paper, we report O-1s absorption edge measurements in the system $Pb_2Sr_2Eu_{1-x}Ca_xCu_3O_{8+z}$ superconductors by electron energy-loss spectroscopy. The procedure measures the unoccupied density of states at the O sites having p symmetry. Our measurements clearly shows the presence of a pre-edge peak in the O-1s excitation. The intensity of the this peak increases with the Ca substitution, indicating the increase in hole concentration in the compounds.

EXPERIMENTAL

Samples of the $Pb_2Sr_2Eu_{1-x}Ca_xCu_3O_{8+z}$ (x = 0.0, 0.1, 0.3, 0.5) were prepared from high purity powders of PbO, $SrCo_3$, Eu_2O_3, $CaCO_3$ and CuO. The stoichiometric cuprate precursor of $Sr_2Eu_{1-x}Ca_xCu_3O_7$ was mixed with PbO and the mixture was heated in a flowing gas mixture of 1% oxygen in nitrogen at 850°C for six hours. The samples were found to be single phase as determined by x-ray diffraction. A quantum design SQUID susceptometer was used to measure the susceptibility of samples in powder form.

Samples for electron microscopy were prepared by the method described in the reference (6) except that Al grids were used in place of the copper grids. Energy dispersive x-rays (EDX) and electron energy-loss measurements were made on a VG501 scanning transmission electron microscope. The primary energy of the beam was held constant at 100 keV. The energy resolution of the microscope was kept at 1 eV. In order to increase the signal to noise ratio, several runs were made for the samples and all the data were added for the final analysis. The background of the all the EEL spectra were fitted in terms of a power series and subtracted from the data.

RESULTS AND DISCUSSION

The powder x-ray diffraction results of all the samples were reported by us elsewhere(7,8). The analysis of the pattern indicated that samples with $0 < x < 0.5$ are single phase material with lattice parameter values in agreement with the reported values (1-3). The onset temperature of the superconducting transition was estimated from the temperature at which the susceptibility crosses zero. We obtained values of 47, 64, and 69 K for x = 0.1, 0.3 and 0.5 samples. x = 0.0 sample was non-superconducting.

The spot mode operation was used to perform energy dispersive x-ray measurements (Fig. 1). The area ratio for O/Cu, Sr/Cu, Pb/Cu, Ca/Cu and Eu/Cu are given in Table I. With in experimental errors, the area ratios for O/Cu, Sr/Cu, are found to be constant. A slight decrease in the amount of Pb/Cu was noticed. As expected the area ratios for Ca/Cu and Eu/Cu were found to increase and decrease, respectively.

The electron energy-loss spectra for the x = 0.0, 0.1, 0.3, 0.5 samples are shown in Fig. 2. The solid lines through the data are just a simple guide for the eye. No pre-edge structure is noticed for the $Pb_2Sr_2EuCu_3O_{8+z}$

TABLE I. EDX measurements of the area ratio of the peaks for $Pb_2Sr_2Eu_{(1-x)}Ca_xCu_3O_{8+z}$ superconductors

Samples	Area Ratios				
	[O]/[Cu]	[Sr]/[Cu]	[Pb]/[Cu]	[Ca/Cu]	[Eu]/[Cu]
x=0.0	0.23±.03	0.64±.05	1.63±.1	0.005±.01	0.32±.02
x=0.1	0.21±.03	0.61±.05	1.59±.1	0.012±.01	0.29±.02
x=0.3	0.22±.03	0.62±.05	1.43±.1	0.090±.01	0.27±.02
x=0.5	0.21±.03	0.69±.05	1.42±.1	0.095±.01	0.23±.02

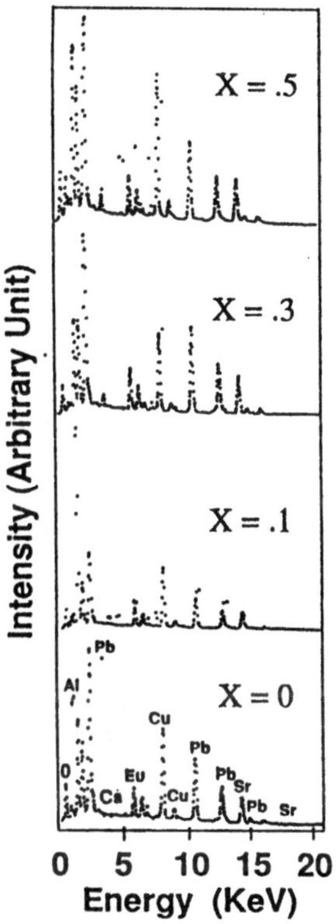

Fig. 1. Energy dispersive x-ray spectra for $Pb_2Sr_2Eu_{1-x}Ca_xCu_3O_{8+z}$ (x=0.0, 0.1, 0.3, 0.5) superconductors.

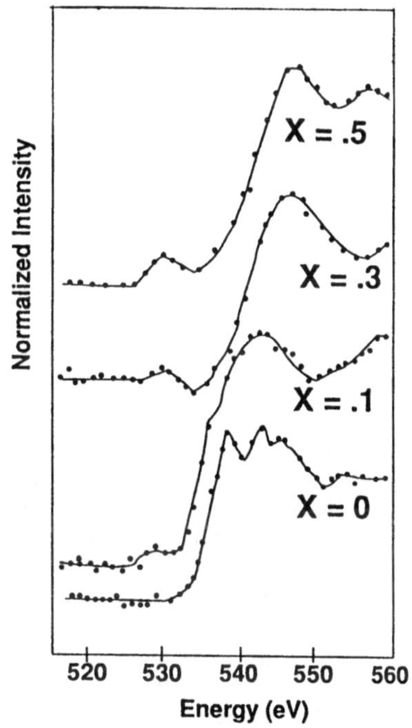

Fig. 2. Electron energy-loss spectra for $Pb_2Sr_2Eu_{1-x}Ca_xCu_3O_{8+z}$ (x=0.0, 0.1, 0.3, 0.5) superconductor

(x= 0.0) sample. This sample is non-superconducting, which is in agreement with other studies with Y atoms. For the remaining three samples, a peak in the pre-edge region at 528 eV is observed. The intensity of this peak is found to increase with the Ca concentration. By Mossbauer spectroscopy measurements(9), it has been found that the Eu is in the trivalent state in the materials. Thus the replacement of trivalent Eu in the lattice by divalent Ca is essentially introduces holes in the lattice. Ohta and Maekava et al(10) have examined $Pb_2Sr_2Y_{1-x}Ca_xCu_3O_{8+z}$ superconductors through the calculations of Madelung energy in the ionic model. It was found that the extra hole(s) in the x = 0.1, 0.3, and 0.5 samples are introduced in the CuO_2 planes and therefore the compounds show the superconducting behavior. These results are in agreement with other studies(1-3).

It is interesting and instructive to contrast results of the present study with those of two other studies. Nucker, et al (5) studied $YBa_2Cu_3O_{6+y}$ while Vaishnava, et al (6) studied 1:2:3 superconductors doped with Fe. In the former study the initial holes introduced by oxidation of the samples (increasing y values) are captured by the charge reservoirs (chains) which produces a non-superconducting state of the material. At the stoichiometry y = 0.5, the holes start flowing into the CuO_2 planes which produces both superconductivity and the pre-edge feature in the O-1s excitation. In the latter study, the Fe doping of the 1:2:3 superconductors introduced structural disorder as well as a structural transition. These structural effects decreased the transition temperature and produced systematic changes in the O pre-edge structure. The results on the present study on $Pb_2Sr_2Eu_{1-x}Ca_xCu_3O_{8+z}$ are similar to those of Nucker, et al (5) when y 0.5. The Ca doping introduces holes which go to CuO_2 planes; the transition temperature and the O pre-edge structure increase with Ca doping up to x = 0.5. The picture is not very clear for the x values greater than 0.5 as the samples are not single phase materials (1,2,7,8). A simplistic but essentially correct description of our results is the claim that x in $Pb_2Sr_2Eu_{1-x}Ca_xCu_3O_{8+z}$ and y in $YBa_2Cu_3O_{6+y}$ produce mirror image behaviors.

In summary, from the electron energy-loss spectroscopy on $Pb_2Sr_2Eu_{1-x}Ca_xCu_3O_{8+z}$ samples, a peak in the pre-edge region of O-1s excitation has been observed. For x = 0.1, 0.3, 0.5 samples, the intensity of this peak is directly related with Ca concentration in the samples. With increasing Ca concentration, the holes are localized on the CuO_2 planes which also explains the superconducting properties of the three samples.

Prem P. Vaishnava would like to thank Center for Fundamental Materials Research at Michigan State University for the award of academic affiliate fellowship to perform this work.

REFERENCES

1. Cava RJ, Batlogg B, Krajewski JJ, Rupp LW, Schneemeyer LF, Siegrist T,vanDover RB, Marsh P, Peck WF Jr., Gallgher PK, Glarum SH, Marshall JH, Farrow RC, Waszczak J, Hull R, and Trevor P (1988) Nature 336: 211
2. Cava RJ, Marezio M, Krajewski JJ, Peck WF Jr, Santoro A, and Beech F. (1989) Physica C 157: 272
3. Subramanian MA, Gopalakrishnan J, Torardi CC, Gai PL, Boyes ED, Askew TR, Flippen RB, Farneth WE, and Sleight AW (1989) Physica C 157: 124
4. Rao CNR, Bhat V, Nagarajan R, Ranga Rao G, and Sankar G (1989) Phys. Rev. B 39: 9621
5. Nucker N, Fink J, Fuggle JC, Durham PJ and Timmerman WM (1988) Phys. Rev. B 37: 5158
6. Vaishnava PP, Taylor CA Jr, and Foiles CL (1990) Phys. Rev B 41: 4195
7. Nelson CD, Pratt WP Jr, and Vaishnava PP (1990) In: Christen D, Narayan J, and Schneemeyer L (eds) in High-Temperature Superconductors: Fundamental Properties and Novel Materials Processing, Materials Research Society, Pittsburg, pp 1077-1080
8. Vaishnava PP, Nelson CD and Pratt WP Jr (1990) In: Christen D, Narayan J, and Schneemeyer L (eds) in High-Temperature Superconductors: Fundamental Properties and Novel Materials Processing, Materials Research Society, Pittsburgh, pp 1021-1024
9. Vaishnava PP, Nelson CD, Pratt WP Jr, and Boolchand P (1990) Physica B 165 & 166: 1325
10. Ohta Y, and Maekawa S (1990) Phy. Rev. B 41: 6524

Changes of Electronic States of Superconducting Oxides by Various Surface Treatments Studied by XPS and SEM

MASAYASU NAGOSHI[1], TERUO SUZUKI[1,*], YASUO FUKUDA[2], KATSUYOSHI OH-ISHI[3], KAZUHIRO UEKI[3], AYAKO TOKIWA[3], MASAE KIKUCHI[3], YASUHIKO SYONO[3], and MASASHI TACHIKI[3]

[1] Steel Research Center, NKK Corporation, Kawasaki, Kanagawa, 210 Japan
[2] Research Institute of Electronics, Shizuoka University, Hamamatsu, Shizuoka, 432 Japan
[3] Institute for Materials Research, Tohoku University, Sendai, 980 Japan

ABSTRACT

Changes of electronic states and surface morphology by scraping and fracturing have been investigated for the Nd-Ce-Cu-O and Ba-K-Bi-O superconductors using x-ray photoelectron spectroscopy (XPS) and scanning electron microscopy (SEM) in order to check the suitability of the surface treatments for photoemission studies. Scraping the samples with a file considerably changes the electronic states at the surfaces : the spectral intensity at the Fermi level is decreased and the line shape of core levels varied. On the other hand, the clear Fermi edge is observed for the fractured surfaces. These results are discussed in terms of oxygen defects and structural changes in the surface region.

KEY WORDS : photoemission, surface treatment, Nd-Ce-Cu-O superconductor, Ba-K-Bi-O superconductor

INTRODUCTION

Photoemission is one of the most powerful tools for the investigation of the electronic states of high - T_c superconducting oxides. Actually, a number of the studies have been carried out using x-ray photoelectron spectroscopy (XPS) and ultra-violet photoelectron spectroscopy (UPS) as well as other electron spectroscopic techniques. However, the contradictory results have been obtained for the surfaces prepared by various methods because of strong surface sensitivity of the photoemission due to short mean-free-paths of photoelectrons in solids [1]. This has prevented us from understanding the real electronic structures of solids. In order to obtain the surfaces which reflect the bulk properties of the oxides, several *in situ* techniques (ion sputtering, fracturing, scraping, annealing in O_2, and so on [2]) have been used. Although scraping and fracturing, especially the former, have been widely used for cleaning sintered samples, few detailed studies on their suitabilities have been reported.

In this paper, we present the changes of the electronic states in the surface region of superconducting oxides by scraping and fracturing studied using XPS and SEM to check the suitability of the surface treatments for photoemission experiments. $Nd_{1.85}Ce_{0.15}CuO_{4-y}$ (NCCO) and $Ba_{0.5}K_{0.5}BiO_{3-\delta}$ (BKBO) were chosen as the samples which are representatives of the superconductors with and without Cu - O_2 planes, respectively.

EXPERIMENTAL

XPS measurements were carried out at room temperature using a SSX-100 electron analyzer (Surface Science Instrument) with a monochromatized AlKα radiation. The incident x-ray beam was focused at about 300 μm in diameter on the sample surfaces to select clean area as possible. The typical energy resolution of the system was 0.36 eV.

Polycrystalline superconductors, $Nd_{1.85}Ce_{0.15}CuO_{4-y}$ (T_c = 23 K) [3] and $Ba_{0.5}K_{0.5}BiO_{3-\delta}$ (T_c = 20 K) [4], were synthesized by the solid-state reaction and the former was annealed at 900 °C for 12 h in vacuum. The samples were confirmed to have a single phase by x-ray powder diffraction. The surfaces studied here were prepared by fracturing or scraping with a stainless file just before the XPS measurements under a pressure of about 1 x 10^{-8} Pa at room temperature. The morphologies of the fractured and the scraped surfaces were observed by SEM (JSM T - 330) after the XPS measurements.

RESULTS AND DISCUSSION

The morphologies of the fractured and the scraped surfaces are shown in Fig.1 for NCCO, in which they are completely different. The area of trans-granular fractures is about 90 % of the fractured surface and the remaining part is composed of voids and grain boundaries. The surface is smoothed and new surfaces appear by scraping. No voids and no grain boundaries are distinguished for the scraped surface. Similar results were obtained for BKBO.

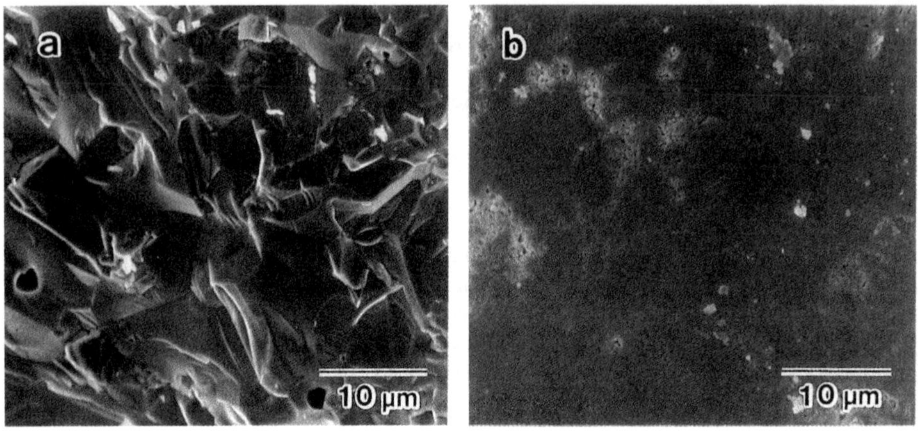

Fig.1 Scanning electron micrographs of the fractured (a) and the scraped (b) surfaces for $Nd_{1.85}Ce_{0.15}CuO_{4-y}$ (reduced by annealing in vacuum).

Figure 2 shows the valence band spectra for the scraped and the fractured surfaces of NCCO and BKBO. The valence bands of the former and the latter are mainly composed of the Cu3d - O2p and the Bi6s - O2p mixed bands, respectively [5]. The Fermi edge is clearly observed for the fractured surfaces in both systems, which is in good agreement with the metallic property of these materials. For the scraped surface, on the other hand, the Fermi edge is considerably dulled for NCCO. This shows that the surface property is changed from metallic to non - metallic. Actually, electric resistance for the scraped surface was about three orders of magnitude larger than that for an unscraped surface. The main band at about 4 eV, which is attributed to mainly $3d^9\underline{L}$ ($\underline{L}$ denotes a ligand hole) final state [3], is sharpened by scraping : the line width decreases from 5.0 to 4.3 eV. This suggests that structural changes occur at the surface. For BKBO, the spectral intensity at the Fermi level is remarkably decreased by scraping, although the line shape of the main band at about 4 eV is unchanged. These results imply that the valence states in the surface region of NCCO and BKBO are considerably modified by scraping.

In order to investigate the alteration of the electronic states, core level photoemission spectra were measured. Figure 3 shows the Cu2p spectra for the fractured and subsequently scraped surfaces of NCCO. By scraping, the main line ($\underline{2p}3d^{10}\underline{L}$; $\underline{2p}$ denotes a 2p hole) at about 933 eV is sharpened and shifted to the low-binding-energy side and the satellite line ($\underline{2p}3d^9$) at about 943 eV is decreased in intensity. The intensity ratio of the satellite to the main line is 0.30 for the fractured surface and 0.11 for the scraped. The binding energy of $Cu2p_{3/2}$ for the scraped surface (932.3 eV) is close to that for Cu_2O (932.0 eV). These results mean an increase of Cu^+ [6], i.e., the reduction of Cu^{2+} ions.

Figure 4 shows the $Bi4f_{7/2}$ spectra for the fractured and the scraped surfaces of BKBO, in which the broad lines suggest the existence of more than one chemical state for Bi atoms. The lines are divided into four peaks which correspond to Bi^{3+}, Bi^{5+}, Bi^{4+} (Bi^{4+} is the Bi state which is dominant in $BaBiO_{3-\delta}$), and energy loss components [7]. The Bi^{3+} component is dominant for the scraped surface. On the other hand, the Bi^{5+} and Bi^{4+} components have larger peak area of the line for the fractured surface than those for the scraped surface. These results suggest the reduction of Bi ions from Bi^{4+} and Bi^{5+} to Bi^{3+} by scraping.

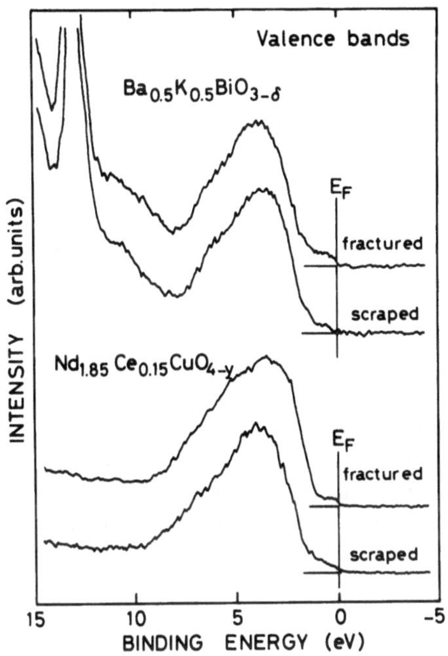

Fig.2 Valence band spectra for the fractured and the scraped surfaces of $Nd_{1.85}Ce_{0.15}CuO_{4-y}$ and $Ba_{0.5}K_{0.5}BiO_{3-\delta}$. Clear Fermi edge is observed for only the fractured surfaces in both systems.

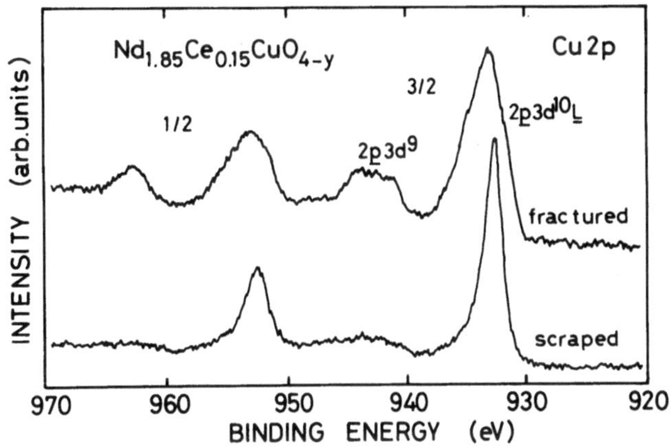

Fig.3 Cu2p spectra for the fractured and the scraped surfaces of $Nd_{1.85}Ce_{0.15}CuO_{4-y}$.

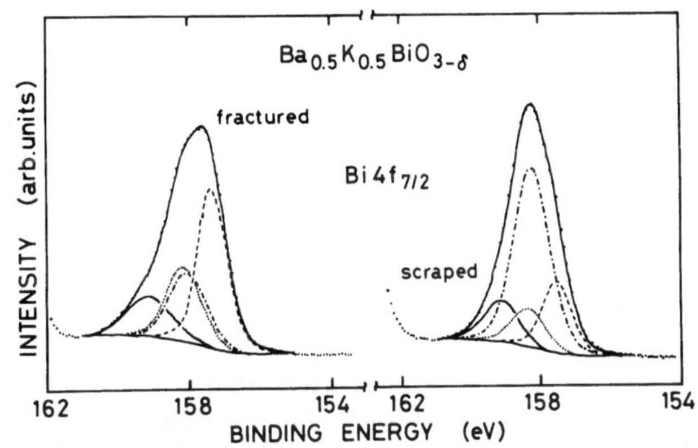

Fig.4 $Bi4f_{7/2}$ spectra for the fractured and the scraped surfaces of $Ba_{0.5}K_{0.5}BiO_{3-\delta}$. The lines are divided into four peaks : Bi^{3+} (-·-·-), Bi^{5+} (······), Bi^{4+} (- - -), and energy loss (—) components.

Table 1 Full width at half maximum (FWHM) of the O1s and $Ba4d_{5/2}$ lines for the scraped and fracture surfaces of $Nd_{1.85}Ce_{0.15}CuO_{4-y}$ and $Ba_{0.5}K_{0.5}BiO_{3-\delta}$. The precision in the FWHM is within ± 0.02 eV.

material	surface	FWHM (eV) O1s	FWHM (eV) $Ba4d_{5/2}$
$Nd_{1.85}Ce_{0.15}CuO_{4-y}$	scraped	1.88	---
	fractured	1.56	---
$Ba_{0.5}K_{0.5}BiO_{3-\delta}$	scraped	1.44	1.39
	fractured	1.38	1.28

It is found that the O1s and Ba4d lines are considerably broadened by scraping as summarized in Table 1. The broadening of core lines was reported for the ion sputtered-surfaces of several oxides such as SiO_2 [8 - 10] and Al_2O_3 [10], in which no reduction of metallic ions is accompanied. This was attributed to structural changes by strains and defects [8 - 10]. For NCCO, the broadening of x-ray diffraction peaks were observed for the scraped surface [3]. Therefore, the broadening of the core lines can be explained by structural changes and oxygen defects. We should emphasize that the surface degradation due to structural changes and oxygen defects modifies the valence states of the scraped surfaces (Fig.2).

These results are quite different from those for the Bi-Sr-Ca-Cu-O superconductors : the scraped surfaces consist of the stable Bi - O layers even for the sintered samples because of the cleavage between the double Bi - O layers [11]. The cleavage, which is due to weak interactions of the $6s^2$ lone pairs localized at Bi atoms [11], can relax scraping-induced stress by the slide between the layers facing each other [12]. On the other hand, for NCCO and BKBO with higher three- dimensional structures, it may be difficult to relax the stress by the slide of atomic layers; therefore, strains and defects would be induced in the surface region by scraping.

We found weak C1s lines for the fractured surfaces in both systems, whereas no C1s lines were observed for the scraped surfaces (the spectra are not shown here). These lines for the fractured surfaces may be attributed to adsorbates and contaminants at pore surfaces or grain boundaries (Fig.1 (a)). These pores and grain boundaries are broken and new surfaces appear by scraping (Fig.1 (b)), therefore they disappeared. However, the electronic states of the scraped surfaces are considerably modified as shown in Fig.2. It should be emphasized that the electronic states of the surfaces are little altered by fracturing, even though there are small amounts of carbon contaminations. This concludes that the fractured surfaces are suitable for the photoemission study of sintered NCCO and BKBO.

ACKNOWLEDGEMENT

This work was partly supported (Y.F., Y.S., and M.T.) by the Grant of Aid from the Ministry of Education, Science, and Culture of Japan.

* present address : Electronic Research Center,.NKK Corporation, Toyama 934, Japan

REFERENCES

1. Seah MP, Dench WA (1979) Surf Interf Anal 1: 2 - 11
2. Egdell RG, Flavell WR, Golden MS (1990) Supercond Sci Technol 3: 8 - 19
3. Suzuki T, Nagoshi M, Fukuda Y, Oh-ish K, Syono Y, Tachiki M (1990) Phys Rev B 42: 4263 - 4271
4. Ueki K, Tokiwa A, Kikuchi M, Suzuki T, Nagoshi M, Suzuki R, Kobayashi N, Syono Y (1989) In: Ishiguro T, Kajimura K (eds) Advances in Superconductivity II, Proceedings of the 2nd International Symposium on Superconductivity (ISS'89). Springer - Verlag, Tokyo, pp 489 - 492
5. B.Batrogg (1989) In: Kamimura H, Oshiyama A (eds) Mechanisms of High Temperature Superconductivity, Springer Series in Materials Science Vol.21. Springer - Verlag, Belrin, pp 324 - 331
6. Panzner G, Egert B, Schmidt HP (1985) Surface Science 151: 400 - 407
7. Nagoshi M, Fukuda Y, Suzuki T, Ueki K, Tokiwa A, Syono Y, Tachiki M (submitted) Phys Rev B
8. Hofmann S, Thomas III JH (1983) J Vac Sci Technol B 1: 43 - 47
9. Ajioka T, Ushio S (1986) Appl Phys Lett 48: 1398 - 1402
10. Hashimoto H, Hirokawa K, Fukuda Y, Suzuki K, Suzuki T, Usuki N, Gennai N, Yoshida S, Koda M, Sezaki H, Horie H, Tanaka A, Ohtubo T (in press) Anal Chem
11. Syono Y, Kikuchi M, Nakajima S, Oku T, Hiraga K, Kobayashi N, Iwasaki H, Muto Y (1989) In: Jorgensen JD, Kitazawa K, Tarascon JM, Thompson MP, Torrance JB (eds) High Temperature Superconductivity, Relationship between Properties, Structures, and Solid State Chemistry, MRS Symp Proc Vol.156. Materials Research Society, Pittsburg, pp 229 - 238
12. Syono Y, Nagoshi M, Kikuchi M, Tokiwa A, Aoyagi E, Suzuki T, Kusaba K, Fukuoka K (1989) In: Schmidt SC, Johnson JN, Davison LW (eds) Shock Compression of Condensed Matter - 1989. Elsevier Science Publishers, B.V., pp 579 - 582

The Effect of Oxygen Nonstoichiometry on the Structure of the Low T_c Phase in Bi-Sr-Ca-Cu-O Superconductors

SHIGEKI KOBAYASHI, KAZUHIKO DOHMAE, and NORITOMO SUZUKI

Toyota Central Research and Development Laboratories, Inc., Nagakute, Aichi, 480-11 Japan

ABSTRACT

The oxygen content of the low Tc phase was varied by annealing in N_2 and O_2 at 500–750 °C. The oxygen unloading increased the Tc up to 100 K and the length of c–axis by 0.01nm. The associated structural change was examined by X–ray diffraction, XPS, and HRTEM. XPS measurements were performed on the clean fracture surface. No major difference was detected on Bi and Cu spectra , indicating that the change of oxygen content was too small to cause detectable change in the valence of Bi and Cu ions. High resolution TEM showed that the average period of modulation in Bi–O layer was unvaried but the modulation was locally disordered by oxygen unloading. These results strongly indicate that oxygen unloading takes place from the Bi–O layer and the Tc of low Tc phase is mainly controlled by the oxygen in Bi–O layer. Preliminary EXAFS data were also consistent with this idea.

KEY WORDS: BSCCO, oxygen nonstoichiometry, TEM, XPS, structural modulation

INTRODUCTION

Oxygen nonstoichiometry is a key parameter to control the properties of the oxide superconductors such as YBCO[1], BSCCO[2], and TBCCO[3]. In Bi–Sr–Ca–Cu–O(BSCCO) system[4,5], the critical temperature of the low Tc phase was affected by annealing at T > 200 °C (Fig. 1) and annealing in nitrogen at 750 °C increased the Tc up to 100 K. Since the low Tc phase of BSCCO is of an over–doped system, the Tc is increased by annealing in nitrogen(reducing the oxygen content), which is opposite to the behavior observed in YBCO. The change in the stoichiometric parameter δ in $Bi_2Sr_2CaCu_2O_{8+\delta}$ was as small as 0.1–0.2[6]. The similar annealing effect was also observed in the high Tc phase of BSCCO but its effect was much smaller than that observed in the low Tc phase. In this paper, we focused on the low Tc phase of BSCCO and examined the effect of oxygen content on its structure by X–ray diffraction, high resolution TEM and XPS.

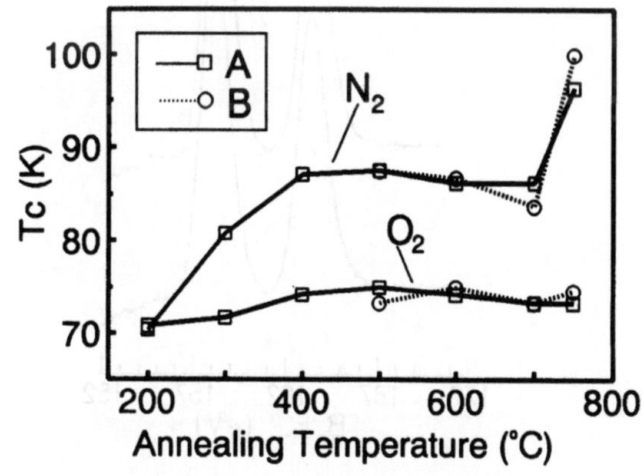

Fig. 1. The effect of annealing on the Tc of the low Tc phase: specimen A[4], B[5].

EXPERIMENTAL

Details on specimen preparation were described in the previous papers[4,5]. The starting powder was solid–state–reacted one of $Bi_2Sr_2CaCu_2O_x$ composition. It was confirmed by X–ray diffraction that sintered specimens contained mainly the low Tc phase. Specimens were annealed in oxygen and nitrogen at 500–750 °C for 2 hr, followed by furnace–cooling with a rate of about 1 °C/min. XPS measurements were performed on clean fracture surface of rod specimens which were fractured under vacuum inside the instrument. TEM foils were prepared by ion–milling disc samples or by crushing bulk samples. Foils were examined at 200 kV with a JEOL 2000EX II electron microscope.

RESULTS AND DISCUSSION

Using specimens containing only the low Tc phase, structural change due to annealing was examined by X–ray diffraction. For the specimens annealed at T < 700 °C, the weight change during annealing was very small (0.1–0.3 %) and no drastic change of the structure was detected but the length of c–axis was increased by 0.01nm with oxygen unloading [5,6]. During annealing in nitrogen at higher temperature, reducing reaction caused large weight reduction and produced second phases. However, no evidence was noticed in this study that major portion of the specimens transform into the high Tc phase.

As shown in Fig. 1[4,5], oxygen unloading (annealing in nitrogen) increased the Tc of low Tc phase in BSCCO system. This is opposite to the behavior observed in YBCO system where

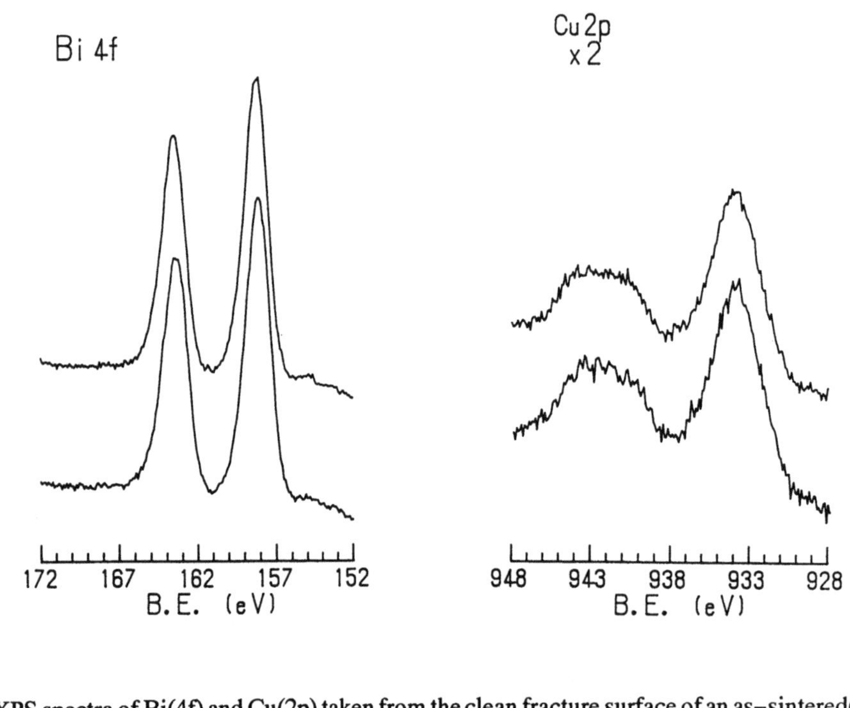

Fig. 2. The XPS spectra of Bi(4f) and Cu(2p) taken from the clean fracture surface of an as–sintered(untreated) specimen(upper curve) and a specimen annealed in nitrogen(lower curve).

oxygen loading increases Tc. This can be explained by the fact that the low Tc phase in BSCCO is of over-doped system[7]. In BSCCO system, the Bi ions as well as the Cu ions can take on multivalencies. It is then of great interest to know whether oxygen atoms are removed from Bi–O or Cu–O layer. However, as already mentioned, the variation of the oxygen content in the low Tc phase of BSCCO was very small compared to the YBCO system. Also the presence of the modulation in the Bi–O layer and the solid solution among Bi, Sr, Ca atoms make the structural analysis very difficult.

XPS measurements were performed on clean surface of the low Tc phase. No significant changes were observed on the spectra of Bi(4f), Sr(3d), Ca(2p) and Cu(2p). The spectra of Bi and Cu were shown in Fig.2. These measurements indicate that the variation of oxygen content was too small to cause any detectable change in the valence of Cu or Bi atoms.

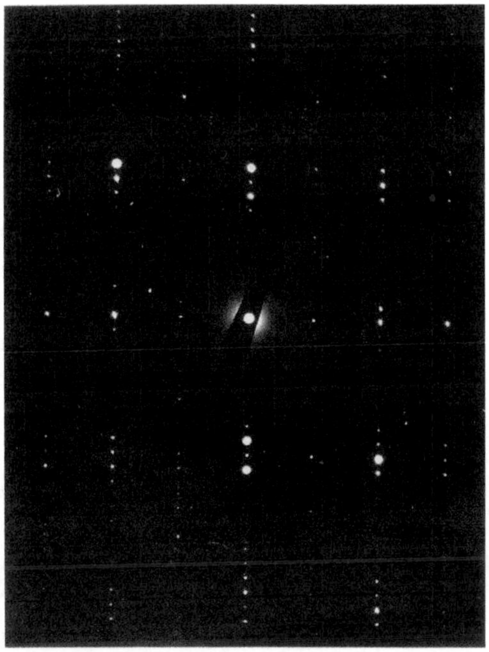

Fig. 3. Electron diffraction pattern corresponding to Fig 4a for a specimen annealed in nitrogen; beam normal to (001).

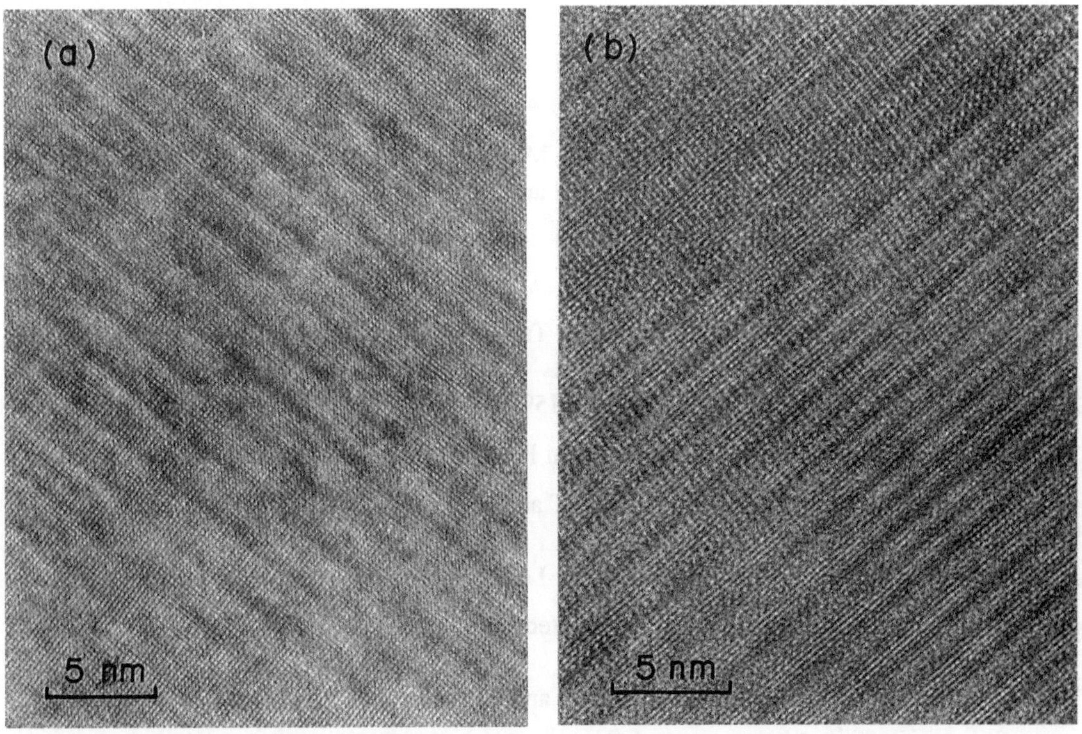

Fig. 4. High resolution TEM micrographs for specimens annealed in (a) nitrogen and (b) oxygen. Pictures were taken with an electron beam parallel to c–axis.

The presence of incommensurate structural modulation in Bi–O layer is now well known. The cause of this modulation is thought to be due to the excess oxygen atoms in Bi–O layer. The addition of Pb into the BSCCO superconductors caused the change in the modulation of Bi–O layer[8]. Similar change of modulation is also expected in the low Tc phase by oxygen unloading due to annealing. High resolution TEM was attempted to look for the possible local disordering of modulation caused by annealing. The analyses of (001) electron diffraction patterns showed that the average period of modulation was unchanged and was 2.7nm(Fig. 3). However, high resolution TEM indicated that oxygen unloading caused local disordering of modulation as shown in Figs. 4a and b, which were taken with an incident beam parallel to the c axis. Figure 4a shows that the image corresponding to the modulation was wavy and discontinuous in the specimen annealed in nitrogen. These features were clearly different from those for the specimens untreated or annealed in O_2(Fig. 4b).

The change in the c–axis length due to annealing indicates that some structural modification takes place during annealing. However, results of X–ray diffraction and XPS showed no significant difference of structure between oxygen loaded and unloaded specimens, suggesting that structural modification due to oxygen unloading is small. From the TEM observations, it was found that the average period of modulation in Bi–O layer was unvaried but local arrangement was disordered by oxygen unloading. These results strongly suggest that oxygen unloading takes place from the Bi–O layer and the Tc of low Tc phase is mainly controlled by the oxygen in Bi–O layer. EXAFS study on this matter is under way. Preliminary results[9] measured at the Cu K–edge showed no difference between sample annealed in nitrogen and untreated sample. This is consistent with the above conclusion.

ACKNOWLEDGEMENTS

The authors would like to thank Dr. T. Hioki, Mr. N. Suzuki and Mr. M. Ohkubo, members of superconductor research group , for their interest and discussion.

REFERENCES

1. Takayama–Muromachi E, Uchida Y, Ishii M, Tanaka T, Kato K (1987) Jpn. J. Appl. Phys. 26: L1156–58.

2. Groen WA and Leeuw DM (1989) Physica C159: 417–421.

3. Shimakawa Y, Kubo Y, Manako T and Igarashi H (1989) Phys. Rev. B40: 11400–11402.

4. Kobayashi S and Wada S (1990)In: Ishiguro T and Kajimura K (eds) Advaces in Superconductivity II Springer–Verlag, Tokyo, pp 125–128.

5. Kobayashi S and Wada S (1991) J. Mater. Sci. Lett. 10: L73–76.

6. Kobayashi S In: Kajimura and Hayakawa H (eds) Advaces in Superconductivity III Springer–Verlag, Tokyo, pp 411–414.

7. Koike Y, Iwabuchi Y, Hosoya S, Kobayashi N and Fukase T (1989) Physica C159: 105–110.

8. Matsui Y, Maeda A, Uchinokura K and Takekawa S (1990) Jpn. J. Appl. Phys. 29: L273–276.

9. Sakurai K and Kobayashi S : unpublished work.

2.3 Transport and Thermal Properties

The Effect of Nd Substitution on Crystal Structure and Resistivity in $La_{2-y}Sr_yCuO_{4-\delta}$

Y. GOTO, T. KAMIYAMA, T. OHSHIMA, and H. ASANO

Institute of Materials Science, University of Tsukuba, Tsukuba, Ibaraki, 305 Japan

ABSTRACT

The effect of Nd substitution for La on crystal structure and electrical resistivity in $La_{2-y}Sr_yCuO_{4-\delta}$ is investigated. Single phase was obtained for x=0–0.4, y=0; x=0–0.8, y=0.1; x=0–0.8, y=0.2; x=0–1.0, y=0.3 in $La_{2-x-y}Nd_xSr_yCuO_{4-\delta}$. The Nd substitution for La induces an increase in the orthorhombicity. This is explained by the bond-length mismatch between LaO and CuO_2 layers because the Nd ion is smaller than the La ion. It was shown that the phase transition to low-temperature tetragonal (LTT) phase like that in $La_{2-z}Ba_zCuO_{4-\delta}$ (z=0.125) occurs in rare-earth doped $La_{1.85}Sr_{0.15}CuO_4$. The phase transition explains the suppression of T_c by Nd substitution for La and various behavior of electrical resistivity.

KEY WORDS: $La_{2-x-y}Nd_xSr_yCuO_{4-\delta}$, $La_{2-z}Ba_zCuO_{4-\delta}$, crystal structure, electrical resistivity

INTRODUCTION

It has been pointed out that the suppression of superconductivity by lanthanide substitution for La in $La_{2-y}Sr_yCuO_{4-\delta}$ is much larger than that for Y in $YBa_2Cu_3O_{7-\delta}$ [1]. As the reason for the depression of T_c in $La_{2-y}Sr_yCuO_{4-\delta}$, several mechanisms have been proposed. Tarascon *et al.* have suggested the unit-cell volume contraction due to a size effect of rare-earth ion rather than magnetic interaction between Cu and the rare-earth atom [2]. However, Takagi *et al.* have speculated on local strain introduced by the difference in ionic radii or the oxygen vacancy [1].

On the other hand, some signs on structural instability comparable to that in $La_{2-z}Ba_zCuO_{4-\delta}$ (z=0.125) were reported by several authors [3-5]: a slight depression in T_c also around y=0.125 [4], and lattice stiffening and a small attenuation peak at $T\approx 10$ K in ultrasound measurements on the sample of $y\approx 0.12$ [5]. These phenomena have been regarded as an incipient phase transition at low temperature.

Recently, it was shown that the low-temperature phase transition like that in $La_{2-z}Ba_zCuO_{4-\delta}$ (z=0.125) occurs in rare-earth doped $La_{1.85}Sr_{0.15}CuO_4$ [6]. This phase transition occurs in Nd-, Gd-, Eu-, Tb- and Dy-doped systems but not in Pr-doped one. In $La_{1.25}Nd_{0.60}Sr_{0.15}CuO_4$, the temperature dependence of the neutron powder-diffraction peaks gives strong evidence for a discontinuous structural transformation to low temperature tetragonal (LTT) phase.

In order to investigate the effect of Nd doping, we have synthesized $La_{2-x-y}Nd_xSr_yCuO_{4-\delta}$ ($0<x<1.4$, $0<y<0.4$), and studied the structure, electrical resistivity and oxygen content for various x and y.

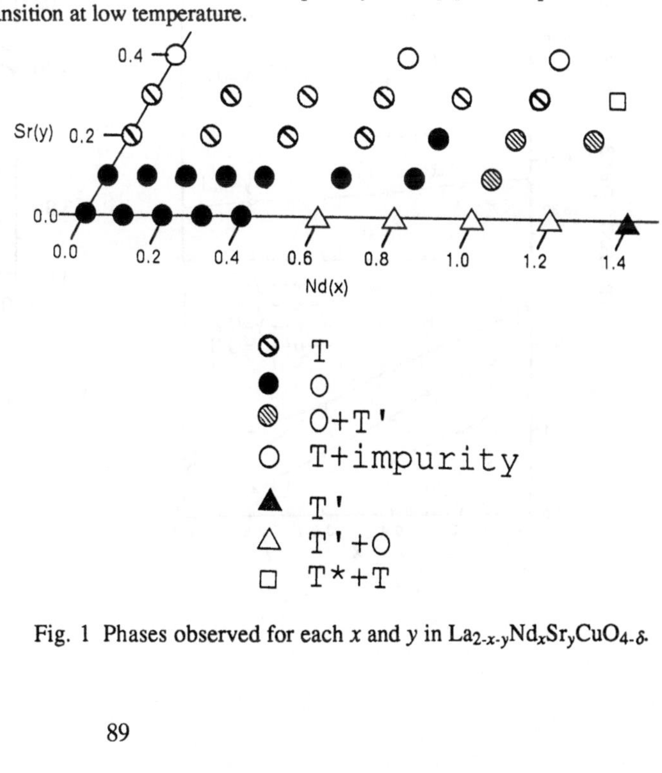

Fig. 1 Phases observed for each x and y in $La_{2-x-y}Nd_xSr_yCuO_{4-\delta}$.

EXPERIMENTAL

Samples used in this study were prepared by the solid-state reaction method. La_2O_3, Nd_2O_3, $SrCO_3$ and CuO powders were mixed in a ratio of a chemical formula $La_{2-x-y}Nd_xSr_yCuO_{4-\delta}$ ($0< x <1.4$, $0< y <0.4$), and calcined at 900°C for 3 hours in air. They were ground, pelletized and sintered at 1130°C for 16 hours in air, followed by furnace cooling.

The obtained samples were characterized by X-ray powder diffraction on a Rigaku RAD-C diffractometer. Phases which appeared for each x and y are shown in Fig.1. Orthorhombic single phase was obtained for x=0–0.4, y=0; x=0–0.8, y=0.1; x=0.8, y=0.2; and tetragonal one for x=0–0.6, y=0.2; x=0–1.0, y=0.3. For single-phase samples, the structural parameters were determined by Rietveld method using a program RIETAN [7].

The oxygen content of the samples was determined by an iodometric titration technique.

The electrical resistivity of the samples was measured by the usual four-probe method from room temperature down to about 10 K with a current of 1 to 5 mA. The T_C was defined from the midpoint of the resistivity transition.

RESULTS AND DISCUSSION

Figure 2 shows the variation of the lattice parameters of single-phase $La_{2-x-y}Nd_xSr_yCuO_{4-\delta}$ at room temperature. In order to compare the lattice parameters of tetragonal phase with those of orthorhombic one, the lattice parameter a of tetragonal phase was multiplied by $\sqrt{2}$. The lattice parameter b decreases considerably with x by a size effect of the Nd ion which is smaller than the La ion. On the other hand, the difference between the lattice parameters a and c, or the orthorhombic distortion, increases with x. In order to explain this result, we introduce the tolerance factor [8], which is a measure of the bond-length matching, by

$t=l(\text{R–O})/\sqrt{2}l(\text{Cu–O})$,

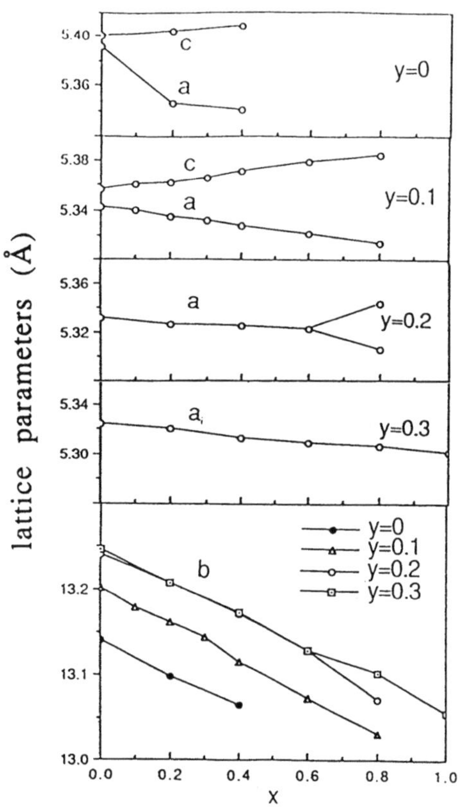

Fig. 2 Lattice parameters of $La_{2-x-y}Nd_xSr_yCuO_{4-\delta}$ at room temperature.

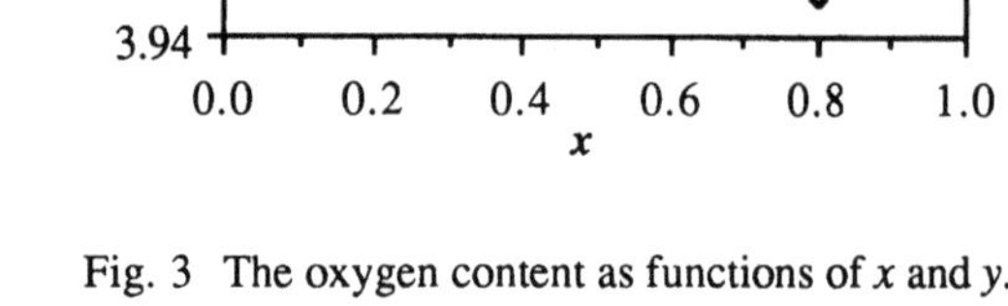

Fig. 3 The oxygen content as functions of x and y.

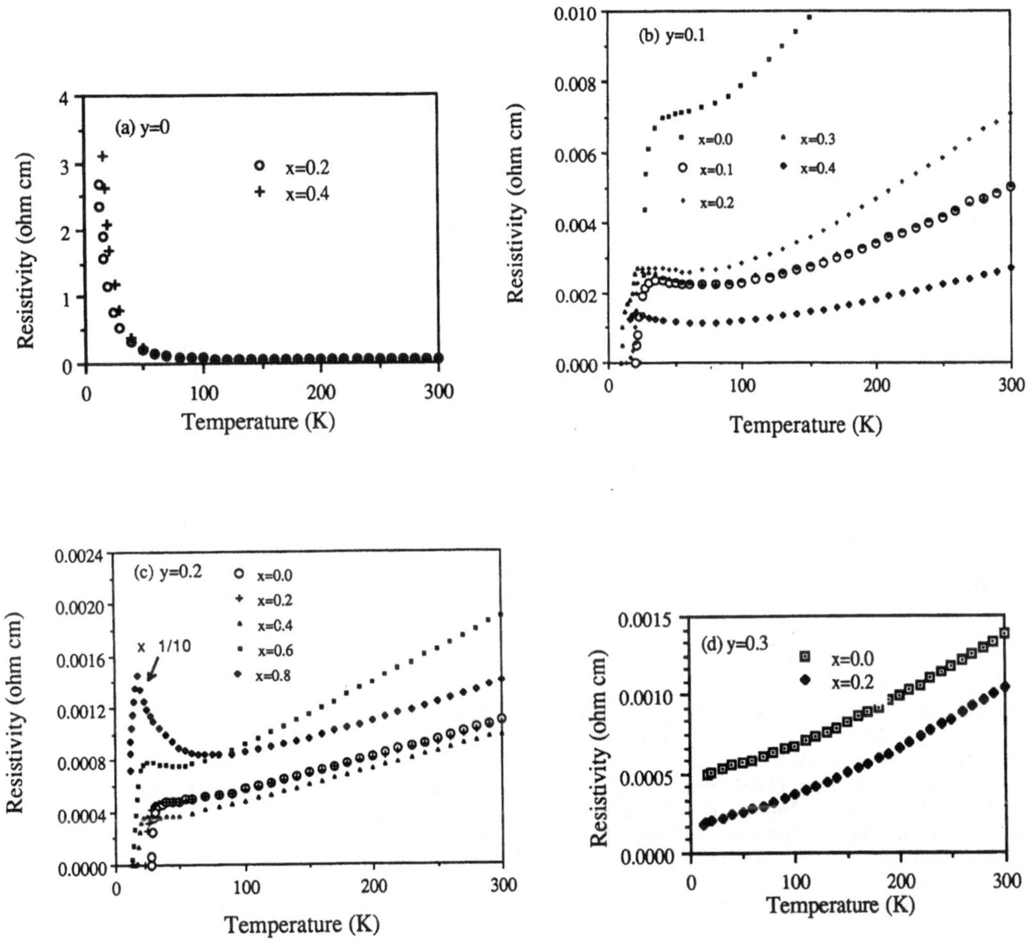

Fig. 4 Temperature dependence of the electrical resistivity for $La_{2-x-y}Nd_xSr_yCuO_{4-\delta}$:(a) y=0, (b) y=0.1 (c) y=0.2 and (d) y=0.3.

where R is the rare-earth ion, and the bond lengths l(R–O) and l(Cu–O) are taken as the sums of the ionic radii [9]. The perfect matching of the bond lengths occurs for t=1. However, $t\approx0.9$ in the case of La_2CuO_4. Therefore, the CuO_2 plane is under compression and the LaO plane under tension. The stress between two layers may be relieved by a tilting of the CuO_6 octahedra making Cu–Cu distance shorter, when the crystal structure is distorted from tetragonal to orthorhombic [8,10]. If La^{3+} ions are substituted by smaller Nd^{3+} ions, the tolerance factor becomes smaller and the bond-length mismatch is enhanced. This explains the increase of orthorhombicity with x [10].

Figure 3 shows the oxygen content as functions of x and y. For y=0, it increases with increasing x. This is easily understood because introduction of excess oxygen between LaO layers expands the layer and relax the bond-length mismatch between LaO and CuO_2 layers [10]. This tendency also holds for y=0.1 and 0.2, and x<0.2. For y=0.1 and 0.2, and x>0.4, however, oxygen content gradually decreases with x. For y=0.3, oxygen content is less than four because the oxygen site in the CuO_2 plane easily becomes deficient. Namely, two tendencies coexist and thus the data scattered. Samples with y=0.3 should be prepared under high oxygen pressure [10].

Figure 4 shows the temperature dependence of the electrical resistivity for $La_{2-x-y}Nd_xSr_yCuO_{4-\delta}$. For y=0, the electrical resistivity shows semiconducting behavior over the measured temperature range. With increasing y, the resistivity drastically decreases and T-linear resistivity dominates at high temperature. For y=0.1, normal state resistivity decreases with increasing x, which may be explained by the improvement of crystality by Nd doping. T_c of the samples with y=0.1 and 0.2 decreases with increasing x, as was observed in the sample with y=0.15 in previous reports [1,6,11]. For y=0.3, resistivity decreases monotonically with decreasing temperature but does not show superconductivity above 16 K (x=0) and 12 K (x=0.2). These compounds are non-superconducting metals.

The resistivity minimum observed in $La_{1.875}Sr_yBa_{0.125-y}CuO_4$ ($0 \leq y \leq 0.1$) around $T_m \approx 60$ K [3] appears also in the present samples with y=0.1 and 0.2. Similar phenomenon was observed by Tarascon *et al.* in $La_{1.6}Ln_{0.2}Sr_{0.2}CuO_4$ (Ln=Gd, Eu) [2]. They ascribed it to magnetic scattering such as , for example, the Kondo effect. Maeno *et al.* speculated that T_m characterizes an instability in electronic state, which precedes the structural transition [3]. In $La_{1.875}Sr_yBa_{0.125-y}CuO_4$, T_m decreases slightly from 60 K to 40 K with increasing y ($0 \leq y \leq 0.1$) [3]. In $La_{2-x-y}Nd_xSr_yCuO_{4-\delta}$, T_m stays almost constant ($0 \leq x \leq 0.4$, y=0.1) or increases slightly with x ($0 \leq x \leq 0.8$), and reaches $T_m \approx 75$ K in x=0.8 and y=0.2. The magnitude of the anomaly at T_m, which is defined by the ratio of the maximum resistivity around T≈20 K to the minimum one at T_m, becomes larger in larger x and reaches the largest for x=0.8 and y=0.2. This result is in contrast with that of $La_{1.875}Sr_yBa_{0.125-y}CuO_4$ ($y \leq 0.025$), where the magnitude of the anomaly at T_m does not change much [3]. Two mechanisms which Tarascon *et al.* and Maeno *et al.* have proposed may interplay and explain the difference. However, according to the result of Büchner *et al.* [6] on $La_{2-x-y}Nd_xSr_yCuO_{4-\delta}$ (y=0.15), fraction of the LTT phase increases with increasing x. Therefore, the difference between two results mentioned above is also ascribable to the change in fraction of the LTT phase by Nd doping.

The two-step resistivity transition reported in $La_{2-z}Ba_zCuO_{4-\delta}$ (z=0.125) and $La_{1.875}Sr_yBa_{0.125-y}CuO_4$ is also observed clearly in the sample with x=0.3 and y=0.1, but not observed in y=0.2.

REFERENCES

1. H. Takagi, S. Uchida, H. Eisaki, S. Tanaka (1988) J. Appl. Phys. 63: 4009
2. J. M. Tarascon, L. H. Greene, W. R. Mckinnon, G. W. Hull (1987) Solid State. Commun. 63: 499
3. Y. Maeno, A. Odagawa, N. Kakehi, T. Suzuki, T. Fujita (1991) Physica C 173: 322
4. H. Takagi, T. Ido, S. Ishibashi, M. Uota, S. Uchida, Y. Tokura (1989) Phys. Rev. B 40: 2254
5. T. Fukase, T. Nomoto, T. Hanaguri, T. Goto, Y. Koike (1990) Physica B 165&166: 1289
6. B. Büchner, M. Braden, M. Cramm, W. Schlabitz, W. Schnelle, O. Hoffels, W. Braunisch, R. Müller, G. Heger, D. Wohlleben (1991) Proc. of M^{2}S-HTSC, Kanazawa, Physica C (to be published)
7. F. Izumi, H. Asano, H. Murata, N. Watanabe (1987) J. Appl. Crystallogr. 20: 411
8. J. B. Goodenough, A. Manthiram (1990) J. Solid State Chem. 88: 115
9. R. D. Shannon (1976) Acta Crystallogr. Sect A32: 751
10. Y. Takeda, K. Yoshikawa, O. Yamamoto, M. Takano (1991) J. Solid State Chem. 92: 241
11. B. Chevalier, A. Tressaud, P. Caracino, C. Robin, J. Etourneau (1991) Physica C 175: 435

Influence of Magnetic Ordering on Transport Properties of $Pr_{2-x}Ce_xCuO_4$

NATHANAEL A. FORTUNE[1], KEIZO MURATA[1], FUMITOSHI IGA[1], YUKO YOKOYAMA[1], YOSHIKAZU NISHIHARA[1], and YOSHIKAZU HIDAKA[2]

[1] Electrotechnical Laboratory, Tsukuba, Ibaraki, 305 Japan
[2] NTT Interdisciplinary Research Laboratories, Tokai, Ibaraki, 319-11 Japan

ABSTRACT

In the 'electron-doped' high T_c copper oxides, the combination of chemical substitution and reduction transforms the anti-ferromagnetic parent compounds into metallic superconductors. To study the influence of chemical substitution and resulting suppression of anti-ferromagnetic ordering on the normal state properties, we have measured the temperature dependent Hall coefficient R_H(T) and resistivity ρ(T) on as-grown non-superconducting $Pr_{2-x}Ce_xCuO_4$ single crystals. We find that the onset of anti-ferromagnetic order at the Cu sites in the Cu-O_2 planes (below T_{Neel}) and the appearance of a weak induced moment on the rare-earth sites induces corresponding changes in R_H(T) and ρ(T). Suppression of anti-ferromagnetic order is necessary but not sufficient for inducing superconductivity.

KEY WORDS: electron superconductors, anti-ferromagnetism, induced moment, Hall coefficient, resistivity

EXPERIMENTAL METHOD

Single crystals of $Pr_{2-x}Ce_xCuO_4$ were grown using an excess of CuO flux, as described in detail elsewhere [1,2]. Crystals with cerium doping levels $x = 0$ and $x= 0.08$ were large enough to determine the orientation of the in-Cu-O_2 plane **a** axes by x-ray diffraction, thinned to 150 μm along the **c** axis (normal to the Cu-O2 planes), then cut into a 6 contact Hall bar pattern. For transport measurements, the electrical current was applied along the in-plane **a** axis. Magnetic fields were applied along the **c** axis. Samples with higher cerium doping levels (x=0.146) were used as is (typical thickness 25 μm) with magnetic fields applied along the **c** axis, as before, but with the electrical current applied in an arbitrary direction within the Cu-O_2 plane. Ohmic contacts with contact resistances on the order of 1 ohm (1-10 ohm for $x = 0.146$) were made using a 90% In,10% Ag solder alloy. Transport measurements were performed continuously as the temperature of the samples was slowly varied (5-10 K/hr) using a 5 tesla superconducting magnet. Spurious contributions to the measured DC voltage across the Hall terminals due to longitudinal resistance components, thermoelectric effects, etc were eliminated using field reversal and the ohmicity of the Hall voltage.

EXPERIMENTAL RESULTS

The temperature dependence of the resistivity ρ(T), relative to that at 300 K ρ(300 K), is shown in Fig. 1 for as-grown (unreduced) crystals with different Ce doping levels.The solid circles, solid line and dashed line correspond to $x = 0$, 0.08 and 0.146, respectively. The nominal electron carrier concentration is expected to increase proportionally with x. The change in the slope of the resistivity with respect to temperature at 300 K from negative to positive with increasing x is consistent with increasing delocalization of the doped carriers with increasing x as the system approaches a metallic state. Fox x= 0, ρ(T) is linear on a log-log plot, consistent with a single power-law temperature dependence over the measured range. The $x = 0.08$ crystal, however, exhibits a change in slope between T > 200 K and T < 100 K. A plot of the temperature dependence of the differential resistivity dρ/dT for this crystal reveals that the change in slope of the resistivity that occurs just below 170 K. As will be discussed below, 170 K corresponds to T_{Neel}, the onset of anti-ferromagnetic ordering of the Cu^{2+} moments in the CuO_2 planes [3,4]. Above 150 K, ρ(T) for the x=0.146 as grown crystal exhibits the positive slope characteristic of the metallic copper-oxide superconductors. At lower temperatures, however, we find a weakly increasing resistivity with decreasing temperature instead of superconductivity. Similar behavior in $Nd_{2-x}Ce_xCuO_4$ thin films has been attributed to weak localization due to oxygen defects [5]. As will be discussed in a separate publication, heat treatment in a sufficiently reducing atmosphere induces superconductivity in x=0.146 crystals but not $x = 0.080$ crystals, as expected.

The temperature dependence of the Hall coefficient $R_H(T)$ is shown in Fig. 2. for $x = 0.08$ and $x = 0.146$. The low intrinsic (undoped) carrier density and large, strongly temperature dependent longitudinal resistivity has to date precluded accurate measurement of $R_H(T)$ for for $x = 0$. The as-grown $x = 0.146$ crystal, which is not known to order anti-ferromagnetically, exhibits a smoothly varying temperature dependent negative Hall coefficient. In contrast, the as-grown $x = 0.08$ crystal, which anti-ferromagnetically orders below $T_{Neel} = 170$ K, exhibits a smooth change in slope below T_{Neel} and a second, more abrupt change in slope below 50 K.

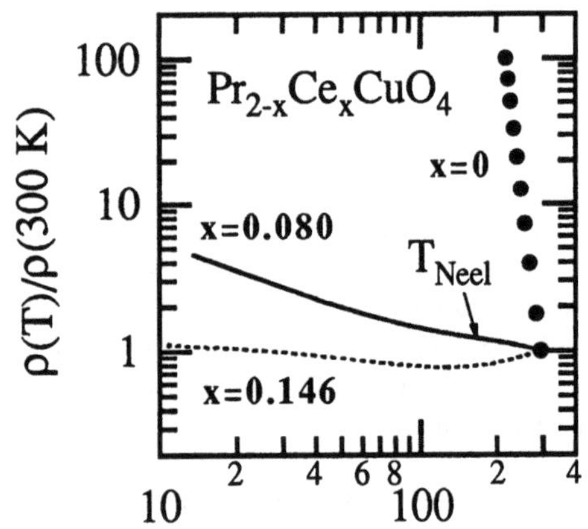

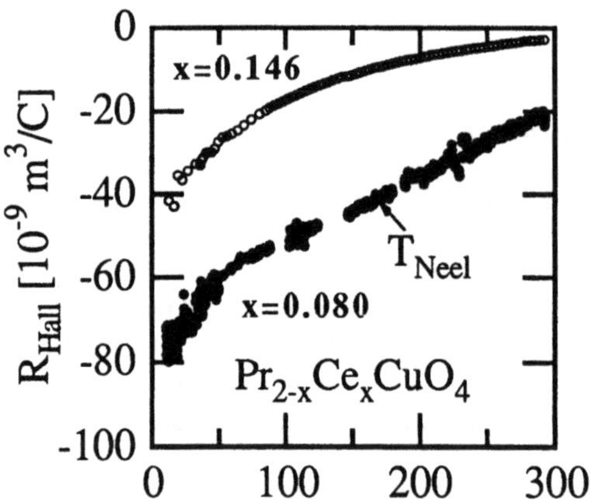

Fig. 1 (left) Temperature dependence of the resistivity relative to that at 300 K, $\rho(T)/\rho_0$, for as-grown $Pr_{2-x}Ce_xCuO_4$ single crystals. The solid circles, solid line and dashed line correspond to $x = 0$, 0.08 and 0.146, respectively.

Fig. 2 (right) : Temperature dependence of the Hall coefficient $R_H(T)$ for two as-grown $Pr_{2-x}Ce_xCuO_4$ single crystals. Solid circles and hollow circles correspond to x =0.08 and 0.146, respectively.

DISCUSSION

Independent neutron diffraction measurements on crystals from the same batches (for $x = 0$ and $x = 0.08$) as those used in the present experiment indicate that anti-ferromagnetic ordering of the Cu^{2+} moments in the Cu-O_2 plane occurs at $T_{Neel} = 255$ K, 170 K and 120 K for $x = 0$, 0.08 and 0.125, respectively [3,4]. For $x = 0$, the crystal-field induced singlet ground state of the Pr^{3+} ion [6] would be expected to quench the Pr^{3+} moment. Second nearest neighbor coupling of the rare-earth ions to the Cu^{2+} moments leads to a small admixture of higher-lying states and a net magnetic moment [4]. In a mean field model, the temperature dependent induced 3D ordered magnetic moment at the rare-earth site $\langle M_{RE}(T)\rangle$ is given by [4]

$$\langle M_{RE}(T)\rangle \propto \chi^*_{||ab}(T) \cdot J \cdot \langle M_{Cu}(T)\rangle \qquad (1)$$

where $\chi^*_{||ab}(T)$ is the single-ion paramagnetic susceptibility of the rare-earth ions within the Cu-O_2 plane, J is the Cu^{2+}-rare-earth interaction energy and $\langle M_{Cu}(T)\rangle$ is the Cu^{2+} 3D ordered magnetic moment [4].

Because of the strong anti-ferromagnetic correlation of the Cu^{2+} moments, the rare-earth ions (rather than the Cu^{2+} moments) dominate the measured bulk susceptibility $\chi_{||ab}(T)$ for $x = 0$ [6] and $x = 0.08$ [7]. Hence $\chi^*_{||ab}(T) \approx \chi_{||ab}(T)$(parallel to the Cu-$O_2$ planes) and we may write

$$\langle M_{RE}(T)\rangle \propto \chi_{\|ab}(T)\cdot J \cdot \langle M_{Cu}(T)\rangle \qquad (2)$$

in place of Eq. 1. Since for $x = 0.08$, structure is observed in $\rho(T)$ and $R_{Hall}(T)$ at temperatures corresponding to the onset (170 K) and saturation ($\approx$ 50 K) of anti-ferromagnetic and induced magnetic order while no corresponding features are observed for $x = 0.146$, in which magnetic ordering is absent, we may surmise that the anti-ferromagnetic ordering is responsible for the observed structure. In a manner analogous to Kondo scattering by magnetic impurities,we suggest that the induced magnetic moment on the rare-earth sites may be responsible for the observed magnetic scattering.

In the case of Kondo scattering from independent magnetic impurities or from a Kondo lattice of independent magnetic moments, the measured Hall coefficient is the sum of an "ordinary" Hall coefficient $R_H^0(T)$ and an "anomalous" magnetic contribution $R_H^{mag}(T)$. That is,

$$R_{Hall}(T) = R_H^0(T) + R_H^{mag}(T), \qquad (3)$$

where $R_H^{mag}(T)$ is given by [8]

$$R_H^{mag}(T) \propto \rho_{mag} \cdot \chi_{Kondo}(T) \cdot [\,1 - aT\cdot\chi_{Kondo}(T)]. \qquad (4)$$

Here ρ_{mag} is the magnetic contribution to the total resistivity, $\chi_{Kondo}(T)$ is the (isotropic) susceptibility due to the magnetic impurity and a is a constant [8]. If the induced rare-earth magnetic moment takes the place of the magnetic impurities in the Kondo magnetic skew scattering model above, then

$$\chi_{Kondo}(T) = \langle M_{RE}\rangle/H \propto (\chi_{\|ab}(T)/H)\cdot J \cdot \langle M_{Cu}(T)\rangle \qquad (5)$$

and by substitution,

$$R_H^{mag}(T) \propto \rho_{mag} \cdot (\chi_{\|ab}(T)/H)\cdot J \cdot \langle M_{Cu}(T)\rangle \cdot [\,1 - aT\cdot(\chi_{\|ab}(T)/H)\cdot J \cdot \langle M_{Cu}(T)\rangle]. \qquad (6)$$

Calculation of $R_H^{mag}(T)$ from $R_{Hall}(T)$ requires knowledge of$R_H^0(T)$, which is not currently theoretically understood, and separation of the "ordinary" and magnetic contributions to $\rho(T)$. Nevertheless we can easily see within this model that temperature dependent changes in $\langle M_{RE}\rangle$ will be reflected in $R_{Hall}(T)$. For example, since $\langle M_{Cu}(T)\rangle$ and $\langle M_{RE}\rangle = 0$ for $T > T_{Neel}$, $R_H^{mag}(T) = 0$ for $T > T_{Neel}$. As T decreases below T_{Neel}, and the 3D induced ordered magnetic moment on the rare-earth sites begins to develop, the temperature dependence of $R_{Hall}(T)$ will begin to deviate from that for $T > T_{Neel}$, as observed. Likewise, $\chi_{\|ab}(T)$, $\langle M_{Cu}(T)\rangle$ (and hence $\langle M_{RE}\rangle$) all saturate at low temperature, becoming temperature independent below roughly 50 K [3,4]. From Eq. 6, we can therefore predict that below 50 K, the temperature dependence of $R_H^{mag}(T)$ will also diminish. As noted in Eq. 3, $R_{Hall}(T)$ is the sum of $R_H^{mag}(T)$ and $R_H^0(T)$. Assuming that $R_H^0(T)$ varies smoothly with temperature, an abrupt change in the temperature dependence of $R_H^{mag}(T)$ at 50 K will appear as an abrupt change in the temperature dependence of $R_{Hall}(T)$, as observed. The smoothly varying background well above T_{Neel}, and well below 50 K will be primarily due to $R_H^0(T)$.

For the $x = 0$ crystal (Pr_2CuO_4), T_{Neel} is too close to room temperature (255 K) and the intrinsic (undoped) carrier density too small to clearly distinguish the onset of magnetic behavior. If, however, we assume that $\rho(T)$ is dominated by the "ordinary", non-magnetic temperature dependent resistivity $\rho_0(T)$, we may infer that $\rho_0(T)$ has an approximately power-law dependence on temperature over the measured temperature range. Measurements of $\rho(T)$ for both $x = 0$ and $x = 0.08$ at elevated temperatures would be useful to verify this power-law dependence.

In closing, we note that the suppression of anti-ferromagnetic order by chemical substitution of Ce for Pr is a necessary but not a sufficient condition for inducing superconductivity, as can be seen from our results for an $x = 0.146$ as-grown single crystal. In the case of $Nd_{2-x}Ce_xCuO_4$ polycrystals, we have already demonstrated the need for heat treatment in a reducing atmosphere in order to delocalize the chemically doped carriers by removing or redistributing oxygen defects in the as-grown material [9,10]. The optimization of superconductivity by chemical doping and reduction and the effects on magnetic and transport properties of $Pr_{2-x}Ce_xCuO_4$ single crystals will be presented in a subsequent publication.

REFERENCES

[1] Hidaka Y, Suzuki M (1989) Nature 388: 635-638

[2] Matsuda M, Endoh Y, Hidaka Y (1991) Physica C 179: 347-352

[3] Thurston TR, Matsuda M, Kakurai K, Yamada K, Endoh Y, Birgeneau RJ, Gehring PM, Hidaka Y, Kastner MA, Murakami T, Shirane G (1990) Phys. Rev. Lett. 65: 263-266

[4] Matsuda M, Yamada K, Kakurai K, Kadowaki H, Thurston TR, Endoh Y, Hidaka Y, Birgeneau RJ, Kastner MA, Gehring PM, Moudden AH, Shirane G (1990) Phys. Rev. 42: 10098-10107

[5] Tanda S, Honma M, Nakayama T (1991) Phys. Rev B. 43: 8725-8728

[6] Allenspach P, Cheong S-W, Dommann A, Fischer P, Fisk Z, Furrer A, Ott HR, Rupp B (1989) Z. Phys. B 77: 185-191

[7] Fortune NA, Murata K, Hidaka Y, Ishibashi M, Yokoyama Y, Iga F, Nishihara Y (in press) Physica C

[8] Coleman P, Anderson PW, Ramakrishnan TV (1985) Phys. Rev. Lett 55: 414

[9] Fortune NA, Murata K, Ishibashi M, Yokoyama Y, Nishihara Y (1991) Phy. Rev. B 43: 12930-12934

[10] Fortune NA, Murata K, Yokoyama Y, Ishibashi M, Nishihara Y (1991) Physica C 178:437-444

Quantum Phase Fluctuation and Conduction Properties of Highly Resistive Superconducting Weak Links

NOBUYUKI YOSHIKAWA, XIAO CHEN, and MASANORI SUGAHARA

Faculty of Engineering, Yokohama National University, Yokohama, 240 Japan

ABSTRACT

We have studied the conduction properties of small superconducting weak links with high normal resistance and inductance in order to study the relation between the macroscopic quantum effect in the weak links and junction parameters. Experimental evidence of a critical resistance which determines the existence of the Josephson effect in the weak links is presented. Unusual nonlinearity which is seemingly dual to the Josephson effect is observed in the junctions with resistance much larger than the critical value.

KEY WORDS: superconducting weak links, phase fluctuation, macroscopic quantum effect, PQT effect, SET effect

INTRODUCTION

Recent theoretical and experimental works on very small tunnel junctions [1-3] predict a charge quantization effect in junctions with very low capacitance and high resistance. Since the charging energy made even by a single electron tunneling is not negligible in such junctions, we can observe a single electron tunneling from the macroscopic measurement of the junction property. Because of the uncertainty principle between the phase and the electron number, the definiteness of the electron number destroys the phase coherency in a small superconductor junction and smears its Josephson effect. In such charge definite state, the junction characteristics assume some duality to the Josephson effect, i.e. the periodic dependence of the junction voltage on the junction charge and the voltage oscillation in a constant current biassed state. This effect associated with the charge quantization has been clearly observed in the case of normal tunnel junctions [4,5].

On the other hand, we have been studying the macroscopic quantum effect (MQE) in the both types of small junction, i.e. tunnel junctions and bridge junctions considering the quantum electrodynamic fluctuation in the junction [6,7]. Our theoretical investigation indicates that each of the two types of junction can have two kinds of MQE: the Josephson effect and the new effect completely dual to the Josephson effect. We are calling the new MQE phase quantum tunneling (PQT) effect. The theory also predicts that the duality of the Josephson effect in the tunnel junction should be observed in the highly resistive bridge junction, while the duality of the Josephson effect in the bridge junction should be observed in the highly resistive tunnel junction.

In our stand point, MQE in the weak links is characterized by the kinetic inductance L_J and the resistance R_J of the junction part. In the junction with small L_J and R_J, the energy necessary for the flux-quantum-tunnel across the weak link is much larger than the electronic quantum fluctuation energy. Since the phase difference between two electrodes is well defined there, the Josephson effect can be expected in such junctions where electrodes are superconductive. On the other hand, in the junction with very large L_J and R_J, the energy for the flux tunneling is so small compared with the electronic quantum fluctuation energy that the large phase fluctuation takes place in the junction.

Because the phase difference is indefinite in such junctions, the Josephson effect is faded out even at low temperatures. As an extreme case, if we assume the large coherent fluctuation of flux across the junction, we can expect the new MQE completely dual to the Josephson effect considering the uncertainty relation, where charge quantization is arising from the large coherent flux fluctuations.

The objective of this study is to experimentally observe the new MQE dual to the Josephson effect in the superconducting weak links. In this report, we represent the preparatory study of weak links with large R_J and L_J where the Josephson effect is smeared and unusual junction characteristics are observed.

EXPERIMENTAL

From the discussion stated in the previous section, in order to increase the junction resistance $R_J=\rho d/S$ and the kinetic inductance $L_J=\mu_0\lambda d/S$, where d and S are the length and the cross section of the weak links and ρ and λ are the resistivity and the magnetic penetration depth respectively, we must fabricate the small weak links with extremely small cross section S. Besides, the length of the bridge junctions should be the order of the coherence length to guarantee the quantum-mechanical coupling between two electrodes. To this end, we fabricated Nb variable thickness bridge junction with narrow width and short length using step edge defined process [8]. First, a SiO step of 20nm height was formed on a Si substrate, and a Nb film of about 20-40 nm thick was deposited on the substrate. After that a resist stencil with narrow opening line of width ~100nm was formed crossing the SiO step by the electron beam lithography. Then, Nb of the opening part was etched by a reactive ion beam at the angle of 45° from the substrate normal so that the SiO step may shadow a narrow Nb filament. In this process, the bridge width and thickness was defined by the height of the SiO step and the length was defined by the width of the narrow opening line made by electron beam lithography. A resultant junction size was about 20nm in width and 100nm in length.

Current-voltage characteristics of the Nb bridge junctions made by this process were measured by the four terminal method at temperatures down to 4.2K. These junctions had relatively large normal resistance (1-50Ω) and exhibited the Josephson effect clearly where the microwave induced constant voltage steps were definitely observed [8]. After the measurement, the bridge part was slightly etched over and again by the reactive ion beam and the changes of the current-voltage characteristics were measured.

RESULT AND DISCUSSION

Figure 1 shows the dependence of the critical current I_c of the bridge junctions at $T=0.8T_c$ (T_c: the critical temperature) on the junction normal resistance R_J which is determined by the linear part of the current-voltage characteristics. With increase of R_J by further ion beam etching, I_c tends to decrease gradually because the area of the cross section of the bridge part is reduced and the superconducting properties of the linked part are degraded by the ion beam bombardment. Besides the abrupt decrease of I_c is found at about $R_J\approx100\Omega$ in these samples. Since the estimated value of the thermal noise current at 4.2K is about 0.1μA, we believe this is not due to the thermal fluctuation.

In Fig. 2 we display the value of the junction normal resistance R_J for the samples measured, where each vertical points correspond to the data given by successive etchings of the same sample. Solid squares indicate the value of R_J where the Josephson effect is observed, and open squares are the data where the Josephson effect is not observed. Our criterion of the Josephson effect is the existence

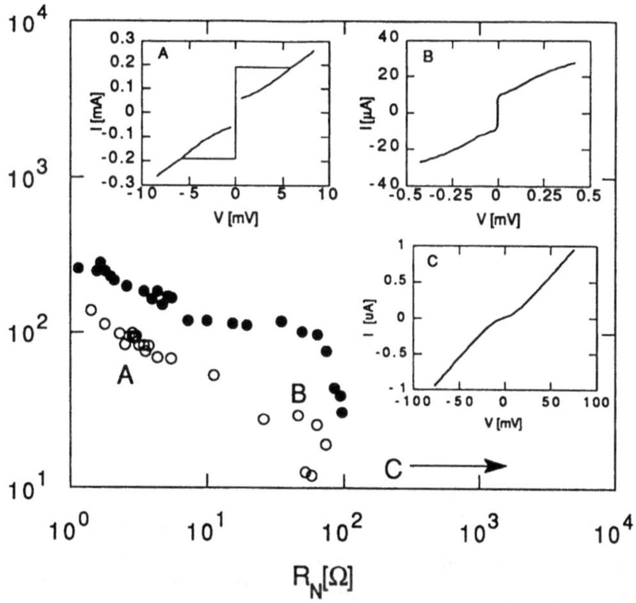

Fig. 1 The dependence of the critical current I_c of the weak links at $T=0.8T_c$ on the junction normal resistance R_N. The inset shows the current-voltage characteristics of junctions measured at T=4.2K corresponding to the data indicated in the main part of the figure.

of the zero-voltage current and the microwave induced constant voltage steps. From the Fig.2, we can recognize that some threshold value of R_J exists for the Josephson effect to be observed. This value is about 100Ω in these samples. The theoretical study of the weak links [7] indicates that the critical value of the junction resistance for the Josephson effect is about several hundred Ω for the kinetic inductance $L_J \approx 10^{-11}$[H] of the bridge part which is estimated from a sample geometry. Experimentally observed threshold resistance is consistent with the theoretically predicted value. Although experimental evidence of a critical resistance for the Josephson effect is suggested here, certainly it is not conclusive. In order to confirm the existence of the critical resistance, we should measure the junction properties at much lower temperatures.

In the inset to Fig.1, we also show the typical current-voltage characteristics at T=4.2K of the bridge junctions corresponding to the data point indicated in the main part of Fig.1. When R_J is sufficiently low compared with the critical value (the data points indicated by A and B) , typical current-voltage characteristics of the weakly coupled junctions are observed distinctly. On the other hand, unusual current-voltage characteristics are observed in the junctions which have R_J much larger than the critical value after the successive ion etching (indicated by C). In Fig.3 we show the temperature dependence of the current-voltage characteristics of the junction with $R_J \approx 80$kΩ. Remarkable features of the current-voltage relation of the high R_J junction are the followings.

(i) The nonlinearity is seemingly dual to that of the Josephson effect.

(ii) There is a voltage offset of about ΔV=20mV.

(iii) With increase of the temperature, the nonlinearity is smeared but the value of the voltage offset seems to be not changed.

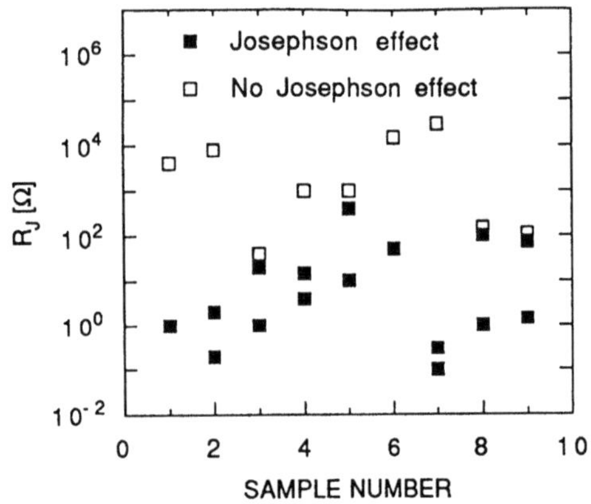

Fig. 2 The value of the junction normal resistance R_J for various samples measured, where each vertical points correspond to the data given by successive etchings of the same sample. Solid squares indicate the value of R_J where the Josephson effect is observed, and open squares are the data where the Josephson effect is not observed

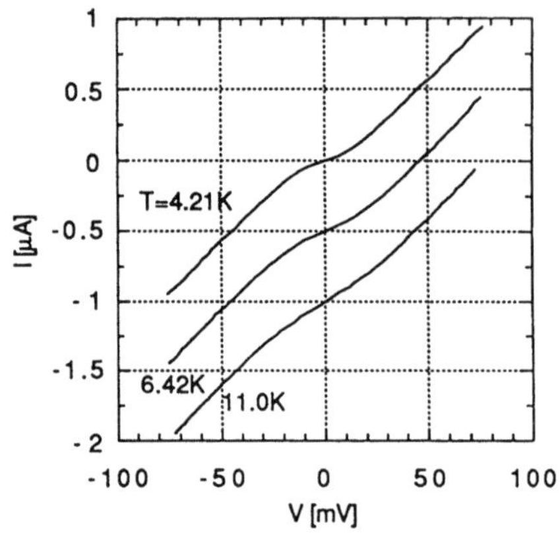

Fig. 3 The temperature dependence of the current-voltage characteristics of the junction with large R_J. Curve of T=6.42K is offset by -0.5μA and curve of T=11.0K by -1μA.

(iv) The nonlinearity is also faded out when the microwave is irradiated slightly to the junction.

There are some possibilities for the origin of these nonlinearity. Firstly the single electron tunneling (SET) effect must be pointed out where the small tunnel junction formed in the grain boundary dominates the junction characteristics. In this case, the estimated junction capacitance is about 0.004fF which corresponds to the junction area of about 5nm. Another possibility is the PQT effect in the weak link itself. To clarify the origin of the nonlinearity, further experimental investigation, such as the measurement of the charge-voltage relation of the junction, is needed.

In summary, the conduction properties of small superconducting weak links with high normal resistance and inductance were measured in order to study the MQE in the weak links. Experimental evidence of the existence of the critical resistance for the Josephson effect was presented in the case of weakly coupled junctions. Unusual nonlinearity seemingly dual to the Josephson effect was also observed in highly resistive junctions.

REFERENCES

1. Widom A, Megaloudis G, Clark T G, Prance H, Prance R J, (1982) J. Phys. A15: 3877
2. Likharev K K, Zorin A B (1985) J. Low Temp. Phys. 59: 347
3. Averin D V, Likharev K K (1986) J. Low Temp. Phys. 62: 345
4. Fulton T A, Dolan G J (1987) Phys. Rev. Lett. 59: 109
5. Delsing P, Likharev K K, Kuz'min L S, Claeson T(1989) Phys. Rev. Lett. 63: 1861
6. Sugahara M, Yoshikawa N (1989) Jpn. J. Appl. Phys. 28: 1816
7. Sugahara M, Yoshikawa N (1991) Jpn. J. Appl. Phys. 30: 263
8. Sugahara M, Yoshikawa N, Furuoya S (1991) IEEE Trans. Magn. MAG-27: 1328

The Properties of $Tl_2Sr_2CuO_z$

KAZUYUKI HAMADA, TETSUYUKI KANEKO, SEIJI ADACHI, and H. YAMAUCHI

Superconductivity Research Laboratory, ISTEC, Koto-ku, Tokyo, 135 Japan

We successfully synthesized nearly single-phase samples of the nominal composition of $Tl_2Sr_2CuO_z$. The temperature dependence of electrical resistivity for the sample showed that this was a normal metal. On the other hand, $Tl_2Ba_2CuO_z$ showed a transition from a superconductor to a nonsuperconducting normal metal, as the excess oxygen content z increased from 6.0 to 6.1. The properties of $Tl_2Sr_2CuO_z$ are disscussed in comparison with those of $Tl_2Ba_2CuO_z$.

KEY WORDS: Tl-based superconductors, substitution effects, superconducting transition

INTORODUCTION

Since the discovery of high temperature superconductivity there have been many similar superconductors discovered. Several of these compounds belong to the family; $Tl_2Ba_2Ca_{n-1}Cu_nO_{2n+4}$ where n=1 to 5. The end member of this group with n=1, i.e. $Tl_2Ba_2CuO_z$ (Tl2201 phase) has the simplest structure in this family. Figure 1 shows the crystal structure of Tl-2201 phase. Shimakawa et al.[1] reported that this compound transformed from normal metallic state to superconducting one by annealing treatment in reducing atmosphere. The nature of the transition was referred to oxygen content. The sample with excess oxygen has metallic characteristies. On the other hand, the sample at z=6.00 is a superconductor with Tc of around 85K.[2] The decrease in the hole concentration due to removal of excess oxygen was previously confirmed by the measurements of Hall coeffcient and electrical resistivity.

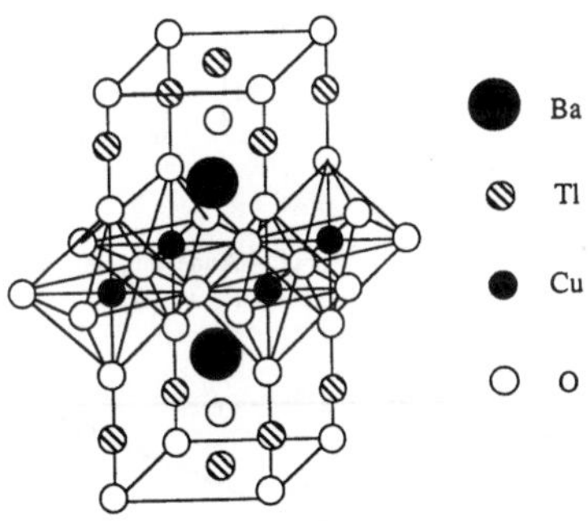

Fig.1 Crystal structure of $Tl_2Ba_2CuO_z$

It is important to investigate substitution effects for understanding the normal metal-superconducor transition in the Tl-2201. Ganguli et al.[3] investigated Sr substitution for Ba. They reported that single-phase $Tl_2Ba_{2-x}Sr_xCuO_z$ of the Tl-2201 structure were obtained in the range of x=0.0 to 1.2. We studied the similar substitution independently and successfully synthesized nearly single-phase full-substituted samples, $Tl_2Sr_2CuO_z$, of theTl-2201 structure. In this paper, we report the synthesis of $Tl_2Sr_2CuO_z$ and its properties in comparison with those of $Tl_2Ba_2CuO_z$.

EXPERIMENTAL

$Tl_2Sr_2CuO_z$ samples were prepared by solid state reaction. As starting materials were used highly pure (99.9%) powders of Tl_2O_3, $SrCO_3$ and CuO. At first a precursor, Sr_2CuO_3, was prepared. $SrCO_3$ and CuO powder mixture at a molar ratio of 2:1 were fired at 910℃ for 10h in air. Powder x-ray diffraction revealed that single-phase Sr_2CuO_3 was obtained. This precursor and Tl_2O_3 were well mixed at a nominal composition of $Tl_2Sr_2CuO_z$. The mixture was pressed into a pellet. It was wrapped with a gold foil and then sintered at 880℃ for 1h in flowing oxygen. For synthesis of $Tl_2Ba_2CuO_z$, Tl_2O_3, BaO_2 and CuO powders were used as starting materials. They were mixed at a nominal composition of $Tl_2Ba_2CuO_z$. A press-formed pellet which was wrapped with a gold foil was sintered at 850℃ for 1h in oxygen. Annealing treatments at 500℃ for 1-5h in argon gas were also carried out for the obtained $Tl_2Sr_2CuO_z$ and $Tl_2Ba_2CuO_z$.

The phases present and the lattice constants were determined by x-ray powder diffraction using Cu Kα radiation. The lattice constants were determined by the least-squares method. Electrical resistivity was measured by a conventional four-probe method. Thermoelectric power was also measured.

RESULTS

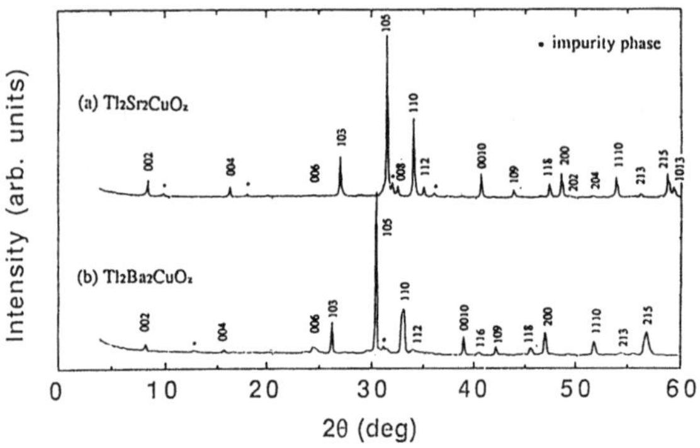

Fig.2 Powder x-ray diffraction patterns for as-sintered (a)$Tl_2Sr_2CuO_z$ and (b)$Tl_2Ba_2CuO_z$

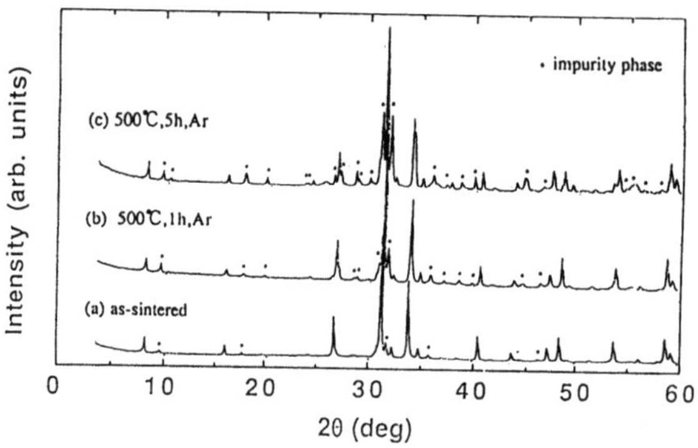

Fig. 3 Powder x-ray diffraction patterns for $Tl_2Sr_2CuO_z$ (a)as-sintered, (b)annealing at 500℃ for 1h in argon and (c)annealing at 500℃ for 5h in argon

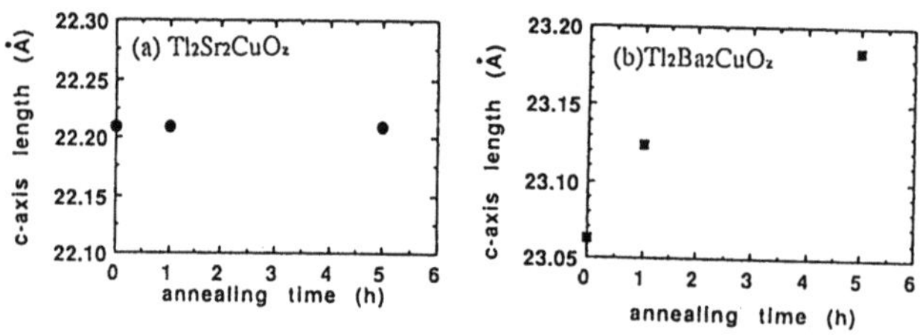

Fig. 4 Annealing time vs. c-axis length of (a) $Tl_2Sr_2CuO_z$ and (b) $Tl_2Ba_2CuO_z$

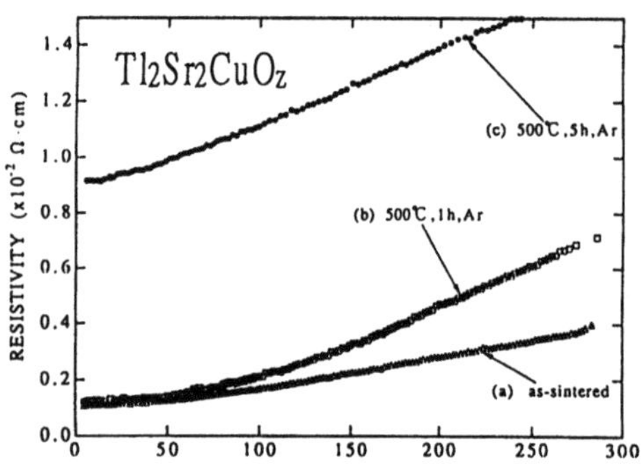

Fig. 5 Temperature dependence of electrical resistivity for $Tl_2Sr_2CuO_z$ (a) as-sintered (b) annealing at 500℃ for 1h in argon and (c) annealing at 500℃ for 5h in argon

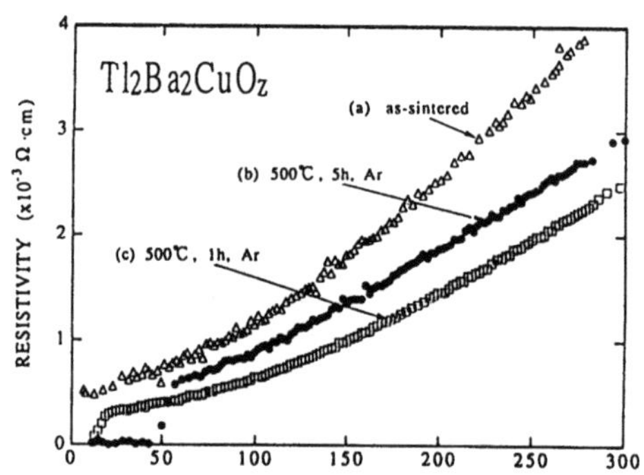

Fig. 6 Temperature dependence of electrical resistivity for $Tl_2Ba_2CuO_z$ (a) as-sintered (b) annealing at 500℃ or 1h in argon (c) annealing at 500℃ for 5h in argon

1. $Tl_2Sr_2CuO_z$

Figure 2(a) shows the powder x-ray diffraction pattern for as-sintered $Tl_2Sr_2CuO_z$. This sample was nearly single phase of the Tl-2201 structure.

Almost all peaks were successfuly indexed for a tetragonal unit cell with the lattice constants of a=5.305Å, c=22.195Å. Small impurity peaks in the figure were assigned to $SrTl_2O_4$.[4]. Figure 3 shows the diffraction patterns for (a) as-sintered and (b) and (c) annealed samples. Although impurity peaks appeared, the main phases had the Tl-2201 structure. The longer the annealing duration, the stronger the intensity of the impurity peaks. It suggested that $Tl_2Sr_2CuO_z$ was unstable at 500℃ in argon. The change in the c-axis length of the Tl-2201 phases is shown in Fig.4(a). No change was observed. In the case of $Tl_2Ba_2CuO_z$, the oxygen loss by annealing treatment in reducing atmosphere was related to the elongation of the c-axis length.[1] Therefore, it is considered that oxygen loss did not occur in our experiment of $Tl_2Sr_2CuO_z$. Figure 5 shows the temperature dependence of electrical resistivity for the samples as-sintered and annealed. All the samples have metallic behaviors.

2. $Tl_2Ba_2CuO_z$

Nearly single-phase samples of $Tl_2Ba_2CuO_z$ were obtained as shown in Fig.2(b). The obtained Tl-2201 phase had a tetragonal unit cell with a=5.475Å and c=23.062Å. The annealed samples were also nealy single phase. The c-axis length for the Tl-2201 structure as a function of annealing time is shown in Fig.4(b). An elongation of the c-axis length was observed as the annealing time increased. The resistivity-vs.-temperature curve is shown in Fig.6. A metal-superconductor transition was ovserved. The obtained results were parallel to the previously reported data by Shimakawa et al.[1].

DISSCUSION

In the case of $Tl_2Ba_2CuO_z$, the 85K superconducting transition is obtained by removal of excess oxygen. We speculated that $Tl_2Sr_2CuO_z$ is analogous to $Tl_2Ba_2CuO_z$. None the less annealing treatments for $Tl_2Sr_2CuO_z$ in reducing atmosphere resulted in decomposition of the Tl-2201 phase. The removal of oxygen to exhibit superconductivity was not achieved.

Ganguli et al.[3] studied $Tl_2Ba_{1-x}Sr_xCuO_z$ for various values of x. They reported that Tc decreased gradually from 92K(at x=0.0) to 77K(at x=0.8). Superconductivity was lost beyond x=1.0. The x=1.2 sample was a normal metal. They interpreted the Tc decrease in terms of a change in electronic structure. They pointed out that the substitution of less electropositive Sr for more electropositive Ba could bring down the Tl6s level and should increase the overlap of Tl6s with $Cu3dx^2-y^2$. This overlap is crucially important to obtain superconductivity in Tl-2201 compounds.[5] The substitution for Ba by Sr is likely to increase the hole concentration because the lattice constants are on the decrease.[6] Subsequentially Tc decreases with increasing Sr content. Therefore we speculate that $Tl_2Sr_2CuO_z$ is more over-doped by exess oxygen than $Tl_2Ba_2CuO_z$. It is well-known that the higher hole concentration corresponds to the lower Seebeck coeficient.[7] Figure 7 shows the tempera-

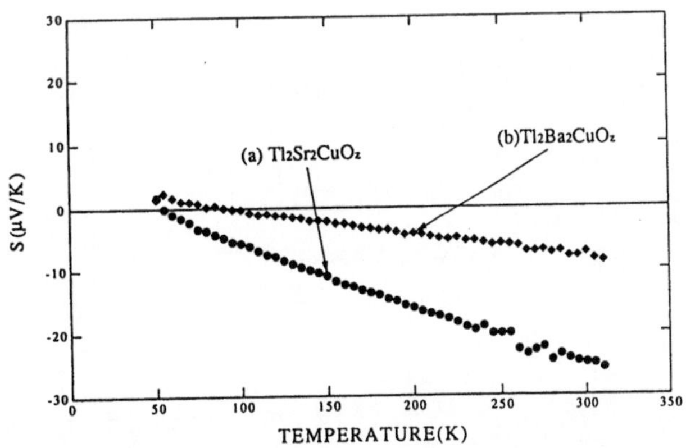

Fig.7 Temperature dependence of Seebeck coefficient for as-sintered (a) $Tl_2Sr_2CuO_z$ and $Tl_2Ba_2CuO_z$

ture dependence of Seebeck coeficient, S, for as-sintered $Tl_2Sr_2CuO_x$ and $Tl_2Ba_2CuO_x$ in our experiment. The S value for $Tl_2Sr_2CuO_x$ was lower than that for $Tl_2Ba_2CuO_x$ in the temperature range of 50-300K. This result supports our speculation.

CONCLUSION

We successfully synthesized nearly single-phase samples of the nominal composition of $Tl_2Sr_2CuO_x$(Tl2201 phase). In the case of $Tl_2Ba_2CuO_x$, the 85K superconducting transition is obtained by removal of excess oxygen. On the other hand, in the case of $Tl_2Sr_2CuO_x$, annealing treatments in reducing atmoshere resulted in decomposition of Tl-2201 phase. The removal of oxygen to exhibit superconductivity was not achieved.

ACKNOWLEDGE

This work was supported by NEDO in program for R&D of Basic Technology for Future Industries.

REFERENCE

1. Y Simakawa, Y. Kubo, T.Manako and H.Igarashi Phys.Rev.B,Vol.40,No.16,40 (1989)
2. C.C.Torardi, M.A.Subramanian, J.C.Calabrese, J.Gopalakrishnan, E.M.McCarron, K.J.Morrissey, T.R.Askew, R.B.Flippen, U.Chowdhry, and A.W.Sleight Phys.Rev.B 38,225,(1988)
3. A.K.Ganguli and M.A.Sabramanian J.Solid State Chemistry 90, 382 (1991)
4. C.Michel, M.Hervieu, B.Raveau, S.Li, M.Greaney, S.Fine, J.Potenza, and M.Greenblatt Mat.Res.Bull. 26, 123 (1991)
5. M.Kikuchi, S.Nakajima, Y.Shono, K.nagase, R.Suzuki, T.Kajitani, N.kobayashi and Y.muto Physica C 166, 497 (1990)
6. M.G.Smith, A.Manthiram, J.Zhou, J.B.Goodenough and J.T.Markert NATURE 549 (1991)
7. N.Mitra, J.Trefny, B.Yarar, G.Pine, Z.Z.Sheng and M.A.Harmann Phys. Rev. B 38,7064 (1988)

Thermoelectric Power of Cation Substituted Bi-232 Superconductor

K. MATSUURA, F. MUNAKATA, T. KAWANO, K. KUBO, and H. YAMAUCHI

Superconductivity Research Laboratory, ISTEC, Koto-ku, Tokyo, 135 Japan

ABSTRACT

The Seebeck coefficients were measured for several cation substituted "Bi-232" or "Bi-2212" type superconductors. A clear difference between the effects of the La for Sr and the Y for Ca substitutions on the Cu valence and also on the Seebeck coefficient was observed. The value of the Seebeck coefficient changed when substituting Pb for Bi. Furthermore, with increasing Pb content the system underwent a metal-to-semiconductor transition accompanied with a change in the structural modulation. The results suggested that the thermoelectric power was affected by changes in the electronic state as well as in diffusion process due to these cation substitutions.

KEY WORDS: thermoelectric power, Bi cuprate superconductor, cation substitution effect

INTRODUCTION

Thermoelectric power studies allow to probe electronic properties of conductors because of the sensitivity of the method to electron energy-band structures and also electron-excitation interactions. Thus, these studies may contribute to the development of a model for high-Tc superconductivity. It is generally known that the $Bi_2(Sr,Ca)_3Cu_2O_y$ ("Bi-232" or "Bi-2212") superconductors have a structural modulation, which is furthermore demonstrated to disappear when replacing more than 20 % of the Bi ions by Pb.[1] Both the superconducting transition temperature and the crystal structure of "Bi-232" are also known to vary with the substitution of rare earth elements for the (Sr, Ca) sites.[2-4] Sr-free "Bi-232" superconductors, i.e. $Bi_2Ca_{2.5}Nd_{0.5}Cu_2O_y$ and $Bi_2Ca_{2.5}La_{0.25}Pr_{0.25}Cu_2O_y$, have tetragonal crystal structures and commensurate modulations with the period equal to 9b.[5] The superconducting properties of these phases depend on the oxygen partial pressure during the sintering, but the modulation period remains unchanged. Additionally, a correlation between the mean ionic radius of the rare earth element and the average valence of copper (or Tc) is observed. In this paper, we report the effects of various cation substitutions on the thermoelectric power of "Bi-232" superconductors.

EXPERIMENTAL

The cation substituted "Bi-232" samples, i.e. $Bi_{2-x}Pb_xSr_2CaCu_2O_y$ (x= 0.0, 0.2, 0.3, 0.5), $Bi_2Sr_2Ca_{1-x}Y_xCu_2O_y$ (x=0.0, 0.2, and 0.4), $Bi_2Sr_{1-x}La_xCaCu_2O_y$ (x=0,0, and 0.5), $Bi_2Ca_{2.5}Nd_{0.5}Cu_2O_y$, and $Bi_2Ca_{2.5}La_{0.25}Pr_{0.25}Cu_2O_y$ were synthesized by a conventional solid-state reaction method using Bi_2O_3, PbO, $SrCO_3$, $CaCO_3$, Y_2O_3, La_2O_3, Nd_2O_3, Pr_6O_{11}, and CuO as starting materials. After calcined at 1033 - 1173 K for 10 hours in air, the powders were ground and pressed into parallelepiped bars of 3 x 3 x 20 mm^3. Sintering of the $(Bi,Pb)_2Sr_2CaCu_2O_y$ samples was performed at 1043 - 1093 K for 30 hours in N_2

and the bars of $Bi_2(Sr,La)_2(Ca,Y)_1Cu_2O_y$ were sintered at 1133 - 1203 K for 30 - 35 hours in air and cooled down in approximately 5 hours. Finally, the bars of $Bi_2Ca_{2.5}Nd_{0.5}Cu_2O_y$ and $Bi_2Ca_{2.5}La_{0.25}Pr_{0.25}Cu_2O_y$ were sintered at 1113-1133 K for 35 hours in O_2 and cooled to 773 K at a rate of 60 K/hour. Then the samples were annealed at 773 K for 10 hours in O_2 and cooled to room temperature at the same rate. The thermoelectric power was studied by measuring Seebeck coefficients. The Seebeck coefficients were evaluated by a dc method using a 10 nV-resolution digital voltmeter at temperatures between 30 K and 310 K and applying a temperature difference of about 0.3 K to the sample. The temperature difference was determined using two pairs of Cu-Constantan thermocouples and the temperature was measured by means of a calibrated Si diode sensor. The two pairs of thermocouples were contacted on hot and cold sides of the sample. Electrical resistivity measurements were carried out at temperatures below room temperature using a standard four-probe method. The superconducting transition temperatures (Tc) reported are onset temperatures obtained from the ac magnetic susceptibility data. The average Cu valences (p) were determined by a coulometric titration technique.[6]

RESULTS AND DISCUSSION

The temperature dependence of Seebeck coefficient for air-sintered $Bi_2(Sr,La)_2(Ca,Y)Cu_2O_y$ samples, i.e. $Bi_2Sr_2CaCu_2O_y$, $Bi_2Sr_{1.5}La_{0.5}CaCu_2O_y$, $Bi_2Sr_2Ca_{0.8}Y_{0.2}Cu_2O_y$, and $Bi_2Sr_2Ca_{0.6}Y_{0.4}Cu_2O_y$, is shown in Fig. 1. $Bi_2Sr_2CaCu_2O_y$ appears to be over-doped with holes, as its Seebeck coefficient is negative and superconducting transition temperature (Tc=67 K) is lower than the maximum value (Tc~ 90 K) for "Bi-232" superconductors. The Seebeck coefficient became positive and its value increased when extra electrons were introduced due to the substitution of trivalent La or Y for the divalent (Sr, Ca) sites. A similar behaviour was also observed when the oxygen content of the under-doped "Bi-232" was reduced.[7] The Tc values for $Bi_2Sr_{1.5}La_{0.5}CaCu_2O_y$ and $Bi_2Sr_2Ca_{0.8}Y_{0.2}Cu_2O_y$ were higher than the value obtained for $Bi_2Sr_2CaCu_2O_y$ indicating that the optimum hole concentration (n_{opt}) corresponding to the maximum Tc was achieved by these cation substitutions. Furthermore, the Tc of $Bi_2Sr_2Ca_{0.6}Y_{0.4}Cu_2O_y$ was lower than the Tc of $Bi_2Sr_2Ca_{0.8}Y_{0.2}Cu_2O_y$, which suggests that its hole concentration is smaller than n_{opt} due to over-doping of Y for Ca. For the $Bi_2(Sr,La)_2(Ca,Y)Cu_2O_y$ samples the value of Seebeck coefficient increased as the Cu valence (p) decreased, in the same way as in $YBa_2Cu_3O_{y-\delta}$ due to decreasing oxygen con-

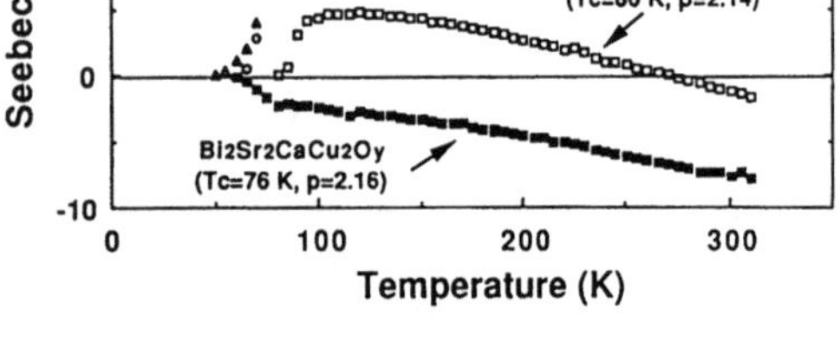

Fig. 1. Temperature dependence of the Seebeck coefficient for the $Bi_2(Sr,La)_2(Ca,Y)Cu_2O_y$ samples.

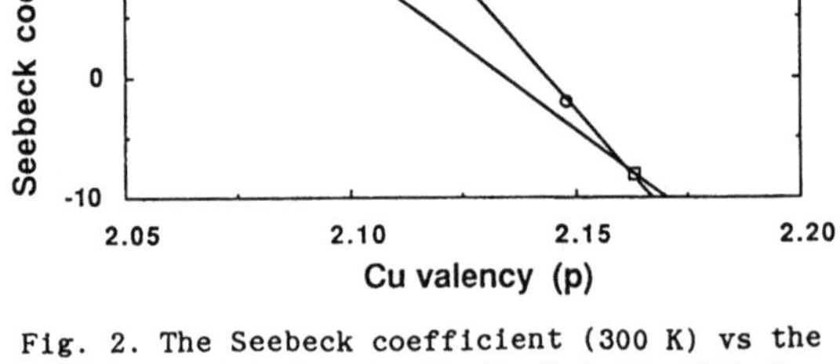

Fig. 2. The Seebeck coefficient (300 K) vs the Cu valence (p) for the $Bi_2(Sr,La)_2(Ca,Y)Cu_2O_y$ samples.

tent.[8] The Seebeck coefficient at 300K (S_{300K}) versus the Cu valence (p) for $Bi_2(Sr,La)_2CaCu_2O_y$ and $Bi_2Sr_2(Ca,Y)Cu_2O_y$ is shown in Fig. 2. It is clearly seen that the dependence of the S_{300K} on p is different for $Bi_2(Sr,La)_2CaCu_2O_y$ and $Bi_2Sr_2(Ca,Y)Cu_2O_y$. Since the apical oxygen is surrounded by Sr atoms, the most probable consequence is that the $d_{3z^2-r^2}$ orbital is affected more strongly by the substitution of La for Sr than by the substitution of Y for Ca. On the other hand, as the CuO_2 planes are separated by Ca atoms, the $d_{x^2-y^2}$ orbital may be more sensitive to the Y for Ca substitution than to the La for Sr. According to the above assumptions the substitution effect on the electronic state is different for the Sr and Ca sites. The temperature dependence of Seebeck coefficient for the $Bi_{2-x}Pb_xSr_2CaCu_2O_y$ samples with x=0.0, 0.2, 0.3 and 0.5 is shown in Fig. 3. These samples have nearly the same superconducting transition temperature (Tc= 87-83 K) and Cu valence (p= 2.08-2.11). Although the curves presented are quite similar, the samples were divided into two groups, i.e. samples with x= 0.0 and 0.2 and the samples with x= 0.3 and 0.5. The former group has a modulation period around 4.9b, while the latter has a modulation period either around 9.5b (x= 0.3) or no structural modulation at all (x= 0.5). The difference in the Seebeck coefficients between these two groups was about 1μ V/K. Thermoelectric power (S) in a metallic system is given by the well-known Mott formula,

$$S = -\frac{\pi^2 k_B^2}{3|e|} T \left(\frac{\partial \ln A}{\partial \varepsilon} + \frac{\partial \ln \Lambda}{\partial \varepsilon} \right)_{\varepsilon = \varepsilon_F} \quad (1)$$

where any contributions from phonon and magnon drags have been ignored, and A is the area of the Fermi surface for energy ε and Λ is the mean free path. Since the first term in (1) is related to the carrier concentration, it must be essentially equal for all the samples. Therefore, the difference in the thermoelectric power between the two groups was thought to be caused mainly by the differences in the second term of (1). These differences may be due to the changes in the diffusion process or in the crystal structures. Figure 4 presents the temperature dependence of electrical resistivity for $Bi_{2-x}Pb_xSr_2CaCu_2O_y$ (x=0.2 and 0.5). The sample with x= 0.2 shows metallic behaviour, while the sample with x= 0.5 behaves like a semiconductor. The

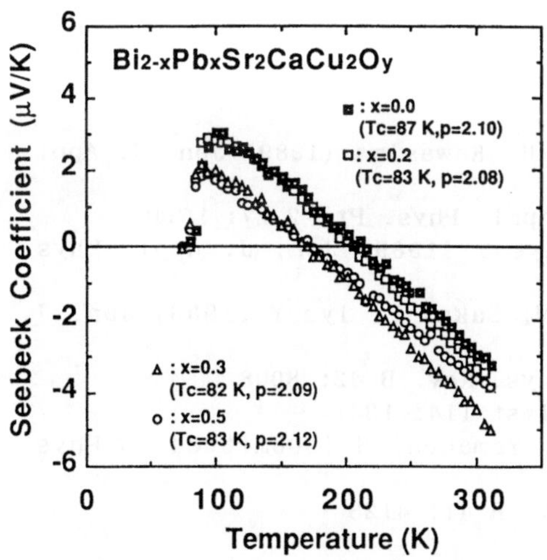

Fig. 3. Temperature dependence of the Seebeck coefficient for the $Bi_{2-x}Pb_xSr_2CaCu_2O_y$ samples.

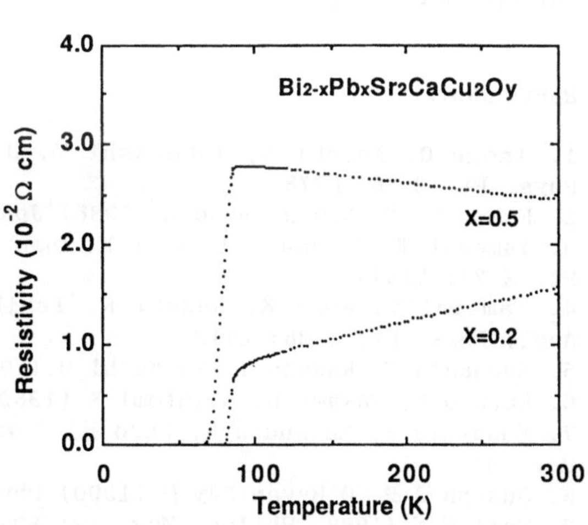

Fig. 4. Temperature dependence of the electrical resistivity for some of the $Bi_{2-x}Pb_xSr_2CaCu_2O_y$ samples.

semiconductive behaviour of the sample with x= 0.5 suggests that the carriers are localized by a certain mechanism such as Mott model[9] or Anderson model[10]. This resistivity data supports the above-mentioned interpretation of the Seebeck coefficient data for $Bi_{2-x}Pb_xSr_2CaCu_2O_y$. The temperature dependence of Seebeck coefficient for the Sr-free "Bi-232" samples, i.e. $Bi_2Ca_{2.5}Nd_{0.5}Cu_2O_y$ and $Bi_2Ca_{2.5}La_{0.25}Pr_{0.25}Cu_2O_y$, is shown in Fig. 5. The sign of the Seebeck coefficient is positive for both the samples. For $Bi_2Ca_{2.5}Nd_{0.5}Cu_2O_y$, the value of the Seebeck coefficient is larger and the superconducting transition temperature (Tc) and Cu valence (p) are smaller than for $Bi_2Ca_{2.5}La_{0.25}Pr_{0.25}Cu_2O_y$. This relationship between the Seebeck coefficient and p (or Tc) is similar to that obtained for the $Bi_2Sr_2(Ca,Y)Cu_2O_y$ samples with an under-doped state indicating that $Bi_2Ca_{2.5}Nd_{0.5}Cu_2O_y$ and $Bi_2Ca_{2.5}La_{0.25}Pr_{0.25}Cu_2O_y$ were also under-doped with holes.

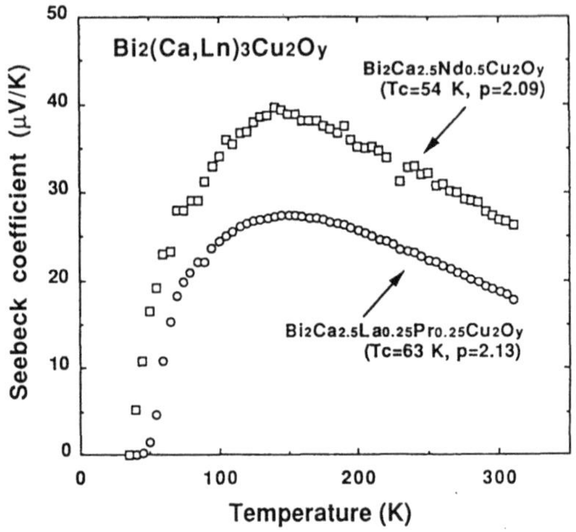

Fig. 5. Temperature dependence of the Seebeck coefficient for $Bi_2Ca_{2.5}Nd_{0.5}Cu_2O_y$ and $Bi_2Ca_{2.5}La_{0.25}Pr_{0.25}Cu_2O_y$.

CONCLUSION

The effects of various cation substitution on thermoelectric power of "Bi-232" or "Bi-2212" type superconductors were investigated in the first work. A clear difference between the effects of the La for Sr and the Y for Ca substitutions on the Cu valence and also on the Seebeck coefficient was observed. The value of the Seebeck coefficient changed when substituting Pb for Bi. Furthermore, with increasing Pb content the system underwent a metal-to-semiconductor transition accompanied with a change in the structural modulation. The results suggested that the thermoelectric power was affected by changes in the electronic state as well as in diffusion process due to these cation substitutions.

REFERENCES

1. Inoue O, Adachi S, Takahashi Y, Hirano H, Kawasima (1989) Jpn. J. Appl. Phys. Pt. 2 28: L778
2. Fukusima N. Niu H, Ando K (1988) Jpn. J. Appl. Phys. Pt. 2 27: L790
3. Tamegai T. Watanabe A, Koga K, Ogura I, Iye Y (1989) Jpn. J. Appl. Phys. Pt. 2 27: L1074
4. Tamegai T, Koga K, Suzuki K, Ichihara M, Sakai F, Iye Y (1989) Jpn. J. Appl. Phys. Pt. 2 28: L112
5. Munakata F, Kawano T, Yamauchi H (1990) Phys. Rev. B 42: 8008
6. Kurusu K, Takami H, Shintomi K (1989) Analyst 114: 1341
7. Munakata F, Matsuura K, Kubo K, Kawano T, Yamauchi H (submitted to Phys. Rev. B)
8. Ouseph J P, O'Bryan Ray M (1990) Phys. Rev. B 41: 4123
9. Mott F N (1968) Philos. Mag. 19: 835
10. Anderson W P (1958) Phys. Rev. 109: 1492

2.4 Magnetic Resonance and Mössbauer Studies

Tin Doping in High-T_c Superconductor $YBa_2Cu_3O_{7-y}$

TATSUYA MURAKI, MASAFUMI TANIWAKI, and KÔICHI MUKASA

Department of Electronic Engineering, Faculty of Engineering, Hokkaido University, Sapporo, 060 Japan

ABSTRACT

This work was performed to obtain the solubility of Sn in $YBa_2Cu_3O_{7-y}$ and to establish the heat treatment for preparing $YBa_2(Cu,Sn)_3O_{7-y}$. For the sample of the nominal composition $YBa_2(Cu_{1-x}M_x)_3O_{7-y}$, the lattice parameters in 1-2-3 phase, and the impurity phases were examined as a function of tin content x and the firing temperature by powder X-ray diffraction. From the dependence of the impurity phase peak intensity on x, the Sn solubility was obtained as a funtion of firing temperature. Tin substitutes for 0.09, 0.92 and 1.11 at%Cu by firing at 900, 940 and 960 ℃, respectively. In the samples fired at high temperatures, the lattice parameters change with x at higher Sn content than the solubility obtained above, which means that Sn dissolves slightly. It is considered that tin atoms occupy the Ba or Y sites which are vacant due to the lack of Ba with the increase of the impurity phases.

KEY WORDS: high-*Tc* superconductor, tin solubility, crystal structure, X-ray diffraction

INTRODUCTION

The element substitution has been used to investigate the mechanism of high-*Tc* superconductivity. Magnetic or non-magnetic metal doping have been tried by many reserchers. Especially, many studies on structural, magnetic and lattice dynamics properties of the $RBa_2(Cu_{1-x}M_x)_3O_{7-y}$ have been performed (M = Fe, Co, Al, Zn, Ni, Ag, Li, Ga, Pt, Cr, V, etc)[1-4]. In Mössbauer spectroscopy (MS) technique, which is very useful method to investigate the mechanism of high-*Tc* superconductivity, the dopants such as ^{119}Sn and ^{57}Fe are used as a probe[5-12]. In this technique, it is important to clarify the site occupied by the doped element. Through many ^{57}Fe MS studies of the 1-2-3 phase high-*Tc* superconductor, it has been established that Fe impurities are almost subsutituted for Cu site in the Cu-O chains and its solubity for Cu site is more than 20 at%. The crystal structure of the 1-2-3 phase transforms from orthorhombic to tetragonal as x increases ($x \simeq 0.03$)[13]. To the contrary, the results of ^{119}Sn MS studies are confusing[8-12]. For example, the reported ^{119}Sn spectra show various shapes. For Sn doped 1-2-3 system, one proup reported that the spectrum consisted of a strong singlet and a weak doublet[10,11], and other group did that the spectrum consisted of only one doublet[8]. Another research group approximated the measured spectrum with a singlet[12]. From such a confusion, it is considered that Sn atoms may not be substituted for Cu site in high-Tc oxide $YBa_2Cu_3O_{7-y}$. Therefore, we tried to establish the heat treatment for substitution of Sn for Cu in high-*Tc* superconductor $YBa_2Cu_3O_{7-y}$. In this work the ordinary method, i.e., firing of the mixture of $2BaCO_3$, $1/2Y_2O_3$, $3(1-x)CuO$ and $3xSnO_2$ was performed. The crystal structure and the impurity phases in the fired samples were examined by X-ray diffraction.

EXPERIMENTAL

Starting materials were prepared so that their compositions became $YBa_2(Cu_{1-x}Sn_x)_3O_{7-y}$ (x = 0.01, 0.02, 0.03, 0.04 and 0.05). The thoroughly ground mixtures of Y_2O_3, $BaCO_3$, CuO and SnO_2 were fired at 900 ~ 970 ℃ in air for 40 hours. These samples were annealed at 460 ℃ for 8 hours and cooled down to room temperature slowly in flowing oxgen. The structure in obtained oxides was estimated by powder X-ray diffraction (Cu-K α line) and the impurity phases containing Sn were identified in detail. The solubility of Sn in $YBa_2Cu_3O_{7-y}$ was estimated from the ratio of the sum of the maximum reflection line intensities of the impurity phases containing Sn to the sum of 013, 110 and 103 line intensities in 1-2-3 phase.

RESULTS AND DISCUSSION

(1) Characterization

The X-ray diffraction pattern of the $YBa_2(Cu_{0.96}Sn_{0.04})_3O_{7-y}$ sample fired at 900℃ is shown in Fig.1. The major reflection of this pattern indicates the 1-2-3 phase. There are reflections of

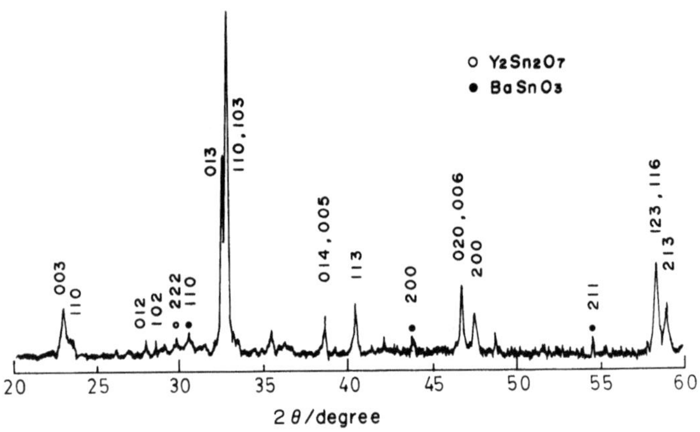

Fig.1
Powder X-ray diffraction pattern from $YBa_2(Cu_{0.95}Sn_{0.05})_3O_{7-y}$ containing impurities. ●;$BaSnO_3$, ○;$Y_2Sn_2O_7$

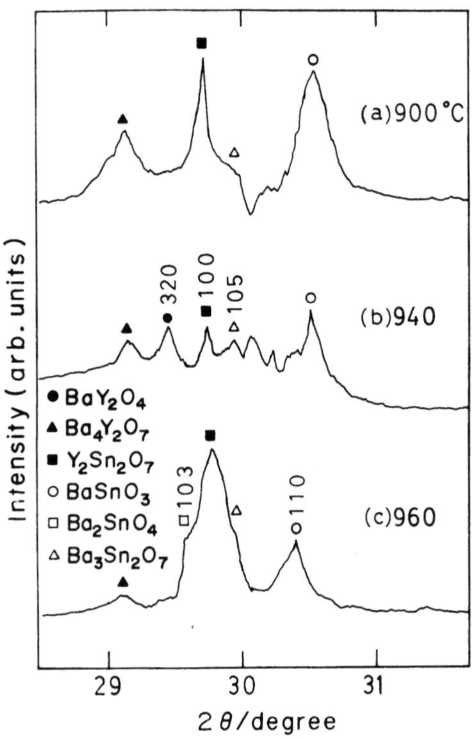

Fig.2
Powder X-ray diffraction pattern from impurity phases in $YBa_2(Cu_{0.96}Sn_{0.04})_3O_{7-y}$ fired at 900, 940 and 960℃.

impurity phases containing Sn such as $BaSnO_3$ and $Y_2Sn_2O_7$. The strongest lines of these impurity phases appear near (013), (103) and (110) reflections of 1-2-3 phase. These impurity phases were different with firing temperature and their line intensities increased with increasing Sn content, x. Figue 2-(a), (b) and (c) show the X-ray diffraction patterns of impurity phases containing Sn in the samples fired at 900, 940 and 960 ℃, respectively. All of patterns contained $Ba_4Y_2O_7$ [2θ =30.9°, 30.1°], $Y_2Sn_2O_7$[29.8°], $BaSnO_3$[30.68°] phases. In the sample fired at 940 ℃, many small peaks appeared. It is considered that those peaks correspond to the thermally stable phases[14]. By firing at 960 ℃, the (110) peak of $Y_2Sn_2O_7$ seemed to be broadened, however, this broadening is due to the appearance of Ba_2SnO_4 and $Ba_3Sn_2O_7$ phases. The $BaSnO_3$ phase decreased.

(2) Solubility of Sn in $YBa_2Cu_3O_{7-y}$

Figue 3 shows the integrated area intensity of Sn-containing impurity phases against Sn content, x, as a function of firing temperature. From the extrapolation to zero line intensity, the soluble Sn content x in $YBa_2Cu_3O_{7-y}$ was estimated for each firing temperature. Figue 4 shows the soluble Sn content x against firing temperature. Tin substitutes for 0.09 at%Cu, 0.92 at%Cu and 1.1 at%Cu by firing at 900 ℃, 940 ℃ and 960 ℃, respectively. This serves as a guideline to synthesize 1-2-3 phase without any impurity compound.

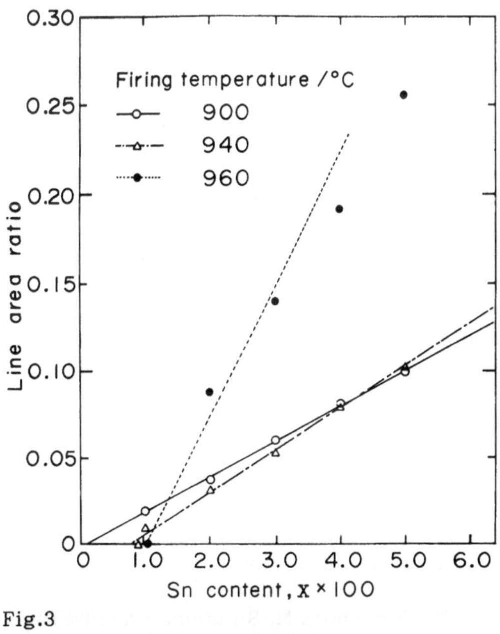

Fig.3
The line area ratio of impurities containing Sn to the total area of 013, 110 and 103 is shown against Sn contant x in $YBa_2(Cu_{1-x}Sn_x)_3O_{7-y}$. The solubility is obtained by extraporation.

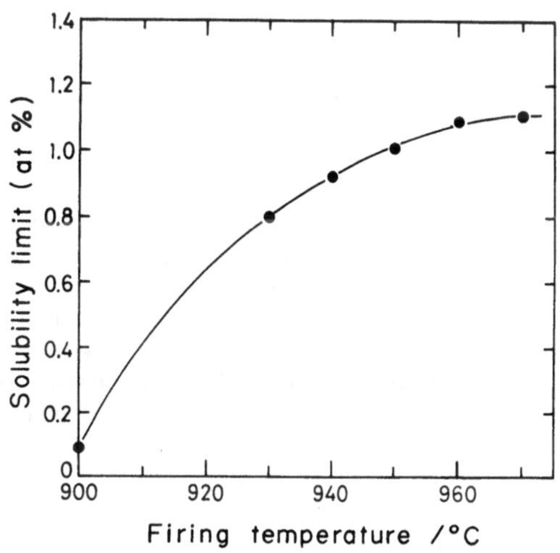

Fig.4
Firing temperature dependence of Sn solubility. The saturation curve is shown and the maximum solubility is about 1.1%.

(3) Lattice parameters of $YBa_2(Cu_{1-x}Sn_x)_3O_{7-y}$

The lattice parameters and the unit-cell volume, *V*, are shown as a function of doping concentration in Fig.5 and 6, respectively. In the sample fired at 900 ℃, *a*-, *b*- and *c*-axis lengths didn't depend on Sn content, x. In the samples of x<0.03 fired at 940 and 960 ℃, the lattice parameter, *a*, increased linearly with increasing x. The slope of the straight line of *a*-axis length vs. x was enlarged with increasing firing temperature. The *a*-axis length in the samples fired at 940 and 960 ℃ did not change at x>0.03. The unit-cell volume, *V*, scarcely changed with x in the samples fired at 900 ℃. It increased monotonously with x for $x \leqq 0.03$ and did not change with x for x>0.03 in the samples fired at 940, 960 ℃.

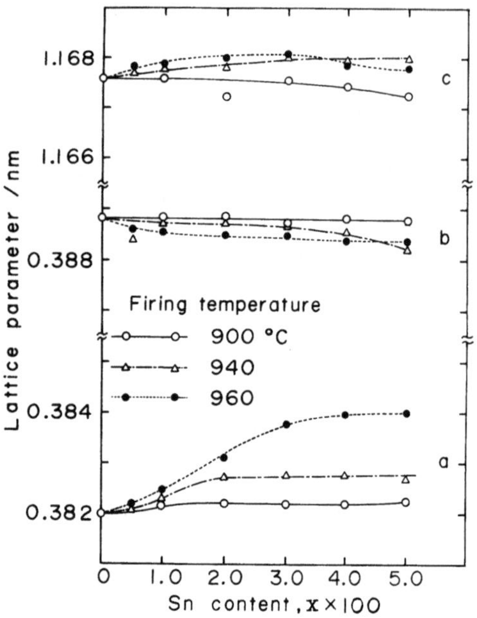

Fig.5
The lattice parameter of $YBa_2(Cu_{1-x}Sn_x)_3O_{7-y}$ is shown as a function of the Sn content x.

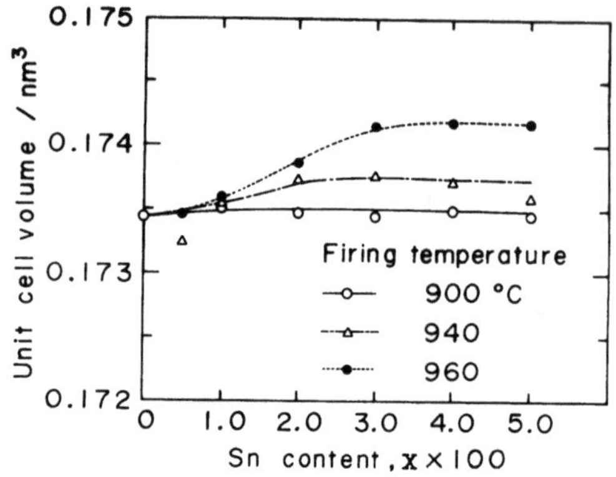

Fig.6
The unit-cell volume of $YBa_2(Cu_{1-x}Sn_x)_3O_{7-y}$ is shown as a function of Sn content x.

Previously it was shown that Sn-solubility was 0.0092 and 0.011 in the samples fired at 940 and 960 ℃, respectively. However, the lattice parameters of 1-2-3 phase changed with x at higher concentration than these values. It is considered that Sn dissolves into 1-2-3 matrix, yielding impurity phases which contain Sn at high x. The impurity phases such as $Ba_4Y_2O_7$, $Y_2Sn_2O_7$, $BaSnO_3$ increased with increasing x. The increase in these impurity phase causes the lack of Ba in 1-2-3 matrix, because the impurity phases are mainly Ba-Sn-O systems. Therefore, tin atoms probably substitute Ba or Y sites, at x>0.0092 in the sample fired at 940℃, and at x>0.011 in that fired at 960℃.

REFERENCES

(1) Strobel P, Paulsen C, Tholence JL (1988) Solid State Commun. **65**:585.

(2) Xiao G, Streitz FH, Gavrin A, Du YW, Chien CL (1987) Phys. Rev. B35:8782.

(3) Kistenmacher TJ (1988) Phys. Rev. **B38**:8862.

(4) Tarascon JM, Barboux P, Miceli PF, Greene LH, Hull GW, Eibschutz M, Sunshine SA (1988) Phys. Rev. **B37**:7458.

(5) Nishihara Y, Obara H, Katayama T, Oyanagi H, Murata K, Kimura Y, Kajimura K (1989) J. Phys. Soc. Jpn. **58**:2925.

(6) Hechel D, Nowik I, Bauminger ER, Felner I (1990) Phys. Rev. **B42**:2166.

(7) Simopoulos A, Niarchos D (1988) Phys. Rev. **B38**:8931.

(8) Yuen T, Lin CL, Crow JE, Myer GN, Salomon RE, Schlottmann P, Bykovetz N, Herman WN (1988) Phys. Rev. **B37**:3770.

(9) Zhang H, Wang GM, Zhang QR (1989) Phys. Lett. **A138**:517.

(10) Boolchand P, Enzweiler RN, Zitkovsky I, Wells J, Bresser W, McDaniel D, Meng RL, Hor PH, Chu CW, Huang CY (1988) Phys. Rev. **B37**:3766.

(11) Nishida T, Katada M, Matsumoto Y (1990) Jap. J. Appl. Phys. **29**:259

(12) Wang G, Chen LT, Chang HL, Chu ML, Hsu FW, Chen TM, Yang TJ, Gou YS, Uen TM (1989) Jap. J. Appl. Phys. **28**:2459

(13) Xu Y, Suenaga M, Tafto J, Sabatini RL, Moodenbaugh AR, Zolliker P (1989) Phys. Rev. **B39**:6667.

(14) Roth RS, Dennis JR, Davis KC (1987) Adv. Ceramic Mat. **2**:303.

^{63}Cu and ^{17}O Nuclei Spin-Lattice Relaxation in $Y_{1-x}Ca_xBa_2Cu_4O_8$

TAKATO MACHI[1], IZUMI TOMENO[1], TAKAYUKI MIYATAKE[1], KEISHI TAI[1], NAOKI KOSHIZUKA[1], AKIHIKO HAYASHI[2], YUTAKA UEDA[2], TAKASHI IMAI[3], and HIROSHI YASUOKA[2]

[1] Superconductivity Research Laboratory, ISTEC, Koto-ku, Tokyo, 135 Japan
[2] The Institute for Solid State Physics, The University of Tokyo, Minato-ku, Tokyo, 106 Japan
[3] Department of Physics and Materials Research Laboratory, University of Illinois at Urbana-Champaign, Urbana, IL 61801, USA

ABSTRACT

The temperature dependences of the ^{63}Cu and ^{17}O nuclei spin-lattice relaxation rate $^{63}(1/T_1)$ and $^{17}(1/T_1)$ in $YBa_2Cu_4O_8$ have been measured by the nuclear quadrupole resonance(NQR) and nuclear magnetic resonance(NMR) techniques, respectively. The frequency spectra in $Y_{1-x}Ca_xBa_2Cu_4O_8$ (x=0, 0.025, 0.075) and $^{63}(1/T_1)$ in the case of x=0, 0.075 were also measured. From the frequency spectra, we conclude that the Ca ions substitute for Y sites rather than Ba sites. The antiferromagnetic correlation derived from the Ca concentration dependence of $^{63}(1/T_1)$ is reduced with increasing Ca content. The absolute value of $^{17}T_1/^{63}T_1$ is located between that of $YBa_2Cu_3O_7$ and $YBa_2Cu_3O_{6.63}$. Therefore, it seems that the transition temperature T_c depends on the value of antiferromagnetic correlation length ξ that indicates a degree of breaking down of the correlation by introduction of itinerant carriers.

KEY WORDS: NMR, NQR, $YBa_2Cu_4O_8$, spin-lattice relaxation rate $(1/T_1)$

INTRODUCTION

One can obtain static- and dynamic- microscopic information about different atoms and different crystallographic sites by nuclear magnetic resonance(NMR) and nuclear quadrupole resonance(NQR). Especially, the spin-lattice relaxation rate $(1/T_1)$ has contributed to understanding of the dynamics of the conventional BCS superconductors. In the case of the high Tc superconductors, $(1/T_1)$ has been measured for the same reasons, for example $^{63,65}Cu$[1-11], ^{17}O[12-17] and ^{89}Y[18,19] nuclei in $YBa_2Cu_3O_{7-\delta}$. The $(1/T_1)$ in high T_c superconductors below T_c has no Hebel-Slichter peak[20] representing the influence of the coherence factor. In the normal state, namely $T>T_c$, the temperature dependence of $^{63}(1/T_1)$ at the planar Cu sites shows the common feature which is reflected in the antiferromagnetic spin fluctuation of *3d* electrons at least in the low doping superconductors. $^{63}(1/T_1)$ increases rapidly with temperature from T_c up to some temperature and follows an approximately T-linear behavior at high temperatures. On the other hand, $^{17}(1/T_1)$ at the planar O(2,3) sites and $^{89}(1/T_1)$ at Y sites shows the behavior like a Fermi liquid, since the spin fluctuation is geometrically cancelled out at O(2,3) and Y sites. A phenomenalogical model of the antiferromagnetic spin fluctuation was proposed by Millis, Monien and Pines(MMP)[21] based on a single spin degree of freedom, which was motivated by the result of the proportional relationship between $^{17}(1/T_1)$ and the spin susceptibility. The dynamic susceptibility $\chi(q,\omega)$ in the MMP model coincides with a formula of the self-consistent re-normalization(SCR) theory applied to a nearly antiferromagnetic metal of the two dimensional electron system by Moriya *et al.*[22], where q is the wave vector and ω is the nuclear Larmor frequency. The temperature dependence of $\mathrm{Im}\chi(q,\omega)$ in the recent neutron scattering experiments are in agreement with the result of NMR[23,24]. Therefore, it seems that the credibility of the MMP model is considerable high.

$YBa_2Cu_4O_8$ with T_c=80K has double Cu(1)-O chains and two Cu(2)-O_2 planes in the unit cell[25,26]. $YBa_2Cu_4O_8$ is regarded as a under doping superconductor like the 60K phase of $YBa_2Cu_3O_{7-\delta}$, because the uniform susceptibility decreases with decreasing temperature and T_c increases with Ca substitution[27] or pressure[28]. In addition, the oxygen content in $YBa_2Cu_4O_8$ is more stable[29] than that of $YBa_2Cu_3O_{7-\delta}$ and therefore the lineshape of the NQR signal is very sharp. Consequently, $YBa_2Cu_4O_8$ is a suitable material to confirm any model by measurement of $^{63}(1/T_1)$ at the planar Cu(2) sites and $^{17}(1/T_1)$ at O(2,3) sites. Therefore, we measured the $^{63}(1/T_1)$ and $^{17}(1/T_1)$ in $YBa_2Cu_4O_8$ and report the results in this paper. Furthermore, since it is expected that the increase of Tc with Ca substitution has a relationship to the change in the spin fluctuations, we also report the temperature dependence of $^{63}(1/T_1)$ at the planar Cu(2) sites in $Y_{0.925}Ca_{0.075}Ba_2Cu_4O_8$. Because it is not yet clear which sites is substituted by Ca ions, that is the Y sites[27] or at the Ba sites[30], we discuss this point using the frequency spectra in several Ca compositions for both Cu(1) and Cu(2) sites.

EXPERIMENTAL

The polycrystalline samples of $Y_{1-x}Ca_xBa_2Cu_4O_8$ (x=0, 0.025, 0.075) were prepared by the solid state reaction method with a hot isostatic press(HIP) treatment in an oxygen atmosphere[27]. It was confirmed that these were almost entirely single phase samples by X-ray diffraction studies. T_c was determined by measuring the temperature dependence of the susceptibility. $^{63}(1/T_1)$ and the frequency spectra were measured by the NQR technique. The frequency spectra were obtained by measuring the spin echo intensities for several frequencies at 4.2K. All NQR measurements were carried out for powder samples at zero magnetic field using a handmade coherent pulsed NMR spectrometer. $^{17}(1/T_1)$ for the planar O(2,3) sites in $YBa_2Cu_4O_8$ was measured for an ^{17}O enriched and magnetically oriented powder sample at 33.0MHz between T_c and 220K. $(1/T_1)$ was measured by the ordinary saturation recovery method. Since the recovery curve of $^{63}(1/T_1)$ is a single exponential, $^{63}(1/T_1)$ was obtained from the slope of the logarithmic recovery curve. The recovery curve at O(2,3) sites was fitted to the theoretical expression given by Narath[31] for the (1/2,3/2) transition in the case of the nuclear spin I=5/2. It should be noted that we defined $(1/T_1)$ as $2W$ in this paper, where W is the fundamental relaxation rate.

RESULTS AND DISCUSSION

SUBSTITUTED SITES OF Ca IONS

The frequency spectra of the Cu(1) and Cu(2) sites are shown in Fig. 1 and Fig. 2, respectively. The linewidth for both sites broadened by the Ca-random substitution effect with increasing Ca content. Since the ratios of the central frequencies for the two isotopes at both sites, $^{63}\nu_Q(i)/^{65}\nu_Q(i)$ (i=1,2), are nearly equal to 1.08, it is clear that any magnetic order is absent. $^{63}\nu_Q(1)$ for the Cu(1) sites increases slightly with Ca content. $^{63}\nu_Q(2)$ for the planar Cu(2) sites is shifted about 120kHz higher frequency with 2.5% Ca substitution. The change of $^{63}\nu_Q(i)$ under high pressures was reported by Zimmermann *et al.*[32] and more recently by Zheng *et al.*[33]. According to their results, $^{63}\nu_Q(1)$ decreases and $^{63}\nu_Q(2)$ increases with increasing pressure. Therefore, the increase of T_c with Ca substitution does not arise from chemical pressure effect. The carrier number in the Hall measurement increases with Ca content as reported by Tai *et al.*[34]. If Ca ions occupy Ba sites, the charge valance does not change and then the carrier number cannot increase. In the above point of view, we think that Ca ions substitute for Y sites rather than Ba sites. Futhermore, using an asumption that the change of $^{63}\nu_Q(i)$ is proportional to the number of the conducting carriers, it is apparent that the additional hole from Ca substitution goes mainly into the planes and barely into the chains.

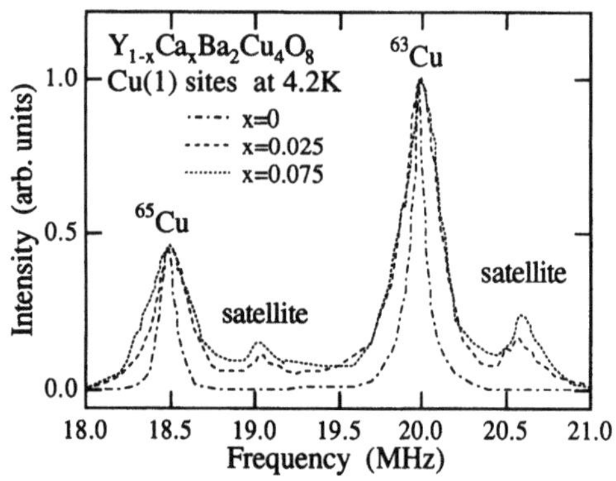

Fig. 1 The frequency spectra for Cu(1) sites at 4.2K.

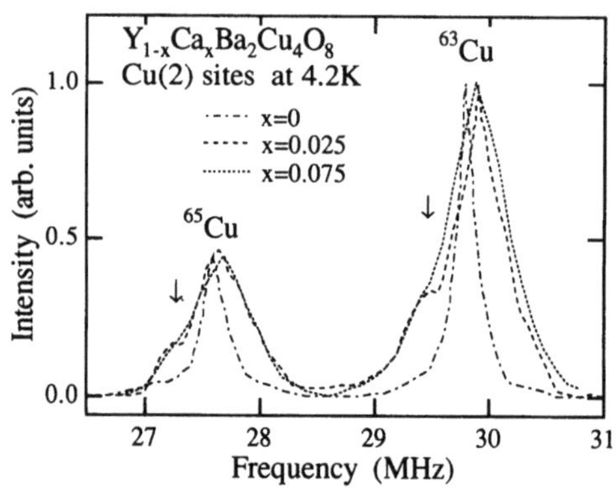

Fig. 2 The frequency spectra for Cu(2) sites at 4.2K.

Other authors claimed that the Ca ions substitute for Y sites. X-ray studies[35-37] show that the volume of the unit cell increases with Ca content. However, Wada *et al.*[38] reported that T_c is nearly constant for Sr substitution of the Ba sites in the $YBa_2Cu_4O_8$, although the volume of the unit cell monotonically decreases with increasing Sr content. Because there is a relation between the ion radii r of Y, Ca, Sr and Ba, that is $r(Y^{3+}) < r(Ca^{2+}) < r(Sr^{2+}) < r(Ba^{2+})$, it is appropriate that the Ca ions substitute Y with smaller radius than that of Ba. The same conclusion was reached by Raman scattering[39].

We think the satellite peak at the Cu(1) sites arises from the stacking faults, where a stacking faults is the local replacement of $(Cu\text{-}O)_2$ double chains by a (Cu-O) single chain. Yamaguchi *et al.*[40] found the existence of the stacking faults by high-resolution transmission electron microscopy(HRTEM). Recently, Matsui *et al.*[41] reported that the density of the stacking faults increases with increasing Ca content. The growth of the satellite peak with Ca content is consistent with these HRTEM results. In addition, Yoshida *et al.*[39] reported that the Raman peak of the vibration of the oxygen in the Cu-O chain becomes broader with Ca substitution, which is caused by the distortion of the Cu-O bond arising from the stacking faults. Therefore, we regard the satellite peak at Cu(1) sites as the expression of the stacking faults. On the other hand, a shoulder is observed for the planar Cu(2) sites on the lower frequency side of

$\nu_Q(2)$ as shown in Fig. 2. We suggest that the origin of the peak is ascribed to the Cu(2)-O_2 planes adjacent to Ca substituted Y sites.

$^{63}(1/T_1)$ IN $YBa_2Cu_4O_8$ AND $Y_{0.925}Ca_{0.075}Ba_2Cu_4O_8$

The temperature dependences of $^{63}(1/T_1)$ for the planar Cu(2) sites in $YBa_2Cu_4O_8$ and $Y_{0.925}Ca_{0.075}Ba_2Cu_4O_8$ are shown in Fig. 3. In the superconducting state, $^{63}(1/T_1)$ for neither compounds has a the Hebel-Slichter peak[20]. In the normal state, $^{63}(1/T_1)$ increase rapidly with temperature below T_c up to about 250K and 180K in the case of x=0 and x=0.075, respectively. Subsequently, $^{63}(1/T_1)$ increases approximately linearly with temperature. Therefore, we think again that this behavior of $^{63}(1/T_1)$ is a common feature to at least low doping high-T_c superconductors.

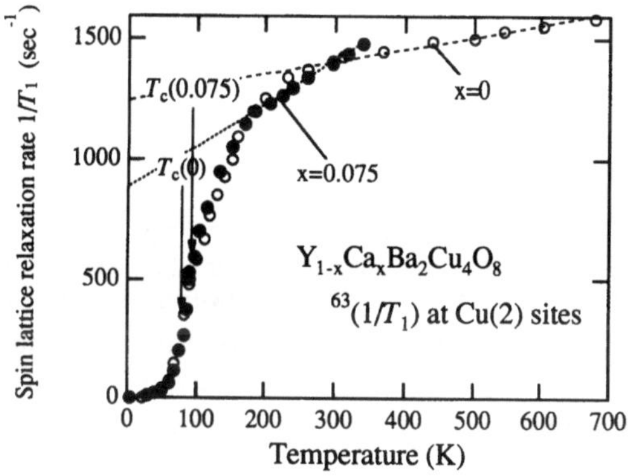

Fig. 3 The temperature dependences of $^{63}(1/T_1)$ for the planar Cu(2) sites in $YBa_2Cu_4O_8$ and $Y_{0.925}Ca_{0.075}Ba_2Cu_4O_8$

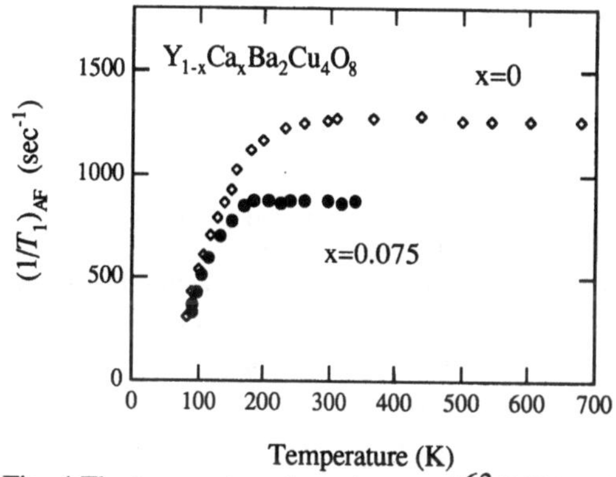

Fig. 4 The temperature dependences of $^{63}(1/T_1)_{AF}$ in $YBa_2Cu_4O_8$ and $Y_{0.925}Ca_{0.075}Ba_2Cu_4O_8$

Applying the analysis of Imai[41] to interpret the temperature dependence of $^{63}(1/T_1)$ for the Cu(2) sites in the 1-2-4 compounds, $^{63}(1/T_1)$ in the normal state can be divided into $^{63}(1/T_1)_0$ and $^{63}(1/T_1)_{AF}$, where $^{63}(1/T_1)_0$ and $^{63}(1/T_1)_{AF}$ indicate the contribution to $^{63}(1/T_1)$ from the zone center and the zone corner, respectively. Because $^{63}(1/T_1)_{AF}$ reflects the contribution of the spin fluctuations, it would be expected that $^{63}(1/T_1)_{AF}$ is reduced by increasing the carrier number and there by decreasing the microscopic antiferromagnetic spin correlations. Actually, it is known that $^{63}(1/T_1)_{AF}$ in $YBa_2Cu_3O_{7-\delta}$ decreases as the hole concentration increases[9].

One can obtain the value of $^{63}(1/T_1)_{AF}$ by an assumption that the approximately T-linear component of $^{63}(1/T_1)$ represents $^{63}(1/T_1)_0$. Namely, $^{63}(1/T_1)_{AF}$ is defined by subtracting the T-linear component from the raw $1/T_1$ data in the normal state. The slopes of the T-linear term in $YBa_2Cu_4O_8$ and $Y_{0.925}Ca_{0.075}Ba_2Cu_4O_8$, namely $^{63}(1/T_1T)_0$, are about 0.52 $sec^{-1}K^{-1}$ and 1.78 $sec^{-1}K^{-1}$, respectively. The temperature dependence of $^{63}(1/T_1)_{AF}$ obtained by the above procedure for $Y_{1-x}Ca_xBa_2Cu_4O_8$ are shown in Fig. 4. The values of $^{63}(1/T_1)_{AF}$ in $Y_{0.925}Ca_{0.075}Ba_2Cu_4O_8$ is smaller than that of $YBa_2Cu_4O_8$. Therefore, it seems that the contribution of spin fluctuations to $1/T_1$ for the planar Cu(2) sites is inversely proportional to the carrier concentration similar to that observed for the 1-2-3 compounds as discussed above.

$^{17}(1/T_1)$ IN $YBa_2Cu_4O_8$

The temperature dependence of $^{17}(1/T_1)$ at the planar O(2,3) sites in $YBa_2Cu_4O_8$ between T_c and 220K is shown in Fig. 3. $^{17}(1/T_1)$ varies approximately linearly with temperature in quasiparticle like behavior. The slope of the T-linear term, $^{17}(1/T_1T)$, is about 0.27 $sec^{-1}K^{-1}$. The ratio of $^{63}(1/T_1T)_0/^{17}(1/T_1T)$ is about 1.9, that is, the same order as in $YBa_2Cu_3O_7$, 3.9±0.4[42]. Using a Mila-Rice Hamiltonian[43], the relation between $^{63}(1/T_1)_0$ and $^{17}(1/T_1)$ represents as follows[42],

$$^{63}(1/T_1)_0 = 2\frac{A^2+4B^2}{(A+4B)^2}\left(\frac{^{63}\gamma}{^{17}\gamma}\right)\left(\frac{^{63}K}{^{17}K}\right)^{17}(1/T_1)$$

where K is Knight shift; A and B indicate the on-site and isotropic transferred hyperfine field, respectively. Since the temperature dependence of ^{63}K[44,11,45] and ^{17}K[46] in $YBa_2Cu_4O_8$ is similar to that of $YBa_2Cu_3O_{7-\delta}$ and the ratio of $^{63}(1/T_1T)_0/^{17}(1/T_1T)$ is nearly of the same order as in $YBa_2Cu_3O_7$, we think that the hyperfine field is quite similar in the two compounds.

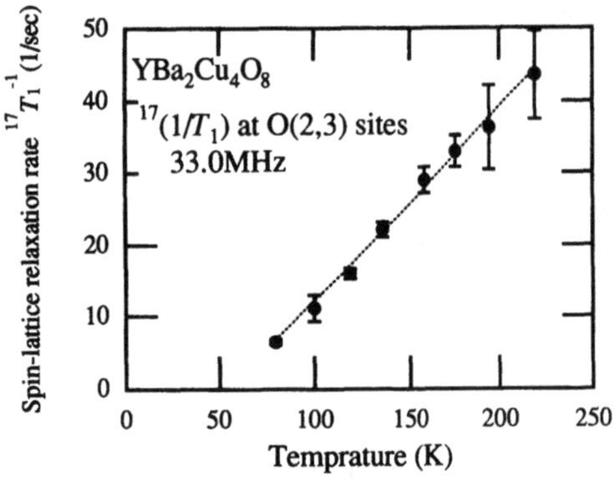

Fig. 5 The temperature dependence of $^{17}(1/T_1)$ at the planar O(2,3) sites in $YBa_2Cu_4O_8$.

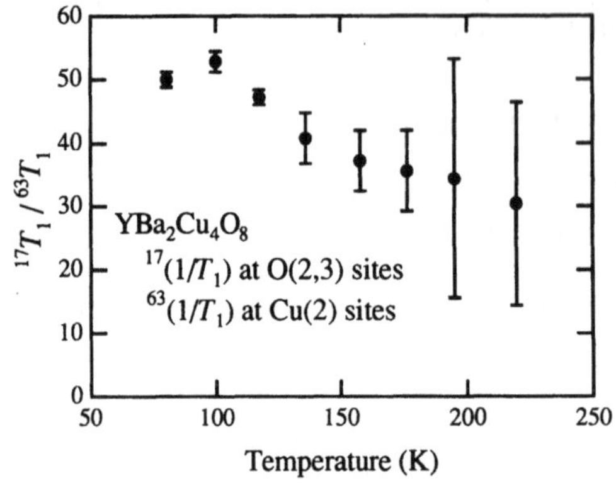

Fig. 6 The temperature dependence of the ratio of two relaxation rates $^{17}T_1/^{63}T_1$ in $YBa_2Cu_4O_8$.

It was reported the temperature dependence of the Knight shift ^{17}K for the transition (±1/2,±3/2) at the planar O(2,3) sites, elsewhere[46]. It seems that $^{17}(1/T_1T)$, ^{17}K and the uniform susceptibility scale with one another. Therefore, we can regard the MMP model as applicable to the case of $YBa_2Cu_4O_8$. According to the MMP model and the assumption that the wave vector q is near (π,π), $^{17}T_1/^{63}T_1$ is proportional to $1+\xi^2$, where ξ represents the antiferromagnetic correlation length. The temperature dependence of the ratio of the two relaxation rates $^{17}T_1/^{63}T_1$ is shown in Fig. 5. The behavior of $^{17}T_1/^{63}T_1$ is similar to the case of $YBa_2Cu_3O_{7-\delta}$, and the absolute value is located between that of $YBa_2Cu_3O_7$ and $YBa_2Cu_3O_{6.63}$[16]. Therefore, the antiferromagnetic correlation length ξ in $YBa_2Cu_4O_8$ is midway between the value for the 90K phase and the 60K phase of 1-2-3 compounds. Roughly speaking, we think that the T_c depends on the value of ξ that indicates a degree of breaking down of the antiferromagnetic correlations by the introduction of itinerant carriers.

ACKNOWLEDGEMENT

We thank Dr. J. O. Willis for helpful discussions and a critical reading of the manuscript.

REFERENCES

[1] M.Mali, *et al.*; Phys.Lett.A 124(1987)112.
[2] W.W.Warren,Jr., *et al.*; Phys.Rev.Lett. 59 (1987) 1860.
[3] T.Imai, *et al.*; J.Phys.Soc.Jpn. 57(1988)1771.
[4] C.H.Pennington, *et al.*; Phys.Rev. B 39 (1989) 2902.
[5] M.Horvatic, *et al.*; Phys.Rev.B 39(1989)7332.
[6] K.Ishida, *et al.*; J.Phys.Soc.Jpn. 58(1989)36.
[7] T.Oashi, *et al.*; Physica C 157 (1989)315.
[8] R.E.Walstedt, *et al.*; Phys.Rev.B 40(1989)2572.
[9] T.Imai, *et al.*; Physica C 162-164(1989)169.
[10] H.Zimmermann, *et al.*; PhysicaC 159(1989)681.
[11] T.Machi, *et al.*; Physica C 173(1991)32.
[12] K.Ishida, *et al.*; J.Phys.Soc.Jpn. 57(1988)2897.
[13] P.C.Hammel, *et al.*; Phys.Rev.Lett. 63(1989) 1992.
[14] P.Wzietek, *et al.*; Europhys.Lett. 8(4) (1989)363.
[15] M.Horovatic, *et al.*; Physica C 159(1989)689.
[16] M.Takigawa, *et al.*; Phys.Rev. B 43(1990)247.
[17] Y.Yoshinari, *et al.*; J.Phys.Soc.Jpn. 59(1990) 3698.
[18] H.Alloul, *et al.*; Phys.Rev.Lett. 61(1988)746.
[19] T.Ohno, *et al.*; J.Phys.Soc.Jpn. 59(1990)1139.
[20] L.C.Hebel, *et al.*; Phys.Rev. 113(1957)1504.
[21] A.J.Millis, *et al.*; Phys.Rev. B42(1990)167.
[22] T.Moriya, *et al.*; J.Phys.Soc.Jpn. 59(1990)2905.
[23] C.Vettier, *et al.*; Proceedings of ISS90.
[24] J.Rossat-Mignod, *et al.*; Proceedings of ICM91.
[25] P.Marsh, *et al.*; Nature 341(1989)41.
[26] P.Fisher, *et al.*; Solid State Comm.69(1989) 531.
[27] T.Miyatake, *et al.*; Nature 314(1989)41.
[28] B.Bucher, *et al.*; Physica C 157(1989)478.
[29] J.Karpinski, *et al.*; Nature 336(1988)660.
[30] I.Mangelschots, *et al.*; Physica C 172(1990)57.
[31] A.Natath; Phys.Rev. 162(1967)162.
[32] H.Brinkmann, *et al.*; Proceedings of EMRS 90.
[33] G.-Q.Zheng, *et al.*; Solid State Comm.79(1991) 51.
[34] K.Tai, *et al.*; unpublished data
[35] T.Miyatake, *et al.*; Proceedings of ISS89, p57.
[36] R.G.Buckley, *et al.*; Physica C165(1990)391.
[37] T.Wada, *et al.*; J.Materials Research 6(1991)18.
[38] T.Wada, *et al.*; Phys.Rev. B41(1990)11209.
[39] M.Yoshida, *et al.*; Jpn.J.Appl.Phys. (to be published)
[40] K.Yamaguchi, *et al.*; Jpn.J.Appl.Phys. 28(1989) L1942.
[41] Y.Matsui, *et al.*; J.Electron Microsc. 40(1991) 221.
[42] T.Imai; J.Phys.Soc.Jpn. 59(1990)2508.
[43] F.Mila, *et al.*; Phys.Rev. B40(1989)11382.
[44] H.Zimmermann, *et al.*; Proceedings of EMRS 90.
[45] H.Zimmermann, *et al.*; Proceedings of M2S 91.
[46] I.Tomeno, *et al.*; Proceedings of ISS91.

Oxygen NMR Study of Magnetically Oriented Powder Samples of $YBa_2Cu_4O_8$

I. Tomeno[1], T. Machi[1], K. Tai[1], J.O. Willis[1,*], T. Miyatake[1], M. Kosuge[1], N. Koshizuka[1], A. Hayashi[2], Y. Ueda[2], and H. Yasuoka[2]

[1] Superconductivity Research Laboratory, ISTEC, Koto-ku, Tokyo, 135 Japan
[2] Institute for Solid State Physics, The University of Tokyo, Minato-ku, Tokyo, 106 Japan

ABSTRACT

The Knight-shift K and the spin-lattice relaxation rate T_1^{-1} at the planar oxygen sites are presented for magnetically aligned polycrystalline $YBa_2Cu_4O_8$. Both the ^{17}O Knight-shift K_c along the c axis and the normal-state bulk susceptibility χ decrease gradually with decreasing temperature. This indicates that the oxygen spin correlates with the copper spin. The quantity $(^{17}T_1T)^{-1}$ is nearly proportional to $^{17}K_c$, in marked contrast to the planar Cu result for $(^{63}T_1T)^{-1}$. The oxygen NMR features for $YBa_2Cu_4O_8$ are similar to those for the oxygen deficient $YBa_2Cu_3O_x$.

KEY WORDS: $YBa_2Cu_4O_8$, Oxygen NMR

INTRODUCTION

The 80K superconductor $YBa_2Cu_4O_8$ (Y124) has the crystal structure consisting of a sequence of double CuO_2 planes and double CuO chains. The oxygen stoichiometry in the chain sites of Y124 remains unchanged up to 800 °C, in contrast to the $YBa_2Cu_3O_x$ (Y123) case. The recent band structure calculations [1,2] attributed the oxygen stability to an energy-lowering of the oxygen 2p-states at the chain and apex sites. A pressure dependence of Tc for Y124 is significantly larger than for Y123.[3] This suggests some differences in the superconducting properties between Y123 and Y124, in spite of the similarity of the crystal structures.

Nuclear magnetic resonance (NMR) studies of Y123 have been intensively made to clarify the microscopic magnetic nature of the CuO_2 planes.[4] The oxygen content x in Y123 has a marked influence on both the Knight-shift K and the relaxation-rate T_1^{-1} of the Cu and O nuclei. For the fully oxygenated Y123, $^{17}K_c$ and $(^{17}T_1T)^{-1}$ for the planar oxygen sites are temperature independent, while for the oxygen deficient Y123 these quantities decrease with decreasing temperature. A similar temperature dependence is found for the planar Cu Knight-shift K_{ab} perpendicular to the c axis in the Y123 system. On the other hand, $(^{63}T_1T)^{-1}$ at the planar Cu site exhibits a maximum in the range of 100 and 150 K. Different relaxation behavior between these nuclei can be understood based on the fact that the antiferromagnetic Cu spin fluctuations cancel out at the oxygen sites.

We carried out oxygen NMR experiments at the plane sites of Y124 in order to compare the present results with Y123 from the microscopic view point.

---

*Permanent address: Exploratory Research and Development Center, Los Alamos National Laboratory, Los Alamos, NM 87545, USA

SAMPLE PREPARATION AND CHARACTERIZATION

The Y124 materials were synthesized by a high-oxygen-pressure technique, as described previously.[5] Magnetization measurements were carried out using a SQUID magnetometer. The T_c value for our sample is 80 K at 10 Oe. Figure 1 shows that the normal-state magnetic susceptibility χ in Y124 decreases gradually with decreasing temperature.

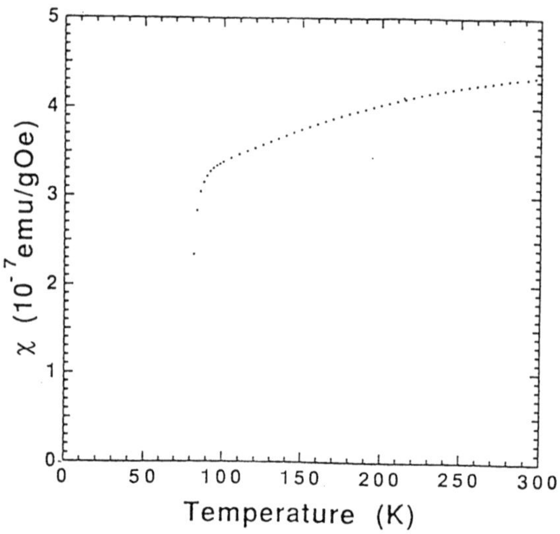

Figure 1 The bulk susceptibility χ in Y124 as a function of temperature.

The substitution of ^{17}O for ^{16}O in the Y124 powder was made by a gas-exchange technique at 700 °C and one atmosphere. A mixture of ^{17}O enriched sample and Epoxy (Stycast 1226) was cured in an 8T magnetic field at room temperature. X-ray diffraction indicated that the sample is aligned with the c axis parallel to the field under these conditions.
A spin-echo NMR spectrometer was used to determine the oxygen Knight-shift and spin-lattice relaxation rate at the plane sites.

RESULTS AND DISCUSSION

A. KNIGHT-SHIFT

As illustrated in Fig.2, there are four distinct oxygen sites in Y124. Figure 3 shows the ^{17}O NMR spectrum at 98 K and 33 MHz with magnetic field along the c axis. The electronic environment at each atomic site in Y124 is expected to be similar to that for Y123. In view of the Y123 results, we assign the peaks in the spectrum as follows. The main peak b0 is due to the central transition($1/2 \leftrightarrow -1/2$) at the O(2,3) sites. The first satellite transitions ($\pm 1/2 \leftrightarrow \pm 3/2$) at the O(2,3) sites correspond to the b1 and b2 peak pairs, and the second satellite transitions ($\pm 5/2 \leftrightarrow \pm 3/2$) appear at the b3 and b4 peaks. Poorly oriented grains may contribute to a broad background centered at 57 kOe. The weak peaks ai and ci (i=0,1 and 2) originate from the O(1) and O(4) sites.

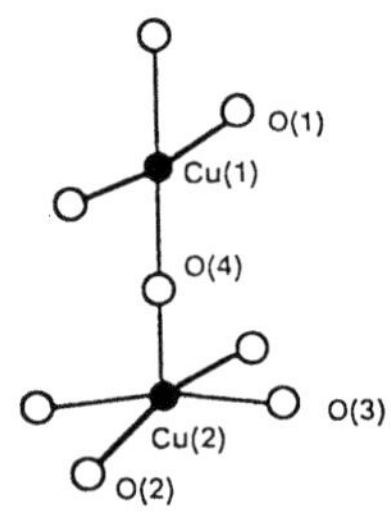

Figure 2 The oxygen and copper sites for Y124.

We determine $^{17}K_c$ for the O(2,3) sites using the b1 and b2 positions. The temperature dependence of $^{17}K_c$ is shown in Fig.4. We compare the present result with the bulk susceptibility χ. Generally χ consists of spin, orbital and isotropic diamagnetic core contributions. The latter two parts are expected to be temperature independent. A similar classification is applicable to the Knight shift. Below Tc, $^{17}K_c$ decreases with decreasing temperature. This indicates that the creation of Cooper pairs suppresses the spin part K_s of $^{17}K_c$ below Tc. The small negative $^{17}K_c$ value at 4.2 K suggests that the Knight shift minus the spin part K_s is also negative. We assume that the temperature independent part of $^{17}K_c$ is -0.2 % at 4.2 K. The $^{17}K_c$ vs χ relationship is plotted in Fig.4. $^{17}K_c$ is proportional to χ in the range between Tc and 140 K, and then appears to deviate from the linearity above 200 K. The spin part of χ is mainly governed by the Cu spins at the plane site. The present result therefore suggests that the oxygen spin correlates with the copper spin below 140 K. We estimate the temperature independent part of χ as 2×10^{-7} emu/gOe, assuming that Ks=Kc+0.2%. In the normal state, $^{17}K_c$ increases gradually with increasing temperature and then becomes constant. A similar temperature dependence is observed for $^{17}K_c$ in the oxygen-deficient Y123.[4]

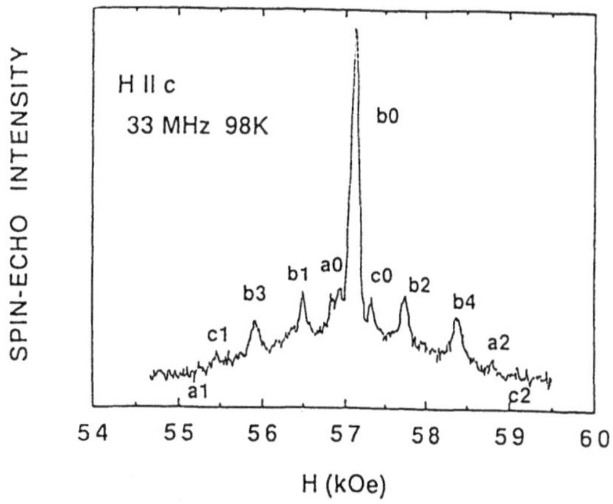

Figure 3 The oxygen NMR spectrum of aligned crystalline Y124 at 33 MHz and 98K.

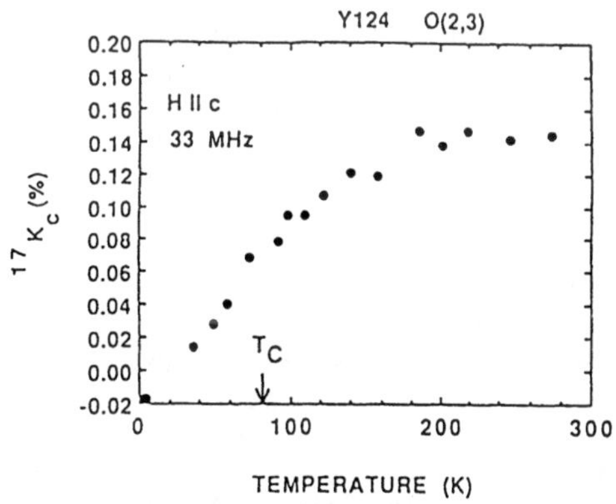

Figure 4 The Knight shift $^{17}K_c$ for the O(2,3) site as a function of temperature.

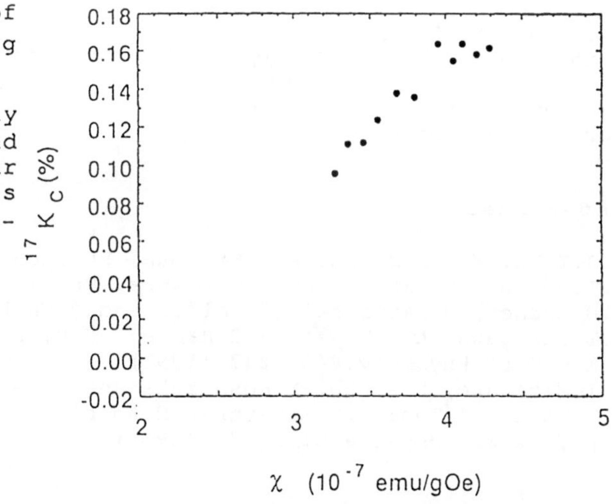

Figure 5 The $^{17}K_c$ vs χ relationship for Y124.

B. SPIN-LATTICE RELAXATION

We determined T_1 by the spin-echo recovery technique for the first satellite at the O(2,3) sites. Figure 6 shows that both $(^{17}T_1T)^{-1}$ and $^{17}K_s$ have a similar temperature dependence, indicating that $(^{17}T_1T)^{-1}$ depends on the spin susceptibility. At the planar Cu sites $(^{63}T_1T)^{-1}$ [6] is significantly larger than $(^{17}T_1T)^{-1}$, indicating that the spin density at E_F is located at the Cu site. A combination of the uniform and the antiferromagnetic susceptibilities contributes to $^{63}T_1$, while the uniform part is dominant at $^{17}T_1$ due to the crystal symmetry. The relation $(^{17}T_1T)^{-1} \propto \chi$ implies that the oxygen spin fluctuation is influenced by the Cu spin at q=(0,0), although the oxygen spin does not directly couple with the antiferromagnetic Cu spin fluctuations at q=(π,π). Korringa relation predicts that $(^{17}T_1T^{17}K_s^2)^{-1}$ is constant. The oxygen NMR result for the Y123 with x=7 is consistent with this relation, while $(^{17}T_1T^{17}K_s^2)^{-1}$ for the oxygen-deficient Y123 decreases with increasing temperature.[4] The plot in Fig.7 shows that Y124 is similar in behavior to the oxygen-deficient Y123 case. Comparison with this Y123 result suggests that the antiferromagnetic spin fluctuations persist in the normal state of Y124.

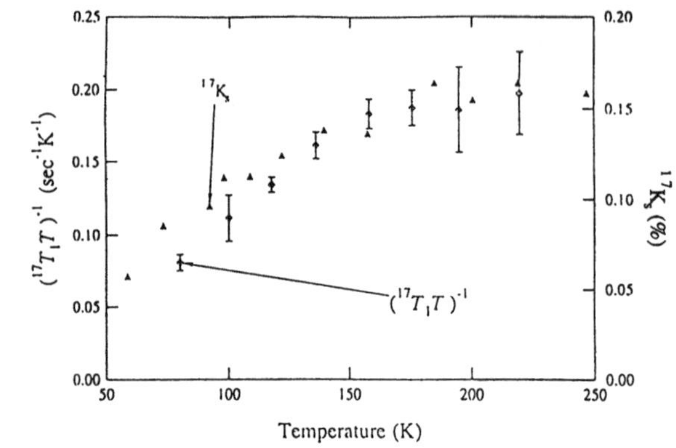

Figure 6 The temperature dependence of $(^{17}T_1T)^{-1}$ for Y124. For comparison, the spin part $^{17}K_s$ of Knight-shift is also plotted.

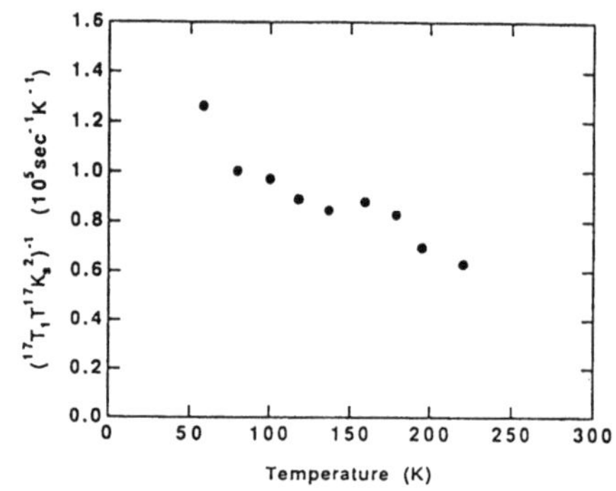

Figure 7 The temperature dependence of $(^{17}T_1T^{17}K_s^2)^{-1}$ for Y124.

References

1. J.Y.Yu, K.T.Park and A.J.Freeman, Physica C172, 467 (1991).
2. T.Oguchi, T.Sasaki and K.Terakura, Physica C172, 277 (1991).
3. B.Bucher, J.Karpinski, E.Kaldis and P.Wachter, Physica C157,478 (1989).
4. M.Takigawa, A.P.Reyes, P.C.Hammel, J.D.Thompson, R.H.Heffner, Z.Fisk and K.C.Ott, Phys.Rev.B43, 247 (1991).
5. T.Miyatake, S.Gotoh, N.Koshizuka and S.Tanaka, Nature 341, 41 (1989).
6. T.Machi, I.Tomeno,T.Miyatake, N.Koshizuka , S.Tanaka, T.Imai and H.Yasuoka, Physica C173, 32 (1990).

^{151}Eu Mössbauer Analysis of T'-Phase Superconductor $Eu_{1.85}Ce_{0.15}CuO_4$

M. TANIWAKI[1], H. ITOH[2], and M. KUSUHASHI[2]

[1] Department of Electronics, Faculty of Engineering, Hokkaido University, Sapporo, 060 Japan
[2] Muroran Research Laboratory, The Japan Steel Works Ltd., Muroran, Hokkaido, 051 Japan

The electronic state and the lattice vibration of Eu in T'-phase $Eu_{1.85}Ce_{0.15}CuO_4$ are investigated by ^{151}Eu Mössbauer effect. The superconductivity appears in samples fired at 1100℃ and annealed in argon, in which the line shift of ^{151}Eu is 0.76-0.79mm/s and the Debye temperature is 165-179K. These values are lower than those in T*-phse $(La,Eu)_{2-y}Sr_yCuO_4$ and higher than those in T'-phase $(La,Eu)_{2-y}Ce_yCuO_4$. The sample which was sufficiently doped with Ce^{4+} shows the low isomer shift. Due to doping of the Ce^{4+}, the Eu-O distance increases and as a result, the s-electron density of Eu decreases. The isomer shift and the Debye temperature of T'-phase are discussed with those in other high Tc superconductors. It is suggested that the transition temperature is concerned with the lattice vibration.

KEY WORDS: T'-phase superconductor, Mössbauer effect, Debye temperature, Tc, lattice vibration

1. INTRODUCTION

The present authors, and one of them and others have studied the microscopic states such as the electronic state and the lattice vibration of Eu in various high Tc superconductors[1-7]. From these studies, it was shown that the s electron density and the Debye temperature of Eu with large Eu-O distance are low. The ^{151}Eu Mössbauer effect of electron doped superconductor T'-phase $(La,Eu)_{2-y}Ce_yCuO_4$ was studied by the present authors. It was shown that the s-electron density of Eu in this superconductor is lower than that in T*-phase $(La,Eu)_{2-y}Sr_yCuO_4$[3,4] and 123-phase $Ba_2EuCu_3O_7$ and the lattice vibration in T'-phase is softer than in the others, because of the longer Eu-O distance in the T'-phase[2,3,4,7]. Furtheremore a correlation between the Debye temperature and the superconducting transition temperature was obtained. These results on the T'-phase superconductor, however, does not necessarily show the state of all lanthanides in T'-phase, because we measured only one of two major lanthanides in this phase. In the present work, the ^{151}Eu Mössbauer effect of T'-phase $Eu_{1.85}Ce_{0.15}CuO_4$ is observed in order to clarify the microscopic states of lanthanide in T'-phase and examine the correlation between these states and the superconductivity. For comparison, the ^{151}Eu Mössbauer effect of non-superconductive T'-phase Eu_2CuO_4 is also measured. Combining these results with those of the previous works, relations between the isomer shift of Eu and the Eu-O distance, and between the Debye temperature and the superconducting transition temperature are discussed.

2. EXPERIMENTAL PROCEDURES

The T'-phase $Eu_{1.85}Ce_{0.15}CuO_4$ and Eu_2CuO_4 samples were prepared by solid state reaction from mixtures of Eu_2O_3, Ce_2O_3 and CuO. The mixed powder was calcined at 900℃ for 20hrs in air, then, sintered at a temperature of 1000-1100℃ for 10-16hrs. After that, it was annealed in flowing argon at 875 or 885 ℃ for 12hrs and cooled(or quenched) to room temperature. The samples and heir heattreatment are tabulated in table 1. The ^{151}Eu Mössbauer effect was observed using a $^{151}SmF_3$ γ-ray source, where the σn was 3 in all samples(σ : the Mössbauer cross-section, n: the area density of the Mössbauer nucleus). The spectrum was standardized by EuF_3 and Fe at room temperature.

3. RESULTS

All obtained Mössbauer spectra were analyzed by non-linear least square method using a single Lorentzian as shown in Fig. 1. The Mössbauer line parameters and the Debye temperature, Θ_D are tabulated in table 1, with the lattice constant and the superconductive characteristics, where Θ_D was estimated from the line area. The samples which showed superconductivity above 4.2K are EC5 and EC6. These were the $Eu_{1.85}Ce_{0.15}CuO_4$ samples fired at 1100℃ and annealed in argon. Their δ and Θ_D were (0.764mm/s, 165K) and (0.787mm/s, 179K).

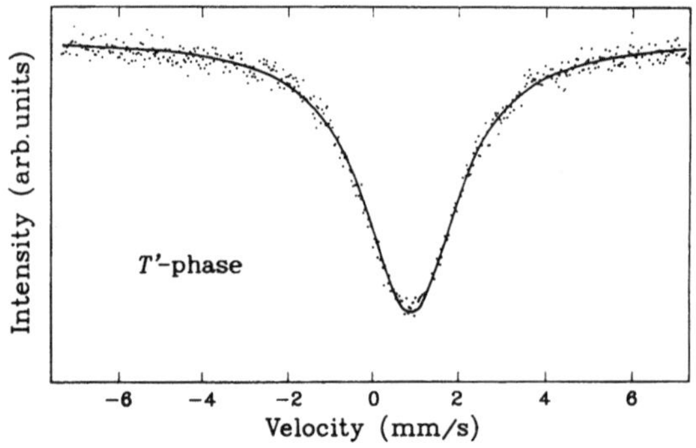

Fig. 1. The ^{151}Eu Mössbauer spectrum of T'-phase superconductor at room temperature. The velocity axis is standardized by EuF_3.

Table 1
The heat treatment of $Eu_{1.85}Ce_{0.15}CuO_4$(EC1-EC6) and Eu_2CuO_4(E1 and E2) after calcination at 900℃ for 20hrs and their data on structure, superconductivity and the Mössbauer effect. M.V.; Meissner volume, Θ_D: The Debye temperature of Eu estimated from the line area.

Sample	Sintering ℃, hrs	Annealing ℃, hrs	Lattice constant a nm, c nm	*Tc*, M.V. K %	Line shift mm/s	Line width mm/s	Θ_D K
EC1	-	-	0.3902, 1.1896	-	0.832 ± 0.018	2.33 ± 0.07	172
EC2	1100, 16	-	0.3906, 1.1860	-	0.870 ± 0.025	2.23 ± 0.08	160
EC3	1000, 10	875, 12	0.3906, 1.1867	-	0.840 ± 0.019	2.38 ± 0.07	163
EC4	1050, 16	885, 12	0.3903, 1.1856	-	0.814 ± 0.020	2.40 ± 0.07	166
EC5	1100, 16	875, 12	0.3903, 1.1841	11.7, 2.8	0.764 ± 0.019	2.38 ± 0.07	165
EC6	1100, 16	885, 12	0.3905, 1.1848	12.9, 3.4	0.787 ± 0.018	2.40 ± 0.06	179
E1	1000, 16	875, 12	0.3903, 1.1903	-	0.819 ± 0.021	2.23 ± 0.08	172
E2	1100, 16	885, 12	0.3903, 1.1904	-	0.847 ± 0.008	2.41 ± 0.04	195

4. DISCUSSION

4. 1. Isomer shift and Eu-O distance

The $Eu_{1.85}Ce_{0.15}CuO_4$ samples which show superconductivity are EC5 and EC6 which were fired at high temperature, 1100℃ and annealed in argon. The other $Eu_{1.85}Ce_{0.15}CuO_4$ samples, EC1 - EC4, which were fired below 1100℃ or were not annealed in argon, were not superconductive above 4.2K. The firing at a high temperature and annealing in argon is necessary for electron doping. The former is for deoxydization and the latter is for dissolving Ce^{4+} as reported by Itoh and Kusuhashi. The solution of Ce shorten the lattice constant, especially the c-axis as shown in Fig. 2. Here we will examine the relation between the c-axis length and Mössbauer isomer shift, δ. The isomer shift, δ, is plotted against the c-axis length in Fig. 3. The isomer shift, δ, decreased with shortening c-axis. Since the δ of ^{151}Eu(i. e., the s-electron density of Eu) in Eu_2O_3 increases with decreasing Eu-O distance[8], we can conclude that the Eu-O distance increases with doping Ce^{4+} in T'-phase Eu_2CuO_4, in spite of the decrease in the c-axis length. Probably the Ce^{4+} substituted for Eu^{3+} attracts oxygen ions to expand Eu-O bond.

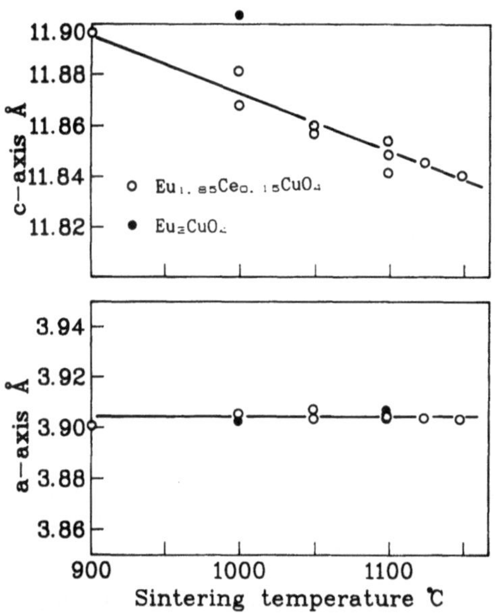

Fig. 2. The dependence of the lattice parameters on the sintering temperature in T'-phase $Eu_{1.85}Ce_{0.15}CuO_4$ and Eu_2cuO_4.

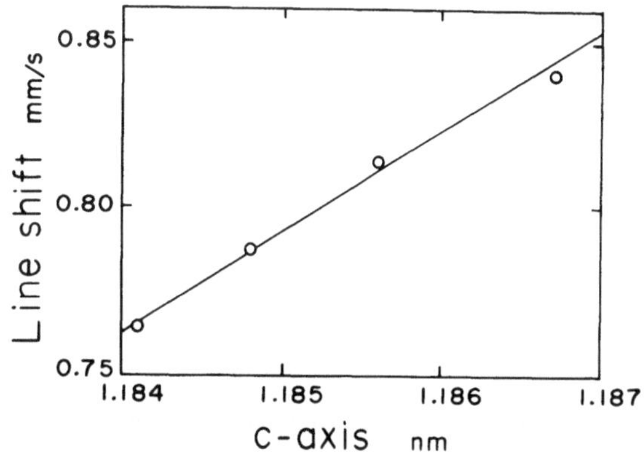

Fig. 3. The relation between the ^{151}Eu isomer shift and the c-axis in T'-phase $Eu_{1.85}$-$Ce_{0.15}CuO_4$ annealed in argon.

4. 2. Comparison of isomer shift between T'- and T*-phases

The isomer shifts, δ s, of ^{151}Eu in T'-phase $Eu_{1.85}Ce_{0.15}CuO_4$, Eu_2CuO_4 and $(La,Eu)_{2-y}Ce_yCuO_4$, and T*-phase $(La,Eu)_{2-y}Sr_yCuO_4$ are plotted against the unit cell volume in Fig. 4. We find the rough tendency that the δ decreases with increasing unit cell volume. The large unit cell volume generally means the long Eu-O distance. Because of it this tendency is explained by the dependence of δ on Eu-O distance, however, the following fact cannot be explained only by the idea. The unit cell volumes of $Eu_{1.85}Ce_{0.15}CuO_4$ and Eu_2CuO_4 are about 1.8 nm^3, which are much smaller than those of $(La,Eu)_{2-y}Ce_yCuO_4$ and $(La,Eu)_{2-y}Sr_yCuO_4$. From this, the δ s in $Eu_{1.85}Ce_{0.15}CuO_4$ and Eu_2CuO_4 are predicted to be higher than those in $(La,Eu)_{2-y}Ce_yCuO_4$ and $(La,Eu)_{2-y}Sr_yCuO_4$. The δ (0.764, 0.787 mm/s) in superconductive $Eu_{1.85}Ce_{0.15}CuO_4$ is, however, between the δ in the T*-phase $(La,Eu)_{2-y}Sr_yCuO_4$ and that in the T'-phase $(La,Eu)_{2-y}Ce_yCuO_4$. As described in the former section,

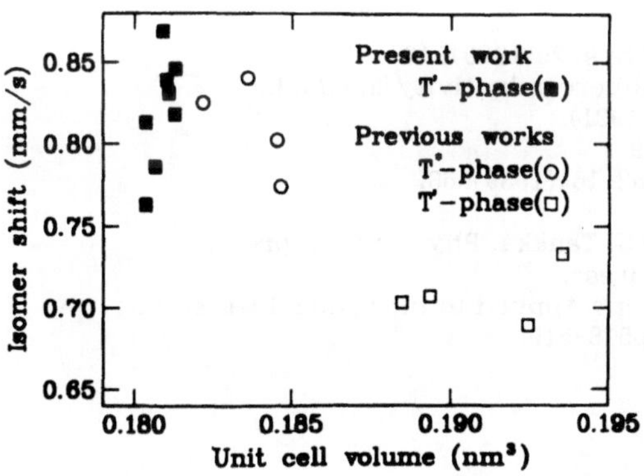

Fig. 4. The plot of the ^{151}Eu isomer shift against the unit cell volume in high Tc superconductors.

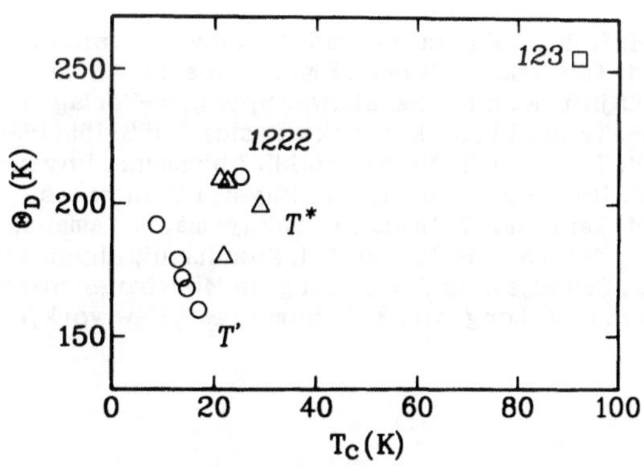

Fig. 5. The relation between the Debye temperature and the transition temperature in high Tc superconductors.

the effect of the doped ions must be considered. In the T*-phase $(La,Eu)_{2-y}Sr_yCuO_4$, the Eu-O distance must be smaller than that predicted from the unit cell volume, because the binding force between the doped Sr^{2+} and oxygen ions is less than that between Eu^{3+} and oxygen ions, while in the superconductive T'-phase the Ce^{4+} attracts oxygen ions and enlarges the Eu-O distance. In fact, the δ is high in the T'-phase Eu_2CuO_4 and nonsuperconductive T'-phase $Eu_{1.85}Ce_{0.15}CuO_4$ which scarcely contains Ce^{4+}, since the Eu-O distances in these are not enlarged.

4. 3. Debye temperature and Tc

The Debye temperatures, Θ_Ds in high Tc superconductors(123-, 1222-, T*- and T'-phases) obtained by the present authors, and one of them and others) are plotted against Tc in Fig. 5. These data are classified into three groups, T'-phase, T*-phase and 1222-phase, and 123-phase. The Θ_D(255K) and Tc(92K) in 123-phase superconductor $Ba_2EuCu_3O_7$ is the highest of all. The Θ_D(180-209K) and Tc(21-29K) in T*- and 1222-phase are respectively between the value of 123-phase and that of T'-phase. The T'-phase superconductors have the lowest Θ_D(160-192K) and Tc(16-20K) of all. From these facts, we conclude that the Θ_D of Eu in the phase having higher Tc shows the higher value.

In contrast with the above discussion, the Tc in T'-phase increases with increaing Θ_D as pointed out in a previous paper by the present authors[1]. Such a tendency seems to be observed also in the T*-phase superconductors. We may consider that the lattice softening enhances the superconductivity in a series of superconductors having the same structure.

In this section we examined correlations between the Tc and the Θ_D of Eu in high Tc superconductors. The experimental facts seem to indicate that the lattice vibration has an important role in high Tc superconductivity. As it is well known, Eu does not contribute to the superconductivity. However, the lattice vibration of the matrix will be reflected on the Θ_D of Eu. Especially that of superconducting plane(Cu-O plane) must affect the Θ_D of Eu, because Eu atoms are in the neighboring sites of the plane.

5. SUMMARY

In this work, the ^{151}Eu Mössbauer effect in T'-phase Eu_2CuO_4 doped with Ce was studied. In this system, the sample with the higher isomer shift shows the higher transition temperature. By doping Ce^{4+}, the Eu-O distance in the Eu_2CuO_4 is enlarged because Ce^{4+} attracts oxygen ions. As a result, the s-electron density increases, i. e., the isomer shift increases. For doping Ce^{4+} sufficiently in T'-phase Eu_2CuO_4 system, the firing at high temperature($\geqq$ 1100℃) and the annealing in argon are necessary. The isomer shift was 0.787mm/s in the sample showing the highest Tc, 12.9K. The correlation between the isomer shift and the Eu-O distance, and that between the Debye temperature and the transition temperature in various high Tc superconductors were discussed. The Eu-O distance in T'- and T*-phases does not depend only on the unit cell volume but also depend on doped ions. The Debye temperature of Eu in the phase with higher Tc is higher. In a series of the superconductors of the same phase, the Debye temperature in the sample with higer Tc is lower. These results suggest that the lattice vibration is concerned with the high Tc superconductivity.

REFERENCES

1. H. Itoh, M. Kusuhashi and M. Taniwaki, Jpn. J. Appl. Phys. 29(1990)L1604.
2. M. Taniwaki, H. Itoh and M. Kusuhashi, Advances in Superconductivity III, eds. K. Kajimura and H. Hayakawa,(Springer-Verlag, Tokyo, 1991).
3. M. Taniwaki and K. Sasaki, Pysica C 153-155(1988)1549.
4. M. Taniwaki, T. Muraki and K. Shiramine, Physica C 162-164(1989)568.
5. H. Itoh and M. Kusuhashi, Physica C, in press.
6. M. Taniwaki, T. Maeda, K. Sakuyama, H. Yamauchi and S. Tanaka, Physica C, in press.
7. M. Taniwaki, H. Itoh and M. Kusuhashi, Physica C, in press.
8. F. Grandjean and G. J. Long, in Mössbauer Spectroscopy Applied to Inorganic Chemistry, ed. G. J. Long, Vol. 3(Plenum Press, New york,1990)pp518-519.

2.5 Optical Properties

Raman Spectroscopy as a Probe of Electronic Structure. Superconducting Gaps, and Electron-Phonon Coupling in HTSC

ERK THORSTEN HEYEN

Max-Planck-Institut für Festkörperforschung, 7000 Stuttgart, Germany

ABSTRACT

In order to investigate the electronic structure in the normal state as well as the mechanism of Raman activity, we have measured and calculated the dependence of phononic Raman intensities on laser wavelength for $YBa_2Cu_3O_7$ and $YBa_2Cu_3O_6$. Under resonance conditions, infrared-active modes are observed in $YBa_2Cu_3O_6$. We also report measurements of the superconducting gaps and the electron-phonon coupling in untwinned $YBa_2Cu_4O_8$ (YBCO-124) single crystals. We find *two* critical gap values. Finally, we demonstrate that a double peak observed near 300 cm^{-1} at low temperature in the Raman and inelastic neutron scattering spectra of $NdBa_2Cu_3O_7$ is due to a coupled (and therefore mixed) excitation involving renormalized Nd^{3+} $4f$ crystal-field excitations and phonons.

KEY WORDS: Electronic Structure, Gap Anisotropy, Electron-Phonon-Coupling, Crystal Field Excitations, Phonons

INTRODUCTION

Since the discovery of the Cuprate Superconductors, much interest has been focused on their electronic structure in the normal as well as in the superconducting state. The complex crystal structure and correlation effects make it very difficult to describe correctly the electronic structure. The distinct anisotropy of these materials and up to four Fermi-surface (FS) sheets make the interpretation of experiments attempting the determination of the gap functions $\Delta_i(\mathbf{k})$ difficult.

Recently, absolute Raman efficiencies of $YBa_2Cu_3O_7$ have been measured and compared favorably with results of LDA "ab initio" calculations [1], indicating that this theory yields a good description of the normal state.

Similar measurements for insulating $YBa_2Cu_3O_6$ helped to clarify its electronic structure which is strongly affected by electron correlations [2]. In this material, also infrared-active, Raman-forbidden longitudinal (LO) phonons have been observed under resonant conditions when they become Raman-quadrupole-allowed by the Fröhlich interaction.

All high T_c superconductors (HTSC) show a strong electronic continuum which is almost independent of temperature and energy [3]. Although its origin is still controversial [4–7], the change of the spectra below T_c can be used to determine the superconducting gaps [8]. For $YBa_2Cu_4O_8$, these measurements suggest the existence of varying gap values on the different FS sheets [9].

The superconducting gaps can also be investigated by the measurement of phonon anomalies at T_c [10], yielding not only values for the superconducting gaps at least at some part of the FS as well as electron-phonon coupling constants [11], but also confirm the existence of two gaps in $YBa_2Cu_4O_8$ [9].

In this material, also anomalies of the phonon intensities have been detected and interpreted as a resonance between the phonon and the superconducting gaps [9].

Scattering by strongly coupled excitations involving a phonon and Nd $4f$-electron crystal field (CF) excitations has been reported for $NdBa_2Cu_3O_7$[12,13]. The dependence of the coupled modes on temperature and oxygen isotope has been successfully described by a simple model [13].

RESONANT RAMAN SCATTERING IN $YBa_2Cu_3O_7$

Raman scattering spectra for $YBa_2Cu_3O_7$ usually show peaks near 115, 150, 340, 435, and 500 cm^{-1} which are due to the five A_g phonons corresponding to a motion of the Ba, Cu(2) (plane copper), O(2)-O(3) (plane oxygen), O(2)+O(3), and O(4) (apical oxygen) atoms in z-direction. Only modes having even parity with respect to the inversion can be seen in Raman spectra. In this material the vibrations of the Y, Cu(1) (chain copper), and O(1) (chain oxygen) do not yield Raman-active modes since they have necessarily odd parity.

The scattering intensity is very directly influenced by the electronic structure and the way the electronic states are modulated by the phonon. The scattering efficiency S for scattering by a certain phonon can be approximated as [1]

$$S(\omega) = \frac{V}{(4\pi)^2}\left(\frac{\omega}{c}\right)^4 \left|\mathbf{e}_i \cdot \frac{d\tilde{\epsilon}(\omega)}{dQ} \cdot \mathbf{e}_s\right|^2, \qquad (1)$$

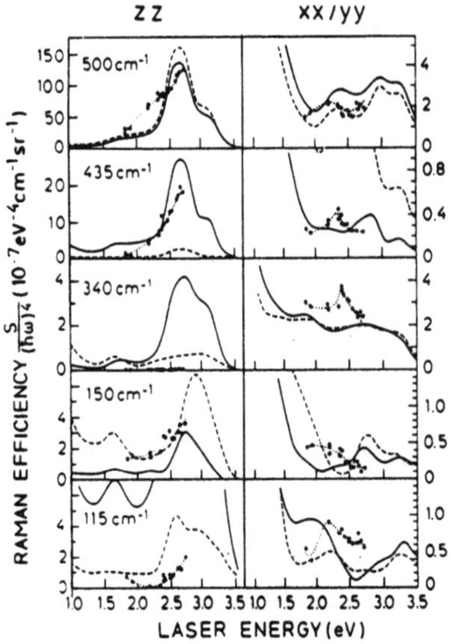

Figure 1: Raman efficiencies in $YBa_2Cu_3O_7$ for z- or x-polarized laser and Stokes light as a function of laser photon energy in absolute units, compared with LDA calculations (also absolute!). The dashed lines represent calculations for the unmixed ionic motion of the Ba, Cu(2), O(2)–O(3), O(2)+O(3), and O(4) ions, the solid lines calculations for modes mixed according to [16]. Note that the 435 cm^{-1} mode derives its strength from the O(4) mode admixture while the Ba and Cu(2) modes (115 and 150 cm^{-1}) are essentially unmixed [1].

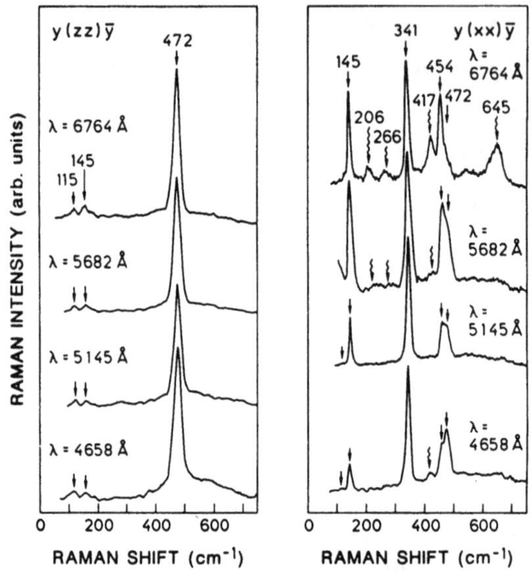

Figure 2: Raman spectra of the insulating, antiferromagnetic compound $YBa_2Cu_3O_6$ at 10 K for various laser wavelengths showing the five Raman-allowed phonons (arrows) and the Raman-forbidden infrared-active E_u phonons which become Raman-active by Fröhlich interaction when the laser photon energy approaches an electronic resonance [2].

where $\hbar\omega$ is the photon energy of the laser light, $\tilde{\epsilon}(\omega)$ the dielectric tensor, Q the phonon normal coordinate, $\mathbf{e}_i$ and $\mathbf{e}_s$ the polarization vectors of incident and scattered light, respectively, V the volume of the unit cell, and c the velocity of light. The actual elongation q_n of the n-th atom is the given by $q_n = Q\epsilon_n\sqrt{\hbar/(2M_n\omega_{ph})}$, where ω_{ph} is the phonon frequency, ϵ_n the phonon eigenvector and M_n the mass of the respective atom. We have measured the scattering efficiencies for all 5 A_g phonons as a function of laser photon energy. This technique is called Resonant Raman Scattering (RRS). After correcting for varying penetration depths, reflectivities etc. by using ellipsometric data [14], we obtained the results in absolute units (without an arbitrary scaling factor) by comparison of the measured efficiencies with those of the BaF_2 phonon which are known (see [1]). The results are displayed in Fig. 1. We notice that the efficiencies vary by three orders of magnitude for different phonons and polarizations. In particular, the zz-polarized spectra show a strong resonance towards 2.8 eV.

In spite of doubts cast on the relevance of local-density approximation (LDA) band theory for high temperature superconductors, it is so far the only *ab initio* theory which has yielded quantitatively correct results for the shape of the FS, phonon frequencies, electron-phonon coupling constants, and the dielectric function [15]. This motivated us to interpret our results in terms of LDA bands. We therefore calculated the electronic band structure using the Linear Muffin-Tin Orbital (LMTO) method and subsequently the dielectric function not only for the undistorted structure, but also for a distortion corresponding to one of the five phonons mentioned above. From these dielectric functions, we obtained the Raman efficiency $S(\omega)$ from Eq.(1) by numerical differentiation. These results are plotted as dashed lines in Fig. 1. We also calculated the intensities for phonon eigenvectors proposed by full potential total energy calculations (solid lines) [16]. These eigenvectors differ from the "pure phonons" mainly in that they suggest a strong mixing of the Cu(2) and Ba motion as well as of the three oxygen modes.

First, we notice that the average intensities of all modes and polarizations are well described by our calculations, which have used as only variable parameter a reasonable lifetime broadening corresponding to a Lorentzian half width of 0.12 eV. Also the shape of most resonance profiles is well reproduced. In particular, we can now understand the origin of the strong Raman intensity of the 500 cm^{-1} mode in zz-polarization: The "chain band", corresponding mainly to Cu(1) $d_{z^2-y^2}$, O(4) p_z, and O(1) p_y orbitals, acts as final state for a transition from an occupied band having O(1), O(2), O(3), O(4) p_z, and Cu(2) $d_{3z^2-r^2}$ character (for

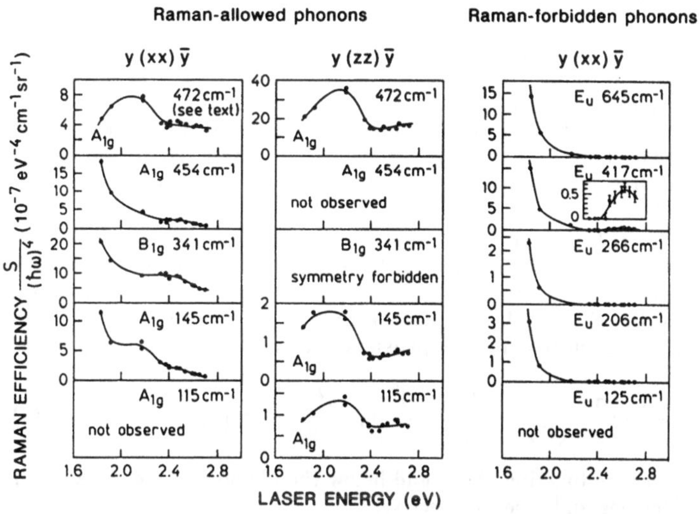

Figure 3: Raman efficiencies in insulating $YBa_2Cu_3O_6$ for x-(left panel) or z-polarized (middle panel) laser and Stokes light as a function of laser photon energy in absolute units. For xx-polarization, resonances only for the plane atoms are observed near the charge-transfer excitations at 1.7 and 2.5 eV, while the zz-intensity, showing only a weak resonant enhancement near 2.1 eV, is caused by the huge deformation potential of the occupied "chain band". Near the 1.7 eV resonance, the infrared-active E_u phonons (right panel) gain Raman-intensity by Fröhlich interaction [2].

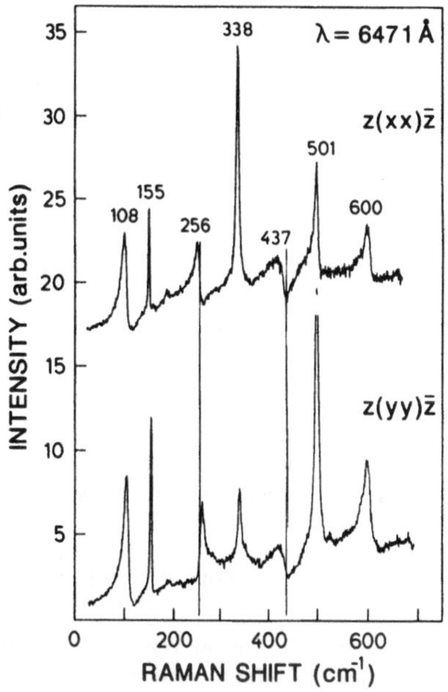

Figure 4: Raman spectra of an untwinned $YBa_2Cu_4O_8$ single crystal, taken at 10 K with a λ=6471 Å laser. Strong electron-phonon coupling for practically all modes can be observed via the asymmetric Fano line shapes. Also phonons corresponding to chain copper and oxyge vibrations are detected at 256 and 600 cm^{-1} [9,18].

a review of LDA band structure calculations, see [17]). This z-polarized transition may be considered as a plane-to-chain charge transfer. Its spectrum has a strong peak at 2.8 eV, which is clearly observed in measurements of the dielectric function [14]. The chain band is very strongly modulated by the O(4) vibration (−45 meV for 0.01 Å O(4) elongation), so that the derivative of the dielectric function and hence $S(\omega)$ (Eq.(1)) is very large, as observed experimentally. Also we find that the zz-intensity of the mode at 435 cm^{-1} (which has previously been characterized as corresponding mainly to O(2)+O(3) motion) is entirely due to a small admixture of O(4) motion (solid line). Without this admixture (dashed line) this intensity would be much smaller. The 340 cm^{-1} mode, however, has clearly no such admixture, since it has experimentally zero zz-intensity. Finally, we find good agreement between theory and experiment only if we assume that the Ba and Cu(2) modes at 115 and 150 cm^{-1}, respectively, do essentially not mix (dashed lines).

RRS spectra thus yield more evidence that LDA band theory yields a good description of the normal state electronic structure of the HTSC. Comparing theory and experiment, we can now understand the origin of Raman intensities which vary strongly for different phonons and polarizations. More detailed information is given in [1].

RESONANT RAMAN SCATTERING IN $YBa_2Cu_3O_6$

Similar investigations have also been performed for single crystals of the insulating, antiferromagnetic compound $YBa_2Cu_3O_6$ [2]. The results are shown in Figs. 2 and 3. This material, which can be characterized as a charge-transfer insulator, has its lowest (x-polarized) absorption near the charge-transfer gap $E_g \approx 1.7$ eV, corresponding to a transition from the plane oxygen p bands to the Cu(2)-derived Upper Hubbard Band. Its electronic structure cannot be adequately described by LDA theory since correlation effects are essential in this material.

The xx-polarized Raman spectra (Fig.(2) show that only the three phonons corresponding to vibrations of the plane atoms can be observed while the O(4) and Ba phonons have vanishing intensity. The three plane phonons also show a strong resonance (Fig.(3) towards 1.7 eV and another resonance at 2.5 eV, which has also been observed as a structure in in the dielectric function as seen by ellipsometric measurements. These results indicate (Eq.(1)) that only plane-derived bands contribute significantly to the xx-component of the dielectric function, thus strongly supporting the assumptions of the 3-band Hubbard model that only Cu(2) $d_{x^2-y^2}$, O(2) p_x, and O(3) p_y orbitals are relevant. (We stress that this is not the case for superconducting $YBa_2Cu_3O_7$, as shown above).

The zz-polarized RRS profiles show no strong resonant enhancement as was observed for $YBa_2Cu_3O_7$. This can be easily understood since the chain band of $YBa_2Cu_3O_7$, acting as final state for a very strong zz-polarized transition, is for $YBa_2Cu_3O_6$ completely occupied. Thus all zz-efficiencies are smaller than for $YBa_2Cu_3O_7$, but still rather high. However, the relative zz-intensities are similar to the $YBa_2Cu_3O_7$ case. This observation shows, that the occupied chain band, which is presumably still strongly modulated by the O(4) phonon now serves as initial state for other weak transitions, e.g. to the Upper Hubbard Band which could induce the small and broad resonance near 2.1 eV [2].

Intrestingly, four additional peaks at 206, 266, 417, and 645 cm^{-1} are observed as soon as the laser energy approaches the 1.7 eV resonance [2]. These four frequencies can readily be identified as those of the infrared-active longitudinal phonons of E_u symmetry, which are normally Raman-forbidden. It is, however, well-known that these phonons can gain intensity due to the Fröhlich LO-phonon–electron coupling Hamiltonian which diverges for momentum transfers $q \rightarrow 0$. In a simple effective mass approximation, one obtains for this process:

$$S(LO,\omega) \sim \omega_{ph}\left(\frac{1}{\epsilon_-}-\frac{1}{\epsilon_+}\right)\cdot\left|\frac{\mathbf{q}}{q}\cdot\left(\frac{1}{\tilde{m}_e}-\frac{1}{\tilde{m}_h}\right)\cdot\mathbf{q}\right|^2\left|\mathbf{e}_i\cdot\frac{d^2\tilde{\epsilon}(\omega)}{d\omega^2}\cdot\mathbf{e}_s\right|^2, \qquad (2)$$

where ϵ_- and ϵ_+ are the real parts of the dielectric function above and below the phonon frequency, and $\tilde{m}_e$ and $\tilde{m}_h$ are the effective mass tensors for electrons and holes, respectively.

We were able to show that our experimental results behave exactly as expected from Eq.(2) [2]. In particular, the relative phonon intensities for the four peaks which depend, according to Eq.(2), exclusively on $1/\epsilon_- - 1/\epsilon_+$ (known from infrared reflectivity measurements), is well reproduced. This effect becomes only significant in case of resonance where $d^2\tilde{\epsilon}/d\omega^2$ is large. The A_{2u} modes having $q \parallel z$ cannot be observed since the zz-component of the effective-mass tensors is expected to be very large in these almost two-dimensional materials.

SUPERCONDUCTING GAPS AND ELECTRON-PHONON-COUPLING IN $YBa_2Cu_4O_8$

The 80 K superconductor $YBa_2Cu_4O_8$ differs from the well-known $YBa_2Cu_3O_7$ mainly by having two chains stacked onto each other instead of one. Consequently, also the chain atoms yield two Raman-active A_g phonons at 250 and 600 cm^{-1}, corresponding to zz-polarized motion of Cu(1) and O(1), respectively [18]. All other vibrations remain practically unchanged. Each of the chains is now strongly zig-zagged, different from $YBa_2Cu_3O_7$ where they are straight. Similar to $YBa_2Cu_3O_7$, the Raman spectrum of untwinned $YBa_2Cu_4O_8$ single crystals (Fig. 4) consists of an electronic continuum, which is almost independent of energy as well as of temperature (for $T > T_c$), and the 7 phonon peaks. After having subtracted the phonon part, we obtained the pure electronic contribution for various temperatures (Fig. 5).

For temperatures lower than T_c, the intensity decreases considerably below a certain threshold energy, which is $2\Delta_2 \approx 325$ cm^{-1} for xx- and $2\Delta_1 \approx 115$ cm^{-1} for yy-polarization. These results suggest the existence of two critical values for the superconducting gaps.
This conclusion is strongly confirmed by the observation of anomalies of phonon frequencies and linewidths at T_c for almost all phonons:

Phonon		Frequency	Width	Intensity
Ba	108 cm^{-1}	-	+	+
Cu(2)	155 cm^{-1}	?	+	0
Cu(1)	256 cm^{-1}	+	0	0
O(2)/O(3)	338 cm^{-1}	-	0	+
O(2)/O(3)	438 cm^{-1}	+	?	0

The three rightmost columns denote the sign of the anomaly when the sample is cooled from the normal state to the superconducting state. According to the model of Zeyher and Zwicknagel phonon frequencies soften at T_c, when the phonon energy is lower than 2Δ, and harden for $\omega_{ph} > 2\Delta$. Similarly the phonon linewidth decreases for $\omega_{ph} < 2\Delta$ and increases otherwise. These effects are all proportional to the electron-phonon coupling constant, we can thus conclude that all phonons couple significantly to the conduction electrons, not just the phonons corresponding to vibrations of the plane atoms. Intensity anomalies occur at T_c if the

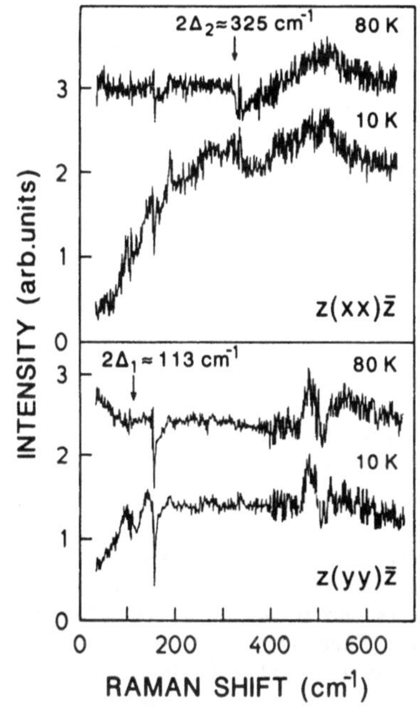

Figure 5: Electronic Raman spectra at 10 and 80 K for a laser wavelength λ=6764 Å after all phonon structures have been subtracted clearly showing the opening of superconducting gaps below $2\Delta_1$=113 cm^{-1} for *yy*-polarization and below $2\Delta_2$=325 cm^{-1} for *xx*-polarization. Residual structures are artifacts of the phonon subtraction. Intensity units are the same as in Fig.4. The 80 K curves are offset by one unit with respect to the 10 K curves [9].

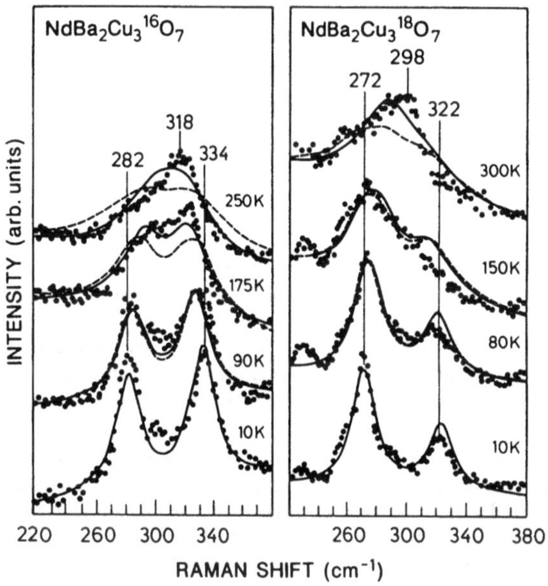

Figure 6: Mixed phonon–CF-excitation coupled modes in the Raman spectra of $NdBa_2Cu_3O_7$ for ^{16}O and ^{18}O isotopes. The dots are experimental data; the solid and dashed lines represent a simple temperature dependent theory with and without anharmonic corrections [12,13].

phonon energy is similar to the superconducting gap, as can be directly seen from third order pertubation theory [9].

A quantitative analysis of the results shown in the table indicates that the lower three phonons couple strongly to a part of a FS which opens a gap of roughly 115 cm^{-1} , while the two O(2)/O(3) phonons indicate a gap near 325 cm^{-1} .

$YBa_2Cu_4O_8$ as well as $YBa_2Cu_3O_7$ have one FS for each of the planes and each of the chains, i.e. 4 or 3 FS sheets, respectively. In the superconducting state, a gap must open on each FS. There is no reason why the gaps on the plane and chain FS should be the same. By comparison with the results for $YBa_2Cu_3O_7$ (where only the upper gap was observed by Raman scattering; however, the lower gap was also observed with other techniques) we conclude that the lower gap corresponds to the chain bands and the upper gap to the plane bands, since the double chains are the only significant difference between both materials. $2\Delta_2$ is approximately the same as in YBCO-123 although T_c is 20% lower, indicating that this ratio is no appropriate variable for the description of HTSC. Both gap functions seem to be anisotropic in **k**-space as indicated by the fact that the decrease in intensity below the respective 2Δ is not abrupt but rather very broad. $2\Delta/kT_c$-ratios larger than 3.5 may be the consequence of multiple gaps or gap anisotropy rather than strong coupling, since only $2\bar{\Delta}/kT_c$, with $\bar{\Delta}$ being the average gap over all Fermi surfaces, is a significant measure for strong or weak coupling. With every experimental technique– it be infrared-reflectance, tunneling or Raman spectroscopy– one only obtains a weighted average of the gap functions. Usually, however, one does unfortunately not know how the weighting is performed, it can be different for each method.

The details of this investigation have been published in [9].

COUPLING OF PHONONS AND CRYSTAL FIELD EXCITATIONS IN $NdBa_2Cu_3O_7$

The rare earth ions of the HTSC have, with the exception of La and Y, strongly localized $4f$-electrons whose ionic levels are split by spin-orbit interaction into angular-momentum eigenstates, which are further split by the influence of the crystal electric field. For $NdBa_2Cu_3O_7$, one expects five Kramers doublets (symmetry in the tetragonal group D_{4h}: $3M_6 + 2M_7$). Transitions between these levels, which are called crystal field (CF) excitations, have been measured by inelastic neutron scattering [19]. CF excitations are, in principle, also Raman-active, having A_{1g} ($M_6 \rightarrow M_6$ or $M_7 \rightarrow M_7$) or B_{1g} symmetry ($M_6 \rightarrow M_7$ or $M_7 \rightarrow M_6$); in reality, however, their intensity is too low to be detectable in these materials due to the localized nature of the $4f$-electrons.

In $NdBa_2Cu_3O_7$ (isostructural to $YBa_2Cu_3O_7$), the CF excitation from the ground state (M_6) to the fourth state (M_7) has a similar energy and the same symmetry as the B_{1g} phonon near 305 cm^{-1} , corresponding to O(2)/O(3) out-of-phase vibration (which has a frequency of 340 cm^{-1} in $YBa_2Cu_3O_7$). Since the phonon modulates the CF and thus the CF excitation energies, both elementary excitations may couple and mix. In this case, one obtains two elementary excitations which both have phonon as well as CF character. Although Raman scattering is practically only sensitive to phonons and not to CF excitations, one observes two peaks (Fig. 6)[12,13] since both have phonon content. A theoretical model [12,13] describes the spectrum well if one assumes frequencies for the unperturbed phonon and CF excitation of ω_{ph}=308 cm^{-1} , ω_{CF}=304 cm^{-1} , and a coupling matrix element V=35 cm^{-1} . If ^{16}O is substituted by ^{18}O, the unperturbed phonon frequency decreases by 6% as phonon frequencies are inversely proportional to the square root of the atomic masses. Since phonon and CF excitations are strongly mixed, one finds, however, that each of the two peaks shift only by 3%, which confirms our interpretation.

Additional support is obtained from the temperature dependence: The coupling is only possible if the Nd ions are in their electronic ground state in all unit cells. At non-zero temperatures, however, also the four excited CF states are populated according to Boltzmann statistics and subsequently, the effective coupling decreases. After the three parameters ω_{ph},ω_{CF}, and V have been fixed by the ^{16}O spectrum at 10 K, our model describes very well the spectra for both isotopes and elevated temperatures. One finds $\omega_{ph} < \omega_{CF}$ for ^{18}O and $\omega_{ph} > \omega_{CF}$ for ^{16}O, which induces the qualitatively different temperature dependence for the ^{16}O and ^{18}O cases.

Recently, the magnitude of V has been reproduced by a parameter-free first-principles calculation within 5% accuracy.

This research would not have been possible without the invaluable contributions of M. Cardona, O.K. Andersen, S.N. Rashkeev, I.I. Mazin, and R. Wegerer. E. Kaldis, J. Karpinski, S. Rusiecki, P. Murugaraj, and B. Nick provided high quality samples.

REFERENCES

[1] Heyen ET, Rashkeev SN, Mazin II, Andersen OK, Liu R, Cardona M, Jepsen O (1990) Phys Rev Lett 65: 3048

[2] Heyen ET, Kircher J, Cardona M (1991) Phys Rev B, to be published

[3] Staufer T, Hackl R, Müller P (1990) Solid State Commun 75: 975

[4] Cardona M, (1991) Proc. M²S HTSC III, Kanazawa, Physica C, to be published

[5] Virosztek A, Ruvalds J, to be published; and Ruvalds J, Virosztek A (1990) Phys Rev B 43: 4064

[6] Varma CM et al., (1989) Phys Rev Lett 63: 1996

[7] Reznik D etb al., (1991) Proc. M²S HTSC III, Kanazawa, Physica C, to be published

[8] Cooper SL, Klein MV, Pazol BG, Rice JP, Ginsberg DM, (1988) Phys Rev B 37: 5920

[9] Heyen ET, Cardona M, Karpinski J, Kaldis E, Rusiecki S, (1991) Phys Rev B 43: 12958

[10] Macfarlane RM, Rosen H, Seki H, (1987) Solid State Commun 63: 831

[11] Friedl B, Thomsen C, Cardona M, (1990) Phys Rev Lett 65: 915

[12] Heyen ET, Wegerer R, Cardona M, (1991) Phys Rev Lett 67: 144

[13] Heyen ET, Wegerer R, Schönherr E, Cardona M, (1991) Phys Rev B 44: to be published

[14] Kircher J, Kelly MK, Rashkeev SN, Alouani M, Fuchs D, Cardona M (1991) Phys Rev B 44: 217

[15] Andersen OK et al, (1991) Proc. M²S HTSC III, Kanazawa, Physica C, to be published

[16] Rodriguez CO, Liechtenstein AI, Mazin II, Jepsen O, Andersen OK, Methfessel M, (1990) Phys Rev B 42: 2692

[17] W. Pickett, (1989) Rev Mod Phys 61: 433

[18] Heyen ET et al., (1990) Phys Rev B 41: 11058

[19] Allenspach P, Furrer A, Bruesch P, Unternaehrer P, (1989) Physica B 156-157: 864

Lattice Vibrational Modes in 1222 Phase of Pb-Sr-La-Ln-Cu-O System (Ln; Gd, Dy, Ho, Er)

YOSHIKAZU HAYASHI, YOSHIYUKI HARADA, HIROYUKI SASAKURA, and KIYOTAKA NAKAHIGASHI

Materials Science, School of Integrated Arts and Sciences, University of Osaka Prefecture, Sakai, Osaka, 591 Japan

ABSTRACT

Measurements of infrared absorption in insulating compounds of the 1222 phase of Pb-Sr-La-Ln-Cu-O system have been made. Lanthanoid(Ln) ions of Er, Ho, Dy and Gd have been used. Absorption peaks near 600 cm^{-1} and 530 cm^{-1} in a crystal containing with Gd ions split or shift in other crystals with Dy, Ho and Er. Results are analyzed by comparing vibrational frequencies with interatomic distances between constituent ions precisely deduced from x-ray structural analysis. Vibrational frequencies of some of the phonon modes are found to increase with the increase of lattice constant contrary to the most of the other modes. The result may be useful to identify phonon modes.

KEY WORDS: 1222 phase, Pb-Sr-La-Ln-Cu-O, phonon spectra, infrared absorption

INTRODUCTION

Synthesis of a new kind of Pb-based compound with so called 1222 phase have been successfully made recently[1-3]. Although no superconductive properties have yet been observed in Pb-Sr-La-Ln-Cu-O system used in the present experiment, it is hopeful to get superconductors if an appropreate process, such as sintering under high oxygen partial pressure, is employed[3]. An x-ray analysis for the series of these compounds were made[3]. Lattice constants for both a- and c-axes are found to increase with the increase of the ionic radius of Ln contained in each crystal. Therefore, decrease of the frequency of each vibrational mode is expected with the increase of the ionic radius of Ln, namely, from the highest for Er to the lowest for Gd. However, according to the detailed x-ray analysis, interatomic distances between some kinds of ions along c-axis are found to decrease with the increase of the ionic radius of Ln. The vibrational frequecy of the related phonon mode may increase with the increase of the ionic radius of the Ln, contrary to those for the most of the other modes, for instance, originating from the in-plane vibration. It is the purpose of the present paper to investigate the relation between the lattice vibrational frequency and the interatomic distance. The result is useful to discriminate some of the vibrational modes parallel to the c-axis.

EXPERIMENTAL METHOD AND RESULTS

Various compounds with the composition of $(Pb_{0.5}Cu_{0.5})Sr_xLa_yLn_{4-x-y}Cu_2O_z$ were prepared by solid state reaction using high purity powders of Sr_2CuO_3, Ln_2O_3, PbO and CuO[3]. Single phase compounds were obtained in relatively narrow range of the compositions for Sr, La and Ln. The optimum composition for each sample depends on the kind of Ln ions contained in the sample. Table 1 lists x and y values for single phase compounds used in the present experiment.

Table 1 Typical x and y values for single phase compounds with different Ln.

Ln	Er	Ho	Dy	Gd
x	1.65	1.6	1.55	1.4
y	0.6	0.65	0.65	0.65

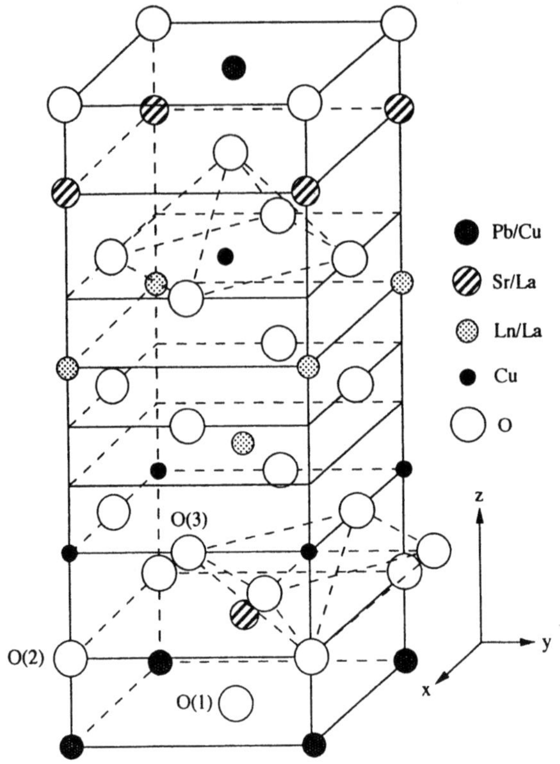

Fig.1 Schematic drawing of a crystal structure of Pb-Sr-La-Ln-Cu-O system.

Figure 1 shows a schematic drawing of one half of a unit cell of the crystal. Lattice constants along a- and c-axes against ionic radius are shown in Fig.2. It may be natural that the lattice constants becomes larger by introducing the larger Ln ions although no simple relation between the lattice constants and the radius of Ln ions is observed. However, a detailed Rietveld analysis of x-ray diffraction has revealed that almost linear dependence exists between ionic radius of Ln and the interatomic distance for every pair of one of the O ions and the neighboring Cu or Pb/Cu or Ln/La or Sr/Ln[3]. Typical examples are shown in Fig.3. It is interesting to note that interatomic distances for Pb/Cu-O(2) and Cu-O(2) show relatively stronger dependence on the ionic radius of Ln: The former decreases while the latter increases.

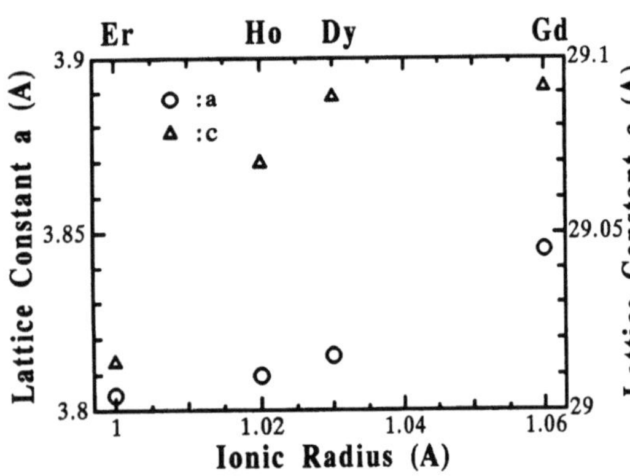

Fig.2 Lattice constants along a- and c-axes vs ionic radius of Ln contained in each compound.

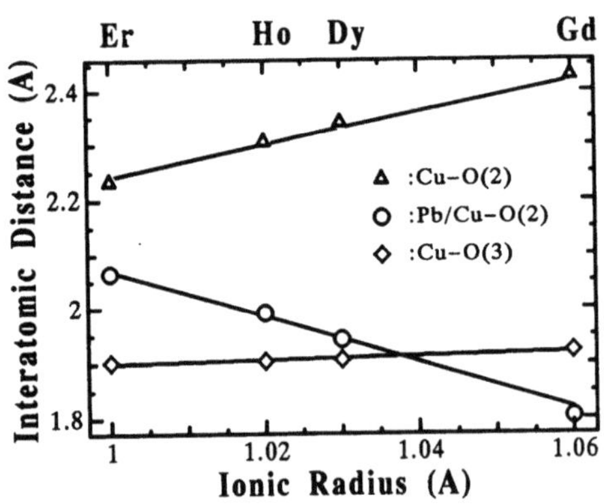

Fig.3 Interatomic distance between Cu-O(3), Cu-O(2) and Pb/Cu-O(2) dependent on ionic radius of Ln.

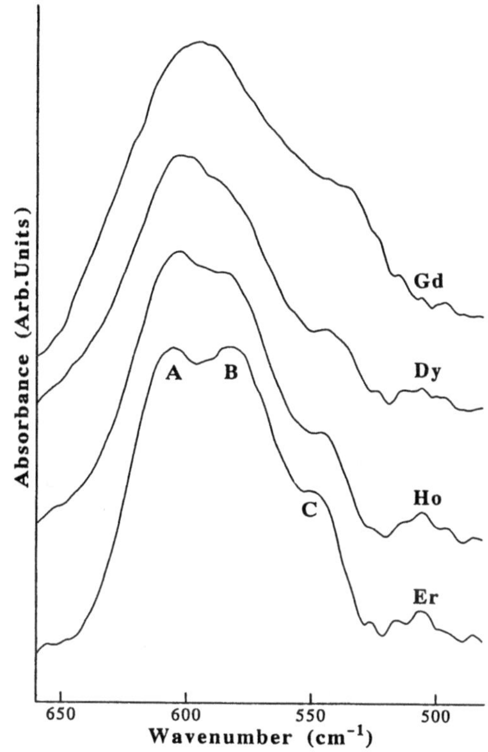

Fig.4 Infrared absorption spectra for 4 kinds of Pb-based oxides of high-Tc family containing Er, Ho, Dy and Gd.

We have made infrared absorption measurements on these compounds. The method of measurements is the same as those employed previously for measurements in Bi-based high-Tc superconductors[4,5]. A Nicolet 60SXR type Fourier transform infrared (FTIR) spectrometer was used. Very small amount (less than 1 mg) of powdered compound is mixed with CsI powder of about 400 mg and pressed to form a disk with 2 cm diameter for use as a sample. Infrared absorption spectra between 660 cm^{-1} and 480 cm^{-1} are shown in Fig.4. The symbols of Lanthanoid elements labelled above each curve, namely, Gd, Dy, Ho and Er, indicate the compound containing each Ln ion. Curves are arranged from top to bottom in order of ionic radius from larger to smaller. Three kinds of peaks marked as A, B and C under the lowest curve labelled Er, appear. Two of them shift to lower frequency side as the increase of the ionic radius but the peak marked as B shifts to the higher frequency side. These results can be recognized as a consequence of the change of interatomic distances as will be discussed in the next section.

DISCUSSION

The vibrational frequency of the harmonic oscillator is proportional to the square root of k/M, where k is the force constant and M is the mass. In the present case, the vibrational motions of oxygen ions are considered to be mainly responsible for the phonon spectra. Referring to Fig.1, interactions between O-Cu and O-Pb/Cu may be important. Although different values of the reduced mass should be taken as M for different vibrational modes, here we assume for simplisity, the same value of M for every mode. In addition, four different crystals are assumed to have essentially the same characteristics except for interatomic distances. Hence, only the interatomic force between O-Cu and O-Pb/Cu may be effective for vibrational frequencies.

According to the simple argument, the vibrational frequency in ionic crystals should be proportional to $d^{-3/2}$, where, d is the interatomic distance[6].

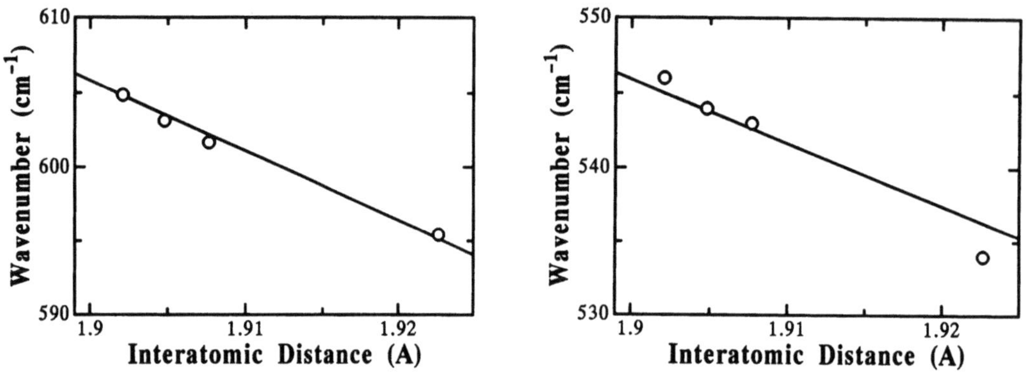

Fig.5 Frequencies corresponding to peaks A and B as a function of the distance between Cu-O(3). The left and the right figures show the frequencies for peaks marked as A and C in Fig.4, respectively. Solid line in each figure indicates -3/2 power dependence of the vibrational frequency to the interatomic distance.

Interatomic distance of Cu-O(3) increases almost linearly with the increse of the ionic radius of Ln as shown already in Fig.3. In Fig.5, frequencies corresponding to peaks marked as A and C are plotted against interatomic distance for Cu-O(3). Fittings to the simple -3/2 power dependence are good. Similar measurements have been made by Tajima et al, in several materials of high-Tc family[7]. They found considerablly stronger power dependence for the highest stretching mode. In the present case, since no significant deviation from -3/2 power dependence is observed, the highest stretching mode seems to have not been observed.

The peak marked as B shows a shift to lower frequency side. Decrease of the effective interatomic distance with the increase of the ionic radius of Ln may be responsible for this shift. Interatomic distance between Pb/Cu and O(2) decreases rather strongly as the ionic radius of Ln increases as shown in Fig.3. On the other hand, interatomic distance of Cu-O(2) increases. We can calculate the relative vibrational frequency by assuming that force constants from two neighboring ions of O(2), i.e. Cu and Pb/Cu, are additive and the same -3/2 power dependence also holds between Cu-O(2) and Pb/Cu-O(2). The calculated results agree well to the experiment. Hence, the peak marked as B may possibly be due to the vibration of O(2) ions along c-axis.

ACKNOWLEDGEMENTS

The authors express their thanks to H. Teraoka, K. Watari, S. Minamigawa for helpful collaboration in sample preparation and x-ray analysis, M. Nagayama for the infrared measurements. We also thank to Prof. A. Yanase and all the other members of Materials Science for their valuable discussions.

REFERENCES

1. Adachi S, Inoue O, Kawashima S, Adachi H, Ichikawa Y, Setsune K, and Wasa K (1990) Physica C 168: 1
2. Maeda T, Sakuyama K, Koriyama S, Ichinose A, Yamauchi H, and Tanaka S (1990) Physica C 169: 133
3. Sasakura H, Nakahigashi K, Teraoka H, Watari K and Minamigawa S (1991) presented at the Spring meeting of the Phys.Soc.Jpn.(in Japanese)
4. Hayashi Y, Fukuoka N, Harada Y and Nakanishi S (1990) J.Phys.Soc.Jpn. 59: 4192
5. Hayashi Y, Harada Y, Sasakura H and Teraoka H (in press) In: Muto Y and Tachiki M (eds) in Proceeding of M^2S-HTSC III
6. Keyes R.W. (1962) J.Appl.Phys. 33: 3371
7. Tajima S, Ido T, Ishibashi S, Itoh T, Eisaki H, Mizuo Y, Arima T, Takagi H and Uchida S (1991) Phys.Rev. 43: 19496

Infrared Refrection and Raman Scattering in $Ba_{1-x}Rb_xBiO_3$

A. Inoue[1], A. Iyo[1], H. Uwe[1], T. Sakudo[1], Y. Tanaka[2], and M. Tokumoto[2]

[1] Institute of Applied Physics, University of Tsukuba, Tsukuba, Ibaraki, 305 Japan
[2] Electrotechnical Laboratory, Tsukuba, Ibaraki, 305 Japan

ABSTRACT

We have measured Raman scattering and infrared (IR) reflectivity on $Ba_{1-x}Rb_xBiO_3$ at room temperature. The oscillator strength of Bi-Bi stretching mode at $120cm^{-1}$ which becomes IR-active by the charge disproportionation, decreases with increasing Rb concentration, while the frequency change is small. The phonons observed by Raman scattering have the same frequencies as those of longitudinal optical phonons, although the system has the center of symmetry.

KEY WORDS: $Ba_{1-x}Rb_xBiO_3$, the semiconductor metal transition, charge disproportionation, infrared reflectivity, oscillator strength, Raman scattering

INTRODUCTION

After the discovery of superconductivity near 30K in $Ba_{1-x}K_xBiO_3$ (BKB) [1,2] and $Ba_{1-x}Rb_xBiO_3$ (BRB) [3], many theoretical and experimental efforts have been done to clarify the mechanism of the superconductivity and the semiconductor metal transition. $BaBiO_3$ (BBO) which is the end material of those superconductors exhibits semiconducting behavior with about 2eV energy gap, although metallic behavior is expected by a conventional band calculation [4]. By the neutron diffraction experiment, the small distortion of oxygen breathing mode was found [5]. This distortion is attributed to the effect of charge disproportionation of Bi^{4+} to Bi^{3+} and Bi^{5+} which will form the charge density wave (CDW). The energy gap is introduced at Fermi surface of half filled Bi^{4+} band by CDW. A certain amount of substitution of K or Rb for Ba or Pb for Bi on BBO causes the semiconductor metal transition. In $BaPb_{1-x}Bi_xO_3$ (BPB), the Pb dependence of energy gap was measured by reflective spectrum, and the gap was found to vanish in the metallic phase [6]. In the IR spectrum of BBO, the Bi-Bi stretching mode which is R_{25}' mode in the cubic phase was found. This mode becomes IR active because of the charge disproportionation. When the BPB system goes to metallic phase, the oscillator strength of Bi-Bi stretching mode decreases rapidly while the frequency of this phonon is not so much changed. This means that the coherence length of CDW decreases without decreasing the amplitude of CDW. In the case of BKB and BRB, K and Rb are substituted for Ba, respectively. Therefor one can study the effect of carrier which is doped by K or Rb on the semiconductor metal transition without introducing disorder at Bi sites.
In this paper we show the results of IR reflectivity and Raman scattering in semiconducting region of BRB which will be hole-doped CDW system and discuss the semiconductor metal transition [7].

EXPERIMENT

We synthesized the BRB by the atmosphere exchange method [8] which is a modified two-step synthesis procedure by Hinks et. al [1]. Stoichiometric mixtures of BaO_2, Rb_2O or Rb_2CO_3 and Bi_2O_3 ground in dry He gas were sintered at 900~800°C in flowing N_2 gas.

After the 2 or 3 hours firing, on the cooling process, flowing gas was exchanged from N_2 to O_2 at 650~600°C and the temperature was cooled rapidly to avoid the phase separation to about 400°C at which the sample was annealed for 3 hours. The firing and gas exchange temperatures depend on Rb concentration x. The surfaces of the samples were polished to have sufficient smoothness for the optical measurement. The IR reflective spectrum measurement was performed from 50 cm^{-1} to 1000 cm^{-1} with BOMEM DA3 Fourier transform spectrometer at room temperature. The raw reflective spectrum from an Au mirror was used as reference. The Raman spectra were also measured for the back scattering geometry at room temperature with SPEX 1404 double Monochromator and the 5145 Å Ar gas Laser.

EXPERIMENTAL RESULTS AND DISCUSSION

Results from the IR reflective spectra of BRB by using the K-K transformation are shown in Fig.1, (a) for the imaginary part of dielectric constant Im(ε) and (b) for the imaginary inverse dielectric constant Im(-1/ε). The frequencies of the transverse optical (TO) and the longitudinal optical (LO) phonons were determined from the peaks of the Im(ε) and Im(-1/ε) spectrum, respectively. Four TO and LO phonons can be seen to remain until 30% Rb concentration. No other phonon appears in IR spectrum, which is different from the case of BPB. The frequencies of those phonons of BBO are consistent with the previous reports [6]. The Rb concentration x dependence of those phonon frequencies is shown in Fig.2. The frequencies change little with increasing x. Three TO phonons (Ω_1,Ω_2,Ω_4) originate from the cubic Γ-point phonons. The Ω_3 phonon comes from the zone-boundary R_{25}' mode in the cubic phase as shown by Uchida et. al in BPB [6].

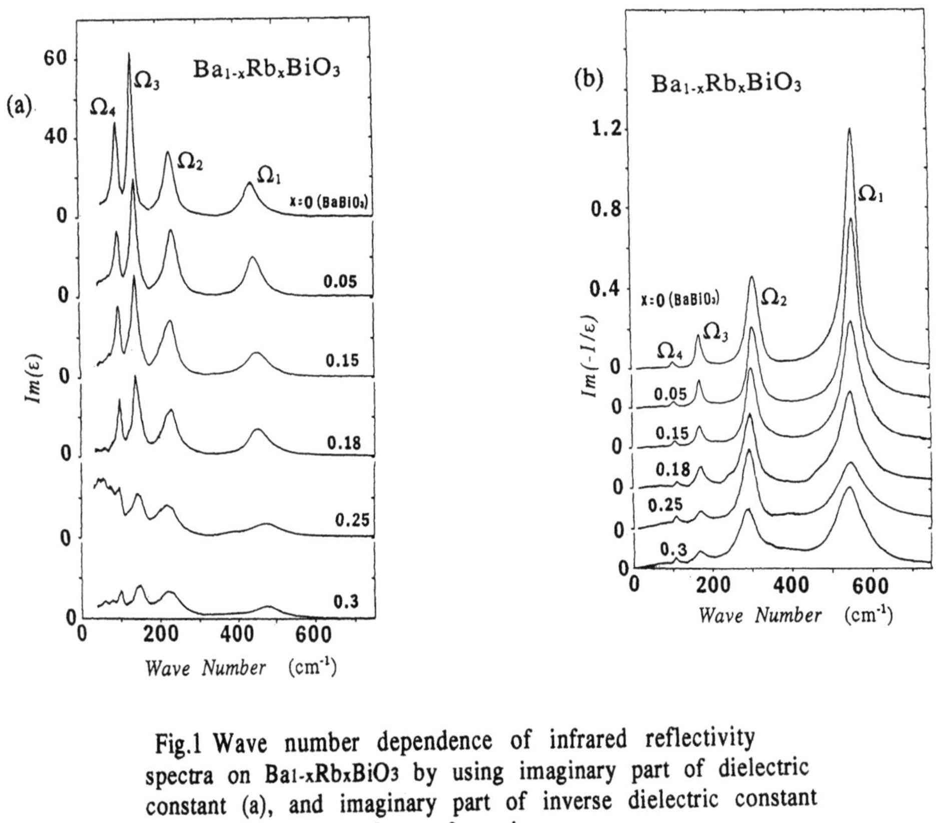

Fig.1 Wave number dependence of infrared reflectivity spectra on $Ba_{1-x}Rb_xBiO_3$ by using imaginary part of dielectric constant (a), and imaginary part of inverse dielectric constant (b) calculated by K-K transformation.

The reflective spectra were also fitted by a damped-harmonic-oscillator model which gives the complex dielectric constant as

$$\varepsilon(\omega) = \varepsilon_\infty + \Sigma\, \omega_j^2 S_j / (\omega_j^2 - \omega^2 - i\, \omega\gamma_j)\ .$$

ε_∞ is the dielectric constant at the frequency well above all phonon frequencies and ω_j, γ_j and S_j are TO phonon frequency, damping constant and oscillator strength of TO phonon, respectively. In Fig.3, S_j are plotted to x. As in the case of BPB, the oscillator strength S_3 decreases with the increase of x with zero at about x=0.35 which is the critical concentration of the semiconductor metal transition [2,3]. Since the frequencies of TO phonons, especially Ω_3, are almost independent of x, ionic charges which associate with the vibrations will not change drastically with x. The Ω_3 mode is the R_{25}' mode in the cubic phase which involves only the Bi-Bi antiparallel and equal-amplitude motion. The oscillator strength S_3 in the multiple-cell structure comes mainly from the charge disproportionation which induces dipoles. The Rb substitution for Ba will change Bi^{3+} to Bi^{4+} which plays the roll of Pb^{4+} in BPB. When the pairs of the Bi^{4+}-Bi^{4+} appear by increasing the Bi^{4+}, the pairs change to the pairs of the Bi^{3+}-Bi^{5+} by the charge disproportionation. With increasing x, the numbers of Bi^{4+} and of Bi^{5+} increase and that of the IR active Bi^{3+}-Bi^{5+} pair decrease with x. The Bi^{5+}-Bi^{5+}, Bi^{3+}-Bi^{4+} and Bi^{3+}-Bi^{4+} stretching mode will have smaller oscillator strength. The CDW which is the ordered array of Bi^{3+} and Bi^{5+} will shorten the coherence length. These effects result in the decrease of S_3. The increase of the Bi^{4+} ions will turn the localized nature of the carrier to the itinerant one at the semiconductor metal transition.

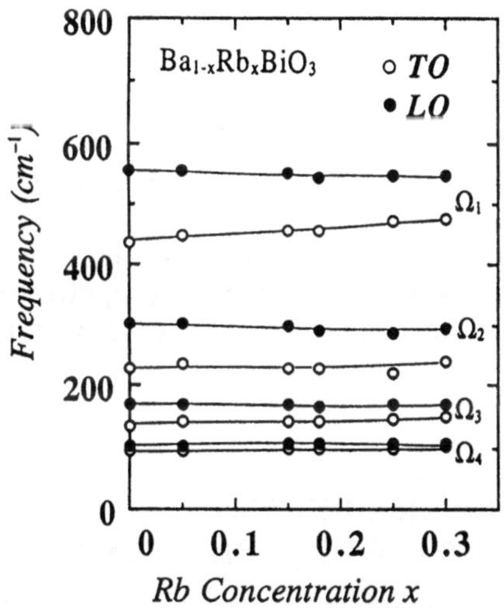

Fig.2 Rb concentration dependence of TO-and LO-phonon frequency.

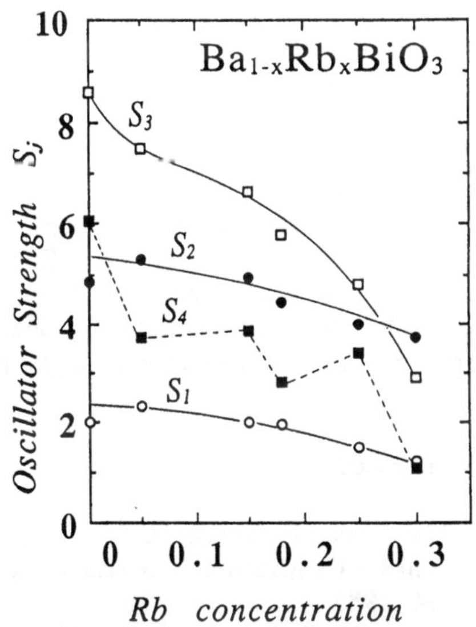

Fig. 3 Rb concentration dependence of oscillator strength of TO-phonon.

The Raman spectra are shown in Fig.4. The frequencies determined from Raman spectra at about 580cm^{-1} and 300cm^{-1} nearly equal the highest and the second highest LO phonon frequencies, Ω_1 mode and Ω_2 mode, respectively. These LO phonons have the highest and second highest intensities in Im(-1/ε) spectra as shown in Fig.1 (b). Ω_1 mode and Ω_2 mode are both IR and Raman active, though this system has center of symmetry in the structure analysis. The surface depletion layer might made the modes Raman active, or the coherence length of CDW might be shorter than 5145Å which is incident light of Raman scattering.

Fig.4 Raman spectra on $Ba_{1-x}Rb_xBiO_3$.

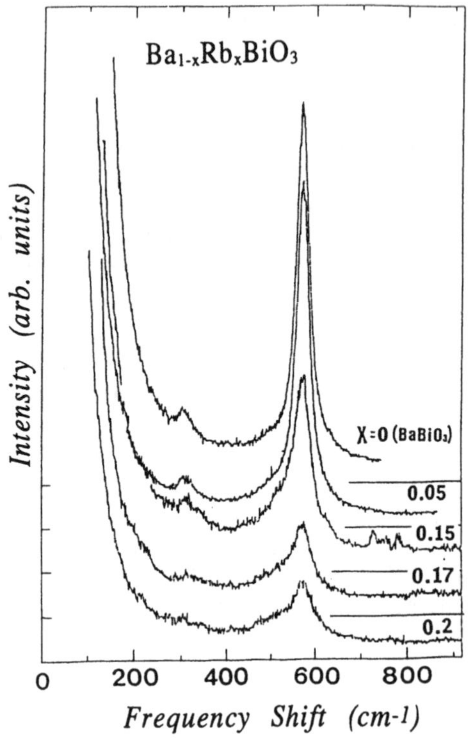

CONCLUSION

The infrared and Raman spectra for $Ba_{1-x}Rb_xBiO_3$ were measured and analyzed for x=0~0.3 at room temperature. Four TO and LO phonons, the frequency, the damping constant, and transverse oscillator strength are determined. The infrared activeness of the Bi-Bi stretching mode is one of the evidence for the charge disproportionation. Decreasing of the oscillator strength for the Rb substitution is found as in the case of BPB.

REFERENCES

1. Mattheiss LF, Gyorgy EM, Jonson Jr. DW (1988) Phys. Rev. B37: 3745
2. Hinks DG, Mitchiell AW, Zheng Y, Richards DR, Dabroski B (1989) Appl. Phys. Lett. 54: 1985
   Cava RJ, Batlogg B, Krajewski JJ, Farrow R, Rupp Jr LW, White AE, Short K, Peck Jr WF, Kometani T (1988) Nature 332: 814
3. Tomeno I, Ando K (1989) Phys. Rev. B40: 2690
4. Cox DE, Sleight AW (1976) Solid State Commun. 19: 969
   Cox DE, Sleight AW (1976) Acta. Crystallogr. Sect. B35: 1
5. Tajima S, Uchida S, Masaki A, Takagi H, Kitazawa K, Tanaka S, Katsui A (1985) Phys. Rev. B32: 6302
6. Uchida S, Tajima S, Masaki A, Sugai S, Kitazsawa K, Tanaka S (1985) J. Phys. Soc. Jpn. 54:4395
7. Inoue A , Iyo A, Oiji Y, Tanaka Y, Tokumoto M, Uwe H, Sakudo T (1991) Proceedings of M^2S-HTSC III, Kanazawa
8. Iyo A, Yamada A, Oiji Y, Takahashi A, Inoue A, Watanabe H, Uwe H, Sakudo T, Murata K (1990) Proc. of 3rd Internat. Symp. Superconduct. (ISS'90) Sendai:113
9. Cowley RA (1964) Phys. Rev. 134 (1964):1981
10. Khann Y, Nahm K, Rosenberg M, Willner H (1977) Phys. Status. Solidi a39: 79

Photoinduced Effects of $YBa_2Cu_3O_y$ Thin Films

K. KAWAMOTO[1], H. MASUDA[1], K. NATSUME[3], N. HONMA[2], I. HIRABAYASHI[1], T. MORISHITA[2], and S. TANAKA[1]

[1] Superconductivity Research Laboratory, Nagoya Division, ISTEC, Nagoya, 456 Japan
[2] Division V, Superconductivity Research Laboratory, ISTEC, Koto-ku, Tokyo, 135 Japan
[3] Department of Electrical and Computer Engineering, Nagoya Institute of Technology, Nagoya, 466 Japan

ABSTRACT

We report the photo-irradiation effect on $YBa_2Cu_3O_y$ thin films. After high irradiation by an Ar laser at low temperature (below 100K), the resistivity of the YBCO films decreases. Particularly, for oxygen deficient films with y=6.4-6.5, the resistivity change is completely recovered by thermal annealing at ambient temperature. In order to put the recovery process, we performed isothermal annealing on the irradiated samples at around 300K and obtained activation energy as 0.41 eV. This value is similar to that of oxygen diffusion in the layer of Cu-O chains. It suggests that this phenomenon is due to a local rearrangement of oxygen in the layer of Cu-O chains by irradiation of light.

key words : photo-irradiation effect, oxygen ordering, $YBa_2Cu_3O_y$ thin film

INTRODUCTION

Optical effects in the superconducting materials have been investigated widely from the viewpoint of both application and basic physics [1,2,3]. Among them the optical response which means a transient resistivity change by irradiation through the nonequilibrium effect on the weak superconductivity has become one of the promising candidate for application to high sensitive ultrafast IR detectors. In order to apply this effect into actual use, it is necessary to resolve the problems on the stability of the materials.
Recently we have investigated the optical response of YBCO films at low temperature and found some change in the signal response after the high level laser irradiation even at liquid helium temperature. This effect is reversible by the thermal annealing at around room temperature. It suggests some relationship to the recent topic of the oxygen ordering effect of the $YBa_2Cu_3O_y$ at the boundary of metal-insulator transition. Jorgensen et al [4,5] found that the samples quenched at high temperature which realizes $y \cong 6.5$ shows an anomalous annealing effect at room temperature by the powder neutron scattering and that it relates to the oxygen ordering. After that many researchers discuss the role of oxygen ordering [6,7,8] and they obtained a common understanding on the importance of entropy term in the discussion of the structure of YBCO even at ambient temperature. Very recently Kudinov et al [9,10] also found an interesting photo-irradiation effect on the YBCO of y=6.41. They made a transformation of insulating samples into superconductor only by laser irradiation at low temperature, and they observed a growth of the diamagnetic moment as well as the zero resistance for the samples which irradiated by the Ar laser up to the total photon dose of 10^{23} cm^{-2}. This effect is thermally reversible. This photo-irradiated effect is anomalous because the photo-irradiated state seems to be a more ordered phase which corresponds to the lower temperature phase.
In this paper, in order to examine the relationship between the degradation of the optical response signal and the above novel phenomena, we will report on the photo-irradiation and thermal annealing effects below room temperature for the oxygen deficient YBCO films of $y \cong 6.5$.

EXPERIMENTAL

The samples used in the experiment are $YBa_2Cu_3O_y$ thin films prepared by the metalorganic chemical vapor deposition (MOCVD) technique. [11] The source materials used were β-diketone of $Y(thd)_3$, $Ba(thd)_2$, and $Cu(thd)_2$. The films were grown on the polished MgO(100) substrates at the temperature of 700°C. The structure of the films was examined by the X-ray diffraction pattern. The surface morphology and cross section were observed by the scanning electron microscopy. The thickness of the samples was approximately 400nm. The films were etched by an excimer laser (operated with KrF) to 0.5mm in width and 1mm in length, and were isochronally (30 min for each temperatures) annealed in vacuum at 200-270°C to reduce the oxygen content over the boundary of the metal insulator transition around y≅6.5. The thin films were irradiated in a cryostat by Ar laser (wavelength of 514.5nm) under low temperature (9K) . The power density of Ar laser was $15W/cm^2$ and laser light was introduced into the sample chamber controlling the temperature by He gas flow. The sample temperature rose up during irradiation to be about 40-50K which was estimated from the resistivity. In order to examine roughly the recovery process, isochronal annealing was carried out for the photo-irradiated films and to estimate the activation energy, an isothermal annealing schedule was applied at several point near the room temperature.

RESULTS AND DISCUSSION

Figure.1 shows the annealing temperature dependence of resistivity of the film by vacuum annealing. The film before vacuum annealing showed metallic behavior at the normal state and Tc(zero) was 70K. As increasing the temperature of vacuum annealing, unit cell length of the c-axis increased and Tc decreased. After 270°C annealing, the film changed into a semiconducting behavior in resistivity, and Tc disappeared. In this case, the oxygen content is considered as y=6.4-6.5 from the unit cell length of the c-axis.

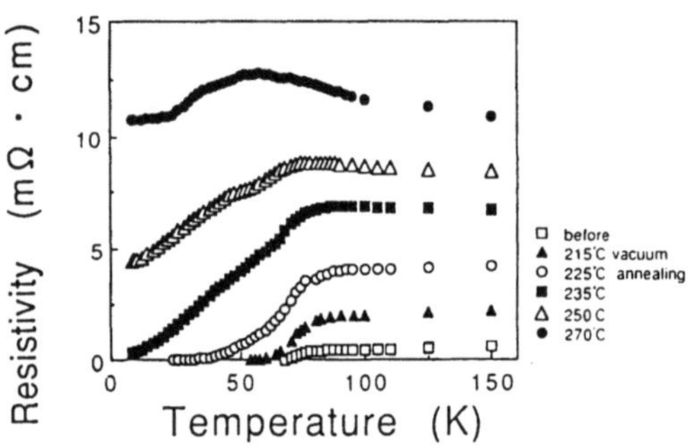

Fig.1 Resistivity change by vacuum annealing.

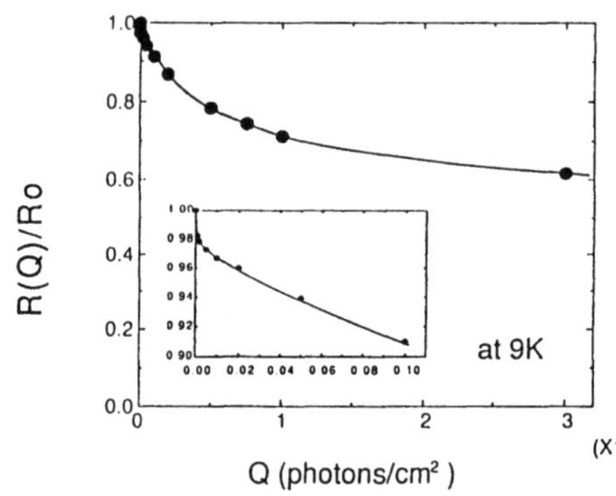

Fig. 2 Resistivity change measured at 9K as a function of cumulative photon dose.

By the photo-irradiation on the as-grown superconducting films up to $1.5X10^{24}$ photons/cm^2, we found a slight normal resistivity decrease of 3% at 100K, but could not assert the increase of Tc considering precision of the experiment.

When the $YBa_2Cu_3O_y$ film annealed in vacuum at 270°C was irradiated, resistivity of the film decreased more remarkably than before vacuum annealing. (figure.3). A typical result as a function of cumulative photon dose is shown in figure.2. There are two regimes in the observed photoinduced resistivity change corresponding to the total photon dose Q. The first regime of $Q < 10^{23}$ photons/cm^2, the relative resistivity R/R_0 decreases linearly with Q. Sometimes we observed two different decreasing rates in this regime depending on the samples. This rapid decrease may relate to the structurally weak part such as grain boundaries or twining plane in the samples. In the high dose regime of $Q > 10^{23}$ photons/cm^2, the resistivity decrease shows another behavior which is well described by the relations $\Delta R = (R(Q) - R(0))$ is proportional to logQ or exp((-α/Q)^{β}); such functional forms are observed in the description of the relaxation in the complex systems such as flux creep and the phenomena occurring in the disordered systems. After irradiation of $3X10^{24}$ photons/cm^2, the resistivity of the film decreased 60% to compare with that before irradiation as shown in Figure.3. This low resistivity (R) state remains for a long time as long as the samples are kept at 9K. The low R state can be annealed to be the high R state which is thermally stable at ambient temperature. The samples are completely reversible by this irradiation-annealing cycle.

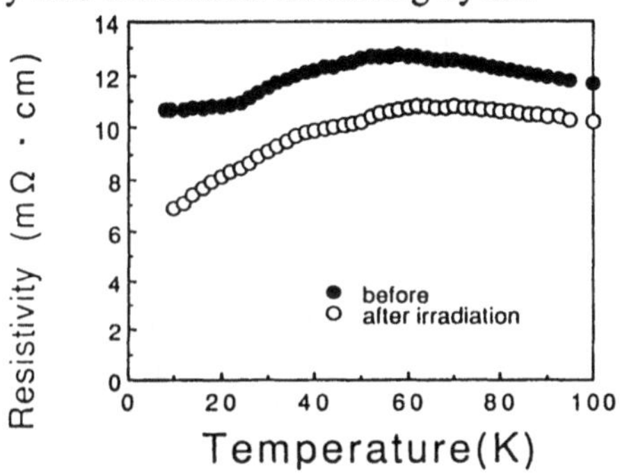

Fig.3 Temperature dependence of the resistivity before and after irradiation.

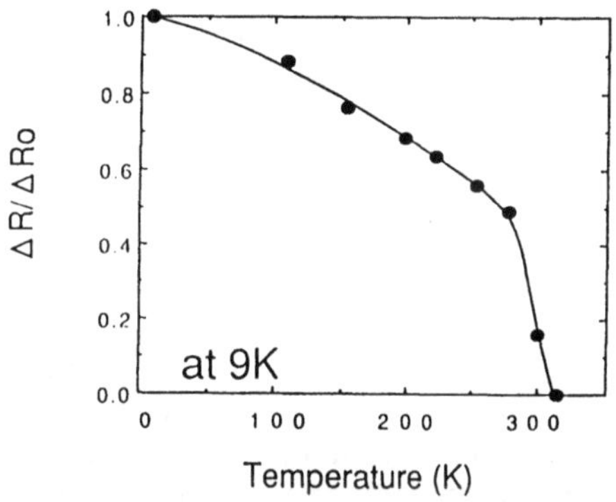

Fig. 4 Recovery of the resistivity change by isochronal annealing below 315K.

Figure.4 shows the recovery curve of the resistivity by isochronal annealing. Monitoring of the resistivity was carried out at 9K after 30 min annealing at relevant temperatures. There is a large variation in the annealing curve around room temperature. In order to estimate the activation energy for the recovery process, we measure the isothermal annealing at several points around 300K. Assuming the simple monomolecular process, activation energy (ΔE) for this process can be estimated as 0.41 eV. This value is in good agreement to the activation energy of oxygen diffusion within the layer of CuO chains for $YBa_2Cu_3O_{6.62}$ estimated by Tu et.al.[12].

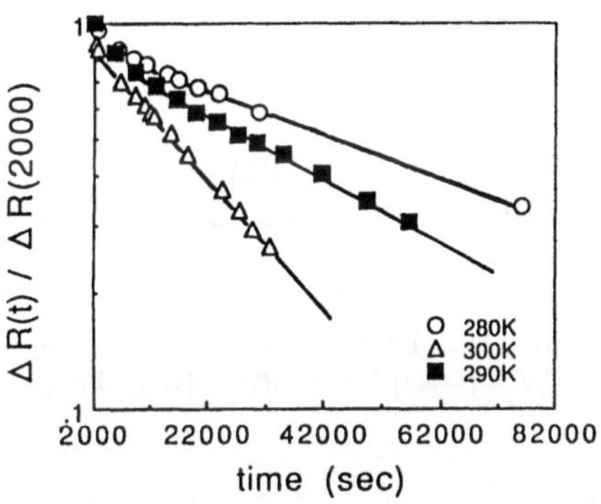

Fig.5 Recovery of resistivity change by isothermal annealing.

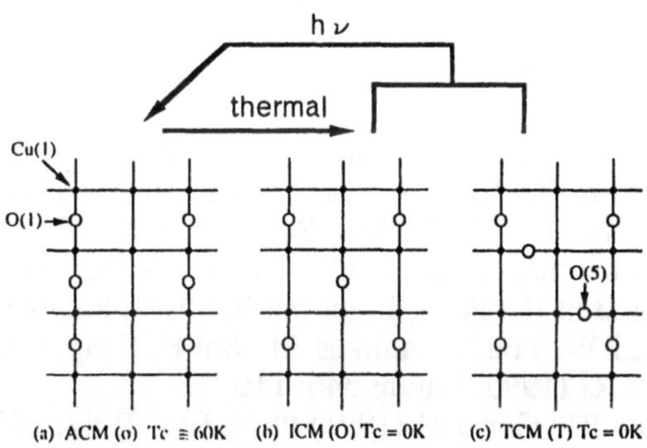

Fig.6 Oxygen ordering in the layer of Cu-O chains.

Here we would like to consider a mechanism of the reversible photo-irradiation effect. Before entering the detailed discussion, it may be worth remarking on the similar reversible photo-irradiation effect observed in the non-superconductors. We can classify the mechanism into two cases. The first one is a purely electronic mechanism. The persistent photoconductivity observed in the deep levels [13] or nipi superlattice [14] structure. In this case, the photogenerated carriers are spatially separated and thus recombination is inhibited if the sample is kept at the low temperature. This state is annealed at elevating the temperature. The other model involves the photo-structural change in the flexible structure such as low dimensional[15] and amorphous [16] materials. In the Staebler-Wronski effect which is observed in a- Si:H, the conductivity decreases by irradiation and recovered by thermal annealing. It is the converted situation to the present experiment. This effect has been discussed for a long time because of its importance for the solar cell application and several models relating to the mobile property of hydrogen are proposed.

Recently it has been commonly accepted that since the structure of $YBa_2Cu_3O_y$ is dependent on the very complicated problem of the ordering of oxygen, the entropy effect should be taken into account to explain phenomena occurring even at ambient temperature. Especially near the metal-insulator transition of y=6.4-6.5, it is known that the superconducting properties are remarkably dependent on the local, short range order of the oxygen atoms in the layer of the CuO chains. According to a recent calculation by Gupta et.al.[7], the structure of the Alternate Chain Model (ACM in Fig.6(a)) involving parallel full-chain/empty-chain becomes orthorhombic and shows superconductivity. On the other hand, the tetragonal chain model (TCM in Fig.6(c)) with random oxygen sites and another model having the identical half filled chain (Identical Chain Model, ICM in Fig.6(b)) do not undergo charge transfer from the layer of CuO chains to the CuO_2 layer and thus not superconducting. Therefore if we assume that the photo-irradiation causes the higher ordered state having low entropy such as ACM and that thermal anealing effect at around room temperature is enough to prefer a partially disordered state involving local structure of TCM or ICM, our observation can be explained by this oxygen ordering model considering the importance of entropy effect at room temperature. Actually, the agreement of activation energy of recovery for the photo-irradiated samples with that of the oxygen diffusion in the layer of the CuO chains seems to support the above speculation. On the other hand, Kudinov et.al.[9] explained their observation by the photodoping model that the photogenerated electrons are trapped at the sites of oxygen vacancy in the CuO chains and the resultant holes are accommodated in the CuO_2 planes. This model assumes the complete electron hole charge separation between adjacent crystalline layers. The material y=6.4-6.5 seems to be neither so highly insulating nor disordered as observed in the superlattice or amorphous semiconductors. We think it natural to consider some stabilization mechanisms involving atomic displacement such as the switching of oxygen between the sites in the layer of CuO chains. They observed very large change in resistivity and magnetic properties even with 1/100 photon dose than ours. There may be some differences in the samples and also in the experimental conditions. It should be made clear by further experiments.

REFERENCES

1. Tanabe K, Kubo S, Enomoto Y, Asano H, Yamaji A, (1991) Jpn. J. Appl. Phys. 30: L110
2. Brorson SD, Kazeroonian A, Moodera JS, Face DW, Cheng TK, Ippen EP, Dresselhaus MS, Dresselhaus G, (1990) Phys. Rev. Lett. 64: 2172
3. Kleinhamnes A, Chang CL, Moulton WG, Testardi LR, (1991) Phys. Rev. B44: 2313
4. Jorgensen JD, Pei S, Lightfoot P, Shi H, Paulikas AP, Veal BW, (1990) Physica C167:571
5. Veal BW, You H, Paulikas AP, Shi H, Fang Y, Downey JW, (1990) Phys. Rev. B42: 4770
6. Cava RJ (1990) Nature 346: 110
7. Gupta RP, Gupta M (1991) Phys. Rev. B 44: 2739
8. Moran-Lopez JL, Schuller IK ed. "Oxygen disorder effect in high Tc superconductors"
9. Kudinov VI, I.Kirilyuk A, Kreines NM (1990) Phys. Lett. A 151: 358
10. Kudinov VI, Chaplygin IL, Kirilyk AI, Kreines NM, (1991) Proc. of M^2S-HTSCIII
11. Higashiyama K, Ushida T, Higa H, Hirabayashi I, Tanaka S (1991) JJAP 30: 1209
12. Tu KN, Yeh NC, Park SI, Tsuei CC, (1989) Phys. Rev. B39: 304
13. Sheinkman MK, Shik AY, (1976) Sov. Phys. Semicond. 12: 816
14. Hundhausen M, Ley L, (1985) J. Non-Cryst. Solids 77&78: 1051
15. Hirabayashi I. Morigaki K, Yamanaka S, (1983) J. Phys. Soc. Jpn 52: 671
16. Staebler DL, Wronski CR, (1977) Appl. Phys. Lett 31: 292

This work was supported by New Energy and Industrial Technology Development Organization under the management of the R & D Basic Technonogy for Future Industries.

Doping Effect on the Magnon Scattering Spectrum of $(Eu,Ce)_2(Ba,Eu)_2Cu_3O_{8+z}$

Y. MIZUO[1], S. TAJIMA[2], M. YOSHIDA[2], T. WADA[2,*], Y. YAEGASHI[2], H. YAMAUCHI[2], and N. KOSHIZUKA[2]

[1] Department of Physics, Tokai University, Hiratsuka, Kanagawa, 259-12 Japan
[2] Superconductivity Research Laboratory, ISTEC, Koto-ku, Tokyo, 135 Japan

ABSTRACT

We have investigated the carrier concentration dependence of the Raman scattering spectra, using the single crystals of $(Eu,Ce)_2(Ba,Eu)_2Cu_3O_{8+z}$. We prepared several samples with various oxygen contents by annealing fully oxygenated samples with T_c=30K in N_2-atmosphere at various temperatures. The two magnon peak was clearly observed in the lowest oxygen content sample. With increasing oxygen content, the magnon peak gradually disappeared and the spectral weight moved towards lower energy. This magnon behavior is the same as that observed in the other hole-doped high-T_c superconducting cuprates, and is clearly different from the electron-doped one.

KEY WORDS: $(Eu,Ce)_2(Ba,Eu)_2Cu_3O_{8+z}$, Raman scattering spectra, two magnon peak

INTRODUCTION

In the Raman spectra of the parent insulators of high-T_c superconducting cuprates, the zone-boundary magnon pair excitations within the CuO_2-plane are observed as broad peaks around 2600-3200cm^{-1}, which is an evidence of two-dimensional antiferromagnetic order.[1] The doping effect on the magnon scattering spectra was investigated for several materials. In $La_{2-x}Sr_xCuO_4$, the two magnon peak gradually disappears, its spectral weight shifting towards lower energy, with increasing Sr-content x.[2] The shift of the magnon peak energy was also observed in $YBa_2Cu_3O_{6.6}$ with Tc=60K.[3] On the other hand, in the electron-doped superconductors $L_{2-x}Ce_xCuO_4$ (L=Nd, Pr and Sm) the magnon peak broadens and is smeared out without softening of the peak energy.[4] To understand the change of spin-spin correlation within the CuO_2-plane versus doping, the composition dependence of the magnon spectra should be determined.

In the present work, we investigated the carrier concentration dependence of the magnon spectra for $(Eu,Ce)_2(Ba,Eu)_2Cu_3O_{8+z}$ which is one of the hole-doped superconducting cuprates. $(Eu,Ce)_2(Ba,Eu)_2Cu_3O_{8+z}$ (EuEu-223) with T_c=30K consists of a $YBa_2Cu_3O_{6+z}$-type block i.e.$(Eu,Ce)(Ba,Eu)_2Cu_3O_{6+z}$ and a fluorite block i.e. $(Eu,Ce)O_2$.[5] The oxygen sites around the Cu(1)-site at the center of the 123-type block are randomly occupied with occupancy of about 0.5, and thus the crystal structure has a tetragonal symmetry. The hole concentration in the CuO_2-plane can be controlled by changing the oxygen content around the Cu(1)-site.

We have prepared several samples with various oxygen contents, by annealing the fully oxygenated single crystals in N_2-atmosphere at various temperatures, and have measured the Raman scattering spectra of these samples.

EXPERIMENTAL

The single crystals were grown by the CuO-flux method. The details of crystal growth are described elsewhere.[6] The average sample size is 0.5 × 0.5 × 0.05mm^3. After post annealing at 600°C under high oxygen partial pressure of 400atm by HIP-technique, the crystals exhibited superconductivity with T_c=28 K. For this fully oxygenated sample, we performed thermal gravimetry analysis (TGA) in N_2-atmosphere and measured the change of its weight versus

* Present address: Matsushita Central Reserach Laboratory, Yakumonakamachi 3-15, Moriguchi-shi, Osaka 570, JAPAN

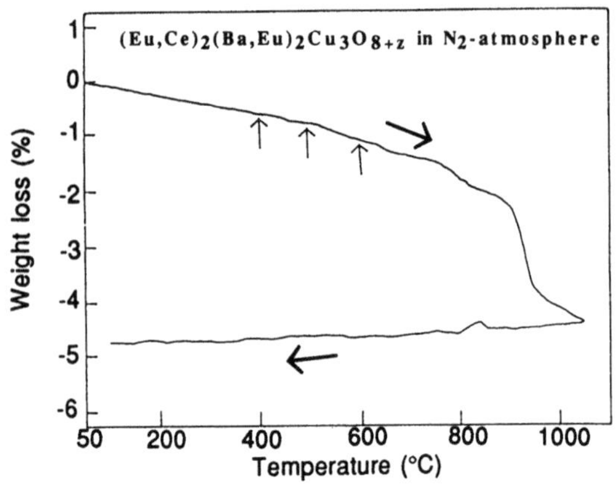

Fig.1 The change in weight of $(Eu,Ce)_2(Ba,Eu)_2Cu_3O_{8+z}$ single crystals, measured by Thermal gravimetry analysis (TGA) in N_2-atmosphere at the rate of 5°C/min. The starting sample weight is 8.0mg. The temperature at which we reduced the samples for the measurement of Raman spectrum are shown by arrows.

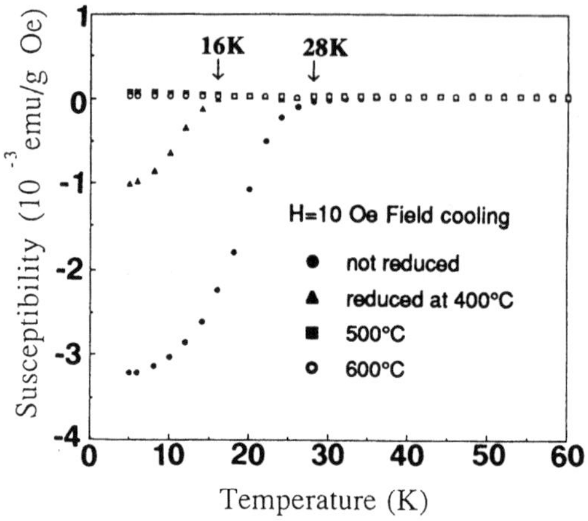

Fig.2 Temperature dependence of magnetic susceptibility for $(Eu,Ce)_2(Ba,Eu)_2Cu_3O_{8+z}$ single crystals annealed in N_2-atmosphere at various temperature. These were measured by SQUID magnetometer in the field cooling condition at a field of H=10Oe.

temperature. The result of the TGA is shown in Fig.1. The value of z for the fully oxygenated 223-compound was estimated to be about 1.0.[7] After annealing the samples in N_2 up to 600°C at the rate of 5°C/min, the sample weight decreased 1.17%, which corresponds to the reduction of oxygen content from z=1.0 to z=0.33. As shown in Fig.1, there is no phase transition before the sample melted around 900°C. This means that in the 223-compound the oxygen content and presumably the hole concentration can be controlled continuously without any phase transition by changing the annealing temperature. Based on the result in Fig.1, we set the temperature from 400°C to 600°C for reduction, and annealed the fully oxygenated samples in N_2-atmosphere at each temperature for 48 hours.

The superconducting transition temperature T_c was measured by a superconducting quantum interference device (SQUID) magnetometer, in the magnetic field of H=10Oe. To get enough signal strength, many crystals (total mass of 10mg) were set in the sample holder. As shown in Fig.2, the onset T_c value decreases as the annealing temperature increases, and finally superconductivity disappears for the samples reduced at 500°C and 600°C. From this systematic change in T_c, we can confirm that the oxygen content in the sample changes as expected.

The Raman scattering measurements were carried out in the back scattering configuration at room temperature. The incident light was 5145Å-beam from an argon-ion laser with the power of 50mW. For the measurement of two magnon scattering spectra with B_{1g}-symmetry, the polarization of the incident and scattering light was set perpendicular to each other ([x',y']-configuration).

RESULTS & DISCUSSION

Fig.3 (a)-(d) shows the B_{1g}-Raman spectra of EuEu-223 for various reduction temperatures. In the low energy region, we can observe the phonon peaks which are characteristic of the (223)-compounds.[8] For the non-superconducting sample reduced at 600°C, a broad peak is observed at 2580cm^{-1}, which is assigned to the two magnon scattering peak resulting from two-dimensional antiferromagnetic order, as commonly observed in the other insulating phases of high-T_c superconducting cuprates. The peak energy is almost the same as that reported in $YBa_2Cu_3O_6$,[3] $Bi_2Sr_2Ca_{0.5}Y_{0.5}Cu_2O_{8+y}$[9] and Nd_2CuO_4.[4] Although the 2580cm^{-1} peak completely disappears in the highest oxygen content sample, a finite scattering intensity is observed up to high energy. This electronic continuum in the Raman spectra is commonly

observed in all the high-T_c superconducting cuprates, but its origin is not yet clear. As the reduction temperature decreases, that is, the oxygen content increases, the magnon peak gradually disappears while the spectral weight shifts towards lower energy. The continuous spectral change in Fig.3 indicates that the oxygen content in the samples changes continuously and homogeneously. This fact as well as the continuous change in T_c shown in Fig.2 indicates that the electronic states within the CuO_2-plane changes continuously in these samples. A similar doping effect on the Raman spectra is observed in $La_{2-x}Sr_xCuO_4$[2] and $YBa_2Cu_3O_{6+y}$[3], but not in the electron-doped systems such as $Pr_{2-x}Ce_xCuO_4$[4] where the two magnon peak gradually becomes broad and weak without softening of its peak energy. Thus, we may conclude that the doping effect on the magnon spectrum observed in EuEu-223 is characteristic of the hole-doped system. In the hole-doped high-T_c superconductors, the holes are considered to be predominantly introduced into the O2p-orbital.
According to the model of Aharony et al,[10] O^{1-}-spins are coupled with the nearest neighbor Cu^{2+}spins, which induces the ferromagnetic interaction between Cu-Cu spins, resulting in the reduction of the exchange energy J. On the other hand, in the electron-doped systems, the doped electrons may be introduced into the Cu-orbital and dilute the antiferromagnetic coupling between Cu-spins. The difference in the doping effect on the magnon spectrum between these systems may be attributed to the difference between dilution and frustration of the magnetic ordering.

Concerning the Raman scattering extending to higher energy in the metallic composition range, there are some controversial opinions whether it indicates some charge excitation or an over-damped magnon. Although the introduction of mobile carriers gives rise to electronic Raman scattering in the low energy region, the higher energy continuum is not necessarily related to the itinerant hole excitation. Judging from the continuous change in the magnon spectra in Fig.3, we can not rule out the possibility of the over-damped magnon or some kind of spin fluctuation as the origin of the Raman continuum.

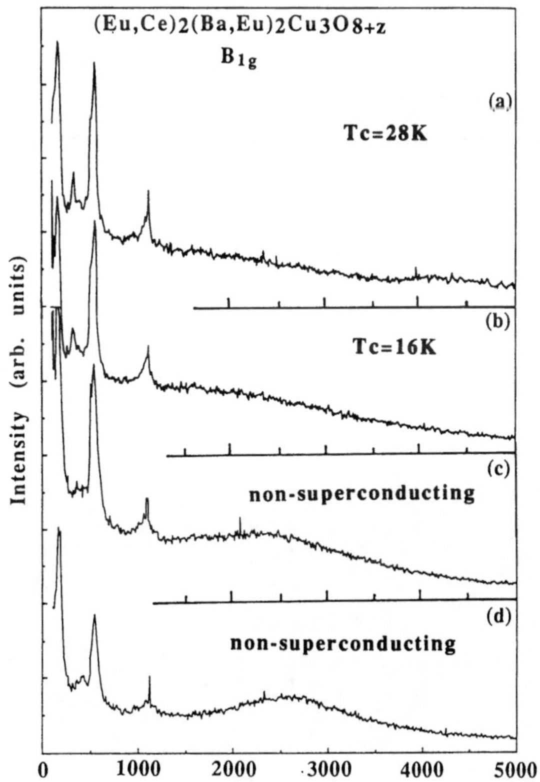

Fig.3 B_{1g}-Raman scattering spectra of $(Eu,Ce)_2(Ba,Eu)_2Cu_3O_{8+z}$ at room temperature for the samples not reduced (a), and annealed in N_2-atmosphere at 400°C (b), 500°C (c), and 600°C (d), respectively.

CONCLUSION

We have investigated the B_{1g}-Raman scattering spectra of EuEu-223 single crystals with various oxygen contents. A two magnon scattering peak is observed around $2580cm^{-1}$ in the lowest oxygen content sample. With increasing oxygen content, the peak gradually disappears, the spectral weight shifting towards lower energy. This behavior is also observed in the other hole-doped high-T_c superconducting cuprates, but is different from that in electron-doped ones. Thus, it is concluded that this magnon behavior is characteristic of hole doping system, which can be explained by the model of magnon frustration due to the hole doping.

ACKNOWLEDGEMENT

The authors would like to thank T.Sakurai at Superconductivity Reserach Laboratory for his help in performing the TGA measurements.

REFERENCES

1. Lyons KB, Fleury PA, Remeika JP, Cooper AS, and Negran TJ, (1988) Phys. Rev. B, Vol.37, No.4: 2353.
2. Sugai S, Sato M, and Shamoto S, (1990) Solid State Commun, Vol.76, No.3: 365.
3. Lyons KB, Fleury PA, Schneemeyer LF, and Waszczak JV, (1988) Phys. Rev. Lett. Vol.60, No.8: 732.
4. Tomeno I, Yoshida M, Ikeda K, Tai K, Takamuku K, Koshizuka N and Tanaka S, (1991) Phys. Rev. B, Vol.43, No.4: 300.
5. Sawa H, Obara K, Akimitsu J, Matsui Y and Horiuchi S (1989) J. Phys. Soc. Jpn. Vol58: 2252.
6. Tajima S, Mizuo Y, Wada T, Yaegashi Y, Ichinose A, Takata T, Koshizuka N, and Yamauchi H (in press) J. Crystal Growth
7. Wada T, Ichinose A, Yaegashi Y, Yamauchi H, and Tanaka S (1990) Jpn. J. Appy. Phys. Vol.29: L266.
8. Yoshida M, Tajima S, Mizuo S, Wada T, Ichinose A, Yaegashi Y, Koshizuka N, Yamauchi H, and S.Tanaka (1991) Phys.Rev. B, Vol.44, No.2: 782.
9. Sugai S and Sato M (1989) Phys. Rev. B, Vol. 40, No.13: 9292.
10. Aharony A, Birgeneau RJ, Coniglio A, Kastner MA, and Stanley HE (1988) Phys. Rev. Lett. Vol.60, No.13: 1330

2.6 Magnetic Properties

AC Hysteresis Loss, Lower Critical Field, and Critical Current Density at 77 K in Melt Processed $YBa_2Cu_3O_{7-\delta}$ Wires and Rods

E.W. Collings[1], M.D. Sumption[1], K.R. Marken, Jr.[1], Zhou Lian[2], M.V. Parish[3], and W.J. Carr, Jr.[4]

[1] Battelle Memorial Institute, Columbus, OH 43201, USA
[2] Northwest Institute for Nonferrous Metal Research, Baoji, China
[3] CPS Superconductor, Inc., Milford, MA 01757, USA
[4] Pittsburgh, PA 15235, USA

ABSTRACT
The AC hysteretic loss at 77 K in three samples of melt-processed $YBa_2Cu_3O_{7-\delta}$ (YBCO) rods and wires has been measured over a wide range of applied field sweep amplitudes, H_m, using vibrating sample magnetometry. The results are discussed in two amplitude ranges, $H_m < H^*$, and $H_m > H^*$, the latter being the penetration field. A magnetic method of ascribing an effective radius to cylinders of imperfect roundness is employed. From the heights of the hysteresis loops, the magnetic critical current density, J_c, is determined as a function of field strength. The per-cycle, per unit volume, hysteretic losses, Q_h, are intercompared after having been normalized to a common reference sample radius, $R_{ref.}$. Loss differences that remain are ascribed primarily to differences in J_c. The hysteretic method is applied in determining lower critical field, H_{c1}. It was found that samples prepared using two variants of the melt-processing approach had H_{c1}s of 225 and 163 Oe. Two samples prepared independently by the *same* melt process had H_{c1}s of 163 and 162 Oe.

KEY WORDS: AC loss, critical current density, lower critical field, melt processed $YBa_2Cu_3O_{7-\delta}$.

INTRODUCTION

The purpose of the investigation was to measure AC hysteretic loss in three samples of melt-processed $YBa_2Cu_3O_{7-\delta}$ (YBCO) superconductor, over a wide range of field sweep amplitudes, H_m, and to use the results of the measurements: (1) to intercompare, independently of sample radius, R_0, the the hysteretic losses of various samples both below and above the penetration field, H^*; (2) to extract critical current density, J_c, from the heights of the fully penetrated hysteresis loops; (3) using the $H_m < H^*$ data sets, to employ a hysteretic method to measure the lower critical fields, H_{c1}.

EXPERIMENTAL

Measurements

Magnetization measurements were performed using a PAR EG&G vibrating sample magnetometer. Samples 6~12 mm long were exposed to the transverse field of an electromagnet energized by a bipolar power supply with a maximum current capability of ±65 A. Prior to starting any series of magnetization (M) *versus* field (H) measurements, the sample was always cooled to the operating temperature, 77 K, in zero field. Two sets of data were taken, representing: (1) the initial magnetization curve of a virgin sample; (2) a series of hysteresis loops taken at field-amplitudes, H_m, of about 10 Oe to 16 kOe without intermediate demagnetization.

Materials

Melt-processed YBCO samples were prepared and supplied by the Northwest Institute for Nonferrous Metals (NIN), Baoji, China, and CPS Superconductor (CPSS), Milford, MA. The NIN "powder-melt-processed" samples[1,2], herein designated $DC1_A$ (diam., $8.97x10^2$cm) and $DC1_{AB}$ (diam., $8.97x10^{-2}$cm), were prepared to approximately cylindrical shapes from a single stick of material by hand grinding. Two magnetometry samples were also prepared from CPSS fiber or fiber-bundles[2,3], viz.: (1) CPS3, a pair of individual Ag-coated fibers (diam., $3.60x10^{-2}$cm); (2) CPS4, an assemblage of five Ag-coated fibers (av. fib. diam., $1.91x10^{-2}$cm) imbedded in extra Ag. Optical metallography and computerized image analysis performed on DC1, CPS3, and CPS4 indicated that all three materials supported dense arrays of 211-phase precipitates. Some geometrical properties of the precipitates are listed in Table I.

Table I Image Analysis of 211-Phase Precipitates in Melt-Processed $YBa_2Cu_3O_{7-\delta}$

Sample Code		DC1	CPS3	CPS4
Occupied Area,	%	38.6	24.4	22.8
Length (Max. Diameter),	µm	1.96 ± 1.13	1.95 ± 1.31	1.92 ± 1.27
Width (Perp. to above),	µm	1.30 ± 0.74	1.10 ± 0.60	1.14 ± 0.63
Mean Diameter,	µm	1.69 ± 0.93	1.62 ± 0.99	1.61 ± 0.99
Aspect Ratio		1.54 ± 0.37	1.86 ± 1.02	1.72 ± 0.58

INITIAL MAGNETIZATION

For an infinite right circular cylinder in a transverse magnetic field, H, the susceptibility, obtained from the slope of the initial magnetization line, is $dM_v/dH = -1/2\pi$ (where M_v is the volume-normalized moment). However, for relatively short cylinders a correction factor, η, may need to be applied[4] to allow for field encroachment into the ends. This factor decreases below 0.98 as the length/diameter ratio drops below about 16.[5] Although the sample's length can be measured quite accurately not so the radius, R_0, especially if the rod is thin or slightly irregular. In such cases an effective radius can be derived from the sample's "magnetic volume", $V^m \equiv (-dM_m/dH)(2\pi/\eta)$, where M_m is the measured moment. Whenever it was dictated by geometrical considerations, magnetic volumes and radii were used in analyzing the magnetization results.

AC HYSTERETIC LOSS

AC Hysteretic Loss Below H*

As we have recently been reminded by Carr[2,5,6,7] (see also Campbell and Evetts[8]), when H_m exceeds H_{c1} and flux begins to enter the superconductor, an electric field is produced which acts on both the body current (Bean current[9]) and the surface Meissner current, which below H_{c1} is not itself responsible for any loss. In general for $H_m < H^*$, the penetration field, the former leads to the well-known incomplete-penetration loss $Q_{hB} \propto H_m^3$. The Meissner current provides a contribution, Q_{hM}, that turns out to be proportional to $H_{c1}(H_m-H_{c1})^2$. Taking these two sources of loss into account, Carr[2,5,6,7] has developed for the first time an expression for the energy loss per cycle per unit volume of a *transverse-field-mounted* cylinder

$$Q_h = (10/3\pi^2 J_c R_0)\{(3/2)h_{c1}(H_m-h_{c1})^2 + (H_m-h_{c1})^3\} \qquad (1)$$

in which h_{c1} is the field-at-the-magnet needed to supply H_{c1} to the sides of the cylinder (hence $H_{c1} = 2h_{c1}$). With J_c in A/cm^2, R_0 in cm, and H in Oe, the units of Q_h are erg/cm^3; We refer to Eqn. (1) as the "2-3 Equation", and to the part in {}-brackets as the "2-3 Kernel". It has been shown elsewhere[5] that Q_{hM} dominates the loss when H_m is near h_{c1}, but that as it approaches H*, although $Q_h-Q_{hB} = Q_{hM}$ continues to increase in the absolute sense, since $Q_{hM}/Q_{hB} \approx (3/2)h_{c1}/H_m$, it becomes *relatively* negligible. Equation (1) indicates that the pre-penetration loss is proportional to $1/R_0$, which provides us with a way of eliminating the extrinsic variable, thickness, when comparing the losses of different size samples. In what follows we choose to normalize the loss to sample-radius $R_{ref.} = 100$ µm. Any differences in Q_h that remain must be attributable to the *intrinsic* materials variables, J_c and h_{c1}. The former will operate over the entire H_m range, but h_{c1} will be important only in the Q_{hM}-dominant region.

The radius-normalized volumetric per-cycle loss ($Q_{h,norm.} = Q_h \cdot R_0/R_{ref.}$) of $DC1_{AB}$, CPS3, and CPS4, are compared in Fig. 1. The low-H_m scatter is undoubtedly due to variations in the surface-roughness loss -- $DC1_{AB}$, being hand-shaped, would have the largest surface loss. We attribute the high-H_m scatter to breakdown of Eqn. (1) as H_m

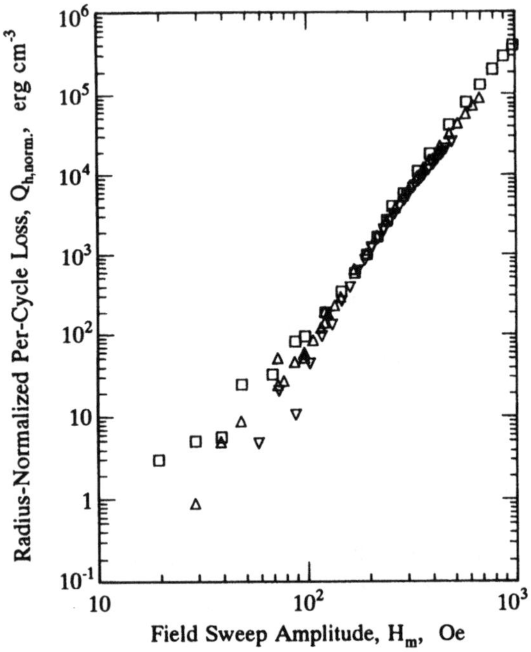

Fig. 1. Low-field (H<H*) radius-normalized hysteretic loss *versus* field-sweep amplitude for $DC1_{AB}$ (□), CPS3 (△), and CPS4 (▽) at 77 K.[5]

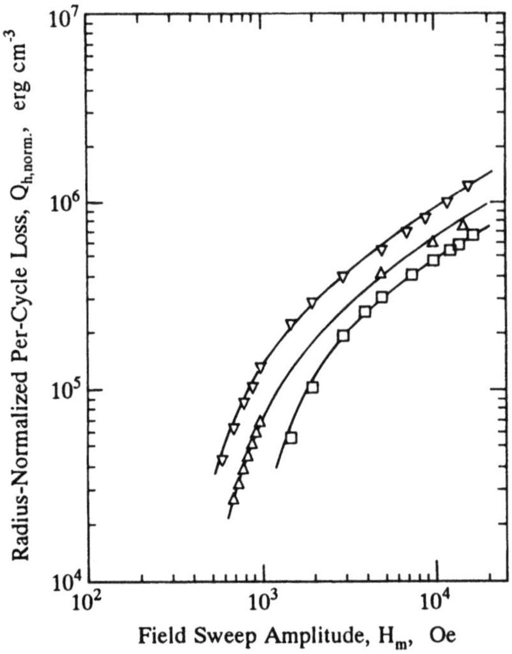

Fig. 2. High-field (H>H*) radius-normalized hysteretic loss *versus* field-sweep amplitude for $DC1_{AB}$ (□), CPS3 (△), and CPS4 (▽) at 77 K.

approaches H*, which varies from sample to sample according to $H^* = 0.8J_cR_0$.[10,p.84] On the logarithmic scale of the figure the mid-range losses appear to coincide. However, when presented in greater detail some differences do emerge, traceable primarily to differences in J_c.[5]

AC Hysteretic Loss Above H*

According to Carr[10,p.67] and others, the hysteretic loss per cycle per unit volume of a fully penetrated superconductor in a transverse magnetic field is given by

$$Q_h = (1.6/3\pi)(R_0J_c)H_m \quad (2)$$

Since the Meissner contribution was shown to be negligible in the pre-penetrated superconductor close to H*, it will certainly be so in the fully penetrated case. This time we note that Q_h is *proportional* to R_0, so that radius normalization should be introduced through the factor $R_{ref.}/R_0$. The radius-normalized high-H_m losses are depicted in Fig. 2. In it we see a separation of three data sets, which we will demonstrate below to be clearly a consequence of J_c variation.

MAGNETIC CRITICAL CURRENT DENSITY

The Bean critical state model for a cylinder in a transverse applied field, H, relates J_c to the total height, $\Delta M_v(H)$, of the volume-normalized hysteresis loop (measured between the shielding and trapping branches) by way of[11]

$$\Delta M_v(H) = (0.8/3\pi)J_cR_0 \quad (3)$$

Bean's model was of course based on the premise that J_c was independent of field; nevertheless even Bean[9] and many others to follow employed Eqn. (3) to determine the *field-dependence* of J_c. A more rigorous treatment would be to introduce field dependence at the outset, or to employ more advanced methods for extracting J_c from the hysteresis loop.[12] Close examination of the low-field M(H) loops showed no evidence of intergrain weak linking. Thus we are able to claim that the sample diameter, $2R_0$, is a true measure of the size of the magnetization current loops. Assuming the validity of Eqn. (3) the magnetic J_cs of $DC1_A$, CPS3, and CPS4, in two magnetic field ranges, have been computed. The results are presented in Figs. 3 and 4.

HYSTERETIC LOSS AND MAGNETIC J_c

As indicated above, we attribute the differences in radius-normalized hysteretic loss to differences in J_c. But since J_c is field dependent, when assigning a $J_c(H')$ to a particular $Q_h(H_m)$, the question arises as to how to go about selecting a suitable H'. Certainly $H' = H_m$ will not yield a satisfactory average J_c. But the issue has already been addressed in connection with a comparison of the AC losses of a pair of NbTi/Cu multifilamentary composite strands.[13] From Eqns. (2) and (3) we can write

$$Q_h(H_m)/\Delta M_v(H') = 2H_m \quad (4)$$

provided, under the present "semi-Bean conditions", that H' is suitably selected. It was demonstrated, at least for commercial Cu/NbTi, that the weighting of J_c towards low fields could be taken into account, and Eqn. (4) reasonably

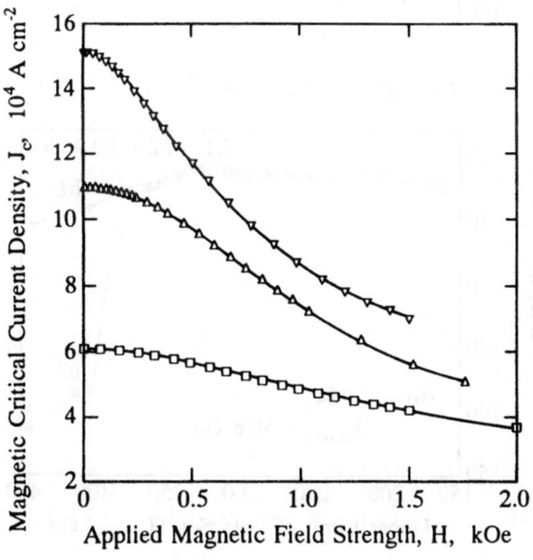

Fig. 3. Magnetic critical current density in the low-field range for $DC1_{AB}$ (□), CPS3 (△), and CPS4 (▽).

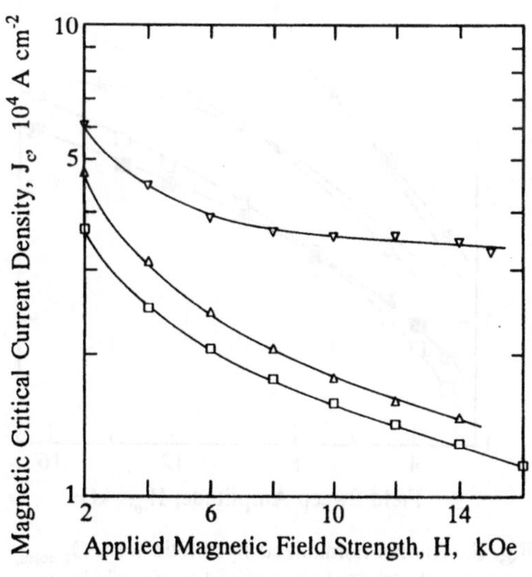

Fig. 4. Magnetic critical current density in the high-field range for $DC1_{AB}$ (□), CPS3 (△), and CPS4 (▽).

well satisfied, by setting H′ equal to $H_m/3$. We will assume that the same approach to magnetic-J_c averaging over the cycle also holds for melt-processed YBCO. Thus in Fig. 5 we plot concurrently $Q_h(H_m)$ and $2H_m\Delta M_v(H_m/3)$ *versus* H_m, the ΔM_v of course having also been normalized to $R_{ref.} = 100$ μm. The pairing up of the curves for $DC1_{A,AB}$, CPS3, and CPS4, indicates that the differences in Q_h that remain after radius differences have been compensated for can be justified in terms of magnetic J_c. In principle, the Fig. 5 result is entirely to be expected; a higher level of Q_h,J_c comparison would require *transport* measurements of J_c.

LOWER CRITICAL FIELD, H_{c1}, BY THE COEFFICIENT-RATIO METHOD

The right hand side of Eqn. (1) is the product of two terms. The the first term, is a function of J_cR_0, the properties of which have been considered above. The second term, which we designate Q_{hK}, is a function of H_m and h_{c1}. It has been demonstrated elsewhere[2,5,6,7] how the lower critical field, $H_{c1} = 2h_{c1}$ can be derived from Q_{hK} when expressed in the form:

$$Q_{hK} = (1/2)h_{c1}^3 - (3/2)h_{c1}H_m^2 + H_m^3 \quad (5)$$

$$\equiv A_0 + A_2H_m^2 + H_m^3 \quad (6)$$

The Coefficient-Ratio Method is used in which a pair of "interim critical fields", h1 and h2, are defined in terms of the ratios of coefficients taken in pairs: $h1 \equiv -(2/3)(A_2/A_3)$ and $h2 \equiv \sqrt[3]{(2A_0/A_3)}$. When h1 = h2, Eqns. (5) and (6) have the same functional form, and $h1 = h2 = h_{c1}$. To provide scope for this equality to take place, Eqn. (6) is fitted to successively truncated (Q_h,H_m) data sets. The lowest point,$(Q_h,H_{m,min})$ is fixed, while the upper point, $(Q_h,H_{m,max})$, is successively reduced by truncation; h_{c1} is found at the point(s) of intersection of h1 and h2, respectively, *versus* $H_{m,max}$. Figure 6 presents the results of Coefficient-Ratio Method determinations of h_{c1} for the two melt-processed samples CPS3 and CPS4; the results for $DC1_{AB}$ have been presented elswhere.[6,7] Close examination of the plots yields values for $h1 = h2 = h_{c1}$ and hence for the actual lower critical fields, $H_{c1} = 2h_{c1}$, which turn out to be: (i) for $DC1_{AB}$, $\langle H_{c1}\rangle_{av.} = 224.9$ Oe; (ii) for CPS3, $H_{c1} = 163.1$ Oe; (iii) for CPS4, $H_{c1} = 161.8$ Oe. The difference between (i) and (ii) is attributable to differences microstructure (see Table I) arising from differences in the melt-processing procedures to which the NIN and CPSS samples had been subjected. Conversely, the agreement between (ii) and (iii) emphasizes the reproducibility of the CPSS preparation procedure.

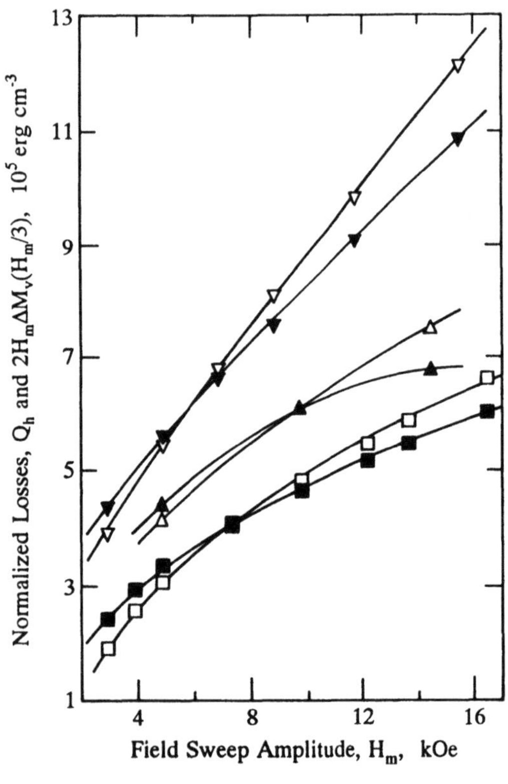

Fig. 5. Radius-normalized hysteretic loss, $Q_{h,norm.}$ (open symbols), and normalized hysteresis-loop height, ΔM_v, (filled symbols) for $DC1_{A,AB}$ (□■), CPS3 (△▲), and CPS4 (▽▼).

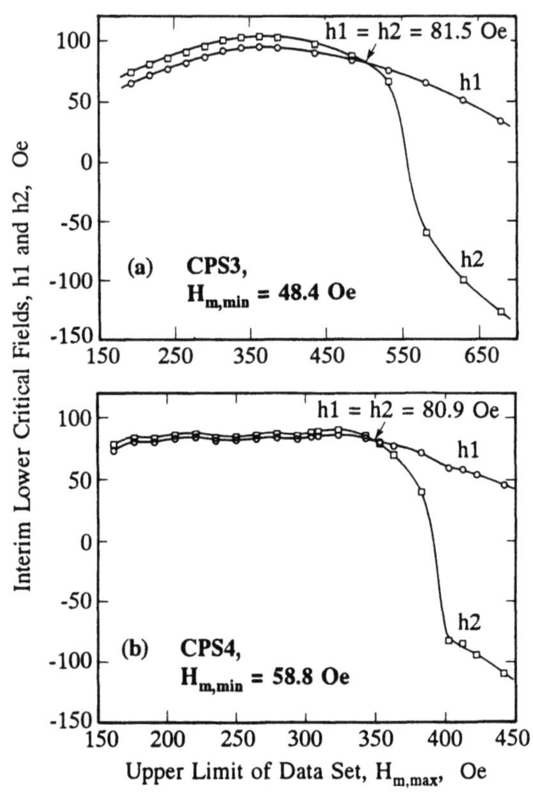

Fig. 6. Interim critical fields h1 and h2 *versus* $H_{m,max}$ at fixed $H_{m,min}$ for (a) CPS3, and (b) CPS4.[2,5]

CONCLUDING SUMMARY

Magnetization measurements of three samples of melt-processed YBCO have been described. A magnetic method, which utilizes the initial magnetization curve of a virgin sample, has been described for assigning an effective radius to cylinders of imperfect roundness. The results of a series of magnetic hysteresis measurements, of gradually increasing field-sweep amplitude and without intermediate demagnetization, have been utilized in three ways: (1) After having been normalized to a common reference radius the hysteretic losses, Q_h, of the three samples have been intercompared, the remaining differences being attributed primarily to differences in J_c. (2) From a few selected full-penetration hysteresis loops the magnetic J_cs of the samples were derived as function of applied field strength after it was first ascertained that the current loops were constrained by sample diameter rather than weak-link-defined grain boundaries. Using these J_cs, their responsibility for the separation the $Q_{h,norm.}$ *versus* H_m curves for the three samples was confirmed. (3) Application of pre-penetration Q_h *versus* H_m data to the measurement of lower critical field, by way of the Coefficient-Ratio Method was demonstrated. It was pointed out that samples prepared using two variants of the melt-processing approach had H_{c1}s of 225 and 163 Oe; and that the two samples that had been independently prepared by the *same* melt process had similar H_{c1}s, viz. 162 and 163 Oe.

ACKNOWLEDGEMENTS

The optical microscopy and image analysis referred to in the text were carried out by R. L. Pratt and P. M. Schumacher, respectively. The research at Battelle was supported by the Electric Power Research Institute (**EPRI**) under Contract No. RP8009-06; sample preparation at CPSS was supported by DARPA under Contract No. N00014-88-C-0512.

REFERENCES

1. Zhou Lian, Zhang Pingxiang, Ji Ping, *et al.*, The properties of YBCO superconductors prepared by a new approach: the 'powder melting process'", *Supercond. Sci. Technol.* **3**, 490-492 (1990).

2. M. D. Sumption, E. W. Collings, M. V. Parish, *et al.*, Lower critical field determination in thick rods of high-J_c melt-processed YBCO, *Fifth Ann. Conf. on Supercond. and Appl.*, Buffalo, NY, Sept. 24-26, 1991 -- submitted for publication.

3. J. D. Hodge, L. J. Klemptner, and M. V. Parish, Production of continuous filaments of melt-textured $YBa_2Cu_3O_{7-\delta}$ superconductor, *J. Mater. Sci. Let.* -- in press.

4. R. B. Goldfarb and J. A. Brug, Demagnetizing factors for cylinders, *Paper K26 12, presented at APS March Meeting*, Cincinnati, OH, March 18-22, 1991; also R. B. Goldfarb, personal communication.

5. E. W. Collings and W. J. Carr, Jr., Surface and bulk contributions to low-field hysteresis loss and H_{c1} in melt-processed YBCO, *The 2nd Tokai Univ. Int. Wkshp. on Supercond.*, Honolulu, HI, Oct. 9-12, 1991 -- submitted for publication.

6. E. W. Collings, K. R. Marken, Jr., M. D. Sumption, *et al.*, Low field AC loss and lower critical field in PMP high-T_c superconductors, *CEC/ICMC Conference*, Huntsville, AL, June 10-14, 1991 -- to be published.

7. E. W. Collings, M. D. Sumption, and W. J. Carr, Jr., Hysteretic method of lower critical field determination in high-T_c superconductors, *6th Int. Wkshp. on Crit. Currents*, Cambridge, UK, July 8-11, 1991 -- to be published.

8. A. M. Campbell and J. E. Evetts, **Critical Currents in Superconductors**, Taylor and Francis Ltd, London; Barnes and Noble Books, New York, 1972, pp. 75-76.

9. C. P. Bean, Magnetization of high-field superconductors, *Rev. Mod. Phys.* **36**, 31-39 (1964).

10. W. J. Carr, Jr., **AC Loss and Macroscopic Theory of Superconductors**, Gordon and Breach, New York, 1983.

11. W. J. Carr, Jr., and G. R. Wagner, Hysteresis in a fine filament NbTi composite, *Adv. Cryo. Eng. (Materials)* **30**, 923-930 (1984).

12. K. Yamafuji and Y. Mawatari, Estimation of critical current densities from magnetization measurements in a single crystal of a high-T_c superconductor -- 1. One-dimensional flux penetration, *6th Int. Wkshp. on Crit. Currents*, Cambridge, UK, July 8-11, 1991 -- to be published.

13. E. W. Collings, K. R. Marken, Jr., M. D. Sumption, *et al.*, AC loss measurements of two multifilamentary NbTi composite strands, *Adv. Cryo. Eng. (Materials)* **36**, 169-176 (1990).

A New Method of Determining $T_c(H)$ of Single-Crystal $La_{2-x}Sr_xCuO_4$ from the Field-Dependent Susceptibility Above T_c

MASASHI HASE[1], ICHIRO TERASAKI[1], ATSUTAKA MAEDA[1], KUNIMITSU UCHINOKURA[1], ISAO TANAKA[2], and HIRONAO KOJIMA[2]

[1] Department of Applied Physics, The University of Tokyo, Bunkyo-ku, Tokyo, 113 Japan
[2] Institute of Inorganic Synthesis, Yamanashi University, Kofu, Yamanashi, 400 Japan

ABSTRACT

Magnetic susceptibility of superconducting single-crystal $La_{2-x}Sr_xCuO_4$ ($x = 0.12$, 0.14 and 0.19) was measured with a SQUID magnetometer from 30 to 300 K under the magnetic fields (H) of 1 $\sim$ 5 T. The field-dependent susceptibility is evident below 40 K under H perpendicular to the CuO_2 planes and is assigned to the superconducting fluctuation-induced diamagnetism (SFD). The field-independent part was subtracted by taking the difference of the susceptibilities for the different H. The field-dependent part was fitted by using the two-dimensional form of the Lawrence-Doniach model for SFD. The magnetic-field dependence of the superconducting-transition temperature was obtained by this analysis. The values of the in-plane coherence length at 0 K are determined as 28.3, 29.2 and 32.5 Å for the samples of $x = 0.12, 0.14$ and 0.19, respectively.

KEY WORDS : Field-Dependent Susceptibility, Superconducting Fluctuation, Lawrence-Doniach Model, Superconducting Transition Temperature, Coherence Length

INTRODUCTION

The Ginzburg-Landau coherence length (ξ_{GL}) is an important microscopic parameter predominantly characterizing the superconducting properties. Several studies for determining ξ_{GL} have been done for the high-T_c cuprates. These studies are roughly divided into two groups; one method uses the data below the superconducting-transition temperature (T_c) [1,2] and the other uses those above T_c [3-9].

In the former method, the value of the upper critical field ($H_{c2}(T)$) at 0 K ($H_{c2}(0)$) is estimated from the magnetic-field (H) dependence of T_c ($T_c(H)$). The value of ξ_{GL} at 0 K ($\xi_{GL}(0)$) is obtained from the value of $H_{c2}(0)$ using the Ginzburg-Landau (GL) theory. The value of $T_c(H)$ is derived from the data of the magnetoresistivity [1] or the magnetic susceptibility [2]. A conventional definition of $T_c(H)$, *e.g.*, the definition as the mid-point temperature of the resistivity curve under H, is inappropriate owing to the so-called broadening of the superconducting transition seen in most of the high-T_c cuprates [1]. Welp *et al.* [2] have determined $T_c(H)$ of single-crystal $YBa_2Cu_3O_{7-\delta}$ from the data of the magnetization (M) which shows a $1 - T/T_c(H)$ behavior in a so-called reversible region. Their data for $\partial M/\partial H$, however, depend on H, which disagrees with the GL theory. Below T_c we cannot neglect the effect of the magnetic-flux motion, which gives ambiguity to the determination of $T_c(H)$.

In the latter method, the value of $\xi_{GL}(0)$ is determined as one of the fitting parameters by estimating the superconducting fluctuation above T_c, where there is no quantization of magnetic flux. For this purpose, the specific heat [3], the resistivity [4] and the susceptibility [5,6] have been measured. In their analyses, the normal-state behaviors of each quantities are assumed, which have not been firmly established. Another method is the measurements of the magnetoresistivity [7-9], which are explained by the Aronov-Hikami-Larkin theory [10] or the Bieri-Maki theory [11]. The analyses, however, recently turned out to have some confusion [12].

As was described above, previous methods of determining $T_c(H)$ or ξ_{GL} have problems. This circumstance requires a new experimental method. In the present paper, we propose a new method of determining $T_c(H)$ and subsequently ξ_{GL} by measuring superconducting fluctuation-induced diamagnetism (SFD) above T_c, which does not require any assumption of the form of the field-independent part of the susceptibility.

EXPERIMENT

Single-crystal $La_{2-x}Sr_xCuO_4$ ($x = 0.12, 0.14$ and 0.19) were prepared by a traveling solvent floating zone (TSFZ) technique. The details of the procedure and the characterization of the samples have been reported [13]. Each sample is a rectangular slab with a typical dimension of about $4 \times 4 \times 1$ mm^3. The widest plane of the sample is parallel [perpendicular] to the c axis [CuO_2 plane].

Magnetic susceptibility was measured from 30 to 300 K under several magnetic fields of $1 \sim 5$ T by an SHE model 905 SQUID magnetometer. Magnetic fields were applied parallel (susceptibility in this configuration is defined as $\chi_{\|}(H,T)$) or perpendicular ($\chi_{\perp}(H,T)$) to the CuO_2 planes in the widest plane in order to avoid the influence of the demagnetization effect. The superconducting-transition temperature in a weak magnetic-field limit (T_{c0}) was determined to be about 34 K for each sample from the data of $\chi_{\perp}(H,T)$ under 0.4 mT. Anisotropic susceptibility in the normal state of the same samples of $La_{2-x}Sr_xCuO_4$ with various x has been briefly reported [14].

RESULTS AND ANALYSES

The measured susceptibility under several H is shown in Figs. 1(a)-(b). The field dependence of $\chi_{\perp}(H,T)$ becomes evident below 40 K. On the other hand, that of $\chi_{\|}(H,T)$ becomes evident below 35 K. The field dependence of $\chi_{\perp}(H,T)$ is stronger than that of $\chi_{\|}(H,T)$ for each sample. The field-dependent susceptibility may be attributed to the SFD and/or inhomogeneity of the samples. The inhomogeneity, however, cannot be the sole source of the field dependence, because we can clearly observe it in $\chi_{\perp}(H,T)$ at 40 K, which is higher than the highest T_c of $La_{2-x}Sr_xCuO_4$. On the other hand, the strong two-dimensional nature [7-9] of the high-T_c superconductors induces large superconducting fluctuation. Therefore we will attribute the field-dependent susceptibility to SFD and analyze our experimental results, and later we will show that the result of analyses is consistent with this assumption. In this paper, only the field-dependent part of $\chi_{\perp}(H,T)$ will be analyzed.

Before analyzing SFD, the contribution of the field-independent part of the susceptibility ($\chi_0(T)$) must be subtracted. For this purpose, one method is to assume the form of $\chi_0(T)$ and to subtract it from the experimental value to obtain SFD [5,6]. The resulting SFD depends on the assumed form of $\chi_0(T)$, which has not been firmly established. To avoid this ambiguity we have taken the difference of the susceptibilities at different H, which does not include $\chi_0(T)$. This method is free from the particular assumption of the form of $\chi_0(T)$. We define $\delta\chi_{\perp}(H_1,H_2,T)$ as follows;

$$\delta\chi_{\perp}(H_1,H_2,T) \equiv \chi_{\perp}(H_1,T) - \chi_{\perp}(H_2,T), \quad H_1 > H_2. \tag{1}$$

We apply the Lawrence-Doniach (LD) model [15] for SFD to the analysis of $\delta\chi_{\perp}(H_1,H_2,T)$ because of the following reason. The LD model treats the SFD of multilayered superconductors. The high-T_c cuprates can be regarded as multilayered superconductors because they consist of the alternate stacking between the superconducting CuO_2 planes and the other layers and the two-dimensional nature is observed [7-9]. Therefore we assume that the two-dimensional form of the LD model can be used for analyzing our data. The expression of SFD of this model is written as follows;

$$\chi_{Fl.}(H,T) = -AT\{\frac{T_c(H)}{T - T_c(H)}\}, \quad A = \frac{1}{3}\pi k_B \Phi_0^{-2}\frac{\xi^2(0)}{s}, \tag{2}$$

where k_B the Boltzmann constant, Φ_0 the flux quantum, $\xi(0)$ the coherence length at 0 K and s the repeat distance of the superconducting layers. Thus $\delta\chi_{\perp}(H_1,H_2,T)$ is given by the following equation,

$$\begin{aligned}\delta\chi_{\perp}(H_1,H_2,T) &\equiv \chi_{\perp}(H_1,T) - \chi_{\perp}(H_2,T) \\ &= \chi_{\perp Fl.}(H_1,T) - \chi_{\perp Fl.}(H_2,T) \\ &= AT\{\frac{T_c(H_2)}{T - T_c(H_2)} - \frac{T_c(H_1)}{T - T_c(H_1)}\}, \quad H_1 > H_2.\end{aligned} \tag{3}$$

The values of $T_c(H)$ were uniquely determined by fitting Eq. (3) to the data and the value of A was also determined as a fitting parameter. The results are shown in Fig. 2. The closed circles and the solid curves represent the experimental results of $\delta\chi_{\perp}(H_1,H_2,T)$ and the theoretical fits, respectively. Fitting is satisfactory, and we conclude that $\delta\chi_{\perp}(H_1,H_2,T)$ can be well explained by Eq. (3) in spite of its simple formula.

The value of $T_c(H)$ obtained from the fits is shown in Fig. 3. Closed and open symbols correspond to the values of $T_c(H)$ obtained by the fits and those of T_{c0}, respectively. It decreases linearly with increasing H above 1 T. However, the upper critical field ($H_{c2\perp}(T)$) shows a positive curvature near T_{c0} for each sample, that is, the value of T_{c0} is larger than that of T^*_{c0} defined as the extrapolated temperature of $T_c(H)$ to $H = 0$. The values of T_{c0} and T^*_{c0} are listed in Table I. Although this phenomenon near T_{c0} is observed in a conventional superconductor [16] and in the high-T_c cuprates [1], its origin is not clear.

We discuss the x dependence of the value of $H_{c2\perp}(T)$ at 0 K ($H_{c2\perp}(0)$) and the in-plane coherence length at 0 K ($\xi_{ab}(0)$). As is seen in Table I, the value of $|dH_{c2\perp}(T)/dT|$ of the sample with $x = 0.19$ is smaller than that with $x = 0.12$ or 0.14. The values of $H_{c2\perp}(0)$ and $\xi_{ab}(0)$ can be estimated by assuming the WHH formula [17] and the anisotropic GL theory. The results are listed in Table I. The values of $H_{c2\perp}(0)$ are 41.2, 38.6 and 29.2 T and those of $\xi_{ab}(0)$ are 28.3, 29.2 and 32.5 Å for x=0.12, 0.14 and 0.19, respectively. The value of $\xi_{ab}(0)$ increases with increasing x. This result is consistent with the results in ref. [7].

Here we discuss the magnitude of SFD. The value of $A^{(1)}$ in Table I is defined as that of A obtained from the fits. The value of $A^{(2)}$ in Table I is calculated by Eq. (2) with using the obtained $\xi_{ab}(0)$ and half the lattice constant along the c axis as $\xi(0)$ and s, respectively. The value of $A^{(1)}$ agrees well with that of $A^{(2)}$ for each sample, that is, the LD model can quantitatively explain the SFD of the high-T_c cuprates.

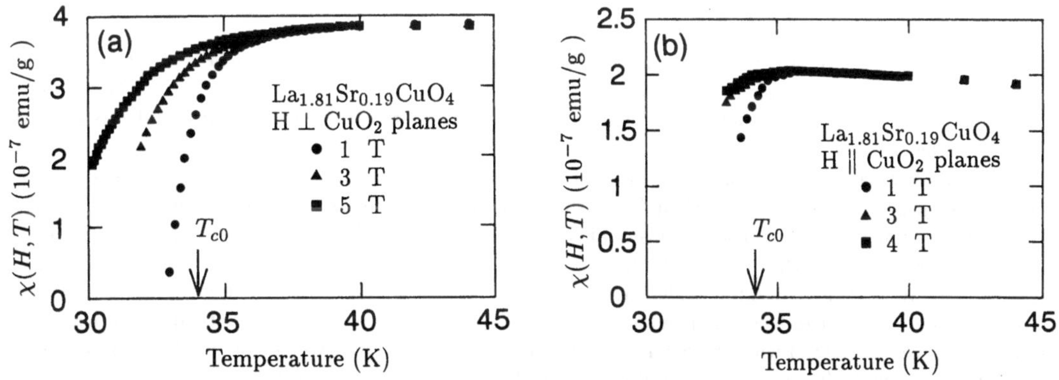

Fig. 1. Temperature dependence of susceptibility. (a) x=0.19, $\chi_\perp(H,T)$ and (b) x=0.19, $\chi_\parallel(H,T)$.

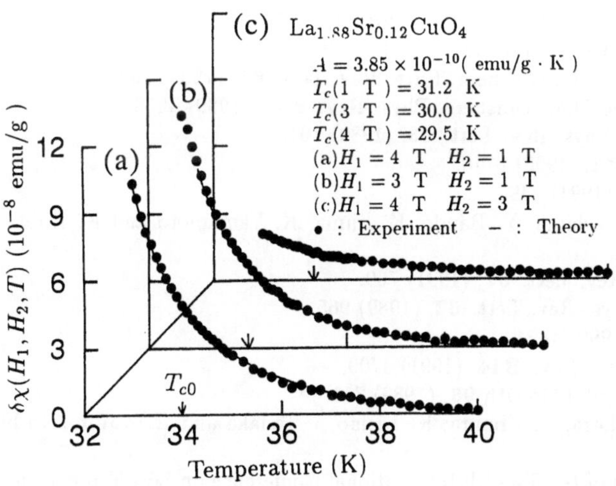

Fig. 2. Temperature dependence of $\delta\chi_\perp(H_1, H_2, T)$.

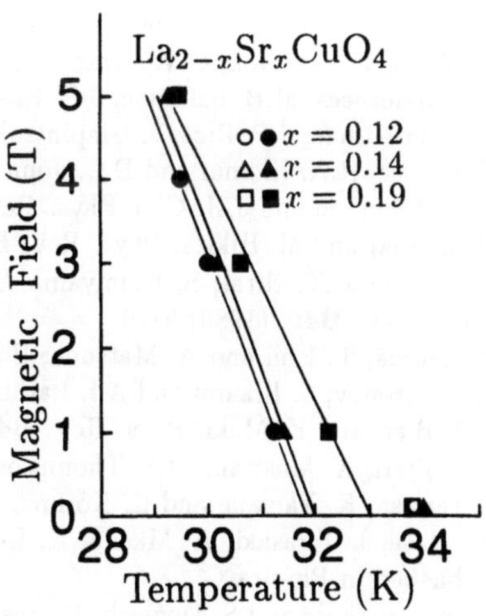

Fig. 3. $H_{c2\perp}(T)$ vs. T.

Table I The results of the fits are listed.

		$x = 0.12$	$x = 0.14$	$x = 0.19$
$-dH_{c2\perp}(T)/dT$	(T/K)	1.85	1.72	1.35
$H_{c2\perp}(0)$	(T)	41.2	38.6	29.2
$\xi_{ab}(0)$	(Å)	28.3	29.2	32.5
$A^{(1)}$	(10^{-9} emu cm^{-3} K^{-1})	2.70	2.32	2.28
$A^{(2)}$	(10^{-9} emu cm^{-3} K^{-1})	4.09	4.36	5.42
T_{c0}	(K)	34.0	34.0	34.0
T^*_{c0}	(K)	31.8	32.0	33.0

At the beginning of the analysis, we have assumed that the field-dependent susceptibility is due to SFD and not due to the inhomogeneity of the samples and also assumed that it can be described by the two-dimensional LD model for SFD. The result of the analysis shows good fit of the data to LD model and the behavior of $T_c(H)$, the values of $\xi_{ab}(0)$ and the magnitude of A obtained are quite reasonable as was described above. Therefore we conclude that our assumptions are valid and our method of determining $T_c(H)$ can be applied to our samples and also may be applied to the high-T_c superconductors in general.

SUMMARY

Magnetic susceptibility of superconducting single-crystal $La_{2-x}Sr_xCuO_4$ ($x = 0.12$, 0.14 and 0.19) was measured with a SQUID magnetometer from 30 to 300 K under the magnetic fields of 1 ~ 5 T. The field-dependent susceptibility was clearly observed below 40 K under H perpendicular to the CuO_2 planes, and was analyzed as the effect of the superconducting fluctuation-induced diamagnetism (SFD) with using the Lawrence-Doniach (LD) model for SFD based on the Ginzburg-Landau theory. The contribution of SFD was obtained from the difference of the susceptibilities measured under the different magnetic fields. This diamagnetic susceptibility is quantitatively explained by the two-dimensional form of the LD model. We succeeded in determining the magnetic-field dependence of the superconducting-transition temperature, *i.e.*, $T_c(H)$. The advantages of our method are as follows; (i) the field-independent part of the susceptibility can be excluded without any particular assumption of the form of the field-independent susceptibility, (ii) there is no flux quantization above T_c and therefore no complexity related with the magnetic-flux motion and (iii) only two parameters ($T_c(H)$ and A) are needed. The values of $\xi_{ab}(0)$ are 28.3, 29.2 and 32.5 Å for x=0.12, 0.14 and 0.19, respectively, and become longer with the increasing x.

REFERENCE

1. Y. Iye, T. Tamegai, H. Takeya and H. Takei, Jpn. J. Appl. Phys. **26**, (1987) L1057.
2. U. Welp, W.K. Kwok, G.W. Crabtree, K.G. Vandervoort and J.Z. Liu, Phys. Rev. Lett. **62**, (1989) 1908.
3. S.E. Inderhees, M.B. Salamon, J.P. Rice and D.M. Ginsberg, Phys. Rev. Lett. **66**, (1991) 232.
4. T.A. Friedman, J.P. Rice, J. Giapintzakis and D.M. Ginsberg, Phys. Rev. **B39**, (1989) 4258.
5. W.C. Lee, R.A. Klemm and D.C. Johnston, Phys. Rev. Lett. **63**, (1989) 1012.
6. D.C. Johnston and J.H. Cho, Phys. Rev. **B42**, (1990) 8710.
7. M. Suzuki and M. Hikita, Phys. Rev. **B44**, (1991) 249.
8. Y. Matsuda, T. Hirai, S. Komiyama, T. Terashima, Y. Bando, K. Iijima, K. Yamamoto and K. Hirata, Phys. Rev. **B40**, (1989) 5176.
9. K. Semba, T. Ishii and A. Matsuda, Phys. Rev. Lett. **67**, (1991) 769.
10. A.G. Aronov, S. Hikami and A.I. Larkin, Phys. Rev. Lett. **62**, (1989) 965.
11. J.B. Bieri and K. Maki, Phys. Rev. **B42**, (1990) 4854.
12. J.B. Bieri, K. Maki and R.S. Thompson, Phys. Rev. **B44**, (1991) 4709.
13. I. Tanaka, K. Yamane and H. Kojima, J. Crystal Growth **96**, (1989) 711.
14. M. Hase, I. Terasaki, A. Maeda, K. Uchinokura, T. Kimura, K. Kishio, I. Tanaka and H. Kojima, to be published in Physica C.
15. W.E. Lawrence and S. Doniach, Proceedings of the Twelfth International Conference on Low Temperature Physics, Kyoto, Japan, 1970, edited by E. Kanda (Academic Press of Japan, 1971), p. 361.
16. T.W. Haywood and D.G. Ast, Phys. Rev. **B18**, (1978) 2225.
17. N.R. Werthamer, E. Helfand and P.C. Hohenberg, Phys. Rev. **147**, (1966) 295.

Anomalous Field Dependence of Magnetic Susceptibility in the Normal State of High T_c Superconducting Cuprates

KOHICHI KUBO, FUMIO MUNAKATA, KIYOTAKA MATSUURA, and H. YAMAUCHI

Superconductivity Research Laboratory, ISTEC, Koto-ku, Tokyo, 135 Japan

ABSTRACT

The field dependence of magnetic susceptibility in the normal state of $Bi_2Sr_2Ca_{1-X}Y_XCu_2O_z$ was studied by SQUID magnetometry. The results suggested that there existed a weak-ferromagnetic interaction in the electronic system at the samples with X=0.0,0.2,0.5 and 0.8, and a certain relation between such an interaction and superconductivity.

KEY WORDS: Bi-2212 system, magnetic susceptibility, normal state, weak ferromagnetism

INTRODUCTION

To reveal the nature of the high Tc superconductivity is an important subject, because it is an unsolved problem in physics and indispensable for the design of new high Tc superconductors. Nonetheless, consistency between the theories and experimental results has not reached any satisfactory levels. This may be due to many possible causes and the superconducting state in which there exists a strong screening of electromagnetic interaction tends to prevent us from taking direct information for the superconducting mechanism. So, we need to obtain knowledge from precursory phenomena or relavant effects induced by superconductivity. We consider that there are included keys which cause instability to superconductivity.

It has been so far reported that the normal state of the electronic system of the high Tc superconductor is mysterious. The transport properties exhibited peculier behaviors which were neither metallic nor insulating and still remain puzzling. The resistivity shows a linear temerature dependence throughout an unusually wide range of temperature. The thermoelectric power, S, and Hall coefficient, R_H, also exhibited characteristic and involved features. The S-vs-T and R_H-vs-T curves posessed broad maxima in terms of temperature around 50 K to 200 K. The characteristic feature for the thermoelectric power has been extensively believed to be caused by phonon or magnon drugs. The idea that the magnetic interaction was a cause for the anomalous features has also been existing. In the case of Hall coefficient, there also exist some possible models. A skew scattering mechanism by magnetism is one of the candidates for the anomalous temperature dependence[1]. If it is mainly due to a skew scattering, the Hall resistivity must include a saturating term at high fields. However such a behaviour has not been reported by now. It seems to be commonly observed that if the carrier density exceeded an optimum value, the anomalous phenomena tended to be annihilated and turned to show a normal metallic behaviour. This interpretation for the normal state of the electronic system seems to be valid for the temperature dependence of magnetic susceptibility showing a feature of a broad maximum whose magnitude is also unusual[2]. According to Johnston[3], the temperature dependence of the magnetic susceptibility for the La-"214" phase could reasonably fit to the theory for a 2-dimensional Heisenberg system in the low carrier region. This suggests that the spin correlation among electrons decreased by increasing the hole concentration. In other words, superconductivity seemed to appear between a nearly localized electron system and a delocalized and itinerant electron system. The magnetic correlations among electrons may play an important role for high Tc superconductivity. Thus, we performed further experiments of magnetic susceptibility in the normal state of high Tc superconductors. The measurement of magnetic susceptibility includes many possible terms which are not essencial. Especially, refined experiments are required to take these possibilities into account. The anti-ferromagnetism in the CuO_2 plane of the Bi-Sr-(Ca,Y)-Cu-O ("2212"phase) system which still possiblly exist up to the superconducting region was not detactable by DC SQUID measurements, because the orientation of the CuO_6 octahedra did not deviate from the normal axis as extensively reported. This is why the Bi-"2212" has an advantage for the refinement of the magnetic susceptibility data. By this reason, we adopted the Bi-"2212" system for the measurement of magnetic susceptibility in the normal state.

EXPERIMENTAL

The policrystalline samples of $Bi_2Sr_2Ca_{1-X}Y_XO_Z$ with X=0.0,0.2,0.5 and 0.8 were successfully sinthesized and prepared by a conventional solid state reaction. Bi_2O_3, $CaCO_3$, Y_2O_3, $SrCO_3$ and CuO powders of high purity were mixed and ground and pressed into cylindrical pellet under a pressure of about 1ton/cm^2. The resulting mixtures being calcined in air at a temperature arround 790℃ for 10 hours and this procedure was repeated twice. The samples were then reground and pressed into bar-shape and sintered in air at a temperature between 850℃ and 920℃ for 30 hours. A microcomputer-controlled diffractometer equipped with a copper target and a graphite monochromater for CuKα radiation was used to obtain powder X-ray diffraction (XRD) patterns. All the samples were single phase. The magnetization data were taken in an automatic superconducting quantum interference device (SQUID) magnetometer (Quantum Design).

We performed measurements of temperature dependence of para-magnetization in the temperature range of 100 K to 200K with the step width of 1 K. The magnetic fields H of 1T and 5.5T were utilized in the measurement.

Secondly, the field dependence of magnetic moment was measured in the range of the field of 5.5T to -5.5T with the step-width of 0.1T, and was measured between -5.5T to 5.5T with increasing the field strength, while the temperature was kept at 120K and 200K. A 16-time averaging was carried out at each temperature step and we carefully changed the field strength not to overshoot in order to perform a highly refined measurement.

RESULTS AND DISCUSSION

Figure 1 shows the temperature dependence of the magnetic susceptibility in the normal state. This is considered to be an essencial behavior for high Tc superconducting cuprates, since the result is surprisingly close to the La-"214" system with respect to not only the temperature dependence but also to the carrier contents, apart from the superconducting transition temperature Tc and carrier content at optimized Tc.

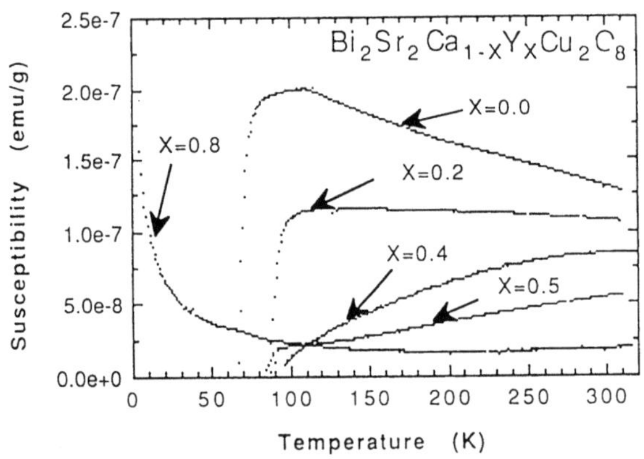

Figure 1 Temperature dependence of magnetic susceptibility in normal state for $Bi_2Sr_2Ca_{1-x}Y_xCu_2O_z$ (x=0.0,0.2,0.4,0.5,0.8)

Figure 2 shows the temperature dependence of the magnetic susceptibility for $Bi_2Sr_2Ca_{0.8}Y_{0.2}Cu_2O_z$ at fields of 1T and 5.5T. This change was observed not only for this sample but also for all other compositions of this system.

All the curves commonly decreased in their magnitude by increasing field strength. Thus, we performed refined measurements of the magnetic moment by changing the field strength, in order to know which part of the field strength the nonlinearity occurred. The results are shown in figs. 3 and 4.

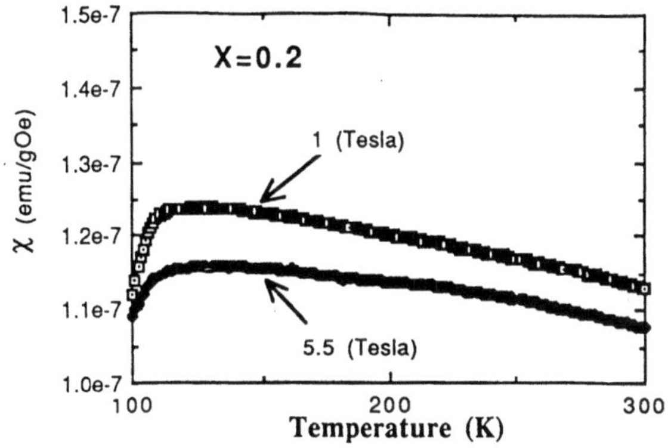

Figure 2 Dependency of magnetic susceptibility on field strength for $Bi_2Sr_2Ca_{0.8}Y_{0.2}Cu_2O_z$ (1 T and 5.5T).

A weak ferromagnetism was observed around H=0, The M-vs-H curve was almost linear in the field region higher than 0.2T. Therefore, we extrapolated the data using the following equation to the zero field limit from the high field portion, and calculted Mo:

$$M = M_0 + \chi H \qquad (1)$$

Here Mo was found to depend on the dopant content and temperature as shown in Fig. 5. This behavior is quite parallel to the dependence of Tc on x.

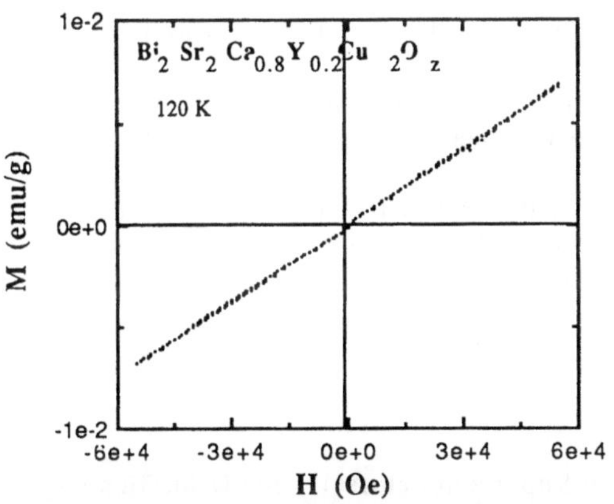

Figure 3 Magnetic moment vs magnetic field for $Bi_2Sr_2Ca_{0.8}Y_{0.2}Cu_2O_z$ in field range of -5.5 T to 5.5 T

Nonlinearlity is not seen in this scale

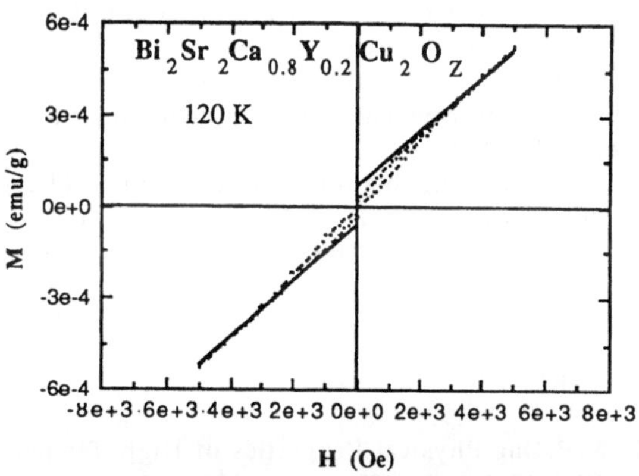

Figure 4 Magnetic moment vs magnetic field for $Bi_2Sr_2Ca_{0.8}Y_{0.2}Cu_2O_z$ in field range of -0.6 T to 0.6 T

Anomaly was observed around at 0 Tesla

For the La-"214" system, not only an anomaly in the high field region which was caused by the Dyaloshinsky-Moriya interaction but also a similar anomaly in the low field region have been reported. The present experimental data must be carefully studied because a variety of cases such as those listed in the following are existing. Firstly, if magnetic impurities were included in the sample, only the amount of 0.1ppm order of iron atoms may cause a weak-ferromagnetism, if the atoms wre clustered. Nonetheless, annealing or sintering of such samples at a high temperature around 800 ℃ can eliminate such a weak-ferromagnetism.

Recently, Eto et.al.[4] showed using a band calculation technique for the La-"214" system that holes occupied the Cu dz^2 band with the spins of the band antiparallel to those of the localized Cu dx^2-y^2 holes for the Sr concentration less than that which corresponds to the onset of superconductivity, while the holes occupy the Cu dz^2 band with their spins parallel with those of Cu dx^2-y^2 holes in the superconducting state. If such a high spin state shows, there exists a ferromagnec interaction between Cu $dz^2$and O $2p_x$ by-passed through Cu dx^2-y^2 due to the orthogonality between these orbitals, when the carrier concentration is below that which corresponds to the onset of superconductivity and the holes still remained in the O $2p_x$ orbital.

But another possibilities for this weak-ferromagnetism may be required to be taken into account. A parasitic ferromagnetism due to distorted antiferromagnetic Cu-O_2 planes can also be a possible cause.

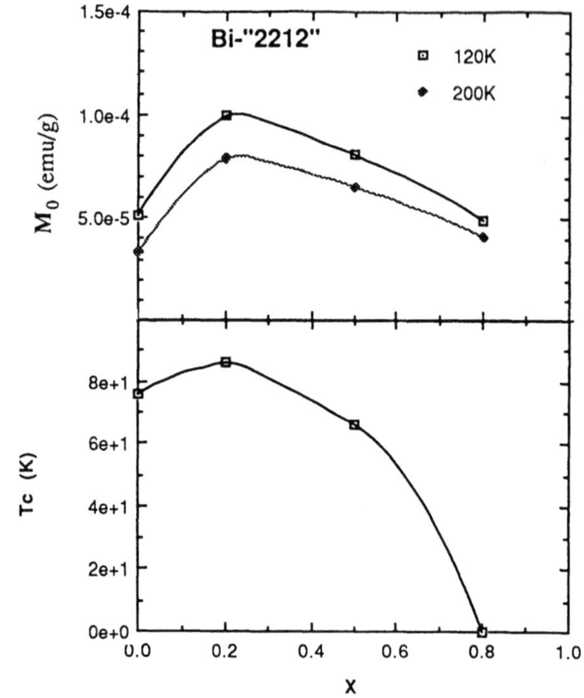

Figure 5 Extrapolated moment M0 vs dopant content X (upper viewgraph)
Tc vs dopant content (lower viewgraph)
There is some correlation between M0 and Tc

SUMMARY

1. The normal-state DC magnetic susceptibility for the Bi-"2212" compounds was found to depend on the field strength.

2. The M-H curve suggested that the magnetic moment M consisted of a nonlinear term which was ferromagnetic and another linear term. The ferromagnetic term tended to level off around 1000 Oe.

3. A linear extrapolation from the high field region of the M-H curve intercepted M-axis at M_0 (>0)(when H=0). It was found that the value of M_0 depended on the dopant concentration,X, in a manner parallel to the Tc-vs-X curve.

REFERENCE

1) N.P.Ong Physical Properties of High-Temperature Superconductors II, ed. D.M.Ginsberg (WorldScientific, Singapore 1990)

2) J.B.Torrance, A.Bezinge, A.I.Nazzal, T.C.Huang and S.S.P.Parkin et.al. Phys. Rev. B. V.40 N.13 (1989) 8872

3) D.C.Johnston, Phys. Rev. Let., V.62 N.8 (1989) 957

4) M.Eto and H.Kamimura, J. Phys. Soc. Jpn. 60 (1991) No.7.

Nonlinear Microwave Electrodynamics of Superconducting NbN and Nb Thin Films

C.C. CHIN[1,5,*], D.E. OATES[4], G. DRESSELHAUS[3,5], and M.S. DRESSELHAUS[1,2]

[1] Department of Physics, [2] Department of Electrical Engineering and Computer Science, [3] Francis Bitter National Magnet Laboratory, Massachusetts Institute of Technology, Cambridge, MA 02139, USA
[4] Lincoln Laboratory, Massachusetts Institute of Technology, Lexington, MA 02173, USA
[5] Rome Laboratory, USAF, Hanscom AFB, MA 01731, USA

ABSTRACT

The surface resistance R_s of NbN and Nb thin films with particular reference to nonlinear effects is reported at high values of the microwave magnetic field $H_{rf} \leq 500$ Oe using a stripline resonator in the frequency range $0.6 \leq f \leq 12$ GHz. For Nb thin films, R_s does not increase with H_{rf} for low H_{rf} values, as expected from weak-link theory. However, for NbN thin films, R_s at temperature T follows a $R_s(T,f) = R_{s0}(T,f) + S(T,f)H_{rf}$ dependence for $f < 6$ GHz, where $R_{s0}(T,f)$ is the surface resistance at zero H_{rf}, and the slope $S(T,f)$ is proportional to $f^{2.3}$.

KEY WORDS: Nb, NbN, Microwaves, Stripline, Weak links

1 INTRODUCTION

The measurement of nonlinear electrodynamic effects in superconductors [1] is of interest for both practical applications and fundamental studies. In any practical electronic application, the designer must know how much power the conductors can handle and at which power level nonlinear effects such as harmonic generation and intermodulation (IM) distortion become appreciable. Therefore the magnitude and detailed nature of the nonlinear effects must be measured and understood in order to facilitate widespread application of superconductors in microwave frequency electronics.

Nonlinear effects can also be used to study the fundamental properties of the superconducting materials. One of the sources of nonlinear electrodynamics in type II superconductors can be vortex formation and motion. Measurements at microwave frequencies should yield information about the nature of pinning forces and the nucleation time of vortices.

In this work, we present results for the surface impedance Z_s of thin superconducting films of NbN and Nb at high values of the microwave magnetic field H_{rf} using a stripline resonator. We find that the nonlinearity of Z_s has a different behavior in two regions of H_{rf}: one of weak nonlinearity and one of strong nonlinearity. Weak nonlinearity is characterized by an increase in the surface resistance, and the resonance curves remain Lorentzian and symmetric about the center frequency. In the region of strong nonlinearity, the resonance curves become asymmetric and show hysteresis, R_s increases sharply, and the insertion loss increases (see Fig. 1). The shape of the resonance curves resembles that calculated for a nonlinear harmonic oscillator [2, 3]. The modeling of these curves will be published later [4]. Often the strongly nonlinear regions exhibit a thermal instability.

NbN is of great interest because of its similarities to high-T_c materials; for example, NbN is composed of oriented grains connected by weak links. Furthermore, NbN is strongly type II with a large upper critical field and a high critical current density. Also, the mechanism of superconductivity in NbN is understood and all relevant BCS parameters have been measured. We also include Nb in the study because of the availability of high quality, well-characterized Nb films.

The surface resistance R_s of NbN increases on increasing the magnitude of H_{rf}. This effect is believed to be due to the granularity of the NbN. Halbritter [5] calculated the change of Z_s as a function of H_{rf} due to the high frequency vortex penetration into the intergranular weak links. Comparison of the Z_s vs H_{rf} data of granular NbN films with Halbritter's theory may provide important insights for the high-T_c data which also show a power dependence of Z_s [6]. In contrast, our Nb sample shows no power dependence of Z_s, perhaps because of lesser granularity and a low nucleation frequency of intergranular vortices.

When H_{rf} reaches a critical value H_{rf}^c, the resonance curves for the NbN and Nb resonators become asymmetric about the resonant frequency, with the peak resonant frequency shifted to lower frequency, as shown in Fig. 1(a) for NbN and 1(b) for Nb. Recent studies on high quality $YBa_2Cu_3O_{7-x}$ films by the stripline resonator method show similar phenomena. The data is reported elsewhere [4]. Halbritter [2] argues that the asymmetry of the resonance curve is caused by the power dependence of the surface reactance X_s of the superconductor, which may be due to pair breaking, nucleation of high frequency vortices, or other nonlinear effects. By measuring the temperature and frequency dependence of H_{rf}^c for both NbN and Nb, we are able to infer the physical reasons for asymmetry of the resonance curves.

*Present address: Superconductivity Research Laboratory, Koto-ku, Tokyo 135, Japan

2 NONLINEAR SURFACE IMPEDANCE Z_s

The complex surface impedance $Z_s(T,f)$ was measured using the stripline resonator method of determining $R_s(T)$ and $\lambda(T)$ as described elsewhere [7, 8]. To determine the superconducting penetration depth $\lambda(T)$, we measured the center frequency f_0 of the fundamental resonance as a function of T from low T (4.2 K) to T_c (or as close to T_c as possible). We then fitted $f_0(T)$ to a two-fluid or Mattis-Bardeen model expression for $\lambda(T)$ by using the calculated current distributions [9] to model the dependence of the stripline inductance on $\lambda(T)$. To determine R_s, we find the Q of the resonator by measuring the 3 dB points of the transmitted-power-vs-frequency curve. The capacitive coupling gaps at the ends of the stripline are adjusted to yield an under-coupled condition at resonance so that the measured Q and the unloaded Q_0 are almost equal. The Q is measured as a function of T and input power at the fundamental frequency of 0.6 GHz and at all of the overtone frequencies up to 12 GHz. The calculated current distribution, which depends on $t/\lambda(T)$ where t is the thickness of the film, is used to extract the surface resistance values from the measured Q (see Ref. [9]).

Figure 2 shows examples of the H_{rf} dependence of R_s of NbN for several low-order modes up to 6 GHz measured at 4.2 K, i.e., we plot R_s vs the peak H_{rf} at the edges of the transmission line. At low H_{rf} (< 30 Oe), the resonance curve is a Lorentzian and thus is symmetrical about the resonant frequency. When the incident power is increased to produce H_{rf} larger than about 30 Oe in NbN, the resonance curves become asymmetric and much broader, indicating a larger increase in R_s than that observed below 30 Oe (see, for example, the resonance curves in Fig. 1(a) for the mode at 1.14 GHz). Figure 2 shows that, at low fields, R_s increases linearly with H_{rf} (or I_{rf}) and is given by

$$R_s(T,f) = R_{s0}(T,f) + S(T,f)H_{rf},$$

where $S(T,f)$ is the slope of the linear increase and $R_{s0}(T,f)$ is the surface resistance at zero H_{rf}. Both S and R_{s0} are functions of frequency and temperature. Figure 2 also shows that R_s vs H_{rf} is divided into two regions: one of weak nonlinearity at low fields, showing a linear increase in R_s, and one of strong nonlinearity at high fields in which R_s increases more sharply. For $f < 6$ GHz all modes show a similar linear dependence of R_s with H_{rf} at 4.2 K. The value of $R_{s0}(4.2\text{ K},f)$ is found by extrapolating the linear least squares fit to $H_{rf} = 0$ for each of the f values measured, showing that $R_{s0}(4.2\text{ K},f)$ is proportional to f^2, the expected result [8]. Above 6 GHz, a new feature appears. When H_{rf} reaches a critical value, a plateau in R_s (a region of nearly zero slope) is observed at all frequencies above 6 GHz and up to 10 GHz. Below the onset of the plateau $R_s(H_{rf})$ shows a linear increase with H_{rf}, as is also found in the lower frequency modes. The resonance curve remains Lorentzian both for H_{rf} values corresponding to the plateau region and below. Figure 3 shows data for the slope $S(4.2\text{ K},f)$ vs f on a log-log scale obtained by a least squares fit to $R_s(f)$ for $0.5\text{ GHz} < f < 10$ GHz, using data for the low field, linear regions of Fig. 2. A least squares fit to $S(4.2\text{ K},f) = Af^{\nu}$ yields $S(4.2\text{ K},f)$ proportional to $f^{2.3\pm0.3}$, as shown in Fig. 3, in good agreement with Halbritter's weak-link theory [5] which predicts an f^2 dependence of $S(T,f)$, resulting from flux penetration into the weak links between the grains.

The temperature dependence of H_{rf}^c for NbN was measured at 1.14 GHz (see Fig. 4) and the data were fit to the T dependence of the dc lower critical field $H_{c1}(T)$ given by [10]

$$H_{c1}(T) = \frac{\Phi_0}{4\pi\mu_0\lambda^2(T)}\ln\kappa,$$

where Φ_0 is the magnetic flux quantum, κ is the Ginzburg-Landau parameter, and $\lambda(T)$ is assumed to have a two-fluid model temperature dependence. Since $H_{c1}(T)$ is essentially proportional to $[1-(T/T_c)^4]$ as shown in Fig. 4. One sees a large discrepancy between the T dependence of $H_{rf}^c(T)$ and $H_{c1}(T)$ for NbN. H_{c1} of the NbN thin film as measured by a SQUID at 4.8 K is in the range 350 to 500 Oe, which should be compared to the H_{rf}^c value of 40 Oe. In order to see whether such a discrepancy is due to strong fluxoid pinning sites, the T dependence of the dc critical current J_c was measured and was found to vary as $1-(T/T_c)^2$, which rules out flux pinning as the explanation of the H_{rf}^c vs T curve.

3 DISCUSSION

We conclude that the nonlinearities in $R_s(H_{rf})$ may be divided into weak and strong regions. The weak nonlinear regimes for NbN show a linear increase in R_s with H_{rf}, but in Nb, no increase in R_s is observed. The strong nonlinear regimes in both NbN and Nb are characterized by asymmetric resonance curves and sharp increases in $R_s(H_{rf})$, resembling critical behavior. The linear increase of the R_s of NbN with H_{rf} is consistent with flux penetration in the intergranular material of the NbN films. The slope at $T = 4.2$ K agrees within experimental error with the $S(T,f) \propto f^2$ prediction of Halbritter's granularity theory [5].

For Nb no increase of the R_s with H_{rf} is observed. The weak links of Nb are attributed to oxide-filled fissures caused by oxidation [11], and $R_s \propto H_{rf}$ at 0.1 GHz [12]. The increase of R_s was found to disappear with increasing frequency [5], showing that the nucleation frequency of the vortices in Nb is ~ 0.5 GHz [5]. The Nb films are also granular, but the behavior of the links between the grains is not as weakened as that in NbN.

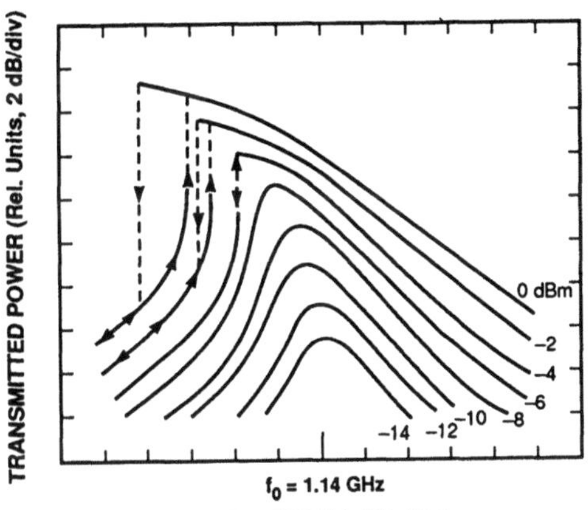

(a)

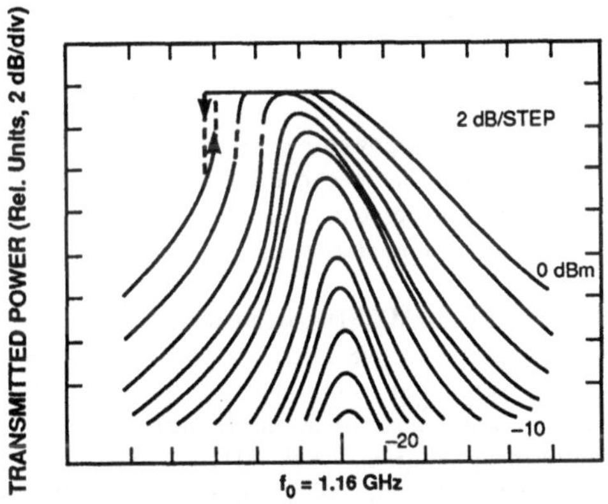

(b)

Figure 1: (a) Lineshape of the transmitted power resonance curves of NbN at strong microwave magnetic field H_{rf}. Each curve is labeled by the input power level in increments of 2 dBm. The asymmetry of the resonance curves at high power is shown, and these curves show hysteresis at an input power greater than -2 dBm. (b) Lineshape of the transmitted resonance curves of Nb at strong H_{rf}. The resonance becomes asymmetrical at input power of -10 dBm and exhibits hysteresis at 0 dBm input power. The flat top of the resonance curve may be due to the instability of the resonance at very high input power levels.

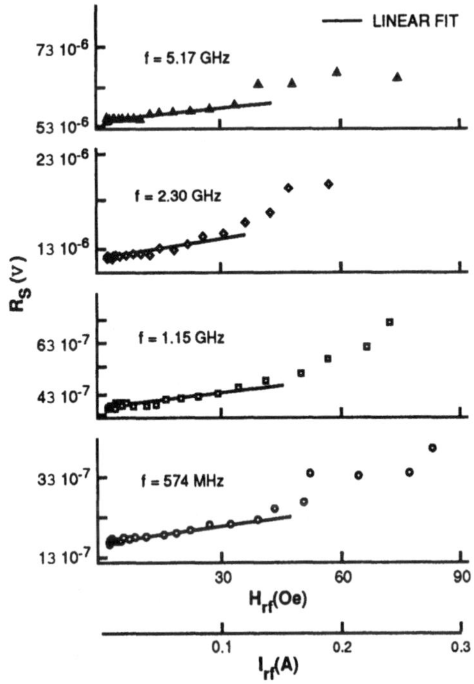

Figure 2: Surface resistance R_s of NbN vs H_{rf} or I_{rf} measured at 4.2 K for different frequencies up to 6 GHz. Note the change of the R_s scale for the various frequencies.

An important finding is that in both Nb and NbN the measured H_{rf}^c is significantly smaller than the dc H_{c1} for stripline samples fabricated by the same methods as films used in resonator experiments [1]. Thus H_{rf}^c may be the field value at which flux penetration into the grains occurs. Halbritter has suggested [13] that the flux penetration takes place at microscopic defects and regions of "weakened superconductivity." This theory does not, however, explain why the dc H_{c1} is higher than the H_{rf}^c, nor does the theory explain the effects of the defects. If flux is penetrating into the grains, the effects should have been seen in the magnetization curve. Perhaps the R_s measurements are more sensitive to flux penetration at the defects than dc measurements since R_s is very small so that changes in R_s are more easily observed. Using the I_{rf} value where H_{rf}^c is reached for Nb, we compute an average critical current density $J_{rf}^{ave.} \approx 1.1 \times 10^6/\mathrm{cm}^2$ which is very close to the dc J_c measured using the same stripline geometry as was used for the rf measurements. We have verified this relationship for two different widths of the stripline. For NbN, however, $J_c = 2.5 \times 10^6$ A/cm^2 and $J_{rf}^{ave} = 1.0 \times 10^5$ A/cm^2. This difference in rf and dc J_c is consistent with the measurements of $H_{rf}^c(T)$, which indicates that in NbN the H_{rf}^c is not directly related to J_c.

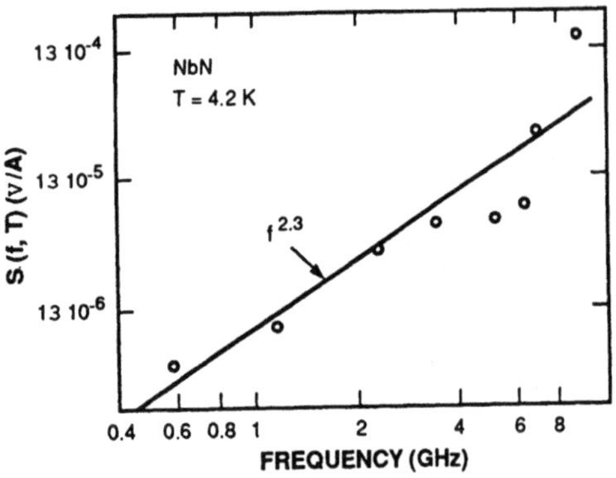

Figure 3: Plot of the slope $S(T,f)$ vs f at T = 4.2 K. The points are experimental data and the line is a least squares fit of the data yielding an $f^{2.3\pm0.3}$ dependence.

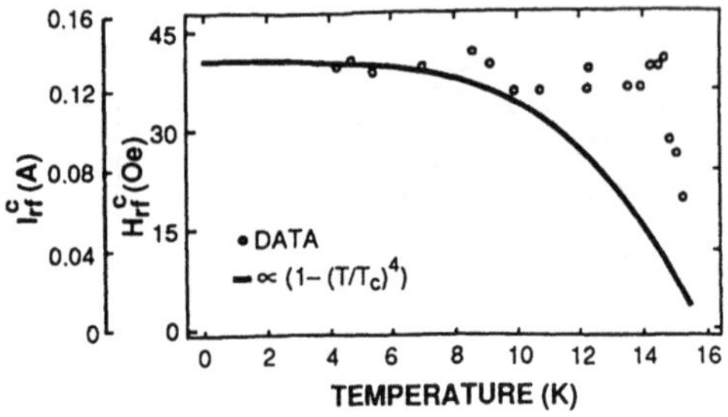

Figure 4: The temperature dependence of $H_{\rm rf}^{\rm c}(T)$ and $I_{\rm rf}^{\rm c}(T)$ of NbN. The solid curve denotes a dependence proportional to $[1-(T/T_c)^4]$ for reference.

In summary, we have shown that the nonlinear behavior of R_s of NbN at high power is consistent with Halbritter's weak-link model. Increases in R_s result from flux penetration at weak links that occur at grain boundaries. This nonlinear behavior has also been seen in granular films of the high-T_c oxide superconductors [4]. These results thus imply that it is necessary to employ single crystal films which minimize grain boundaries and hence minimize R_s for high power device applications.

This paper is based in part on the Ph.D. thesis of C.C.C. submitted to the MIT Department of Physics. C.C.C. acknowledges the hospitality of the MIT Lincoln Laboratory where the experiments were carried out. A more complete report of these results will be published [1]. The research at MIT was supported in part by the Air Force Office of Scientific Research AFOSR–90–0123. The work at the Rome Laboratory was sponsored by the Department of the Air Force where G.D. was supported by the AFOSR University Research Resident Program. At the MIT Lincoln Laboratory the work was supported in part by the Department of the Air Force and in part by the Defense Advanced Research Projects Agency under the auspices of the Consortium for Superconducting Electronics.

We acknowledge fruitful discussions with and suggestions by Dr. J. Halbritter concerning rf losses, flux penetration, and the application of nonlinear oscillator theory. The authors thank Dr. Alfredo Anderson for providing the NbN samples, R.P. Konieczka for resonator fabrication, and W.T. Brogan and E E. Macedo for sample preparation.

References

[1] C. C. Chin, D. E. Oates, G. Dresselhaus, and M. S. Dresselhaus, submitted for publication.

[2] J. Halbritter, J. Appl. Phys. **41**, 4581 (1969).

[3] C. Hayashi, *Nonlinear Oscillations in Physical Systems* (Princeton University, Princeton, NJ, 1985).

[4] P. Nguyen, D. Oates, C.C. Chin, G. Dresselhaus, and M.S. Dresselhaus, unpublished.

[5] J. Halbritter, J. Appl. Phys **68**, 6315 (1990).

[6] D.E. Oates, A.C. Anderson, D. M. Sheen, and S. M. Ali, IEEE Trans. Microwave Theory Tech. **39**, 1522 (1991).

[7] D.E. Oates, A.C. Anderson, and P.M. Mankiewich, J. Supercond. **3**, 251 (1990).

[8] D.E. Oates, A.C. Anderson, C.C. Chin, J.S. Derov, G. Dresselhaus, and M.S. Dresselhaus, Phys. Rev. B **43**, 7655 (1991).

[9] D. Sheen, S. Ali, D.E. Oates, K.A. Kong, and R.S. Withers, IEEE Trans. Appl. Superconduct. **1**, 108 (1991).

[10] T. Van Duzer and C.W. Turner, *Principles of Superconductive Devices and Circuits* (Elsevier, New York, 1981).

[11] J. Halbritter, Appl. Phys. A **43**, 1 (1987).

[12] B. Piosczyk, P. Kneisel, O. Stoltz, and J. Halbritter, IEEE Trans. Nucl. Sci. **NS-20**, 108 (1973).

[13] J. Halbritter, private communication.

3 Chemistry

3.1 Synthesis

Preparation of "124" Superconductors Containing Ca from Alkoxides

SHIN-ICHI KORIYAMA, TAKAAKI IKEMACHI, and H. YAMAUCHI

Superconductivity Research Laboratory, ISTEC, Koto-ku, Tokyo, 135 Japan

ABSTRACT

The samples of "124" superconductors containing Ca, i.e., $(Y_{1-x}Ca_x)Ba_2Cu_4O_y$, are synthesized in flowing O_2 gas of $P(O_2)=1$ atm by means of a sol-gel method using only alkoxides. The sample with the Ca content, x, equal to 0.05 shows sharp x-ray diffraction peaks for the "124" phase and sharp superconducting transitions in both electrical resistivity ($Tc^{(on)}$=89K, $Tc^{(R=0)}$=84K) and magnetic susceptibility (Tc $mag^{(on)}$=90K). The solubility limit, x, for Ca in the "124" phase when it was synthesized in flowing O_2 gas is less than 0.1 as previously reported for high-pressure syntheses. The "124" particles in the synthesized samples are flakes with flat c planes.

KEY WORDS: 124 superconductor, sol-gel method, low temperature synthesis, Y-Ba-Cu-O , Y-Ba-Ca-Cu-O

INTRODUCTION

The "124" phase ($YBa_2Cu_4O_8$) was discovered as lattice defects in the "123" phase (YBa_2Cu_3Oy)[1]. Later, bulk "124" samples were synthesized by a high oxygen pressure technique. The "124" phase was classified to be a 80K superconductor[2],[3] and found to be thermally stable up to 800°C[3]. The "124" superconductor should be useful because its Tc is little depressed even when processed, but it has not attracted much attention because the synthesis requires a high pressure process and its Tc value is not enough high to be used at liquid nitrogen temperature. An increase in Tc to 90K for a "124" compound with Ca doped[4] was achieved. The "124" compounds which did not contain Ca have been synthesized in flowing O_2 gas flow of $P(O_2)=1$ atm using a catalyst[5] and by sol-gel methods[6]. Recently, the "124" compounds containing Ca were synthesized even at relatively low oxygen pressures e.g. $P(O_2)=3$ atm[7] and $P(O_2)=1$ atm[8] but the quality of such samples was poor compared with that of the samples synthesized at high oxygen pressures[9]. To prepare high-quality "124" compounds containing Ca in flowing O_2 gas of $P(O_2)=1$ atm, preparation of highly homogeneous precursors and syntheses at low temperature are thought to be the key points. The sol-gel method has advantages for a low temperature synthesis of ceramics. However, to date, Cu-alkoxides which are necessary for a sol-gel synthesis of cuprates have not been available.

In this paper, we synthesize a Cu-alkoxide and develop a low temperature technique for the synthesis of cuprate superconductors. Using the Cu-alkoxide together with other metal-alkoxides, we demonstrate that "124" samples containing Ca are prepared in flowing O_2 gas of $P(O_2)=1$ atm when fired at 750°C.

EXPERIMENTAL

Preparations of metal-butoxide

The Ca- and Cu-secondary-butoxides were prepared by using reactions between metal chlorides and sodium-secondary-butoxide in liquid phase i.e. using the following reaction:

$MCl_2+2Na(s\text{-}OBu)\rightarrow M(s\text{-}OBu)_2+2NaCl\downarrow$ (M=Ca and Cu).

First, MCl_2 and Na(metal) were separately dissolved into secondary-butanol solvents (s-BuOH). Then, the two solutions were mixed and stirred in flowing Ar gas. The resultant suspensions were centrifuged at 4000G for 10 min. The obtained solutions were analyzed by inductively coupled plasma (ICP) technique and were tested to determine if a each solute in the solutions had hydrolyzed by adding water.

The Ba-secondary-butoxide was prepared by dissolving Ba-metal into secondary-butanol solvent i.e. using the following reaction:

$Ba+2s\text{-}BuOH\rightarrow Ba(s\text{-}OBu)_2+H_2\uparrow$.

The solution was tested to determine if a solute in the solution had hydrolyzed by adding water.

The Y-normal-butoxide/xylene solution was purchased from Nihon Kagaku Sangyo Co. Ltd, Tokyo.
The solution was tested to determine if a solute in the solution had hydrolyzed by adding water.

Preparation of "124" compounds

0.43M(mol/l) $Y(n\text{-}OBu)_3$ /xylene, 0.1M $Ca(s\text{-}OBu)_2$ /s-BuOH, 0.1M $Ba(s\text{-}OBu)_2$ /s-BuOH and 0.05M $Cu(s\text{-}OBu)_2$ /s-BuOH were used as starting solutions. The starting solutions were mixed for the composition of Y/Ca/Ba/Cu =1-

x/x/2/4 (x=0, 0.05 and 0.1) and stirred in flowing Ar gas at 70°C for 10 hours. Ambient air was sent to the solutions being stirred at the rate of 50 ml/min for 30 min in order to be hydrolyzed the solutes in the solutions. The hydrolyzed suspensions were evaporated to obtain gel powders. The gel powders were vacuum dried at 500°C for 1 hour. The dried powders were die-pressed and heated at 750°C or 800°C for 40 hours in flowing O_2 gas of $P(O_2)$=1 atm.

The synthesized gel powders were examined by x-ray diffractometry (XRD), scanning electron microscopy (SEM) and energy dispersive x-ray spectrometry (EDX).
The fired samples were examined by XRD. The temperature dependences of electrical resistivity and magnetic susceptibility were measured using a conventional four probe method and dc magnetometer (SQUID), respectively. Fractured surfaces of the samples were observed by SEM.

RESULTS AND DISCUSSION

Metal-butoxides

According to ICP analysis, the Ca concentration in the centrifuged Ca-butoxide solution was 0.1M, and the Cu concentration in the centrifuged Cu-butoxide solution was 0.05M. The Cu concentration of 0.05M was sufficient for the syntheses of cuprates. Ba-metal of 0.01 mol was dissolved into s-BuOH solvent of 0.1 ℓ. All the prepared metal-alkoxide solutions and purchased Y-butoxide solution were separated into transparent solvents and precipitates by adding water. These results suggested that each metal component in the solutions was in a metal-alkoxide form i.e. $Y(n\text{-}OBu)_3$, $Ca(s\text{-}OBu)_2$, $Ba(s\text{-}OBu)_2$ and $Cu(s\text{-}OBu)_2$ because the each solute in the solution was hydrolyzable.

"124" compounds

Figure 1 shows (a) a scanning electronmicrograph and (b) EDX spectra for the gel powder particles in the "124" precursor of containing Ca with x=0.05. The particles were of round shape with the diameter less than 1 μm. The EDX spectra show that both particles A and B with relatively large sizes contain the four elements Y, Ca, Ba and Cu. The spectra of Ca for the particles A and B are very weak because the Ca content, x, of the gel powder is very small at 0.05. The spectra of Y, Ba and Cu are weaker in particles C and D because of the smaller size of these particles relative to particles A and B. No observation of spectra for the Ca in the particles C and D were supposed that to be caused the spectra for the Ca were as low as noise level. Figure 1 (c) shows an XRD pattern for the gel powder. The gel powder was amorphous because no XRD peaks were observed except those from NaCl which was produced during reaction. Therefore, it was thought that the prepared gel powder was microscopically homogeneous.

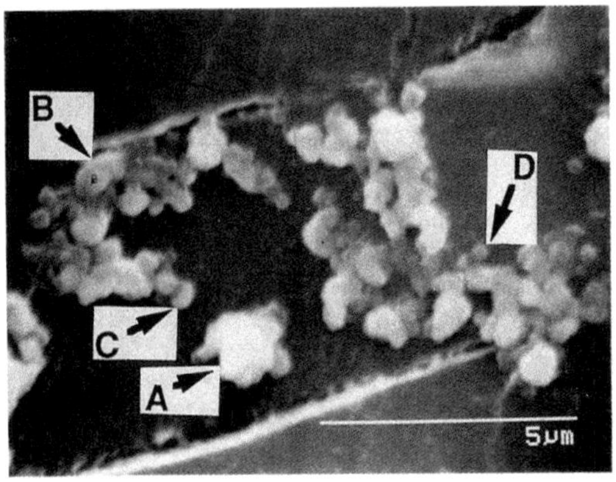

(a)

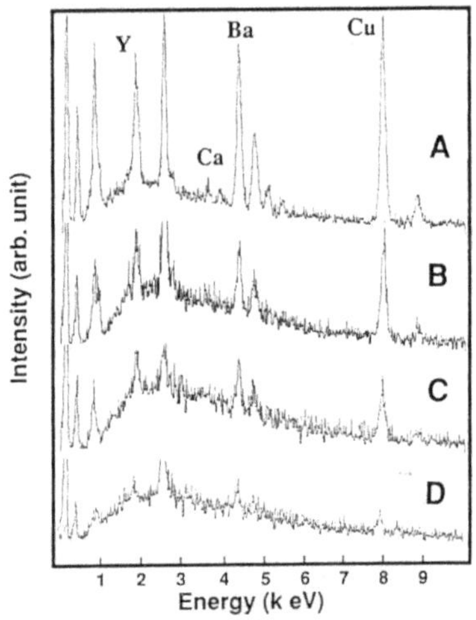

(b)

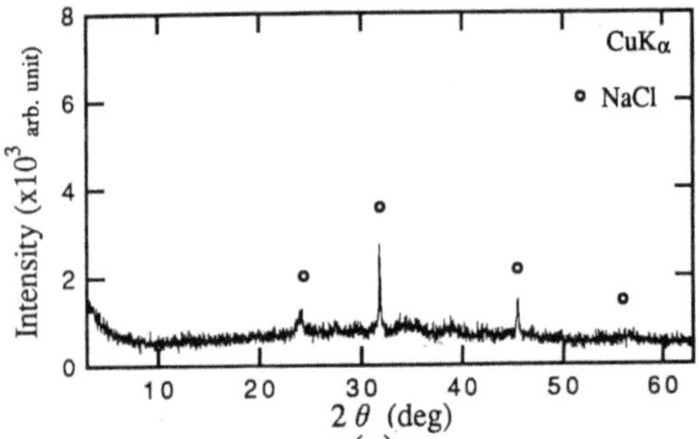

(c)

Fig.1 (a) a scanning electronmicrograph, (b) EDX spectra and (c) an XRD pattern for the gel powder of "124" precursor with x=0.05.

Figure 2 shows (a) XRD patterns for the samples with x=0.05 fired at 750°C and 800°C and (b) the temperature dependences of electrical resistivity for the two fired samples. The sample fired at 750°C showed sharp XRD peaks characteristic to the "124" phase but no other intense peaks. The sample fired at 800°C showed peaks characteristic to the "124" phase and those for CuO. This means that the "124" compound containing Ca partially decomposed at 800°C. This agrees with the results of electrical resistivity measurement. The superconducting transition of the sample fired at 750°C was sharper than that of the sample fired at 800°C.

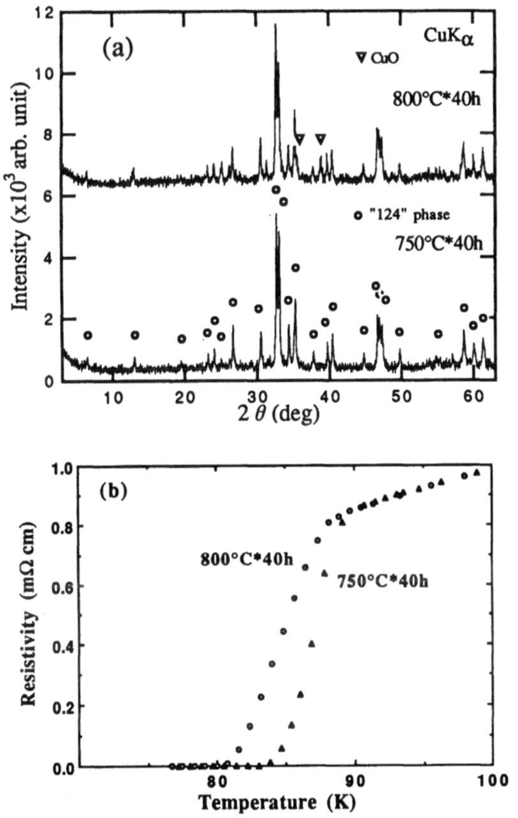

Fig.2 (a) XRD patterns and (b) the temperature dependences of electrical resistivity for the firered samples with x=0.05.

Figure 3 shows (a) XRD patterns, (b) the temperature dependences of electrical resistivity and (c) the temperature dependences of magnetic susceptibility for three samples with x=0, 0.05 and 0.1 fired at 750°C. All the three samples showed sharp XRD peaks characteristic to the "124" phase and sharp superconducting transition in both electrical resistivity and magnetic susceptibility. The sample with x=0.1 showed the XRD peaks from CuO and a unknown phase in addition to those from the "124" phase. Therefore, it is supposed that the solubility limit of Ca in the "124" phase when it was synthesized under the previously mentioned condition is less than 0.1. This agrees with the results of electrical resistivity measurement and magnetic susceptibility measurement. In electrical resistivity measurement, the $Tc^{(on)}$ for the sample with Ca content x=0.1 was higher only 1K than that for the sample with x=0.05 although the $Tc^{(on)}$ for the sample with x=0.05 was higher 7K than that for the sample with x=0. In magnetic susceptibility measurement, the volume fraction of superconductivity at 10K for the sample with x=0.1 was smaller than that for the sample with x=0 or 0.05. On the other hand, the sample with x=0.05 was thought nearly single phase of the "124" phase because it showed no XRD peaks from impurity phases and showed a Meissner signal at 10K as high as that with x=0. The nearly single phase sample with Ca content x=0.05, showed superconducting transition in electrical resistivity at the onset temperature ($Tc^{(on)}$) of 89K, the zero-resistivity temperature ($Tc^{(R=0)}$) of 84K and in the magnetic susceptibility at the onset temperature (Tc $mag^{(on)}$) of 90K.

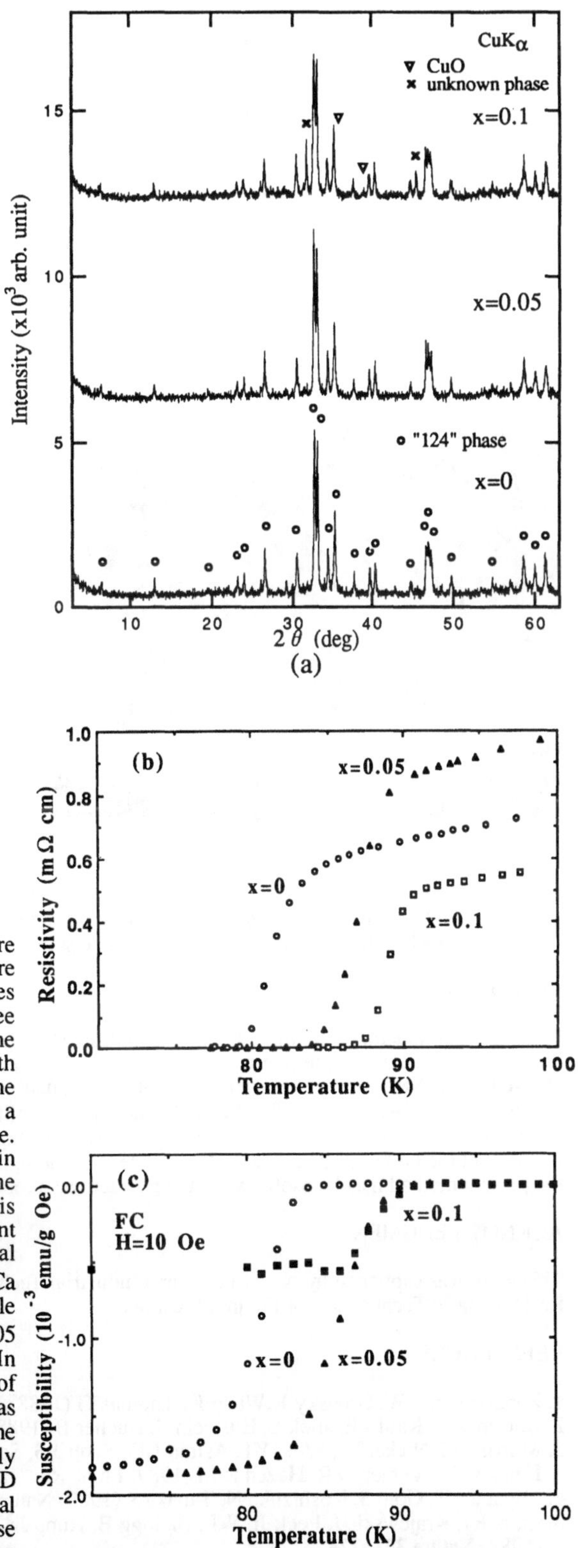

Fig.3 (a) XRD patterns, (b) the temperature dependences of electrical resistivity and (c) the temperature dependences of magnetic susceptibility for the samples fired at 750°C.

The scanning electronmicrograph for a fractured surface of the sample with x=0.05 fired at 750°C is given in Figure 4 (a). In Figure 4 (b), the XRD pattern for the powder of the same sample, which was strongly pressed onto a sample holder is shown. The synthesized "124" grains were flaky. The (00ℓ) peaks characteristic to the "124" phase were intensified compared with powder XRD pattern (Fig. 3). It was found that the "124" grains were flakes with flat c planes. This flake with flat c plane is similar to that for Bi-based superconducting materials in a shape. For most of the forming wires or tapes of the high temperature superconducting materials, Bi-based superconductors have been used. Bi-based superconductors are useful for the forming wires or tapes because the flaky particles for those compounds are easily oriented when those are processed and the Tc for those compounds are a little depressed during the process. The present "124" superconductors had the similar characters to Bi-based superconductors. Such "124" superconductors may be useful for the forming superconducting wires and tapes.

(a)

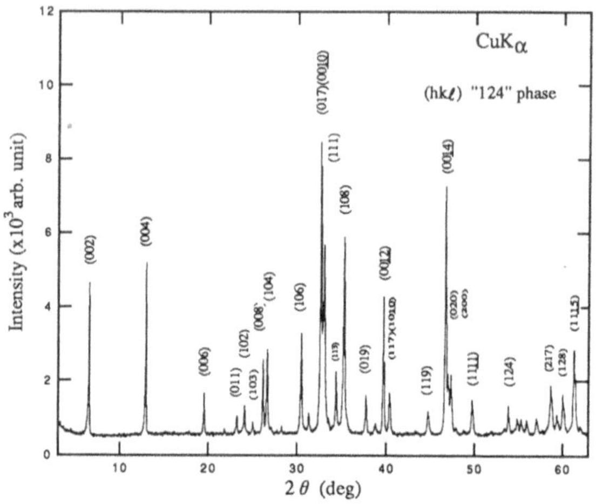

(b)

Fig.4 (a)The scanning electronmicrograph for a fractured surface of the sample with x=0.05 fired at 750°C and (b) the XRD pattern for the powder of the same sample, which was strongly pressed onto sample holder

CONCLUSION

"124" superconductors containing Ca were synthesized by firing in flowing O_2 gas of $P(O_2)$=1 atm at 750°C for only 40 hours. The nearly single phase sample for the "124" phase with Ca content, x=0.05, i.e. $(Y_{0.95}Ca_{0.05})Ba_2Cu_4O_y$, showed sharp XRD peaks characteristic to the "124" phase and sharp superconducting transitions in both electrical resistivity ($Tc^{(on)}$=89K, $Tc^{(R=0)}$=84K) and magnetic susceptibility (Tc $mag^{(on)}$=90K). The solubility limit of Ca in the "124" phase was less than 0.1 as previously reported for high pressure syntheses. The "124" grains synthesized by the present method were flakes with flat c planes. Because of this shape of grains, the "124" powders synthesized by the present method may be useful when the powder was processed into a wire or a tape shape.

ACKNOWLEDGMENT

This work was supported by New Energy and Industrial Technology Development Organization in the programme for R&D of Basic Technology for Future Industries.

REFERENCES

1. Zandbergen HW, Gronsky R,Wang K, Thomas G (1988).Nature 331: 596
2. Karpinski J, Kaldis E, Jilek E, Rusiecki S, Bucher B (1988) Nature 336: 660
3. Morris DE, Nickel JH, Wei JYT, Asmar NG, Scott J S, Scheven U M, Hultgren C T, Markelz A G, Post J E, Heaney P J, Veblen D R, Hazen R M (1989) Phys. Rev. B 39: 7347
4. Miyatake T, Goto S, Koshizuka N, Tanaka S (1989) Nature 341: 41
5. Cava RJ, Krajewski JJ, Peck Jr. WF, Batlogg B, Rupp Jr. LW, Fleming RM, Lames ACWP, Marsh P (1989) Nature 338: 328
6. Murakami H, Yaegashi S, Nishino J, Shiohara Y, Tanaka S (1990) Jpn. J. Appl. Phys. 29: L445
7. Wada T, Suzuki N, Yamaguchi K, Ichinose A, Yaegashi Y, Yamauchi H, Koshizuka N, Tanaka S (1991) J. Mater. Res. 6: 18
8. Kellner K, Wang XZ, Gritzner G, Bäuerle D (1990) Physica C 173: 208
9. Wada T, Suzuki N, Ichinose A, Yaegashi Y, Yamauchi H, Tanaka S (1990) Appl. Phys. Lett. 57: 81

Synthesis and Properties of Tl-Based "2223" Superconductor with $T_c^{R=0}$=127 K

TETSUYUKI KANEKO, KAZUYUKI HAMADA, SEIJI ADACHI, H. YAMAUCHI, and SHOJI TANAKA

Superconductivity Research Laboratory, ISTEC, Koto-ku, Tokyo, 135 Japan

ABSTRACT

Single phase superconducting Tl-Based "2223" material with a T_c higher than 125 K, i.e. $T_c^{R=0}$ (zero-resistance temperature) = 127 K and T_c^{mag} (diamagnetic onset temperature) = 130 K, was obtained using Ca-rich nominal composition of $Tl_{1.7}Ba_2Ca_{2.3}Cu_3O_z$. After sintering at 890 ℃ the critical temperature of the samples were around 120 K, but post-annealing with an encapsulation in an evacuated quartz tube effectively improved the T_c values. The critical temperature correlated with the post-annealing time, and for the samples annealed at 750 ℃ for longer than 80 h, the T_c values exceeded 125 K. As a structure change connected with the improvement of the T_c an increase in the c - axis was observed while the a - axis was not affected by post-annealing.

KEY WORDS: $Tl_2Ba_2Ca_2Cu_3O_{10}$ structure, Post-annealing, Encapsulation.

INTRODUCTION

In 1988, Sheng et al. discovered superconductivity in Tl-Ba-Ca-Cu-O system [1] and Parkin et al. confirmed that a Tl-"2223" compound, i.e. $Tl_2Ba_2Ca_2Cu_3O_{10}$, had a zero resistance temperature as high as 125 K [2]. Since that have been many attempts to increase the T_c of this compound, but no reproducible data have been reported about such studies. However, it should be noted that Tl-Ba-Ca-Cu-O superconductors with T_c^{mag}'s up to 127 K were successfully synthesized by Adachi et al., though their $T_c^{R=0}$'s did not exceed 125 K [3]. In their preparation method, encapsulation of the sample during sintering and a low temperature heat treatment at around 700 ℃ for a long period of time seemed to be the most important factors to improve T_c. In this work, we first prepared single-phase "2223" samples by sintering in flowing oxygen and then investigated the effect of post-annealing at a relatively low temperature in an evacuated quartz tube.

EXPERIMENTAL

Samples were prepared by a solid state reaction using high-purity powders of Tl_2O_3, BaO_2, CaO and CuO as starting materials. Appropriate amounts of the powders were mixed to obtain a composition of $Tl_{1.7}Ba_2Ca_{2.3}Cu_3O_z$. The choice of this Ca-rich "2223" stoichiometry was based on our preliminary experiments as well as previously reported results [4]. Before sintering the mixed powders were pressed into pellets. The pellets were wrapped in gold foil and sintered at 890 ℃ for 5 h in flowing oxygen. Some of the as-sintered samples were wrapped again with and encapsulated in evacuated (~ 10^{-4} Torr) quartz tubes. Both the encapsulated and unencapsulated samples were post-annealed at 750 ℃ for various periods of time (2 - 250 h). Structural properties of the samples were studied by X-ray powder diffraction (XRD) using Cu-Kα radiation. Electrical resistivity was measured using a standard four-probe method and DC magnetic susceptibility measurements were carried out with a SQUID magnetometer (Quantum Design: Model MPMS) collecting the data in the field cooling mode under a DC magnetic field of 10 Oe.

RESULTS AND DISCUSSION

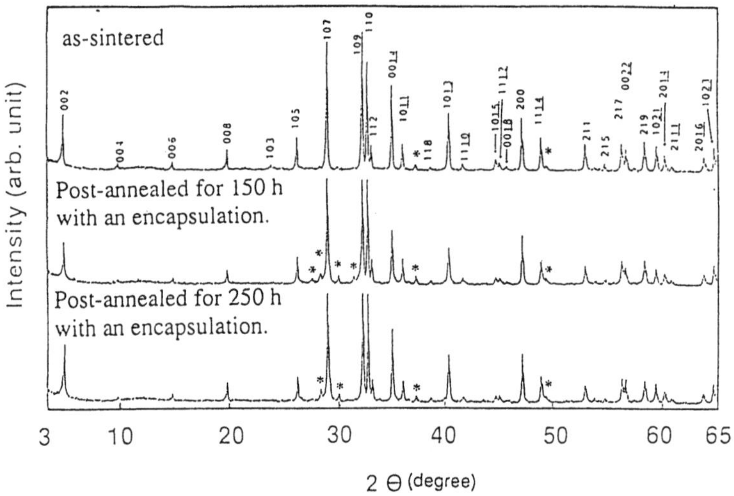

Fig.1 Powder X-ray diffraction patterns for the samples. (Peaks marked with * are due to impurity phases.)

The XRD pattern for the as-sintered sample is shown in fig. 1. Almost all peaks can be indexed for the tetragonal "2223" structure with the lattice constants of a = 3.849 A and c = 35.625 A. Figure 2 presents the temperature dependence of resistivity for the same sample and the temperature dependence of dc magnetic susceptibility is shown in fig. 3. From these results it is concluded that the as-sintered sample has a T_c around 120 K as the observed value for the zero resistance and magnetic onset temperatures were 118 K and 122 K, respectively.

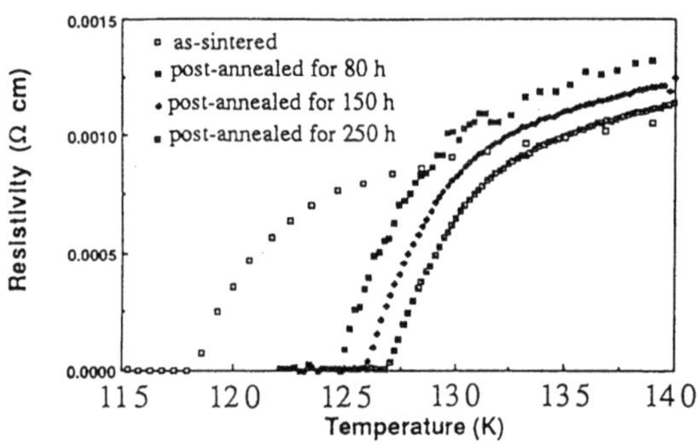

Fig. 2. Temperature dependence of resistivitity for the samples as-sintered and post-annealed with an encapsulation.

The XRD patterns for the samples post-annealed at 750 ℃ in an evacuated quartz tube for 150 h and 250 h are also shown in fig. 1. In both patterns, the number of extra peaks being due to impurity phases (marked by *) slightly increased compared with that of the as-sintered sample. Obviously the "2223" phase partly decomposed by post-annealing. However, almost initial "2223" phase has been left in the sample in spite of very long period of post-annealing such as 250 h. As seen in fig.4, if the sample was post-annealed without an encapsulation, the "2223" phase decomposed completely due to the evaporation of Tl and the sample did not contain any superconducting phase. The temperature dependence of resistivity for this sample resembled that of an insulator (in fig. 5). The above results demonstrates that the encapsulation technique prevents the decomposition of the "2223" phase during post-annealing.

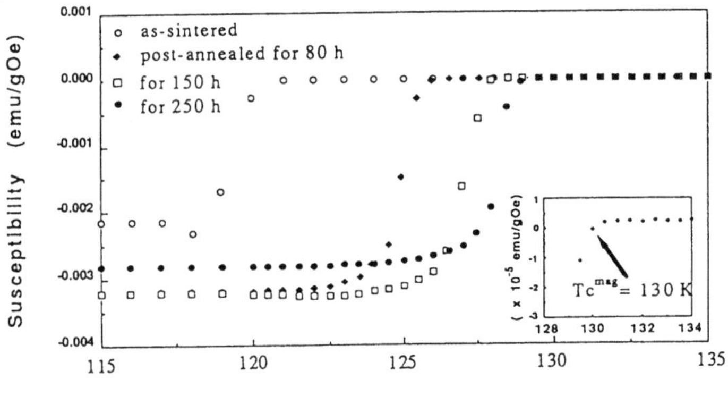

Fig. 3 Temperature dependence of DC magnetic susceptibility for the samples as-sintered and post-annealed with an encapsulation. The data for the highest one between 128 and 134 K are enlarged in the inset.

Table 1 lists the observed $T_c^{R=0}$ and T_c^{mag} values for the samples which were post-annealed at 750 ℃ for 0 to 250 h with encapsulation. The temperature dependence of resistivity as well as the magnetic suscep-

Table I

Zero-resistance temperatures, $T_c^{R=0}$, and onset temperatures of diamagnetic signal, T_c^{mag}, of encapsulated samples during post annealing at 750 ℃ for 0 - 250 hours.

Sample	Post-annealing time (h)	$T_c^{R=0}$ (K)	T_c^{mag} (K)
1	0 (as-sinterd)	117.9± 0.1	122.0± 0.5
2	2	120.6± 0.1	124.5± 0.5
3	10	122.9± 0.1	126.0± 0.5
4	80	124.8± 0.1	127.0± 0.5
5	150	125.8± 0.1	128.5± 0.5
6	250	126.9± 0.1	130.0± 0.5

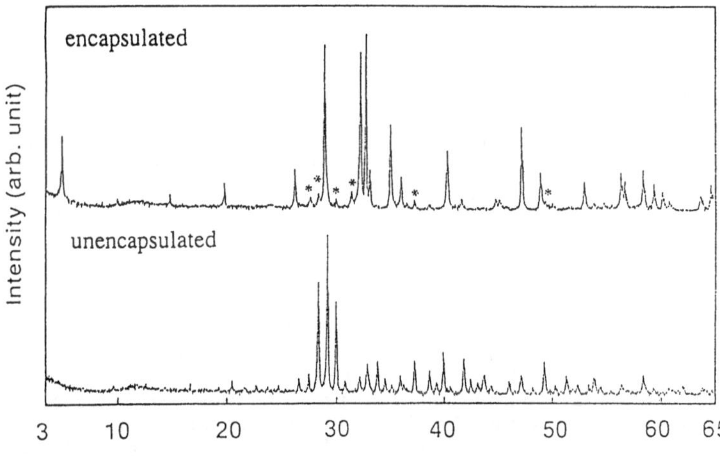

Fig. 4 Powder X-ray diffraction patterns for the encapsulated and unencapsulated samples when post-annealed for 80 h.

tibility curves for some of the samples are also shown in figs. 2 and 3, respectively. Both T_c values increased as the post-annealing time was increased and the highest zero resistance temperature of 126.9 K and the T_c^{mag} of 130.0 K were obtained when the sample was post-annealed for 250 h. On the other hand, the magnitude of diamagnetic signal remained almost unchanged among the samples studied indicating that, (although the critical temperature was effectively improved by using longer annealing periods) the volume (fraction) of the superconducting phase did not depend on the length of the post-annealing.

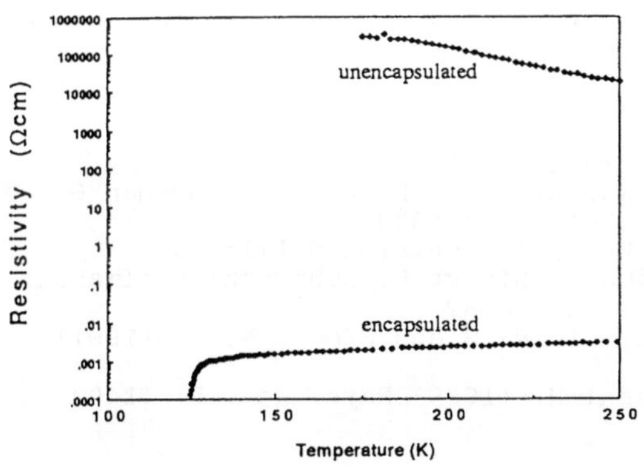

Fig. 5 Comparison of the temperature dependences of resistivity for the encapsulated and unencapsulated samples post-annealed for 80 h .

The relationship between the lattice constants, i.e. a and c - axes, and the period of post-annealing is presented in fig. 6. The c-axis increased with increasing post-annealing time. While the a - axis did not significantly change. This behavior that both the c - axis and T_c increased at the same time is contrary to the results previously reported by Presland et al.[5]. According to their observations introduction of oxygen by a low temperature post-annealing at 400 ℃ resulted in a raise of T_c and a simultaneous decrease in c - axis.

A similar correlation between the c - axis and the oxygen content of the sample was also observed for the "2201" phase [6]. However, the T_c of the "2201" phase increases when oxygen content decreases,

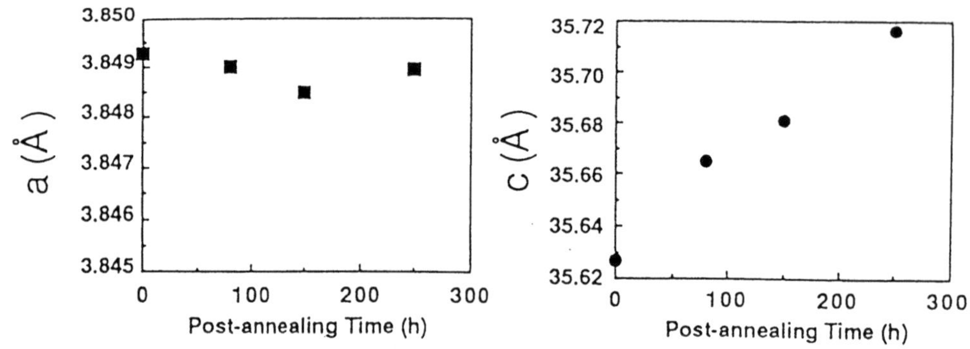

Fig. 6 Relationships between post-annealing time and the lattice constants.

because the "2201" phase sintered in ambient oxygen pressure is in an over-doped state. Obviously our as-sintered "2223" sample was in an under-doped state, but the increase in the carrier concentration and thus the improvement of the T_c were not due to an increase in oxygen content. According to our proposition the change in the carrier concentration was caused by a loss of Tl during the post-annealing. Actually, some white crystals were grown on the inside wall of the quartz tube after the post-annealing indicating that the evaporation of Tl and the recrystallization of some Tl compound most probably occurred.

SUMMARY

A single phase "2223" superconductor was obtained with a nominal composition of $Tl_{1.7}Ba_2Ca_{2.3}Cu_3O_z$ when the sample sintered in oxygen. The as-sintered sample had a T_c of about 120 K ($T_c^{R=0}$ = 118 K and T_c^{mag} = 122 K). When the sample was encapsulated in an evacuated quartz tube and post-annealed at 750 ℃ the T_c increased with increasing the post-annealing time and exceed 125 K. The highest recorded values for $T_c^{R=0}$ and T_c^{mag} were 127 K and 130 K, respectively. The benefit of the sample encapsulation technique was apparent when long post-annealing times, i.e. longer than 80 h were used. When the sample was post-annealed, the lattice constant of c - axis enlarged, but the lattice constant of a - axis scarcely changed by post-annealing.

ACKNOWLEDGMENT

Part of this work was supported by New Energy and Industrial Technology Development Organization in the Program for R&D of Basic Technology for Future Industries.

REFERENCES

1. Sheng Z. Z and Hermann A. M, (1988) Nature 332: 138
2. Parkin S. S. P, Lee V. Y, Engler E. M, Nazzal A. I, Huang T. C, Gorman G, Savoy R and Beyers R, (1988) Phys. Rev. Lett. 60: 2539
3. Adachi S, Mizuno K, Setsune K and Wasa K, (1990) Physica C 171: 543
4. Kikuchi M, Kajitani T, Suzuki T, Nakajima S, Hiraga K, Kobayashi N, Iwasaki H and Syono Y, (1989) Jpn. J. Appl. Phys. 28: L382
5. Presland M. R, Tallon J. L, Buckey R. G, Liu R. S and Flower N. E, (1991) Physica C 176: 95
6. Shimakawa Y, Kubo Y, Manako T and Igarashi H, (1989) Phys. Rev. B: 11400

T′ – and Impurity Phases Appeared in $Nd_{2-x}Ce_xCuO_4$ and $Eu_{2-x}Ce_xCuO_4$ Systems

HIDEAKI ITOH and MIKIO KUSUHASHI

Muroran Research Laboratory, The Japan Steel Works, Ltd., Muroran, Hokkaido, 051 Japan

ABSTRACT

T'-phase and the impurity phases in $Nd_{2-x}Ce_xCuO_4$ and $Eu_{2-x}Ce_xCuO_4$ ($0.1 \leq X \leq 0.2$) prepared by solid-state reaction were examined by X-ray diffraction and EPMA. The impurity phases are believed to be cubic $(Ln,Ce)_2O_{3+\delta}$ and $(Ln,Ce)CuO_2$ (Ln=Nd,Eu). Although amount of impurity phases and c-axis lattice constant of T'-phase decrease and Ce can dissolve in T'-phase with increasing sintering temperature in both systems, solubility limit exists near X=0.15 in $Eu_{2-x}Ce_xCuO_4$ system at 1100°C. In $Eu_{2-x}Ce_xCuO_4$ system, we could not obtain completely single T'-phase after heat treatment under flowing argon gas and the highest Tc was observed just beyond the solubility limit.

KEY WORDS: $Nd_{2-x}Ce_xCuO_4$, $Eu_{2-x}Ce_xCuO_4$, T'-phase, impurity phase, solubility limit

INTRODUCTION

After discovery of the n-type superconductor $Nd_{2-x}Ce_xCuO_4$ by Tokura et al. [1], many superconducting T'-phase materials, such as $(La,Eu)_{2-x}Ce_xCuO_4$ [2] and $Eu_{2-x}Ce_xCuO_4$ [3], were found and lots of studies have been carried out. It is generally known that the single T'-phase is hard to be obtained. In our previous work [4] on the Mössbauer effect, $(La,Eu)_{2-x}Ce_xCuO_4$ was not obtained as the single T'-phase. Is it possible to get single T'-phase by solid state reaction? We aim to get single T'-phase especially in $Eu_{2-x}Ce_xCuO_4$ because Eu is one of the elements suitable for Mössbauer measurement. This paper reports the effect of sintering temperature on the impurity phases and on T'-phase in $Nd_{2-x}Ce_xCuO_4$ and $Eu_{2-x}Ce_xCuO_4$ systems ($0.10 \leq X \leq 0.20$). Cava [5] mentioned for $Nd_{2-x}Ce_xCuO_4$ system that single T'-phase was obtained by the long time sintering and the subsequent heat treatment in argon gas made no change on T'-phase. We applied the same procedure as Cava did for $Eu_{2-x}Ce_xCuO_4$ system to get single T'-phase.

EXPERIMENT

1.Effect of the Sintering Temperature

The samples were prepared by solid state reaction from Nd_2O_3, Eu_2O_3, CeO_2 and CuO with 99.9% purity, which mixtures were calcined in air at 900°C for 20 hours, ground, calcined again with the same condition, and then ground again, pressed into pellets and sintered in air at 950-1150°C for 16 hours. After sintering, the samples were rapidly cooled to room temperature, and then subjected to X-ray powder diffraction analysis(XRD) using Cu target and EPMA. The samples were then heat treated under flowing argon gas or under vacuum at 885°C for 12 hours. After heat treatment, the samples were cooled to room temperature at a cooling rate of 100°C/h in the same atmosphere, and subjected to XRD, EPMA and measurements of the magnetization using conventional AC inductance method.

2.Effect of the Long Time Sintering for $Eu_{2-x}Ce_xCuO_4$ System

The same starting mixtures were calcined twice at 1000°C for 20 hours. The long time sintering was made at 1100°C for 144 hours and the subsequent heat treatment in flowing argon gas was made at 885°C for 20 hours. The samples prepared by this procedure were subjected to measurements of resistivity in addition to the analyses and measurements as mentioned above.

RESULTS AND DISCUSSIONS

1.Effect of Sintering Temperature

Figure 1 shows X-ray powder diffraction patterns of the samples sintered at 1000°C and 1100°C. These patterns consist of the major peaks of T'-phase and other minor peaks. The most intensive peak of the minor peaks appears at approximately 28.2 degree for $Nd_{2-X}Ce_XCuO_4$ and 28.4 degree for $Eu_{2-X}Ce_XCuO_4$. From JCPDS cards, most intensive peak of cubic Nd_2O_3, Eu_2O_3, CeO_2 and $Nd_2Ce_2O_7$ appears at 27.8, 28.4, 28.5 and 28.1 degree respectively. So, from these values and the patterns of the impurity peaks marked with open circle in Fig.1, the impurity phases are considered to be cubic $(Ln,Ce)_2O_{3+\delta}$ (Ln=Nd and Eu). At lower sintering temperatures, other impurity peaks (35.5 and 38.7 degree) are observed. These peaks marked with open square in Fig.1 coincide with CuO. Impurity peaks are reduced with increasing sintering temperature. The samples partially melted during sintering above 1075°C for $Nd_{2-X}Ce_XCuO_4$ and 1100°C for $Eu_{2-X}Ce_XCuO_4$. But some of impurity peaks still remain even at high sintering temperature such as partial melting occurred.

Figure 2 shows the peak intensity ratio of (222) peak of cubic $(Ln,Ce)_2O_{3+\delta}$ phase to (101) peak of T'-phase obtained by sintering at 1000°C and 1100°C. When the samples were sintered at 1000°C, the ratio increases with increasing Ce content(X). But at 1100°C, the ratio remains below 0.1 with Ce content between 0.10 through 0.20 for $Nd_{2-X}Ce_XCuO_4$, and for $Eu_{2-X}Ce_XCuO_4$, it abruptly begins to increase at X=0.15. It seems that Ce can easily dissolve in Nd_2CuO_4, but its solubility limit exists near X=0.15 for $Eu_{2-X}Ce_XCuO_4$.

From EPMA, CuO and $(Ln,Ce)_2O_{3+\delta}$ particles are observed at lower sintering temperature. But at higher temperature, CuO disappears and $(Ln,Ce)CuO_2$ particles appear instead. $(Ln,Ce)_2O_{3+\delta}$ is not observed in XRD because the peaks of this phase are hidden by the peaks of T'-phase.

Jorgensen et al.[6] reported another phase separation in $Nd_{2-X}Ce_XCuO_4$ system based on neutron powder diffraction data. They found shoulders on some of the Bragg peaks of T'-phase and concluded that these two phases have essentially the same structure and are so similar in lattice constants. In this work on X-ray powder diffraction, we can not obtain the suggestion for the phase separation as reported by Jorgensen et al.

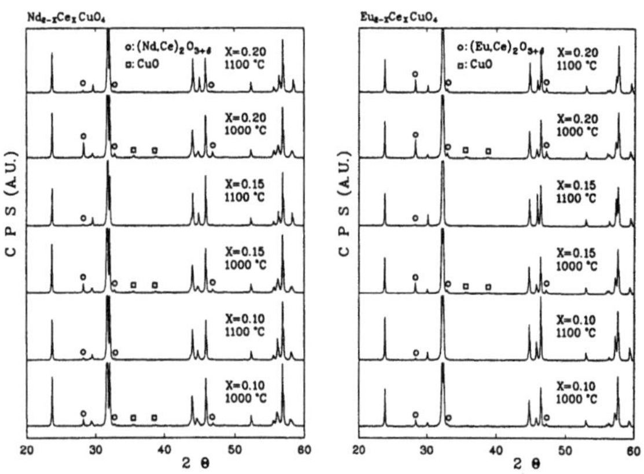

Fig.1 X-ray diffraction patterns of $Nd_{2-X}Ce_XCuO_4$ and $Eu_{2-X}Ce_XCuO_4$ sintered at 1000°C and 1100°C.

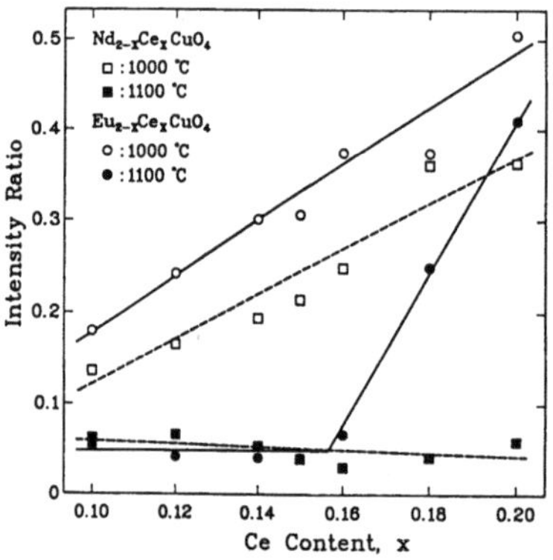

Fig.2 XRD intensity ratio of (222) peak of cubic impurity phase to (101) peak of T'-phase.

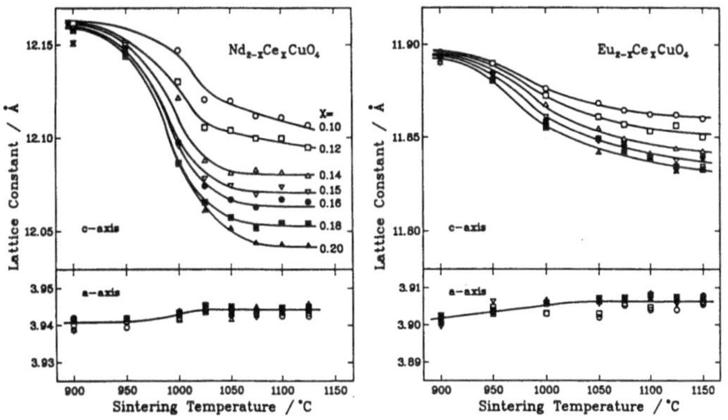

Fig.3 Temperature dependence of the lattice constant of $Nd_{2-x}Ce_xCuO_4$ and $Eu_{2-x}Ce_xCuO_4$.

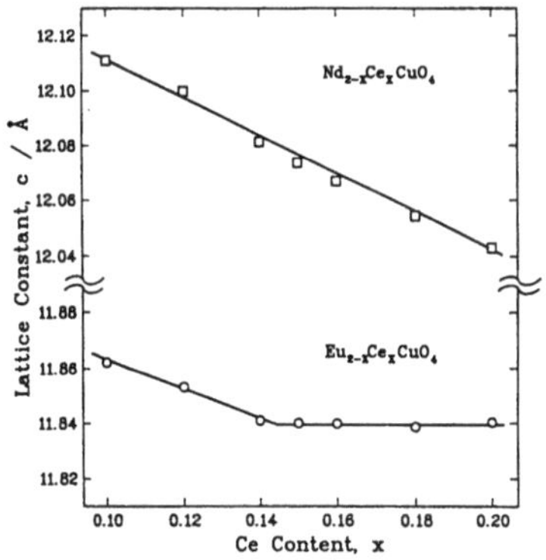

Fig.4 Lattice constant change of T'-phase sintered at 1100°C.

Figure 3 shows the temperature dependence of the lattice constants of $Nd_{2-x}Ce_xCuO_4$ and $Eu_{2-x}Ce_xCuO_4$. The constant of a-axis slightly increases with increasing sintering temperature in both systems. The constant of c-axis decreases with increasing sintering temperature and a drastic decrease is seen in $Nd_{2-x}Ce_xCuO_4$ system at sintering temperature range from 950 to 1050°C. But the change of the lattice constant is smaller for $Eu_{2-x}Ce_xCuO_4$ system due to the smaller ionic radius of Eu^{3+}. Above 1050°C, the constant of c-axis is only slightly changed with increasing sintering temperature in both systems. From these temperature dependence, it is confirmed that higher sintering temperature than 1050°C are needed for dissolution of Ce into T'-phase.

It has been understood for $Nd_{2-x}Ce_xCuO_4$ system that the c-axis is shortened with increasing Ce content [7,8]. Figure 4 shows such dependence of lattice constant on Ce content sintered at 1100°C. For $Nd_{2-x}Ce_xCuO_4$, the c-axis shortens lineally with Ce content, but for $Eu_{2-x}Ce_xCuO_4$, it saturates at X=0.15. This phenomenon indicates again that solubility limit exists near X=0.15.

The heat treatment after sintering makes no change to the lattice constants of T'-phase, but makes transformation CuO existed at lower sintering temperature into Cu_2O, which is detected by both X-ray diffraction and EPMA. The X-ray diffraction patterns before and after heat treatment look almost unchanged. For $Nd_{2-x}Ce_xCuO_4$ system, the samples sintered below 1000°C show no superconducting transition. Meissner magnetization is observed with Ce content range of 0.10-0.20 in this work when the samples were sintered at 1025°C and at the range of Ce content of $0.14 \leq X \leq 0.18$ in case of sintering above 1075°C. Figure 5 shows the relative intensity of the Meissner signal to ideal Meissner effect at 15K for $Nd_{2-x}Ce_xCuO_4$. Meissner signal is most intensive at X=0.15. At higher Ce content, Meissner volume shows the peak which shifts to lower sintering temperature with increasing Ce content. It reveals that suitable carrier doping for superconductivity is completed at X=0.15 ordinarily, but in case of higher Ce content, it will be performed by lower temperature sintering because Ce does not fully dissolve. For $Eu_{2-x}Ce_xCuO_4$ system, superconductivity appears only at $0.15 \leq X \leq 0.20$ and then sintered above 1050°C. Meissner volume becomes maximum at 1100°C for any Ce content range.

2.Effect of the Long Time Sintering for $Eu_{2-x}Ce_xCuO_4$ System

From X-ray powder diffraction patterns, no impurity peak is observed after long time sintering at X=0.15 or below. $(Ln,Ce)_2O_{3+\delta}$ peaks appear after subsequent heat treatment under flowing argon gas. From EPMA, same impurity phases as appeared in 16hr sintered materials still remain after long time sintering even in samples with X=0.15 or below. Figure 6 shows the change of the lattice constant with Ce content. Long time sintering decreases the c-

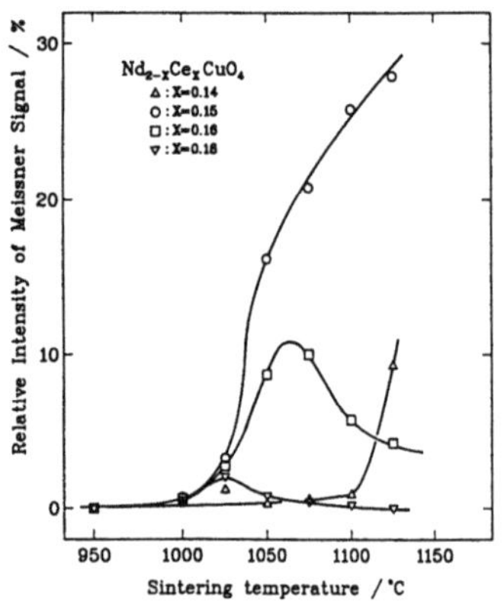

Fig.5 Relative intensity of the Meissner signal to ideal Meissner effect at 15K for $Nd_{2-x}Ce_xCuO_4$.

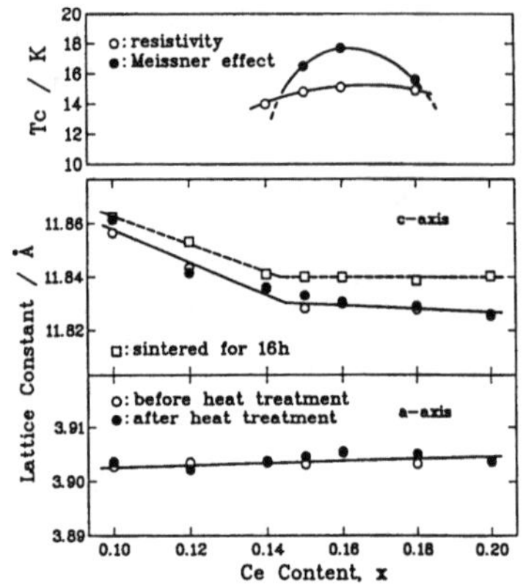

Fig.6 Change of the lattice constants and Tc of T'-phase in $Eu_{2-x}Ce_xCuO_4$ sintered at 1100°C for 114 hours.

axis lattice constant. It seems that much more amount of Ce can dissolve into T'-phase during long time sintering, but solubility limit exists at the same Ce content as for 16hr sintered materials. After heat treatment, no change of lattice constant is observed in spite of the partial decomposition of T'-phase.

Tc determined by resistivity and Meissner effect measurement is shown in Fig.6 together. The difference of Tc between two methods is large, but both curves show same behavior. The Tc becomes the maximum at x=0.16, which Ce content is just beyond the solubility limit. We consider the mechanism of this phenomenon as follows. At solubility limit near x=0.15, carrier doping for superconductivity is incomplete. In the range of Ce content more than the solubility limit, Ce concentration in T'-phase will slightly change according to Ce content, X. This concentration change is so small that lattice constant of T'-phase scarcely change, but superconductivity is affected.

The effect of sintering temperature on impurity phases and T'-phase in $Nd_{2-x}Ce_xCuO_4$ and $Eu_{2-x}Ce_xCuO_4$ ($0.10 \leq x \leq 0.20$) has been studied. 1) The impurity phases are believed to be cubic $(Ln,Ce)_2O_{3+\delta}$ and $(Ln,Ce)CuO_2$ (Ln=Nd,Eu). 2) Although amount of impurity phases decreases with increasing sintering temperature, single T'-phase is not obtained in both systems when the samples are sintered too for 16 hours. 3) For $Eu_{2-x}Ce_xCuO_4$ system, single T'-phase is not obtained too, in case of sintering at 1100°C for 144 hours. 4) For $Eu_{2-x}Ce_xCuO_4$ system, solubility limit of Ce is found near X=0.15 at 1100°C. 5) For $Eu_{2-x}Ce_xCuO_4$ system, the Tc becomes the maximum at x=0.16, which just beyond the solubility limit.

REFERENCES

1. Tokura Y, Takagi H, Uchida S (1989) Nature 337: 345
2. Yoshimoto M, Hashimoto T, Takaya T, Koinuma H (1989) Jpn J Appl Phys 28: L1115
3. Markert JT, Early EA, Bjornhorm T, Ghamaty S, Lee BW, Neumeier JJ , rice RD Seaman CL, Maple MB (1989) Physica C158: 178
4. Itoh H, Kusuhashi M, Taniwaki M (1990) Jpn J Appl Phys 29: L1604
5. Cava RJ (1991) Proc M²S-HTSC III, Kanazawa, Japan. Physica C (to be published)
6. Jorgensen JD, Lightfoot P, Pei S, Dabrowski B, Richards DR,Hinks DG (1991) Proc 3rd Int Symp Superconductivity(ISS'90),Sendai,1990. Springer-Verlag. Tokyo, pp 337-342
7. Uji S, Aoki H, Matsumoto T (1989) Jpn J Appl Phys 28: L563
8. Huang TC, Moran E, Nazzal AI, Torrance JB (1989) Physica C158: 148

Preparation and Properties of $(Pb,Cu)(Sr,La)_2(Y,Ca)Cu_2O_z$

KIYOSHI YAMAMOTO, TOSHIHIKO MAEDA*, and H. YAMAUCHI

Superconductivity Research Laboratory, ISTEC, Koto-ku, Tokyo, 135 Japan

ABSTRACT

Lead based copper oxide superconductors which are nominally represented by $(Pb_{(1+x)/2}Cu_{(1-x)/2})(Sr_{2-y}La_y)(Y_{1-(x+y)}Ca_{x+y})Cu_2O_z$, (Pb/Cu)-1212 are synthesized and characterized. Nearly single-phase samples were obtained in the range of x+y<0.5, i.e., Ca/(Ca+Y)<0.5. As-sintered samples without post-anealing and quenching procedure exhibited superconductivity in the range of 2x+y>0.6, and T_c^{on} of multi-phase sample with x=0.5, y=0.2 was 65K. T_c decreased as La content,y, increased in Ca/(Ca+Y)=constant (x+y=constant). The increase in amount of positive charge in the (Pb/Cu) site and/or the decrease in amount of positive charge in the Sr site were thought to enhance T_c.

KEY WORDS: Pb-based copper oxide, 1212 structure, substitution effect

INTRODUCTION

Since the superconducting lead cuprates, $Pb_2Sr_2(Y,Ca)Cu_3O_8$, were discovered by Cava et al. [1], a variety of new lead cuprates have been synthesized [2-7]. In such lead cuprates, $(Pb,Cu)Sr_2(Y,Ca)Cu_2O_z$ phase, (Pb/Cu)-1212, were independently discovered by Cava et al. [1], Subramanian et al. [2] and Lee et al. [3]. The (Pb/Cu)-1212 phase is thought to be a candidate for new superconductors which have T_c's higher than 100K, since the crystallographic structure of this phase is similar to those of $YBa_2Cu_3O_z$ and $TlBa_2CaCu_2O_7$.

Maeda et al. [7] reported that the chemical composition of the (Pb/Cu)-1212 phase was given by the formula, $(Pb_{(1+x)/2}Cu_{(1-x)/2})Sr_2(Y_{1-x}Ca_x)Cu_2O_z$ with x in the range of 0 - 0.35 and that Tc of this phase tended to increase as the Ca content, x, increased. The decrease in the amount of positive charge due to substitution of Ca^{2+} for Y^{3+} was completely offset by the increase in the Pb^{4+}/Cu^{2+} ratio in the (Pb,Cu)-O layers. Furthermore, Koriyama et al. [8] substituted Ba for Sr in the Pb-based layered copper oxides, $(Pb_{0.5}Cu_{0.5})Sr_2(Y_{1-x}Ca_x)Cu_2O_z$ (x=0 and 0.25). This Ba-substituted (Pb/Cu)-1212 compounds exhibited superconductivity.

In the present work, (Pb/Cu)-1212 lead cuprates which were nominally represented by $(Pb_{(1+x)/2}Cu_{(1-x)/2})(Sr_{2-y}La_y)(Y_{1-(x+y)}Ca_{x+y})Cu_2O_z$ were synthesized and characterized. When the Ca/Y ratio was constant for example, x+y=0.5, The effect of simultaneous partial substitution of La and Pb respectively for Sr and Cu were also examined.

EXPERIMENTAL

The samples were prepared by a solid-state reaction method using highly pure (>99.9%) powders of PbO, $SrCO_3$, La_2O_3, Y_2O_3, $CaCO_3$ and CuO. These powders were mixed into nominal compositions of $(Pb_{(1+x)/2}Cu_{(1-x)/2})(Sr_{2-y}La_y)(Y_{1-(x+y)}Ca_{x+y})Cu_2O_z$. The mixed powders were calcined at 850°C for 10 h in air, and then pressed into parallelepiped bars of 3x3x20mm^3. The bars were fired at 1040-1060°C for 1 h in a flowing O_2 gas and then slowly cooled

* Permanent address: Yokohama R&D Laboratories, The Furukawa Electric Co.,Ltd., 4-3,Okano 2-chome, Nishi-ku, Yokohama 220, JAPAN

down to room temperature at a rate of $1^{o}C$/min. In this study, post-annealing and quenching procedure [7] were not performed. Samples were characterized by X-ray powder diffractmetry (XRD) using CuKa radiation with a graphite monochromater. The temperature dependence of electrical resistivity was measured by a conventional four-probe method in the temperature range from room temperature down to 4.2K. The measurement of DC magnetic susceptibility with respect to temperature was made using a SQUID magnetomater.

RESULTS AND DISCUSSION

$(Pb_{0.5}Cu_{0.5})(Sr_{2-y}La_y)(Y_{1-y}Ca_y)Cu_2O_z$: x=0

Firstly, we investigated the effect of La substitution for Sr, when the Pb/Cu ratio was constant with x=0. For these samples, the increase in the amount of positive charge due to La substitution was offset by the decrease due to Ca substitution for Y.

Figure 1 shows the powder XRD patterns. The samples with y<0.6 were of single-phase, whose diffraction lines were indexed for a tetragonal unit cell with **a**~0.38nm and **c**~1.18nm. On the other hand, the samples with y>0.6 contained impurities which were identified to be $SrCuO_2$ and $Sr_{5-x}Pb_{3+x}Cu_yO_z$ [9]. Figure 2 shows the La-content, y, dependence of the lattice constants, **a** and **c**/3. Although impurities were contained in the samples with y>0.6, the value of **a** monotonically increased as y increased. The value of **c** decreased as y increased up to 0.5 and then started to increase when y exceeded 0.6. This suggested that the present (Pb/Cu)-1212 compounds with y<0.5 were single phase and had nearly the same compositions as the nominal one.

Figure 3 shows the temperature dependence of electrical resistivity for $(Pb_{0.5}Cu_{0.5})(Sr_{2-y}La_y)(Y_{1-y}Ca_y)Cu_2O_z$. The normal-state electrical resistivity decreased with increasing La content, y. The samples with y=0.6 and 0.8 showed superconductivity with T_c^{on}'s of 11K and 41K, respectively, although the samples contained impurities. The compositions of (Pb/Cu)-1212 phase, which exhibited superF-conductivity , were thought to be diffrent from the nominal compositions. The magnitudes of ionic radius were Sr>La>Ca>Y. The partial subtitution of La and Ca respectively for Y and Sr were thought to be made. The electrical resistivity of samples with y>0.6 increased with increasing La content, y, probably due to the existance of impurities. The temperature dependence of electrical resistivity for the sample with y=1.0 was semiconductive and the electrical resitivity at room temperature was too high to be be plotted in the scale of Fig.3.

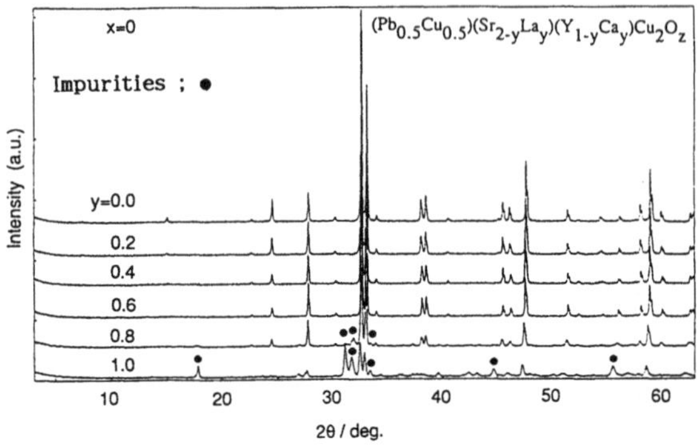

Fig.1 The powder XRD patterns of $(Pb_{0.5}Cu_{0.5})(Sr_{2-y}La_y)(Y_{1-y}Ca_y)Cu_2O_z$.

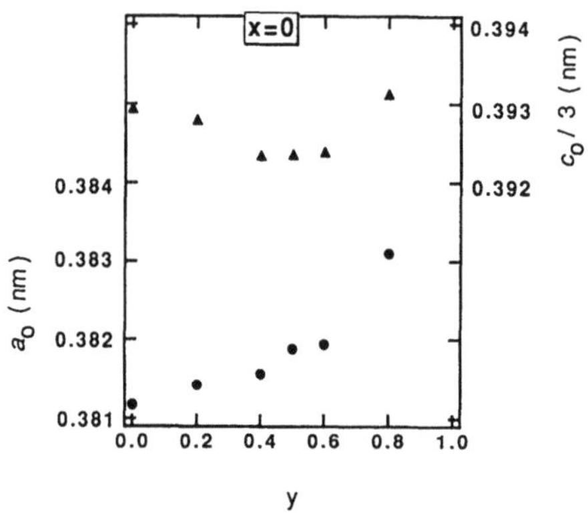

Fig.2 La-content, y, dependence of the lattice constants, **a** and **c**/3.

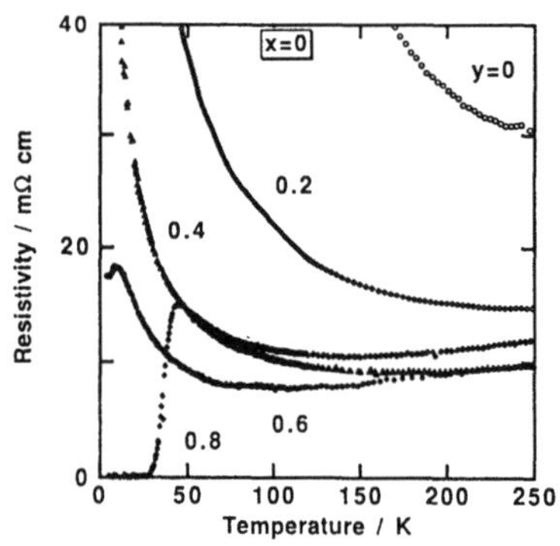

Fig.3 The temperature dependence of electrical resistivity.

$(Pb_{(1+x)/2}Cu_{(1-x)/2})(Sr_{2-y}La_y)(Y_{0.5}Ca_{0.5})Cu_2O_z$: x+y=0.5

Secondarily, the effects of La substitution for Sr when total amount of Y and Ca, x+y, was fixed at 0.5. For this case, the increase in the amount of positive charge was offset by the decrease in the Pb/Cu ratio.

Figure 4 shows powder XRD patterns for $(Pb_{(1+x)/2}Cu_{(1-x)/2})(Sr_{2-y}La_y)(Y_{0.5}Ca_{0.5})Cu_2O_z$. The XRD analysis indicated that samples with x+y=0.5 were single phase except for a nearly single-phase sample with x=0.5. The solubility of Ca in this (Pb/Cu)-1212 was thought to be be expanded by La substitution for Sr, as the solubility of Ca was expanded by Ba substitution for Sr [8]. If samples with x=0.5 are synthesized in the most suitable condition, those samples are thought to be single phase. Figure 5 shows the lattice constants, **a** and **c**/3, with respect to the La content, y. The value of **a** was independent of y, but the value of **c** monotonically decresed with increasing La content,y. This is similar to the case [7] of the Ca substitution for Y sites in (Pb/Cu)-1212 phase. This suggests that the value of **c** decreased by the decrease in the average ionic radius between Sr and La in the (Sr/La) site.

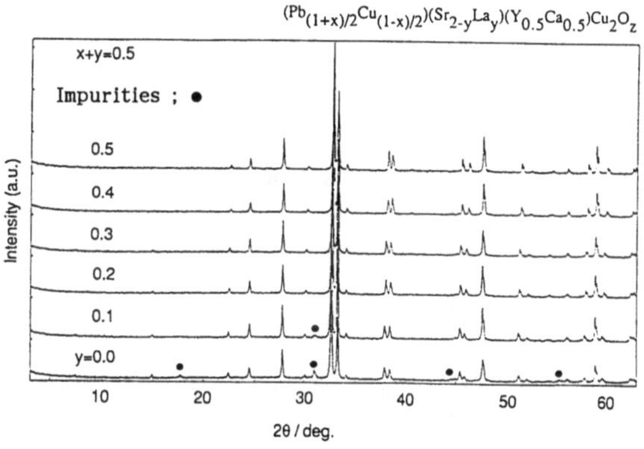

Fig.4 The powder XRD patterns of $(Pb_{(1+x)/2}Cu_{(1-x)/2})(Sr_{2-y}La_y)(Y_{0.5}Ca_{0.5})Cu_2O_z$.

Figure 6 shows the temperature dependence of electrical resistivity for $(Pb_{(1+x)/2}Cu_{(1-x)/2})(Sr_{2-y}La_y)(Y_{0.5}Ca_{0.5})Cu_2O_z$. T_c decreased as the La content, y, increased. This suggested that the increase in the amount of positive charge at the (Pb/Cu) site and/or the decrease at the Sr site enhanced T_c.

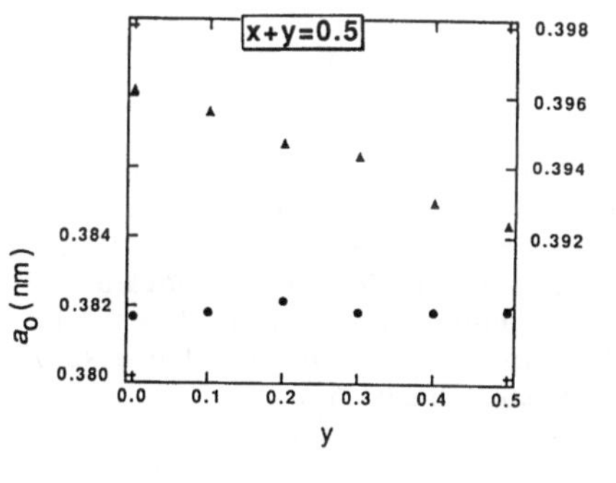

Fig.5 La-content, y, dependence of the lattice constants, **a** and **c**/3.

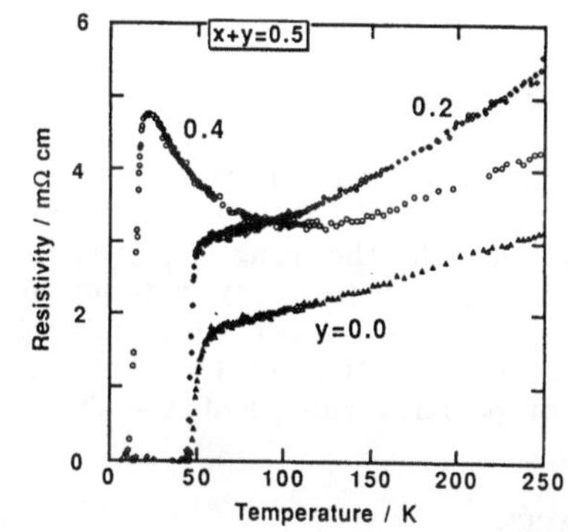

Fig.6 The temperature dependence of electrical resistivity.

$(Pb_{(1+x)/2}Cu_{(1-x)/2})(Sr_{2-y}La_y)(Y_{1-(x+y)}Ca_{x+y})Cu_2O_z$

Figure 7 shows the relationship between superconducting property and the nominal compositions. Single-phase samples were obtained in the range of x+y≤0.5, i.e., Ca/(Ca+Y)≤0.5. The samples cotaining impurities exhibited superconductivity. The samples with x=0.5 and y=0.2

exhibited superconductivity with T_c^{on} of 65K and T_c^{zero} of 61K as shown in Fig.8(a). Figure 8(b) shows the temperature dependrnce of the magnetic susceptibility. The onset temperature of the diamagnetic signal was 62K.

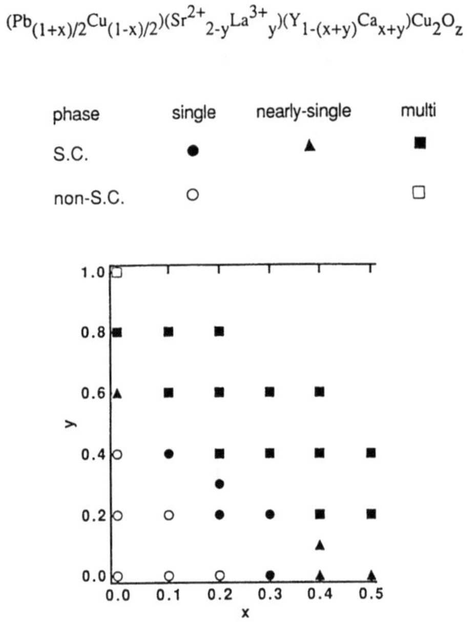

Fig.7 The relationship between superconducting property and the nominal compositions (S.C. : superconducting).

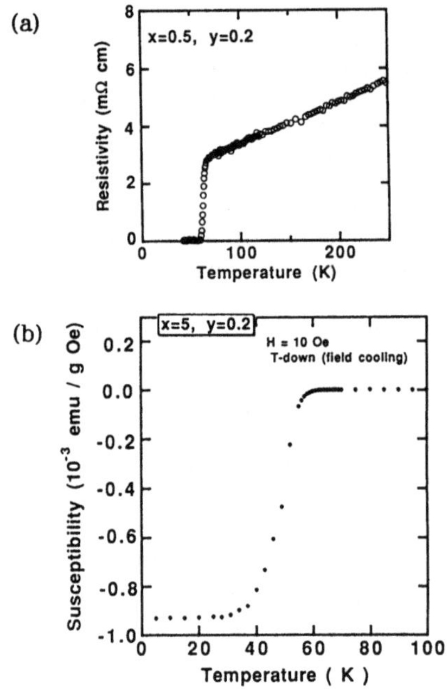

Fig.8 The temperature dependence of electrical resistivity (a) and magnetic susceptibility (b) (x=0.5, y=0.2).

CONCLUSION

The (Pb/Cu)-1212 lead cuprates which were nominally represented by $(Pb_{(1+x)/2}Cu_{(1-x)/2})(Sr_{2-y}La_y)(Y_{1-(x+y)}Ca_{x+y})Cu_2O_z$ were synthesized and characterized. Nearly single-phase samples were obtained in the range of $x+y \leq 0.5$. Nearly single-phase samples of (Pb/Cu)-1212, which exhibited superconductivity without post-annealing and quenching, were obtained. The multi-phase samples with x=0.5 and y=0.2 exhibit superconductivity with T_c^{on} at 65K. The increase in the amount of positive charge at the (Pb/Cu) site and/or the decrease in the amount of positive charge at the Sr site were thought to enhance T_c.

REFERENCES

1. Cava RJ, Batlogg B, Krajewski JJ, Rupp LW, Schneemeyer LF, Siegrist T, van Dover RB,Marsh P, Peck WF, Jr, Gallagher PK, Glarum SH, Marshall JH, Ferrow RC, Waszczak JV, Hull R, Trevor P (1988) Nature **336**: 211
2. Subramanian MA, Goparakrishnan J, Torardi CC, Gai PL, Boyes ED, Askew TR, Flippen RB,Farneth WE, Sleight AW (1989) Phsica C **159**: 124
3. Lee JY, Swinna JS, Steinfink H (1989) J. Mater. Res. **4**: 763
4. Inoue O, Adachi S, Takahashi Y, Hirano H, Kawashima S (1989) Jpn. J. Appl. Phys. **28**: L60
5. Rouillon T, Retoux R, Groult D, Michel C, Harvieu M, Provost, Raveau B (1989) J. Solid State Chem. **78**:322
6. Tokiwa A, Oku T, Nagoshi M, Kikuchi M,Hiraga K, Syono Y (1989)Physica C **161**:459
7. Maeda T, Sakuyama S, Yamauchi H, Tanaka S (1991) Phys. Rev. B **39**: 7866
8. Koriyama S, Sakuyama K, Maeda T, Yamauchi H, Tanaka S (1990) Physica C **166**: 413
9. Kim JS, Tang XX, Manthiram A, Swinnea JS, Steinfink H (1990) J. Solid State Chem. **85**: 44

Synthesis of Superconducting $(La,Sr)_2(Ca,Y)Cu_2O_6$ Ceramics

SEIJI ADACHI, TAKESHI SAKURAI, YUJI YAEGASHI, H. YAMAUCHI, and SHOJI TANAKA

Superconductivity Research Laboratory, ISTEC, Koto-ku, Tokyo, 135 Japan

ABSTRACT

The co-substitution effect of Sr for La and Ca for Y in $La_{1.8}Sr_{0.2}CaCu_2O_6$ superconductor was investigated. Ceramic samples with nominal compositions of $La_{1.8-x}Sr_{0.2+x}Ca_{1-x}Y_xCu_2O_6$ (x=0.0 to 0.5) were prepared by using an O_2-HIP (Hot Isostatic Pressing) technique. Single-phase samples were obtained over the whole range of x. The value of T_c gradually decreased with increasing x.

KEY WORDS: oxide superconductor, High-T_c superconductor, "326" structure, substitution effect, O_2-HIP process, The Rietveld analysis

INTRODUCTION

Since the discovery of superconducting La-Ba-Cu-O, many variations of cuprate superconductors have been found. Several classifications of cuprate superconductors were proposed in terms of their crystal structures.[1-4] It is well known that all the p-type superconducting cuprates are characterized by two-dimensional Cu-O_5 pyramidal sheets in their crystal structures. Focusing on the superconducting cuprates containing two Cu-O_5 pyramid blocks with an intermediate cationic layer, its crystal structure may be devided into following four parts:

1) charge reservoir block
2) site next to apical oxygen
3) Cu-O_5 pyramid
4) intermediate cationic layer

We can control their superconducting properties by modifying some of the four parts.

The $La_{1-x}Sr_xCaCu_2O_6$ system has the simplest crystallographic structure among those which have two Cu-O_5 pyramids blocks with an intermediate cationic layer. This structure (denoted by "326" structure) is composed of parts of 2), 3) and 4) as shown in Fig.1. It has no charge reservoir blocks. Positive charges essential for the occurance of superconductivity are supplied from the parts 2) and 4) into the pyramidal sheets. Modifications in both parts

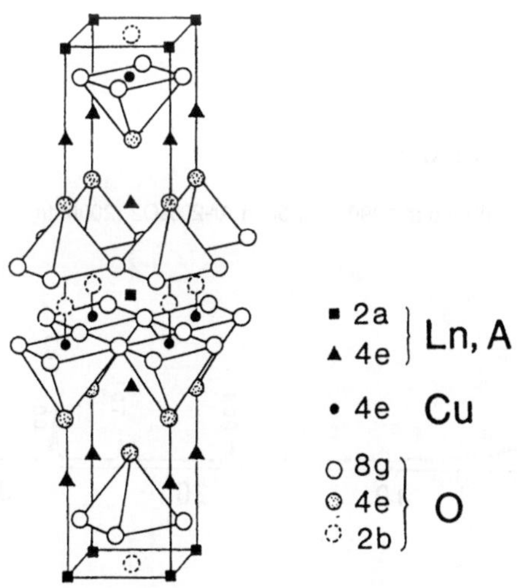

Fig.1. Schematic representation of "326" structure.

must result in a change in the charge doping feature. It is interesting to investigate the effect of site-selective substitution on superconducting properties in this simple "326" structure. In this paper, we introduce Y atoms into La-Sr-Ca-Cu-O system, which likely to occupy the 2a site preferentially, and investigate the co-doping characteristics in $(La,Sr)_2(Ca,Y)Cu_2O_6$ superconductors.

Superconductivity of the "326" cuprate has been obtained by Cava et al.[5] They reported that $La_{1.6}Sr_{0.4}CaCu_2O_6$ exhibited superconductivity around 60K. Kinoshita et al.[6] and Sakurai et al.[7-9] demonstrated the synthesis of "326" superconductors with high reproducibility by using an O_2-HIP (Hot Isostatic Pressing) technique which is commercially available. In this work, we used the HIP technique for making "326" materials superconductive or "superconductor-izing" them.

EXPERIMENTAL

Ceramic samples were prepared by solid state reaction. La_2O_3, $SrCO_3$, $CaCO_3$, Y_2O_3 and CuO powders were used as starting materials. They were mixed in nominal compositions of $La_{1.8-x}Sr_{0.2+x}Ca_{1-x}Y_xCu_2O_6$ (x=0.0 to 0.5). The x=0.0 material was previously well-studied by Sakurai et al.[7-9] This chemical formula was adopted in order to keep the total valency of cations except Cu constant even when x was varied. The powder mixture was calcined at 1000°C for 10h in air, pulverized and pressed into rectangular bars. They were sintered at 1060°C for 5h. Sintered samples were annealed at 1090°C for 5h in Ar-20%O_2 of 2000 atm by using an HIP apparatus.

For the detection of superconductivity, the resistivity was measured by a standard four-probe method. The oxygen content was analyzed by coulometric titration. Powder X-ray diffraction (XRD: using CuKα radiation) was carried out for structural analyses. The Rietveld analyses of XRD data were performed by using a computer program, RIETAN.[10]

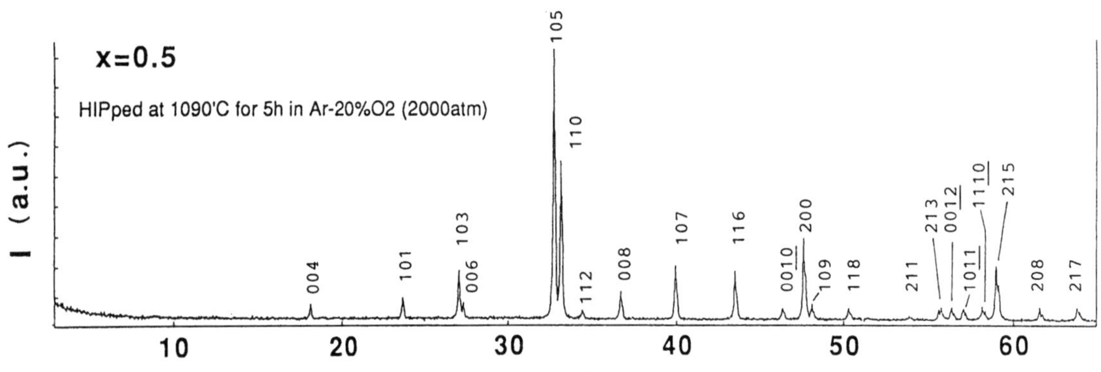

Fig.2. Powder XRD pattern for the x=0.5 sample.

RESULTS AND DISCUSSION

Single-phase ceramic samples having tetragonal "326" structure were obtained over all the range of x. Figure 2 shows the powder XRD pattern of the x=0.5 sample. The lattice constants are listed in Table I and plotted against x in Fig.3. The a-axis length decreased and the c-axis increased monotonically with increasing x. The cell volume slightly decreased with increasing x. The oxygen contents of the x=0.0 and 0.5 samples were analyzed to be 6.01 and 5.95, respectively. The shrinkage of unit cell corresponded to less oxygen content.

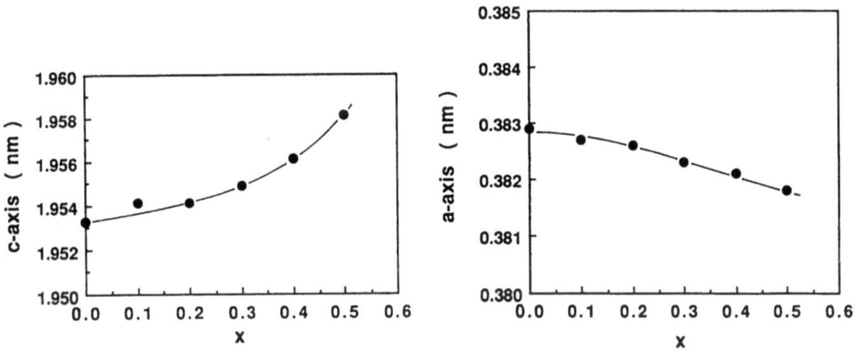

Fig.3. Lattice constants as a function of x value.

Table I. Lattice parameters, T_c values and oxygen content.

x	a [nm]	c [nm]	V [nm^3]	T_c [K] (onset)	T_c [K] (zero)	oxygen content
0.0	0.3829	1.9533	0.2864	60	36	6.01
0.1	0.3827	1.9541	0.2862	52	32	
0.2	0.3826	1.9541	0.2860	47	27	
0.3	0.3823	1.9549	0.2857	44	24	
0.4	0.3821	1.9561	0.2856	42	23	
0.5	0.3818	1.9581	0.2853	39	20	5.95

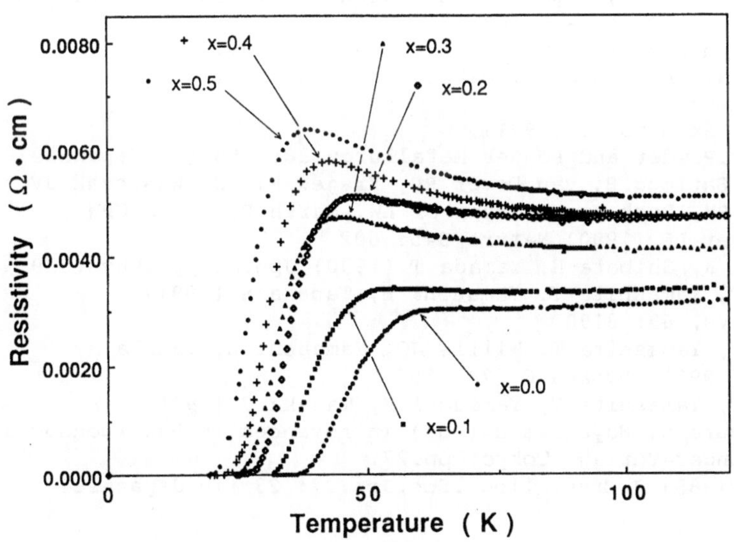

Fig.4. Temperature dependences of electrical resistivity.

Figure 4 shows the temperature dependence of electrical resistivity. T_c values are summarized in Table I. T_c decreased with increasing x. The observed depression in T_c seemed to be caused by following reasons. 1) The effective number of holes responsible for superconductivity decreased due to oxygen vacancies. 2) The shrinkage in unit cell was not preferable to high T_c. 3) The introduced Y atoms brought about local perturbations to deteriorate superconductivity.

It is likely that Sr and Y atoms are preferentially located at 4e and 2a sites, respectively. Assuming such cationic distribution, the Rietveld analysis of powder XRD data for the x=0.5 sample was performed. It was assumed that oxygen atoms at 4e sites were partially vacant. The space group of this "326" structure was tentatively assumed to be I4/mmm. The isotropic thermal parameters for cations and oxygen were fixed at $0.005 nm^2$ and $0.010 nm^2$, respectively. The obtained R-factors were :R_{wp}=8.13%, R_p=5.54%, R_e=2.27%, R_I=9.09% and R_F=5.81%. After several trial runs for a variety of possible distributions of cations, we found that the exchange of small amounts of La and Ca atoms made the R-factors lower. The lowest R-factors we obtained were :R_{wp}=7.18%, R_p=4.81%, R_e=2.27%, R_I=7.52% and R_F=4.52%. The final structure parameters are listed in Table II. The refined lattice parameters were a=0.381748(6)nm and c=1.95839(4)nm. It is difficult to precisely determine the distribution of four elements in this structure. However, the obtained results suggested that a cationic ordering as our initial speculation was not achieved.

Table II. Refined structural parameters of the x=0.5 samples.

Atom	Site	x	y	z	occupation
La/Sr/Ca/Y (1)	4e	0	0	0.1777(3)	0.60/0.35/0.05/0.00
La/Sr/Ca/Y (2)	2a	0	0	0	0.10/0.00/0.40/0.50
Cu	4e	0	0	0.4139(5)	1.00
O (1)	8g	0	0.5	0.0764(17)	1.00
O (2)	4e	0	0	0.3027(22)	0.975

REFERENCES

1. Tokura Y, Arima T (1990) Jpn.J.Appl.Phys. 29: 2388
2. Adachi S, Inoue O, Kawashima S, Adachi H, Ichikawa Y, Setsune K, Wasa K (1990) Physica C 168: 1
3. Wada T, Ichinose A, Yamauchi H, Tanaka S (1991) Nihon Seramikkusu Kyoukai Gakujutsu Ronbun-shi 99: 435 [in Japanese]
4. Kokubu M, Akimitsu J (1991) J.Jpn.Soc.Powder and Powder Metallurgy 38: 195 [in Japanese]
5. Cava RJ, Batlogg B, van Dover RB, Krajewski JJ, Waszczak JV, Flemming RM, Peck Jr WF, Rupp Jr LW, Marsh P, James CPWP, Schneemeyer LF (1990) Nature 345: 602
6. Kinoshita K, Shibata H, Yamada T (1990) Jpn.J.Appl.Phys. 29: L1632
7. Sakurai T, Yamashita T, Yamauchi H, Tanaka S (1991) J.Appl.Phys. 69: 3190
8. Sakurai T, Yamashita T, Willis JO, Yamauchi H, Tanaka S, Kwei GH (1991) Physaca C 174: 187
9. Sakurai T, Yamashita T, Yamauchi H, Tanaka S (1991) In: Kajimura K, Hayakawa H (eds) in Advances in Superconductivity III, Springer-Verlag, Tokyo, pp.271
10. Izumi F (1985) J.Crystallogr.Soc.Jpn.27: 23 [in Japanase]

This work was supported by NEDO in the program for R&D of Basic Technology for Future Industries.

The Critical Annealing Condition to Superconductor-ize $La_{1.8}Sr_{0.2}CaCu_2O_6$(La-326) Phase

TAKESHI SAKURAI, TORU YAMASHITA*, H. YAMAUCHI, and SHOJI TANAKA

Superconductivity Research Laboratory, ISTEC, Koto-ku, Tokyo, 135 Japan

ABSTRACT

The critical annealing conditions to superconductor-ize (obtaining superconductivity) La-326 compound was studied in the field where La-326 compound was stable. The phase stability was determined by the studies of the phase decomposition at high temperatures and phase forming under the high pressures. The superconducting La-326 phase was stable up to 800 ℃ under the ambient pressure and up to 1000 ℃ under P_{O2}=10*atm*. The many conditions, which were compositions of the mixture of Ar gas and oxygen gas and temperatures, were examined in order to optimize the condition. It was concluded that oxygen pressure, P_{O2} was essentially important to superconductor-ize La-326, regardless total pressure; P_T.

KEY WORDS: La-326 compound, partial oxygen pressure, phase stability, superconductor-ization

INTRODUCTION

The La-326 compound (Fig.1) was superconductor ized firstly at the composition of the $La_{1.6}Sr_{0.4}CaCu_2O_6$ by Cava, et. al [1]. It have been seemed to be some reasons why the La-326 compound did not show the superconductivity. One is the oxygen content, that must be controlled to 6 in order to dope the sufficient amount of hole carrier [2, 3]. The other is the vacancy site occupied by an oxygen atom. When the oxygen vacancy site, O(3), between the CuO_2 planes is occupied by an oxygen atom, it makes oxygen defect easily at the other site which is thought essentially important to obtaining superconductivity [4, 5, 6]. Therefore, it had been difficult to make the La-326 compound superconductor.

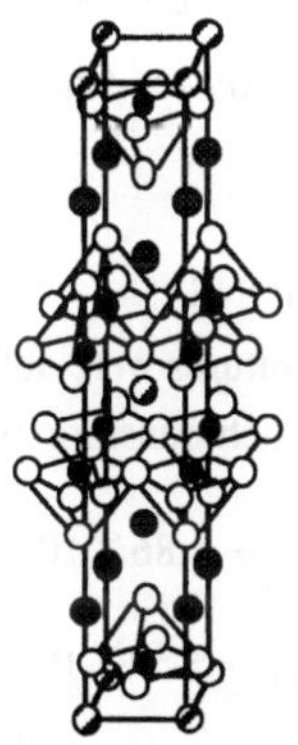

Fig.1 Schematic crystal structure of the La-326 compound ($La_{1.8}Sr_{0.2}CaCu_2O_6$).

The superconductivity was obtained in the La-Sr-Ca-Cu-O [1, 7, 8] and La-Ca-Cu-O [9] systems by means of the annealing under the high oxygen pressure. In these systems, Kinoshita, et. al. reported [8] that superconducting Tc^{onset} temperatures depend on the partial oxygen pressure; P_{O2}, during the annealing. However, in the La-Sr-Ca-Cu-O system, superconductivity was not obtained when the sample was post-annealed at the low temperature and high oxygen pressure [10]. We also reported that post-annealing temperature; *Ta*, was a key parameter to obtaining superconductivity [7]. We must consider the effects of the total pressure; P_T, and the oxygen partial pressure; P_{O2}, on the superconducting properties, when the mixture gas is used to be the pressure media during the post-annealing like an O_2-HIP method. So, there are three parameters, at least, to control the superconductivity in the La-326 compound. They are post-annealing temperature; *Ta*, P_T, and P_{O2}.

* Present Address: Electron Microscope Centre, University of Queensland, St. Lucia Brisbane, Queensland 4072, Australia

EXPERIMENTAL PROCEDURE

First, the phase decomposition of the superconducting La-326 was studied in order to determine the upper limit of the superconducting phase. Second, the phase formation of the superconducting La-326 was examined at a variety of combinations of P_{O2} and P_T at the optimized annealing temperature, *Ta*.

Samples with the composition of $La_{1.8}Sr_{0.2}CaCu_2O_6$ were prepared by the conventional solid state reaction method and annealing under the high pressure [7]. They exhibited Tc^{onset}=58K and $Tc^{R=0}$=37K by the electrical resistivity measurement, and Tc^{onset}=57K by the susceptibility measurement. Formed phase was confirmed as the single phase of La-326 structure whose lattice constants were a=3.8207 and c=19.535. Of course, the reproducibility was also confirmed. Moreover, the cation ordering between A(2a) and A(4e) sites was obtained in the superconducting sample by the crystal structure refinement using a joint powder x-ray/neutron diffraction method [11]. For the phase decomposition, superconducting samples was measured by TG-DTA (thermo-gravimetory and differential thermal analysis) method, which was performed from the room temperature up to the partial melting point under the controlled oxygen partial pressures, that is, P_{O2}=0.1, 1, 5 and 10*atm*. The apparatus has been remodeled to the pressure containing mechanism for the high pressure measurement up to 10*atm*. After the phase transformation points were determined by the TG-DTA, formed phases and superconducting properties were examined at the each region by quenching the sample from the temperatures.

The phase formation was studied by the annealing under the many combinations of P_T and P_{O2} at the optimized *Ta*. As this compound has similar chemical composition to the La-214 superconductor, following criteria were used to distinguish between La-326 superconductor and La-214 superconductor. When the sample was satisfied following terms, the sample was identified La-326 superconductor. That is, the La-326 compound was identified by the x-ray diffraction method, Tc^{onset} exceeds 40K measured by electrical resistivity and Tc^{onset} exceeds 40K also measured by magnetic susceptibility.

RESULTS AND DISCUSSIONS

The transition temperatures were able to be distinguished each other using the both TG curves and DTA curves. The P_{O2} dependence of the each transition was concluded following.

transition.*A*; $\mathrm{Log}P_{O2}=-4.86\times 10^4/T+33.97$ -(eq.1), La-326'→liquid,solidus line

transition.*C*; $\mathrm{Log}P_{O2}=-3.30\times 10^4/T+24.62$ -(eq.2), La-326'→214+others

transition.*D*; $\mathrm{Log}P_{O2}=-2.84\times 10^4/T+21.69$ -(eq.3), La-326→La-326'

When a sample was heated, the superconducting La-326 phase was transformed at *D* into the other La-326' phase. At the temperature of C, La-326' phase was decomposed into 214 phase and others. Finally, it started to melt the non-reacted La-326' at the temperature, A. Solidus temperature at P_{O2}=200*atm* is calculated to be 1260 ℃, which is consistent to the actual post-annealing data using an O_2-HIP technique.

It is obvious that $Tc^{R=0}$ temperatures were improved up to 46.9K by the post-annealing at 800 ℃ and quenched under the P_{O2}=1*atm* (Fig.2). Such an improvement of superconducting properties were proceeded in the region under the phase decomposition, *D*. The reason is suspected to be the release of the excess oxygen. Oxygen gas interaction at the phase transitions are also detected by means of TG-curves. As the initial oxygen content is confirmed 6.01± 0.01 by the coulometric titration technique [12], total oxygen content at any temperatures are able to be estimated. The O(3) site located between the facing CuO_2 planes would be occupied by an excess oxygen atom. Though O(3) is a vacancy site in the La-326 structure, the excess oxygen atoms intrude this site in order to balance the vacancy density under the high pressure of the O2-HIP annealing, or balance the electrical neutrality, because A(2) site between the facing CuO_2 phases was occupied by tri-valent ion [11]. $Tc^{R=0}$ temperature was deteriorated when the sample was annealed at 980 ℃, where the La-326' phase was observed. This reason is suspected that the oxygen gas was released irreversively from the O(1) or O(2) sites

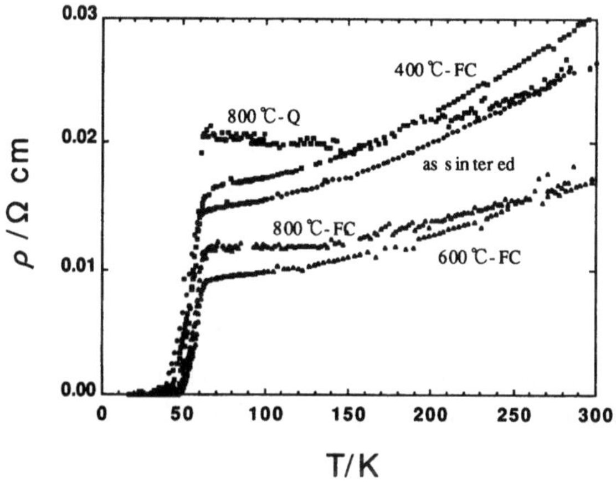

Fig.2 Electrical resistivity -vs- temperature curves for the samples post-annealed or quenched at the various temperature lower than the phase transformation temperature, *D.* Post-annealed and furnace cooled samples were marked "FC". Quenched sample was marked "Q".

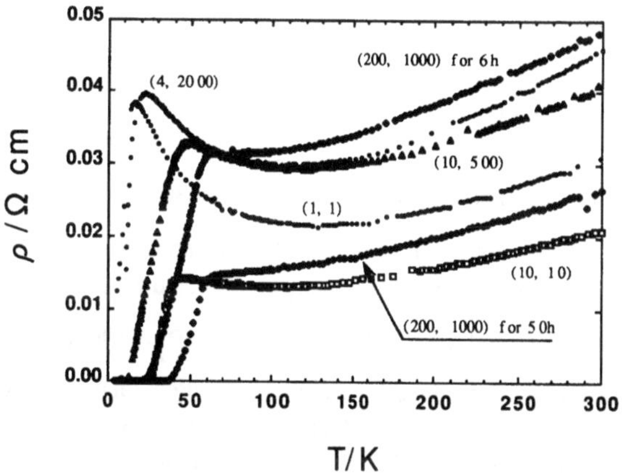

Fig.3 Electrical resistivity -vs- temperature curves for the samples which were annealed at various combinations of (P_{O2}, P_T).

Phase formation was studied in the field, whose boundary was determined by former experiment. Superconductor-ized La-326 samples were observed when $P_{O2} \geqq 10 atm$, regardless P_T. Moreover, 326 single phase was obtained in the field, which was enclosed in the (P_{O2}, P_T) combinations, (10, 10), (10, 500) and (200, 1000). Superconducting properties seems to depend on the combinations of (P_{O2}, P_T) (Fig.3). Figure.4 shows the electrical resistivity-vs-temperaure curves for the samples annealed at various combinations of (P_{O2}, P_T). Higher $Tc^{R=0}$ tends to be observed when relatively high oxygen partial pressure is achieved comparing to the total pressure; P_T. In the field of $P_{O2} < 10 atm$, no sample was superconductor-ized in any combinations of (P_{O2}, P_T).

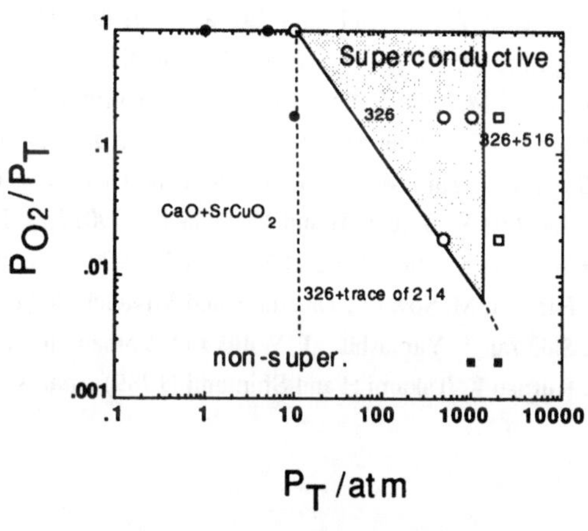

Fig.4 P_{O2}, -vs- P_T diagram. Superconducting La-326 samples with single phase were observed in the shadowed area. When P_T exceeds 1000*atm*, impurities were formed, 516 is $La_5SrCu_6O_{15}$ structure and 214 is La_2CuO_4 structure.

Tc^{onset} temperatures are slightly increased with the conditions of high pressure annealing, but they can't be regarded as La-326 superconductor due to the criteria described previously. So, we suspect P_{O2} is essentially important to superconductor-ize La-326 compound. The reason of the superconductor-ization is considered that P_{O2} controls not only to achieve the oxygen content to be stoichiometory but also the diffusion coefficients of oxygen atom in the lattice.

CONCLUSION

The optimum annealing condition to superconductor-ize La-326 compound was examined by the studies of the phase decomposition and phase formation at the composition of $La_{1.8}Sr_{0.2}CaCu_2O_6$. It was concluded that P_{O2} was essentially important parameter to form and superconductor-ize 326 compound. Total oxygen content and atomic site which was occupied by oxygen were related to the P_{O2}. Superconducting La-326 phase was formed by annealing under the condition of $P_{O2} \geqq 10atm$ and $P_T \leqq 1000atm$. Though superconducting phase could be formed under the mixture gas including oxygen, high pressure with pure oxygen gas condition is recommended to get a high quality in ρ -T and χ -T properties. And, superconducting phase which have been formed once is stable up to 800 ℃ under the P_{O2}=1atm and 1000 ℃ under the P_{O2}=10atm.

ACKNOWLEDGEMENT

This work was supported by New Energy and Industrial Technology Development Organization as a part of program for R&D of Basic Technology for Future Industries.

REFERENCES

1. Cava R J, Batlogg B, Van Dover R B, Krajewski J J, Waszczak J W, Fleming R M, Peck Jr W F, Lupp Jr L W, Marsh P, James A C W P and Schneeneyer L F (1990) Nature 345: 602
2. Izumi F, Takayama-Muromachi E, Nakai Y and Asano H (1989) Physica C. 157, 89
3. Caignaert V, Nguyen N and Raveau B (01990) Mat. Res. Bull. 25: 199
4. Lightfoot P, Pei A, Jorgensrn J D, Tang X X, Manthiram A and Goodenough J B (1990) Physica C. 169: 464
5. Hiratani M, Saito S, Suga M and Sowa T (1990) Solid. state. comm. 75: 425
6. Currie D B, Weller M T, Rowles S and Gregory D H (1990) Mat. Res. Bull. 25: 1279
7. Sakurai T, Yamashita T, Yamauchi H and Tanaka S (1991) J. Appl. Phys. 69: 3190
8. Kinoshita K, Shibata H and Yamada T (1990) Jpn. J. appl. phys. 29: L1632
9. Kinoshita K, Shibata H and Yamada T (1990) Physica C. 171: 523
10. Hiratani M, Sowa T, Takeda Y and Miyauchi K (1989) Solid. state. comm. 72: 541
11. Sakurai T, Yamashita T, Willis J O, Yamauchi H, Tanaka S and Kwei G H (1991) Physica C 174: 187
12. Kurusu K, Takami H and Shintomi (1989) Analyst 114: 1341

New Organic Superconductor, $(BEDT\text{-}TTF)_4Pt(CN)_4H_2O$

HATSUMI MORI[1], IZUMI HIRABAYASHI[1], SHOJI TANAKA[1], TAKEHIKO MORI[2], YUSEI MARUYAMA[2], and HIROO INOKUCHI[2]

[1] ISTEC, Nagoya, 456 Japan
[2] Institute for Molecular Science, Okazaki, Aichi, 444 Japan

ABSTRACT

New organic superconductor, $(BEDT\text{-}TTF)_4Pt(CN)_4H_2O$ (BEDT-TTF:bis(ethylenedithiolo)tetrathiafulvalene), was recently found. Under an ambient pressure the M=Pt salt shows broad metal-insulator transition around 100 K, but by suppressing the transition, the superconductivity was observed at 2K under 6.5 kbar. The crystal structures and electrical resistivities of the Pt-, Pd-, and Ni-cyano salts based upon BEDT-TTF are reported.

KEY WORDS:organic superconductor, BEDT-TTF, Pt-cyano complex, Pd-cyano complex, Ni-cyano complex

INTRODUCTION

So far 42 kinds of organic superconductors have been found and the highest Tc of 30K for Rb_3C_{60} reaches the T_c of La-Ba-Cu-O system, which is the starting material of oxide-fever. However, doped C_{60} comlexes are not stable in the air, so that it is still necessary to cultivate the pseudo-one or two-dimensional organic superconductors so far studied, which are stable H_2O and O_2 for a long time. In this work, Pt, Pd, and Ni-cyano salts based upon BEDT-TTF were investigated. It is interesting to observe how the anion controls the donor arrangement, which determines the electronic states of the salt. This paper reports preparations, electrical resistivities, and crystal structures of $(BEDT\text{-}TTF)_4M(CN)_4H_2O$ (M=Pt, Pd, and Ni), $(BEDT\text{-}TTF)_4Pd(CN)_4$, and $(BEDT\text{-}TTF)Pd(CN)_2$. Among them, $(BEDT\text{-}TTF)_4Pt(CN)_4H_2O$ exhibits a superconductivity at 2K under 6.5kbar.[1)]

EXPERIMENTAL

The single crystals were prepared by a standard electrochemical technique under a current of 0.5 μA. The black plates of Ni-, Pd-, and Pt-cyano salts based upon BEDT-TTF were grown by using $BaPt(CN)_4$ ($K_2Ni(CN)_4$, or $Pd(CN)_2$), KCN, and 18-crown-6 ether as supporting electrolyte in 1,1,2-trichloroethane and 10% vol. ethanol. Crystal data are shown in Table 1. $(BEDT\text{-}TTF)_4$-$M(CN)_4H_2O$ (M=Ni, Pd and Pt) are isostructural to each other. The previously reported $(BEDT\text{-}TTF)_4M(CN)_4$ (M=Ni[2)] and Pt[3)]) are also isostructural. Unlike these salts, our salts contain H_2O. Though M=Pt and Ni salts with and without H_2O afford slightly different lattice parameters, these salts show quite different physical properties. Intensities of x-ray reflections were measured by the ω-2θ technique on a Rigaku four-circle diffractometer with graphite monochromated Mo radiation ($2\theta<60°$) and were corrected for usual Lorentz and polarization effects, but not for the absorption. The structures were solved by the Patterson method and refined by the block-diagonal least-squares procedure ($|F_0|>3\sigma(|F_0|)$). The number of used reflection data and final R values are shown in Table 1. The non-hydrogen atoms were refined anisotropically.
The resistivity measurements were carried out by the usual four-probe technique with applying low-frequency (100Hz) ac current. 25 μm gold wires are pasted to a crystal with a gold paste.
Pressure cell of clamp type was used with an oil (Daphne #7373) as a pressure medium. The pressure was determined from the resistance of a manganin wire at room temperature, and from the T_c of Sn at low temperature. The pressure was applied up to at least 6.5 kbar on two samples of $(BEDT\text{-}TTF)_4Pt(CN)_4H_2O$.

Table 1. The color, shape, electrical conductivity, crystallographic data, and other properties for metal-cyano salts (M=Ni, Pd, and Pt) of BEDT-TTF (ET).

M=	Ni		Pd	
	$(ET)_4Ni(CN)_4H_2O$	$(ET)_4Ni(CN)_4$	$(ET)_4Pd(CN)_4H_2O$	$(ET)_4Pd(CN)_4$
color	black	black	black	black
shape	plate	plate	plate	plate
electrical conductivity (Scm^{-1}(rt))	70-110 T_{MI}=100K	20 T_{MI}=230K	- -	- -
F.W.	1719.4	1701.4	-	-
system	triclinic	triclinic	triclinic	triclinic
space group	$P\bar{1}$	$P\bar{1}$	-	-
a/Å	11.493(3)	10.994(2)	11.488(2)	10.968
b	16.693(6)	16.443(2)	16.634(2)	16.461
c	9.543(2)	9.703(1)	5.536(2)	9.705
α/°	91.28(2)	97.85(1)	91.24(1)	98.00
β	112.47(1)	115.15(1)	112.26(1)	115.28
γ	75.57(2)	95.94(1)	75.65(1)	95.56
V/Å^3	1632.9(8)	1547.2	1628.6(4)	1542.7
Z	1	1	-	-
Dc	1.75	1.83	-	-
No of data	5243	4268	-	-
R	0.069	0.067	-	-
Reference	this work	2	this work	this work

M=	Pd	Pt	
	(ET)-$Pd(CN)_2$	$(ET)_4Pt(CN)_4H_2O$	$(ET)_4Pt(CN)_4$
color	black	black	black
shape	plate	plate	plate
electrical conductivity (Scm^{-1}(rt))	70 semiconductor	130-280 T_{MI}=120K (0kbar) superconductivity 2K(6.5kbar)	20 T_{MI}=250K
F.W.	-	1855.8	1837.8
system	monoclinic	triclinic	triclinic
space group	-	$P\bar{1}$	$P\bar{1}$
a/Å	34.913(11)	11.514(2)	9.721
b	13.470(4)	16.704(7)	11.127
c	15.010(6)	9.563(2)	16.552
α/°		91.22(4)	76.90
β	108.50(2)	112.30(2)	81.52
γ		75.48(2)	62.88
V/Å^3	6694.3(39)	1641.6(9)	1550
Z	-	1	1
Dc	-	1.88	1.97
No of data	-	6037	6069
R	-	0.041	0.032
Reference	this work	this work	3

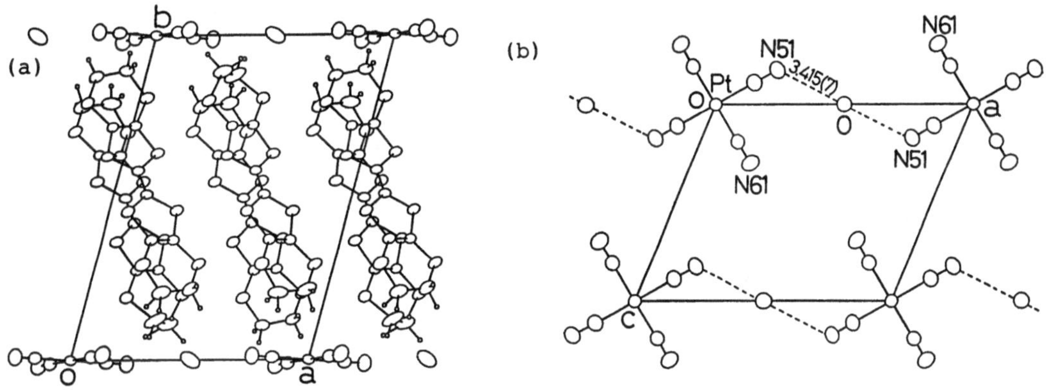

Fig. 1. (a) Crystal structure and (b) donor arrangement of (BEDT-TTF)$_4$-Pt(CN)$_4$H$_2$O.

RESULTS AND DISCUSSION

The crystal structure of (BEDT-TTF)$_4$Pt(CN)$_4$H$_2$O is shown in Fig. 1(a). The conducting BEDT-TTF sheet and the insulating anion sheet containing [Pt(CN)$_4$]$^{2-}$ and H$_2$O stack alternately along the b-axis. In the two-dimensional conducting sheet, the donor··donor interactions exist not only in the stacking direction, but also in two kinds of side-by-side directions. This packing of donors is similar to that of (BEDT-TTF)$_3$(ClO$_4$)$_2$.[4] BEDT-TTF salts of ClO$_4$-type packing usually undergo metal-insulator transition at 50-250 K under an ambient pressure. Among them, only (BEDT-TTF)$_3$Cl$_2$2H$_2$O exhibits superconductivity at 2 K under 16 kbar.[5]
Figure 1(b) shows the anion arrangement of (BEDT-TTF)$_4$Pt(CN)$_4$H$_2$O. [Pt(CN)$_4$]$^{2-}$, an anion with the planar square structure, is located on an inversion center, while O atom of H$_2$O is also on another inversion center. Since the distance between N51 and O atoms (Fig. 1(b)) is 3.415(7) Å, which is shorter than the calculated value of N···H-O (2.79+0.957=3.747 Å), an anion ([Pt(CN)$_4$]$^{2-}$) and H$_2$O are reasonably considered to construct a one-dimensional polymer through a hydrogen bond like $\{$NC-Pt-CN··H-O-H$\}_n$.
The electrical resistivity of (BEDT-TTF)$_4$Pt(CN)$_4$H$_2$O is depicted in Fig. 2. The conductivity at room temperature is 130-280 S cm^{-1} which is rather high among BEDT-TTF salts; Usually a conductivity of even a metallic BEDT-TTF salt is <100 Scm^{-1}. Under an ambient pressure the resistivity is metallic to about 120 K and increases slowly at low temperatures. Below 10 K, the resistivity goes up rapidly, but the resistivity at 1.5 K is only five times as large as that at room temperature (R(1.5 K)/R(rt)≅5). The measurement of thermoelectric power indicates the semimetallic behavior even below the broad transition of electrical resistivity around 100 K.
Table 1 also shows that the electrical conductivity of (BEDT-TTF)$_4$Ni(CN)$_4$H$_2$O at room temperature is 70-110 Scm^{-1}, which is a little smaller than that of

Fig. 2. Temperature dependence of electrical resistivity at various pressure for (BEDT-TTF)$_4$Pt(CN)$_4$H$_2$O (a) sample 1 and (b) sample 2.

$(BEDT-TTF)_4Pt(CN)_4H_2O$. The resistivity is metallic down to about 100 K with a small hump around 170 K. This anomaly was observed on three samples, the hight of the hump being dependent on the sample. After the metal-semiconductor transition, the resistivity increases. But the value of the resistivity at 1.5 K is only one order of magnitude higher than that at room temperature.

$(BEDT-TTF)_4M(CN)_4$ (without H_2O, M=Pt, Pd and Ni) have the analogous lattice parameters to $(BEDT-TTF)_4M(CN)_4H_2O$ (with H_2O, M=Pt, Pd and Ni) (Table 1) and the donor arrangements of both salts (with or without H_2O) are ClO_4-type. $(BEDT-TTF)_4Pt(CN)_4$ undergoes a metal-insulator transition at 250 K,[3], whereas $(BEDT-TTF)_4Ni(CN)_4$ exhibits a broad metal-insulator transition around 230 K and the resistivity reaches 10^4 Ωcm at 20 K.[2] It is noteworthy that the slight difference of the crystal structures, which is caused by the inclusion of water, result in the significant difference of the physical properties.

As shown in Table 1, $(BEDT-TTF)-Pd(CN)_2$ has lattice constants quite different from those of $(BEDT-TTF)_4M(CN)_4H_2O$ (M=Pt and Ni). The crystal structure analysis is under way. The conductivity at room temperature is 70 Scm^{-1} and a semiconductive behavior was observed.

In order to suppress the metal-semiconductor transition, a pressure was applied on a single crystal of $(BEDT-TTF)_4Pt(CN)_4H_2O$.

The temperature dependence of the resistivity at various pressures is illustrated in Fig. 2. By applying pressure, the temperature of metal-semiconductor transition is lowered. Under the pressure of 6 kbar, the metallic state remains down to 60 K, then the resistivity increases twice of the minimum resistivity, and a small resistance drop appeared at 1.6 K. Under 8 kbar, after the resistivity minimum around 50 K the resistance drop occurs (sample 1). The pressure dependence of resistivity at low temperatures (sample 2) is shown in Fig. 2(b). The sample 2 has the same resistivity behavior at high temperatures as that of the sample 1. However, the resistivity drop of the sample 2 appeared at lower pressure probably due to a better quality of the sample. Under 5 kbar a partial resistance drop appeared, under 6 kbar 90 % of a resistance drop occurred, and the zero resistance was observed under the pressure of 6.5 kbar, indicating the superconductivity. The onset temperature is 3.5 K, the midpoint is 2.2 K, and the offset is 1.5 K. The large width of the transition region might be caused by pressure inhomogeneity in a sample.

In conclusion, we discovered a new organic superconductor, $(BEDT-TTF)_4$-$Pt(CN)_4H_2O$ with T_c at 2 K under 6.5 kbar, which is the first (2-) anion organic superconductor. The donor arrangement of $(BEDT-TTF)_4Pt(CN)_4H_2O$ is ClO_4-type which so far gave only one organic superconductor, $(BEDT-TTF)_3Cl_2 2H_2O$[5]. The isostructural $(BEDT-TTF)_4Ni(CN)_4H_2O$ has the temperature dependence of the resistivity similar to the M=Pt salt under an ambient pressure. The investigation of the pressure dependent properties of this salt is in progress.

REFERENCES

1) H. Mori, I. Hirabayashi, S. Tanaka, T. Mori, Y. Maruyama, and H. Inokuchi, Solid State Commun., in press.
2) M. Tanaka, H. Takeuchi, M. Sano, T. Enoki, K. Suzuki, and K. Imaeda, Bull. Chem. Soc. Jpn., 62, 1432(1989); M. Tanaka, H. Takeuchi, A. Kawamoto, J. Tanaka, T. Enoki, K. Suzuki, K. Imaeda, and H. Inokuchi, The Physics and Chemistry of Organic Superconductors, ed. G. Saito and S. Kagoshima (Springer 1990) p298; L. Ouahab, J. Padiou, D. Grandjean, C. Garrigou-Lagrange, P. Delhaes, and M. Bencharif, J. Chem. Soc., Chem. Commun., 1989, 1038.
3) R. P. Shibaeva, R. M. Lobkovskaya, V. E. Korotkov, N. D. Kushch, E. B. Yagubskii, and M. K. Makova, Synthetic Metals, 27, A457(1988); L. Ouahab, J. Padiou, D. Grandjean, C. Garrigou-Lagrange, P. Delhaes, and M. Bencharif, J. Chem. Soc., Chem. Commun., 1989, 1038; R. M. Lobkovskaya, N. D. Kushch, R. P. Shibaeva, E. B. Yagubskii, and M. A. Simonov, Kristallografiya, 34, 1158(1989).
4) H. Kobayashi, R. Kato, T. Mori, A. Kobayashi, Y. Sasaki, G. Saito, T. Enoki, and H. Inokuchi, Chem. Lett., 1984, 179.
5) T. Mori and H. Inokuchi, Chem. Lett., 1987, 1657; T. Mori and H. Inokuchi, Solid State Coummun., 64, 335(1987).

A New Type of Molecular Superconductor Based on C60 Doped with Cesium and Rubidium

KATSUMI TANIGAKI, THOMAS W. EBBESEN, SUSUMU SAITO, JUNICHIRO MIZUKI, JAW-SHEN TSAI, YOSHIMI KUBO, and SADANORI KUROSHIMA

Fundamental Research Laboratories, NEC Corporation, Tsukuba, Ibaraki, 305 Japan

ABSTRACT

A new type of C60 superconductor with a binary dopant system of cesium and rubidium is reported, which has a critical transition temperature of Tc=33 K. In the binary-metal doped C60's with a nominal composition of CsxRbyC60 (where x+y=3), Tc increases with the content of Cs up to x=2 and the superconducting phase disappears in the region of x>2. The increase in Tc with the ionic radius of alkali-metals indicates that the lattice parameter controls Tc, this being interpreted in terms of the change in the density of states at the Fermi level.

KEY WORDS: C60, Fullerenes, Superconductor

INTRODUCTION

The success in macroscopic-scale production [1] of fullerenes [2] (C60, C70 ...) resulted in a surge of interest in fundamental properties of these new class of molecules. Among the properties, the discovery of conductivity [3] in MxC60 and MxC70 (where M=Li, Na, K, Rb, Cs) and the superconductivity in KxC60 (transition temperature Tc=18 K) [4] has stimulated broad interest in this material. The quantitative preparation of KxC60 (x=0-6) [5] shows that the superconducting phase is K_3C60, as was originally proposed [4]. The crystal structure of the superconducting phase is determined to be face-centered cubic (fcc) by X-ray diffractometry [6], where K-atoms occupy both interstitial tetrahedral and octahedral sites.

Very soon after finding the K-doped C60 superconductor [4], Rb-doped C60 (Rb_3C60) was reported to show a higher Tc (ca. 28 K) [5, 7]. However, Cs_3C60 was not a superconductor [5], although a much higher Tc was expected from the trend in the Tc [Tc(K_3C60) < Tc(Rb_3C60)]. In this paper we report a new type of C60 superconductor intercalated with a Cs/Rb binary dopant system, which shows the highest Tc=33K among the alkali-metal doped C60's when the nominal composition is Cs_2Rb_1C60 [8]. The superconducting properties of these materials are discussed with emphasis on the important role of the lattice parameter for controlling the Tc's.

EXPERIMENTAL

The soot containing C60 was produced with a arc-discharge method using pure carbon rods according to the literature [1]. After diethyl ether treatment to remove hydrogenated molecules, C60 was chromatographically isolated using a neutral alumina column (activity I, Mark) with chromatographic grade toluene/hexane elutants. The purity of the C60 was confirmed to be at least 99.9 % by infrared, ultraviolet and mass spectroscopies.

Typically, the C60 (10 mg) was placed in a quartz tube (5mm in diameter) together with a desired amount of Rb and/or Cs. The small amounts of alkali-metal were measured by using metal-filled capillary tubes (0.5mm in diameter), which were cut and handled under a helium atmosphere in a glove box. The quartz tubes prepared in this fashion were degassed to 10^{-2} torr and were refilled with helium (700 torr). The degassing /refilling process was repeated twice before sealing the tubes. These were then heated at around 400 °C for 74 hours to complete the doping process.

The temperature dependence of the dc magnetization was measured with a SQUID magnetometer (Quantum Design MPMS). In zero field cooling process to measure magnetic field shielding, the samples were cooled to 4 K without magnetic field and diamagnetic susceptibilities were measured on warming under the applied filed of 10 Oe. The magnetic field repulsions (Meissner-Ochsenfeld effect) were also measured as a function of temperature in an applied field cooling mode with 10 Oe.

RESULTS AND DISCUSSION

Cs-doped C60 with a nominal composition of Cs_3C60 was reported to show nonsuperconducting features [5], even though the stoichiometric study using potassium indicates that x=3 is essentially important [6] in achieving superconducting properties for MxC60 (where, M=K or Rb). Expected increase in Tc for Cs-doped C60 from the tendency of Tc=28 K (M=Rb) > Tc=18 K (M=K) has given us further impetus to study this doped system. We repeated dc magnetization measurements for different nominal compositions of Cs to C60 from 0.5 to 6, but no superconducting features were observed. We presumed that this would be due to a crystal phase change, because there would be little difference in electron transferring from M to C60 among K, Rb and Cs.

Improved technique of making a single-phase superconducting K_3C60 showed that the superconducting phase is face-centered cubic (fcc) through X-ray analyses [6]. The fcc structure has two kinds of spaces, one octahedral and two tetrahedral sites, for doping metals as shown in Fig.1. We considered that Cs could be doped into a superconducting phase using a binary dopant system of Cs and Rb so that the two kinds of interstitial sites could be shared by the two different types of metals in such a way as to keep the fcc structure. The Rb element first introduced into the fcc structure will enlarge the lattice and Cs could be introduced without any destruction of the structure, resulting in further lattice expansion.

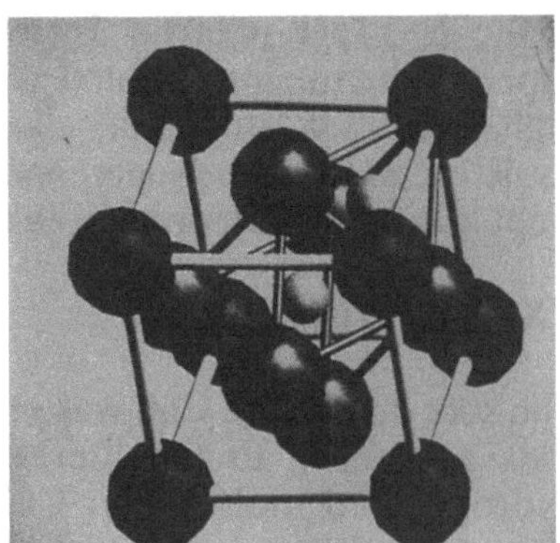

Fig.1 The fcc crystal structure, where black ball represents C60. The larger octahedral site is shown in the center and the smaller tetrahedral site is in the upper right. The bright balls are metals.

Figure 2 shows the diamagnetic shielding signals against temperature for the samples of CsxRbyC60's. RbyC60's with y=2,3,4 showed magnetic field shielding below 29K in zero-filed cooling (ZFC) measurements. These data are in very good agreement with the previous reports [5, 7]. Tc was always the same value regardless of y, although the degree of the shielding is dependent on the nominal composition. This would be due to the fact that only fcc Rb_3C60 is superconducting among the existing phases as reported for KxC60, where fcc K_3C60 [6], bct K_4C60 [9] and bcc K_6C60 [10] phases are confirmed, and the K_1C60 and K_2C60 phases do not exist [11]. When the content of Cs is increased for $CsxRb_{3-x}C60$ (where x=0,1,2), Tc became higher with the Cs content and Tc=33 K was obtained for Cs_2Rb_1C60. The field cooling (FC) measurements for Cs_2Rb_1C60 clearly showed the magnetic field flux expulsion, namely the Meissner-Ochsenfeld effect, with the same onset of 33K as shown in Fig.3. The FC data of equal or lower magnitude than the ZFC data excludes any other known phenomena besides superconductivity. The Tc is the highest among those so far reported for alkali-metal doped C60 superconductor families. The magnitude of the ZFC shielding signal also exceeds 60 % of that expected for a homogeneous superconductor.

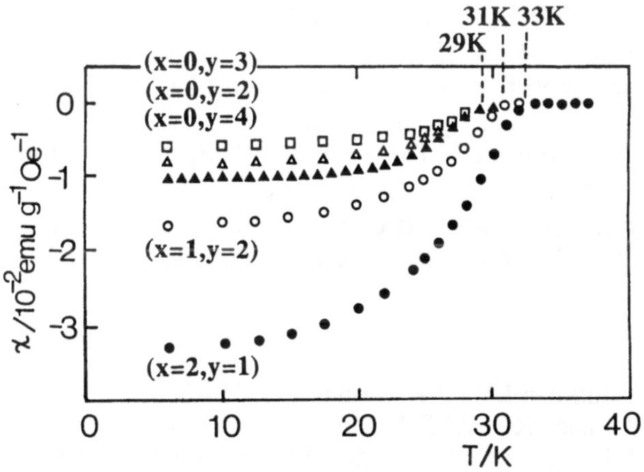

Fig.2 Magnetic susceptibilities measured for CsxRbyC60. Tc=33 K is observed for Cs_2Rb_1C60.

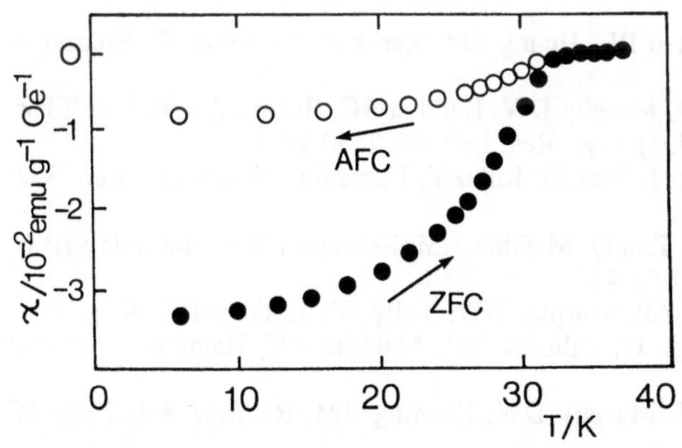

Fig.3 Meissner-Ochsenfeld effect observed for Cs_2Rb_1C60 on FC measurements.

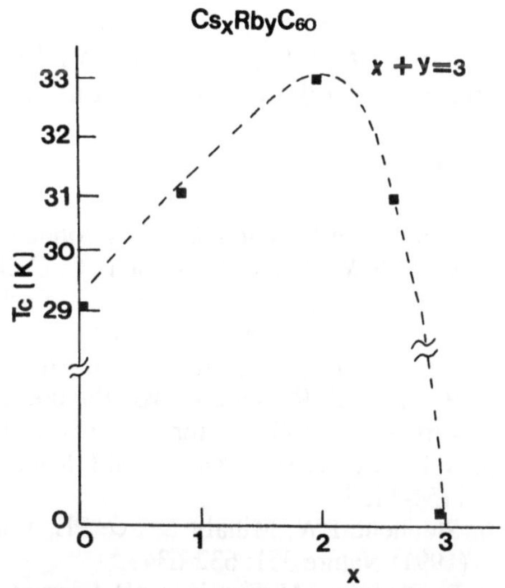

Fig.4 The dependence of Tc's on the composition of x for CsxRbyC60, where x+y=3.

The relationship between Tc and the Cs content for CsxRbyC60's (where x+y=3) has been studied in further detail. As shown in Fig.4, Tc [Tc=29 K: Rb_3C60] was found to increase gradually with the Cs content [Tc=31 K: Cs_1Rb_2C60] upto the composition of x=2 [Tc=33 K:Cs_2Rb_1C60] and to drop very rapidly at the regions of x>2, transforming to a nonsuperconducting phase. From the viewpoint of the two types of interstitial sites described in the early part of this manuscript, Cs_1Rb_2C60 is thought to be a reasonable composition as a single phase. However, the situation is more profound than envisaged. The experimental results indicated an interesting phenomenon that the superconducting fcc phase can be maintained upto the nominal compositions of Cs_2Rb_1C60. This seems to imply that the both alkali-metals occupy tetrahedral and octahedral sites in a random fashion, although the larger octahedral site may be occupied by the larger Cs element. Further insight of the structure and mechanism of this new binary system superconductor will require more detailed structural information.

Lattice parameter analyses [12] recently carried out for several alkali-metal doped C60's indicated that the lattice parameter a_0(Å) increases monotonically with the Tc's in the order of K_3C60 (a_0=14.24, Tc=18 K) < K_2Rb_1C60 (a_0=14.30, Tc=22 K) < K_1Rb_2C60 (a_0=14.36, Tc=26 K) < Rb_3C60 (a_0=14.44, Tc=29 K) < Cs_1Rb_2C60 (a_0=14.49, Tc=31 K) < Cs_2Rb_1C60 (a_0=14.51, Tc=33 K). As is discussed in ref.12, the superconducting properties obtained for alkali-metal doped C60's would be resulted from the fcc isostructural series. The very good linear relationship between Tc and a_0 would mean that the change in Tc observed for the M_3C60 superconductors (where M=K, Rb, Cs or a binary system among these) is explained by the change in density of states at the Fermi level and there are little variations in the phonon frequency and electron-phonon coupling in the frame work of BCS theory.

REFERENCES

1. Kratschmer W, Lamb LD, Fostiropoulos K, Huffman DR (1990) Nature 347: 354-358
2. Kroto HW, Heath JR, O'Brien SC, Curl F, Smalley RE (1985) Nature 318: 162-164
3. Haddon RC, Hebard AF, Rosseinsky MJ, Murphy DW, Duclos SJ, Lyons KB, Miller B, Rosamilia JM, Fleming RM, Kortan AR, Glarum SH, Makhija AV, Muller AJ, Eick RH, Zahurak SM, Tycko R, Dabbagh G, Thiel FA (1991) Natuare 350: 320-322
4. Hebard AF, Rosseinsky MJ, Haddon RC, Murphy DW, Glarum SH, Palstra TTM, Ramirez AP, Kortan AR, (1991) Nature 350: 600-601
5. Holczer K, Klein O, Huang SM, Kaner RB, Fu KJ, Whetten RL, Diederich F (1991) Science 252: 1154-1157
6. Stephene PW, Mihaly L, Lee PL, Whetten RL, Huang SM, Kaner R, Diederich F, Holczer K (1991) Nature 351: 632-634
7. Rosseinsky MJ, Ramirez AP, Glarum SH, Murphy DW, Haddon RC, Hebard AF, Palstra TTM, Kortan AR, Zahurak SM, Makhija AV (1991) Phys. Rev. Lett. 66: 2830-2832
8. Tanigaki k, Ebbesen TW, Saito S, Mizuki J, Tsai JS, Kubo Y, Kuroshima S (1991) Nature 352: 222-223
9. Zhou O, Fischer JE, Coustel N, Kycia S, Zhu Q, McGhie AM, Romanow WJ, McCauley JP Jr, Smith AB III, Cox DE (1991) Nature 351: 462-464
10. Fleming RM, Rosseinsky MJ, Rmirez AP, Murphy DW, Tully JC, Haddon RC, Siegrist T, Tycko R, Glarum SH, Marsh P, Dabbagh G, Zahurak SM, Makhjia AV, Hampton C (1991) Nature 352: 701-703
11. Tycko R, Dabbagh G, Rosseinsky MJ, Murphy DW, Fleming RM, Ramirez AP, Tully JC (1991) Science 253:884-886
12. Fleming RM, Ramirez AP, Rosseinsky MJ, Murphy DW, Haddon RC, Zahurak SM, Makhja AV (1991) Nature 352: 787-788.

Preparation and Properties of Alkali-Fulleride Superconductors

MASATO KATO, HIDEYUKI FUNASAKA, KOJI SAKURAI, KAZUNORI YAMAMOTO, YOSHIHARU KANO, TOSHIAKI ISHIGURO, YUKIO WADA, and YOICHIRO KISHIMOTO

Power Reactor and Nuclear Fuel Development Corp., Tokai-mura, Naka-gun, Ibaraki, 319-11 Japan

ABSTRACT

We report on the properties of alkali-fulleride superconductors prepared by the reactions of C_{60} crystallized in several organic solvents (toluene, CS_2, CH_2Cl_2) with alkali-metals. X-ray diffraction experiments revealed that the structure of solid C_{60} was related to the solvent for crystallizing it. The structure of solid C_{60} grown from CH_2Cl_2 solution, which has not been reported yet, was not identified as fcc or hcp structure. The alkali-fullerides for nominal composition of K_3C_{60} and Rb_3C_{60} were prepared using solid C_{60} grown from CH_2Cl_2 solution. Their X-ray diffraction patterns and superconducting transition temperatures were different from those of other alkali- fulleride samples.

KEY WORDS : C_{60}, superconductor, alkali-fulleride, alkali-metal doped C_{60}

INTRODUCTION

The production of bulk quantities of the fullerene C_{60} has stimulated considerable interest in the synthesis and properties of this form of carbon.[1, 2] The crystal of solid C_{60} grown from solutions of aromatic solvents (such as benzene or toluene) has the fcc or hcp structure [3,4], and that grown from CS_2 or pentane-benzene solution has the orthorhombic [5] or monoclinic lattice.[6] The crystal structure of solid C_{60} is related to the solvent used for crystallization. Several reported syntheses of alkali-metal doped C_{60} superconductors have relied on the direct reaction of alkali-metal vapour with solid C_{60}.[7,8] The superconductivities for K_3C_{60} and Rb_3C_{60} which were synthesized from fcc C_{60} showed superconducting transition at 19.3 K and 30 K, respectively[9,10], and K_xC_{60} and Rb_xC_{60} synthesized with orthorhombic C_{60} have transition temperatures of 19.4 K and 28.5K, respectively.[11,12] It was found that solid C_{60} grown from CH_2Cl_2 solution had a new crystal structure, which has not been reported yet. In this study we prepared three types of solid C_{60} obtained from toluene, CS_2 and CH_2Cl_2 solutions and synthesized superconductor with them by alkali-metal doping. Some experiments were performed to know their properties.

EXPERIMENTAL

C_{60} was synthesized in a stainless-steel chamber using high-purity graphite electrodes(99.997%), as described in ref 2. The carbon soot produced from an arc discharge in 50-150 torr of helium was extracted by toluene, and pure C_{60} was obtained from this solution by chromatography on neutral alumina. The purity of the C_{60} was estimated to be at least 99.7% by HPLC. The solid C_{60} was produced from the eluant solution (5% toluene in hexane) by solvent evaporation, and then was dissolved again in toluene, CS_2 and CH_2Cl_2, respectively, and re-crystallized from each solution. The solid C_{60} was dried under vacuum for 4 hours at 250℃, and then was synthesized to superconductor by alkali-metal doping. The samples of nominal composition K_3C_{60} and Rb_3C_{60} were prepared by reaction of each C_{60} with potassium or rubidium. The C_{60} was placed in a pyrex tube together with alkali-metal in a glove box. The tube was sealed under Ar atmosphere and then heated for 8-150 hours at 200-450 ℃. The magnetization of each sample was determined as a function of reaction time with the use of superconducting quantum interference device magnetometer(MPMS2, Quantum Design). Crystal of undoped and doped C_{60} were characterized by powder X-ray diffractometer.

RESULTS AND DISCUSSION

The X-ray powder diffraction patterns for solid C_{60} grown from toluene, CS_2 and CH_2Cl_2 solutions were shown in Fig.1. The diffraction pattern of solid C_{60} (CH_2Cl_2) was different from those of C_{60}(toluene) and C_{60}(CS_2) which are fcc and orthorhombic lattice, respectively. The crystal of C_{60}(CH_2Cl_2) was observed as an acicular crystal under the optical microscope. This solid may be the cocrystallized molecular crystal of the type $C_{60}\cdot CH_2Cl_2$, although it has not been identified yet.

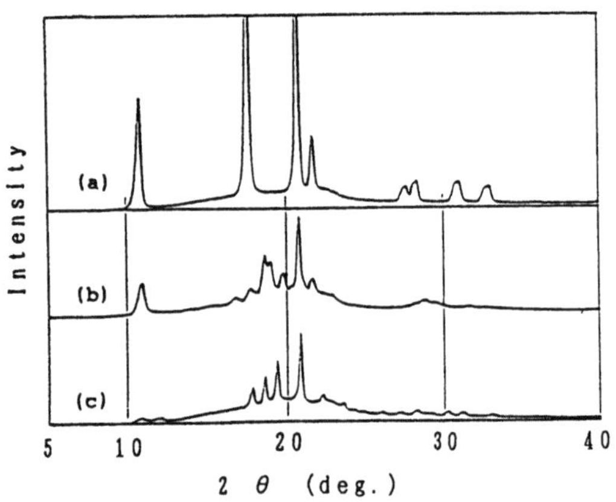

Fig.1 X-ray powder diffraction patterns for C_{60}, which were grown in toluene(a), CS_2(b) and CH_2Cl_2(c).

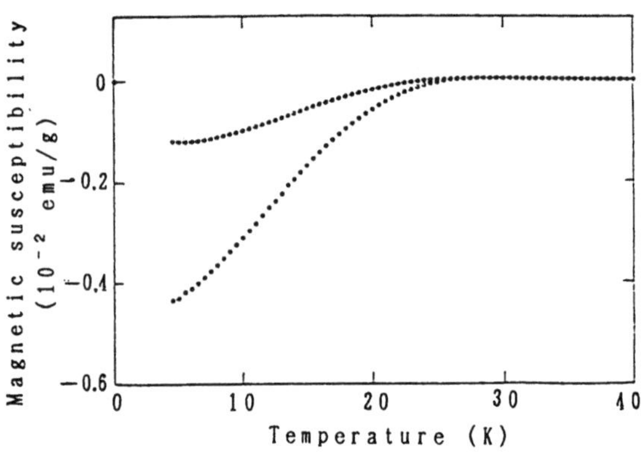

Fig.2 Temperature dependence of the magnetization obtained for a Rb_3C_{60} (CH_2Cl_2) after heated for 150 hours at 400℃.

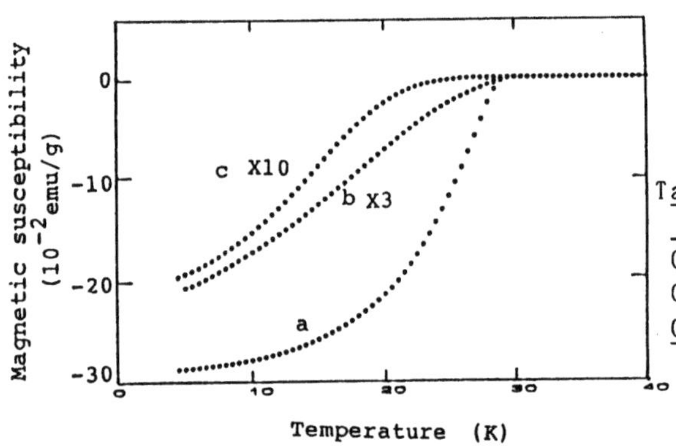

Fig.3 Temperature dependence of the magnetization obtained for Rb_3C_{60} (a:toluene, b:CS_2 and c:CH_2Cl_2) after heated for 150 hours at 400℃.

Table 1 Onset of Tc for various samples

type of solid C_{60}	K_3C_{60}	Rb_3C_{60}
C_{60} grown in toluene	19 K	30 K
C_{60} grown in CS_2	18.5 K	30 K
C_{60} grown in CH_2Cl_2	17.5 K	27.5K

Temperature-dependent magnetic susceptibility was obtained for Rb_3C_{60} which was prepared by reaction of $C_{60}(CH_2Cl_2)$ with rubidium(Fig.2). The sample was first cooled to 5 K in zero field and was taken in a 10 Oe field on warming the sample to 40 K. The sample was then cooled in the same field down to the base temperature. The Meissner effect showed clear transition at 27.5 K. The same measurements were performed on Rb_3C_{60}(toluene) and $Rb_3C_{60}(CS_2)$ of which Meissner effect of both samples were observed with Tc of 30 K (Fig.3). These values are consistent with the published values . [9,11] The coincidence of superconducting transition temperature of $Rb_3C_{60}(CS_2)$ with Rb_3C_{60}(toluene) was elucidated by the phase transition from orthorhombic C_{60} to fcc C_{60}.[5] This phase tansition was confirmed by the X-ray diffraction experiments of both undoped and alkali-metal doped C_{60}. On the other hand the superconducting transition temperature of $Rb_3C_{60}(CH_2Cl_2)$ was 2.5 K lower than that of other samples. X-ray diffraction pattern of $Rb_3C_{60}(CH_2Cl_2)$ has extra peaks comparing with fcc Rb_3C_{60}(toluene).

The onset of Tc for various samples of K_3C_{60} and Rb_3C_{60} were shown in Table 1. The Meissner effects for K_3C_{60}(toluene, CS_2 and CH_2Cl_2) were observed with Tc of 19 K, 18.5 K and 17.5 K, respectively. The superconducting transition temperature of $K_3C_{60}(CH_2Cl_2)$ was also lower than that of other two samples. X-ray diffraction experimants of K_3C_{60} showed similar results as Rb_3C_{60}.

It seems to suggest that the differences of superconductivity for various samples were due to crystal structures of doped C_{60}. The fractional diamagnetic shielding for alkali-fulleride(CH_2Cl_2) reached to the saturation point rapidly, which may be related to the reaction time in diffusing alkali-metal into the crystal lattice of solid C_{60}. It is indicated that alkali-metal atom may diffuse into structure of $C_{60}(CH_2Cl_2)$ easily since the packing factor of $C_{60}(CH_2Cl_2)$ is smaller than fcc C_{60}. The lower Tc of alkali-fulleride(CH_2Cl_2) may be involved in structure of $C_{60}(CH_2Cl_2)$, such as the size and the number of interstitial sites.

In summary, we have observed the new crystal structure of C_{60} which was grown from CH_2Cl_2 solution. The superconductivity of alkali-metal doped $C_{60}(CH_2Cl_2)$ was demonstrated by a Meissner effect. It seems that the superconductor of alkali-metal doped $C_{60}(CH_2Cl_2)$ differs from that of other samples in crystal structure because those show different superconducting transition temperature and X-ray diffraction. It is expected that optimization of composition and crystallinity will lead to futher improvement of superconducting properties.

REFERENCES

1. H.W.Kroto, J.R.Heath, S.C.O'Brien, R.F.Curl, R.E.Smalley (1985) Nature 318: 162
2. W.Kratschmer, L.D.Lamb, K.Fostiropouos, D.R.Haffman (1990) Nature 347:354
3. J.E.Fischer, P.A.Heiney, A.R.McGhie, W.J.Romanow, A.M.Denenstein,J.P.McCaule A.B.Smith (1991) Sience 252:1288
4. P.A.Heiney, J.E.Fischer, A.R.McGhie, W.J.Romanow, A.M.Denenstein,J.P.McCaule A.B.Smith (1991) Prys. Rev. Lett. 66:2911
5. K.Kikuchi, H.Saito, S.Suzuki, I.Ikemoto, Y.Achiba, (1991) Conference on C_{60}. (Tokyo)
6. R.M.Fleming, A.R.Kortan, B.Hessen, T.Siegrist, F.A.Thiel, R.C.Haddon, R.Tycho, G.Dabbagh,M.L.Kaplan, A.M.Mujsce (1991) Phys.Rev.B 44:888
7. A.F.Hebard, M.J.Rosseinsky, R.C.Haddon, D.W.Muphy, S.H.Glarum, T.T.M.Palstray, A.P.Ramirez, A.R.Kortan (1991) Nature 350:600
8. M.J.Rosseinsky, A.P.Ramirez, S.H.Glarum, D.W.Muphy, R.C.Haddon, A.F.Hebard y, T.T.M.Palstra, A.R.Kortan, S.M.Zahurak, A.V.Makhija (1991) Phy.Rev.Letter 66:2830

9. K.Holczer, O.Klein, S.M.Huang, Richerd, B.Kaner, K.J.Fu, R.L.Whetten, F.Diederich (1991) Sience 252:1154
10. C.C.Chen, S.P.Kelty, C.M.Lieber (1991) Sience, 253:886
11. A.Ugawa, H.Yamakado, K.Kikuchi, S.Suzuki, K.Saito, Y.Achiba, I.Ikemoto (1991) Conference on C_{60} (Tokyo)
12. Y.Maruyama, Y.Achiba (1991) M^2S-HTSC III (Kanazawa)

Preparation of $Ba_{1-x}A_xBiO_3$(A=K,Rb) and Phase Separation in Nitrogen Atmosphere

AKIRA IYO, HIROMOTO UWE, and TUNETARO SAKUDO

Institute of Applied Physics, University of Tsukuba, Tsukuba, Ibaraki, 305 Japan

ABSTRACT

In preparation of $Ba_{1-x}K_xBiO_3$ and $Ba_{1-x}Rb_xBiO_3$ in atmosphere of nitrogen gas, the compounds in a certain range of x are found to be stable only above a certain temperature T_N, below which the compounds separate into two phases. The compounds were sintered in nitrogen, and on cooling process the atmosphere was exchanged from nitrogen to oxygen. From the powder X-ray diffraction at room temperature, and by using the relation between the lattice parameter and the composition, we obtained phase diagrams. $Ba_{1-x}K_xBiO_3$ ($Ba_{1-x}Rb_xBiO_3$) undergoes the phase separation between x=0.08 (0.15) and x=0.28 (0.44) composition, and T_N at x=0.2 (0.3) is 760°C (640°C).

KEY WORDS: $Ba_{1-x}K_xBiO_3$, $Ba_{1-x}Rb_xBiO_3$, phase separation, preparation method

INTRODUCTION

Since the discovery of superconductivity near 30K in $Ba_{1-x}K_xBiO_3$ [1, 2], this material has evoked researches not only for physical properties but also for applications for devices. The distinctive features of this system are the absence of magnetic order and two dimensional conduction planes which are believed to cause the high-Tc in the copper-oxide superconductors. Studies of superconductivity of this $Ba_{1-x}K_xBiO_3$ system may give hints of the superconducting mechanism of the high-Tc copper-oxide superconductors. From this aspects, many kinds of experiments have been carried out.

It is important to approach from semiconducting phase for a proper understanding of the origin of superconductivity in this system, as is done in high-Tc cuprates superconductors. However, systematic studies from semiconducting phase to superconducting phase have not carry out sufficiently. Furthermore, the experimental results have not always agree with one another. This seems to be due to the difficulties of sample preparation in this system. Though some preparation methods have been reported at present [2, 3, 4], it is necessary to reconsider the method to obtain good quality samples. In our previous report, we found the phase separation in nitrogen atmosphere in $Ba_{1-x}Rb_xBiO_3$ [5]. In this study, we have investigated the system of $Ba_{1-x}K_xBiO_3$ (BKB) and also $Ba_{1-x}Rb_xBiO_3$ (BRB) in more detail. Both systems of BKB and BRB are found to undergo phase separation in nitrogen atmosphere in a certain range of composition x.

EXPERIMENTAL

Stoichiometric mixture of BaO_2, KO_2 (Rb_2O), and Bi_2O_3, which was ground in dry helium gas, was placed in an alumina crucible and sintered at a temperature from 750°C to 950°C (=950-500x) for 2 hours in flowing nitrogen. On cooling process after the firing, the temperature was held at a certain temperature for several hours, and after the atmosphere was exchanged from nitrogen to oxygen, the sample was cooled in the furnace to room temperature. Containing phases in the samples were supposed to be fixed when oxygen gas was introduced, and the sample was analyzed by the powder X-ray diffraction (XRD) method at room temperature. The temperature at which the atmosphere was exchanged from nitrogen to oxygen was varied.

In oxygen atmosphere, samples of high x composition could be oxidized easily at low temperature (ex. 320 °C for x=0.4 in BKB) but samples of low x composition had to be oxidized at higher temperature (ex. 480 °C for x=0.05 in BKB). Therefore, for the case of the gas-exchange temperature too low for the oxidation, after taking XRD pattern for the phase oxidized already, samples were annealed in oxygen gas at an appropriate higher temperature and analyzed by powder XRD again.

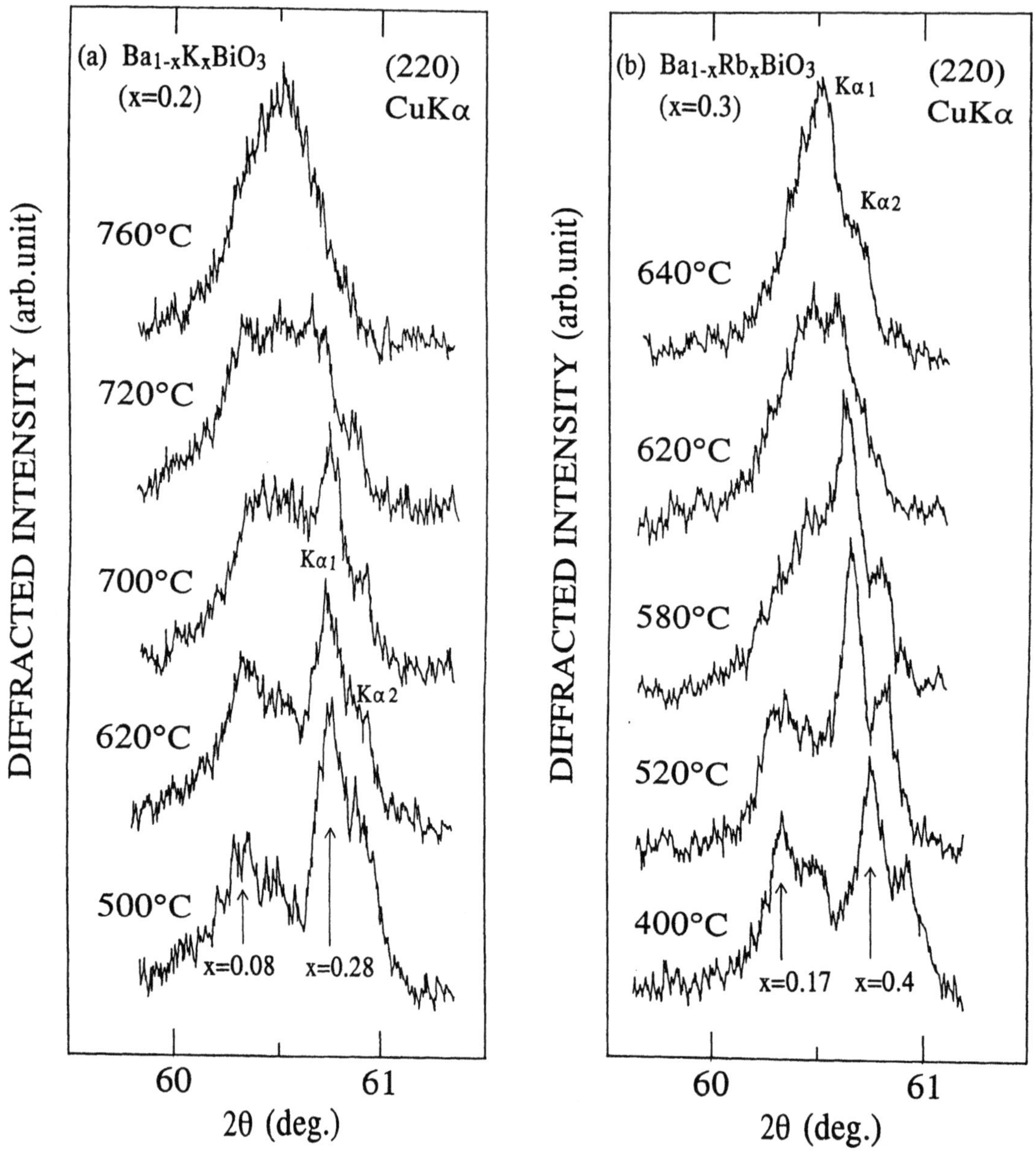

Fig. 1(a) and (b). The X-ray diffraction peaks for the (220) plane in the pseudo cubic phase as a function of the temperature at which the atmosphere was replaced from nitrogen to oxygen for $Ba_{0.8}K_{0.2}BiO_3$ and $Ba_{0.7}Rb_{0.3}BiO_3$.

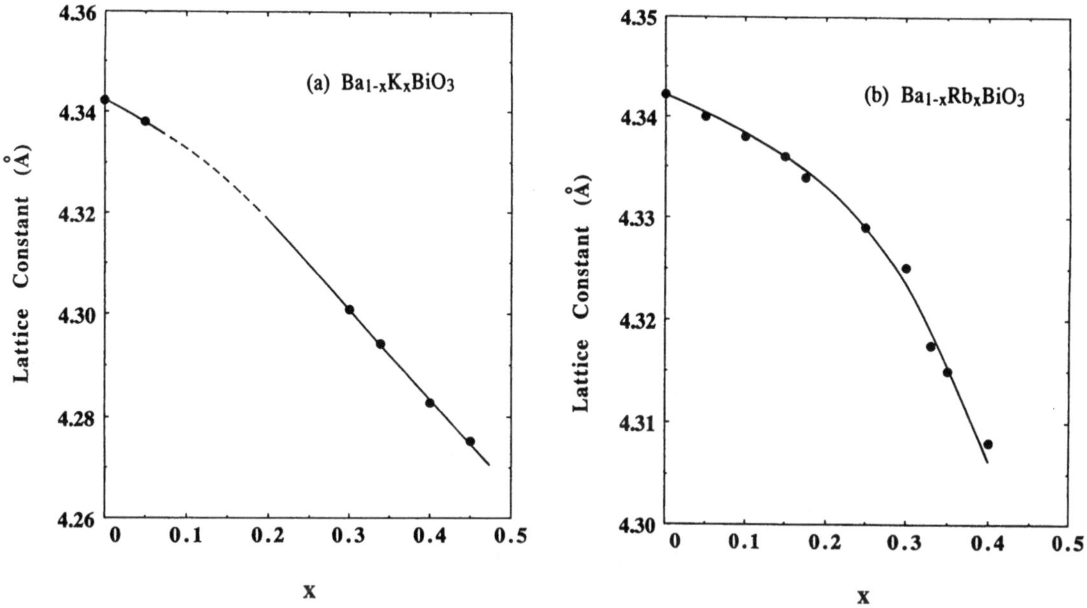

Fig. 2(a) and (b). The relation between the lattice parameter and the composition x for $Ba_{1-x}K_xBiO_3$ and $Ba_{1-x}Rb_xBiO_3$, respecively. The date points were determined by the highest intensity X-ray diffraction peak for the (220) planes in the pseudo cubic phase.

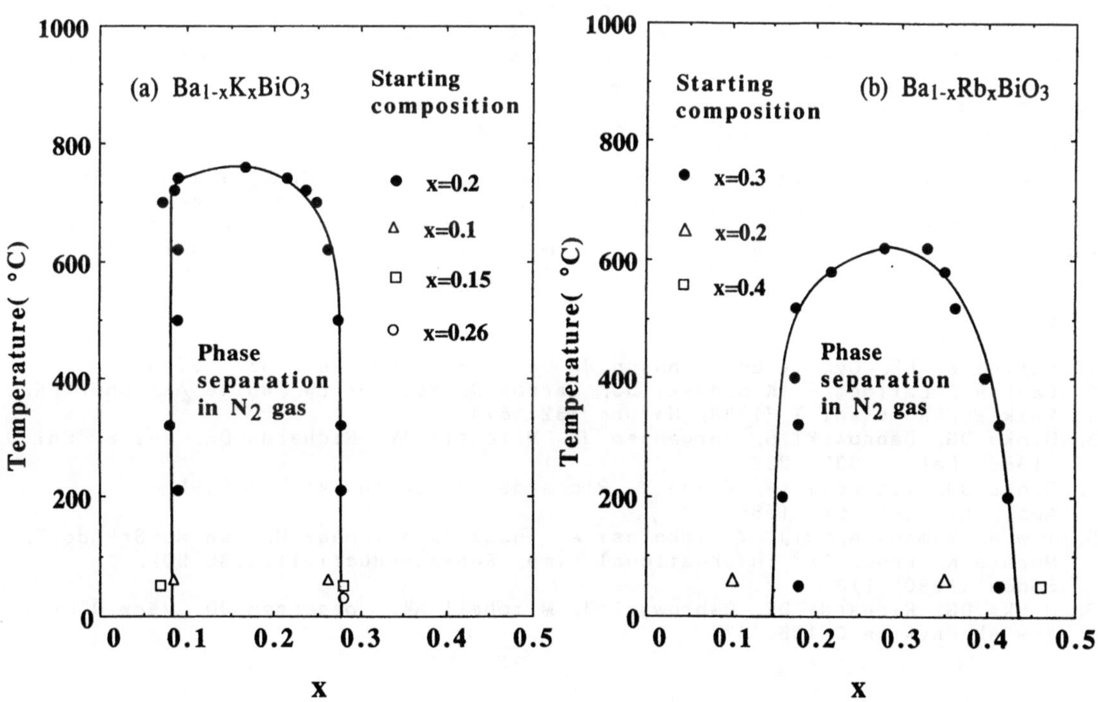

Fig. 3(a) and (b). The phase diagrams of $Ba_{1-x}K_xBiO_3$ and $Ba_{1-x}Rb_xBiO_3$ in nitrogen atmosphere.

RESULTS AND DISCUSSION

If a sample consists of multiphases, the powder X-ray diffraction exhibits a superposed pattern from different compositions x. In fig. 1(a) and (b), the XRD peaks for the (220) planes in the pseudo cubic phase are shown for various samples as a function of the temperature at which the atmosphere was replaced from nitrogen to oxygen for $Ba_{0.8}K_{0.2}BiO_3$ and $Ba_{0.7}Rb_{0.3}BiO_3$, respectively. As the temperature is lowered, peaks are found to be split into two parts, which exhibits the phase separation.

By using the relation between the lattice parameter and the composition x shown in Figs. 2(a) and(b), the composition x of each separated phases was determined. For $Ba_{1-x}K_xBiO_3$, this relation was obtained from an interpolation and the data for samples for composition x which did not undergo the phase separation even at room temperature. For $Ba_{1-x}Rb_xBiO_3$ the relation was obtained from samples which were prepared by quenching from T_N to room temperature after atmosphere was replaced by oxygen as will be mentioned later.

Figure 3(a) and (b) show the phase diagrams thus obtained for BKB and BRB in nitrogen atmosphere, respectively. For BKB, $Ba_{0.8}K_{0.2}BiO_3$ can exist as a single phase only above 760°C in nitrogen gas, below this temperature the sample undergoes phase separation into x=0.08 and x=0.28 compositions. Samples for composition x=0.1, 0.15, 0.26 also underwent phase separation into about x=0.08 and x=0.28 compositions in nitrogen gas at near room temperature, (Fig. 3(a)) and the samples for composition x=0, 0.05, 0.3, 0.34, 0.4, 0.46 did not undergo phase separation till room temperature in nitrogen gas. So it is easy to obtain single-phase samples for 0<x<0.08, 0.28<x<0.5 compositions by the two-step synthesis method [3]. But it is difficult to prepare the composition for 0.08<x<0.28.

For BRB, $Ba_{0.7}Rb_{0.3}BiO_3$ found to be stable only above 640°C, below which compounds separate into x=0.15 and x=0.44 compositions at room temperature. BRB undergoes the phase separation for a wider range of composition x than BKB. In order to obtain single phase samples which would undergo phase separation in nitrogen atmosphere, we used a following method. On cooling process after firing in nitrogen gas, the atmosphere is replaced by oxygen gas at the temperature before the phase separation causes, and then the sample is quenched to room temperature in oxygen gas, because another kind of phase separation undergoes at high temperature in oxygen gas. [6]

In summary, we have determined phase diagrams of $Ba_{1-x}K_xBiO_3$ and $Ba_{1-x}Rb_xBiO_3$ in nitrogen gas. $Ba_{1-x}K_xBiO_3$ undergoes the phase separation between x=0.08 and x=0.28 composition, and T_N at x=0.2 is 760°C. $Ba_{1-x}Rb_xBiO_3$ undergoes the phase separation between x=0.15 and x=0.44 composition, and T_N at x=0.3 is 640°C.

REFERENCES

1. Mattheiss LF, Gyorgy EM, Johnson Jr DW (1988) Phys.Rev. B37: 3745
2. Cava RJ, Batlogg B, Krajewski JJ, Farrow R, Rupp Jr LW, White AE, Shork K, Peak WF, Kometani T (1988) Nature 332: 814
3. Hinks DG, Dabrowski B, Jorgensen JK, Mitchell AW, Richards DR, Pei S, Shi D (1988) Nature 333: 836
4. Hinks DJ, Mitchell AW, Zheng Y, Richards DR, Dabrowski B (1989) Appl.Phys.Lett 54: 1585
5. Iyo A, Yamada A, Oiji Y, Takahasi A, Inoue A, Watanabe H, Uwe H, Sakudo T, Murata K, Proc. 3rd International Symp. Superconductivity(ISS'90), Sendai(1990):113
6. Hinks DG, Richards DR, Dabrowski B, Mitchell AW, Jorgensen JD, Marx DT (1988) Physica C 156: 477

3.2 Substitution and Doping Effects

Superconductivity in Pb-Based "1222" Copper Oxides

NAOMICHI SAKAI[1], TOSHIHIKO MAEDA[1,*], FUJIO IZUMI[2], HAJIME ASANO[3], and H. YAMAUCHI[1]

[1] Superconductivity Research Laboratory, ISTEC, Koto-ku, Tokyo, 135 Japan
[2] National Institute for Research in Inorganic Materials, Tsukuba, Ibaraki, 305 Japan
[3] Institute of Materials Science, The University of Tsukuba, Tsukuba, Ibaraki, 305 Japan

ABSTRACT

The superconducting properties of the Pb-based 1222 copper oxides, $(Pb,Cu)(Ln,Sr)_2(Ln,Ce)_2Cu_2O_Z$ (Ln = Sm, Eu, Gd, Dy and Ho), have been found to strongly depend on the oxygen content as well as on the species of Ln ions. Although the samples with Ln= Sm, Eu, Gd, and Dy synthesized at ambient pressure exhibited superconductivity, the sample with Ln=Ho required post-annealing in an oxygen gas of high pressure to make it superconducting. We explain this in terms of the average ionic radius and the oxygen content. In addition, the structure parameters of superconducting $(Pb,Cu)(Sr,Nd)_2(Ho,Ce)_2Cu_2O_Z$ have been refined by the Rietveld analysis of neutron powder diffraction data.

KEY WORDS: Superconducting lead cuprates, "1222" structure, oxidizing atmosphere, ionic radius, Rietvelt analysis

INTRODUCTION

Since the first superconducting lead cuprate, Pb2Sr2(Y,Ca)Cu3O8 (the (Pb/Cu)-3212 phase), was discovered by Cava et al. [1], a variety of lead cuprates have been synthesized [2-12]. Adachi et al. synthesized $(Pb,Cu)(Sr,Pr)_2Pr_2Cu_2O_9$ (the (Pb/Cu)-1222 phase) which was the first lead cuprate containing fluorite blocks [11]. However, this compound was nonsuperconducting. Crystallographically, this phase was isomorphous with $TlSr_2(Nd,Ce)_2Cu_2O_9$ (the Tl-1222 phase). That is, the (Pb/Cu)-1222 structure contains (Pb,Cu)O monolayers instead of TlO monolayers of the Tl-1222 structure. Therefore, the crystal structure of the (Pb/Cu)-1222 phase contains two $[(Pb/Cu)_{0.5}(Sr,Pr)_{0.5}O]$ rock-salt blocks, $(Sr,Pr)CuO_{2.5}$ oxygen-deficient perovskite block and PrO_2 fluorite block. As Maeda et al. [13] proposed for the (Pb/Cu)-1212 phase, Pb and Cu ions in the (Pb,Cu)O monolayer of the (Pb/Cu)-1222 phase are expected to be approximately tetravalent and divalent, respectively.

Recently, Maeda et al. [14] have synthesized (Pb/Cu)-1222 compounds, $(Pb,Cu)(Sr,Eu)_2(Eu,Ce)_2Cu_2O_z$, which had a superconductivity transition temperature (T_c) around 25 K. Since then, many (Pb/Cu)-1222 superconductors were synthesized using various kinds of rare-earth elements [15]. In the present study, we investigated the dependence of superconducting properties on the oxygen content for the (Pb/Cu)-1222 compounds.

EXPERIMENTAL

Samples were prepared by a conventional solid-state-reaction method using highly pure powders of oxides and carbonates of the metallic elements. The powders were

*Present address: Yokohama R & D Laboratories, Furukawa Electric Co.,Ltd., 2-4-3, Okano, Nishi-ku, Yokohama 220, Japan

mixed so as to yield nominal compositions of $(Pb_{0.5}Cu_{0.5})(Sr_{1-x}Ln_{1}x)_2(Ln_{0.75}Ce_{0.25})_2Cu_2O_z$ (Ln= Sm, Eu, Gd, Dy and Ho; x= 0.1- 0.2) The mixed powders were calcined at 850°C for 10 h in air and then pressed into pellets. The pellets were fired at 1080°C for 1 h in a flowing O_2 gas and then cooled down to room temperature with a rate of 5°C/min. The samples were characterized by means of x-ray powder diffractometry. The temperature dependence of electrical resistivity was measured by a conventional four-probe method. The oxygen content, z, was measured by a coulometric titration technique [16]. The structure parameters of $(Pb,Cu)(Sr,Nd)_2(Ho,Ce)_2Cu_2O_z$ which was made superconducting by a high-pressure O_2 treatment were refined by the Rietveld method using time-of-flight neutron powder diffraction data.

RESULTS AND DISCUSSION

Figure 1 schematically shows the (Pb/Cu)-1222 structure which characteristically contains (Pb,Cu)O monolayers and fluorite blocks. A structure analysis using the neutron diffraction data for $(Pb_{0.5}Cu_{0.5})(Sr_{0.875}Nd_{0.125})_2(Ho_{0.69}Ce_{0.31})_2Cu_2O_z$ indicated that oxygen at the apical site of a $[CuO_5]$ pyramid was 14% deficient and that the (Pb,Cu)O monolayer was highly disordered and defective [17]. We previously reported that the superconducting properties of the (Pb/Cu)-1222 phase were strongly dependent on the effective ionic radius, r, of the (Ln,Ce) site [15]. That is, only samples of $(Pb_{0.5}Cu_{0.5})(Sr_{0.875}Ln_{0.125})_2(Ln_{0.75}Ce_{0.25})_2Cu_2O_z$ with Ln= Sm, Eu, Gd and Dy, which have r values close to 1.05 Å, showed superconductivity. The samples with r values for (Sr, Ln) less than 1.01 Å, for example Ln= Ho, did not exhibit superconductivity; such samples were hard to form the (Pb/Cu)-1222 phase.

Figure 2 shows the X-ray powder diffraction patterns for $(Pb_{0.5}Cu_{0.5})(Sr_{1-x}Ln_x)_2(Ln_{0.75}Ce_{0.25})_2Cu_2O_z$ with Ln= Sm, Eu, Gd, Dy and Ho. The Ln content x, was determined so as to make the r value for (Sr,Ln) to be 1.285 Å and to obtain single-phase samples. However the samples containing Dy and Ho included secondary phases, while the samples containing Sm, Eu and Gd were nearly pure. Figure 3 shows the lattice constants as a function of r for (Ln,Ce) in the fluorite block.

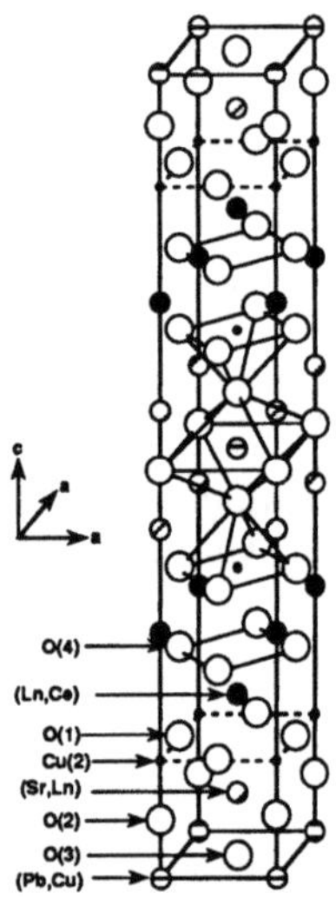

Fig.1 Tetragonal unit cell of $(Pb,Cu)(Sr,Eu)_2(Eu,Ce)_2Cu_2O_9$.

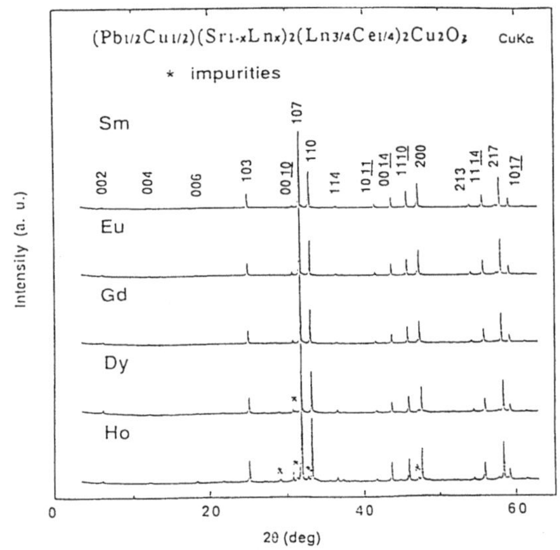

Fig.2 XRD patterns for $(Pb_{0.5}Cu_{0.5})(Sr_{1-x}Ln_x)_2(Ln_{0.75}Ce_{0.25})_2Cu_2O_z$ with Ln=Sm, Eu, Gd, Dy and Ho.

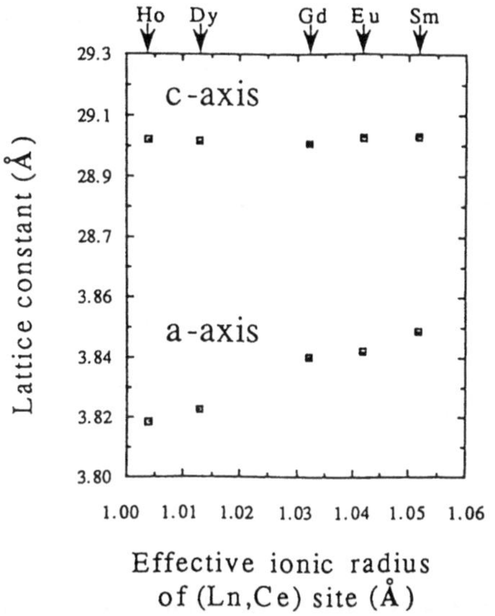

Fig.3 Lattice constants, a and c, vs effective ionic radii for the (Ln,Ce) site for $(Pb_{0.5}Cu_{0.5})(Sr_{1-x}Ln_x)_2(Ln_{0.75}Ce_{0.25})_2Cu_2O_z$ with Ln=Sm,Eu,Gd,Dy and Ho.

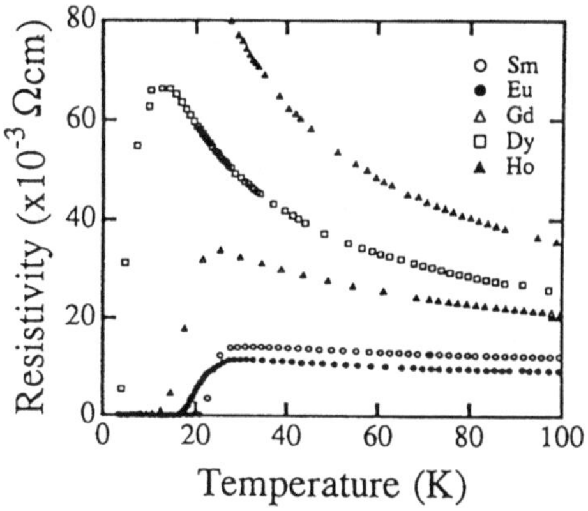

Fig.4(a) Temperature dependence of electrical resistivity for $(Pb_{0.5}Cu_{0.5})(Sr_{1-x}Ln_x)_2(Ln_{0.75}Ce_{0.25})_2Cu_2O_z$ containing Sm, Eu, Gd, Dy and Ho.

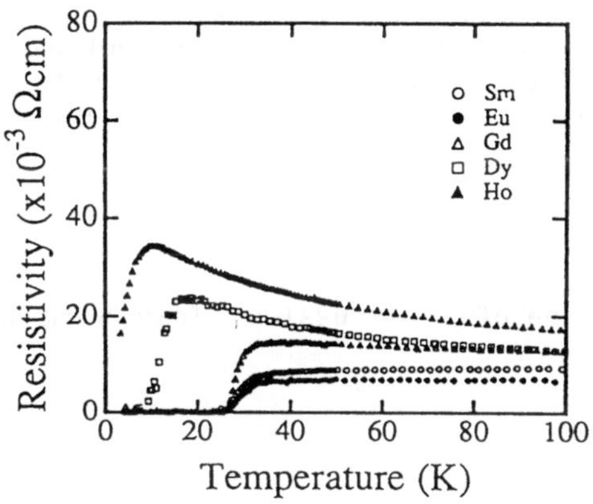

Fig.4(b) Temperature dependence of electrical resistivity for $(Pb_{0.5}Cu_{0.5})(Sr_{1-x}Ln_x)_2(Ln_{0.75}Ce_{0.25})_2Cu_2O_z$ containing Sm, Eu, Gd, Dy and Ho, annealed in 90 atm oxygen atmosphere.

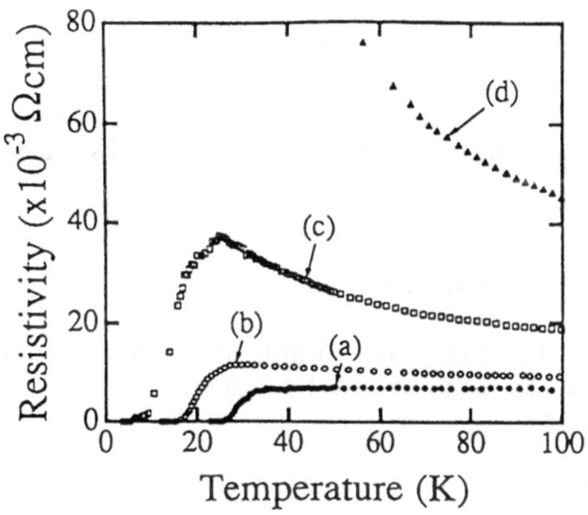

Fig.5 Temperature dependence of electrical resistivity for samples (a) annealed at 500℃ for 4 h in 90 atm oxygen (b) as sintered (c) annealed at 400℃ for 10 h in air and (d) annealed at 700℃ for 10 h in a nitrogen atmosphere.

The length of an a-axis was proportional to the r value for (Ln,Ce). In superconductors containing fluorite blocks, the extent of oxygen deficiency increases with decreasing a value [18]. Therefore, the annihilation of superconductivity in the sample containing Ho is probably due to the considerable oxygen deficiency. In order to introduce oxygen as much as possible, post annealing in an O_2 gas of high pressure was carried out.

Figure 4 shows the temperature dependence of electrical resistivity for (a) as-sintered samples and (b) samples annealed at 500°C for 4 h in 90 atm O_2 gas. The superconducting properties of all the samples were found to be fairly improved after the post-annealing. Especially, the as-sintered sample with Ln= Ho, which did not exhibit superconductivity, became superconducting after the post-annealing. Figure 5 shows the temperature dependeces of electrical resistivity for the samples containing Ln= Eu, which were annealed at various O_2 partial pressures. Table 1 lists oxygen contents, z, determined by the coulometric titration method and the lattice constants, a and c. As z was increased, T_c rose and the lengths of the a- and c-axis decreased and increased, respectively. This is, the superconducting properties of the (Pb/Cu)-1222 phase proved to be strongly dependent on the oxygen content.

Conditions	Z	T_c (K)	a(Å)	c(Å)
O_2, 500℃, 90 atm, 40hr	9.30	32	3.845_3	29.01_1
As sintered	9.23	24	3.842_3	29.02_6
N_2, 700℃, 10hr	9.02	Non	3.840_6	29.06_4

$(Pb_{1/2}Cu_{1/2})(Sr_{1-x}Eu_x)_2(Eu_{3/4}Ce_{1/4})_2Cu_2O_z$ x=0.1025

Table.1 Oxygen content, superconducting transition temperature T_c(onset) and lattice constants for $(Pb_{0.5}Cu_{0.5})(Sr_{0.8975}Eu_{0.1025})_2(Eu_{0.75}Ce_{0.25})_2Cu_2O_z$ annealed under varius conditions.

SUMMARY

The superconducting properties of the (Pb/Cu)-1222 compounds, $(Pb_{0.5}Cu_{0.5})(Sr_{1-x}Ln_x)_2(Ln_{0.75}Ce_{0.25})_2Cu_2O_z$ (Ln= Sm, Eu, Gd, Dy, Ho), were investigated. T_c's were strongly dependent on both the oxygen content and the effective ionic radius for the (Ln,Ce) site. Although the samples containing Sm, Eu, Gd and Dy synthesized in oxidizing atmospheres exhibited superconductivity, the sample containing Ho required post-annealing in the O_2 gas of high pressure to make superconducting. The Rietveld analysis using neutron powder diffraction data showed that the apical oxygen site in the CuO_5 pyramid in superconducting $(Pb,Cu)(Sr,Nd)_2(Ho,Ce)_2Cu_2O_z$ is considerably deficient. Therefore, the annihilation of superconductivity in this sample is probably due to the oxygen deficiency.

ACKNOWLEDGEMENTS

We would like to thank K. Sakuyama for his helpful discussion. This work was partially supported by NEDO.

REFERENCES

[1] R. J. Cava et al., Nature 336 (1988) 211.
[2] M. A. Subramanian et al., Physica C 159 (1989) 124.
[3] O. Inoue et al., Jpn. J. Appl. Phys. 28 (1989) L60.
[4] T. Rouillon et al., J. Solid State Chem. 78 (1989) 322.
[5] J. Y. Lee et al., J. Mater. Res. 4 (1989) 763.
[6] H. W. Zandbergen et al., Physica C 159 (1989) 81.
[7] T. Rouillon et al., Physica C 159 (1989) 201.
[8] A. Tokiwa et al., Physica C 161 (1989) 459.
[9] T. Rouillon et al., Physica C 167 (1990) 107.
[10] R. J. Cava et al., Physica C 167 (1990) 67
[11] S. Adachi et al., Physica C 168 (1990) 1.
[12] S. Adachi et al., Jpn. J. Appl. Phys. 29 (1990) L890.
[13] T. Maeda et al., Phys. Rev. B 43 (1991) 7866.
[14] T. Maeda et al., Physica C 169 (1990) 133.
[15] T. Maeda et al., Physica C 177 (1991) 337.
[16] K. Kurusu et al., Analyst 114 (1989) 1341.
[17] T. Maeda et al., to be submitted.
[18] A. Tokiwa et al., Solid State Phys. 25 (1990) 995.

Cation Substitution Effect on Oxygen Nonstoichiometry in Bi-232 Cuprate Superconductors

F. MUNAKATA, T. KAWANO, M. WAKATA, S. TAKANO, and H. YAMAUCHI

Superconductivity Research Laboratory, ISTEC, Koto-ku, Tokyo, 135 Japan

ABSTRACT

The effects of oxygen nonstoichiometry on structural modulation and superconducting properties of Bi cuprate superconductors were investigated. The results suggested that the superconducting properties depended on the Cu valency and the modulation period was correlated with the Bi valency. For $Bi_{2-x}Pb_xSr_2CaCu_2O_{8+y}$ samples of nearly the same $Tc^{on\ set}$'s, it was found that the normal state conduction changed from metallic to semiconductive with increasing Pb content. This suggested that the cation substitution affected oxygen nonstoichiometry and the conduction mechanism.

KEY WORDS: Bi cuprate superconductor, Cation substitution effect, Oxygen nonstoichiometry

INTRODUCTION

Since the discovery of 105 K Bi superconductors,[1] a number of studies have been carried out on $Bi_2Sr_2Ca_{n-1}Cu_nO_{2n+4}$(n=1,2,3) phase.[1,2] The partial substitution(i.e. x > 0.6) of Ln(=Y, Nd, Eu or Gd) for Ca in superconducting $Bi_2Sr_2CaCu_2O_y$ resulted in an antiferromagnetic insulator.[3-5] The structual modulation along the b-axis in $Bi_2Sr_2Ca_{1-x}Y_xCu_2O_y$ was also investigated.[5,6] In this system, the incommensurate modulation period of 4.8b changed linearly to a commensurate one with a value equal to 4.0b as the Y content was increased from x = 0.0 to x = 1.0. On the other hand, partial replacement of Bi by Pb improved the superconducting temperatire as well as increased the periodicity of the modulation resulting finally in a disappearance of the modulation in a highly substituted $Bi_{2-x}Pb_xSr_2CaCu_2O_y$.[7] Recently, we reported that $Bi_2Ca_{3-x}Ln_xCu_2O_z$(Ln=Pr and Nd)[8] and $Bi_2Ca_{3-x}(La_yLn_{1-y})_xCu_2O_z$(Ln=Y or a rare earth element except Lu and Yb)[9] exhibited superconductivity. The modulation in $Bi_2Ca_{2.5}Nd_{0.5}Cu_2O_y$ and $Bi_2Ca_{2.5}La_{0.25}Pr_{0.25}Cu_2O_y$ is nearly commensurate with a period equal to 9 b.[10]

In this paper, we report the effects of oxygen nonstoichiometry on superconducting properties and structural modulation in Bi cuprate superconductors , and discuss the relationship between the cation substitution effect on oxygen nonstoichiometry in Bi-232 cuprate superconductors and superconducting properties.

EXPERRIMENTAL

The samples of $Bi_2Sr_2Ca_{1-x}Y_xCu_2O_{8+y}$, $Bi_{2-x}Pb_xSr_2CaCu_2O_{8+y}$, and $Bi_2Ca_{2.5}(Ln_{0.5}Ln^*_{0.5})_{0.5}Cu_2O_{8+y}$(Ln, Ln* = La, Nd, Sm, Eu, Gd, Y, and Pr) were synthesized by a conventional solid-state reaction method. After being calcined in air at 1073 K and 1033K for 10 hours, the powders were ground and pressed into parallelepiped bars of 3 x 3 x 20 mm^3. The bars of $Bi_2Sr_2Ca_{1-x}Y_xCu_2O_{8+y}$ were then sintered for 35 hours in air : the x = 0.0 sample at 1143K and the others at 1203K. The sintering of the $Bi_{2-x}Pb_xSr_2CaCu_2O_{8+y}$ samples was performed in N_2 at temperature between 1043K and 1093K for 30 hours. Finally, the bars of $Bi_2Ca_{2.5}(Ln_{0.5}Ln^*_{0.5})_{0.5}Cu_2O_{8+y}$ were sintered at a temperature range from 1113K to 1133K for 35 hours in O_2 and cooled to 773K at a rate of 60K/hour. At this temperature Then the samples were O_2-annealed for 10 hours and then cooled down to room temperature at the same rate. The structure analysis were carried out by x-ray powder diffraction(XRD). The oxygen content and the valencies of Cu and Bi determined by a coulometric titoration.[11]

Superconducting transition onset temperature were determined by AC magnetization experiments and electrical resistivity was measured at temperatures below room temperature using a standard four-probe method. The modulattion structures were sintered by TEM(Hitachi H-800). The TEM specimens, i.e. ceramic flakes, were prepared by crushing the sample bars.

RESULTS AND DISCUSSION

Figure 1 shows the dependences of Tc $^{on\ set}$, the oxygen content(8+y), the average valencies of both the Cu(p(Cu)) and Bi(q(Bi)) ions as well as the modulation period(m) on the Y content(x) for $Bi_2Sr_2Ca_{1-x}Y_xCu_2O_{8+y}$ samples. The data suggests that superconductivity occurs only when the hole number(p for $(Cu\text{-}O)^{+p}$) exceeds 0.1. The average valency of Bi first monotonically decreases from 3.12 to 2.99 as x increases and remains constant for x greater than 0.8. The change in the modulation period with increasing x appears to follow the same tendency than the Bi valency.

From the data given in Figure 2 it is conclude that in the $Bi_{2-x}Pb_xSr_2CaCu_2O_{8+y}$ system. The Cu valency is almost independent of the Pb content while the (Bi,Pb) valency monotonically decreases as the Pb content increased. On the contrary, the modulation period first increases and then disappears for the samples with x > 0.4. This result is explained by assuming excess oxygen ions in the Bi(Pb)-O layers.[12]

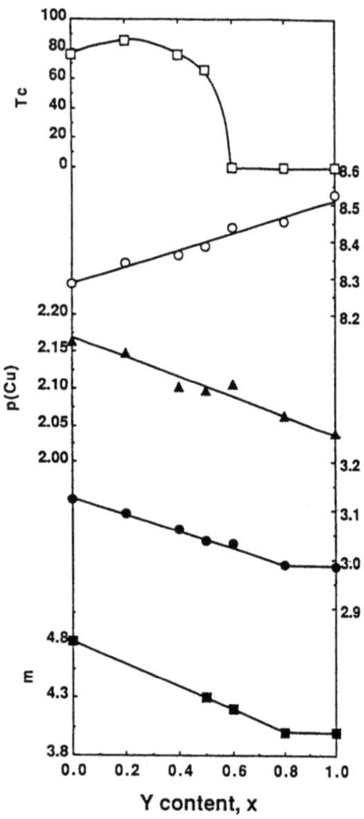

Fig. 1. Tc $^{on\ set}$, 8+y, p(Cu), q(Bi), and m vs Y content(x) for $Bi_2Sr_2Ca_{1-x}Y_xCu_2O_{8+y}$.

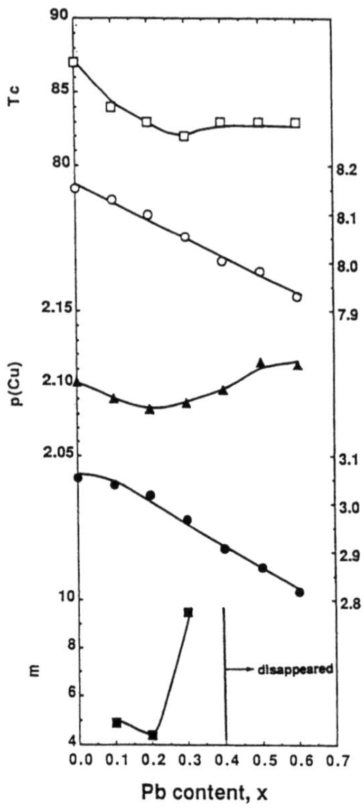

Fig. 2. Tc $^{on\ set}$, 8+y, p(Cu), q(Bi), and m vs Pb content(x) for $Bi_{2-x}Pb_xSr_2CaCu_2O_{8+y}$.

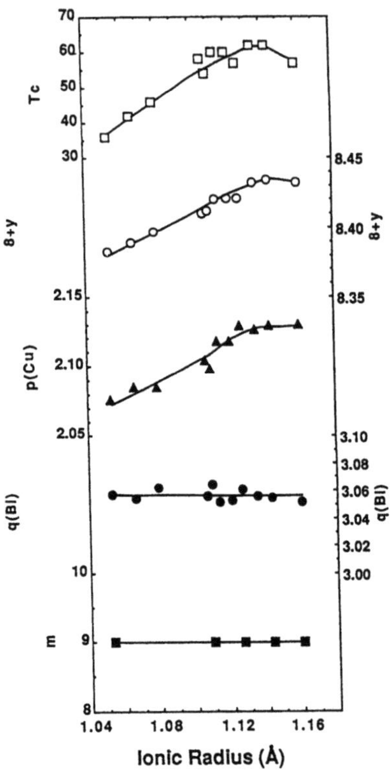

Fig. 3. Tc $^{on\ set}$, 8+y, p(Cu), q(Bi), and m vs mean ionic radius of rare-earth elements for $Bi_2Ca_{2.5}(Ln_{0.5}Ln^*_{0.5})_{0.5}Cu_2O_{8+y}$

Figure 3 summarizes the properties of $Bi_2Ca_{2.5}(Ln_{0.5}Ln^*_{0.5})_{0.5}Cu_2O_{8+y}$ as a function of the mean ionic radius of the rare-earth elements. As the mean ionic radius increases the oxygen contet of the sample increased, and the valency of Cu behaves in a same way. Also, the $T_c^{on\ set}$ values show the same tendecy the only exception being the sample with Ln*=La. On the other hand, the average valency of Bi (q(Bi)) does not depend on the ionic radius. These results suggest that the change in Tc is closely related to the valency of Cu(or the oxygen content) and is independent of the valency of Bi. Therefore, it is conclued from the results presented in Figures 1, 2, and 3 that the superconducting properties depend on the Cu valency and the modulation period correlates the Bi valency. This conclusion suggests that cation substitution affects the oxygen nonstoichiometry of the Bi and Cu ions as the same time, and results the change of the superconducting properties and crystal structure.

The temperature dependence of the resistivity in the $Bi_2Sr_2Ca_{1-x}Y_xCu_2O_{8+y}$ and $Bi_2Ca_{2.5}(Ln_{0.5}Ln^*_{0.5})_{0.5}Cu_2O_{8+y}$ systems was previouly reported.[8,10] According to these results Tc $^{on\ set}$ decreased with decreasing of Cu valency and along with the depression of T_c the metal-to-semiconductor transition in the normal state conduction resulted. However, there are only few reports on the normal state conduction properties of $Bi_{2-x}Pb_xSr_2CaCu_2O_{8+y}$. To investigate the effect of Pb substitution on the normal state conduction of $Bi_2Sr_2CaCu_2O_{8+y}$, the temperature dependence of the resistivity for $Bi_{2-x}Pb_xSr_2CaCu_2O_{8+y}$ samples was measured. The results of these measuraments are shown in Figure 4. The values for the superconducting transition temperature determined both by ac magnetization and the resistivity measuraments as well as zero-resistivity temperatures are also summarized in Table 1 for $Bi_{2-x}Pb_xSr_2CaCu_2O_{8+y}$. The superconducting transition temperature is nealy the same for each sample studied. However, the temperature dependences of the resistivity in $Bi_{2-x}Pb_xSr_2CaCu_2O_{8+y}$ samples show metallic-like behaviour in the compositional region for x < 0.4, and the semiconductive behavior for x > 0.4. The crystal structure of $Bi_{2-x}Pb_xSr_2CaCu_2O_{8+y}$ also depends on the Pb content exhibiting the tetragonal-orthorhombic -transition at x = 0.2, and the orthorhobic-monoclinic-transition at x = 0.4. The modulation period first increases from 4.8b (x=0.0) to 9.5b (x=0.3) with increasing Pb content but for the monoclinic samples with x > 0.4 no modulation is observed. The above results suggest that the Pb substitution affects the oxygen nonstoichiometry and the electronic structure.

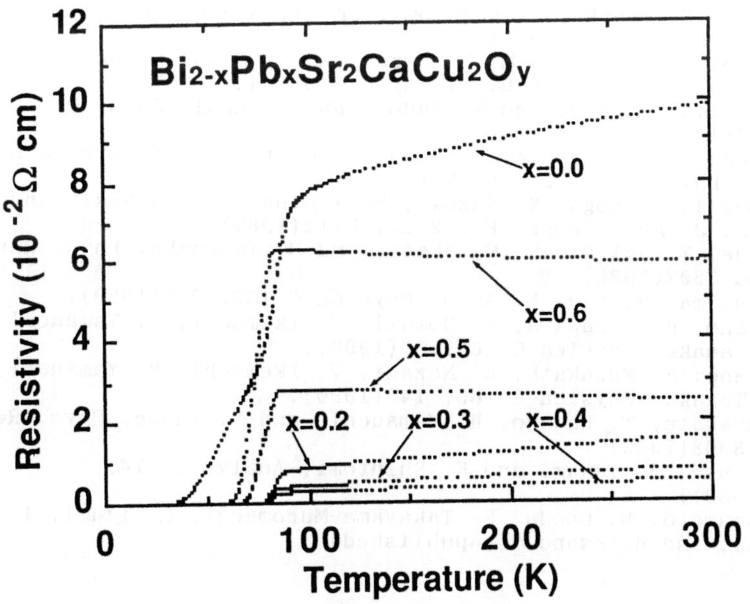

Fig. 4. Temperature dependence of the resistivity for $Bi_{2-x}Pb_xSr_2CaCu_2O_{8+y}$.

Table 1. Tc on set determined by AC magnetization ($T_c^{on\ set\ AC}$), Tc on set determined by resistivity ($T_c^{on\ set\ R}$), and zero-resistance temperature ($T_c^{zero\ R}$) for $Bi_{2-x}Pb_xSr_2CaCu_2O_{8+y}$.

x	0.0	0.2	0.3	0.4	0.5	0.6
$T_c^{on\ set\ AC}$	87K	83K	82K	83K	83K	83K
$T_c^{on\ set\ R}$	89K	85K	83K	84K	86K	84K
$T_c^{zero\ R}$	33K	79K	77K	78K	64K	60K

CONCLUSION

The effects of oxygen nonstoichiometry on structural modulation and superconducting properties of Bi cuprate superconductors were investigated. The results suggested that the superconducting properties depend on the Cu valency and the modulation period is correlated with the Bi valency. In $Bi_{2-x}Pb_xSr_2CaCu_2O_{8+y}$ having nearly the same Tc $^{on\ set}$, it was found that the normal state conduction properties changed from metallic to semiconductive with increasing Pb content. This suggested that the Pb substitution which changed the crystal structure affected the conduction mechanism.

ACKNOWLEDGMENTS

The authors would like to thank Dr. K. Matuura of ISTEC-SRL for his kind help in the resistivity measuraments. They are also grateful to Professor S. Tanaka for helpful discussions.

REFERENCES

1. H. Maeda, Y. Tanaka, M. Furutomi, and T. Asano, Jpn. J. Appl. Phys. Pt. 2 **27**, L209(1988).
2. C. Michel, M. Hervieu, M. M. Borel, A. Grandin, F. Deslandes, J. Provost, and B. Raveau, Z. Phys. B **68**, 421(1987).
3. N. Fukusima, H. Niu, and K. Ando, Jpn. J. Appl. Phys. Pt. 2 **27**, L790(1988).
4. T. Tamegai, A. Watanabe, K. Koga, I. Ogura, and Y. Iye, Jpn. J. Appl. Phys. Pt. 2 **27**, L1074(1989).
5. T. Tamegai, K. Koga, K. Suzuki, M. Ichihara, F. Sakaki, and Y. Iye, Jpn. J. Appl. Phys. Pt. 2 **28**, L112(1989).
6. Y. Inoue, Y. Shichi, F. Munakata, and M. Yamanaka, Phys. Rev. B **40**, pp.7307(1989).
7. N. Fukusima, H. Niu, K. Ando, Physica C **159**, 777(1989).
8. T. Kawano, F. Munakata, A. Nozaki, T. Ikemachi, H. Yamauchi, and S. Tanaka, Physica C **165**, 67(1990).
9. T. Kawano, F. Munakata, A. Nozaki, T. Ikemachi, H. Yamauchi, and S. Tanaka, Physica C **165**, 143(1990).
10. F. Munakata, T. Kawano, H. Yamauchi, and Y. Inoue, Phys. Rev. B, **42**, 8008(1990).
11. K. Kurusu, H. Takami and K. Shintomi, Analyst, **114**, L1341(1989).
12. A. Yamamoto, M. Onoda, E. Takayama-Muromachi, F. Izumi, T. Ishigaki, and H. Asano, (unpublished).

Effects of Li-Doping on the Superconductivity of $Bi_2Sr_2Ca_2Cu_3O_{10}$ Whiskers

ICHIRO MATSUBARA[1], TORU OGURA[1], HIROSHI YAMASHITA[1], MAKOTO KINOSHITA[1], and TOMOJI KAWAI[2]

[1] Government Industrial Research Institute Osaka, Ikeda, Osaka, 563 Japan
[2] The Institute of Scientific and Industrial Research, Osaka University, Ibaraki, Osaka, 567 Japan

ABSTRACT

Li-doped $Bi_2Sr_2Ca_2Cu_3O_{10}$ (2223) whiskers of single crystals with several millimeters length have been successfully prepared. It is revealed that Li-doping is effective for increasing critical temperature (Tc) of the 2223 phase. The whiskers are prepared by annealing $Bi_2Sr_2CaCu_2O_8$ (2212) whiskers in a calcined powder with the composition of $Bi_2Sr_2Ca_4Cu_6Pb_{0.5}Li_xO_y$ (x=0.2, 0.3, 0.4,). The 2212 whiskers are successfully converted into the 2223 phase only within a narrow temperature region around 840°C, which depends on the content of Li in the calcined powder. The Li content of x=0.5 is too high for the complete conversion into the 2223 phase. The Tc value of the Li-doped 2223 whiskers is reproducibly 1.1-1.3K higher than that of undoped 2223 whiskers.

KEY WORDS: oxide superconductor, $Bi_2Sr_2Ca_2Cu_3O_{10}$ phase, whisker, Li-doping,

INTRODUCTION

Li-doping into the $Bi_2Sr_2CaCu_2O_8$ (2212) compounds has recently attracted much attention because it is effective for increasing the critical temperature (Tc) for the 2212 phase. The Li-doping has been achieved by various methods, for example, conventional powder reaction method[1], electrochemical method[2], melt processing[3], recrystarization from glass[4], and floating zone method[5]. All of these methods have confirmed the effectiveness of the Li-doping. However, the mechanism of the enhancement of Tc has not been so clear. Several mechanisms have been proposed for the increasing of Tc such as the optimization of hole concentration, the change of atomic ratio of alkali earth metals, or change of electronic structure. The effects of Li-doping on the $Bi_2Sr_2Ca_2Cu_3O_{10}$ (2223) phase will give us another important insight into the role of Li in the Bi-based superconductors. Recently, we have succeeded to prepare the Li-doped 2223 whiskers of single crystals by Conversion by Annealing in Powder (CAP) method[6]. The original 2212 whiskers are converted into the 2223 whiskers by annealing in a Ca and Cu rich calcined powder. By using $Bi_2Sr_2Ca_4Cu_6Pb_{0.5}Li_xO_y$ (x=0.2, 0.3) powder as the calcined powder, the phase conversion and Li-doping have been spontaneously achieved. It has been reported that the addition of small amounts of lead enhances the growth of the 2223 phase by promoting the nucleation of the 2223 phase and the diffusivity of calcium and copper[7]. We added small amounts of lead to the calcined powder and confirmed the incorporation of lead into the whiskers by EPMA. The obtained Li-doped 2223 whiskers have higher Tc value compared with the undoped 2223 whiskers. In this paper, we report on the preparing conditions and properties of the Li-doped whiskers when we use the highly Li containing $Bi_2Sr_2Ca_4Cu_6Pb_{0.5}Li_xO_y$ (x=0.4, 0.5) powder in the CAP treatment.

EXPERIMENTAL

The 2223 whiskers of single crystals have been prepared by the CAP method[8]. The starting materials of the 2212 whiskers were prepared by annealing glass plates in a stream of O_2 gas at 860°C for 120h as described in our previous papers[9, 10]. Each whisker is composed of several plate-like single crystals which are stacked in a layered structure. The 2212 whiskers with such a structure were placed in a powder with the composition of $Bi_2Sr_2Ca_4Cu_6Pb_{0.5}Li_xO_y$ (x=0.4, 0.5) which had previously been calcined at

780°C for 20h and 820°C for 20h with intermediate grinding. The powder with whiskers was set between two sintered pellets with the composition of $Bi_2Sr_2Ca_4Cu_6Pb_{0.5}O_y$ as shown in Fig. 1, and annealed at 820-840°C for 150h. After cooling to a room temperature, the whiskers were removed mechanically from the powder. The obtained Li-doped 2223 whiskers have the dimensions of 1-5um thick, 10-100um wide, and 1-3mm long. The susceptibility was measured with a superconducting quantum interference device (SQUID) in a magnetic field of 30e from 5k to 120k (field cooling). The Li content was determined by atomic absorption analysis. The composition was determined by EPMA.

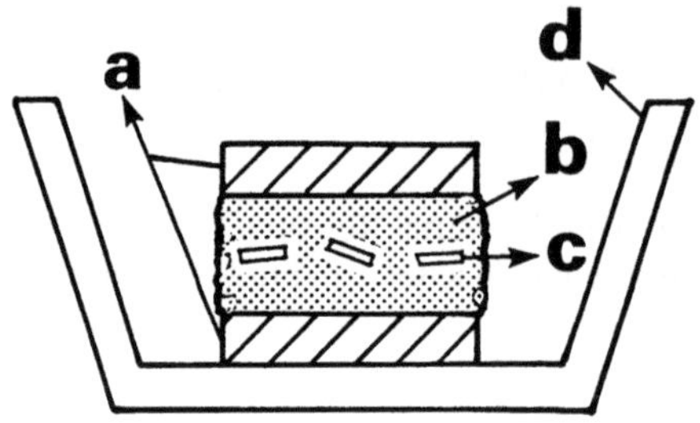

Fig. 1. Schematic representation of the CAP treatment. a: sintered pellets with a composition of $Bi_2Sr_2Ca_4Cu_6Pb_{0.5}O_y$ b: calcined powder with a composition of $Bi_2Sr_2Ca_4Cu_6Pb_{0.5}Li_xO_y$ (x=0.4, 0.5) c: whiskers d: alumina boat.

RESULTS AND DISCUSSION

Figure 2 shows the temperature dependence of susceptibility for the starting 2212 whiskers and the CAP treated whiskers at various temperatures for the (a) x=0.4 and (b) x=0.5 powders. When the CAP treatment is carried out at 840°C for the x=0.4 powder and at 838°C for the x=0.5 powder, we do not find the whiskers in the powder any more. This might have occurred through the erosion by liquid phase generated in the powder during annealing because the annealing temperature is in the range of partial melting of the powder. It is evident from the susceptibility signal around 75K that the unconverted 2212 phase is remained in the whiskers annealed lower than 835°C in the case of x=0.4 powder. The phase conversion is completed for the CAP treatment temperature of 838°C. Therefore, the annealing temperature should be controlled within the accuracy of about 2 degrees to obtain the pure 2223 whiskers. On the other

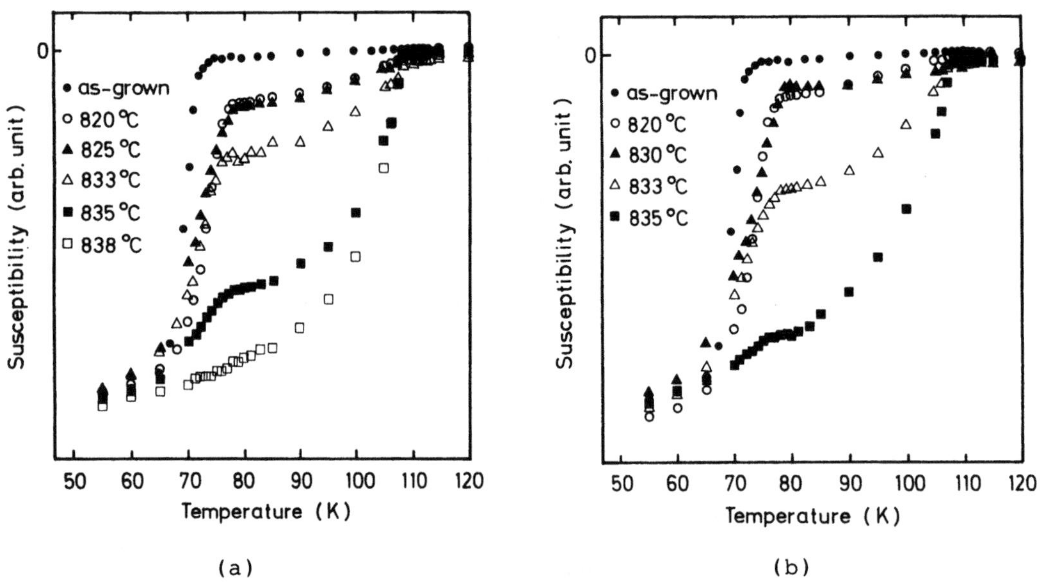

Fig. 2. Temperature dependence of susceptibility for the whiskers CAP treated at various temperatures by using the (a) x=0.4, (b) x=0.5 calcined powders.

hand, the complete phase conversion has not been observed in the case of the x=0.5 powder (Fig. 2(b)). In the whiskers annealed at 835°C, a small amount of the original 2212 phase is detected in the susceptibility signal. The appropriate annealing temperature for the complete phase conversion from the 2212 phase to the 2223 phase does not exist when we use the x=0.5 powder in the CAP treatment.

The relative atomic ratio for the CAP treated whiskers is shown in Fig. 3. The composition is calculated as the sum of Bi and Pb contents is fix to two. The composition of Cu, Ca, and Sr steeply increases with increasing the annealing temperature higher than 833°C for the x=0.4 powder(Fig. 3(a)). The composition of the whiskers annealed at 838°C is $(Bi_{1.84}Pb_{0.16})Sr_{1.36}Ca_{1.57}Cu_{2.82}O_y$. The content of Li is calculated to be 0.06 written in the chemical formula from the measurement of atomic absorption analysis. In the case of the powder with x=0.5 for the CAP treatment, although the phase conversion is not completed, the content of Cu, Ca, and Sr increases with increasing the conversion ratio from the 2212 phase to the 2223 phase at the annealing temperature higher than 830°C. The composition of the whiskers annealed at 835°C is $(Bi_{1.83}Pb_{0.17})Sr_{1.27}Ca_{1.47}Cu_{2.72}O_y$. The lack of Sr, Ca, and Cu content compared with the case of x=0.4 powder reflects the remaining of the original 2212 phase in the whiskers.

The most appropriate annealing temperature (Ta) in the CAP treatment and the Tc value for the whiskers obtained by using the calcined power with different Li content are summarized in table I. The Ta value decreases by 10-15°C with increasing the Li content in the calcined powder. In the DTA curve of calcined powder, a large endothermic peak due to the generation of liquid phase is observed. The peak appears at lower temperature for the power with higher Li content. The difference of Ta value for each powder shows a good correlation with the generation of liquid in the calcined powder, strongly suggesting the necessity of liquid phase for the phase conversion. When the Li content is increased in the calcined powder up to x=0.5 the erosion of the whiskers due to the liquid phase occurs before the complete phase conversion. Therefore the pure Li-doped 2223 whiskers are not obtained for the x=0.5 powder.

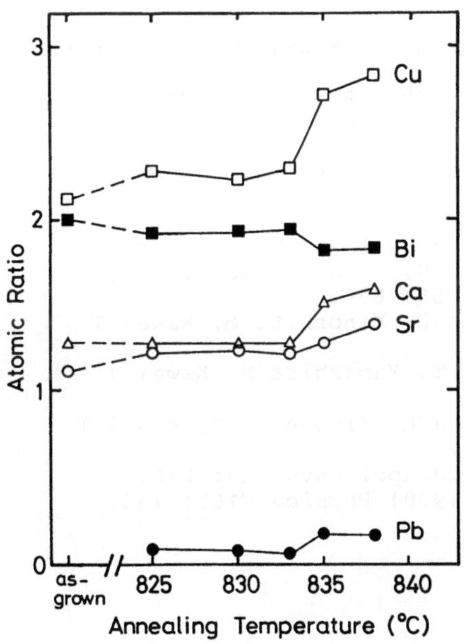

(a)

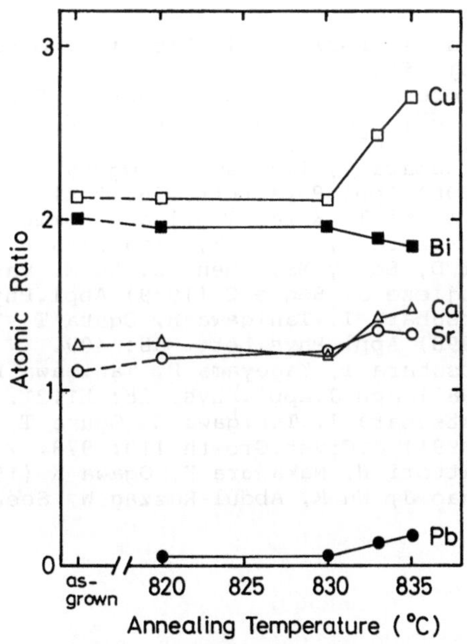

(b)

Fig. 3. Change of relative atomic ratio for the CAP treated whiskers using the (a) x=0.4, (b) x=0.5 calcined powder. The ratio is calculated as the sum of Bi and Pb contents is fixed to 2.

Table I. The most appropriate annealing temperature (Ta) in the CAP treatment and the Tc value for the whiskers obtained by using the $Bi_2Sr_2Ca_4Cu_6Pb_{0.5}Li_xO_y$ (x=0.0, 0.2, 0.3, 0.4, 0.5) calcined powder.

x[a]	Ta[b], °C	Tc[c], K	ref.
0.0	853	107.0	[6]
0.2	843	108.1	[6]
0.3	840	108.3	[6]
0.4	838	108.2	Present
0.5	---	108.3[d]	Present

a) Content of Li in $Bi_2Sr_2Ca_4Cu_6Pb_{0.5}Li_xO_y$ calcined powder.
b) The most appropriate annealing temperature for the phase conversion.
c) Determined by susceptibility measurement.
d) Data for the whiskers CAP treated at 835°C.

The Tc value of the Li-doped 2223 whiskers is enhanced compared with the undoped 2223 whiskers (x=0.0). The Tc value exceeding 108K is obtained for the Li-doped 2223 whiskers. The undoped whiskers have been prepared by the same method using a calcined powder without Li. The Tc value for the undoped whiskers(107.0K) is 1.1-1.3K lower than that for the Li-doped whiskers. This reproducible result indicates that the Li-doping is effective not only for the 2212 phase but also for the 2223 phase. The systematic and reproducible improvement of Tc by the doping has not been observed for the 2223 phase. In many cases, the doping of foreign atoms, such as In or Sb, makes the second phase. Some reasons for the effect of Li-doping are reported as described above. Present result suggests that the role of Li is not the optimization of hole concentration if the Li-doping acts as the same manner for both the 2212 and 2223 phase. The Tc of the 2223 phase is commonly accepted to be lowed by the reduction of hole concentration [11,12], that is the opposite situation to the case of 2212 phase.

REFERENCES

1. Kawai T, Horiuchi T, Mitsui K, Ogura K, Takagi S, Kawai S (1989) Physica C161: 561.
2. Strobel P, Bonnet B, Mouget C, Souletie B (1990) Physica C172: 193.
3. Dou SX, Lie HK, Wu WM, Wang WX, Sorrell CC, Winn R, Savvides N (1990) Physica C172: 295.
4. Matsubara I, Tanigawa H, Ogura T, Yamashita H, Kinoshita M, Kawai T (1990) Appl.Phys.Lett. 56: 2141.
5. Horiuchi T, Kawai T private communication.
6. Matsubara I, Ogura T, Yamashita H, Kinoshita M, Kawai T unpublished.
7. Shi D, Boley MS, Chen JG, Xu M, Vandervoort K, Liao YX, Zangvil A, Akujieze J, Segre C (1989) Appl.Phys.Lett. 55: 699.
8. Matsubara I, Tanigawa H, Ogura T, Yamashita H, Kinoshita M, Kawai T (1990) Appl.Phys.Lett. 58: 409.
9. Matsubara I, Kageyama H, Tanigawa H, Ogura T, Yamashita H, Kawai T (1989) Jpn.J.Appl.Phys. 28: L1121.
10. Matsubara I, Tanigawa H, Ogura T, Yamashita H, Kinoshita M, Kawai T (1991) J.Cryst.Growth 110: 973.
11. Hattori H, Nakanura K, Ogawa K (1990) Jpn.J.Appl.Phys. 29: L36.
12. Zhao J, Wu M, Abdul-Razzag W, Seehra MS (1990) Physica C165: 135.

Effects of Ti Addition for $Bi_{1.6}Pb_{0.4}Sr_2Ca_2Cu_3O_x$

KAZUO NOZAKI, SHINICHI ICHIMURA, TAKUYA KUROTAKI, KAZUYA OGURI, and YOSHITAKE NISHI

Department of Materials Science, Tokai University, Hiratsuka, Kanagawa, 259-12 Japan

ABSTRACT

Influence of Ti addition on T_c is studied for high T_c $Bi_{1.6}Pb_{0.4}Sr_2Ca_2Cu_3O_x$. Although the large amount of Ti is added, the high T_c (above 100 K) samples are found. In order to explain the T_c changes, electronic and morphological discussions are attempted.

KEY WORDS : Bi system superconductor, T_c, addition, titanium

INTRODUCTION

High T_c has been reported above 100 K for Bi-Pb-Sr-Ca-Cu-O [1-3]. It is important to know effects of solutes on T_c. Although we have recently found that the small addition (m < 0.01) of Ti [4], Ag [5] and Pd [6] don't largely decrease T_c of $YBa_2(Cu_{1-m}M_m)_3O_{7-y}$, influences of the solutes on T_c have never been investigated for Bi-Pb-Sr-Ca-Cu-O, yet. Thus, we undertake to obtain T_c informations by the Ti addition (substitution for Bi, Pb, Sr, Ca and Cu) for $Bi_{1.6}Pb_{0.4}Sr_2Ca_2Cu_3O_x$.

EXPERIMENTAL PROCEDURE

Samples with nominal composition were prepared from high-purity powders of CuO (99.9 mass%), TiO_2 (99.9 mass%), Bi_2O_3 (99.99 mass%), PbO (99.9 mass%), $SrCO_3$(99.9 mass%) and $CaCO_3$ (99.9 mass%). The powders were mixed and reacted in air at 1023 K for 6 h and then air-cooled. The reacted powders were resintered in air at 1073 K for 18 h and then air-cooled. After crushing, a pelletized tablet, 1 mm thick and 13 mm diameter, was sintered at 1123 K for 100 h and then furnace cooled in air.

The electrical resistivity [2,3] was measured in a cryostat at equilibrium temperatures. The electrical resistivity was measured using a standard 4-probe technique and a Keithley 181 nanovoltmeter. The temperature was measured by Au*Fe-chromel thermocouple attached to the specimen [2,3]. Figure 1 shows the temperature dependence of electrical resistance. The deviated point is taken at $d(R/R_{300K})/dT = 0.004$, where $d(R/R_{300K})/dT$ and R_{300K} are slope reduced electrical resistivity (R/R_{300K}) against the temperature graph and the resistivity at 300 K, respectively. The point of electrical resistivity is R^{on}. T_c^{mid} is designated the temperature of 1/2 R^{on}.

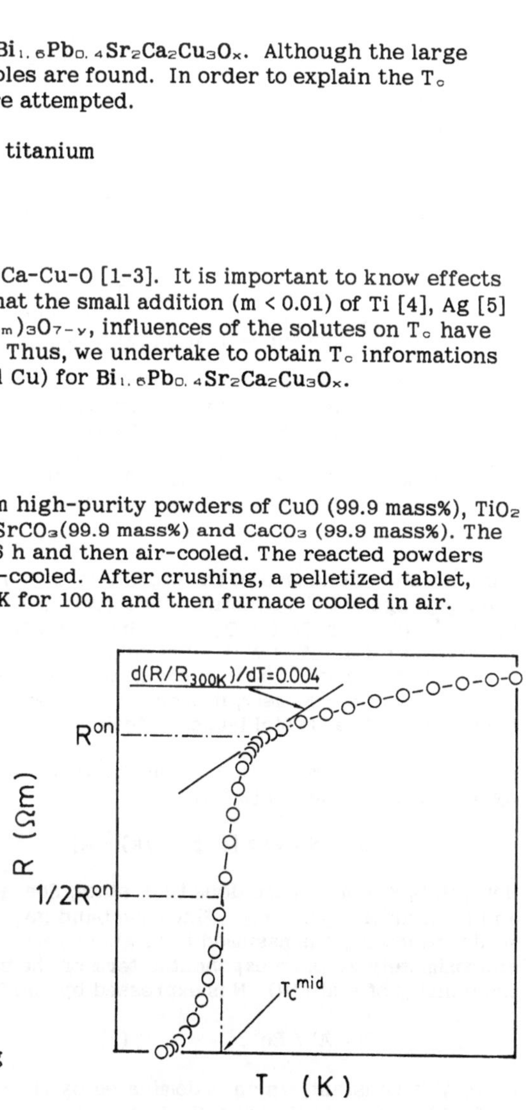

Fig.1 Change in electrical resistivity (R) versus temperature (T).

The structure of the sample was examined by means of x-ray diffraction (RU-200B, Rigaku Denki, Tokyo). The diffraction was performed under a step scanning method (Cu target, 40 kV, 50 mA, 0.01 deg/step, 10 s/step).

RESULTS AND DISCUSSION

Figure 2 shows changes in T_c^{mid} at 1 mA/mm^2 against the titanium concentration z for the $Bi_{1.6}Pb_{0.4}Sr_2Ca_2Cu_{3-z}Ti_zO_x$, $Bi_{1.6-z}Ti_zPb_{0.4}Sr_2Ca_2Cu_3O_x$, $Bi_{1.6}Pb_{0.4-z}Ti_zSr_2Ca_2Cu_3O_x$, $Bi_{1.6}Pb_{0.4}Sr_{2-z}Ti_zCa_2Cu_3O_x$ and $Bi_{1.6}Pb_{0.4}Sr_2Ca_{2-z}Ti_zCu_3O_x$. This is the first observation that adding the such a large amount of solutes does not largely decrease T_c for the high T_c oxides above

90 K of T_c, although the high T_c phase are found for a small addition below 0.01 of m in $YBa_2(Cu_{1-m}M_m)_3O_{7-y}$ [4-6]. Here, M is for Ti [4], Ag [5] and Pd [6], respectively. In order to explain the T_c changes, electronic and morphological discussions are attempted.

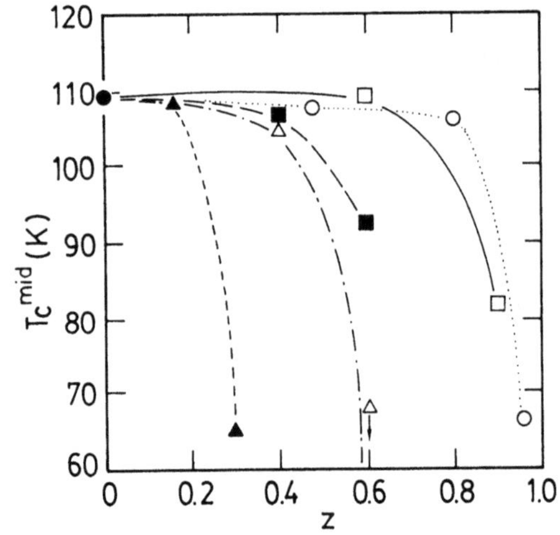

Fig.2 Changes in temperature (T_c^{mid}) of superconducting transition versus titanium concentration (z).

● $Bi_{1.6}Pb_{0.4}Sr_2Ca_2Cu_3O_x$
·····○····· $Bi_{1.6-z}Ti_zPb_{0.4}Sr_2Ca_2Cu_3O_x$
– –▲– – $Bi_{1.6}Pb_{0.4-z}Ti_zSr_2Ca_2Cu_3O_x$
—·–△–·— $Bi_{1.6}Pb_{0.4}Sr_{2-z}Ti_zCa_2Cu_3O_x$
– —■— - $Bi_{1.6}Pb_{0.4}Sr_2Ca_{2-z}Ti_zCu_3O_x$
——□—— $Bi_{1.6}Pb_{0.4}Sr_2Ca_2Cu_{3-z}Ti_zO_x$

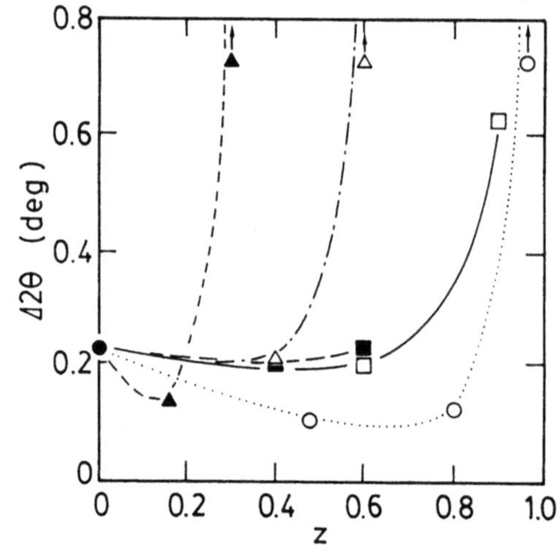

Fig.3 Changes in $\Delta 2\theta$ for high-T_c phase (triple CuO layer) of (002) plane against titanium concentration (z).

● $Bi_{1.6}Pb_{0.4}Sr_2Ca_2Cu_3O_x$
·····○····· $Bi_{1.6-z}Ti_zPb_{0.4}Sr_2Ca_2Cu_3O_x$
– –▲– – $Bi_{1.6}Pb_{0.4-z}Ti_zSr_2Ca_2Cu_3O_x$
—·–△–·— $Bi_{1.6}Pb_{0.4}Sr_{2-z}Ti_zCa_2Cu_3O_x$
– —■— - $Bi_{1.6}Pb_{0.4}Sr_2Ca_{2-z}Ti_zCu_3O_x$
——□—— $Bi_{1.6}Pb_{0.4}Sr_2Ca_2Cu_{3-z}Ti_zO_x$

Figure 3 shows changes in the width ($\Delta 2\theta$) of (002) plane of the high-T_c phase (triple Cu-O layer [7]) against the titanium concentration z for the $Bi_{1.6}Pb_{0.4}Sr_2Ca_2Cu_{3-z}Ti_zO_x$, $Bi_{1.6-z}Ti_zPb_{0.4}Sr_2Ca_2Cu_3O_x$, $Bi_{1.6}Pb_{0.4-z}Ti_zSr_2Ca_2Cu_3O_x$, $Bi_{1.6}Pb_{0.4}Sr_{2-z}Ti_zCa_2Cu_3O_x$ and $Bi_{1.6}Pb_{0.4}Sr_2Ca_{2-z}Ti_zCu_3O_x$. The high T_c (above 100 K) samples approximately show small $\Delta 2\theta$ values of the peaks. The changes in the $\Delta 2\theta$ values approximately agree with the changes in the T_c^{mid} values. Namely, it is one of dominated factors that the Ti additions keep the crystal perfection of the crystal lattice ordering.

A carrier concentration (c^*) is one of the dominant factors to control the T_c. c^* can be expressed by a following equation [2,8,9] :

$$c^* = N / \exp[(\pm eQ/k) - A]. \qquad (1)$$

Here, N, Q, k, e and A are density of state, Seebeck coefficient, Boltzman constant, electron charge and constant,respectively. Since the band gaps for the high T_c oxides are smaller than those of semiconductors, A is assumed to be about zero [8,9]. Since the Q value of high T_c oxides are approximately zero, an exponential term of the samples are about 1. Therefore, c^* is dominated by the density of state (N). N is expressed by the following equation [10] :

$$N = A' / Re^{1/2} \qquad (2)$$

Here, A' is constant, which is dominated by the electron mass, charge, Planck constant, mean free path. If we assume that the Ti addition does not largely change the mean free path, the N value is inversely proportional to the $Re^{1/2}$ value. Figure 4 shows changes in the electrical resistivity (R) at 200 K against the titanium concentration z for the $Bi_{1.6}Pb_{0.4}Sr_2Ca_2Cu_{3-z}Ti_zO_x$, $Bi_{1.6-z}Ti_zPb_{0.4}Sr_2Ca_2Cu_3O_x$, $Bi_{1.6}Pb_{0.4-z}Ti_zSr_2Ca_2Cu_3O_x$, $Bi_{1.6}Pb_{0.4}Sr_{2-z}Ti_zCa_2Cu_3O_x$ and $Bi_{1.6}Pb_{0.4}Sr_2Ca_{2-z}Ti_zCu_3O_x$. The changes in the 1/R values approximately agree with the changes in the T_c^{mid} values. Therefore, the lower the R values (see Fig. 4), the larger the N and c^* values become, resulting in higher T_c (see Fig. 2).

Although the both crystal perfection and electrical resistivity do not dominate the T_c change perfectly, it is partly controlled by them.

CONCLUSION

In summary, a large amount of Ti addition doesn't decrease T_c^{mid} for $Bi_{1.6}Pb_{0.4}Sr_2Ca_2Cu_3O_x$.
The high crystal perfection (low $\Delta 2\theta$ of x-ray diffraction) of the triple Cu-O layer phase is found in the high T_c (above 100 K)samples. Furthermore, low electrical resistivity (the high density of state and high carrier concentration) is found in the high T_c samples.

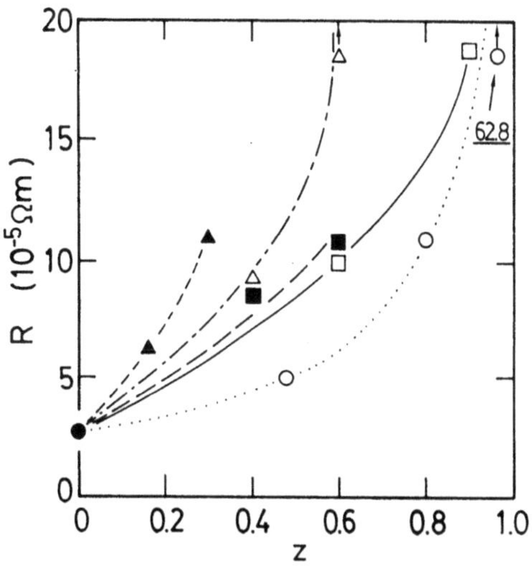

Fig.4 Changes in electrical resistivity (R) at 200 K against titanium concentration (z).

● $Bi_{1.6}Pb_{0.4}Sr_2Ca_2Cu_3O_x$
····○···· $Bi_{1.6-z}Ti_zPb_{0.4}Sr_2Ca_2Cu_3O_x$
– – ▲ – – $Bi_{1.6}Pb_{0.4-z}Ti_zSr_2Ca_2Cu_3O_x$
–·–△–·– $Bi_{1.6}Pb_{0.4}Sr_{2-z}Ti_zCa_2Cu_3O_x$
–—■—– $Bi_{1.6}Pb_{0.4}Sr_2Ca_{2-z}Ti_zCu_3O_x$
——□—— $Bi_{1.6}Pb_{0.4}Sr_2Ca_2Cu_{3-z}Ti_zO_x$

REFERENCES

1. Nishi Y, Moriya S, Manabe T (1989) J.Appl.Phys.65: 2389
2. Nishi Y, Oguri K, Ohinata H, Tanioka K, Kita Y, Ninomiya N (1990) Phys.Rev.B 41: 6520
3. Nishi Y, Kita Y, Tanioka K (1990) J.Mater.Sci.25: 5105
4. Nishi Y, Moriya S, Tokunaga S (1988) J.Mater.Sci.lett.7: 359
5. Nishi Y, Moriya S, Tokunaga S (1988) J.Mater.Sci.lett.7: 596
6. Nishi Y, Moriya S, Tokunaga S (1987) Phys.Lett.A 126: 55
7. Endo U, Koyama S, Kawai T (1988) Jpn.J.Appl.Phys.27: L1476
8. N.M.Tallan, and I.Bransky (1971) J.Electrochem.Soc.118: 345
9. Brebrick RF (1975) in Defects in Solids, Vol.2 of Treatise on Solis State Chemistry,edited by N.B.Hannay (Plenum, New York,), pp. 357-362.
10. Mott NF (1969) Philos.Mag.19: 835

Structural and Superconducting Properties of Iodine Intercalated $Bi_2Sr_2Ca_{n-1}Cu_nO_y$ (n=1,2,3)

D. Pooke*, K. Kishio, N. Motohira, T. Tamura, Y. Tomioka, T. Koga, K. Kitazawa, and K. Yamafuji

Department of Industrial Chemistry, The University of Tokyo, Bunkyo-ku, Tokyo, 113 Japan

ABSTRACT

We examine the structural and electronic properties of the $Bi_2Sr_2Ca_{n-1}Cu_nO_y$ (n=1,2, and 3) family of superconductors intercalated with iodine. Intercalation leads to a c-axis expansion of 0.7nm in all 3 members, suggesting that iodine is inserted into the common structural unit, the BiO-BiO double layers. Resistivity and susceptibility measurements on n=2 and 3 materials indicate changes in Tc which depend on the oxygen stoichiometry and cation ratios of the parent material. The magnetic field dependence of resistivity for a single crystal n=2 intercalated specimen is presented for the first time. There is evidence for an enhanced flux pinning potential at low fields, while it shifts to a reduced potential at higher fields.

KEY WORDS: Bi system superconductor, intercalation, structure, Tc, flux pinning.

INTRODUCTION

The intercalation reaction offers interesting possibilities for the control of coupling between CuO_2 sheets and for carrier doping in the high-Tc cuprates. Both these factors are undoubtedly crucial to the superconducting properties, as witnessed by the sensitivity of Tc to doping, including oxygen stoichiometry, and by the central role played by anisotropy in the transport properties, especially with regard to fluctuation effects and flux dynamics. Intercalation of appropriate species into the high-Tc superconductors may thus help elucidate many of the current issues in this field.

Such a reaction was recently demonstrated [1] in the $Bi_2Sr_2CaCu_2O_{8+\delta}$ system (n=2) with the intercalation of iodine. A stage one compound, corresponding to the insertion of the guest species between each layer with an associated 0.7nm increase in c-axis, was obtained. An approximate 10K decrease in Tc was observed, despite the large increase in c-axis. The successful intercalation of the n=2 member of the series of compounds $Bi_2Sr_2Ca_{n-1}Cu_nO_y$ has been extended to the n=1 and 3 members as well [2,3]. Our initial results [3] on changes in Tc with intercalation indicated some variability between samples. For instance, n=2 single crystals displayed decreases of zero-resistance Tc in the range 15-25K, while sintered polycrystalline specimens showed much larger negative shifts in Tc, up to 40K as marked by zero resistivity. In this paper we present a more detailed study of intercalation of iodine into the Bi system superconductors, using single crystal specimens of both the n=1 and 2 members, and sintered pellets of n=1,2 and 3. Tc is investigated at various oxygen contents, and resistivity measurements in magnetic fields are presented.

EXPERIMENTAL

Single crystals of n=1 and 2 were prepared by a floating zone method described previously [4]. The n=2 crystals were grown from a feed rod of stoichiometry $Bi_{2.2}Sr_{1.8}CaCu_2O_{8+\delta}$. Small n=1 single crystals were grown from a starting composition of $Bi_2Sr_{1.6}La_{0.4}CuO_{6+\delta}$. Sintered polycrystalline pellets of n=2 were prepared, both with and without Pb substitution, by direct reaction of oxides of Bi, Pb, and Cu, and the nitrates of Sr and Ca. Samples of n=3, of stoichiometry $Bi_{1.84}Pb_{0.34}Sr_{1.91}Ca_{2.03}Cu_{3.06}O_{10+\delta}$, were prepared following Ref. 5, reacting in air at 845-855°C.

Iodine intercalation was achieved by sealing samples with elemental iodine in Pyrex ampoules, as described earlier [3]. Polycrystalline samples can be completely intercalated in as little as 1h at 180°C using a saturated vapour pressure of iodine; single crystals take less than 24h under the same conditions. Completion of the reaction can be judged both by mass change (corresponding to one I per two Bi in the unit formula), and by x-ray diffraction.

* Permanent address, DSIR Physical Sciences, Lower Hutt, New Zealand

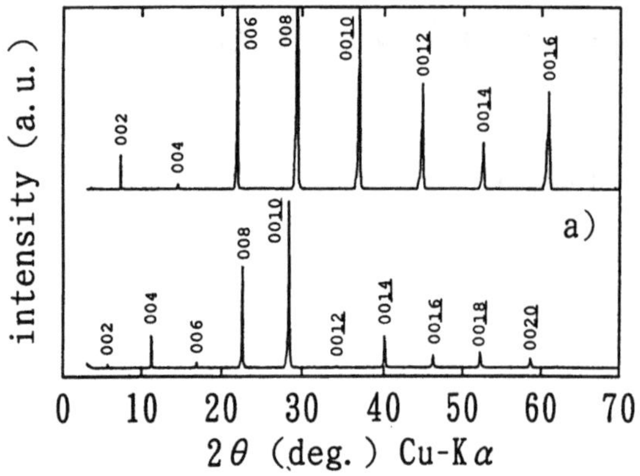

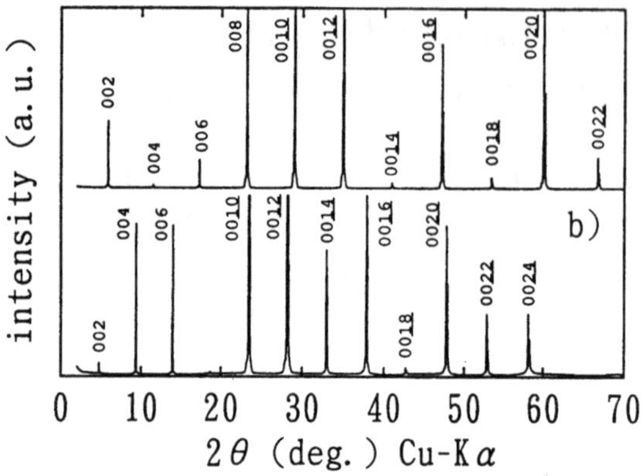

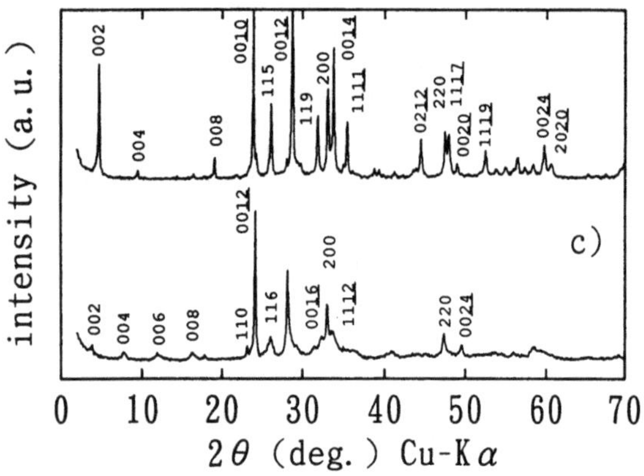

Fig. 1 X-ray diffraction patterns for the pure and intercalated Bi superconducting phases a) :n=1 single crystal, b) :n=2 single crystal, and c) :n=3 powder. The intercalated material is the lower trace in each case.

RESULTS AND DISCUSSION

In Fig. 1 single crystal XRD patterns for n=1 and 2, and powder patterns for n=3 are shown for both the parent and intercalated materials. From single crystal patterns, we find c-axis lattice expansions of 0.71nm for n=1 and 2, while powder patterns for n=3 give a 0.70nm expansion on intercalation. Results for sintered specimens of n=1 and 2 give similar results, 0.74 and 0.72nm respectively. We have found negligible change in the a- and b-axis lattice parameters in all cases. These results are in good agreement with previously reported values [1,2]. The equivalence of the c-axis expansion in all 3 phases suggests the most likely site for the iodine to enter the structure is between the BiO bi-layers, this being the common structural unit in the series. A high resolution electron microscopy study [6] has provided more direct evidence for the intra- BiO layer site.

Figure 2 displays the diamagnetic transitions, measured by AC susceptibility, for a $Bi_{1.84}Pb_{0.34}Sr_{1.91}Ca_{2.03}Cu_{3.06}O_{10+\delta}$ sample before and after reaction in iodine. Two critical temperatures are visible in each of the transitions; the higher transition is understood [7] to be the intrinsic critical temperature, with the lower transition characteristic of coupling between grains, and marking the zero of resistivity. The sintered pellets tend to absorb more iodine than is required for intercalation which is evidenced in the marked depression of the lower transition. Tails in the resistivity are sometimes observed, but are absent in single crystals.

To understand the variability in Tc seen in some samples we have probed this as a function of oxygen stoichiometry of the parent material, for both the n=2 and n=3 phases. The oxygen content of a sample can be reproducibly controlled by rapid quenching from a known anneal temperature and oxygen partial pressure [8]. The results are summarised in Fig. 3, where Tc's are plotted as a function of anneal temperature in air, for both n=2 and 3 phases before and after intercalation at 180°C. The points at 300°C represent samples slowly cooled in oxygen, rather than air annealed. Any change in Tc will result from interaction between charge transfer associated with the insertion of iodine itself, and a change in coupling as the c-axis is increased. Concentrating on the low oxygen content data, for n=2 the results are consistent with a modest transfer of holes to the CuO_2 planes. This was supported by a single crystal Hall coefficient measurement reported recently [3]. DC magnetization measure-

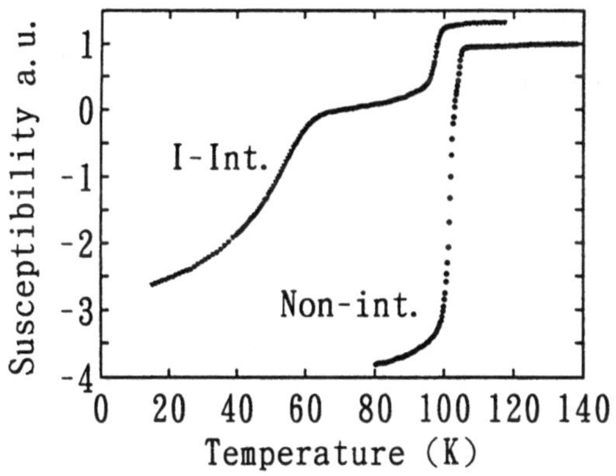

Fig. 2 Diamagnetic transitions for a Bi(n=3) sample annealed at 700°C before and after intercalation with iodine.

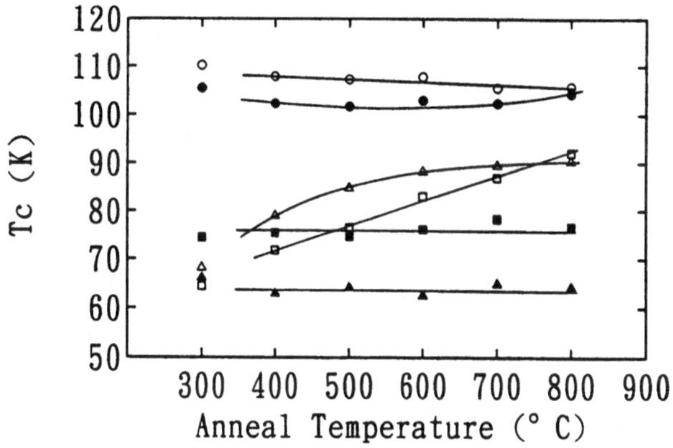

Fig. 3 Tc of Bi system superconductors as a function of anneal temperature in air, before (open symbols) and after (closed symbols) intercalation; ●: n=3, ▲: n=2, no Pb, ■: n=2 Pb=0.2. The data points at 300°C are for slow cooling in oxygen.

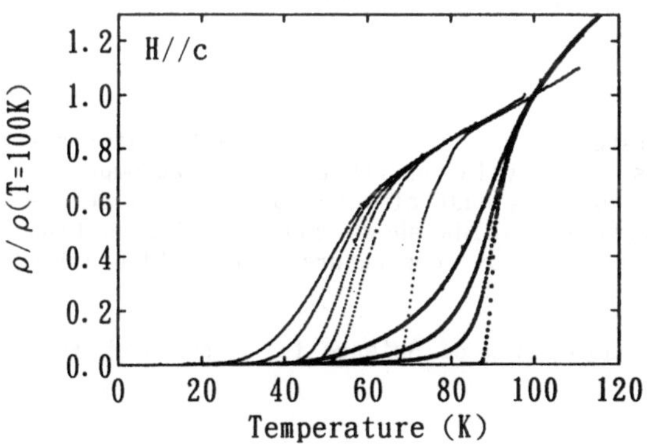

Fig. 4 Broadening of the resistivity transitions for an intercalated Bi n=2 single crystal (light points, at magnetic fields of 6,4,2,1,0.5 and 0 Tesla from left to right), and a non-intercalated crystal (dark symbols, at fields 6,2,0.5 and 0 Tesla from left to right). The resistivities are normalised to the respective T=100K values.

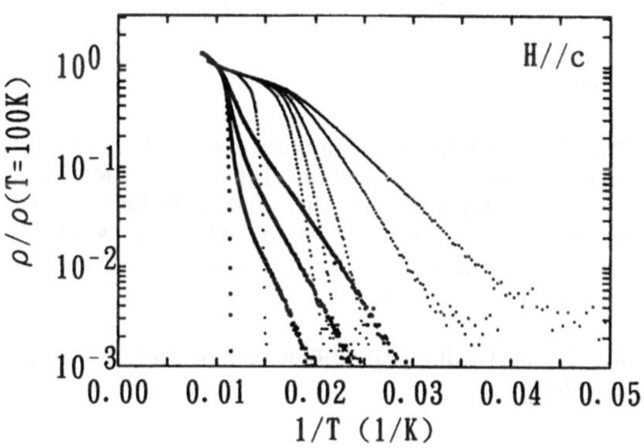

Fig. 5 Arrhenius plot for the data of Fig. 4. Symbols are as previously stated

ments of intercalated single crystals also show Tc's close to 75K, though it is difficult to vary oxygen content over a wide range in this case. The larger decrease in Tc for the Pb free n=2 sample is not presently understood. The n=3 results are also consistent with some charge transfer; indeed, a small Tc increase is observed in samples unloaded at 765°C in 2.5% O_2. The Tc - hole concentration phase diagram is steepest here and any change in doping will have its greatest affect on Tc.

Tc in the intercalated material seems largely unaffected by increasing the oxygen content in the parent phase, with Tc apparently increasing for Pb substituted n=2 samples of high oxygen content. However, we find that oxygen unloads from the samples during the reaction; this occurs even without iodine in the ampoule, for which Tc increases by a larger extent but displays a broad. transition. The labile oxygen is believed to reside between the BiO bi-layers, and the expansion of the lattice at this point by iodine intercalation may facilitate such diffusion. It is noteworthy that oxygen can unload at such low temperatures.

In Fig. 4 we present data for the broadening of the resistivity transition in magnetic fields applied parallel to the c-axis of an intercalated n=2 single crystal. With a large increase in the c-axis on intercalation, the intuitive expectation is to observe a more two dimensional, or anisotropic, behaviour in the transport properties. Anisotropy is a key issue in the high Tc cuprates, reflected in broadening of the resistivity, especially with H//c. The Bi system, being much more anisotropic than the Y-Ba-Cu system, correspondingly shows this to a larger extent. The mechanism has been discussed in terms of fluctuation effects [9], flux flow [10], and at lower temperatures, flux creep [11]. In the high temperature region, just below the onset of the resistive transition, fluctuation or flux flow effects are expected to be dominant; the low temperature tail of the resistivity is related to flux creep. At least two models, then, are needed to explain the complete transition. Examining the high temperature region, there is a significant increase of broadening of the transition compared to the native material (also shown); in the spirit of a fluctuation model, the result is consistent with a much enhanced anisotropy, associated with a decrease in coherence length along the c-axis.

To analyse the low temperature regime, where flux creep is a good description for the vortex dynamics, the same data is plotted in Fig. 5 in Arrhenius form. The activated behaviour is clear. In the flux creep model the activation energy from such data is equated to the flux pinning potential. A large difference between the intercalated and parent specimens is obvious. At low fields the pinning potential is enhanced over the non-intercalated material, but at higher fields the opposite is true, with a much reduced potential. The enhancement at low field is consistent with DC magnetization at 10 Oe, which show a significantly small signal in the field cooled measurements. In the intercalated material, the enhancement of the low field pinning potential poses a question on interpretation of the pinning mechanism. A detailed analysis of the resistive curve in the high temperature region may provide some clues in understanding the two-dimensional character of the Bi superconductors.

SUMMARY

The Bi system superconductors are easily intercalated with iodine, giving stable materials which remain superconducting, with increased c-axes. Intercalation appears to be accompanied by some charge transfer, and a reduction in c-axis coupling must also be important in determining Tc. Broadening of the resistivity in magnetic fields shows enhanced fluctuation effects, consistent with lower c-axis coherency, while the low temperature, low field pinning potential is enhanced.

D. Pooke wishes to thank the Japan Society for the Promotion of Science for a research fellowship.

REFERENCES

1. Xiang X-D, McKernan S, Vareka WA, Zettl A, Corkill JL, Barbee III TW, Cohen ML (1990) Nature (London) 348: 145
2. Xiang X-D, Zettl A, Vareka WA, Corkill JL, Barbee III TW, Cohen ML (1991) Phys.Rev.B 43: 11496
3. Nakashima H, Pooke DM, Motohira N, Tamura T, Nakayama Y, Kishio K, Kitazawa K, Yamafuji K (1991) Proceedings of M^2S-HTSC III, Kanazawa, Japan. Physica C (in press)
4. Motohira N, Kuwahara K, Hasegawa T, Kishio K, Kitazawa K (1989) J.Ceram.Soc.Jpn. Intl.Ed. 97: 994
5. Endo U, Koyama S, Kawai T (1988) Jpn.J.Appl.Phys. 27: L2040
6. Chenevier B, Ikeda S, Kadowaki K (1991) Proceedings of M^2S-HTSC III, Kanazawa, Japan. Physica C (in press)
7. Goldfarb RB, Lelental M, Thompson CA (1991) presented at Office of Naval Research Workshop on Magnetic Susceptibility and Other Spin Systems, Berkeley Springs, West Virginia
8. Tallon JL, Buckley RG, Gilberd PW, Presland MR (1989) Physica C 158: 247
9. Kambe S, Kitazawa K, Naito M, Tanaka I, Kojima H (1989) Jpn.J.Appl.Phys. 28: L555
10. Matsushita T, Ni B (1990) Physica C 166: 423
11. Palstra TTM, Batlogg B, Schneemeyer LF, Waszczak JV (1988) Phys.Rev.Lett. 61: 1662

Superconducting Properties of Pb-Based Layered Copper Oxides, $(Pb_{(1+x)/2}Cu_{(1-x)/2})(Sr_{1-y}Ca_y)_2(Y_{1-x}Ca_x)Cu_2O_{7+\delta}$

K. ISAWA, T. MAEDA*, K. SAKUYAMA**, and H. YAMAUCHI

Superconductivity Research Laboratory, ISTEC, Koto-ku, Tokyo, 135 Japan

ABSTRACT

We previously reported that single-phase superconducting lead cuprates given by the formula, $(Pb_{(1+x)/2}Cu_{(1-x)/2})Sr_2(Y_{1-x}Ca_x)Cu_2O_{7+\delta}$ ($0 \leqq x < 0.4$) were successfully synthesized. Although the samples with $x \geqq 0.4$ were multiphase, Tc continued to increase as x increased up to 0.5. In the present study, part of Sr is substituted by Ca following the formula, $(Pb_{(1+x)/2}Cu_{(1-x)/2})(Sr_{1-y}Ca_y)_2(Y_{1-x}Ca_x)Cu_2O_{7+\delta}$ ($0 \leqq x \leqq 0.6$, $0 \leqq y \leqq 0.2$). The compositional range in which single-phase samples are obtained is determined in terms of x and y. We optimize the Tc value in terms of the temperature (Tq) at which the sample is post-annealed and subsequently quenched in flowing O_2 gas. We obtain a nearly single-phase superconductor with x=0.5 and y=0.1 which exhibits Tc=60 K as-sintered and Tc=80 K when post-annealed and quenched at Tq=800℃.

KEY WORDS : high temperature superconductivity, superconducting lead cuprates, "1212" structure

INTRODUCTION

Since the discovery of a superconducting Pb-based copper oxide by Cava et al.[1], a variety of new lead cuprates have been synthesized and characterized. $(Pb,Cu)Sr_2(Y,Ca)Cu_2O_z$ compounds of the (Pb/Cu)-1212 phase were found by Subramanian et al.[2]. The structure of these compounds was isomorphous to that of $TlBa_2CaCu_2O_7$ (the Tl-1212 phase) [2, 3]. However, these compounds did not exhibit superconductivity. Rouillon et al. [4] observed superconductivity for the samples with a nominal composition of $(Pb_{0.5}Sr_{0.5})Sr_2(Y_{0.5}Ca_{0.5})Cu_2O_7$. Koriyama et al. [5] confirmed superconductivity in the (Pb/Cu)-1212 compounds. They observed superconductivity transition around 70 K for Ba-substituted samples. Recently, Maeda et al. [6] pointed out that the chemical composition of the (Pb/Cu)-1212 phase was represented by the formula, $(Pb_{(1+x)/2}Cu_{(1-x)/2})Sr_2(Y_{1-x}Ca_x)Cu_2O_z$ ($x=0 - 0.35$, $z \sim 7$) and reported that the samples with $x \geqq 0.4$ were multi-phase, while Tc increased as x increased [7]. This high-Tc compound was expected to be a (Pb/Cu)- 1212 compound in which the Sr site was partially substituted by Ca.

In the present study, substitution effects of Ca for Sr in this phase were investigated. We made the Sr sites partially substituted by Ca according to the formula, $(Pb_{(1+x)/2}Cu_{(1-x)/2})(Sr_{1-y}Ca_y)_2(Y_{1-x}Ca_x)Cu_2O_{7+\delta}$ ($0 \leqq x \leqq 0.6, 0 \leqq y \leqq 0.2$). The compositional range in which single-phase sample was obtained is determined in terms of x and y.

EXPERIMENTAL

The samples were prepared by a conventional solid-state reaction method using high purity powders of PbO, $SrCO_3$, $CaCO_3$, Y_2O_3 and CuO. These powders were mixed and calcined at 850 ℃ for 10 h in air. After grinding, these mixtures were fired at 1060℃ for 1 h in a flowing O_2 gas and then slowly cooled down to room temperature at a rate of 60 ℃ / h. The as-sintered samples were post-annealed at 800℃ for 1 h in a flowing O_2 gas and then quenched to room temperature in the same atmosphere. Especially, samples with x=0.35 and y=0 were post-annealed at 700℃, 800 ℃, 900℃ and 1000℃ and subsequently quenched. The obtained samples were characterized by X-ray powder diffraction (XRD). The temperature dependences of electrical resistivity and dc magnetic susceptibility were measured by a conventional four-probe method and by a SQUID magnetometer, respectively.

Present address: * Yokohama R&D Laboratories, Furukawa Electric Co., Ltd.,2-4-3, Okano, Nishi-ku, Yokohama-shi, Kanagawa 220, Japan.

** Tohoku Electric Power Co., Ltd., 3-7-2, Ichiban-cho, Aoba-ku, Sendai-shi, Miyagi 980, Japan.

RESULTS AND DISCUSSION

Figure 1 schematically shows the crystallographic structure of a (Pb/Cu)-1212 compound. It is similar to the structure of a "123" compound, but contains no Cu-O chains[8]. Figure 2 shows powder XRD patterns for the samples with x=0.3 and y=0, 0.05, 0.1, 0.15 and 0.2. Almost all of the diffraction lines were able to be indexed for a tetragonal unit cell with the lattice constants, a ~ 3.8 Å and c ~ 11.9 Å. Diffraction lines due to impurities are marked by ● only for the patterns for y=0.15 and y=0.2.

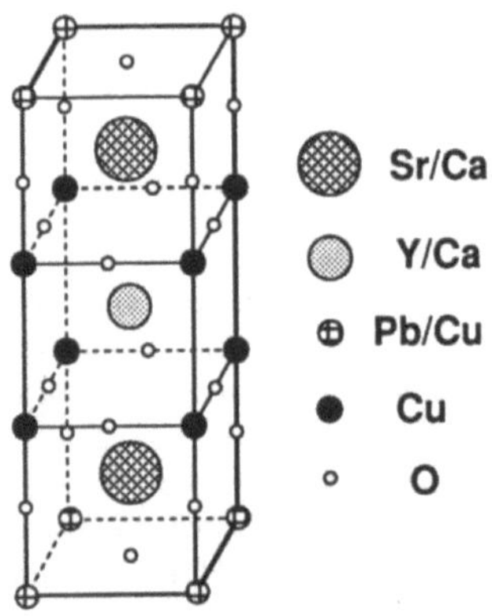

Fig. 1. Schematic crystallographic structure of the (Pb/Cu)-1212 phase.

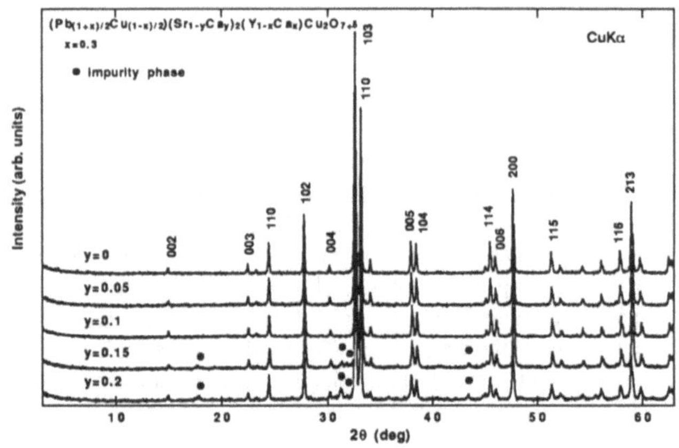

Fig. 2. Powder XRD patterns for the samples with x=0.3.

The compositional area in terms of x and y in the formula, $(Pb_{(1+x)/2}Cu_{(1-x)/2})(Sr_{1-y}Ca_y)_2(Y_{1-x}Ca_x)Cu_2O_{7+\delta}$ ($0 \leqq x \leqq 0.6$, $0 \leqq y \leqq 0.2$), in which single-phase samples are obtained, according to the XRD results are mapped in Fig. 3. The area where superconducting samples is also indicated. The compositions which yielded single-phase samples are indicated by ◎ and ○, which represent superconducting and nonsuperconducting samples, respectively. The single phase area was found to extend up to x ~ 0.4 when y < 0.5x. Thus, the solubility of Ca in the Sr sites increased as x increased. In the cases of x=0.35 and x=0.4, superconducting single-phase (or nearly single-phase) samples were obtained up to y ~ 0.2. This result indicates that the partial substitution of Ca for Sr stabilized the (Pb/Cu) -1212 compounds. Moreover superconducting properties of the compounds are strongly related to the Ca concentration. In other words, the Ca concentration in the Sr sites as well as that in the Y sites influences the superconducting property . Especially, no superconducting samples were obtained for x=0 up to y=0.2 even after post-annealing at 800℃ in flowing O_2 gas.

The temperature dependence of the dc magnetic susceptibility for the samples with x=0.4 and y=0 ~ 0.2 is shown in Fig. 4. The superconductivity transition temperature (Tc) decreased monotonically as y increased.

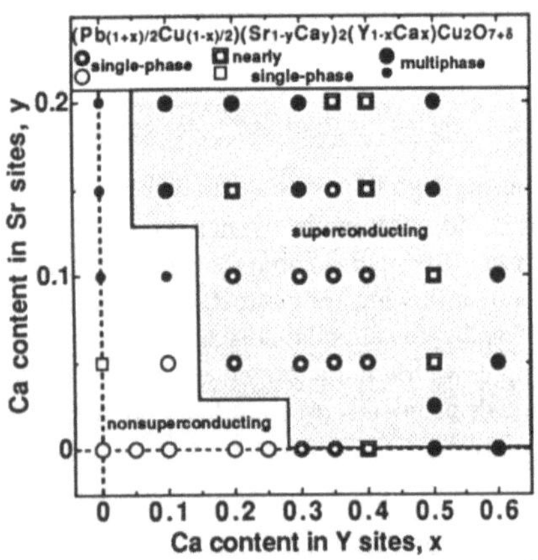

Fig. 3. Compositional field for single-, nearly single- and multi-phase samples.

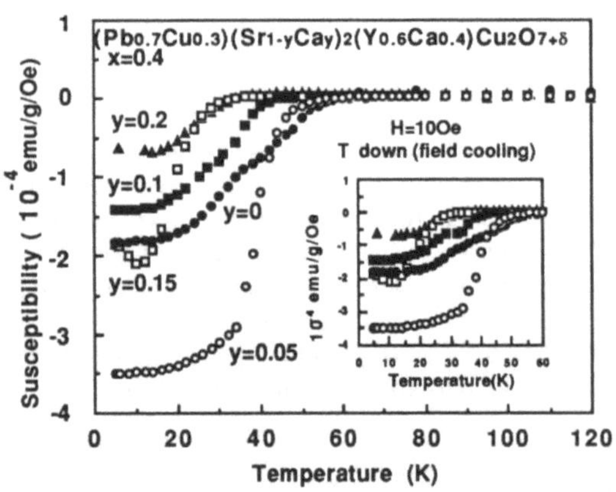

Fig.4. Temperature dependence of dc magnetic susceptibility for the samples with x=0.4 and y=0, 0.05, 0.1, 0.15, and 0.2.

The post-annealing was performed at various temperatures (Tq) to find the optimal value for Tq (from which samples were quenched in flowing O_2 gas) to obtain the highest Tc. The dependence of Tc on Tq was investigated for the samples with x=0.35 and y=0 because this composition exhibited the highest Tc without Ca in the Sr sites. Figure 5 shows the temperature dependence of electrical resistivity after quenching from Tq= 700℃, 800℃, 900℃ and 1000℃. Up to Tq=900℃, as Tq increased, the normal-state resistivity decreased and Tc increased. Tc (onset) and Tc (zero resistance) for the sample quenched from Tq= 900℃ were 45 K and 38 K, respectively. However, the temperature dependence of the electrical resistivity for the sample quenched from Tq=1000℃ which contained only a little amount of impurity was semiconducting and its normal-state resistivity was about 250 times higher than those for the samples. The temperature dependences of dc magnetic susceptibility for the sample quenched from Tq=700℃, 800℃ and 900℃ are shown in Fig. 6. All the samples were confirmed to be bulk superconductors, and Tc's of the samples were magnetically determined to be 26 K, 30 K and 45 K for Tq=700, 800 and 900℃, respectively.

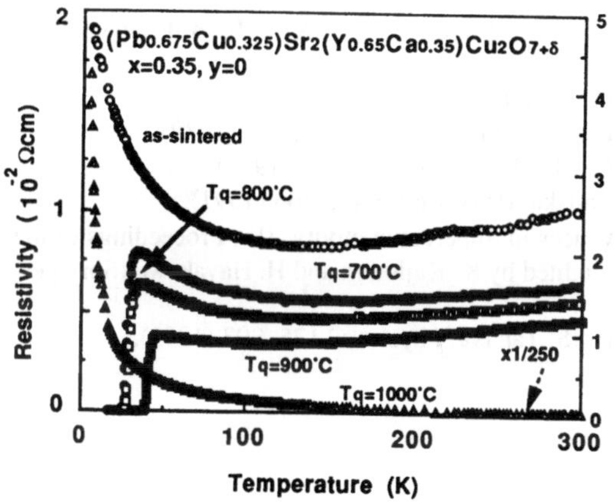

Fig. 5. Temperature dependences of electrical resistivity for samples with x=0.35 and y=0 with various Tq's.

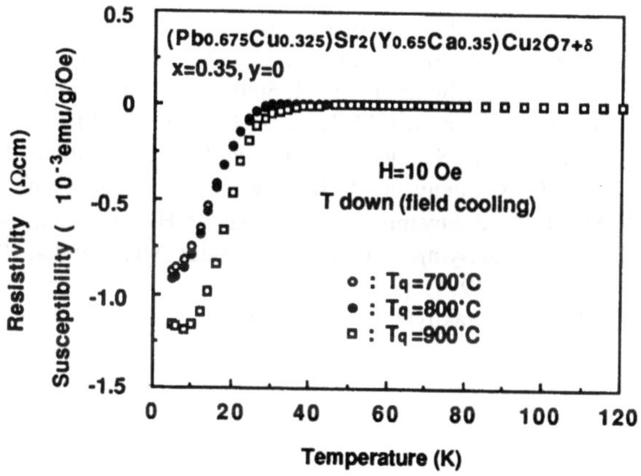

Fig. 6. Temperature dependence of dc magnetic susceptibility for the sample with x=0.35 and y=0 when quenched from three different Tq's.

The dependence of the lattice constants, *a* and *c*/3, on Tq for the samples with x=0.35, y=0 is shown in Fig. 7. The parameter *a* was nearly independent of Tq, but became shorter after quenching because of the removal of excess oxygen [8]. Contrary to this, parameter *c* remained almost the same even after quenching, except for the case of the sample quenched from Tq= 1000℃.

Figure. 8 shows the resistivity -vs- temperature curves for the as-sintered sample and the sample quenched from Tq=800℃, both of which were nearly single phase and of the composition with x=0.5 and y=0. Both samples exhibited superconductivity. Tc (onset) increased from 60 K to 80 K by quenching.

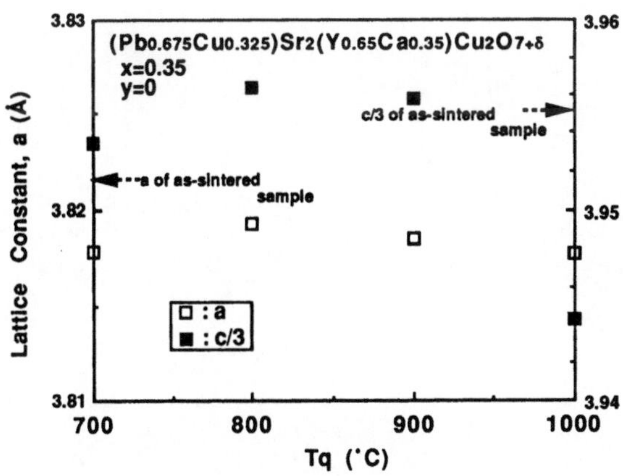

Fig. 7. Relationship between the lattice constants,*a* and *c*/3, and Tq.

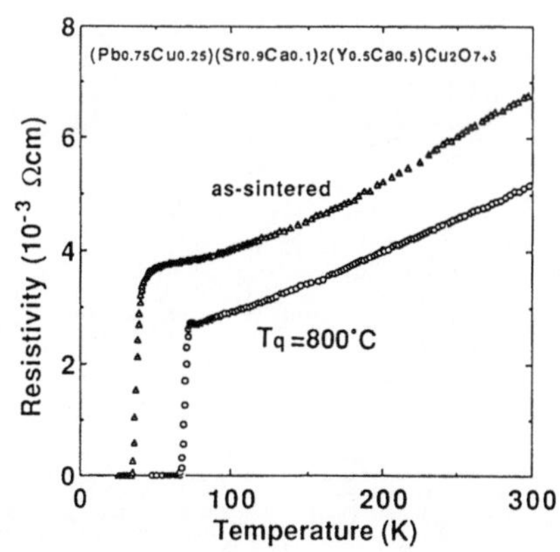

Fig. 8. Temperature dependence of electrical resistivity for the as-sintered and quenched samples with x=0.5 and y=0.1.

In summary, the substitution effects of Ca for Sr in $(Pb_{(1+x)/2}Cu_{(1-x)/2})(Sr_{1-y}Ca_y)_2(Y_{1-x}Ca_x)Cu_2O_{7+\delta}$ ($0 \leqq x \leqq 0.6$, $0 \leqq y \leqq 0.2$) were investigated. We found that partial substitution of Ca for Sr expanded the solubility of Ca in the Y site. For the samples with x=0.35 and y=0, Tc was dependant on Tq, and the highest Tc of 45 K was obtained for the sample quenched from Tq=900℃. The sample with x=0.5 and y=0.1 quenched from Tq=800℃ exhibited a superconductivity transition around 80 K.

REFERENCES

1.R. J. Cava, B. Batlogg, J. J. Krajewski, L. W. Rupp, L. F. Schneemeyer, T. Siegrist, R. B. van Dover, P. Marsh, W. F. Peck, Jr., P. K. Gallagher, S. H. Glaum, J. H. Marshall, R. C. Farrow, J.V. Waszczak, R. Hull and P. Trevor, Nature 336, 211 (1988).
2.M. A. Subramanian, J. Goparakrishnan, C. C. Trardi, P. L. Gai, E. D. Boyes, T. R. Askew, R. B. Flippen, W.E. Farneth and A. W. Sleight, Physica C 159, 124 (1989).
3. J. Y. Lee, J. S. Swinnea and H. Steinfink, J. Mater. Res. 4, 763 (1989).
4. T. Rouillon, J. Provost, M. Hervieu, D. Grout, C. Michel and B. Raveau, Physica C 159, 201 (1989).
5. S. Koriyama, K. Sakuyama, T. Maeda, H. Yamauchi and S. Tanaka, Physica C 166, 413 (1990).
6. T. Maeda, K. Sakuyama, S. Koriyama, H. Yamauchi and S. Tanaka, Phys. Rev. B43,7866 (1991).
7.T. Maeda, K. Sakuyama, S. Koriyama and H. Yamauchi, Advances in Superconductivity III , Proceedings of the 3rd International Symposium on Superconductivity, Sendai, 1990, edited by K. Kajimura and H. Hayakawa (Springer-Verlag, Tokyo, 1991), pp. 261-264.
8. T. Maeda, K. Sakuyama, F. Izumi, H. Yamauchi, H. Asano and S. Tanaka, Physica C 175, 393 (1991).

Magnetic and Energy Dispersive X-Ray Characterization of Silver Substitution in $Pb_2Sr_2Y_{0.5}Ca_{0.5}Cu_3O_{8+z}$ Superconductors

P.P. VAISHNAVA and D. KISSEL

GMI Engineering & Management Institute, Flint, MI 48504, USA

ABSTRACT

The present investigation was performed with a purpose to study the effect of silver substitution and addition on the transition temperature, the Meissner fraction and the diamagnetic shielding of $Pb_2Sr_2Y_{0.5}Ca_{0.5}Cu_3O_{8+z}$ superconductor. Samples were prepared by substitution (Cu depletion) and addition of silver to the title superconductor. X-ray diffraction measurements indicated that all the samples were single phase materials. Energy dispersive X-ray measurements were performed to determine exact amounts of the elements in the samples. From magnetic measurements, which were performed between 10 and 300 K using a SQUID magnetometer, it was found that both diamagnetic shielding and the Meissner fraction decreased by substitution and addition of silver to the superconductor. Superconducting transition temperature and the lattice parameters were not affected.

KEY WORDS: PbSrYCaCuO system superconductors, Diamagnetic shielding, Meissner fraction

INTRODUCTION

Silver has been used extensively as dopant to alter the superconducting properties of oxide ceramics such as Y-Ba-Cu-O(1-3), Bi-Sr-Ca-Cu-O(4), and Tl-Ba-Ca-Cu-O(5). Various effects have been observed as a result of doping silver in these superconductors. They include i) the promotion of crystallization together with c-axis orientation, ii) a high temperature shift of the point of zero resistivity, and iii) an increase in the critical current density (J_c).

In this paper, results of substitution and addition of silver to a relatively new family of superconductors are presented. This new family of superconductors with a formula $Pb_2Sr_2ACu_3O_{8+z}$ (where A is a lanthanide or a mixture of Ln + Sr or Ca) was discovered by Cava et al(6) and independently by Subramanian et al(7). These superconductors have attracted much attention recently because of their unique superconducting properties. Some of the properties are: a) this family displays the same kind of substitutional chemistry as displayed by Y-Ba-Cu-O. b) The average valency of Cu in this family is less than 2+. In Y-Ba-Cu-O superconductors, the average valency is always greater than 2+. c) These superconductors have usual double CuO_5 pyramidal layers that are separated by SrO-PbO-CuO_z-PbO-SrO layers. Such layers in 1:2:3 materials are separated by BaO-CuO_y-BaO layers. Thus the distance between the charge reservoir blocks (chain layers or oxygen deficient layers) is much greater in this family.

EXPERIMENTAL

Samples of $Pb_2Sr_2Y_{0.5}Ca_{0.5}Cu_{3-x}Ag_xO_{8+z}$ (x=0.1, 0.3) for Cu depletion or silver substitution and $Pb_2Sr_2Y_{0.5}Ca_{0.5}Cu_3Ag_yO_{8+z}$ (y = 0.1, 0.3) for silver addition were prepared from high purity powders of PbO, $SrCo_3$, Eu_2O_3, $CaCo_3$, and CuO. The stoichiometric cuprate precursor of $Sr_2Y_{0.5}Ca_{0.5}Cu_3O_7$ was mixed with PbO and silver oxide and the mixture was heated in a flowing gas mixture of 1% oxygen in nitrogen at 850° C for 6 hours. The samples were found to be single phase as determined by X-ray diffraction. A Quantum design susceptometer was used to measure the susceptibility of samples in powder form. Energy dispersive X-ray measurements were performed on a Princeton Gamma-Tech scanning transmission electron microscope.

RESULTS AND DISCUSSION

The X-ray diffraction pattern for silver substituted (x=0.0, 0.1, 0.3) samples are shown in Fig.1. Similar patterns were obtained for silver addition samples. The analysis of the spectra indicated that all the samples were single phase materials. The lattice parameter values were found to be in agreement with the reported values (6,7). From the energy dispersive X-ray diffraction spectra (Fig. 2) the area ratios for Pb/Cu, Sr/Cu, Y/Cu, Ca/Cu were found to remain the same for all the samples with substitution and addition. The increase in the area ratio for Ag/Cu was found to be in agreement with the stoichiometry of the samples.

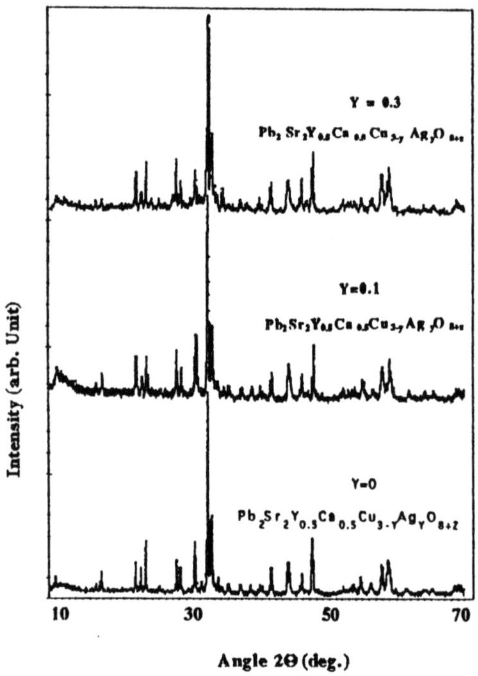

Fig. 1. X-ray diffraction patterns for $Pb_2Sr_2Y_{0.5}Ca_{0.5}Cu_{3-y}Ag_yO_{8+z}$ (y=0.0, 0.1,0.3) superconductors

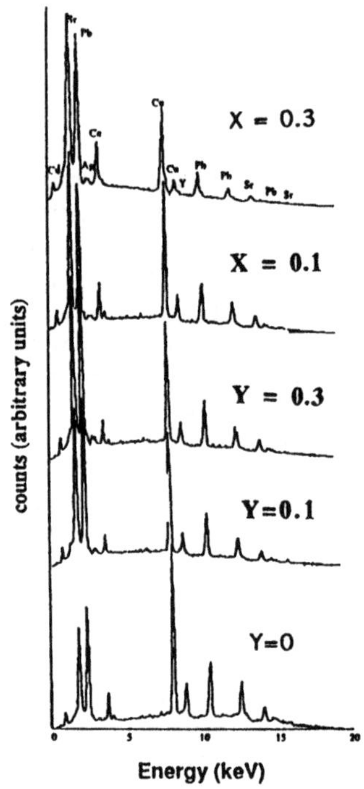

Fig. 2. Energy dispersive x-ray spectra for $Pb_2Sr_2Y_{0.5}Ca_{0.5}Cu_3Ag_xO_{8+z}$ and $Pb_2Sr_2Y_{0.5}Ca_{0.5}Cu_{3-y}Ag_yO_{8+z}$ (y=0.0, 0.1, 0.5) supercomductors

Figs. 3, 4, 5 show the magnetic susceptibility verses temperature for the samples with y=0 (x=0), y=0.1 and y =0.3, respectively. In these samples the Cu has been replaced by silver. The spectra were recorded in the field-cooled (FC) and zero field-cooled (ZFC) modes. In the ZFC mode, the samples were cooled to 4.5 K in

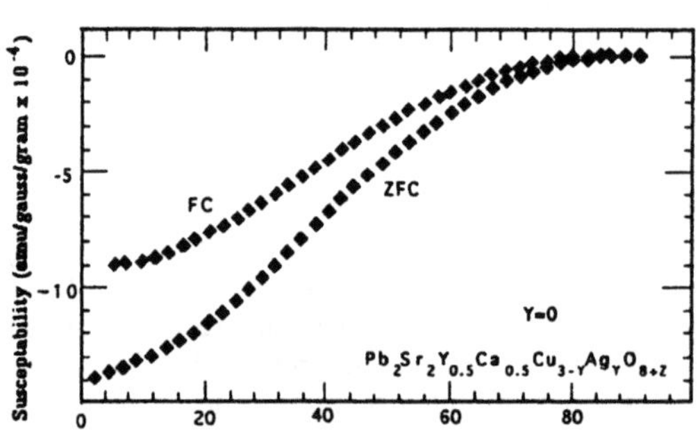

Fig. 3. Magnetic susceptibility vs temperature for $Pb_2Sr_2Y_{0.5}Ca_{0.5}Cu_3O_8$ in field-cooled and zero field-cooled modes

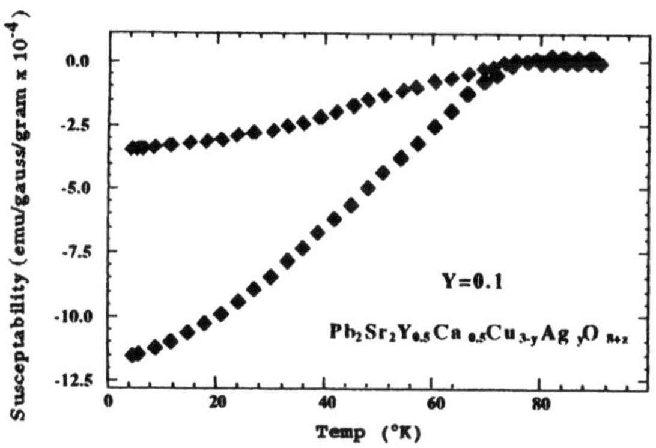

Fig. 4. Magnetic susceptibility vs temperature for $Pb_2Sr_2Y_{0.5}Ca_{0.5}Cu_{3-y}Ag_yO_{8+z}$ (y=0.1) superconductor

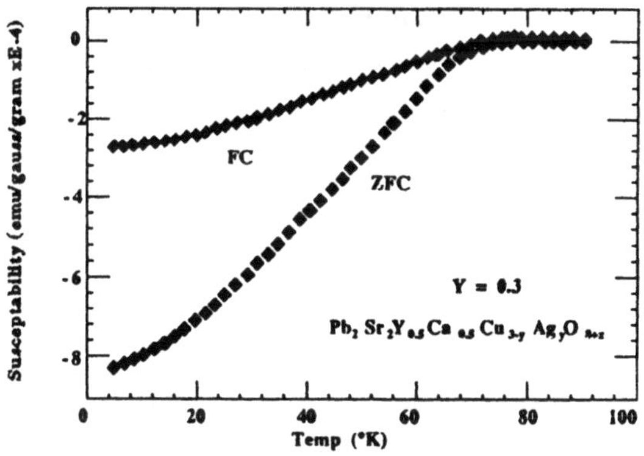

Fig. 5. Magnetic susceptibility vs temperature for $Pb_2Sr_2Y_{0.5}Ca_{0.5}Cu_{3-y}Ag_yO_{8+z}$ (y=0.3) superconductor

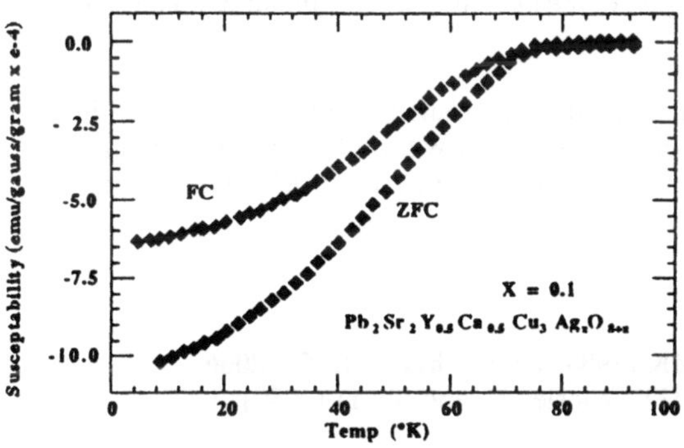

Fig. 6. Magnetic susceptibility vs temperature for $Pb_2Sr_2Y_{0.5}Ca_{0.5}Cu_3Ag_xO_{8+z}$ (x=0.1) superconductor

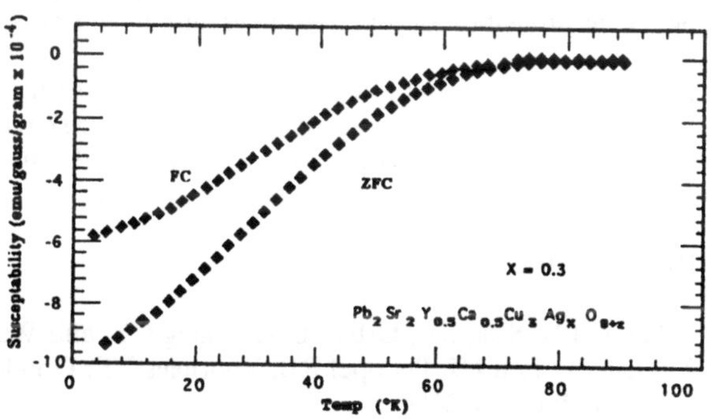

Fig. 7. Magnetic susceptibility vs temperature for $Pb_2Sr_2Y_{0.5}Ca_{0.5}Cu_3Ag_xO_{8+z}$(x=0.3) superconductor

a 1 Oe residual field, a 20 Oe field was then applied and susceptibility was measured with increasing temperature. In the FC mode, the 20 Oe field was applied at 90 K and then measurements were made with decreasing temperature. Figs. 6, 7 show similar measurements for x=0.1, 0.3 samples. The values of the onset temperature, diamagnetic shielding, and the Meissner fraction for all the samples are listed in Table I. The onset temperature of the superconducting transition was measured from the temperature at which susceptibility crossed zero. This temperature (72 K) was found to be approximately the same for all the samples. For the ZFC measurements, the fractional diamagnetic shielding effect was 5, 4.3, 3% (for the substitution samples y =0, .1, .3) and 3.8, 3.5% (for the addition x=0.1, 0.3 samples) of the ideal susceptibility value ($-1/4\pi$). Likewise, for the FC measurements, the respective fractional Meissner effect was 3.2, 1.3, 1%, and 2.2 and 2.1% for the substitution and addition samples, respectively. These low Meissner fraction values are similar to the values observed by Mohamed et al(8) in ploycrystalline Y-Ba-Cu-O and indicate the fact that an appreciable amount of flux trapping should be occurring in our samples. It is clear from Table I that the diamagnetic shielding as well as the Meissner fraction decrease by silver substitution and addition. Similar results are obtained by Meisner and Taylor (9) when Cu was depleted and added in Y-Ba-Cu-O samples. Since there is no effect on the transition temperature, it can be presumed that silver did not substitute for Cu in the lattice. It is most likely that the silver went in the grain boundaries in all the samples.

TABLE I. Onset temperature, diamagetic shielding, and Meissner fraction for $Pb_2Sr_2Y_{0.5}Ca_{0.5}Cu_{3-y}Ag_yO_{8+z}$ and $Pb_2Sr_2Y_{0.5}Ca_{0.5}Cu_3Ag_xO_{8+z}$ superconductors

	Samples				
	x=0, y=0	**x=0.1**	**x=0.3**	**y=0.1**	**y=0.3**
Onset Temperature (K)	72	72	72	72	72
Diamagnetic Shielding	5%	3.8%	3.5%	4.3%	3%
Meissner Fraction	3.2%	2.2%	2.2%	2.1%	1%

In summary, the magnetization data on our samples indicate that the superconductivity is mostly determined by the copper contents. The transition temperature and the lattice parameters of the samples are not affected by substitution or addition of silver to the compound. A minimal decrease in the values of diamagnetic shielding and the Meissner fraction in our samples can attributed to the increased intergranular coupling caused by the accumulation of silver at the grain boundaries.

Prem P. Vaishnava would like to thank the Center for Fundamental Materials Research at Michigan State University for the award of academic affiliate fellowship to perform this work. We would also like to thank Carl Nelson and Professor Bill Pratt for their help in the susceptibility measurements and useful discussions.

REFERENCES

1. Peters PN, Sisk, RC, Urban EW, Huang CY, and Wu MK (1988) Appl. Phys. Lett. 52: 2066
2. Singh JP, Leu HJ, Poeppel RB, Voorhees VE, Goudey GT, Winsley K, and Shi D (1989) J.App. Phys. 66: 3154
3. Tsuchida K, Miura Y, Tsudo H and Kato A (1989) J. Less-Common Metals 146: L19
4. Matsumoto Y, Hombo J, Yamaguchi Y, and Mitsunaga T (1989) Mater. Res. Bull. 24: 1469
5. Matsumoto Y, Hombo J, Yamaguchi Y (unpublished).
6. Cava RJ, Batlogg B, Krajewski JJ, Rupp LW, Scheenmeyer LF, Siegrist T, vanDover RB, Marsh P. Peck WF, Gallagher PK, Glarum SH, Marshall JF, Farrow RC, Waszxzak JV, Hull R, and Trevor P (1988) Nature 336: 211
7. Subramanian MA, Gopalakrishnan J, Torardi CC, Gai PL, oyes ED, Askew TR, Flippen RBN, Farneth WE, and Sleight AW (1989) Physica C 157: 124
8. Mohamed MAK, Jung J, and Franck JP (1989) Phys. Rev B 39: 9614
9. Meisner GP, and Taylor CA (1991) J. Appl. Phys. 69: 5370

Ca Doping Effects in $YBa_2Cu_4O_8$ Prepared by the Citrate Pyrolysis Method

KUNIYUKI KOYAMA*, ALAIN JUNOD, THOMAS GRAF, GILLES TRISCONE, and JEAN MULLER

Département de Physique de la Matière Condensée, Université de Genève, CH-1211 Genève 4, Switzerland

ABSTRACT

Ca doping effects in $YBa_2Cu_4O_8$ (124) phase are investigated by the citrate pyrolysis method under both atmospheric and high oxygen pressure. The decomposition temperature of (124) to the $Y_2Ba_4Cu_7O_{15-\delta}$ (247) and the $YBa_2Cu_3O_{7-\delta}$ (123) phases decreases with Ca doping. The onset T_c measured by DC susceptibility increases linearly from 81K to 89K with increasing Ca content, x, from x=0 to x=0.15 in nominal compositions $Y_{1-x}Ca_xBa_2Cu_4O_8$. The orthorhombicity; 2(b-a)/(b+a) decreases linearly with increasing x. EDAX analysis for individual grains convinced us that Ca is entering the structure, but its substitution for Y in the (124) phase seems rather unlikely.

KEYWORDS: (124) phase superconductor, Ca-doping, magnetic susceptibility, EDAX, Meissner effect

INTRODUCTION

$YBa_2Cu_4O_8$ (124), of interest because of its stable oxygen content, is easily obtained by sintering under high oxygen pressure [1,2]. More recently, the successful synthesis of the (124) phase was also achieved under ambient pressure [3-6]. The T_c is enhanced from 80K to 90K in the Ca doped (124) system [7-9]. The T_c is also enhanced greatly from 80K to 108K by applying high pressure of about 10 GPa [10]. It is not clear to what extent Ca is substituted at the Y-site. The possibility that Ca ions occupy the Ba-sites in the (124) phase was found by NQR experiments [11].

We examined the Ca doping effects in the (124) phase synthesized from sub-micron powders prepared by the citrate pyrolysis method followed by sintering under both atmospheric and high pressure oxygen.

SAMPLE PREPARATION

The well mixed powders, consisting of sub-micron particles, are obtained by the citrate pyrolysis method [6,12]. The starting products, CuO, Y_2O_3, $BaCO_3$ (all 99.999%) and CaO (99.95%) were weighed in the stoichiometric ratio $Y_{1-x}Ca_xBa_2Cu_4O_8$ and dissolved in hot nitric acid. After mixing, citric acid was added. The blue solution was then neutralized with ammonia, monitoring the pH. The solution was evaporated on a hot plate until spontaneous combustion occurred. The resulting dark brown expanded foam consists of well mixed sub-micron particles of Cu (elementary), BaO, CaO, Y_2O_3 and organic contaminants [6]. The particle size was in the 500-1000Å range. The pyrolysis powders were calcined at 780℃ in flowing O_2 during 100 hours [6]. The fired powders were crushed, mixed and pressed into pellets, and sintered at 780℃ for 100 hours. The heating and cooling rate was 60℃ per hour in both processes. Then, pellets were annealed under 90 bar oxygen pressure at 850℃-1000℃. The calcining and sintering temperature dependence under atmospheric oxygen was also examined precisely for the samples of x=0 and x=0.1, as indicated below.

*Permanent Address: *Technology Development Center, Toyohashi University of Technology, Aza-Hibarigaoka 1-1, Tempaku-cho, Toyohashi, Aichi, 441 Japan*

RESULTS AND DISCUSSION

After each particular heat treatment, the phase composition and crystal structure were studied by powder X-ray diffraction (XRD) using a Guinier camera (Cu-Kα line). Silicon was added as an internal standard and the lattice parameters were determined by least-squares fitting.

Figure 1 shows the change of lattice constants as a function of sintering temperature (1 bar and 90 bar) for the samples with (a) x=0 and (b) x=0.1. The decomposition temperature of the (124) to the (247) phase at 1 bar O_2 atmosphere is about 820℃ for the sample with x=0 [6]. This decomposition temperature decreases to 800℃ for the sample with x=0.1. The window for the formation of the (247) is very narrow for the Ca-doped system under atmospheric pressure. The optimum annealing temperature under 90 bar oxygen is determined as 900℃ for x=0.1, judging from the minimum fraction of CuO second phase in XRD patterns. The optimum annealing temperature decreases linearly with increasing x, from 1000℃ for x=0 to 850℃ for x=0.15 under 90 bar oxygen pressure. Ca addition lowers the decomposition temperature of the (124) phase under not only 1 bar but also 90 bar oxygen pressure. The lattice constants of the high pressure annealed samples coincide with those of the

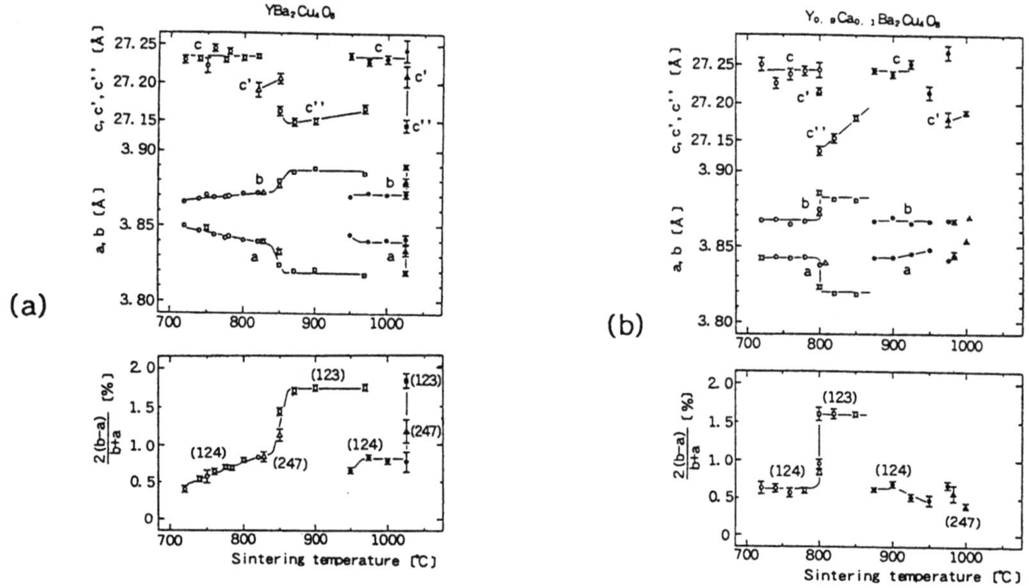

Fig.1 Lattice constants and orthorhombicity as a function of sintering temperature for the sample with (a) x=0 and (b) x=0.1 in $Y_{1-x}Ca_xBa_2Cu_4O_8$. Solid symbols indicate the data for high pressure annealed samples. $c'=(c/2)+\Delta$ for the (247) and $c''=2\cdot(c+\Delta)$ for the (123) phases, where $\Delta=1.9$Å.

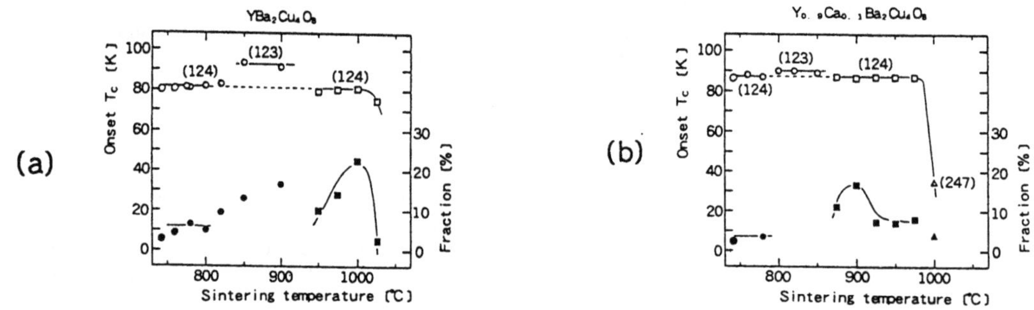

Fig.2 Onset T_c (open symbols) and Meissner fractions (solid symbols) determined from DC susceptibility as a function of sintering temperature for the sample with (a) x=0 and (b) x=0.1 in $Y_{1-x}Ca_xBa_2Cu_4O_8$. Circles indicate the data of atmospheric sintered samples, and squares that of the high pressure annealed ones.

atmospheric sintered samples.

The onset T_c values determined from DC-χ are summarized as a function of sintering temperature for the sample with (a) x=0 and (b) x=0.1 as shown in Fig.2. The intrinsic value of T_c for x=0 is 81K, and that for x=0.1 is 87K. The higher values than those given above are not reliable because of the decomposition of the (124) phase to the (247) and (123) phases as shown in Fig.1, a difficulty that was generally not considered in the literature. The onset T_c in the (247) phase for the high pressure annealed sample decreases strongly due to oxygen deficiency by quenching after the high pressure annealing. The Meissner volume fractions show a maximum for the high pressure annealed samples; 1000℃ for x=0 and 900℃ for x=0.1 under 90 bar oxygen pressure. These annealing conditions coincide with those for the maximum distortions and those for the minimum intensities of the CuO second phase peak in the X-ray diffraction patterns.

The DC Meissner susceptibility was measured during cooling in a constant field of 20 Oe with a SQUID magnetometer (Quantum Design). Figure 3 shows the DC susceptibility (χ) for the various x samples annealed under 90 bar oxygen at

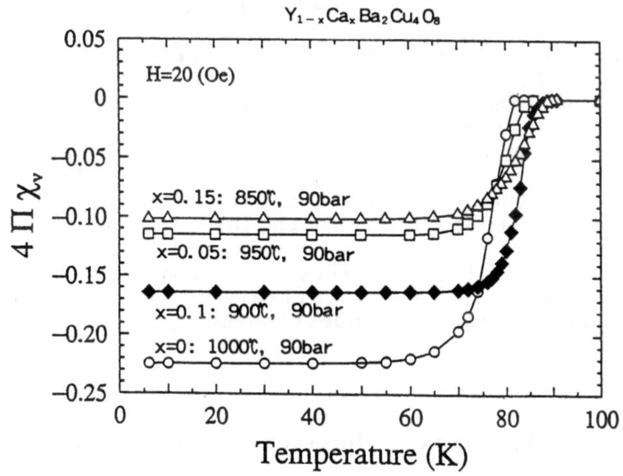

Fig.3 Temperature dependence of DC susceptibility of the various x samples annealed at each optimum annealing temperature under 90 bar oxygen atmosphere.

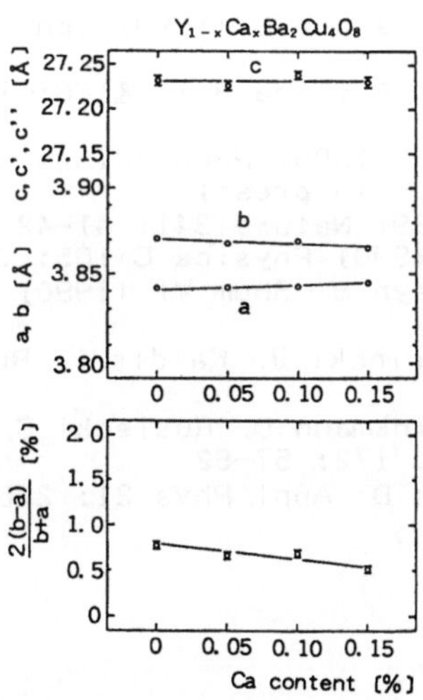

Fig.4 Lattice constants and orthorhombicity as a function of the nominal Ca composition x in $Y_{1-x}Ca_xBa_2Cu_4O_8$.

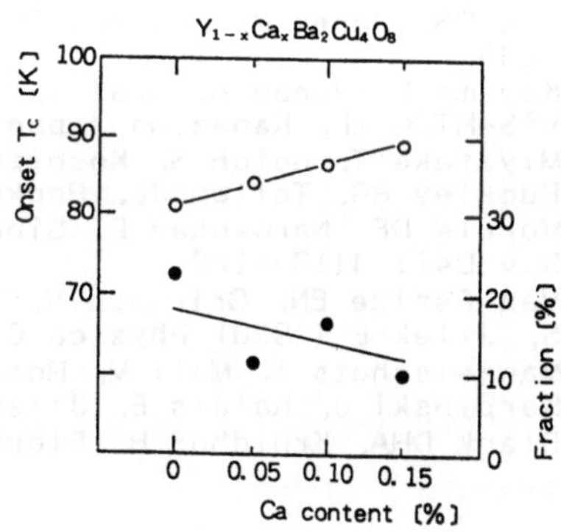

Fig.5 Onset T_c's and Meissner fractions determined from DC-χ as a function of the nominal Ca composition x in $Y_{1-x}Ca_xBa_2Cu_4O_8$. (high pressure annealed samples at each optimum temperature under 90 bar oxygen atmosphere)

each optimum annealing temperature as a function of temperature. The onset T_c is clearly enhanced by the Ca-doping for the samples which contain no (247) and (123) phases in the X-ray diffraction patterns.

Lattice constants and onset T_c values determined from DC-χ are summarized as a function of x in Figs. 4 and 5, respectively. The orthorhombicity 2(b-a)/(b+a) decreases linearly with increasing x. The lattice constant c remains constant with x. The onset T_c determined by DC-χ increases linearly from 81K to 89K with increasing Ca content from x=0 to x=0.15. Lattice constants and onset T_c for the atmospheric sintered samples are nearly the same as those of the high pressure annealed samples.

Systematic microstructural investigations were performed after polishing the samples using diamond paste. They were examined in a light microscope and in a high resolution scanning electron microscope. The most obvious difference between (124) specimens prepared at normal and high pressure is found in the grain size. While we observed an average grain size below 1μm in the former case, it increases to more than 10μm by the high pressure annealing.

Energy-dispersive analysis of X-rays (EDAX) was performed on the L lines of Ba and Y, and the K lines of Cu and Ca. Atomic compositions were determined by EDAX for each grain and globally for the samples with x=0, 0.1 and 0.15 prepared at atmospheric and high pressure oxygen. There exists nearly the same amount of Ca as the nominal composition in the global analysis for all samples including the samples which decompose to the (247) and the (123) phases. EDAX analyses in the (124) phase grains for the samples prepared at high pressure oxygen show that Ca is entering the structure as given by the nominal concentration. On the other hand, the composition ratio of Y in the (124) grains is larger than that of the global analysis by nearly the same amount as the nominal Ca content for both the samples with x=0.1 and x=0.15. If Ca had substituted for Y, the Y composition ratio in the (124) grains should coincide with the global analysis. From these results, it appears that Ca substitution for Y is rather unlikely. However, the doping effect on T_c within the (124) phase cannot be questioned.

REFERENCES

1. Karpinski J, Kaldis E, Jilek E, Rusiecki S, Bucher B (1988) Nature 336: 660-662
2. Morris DE, Asmar NG, Nickel JH, Sid RJ, Wei JYT (1989) Physica C 159: 287-294
3. Cava RJ, Krajewski JJ, Peck Jr. WF, Batlogg B, Rupp Jr. LW, Fleming RM, James ACWP, Marsh P, (1989) Nature 338: 328-330
4. Murakami H, Yaegashi S, Nishino J, Shiohara Y, Tanaka S (1990) Jpn J Appl Phys 29: L445-L446
5. Liu RS, Janes R, Edwards PP, Campbell AM (1990) Physica B 165&166: 1667-1668
6. Koyama K, Junod A, Graf T, Triscone G, Muller J (1990) Proceedings of M^2S-HTSC III, Kanazawa Japan. Physica C 182-184: (in press)
7. Miyatake T, Gotoh S, Koshizuka N, Tanaka S (1989) Nature 341: 41-42
8. Buckley RG, Tallon JL, Pooke DM, Presland MR (1990) Physica C 165: 391-396
9. Morris DE, Narwankar P, Sinha APB, Takano K, Fayn B, Shum VT (1990) Phys Rev B41: 4118-4122
10. Van Eenige EN, Griessen R, Wijngaarden RJ, Karpinski J, Kaldis E, Rusiecki S, Jilek E (1990) Physica C 168: 482-488
11. Mangelschots I, Mali M, Roos J, Zimmermann, Brinkmann D, Rusiecki S, Karpinski J, Kaldis E, Jilek E (1990) Physica C 172: 57-62
12. Blank DHA, Kruidhof H, Flokstra J (1988) J Phys D: Appl Phys 21: 226-227

3.3 Structures and Properties

Infrared and Raman Spectra of Cuprate Superconductors

JIRO TANAKA, MICHIHIRO SHIMADA, KOJI KAMIYA, and MASAAKI SHIMIZU

Department of Chemistry, Nagoya University, Nagoya, 464-01 Japan

Abstract

The electronic excitations between the filled and unoccupied O_{2p} molecular orbitals of CuO_4^{6-} cluster in the CuO_2 plane are most important among possible elementary excitations in cuprate superconductors. The scattering rates inferred from the Raman and infrared spectra are used to estimate Eliashberg function. We found that the A_{1g} exciton state of the above-mentioned excited state is most important for the Cooper pair formation. The transition temperatures are evaluated with the gap equation and reasonable agreement is found between the calculated and observed values.

Introduction

Electronic spectra of cuprate superconductor are of particular importance to find the mechanism of Cooper pair formation. In fact, both the infrared spectra and the electronic Raman spectra showed that the low energy electronic excitation dominates from near infrared even to phonon spectral region. The mid IR band show interference effect with the vibrational modes and also overlap with Drude reflectance of free carrier.

Under this situation, the spectra show complicated pattern. We measured infrared reflection spectra carefully on good quality single crystals and analyze them into detail. We also measured Raman spectra on single crystals of $La_{2-x}Sr_xCuO_4$ and $Bi_2Sr_2CaCu_2O_{8+y}$ at room temperature. We calcualted the optical dielectric functions by Kramers-Kronig transformation and estimated the Eliashberg function. The pair potentials are calculated by the Eliashberg equation and transition temperatures are estimated by solving the equation at finite temperature.

Experimental

The single crystals of $La_{2-x}Sr_xCuO_4$ with varying x contents were gifted from Prof. Kojima and Dr. Tanaka of Yamanashi University. The single crystals of $Bi_2Sr_2CaCu_2O_{8+y}$ were grown in our laboratory and large single crystals were gifted from Prof. Kawai and Dr. Kitahama of Osaka University. The reflection spectra of these crystals were measured as shown in Figs.1 and 2. Three microspectrophotometers were used from far-infrared to ultraviolet regions. Specially designed microscopes were used with a large aperture

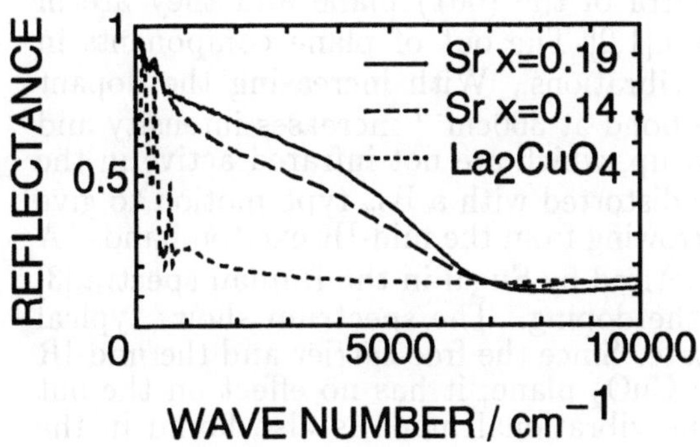

Fig.1 Reflection Spectra of $La_{2-x}Sr_xCuO_4$

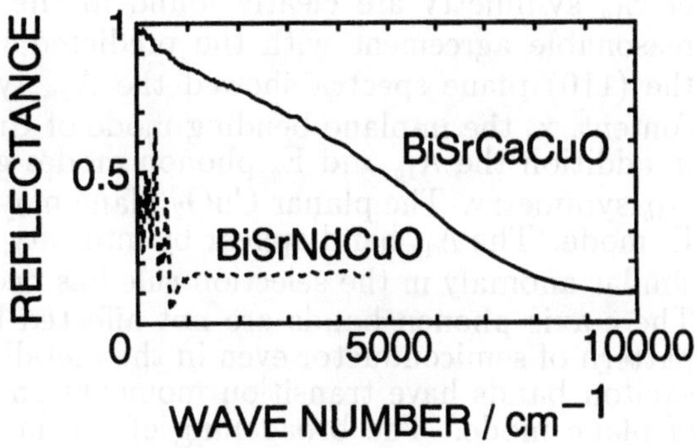

Fig.2 Reflection Spectra of $Bi_2Sr_2XCu_2O_{8+y}$(X=Ca,Nd)

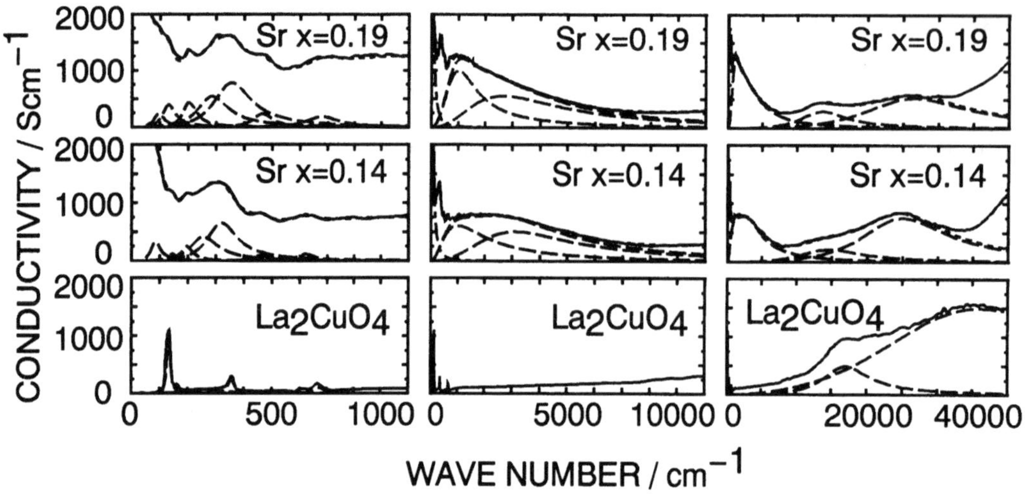

Fig.3 Optical Conductivities of $La_{2-x}Sr_xCuO_4$ (x=0,0.14,0.19)

(———) observed curve, (-·-·-·-) Drude term, (- - - - -) each bands, (··········) total fit

objective lens for far infrared and infrared measurement. The Raman spectra were recorded by the 355nm excitation of YAG laser with a JASCO monochromator and a multichannel plate detector of PAR.

The optical dielectric functions are calculated as shown in Figs.3 and 4. The peaks of optical conductivity spectra were assigned to each elementary excitations. The A_{1g}, B_{1g} and B_{2g} components of electronic Raman spectra were obtained by the polarization measurement with oriented single crystal surfaces. (Figs.5 and 6)

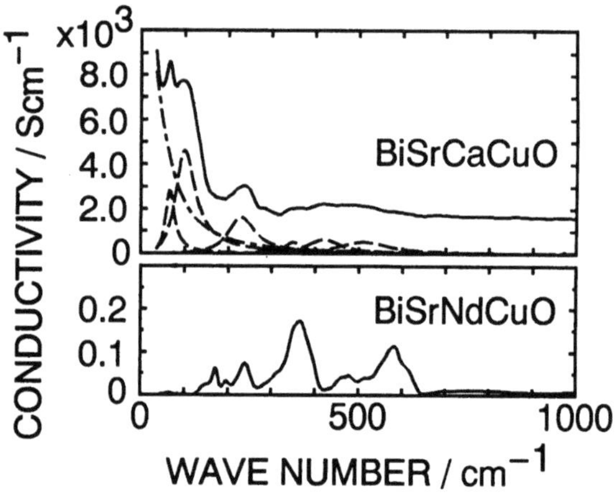

Fig.4 Optical Conductivities of $Bi_2Sr_2XCu_2O_{8+y}$ (X=Ca,Nd). All curves have the same meanings as in Fig.3.

Phonon Bands

In the insulating phase of La_2CuO_4 and $Bi_2Sr_2YCu_2O_{8+y}$, the optical phonon bands of E_{1u} symmetry are clearly found in the spectra of the (001) plane and they are in reasonable agreement with the predicted values.[1,2] The out of plane components in the (110) plane spectra showed the A_{2u} type vibrations. With increasing the dopant content, x, the in-plane bending mode of Cu-O bond at 360cm^{-1} increases intensity and in addition the A_{1g} and E_g phonon modes grow up, which are not infrared active in the D_{4h} symmetry. The planar CuO_2 plane may be distorted with a B_{2u} type motion to give E_g mode. The A_{1g} band appear by intensity borrowing from the mid-IR exciton bands. A similar anomaly in the selection rule has been noticed by Sugai in the Raman spectra.[3] The c-axis phonon bands are not affected by the doping. The spectrum shows typical pattern of semiconductor even in the metalic state. Since the free carrier and the mid-IR exciton bands have transition moments on the CuO_2 plane, it has no effect on the out of plane mode. The broadening of the in plane vibrational mode is also found in the $Bi_2Sr_2CaCu_2O_{8+y}$ phonon bands.

Drude Electron and Scattering Rate

The dielectric functions are expressed in the harmonic oscillator model by

$$\varepsilon(\omega) = \varepsilon_\infty - \frac{\omega_p^2}{\omega^2 + i\omega/\tau_0} + \sum_{j=1}^{n} \frac{\Omega_0^2 f_j}{\omega_j^2 - \omega^2 - i\omega\gamma_j} \tag{1}$$

The Drude component is the second term and the phonon, the Mid-IR exciton and the charge transfer bands are given in the third term for each excited states. Here ω_p is the plasma frequency and $1/\tau_0$ is the scattering rate of the free carrier. The f_j is the oscillator strength of the j-th transition and the value of Ω_0 is taken as number of Cu atoms in a unit volume.[4]

We fitted the $\sigma(\omega)$ spectra and $\varepsilon_1(\omega)$ spectra by selecting parameters. The fitting curves in the case of $La_{2-x}Sr_xCuO_4$ is shown in Figs.3 and 4.

Alternative approach to analyze the dielectric function is to use only the Drude term to represent all functions by variation of $\tilde{\tau}(\omega)$ and $\Omega_p(\omega)$ $(= \sqrt{m/m*(\omega)}\omega_p)$

$$\varepsilon(\omega) = \varepsilon_\infty - \frac{\Omega_p^2(\omega)}{\omega^2 + i\omega/\tilde{\tau}(\omega)} \tag{2}$$

This method is differnt from the previous one, but it may give some insight into the scattering mechanism and the mass enhancement of the free carier. We followed the method of Allen and Sievers [5,6] and take $\tilde{\tau}(\omega)$ as complex quantity by

$$\frac{1}{\tilde{\tau}(\omega)} = R(\omega) + iI(\omega) \tag{3}$$

The real part of scattering, $R(\omega)$, is used to find the contribution of the optical modes to the Eliashberg function.[7]

Mid IR Exciton Band and Other Excited States

We have discussed the origin of the mid IR exciton band by the excitaion of electron in the molecular orbitals composed of the $O_{2p\pi}$ and the $O_{2p\sigma}$ atomic orbitals.[8] The holes in the CuO_2 plane exist in these orbitals. At least two excited states exist in the 0.1 $\sim$ 0.5 eV region are observed in the ab plane. The intensities are significantly large to give large excitonic dispersion effect. These are transverse mode of the exciton measured on the (001) plane. In addition, the longitudinal mode is important to give the peak of the dielectric loss function at 1 eV region. The functions are calculated by using the parameters [4]; it turns out that the main part of $\mathrm{Im}(1/\epsilon(\omega))$ originates from the mid IR exciton bands.[7]

The next excited state appears at 12000 cm^{-1} region in $La_{2-x}Sr_xCuO_4$, which were first found by Suzuki.[9] We assigned this band as the transition from the deeper $O_{2p\sigma}$ orbital to the holes of the CuO_4 group orbital. The strong charge transfer band from O_{2p} to Cu_{3d} vacant orbital appears at 20000 $\sim$ 25000 cm^{-1} region, and their energies show blue shift by increasing the hole concentration in the O_{2p} group orbital; the tendency is reasonable since the holes in the O_{2p} orbital may stabilize the level relative to the Cu_{3d} vacant orbital.

Raman Spectra

The electronic Raman spectra of the single crystals are measured on the (001) face with the exciting light polarized along the Cu-O bond (x,y) and the axes rotated by 45 degrees (x',y'). The A_{1g}, B_{1g} and B_{2g} Raman bands are obtained by the configuration of y'y'-yx, y'x' and yx of the exciting and the outgoing light. The spectra for $La_{1.81}Sr_{0.19}CuO_4$ and $Bi_2Sr_2CaCu_2O_{8+y}$ are shown in Figs.5 and 6.

The characteristic of the A_{1g} electronic Raman spectra is that the maximum is found

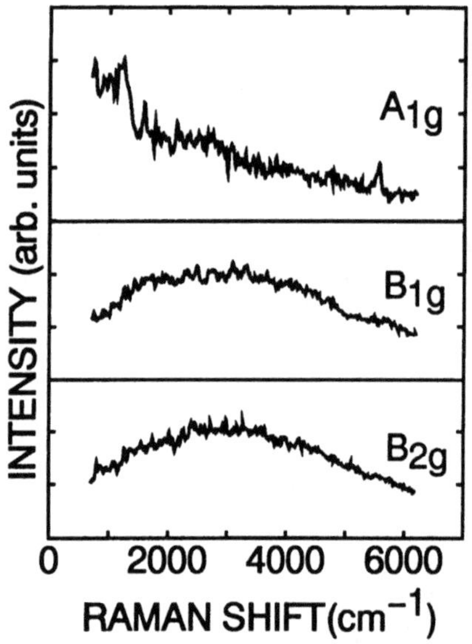

Fig.5 Raman Spectra of $La_{2-x}Sr_xCuO_4$:x=0.19

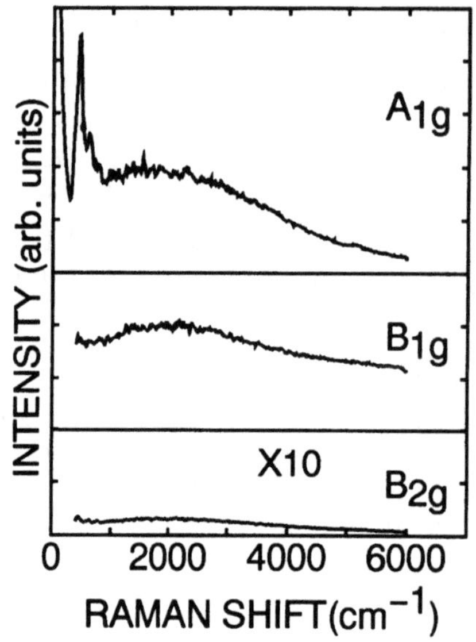

Fig.6 Raman Spectra of $Bi_2Sr_2CaCu_2O_{8+y}$

at $\omega \rightarrow 0$ and other peak is found at 3000 cm^{-1} region. However, the B_{1g} and the B_{2g} bands show single broad maximum at 2000 - 3000 cm^{-1} with the ordering of the level $B_{1g} < B_{2g}$.

Dispersion of the Exciton States

The mid IR exciton state of the CuO_4^{6-} cluster has a large transition moment, and the exciton band splitting will be very large. We calculated the k-dependent band splitting for the transverse and the longitudinal exciton states of the E_{1u} symmetry by using the dipole-dipole interaction of the transition moment. We put the transition moments on the central Cu atom in the CuO_4 cluster. The band splitting is dependent on the magnitude of the transition moment and we scale it with the splitting of the TO and the LO exciton bands and the k-dependence of EELS spectra.

The Raman active exciton states are also calculated with the arrangement of moments as shown in Fig.7. The electron with a center of symmetry (A_{1g} symmetry) will be

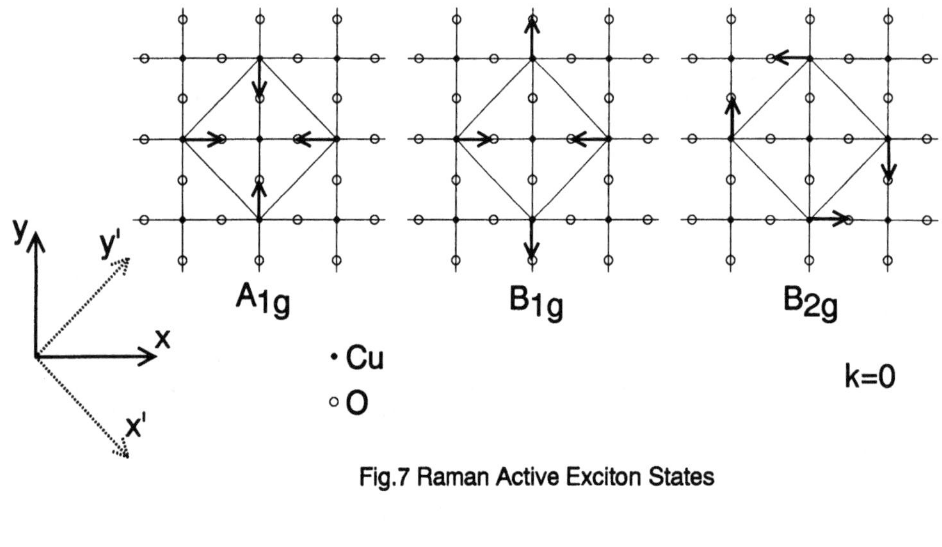

Fig.7 Raman Active Exciton States

stabilized by the polarization field of the A_{1g} exciton state. It gives the electronic Raman band of A_{1g} symmetry at $\omega \to 0$ region. The coupled A_{1g} exciton state appears at 3000 cm^{-1} region. The energy level of the B_{1g} state is lower than the B_{2g} state at k = 0; this is consistent with experiment. The important result from this analysis is that the electron coupled with the A_{1g} exciton state shows significant stabilization to $\omega \to 0$. In fact, the A_{1g} Raman spectra show finite spectral density, S(ω), near $\omega \to 0$.

Pair Potnetial

The Eliashberg function for the kernel of integral equation is connected to the scattering rate of the free carrier, which is given by

$$\alpha^2(\omega)F(\omega) = S(\omega) - \frac{1}{2\pi}\frac{\partial}{\partial\omega}(\frac{\omega_p^2}{\omega}Im(\frac{1}{\epsilon(\omega)})) \tag{4}$$

The Eliashberg function represents the coupling of the carrier with the elementary excitations. We take two couplings; one is the elcectron coupling with A_{1g} type exciton state, which is taken to be proportional to ths spectral density $S(\omega)$ of the A_{1g} Raman spectra. This coupling is important since the s-wave Cooper pair has symmetry of A_{1g} and the Hamiltonian will be totally symmetric. The second term is due to the LO type E_{1u} excited states of the phonon and the exciton.

The function calculated by this method shows significant positive peaks at phonon, and exciton regions and some negative regions. It should be taken positive for all frequency range. We normalize $S(\omega)$ to satisfy this condition. We use the well known gap equations to calculate $\Delta(\omega)$ with the $\alpha^2F(\omega)$ and the values of $\mu*$ at 0.3 ~ 0.4.

$$\Delta(\omega) = [Z(\omega)]^{-1}\int_0^{\omega_c} d\omega' Re\frac{\Delta(\omega')}{[\omega'^2 - \Delta^2(\omega')]^{1/2}}[K_+(\omega,\omega') - \mu*] \tag{5}$$

$$[1 - Z(\omega)]\omega = \int_0^{\omega_c} d\omega' Re\frac{|\Delta(\omega')|}{[\omega'^2 - \Delta^2(\omega')]^{1/2}}K_-(\omega,\omega')$$

$$K_\pm(\omega,\omega') = \int_0^{\omega_c} d\Omega\alpha^2(\Omega)F(\Omega)[(\frac{f(-\omega') + n(\Omega)}{\omega' + \omega + \Omega + i\delta} \pm \frac{f(-\omega') + n(\Omega)}{\omega' - \omega + \Omega - i\delta})$$

$$\mp(\frac{f(\omega') + n(\Omega)}{-\omega' + \omega + \Omega + i\delta} \pm \frac{f(\omega') + n(\Omega)}{-\omega' - \omega + \Omega - i\delta})]$$

where $f(\omega)$ and $n(\omega)$ are Fermi and Bose distribution functions. The values of $\mu*$ are about 0.3 -0.4 for these crystals.

Table 1 Energy Gap and Transition Temperatures

Crystals	Δ(calc.)/meV	Δ(exp.)/meV	T($\Delta(\omega)$=0)/K	T_c/K	μ^*
$La_{1.86}Sr_{0.14}CuO_4$	8	8	45	31	0.32
$La_{1.81}Sr_{0.19}CuO_4$	9	8	51	33	0.33
$Bi_2Sr_2CaCu_2O_{8+y}$	21	21	118	85	0.36

Table 1 shows the calculated $\Delta(0)$ and the temperature at which $\Delta(\omega)$ converges to 0. In spite of simple approach, the results are in reasonable agreement with experimental values of $\Delta(0)$ and Tc. Moreover the phonon structures are found in the superconducting density of states which agree with tunneling spectral data.[10,11]

Coupled Mode for Pairing

The main excitation which couples with the electron is the A_{1g} exciton state. We have to revise our previous argument that the LO exciton is most important. (In that paper we made a mistake that the $c\omega$ term is small compared to the LO term in representing Fig.10, however, numerical calculation was correct.[7])
We analyze the contribution of the phonon and the LO exciton separately by decomposing the Eliashberg function into part. The phonon and the E_u exciton contributions are separately estimated by calculating λ by the following integration.

$$\lambda = \int_0^{\omega_c} \frac{\alpha^2(\omega)F(\omega)}{\omega} d\omega \qquad (6)$$

The A_{1g} contribution is estimated by the Raman intensity $S(\omega)$.
A quantitative estimate for these three contribitions are as follows.

Table 2 Contribution of each Excitation to Coupling Constant

Crystals	λ(total)	A_{1g} exciton	LO-phonon	E_u exciton (LO)
$La_{1.86}Sr_{0.14}CuO_4$	1.47	1.06 (0.72)	0.39 (0.26)	0.02 (0.02)
$Bi_2Sr_2CaCu_2O_{8+y}$	2.45	2.05 (0.84)	0.28 (0.11)	0.11 (0.05)

The values in parenthesis are percent of the contribution.

This analysis shows that presence of the mid IR exciton state is most effective for increasing transition temperatures of cuprate superconductors.

Acknowledgement
The authors thank to Profs.H.Kojima and T.Kawai and Drs.I.Tanaka and Y.Kitajima for gift of crystals. They thank to Prof.Y.Kuroda of Nagoya University and Dr.Y.Tanaka of Niigata University for helpful comment and also to Mr. S.Yogo of Earth Science Department of our faculty for preparing clean crystalline surface.

References

[1]F.E.Bates and J.E.Eldrigde, Solid State Comm. 72 (1989) 187
[2]J.Prade, A.D.Kulkarni, F.W.de Wette, W.Kress, M.Cardona, R.Reiger and U.Schröder, Solid State Comm. 64 (1987) 1267
[3]S.Sugai, S.Shamoto, M.Sato, T.Ido, H.Takagi and S.Uchida, Solid State Comm. 76 (1990) 371
[4]M.Shimada, M.Shimizu, J.Tanaka, I.Tanaka and H.Kojima, to be published
[5]J.W.Allen and J.C.Mikkelsen, Phys. Rev. B 15 (1977) 2952
[6]B.C.Webb, A.J.Sievers and T.Mihalisin, Phys. Rev. Lett. 57 (1986) 1951
[7]J.Tanaka, M.Shimada, M.Shimizu and Y.Tanaka, Physica C 176 (1991) 170
[8]J.Tanaka, K.Kamiya and C.Tanaka, Physica C 161 (1989) 451
[9]M.Suzuki, Phys. Rev. B 39 (1989) 2313
[10]N.Miyakawa, D.Shimada, T.Kido and N.Tsuda, J. Phys. Soc. Jpn. 59 (1990) 2473
[11]T.Ekino, to be published in the Soviet Journal of Low-Temperature Physics

Formation of $Y_2Ba_4Cu_7O_y$ at Ambient Pressure

YUTAKA IWAI and HIROSHI SAITO

Nagaoka College of Technology, Nagaoka, Niigata, 940 Japan

Preparation of $Y_2Ba_4Cu_7O_y$ (247 phase) was examined at 1 atm of oxygen partial pressure under the following conditions; (1)using $NaNO_3$ as a reaction accelerator, (2)without reaction accelerator and without seed and (3)using the seeding method, respectively. They were pressed and fired at 850-900°C under oxygen gas flow. The 247 phase was obtained in the temperature range from 860°C to 890°C using $NaNO_3$. Under the condition(2), however, the 247 phase could not be formed in the same temperature range. But, the seeding method was effective; the 247 phase formed at the temperature range from 870°C to 890°C. The sample of the 247 phase prepared at 870°C under the condition(1) exhibited superconductivity with zero resistive point at 93K.

Keywords: 247 phase, seeding method, ambient pressure

1. Introduction

$Y_2Ba_4Cu_7O_y$ superconducting oxide (hereafter refer to as the 247 phase) has the characteristic structure consists of alternating blocks of $YBa_2Cu_4O_8$(124 phase) and $YBa_2Cu_3O_{7-x}$(123 phase). The 247 phase, as well as the 124 phase, have been prepared using the high pressure condition[1-4]. Since Cava et al. first succeeded in the bulk synthesis of the 124 phase at one atmosphere of oxygen partial pressure using alkali metal carbonates as a reaction accelerator[5], several researches have been performed to prepare these phases at ambient pressure; Pooke et al.[6] and Buckley et al.[7] prepared the 124 and 247 phase using alkali nitrate.

In the fields of ceramics, seeding method has been popular technique for the solid state reaction[8]. It is effective when an activation energy of nucleation is high and therefore the transformation is sluggish and sometimes yields the metastable phase. In our previous research, the seeding method was applied to synthesis of the 124 phase and was found to be effective[9].

In this report, we described on the preparation of the 247 phase under one atmosphere of oxygen partial pressure by the following condition; (1)using $NaNO_3$ as a reaction accelerator, (2)without $NaNO_3$ and without seed and (3)applying the seeding method.

2. Experimental

Y_2O_3(99.9%), $BaCO_3$(99.9%) and Cu(99.99%) were used as the starting materials and they were dissolved in nitric acid. The atomic ratio of Y:Ba:Cu was adjusted to be 2:4:7. The solution was converted into nitrate salt followed by decomposition at 750°C for 5h under the oxygen gas flow; the resultant material was used as a raw material. (1)For preparation using the reaction accelerator, the raw material was mixed with 20mol% of $NaNO_3$(99.9%). (2)For direct preparation, the raw material was used without any additives. (3)For seeding preparation, the raw material was mixed with 20wt% of seed of the 247 phase, which was, in the present research, prepared by utilizing $NaNO_3$ accelerator describing later. The raw materials and additives were ground and mixed using agate mortar and pestle. The resultant powders were pressed into pellets at 1000 kgf/cm^2 and heated at the temperature from 860° to 900°C for 20h under the oxygen gas flow and cooled in furnace. The electric resistivity of the samples was measured by a four-probe method. The phases in the samples were identified by the X-ray powder diffraction(XRD) method with CuKα radiation.

3. Results and Discussions

The XRD patterns for the as-prepared samples are shown in Figs.1-3. Some main diffraction peaks of the 123, 247 and 124 phases were overlapped each other, then peaks which can be assigned to only one of the phases were shown using the following notations, ▲(123 phase), O(247 phase) and ●(124 phase). Fig.1 shows the series of samples fired at 850° - 900°C for 20h under the oxygen gas flow with $NaNO_3$ reaction accelerator. Well crystallized nearly single 247 phase were observed in the XRD patterns of samples prepared at 860°-890°C. Favorable sample was obtained by firing at 870°C. $NaNO_3$ in the sample was almost vaporized during the firing process. From the flame spectrochemical analysis, the concentration of residual Na^+ is approx. 200mol·ppm in the sample.

Fig.2 shows the samples prepared under the condition(2) by firing raw material without $NaNO_3$ accelerator. The firing condition was the same in the case of Fig.1. The 124 phase was formed at 850°C and the 123 phase was at 860° - 900°C, but the 247 phase was not observed clearly. It suggests the difficulty of direct preparation of the 247 phase under the condition. For the reason, we consider that the activation energy of the nucleation of the 247 phase is very large, then the nucleation process was very sluggish and sometimes produce the 123 phase as a metastable phase[8].

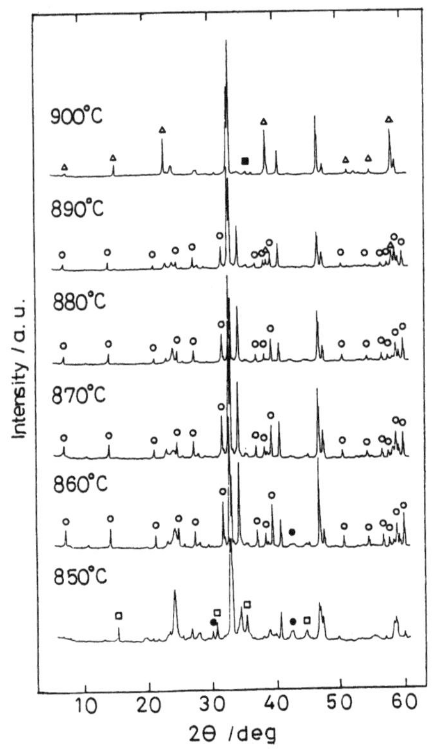

Fig. 1 XRD patterns using CuKα radiation for samples prepared using $NaNO_3$ as reaction accelerator. The sample was obtained by firing at 850°-900°C for 20h in oxygen gas flow. ▲, O, □ and ■ denote to be the diffraction peaks of the 123, 247, 124 phase and CuO. ● denoted to be unidentified phase.

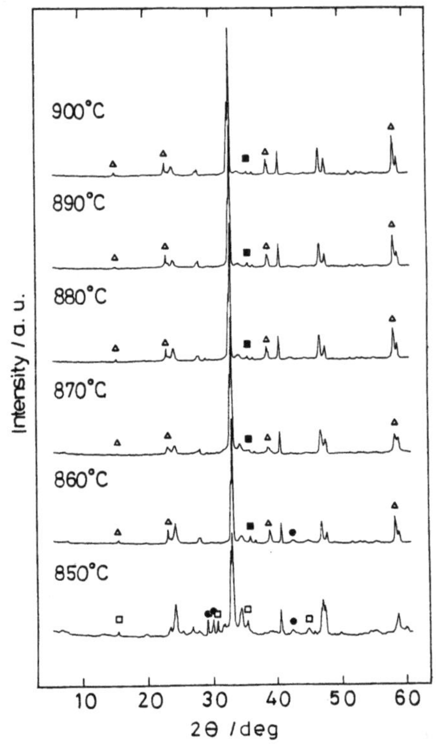

Fig. 2 XRD patterns using CuKα radiation for samples prepared without $NaNO_3$. The samples were obtained by firing the raw material at 850°-900°C for 20h in oxygen gas flow. ▲, O, □ and ■ denote to be the diffraction peaks of the 123, 247, 124 phase and CuO. ● denoted to be unidentified phase.

Fig.3 shows the samples prepared from raw materials with 20wt% of seed. The diffraction peaks of the 247 phase were observed in samples prepared at 870° - 890°C. The trace of the 247 phase was observed in samples prepared at 860°C and 900°C, respectively, but disappeared after repeating the firing with intermediate pulverization. The very sluggish process of the nucleation can be skipped in such system, and the 247 phase can be grown around the seed. The result shows that the seeding is effective to prepare the 247 phase as well as the 124 phase[9].

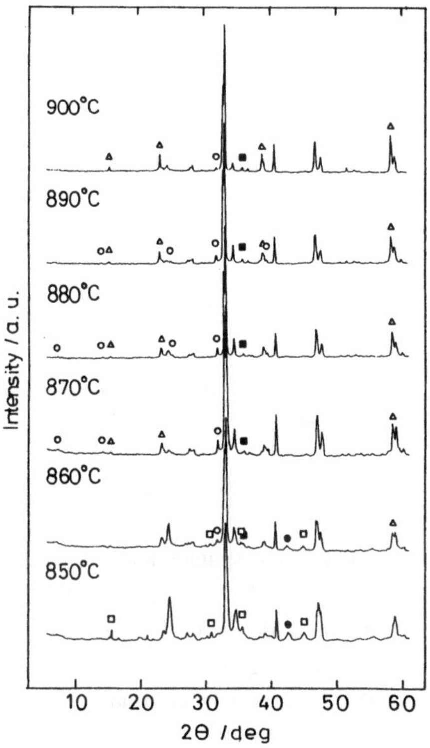

Fig. 3 XRD patterns using CuKα radiation for seeded samples. The samples were obtained by firing the raw material with 20wt% of seed at 850°-900°C for 20h in oxygen gas flow. ▵, O, □ and ■ denote to be the diffraction peaks of the 123, 247, 124 phase and CuO. ● denoted to be unidentified phase.

In Fig.4, the XRD patterns of the seeded samples which were fired at 870°C; (1) shows the as-prepared sample and (2) shows the sample fired twice with intermediate pulverization. Diffraction peaks with indices are of the 247 phase, and ▵ with indices and ■ denoted to be the 123 phase and CuO, respectively. The 123 phase was coexisted with the 247 phase in the samples, but the fraction of the 123 phase decreased with repeating the firing process, suggesting that the chemical reaction of

$$2YBa_2Cu_3O_{7-d} + CuO + xO_2 \rightarrow Y_2Ba_4Cu_7O_y$$

proceeded during the process. The 247 phase will grow in the sample through the prolonged firing.

To apply the seeding method, the pure 247 phase without $NaNO_3$ can be prepared more easily. We consider that the reaction period can be reduced and quality of the 247 phase samples can be improved by simultaneous use of alkoxide or other organometallic methods.

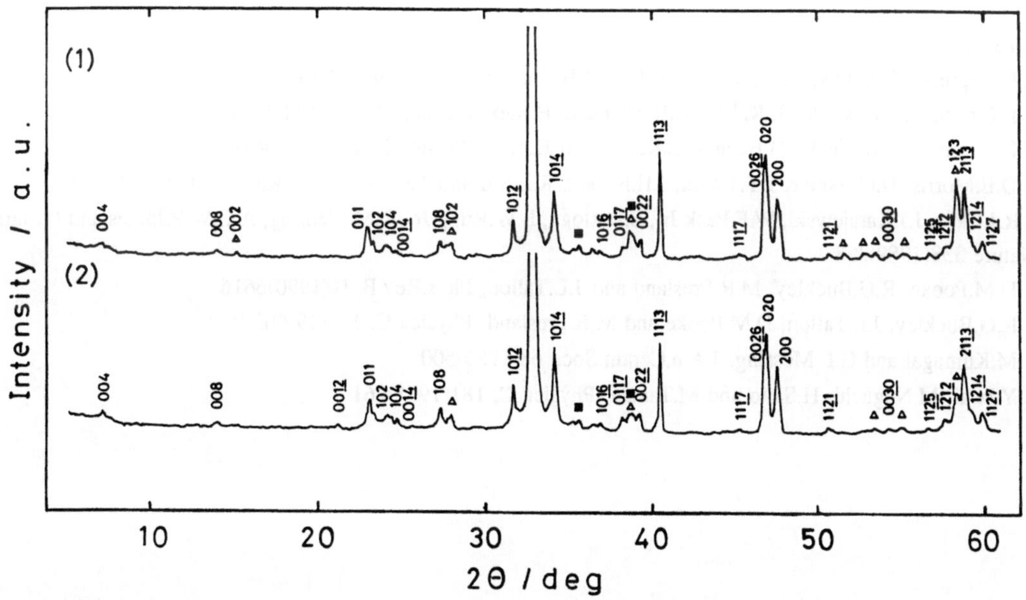

Fig.4 XRD patterns of samples prepared by seeding at 870°C. (1) show the sample as prepared. (2) shows the sample after twice firing with intermediate pulverization. ▵ with index denote to be the diffraction peak of the 123 phase. ■ denote to be the diffraction peak of CuO.

Fig.5 shows the electric resistivity versus temperature curve for the samples which was prepared at using $NaNO_3$ and for the sample which by seeding method. They were fired at 870°C for 20h and cooled in furnace in oxygen gas flow. The electric resistivity of both samples decreased with decreasing temperature. The sample prepared using $NaNO_3$ exhibited a sharp transition and the zero-resistivity condition at 93K. The sample prepared by seeding method exhibited zero-resistivity condition at 92K.

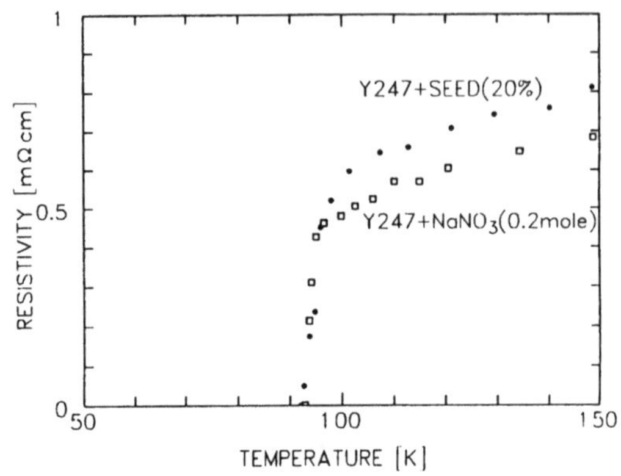

Fig.5 Temperature dependence of the electric resistivity of sample prepared at 870°.

4. Summary

The preparation of the 247 phase superconducting oxides was examined at the temperature from 850° to 900°C under P_{O2}=1atm. From the direct preparation using $NaNO_3$ accelerator, the single 247 phase was obtained at 860° to 890°C. The seeding method was effective when the firing temperature is 870°-890°C. From unseeded raw material, the 247 phase could not be formed under the conditions examined in the present research. The sample of the 247 phase which was prepared at 870°C using $NaNO_3$ exhibited the zero-resistivity condition at 93K.

Acknowledgements:

The authors thank Mr.N.Tanaka and Mr.N.Takasugi in Nagaoka College of Technology for their assistance in experiments.

References:

[1] I.Karpinski, E.Kaldis, E.Jilek, S.Rusiecki and B.Bucher, Nature 336(1988)660
[2] I.Karpinski, S.Rusiecki, E.Kaldis, B.Bucher and E.Jilek, Physica C, 160(1989)449
[3] I.Karpinski, S.Rusiecki, B.Bucher, E.Kaldis and E.Jilek, Physica C, 161(1989)618
[4] D.E.Morris, N.G.Asmar, J.Y.T.Wei, J.H.Nickel, R.L.Sid and J.S.Scott, Phys.Rev.B 40(1989)11406
[5] R.J.Cava, J.J.Karajewski, W.F.Peck Jr., B.Batlogg, L.W.Rupp Jr., R.M.Fleming, A.C.W.P.James and P.Marsh, Nature 338(1989)328
[6] D.M.Pooke, R.G.Buckley, M.R.Presland and J.L.Tallon, Phys.Rev.B 41(1990)6616
[7] R.G.Buckley, J.L.Tallon, D.M.Pooke and M.R.Presland, Physica C, 165(1990)391
[8] M.Kumagai and G.L.Messing, J.Am.Ceram.Soc., 68(1985)500
[9] Y.Iwai, M.Noguchi, H.Saito and M.Takata, Physica C, 181(1991)261

Defect Chemistry of High T_c Superconducting Oxides

KAZUO FUEKI and YASUSHI IDEMOTO

Science University of Tokyo, Noda, Chiba, 278 Japan

ABSTRACT

It was found that the average valence of Pb+Bi in $BaPb_{1-x}Bi_xO_3$ is higher than +4 in a compositional range $0.1 \leq x \leq 0.3$ where superconductivity appears at low temperatures. The carrier concentration calculated with the assumption that one excess oxygen atom creates two holes agreed well with those determined from the Hall coefficient. The density measurement revealed that $BaPb_{1-x}Bi_xO_3$ possesses high concentration vacancies of Schottky type. The excess oxygen was concluded to be incorporated into anion vacancy. Close relationship between oxygen content and Tc was confirmed for more than ten kinds of superconducting oxides.

KEY WORDS: barium lead bismuth oxide, oxygen content, critical temperature

1. INTRODUCTION

High Tc superconducting oxides possess the perovskite-related structures in which B site is occupied by copper ion, as the skeleton of crystal. Generally, the perovskite type oxides with transition metal ions on B sites have large oxygen nonstoichiometry that controls the concentrations of electronic and ionic defects. Since the transport properties such as electronic conductivity and oxide ion diffusion coefficient depend on the defect concentration, the superconducting properties of oxide superconductors are expected to be dependent of oxygen content.

2. DEFECTS IN $BaPb_{1-x}Bi_xO_3$ AND Tc[1]

$BaPb_{1-x}Bi_xO_3$ is a superconducting oxide of perovskite structure and the superconductivity appears when x ranges from 0.1 to 0.3, as shown in Fig.1. Many studies have been made on the superconductivity of the oxide, but the reason why the superconduction appears in this compositional range has not been mentioned. The authors and their coinvestigators have determined the average valence of Pb+Bi by iodometry and found the result given in Fig.2. The average valence of Pb+Bi is higher than +4 in the range $0.1 \leq x \leq 0.3$ where the superconduction appears. Therefore, the exact chemical formula has to be represented not by $BaPb_{1-x}Bi_xO_3$ but by $BaPb_{1-x}Bi_xO_{3+\delta}$. Comparison of Fig.1 with Fig.2 indicates that Tc has a chose relationship with the excess oxygen content δ. The determination of Tc as a function of δ for the oxide with x=0.28 has shown that Tc increases with increasing δ. If one excess oxygen atom is assumed to create two holes, one can obtain the carrier

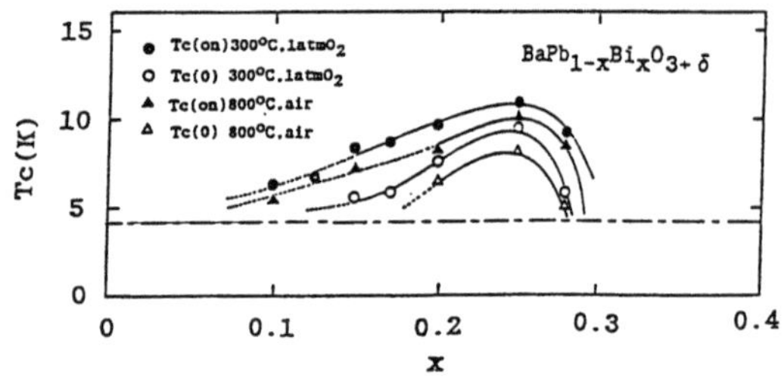

Fig.1 Relationship between Tc and x ($BaPb_{1-x}Bi_xO_3$).

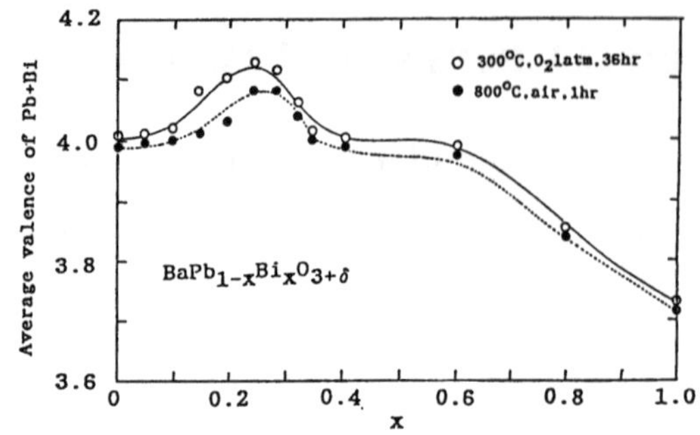

Fig.2 Relationship between average valence of Pb+Bi and x ($BaPb_{1-x}Bi_xO_{3+\delta}$).

concentration, which is drawn by solid lines in Fig.3. For comparison, the carrier concentrations determined by the Hall coefficient measurement are given by dotted lines. Fair agreement is seen between the present data and the data by Moiseev and his coinvestigator[2].

In order to clarify whether the excess oxygen occupies the interstitial site or oxide ion vacancy, the density was determined by the pycnometric and X-ray diffraction methods. It was found that $BaPb_{1-x}Bi_xO_3$ possess cationic and anionic vacancies of large concentration, essentially of Schottky type. Figure 4 gives the oxygen deficiency in unit cell. The highest percentage of oxygen deficiency is 13%, which is close to that of molten salts. It is concluded that the excess oxygen occupies oxide ion vacancies.

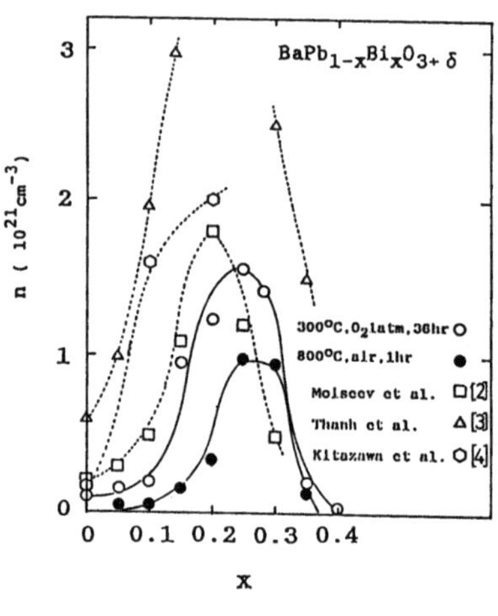

Fig.3 Carrier concentration vs. x ($BaPb_{1-x}Bi_xO_{3+\delta}$).

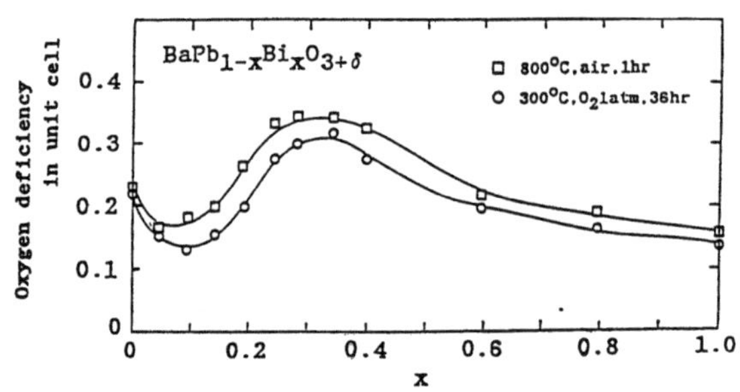

Fig.4 Oxygen deficiency in unit cell as a function of x ($BaPb_{1-x}Bi_xO_{3+\delta}$).

3. OXYGEN CONTENT AND Tc

It is well known that the critical temperatures of $(La_{1-x}Sr_x)_2CuO_4$ and $LnBa_2Cu_3O_{6+\delta}$ are closely related to their oxygen content and copper valence. A series of studies were carried out to elucidate whether other kinds of superconductors have the similar relationship between Tc and oxygen content. Figure 5 shows the plot of Tc against bismuth valence (and oxygen content $3-\delta$) for $Ba_{0.6}K_{0.4}BiO_{3-\delta}$. Superconductivity appears when the bismuth valence is higher than 4.20. Figures 6 and 7 give the Tc-oxygen content relationships for superconducting oxides of T[2] and T* structures[2,3]. The relationships for the 2201, 2212[3] and 2223[4] phases of Bi system are shown in Figs.8 to 10, respectively. A maximum is observed for the 2201 phase. In the case of

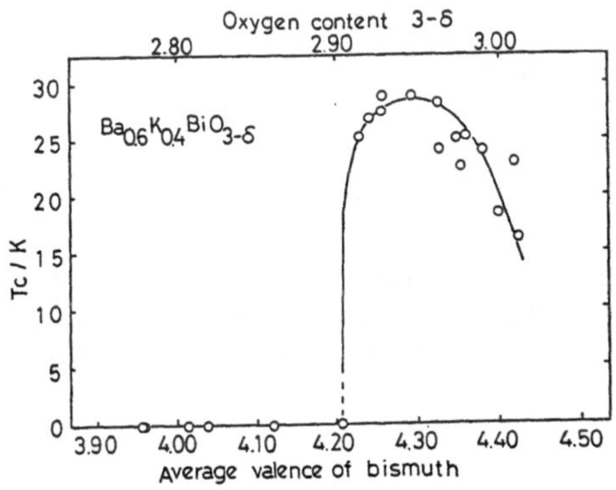

Fig.5 Relationship between Tc and bismuth valence ($Ba_{0.6}K_{0.4}BiO_{3-\delta}$).

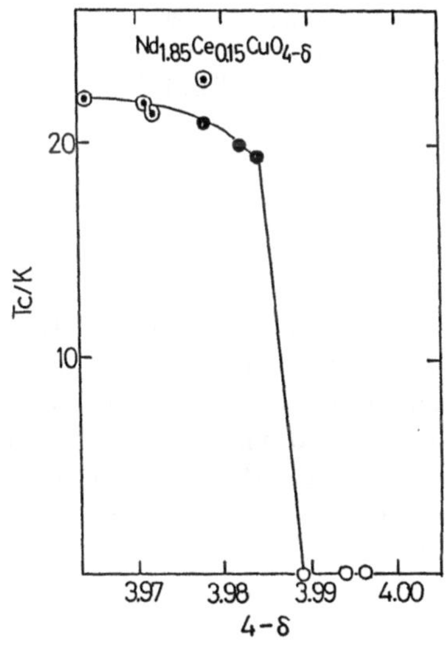

Fig.6 Relationship between Tc and oxygen content ($Nd_{1.85}Ce_{0.15}CuO_{4-\delta}$).

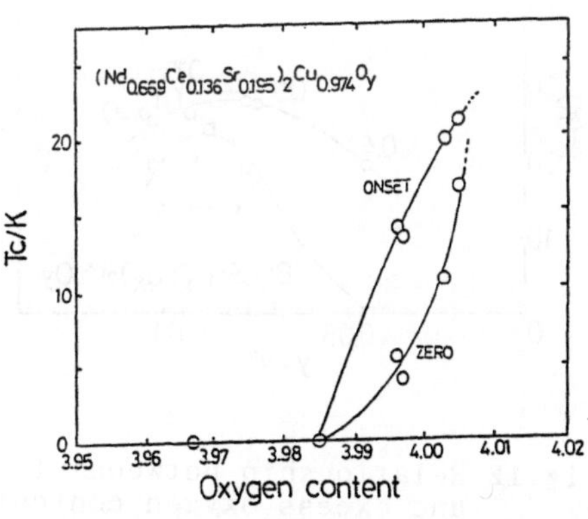

Fig.7 Relationship between Tc and oxygen content($(Nd_{0.669}Ce_{0.136}Sr_{0.195})_2Cu_{0.974}O_y$).

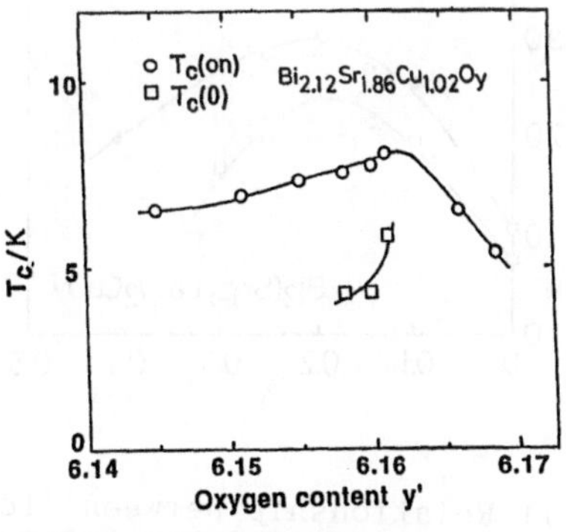

Fig.8 Relationship between Tc and oxygen content y' ($Bi_{2.12}Sr_{1.86}Cu_{1.02}O_{y'}$).

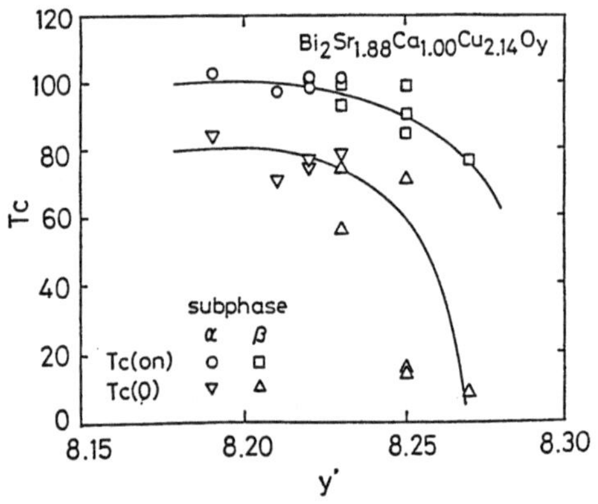

Fig.9 Relationship between Tc and oxygen content y' ($Bi_2Sr_{1.88}CaCu_{2.14}O_{y'}$).

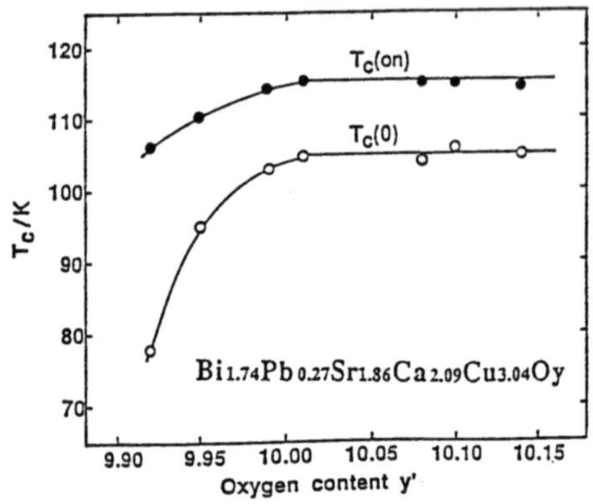

Fig.10 Relationship between Tc and oxygen content y' ($Bi_{1.74}Pb_{0.27}Sr_{1.86}Ca_{2.09}Cu_{3.04}O_{y'}$).

the 2212 phase, Tc decreases with increasing oxygen content with the increase in content, whereas it increases and reaches a constant value in the case of the 2223 phase. Strontium in $Bi_2Sr_2CuO_y$ was replaced by La to form $Bi_2(Sr_{1-x}La_x)_2CuO_y$ and the critical temperature was determined as a function of x. The Tc-x relationship was found to have a maximum at x=0.2, as shown in Fig.11[5]. The oxygen content y was determined by iodometry and the oxygen content y' was calculated with the assumption that the valences of bismuth and copper are +3 and +2, respectively. Figure 12 shows the relationship between Tc and y-y' .

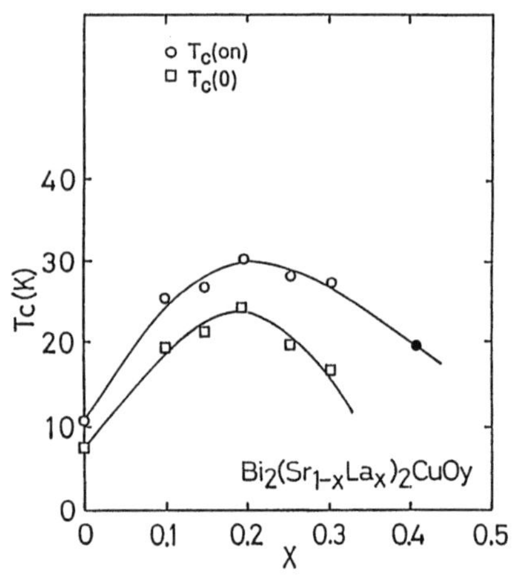

Fig.11 Relationship between Tc and x($Bi_2(Sr_{1-x}La_x)_2CuO_y$).

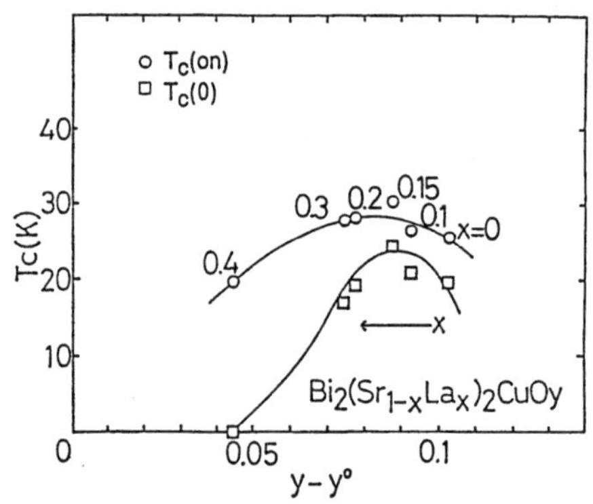

Fig.12 Relationship between Tc and excess oxygen content y-y' ($Bi_2(Sr_{1-x}La_x)_2CuO_y$).

As y-y˚ increases x decreases, and Tc has a maximum at y-y˚ =0.08 corresponding to x=0.2. This result indicates that the cationic substitution results in the change in Tc through the change in oxygen content. Simultaneous replacement of Sr by La and Bi by Pb in $Bi_2Sr_2CuO_y$ was made to form $(Bi_{1-x}Pb_x)_2(Sr_{1-x}La_x)_2CuO_y$. The plot of Tc against y-y˚ is shown in Fig.13[5]. With the increase in x, y-y˚ increases and the maximum Tc appears at around y-y˚ =0.09. Simultaneous substitution of Y for Ca and Pb for Bi in $Bi_2Sr_2CaCu_2O_y$ was made to form $(Bi_{1-x}Pb_x)_2Sr_2(Ca_{1-x}Y_x)Cu_2O_y$. Figure 14 shows the plot of Tc against y-y˚ [5]. y-y˚ increases with decreasing x, and Tc has a maximum at y-y˚ =0.14. A close relationship between Tc and oxygen content is concluded for all superconducting oxides studied.

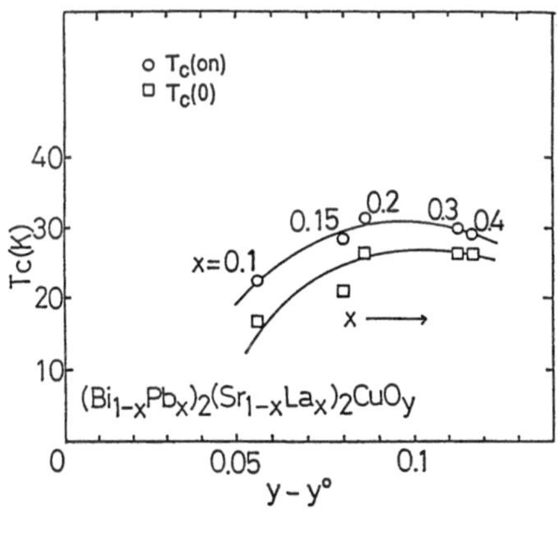

Fig.13 Relationship between Tc and y-y˚ ($(Bi_{1-x}Pb_x)_2(Sr_{1-x}La_x)_2CuO_y$).

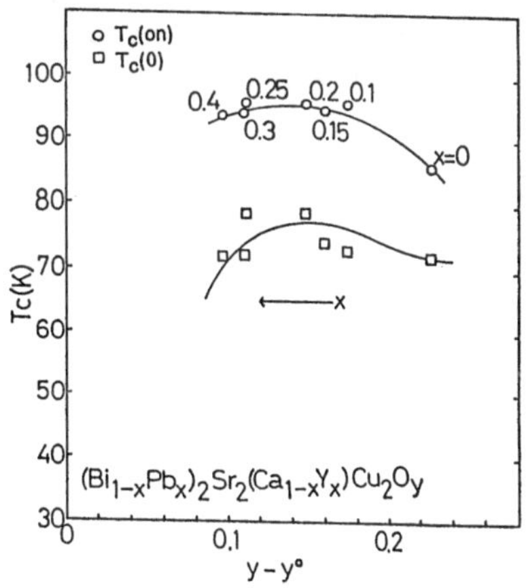

Fig.14 Relationship between Tc and y-y˚ ($(Bi_{1-x}Pb_x)_2Sr_2(Ca_{1-x}Y_x)Cu_2O_y$).

4. CONCLUSION

(1) The average valence of Pb+Bi in $BaPb_{1-x}Bi_xO_3$ was found to be higher than +4 in a range $0.1 \leq x \leq 0.3$ where superconductivity was observed at low temperatures, and a close relationship between Tc and excess oxygen content was concluded.

(2) The carrier concentration was calculated with the assumption that one excess oxygen atom creates two holes in BPBO. The calculated carrier concentration was found to agree well with those determined from Hall coefficient data.

(3) High vacancy concentration of Schottky type was found in BPBO. The concentration was extremely high in a range $0.1 \leq x \leq 0.3$. The highest concentration reached as high as 13% which is very close to the concentration in liquid.

(4) Close relationship was found between Tc and oxygen content for various kinds of superconducting oxides.

(5) Strontium and bithmuth in $Bi_2Sr_2CuO_y$ were replaced by La and Pb, respectively, to form $Bi_2(Sr_{1-x}La_x)_2CuO_y$ and $(Bi_{1-x}Pb_x)_2$ $(Sr_{1-x}La_x)_2CuO_y$, and Y and Pb in $Bi_2Sr_2CaCu_2O_y$ to were substituted for Ca and Bi, respectively, in form $(Bi_{1-x}Pb_x)_2Sr_2(Ca_{1-x}Y_x)Cu_2O_y$. It was found that the cationic replacement causes the change in Tc through the oxygen content change.

REFERENCES

1. Fueki K, Idemoto Y, Yamauchi T Physica C (in submission).
2. Moiseev DP, Uranova SK, Fenik MB Sov. Phys. Solid State 23 (1981) 1371.
3. Thanh TD, Koma A, Tanaka S Appl. Phys. 22 (1980) 205.
4. Kitazawa K, Katsui A, Tanaka S Solid State Commun. 52 (1984) 459.
5. Idemoto Y, Fueki K, Shimbo T Physica C 166 (1990) 153.
6. Idemoto Y, Fueki K Physica C 168 (1990) 167.
7. Idemoto Y, Ichikawa S, Fueki K Physica C 181 (1991) 171.
8. Fueki K, Idemoto Y, Tokunaga H proceedings of M^2S-HTSC Knazawa(1991) (to be published).

Changes in the Bi and Cu Valences and Structure of Bi 2212 Phase by Ar Annealing

SHIRO KAMBE, SHIGETOSHI OHSHIMA, and KATSURO OKUYAMA

Department of Electronic and Information Engineering, Faculty of Engineering, Yamagata University, Yonezawa, Yamagata, 992 Japan

Abstract

For Bi 2212 phase, the movement of the apical oxygen in the CuO_2 plane causes a change in the distribution of the excess charge between the BiO and CuO_2 layer, which may increase the Tc. For N_2 (or Ar)-annealed samples, the shrinkage of the d(O-Cu-O) leads to the increase in the Tc values. The decrease in the Tc values for air-sintered samples with d(O-Cu-O) < 0.383nm might be correlated with the change in the apical oxygen position in the CuO_2 plane.

KEY WORDS: Bi-based superconductor (2212 phase), N_2 and Ar annealing, Bi and Cu valences, Cu-O distance, modulation structure

Introduction

It has already been reported that a maximum Tc of 98.5K is attainable for the 2212 phase of the Pb-substituted Bi system after annealing in N_2 or Ar and 39.5K for the 2201 phase [1,2]. These Tc values are 15-20K higher than those prepared in ambient oxygen. The origin of these extremely high Tc values cannot be explained simply in terms of the hole concentration of the whole system [1,2]. For the 2201 phase, the high Tc values are correlated with the change in the charge distribution [3] caused by the transformation of the apical oxygen position in the CuO_2 plane [4]. In this paper, we report the change in the Bi, Cu valences by Ar-annealing for the Y-substituted 2212 phase and the dependence of the Tc and the modulation period on the distance of O-Cu-O in plane. From these results we discuss what is the determining factor of the Tc and the modulation period.

EXPERIMENTAL

Starting powders were prepared by dissolving $Bi(NO_3)_3*5H_2O$, $Pb(NO_3)_2$, $Sr(NO_3)_2$, and $Ca(NO_3)_2*4H_2O$ in water, drying by evaporation, and calcining in air at 1073K for 2 hours. These powders were ground, pressed into pellets, sintered in air at 1133K for 80 hours. Some of the pellets were annealed in Ar at 973K for 8 hours after sintering. Five kinds of samples, x=0, 0.10, 0.20, 0.30 and 0.40 in $Bi_{2.0}Sr_{1.8}(Ca_{1-x}Y_x)_{1.2}Cu_{2.0}O_y$ were prepared. The superconductive properties were determined by measuring the temperature dependence of ac magnetic susceptibility, by the field-cooling method (Meissner effect). Each sample was reground and pelletized prior to the ac susceptibility measurement. The cell parameters were determined by powder x-ray diffraction (XRD) using Cu Kα line. The peaks were assigned according to those indexed by Onoda et al. [5]. All the peaks appeared in XRD were indexed as those in the 2212 phase. The excess charge was determined by the iodometric titration method [1], where the value is defined as (Bi valence - 3) + (Cu valence - 2). The valence of Bi was determined by the permanganate titration method [3]. The valence of Cu was estimated by subtracting the Bi valence from (excess charge + 5).

RESULTS AND DISCUSSION

In Fig. 1, the Tc dependence on δ is illustrated. For air annealed samples (○), the maximum Tc of 89.4K was observed with δ=0.15. On the other hand, after annealing in Ar, the optimum δ for the highest Tc (84.4K) shifted downwards to 0.08, indicating that the Tc values cannot be treated only as a function of the excess charge [1]. In contrast to the previous results [1], the Tc values did not increase by Ar-annealing. This may be due to the difference of the Sr/Ca ratio. It has been reported that when the Sr/Ca ratio becomes smaller, annealing in the absence of oxygen lead to the decrease in the Tc values [6].

In order to examine the relation between the Tc values and the change in the charge distribution by Ar-annealng, the Bi and Cu valences were determined separately. As illustrated in Fig. 2, an increase in the Y content (x) lead to the decrease in the Bi valence from 3.16 to 3.04 linearly. Although annealing in Ar reduced the Bi valence by 0.02-0.04, dependence of the Bi valence on the x did not change. Dependence of the Cu valence on the x was changed by annealing in Ar. For air-sintered samples, Cu valence was varied from 2.08 to 1.94. As the Y content became larger, the Cu valence decreased. After annealing in Ar, the Cu valence was fixed at 2.0. It did not change as x varied from 0 to 0.4.

These results indicate that the distribution of the charge between the CuO_2 and the BiO planes changes after annealing in Ar as it was for the 2201 phase [3]. From the fact that the charge distribution between the BiO and the CuO_2 planes dramatically changes after annealing in the absence of O_2, a structural change due to the oxygen positions, particularly due to the apical and in-plane oxygen atoms of the CuO_2 plane may occur at the same time as it did for the Pb-substituted 2212 phase [7] and for the 2201 phase [4].

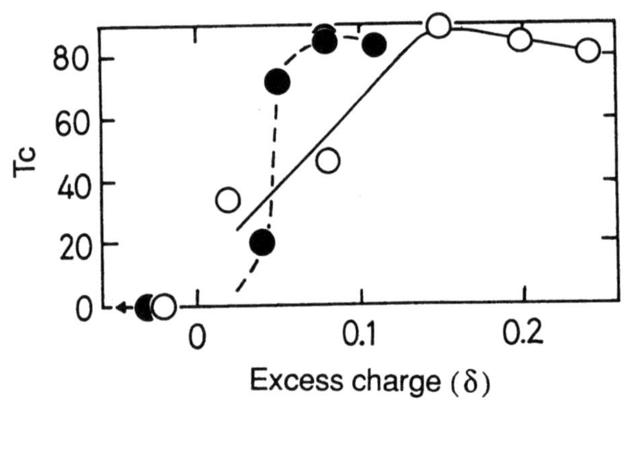

Fig. 1 Tc dependence on the excess charge. Those for air-sintered samples are represented by ○ and those for the samples annealed in the absence of O_2 are represented by ●.

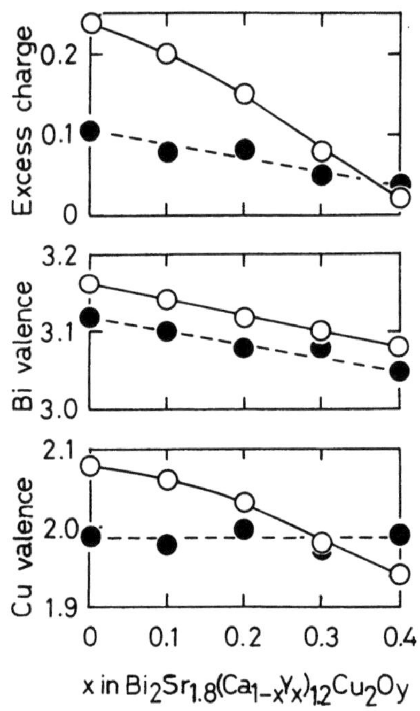

Fig. 2. Excess charge, Bi and Cu valences against Y content for $Bi_2Sr_{1.8}(Ca_{1-x}Y_x)_{1.2}Cu_2O_{8+y}$. Those for air-sintered samples are represented by ○ and those for the samples annealed in the absence of O_2 are represented by ●.

It has been known that the distance of O-Cu-O in plane is correlated with the Tc values [8]. In order to elucidate whether the change in the Tc values by N_2 (or Ar)-annealing can be explained by the change in the d(O-Cu-O), Tc dependence on the d(O-Cu-O) was examined (Fig. 3). For air-annealed samples (○), the optimum d(O-Cu-O) around 0.383nm for the highest Tc was obtained, approximately consistent with the results by Whangbo et al. [8], On the other hand, for N_2 (or Ar)-annealed samples (●), the shrinkage of the d(O-Cu-O) monotonously lead to the increase of the Tc value, resulting that N_2 (or Ar) annealing changes the Tc dependence on the d(O-Cu-O). As the shrinkage of the d(O-Cu-O) gives a rise of the hole concentration in the x^2-y^2 orbital [8,9], the change in the hole concentration may be correlated with the increase of the Tc values for N_2 (or Ar)-annealed samples. On the other hand, the decrease in the Tc values for air-sintered samples with d(O-Cu-O) < 0.383nm cannot be explained simply by the change in the d(O-Cu-O). In the range of d(O-Cu-O) < 0.383nm, the apical oxygen position of the CuO_2 layer may be correlated with the change in the Tc values [4].

In Fig. 4, the modulation period (s) is plotted against the d(O-Cu-O). For N_2 (or Ar)-annealed samples (●), the shrinkage of the d(O-Cu-O) from 0.3846nm to 0.3806nm lead to the increase of the s from 4.25 to non-modulation, showing that the s value is determined by d(O-Cu-O). The trend of s vs. d(O-Cu-O) for air-annealed samples (○) was quite similar to that for Ar-annealed samples (●). But the s values for d(O-Cu-O) < 0.382nm split into 3.9-4.2 and 5.5-6.4, resulting that the modulated structure is disordered. From this reason, no plot was carried out for the air-annealed samples with d(O-Cu-O) < 0.382nm. As the a-axis length of the usual Bi complex oxides with tetragonal or pseudotetragonal symmetry is close to 0.380nm [12], the d(O-Cu-O) of the Bi 2212 phase is 0.2-1.2% mismatched. Larger misfit causes shorter modulation period, showing that the modulation relaxes the mismatch between the BiO and CuO_2 layers as pointed by Fukushima et al. [10]. As the modulation is caused by the extra oxygen in the BiO layer [11], splitting of the modulation in air-annealed samples for d(O-Cu-O) < 0.382nm suggests the existence of disorder of the oxygen concentration. The disorder might be correlated with the change in the apical oxygen position in the CuO_2 plane.

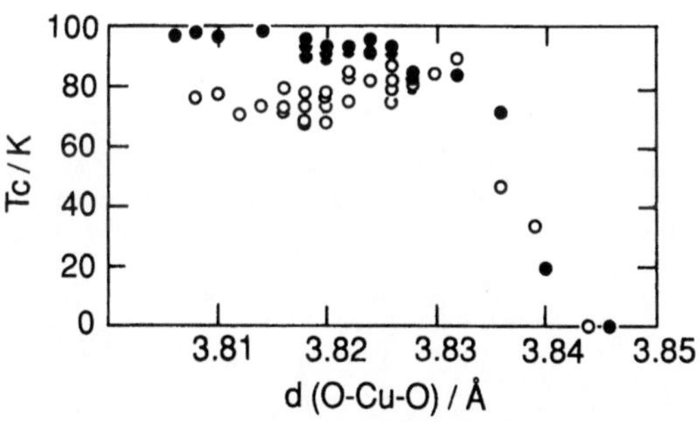

Fig. 3. Tc against the distance of O-Cu-O in plane for Bi 2212 phase. The trend of Tc dependence for the samples prepared in N_2 (or Ar)-annealing (●) is different from that of the samples prepared in ambient oxygen (○)

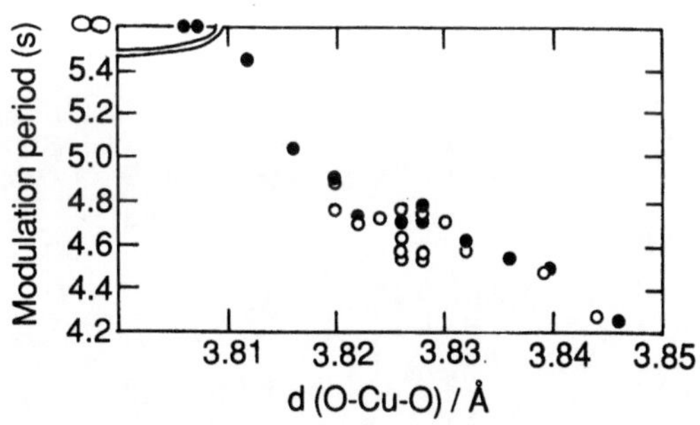

Fig. 4. Modulation period dependence on the distance of O-Cu-O. ○ for air-annealed samples and ● for N_2 (or Ar)-annealed samples.

CONCLUSION

The change in the Tc values by annealing in the absence of oxygen cannot be explained simply in terms of the hole concentration for the 2212 phase. Distribution of the charge between the BiO and the CuO_2 planes dramatically changes after annealing in N_2 or Ar, suggesting that N_2 or Ar annealing causes the structural change including the oxygen positions around Cu so as to affect the distribution of the excess charge between BiO and CuO_2 planes, which may be correlated with Tc.
The shrinkage of the d(O-Cu-O) leads to the increase of the Tc values. The decrease of the Tc values for air-annealed samples with d(O-Cu-O) < 0.383nm might be correlated with the change in the apical oxygen position in the CuO_2 plane.

Acknowledgement

We are grateful to Ms. R. Sekine of Tokyo Institute of Technology and Dr. M. Kawai of the Institute of Physical ahd Chemical Research for useful discussion. We are also indebted to Mr. K. Yamada of Yamagata University for sample preparation.

References

[1] S. Kambe, T. Matsuoka, M. Takahasi, M. Kawai and T. Kawai, Phys. Rev. B **42** (1990) 2669.
[2] S. Kambe, M. Kawai and T. Kawai, Solid State Comm. **75** (1990) 435
[5] M. Onoda, A. Yamamoto, E. Takayama-Muromachi and S. Takekawa, Jpn. J. Appl. Phys. **28** (1988) L833.
[3] Y. Murakoshi, S. Kambe and M. Kawai, Physica C, 178 (1991) 71.
[4] R. Sekine, S. Kambe, M. Kawai and Y. Murakoshi, Physica C (1991) in press.
[9] R. Sekine, M. Kawai and H. Adachi, Physica C 19 (1989) 161.
[6] Yamazoe et al., preprint.
[7] N. Kuriyama, H. Nasu and K. Kamiya, Jpn. J. Appl. Phys. 30 (1991) L1462
[8] M. -H. Whangbo, D. B. Kang and C. C. Torardi, Physica C 158 (1989) 371
[10] H. Niu, N. Fukushima, S. Takeno, S. Shimakawa and K. Ando, Jpn. J. Appl. Phys. **28** (1989) L784.
[11] A. Yamamoto, M. Onoda, E. Takayama-Muromachi, and F. Izumi, PHys. Rev. B **42** (1990) 4228.
[12] A. F. Wells (1989) Structural Inorganic Chemistry. Oxford Science Publications, Oxford, pp 894.

The Determination of the Oxygen Content in $YBa_2Cu_3O_{7-\partial}$ Superconductors Using Thermal Analysis

H. VLAEMINCK, D. DEPLA, and S. HOSTE

Laboratory of General and Inorganic Chemistry, Rijksuniversiteit Gent, 9000 Ghent, Belgium

ABSTRACT

Careful analysis of the experimental parameters in applying thermal analysis for the determination of the oxygen content in $YBa_2Cu_3O_{7-\partial}$ leads to a procedure yielding a reproducibility of 0,03. Only after the influences of the dynamic pressure effects occuring in the thermogravimeter were assessed, could this value been obtained. The reduction curves furthermore display at least four separate step processes, the first of which represents loss of oxygen from the Cu-O chain structures and the appearance of a tetragonal phase. Full reduction to metallic Cu, without any interference of the accompanying oxides of Ba and Y can be obtained by using a H_2/N_2 5 volume % mixture in a thermal reduction between room temperature and 1248K.

KEY WORDS: Oxygen determination, iodometry, thermogravimetry

INTRODUCTION

The critical transition temperature and the properties of the high Tc-superconductor $YBa_2Cu_3O_{7-\delta}$ are greatly affected by the oxygen deficiency δ. Therefore many methods have been devised to establish the value of δ with the highest accuracy and reproducibility.(1-13) In this work, a method based on thermal analysis in a reducing atmosphere, has been improved for the determination of the oxygen content in $YBa_2Cu_3O_{7-\delta}$ yielding a reproducibility of 0,03. Several intermediate steps found on the thermogravimetric reduction curve and related to the formation of different species during the reduction process, have been characterized using X-ray diffraction and differential thermal analysis.

EXPERIMENTAL

The $YBa_2Cu_3O_{7-\delta}$ samples were prepared from spray dried nitrate precursor materials.(14) This precursor material was calcined for 10 hours at 1053K in air followed by ball milling for 3 hours. Eleven samples with different oxygen content ($0,03 < \delta < 0,95$) have been prepared(8) by varying the sintering and annealing conditions.
A sample of Y_2BaCuO_5 was prepared from a stoichiometric mixture of $BaCO_3$, Y_2O_3 and CuO. It was heated in air at 773K for 18 hours, at 1073K for 20 hours and at 1273K for 20 hours with intermittent grinding The structure of the final powder, characterized with X-ray diffraction, corresponds with the structure of Y_2BaCuO_5 as determined by C. Michel and B. Raveau.(17)

The thermal studies were carried out using a Simultaneous Thermal Analyzer (Stanton Redcroft STA 1500) for thermogravimetric (TG) and differential thermal analysis (DTA) modified to accept samples of 50-150 mg. rather than the conventional 10 mg. Thermogravimetry samples weighing about 135 mg. were obtained by grinding the $YBa_2Cu_3O_{7-\delta}$ superconducting pellet or powder in an agate mortar in order to maximize the homogeneity of the powder. TG-DTA runs were recorded in Al_2O_3 crucibles, using Al_2O_3 as inert reference material for DTA, from room temperature up to 1248K at a constant rate of 5K/min. in 58 cm^3/min 5 volume percent H_2/N_2 reduction mixture and a side branch flow of N_2 (1,5 cm^3/min.).
Microbalances are affected by a variety of errors. In order to obtain the highest quality for our experimental data, a careful study of the effects of buoyancy, underflow and side branch flow was undertaken. The first correction which we ascertain is for the buoyancy effect which results

in an apparent gain in weight corresponding with the mass of displaced gas. We can determine the change in mass of displaced gas, when heating a superconductor from 293K to 1273K, with the equation:

$$\Delta\delta = \frac{\Delta m}{V} \tag{1}$$

where $\Delta\delta$ (N_2) ~ $\Delta\delta$ (5 volume percent H_2/N_2) = 0,908 mg/cm^3 and V is the total volume of the part of the hangdown in the sample room. This total volume includes the volume Pt-13%Rh/Pt (0,032 cm^3), the volume Al_2O_3 (0,073 cm^3) and the volume of 135 mg. superconductor (0,021 cm^3). The change in mass of the displaced gas is

$$0{,}908 \text{ mg./cm}^3 \, . \, 0{,}126 \text{ cm}^3 = +0{,}11 \text{ mg.}$$

A second correction we have to make is for the side branch flow of an inert gas in order to protect the balance mechanism for corrosive gasses or vapors. As a consequence of this, a dynamic pressure arises in the sample room resulting in a downward force and in an apparent gain in weight. Therefore we have determined the weight change as function of the side branch flow from 0 to 1,5 cm^3/min at 293K, 573K, 973K and 1273K for N_2 and for a 5 volume percent H_2/N_2 mixture with constant underflow. The linear dependence noted here, produces an interpolated apparent weight gain at 1273K of + 0,03 mg, for both atmospheres. At 973 K however, the weight changes for N_2 and for a 5 volume percent H_2/N_2 gas mixture are slightly different. The formation of hydrides and adsorption-diffusion dynamics of H_2 in alumina may be responsible for this behavior and this effect warrants further study.

The underflow (third correction) causes an upward force resulting in a weight loss at higher temperatures. The procedure followed here was similar to that used in the determination of the sidebranch effect. When changing the underflow from 0 to 58 cm^3/min (for N_2 and for a 5 volume percent H_2/N_2 gas mixture as underflow), we find a weight loss of -0,10mg. at 1273K. Again, the slightly different behavior between N_2 and the reducing mixture was noted around 973K.

Finally, the total experimental correction, necessary to compensate for buoyancy effect and other aerodynamic forces is -0,04 mg. for 135 mg. superconductor resulting in an underestimation of 0,015 on the oxygen content of $YBa_2Cu_3O_{7-\delta}$.

RESULTS AND DISCUSSION

The reduction curve typically obtained for a superconducting sample with δ =0,10 is shown in figure 1 where the DTA curve is superimposed on the TG curve. We calculated the oxygen content based on the overall reaction given by

$$2\, YBa_2Cu_3O_{7-\delta} + (7\text{-}2\delta)\, H_2 \longrightarrow Y_2O_3 + 4BaO + 6Cu + (7\text{-}2\delta)\, H_2O \tag{2}$$

from the weight loss between room temperature and 1248K. The final temperature reported in the literature for the calculation of the oxygen content, is subject to debate. Cava et al.(15) determined the oxygen content by measuring the weight loss on reduction at 1173K in H_2 for 2h. while Syono et al.(16) proposed to use 973K. From TG data, published by Gallagher et al.(7), we can see that the weight still changes slightly above 1173K. Mitzi et al.(17) have determined the oxygen content from the weight loss between 293K and 1253K. Takayama et al.(18) warns against the fact that copper can be lost by vaporization at the higher temperatures. On the basis of all the experimental evidence described above, we have chosen 1248K as final temperature. From X-ray diffraction data (7) in the final stage and from thermal analysis data of Y_2O_3, and CuO in a 5 volume % H_2/N_2 reduction mixture, we conclude that only Cu, BaO and Y_2O_3 are present at 1248K. The reduction of pure CuO to metallic Cu in a reducing atmosphere occurs between 473K and 673K and is a strongly exothermic, single step process yielding 99,7% of theoretical weight loss. No change in weight of pure Y_2O_3 was observed at temperatures to 1273K in the reducing atmosphere.

The study of BaO in reducing atmosphere is more difficult because BaO is very hygroscopic [formation of $Ba(OH)_2$ and $BaCO_3$]. Instead of using p.a. BaO, we started from $Ba(NO_3)_2$ to examine the reduction of BaO (figure 2). Although the total weight loss noticed in 5% H_2/N_2 between 298K and 1023K is larger than expected for the sole decomposition of $Ba(NO_3)_2$ to BaO , the reduction curve was very similar to the decomposition thermogram under an inert atmosphere (pure N_2). From a detailed study of the latter we learned that $Ba(NO_3)_2$ decomposes to a mixture of BaO and BaO_2 (first plateau) in figure 2. When the heating rate is sufficiently slow the mixture reacts with N_2 to form Ba_3N_2[23]which sublimates without further decomposition and is thus removed from our sample. The residue, with constant weight, obtained above 1023K can not be BaO_2 which decomposes above 873K [19]. Furthermore, being a good oxidizer, BaO_2 is expected to react readily with the strongly reducing H_2/N_2 atmosphere. Thus, the residue must consist of BaO and its stability over the 873K to 1273K region proves its inertness in a reducing atmosphere.
Furthermore this is confirmed by the observation that the respective reduction E° potentials for Y^{3+} (-2,91V) and for Ba^{2+} (-2,37V) are much lower than for Cu^{2+} ions (+0,34V).
Turning our attention back to the superconductor reduction curve (fig.1) we find that the reduction proceeds in 4 steps related to different species present at these stages of the reduction process. In order to identify the 4 intermediate steps , we have compared the reduction traces of 11 samples with different oxygen content.[8]

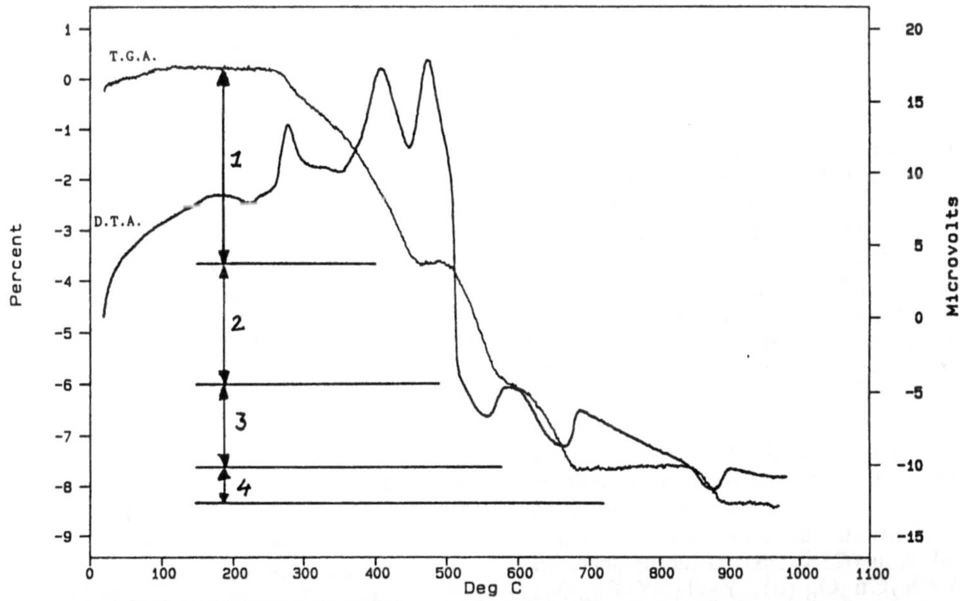

Figure 1: Thermogram of $YBa_2Cu_3O_{7-\partial}$ with δ=0,10 in 5 volume percent H_2/N_2. Weight change (TG) and differential analysis curve (DTA) both show exhibit clear features indicating that oxygen removal from the Cu-O chains occurs between 523K and 623K

The oxygen content of these 11 samples was determined with iodometric titration [8] and thermal analysis. Our results for 11 samples of $YBa_2Cu_3O_{7-\partial}$ with δ ranging from 0,03 to 0,95 show a remarkable and gratifying correlation between both iodometry and thermal analysis (correlation coefficient squared = 0,997) for 9 points as seen in figure 2. Despite the fact that the exact distribution of charge between Cu and O in these materials remains a debated question, this problem does not directly affect the explanation of the reduction process of $YBa_2Cu_3O_{7-\partial}$, which, for convenience, is described in terms of a Cu valence state related to distinct crystallographic sites rather than in terms of an average Cu valence.

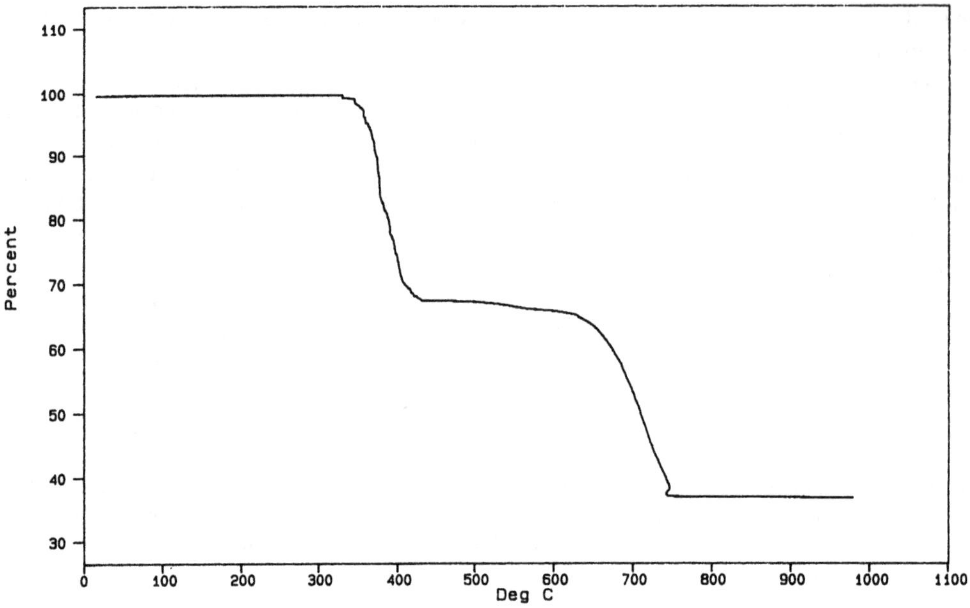

Figure 2: Thermogram of $Ba(NO_3)_2$ in 5 volume percent H_2/N_2 reducing atmosphere

We have a strong indication that the first reduction step in the TG curve occuring between room temperature and 773 K involves the release of oxygen from the (Cu-O) chains with the formation of the tetragonal $YBa_2Cu_3O_6$ phase. This is substantiated by the fact that the weight changes associated with step 1 in figure 1 account for most the variation observed in the oxygen content of the 11 samples as shown in figure 3.

In step 1 (fig.1 , 523K -773K), we can see a small distinctive feature between 523K and 623K accompanied by a exothermic peak in the DTA curve. A X-ray diffraction pattern of a sample taken at 623K, reveals the presence of $YBa_2Cu_3O_6$ (o), Y_2O_3, $Y_2Ba_4O_7$, Y_2BaO_4, Ba_2CuO_3, CuO and Cu_2O (figure 4) as determined by simulating the experimental X-ray diffraction pattern from a linear combination of pure reference spectra..

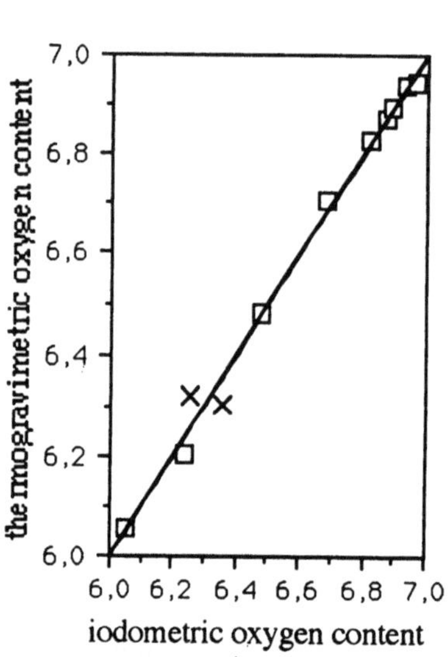

figure 2: correlation observed between oxygen content independently measured using thermal and iodometric methods (x= very inhomogeneous samples)

From these XRD-data, it thus follows that oxygen is released from the (Cu-O) chains with formation of Cu(II) compounds and $Cu(I)_2O$. Free BaO, in contrast to Y_2O_3, isn't formed at 623K but it appears that two Y-Ba compounds, Y_2BaO_4 and $Y_2Ba_4O_7$, are present. X-ray data of these two compounds are given in detail by W. Kwestroo et al. [20] and D.M. De Leeuw et al. [21].

We have also compared the TG reduction curves of $YBa_2Cu_3O_{7-\partial}$, which is known to contain (Cu-O) chains, with that of Y_2BaCuO_5 (green phase/no chains).

We can calculate the weight loss between room temperature and 1248K based on the following reduction reaction

$$Y_2BaCuO_5 + H_2 \longrightarrow Y_2O_3 + BaO + Cu + H_2O \quad (3)$$

and find an experimental weight loss of 3,51% in complete agreement with the theoretical weight loss of 3,49%.

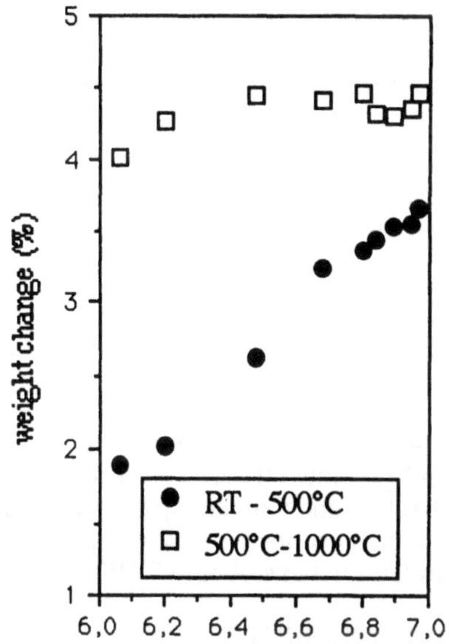

figure 3: thermogravimetric weight change (%) as function of oxygen content for two separate parts of the reduction curve

In contrast to the superconducting oxides, his experimental TG curve does not exhibit any weight loss below 673K. This observation clearly proves, that the feature noted in step 1 of the superconductor reduction curve, is indeed related to escape of oxygen from the (Cu-O) chains.

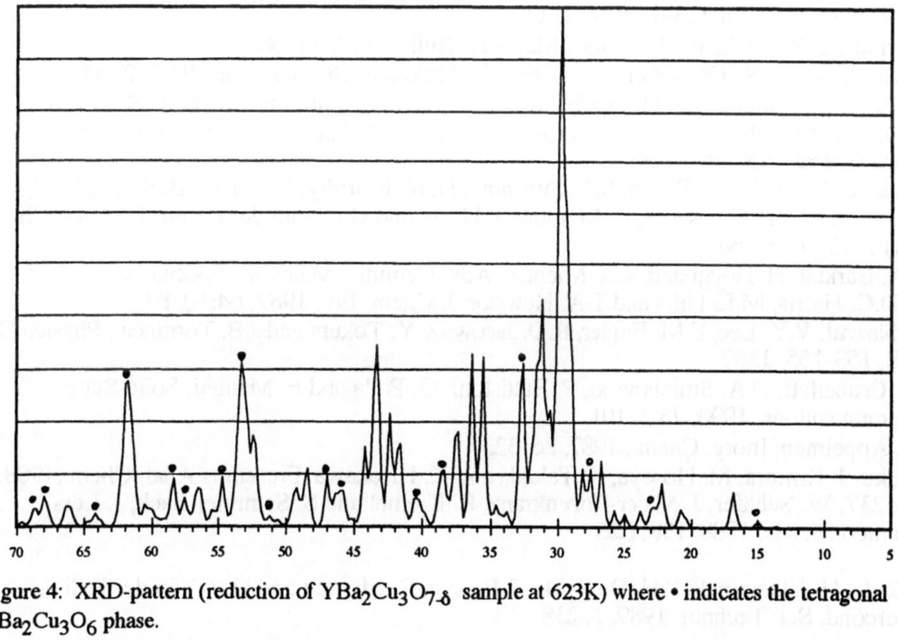

Figure 4: XRD-pattern (reduction of $YBa_2Cu_3O_{7-\delta}$ sample at 623K) where • indicates the tetragonal $YBa_2Cu_3O_6$ phase.

The plateau at 773K in the TG curve in fig.1 is associated with Y_2O_3, $Y_2Ba_4O_7$, CuO and Cu_2O as indicated found from the simulated XRD-pattern. At 773K the original orthorhombic or tetragonal $Y_1Ba_2Cu_3O_{7-\delta}$ structures have disappeared completely. The last plateau (between 953K and 1133K) in fig.1 must correspond to the remarkably stable Cu^{1+} ionic species in the remaining Y_2O_3-BaO framework. This is in stark contrast to the reduction of pure CuO under identical circumstances, where complete and strongly exothermic reduction to metallic Cu occurs in a single step between 473K and 673K.

CONCLUSIONS

It has been shown that the oxygen content in samples of 135 mg. $YBa_2Cu_3O_{7-\delta}$ can be routinely obtained from the total weight loss in the TG curve between room temperature and 1248K, in a reducing atmosphere with a reproducibility of 0,03 for 3 measurements. In order to obtain the highest precision, care must be taken to include the necessary corrections for buoyancy and gas flow effects. This amounts to an underestimation of the apparent oxygen content with 0,015 units in our experiments.
Thermal reduction occurs in discrete steps, the first of which corresponds to the removal of oxygen from Cu-O chains. Above 773K, the $YBa_2Cu_3O_{7-\delta}$ -structure is completely destroyed.

ACKNOWLEDGEMENTS

The authors would like to thank the Belgian Government through the National Impulse Programme for High Temperature Superconductors and the University's Research Funding Committee. We gratefully acknowledge the help of Ing. R. Mouton with the recording of the thermograms.

REFERENCES

[1] E. Takayama, Y. Uchida, M. Ishii, T. Tanaka and K. Kato, Jap. J. Appl. Phys., 1987, 26:7, L1156
[2] H.M. Rietveld, J. Appl. Cryst., 1969, 2, 65
[3] K. Conder, S. Rusiecki and E. Kaldis, Mat. Res. Bull., 1989, 24, 581
[4] G.S. Grader and P.K. Gallagher, Adv. Ceramic. Materials, Special Issue, 1987, 2, 649
[5] P. Kishan, L.K. Nagpaul and S.N. Chatterjee, Solid State Communications, 1988, 65, 1019
[6] N. Brnicevic M. Paljevic, Z. Ruzic-Toros, M. Tonkovic A. Kashta, M. Prester, E. Babic, Physica C, 153-155, 820 (1988)
[7] P.K. Gallagher, H.M. O' Bryan, S.A. Sunshine, D.W. Murphy, Mat. Res. Bull.,22 (1987) 995
[8] H.Vlaeminck, H.H. Goossens, R. Mouton, S. Hoste and G.P. Van der Kelen, J. Mater. Chem. (1991), 1(5), 863-866
[9] e.g.: A. Barkatt, H. Hojaji and K.A. Michael, Adv. Ceramic. Materials, Special Issue, 1987, 2, 701; D.C. Harris, M.E. Hills and T.A. Hewston J. Chem. Ed., 1987, 64:10, 847
[10] A.I. Nazzal, V.Y. Lee, E.M. Engler, R.D. Jacowitz, Y. Tokura and J.B. Torrance, Physica C., 1988, 153-155, 1367
[11] R.O. Crubellati, D.A. Smichowski, P. Battistoni, G. Polla and E. Manghi, Solid State Communications, 1990, 75:2, 101
[12] E.H. Appelman, Inorg. Chem., 1987, 26, 3237
[13] M. Oku, J. Kimura, M. Hosoya, K. Takada and K. Hirokawa, Fresenius Anal. Chem.;1988, 332, 237; W. Schäfer, J. Maier-Rozenkranz, R. Kiemel and S. Semmler-Sack, J. Less Common Metals, 1988, 138, L25
[14] S. Hoste, H. Vlaeminck, P.H. De Ryck, F. Persyn, R. Mouton and G.P. Van der Kelen, Supercond. Sci. Technol. 1989, 1, 238
[15] R.J. Cava, B. Batlogg, R.B. van Dover, D.W. Murphy, S. Sunshine, T. Siegrist, J.P. Remeika, E.A. Rietman, S. Zahurak and G.P. Espinosa, Phys. Rev. Lett. 58 (1987) 1676
[16] Y. Syono, M. Kikuchi, K. Oh-ishi, K. Hiraga, H. Arai, Y. Matsui, N. Kobayashi, T. Sasaoka and Y. Muto, Jpn. J. Appl. Phys. 26(1987) L498

[17] D.B. Mitzi, P.T. Feffer, J.M. Newsam, D.J. Webb, P. Klavins, A.J. Jacobson, A. Kapitulnik Phys. Rev.B, vol.38, n°10,6667-6676 (1988)

[18] E. Takayama, Y Uchida, M. Ishii, T Tanaka, K. Kato , Jpn. J. Appl. Phys. 26 (1987) L1156-L1158

[19] Progress in Inorganic Chemistry volume 4", F. Albert Cotton, Interscience Publicers, 1962, p. 145-146

[20] W. Kwestroo, H.A.M. van Hal, C. Langereis, Mat. Res. Bull., Vol.9, pp 1631-1638,1974

[21] D.M. De Leeuw, C.A.H.A. Mutsaers, C. Langereis, H.C.A. Smoorenburg and P.J.Rommers, Physica C., 152 (1988), 39-49

[22] C.Michel and B. Raveau, J. Solid State Chem., 43, 73 (1982)

[23] Atlas Der Physikalischen Und Anorganischen Chemie, die Eigenschaften Der Elementen Und Ihrer Verbindingen In Graphinsher Darstellung Auf 2 , Von Antropoff, Von Stackelberg , 1929 , Verslag Chemie / Gmbh / Berlin , p57

Low Temperature Phases in the $(Ba_{1-x}La_x)CuO_y$ System

NOBUYUKI SUGII, KOICHI KUBO, and H. YAMAUCHI

Superconductivity Research Laboratory, ISTEC, Koto-ku, Tokyo, 135 Japan

ABSTRACT

$BaCuO_{2.5}$ and $(Ba_{1-x}La_x)CuO_y$ samples are synthesized by a spray-drying method or an O_2 hot isostatic pressing technique. $BaCuO_{2.5}$ can be synthesized at 650 ℃ at 1 atm oxygen. It has an orthorhombic unit cell (a=7.40 Å, b=4.23 Å and c=11.40 Å), and is similar to $Ba_2Cu_3O_5$. The magnetic susceptibility of $BaCuO_{2.5}$ shows a Curie-Weiss behavior. In the $(Ba_{1-x}La_x)CuO_y$ system, no solid solution fields form, but mixtures of $BaCuO_{2.5}$ and $(La,Ba)_2CuO_4$ are obtained. These two phases are dispersed in a micro-domain. The state of mixture of $BaCuO_{2.5}$ and $(La,Ba)_2CuO_4$ in $(Ba_{0.5}La_{0.5})CuO_y$ is unstable at high temperatures, and changes into $La_3Ba_3Cu_6O_{14}$ above 750 ℃.

KEY WORDS: $BaCuO_{2.5}$, $(Ba,La)CuO_y$, low temperature synthesis, spray-drying method

INTRODUCTION

It has been reported that $BaCuO_{2.5}$ has a high oxidation state of Cu ion (i.e. +3), and is stable at low temperatures below 650 ℃ in air and/or under oxygen pressures as high as 400 atm at temperatures as high as 700 ℃ [1]. However, its crystal structure is unknown until now. Thompson et al. denied the existence of this phase and rather identified a new phase, $Ba_2Cu_3O_5$, instead [2]. Recently, Machida et al.[3] reported the formation and decomposition of $BaCuO_{2.5}$, and they also reported that the mean oxidation state of Cu was +3. It is thus worthwhile to determine the crystallographic structure of this compound and to study the physical properties. It is also useful to search for new phases based on this compound because it is thought to be a candidate for high-T_c superconductor-related materials.

In this study, we synthesized $BaCuO_{2.5}$ and related $(Ba_{1-x}La_x)CuO_y$ by a spray-drying method and/or an O_2 hot isostatic pressing (HIP) technique. The crystallographic structure and micro-structures of these materials were investigated. The magnetic properties of $BaCuO_{2.5}$ were also studied.

EXPERIMENTAL

$BaCuO_{2.5}$ and $(Ba_{1-x}La_x)CuO_y$ were synthesized by a spray-drying method. An adequate mixture of aqueous solutions of $Ba(NO_3)_2$, $Cu(NO_3)_2$ and $La(NO_3)_3$ was prepared. Light-blue fine powder was obtained from the solution when spray-dried at 150 ℃. Then the powder was heated at 650 ℃ in oxygen gas of 1 atm for 72 h to 220 h.

Other samples were synthesized using an O_2 HIP technique. Sintered or pressed pellets from the powder previously prepared were hot isostatically pressed at 750 ℃ in 400 atm oxygen for 6 h. An Ar/O_2 mixed gas (20% O_2) was used so that the total pressure in this HIP process was 2000 atm.

The crystallographic structure was determined by X-ray diffraction (XRD), electron diffraction (ED) and transmission electron microscopy (TEM). ED and TEM were conducted using JEM 4000-EX (JEOL) transmission electron microscope operated at 400 KV. The magnetic susceptibility of $BaCuO_{2.5}$ was measured using a SQUID magnetometer (Quantum Design MPMS).

RESULTS AND DISCUSSION

1. Synthesis and characterization of $BaCuO_{2.5}$

$BaCuO_{2.5}$ samples were synthesized at 650 ℃ at 1 atm oxygen. The result of thermogravimetry and differential thermal analysis of the spray-dried powder (stoichiometric mixture of $Ba(NO_3)_2 \cdot nH_2O$ and $Cu(NO_3)_2 \cdot nH_2O$) suggested that the decomposition of nitrates occurred between 550 ℃ and 650 ℃. Furthermore, another weightloss was observed above 700 ℃. This is thought to correspond to the decomposition of $BaCuO_{2.5}$. Therefore, the temperature range where $BaCuO_{2.5}$ can be synthesized is narrow at 1 atm oxygen. These results are consistent with the data reported by Machida et al. [3]. XRD patterns of $BaCuO_{2.5}$ prepared under various conditions are shown in Fig. 1. The sample whose reaction time was 72 h (sample a) exhibited broad XRD peaks. The sample reacted for 220 h (sample b) had sharper XRD peaks and peaks due to unidentified impurity phases were also observed. The XRD pattern for a HIP-processed sample (sample c) suggested that $BaCuO_{2.5}$ structure remained up to 750 ℃ at 400 atm oxygen. The pattern for $BaCuO_2$ shown in Fig.1-d is not included in all the samples, a,b and c.

Electron diffraction patterns of $BaCuO_{2.5}$ are shown in Fig. 2. The diffraction pattern collected along the [001] zone axis suggests that $BaCuO_{2.5}$ might have a hexagonal unit cell, however, another diffraction pattern collected along $[0\bar{3}1]$ suggests that it has an orthorhombic symmetry rather than a hexagonal one. The XRD pattern for $BaCuO_{2.5}$ given in Fig. 1 can also be indexed by an orthorhombic unit cell. Unit cell parameters are as follows: a=7.40 Å, b=4.23 Å and c=11.40 Å. These unit cell parameters are similar to those reported for $Ba_2Cu_3O_5$ by Thompson et al. [2]. It is thought that the coordination of a Ba atom is similar to that in $Ba_2Cu_3O_5$, and the coordinations of a Cu atom and an O atom are different. A detailed structural analysis is in progress.

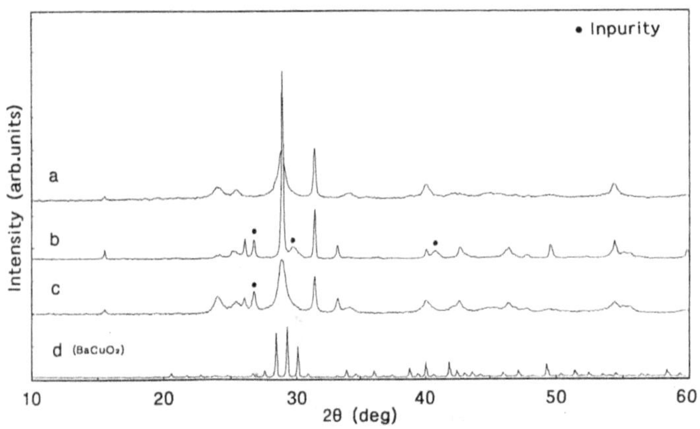

Fig. 1 XRD patterns for $BaCuO_{2.5}$
a; 650 ℃ 72 h, b; 650 ℃ 220 h,
c; 650 ℃ 220 h + 750 ℃ 6 h at 400 atm O_2
d; $BaCuO_2$

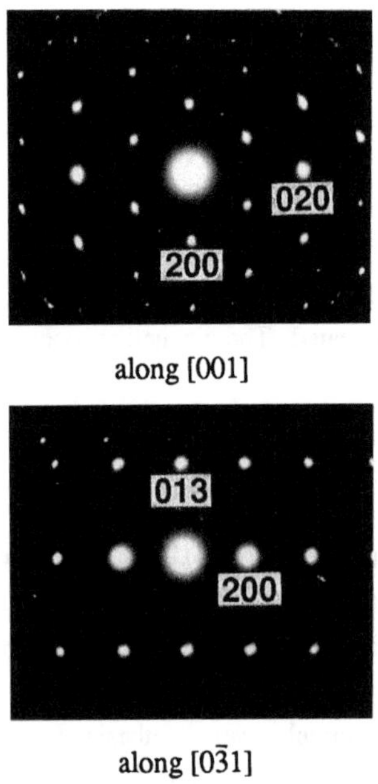

along [001]

along $[0\bar{3}1]$

Fig. 2 ED patterns for $BaCuO_{2.5}$

Figure 3 shows the temperature dependence of the inverse magnetic susceptibility for $BaCuO_{2.5}$ (sample a). It obeys the Curie-Weiss law. From the Curie's constant, the effective Bohr magneton, *p*, was calculated to be 0.69. This value was not changed even when the sample was prepared under different preparation condition (sample c). Therefore, this value for *p* is thought to represent the nature of the Cu ion in $BaCuO_{2.5}$ and not to be due to impurities.

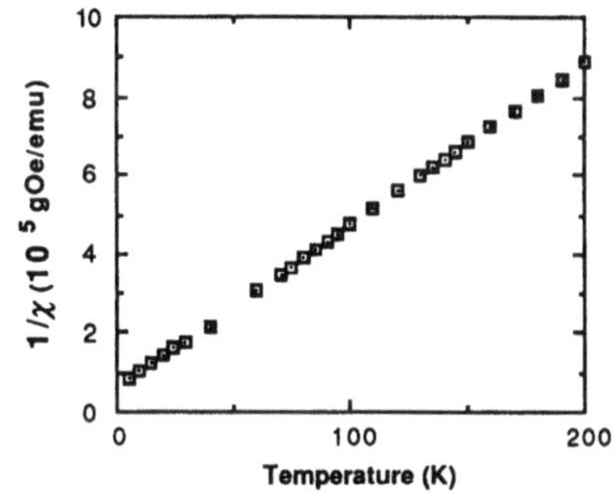

Fig. 3 Temperature dependence of inverse magnetic susceptibility for $BaCuO_{2.5}$

2. Synthesis and characterization of $(Ba_{1-x}La_x)CuO_y$

The samples were prepared by the same procedure as for $BaCuO_{2.5}$. The spray-dried powder was quickly heated to 650 ℃ to avoid a segregation due to the different decomposition temperatures of nitrates.

Figure 4 shows XRD patterns for $(Ba_{1-x}La_x)CuO_y$. As the La content, x, increased the amount of $(La,Ba)_2CuO_4$ phase having K_2NiF_4 structure increased. In the whole range of x, $(La,Ba)_2CuO_4$ phase coexisted with $BaCuO_{2.5}$, and no solid solution was formed. The lattice constants of the $(La,Ba)_2CuO_4$ phase are shown as a function of x in Fig. 5. Both lattice constants changed monotonically as x increased. The Ba content, w, in $La_{2-w}Ba_wCuO_4$ was estimated from the c-axis length [4]. The value w was 0.20 for the sample whose x was 0.1, and w decreased to zero as the La content x in $(Ba_{1-x}La_x)CuO_y$ increased. These data suggest the solubility of Ba in $(La,Ba)_2CuO_4$ in the state of mixture of $BaCuO_{2.5}$ and $(La,Ba)_2CuO_4$ is at most 10 % even when an excess amount of Ba exists around the $(La,Ba)_2CuO_4$ particles. For the samples prepared by a HIP process, no remarkable changes in the phase relationships were observed.

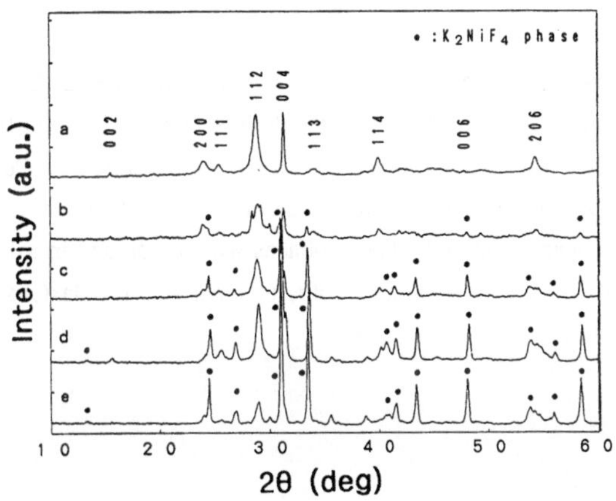

Fig. 4 XRD patterns for $(Ba_{1-x}La_x)CuO_y$
a; x=0, b; x=0.1, c; x=0.3
d; x=0.5, e; x=0.7

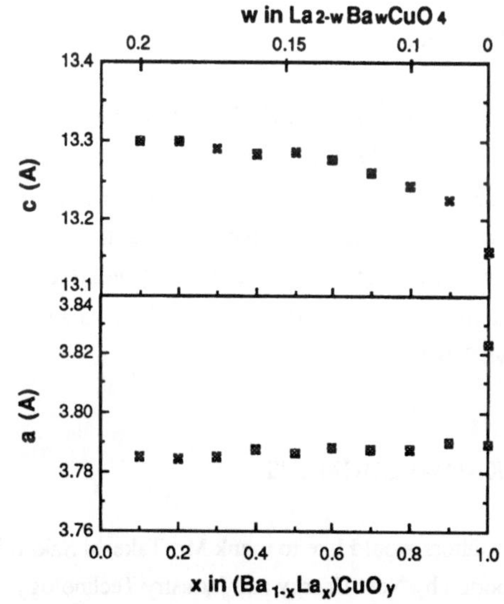

Fig. 5 Lattice constants of the $(La,Ba)_2CuO_4$ phase

The micro-structure of $(Ba_{0.5}La_{0.5})CuO_y$ was investigated. A transmission electron micrographs of this sample are shown in Fig. 6. The $BaCuO_{2.5}$ and $(La,Ba)_2CuO_4$ phases were dispersed in micro-domains whose sizes were between 10 nm and 50 nm.

The stability of the state of mixture of $BaCuO_{2.5}$ and $(La,Ba)_2CuO_4$ was investigated for $(Ba_{0.5}La_{0.5})CuO_y$. The sample was heated at 750 ℃ and 800 ℃ each for 72 h. Figure 7 shows XRD patterns of these samples. Above 750 ℃, both $BaCuO_{2.5}$ and $(La,Ba)_2CuO_4$ phases were partially decomposed, and the $La_3Ba_3Cu_6O_{14}$ phase formed.

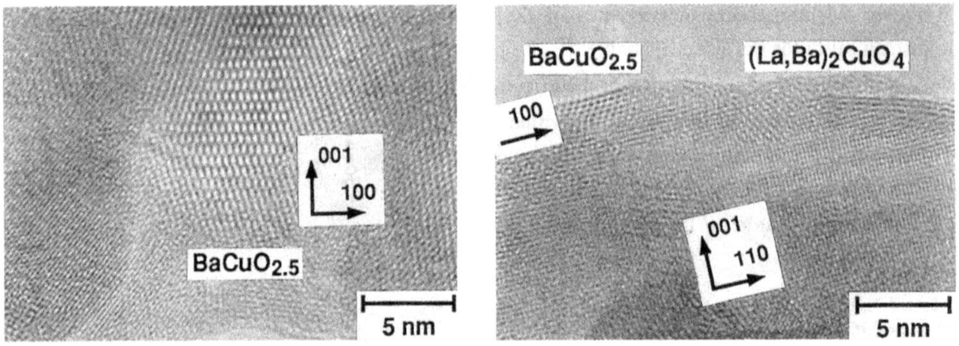

Fig. 6 Transmission electron micrographs for $(Ba_{0.5}La_{0.5})CuO_y$

Above 800 ℃, the decomposition of $BaCuO_{2.5}$ and $(La,Ba)_2CuO_4$ phases was almost completed. This result suggests that the $La_3Ba_3Cu_6O_{14}$ phase which is more uniform than the state of mixture of $BaCuO_{2.5}$ and $(La,Ba)_2CuO_4$ is stable at temper-atures as high as 800 ℃, and that the state of mixture is quasi-stable. It is thought that the reaction kinetics is rather dominant than thermo-dynamics at temperatures as low as 650 ℃.

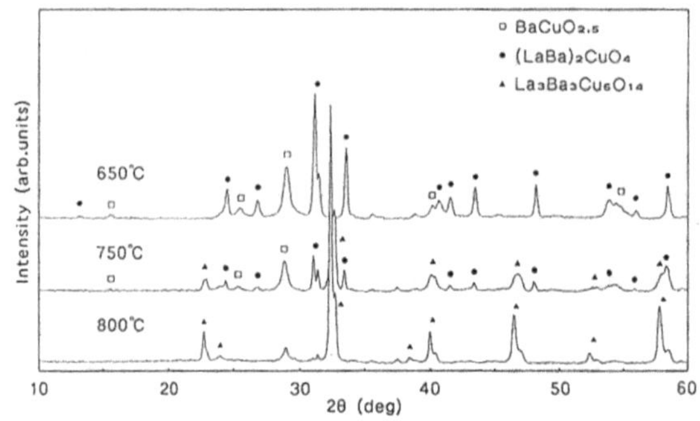

Fig. 7 XRD patterns for $(Ba_{0.5}La_{0.5})CuO_y$ heat-treated at different temperatures

SUMMARY

$BaCuO_{2.5}$ and $(Ba_{1-x}La_x)CuO_y$ were synthesized by a spray-drying method and/or an O_2 HIP technique. $BaCuO_{2.5}$ were successfully synthesized at 650 ℃ at 1 atm oxygen. $BaCuO_{2.5}$ had an orthorhombic unit cell (a=7.40 Å, b=4.23 Å and c=11.40 Å), and was similar to $Ba_2Cu_3O_5$. The magnetic susceptibility of $BaCuO_{2.5}$ showed a Curie-Weiss behavior, and effective Bohr magneton, p=0.69. In the $(Ba_{1-x}La_x)CuO_y$ system, no solid solution fields formed, instead mixtures of $BaCuO_{2.5}$ and $(La,Ba)_2CuO_4$ were obtained. These phases were dispersed in micro-domains. The state of mixture of $BaCuO_{2.5}$ and $(La,Ba)_2CuO_4$ in $(Ba_{0.5}La_{0.5})CuO_y$ changes into $La_3Ba_3Cu_6O_{14}$ above 750 ℃.

ACKNOWLEDGMENT

The authors would like to thank Mr. Takeshi Sakurai of SRL-ISTEC for his help in HIP processing. This work was supported by New Energy and Industry Technology Development Organization as a part of the program for Research and Development of Basic Technology for Future Industries.

REFERENCES

1. Arjomand M, Machin DJ (1975) J Chem Soc Dalton Trans: 1061
2. Thompson JG, Fitz Gerald JD, Withers RL, Barlow PJ, Anderson JS (1989) Mater Res Bull 24: 505
3. Machida M, Yasuoka K, Eguchi K, Arai H (1991) J Solid State Chem 91: 176
4. Uemura T, Yoshimura K, Kosuge K (1991) J Phys Soc Japan 60: 1300

New Layered Perovskite Vanadium Oxide with Negative Magnetization

NOBURU FUKUSHIMA, SHIRO TAKENO, SHIGENORI TANAKA, and KEN ANDO

Advanced Research Laboratory, Research & Development Center, Toshiba Corporation, Kawasaki, Kanagawa, 210 Japan

ABSTRACT

A new layered perovskite vanadium oxide $Sr_4V_3O_{8-d}$ was synthesized and its crystal structure and magnetic properties were studied. $Sr_4V_3O_{8-d}$ has a novel crystal structure, containing a V-O octahedron and a two-fold V-O fence in a unit cell. It exhibits negative magnetization below 30 K, when cooled in a magnetic field. This peculiar magnetism is attributed to N-type parasitic ferrimagnetism.

KEY WORDS:vanadium oxide, negative magnetization, parasitic ferrimagnetism

INTRODUCTION

Layered perovskite vanadates $Sr_{n+1}V_nO_{3n+1}$ [1,2] are attracting attention because of such similarities to the cuprate superconductors as 1) the existence of two-dimensional metal-oxygen sheets, 2) the expected existence of localized spin 1/2 on transition metal sites in the insulating phases, and 3) the existence of metal-insulator transitions[3]. It is important to elucidate the similarities and dissimilarities in physical properties between the cuprates and the vanadates, such as electronic structures and magnetic properties. Nevertheless, the magnetic order, which would be a clue to clarify the mechanism of the high Tc superconductivity in Cu-O superconductors, has never been confirmed in these layered V oxides; the cusps of the temperature dependence of the magnetic susceptibility[1,2] seen in these system may be attributed to the disordered magnetism such as the spin freezing which was observed in the oxygen deficient $Sr_3V_2O_{7-d}$[4].

We synthesized a new phase in the layered perovskite vanadate Sr-V-O system, $Sr_4V_3O_{8-d}$, and studied the crystal structure and the magnetic properties. This compound has a novel crystal structure containing a V-O octahedron and a two-fold V-O fence in a unit cell, and exhibits negative magnetization when it is cooled in a magnetic field. We attributed this magnetic property to a kind of parasitic ferrimagnetism[5]. This result implies the first confirmation of the existence of the underlying magnetic order in these layered perovskite vanadates, though the valence of V ions is no longer tetravalent (s=1/2).

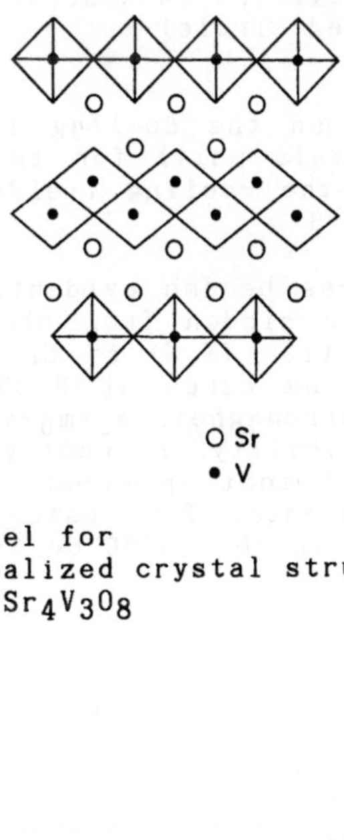

FIG. 1 Model for idealized crystal structure of $Sr_4V_3O_8$

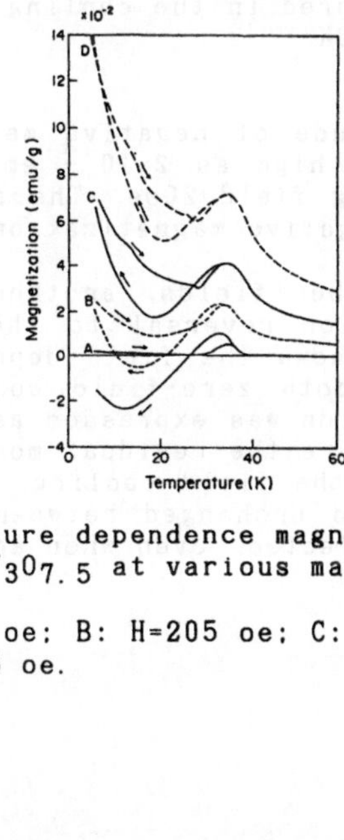

FIG. 2 Temperature dependence magnetization for $Sr_4V_3O_{7.5}$ at various magnetic fields.
A: H=19 oe; B: H=205 oe; C: H=490 oe; D: H=965 oe.

EXPERIMENTAL

$Sr_4V_3O_{8-d}$ samples were synthesized by heating an appropriate mixture of $SrCO_3$ and V_2O_3 at 1250°C in H_2 flow. The structural features were analyzed by means of transmission electron microscopy (**TEM**) and powder X-ray diffraction. From the high resolution **TEM** images, the electron diffraction patterns and the powder X-ray diffraction patterns, the crystal structure of this compound was found to be stacked along the **c** axis, with a single V-O octahedron layer and a two-fold V-O fence with a two-fold Sr-O rock salt layer in between. Since the V-V distance along the **c** axis in the two-fold V-O fence is as short as 2 **A**, this layer is thought to form such a double fence structure, as can be seen in $YBa_2Cu_4O_8$. Figure 1 shows a crystal structure model for this compound. Estimated lattice constants for the tetragonal unit cell are **a**=3.8 **A** and **c**=14.5 **A**. The oxygen contents, obtained by thermo-gravimetric measurements, were 7.5-8.0, which shows a good agreement with the structural model proposed above.

In addition, this compound indicates a superstructure along the **a** axis at a period double the length of lattice parameter **a**; this periodicity changes with a change in the oxygen contents. This superstructure is now expected to arise from the alternating directions of the V-O fences. Precise structural features will be reported elsewhere[6].

Magnetization measurements were carried out on a $Sr_4V_3O_{7.5}$ powder sample by means of a SQUID magnetometer (HOXAN HSM-2000). Figure 2 shows the temperature dependence of magnetization, both for the measurements after zero-field cooling (ZFC) and for the field cooling (FC) measurements at various magnetic fields. In the ZFC measurements, cusp-like peaks were observed at about 35 K, while the magnetization showed negative values at temperatures below the cusps in the FC measurements at lower magnetic fields.

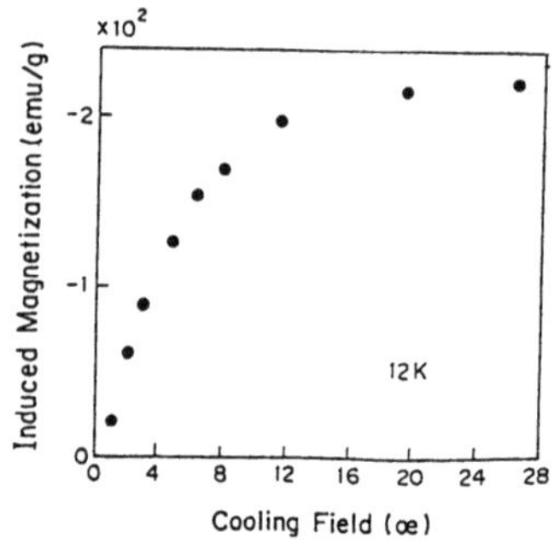

FIG. 3 Relationship between induced negative magnetization and cooling fields for $Sr_4V_3O_{7.5}$, measured in the cooling field at 12K.

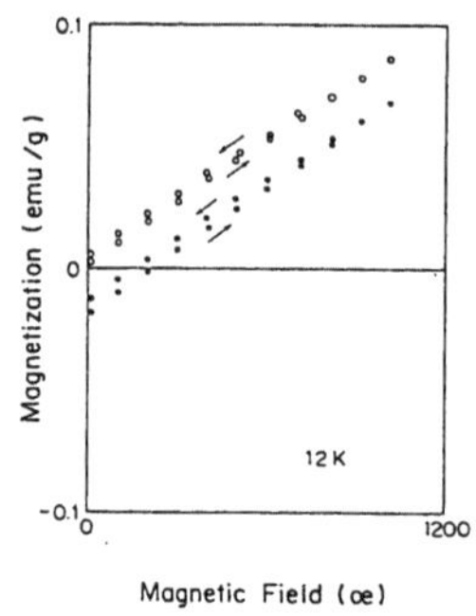

FIG. 4 Magnetization curves for Sr4V3O7.5, after ZFC (open circle), and after FC in H=20 oe (closed circle)

The magnitude of negative magnetization depended on the cooling fields and reached as high as $2x10^{-2}$ emu/g ($2.2x10^{-3}$ μ_B/formula unit) for the case of the cooling field=20oe. The relationship between the cooling fields and the induced negative magnetization is displayed in Fig. 3.

At the higher fields, another magnetization process became evident, and the magnetization reversal to the negative values was hidden from observation. Figure 4 shows the field dependences of magnetization (M-H) of $Sr_4V_3O_{8-d}$ at 12 K for both zero-field cooling and field cooling cases at H=20 oe. The magnetization was expressed as that of the weak ferromagnet, $m_c=m_0+x_cH$, where m_0 and x_c are the residual moment and the susceptibility, respectively. As a result of the field cooling, a negative residual moment appeared, while the x_c remained unchanged between ZFC and FC measurements. The residual moment was not affected, even when applying the field up to the 10000 oe.

DISCUSSION

This compound exhibits semiconducting behavior in electric conduction. The magnetic properties can therefore be interpreted in the localized spin picture. Though magnetization reversals to the negative value were observed also in Neel's type-N ferrimagnets[7], the thermo-irreversible nature of the magnetization in this compound cannot be explained in the framework of ordinary ferrimagnetism. In addition, we can rule out the possibility of the disordered magnetism, such as spin-freezing, as the origin of this peculiar magnetism, because the characteristic temperatures, such as the magnetization peak temperature, did not shift with change in the magnetic field.

The difference between the zero-field cooling and the field-cooling measurements is rather similar to the parasitic ferrimagnetism reported for YFe_2O_4[5]. In YFe_2O_4, all iron ion sites are equivalent, and the antiferromagnetic ordering along the c axis is observed below 205 K[8]. The field-cooling induced positive weak ferromagnetic nature was attributed to the non-equivalent occupation of Fe^{2+} and Fe^{3+} on the antiferromagnetic sublattices.

For $Sr_4V_3O_{8-d}$, we now presume that the magnetization reversal observed in the FC measurements can be attributed to N-type parasitic ferrimagnetism. At first, we shall discuss the crystallographic sites where the magnetic coupling takes place, which corresponds to the question whether this magnetism basically belongs to a kind of ferrimagnetism, or antiferromagnetism in the same manner as the parasitic ferrimagnetism in YFe_2O_4. There are two kinds of crystallographically non-equivalent sites for vanadium ions in $Sr_4V_3O_{8-d}$. The magnetic coupling between these non-equivalent sites is unlikely, because the superexchange interaction is too weak to be effective via such long distances and intervening Sr ions between two sites. Thus, magnetic coupling between crystallographically equivalent sites is more plausible for this case, and the magnetism of $Sr_4V_3O_{8-d}$ can be treated as a kind of antiferromagnetism.

Next, we consider the magnetization below the antiferromagnetic ordering temperature, based upon the parasitic ferrimagnetism model. The valence of the vanadium ions is expected to be divalent and trivalent from the estimated value of the oxygen content. In the zero-field cooling, $Sr_4V_3O_{8-d}$ is an antiferromagnet, and individual antiferromagnetic sublattices, occupied by V^{2+} (s=3/2) and V^{3+} (s=1) randomly and homogeneously, are magnetically equivalent. On the other hand, V^{2+} and V^{3+} tend to locate on different antiferromagnetic sublattices in the field cooling; V^{2+} ions occupy preferentially the sublattices parallel to the applied field and V^{3+} ions prefer those opposite to the field. Thus, the magnetic moments on the individual antiferromagnetic sublattice are non-equivalent. This tendency is interpreted to arise from the competition between the Zeeman energy sum and the ion mixing energy.

Supposing that the temperature dependences of partial magnetization of individual sublattices are different, and supposing that the magnetization of the sublattice opposite to the field overcomes that of the sublattice parallel to the field over a certain temperature range, negative net magnetization can be expected as well as for the case of N-type ferrimagnets.

This non-equivalent occupation of V^{2+} and V^{3+} takes place via a thermally activated electron exchange process. As this process seems inactive at lower temperatures, the non-equivalence of V^{2+} and V^{3+} distribution is fixed after field cooling, and the negative residual moment cannot be reversed by applying a high magnetic field.

Though further interpretation requires the molecular- magnetic field calculation as carried out for YFe_2O_4[5], the derivation of negative magnetization will be achieved, because the molecular-field calculation for the case of the magnetic coupling between crystallographically equivalent sites can be treated in the same way as for the case of non-equivalent sites, in which the negative magnetization of the Neel's type-N ferrimagnets is reproducible[9].

As mentioned above, the magnetism observed in $Sr_4V_3O_{8-d}$ can be treated as N-type parasitic ferrimagnetism, which is based on the antiferromagnetic order and the electron exchange between V ions on the equivalent sites. Furthermore, weak remanences of magnetization, comparable to the disaccommodation by the electron exchange between Fe^{2+}-Fe^{3+} in ferrites, were observed at various temperatures below 35 K. These remanences also imply that this peculiar magnetism can be attributed to the charge re-arrangement between the V ions on the antiferromagnetic sublattices. In addition, as the thermally activated transfer between the crystallographically equivalent sites is expected to be easier than electron exchange between the non-equivalent sites, these remanences also confirm the relevance of the parasitic ferrimagnetism model based on the underlying antiferromagnetism in the crystallographically equivalent sites.

In summary, we discovered a new layered perovskite vanadium oxide $Sr_4V_3O_{8-d}$ and presented a structural model. It exhibits negative magnetization, when it is cooled in magnetic fields. This magnetism is attributed to the N-type parasitic ferrimagnetism. Our finding of this peculiar magnetism implies the first confirmation of the underlying antiferromagnetic order in these layered perovskite vanadates. In addition, we remark that the anomalous magnetization process in Co_2VO_4[10], Co_2TiO_4[11] and $LaVO_3$[12], in which a magnetization reversal by field cooling was observed, may be ascribed to N-type ferrimagnetism as well as discovered in the present compound $Sr_4V_3O_{8-d}$.

REFERENCES

1 M. Cyrot, B. Lambert-Andron, J. L. Soubeyroux, M. J. Rey, Ph. Dehauht, F. Cyrot-Lackman, G. Fourcaudot, J. Beille and J. L. Tholence, J. Solid. State Chem. 85, 321(1990)
2 Ayumi Nozaki, Hiroshi Yoshikawa, Takahiro Wada, H. Yamauchi and Shoji Tanaka, Phys. Rev. B43, 181(1991)
3 N. Fukushima, S. Tanaka, H. Niu and K. Ando, Jpn. J. Appl. Phys. 29, L2190(1990)
4 Hiromi Niu, Noburu Fukushima, and Ken Ando, Phys. Rev. B
5 T. Sugihara, K. Siratori, I. Shindo and T. Katsura, J. Phys. Soc. Jpn. 45, 1191 (1978)
6 S. Takeno, S. Nakamura, T. Nomaki, N. Fukushima and K. Ando, to appear in J. Solid State Chem.
7 E. W. Gorter, Philips Res. Rep. 9, 403(1954)
8 J. Akimitsu, Y. Inada, K. Siratori, I. Shindo and N. Kimizuka Solid State Commun. 32, 1065 (1979)
9 N. Miura, I. Oguro and S. Chikazumi, J. Phys. Soc. Jpn, 45, 1534(1978)
10 N. Menyuk, K. Dwight and D. G. Wickham, Phys. Rev. Lett. 4, 119(1960)
11 N. Sakamoto and Y. Tamaguchi, J. Phys. Soc. Jpn., 17 Suppl. B1, 276(1962)
12 N. Shirakawa and M. Ishikawa, Jpn. J. Appl. Phys. 30, L755(1991)

Syntheses and Physical Properties for Tl-Sr-V-O System

KAZUO SAKAI[1], SHINJI MIGITA[1], HIDEKI FUJII[1], TAISUKE SHINDO[1], and RYOZO AOKI[2]

[1] Department of Electrical Engineering, [2] Research Center for Superconducting Materials and Electronics, Osaka University, Suita, Osaka, 565 Japan

ABSTRACT

Tl-Sr-V-O system was investigated to search for new non-Cu and n-type superconductors. A ferromagnetic behavior, however, was observed in Tl-Sr-V-O system and a remarkable hysteresis between ZFC and FC processes was found. A new peak at $2\theta=6.8°$ which suggests a long periodical structure was found.

[Key words] superconducting oxide, ferromagnetic ordering, $SrVO_3$, Tl-Sr-V-O compound, magnetic anomaly

1. INTRODUCTION

The high-T_c superconducting oxide is recognized, for the present stage, as having the following characteristics;(i) it must be a copper-oxide compound, (ii)it has a layered structure including a two-dimensional network structure of $[CuO_2]_n$, and (iii)the superconducting phenomenon takes place by doping usually the hole carrier in the $[CuO_2]$-layer, except for e.g. $(Nd_{1-x}Ce_x)_2CuO_4$. If the superconducting mechanism in the high-T_c oxide superconductor essentially limited to the hybridization of the Cu_{3d} and O_{2p} orbitals, one would have a severe condition for the search of a new type of high-T_c superconductor. However, various different models for the high T_c superconducting mechanism leaves more wide selection of the structure and compositions other than $[CuO_2]_n$ and really similar kind of the transition element Nb has been reported as bringing the oxide superconductivity[1]. Accordingly, we intend to search for a new non-Cu and n-type superconducting oxide.

We adopted the vanadium system by the following consideration. The vanadium element locates at the less than half site of the 3d series, while the copper element at the end of the more than half site and both are in a complimentary situation. Accordingly when we consider the electronic state in V^{4+} ion, it will just correspond to the hole state of the Cu^{2+} ion. Furthermore, vanadium element can take a various valence state changing successively from 2+ to 5+ by the condition of the surrounding ions. If a valence fluctuation mechanism would play an important role for the high-T_c superconducting mechanism, V-compound may have an advantage to a n-type superconductor. In addition, some evidence of superconductivity in this Tl-Sr-V-O system was once reported by Matsuda et al.[2], although not yet has been reconfirmed by any other investigators. Thus we try to ascertain any possibilities of realizing superconductivity in this Tl-Sr-V-O system and to investigate the chemical reaction conditions of Tl_2O_3 and $SrVO_3$.

2. EXPERIMENT

2-1. SAMPLE PREPARATION

Tl-Sr-V-O system was prepared by a solid state reaction. The mother compound $SrVO_3$ was prepared from $SrCO_3$ and V_2O_5. These starting materials were mixed in ethanol for 2 hrs and calcined at 800℃ in dry air flow. After regrounding and in pellet with 1000kg/cm^2 press, it was sintered at 1100℃ with H_2 gas flow. For Tl-Sr-V-O systems it was mixed with Tl_2O_3 in any arbitrary compositional ratios, pelletized under 2000kg/cm^2, and covered with gold foil in order to avoid the Tl vaporization, then the pellet was sintered at 650℃ for 4hrs and at 800～900℃ for 1～12 hrs under gas flow of the H_2 and Ar mixture which ratios were selected to choose the reduction conditions.

2-2. MEASUREMENT

Temperature dependence of the resistivity and susceptibility for the obtained samples were measured by the dc-four-probe method and the SQUID susceptometer, respectively. The moment for the specimen was measured via two following processes; first the specimen was cooled to 4.2K without a magnetic field and after applied the magnetic field with 100 Oe, the moment was measured raising gradually up to 150K at the speed of 5K/min (zero field cooling). It was measured lowering at a similar speed to 4.2K with the applied magnetic field of 100 Oe(field cooling). The structure was examined by a X-ray powder diffraction pattern(CuKα). The compositional ratios of the obtained specimens were evaluated by ICP analysis.

3. RESULTS AND DISCUSSION

Fig.1 shows the change of X-ray peak patterns in the specimens obtained under H_2:Ar=10:90(10%H_2) with Tl content (x) in the Tl-Sr-V-O system. Two main phases of $Sr_3V_2O_8$(marked by the circle in figure) and the unreacted $SrVO_3$ (the triangle) exhibit for x<0.33. Thus the reaction time was increased to 12hrs due to obtain a single phase, but $SrVO_3$ still remained. Most of Tl_2O_3 are reduced to Tl metal(the square) whose spherical solid precipitated on the specimen and little Tl is soluble in $SrVO_3$. The ICP analysis showed without Tl vaporization that Tl content in the specimens almost corresponds to the nominal composition. On the other hand, for x>0.44 only $Sr_3V_2O_8$ appeared dominantly. It means that $SrVO_3$ transfers to $Sr_3V_2O_8$ phase with the increment of Tl content. In this x>0.44 case, a little amount of several unknown peaks besides a main phase appeared in the region from 2θ = 25 to 31°. These peaks may be caused by a series of $Sr_{n+1}V_nO_{3n+1}$[3,4] or Tl-Sr-V-O compounds. The results under 4%H_2 are similar to those under 10%H_2, only difference in any unknown peaks appearance.

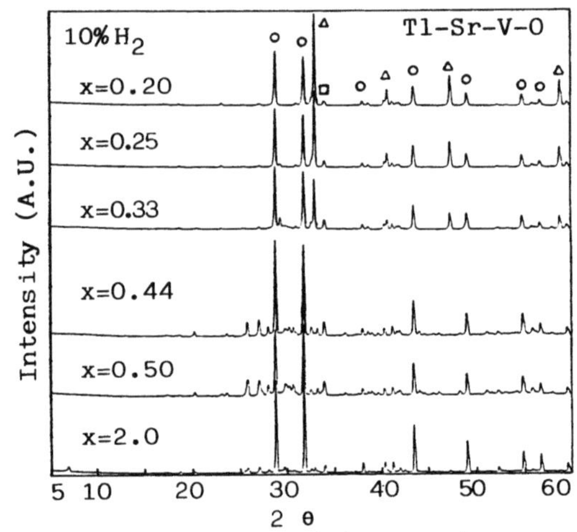

Fig.1 Tl content(x) dependence of the X-ray diffraction patterns under 10%H_2 in the Tl-Sr-V-O systems. The marks of O, Δ and □ indicate the peaks come from $Sr_3V_2O_{8-\delta}$, $SrVO_3$ and Tl, respectively.

Figs.2 and 3 show temperature dependence of the susceptibility(χ -T) and the resistivity(ρ -T) with Tl content changes, respectively. A ferromagnetic ordering was observed in the whole specimens. This magnetism probably originates in the spin arrangement of V ion, because Tl has a diamagnetic character (-2.4×10^{-7} emu/g・Oe). The $SrVO_3$, whose V^{4+} ion has one spin state basically, showed a Pauli-paramagnetic behavior and a metallic conduction. The $Sr_3V_2O_8$, the other main phase, was investigated by synthesis and it showed a paramagnetic and insulating behavior. Really the stoichiometric $Sr_3V_2O_8$ should show a non-magnetic behavior by taking account of the V^{5+} electronic state. Nevertheless, the specimen obtained here showed remarkable ferromagnetic character. Hence it is presumed that the paramagnetic characteristics would be appeared for lack of the oxygen from the $Sr_3V_2O_8$. Furthermore, when the oxygen gets out of $Sr_3V_2O_8$ more strongly with the help of the reduction by Tl metal, the oxygen-deficient $Sr_3V_2O_{8-\delta}$ would bring the ferromagnetic characteristics in the present Tl-Sr-V-O systems. In this system, as Tl content x increases the resistivity character ρ -T transfers from the metallic to the semiconducting as shown in Fig.3. On the other hand, the susceptibility character χ -T has some complicated features with x increase. First χ at low temperatures increases with x and shows maximum at x=0.33 and again decrease until x=2.0. This χ -T characteristics was found as corresponding to the structure changes. As shown in Fig.1 at x=0.4 the component $SrVO_3$ is distinguished and relatively $Sr_3V_2O_{8-\delta}$ grows up.

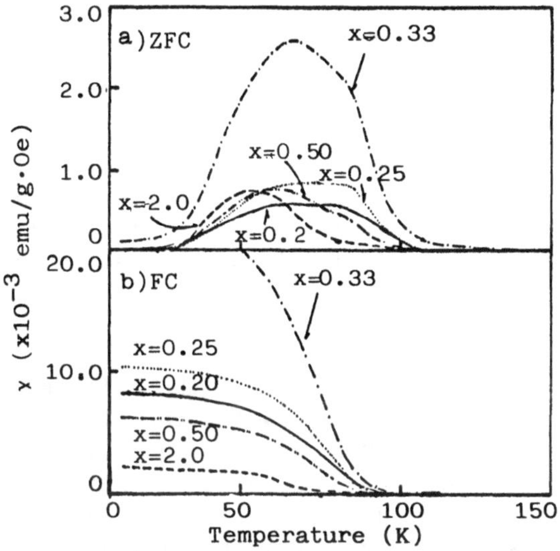

Fig.2 Temperature dependence of susceptibility in the Tl-Sr-V-O systems(Tl content x) under 10%H_2.
a)Zero field cooling,
b) field cooling

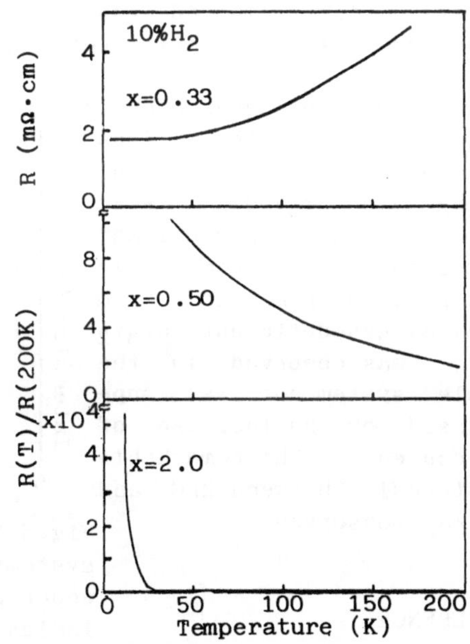

Fig.3 Temperature dependence of resistivity in the Tl-Sr-V-O systems (Tl content x) under 10%H_2.

The susceptibilities under zero magnetic field cooling(ZFC) and the applied magnetic field cooling(FC) were shown in Figs. 2-a) and b), respectively. Remarkable hysteresis of χ -T curve between ZFC and FC processes was found over the whole Tl-Sr-V-O systems. The χ_{max} in FC is by one order of magnitude larger than to that in ZFC. These kinds of hysteresis in the magnetization curves are usually observed in any ferromagnetic substances with the magnetic domain structures. In this cases the main cause of the hysteresis comes from the magnetic domain wall pinning. However, in the present measurement, the extent is significantly large. It is to be noted that the increase of Tl content suppressed this anomalous hysteresis at x>0.33.

When the reduction atmosphere was changed during sintering, the properties of the prepared sample were changed. In case of 50%H_2 and 100%H_2, X-ray peak patterns were modified as shown in Fig.4. New peaks appeared in the region of 28 to 32° under 100%H_2, while at 2θ=30.0° under 50%H_2 in spite of the Tl content in the region of 0.67<x<2.0. More to be noted is that a low angle peak which suggests the formation of a long periodical structure was appeared at 2θ = 6.8° (d=12.8 A), although it has very little intensity in this stage. The ferro-magnetic behavior is suppressed in the specimens obtained under both these conditions.

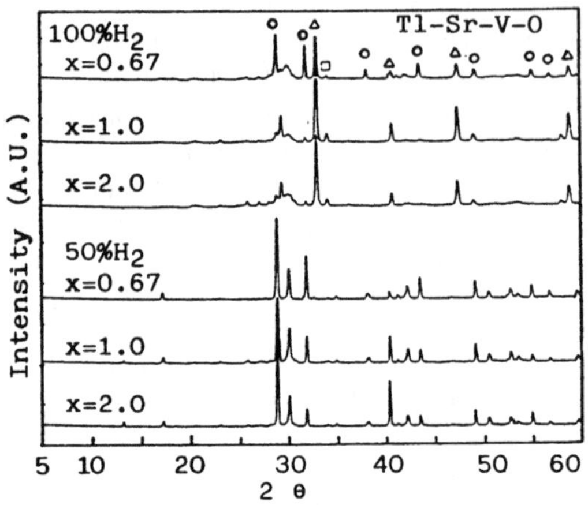

Fig.4 Tl content(x) dependence of the X-ray diffraction patterns for the Tl-Sr-V-O systems under 50%H_2 and 100%H_2, respectively. The marks of ○, △ and □ indicates the peaks come from $Sr_3V_2O_{8-\delta}$, $SrVO_3$ and Tl, respectively.

From these results the favorable sintering conditions for the formation of Tl-Sr-V-O compound, especially for the formation of a long periodical structure are summarized; (i) under 50%H_2, (ii) not so long time, and (iii) at rather lower temperature. Thus we adopted to form the appropriate samples in a condition at 800℃ for 1h under 50%H_2, then the X-ray patterns for the specimens were obtained as shown in Fig.5. The new peaks besides $Sr_3V_2O_8$ were observed from 2θ=26 to 34° and at 6.8° in both x=0.67 and 1.0. However, the compound was an ferromagnetic insulator, in spite of the appearance of a peak corresponding to a long periodical structure.

In summary, the superconductivity could not be observed in Tl-Sr-V-O systems prepared under the various reduction conditions. Most part of Tl_2O_3 were reduced to Tl metal, and $SrVO_3$ and $Sr_3V_2O_{8-\delta}$ were formed as predominant compounds. The ferromagnetic ordering which may come from the oxygen-deficient $Sr_3V_2O_{8-\delta}$ was observed in the whole system and was suppressed by the increase of Tl content. The remarkable hysteresis between ZFC and FC was observed.

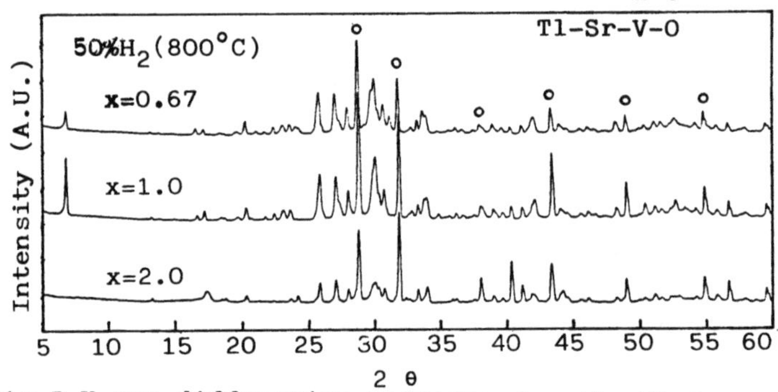

Fig.5 X-ray diffraction patterns for the Tl-Sr-V-O systems(Tl content x) obtained after sintering under 50%H_2 in 800℃ for 1hr. The mark of ○ indicates the peaks come from $Sr_3V_2O_{8-\delta}$.

REFERENCE

[1]Akimitsu J, Amano J,Sawa H,Nagase O,Gyoda K, Kogai M(1991)Jpn J Appl Phys 30:L1155-1156

[2]Matsuda S, Takeuchi S, Soeta A, Doi T, Aihara K, Kamo T(1990)Jpn J Appl Phys 29:L1781-1784,

[3]Nozaki A, Yoshikawa H, Wada T, Yamauchi H, Tanaka S(1991)Phys Rev 43:181-185

[4]Itoh M, Shikano M, Liang R, Kawaji H, Nakamura T(1990)J Solid State Chem 88:597-600

Effects of Aging on T_c and J_c of $YBa_2Cu_3O_{7-y}$ Irradiated by Electron Beam

K. ITOH, S. TAKAGI, K. YASUDA, K. KATORI, N. KAIDA, and Y. NISHI

Department of Materials Science, Tokai University, Hiratsuka, Kanagawa, 259-12 Japan

ABSTRACT

Effect of aging on T_c and J_c are investigate for $YBa_2Cu_3O_{7-y}$ sample which T_c^{mid} values are from 90 to 92 K. Although the T_c change cannot be found of the $YBa_2Cu_3O_{7-y}$ sample just after the electron beam(EB) irradiation, the aging increases the T_c and J_c values.

INTRODUCTION

Since high T_c above the boiling point of N_2 had been discovered for the $YBa_2Cu_3O_{7-y}$, many HTSC works have been reported. Although slight increases of the T_c were found by means of small addition of solute[1] and compressive stress[2], we would like to obtain the process that largely enhances T_c. We found large T_c increases of the EB irradiated $YBa_2Cu_3O_{7-y}$ (which T_c is from 94 to 96 K) by the aging at room tmperature[3]. Since the maximum T_c depends on the the T_c before the EB irradiation, the effect of the treatment on the large T_c increasing cannot be found in the YBCO sample which T_c^{mid} is below 92 K before EB irradiation. Thus, the influence of the EB irradiation on T_c and J_c were investigated for the $YBa_2Cu_3O_{7-y}$ which T_c^{mid} is below 92 K before EB irradiation.

EXPERIMENTAL

Samples with nominal composition $YBa_2Cu_3O_{7-y}$ were prepared from high purity powders of CuO, $BaCO_3$ and Y_2O_3. The powders were mixed and reacted at 1210 K for 2 hours and then cooled under pure oxygen. After crushing, a pelletized tablet, 1 mm thick, 13 mm diameter, was sintered under pure oxygen at 1210 K for 8 hours and furnace-cooled. The cooling rate was varied by controlling heat capacity of the furnaces[4].

All the measurements of T_c and J_c were measured by means of a standard 4-prove technique and a Keithley 181 nanovoltmeter[2]. The distance between the voltage probes is 4 mm. The temperature was measured by Au*Fe-chromel thermocouple attached to the specimen in a cryostat at equilibrium temperature. It was calibrated by standard Pt resistance thermometer (SCIENTIFIC INSTRUMENTS, INC. 4400 W. Tiffany Drive Mangonia Park, FL 33407 : MODEL P3AB, SERIAL # 81788) at equilibrium temperature. The structure of the samples was examined by means of x-ray diffraction.

The electron beam (EB) irradiation was homogeneously performed by Electrocurtain processer (Type; CB175/15/180L, ENERGY SCIENCE INC. 8 GILL STREET WOBURN, MA 01801)[3]. Fig. 1 shows schematic apparatus. The acceleration potential and the irradiating current were 175 kV and 0.15 mA/cm^2, respectively. The EB treatment is not continuously performed. Irradiation time was kept constant at 0.23 s in order to control the temperature. The temperature of the sample was below 323 K just after the EB irradiation. The integrated irradiation time was 1.1 s (83 Mrad of dose). Here, the total amount of absorbed dose value was converted by the absorbed dose of the distillation water. Although the other EB treatments are usually in vacuum, the irradiated specimen was under atmospheric pressure in the apparatus. It probably prevented removal of oxygen atoms from the $YBa_2Cu_3O_{7-y}$ sample surface.

RESULTS AND DISCUSSION

Figure 2 shows the T_c^{mid} change against the aging time (t_a) of the EB irradiated $YBa_2Cu_3O_{7-y}$ aged at room temperature. The electorical resistivity of the high-T_c $YBa_2Cu_3O_{7-y}$ shows a metallic decreases with temperature. Onset of the transition is defined as the temperature at $d(R/R_{300K})/dT=0.03$, where $d(R/R_{300K})/dT$ and R_{300K} are the slope of the reduced electorical resistivity (R/R_{300K}) versus temperature and the resistivity at 300K, respectively.

The transition temperature (T_c^{mid}) is defined as the temperature where the electrical resistivity is 1/2 its value at the onset of the transition. The aging slightly increases T_c^{mid}. The maximum

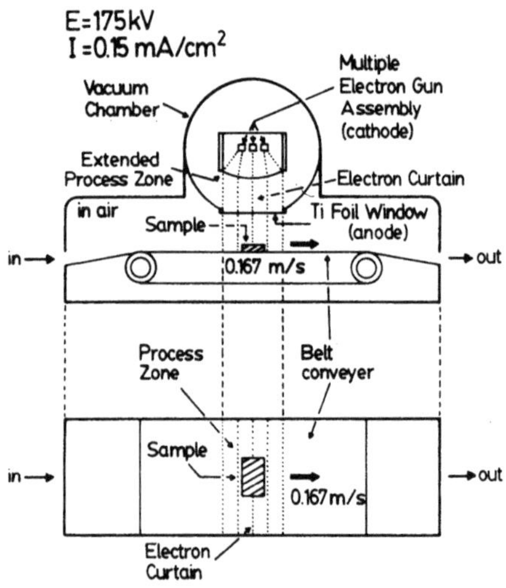

Fig. 1. Schematic diagram of electrocurtain processor.

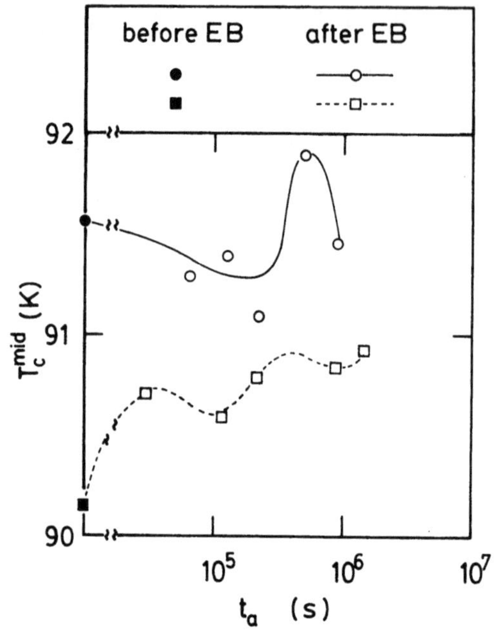

Fig. 2. Change in T_c^{mid} of superconducting transition against aging time (t_a) at room temperature of $YBa_2Cu_3O_{7-y}$.

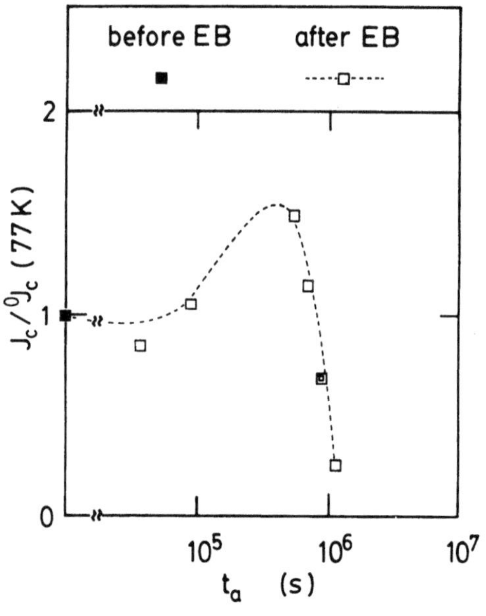

Fig. 3. Change in critical current densities ($J_c/^oJ_c$) at 70 and 75 K against the aging time (t_a). The mark ▣ is for the another sample which confirm reproductively.

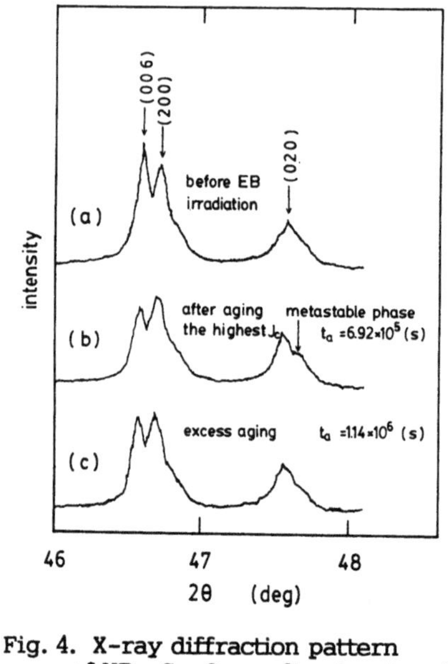

Fig. 4. X-ray diffraction pattern of $YBa_2Cu_3O_{7-y}$ after treatment (EB irradiation and aging), together with untreated sample before EB irradiation.

T_c^{mid} is 91.9 K for 5.146 x 10^5 s of the aging time.

Figure 3 shows J_c changes against the aging time (t_a) of the EB irradiated $YBa_2Cu_3O_{7-y}$ aged at room temperature. J_c was determined from the onset of voltage at 1 μ V level. The solid lines is for the reduced critical current density ($J_c/^oJ_c$) at 77 K. ^oJ_c is the J_c before EB irradiaton. The marks is for the another sample which confirms reproducibly. The maximum $J_c/^oJ_c$ is 1.5 for 5.30x10^5 s of the aging time. The excess aging (see Fig. 3) decreases J_c. The both T_c and J_c increase. The oxygen rich high T_c zone is assumed, whereas, the oxygen poor low T_c zone may exist,simultaneously. The J_c also relates to the the pinning site. If the oxygen poor low T_c zone acts as an additional pinning site on superconducting current, it probably enhances the J_c value.

Figure 4 shows x-ray diffraction peaks of the treated (EB irradiated at 83 Mrad of dose and aged for 6.92×10^5 s) $YBa_2Cu_3O_{7-y}$, together with the untreated sample [see Fig. 4(a)]. Peaks can be identified as orthorombic structure, which corresponds to oxygen-deficient perovskite. It corresponds to the T_c and J_c changes, as shown in Fig. 1 and 2. As the T_c and J_c increases, the metastable phase appears as shown in Fig. 4(b). The atom site distribution in the orthorombic structure is not most stable from the thermodynamics point of view. Thus, the excess aging approaches to the equilibrium. Therefore, the angles of the x-ray diffraction peaks after the excess aging [see Fig. 4(c)] approach that before the EB irradiation, because the structure seems to be similar to that before the irradiation [see Fig. 3 and 4(a)].

CONCLUSION

In summary, the aging increases the T_c and J_c values of $YBa_2Cu_3O_{7-y}$ ($T_c^{mid} < 92$ K) irradiated by the electron beam. The maximum J_c at 77 K is about 1.5 times larger than that before the EB irradiation. The aging induces metastable oxygen rich high T_c zone and oxygen poor low T_c zone, simultaneously. If the oxygen poor zone acts as an additional pinning site, the phenomena can be explained.

REFERENCES

1. Nishi Y, Moriya S and Tokunaga S (1988) J. Mater. Sci. Lett. 7:359
2. Nishi Y, Ninomiya N, Oguri K, and Tokunaga S (1986) J. Appl. Phys. 65: 3972.
3. Nishi Y, Takagi S, Yasuda K, Itoh K, (1991) J. Appl. Phys 70: L367.
4. Nishi Y, Kita Y, Igarashi A and Tokunaga S (1988) J. Mater. Sci. Lett. 7: 279.

Rapid Rate of He Ion Etching for the High T_c $YBa_2Cu_3O_{7-y}$

YASUO TAKAHASHI, NORIYUKI INOUE, SEIICHIRO TAKAGI, TAKASHI SUDO, KAGEYOSHI SAKAMOTO, TOSHIO SUZUKI, TOSHIO SHIMA, and YOSHITAKE NISHI

Department of Materials Science, Tokai University, Hiratsuka, Kanagawa, 259-12 Japan

Abstract

A rapid rate (R_s) is observed by a helium ion etching. R_s depends on the etching time. T_c change by the etching is studied for a high T_c $YBa_2Cu_3O_{7-y}$. A tremendous T_c decrease is caused by a excess helium ion etching.

KEY WORDS: High Tc $YBa_2Cu_3O_{7-y}$, Ion etching rate, Helium ion etching

Introduction

An ion etching is able to form fragile materials without fracture. From an engineering point of view, it is important to know an influence of the ion etching on a superconducting transition temperature for a high T_c $YBa_2Cu_3O_{7-y}$ [1]. We found a rapid rate by an argon ion etching [2] and a polishing effect by helium ion etching [3]. Thus, we have started the present study to observe the etching rate and T_c change by the helium ion etching for the $YBa_2Cu_3O_{7-y}$.

Experimental procedure

Sample preparation

Samples with nominal composition $YBa_2Cu_3O_{7-y}$ were prepared from high purity powders of CuO (99.9%), $BaCO_3$ (99.9%) and Y_2O_3 (99.99%). The powders were mixed and reacted at 1210 K for 2 h and then air-cooled under pure oxygen. After crushing, a pelletized tablet, 0.8 mm in thick 13 mm in diameter, was sintered under pure oxygen at 1210 K for 8 h and furnace-cooled. The cooling rate was 4.2 K/min at 973 K.

Ion etching

The helium ion etching was performed under 4 x 10^{-4} Torr helium atmosphere just above the sample by Kaufman type apparatus, which is shown schematically in figure 1 (ISM-S,Elionix, Tokyo)[4]. Kaufman source of this type is operated with diameters of 70 mm for the use in ion etching stations. The energy for the ion acceleration and the ion current density were 1.0 keV and 0.6 mA/cm^2, respectively. The ion beam was perpendicular to the sample surface.

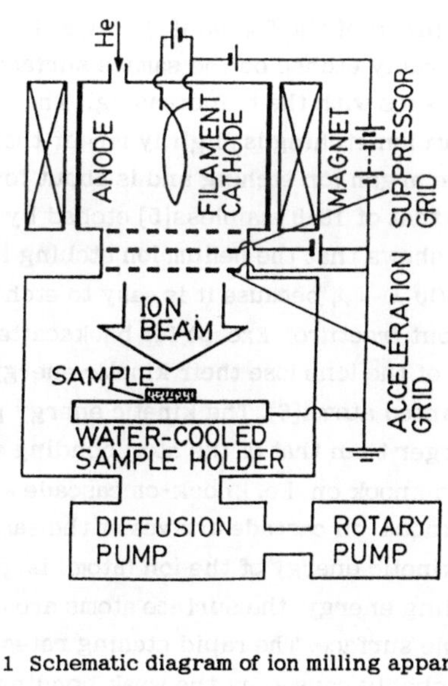

Fig. 1 Schematic diagram of ion milling apparatus

Etching rate, R_s was measured with micrometer and surface roughness meter. These resolutions were below 1 m and 0.1 nm, respectively.

Sample characterization

The T_c was measured using a standard 4-prove technique and a Keithley 181 nanovoltmater. The temperature was measured by an Au·Fe-chromel thermocouple attached to the specimen in a cryostat. Figure 2 shows the temperature dependence of electrical resistivity of the $YBa_2Cu_3O_{7-y}$. The deviated point is taken at $dR/dT=0.03R_{300\ K}$ in the present work, where dR/dT and R_{300K} are the slope of R-T and resistivity at 300 K, respectively. The midpoint of the transition is designated T_c^{mid}.

The structures of a samples were examined by means of x-ray diffraction (Cu k).

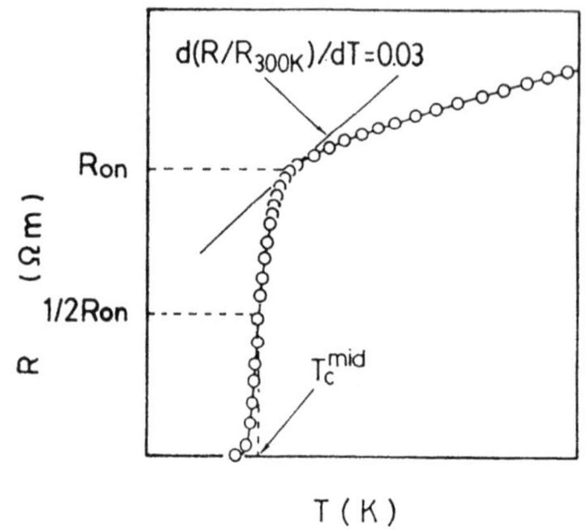

Fig. 2 Changes in electrical resistivity (R) with temperature (T) of $YBa_2Cu_3O_{7-y}$.

Results and Discussions

Ion etching rate (Rs)

Figure 3 shows change in the etching rate(R_s) against the etching time (t_e) for the $YBa_2Cu_3O_{7-y}$. The R_s depends on the etching time (t_e). The larger t_e, the smaller the R_s becomes. The helium etching reduces the spread of the R_m data[3]. Since the convex part is preferably etched on the sample surface, the R_s decreases with the t_e increasing. The R_s of the helium ion etching is slightly larger than that of the argon ion etching and is about four times larger than that of 18-8 stainless[5] etched by the argon ion. This shows that the helium ion etching is useful for the $YBa_2Cu_3O_{7-y}$, because it is easy to etch it rapidly without fracture. Except for backscattering ions, most of the ions lose their kinetic energy and become implanted atoms[6]. The kinetic energy gained by an atom is larger than that of the atom bonding energy. The atoms knock on, i.e. knock-on cascade are formed[7]. If the knock-on cascade arrives at the sample surface and the kinetic energy of the ion (atom) is greater than the bonding energy, the surface atoms are ejected from the sample surface. The rapid etching rates of the $YBa_2Cu_3O_{7-y}$ is probably caused by the weak bonding energy of the Cu-O bonding.

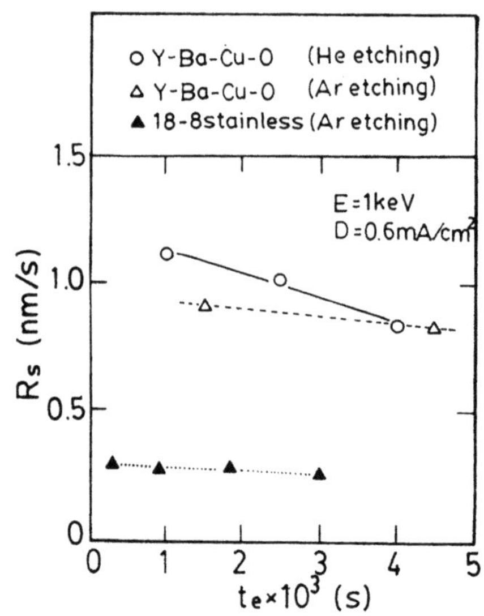

Fig. 3 Change in etching rate(R_s) against etching time (t_e).
Solid, dotted and broken line are for helium etching, argon etching (Ref.2) of $YBa_2Cu_3O_{7-y}$ and 18-8 stainless (Ref.5), respectively.

T_c change by etching

Figure 4 shows the T_c^{mid} change against the etching time (t_e) of the $YBa_2Cu_3O_{7-y}$. Solid and broken lines are for the helium ion etching and for the argon ion etching[1], respectively (at 1 keV of the ion energy). Although the argon ion etching tremendously changes the T_c value above 600 s, the helium ion etching does not largely change the T_c value up to 2000 s. The excess helium ion etching at 3000 s decreases the T_c value below 4.2 K.

x-ray diffraction pattern

Figure 5 shows x-ray diffraction patterns of the samples before and after the helium ion etching (t_e= 0 s and 3000 s) Most peaks can be identified as an orthorombic structure, which corresponds to an oxygen-deficient perovskite. The excess etching decreases the angle value of the peak and increases the peak width. The larger the peak width, the lower the crystal perfection becomes. It decreases the T_c value.

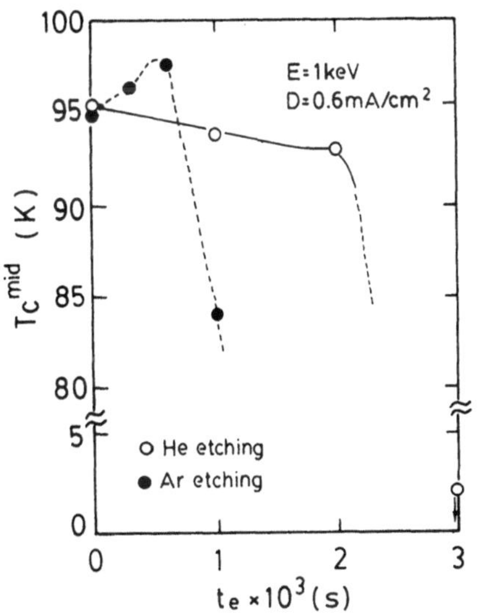

Fig. 4 Changes in superconducting transition temperature(T_c) against etching time (t_e) for $YBa_2Cu_3O_{7-y}$. Solid and broken line are for helium etching and argon etching, respectively.

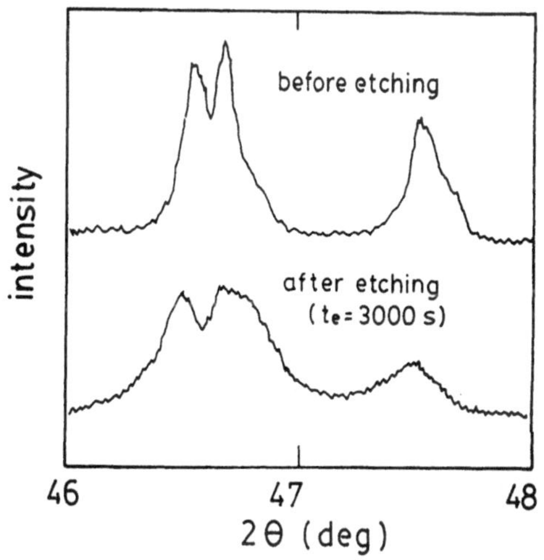

Fig. 5 x-ray diffraction pattern of the $YBa_2Cu_3O_{7-y}$. Samples are taken before and after helium ion etching(t_e = 0 s, 3000 s).

Conclusion

In summary, the high rate (R_s) of helium ion etching was observed of the high T_c $YBa_2Cu_3O_{7-y}$. The etching time to keep the high T_c of a helium ion etching is three times longer than that of argon ion etching. The excess helium ion etching at 3000 s decreases the T_c value below 4.2 K.

References

1.Y.Nishi, S.Moriya, N.Inoue, S.Tokunaga and T.Shima, (1988) J.Mater.Sci.Lett.7:997.
2.Y.Nishi, S.Moriya, N.Inoue, S.Tokunaga and T.Shima, (1988)J.Mater.Sci.Lett. 7:281.
3.N.Inoue, Y.Takahashi, T.Sudo, T.Shima and Y.Nishi., (1991) J.Appl.Phys., January in press.
4.P.G.Gloerson, (1975) J.Vac.Sci.Technol. 12:28.
5.Y.Nishi, N.Inoue, K.Watanabe, T.Morishita and T.Shima., (1987)J.Mater.Sci.Lett. 6:63
6.T.Ishitani, and R.Shimizu, (1974) Phys.Lett. 46:487.
7.P.Sigmond,(1969) Phys.Rev. 184:383.

STM Observation on Bi-O Surfaces of $Bi_2Sr_2Ca_{1-x}Y_xCu_2O_y$ Single-Crystals

KAZUTO IKEDA, KENSHI TAKAMUKU, KOJI YAMAGUCHI, RITTAPORN ITTI, NAOKI KOSHIZUKA, and SHOJI TANAKA

Superconductivity Research Laboratory, ISTEC, Koto-ku, Tokyo, 135 Japan

ABSTRACT

We observed STM images of the Bi-O surfaces of cleaved Bi2212 single crystals with and without substituted Y ions. A series of STM images obtained showed that the period of the superstructure decreased with increasing Y substitution ratio. We also measured the tunneling spectra of these crystals. The electronic states on the Bi-O surfaces are shown to have a large (semiconducting) energy gap in the non-superconductive crystals.

KEY WORDS: $Bi_2Sr_2Ca_{1-x}Y_xCu_2O_y$, Bi-O surface, STM

I. INTRODUCTION

The superstructures in the Bi-O planes of the $Bi_2Sr_2CaCu_2O_y$ superconductor have been studied by scanning tunneling microscopy (STM) in the interest of correlating the modulation structure and high-T_c superconductivity.[1-5] Electron diffraction observations revealed that the period of the superstructure in the Bi2212 compounds changs with Y substitution.[6] On the other hand, a photoelectron spectroscopic study showed that the Bi-O bond becomes ionic with Y substitution. This is in good agreement with the change of the Bi oxidation state calculated on the basis of the bond valence sum (BVS) method.[7] We report here STM images of *in situ* cleaved Bi-O surfaces of single crystals which differ in Y substitution ratio. We also discuss the tunneling spectra measured on these surfaces.

II. EXPERIMENTAL

Three compositions of Bi2212 single crystals were grown by the floating zone (FZ) method. The first kind was unsubstituted, and the second and third crystals were Y substituted ones made from the powders mixed in the ratios of Bi:Sr:Ca:Y:Cu=2:2:1:0:2, 2:2:0.5:0.5:2 and 2:2:0:1:2, respectively. The first kind showed superconductivity, but the latter two were not superconductive. T_c of the first one measured by a SQUID magnetometer was 89 K. The composition of this crystal was determined to be Bi:Sr:Ca:Y:Cu=2.1:1.7:1.0:2 (normalized with Cu=2) by electron probe micro analysis (EPMA). It seems that the reduction the number of hole by the compositional difference from 2:2:1:2 increases T_c because the hole density approaches its optimum value. Compositions of the second and the third crystals were determined to be by EPMA Bi:Sr:Ca:Y:Cu=2.0:1.8:0.5:0.6:2 and 1.9:2.0:0:0.7:2, respectively. Figure 1 shows the

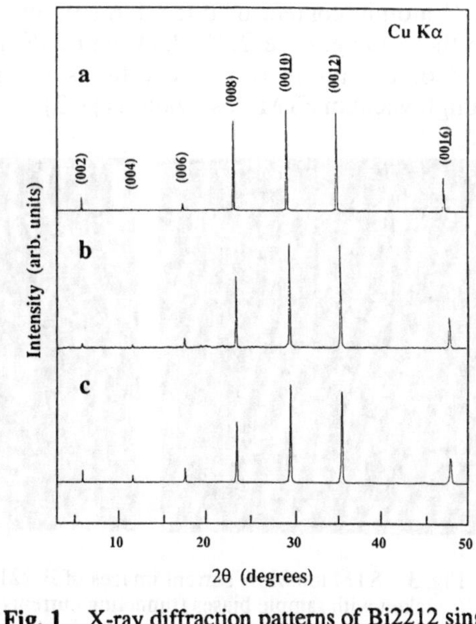

Fig. 1 X-ray diffraction patterns of Bi2212 single crystals with Y content of 0 (a), 0.6 (b), and 0.7 (c).

X-ray diffraction patterns of three compositions. This indicates that the c-axis lattice parameter decreases with increasing Y substitution ratio. STM observations were performed at room temperature under a pressure of $<2\times10^{-8}$ Pa using a JEOL STM system (JSTM-4000XV) equipped with a scanning electron microscope (SEM). We used electrochemically etched tungsten and platinum-iridium (10%) tips; Pt-Ir tips were made by Material Analytical Services. Tips were *in situ* heated and treated by placing them close (<1 mm) to the Si heater (~1250°C) before observation. Crystals were cleaved at room temperature under ultrahigh-vacuum condition with a pressure of 3-4×10^{-8} Pa in the sample treatment chamber by using adhesive tape. STM observations were performed after confirming by SEM that the tip was in close proximity to the clean cleaved surfaces as shown in Fig. 2. Tunneling spectroscopic measurements were made by moving the STM tip to the bright (projecting) stripes on the image when the superstructure could be observed under a stable condition. Sample bias voltage was scanned from +5 V to -5 V 16 times at each point.

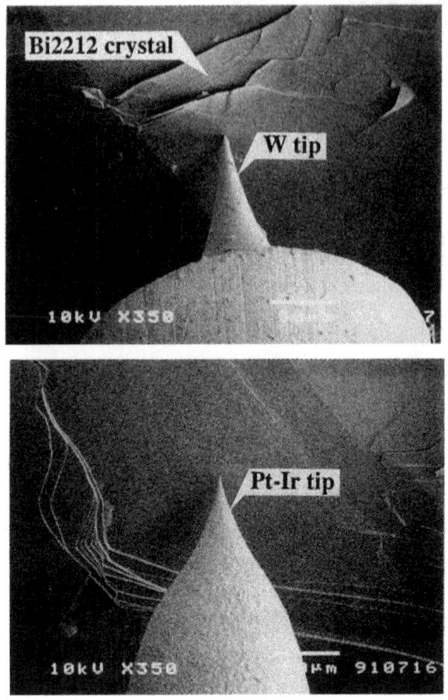

Fig. 2 SEM images of the STM tips and cleaved surfaces of Bi2212 crystals.

III. RESULTS AND DISCUSSION

a. Superstructures of Y substituted Bi2212 single crystals

Several groups have reported that the period of the superstructure of the Bi2212 compound changes by Y substitution for Ca [6] and Pb substitution for Bi [8-12] by using electron diffraction (ED). These changes are discussed in relation to the problem of excess oxygens in the Bi-O planes. Figure 3 shows STM current images (45 nm by 45 nm) of surfaces of the crystals with Y atomic content of 0 (a), 0.6 (b), and 0.7 (c). The periods of the superstructures obtained from these images are 2.71, 2.44, and 2.40 nm for Y=0, 0.6, and 0.7 crystals, respectively. The period of the unsubstituted crystal is in agreement with the values obtained from previous ultrahigh vacuum STM observations.[1-3]

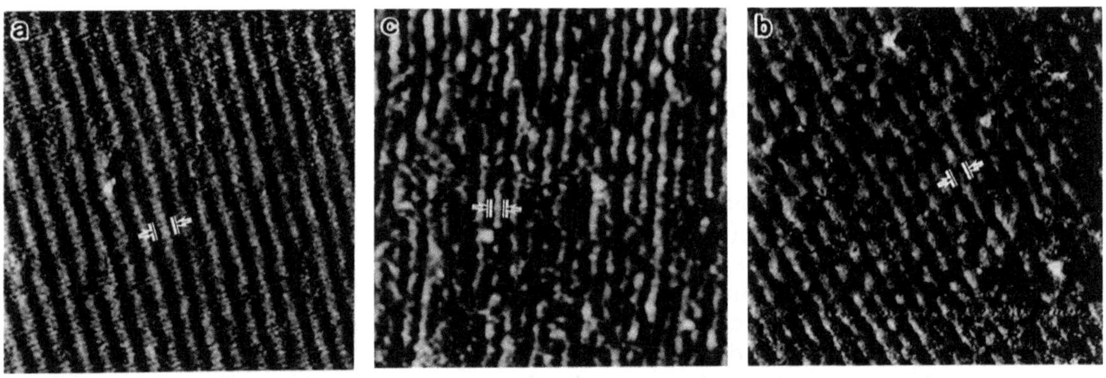

Fig. 3 STM top-view current images of Bi2212 single crystals with Y contents of 0 (a), 0.6 (b), and 0.7 (c) taken with sample biases (tunneling currents) of 1.80 V (0.2 nA), 3.50 V (0.2 nA), and 3.00 V (0.2 nA), respectively. All images are 45 nm by 45 nm.

The superstructure reflections have been observed in the [001] ED pattern.[13-17] Figure 4 shows ED patterns of the same crystals measured by STM. We calculated the periods of the superstructure from the spacings between spots marked by arrows around the (060) reflection, which corresponds to the second spots of the superstructure. The periods obtained from ED patterns are 2.54, 2.40, and 2.29 nm for the Y=0, 0.6, and 0.7 crystals, respectively. Thus the decrease in the period with increasing Y substitution ratio observed by STM is consistent with that obtained from the ED patterns. However, the periods obtained by STM are larger than that by ED . We speculate the reason as follows; STM images exhibit the surface structure but this structure is different from that in bulk. Kirk *et al.* reported that the superstructure originates in the expansion of Bi-O bonds to fit the larger Cu-O bonds.[1] The expansion energy of the surfaces must then be smaller than that in the bulk and the stress of the surface may be a little less.

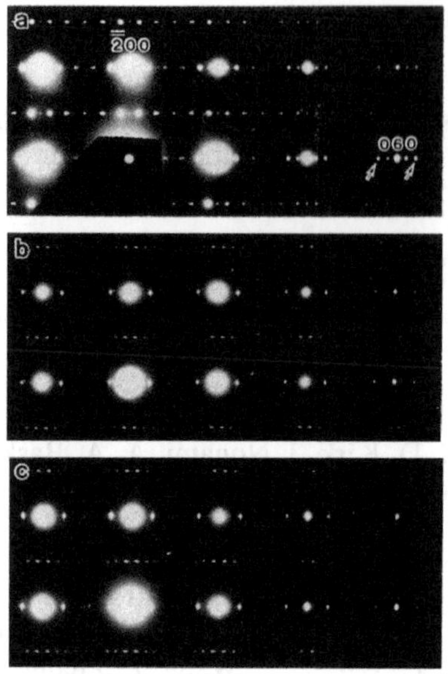

Fig. 4 [001] electron diffraction patterns of Bi2212 single crystals with Y content of 0 (a), 0.6 (b), and 0.7 (c).

b. Tunneling spectroscopic studies on Bi-O surfaces

Figure 5 shows tunneling spectra on Bi-O surfaces of the Bi2212 crystals with Y content of 0 (a), 0.6 (b), and 0.7 (c). These spectra are the averaged ones for more than ten different points. These spectra were measured under the conditions that the STM images were observed at 2.00 V, 3.40 V, and 3.00 V sample biases and 0.2 nA tunneling currents for Y=0, 0.6, and 0.7 crystals, respectively. The spectrum of the superconductive Y=0 crystal has structures in the voltage beyond ±0.5 V. This is consistent with the results of the previous studies; there are no electronic states near zero sample bias voltage.[2] On the other hand, no structure exists between -1 and +1 V in the spectra of non-superconductive Y=0.6 and 0.7 crystals. The photoelectron spectroscopic study by Itti *et al.* [7] revealed that the Bi-O bonding becomes ionic with the change from the superconductive to the non-superconductive state for Bi2212 compound by Y substitution. Itti *et al.* concluded that the shifts of the Bi core-level spectra with Y substitution indicates an increase of the Bi oxidation state. They confirmed this by the bond valence sum (BVS) consideration [18] which gives information on the oxidation state of a cation using the distances between the cation and the surrounding anions. The result of our tunneling spectroscopic studies shows that the semiconductive gap energies of the Bi-O surfaces increase with the change to the non-superconductive state for Bi2212 crystals, and this is consistent with the the result of the photoemission study.

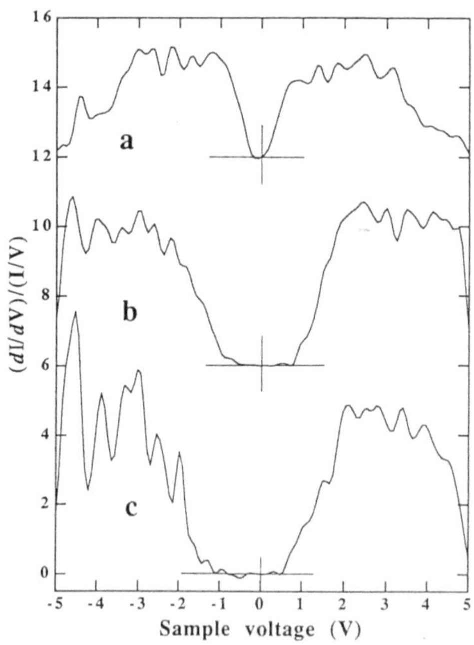

Fig. 5 Normalized conductivity vs sample voltage for Bi-O surfaces of Bi2212 single crystals with Y content of 0 (a), 0.6 (b), and 0.7 (c).

IV. CONCLUSION

By STM observation, we found that the period of the superstructure of the Bi-O surfaces decreased with increasing Y substitution ratio. The tunneling spectra show that the gap energy of the Bi-O surface of the superconductive Bi2212 crystal is smaller than that of the non-superconductive Y substituted crystals.

ACKNOWLEDGMENT

This work was supported by New Energy and Industrial Technology Development Organization under the management of the R & D of Basic Technology for Future Industries.

REFERENCES

[1] M. D. Kirk, J. Nogami, A. A. Baski, D. B. Mitzi, A. Kapitulnik,T. H. Geballe, and C. F. Quate, Science **242** (1988) 1673.
[2] C. K. Shih, R. M. Feenstra, J. R. Kirtley, and G. V. Chandrashekhar, Phys. Rev. B **40** (1989) 2682.
[3] C. K. Shih, R. M. Feenstra, and G. V. Chandrashekhar, Phys. Rev. B **43** (1991) 7913 .
[4] X. L. Wu, Z. Zhang, Y. Wang, and C. M. Lieber, Science **248** (1990) 1211.
[5] X. L. Wu, Y. L. Wang, Z. Zhang, and C. M. Lieber, Phys. Rev. B **43** (1991) 8729.
[6] T. Tamegai, K. Koga, K. Suzuki, M. Ichihara, F. Sakaki, and Y. Iye, Jpn. J. Appl. Phys. **27** (1989) L112.
[7] R. Itti, F. Munakata, K. Ikeda, H. Yamauchi, N. Koshizuka, and S. Tanaka, Phys. Rev. B **43** (1991) 6249.
[8] C. H. Chen, D. J. Werder, G. P. Espinosa, and A. S. Cooper, Phys. Rev. B **38** (1989) 4686; D. J. Werder, C. H. Chen, S. Jin, and R. C. Sherwood, J. Matter. Res. **4** (19890 748.
[9] J. Schneck, L. Pierre, J. C. Toledano, and C. Daguer, Phys. Rev. B **39** (1989) 9624.
[10] P. Border *et al.*, Studies High Temo. Supercond. **2** (1989) 171.
[11] H. W. Zandbergen, W. A. Groen, F. C. Mijlhoff, G. van Tendeloo, and S. Amelinckx, Physica C **156** (1988) 325.
[12] R. Ramesh, G. van Tendeloo, G. Thomas, S. M. Green, and H. L. Luo, Appl. Phys. Lett. **53** (1988) 2220; K. K. Fung, C. Y. Yang, and Y. F. Yan, *ibid.* **55** (1989) 280.
[13] T. M. Shaw, S. A. Shivashankar, S. J. LaPlaca, J. J. Cuomo, T. R. McGuire, R. A. Roy, K. H. Kelleher, and D. S. Yee, Phys. Rev. B**37** (1988) 9856.
[14] S. Horiuchi, H. Maeda, Y. Tanaka, and Y. Matsui, Jpn. J. Appl. Phys. **27** (1988) L1172.
[15] E. A. Hewat, M. Dupuy, P. Bordet, J. J. Capponi, C. Chaillout, J. L. Hodeau, and M. Marezio, Nature **333** (1988) 53.
[16] P. L. Gai and P. Day, Physica C **152** (1988) 335.
[17] Z. Hiroi, Y. Ikeda, M. Takano, and Y. Bando, J. Mater. Res. Vol. **6** (1991) 435.
[18] I. D. Brown and D. Altermatt, Acta Crystallogr. Sect. B**41** (1985) 244.

3.4 Crystal Growth

Crystal Growth of $YBa_2Cu_3O_x$ Phase by Crystal Pulling

Y. YAMADA, M. NAKAGAWA, K. ISHIGE, and Y. SHIOHARA

Superconductivity Research Laboratory, ISTEC, Koto-ku, Tokyo, 135 Japan

ABSTRACT

Single crystals of oxide superconductor of YBa2Cu3Ox were grown directly from the melt by crystal pulling method. Influences of crucibles made of various different materials were investigated. Pt, yttria stabilized zircomia, calcia stabilized zirconia, and calcia were not suitable for this experiment. Alumia and magnesia crucibles could have possibility to maintain a melt as containers. Barium cuprate was selected as a solvent since there are much informations of the phase diagrams of this compounds. Conditions for controlling continuous growth have not been determined yet.

KEY WORDS: Y system superconductor, crystal growth, crystal pulling

INTRODUCTION

A large number of studies on single crystal growth of high Tc oxide superconductors have been reported[1,2,3,4]. One of the most successful and the easiest growth methods is a self-flux method. This is a convenient process for preparing single crystals. However, it is difficult to control crystal growth. Obtained crystals are small up to about a few cubic millimeter or about 50 micrometer thick. MPMG(melt-quench powdering and melt growth) method, more which is a sophisticated process, can offer large crystals. But it has some inclusions and pores since its formation reaction is peritectic. It also has a restriction in size and shape. For further physical investigation and future application, it is very important to develop other promising methods in order to grow a large crystal continuously with high quality. Not only a single crystal but also oriented bulk polycrystals are available for power application. For this purpose, we have tried to grow YBa2Cu3Ox(Y123) crystals by a crystal pulling method using seed crystals and molten barium cuprate with yttrium oxide.

The advantages of this technique are following. It is easy to control crystal size. The crystals have less inclusions and less imperfections. The most disadvantage is that a crucible is required to maintain the melt for a long time. The melt reacts and corrodes the crucibles and dissolve the material in to the melt. This might be possible not only to destroy the crucible but also to change the nature of the melt and degrade the quality of the crystals. In this paper we present the feasibility of some crucibles for performing crystal pulling and describe the crystals obtained.

EXPERIMENTS

In order to grow crystals by crystal pulling from a melt, it is necessary to prepare the melt and a seed. Y123 decomposes to Y2BaCuO5(Y211) and liquid of barium cuprate with yttrium oxide at a temperature of about 1000°C. Therefore, a solvent is required to crystalize Y123 as a primary phase from the melt. Barium cuprate was selected as the solvent, because Y123 exists as the primary phase in the system and other solvent has much influence on the Y123 crystals by contaminations. The starting materials of the solvent were prepared by a mixture of high purity(3N) BaCO3 and CuO. The composition was selected at Ba:Cu=28.5:71.5 which is the eutectic composition in the binary system[5,,6]. The mixed powders were presintered at 880°C for 24 hrs in air. This solvent and the solute of Y123, which was also made of high purity Y2O3, BaCO3 and CuO, were mixed and melted at above 900°C in a crucible.

The crucibles, which have less solubility and less reactivity to the molten, are suitable for crystal growth. The melt was, however, so reactive that we had to investigate the influences of various materials of crucibles made of Pt, YSZ(yttria stabilized zirconia), CSZ(calcia stabilized zirconia), calcia, alumina and magnesia. If the materials of the crucibles were dissolved into the melt, the phase diagram would change to a new one and the solubility of Y123 would change depending on the dissolved amount of the elements. Furthermore, other solid phases would be

Table 1 Change of viscosity, separated solid in the case of using some crucibles and their erosion resistance.

crucibles	corrosion resistance	viscosity	separated solid on the surface
alumina	good	gradually decreasing	Y123
magnesia	poor	slightly decreasing	Y123
Pt	good	gradually increasing	BaPtO2.38
YSZ	good	gradually increasing	ZrBaO3
CSZ	very poor	gradually decreasing	----
calcia	very poor	decreasing	----

possibly separated on the surface where the temperature is the lowest in the melt. Complete understanding of these effects needs details of these phase diagrams of the compounds. It is difficult to investigate the phase diagrams, so we tried more convenient procedures by means of investigating a change of viscosity of the melt, whether or not Y123 was solidified from the molten and also corrosion resistance of the crucibles while the molten was maintained in the crucibles at a high temperature. Since these three subjects are the most important factors for crystal pulling, and it is supposed that the viscosity reflect the solubility of Y123. By observing the stirring behavior, the melt was investigated whether or not the excess Y123 was dissolved in a long time and whether Y123 or other phases was separated on the surface examined by X-ray diffraction.

The seeds which were used for crystal pulling were made of SmBa2Cu3Ox by MPMG method because Sm123 has a much higher decomposition temperature(1060°C) than Y123 and the Sm123 grains of the samples were considerably oriented and large enough for the seed.

The growth temperatures were within the primary phase range up to 1010°C. The temperature gradient at the center of the molten surface was about 10°C/cm perpendicular to the surface. Seed rotation rates were 6 to 50 rpm and pulling rates were 0.1 to 1.0 mm/h.

RESULTS AND DISCUSSION

The influences of the crucible materials to viscosity of the melt and separated solids on the surface are listed in table 1 with corrosion resistance of the crucibles. Using alumina and magnesia crucibles, excess Y123 solid phase in the melt was gradually dissolved at the pulling temperature. The melt slightly permeated through the wall of the magnesia crucible. Calcia and CSZ crucibles were also permeated and they could not maintain the liquid. No melt permeated through Pt and YSZ crucibles. However, the viscosity of the melt increased gradually and other solid phases than Y123 separated in it. They were not suitable on a small scale experiment like we did. Although the crucibles investigated have not met the requirement for crystal growth perfectly, grown Y123 crystals were obtained in alumina and magnesia crucibles.

During the crystal pulling experiments, it often occurred that the surface of the melt was covered by the solid Y123. This means that the melt was supersaturated at the surface where the temperature is the lowest in the melt. Until the surface was completely covered, crystal pulling could be performed. Figs. 1 and 2 show the photographs of the specimens pulled by this method. It is obvious that the solidification started at the seed. Fig. 3 shows the longitudinal section of the sample shown in fig.2. Y123 crystals and the seed are indicated as black areas. Gray-looked region shows other phases solidified from the solvent. Observing through a polarized microscopy, Y123 crystals were grown from the Sm123 seed with almost the same orientation. However, there were some crystals which had different orientations. It is supposed that these crystals were nucleated independently of the seed. In order to grow continuously from the seed and prevent further nucleation, it is necessary to decrease supersaturation. There are two ways of decreasing supersaturation. One is increasing temperature and another is increasing the amount of the solvent.

In conclusions, we have successfully grown Y123 crystals from the melt by the crystal pulling technique. Some of Y123 crystals were considered to be grown epitaxialy to the seed made of Sm123. Alumina and magnesia crucibles were selected because they could maintain the melt and Y123 crystals could be obtained in these crucibles. It is clear that the possibility of continuous crystal growth is promising by crystal pulling.

ACKNOWLEDGMENT

This work was supported by New Energy and Industrial Technology Development Organization under the management of the R&D of Basic Technology for Future Industries.

REFERENCES

1. Szbo Gy, Juhasz z, Keszei b, Gosztonyi L, Vandlik J, Battisting G, Sandor S (1988) Physica C 153-155: 415
2. Gencer F, Abell J S, (1991) J. Crystal Growth 112: 337
3. Yaoshui Wang, Bemmema P, Van Der Linden P (1990) J. Crystal Growth 106: 483
4. Sceel H J (1988) Physica C 153-155: 44
5. Licci F (1991) J. Crystal Growth 112: 600
6. Wei Zhong, Osamura K, Ochiai S (1990) J. Am. Cerm. Soc. 73: 1958

Fig. 1 The photograph of the seed holder, seed and pulled crystals

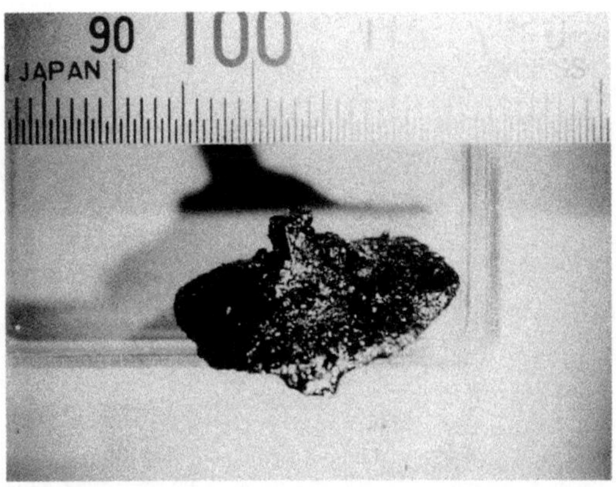

Fig. 2 The photograph of the pulled crystals

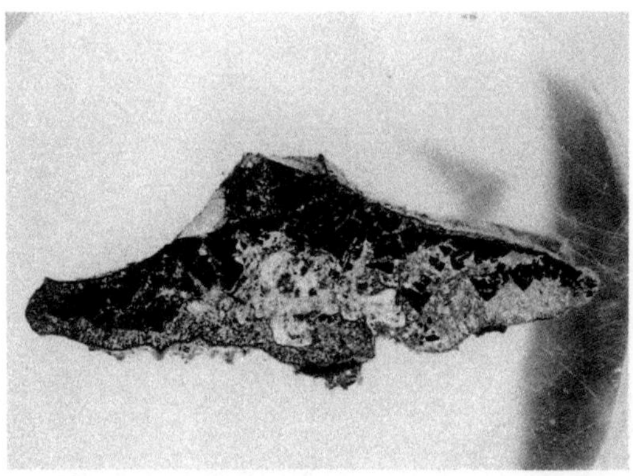

Fig. 3 The longitudinal section of the pulled crystals shown in fig. 2

Use of a Thermal-Gradient Furnace to Study the Effects of Liquid Phase on the Formation of Bi-HTSC

PETER E.D. MORGAN, JONATHAN D. PICHÉ, and ROBERT M. HOUSLEY*

Rockwell International Science Center, Thousand Oaks, CA 91360, USA

ABSTRACT

Firing a long specimen in a thermal gradient furnace is an efficient way of looking for the fine details of synthesis/morphology behavior in Ag-"doped" (Bi,Pb)-HTSC as a function of temperature. "2223" type (110K phase) begins to form only from a eutectic-liquid at ~822°C. Ca_2PbO_4 dissolves in this liquid and, on cooling, some "2201" precipitates. The accompanying morphological changes are readily seen and interpreted by SEM/EDS of fractured surfaces

KEY WORDS: Bi-HTSC, liquid phase effects, eutectic temperature, morphology, thermal-gradient.

INTRODUCTION

Liquid plays a crucial role in the synthesis of all 3-Cu-layer superconductors, "2223", [1**]. Although the 2-Cu-layer analogues, "2212", seem able to form in the solid state, "2223" appears more slowly, apparently only by solution-reprecipitation (S-R), from a liquid. This encourages grain-growth which can improve the weak-link properties but what effect this may have upon flux pinning is not precisely known. The need for liquid limits the purity of the final product, because, on cooling, the liquid inevitably solidifies into generally undesirable phases which are not by themselves flux pinning sites; much finer scale defects [3], perhaps *associated* with second phases, must be contemplated. It will be best to carry out the synthesis so that, on cooling, the amount of liquid remaining to crystallize will be minimal. This requires a precise knowledge of the lowest eutectic temperature, which is difficult, or impossible, to measure by other methods such as DTA.

Using a thermal-gradient (T-G) furnace and observing the morphological changes as a precise function of temperature, it is possible accurately to determine this important temperature. Moreover, changes in the eutectic temperature and the nature of the eutectic liquid with additives can also be detected. Additives in low amounts may radically effect such a eutectic [4], even moving it to a different region of the phase diagram. As only a percent or so of a liquid can be effective, fractions of a percent of an additive (e.g. Ag, perhaps Sb, etc.) may produce observable changes. Although Ag hardly enters the lattices of these superconductors, it does enter the eutectic liquid [1,4] depending upon the partial pressure of oxygen.

In the reacted Tl-2223-case, the first liquid to form is Ba-Cu rich at ~890°C [5] whereas, on adding silver, we believe the first liquid is Cu-Ag rich at ~830°C [4]; in the nominal Pb-doped Bi-2223-case, with Ag also present, the lowest liquid is Pb-Bi-Ag rich [1] and in the Tl-Pb-Ba-Sr-Ca-Cu-1223 situation the first liquid is probably Ba-Pb rich at ~880°C [6]. Eutectic liquids containing silver deposit metallic silver [4] on cooling and this has frequently been misunderstood as evidence that Ag is inert [e.g. 7].

We anticipate that processes that involve large scale melting (the hare!) will eventually make way for more controlled methods where only low levels of liquid will be utilized (the tortoise!). Yet much experience, data and understanding needs to be acquired for such control.

---

*Also visiting professor, California Institute of Technology, Pasadena, CA.

**Small liquid droplets [1] have now been seen for the Tl-case also [2].

EXPERIMENTAL

Precursor powders were made from all-nitrates after we positively confirmed that the presence of low levels of carbonates, as in mixtures made from oxides and carbonates, was extremely deleterious to the formation of desired products at the lowest possible temperatures. Mixed solid powdered nitrates of Bi, Pb, Sr, Ca, Cu and Ag, for the nominal stoichiometry of $Bi_{1.84}Pb_{0.34}Sr_{1.9}Ca_{2.0}Cu_{3.1}O_x/Ag_{0.02}$ [8], were blended for a short time in a very thick slurry with concentrated nitric acid. This slurry was placed in a pyrex dish on a heater plate and slowly heated up while gentle hand-stirring with a glass rod. The solid substantially melted, turning from blue to green, as nitric acid and water-of-crystallization evaporated off. After the greenish mix had resolidified (~170°C) it was placed in a vacuum oven at up to ~220°C where, while remaining solid, it further lost nitrate becoming a brownish/black color. At this stage a few per cent of copper was lost as a result of the well known volatility of copper nitrate [9], not commented upon by most previous observers using nitrate mixtures. The blue/green deposits seen in the outlet tube from the vacuum-oven and black deposits around the rim of the glass panel door showed predominantly copper with low calcium but no other elements by EDS. The powder was lightly milled and prefired in air (no evidence of significant carbonate pick-up) to 600-800°C overnight. With prefiring at 600°C substantial nitrate remained and caused problems. At 700°C some nitrate still remained but this was a good powder for finding the lowest temperature at which "2212" compound formed. The 800°C powder was chosen for the careful T-G work; it contained "2212" as a major phase with minor Ca_2PbO_4 and very minor Ca, Sr, Cu oxides, but no "2201" (attesting to the high activity of the powder) or "2223" by XRD.

The lightly hand-ground powder was tamped into one-ended MgO tubes that were 8mm in inner diameter and 40cm long (Al_2O_3 tubes were tried and found to chemically react!). With the open ends toward the cooler end of the furnace, overlapping tubes were placed along the length of a T-G (~2°C/cm) furnace. For the regions of key interest the furnace was subsequently calibrated with thermocouples precisely placed at the important points. Samples were heated to operating temperature-gradient over a 24h period and then held for 100h at constant-gradient, ~900°C at the hot end, ~750°C at the cooler end (temperatures are estimated to be well within ±2°C). After slow cooling, due to the large thermal inertia of the furnace, the tubes were marked, to indicate the placement in the furnace, then broken open carefully and the specimens extracted. The inside of each specimen was first powdered and XRD, using CuK_α radiation, was carried out. Emphasis was placed on the (002) lines of the "2212" (at 5.8° 2Θ) and the "2223" (at 4.9° 2Θ). Fractured surfaces were examined by SEM/EDS whereby the morphological changes, manifesting the presence or absence of liquid, were observed on these low density materials.

RESULTS AND DISCUSSION

Fig.1 shows the ratios of the (background subtracted) heights of the "2223" and "2212" X-ray peaks as a function of temperature; a ratio is used to minimize possible variations in preferred orientation of the platelet powder on the X-ray specimen, which should proportionately effect the strengths of the (002) peaks in a like-manner. The ratio rises rapidly from zero above ~820°C. "2212" is already well formed below 750°C in all-nitrate mixes (lower than reported in other work, [e.g. 10,11[3]]).

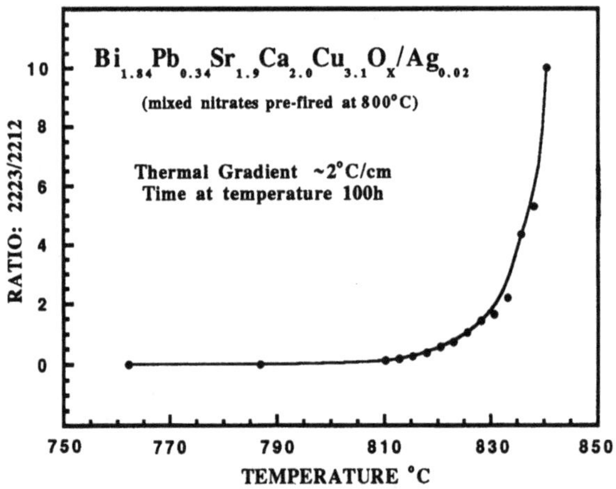

Fig. 1 Height of "2223" (002)/ height of "2212" (002) vs firing temperature, line for eye guidance only.

At this point we provisionally assumed (in light of earlier work [1]) that eutectic liquid formation, aside from perhaps low temperature

[3] In this ref., "At temperature above 850°C, the 37Å phase, $Bi_{2+x}(Sr,Ca)_{4-y}Cu_3O_{10}$ (aka "2223") *becomes stable* (my italics) for n≥3". They surely mean they see it *appear* at 850°C. Many papers, too numerous to mention, on phase studies make similar misleading statements. Long term anneals at lower temperature *might* reveal the lower temperature stability of "2223".

transient liquid, commenced at ~825°C. SEM pictures confirm this supposition as in Fig.2: the 821°C region of the specimen is uniform, friable and unreacted much as we would expect of an unaffected lightly-ground-up 800°C prefired powder with indistinct (presumably crushed) "2212" platelets. At

Fig. 2 Microstructural changes as a function of temperature over the eutectic region. Visible onset of "2223" crystallization, often as overgrowths, commences by 823°C and then continues.

823°C we are at a transitional point where some platelets are growing while other regions remain somewhat unreacted. At 826°C, the platelets are uniformly better formed. At 828°C the platelets are distinct, and droplet-related features [1] common, and occasionally there are larger platelets (but not much thicker!) in certain regions. At 831°C, while the XRD indicates increasing "2223", the microstructure now changes less, this must reflect more S-R than grain-growth. A diligent search now only occasionally turned up noticeable regions of concentrated silver, e.g. in the 826°C sample, as might have been present in liquid droplets [1]; we believe this to be due to the slow cooling, allowing more diffusional annealing than in the earlier sample [1] which had been rapidly cooled by removal from the hot furnace to ambient. In the present samples we also don't see such distinctive and obvious frozen droplet features as clearly as before [1] and perhaps Ag has been able partially to diffuse away and maybe enter the Bi layers of the Bi-HTSC. Higher temperatures were not pursued because chemical reaction between the increasing liquid and the MgO tube became evident by ~850°C.

That such a major change occurs between 821°C and 826°C, i.e. well within 5°C, although not as sharp as the <<1°C for the Tl-case [4], implicates a liquid. We believe that the first liquid appeared at ~822°C although this might be only in special regions where perhaps e.g. Pb, Bi, and Ag were more concentrated; accelerated platelet growth and formation of "2223", initially probably as overgrowths, then commences. This is slightly below where "2223" has been seen to form before at 830°C where there was no silver [12] and also for no Pb [10]. Liquid is widespread at low levels, at 825°C. At higher temperatures average crystal growth proceeds but mostly by extension of the larger grain regions at the expense of the smaller ones rather than overall growth. As the platelets do not so much thicken, growth of the new generation is largely by extension of some grains in the a/b direction, with the necessary disappearance of others.

CONCLUSIONS

The presence of eutectic liquid in the Ag-doped Bi,Pb-2223 system is easily spotted at >822°C by SEM studies of the morphology of non-dense specimens fired in a T-G. "2212" converts to "2223" with Ca_2PbO_4 rather obviously dissolving in the liquid. This coincides with accelerated platelet growth, which is more gradual in the Bi-case than for the Tl-case, even after 100h firing, extending over ~20°C. This could be a manifestation of the sluggish removal of "2212" which has already been overgrown by "2223" [1]. Perhaps, given longer times, the change would appear more sharp. "2201" and other phases precipitate from the liquid on cooling. As it must dissolve and convey all the main 5 elements for S-R, the precise nature, elemental content etc., of the low-level liquid, will very sensitively effect the kinetics of "2223" formation.

REFERENCES

1. Morgan PED, Housley RM, Ratto JJ (1991) Physica C 176: 279
2. Nabatame T. Hitachi research lab., personal communication.
3. Hylton TL, Beasley MR(1990) Phys. Rev. B 41: 11669.
4. Morgan PED, Okada M, Matsumoto T, Soeta A (1990) In: Ishiguro T, Kajimura K (eds) in Advances in Superconductivity II. Springer-Verlag, Tokyo, pp 435-438
5. Ratto JJ, Porter JR, Housley RM, Morgan PED (1990) Jpn. J. Appl. Phys. 29: 244
6. Doi T, Morgan PED, Yuasa T, Porter JR, work in progress.
7. Kim CJ, Hahn MS, Suhr DS, Kim KB, Lee HJ, Lee HG, Hong GW, Wong DY (1991) Matls. Lett. 11: 79
8. Kanai T, Kamo T, Matsuda S-P (1991) In: Kajimura K, Hayakawa H (eds) in Advances in Superconductivity III. Springer-Verlag, Tokyo, pp. 547-50
9. Stern KH (1972) J. Phys. Chem. Ref. Data 1: 747
10. Schulze K, Majewski KP, Hettich B, Petzow G (1990) Z. Metallkde. 81: 836
11. De Guire MR, Bansal NP, Farrell DE, Finan V, Kim CJ, Hills BJ, Allen CJ (1991) Physica C 179: 333
12. Oh SS, Osamura K (1991) Supercond. Sci. Tech. 4: 239

HISTORICAL POSTSCRIPT

Lest it be supposed that using a thermal gradient is new, this standard technique has been used for many years for a variety of materials: alloys, glasses, ceramics etc. [13-18].

13. Tammann G (1914) Z. Anorg. Chem. 87: 248
14. Tesche O (1928) Z. tech. Physik. 9: 419
15. Steger W (1936) Ber. deut. keram Ges. 17: 177
16. Darken LS, Gurry RW (1945) J. Am. Chem. Soc. 67: 1398
17. Stone RL (1953) J. Am. Ceram.Soc. 36:140
18. Welterlen JD, DuFresne RS (1957) J. Am. Ceram.Soc. 40: 457

4 Flux Pinning

4.1 Flux Pinning Properties

Flux Pinning Behavior in Twin Boundaries of Single Crystal $YBa_2Cu_3O_{7-\delta}$

WAI-KWONG KWOK, STEVE FLESHLER, ULRICH WELP, VALERII M. VINOKUR, JOHN DOWNEY, and GEORGE W. CRABTREE

Materials Science Division and Science and Technology Center for Superconductivity, Argonne National Laboratory, Argonne, IL 60439, USA

ABSTRACT

We present magnetoresistance measurements up to H = 6 Tesla using a crossed magnetic field technique to demonstrate pinning by twin boundaries (TB) in $YBa_2Cu_3O_{7-\delta}$. We investigate geometries for H ‖ **c** ‖ TB, H ‖ **ab** ‖ TB and for H rotated within a twin boundary plane. We find the critical angle θ^* for TB pinning to be field dependent and field independent for H ‖ **c** ‖ TB and H ‖ **ab** ‖ TB respectively in fields up to 6 Tesla and temperatures near T_c. In addition, we report on the full phase diagram for the onset of TB pinning. The TB pinning onset temperature coincides with the 'shoulder' feature observed in the resistive transition in magnetic fields in these crystals.

KEY WORDS: $YBa_2Cu_3O_{7-\delta}$ superconductor, transport, pinning, twin boundaries

INTRODUCTION

The resistive transition in magnetic field of the high T_c superconductors displays a distinct 'fan shaped' broadening well below the onset of superconductivity [1]. This broadening has generally been attributed to large fluctuations [2] near the onset T_c and to dissipation by vortex motion [3] in the tail of the resistive transition. Defects in the crystalline structure or inhomogeneity in the material may inhibit this vortex motion, leading to zero resistance at lower temperatures. Among these defects, the most salient feature in $YBa_2Cu_3O_{7-\delta}$ is twin boundaries which are formed in the <110> and <1$\bar{1}$0> crystallographic directions to accomodate the strains induced as the material is cooled through its tetragonal to orthorhombic phase near 700°C during its synthesis. The twin boundaries are approximately 10-30Å wide [4] nonorthorhombic regions where oxygen atoms may be displaced from their ideal positions due to strain fields. The twin boundaries are probable locations of atomic defects such as oxygen vacancies and Cu atom displacements which may act as pinning centers. These boundaries separate regions of the crystal characterized by the alternation of the **a/b** axis in the basal plane of the crystal which can be clearly observed under a polarized light microscope. In this paper, we report on the direct measurement of the anisotropic flux pinning by twin boundaries in a high quality single crystal of $YBa_2Cu_3O_{7-\delta}$. We employ a novel crossed magnetic field technique where the resultant magnetic field vector of two orthogonal superconducting magnets is rotated to obtain angular dependent magnetoresistance. We observe a critical angle θ^* for twin boundary pinning characterized by a sharp drop in resistance as the magnetic field is rotated across a twin boundary plane. In addition we also extract a phase diagram for the onset of anisotropic twin boundary pinning. For magnetic fields in the c-axis, we can correlate the onset of twin boundary pinning with the 'shoulder' feature commonly observed in the resistive transition curve in magnetic field.

EXPERIMENT

The sample was prepared by a self-flux method described elsewhere [5] which yielded platelet single crystals. From a large batch, a crystal with a single domain of twin boundaries oriented only in the <110> direction was selected. The crystal edge corresponded to either the **a** or **b** crystallographic axis such that twin boundaries intersected the crystal edge at a 45° angle. The crystal was cleaved along the c-axis, such that the twin boundaries intersected the edge of the crystal at nearly 90°. The final dimensions of the crystal were 0.7 x 0.2 x 0.02 mm^3. AC resistivity was measured by the standard four probe method with silver conducting pads sintered onto the crystal for low contact resistances. Typical measuring current density was ~0.7 A/cm^2 at 17 Hz in the **ab** plane of the crystal, directed perpendicular to the twin boundaries. This particular crystal showed zero field critical temperature T_c of 90.3K with transition width ΔT_c<200mK. The crystal was placed in the bore of two superconducting magnets: a 1.5 Tesla split coil magnet used to generate a transverse magnetic field (H_T), which resided in the bore of an 8 Tesla solenoid magnet contributing a longitudinal magnetic field (H_L). A resultant magnetic field at an arbitrary angle with respect to the sample is obtained by separately energizing these two orthogonal magnets. Extremely high angular resolution of $\Delta\theta$~0.005^o may be obtained by adjusting the two magnet currents. This technique was recently used to study intrinsic pinning in these materials [6].

We investigated three geometries of the magnetic field orientation with respect to the twin boundaries with this setup as shown in Fig. 1. In the first geometry, the sample was first mounted such that the transverse magnetic field was parallel to the **ab** plane of the crystal. The sample was then rotated mechanically about the **c** axis. The angular dependence of the resistance for H=1.5 Tesla and T=89.44K is shown in Fig. 2. We see a clear $\sin^2\theta$ modulation of the resistance due to

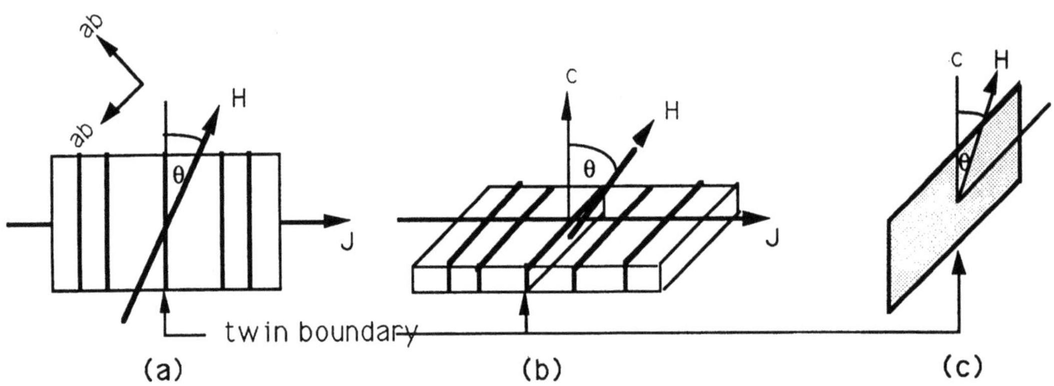

Fig. 1 Orientation of magnetic field and twin boundary for three geometries (a) H || **ab** and (b) H || **c** and (c) H rotated within a twin boundary plane. The measuring current is in the **ab** plane of the crystal.

flux flow effects originating from the Lorentz force induced flux motion as reported in our earlier work [7]. A very sharp drop in resistivity is observed close to θ~0° when the magnetic field is aligned parallel to the twin boundaries. For as-grown crystals in our previous studies, this drop in resistivity was observed at θ~45°. However, due to the orientation of the twins in this sample, the drop occurs near the point of maximum Lorentz force. This makes the comparison of the anisotropy of twin boundary pinning for H || **ab** and H || **c** more consistent since in both geometries we consider the case for $H \perp I$ (ie. maximum Lorentz force). The Lorentz force for H || **ab** is directed in the twin boundary plane along the crystallographic **c**-axis. The sharp drop in resistance may be analyzed in terms of a critical angle θ^* for twin boundary pinning, which describes the field misalignment which can be tolerated without causing the twin boundary pinning to disappear. For slightly misaligned fields, the vortex can take advantage of the pinning energy $\Delta\varepsilon$ at the cost $1/2\varepsilon_t\theta^2$ of bending away from the field direction by an angle θ where ε_t is the tilt modulus for a single vortex line. The critical angle θ^* will then be proportional to $\sqrt{2\Delta\varepsilon/\varepsilon_t}$. With estimates for the pinning energy, twin density, and vortex-vortex interaction, Blatter, Rhyner and Vinokur [8] obtained a critical angle of ~ 5° at H=1 T, in close agreement with our measured critical angle of θ^*~3.5^o.

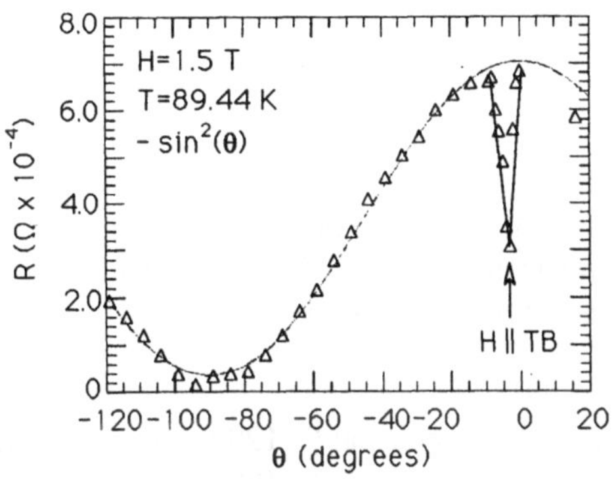

Fig. 2 Angular dependence of the resistance of $YBa_2Cu_3O_{7-\delta}$ at fixed field and temperature for H rotated in the **ab** plane.

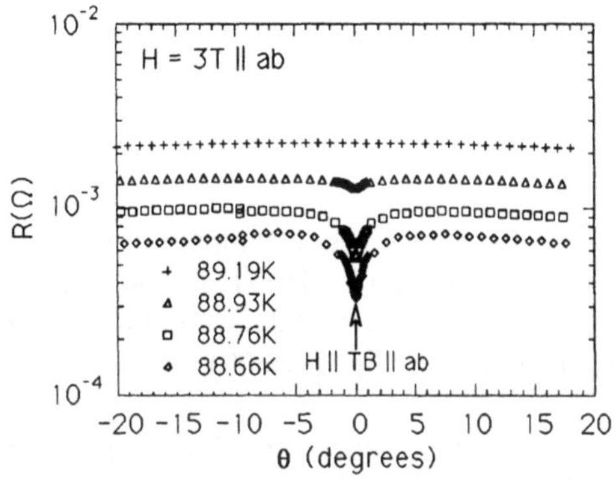

Fig. 3 Angular dependence of the resistance near H || TB || **ab** at 3T and different temperatures.

Using the crossed magnetic field technique, we investigated the temperature dependence of the critical angle and the sharp drop in resistance. We remounted the crystal such that the longitudinal magnetic field was in the **ab** plane, parallel to the twin boundaries but perpendicular to the measuring current, and the transverse magnetic field H_L was also in the **ab** plane but parallel with the measuring current. By energizing both magnets separately, the resultant magnetic field was swept through small angles about the twin boundary plane. Angular dependent magnetoresistance plotted as log R versus the angle (θ) between the applied magnetic field and the twin boundary for H=3T in the **ab** plane for various temperatures is shown in Fig. 3. At temperatures just below T_c=90.3K, we do not observe any pinning by twin boundaries. However, with gradually decreasing temperatures, we see the onset of twin boundary pinning for temperatures below T=89.19K. The critical angle θ^* defined as half the angular distance between the two shoulders on either side of the sharp resistance drop, remains virtually constant. However, the strength of the pinning increases with decreasing temperatures quite rapidly, as shown by the rapid growth of the valley at $\theta=0^o$. With this method we determined the onset of twin boundary pinning for H || **ab** || TB.

An alternate method to determine the onset point is to measure the temperature dependence of the resistivity at fixed field and angle at H || TB ($\theta=4^o$ in Fig. 2) and at H misaligned from the twin boundary by more than the critical angle θ^*($\theta=0^o$ in Fig. 2) and determine the temperature at which the resistive curves deviate. This point is shown in Fig. 4 where we have expanded the region of the tail of the resistive transition in a field of 3 Tesla. The onset temperature of twin boundary pinning determined with this method is approximately 89.2K, in close agreement with the value obtained using the crossed magnetic field technique.

For the second geometry of Fig. 1, the crystal was mounted such that H_L || c || TB and H_T || ab || I. By energizing the two magnets, the resultant magnetic field was tilted from the **c** axis parallel to TB towards the **ab** plane. The angular dependence of the resistance at H=3T for several fixed temperatures is shown in Fig. 5. We observe a much larger critical angle θ^* for this field direction in addition to a weaker temperature dependence R(TB)/R(θ^*). We also note a large critical angle dependence on the magnetic field for this field direction as demonstrated in Fig. 6, where the angular

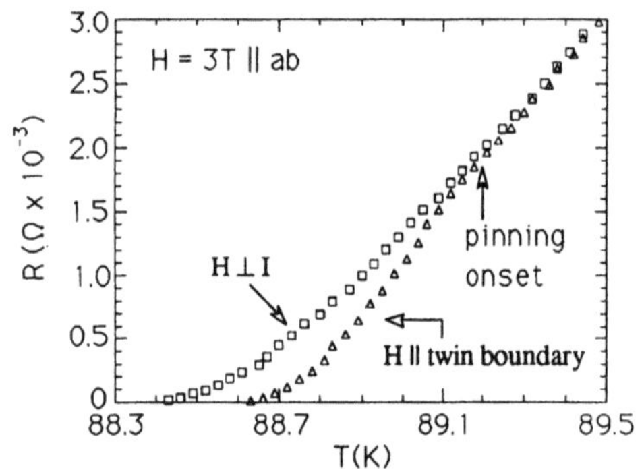

Fig. 4 Resistance versus temperature near the foot of the transition for H=3T, for two different geometries, H || twin boundaries and H ⊥ I (max. resistance point in Fig. 2, $\theta=0^o$).

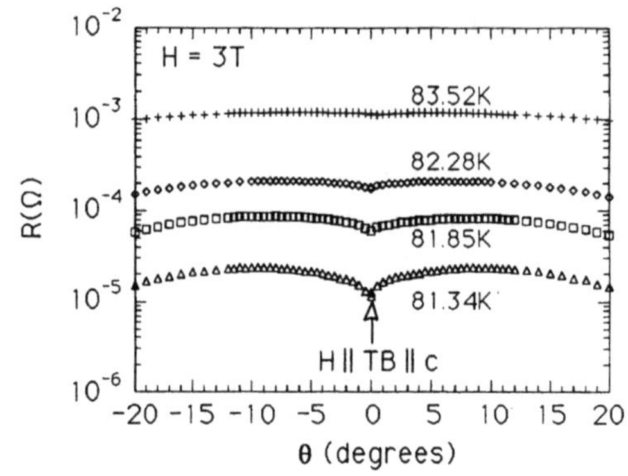

Fig. 5 Angular dependence of the resistance near H || TB || **c** at 3T for several temperatures.

dependent resistance curves for H=1T and 4T normalized to their values at $\theta=12^o$ are shown. This can qualitatively be explained due to the increase in tilt modulus ε_t with magnetic field strength at a fixed angle while the pinning energy remains constant, at least in the narrow temperature range near T_c under consideration and hence the critical angle decreases accordingly. The larger critical angle for H || **c** may be qualitatively understood in light of the difference in the tilt modulus of the vortices for H || **ab** and H || **c**. For an anisotropic effective mass tensor, the angular dependence of the vortex line energy for a single vortex is given by [10]:

$$\varepsilon(\phi) = \varepsilon_o\left(\frac{\sin^2(\phi)}{\Gamma} + \cos^2(\phi)\right)^{1/2} \qquad (1)$$

where $\varepsilon_o=[\Phi^2/(4\pi\lambda)^2]\ln\kappa$, Γ is the effective mass ratio, and ϕ is the angle tilted from the **c** axis. For H || **ab**, $\phi=\pi/2$ and $\varepsilon(\pi/2) = \varepsilon_o/\sqrt{\Gamma}$. The energy to tilt a single vortex line within the **ab** plane can be expressed as:

$$\frac{1}{2}\varepsilon_t\theta^2 = \varepsilon(\phi)(1+\theta^2)^{1/2} - \varepsilon(\phi) \qquad (2)$$

Using $(1+\theta^2)^{1/2} \sim 1+\frac{\theta^2}{2}$, we obtain $\varepsilon_t \sim \varepsilon_o/\sqrt{\Gamma}$. On the other hand, for H || **c**, a Taylor expansion about $\phi=0$ yields $\varepsilon_o(1 - \frac{\theta^2}{2} + \frac{\theta^2}{2\Gamma})$. Substituting into equation (2), we obtain $\varepsilon_t \sim \varepsilon_o/\Gamma$ for a small angle tilt about the **c** axis. Thus if we assume from our simple model, θ^* is proportional to $\sqrt{2\Delta\varepsilon/\varepsilon_t}$, then we would expect a smaller critical angle for H || **ab** compared with H || **c**. However, the absence of the field dependence of the critical angle for H || **ab** suggest that vortex-vortex interaction energy may also have to be taken into consideration due to the anisotropy of the penetration depth λ which is larger for H || **ab** than H || **c**, and to the rapid change in the ratio a_o/d with magnetic field. Further refinement of the model is required to explain these results.

In the third geometry (see Fig. 1), the sample was mounted such that H_L || **c** || TB and H_T || **ab** || TB The resultant magnetic field was rotated within a twin boundary plane as shown in Fig. 7 for H=4T and T=80.23K. The angular sweep for H crossing perpendicular with the twin boundary at H || **c** is also shown for reference. For magnetic field rotations in the twin boundary plane, the resistance always remains lower than field rotations crossing a twin boundary, suggesting that the vortices remain pinned as long as the entire length of the vortex lies within the twin boundary plane regardless of the orientation.

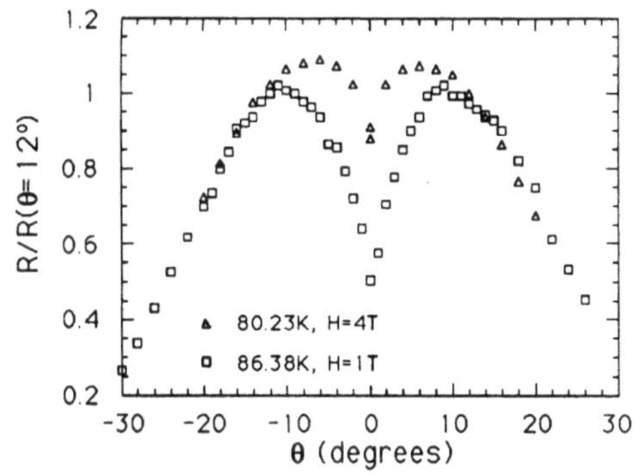

Fig. 6 Magnetic field dependence of the critical angle for H ll c for 1T and 4T.

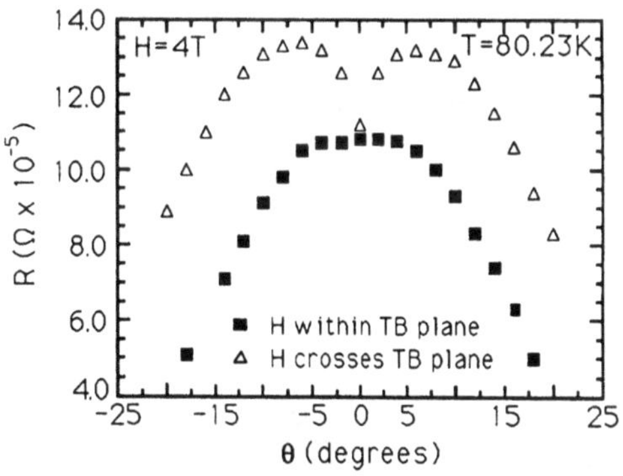

Fig. 7 Angular dependence of the resistance for (closed squares) H rotated within a twin boundary plane from **c** to **ab**, and (open triangles) H rotated from **c** to **ab**, crossing a twin boundary, as in Fig. 6.

Using the crossed magnetic field technique as shown above in Figs. 4 - 6, we determined the phase diagram for the onset of twin boundary pinning for the two geometries, H ll **ab** ll TB and H ll **c** ll TB as shown in Fig. 8. The zero resistance points of the resistive transition at fixed fields for H ll **c** which have been associated with the irreversibility or depinning line are also shown for comparison and lie much below the twin boundary pinning onset curve. In high quality single crystals of $YBa_2Cu_3O_{7-\delta}$, several unique features in the resistive transition in magnetic field have been reported. Among them, a shoulder near the tail of the resistive transition characterized by a nearly universal behavior above this shoulder and different resistive behavior below the shoulder between twinned and untwinned crystals has been observed [11]. Remarkably, the onset of twin boundary pinning can be correlated with the inflection point in the resistance versus temperature curve occurring at the 'shoulder' observed in all high quality single crystals of $YBa_2Cu_3O_{7-\delta}$. These points are shown as arrows in Fig. 9.

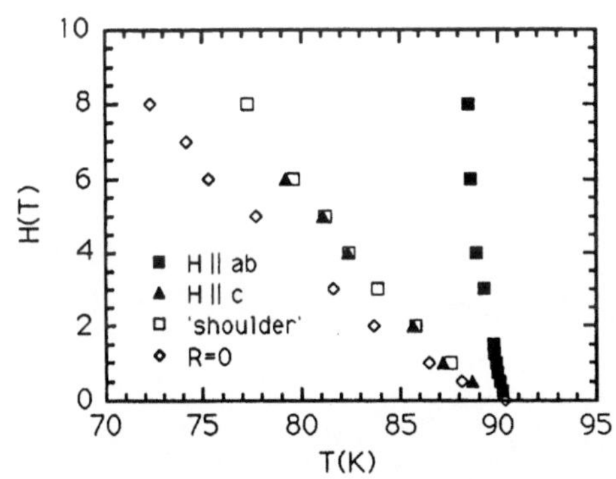

Fig. 8 Phase diagram for the onset of twin boundary pinning for H ll **ab** ll TB and H ll **c** ll TB. Also shown is the inflection point of the 'shoulder' from the R vs T(H) curves of Fig. 9 and the point of zero resistance for H ll **c**.

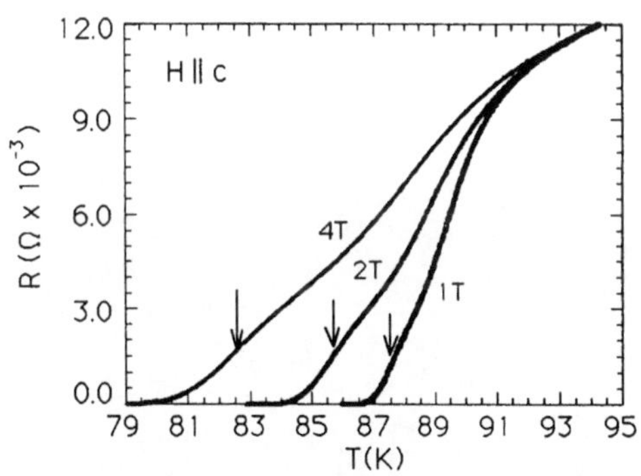

Fig. 9 Resistance versus temperature for Hll **c**. The arrows depict an inflection point due to the 'bump' in the resistive transition.

In summary, we have shown evidence of strong twin boundary pinning close to T_c in $YBa_2Cu_3O_{7-\delta}$ whenever the magnetic field is aligned within a critical angle θ^* of the twin boundary plane. The critical angle for twin boundary pinning is larger for H ll **c** than for H ll **ab**, and field dependence was observed only in the former case. Using a crossed magnetic field technique, we determined the phase diagram for the onset of twin boundary pinning. For H ll **c**, the onset temperature coincides with the downturn at the 'shoulder' observed in the resistive transition in magnetic field, suggesting that vortex pinning mechanisms set in at temperatures above the irreversibility line.

ACKNOWLEDGEMENT

This work was supported by the U. S. Department of Energy, BES--Materials Science under contract #W-31-109-ENG-38 (WKK, VMV, SF, JD, GWC) and the NSF-Office of Science and Technology Centers under contract #STC8809854 (UW) Science and Technology Center for Superconductivity.

REFERENCES

1. Iye Y, Tamegai T, Takeya H, Takei H (1988)Technical Report of the Institute for Solid State Physics, Univ. of Tokyo, Ser. A, No. 1919
2. Tsuneto T (1988) J. Phys. Soc. Jpn. 57: 3499
3. Palstra T T M, Batlogg B, van Dover, R B, Schneemeyer L F, Waszczak J V (1988) Phys. Rev. Lett. 61: 1662
4. Zhu Y, Suenaga M, Xu Y, Sabatini R L, Moodenbaugh A R, (1989) Appl. Phys. Lett. 54, 374
5. Kaiser D L, Holtzberg F, Chisholm M F, Worthington T K (1987) J. Cryst. Growth 85: 593
6. Kwok W K, Welp U, Vinokur V M, Fleshler S, Downey J, Crabtree G W, (1991) Phys. Rev. Lett. 67: 390
7. Kwok W K, Welp U, Crabtree G W, Vandervoort K G, Hulscher R, Liu J Z (1990) Phys. Rev. Lett. 64: 966
8. Blatter G, Rhyner J, Vinokur V M (1991) Phys. Rev. B43: 7826
9. Farrell D E, Rice J P, Ginsberg D M, Liu J Z (1990) Phys. Rev. Lett. 64:1573
10. Kogan V G (1981) Phys. Rev. B24: 1572
11. Crabtree G W, Kwok W K, Welp U, Downey J, Fleshler S, Vandervoort K G, Liu J Z (1991) Proceedings of M^2S-HTSC, Kanazawa, Japan. Physica C (to be published)

Flux Pinning and Flux Creep in Neutron Irradiated $(Y,Gd)Ba_2Cu_3O_x$

J.O. Willis[1,*], K.E. Sickafus[2], and D.E. Peterson[2]

[1] Superconductivity Research Laboratory, ISTEC, Koto-ku, Tokyo, 135 Japan
[2] Los Alamos National Laboratory, Los Alamos, NM 87545, USA

ABSTRACT

Powder samples of $Y_{0.9}Gd_{0.1}Ba_2Cu_3O_x$ were irradiated with mixed spectrum (~50% E<0.5eV, 50% E>0.5eV) neutrons with most interactions expected to occur at the Gd site. As a function of fluence the samples showed increased (~X3-X8) magnetically measured critical current densities J_c at low fluences, falling off at the highest values. An analysis of magnetic relaxation data, which allows for a nonlinear pinning potential U vs J relationship, revealed substantial increases in U at constant J, indicating that the irradiation introduced more effective pinning centers than those originally present.

KEY WORDS: Y123, neutron irradiation, flux creep, flux pinning, pinning potential

INTRODUCTION

The anomalously large magnetic relaxation properties of the high temperature superconductors are thought to be a result of the very small size of the coherence length ξ_o, and thus the volume of the vortex core, which limits the size of the pinning energy and of the generally higher thermal fluctuation energies at the temperatures at which measurements are made. This phenomenon has most often been treated as thermally activated flux creep, but models of flux lattice melting at high temperatures and fields have also been proposed. For both fundamental and practical reasons, the question of whether the pinning potential can be varied by processing, doping, or by introducing controlled damage is important. It is expected to restrict the range of theories and to suggest routes for the development of applications.

Radiation damage, particularly with neutrons, is known to be an effective method for varying the number (and type) of defects in superconductors[1]. For fast neutrons the source of the change in J_c is thought to be pinning centers produced in the form of collision cascades which consist of crystalline defects of 5 to 10 nm in size[1,2]. These are close to the optimum in size to produce strong pinning because they are about the same size as the diameter of a vortex core which has an effective diameter of the superconducting coherence length ξ_o. In a Gd doped sample [$Y_{0.9}Gd_{0.1}Ba_2Cu_3O_x$, hereafter called (Y,Gd)123], thermal neutron collisions are expected to occur almost exclusively with the Gd nucleus because of its extremely large capture cross section, $<\sigma_{th}>=5x10^4$ barns. This is also much larger than typical fast neutron cross sections (1~10 barns) so that a large number of the interactions produced by a mixed energy spectrum, such as will be employed here, will be with the Gd nucleus. This nucleus decays by gamma ray emission and recoils with relatively low enough energy that, while it can be ejected from its lattice site, there is generally insufficient energy to produce secondary displacements. Here we report the effects of these mixed spectrum neutron irradiations on the superconducting properties, primarily J_c, magnetic relaxation, and the effective pinning potential U of (Y,Gd)123. Results of the microstructural examination by SEM and TEM are to be reported separately[3].

EXPERIMENTAL

The $Y_{0.9}Gd_{0.1}Ba_2Cu_3O_x$ sample was prepared by a standard solid state reaction route. It was then ground to a fine powder (average particle diameter 1.46μm by SEM observations) to allow for uniform irradiation by the neutrons. The material was divided into eight lots, each approximately 200 mg, and sealed into individual Al holders with the powder constrained in the form of a thin layer. These were irradiated for specific lengths of time ranging from 0 to 35 hours in the Omega West Reactor at the Los Alamos National Laboratory at a temperature of about 80 C. The neutron spectrum was mixed with about half of the intensity above an energy E of 0.5 eV and the remainder at lower energies.

The superconducting properties were determined by SQUID magnetometry (Quantum Design Model MPMS). The superconducting transition temperature T_c was determined from the onset of diamagnetism of the 10 Oe zero field cooled (ZFC) susceptibility. The midpoint is the temperature T_{mid} for which the susceptibility reaches 50% of its low T (7 K) value. The ZFC susceptibility remained constant at about 80% of perfect diamagnetism (after correction for the demagnetizing factor of a sphere) and the Meissner (field-cooled) susceptibility at about 30% until the fluence reached $2x10^{18}ncm^{-2}$, at which point the values began to drop, reaching 60% and 15%, respectively, at the maximum fluence. Clearly, the material is beginning to show the effects of significant damage at these higher fluences.

Magnetization hysteresis loops were measured at 7, 25, 45, and 75 K to a maximum field of 50 kOe on each sample using the "fast hysteresis mode." The critical current density values J_c were determined from these loops assuming the critical state model using the Bean formula for a cylinder $\Delta M=J_c d/30$, where ΔM is the full width of the hysteresis loop in emu/cm^3, d is the sample diameter in cm, and J_c is in A/cm^2. Relaxation measurements were performed by first zero field cooling the sample to the measurement temperature, raising the field to 10 kOe in the "no overshoot mode" and then recording the sample magnetization periodically for approximately 2 hours. A sample scan length of 3 cm was used and all samples were in the fully penetrated state at 10 kOe even at the lowest measurement temperature of 5 K, for which the peak in the magnetization was reached at about 4 kOe. Measurements were typically made at temperatures of 5, 10, 15, 20, 25 and 30 K. The data were corrected for the equilibrium and the larger paramagnetic magnetization, which were calculated as the average value of M(H) for H increasing and H decreasing.

RESULTS AND DISCUSSION

A list of parameters characterizing the samples is given in Table I. T_c is almost constant at the unirradiated value of 91 K until the highest dose is reached; in contrast, T_{mid} drops somewhat more gradually, but it too shows a sharp drop for the most highly irradiated sample. The maximum enhancements in J_c are a factor of 3.5 at 7 K and nearly 8 at 75 K. This is very similar to results observed previously for fast neutron irradiated polycrystalline material. The enhancement is more striking and the plateau narrower for the 75K data presumably because of its smaller absolute initial value and a sharper cut-off as T_c drops rapidly at high fluence.

The magnetization data show an approximately logarithmical time dependence away from the initial value imposed upon the ZFC sample by the application of the 10 kOe magnetic field. The standard model for flux creep in superconductors proposed by Anderson[4] and extended by Beasley, Labusch, and Webb[5] (BLW) has been extensively used to analyze data on high temperature superconductors[1,6]. The model assumes pinning of a flux vortex by trapping in a defect that has a pinning potential U(J, T). In the presence of the Lorentz driving force $F = J \times B$ and at finite temperatures, the magnetization creeps away from the value corresponding to J_c. In the simplest model, applicable for $U/kT >>1$, the approximation that U is linear in J can be made resulting in the familiar equation $Ueff(T) = kT/S(T)$, where $S(T)=dlnM/dlnt$, the normalized logarithmic relaxation rate, and t is time. Figure 1 is a plot of U_{eff} vs. T for all of the samples as derived from this linear approximation. The data below 25K for the unirradiated sample show approximately T-independent behavior, whereas the remainder of the data exhibit increases in U_{eff} with T and also with fluence. The $12x10^{18}$ n/cm^2 sample shows a different response at the higher temperatures, possibly related to its

Table I. Sample parameters of neutron irradiated (Y,Gd)123.

Fluence 10^{18} n/cm^2	T_c (ZFC 10 Oe) (K)	T_{mid} (K)	J_c (7K, 10 kOe) (10^6A/cm^2)	J_c (25K, 10 kOe) (10^6A/cm^2)	J_c (45K, 10 kOe) (10^6A/cm^2)	J_c (75K, 10 kOe) (10^3A/cm^2)
0	90.9	85.0	9	1.7	0.56	50
0.13	90.9	85.4	11	3.0	1.2	130
0.27	90.9	85.2	15	4.2	1.7	190
0.53	91.0	84.7	29	9.1	3.9	390
1.1	90.9	84.5	26	9.6	3.7	370
2.0	90.9	83.2	32	10.5	4.3	350
4.3	88.9	80.2	31	9.5	3.6	210
12	78.0	61.5	15	4.2	0.98	70

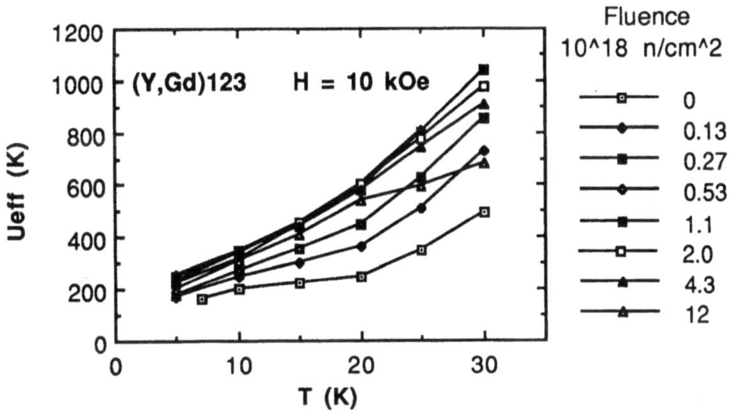

Fig. 1. The effective pinning potential U_{eff} of (Y,Gd)123 as obtained from the linear flux creep equation as a function of temperature T for the neutron fluences shown.

much lower T_c. However, because near J_c U_{eff} is very small ($U_{eff} = 0$ at J_c by definition), the linear approximation (U/kT>>1) is in general not correct, and unphysical results (U increasing strongly with T as seen here) are often obtained[6]. Those authors have also suggested that the linear approximation is incorrect here. Malozemoff and Fisher[7] have proposed a "universal" T-independent relaxation rate (equivalent to U_{eff} ~kT), S~0.03, for Y123 based on measurements on many different forms of Y123. The present data plotted in Fig. 2, gives a similar value after small amounts of radiation damage are introduced. It might be argued that the somewhat low values of J_c and U for the unirradiated sample have been enhanced to values characteristic of single crystals or thin films.

Alternative methods of analysis include models with a distribution of pinning energies[8] U(E) and a model which assumes a single pinning energy but allows for a nonlinear U(J) relation. We use the latter model due to Maley et al.[9] to further analyze the data. Using the rate equation of BLW: $dM/dt=(B\omega a/\pi d)\exp(-U_{eff}/kT)$ where ω and a are the hop frequency and distance for a flux bundle, and d is the grain diameter. This leads to the expression $U_{eff}/k = -T[\ln(dM/dt)-\ln(B\omega a/\pi d)]$ for the effective pinning potential. Here we have estimated the value of the constant $A= \ln(B\omega a/\pi d)$ empirically by plotting the data as U_{eff}/k vs M at each temperature for all of the samples and adjusting A to obtain the smoothest curve. The value A=20 (±3) implies physically reasonable values for the parameter ωa of 22 cm/s (i.e., for an a of 10 nm, ω becomes 2.2×10^7 Hz) and is in agreement with previous results[9,10]. If we include the implicit dependence of U on T using the physically reasonable approximation for U(T) given by Tinkham[11], $U(T) = U_0 g(T)$ where $g(T)=[1-(T/T_c)^2]/[1-(T/T_c)^4]^{-1/2}$, the maximum correction is about 10% for the highest temperature data. Including this implicit T dependence, Fig. 3 is a plot of U_{eff}

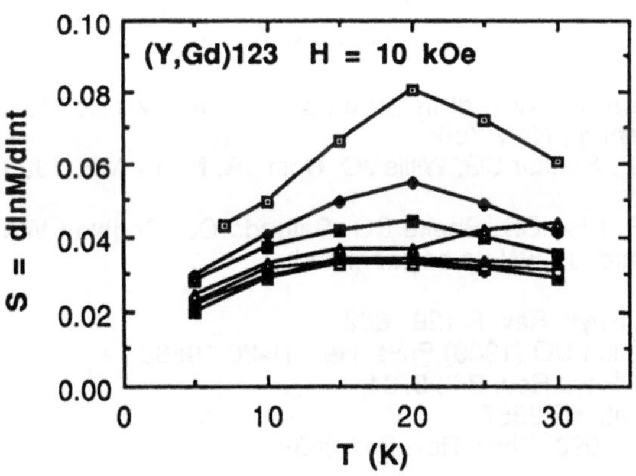

Fig. 2. The normalized logarithmic relaxation rate S(T)=dlnM/dlnt of (Y,Gd)123 plotted versus temperature T for the neutron fluences shown. The symbols used are the same as in Fig. 1.

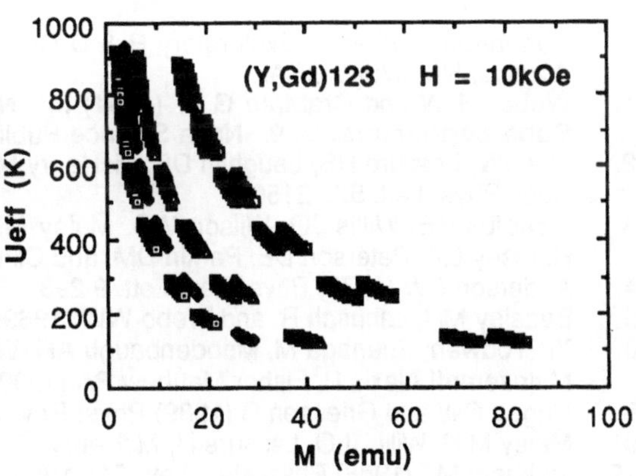

Fig. 3. The effective pinning potential U_{eff} of (Y,Gd)123 as obtained from the nonlinear equations for flux creep versus the magnetization M for the neutron fluences shown. The symbols used are the same as in Fig. 1.

versus M (or J). The temperature is an implicit parameter that allows different values of M to be probed through the temperature dependence of J_c. The results demonstrate the explicit nonlinear dependence of U_{eff} on M that make the use of the linear model questionable without careful analysis. They also show an enhancement of the pinning potential U as a function of neutron fluence for low and intermediate values, and a decrease at the highest values (160, 500, and 250K, respectively, at 30 emu/cc). The functional form of U(M) is approximately logarithmic, as observed by Zeldov et al.[12] in their experiments near T_c, at large values of M but deviates from this dependence at lower values of M. A fit to the data yields values for U_0 from 300 to 450 dropping back to 380K and M_c (corresponding to J_c) from 40 to 100 and then back down to 55 emu/cc as a function of fluence. A somewhat better fit to a power law was obtained with exponents ($U=U_0(M_c/M)^a$) varying from 1.3 to 1.7 and dropping to 1.36 as a function of fluence. It should be noted that both these dependences yield a U that goes to infinity as J goes to zero.

These results are in good qualitative agreement with those of Ref. 9 and with many of the results of Ref. 1. The work of Schindler[13] on fast neutron irradiation of Y123 single crystals present somewhat different results in that U_{eff} increases with irradiation independent of an increase or decrease in Jc. These differences are thought to be related to differing initial defect type and densities in the crystals and are not in conflict with flux creep theory.

CONCLUSIONS

Polycrystalline samples of $Y_{0.9}Gd_{0.1}Ba_2Cu_3O_x$ have been irradiated near room temperature with a mixed spectrum of neutrons with a median energy of 0.5 eV. The T_c, J_c, and U the pinning energy (via magnetic relaxation measurements) were measured. The linear flux creep model is shown to give unphysical results for U_{eff} as has also been observe in many other experiments. However, the normalized flux creep rate S(T) is in agreement with the model of Malozemoff and Fisher suggesting a universal mechanism for flux creep in all Y123 samples. The samples show clear enhancements in response to the irradiations for both J_c and U, when the latter is analyzed in terms of a nonlinear U(J) relation.

ACKNOWLEDGEMENTS

Work at Los Alamos National Laboratory was performed under the auspices of the US Dept. of Energy. JOW acknowledges partial support by the USDOE through the Exploratory Research and Development Center at Los Alamos.

REFERENCES

*Permanent Address: Exploratory R & D Center; MS K763, Los Alamos National Laboratory, Los Alamos, NM 87545 USA.

1. Weber H W and Crabtree G W (1991) In: Narlikar AV (ed) in **Studies of High Temperature Superconductors**, V. 9. Nova Science Publishers, New York.
2. Lee JW, Lessure HS, Laughlin DE, McHenry ME, Sankar SG, Willis JO, Cost JR, Maley MP (1990) Appl. Phys. Lett. 57: 2150
3. Sickafus KE, Willis JO, Wilson WB, Maley MP, Kise CD, Boeke SG, Salgado CJ, Coghlan WA, Ramsey GF, Peterson DE, Parkin DM, and Clinard, Jr. FW (to be submitted)
4. Anderson PW (1972) Phys. Rev. Lett. 9 293
5. Beasley MR, Labusch R, and Webb WW (1969) Phys. Rev. B 129: 682
6. Xu Youwen, Suenaga M, Moodenbaugh AR, Welch DO (1989) Phys. Rev. B420 10882.
7. Malozemoff Alexis P, Fisher Matthew PA (1990) Phys Rev. B42 6784.
8. Hagen CW and Griessen R (1989) Phys. Rev. Lett. 62 2857
9. Maley M P, Willis J O, Lessure H, McHenry M E (1990) Phys. Rev. B42 2639
10. Tinkham MT (1988) Phys. Rev. Lett. 61: 1658
11. McHenry ME, Simizu S, Lessure H, Maley MP, Coulter JY, Tanaka I, and Kojima H (in press) Phys. Rev. B
12. Zeldov E, Amer NM, Koren G, Gupta A, McElfresh MW, and Gambino RJ (1990) Appl. Phys. Lett. 56: 680
13. Schindler W (1991) J. Appl. Phys. 70 1877

Flux Pinning Study of $Y_xRE_{1-x}Ba_2Cu_3O_{7-\delta}$ Superconductors

D.A. GARVIE, T.R. FINLAYSON, and Z. PRZELOZNY

Department of Physics, Monash University, Australia, 3168

ABSTRACT

A series of substituted compounds has been made from powders produced by the freeze-drying process followed by pressing and sintering. The rare earths, Gd, Ho, Yb and Dy, have been substituted for the Y in various percentages. AC susceptibility and Vibrating Sample Magnetometry measurements have been made and the critical state model used to obtain a critical current and its temperature dependence for each of the compositions. For all substitutions an enhancement of flux pinning has been achieved. To date the absence of a specific trend with alloy concentrations, coupled with microstructural observations, suggest a secondary effect on flux pinning, mostly from the influence on sintering behaviour caused by the rare-earth addition.

KEY WORDS: AC susceptibility, rare earth substitution, flux pinning

INTRODUCTION

Most of the $REBa_2Cu_3O_{7-\delta}$ compounds have critical temperatures in the vicinity of 90K [1]. This suggests that the interaction between the superconducting electrons and the magnetic moments of the rare earth atoms, is very weak. The superconducting electrons are therefore expected to be primarily associated with the Cu–O polyhedral clusters [2].

As high as the critical temperatures may be for these compounds they will continue to be limited in their applicability until their critical current densities (J_c) can be significantly improved. Critical current densities of the order 10^6 A/cm^2 have been successfully achieved using thin film techniques, such as magnetron sputtering, and in single crystals. Polycrystalline samples of high critical current densities are very difficult to prepare. The granular, weak link nature and anisotropy effects [3] are contributing factors to this problem.

The aim of the experiments presented here, was not to develop a material with the highest J_c obtained, but rather to investigate, systematically, a way in which polycrystalline materials may be improved regardless of the processing technique involved. A disadvantage of powder metallurgy techniques is a disruption of the starting stoichiometry by mechanical separation during powder mixing or through the trapping of powder particles in the pores of the mortar and pestle. In the freeze–drying process a nitrate solution is formed and the materials are mixed as a solution, thus removing these problems. As a result, the freeze–dried powders are extremely homogeneous.

Stronger intergranular coupling strength or an enhancement of flux pinning are possible ways to increase critical current densities. The possibility of a magnetic interaction between flux lines and rare earth paramagnetic moments gives rise to a systematic investigation of the influence of rare earth substitution, using atoms having different magnetic moments, on flux pinning. Magnetisation measurement techniques such as AC susceptibility and Vibrating Sample Magnetometry, yield information for both the inter– and intra–granular superconducting mechanisms. The effect of the rare earth addition can thus be monitored using these techniques.

EXPERIMENTAL METHODS

Each compound was produced the same way (freeze drying and sintering) so as to permit an unambiguous comparison between them all. Ten compositions, which can be summarised by the following formula, were made: $Y_xRE_{1-x}Ba_2Cu_3O_{7-\delta}$ where RE = Dy, Gd and Ho and x = 0.2, 0.5, 0.7 and 1.0. Appropriate amounts of Y_2O_3, Dy_2O_3, Gd_2O_3, Ho_2O_3, $BaCO_3$ or CuO were weighed into a beaker with approximately 150ml deionised water and the beaker placed onto a hot plate. When the mixture reached an equilibrium temperature between 70 and 80°C , HNO_3 was added drop by drop until the amount of HNO_3 (plus ≈ 5% more) required to convert all the oxides and carbonates to nitrates had been reached. At this point the mixture was left stirring until no powder remained (≈ 3hrs). The solution was then sucked through a nozzle and sprayed at a bowl immersed in liquid nitrogen, which was rotating on a platform so as to keep the sprayed layer of a reasonably uniform thickness. The bowl containing the now frozen nitrate solution was placed inside a freeze drier at -15°C and under a vacuum of 20 μmHg for three days to sublime all the water out of the frozen layer. The resulting nitrate powder was then converted to oxides by decomposing it in a furnace for 1 hour at 800°C. After decomposition the oxides were mixed thoroughly and pellets produced using a uniaxial press of 0.5 ton followed by isostatic pressing to 30,000 psi. The pellets were heat treated at 940°C for 20 hours and 440°C for 6 hours, both in flowing O_2. One further material, $Y_{0.5}Yb_{0.5}Ba_2Cu_3O_{7-\delta}$, was produced using the same procedure but with Yb_2O_3 as a raw material.

AC susceptibility measurements were carried out on rectangular slabs of the approximate dimensions 9 x 3 x 1.5 mm. AC field amplitudes (h_m) in the range 0.04→6 Oe and at 270 Hz, were applied to each sample. For one of the compositions constant DC fields of 12 and 20 Oe were also applied for a subset of the range of AC fields. The AC susceptibility rig has been explained in detail elsewhere [4].

The standard lock-in amplification system enables the separation of signals proportional to the real (χ') and imaginary (χ'') parts of the susceptibility. The real component yields the temperature at which the sample goes diamagnetic. At low frequencies the imaginary component is related to the loss associated with the pinning of magnetic flux lines as the sample is taken through an AC cycle. The χ''(T) curve is bell shaped and peaks at temperature, T^*, at which the flux fronts meet at the center of the sample, d/2, where d is the sample thickness. The field that corresponds to this T^* is referred to as $H_l^* = h_m(T^*)$. The peak at T^* is shifted to lower temperatures as h_m is increased. Assuming the critical state model, one can derive:

$$J_c = \frac{2H_l^*}{d}. \tag{1}$$

Plotting the J_c obtained using Eqn. 1 against the intergranular loss-peak temperature (T^*) gives J_c as a function of temperature.

High-field magnetisation loops were performed for three of the samples at different temperatures ranging from 5-85K. Flux creep measurements were also performed at 1 kOe at 10K intervals from 5-85K, on the sample with the highest critical current density.

RESULTS AND DISCUSSION

Fig. 1 is a plot of J_c (from Eqn. 1) versus temperature, for each of the Dy-substituted compounds. Similiar data, not presented here, have been obtained for each of the Ho- and Gd- substituted samples. It is seen that the substitution of any percentage of rare earth, increases the critical current density by comparison with that of $YBa_2Cu_3O_{7-\delta}$. The critical state appears to be the same for each composition. To date the absence of a specific trend with alloy concentrations, coupled with microstructural observations, suggests a secondary effect on flux pinning, mostly from the influence on sintering behaviour of the rare-earth addition.

Yb was chosen as a substitute because of its small magnetic moment, compared to those of the other substitutes chosen. To date no systematic influence with magnetic moment concentration has been obtained. As can be seen from Fig. 1, the Yb substituted sample has a higher J_c than the 100% YBCO sample but its J_c is still lower than those for the other substituted samples.

SEM work has shown that the Yb compound is less porous than the YBCO and therefore would be expected to have a higher grain coupling density. However, this sample is comparable microstructurally to the sample with the highest J_c (80% Dy). This suggests that the rare earth chemistry as well as the magnetic moment are helping to improve the flux pinning. Much work is still required to test the reproducibility of these results. The onset temperature for each of the compounds was found to be approximately 90K in each case.

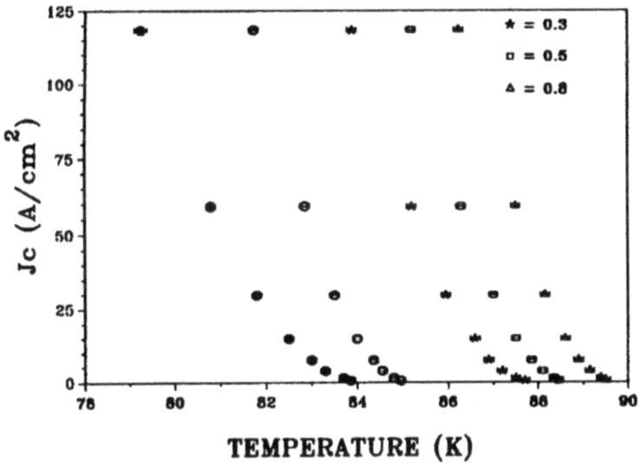

Fig. 1 : Critical current densities for Dy–doped materials, by comparison with the $Y_{0.5}Yb_{0.5}$ sample (○) and the undoped materials (●).

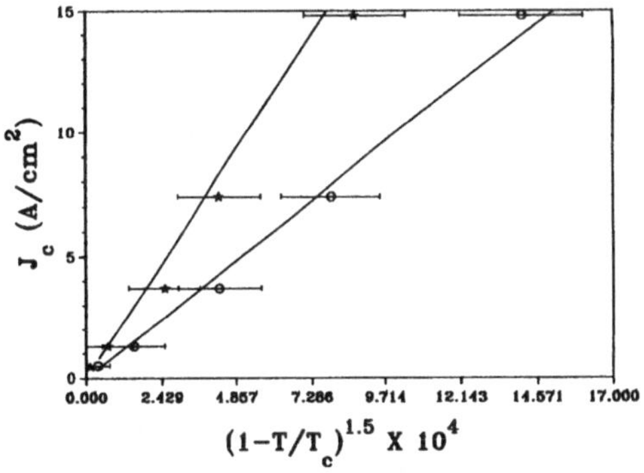

Fig. 2 : Analysis of the critical current density data for $Y_{0.7}Ho_{0.3}$ for applied DC fields of 0 (✱) and 12 (○) Oe.

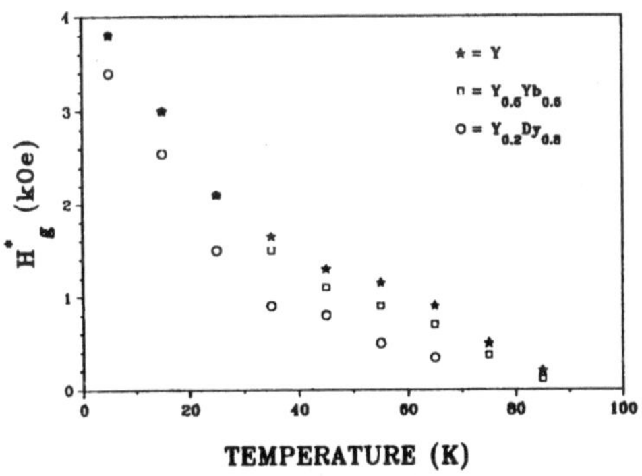

Fig. 3 : H* versus temperature for three of the sample compositions shown in the key.

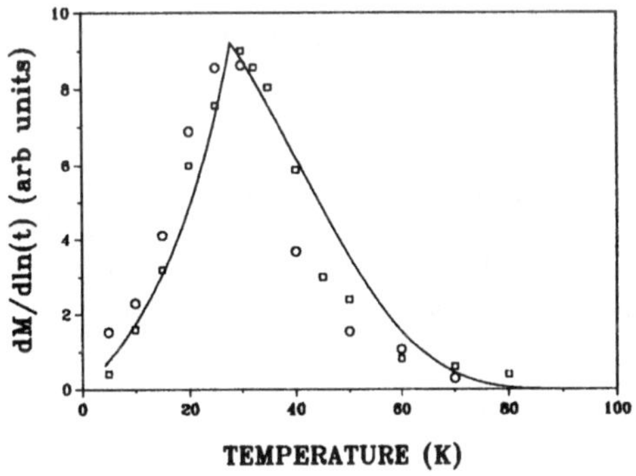

Fig. 4 : Flux creep data for $Y_{0.2}Dy_{0.8}$ (○) and a plasma sprayed Y sample (□).

In general the temperature dependence of the critical current density, J_c, is of the form:

$$J_c \sim \left(1 - \frac{T}{T_{c_J}}\right)^{\alpha} \tag{2}$$

where T_{c_J} [5] is the decoupling temperature and α is related to the coupling mechanism between superconducting grains and the critical state model. Weak links between grains are responsible for magnetic irreversibility at fields lower than the first penetration field of the grains, H_{c1_g}. The first penetration field of the links is known as the Josephson first penetration field, H_{c1_J}. Josephson vortices are able to enter the junction for fields higher than H_{c1_J}. The absence of a normal core for a Josephson vortex, means that any pinning force, f_p, must evolve from a magnetic interaction of the form $\phi_0 M_{rev}/\lambda_J$, where ϕ_0 is the flux quantum, M_{rev} the reversible magnetisation and λ_J the Josephson penetration depth. For fields close to H_{c1_J}, $M_{rev} \approx H_{c1_J}$ and it can be shown for grains of average radius less than the Josephson penetration depth (i.e. $r_g < \lambda_J$) that

$$f_p = c j_m^{3/2} \tag{3}$$

where $c = r_g^{3/2}\phi_0^{1/2}/\ln(\lambda_J/r_g)$ and j_m is the maximum tunneling current through the weak links. The addition of a DC field leads to the expression [4]:

$$f_p = \frac{c}{2} j_m^{3/2} ln\left[\frac{\phi_0}{4Br_g^2}\right]. \tag{4}$$

Scanning electron microscopy work has shown that the grain sizes for each of the compositions is small (5 μm or less) and therefore these materials should be expected to be well described by the critical state model presented.

To test this further, the temperature dependence was investigated for one of the compositions in the presence of zero and higher DC fields (Fig. 2). The solid lines in Fig. 2 are fits to the data. As follows from the temperature dependence (not presented but obtained from results similiar to those presented in Fig. 1) , the exponent α from Eqn. 2 is equal to 3/2 which, using Eqn. 3 leads to the condition that j_m is a linear function of reduced temperature as is the case for an SIS (Superconductor–Insulator–Superconductor) junction [6]. The constant c (from Eqn. 3) from the fits in Fig. 2 was calculated to be 2×10^4 for zero DC field and 9.9×10^3 for 12 Oe. This again is consistent with the model in which the constant halves (refer to Eqn. 4) with the application of a DC field.

The application of high DC magnetic fields, using VSM, is another useful probe, of the rare earth substitution, on the irreversible magnetic behaviour of the materials. The temperature dependence of the intragranular fields, H_g^* (analogous to H_l^* but corresponding to flux fronts meeting at the centre of an average sized grain), over the range 5 to 85 K is shown in Fig. 3. These data have been found from the measurements of magnetic losses, W ($= \oint B\, dH$), and H_g^* corresponds to the maximum of W/H_m^2, where H_m denotes the amplitude of the applied DC magnetic field. We found that W/H_m^2 is a linear function of H_m, for $H_m < H_g^*$. This indicates that the intragranular critical state is field independent [4]. The same shape of the $H_g^*(T)$ from Fig. 3 for different compositions supports the conclusion, that the magnetic moment of the rare earth substitutions has no influence on the pinning mechanism. Since the elementary magnetic pinning force, p_m, is equal to $\phi_0 H_{c1}/\lambda$, and the vortex core interaction force, $p_c = H_c^2\xi$, then one has $p_c/p_m \propto \kappa$, where $\kappa = \lambda/\xi$. As a result, for the high κ superconductors like $YBa_2Cu_3O_{7-\delta}$, the magnetic interaction has no effect on the critical current density, as was found experimentally.

In Fig. 4 the creep rate data are plotted versus temperature at 1 kOe applied DC field, for $YBa_2Cu_3O_{7-\delta}$ and $Y_{0.2}Dy_{0.8}Ba_2Cu_3O_{7-\delta}$ samples. The same temperature dependence of the creep rate indicates the same temperature dependence of activation energy, which together with the data from Fig. 3, results in the common interaction mechanism (possible via twin planes) in both compositions. The solid line in Fig. 4 is a fit from the modified conventional flux creep theory [7].

CONCLUSIONS

1. Eleven compositions were successfully prepared with the result being an increase in the intergranular critical current density for all of the substitutions. The pinning force appears to be governed by both the rare earth chemistry and the magnetic moment of the substituted rare earth element.

2. The samples appear to have the same pinning mechanics and each sample can be well described by the critical state model presented.

3. VSM measurements have shown that the magnetic moment of the substituted rare earth does not have any effect on the intragranular flux pinning mechanics.

ACKNOWLEDGEMENT

This work has been supported by a GIRD Grant from the Industrial Research and Development Board under Project No 15011. DAG acknowledges support under an Australian Postgraduate Research Award.

REFERENCES

1. Hosoya S, Shamoto S, Masashigi O, Masatoshi S (1987) Jpn. J. Appl. Phys 26: 325–456
2. Yang KN, Dalichaouch Y, Ferreira JM, Lee BW, Neumeier JJ, Torikachvili MS, Zhou H, Maple MB, Hake RR (1987) Solid State Communications 63: 515–519
3. Narlikar AV, Agarwal SK (1988) Current Science 57: 753–760
4. Finlayson TR, Garvie DA, Lowe DB, Przelozny Z, Smith TF (1990) Cryogenics 30: 467–473.
5. Clem JR (1988) Physica C 153-155: 50–55
6. De Gennes PG (1966) Superconductivity of metals and alloys. W. A. Benjamin Inc. Amsterdam, pp 234–246
7. Przelozny Z, Finlayson TR (1991) In press

The Pinning Centers in MPMG-Processed YBCO Observed by Using Bitter Pattern Technique

Y. Higashida[1], K. Michishita[1], Y. Kubo[1], A, Saji[2], T. Inoue[2], H. Yoshida[2], M. Murakami[3], H. Fujimoto[3], K. Yamaguchi[3], T. Takata[3], A. Kondoh[3], N. Nakamura[3], and N. Koshizuka[3]

[1] Research and Development Laboratory, Japan Fine Ceramics Center, Nagoya, 456 Japan
[2] Chubu Electric Power Co., Inc., Nagoya, 459 Japan
[3] Superconductivity Research Laboratory, ISTEC, Koto-ku, Tokyo, 135 Japan

ABSTRACT

The effect of a magnetic field on the distribution of fluxoids trapped at Y2BaCuO5 inclusions in MPMG-processed YBCO was investigated using the Bitter pattern technique. The decorated pattern exhibited a very different distribution according to whether or not the magnetic field is applied. When it is applied, the ferromagnetic particles are attracted to the fluxoids which penetrate the superconductor.

KEY WORDS: Bittert pattern technique, decoration method, plux pinning, flux line lattice,

INTRODUCTION

In order to obtain a high critical current density, it is essential to prevent any magnetic flux motion in a superconductor. Any inhomogeneiety in a superconductor may work as a pinning center against such magnetic flux motion.

The MPMG-processed YBCO dispersed with fine Y2BaCuO5 (211) inclusions possesses a very large critical current density [1]. The reason has been attributed to the presence of 211 inclusions which may act as dominant pinning centers. This assertion is based upon the experimental fact that the critical current density of the MPMG-processed YBCO depends strongly upon the density and average size of the 211 inclusions [2]. There seems to be some confusion as to whether or not the size of the pinning centers must be of the same order of coherence length of a superconductor. There is strong evidence that large precipitates contribute to flux pinning in conventional superconductors [3,4].

In order to verify whether or not 211 inclusions act as pinning centers, the direct observation of the fluxoid trapped by MPMG-processed YBCO was carried out using the decoration method of ferromagnetic particles [5].

It it is important to distinguish whether the decorated pattern is truly created as a result of intrinsic magnetic decoration or is simply due to the non-magnetic adsorption of fine ferromagnetic particles. Ultra fine particles are very active and easily adsorbed on the material surface which is kept cooled at temperatures as low as 10 K.

In this paper, the effects of a magnetic field on the decorated pattern are presented.

EXPERIMENTAL

The sample thickness was about 0.05 mm. The surface was polished with 1 um diamond paste. One sample was cooled in the magnetic field H = 0 Oe down to 10 K. Another sample was field-cooled to the same temperature at H = 10 Oe. These samples were subject to decoration. The experimental apparatus was illustrated in Fig. 1. The ferromagnetic particles were nucleated by evaporating 20 mg of nickel wire in a helium gas atmosphere. The helium gas pressure was 0.1 Torr, which is considered most suitable for nucleating fine nickel particles. The helium carrier gas, which was flushed through the chamber, drove the "smoke " of the nickel particles to the vicinity of the sample surface. The decorated pattern was observed using a polarized microscope.

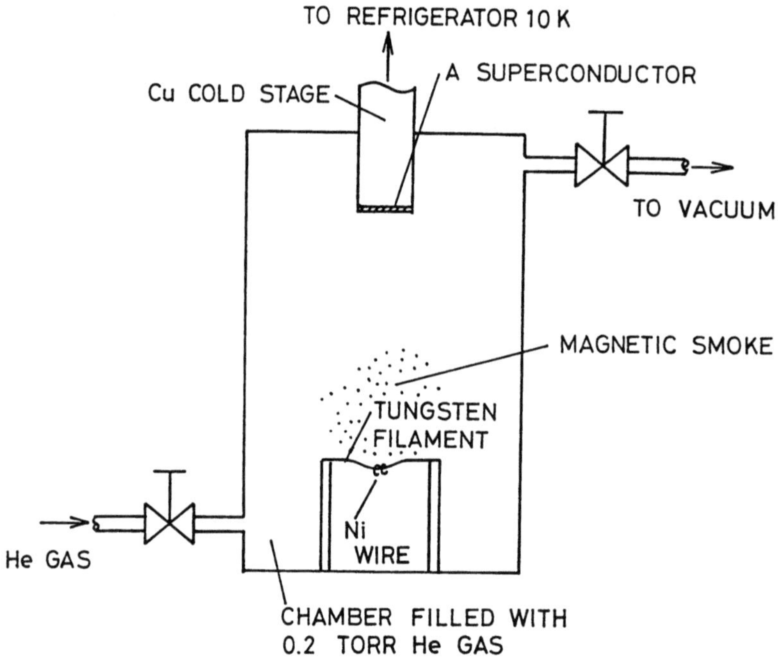

Fig.1 Schematic illustration of the apparatus for decoration experiment of ferromagnetic particles.

RESULTS AND DISCUSSIONS

Fig.2 shows the decorated patterns for MPMG-processed samples dispersed with 211 inclusions. The pattern (a) corresponds to the sample which was cooled in a zero magnetic field, and the pattern (b) corresponds to the one which was field-cooled at H = 10 Oe and the magnetic field was turned off before decoration. The grey islands in (a) and the white ones in (b) are 211 inclusions. The black dots are the deposited nickel particles. Any one dot may be composed of aggregate of numerous nickel particles. These dots are mainly present on and around 211 inclusions in Fig.2 (b).

However, as shown in Fig.2 (a), non-magnetic decoration occurred even when no magnetic field was applied to the sample. The decorated patterns include nickel particles adsorbed both due to simple trapping and due to magnetic attraction. So that, we must be careful when interpreting the decorated results.

Fortunately a detailed inspection of Fig.2 (a) and (b) indicates that each decorated nickel pattern has somewhat different features:

1)The pattern (a) has a rather broad size distribution of nickel particles, while pattern (b) has a definite size.
2)The pattern (a) has random spatial distribution of nickel particles , while pattern (b) has the one localized near and around the 211 inclusions.
3)The pattern (a) has rather irregular shape of nickel particles, while the pattern (b) has circular shapes.

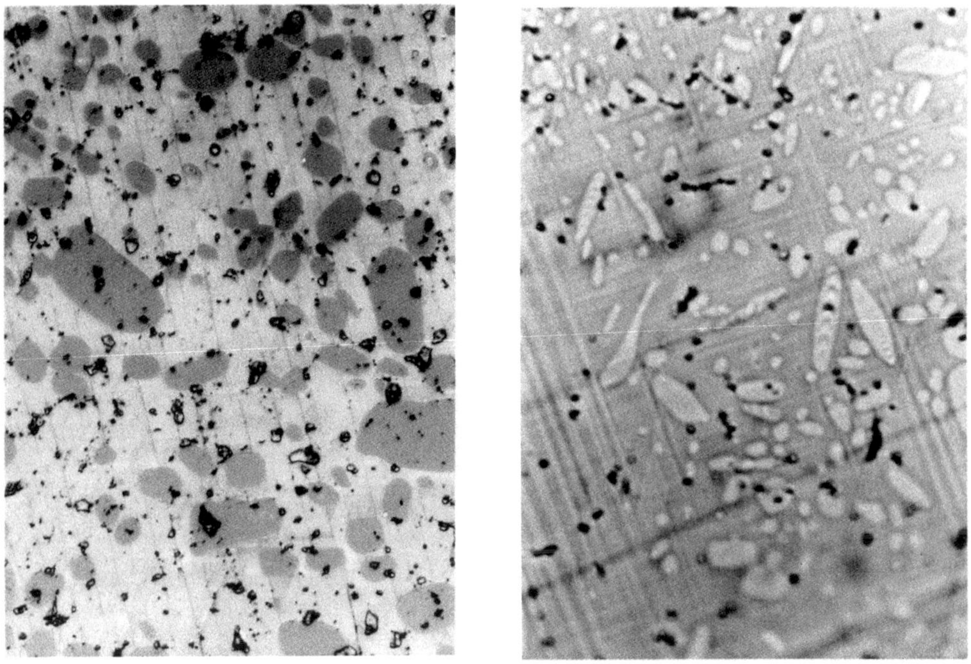

(a) H = 0 Oe. (b) H = 10 Oe.
Fig.2 The polarization micrographs for the decorated patterns of Ni particles on the MPMG-processed YBa2Cu3O7 containing Y2BaCuO5. Temperature is 10 K.

Thus we are confident that the pattern (b) is formed as a result of magnetic attraction of fine nickel particles into the fluxoids in a superconductor. It is probable that the magnetic flux in a superconductor is quantized and trapped near the non-superconducting 211 inclusions in pattern (b), which is not the case with pattern (a) where no magnetic field was applied.

The experiments on the temperature dependence of the decorated pattern is under investigation and will be presented elsewhere.

CONCLUSIONS

The decorated pattern of the superconductor by ferromagnetic particles exhibits very different distributions according to whether or not the magnetic field is applied. When the magnetic field is applied, the ferromagnetic particles are attracted to the superconductor into which fluxoids penetrate.

ACKNOWLEDGEMENT

A part of this work is supported by NEDO. The apparatus for decoration experiments was fabricated with the financial support of the Japan Keirin Association.

REFERENCES

1. H. Fujimoto, M. Murakami, S. Gotoh, N. Koshizuka, T. Oyama, Y. Shiohara, and S. Tanaka, Advances in Superconductivity, Proc. International Conf. on Superconductivity, Tsukuba, Japan, 1989, ed. K. Kajimura and T. Ishiguro, Springer, 1989.
2. M. Murakami et al., Physica C in press, proc. M2S-HTSC Conf.,Kanazawa, July 22-26, 1991.
3. K. Yamafuji et al., Cryogenics 31 (1991) 431.
4. R. I. Coote, J. E. Evetts and A. M. Campbell, Canad. J. Phys. 50 (1972) 421.
5. Y. Higashida et al., Physica C in press, Proc. M2S-HTSC Conf., Kanazawa, July 22- 26,1991.

Microstructures of MPMG Processed YBaCuO

K. Yamaguchi, M. Murakami, H. Fujimoto, and N. Koshizuka

Superconductivity Research Laboratory, ISTEC, Koto-ku, Tokyo, 135 Japan

ABSTRACT

Microstructures of high-J_c YBaCuO material fabricated by Melt-Powder-Melt-Growth process were studied by transmission electron microscopy to elucidate the pinning mechanism in this material. It was found that the interface between the $YBa_2Cu_3O_7$ matrix and the Y_2BaCuO_5 inclusion is sharp. The fact that the gradient of the superconducting order parameter at a sharp interface can be steep, may lead to a strong pinning force and large J_c value in the material.

KEY WORDS : melt process, interface, $YBa_2Cu_3O_7$, Y_2BaCuO_5, transmission electron microscopy

INTRODUCTION

For the practical use of high-T_c superconductors, the developments of high-J_c materials are required. Several groups reported that bulk YBaCuO materials with high-J_c could be fabricated by melt processes.[1-5] We have also developed a kind of melt process named MPMG (Melt-Powder-Melt-Growth) for high-J_c bulk YBaCuO material.[6]

Using the MPMG process, bulk YBaCuO with finely dispersed nonsuperconducting Y_2BaCuO_5 (211) inclusions is obtained. J_c of the material increases with increasing number of inclusions. The sample with 30 % volume of 211 inclusions exhibited the J_c value of $3x10^4$ A/cm^2 at 77 K and 1 T (parallel to the c-axis), while the sample without such inclusions a relatively small value of $5x10^3$ A/cm^2.[7] It suggests that the flux pinning centers are introduced by introducing the 211 inclusions.

In order to understand the pinning mechanism in this system, microstructures of the MPMG processed materials were investigated by transmission electron microscopy (TEM). Two dimensional discrete Fourier transformations (2d-FT) of the TEM images were also conducted to evaluate the strain in the 123 matrix.

EXPERIMENTAL

Both the samples with 30 volume % of 211 inclusions (J_c=$3x10^4$ A/cm^2), and without 211 inclusion (J_c=$5x10^3$ A/cm^2) were used for the experiments.

The samples for TEM observations were prepared by Ar^+ ion milling after mechanically thinning to 50 microns thick: details of the preparation are described in our previous reports.[8,9] All the TEM observations here were performed by a microscope operated at 400 kV and having a point resolution of 0.17 nm.

To get information on the strain in the 123, the 2d-FT was conducted using a digital image processor. A crystal lattice image of the 123 was taken by a TV camera. Before the 2d-FT, the image was masked by a circular aperture to restrict the region to be transformed and avoid artifacts caused by 2d-FT: the intensity of pixels outside of the aperture was set to zero. As the range of intensities of the 2d-FT pattern was very large, a logarithmic transformation was performed to resolve small details. Then the positions of the spots in the two transformed patterns, one next to and the other 50 nm distant from the interface, were compared to discriminate any difference between the lattice parameters, which may be caused by strain.

RESULTS

Figure 1 shows TEM micrographs of the materials (a) with and (b) without the inclusions viewed perpendicular to the c-axis of the 123. Many stacking faults are observed as lines in the 123 matrix for both materials. The density of the stacking faults does not increase with the introduction the 211 inclusions.

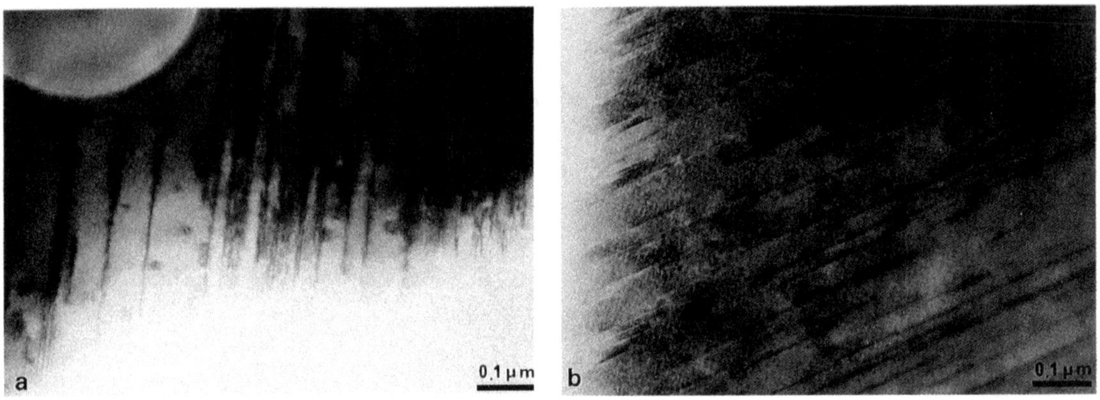

Fig.1. TEM images of the material (a) with and (b) without 211 inclusions viewed perpendicular to the c-axis of the 123. Lines correspond to stacking faults

Figure 2 is a micrograph in the vicinity of the 211 inclusion viewed along the c-axis of the 123. In this region, two crystals of 123 and 211 are superimposed and moire fringes are generated. These fringes are not distorted except where they are bent according to the twin boundaries of the 123 matrix. This means that the crystal of the 123 has no crystallographic defects and the twins run up to the interface.

Crystal lattice images of the 123 in the vicinity of the inclusion viewed along and perpendicular to the c-axis of the 123 are shown in Fig. 3-a and -b, respectively. In both micrographs, no distortion of the lattice images of the 123 can be recognized even at the interface, which suggest the 123 crystal lattice is not distorted.

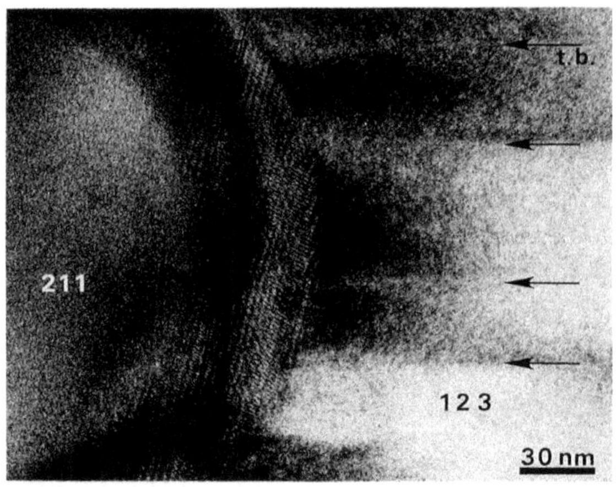

Fig.2. TEM image in the vicinity of a 211 inclusion viewed along the c-axis of the 123. Twin boundaries in the 123 are marked by arrows.

Figures 4-a and -b are crystal lattice images next to and 50 nm distant from the interface viewed along the c-axis of the 123 and the respective masked images using circular apertures are shown in Figs.4-c and -d. These two images are transformed by 2d-FT and the logarithmic transformation as shown in Figs. 4-e and -f. These spots can be indexed as the [001] zone-axis electron diffraction pattern. There is no difference in the position of the spots between the two transformed patterns. The distance between the 000 and the 200 spots is 102 pixels and that between the 000 and the 020 spots is 101. Therefore the crystal lattice constants a and b at the interface are not different from those of the region at 50 nm from the interface within an error of 1 %.

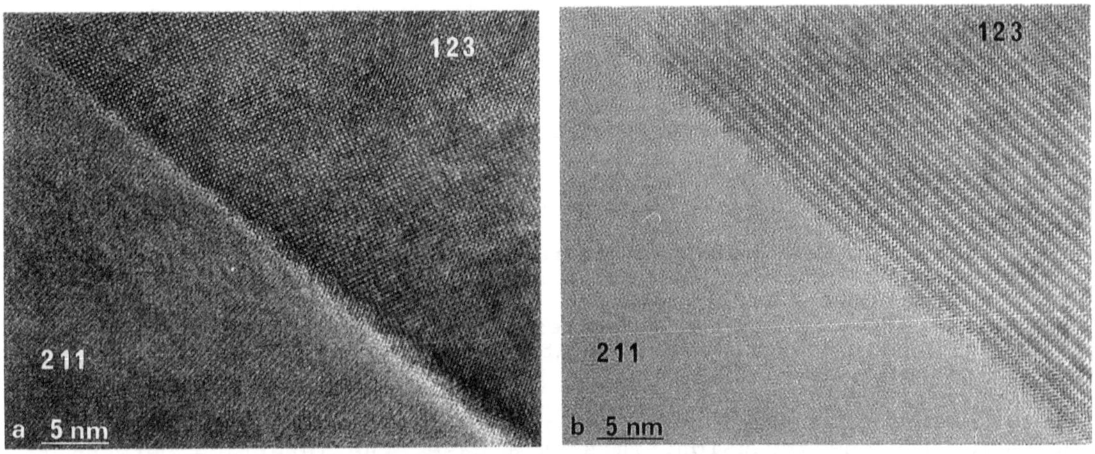

Fig.3. Crystal lattice images in the vicinity of the inclusion viewed (a) along and (b) perpendicular to the c-axis of the 123.

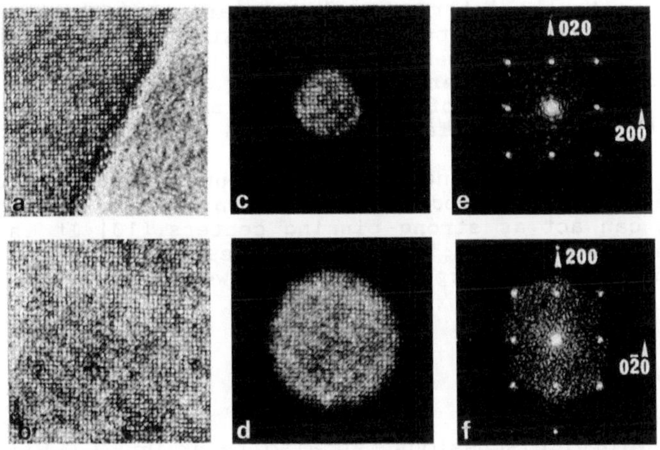

Fig.4. Crystal lattice image of the 123 (a) next to and (b) 50 nm distant from the interface viewed along the c-axis, and the respective masked images (c), (d), and Fourier transformed patterns (e), (f).

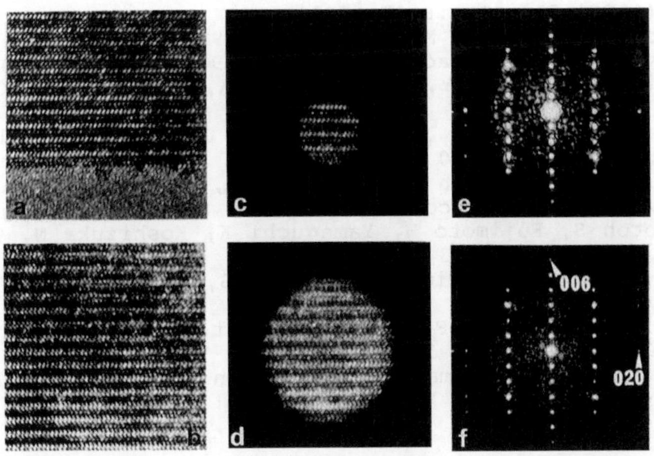

Fig.5. Crystal lattice images of the 123 (a) next to and (b) 50 nm distant from the interface viewed perpendicular to the c-axis, and the respective masked images (c), (d), and Fourier transforms (e), (f).

Figure 5 shows the results of 2d-FT of the crystal lattice images viewed perpendicular to the c-axis as in Fig.4. The transformed patterns, Figs.5-e and -f, correspond to the [100] zone-axis electron diffraction pattern. The distance between the 000 and the 020 spots is 102 pixels and that between the 000 and the 006 spots is 101 pixels. Thus, the variation in the lattice parameters b and c between the two regions is zero within an error of 1 %.

DISCUSSION

In the MPMG processed materials, there are many stacking faults in the 123 matrix. Although they may act as pinning centers, the density of these stacking faults does not increase by introducing the 211 inclusions. Thus, the difference in J_c between the materials with and without the inclusions can not be caused by the stacking faults, so it should be ascribed to the 211 inclusions themselves.

The mean size of the 211 inclusions is about 1 micron, so many fluxoids can interact with one inclusion. In such a case, the interface between the superconducting matrix and the nonsuperconducting inclusion is important. From the TEM observations and 2d-FT, the interface was confirmed to be sharp. The crystal lattice of the 123 matrix seems to have no crystallographic defects or distortions even at the interface. Furthermore, twins in the 123 matrix are running to the interface, which means that the crystal of the 123 is orthorhmbic at the interface. At such sharp interfaces, the superconducting order parameter in the 123 matrix falls to zero rapidly within a width of the coherence length. Since the pinning force is proportional to the gradient of the order parameter, a strong pinning force is expected at the sharp interface in the MPMG processed material.

The magnitudes of flux pinning by the nonsuperconducting inclusions with sharp interfaces have been modeled theoretically and this points out that such inclusions can act as strong pinning centers.[10] It is concluded that the strong pinning forces and the large values of J_c in the MPMG processed materials are due to the sharp interface between the 211 inclusion and the 123 matrix.

ACKNOWLEDGEMENTS

The authors are grateful to Mr. S.Tokuono of Superconductivity Research Laboratory for helpful discussions. We also thank Dr. J.O.Willis for helpful discussions and a critical reading of the manuscript.

REFERENCES

1.Jin S, Tiefel TH, Sherwood RC, van Dover RV, Davis ME, Kammlot G.W, Fastnacht RA (1988) Phys.Rev.B 37: 7850
2.Murakami M, Morita M, Doi K, Miyamoto K (1989) Jpn.J.Appl.Phys. 28: 1189
3.Salama K, Selvamanickam V, Gao L, Sum K (1989) Appl.Phys.Lett. 54: 2352
4.Meng RL, Kinalidis C, Sun YY, Gao L, Tao YK, Hor PH, Chu CW (1990) Naure 345: 326
5.Lian Z, Pingxiang Z, Ping J, Keguang W, Jingrong W, Xiaozu W (1990) Supercond.Sci.Technol. 3: 490
6.Fujimoto H, Murakami M, Gotoh S, Koshizuka N, Oyama T, Shiohara Y, Tanaka S (1990) In: Advances in Superconductivity II. Springer-Verlag, pp285
7.Murakami M, Gotoh S, Fujimoto H, Yamaguchi K, Koshizuka N, Tanaka S (1991) Supercond.Sci.Technol. 4: S43
8.Yamaguchi K, Murakami M, Fujimoto H, Gotoh S, Koshizuka N, Tanaka S (1990) Jpn.J.Appl.Phys. 29: L1428
9.Yamaguchi K, Murakami M, Fujimoto H, Gotoh S, Oyama T, Shiohara Y, Koshizuka N, Tanaka S (1991) J.Mater.Res. 6: 1404
10.Tokuono S, Tanaka S, Fukuyama H to appear in Physica C (Proc.M^2S-HTSC III)

Critical Current Properties of YBCO Prepared by Melt Processes

HIROYUKI FUJIMOTO, MASATO MURAKAMI, NAOMICHI NAKAMURA, SATOSHI GOTOH, AKIHIRO KONDOH, NAOKI KOSHIZUKA, and SHOJI TANAKA

Superconductivity Research Laboratory, ISTEC, Koto-ku, Tokyo, 135 Japan

ABSTRACT

YBaCuO prepared by the Melt-Powder-Melt-Growth (MPMG) process has a high critical current density. It has been found that non-superconducting Y_2BaCuO_5 (211) particles dispersed in the 123 superconducting phase work as pinning centers, resulting in high Jc values. The direct summation model predicts that the critical current density is proportional to the ratio of the volume fraction of non-superconducting particles to the size of those particles. The measured values of the critical current density increased with increasing volume fraction of the 211 particles and decreasing size of those particles. Therefore, the critical current density is proportional to the ratio of the volume fraction of the 211 particles to the size of those particles, which is in good agreement with the model.

KEY WORDS : $YBa_2Cu_3O_x$, Y_2BaCuO_5, melt process, critical current density, flux pinning

INTRODUCTION

Since the discovery of the Y-Ba-Cu-O superconductor with a critical temperature above the boiling point of liquid nitrogen in 1987 [1], considerable efforts have been made to increase the critical current density, Jc, which is very important for applications, and Jc has been increased remarkably. Samples prepared by melt processes such as MTG (Melt Textured Growth) [2,3], QMG (Quench and Melt Growth) [4,5] and MPMG [6,7] have large critical current densities exceeding 10^4 A/cm^2 at 77 K. Grain alignment in the MTG process has overcome the weak-link problem and further non-superconducting Y_2BaCuO_5 (211) particles dispersed in the 123 superconducting phase have enhanced flux pinning.

YBaCuO prepared by the MPMG process has a high critical current density of over 30,000 A/cm^2 at 77K and 1T. It has been found that non-superconducting 211 particles, which can be dispersed in the 123 matrix in this process, can work as pinning centers, resulting in higher Jc values. These results have been supported by the facts that (1) the larger the volume fraction of the 211 phase, the higher the critical current density, (2) direct observation of flux density using the magneto-optical effect of iron garnet films showed that the 211 phase was the pinning site[8], and (3) observation of the flux line lattice using a decoration technique showed that flux lines were trapped at the 211 particles[9].

In this paper we report microstructures of the samples prepared by various melt processes and the relation between the size and volume fraction of the 211 phase and Jc values which support the fact that the 211 particles contribute to flux pinning.

EXPERIMENTAL

Samples were prepared by the MPMG process [6] and the modified MTG process [3]. In the MPMG process (samples A ~ E) appropriate amounts of mixtures of Y_2O_3, $BaCO_3$ and CuO were calcined at 900 °C; then the powders were melted at over 1200 °C and rapidly cooled down to room temperature. After the melted oxides were pulverized and pressed into pellets, the samples were reheated to 1100 °C and rapidly cooled down to 1000 °C or the peritectic temperature and cooled slowly to room temperature at a rate of 1 °C/h through the peritectic reaction. At the final stage, the samples were annealed at 600 °C and cooled down to room temperature in flowing oxygen. Changing the starting compositions makes it possible to obtain samples with 211 particles of different volume fractions.

In the modified MTG process (samples F and G), the mixtures of sintered $YBa_2Cu_3O_x$ and Y_2BaCuO_5 or Y_2O_3, $BaCuO_2$ and CuO were pressed into pellets. Then the samples were heated to over 1000 °C, which is a relatively low melt processing temperature, and rapidly cooled down to 1000 °C or the peritectic temperature and cooled slowly to room temperature at a rate of 1°C/h through the peritectic reaction. The samples were annealed at 600 °C and cooled down to room temperature in flowing oxygen after the melt growth.

Magnetization properties were measured using a vibrating sample magnetometer (VSM), a SQUID magnetometer and an AC susceptometer. Microstructure was observed using an optical microscope, a transmission electron microscope (TEM), a scanning electron microscope (SEM), and a wavelength dispersive X-ray spectroscope (WDX).

RESULTS AND DISCUSSION

Microstructure

Sample name, process and specification are shown in Table 1. Figure 1 (a) ~ (d) show optical micrographs of the samples A,C,E and G, respectively. Sample A, C and E were prepared by the MPMG process. In sample A, 211 particles, which are observed as dark spots, are homogeneously and finely distributed in the 123 superconducting phase as shown in Figure 1 (a). The size of the 211 phase particles is around 1 μm and the volume fraction of 211 phase is 25 % in this sample. The volume fraction was estimated using an image analysis assuming that the line fraction of 211 particles is the same as the volume fraction of 211 particles. Figure 1 (b) shows a micrograph of sample C. The size of the 211 phase particles is the same as that of sample A, but the volume fraction is 10 % which is different from sample A. Sample C was prepared reducing the excess amount of 211 phase of starting composition, e.g. $1.2Y_2O_3:2.1BaCO_3:3.1CuO$. Sample E was also prepared by the MPMG process, but the size of the 211 particles is larger than that of sample A and C as shown in Figure 1 (c). In this process for sample E, the melting and quenching at the first step was conducted at a low melt processing temperature. Figure 1 (d) shows a micrograph of sample G which was prepared by the modified MTG process. The size of the 211 particles which are represented as black particles is larger than that of sample A and C.

Critical Current Density

It has been reported that non-superconducting particles can work as pinning centers in conventional superconductors [10] and also in high Tc superconductors [11]. For the melt processed YBaCuO, the 211 particles are considered to work as the main pinning centers. Since the size of the 211 particles is larger than the coherence length, the interface between the 211 particles and the 123 matrix contributes to flux pinning [12~14]. Since the flux line lattice (FLL) of high temperature superconductors is plastic, or the elastic constant for shear, C_{66}, is very small due to a large $\kappa(=\lambda/\xi)$, the critical current density can be obtained from the direct summation of the elementary pinning force as follows,

$$Jc=Npd^2\xi Bc^2(1-B/Bc_2)/(\phi_0 B)^{1/2}, \qquad (1)$$

where Np and d are the density and size of the 211 particles respectively, Bc is the thermodynamical critical field, Bc_2 the upper critical field, ξ the coherence length and ϕ_0 the flux quantum [15]. At low field, $(1-B/Bc_2)=1$, and the volume fraction of the 211 particles, $V_f=Np\times d^3$, therefore Jc is proportional to V_f/d.

$$Jc=(V_f/d)\xi Bc^2/(\phi_0 B)^{1/2} \qquad (2)$$

Table I Process, Heat treatment temperatures, and volume fraction and size of 211 phase of samples.

Sample	Process	First melting temperature, °C	Second melting temperature, °C	size of 211 μm	volume fraction of 211, %
A	MPMG	1400	1100	1.0	25
B	MPMG	1400	1100	1.1	24
C	MPMG	1400	1100	1.0	10
D	MPMG	1400	1100	1.3	17
E	MPMG	1250	1100	9.0	42
F	MTG	-	1050	3.0	20
G	MTG	-	1050	3.5	24

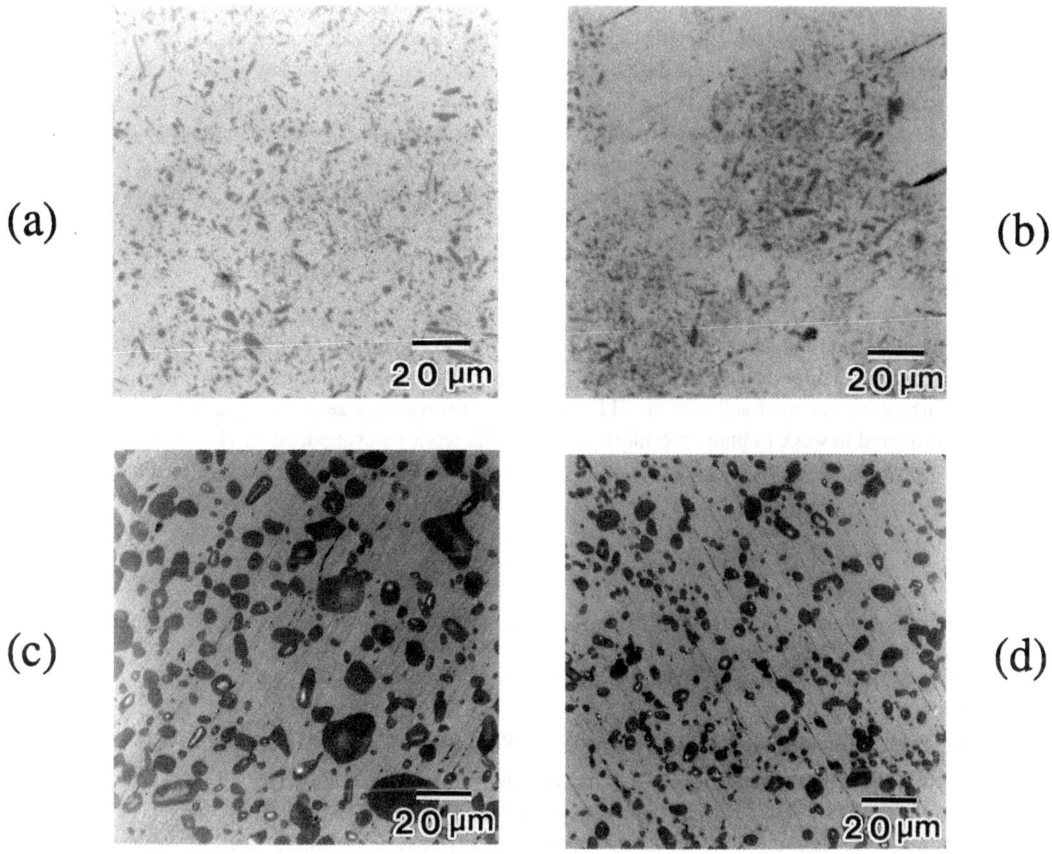

Figure 1 Optical micrographs of samples, (a) sample A prepared by the MPMG process, (b) sample C prepared by the MPMG process, (c) sample E prepared by the MPMG process, (d) sample G prepared by the modified MTG process.

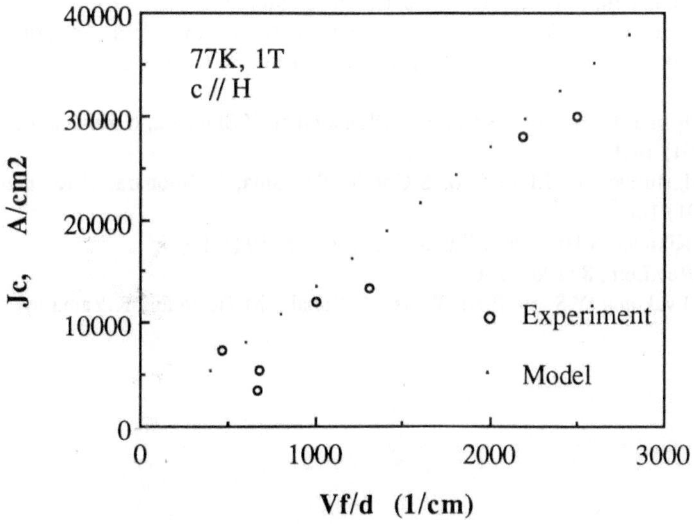

Figure 2 Relation between the ratio of the volume fraction to the size of the 211 particles (V_f/d) and the critical current density at 77K and 1T for the magnetic field parallel to the c-axis.

Figure 2 shows the relation between the ratio of the volume fraction to the size of the 211 particles (V_f/d) and the critical current density at 77K and 1T for the magnetic field parallel to the c-axis. Critical current density was calculated from magnetization hysteresis based on the critical state model [16]. The volume fraction and size of the 211 particles were estimated using an image analysis. It can be seen from Fig. 2 that the critical current density increases with increasing ratio of the volume fraction of the 211 particles to the size of those particles. A dotted line indicates the values which were calculated using equation (2). For calculating critical current densities, the values ξ=2nm and Bc=0.22T were used [17]. As a result, the estimation of Jc agrees very well with the experimental results.

CONCLUSIONS

The size and volume fraction of non-superconducting Y_2BaCuO_5 particles, which are distributed in the 123 superconducting matrix, can be controlled in the melt-processed YBaCuO. The value of the critical current density increased with increasing volume fraction of the 211 particles and decreasing size of those particles. Therefore, the 211 particles are considered to work as pinning centers. A part of this work was supported by New Energy and Industrial Technology Development Organization under the management of the R&D of Basic Technology for Future Industries.

REFERENCES

[1] M.K.Wu, J.R.Ashburn, C.J.Torng, P.H.Hor, R.L.Meng, L.Gao, Z.J.Huang, Y.Q.Wang and C.W.Chu : Phys.Rev.Lett., **58** (1987) 908.

[2] S.Jin, T.H.Tiefel, R.C.Sherwood, R.B.van Dover, M.E.Davis, G.W.Kammlott and R.A.Fastnacht : Phys.Rev.B, **37** (1988) 7850.

[3] K.Salama, V.Selvamanickam, L.Gao and K.Sun : Appl.Phys.Lett. **54** (1989) 2352.

[4] M.Murakami, M.Morita, K.Doi, K.Miyamoto and H.Hamada : Jpn.J.Appl.Phys., **28** (1989) L399.

[5] M.Murakami, M.Morita, K.Doi and K.Miyamoto:Jpn.J.Appl.Phys., **28** (1989) 1189.

[6] H.Fujimoto, M.Murakami, S.Gotoh, N.Koshizuka, T.Oyama, Y.Shiohara and S.Tanaka:Advances in Superconductivity II (Springer-Verlag, Tokyo, 1990) p.285.

[7] M.Murakami : Mod.Phys.Lett. B, **4** (1990) 163.

[8] S.Gotoh, N.Koshizuka, M.Yoshida, M.Murakami, H.Fujimoto and S.Tanaka : Jpn.J.Appl.Phys. **29** (1990) L1083.

[9] Y.Higashida, Y.Kubo, M.Murakami, H.Fujimoto, K.Yamaguchi, T.Takata, A.Kondoh and N.Koshizuka : to be published in Physica C.

[10] R.I. Coote, A.M.Campbell, and J.E.Evetts : Canad.J.Phys., **50** (1972) 421.

[11] T. Matsushita : Progress in High Temperature Superconductivity, vol. 15, (World Scientific, 1988) p.49.

[12] M.Murakami, S.Gotoh, N.Koshizuka, S.Tanaka, T.Matsushita, S.Kambe and K.Kitazawa : Cryogenics **30** (1990) 390.

[13] M.Murakami, T.Oyama, H.Fujimoto, S.Gotoh, K.Yamaguchi, Y.Shiohara, N.Koshizuka and S.Tanaka : IEEE tran. mag., **27** (1991) 1479.

[14] K.Yamaguchi, M.Murakami, H.Fujimoto, S.Gotoh, T.Oyama, Y.Shiohara, N.Koshizuka and S.Tanaka : J.Mat.Res., **6** (1991) 1404.

[15] K.Matsumoto, H.Kikuchi, N.Uno and Y.Tanaka : Cryogenics **30** (1990) 5.

[16] C.P.Bean : Phys.Rev.Lett., **8** (1962) 250.

[17] T.Matsushita, M.Iwakuma, Y.Sudo, B.Ni, T.Kisu, K.Funaki, M.Takeo and K.Yamafuji : Jpn.J.Appl.Phys. **26** (1987) L1524.

Effects of MgO Doping on Bi-Containing Superconductors Prepared by a Partial Melting Process

T.H. SUNG, T. IZUMI, Y. NAKAMURA, T. SUGAWARA, and Y. SHIOHARA

Superconductivity Research Laboratory, ISTEC, Koto-ku, Tokyo, 135 Japan

ABSTRACT

The uniaxial pressing in a partially molten state was attempted on MgO doped Bi-containing superconductors in order to realize the high Jc. The samples with the composition of Bi:Sr:Ca:Cu:Mg=2:2:1:2:0.25 were heated to 900°C and held for 1h and 3h in platinum crucible and quenched to investigate the location of the MgO particles. In the uniaxial hot-press method, the powders were pressed into pellets (20mm in diameter; 1mm in thickness) and heated to predetermined temperatures and held and then pressed to one half of the original thickness. The particle size of MgO was almost independent of holding time at 900°C and of crystallization from the partially molten state to solid state. The fine MgO particles were dispersed in both the 2212 and 2201 crystals. These results indicate that the MgO particles wet with these phases. And the fine particles of the Sr-Ca-Cu-O compound were dispersed in 2212 and 2201 crystals by repetition of pulverizing, molding, annealing and quenching processes. The microstructures of the sample was well oriented under the condition of annealing at 900°C for 1h and cooling to 880°C with the rate of 10°C/min and pressing by applying 900kgf after holding for 1h at 880°C.

KEYWORD:Bi-Sr-Ca-Cu-O, MgO, partial melting, precipitate, uniaxial pressing

INTRODUCTION

Since the discovery of high Tc superconductors, many works have been attempted to obtain the high critical current density. However, the critical current density in bulk materials have so far been below the required value. Such a characteristic in critical current density was caused by the followings;(1)weak coupling between grains[1], (2)the absence of pinning centers, (3)random crystal orientation etc.. These problems should be solved to improve critical current density. Therefore, it is necessary to eliminate weak links from bulk materials. A melt and solidification process such as MTG[2,3] or QMG[4-7] and MPMG[8] is able to avoid weak links. On the other hand, it is known that the presence of effective pinning centers is able to achieve high Jc values. In the $YBa_2Cu_3O_7$ superconductor, oxygen defect[9] and Y_2BaCuO_5(211)[10] inclusions have been proposed to serve as the pinning centers. However, it is still unclear what constitutes an effective pinning center. Such pinning behaviors are more serious in the Bi-Sr-Ca-Cu-O system than the Y-Ba-Cu-O system. In the Bi-Sr-Ca-Cu-O system, it was also reported that Ca_2CuO_3 precipitates were shown the possibility to realize the high Jc by a splat quenching method[11] and melt process[12], enhancing the pinning force. However, in the later method, Sr-Ca-Cu-O precipitates existed mainly at the grain boundary with several tens of μm. Therefore, we propose to introduce two kinds of methods as pinning centers in grains; one is a mechanical method, in which MgO, non-reacted with Bi-containing superconductor, was introduced by partial melting, and the other is a chemical method, in which Sr-Ca-Cu-O compound(designated as a gold crystal from the color in this paper) was expected to serve the peritectic reaction for Bi-containing superconductor and to be fine due to repetition of pulverizing, molding, annealing and quenching processes.

EXPERIMENTAL

The appropriate amounts of Bi_2O_3, $SrCO_3$, $CaCO_3$, CuO and MgO with the composition of $Bi_2Sr_2CaCu_2Mg_{0.25}O_x$ were mixed thoroughly in an agate mortar and pestle using ethanol and calcined at 800°C for 12h in order to decompose the carbonate. After pulverizing, sample bars, 2.5mmø x 10mm were prepared by cold isostatic pressing at about 2ton for 1min. The bar sample was heated at 880°C-1400°C for 0.5-1h in a Pt crucible in order to investigate the distribution of MgO in molten and partially molten state. For the fine precipitates of Sr-Ca-Cu-O compounds sample, pellets were formed 20mm in diameter and about 2mm in thickness by applying 1.6ton/cm^2 pressure, and heated at 900°C for 1h onto Pt plate and quenching with metal copper plate and 3 times as the same treatment. The samples were heated at 900°C for 1h and cooled to 890°C-860°C with the rate of 10°C/min and quenched with Cu plate after holding for 1h, respectively, in order to investigate the appropriate temperature of the partially molten state at which pressing is applied. The sample was put a sandwich style with Pt plates between the upper and lower MgO plates[13]. The sample was heated at 900°C for 1h and cooled to 880°C for 2min and pressed to 900kgf with the rate of 1kgf/sec after the holding for 1h at 880°C. The crystalline phases were analyzed using X-ray powder diffraction. The microstructures were observed by SEM and OM(optical microscope) and analyzed by EPMA

RESULTS AND DISSCUSSION

One of the two methods which are the mechanical method and the chemical method for introducing pinning centers, the former process was studied by using partial melting method, because MgO particles were separated and floated with in the liquid phase(Fig. 1). Therefore, the samples were treated to be partially molten state at 900°C in order to obtain fine dispersion of MgO in 2212 crystals. The fine MgO particles were dispersed in the 2212, 2201 crystals and gold crystals and the particle size of MgO was almost independent of holding time at 900°C and even after crystallization of 2212 and 2201 phases(Fig. 2). These results indicate that the MgO particles wet with these phases. In the chemical process with Sr-Ca-Cu-O compound, the optical micrographs of the surface for the samples show the results that the gold phase was finely dispersed by the repetition of the processes, during 3 times treatments using the previous procedures(Fig.3). These trends were shown in the both samples with and without MgO in 2212 composition. In the particle size of the gold crystal, the samples with addition of MgO in 2212(Fig. 3 (a)(c) and ((e)) were constituted of smaller size than the samples without MgO in 2212. The reason why MgO increased the volume fraction of solid in the sample, it decreased the diffusivity for the crystal growth of Sr-Ca-Cu-O compound. The uniaxial pressing method in the partially molten state was attempted using these samples which treated 3 times by the repetition processes. Figure 4 shows the X-ray powder diffraction patterns with the samples which were annealed at 900°C for 1h and cooled to 890°C, 880°C, 870°C and 860°C and quenched with Cu plate after holding for 1h, respectively, in order to investigate the adequate temperature of the partially molten state which will be pressed. From the results of the X-ray powder diffractometry, Figure 5 shows the plot of the area ratio of 2212(115)/2201(113)which indicates relative amounts of 2212 formation and the area ratio of gold crystal/2212(113) for gold crystal in the samples as a function of annealing temperature. In the view of quantitative analysis, amount of gold crystals decreased but the amount of 2212 crystals increased with decreasing the temperature. The samples were pressed at 880°C, the reason why the sample was constituted of crystal of 2212, liquid and gold crystal at that temperature. Figure 6 shows the fractured surface of the sample which was annealed at 900°C for 1h and cooled to 880°C and held for 1h and then loaded by applying 900kgf. Figure 6 shows the oriented array in a vertical direction to the c-axis of 2212 crystal.

ACKNOWLEDGEMENTS

This study has been supported by the New Energy and Industrial Technology Development Organization(NEDO) under the management of the R&D of Basic Technology for Future Industries.

REFERENCES

1. D.Dimos, P. Chaudhari, J. Mannhart and F.K. LeGoues: Phys. Rev. Lett., 61(1988)219.
2. S. Jin, T. Tiefel, R. Sherwood, R. van Dover, M. Davis, G. Kammlott and R. Fsatnacht: Phys. Rev., B37(1988)7850.
3. S. Jin, T. Tiefel, R. Sherwood, M. Davis, R. van Dover, G. Kammlott and R. Fsatnacht and H. Keith: Appl. Phys. Lett., 52(1988) 2074.
4. M. Murakami, M. Morita, K. Doi, K. Miyamoto and H. Hamada: Jpn. J. Appl. Phys., 28(1989) L399.
5. M. Murakami, M. Morita, K. Doi and K. Miyamoto: Jpn. J. Appl. Phys., 28(1989) L1189.
6. M. Murakami, M. Morita and K. Miyamoto: Proc. Osaka Univ. Int. Symp. on New Development in Applied Superconductivity, Suita, 1988 (World Scientific, Singapore, 1989) p.95.
7. M. Murakami and M. Morita: Proc. Mater. Res. Soc. Meeting, 1988 (MRS, Tokyo, 1989) p.641.
8. H. Fujimoto, M. Murakami, S. Gotoh, N. Koshizuka, T. Oyama, Y. Shiohara and S. Tanaka: Advances in Superconductivity2(Springer-Verlag, Tokyo, 1990) p.285.
9. P. Kes, A. Pruymboom, J. van den Berg and A. Mydosh: Cryogenics, 29 (1989) 228.
10. M. Murakami, M. Morita, K. Doi and K. Miyamoto:Jpn. J. Appl. Phys., 28(1989) L1189.

11. D. Shi, M.S. Boley, U. Welp, J.G. Chen and Y. Liao: Phys. Rev., B40 (1989) 5255.
12. K. Egawa, T. Umemura, S. Kinouchi, M. Wakata and S. Utsunomiya: Jpn. J. Appl. Phys.,30 (1991) L1160.
13. T. Noda and Y. Shiohara: Advances in Superconductivity3 (Springer-Verlag, Tokyo, 1991) p.777.

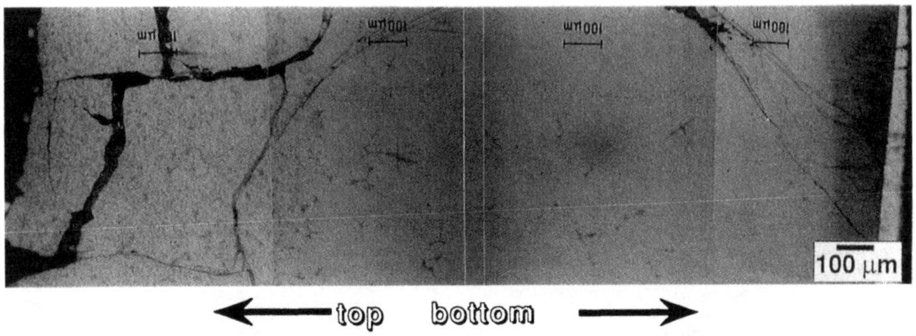

Figure 1. The optical microphotograph of the ground cross section of the sample which was heated at 1200°C for 1h and quenched in water.

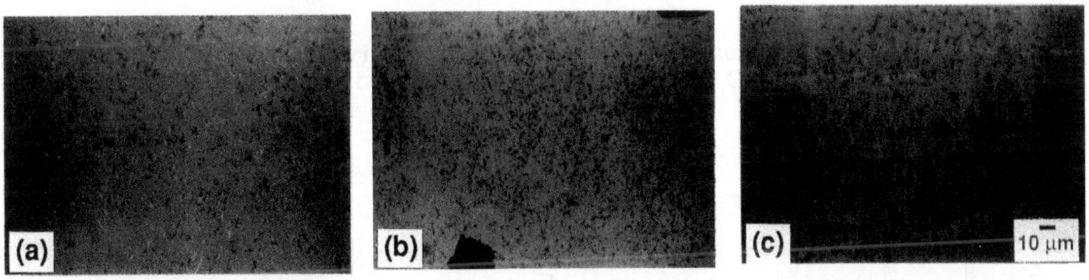

Figure 2. The optical microphotograph of the ground cross section of the sample which was heated at 900°C for 1h(a), at 900°C for 3h(b) and cooled from 900°C to 800°C with the rate of 2°C/h and to room temperature with the rate of 200°C/h(c).

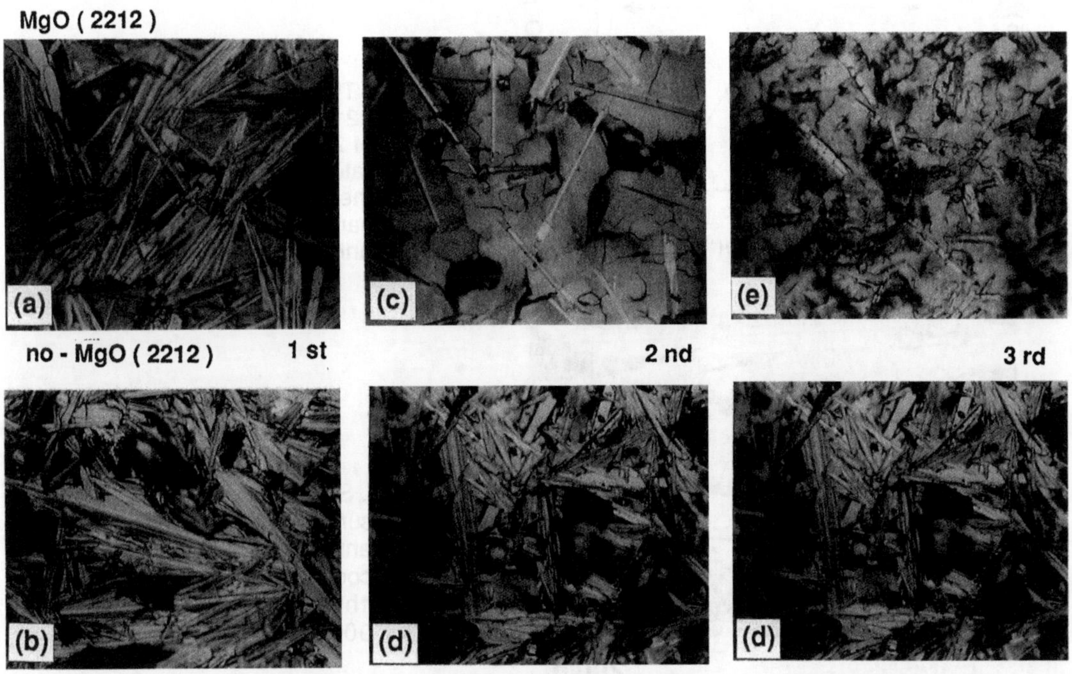

Figure 3. The optical microphotograph of the surface of the samples with addition of MgO(a),(c) and (e), without MgO (b), (d) and (f) which was heated at 900°C for 1h(a),(b), and 2nd heated at 900°C for 1h(c),(d) and 3rd heated at 900°C for 1h(e),(f).

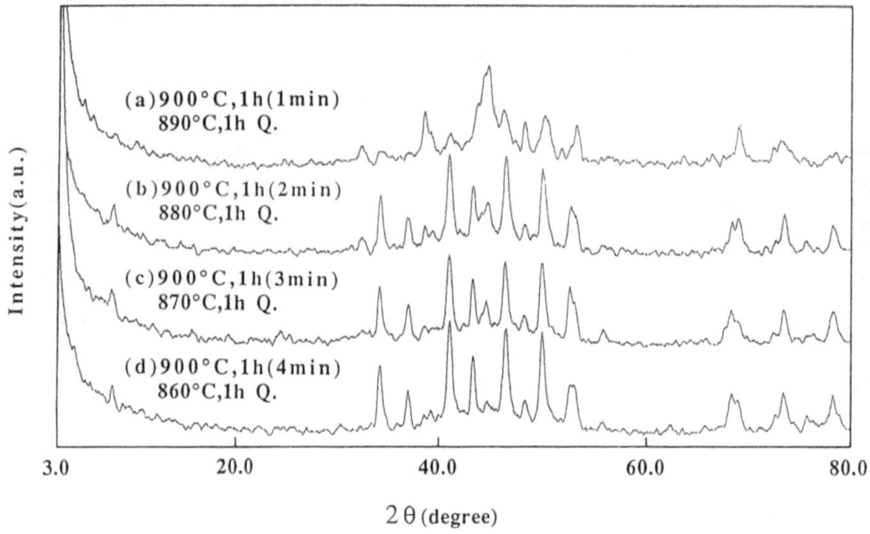

Figure 4. The X-ray powder diffraction patterns with the samples which were annealed at 900°C for 1h and cooled to 890°C, 880°C, 870°C and 860°C and quenched with Cu plate after holding for 1h, respectively.

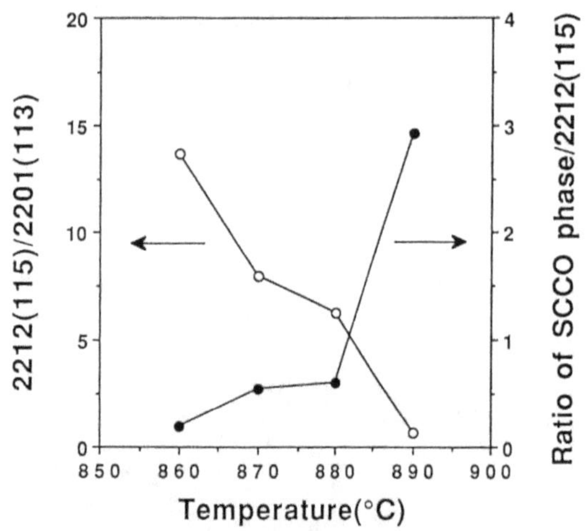

Figure 5. The plot of the area ratio of 2212(115)/2201(113) as amount of 2212 formation and the area ratio of gold crystal/2212(113) as the amount of gold crystal in the samples as the function of annealing temperature.

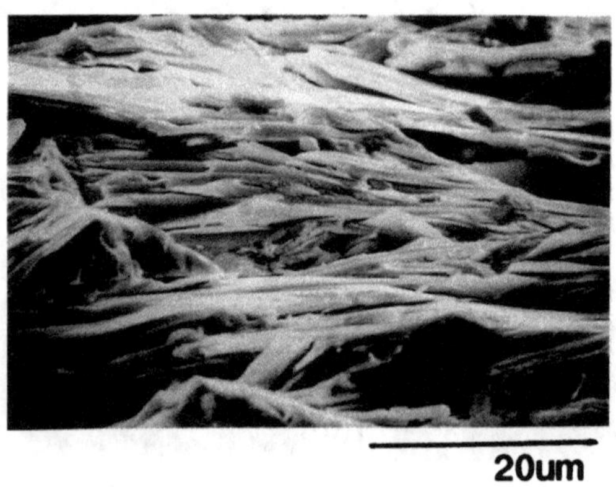

Figure 6. SEM photoes of the fractured surface of the sample which was annealed at 900°C for 1h and cooled to 880°C and held for 1h and then loaded by applying 900kgf.

Effects of Cation Substitution on the Magnetization for Bi-2212 Phase

M. WAKATA, S. TAKANO, F. MUNAKATA, and H. YAMAUCHI

Superconductivity Research Laboratory, ISTEC, Koto-ku, Tokyo, 135 Japan

ABSTRACT

Effects of various cation substitutions on the pinning properties of the Bi-2212 phase have been investigated. It has been found that the substitution of Y or rare earth element for (Sr, Ca) or Pb for Bi improves the pinning properties, while changes in oxygen content influence only slightly. The former substitution does not, however, change the irreversibility line, suggesting that the enhancement of the pinning is due to point like defects as in the case of the background pinning. The latter substitution improves the irreversibility line at high temperatures. In this case the pinning centers introduced are considered to be larger in size and possibly caused by the structural transformation between orthorhombic and monoclinic phases giving rise to a some kind of lattice distortion.

KEY WORD: Bi-2212 System, Cation Substitution, Flux Pinning, Irreversibility Line

INTRODUCTION

As compared to the 123-type superconductors the intra-grain critical current densities (J_c) for the Bi-based superconductors show stronger dependences on magnetic-field and temperature. According to a reasonable explanation the pinning centers in the Bi-system resemble point like defects, but the detailed nature of these centers is not yet clear.

Substitutional effects on the properties of the $Bi_2Sr_2CaCu_2O_\delta$ (Bi-2212) have been widely studied. The critical temperature (T_c), for example, behaves almost in a same way both in $Bi_2Sr_{2-x}La_xCaCu_2O_\delta$ and $Bi_2Sr_2Ca_{1-y}Ln_yCu_2O_\delta$ (Ln=Y, or rare earth element except La, Ce and Pr) systems. For the samples synthesized in air the T_c first increases up to about 90K with increasing x or y (x, y$\leqq$0.2). Further increase in x or y decreases the T_c and finally the superconductivity disappears. A similar behavior of the T_c is also observed for non-substituted samples when decreasing the oxygen content, δ. On the other hand single phase $Bi_{2-z}Pb_zSr_2CaCu_2O_\delta$ samples can be synthesized in a wide range of z in a reducing atmosphere. In this system the oxygen content decreases with increasing Pb content, but the T_c remains almost constant.[1].

The reported substitutional studies are mainly concentrated on the carrier density, the T_c as well as on the lattice and the modulation structures but the reports concerning the effects on the pinning are not frequent. In this study we have investigated the effects of various substitutions on the magnetization in order to clarify the origin of the background pinning in the Bi-2212 system.

EXPERIMENTAL

All samples were prepared by a conventional solid state reaction method using oxides and carbonates as starting materials. The $Bi_2Sr_{2-x}La_xCa_{1-y}Ln_yCu_2O_\delta$ (x+y=0.2) were calcined twice in air at 1063K for 10hr and sintered in air at 1133K for 30hr. In the case of $Bi_{2-z}Pb_zSr_2CaCu_2O_\delta$ (z=0, 0.1, 0.2, 0.3, 0.4, 0.5, 0.6, 0.8) the calcination was performed twice at 1033K for 10hr in air and the sintering was carried out at 1043-1093K for 30hr in a nitrogen atmosphere.
Structures and T_c^{on}'s for the samples were determined by X-ray diffraction (XRD) analyses and ac-susceptibility measurements, respectively. Magnetization studies were carried out using a SQUID magnetometer. For the determination of the irreversibility line each sample was ground and

pressed into a disc shaped pellet at 7000kgf/cm^2. The magnetization curve for these pellets was measured at a temperature (T), where the magnetic field was parallel to the axis of the sample disc.

RESULTS AND DISCUSSION

Effects of the Substitution of Y or Rare Earth Element for (Sr, Ca): The evaluation of the substitutional effects on the pinning properties was hampered by the fact that the rare earth doping affects the T_c also. To overcome this difficulty the magnetization curves for the substituted samples were compared not only with the curve of the non-doped and air-sintered sample but also with the data obtained for the non-doped and N_2-sintered sample (with nearly the same T_c as the substituted sample had). The compositions and the T_c values for these samples are shown in Table 1. The given values for the oxygen content were determined by a coulometric titration[1]. XRD analyses revealed that all the samples were single phase with a tetragonal structure. The hysteresis of the magnetization (ΔM) for each sample is shown in Fig. 1 as a function of the magnetic field. As there do not exist distinct differences in grain sizes or in crystal orientations, it can be said that ΔM is proportional to the "mean" value of the J_c's. As seen from the curves in figures, ΔM for the sample [Air] is larger in low fields and smaller in high fields than for the sample [N_2], which may be due to the fact that the carrier density is higher and the T_c (and therefore the B_{c2}) is lower in the air-sintered sample. However, the difference is relatively small in spite of the large difference in T_c. Therefore, it is assumed that the change in the oxygen content affects only slightly the pinning properties. On the other hand, ΔM's for La and/or Y substituted samples are higher at high fields than those obtained for [Air] and [N_2]. Such improvement of the pinning properties is more effective in the case of a simultaneous substitution of La and Y for Sr and Ca than in the case of La or Y substitution.

Table 1 The characteristics for the $Bi_2Sr_{2-x}La_xCa_{1-y}Y_yCu_2O_\delta$ samples sintered in air ([Air], [0.2La], [0.2Y] and [0.1La0.1Y]) and in N_2 ([N_2])[1].

Sample Name	Air	N_2	0.2La	0.2Y	0.1La0.1Y
x, y	0, 0	0, 0	0.2, 0	0, 0.2	0.1, 0.1
δ	8.284	8.154	8.337	8.341	-
T_c^{on} (K)	76	87	90	90	90

Pinning is based on the existence of some kind of inhomogeneities inside the grains. The inhomogeneity introduced by the above substitution ($M^{2+} \rightarrow Ln^{3+}$) may be (A) a random distribution of the charge at the lattice site and (B) a local lattice distortion. In order to investigate these effects, ΔM at 20K for the $Bi_2Sr_2Ca_{0.8}Ln_{0.2}Cu_2O_\delta$ samples were also measured. These samples are also single phase with a tetragonal structure. As shown in Fig. 2 the lattice parameters, a and c, are decreased with decreasing ionic radius of Ln^{3+}), which has already been reported by Yoshizaki et al.[2]. Figure 3 shows ΔM value as a function of the ionic radius of Ln^{3+}. As can be seen, ΔM is increased at each field with decreasing ionic radius. This tendency suggests that the point (B) mentioned above is one of the origins for the pinning because a larger difference in the ionic radius between Ca^{2+} (1.12Å) and Ln^{3+} induces a larger local lattice distortion. The linear extrapolations of ΔM's to the ionic radius of 1.12Å are shown by solid lines in Fig. 3. However, such values are still higher than those for [N_2], which are also plotted in Fig. 3 at the ionic radius of 1.12Å. This fact suggests that the point (A) is also enhancing the pinning properties of the cation substituted samples.

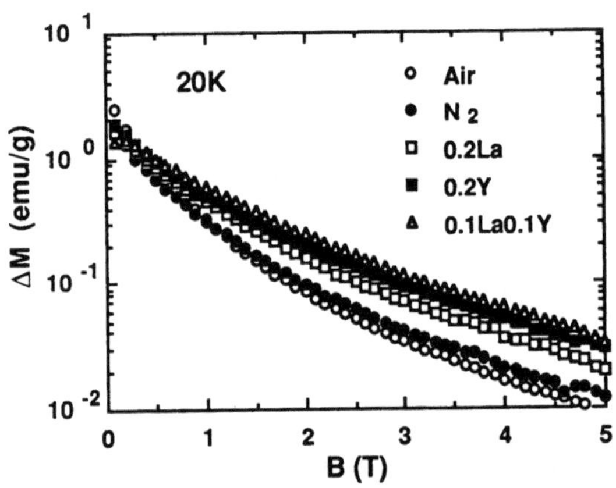

Fig. 1 The magnetic field dependence of the ΔM at 20K for the samples of [Air], [N_2], [0.2La], [0.2Y], and [0.1La0.1Y].

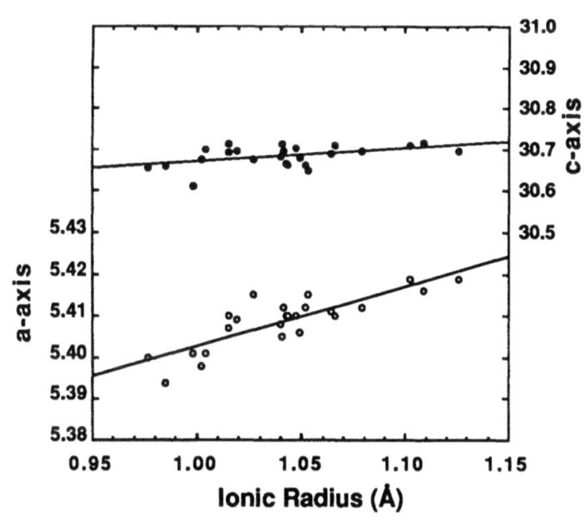

Fig. 2 The lattice parameters as a function of the ionic radius of La^{3+} for $Bi_2Sr_2Ca_{0.8}Ln_{0.2}Cu_2O_\delta$.

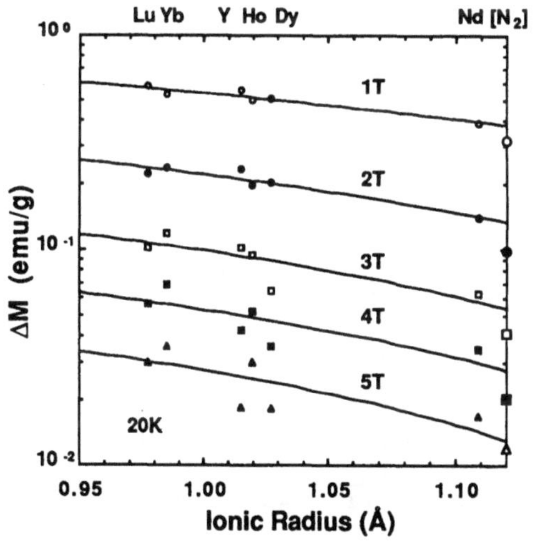

Fig. 3 The ΔM(20K) values as a function of the ionic radius of La^{3+} for $Bi_2Sr_2Ca_{0.8}Ln_{0.2}Cu_2O_\delta$.

Effects of the Substitution of Pb for Bi: Table 2 summarizes the compositions , the T_c values and the modulation periods for the $Bi_{2-z}Pb_zSr_2CaCu_2O_\delta$ samples sintered in N_2[1,3]. In this system the oxygen content decreases with increasing Pb content, however, the T_c remains almost unchanged as the average valency of (Bi, Pb) is simultaneously decreased[1]. The lattice structure is tetragonal for z<0.2, orthorhombic for z<0.4 and monoclinic for $z \geqq 0.4$. The presence of a Ca_2PbO_4 impurity phase was observed for the samples with $z \geqq 8$. The modulation period increased as the Pb content was increased. Pinning properties of these samples are summarized in Fig. 4 as a function of z.

As shown in Fig. 4, ΔM increases with increasing Pb content (z) in the region (z<0.4), where the structure is tetragonal or orthorhombic. The maximum value for ΔM is at a point around z=0.4, which is near the phase boundary between the orthorhombic and monoclinic structures. Above this point, the structure becomes the monoclinic and ΔM decreases with increasing z. In this region the modulation structure also disappears. These results suggest that some kind of inhomogeneous structure exists near the phase boundary between the orthorhombic and monoclinic structures and it is effective for the pinning. Obviously, the substitution of Pb for Bi introduces the intra-grain inhomogeneities like points (A) and (B) mentioned above. However, another inhomogeneity: (C) a some kind of distorted structure caused by the structure transformation, is thought to be more essential for the pinning in this system. In order to obtain more information the irreversibility line has been evaluated.

As the Bi-based superconductors exhibits strong two dimensional anisotropy in electromagnetic properties, it is difficult to obtain reliable data on the irreversibility line, $B_{irr}(T)$, by using a polycrystalline sample with randomly oriented grains if $B_{irr}(T)$ is also strongly anisotropic. Therefore, we used pressed samples and applied magnetic fields perpendicular to the surface. XRD measurements confirmed that the crystal-axes were well aligned in all the samples, certifying that the angles between the field and c-axis were small in each grain. The irreversibility line, $B_{irr}(T)$ is the field where ΔM(B,T)=0. In this experiment it was defined as the field where $\Delta M(B,T)=10^{-3}$emu/g.

Table 2 The characteristics for the $Bi_{2-z}Pb_zSr_2CaCu_2O_\delta$ samples sintered in N_2.

Sample Name	N_2	0.1Pb	0.2Pb	0.3Pb	0.4Pb	0.5Pb	0.6Pb
Pb Content, z	0	0.1	0.2	0.3	0.4	0.5	0.6
Oxygen Content, δ	8.154	8.130	8.100	8.054	8.003	7.983	7.931
T_c^{on} (K)	87	84	83	82	83	83	83
Modulation Period	-	4.9b	4.4b	8.6-10.4b	∞	∞	∞

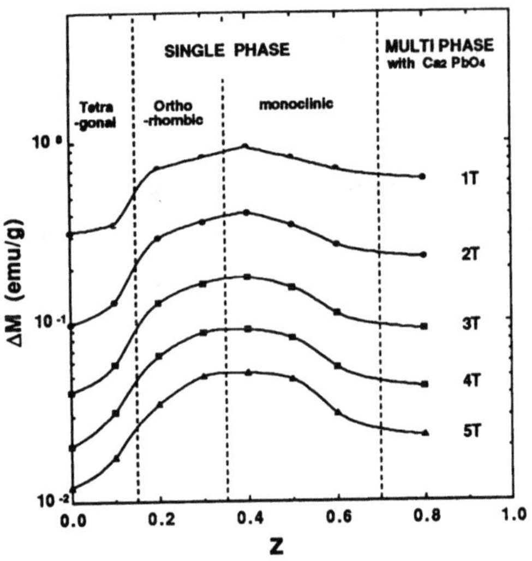

Fig. 4 ΔM at 20K as a function of z for the $Bi_{2-z}Pb_zSr_2CaCu_2O_\delta$ samples

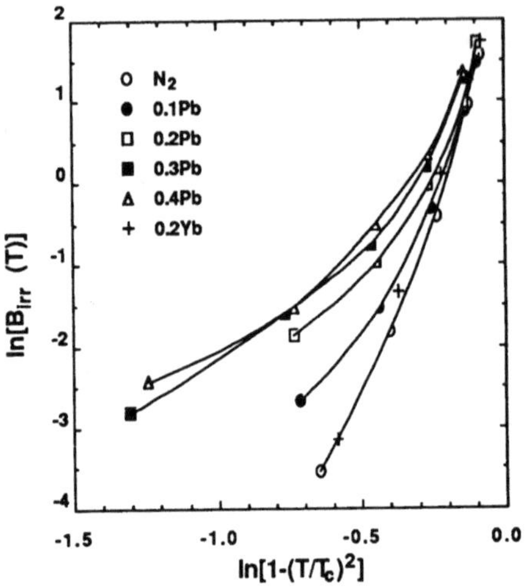

Fig. 5 The irreversibility line, $B_{irr}(T)$ for $Bi_{2-z}Pb_zSr_2CaCu_2O_\delta$ and $Bi_2Sr_2Ca_{0.8}Yb_{0.2}Cu_2O_\delta$

Figure 5 shows $\ln[B_{irr}(T)]$ as a function of $\ln[1-(T/T_c)^2]$ for six samples. It is known that this relation is linear near T_c and the slope (exponent: n) is about 1.5 for the Y-system and 3-4 for the Bi-system[4]. Although it is impossible to calculate exact values for the exponents from our data as the temperature is far from the T_c, the trends are clearly seen in Fig. 5. At low temperatures B_{irr} is almost independent of the sample composition. However, the temperature dependence is improved by the Pb substitution. The minimum slopes are 7.0 and 1.8 for the samples, [N_2] and [0.4Pb], respectively. The curvature of each line in Fig. 5 is positive, indicating that the exponents are smaller than the observed slopes. In this figure the data for the sample [0.2Yb] are also plotted (mark: +). No improvement in the normalized temperature dependence of B_{irr} is observed, suggesting that the pinning centers introduced by the substitution of Yb for Ca are point like defects as in the case of the background pinning. On the other hand, when Pb is substituted for Bi the pinning centers introduced are larger in size and B_{irr} is improved. This is consistent with the above discussion on another inhomogeneity (C): a some kind of distorted structure caused by the structural transformation between the orthorhombic and monoclinic phases.

Independent of this work Paulose et al.[5] have reported that the Pb substitution enhances the pinning at Pb contents of z=0.3 and 0.4 (but the sample with z=0.2 showed rather weak pinning). According to their considerations, the pinning centers may be due to the Ca_2PbO_4 impurity phase the concentration of which is about 3% in their samples. However, such a impurity phase was not observed in our samples with $z \leqq 0.6$, and the normal state resistivity had the minimum at z=0.4[3]. Furthermore, as shown in Table 2, the modulation period increases with increasing z, when approaching the phase boundary between the orthorhombic and monoclinic structures. and modulation disappears in the monoclinic structure region[3]. These facts suggest that the pinning is due to a some kind of distorted structure rather than an impurity phase.

ACKNOWLEDGEMENTS

A part of this work was supported by NEDO for R&D of Basic Technology for Future Industries.

REFERENCES

1. T. Kawano et al., in preparation for publication in Physica C
2. R. Yoshizaki et al., Physica C 171 (1990) 315
3. F. Munakata et al., Proc. 4th Int. Symp. on Superconductivity (Tokyo, 1991)
5. P. L. Paulose et al., Proc. 3rd M^2S-HTSC (Kanazawa, 1991)

Uniform Distribution of Precipitates in Bi-Base Low T_c Single Crystal-like Samples

SHIN-ICHI KINOUCHI, KUNIHIKO EGAWA, TOSHIO UMEMURA, AYUMI NOZAKI, and SHIN UTSUNOMIYA

Materials & Electronic Devices Laboratories, Mitsubishi Electric Corporation, Sagamihara, Kanagawa, 229 Japan

ABSTRACT

Fine precipitates may be one of candidates of pinning centers in oxide superconductors. The process of the crystal growth giving fine distributions of precipitates in $Bi_2Sr_2CaCu_2O_y$(2212) crystals was investigated. Pt powders were mixed with glassy oxide powders and the mixtures were heated into partially melting state and slowly cooled down for crystallization. This procedure has made the precipitate distribution more uniformly and the size of precpitates smaller than in the case of no Pt addition. To clear the role of Pt elements ,the grain structures in the cool down process were observed and analyzed.

KEY WORDS: Bi system superconductor, pinning, precipitate, Pt

INTRODUCTION

For energy applications of high T_c superconductors, many studies suggest it to be necessary to bring strong pinning centers in grains. It has been reported for the Y-system that the introduction of pinning centers into them could be done by bringing fine precipitates in grains [1]. This suggests the possibility for fine precipitate to be one of pinning centers. On this point of view, we have been studying the uniform distribution of fine particles in (2212) grains, using a process which goes through glass state and partial melting state [2]. That is, after melting and rapidly cooling, glass-like samples are heated into partial melting state and slowly cooled down to room temperature for crystallization. The (2212) crystals have grown 100μm order in size and included some kinds of precipitates about 10μm in diameter. But the size of precipitates could not have been controlled. Although the size become smaller in the Ca-poor sample, the amount of precipitates decreased in those. Recently, we have found that the addition of Pt elements makes the precipitates distribution more uniformly and the size finer. So as to distribute those more effectively, this effect of Pt elements have to be cleared. We observed the variation of the phase structures due to the addition quantity of the Pt element in the growth process. In particular, our attention was put on the correlation between Pt-compounds and (Sr,Ca)-compound.

EXPERIMENTAL

Starting materials were glassy oxides, which were obtained by melting the polycrystalline samples with the compositions of Bi:Sr:Ca:Cu=2:1.9:1.4:2.2 in a Pt-crucible at 1500°C and quickly cooling. By mixing Pt powders in the course of grinding those, three samples containing Pt- 0, 3 and 30(wt%)[calling P0,P3 and P30, respectively] have been got. Those were pressed into the discs of 20ϕ*3mm and inserted into the furnace of 915°C . After the temperature of the discs got stable at 915°C, it was cooled down to 885°C at the rate of 5°C/hr. At 885°C those were pulled out and quickly put into liquid N_2. From previous studies it was known that 885°C is the temperature at which two kinds of precipitates grow large in the sample without Pt elements [3]. The structure observations were done by EPMA(Electron Probe Microscopic Analysis),SEM(Scanning Electron Microscope),XRD(X-RAY Diffraction) and optical microscope.

RESULTS AND DISCUSSION

XRD results are shown in Fig.1. In the P0, major peaks were assigned to those of the (2212) phase and some peaks were assigned to those of $Bi_2Sr_2CuO_w$ (2201) phase. This result differs from that of which sample was annealed without powdering. In such the specimen, the (2212) phase did not be observed above 870°C. This indicates that the powdering process makes the (2212) phase boundary in the crystal phase and the temperature plane shift to the high temperature. The result of the P3 was similar to that of the P0, and the peaks of both the (2212)-

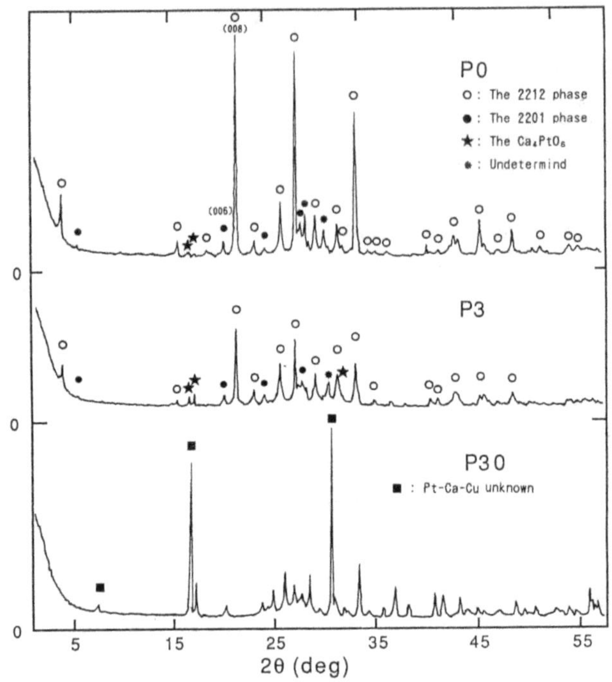

Fig.1. XRD results of the P0, P3, P30, including Pt-elements, 0, 3, 30(wt%), respectively.

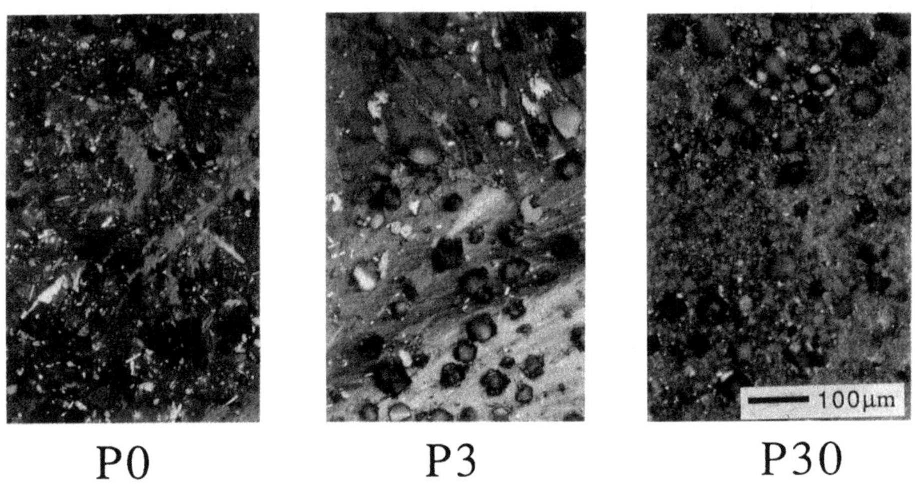

Fig.2. Polarized microscopic pictures of the polished surfaces.

and the (2201)-phasewere observed. However, the intensities of the peaks increased remarkably, which seemed to be assigned to Ca_4PtO_6, but their lattice constants became longer. Perhaps, this would be because some Sr ions replace the Ca ions. These peaks could be observed also in the P0, but were small in magnitude. The results of the P30 are quite different from the others. There were no (2212) and (2201) peaks. The ratio, R, of the amount of the (2201) to the (2212) was defined as R=[(006)peak intensity of (2201)]/[(008) of (2212)]. As shown in Figure 1, R=0.1 for the P0 and R=0.23 for the P3, and, it was remarkable that the (008)peak intensity in the P3 was much smaller than in the P0.

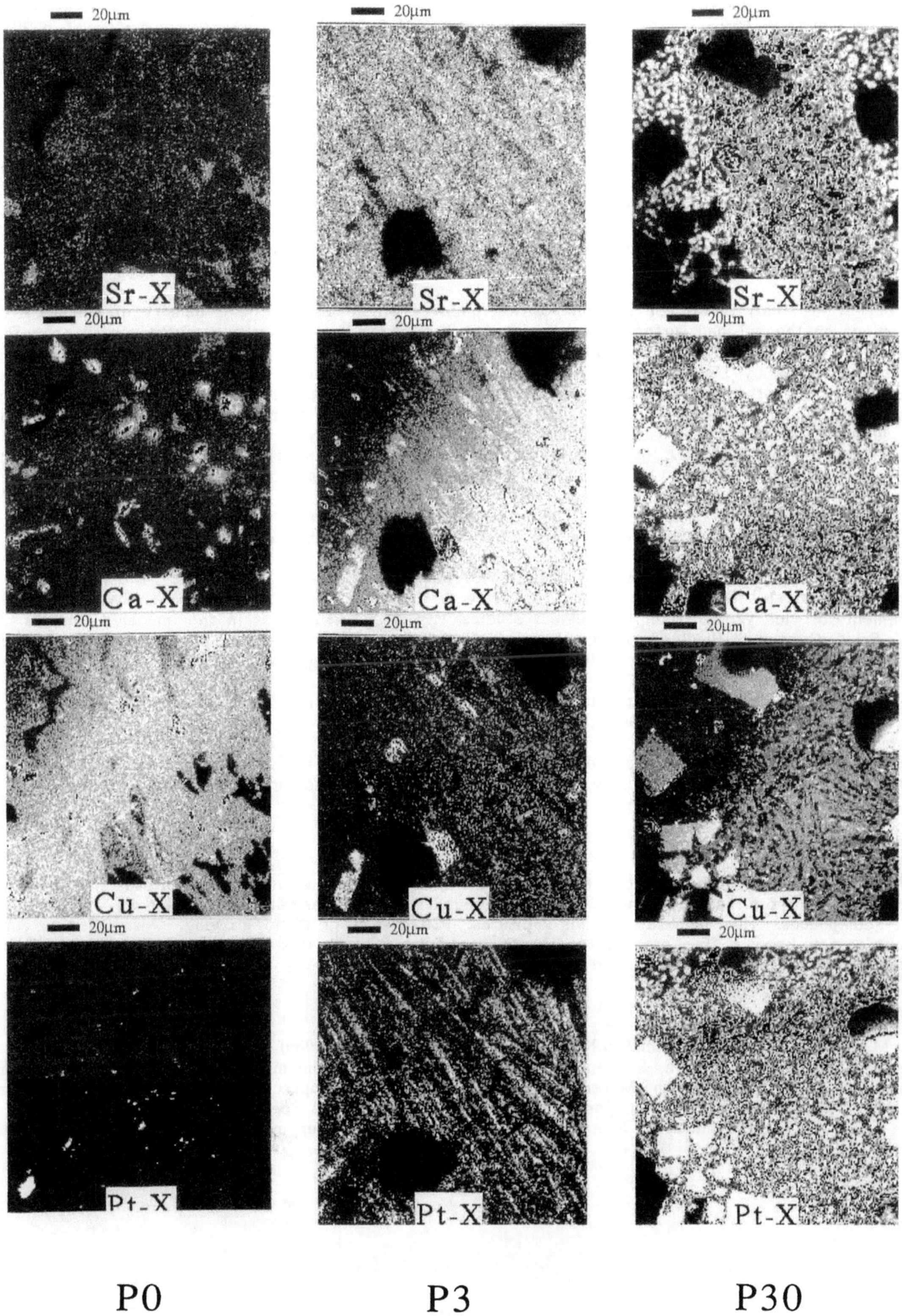

FIG.3. Composition pictures from EPMA analyses, from the left, the P0, p3, P30. With brightening, the element is denser.

The XRD results indicate two effects by Pt elements. One is to increase the Ca_4PtO_6 compound. The second is to restrain the (2212) phase from increasing at 885°C. However, an excessive addition of Pt elements prevented the (2212) and the (2201) crystals from growing.

As shown in Fig.2., optical microscopic observation of the P0 shows there grew large crystals. From XRD results, those were the (2212) phase, in which there were formless precipitates below 50μm. From previous studies, these precipitates were expected to be $(Sr,Ca)_2CuO_z$. In the P3, the crystals of a stripe form were observed, which were often observed in the case that the (2201) phase was detected by the XRD. And also, in the P0 there were some precipitates below 30μm with no special forms. The P30's structure was quite different from the above two. The matrix seemed not to be completely crystallized, however, in which precipitates 20-30μm in size were observed. Those precipitates crystallized in rectangular-like forms and did not appear in the P0 and P3. This fact was consistent with the result of the XRD. The major peaks of the P30's result of the XRD were those of rectangular precipitates.

From EPMA analyses we got the composition pictures which represented a density of the specified element on the polished surface, as shown in Fig.3. Comparing composition pictures with the SEM picture , principal compositins of precipitates and the matix could be cleared. On the P0, the precipitates observed by the optical microscope were classified into the two kinds, one was Ca rich compound, the other is Cu rich compound. Compairing with our previous studies, each should correspond to $(Ca,Sr)_2CuO_3$, $(Sr,Ca)_{3.0}Cu_{5.0}O_x$, respectively. And there was also another compound including Pt elements, which might come from Pt crucible. This would Ca_4PtO_6 , because Ca_4PtO_6 was observed by the XRD. In contrast with previous studies, $Bi_{2.0}Sr_{2.7}Ca_{1.0}O_y$ and Sr-Pt-O compound could not be found. On the P3, as in the P0, Ca- rich compounds, Cu- rich compounds, and Pt-compounds are found in the composition pictures. Compairing with the result observed in the P0, Pt compounds were in large amount. The rectangular-like precipitates observed by the optical microscope in the P30 were found to be Pt and Ca rich compounds from the composition pictures.

From those composition pictures, Pt and Ca rich compounds increased with the Pt amount increasing. This indicates that with increasing Pt elements the more Ca elements have to be bound to Pt elements or vice versa. From the previous studies, the final precipitates in the annealed samples are $(Ca,Sr)_2CuO_3$ and Sr-Pt-O compounds. It needs Ca elements for the former precipitates to grow. So the more Pt-compounds increase, with the $(Ca,Sr)_2CuO_3$ decreasing.

On the other hand, the fine ditribution of Pt-compounds is worthy of notice, since too big precipitates are obstacles for superconductivity. Large Pt-compound about 40μm in size was observed in the P3, however, in Fig.3. almost Pt-compounds appeared to be about 1μm in size ,along a growth direction. The former big compound would be introduced into, when Pt powders was added in ,because of its form, like a needle. So, the fine Pt particles are necessary to get fine Pt compounds.

CONCLUSIONS

We have investigated the role of Pt elements for fine distributions of the precipitates in the (2212) crystals. We guess that Pt elements preferably couple with Ca elements and become Pt-Ca compounds. This prevents the precipitates, $(Ca,Sr)_2CuO_3$, from growing. Also, Pt-Ca compounds may become fine precipitates if not too much. So, it is necessary to find most suitable contents of Pt elements to get fine distributions of the precipitates.

REFERENCES

1. M. Murakami, M. Morita, and N.Koyama, Jpn.J.Appl.Phys. 28 (1989) L1754
2. T. Umemura, K. Egawa, S. Kinouchi, A. Nozaki, and S.Utsunomiya, Proceedings of the Third International Conference on Materials and Mechanisms of Superconductivity on High-Temperature Superconductors, Kanazawa, Japan, 1991, to be published as a Special Issue of Physica C.
3. K. Egawa, T. Umemura, S. Kinouchi, W. Wakata, and S. Utsunomiya, Jpn.J.Appl.Phys. 30 (1991) L1160

Flux Pinning Characteristics in Superconducting Bi-(2212) Single Grains with Normal Particles

T. UMEMURA[1], K. EGAWA[1], S. KINOUCHI[1], A. NOZAKI[1], S. UTSUNOMIYA[1], B. NI[2], E.S. OTABE[3], and T. MATSUSHITA[3]

[1] Materials & Electronic Devices Laboratories, Mitsubishi Electric Corporation, Sagamihara, Kanagawa, 229 Japan
[2] Department of Electronics, Kyushu University, Fukuoka, 812 Japan
[3] Department of Computer Science and Electronics, Kyushu Institute of Technology, Iizuka, Fukuoka, 820 Japan

ABSTRACT

The temperature and magnetic field dependence of the critical current density, J_c, in the melt processed $Bi_2Sr_2Ca_1Cu_2O_y$ superconductors, (2212), with normal particles was measured. Although J_c at 4.2K was about one order of magnitude larger than in the polycrystalline $(Bi,Pb)_2Sr_2Ca_2Cu_3O_z$, its temperature dependence seemed to be enhanced by the existance of weak links. The irreversibility magnetic fields were represented as $[1-(T/T_c)^2]^n$ with the exponent n=5.44 at low temperatures and n=1.5 in high temperatures. Discussion is given on the dominant pinning centers by comparing the results and the theoretical predictions based on the flux creep.

KEYWORDS: Bi-based superconductor, flux pinning, critical current density, irreversibility line, flux creep, weak link,

INTRODUCTION

Bad influence of connections at grain boundaries on the critical current density in the Bi-based superconductors can be reduced by crystal-axis alinement, stronger grain coupling and decrease of grain boundaries. Therefore, the superconducting wires made of the Bi-system are commonly fabricated by the powder -in-tube method. Many measurements have been done for the wires on the temperature and magnetic field dependence of the critical current density and its angle dependence between the wire and the magnetic field. Although the effect of weak links at grain boundaries can be efficiently reduced by the mechanical pressing and so on, the intragrain critical current density still shows a drastic decrease due to the flux creep with increasing the temperature and the magnetic field.

We have been investigating the characteristics of the flux pinning in the high-T_c oxide superconductors through the comparision between the experimental results and the theoretical predictions based on the flux pinning, and have pointed out that from the temperature dependence of the pinning force density in the polycrystalline $(Bi,Pb)_2Sr_2Ca_2Cu_3O_z$ superconductors a collective pinning by point defects seem to be dominant [1]. The irreversibility line, i.e. a boundary between zero and nonzero critical current density on the temperature vs magnetic field plane, is far from the phase boundary $B_{c2}(T)$ in these superconductors. This suggests that their application may be restricted only within low temperature and low magnetic field. On the other hand, it is well known that the macroscopic pinning force density in Y-Ba-Cu-O with fine normal particles prepared by the so-called MPMG method is significantly enhanced in comparision with that in sintered Y-Ba-Cu-O. Existence of such fine normal particles was also reported for Bi-systems[2]. Therefore, it is important to distinctively clear the contribution to the flux pinning from each kind of pinning center such as lattice defects or fine normal particles in grains in these superconductors. For the improvement of the irreversibility line, it is needed to introduce the pinning centers which effectively strengthen the macroscopic pinning force even at high temperatures and high magnetic fields.

In this study, the Bi-based superconductors with fine particles were synthesized by melt processing, and what is the cause of the observed pinning characteristics was discussed.

EXPERIMENTS

The mixtures of Bi_2O_3, $SrCO_3$, $CaCO_3$ and CuO powders in Bi: Sr: Ca: Cu=2: 2: 1.2: 2.1 were calcined at 1073K for 12 hours in air. After powdered and pressed into discs of $5mm^t$, the powder was reacted at 1073K for 48hours in air. The reduced chips from the discs were melted in Pt crucible at 1775K, remained for 5minitues, and quickly quenched to the plate not exceeding $1mm^t$ between the copper-plates. After powdered below $0.1mm^d$ with an addition of PtO-0.5wt%, the powdered were pressed into the disc of $3mm^t$. Figure 1 shows the procedure of heat treatment for crystallization. The disc was heated on Pt nets up to 1195K at the rate of 2deg/hour, and remained in the partially melting state at the temperature, and then crystallized by cooling down to room temperature at the rate of -1deg/hour in flowing O_2. Such a gradual temperature increase before reaching the highest temperature was effective for decreasing voids in the disc in crystallization. As a result, the obtained dense disc contained some single grains with the size larger than 1mm in dimension[3]. Since the single grains were cleavable, it was easy to obtain the thin slab with the wide surface normal to the c-axis. The measured specimen for the ac inductive measurement and the magnetization were of $1.5x1.5x0.15mm^3$ and $1.5x1.5x0.20mm^3$ in size, respectively.

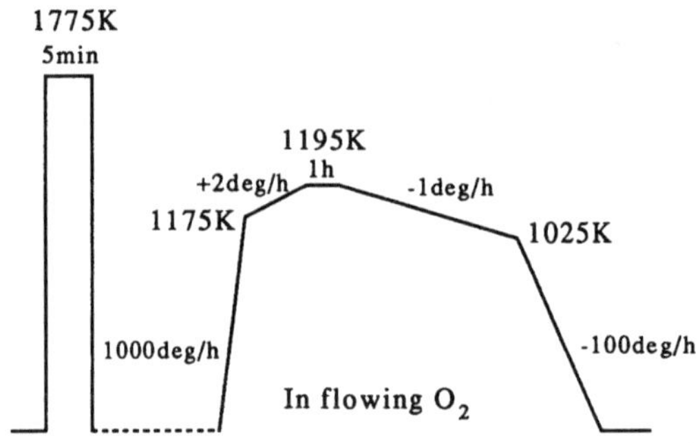

Fig. 1 Procedure of heat treatments of the melt processed Bi-(2212).

The superconducting transition temperature, T_c, was 83K resistively. The critical current density was measured ac inductively at various temperature in the deware with a temperature control facility. The temperature stability during the measurement was within 0.3K. In the ac inductive measurement, the dc and superposed small ac magnetic fields were applied parallel to the wide surface of the specimen, that is, normal to the crystal c-axis. The frequency of the ac magnetic field was 35Hz. Critical current densities at each dc magnetic field have been estimated from the magnetic flux going in and out the specimen measured as a function of the amplitude of the ac magnetic field[4]. The magnetization in the magnetic field normal to the c-axis was measured to estimate the irreversibility magnetic field, B_i. B_i at each temperature was determined under the criterion of $J_c=10^6 A/m^2$ estimated using the critical state model for the magnetization.

RESULTS AND DISCUSSION

The magnetic field dependence of the critical current density measured by the ac inductive method is shown in Fig. 2. J_c at 4.2K was about one order of magnitude larger than in the polycrystalline $(Bi,Pb)_2Sr_2Ca_2Cu_3O_z$ superconductors, (2223), but these are almost the same at the temperature of $T_c/2$, that is, 53K for the (2223) and 41.5K for the (2212), respectively[1].

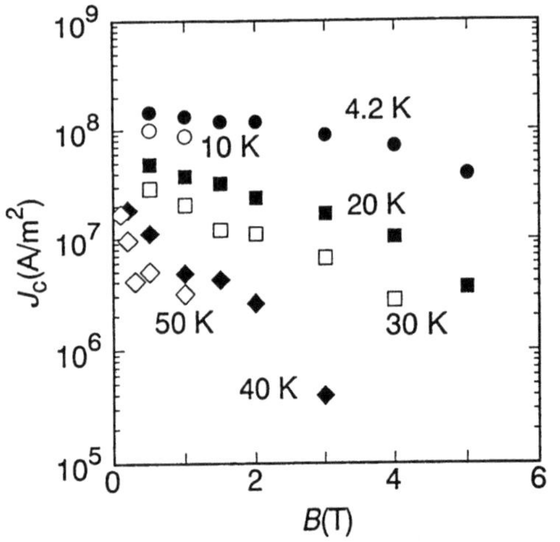

Fig. 2 Critical current density vs magnetic field at various temperatures.

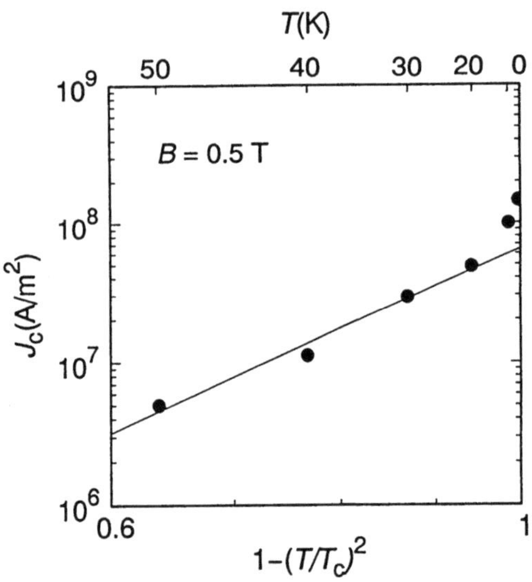

Fig. 3 Double logarithmic plots of critical current density vs $1 - (T/T_c)^2$.

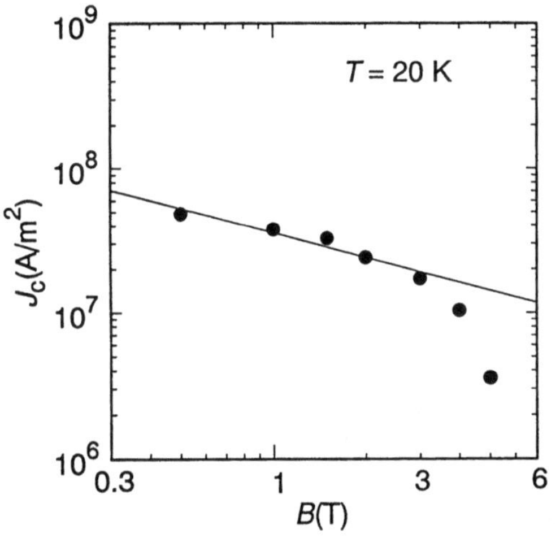

Fig. 4 Double logarithmic plots of critical current density vs B.

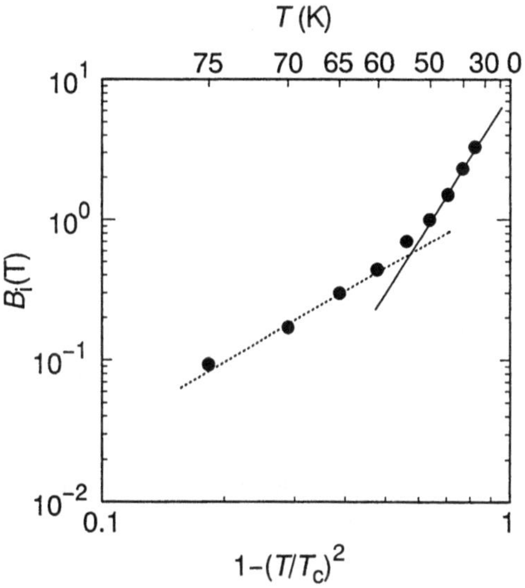

Fig. 5 Temperature dependence of the irreversibility field as a function of $1 - (T/T_c)^2$. Solid and broken lines represent the temperature dependences given by Eq. (3) with n=5.44 and n=1.5, respectively.

The temperature dependence of the J_c at the magnetic field of B=0.5T is replotted in Fig. 3 as a function of $1- (T/T_c)^2$. The cause for the abrupt decrease of the Jc at very low temperatures seems to be attributed to weak links on the analogy with the results in Ag theathed tape wires [5]. The temperature dependence of the J_c in the region of the temperature above 20K is approximately represented by the straight line with the exponent m=5.84, where m is the scaling parameter of the ideal pinning force density free from the flux creep at low magnetic field represented as

$$F_p(B,T)=AB_{c2}{}^m(T)B^{\gamma}. \qquad (1)$$

Figure 4 shows the magnetic field dependence of J_c at 20K where it seems not to be efficiently affected with weak links. The slope of the present straight line gives γ=0.43. The rapid decrease of the J_c at high magnetic fields seems to be attributed to the flux creep, since the applied magnetic field is close to the irreversibility field.

From the microstructure in the grain, the J_c due to the flux pinning by normal particles can be estimated theoretically. The present specimens contained the normal particles about 5μm^d and the volume fraction, ζ, about 10% in grains. The contribution to the J_c can be estimated at low temperature and low magnetic field by

$$J_c \simeq \frac{\pi B_c^2 \zeta \xi_{//}}{4\mu_0 B a_f D} \qquad (2)$$

where B_c is the thermodynamic critical field, a_f the flux lattice spacing, $\xi_{//}$ the coherence length in the direction parallel to the a-b plane and D the mean size of the normal particles. At T=20K and B=1T, substituting B_c=0.5T, $\xi_{//}$ $=2.0\times10^{-9}$m and $a_f=4.89\times10^{-8}$m, we have $J_c=1.28\times10^8$ A/m^2[6].

On the other hand, from the relating of $B_c \propto B_{c2}$ and $\xi \propto B_{c2}{}^{-1/2}$ for the temperature dependence, the temperature dependence of the J_c expressed by Eq.(2) is represented by the exponent m=1.5. Since the observed m value, 5.84, is much larger than this, the observed Jc does not originate from the normal particles. Such a large m value suggests that a kind of point defects of some kind are appropriate candidates for pinning centers at low temperature. The observed J_c was 3.8×10^7A/m^2 in the same condition, and about one third of the expectation value due to the normal particles. The results mentioned above suggests that both the contributions from normal particles and point defects are much reduced by the common cause, e.g. weak links.

In the optical observation of the melt processed single grains, it was often observed that the large single grain consisted of superposed platelike crystal domains, which likely brought the small misalignment of the c-axis normal to the plate-like domain. Under the condition for the J_c measurement, since supercurrents flow across interfaces of misaligned domains, such the domain boundaries may be considered as a candidate for weak links.

Figure 5 shows the irreversibility field, B_i, determined using the J_c evaluated from the magnetization. The tendency of the temperature dependence is different between the low and high temperature regions. Supposing the temperature dependence of the B_i as

$$B_i \propto [\, 1 - (T/T_c)^2]^n, \tag{3}$$

we have $n=2m\,/\,(3-2\gamma)$ from the flux creep theory[7]. Substituting the values of m and γ obtained from the J_c properties shown in Fig. 3, 4, the solid straight line of the exponent n=5.44 as shown in the figure is obtained. The observed temperature dependence of the B_i agrees qualitatively with the prediction from the results of the J_c measured inductively below 50K.

In order to compare quantitatively the results observed at low temperatures and the theoretical predictions, we evaluated the size of the flux bundle. In the elastic limit of the flux lattice, the number of fluxoids in the flux bundle is represented by g^2 which is given by

$$g = \left(\frac{C_{66}}{2\pi J_c B a_f} \right)^{\frac{1}{2}} \tag{4}$$

In the above $C_{66}=B_c{}^2B/4\mu_0 B_{c2}$ is the shear modulus at low fields. At T=20K and B=1T, supposing B_c=0.5T and $B_{c2//}$ = 100T, we have $C_{66}=4.97\times10^2 N/m^2$ and g=6.5. On the other hand, if the observed results of B_i=1.0T at 50K and Tc=83K are substituted into Eq. (4), we have g=10.5. This is almost the same as the theoretical prediction.

The temperature dependence of the B_i in the high temperature region reprecented by the dotted line in Fig. 5 gives the exponent n=1.5. This is just the same with the theoretical prediction assuming the flux pinning due to the normal particles. This is similar to the results in the Y-Ba-Cu-O specimen prepared by the MPMG process, and shows the decrease of B_i with elevating the temperature becoming small in the high temperature region. Therefore we guess this comes from the flux pinning due to the normal particles.

In conclusion, by comparing the observed results and the theoretical prediction based on the flux creep, it was pointed out that the behavior of the flux pinning was attributed to point defects in the low temperature region and the normal particles in the high temperature region. For the improvement of the B_i, the specification of the pinning centers and the elucidation of the pinning mechanism are necessary.

REFERENCES

1. Matsushita T, Otabe E. S, Nakatani T, Ni B, Yamafuji K, Umemura T, Egawa K, Wakata M, Utsunomiya S (1991) Proceedings of International Cryogenic Materials Conference, Huntsville. (to be published)
2. Hikata T, Ueyama M, Mukai H, Shibuta N, Kato T, Sato K (1991) in Kajimura K, Hayakawa H (eds) Adv. Superconductivity III, Springer-Verlag, Tokyo, pp 515- 518
3. Egawa K, Umemura T, Kinouchi S, Wakata M, Utsunomiya S (1991) Proceedings of International Cryogenic Materials Conference, Huntsville. (to be published)
4. Ni B, Munakata T, Matsushita T, Iwakuma M, Funaki K, Takeo M, Yamafuji K (1988) Jpn. J. Appl. Phys. 27: 1659
5. Hikata T, Sato K, Ni B, Otabe E.S, Matsushita T, Takeo M, Yamafuji K (1991) in Kajimura K, Hayakawa H (eds) Adv. Superconductivity III, Springer-Verlag, Tokyo, pp 511- 514
6. Ni B, Yamafuji K, Mori N, Ogi K, Matsushita T, Funaki K (1991) in Kajimura K, Hayakawa H (eds) Adv. Superconductivity III, Springer-Verlag, Tokyo, pp 503 - 506
7. Matsushita T, Fujiyoshi T, Toko K, Yamafuji K (1990) Appl. Phys. Lett. 56 : 2039

High Resolution Electron Microscopy of Sintered $YBa_2Cu_3O_{7-x}$

YOSHITSUGU TOMOKIYO[1], TAKASHI YOSHINO[2], YOKO SUYAMA[3], TAKESHI MANABE[1], and EISHI TANAKA[1]

[1] Research Laboratory of High Voltage Electron Microscope, Kyushu University, Fukuoka, 812 Japan
[2] NGK Insulators Ltd., Nagoya, 467 Japan
[3] Japan Fine Ceramics Center, Nagoya, 456 Japan

ABSTRACT

Sintered specimens have been observed through high resolution electron microscopy. Lattices of orthorhombic structure are slightly distorted in the vicinity of a twin boundary. Three types of grain boundaries (GBs) are present: (1) Some grain boundaries are parallel to a C-plane of one grain where a lot of stacking faults is concentrated along the GB. (2) The others have amorphous layers of 1~4 nm in thickness at the boundaries. Lattice distortion and stacking faults were not found near the GBs. (3) A coincidence tilt boundary accompanies neither stacking fault nor amorphous layer, but an axial length varys near the GB. Cracks parallel to a C-plane are frequently presnt. A width of crack is in the range of 1 nm ~ several tens of nm. A stacking fault present in an interior of grain consists of extra planes of Ba layer and Cu-O chain. The lattice distortion at a twin boundary, a nano-crack and an isolated stacking fault in the internal region of a grain may act as an effective pinning center but most of the structures mentioned above seem to be responsible for low Jc value.

KEY WORDS: micro structure, twin boundary, grain boundary, stacking fault, crack, pinning center

INTRODUCTION

High Tc superconductor $YBa_2Cu_3O_{7-x}$ has been extensively studied from various points of view, but critical current density Jc is insufficiently low for a practical application, especially in a sintered sample. Various attempts have been made so far to improve Jc and its value has become nearly the required level [1~4]. Weak link at GBs, effective pinning centers and flux creep are considered to be important factors that determine the Jc. It is quite important for investigation of pinning mechanism to observe nano-structures since the coherence length in the high Tc oxide superconductor has been estimated to be as small as 1 nm. In this paper we examine sintered $YBa_2Cu_3O_7$, by means of high resolution electron microscopy (HREM), concentrating on nano-structures which seem relevant to Jc. Physical properties and structure parameters of specimens used in the present study are listed in Table 1. Thin foil specimens were prepared by an usual ion milling technique. Electron microscopy was carried out by using a high resolution electron microscope JEM-4000EX/T at 400 kV.

RESULTS AND DISCUSSION

1. Twin boundary

An orthorhombic phase of $YBa_2Cu_3O_{7-x}$ usually consists of twin lamellas to reduce total lattice strains introduced upon phase transision from a tetragonal phase. An average width of twin lamella and distribution of widths depnd on oxygen content 7-x. Crystal lattices are distorted at a twin boundary on account of orthorhombicity [5]. One of the typical examples of HREM images of twin boundaries is shown in Fig.1a. Dark region along the boundary indicates slight lattice distortion. Axial lengths and orthorhombicity b/a measured from the HREM image is shown in Fig.1 b and c as a function of distance from the boundary. The variation of b/a near the boundary seems to be due to lattice distortion produced to accomodate stress originated from a difference in axial lengths at the boundary.

Table 1 Values for physical properties of specimens used.

Specimen	Oxygen Content 7 - x	Critical temp. Tc (K)	Critical current Jc (A/cm²)	Lattice parameters a	b	c (nm)
A	6.864	93.8	243 (at 77 K)	0.382	0.388	1.168
B	6.537	59	403 (at 39 K)	0.383	0.388	1.173
C	6.220	< 14	—	0.386	0.387	1.181

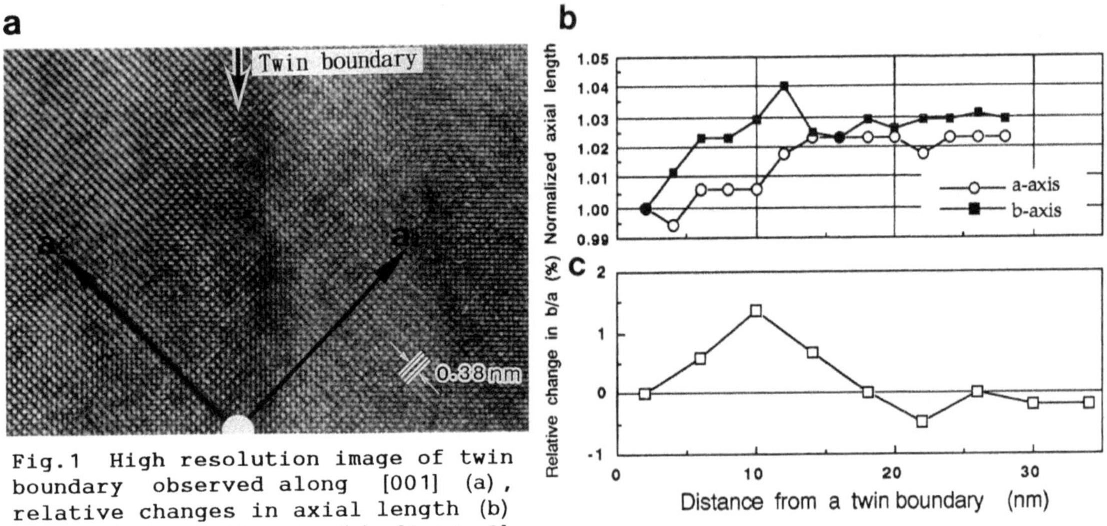

Fig.1 High resolution image of twin boundary observed along [001] (a), relative changes in axial length (b) and orthorhomicity b/a (c). [Spec.:A].

2. Grain boundary

Figures 2~5 show typical examples of GBs observed. Intergranular phases or voids have not been observed in our specimens. There exist three types of GBs; ① A GB has a thin amprphous layer of 1~4 nm in thickness. Neither lattice distortion nor stacking fault (SF) is present near the boundary as apparent in Fig.2a. ② A GB is parallel to a C-plane of one grain where a lot of SFs is present as apparent from wavy contrast parallel to a C-plane. The contrast also indicates the presence of dislocation loops (Fig.3). This GB corresponds to type I of classification by Nakahara et al. [6] Local change in composition may occur at the heavily faulted region because the stacking faults occasionally accompany extra layers of heavy elements. ③ A coincidence tilt GB is frequently observed. In Fig.4a C-planes of both grains are tilted by about 20°. Inter-atomic distances measured from Fig.4a are ploted in Fig.4b as a function of distance from the GB. A length of C-axis varies in a narrow region of about 20 nm along the GB. Figure 5 shows another example. The GB is parallel to a C-axis of one grain where lengths of a- and b- axes vary near the GB. These changes in axial lengths probably take place as a result of relaxation and rearrangement of atoms near the GB to maintain charge neutrarity. In the case of coincidence tilt GB of a-b type, Jc value is very sensitive to a misorientation angle [7,8]. However, the GB of c-c type shown in Fig.2c seems not to affect Jc value. Above results suggest that the effect of GB on Jc is different depending on character of the GB even if the boundary has neither voids nor precipitates of other phases. All the GBs may not decrease Jc.

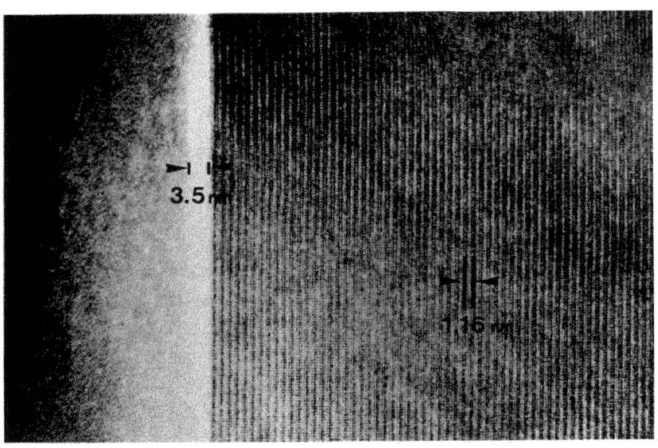

Fig.2 Typical grain boundary with thin amorphous layer. [Spec.:A].

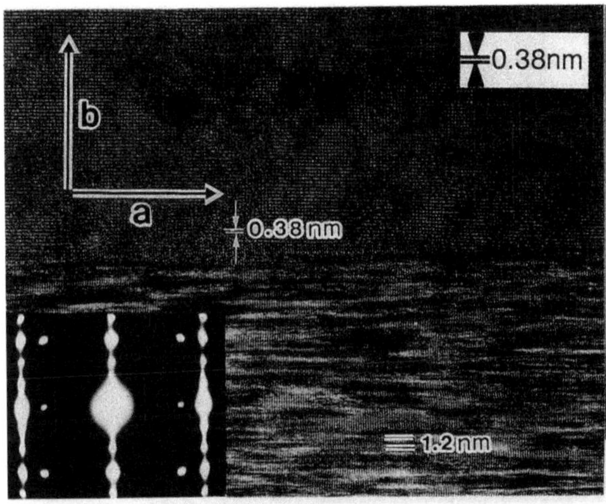

Fig.3 Example of grain boundary parallel to (001) plane of one grasin where stacking faults are densely concentrated. [Spec.:A].

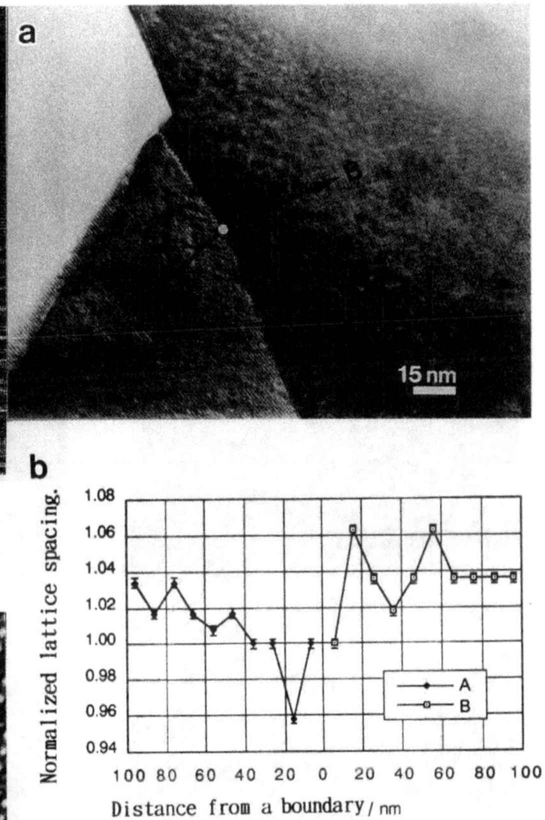

Fig.4 (a) Coincidence tilt boundary formed by two grains whose C-axes are parallel to the micrograph. (b) Relative change in lattice spacing of (001) planes measured as a function of distance from the boundary. [Spec.:C].

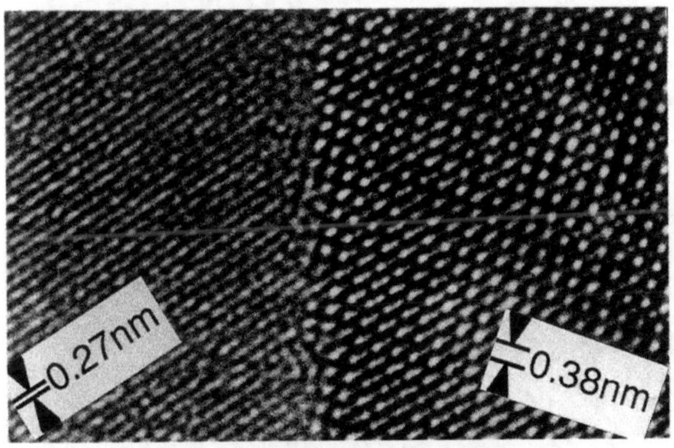

Fig.5 Coincidence tilt boundary. C-axis of right hand grain is perpendicular to the micrograph. [Spec.A]

3. Crack

Cracks observed were always parallel to a C-plane, suggesting the weakness of bonding along c-axis. A width of crack is 1 nm for smaller case and several tens of nm for larger case (Figs.6a and b). Some cracks are considered to be produced during specimen preparation, namely before final sintering, because many cracks are not empty but filled with amorphous phase or crystallites. $YBa_2Cu_3O_{7-x}$ has a tendency of cleavage along a C-plane because it has a layer structure along a C-axis. Furthermore lattice shrinkage during absorption of oxygen or upon cooling is larger along a C-axis than along other two axes. These are the origins of cracks parallel to a C-plane [6]. Microcracks cause flux creep [9] and reduce the Jc value, although nano-craks may hardly affect Jc.

4. Stacking fault

Stacking fault (SF) is one of the most typical lattice defects in $YBa_2Cu_3O_{7-x}$ and is found on a (001) plane (Fig.5a). SFs are frequently present in an internal region of grain as well as localized at certain GB as described above. SFs terminate at GBs or at partial dislocations. One of the typical examples of SFs is shown in Fig.5b where we can clearly see extra layers of Ba and Cu-O planes. This kind of SF seems to act as an effective pinning center, while SFs concentrated near a GB may reduce current transport.

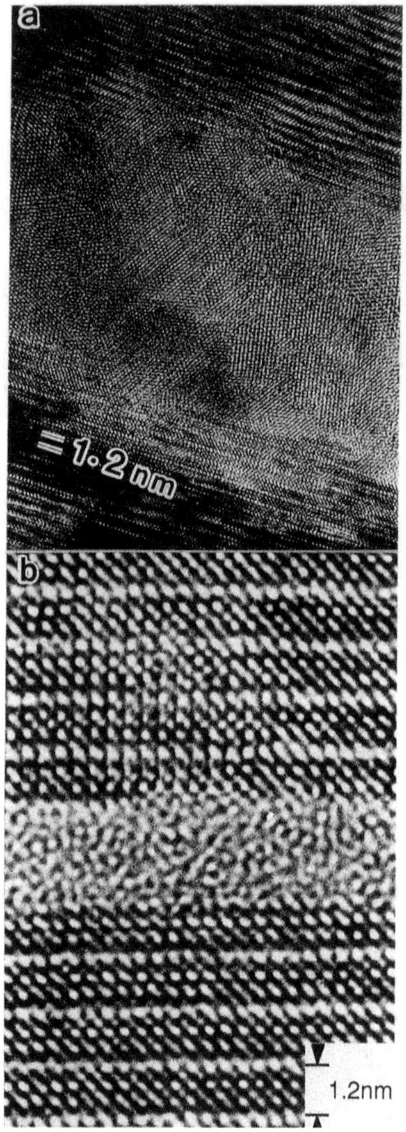

Fig.6 Microcrack filled with crystallites (a), [Spec.:B] and nanocrack filled with amorphous phase (b), [Spec.:A].

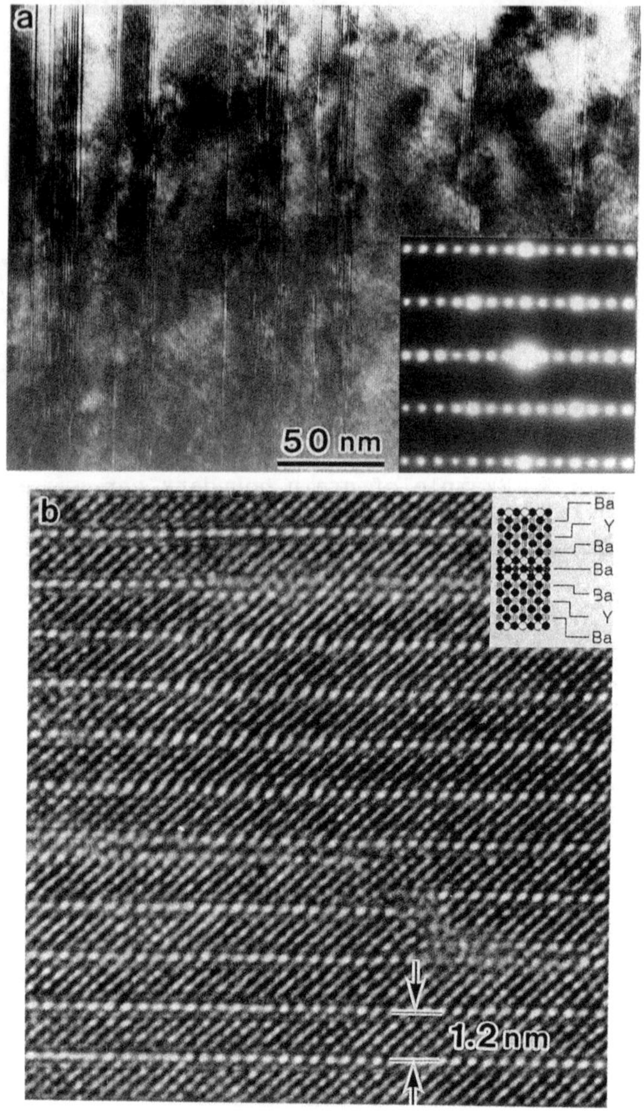

Fig.7 Stacking faults present in the internal region of grain.
(a) Low magnification image, [Spec.:B]
(b) High resolution images observed along b-axis. Extra plane of Ba layer is seen at the fault, [Spec.:A].

References

[1] S.Jin,T.H.Tiefel, R.C.Sherwood, R.V. vasn Dover, M.E.Davis, G.W.Kammlott and R.A.Fastnacht:Phys. Rev., B37 (1988) 7850
[2] M.Murakami, M.Morita, K.Doi and K.Miyamoto:Jpn.J.Appl.Phys.,28 (1989) 1189
[3] KSalama,V.Selvamanickam, L.Gao and K.Sun:Appl.Phys.Lett.54 (1989) 2352
[4] R.L.Meng,C.Kinalidis, Y.Y.Sun, L.Gao, Y.K.Tao, P.H.Hor and C.W.Chu:Nature 345 (1990) 326
[5] T.Yoshino, Y.Tomokiyo and Y.Suyama:Proc. 3rd Int.Sympo. on Superconductivity (ISS'90), Sendai, (1990) p.355.
[6] S.Nakahara, G.J.Fisanick, M.F.Yan, R.B.van Dover, T.Boone and R.Moore:J.Crystal Growth, 85 (1987) 639
[7] D.Dimos, P.Chaudhari, J.Manhart and F.K.LeGoues:Phys. Rev. Lett.,61 (1988) 219
[8] K.Jagannadham and J.Narayan:Phil. Mag.,B61 (1990) 129
[9] H.Dersch and G.Blatter:Phys. Rev.,B38 (1988) 11391

4.2 Critical Current Characteristics

J_c Derived from Pinning Potential on Epitaxial YBCO Thin Film

TAKASHI MATSUURA and HIDEO ITOZAKI

Basic High Technology Laboratories, Sumitomo Electric Ind., Ltd., Itami, Hyogo, 664 Japan

ABSTRACT

The relation between the critical current density and the pinning potential of an epitaxial $Y_1Ba_2Cu_3O_{7-x}$ thin film was investigated. The pinning potential was evaluated from the temperature dependence of the resistivity in various values of a magnetic field. Critical current density of the thin film was calculated using the obtained pinning potential. This calculated result was consistent with the critical current density measured by a four-probe method.

KEY WORDS:Y system superconductor, pinning potential,critical current density ,resistive transition

INTRODUCTION

It has been reported that critical current density (J_c) of more than $1x10^6$ A/cm^2 at 77 K for a $Y_1Ba_2Cu_3O_{7-x}$(YBCO) superconducting thin film is sustained in a magnetic field of 1 Tesla[1]. To explain the high J_c of the thin film in a high magnetic field, a mechanism of flux pinning in the film should be investigated. Evaluation of pinning potential in the thin film has been reported[2,3,4]. However, it was not clear whether the critical current density could be explained by the pinning potential.

In this paper, we report on the quantitative relation between the pinning potential and the critical current density of the film. We show that both the magnetic field dependence and the anisotropy of the critical current density can be explained by the pinning potential. We also show that the Jc is determined by the portion of a narrow cone-like well of the pinninng potential well.

PROCEDURE

We prepared an epitaxially grown c-axis oriented YBCO thin film. The film was deposited on a MgO(001) single crystal by RF magnetron sputtering[1]. The thickness of the film was 600 nm. The critical temperature was 86 K and the critical current density at 77.3 K was 1.9 x 10^6 A/cm^2 in a zero magnetic field.

The pinning potential of the YBCO epitaxial thin film was estimated by a resistive method. This method itself has been already used for several measurements[2,5,6]. According to the flux creep model, the induced resistivity ρ near critical temperature is represented by

$$\rho/\rho_0=\exp(-U_e/kT) \qquad (1)$$

where ρ_0 is the prefactor, U_e is the effective pinning potential and k is the Boltzman constant[7]. Here the effective pinning potential U_e depends on the current density, temperature and the magnetic field[6]. If the resistivities are measured at a high current density, the effective pinning potential decreases by the Lorentz force. As the Lorentz force is very small at a low current density, U_e is almost the same as the unperturbed pinning potential U_0. This U_0 depends on both the temperature and the magnetic field, as proposed by Tinkam[8]. The U_0 is described by

$$U_0 = \alpha H_c^2 \zeta a f^2$$
$$= C(1-T/T_c)^{1.5}/H \quad (2)$$

where H_c is the thermodynamic critical field, ζ is the coherence length along the direction of the field direction, af is the flux lattice spacing and H is the applied field. α and C include all numerical factors.

RESULTS AND DISCUSSION

We evaluated the current density dependence of the pinning potential. Figure 1(a) shows the resistive transition at a magnetic field of 6 Tesla which is perpendicular to the a-b plane for several current densities. These data were plotted in log ρ vs $(1-T/T_c)^{1.5}/T$ in Fig. 1(b). These plotted lines were also linear below resistivity of about 10 $\mu\Omega\cdot$cm. The slopes of the straight lines represent the effective pinning potential U_e. Figure 2 shows U_e at 77 K as a function of the current density. U_e is constant at the current density below about 100 A/cm^2. This indicates that U_e is almost the same as U_0 at the current density below 100 A/cm^2. We thus write U_e at the current density above 100 A/cm^2 as

$$U_e(T,H,J) = KU_0(T,H)\ln(J_o/J) \quad (3)$$

where J is the current density and U_0 is the unperturbed pinning potential. K and J_o are constant. This logarithmic dependence indicates that the model of the potential well is curved, as illustrated in Fig. 3[7]. It is easy to image that the flux tend to be localized in a narrow cone-like well potential. The critical current density is mainly determined by the steep portion of the slope of this narrow cone-like well.

It has the maximum restoring force. The pinning potential of the narrow cone is presented as KU_0. The ratio of radius R/r is presented as $\exp[(1-K)/K)]$, where r is the radius of the narrow cone and R is the radius of the pinning potential well. This relation shows that the pinning force is KU_0/r. Since K is determined to be 0.14 from Fig. 2, R/r is estimated to be 400. Therefore the shape of the pinning potential well is presented as Fig. 4.

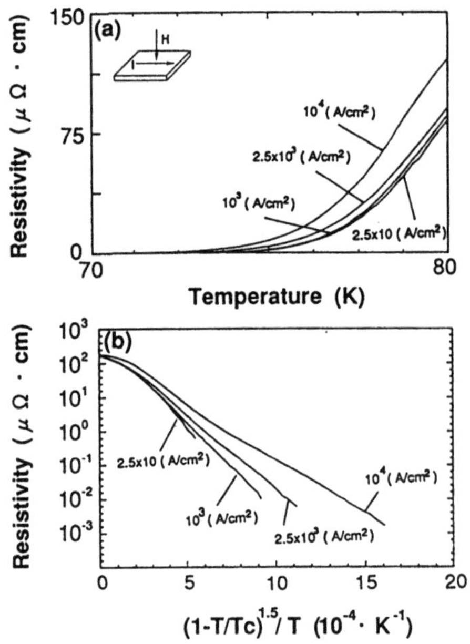

Fig.1 (a)Temperature dependence of resistivity of YBCO epitaxial thin film at H=6 Tesla and various current densities.
(b)log ρ vs $(1-T/T_c)^{1.5}/T$ at H=6 Tesla and various current densities.

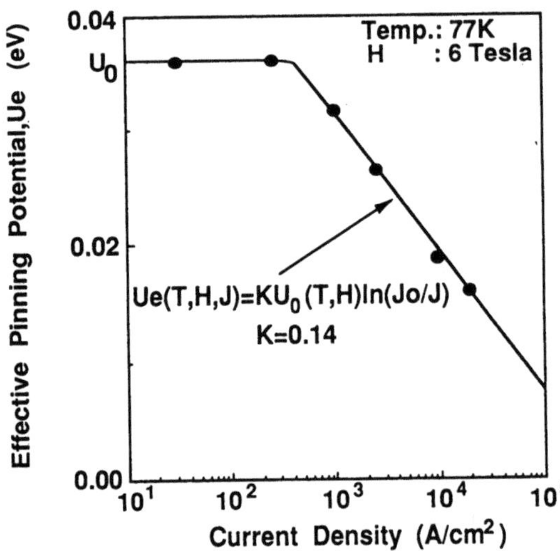

Fig.2 Current dependence of the effective pinning potential as derived from Fig.1.

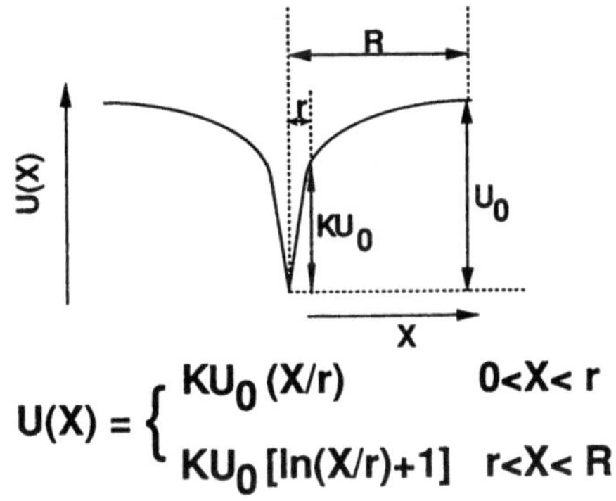

Fig.3 Model of pinning potential well.

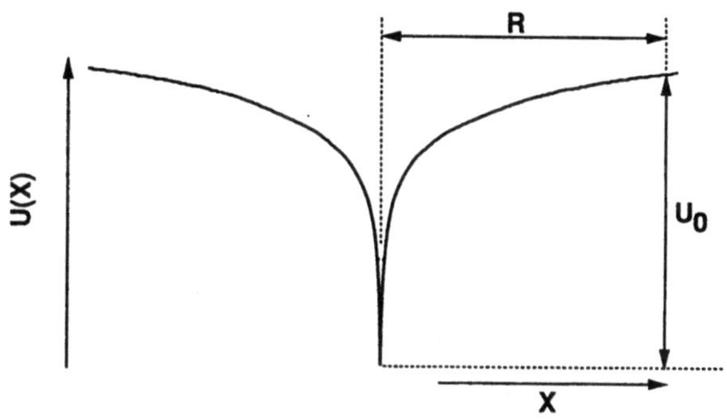

Fig.4 Shape of pinning potential well in YBCO thin film. It is derived from the analysis of Fig.2.

We also evaluated the magnetic field dependence of the pinning potential. The current density of 10 A/cm^2 was chosen to evaluate the unperturbed pinning potential. These data were shown in Ref.9. Figure 5 shows the pinning potential as a function of the magnetic field H at 77.3 K. The pinning potential is in inverse proportion to magnetic field H. This is consistent with Eq.(1). The constant C is 31 eV•Tesla in the parallel field and 5 eV•Tesla in the perpendicular field.

The critical current density can be derived from the obtained pinning potential. We used the following assumption: each flux moves independently when the Lorentz force acts on the flux. This is because , in high k material like a high Tc superconductor, the shearing correlation length, l_{66} is smaller than a_f (flux lattice spacing)[10]. The pinning force on one flux is equal to the Lorentz force on the flux at the critical current under this assumption. We also supposed that K is independent on the magnetic field. As the shape of the pinning potential well is represented in Fig. 3, the pinning force on the flux is dominated by the cone of the pinning potential well as mentioned above. Thus the pinning force on one flux is described by KU_0/r. The Lorentz force acting on the flux of the effective length L is $J_c \phi L$ at the critical current. Therefore these two forces are equal as in the following equation

$$KU_0/r = J_c \phi L \quad (4)$$

where J_c and ϕ are the critical current density and the quantum flux,respectively. The parameter L is expected to be the interval of the pinning center. Using Eq. (2), we obtain the critical current as follows

$$J_c = KC(1-T/Tc)^{1.5}/(H \phi Lr) \quad (5)$$

We measured the critical current density of the film by a four probe method in different directions of the magnetic field at 77 K. In order to fit the calculated critical current density with those measured by the four-probe method, 900 nm^2 was chosen as the value of Lr. The result of this fitting is shown in Fig. 6. The anisotropy and the magnetic field dependence of the Jc calculated from the pinning potential are consistent with those of the J_c measured by the four-probe method. This result suggests that the critical current density is almost dominated by the pinning potential.

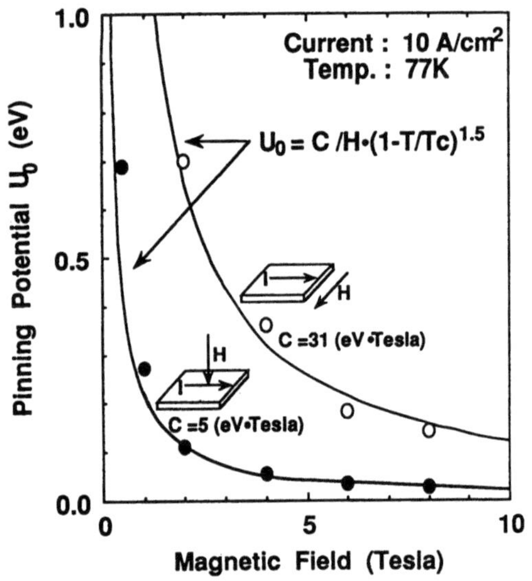

Fig.5 Magnetic field dependence of pinning potential in fields parallel to the a-b plane (open circle) and perpendicular to the a-b plane (closed circle).

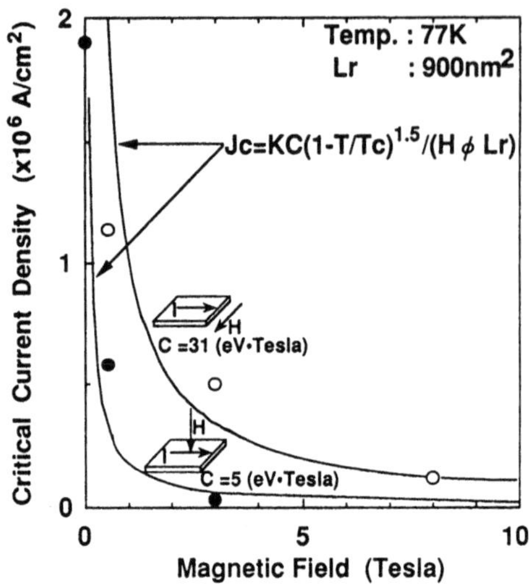

Fig.6 Calculation of critical current density of YBCO epitaxial thin film in various magnetic fields parallel to the a-b plane(open circle) and perpendicular to the a-b plane (closed circle).

SUMMARY

We calculated the critical current density of YBCO thin film based on the pinning potential estimated by the resistive method. The radius of the narrow cone-like well is determined to be 1/400 of the radius of the pinning potential well from the slope of the logarithmic dependence of the pinning potential. The pinning potential is inversely proportional to the applied magnetic field and has the anisotropy in the direction of the magnetic field. The critical current density calculated with the assumption of the isolated flux is consistent with that measured by the four probe method. This result shows that the critical current can be explained by the pinning model.

ACKNOWLEDGEMENTS

We thank Dr. Saburo Tanaka for his co-operation in the setup of the equipment and helpful discussions, Mr. Keizo Harada for providing the sputtered films and Mr. S. H. Thomsen for the assistance with the measurement.

REFERENCES

[1]S.Tanaka , H.Itozaki (1988) Jpn. J. Appl. Phys. 27: L622
[2]J.Z.Sun,K.Char,M.R.Hahn,T.H.Geballe , A.Kapitulnik (1989) Appl. Phys. Lett. 54: 663
[3]P.Fischer,H.-W.Nemuller,B.Roas,H.F.Braun,G.Saemann-Ischenko (1989) Solid State Comun. 72: 871
[4]K.Enpuku,T.Kisu,R.Sako,K.Yoshida,M.Takeo , K.Yamafuji (1989) Jpn. J. Appl. Phys. 28: L991
[5]T.T.M.Palstra,B.Batlogg,R.B.van Dover,L.F.Schneemeyer,J.V.Waszczak (1989) Appl. Phys. Lett. 54: 763
[6]E.Zeldov,N.M.Amer,G.Koren,A.Gupta,M.W.McElfresh, R.J.Gambino (1990) Appl. Phys. Lett. 56: 680
[7]Y.B.Kim,C.F.Hempstead , A.R.Strnad (1963) Phys. Rev. 131: 2486
[8]M.Tinkam (1988) Phys. Rev. Lett. 61: 1658
[9]T.Matsuura,S.Tanaka,K.Harada,H.Itozaki, S.Yazu (1990) In: T.Ishiguro, K.Kajimura (eds) in Advances in Superconductivity II. Springer,Tokyo,pp.687
[10]K.Yamafuji,T.Fujiyoshi,K.Toko,T.Matsushita (1989) Physica C 159: 743

Decay Rate Measurement of Magnetization and Transport Properties in Bi-2212 Single Crystal

SEIJI HAYASHI[1], KAZUYUKI SHIBUTANI[1], TOSHIO EGI[1], ICHIRO SHIGAKI[1], RIKUO OGAWA[1], YOSHIO KAWATE[1], VINCENT BOUCHIAT[2], KATSUKI KITAHAMA[3], and SHICHIO KAWAI[3]

[1] Superconducting and Cryogenic Technology Center, Kobe Steel, Ltd., Kobe, 651-22 Japan
[2] School of Industrial Physics and Chemistry of Paris, France
[3] The Institute of Scientific and Industrial Research, Osaka University, Osaka, 567 Japan

ABSTRACT

Measurement of decay rate of magnetization was performed as a function of temperature for Bi-2212 single crystal which was grown by TSFZ method. Normalized decay rate increases monotonically with increasing temperature, and does not have such plateau region as reported for YBCO. According to the Hagen and Griessen's inversion scheme, we calculated the distribution of pinning potential, and obtained a narrow distribution centered around 40 meV. This result can explain consistently other properties of Bi-2212 single crystal.

KEY WORDS: Bi-2212 single crystal, decay rate of magnetization, pinning potential, distribution of pinning energy

INTRODUCTION

Extensive efforts have been concentrated on the study of critical current density (Jc) of the high-temperature superconductors (HTSCs) because of the technological and scientific importance. Because of the extremely short coherence length in HTSCs, weak-link has been the major obstacle for increasing Jc in HTSC-materials.[1] In fact, higher Jc has been achieved at 4.2K for the samples which have better crystallographic quality.[2,3,4] However, Jc decreased rapidly with increasing temperature in these materials. It is now of common knowledge that it is necessary to introduce artificial pinning centers to make possible applications at 77K.

Many researchers are continuing to make their effort to find out available pinning centers. Murakami et al. succeeded in introducing artificial pinning center in YBCO by the MPMG process.[5] Pinning energy can be estimated by the measurement of magnetization decay. The literature on magnetic relaxation of high-temperature superconductors is extensive, and most of all these experiments were done for YBCO materials.[6,7,8] According to the normalized decay rate analysis, Malozemoff asserted that the flux-creep relaxation-rate data could be classified into four temperature regimes.[7]

In case of Bi-based materials, report on magnetic relaxation is not so extensive as compared to Y-based materials. Present authors observed very high critical current density at 4.2K and obtained small pinning potential about 10 meV for Bi-2212 single crystal.[9] Based upon the flux-annealing method experiment, small distribution of pinning potential was infered. Introduction of artificial pinning centers to Bi-based materials will be the important subject for the future investigation. Therefore, it is important to investigate pinning characteristics for bare materials which have no intentional pinning. In the present paper, we report on the measurement of magnetization-decay, and discuss the distribution of pinning potential in Bi-2212 single crystal.

EXPERIMENTAL PROCEDURES

The Bi-2212 single crystal were grown by the travelling solvent floating zone (TSFZ) method which is described in Ref.10 in details.[10] The sample is a plane with thickness about 30μm along the c-axis and area about 2.4mm x 4.6mm parallel to the a-b plane that was cut off from the bulk crystal. The critical temperature of this sample was 86K.

Magnetization was measured using a vibrating sample magnetometer (VSM). In this apparatus, maximum magnetic field 6T is applicable, and temperature can be changed from 4.2K to room temperature. In case of the present measurement, temperature was changed from 4.2K to about 30K within the accuracy of 0.1K. Since the Jc decreases rapidly with increasing temperature and then temperature-instability affects very much to the decay rate, this temperature stability during the measurement is very important condition. Although the angle between magnetic field and sample can be changed in our system, magnetic field was applied perpendicular to a-b plane of the crystal. The angle can be controlled within the accuracy of 1 degree. Two types of measurements were performed as; decay rate measurement at various temperatures and so-called flux annealing experiment.

RESULTS AND DISCUSSION

Results of decay of magnetization at several different temperatures are shown in Fig.1. These measurements were performed as follows. After stabilizing temperature to desired value, we applied magnetic field higher than the twice of saturation field, then decreased to zero field. Just after the magnetic field decreased to zero, decay of the magnetization started to measure. These procedures assure that the present decay corresponds the decay exactly from the critical state. As shown in this figure, decay rate increases with increasing temperature. The almost linear dependence of normalized magnetization on to logarithm of time suggests the activation of flux lines from pinning sites.

Figure 2 shows the normalized decay rate at t=100s as a function of temperature. As shown in this figure, there is no plateau region that was reported for YBCO as suggesting vortex glass state. And whole picture can be interpreted as shifted to lower temperatues compared to YBCO.[7] If we assume that each pinning center has the same pinning energy U(T) and flux moves by thermal activation, normalized decay rate S is given by,

$$S = \frac{d\ln M}{d\ln t} = \frac{dM}{M d\ln t} = -\left[\frac{U(T)}{kT} - \ln\left(\frac{t}{t_{eff}}\right)\right]^{-1}$$

where t is the observation time and t_{eff} is a effective attempt time for flux movement.[7] Though this equation well explains time dependence of magnetization decay, it does not follow the expected temperature dependence.

Figure 3 shows the result of flux annealing measurement. In this case, measurement was performed as follows; the sample was cooled to 5.8K in zero magnetic field, then magnetic field was increased to 2T, decay measurement was started right after magnetic field reached to maximum. Continuing decay measurement, sample was quenched to 4.2K at t=550s. It took about 30s for quenching. As seen in the figure, slope of magnetization decay rapidly changes. This is based on the idea as follows. If the distribution of pinning potential exists, decay after the quench will be different from that of without quench. Because, in case of quench from higher temperature to 4.2K, weak pinning will be already relaxed before the quench due to higher thermal activation, and on the other hand, weak pinning, will remain in case of the temperature fixed at 4.2K.

Figure 4 is the expanded graph of Fig.3, where time scale is transfered as t'=t+7050, and normalized magnetization was multiplied by 0.897 taking into account the temperature dependence and time dependence of magnetization.[9] As shown in Fig. 4, Quenched datum shows somewhat smaller slope than continued decay-line of 4.2K without quench (denoted by broken line). This result suggests the presence of distribution of pinning potential.

Hagen and Griessen have given the expression of magnetization for a superconductor consisting of a collection of regions with different activation energies as follows.[8]

$$M(t,T)=M_0\frac{b(T)}{a(T)}\int_{E_0^*(t,T)}^{\infty} m(E^*)\left[1-\frac{kT}{E^*b(T)}\ln\left(1+\frac{t}{t_{eff}}\right)\right]dE^*$$

Inverting this equation, the distribution function m(E) is expressed as follows.

$$m(E_0^*(t_b,T)) = \left[\frac{d}{dT}\left\{\frac{a(T)}{M_0 b(T)}\left[\ln\left[\frac{t_b}{t_{eff}}\right]\left[\frac{dM}{d\ln t}\right]-M\right]\right\}\right]\left[\ln\left[\frac{t_b}{t_{eff}}\right]\frac{d}{dT}\left[\frac{kT}{b(T)}\right]\right]^{-1}$$

Since our experiment was done far below the Tc, we can take a(T) and b(T) to be unity in good approximation.[8] The calculated distribution is shown in Fig.5 by solid curve. For the comparison, the data for YBCO single crystal which is the quotation from Hagen and Griessen's paper is also shown by broken line. The prominent features in Bi-2212 single crystal are the facts that pinning energies are concentrated to low energy and have narrow distribution. It has a non-negligible weight at low energies and has no weight at higher energies than 80 meV. This is probably due to the extremely small coherence length and quasi two-dimensional character of Bi-2212 single crystal. This distribution can consistently explain such experimental results as; rapid decrease of Jc with increasing temperature, behaviour of the normalized decay rate as a function of temperature, temperature dependence of pinning potential. This distribution will also be able to explain very high transport critical current density and its temperature dependence.[9,11]

In conclusion, we have measured the decay of magnetization for Bi-2212 single crystal at various temperatures. According to the inversion scheme proposed by Hagen and Griessen, we deduced the distribution of activation energy. Obtained result showed narrow distribution centered around 40 meV. This distribution is consistent with other experimental results reported before.

REFERENCES

1. Gao L, Meng R.L, Xue Y.Y, Hor and Chu C.W, (1991) Appl. Phys. Lett. 58: 92.
2. Dinger T.R, Worthington T.K, Gallagher W.J, and Sandstrom R.L, (1987) Phys. Rev. Lett. 58: 2687.
3. van Dover R.B, Schneemeyer L.F, Gyorgy E.M, and Waszczak J.V, (1988) Appl. Phys. Lett. 52: 1910.
4. Hayashi S, Shibutani K, Shigaki I, Ogawa R, Kawate Y, Maret V, Kitahama K, and Kawai S, (1991) Physica C 174: 329.
5. Murakami M, Oyama T, Fujimoto H, Taguchi T, Gotoh S, Shiohara Y, Koshizuka N and Tanaka S, (1990) Jpn. J. Appl. Phys. 29: L1991.
6. Yeshuran Y and Malozemoff A.P, (1988) Phys. Rev. Lett. 60: 2202.
7. Malozemoff A.P, (1990) Proc. A.I.P.Conf. ed. Y.H.Kas et al. p89.
8. Hagen C.W and Griessen R, (1989) Phys. Rev. Lett. 62: 2857.
9. Shibutani K, Hayashi S, Egi T, Shigaki I, Ogawa R, Kawate Y, Kitahama K and Kawai S, (1991) Proceedings of ICMC'91.
10. Shigaki I, Kitahama K, Shibutani K, Hayashi S, Ogawa R, Kawate Y, Kawai T, Kawai S, Matsumoto M and Shirafuji J, (1990) Jpn. J. Appl. Phys. 29: L2013.
11. Shibutani K, Hayashi S, Egi T, Shigaki I, Ogawa R, Kawate Y, Kitahama K and Kawai S, (1991) Proc. M2S-HTSC3.

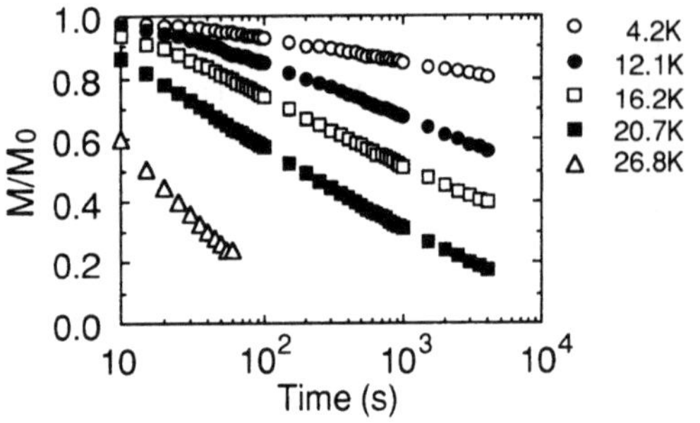

Fig.1 Relaxation of the normalized magnetization at various temperatures. Magnetic fluxes which are perpendicular to a-b plane of the crystal are trapped.

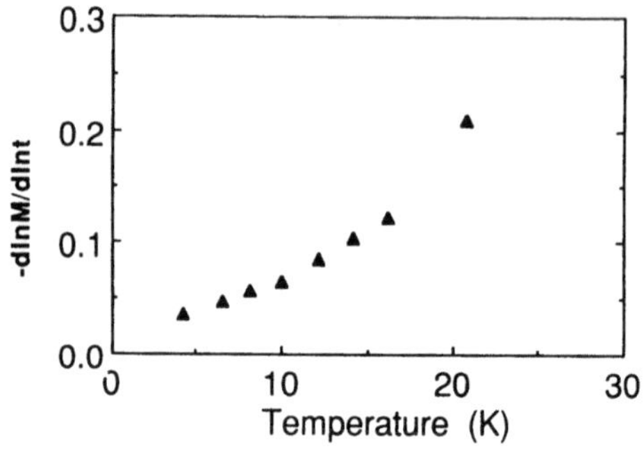

Fig.2 Normalized decay rate at t=100s. The decay rate monotonically increases with temperature and does not have plateau region.

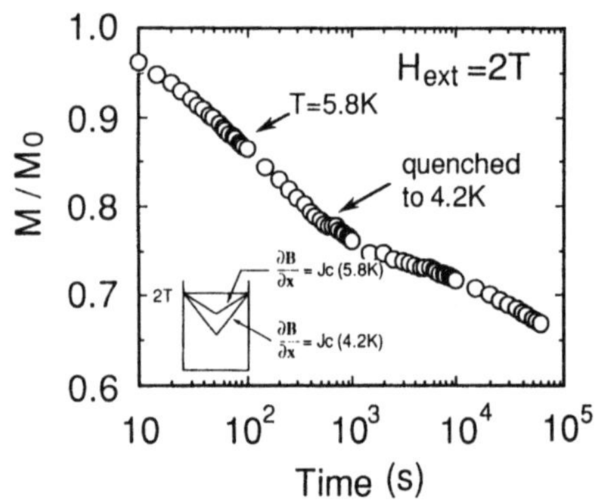

Fig.3 Decay of the magnetization. The sample was quenched from 5.8K to 4.2K at t=550s. By quenching, decay rate rapidly changes. Inset shows the induction profile at t=0s for T=5.8K and T=4.2K.

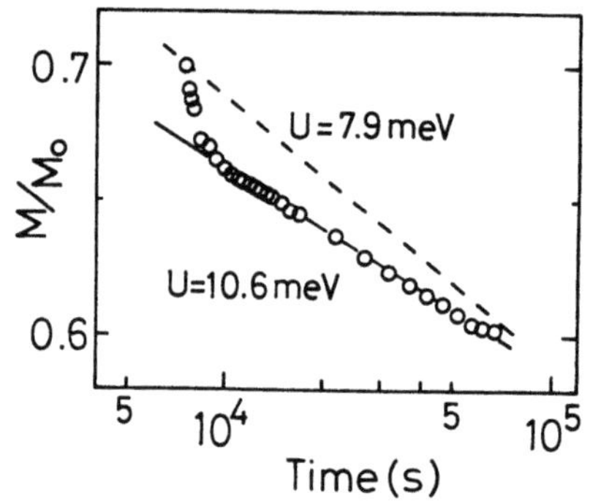

Fig.4 Expansion of the datum which corresponds to t>550s in Fig.3. Time scale is added 7050s and magnetization is multiplied by 0.897 compared to Fig.3. Broken line corresponds to the continued decay at 4.2K without quench.

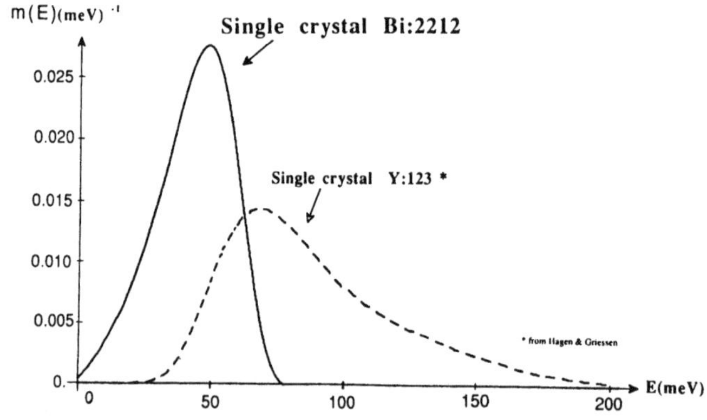

Fig.5 Distribution of pinning potential. Solid curve corresponds to the Bi-2212 single crystal. The data for YBCO single crystal which is the quotation from Hagen and Griessen's paper are also shown by the broken line for the comparison.

Irreversibility Lines and Doping Effect in $(Bi,Pb)_2Sr_2Ca_2Cu_3O_y$

HIROSHI IKEDA[1,2], RYOZO YOSHIZAKI[1,2], MASAYA WATAHIKI[1], KOZO YOSHIKAWA[3], and NOBUYOSHI TOMITA[4]

[1] Institute of Applied Physics, [2] Cryogenics Center, University of Tsukuba, Tsukuba, Ibaraki, 305 Japan
[3] Takasago Research and Development Center, Mitsubishi Heavy Industries Ltd., Takasago, Hyogo, 676 Japan
[4] Kobe Shipyard and Machinery Works, Mitsubishi Heavy Industries Ltd., Kobe, 652 Japan

ABSTRACT

Irreversibility lines have been studied in the pure and doped $(Bi,Pb)_2Sr_2Ca_2Cu_3O_y$ oxide superconductors. We obtained irreversibility temperature $T_r(H)$ and irreversibility field $H_r(T)$ from magnetic measurement and irreversibility field B^*_{c2} from resistivity measurement. The magnetic irreversibility lines are consistent with each other, while the transport one locates at the higher temperatures. The doping effect on the irreversibility lines are investigated by substituting La or Gd for 2% Ca.

KEY WORDS: copper-oxide superconductors, Bi(2223), irreversibility lines, doping effect

INTRODUCTION

Since the first discovery of high-Tc superconductors, the behaviors of flux in the mixed state have attracted the attention of many researchers [1,2]; they are characterized by time decay of the superconducting diamagnetism at low temperature and by extended region of flux-reversible states below Tc. Although the microscopic mechanism of an irreversibility line has not been settled yet and there are many proposed models [2], it is important to investigate the irreversibility lines not only from the physical point of view but also for the practical purposes. If the irreversibility lines are changeable against introducing some defects into the superconductors, the ways would open for the possible application of high-Tc superconductors. In the present paper we have studied the irreversibility lines for the $(Bi,Pb)_2Sr_2Ca_2Cu_3O_y$ (hereafter abbreviated to Bi(2223)) compounds by means of magnetic and transport measurements. Doping effect on the irreversibility lines was also studied with replacing 2% Ca by La or Gd.

EXPERIMENTAL

The starting composition of the powder was chosen so as to be $Bi_{1.84}Pb_{0.34}Sr_{2.03}Ca_{1.9}Cu_{3.06}O_y$. The compounds were prepared by a solid-state reaction. After the ordinary sintering process, the samples were pressed with uniaxial stress at elevated temperature. The details of the preparation method and the fundamental properties of the samples were reported elsewhere [3,4]. The doped samples were provided by substituting La or Gd for 2% Ca with maintaining a single phase of Bi(2223). According to secondary electron microscope (SEM) analysis of the sample, large grains (5-10 µm in ab-plane) stack closely with almost aligned c-axis (called c-axis in this paper) due to the uniaxial hot-pressing. One of the most characteristic features of the hot-pressed samples is the less number of weak links among the grains [4]. As a result, the sample can bear a transport current density of about $1x10^4$ A/cm^2 at 77.3 K and 0 T in a bulk state. We measured the non-doped sample and doped one with high resolution transmission electron microscopy (TEM).

The magnetic properties were measured by using a SQUID magnetometer (Quantum Design, Inc.). In zero-field cooling (ZFC) process a sample was cooled in zero-magnetic field of about 20 µT, and a magnetic field was applied at 5 K in a non-overshoot mode. Magneto-transport measurement was carried out with using conventional four-probe method in the transverse field geometry, where the magnetic field was applied along the c-axis.

RESULTS AND DISCUSSION

The TEM pictures for the crystal ac-plane are shown in Fig. 1; (a) for the non-doped sample and (b) for the Gd-doped one. It should be stressed that the unit layers with the thickness of about 18 Å stack over 50 layers without any intergrowth layer for the non-doped samples. On the other hand, a small amount of intergrowth of other layers such as 2212 was observed in the doped samples, and the quality is still appreciable as a fairly good Bi(2223) compound.

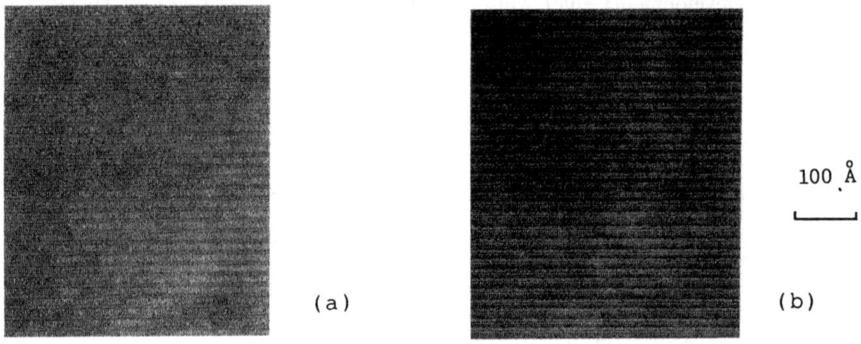

Fig. 1 High resolution TEM pictures for (a) non-doped and (b) Gd-doped Bi(2223) compounds.

The irreversibility temperature, $T_r(H)$, was determined from the merging temperature of the magnetic moment observed in the ZFC process into the one in field cooling (FC) process as shown in Fig. 2. It is noted in this figure that the Meissner effect in FC process indicates two-step flux expulsion with decreasing temperature: An ordinary Meissner effect is seen below Tc and the second flux expulsion occurs around 30 K. This behavior becomes more remarkable as increasing external magnetic field, whereas single step expulsion is observed for the magnetic field less than 0.5 T [5]. Similar two-step like behaviors of the Meissner signals are observed in other high-Tc superconductors, sintered LSCO [6] and Bi(2212) [7-9]. Figure 3 shows hysteretic curves of the magnetization for the non-doped sample. The irreversibility field, $H_r(T)$, was determined from the lowest magnetic field where the magnetization was reversible.

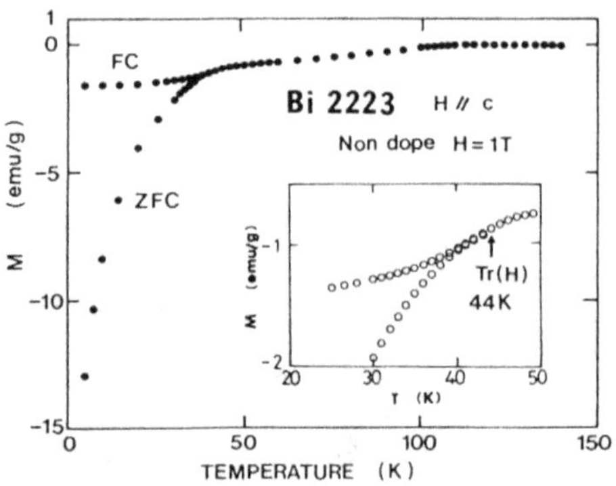

Fig. 2 Temperature dependence of magnetization for non-doped Bi(2223). The $T_r(H)$ is indicated in the inset.

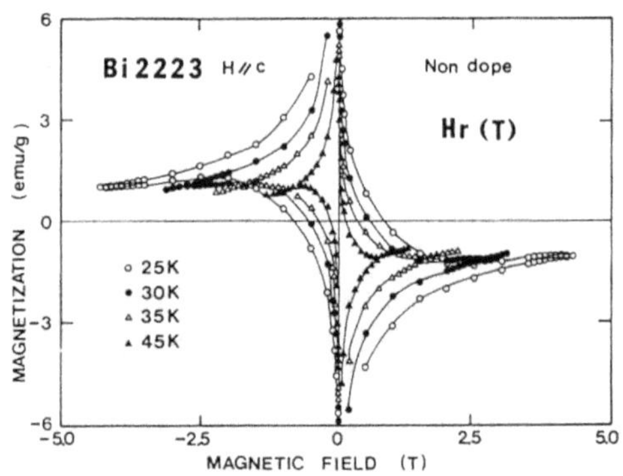

Fig. 3 Hysteretic curves of Bi(2223) samples at various temperatures for obtaining $H_r(T)$.

We can estimate the irreversibility line from the transport effect of the superconductors. We define the transport irreversibility line from the temperature at zero-resistivity in the presence of magnetic field as shown in Fig. 4. The irreversibility field B^*_{c2} was estimated from $\rho = 1\times10^{-6}\ \Omega\cdot cm$ with the transport current of 10 mA ($J = 0.25\ A/cm^2$) for the non-doped sample. Those results of the irreversibility lines are plotted in H-T plane in Fig. 5, (a) from the magnetic and (b) the transport measurement. It is noted that the results of the two magnetic measurements coincide with each other. The irreversibility line for the doped sample locates at the lower temperature side compared to the non-doped one. In contrast, the irreversibility line obtained from B^*_{c2} locates at the higher temperature side.

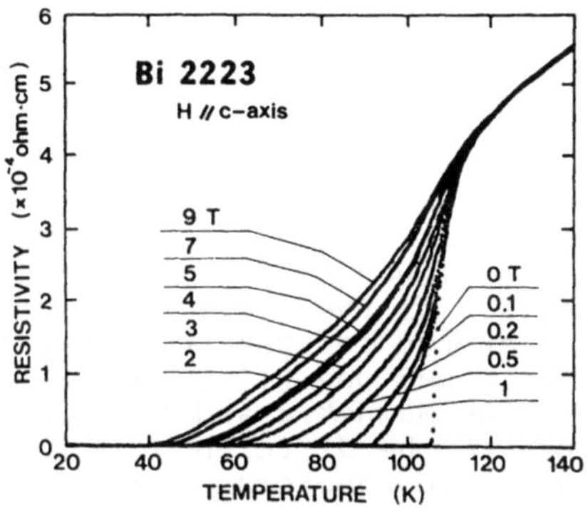

Fig. 4 Resistive transition curves in various magnetic field with transport current of 10 mA.

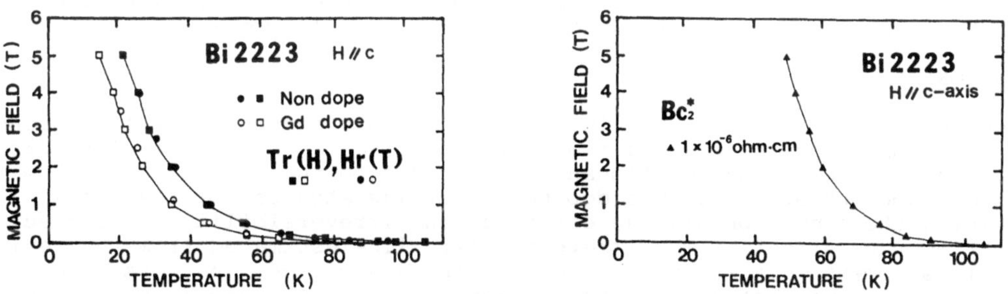

Fig. 5 Irreversibility lines in H-T plane. The results are obtained from (a) the magnetic measurement and (b) the transport one. The square symbols are the data from $T_r(H)$ and the circles are from $H_r(T)$. The solid marks represent the results for the non-doped sample and the open marls are for the doped one.

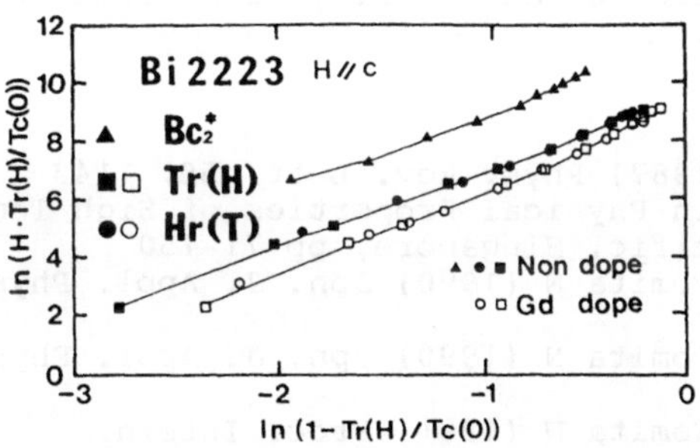

Fig. 6 The irreversibility lines plotted in the logarithmic scale of Ht vs [1-t]. The symbols are the same with those in Fig. 5.

According to a flux-depinning model [2], the irreversibility lines are expected to follow the relationship,

$$H = a[1-t]^n/T, \qquad (1)$$

where a is a constant and $t=T_r(H)/Tc(0)$ is reduced temperature. Then we plotted the present results in logarithmic scale in Fig. 6. First, we can see the linear dependence of the results, which indicates the irreversibility lines for the Bi(2223) compounds follows the eq. (1) in the wide range of temperature. Secondly, the slope n is about 2 in the higher temperature region and seems to change to 2.5 in the lower temperatures. Thirdly, the doping effect causes the shift of the irreversibility lines toward lower temperature side. Finally, the slope is almost independent of the measurements although the magnitude differs with each other.

As for the slope n, its value is obtained from the superconducting cohesive energy and the coherence length along the magnetic field based on the mean field theory [2]. In the present case of Bi(2223) and H//c-axis, it is probable that the coherence length along the c-axis becomes shorter than the thickness of three CuO_2 superconducting sheets and must be replace by it for estimating the flux pinning energy. Then the power exponent n in eq. (1) is expected to become not 1.5 but 2. Concerning to the doping effect, single phase was maintained and no appreciable impurity phase was observed in TEM analysis for the present doped samples. However, lattice mismatches such as intergrowth of other layers and low-angle grain-boundaries are recognized in the crystallets of the doped samples, although the amount is small. Those lattice imperfections may act as weakening the flux pinning and lowering the irreversibility lines. This tendency is consistent with the fact of the distribution energy measurement of pinning potentials that the peak energy for the non-doped sample is about 25 meV and less than 20 meV for the doped one [10]. As for the difference of the irreversibility lines between the magnetic and transport measurements, one of the reasons may be ascribed to the difference of the magnetic Jc and the transport Jc for the sintered superconductors. Since Jc(Mag)>Jc(Trans) in oxide superconductors and the Lorentz force is proportional to Jc, it is probable that the transport irreversibility line lies at the higher temperatures than the magnetic irreversibility ones.

CONCLUSION

We observed the irreversibility lines for the single phase Bi(2223) compound by means of magnetic and transport measurements. We found that the irreversibility lines followed the relationship expected from the depinning model with the power exponent of 2-2.5. The doping effect was studied for the sample with maintaining the single phase. The effect on the irreversibility lines resulted in lowering them slightly. The irreversibility line estimated from the resistivity measurement located at the higher temperature side. The crystallographical difference between the doped sample and the non-doped one was investigated by a high resolution TEM; the difference was quite small except the presence of intergrowth and low angle grain-boundaries in doped sample although the amount is small.

ACKNOWLEDGEMENTS

The authors are grateful to J.P. Bird for helpful discussion. This work is supported by Grant-in-Aid for Scientific Research on Priority Areas from the Ministry of Education, Science and Culture.

REFERENCES

1. Müller KA, Takashige M and Bednorz JG (1987) Phys. Rev. Lett. **58:** 1143
2. Malozemoff AP (1988) In: Ginsberg (ed) in Physical Properties of High Temperature Superconductors I. World Scientific, Singapore, pp 71-150
3. Ikeda H, Yoshizaki R, Yoshikawa K, and Tomita N (1990) Jpn. J. Appl. Phys. **29:** L430
4. Yoshizaki R, Ikeda H, Yoshikawa K, and Tomita N (1990) Jpn. J. Appl. Phys. **29:** L753
5. Yoshizaki R, Ikeda H, Yoshikawa K, and Tomita N (1991) Proc. Intern. Workshop on Electrical Properties and Mechanisms in High-Tc Superconductors, Tsukuba, (to be published)
6. Sato M, Shamoto S, Sera M and Fujishita H (1989) Solid State Commun. **72:** 689
7. Maniwa Y, Sato H, Mituhashi T, Mizoguchi K, Shiozaki I and Kume K (1991) Jpn. J. Appl. Phys. **30:** L572
8. Kritscha W, Sauerzopf FM, Weber HW, Crabtree GW, Chang YC and Jiang PZ (1991) Physica C **171:** 59
9. Kadowaki K (1991) Proc. M^2S-HTSC III, Kanazawa, (to be published)
10. Yoshizaki R, Ikeda H, Watahiki M, Yoshikawa K and Tomita N (1991) In: Kajimura and Hayakawa (eds) in Advances in Superconductivity III. Springer-Verlag, Tokyo, pp 483-486

AC Magnetic Susceptibility Measurement and Irreversibility Line in $(Bi,Pb)_2Sr_2Ca_2Cu_3O_y$

MASAYA WATAHIKI[1], RYOZO YOSHIZAKI[1,2], HIROSHI IKEDA[1,2], KOZO YOSHIKAWA[3], and NOBUYOSHI TOMITA[4]

[1] Institute of Applied Physics, [2] Cryogenics Center, University of Tsukuba, Tsukuba, Ibaraki, 305 Japan
[3] Takasago Research and Development Center, Mitsubishi Heavy Industries, Ltd., Takasago, Hyogo, 676 Japan
[4] Kobe Shipyard and Machinery Work, Mitsubishi Heavy Industries, Ltd., Kobe, 652 Japan

ABSTRACT

Flux motion in a copper-oxide superconductor Bi(2223) was investigated by means of ac magnetic susceptibility measurement. Complex susceptibility was observed as functions of various temperature, frequency, amplitude of ac magnetic field and external dc magnetic field. We obtained irreversibility line, penetration depth and flux-pinning energies.

Key Words: copper-oxide superconductors, Bi(2223), ac susceptibility, irreversibility line, pining potential

INTRODUCTION

Since the discovery of high-temperature superconductors, extensive research has been conducted to require superconductors with high-critical currents. In the mixed state between H_{c1} and H_{c2} of the type-II superconductors an array of flux line permeates into the crystal and forms a triangular lattice. The detailed knowledge of the microscopic of the flux line in the mixed state is of importance for application of high-T_c superconductors, because the critical current is closely related to flux motion. In relation to flux motion an representative concept in high-T_c superconductors is an irreversibility line which is clearly observed as the boundary of the flux reversible and irreversible region, although the attribution of its physical origin is still controversial [1]. Measurements of ac susceptibility of the type-II superconductors are an extremely powerful method for investigating the motion of flux line: e.g., The irreversibility temperatures for YBCO single crystals depend strongly on the frequency ω of the applied ac field as well as on the external magnetic field [2].

In this paper we report measurements of the complex ac susceptibility for single-phase $(Bi,Pb)_2Sr_2Ca_2Cu_3O_y$ compounds in an applied static magnetic field. We estimated pinning potentials by use of a depin model and penetration depth by use of a Campbell method [3,4].

EXPERIMENTS

We synthesized high-quality $(Bi,Pb)_2Sr_2Ca_2Cu_3O_y$ compounds by a uniaxial hot-press method. The single phase was confirmed from a Rietveld analysis of the X-ray powder diffraction spectra and an electron microscopy. The sample size was typically 2x2x3 mm^3. The lateral length along the c-axis was the longest. Complex ac susceptibility was observed by a conventional Hartshorn bridge-type susceptometer. The direction of the external static field, H_0, was parallel to the ac magnetic filed and to the aligned c-axis. The demagnetization effect of the sample was neglected.

RESULTS AND DISCUSSION

We measured temperature dependence of ac susceptibility for various amplitudes and frequencies of the ac magnetic field under H_0 up to 5 T. Figure 1 shows temperature dependence of χ' and χ'' for f=400 Hz and H_0=0 T. The onset of diamagnetism χ' and the energy loss peak χ'' start at the same temperature of about 104 K which is in agreement with the result (106 K) in dc susceptibility measurements. The transition width extends with increasing the amplitude of ac field. As for the χ" spectra, only single peak was observed, and the the peak shifted toward low temperature side with growing its peak height, increasing ac amplitude. Figure 2 shows the static field dependence of ac susceptibility spectra for f=49.6 Hz and ac field of 5 Oe. Increasing the external field make the χ'and χ" spectra shift with slight increasing of the transition width.

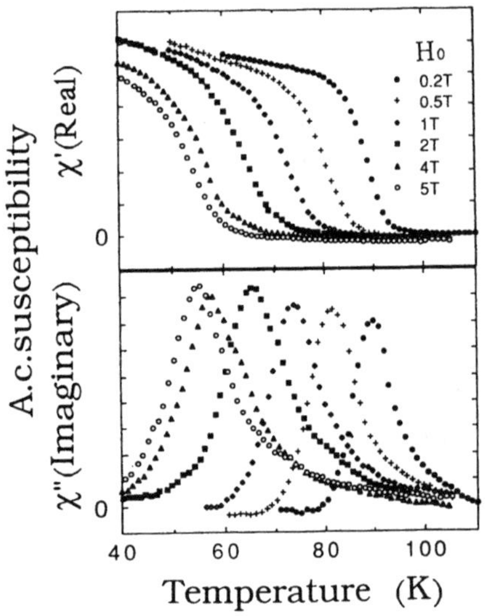

Fig. 1 Temperature dependence of ac susceptibility for f=400 Hz and H_0=0 T. The real part is in top and the imaginary part in the bottom.

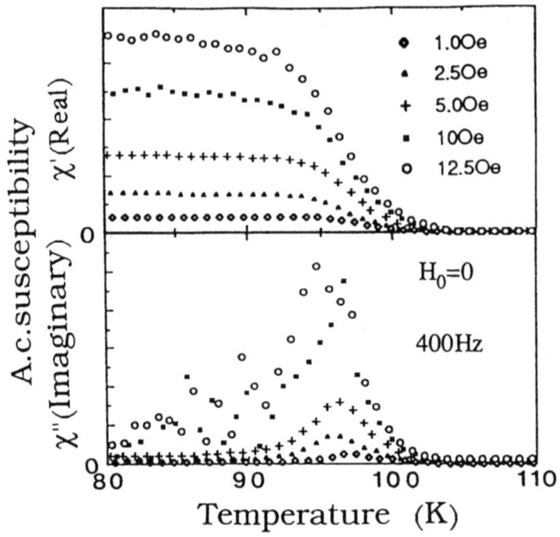

Fig. 2 The static field dependence of ac susceptibility spectra for f=49.6 Hz and ac field of 5 Oe.

We measured the frequency dependence of the χ" peak temperature T^* in the frequency range from 110 Hz to 101 kHz. The result plotted in H-T plane give us the irreversibility lines, which is shown in Fig. 3. The irreversibility lines depends on the frequency; they shift toward lower temperature side with lowering the frequency. The irreversibility lines obtained in the present ac susceptibility measurement and the dc magnetic measurement [5] are represented in Fig. 4 for comparison.

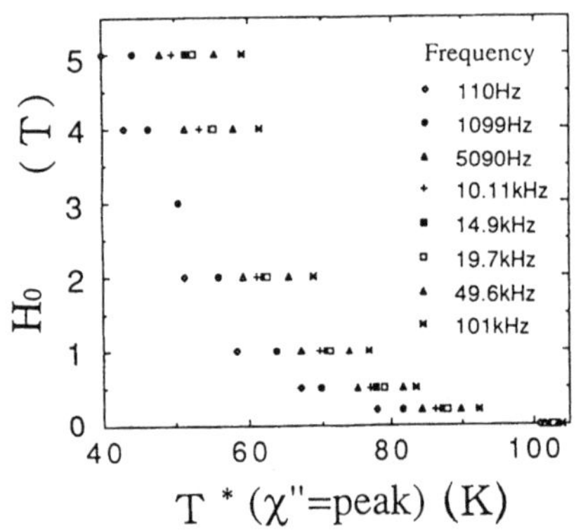

Fig. 3 The frequency dependence of the χ" peak temperature T^* for ac field of 5 Oe.

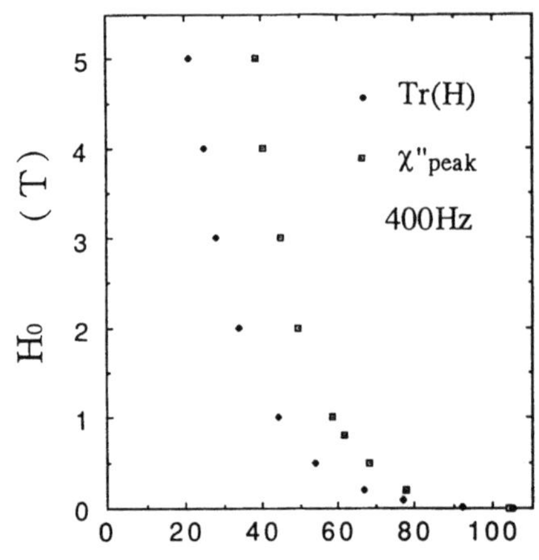

Fig. 4 The irreversibility lines obtained form the χ" peak and the irreversibility temperature in dc magnetic measurement (Ref. 5).

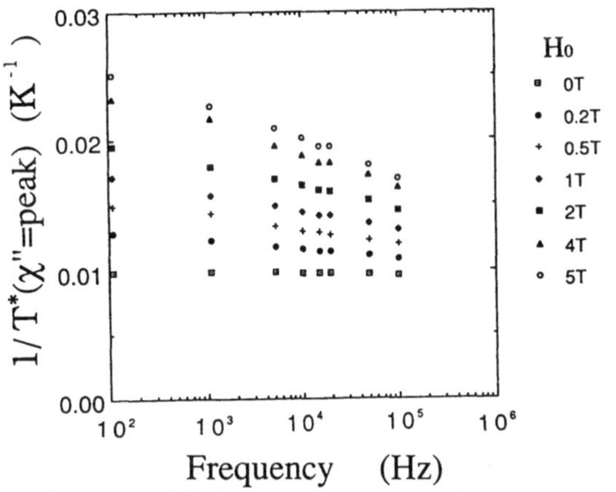

Fig. 5 The irreversibility temperature obtained from the χ" peak temperature is plotted as a function of frequency of the ac field.

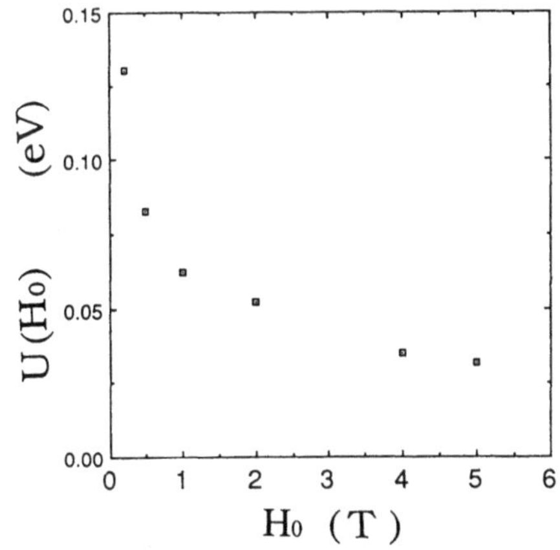

Fig. 6 Pinning potential energy U vs. external dc magnetic field H_0 derived from a depin model

In order to estimate the pinning energy strength we use a flux creep model [1]. The predicted equation for the irreversibility temperature T^* in this model is [3,4]

$$U/kT^* = -\ln(\omega/\omega_0),$$

where w is the measurement frequency, ω_0 is a attempt frequency for flux lines to escape from a pinning site with a pinning potential U. The frequency dependence of T^* is shown in Fig. 5 for a fixed ac field (5 Oe) in various H_0. One finds linear relation between $1/T^*$ and ln ω as expected from the theory. By use of the equation we determined a pinning potential U, which is shown in Fig. 6 for various magnetic field. U distributes from 30 meV at strong H_0 to 130 meV at weak H_0. This distribution of U is in good agreement with the results determined from the decay measurement of the remanent moment in Bi(2223) where the distribution the pinning potential energy has a peak at about 30 meV and a trailed tail extended up to 160 meV [6].

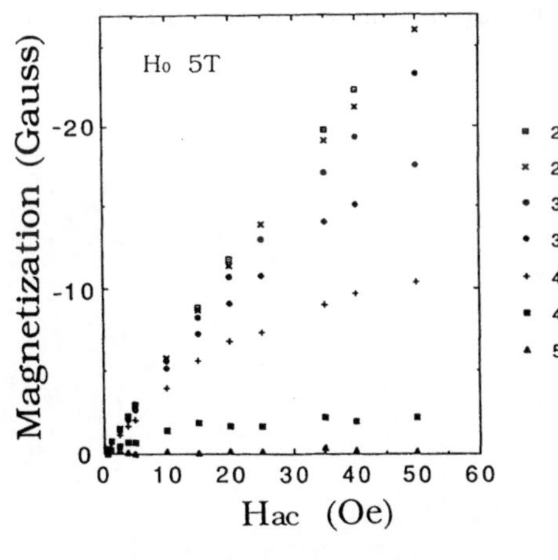

Fig. 7 The ac-field amplitude dependence of the magnetization for 1099 Hz and H_0=5 T.

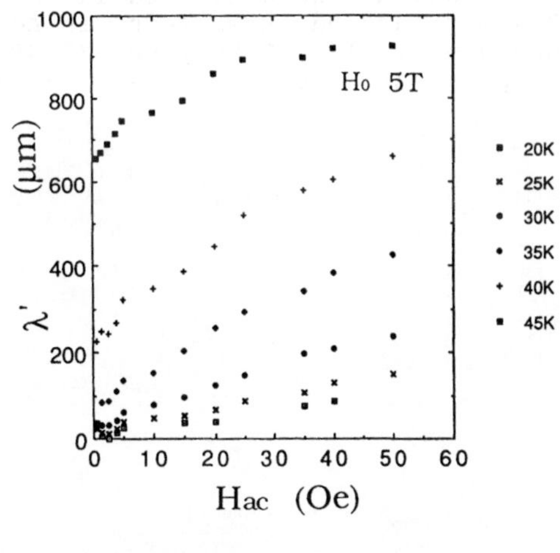

Fig. 8 The penetration depth of the ac field vs. amplitude of the ac field.

Figure 7 shows the ac field amplitude dependence of magnetization at various temperatures for f=1099 Hz and H_0=5 T. The magnetization increases almost linearly with the ac field amplitude, H_{ac} at low temperatures. With increasing the temperature, however, the magnetization deviates from the linear dependence and becomes to saturate at high H_{ac}. The increase of the deviation from the linear dependence implies the increased flux that is penetrated into the sample. Then one can deduces the penetration depths λ' of the ac field at various temperatures [7] The results are shown in Fig. 8, where the thickness of the sample perpendicular to the magnetic field is 2 mm. The present sample behaves as a hard superconductor below 40 K, while the penetration rapidly increases at 45 K. This is consistent with the result in Fig. 3 that the χ'' peak locates at a little lower than 45 K.

CONCLUSION

We observed ac susceptibilities of the single phase Bi(2223) compounds with various parameters of amplitude and frequency of the ac field, external dc magnetic field and temperature. We obtained irreversibility lines from the peak temperature of χ'' and estimated a pinning potential from the frequency dependence of the irreversibility lines. Those results are compared with the ones derived from the dc magnetic measurement and we find that the both results are consistent with each other. We estimated the penetration depth of the ac field with the Campbell method.

ACKNOWLEDGEMENTS

This work is supported by Grant-in-Aid for Scientific Research on Priority Areas from the Ministry of Education Science and Culture.

REFERENCES

1. Malozemoff AP (1989) In: Ginsberg (ed) in Physical properties of High Temperature Superconductors I. World Scientific, Singapore, pp 71-150
2. Malozemoff AP, Worthington TK, Yeshurun Y, Holtzberg F and Kes PH (1988) Phys. Rev. B **38:** 7203
3. Sagdahl LT, Laegreid TL, Fossheim K and Murakami M (1991) Physica C **172:** 495
4. Ni B, Munakata T, Matsushita T, Iwakuma M, Funaki K, Takeo M and Yamafuji K (1988) Jpn. J. Appl. Phys. **27:** 1658
5. Ikeda H, Yoshizaki R, Watahiki M, Yoshikawa K and Tomita N (1991) submitted to ISS'91
6. Yoshizaki R, Ikeda H, Watahiki M, Yoshikawa K and Tomita N (1991) In: Kajimura and Hayakawa (eds) in Advances in Superconductivity III (1991) Springer-Verlag, Tokyo, pp 483-486
7. Campbell AM (1969) J. Phys. C **2:** 1492

The Irreversibility Line of MPMG Processed High-J_c $YBa_2Cu_3O_x$

NAOMICHI NAKAMURA, MASATO MURAKAMI, HIROYUKI FUJIMOTO, and
NAOKI KOSHIZUKA

Superconductivity Research Laboratory, ISTEC, Koto-ku, Tokyo, 135 Japan

ABSTRACT

The magnetic irreversibility line $B_{irr}(T)$ and the critical current density J_c of melt-powder-melt-growth (MPMG) processed 123 samples are measured to investigate the relation with the amounts and sizes of 211 precipitates. Places of $B_{irr}(T)$ of the samples show no correlation with J_c values at 77 K in 1 T. Analysis of J_c-B behaviors at 77 K indicates that flux pinning by the 211 precipitates is dominant in fields below 1 T but not always in higher fields.

KEY WORDS: MPMG material, irreversibility line, critical current, flux pinning

INTRODUCTION

It has been pointed out that there is a magnetically reversible region on the magnetic phase diagram below the upper critical field in bulk oxide superconductors [1]. The boundary of this reversible range is called the irreversibility line. The line is one of the most interesting subjects to study for basic understanding of the flux dynamics which correlates closely with the critical current properties most important for applications.

Several models have been proposed to describe the irreversibility line such as the superconducting glass model [1], thermal depinning within the frame work of conventional flux-creep [2,3], melting of the vortex lattice [4], and a transition between vortex-glass and vortex-liquid phases [5]. In the ongoing effort to understand the flux dynamics it seems essential to establish the basic experimental features of the irreversibility line for different samples with different defect densities, anisotropies and types of pinning centers. Xu and Suenaga [3] have reported a systematic shift of the irreversibility line in the direction of the high temperature region with an increase of J_c in the c-axis aligned polycrystalline $YBa_2Cu_3O_x$ (123) system, suggesting a picture such that the reversible region becomes narrower as pinning becomes stronger. On the other hand, some authors have shown that the irreversibility line of 123 single crystals did not shift by irradiation of protons [6] or electrons [7] in spite of a significant increase of J_c, making it complicated to understand the relation between the strength of flux pinning and the irreversibility line.

Recent work has shown that 123 materials prepared by a melt-powder-melt-growth (MPMG) process have notably high J_c values [8]. Furthermore, Wacenovsky et al. [9] and Sagdahl et al. [10] have reported that the irreversibility lines of MPMG processed samples occurred in a significantly higher temperature region compared to that of single crystal 123, in agreement with the picture of Ref. 3. Though fine Y_2BaCuO_5 (211) precipitates introduced through the MPMG process are considered to act as pinning centers, the influence of the size and the amount of 211 precipitates on the irreversibility line has not been sufficiently investigated so far; the relation between 211 precipitates, the strength of flux pinning, and the irreversibility line is still uncertain. In this paper the results of J_c measurements and the irreversibility lines for MPMG samples with different microstructures are reported. The role of the 211-phase is discussed within a model of pinning by normal precipitates.

EXPERIMENTAL

Samples were prepared by the MPMG process whose details have been described elsewhere [8]. Powders of Y_2O_3, $BaCO_3$ and CuO were mixed in compositions of Y:Ba:Cu = 1.2:2.1:3.1 and Y:Ba:Cu = 1.8:2.4:3.4 denoted as Y1.2 and Y1.8 for the purpose of doping Y_2BaCuO_5 (211) precipitates of 10 vol.% and 40 vol.%, respectively. The volume fractions, V_f, and the average grain sizes, d, of the 211 precipitates were varied by controlling the process parameters and determined as listed in table 1 by line analysis on optical micrographs of polished surfaces. The difference of the V_f values from the expected ones are presumably due to loss of the liquid phase during the melting process resulting in the deviation from the starting composition. The superconducting transition temperatures, T_c, determined magnetically were 92 K for all the samples.

A commercial SQUID magnetometer (Quantum Design Inc.) was used for measurements of the magnetization curves over the temperature range above 77 K. Magnetic fields up to 5.5 T were applied parallel to the c-axis of the samples in all the measurements. J_c values were derived using Bean's model [11] of $J_c = 2(M_{dec}-M_{inc})/\mu_0 t$ where t is the thickness of a slab-shaped sample and μ_0 is the permeability of free space. M_{dec} and M_{inc} are magnetic moments for decreasing and increasing portion of the cycled applied fields, respectively. J_c values at 77 K in a field of 1 T are also listed in table 1. The irreversibility fields B_{irr} were determined as the field at which the ratio M_{dec}/M_{inc} became unity.

RESULTS AND DISCUSSION

The results for the irreversibility lines for the five samples in table 1 are shown in Fig 1. It should be noted that the positions of the irreversibility lines are clearly sample-dependent, and do not correspond to the J_c values. B_{irr} of the sample Y1.2-1 (J_c of 2.70×10^8 A/m^2) and Y1.2-3 (1.42×10^8 A/m^2) are almost the same and the highest of the five samples. The samples Y1.8-1 (2.80×10^8 A/m^2) and Y1.8-2 (0.73×10^8 A/m^2), having the highest and the lowest J_c value, respectively, have almost the same H_{irr} values which are the lowest of the samples. This result is not in agreement with the picture that the reversible region is narrower in samples with stronger pinning [3,9,10].

The field dependences of J_c at 77 K for samples Y1.2-1 and Y1.8-1, which show the highest and the lowest B_{irr} values, respectively, but have almost the same J_c

table 1. Properties of the samples measured.

Sample	V_f(%)	d(μm)	J_c($\times10^8$ A/m^2)	J_c^{cal}($\times10^8$ A/m^2)
Y1.2-1	25	1.0	2.70	3.3
Y1.2-2	17	1.3	1.33	1.7
Y1.2-3	22	1.2	1.42	2.4
Y1.8-1	24	1.1	2.80	2.9
Y1.8-2	42	9.0	0.73	0.6

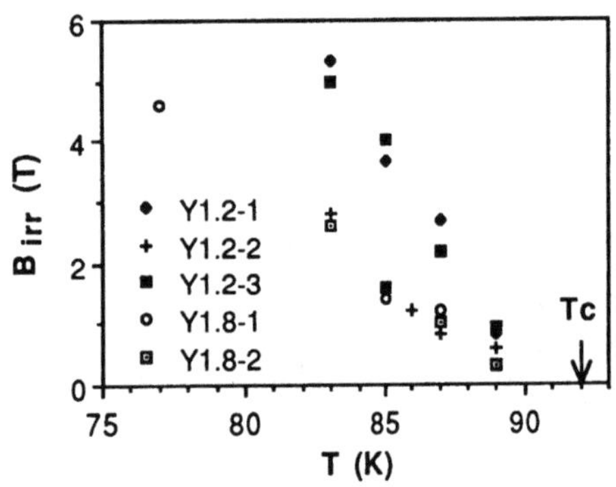

Fig.1 Irreversibility lines for the five samples measured.

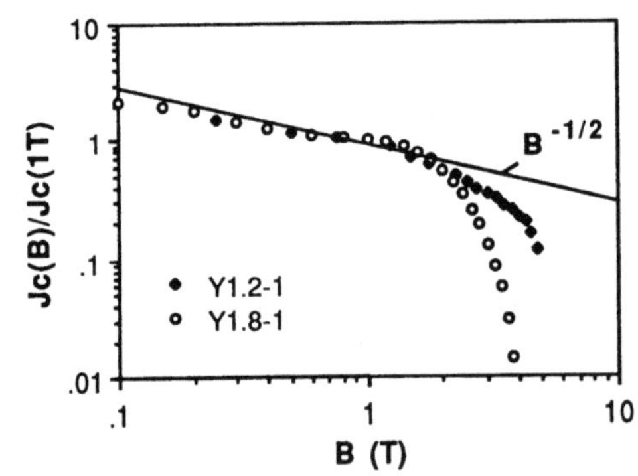

Fig.2 Normalized J_c versus B at 77 K on a log-log scale.

values at 1 T, are shown in Fig. 2 on a log-log scale. The J_c values are normalized by values at 1 T. In fields below 1 T, the field dependences of J_c are almost the same for these samples. On the contrary, the behavior becomes sample dependent in higher fields. The results indicate that the property of the flux pinning changes around 1 T. Therefore, it is reasonable to consider that the J_c values around 1 T are dominated by the pinning properties in the low field region while B_{irr} is determined by those in the high field region, resulting in the absence of the correlation between them as shown in Fig. 1.

We now estimate the J_c-B behavior assuming the flux pinning by 211 precipitates modifying the model used in Ref. 8. Flux lines can be pinned at normal precipitates because of the condensation energy stored within the volume where a flux line interacts with a normal precipitate. The difference in free energy between the superconducting and normal state per unit volume of $B_c^2/2\mu_0$, where B_c is the thermodynamic critical field, gives the maximum condensation energy. At high fields the available condensation energy goes down as $(1-B/B^*)$ [12] where B^* denotes a field at which the pinning force becomes zero, and we here regard B^* as B_{irr}. The pinning energy, U_p, is then given as $(B_c^2/2\mu_0)\pi\xi^2 d(1-B/B_{irr})$, where $\pi\xi^2 d$ gives the volume in which a flux line interacts with a normal precipitate with a size of d. The corresponding elementary pinning force, f_p, is then given by $U_p/2\xi$. We here assume that the density of the bulk pinning force, F_p, is given by direct summation of the elementary pinning, i.e. $F_p = Nf_p$ where N is the number of interactions between pinning centers and flux lines per unit volume. Calculations under the assumption have shown good agreements with experimental data in conventional high-κ materials [13]. For 211 precipitates with the number density of N_p, N can be replaced by $N_p d/a_f$ where $a_f = 1.07(\phi_0/B)^{1/2}$ is the spacing of the flux lines. Then from the relation of $F_p = J_c B$, the expression of J_c can be obtained as

$$J_c = (B_c^2/2\mu_0)(V_f/d)\pi\xi(\phi_0 B)^{-1/2}(1-B/B_{irr}). \qquad (1)$$

First the case of the low field region is considered. Calculated J_c values from eq. (1) with the parameters of ξ = 2 nm, B = 1 T, B_c = 0.22 T [14], and V_f and d from table 1, and under the approximation that $B/B_{irr} \ll 1$, are shown in table 1. The measured and the calculated values agree within a factor of 2. Moreover, the field dependence of J_c is closely proportional to $B^{-1/2}$ in low fields for all the samples as shown in Fig. 2, showing a good agreement with the prediction of eq. (1). These results suggest that 211 precipitates can work as pinning centers at least in the relatively low field region.

Figures 3-(a) and -(b) show normalized J_c versus B plots for the samples Y1.2-1 and Y1.8-1, respectively, with the best fits to the equation of $J_c = J_0(1-B/B^*)/B^{1/2}$ by parameterizing J_0 and B^* according to eq. (1). It is seen that the data of sample Y1.2-1 are well fitted by eq. (1) indicating that the pinning by

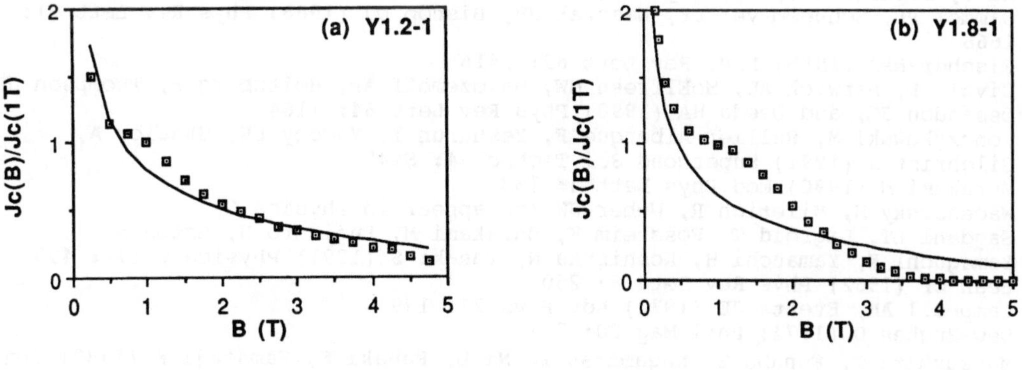

Fig.3 J_c vs B plots for (a) Y1.2-1 and (b) Y1.8-1 at 77 K. Solid lines represent the best fits to eq. (1).

211 precipitates is valid in rather high fields. On the contrary, the data of the sample Y1.8-1 cannot be fitted well. The data shows a significant shoulder structure, which should not exist if eq. (1) is valid, between fields of 1-2 T and then rapidly falls. The shoulder indicates that another pinning mechanism than pinning by 211 precipitates increases with increasing field in the sample.

A possible pinning mechanism increasing with field has been proposed by Daeumling et al. [15]. They considered that oxygen-deficient regions in 123 act as pinning centers in finite fields because the difference of the critical field between oxygen-deficient and fully-oxidized regions, resulting from the suppression of the superconductivity of the former by the applied field, increases the pinning force. Such a pinning mechanism is not effective in low field region because of proximity-effect coupling. In the high field region, the critical current density falls rapidly because of granular properties of the oxygen-deficient regions.

Then the following picture would explain the behavior of J_c in Fig.3-(b). In the low field region, flux pinning by 211 precipitates is dominant as shown in Fig. 2 and table 1. Pinning by oxygen-deficiencies in the 123-phase becomes effective in increasing field, resulting in the shoulder in the J_c-B curve. On the other hand, oxygen-deficiencies at boundaries of 123- and 211-phases broaden the width of the boundary regions across which the material changes from a superconductor to a non-superconductor. Pinning of the vortex core vanishes when this distance becomes much larger than ξ [16]. For 123 materials, the average distance d_o between ordered O vacancies is $\sim a/(7-x)^{1/2}$ where a is the lattice parameter of the basal plane [15]; for example, $d_o \sim \xi$ when $7-x = 0.05$. Therefore, it is possible that the boundary width is broadened beyond ξ by the suppression of the superconductivity in oxygen-deficient regions at the boundary in increasing field, resulting in the loss of the pinning ability of 211 precipitates. In order to confirm the above idea, experiments of oxygen-annealing are now on progress.

In summary, the field dependence of J_c of MPMG samples is well described with a model of flux pinning by 211 precipitates at least in the relatively low field region. In high field region, it is possible that the dominant pinning mechanism is different from that in the low field region. Therefore it is not adequate to correlate the irreversibility lines and J_c values measured in the low field region.

The authors thank N. Sakai and A. Kondoh for sample preparation and Dr. J. O. Willis for helpful discussions and a critical reading of the manuscript. This work was partly supported by NEDO under the management of the R&D of Basic Technology for Future Industries.

REFERENCES

1. Müller KA, Takashige M, Bednorz JG (1987) Phys Rev Lett 58: 408
2. Yeshurun Y, Malozemoff AP (1988) Phys Rev Lett 60: 2202
3. Xu Y, Suenaga M (1990) Phys Rev B 43: 5516
4. Gammel PL, Schneemeyer LF, Wasczak JV, Bishop DJ (1988) Phys Rev Lett 61: 1666
5. Fischer MPA (1989) Phys Rev Lett 62: 1416
6. Civale L, Marwick AD, McElfresh MW, Malozemoff AP, Holtzberg F, Thompson JR, Ossandon JG, and Deeds HA (1990) Phys Rev Lett 64: 1164
7. Konczykowski M, Rullier-Albenque F, Yeshurun Y, Yacoby ER, Shauiov A, Gilchrist J (1991) Supercond Sci Technol 4: S445
8. Murakami M (1990) Mod Phys Lett 4: 163
9. Wacenovsky M, Miletich R, Weber HW (to appear in Physica C)
10. Sagdahl LT, Lægreid T, Fossheim K, Murakami M, Fujimoto H, Gotoh S, Yamaguchi K, Yamauchi H, Koshizuka N, Tanaka S (1991) Physica C 172: 495
11. Bean CP (1962) Phys Rev Lett 8: 250
12. Campbell AM, Evetts JE (1972) Adv Phys 21: 199
13. Dew-Hughes D (1974) Phil Mag 30: 293
14. Matsushita T, Funaba S, Nagamatsu Y, Ni B, Funaki K, Yamafuji K (1989) Jpn J Appl Phys 28: L1508
15. Daeumling M, Seuntjens KM, Larbalestier DC (1990) Nature 346: 332
16. Campbell AM, Evetts JE, Dew-Hughes D (1968) Phil Mag 18: 313

High Critical Current Densities at 77 K Under High Magnetic Fields Observed for BiSrCaCuO Thin Films Prepared by an MOCVD Method

HIROFUMI YAMASAKI, KAZUHIRO ENDO, YOSHIHIKO NAKAGAWA, MASAICHI UMEDA, SHIN KOSAKA, SHUNJI MISAWA, and SADAFUMI YOSHIDA

Electrotechnical Laboratory, Tsukuba, Ibaraki, 305 Japan

ABSTRACT

High critical current densities (J_C) were observed for BiSrCaCuO thin films at 77 K under high magnetic fields. The best values are 3.3×10^9 A/m^2 (1 T) and 9.1×10^8 A/m^2 (8 T), which are the highest J_C for Bi-oxide thin films. These as-deposited films prepared by an MOCVD method consist of c-axis oriented $Bi_2Sr_2Ca_2Cu_3O_x$ phase with T_C values as high as 97 K. There were no signs of weak-links in the $J_C(H)$ behavior. J_C dropped abruptly when the applied field deviates from the ab-plane, which evidently supports the idea of intrinsic pinning proposed by Tachiki and Takahashi.

KEY WORDS: Bi-oxide thin film, MOCVD, transport critical current density, weak-link, intrinsic pinning

INTRODUCTION

Since its first discovery by Maeda et al. [1], BiSrCaCuO (BSCCO) system has attracted much attention for its higher T_C and better chemical stability, compared with the $YBa_2Cu_3O_7$ compound. It is well known that there are many superconducting phases showing various T_C's: the high-T_C phase $Bi_2Sr_2Ca_2Cu_3O_x$ with T_C~110 K and the low-T_C phase $Bi_2Sr_2CaCu_2O_y$ with T_C~80 K are most commonly synthesized. In general it is difficult to obtain the 2223-high-T_C phase. In the case of bulk synthesis, Pb addition is necessary to obtain the 2223-high-T_C single phase samples. The situation is the same for thin film preparation. The Bi-2223 films with high T_C (>90 K) have not been prepared without post annealing, using any growth techniques, and most of the films reported consist of the 2212-low-T_C phase in an as-deposit state. Accordingly, there have been very few reports describing critical current densities (J_C) at 77 K [2, 3].

We have succeeded in preparing as-deposited Bi-2223 thin films without post annealing, having typical T_C values of 92-97 K (defined by zero-resistivity), by metalorganic chemical vapor deposition (MOCVD) [4]. These films had the highest J_C at 77 K under magnetic fields (1-8 T) ever reported for the BSCCO system. In this paper we report our investigation on their J_C properties.

EXPERIMENTAL

The details of the sample preparation procedures are reported elsewhere [4]. The outline is as follows. Bi-oxide thin films were prepared by a cold-wall type MOCVD apparatus, using the source materials of triphenyl bismuth $Bi(C_6H_5)_3$, $Sr(DPM)_2$, $Ca(DPM)_2$ and $Cu(DPM)_2$. Substrates of $LaAlO_3$(100) single crystals having very small lattice mismatches with BSCCO, 5 × 10 × 0.5 mm^3 in size, were placed on an inconel susceptor and then inductively heated to 800°C. The growth rate was kept very slow, around 1.2 nm/h, for there was a clear inverse correlation between T_C and the growth rate [4]. The thickness of the films used in this study was 24-96 nm. The films were characterized by an X-ray diffractometer. The surface morphology was examined by scanning electron microscopy (SEM), and the compositions were confirmed by inductively coupled plasma spectroscopy. Resistance was measured as a function of temperature by a conventional four-probe DC method.

Transport critical currents (I_C) were measured by a conventional four-probe DC method using evaporated Au film pads as electrodes and In foils as current leads. We measured I_C only at 77.3 K in a liquid nitrogen bath without making bridge patterns in the films, in order to avoid chemical changes that might occur by the contact with the resist material or the etching solution. The I_C measurements with a pulse current supply were also performed to confirm that the heating effect at the electrodes are negligible. Critical currents are defined by a criterion of 2 μV/cm electric field for the voltage tap distance of 3 mm, and critical current densities are calculated by dividing I_C with the cross-sectional area.

RESULTS AND DISCUSSION

For all the as-deposited films used in this study, all the X-ray diffraction peaks were indexed as (00n) of the Bi-2223 phase, which means that they consist of single phase $Bi_2Sr_2Ca_2Cu_3O_x$ with the c-axis oriented normal to the substrate.

The *I-V* traces for the I_C measurements for one of the best J_C samples are shown in Fig. 1. The critical current was also measured with a pulse current, whose pulse width was 4 ms with the interval of 0.5 s. Although poor S/N ratio hindered the exact I_C determination in the case of the pulse current, it can be concluded that the heating effect was negligible when the DC current was used for the I_C measurements. Then the critical currents were measured in magnetic fields parallel to the film surface (H ∥ ab-plane) and perpendicular to the transport current. As the applied magnetic fields increased, I_C decreased and the transition became broader.

Critical current densities were measured at 77.3 K for many samples in magnetic fields parallel to the film surface and perpendicular to the transport current (Fig. 2). In Fig. 2, filled symbols show the J_C data for 96 nm-thick films, open symbols are for 42 nm-thick films and crosses (x) are for a 24 nm-thick film. It can be seen that the J_C values greater than 10^9 A/m^2 at 1 T were

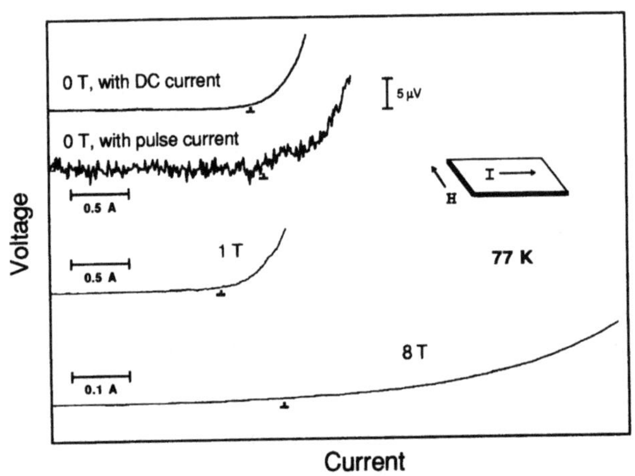

Fig. 1 Current-voltage characteristics for one of the best J_C samples.

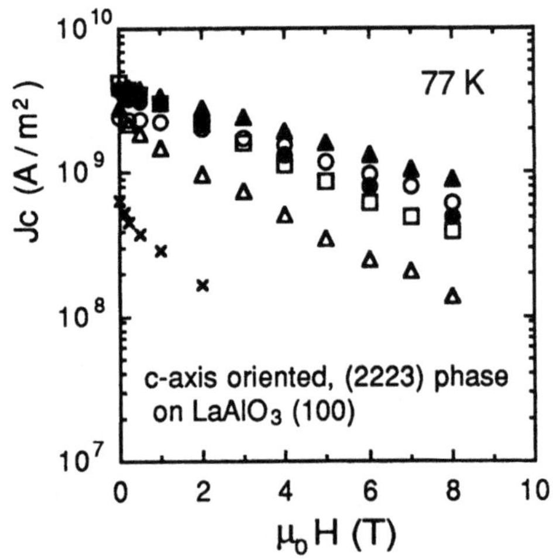

Fig. 2 J_C at 77.3 K for the Bi-2223 phase films (H ∥ ab-plane).

reproducibly observed for the films with the thickness of 42-96 nm. The best values are 3.3×10^9 A/m^2 at 1 T and 9.1×10^8 A/m^2 at 8 T for a 96 nm-thick film. The transition temperature (T_c) of this best J_c sample was 97 K. Although J_c values higher than 10^{10} A/m^2 are claimed for the post-annealed Bi-oxide thin films in ref. 2, they measured I_c for micro-bridges of 20 μm in length and 20 μm in width, without mentioning the I_c criterion. It seems that they used much looser criterion than that used in the present study (2 μV/cm). Therefore, it is reasonable for us to claim that our J_c values are the highest values for Bi-oxide films, observed at 77.3 K under magnetic fields greater than 1 T. The remarkable feature in Fig. 2 is that J_c declines very gradually with increasing magnetic field; there is no abrupt drop of J_c in low-field range (<0.1 T). It is also to be noted that the $J_c(H)$ behavior was reversible in field and no hysteresis was observed. It is well known that the $J_c(H)$ behavior of a sample consisting of a weak-link network shows an abrupt J_c drop in the low-field range (<0.01-0.1 T, depending on the sample) and a remarkable hysteresis [5-8]. Therefore, it is concluded that there are no signs of weak-links in the $J_c(H)$ behavior of the films in the present study.

Figure 3 shows the surface morphology of the films, for which J_c data are shown in Fig. 2 ((a) corresponds to filled triangles and (b) to open circles). We can see irregularly-shaped small particles which are considered to be second phase precipitates, and a few square particles which are assumed to be BSCCO crystals just grown from the film base. It is to be noted that there are no apparent grain boundaries in the film base. This SEM observation also supports the idea that these films do not consist of a weak-link network.

Another evidence for the weak-link-free J_c behavior is obtained from the J_c measurements in the magnetic field parallel to the transport current, which corresponds to the Lorenze-force-free configuration. Critical current densities measured in magnetic fields parallel to the current, $J_c(\parallel)$, are compared with the usual J_c values measured in magnetic fields perpendicular to the current, $J_c(\perp)$. The Jc enhancement of $1 \leq J_c(\parallel)/J_c(\perp) \leq 2$ was observed in the fields up to 14 T, as shown in Fig. 4. This feature supports the assumption that the critical current is determined by the balance between the pinning strength and the Lorentz force, and that this system does not consist of a weak-link network. It was reported that $J_c(\parallel)$ is essentially equal to $J_c(\perp)$ for textured $Bi_2Sr_2CaCu_2O_y$ tapes with the J_c values over 10^9 A/m^2 at 4.2 K under high magnetic fields (>20 T) [9]. They interpreted the results as follows: the microscopic superconducting current takes a zig zag path because of the random orientation

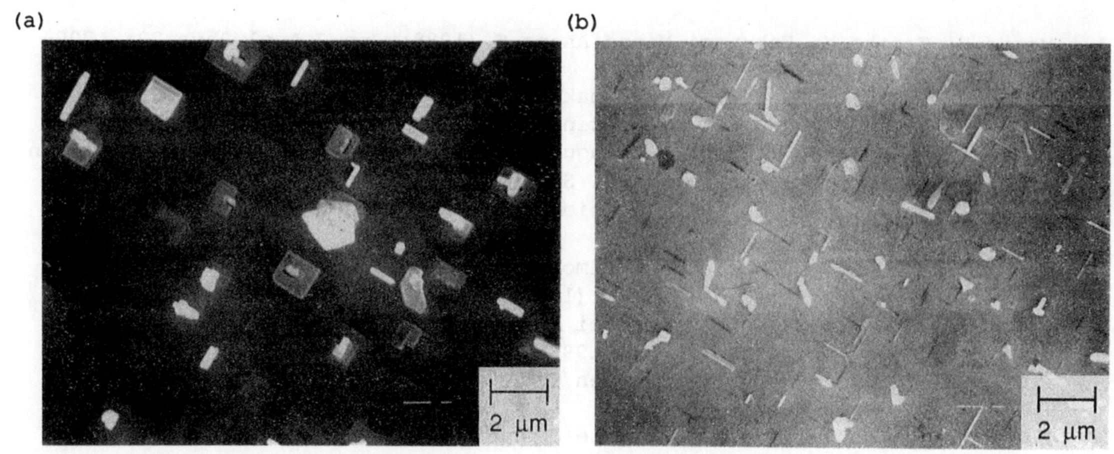

Fig. 3 SEM photographs showing the surface morphology of the Bi-2223 phase films. (a) film with the thickness of 96 nm, and (b) film with t = 42 nm.

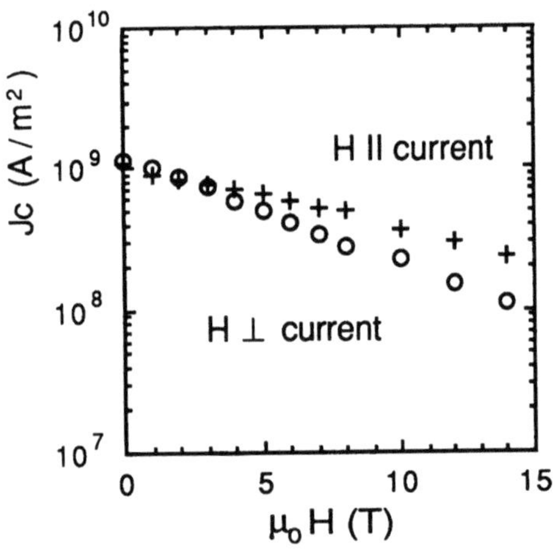

Fig. 4 Effect of the current direction relative to the applied magnetic field, upon J_C.

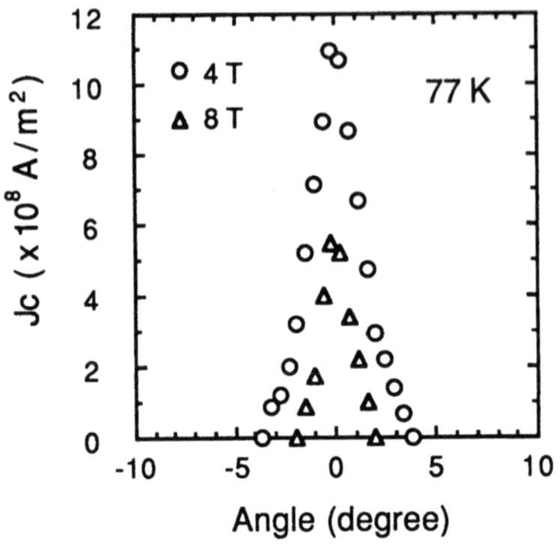

Fig. 5 The effect of the angle between the applied filed and the film surface (ab-plane) for a 42 nm thick film.

of the grains within the plane perpendicular to the c axis, and the volume pinning force is independent of the field direction.

Record-high J_C values have been observed for as-deposited $Bi_2Sr_2Ca_2Cu_3O_x$ thin films prepared by an MOCVD method, and there is no sign of weak-links. Then what are the primary flux pinning centers in these films? In order to obtain more insights into this issue, we measured J_C as a function of the angle between the applied magnetic field and the film surface (ab-plane). J_C decreased sharply as the magnetic field deviates from the ab-plane (Fig. 5). This result implies that there is a very strong pinning for the magnetic fluxes parallel to the ab-plane and that pinning strength is very weak for the magnetic component parallel to the c-axis. This interpretation is consistent with the model of the step-like inclusion of magnetic fluxes [10], based on the intrinsic pinning mechanism proposed by Tachiki and Takahashi [11] which assumes that insulating block layers imbedded among superconducting CuO_2 layers become strong pinning centers against magnetic flux motion.

REFERENCES

1. Maeda H, Tanaka Y, Fukutomi M and Asano T (1988) Jpn J Appl Phys 27: L209-L210
2. Itozaki H, Higaki K, Harada K, Tanaka S, Fujimori N and Yazu S (1989) In: Advances in Superconductivity. Springer-Verlag, Tokyo, pp 599-604
3. Hayashi N, Okuda S, Takano S, Minaguchi D, Terai S and Hasegawa K (1991) In: Advances in Superconductivity III. Springer-Verlag, Tokyo, pp 1037-1040
4. Endo K, Yamasaki H, Misawa S, Yoshida S and Kajimura K (1991) Nature (in press)
5. Watanabe K, Noto K, Morita H, Fujimori H, Mizuno K, Aomine T, Ni B, Matsushita T, Yamafuji and Muto Y (1989) Cryogenics 29: 263-267
6. Ekin JW, Hart HR, Jr. and Gaddipati AR (1990) J Appl Phys 68: 2285-2295
7. Yang Y, Beduz C and Ashworth SP (1990) Cryogenics 30: 618-622
8. Chang H-L, Juang J-Y, Liou S-H, Uen T-M and Gou Y-S (1990) Jpn J Appl Phys 29: L2037-L2040
9. Kumakura H, Togano K, Maeda H, Kase J and Morimoto T (1991) Appl Phys Lett 58: 2830-2832
10. Iye Y, Tamegai T and Nakamura S (1991) Physica C174: 227-232
11. Tachiki M and Takahashi S (1989) Solid State Commun 70: 291-295

Anisotropy of Irreversibility Line in QMG Processed Y-Ba-Cu-O

TERUO MATSUSHITA[1], EDMUND SOJI OTABE[1], BAORONG NI[2], KAORU YAMAFUJI[2], KEIICHI KIMURA[3], MITSURU MORITA[3], MASAMOTO TANAKA[3], SEIKI TAKEBAYASHI[3], KATSUYOSHI MIYAMOTO[3], and MISAO HASHIMOTO[3]

[1] Department of Computer Science and Electronics, Kyushu Institute of Technology, Iizuka, Fukuoka, 820 Japan
[2] Department of Electronics, Kyushu University, Fukuoka, 812 Japan
[3] Advanced Materials & Technology Research Laboratories, Nippon Steel Corporation, Kawasaki, Kanagawa, 211 Japan

ABSTRACT

The anisotropy of the irreversibility field was measured for QMG processed Y-Ba-Cu-O and the irreversibility field in the a-b plane was larger by a factor of 4.2 than that in the c-axis at 85 K. The result was compared with the collective flux creep theory and a good agreement was obtained. It is concluded that large values of the irreversibility field in this material is determined by the flux pinning strength but its anisotropy comes from the anisotropic upper critical field.

KEYWORDS: Y-Ba-Cu-O, quench and melt growth, irreversibility line, anisotropy, flux creep, flux pinning

INTRODUCTION

There exists a reversible region in the magnetic field vs temperature plane where the critical current density is zero at high temperatures due to the thermally activated flux creep. The boundary between the reversible region and the irreversible one with nonzero critical current density is called the irreversibility line. This characteristic line is relatively close to the phase boundary, $B_{c2}(T)$, in Y-based superconductors with strong pinning force, while it is far from $B_{c2}(T)$ in weakly pinned Bi-based ones. Therefore, it is desirable to reduce the effect of the flux creep so as to make the reversible region narrow by increasing the pinning strength.

The irreversibility line is anisotropic with respect to the direction of the magnetic field. There exists a question on this anisotropy. That is, the anisotropy is larger in Y-based superconductors [1] with less significant anisotropy in B_{c2} than in two-dimensional Bi-based ones [2]. This fact suggests that the anisotropy in the irreversibility line is not influenced only by the anisotropic B_{c2}. In order to clarify the mechanism of determining the irreversibility line, its anisotropy was measured for melt-processed bulk Y-Ba-Cu-O and the result was compared with the collective flux creep theory.

EXPERIMENTS

The specimen used for the measurements of the irreversibility line was prepared by the quench and melt growth (QMG) method. The critical temperature of the specimen was 91.7 K. Its size was 2.66 mm wide, 0.66 mm thick and 11.5 mm in length. The c-axis was directed normal to the wide surface. The critical current density, J_c, was measured resistively. The distance between voltage terminals was 2.8 mm and the voltage criterion of 1 μV was used for measurements of J_c. The magnetic field was applied in two directions; in the a-b plane and along the c-axis. It was directed always perpendicular to the current. The irreversibility field, $B_i(T)$, was defined by the magnetic field at which J_c was reduced to 1×10^4 A/m^2 at given temperature, T. The measurement was carried out in the field range of 0 to 5 T.

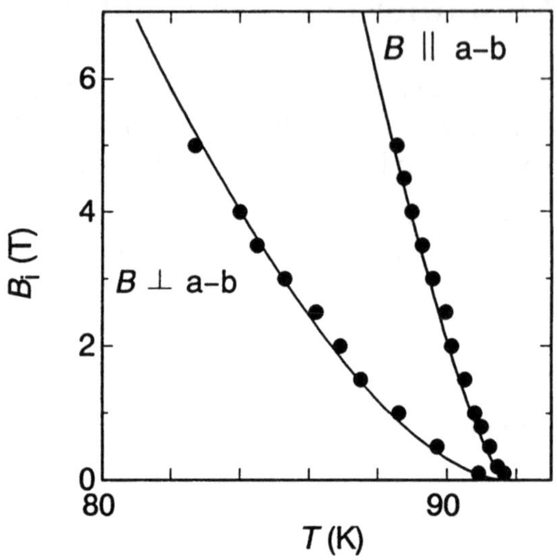

Fig. 1. Observed anisotropy of the irreversibility field in QMG processed Y-Ba-Cu-O.

The obtained results are shown in Fig. 1. If the results are expressed as

$$B_i(T) = B_i(0)(1 - t^2)^n \tag{1}$$

with t denoting the reduced temperature, T/T_c, $B_i(0)$ = 203 T and n = 1.39 are obtained for the field in the a-b plane, and $B_i(0)$ = 93.2 T and n = 1.72 are obtained for the field along the c-axis. These relations are shown in the figures. Hence, the irreversibility field in the a-b plane is estimated to be larger by a factor of 4.2 than that in the c-axis at 85 K.

DISCUSSION

According to the collective flux creep theory [3], the irreversibility field is given by

$$B_i^{(3-2\gamma)/2} = \left(\frac{K}{T_c}\right)^2 (1-t^2)^m \left(1 - \frac{B_i}{B_{c2}}\right)^\delta, \tag{2}$$

where m, γ and δ are dimensionless parameters representing temperature and magnetic field dependences of the ideal pinning force density in the absence of the flux creep:

$$F_{p0} = AB_{c2}^m(T)B^\gamma \left(1 - \frac{B}{B_{c2}}\right)^\delta. \tag{3}$$

In Eq. (2) the upper critical field, B_{c2} is assumed to vary with the temperature as

$$B_{c2}(T) \propto 1 - t^2. \tag{4}$$

A constant K in Eq. (2) is given by

$$K = \frac{0.835g^2[AB_{c2}^m(0)]^{1/2}}{\zeta^{3/2}\ln(Ba_f\Omega/E_c)}, \tag{5}$$

where g^2 is a number of fluxoids in the flux bundle, a_f is the fluxoid spacing, Ω is an oscillation frequency of the flux bundle and E_c is an electric field criterion for determination of J_c. Substitution of these typical values into the logarithmic term leads to $\ln(Ba_f\Omega/E_c) \simeq 14$ [3]. The number ζ is connected to the half size of the pinning potential called the interaction distance:

$$d_i = a_f/\zeta. \tag{6}$$

For small pointlike defects with high density, we have $\zeta = 2\pi$ [4]. On the other hand, $\zeta = (d_p - a_f)/4(d_p + a_f)$ was derived for strong pinning centers with low concentration [5], where d_p is a mean pin spacing. The present specimen with 211 particles is close to the latter situation. In this case the pin spacing is of the order of several micrometers and is much larger than the fluxoid spacing. Hence, we have $\zeta = 1/4$. In the elastic region of the fluxoid lattice, g is considered to be given by the transverse elastic correlation length divided by a_f:

$$g = \left(\frac{C_{66}d_i}{F_{p0}a_f^2}\right)^{1/2}, \tag{7}$$

where C_{66} is the shear modulus of the fluxoid lattice. The shear modulus is expressed as

$$C_{66} = \frac{B_c^2 B}{4\mu_0 B_{c2}}\left(1 - \frac{B}{B_{c2}}\right)^2, \tag{8}$$

where B_c is the thermodynamic critical field.

Here we treat the case where the particles of $211(Y_2BaCuO_5)$ phase of 2.0 μm in diameter are distributed as pinning centers in the superconductor. Its volume fraction is assumed to be 25%. In this case the pinning parameters predicted theoretically are m=1.5 and γ=0.5 [6]. Using these parameters, the ideal pinning force density at 0 K is estimated to be $AB_{c2}^m(0) = 3.75 \times 10^9$ A/m^2 [7]. On the other hand, the shear modulus C_{66} at 0 K and 1 T is estimated as 4.21×10^3 N/m^2 [7]. Substitution of these values into Eq. (7) leads to $g = 2.4$ at 0 K and 1 T. In the practical case this g value is expected to be smaller than this estimate, since the shear modulus is decreased from Eq. (8) because of existing fluxoid lattice defects. However, the flux bundle size is not smaller than the fluxoid spacing and hence, the minimum value of g is unity. That is, g is expected to vary within the range of 2.4 and 1. For simplicity, therefore, we assume that g is given by a constant 2 in the whole range of the magnetic field and temperature.

Substituting the above values into Eq. (5), we have $K = 1.83 \times 10^3$. With the theoretical parameters of m and γ Eq. (2) is reduced to

$$B_i = 3.98 \times 10^2 \left[1 - \left(\frac{T}{91.7}\right)^2\right]^{1.5}\left(1 - \frac{B_i}{B_{c2}}\right)^\delta. \tag{9}$$

In the case of pinning by large normal particles, the pinning efficiency is not high. Hence, $\delta = 1$ in the strong pinning regime will not be attained in the present specimen. Thus, we assume $\delta = 2$ as is observed in Nb_3Sn. In the case of the magnetic field in the a-b plane, B_{c2} is expected to be as high as 60 T and hence, the factor of $(1 - B/B_{c2})^2$ can be approximated by 1. Hence, we have $B_i = 15.8$ T at T=86.2 K directly from Eq. (9). This value is slightly larger than 10.2 T obtained by extrapolation of the experimental result through Eq. (1).

In the magnetic field along the c-axis, B_{c2} is not so high as in the previous case. According to the experimental result by Sagdal *et al.*[8], B_{c2} along the c-axis at 86.2 K is estimated as 8.9 T. Hence, B_i at this temperature can be obtained from

$$B_i = 15.8\left(1 - \frac{B_i}{8.9}\right)^2 . \qquad (10)$$

This equation is easily solved and we have B_i=4.27 T. This value is also larger than the experimental result of 2.5 T.

Although there are quantitative difference in B_i values in the two field directions, the theoretically predicted anisotropy of B_i is 3.70, while the experimental one is 4.08. The slightly larger theoretical values of B_i may come from the error in choice of g value or in assumption of logarithmic term in Eq. (5). Figure 2 represents the present theoretical results on the anisotropic irreversibility fields, where both of them are divided by 1.6 so as to get a global agreement between the theoretical and experimental results. Hence, the anisotropy can be explained well by the collective flux creep theory.

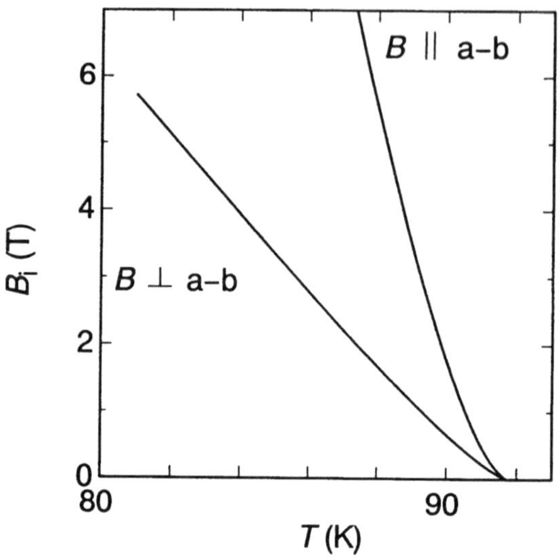

Fig. 2. Theoretical result of anisotropy of the irreversibility field. The results are divided by 1.6 so as to get a good agreement with experiments.

In conclusion, the large irreversibility fields in melt-processed Y-Ba-Cu-O in comparison with Bi-based superconductors can be attributed to its strong pinning strength. Since B_i is of a comparable order of magnitude to B_{c2} for the field direction along the c-axis, B_i is suppressed by B_{c2}, resulting in the large anisotropy in B_i.

REFERENCES

1 Slaski M, Sagdahl LT, Heill LK, Fossheim K, Murakami M, Fujimoto H, Koshizuka N and Tanaka S (1991) Proc. M²S-HTSC III, Kanazawa, Physica C (to be published)

2 Matsushita T, Otabe ES, Nakatani T, Ni B, Umemura T, Egawa K, Wakata M and Utsunomiya S (1991) Jpn. J. Appl. Phys. 30: (to be published)

3 Matsushita T, Fujiyoshi T, Toko K and Yamafuji K (1990) Appl. Phys. Lett. 56: 2039

4 Campbell AM, Küpfer K and Meier-Hirmer R (1985) *Proc. Int. Symp. Flux Pinning and Electromagnetic Properties in Superconductors*, Fukuoka. Matsukuma, Fukuoka, p. 54

5 Matsushita T (1981) Jpn. J. Appl. Phys. 20: 1955

6 Matsushita T, Otabe ES, Ni B, Kimura K, M)*Pc*KM, Tanaka M, Kimura M, Miyamoto K and Sawano K (1991) Jpn. J. Appl. Phys. 30: L342

7 Matsushita T, Otabe ES and Ni B (1991) *Proc. 6th Workshop on Critical Currents*, Cambridge, Supercond. Sci. Technol. (to be published)

8 Sagdahl LT, Lægreid T, Fossheim K, Murakami M, Fujimoto H, Gotoh S, Yamaguchi K, Yamauchi H, Koshizuka N and Tanaka S (1991) Physica C 172: 495

4.3 Flux Motion

Unusual Vortex States in High-T_c Superconductors – What Determines J_c-Values? –

KAZUO KADOWAKI

National Research Institute for Metals, Tsukuba, Ibaraki, 305 Japan

ABSTRACT

Experimental results in the unusual vortex state in highly anisotropic high-T_c superconductor $Bi_2Sr_2CaCu_2O_{8+\delta}$ are discussed in terms of the two dimensional pancake vortex model. The characteristic resistivity broadening and the temperature dependence of superconducting diamagnetic moment are argued by thermal fluctuations of the order parameter of superconductors in the reversible vortex state. The irreversibility line was determined by static magnetic measurements and was argued with a decoupled superconducting layer model. In addition, it is observed that the unceasing relaxation of magnetization occurs even at 55 mK, suggesting evidence of a new relaxation mechanism such as quantum creep of the vortex. All these features are very intriguing properties by themselves but are not favorable for applications, since they degrade and limit the J_c-value.

Abrikosov vortex state, irreversibility line, flux-creep, flux-flow, resistivity broadening

INTRODUCTION

There has long been a controversy concerning the unusual vortex state near T_c in high-T_c superconductors, since the phenomena observed markedly contrast with the ones in the conventional superconductors. A most striking one is the resistivity broadening phenomenon, which indicates dissipation of the super-current flow in magnetic fields in the superconducting states. Another feature is a very fast relaxation phenomenon even at very low temperatures, which is associated with the fast decay of the persistent current in the sample, suggesting the existence of a new dissipation mechanism in the superconducting state. Until recently, those two typical phenomena have been treated by the traditional models: flux creep and flux flow of the vortex or somewhat modified versions of them. The fundamental idea there is related to motion of vortices due to the electromagnetic Lorentz force. Although such models are simple and they may be plausible for conventional superconductors, a serious disagreement with them has been found in the results of the precise measurements of the angular dependence of the resistivity on high quality single crystals for compounds such as $YBa_2Cu_3O_{7-\delta}$, $La_{2-x}Sr_xCuO_{4-\delta}$ as well as $Bi_2Sr_2CaCu_2O_{8+\delta}$. Most distinct contradictions are the Lorentz force independence of the resistivity broadening in magnetic fields and the fast relaxation of the diamagnetic magnetization at very low temperatures, where all the thermally activated phenomena should be frozen out. Those phenomena cannot be explained by the traditional models.

Here we clarify the origin of the dissipation in the resistivity in fields and show the results of the recent magnetization study near and below T_c as evidence of the most prominent effect of the superconducting fluctuations. A model is proposed, in which pancake vortices instead of conventional vortex lines are moving around like gas molecules and cause dissipation. It is shown that this model is compatible with them, and moreover, it can comprise the traditional approaches. As a consequence, this leads us to a more general understanding of the vortex state in these high T_c superconductors. A peculiar behavior in highly anisotropic systems is the existence of a boundary which separates the vortex state into two states: reversible and irreversible states. Below this boundary, superconducting properties are more or less like traditional ones, whereas above that most of them are unusual. It appears that the boundary enters more deeply into the vortex state in the B–T phase diagram in higher anisotropic superconductors. It is shown that these unusual behaviors originate essentially from fluctuations of motion of vortex pancakes floating around over the CuO_2 plane, which process is intrinsically dissipative in nature even in the superconducting state. Since this intrinsic resistive state practically limits attainable values of J_c in the system, it is very important to understand this phenomenon for applications.

RESISTIVITY BROADENING PHENOMENA AND STRONG SUPERCONDUCTING FLUCTUATION EFFECTS

The resistivity broadening phenomena for the ρ_a and ρ_c of single crystalline $Bi_2Sr_2CaCu_2O_{8+\delta}$ are shown in Figs. 1 and 2. These are especially chosen as typical examples of the resistivity broadening phenomena in extremely anisotropic systems. The results of the more extended study have already been published[1,2]. The most prominent fact in this case is that there is no observable Lorentz force effect on the resistivity broadening for any configurations of the measuring current and the applied magnetic field directions. This has been a key issue concerning the

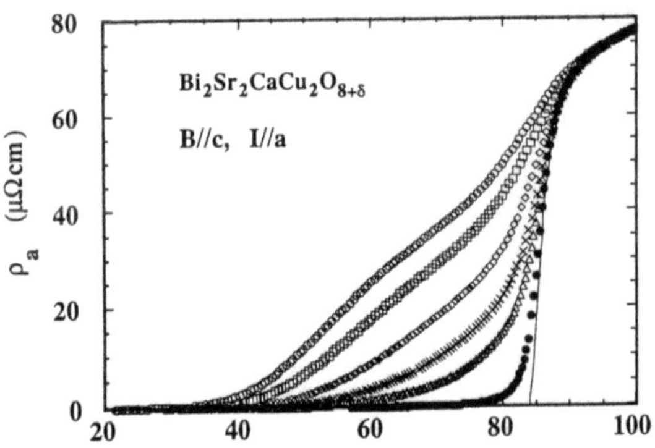

Fig. 1. Temperature dependence of the a-axis resistivity curves of single crystalline $Bi_2Sr_2CaCu_2O_{8+\delta}$ in various magnetic fields of 0 T (–), 0.1 T (●), 0.5 T (△), 1 T (×), 2 T (◊), 4 T (□) and 6 T (○) applied along the c-axis.

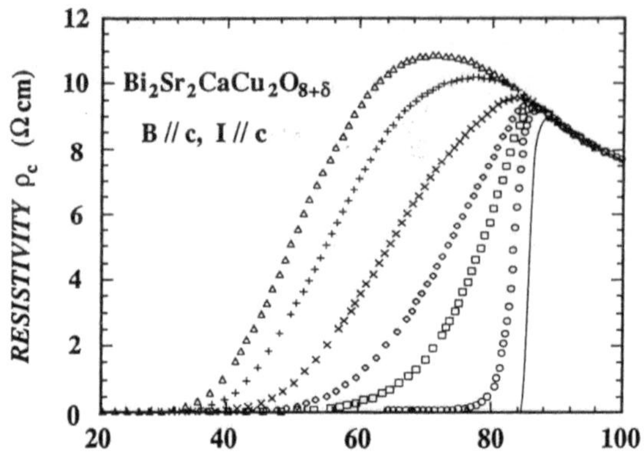

Fig. 2. Temperature dependence of the c-axis resistivity curves of single crystalline $Bi_2Sr_2CaCu_2O_{8+\delta}$ in various magnetic fields of 0 T (–), 0.1 T (○), 0.5 T (□), 1 T (◊), 2 T (×), 4 T (+) and 6 T (△) applied along the c-axis.

dissipation mechanism in the vortex state. Contrasting with this, it is noted that in $YBa_2Cu_3O_{7-\delta}$ the effect of the Lorentz force has indeed been observed in addition to the Lorentz force independent background[3,4]. The Lorentz force dependent part can be interpreted by the flux-creep and the flux-flow model quite well[3]. The fact that the effect of the Lorentz force could not be detected in any possible current and magnetic field configurations in the $Bi_2Sr_2CaCu_2O_{8+\delta}$ system leads us to a conclusion that the Lorentz force driving the vortex motion is somehow inactive and the other dissipation mechanism must be dominant.

We have proposed a simple model in previous publications[2,5], which can account for the Lorentz force independent resistivity. This is essentially based on the thermal fluctuations of the superconducting order parameter due to the short coherence length as well as the low dimensionality of the electronic structure of the system and can explain the experimental facts qualitatively. The model assumes two dimensional pancakes which move around with a velocity of v in the decoupled CuO_2 plane. The fluctuating component of the velocity $<v>_{ave}$ is induced by a current I due to the work $2\pi\hbar I/2e$ done in the system. When the number of pancakes n is in a distance L between the two points A and B, the resulting fluctuating flux in a unit time amounts to $n<v>_{ave}$. Therefore, the phase change induced by the fluctuating pancakes is $2\pi n<v>_{ave}$. As a consequence, the induced voltage between the two points according to the Josephson effect can be written as

$$V = 2\pi n(\hbar/2e)<v>_{ave}L \quad (1)$$

$$= (\phi_0 n/c)<v>_{ave}L \quad (2)$$

$$= (B\perp/c)<v>_{ave}L, \quad (3)$$

where $B\perp = n\phi_0$. Since $<v>_{ave}$ is induced by the applied current, $<v>_{ave}$ must be proportional to the current below a certain level I_c. This can be proved more rigorously using a similar argument given by McCumber[6], although it is rather complicated to calculate $<v>_{ave}$ explicitly as was performed by McCumber and Halpelin[7]. The linear I-V characteristic can also be expected in the low current limit but the deviation may occur at higher current levels near I_c. The thermal fluctuation process must be dominant, as long as the relaxation time τ is much shorter than $l/<v>_{ave}$, where l is the mean free path of a pancake.

It must be noted that by this formulation the fluctuation regime can continuously be connected to the flux-flow regime, when the fluctuation becomes less dominant, *i. e.*, $\tau \geq l/<v>_{ave}$. In this regime, the motion of pancakes becomes similar to that of ordinal vortices, so that the traditional Lorentz force effect on the vortex motion can be seen. Therefore, the velocity $<v>_{ave}$ can be written in a more general fashion and can be divided into two contributions: $<v>_{ave} = <v>_{fluct} + <v>_{flow}$, where $<v>_{fluct}$ is the averaged velocity due to fluctuations and $<v>_{flow}$ is that due to vortex motion. As a consequence, it is naturally expected in isotropic systems with long coherence length or in higher current limit near the critical current I_c even in anisotropic systems that the contribution from the conventional Lorentz force can be observable, due to the increasing contribution from $<v>_{flow}$ to the resistivity. Using the simplest thermodynamical approach of the pancake motion, $<v>_{ave}$ can roughly be approximated by $\{3k_B(T-T_s)/m^*\}^{1/2}$. Therefore,

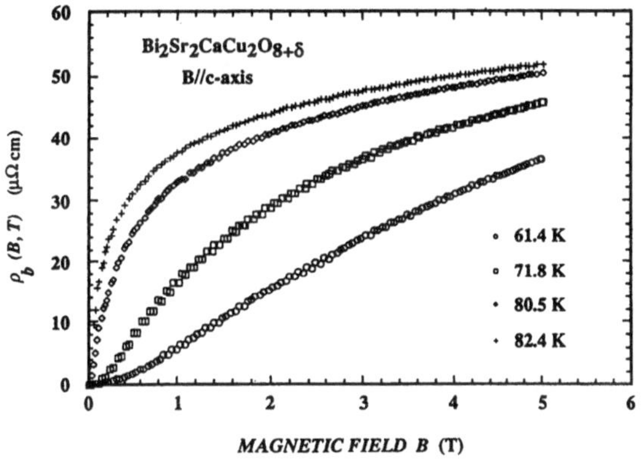

Fig. 3. Magnetoresistance curves of the b–axis resistivity of single crystalline $Bi_2Sr_2CaCu_2O_{8+\delta}$ at various temperatures below T_c.

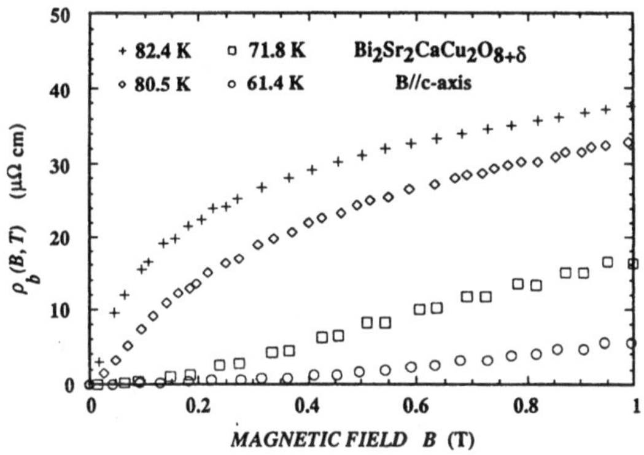

Fig. 4. Magnetoresistance curves of the b–axis resistivity of single crystalline $Bi_2Sr_2CaCu_2O_{8+\delta}$ at various temperatures below T_c in low fields.

$$V = (B\perp/c)L\{3k_B(T-T_{irr})/m^*\}^{1/2}, \tag{4}$$

where T_{irr} is the irreversibility temperature and m^* the effective mass of the pancake as a interacting particle. The consequence of the above approach, although the approximation is so primitive, gives rather satisfactory agreement with the experimental results. In Figs. 3 and 4, the magnetoresistance of ρ_b is shown at a fixed temperature. At low temperatures near T_{irr} and in the low field region the field effect is nearly linear, in agreement with the prediction. At higher temperatures near T_c, the field effect is much stronger than the expected and has a tendency to saturates, indicating importance of other contributions such as critical fluctuations near T_c. The expected temperature dependence of the resistivity does not seem to follow $T^{1/2}$, in particular, at low temperatures. The discrepancy may be due to the fact that the viscosity associated with the interaction between the pancakes becomes important as the temperature approaches T_{irr}, where the motion of the vortex pancakes are frozen out and they form a vortex lattice. Apparently, such an effect is not taken into consideration in the analysis and it is certainly necessary to have more detailed approaches to account for it. It is worthwhile mentioning the relation to the Tinkham's approach[8]. Although the pancake vortex state due to the low dimensionality of the system is not emphasized there, the phase slippage process as an essence of dissipation mechanism is the same origin as the one described here. However, he sought the origin of the phase slippage to the weak pinning force which causes the essentially zero critical current. On the contrary, the origin of the phase fluctuation in this case is attributed to the thermally activated pancake motion.

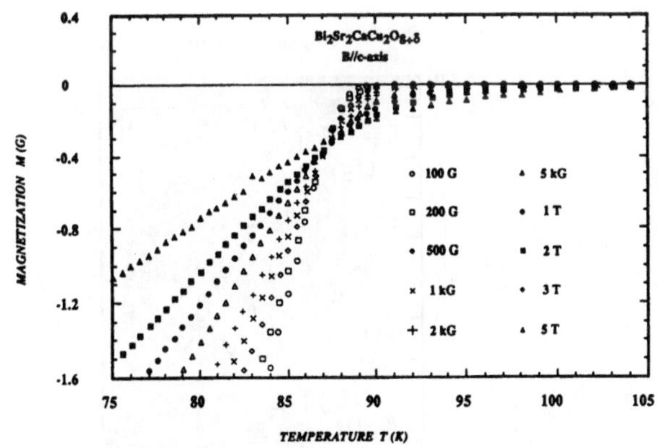

Fig. 5. Temperature dependence of magnetization curves in single crystalline $Bi_2Sr_2CaCu_2O_{8+\delta}$ in various magnetic fields applied to the c–axis in the vicinity of T_c.

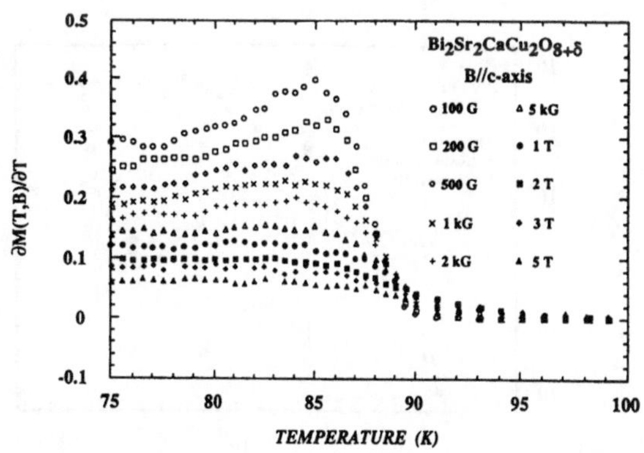

Fig. 6. Temperature dependence of the derivative of magnetization of single crystalline $Bi_2Sr_2CaCu_2O_{8+\delta}$ in various fields applied along the c–axis in the vicinity of T_c.

SUPERCONDUCTING DIAMAGNETIC MOMENT

In order to study more directly the superconducting fluctuation effect in the vortex state, the temperature dependence of the superconducting diamagnetic moment has been measured in various magnetic fields. The results for the single crystalline $Bi_2Sr_2CaCu_2O_{8+\delta}$ are presented in Fig. 5 only near T_c. It is clearly seen that the superconducting transition becomes wider and wider with increasing the field intensity. Note that the superconducting transition in a magnetic field of 1×10^{-4} T is stepwise within the width of 0.5 K at $T_c = 88.5$ K $\pm$ 0.3 K. In order to extract the fluctuation contribution only, the magnetization curves were normalized at 130 K. A remarkable feature here is a significant contribution to the diamagnetism well above T_c, which increases with increasing fields. At higher fields it is no longer possible to determine the superconducting transition point without defining properly the transition temperature. Although a construction extrapolating the linear part of magnetization just below T_c towards the zero magnetization point has been used for the determination of $B_{c2}(T)$ line in fields in $YBa_2Cu_3O_{7-\delta}$[9], it is obvious that this method is no longer effective in this case. If one still clings to apply this method to deduce $B_{c2}(T)$ line, one would obtain even a positive slope for the $dB_{c2}(T)/dT$–value. One reason for this can be seen by looking at Fig. 6, in which the temperature derivative of the magnetization curves $\partial M(B,T)/\partial T$ is shown. The value $\partial M(B,T)/\partial T$ is strongly field dependent as well as temperature dependent, in strong contradiction to the ones in the Abrikosov vortex state in traditional superconductors. This fact implies that the simple linear extrapolation to deduce the field dependence of $B_{c2}(B,T)$ cannot be applied. As seen in Fig. 5, the fact that diamagnetism above T_c gradually increases also suggests existence of strong superconducting fluctuations. Such anomalous behaviors cannot be described by the standard Ginzburg`Landau formalism without taking into account the strong superconducting fluctuation effects. The analysis of magnetization near T_c was tried using the scaling law suggested by Welp *et al.*[10]. However, the result is not successful in this particular case, in strong contrast to $YBa_2Cu_3O_{7-\delta}$[10]. The reason is not understood well, but it may be related to the highly two dimensional nature of superconductivity, which is only the considerable difference in the two systems and may be beyond approximation for $Bi_2Sr_2CaCu_2O_{8+\delta}$. Remember that the anisotropic effective mass ratio of $\gamma = (m_c/m_{ab})^{1/2} \approx 95$ –250 at least in $Bi_2Sr_2CaCu_2O_{8+\delta}$[11], which is significantly higher than that of 55[12] obtained previously and is compared with the value of 8 in $YBa_2Cu_3O_{7-\delta}$[13]. It seems that the $Bi_2Sr_2CaCu_2O_{8+\delta}$ system violates the range of validity of the approximation used to deduce the scaling low.

THE IRREVERSIBILITY LINE

By definition, the irreversibility line has been obtained by measuring the difference between the zero field cooled and field cooled magnetization in various magnetic fields. The result is presented in Fig. 7. The irreversibility line for $YBa_2Cu_3O_{7-\delta}$ obtained by similar method by Yeshurum and Malozemoff[14] as well as obtained by the resistivity onset and ac–susceptibility[15] are also included. The power low $\Delta T \propto B^{2/3}$, which is identical as the scaling law for the anisotropic 3D system, is evident for $YBa_2Cu_3O_{7-\delta}$[14], whereas the power law behavior with an exponent of approximately 1.1, indicating a close value to the linear relation to $1-t$, was observed only at the low field limit. It seems that this temperature dependence of B_{irr} is neither consistent with the scaling law for the 3D system nor even that for the 2D system. Contrasting to the low field region, at higher fields above about 0.01 T, the irreversibility field B_{irr} takes off nearly vertically at around 20 K. Such an abrupt increase of B_{irr} has not been observed in $YBa_2Cu_3O_{7-\delta}$ below 6 T, but inconsistency met in the analysis of $B_{c2}(T)$ line may indicate a similar increasing behavior of B_{irr}[16]. Moreover, the proportionality constant to $1-t$ is approximately two to three orders

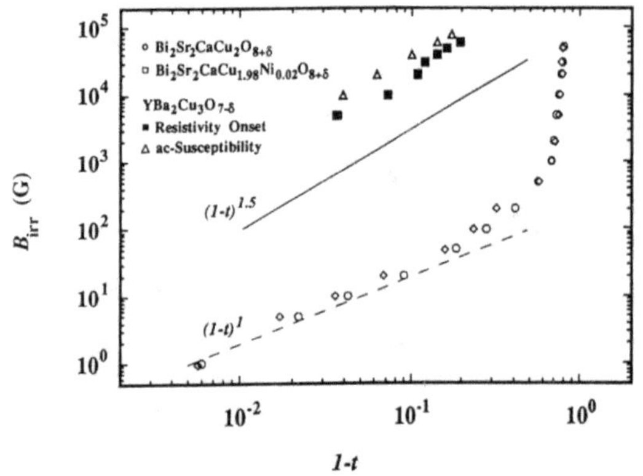

Fig. 7. The irreversibility line for single crystalline $Bi_2Sr_2CaCu_2O_{8+\delta}$ as well as $Bi_2Sr_2CaCu_{1.98}Ni_{0.02}O_{8+\delta}$ measured by the SQUID magnetometer.

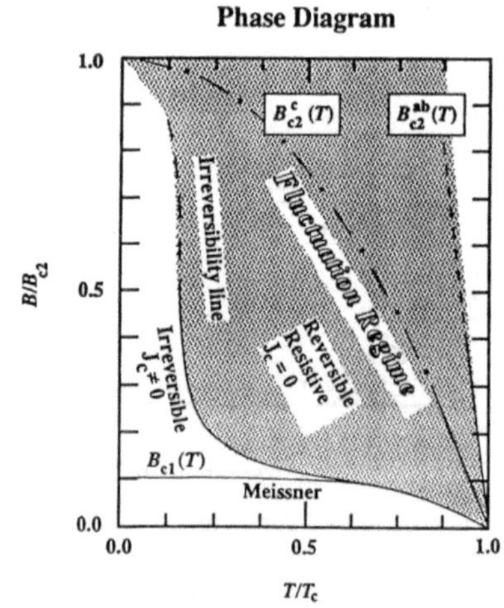

Fig. 8. The schematic phase diagram proposed for highly anisotropic layered superconductors.

of magnitude larger for $YBa_2Cu_3O_{7-\delta}$ than that for $Bi_2Sr_2CaCu_2O_{8+\delta}$. It is emphasized that the $Bi_2Sr_2CaCu_{1.98}Ni_{0.02}O_{8+\delta}$ sample does not show appreciable difference in the irreversibility line, indicating inactivity of the Ni atoms in the vortex dynamics. However, it should be mentioned that the results for the polycrystalline samples as well as the irradiated sample seem to differ from that presented here. For example, experimental results reported seem to indicate that the irreversibility line is significantly affected by irradiation which introduces various types of damages in the system[17–20]. Although some report only an increase of the hysteresis of magnetization at low temperatures without influencing to the irreversibility line[21], it is interesting that the nature of the pinning strength is certainly changed by the irradiation damage. Since this is only positive evidence for improving the critical current density in high-T_c materials, it is highly desirable to clarify this point in order for the applications. In Fig. 8 the proposed phase diagram which is constructed by summarizing the argument given above is presented. According to the low field magnetization measurements, the Meissner state, in which the perfect diamagnetism is observed, exists below B_{c1}. The B_{c1} curve at high temperatures can uniquely be determined by taking a point at which the magnetization curve changes sharply similar to a typical type II superconductor, though it becomes difficult to determine it below about 40 K, where the pinning effect seems to come into a play. Nevertheless, the most interesting phenomena are an uniqueness of B_{c1} as well as the initial susceptibility value of $-1/4\pi$ above about 40 K, irrespective of the shape of the sample. This shape independence cannot be explained by a simple perfect diamagnetism of a bulk sample even with no pinning effect and can only be accounted for the decoupled superconducting layers. This speculative conclusion leads us to a picture that the vortex state above B_{c1} and T_{irr} must be considered as a decoupled vortex state, whereas below T_{irr} the vortex state is rather conventional as depicted schematically in Fig. 9. Following this picture, the shaded region in Fig. 8 is a decoupled state, where the vortex becomes a pancake state and is floating around over the CuO_2 plane. As described above, most of anomalies occur in this regime. This region is tentatively called as a fluctuation regime, since the resistivity and the magnetization data show strong superconducting fluctuation effects. However, the physical nature of the superconducting fluctuations has not well been characterized yet and still remains unsolved.

RELAXATION PHENOMENA AT LOW TEMPERATURES

One of the remarkable difference between the conventional superconductors and highly anisotropic superconductors is fast relaxation phenomenon even at very low temperatures. This has been speculated by extrapolating the relaxation rate $S = -\partial \boldsymbol{M}(t)/\boldsymbol{M}(0)\partial \ln t$ from high temperatures, which does not seem to become zero but remains at finite even at T = 0 K[22]. This was indeed confirmed by the relaxation measurement of magnetization in $Bi_2Sr_2CaCu_2O_{8+\delta}$ performed at the lowest temperature of 55 mK, which is shown in Fig. 10. The temperature dependence of the relaxation can well be described by the logarithmic decay, although a slight deviation at the long time limit was observed. The decay rate S in this case is about 0.03, which is approximately three to five times larger than that of $YBa_2Cu_3O_{7-\delta}$[23]. Moreover, it is observed that the value of the relaxation rate is unchanged below about 1 K but depends on the applied magnetic field, which seems to indicate that the relaxation rate depends on the initial state of the magnetization value. In addition, it is noted that at low temperatures below 1 K, sudden jumps of the magnetization were periodically observed while the magnetic field was ramped up and down. Similar phenomenon which is reminiscent of flux-jump has been reported in single crystal $YBa_2Cu_3O_{7-\delta}$ at low temperatures[24]. It is noted that the maximum magnetization value obtained is limited by this phenomenon, where the relaxation measurement was performed.

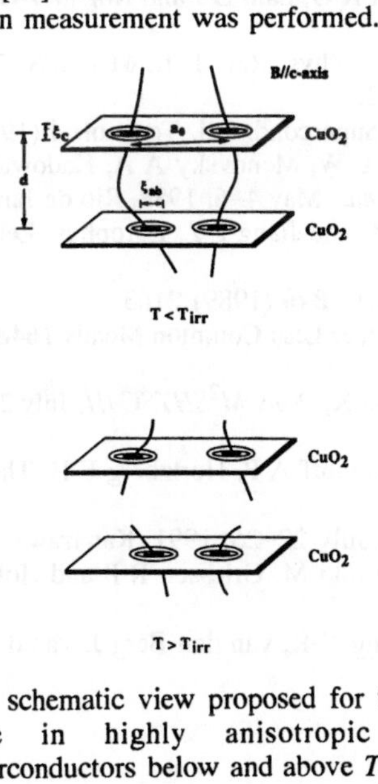

Fig. 9. The schematic view proposed for the vortex state in highly anisotropic layered superconductors below and above T_{irr}.

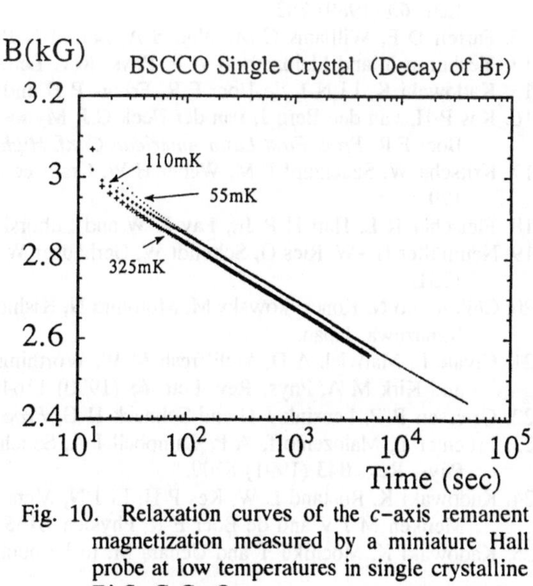

Fig. 10. Relaxation curves of the c-axis remanent magnetization measured by a miniature Hall probe at low temperatures in single crystalline $Bi_2Sr_2CaCu_2O_{8+\delta}$.

It is obvious that the temperature independent relaxation phenomenon excludes a possibility that the relaxation mechanism is thermally activated type, which may be the case at high temperatures. Although the distribution of the pinning potential to lower energies is argued to account for the phenomenon, the temperature independent relaxation rate can not be explained by thermally activated processes only. Therefore, it is necessary to seek other mechanism as an origin of the relaxation. Recently, the quantum tunneling has been proposed as a mechanism[23] and may be relevant due probably to the short coherence volume. Along with this line, the decay of magnetic moment is calculated by the simple WKB approximation and the results agrees well with the experiments[25]. It is noted that the relaxation of the deformed flux-line lattice in the case of collective pinning may also be possible due to the quantum jump due likely to the zero point motion of the vortex. This relaxation process, which may be nomenclated as a quantum creep phenomenon, is also dissipative, causing a serious problem for applications. The unified treatment of the relaxation phenomena will be given elsewhere[25].

CONCLUDING REMARKS

So far, we have discussed several points, which are the most characteristic features in highly anisotropic superconductors such as $Bi_2Sr_2CaCu_2O_{8+\delta}$. These have in turn also serious problems for applications, because all these phenomena degrade, limit and reduce the J_c-values. As was described, those are not caused by the weak pinning force but more fundamental problems related to the nature of the superconducting state is essentially involved. It is certainly necessary to overcome these points in order for applications. It seems that in addition to the increase of pinning potential, which is always important, it is highly desired to modify the fundamental nature of the superconducting properties by introducing impurities, vacancies, by changing stoichiometry or by alloying in order to realize less anisotropic superconducting state with the rigid flux-line lattice. A part of countermeasure is currently progressing and some are still left for the future problems.

REFERENCES

1. Kadowaki K, Li J N and Franse J J M, *Advances in Superconductivity III*, eds. by Kajimura K and Hayakawa H, (Springer Verlag) 1991 p. 145.
2. Kadowaki K, *Proc. Int. Conf. Elec. Prop. and Mech. High-T_c Supercond.*, July 29–31, 1991, Tsukuba, Japan.
3. Kwok W K, Welp U, Crabtree G W, Vandervoort K G, Hulscher R and Liu J Z, Phys. Rev. Lett. **64** (1990) 966.
4. Kadowaki K, Li J N and Franse J J M, Physica **C170** (1990) 298.
5. Kadowaki K, *Proc. Int. Conf. Elec. Prop. and Mech. High-T_c Supercond.*, July 29–31, 1991, Tsukuba, Japan.
6. McCumber D E, Phys. Rev. **172** (1968) 427.
7. McCumber D E and Halperin B I, Phys. Rev. **B1** (1970) 1054.
8. Tinkham M, Phys. Rev. Lett. **61** (1988) 1658.
9. Welp U, Kwok W K, Crabtree G W, Vandervoort K G and Liu J Z, Phys. Rev. Lett. **62** (1989) 1908.
10. Welp U, Fleshler S, Kwok W K, Klemm R A, Vinokur V M, Downey J and Crabtree G W, Preprint.
11. Okuda K, Kawamata S, Noguchi S, Itoh N and Kadowaki K, to be published in J. Phys. Soc. Jpn.
12. Farrell D E, Bonham S, Foster J, Chang Y C, Jiang P Z, Vandervoort K G, Lam D J and Kogan V G, Phys. Rev. Lett. **63** (1989) 782.
13. Farrell D E, Williams C M, Wolf S A, Bansal N P and Kogan V G, Phys. Rev. Lett. **61** (1988) 2805.
14. Yeshurun Y and Malozemoff A P, Phys. Rev. Lett. **60** (1988) 2202.
15. Kadowaki K, Li N J, de Boer F R, Frings P H and Franse J J M, Supercond. Sci. Technol. **4** (1991) S88.
16. Kes P H, van den Berg J, van der Beek C J, Mydosh J A, Roeland L W, Menovsky A A, Kadowaki K and de Boer F R, *Proc. First Latin American Conf. High Temp. Supercond.*, May 4–6, 1988, Rio de Janeiro, Brazil.
17. Kritscha W, Sauerzopf F M, Weber H W, Crabtree G W, Chang Y C and Jiang P Z, Europhys. Lett. **12** (1990) 179.
18. Fleischer R L, Hart H R Jr., Lay K W and Luborsky F E, Phys. Rev. **B40** (1989) 2163.
19. Neumüller H -W, Ries G, Schmidt W, Gerhäuser W and Klaumünzer, J. Less Common Metals **164&165** (1990) 1351.
20. Chikumoto N, Konczykowsky M, Motohira N, Kishio K and Kitazawa K, *Proc. M^2SHTSC III*, July 22–26, 1991, Kanazawa, Japan.
21. Civale L, Marwick A D, McElfresh M W, Worthington T K, Malozemoff A P, Hortzberg F H, Thompson J R and Kirk M A, Phys. Rev. Lett. **65** (1990) 1164.
22. Griessen R P, Lensink J G and Schnack H G, *Proc. M^2SHTSC III*, July 22–26, 1991, Kanazawa, Japan.
23. Fruchter L, Malozemoff, A P, Campbell I A, Sanchez J, Konczykowski M, Griessen R P and Holtzberg F H, Phys. Rev. **B43** (1991) 8709.
24. Kadowaki K, Roeland L W, Kes P H, Li J N, Menovsky A A, Huang Y K, van den Berg J, van der Beek C J, Menken M J V and de Boer F R, Physica **B155** (1989) 136.
25. Kadowaki K, Mochiku T and Uehara M, to be published.

4.4 Flux Distribution

Polarized Neutron Studies on Magnetic Flux in Superconductors

YASUO ENDOH[1], T. WATANABE[1], and S. ITOH[2]

[1] Department of Physics, Tohoku University, Sendai, 980 Japan
[2] National Laboratory for High Energy Physics, Tsukuba, Ibaraki, 311 Japan

ABSTRACT

Neutron depolarization studies have been performed by using single crystals of the high temperature superconductors in order to elucidate the bulk magnetic state. The quantitative analysis of neutron wavelength dependence of the polarization yielded the determination of the inportant quantities such as lower critical field, H_{c1} and the penetration depth Λ.

KEY WORDS Neutron Depolarization. $La_{2-x}Sr_xCuO_4$, Magnetic flux, Lower critical field, Penetration depth

NEUTRONDEPOLARIZATION

It is well recognized that neutrons are easily depolarized when neutrons pass through a superconducting sheet under the large magnetic field. It is obvious from this fact that the neutron polarization roates around the magntic induction. Since neutrons are not depolarized at all in the Meissner state below the lower critical field, H_{c1}, the depolarization shows that the magnetic induction is trapped in the superconductor of the mixed state. The basic concept is that a polarized neutron precesses around themagnetic field with the Larmor frequency following the equation of motion for the neutron polarization, **P**.

$$d\mathbf{P}/dt = \gamma [\, \mathbf{P} \times \mathbf{H} \,] \quad , \qquad (1)$$

where γ represents the gyromagnetic ratio of the neutron magnetic moment. In fact, the polarization of the transmitted neutron through the superconductors above the lower critical field, H_{c1} shows the cosine oscillation with respect to the neutron wavelength as shown in Fig.1, when the experimental condition is set up in such a way that **P** is perpendicular to the mangetic flux. This oscillation is readily obtained by the integration of the above differential equation. This wavelength dependence can readily be observed by using pulsed neutron source, because the polychromatic neutron beam is delivered to the experimental position, which was described in the previous publication in detail [1]. Though the depolarization method is an old experimental technique, the wavelength dependent measurement is quite unique and novel, which is very useful for the detection of the magnetic induction penetrated in the bulk superconducting material [2].

EXPERIMENT

The neutron depolarization measurements were performed on the TOP spectrometer installed at the pulsed neutron facility at the National Laboratory for High Energy Physics

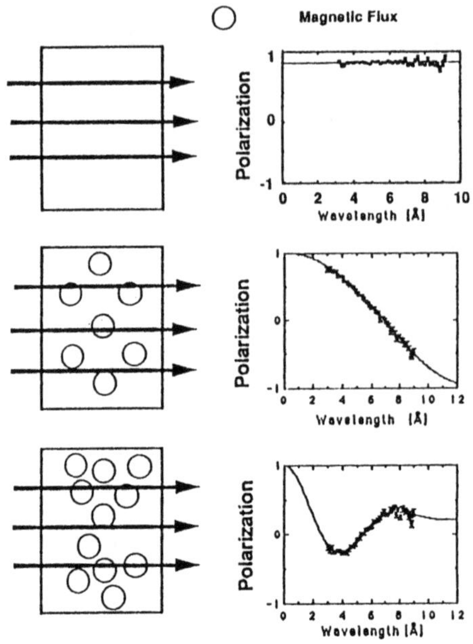

Fig.1 Conceptual illustration of neutron depolarization by the trapped magnetic flux. Polarized neutrons whose polarization direction is shown in the figure, pass through the superconductor from the left hand side. The fluxes shown by the circles penetrate the sample. The upper figure shows the Meissner state below Hc1, the middle corresponds to the case H>Hc1, but the flux density is low and the bottom is the higher flux density.

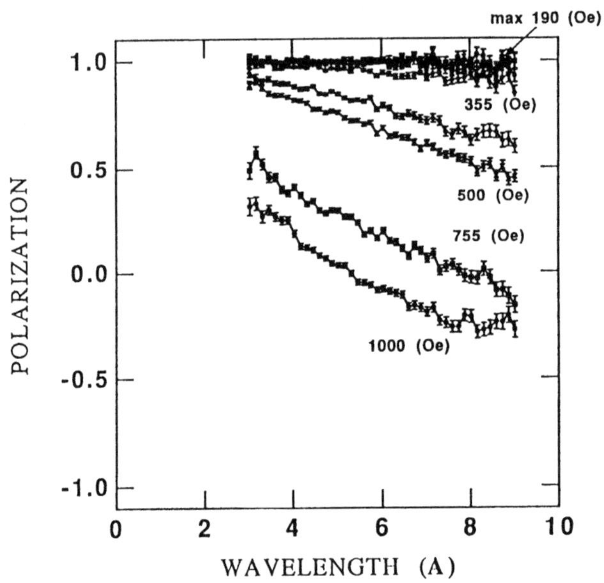

Fig.2 Typical example of the neutron wavelength dependence of depolarization. Depolarization was measured at 10K at 7 Oe after zero field cooling. The field value in the figure is the external field applied before the measurement started at 10K.

(KEK). The depolarization was always measured at the external field of 7 Oe while changing either the measuring temperature or the cooling field at the samples. The cooling speed through the superconducting transition temperature Tc was approximately kept constant. Single crystals used in this experiment of $La_{1.85}Sr_{0.15}CuO_4$ were grown by the TSFZ method [3]. The circular plates several mm thick, where the crystalline c axis in the tetragonal notation is oriented in the plate were prepared. The neutron depolarization does not occur for the sample of zero field cooling, which is a clear evidence that the single crystal using this experiment realizes a perfect Meissner state. Then neutrons were depolarized once it was passed in the field above a certain threshold fild at the measured temperature as shown in Fig.1. This threshold field to start depolarization was determined to be the lower critical field, H_{c1}. After the series of the measurement at various temperatures we have drawn the curves of H_{c1} with respect to temperature, which are in a fairly good agreement with the magnetization results [4]. Note that the latter result is based on the Bean's model [5], but the determination from the present experiment is rather straightforward.

The depolarization data was further analyzed by introducing an empirical formula of the damped cosine oscillation with respect to the neutron wavelength [2].

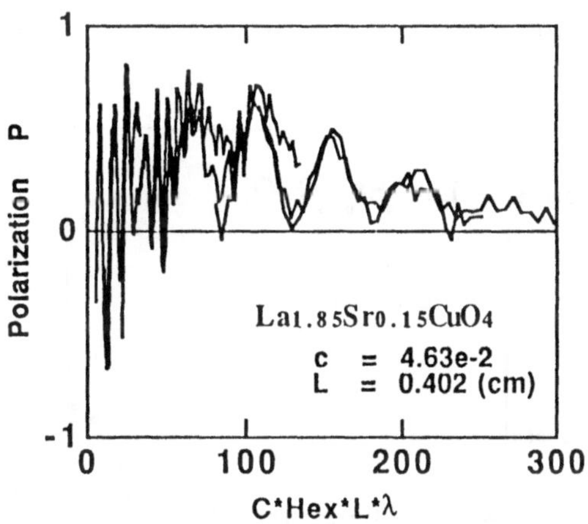

Fig.3 The polarization as the function of CH*Lλ (H* represents the cooling field, L and λ are respectively the sample thickness and the neutron wavelength.) Neutron depolarization is shown to be scaled by H*Lλ.

$$P(\lambda) = Po + (1-Po) \exp(-\alpha\lambda^2) \cos(I\lambda + \phi) . \qquad (2)$$

Where Po is related to the direction of neutron polarization, which is 0 when neutron polarization is perfectly perpendicular to the magnetic induction. I and α are related to the magnetic induction in the sample as follows.

$$I = CLM, \qquad (3)$$

$$\alpha = 1/\sigma . \qquad (4)$$

C is a physical constant related to the gyromagnetic ratio. L and M are respectively the neutron path length or the sample thickness in this case and the magnetic induction. σ is defined as the field distribution in the bulk sample which will be discussed later in this section.

After collecting data from the series of the scan changing the cooling field as well as the measuring temperature, we found an important fact that the polarization is scaled by the function of the cooling field value integrated over the sample thickness times the neutron wavelength as shown in Fig.3. This novel result of 'scaling' suggests that the magnetic flux density is proportional to the cooling field at least for smaller external field. Therefore it is experimentally confirmed that each magnetic flux trapped in the single crystal of $La_{1.85}Sr_{0.15}CuO_4$ above H_{c1} is approximately proportional to the cooling field.

The average magnetic induction in the sample was estimated from the period of the cosine oscillation in the polarization. The value is equal to the residual magnetization at the measured field of 7 Oe after field cooling, which is proportional to the cooling field. Therefore it is easily concluded that the magnetic induction in the high temperature super-conductor is induced by the trapped magnetic flux. It is important that the present depolarization results without using any model accord with the bulk magnetization results, which are analyzed by the Bean's model [5]. It is also equally important that the perfect Meissner effect is veri-fied by the depolarization method since neutrons penetrate deep in the sample. On the other hand, the usual magnetization measurements cannot be ignored the effect of the exclusion of the magnetic induction by the superconducting surface skin.

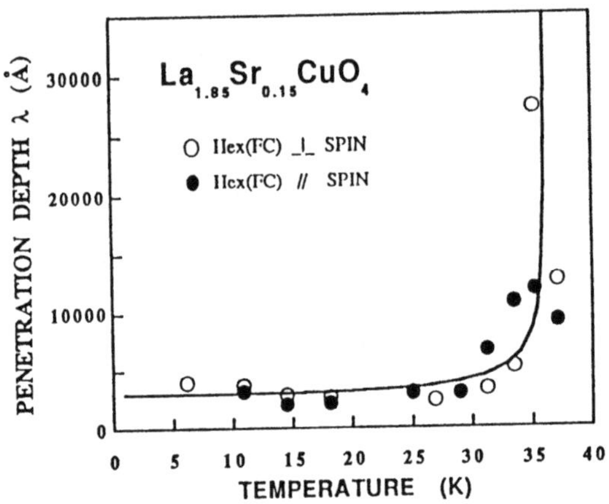

Fig.4 The penetration depth determined by the depolarization measurements for the cooling field of 400 Oe as the function of temperature. The data was fitted to the Ginzburg Landau model.

The damping in polarization with respect to the neutron wavelength tells us that the dephasing of the polarization is dominated by the spatial distribution of the magnetic field in the bulk superconductor. The spatial distribution of the magnetic field was simply assumed to be originated by the magnetic flux which forms vortex lattice penetrated in the superconductor. In our experimental condition with the lower magnetic field, the number of the magnetic vortexes can be estimated by the magnetic induction value described above, which turned out to be well separated with each other. Then we postulate that the damping of neutron polarization, which is expressed by the exponential form as the function of neutron wavelength, is proportional to the mean square deviation of the magnetization density. This value is further converted to the penetration depth by solving the London equation [6].

$$\sqrt{\Delta H^2} \sim \Phi / 4\pi^{1/2}\Lambda^2 , \qquad (5)$$

where Φ and Λ are respectively the magnetic flux and the penetration depth. This assumption is not unreasonable, because as mentioned just above, the depolarization function is scaled by the sample thickness. Furthermore, the ratio of Λ/ξ (ξ : coherence length) is large enough in the high temperature superconductor to simplify that the dephasing of the neutron polarization occurs mostly in the nonuniform area of the magnetic field defined by Λ. But neutron polarization rotates rapidly in the narrow region inside the magnetic flux core defined by ξ. The experimentally determined penetration depth Λ was then summarized as the functions of both cooling field and temperature, which is shown in Fig.4. Although the more precise spatial distribution of the magnetic flux was not studied by the fact that the effect of the demagnetization was not treated properly in the current experimental set up. Thus our experiment is not complete yet, but it demonstrates that the neutron depolarization is the unique tool to study the physics of magnetic flux state, because it provides the bulk magnetization value. In particular if the neutron depolarization method is combined with other neat tech-niques such as controlling the accurate neutron polarization as well as analyzing the time dependent magnetization after the current is put on, the flux creeping or the relaxation mechanism of the magnetic flux will be able to study. We consider such experiments in the future.

Acknowledgement
Authors thank H.Kojima, I.Tanaka, N.Toyota and K.Yamada for their contributions in the present experiment. The work has been supported by the Grant in Aid for the Scientific Research from the Ministry of Education, Science and Culture.

References
1 S.Mitsuda and Y.Endoh, J. Phys. Soc. Jpn., 54 (1985) 1570
2 Y.Endoh, Prog. Theor. Phys. Suppl
3 I.Tanaka, K.Yamane and H.Kojima, J. Crys. Growth, 96 (1989) 711
4 M.Naito and A.Matsuda, phys.Rev. B41 (1990) 4823
5 C.P.Bean, Rev.Mod.Phys., 36 (1964) 31
6 P.Pincus et al., Phys.Lett., 13 (1964) 21

Magnetization Study of High-T_c Oxide Superconductors by Magneto-Optical Observations

SATOSHI GOTOH*, NAOMICHI NAKAMURA, NAOKI KOSHIZUKA, KAZUMI OHATA, HIROHIKO IZUMI**, and TADATAKA MORISHITA

Superconductivity Research Laboratory, ISTEC, Koto-ku, Tokyo, 135 Japan

ABSTRACT

Magnetization and the spatial variation of magnetic flux in thin film samples of $YBa_2Cu_3O_x$ were studied by using a high contrast magneto-optical observation method(HICOM). The demagnetizing effect in the mixed state was also investigated in comparison with ferromagnetic materials. The magnetic flux distribution conformed to the critical state model in a $YBa_2Cu_3O_x$ thin film made by laser deposition. The trapped flux in the sample with high J_c showed the nature of permanent magnet. In particular, a thin film magnet with perpendicular magnetization could be realized because the flux density in the center of the sample was not reduced by the demagnetizing field. However, the flux gradients in the edges of the sample were changed following the critical state determined by the demagnetizing field.

KEY WORDS: $YBa_2Cu_3O_x$ thin film, magneto-optical observation, iron garnet, flux distribution, flux pinning

INTRODUCTION

Magnetization measurements are very useful for estimation of the critical current density, J_c, which is usually analyzed by the critical state model[1]. In melt processed $YBa_2Cu_3O_x$(YBCO), the critical state is well established and the magnetic and transport critical currents are in good agreement[2]. In YBCO thin films, the transport critical currents also agree with the magnetically measured values at low temperatures[3]. However, magnetization measurements show the averaged values for the total sample volume. The magnetic flux distribution and any local pinning variations, etc., in the sample are therefore not directly observable by magnetization measurements.

We have recently developed a new technique named HICOM(high contrast observation method for magnetic flux density)[4-6] using the magneto-optical Faraday effect of ferrimagnetic iron garnet films. In this method, high-contrast images of the flux distribution under increasing and decreasing field conditions are directly obtained without an image processing system. The principle of the method is the observation of the image of the stripe magnetic domain patterns in an iron garnet film, which reflects the flux density in the sample. In this paper, we report on the observations of flux penetration and trapping in YBCO thin films prepared by laser ablation.

EXPERIMENTAL

Thin film samples of $YBa_2Cu_3O_x$ on MgO(100) substrates with a critical temperature of 83K were prepared by excimer laser ablation using 193 nm ArF[7]. Epitaxial growth and c-axis orientation were achieved at a substrate temperature of 973 K and 5 mTorr oxygen partial pressure. Rectangular c-axis oriented specimens with dimensions approximately 250nmx3mmx3mm were cut from the larger initial sample.

The magneto-optical thin film of $(BiY)_3(GaFe)_5O_{12}$ for optical detection was prepared on a transparent gadolinium gallium garnet (GGG) substrate[8]. The thickness of the iron garnet film and the GGG substrate were 4μm and 0.4mm, respectively. An optical observation system with an electromagnet, an optical cryostat and a polarization microscope was used[5]. The YBCO film was fixed on a copper cooling holder inside the cryostat and then the film/substrate was placed on the sample surface film-side down and in good thermal contact with the sample. The temperature is variable from 7.7 K to 300 K by controlling both the flow of cooled helium gas and the power of a heater inside the copper sample holder. The polarization microscope (OLYMPUS, BHT-P) has a resolution of about 1.3μm and a working distance of up to 8mm. The electromagnet was placed under the cryostat to apply a magnetic field perpendicular to the sample surface.

Magnetization properties were measured using a SQUID magnetometer (Quantum Design,Inc.). The applied field was perpendicular to the film surface (c//H) in all cases with the same magnetic field orientation as in the HICOM observations. A large demagnetizing effect was brought about by this configuration; the demagnetizing factor, N, is given by $N = 1-\pi t/(2d)$ for $t \ll d$ where t is the film thickness and d is the effective sample diameter[9]. The demagnetizing factor was thereby determined from the sample shape to be 0.9998. This value was in good agreement with that estimated from the slope of the zero-field cooled initial magnetization curve, if the latter was assumed to show perfect diamagnetism.

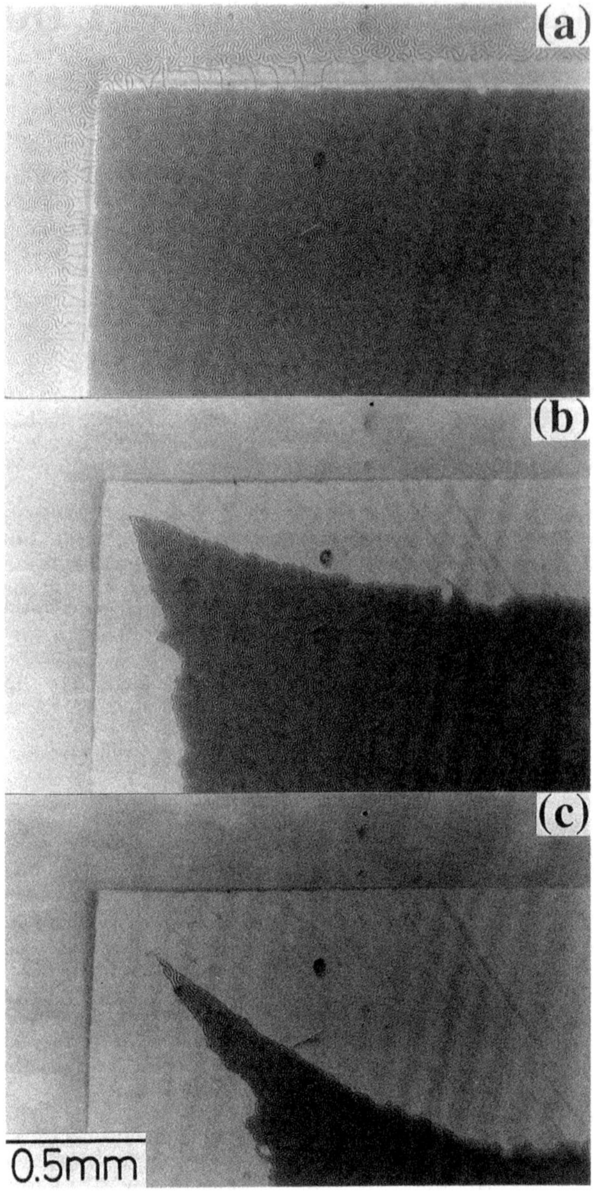

Fig. 1 Changes in the stripe domain patterns on a YBCO thin film at 10 K with increasing fields after zero-field cooling. (a) Ha=52 Oe, (b)Ha=347 Oe, (c) Ha=607 Oe.

RESULTS AND DISCUSSION

The changes in the stripe domain patterns of the iron garnet film on the YBCO film zero-field cooled to 10 K and measured at 10 K with increasing applied fields are shown in Fig. 1. Assuming the lower critical field, H_{c1}, of the YBCO film to be about 500 Oe at 10 K, the effective H_{c1} becomes about 0.05 Oe because of the demagnetizing factor. We could not thereby confirm the diamagnetism at lower applied fields by HICOM. In fact, the applied field penetrated the sample edges at a relatively lower field of 52 Oe as shown in Fig. 1(a). When the applied field is increased further, the flux penetration starts smoothly with bow-shaped flux fronts from the sample edges according to the critical state (Fig. 1(b)). This is in contrast with a rectangular flux front reflecting the sample shape as observed in a MPMG-processed YBCO sample[5]. The bow-shaped flux front can be qualitatively explained by the large demagnetizing effect. In a rectangular specimen, the demagnetizing factor becomes smallest in the corners and becomes largest in the center of the edges. In the diamagnetic superconducting state, the direction of the demagnetizing field is the same as the applied field so that the effective field inside the sample is enhanced although it reduces the applied field in the usual ferromagnetic materials. Accordingly, the flux penetration starts from the center of the edge with the largest demagnetizing factor, and the flux front, as a result, bows into the sample. A more quantitative discussion using the shielding current distribution will be reported elsewhere. Since the bow-shaped flux front symmetrically moves to the center of the sample as shown in Fig. 1(c), the magnetic behavior is homogeneous and thus the pinning variation is small throughout the sample. Figure 2 shows the domain patterns at 10 K after field-cooling in a magnetic field of 140 Oe. Fig. 2(a) shows no change of the domain structure neither inside nor outside the sample. This indicates that almost all of the applied field is trapped in the sample due to large pinning forces. When the applied field was removed, the trapped flux appeared as a positive remanent magnetization. This trapped flux is shown in Fig. 3. The remanent magnetization, $4\pi M_r$ is an extremely large value, $60x10^4$ G, because of the large demagnetizing effect. If there were no demagnetizing field in the sample, the remanent magnetization, M_{ro}, having the flux distribution determined by J_c is obtained as shown in Fig. 4(a). In fact, when the sample is isolated in space, M_{ro} induces a demagnetizing field, H_d. In the critical state of type II superconductors with pinning, the critical state is changed to that determined by the effective field, H_{eff}, induced by H_d as shown in Fig. 4(b). The value of the measured magnetization, M_{rd}, is therefore larger than that of M_{ro}. In addition, the surface magnetic field in the center of the sample is unchanged but field of the opposite sign occurs near the edge of the sample as shown in Fig. 4(b). In fact, at zero field, as shown in Fig. 2(b), trapped flux remains in a large area inside the sample with the same domain structure, reflecting the applied field of 140 Oe. Furthermore, the black regions of the edges of the sample and surrounding the sample indicate that the field here has the opposite direction as the field inside the sample. In this way, the YBCO thin film can trap magnetic flux perpendicular to the film plane at low temperatures. The trapped flux in the sample exhibits hard ferromagnetic behavior, which means the sample can act as a permanent magnet[10-12]. This indicates that we can use the film as a thin film magnet with the field direction perpendicular to the plane.

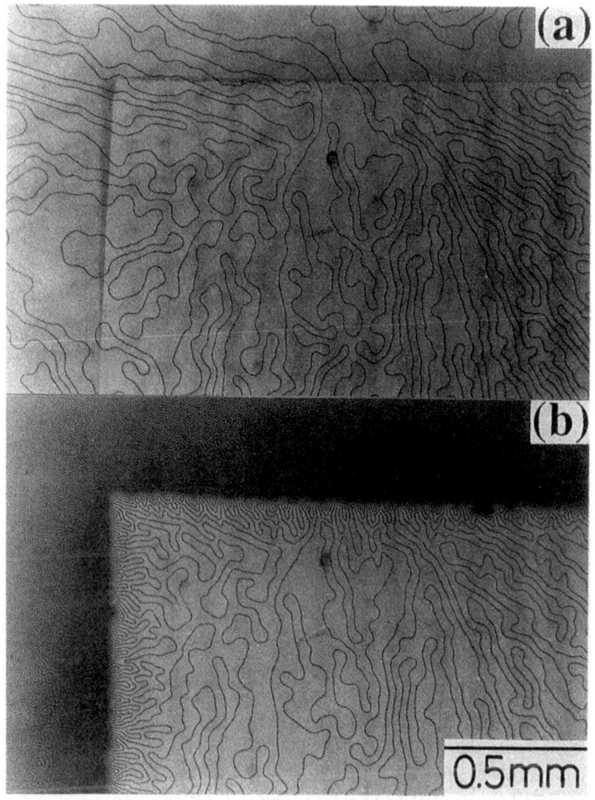

Fig. 2 Changes in the stripe domain patterns on a YBCO thin film at 10 K after field-cooling in a field of 140 Oe. (a)Ha=140 Oe, (b)Ha=0 Oe (remanent state).

CONCLUSIONS

Dynamic flux behavior was directly observed in a YBCO rectangular thin film using the Faraday effect of an iron garnet film with stripe domains. With increasing applied fields, the flux gradually penetrates into the sample according to the critical state with a bow-shaped flux front because of the large non-uniform demagnetizing effect. It is also found that the sample can behave as a thin film magnet at low temperatures with magnetization perpendicular to the film plane.

ACKNOWLEDGMENTS

We would like to thank Dr. Y. Yokoyama of Electrotechnical Laboratory for the preparation of the iron garnet films used. We also thank Dr. J. O. Willis for helpful discussions. This work was supported by New Energy and Industrial Technology Development Organization under the management of the R&D of Basic Technology for Future Industries.

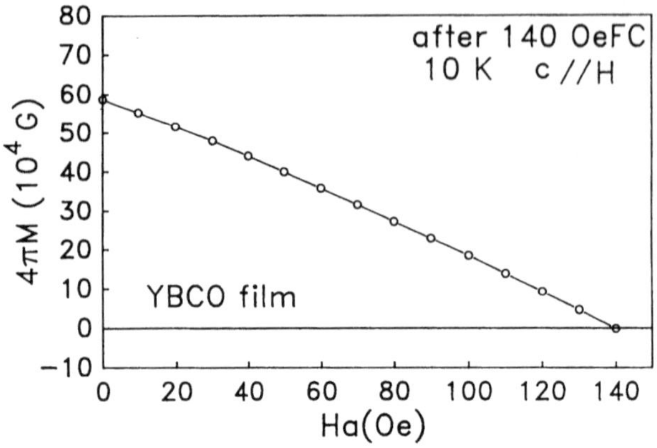

Fig. 3 Magnetization curve of the YBCO thin film at 10 K after field-cooling in a field of 140 Oe with no demagnetizing correction.

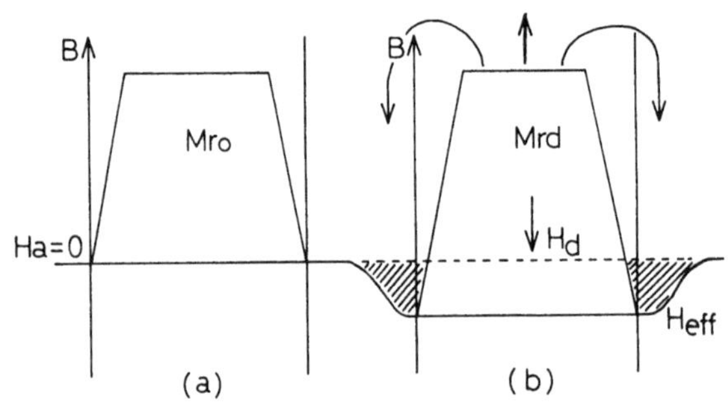

Fig. 4 Schematic illustration of the flux distribution based on the Bean model for the remanent state without(a) and with(b) a demagnetizing field. H_d is the demagnetizing field and H_{eff} is the effective field inside the sample.

REFERENCES

1. Bean CP (1962) Phys. Rev. Lett. 8: 250
2. Murakami M, Gotoh S, Koshizuka N, Tanaka S, Matsushita T, Kambe S, Kitazawa K (1990) cryogenics 30: 390
3. McGuire TR, Dimos D, Koch RH, Laibowitz RB, Mannhart J (1989) IEEE Trans. Magn. MAG-25: 3218
4. Gotoh S, Koshizuka N, Yoshida M, Murakami M, Fujimoto H, Tanaka S (1990) Jpn. J. Appl.Phys. 29: L1083
5. Gotoh S, Koshizuka N (1991) Physica C 176: 300
6. Indenbom MV, Kolesnikov NN, Kulakov MP, Naumenko IG, Nikitenko I, Polyanskii AA, Vershinin NF and Vlasko-Vlasov VK (1990) Physica C 166: 486
7. Ohata K, Sawada T, Izumi H, Suzuki K, Morishita T, Tanaka S (1991) In: Advances in Superconductivity III, Kajimura, Hayakawa (eds) the 3nd International Symposium on Superconductivity(ISS'90) Springer-Verlag,Tokyo, pp 913-916
8. Yokoyama Y, Koshizuka N, Takeda N (1985) IEEE Trans. Magn. MAG-21: 1666
9. Stoner EC (1945) Phil. Mag. 36: 803
10. Wipf SL, Laquer HL (1989) IEEE Trans. Magn. MAG-25: 1877
11. Gotoh S, Murakami M, Koshizuka N, Tanaka S (1990) Physica B 165&166: 1379
12. Gotoh S, Murakami M, Fujimoto H, Koshizuka N (1991) to be published in Physica C

*Present Address: Kawasaki Steel Corporation, 1, Kawasaki-cho, Chiba 260,Japan
**Present Address: Ulvac Japan, Ltd., 5-9-7, Tohkohdai, Tsukuba 300-26, Japan

Flux Density Distribution in $Bi_2Sr_2CaCu_2O_x$ Sintered Specimens

YUKO YOKOYAMA[1], TOMOYA KUBO[2], KAZUMASA INO[2], TOSHIO SUGIMOTO[3], HITOSHI SAKAI[4], YOSHISHIGE SUZUKI[1], and SADAFUMI YOSHIDA[1]

[1] Electrotechnical Laboratory, Tsukuba, Ibaraki, 305 Japan
[2] Faculty of Science & Technology, Meiji University, Kawasaki, Kanagawa, 214 Japan
[3] Faculty of Science & Technology, Nihon University, Funabashi, Chiba, 274 Japan
[4] NGK Insulators, Ltd., Nagoya, 467 Japan

ABSTRACT

We investigated magnetic flux penetration and trapping processes of $Bi_2Sr_2CaCu_2O_x$ sintered specimens ($T_c \sim 90K$) by observing the change of stripe domains of iron garnet films for various temperatures. SEM image showed that sintered specimens consist of the assemblage of randomly oriented frake like grains with the diameters of 50~200 μm. The flux penetration and trapping processes were inhomogeneous. Averaged critical current Jc estimated from the flux density distribution using Bean's model was the order of 10^4 A/cm^2 at low temperatures,which are consistent with the results of the magnetization measurements.

KEY WORDS: Bi system superconductor, flux density distribution, critical current, Bean's model

INTRODUCTION

Since superconductivity in Bi-Sr-Ca-Cu-O system was at first reported by Maeda et al. [1], extensive researchs for the materials have been carried out on this systems. However, there exist some problems to solve in order to apply Bi-Sr-Ca-Cu-O superconductor to shielding, magnet and electronic devices.

The observation of flux penetration and trapping processes is useful to investigate the properties as a shielding material. Flux density distribution is observed using the Faraday effect of an iron garnet film placed on the superconductor. In the previous work, we estimated the critical current density profiles for $YBa_2Cu_3O_x$ thin films from the flux density distribution [2]. The results showed that critical current density Jc is nearly constant in the flux penetration region at low effective fields. This suggests the validity of the Bean's model [3]. In this paper, the flux density distribution of $Bi_2Sr_2CaCu_2O_x$ sintered specimens was observed using iron garnet films and homogeneity, crystalline orientation and critical current density were discussed from the results of the flux density distribution and SEM images.

EXPERIMENT

Appropriate amounts of Bi_2O_3, PbO, $SrCO_3$, $CaCO_3$, and CuO powders with 4 to 5N purity were mixed, sintered and ground. The ground powder of the composition of $Bi_2Sr_2CaCu_2O_x$ was pressed into pellets, partially melted for 30 min at 890℃ in oxygen atmosphere, annealed at 830℃ for 15 min and then quenched to room temperature. The pellets were cut into the disk-shape with the radious R of 1 mm and the thickness d of 0.2 mm.

A polarization optical microscope with an electromagnet and a cryostat was used to observe the domain patterns of the iron garnet films on superconducting specimens. Bi- and Ga-substituted iron garnet films with saturation magnetization of 280 G and thickness of 4 μm ware used in this study. Their saturation field H_s and nucleation field H_n at 15 K were 60 Oe and 44 Oe, respectively. In order to observe precisely the change of domain patterns, Ag films were coated on the surface of the iron garnet films. Magnetization was

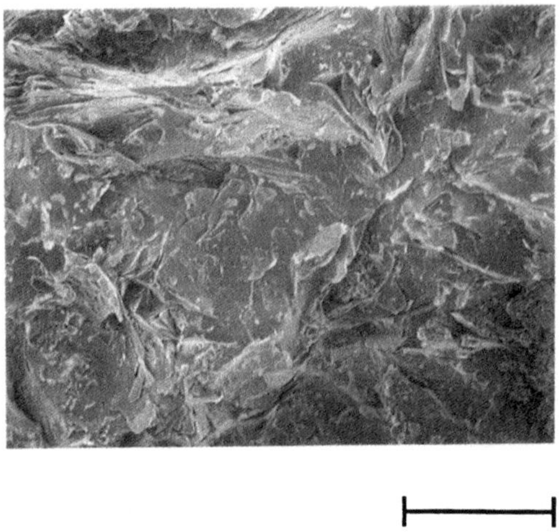

Fig.1. The SEM image of the cross section of $Bi_2Sr_2CaCu_2O_x$ sintered specimen.

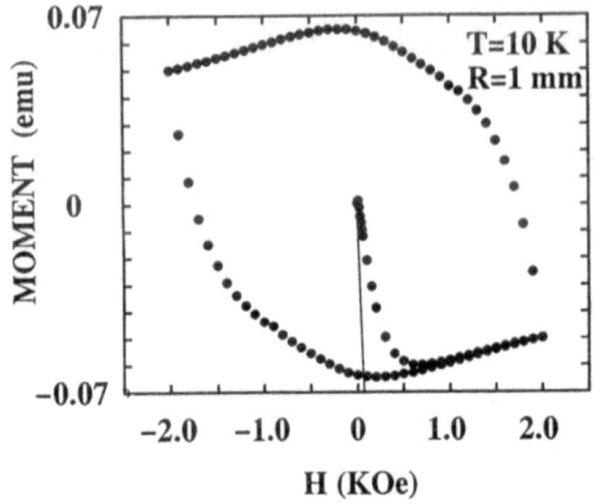

Fig.2. Magnetization curve at 10 K of disk-shaped $Bi_2Sr_2CaCu_2O_x$ sintered specimen with a radius of 1 mm and thickness of 0.2 mm. The solid line shows the initial magnetization process calculated assuming perfect diamagnetism.

measured using a SQUID magnetometer. Magnetic field was applied normal to the disk surface for obserbation of the flux density distribution and magnetization measurements.

RESULTS

Figure 1 is the SEM image of the cross section of a $Bi_2Sr_2CaCu_2O_x$ sintered specimen. The figure shows that the specimen consists of the assemblage of flake like grains. The diameters are 50~200 μm and the directions are random.

The magnetic field dependence of the magnetization at 10 K is shown in Fig.2. The solid line shows the initial magnetizing process calculated assuming that the disk shaped specimen has perfect diamagnetism. The field dependence suggests that the specimen have a strong trapping force of the magnetic flux.

The specimen was cooled to 15 K at zero-field and the domain pattern of the garnet film on the specimen was observed increasing the magnetic field as shown in Fig.3. At a low applied field, the specimen excludes almost all the magnetic flux. As increasing the field, the flux penetrates from the edge of the specimen. The penetration is inhomogeneous. The inhomoginiety is considered to be due to the inhomogeneous inner structures, such as crystalline orientation and superconducting phases, and not due to the surface roughness, because the roughness was less than 10 μm.

Figure 4 displays the domain patterns in the remanent state after applying a magnetic field of 3 kOe, which is enough to saturate the magnetization of the specimen. The inner magnetic flux flows out from the edge of the specimen as decreasing the field, and a certain amount of the flux is trapped at the pinning center in remanent state. The flux density distribution is inhomogeneous along the radius as the flux penetration. Figures 3 and 4 show that the region where the flux is easy to penetrate is same as that where the flux is easy to flow out, as shown in the arrow in Figs. 3 (c) and 4 (a) for a example. The flux penetration through the crystal boundary with superconducting weak-link has been reported in $YBa_2Cu_3O_x$ sintered specimens [4]. In this experiment using the garnet film with saturation field of 60 Oe, we did not observe such a flux penetration.

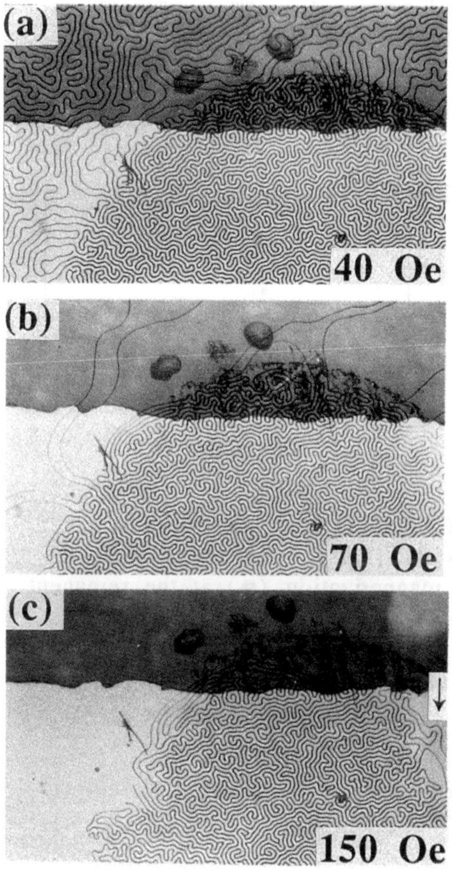

Fig.3. stripe domain structures of iron garnet film on superconducting disk-shaped $Bi_2Sr_2CaCu_2O_x$ specimen at 15 K with a radius of 1 mm and thickness of 0.2 mm. Ag is not evaporated at the upper part of the photograph.

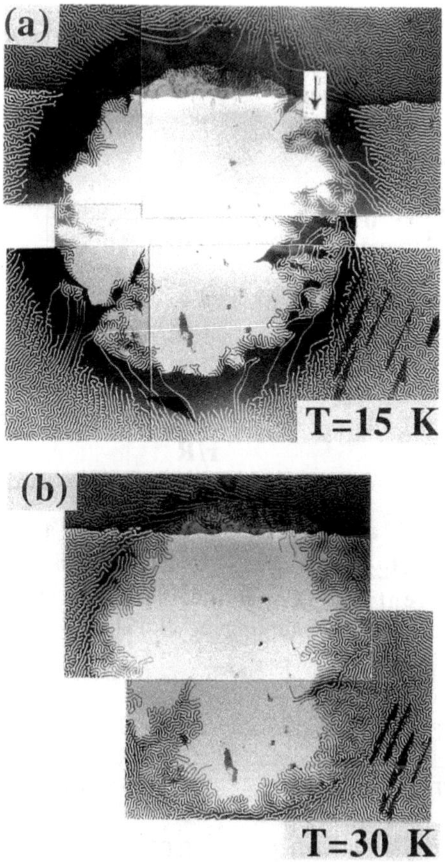

Fig.4. Stripe domain structures of iron garnet film on $Bi_2Sr_2CaCu_2O_x$ specimen at remanent state.

DISCUSSION

At the remanent state of superconductor, critical current J_c flows all over the superconductor. The flux density distribution at the remanence state is equal to the distribution of the shielding field created by the current J_c. Figure 5 shows the flux density distribution at the remanent state of a disk-shaped superconductor calculated by Daumling et al. using Bean's model [5]. In the figure, flux density distribution and the distance from the center are normalized by $J_c \cdot d$ and R, respectivery.

We can estimate the Jc from the flux density distribution of the remanent state using Fig.5. In Fig.4 (a) the magnetization of the garnet film in the white regions near the center saturates to a direction of the applied field and that in the black ones at the outside of the edge saturates to an opposite direction. The field at the boundary between the white region and the stripe domain is equal to a nucleation field H_n. The average distances r of the boundary from the center are 0.83 mm at 15 K and 0.68 mm at 30 K as estimated from Fig.4, and 0.43 mm at 50 K. In Table 1, r/R is a normalized averaged distance of the nucleation field H_n. The value of $H_z/J_c \cdot d$ at the value of the r/R were estimated from Fig.4. The values of J_c are estimated from the values of $H_z/J_c \cdot d$ ($H_z=H_n$) , H_n and d.

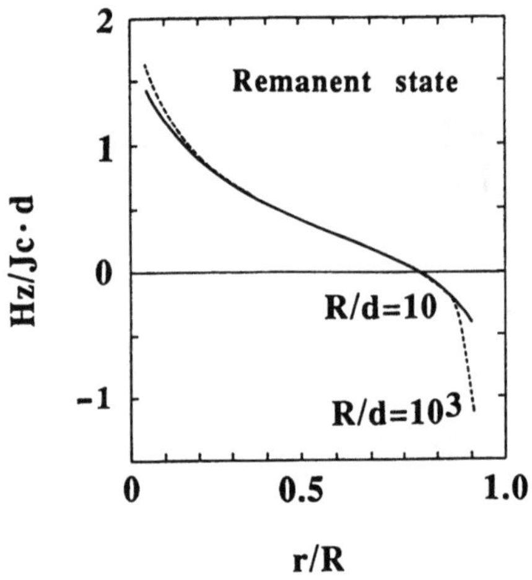

Fig.5. Radius dependence of the axial field H_z normalized by $J_c \cdot d$ for two different disk at remanent state calculated by Daunling et al [5].

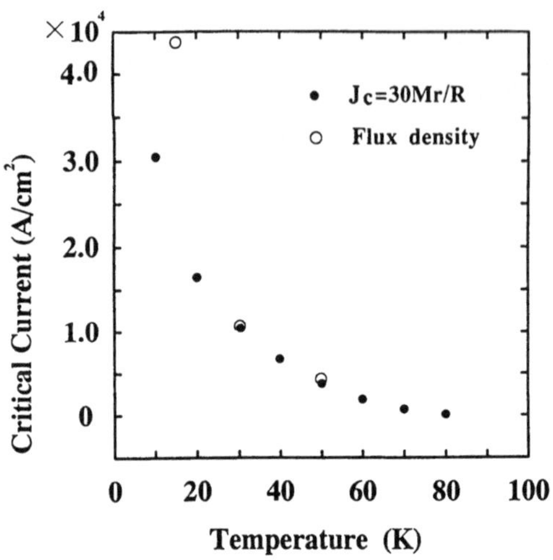

Fig.6. Critical current desity estimated from the flux density distribution (○) and magnetization (●) at the remanent state.

Table 1

Temp. (K)	r/R	$H_z/J_c{\cdot}d$	H_n (Oe)	J_c (10^4A/cm^2)
15	0.83	0.05	44	4.4
30	0.68	0.23	49	1.1
50	0.43	0.51	53	0.52

Figure 6 shows the temperature dependence of the value of J_c estimated from flux density distribution together with the J_c estimated from the remanent magnetization Mr using Bean's model ($J_c=30M_r/R$). The figure suggests the values of J_c estimated from flux density distribution are nearly equal to the values estimated from the magnetization measurements.

CONCLUSION

We observed magnetic flux penetration and trapping processes of disk-shaped $Bi_2Sr_2CaCu_2O_x$ sintered specimens using the iron garnet films. In the specimens, flux penetration and trapping were inhomogeneous, which is not due to surface flatness. The critical current J_c estimated from flux density distribution was 4.4×10^4 A/cm^2 at 15 K.

REFERENCES

1. Maeda H, Tanaka Y, Fukutomi M and Asano T (1988) Jpn. J. Appl. phys. **27** : L209
2. Yokoyama Y, Suzuki Y, Obara H, Ino K, Kubo T and Yoshida S (1991) to be published in proceedings of the 4nd international symposium on superconductivity (ISS'91)
3. Bean C.P. (1962) Phys. Rev. Lett. **8** : 250
4. Gotoh S, Koshizuka N, Yoshida M, Fujimoto H,Murakami M, Izumi H and Ohata K (1991) J. Mag. Soc. Japan **15** : 601 (in Japanese)
5. Daumling M and Larbalestier D C (1989) Phy. Rev. B **40** : 9350

Superconducting Current Density Profiles Estimated from the Flux Density Distribution in $YBa_2Cu_3O_x$ Thin Films

YUKO YOKOYAMA[1], YOSHISHIGE SUZUKI[1], HARUHIKO OBARA[1], KAZUMASA INO[2], TOMOYA KUBO[2], and SADAFUMI YOSHIDA[1]

[1] Electrotechnical Laboratory, Tsukuba, Ibaraki, 305 Japan
[2] Faculty of Science & Technology, Meiji University, Kawasaki, Kanagawa, 214 Japan

ABSTRACT

The flux penetration into superconducting thin films has been observed by means of the magneto-optical Faraday effect in iron garnet (YIG) films. Bi- and Ga-substituted YIG films were used for the observation of the flux density distribution for the $YBa_2Cu_3O_x$ films formed to disk. We proposed an estimation method of the critical current density Jc from the whole profile of the flux density distribution without using Bean's model. The estimated value of Jc is about 6×10^6 A/cm^2 at 30 K which is nearly equal to one estimated from the magnetization measurements at the same temperature.

KEY WORDS : Y-Ba-Cu-O, critical current density, garnet film, Bean's model

INTRODUCTION

Flux density distribution in a mixed state of high temperature oxide superconductor is closely connected with critical current density which is one of the most important parameters for the applications such as electronic devices and magnets. Several methods have been proposed to obtain the information on the flux density distribution. They are powder (or decoration) method using magnetic fine particles [1,2], electrical method using a Hole probe [3,4], magneto-optical method using a magneto-optical material [5-8], and so on.

The superconducting current density J can be estimated from the flux density distribution using the equation $\mathbf{J}\ (A/cm^2)=(10/4\pi)\cdot\mathrm{rot}\ \mathbf{H}(Oe)$. The value of J of the films when the magnetic field is applied normal to the surface of the film (z axis), however, can not be directly estimated from the field gradient (dH_z/dr) because of the effect of the large axial gradient of radial field (dH_r/dz). Therefore, it is necessary to calculate the J from the whole profile of the flux density distribution.

In the previous papers [9.10], we measured the flux density distributions for various applied fields and proposed the method to estimate the critical current density J_c from the flux penetration distance or the field gradient of inflection point using Bean's model [11]. In the present work, we estimate the superconducting current density profile from the flux density distribution and show that the superconducting current density J in the flux penetration region is constant and almost equal to the value estimated from the remanent magnetization using Bean's model.

EXPERIMENTAL PROCEDURE

$YBa_2Cu_3O_x$ films with the c axis normal to the surface were prepared on MgO (100) substrates by molecular beam epitaxy (MBE) method with an ionized oxygen source [12]. After the deposition, $YBa_2Cu_3O_x$ films with the thickness of 1800 Å were formed to a circle shape by chemical etching in order to make easy the calculation of the shielding field yielded by superconducting current flowing in the films.

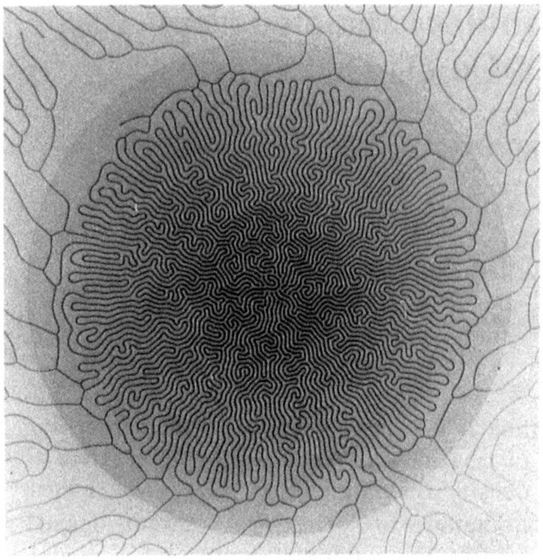

Fig. 1 Stripe domain structures of the garnet film on a disk-shaped $YBa_2Cu_3O_x$ film with a radius of 0.5mm and thickness of 1800 Å at 30 K. The film was coold to 30 K in zero field and then the field of 130 Oe was applied.

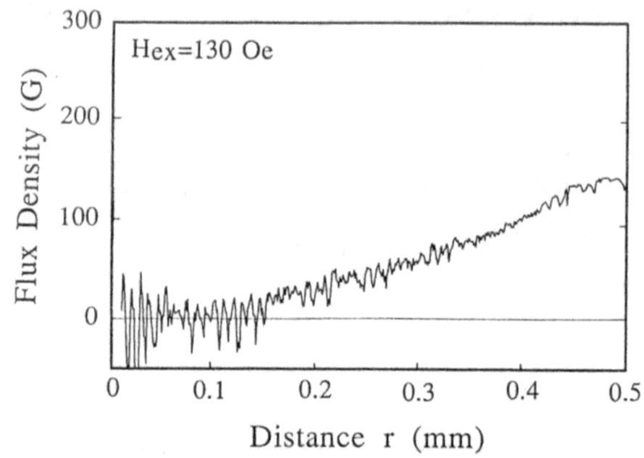

Fig.2 Averaged flux density distribution of Fig.1 analized using image processing technique from the magnetic field dependence of the domain pattern of the garnet film.

Bi and Ga-substituted ytrium iron garnet films were grown on GGG ($Gd_3Ga_5O_{12}$) substrates by liquid phase epitaxy (LPE) method. The temperature dependence of the saturation field, demagnetizing domain width and Faraday rotation and the magnetic field dependence of the domain patterns were measured [9].

In magneto-optical method using an iron garnet film, the film is placed on a superconducting specimen. The iron garnet film shows the domains with up and down directions of the magnetization. The domain patterns change with the magnetic field. Therfore the flux motion, homogeneity and pinning behavior of superconducting specimens are detected through the change of domain patterns.

A polarization optical microscope with an electromagnet and a cryostat was used to observe the change of domain patterns of iron garnet films. A $YBa_2Cu_3O_x$ films contacted with an iron garnet film was fixed on a cooling holder using the grease for low temperature. The temperature difference between the superconducting film surface and the cooling holder was estimated about 20 K when the temperature of the cooling holder was 10 K. The difference decreased with increasing the temperature of the cooling holder. The domain patterns showing the direction of magnetization were photographed and the flux density distribution was estimated from the photograph by image processing analysis. Superconducting current density distribution was calculated directly from the flux density distribution using Maxwell's equations.

RESULTS

We investigated the applied field dependence of the flux density distribution in $YBa_2Cu_3O_x$ films formed with a disk shape of a radius R of 0.5mm and a thickness d of 1800 Å. First, the superconducting film was cooled to 30 K in zero field, and then magnetic field normal to the sample surface was applied.

Figure 1 shows the domain patterns of the garnet film on a $YBa_2Cu_3O_x$ film for the applied field of 130 Oe at 30 K. In order to estimate the flux density distribution, the

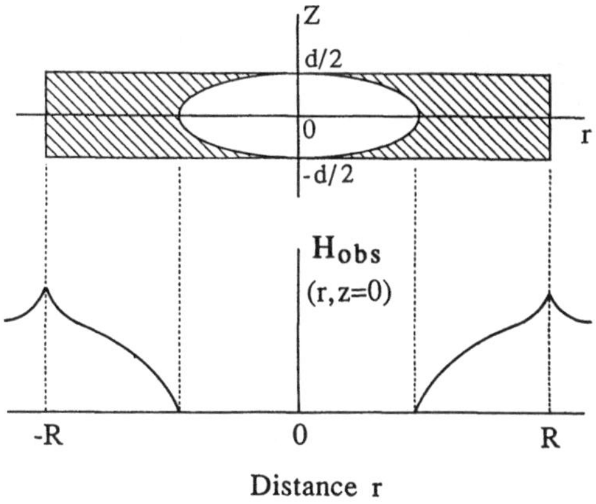

Fig.3 Schematic drawing of the flux penetration of a disk-shaped superconductor with radius R and thickness d, and flux density distribution at the center of the thickness, z=0. The flux penetration region is shown by hatched area.

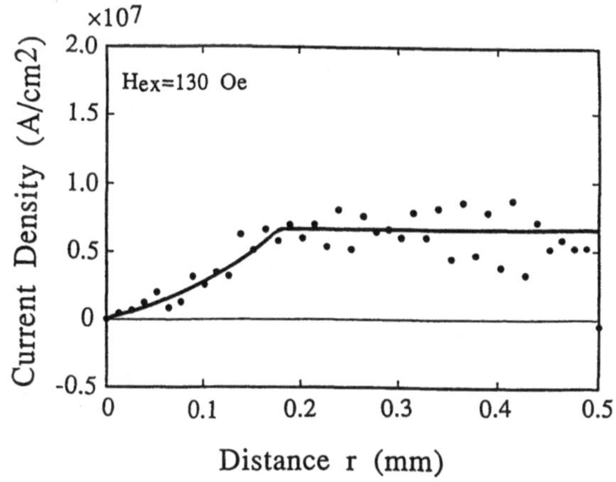

Fig.4 The superconducting current density profile estimated from the flux density distribution of Fig.2 using the model of Fig.3.

domain patterns were analyzed by image processing technique. The procedure is as follows. We divided the image to 514× 514 segments and calculated the ratio of black segments on the circle with the same radius. The ratio was replaced with flux density using the field dependence of the domain patterns as shown in Fig.2. In the figure the flux density distribution largely fluctuates. It is considered that the fluctuation does not reflect the actual flux density distribution, because large fluctuation was observed even at no flux penetration regions. The fluctuation increases with decreasing of the distance because of the decreasing of the measured area. We omitted the result of the flux density distribution near the center which displayed very large fluctuation and assumed it to zero for the following analysis.

DISCUSSION

When magnetic field is applied normal to a disk-shaped type II superconductor, magnetic flux gathers around the superconductor because shielding field is yielded by superconducting current flowing in the superconductor. When the magnetic flux at the edge of a superconductor is beyond the upper critical field H_{c1}, the flux begins to penetrate from the edge. For infinitely long cylinder specimens, H_{c1} equals to applied field. However, for thin films, magnetic flux begins to penetrate at the applied field much smaller than H_{c1} because of very large shielding field at the edge of the film.

The flux penetration profile of a disk-shaped superconductor and its flux density distribution H_{obs} at the center of the thickness (z=0) is shown as Fig.3. The flux penetration region is shown as hatched one and the rest keeps perfect diamagnetism. In the figure, the flux density H_{obs} is equal to the sum of applied field H_{ex} and shielding field H_{sh} which is yielded by the superconducting current J(r) flowing in hatched region as:

$$H_{obs}=H_{ex} + H_{sh}(J(r)) \quad (1).$$

We estimated the superconducting current profiles J(r) from the flux density distribution using eq.(1) as shown in Fig.2. For the calculation of J(r), the disk was divided into forty doughnut-shaped segments with the width of one fortyth of the radius and thickness of d. J(r) is the averaged value over the film thickness d and H_{sh} is calculated using Biot-Savart's law.

Figure 4 shows the values of J(r) estimated from the flux distribution shown in Fig.2. The critical current density is equal to the value of the J(r) in flux penetration region. This figure shows that in the flux penetration region J(r) is almost constant. This fact supports the Bean's model, that is, J(r) is independent of magnetic field. The constant value of J(r), $6\times 10^6 A/cm^2$ corresponds to the critical current density. This value is nearly equal to one obtained using Bean's model from the measurements of magnetization [12].

Our method to estimate the critical current profile from a whole of the flux density distribution is analogous to the one estimated by Frankel [5]. In Frankel's method, the thickness of the specimen in which critical current J_c is flowing decreases with decreasing radius when the flux does not penetrate to the center of the disk and Jc is estimated as a function of the essential flux density H_{obs}, $J_c=\alpha/H_{obs} + H_0$ (α and H_0 are constant). In our method, the superconducting current J(r) is averaged over the film thickness and the distance dependence of J is estimated. Critical current density is determined from the value of J(r) at the flux penetration region.

CONCLUSION

We obtained the flux density distribution of disk-shaped superconducting $YBa_2Cu_3O_x$ films by observing the change of the domain patterns of iron garnet films placed on the superconducting films. We directly estimated the critical current density from the flux density distribution without using the Bean's model. Estimated critical current density was nearly equal to one obtained using Bean's model from the measurements of magnetization.

ACKNOWLEDGEMENT

The authors wish to thank Drs. N. Koshizuka, S. Gotoh and Y. Hasumi for the cooperation on the observation of the domain patterns at low temperatures.

REFERENCES

1. Traüble H and Essman U (1968) Phys. Status Solidi **25** : 395
2. Gammel P L, Bishop D J, Dolan G J, Kwo J R, Murray C A, Schneemeyer L F and Waszczak J V (1987) Phys. Rev. B **36** : 2592
3. Weber H W and Riegler R (1973) Solid State Commun. 12 : 12
4. Kirchner H (1971) Phys. Status Solidi a **4** : 531
5. Frankel D J (1979) J. Appl. Phys. 50 : 5402
6. Koblischka M R, Moser N, Gegenheimer B and Kronmüller H (1990) Physica C **166** : 36
7. Indenbom M V, Kolesnikov N N, Kulakov M P, Naumenko I G, Nikitenko V I, Polyanskii A A, Vershinin N F and Vlasko-Vlasov V K (1990) Physica C **166** : 486
8. Gotoh S, Koshizuka N, Yoshida N, Murakami M and Tanaka S (1990) Jpn. J. Appl. Phys. **29** :L1083
9. Yokoyama Y, Hasumi Y, ObaraH, Suzuki Y, Katayama T, Gotoh S and Koshizuka N (1991) Jpn. J. Appl. Phys. **30** : L714
10. Yokoyama Y, Suzuki Y , Hasumi Y, Obara H, Yoshimi T, Kita Y, Kosaka and Yoshida S (1991) to be published in proceeding HTSC III (Kanazawa)
11. Bean C P (1962) Phys. Rev. Lett. **8** : 250
12. Obara H, Kosaka S, Yokoyama Y, Umeda M and Kimura Y (1991) Phys. Rev. B **44** : 4532

Simulation of Magnetization Curves of Oxide Superconductors and Shape Effects

TAKAHIRO TAGUCHI, MASATO MURAKAMI, HIROYUKI FUJIMOTO, SATOSHI GOTOH, NAOKI KOSHIZUKA, and SHOJI TANAKA

Superconductivity Research Laboratory, ISTEC, Koto-ku, Tokyo, 135 Japan

ABSTRACT

Magnetization of non-ideal type II superconductors can be analyzed by using the critical state model on the assumptions that the critical current density is independent of the external field and that a sample is an infinite thin flat plate. However, in a real case, the critical current density is considered to be dependent on the magnetic field and a sample is not an infinite thin flat plate. Therefore, we calculated magnetization curves considering the two dimensionality of the sample and the magnetic field dependence of J_c.

KEY WORDS: critical state model, magnetization, oxide superconductor, shape effect, J_c

INTRODUCTION

Magnetization measurements are informative for understanding the critical current characteristics of type II superconductors [1-9]. In general, the critical current density is estimated by using the critical state model. Difference in magnetization ($\Delta M = M^+ - M^-$) between the increasing and decreasing field branches is related to sample thickness (d) and the critical current density (J_c) by the following equation in SI units: $\mu_0 M$ in Tesla, J_c in A/m^2 and d in m,

$$\Delta M = \frac{1}{2} J_c d \quad (1)$$

In a practical case, J_c is dependent on the applied magnetic field, though J_c is constant in the Bean model. And a sample is not an infinite thin plate. If we consider such two dimensional shape effect, the hysteresis of simulated curves becomes smaller [10]. We therefore need to take account of the two dimensional effects and the magnetic field dependence of J_c for determining J_c from M-H loops.

In the present study, We analyzed magnetization curves of oxide superconductors using the critical state model considering two dimensional shape effect, and compared them with the results of one dimensional model analysis.

EXPERIMENTAL

We calculated magnetization using the Irie-Yamafuji model, $J_c=KB^{-n}$, where K and n are material parameters [7] and Maxwell's law, $\nabla \times B = \mu_0 J_c(B)$, with a field parallel to the surface of a sample. The field within a specimen decreases with distances as $dB/dx=\mu_0 KB^{-n}$ and n=0 corresponds to the simple Bean critical state model. Here, the parameter K corresponds to J_c at 1 Tesla and n is the parameter that is related to the kinds of pinning centers. By definition, magnetization is given by

$$\mu_0 M = B - \mu_0 H_a, \tag{2}$$

where H_a is the applied external field and μ_0 is the permeability of a vacuum. The mean magnetization of a specimen can be obtained by averaging the entire volume,

$$\mu_0 M = \frac{\int (B - \mu_0 H_a) dv}{\int dv}. \tag{3}$$

In the case of an infinite thin plate, this equation becomes

$$\mu_0 M = \frac{\int_0^d (B - \mu_0 H_a) dx}{d}. \tag{4}$$

In a real case, two dimensionality must be taken into consideration, therefore the equation becomes

$$\mu_0 M = \frac{\int_0^a \int_0^b (B - \mu_0 H_a) dx dy}{S}, \tag{5}$$

where S is the area of a cross section.

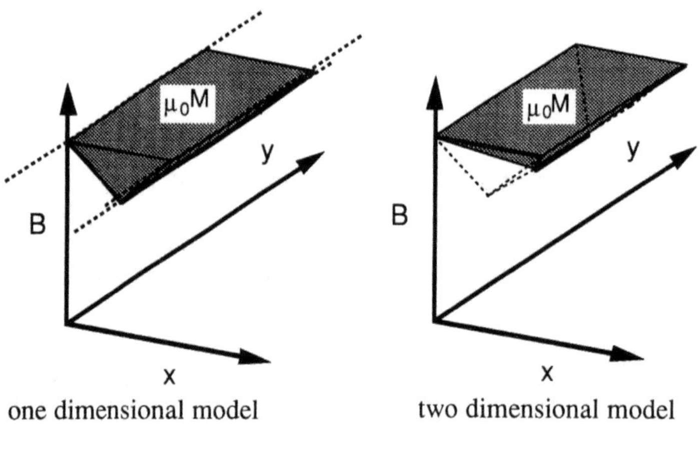

Figure 1. Schematic illustration of the field penetration for one dimensional (left) and two dimensional model (right).

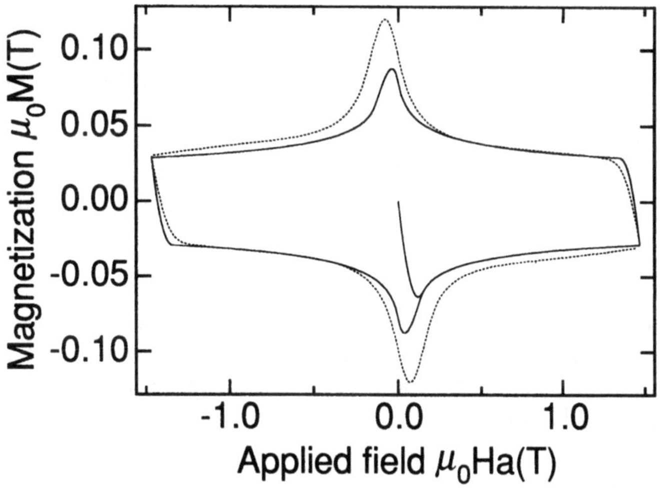

Figure 2 Magnetization loops for MPMG-YBaCuO sample at 77K. The dotted curve is experimental, and the solid curve is calculated using the critical state model with parameters: K=1.38×10^{8}A/m^2 and n = 0.277.

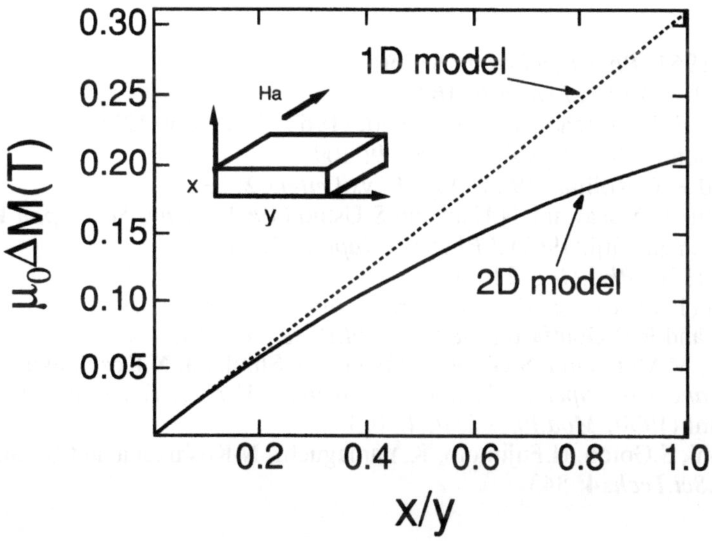

Figure 3 ΔM versus the aspect ratios of a sample. The dotted curve is calculated based on one dimensional model. The solid curve is calculated based on two dimensional model.

YBaCuO samples were prepared by the MPMG process [11,12,13]. The sample contained $Y_2B_aC_uO_5$(211) inclusions of 25% in volume embedded in the $YBa_2Cu_3O_x$ matrix. A rectangular sample with dimensions of 3.64×3.55×0.79 mm^3 was cut from an MPMG processed pellet so that the surface plane of the plate becomes parallel to the c axis. Magnetic field was applied parallel to the c axis. Magnetization measurements were conducted at 77K using a vibrating sample magnetometer with the maximum applied field of 1.5 Tesla.

RESULTS AND DISCUSSION

Figure 2 shows magnetization loops. The dotted line and the solid line show experimental data and two dimensional simulation data, respectively. First, the parameters were obtained by fitting the curves to the experimental data. In the present analysis, a curve of the form $J_c(B)=KB^{-n}$ was fitted to the curve at higher fields (0.4T-1.5T). The material parameters were found to be $K = 1.38\times10^8$ A/m^2 and $n = 0.277$ from the present data. Then, we performed simulation of M-H loops considering two dimensionality. It was found that ΔM of the simulated curves was smaller than the experimental data, when two dimensionality was considered.

Figure 3 shows calculated ΔM using equation (4) and (5). It is clear that when we consider two dimensionality, the hysteresis of simulated curves becomes smaller as the shape of the specimen deviates from an ideal thin plate. We therefore need to take account of the two dimensional effects for understanding M-H loops.

CONCLUSION

If we consider two dimensional shape effect, the hysteresis of simulated curves becomes smaller. We therefore need to take account of the two dimensional effects for determining J_c from M-H loops. However, we may need to use more detailed corrections such as demagnetizing factor, the effect of H_{c1} and surface barrier for simulation of the curves at low fields.

REFERENCES

1. C.P.Bean (1962) *Phys.Rev.Letters* **8**: 250
2. H.London (1963) *Phys.Letters* **6**: 162
3. Y.B.Kim, C.F.Hempstead and A.R.Strnad. (1963) *Phys.Rev.* **129**: 528
4. P.W.Anderson (1962) *Phys.Rev.Letters* **9**: 309
5. J.Silcox and R.W.Rollins (1963) *Appl.Phys.Letters* **2**: 231
6. K.Yasukochi, T.Ogasawara, N.Usui and S.Ushio (1964) *J.Phys.Soc.Japan* **19**: 1649
7. F.Irie and K.Yamafuji.(1967) *J.Phys.Soc.Japan* **23**: 255
8. G.R.Love (1970) *Phil.Mag.* **21**: 1003
9. T.Hirasawa (1973) *Cryogenic Engineering, Japan* **8**: 154
10. D.-X.Chen and R.B.Goldfarb (1989) *J.Appl.Phys.* **66**: 2489
11. H.Fujimoto, M.Murakami, S.Gotoh, T.Oyama, Y.Shiohara, N.Koshizuka and S.Tanaka (1990) *Advances in Superconductivity II. Springer-Verlag, Tokyo,* p 285
12. M.Murakami (1990) *Mod.Phys.Lett.* **4**: 163
13. M.Murakami, S.Gotoh, H.Fujimoto, K.Yamaguchi, N.Koshizuka and S.Tanaka (1991) *Supercond.Sci.Tech.* **4**: S43

5 Wires, Tapes and Bulk

5.1 Bulk — (1) YBCO

Growth Mechanism During Directional Solidification on Y-System Superconductors

TERUO IZUMI, YUICHI NAKAMURA, TAE HYUN SUNG, and YUH SHIOHARA

Superconductivity Research Laboratory, ISTEC, Koto-ku, Tokyo, 135 Japan

Abstract

The unidirectional solidification, which is a zone melt method, was performed on Y-system superconductors. The condition for continuous growth of $YBa_2Cu_3O_y$ crystals was confirmed to be larger G/R than 300°Ch/cm^2 where G is temperature gradient and R is growth rate. The faceted interface was recognized in the sample grown at 1 and 2mm/h. The necessary solute diffusion for peritectic reaction was clarified to be provided through liquid. The solidification model based on the liquid diffusion control was developed and it agrees qualitatively with the experimental results.

KEY WORD: Y-system superconductors, zone melt method, facet growth, peritectic reaction, solidification model

Introduction

Since the high Tc oxide superconductors were discovered, lots of efforts have been formed to obtain higher critical current densities (J_c). Recently, high Jc values have been reported in the samples prepared by several solidification processes[1-3]. However the growth mechanism has not been clarified yet. In this paper, the effects of process parameters on the structure and the proposed solidification model based on the experimental results were also discussed.

Experimental

The growth apparatus used in this study was shown in Figure 1 which consists of two parts. An outer heater coiled on a ceramic tube was arranged and the other filament loop was located inside the outer heater to make a partial melting zone. The maximum temperature in the zone was controlled at 1040°C which is above the peritectic temperature. The partial melting zone which is over 1000°C was kept about 10mm in length. The temperature gradient (G) at 1000°C was about 180°C/cm. The precursors were prepared by the following procedures; the mixtured powders with the ratio of Y:Ba:Cu=1.2:2.1:3.1 were preheated at 900°C for 12h and repeated this heating twice. The preheated powders were pressed into rods of 2mm in diameter and 120mm in length by Cold Isostatic Press (CIP) using a rubber tube. The rods were suspended by clip and inserted into the furnace at high speed initially. Then, the samples were pulled upward at several growth rates (R) which were 1,2,3,6 and 10 mm/h. The sample dropped into a quench tank by releasing the clip during steady state solidification to observe the solidification interface morphology of the 123 crystals. The microstructures were investigated by optical microscopy and EPMA.

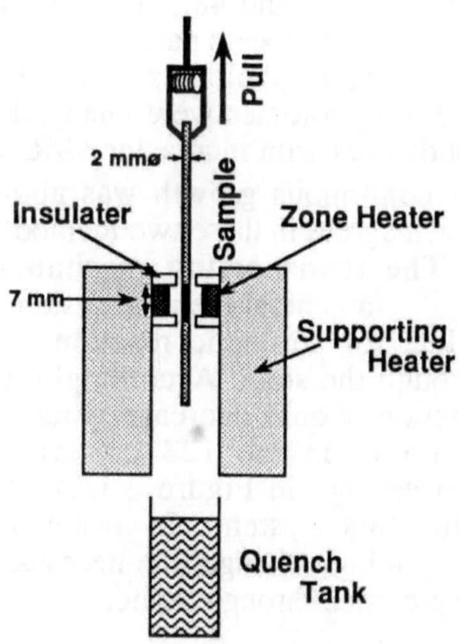

Figure 1 Schematic illustrations of the experimental apparatus for zone melt method.

Results

The observation of the interface morphology leads to the following results. The clear solid/liquid interface was not recognized (mushy) in the sample grown at 10mm/h. The sample grown at growth rates lower than 6mm/h showed continuously grown structure. However, the entrapped liquids existed in the 123 crystals in the samples of R=6 and 3 mm/h. In contrast with these samples, no entrapped area was seen and the faceted interfaces were observed in the samples of R=1 and 2 mm/h. The facetted grains of R=1mm/h were larger than those of R=2mm/h. The typical microstructeres were shown in Figure 2. In this figure, the grains in white are 123 crystals with dispersion of 211 particles and the dark regions are liquids before quenching or the unreacted phase also including 211 particles due to EPMA. Furthermore, Figure 3 shows the yttrium concentration mapping around the interface of the sample grown at 1mm/h. The lighter color indicates higher concentration in yttrium. Therefore, white particles, gray grains and black matrix expressed 211, 123 and liquid phases, respectively. The sharp interface of the 123 crystals were observed and the volume fraction of the 211 phase was changed drastically from in the liquid to in the 123 crystal.

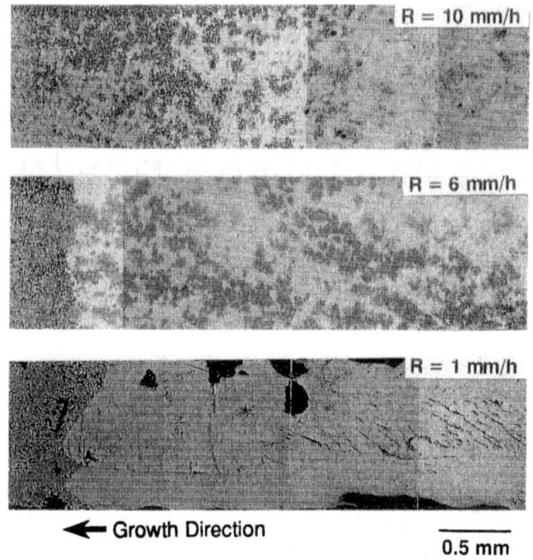

Figure 2 Microstructures of the longitudinal sections of the samples grown at different growth rates.

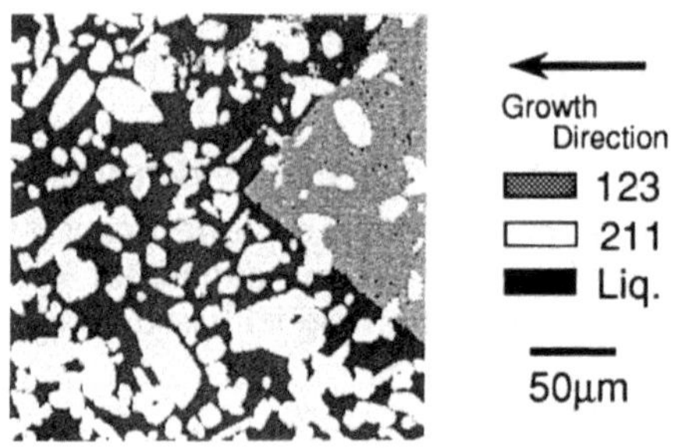

Figure 3 The compositional mapping image at the interface of the sample grown at 1mm/h.

Discussions

1. The condition for continuous growth

As seen in the experimental results, the continuous growth of the 123 crystals were recognized in the samples grown at growth rates lower than 3mm/h. On the view of the constitutional supercooling theory, the lower limit of G/R condition for continuous growth was estimated around 300°Ch/cm^2. In the previous results of the Bridgman method, the limit for continuous growth was confirmed to exist between 301 and 482 °Ch/cm^2. Figure 4 shows the effect of process parameters (G and R) on the interface morphologies in these two processes. The morphologies were changed from mushy to continuous with increasing G/R. The lower limit for continuous growth was about 300°Ch/cm^2 which agrees in these two method.

2. The transporting mechanism

In general metallic systems, the necessary solute for peritectic reaction were transported through the solid. Accordingly, the 211 volume fraction should decrease gradually from in the liquid to in the 123 crystals. However, the morphology in Figure 3 was not in agreement with this system. In order to explain the morphology change, the necessary solute should be provided through liquid.

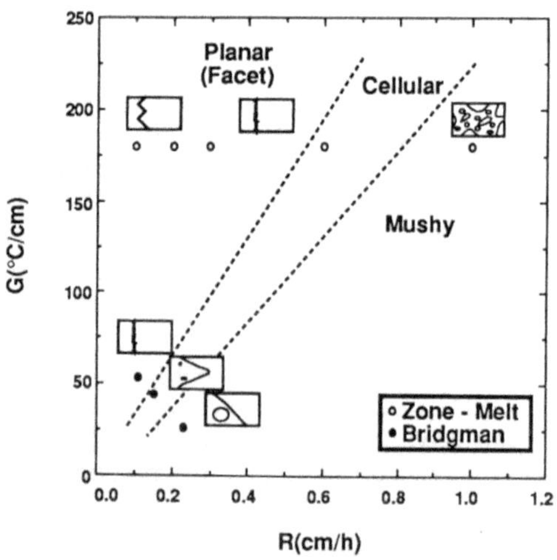

Figure 4 Effect of the process parameters, G and R, on the morphology of the solidification front.

Solidification Model

Though details about the solidification model can be found in other publications[4], the brief summary is shown in the followings. Figure 5 shows the principle of the model based on the assumption that rate limiting for the 123 phase crystal is solute diffusion in the liquid. Yttrium diffusion flux for the 123 growth from the 211 particles was estimated according to the phase diagram. This flux was considered from the composition difference in the liquid at the interfaces of liquid/211 and liquid/123. This composition difference is due to the change of the chemical potential caused by the curvature of the 211 particles. The change relates to the undercooling, ΔT, derived from the Gibbs-Thomson's relationship as;

$$\Delta T = 2\Gamma / r \quad (1)$$

where Γ is Gibbs-Thomson's coefficient $(\sigma/\Delta s)$,

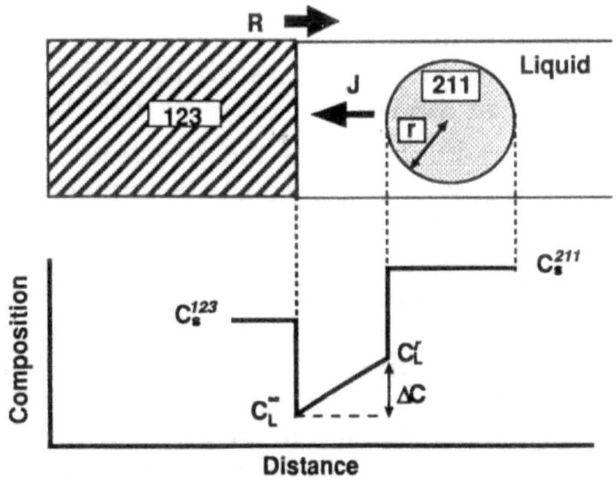

Figure 5 Sketch of the principle of the proposed solidification model.

σ is the interfacial energy between the 211 and the liquid, ΔS is volumetric entropy of fusion, and r is the radius of the 211 particle. From this equation, the composition difference which is considered to be a driving force for diffusion, ΔC, is expressed using the slope of the liquidus from the phase diagram;

$$\Delta C = 2\Gamma / rm \quad (2)$$

where m is the liquidus slope of the 211 phase. The flux from one 211 particle in a radius, r, at a distance from the 123 interface, z, was calculated as follows;

$$J = (2\pi D\Gamma/m)(r/z) \quad (3)$$

where D is liquid diffusion coefficient. Here the diffusion zone for growing 123 crystals was assumed to be within the maximum effective length, δc, and the minimum effective length, δmin. The δc was determined as a characteristic length during planar steady state growth, D/R. The dmin was defined as a following idea. The flux increases with decreasing z and r due to the equation (3). When the flux reaches the necessary flux for the 123 steady state growth, the distance was determined as dmin. The 211 particles are expected to diminish its own volume providing the diffusion flux with approaching to the 123 interface in the diffusion zone. If we give the size distribution of the 211 particles in the outer zone, the total flux, J_{211}, from the 211 particles was estimated by the integration of the flux from the individual 211 particles, J, by z and r as follows;

$$J_{211} = \frac{3D\Gamma V_f^{211}}{2m} \frac{\int_0^\infty r \left(\int_{\delta_{min}}^{\delta c} \frac{\Psi(r_0(r,z))}{z} dz \right) dr}{\int_0^\infty r_0^3 \Psi(r_0) dr_0} \quad (4)$$

where V_f^{211} is the volume fraction of the 211 phase in outer zone, which is calculated from the initial composition of the alloy, r_0 is the radius of the 211 particle at δc, $\Psi(r_0)$ is the size distribution function of the 211 particles in the outer zone, $\Psi(r_0(r,z))$is the size distribution function of 211 particles in diffusion zone related with $\Psi(r_0)$. On the other hand, the necessary flux for growing of the 123 crystal is calculated as follows;

$$J_{123} = (C_s^{123} - C_L^\infty)R \quad (5)$$

where C_s^{123} and C_L are the yttrium concentrations of 123 crystal and the liquid at the interface, respectively, and R is the growth rate of the 123 crystal. The condition for the steady state growth of the 123 crystal lead to the relation equating Eq.4 and Eq.5;

$$(C_s^{123} - C_L^\infty)R = \frac{3D\Gamma V_f^{211}}{2m} \frac{\int_0^\infty r \left(\int_{\delta_{min}}^{\delta c} \frac{\Psi(r_0(r,z))}{z} dz \right) dr}{\int_0^\infty r_0^3 \Psi(r_0) dr_0} \quad (6)$$

Comparison

If we give a typical size distribution of 211 particles in the outer zone which has the mean particle radius, m, and the deviation, σ, the balanced growth velocity could be estimated from this model using some appropriate thermo physical properties. Figure 7 shows the calculated balanced velocities for the continuous steady state growthof the 123 crystal, indicating that the larger balanced velosities can be applied with smaller average particle size and deviation. The order of the calculated balanced growth velosity is in agreement with the experimentally confirmed one for continuous and large planar front growth.

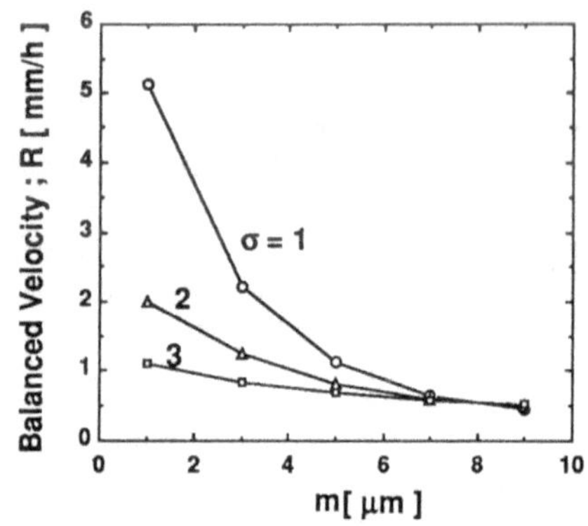

Figure 6 Calculated balanced velocity for the continuous steady state growth of the 123 crystal, with different size distributions of initial 211 particles.

In order to evaluate propriety of the model, the experimental results were compared with the calculated results. Figure 8(a) shows the micro-structure in the outer zone of the sample grown at 1mm/h. The distribution close to that in Figure 7(a) was drawn as open circles in Figure 7(c). The balanced velocity calculated using this distribution, about 1mm/h, is almost same as the actual growth rate. Furthermore, the estimated size distribution in the 123 crystal derived from this model is expressed by filled circles in Figure 7(c). This calculated distribution agrees well at least qualitatively with the experimental results shown in Figure 7(b).

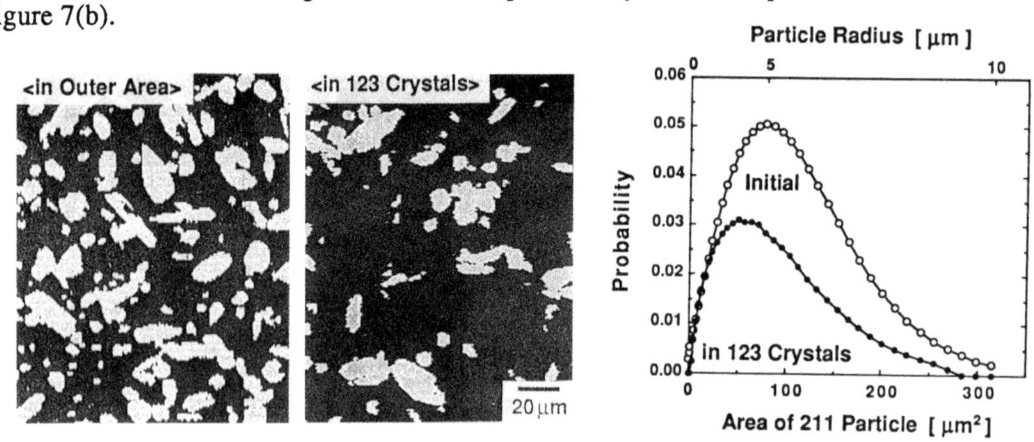

Figure 7 Change of the size distributions of the 211 particles before and after the growth of the 123 crystal at 1mm/h, both microstructures and calculated distributions are shown.

Acknowledgement

The authors would like to thank to Profs. M. C. Flemings of M.I.T., W. Kurz of EPFL in Lausanne, and T. Umeda and T. Suzuki of University of Tokyo for helpful discussions.

This work was supported by New Energy and Industrial Technology Development Organization under the management of the R&D of Basic Technology for Future Industries.

References

1) S.Jin, T.Tiefel, R.Sherwood, R.van Dover, M.Davis, G.Kammlott, and R.Fatnacht; Phys. Rev. B 37 (1988) 7850
2) M.Murakami, M.Morita, K.Doi, and K.Miyamoto; Jpn. J. Appl. Phys. 28 (1989) 1189
3) M.Murakami, H.Fujimoto, T.Oyama, S.Gotoh, Y.Yoshida, N.Koshizuka, and S.Tanaka; ICMC"90 Materials Aspect, in press
4) T.Izumi, Y.Nakamura, and Y.Shiohara submitted for publication

Process and Properties of QMG Crystals

M. Morita, M. Tanaka, K. Kimura, T. Sasaki, L. Trouilleux, S. Takebayashi, K. Miyamoto, M. Hashimoto, and K. Sawano

Advanced Materials & Technology Research Laboratories, Nippon Steel Corporation, Kawasaki, Kanagawa, 211 Japan

Abstract

Recent developments of QMG (Quench and Melt Growth) process and properties of large QMG crystals as bulk magnets are reported. QMG crystals have two prominent features. One is the finely dispersed RE_2BaCuO_5(211) phase (about 1 micrometer). The other is that the $REBa_2Cu_3O_{7-x}$ (123) matrix is a pseudo single crystal which has no high angle grain boundary acting as weak -link.

In QMG process, it is found that fine dispersion of 211 phases is brought about by Pt addition rather than melt-quenching effect. The size of 211 phase was observed to decrease when Pt was added to the precursor. In order to investigate grain growth of 211 phase, time dependence of 211 particles size was studied within BaCu-oxide liquid phase and also in the sintering process. It seems that Pt element prevents the grain growth of fine 211 particles.Melt-quench treatments using Pt crucibles could be omitted. Thus QMG process was simplified by the Pt addition technique.

On the other hand, large QMG crystals made by seeding and RE compositional gradient techniques were able to trap high magnetic field in excess of 1T at 63K and 77K. It is found that QMG crystals can operate as bulk magnets without stabilization unlike conventional metal superconductors at 4.2K.

KEY WORDS: melt-process, Pt addition, coarsening, bulk magnet

I. Introduction

It is known that the QMG process can produce a bulk $YBa_2Cu_3O_{7-x}$ superconductor with high critical current density (Jc) even at 77K and a high magnetic field[1,2]. The QMG crystal is a pseudo single crystal which contains a finely dispersed Y_2BaCuO_5 phase in the $YBa_2Cu_3O_{7-x}$ matrix. QMG process is one of the melt processes to make bulk superconducting materials. Recently, it is known that QMG crystals can be made by other melt processes[3-5]. The high Jc exhibited by the QMG materials is related to the nature of the dispersion of 211 phase. Thus, it is important to elucidate the growth mechanism of the 211 phase in the 123 matrix. Futhermore, QMG crystals are interesting materials in terms of applications. Because of the excellent Jc properties, QMG crystals can be used as bulk superconducting magnets, superconducting bearings, current leads and so on.

In this paper, the authors report the effect of Pt addition on the dispersion of 211 phase in melt process and the properties of bulk magnets which consist of large QMG crystals.

II. Effect of Pt addition on 211 dispersion

II.1 Analysis of Melt -Quenched 123 [6]

In order to investigate the composition of the melt-quenched 123 phase, we firstly analysed melt-quenched material by X-ray diffraction technique. It is known that melt-quenched materials consist of Y_2O_3, $BaCuO_2$ and an unknown phase whose main diffraction peak is $2\theta=36^o$ for Cu(K_α), as can be seen in figure 1. It was believed that this unknown phase was responsible for the formation of the fine 211 phase. This phase has been reported by C. L. Teske and H.K.Muller [7]. This unknown phase was found to be $BaCu_2O_2$ which can be obtained under reducing atmosphere at about 800^oC.

Based on the idea that $BaCu_2O_2$ plays an important role in the fine dispersion of 211 phase in QMG process, we prepared a precursor which had the same X-ray diffraction pattern as the quench-melt one by mixing Y_2O_3, $BaCuO_2$ and $BaCu_2O_2$ .The $BaCu_2O_2$ phase was synthesized by calcining the mixture of an appropriate amount of $BaCO_3$ and CuO at 820°C for 20 h in nitrogen atmosphere.

The mixture was then formed into pellets as a QMG precursor. This precursor was partially melted at 1150°C for 30 minutes and subsequently cooled down at a rate of 1 °C/h between 1005°C and 975°C in order for the growth of 123 to proceed. Microstructural observation of the polished surface of the sample by a polarized-light optical microscope revealed that 5 to 10 micron-sized 211 inclusions were dispersed in the 123 matrix. These inclusions are much larger than those in QMG materials prepared using the melt-quenched precursor. From this result, we conclude that $BaCu_2O_2$ is not responsible for fine dispersion of 211 phase .

In the next experiment, we checked the contamination from Pt crucibles. Elemental analysis indicated that the melt-quenched precursor contains about 0.3 - 0.5wt% of Pt. Therefore, a small amount of Pt was suspected to contribute to the fine dispersion of 211.

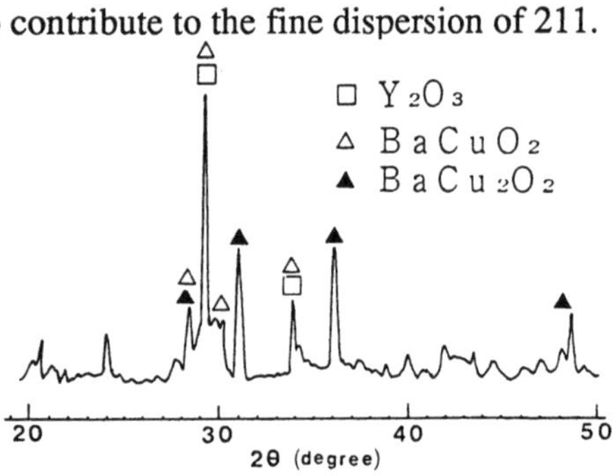

Fig.1 : X-ray diffraction pattern of a sample quenched from 1400°C to room temperature

II.2 Pt addition effect on the size of 211 particles

To elucidate the effect of Pt , various amounts of Pt were added to the precursor. The Pt amount was varied at 0.00, 0.01, 0.05, 0.1, 0.5, 1.0, 5.0 and 10wt%. PtO_2 powder was used. The starting composition of Y:Ba: Cu was determined so that the excess amount of 211 is 30 mol% in the 123 matrix. Y_2O_3, $BaCuO_2$ and $BaCu_2O_2$ powder were mixed at the appropriate composition. The precursors were partially melted at 1150°C for 30 minutes followed by a slow cooling from 1005°C to 970°C at a cooling rate of 1°C/h. The materials were then annealed in oxygen at 600°C and subsequently cooled slowly to room temperature. To estimate Jc for each sample, magnetization measurements were performed using a vibrating sample magnetometer.

Figure 2 shows the optical micrographs of the final microstructures with Pt 0.0, 0.1 and 0.5wt%, respectively. Additions smaller than 0.05wt% had almost no effect on the size of 211 inclusions. The Jc was about $5x10^3 A/cm^2$ at 77K and 1T. The effect appeared above 0.1 wt%. Most of the 211 inclusions were about 1 micrometer in diameter when more than 0.5wt% Pt was added. Jc exceeded $10^4 A/cm^2$ with 1.0wt% Pt (figure 3). As shown in Figure 4, relatively large-sized complex oxide particles of Ba and Pt were observed in the 5.0wt% and 10wt% Pt added samples.

II .3 Mechanisms and prevention of 211 grain growth - Pt effect [8]

We have considered two mechanisms to describe the Pt effect on the growth of 211 :

1 : Pt element may act as a nucleation site when 211 phase is produced from the reaction between Y_2O_3 and the liquid phase, generating a fine dispersion of 211 particles (then in this case, coarsening of fine 211 should not occure).

2 : Pt element prevents coarsening of fine 211 phase. (with or without promoting 211 phase nucleation)

In order to verify this second hypothesis , following experiments were performed. Powders of fine 211 were prepared using a ball mill. The powders of 211, BaO_2 and CuO were mixed to the molar ratio Y:Ba:Cu=14:16:22. This composition corresponds to a ratio of 123: 211=6:4. 1wt% Pt doped pellets and non doped pellets were made from the mixed powders. To prevent generation of 123 phase around 1000°C, these pellets were inserted into the furnace which had been previously heated up to 1112°C. After 5, 10, 20, 40 and 60 minutes, these

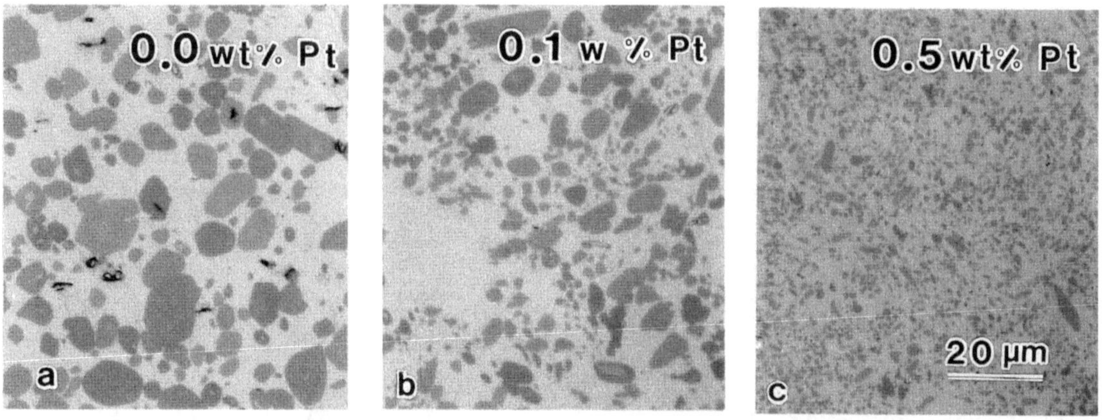

Fig.2 : Polarized-light optical micrographs of Pt-added QMG samples.

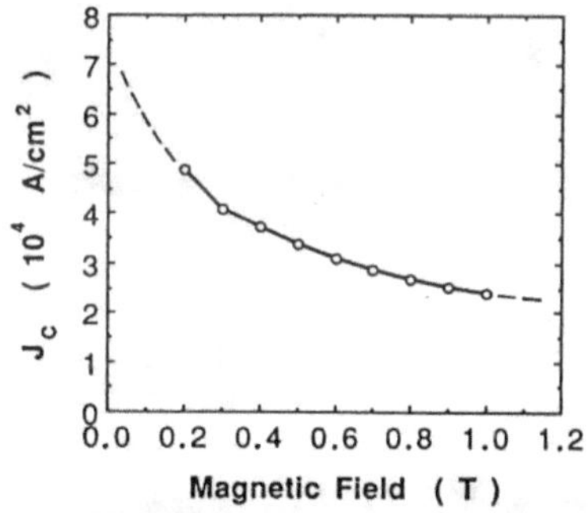

Fig.3: Magnetic field dependence of Jc at 77K

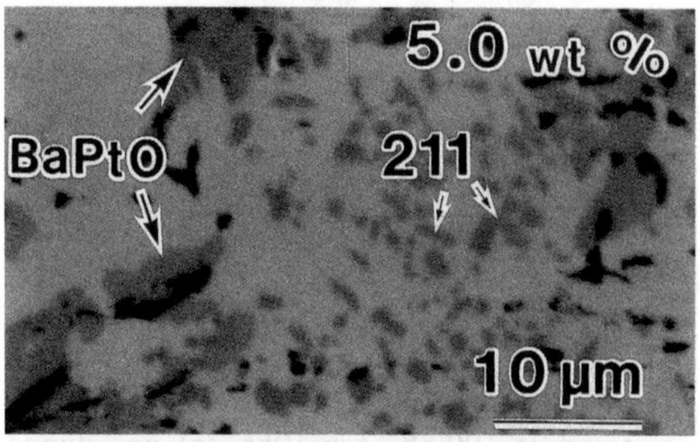

Fig.4 : Observation of BaPt-complex oxides in a 5 wt% Pt added sample

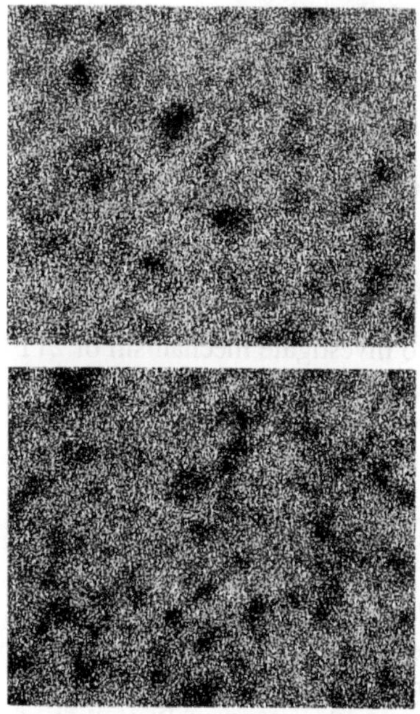

non-doped

1wt%Pt

(a) Sample kept for 5 minutes at 1112°C

Fig.5 : Compositional mapping of yttrium in materials quenched from the partial melt

pellets were quenched into water. Figure 5 shows the compositional mapping for yttrium element. In the case of the samples kept for 5 minutes at 1112°C, there were no differences in the distributions of yttrium between the non-doped sample and Pt doped sample, as shown in figure 5 (a). Y element is rather homogeneously distributed. Figure 5 (b) shows the compositional mapping for a sample kept for 10 minutes. For non- doped sample, yttrium element distributes like islands whose mean diameter is about 20 micrometer. It seems that these yttrium-rich islands correspond tothe coarsed 211 phases. In the case of the Pt doped sample, yttrium element is still homogeneous. For a keeping time longer than 10 minutes (20, 40 and 60 minutes), the results are almost the same. the experiments performed with non-dopped precursors clearly show that the fine 211 particles quickly coarsen to about 20 micron within BaCu-oxide liquid phase. Thus, we ofconclude that Pt element prevents the coarsening of the 211 phase.

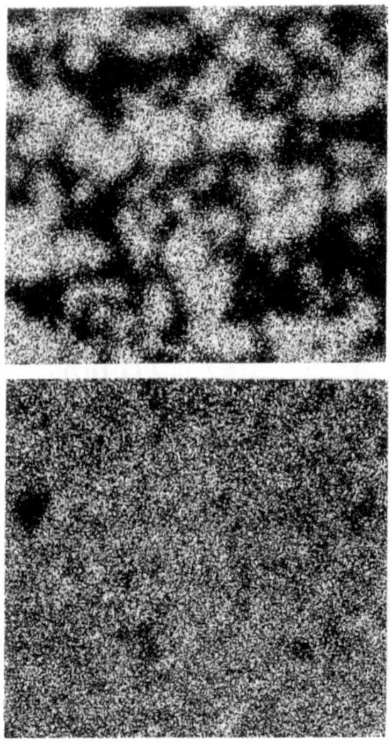

(b) Sample kept for 10 minutes at 1112°C

Furthermore, to investigate mechanism of 211 grain growth, we performed sintering of fine 211 phase. Five kind of powders (nondoped 211, Pt 1wt% doped 211, Pt 10 wt% doped 211 and also Pd 1wt% doped 211 and Pd 10wt% doped 211) were prepared. Initial particle size of 211 phase was 2 micron. PtO_2 and PdO powders with particle size of about 1 micrometer were used. Since additives may prevent sintering itself, Pd element was also adopted to compare the effect of Pt to that of Pd. These powders were pressed into pellets whose diameter is 12.5 mm. These pellets were heated up to 1000°C in 3hr, kept 10 hr and cooled to room temperature. Using this treatment, four pellets shrinked to 10.5 mm except the 10wt% Pt doped pellet for which the decrease of diameter was 1.0mm. It seems that the composition of 1wt% Pt was not high enough for any effect to be seen after sintering unlike within BaCu-oxide liquid phase as we have shown above. Figure 6 shows SEM micrographs for non-doped (a), 10wt% Pt (b) and 10wt% Pd (c). Figure 7 shows optical micrographs for 1wt% Pd (a) and 10wt% Pd (b) and 10wt% Pt (c) . If we suppose that additives remain in the pellets and change to

Fig. 6 : SEM photograph of 211 materials sintered at 1000°C for 10hr. (a) non-doped, (b) Pd 10%wt and (c) Pt 10%wt.

metal state, the ratio of volume fraction of additives, calculated from the density of the metals and initial compositions, are about 1%Pd:10%Pd:10%Pt = 1:10:5. From Figure 7, it seems that the volume fraction of metallic Pt in 10%Pt doped sample was very small compared to that of Pd doped sample. It is believed that Pt diffuses to the surface of the 211 particules. This would result in a change of surface energy of 211 phase, so that coarsening is prevented. So far however, we have not seen evidence whether platinum acts as a nucleation site for 211 phase or not.

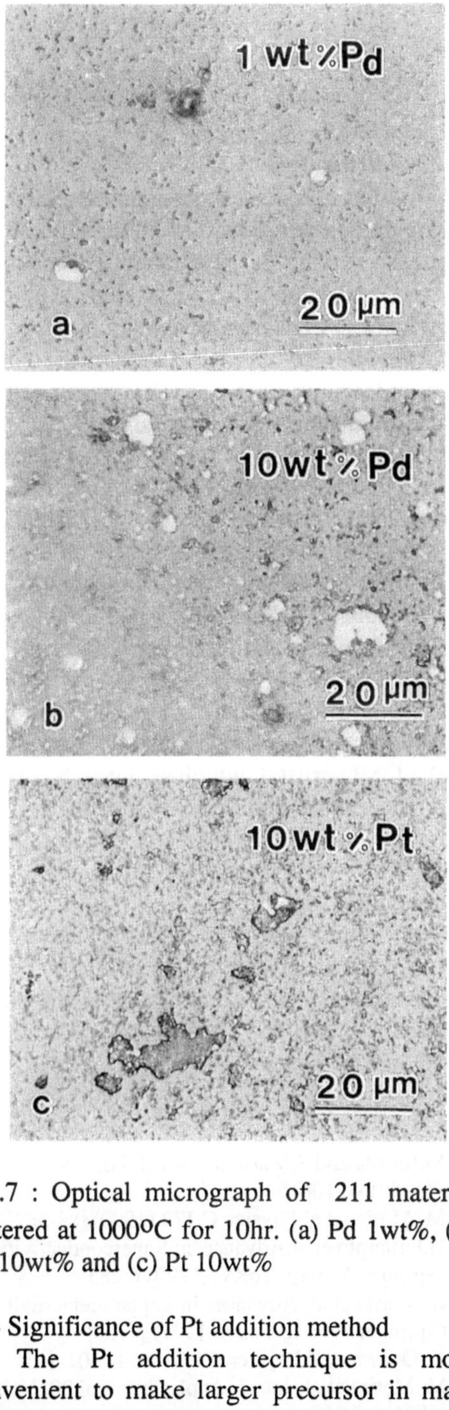

Fig.7 : Optical micrograph of 211 material sintered at 1000°C for 10hr. (a) Pd 1wt%, (b) Pd 10wt% and (c) Pt 10wt%

II.4 Significance of Pt addition method

The Pt addition technique is more convenient to make larger precursor in mass production than the previous melt-quench technique and allows to shape the precursor in an easier way. Obtainment of precursor by melt-quench using Pt crucible could be omitted. Thus QMG process can be simplified by the Pt addtion technique.

III. Properties of Bulk magnets by Pt addition technique[9]

III.1 Preparation of large QMG crystals

Large precursors were easily made by Pt addition technique. Large QMG crystals were prepared by seeding and RE compositional gradient techniques in the precursor. Figure 8 shows an enlarged QMG crystal which has no high angle grain boundary. Traces of facets were observed on the surface. On the basis of these traces, the c-axis of the crystal is normal to the surface.

Fig.8 : As-grown QMG crystal

III.2 Measurements of flux trapping and creep

Magnetic flux densities were measured at 63K, 77K and 84K. 77K is the boiling point of nitrogen under atmospheric pressure. 63K is the triplet point of nitrogen, which was made by evacuation of nitrogen into a vessel. 84K is the boiling point of nitrogen under a pressure of two atmosphere which was obtained by compression of nitrogen into the vessel.

Several Hall probes were attached on the sample. A field was applied, the maximum value of which is limited to be 1.75T due to the capacity of the magnet. Field cool was performed at the different temperatures mentioned above. Then, the external field was removed and flux densities were measured at four different locations from the center to the edge of the sample .

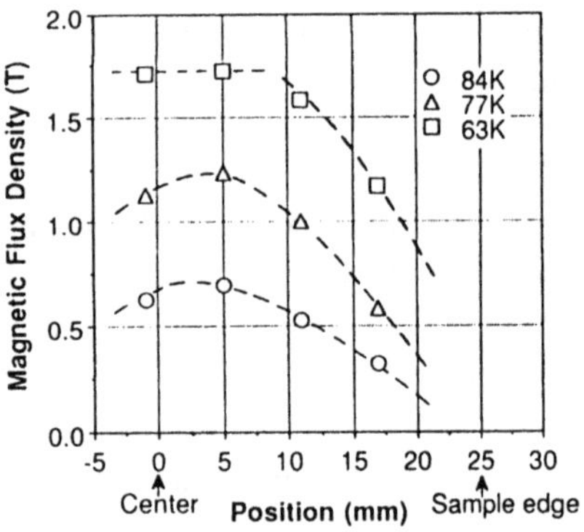

Fig.9: Distribution of trapped field in a QMG sample (50 mm in diameter and 24 mm high)

Figure 9 shows the distribution of magnetic flux densities measured 100s after external field was removed and for three different temperatures. This figure shows that the peak of the distribution does not coincide with the center of the sample. The QMG crystal may have cracks inside. At 84K and 77K, it seems that superconducting current flows in the whole sample according to the flux distribution. The data at 63K shows that around the center of the sample, the distribution of the measured field is flat and almost that of the field initially applied .It seems that the central region of the sample was not in critical state in this latter case only. Table 1 shows the average of flux creep rate (R) and apparent pinning potential (Uo*) determined at three different temperatures. R and Uo* were calculated from the time dependence of flux densities using the following formula :

$$B(t) = B(t_0)\,[1 - R\,\ln(t/t_0)]$$

$$Uo^* = kT / R$$

Here, we considered only the first and the second probe from the sample egde where critical state is achieved, even at 63K. Creep rate decreases with decreasing temperature. An increase of Uo* was also expected, however the values at 77K and 84K were almost the same. Uo*, as measured, is the apparent pinning potential at critical state. The true pinning potential should increase when the temperature decreases[10].

Table 1

Temperature(K)	84	77	63
R $\times 10^{-3}$	2.4	2.2	0.9
Uo*(eV)	0.3	0.3	0.6

IV. Conclusions

1) In QMG process, melt-quench treatment of the precursor produces a Pt contamination from the Pt crucible. Pt element is responsible for the fine dispersion of 211 phases (about 1 micrometer) even in a quantity as low as 0.5wt% Pt.

In order to investigate grain growth of 211 phase, change of microstructure of 211 particles was observed when the precursors were heat treated within BaCu-oxide liquid phase. Other experiments involving sintering of 211 phase were also performed. Both experiments have shown evidence that Pt element prevents coarsening of fine 211 phase

Melt-quench treatments using Pt crucible could be omitted. Hence, the preparation of the precursor was simplified by the Pt addition technique.

2) a QMG crystal, which size was 50mm in diameter and 24 mm high can trap high magnetic field exceeding 1T at 63K and 77K. Flux jump was not observed at these temperatures.

Flux creep rate was decreased by decreasing the temperature from 77K to 63K.

At present, the maximum value of trapped flux density of QMG crystal is 1.72T at 63K. and 1.27T at 77K.

References

1. M.Morita and S.Matsuda, NSMF NEWS (1988)No.10p15
2. M. Morita et al,Physica C **172** (1990)383-387
3. H.Fujimoto et al,Advancesin Superconductivity II (Springer Verlag, ToKyo,1990) p.285
4. M. Morita et al; Advances in Superconduc-tivity III (Springer Veriag,Tokyo,1991) p.733
5. N. Ogawa et al,Physica C177(1991) 101-105
6. M. Morita et al; Jpn. J. Appl. Phys. vol.30, No.5A (1991) p.L813
7. C.L.Teske and H.K.Muller-Buschbaum, Z.Naturtorsch **27b**(1972) 296.
8. K.Kimura et al, submitted to Proc.ISS'91
9. M Morita et al, submitted to Proc. ISS'91
10. M. R. Beasley, R. Labusch and W. W. Webb, Phys. rev., vol.181 (1969) No. 2, p.682-7

Microstructural Control of QMG Crystals

KEIICHI KIMURA, MITSURU MORITA, LIONEL TROUILLEUX, KIYOSHI SAWANO, KATSUYOSHI MIYAMOTO, and MISAO HASHIMOTO

Advanced Materials & Technology Research Laboratories, Nippon Steel Corporation, Kawasaki, Kanagawa, 211 Japan

ABSTRACT

We investigated the role of platinum in the dispersion of Y_2BaCuO_5 phase in the $YBa_2Cu_3O_x$ bulk superconductor. It was found that the addition of platinum suppressed the coarsening of the Y_2BaCuO_5 phase in Ba-Cu-O liquid. The Y_2BaCuO_5 phase grew very fast in the case that platinum was not contained. The addition of platinum and Y_2BaCuO_5 powder as the raw materials make it possible to obtain finely dispersed Y_2BaCuO_5 phase in the $YBa_2Cu_3O_x$ superconducting phase. By applying this result, the bulk superconductors with the critical current density higher than 3 x 10^4 A/cm^4 at 77K and 1T could be obtained.

KEY WORDS : $YBa_2Cu_3O_x$, Y_2BaCuO_5, melt -process, addition of platinum, coarsening

INTRODUCTION

QMG (quench and melt growth) process is one of the melt process [1]. In this technique, the superconducting $YBa_2Cu_3O_x$ phase are formed by the peritectic reaction from the region that Y_2BaCuO_5 solid phase and Ba-Cu-O liquid phase coexist. Now, a bulk superconductor larger than 20 cm^3 without any grain boundary can be obtained by the QMG process with the seeding technique [2]. The bulk superconductor made by QMG process has the high critical current density (*Jc*) whose value is 1-4 x 10^4 A/cm^2 at 77K and 1T [3].

The feature of QMG crystal is that Y_2BaCuO_5 phase is finely dispersed in the $YBa_2Cu_3O_x$ matrix. It is assumed that the Y_2BaCuO_5 phase plays an important role of increasing *Jc* since these phase suppress formation and/or extension of microcracks followed by tetragonal-orthorhombic transformation. Furthermore, it was suggested that the Y_2BaCuO_5 phase were pinning sites at the high temperature region [4]. At the beginning, we thought that the fine dispersion of Y_2BaCuO_5 phase was caused by the quenching as the preliminary heat treatment. But it was found that this was due to platinum contamination from the crucibles [5]. Independently, other group found the same using platinum and rhodium [6]. Therefore, it now becomes possible to obtain the finely dispersed Y_2BaCuO_5 phase without quenching treatment.

However, the mechanism of refinement of Y_2BaCuO_5 phase is not yet clear. There are mainly two reasons about the role of platinum. One is that platinum provides the nucleation sites for Y_2BaCuO_5 from Y_2O_3 when heating. And the other is that it suppress the growth of Y_2BaCuO_5 phase at the region of Y_2BaCuO_5 solid and Ba-Cu-rich liquid. Ogawa et al. showed that the doping of platinum caused heterogeneous nucleation of Y_2BaCuO_5 phase [6]. Furthermore, they suggested that $Ba_4CuPt_2O_9$ could be the nucleation sites of Y_2BaCuO_5 phase [7]. On the other hands, we found that platinum suppressed the coarsening of Y_2BaCuO_5 phase through this study.

EXPERIMENTAL

Y_2BaCuO_5, BaO_2 and CuO powders were used as the raw materials in order to avoid the nucleation of the Y_2BaCuO_5 phase from Y_2O_3. The median size of the Y_2BaCuO_5 powder was about 2μm. These materials were mixed so that the ratio of Y, Ba, Cu was 1.8 : 2.4 : 3.4. This contains 40 mol% excess Y_2BaCuO_5 over stoichiometric composition of $YBa_2Cu_3O_x$. Then they were formed into the pellets. These pellets were heated up to 1112°C within 5 min and quenched into the water after keeping there for 5 to 60 min. The specimens prepared by the above procedure were used as the reference. The specimen including 0.95 mass% PtO_2 were also prepared in order to elucidate platinum effect. The microstructure of these specimens were analyzed using SEM and EPMA.

Next, the size of Y_2BaCuO_5 powder as the raw material was changed. After forming into the pellets using the same conditions described above, the pellets were heated up to 1150°C and kept there for 1 hour. $YBa_2Cu_3O_x$ phase was grown by cooling slowly from 1000°C. The cooling rate was 1°C / hour. The seeding was performed at 1050°C to obtain the large crystals. The structures of the $YBa_2Cu_3O_x$ bulk superconductors formed after melt-growth were observed and the *Jc* was measured at 77K by VSM.

RESULTS AND DISCUSSION

As mentioned above, Y_2BaCuO_5 phase and Ba-Cu-rich liquid phase coexist at 1112°C. However it is difficult to observe this structure even when the specimen was quenched from this temperature since a part of the quenched specimen , original Ba-Cu-O liquid phase, is so reactive with the air. Therefore, it was taken the images of the composition by the back scattering electron and X-ray. The gold evaporated immediately after polishing the surface in order to avoid the reaction with the air. Figure 1 shows the back scattering electron images of the specimens kept for 1 hour at 1112°C. The white area was found to be Y_2BaCuO_5 phase by EPMA. It can be seen from this figure that the Y_2BaCuO_5 phase of the specimen without platinum (a) grew largely to compare with that of the specimen containing platinum (b).

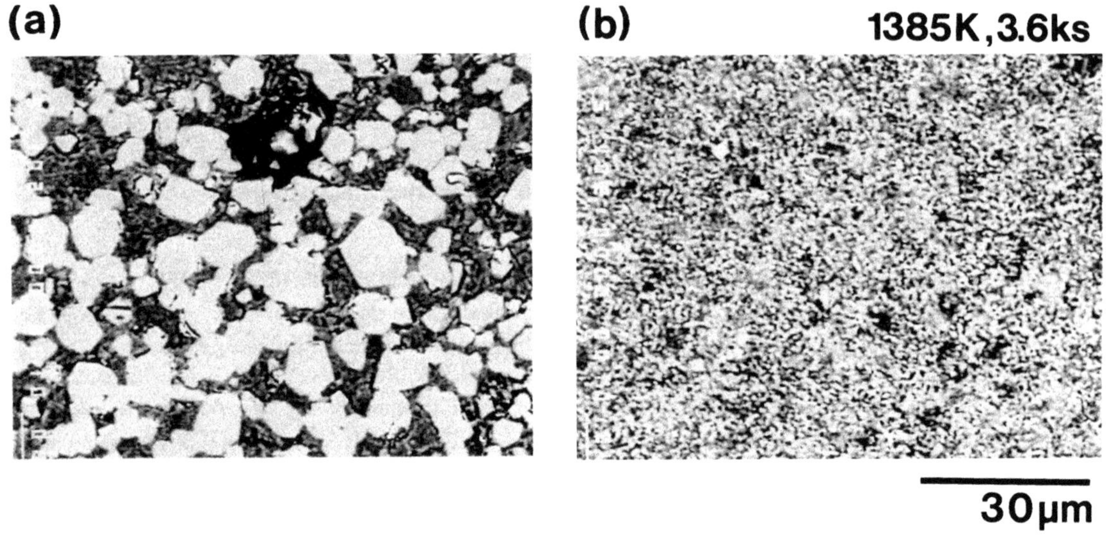

Fig.1 The back scattering electron images of the specimens quenched from the 1112°C after keeping there for 60 min. (a) platinum free, (b) 0.95mass% PtO_2

Figure 2 shows that X-ray images of yttrium in the specimens heat-treated at 1112°C for various time. The yttrium images were good indication to observe the distribution of the Y_2BaCuO_5 phase because yttrium existed mostly in the Y_2BaCuO_5 phase. Yttrium was distributed evenly in the both specimens when the keeping time was 5 min. This shows that Y_2BaCuO_5 phase dispersed finely. It was found that the Y_2BaCuO_5 phase grew rapidly in the specimen without platinum when the keeping time was 10 min. On the other hand, in the case of platinum addition, the Y_2BaCuO_5 phase was still fine even after 60 min. These observations suggest that the addition of platinum suppressed the coarsening of Y_2BaCuO_5 phase in the region where the Y_2BaCuO_5 solid and Ba-Cu-O liquid coexist.

It is assumed that the coarsening occurred simply by the coalescence of the Y2BaCuO5 particles , not by the diffusion through the liquid phase. This assumption could be experimentally verified by the fact that the

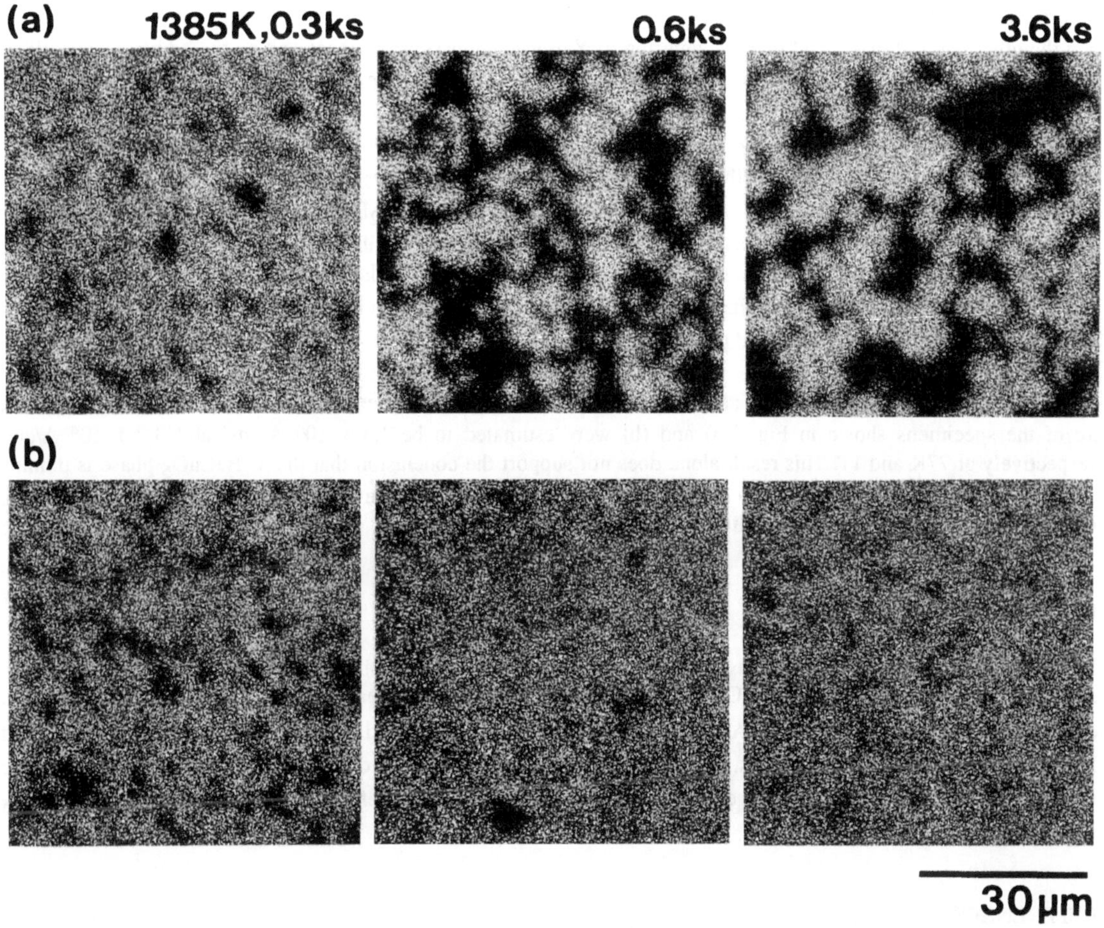

Fig.2 The X-ray images of yttrium of the specimens for the various keeping time at 1112°C. (a) platinum free, (b) 0.95mass% PtO_2

Fig.3 The optical micrograph of the $YBa_2Cu_3O_x$ made by melt process. Y_2BaCuO_5, BaO_2, CuO and PtO_2 powders were used as the starting materials. The median size of the Y_2BaCuO_5 phase was 1.5μm (a) and 6.5μm (b).

intermediate stage of the coalescence of two Y_2BaCuO_5 particles were observed in the specimen without platinum. On the other hand, it was found that the sintering of Y_2BaCuO_5 phase was suppressed by the addition of platinum [8]. This result is consistent with the assumption. However, the mechanism of the effect of platinum is not clear.

Next, the different size of Y_2BaCuO_5 powder were prepared as one of the starting materials in order to see the structure dependence after melt-growth on the size of the original Y_2BaCuO_5 powder. Figure 3 (a) and (b) show the microstructures of the $YBa_2Cu_3O_x$ made from the different size of Y_2BaCuO_5 powders, 1.5μm and 6.5μm respectively as the one of the raw material. It can be seen from these figures suggests that the condition of the dispersed Y_2BaCuO_5 phase in the $YBa_2Cu_3O_x$ matrix is depended on the distribution of the Y_2BaCuO_5 powder as the raw materials. It is assumed that the distribution of Y_2BaCuO_5 phase can be controlled during the QMG process by direct addition of Y_2BaCuO_5 and platinum.

The *Jc* value of these specimens was measured by VSM. The applied magnetic field was parallel to the c-axis. The *Jc* of the specimens shown in Fig. 3(a) and (b) were estimated to be 1.3 x 10^4 A/cm^2 and 3.2 x 10^4 A/cm^2 respectively at 77K and 1T. This result alone does not support the conclusion that the Y_2BaCuO_5 phase is pinning sites, because other factors, especially cracks, are not considered here. However, there is still possibility that the Jc could increase by applying these results.

CONCLUSION

We investigated the role of platinum in the melt(QMG) process. It was observed that the addition of platinum suppressed the coarsening of Y_2BaCuO_5 phase in Ba-Cu-O liquid phase. The bulk superconductors with the critical current densities higher than 3 x 10^4 A/cm^2 at 77K and 1T could be obtained by using Y_2BaCuO_5, BaO_2, CuO and PtO_2 powders as the starting materials. These results suggest the possibility to increase the *Jc* further. However, the more detailed studies are necessary to elucidate the mechanism of platinum effect on the suppression of Y_2BaCuO_5 coarsening.

REFERENCES

1) M.Morita, K.Miyamoto, K.Doi, M.Murakami, K.Sawano and S.Matsuda : Physica C,172(1990),383.
2) M.Morita, S.Takebayashi, M.Tanaka, K.Kimura, K.Miyamoto and K.Sawano : Proceedings of ISS'90,(1990),733.
3) S.Sawano, M/Morita, M.Tanaka, K.Kimura, S.Takebayashi, M.Kimura and K.Miyamoto : Proceedings of ISS'90,(1990),715.
4) T.Matsushita, E.S.Otabe, B.Ni, K.Kimura, M.Morita, M.Tanaka, M.Kimura, K.Miyamoto and S.Sawano: Jpn.J.Appl.Phys.,30(1991),L813
5) M.Morita, M.Tanaka, S.Takebayashi, K.Kimura, K.Miyamoto and K,Sawano : Jpn.J.Appl.Phys.,30(1991), L813.
6) N.Ogawa, I.Hirabayashi and S.Tanaka : Physica C ,177(1991),101.
7) Y.Yoshida, N.Ogawa, I.Hirabayashi and S.Tanaka "Effects of the platinum group element addition on preparation of YBCO superconductor by melt growth method" : Submitted to Physica C.
8) M.Morita, M.Tanaka, K.Kimura, T.Sasaki, L.Trouillex, S.Takebayashi, K.Miyamoto and M.Hashimoto, "Process and properties of QMG crystal : Submitted to Proceeding ISS'91,(1991).

Flux Penetration and Shielding Properties of QMG Crystals

T. SASAKI, M. TANAKA, M. MORITA, K. MIYAMOTO, K. KIMURA, S. TAKEBAYASHI, L. TROUILLEUX, I. ITO, M. HASHIMOTO, and K. SAWANO

Advanced Materials & Technology Research Laboratories, Nippon Steel Corporation, Kawasaki, Kanagawa, 211 Japan

ABSTRACT

Magnetic shielding and flux penetration characteristics of Quench and Melt Growth(QMG) processed Y-Ba-Cu-O crystals have been investigated at 77K and 4.2K. Overall critical current densities in "large single-grained" specimens were estimated to be about 10^4A/cm^2 based on the Bean model. Comparison between shielding characteristics of "singlegrain" and those of "polygrain" demonstrated that weak-linking large-angle tilt grain boundaries should be eliminated in order to apply YBCO bulk to high field shielding. Lorentz force which acts on specimens mechanically during shield measurements, flux creep, and flux jump are discussed.

KEY WORDS:Y-Ba-Cu-O superconductor, melt process, critical current density, magnetic shielding

INTRODUCTION

One of the most possible applications of high T_c oxide superconductors(HTSC) is high magnetic field shielding : the strength of the field is larger than lower critical field of superconductor, H_{c1}, and is ranging from about 0.1T to a few T or higher. It is obvious that critical current density(J_c) characteristics is important for superconducting high field shield as well as for other applications such as bulk magnet, bearing, current lead etc..

Since $YBa_2Cu_3O_{7-x}$ prepared by the modified Quench and Melt Growth(QMG) process is known to exhibit high J_c even at 77K and a high magnetic field[1], QMG crystals have a potential and are promising to be applied for practical superconducting shield operating at liquid nitrogen temperature, 77K. This high J_c value is, however, measured for small sample, for example, 2x5x1mm^3[1]. In order to examine the applicability of QMG crystals as magnetic shields, we should non-destructively evaluate overall J_c in large bulk, *e.g.*, 45mmϕ and 15mm thick, in order to predict shield effect.

For practical superconducting shield application, it is important to establish process and design technique. In this paper, we focus on the latter and discuss what should be taken into consideration. J_c estimation is one of them. The present paper discusses, (1) an estimation method of overall J_c from flux trapping, (2) effect of grain boundary, (3) flux creep which causes decrease of shield effect with time, (4) necessity of stabilization against flux jump, and (5) Lorentz force which acts on specimens mechanically during flux penetration and shielding measurements–it was observed for the first time.

EXPERIMENTAL

Disc-shaped samples were prepared by the modified QMG process[2]. The starting compositions of Y:Ba:Cu were chosen to include appropriate mol% Y_2BaCuO_5(211) inclusions in $YBa_2Cu_3O_{7-x}$(Y-123): 30 mol% 211 for both 50mmϕ and 23mm thick and 16mmϕ and 10mm thick specimens; 20 mol% 211 for 45mmϕ and 15mm thick ones. Platinum was added to precursors so as to have a weight fraction of 0.5%. Samples were unidirectionally grown to be single-grained(SG'ed) and *c*-axis perpendicular to the disc plane with using Sm-123 seed crystals, which have higher 123 formation temperature than Y-123. For comparison, non-seeded-processed samples, *i.e.* , polygrain(PG) ones were also prepared in order to observe flux motion through grain boundaries. Here we use the term "single grain"(SG) for convenience' sake, because QMG crystals are not single crystals but what have finely dispersed 211 inclusions and no high-angle grain boundary. The term "poly grain"(PG) denotes polycrystalline samples which have high-angle grain boundaries that can be easily seen.

The experimental procedures for flux penetration (FP) and magnetic shielding(SH) measurements are as follows. After zero-field-cooled (ZFC), magnetic field H was applied perpendicular to the disc plane by superconducting magnets or an Iron-cored electromagnet. For SG specimens, it means that applied field was along the c-axis. Magnetic flux density at the surface B_s or inside the cylinder B_i was measured using Hall probes. Detected flux density was the component parallel to the applied field. Since the difference between FP and SH is whether shielded free space is the subject of the observation or not [3,4], here measurement for disc without hole is called FP; that for disc with hole SH, . A hole was made at the center of the disc by grinding. Small samples (*ex.* 3x3x1mm^3) for magnetization measurement using a vibrating sample magnetometer(VSM) were cut from the large crystals after FP or SH experiments.

RESULTS AND DISCUSSION

ESTIMATION OF J_C FOR LARGE CRYSTAL

Figure 1 shows flux density profiles measured at the surface and 77K after various hystereses. Enlarging SG sample leads to high magnetic field trapping. The shape of flux-trapping profile for 45mmϕ after field-cooled(FC) indicates that super-current rotatores in the whole of the sample. Though the profile for FP at 1T means the same, it is not clear for FP whether all part of the sample is in mixed state or not, *i.e.* it is not sure that super-current flows in the whole of the sample. We, therefore, adopt trapping profiles after FC to estimate overall J_c. Assuming constant current density J(φ component) as in Bean model[5], conic profile is obtained(Fig. 2). By fitting the value of the center, $B_{surf,c}$, J_c and/or decayed current density can be estimated using Biot-Savart law from

$$B_{surf,c}=\mu_0 J\, D \ln \frac{1+\sqrt{1+4x^2}}{2x} \quad (1),$$

where $2D$ is thickness, x is D/R and R is radius. For the sample, which trapped 1.1T right after external field was reduced to 0, whose size is R=22.5mm and D=15mm, J_c=9800A/cm^2 was calculated. However, the smaller sample cut from the same sample has J_c about 2x10^{4}A/cm^2 at 77K and 1T. This value is higher than the estimated overall J_c. This tendency was seen for other samples and is probably attributed to inhomogeneities and/or weak links which are more likely to be present in the large bulk than in the small one.

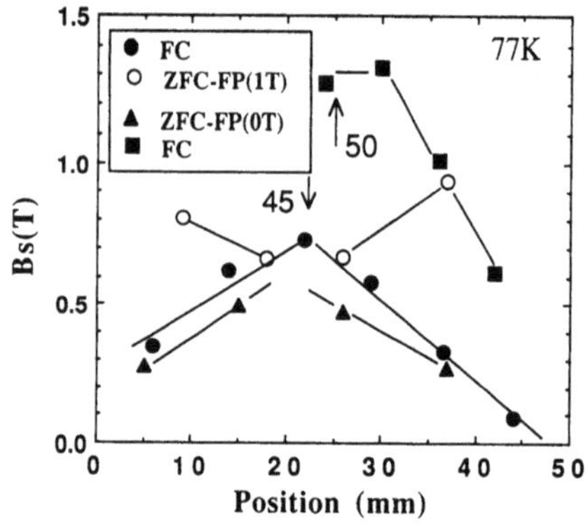

Fig. 1. Flux density profiles for SG discs at the surface and 77K. Center of a disc is indicated by an arrow with corresponding diameter(in mm):the closed circles for trapped flux by 45mmϕ disc after FC; profiles of FP when 1T was applied and field was reduced to 0 are for ZFC 45mmϕ sample; the closed squares for 50mmϕ sample. For FC profiles, 10^{3}sec passed after field was reduced to 0.

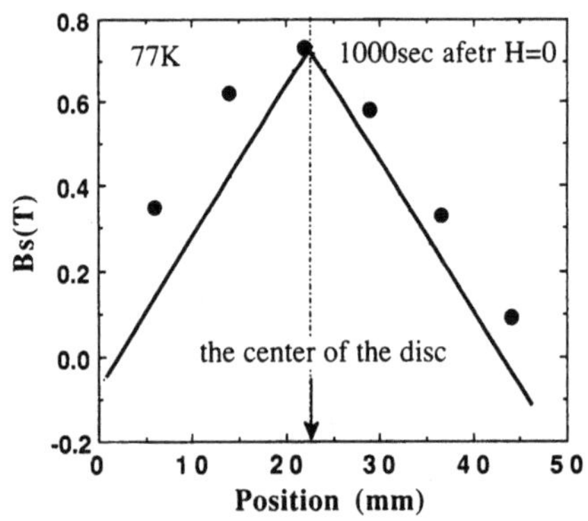

Fig. 2. Measured(points) and calculated(line) flux density profiles for 45mmϕ and 15mm thick SG disc at the surface and 77K. 10^{3}sec passed after field was reduced to 0. Constant current density, 6.1x10^3 A/cm^2 was assumed to fit the value of the center.

FLUX PENETRATION AND SHIELDING

SH curves are compared between SG specimen(16mmϕ, 20mm high with 7mmϕ hole) and PG one (45mmϕ, 15mm high with 10mmϕ hole) at 77K in Fig. 3. In the case of SG, B_i takes off at about 0.2T, *i.e.*, external field was shielded by SG cylinder up to 0.2T with high efficiency. PG sample shows much less shield effect compared to SG, because flux can easily penetrate into the inside of the cylinder through grain boundaries.

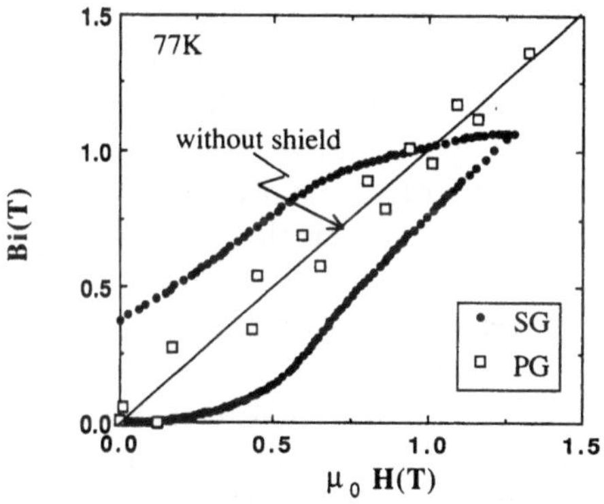

Fig. 3. Comparison with magnetic shielding curves at 77K for 16mmϕ SG cylinder(closed circles) and 45mmϕ PG (open squares). Measured point was the center of the hole for each sample. The straight line shows the behavior in the normal state.

Polycrystalline HTSC is, however, possibly suitable for weak field shielding application, such as shield for SQUID and/or biomagnetic measurement. According to the shielding measurements for the polycrystal samples reported before [6, 7, 8], HTSC could shield no more than 0.01T external field. For high field shielding, polycrystal is not useful but "single grain" as demonstrated in Fig. 3 since high-angle grain boundaries act as weak link.

Figure 4 shows flux jumps observed at 4.2K caused by thermal instabilities. In our measurement, QMG crystals are stable at 77K. Based on adiabatic theory of flux jumping, critical sizes were calculated to be 1.2×10^{-1}m and 7.0×10^{-4}m at 77K($J_c=10^4$A/cm^2) and at 4.2k($J_c=10^5$A/cm^2), respectively, using cryophysical property data tabulated in ref.[9]. QMG crystals were stable at 77K considering the sizes and the sweep rates (about 10mT/s) we have dealt with. The conditions fulfilled the adiabatic criterion.

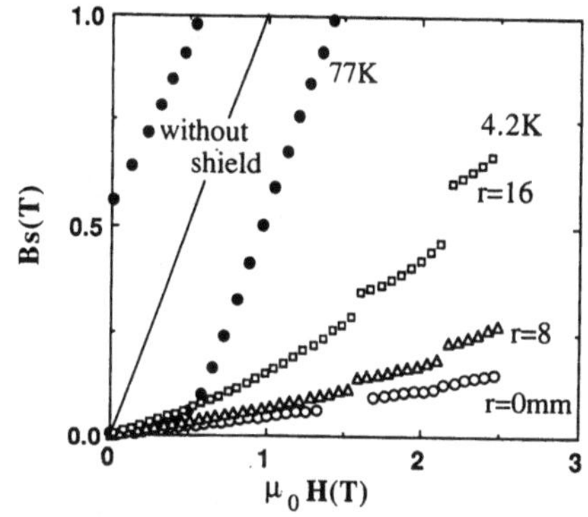

Fig. 4. Flux penetration curves for 45mmϕ and 15mm thick SG disc at 4.2K(open symbols) and at 77K(closed symbols). Distances from the center was represented by "*r*" for data at 4.2K. Note that flux jump was observed only at 4.2K.

At the same experiment where flux jumps occurred at 4.2K, the sample was broken when field was raised to about 6T probably due to Lorentz force(Fig. 5). Such force which acts on samples mechanically was also observed during SH and FP measurements at 77K. This phenomenon was observed for the first time and probably caused by asymmetric flux penetration due to flux jumps and/or inhomogeneities of the sample, because axisymmetric penetration couldn't cause such force[10].

Another problem for designing high field shield is flux creep. Although flux creep rate can be reduced to the level below the detection limit of Hall probe[11], flux creep should be still taken into consideration when it is difficult to adopt methods in ref. [11]. Figure 6 shows the flux density dependence of pinning potential, U_0, and U_0^* obtained as below. Assuming constant current density J rotatorying, U_0^* was calculated from

$$B_{surf,c}=B_0\left[1-\frac{k_BT}{U_0^*}\ln\frac{t}{t_0}\right] \quad (2),$$

where B_0 and t_0 denote flux density and time at which logarithmic relaxation started, respectively; k_B is Boltzmann constant. U_0 increases as B_0 increases as expected from a case of the theory[12] where the sharing correlation length is much larger than a half of

Fig. 5. Broken sample after an external field of about 6T was applied at 4.2K. Note that GFRP sample holder was broken at the same time. It indicates that Lorentz force acted on the sample mechanically.

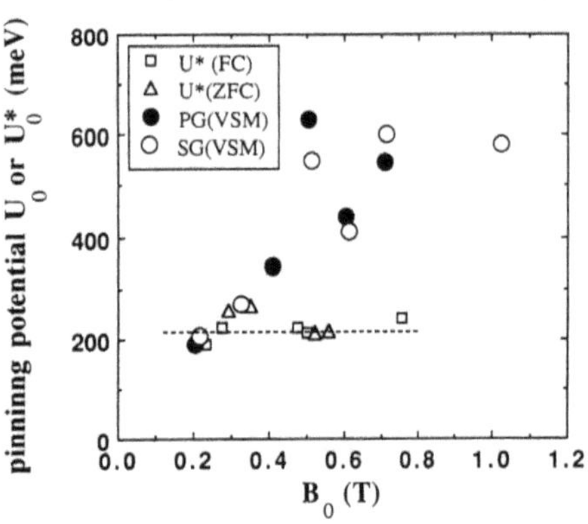

Fig.6. Pinning potential(U_0,circles) and apparent pinning potential(U_0*) plotted against B_0 at which logarithmic decay started.

the fluxoid spacing. U_0* shows much smaller change compared to U_0 and this means that J, flowing rotatively in whole the sample, decays at nearly the same rate. From the phenomenonlogical point of view, this B_0 dependence of U_0* support the constant current density model for estimating overall J, in other words, Bean type model is considered to be fairly good as the first approximation.

CONCLUSIONS

In summary, magnetic shielding and flux penetration characteristics of QMG crystals have been investigated at 77K and 4.2K. Overall critical current density(J_c) of large QMG crystal, whose size was in the magnitude of 10^4mm^3, was estimated to be about 10^4A/cm^2 by constant current density model. Improvement of process to obtain bulk with higher overall Jc is still necessary because the values of overall Jc for large bulk were less than those of small samples. The pinning potential U_0* obtained from flux penetration and shielding measurements was about 0.2eV and showed smaller change compared to the pinning potential U_0. This may support the constant current density model.

Single grain cylinder could shield inside up to about 0.2T external field at 77K with high efficiency. Comparison between singlegrain and polygrain demonstrated that weak-linking high-angle grain boundaries should be eliminated in order to apply YBCO bulk as a high field magnetic shield. Since flux jump was observed only at 4.2k and never at 77k, it is not necessary to stabilize considering the size and the sweep rate we have dealt with. Lorentz force which acted on the specimens mechanically during flux penetration and shielding measurements was observed for the first time and considered to be caused by asymmetric flux penetration. Therefore, not only flux creep but also mechanical Lorentz force should be taken into consideration in high field shield designing.

ACKNOWLEDGEMENT

We would like to thank Prof. T. Matsushita of Kyushu Institute of Technology for fruitful discussion. One of the authors(TS) would like to thank Mr. Iwata for his assistance in A-φ method numerical calculation.

REFERENCES

1. M.Morita et al.,Physica C**172**(1990)383
2. M.Morita et al.,Jpn.J.Appl.Phys.**30**(1991)L813
3. R.B.Harrison et al.,Can.J.Phys.**52**(1974)1107
4. R.B.Harrison et al.,J.Appl.Phys.**45**(1974)403
5. C.P.Bean,Phys.Rev.Lett.**8**(1962)250
6. T.Okada et al.,IEEE Trans. Mag.**25**(1989)2270
7. J.W.Purpura and T.R.Clem,ibid. 2506
8. O.G.Symko et al.,J.Appl.Phys.**65**(1989)2142
9. E.W.Collings,Advances in Superconductivity II (Springer-Verlag,1989)p.327
10. T.Sasaki et al.,submitted to Jpn.J.Appl.Phys.
11. M.Morita et al.,submitted to Proc. ISS'91
12. K.Yamafuji et al.,Physica C**159**(1989)743

Magnets Made of QMG Crystals

MITSURU MORITA, MASAMOTO TANAKA, TSUTOMU SASAKI, MISAO HASHIMOTO, and KIYOSHI SAWANO

Advanced Materials & Technology Research Laboratories, Nippon Steel Corporation, Kawasaki, Kanagawa, 211 Japan

Abstract
Magnetic flux density trapped in large QMG (Quench and Melt Growth) crystal was measured at 85K, 77K and 63K. QMG crystals which had no high-angle grain boundaries were made. A sample which size was 50mm in diameter 24mm thick was used. Distribution of magnetic flux density was measured on the suface of the QMG crystal using several Hall probes. Field cooling was performed to investigate trapping of the magnetic field. Temperature of coolant was varied by controlling the pressure over the nitrogen vessel. At 63K, flux density was roughly double of that at 77K. Maximum value of trapped field was 1.72T. Flux creep rates were $2.4x10^{-3}$, $2.2x10^{-3}$ and $0.9x10^{-3}$ at 84K, 77K and 63K respectively. Since flux creep is still not negligeable at these high temperatures, we tried to prevent it by cooling the QMG material down to 77K after trapping at 84K and also cooling down to 63K after trapping at 77K the flux allows to save a lot of time for the field stability to be achieved.
KEY WORDS: melt-process, bulk magnet, flux trapping, triplet point

1.Introduction

It is known that the QMG process can produce a bulk $YBa_2Cu_3O_{7-x}$ superconductor with high critical current density (Jc) even at 77K and in high magnetic fields[1]. The QMG crystal is a pseudo single crystal which contains a finely dispersed Y_2BaCuO_5 phases in the $YBa_2Cu_3O_{7-x}$ matrix. Because of excellent Jc properties, QMG crystals can be used as bulk superconducting magnets, superconducting bearings, current leads, and so on.

Large QMG crystals are able to be used as bulk magnets which produce high magnetic fields, because of high Jc properties around the boiling point of nitrogen[2]. It is known that QMG bulk magnets can trap a magnetic field exceeding 1T at 77K and 63K[3]. This latter temperature can conveniently be achieved by controlling the pressure over the coolant. In this paper, authors report that magnetic flux trapping is increased by decreasing the temperature down to 63K and the associated flux creep is decreased. Furthermore, we present a convenient way to prevent flux creep at 77K and 63K which allows a good stability of trapped flux .

2. Experimental

2.1 Sample preparation

A mixed Y_2O_3, $BaCuO_2$ and CuO powder containing 0.5wt% Pt powder was uniaxially pressed by dies[4]. Samples of two different sizes were formed. One is 50mm in diameter and 17mm high . The other one is 60mm in diameter and 25 mm high. These precursors were heated at 1100°C for 30 minutes. Seeding was performed at 1030°C by using Sm based QMG crystals[5]. QMG crystal grew in the precursor by slow cooling from 1000°C to 970°C. Using this heat treatment, the samples shrinked to 43mm in diameter and 15mm high for the first one and 50mm in diameter and 24mm high for the second one. Subsequently, the sample were slowly cooled from 600°C to 300°C in one atmosphere of pure O_2. Fig.1 shows a QMG crystal which has no high angle grain boundary. Traces of facets were observed on the surface. On the basis of this traces the c-axis of the crystal is normal to the surface.

Fig.1 Photograph of an as-grown QMG crystal

2.2 Measurements of flux trapping and creep

Magnetic flux densities were measured at 63K, 77K and 84K. 77K is the boiling point of nitrogen under atmospheric pressure. 63K is the triplet point of nitrogen, which was reached by evacuating nitrogen in the vessel as shown in fig.2. 84K is the boiling point of nitrogen under a pressure of two atmosphere which was obtained by compression of nitrogen in the vessel.

Several Hall probes were attached on the sample. First, an external magnetic field was applied and the sample was cooled to the different temperatures mentioned above. Maximum value of applied field was limited to 1.75T due to the property of the magnet. Then, the external field was removed and flux densities were measured at four different locations from the center to the edge of the sample.

3. Results and discussion

3.1 Flux trapping and creep

Fig.3 shows the distribution of magnetic flux densities measured 100s after external field was removed and for three different temperatures.

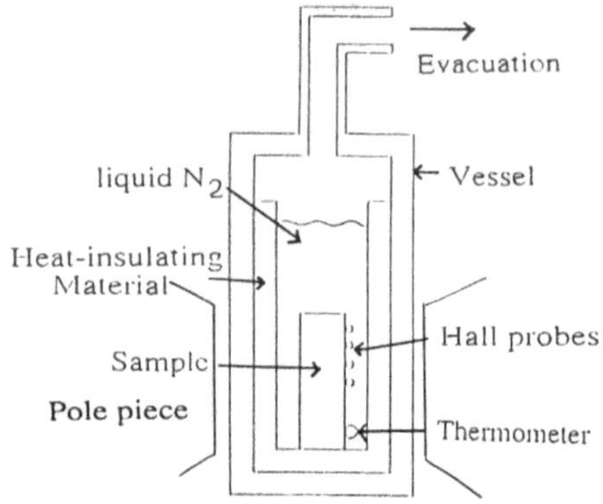

Fig.2 Schematical explanation of experimental device

This figure shows that the peak of the distribution does not coincide with the center of the sample. The QMG crystal may have cracks inside. At 84K and 77K, it seems that superconducting current flows in the whole sample according to the flux distribution. The data at 63K shows that the distribution of the measured field is flat around the center of the sample and almost that of the field initially applied . It seems that the central region of the sample was not in critical state at 63K only. If initial applied field were high enough, the sample would certainly trap a field higher than 2T.

To verify the change of flux distribution at 63K compared to the one measured at 77K, a similar experiment was performed with a smaller sample, for which the critical state was easily achieved. Fig.4 shows that trapped flux densities at 63K were roughly two times higher than the ones measured at 77K.

Assuming that Jc obeys the following formula above 60K in QMG materials[6], that is:

$$Jc(T) \propto [1 - (T/Tc)^2]^{1.5}$$

we can calculate the ratio of Jc(T) to Jc(77). Hence, the ratio Jc(84) : Jc(77) : Jc(63) are 0.44 :1 : 2.39. We conclude that the trapped flux densities measured here (figure 3 and 4) agree reasonably with this data.

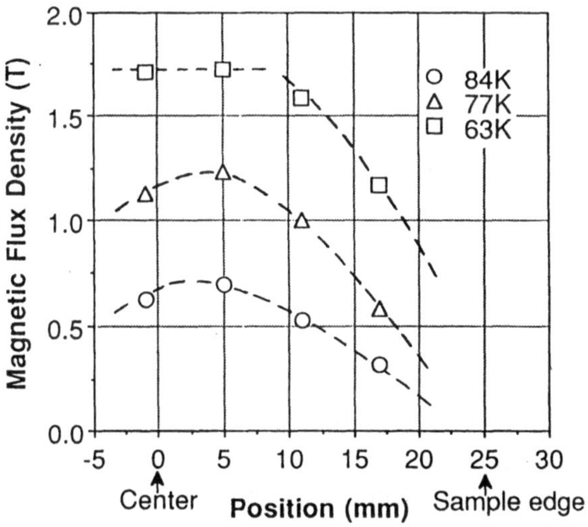

Fig.3 Distribution of trapped field in a QMG sample (50 mm in diameter and 24 mm high)

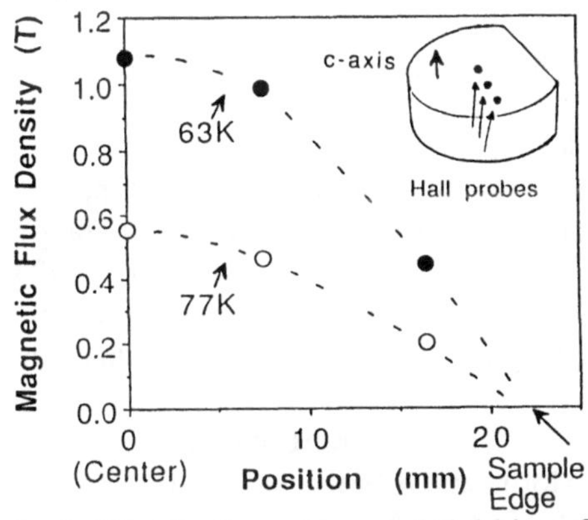

Fig.4 Distribution of trapped field in a QMG sample (43 mm in diameter and 15 mm high)

The advantage of cooling at 63K is pronounced. At such a temperature a higher magnetic field can be trapped by a bigger QMG crystal. Fig.5 shows the field dependence of Jc at 60K, 70K and 77K, as previously determined [7]. At 77K, Jc decreases rapidly over 5T. On the other hand, Jc does not decrease at 60K for a field up to at least 10T. If a QMG crystal is mechanically strong enough, a bulk magnet which trappes high magnetic field exceeding 10T can be made.

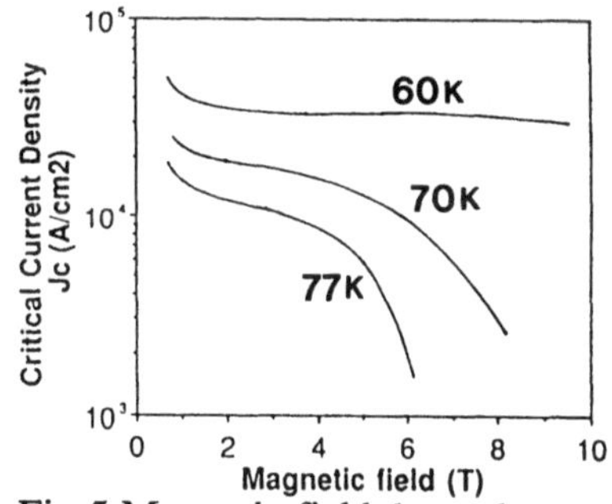

Fig.5 Magnetic field dependence of Jc in QMG materials [8]

3.2 Flux creep

Table 1 shows average of flux creep rate (R) and apparent pinning potential (Uo*) determined at three different temperatures. R and Uo* were calculated from the time dependence of flux densities using the following formula :

$$B(t) = B(t_0)\,(1 - R \ln(t/t_0))$$

$$Uo^* = kT / R$$

Here, we considered only the first and the second probe from the sample egde where critical state is achieved , even at a temperature as low as 63K. Creep rate decreases with decreasing temperature. An increase of Uo*was also expected, however the values at 77K and 84K were almost the same. Uo*, as measured, is the apparent pinning potential at critical state. The true pinning potential should increase when the temperature decreases [8].

Table 1

Temperature(K)	84	77	63
R $\times 10^{-3}$	2.4	2.2	0.9
Uo*(eV)	0.3	0.3	0.6

3.3 Prevention of flux creep

At relatively high temperatures, above 60K, the trapped flux sensibly decreases with time in critical state. In order to prevent such a flux creep, we first cooled a sample down to the temperature T1, where the critical state is established by applying a field of sufficient strenght. Then, the sample is cooled to the temperature T_2. A schematic explanation is shown in fig.6. Fig.6(a) shows the flux trapped at a temperatureT_1, where the pinning centers are associated with a pinning potential U_1. Cooling down to the temperature T_2 from temperature T_1 allows the remaining flux to be trapped in a dipper pinning potential U_2 as shown in fig.6(b). This procedure allows to get a stable trapped field in a very short time.

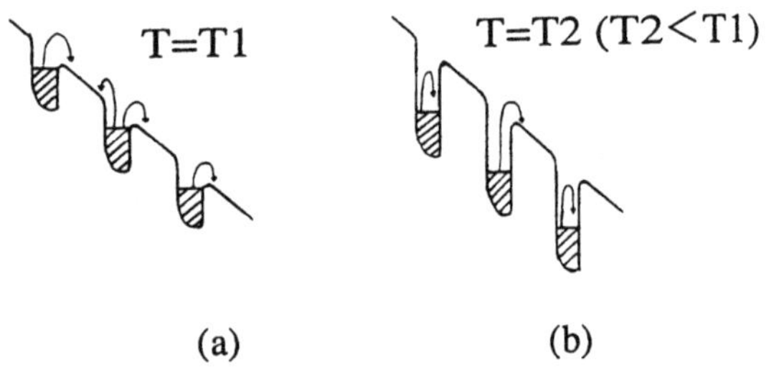

Fig.6 (a) trapped flux at T1 where a high magnetization decay rate is observed and (b) remaining flux at T2 after cooling from T1.

This type of experiment was performed at 77K and 63K after an initial trapping at 84K and 77K respectively.In both cases, after the field is removed, the samples are cooled down as soon and fast as possible because the magnetization decay is high. The results of flux creep prevention by this method are presented in figure 7. Also shown for measured at comparision are the magnetization decay 84K and 77K using the third probe from the sample edge. When cooling down further after trapping, flux creep is no more observed, within experimental uncertainty, at either temperatures. This procedure allows to get a great stability of the trapped field.

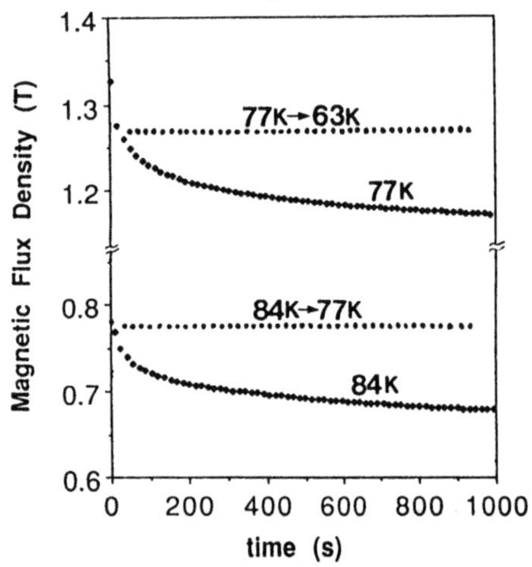

Fig.7 Prevention of flux creep at 77K and 63K after an initial trapping at 84K and 77K respectively.

4. Conclusions

The temperature of liquid nitrogen was easily controlled by changing the pressure over the vessel. At 63K (triplet point), 77K (boiling point under 1atm) and 84K (boiling point under 2atm) flux trappping and creep were measured for a large QMG crystal.

1) The ratio of trapped flux density at each temperature was approximatively B(84K) :B(77K) : B(63K) = 2 : 1 : 0.5.
2) The ratio of creep rate (R) was R(84K) : R(77K) : R(63K)= 24 : 22 :9
3) Flux creep can be prevented by decreasing the temperature of the coolant after the flux is being trapped.

At present, the maximum value of trapped flux density in a QMG crystal is 1.72T. However, larger QMG crystal will be used as a bulk magnet which can hold more than few T.

References

1. M. Morita and S.Matsuda, NSMF NEWS (1988)No.10 p15
2. M. Morita et al;Physica C172 (1990) p.383-387
2. Stefan L. Wipf and Henry L. Laquer; 1989 Applied Superconductivity Conference, paper LN-10. to be published in IEE Trans, MAG-25(1989) p.1877
3. M. Morita et al,(1991)Proceeding of M2-s, kanazawa,PhysicaC(to be published)
4. M. Morita et al; Jpn. J. Appl. Phys. vol.30, No.5A (1991) p.L813
5. M.Morita et al; Advances in Superconduc-tivityIII(Springer-Veriag.Tokyo,1991)p.733
6. T. Matsushita et al, Jpn. J. Appl. Phys. vol. 30, No.3A (1991) p.L342
7. M. Morita et al, Jounal of the cryogenic Soiety of Japan Vol.25,No2(1990) p40
8. M. R. Beasley, R. Labusch and W. W. Webb, Phys. rev., vol.181, No .2 (1969) p.682-700

A Comparison of Melt Processes to Prepare YBaCuO with High J_c

M. MURAKAMI, A. KONDOH, H. FUJIMOTO, K. TAKAMUKU, N. NAKAMURA, and N. KOSHIZUKA

Superconductivity Research Laboratory, ISTEC, Koto-ku, Tokyo, 135 Japan

Abstract

There are several melt processes which promise high J_c in bulk YBaCuO. A comparison of various melt processes such as MTG and MPMG was performed in this study, and it was found that size control of the Y_2BaCuO_5 inclusions is the key to the improvement in J_c values. Large J_c values in the MPMG YBaCuO are attributed to the size reduction of the 211 inclusions.

Keywords: melt process, YBaCuO, microstructure, critical current density

1. Introduction

It has been reported that large J_c values are achievable in melt processed YBaCuO. Jin et al.[1] proposed that well oriented structure is responsible for large J_c values. However, it is known that the introduction of pinning centers are required to obtain large J_c in a magnetic field. Murakami et al. [2] suggested that the Y_2BaCuO_5 (211) inclusions can act as pinning centers in melt grown YBaCuO, although their effectiveness is still controvertial[3,4]. According to transmission electron mciroscopic observation, melt grown YBaCuO crystals contain various defects such as dislocations, stacking faults, and microcracks. Since any inhomogeneity can work as a pinning site, these defects can contribute to flux pinning. Twin planes are also candidates for pinning centers.

In order to find which defect is a dominant pinning center, we need to compare J_c values among melt processed YBaCuO samples whose microstructure is well characterized. It should also be noted that J_c values are dependent on the crystal orientation due to a large anisotropy of superconducting parameters in oxide superconductors. This anisotropy complicates the analysis and occasionally leads to misunderstanding.

It is also important to realize that J_c values strongly depend on the method to measure J_c and the field direction. In most cases, J_c values for the same sample are in the order of transport J_c (H//*ab* plane) >> magnetization J_c (H//*c* axis) = transport J_c (H// *c* axis) >> magnetization J_c (H//*ab* plane) > transport J_c along the *c* axis. In many papers, these J_c values are mixed up and compared without care, which causes great confusion.

As proposed by Murakami[5], it is critical to use transport or magnetization J_c values with H//*c* axis for making clear the pinning effect of the 211 inclusions. According to the pinning model due to the inclusions, J_c values are proportional to V_f/d where V_f is the volume fraction and d is the size of the inclusions. Recent analysis has demonstrated that J_c values (H//*c* axis) in melt grown YBaCuO agreed very well with the above relation, and now it is generally accepted that the key to the improvement in J_c values for melt grown YBaCuO is the size control of the 211 inclusions. Ogawa et al.[6] have found that the size of the 211 can be reduced by small addition of Pt in MTG processed YBaCuO and have succeeded in raising J_c values.

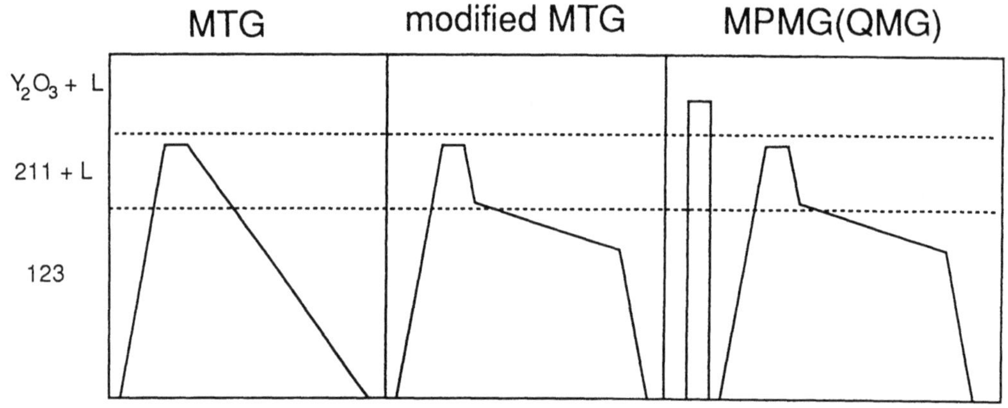

Fig. 1. Schematic illustration of melt processes used in this study. For all processes, a heating condition in the 211 + L region was 1100°C for 20 minutes. For comparison, the temperature was reduced down to 1030°C

In this paper, we observed microstructure of YBaCuO samples prepared by several melt processes with an emphasis placed on the size and distribution of the 211 inclusions.

2. Experimental

YBaCuO samples were prepared by MTG (Jin's process), modified MTG (Salama's process[7]) and MPMG processes. The processes are schematically illustrated in Fig. 1. Although Jin and Salama started with the composition of Y:Ba:Cu = 1:2:3, we started with 1.8:2.4:3.4 (Y1.8) in order to disperse the 211 inclusions. The powders of Y_2O_3, $BaCO_3$ and CuO were mixed and calcined at 900°C for 24 hours and then subjected to the melt processes. An addition of Pt powders was also conducted for some samples.

Microstructure was observed using a polarized optical microscope. J_c values were obtained from magnetization hysteresis using the relationship: $J_c = 2(M^+ - M^-)/d$. As mentioned above, a comparison was made for J_c values with the field parallel to the c axis.

3. Results and discussion

Figure 2 shows typical microstructures of modified MTG and MPMG processed Y1.8. The J_c values(H//c axis) are 5000 and 25000 A/cm^2 at 77K and 1T, respectively. It is clear that J_c becomes larger as the size of the 211 inclusions becomes smaller. It should also be noted that the shape of the inclusions is sphere-like in the MTG processes, while it is needle-like in the MPMG process.

Here we can raise the questions: if it is possible to reduce the size of the 211 inclusions in MTG process by modifying the processing conditions; and what is the beneficial effect of the MPMG process in reducing the size of the 211 inclusions.

Figure 3 shows the microstructure of the modified MTG Y1.8 with the maximum temperature being 1030°C. Apparently, the size of the 211 inclusions can be reduced and a J_c value can be increased from 5000 to 8000 A/cm^2 at 77K and 1T. We have found that as we reduced the temperature in the 211+L region, the size of the 211 could be reduced. However, the density and the grain size were also reduced, which will be harmful for practical applications.

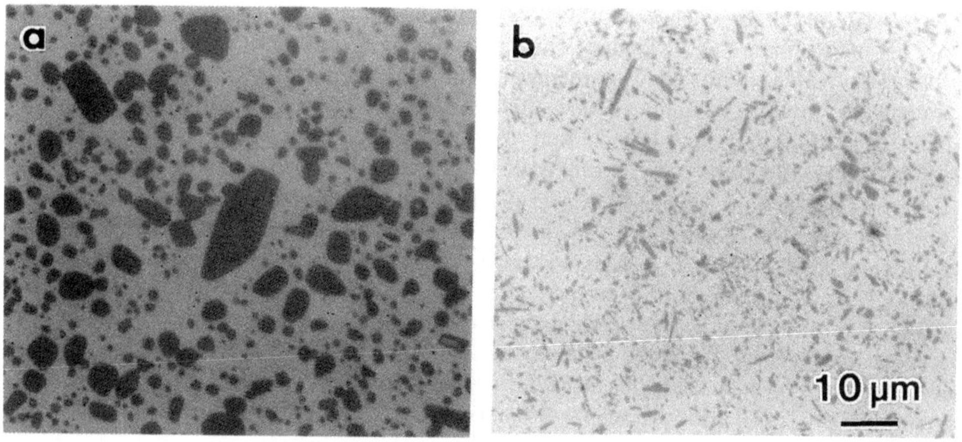

Fig. 2. Optical micrographs of the Y1.8 samples prepared by (a) modified MTG and (b) MPMG processes. Note that the size of the 211 inclusions are much smaller in the MPMG processed sample. J_c values (77K and 1T with H//c axis) are 5000 A/cm^2 and 25000 A/cm^2 for modified MTG Y1.8 and MPMG Y1.8, respectively.

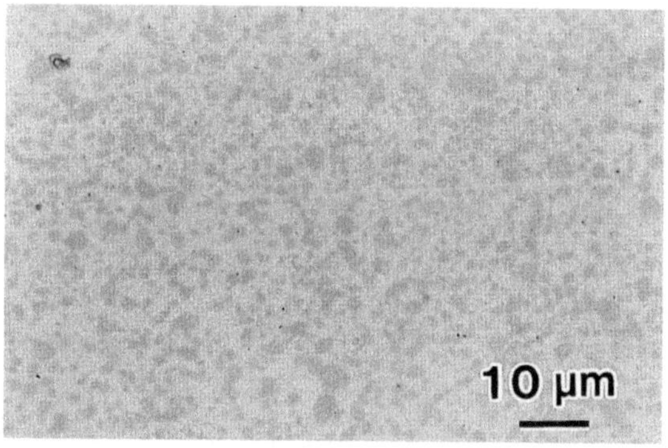

Fig. 3. Optical micrographs of the modified MTG Y1.8. The maximum temperature reached was lowered down to 1030°C. Compared to the sample heated to 1100°C (see Fig. 2 (a)), the size of the 211 inclusions is smaller.

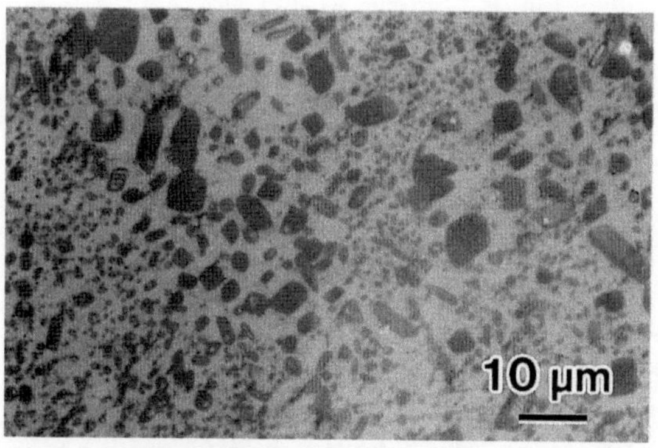

Fig. 4. Optical micrograph of the MPMG Y1.8 which was processed without using a Pt crucible. Note that the size of the 211 inclusions is not as small as the sample fabricated using a Pt crucible, although the size is small compared with the MTG processed Y1.8.

As first reported by Ogawa et al.[6], Pt addition is also effective in reducing the size of the 211 inclusions. A J_c value could be raised from 2000 up to 20000 A/cm^2 at 77K and 1T, which is in good agreement with the idea that the interface of the 211 inclusions provide flux pinning. Since such a high J_c value is comparable to that of the MPMG YBaCuO, they proposed that refinement of the 211 inclusions may be attributed to Pt contamination in the MPMG process because a Pt crucible is used in the first melt quenching process. Careful analysis of the MPMG sample revealed that the sample contains about 0.3-0.5 wt% Pt. Then we performed melt quenching without using a Pt crucible. Fig. 4 shows the final structure of the MPMG sample without Pt. It is clear that the size of the 211 is larger, indicating that Pt addition is partly responsible for the size reduction even in the MPMG process. However, as reported by Nagaya et al.[8], the size reduction of the 211 is possible without using Pt addition, which suggests that the MPMG process involves another mechanism of grain refinement of the 211, although it is still unclear and should be clarified.

4. Conclusions

From a comparative study of J_c values and microstructures of several melt processes for YBaCuO, it was found that the smaller is the size of the 211 inclusions, the flux pinning is larger. This result supports the fact that the interface of 211/123 provides flux pinning.

References

1) S. Jin et al.: *Appl. Phys. Lett.* **52** (1988) 2974.
2) M. Murakami et al.: *Jpn. J. Appl. Phys.* **28** (1989) 1189.
3) S. Jin et al.: *Appl. Phys. Lett.* **59** (1991) to be published.
4) P. J. McGinn et al.: *Appl. Phys. Lett.* **59** (1991) 120.
5) M. Murakami: Proc. 2nd Los Alamos Symposium (1991)
6) N. Ogawa et al.: *Physica C* **177** (1991) 101.
7) K. Salama et al.: *Appl. Phys. Lett.* **54** (1989) 2352.
8) S. Nagaya et al.: *IEEE Trans. Mag* **27** (1991) 1487.

Cerium Oxide Doped YBCO Superconductor by Melt Growth Method

NAOYUKI OGAWA and HITOSHI YOSHIDA

NGK Insulators, LTD., Nagoya, 467 Japan

ABSTRACT

We report on a study of the cerium oxide doped YBCO superconductor prepared by the melt growth method. The sample consists mainly of the large grown $YBa_2Cu_3O_y$ crystals including the finely dispersed Y_2BaCuO_5 inclusions which work as effective pinning centers. These samples show the large magnetic hysteresis loops. The typical critical current density value exceeded 15000 A/cm^2 at 77K and 1T. The almost no doped cerium widespread into $YBa_2Cu_3O_y$ matrix. The cerium barium oxide compound which was segregated away from matrix was found in our samples. We consider cerium compound acts as a grain growth inhibitor for Y_2BaCuO_5 inclusion.

KEY WORDS : YBCO superconductor, melt growth, cerium oxide doping

INTRODUCTION

For the YBCO superconductor, the control of microstructure is very important in order to realize the high critical current density under the magnetic field. Melt growth method obtains the characteristic microstructure having the large grown $YBa_2Cu_3O_y$ crystal embedded small Y_2BaCuO_5 inclusions [1-4]. In the conventional melt growth method such as Melt-Texture-Growth (MTG), relatively large Y_2BaCuO_5 inclusions which are about 20μm in size remain in the grown $YBa_2Cu_3O_y$ crystal [1]. On the other hand, Quench and Melt Growth (QMG) or Platinum Doped Melt Growth (PDMG) method realized a fine dispersion of Y_2BaCuO_5 inclusions [3,4]. In PDMG method, doped platinum or rhodium behaves as an effective grain growth inhibitor for Y_2BaCuO_5 inclusion. The critical current density of the PDMG sample is a few times larger than that of the undoped melt growth sample because the fine Y_2BaCuO_5 inclusions act as the strong pinning center.

Recently we have found another excellent dopant for melt growth method controlling the microstructure and attaining the high critical current density in YBCO superconductor. The dopant is cerium oxide. A few micrometer Y_2BaCuO_5 inclusions disperse homogeneously in matrix even with less than 1 wt% cerium oxide doping.

EXPERIMENTAL

The mixed powder was prepared using Y_2O_3, $BaCO_3$, CuO and CeO_2. Appropriate amount of these powders were wet milled in distilled water with ZrO_2 balls, followed by drying and calcined at 950°C - 1000°C in air. The calcined powder was ground with an agate mortar and pestle. The powder was pressed into disks and sintered. The sample was heated at 1100°C and kept for an hour and slowly cooled at the rate of 1°C/h in the temperature from 1000°C to 920°C. Then it was annealed at 600°C for 48 hours in flowing oxygen gas.
The magnetic property was measured by a SQUID magnetometer. The microstructure was observed by using an optical microscope and a scanning electron microscope (SEM) and

analyzed by an electron probe microanalyzer (EPMA) and an energy disperse X-ray analyzer (EDX). The crystal phase was identified by an X-ray powder diffractometer (XRD).

RESULTS AND DISCUSSION

Shown in figure 1 is the micrograph of the polished surface for 1 wt% cerium oxide doped YBCO superconductor. The matrix is large grown $YBa_2Cu_3O_y$ crystal and the small dark points are Y_2BaCuO_5 inclusions which were trapped as the result of an incomplete peritectic reaction. The fine Y_2BaCuO_5 inclusions disperse homogeneously. In the case of undoped sample, relatively large Y_2BaCuO_5 inclusions remained in $YBa_2Cu_3O_y$ crystal. Cerium compound behaves as a grain growth inhibitor for Y_2BaCuO_5 phase. It is suggested that cerium is one of rare earth elements and stable for tetravalent. On the other hand, all the other rare earth elements are stable in the state of trivalent. Addition of these elements causes the substitution of yttrium elements. The effect of rare earth substitution in QMG process has been reported by Sawano et. al [6]. They observed the growth of $YBa_2Cu_3O_y$ crystal in Sm, Eu, Gd, Ho, Er and Yb substitution as same as Y-system and reported the variation of the $YBa_2Cu_3O_y$ formation temperature. In the case of up to 1 wt% cerium oxide doping, we observed no difference between the doped sample and undoped sample for the $YBa_2Cu_3O_y$ formation temperature. No widespread diffusion of cerium into $YBa_2Cu_3O_y$ matrix was detected by an EPMA. Figure 2 shows the micrograph of the polished surface for 1 wt% cerium oxide doped sample by using an SEM. We confirmed the relatively light gray region marked *A* is Y_2BaCuO_5 composition and the white region marked *B* is barium and cerium rich composition by an EPMA and an EDX analysis of this microstracture. We consider doped cerium exists as the barium cerium oxide compound which segregated away from $YBa_2Cu_3O_y$ lattice in our sample.

Because of no diffusion of cerium element into $YBa_2Cu_3O_y$ matrix, superconducting transition temperature did not degrade in our samples. Figure 3 shows the temperature dependence of susceptibility in zero field cooled (ZFC) and field cooled (FC) conditions. In the FC condition, we obtained very small signal which shows the existence of strong pinning center [5]. We measured the magnetic hysteresis loop by a SQUID magnetometer and estimated a critical current density by using Beans' critical state model. The typical Jc value exceeds 15000 A/cm^2 at 77K and 1T. This value is twice larger than that of undoped sample.

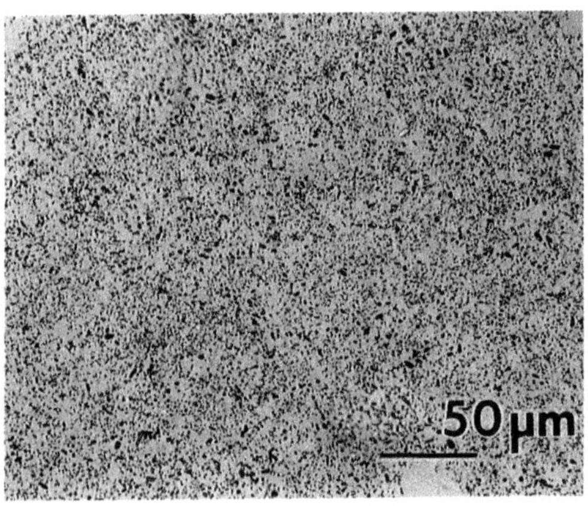

Fig. 1 Optical micrograph of polished surface for 1 wt% cerium doped sample

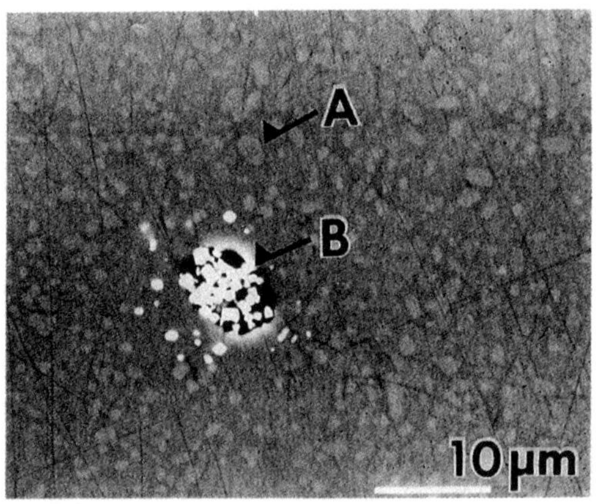

Fig. 2 Scanning electron micrograph of polished surface for 1 wt% cerium doped sample. (A) 211 phase, (B) barium and cerium rich phase

The mechanism of the fine dispersion of 211 inclusion has been reported in QMG process and PDNG process. In the QMG process, the reaction between Y_2O_3 and liquid phase produces the fine Y_2BaCuO_5 whiskers which is significant for the fine Y_2BaCuO_5 dispersion of the final microstructure. In the PDMG process, platinum or platinum related compound behaves as the nucleation site of Y_2BaCuO_5 phase [4]. Therefor, there is no effect of platinum doping for the Y_2BaCuO_5 inclusion which decomposed from $YBa_2Cu_3O_y$ crystal which have partially melted. In order to investigate the mechanism of a fine dispersion of Y_2BaCuO_5 inclusion in the cerium oxide doped sample, we observed the size and the distribution of Y_2BaCuO_5 inclusions at the initial stage of peritectic reaction. The quenched sample was prepared by the following procedure. The pellet pressed into disk was heated at 1100°C and kept for an hour. And the sample was cooled to 1000°C and quenched to room temperature. Figure 4 shows the micrograph of the fractured surface of the quenched sample. We observed a fine Y_2BaCuO_5 inclusions and solidified liquid phase surrounding them. For the cerium oxide doped sample, it is important that a fine dispersion of Y_2BaCuO_5 inclusion must achieve at the initial stage of peritectic reaction. The similar phenomena have been observed in PDMG process [4]. However, there is a difference in the fine dispersion mechanism.

We have identified the crystal phase of the calcined powder and the quenched sample. XRD patterns identified $YBa_2Cu_3O_y$ phase and Y_2BaCuO_5 phase for calcined powder and identified only Y_2BaCuO_5 powder for quenched sample. Thus, incongruent melt occurs following equation in our sample.

$$YBa_2Cu_3O_y + Y_2BaCuO_{5\,cal} \;{-}{-}{-}{-}{>}\; Y_2BaCuO_{5\,cal} + Y_2BaCuO_{5\,dec} + Liq.$$

where $Y_2BaCuO_{5\,cal}$ stands for Y_2BaCuO_5 produced by the calcination and $Y_2BaCuO_{5\,dec}$ stands for Y_2BaCuO_5 produced by the decomposition. About 56-67 percent of Y_2BaCuO_5 inclusion was produced by the decomposition from $YBa_2Cu_3O_y$ crystal estimating from the starting composition of our sample. If cerium compound behaves as a nucleation site like platinum or rhodium in PDMG process, two kinds of Y_2BaCuO_5 would be observed. One is a fine Y_2BaCuO_5 inclusion which has nucleates at the site. And the other is relatively large

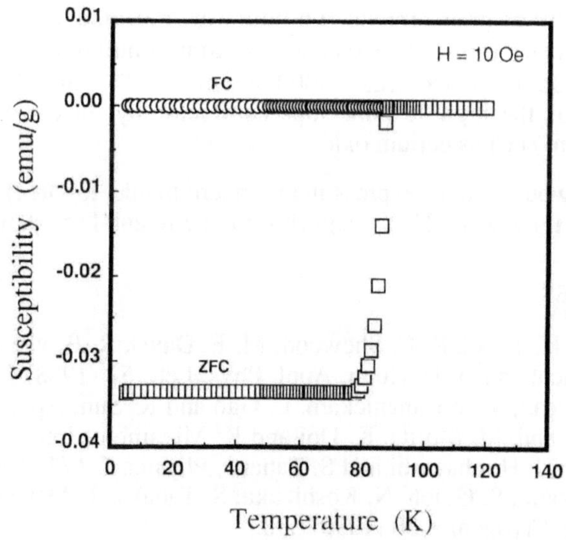

Fig. 3 Temperature dependence of magnetic susceptibility

Y_2BaCuO_5 inclusion which decomposed from $YBa_2Cu_3O_y$ crystal as the result of incongruent melt. As shown in figures 1 and 4, we can not distinguish between Y_2BaCuO_{5cal} and Y_2BaCuO_{5dec} in cerium oxide doped $YBa_2Cu_3O_y$ samples. And we found only grain-shaped Y_2BaCuO_5 inclusion in the quenched sample and could not find the whitsker-shaped Y_2BaCuO_5 inclusion observing in QMG process. Judging from these results, cerium affects directly to the grain growth of Y_2BaCuO_5 inclusions such as suppressing grain boundary diffusion or surface energy. In the case of undoped sample, Y_2BaCuO_5 inclusions coarsen in liquid phase above the partial melting temperature. In the case of cerium oxide doped sample, suppressing of coarsening realizes a fine Y_2BaCuO_5 dispersion in the final microstructure.

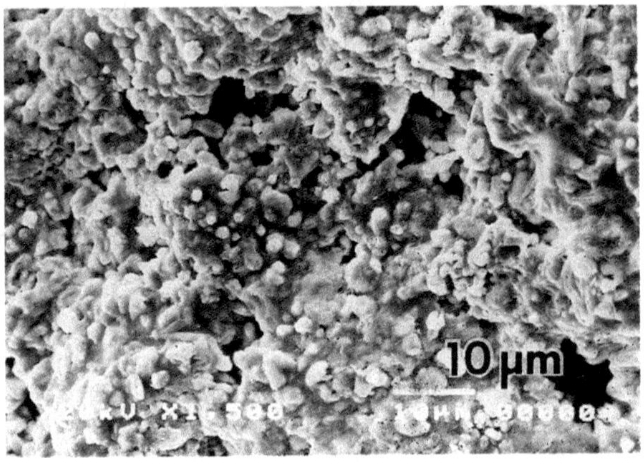

Fig. 4 Scanning electron micrograph of fractured surface for 1 wt% cerium doped sample

SUMMARY

In YBCO superconductor, we have found another effective dopant which improves a critical current density. The cerium oxide doped sample consists of the large $YBa_2Cu_3O_y$ crystal embeded a fine Y_2BaCuO_5 inclusions which disperse homogeneously. We consider the cerium compound acts as grain growth inhibitor for Y_2BaCuO_5 inclusions. The measurement of magnetic property shows the existence of strong pinning center in cerium oxide doped sample. Our typical Jc value exceeds 15000 A/cm^2 at 77K and 1T. It is attractive to realize the preparation of the high Jc bulk superconductor by means of melt growth method with the cheaper dopant such as cerium oxide.

The authors would like to express their sincere thanks to Dr. H. Sakai for his helpful discussion and Messrs. H. Baba and H. Yamaguch for experimental assistance.

REFERENCES

1. S. Jin, T. H. Tiefel, R. C. Shewood, M. E. Davis, R. B. van Dover, G. W. Kammalott, R. A. Fastnacht and H. D. Keith, Appl. Phys. Lett. **52** (1988) 2074
2. K. Salamman, V.Selvamanickam, L. Gao and K. Sun, Appl. Phys. Lett. **54** (1989) 2352.
3. M. Murakami, M. Morita, K. Doi and K. Miyamoto, Jpn. J. Appl. Phys. **28** (1989) 1189.
4. N. Ogawa, I. Hirabayashi and S. Tanaka, Physica C **177** (1991) 101.
5. M. Murakami, S. Gotoh, N. Koshizuka, S. Tanaka, T. Matsushits, S. Kambe and K. Kitazawa, Cryogenics **30** (1990) 390.
6. K. Sawano, M. Morita, K. Doi, A. Hayashi, K. Kimura, K. Miyamoto and S. Matsuda, Proceedings of Advances in Superconductivity II, (1989) 277.

Enhanced Melt Growth of $YBa_2Cu_3O_{7-\delta}$ Superconductor by High-Energy Ball Milling

JUNG-HO AHN, KOOK-HYUN HA, JAE-WOONG KO, HAE-DOO KIM, and HYUNGSIK CHUNG

Korea Institute of Machinery & Metals, Changwon-shi, 641-010 Korea

ABSTRACT

The high-energy ball milling of Y1-2-3, Y_2O_3, or Y2-1-1 powders was investigated to see the effect of refined and highly energized powders on melt growth of Y1-2-3 superconductor. The use of highly ball-milled particles as starting powders promoted the uniform distribution of fine Y1-2-3 particles in the liquid phase above the peritectic temperature, resulting in more complete reaction and more enhanced melt growth as compared to the case of unmilled powders. The high-energy ball milling can reduce thus the problems of melt growth process, extremely slow rate of reaction and difficulty in microstructural control.

KEY WORDS : High Tc superconductor, $YBa_2Cu_3O_{7-\delta}$ superconductor, Mechanical alloying, Melt texuring

INTRODUCTION

The application of bulk high Tc superconductor has been hindered by their low critical current density (Jc) which is caused by a number of problems such as weak-links at grain boundaries or flux creep. It has been well studied that the values of Jc depend stronly on the microstructure of high Tc superconductor, that is, a significant improvement can be made by aligning and growing grains, and also employing effective pinning sites to hinder flux creep.

In $YBa_2Cu_3O_{7-\delta}$ (Y1-2-3) superconductor, fairly high values of Jc have been reported by a variety of melt texturing techniques (e.g.1-3) where well grown and aligned Y1-2-3 grains develop parallel to a-b (CuO_2) plane via the peritectic reaction, " solid 2-1-1 (Y_2BaCuO_5 phase) + liquid ($3BaCuO_2$ + 2CuO) --> 2 $YBa_2Cu_3O_{7-\delta}$" at ~1010°C in air. However, a proper control of microstructure by means of the peritectic reaction is generally difficult, because the solidification reaction occurs from a partial melt at solid-liquid boundaries. The characteristic microstructure obtained after the peritectic reaction contains thus unreacted liquid ($BaCuO_2$ and CuO) and trapped solid (Y2-1-1) phases with an incomplete growth of 1-2-3 phase. To avoid such problem, N. Murakami and his co-workers (2) have employed Melt-Powder-Melt-Growth (MPMG) process. In this process, the grinding of quenched melt (Y_2O_3 + Liq.) prior to melt growth treatment has been employed to achieve the fine and uniform dispersion of Y2-1-1 phase in the liquid, which promotes the peritectic reaction.

In the present work, we have investigated the effect of high-energy ball milling, or mechanical alloying, on the melt growth process. Unlike Murakami and his co-workers' work (2) where the quenched melt (Y_2O_3 + Liq) was mechanically ground, the starting powders of Y1-2-3, Y_2O_3 + BaO + CuO, or Y2-1-1 + $BaCuO_2$ + CuO were directly ball-milled with high mechanical energy prior to melt growth treatment. This method reduces processing procedures, and neither intermediate heating nor the grinding of quenched melt are needed. The investigated compositions were Y1-2-3, and Y1-2-3 + 10 and 20% excess Y2-1-1 to enhance flux pinning effect and to optimize complete reaction. The purpose of high-energy ball milling was not only to accomplish the particle refinement which can increase the reaction area of solid/liquid interface, but also to see the effect of highly energized powders on the subsequent growth behaviour.

EXPERIMENTS

Three different powders were ball milled : Y1-2-3 powders, mixed powders of Y_2O_3 + (BaO + CuO), or Y2-1-1 + ($BaCuO_2$ + CuO). In each case, the powders were prepared to give final composition of $YBa_2Cu_3O_{7-\delta}$, $YBa_2Cu_3O_{7-\delta}$+10 % and 20 % Y2-1-1 (wt.%). Both dry and wet ball milling (with ethanol) with stainless steel or alumina vial and balls were employed in air. The high-energy ball milling was carried out in a planetary ball mill (Retsch PM-4 type) which revolves and rotates at the speed of 150 r.p.m. with changing directions after every 70 seconds. The ball milled powders were removed from the vial after given intervals of 1/2, 20 and 40 hours, and then compacted for heat treatments. Some compacts was subjected to preliminary quenching experiments from 1100°C to see the morphology and grain refinement of Y2-1-1 in liquid phase matrix. The melt growth treatment was carried out in a vertical and float-zone fashion without any thermal gradient along the sample : the sample was rapidly heated to 1100°C for 15 minutes, and then slowly cooled from 1020°C to 980°C in air at cooling rate of 1°C/hr followed by O_2 annealing at 600°C for 24 hours. After heat treatment, a comparison was made on the different samples which were ball-milled for different durations.

RESULTS and DISCUSSION

The result of different milling methods showed that wet ball milling is more effective for the refinement of superconducting powders than dry milling. However, the powders were severely contaminated by Fe and Ni when employing wet milling with stainless steel balls and vial. Ball milling with ceramic (alumina) balls and vial also resulted in the contamination, caused by small amount of brittle fracture on the ceramic ball surface during the both wet and dry ball milling for prolonged durations. The critical temperature (Tc) dropped to 60K for Y1-2-3 powders ball-milled for 40 hours with alumina balls, indicating Al_2O_3 contamination according to EDS analysis. The problem of contamination could be markedly reduced when employing dry milling with stainless steel balls and vial: only a very small amount of contamination (0.029% Fe, 0.034% Ni, and 0.008% wt.% Cr) was detected after 40 hrs milling without any apparent degradation of superconductivity, although refining effect was moderate as compared to the case of wet milling. Figure 1 shows the morphology of Y1-2-3 powders before and after dry milling, showing that the initially large agglomerated particles of Y1-2-3 (> 10 μm) was well refined after 40 hrs milling to small particles containing many submicron particles.

Microstructures obtained after heat treatment at 940°C, i.e. the temperature below the peritectic point, show that the grain size of the 40 hrs ball-milled sample is slightly larger than that of the unmilled sample (Figure 2).

Fig.1 : SEM morphology of Y1-2-3 powders : (a) unmilled, and (b) high-energy ball milled for 40 hrs.

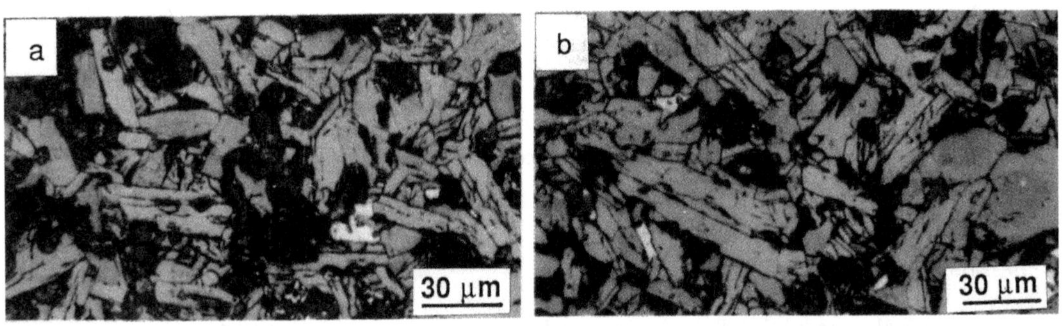

Fig.2: Optical micrographs of Y1-2-3 samples, made of (a) unmilled powders, and (b) 40 hrs ball-milled powders, sintered at 940°C for 4 hours.

Microstructural difference between the unmilled and high-energy ball-milled powders was clearly observed after melt growth treatment. Figure 3 shows the effect of high-energy ball milling on the microstructures of Y1-2-3 samples produced after slow cooling from 1020°C to 980°C in air at the cooling rate of 1°C/hr without any thermal gradient along the sample. Although two samples were heat treated under the same working condition, the grain size of the unmilled powder sample is only 1 mm or less containing some trapped and unreacted (dark) regions of Y2-1-1 + liquid phases between grain boundaries, while the large grains (>10 mm) were well developed in the case of 40 hrs ball-milled powder sample. Such beneficial effect of high-energy ball milling on the melt growth was more pronounced especially for the samples of Y1-2-3 + excess 211 as compared to those of stoichiometric Y1-2-3. As shown in figure 4, a well-aligned microstructure with uniform distribution of Y2-1-1 particles (less than 10 μm in size) were produced during the melt growth for the sample of 40 hrs ball-milled Y1-2-3 + 10% 211 powders, while relatively inhomogeneous microstructure produced in the case of unmilled powder sample. It has been known that the excess Y2-1-1 is beneficial both for the flux pinning and the minimization of unreacted products during the peritectic reaction. The present work confirms the beneficial effect of Y2-1-1 for more complete formation of Y1-2-3 during peritectic reaction, but the effect of high-energy ball milling appears more dominant for the complete reaction and grain growth. No particular structural difference after melt growth was observed on the samples

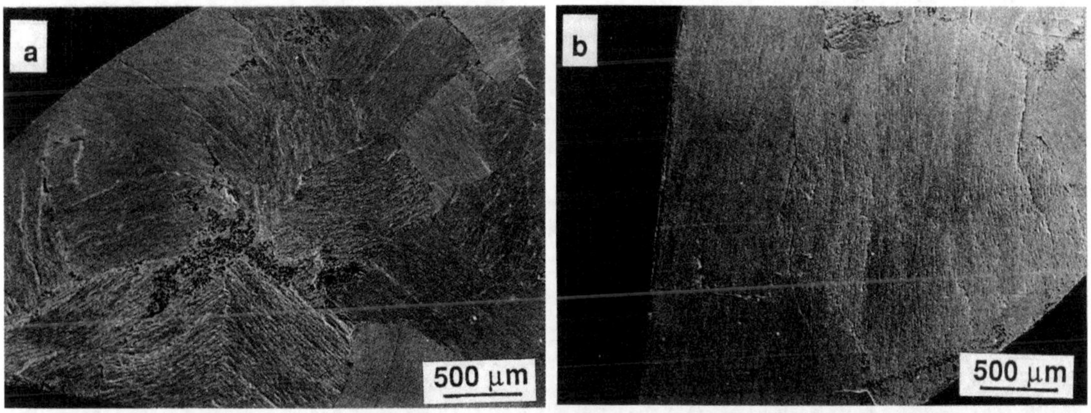

Fig.3: SEM micrographs of melt grown Y1-2-3 sample, using (a) unmilled powders, and (b) 40 hrs ball-milled powders, after slow cooling from 1020°C to 980°C in air at the cooling rate of 1°C/hr without thermal gradient along the sample.

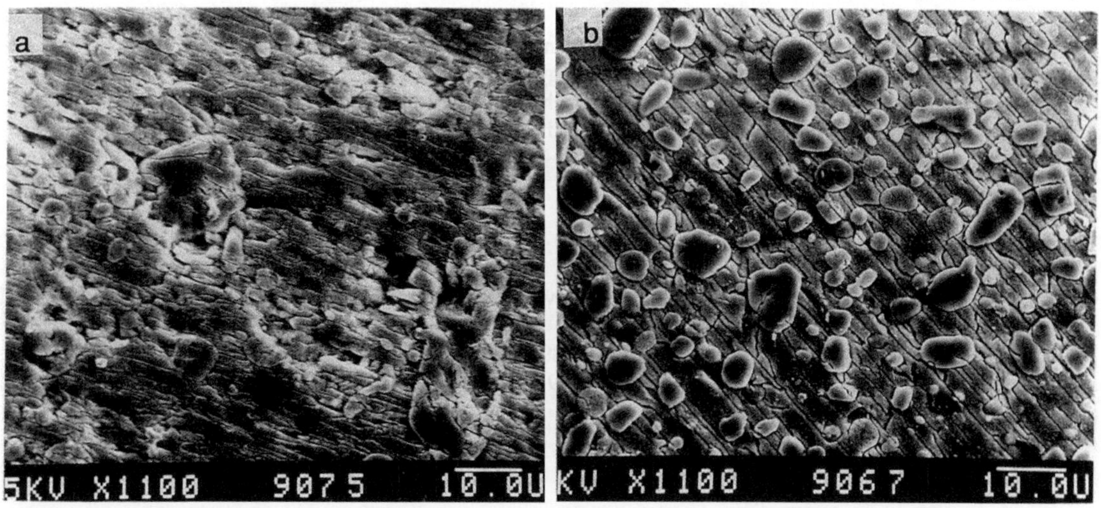

Fig.4: SEM micrographs of melt grown Y1-2-3 + 10 % 211 sample, using (a) unmilled powders, and (b) 40 hrs ball-milled powders after slow cooling from 1020°C to 980°C in air at the cooling rate of 1°C/hr.

whose starting powders were different (Y1-2-3, Y_2O_3 +Liq, or Y2-1-1 + Liq). The Jc value was measured, using the four-probe method, for 40 hrs ball-milled and subsequently melt grown sample; no voltage drop was observed during the continuous D.C. measurement until 20 A (Jc > 2000 A/cm^2) where the contact wire of Ag could not persist due to resistant heating.

It is considered that the enhanced melt growth when using high-energy ball-milled powders results from the uniform and fine distribution of Y2-1-1 particles in the liquid matrix at the temperature range of Y2-1-1 + Liq phases. The fact is shown in Figure 5 where one can see the difference in microstructures between unmilled and 40 hrs ball-milled powder samples, produced after quenching from 1100°C. In these micrographs, the ball-milled powder sample shows quite uniform distribution of fine needle-shaped Y2-1-1 particles in the liquid matrix as compared to the inhomogeneous microstructure in the case of the unmilled powders. It is evident that such fine and uniform distribution of Y2-1-1 is favorable for "solid (2-1-1) + liquid --> solid(1-2-3)" peritectic reaction, and promotes melt texturing.

Fig.5 : SEM micrographs of Y1-2-3 samples, heated at 1100°C for 10 minutes and quenched to room temperature : (a) unmilled sample, and (b) 40 hrs ball-milled samples.

CONCLUSIONS

The highly ball-milled powders resulted in more homogeneous and refined Y2-1-1 particles in liquid phase than the unmilled powders. As a result, the use of high-energy ball-milled particles as starting powders promoted the peritectic reaction, the grain alignment and growth. The high-energy ball milling could be thus effective method to reduce the problems of the melt growth process via peritectic reaction, i.e. both extremely slow reaction rate and difficulty in microstructural control.

REFERENCES

1. Jin S, Sherwood RC, Gyorgy EM, Tiefel TH, Van Dover RB, Nakahara S, Schneemeyer LF, Fastnach RA, and Davis RB (1989) Appl Phys Lett 54(6) : 584-586
2. Murakami M, Moria M, Doi K, and Miyamoto K (1989) J Appl Phys 28 (7) : 1189-1194
3. Salama K, Selvamanickam V, Gao L, and Sun K (1989) Appl Phys Lett 54 (23): 2352-2354

Solidification of High-T_c Oxide Superconductors by a Laser Zone Melting Method

T. NAKAMOTO, M. MIYAJIMA, I. HIRABAYASHI, and Y. SHIOHARA

Superconductivity Research Laboratory, Nagoya Division, ISTEC, Nagoya, 456 Japan

ABSTRACT

Nonsuperconducting precursor thick films of YBCO system were reactively remelted and directionally solidified by a laser zone melting method with a high temperature gradient (G) and a high growth rate (R). A CW Nd:YAG sheet beam laser was focused and scanned onto the films of about 20μm thick on Ag substrates. Scanning velocity (which corresponds to R) dependences of relative peak intensities of X-ray diffraction for the 123 phase, the 211 phase, and Y_2O_3 were investigated. The interplay of the 211 and the 123 phases in the lower R region is understood as a peritectic reaction occurring through the 211 phase and the BaO-CuO liquid phase, whereas that in the higher R region suggests that the direct crystallization of the 123 phase from the melt of the 123 composition due to a high interfacial undercooling introduced by the rapid solidification.

KEY WORDS : YBCO superconductor, laser zone melting, rapid solidification

INTRODUCTION

Since the discovery of high-Tc superconductors, tremendous efforts have been made to produce materials with high critical current density (Jc). In particular, the melting-solidification process has been investigatived as a promising process to control the crystal orientation and improve the grain boundary characteristics (1) (2). By the convensional melt process through the equilibrium phase transformation, the 123 phase is produced by the peritectic reaction through Y_2BaCuO_5 (211) and the liquid. This reaction is very slow because it includes diffusion in the solid phase, so that lengthy reaction time and slow cooling rate are necessary to make the reaction completely.

Recently Nagaya et al. (3) discovered direct crystallization of the 123 under nonequilibrium phase transformation by a rapid solidification using laser zone melting method. This process was achieved by high interfacial undercooling which brought high temperature gradient and high growth rate. In the previous paper, however, they reported the result only in the high R region in which the composition ratio of the 123 phase increased with increasing scanning velocity. Moreover the targets for their treatment were the 123 sinterd samples. It has not necessarily been able to exclude the doubt that the 123 phase remains unmelted because of the high R scan.

Now our question is how is the crossover region between the high R nonequilibrium region which is dominant direct crystallization of the 123 phase and the low R equillibrium phase transformation region in which the 123 phase occurs through peritectic reaction between the 211 phase and the liquid phase of the rest.

In the present paper, we report the experimental result of wide range of R to observe the crossover from the low R to high R regions for the irradiation onto the nonsuperconducting thick films.

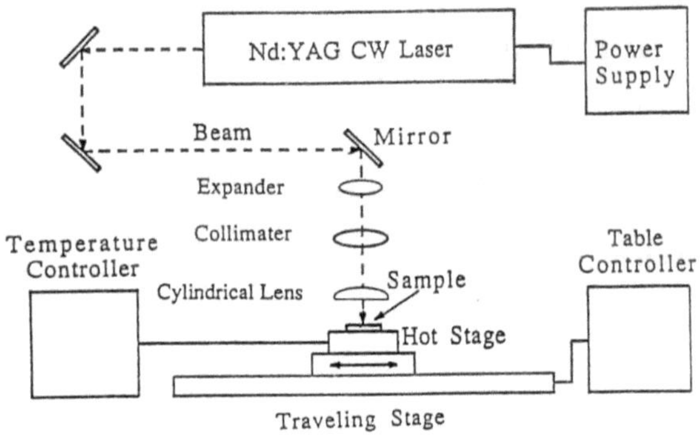

Fig.1 Schematic of the laser irradiation system

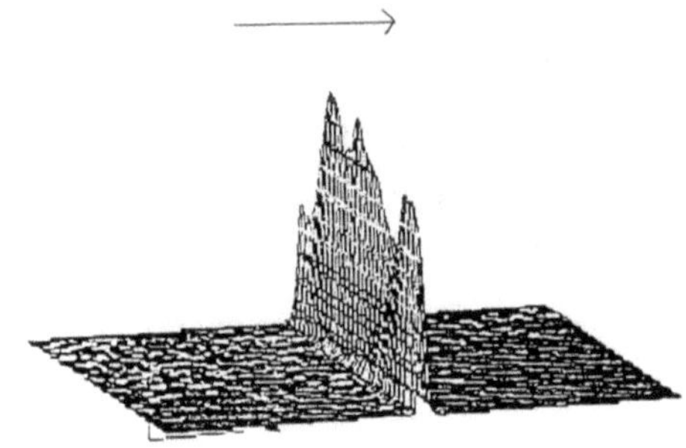

Fig.2 Beam profile

EXPERIMENTAL

Figure 1 shows a schematic illustration of our laser zone melting apparatus. The laser used in this work was a multi-mode CW Nd:YAG laser (λ=1.06μm). Samples were placed on a hot stage which was installed on a travelling stage, and traversed under the focused laser beam. The beam was focused on the sample surface with a focusing system composed of a beam expander, a collimator, and a cylindrical lens. The focused beam shape was about 12mm length and 0.2mm wide. Laser scanning velocity, average laser power, substrate temperature, and focusing condition were controlled as experimental parameters. But in the present experiment, the parameters were fixed except for laser scanning velocity, i.e., laser power density was 2000W/cm^2, focusing distance was adjusted as just in focus, and substrate temperature was room temperature. Laser scanning velocity were changed from 0.8mm/hr to 1000mm/hr.

A typical focused sheet beam profile observed by laser beam analyzer is shown in Fig. 2. The beam intensity was decreased by an array of beam splitter plates when the beam analyzer was used. The sample was scanned toward the arrow-marked direction in order that temperature gradient should orient parallel to the scanning direction.

The target samples were nonsuperconducting films of about 20μm thick on Ag substrates. They were prepared by spincoating with appropriate portion of each salt of naphthenic acid and calcining at 500°C for 15min. Coating and calcining were repeated 15-20 times (4).

Crystal structure and surface morphology of the sample were characterized by X-ray diffraction (XRD) and scanning electron microscopy (SEM), respectively.

RESULT AND DISCUSSION

Figure 3 shows the X-ray diffraction patterns for the unirradiated sample (a), the furnace-annealing sample (b), and the laser irradiated sample (c). To confirm the composition and quality of the unirradiated sample, we also annealed the sample in the furnace at 920°C for 2 hours in flowing oxygen. Scanning velocity for sample (c) was 22.5mm/hr.

In the X-ray diffraction pattern for the calcined sample before irradiatation (a), there are traces of Y_2O_3, $BaCO_3$, CuO, Ag (substrate), and unknown phase. It does not show perovskite structure, because heating temperature (500°C) was too low to form perovskite structure. The furnace-annealing sample (b) shows the c-axis oriented 123 structure and shows Tc=90K by AC susceptibility. In the diffraction pattern of the irradiated sample (c), there are peaks of the 123 phase, the 211 phase, and Y_2O_3. By only one laser scanning, the precursor film reacted and shows the traces of compounds which have perovskite structure. For the sample treated at 22.5mm/hr, the 211 phase remains. It is because peritectic reaction is so slow that it is not enough time for the 211 phase to react with BaO-CuO liquid during the time scale of this irradiation condition. Sample (c) shows a trace of the 123 phase but it is a nonsuperconducting tetragonal phase. Sufficient inclusion of oxygen is necessary for the 123 phase to become orthorombic superconducting phase. It is explicable if considering as follows. Before irradiation the film was not perovskite structure, and there is not sufficient oxygen in the precursor itself to form superconductor. This oxygen deficiency not supplemented by intake of oxygen from an atmosphere during such short time of irradiation. In fact, if the $BaCO_3$ is assumed to be decomposed into CO_2 and BaO and the valence of Cu in the precursor is 2+, the oxygen content should be 6.5 in the formula of $YBa_2Cu_3O_x$. But if the unirradiated sample includes oxygen enough to become a superconductor as in the previous case in which the orthrombic 123 sintered sample was used as the target (3), the irradiated sample become orthoromic superconductor also, because oxygen cannot go out from the sample.

Figure 4 shows the laser scanning velocity dependences of the relative intensities of X-ray diffraction patterns for the 123 phase, the 211 phase, and Y_2O_3. Taking notice of the ratio of the 123 phase and the 211 phase, there are two region, one is low R region below R=3mm/hr, in which the peak intensity ratio of the 123 phase decreases with increasing R, and the other is high R region in which that of the 123 phase increases with increasing R. On the other hand the peak intensity ratio of the 211 phase is almost complementary to that of the 123 phase, i.e., it increases with increasing R in the low R region and decreases with increasing R in the high R region and takes a maximum also at around R=3mm/hr.

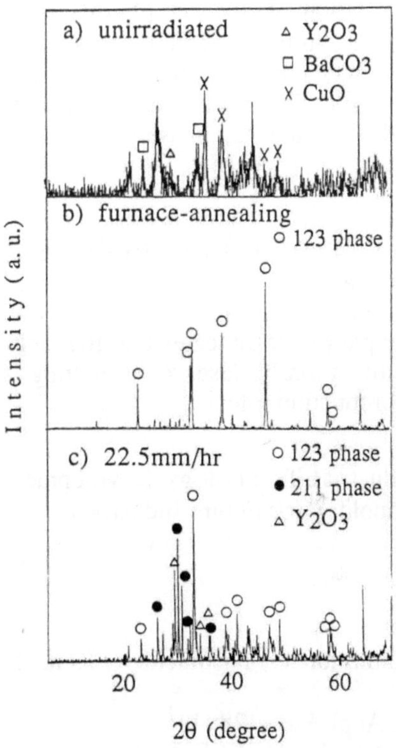

Fig.3 X-ray diffraction patterns

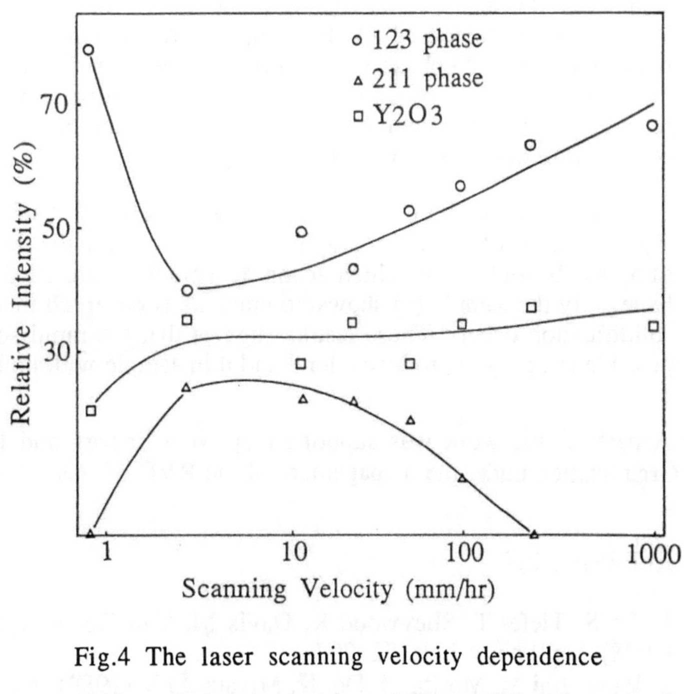

Fig.4 The laser scanning velocity dependence of the relative intensities of X-ray diffraction.

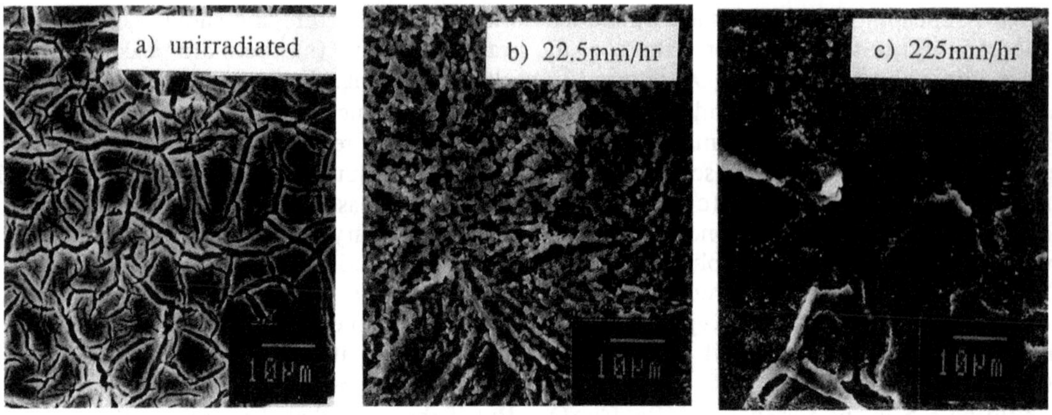

Fig.5 SEM micrograph of the surface

We consider the above result as follows. The low R region is the peritectic reaction occurring through the 211 phase and the BaO-CuO liquid phase, i.e., at first stage of cooling, in the molten film by laser irradiation, Y_2O_3 forms and subsequently the 211 phase grows on the result of the peritectic reaction between Y_2O_3 and the rest liquid. After that, in the 2nd stage of the peritectic reaction the 211 phase react with the liquid phase to form the 123 phase. This peritectic reaction requires much time since the reaction includes diffusion process in solids phase, so that the ratio of the 123 phase increases with decreasing scanning velocity.

On the other hand, the result for high R region is the evidence for the direct crystallization of the 123 phase from the melt of the 123 composition. If crystallization of the 211 phase can be suppressed during the rapid cooling process, an undercooling state can be realized in this system. When it is undercooled to the temperature at which the newly emerged liquidus lines is intercepted, the 123 phase can be directly crystallized from its melt. The result is essentially the same to the previous result (3) except that the starting material is not the 123 phase but the non-reacted precursorial material and that the instantaneous reaction process to form the 123 composition materials occurs at the early stage of irradiation.

To confirm the melting of the sample, we made SEM observation for surface microstructure. Figure 5 shows SEM micrograph of the surface, for the unirradiated sample (a), the irradiated samples (b) and (c) of which scanning velocities are 22.5mm/hr and 225mm/hr, respectively. Especially the sample (c) shows no crack as is observed in sample (a). It indicates that the melt-solidification occurs. These results suggest that the rapid solidification by laser zone melting is possible to apply to produce a long and thin sample with high production rate.

A part of this work was supported by New Energy and Industrial Technology Development Organization under the management of the R&D of Basic Technology for Future Industries.

REFERENCES

1. Jin S, Tiefel T, Shevwood R, Davis M, Van Dover R, Kammlott G, Fastnacht R, Keith H (1988) Appl. phys. Lett. 52: 2074
2. Murakami M, Morita M, Doi K, Miyamoto K (1989) Jpn. J. Appl. Phys. 28: 1189
3. Nagaya S, Miyajima M, Hirabayashi I, Shiohara Y, Tanaka S (1991) IEEE Transactions on Magnetics, 27: 1487
4. Shiono T, Nakamoto T, Nasu H (1990) Trans. IEE of Japan 110-A 7:414

Stress Dependent T_c for Y-Ba-Cu-O System

KAZUYA OGURI, TAKAKAZU MOCHIDA, KOUJI ISHIZUKA, MASAHIKO IMAGAWA, KOUICHI WATANABE, NOBUYUKI NINOMIYA, and YOSHITAKE NISHI

Department of Materials Science, Tokai University, Hiratsuka, Kanagawa, 259-12 Japan

ABSTRACT

Effects of small compressive stress on superconducting transition temperatures are studied for Y-Ba-Cu-O samples. The small stress increases the temperatures. Based on the results of X-ray diffraction and electrical resistivity, we conclude that the crystal perfection of the lattice ordering and the high carrier concentration enhance the temperatures.

key words: superconductor, small stress, Y-Ba-Cu-O

INTRODUCTION

Many workers have reported superconducting property with pressure [1-3]. Since high T_c superconductors have been discovered and the remarkable T_c increasing with pressure has been reported for La system [4]. Thus, we undertake to obtain an information of the influence of the small compressive stress on the T_c value for the high-T_c $YBa_2Cu_3O_{7-y}$.

EXPERIMENTAL PROCEDURE

Samples with nominal composition $YBa_2Cu_3O_{7-y}$ were prepared by high-purity powders of Y_2O_3(99.99 %), $BaCO_3$(99.9 %), and CuO(99.9 %). The powders were mixed and sintered in air at 1200 K for 2h and then air-cooled. After crushing, the pelletized tablet, 2 mm thick and 13 mm diameter, was sintered in air at 1200 K for 8 h and furnace-cooled. The cooling rate is 0.07 K/s at 1000 K [5]. The temperature was measured by Au*Fe-chromel thermocouple attached to the specimen in a cryostat. The electrical resistivity was measured using a standard four-probe technique at equilibrium temperatures.

The structure of the samples was examined by means of x-ray diffraction used by Cu-K α. The small compressive stress was provided by a mechanical tester(see. fig. 1). The samples do not be loaded after sintered.

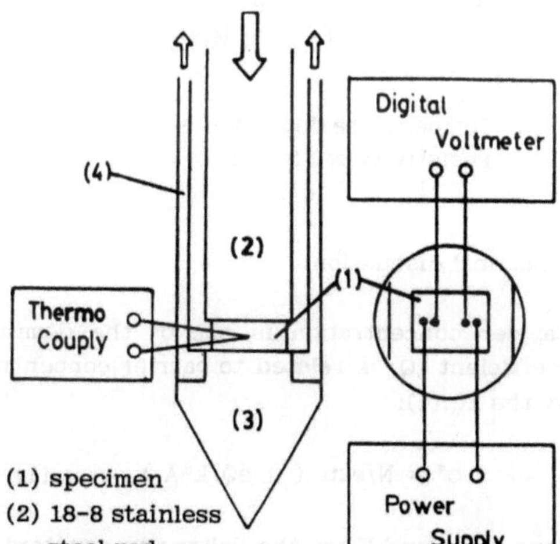

(1) specimen
(2) 18-8 stainless steel rod
(3) Cu block
(4) 18-8 stainless steel pipe

Fig.1 Schematic drawing of compressive stress machine.

RESULTS AND DISCUSSION

T_c measurement

Figure 2 shows the temperature dependence of electrical resistivity (R) of the unloaded and loaded $YBa_2Cu_3O_{7-y}$. The lower the temperature, the lower the electrical resistivity becomes. The deviated point is taken at $d(R/R^{300K})/dT=0.03$ of $YBa_2Cu_3O_{7-y}$, where $d(R/R^{300K})/dT$ and R^{300K} are the slope of the reduced electrical resistivity (R/R^{300K}) against the temperature graph and the resistivity at 300 K, respectively. T_c^{mid} is defined as the temperature corresponding to $R^{on}/2$, and R^{on} is resistivity of the deviated point.

Figure 3 shows change in T_c^{mid} of $YBa_2Cu_3O_{7-y}$ against compressive stress. The small stress enhances T_c^{mid}. The excess stress lowers T_c^{mid}. The maximum values of T_c^{mid} is found 98.4 K at 16.5 MPa on $YBa_2Cu_3O_{7-y}$.

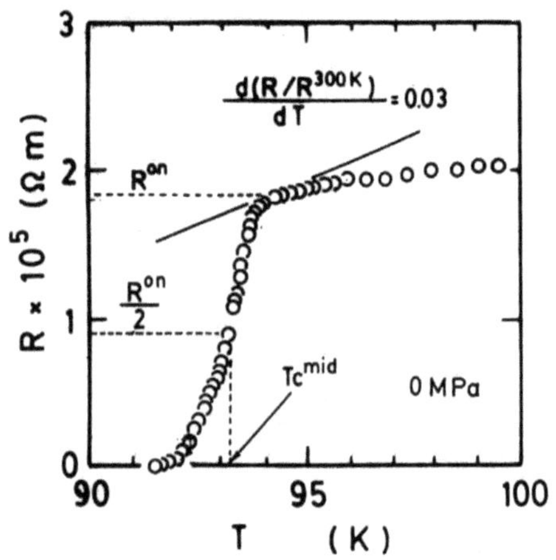

Fig.2 Temperature dependence of electrical resistivity of $YBa_2Cu_3O_{7-y}$.

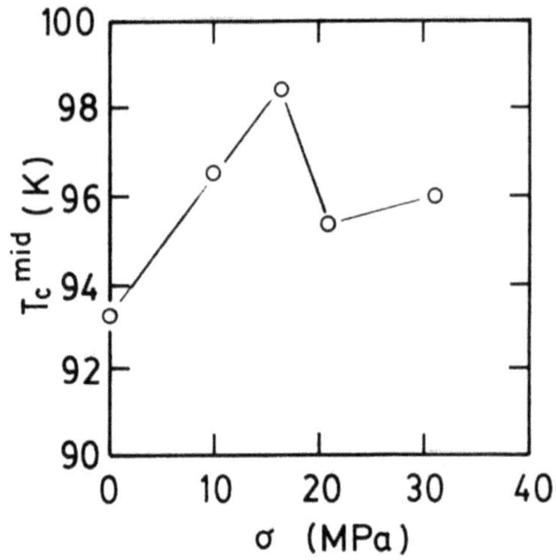

Fig.3 Changes in T_c^{mid} of $YBa_2Cu_3O_{7-y}$ against small compressive stress.

Electrical discussion

Carrier concentration is one of the dominant factors that control the T_c. If the Seebeck coefficient (Q) is related to carrier concentration [6,7], carrier concentration (c^*) is expressed by the Eq.(1):

$$c^* = N/\exp\ (\pm eQ/k - A) \qquad (1)$$

Here, k, e, and N are the Boltzmann constant, electron charge and density of state, respectively; A is a constant. The Seebeck coefficient of the high-T_c $YBa_2Cu_3O_{7-y}$ is small value. So it do not largely affect the carrier concentration. The density of state is the most dominant factor in the Eq.(1). To obtain c*, the density of state is calculated by the use of the electrical resistivity (Re). N is expressed by the Eq. (2)[8]:

$$N = A' / Re^{1/2} \qquad (2)$$

Here, A' is a constant that is dominated by the electron mass, charge, Plank constant, and mean free path. If the stress does not largely change their values, the N value is inversely proportional to the $Re^{1/2}$ value.

Figure 4 shows the change in the electrical resistivity against the stress. the small compressive stress decreases the resistivity and increases the density. Here, the electron mean free path is assumed to be constant and to be equal to the lattice constant of the c axis. Namely, the larger the stress, the higher the density of state becomes, resulting in high T_c. On the other hand, since the excess stress slightly decreases the density, it decreases the T_c.

Figure 5 shows the c* change against the compressive stress. The c* change approximately agrees with the T_c^{mid} change. Namely, the higher the carrier concentration, the higher the T_c^{mid} becomes for the loaded $YBa_2Cu_3O_{7-y}$.

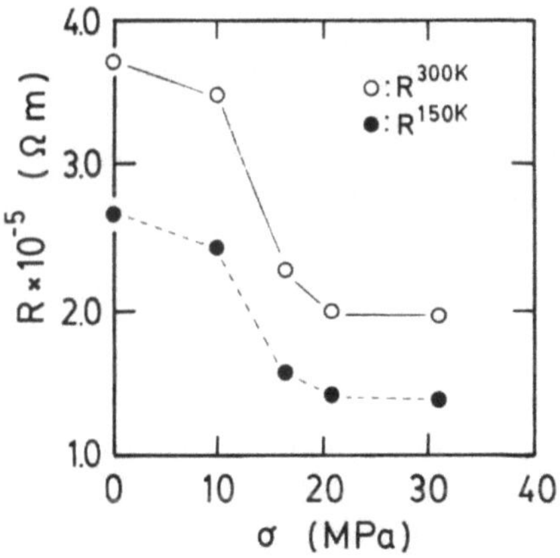

Fig.4 Changes in electrical resistivity against compressive stress of $YBa_2Cu_3O_{7-y}$.

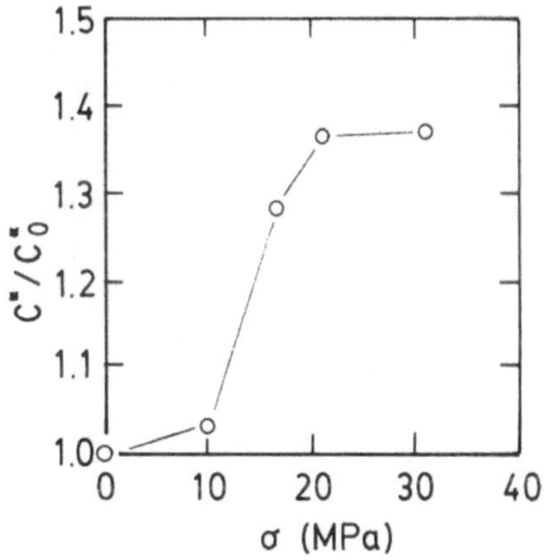

Fig.5 Change in Carrier concentration against compressive stress of $YBa_2Cu_3O_{7-y}$.

X-ray diffraction

Figure 6 shows X-ray diffraction pattern of unloaded $YBa_2Cu_3O_{7-y}$ before and after loading. The excess stress increases the lattice distance and FWHM of the unloaded crystal plane, although the lattice distances and FWHM do not depend on the stress below the endurance limit (σ_c). Namely, remarkable changes in the x-ray pattern cannot be found within the elastic deformation below the endurance limit. On the other hand, the excess compressive stress which decreases T_c^{mid}, because some disaccommodations (such as microcracks) exist.

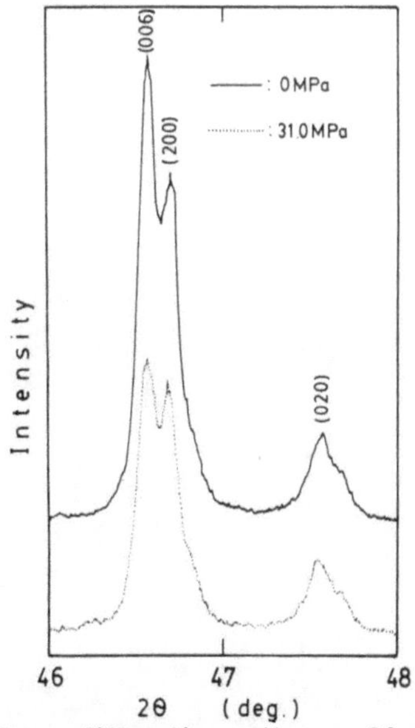

Fig.6 X-ray diffraction patterns of before and unloaded $YBa_2Cu_3O_{7-y}$ after loading.

CONCLUSION

The effect of small compressive stress on T_c is studied for $YBa_2Cu_3O_{7-y}$. The small compressive stress increases T_c, although the excess stress decreases T_c of $YBa_2Cu_3O_{7-y}$. The maximum T_c^{mid} is 98.4 K at 16.5 MPa of $YBa_2Cu_3O_{7-y}$ (T_c = 93.2 K before the loading). Based on the results of the x-ray diffraction and electrical resistivity, we conclude that the high carrier concentration enhance the T_c value and the excess compressive stress decrease T_c by some disaccommodations.

REFERENCES

1. Wu M.k., Ashburn J.R., Torng C.J,. Hor P.H., Meng R.L., Gao L., Haung Z.J., Wang Y.Q., Chu C.W. (1987) Phys. Rev. Lett. 58: 908.
2. Akihama Y, Endo S, Noguchi S, and Okuda (1987) Jpn. J. Appl. Phys. 26: L871.
3. Yomo S, Murayama C, Utsumi W, Takahashi H, Yagi T, Mori N, Tamegai T, Watanabe A, Iye Y (1987) Jpn. J. Appl. Phys. Suppl. 26: 1107.
4. Chu C.W., Hor P.H., Meng R.L., Gao L., Huang Z.J., Wang Y.Q. (1987) Phys. Rev. Lett. 58: 405.
5. Nishi Y, Kita Y, Igarashi A, Tokunaga S (1989) J. Mater. Sci. Lett. 8: 503.
6. Tallan N.M., Bransky I., (1971) J. Electrochem. Soc. 118: 345.
7. Brebrick R.F. in Defects in Solids, Vol. 2 of Treatise on Solid State Chemistry, edited by Hannay N.B. (Plenum, New York, 1975), pp. 357-362.
8. Mott N.F. (1969) Philos. Mag. 19: 835.

Heat Treatments After High Temperature Deformation in Oxide Superconductors

KENJI HIGASHIDA, NOBUTAKA NARITA, YUHKI YASUDA, and KOHJI FUJITA

Department of Metallurgy, Kyoto University, Kyoto, 606 Japan

ABSTRACT

Annealing behavior in O_2 flow after high temperature deformation was investigated using $YBa_2Cu_3O_y$ (YBCO) and $Bi_{0.7}Pb_{0.3}SrCaCu_{1.8}O_y$ (BPSCCO) ceramics. High temperature deformation causes the degradation of transport properties. In deformed YBCO samples, when they are directly subjected to annealing at 900° C, a small amount of CuO takes place together with oxygen release. Generation of such additional phase can arise from structural instability due to the damage introduced by deformation. To stabilize the YBCO phase, deformed specimens were intermediately annealed at around 500° C for 3 hours on a way of heating up to 900° C. This intermediate annealing served not only the prevention of mass decrease due to high temperature annealing but also the suppression of structural degradation. Annealing behavior of BPSCCO samples is also reported.

KEY WORDS: oxide superconductor, high temperature deformation, thermo-gravimetry, texture, heat treatment

INTRODUCTION

In a previous work [1], we showed plastic deformability and texture formation in oxide superconductors using YBCO and BPSCCO ceramics. These ceramics were capable of compressive deformation over 50% in strain at high temperatures, and the compressive deformation led to the development of (001)-oriented textures in those ceramics. However, their superconducting properties were degraded by the high temperature deformation. Such degradation may arise from the accumulation of residual strain as well as crystal defects introduced by the deformation. Therefore, in order to remove such damage, deformed ceramics must be subjected to appropriate annealing. The purpose of the present study is to find a suitable annealing condition for the heat treatments by means of thermo-gravimetric measurements using deformed YBCO and BPSCCO specimens.

EXPERIMENTAL

YBCO and BPSCCO specimens (13mmφ × 5mm) were prepared by a conventional sintering method described previously [2-4]. Compression tests for YBCO and BPSCCO specimens were made in air at temperatures above 840° C and 720° C, respectively. The strain rate was 3×10^{-5}/sec. After the compression tests, specimens were slowly cooled down with the rate of 1.7° C/min. Structural characterization was made by X-ray measurements using Cu-Kα and Co-Kα radiation. Thermo-gravimetric (TG) measurements were carried out in O_2 flow. In the TG experiments, the specimens were heated up to desired temperatures with the rate of 3° C/min and then were held for 10 hours at the temperature, followed by cooling to room temperature with the rate of 1.7° C/min.

RESULTS AND DISCUSSION

Figure 1 shows the temperature dependence of electrical resistivity for a YBCO sample before and after deformation of 50% in compressive strain at 840° C. As is seen from this figure, superconducting transition sharply takes place before deformation (original), while it occurs gradually after deformation. Further, the slope of the resistivity curve in normal conductive state becomes small after deformation. Such degradation of transport properties is supposed to result from the accumulation of residual strain as well as crystal defects introduced by the high temperature deformation. Therefore, in order to remove such damage due to the deformation, we tried to find the suitable annealing condition by means of thermo-gravimetric (TG) measurements in O_2.

In TG measurements of undeformed YBCO samples, their mass was markedly decreased during heating above 500°C and increased during cooling. The mass after cooling down to room temperature was almost completely recovered to that before heating, as already reported by other investigators [5,6]. On one hand, deformed samples exhibited different annealing behavior. Figure 2 shows TG curves of YBCO samples deformed at 840°C. The samples indicated by solid curves were heated up to 400°C, 500°C, 600°C and 900°C, and then held for 10 hours at each temperature, followed by cooling down to room temperature. The dashed curve in the figure is the result of two step annealing, the detail of which is discussed later. During heating of the YBCO samples, their mass is a little increased in the temperature range 350-450°C, resulting from oxygen absorption. Such increase was not observed in undeformed samples. Since high temperature deformation was made in air, some amount of oxygen can be released during the deformation so that the deformed samples are more deficient in oxygen content than undeformed samples are. Therefore, they can absorb oxygen during heating in O_2 flow before large oxygen release at temperatures above 500°C. As is seen from TG curves above 500°C for deformed samples, their mass was markedly decreased during heating while it was increased during cooling in a similar way as that of undeformed samples. However, their mass after cooling down to room temperature became larger compared with that before heating. Such weight gain, which is not observed in undeformed samples, is ascribable to the mass increase in the heating process at temperatures 350-450°C.

In a deformed sample heated up to 900°C (the top solid curve in Fig.2), the mass was slightly decreased during keeping the temperature at 900°C. Such mass decrease at 900°C was not observed in the undeformed sample. In order to make clear the cause of oxygen release at 900°C, X-ray measurements were made. Figure 3 shows the X-ray diffraction pattern of a sample directly subjected to the annealing at 900°C after the deformation at 860°C. In the figure, a small peak of CuO is observed, indicating the partial decomposition of YBCO phase induced by the annealing at 900°C. Such phase decomposition must be harmful to transport properties so that it must be suppressed during the annealing. As is understood from the TG measurements, a marked release of oxygen occurs above 500°C. Such large release of oxygen may be one of the factors which reduce the stability of YBCO phase. In addition, the presence of structural damage such as crystal defects is also one of the factors to reduce the phase stability.

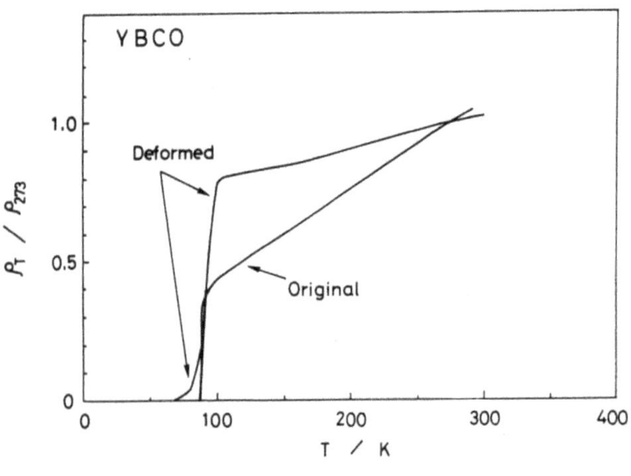

Fig.1 Temperature dependence of electrical resistivity in a YBCO sample before and after deformation.

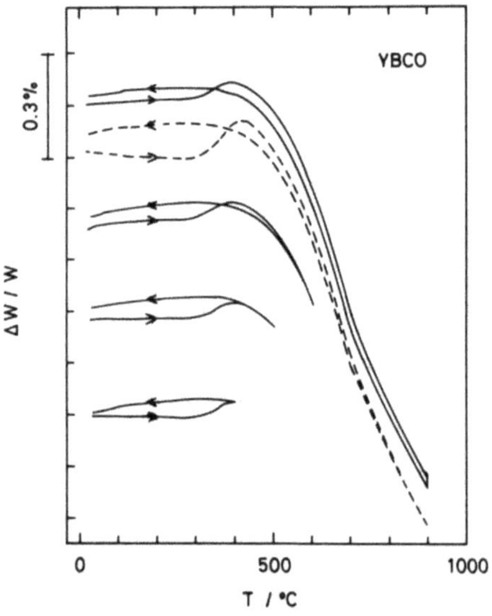

Fig.2 TG curves of deformed YBCO samples.

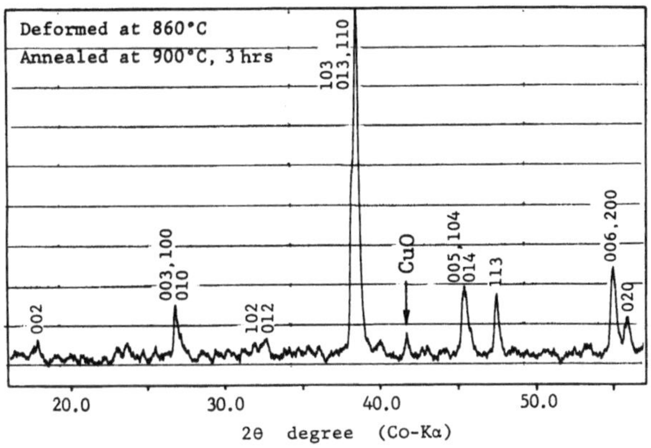

Fig.3 X-ray diffraction pattern of YBCO sample.

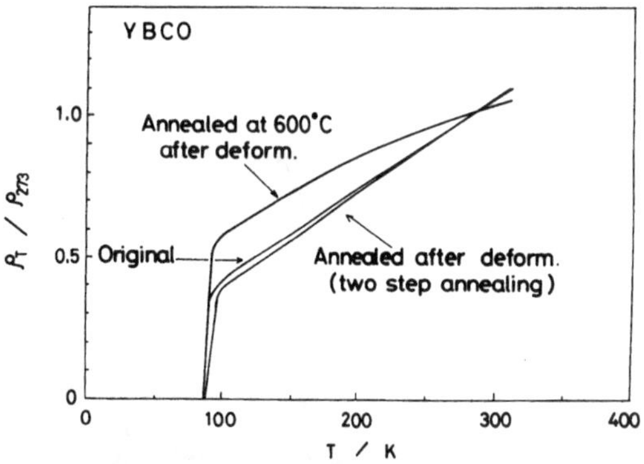

Fig.4 Temperature dependence of electrical resistivity.

In particular, accumulation of damage near grain boundaries may promote the generation of additional phase such as CuO at boundaries during annealing at high temperatures. Therefore, such damage should be reduced by heat treatments at intermediate temperature before marked release of oxygen at higher temperatures. Upon this consideration, a deformed sample was subjected to intermediate annealing at 500° C for 3 hours on a way of heating up to 900° C in order to suppress the structural degradation. The result is shown by the dashed TG curve in Fig.2. In this TG curve, the mass decrease during annealing at 900° C was not observed, suggesting that the intermediate annealing is effective in suppressing the oxygen release at 900° C. Further, a X-ray measurement for the sample subjected to two step annealing showed no diffraction peak of CuO. Figure 4 shows temperature dependence of electrical resistivity for a YBCO sample subjected to the two step annealing after deformation. In the figure, the results for an original sample and a sample annealed at 600° C for 30 hours after deformation are also shown. The transport properties of a deformed sample were improved after the two step annealing, while those in the sample annealed at 600° C is insufficient. Thus, the full annealing at 900° C as well as intermediate annealing at 500° C are necessary in order to improve the transport properties of YBCO deformed at high temperatures.

To examine the effect of annealing on texture formation, X-ray diffraction analyses were made using YBCO samples. The results are shown in Fig.5. The upper pattern in the figure corresponds to that for a original sample. The middle and lower patterns are those as deformed and after two step annealing, respectively. In this figure, the intensity ratio of a (00n) peak (n=1,2,3,5,6,7) to a (020) peak, $I/I_{(020)}$, are also shown below each pattern. The analyses suggest that (001) preferred orientation is developed by the compressive deformation at high temperatures and the developed textures are kept after the annealing. Thus, the annealing process does not hinder the texture formation.

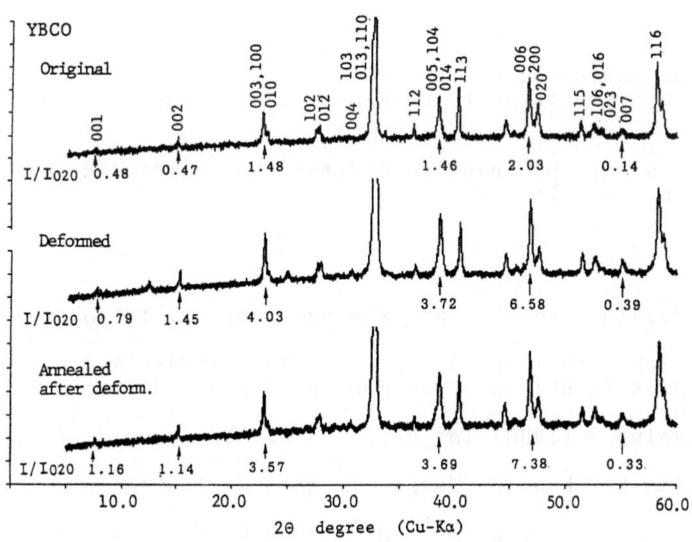

Fig.5 X-ray diffraction patterns of YBCO.

In BPSCCO, (001)-texture was also developed by compressive deformation at high temperatures, as

Fig.6

SEM photograph of BPSCCO; (a) undeformed, (b) deformed at 720°C.

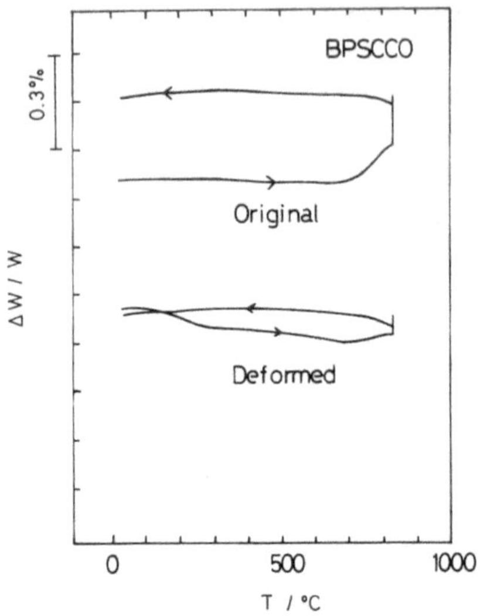

Fig.7 TG curves of BPSCCO

already reported [1]. Figure 6 shows the SEM photographs of BPSCCO samples before and after compression tests at 720°C. The observation was made on the fracture plane parallel to the compression axis. By the compressive deformation, the microstructure was changed from the aggregation of flake-like grains to layered structure densely packed. However, superconducting properties were markedly degraded by the deformation. To predict a suitable annealing atmosphere, we took into account the relation between the number of holes and oxygen deficiency. The number of carriers which are also holes in BPSCCO may be increased with decreasing oxygen deficiency. The deficiency must be decreased by the increase of oxygen pressure during aging at high temperature [4]. From this reason, the flowing O_2 was employed as an annealing atmosphere, although BPSCCO was sintered at 1/13 atm of oxygen pressure to form the structure of high T_c phase. Figure 7 shows TG curves of BPSCCO samples before and after deformation at 720°C. In an undeformed sample, the mass was markedly increased during heating above 700°C and also during annealing at 830°C. In a deformed sample, however, such marked mass increase was not observed. Transport properties of deformed samples were not much improved by such annealing.

In conclusion, the two step annealing in O_2 flow for deformed YBCO ceramics is quite effective to improve their superconducting properties. The oxygen absorption and desorption during annealing for BPSCCO ceramics before and after deformation were observed, but the trial to seek a suitable annealing condition did not succeed.

This work has been supported in part by Grant-in-Aid for Scientific Research from the Ministry of Education and Culture.

REFERENCES

1. Higashida K, Narita N (1990) In: Proc ISS'90. Springer-Verlag, Tokyo, pp 805-808
2. Leng S, Narita N, Higashida K (1987) Jpn J Appl Phys 26: L1394-L1397
3. Narita N, Leng S, Higashida K (1989) In: Proc MRS Int Mtg on Adv Mat Vol 6. MRS, Pittsburgh, pp 403-408
4. Narita N, Higashida K, Mishina S (1989) In: Proc ISS'89. Springer-Verlag, Tokyo, pp 109-112
5. Takeyama-Muromachi E, Uchida Y, Ukino K, Tanaka T, Kato K (1987) Jpn J Appl Phys 26: L665-667.
6. Kikuchi M, Syono Y, Tokiwa A, Oh-ishi K, Arai H, Hiraga K, Kobayashi N, Sasaoka T, Muto Y (1987) Jpn J Appl Phys 26: L1066-1069.

5.1 Bulk — (2) BSCCO

Crystal Growth of the High Quality 80K-Phase Bi-Sr-Ca-Cu-O Bulk Single Crystals

SATORU KISHIDA, HEIZO TOKUTAKA, MAKOTO KATAYAMA, MAKOTO CHIHAYA, KATSUMI NISHIMORI, and NAGANORI ISHIHARA

Department of Electrical and Electronic Engineering, Tottori University, Tottori, 680 Japan

ABSTRACT

We prepared high quality 80K-phase Bi-Sr-Ca-Cu-O (Bi-based) single crystals using the crucible with a cap by a self-flux method, changing the compositions of the starting materials. From the results, we found that the optimum composition of the starting materials was around Bi:Sr:Ca:Cu=2:2:1:2 and that the prepared single crystals showed good superconductivity. The optimum composition, Bi:(Sr,Ca):Cu=2:3:2 (Bi:Sr:Ca:Cu=2:2:1:2) was different from that obtained by using the crucible without a cap, Bi:(Sr,Ca):Cu=3:3:3. This may indicate that excess Bi and Cu in the starting materials evaporate during the crystal growth in the case prepared using the crucible without a cap.

KEYWORDS : Bi-Sr-Ca-Cu-O, single crystal, crystal growth, self-flux method

INTRODUCTION

In order to clarify the physical properties of Bi-Sr-Ca-Cu-O (Bi-based) superconductors, it is necessary to prepare a large and high quality Bi-based single crystal. We previously prepared the Bi-based single crystals using the crucible without a cap by a self-flux method, changing the compositions of the starting materials[1] or the temperature processes of crystal growth[2,3]. The X-ray diffraction patterns showed that the single crystals did not contain any intergrowth with a 7K-phase Bi-based superconductor. However, the single crystals showed the zero-resistance about 73K. In this study, we prepared the high quality 80K-phase Bi-based single crystals using an alumina crucible with a cap by a self-flux method, changing the compositions of starting materials. The resistance-temperature (R-T) characteristics and X-ray diffraction (XRD) of the prepared single crystals were measured. The results were compared with those obtained by using the crucible without a cap.

EXPERIMENTAL

The Bi-based single crystals were prepared using the crucible with a cap by a self-flux method. The conditions were as follows : the melting temperature ; 1000°C, the melting period ; 5h, the cooling rate ; 1°C/h, the starting temperature of cooling in a furnace ; 830°C. The compositions of the starting materials used in this work were (1) Bi:Sr:Ca:Cu=x:2:1:2(Bi ; x=1.25 - 2.5), (2) Bi:Sr:Ca:Cu=y:2:1:z(Bi ; y=1.75, 2, 2.25 : Cu ; z=1.5, 2, 2.5), (3) Bi:(Sr,Ca):Cu=2:w:2((Sr,Ca) ; w= 2 - 4, Sr:Ca=2:1).

The R-T characteristics were measured by a conventional four-point-probe method. Au electrodes were formed by a resistive heating method on the crystal surfaces and then, Cu wires were fixed using Ag-paste on the Au electrodes. The X-ray diffraction (XRD) measurements were carried out on the single crystals cleaved in air using an adhesive tape.

RESULTS AND DISCUSSION

Table 1 shows the typical size, the superconducting phase determined from the XRD patterns and the critical transition temperature Tc(R=0) of the single

Table 1 The characterizations of the Bi-based single crystals prepared using the starting materials of Bi:Sr:Ca:Cu=x:1:1:2 (Bi ; x=2.5 - 1.25).

Composition Bi : Sr : Ca : Cu	Typical size (mm^2)	Superconducting phase	Tc (R=0)
2.5 : 2 : 1 : 2	< 1 × 1	–	–
2.25 : 2 : 1 : 2	5 × 2	80 K	83 K
2 : 2 : 1 : 2	6 × 2	80 K	80 K
1.75 : 2 : 1 : 2	4 × 2	80 K	89 K
1.5 : 2 : 1 : 2	< 1 × 1	–	–
1.25 : 2 : 1 : 2	< 1 × 1	–	–

crystals, which were prepared using the starting materials of (1) Bi:Sr:Ca:Cu=x:2:1:2(Bi ; x=1.25 - 2.5). The single crystals prepared using x=2.25, 2 and 1.75 were plate-like with the sizes of 5 X 2, 6 X 2 and 4 X 2 mm^2, respectively. The sizes of the single crystals using other Bi compositions x were less than 1 X 1 mm^2. From the results, we found that the optimum Bi composition x was about 2. Therefore, the optimum composition was the same with the ideal composition of an 80K-phase Bi-based superconductor, Bi:(Sr,Ca):Cu=2:3:2 (Bi:Sr:Ca:Cu=2:2:1:2). However, Bi:(Sr,Ca):Cu=3:3:3 was the optimum composition of the starting materials obtained by using the crucible without a cap[1]. The result may indicate that excess Bi and Cu in the starting materials evaporate during crystal growth in the case prepared using the crucible without a cap.

Figure 1 shows the XRD patterns of the single crystal prepared using x=1.75. The prepared single crystals were easily cleaved in air. As shown in Fig.1, the cleaved surface of the single crystal prepared using x=1.75 was an a-b face, and the c-axis unit cell length was determined to be about 31A from the XRD pattern. Although the XRD patterns of the single crystal prepared using x=2 or 2.25 were not shown, the results were the same with the XRD pattern shown in Fig.1. From the results, we found that the single crystal prepared using x=1.75, 2 or 2.25 was an 80K single-phase and did not contain any intergrowth with a 7K-phase Bi-based superconductor.

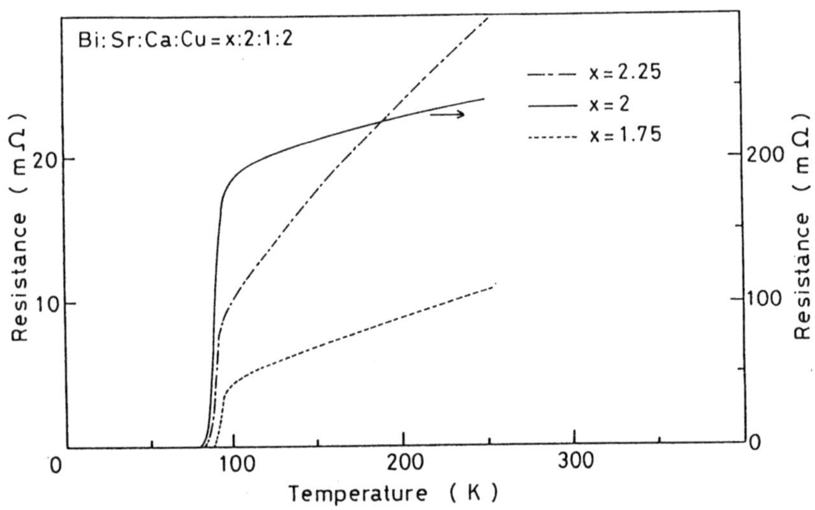

Fig. 1 The X-ray diffraction pattern the Bi-based single crystal, which was prepared using the starting material of Bi:Sr:Ca:Cu=1.75:1:1:2.

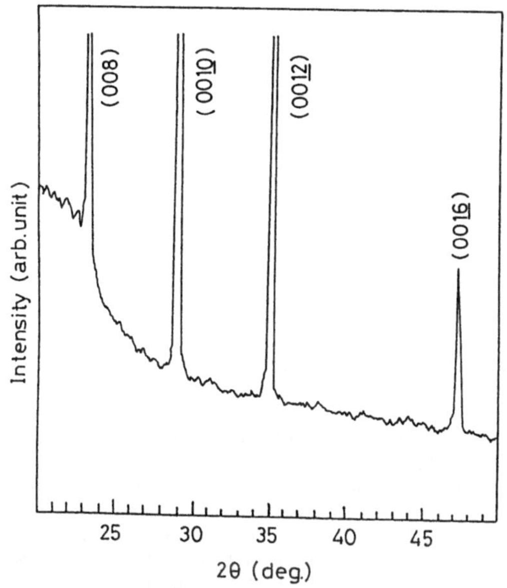

Fig. 2 The resistance-temperature characteristics of the Bi-based single crystals, which were prepared using the starting materials of Bi:Sr:Ca:Cu=x:1:1:2 (Bi ; x=1.75, 2, 2.25).

Figure 2 shows the R-T characteristics of the single crystal prepared using x=1.75, 2 or 2.25. The single crystals showed a metallic temperature dependence of the resistance in the normal state. Tc (R=0) of the single crystals prepared using x=1.75, 2 and 2.25 were 89, 80 and 83K, respectively. Tc of the single crystals using the crucible without a cap were about 73K[1,2]. The results indicated that Tc of the single crystals prepared using the crucible with a cap were higher than those using the crucible without a cap. This may be due that the temperature in the crucible during crystal growth becomes stable by using the crucible with a cap and that it is possible to prepare a high quality single crystal.

Table 2 shows the typical size, the superconducting phase determined from the XRD patterns and Tc(R=0) of the single crystals, which were prepared using

Table 2 The characterizations of the Bi-based single crystals prepared using the starting materials of Bi:Sr:Ca:Cu=y:1:1:z (Bi ; y=1.75, 2, 2.25 : Cu ; z=1.5, 2, 2.5).

Composition Bi : Sr : Ca : Cu	Typical size (mm^2)	Superconducting phase	Tc (R=0)
2.25 : 2 : 1 : 2.5	< 1 × 1	–	–
2 : 2 : 1 : 2.5	< 1 × 1	–	–
1.75 : 2 : 1 : 2.5	< 1 × 1	–	–
2 : 2 : 1 : 2	6 × 2	80 K	80 K
2.25 : 2 : 1 : 1.5	5 × 2	80 K	87 K
2 : 2 : 1 : 1.5	4 × 2	80 K	87 K
1.75 : 2 : 1 : 1.5	< 1 × 1	–	–

the starting materials of (2) Bi:Sr:Ca:Cu=y:2:1:z(Bi ; y=1.75, 2, 2.25 : Cu ; z=1.5, 2, 2.5). The single crystals prepared using (y,z)=(2,2), (2.25,1.5) and (2,1.5) were plate-like with the sizes of 6 X 2, 5 X 2 and 4 X 2 mm^2, respectively. The sizes of the single crystals using z=2.5 were less than 1 X 1 mm^2. This may indicate that excess Cu interrupts the growth of a large 80K-phase Bi-based single crystal. From the XRD patterns, we found that the single crystals showed an 80K single-phase. Tc of the single crystals prepared using (y,z)=(2,2), (2.25,1.5) and (2,1.5) were 80, 87 and 87K, respectively. Tc of the single crystals were also higher than those using the crucible without a cap. In addition, Tc of the single crystal prepared using (x,y)=(2.25,1.5) or (2,1.5), which is the Cu deficient composition in comparison with the ideal composition of an 80K-phase Bi-based superconductor, seems to be higher than that using (y,z)=(2,2).

We also prepared the Bi-based single crystals under the same conditions using the starting materials of Bi:Sr:Ca:Cu= 2:3-v:v:2(Ca ; v=0.5 - 2.5). Although the results were not shown, the single crystal prepared using v=1 was plate-like with the size of 6 X 2 mm^2 and the sizes of the single crystals prepared using other Ca compositions v were less than 1 X 1 mm^2. Therefore, we found that the optimum Ca composition v was 1, that is, Sr:Ca=2:1. The optimum ratio of Sr to Ca was different from that obtained by using the crucible without a cap, Sr:Ca=1:1[1]. Next, we prepared the single crystals using the starting materials of (3) Bi:(Sr,Ca):Cu=2:w:2((Sr,Ca) ; w=2-4, Sr:Ca=2:1). The results indicated that the optimum (Sr,Ca) composition w was 3.

SUMMARY

We prepared 80K-phase Bi-based single crystals using the crucible with a cap by a self-flux method, changing the compositions of the starting materials. The results were compared with those obtained by using the crucible without a cap. From the results, we found that the optimum composition of the starting materials was around Bi:Sr:Ca:Cu=2:2:1:2 under the condition used in this work and that the prepared single crystals showed good superconductivity. The optimum composition, Bi:(Sr,Ca):Cu=2:3:2 (Bi:Sr:Ca:Cu=2:2:1:2) was different from that obtained by using the crucible without a cap, Bi:(Sr,Ca):Cu=3:3:3. This may indicate that excess Bi and Cu in the starting materials evaporates during crystal growth in the case prepared using the crucible without a cap.

ACKNOWLEDGEMENTS

This research was supported by a Grant-in-Acid for Scientific Research on Priority Areas from the Ministry of Education Science and Culture, Japan.

REFERENCES

1. Kishida S, Tokutaka H, Nakanishi S, Fujimoto H, Nishimori K, Ishihara N, Watanabe Y, Futo W (1990) J. Crystal Growth 99: 937-941.
2. Futo W, Kishida S, Tokutaka H, Chihaya M, Nishimori K, Ishihara N (1991) In: Kajimura K, Hayakawa H (eds) in Advances in Superconductivity III. Springer-Verlag, Tokyo, pp.781-784
3. Kishida S, Tokutaka H, Katayama M, Chihaya M, Nishimori K, Ishihara N (1991) Physica C (to be accepted)

Effects of Doping Impurities into the Bi-Sr-Ca-Cu-O Ceramics on Superconducting Properties

HIROYA IMAO[1], SATORU KISHIDA[2], HEIZO TOKUTAKA[2], TOURU IKEUCHI[2], and YUICHI MATSUO[2]

[1] Department of Control Engineering, Matsue College of Technology, Matsue, 690 Japan
[2] Department of Electrical and Electronic Engineering, Tottori University, Tottori, 680 Japan

ABSTRACT

We investigated the effects of doping the impurities of Y_2O_3 and $YBa_2Cu_3O_{7-y}$ into the Bi-Sr-Ca-Cu-O (Bi-based) superconducting ceramics on the critical current density (Jc) and the critical temperature (Tc). Jc of the samples decreased by doping Y_2O_3 into Bi-based ceramics. This may be due that Y_2O_3 exists between the Bi-based superconducting grains and that the electrical coupling decreased. Jc of the samples increased by doping $YBa_2Cu_3O_{7-y}$. This may indicate that doping $YBa_2Cu_3O_{7-y}$ into the Bi-based superconductor is effective on the increase of Jc.

KEY WORDS : Bi-based superconducting ceramics, impurity doping, critical current density, critical temperature

INTRODUCTION

In order to apply superconducting oxides in the field of power electronics, it is important to prepare them with a high critical current density (Jc). We previously reported the effect of pressure before sintering on Jc and critical temperature (Tc) of the (Bi,Pb)-(Sr,Ca)-Cu-O (Bi-based) superconducting ceramics [1]. In this study, we doped the impurities of Y_2O_3 and $YBa_2Cu_3O_{7-y}$ into the Bi-based ceramics and investigated the effects of the sintering temperature and the impurity concentration on Jc and Tc.

EXPERIMENTAL

The 110K-phase Bi-based superconducting ceramics were prepared by a conventional solid-state reaction method [2]. The starting materials with the metal ratio of Bi:Pb:Sr:Ca:Cu=1.6:0.4:2:2:3 were mixed and calcined for 44h at 800°C in air. The products were again ground, mixed, pressed and sintered for 89h at 850°C in an atmosphere of $Ar:O_2$=12:1. Then, Y_2O_3 and $YBa_2Cu_3O_{7-y}$ were doped into the products.

The $YBa_2Cu_3O_{7-y}$ ceramics were prepared by a solid-state reaction method. The starting materials with the metal ratio of Y:Ba:Cu=1:2:3 were mixed and calcined for 12h at 900°C in air. The products were again ground, mixed, pressed and sintered for 12h at 900°C in air.

The 110K-phase Bi-based ceramics and impurities (Y_2O_3 and $YBa_2Cu_3O_{7-y}$) were mixed. The concentration of impurity was changed from 1wt% to 50wt% in Y_2O_3, or from 1wt% to 10wt% in $YBa_2Cu_3O_{7-y}$. The products were pressed and sintered at the temperatures from 600°C to 860°C for 5h in air.

The resistance-temperature (R-T) and current-voltage (I-V) characteristics of the samples were measured by a four-point-probe method. Figure 1 shows the typical R-T characteristics of 1wt% Y_2O_3-doped sample, which was sintered at 850°C. Tc and Tc_{on} were defined as shown in Fig.1.

RESULTS AND DISCUSSION

Figure 2 shows Tc of the 1wt% Y_2O_3-doped samples as a function of sintering temperatures. The samples sintered at temperatures from 600°C to 830°C did

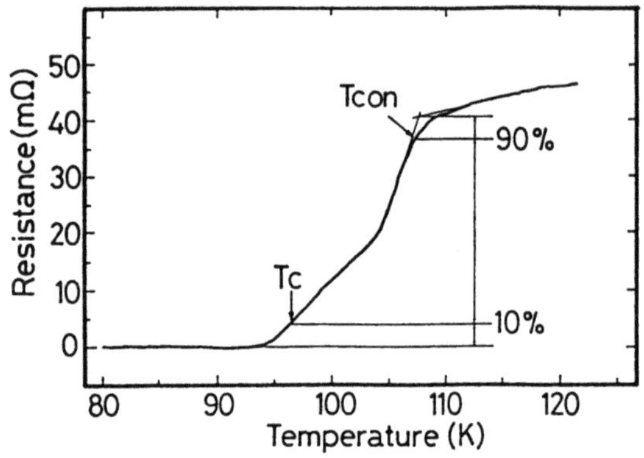

Fig.1 The R-T characteristics of the 1wt% Y_2O_3-doped sample, which was sintered for 5h at 850°C in air.

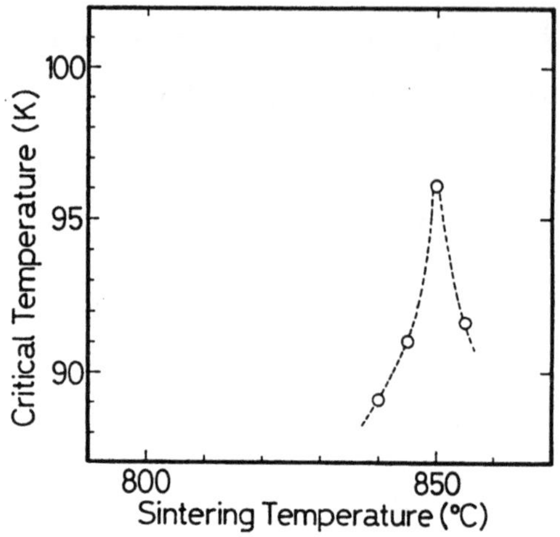

Fig.2 Tc of the 1wt% Y_2O_3-doped samples as a function of sintering temperatures. The samples were sintered for 5h in air.

not show zero-resistance. Tc of the sample sintered at 850°C was 96.1K, and it was the highest value of the 1wt% Y_2O_3-doped samples. However, the sample showed the low Jc of 0.67A/cm^2. Jc of the sample sintered at 845°C or 855°C was less than 0.01A/cm^2. From the results, we found that Tc and Jc of the samples were highly sensitive to sintering temperatures around 850°C.

When the concentrations of Y_2O_3 were changed from 5wt% to 50wt%, the samples showed Tc_{on}, but did not show the zero-resistance at the temperature above 77K. The resistances at 120K (R_{120}) of the samples increased by doping with Y_2O_3.

The AC magnetization-temperature characteristics of the samples were measured. The results showed that 110K- and 80k-phase Bi-based superconductors existed in the samples doped with Y_2O_3 of 1-50wt%. Although there was the superconducting phase in the samples, they did not show the zero-resistance. Therefore, Y_2O_3 is thought to exist between the Bi-based superconducting grains.

Figure 3 shows Tc of the 1wt% $YBa_2Cu_3O_{7-y}$-doped samples as a function of sintering temperatures. The samples sintered at the temperatures from 800°C to 855°C showed zero-resistance at temperatures above 77K. The samples sintered at 600°C and 700°C did not show zero-resistance, although Tc_{on} of the samples were observed. When the sintering temperatures were raised from 800°C to 830°C, Tc of the samples increased and then, saturated at the temperatures above 830°C. The saturated Tc value was about 103K.

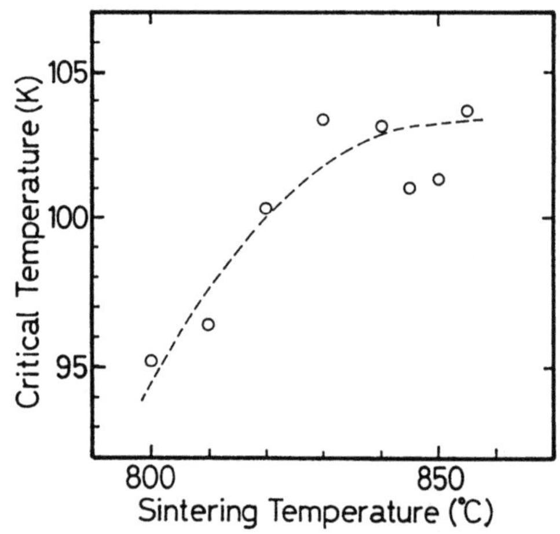

Fig.3 Tc of the 1wt% $YBa_2Cu_3O_{7-y}$-doped samples as a function of sintering temperatures. The samples were sintered for 5h in air.

Figures 4 and 5 show Jc and R_{120} of the 1wt% $YBa_2Cu_3O_{7-y}$-doped samples as a function of the sintering temperatures, respectively. Jc of the samples increased with increasing the sintering temperatures and then, saturated at the temperatures around 850°C as shown in Fig.4. Jc of the sample sintered at 850°C was 231A/cm^{2}. R_{120} of the samples decreased with increasing the sintering temperatures as shown in Fig.5.

Figures 6 and 7 show Tc and Jc of the samples sintered at 850°C as a function of the concentrations of $YBa_2Cu_3O_{7-y}$, respectively. The zero-resistance was also observed at 93.8K in the 10wt% $YBa_2Cu_3O_{7-y}$-doped sample as shown in

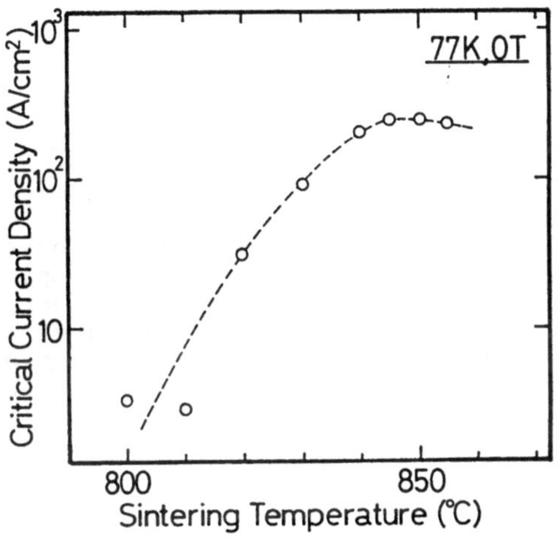

Fig.4 Jc of the 1wt% $YBa_2Cu_3O_{7-y}$-doped samples as a function of sintering temperatures. The samples were sintered for 5h in air.

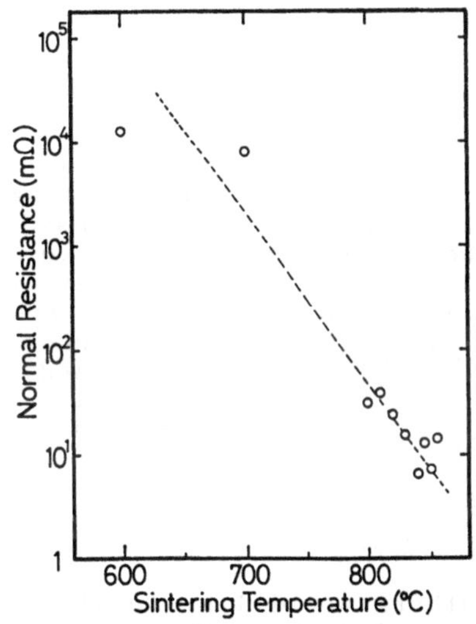

Fig.5 The resistances at 120K (R_{120}) of the 1wt% $YBa_2Cu_3O_{7-y}$-doped samples as a function of sintering temperatures. The samples were sintered for 5h in air.

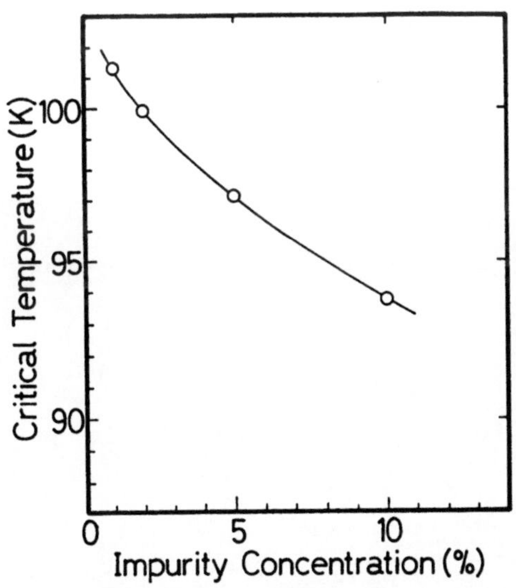

Fig.6 Tc of the samples as a function of the doping concentrations of $YBa_2Cu_3O_{7-y}$. The samples were sintered for 5h at 850°C in air.

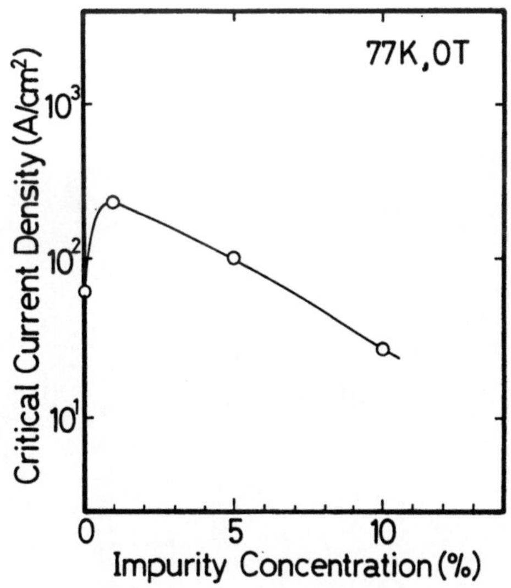

Fig.7 Jc of the samples as a function of the doping concentrations of $YBa_2Cu_3O_{7-y}$. The samples were sintered for 5h at 850°C in air.

Fig.6. The 1wt% $YBa_2Cu_3O_{7-y}$-doped sample showed the maximum Tc at 101.3K. When the concentrations of $YBa_2Cu_3O_{7-y}$ increased from 1wt% to 10wt%, Tc of the samples gradually decreased. Jc of the samples decreased with increasing the concentrations of $YBa_2Cu_3O_{7-y}$ as shown in Fig.7. The sample doped with 1wt% $YBa_2Cu_3O_{7-y}$ showed the maximum Jc of 231A/cm^2. This value was 4 times greater than that of the non-doped sample. Jc of the 10wt% $YBa_2Cu_3O_{7-y}$-doped sample showed 27A/cm^2. The value was as low as that of the non-doped sample.

Jc of the samples increased by doping 1wt% $YBa_2Cu_3O_{7-y}$ into the Bi-based ceramics. This may be due that $YBa_2Cu_3O_{7-y}$ improves the electrical coupling between grains, or acts as a pinning center in the samples.

CONCLUSIONS

We investigated the effects of doping the impurities of Y_2O_3 and $YBa_2Cu_3O_{7-y}$ into the Bi-based superconducting ceramics on superconducting properties. Jc and Tc of the samples decreased by doping Y_2O_3 into the Bi-based ceramics, and they were highly sensitive to the sintering temperatures around 850°C. The decrease of Jc and Tc may be due that Y_2O_3 exists between the Bi-based superconducting grains. Jc and Tc of the samples were increased by doping $YBa_2Cu_3O_{7-y}$ into the Bi-based ceramics. Jc of the 1wt% $YBa_2Cu_3O_{7-y}$-doped sample sintered at 850°C showed 247A/cm^2. The value was 4 times as large as that of the non-doped sample. This may indicate that the doping of $YBa_2Cu_3O_{7-y}$ into the Bi-based superconductor is effective on the increase of Jc.

ACKNOWLEDGEMENTS

The authors would like to thank Dr. Katsumi Nishimori, and to Mr. Naganori Ishihara of Tottori University for many useful discussions. This research was supported by a Grant-in-Aid for Scientific Research on Priority Areas from the Ministry of Education, Science and Culture, Japan.

REFERENCES

1. Imao H, Kishida S, Tokutaka H, Ikeuchi T : Physica C, (to be accepted)
2. Kishida S, Tokutaka H, Futo W, Chihaya M, Nishimori K, Ishihara N (1991) Phys. Stat. Sol. (a) 124: K117-K121

Properties and Application of High Ic Bi Based Oxide Superconductors Prepared by Laser Pedestal Growth Method

HISAO NONOYAMA[1], KAZUHIKO HAYASHI[1], MASAYUKI NAGATA[1], and KEN-ICHI TAKAHASHI[2]

[1] Osaka Research Laboratories, Sumitomo Electric Industries, Ltd., Osaka, 554 Japan
[2] Super-GM, Osaka, 530 Japan

ABSTRACT

$Bi_2Sr_2Ca_1Cu_2Ox$ Superconductors were prepared by laser pedestal growth method. Annealing conditions were investigated to increase Jc and the diameter dependence of Ic was studied to increase Ic. The best annealing condition was 840°C*more than 500 hours. When the diameter was 3.0mm, Ic=500A was obtained. Also, electric current was applied to a sample in liquid helium or with the one electrode of a sample in liquid helium and another over liquid helium as a current lead situation. Oxide superconductor will be thought to be useful for current lead on a superconducting magnet in liquid helium. The possibility of application for it was investigated. In liquid helium, 1000A current could be applied without quenching. But in current lead situation, there was increasing of the generated voltage at the upper electrode with more than 450A current.

KEY WORDS: $Bi_2Sr_2Ca_1Cu_2Ox$ superconductor, laser pedestal growth method, high Ic

INTRODUCTION

Since the discovery of high Tc superconducting Bi-Sr-Ca-Cu-O by Maeda et. al.[1], many researchers have studied for preparing high Jc superconductor for wire. Laser pedestal growth method is one of the directional solidification method and large temperature gradient at liquid-solid interface is easy to obtain [2][3]. So, a well aligned structure with high Jc can be achieved. When the nominal diameter of grown fiber was 0.5mm and the growing speed was 30mm/h, Jc of 17,500A/cm^2 (77.3K,0T) was obtained.[2]
The rod prepared by this method is suitable for a current lead. The first reason is that the thermal conductivity of the rod is much smaller than a rod with metaric composite material. The second reason is that though its Jc becomes higher with smaller diameter, it has a well aligned structure and high Ic can be obtained even in a comparatively large diameter rod. Until now, some high Ic superconductors have been prepared[4][5].

EXPERIMENTAL

Bi_2O3, $SrCO_3$, $CaCO_3$, and CuO were mixed in a rate of 2:2:1:2 (Bi:Sr:Ca:Cu), then calcined, pressed and sintered to form a feed material in any size. Fig.1 is a schematic of the laser pedestal growth method for growing a rod. The feed material that is powder pressed and sintered or previously grown rod is first placed in the chamber. CO_2 lasers are aimed from two directions at the tip of the feed material, which causes the formation of a molten zone at the tip. The molten zone is slowly pulled up and solidified rod is formed continuously at the desired growth rate while the feed material is pushed into the molten zone at any rate. Any sized rods can be prepared, which are depend on feed material size, feed material supplying speed and rod growing speed. In this study, the growing speed was 30mm/h only. All rods after growing were annealed in air in any conditions. Superconducting properties (Ic,Jc) were measured by four terminal method with a pulsed current. Current flow direction was parallel to the growing direction.

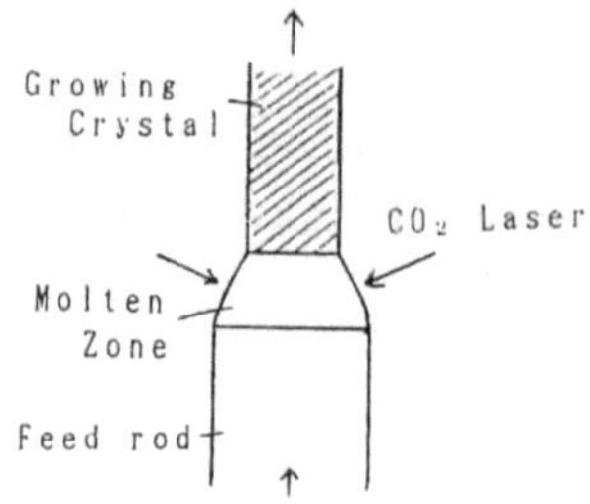

Fig.1 Schematic illustration of the laser pedestal growth method.

In the measurement of temperature dependence of Ic, the sample temperature was controlled as that, sample covered by aluminum thin plate with rolled heater was placed over liquid helium and sample temperature was controlled by operating the heater.
In the current lead situation, the changes of the temperature of some parts on the sample and of generated voltage at the electric terminal were measured.

RESULTS AND DISCUSSION

(1)Annealing condition
Fig.2 shows the annealing condition dependence of Jc. The best annealing temperature is 840°C, and with the longer annealing the Jc becomes higher until 500h. In a post annealing sample there are some kind of phases mainly 0112 and 2201 (or Bi-rich phase) (see photo 1). 2212 is formed by the reaction between the two phases, and advance the unification for 2212 with longer annealing (see Photo 2,3). Thus, in the case of the growing speed 30mm/h, the reaction time between the two phases is about 500h. This time will depend on the size of each crystal. So, if the two phases can be formed smaller in post annealing state, the shorter annealing time will be enough.

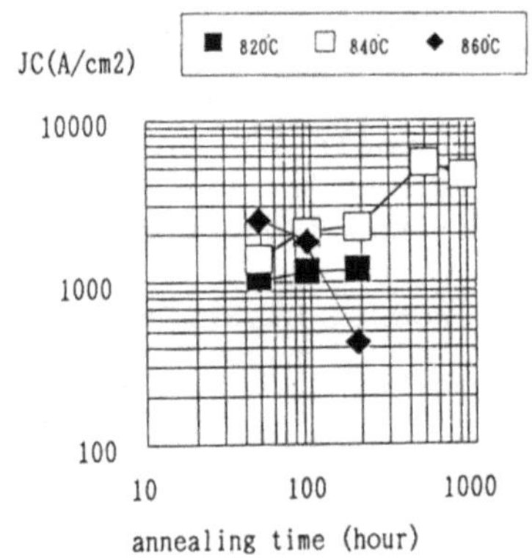

Fig.2 Annealing condition dependence of Jc. (77.3K,0T)

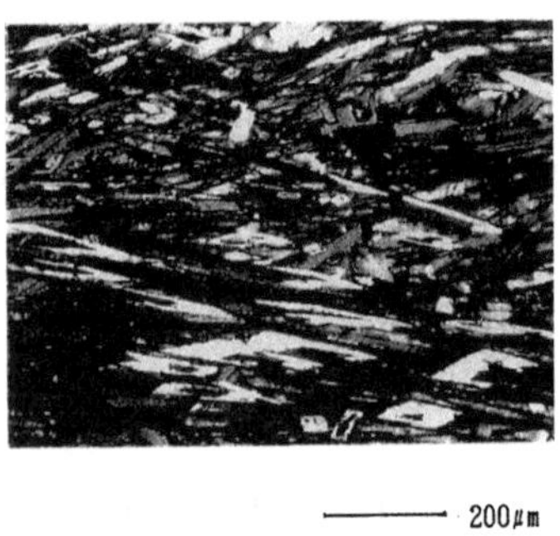

Photo 1 Longitudinal cross-section of as grown sample.

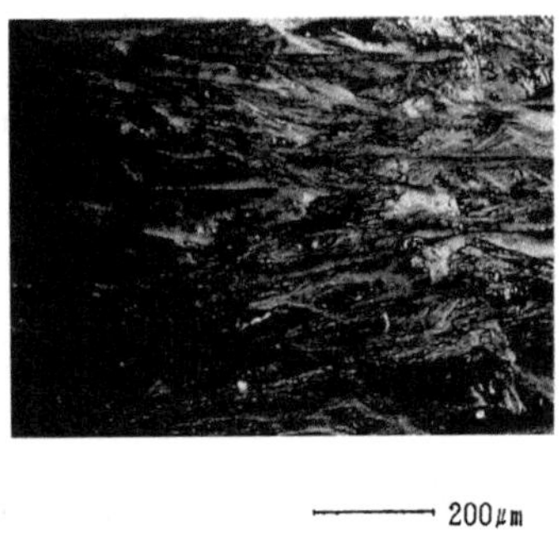

Photo 2 Longitudinal cross-section of 840C*192h sample.

Photo 3 Longitudinal cross-section of 840C*840h sample.

(2)Diameter

Fig.3,4 shows the diameter dependence of Jc, Ic. Jc is lower with the larger diameter. The reason is thought that in a smaller diameter sample, larger temperature gradient at liquid-solid interface is easy to obtain in sample growing. But as shown in Fig.3, the decreasing of Jc in the range of more than 3.0mm is larger than in the range from 0mm to 3.0mm. Therefor, as shown in Fig.4, Ic is the highest in the diameter of 3.0mm. In this result, Ic=500A (77.3K,0T) is obtained in the diameter of 3.0mm. Photo 2 shows the structure of cross section of a 3.0mm diameter sample. Though the orientation is inferior to a sample which diameter is 0.5mm (compare photo 1 with 2), it still has good orientation.

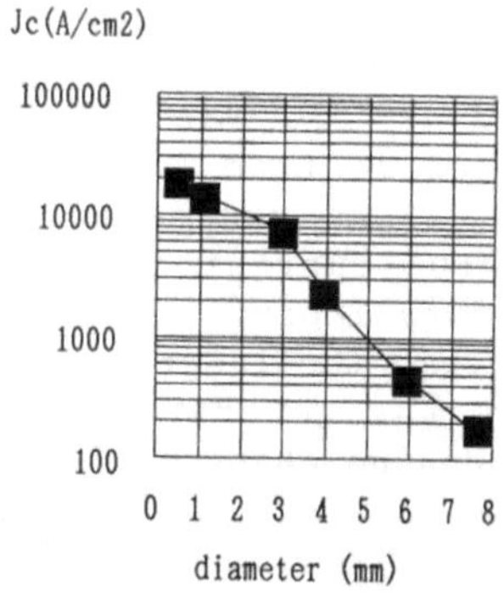

Fig.3 Diameter dependence of Jc. (77.3K,0T)

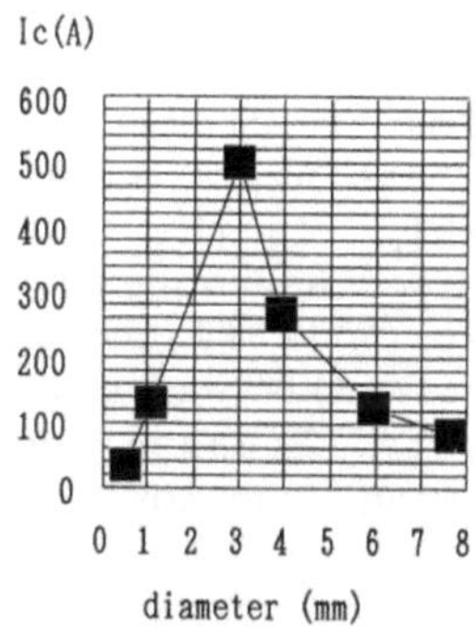

Fig.4 Diameter dependence of Ic. (77.3K,0T)

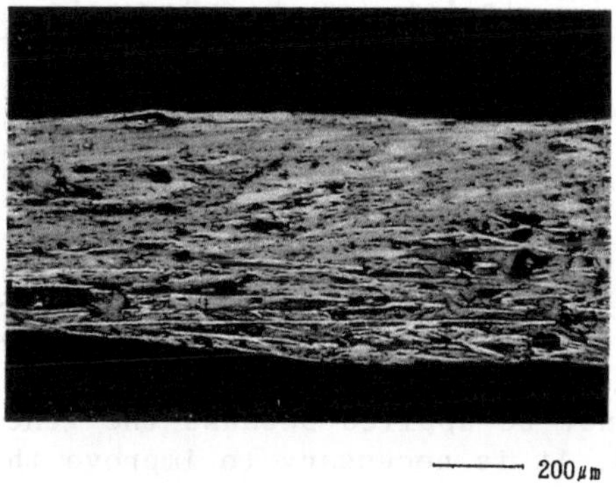

Photo 4 Longitudinal cross-section of 0.5mm diameter sample. (as grown)

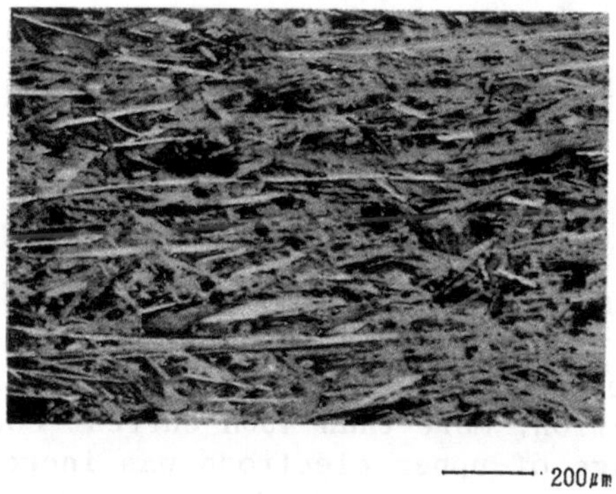

Photo 5 Longitudinal cross-section of 3.0mm diameter sample. (as grown)

(3)Temperature

In the case of applying oxide superconductors to a current lead, it is necessary to make clear the Ic of temperature dependence, because current capacity of a current lead as whole is decided when the extent of temperature which oxide superconductors are placed is limited. Fig.5 shows that result. In this sample, Ic is 35A, Jc is 2,300A/cm² (77.3K, 0T) and Ic is 1,000A, Jc is 65,000A/cm² (4.2K,0T). The ratio of Ic(4.2K) and Ic(77.3K) is more than 25. This value is comparatively larger than that of Y-system or Bi-2223 system (the value is about 7-8). When this result is applied to the aforesaid sample Ic=500A (Fig.2), the Ic in liquid helium is more than 12.5kA and the Ic at 30K is more than 6.8kA.

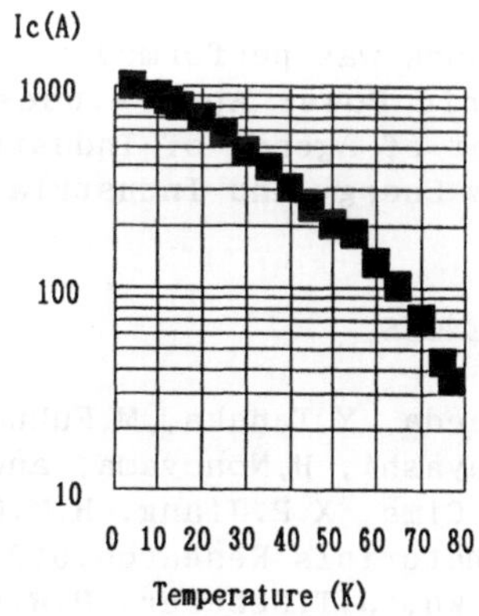

Fig.5 Temperature dependence of Ic. (0T)

(4)Current lead situation
1.0kA constant current was applied to a sample in liquid helium for 12 minutes. The sample size was 1.4mm in diameter and its Ic was 89A (77.3K,0T). There was no change of generated voltage in the sample (=zero) and at the electrodes. Confirmed that result, constant current was applied to another sample with the one electrode of the sample in liquid helium and another over liquid helium (Fig.6) as the situation of current lead. The sample size was 1.7mm in diameter and its Ic was 65A (77.3K,0T). As Fig.7, during the decreasing of liquid helium from A to B, 350A current could be applied stably. In the case of 450A, generated voltage of the upper electrode increased and more than 450A current could not be applied. The results show that (1)it is necessary to improve the way of cooling to apply higher current. (2)it was showed that the laser pedestal sample has low thermal conductivity. Because though the generated voltage of upper electrode was increased, the lower one had no change.

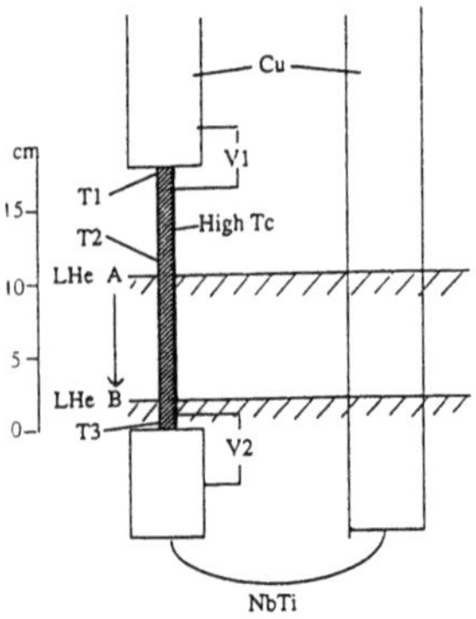

Fig.6 Experimental apparatus drawing

		T1	T2	T3	V1	V2
350A	A ↓	8K	8K	4.2K	0.3mV	0.03mV
	B	8K	8K	4.2K	0.3mV	0.03mV
450A	A ↓	8K	8K	4.2K	1.2mV	0.05mV
	B	8K	8K	4.2K	1.6mV	0.05mV

Fig.7 Change of temperature or voltage at some parts of a sample in Fig.6

In summary, the conditions to prepare high Ic laser pedestal growth sample have been clarified. With 840°C*more than 500h annealing and in the sample diameter of 3.0mm, the highest Ic=500A was obtained. In the current lead situation, more than 450A current could not be applied because the generated voltage of upper electrode was increased. It is necessary to improve the way of cooling to apply higher current.

ACKNOWLEDGMENT

This work was performed as a part of "R & D on Superconducting Technology for Electric Power Apparatuses" as a subject of Super-GM under the Moonlight Project of Agency of Industrial Science and Technology, MITI, being consigned by New Energy and Industrial Development Organization (NEDO).

REFERENCES

1.M.Maeda, Y.Tanaka, M.Fukutomi and T.Asano; Jap.J.Appl.Phys.,27,L209(1988)
2.K.Hayashi, H.Nonoyama, and M.Nagata; the proceedings of ISS'90.
3.M.J.Cima, X.P.Jiang, H.M.Chow, J.S.Haggerty, M.C.Flemings et.al. Journal of Materials Research,5(9),L1834 (1990)
4.J.L.Wu, J.T.Deberer, P.W.Eckels, S,K.Singh, J.R.Hull, R.B.Poepple et.al. IEEE Transaction on Magnetics, 27, No.2, 1991, L1861
5.D.U.Gubser, M.M.Miller, L.Toth, R.Rayne, S.Lawrence; IEEE Transition on Magnetics, 27, No.2, 1991, L1854

Microstructure and J_c of BiSrCaCuO Superconducting Fibers Prepared by Laser Pedestal Growth Method

K. HAYASHI[1], H. NONOYAMA[1], M. NAGATA[1], and K. TAKAHASHI[2]

[1] Osaka Research Laboratories, Sumitomo Electric Industries, Ltd., Osaka, 554 Japan
[2] Super-GM, Osaka, 530 Japan

ABSTRACT

BiSrCaCuO Superconducting fibers were prepared by laser pedestal growth method using a focused CO_2 laser. The relationships among microstructure, Jc-B, and post-growth treatments, were investigated using the fibers prepared at 30mm/h of growing speed. As-grown structures were mainly consisted of aligned 0112 phase and 2201 phase. Aligned 2212 superconducting phase was created along the alignment of 0112 phase after post-annealing. When samples were annealed under tensile force, structures observed by optical microscope became dense and nearly single 2212 phase even at short time annealing, but non-superconducting phases that are sub-micron order size, stacking faults were observed by transmission microscope. Non-superconducting phases were increased when the starting composition was changed from 2212 to 0112. Jc-B characteristics were improved by combination of tension annealing and the change of starting composition.

KEY WORDS : Bi-based superconductor, laser pedestal growth, tension annealing, TEM

INTRODUCTION

Since the discovery of high-Tc oxide superconductors[1], many efforts have been made to prepare high Jc wires for the applications in electric power field. Alignment of microstructure is needed to obtain high Jc in poly crystalline state, because high-Tc oxide superconductors have strong anisotropy of crystal structures and of superconducting properties. It is well-known that melt processed oxide superconductors have high density and highly aligned structures with high Jc. The laser pedestal growth method is also known as the manufacturing process of single crystalline oxide fibers[2] and BiSrCaCuO superconductors[3,4,5]. The laser pedestal growth method is the method of which higher temperature gradient is easy to obtain compared to other melt processes, because the molten zone is smaller as the laser beam is smaller focused. Temperature gradient at liquid-solid interface is the important factor to obtain aligned structure at higher growth rate.[5]
2212 phase of Bi-based oxide superconductor is suitable for laser pedestal growth method. But 2212 phase has inferior Jc-B characteristics at 77.3K to that of 2223 phase and YBaCuO superconductor. For the purpose of practical use of the fibers by laser pedestal growth method, higher growing speed with high Jc and superior Jc-B characteristics are needed. So the mechanism of aligned 2212 phase at higher growing speed condition and the possibility of improvement of Jc-B characteristics were investigated by changing the starting compositions and post annealing treatment. The relationship between Jc-B characteristics and post annealing treatment conditions were also investigated by observation of microstructure.

EXPERIMENTAL PROCEDURE

Powders of Bi_2O_3, $SrCO_3$, $CaCO_3$ and CuO were used as starting materials. Mixing ratio of Bi:Sr:Ca:Cu were mainly 2:2:1:2 and (1-x)(2:2:1:2)+x(0:1:1:2) composition(x is a molecular ratio of 0112.) were also selected. A sintered rod was used for the feed rod. Figure 1 is a schematic illustration of the laser pedestal growth method used in this study. The top of the feed rod was melted by CO_2 laser and the fibers were grown from the molten zone. The seed was the same sintered rod as the feed rod. Fiber diameters were controlled by changing the ratio of growing speed and feed speed, and the diameter of the feed rod. Growing speed was the same 30mm/h for all the samples in this study. All the samples after crystal growth were annealed in air and in part under the tensile force. Critical current densities were measured by the conventional four-terminal method in liquid nitrogen. Current flow direction was parallel to the growing direction and the external magnetic field was applied perpendicular to the current flow direction. The sensitivity of the critical current was 1uV/cm. The microstructure were observed by optical microscopy and transmission microscopy.

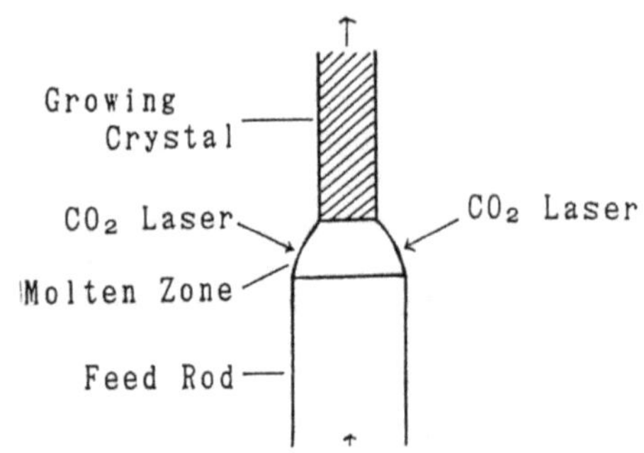

Fig.1 Schematic illustration of laser pedestal growth method.

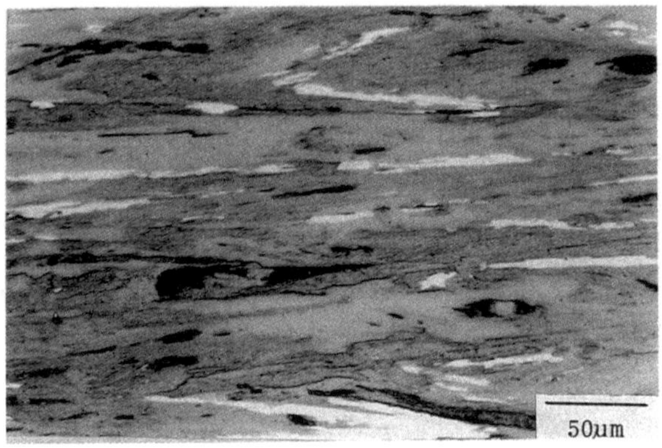

Photo1. Longitudinal cross section of 17,500A/cm2 (77.3K,0T) sample.

RESULTS AND DISCUSSION

The Jc of 17,500A/cm² (77.3K,0T) was obtained for the sample of 0.5mm in average diameter.[5] In this case, superconducting phase was not produced at as-grown state, but produced after post annealing. The micro structure of longitudinal cross section is shown in Photo.1. Aligned 2212 phase was produced along the 0112 phase by the reaction between aligned 0112 phase and 2201 phases, which produced at as-grown state. Other special features of microstructure are the existence of unreacted 0112 phase, compositional fluctuation of 2212 phase between 2201 and 0112, the existence of Kirkendall voids caused by the reaction between 2201 and 0112 phases, and the one dimensional alignment of 2212 phase.

When the tensile force was applied to the samples, which were 0.7mm in diameter, at the time of post annealing simultaneously, Jc and the n-value were increased as shown in figure 2. Samples were plastically deformed, which means the length was stretched and the diameter was decreased. Jc's were increased from 3,000A/cm² to 6,060A/cm² and n-value's were also increased from 10 to 80 as the tension was increased from 0MPa to 2.5MPa, respectively (77.3K,0T). Microstructure of longitudinal cross section of the sample that was annealed under 2.5Pa tension is shown in Photo.2 observed by optical microscope. It is obvious that this sample is dense and nearly single phase. Effects of tension are considered that the Kirkendall voids are decreased by plastic deformation and the numbers of the large non superconducting phase were decreased even in the short time annealing. As a result, current path was improved by tension annealing, so that the n-value and Jc were increased.

Further microscopic structure observation was also done by transmission microscope. Non superconducting phase of the order of sub micron size was observed in the aligned 2212 grains as shown in photo 3. The composition of this phase was Sr-Ca-O determined by EDX analysis. But the ratio of Sr and Ca was changed among the particles. The shape of this phase is oval. The coherency between this particle and the 2212 matrix phase is good by high resolution observation. It is considered that this non superconducting phase is residual unreacted phase caused by 0112 phase that produced at as-grown state. Photo.4

shows the stacking faults in the direction of c-axis and the intragrain amorphous phase in aligned 2212 phases. The length of stacking faults is shorter than normal 2212 phase, so these stacking faults are consisted of deficient of CuO plane. The composition of 2212 phase that produced after annealing had compositional fluctuation as described before. It is considered that this compositional fluctuation is caused by these stacking faults, because the c-axis length of 2201 phase is shorter than that of 2212 phase. Current path is not disturbed by amorphous phases, because these phases exist in the grains. Photo.5 shows the high resolution microscopy of the grain boundaries. There are no grain boundary phases and the coherency of two grains is good, although the angle of two grains is about 10 degree. The structural modulation, which is known to be observed in Bi-based oxide superconductors, is also observed in upper crystal near the grain boundary.

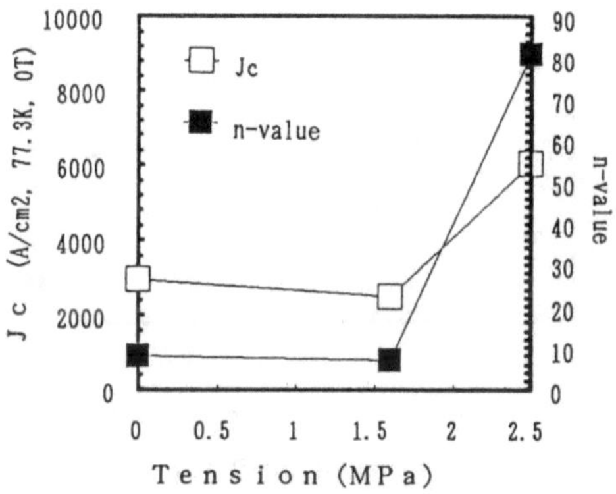

Fig.2 Effect of tension annealing. (2212 composition)

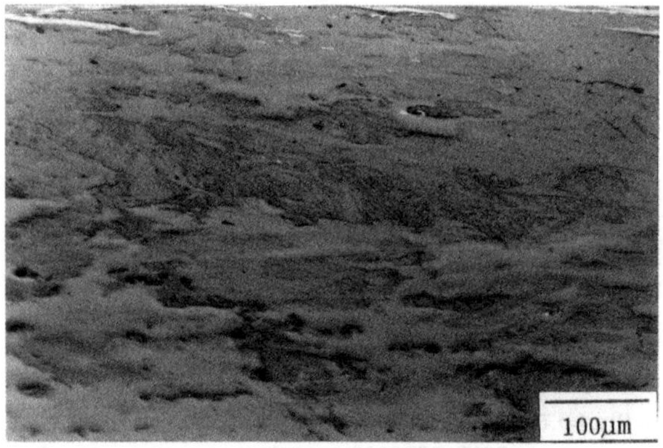

Photo.2 Longitudinal cross section of 2.5MPa tension annealing sample.(2212 composition)

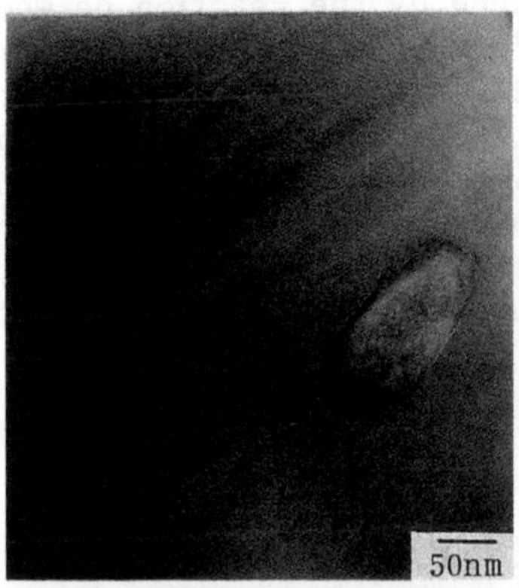

Photo.3 TEM observation of non superconducting phase.

Photo.4 TEM observation of stacking faults and intragrain amorphous phase.

Photo.5 TEM observation of grain boundary between 2212 phase.

As the starting composition was changed from 2212 to 0112 side, unreacted 0112 phases were increased. Photo.6 is the case of 2212-30mol%0112 of starting composition and 0.7mm in diameter after tension annealing. White phases are residual 0112 phase. Jc-B characteristics of 2212 starting composition without tension annealing, 2212 starting composition with tension annealing and 2212-30mol%0112 starting composition with tension annealing are compared in figure 3. Diameters of these samples were around 0.7mm. Jc's were compared among the samples that has Jc of around 5,000A/cm^2 (77.3K,0T) using the normalized Jc divided by zero field Jc. Jc-B characteristics were improved by tension annealing and composition change. These data are superior to that of the 2212 quasi-single crystal samples that are produced directly from the molten zone.[5] It is considered that the improvement of Jc-B characteristics is caused by the improvement of grain connections (decreasing the voids) and dispersing the fine non superconducting phase in the aligned 2212 superconducting phase. But the improvement is limited under the magnetic field of several hundreds' gauss, so the volume of non superconducting phase should be increased and finely dispersed.

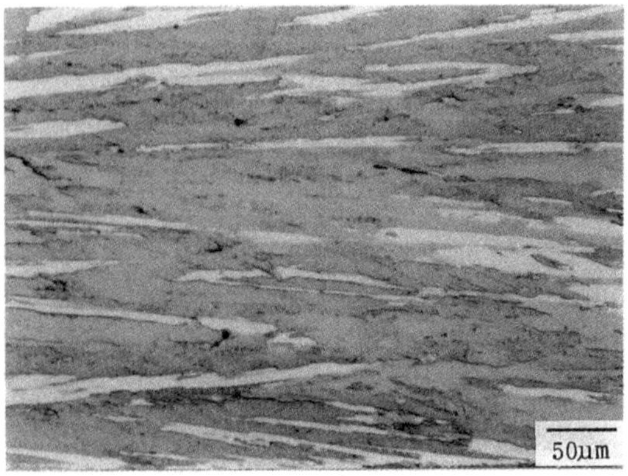

Photo.6 Longitudinal cross section of 2212-30%0112 sample.

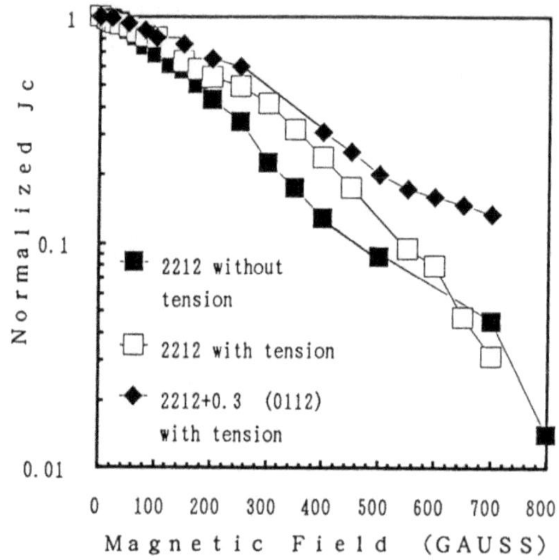

Fig.3 Comparison of Jc-B for some samples.

In conclusion, non superconducting phases, which are expected to act as the pinning center, can be dispersed in aligned 2212 superconducting phases produced by the reaction between 2201 phase and 0112 aligned phase under 30mm/h of growing speed. Kirkendall voids were decreased by tension annealing and the reaction between 0112 phase and 2201 phase was advanced by tension annealing, also. The Jc-B characteristic at 77.3K was improved by the combination of tension annealing and the starting composition change from 2212 to 0112 side. It is expected that the high field Jc-B characteristics at 77.3K will be improved by optimizing the microstructure, such as the volume ratio of fine non superconducting phases. This optimization will be done by optimization of the structure of as-grown state and post growth treatment.

ACKNOWLEDGMENT

The authors are grateful to Japan Fine Ceramics Center for the TEM observation.
This work was performed as a part of "R&D on Superconducting Technology for Electric Power Apparatuses" under the Moonlight Project of Agency of Industrial Science and Technology, MITI, being consigned by New Energy and Industrial Technology Organization (NEDO).

REFERENCES

1.J.G.Bednorz and K.A.Muller, Z.Phys. (1986)B4:189
2.R.S.Feigelson: E.Kaldis(Eds.) Crystal Growth of Electronic Materials, Elsevier Science Publishers (1985):127
3.R.S.Feigelson, D.Gazit, D.K.Fork and T.H.Geball, SCIENCE (1988)240:1642
4.K.Hayashi, H.Nonoyama, M.Nagata, H.Hitotsuyanagi and M.Kawashima: T.Ishiguro, K.Kajimura (Eds.) Advances in Superconductivity , Springer-Verlag Tokyo (1990):309
5.K.Hayashi, H.Nonoyama, M.Nagata, K.Takahashi: T.Ishiguro, K.Kajimura (Eds.) Advances in Superconductivity , Springer-Verlag Tokyo (1991):671

Silver-Coated Fibers Prepared from Superconducting Bi-Sr-Ca-Cu-O FZ-Rods

AKIHITO KUROSAKA[1], MAMORU AOYAGI[1], AKIRA SAJI[2], and TOSHIO INOUE[2]

[1] Fujikura Ltd., Koto-ku, Tokyo, 135 Japan
[2] Chubu Electric Power Co., Inc., Nagoya, 459 Japan

ABSTRACT

Superconducting fibers were prepared by a melt growth technique, i.e. a modified micro-Czochralski method, from $Bi_2Sr_2Ca_1Cu_2O_x$ source crystalline rods, and were then successfully electroplated with a stabilizing element of silver. The source rods were made by the floating zone method from ceramic rods. The maximum length of the fibers obtained was more than 550 mm (0.7 mm diameter). The thickness of silver-coated layer was about 30 μm. After the silver-coated fibers were annealed at 773 K for 6 h in air, the electrical contact of the $Ag/Bi_2Sr_2Ca_1Cu_2O_x$ interface was measured by the three-probe method. Contact resistance of less than 1.3×10^{-2} ohm•cm^2 was obtained at 77 K.

KEY WORDS: Bi system superconductor, melt growth, electroplating

INTRODUCTION

So far many techniques have been employed to fabricate wire of high-Tc oxide superconductors. The dominant methods include powder-in-tube [1], extrusion of a powder/binder mixture [2], and thin-film deposition [3]. These techniques typically employ a noble metal, usually silver, in tube, substrate, or filler metal form to impart flexibility and to aid in fabrication. The noble metal, moreover, can be used as a stabilizing element for high-Tc oxide superconductors. On the other hand, some melt growth techniques [4,5] have been employed to prepare Bi-Sr-Ca-Cu-O superconducting fibers of highly oriented grain structure with the aim of having high critical current density, because Bi-Sr-Ca-Cu-O superconductors have large anisotropy of crystal structure and of superconducting properties. However, the noble metal has not been a component of the superconducting fibers prepared by these methods. In this work, we prepared superconducting fibers by a melt growth method from $Bi_2Sr_2Ca_1Cu_2O_x$ source crystalline rods which were made by the floating zone melting process, and then successfully electroplated the fibers with silver by using argentocyanides solution. Subsequently, we measured some properties of those.

EXPERIMENTAL

Figure 1 is a schematic illustration of the melt growth technique, a modified micro-Czochralski method[5], employed in this study. The tip of a source crystalline rod is preheated by an electric furnace and melted with a heat coil. The molten zone is maintained by wetting of the melt to the heat coil. The crystalline fiber is pulled down from the melt through a pipe. The diameter of the fibers can be specified by the inner diameter of the pipe.

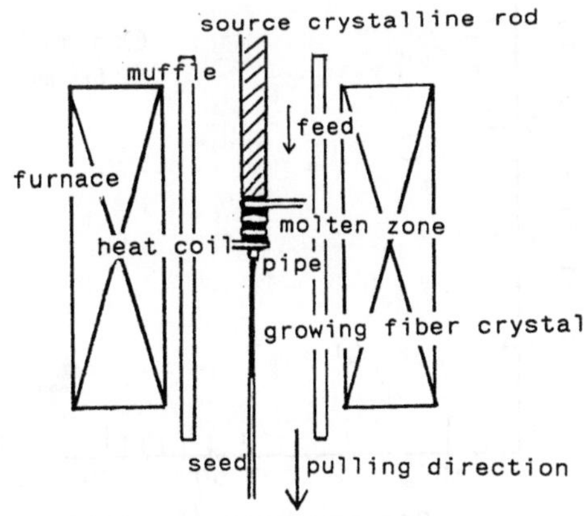

Fig.1. Schematic view of apparatus used for the melt growth technique. [(seed):Pt-wire, (heat coil):Pt-wire, (muffle):silica glass].

The $Bi_2Sr_2Ca_1Cu_2O_x$ source crystalline rods, approximately 3 mm in diameter, were prepared by the floating zone method from ceramic rods. The furnace temperatures were nominally maintained at 700, 750, 800 °C. The molten zone temperature was maintained more than 900 °C by regulating the electric current of the heat coil. The platinum pipes used for setting the fiber diameter had inner diameters of 0.3 mm and 0.6 mm, and fiber growth rates were 5 mm/h and 50 mm/h. Subsequently, the fibers were annealed at 780 °C for 72 hours in air.

The above fibers were electroplated with silver by using the argentocyaides solution, pH=11.5-12, of 33 g/l AgCN, 40 g/l KCN, 37 g/l K_2CO_3 and water. The electrolytic solution temperature was maintained within 23-26 °C, and the current density for the electroplating was 1 A/dm^2. Following the electroplating, the silver-coated fibers were annealed at 500 °C for 6 hours in air.

The temperature gradient of the Pt-pipe/fiber interface was measured by using a Pt-Rh thermocouple. X-ray Powder diffraction was carried out in order to evaluate a fraction of superconducting phase in the fibers, whose microstructures were observed with a optical microscope and a scanning electron microscope. The measurements of contact resistance at the Ag/$Bi_2Sr_2Ca_1Cu_2O_x$ interface and of critical current density in liquid nitrogen were performed by the three-probe method and the four-terminal method, respectively.

RESULT AND DISCUSSION

The fibers of 0.4 mm and 0.7 mm in diameter were prepared through the platinum pipes having inner diameter of 0.3 mm and 0.6 mm, the maximum lengths of those were more than 250 mm and 550 mm, respectively. However, the maximum length of the fibers grown directly from the ceramics rods, applied to the same technique as the present study, was less than 150 mm. We believe that the latter melt including more gas-bubbles compared with the former produces this result because of many voids in the ceramics rods. Hereafter, this paper reports simply on the fibers of 0.4 mm in diameter.

Table 1. Growth rates, nominal temperatures of the furnace and Jc values of the fibers, 0.4 mm in diameter.

Fiber type	Growth rate (mm/h)	Furnace temperature (°C)	Jc, at 77 K (A/cm^2)
1	5	700	98
2	5	750	356
3	5	800	280
4	50	700	0<
5	50	750	69
6	50	800	76

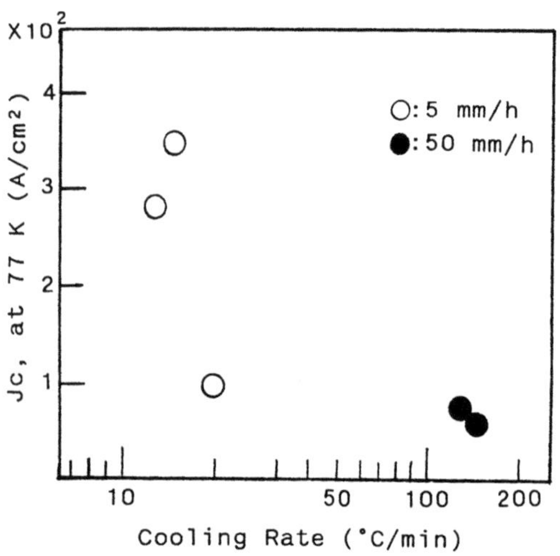

Fig.2. Relationship between the Jc value and cooling rate.

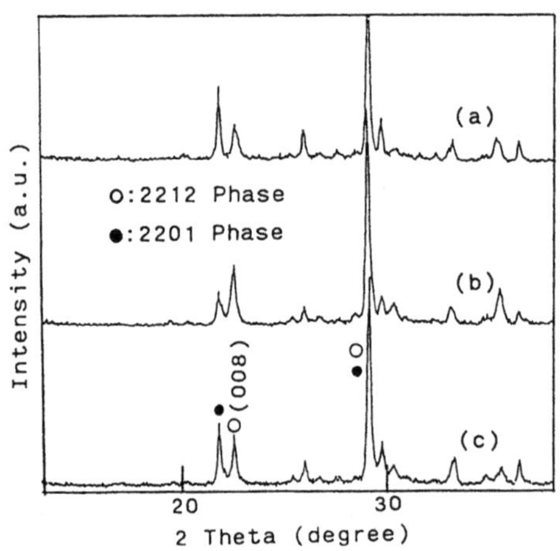

Fig.3. X-ray diffraction patterns of the fibers grown at the growth rate of 5 mm/h. Nominal temperatures of the furnace are (a):700°C, (b):750°C and (c):800°C.

(a)

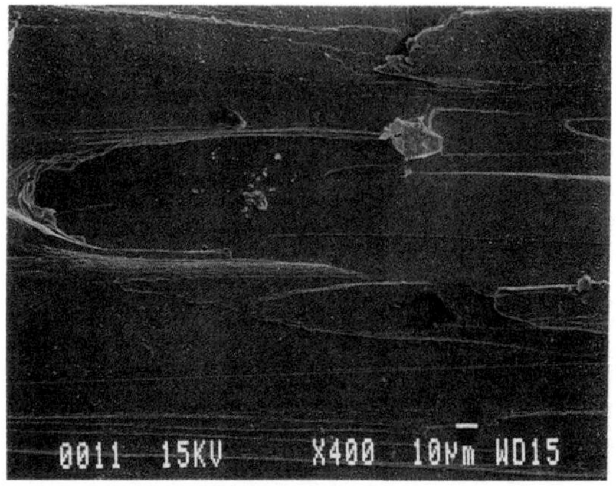

(c)

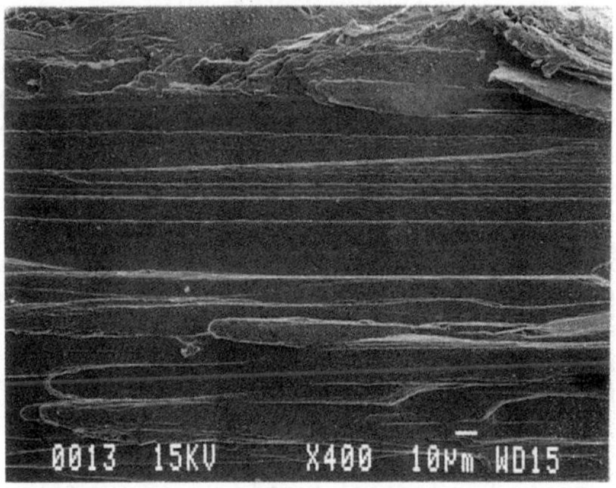

(b)

Fig.4. SEM photographs for surfaces of the fibers grown at the growth rate of 5 mm/h. Nominal temperature of the furnace are (a):700°C, (b): 750°C and (c):800°C.

Table 1 shows the growth rates, the nominal temperature of the furnace and Jc (at 77 K) values of the fibers prepared by the present method. The fibers grown at the growth rate of 5 mm/h, i.e. lower cooling rate, have the higher Jc values. Fig. 2 shows the relationship between the Jc value and cooling rate, which is obtained from the product of the growth rate and the temperature gradient of the Pt-pipe/fiber interface. From this relationship, it can be concluded that Jc values of the fibers depend on cooling rates during its crystal growth. The fibers, made in this work, exhibited considerably lower Jc values compared with the previous papers [4,5]. Therefore we believe that cooling rates and diameter sizes of the fibers are not yet optimized for obtaining higher Jc values. Fig. 3 and Fig. 4 are the X-ray diffraction patterns of the fibers grown at the growth rate of 5 mm/h and the SEM photographs for the surfaces of those, respectively. The (008) peak intensities of 2212-phase can be related to (b)>(a)>(c), as shown in Fig. 3, and the relationship of these intensities corresponds with that of the Jc values. The well-aligned grain structures were observed in all the fibers prepared by the present method, as shown in Fig. 4. Furthermore, the grain sizes of those became smaller with increasing the cooling rates.

Fig.5 shows the cross section of the silver-coated fiber, employed the type-2 shown in Table 1. The fiber was uniformly electroplated with silver, whose layer had about 30 μm in thickness. From the cross section of view, the contact at the $Ag/Bi_2Sr_2Ca_1Cu_2O_x$ interface seems to be well. Fig. 6 shows the contact resistances of that interface below room temperature in the silver-coated fiber, were as-coated and annealed at 500°C for 6 h in air. The contact resistance of the (b) shown in Fig. 6 is extremely low compared with the (a). Presumably, metal component like copper which remain in the Bi system superconducting fiber is diffused to the silver layer by annealing, so the electri-

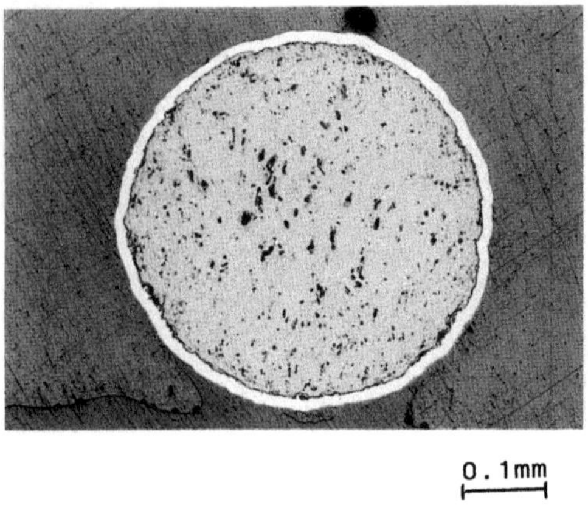

Fig.5. Optical microscope photograph for cross section of the silver-coated fiber.

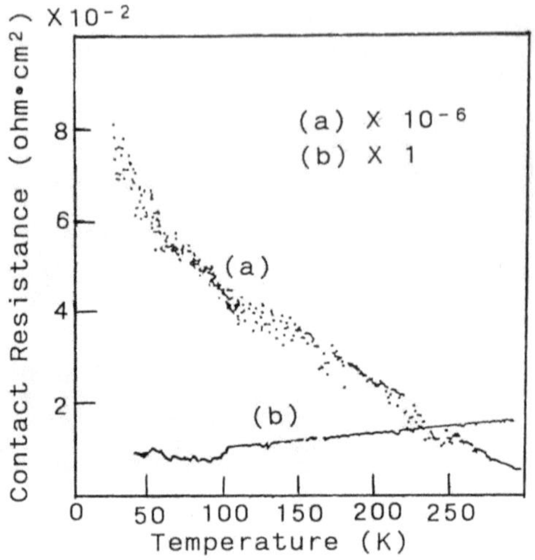

Fig.6. Contact resistances at the $Ag/Bi_2Sr_2Ca_1Cu_2O_x$ interface of (a): as-coated fiber and (b):annealed fiber. Note that the contact resistance in the case of the (a) is magnified 10^{-6}.

cal contact at the $Ag/Bi_2Sr_2Ca_1Cu_2O_x$ interface is improved. The silver-coated fiber, employed the type-2, had the Jc value of 330 A/cm^2. Therefore, we consider that the electroplating with silver does not deteriorate the superconducting property of the fibers.

CONCLUSION

We are aiming to prepare longer silver-coated superconducting oxide fibers with higher Jc values. In the present study, it is ascertained that Bi system superconducting fiber can be electroplated with silver. And furthermore, we believe that Bi system superconducting fibers with higher Jc values can be obtained by optimizing its diameter sizes and cooling rates during crystal growth.

REFERENCES

1. T.Hikata, T.Nishikawa, H.Mukai, K.Sato and H.Hitotsuyanagi (1989) Jpn.J.Appl.Phys. 28: L1204
2. M.T.Lanagan, S.E.Dorris, J.P.Singh, K.C.Goretta, U.Balanchandran, C.A.Youngdahl, J.T.Dusek, J.J.Picciolo and R.B.Poeppel (1990) J. Metals 42: 16
3. J.Zhao, Y.Q.Li, C.S.Chern, P.Norris, B.Gallois, B.Kear and B.W.Wessels (1990) Appl.Phys.Lett. 58: 89
4. R.S.Feigelson, D.Gazit, D.K.Fork and T.H.Geballe (1988) Science 240: 1642
5. A.Kurosaka, M.Aoyagi, H.Tominaga, O.Fukuda and H.Osanai (1989) Appl.Phys.Lett. 55: 390

Textured Bulk Sample of Bi-Sr-Ca-Cu-O

KAZUO MICHISHITA[1], YUMIKO IKUHARA[1], YUKIO KUBO[1], AKIRA SAJI[2], and TOSHIO INOUE[2]

[1] Japan Fine Ceramics Center, Nagoya, 456 Japan
[2] Chubu Electric Power Co., Inc., Nagoya, 459 Japan

ABSTRACT

A three-dimensionally textured structure with the c-axis nearly parallel each other was realized in the bulk sample with $Bi_{2.2}Sr_{1.8}CaCu_2O_y$ composition to which Ag was doped at 10 wt. %. If Ag was'nt doped or Ag-doped $Bi_2Sr_2CaCu_2O_x$ composition was used, this textured structure couldn't be obtained.
DTA analyses suggest that the low-Tc phase is formed by the reaction between a liquid phase and Bi-free insulator phase in $Bi_2Sr_2CaCu_2O_x$ composition, whereas in $Bi_{2.2}Sr_{1.8}CaCu_2O_y$ one the low-Tc phase crystallizes directly from a liquid phase. Both this direct crystallization and Ag-doping seem to play an important role on the formation of the textured structure.

KEY WORDS: Bi system superconductor, Floating zone melting, Magnetic property

INTRODUCTION

The authors have been studying the preparation of Bi-Sr-Ca-Cu-O superconductor by floating zone melting (FZ) method, which is very effective in obtaining a highly aligned structure with the a-b plane of the low-Tc phase parallel to the growth direction[1]. However, the c-axes of low-Tc crystals in the cross-section perpendicular to the growth direction were almost randomly aligned.
Recently, it was found that a textured structure with the c-axis nearly parallel each other as well as a-b plane can be realized. In this paper, we report on this textured bulk sample.

EXPERIMENTAL

Bi-Sr-Ca-Cu-O feed rods and seeds for the FZ experiments were prepared in the following manner: the commercially available oxalate coprecipitates with nominal compositions of $Bi_2Sr_2CaCu_2O_x$ and $Bi_{2.2}Sr_{1.8}CaCu_2O_y$ were calcined and mixed with 0 and 10 wt.% Ag powder (average diameter 2μm)[2,3]. The mixtures were formed into rod shapes with dimensions of about 8mmϕ × 100mmL under an isostatic pressure of 100 MPa and sintered at around 850℃ for 86 hours in air. The growth rate used in the present experiment is 2 mm/h. The detail of the FZ experiments are shown in the literature[1]. Grown samples were cut into rectangular bars for measuring Tc and magnetic properties. Some of rectangular bars were annealed at 840℃ for 3 hours in air.
Differential thermal analysis (DTA) and X-ray diffraction (XRD) analysis were applied to calcined powders and FZ samples, respectively. A polarized light microscopy was used to evaluate the orientation of low-Tc crystals. Tc measurements were performed on samples with dimensions of $0.65 \times 1 \times 18mm^3$ by the standard four terminal method. Magnetization were measured on samples with dimensions of $0.5 \times 1 \times 4mm^3$ using a SQUID magnetometer. Magnetic field was applied parallel to the longitudinal direction (growth direction).

RESULTS AND DISCUSSION

Figure 1(a) and (b) are optical micrographs of FZ rods of $Bi_{2.2}Sr_{1.8}CaCu_2O_y$

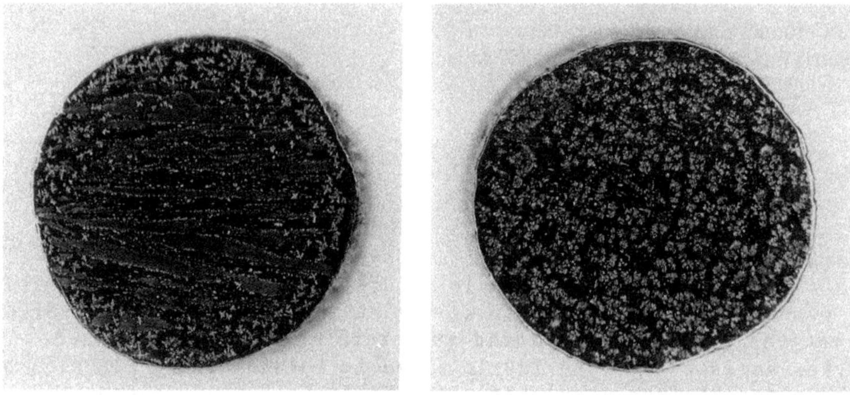

(a) (b)

Fig.1 Optical micrographs of FZ rods of (a) $Bi_{2.2}Sr_{1.8}CaCu_2O_y$ composition doped with Ag 10% and (b) $Bi_2Sr_2CaCu_2O_x$ one doped with Ag 10 %.

composition doped with Ag 10% (sample A) and $Bi_2Sr_2CaCu_2O_x$ one doped with Ag 10% (sample B), respectively. Polished cross-sections perpendicular to the growth direction are shown. In Fig.1(a), it is observed that thin-plate like crystals align nearly parallel each other. In Fig.1(b), Ag grains look white and the low-Tc phase black. XRD analyses showed that the low-Tc phase is predominant for both samples. By a polarized light microscope, it was found that c-axes of the low-Tc crystals for the sample B are almost randomly aligned. Whereas, those of the low-Tc crystals are nearly parallel each other for sample A. In Fig.1(a), c-axes are nearly vertical. Low-Tc crystals for the sample A are larger than those for the sample B. From EPMA analysis of sample B, $Bi_2Sr_2Cu_1O_z$ phase and Bi-free insulator phase were detected. In sample A, Bi-free insulator phase wasn't detected, and the low-Tc phase and $Bi_2Sr_2Cu_1O_z$ phase were predominant.

DTA measurements were performed to investigate the process of crystallization from a melt. While there were two exothermic peaks at 921℃ and 865 ℃ in $Bi_2Sr_2CaCu_2O_x$ composition, there was only peak at 861℃ in $Bi_{2.2}Sr_{1.8}CaCu_2O_y$ one. The 921℃ peak and about 860℃ peak seem to correspond to the solidifying temperature of Bi-free insulator phase and the low-Tc phase, respectively[4,5]. These results suggest that the low-Tc phase is formed by the reaction between a liquid phase and Bi-free insulator phase in $Bi_2Sr_2CaCu_2O_x$ composition, whereas in $Bi_{2.2}Sr_{1.8}CaCu_2O_y$ one the low-Tc phase crystallizes directly from a liquid phase.

Figure 2 shows the temperature dependence of resistivities of sample A and sample B that were annealed at 840℃ for 3 hours in air and slowly cooled. Tc of sample A and sample B are 85K and 86 K, respectively. There is a shoulder about 110K for sample B in contrast to for sample A. As the high-Tc phase may be formed by the reaction between the low-Tc phase and Bi-free insulator phase, it is considered that the high-Tc phase wasn't formed in sample A which contains no Bi-free insulator phase. But when sample B was annealed for 840℃ for 3 hours in air and quenched, the shoulder about 110K wasn't recognized. Therefore, the high-Tc phase might be formed in the cooling process.

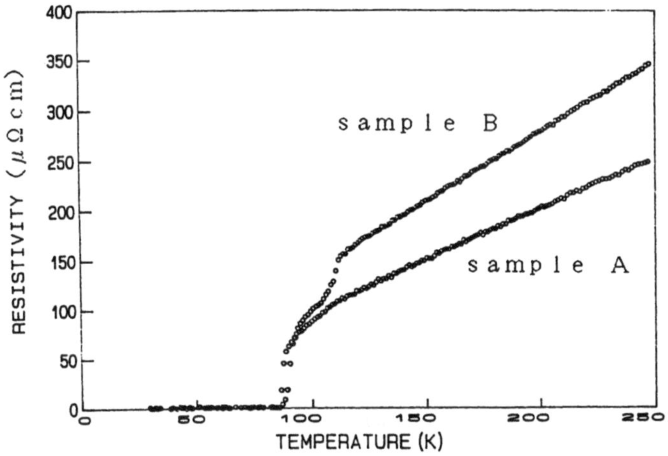

Fig. 2 The temperature dependence of resistivities of sample A and sample B.

In order to investigate the formation mechanism of the textured structure, a sample of Ag-undoped $Bi_{2.2}Sr_{1.8}CaCu_2O_y$ was prepared. This sample contained large-grained low-Tc phase and $Bi_2Sr_2Cu_1O_z$ phase as a main constituents and Bi-free insulator phase wasn't detected as well as for sample A. However, c-axis orientation was much poorer than for sample A. Thus, it is found that Ag-doping together with the direct crystallization of the low-Tc phase seem to play an important role on the formation of the textured structure.

Magnetic hysteresis loops at 77 K for sample A and sample B are shown in Fig.3 (a) and (b), respectively. Both samples are as-grown. It is recognized that ΔM (magnetization difference for increasing and decreasing fields) for sample A is far larger than for sample B. ΔM at zero field are 1.85×10^{-1} emu/cm^3 and 6.2×10^{-3}emu/cm^3, respectively. We are investigating whether this increase is due to the improvement of weak links between low-Tc crystals or the enhancement of pinning force.

In summary, a three-dimensionally textured structure with the c-axis nearly parallel each other was realized in the bulk sample with $Bi_{2.2}Sr_{1.8}CaCu_2O_y$ composition to which Ag powder was doped at 10 wt. %. If Ag isn't doped or Ag-doped $Bi_2Sr_2CaCu_2O_x$ composition is used, this textured structure couldn't be obtained.

DTA measurements suggest that the low-Tc phase in $Bi_{2.2}Sr_{1.8}CaCu_2O_y$ composition crystallizes directly from a liquid phase, while one in $Bi_2Sr_2CaCu_2O_x$ composition is formed by the reaction between a liquid phase and Bi-free insulator phase.
Both this direct crystallization of the low-Tc phase and Ag-doping seem to play an important role on the formation of the textured structure.
This textured sample showed a remarkably increased ΔM compared with randomly aligned one.

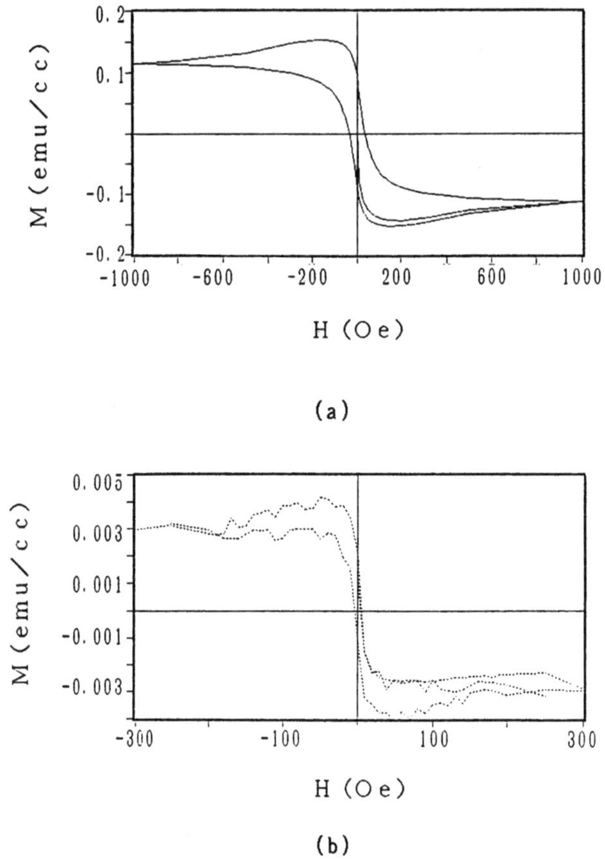

Fig. 3 Magnetic hysteresis loops at 77 K for (a) sample A and (b) sample B.

REFERENCES

1. Y.Kubo, K.Michishita, Y.Higashida, M.Mizuno, H.Yokoyama, N.Shimizu, E.Inukai, N.Kuroda, and H.Yoshida (1989) Jpn.J.Appl.Phys.28: L606
2. Y.Kubo, K.Michishita, N.Shimizu, Y.Higashida, H.Yokoyama, Y.Hayami, E.Inukai, A.Saji, N.Kuroda, and H.Yoshida (1989) Jpn.J.Appl.Phys.28: L1936
3. K.Michishita, N.Shimizu, Y.Higashida, H.Yokoyama, Y.Hayami, T.Tsunooka, E. Inukai, Y.Kubo, A.Saji, N.Kuroda, and H.Yoshida (1989) : T.Ishiguro, K.Kajimura (eds) in Advances in Superconductivity II. Springer-Verlag, Tokyo, pp 273-276
4. Y.Oka, N.Yamamoto, H.Kitaguchi, K.Oda, and J.Takada (1989) Jpn.J.Appl.Phys.28 : L213
5. Y.Oka, N.Yamamoto, Y.Tomii, H.Kitaguchi, K.Oda, and J.Takada (1989) Jpn.J. Appl.Phys.28: L801

The Effects of Process Parameters During Solidification by the Vertical Bridgman Method in Bi-Containing Superconductors

T. SUGAWARA[1], T. IZUMI[1], T.H. SUNG[1], Y. NAKAMURA[1], Z. NAKAGAWA[2], and Y. SHIOHARA[1]

[1] Superconductivity Research Laboratory, ISTEC, Koto-ku, Tokyo, 135 Japan
[2] Tokyo Institute of Technology, Yokohama, 227 Japan

ABSTRACT

The vertical Bridgman directional solidification was performed in order to clarify the groth mechanism from the melt of Bi-contained superconductors. The solidification experiment was carried out by pulling out at the furnace at the predetermined temperature gradient (G) and the several growth rates (R) which are important process parameters. The cell spacings were increased with decreasing growth rate. The interface temperatures of both primary and secoundary phases were increased with increasing growth rates. The microstructures of samples were changed from cellular to discontinuous with decreasing G/R ratio. The superconducting phase of $Bi_2Sr_2CaCu_2O_y$ was obtained as primary phase in the sample grown at 10mm/h.

KEYWORD : Bi-Sr-Ca-Cu-O, Bridgman method, directional solidification, crystal growth mechanism

INTRODUCTION

Since Bi-contained superconductors have been discovered by Maeda et. al.(1) , many reseach investigations have been tried in order to clarify structure(2,3) and characteristics relationship(4-6). For directional solidification method, floating zone(7), Laser-Heated-Pedestal-Growth(8), Czochralski method(9) were reported. Recently, solidification processing of oxide superconductors has been focused because of apparent reduction of weak links in bulk samples. The weak links are related mainly to high-angle grain boundaries which reduce the critical current density of bulk oxide superconductors(10). Therefore, the single crystal like microstructures or the oriented and continuous microstructures are desirable for applications. However, the growth mechanism, which is necessary for further development on the solidification processes, has been hardly reported. The Bridgman method is very visirable for clarifying the growth mechanism, because temperature gradient, G, and growth rate, R, can be controlled independently.

In this paper, the effects of process parameters on the microstructures prepared by the Bridgman method were discussed.

EXPERIMENTAL PROCEDURE

The raw materials were mixed with the ratio of Bi:Sr:Ca:Cu = 42:36:4:18 (following 2201), 32:25:13:30 (following 2212) and calcined twice at 800°C for 12h. The calcined powders were melted at 1400°C for 10 min in Platinum crucible and then quenched. The quenched rumps were crushed and pressed into rods (5mm in diameter and 150mm in length) and loaded into the high purity magnesia crucibles (5mm in inner diameter, 8mm in outer diameter and 300mm in length). The directional solidification was carried out by pulling out from the heated furnace ,controlled at 1400°C, at the several growth rates of 3~300mm/h with a constant temperature gradient. Then the crucible was drawn downward at a high speed during growth in order to observe the solid/liquid interface. The samples were cut along the growth direction and polished. Microstructures of the samples was investigated by EPMA.

RESULTS AND DISCUSSION

1) 2201 COMPOSITON

Photograph 1 shows the composition images at solid/liquid interface in samples grown at different growth rates. As the results of composition analysis, well-textured black phases were Bi-Sr oxide and white phases shown between the cells were 2201 phases. The well-aligned structures were recognized in the samples grown at 3~30mm/h. However, the structures became poor-orientated with increasing growth rates which means decreasing G/R ratios. It was considered that the orientation is deteriorated with decreasing G/R. The cell spacing, λ, became smaller in size with increasing growth rate as shown in this photograph. The mean values of cell spacing were decreased with increasing growth rate as shown in Figure 1. Consequently, it is found that the decreasing growth rate was required for primary crystals to be in a large scale grown.

Photograph.1 shows the positions of the interface is put in order, however, the practical interface temperatures are different . The interface temperature as a function of the growth rates were plotted in Figure 2. The interface temperatures of both primary and secondary phases were lowered with decreasing growth rate. This is explained by solute redistribution in cellular growth as follows. The composition gradient in liquid is predicted to exist because of the temperature gradient (G) within the cells along the growth direction as shown in Figure 3. Solute diffuses

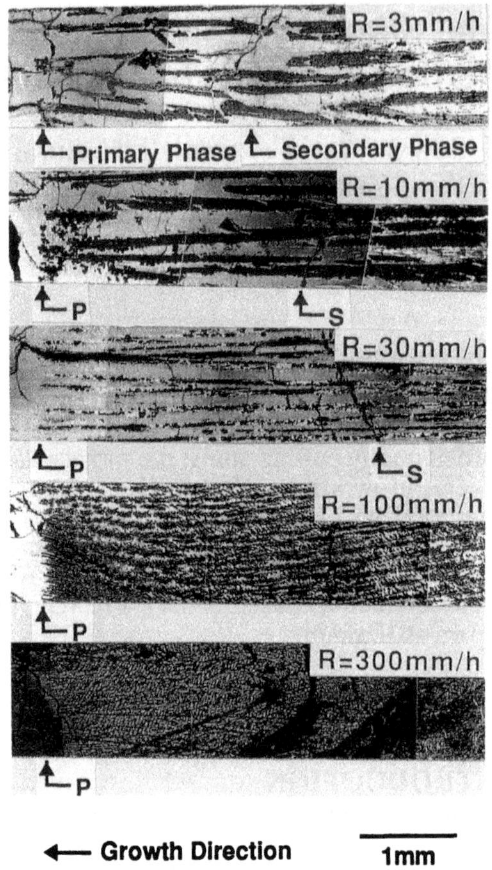

Photo. 1 Backscattered electron images of the samples grown at different growth rates in the starting composition of 2201.

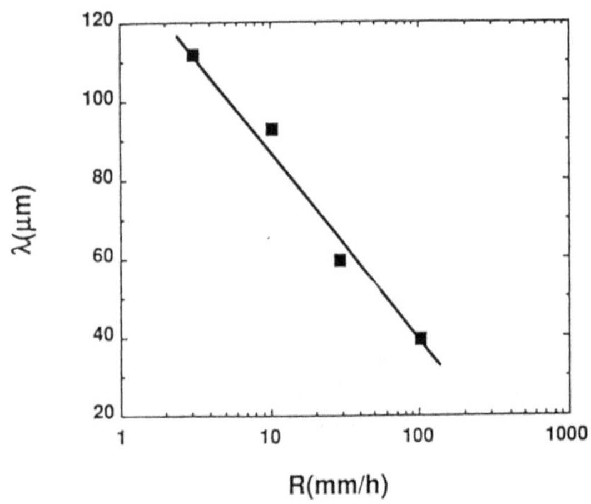

Fig. 1 Growth rate dependence of cell spacings.

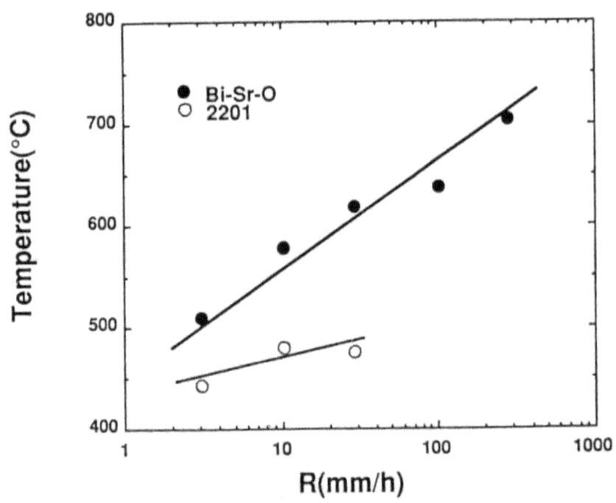

Fig. 2 Growth rate dependence of the interface temperatures of primary phase (Bi-Sr-O) and secondary phase (2201) in the starring composition of 2201.

down this concentration gradient and out ahead of the cell tips. As the result of this diffusion, the concentration of cell tips becomes higher in comparison with the starting composition. At steady state, the requirement of no solute accumulation gives the relation between growth rate(R) and the temperature of the cell tips(T_t) as follows(11);

$$T_t = T_L - D_L G(1/R)$$

where T_L is liquidus temperature and D_L is diffusion coefficient in liquid. Thus temperature of the cell tips becomes lower in proportion to the value of 1/R. Figure 4 shows the relation between the temperature of the cell tips and 1/R. The experimental results agree qualitatively with the above mentioned theory except for the sample grown at 300mm/h. The diffusion seems to be prevented because the microstructure of the sample grown at 300mm/h was dendritic in intercellular spacing. Therefore the decreasing of the tip temperature seems to be no agreement with the theory.

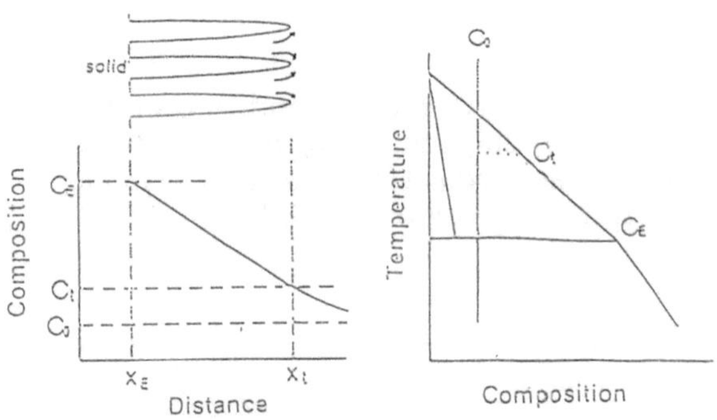

Fig. 3 Sketch of the principle of tip temperature decreasing.

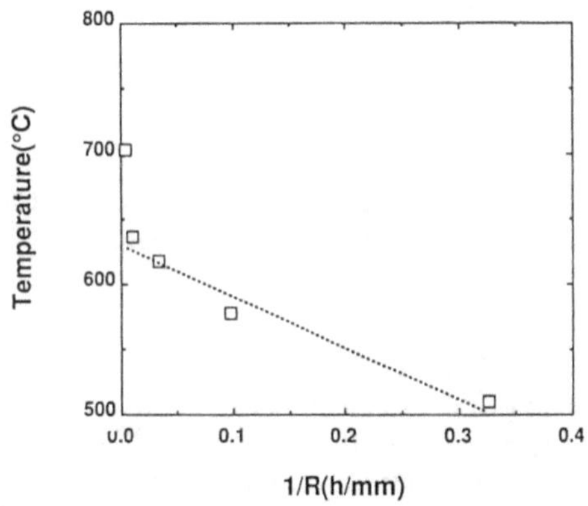

Fig.4 (1/R) dependence of the interface temperature.

2) 2212COMPOSITON

Photograph 2 shows the composition images at solid/liquid interface of the sample grown with different growth rates. The black phase was 2212 and the white phase was 2201. Cell-like and highly oriented structure was recognized in the sample grown at 10mm/h. In the sample of R=30mm/h, the 2201 phase was crystallized as the primary phase and 2212 phase was crystallized in dendritic as a secondary phase. The sharp solid/liquid interface could not be seen and discontinuously grown in the sample of R=100mm/h.

The effect of the process parameters, G and R, on the morphology of the solidification front was shown in Figure 5. As shown in this figure, a higher G/R condition is necessary for continuous solidification and obtaining well aligned structures. Figure 6 shows the 2201 and the 2212 crystallization temperatures against the growth rates. The crystallized temperatures of both 2201 and 2212 were decreased with decreasing growth rates and were equalized at 10mm/h. Then the superconducting phase of 2212 was obtained as primary phase.

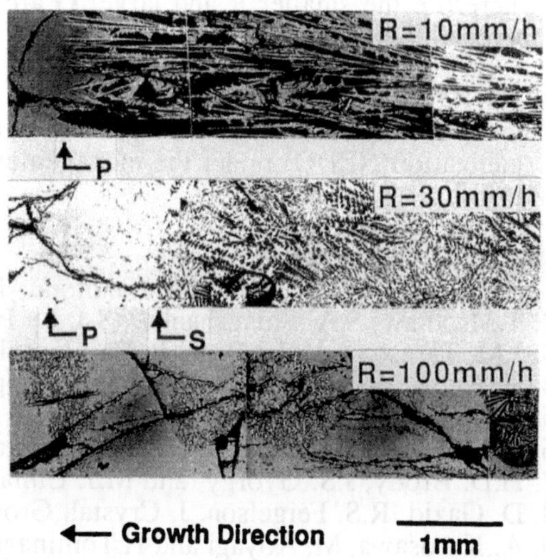

Photo. 2 Backscattered electron images of the samples grown at different growth rates in the starting composition of 2212.

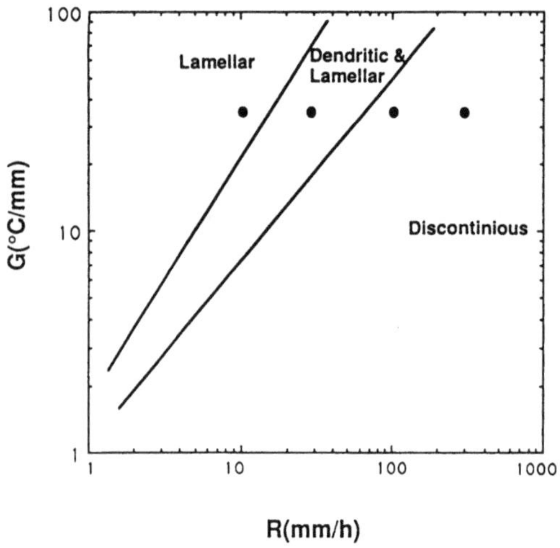

Fig. 5 Effect of process parameters, G and R, on the morphology of the solidification front.

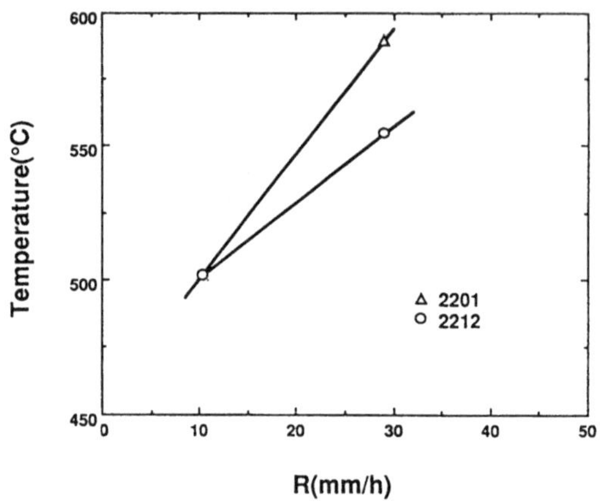

Fig. 6 Growth rate dependence of the interface temperatures of primary phase (2201) and secondary phase (2212) in the starting composition of 2212.

CONCLUSION

The effects of decreasing growth rate were summarized as follows;

in starting composition of 2201

*The orientation was higher.
*Cell spacing was widened. (Crystals are lager in size.)
*Interface temperature decreased.

in starting composition of 2212

*The orientation was higher.
*Interface temperature decreased.
*The 2212 phase was obtained as primary phase at the growth rate of 10 mm/h.

Therefore the smaller R and larger G are required in order to obtain textured and large superconducting crystals.

ACKNOWLEDGEMENTS

This study has been supported by the New Energy and Industrial Technology Development Organization(NEDO) under the management of the R&D of Basic Technology for Future Industries.

REFERENCES

1. H. Maeda, Y. Tanaka and M. Fukutomi, Jpn. J. Appl. Phys. Lett. 27 (1988) 209
2. T.M. Shaw, S.A. Shivashankar, S.J. La Placa, Phys. Rev. B 37 (1988) 9856.
3. J.M. Tarascon, Y. Le Page, P. Barboux, B.G. Bargly, Phys. Rev. B 37 (1988) 9382
4. H. Ueda, Y. asaka, H. Maeda, Jpn. J. Appl. Phys. Lett. 27 (1988) 665
5. K. Tagana, H. Kumakura and H. Maeda, Jpn. J. Appl. Phys. Lett. 27 (1988) 323
6. R. M. Hazen, C.T. Prewitt, R. J. Angel and C.W. Chu, Phys. Rev. Lett. 60 (1988) 1174
7. H.D. Brody, J.S. Gyorgy and M.J. Cima, J.Crystal. Growth. 96 (1988) 225
8. D. Gazid, R.S. Feigelson, J. Crystal. Growth. 91 (1988) 318
9. A. Kurosawa, M. Aoyagi and H.Tominaga, Appl. Phys. Lett. 55 (1989) 390
10. D. Dimos, P. Chaudhari, J. Mannhart and F.K. LeGoues, Phys. Rev. Lett. 61 (1988) 219
11.M.C.Fleming, Solidification Processing (McGraw-Hill, New York, 1974)

Effects of Heat Treatment on T_c of Liquid-Quenched Bi Oxides

S. WATANABE, T. MANABE, K. OGURI, T. AIDA, K. NOZAKI, and Y. NISHI

Department of Materials Science, Tokai University, Hiratsuka, Kanagawa, 259-12 Japan

ABSTRACT

Although T_c cannot be found for liquid-quenched Bi oxides glass. A high T_c^{mid} is found after annealing just below melting point. The maximum T_c^{mid} is lager than that of sintered samples. Based on the electronic and diffractional discussion, the effects of heat treatment on T_c are discussed.

Key words: superconductor, liquid-quenching, Tc, annealing, crystallization

INTRODUCTION

A new class high T_c oxide superconductor of sintered Bi-Pb-Sr-Ca-Cu-O has been discovered[1]. Since the melting point of this oxide is rather low, it is easy to be melted and liquid-quenched. The density of the liquid-quenched sample is apparently larger than that of the sintered one, which may improve the superconducting performance of the sample. However, the T_c can not be found of the glassy Bi system. Thus, we intend to investigate the effects of heat treatments on T_c for the liquid-quenched Bi oxides.

EXPERIMENTAL PROCEDURE

Samples with nominal composition $Bi_{1.6}Pb_{0.4}Sr_2Ca_2Cu_3O_x$ and $Bi_{0.7}Pb_{0.3}SrCaCu_{1.8}O_y$ were prepared from high-purity Bi_2O_3 (99.99%), PbO (99.99%), $SrCO_3$ (99.9%), $CaCO_3$ (99.99%) and CuO (99.99%). The powders were mixed and sintered in air at 1023 K for 6 h and then air cooled. After crushing, pelletized tablets 2 mm thick 13 mm diameter, were sintered in air at 1073 K for 18 h and then air cooled. The pellets were submitted to additional sintering in air at 1123 K for 12h and then furnace-cooled.

Liquid-quenched glassy foils were prepared from chips cut from the sintered tablets using a twin-type piston-anvil apparatus [2-5] as shown in Fig. 1. The twin-type piston-anvil apparatus was constructed to quench the molten sample in a center of quartz tube. The sample was melted by an infrared furnace. After melting of the sample, the sample was quenched. The Cu substrate was moved to touch the sample by air-piston. The speed of the piston was about 0.12 m/s. The liquid-quenched sample was annealed in air and then furnace-cooled.

The electrical resistivity is measured using a standard four-prove technique. Temperature is measured by a Au · Fe-chromel thermocouple attached to the sample in a cryostat at equilibrium temperature. Figure 2 shows the temperature dependence of electrical resistance of the sintered Bi oxides.

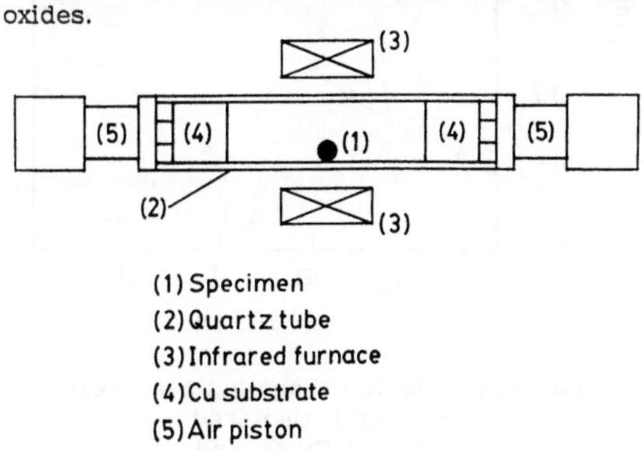

Fig.1 Schematic diagram of twin-type piston-anvil apparatus.

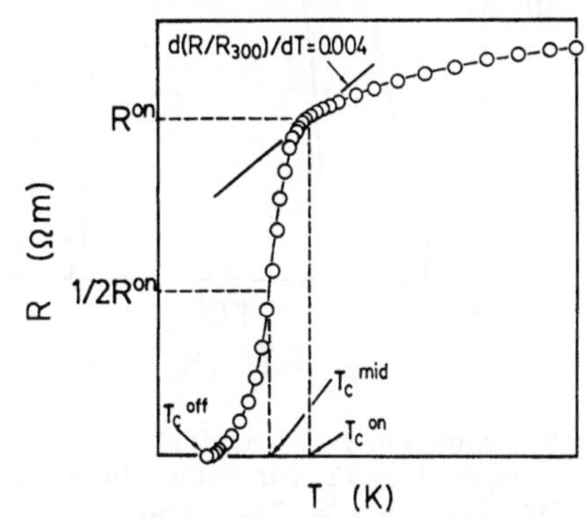

Fig.2 Change in electrical resistivity (R) against temperature (T) for Bi oxides.

The electrical resistivity of the samples shows metallic decreases with temperature. Onset T_c^{on} of the transition is defined as the temperature corresponding to the deviation of the electrical resistivity. The deviated point is taken at $d(R/R_{300})/dT=0.004$ in the present work, where $d(R/R_{300})/dT$ and R_{300} are slope of R-T curve and the resistivity at 300 K, respectively. The midpoint of the transition designated T_c^{mid}.

The structure of the samples were examined by means of x-ray diffraction.

RESULTS AND DISCUSSIONS

[Sintered and liquid-quenched samples.]

We examine the temperature dependence of the electrical resistivity of the sintered samples. The values of T_c^{mid} are obtained 67.2 K for $Bi_{1.6}Pb_{0.4}Sr_2Ca_2Cu_3O_x$ and 69.7 K for $Bi_{0.7}Pb_{0.3}SrCaCu_{1.8}O_y$ respectively. The superconducting transition can not be found above 4.2 K for the liquid-quenched samples. Thus, these samples are annealed.

[Annealing of liquid-quenched samples].

Figure 3 shows changes in T_c^{mid} against the annealing temperature for the liquid-quenched and annealed (for 1 h) $Bi_{1.6}Pb_{0.4}Sr_2Ca_2Cu_3O_x$ and $Bi_{0.7}Pb_{0.3}SrCaCu_{1.8}O_y$. T_c^{mid} are found for the crystallized samples by annealing between 1090 K and 1130 K for $Bi_{1.6}Pb_{0.4}Sr_2Ca_2Cu_3O_x$ and between 1080 K and 1120 K for $Bi_{0.7}Pb_{0.3}SrCaCu_{1.8}O_y$, respectively. The maximum T_c^{mid} are 70.5 K at 1130 K for $Bi_{1.6}Pb_{0.4}Sr_2Ca_2Cu_3O_x$ and 70.2 K at 1120 K for $Bi_{0.7}Pb_{0.3}SrCaCu_{1.8}O_y$, respectively.

Figure 4 shows changes in T_c^{mid} against the annealing time for liquid-quenched and annealed at 1130 K for $Bi_{1.6}Pb_{0.4}Sr_2Ca_2Cu_3O_x$ and at 1120 K for $Bi_{0.7}Pb_{0.3}SrCaCu_{1.8}O_y$, respectively. The maximum T_c^{mid} are 112.5 K for $Bi_{1.6}Pb_{0.4}Sr_2Ca_2Cu_3O_x$ annealed for 18 h and 106.9 K for $Bi_{0.7}Pb_{0.3}SrCaCu_{1.8}O_y$, annealed for 12 h respectively.

[Results of x-ray diffraction.]

In general, as the crystal perfection is high, T_c^{mid} has been known to be high. The higher the crystal perfection, the lower the full width at half maximum ($\Delta 2\theta$) becomes of x-ray diffraction peaks. From these results of x-ray diffraction, we examined $\Delta 2\theta$ of the (117) plane for the high T_c phase. Figure 5 shows changes in the $\Delta 2\theta$ against the annealing time. The results agree with the T_c^{mid} change as shown in Fig.4. Namely,the higher the T_c^{mid} value, the higher the degree of lattice ordering becomes.

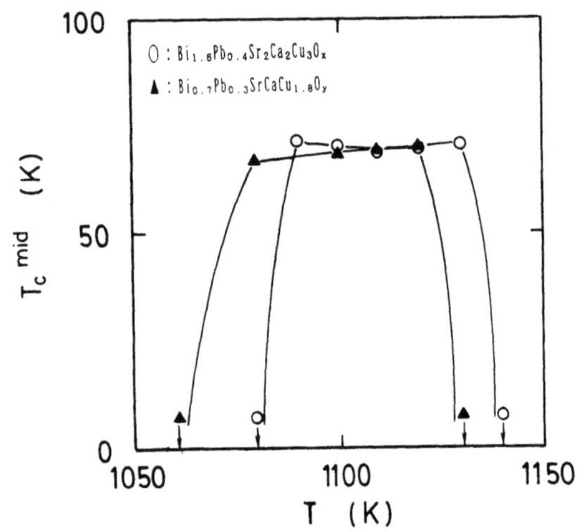

Fig.3 Changes in T_c^{mid} against heat treatment temperature (T) for liquid-quenched $Bi_{1.6}Pb_{0.4}Sr_2Ca_2Cu_3O_x$ and $Bi_{0.7}Pb_{0.3}SrCaCu_{1.8}O_y$.

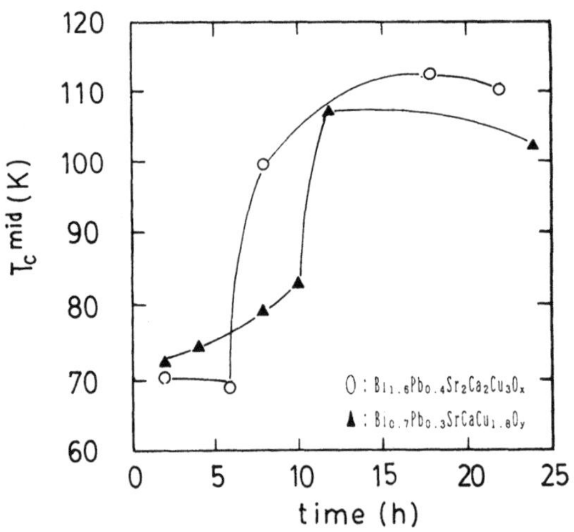

Fig.4 Changes in T_c^{mid} against heat treatment time (t)for liquid-quenched $Bi_{1.6}Pb_{0.4}Sr_2Ca_2Cu_3O_x$ and $Bi_{0.7}Pb_{0.3}SrCaCu_{1.8}O_y$.

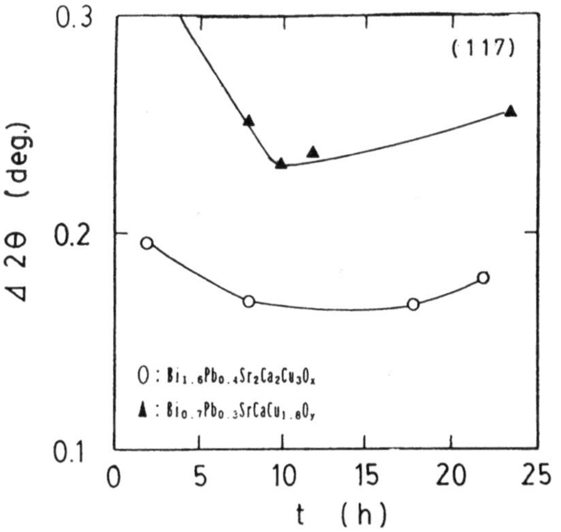

Fig.5 Changes in full width at half maximum ($\triangle 2\theta$) against heat treatment time(t) for liquid-quenched $Bi_{1.6}Pb_{0.4}Sr_2Ca_2Cu_3O_x$ and $Bi_{0.7}Pb_{0.3}SrCaCu_{1.8}O_x$.

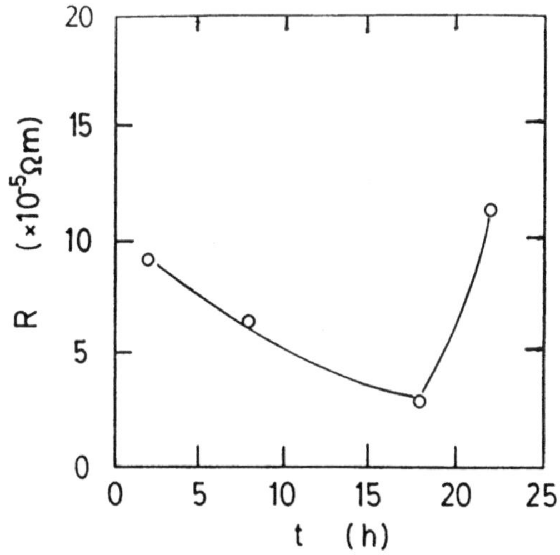

Fig.6 Changes in electrical resistivity at 300 K(R_{300}) against heat treatment time (t) for liquid-quenched $Bi_{1.6}Pb_{0.4}Sr_2Ca_2Cu_3O_x$.

[Electrical discussion.]

Carrier concentration (C) is one of the factors that control the T_c[6,7].

$$C=N/\exp[(\pm eQ/k)-A] \quad (1)$$

Here, Q, k, e and N are the seebeck coefficient, Boltzmann constant, electron change and density of state, respectively; A is a constant. To obtain C, the density of state (N) is calculated by the use of the electrical resistivity (R). N is expressed by the following[8,9].

$$N=A'/R^{1/2} \quad (2)$$

Here, A' is constant that is dominated by the electron mass, charge, Plank constant and mean free path. If the annealing does not largely change their values, N values is inversely proportional to the $R^{1/2}$ values.

Figure 6 shows the changes in the electrical resistivity at 300 K (R_{300}) against the annealing time for liquid-quenched $Bi_{1.6}Pb_{0.4}Sr_2Ca_2Cu_3O_x$. The lower the electrical resistivity, the higher the density of state becomes, resulting in high T_c.

CONCLUSION

In summary, the effects of heat treatments on T_c are investigated of liquid-quenched Bi oxides. The high T_c are found of the liquid-quenched and annealed crystallized samples. Based on the diffractional and electronic discussions, the T_c change can be explained.

REFERENCES

1. Nishi.Y, Kita.Y, Tanioka.K, (1990) J.Mater.Sci. Lett. 25: 5105.
2. Nishi.Y and Harano.H, (1988) J.Appl.Phys. 63: 1141.
3. Nishi.Y and Mikagi.K, (1986) Phys.Rev.B 34: 1792.
4. Nishi.Y, Harano.H, Fukunaga.T and Suzuki.K, (1988) Phys.Rev.B 37: 2855.
5. Nishi.Y, Moriya.S and Manabe.T, (1989) J.Appl.Phys. 65: 2389.
6. Nobumasa.H, Simizu.K, Kitano.Y and Kawai.T, (1988) Jpn.J.Appl.Phys. 27: L1669.
7. Tallan.H.N, and Bransky.I, Electrochem.J, (1971) Soc., 118: 345.
8. R.F.Brebrick, TREATISE ON SOLID STATE CHEMISTRY Vol.2 Defects in Solids Edited by N.B.Hannay [ISBN 0-306-35052-1],(1975) PLENUM PRESS, NewYork, pp.357-362.
9. Mott.F.N, (1969) Philos.Mag., 19: 835.

5.2 Wires and Tapes — (1) YBCO

Development of High Temperature Superconductor Coated Metal Fiber and Multifilamentary Wire

L.D. WOOLF, F.E. ELSNER, W.A. RAGGIO, S.S. PAK, T.L. FIGUEROA, J.D. ROSE, R.B. STEPHENS, R.A. OLSTAD, and T. OHKAWA

Pacific Superconductors Division, General Atomics, San Diego, CA 92186, USA

ABSTRACT

Long superconductor fibers have been continuously produced by electrophoretically depositing $REBa_2Cu_3O_{7-x}$ powder onto a metal substrate fiber and sintering, then eletrophoreticaly depositing silver and sintering. After collecting the coated fiber on a take-up spool, the entire spool is batch-oxygenated to form the 90 K superconducting phase. Multiple fibers are then continuously unspooled and soldered into a copper channel to form the final multifilamentary high temperature superconductor wire. Superconducting fibers over 1000 meters long and multifilamentary wire 450 meters long have been produced.

KEY WORDS: $REBa_2Cu_3O_7$ SUPERCONDUCTOR, FIBER, MULTIFILAMENTARY WIRE CRITICAL CURRENT

Numerous techniques are currently being employed to fabricate wires incorporating the recently discovered high temperature superconducting ceramic materials. The dominant methods include powder-in-tube [1], extrusion of a powder/binder mixture [2] thick film deposition on a substrate [3] and thin film deposition [4]. These techniques typically employ a noble metal [1-3], often Ag, in tube, substrate or filler metal form to impart flexibility and to aid in fabrication. Incorporation of melt-texturing for the first two techniques has led to values for the critical current density J_c exceeding 1000 A/cm^2 at 77 K in an applied magnetic field H of 1 tesla [2,5,6]. Multiple drawing and rolling operations performed using the powder-in-tube technique with BiSrCaBuO-based superconductor followed by extended sintering times of close to 100 hours at temperatures near the melting point, have also led to high values of J_c, particularly at low temperatures even in very high magnetic fields [2,7,8]. Wires produced using these techniques are often short and monofilamentary. In this paper, we report the development of a novel technique incorporating electrophoretic deposition process that has produced kilometer lengths of fully superconducting fiber at rates of 0.3 meters per minute, and a companion method that has been used to continuously incorporate these superconducting fibers into a multifilamentary superconductor wire that is both bend-tolerant and utilizes a minimal amount of noble metal [9].

Superconducting powder of composition $REBa_2Cu_3O_{7-x}$, where RE=Y or a selected rare earth element, is synthesized via solid state reaction of the corresponding RE_2O_3, $BaCO_3$ and CuO. The calcination produces single phase material as determined by differential thermal analysis. A slurry of $REBa_2Cu_3O_{7-x}$ is then prepared by grinding the superconductor powder to a particle size of 0.5-10 microns and suspending it in a non-aqueous liquid.

Superconductor fiber is continuously produced in a single vertical coating system. A nickel alloy substrate is unspooled and fed into a coating chamber containing the slurry of $REBa_2Cu_3O_{7-x}$. Following electrophoretic deposition of the $REBa_2Cu_3O_{7-x}$ particles from the slurry on the metal substrate, the coated metal fiber is passed into a sintering furnace operating in O_2 at approximately 1000°C for about 4 minutes. Subsequent to the sintering of the superconductor coating, the superconductor coated metal fiber enters a chamber containing a slurry of fine Ag powder dispersed in a non-aqueous liquid and is electrophoretically coated with Ag. Following the Ag deposition, the Ag coating is sintered in a furnace operating in O_2 at approximately 900°C for about 1 minute. The resultant fiber is then wound up on a take-up spool. Oxygenation of the entire length of spooled fiber is performed in a batch mode by heating in an oxygen atmosphere at 400°C for 1-3 days.

Next, the oxygenated fiber is continuously processed into a multi-filamentary wire. Twelve spools of oxygenated superconductor fiber are unspooled and combined together to form a parallel set of twelve fibers. These fibers are then laid in the interior of a moving copper channel and soldered in place to form a multifilamentary wire.

A polished cross section of the superconductor fiber is displayed in Figure 1. Clearly shown is the 75 micron diameter metal substrate fiber, with an outer section consisting of a reaction layer with the super-conductor as well as an oxidized outer surface of the metal itself. The 75 micron thick superconductor layer and 20 micron thick Ag layer are also evident. The absence of any significant amount of interdiffusion of the substrate and constituent elements of the coatings is evident in Figure 2. X-ray maps, produced using a scanning electron microscope with an energy dispersive analysis of X-rays attachment, are shown for Ni, the majority component of the substrate, Ba, the most reactive component of the superconductor [10, 11], and Ag, the sole component of the outer coating.

The resistively determined superconducting transition temperatures for these continuously produced fibers are typically in the range of 88-93 K. The critical current density for a representative superconducting fiber is shown in Figure 3. Values for J_c (77K, H=0) vary from 1000-2000 A/cm^2, using a criterion of 1μV/cm, while J_c (50 K, H=0) date are typically in the range of 4000-6000 A/cm^2. The magnetic field dependence of J_c, displayed in Fig. 4 for a fiber at 53K shows typical "weak link" behavior indicative of Josephson-junction coupled grain boundaries [12]. An applied magnetic field of 30-50 gauss reduces J_c by a factor of approximately 2, while a field of 1000 gauss reduces J_c by a factor of 100. Samples taken from a spool containing 1000 meters of continuously-produced and batch-oxygenated superconductor fiber all showed J_c (50 K, H=0) in the range of 4000-6000 A/cm^2.

A polished cross section of a multifilamentary wire containing 12 fibers soldered into a copper channel is shown in Figure 5. That the Ag coating is still visible on each fiber indicates that the soldering process has not appreciably dissolved the Ag. The U-shaped copper channel with the fibers situated above the neutral axis maintains the fibers in compression when the wire is bent so that the fibers are on the inner surface of the bend. We have found that bending this wire in a 7.5 cm radius of curvature gave the same J_c result as for straight wire.

The critical current I_c and critical current density J_c versus temperature in zero applied magnetic field for a 130 meter length of multifilamentary wire containing 1560 meters of superconducting fiber is shown in Fig. 6. Values of I_c are 5A at 77K, increasing to 14A at 50K and 24A at 20K. Corresponding values for J_c at these temperatures are 900 A/cm^2, 2500 A/cm^2 and 4500 A/cm^2.

The magnetic field dependence of I_c for a short piece of the 130 meter wire is shown in Fig. 7. The reduction of I_c with applied magnetic field is consistent with the J_c versus H data for the fibers.

Electric field E versus current density are presented in Fig. 8 for the 130 meter length of wire at 4 different temperatures. Most of the data can be described by a power law $E \alpha J^n$ where n is between 15 and 18. The low E tail of the 73K curve seems to asymptotically approach a slope of 1, corresponding to resistive behavior. This is not due to a break in the fibers since it does not appear at lower temperatures. It is apparently due to thermally activated flux flow at this temperature [13].

The longest length of multifilamentary wire that has been produced is 450 meters, which contains 5400 meters of superconducting fiber. Samples taken from both ends of this wire show values of I_c similar to those reported above.

In conclusion, we have developed continuous processes that rapidly produce long lengths of superconducting fiber and multifilamentary wire incorporating the $REBa_2Cu_3O_{7-x}$ superconductor. Although the transport critical current density is less than has been attained using other techniques, the processes are sufficiently flexible that improved methods or materials may be substituted to increase J_c. This development is currently in progress.

REFERENCES

1. T. Hikata, T. Nishikawa, H. Mukai, K. Sato and H. Hitotsuyanagi, Jap. J. Appl. Phys. 28, L1204 (1989).

2. S. Jin and J. E. Graebner, Mater. Sci. and Eng. 87, 243 (1991).

3. T. H. Tiefel, S. Jin, G. W. Kammlott, J. E. Graebner, R. B. van Dover and N. D. Spencer, Appl. Phys. Lett. 58, 1917 (1991).

4. X.D. Wu, R.E. Muenchausen, S. Foltyn, R.C. Estler, R.C. Dye, A.R. Garcia, N.S. Nogar, P. England, R. Ramesh, D.M. Hwang, T.S. Ravi, C.C. Chang, T. Venkatesan, X.X.Xi, Q. Li and A. Inan, Appl. Phys. Lett. 57, 523 (1990).

5. P.J. McGinn, W. Chen, N. Zhu, U. Balachandran and M.T. Lanagan, Physica C, 165, 480 (1990).

6. M. Okada, R. Nishiwaki, T. Matsumoto, T. Kamo, K. Aihara, S. Matusda and M. Seido, Proc. 1989 Fall Meeting of the Materials Research Society, Boston, MA, Nov. 27 - Dec. 2, 1989.

7. K. Heine, J. Tenbrink and M. Thöner, Appl. Phys. Lett. 55, 2441 (1989).

8. Y. Yamada, K. Jikihara, T. Hasebe, T. Yanagiya, S. Yasuhara, M. Ishihara, T. Asano and Y. Tanaka, Jap. J. Appl. Phys. 29, L 456 (1990).

9. L. D. Woolf, W. A. Raggio, F. E. Elsner, M. V. Fisher, R. B. Stephens, T. L. Figueroa, C. H. Shearer, J. D. Rose, K. M. Schaubel, R. A. Olstad, T. Ohkawa, D. M. Duggan, M. DiMartino and R. L. Fagaly, Appl. Phys. Lett. 58, 534 (1991).

10. C.T. Cheung and E. Ruckenstein, J. Mater. Res. 4, 1 (1989).

11. T. Komatsu, O. Tanaka, K, Matusita, M. Takata and T. Yamashita, Jap. J. Appl. Phys. 27, L1025 (1988).

12. R.B. Stephens, Cryogenics 29, 399 (1989).

13. R.B. Stephens and T.L. Figueroa, to be published in Superconducting Science and Technology.

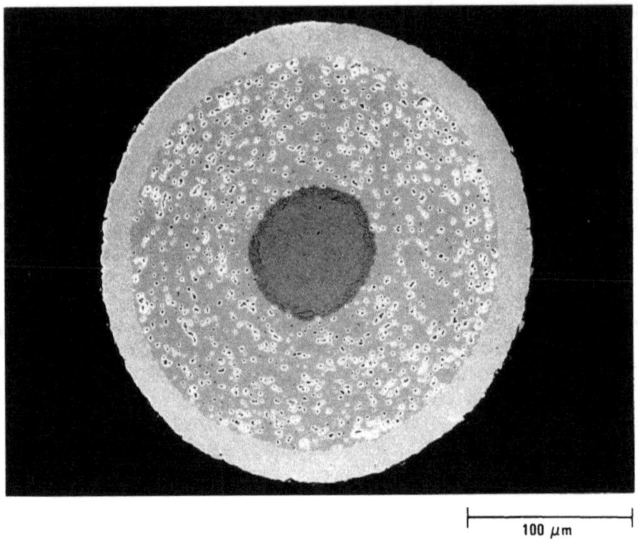

Fig. 1. Scanning electron micrograph of a polished cross section of a silver-coated superconductor-coated metal fiber.

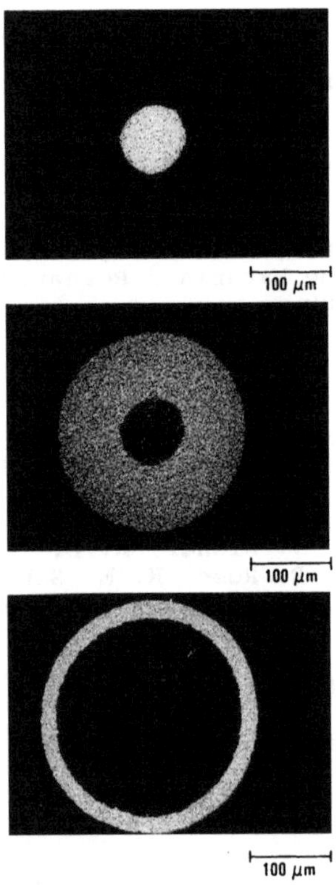

Fig. 2. X-ray map of Ni, Ba and Ag for a silver-coated superconductor-coated metal fiber shown top to bottom respectively

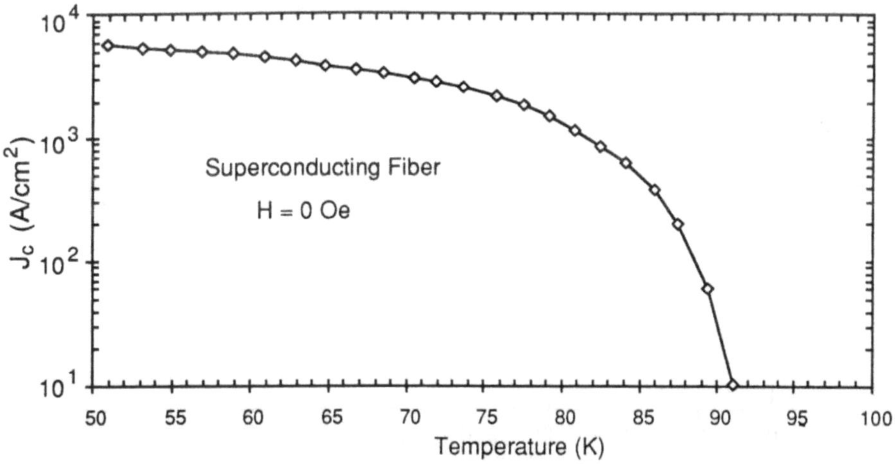

Fig. 3. Critical current density J_c versus temperature T in zero applied magnetic field for a silver-coated superconductor-coated metal fiber.

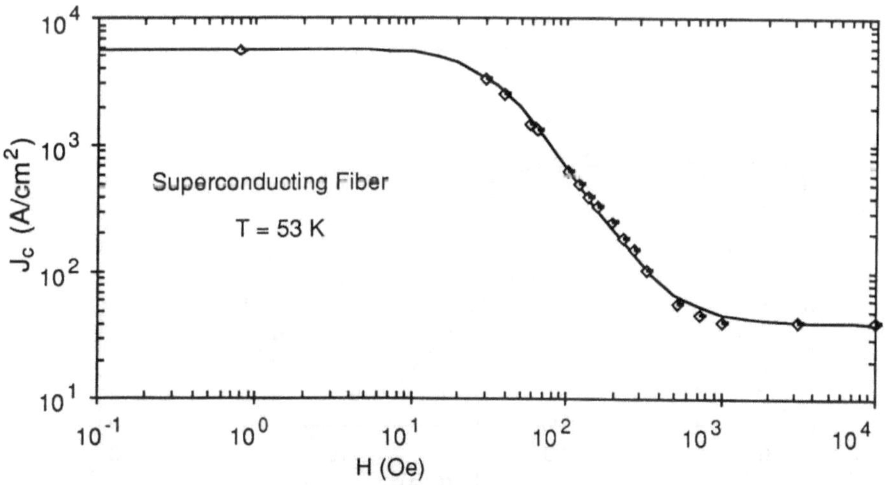

Fig. 4. Critical current density J_c versus applied magnetic field H at a temperature of 53 K for a silver-coated superconductor-coated metal fiber.

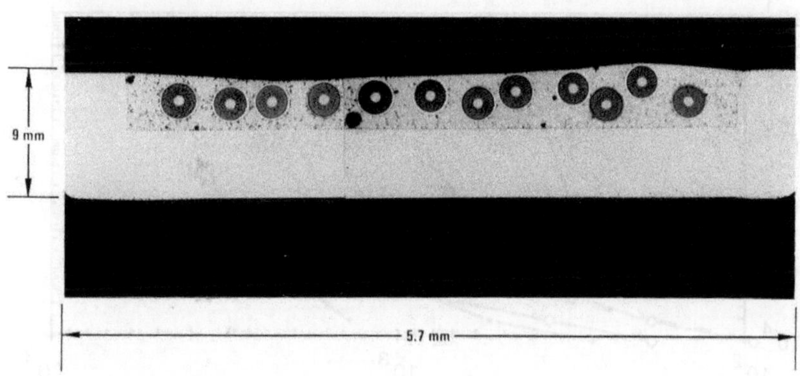

Fig. 5. Polished cross section of a superconducting wire consisting of 12 superconducting fibers soldered into a U-shaped copper channel.

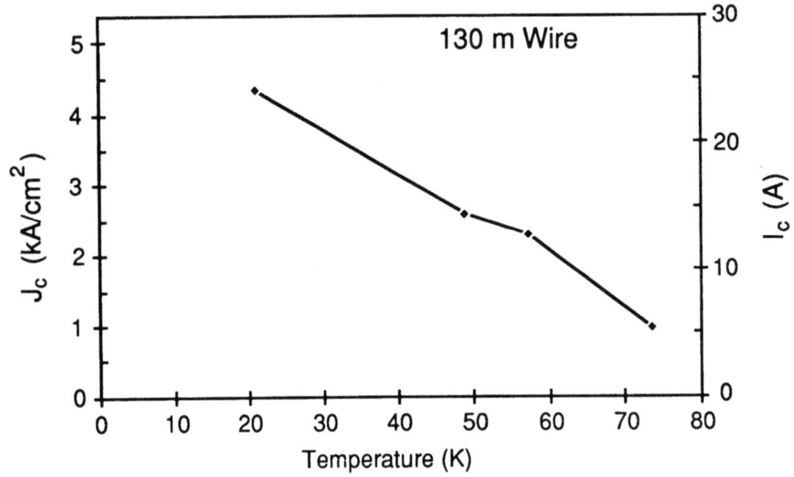

Fig. 6. Critical current I_c and critical current density J_c versus temperature T for a 130 meter length of multifilamentary wire in zero applied magnetic field.

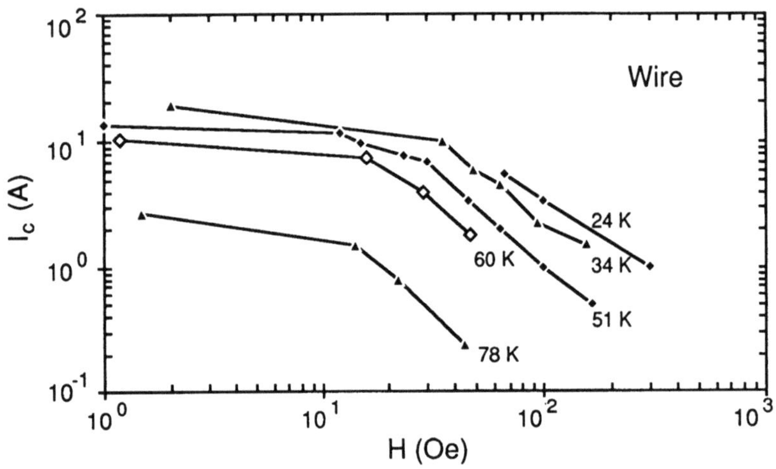

Fig. 7. Critical current I_c versus applied magnetic field H for a short piece of multifilamentary wire.

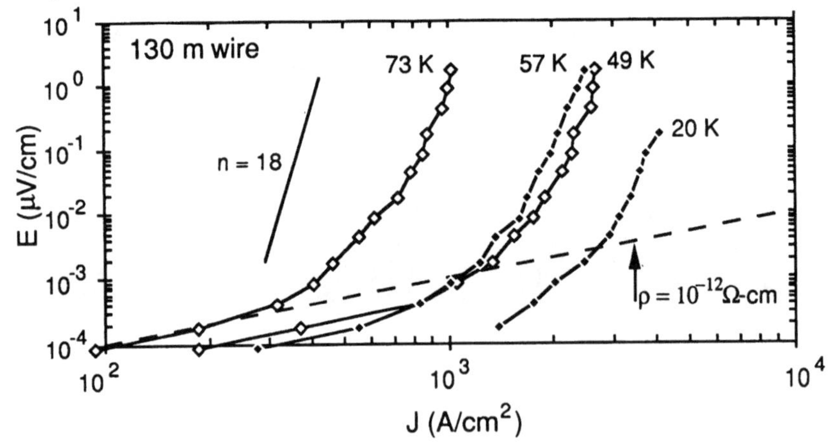

Fig. 8. Electric field E versus current density J at a variety of temperatures in zero applied magnetic field for a 130 meter length of multifilamentary wire.

a–b Plane Aligned $YBa_2Cu_3O_{7-x}$ Thin Film Tapes

Yasuhiro Iijima[1], Kazunori Onabe[1], Naohiro Futaki[1], Nobuo Tanabe[1], Nobuyuki Sadakata[1], Osamu Kohno[1], and Yoshimitsu Ikeno[2]

[1] Materials Research Laboratory, Fujikura Ltd., Koto-ku, Tokyo, 135 Japan
[2] Super-GM, Osaka, 530 Japan

ABSTRACT

A method to reduce intergranular weak links has been developed for application of oxide superconducting thin films, for tape-shaped conductors or other applications using polycrystalline metallic substrates. To avoid weaklinks at the large-angle grain boundaries in the a-b plane, we formed the biaxially oriented YSZ buffer layer on a polycrystalline Ni-based alloy by Ion-Beam-Assisted Deposition (IBAD), and subsequently the a-b plane aligned $YBa_2Cu_3O_{7-x}$ film was deposited by laser ablation. J_c of $2.5x10^5$ A/cm^2 (77 K, 0 T) and $2.2x10^4$ A/cm^2 (77 K, 8.0 T) were obtained on practical polycrystalline substrates.

KEY WORDS: $YBa_2Cu_3O_{7-x}$ thin film, polycrystalline metallic substrate, intergranular weak-link, biaxially alignment, Ion-Beam-Assisted Deposition

INTRODUCTION

For the purpose to fabricate tape-shaped conductors, we had deposited $YBa_2Cu_3O_{7-x}$ films on Ni-based alloy (Hastelloy c276) with thin films as intermediate buffer layers to prevent thermal reactions between superconducting layers and metallic substrates.[4,6] However, all $YBa_2Cu_3O_{7-x}$ films grown on polycrystalline substrates, such as silver, Hastelloy, poly-YSZ, etc., with or without intermediate buffer layers, have c-axis orientation normal to the substrate, but the a- or b- axis distributed randomly. They have J_c of the order from 10^3 to 10^4 A/cm^2 (77 K, 0 T).[1-6] The Jc values degraded drastically in magnetic field, in the behavior similar to the bulk $YBa_2Cu_3O_{7-x}$ samples.[1,5] On the other hand, high quality $YBa_2Cu_3O_{7-x}$ thin films without severe intergranular weak links on suitable single crystalline substrates have epitaxial relations to substrates, c-axis oriented perpendicular to substrates, and the a- or b-axis in-plane aligned on the substrate.
J_c reached above 10^6 A/cm^2 (77 K, 0 T) and the Jc characteristics in magnetic fields were excellent.[10-12,5]

Recently many experiments have concentrated on the weak-links at the grain boundaries of $YBa_2Cu_3O_{7-x}$ misaligned in a-b plane using bi-crystals of $SrTiO_3$, etc.[13-14] They concluded that large-angle grain boundaries in a-b planes had the property of Josephson weak-links. The results indicated that sufficient critical currents can be hardly obtained with the existence of large-angle grain boundaries in the a-b plane of $YBa_2Cu_3O_{7-x}$ films.

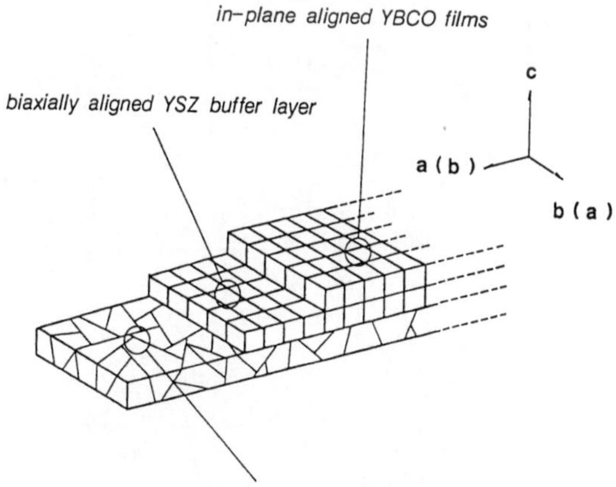

Fig. 1 The structure of the a-b plane aligned $YBa_2Cu_3O_{7-x}$ thin film tape

Figure 1 shows a schematic drawing of a-b plane aligned $YBa_2Cu_3O_{7-x}$ tapes. To align the a-b plane of $YBa_2Cu_3O_{7-x}$, we produced biaxially aligned YSZ films which (100), (010), and (001) axes were uniformly aligned on polycrystalline metallic substrates by Ion-Beam-Assisted Deposition (IBAD); a method to deposit thin films with concurrent energetic particle bombardment.[7] Consequently, the a-b plane aligned $YBa_2Cu_3O_{7-x}$ films were successfully deposited on the YSZ layers by laser ablation.[8-9]

EXPERIMENTAL

Substrates used were mirror-like polished plates of Hastelloy c276 with 18 mm in length, 10 mm in width and 0.5 mm in thickness. The IBAD system with dual ion sources was used to form YSZ films.[7] During the sputtering deposition of YSZ films, an O+Ar ion beam from the assisting ion source concurrently bombarded the films. To vary the incident angle of the assisting ion beam, the substrates were mounted on the holders that could be inclined to the desired position relative to the ion beams. Then, $YBa_2Cu_3O_{7-x}$ films were deposited on Hastelloy c276 with those YSZ films by laser ablation. Sintered $YBa_2Cu_3O_{7-x}$(Y:Ba:Cu = 1:2:3) target was ablated by an Ar-F (wavelength; 193 nm) excimer laser at the repetition rate of 5 per second. The target-substrate distance was 30 mm. The substrate temperature was set at 700 ^oC. The partial pressure of O_2 during the deposition was 200 mTorr. The deposition rate was 0.5 nm/sec. $YBa_2Cu_3O_{7-x}$ films had a thickness from 0.5 to 1.0 micrometers. The crystal structures of both YSZ and $YBa_2Cu_3O_{7-x}$ films were investigated by X-ray diffraction (XRD) and scanning electron microscopy (SEM). In-plane texturings were estimated by X-ray pole figure measurements using Schulz method. Superconducting properties were measured by a d.c. four-probe method. Specimens that have a current path of 0.3 mm in width and 0.5 mm in length was prepared to measure critical current densities, with a 1 microvolt/cm criterion.

RESULTS & DISCUSSIONS

Biaxially aligned YSZ films were formed when the incident angle of the assisting ion beam was from 30 to 60 degrees toward the substrate normal. The film thickness were from 0.3 to 0.8 micrometers. The (100) axis of YSZ was oriented perpendicular to the substrate. The rocking curve full width at half-maximum (FWHM) for (200) peak were from 5 to 7 degrees. Explicit in-plane ordering was recognized from the pole figure of the YSZ (111) peaks. The in-plane texturings was optimized by controlling the incident angle of assisting beam and the thickness of the film. Details of forming conditions for biaxially aligned YSZ films are reported in this volume.[7]

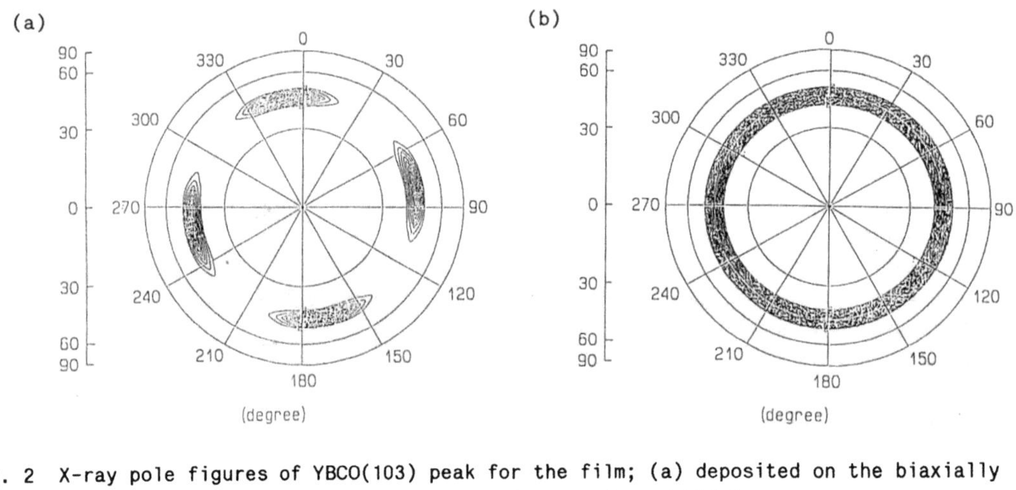

Fig. 2 X-ray pole figures of YBCO(103) peak for the film; (a) deposited on the biaxially aligned YSZ layer, (b) deposited on the conventionally formed YSZ layer.

(a) (b)

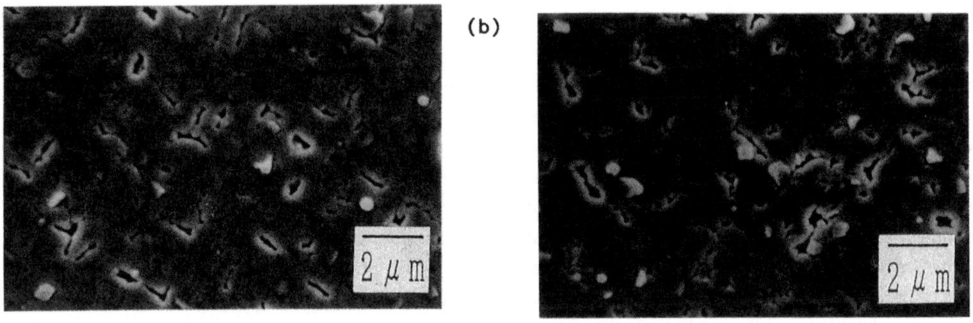

Fig. 3 SEM photographs of the surface of YBCO films; (a) deposited on the biaxially aligned YSZ layer, (b) deposited on the conventionally formed YSZ layer.

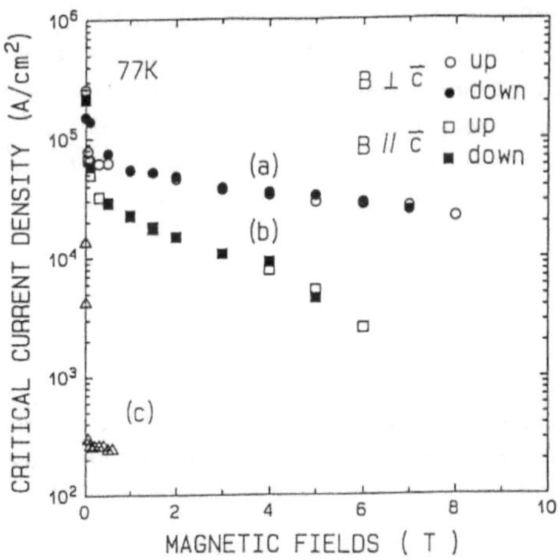

Fig. 4 J_c-B properties of the a-b plane aligned YBCO film;(a) B⊥c, (b) B//c, and the one without in-plane ordering;(c) B⊥c.

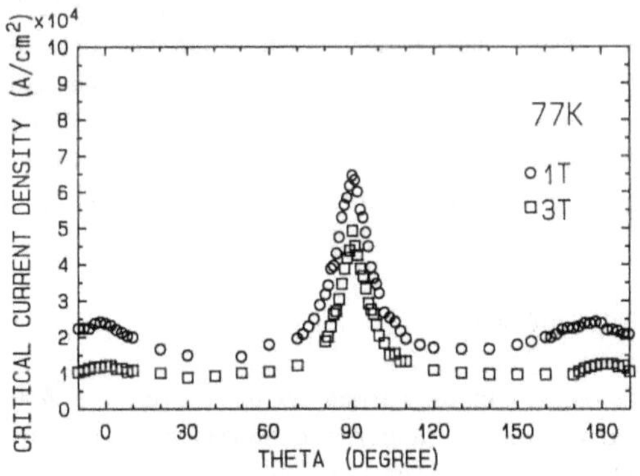

Fig. 5 The angular dependence of J_c of the a-b plane aligned YBCO film in magnetic fields.

Figure 2 (a) shows the pole figure for (103) peaks of the $YBa_2Cu_3O_{7-x}$ film deposited on the biaxially aligned YSZ buffer layer formed with the assisting beam incident angle of 45 degrees. The c-axis was oriented perpendicular to the substrate, and the a- and b-axes were aligned to the YSZ (110) axis on the surface of the substrate. This epitaxial relation is similar to the result for single crystalline (100)YSZ substrates.[12] In comparison, a pole figure for a (103) peak of the $YBa_2Cu_3O_{7-x}$ film deposited on Hastelloy c276 with a YSZ layer formed by RF magnetron sputtering, is shown in figure 2 (b). The YSZ layer was deposited by RF magnetron sputtering. Thickness of The YSZ layers was 0.6 micrometers and YSZ (111) axis was oriented normal to the substrate. The c-axis of the $YBa_2Cu_3O_{7-x}$ film was aligned normal to the substrate, but the other axes distributed randomly. Figure 3 (a) shows an SEM photograph of the surface morphology of the biaxially aligned $YBa_2Cu_3O_{7-x}$ film. Textured grains of $YBa_2Cu_3O_{7-x}$ were observed in the contrast to the ones deposited on conventionally formed buffer layer shown in Figure 3 (b).

Critical temperatures of the films were from 88 K to 90 K at zero-resistivity. By optimizing deposition conditions of IBAD and laser ablation, J_c value of $2.5x10^5$ A/cm^2(77K, 0T) was achieved. Figure 4 shows the J_c in external magnetic fields. $2.2x10^4$ A/cm^2 (77 K, 8 T) was attained under the external magnetic fields applied normal to both c-axis and the current direction as shown in Figure 4 (a). The angular dependence of J_c in magnetic fields is shown in figure 5. The lateral axis indicates the angle between c-axis and magnetic flux. It strongly depended on the orientation of magnetic fields. It was consistent with the pinning behavior of high-quality $YBa_2Cu_3O_{7-x}$ films deposited on suitable single crystalline substrates.[6] The J_c value at 0 T was increased about approximately ten times larger, and the values in magnetic fields were remarkably improved, in comparison to the films deposited on polycrystalline substrates reported so far.[1-6] J_c dependence of the film deposited on conventionally formed YSZ layer is shown in Figure 4 (c). Those results indicates that biaxial alignment of the $YBa_2Cu_3O_{7-x}$ grains decreased the weak links caused by misalignments of a-b plane. However, obtained J_c was still ten times smaller than those on suitable single crystalline substrates. There was a steep decrease and a recognizable hysterisis in the fields less than 0.5 T at 77 K. Dimos et al. have shown that the intergrain transporting current apparently degraded when the misorientation angle exceeded about 5 degrees.[14,15] There should be considerable amount of weak-coupled grain boundaries in the cross section of the $YBa_2Cu_3O_{7-x}$ films. Therefore, J_c could be increased toward the value for single crystalline substrates by further improvement of the alignment of a- and b-axes.

In summary, we succeeded to form biaxially aligned $YBa_2Cu_3O_{7-x}$ films on polycrystalline metallic substrates, using the biaxial YSZ buffer layer deposited by IBAD. J_c value increased more than ten times larger than the one formed by conventional methods on polycrystalline substrates. The results reconfirmed the significance of intergranular weak links arisen from misalignments in a-b plane, for the macroscopic transporting current of $YBa_2Cu_3O_{7-x}$ films. This results suggests a way to apply thin films of oxide superconductors for cables, magnets, magnetic shields, microwave devices, etc., using polycrystalline metallic substrates.

ACKNOWLEDGMENT

This work was performed as a part of " R & D on Superconducting Technology for Electric Power Apparatuses " under the Moonlight Project of Agency of Industrial Science and Technology, MITI, being consigned by NEDO.

REFERENCES

1. Narumi E., Song L. W., Yang F., Patel S., Kao Y. H. & Shaw D. T. (1991) Appl. Phys. Lett., 58 :1202
2. Saitoh J., Fukutomi M., Komari K., Tanaka Y., Asano T., Maeda H. & Takahara H. (1991) Jpn. J. Appl. Phys., 30 :L898
3. Witanachchi S., Patel S., Zhu Y. Z., Kwok H. S. & Shaw D. T. (1990)J. Mater. Res. 5 :717
4. Iijima Y., Hayakawa H., Sadakata N. & Kohno O. (1989) Proc. 11th Int'l Conf. Magnet Tech. (MT-11), Tsukuba, Japan, Aug. 1989 :1142
5. Yoshida N., Kubota M., Okuda S., Takano S., Nagata M., Hara T., Okaniwa K. & Yamamoto T. (1990) Proc. 3rd Int'l Symp. Superconductivity (ISS'90), Sendai, Japan :901
6. Onabe K., Iijima Y., Futaki N., Sadakata N., Kohno O. & Ikeno Y. (1990) Proc. 3rd Int'l Symp. Superconductivity (ISS'90), Sendai, Japan :905
7. Iijima Y., Tanabe N. & Kohno O., "Biaxially aligned YSZ buffer layer on polycrystalline substrates", TFP-19, this volume.
8. Iijima Y., Tanabe N., Kohno O. & Ikeno Y., Proc. 3rd Int'l Conf. M^2S-HTSC, Kanazawa, Japan, July 1991, Special Issue of Physica C, in printing
9. Iijima Y., Tanabe N., Kohno O. & Ikeno Y., submitted to Applied Physics Letters
10. Roas B., Schulz L. & Endres G. (1988) Appl. Phys. Lett., 53 :1557
11. Venkatesan T., Wu X. D., Dutta B., Inam A., Hedge M. S., Hwang D. M., Chang D. M., Nazar L. Wilkens B. (1989) Appl. Phys. Lett., 54 :581
12. Zheng J. P., Dong S. Y. & Kwok H. S. (1991) Appl. Phys. Lett., 58 :540
13. Chaudhari P., Mannhart J., Dimos D., Tsei C., Oprysko C. & Scheuermann M. (1988)Phys. Rev. Lett., 60 :1653
14. Dimos D., Chaudhari P., Mannhart J. & LeGoues F. K. (1988) Phys. Rev. Lett., 61 :219

Enhancement of the Reaction Diffusion in High-T_c Oxide Composites by Additional Elements

KYOJI TACHIKAWA, TAKAYUKI INOUE, KOUICHI ZAMA, and YASUO HIKICHI

Faculty of Engineering, Tokai University, Hiratsuka, Kanagawa, 259-12 Japan

ABSTRACT

The reaction diffusion between a high melting point substrate and a low melting point coating layer produces a thick superconducting layer with an uniform composition. The Ag addition to the coating layer enhances the diffusion both in Y-Ba-Cu-O and Bi-Sr-Ca-Cu-O systems. The Ag is pushed out from the superconducting layer to the surface of the specimen forming a laminated structure composed of a superconducting layer and a Ag layer. This movement of Ag improves the grain orientation in the superconducting layer. The Li substitution for Cu in the coating layer of Bi-Sr-Ca-Cu-O system significantly promotes the diffusion, and a Tc of 85K is obtained after the reaction at 780 ℃ for 30min.

KEY WORDS:Y-Ba-Cu-O, Bi-Sr-Ca-Cu-O, Reaction diffusion, Ag, Li addition

INTRODUCTION

The high-Tc oxides synthesized by a reaction diffusion between a high melting point substrate and a low melting point coating layer can be expected to exhibit a denser and more homogeneous structure than that of the oxides prepared by a conventional sintering process. Moreover, an oriented structure may be formed by the diffusion process without melting the specimen. In the Y-Ba-Cu-O(YBCO) system, a thick, uniform layer of $YBa_2Cu_3O_{7-x}$ was formed by a reaction between a Y_2BaCuO_5 substrate and a superficial $Ba_3Cu_5O_8$ coating layer [1,2]. In the Bi-Sr-Ca-Cu-O(BSCCO) system, a dense and oriented 2212 layer was synthesized by a reaction in a diffusion couple composed of a Sr-base substrate and a Bi-base coating layer [3,4].

In the present study, the effects of the additional element to a diffusion component on the reaction diffusion in the YBCO and BSCCO systems have been investigated. Both in the YBCO and BSCCO systems, the Ag added to the coating layer shows an interesting behavior for the enhancement of the diffusion, and accumulates on the surface of the specimen after the reaction. In the BSCCO system, a partial substitution of Li for Cu in the coating layer significantly promotes the reaction diffusion, and reduces the reaction temperature.

EXPERIMENTAL PROCEDURE

Figure 1 illustrates the principle of the sample preparation. In the preparation of substrate, mixed oxide powders calcined by a conventional process were pressed into rectangular bars 4mm in width, 1.5mm in thickness and 22mm in length. The bars were then sintered at an appropriate temperature. Another mixed oxide powders containing an additional element were suspended in ethyl alcohol to form a slurry, and coated on the surface of sintered bar-shaped substrate to form a composite. The thickness of the coating layer was about 100 µm. The composites were heat treated at different temperatures, and followed by a slow cooling at a rate of 100 ℃ /h to room temperature, all heat treatments were performed in open air.

The composition profiles and crystal grain structure on the cross-section of the specimens were studied by an electron probe microanalyzer (EPMA) and an optical microscope, respectively. X-ray diffraction (XRD) patterns were taken on the polished surface of the reaction layer. The transition temperature Tc and critical current density Jc of the specimens were measured resistively by a DC four-probe method. AC magnetic susceptibility was also measured as a function of temperature. The magnetic hysteresis curves of specimens were measured by a vibrating sample magnetomater (VSM).

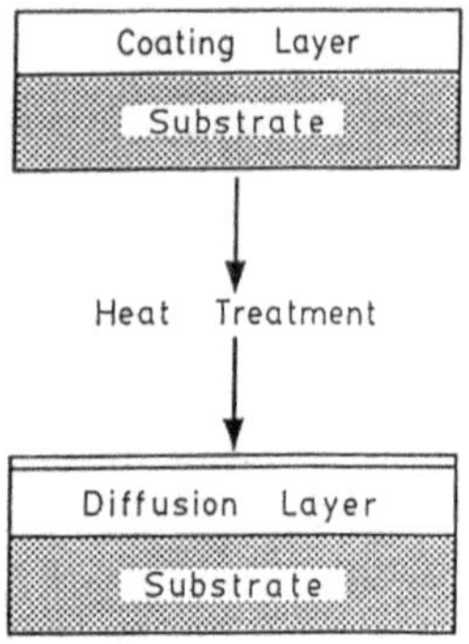

Fig. 1. Procedures of the diffusion process.

Fig. 2. The cross-sectional structure of the 037+30Ag/211 composite reacted at 880 ℃ for 8h.

EXPERIMENTAL RESULTS

Effect of Ag addition to YBCO system

The 211 and 037 oxides were used as the substrate and coating components, respectively. Here, for example, 037 indicates a mixed oxide with a nominal composition of $Ba_3Cu_7O_{10}$, The 211 substrate component was calcined at 900 ℃ for 12h, pressed into a bar-shaped pellet, and then sintered at 950 ℃ for 12h. The 037 coating component was calcined at 800 ℃ for 12h. The 30wt% Ag was added to the 037 oxide using Ag_2O powder. The 037 component with 30wt% Ag addition was calcined at 750 ℃ for 12h. The resulting 037/211 and 037+30Ag/211 composites were reacted at temperatures between 860 ℃ and 960 ℃.

Figure 2 shows the cross-sectional structure of the 037+30Ag/211 composite reacted at 880 ℃ for 8h. A thick reaction layer of about 100 μm in thickness is formed on the substrate. The XRD pattern taken on the polished surface of the reaction layer corresponds to that of the 123 phase. The Ag is pushed out from the diffusion layer to the surface of the specimen opposing the gravity. The Ag significantly enhances the diffusion, and a thick 123 layer is formed after the reaction for only 1h at 880 ℃. In this process, at first the 123 layer is formed at the boundary between the substrate and the coating layer. The Ag pushed out from the 123 layer accumulates at the diffusion front. This accumulation of Ag accelerates the diffusion reaction. Finally, the Ag accumulates on the surface of the specimen, resulting the formation of laminated layers composed of 123 phase and Ag as seen in Fig.2. There was crystal grain orientation in the 123 phase formed neither in 035/211 nor 037/211 diffusion specimen [1]. In the 037+30Ag/211 specimen, a grain orientation is observed near the boundary with Ag layer as seen in Fig.2. The orientation may be caused by the upward movement of Ag during the reaction.

Figure 3 shows the offset Tc versus reaction temperature of the specimen with or without Ag addition. The 037/211 specimen shows offset Tc above 90K by the heat treatment at temperatures 920°-960 ℃, while the 037+30Ag/211 specimen does at a temperature around 880 ℃. Thus, the Ag addition decreases the optimum reaction temperature to obtain high Tc by 50°-60 ℃. This may be due to the enhancement of diffusion by the Ag addition. The transport Jc (77K) of 037/211 diffusion specimen is increased by a factor of three by the Ag addition. This may be caused by the improvement in grain orientation by the Ag addition.

Effect of Ag addition to BSCCO system

In the previous study, the most favorable diffusion couple to synthesize a 2212 phase with uniform composition was found to be 2001/0212 [3]. Here again, for example, 2001 indicates an oxide with a nominal composition of Bi_2CuO_4. In this study, the 0212 oxide was calcined and heat treated at 950 ℃ for 12h. The pressed bar-shaped 0212 substrate was sintered at 950 ℃ for 18h. The 30wt% Ag was added to the 2001 coating layer. The 2001 oxide mixed with Ag_2O was calcined at 650 ℃ for 6h and heat treated at 670 ℃ for 8h. The BSCCO composites were suffered to a two-step heat treatment. The first heat treatment was performed to improve the bonding between the coating layer and the substrate.

Figure 4 shows the EPMA composition mappings taken on the cross-section of the 2001+30Ag/0212 diffusion specimen reacted at 830 ℃ for 20h. An uniform diffusion layer, about 150 μm in thickness, was formed on the substrate. Another thin layer poorer in Bi and richer in Cu is formed adjacent to the substrate. The Ag is pushed out from the diffusion layer, and accumulates on the surface of the specimen quite similar to the YBCO system.

Figure 5 is an XRD pattern taken on the polished surface of the reaction layer in the 2001+30Ag/0212 diffusion specimen. The pattern is composed of the peaks of 2212 phase, and shows an unusually strong (200) peak. This implies that the 2212 phase formed by a diffusion has a strong (100) preferred orientation as reported previously [3,4]. The peak intensity of (200) reflection normalized to (115) reflection in the diffusion specimen is more than ten times stronger than that in the random powder specimen [5]. No reflection peak of Ag is detected in the 2212 layer. Thus, a laminated structure composed of (100) oriented 2212 phase and Ag layer is obtained by the present diffusion process.

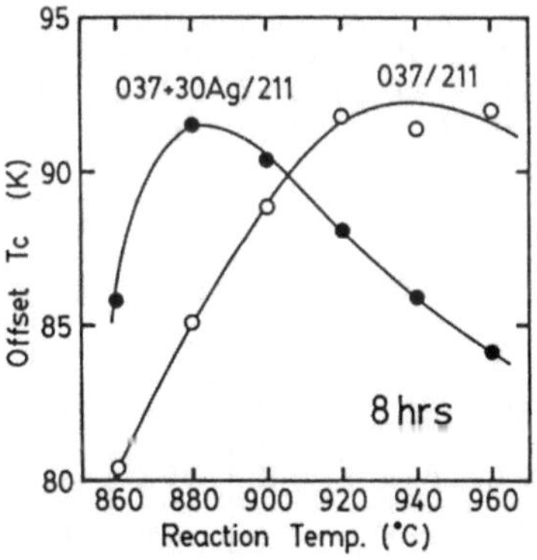

Fig. 3. The offset Tc versus reaction temperature for the specimens with or without 30wt% Ag addition.

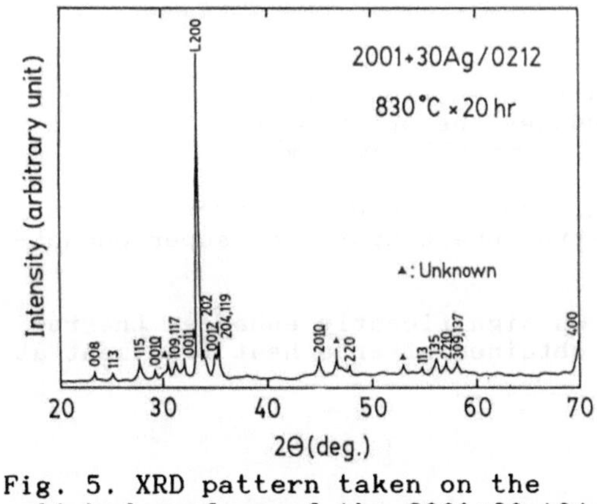

Fig. 5. XRD pattern taken on the polished surface of the 2001+30wt%Ag/0212 diffusion specimen.

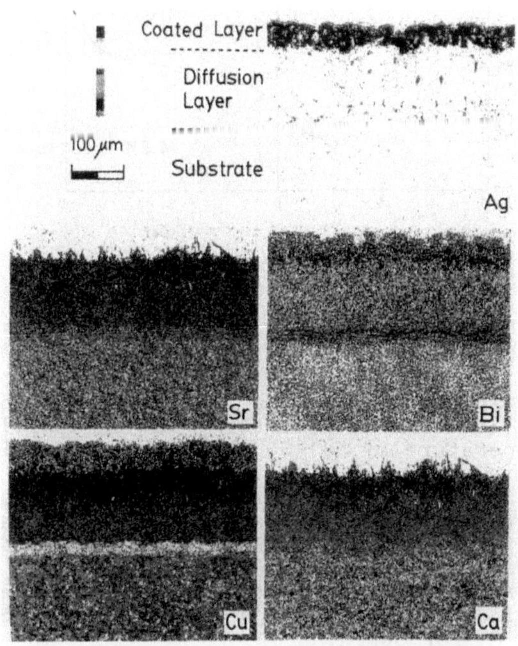

Fig. 4. EPMA composition mappings taken on the cross-section of the 2001+30Ag/0212 specimen reacted at 830 ℃ for 20h. The Ag addition to the coating layer is 30wt%.

The offset Tc of the 2001+30wt%Ag/0212 specimen reacted at 830 ℃ for 8h is 80K. The reaction temperature producing the highest offset Tc is 830 ℃ which is not so much reduced by the Ag addition as in the YBCO system shown in Fig.3. Figure 6 is the magnetic hysteresis curve at 4.2K for the 2001+30Ag/0212 specimen reacted at 830 ℃ for 8h. The magnetic field was applied parallel to the specimen surface. The Jc of the 2212 layer in the specimen calculated from the magnetization is about 7×10^5 A/cm^2 which is fairly large indicating the formation of a dense 2212 structure.

Effect of Li substitution in BSCCO system

The Li substitution for Cu in the coating layer by 25at% was performed. The diffusion composites were prepared by a similar way as described before. The EPMA composition mappings taken on the cross-section of $Bi_2Cu_{0.75}Li_{0.25}O_x$/0212 diffusion specimen reacted at 800 ℃ for 30min reveal that a diffusion layer, 100–150 μm in thickness, is formed after the reaction at 800 ℃ for only 30min. The XRD pattern of the diffusion layer indicates that the diffusion is composed of 2212 phase. The 2212 layer is more homogeneous than that in the specimen with Li substitution for Bi reported previously [4], although the fine dispersion of CuO is observed in the 2212 layer.

Table 1 is the offset Tc of the $Bi_2Cu_{0.75}Li_{0.25}O_x$/0212 diffusion specimen with different heat treatment. After the first heat treatment, the specimen does not show superconductivity. An offset Tc of 85K is obtained after the reaction at 780 ℃ for 30min. Thus the Li substitution significantly promotes the diffusion, and reduces the optimum heat treatment temperature.

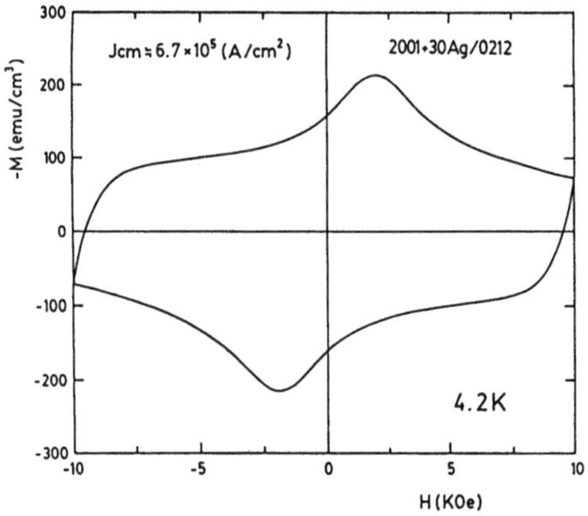

Fig. 6. Magnetization hysteresis at 4.2K of the 2001+30wt%Ag/0212 specimen.

Table 1. The offset Tc of the $Bi_2Cu_{0.75}Li_{0.25}O_x$/0212 diffusion specimen with different heat treatment.

$Bi_2Cu_{0.75}Li_{0.25}O_x$ / 0212		2nd H.T. Temp. (30min)			
		780℃	800℃	820℃	840℃
1st	650℃	65K	71K	78K	80K
H.T.	675℃	85K	84K	86K	86K
Temp.	700℃	81K	85K	85K	84K
(10h)	725℃	78K	76K	77K	75K

CONCLUSION

The Ag addition to the coating layer enhances the reaction diffusion both in the YBCO and BSCCO system. The Ag addition reduces the optimum heat treatment temperature by 50°–60 ℃ and improves the Jc in the YBCO system.

The Ag is pushed out from both the 123 and 2212 superconducting layers to the surface of the specimen forming a laminated structure composed of superconducting oxide and Ag layers.

In the BSCCO system, the Li substitution for Cu significantly enhances the reaction diffusion, and an offset Tc of 85K is obtained after a heat treatment at 780 ℃ for 30min.

REFERENCES

1. K. Tachikawa, N. Sadakata, M. Sugimoto and O. Kohno (1988) Jpn.J.Appl. Phys.27: 1501
2. N. Sadakata, M. Sugimoto, O. Kohno and K. Tachikawa (1989) IEEE Trans. on Magnetics, 25: 2180
3. K. Tachikawa, T. Watanabe and T. Inoue (1990) Supercond. Sci. & Technol.3: 180
4. K. Tachikawa, T. Watanabe, T. Inoue and K. Shirasu (1991) Jpn. J. Appl. Phys. 30: 639
5. M.Onoda, A. Yamamoto, E. Takayama-Muromachi and S. Takekawa (1988) Jpn. J. Appl. Phys. 27: 833

Fabrication of YBCO Thick Films on Ceramic Substrates by the Gas Deposition Method Using Fine Powders

OSAMU OKAMURA, HIROHIKO MURAKAMI, TAMOTSU NODA, and YUH SHIOHARA

Superconductivity Research Laboratory, ISTEC, Koto-ku, Tokyo, 135 Japan

ABSTRACT

Superconducting thick films in 1mm width were fabricated on ceramic substrates ($SrTiO_3$) by the gas deposition method using fine powders of $Y_1Ba_2Cu_3O_{7-x}$ with diameter less than 1μm. After heat treatment in an O_2 gas atmosphere, T_c(onset) value of 93K and T_c(end) of 75K were obtained. In the case of mixed fine powders of Y123 and Ag, T_c(end) increased to 88K.

KEY WORDS : gas deposition, thick film, fine powder, $Y_1Ba_2Cu_3O_{7-x}$, Ag

INTRODUCTION

The gas deposition method is one of the most convenient processes for preparing thick films using fine powders[1, 2, 3]. The fine powders are sprayed onto substrates in the form of aerosol. Materials are not contaminated by impurities such as solvents since this method is a dry process opposite to the doctor blade method or the screen printing method. Moreover, this method is effective to draw a flexible form of superconducting circuits on substrates.

Y123 thick films were prepared on ceramic substrates using $Y_1Ba_2Cu_3O_{7-x}$ fine powders by the gas deposition method and their superconducting characteristics were achieved after various heat treatments in an O_2 gas atmosphere. In addition, Y123 thick films doped with Ag were fabricated by this method using mixed fine powders of Y123 and Ag metal particles. It has been evident that a small content of Ag in a superconductor has an effect for crystal growth and does not degrade superconducting properties[4].

EXPERIMENTAL

The schematic diagram of the gas deposition system used in this investigation is shown in Fig.1. Y123 fine powders with diameter < 1μm stored in the mixing chamber are transported to the deposition chamber with an O_2 gas flow and sprayed through a narrow nozzle. A thick film in the form of a narrow strip 1mm in width and 20~30μm in thickness is deposited on a $SrTiO_3$(100) substrate heated at 500~900℃. The as-deposited films were sintered in a furnace with an O_2 gas atmosphere at a rate of 10℃/min up to 800℃~900℃, held at different temperatures, cooled down at a rate of 2℃/min and held at 500℃ for 5hrs.

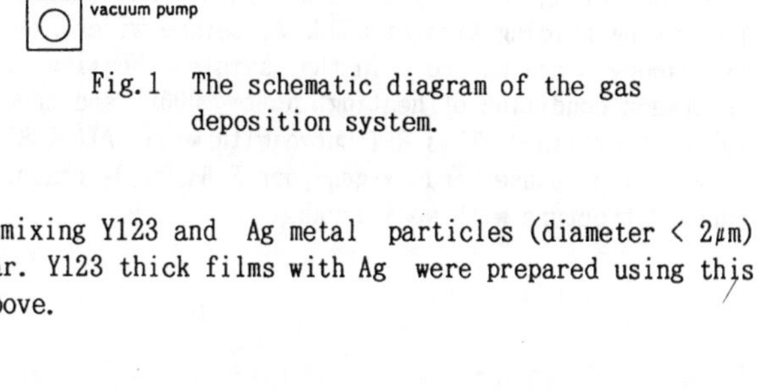

Fig.1 The schematic diagram of the gas deposition system.

Fine powders doped with Ag were prepared by mixing Y123 and Ag metal particles (diameter < 2μm) with various contents (Ag 10~50wt%) in a mortar. Y123 thick films with Ag were prepared using this mixed powders by the same process mentioned above.

Phase development was observed by X-ray diffraction with Cu(Kα) radiation (XRD). Microstructures of the films were investigated by scanning electron microscopy (SEM). Ag-contents in the films were analyzed by inductively coupled plasma chemical analysis (ICP). Temperature dependence of resistance for the samples was measured by a standard four-probe technique.

RESULTS AND DISCUSSIONS

§1 Y123 THICK FILMS

Figure 2 shows SEM micrographs of thick films deposited at various substrate temperatures (500, 700, 900℃) . The condition of after-sintering was heating up to 900℃ and then cooling down in an O_2 atmosphere. The grain size enlarged with incrasing substratre temperature for the as-deposited films. XRD patterns of the samples are shown in Fig.3. All samples indicated random crystal orientation. For the as-deposited samples ,(103) peak was higher than (110) peak opposite to the starting pure fine powders and the sintered sample with $Y_1Ba_2Cu_3O_{7-x}$ ($0 \leq X \leq 0.2$) phase. This result suggests that the as-deposited films transformed to oxygen-poor $Y_1Ba_2Cu_3O_6$ phase by vacuum sintering during deposition and cooling down to room temperature in the vacuum chamber (5×10^{-3}Torr).

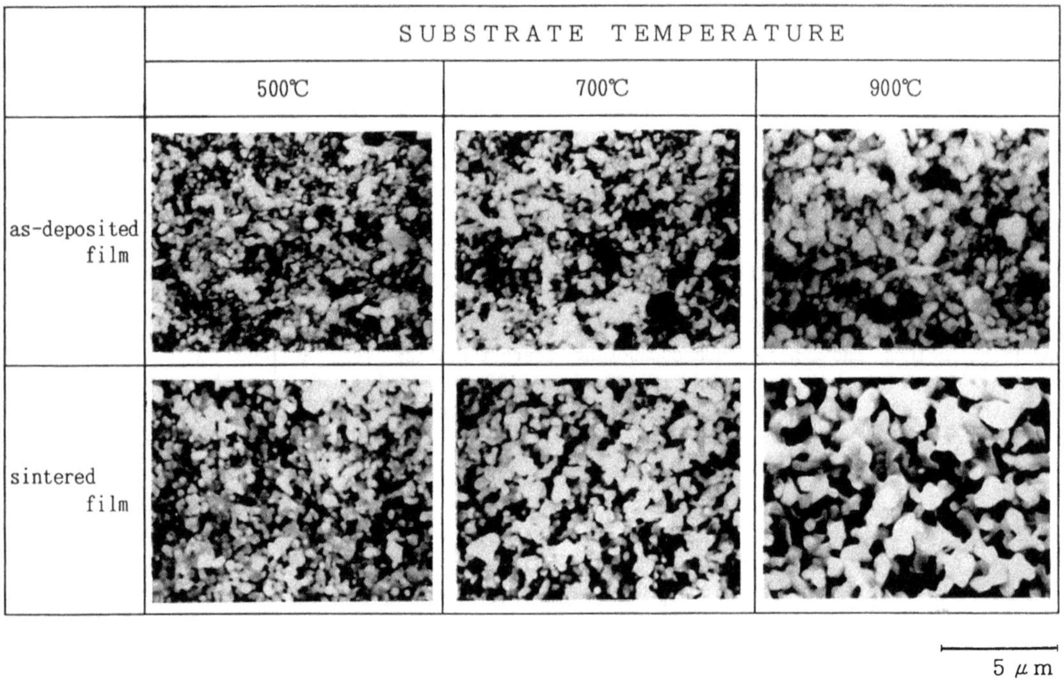

Fig.2 SEM micrographs of thick films deposited at various substrate temperatures.

Figure 4 shows temperature dependence of the electric resistance of the samples with various heat treatments. The films were deposited at the substrate temperature of 900℃. T_c(onset) value increased to 93K as increased sintering temperature up to 900℃. However the film sintered above 950℃ peeled off from the substrate. Heat tratment at 900℃ must be the most suitable to prepare superconducting thick films with oxygen-rich $Y_1Ba_2Cu_3O_{7-x}$ ($0 \leq X \leq 0.2$) phase without peeling off. With increasing holding time at 900℃, ΔT became wider . According to the observation with an optical microscopy , cracks arose in the samples heating for longer than 1hr at 900℃. At the best heat treatment condition of heating up to 900℃ and then cooling down, T_c(onset) of 93K and T_c(end) of 75K were obtained. This R-T curve with wide ΔT(=18K) suggests that this thick film had both oxygen-rich phase and oxygen-poor $Y_1Ba_2Cu_3O_6$ phase. Moreover, the result may be caused by the porous structure with weak-links.

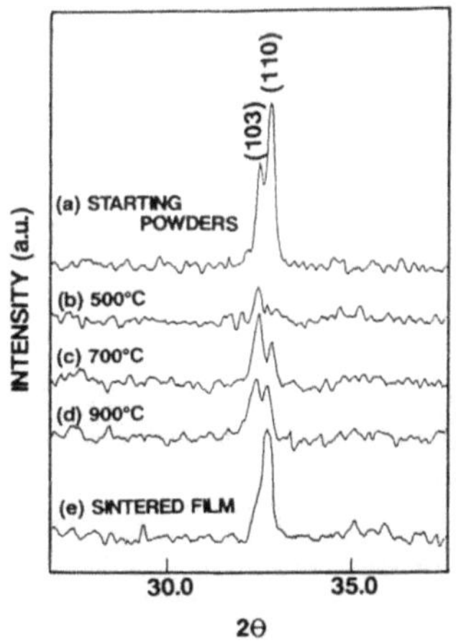

Fig.3 XRD patterns for (a)starting fine powders, (b) as-deposited film with substrate temperature of 500℃, (c)700℃, (d)900℃, (e) sintered film.

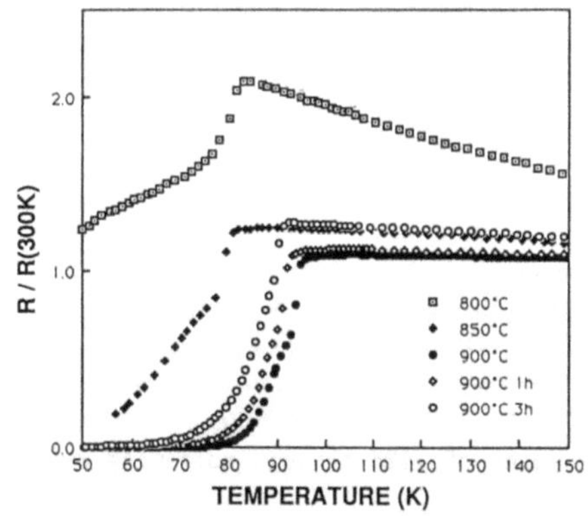

Fig.4 Temperture dependence of the resistance for the sintered thick films.

§2 Y123 THICK FILMS doped with Ag

Figure 5 shows Ag-content of the starting mixed powders of Y123 and Ag versus that of as-deposited thick films behavior analyzed by ICP. The deposited films had different Ag-contents from that of the mixed powders stored in the mixing chamber.

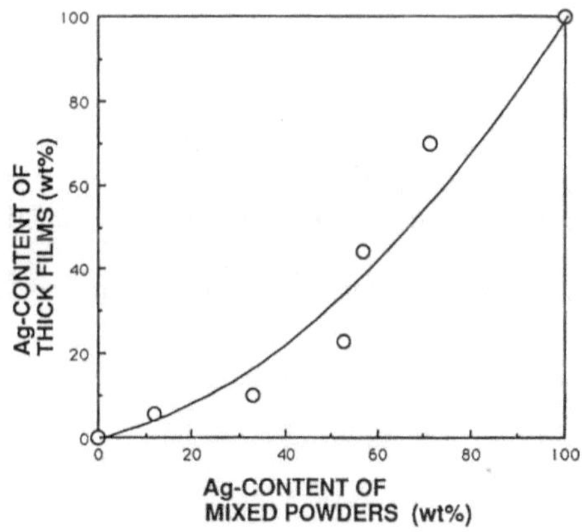

Fig.5 Ag content of the thick films as a function of that of starting powders.

SEM micrographs of the films with different Ag-contents (5, 20wt%) and the temperature dependence of the resistance of the samples are shown in Fig.6 and Fig.7, respectively. The films deposited at the substrate temperature of 900℃ were heated up to 900℃ and then cooled down in an O_2 atmosphere. Grain contacts were improved as increasing Ag-contents and the T_c(end) increased to 88K for the sample with Ag of 20wt%. From these results, we suppose as follows ; it has been

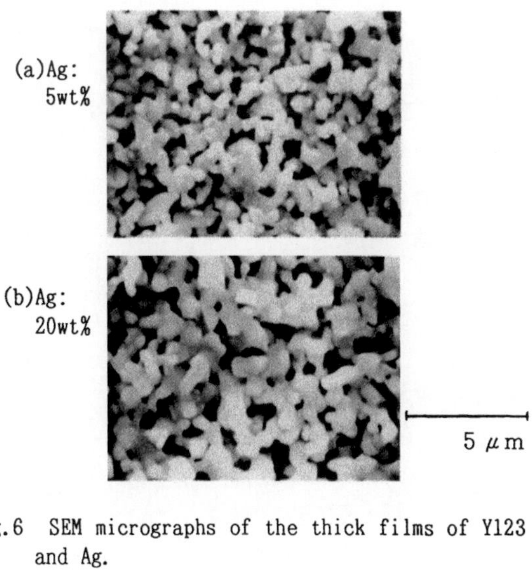

Fig.6 SEM micrographs of the thick films of Y123 and Ag.

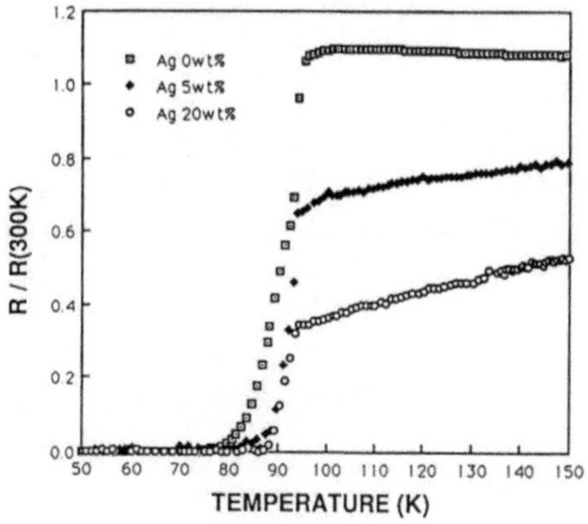

Fig.7 Temperature dependence of the resistance for the samples of Y123 and Ag.

evident that more than 90% of Ag exists in the grainboundaries[4]. Moreover, the melting point of Ag in an O_2 atmosphere decreased to lower than 900℃ due to formation of solid solution impurities into Ag.[3] Y123 grains were linked and coarsened through molten Ag ,resulted in improvement of weak-links.

CONCLUSIONS

Our results are summarized as follows :

(1) Y123 thick films were fabricated on $SrTiO_3$ ceramic substrates by the gas deposition method. The T_c(onset) and T_c(end) of the thick film using Y123 fine powders were 93K and 75K ,restectively. This low value of T_c(end) was probably caused by weak-links of grains.

(2) In the case of mixed fine powders of Y123 and Ag metal particles , T_c(end) increased to 88K by coarsening of Y123 through molten Ag which exists in grain boundaries.

ACKNOWLEGEMENTS

This work was supported by the New Energy and Industrial Technology Development Organization under the management of the R&D of Basic Technology for Future Industries.

REFERENCES

1) S.Kashu, E.Fuchita, T.Manabe and C.Hayashi : Jpn.J.Appl.Phys.23 (1985) L910
2) H.Murakami, J.Nishino, S.Yaegashi, Y.Shiohara and S.Tanaka : Jpn.J.Appl.Phys.30 (1991) L185
3) H.Murakami, O.Okamura, T.Noda, Y.Shiohara and S.Tanaka : Seramikkusu Ronbun-shi 99 (1991) 793
4) F.Mizuno, H.Masuda, I.Hirabayashi and S.Tanaka :Jpn.J.Appl.Phys.28 (1989) L780

Superconducting Y-Ba-Cu-O Tapes Fabricated by Excimer Laser Deposition

N. YOSHIDA[1], K. FUJINO[1], M. KUBOTA[1], S. TAKANO[1], M. NAGATA[1], T. HARA[2], H. ISHII[2], T. YAMAMOTO[2], and K. MAEKAWA[2]

[1] Osaka Research Laboratories, Sumitomo Electric Industries, Osaka, 554 Japan
[2] Engineering Research Center, Tokyo Electric Power Company, Chofu, Tokyo, 182 Japan

ABSTRACT

Thin film Y–Ba–Cu–O superconducting tapes with the length up to 20cm were successfully prepared using a continuous laser deposition technique. Sintered yittria–stabilized zirconia tapes were used as flexible substrates. The as–grown Y–Ba–Cu–O films were highly c–axis oriented over the full length of the tape. The critical current density(J_c) at 77.3K of 3.4×10^4A/cm^2 was obtained for 15cm voltage tap spacing. The highest value of local J_cs was 6.8×10^4A/cm^2 (1cm spacing). The critical current(I_c) of 3.5A was achieved for 15cm voltage tap spacing. The crystal orientations and the morphological feature of the thin film tape are discussed in connection with the J_c values. The effect of bending strain on the J_c were also determined.

key words; Y–Ba–Cu–O film, laser deposition, superconducting tape, bending strain

INTRODUCTION

The discovery of high temperature superconductors has led to numerous efforts for obtaining practical conductors with high J_c. These include superconducting thin film tapes deposited on flexible, long substrates as well as silver sheathed tapes and melt textured fibers. We have shown that the excimer laser deposition makes it possible to form a high quality Y–Ba–Cu–O superconducting film on MgO(001) at a high deposition rate[1]. This method seems applicable to fabricating a high current, long superconductor with high producibility. One of the key technologies to do this is a continuous deposition, where a long tape substrate is sliding on a heater stage during deposition. In this paper we report on our successful continuous deposition of Y–Ba–Cu–O on 20cm long flexible substrates.

EXPERIMENTAL

The fundumental setup for the continuous deposition is shown in Figure 1. A rotating $Y_1Ba_2Cu_3O_x$ target was irradiated by a focused KrF excimer laser with 248nm output. A sintered yittria–stabilized zirconia(YSZ) tape was employed as the subtrate. The tape was 20cm long, 10mm width and 0.1–0.2mm thick. As is well known, YSZ tape of about this thickness can be easily bent elastically, which allows flexibility of the superconducting tape. As illustrated in Figure 1, the tape was sliding on the heater stage during deposition. A deposition rate of 0.05 to 0.1μm/min was chosen. The running speed was between 0.5 to 1mm/min for a film thickness of 1 to 2μm. A stainless steel window was fixed just above the tape substrate. This enabled the film deposition only within the region where process conditions such as the substrate temperature and the ablation plume were proper for formation of high quality films.

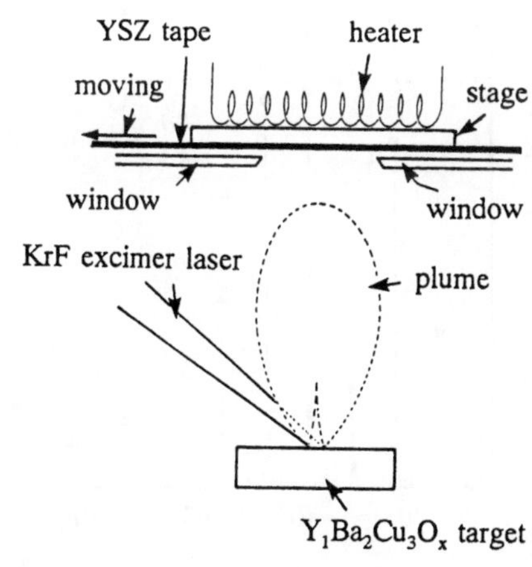

Figure 1. Experimental setup for Y–Ba–Cu–O tape depositions

Previously we reported that the front part of the as–grown Y–Ba–Cu–O tape was found to have a c–axis lattice constant larger than 11.68Å, which is expected for fully oxygenated Y–Ba–Cu–O[1]. This result, corresponding to the existence of oxygen deficiencies in the film, originates from the cooling of the front part of the tape in several hundred-mTorr oxygen ambient during deposition. For oxygenation, Y-Ba-Cu-O tapes were annealed at around 500°C for 1 hour in 1 atm of oxygen gas after deposition. These annealed tapes were investigated in this report. The crystal structure and the morphological feature of the film were examined using a X-ray diffractometer and a high resolution scanning electron microscope(HRSEM, Hitachi S-900).

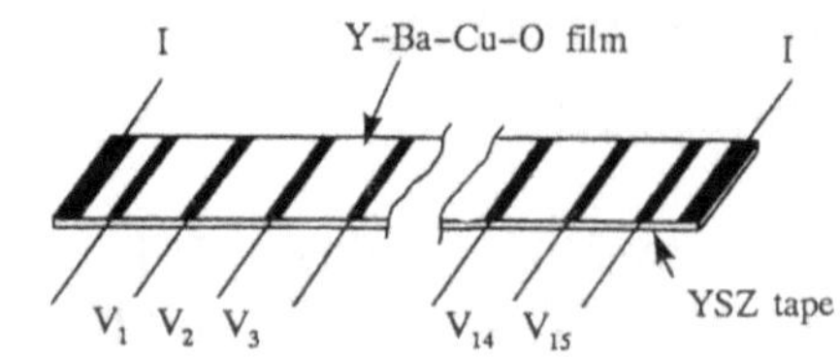

Fifure 2. Configuration of voltage & current taps for Y-Ba-Cu-O tape

The J_c and the I_c measurements were made with the tape immersed in liquid nitrogen using the configuration of voltage and current taps depicted in Figure 2. These taps located at every 10mm are silver thin films that are deposited to reduce the contact resistance[2]. The effect of bending strain on the J_c were also determined. This measurement was done by a conventional four probe method in conditions such as Y-Ba-Cu-O films were bent convexly(tensile strain) and concavely(compressive strain) with the tape attached on a cylindrical base[3]. Several bases with different radius R were used for changing the amount of strain ε applied to the Y-Ba-Cu-O film. ε is approximated as $t_s/2R$, where t_s is the thickness of the tape substrate.

RESULTS AND DISCUSSION

Figure 3 is a comparison of X-ray diffraction patterns for small parts of the two Y-Ba-Cu-O tapes. Each sample has a different J_c value; the lower for (a), the higher for (b). Both are highly oriented with its c-axis perpendicular to the tape substrate. For the lower J_c sample, the peak intensity of a-axis orientation is higher than for the higher J_c sample. Surface SEM micrographs for the same two samples are shown in Figure 4. The lower J_c samples includes more a-axis oriented grains which appear as cross bars than the higher J_c sample. Furthermore denser grains which appear as islands extraneous to the film matrix are observed in the lower J_c sample. These two kinds of grains seem to hinder the transport of supercurrent that flow mainly in the film matrix with c-axis orientation. We believe that

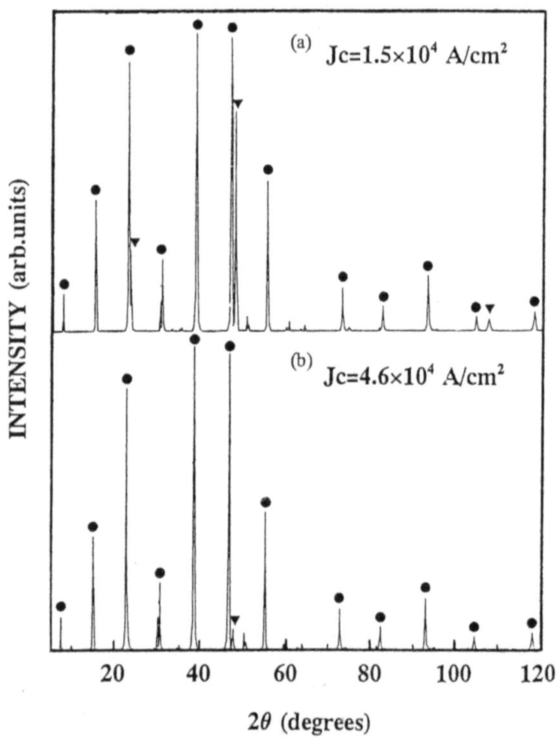

Figure 3. X-ray diffraction patterns for two Y-Ba-Cu-O tapes
(a); J_c=1.5×10^{4}A/cm^2, (b); J_c=4.6×10^{4}A/cm^2
●; (00*l*)
▼; (*h*00)

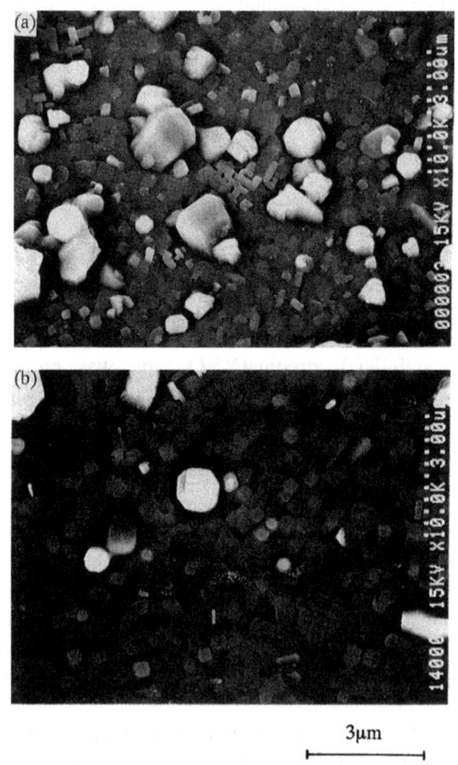

Figure 4. Surface SEM micrographs for the same sample as in Figure 3,
(a); J_c=1.5×10^{4}A/cm^2, (b); J_c=4.6×10^{4}A/cm^2

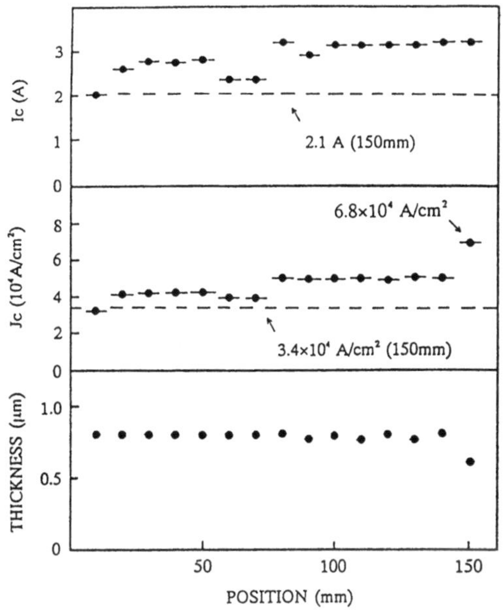

Figure 5. I_c, J_c and thickness distributions for Y-Ba-Cu-O tape with the film thickness of 0.8μm

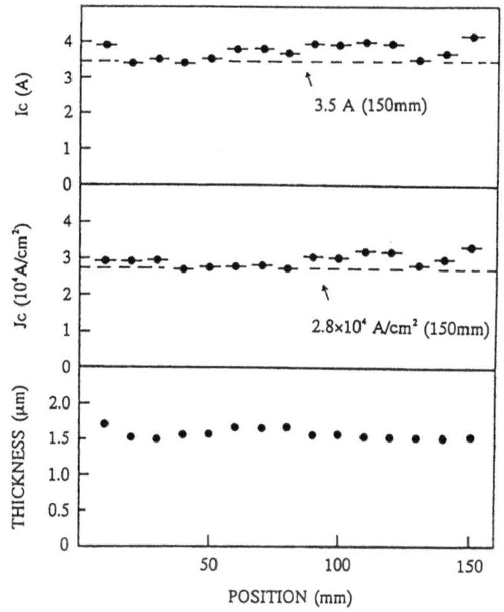

Figure 6. I_c, J_c and thickness distributions for Y-Ba-Cu-O tape with the film thickness of 1.6μm

the formation of these impediment grains to current flow is affected sensitively by the substrate temperature and the plasma plume conditions. So we especially tried to stabilize the plasma plume condition during the tape deposition; The pulsed laser energy was adjusted so as to make constant of the plasma emission intensity from YO and BaO molecules around 600nm in wavelength. We have reported that these emissions are related closely to the formation of high quality films[4].

The resultant I_c, J_c and thickness distributions are plotted in Figure 5 and Figure 6 for two tape samples which have average thicknesses of 0.8μm and 1.6μm respectively. Here we used 1μV/cm criterion to determine the data. The best J_c of 3.4×10^4A/cm^2 was obtained over 15cm voltage tap spacing for the 0.8μm thick sample. The highest value of local J_cs for this sample was 6.8×10^4A/cm^2(1cm spacing). Thickening the Y-Ba-Cu-O film up to 1.6μm was found not to degrade the J_c so much. The I_c of 3.5A over the length of 15cm was obtained for the 1.6μm thick sample. The thickness uniformities of the both samples are excellent. Some scatterings of the I_cs and the J_cs are observed for the 0.8μm thick sample. A dominamt cause of these scatterings is thought to be a surface temperature fluctuation of the tape during deposition, which is due to imperfect contact of the moving tape with the heater stage.

A more detailed observation of the film morphology was done using a HRSEM. A high resolutiuon image of the Y-Ba-Cu-O tape surface is shown in Figure 7. The granularity of the film and the terraced structure of each grain are clearly observed. These terraced structures have been investigated so far only with scanning tunneling microscopy and atomic force microscopy[5],[6]. This is the first observation of these microstructures using HRSEM as far as we notice. The grains have a rectangular shape and adjacent grains are likely to align each other in their a- and b-axis directions. Cornell group has reported that many of the Y-Ba-Cu-O grains deposited on a single grain in a polycrystalline MgO substrate are aligned[7]. Since the grains of the sintered YSZ tape we used is found to have the size of around 0.2μm and to be oriented randomly, we can claim that the Y-Ba-Cu-O grain deposited continuously on the YSZ tape has a tendency to align in the in-plane direction with the adjacent Y-Ba-Cu-O grains regardless of random orientation of the YSZ substrate. However, this fact is inconsistent with the result of X-ray pole figure measurement shown in Figure 8. This figure is the azimuthal distributuion of (102) pole intensity of the Y-Ba-Cu-O film. The data can be interpreted as a statistic of in-plane orientation over the whole sample with the size of around 10mm. Therefore the in-plane alignment observed in Figure 7 is concluded to be valid only in a short range.

Lastly, we refer to the flexibility of our tape samples. The J_c values as a function of the bending strain are shown in Figure 9. Another abscissa represents the bending radius for the tape thickness of 0.16mm. These measurements were done for the voltage tap spacing of 5cm. In the case of tensile bending, the J_c of 2.1×10^4A/cm^2 is gradually

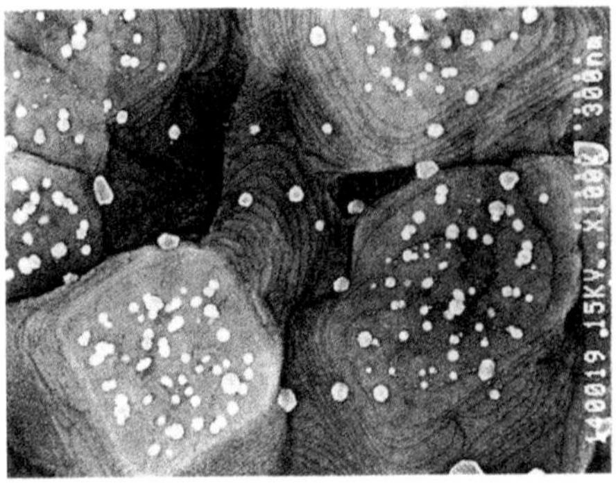

Figure 7. HRSEM image of Y-Ba-Cu-O tape

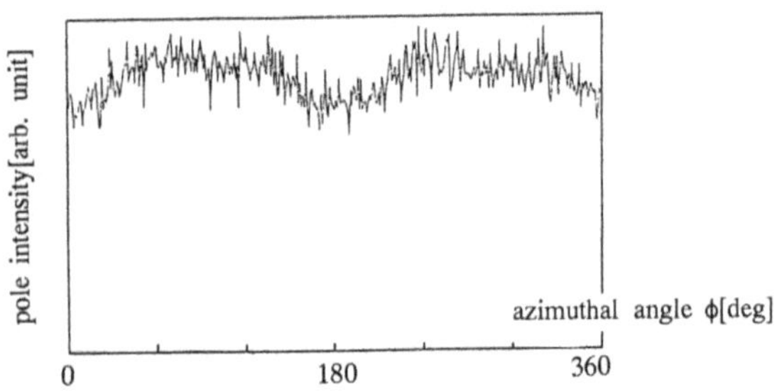

Figure 8. Azimuthal scan of (102) poles of Y-Ba-Cu-O tape

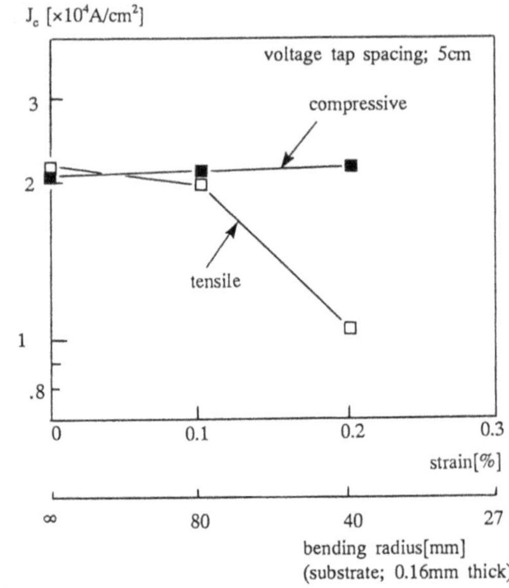

Figure 9. J_c values as a function of bending strain for Y-Ba-Cu-O tape

degraded to around 1×10^4A/cm^2 as the strain is increased up to 0.2%. On the other hand, no degradation in J_c is observed up to the compressive strain of 0.2%. These tendencies are quite similar to those obtained previously for our small samples[8]. The strain of 0.2% is equivalent to the bending radius of 40mm for our tape samples. These results show that our Y-Ba-Cu-O tape has a practical flexibility when the tape is bent concavely.

In summary, we successfully prepared thin film Y-Ba-Cu-O superconducting tapes with the length up to 20cm using the continuous laser deposition technique. The J_c of 3.4×10^4A/cm^2 and the I_c of 3.5A are obtained as the best data over the length of 15cm. The higher J_c sample was found to be free of the extraneous grains which appear as islands and cross bars. The HRSEM image revealed that the Y-Ba-Cu-O grains deposited continuously on the YSZ tape are aligned each other in a short range. With the flexibility of our Y-Ba-Cu-O tape, we confirmed that the J_c is never degraded up to the strain of 0.2% in the compressive bending state. In this paper, we employed the deposition rate up to 0.1μm/min. We have reported a successful high deposition rate of 3.7μm/min using a MgO(001) substrate[1]. So the running speed of several tens mm/min can be expected for our continuous deposition. Though the in-plane alignment of the Y-Ba-Cu-O grains in a short range was confirmed, the alignment of the entire Y-Ba-Cu-O tape was failed. To improve the J_c value of this Y-Ba-Cu-O tape further, some tight locking mechanism to align the grains with each other may be necessary.

REFERENCES

[1] N.Yoshida, M.Kubota, S.Takano, M.Nagata, T.Hara, H.Ishii, T,Yamamoto and K.Maekawa: in *Proc. International Conf. Advanced Materials,* Strassbourg(1991). in press.

[2] J.W.Ekin, A.I.Braginski, A.J.Panson, M.A.Janocko, D.W.Kapone II, N.J.Zaluzec, B.Flandermeyer, O.F.de Lima, M.Hong, J.Kwo and S.H.Lion: *J. Appl. Phys.*, **62**, (1987)p4821.

[3] K.Rokudo, J.Yamaguchi, H.Miyazaki, T.Mitsui, H.Hitotsuyanagi and N.Yoshida: in *Proc. International Meeting on Advanced Matarials*, vol.6, Tokyo, 1988, (Mat. Res. Soc., Pittsburg, 1988)p891.

[4] N.Yoshida, S.Okuda, S.Takano, M.Nagata, T.Yamamoto, T.Hara, K.Okaniwa: *Sumitomo Electric Technical Review*, No.31,(Jan. 1991)p49.

[5] M.Hawley, I.D.Raistrick, J.G.Berry and R.J.Houlton: *Science*, **251**, (1991)p1587.

[6] C.Gerber, D.Anselmetti, J.G.Bednorz, J.Mannhart and D.G.Schlom: *Nature*, **350**, (1991)p279.

[7] B.H.Moeckley, D.K.Lathrop, S.E. Russek, R.A.Buhrman, M.G.Norton and C.E.Carter: *IEEE Trans. Mag.* **27**, (1991)p1017.

[8] N.Yoshida, M.Kubota, S.Okuda, S.Takano, M.Nagata, T.Hara, K.Okaniwa and T.Yamamoto: in *Proc. the 3rd International Symposium on Superconductivity(ISS'90)*, Sendai, (1990)p901.

Fabrication of Y-Ba-Cu-O Superconducting Filament by Solution Spinning

TOMOKO GOTO[1] and ASAICHI KAWANAKA[2]

[1] Department of Materials Science & Engineering, Nagoya Institute of Technology, Nagoya, 466 Japan
[2] Nitivy Co., Ltd., Fujieda, Shizuoka, 426 Japan

ABSTRACT

The Y-Ba-Cu-O superconducting filament has successfully prepared by dry spinning through a homogeneous aqueous solution containing Y, Ba and Cu acetates, polyvinyl alcohol and organic acids. After removing volatile components, the 124 filament is obtained by heating at 780 °C under ambient pressure. A partial melt growth process is used to enhance the J_c of the filaments. By controlling heating condition, dense and highly oriented texture are produced and J_c of more than 10^4 A/cm^2 at 77 K and 0 T is achieved. The phase decomposition from 124 phase to 123 phase and partial melting result in an enhancement of flux pinning.

KEY WORDS: Y-Ba-Cu-O filament, solution spinning, 124 phase, partial melting, enhancement of flux pinning

INTRODUCTION

Since the discovery of high T_c oxide superconductor with critical temperature above liquid nitrogen temperature, energy research investigations have been carried out on high T_c superconducting wire with high J_c. However, bulk high T_c superconductors have a limitation in critical current density (J_c) due to the weak link problem and the insufficient flux pinning.

We have studied the preparation of superconducting oxide of long filaments using a textile fiber spinning technology for the precursor of the oxide and a high transport J_c value of more than 10^4 A/cm^2 at 77 K and 0 T was attained for the melt-processed filament produced by a suspension spinning [1] and a solution spinning [2]. The solution method produces more uniform and finer materials than does the suspension spinning method and introduces a fine flux pinning center into the superconducting phase, such as at elemental level. In this paper we describe a new fabrication process of the Y-Ba-Cu-O superconducting filament by solution spinning.

EXPERIMENTAL

Y, Ba, and Cu acetates were employed for the starting materials. An aqueous solution of PH=4.3 was prepared in advance by dissolving 5.5 wt. % polyvinyl alcohol (PVA) with a degree of polymerization of 1700 and saponification degree of 90.5 mol % and 6 wt. % propionic acid. 1.5 wt.% 2-hydroxy isobutyric acid and 15.4 wt. % of the mixture of the acetates were dissolved in the aqueous solution. The resultant solution was concentrated to obtain a stable viscous homogeneous spinning dope with a viscosity of 500 P. The dope was deaerated at 80 °C for 1 day and was extruded as a filament into a hot air zone and coiled on a winding drum. The as-drawn filament was heated to remove volatile components and to generate the superconducting phase.

Four kinds of filaments with a composition of Y:Ba:Cu =1:2:4 (A), 1:2:3 (B), 1.2:2:3 (C) and 1.5:2.25:3.25 (D) were spun and were heated at 150 °C for 1 h and 500 °C for 1 h in air to remove volatile components. Then the filaments were heated under various heating conditions.

RESULTS AND DISCUSSION

The dry-spinning through the homogeneous aqueous solution of Y, Ba and Cu acetates was performed by applying a spinning technique of PVA and continuous filaments with a diameter of 250 μm were obtained. The as-drawn filament was pyrolyzed to remove volatile components. Then filament A was heated at 780 °C for 3 h in flowing oxygen to generate the 124 phase. The heated filament consisted mostly of a single phase of the 124 structure. [3] A decomposition of the 124 filament is investigated for introducing fine flux-pinning centers into the 123 filament.[4]

The filaments were sintered at 910 °C for 15 min followed by furnace cooling at a cooling rate of 100 °C/h in flowing oxygen. A diameter of filament B was reduced to about 70 μm, in spite of starting from the as-drawn filament of 250 μm in diameter. A partial melt growth process was used to enhance the J_c of filaments B, C and D. Filament B was partially-melted at 1030 °C for 15 min and was cooled to 930 °C at a rate of 100°C/h, subjected to slow cooling to 830 °C at a cooling rate of 30 °C/h, and then furnace-cooled in flowing oxygen.[2] The filaments with diameter of 60 μm were obtained. Filaments C and D were partially-melted under the same condition of filament B. However the filaments did not show the superconductivity. The optimum heating condition was dependent on a composition and diameter of the filament. Filament C was partially-melted at 1000 °C for 15 min and was cooled to 930 °C at a rate of 100 °C/h. After 1 h, the filament was slow-cooled to 830 °C at a cooling rate of 30 °C/h and then furnace-cooled in flowing oxygen. Filament D was partially-melted at 1000°C for 1 h and was cooled as same as for filament C. The diameter, maximum transport J_c at 77 K and 0 T and the structure of the filaments are listed in Table 1.

Table 1 Maximum transport J_c at 77 K and 0 T and structure of the filaments.

Filament	Composition Y : Ba : Cu	Heating treatment	Diameter (μm)	Transport J_c (A/cm^2)	Structure
A	1 : 2 : 4	sintering	70	1100	123+CuO
B	1 : 2 : 3	sintering	70	1900	123
C	1.2:2 : 3	sintering	70	1000	123
D	1.5:2.25:3.25	sintering	110	22	123+211
B	1 : 2 : 3	partial melting	60	17000	123+211+L*
C	1.2:2 : 3	partial melting	60	7300	123+211+L*
D	1.5:2.25:3.25	partial melting	92	3200	123+211+L*

L*: = $CuO+BaCuO_2$

The highest J_c of 17000 A/cm^2 at 77 K and 0 T was attained for partially-melted filament B. The filaments with high J_c were characterized by scanning electron micrography. Figure 1 shows the fracture surface and the polished and etched surface on the longitudinal cross-section of the filament B. The diameter of the filaments was reduced to about 60 μm. The sample shows a dense and well aligned structure consisting of plate like grains.
It was reported that the 211 particles can contribute to increase the J_c and maximum J_c was obtained with a 50 mol% excess of the 211 particle on the bulk QMG processed material and the 211 particles can be regarded as working for the J_c increase due to the reduction in crack [5] or the increasing pinning force.[6] The J_c of the present filament decreases with increasing 211 phase as shown in Table 1. As the filament was very thin, the excess 211 particle may reduce the current path and did not result in a increasing J_c. The fracture surface and longitudinal cross-section of filament C are shown in Fig.2. Filament C had a denser texture than that for filament B. When the sample contains fine 211 precipitates, a strain field is created in the vicinity of every 211 precipitates, which favorably disperses stress and reduces the stress concentration and hence prevent crack formation.[6]

An X-ray transmission Laue pattern of the partial-melted filament with high J_c was taken by a flat cassette camera to examine the orientation of the crystals. [100] and [010] fiber textures were deduced for the partially-melted filament. the fiber axis appears to coincide with the direction perpendicular to the c-axis of the orthorphobic phase in the superconducting layer. Such crystallographic alignment should be beneficial for the flow of supercurrents in this severely anisotropic layer compound. The magnetization curves of these filaments were recorded on a Faraday method susceptibility

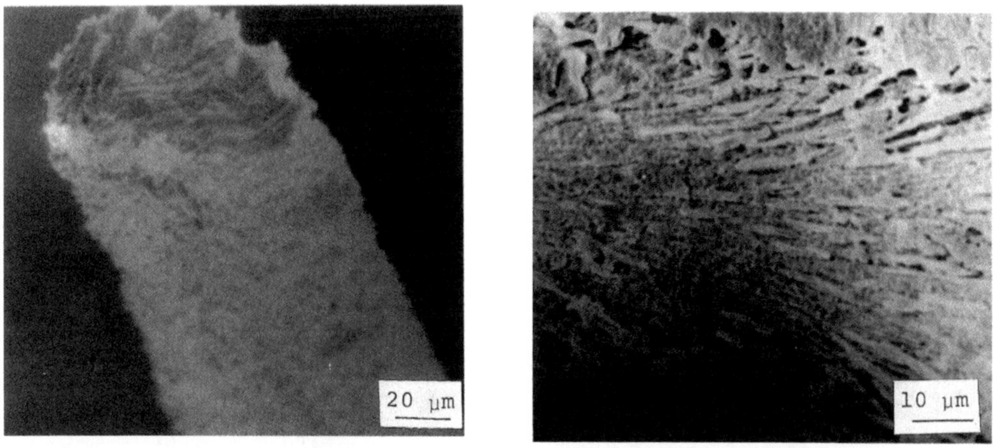

(a) (b)

Fig.1 Scanning electron micrographs of the partially-melted filament B with high Jc. (a) fracture surface (b) longitudinal cross-section.

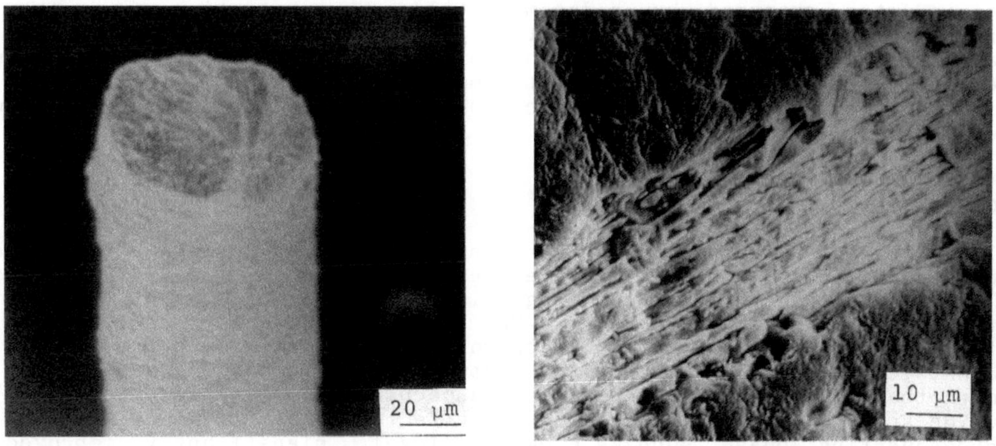

(a) (b)

Fig.2 Scanning electron micrographs of the partially-melted filament C with high Jc. (a) fracture surface (b) longitudinal cross-section

instrument. The specimens of about 5 mm long were placed parallel to the magnetic field which could be varied from 0.001 to 52 kOe. The magnetization curves taken with increasing field and decreasing field were never completely reversible at both temperatures of 77 K and 4.4 K, respectively and showed that the materials are type II superconductors. Although the diamagnetism disappeared at about 2.5 T at 77 K for the suspension-spun Y-system filament,[4] the diamagnetism for the solution-spun filament was maintained even at more than 5.2 T and 77 K.[5] The critical current density had been calculated from the differences ΔM of magnetizations, in the same increasing and decreasing field, using Bean's formula, i.e. $J_c = 30\ \Delta M/d$, where d is the diameter of the sample. The resultants for sintered filament A, sintered filament B and partially-melted filament B are shown in Fig.3. Here, J_c at 77 K is more than two orders of magnitude smaller than J_c at 4.4 K and decreases with increasing field. However it is noticeable that the field dependence of J_c at 77 K for the solution-spun filaments is relatively weak and the J_c at 77 K at 5 T showed more than 10^3 A/cm^2. The relation between the pinning force (F_p) and applied field are shown in Fig.4. The maximum F_p was observed at about 3 T for the partially-melted filament B and sintered filament A. The phase decomposition from 124 phase to 123 phase and partial melting results in a enhancement of the flux pinning.
It was reported that significantly increased J_c was obtained in the Y-123

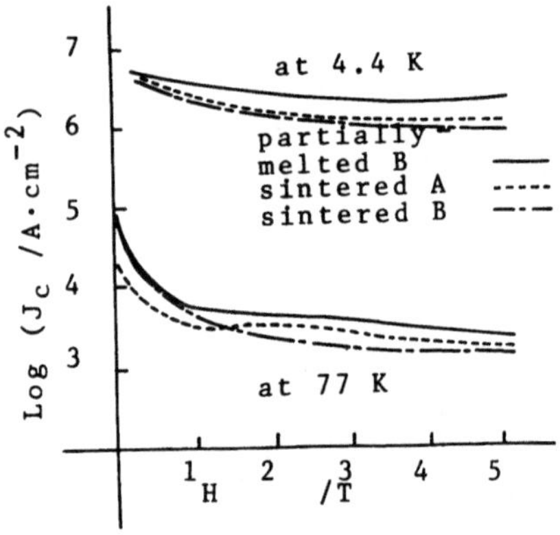

Fig.3 Magnetically determined critical current density (J_c) versus applied field (H) at 4.4 K and 77 K.

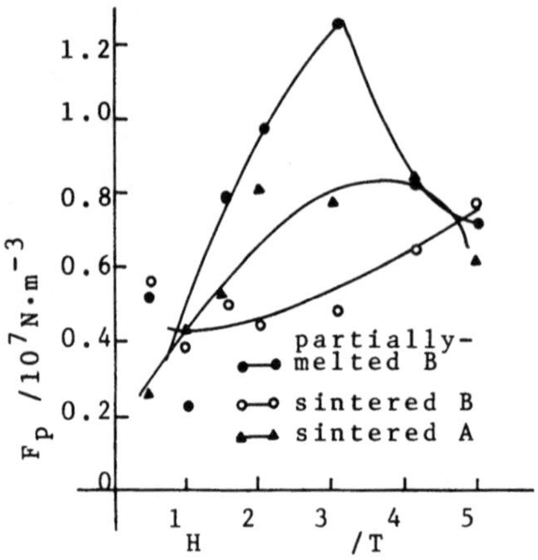

Fig.4 Relation between pinning force (F_p) and applied field (H).

superconductor prepared by a decomposition process starting from the 124 compound precursor. The cause of the J_c improvement was attributed to the enhanced flux pinning by the presence of fine-scale transformation induced defects. [7] For the present filament, the enhanced flux pinning is considered to be due to the presence of fine-scale transformation-induced defect.

CONCLUSION

A long Y-Ba-Cu-O filament was prepared by dry-spinning through a homogeneous aqueous solution containing mixed acetates of Y, Ba and Cu, polyvinyl alcohol and organic acids. After removing volatile components, the filament is sintered or is partially-melted. The 124 phase filament was obtained by heating under ambient pressure. By controlling the melt growth process of the filaments, dense and highly oriented textures were produced, and a maximum J_c of more than 10^4 A/cm^2 at 77 K and 0 T was achieved for the filament. As the solution-spinning is able to introduce a effective fine flux pinning into the superconducnting phase such as elemental level, significantly improved J_c value and the field dependence were achieved. Phase decomposition from 124 phase to 123 phase and partial melting resulted in a enhancement of pinning force.

ACKNOWLEDGMENTS

Thanks are due to the Instrument Center, the Institute for molecular Science for assistance in obtaining the magnetization curves. This work was partly supported by a Grant-in-Aid for Science Research on Chemistry of New Superconductors for the Ministry of Education, Science and Culture (No. 03211216).

REFERENCES

1. Goto T and Kino Y (1990) Jpn.J.Appl.Phys. 29: L2030-L2032
2. Goto T and Sugishita T (1991) Jpn.J.Appl.Phys. 30: L997-L999
3. Goto T and Sugishita T (1990) Physica C 171: 441-443
4. Goto T and Sugishita T (in press) Proc. on the 2nd ISSP International Symposium on Physics and Chemistry of Oxide Superconductors
5. Tanaka M, Morita M, Takebayashi S, Kimura M, Kimura K, Sugiyama M, Miyamoto K and Sawano K (1990) in Kajimura, Hayakawa (eds) Advances in Superconductivity III .Springer-Verlag, Tokyo, pp721-725
6. Murakami M, Gotoh S, Koshizuka N, Tanaka S, Matsushita T, Kambe S and Kitazawa K (1990) Cryogenics 30: 390-396
7. Jin S, Tiefel T.H. Nakahara S, Graebner J.E, Bryan H.M.O, Fastnacht R.A and Kammlott G.W (1990) Appl.Phys.Lett. 56: 1287-1289

Magnetic and Electrical Properties of Ag-Doped Superconducting Y-Ba-Cu-O Tapes

YUMIO YOKOCHI[1], YOICHIRO MASUDA[1], SIGETAKA FUJITA[1], HIDEHIRO SEKI[1], YOSHINORI SAKAMOTO[1], AKIRA TOGASAWA[1], MAKOTO BABA[1], AKIRA BABA[1], and MASAYOSHI MASUDA[2]

[1] Hachinohe Institute of Technology, Hachinohe, Aomori, 031 Japan
[2] Shohnan Institute of Technology, Fujisawa, Kanagawa, 251 Japan

ABSTRACT

Improvement of the critical current density is one of the very important subjects for the high Tc superconducting oxide ceramics. This paper describes the superconducting properties of Ag added Y-Ba-Cu-O tapes. Y-Ba-Cu-O superconducting powder added Ag powder on an Ag substrate was cold-rolled to form a tape and sintered. In this case, a thermal stress generated cracks which deteriorated the superconducting properties. In order to eliminate the generation of cracks, the conditions of heat treatment, formation process and so on were investigated in detail. As the results, the improved crack-free superconducting tapes with magnetizing force M=-2.5 emu/g, Tc=97 K and J_c=2.1X10^4 A/cm^2 (by Bean-London Model) were obtained.

KEY WORDS: Magnetic Susceptibility, Ag-Added Y-Ba-Cu-O Tapes, Oxide Superconductor

1. INTRODUCTION

A number of fundamental and theoretical researches on the oxide superconducting materials with high Tc are performed extensively. And its applied works are also proceeded actively in many fields. The application to the field of power electronics is one of the most important among them and the special attention to developing and establishing the method to obtain wires from oxide materials with high current density is paid. This paper reports the effects of the amount of Ag added to Y-Ba-Cu-O powder and of the improvement of weak combination of grain boundary caused by repeated application of a series of processings consisting of spreading powder, heat-treating and pressing from experimental results.

2. EXPERIMENTAL

2.1 Preparation of Samples

We used as superconducting powders Y-Ba-Cu-O powder without Ag and Y-Ba-Cu-O powder with Ag powder (5, 10, 15 %). The samples prepared were tapes of 0.30 mm in thickness which were produced in the following process (see Fig.1). The powder was spreaded on

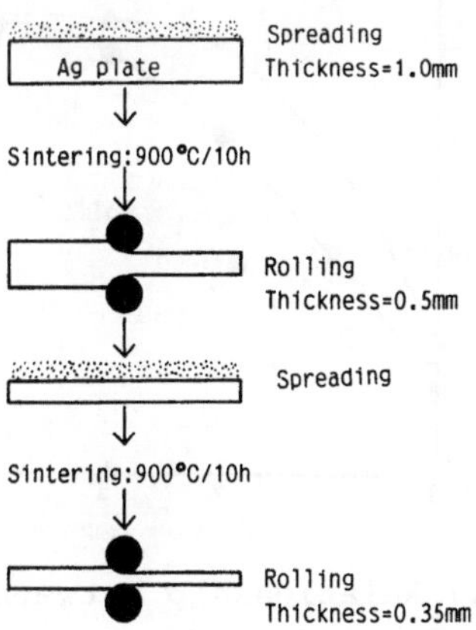

Fig.1 Fabrication process of tape-shaped wire

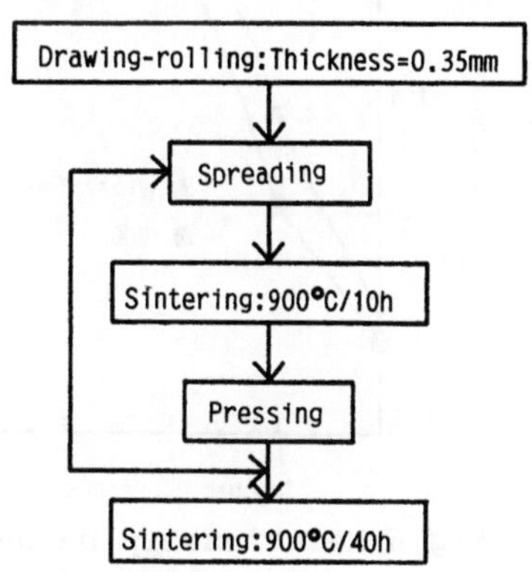

Fig.2 Experimental procedure of processing

the metallic silver plate of 50 x 50 x 1 mm³ of the purity 99.5 % and the plate with the powder was cold-rolled to make a tape of 0.35 mm in thickness. The tape was cut into pieces of the size 40 x 30 mm², then the series of processing which consisted of spreading powder, pressing and heat-treating was applied to the pieces in several times according to the prescribed condition in Fig. 2.

2.2 Methods of Estimation

The micro organization of samples were observed with both the metallurgical microscope and the scanning electron microscopy (SEM in short). The critical temperature Tc was obtained by the four terminal method and the critical current density Jc was calculated by the permanent current method using the circuit as shown in Fig. 3. And the magnetic characteristics were obtained by the vibrating sample magnetometer.

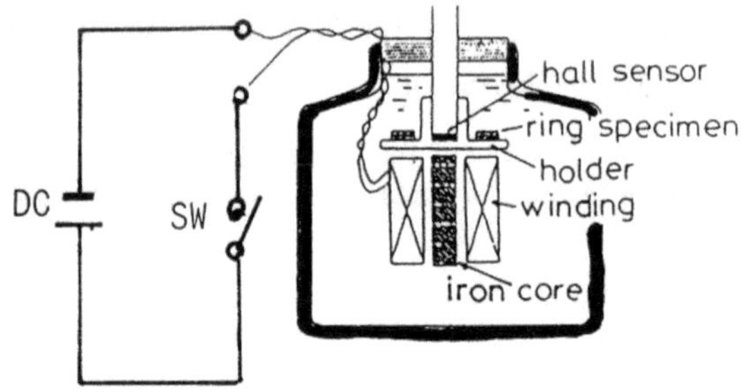

Fig.3 A schematic of the experment

3. RESULTS AND DISCUSSION

Fig. 4 shows that the superconducting layer is controlled from 0.1 to 0.11 mm in thickness. Fig. 5 (a) and (b) are the SEM photographs of the surfaces of the samples, non processing and applying above mentioned process two times, respectively. And it is found from these photographs that the processing causes the crack of the superconducting layer to be reduced and the density of the layer to be increased.

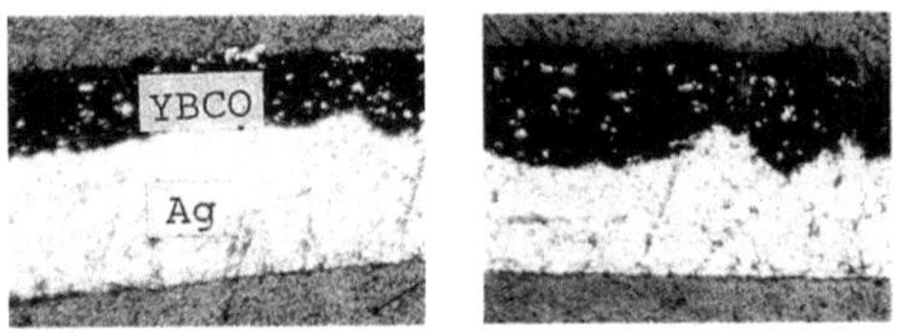

(a) Once (b) 4times 0.1mm

Fig.4 Cross-sectional view of tape processed

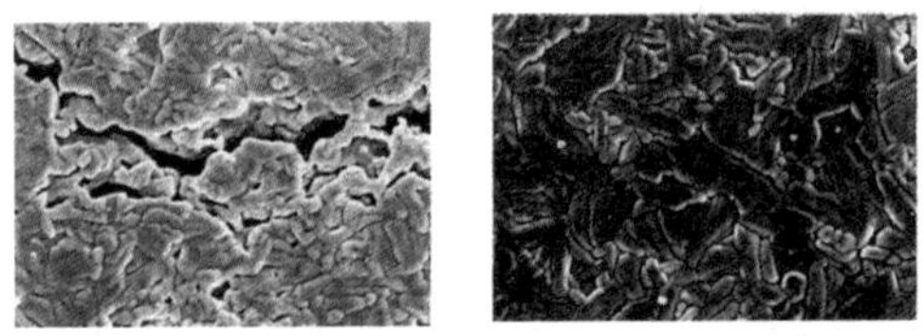

(a)Non Processed (b)Processed 10 μ

Fig.5 Sem Photographs

Fig. 6 shows the relation between the magnetization M and the number n of processing applied. The magnetization M increases as n does. Combining the facts in Figs. 5 and 6, we can conclude that the magnetization M increase depending upon the reduction

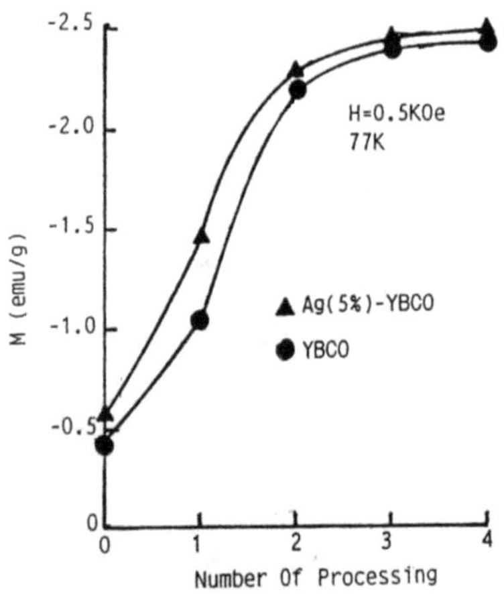

Fig.6 Relationship between M and the number of processing

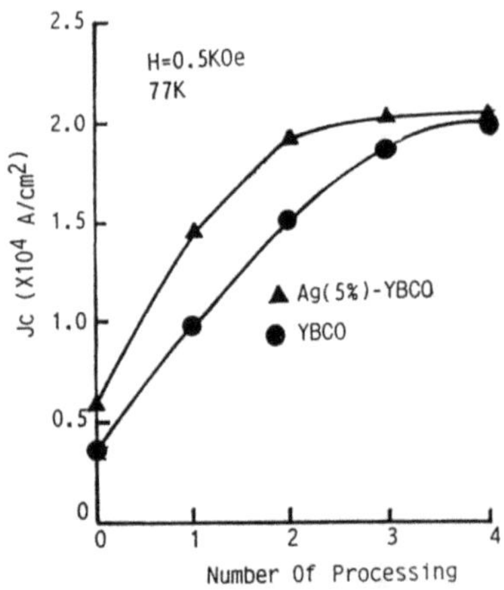

Fig.7 Relationship between Jc and the number of processing (by Bean-London Model)

of cracks in the superconducting layer and the increas of the density of it caused by the repeated processing. The maximum value 2.5 emu/g of M is obtained by using the powder with 5 % Ag and applying the processing four times. Fig. 7 shows the Jc values of the same samples of Fig. 6 and it is found that the changing tendency of Jc is similar to that of M. And the maximum value 2.1 x 10^4 A/cm^2 of J_c is calculated by Bean-London model in the same condition as for M. Fig. 8 shows the relation between the amount of Ag added and the magnetization M. From the figure, it is found that the maximum value 2.5 emu/g of M is lobtained when the amount of Ag is 5 % and that M decreases as the amount of Ag increases over 5 %.

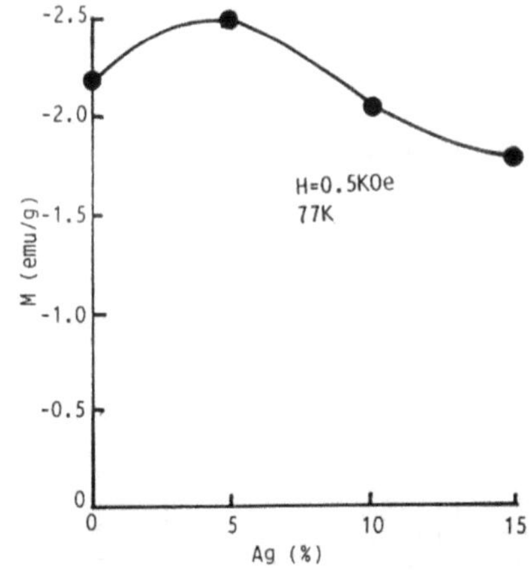

Fig.8 Relationship between M and the Ag added

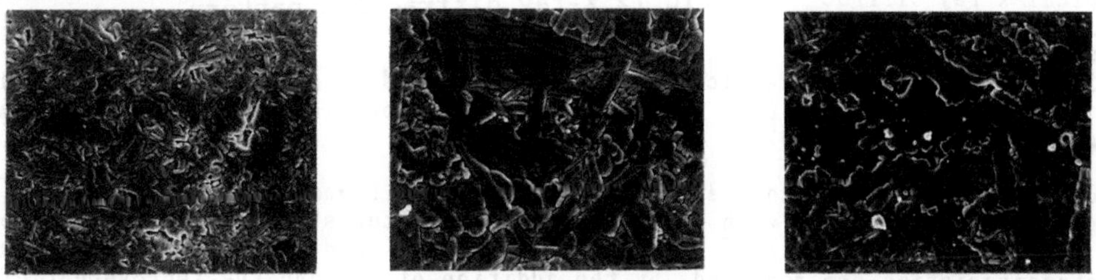
(a) YBCO (b)YBCO with 5% Ag (c) YBCO with 10% Ag

Fig.9 SEM photographs 10 μ

Fig. 9 (a), (b) and (c) are the SEM photographs of the samples Y-Ba-Cu-O, Y-Ba-Cu-O with 5 % Ag and Y-Ba-Cu-O with 10% Ag. By these photographs it can be recognized that in the case of 5 % Ag addition, the crystal grains diffused together to maximum size

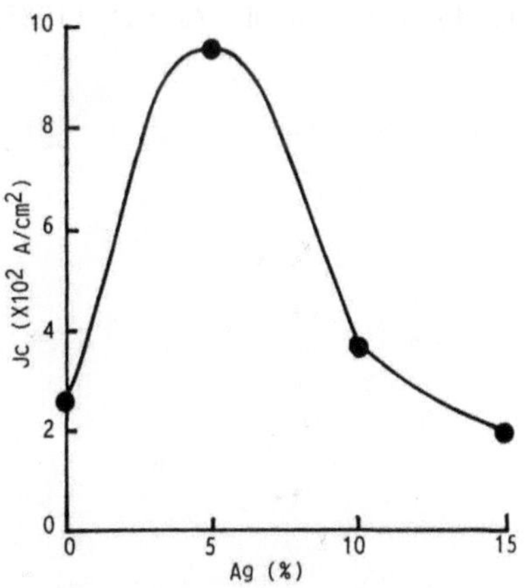

Fig.10 Relationship between Jc and the Ag added (By observation)

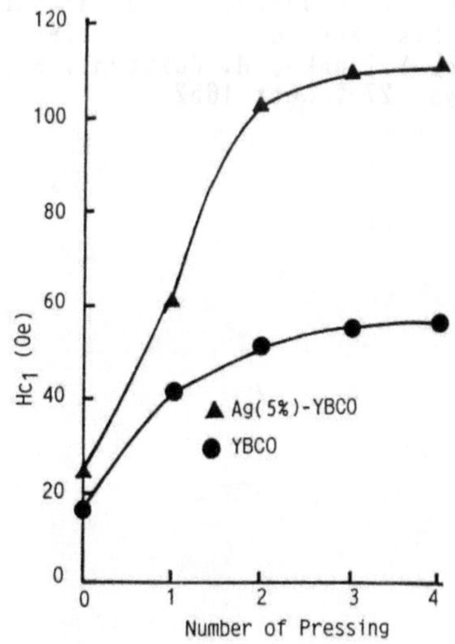

Fig.11 Relationship between Hc_1 and the number of processing.

60 μm, while in the case of 10 % addition, the silver does not diffuse into the layer and stays on the surface. So in the latter case, the characteristics decrease. The critical current density Jc calculated by the permanent current method is shown in Fig. 10 and the maximum value 950 A/cm^2 of J_c is realized, when adding 5 % Ag. This is caused by the fact that Hc_1 is improved by adding Ag. In fact, as shown in Fig. 11, the maximum values 53 Oe and 107 Oe of H_{c1} are obtained at the four times applications of the processing to Y-Ba-Cu-O and Y-Ba-Cu-O with 5 % Ag respectively. The critical temperature Tc of the sample Y-Ba-Cu-O with 5 % Ag is 97 K. The typical X-ray diffraction pattern of the sample Y-Ba-Cu-O with 5% Ag is shown in Fig. 12 and its lattice constants a, b and c obtained by this data are 3.818 A, 3.891 A and 11.695 A, respectively. From these constants, the amount of oxygen calculated was 6.92.

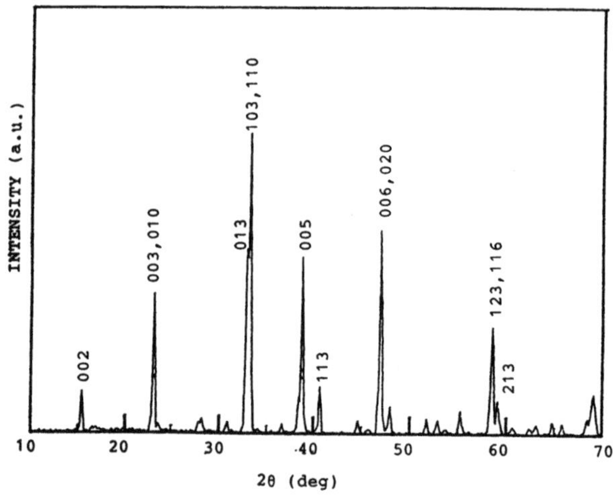

Fig.12 X-ray diffraction pattern

As conclusion, we obtained the best sample by the four times applications of the processing to Y-Ba-Cu-O powder with 5% Ag and its characteristics M, Tc and Jc are 2.5 emu/g, 97 K and 2.1 x 10^4 A/cm^2 by Bean-London Model and 940 A/cm^2 in experiment. These characteristics are realized by the addition of Ag and the improvement of weak combination of grain boundary caused by the repeated applications of the processing.

REFERENCES

1. M.K. Wu, J.R. Ashburn, C.J. Trong, P.H. Hor,R.L. Meng, L.Gao, Z.J. Huang, Y.H.Huang Y.Q. Wang, and C.W.Chu: (1987) Phys. Rev. 58 : Lett. 908
2. K.Watanabe, N. Kobayashi, H. Yamano, H.Kurasawa, T.Hirai, H. Kawabe and Y. Muto: (1989) Jpn. J. Appl.Phys. 28 : Lett 1417
3. D. Dimos, P. Chaudhari. J.Mannart and F.K.Le Goues: (1988) Phys. Rev. 61. : Lett. 219
4. T. Asano, Y.Tanaka, M. Fukutomi, K.Jikihara, J. Machida and H. Maeda: (1988) Jpn.J. Appl.Phys. 27 : Lett 1652

Forming of $Y_1Ba_2Cu_3O_{7-y}$ Superconductive Thick Film on Metal Substrate with Sprayed YSZ Buffer Layer

K. MATSUI, R. KATSUTANI, M. TAGAMI, and Y. NAKANISHI

Joining and Material Processing Department, Research Institute, Ishikawajima-Harima Heavy Industries Co., Ltd., Yokohama, 235 Japan

ABSTRACT

For practical uses of high Tc superconductive ceramics, it is required to form the superconductor film on metal of structural member. The superconductor film directly formed on metal substrate is unstable and it is necessary to make a buffer layer between superconductor film and metal substrate. Accordingly, the authors developed the YSZ buffer layer by spraying method. And on this buffer layer, they formed $Y_1Ba_2Cu_3O_{7-y}$ thick film by the powder coating method with sintering process and confirmed that this film has good Tc characteristics. Moreover, they investigated the effect of CIP treatment on the superconductive characteristics of this thick film formed by the same method.

Key ward: $Y_1Ba_2Cu_3O_{7-y}$, thick film, YSZ buffer layer, metal substrate, powder coating method

INTRODUCTION

Many studies on development of high Tc superconductive ceramics have been carried out. For practical uses of this material, it is required to form a superconductive film on the metal of structural member. However, the superconductive film directly formed on metal substrate is unstable and it is necessary to make a buffer layer between the superconductive film and metal substrate. YSZ(Yttria Stabilized Zirconia) or MgO layers by CVD, PVD or sputtering process have been developed as the buffer layers [1], but their thickness was insufficient for superconductive thick film by sintering process.
Accordingly, we developed the YSZ buffer layer by low pressure plasma spraying method for superconductive thick film. And on this buffer layer, we formed $Y_1Ba_2Cu_3O_{7-y}$(YBCO) superconductive thick film by the powder coating method with sintering in an electric furnace. Moreover, we investigated the effect of CIP (Cold Isostatic Press) treatment on the superconductive characteristics.

EXPERIMENTAL PROCEDURE

Fig.1 shows the process flow of the forming of YBCO superconductive thick film on metal substrate with YSZ buffer layer. The buffer layer, consisting of 92%ZrO_2 and 8%Y_2O_3, was formed by low pressure plasma spraying method on the bond coating layer. The bond coating layer was first laid on metal substrate by air plasma spraying method and this layer consisted of 77%Ni-22%Cr-10%Al-1%Y. The metal substrate was INCONEL 718 alloy and its shape was 10 mm square and 4 mm thick. The thickness of YSZ was changed from 50 to 200 μm. After then, the superconductive thick film was formed by the powder coating method. This method consists of dissolving process of YBCO calcination powder in organic binder, painting process the calcination solution on substrate and sin-

tering process in electric furnace.
Fig.2 shows the sintering procedure. The sintering temperature was 1223 K. The samples were cooled in furnace, whose cooling rate to 923 K was 100 K/ks and 55 K/ks below 923 K.

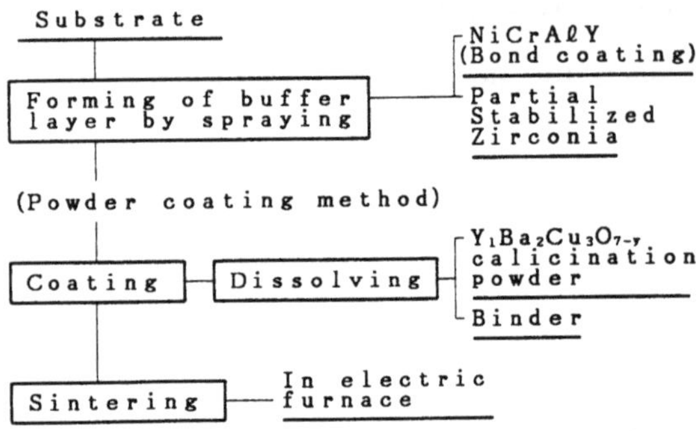

Fig. 1 Process flow

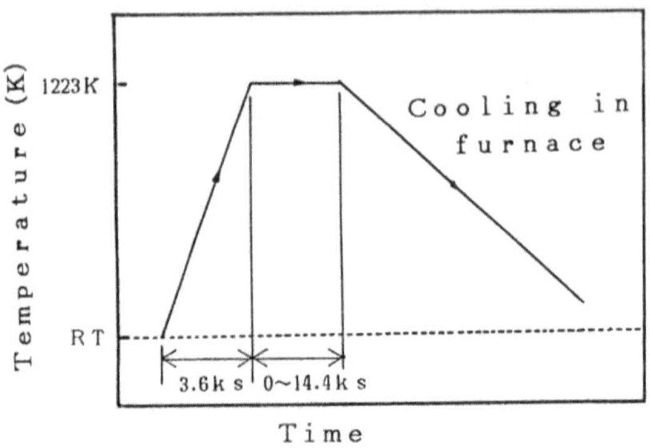

Fig. 2 Sintering procedure

RESULTS AND DISCUSSION

Fig.3 shows the cross section microstructure of the superconductive thick film on the metal substrate with YSZ buffer layer. The thickness of the film was 300 μ m. It was tightly connected to YSZ buffer layer and was very porous.
Fig.4 shows an example of the temperature-resistance curve of this superconductive film that was formed on metal substrate with YSZ buffer layer of 200 μ m thickness. To make a comparison, the temperature-resistance curve of the superconductive thick film formed on YSZ bulky substrate by same process is shown in the same figure. The resistance of the superconductive film on metal substrate didn't change until transition temperature range. The Tc_{end} of this film was 88.5 K and as same level as that of the YBCO bulky material, and higher than that of the same films formed on YSZ bulky substrate.

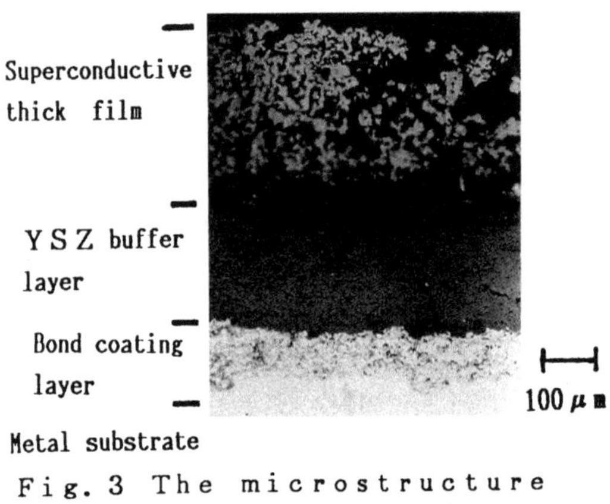

Fig. 3 The microstructure of cross section

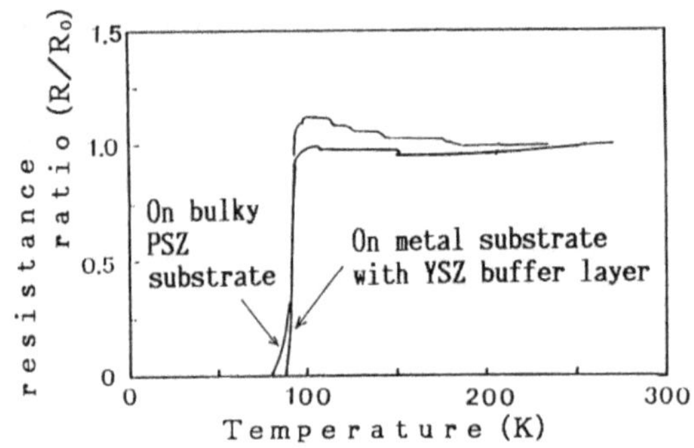

Fig. 4 Temperature-resistance curve of thick film

Fig.5 shows the effect of thickness of YSZ buffer layer on Tc_{end} of the superconductive thick film formed on metal substrate sintered for 7.2 ks at 1223 K. The film that was directly formed on metal substrate is shown as 0 μ m buffer layer thickness in the same figure. This temperature-resistance profile represented the superconductive characteristics, while in higher temperature range showed semiconductor characteristics. Otherwise, in the film with YSZ buffer layer, the Tc_{end} was 85 K, though the YSZ buffer layer was 50 μm thick. The Tc_{end} became slightly higher in the samples with more thick YSZ layer. So that it was recognized that the Tc_{end} could be improved by the YSZ buffer layer of over 50 μm thickness.

Fig.6 shows the influence of holding time in sintering process for Tc_{end} of the superconductive thick film formed on metal substrate with YSZ buffer layer of 200 μm thickness. To make a comparison, the influence of holding time in forming superconductive films on YSZ bulky substrate by same process is also shown in the same figure. The Tc_{end} of the superconductive film on metal substrate was 65 K when sintered without holding time at 1223 K, but exceeded 87 K when the holding time at 1223 K was over 3.6 ks. On the superconductive film on metal substrate with YSZ buffer layer, the most suitable holding time in sintering process was shorter than that of the superconductive thick films formed on YSZ bulky substrate.

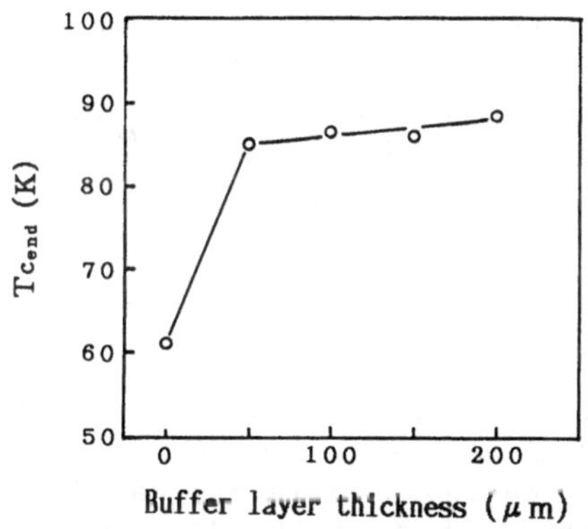

Fig. 5 Relation between buffer layer and Tc_{end}

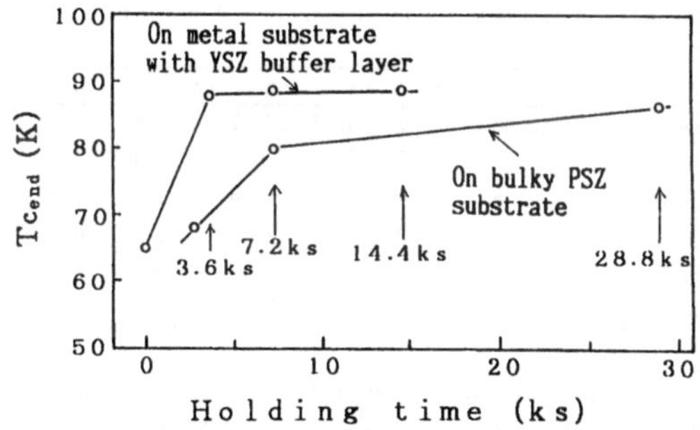

Fig. 6 Relation between holding time and Tc_{end}

Fig.7 shows the X-ray diffraction patterns of superconductive thick film. To make a comparison, the results of the YBCO bulky material and the superconductive thick film directly formed on metal substrate were also shown in the same figure. The peak angles about the crystal structure characteristics of YBCO superconductor were shown in the pattern of the film directly formed on metal substrate. But the diffraction intensity of the film on metal substrate with YSZ buffer layer was similar to bulky material and much higher than that of the films directly formed on metal substrate.

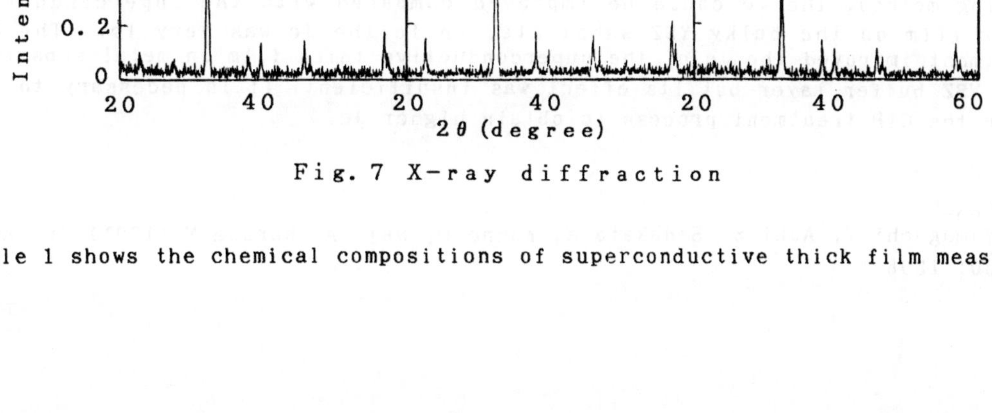

Fig. 7 X-ray diffraction

Table 1 shows the chemical compositions of superconductive thick film measured by EPMA. To make a comparison, the results of the superconductive thick film directly formed on metal substrate and formed on YSZ bulky substrates were also shown in the same table. In all specimens, the ratio between Y, Ba and Cu was approximately 1:2:3 and Zr was not observed. The fundamental composition did not change, so that the influence of Zr on the Tc characteristics was not recognized. In the film directly formed on metal substrate, Ni, Cr and Fe were included as impurities by diffusion. Accordingly it was considered that higher Tc_{end} of the film on metal substrate with the YSZ buffer layer compared with the directly formed film was achieved by the prevention of impurity-element's diffusion from substrate by the buffer layer. On the other hand, because in the thick film formed on YSZ bulky substrate there were no impurity elements diffused from substrate, it was presumed that the YSZ buffer layer has another effect on the superconductive characteristics, for example the oxygen content control.

Table 1 Composition of thick film (EPMA)
(atom%)

	Tc(K)	Y	Ba	Cu	Zr	Ni	Cr	Fe	O
On metal substrate	88.5	7.92	16.76	23.47	–	0.05	–	–	51.80
with YSZ buffer		1.00	2.12	2.96		0.006			
Directly on	61	7.87	15.96	23.20	**	0.07	0.07	0.02	52.81
metal substrate		1.00	2.03	2.95		0.009	0.009	0.003	
On bulky YSZ	80	7.93	16.27	23.42	–	**	**	**	52.38
substrate		1.00	2.05	2.95					

Table 2 Critical current density (Jc) at 77K

Treatment	Jc(A/cm2)
CIP 98MPa	45
CIP 294MPa	58
CIP 490MPa	52
Bulky marerial	196

The Jc of the superconductive thick film directly formed on metal substrate with YSZ buffer layer at 60 K were 10 to 20 A/cm^2, that was very low. As a method to increase the Jc of this film, the CIP treatment was investigated because it can raise the density of superconductive thick film. The CIP was treated after coating of YBCO calcination powder solution.

The CIP treatment consists of sealing process of specimen and pressing process in autoclave at room temperature. The CIP pressure was changed from 98 to 490 MPa, and the pressing time was set to 180 s. After then, the sintering was carried out. The sintering temperature was 1223 K and holding time was 2.4 ks. Though the cooling rate from 1223 K to 923 K was as same as NO-CIP specimen, the rate below 923 K was set at 11 K/ks in furnace, which was smaller than that of NO-CIP specimens.

The Tc_{end} of the superconductive thick film with CIP treatment on metal substrate with YSZ buffer layer exceeded 83 K, 88 K in maximum. Table 2 shows the Jc of the superconductive thick film with CIP treatment on metal substrate with YSZ buffer layer. The Jc of this film at 77 K was 45 to 58 A/cm^2, which was much higher than that of NO-CIP films, but approximately 1/3 of YBCO bulky material. The CIP treatment improved the Jc of the superconductive thick film on metal substrate with YSZ buffer layer but its effect was insufficient. So that, it was judged to be necessary to improve the CIP treatment process to obtain higher Jc characteristics.

In summary, forming of $Y_1Ba_2Cu_3O_{7-y}$ superconductive thick film on metal substrate with YSZ buffer layer by spraying method became possible by the powder coating method. The Tc could be improved compared with the superconductive thick film on the bulky YSZ substrate, while the Jc was very low. The CIP treatment improved the Jc of the superconductive thick film on metal substrate with YSZ buffer layer but its effect was insufficient. It is necessary to improve the CIP treatment process to obtain higher Jc.

REFERENCE

[1] Yamaguchi T, Aoki S, Sadakata N, Kohno O, Saji A, Kuroda N (1991) Jpn Appl Phy 30: 1398

5.2 Wires and Tapes — (2) BSCCO

Fabrication of $Bi_2Sr_2CaCu_2O_8$ Tapes and Coils

HIROAKI KUMAKURA[1], HITOSHI KITAGUCHI[1], KAZUMASA TOGANO[1], HIROSHI MAEDA[1], JUN-ICHI SHIMOYAMA[2], TAKESHI MORIMOTO[2], KATSUMI NOMURA[3], and MASAHIRO SEIDO[3]

[1] National Research Institute for Metals, Tsukuba, Ibaraki, 305 Japan
[2] Research Center, Asahi Glass Co., Ltd., Yokohama, 221 Japan
[3] Advanced Research Center, Hitachi Cable, Ltd., Tsuchiura, Ibaraki, 300 Japan

ABSTRACT

Textured $Bi_2Sr_2CaCu_2O_8$/Ag composite tapes were fabricated by both doctor blade process and dip-coating process applying partial melting and subsequent slow cooling. Texturing significantly improved current transfer across the grain boundaries and increased transport J_c. At 4.2K, J_c values well above 10^5A/cm^2 were obtained up to 25T for the doctor blade processed tapes. Small coils with 64 turns were fabricated using the tapes prepared by the continuous dip-coating process. 6000Gauss was generated at 12T and 4.2K by this coil, which is one of the highest magnetic fields achieved by the coils made by high-T_c oxide superconductors.

KEY WORDS: Bi system superconductor, textured structure, critical current density

INTRODUCTION

The application of superconducting materials to magnets requires the winding of tape or wire conductors with sufficient flexibility and sufficinetly high critical current density J_c into a coil. Since the oxide superconductors are intrinsically brittle, as is the case of A15 intermetallic superconductors, special techniques must be developed for making flexible tape or wire conductors. Among the various techniques, doctor blade process[1] and dip-coating process[2] are promising because flexible tapes can be easily fabricated by these methods.

However, the oxide tape conductors fabricated by these methods usually shows very small J_c due to "weak coupling" between grains which significantly reduces current transfer across the grain boundary. In order to improve the grain coupling, and hence, to increase the J_c, aligning the grain orientation is effective. In our previous works[1,2], we have applied solidification from the partially molten state to the $Bi_2Sr_2CaCu_2O_8$(2212)/Ag composite tapes prepared by the doctor blade process and dip-coating process, and succeeded in fabricating grain oriented 2212/Ag composite tapes with excellent J_c.

In this paper, we summarize our recent results on the textured 2212/Ag tapes prepared by doctor blade method and dip-coating method. Fabrication of small coils and the results of coil test were also reported.

SHORT TAPES BY DOCTOR BLADE PROCESS

Tape Preparation and Microstructure

2212/Ag composite tapes of ~5cm length were prepared by applying the doctor blade casting process and solidification from the partially molten state[3]. Schematic diagram of the tape preparation is shown in Fig. 1. 2212 pellets were prepared by a powder method and the pellets were crashed into fine

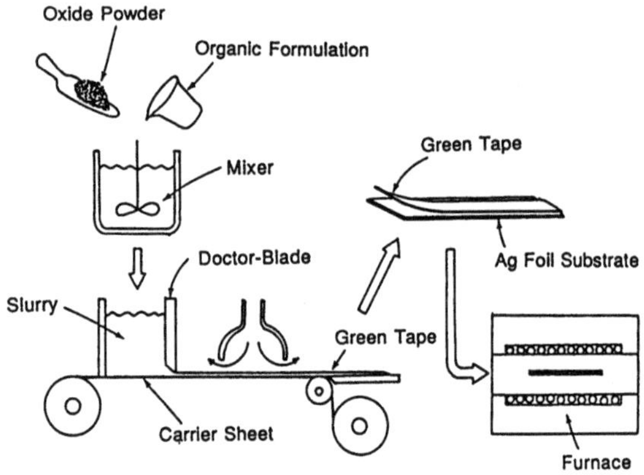

Fig. 1 Schematic diagram of doctor blade process.

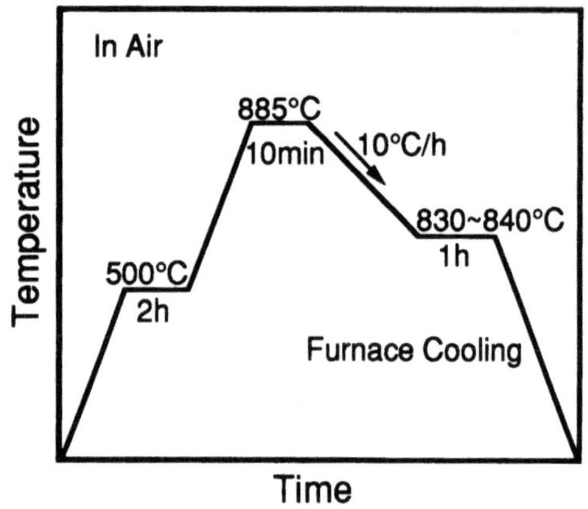

Fig. 2 Optimum heat treatment schedule.

powder. The 2212 powder was mixed with organic materials consisting of solvent(trichloroethlene), binder(polyvinyl butyral) and dispersant(sorditan-trioleate) to form slurry. The slurry was cast under a doctor blade into a green tape of 50μm thickness. Short tapes of ~5cm length were cut from this green tape and mounted on Ag substrates of 50μm thickness, and then heat treated.

During the heat treatment, the thickness of the oxide layer decreases to 1/3 of the green tape thickness. Microstructure and J_c were very sensitive to the heat treatment condition. Optimum heat treatment schedule is shown in Fig. 2. First, the composite tapes was heated at 500°C to remove the organic materials, and the temperature was increased to 885°C, slightly higher than the melting point of the 2212. The tape was then slowly cooled to 830-840°C at 10°C/h and finally furnace cooled to room temperature. A grain oriented microstructure with the c-axis aligned perpendicular to the tape surface was attained by this slow cooling from 885°C.

Besides slow cooling, there are two important factors for the grain alignment. One is the thickness of the oxide layer[3]. The thicker the oxide layer is, the lower the degree of texturing becomes. When the oxide layer thickness after the heat treatment exceeds 30μm, highly textured microstructure was only observed near the surface and oxide/Ag interface. These result suggest that the texturing proceeds from the surface or possibly from the oxide/Ag interface inward during slow cooling from 885°C.

Another important factor for the texturing is Ag substrate. 2212 green tape was heated and slow cooled on a MgO substrate similarly to 2212 on Ag substrate[4]. However, texturing was observed only at the surface and interface even if the thickness was below 30μm. This suggests that the Ag substrate plays an important role in texturing. Ag reacts with 2212 oxide lowering the melting point, although no new compounds between Ag and oxide were observed by scanning electron microscope(SEM) and transmission electron microscope(TEM). However, some portion of Ag substrate is dissolved into the oxide. TEM observation indicates that Ag-rich spherical particles were surrounded with CuO-rich region. This suggests that Ag-CuO eutectic(or peritectic) reaction must occur which lowers the melting point of the oxide. This may promotes the formation of a highly textured microstructure. However, the precise mechanism of the texturing is not yet understood and detailed investigation is now in progress.

Au and Pt were also used for substrate materials. Oxide/Au and oxide/Pt composite tapes were heat treated in a similar way mentioned above. However, no texturing was observed when Au and Pt were used as substrates. Au little reacts with 2212 oxide and melting point does not change. Pt, on the other hand, reacts with 2212 oxide and non superconducting compound was formed.

Superconducting Properties

The transport J_c was measured as a function of temperature and magnetic field by the resistive method in the temperature range from 4.2K to 77K and magnetic fields up to 25T. The magnetic field was applied perpendicular and parallel to the c-axis of grains. Measurements at 4.2K and 18-25T were carried out at the High Field Laboratory for Superconducting Materials at Tohoku University. The criterion for critical current determination was 1μV/cm and the cross-sectional area of the oxide layer was used to calculate J_c.

Transport J_c was significantly improved by the c-axis grain alignment[5]. For example, J_c values well above 20000 A/cm^2 were easily obtained at 77K and zero field. Figure 3(a) and (b) shows the temperature dependence of J_c-B curves in magnetic fields perpendicular and parallel to the c-axis, respectively. J_c decreases rapidly with increasing temperature. Especially for B // c, the magnetic field dependence of J_c is significantly enhanced with increasing temperatures. J_c anisotropy obtained from J_c values in Figs. 3(a) and (b) increases rapidly with increasing temperature. This indicates that at high temperatures flux pinning is effective only for the flux that is perpendicular to the c-axis. At low temperatures, on the other hand, J_c is much less anisotropic. At 4.2K, J_c values higher than 10^5 A/cm^2 and $6x10^4$ A/cm^2 were obtained up to 25T in perpenducular and parallel magnetic fields, respectively[6].

The excellent J_c-B properties at 4.2K of the textured 2212 tapes suggests that the tapes are very promising for the high magnetic field application above 20T for which no superconducting tape or wire of traditional materials is currently available.

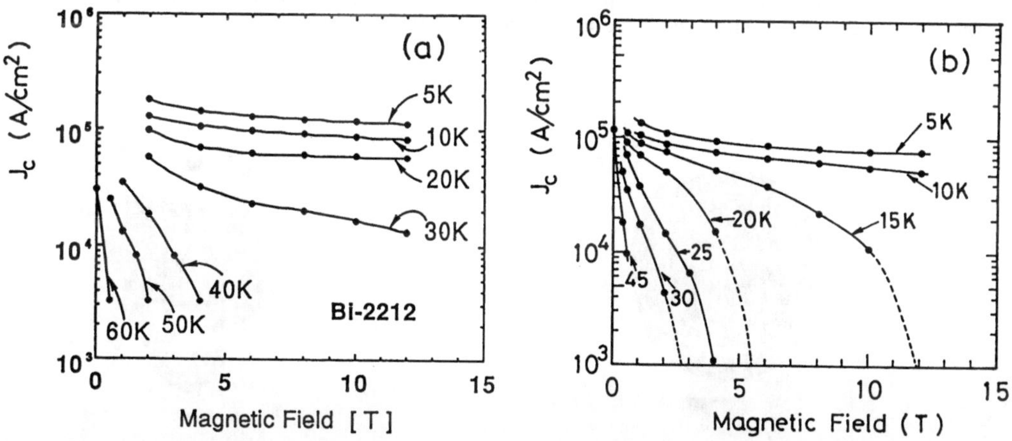

Fig. 3 Temperature dependence of J_c-B curves in magnetic fields (a)perpendicular and (b)parallel to the c-axis.

LONG TAPES AND COILS BY DIP-COATING PROCESS

Tape fabrication and Properties

Long 2212/Ag tapes of about 6m were more easily obtained by continuous dip-coating process than doctor blade process[2]. Schematic drawing of the dip-coating process is shown in Fig. 4. A slurry consisting 2212 powder and organic materials was prepared in a similar way as a slurry for doctor blade process. Ag substrate tape of 6m in length, 13mm in width and 50μm in thickness was continuously coated with slurry on both sides by passing the tape into the slurry as shown in the figure. In order to obtain uniform layer of the slurry on the Ag tape, it is very important to adjust properly the viscosity of slurry. Viscosity of slurry can be controlled by changing the volume fraction of the 2212 oxide in the slurry.

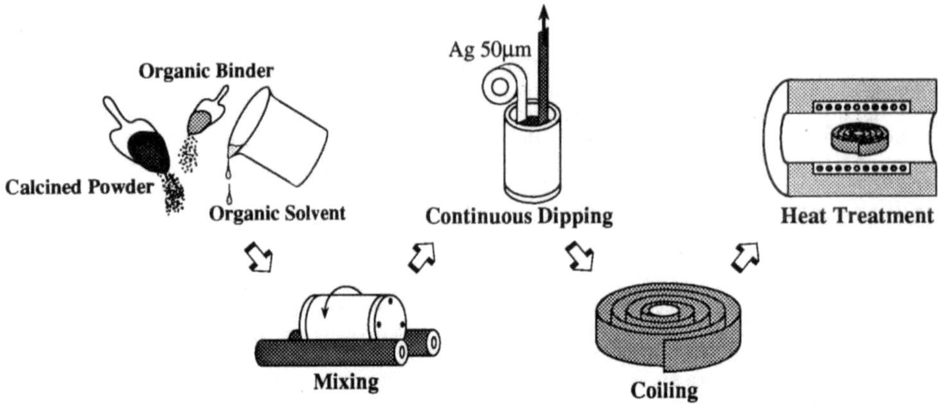

Fig. 4 Schematic drawing of the continuous dip-coating process.

After the solvent in the slurry was completely vaporized, the coated tape was cut into four tapes of 1.5m length. The each tape was wound loosely into a pancake coil with small gap between neighboring tape, and then, heat treated. Heat treatment condition was similar to that of the doctor blade process, consisting of partial melting and subsequent slow cooling. One of the coils after the heat treatment is shown in Fig. 5. After the heat treatment, from two to four coils were combined together coaxially by inserting one coil into a gap of another coil and repeating this process. Myler tapes were also inserted along each gap of the coils for insulation. Ag strips were attached to each coil for the current and voltage leads. The combined coil was wound tightly, and finally the remaining gap was filled with a wax for fixing. Figure 6 shows the final coil composed of four conductors.

Fig. 5 2212 coil after the heat treatment.

Fig. 6 Final coil composed of four conductors shown in Fig. 5.

2212 oxide layers of the coils has a textured microstructure. Figure 7 shows SEM micrograph of the fractured cross section of the 2212 phase. This microstructure is similar to that of the doctor blade processed tapes[3]. The textured microstructure was obtained by partial melting and subsequent slow cooling.

Figure 8 shows J_c-B curves at 4.2K for the pancake coil and short tape prepared by dip-coating process. Short tape was cut from the 1.5m tape. J_c-B curve for the short tape by doctor blade process are also shown for comparison. The dip-coated tape shows excellent J_c values with small magnetic field dependence at 4.2K, indicating the tape is promising for the

Fig. 7 Microstructure of 2212 oxide prepared by dip-coating process.

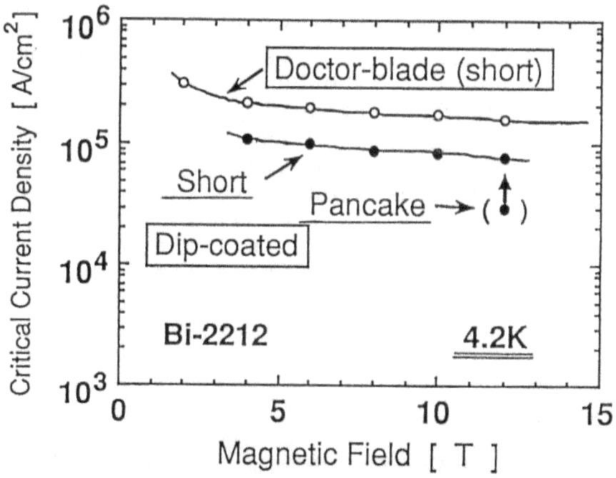

Fig. 8 J_c-B curves at 4.2K for a pancake coil and short tape prepared by the dip-coating process.

high field application. However, J_c values of the dip-coated tape are slightly smaller than the best J_c values of the doctor blade processed tapes. This is because the homogeneity of the 2212 layer for the dip-coated tape is still not perfect, and higher J_c can be obtained by the improvement of the dip-coating process. J_c of the pancake coil was higher than 30000A/cm^2 at 12T, which coresponds to the maximum current available by our power supply.

Coil Test[7]

A coil performance test was carried out at 4.2K using a 12T superconducting magnet for the coil shown in Fig. 6. The parameters of the coil were listed in Table 1. Four conductors with 1.5m length were connected in series using Ag strip, and the total length of the tapes used was 6m. The coil was inserted at the center of the superconducting magnet, and the voltage of the four conductors was monitored simultaneously with increasing applied current in a bias magnetic field of 12T. Magnetic fields induced by the coil was measured by a hole probe set at the center of the coil. Before applying bias field, induced field by the coil was measured as a function of applied current in order to obtain calibration curve of induced field vs. applied current.

Table 1 Prameters of the coil.

Conductor	Dip-Coating Processed Bi-2212/Ag Tape
Conductor Size	Width 13mm, Thickness 0.15mm, Length 1500mm] X4
Number of Turns	64 (16X4)
Number of Pancake	1
Coil Size	Inner Diameter 13mm, Outer Diameter 45mm, Height 13mm

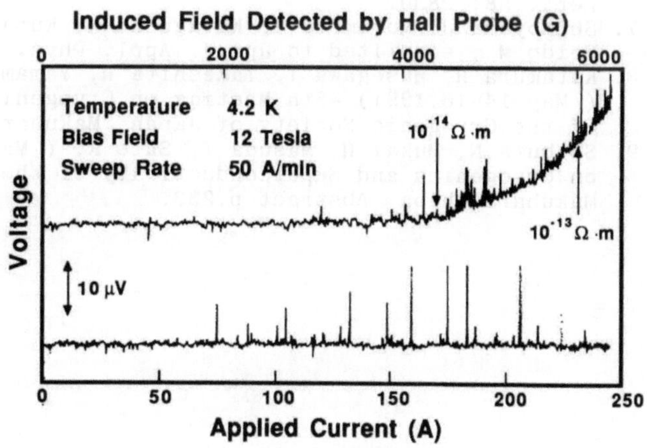

Fig. 9 Voltage-current curves of the coil.

Examples of the coil tests are shown in Fig. 9, where the voltage along the two conductors(1.5m) was shown as a function of applied current. One of the four conductors showed resistive transition, and critical current I_c defined by the $10^{-13}\Omega\cdot m$ criterion was 230A. Other three conductors showed no voltage up to 250A, which is the maximum current available using our power supply. The results indicate that the dip-coated tape has some nonuniformity in I_c values. Calibration curve indicates that 6000Gauss was induced by I_c=230A. This value is almost equal to the magnetic field calculated from I_c and the coil parameters shown in Table 1. This is one of the highest magnetic fields achieved recently by the coils made by oxide high-T_c superconducotrs [8,9]. Magnetic field induced by the coil can be increased by improving the uniformity of the I_c along the tape.

CONCLUSION

$Bi_2Sr_2CaCu_2O_8$/Ag tapes with excellent J_c-B properties were fabricated by the doctor blade process and the continuous dip-coating process. Especially, the dip-coating process is adequate to the fabrication of long tape conductors. Pancake coil of 64 turns was fabricated using this long tape, and succeeded in generating 6000Gauss at 4.2K and bias field of 12T. The result indicates that the 2212 tape conductors fabricated in this experiment is promising for the practical application in high magnetic fields.

ACKNOWLEDGEMENT

The authors are grateful to Profs. Y. Nakagawa and K. Watanabe of Tohoku University for providing them the opportunity to use the high field testing facilities.

REFERENCES

1. Kase J, Togano K, Kumakura H, Dietderich DR, Irisawa N, Morimoto T, Maeda H (1990) Jpn. J. Appl. Phys. 29: L1096.
2. Togano K, Kumakura H, Kadowaki K, Kitaguchi H, Maeda H, Kase J, Shimoyama J, Nomura K, (1991) Proceedings of ICMC '91, Huntsville Arabama(to be published).
3. Kase J, Morimoto T, Togano K, Kumakura H, Dietderich DR, Maeda H (1991) IEEE Trans. Magn. 27: 1254.
4. Dietderich DR, Ullmann B, Freyhardt HC, Kase J, Kumakura H, Togano K, Maeda H (1990) Jpn. J. Appl. Phys. 29: L1100.
5. Kumakura H, Togano K, Kase J, Morimoto T, Maeda H (1990) Cryogenics 30: 919.
6. Kumakura H, Togano K, Maeda H, Kase J, Morimoto T (1991) Appl. Phys. Lett. 58: 2830.
7. Shimoyama J, Morimoto T, Kitaguchi H, Kumakura H, Togano, Maeda, Nomura K, Seido M , submiited to Jpn J. Appl. Phys.
8. Kitamura H, Hasegawa T, Takeshita H, Yamamoto K, Murase S, Ogiwara H, (May 14-16,1991) 45th Meeting on Cryogenics and Superconductivity of the Cryogenic Society of Japan, Makuhari Chiba, Abstract p.231.
9. Shibuta N, Mukai H, Masuda T, Sato K, (May 14-16,1991) 45th Meeting on Cryogenics and Superconductivity of the Cryogenic Society of Japan, Makuhari Chiba, Abstract p.233.

Preparation and Electrical Characterization of Melt Textured 2212 Bi-Sr-Ca-Cu-Oxide Layers on Ag-Tape

NEUMÜLLER HEINZ-WERNER[1], AßMANN HELMUT[2], KREß BERNHARD[2], and RIES GÜNTHER[1]

[1] Research Laboratories, Siemens AG, D-8520 Erlangen, FRG
[2] Power Generation Group, Siemens AG, D-8520 Erlangen, FRG

ABSTRACT

Screen printed pastes and Doctor blade processed green tapes prepared from $Bi_2Sr_2CaCu_2O_8$ powder were subjected to a melt-texturing process on silver foil. Melting and crystallization temperatures were determined using DTA analysis. Highly textured layers were formed by slow cooling from the melting temperature to the crystallization temperature and quenching after a holding period. At 4.2 K and 77 K critical transport current densities j_c as high as 1.3×10^5 A/cm^2 and 1.1×10^4 A/cm^2, respectively were obtained in zero field. These values are in good accordance with critical current densities j_{cm} determined by magnetization measurements. $j_{c\parallel}$ ($\boldsymbol{B} \parallel$ tape) can be related to $j_{c\perp}$ ($\boldsymbol{B} \perp$ tape) assuming that the field component $B \parallel c$-axis alone controls $j_c(B,T)$. The analysis results in an average misorientation of the c-axes of the grains of about six degrees. This is directly confirmed by SEM micrographs.

INTRODUCTION

In parallel to the intense activities on HTSC in basic research great efforts were made to introduce these materials into technical applications. The use of HTSC for magnets and power engineering requires critical current densities $j_c(B)$ exceeding 10^5 A/cm^2 at reasonable fields and the availability of conductor material in long length. In this way the preparation methods mostly directed to the generation of high current densities have also to consider a possible route for a continuous manufacturing of the HTSC material. Comparing the relevant Y- and Bi-based compounds the latter is more promising for a production on a technical scale at the present state. High current densities in HTSC require both intergranular connectivity avoiding the "weak link" problem and a high degree of grain orientation due to the anisotropy of critical current. Both properties are given for the 2223 BPSCCO compound [1, 2] made by the powder in tube method and also for the melt-textured 2212 BSCCO on Ag-tape produced by a melt growth process [3, 4]. Whereas the 2223 PBSCCO has more potential for a technical application at 77 K the 2212 compound is limited to an operation between 4.2 K and 20 K in high field magnets due to the large influence of thermal activated flux flow at elevated temperatures [5]. This paper describes the techniques to generate laminate systems on the basis of pastes [6, 7] and ceramic tapes produced by Doctor blade (DB) technique [8] using the 2212 BSCCO compound. We report on sample preparation starting from different precursor materials and on the results of measurements of transport and magnetization currents. A correlation between the microstructure and the orientation dependence on $j_c(B,T)$ is also established.

PREPARATION AND THERMAL ANALYSIS

The preparation of the precursor material followed two different ways. In the first route we prepared high viscosity pastes (D3, D5, D8) using different starting powders mixed in the cation ratio Bi:Sr:Ca:Cu= 2:2:1:2. The powders of the pastes D5, D8 were prepared by calcination of mixtures of oxides (Bi,Cu) and carbonates (Ca,Sr). For the paste D3 the starting powder was obtained from a bulk melt process using alumina crucibles [9]. Table 1 shows details of powder preparation and characterization. The chemical composition was determined by using ICP-AES. In the second route a green DB cast foil of 50 μm in thickness [8] was used. Majority phase of the calcined powder was the 2201 phase in the case of powder D5 and the 2212 phase in the case of powder D3,D8 and the DB foil. For the pastes, the organic formulation consisted of solvents (terpineol, isooctanol), binder (hydroxypropylcellulose), plasticizer (2-furoic-acid) and dispersant [4]. For the DB tape the following organics were used according to Ref. [8]: solvent (trichloroethylene), binder (polyvinyl butyral) and dispersant (sorbitantrioleate). The pastes were directly deposited on silver foils 25 μm to 100 μm in thickness

Material	Calcination	Stoichiometry Bi Sr Ca Cu	Impurities ppm
D3	melting 1000-1100°C cast ingot tempered 12-60 h/815°C	1.94 1.73 0.93 2.00	Al 2900 Fe 40
D5	2x12h/820°C	2.08 1.77 0.95 2.00	Fe 140
D8	3h/780°C 16h/800°C 16h/850°C	2.37 1.96 1.04 2.00	
DB	2x12h/820°C	2.56 1.97 0.97 2.00	

Table 1: Starting materials for the melt-texturing process.

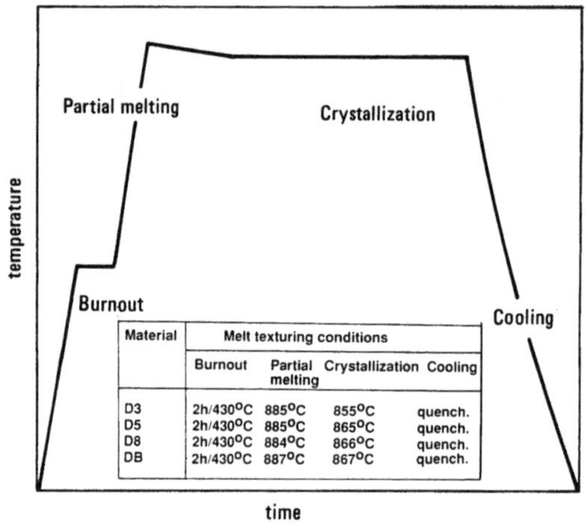

Fig. 1: Typical heat treatment programme for the melt-texturing process. The insert shows the melt-texturing conditions for the different starting materials.

using a screen printing technique [7]. The thickness of the screen printed layers was typically 25–50 μm before burnout and 10–16 μm after the melt-texturing process. With increasing storage time samples prepared from paste of the same batch showed decreasing current densities, while the melt-texturing parameters remained unchanged. This may be attributed to a sedimentation of the coarser particles of the paste. No ageing effects in j_c were observed for the DB tapes.

Determination of processing temperatures by thermal analysis

Highly textured layers were formed by a time-temperature [8] program which consists of the burnout of the organics, heating to a melting temperature, slow cooling to the cristallization temperature and quenching after a holding period (Fig. 1) to room temperature (RT). The key steps of the melt-texturing process are related to changes in enthalpy and could therefore be optimized by DTA. The burnout of the organics is an exothermal process with simultaneous considerable weight loss. TG and DTA measurements show that the burnout is completed after 1 h at 400 °C. The behaviour of the pastes and of the DB tape during burnout and melting was investigated in alumina crucibles in order to allow complete melting. The DTA-curves of all materials showed two major endothermal peaks during heating in the temperature range between 850 °C and 910 °C (Fig. 2). During cooling, three exothermal peaks appeared between 900 °C and 700 °C. The melting temperature for the melt-texturing process was chosen as the turning point of the first endothermic peak between 800 °C and 885 °C during heating in alumina crucibles. It is assumed that a sufficient amount of partial melt is present at this temperature. Silver crucibles were used for the determination of the crystallization temperature. After heating the precursor material to their individual melting temperature, the temperature was reduced with a rate of 10 K/min. During this procedure only one exothermal peak was observed for all the starting materials tested. The individual crystallization temperature was chosen as the first turning point of the exothermal peak during cooling down in silver crucibles (Fig. 2). Thus, for all starting materials, the three important processing temperatures (burnout, partial melting and crystallization) have been determined individually . The set of parameters for melting and crystallization of all the materials is shown in the insert of Fig. 1.

The high j_c material was obtained when the samples were cooled down with 2 K/h from the melting to the crystallization temperature and then held between 30 h and 70 h. Quenching of the samples was necessary in order to achieve high critical temperatures. The use of the calcinates with the 2212 phase as majority phase clearly increases the critical current density. No systematic correlation was found for the j_c dependence on the thickness of the Ag-tape and the reacted BSCCO layer within the values given above. j_c measurements on samples reacted within the same batch showed a variation of j_c of about a factor two. The degradation of j_c is mainly caused by microcracks which arise during the quenching process and the cool down from RT to low temperatures due to the thermal mismatch between silver foil and the oxide layer. The j_c values of powder D3 did always not exceed 4000 A/cm^2. These layers showed smaller grains and a more loose structure

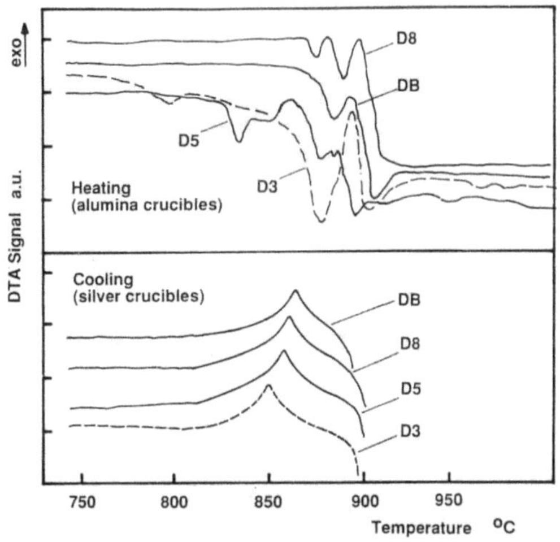

Fig. 2: Comparison of the DTA signals of the different starting materials during heating and cooling.

Fig. 3: SEM micrograph of the fractured 2212 BSCCO layer on Ag-tape (longitudinal cross section).

which is attributed to the high content of Al impurities (cf. Table 1).

MICROSTRUCTURAL ANALYSIS

Plate-like crystals with lengths exceeding 100 μm were observed in the melt-textured samples by SEM. Figure 3 shows the fractured BSCCO layer broken along the longitudinal direction of current flow. The layer is composed of stacks of well aligned plate-like grains. This lamellar structure was displayed on fractured surfaces of samples without aluminium impurities (D5, D8 and DB). XRD patterns of all melt textured samples confirmed the high degree of orientation. Only (00ℓ) reflections of the 2212 BSCCO are observed as shown in Fig. 4 as an example. But, according to Fig. 3 the plates are not perfectly oriented parallel to the surface of the Ag-tape. Some misorientation occurs covering an angle of about $5.5° \leq |\theta| \leq 6.5°$.

Within the stacks several non superconducting phases could be observed. The results of Microprobe analysis (Fig. 5) show plates of the 2201 BSCCO phase within the 2212 BSCCO stacks, which might hinder the current flow. The presence of silver in the sample volume was also found in the 2223 BSCCO Ag-tapes [2]. The silver not only lowers the melting temperature but also takes part in the liquid phase reaction. Besides earth-alkaline cuprates ((CaSr)CuO_2, $CaCuO_2$, CuO) small bismuth-rich copper-free crystallites could be identified. The composition of the lamellae of samples prepared from calcined powder (D5, D8 and DB) was found to be calcium deficient while plates of samples prepared from powder by a melt process (D3) had the theoretical calcium content of the 2212 phase (14 at%). The average calcium content of samples made from D5, D8 and DB was found to be about 12 at%.

MEASUREMENTS AND DISCUSSION

Experimental

For the electrical characterization the samples (typ. $5 \times 25\,mm^2$) were mounted onto a special substrate [5] using a low melting solder. The handling of the samples was done very carefully to avoid the generation of cracks in the BSCCO layer. The electrical measurements were carried out by the conventional four probe technique. The critical transport current density was determined at a field strength $E = 1\,\mu V/cm$. The temperature and field dependent properties were measured in a continuos flow cryostat which was fitted into the warm bore of an 8 T superconducting magnet. The temperature stability during the j_c measurements was better than 0.1 K. For the magnetic measurements samples with a quadratic shape of the BSCCO layer ($2 \times 2\,mm^2$) were fed into the He flow cryostat (4.2 K–300 K) of a vibration sample magnetometer (Oxford Instruments) and measured in fields up to 12 T. The noise base was about 5×10^{-9}Am at a field of 12 T.

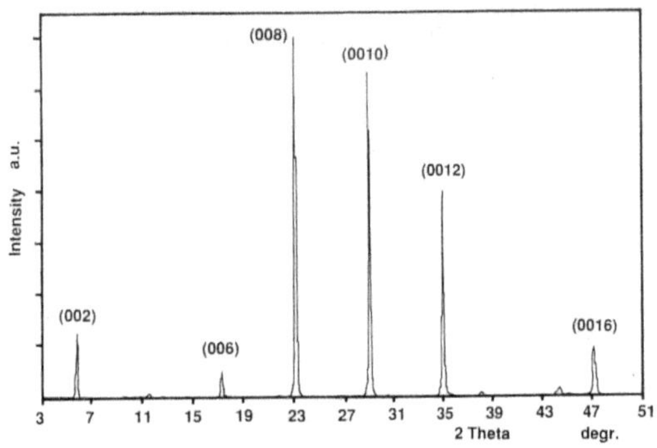

Fig. 4: XRD of a melt textured layer (powder D8).

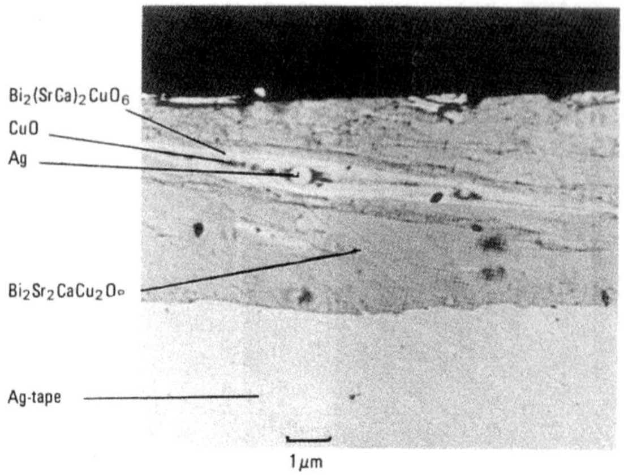

Fig. 5: Backscattering electron image of the melt-textured 2212 BSCCO layer (longitudinal cross section).

Critical temperature

Samples quenched from 860 °C to room temperature generally had values of the critical temperature $T_c(50\%) =$ 90 K and $T_c(\mathrm{off}) = 87.5$ K measured resistively. If the samples were slowly cooled from 500 °C to room temperature the critical temperature lowered to about $T_c(50\%) = 75$ K–77 K. This effect can be correlated to the influence of the temperature dependent partial pressure of oxygen in accordance to the results of Ref. [10]. Due to large surface to volume ratio the oxygen diffusion can rapidly take place; no changes in T_c were observed in thicker bulk 2212 melt samples [5] of dimensions $0.7 \times 0.3\,\mathrm{mm}^2$ for the similar procedure.

Critical transport current measurements

Figure 6 shows the field and temperature dependent transport critical current density j_c as a function of the field orientation $\boldsymbol{B} \parallel$ tape ($j_{c\parallel}$) and $\boldsymbol{B} \perp$ tape ($j_{c\perp}$) respectively. The values in the range 77 K down to 50 K were determined from a DB-sample; the critical current densities at 4.2 K were obtained from a paste sample (D8). Meanwhile, this material could be improved to a maximum j_c value of 10400 A/cm^2 at 77 K in zero field but the drastic decrease of j_c with field is still present, which is one magnitude in a field of 0.1 T even for the favourable orientation $\boldsymbol{B} \parallel$ tape. This reflects the influence of thermal activated flux line depinning in the 2212 BSCCO material [3, 5]. Lowering the temperature j_c rapidly increases and the strong $j_c(B)$ dependence is reduced both for orientation $\boldsymbol{B} \parallel$ tape and $\boldsymbol{B} \perp$ tape. At 4.2 K the critical current is practically independent on field above $B = 1$ T with values $j_{c\parallel} = 40000$ A/cm^2 and $j_{c\perp} = 30000$ A/cm^{2}at 12 T. In case of the DB-tape $j_c = 50000$ A/cm^2 was measured at 4.2 K and 10 T even for the orientation $\boldsymbol{B} \perp$ tape. Figure 6 also shows that the anisotropy ratio $j_{c\parallel}/\ j_{c\perp}$ is drastically reduced with decreasing temperature. At 4.2 K and $B = 12$ T the ratio is only 1.3 and further decreases with increasing field. Thus, the melt textured 2212 BSCCO material represents a more promising candidate for a technical application at 4.2 K. Generally the $j_c(B)$-curves showed some hysteresis in increasing and decreasing field only at $T = 4.2$ K. Contrary to the 2223 BPSCCO powder in tube conductors the hysteresis remained very small and is restrictet to below 7 T.

In Fig. 7 $j_c(B)$ is plotted for orientation $\boldsymbol{B} \parallel$, $\boldsymbol{B} \perp$ tape and an angle of $\theta = 30°$ to the tape surface ($j_{c30°}$). When $j_{c30°}$ is plotted versus $B^* = B \sin\theta$, the field component perpendicular to the ab-planes, these curves nicely coincide with those for orientation $\boldsymbol{B} \perp$ tape both for 70 K and 60 K. In the same way $j_{c\parallel}$ is in good accordance with $j_{c\perp}$ when scaled against $B^* = 0.1B = B \sin\theta$ for all temperatures below 77 K (cf. Fig. 6). As shown in Fig. 3 $\theta = 6°$ is related to the mean angular deviation of the 2212 BSCCO plates found in the microstructure of the layer. Only a small variation of $5.5° \leq |\theta| \leq 6.5°$ was observed in the samples prepared from the different precursors. A similar correlation of the $j_c(B)$ anisotropy with the microstructure was recently found on 2223 BPSCCO Ag-tapes made by the powder in tube technique [2]. The more loosely packed 2223 BPSCCO plate-like grains with a misalignement of $\pm 10°$ were not as well aligned as in case of the melt textured 2212 BSCCO layer in accordance with the observations of Ref. [3]. If the vortex cores pass the

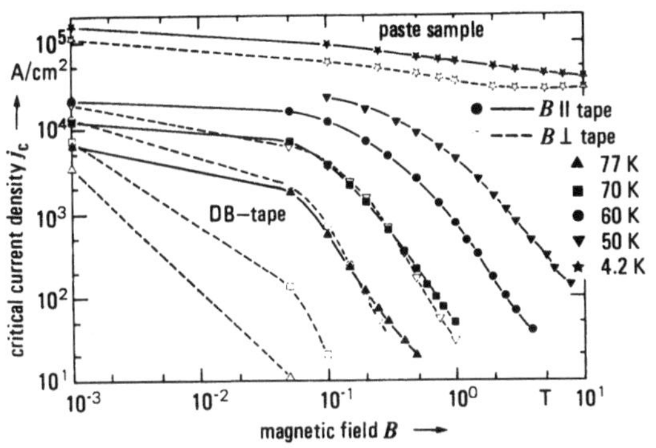

Fig. 6: Critical current densities of highly textured 2212 BSCCO layers on Ag-tapes for $\boldsymbol{B} \parallel$ tape and $\boldsymbol{B} \perp$ tape.

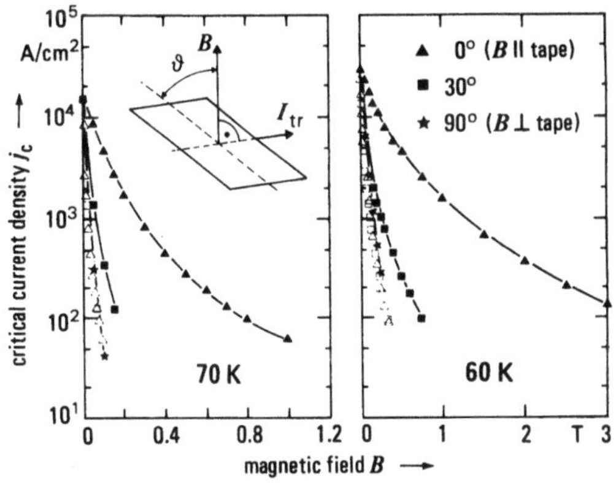

Fig. 7: $j_c(B)$ as function of field orientation to the tape for two temperatures. The open symbols correspond to $j_{c\parallel}(B\sin 6.5°)$ (△) and $j_{c30°}(B\sin 30°)$ (□). $j_{c\parallel} = j_c(B, \theta = 0°)$; $j_{c30°} = j_c(B, \theta = 30°)$.

ab-planes at an angle θ they cross them in form of kinks [11] in order to minimize the energy of the vortex. A vortex density $B\sin\theta$ is generated in each of the double oxide layers, which determines j_c alone. The vortex itself consists of a stack of pancakes interacting with the pinning active defects. This was also confirmed by irradiation experiments recently [12].

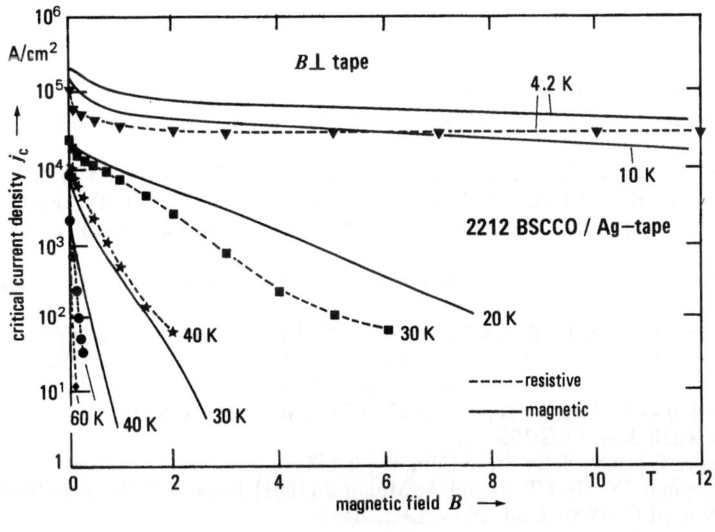

Fig. 8: Critical current density $j_c(B,T)$ determined by magnetic and resistive measurement for orientation $\boldsymbol{B} \perp$ tape.

Magnetization

The magnetization current j_{cm} is plotted in Fig. 8 for a field perpendicular to the tape. j_{cm} is determined from the irreversible magnetization using the relation: $j_c = 3\Delta M/\mu_0 D$ where D is the average sample radius and ΔM is the width of the irreversible magnetization loop. The field ramp rate of the VSM was chosen 0.01 T/s which induced an electric field strength of 0.05 μV/cm on the sample surface. These j_{cm}-values at zero field are in good accordance with the results on j_{ct} derived from the resistive transport measurements for the equivalent field direction $\boldsymbol{B} \parallel$ tape. This is a proof that the BSCCO material shows no granular behaviour as also was confirmed from various measurements in literature [13] on polycrystalline samples.

But, when field is applied, large differences occur between j_{ct} and j_{cm} with increasing field and temperature, which can be attributed to the influence of the Lorentz force and thermal depinning. Whereas j_{cm} is nearly field independent between 4.2 K and 10 K and above 1 T the field depedence is drastically increased when going to 20 K. Obviously the thermal energy is sufficient to depin the flux lines confined within the shallow pinning potentials at very low temperature [14].

CONCLUSIONS

Melt-textured 2212 BSCCO layers on Ag-tape were prepared from organic pastes and green Doctor blade casted tape. Melting and crystallization temperature of the precursors of different compositions were determined using DTA analysis. The critical transport current densities exceeded 10000 A/cm^2 and 10^5 A/cm^2 at 77 K and 4.2 K respectively in zero field. Critical transport and magnetization current showed good agreement in zero field; the differences with increasing field and above 20 K could be attributed to thermal activated depinning. The results of the field orientation dependent $j_c(B)$ measurements could be correlated to the misalignement of the plate-like grains within the layer. In this way the $j_c(B)$- dependence of at least the field orientation $\boldsymbol{B} \parallel$ tape could be clearly improved when the plates could be more perfectly aligned in parallel to the surface of the Ag-tape.

ACKNOWLEDGEMENT

The authors would like to thank Dr. K. Togano, NRIM Tsukuba Laboratories for kindly providing the DB-foil, Mrs. W. Ruppert and Mr. P. Kummeth for the transport and magnetization measurements, Mrs. Urstadt and Mrs. Küsebauch for the Microprobe and SEM analysis. This work was supported by the EC Brite Euram programme and partly by the Bundesministerium für Forschung und Technologie

REFERENCES

1. Sato K, Hikata T, Iwasa Y (1990) Appl. Phys. Lett 57: 1928
2. Wilhelm M, Neumüller HW, Ries G (1991) Proc. of M²-HTSC Conf., Kanazawa Japan, July 22–26. Physica C (to be published)
3. Kumakura H, Togano K, Kase J, Morimoto T, Maeda H (1990) Cryogenics 30: 919
4. Kreß B, Neumüller HW, Aßmann H, Wohlfart M, Kase J, Kumakura H, Togano K (1991) Proc. of M²-HTSC Conf., Kanazawa Japan, July 22–26. Physica C (to be published)
5. Neumüller HW, Ries G (1989) Physica C 160: 471
6. Hoshino K, Takahara H (1988) Proc. 1st Int. Symposium on Superconductivity, Nagoya Japan, Aug. 28–31: 325
7. Aßmann H, Steinmann B, Kreß B, Günther A (1990) Proc. 7th CIMTEC, Trieste Italy, July 2–5, Materials Science Monographs 70, Edit. P. Vincenzini, Elsevier, Amsterdam: 905
8. Togano K, Kumakura H, Dietderich DR, Maeda H, Kase J, Morimoto T, Ullmann B, Freyhardt HC (1991) Proceedings of ICMC 90 Topical Conf. HTSC, Materials Aspects, May 9–11, 1990, Garmisch-Partenkirchen, FRG:165
9. Bock J, Preissler E (1989) Solid State Comm. 72: 453
10. Triscone G, Genoud JY, Graf T, Junod A, Muller J (1991) Physica C (to be published)
11. Feinberg D, Villard C (1990) Mod. Phys. Lett. B4: 9
12. Gerhäuser W, Ries G, Neumüller HW, Schmidt W, Eibl O, Saemann- Ischenko G, Klaumünzer S (1991) Phys. Rev. Lett (to be published)
13. Ries G, Neumüller HW, Schmidt W (1991) Proc. 6th Int. Workshop on Critical Currents, Cambridge GB, July 8-11. Supercond. Sci.& Technol. (to be published)
14. Neumüller HW, Ries G, Bock J, Preisler E (1990) Cryogenics 30: 639

Bismuth Superconducting Wire and Application

KEN-ICHI SATO, NOBUHIRO SHIBUTA, HIDEHITO MUKAI, TAKESHI HIKATA, TAKATO MASUDA, MUNETSUGU UEYAMA, TAKESHI KATO, and JUN FUJIKAMI

Osaka Research Laboratories, Sumitomo Electric Industries, Ltd., Osaka, 554 Japan

ABSTRACT

The continued R&D results of the silver-sheathed bismuth(2223 phase) superconducting wire have suggested that three different fields of application, large current conductors, low field magnets and super-high field magnets will be feasible in the near future.

Maximum Jc at 77.3 K reached 53,700 A/cm² in a zero magnetic field and 12,000 A/cm² at 1 Tesla. Jc and Jc-B enhancements were obtained with finely dispersed nonsuperconducting phases. Prototypes such as 100 m-long wire, 70 m-long multifilamentary wire, a spirally wound 5 m-long conductor and a super-high field magnet were made and tested in detail.

KEY WARDS: bismuth system, wire, critical current, conductor, magnet

INTRODUCTION

Toward the actual application of high-Tc superconductors, especially in the power field such as large current conductors and magnets, we have been focusing on developing a practical wire using the high-Tc(110K) phase of bismuth compound[1,2] and powder-in-tube processing technology[3-14]. It is well known that among bismuth compound superconductors there is a high-Tc(110K) phase and a low-Tc(80K) phase. From the results obtained, the high-Tc phase is most suitable for practical applications at liquid nitrogen temperature. It is necessary to get Jc over 10,000 A/cm² under a magnetic field corresponding to each application, for example 0.1 Tesla for large current conductors and 1 Tesla for magnets. The critical current densities at these typical magnetic fields were over 10,000 A/cm² with these technologies. Figure 1 shows the features of the combination of the high-Tc phase of bismuth compound and the powder-in-tube technique.

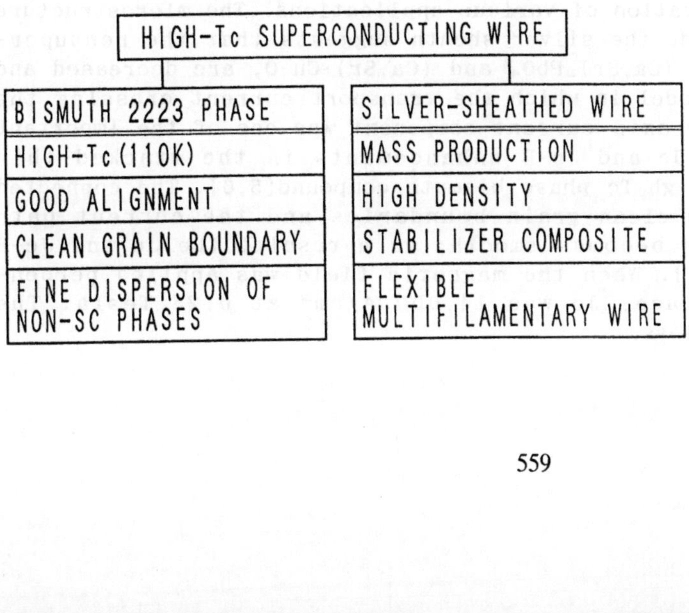

Fig. 1 Features of the combination of high-Tc phase of bismuth and powder-in-tube technique.

There are many hurdles to actual application. The most fundamental ones are long wire fabrication, flexibility and large current techniques. We previously reported that these basic problems could be overcome with powder-in-tube processing, multifilamentary wire and bundle conductor techniques[8-10]. A 60 m-long silver-sheathed bismuth wire was made and proved to have Jc of 2,450 A/cm^2 with Ic of 10.5 A for the whole length at 77.3 K. 1296-multifilamentary wire could tolerate cyclic bending 20 times with 0.33 % strain. We also clarified that the stacking thin-tape type wire technique could make a large current conductor transporting over 1,000 A at 77.3 K.

Apart from liquid nitrogen temperature application, super-high field magnet application at liquid helium temperature is also exciting[7]. Because of the limitation of the upper critical field of metallic superconductors, such as NbTi and Nb_3Sn, we cannot get a super-high field magnet above 20 Tesla at 4.2 K. Bismuth superconducting wire was revealed to have Jc over 10^5 A/cm^2 at 4.2 K and even at 23 Tesla[9]. We can expect to build a super-high field magnet above 20 Tesla with a bismuth superconducting wire.

In this paper, we report on the recent progress in wire development and the results of prototypes.

EXPERIMENTAL

Appropriate amounts of Bi, Pb, Sr, Ca, and Cu oxides and carbonates were mixed, sintered and ground, as described in the previous paper[4]. The ground powder was put into silver tubes, drawn to round wires and then made into tape-shaped wires. These wires were sintered twice and made into high-Jc wires. The typical critical temperature with a cation ratio of Bi:Pb:Sr:Ca:Cu =1.8:0.4:2:2.2:3 was 106 K. Typical dimensions of the wires were 0.1-0.7 mm thick and 2.5-4.5 mm wide.

RESULTS AND DISCUSSION

Wire development

The magnetic field dependence of Jc at 77.3 K of the wire having the highest Jc in a zero magnetic field, 53,700 A/cm^2[11,13], is shown in Fig. 2 with the wires having a zero field Jc of 20,000 A/cm^2 and 5,500 A/cm^2. The applied magnetic field was parallel to the tape-shaped wires. These improvements were attributed to the increase of the irreversibility lines[12]. The maximum Jc values at 77.3 K was 42,300 A/cm^2 at 0.1 Tesla and 12,000 A/cm^2 at 1 Tesla.[11]. From this figure, we can see that the properties of the wire were good for consideration of various applications. The microstructure of the superconductor inside the silver sheath suggests that the nonsuperconducting phases, such as $(Ca,Sr)_2PbO_4$ and (Ca,Sr)-Cu-O, are decreased and dispersed. We proposed a model in which the transport current crossing the grain boundaries with a c-axis current component was one of the important factors when considering Jc and Jc-B enhancements in the stacked thin platelet crystals of the high-Tc phase bismuth compound[5,6]. The connected grain area increases with clean grain boundaries and the current path through these boundaries becomes smooth; as a result, the Jc and Jc-B characteristics improve[11]. When the magnetic field was applied perpendicular to the tape surface, Jc was 14,000 A/cm^2 at 0.25 Tesla. The anisotropy factor is 4 or less.

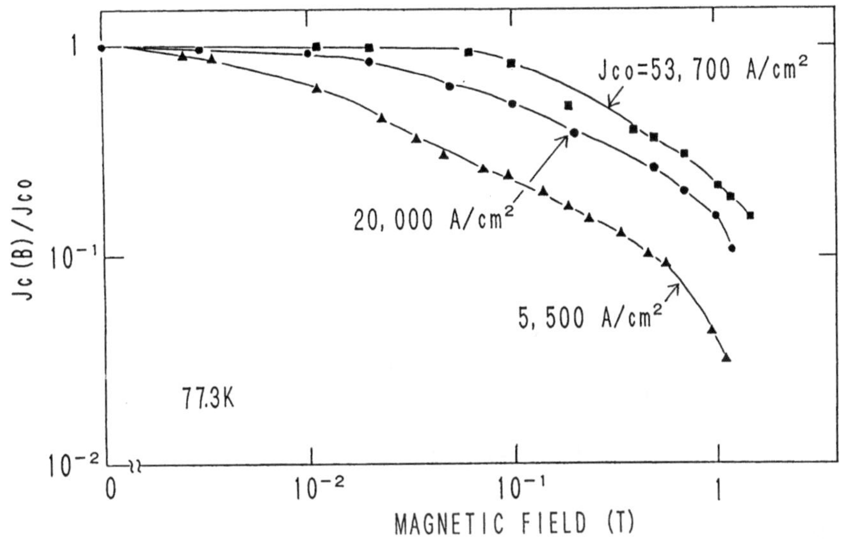

Fig. 2 Magnetic field dependence of Jc at 77.3 K. Jco(A/cm²): ■, 53,700; ●, 20,000; ▲, 5,500.

Regarding the long wire fabrication, a 100 m-long wire was made and evaluated using the 4-probe DC method in liquid nitrogen. Figure 3 shows the relationship of the critical current density of the whole length to the electrical resistivity. Jc defined with $10^{-13}\Omega$ m criterion for 100 m length was 5,250 A/cm² with Ic of 24.3 A. The wire was evaluated in a single pancake coil shape having a 150 mm inner diameter and a 398 mm outer diameter. The self field of this coil was 290 Gauss at its maximum value. From a result of a short evaluation sample cut from the 100 m-long wire, the zero field critical current density of this 100 m-long wire was estimated to be 6,570 A/cm² with Ic of 30.4 A.

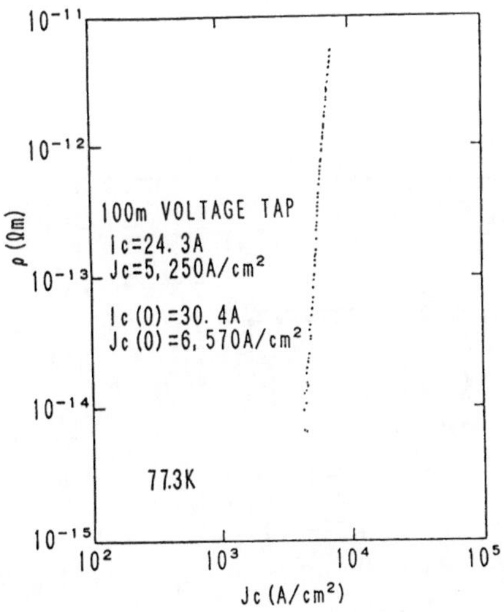

Fig. 3 I-V characteristics of 100 m-long wire.

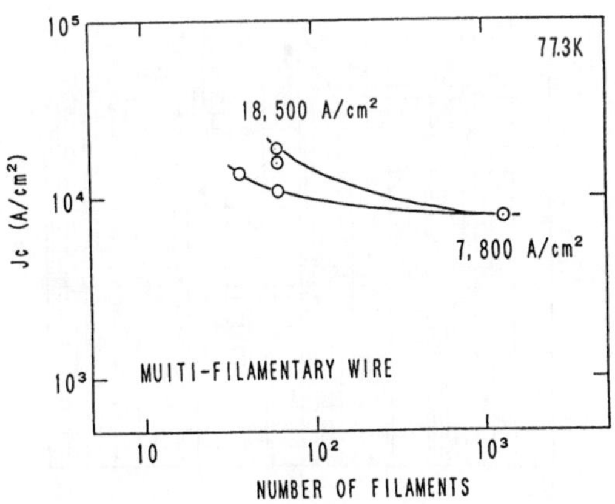

Fig. 4 Jc of multifilamentary wires.

Compared with single-core wire, multifilamentary wires proved to have excellent anti-strain characteristics when bent after sintering[8]. Multifilamentary wires were fabricated using a single- or double-stacking technique of round wires, and also made into tape-shaped wires. Jc of these wires is shown in Fig. 4. Maximum Jc of single-stacked 61-filament wire reached 18,500 A/cm^2 at 77.3 K. After obtaining these results, a 70 m-long wire with 61 filaments was fabricated. This wire was also sintered as a single pancake coil with an inner diameter of 105 mm and an outer diameter of 252 mm. Jc at 77.3 K over the whole length was 4,340 A/cm^2 with Ic of 5.2 A. We can expect a larger zero field Jc because the maximum self field was around 200 Gauss.

Figure 5 shows typical examples of the long wire results, including the prototypes of the conductors as mentioned below. This graph shows the relationship of the critical current densities to Ic*L(L means the length of wire or conductor), which suggests the current status of the availability of silver-sheathed bismuth superconducting wire.

Conductor development

The authors previously reported the results of prototypes of large current conductors, such as 600~ 1,000 A-carrying rigid conductors at 77.3 K[8-10], a 80 A flexible conductor at 77.3 K[8] and 1,500 A-carrying cryogenic current leads[8]. In addition to these prototypes, we made a 5 m-long spirally wound conductor using the above-mentioned 61-filament wire. Two sintered wires of 5 m length were wound on a 25 mm diameter former with 205 mm pitch. This conductor was evaluated at 77.3 K and each wire proved to carry 12.5~ 13.3 A, showing Jc of 10,400~ 11,100 A/cm^2. These values were taken after immersion 10 times in liquid nitrogen. The wires were evaluated after each immersion, as shown in Fig. 6. There was no change of critical current with each immersion. From this result, we can expect to make a spirally wound conductor carrying a large current by increasing the number of wires.

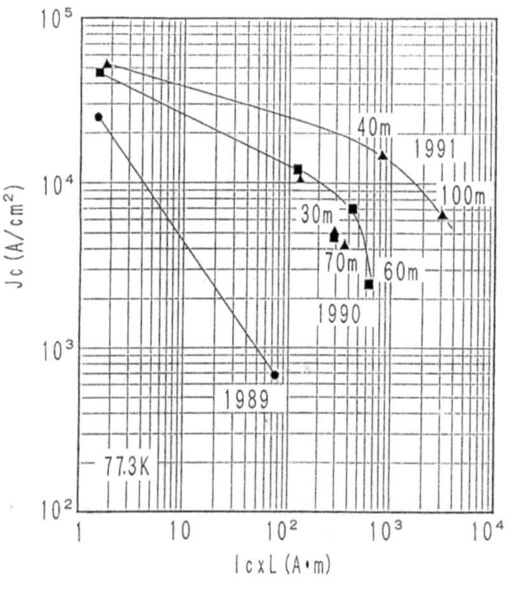

Fig. 5 Relationship between Jc and IcxL of wires and conductors.

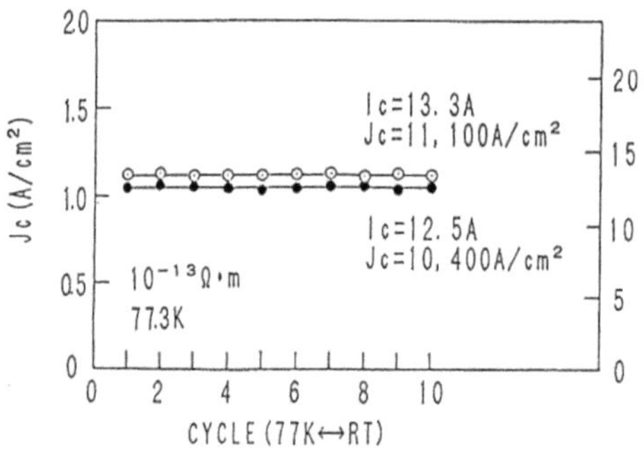

Fig. 6 Change of Ic of 5 m-long conductor during immersion in liq. N_2

Magnet development

At present, magnet application is considered for two different fields. One is the relatively low field application up to 1 Tesla at 77.3 K and the other is super-high field application above 20 Tesla at 4.2 K. In the future, high field application at 77.3 K should be the main target. We reported on the prototypes of a 1,070 Gauss single-pancake coil at 77.3 K, a 1 Tesla single-pancake coil at 4.2 K and a 23.225 Tesla three double-pancake coil at 4.2 K[13]. For building a high or super-high field magnet, it is necessary to suppress the mechanical damage to the superconducting wire due to the large electromagnetic force. One example of this effect was shown previously[13]. The previous data showed that Ic of the three double-pancake coil at 23 Tesla was a half of that at 5 Tesla. In the case of the short wire sample, the change of critical current was 20% from 5 to 23 Tesla. This degradation was considered to be attributable to the high stress, 2,000 kg/cm^2 at 23 Tesla.

In considering these results, an improved technique was chosen, including epoxy resin impregnation and a stainless steel jacket. Figure 7 shows the test results of a newly designed three double-pancake coil, using the Hybrid magnet in HFLSM at Tohoku University. This 84-turn coil was made of a bundle conductor 0.56 mm thick and 3.7 mm wide. The maximum field was 23.314 Tesla with Ic of 95 A at 23 Tesla back-up field.[15] This coil showed little change of the critical current from 5 to 23 Tesla and demonstrated the effectiveness of anti-stress improvements.

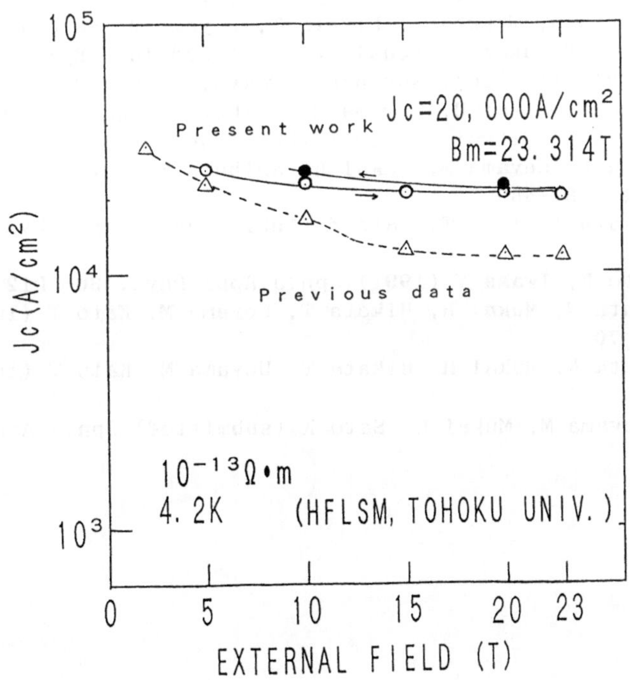

Fig. 7 Jc of three double-pancake coil with a external field.

CONCLUSIONS

Long and high-Jc bismuth superconducting wire was developed with the powder-in-tube technique and the high-Tc phase of bismuth compound. The maximum critical current densities at 77.3 K were 53,700 A/cm^2 in a zero magnetic field, 42,300 A/cm^2 at 0.1 Tesla and 12,000 A/cm^2 at 1 Tesla. Long wires, 70 $\sim$ 100 m, were fabricated and proved to have critical current densities ranging from 4,000 to 7,000 A/cm^2. The result of a 5 m spirally wound conductor revealed the possibilities of large current conductor application. An improved technique against large electromagnetic force showed that it was possible to achieve the expected characteristics of the wire for super-high field magnets.

REFERENCES

1. Maeda H, Tanaka Y, Fukutomi M, Asano T (1988) Jpn.J.Appl.Phys. **27**: L209-L210
2. Takano M, Takada J, Oda K, Kitaguchi H, Miura Y, Ikeda Y, Tomii Y, Mazaki H (1988) Jpn.J.Appl.Phys. **27**: L1041-L1043
3. Sato K, Hikata T, Mukai H, Hitotsuyanagi H (1989) Proc. of ISTEC Workshop on Superconductivity. ISTEC, Tokyo: pp 119-122
4. Hikata T, Nishikawa T, Mukai H, Sato K, Hitotsuyanagi H (1989) Jpn.J.Appl.Phys. **28**: L1204-L1206
5. Sato K, Hikata T, Mukai H, Masuda T, Ueyama M, Hitotsuyanagi H, Mitsui T, Kawashima M (1990) In: Ishiguro T, Kajimura K (eds) Proc. of 2nd Int. Symposium on Superconductivity(ISS'89). Springer, Tokyo, pp 335-340
6. Hikata T, Ueyama M, Mukai H, Sato K (1990) Cryogenics **30**: 924-929
7. Sato K, Hikata T, Iwasa Y (1990) Appl.Phys.Lett. **57**: 1928-1929
8. Mukai H, Shibuta N, Masuda T, Hikata T, Ueyama M, Kato T, Sato K (1990) In: Kajimura K, Hayakawa H (eds) Proc. of 3rd Int. Symposium on Superconductivity(ISS'90). Springer, Tokyo, pp 607-612
9. Sato K, Hikata T, Mukai H, Ueyama M, Shibuta N, Kato T, Masuda T, Nagata M, Iwata K, Mitsui T (1991) IEEE Trans. MAG **27**: 1231-1238
10. Sato K, Hikata K, Ueyama M, Mukai H, Shibuta N, Kato T, Masuda T (1991) Cryogenics **31**: 687-689
11. Ueyama M, Hikata T, Kato T, Sato K (1991) Jpn.J.Appl.Phys. **30**: L1384-L1386
12. Hikata T, Sato K, Iwasa Y (1991) Jpn.J.Appl.Phys. **30**: L1271-L1272
13. Sato K, Shibuta N, Mukai H, Hikata T, Ueyama M, Kato T (In press) J.Appl.Phys. **70**
14. Sato K, Shibuta N, Mukai H, Hikata T, Ueyama M, Kato T (In press) Physica C
15. Shibuta N, Ueyama M, Mukai H, Sato K (submitted) Jpn.J.Appl.Phys.

Microstructure Control and Critical Current Density of Silver Sheathed Bi2223 Tapes

KOZO OSAMURA, SANG SOO OH, and SHOJIRO OCHIAI

Department of Metallurgy, Kyoto University, Kyoto, 606 Japan

ABSTRACT

Silver sheathed Bi2223 tapes were prepared by the powder in tube method. In order to get high critical current density (J_c), the microstructure was controlled through a serial step of handlings. Especially, the addition of other elements, the influence of load for pressing and the degree of reduction of size were in detail investigated. High critical current densities have been achieved relating to the improvement of microstructure by addition of another element, densification, formation of well-textures structure and minimization of work instability.

KEY WORDS: Bi system superconductor, powder-in-tube method, critical current density

INTRODUCTION

The silver sheathed Bi2223 tapes have been usually prepared by the powder in tube method. It comprises a serial step of handlings as listed in Table 1. Each step affects the final performance and so shall be carefully examined to get optimum condition. There is a variety of optimum composition for Bi2223 powder depending on investigators /1,2,3/. Recently residual carbon is reported /2/ to affect the property seriously. It is not easy to understand the deformation mechanism of the composite consisting of ductile metallic sheath and brittle oxide powder. The superconducting critical current increases by repeating heavy cold-working and heat-treatment as reported by various authors /4,5/. The heat treatment condition is sensitive and cooling rate affects the property /3/. The highest J_c at 77 K has been reported to be 50000 A/cm^2 /1/, 33000 A/cm^2 /6/ and 23000 A/cm^2 /7/. These values are, however, not satisfactory for Bi2223 tapes, because of high potentiality of this material. In the present report, optimum conditions for several steps in the powder in tube method have been discussed to improve the critical current density.

EXPERIMENTAL PROCEDURE

The powder with composition of $Bi_{1.6}Pb_{0.4}Sr_{1.6}Ca_{2.0}Cu_{2.8}O_z$ was filled in the silver tube. In order to investigate the effect of thermomechanical treatment on the superconducting properties, the following cold-working and heat treatment was applied for the specimens. For reducing the diameter, the swaging (S), groove rolling (Rg) or drawing (D) techniques was applied. By the first pressing (P), the wire was deformed to a tape shaped specimen. After that, the annealing and pressing were repeated to refine the microstructure. The annealing was performed at temperatures between 1103 and 1113 K for various periods from 36 to 360 ks. The transport critical current measurements have been performed at 77 K with 1 μV/cm criterion and at 4.2K with 0.1 μV/cm. The magnetic field dependence of critical current was measured under magnetic fields parallel to the tape surface and perpendicular to current flow. The critical current density (J_c) is defined as the current divided by the cross-sectional area of the oxide phase. The thickness of oxide layer (t_o) is determined by optical and scanning electron microscopes. The crystallographic orientation of grains was investigated by the X-ray pole figure technique.

EXPERIMENTAL RESULTS AND DISCUSSION

The effect of addition of other elements on J_c has been investigated. Fig. 1 shows the magnetic field dependence of J_c for the specimen including MgO. At zero field, J_c increases in sequence for increasing amount of MgO. But the magnetic field dependence is optimized for the 0.05MgO added specimen. From EPMA analysis /8/, it has been found that fine MgO particles disperse in the matrix and at grain boundaries. Among several additional elements investigated here, BaO is also found to improve both T_c and J_c, but the degradation of J_c occurred by addition of Ga_2O_3 and SnO_2.

Various work techniques were attempted to improve the J_c. By swaging or groove rolling, the powder has been remarkably compacted. When the tapes were reduced in size by groove rolling, J_c was improved comparing with the results by swaging. It was found that a combination of RgP1HT1P2HT2 gives a high J_c value, where after groove rolling, the pressing and heat treatment at 1113 K for 360 ks were two times repeated.

When the tube is deformed to tape shape by pressing, the uniaxial pressure was changed to get tapes with different thickness. As shown in Fig.2, J_c increases with increasing pressure, where the tapes were prepared by a combination of swaging, drawing, pressing and heat treatment. The area ratio of oxide layer to the whole area decreases with increasing pressure. This indicates that the oxide layer is densified during pressing.

J_c is improved by repeating pressing and heat treatment. It has been reported /4/ that three time repetition of pressing and heat treatment gives an optimum value of J_c. When we look at the microstructure of these tapes, the well textured structure is produced along the tape surface. Fig.3 shows a contour map of (0010) plane of Bi2223 phase obtained by the pole figure analysis. Intensity maximum comes at the center, indicating that (00l) plane, that is, a-b plane of superconducting phase is lying parallel to the tape surface. The repetition of pressing and heat treatment is suggested to promote this texturing.

Table 1 Technical key factors for the powder-in-tube method for preparing high performance silver sheathed oxide tapes.

Handling	Key Factor
Preparation of Powder	* Composition * Impurity * Addition of other oxides * Powder Size
Tubing	* Dimension of Tube * Alloying Element * Compaction
Cold-Working for Reduction	* Work Technique swaging groove rolling drawing * Reduction Rate * Degree of Reduction
Cold-Working for Refinement	* Work Technique pressing rolling * Reduction Rate * Load for Work * Degree of Reduction
Heat Treatment	* Temperature and Time * Atmosphere * Cooling Rate

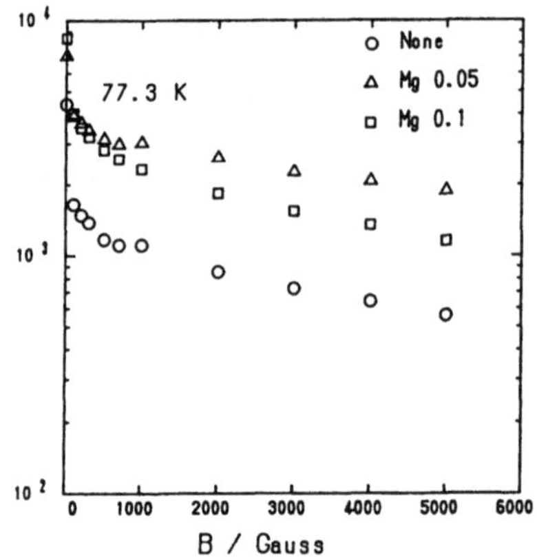

Fig.1 Magnetic field dependence of critical current density for the silver Bi2223 tapes added by MgO.

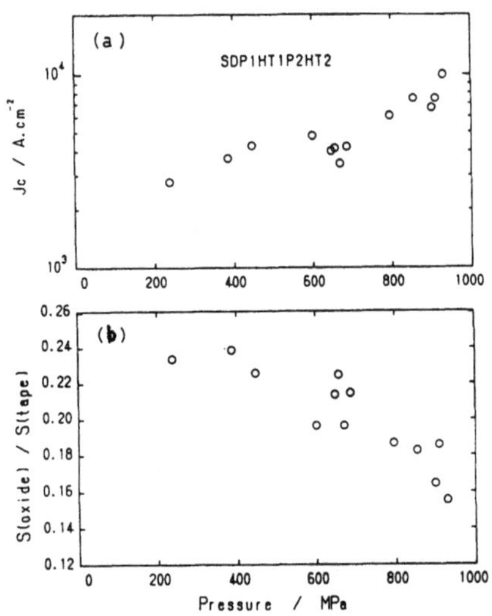

Fig.2 Change of J_c and the area ratio as a function of pressure, when the tape is prpeared by pressing.

In order to examine in detail the influence of cold-work, the following work schedule was applied /7/. Two types of tube filled superconducting powder were reduced in size by groove rolling up to various diameters between

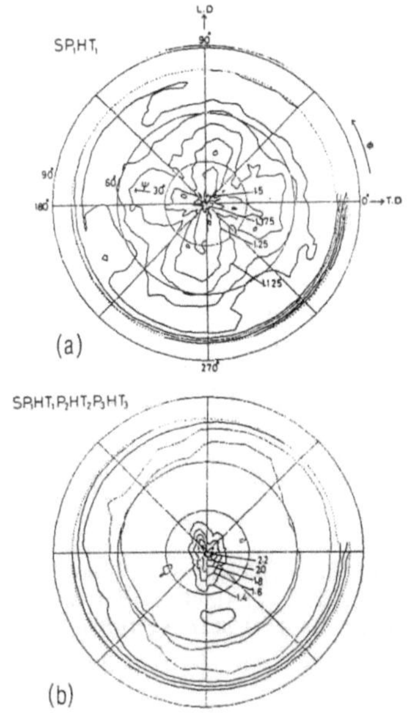

Fig.3 Pole figure of (0010) plane of Bi2223 phase for two specimens prepared by SP1HT1 and SP1HT1P2HT2P3HT3.

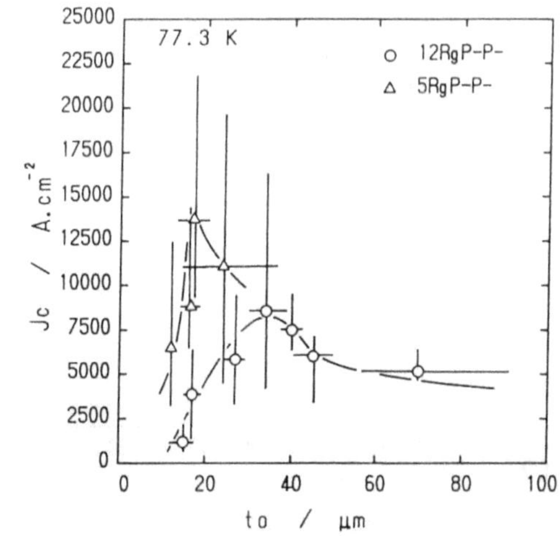

Fig.4 Oxide layer thickness dependence of critical current density at 77 K for the tapes prepared from the different starting tubes.

1.0 and 0.3 mm. Then these wires were pressed to tape shape with a constant load of 900 MPa. Then heat treatment was performed at 1113 K for 360 ks, where pressing and heat treatment were repeated two times. Even though several tapes were prepared with the same condition, their J_c and thickness are scattered very much. The averaged values of J_c and thickness were calculated for tapes of the same group of work condition as shown in Fig.4. The average data give a clear information about the oxide layer thickness dependence of J_c. J_c increases with decreasing thickness and reaches a maximum and then degrades rapidly. The critical thickness is estimated to become about 20 ~ 30 μm. Fig.5(a) shows the magnetic field dependence of J_c at 77 K for the tapes prepared from different wires. It is seen that the maximum H_c at zero magnetic field is 23000 A/cm^2 for 5Rg0.5P-P- and the initial drop at low magnetic field is not so big. The J_c decreases gradually with increasing magnetic field. The highest J_c is found to be 5000 A/cm^2 at 0.5T. At 4.2 K, a high J_c of 3×10^4 A/cm^2 was recorded as shown in Fig.5(b).

As mentioned in the following, the degradation of J_c due to heavy reduction shown in Fig.4 is related to the variation of oxide layer thickness. The variation of thickness was assessed from the photograph of the cross section along the current flow direction. The degree of deviation was evaluated using the coefficient of variation (COV), which is defined as the standard deviation divided by average thickness. Fig.6 indicates that COV becomes large when the oxide layer thickness decreases. Especially COV becomes quite large when the thickness is less than 20 ~ 30 μm. This abnormal increase is called the work instability. The J_c is defined as the critical current divided by the cross sectional area, where the average thickness is usually used. When COV is very large, however the average thickness is not so good measure, because the critical current is determined by the smallest cross section perpendicular to the current flow. Here it is suggested that the average thickness is replaced by t_o - 3 σ , where σ is the standard deviation. The corrected J_c is plotted by using solid signs in Fig. 7. They tend to increase monotonously with decreasing thickness except the small three data. Thus the influence of work instability seems to be corrected. It is clear that a very high J_c value exceeding 10^5 A/cm^2 is obtainable, when the present technique is improved.

SUMMARY

Here the discussion has been done about achieving high J_c. There are two aspects to understand this problem. In the macroscopic viewpoint, several factors listed in Table 1 are important. It could be pointed out that the present Bi2223 tapes has a very high potential for critical current density. Secondly in order to make clear fully the origin of J_c, the pinning mechanism should be investigated. In another word, the pinning center determining

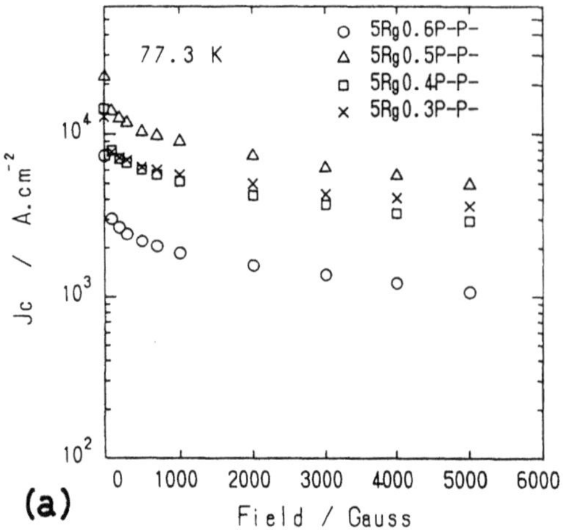

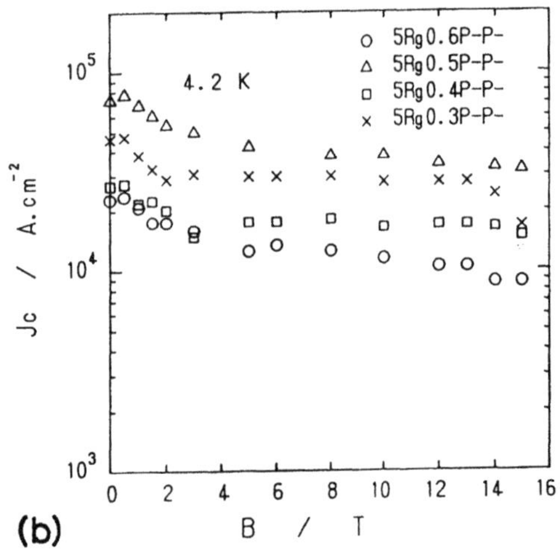

Fig.5 Magnetic field dependence of critical current density for the tapes prepared from the groove-rolled wires with different diameter measured at 77 K (a) and 4.2 K (b).

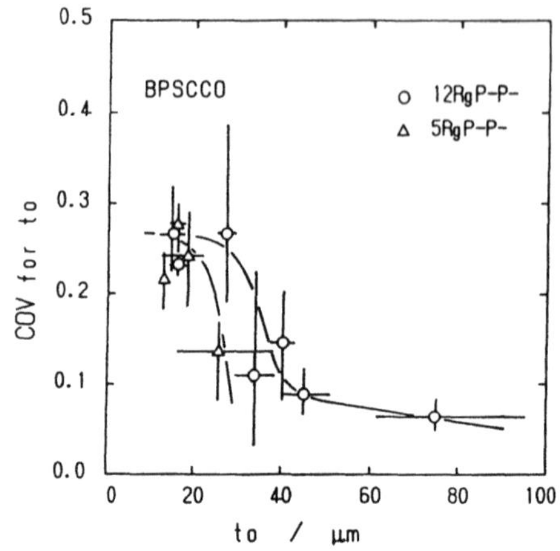

Fig.6 Change of the coefficient of variation (COV) as a function of oxide layer thickness for the tapes prepared from the different starting tubes.

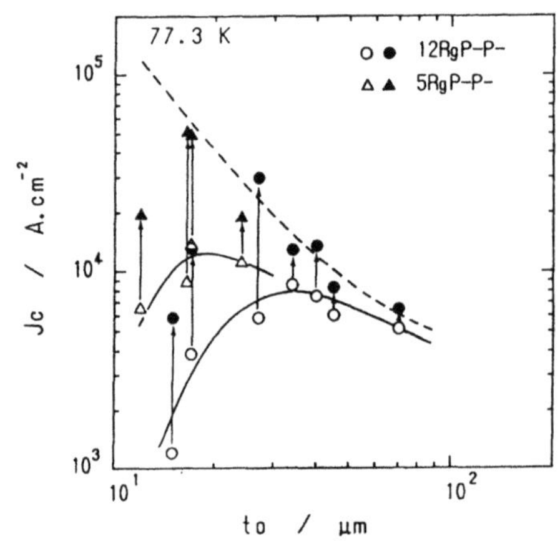

Fig.7 Correction of critical current density by using the smallest oxide layer thickness for the tapes prepared from the different starting tubes.

J_c has to be specified among proposed candidates.

Acknowledgments: This study was made possible by a Scientific Research Grant-in-Aid (Project No 03210220) from the Ministry of Education, Science and Culture, Japan.

REFERENCES

1. K. Sato, T. Hikata, H. Mukai, T. Masuda, M. Ueyama and H. Hitotsuyanagi,T. Mitsui and M. Kawashima; IEEE T rans.Magn. 27:1231(1991).
2. R. Flukiger, A. Jeremie, B. Hensel, E. Seibt, J. Q. Xu and Y. Yamada; ICMC(Huntsville, 1991).
3. S .S . Oh, T. Kubota and K. Osamura; Physica C171(1990)265-270.
4. K. Osamura, S. S. Oh and S. Ochiai; Supercond. Sci. & Tchnol., 3(1990)143.
5. H. Krauth, K. Heine, J. Tenbrink, M. Wilhelm and H. Neumuller; ISS'90 (Sept, 1990).
6. Y. Yamada, B. Obst and R. Flukiger; Supercond. Sci. & Technol. 4(1991)165.
7. K. Osamura, S. S. Oh and S. Ochiai; Supercond. Sci. & Technol, submitted (1991).
8. M. Kamo, S. S. Oh and K. Osamura; Jap. J. Inst. Metals, submitted (1991).

Preparation of Ag-Sheathed Superconducting Wires Based on the Bi(Pb)SrCaCuO Oxide

M. SUZUKI, T. KIMURA, and T. TAKESHITA

Central Research Institute, Mitsubishi Materials Corporation, Omiya, Saitama, 330 Japan

ABSTRACT

Bi(Pb)SrCaCuO superconducting oxides in an Ag tube can be aligned in one of the crystal orientation by deformation. In this report,we studied differential speed rolling (DS rolling) for production of long wires with a high Jc.

Ag tubes with oxides were first reduced in its diameter and then rolled into tape by DS rolling. The Jc of this tape was found to be 19,000 A/cm^2 and is twice as large as that of wires made by conventional rolling. This result indicates that the DS rolling helps to align the crystalline orientation and also strengthen the connection between crystalline grains.

KEY WORDS : superconducting wire, Bi(Pb)SrCaCuO, Ag sheathed wire, differential speed rolling, crystalline orientation

INTRODUCTION

Among Bi system superconductive compounds,$Bi_2Sr_2Ca_2Cu_3$ oxide has highest Tc of 110K.[1] Many research works have been made to produce long wires for coils and cables applications for this system. In order to produce long wires, Ag sheath has been used for solid and liquid phase process.

In solid state process, cold working such as swaging and groove rolling are used to compact powder, and after sintering the compacts, rolling, pressing or cross rolling are used sequentially to produce tape. The highest Jc wire was obtained by pressing alone in stead of these processes. In the case of long wires or coils,lower Jc was reported as compared with this champion data. This may be due to dimensional instability and heterogeneity of core material.

On the other hand, Bi compounds are known to have cleavage plane along to the basal plane and can be oriented easily. Therefore shearing force is thought to be favourable to obtain a prefered orientation. In the roll bite of differential speed rolling(DS rolling) shearing deformation is preceded,and this deformation mechanism is a steady one. Therefore this process is considered to produce good dimensional stability.

EXPERIMENTAL

In $Bi_2Sr_2Ca_2Cu_3$ compound, a partial replacement of Bi with Pb is known to give higher Tc and Jc. Starting materials used were high purity Bi_2O_3,$SrCO_3$, $CaCO_3$,CuO and PbO powders. These powders were mixed in the cation ratio of $Bi_{1.8}Pb_{0.4}Sr_{2.0}Ca_{2.2}Cu_{3.1}$. Superconductive powder was prepared by repeated solid state reaction in the condition of 800℃ × 20hr+ 835℃ × 35hr and milling.

Volumetric mean particle size was 6.5 μm by micro track analysis.

ChemicalComposition of powder by X-ray analysis is shown in Table 1 .

Superconducting wires was made by " powder in tube" process. Superconductive powder was filled into a Ag pipe and was fabricated into a tape. This process is shown in Fig. 1 . Wires were reduced in diameter down to ϕ 1.3mm by groove rolling and swaging, and then rolled to 0.2mm thick tapes by normal and DS rolling respectively. Then sintering for the growth of plate of the compound was first done and cold rollings by both methods above were done in order to obtain prefered grain orientation and finally sintering for intergrain bonding was made.

In DS rolling as shown in Fig. 2 , neutral point of the roll with higher velocity V_1 moves to the exit of contact area in rolling operation and on the roll with slower speed V_2 ,it moves to entrance. In the zone pinched

Table 1 Cemical composition of starting powder

	Bi	Pb	Sr	Ca	Cu
wt %	35.15	7.58	15.96	7.86	18.31
composition ratio	1.85	0.40	2.00	2.15	3.16

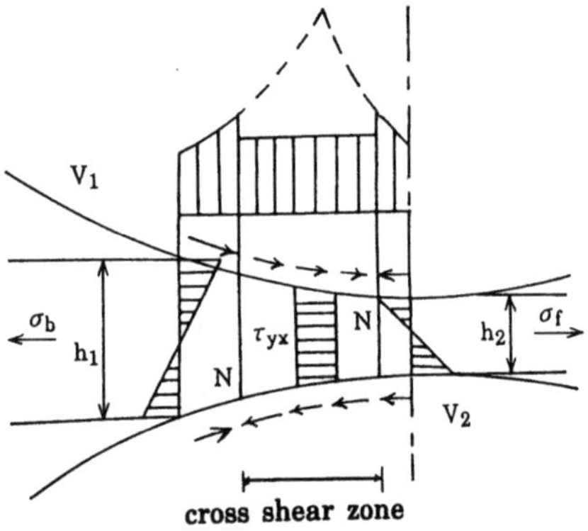

FIG.2 Stress distribution in DS rolling

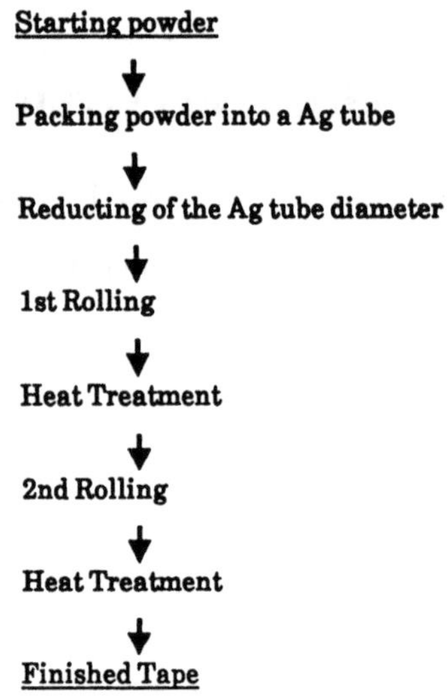

Fig.1 Preparation of Ag sheathed tape

by neutral points, friction force by rolls acts reversely and only shearing deformation precedes.(cross-shear zone)[2] As the velocity ratio V_1/V_2 increases, cross shear zone is enlarged. The critical velocity ratio and front and back tention (σ_f, σ_b) where all contact area come to crossshear zone are calculated respectively to be as follows:

$V_1/V_2 = 1/(1-R)$ $\qquad$ $\sigma_f - \sigma_b = \ln(h_1/h_2)$

where

h_1,h_2: thickness of initial and rolled strip respectively
R: thickness reduction
k: shear strength

In our experiment, front and back tension were not applied, and velocity ratio was kept constant at 1.41 which is sufficient to realize perfect cross shear condition.

DC 4probe method was used to measure Ic at liquid nitrogen temperature without applied magnetic field. Ic was defined as a current which generates 1 μV/cm in the voltage probe. Transvers and longitudinal cross sections of core were examined with optical microscope at several points and the cross secional area was used to calculate Jc value. X-ray diffraction analysis and pole figure measurement were made with CuK α for tapes.

RESULTS AND DISCUSSION

Ic and Jc of samples are listed in Table 2 . In our experiments, DS rolling yield highest Jc values. Dependence of Jc on the reduction ratio in the second rolling is shown in Fig. 3 . Jc value of tapes increase slightly with increasing reduction for normal rolling, while in tapes made by DS rolling Jc's increase steeply with reduction in second rolling. Champion data of Jc is reported tobe 47000A/cm^2 for wires prepared by press method.[3] But in the case of long wires, at least, of several meters which can be wound into coils maximum Jc value is 10200A/cm^2.[4] This difference may be due to the deformation mechanism in conventional processes. Continuous pressing may result in local decrease of cross section or cracking at the edge of anvil.[5]

As rolling deformation proceeds,Ag work hardens readily. But shear strength of the core Bi 2223 is thought to be kept constant. This difference in deformation causes the fluctuation of core thickness along tape length. On the contrary, DS rolling surpresses necking and yields superior homogeneity in the core along its longitudinal axis as shown in Fig. 4 .

Furthermore, the effect of DS rolling is noteworthy on the shape of cross section in transverse direction as clearly shown in Fig. 5 . The width of Ag sheath is about a half of that calculated by experimental formula for the flat rolling of round wire. This indicates that new model is needed to explain

Table 2 Ic and Jc values of tapes

Press

Pressure	Ic (A)	Jc (A/cm²)	S (cm²)
3 t/cm²	19.0	8800	2.16×10^{-3}
5	23.3	10800	2.16×10^{-3}

Rolling

Thickness	Ic (A)	Jc (A/cm²)	S (cm²)
0.18mm	17.5	9700	1.8×10^{-3}
0.15	15.0	9400	1.6×10^{-3}
0.13	16.5	10400	1.58×10^{-3}

Differential speed rolling

Thickness	Ic (A)	Jc (A/cm²)	S (cm²)
0.18mm	21.5	11900	1.8×10^{-3}
0.15	22.5	14100	1.6×10^{-3}
0.13	24.5	18800	1.3×10^{-3}

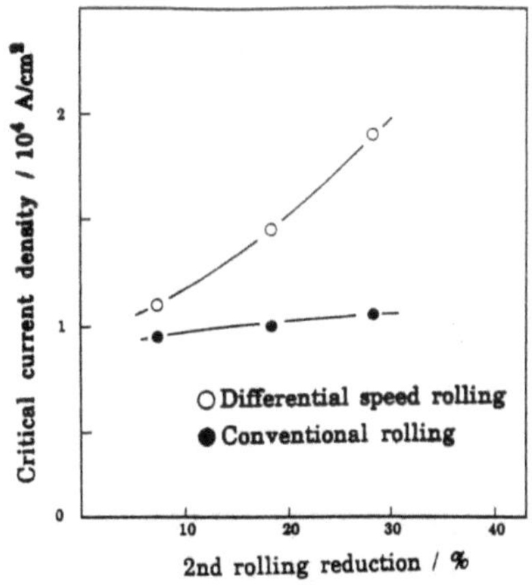

Fig.3 Relation between Jc and 2nd rolling reduction

plastic behavior of this superconducting wire. In these two rolling methods, DS rolling gives spread in width about 13% less than conventional rolling.

Conventionally rolled tapes show apparent edge drop in cores but DS rolling gives uniform core thickness. These indate that the existence of cross shear zone produces uniform pressure along its width and surpresses three dimensional deformation. These difference in macroscopic deformation mode result in different microstructures. Fig. 6 shows improvement in platelet orientation in a DS rolled tape.

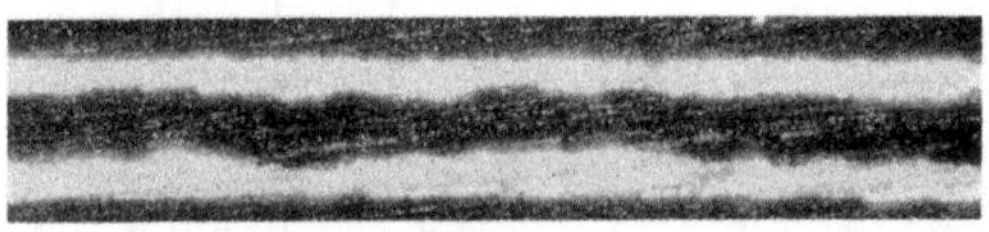

a. conventionalLy rolled

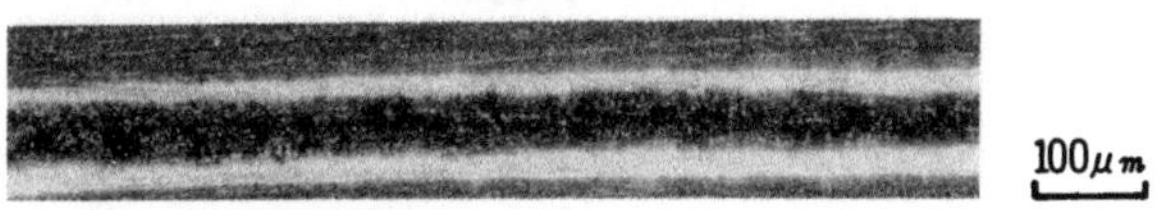

b. DS rolled

Fig.4 Longitudinal cross section of tapes

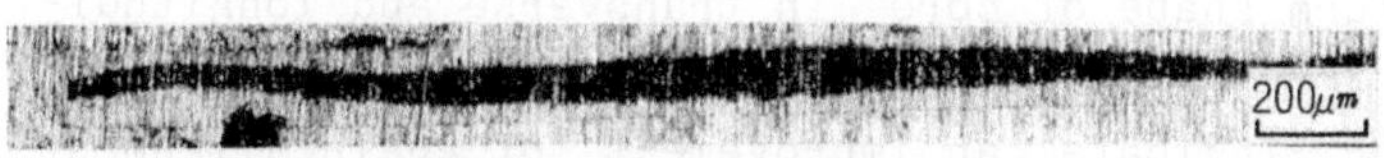

a. conventionally rolled

b. DS rolled

Fig.5 Transverse cross section of tapes

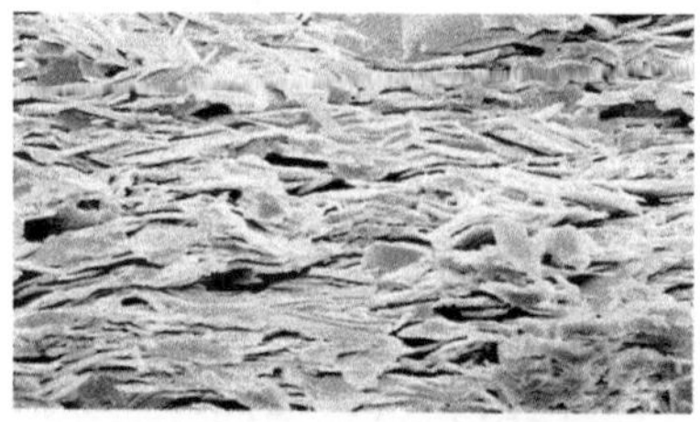

a. conventionally rolled

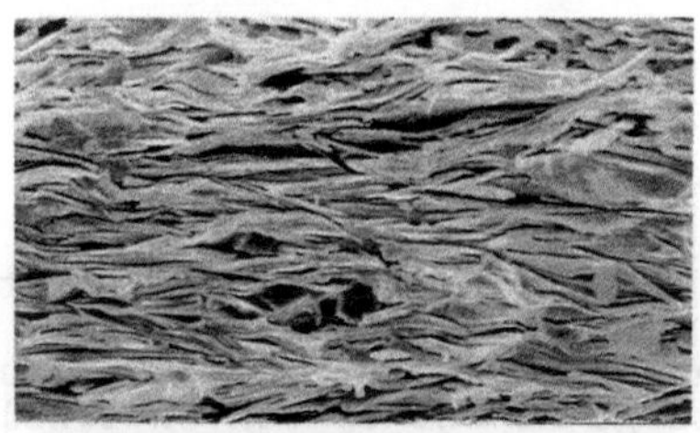

b. DS rollled

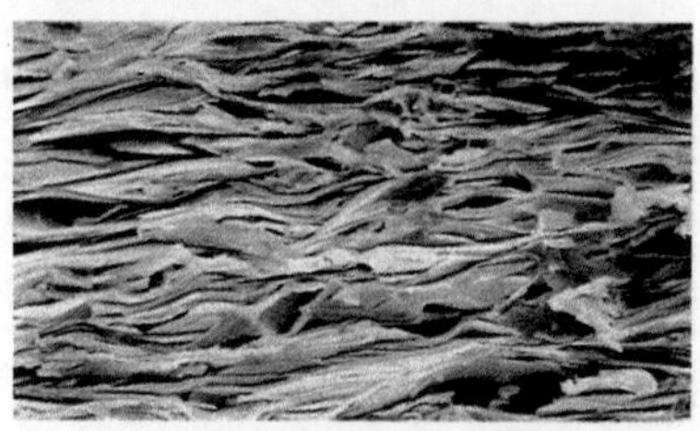

c. pressed

Fig.6 SEM photographs of longitudinal cross section

Fig. 7 shows the pole figure of (0012)plane showing remarkable center polarization. The ratio of X-ray reflectioN from intensity from basal plales ,

$$F = \Sigma I_{00x} / \Sigma I_{hkl}$$

was used as an index of prefered orientation and was correlated with Jc values. F values are over 0.96 and there is only slight difference in line scanned profile as shown in Fig. 8 . There may be improvement of superconductive connection in grain boundary regeons[8] but further investigations are necessary to establish real reason for this improvement.

But this results of our experiments are encouraging to fabricate long superconducting wires with high Jc.

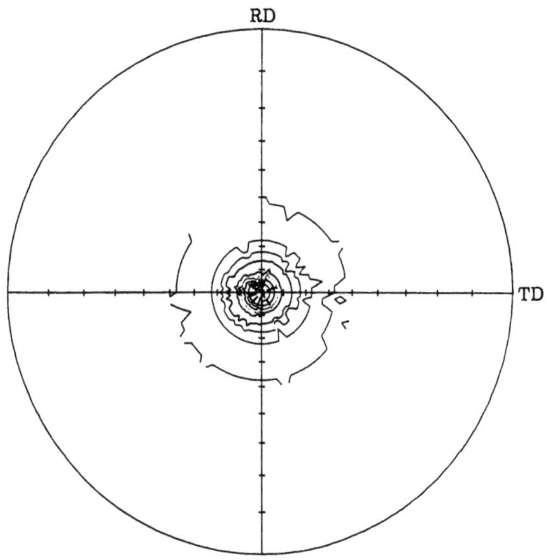

Fig.7 (0012)Pole figure of DS rolled tape

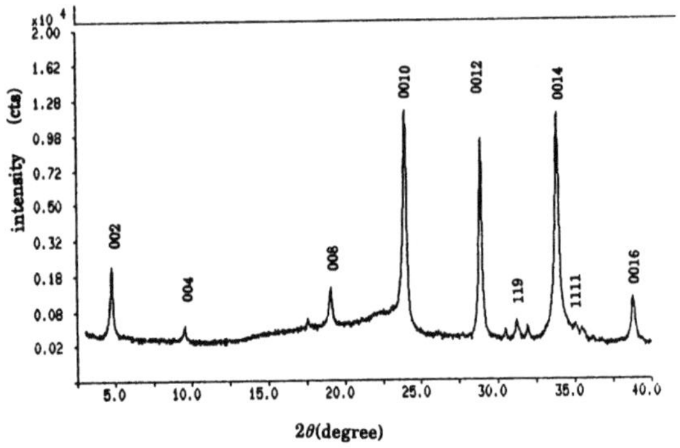

a. conventionally rolled

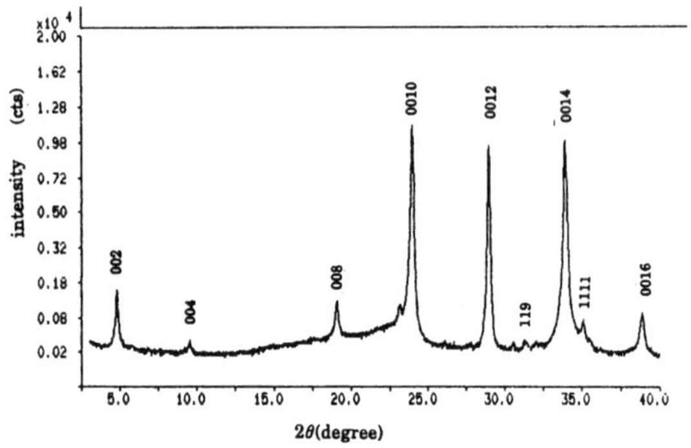

b. DS rolled

Fig.8 X-ray diffraction patterns

CONCLUSION

① Fabricating superconductive Bi 2223 tape by"powder in tube " method using DS rolling with speed ratio 1.41 which is enough to make the whole contact area cross-shear-zone , gives Jc value up to 19000A/cm². This is twice as large as one by conventional rolling. This value is superior to the values reported as long wire.

② DS rolling gives more uniform shape of core in transverse and longitudinal direction. This shows a possibility of producing long tapes with good uniformity.

③ Microstructural observation indicate an improvement of grain alignment but X-ray analysis shows only slight improvement in texture. Therefore this may be due to strengthening of intergrain connection.

REFERENCE

1. H.Maeda et al:Jpn.J.Appl.Phys.,27(1988),L209
2. H.Shiozaki et al:J.Jpn.Soc.Tech.Plast.,24(1983),25
3. M.Ueyama et al:Proc.Jpn.Soc.Cryo.Eng.,(1990.11),5
4. F.Hosono et al:ISS'90(1991),675
5. H.Kikuchi et al:Ref.4,655
6. M.Kobayashi:J.Jpn.Soc.Tech.Plast.,19(1978),630
7. N.Enomoto:Ref.4,625
8. S.S.Oh et al:ISS'89(1990),389

Multilayer Composite Wires of Bi-Based Superconductor Prepared by Doctor-Blade Method

MASANAO MIMURA[1], NAOKI UNO[1], and KUNIO DOI[2]

[1] Yokohama R&D Laboratories, The Furukawa Electric Co., Ltd., Yokohama, 220 Japan
[2] Engineering Research Association for Superconductive Generation Equipment and Materials (Super-GM), Osaka, 530 Japan

ABSTRACT

Fabrication of electric wires with high critical current (Ic) is one of the key technologies for electric power application of high-Tc superconducting materials. For this, the wires must have the composite metals as a stabilizer. Silver is thought to be the most useful metal for the purpose.
In this study, round composite wires with a multilayer structure of Jelly-Roll type are prepared using Bi-based Superconductor by a doctor - blade technique and their superconducting properties are investigated. It was observed that a-b planes of each superconducting grain are highly oriented along the round silver sheets and each laminar grain grew preferably along the axial direction of the wire. Ic of 60A and 30A are obtained for multilayer wires of 3 cm and 100 cm in length, respectively. Moreover, much higher transport current of 270A is obtained for 7-stranded cable. These multilayer composite wires have high Ic and no anisotoropy to the direction of applied magnetic field. Therefore, the composite wires with multilayer structure are one of promising candidates for electric power application.

KEY WORDS: Jelly-Roll, doctor-blade, stranded cable

INTRODUCTION

Since the discovery of high-Tc oxide superconductors with critical temperatures (Tc) above 77k, many studies on the conductor fabrication by a conventional powder metallurgy technique using these materials have been carried out. For tape wires fabricated by this technique, a highly oriented grain structure yielded a high critical current density(Jc)[1],[2]. However, superconducting properties of tape wires with a single core of superconductor was very anisotoropic to magnetic field direction. In addition, critical current (Ic) of such tapes was very small. Furthermore, cold-rolling to tapes would not be suitable for long wire fabrication. Recently, we reported preparation and properties of round shape wires with multilayer structure of Jelly-Roll type[3].
In the present study, long wires and stranded cables with the multilayer structure were fabricated and they were shown to have very high values of Ic.

EXPERIMENTAL

Figure 1 shows a used for fabrication of multilayer composite wire using Jelly-Roll method. A mixture of Bi_2O_3, $SrCO_3$, $CaCO_3$ and CuO powders in a nominal composition($Bi_2Sr_2CaCu_2O_y$) was calcined at 750 - 800 °C for 30h. The calcined powders were mixed with ethanol-toluen mixed solvent, polyvinyl butyral binder, surface active despensant and ethyl phtalate, and then mixed using a ball-mill. The resulting slurry was casted on a carrier sheet by a doctor - blade into 100mm width, 10-400μm thickness. The green sheet were put on a 50-500μm thick silver sheet, and then wound around a silver rod. The composite rod contained 10-20 oxide layers. After the organic formulations were removed by heating at 500-650 °C for 100h. It was put into a silver tube, and then swaged and drawn to a wire with outside diameter of 1-3 mm. The wire was cut into 3-100 cm length. Moreover, using these wires, 7-stranded cables were formed. All of them were made to be superconducting wires or cables by a final sintering at 850 - 950 °C.
The critical current were measured by a four-probe method at 77K and 4.2K in a magnetic field up to 10T which was applied perpendicularly to a transport current. In this measurement Ic was defined as the current which induced 1μV/cm and the Jc was calculated by dividing it by a cross-sectional area of the oxide. For comparison, Ic(Jc) of composite tapes with single core of superconductor were also measured.
The microstructures of wires were characterized by a scanning electron microscope (SEM) and a optical microscope. X-ray diffraction analysis (XRD) were performed for the evaluation of alignment of superconducting grain[2].

RESULTS AND DISCUSSION

Figure 2 shows a cross - sectional view of the mulilayer composite wire with a diameter of 3 mm. A black part reveals 11 layers of $Bi_2Sr_2CaCu_2O_y$ superconducting phase (Bi-2212). The volume fraction of Bi-2212 was controlled by it' thickness and the number of layer. As shown in Fig.3, laminar grains of the Bi-2212 were highly oriented along the round silver sheets. Figure 4 shows the SEM photograph of fractured cross-section view of the outside 1st and 6th superconductor layers. No significant difference was observed for each layer. From the results of XRD, the degree of the c-axis orientation value (F)[2] was estimated to be 92%. Each laminal grain grew preferably along the axial direction of the wires. This result reveals that the both directions of the transport current and crystal grain growth are nearly the same; thus, the multilayer composite wire of Jelly-roll type was expected to have high value of Ic. Actually, Ic of 60A ($Jc=3.6X10^3 A/cm^2$) was obtained for the 3 mm wire of 3 cm in length. Figure 5 shows the magnetic field dependence of Jc at 77K for the multilayer composite wire together with the single core tape. For the single core tape, the anisotoropy in Jc increased with increasing the degree of c-axis orientation[4]. While, the composite wire revealed no anisotoropy as shown in Fig.5. It was due to the high c-axis orientation along the round silver. Figure 6 shows the magnetic field dependence of Jc at 77K and 4.2K for the single core tape. Jc at 77K and 0T was relatively high values of $1.2x10^4$ A/cm^2, which showed that the "weak link" between superconducting grains were much decreased by the increase of c-axis oriented grains. However, Jc drastically decreased at fields of above 0.1T, which was probably due to very weak intragrain pinning. On the contrary, Jc at 4.2K was much less sensitive to the magnetic field; Jc was $2.0x10^5$ A/cm^2 and $9.0x10^4$ A/cm^2 at 1T and 10T, respectively. The temperature dependence of Jc^M at 0.5T for the multilayer wire is shown Fig.7 , which was estimated using Bean's model. For the magnet application, this wire had sufficient Jc at temperatures below 30K. However, Jc at 77 K was thought to be slightly insufficient.

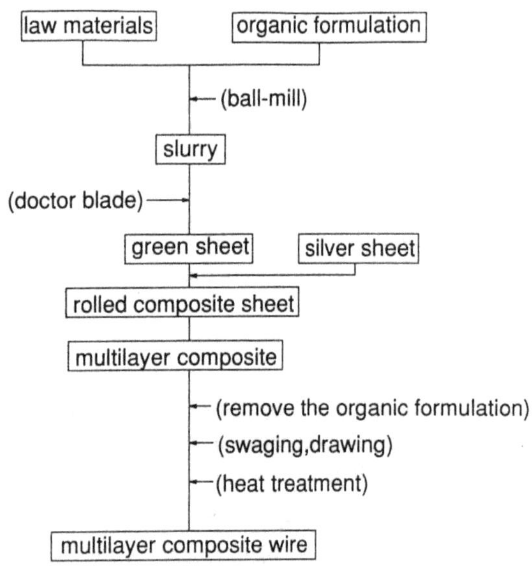

Fig.1 Fabrication process of Bi-based multilayer composite wires using Jelly-Roll method.

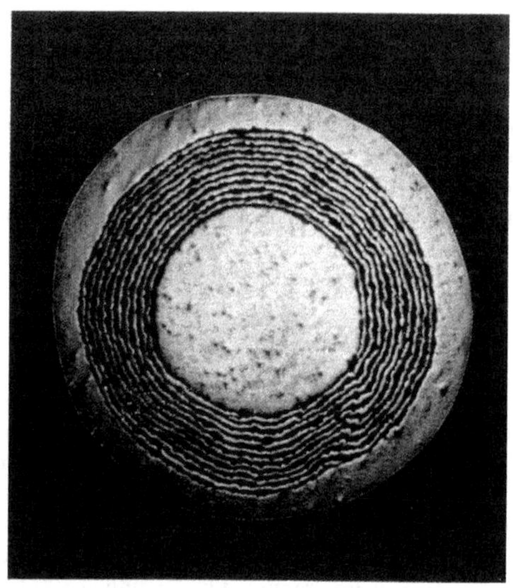

Fig.2 Cross sectional view of the multilayer composite wire with a diameter of 3 mm.

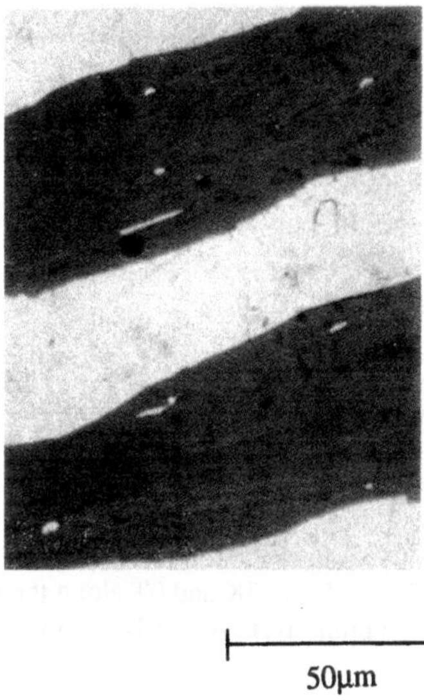

Fig.3 Micrograph of microstructure of the Bi-Based multilayer wire.

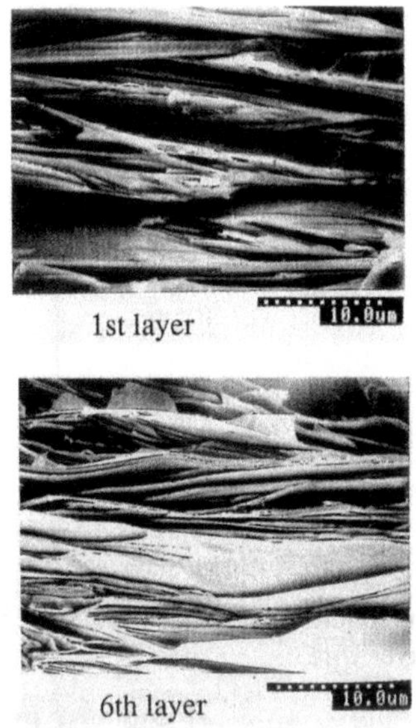

Fig.4 SEM photograph of fractured cross-sectional view of the outside 1st and 6th superconductor layers.

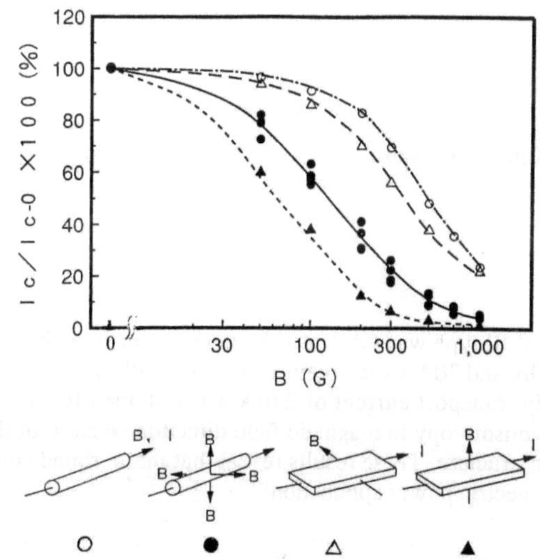

Fig.5 Anisotoropy of Jc to the direction of applied fields for the multilayer wire and the single core tape.

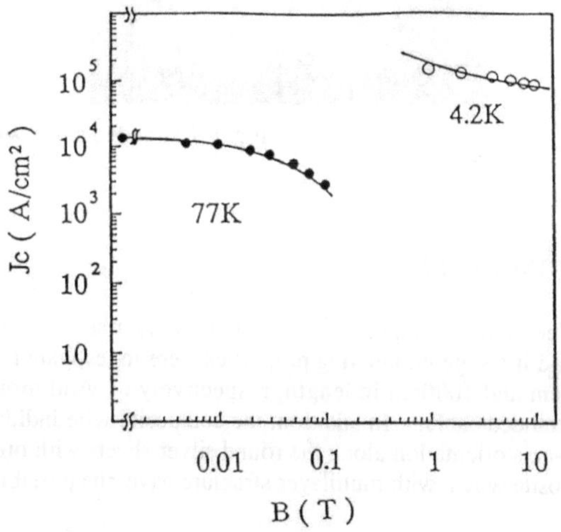

Fig.6 Magnetic field dependence of Jc at 77K and 4.2K for the single core tape.

For practical application, it is necessary for the wires to be highly uniform in superconducting properties along the longitudinal direction. Figure 8 shows the variation of Ic at 77k and 0T along the longitudinal direction, of the multilayer wire of 100cm length and of 3mm diameter. The Ic of the wire was about 40A and was nearly constant along the longitudinal direction. For obtaining more high-Ic wires, we made 7-stranded cables with twist pitch of 20cm, as shown in Fig.9. The Ic of this cable was as high as 270A at 77 K 0 T, which strongly suggested that the stranding of multilayer composite wires has the possibility for electric power application.

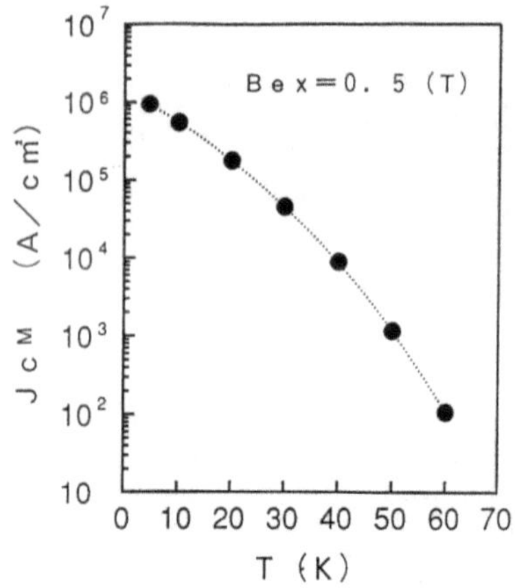

Fig.7 Temperature dependence of Jc^M at 0.5T for the multilayer wire.

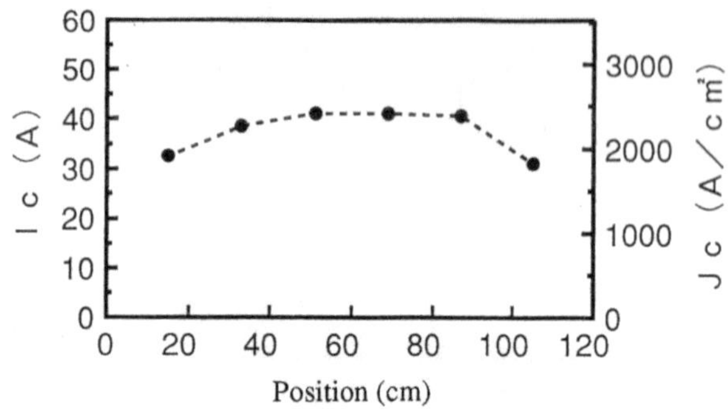

Fig.8 Variation of Ic at 77K and 0T along the longitudinal, direction of the multilayer wire of 100 cm length.

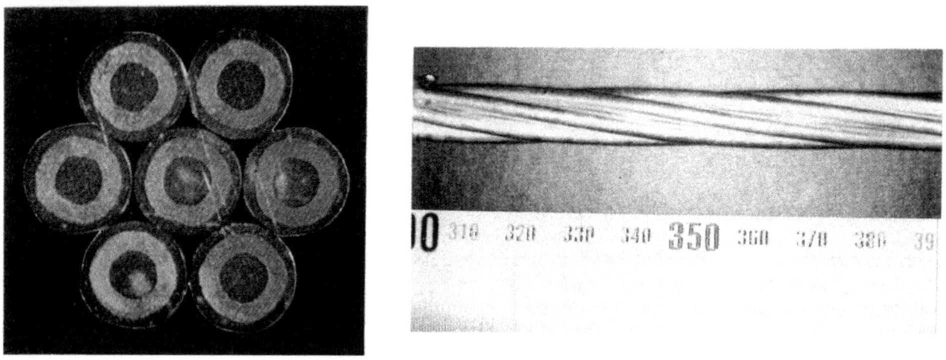

Fig.9 7-stranded cable with twist pitch of 20cm.

CONCLUSION

The high Ic composite wire with multilayer structure of Jelly-Roll type was fabricated by doctor - blade technique and it's superconducting properties were investigated. Ic of 60A and 30A were obtained for the multilayer wires of 3cm and 1000cm in length, respectively . And more highly transport current of 270A was obtained for the 7-stranded cable. In addition, the composite wire indicated no anisotoropy in magnetic field direction because of the c-axis orientation along the round silver sheets with multilayer structure. These results reveal that these round composite wires with multilayer structure have the possibility for electric power application.

ACKNOWLEDGEMENT

This work was performed as a part of "R&D on Superconducting Technology for Electric Power Apparatuses" under the Moonlight Project of Agency of Industrial Science and Technology(AIST), MITI, being consigned by New Energy and Industrial Technology Development Organization(NEDO).

Reference

[1] M. Mimura, H. Kumakura, K. Togano and H. Maeda: Appl. Phys. 54, 582 (1989).
[2] N. Enomoto, H. Kikuchi, M. Mimura, M. Nakajima, N. Uno, T. Hara, K. Okaniwa and T. Yamamoto: " Advances in Superconductivity II" ed. T. Isiguro & K. Kajima, Springer-Verlag, Tokyo, 371 (1991)
[3] N. Enomoto, N, Uno and K. Doi: Applied Superconductivity Conference. Sept. 24-28 (1990).
[4] K. Kumakura, K. Togano, H. Maeda and M. Mimura: J. Appl. Phys. 67. 1 (1990).

Preparation of Multi-Layered Conductors Using Bi-Based Superconducting Tapes

HIROYUKI KIKUCHI[1], NAOKI UNO[1], SATORU TANAKA[2], TSUKUSHI HARA[3], HIDEO ISHII[3], TAKAHIKO YAMAMOTO[3], and KAZUO MAEKAWA[3]

[1] Yokohama R&D Laboratories, The Furukawa Electric Co., Ltd., Yokohama, 220 Japan
[2] Chiba R&D Laboratories, The Furukawa Electric Co., Ltd., Ichihara, Chiba, 290 Japan
[3] Engineering Research Center, Tokyo Electric Power Co., Chofu, Tokyo, 182 Japan

ABSTRACT

In order to apply the high-Tc copper oxides for electric power transmission, we prepared multi-layered electric wires using Bi-Pb-Sr-Ca-Cu-O superconductors. Superconducting tapes prepared by powder-in-tube technique were wound spirally around a silver former. The transport critical current(Ic) at 77K and 0T were measured by four-prove method. Two kinds of criteria, 0.01 μV/cm and 10^{-10} ohm·cm, yielded practically the same value of Ic. Ic's of 60-174A were obtained for the present multi-layered conductors.

KEY WORDS: BPSCCO, multi-layered conductor, critical current, silver sheathed tape

INTRODUCTION

Since the discovery of superconductivity in Bi-Sr-Ca-Cu-O system, a multitude of studies for the new superconductor have been carried out. It is well known that the transport properties of silver-sheathed BiSrCaCuO superconducting wires are much improved by the improvement of grain coupling, grain alignment, and crystal orientation [1][2]. For conventional metal superconductors, their application for electric power transmission was widely studied [3,4]. For high-Tc copper oxides, some research groups [5-7] designed superconducting cables and evaluated AC and DC losses in Bi-based compounds.
The present paper describes the preparation of multi-layered superconducting wires suitable for the transmission lines and their transport properties.

EXPERIMENTAL

Powder-in-tube method was used to prepare silver-sheathed tapes composed of $(BiPb)_2Sr_2Ca_2Cu_3O$ (here after referred to as BPSCCO). Appropriate amounts of oxides and carbonates for the metallic constituents were mixed and calcined at 800-830 °C in air. The calcined powder was formed to highly dense cylindrical or rectangular rods by cold isostatic pressing (CIP) at 300 MPa. These rods were put into a silver tube with 25 mm outer and 15 mm inner diameters or a rectangular silver tube with 20-50x10 mm^2 outer and 15-40x5mm^2 inner sizes. These composites were cold-worked into tapes with 0.1-0.5 mm thickness and 3.5-57 mm width by swaging and rolling. 26 or 29 tapes were wound spirally in four layers around a silver former with 20-30 mm diameter and 0.3 -1 m length. Finally, silver tapes with 0.2 mm thickness were wound around the conductors to stabilize the composite structure. After sintered from 830°C to 840°C for 100-150 hours, the transport critical current(Ic) at 77K and 0T for the single silver-sheathed tapes and multi-layered fconductors were determined by a standard four-probe technique using two kinds of criteria, 0.01 μV/cm and 10^{-10} ohm·cm.

RESULTS AND DISCUSSION

Figure 1 shows the Ic valeus measured for every 1 m parts of single silver-sheathed BPSCCO tapes with 0.1 and 0.15 mm thickness and with 20 cm length. For the same tapes, Figure 2 gives the dependence of Ic on the distance between voltage terminals used for the Ic measurement. Relatively homogeneous distribution of Ic was observed for both tapes. Ic was clearly indicated that, Ic throughout a tape was restricted to that in a short part which was the poorest in Ic.

Figure 3 shows the dependence of the width of BPSCCO core for a silver-sheathed with 0.2 mm thickness, which indicates that Ic is nearly proportional to the width of BPSCCO core. Therefore, for the preparation of high-Tc multi-layered conductors, relatively wide silver-sheathed tapes, i.e., 12 mm width, were used. Their thickness were ranged between 0.2 and 0.36 mm. Figure 4 shows current-voltage curves at 77K and 0T for silver-sheathed tapes with 1 m in length. The Ic's for the tapes with thickness of 0.36 mm and 0.2 mm were 10 A and 3 A, respectively. Figure 5 and 6 show the schematic cross-sectional and general view of the multi-layered superconductor, respectively. The current-voltage curves for the multi-layered conductor formed using 26 tapes with 0.2 mm thickness are shown in Figure7. Total length of the conductor was 1 m, and four voltage terminals A, B, C and D were attached as shown in Figure 7. When terminals A and B were used, Ic was determined to be 80 A, which was in good agreement with the value estimated using Ic of each tape. However, there were lower part in Ic around the center of the conductor where low value of Ic, 60 A was observed.

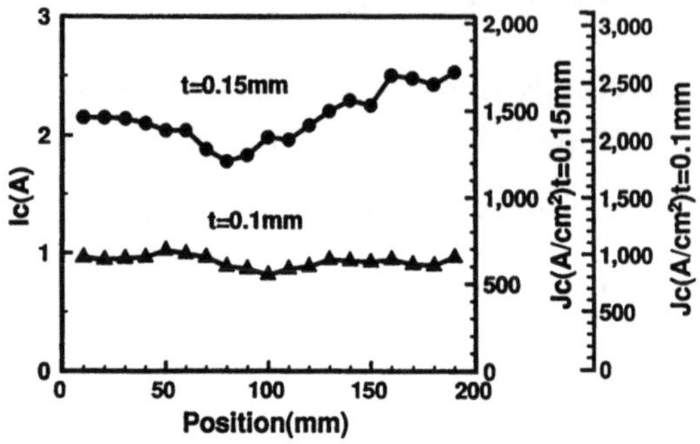

Fig.1 Ic distribution for silver-sheathed BPSCCO tape

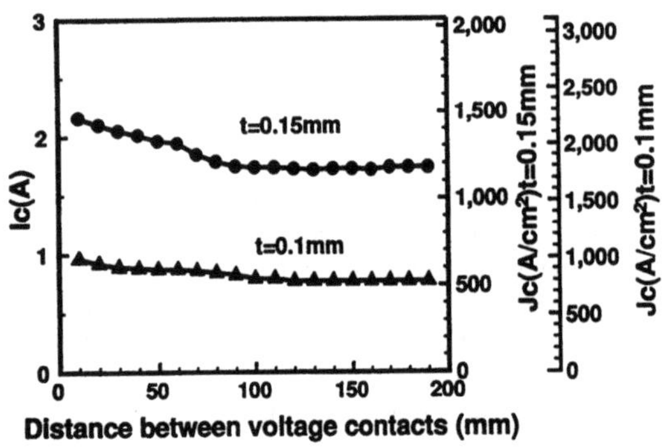

Fig.2 Relationship between Ic and the distance between voltage terminals

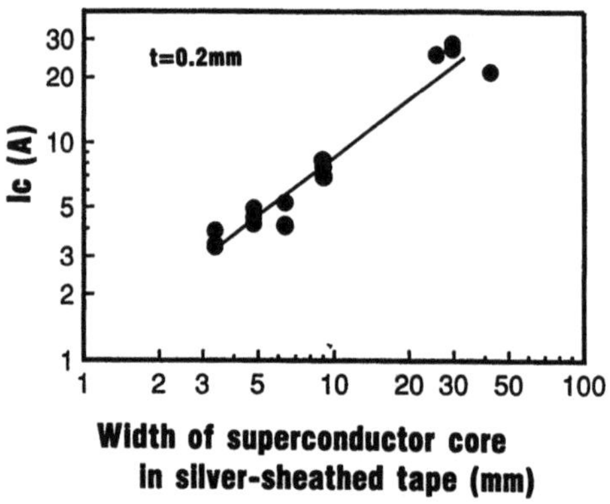

Fig.3 Ic as a function of the width of superconductor core in silver-sheathed tape

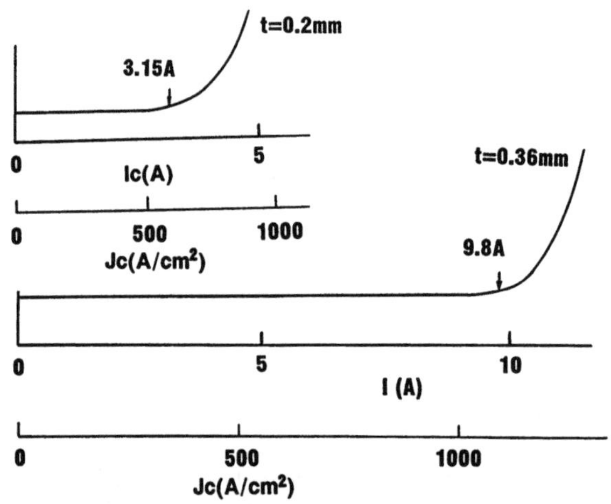

Fig.4 Current-voltage curves for tapes of 1 m in length The distance between voltage contacts is 90 cm

The degradation of Ic was probably due to mechanical damage induced by thermal stress during annealing. Table 1 lists Ic values for the multi-layered conductor measured using four different pairs of voltage terminals. When two kinds of criteria, 10^{-10} ohm·cm and 0.01μV/cm, were used for Ic determination, the two criteria yielded practically the same value of Ic.

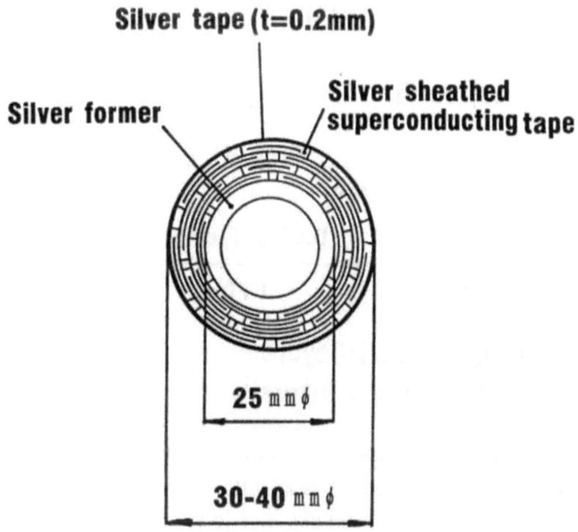

Fig.5 Crosssectional view of Bi-based multi-layered superconductor

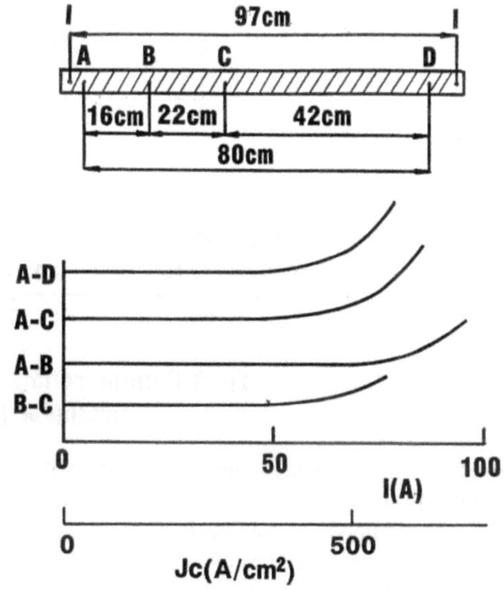

Fig.7 Current-voltage curves for multi-layered conductor (0.2mm^t, 26 tapes , 4 layers)

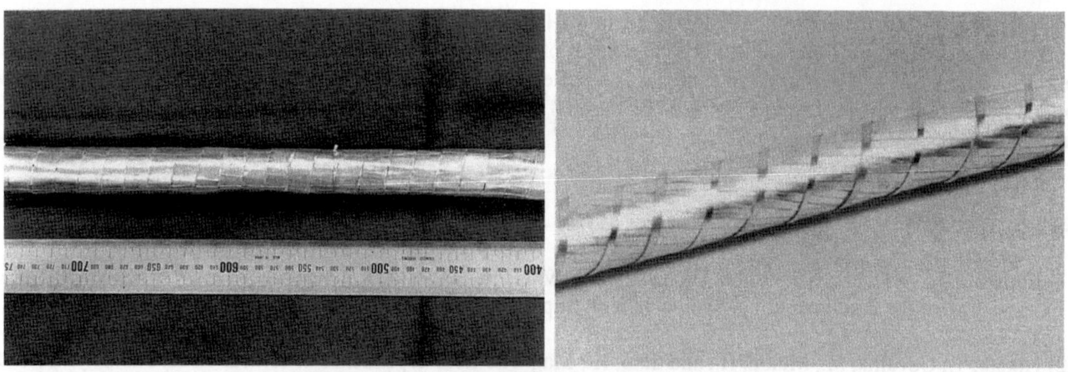

Fig.6 Multi-layered conductor using BPSCCO.

Table1 Ic determined by current-voltage curves shown in Fig.7

Distance between voltage contacts (cm)		Ic (A) Criterion 0.01 (μV/cm)	Ic (A) Criterion 10^{-10} (Ω·cm)	Jc(A/cm²) Criterion 0.01 (μV/cm
A-D	80	57	62	452
A-C	38	59	61	445
A-B	16	80	79	577
B-C	22	60	57	416

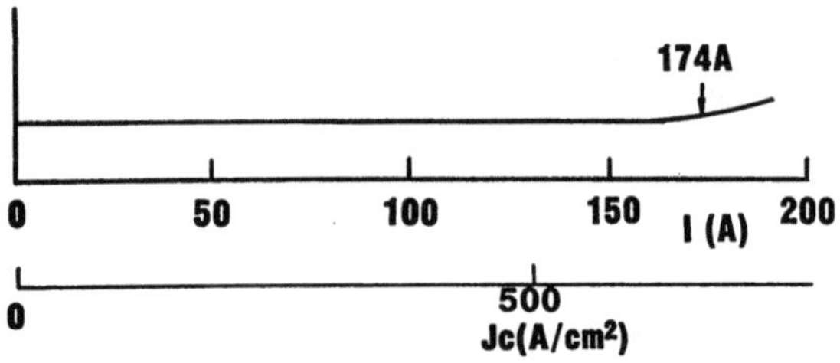

	Ic(A)		Jc(A/cm²)
	Criterion		
Distance between voltage contacts (cm)	0.01 (μV/cm)	10^{-10} (Ω·cm)	0.01 (μV/cm)
20	174	176	678

Fig.8 Current-voltage curve and Ic for multi-layered conductor (0.36mm^t, 29 tapes , 4 layers)

Figure 8 shows the current-voltage curve and value of Ic for another multi-layered conductor formed using the 0.36 mm tapes. In this case, 29 tapes were wound and total length of the conductor was about 30 cm. High value of Ic, 174 A, was obtained. As previously mentioned, a single tape with 0.36 mm thickness had Ic of 10 A, therefore total Ic of the multi-layered conductor was expected to be 290 A. Thus, it was suggested that this conductor also mechanically damage by thermal stress or bending strain.

CONCLUSIONS

In order to apply for power transmission lines, multi-layered conductors were prepared using Bi-based higth-Tc superconducting tapes. The transport critical current of 60-174 A were obtained in the conductors of 0.3 - 1 m in length. The value of Ic around the center was smaller than those at the both sides of the conductor. This was probably due to mechanical damage induced by thermal stress. The Ic's determined using two kinds of criteria, 0.01 μV/cm and 10^{-10} ohm·cm, were practically the same.

REFERENCES

[1] Enomoto, N., Kikuchi, H., Mimura, M., Nakajima, M., Uno, N., Hara, T., Okaniwa, K and Yamamoto, T., Proceeding of the 2nd Internatinal Symposium on Superconductivity (ISS,89) November 14-17, 1989, Tsukuba PP.371-374

[2] Hikata, T., Ueyama, M., Mukai, H., and Sato, K., Magnetic field dependence of critical current density in Bi-Pb-Sr-Ca-Cu-O silver sheathed wire, Cryogenic 1990 Vol 30 November PP.924-929

[3] Cohen, M H.,"Superconductivity in Science and Technology" (University of Chicago Press, Chicago, 1968), PP.77-101

[4] Gregory, W D., Mathews Jr.,W N. and Edelsack E A. "The Science and Technology of Superconductivity" Volume 2, Plenum Press New York-London, 1973, PP.433-458

[5] Hara, T., Okaniwa, K., Tanaka, S. and Ichiyanagi, N. (1990) 1990 IEEE Winter Meeting 075-2PWRD

[6] Hara, T. and Nakamura, M. (1990) IEE of Japan 110-B 58-66

[7] Nakade, M., Hara, T., Yamamoto, T., Hoshino, H., Tanaka, S., Ichiyanagi, N. and Iwata, Z. 3rd Intenational Symposium on Superconductivity (ISS'91) November 6-9, 1990, Sendai, Japan

Preparation and Properties of Ag-Sheathed $Bi_2Sr_2CaCu_2O_{8+\delta}$ Superconducting Tapes

JUN-ICHI KAI, KEISUKE YAMAMOTO, and MAKOTO HIRAOKA

Central Research Laboratory, Mitsubishi Cable Industries, Ltd., Amagasaki, Hyogo, 660 Japan

ABSTRACT

Ag-sheathed $Bi_2Sr_2CaCu_2O_{8+\delta}$ (2212) superconducting tapes have been developed by growing the 2212 phase through a partially melted state. In this paper, the $Bi_2Sr_2Ca_{1-X}Cu_{2-X}O_{8+\delta}$ (X=0,0.18 and 0.36) powder were investigated. With these powders, superconducting tapes about 0.18mm in thickness were prepared by powder-in-tube method and heat-treated under the maximum heating temperature of 890°C to achieve partial melt. The samples were then kept at 860°C for 40 hours. When X was 0.36, the highest critical current density Jc at 4.2K were 1.67×10^5A/cm^2 and 1.06×10^5A/cm^2 at 0T and 3T, respectively. Influence of powder compositions on electromagnetic properties such as Jc and A.C. susceptibility in various magnetic fields will be also discussed.

KEY WORDS: Bi-Sr-Ca-Cu-O, Ag-sheathed superconducting tape, partially melted state, Jc, A.C. susceptibility

INTRODUCTION

High critical current density Jc in Ag-clad wires and thick films on a Ag tape using the $Bi_2Cr_2CaCu_2O_{8+\delta}$(2212) have been reported [1],[2],[3]. These were prepared by growing the 2212 phase through a partially melted state. This method rendered higher Jc values due to the highly oriented 2212 phase, namely c-axis perpendicular to the tape plane. However, when the stoichiometric composition 2212 was used, coarse grains of Sr-Ca-Cu-O non-superconducting phase were found. These grains disturb the orientation of 2212 phase and lower the critical current. This paper reported that coarsening of the Sr-Ca-Cu-O grains was suppressed by optimizing the starting material composition.

EXPERIMENTAL

The atomic composition of $Bi_2Sr_2Ca_{1-X}Cu_{2-X}O_{8+\delta}$ (X=0, 0.18 and 0.36) was investigated. Bi_2O_3, $SrCO_3$, $CaCO_3$ and CuO powders were mixed and calcined at 820°C for 50 hours in air and N_2. The calcined powders were packed into silicone-rubber molds and pressed by CIP (Cold Isostatic Pressing) equipment. By the pressing, green rods 4.9mm in diameter and 100mm in length were obtained. The rods were then inserted into silver tubes 7mm in outer diameter, 5mm in inner diameter and 330mm in length. Both ends of the tubes were filled by Pb-Sn solder. The composite tubes were cold-drawn and then cold-rolled, and tapes about 10m in length, 0.18mm in thickness and 2.8mm in width were obtained. The superconducting core thickness of these tapes was 0.1mm. These tapes were cut in 40mm-long and heat-treated under the maximum heating temperature of 890°C to achieve partial melt. The specimens were then slow-cooled to 860°C at a rate of 2.5°C per hour and kept at 860°C for 40 hours.

The transport critical current Ic at 4.2K in external magnetic D.C. fields H_{dc} varying 0T to 3T was measured by D.C. four-probe method and determined with 1 μV/cm criterion. Jc was calculated on the cross-sectional area of the tapes excluding silver sheath. Microstructure of the specimens were observed by scanning electron microscope (SEM). The composition of the phases was analyzed with energy dispersive X-ray analysis (EDX). Identification of the superconducting phase was carried out using the X-ray diffractometer (XRD). A.C.susceptibility was measured at four different values of applied A.C. fields H_{ac} amplitude between 0.1 Oe and 12.5 Oe at a frequency of 313Hz.

RESULTS AND DISCUSSION

Fig.1 shows composition map indicating the atomic ratio X for Ca and Cu, and Fig. 2 shows magnetic field dependence of the Jc at 4.2K for the specimens using different powders. D.C. magnetic fields H_{dc} up to 3T were applied parallel to the tape surface and current flow. The Jc of these tapes increase drastically with increasing X value, or decreasing Ca_2 and Cu from stoichiometric starting composition 2212. The maximum Jc were $1.67\times10^5 A/cm^2$(Ic=237A) and $1.06\times10^5 A/cm^2$(Ic=157A) at 0T and 3T, respectively, when X is 0.36.

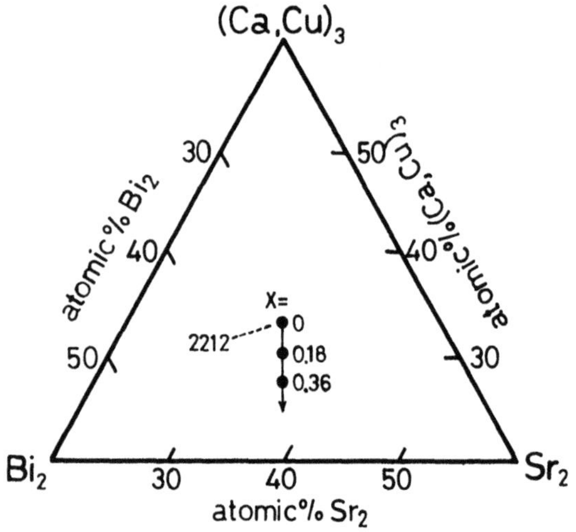

Fig. 1 Composition map indicating the atomic ratio X for Ca and Cu.

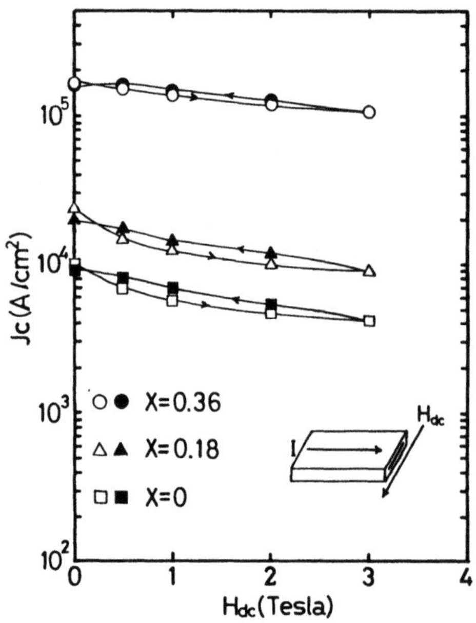

Fig. 2 Magnetic field dependence of Jc at 4.2K for magnetic field parallel to the tape plane.

Fig. 3 shows the back-scattered electron image (BEI) photographs of the polished surface on longitudinal cross-section of the tapes by SEM. When X is 0, or the starting composition is the stoichiometric 2212, coarse grains of the Sr-Ca-Cu-O non-superconducting phase analyzed with EDX are found. These grains disturb the orientation of 2212 phase and lower the critical current due to cuts in the current pass. On the other hand, such coarse Sr-Ca-Cu-O grains are not observed when X were both 0.18 and 0.36. It is indicated that adjusting the starting material composition is very effective for suppressing the formation of Sr-Ca-Cu-O grains. Fig. 4 shows the XRD patterns of surface of tapes, prepared by the partial polishing of silver sheath of the tape plane. In all the tapes, the (00l) peaks of 2212 phase are dominant, indicating that the 2212 phases are oriented with the (00l) plane perpendicular to the tape plane. The (00l) peaks gradually become sharper with increasing X value or optimizing the stating composition. As the results, coarsening of Sr-Ca-Cu-O grains are suppressed and the c-axis of 2212 phases are oriented perpendicular to the tape plane by optimizing the starting composition.

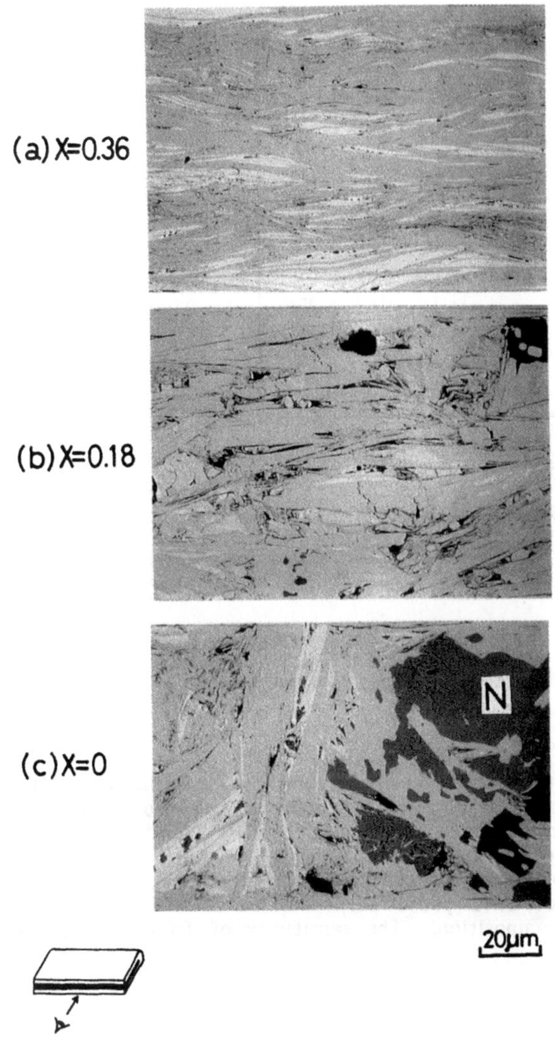

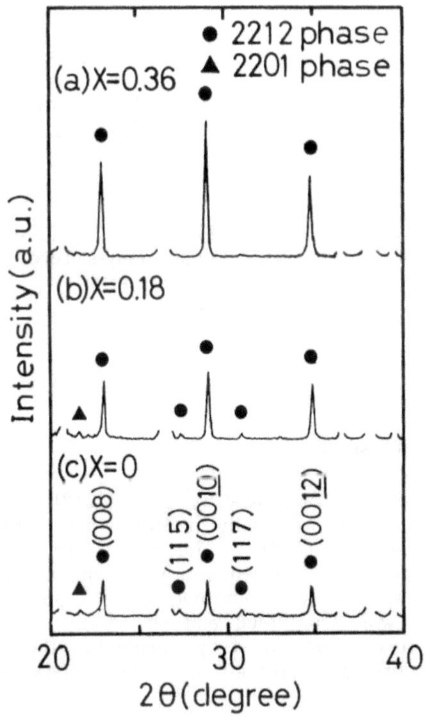

Fig. 3 BEI micrographs of the polished surface on the longitudinal cross-section of the tapes by SEM, where mark N denotes Sr-Ca-Cu-O non-superconductig grain.

Fig. 4 XRD patterns for the tape plane, prepared by the partial polishing of silver sheath of the tape plane.

Temperature dependence of A.C. susceptibility was measured in various applied A.C. fields H_{ac} to investigate weak-links between superconducting grains. A.C.susceptibility χ is sensitive to change in H_{ac}. This phenomenon is due to weak-links between grains [4],[5],[6]. For a given H_{ac} the imaginary component of the observed susceptibility can be maximized at Tp, the peak temperature. H_{ac} dependence of the peak temperature Tp in χ''-T relation for the specimens prepared by various conditions are shown in Fig. 5. It is found that Tp shifts to lower temperatures with increasing H_{ac}, but applied field dependence of Tp decreased with increasing X value. This behavior suggests that weak-links are improved by optimizing the starting composition.

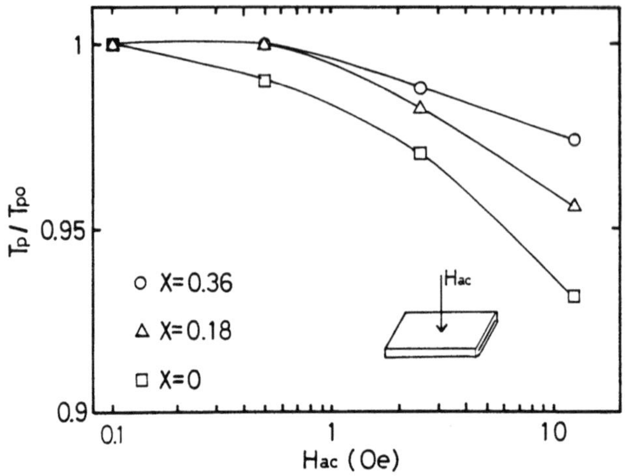

Fig. 5 The A.C. applied field H_{ac} dependence of the peak temperature Tp normalized by Tp_0 at 0.1Oe in χ"-T relation.

CONCLUSION

Ag-sheathed $Bi_2Sr_2CaCu_2O_{8+\delta}$ (2212) superconducting tapes were prepared by growing the 2212 phase through partially melted state. By optimizing starting composition of the tapes, Jc was increased drastically. The highest Jc at 4.2K were $1.67\times10^5 A/cm^2$ and $1.06\times10^5 A/cm^2$ at 0T and 3T, respectively when starting composition was the $Bi_2Sr_2Ca_{1-X}Cu_{2-X}O_{8+\delta}$ (X=0.36). The reason why high Jc was obtained on X=0.36 is that coarsening of the Sr-Ca-Cu-O non-superconducting phase was suppressed and the c-axis orientation of 2212 phase perpendicular to the tape plane was further promoted by optimizing the starting composition. The sensitivity of Tp to change in H_{ac} correlates strongly with the increasing Jc.

ACKNOWLEDGMENT

The authors thank Dr.M.Nunogaki and A.Iwamoto at Institute of Scientific and Industrial Research of OSAKA University for their help in measurements of critical current in D.C. magnetic fields.

REFERENCES

[1]K.Heine, J.Tenbrink and M.Thoner, Appl.Phys.Lett.55,2441(1989)
[2]J.Tenbrink, K.Heine and H.Krauth, Cryogenics 30,423(1990)
[3]J.Kase, K.Togano, H.Kumakura, D.R.Dietderich, N.Irisawa, T.Morimoto and H.Maeda, J.Appl.Phys.29,L1096(1990)
[4]K.Yamamoto, J.Kai and M.Hiraoka, Advances in Superconductivity 2, Proceedings of the 3rd International Symposium on Superconductivity (ISS'90),659(1991)
[5]S.Noguchi, J.Inoue, K.Okuda, Y.Maeno and T.Fujita, J.J.Appl,Phys.27,L390(1988)
[6]H.Mazaki, M.Takano and R.Kanno, J.J.Appl.Phys.26,L780(1987)

Processing and Properties of Superconductor Wires

J.P. SINGH, S.E. DORRIS, M.T. LANAGAN, C.-T. WU, K.C. GORETTA, J. JOO, U. BALACHANDRAN, and R.B. POEPPEL

Materials and Components Technology Division, Argonne National Laboratory, Argonne, IL 60439, USA

ABSTRACT

Numerous variables are encountered in the fabrication of superconductor wires by extrusion and by powder-in-tube processing. The decomposition of plastically extruded YBCO is minimized by firing at reduced total pressure. Sintering the extruded YBCO wires at an oxygen partial pressure of 7.6 mm Hg results in densification and a fine grained microstructure. A reduction in average grain size from 23 to 4 μm increased strength from 83 to 191 MPa. Addition of 15 vol.% Ag particles causes a further increase in strength to 222 MPa due to compressive stresses in the YBCO matrix from the Ag. In powder-in-tube processing, the primary considerations are the mechanical-working and heating schedules. The best properties in Ag-clad Bi-based superconductors are obtained with many small reductions during working and by the use of liquid-phase formation during sintering to promote grain growth.

KEY WORDS: YBCO/BSCCO superconductors, extrusion, powder-in-tube processing, oxygen partial pressure, strength

INTRODUCTION

The high critical temperatures of ceramic superconductors allow the use of relatively simpler cryogenic systems and comparatively inexpensive liquid nitrogen, suggesting the possibility that ceramic superconductors might one day be widely useful and provide large-scale energy conservation. Large-scale applications will require bulk fabrication by which long lengths of continuous superconductor can be produced rapidly. To this end, considerable effort has been devoted to the fabrication of superconducing wires by extrusion [1]. Important considerations in fabricating wire by plastic extrusion include control of atmosphere during removal of organic constituents and optimization of mechanical properties by control of microstructure and by addition of second phases. Long lengths of wire with high strength have been fabricated, but their capacity to carry large currents has been limited [1]. Therefore, the alternative route of wire fabrication by powder-in-tube processing has been also investigated [2] for achieving high critical current densities. In powder-in-tube processing, the primary consideration is the mechanical-working and heat-treatment schedule. In this paper, we will describe the important parameters in extruding $YBa_2Cu_3O_x$ (YBCO) wire and processing BiSrCaCuO (BSCCO) wire by the powder-in-tube technique. The BSCCO powders were used with and without lead (Pb) doping. The effects of these processing parameters on microstructures and the resulting properties will be discussed.

EXPERIMENT

The YBCO powder is made by solid-state reaction of the constituent oxides. Appropriate proportions of Y_2O_3, $BaCO_3$, and CuO are mixed and milled in methanol for ≈12 h. After drying, the mixtures are calcined in flowing oxygen at a reduced total pressure of ≈ 2 mm Hg. This method [3] leads to decomposition of $BaCO_3$ at lower temperatures, avoids the formation of impurity phases, and, in the case of YBCO, produces phase-pure material after a single calcination at 800°C [3]. The $Bi_{1.8}Pb_{0.4}Sr_2Ca_{2.2}Cu_3O_x$ (2223 BSCCO) powder was made by solid state reaction of the constituent oxides. Appropriate amounts of Bi_2O_3, PbO, $SrCO_3$, $CaCO_3$, and CuO are mixed and milled for ≈12 h in isopropyl alcohol with alumina media. The milled slurry is dried and then calcined in two steps. First, the powder mixture is slowly heated to ≈700°C for 6 h in flowing oxygen at a reduced total pressure of ≈5 mm Hg. Subsequently, the atmosphere is changed to ambient air and the temperature is raised to 860°C, where it is held for 100 h. The $Bi_2Sr_{1.7}Ca_1Cu_2O_x$ (2212 BSCCO) powder was made by a similar process, the details of which can be found in Ref. 4.

To extrude wires, the YBCO powder is mixed with appropriate solvent, dispersant, binder, and plasticizer to make a slip that is easily formed into various shapes while retaining satisfactory strength in the green state, i.e., before firing [1]. The solvent provides the fluid medium for the powder and the other additives. Because water is known to attack the YBCO superconductor by preferentially dissolving Ba, organic solvents such as methyl ethyl ketone and xylene are used. In many cases, after the superconductor slip is prepared, silver flakes (mean particle size ≈15 μm) are added to make a YBCO/Ag composite (85 vol.% YBCO/15 vol.% silver). This leads to improved mechanical properties in the superconductor [5], as will be described later. ZrO_2 milling media are then added, and the formulation is vibratory-milled for ≈16 h. Milling reduces the particle size of the YBCO powder, making it more sinterable and more likely to produce a high-density superconductor.

After milling this formulation, some of the solvent is allowed to evaporate. This yields a plastic mass that is forced through a die at high pressure to produce a YBCO/Ag superconducting wire. Superconductor wires have been extruded at Argonne National Laboratory (ANL) with diameters between 0.1 and 3.0 mm and lengths up to 20 m. In the green state, the extruded wires exhibit great flexibility and can be easily formed into coils. Details of the firing schedules for these wires can be found in Refs. 1 and 6.

The heat treatment of wire includes several steps. First, the wire is fired in flowing oxygen (≈1160 mL/min) at a reduced total pressure (≈7.6 mm Hg) to facilitate removal of gaseous species formed during decomposition of organic additives and prevent their reaction with the superconductor. The temperature is slowly increased to ≈260°C at a rate of 0.1°C/min. The wire specimen is held at this temperature for 10 h. After the organics are removed at 260°C, the temperature is raised to the sintering level at 1°C/min. The specimens are sintered at different temperatures and oxygen partial pressures and for different lengths of time to obtain various microstructures. After sintering, the specimens are annealed at 450°C for 12 h in flowing oxygen atmosphere at a total pressure of 760 mm Hg.

For powder-in-tube processing, superconducting powder (BSCCO 2223 or 2212) is loaded in a silver tube (6.35-mm outer diameter, 4.35-mm inner diameter) and the assemblies are compacted by swaging, drawing, or rolling. The Ag sheath provides a barrier that protects the superconductor from damage and environmental degradation. The reactivity of high-T_c superconductors with moisture is well documented [7].

In some of the experiments, each BSCCO/Ag tube was swaged into wires 1.1–1.3 mm in diameter that were used to fabricate the multifilament wires. The swaged 2212 and 2223 wires, which each contained a single oxide filament, were sintered at 840°C in air for 50 h. After this treatment which is designed to induce grain growth, the wires were subjected to a combination of flat rolling and sintering. In the first rolling step, seven pieces of each single-filament wire were cut, packed into a 6.35-mm-diameter Ag tube, swaged to 5.15 mm, and rolled into tape of 1.44 mm total thickness. A second sintering, identical to the first, was then performed. An additional rolling operation decreased each tape thickness to 0.72 mm, and final sintering was done at 840°C for 30 h.

For the monofilaments of 2212/Ag, wires were cold-rolled into tapes. The flattened tapes were heated in air at 890°C for 0.5 h and then cooled to 870°C and held for 10 h. Final rolling produced a tape 0.12 mm thick. The final heat treatment in air was conducted at ≈840°C for 100 h.

The phase composition of the powders was confirmed by X-ray diffraction and differential thermal analysis (DTA). Critical temperatures and current densities were measured by four-probe resistivity. Critical current density (J_c) was measured at 4.2 and 77 K; a 1 μV/cm criterion was used. Strength of the YBCO wires was measured in a three-point bending mode with a support span of 1.825 cm and a crosshead speed of 0.127 cm/min.

RESULTS AND DISCUSSION

Green-state wires produced by extrusion contain ≈10 wt.% organics, which must be completely removed without damaging the superconductor. Incomplete removal of organics can cause decomposition of the superconductor or leave carbon-rich material at the grain boundaries, either of which will degrade superconducting properties. Organics can be removed easily by thermal decomposition in the temperature range of 240–350°C, but if the decomposition proceeds too rapidly, the wires can bloat severely and in some cases even explode. In the firing of any ceramic, it is important to control the rate at which organics are removed, but it is especially important with YBCO, which decomposes in the presence of CO_2 and H_2O [8], two products that are formed during decomposition of organics.

When extruded YBCO is fired at reduced total pressure, the concentrations of CO_2, CO, and H_2O are minimized and decomposition of YBCO is greatly minimized or completely avoided. But when firing is at ambient pressure, the harmful gaseous products accumulate and lead to gross decomposition of YBCO. This was demonstrated in the following experiment: YBCO powder was mixed with the organics that are used in extrusion, then the mixture was fired in flowing oxygen at either ambient or reduced pressure (≈2 mm Hg). Two different powders were used: some produced by solid-state reaction at reduced pressure, and some produced by a liquid-mix technique. Figures 1 and 2 show the schematic X-ray patterns of samples taken from the mixtures at 240, 300, and 350°C. Figure 1 shows that when the YBCO/organic mixtures were fired at ambient pressure, both YBCO samples decomposed (the liquid-mix powder at 240°C and the solid-state powder at 300°C). On the other hand, Fig. 2 shows that YBCO did not decompose when fired at reduced total pressure.

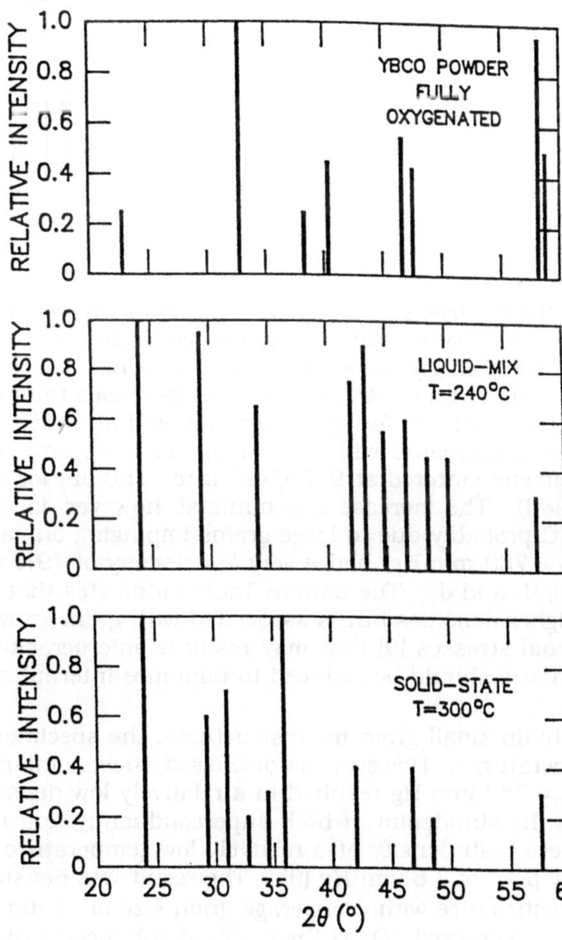

Fig. 1. Schematic X-ray patterns of fully oxygenated YBCO powder and two other YBCO powders fired at ambient pressure in contact with organics used in extrusion. Both powders fired with organics decomposed during firing.

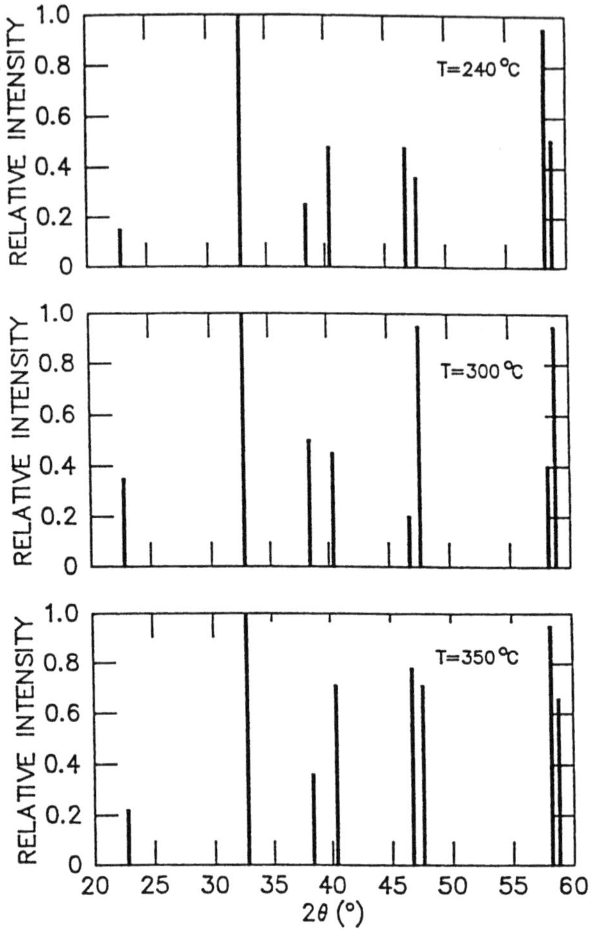

Fig. 2. Schematic X-ray patterns of YBCO powder as a function of temperature. The powder was fired at reduced total pressure in contact with organics used in extrusion and showed no sign of decomposition.

The YBCO wires sintered at various temperatures in flowing oxygen at an oxygen partial pressure (pO_2) of 760 mm Hg ranged in density from 79 to 98% theoretical (Table I). As expected, density increases with increasing sintering temperature and time. Table I shows that average grain size, i.e., length, increases from ≈4 μm for specimens sintered at 910°C for 10 h to 23 μm for specimens sintered at 950°C for 20 h. In addition, maximum grain size also increased with increasing sintering time and temperature, growing from ≈5 μm for specimens sintered at 910°C/10 h to ≈150 μm for specimens sintered at 950°C/20 h (Table I). The increase was minimal, however, for specimens sintered at temperature above 935°C probably due to large grains impinging on each other. Specimens sintered at 910°C (pO_2 = 760 mm Hg) had a very low density of 79% theoretical, and consequently had low strength and J_c. The data in Table I indicates that sintering at higher temperatures results in higher densities but is associated with grain growth. Large grains can cause undesireable internal stresses [9] that may result in microcracking and reduced strength. Therefore, the grain size should be reduced to minimize internal stresses and increase strength.

To obtain small grain microstructures, the specimens must be sintered at relatively low temperatures. However, as discussed above, sintering at lower temperatures (≈910°C) and pO_2 = 760 mm Hg resulted in a relatively low density of 79% theoretical: this is undesirable from the standpoint of both superconducting and mechanical properties. Therefore, to achieve high density at a relatvely low temperature of 910°C, the specimens were sintered at a low pO_2 of 7.6 mm Hg [9]. The result was density of 91% theoretical and a fine microstructure with an average grain size of ≈4 μm. We believe that the increase in density with a decreased pO_2 is the result of enhanced sintering kinetics probably caused by both an increase in defect concentration and a decrease in activation energy of the rate-controlling species [10]. It can be seen in Table I that strength shows an inverse dependence on grain size in specimens with densities of more than over 90% theoretical. Strength increased from 83 to 191 MPa when average grain size decreased from 23 to 4 μm.

Table I. Density, Grain Size, and Strength of YBCO Wires Sintered at Various pO_2 with Different Heat Treatment.

Sintering Temperature and Time (°C/h)	pO_2 (mm Hg)	Density (% theoretical)	Average Grain Size (μm)	Maximum Grain Size (μm)	Strength (MPa)
910/10	760	79%	≈4.0	5	120 ± 10
925/10	760	90%	6.0	20	141 ± 9
935/12	760	96%	16.0	171	91 ± 7
935/20	760	96%	16.5	182	95 ± 6
950/20	760	98%	23.0	150	83 ± 5
910/10	7.6	91%	≈4.0	4	191 ± 7

Some of the YBCO wires with 15 vol.% Ag addition were also sintered at 890°C for 10 h at a pO_2 of 7.6 mm Hg to obtain a density of 91% and an average grain size of 4 μm. These specimens had a strength value of 222 MPa. For the same density and grain size, this represents an increase in strength of YBCO specimens from 191 to 222 MPa as a result of 15 vol.% Ag addition. It is believed that the Ag particles induce compressive stress in the YBCO matrix [11] and pin the propagating crack, resulting in strength improvement. The J_c of YBCO/Ag wires was typically 400–600 A/cm^2 and ranged as high as 2800 A/cm^2. In every case, J_c drops dramatically with applied magnetic field (field of 50 G results in a ≈50% decrease in the zero-field J_c), indicating that the YBCO superconductors are controlled by weak links.

An alternative approach that produces superior superconducting properties is powder-in-tube processing of the BSCCO superconductors. For multifilament 2212/Ag and 2223/Ag wires, earlier efforts [2] focused on swaging and rolling, with use of intermediate heat treatments to promote grain growth. Mulifilament wires with transport J_c values at 4.2 K of >≈3.5 x 10^4 A/cm^2 were produced; J_c values at 77 K were only ≈10^3 A/cm^2.

The present efforts in 2212/Ag wires have incorporated many changes in the processing sequence. First, drawing is now used rather than swaging; reduction per pass is generally 10–15%. Second, only two heat treatments are employed. Third, after the first rolling, the heat treatment induces significant partial melting. The first heat treatment yields relatively large grains that exhibit significant texture. The final rolling and heat treatment improve the texture substantially (Fig. 3). Several intermediate heat-treatment schedules have been attempted, but only those that induced appreciable melting have produced highly aligned microstructures and good superconducting properties.

The J_c values of these newer tapes have been measured only at 4.2 K in zero applied magnetic field, and only short sections have been tested. Measurements in field and at higher temperatures are planned. The transport J_c was 7.3 x 10^3 A/cm^2 after the first rolling and heat treatment and 1.0 x 10^5 A/cm^2 after the final rolling and heat treatment. The critical current of the final tape was 83 A. It is clear that the increase in J_c is related to the improved texture of the final product.

The J_c achieved is probably sufficient for low-temperature applications such as high-field magnets. Toward that end, coils are now being wound from the 2212/Ag tapes. A concern for these coils is mechanical reliability: 2212/Ag tapes have to date been susceptible to mechanical damage and thermal shock. For example, when the tape that carried 83 A was warmed to room temperature and the current leads were disconnected and reconnected, the critical current at 4.2 K was reduced to 45 A. Efforts to improve the reliability of these tapes are underway. Alloying of the core appears to offer the most promise. In addition, the possible problems associated with the BSCCO/Ag interface are under study.

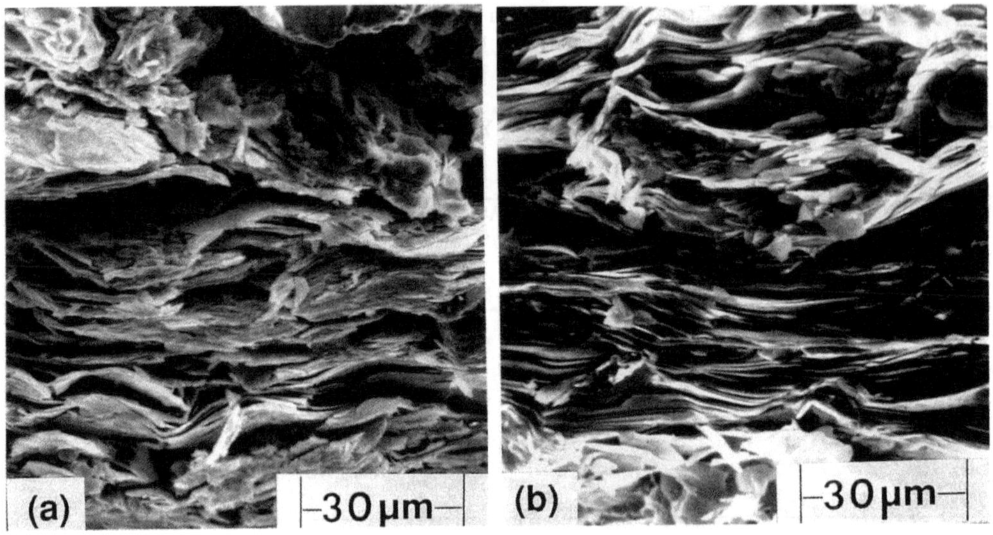

Fig. 3. Scanning electron micrographs of transverse cross sections of 2212 BSCCO/Ag tapes (a) after first rolling and heat treatment at 890°C and (b) after a second rolling and heat treatment at 837°C for 100 h.

SUMMARY

Continuous YBCO wires were produced by plastic extrusion. The strength and reliability of these wires have been improved substantially by control of the heat-treatment atmosphere and by addition of metallic Ag. BSCCO wires were produced in Ag tubes. While J_c values at 4.2 K have reached 10^5 A/cm^2, the mechanical properties of these wires have yet to be optimized.

ACKNOWLEDGMENTS

Work supported by the U.S. Department of Energy (DOE), Office of Utility Technologies, Conservation and Renewable Energy, as a part of a DOE program to develop electric power technology (JPS, SED, MTL, KCG, UB, RBP), under Contract W-31-109-Eng-38; and by the National Science Foundation (DMR 8809854) through the Science and Technology Center for Superconductivity (CTW, JJ).

REFERENCES

1. Dorris S, Dusek JT, Lanagan MT, Picciolo JJ, Singh JP, Creech JE, Poeppel RB (1991) Ceram. Bull. 70: 722
2. Wu CT, Goretta KC, Shi D, Lanagan MT, Poeppel RB (1991) In: Whang SH, DasGupta A, Laibowitz RB (eds) High Temperature Superconducting Compounds III. The Metallurgical Society, Warrendale, PA, USA (in press)
3. Balachandran U, Poeppel RB, Emerson JE, Johnson SA, Lanagan MT, Youngdahl CA, Shi D, Goretta KC, Eror NG (1989) Mater. Lett. 8: 454
4. Karbarz FA, Lacy OD, Goretta KC, Balachandran U, Shi D, Chen JG, Xu M, Hash MC (1990) Mater. Res. Bull. 25: 759
5. Singh JP, Leu HJ, Poeppel RB, Van Voorhees E, Goudey GT, Winsley K, Shi D (1989) J. Appl. Phys. 66: 3154
6. Lanagan MT, Poeppel RB, Singh JP, Dos Santos DI, Degener JK, Dusek JT, Goretta KC (1989) J. Less-Common Met. 149: 305
7. Bansal NP, Sandkuhl AL (1988) Appl. Phys. Lett. 52: 323
8. Gao Y, Merkle KL, Zhang C, Balachandran U, Poeppel RB (1990) J. Mater. Res. 5: 1363
9. Singh JP, Guttschow RA, Kupperman DS, Dusek JT, Poeppel RB (1991) In: Proc. 3rd Symp. on High-Temperature Superconductors: Processing and Microstructure Property Relationships. 1991 TMS Meeting and Exposition, New Orleans, LA, USA, Feb.17-21, 1991 (in press)
10. Chen N, Shi D, Goretta KC (1989) J. Appl. Phys. 66: 2485
11. Kupperman DS, Singh JP, Faber Jr. J, Hitterman RL (1989) J. Appl. Phys. 66: 3396

Effect of Internal Stress in BSCCO/Silver Plates on Critical Current Density

Hitoshi Sakai, Yoshimi Ohzawa, and Hitoshi Yoshida

Corporate Research and Development Group, NGK Insulators, Ltd., Nagoya, 467 Japan

ABSTRACT

The effect of internal stress in BSCCO-Ag-BSCCO plate on Jc was studied. The thermal cycle tests running around room temperature and liquid nitrogen temperature were carried out for BSCCO-Ag-BSCCO samples with various internal stress values. The changes in Jc of BSCCO was measured by four-probe method. The Jc value decreased with increasing the number of thermal cycles, then saturated around 10 cycles. The saturated Jc value was larger for the sample with lower internal stress. It is suggested that the decrease of Jc be due to the microcracks formed during the relaxation of the stress.

KEY WORDS: $Bi_2Sr_2CaCu_2O_y$, internal stress, silver substrate, thermal cycle

INTRODUCTION

Silver is one of the most feasible substrates for the high Tc oxide superconductor thick films or wires. However, the internal stress occurred in the superconductor on the silver substrate is large because of the mismatch of the thermal expansion coefficients and it may cause degradation of the superconducting properties.

This paper presents the effect of internal stress in $Bi_2Sr_2CaCu_2O_y$(BSCCO)-Ag-BSCCO plates on the critical current density(Jc) to clarify the degradation behavior of BSCCO by thermal cycle.

EXPERIMENTAL

Sample Preparation

BSCCO superconducting thick films were formed on both sizes of the silver substrate as drawn schematically in Fig. 1. The thickness of BSCCO was kept constant (0.3mm). The internal stress induced in BSCCO was adjusted by the thickness of the silver substrate. The value of the internal stress in BSCCO was calculated by FEM analysis using the data of the mechanical and the thermal properties of silver and BSCCO such as the thermal expansion coefficient and the stress-strain curve as a function of temperature. Assuming that the BSCCO bonds to silver substrate at 800°C, and is cooled to liquid nitrogen temperature, the internal stress (compressive) in BSCCO increased almost linearly with increasing the thickness of the silver substrate as shown in Fig. 2.

Thermal Cycle Tests

Three cooling-heating cycle rates were applied. (1)Rapid cycle: the sample was dropped into a liquid nitrogen container and taken out to room temperature. (2)Moderate cycle:

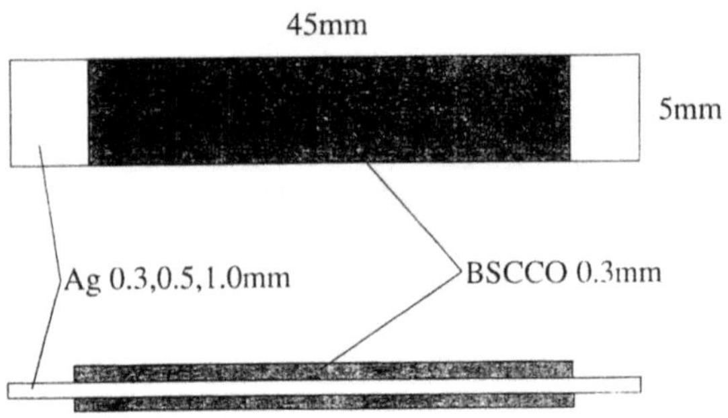

Fig. 1 Schematic of the sample

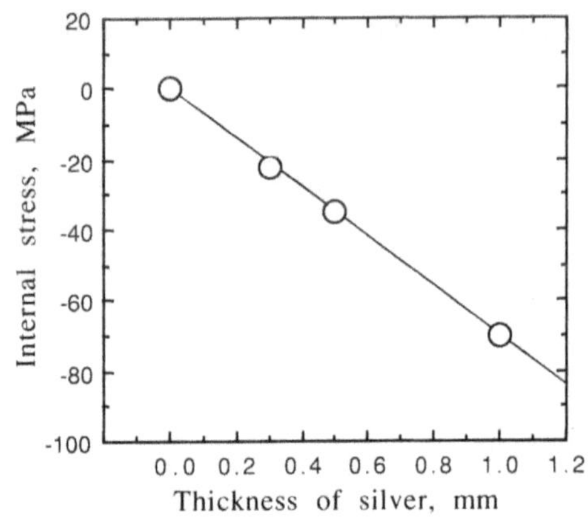

Fig.2 Calculated internal stress of BSCCO (0.3mm) in BSCCO-Ag-BSCCO plate

Table 1 Cooling and heating rates for thermal cycle tests

	cooling rate, °C/min	heating rate, °C/min
Rapid cycle	-4000	1000
Moderate cycle	-50	50
Slow cycle	-5	5

the sample was cooled and heated in the vapor phase on liquid nitrogen. (3)Slow cycle: the sample was covered by a thermal insulator during cooling and heating. The cooling and heating rates measured by the thermocouple attached to the sample were summarized in Table 1.

The thermal cycle test was carried out up to 20 cycles. The Jc of the samples were measured at each cycle by four-probe method. After 20 cycles, the crosssection of the BSCCO samples were observed by a scanning electron microscope (SEM).

RESULTS

The changes in Jc against the number of thermal cycles were shown in Fig. 3. The Jc values were plotted as the relative values(%) normalized by the initial Jc value. The initial Jc values ranged between 800 and 1200A/cm^2. As seen in Fig. 3, the relative Jc

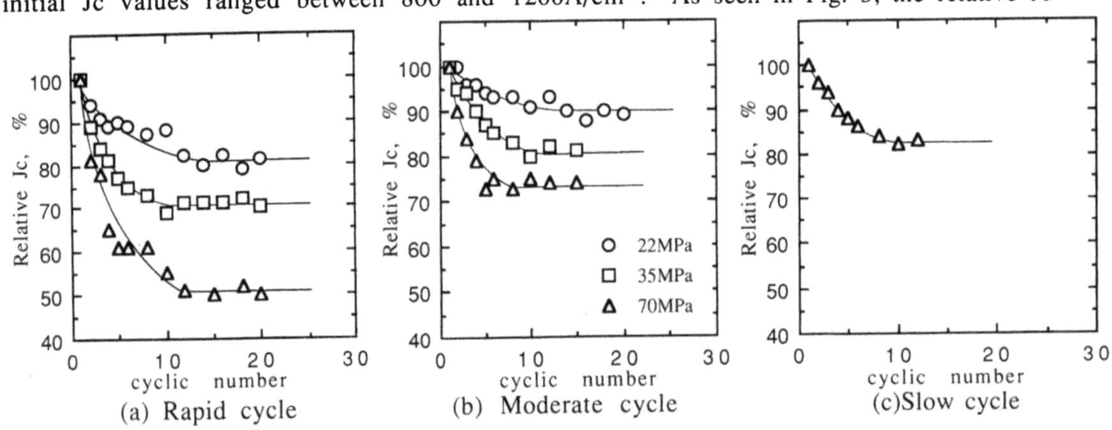

Fig. 3 Changes in Jc against the number of thermal cycles at three cycle rates

values decreased initially with increasing the number of cycles and were saturated around 10 cycles regardless of the values of the internal stress and the cooling-heating cycle rates.

Figure 4 showed the saturated relative Jc values plotted against the internal stress. As seen in Fig. 4 the saturated Jc values decreased with increasing the internal stress and decreased more rapidly by the faster cooling-heating cycle rates.

The SEM crosssectional photographs of the BSCCO samples before and after thermal cycles of rapid cooling-heating cycle were shown in Fig. 5. Microcracks were observed in the sample with the internal stress of 70MPa after 20 times rapid thermal cycle.

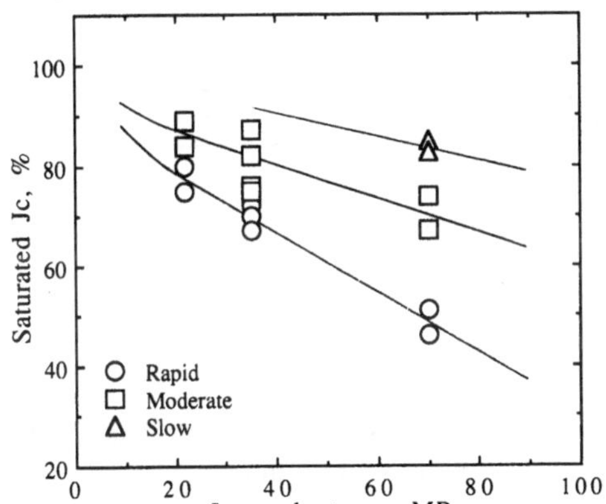

Fig. 4 Saturated Jc as a function of internal stress

(a) before thermal cycle (b)after thermal cycle

Fig. 5 SEM photographs of BSCCO before and after 20 times thermal cycle tests 10μm

DISCUSSION

The internal stress calculated in the BSCCO-Ag-BSCCO plate is the stress due mainly to the mismatch of the thermal expansion coefficients when cooled and heated quasi-statically. Also the stress induced transiently to the samples was due to the cooling-heating cycles at the given rates. A simple model was used to estimate the transient stress by the cooling-heating cycle at the infinite rates as follows. The silver layer was neglected because the thermal conductivity of silver is larger than that of BSCCO by 2 orders of magnitude. The transient stress σ induced on the surface of the sample immediately after the sample at temperature T_i is placed in the environment at a different temperature T is expressed as

$$\sigma=\frac{E\alpha}{1-\mu}(Ti - T),$$

where E is the Young's modulus, α the thermal expansion coefficient and μ the Poisson's ratio. In the case of BSCCO sample rapidly cooled to the liquid nitrogen temperature, the transient stress was calculated to be 55MPa. The transient stress is compressive when heating, and tensile when cooling. When the cooling-heating cycle rates are less than 50°C/min, the transient stress is at most a few MPa which is negligible compared with the internal stress.

The saturated Jc values in Fig. 4 were re-plotted in Fig. 6 as a function of the total stress which is the summation of the internal stress and the transient stress. As seen in Fig. 6, the saturated Jc values decreased almost linearly with increasing the total stress regardless of the cooling-heating cycle rates. Hence it is suggested that the saturated Jc value be determined by the maximum stress applied to BSCCO.

The effect of internal stress on the critical temperature(Tc) has to be considered. The changes of Tc when the isostatic pressure is applied to low Tc phase of BSCCO have been reported by several authors [1]-[5] and summarized in Table 2. The dTc/dP value ranges from 0.06 to 2.4K/GPa. This effect seems to be negligible because the maximum internal stress in this study is 70MPa, which corresponds to the 0.17K increase of Tc or less.

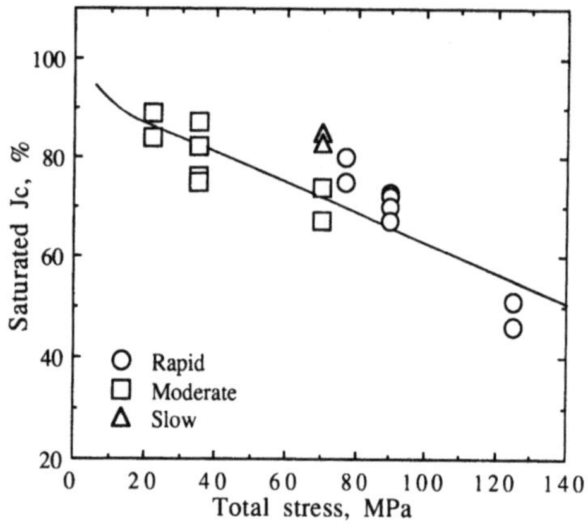

Fig. 6 Saturated Jc as a function of total stress

Table 2 Effect of isostatic pressure on Tc of low Tc phase BSCCO

Tc(K)	dTc/dP(K/GPa)	Ref.
87	+1.4	[1]
85	+2.4	[2]
83.5	+1.2	[3]
83	+0.06	[4]
70.5	+1.0	[5]

The decrease of Jc is considered to be due mainly to the formation of microcracks found in the samples after the thermal cycle. The microcracks develop at each cycle and the local stress is released. When the total stress is large, a large number of microcracks must be formed to release the stress. Hence the saturated Jc value decreased with increasing the total stress. The relaxation of the certain amount of stress may retard the crack formation. Hence it is suggested that the Jc value be saturated around 10 cycles.

SUMMARY

The effect of internal stress of BSCCO-silver-BSCCO plate on Jc after thermal cycle was studied.
(1)Jc decreased with increasing the number of thermal cycles, however, saturated around 10 cycles regardless of the cooling-heating cycle rates.
(2)The saturated Jc value decreased linearly with increasing the total stress which is the summation of the internal stress mainly due to the mismatch of the thermal expansion coefficients and the transient stress induced by the rapid cooling-heating cycle.
(3)It is suggested that the decrease and the saturation of Jc by the thermal cycle be due to the microcracks formed during the relaxation of the total stress.

REFERENCES

[1]Tamura S, Takekawa S, Nozaki H and Umezono A (1988) J.Phys.Soc.Japan 57: 2215.
[2]Schirber JE, Morosin B and Ginley DS (1989) Physica C 157: 237.
[3]Matsumoto T, Aoki H, Matsushita A, Uehara M, Mori N, Takahashi H, Murayama C and Maeda H (1988) Japan.J.Appl.Phys. 27: L600.
[4]Okai B, Kosuge M, Takahashi K and Ohta M (1988) Japan.J.Appl.Phys. 27: L674.
[5]Kamarad J, Arnold Z, Smrcka L, Suski T, Wisniewski P (1989) Supercon.Sci.Technol. 2: 269.

Preparation of Bi-Based High-T_c Superconducting Thick Films Using Combinations of Low-T_c Powder and $PbCa_{1-x}Sr_xCuO_3$ Sols

YOSHIO MASUDA[1], YOSHITO HUKUMOTO[1], RIKURO OGAWA[1], YOSHIO KAWATE[1], and NAOHIRO HARA[2]

[1] Kobe Steel Ltd., Kobe, 651-22 Japan
[2] Kobelco Research Institute Inc., Kobe, 650 Japan

ABSTRACT

Using combinations of Low-Tc powder and $PbCa_{1-x}Sr_xCuO_3$ sols, thick films of High-Tc phase havebeen produced on Ag substrates by casting. High-Tc phase content in the films was increased when the X values of the sols were 0.1 and 0.2. It has been found that the ratios of the Ca_2PbO_4 and the CuO contents to the $CaCu_2O_3$ content and the residual 2201 phase content in the calcined film, and also sintering the film at 850℃ are important in increasing the High-Tc phase content in the film. The Jc property of the films was improved by uni-axial pressure and a pressed film showed a Jc value of 330 A/cm^2.

KEY WORDS: Bi, Superconductor, Thick film, High-Tc, Low-Tc, Sol, casting.

INTRODUCTION

The Bi-based high-Tc superconductor ($(Bi,Pb)_2Sr_2Ca_2Cu_3O_{10}$, known as High-Tc or 2223 phase) has an excellent superconducting property with a superconducting critical temperature (Tc) of above 110 K. This materials is studied intensively in various facilities. However, the mechanism for production of the High-Tc phase is not yet understood sufficiently, and a long reaction time is needed to produce a lot of the phase.[1-4] It was presumed in our previously reported article[4] that the High-Tc phase is produced by the reaction between Low-Tc phase ($Bi_2Sr_2CaCu_2O_8$, named as 2212 phase) and a Pb-Ca-Cu oxide liquid phase produced as a intermediate from Ca_2PbO_4 and CuO. It is hypothesized in this study that by using combinations of powdered Low-Tc phase and sols of the $PbCa_{1-x}Sr_xCuO_3$ composition, the production time of the High-Tc phase might be shortened and the production mechanism of the phase could be understood more clearly. Therefore, thick films of the high-Tc phase have been produced on Ag substrates by the casting technique, using mixtures of powdered Low-Tc phase and sols of the $PbCa_{1-x}Sr_xCuO_3$ composition.

EXPERIMENTAL

Sols of the $PbCa_{1-x}Sr_xCuO_3$ (X=0-1.0) composition were prepared from mixed solutions of the raw pertinent metal acetates. Details of this procedure were described in another paper.[5] The sols were mixed with powdered Low-Tc phase supplied by Todakogyo Ltd. The viscosity of the mixtures was controlled by adding an organic binder. The films were formed on the Ag substrate by casting the mixtures. Film thicknesses ranging from about 10μm to about 80μm were produced by repeating the process of casting-drying-calcination. The crystalline phases deposited in the mixture at various temperatures were studied by X-ray diffraction (XRD). The precursor thick films were dried and calcined at 780°-810℃ and then sintered at 830°-855℃. Uni-axial pressure was applied to some calcined films to increase the density of the film and to improve the superconducting properties. The films were analyzed and characterized by using the XRD technique, scanning electron microscopy (SEM) accompanied by energy dispersion X-ray spectroscopy(EDX), electron probe microanalysis (EPMA), thermal analyses (TG & DTA) and chemical analyses. The contents of the produced crystalline phases and High-Tc phase in the films were calculated by using the XRD peak height of each of the phases. Current properties of the films were measured at 77 or 4.2 K by the normal four-probe technique in a zero field. In some cases the field was changed to 2 T.

RESULTS AND DISCUSSION

When viscosities of the mixtures of the Low-Tc powder and the $PbCa_{1-x}Sr_xCuO_3$ sols were adjusted to be in the range of 1 to 2 Pa.s, smoothly coated films were obtained when the mixtures were cast. The crystalline phases deposited by heating the coating films at various temperatures are shown in Fig. 1. The Low-Tc powder used as a starting material cannot be seen, but calcium acetate as a constituent of the sol is deposited in the film which was dried at 200℃. It is assumed that the powder is dissolved by the $PbCa_{1-x}Sr_xCuO_3$ sol. Bismuth oxides such as β-Bi_2O_3 and $CuBi_2O_4$ are deposited in the first stage.

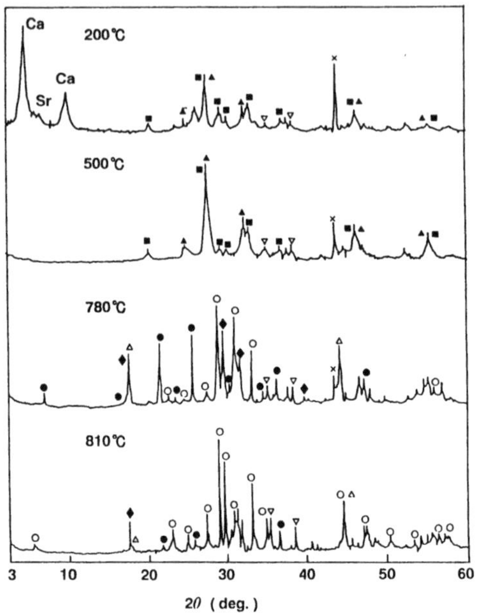

Fig. 1 X-ray diffraction patterns of films (X = 0.1) heated at various temperatures from 200℃ to 810℃.

Sr : Sr $(CH_3COO)_2$ Ca : Ca $(CH_3COO)_2$ ▲ : β-Bi_2O_3
■ : $CuBi_2O_4$ ▽ : CuO ○ : 2212
● : 2201 ◆ : Ca_2PbO_4 △ : $CaCu_2O_3$
× : unknown

When the film is heated at 780℃, 2201($Bi_2Sr_2CuO_6$) phase, Ca_2PbO_4, $CaCu_2O_3$ and a little of Low-Tc phase are deposited in the film. It was revealed by the EDX and EPMA analyses that part of the Ca component in Ca_2PbO_4 and $CaCu_2O_3$ can be replaced by Sr. Two Sr-Ca-Cu oxides phases which are Ca-rich and Ca-poor were reported by Fukushima.[6] When the heating temperature is raised to 810℃, the 2201 phase disappears and the Low-Tc phase is grown as the main phase. When the films casted on Ag substrate were fired directly to a sintering temperature without enough calcination, the films on the Ag substrate were melted and separated from the substrate. Therefore, it may be seen that calcination is an important part of the process in getting a good film. Figure 2 shows that the contents of the crystalline phases produced in the films by calcination at 810℃ depend upon the Sr concentrations (X values) in the $PbCa_{1-x}$-Sr_xCuO_3 sols. Ca_2PbO_4 and CuO contents are increased and the 2201 phase is decreased until the Sr concentration is increased by 0.1 mol, but above this concentration (0.1 mol), Ca_2PbO_4 and CuO contents are decreased. $CaCu_2O_3$ content is increased gradually as the Sr concentration is increased. The peak temperatures of the melting phenomena, which were observed to be near 820℃ in the thermal analyses, are changed depending upon the Ca_2PbO_4 contents.

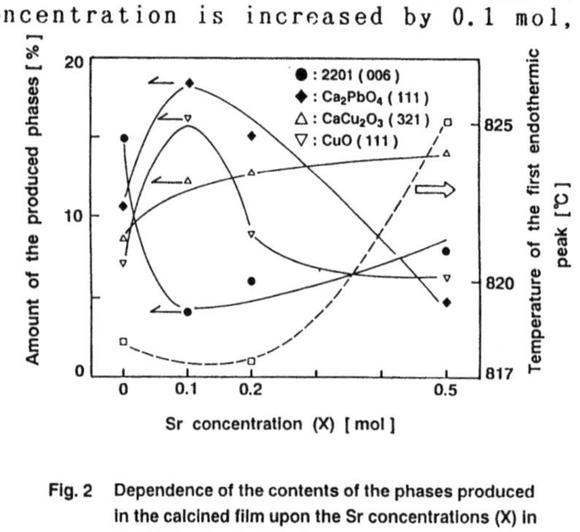

Fig. 2 Dependence of the contents of the phases produced in the calcined film upon the Sr concentrations (X) in the $PbCa_{1-X}Sr_XCuO_3$ sols.

Upon sintering of the coating films at 830°, 840° and 850℃, the films were almost composed of the Low-Tc phase, the High-Tc phase and $CaCu_2O_3$. Ca_2PbO_4 was found to a certain extent in the films sintered at 830° and 840℃. When the film was sintered at 855℃, most of the coated film dropped off the substrate. The contents of the High-Tc phase in the sintered films are shown in Fig. 3. High-Tc phase cannot be produced in large amounts at 830° and 840℃. When the coating film with a Sr concentration of 0.1 mol is sintered at 850℃ for 60 h, the High-Tc phase content is increased to over 70 %. The High-Tc phase production speed of this film was higher than that of the film produced by using the sol containing all pertinent metal elements.[4] However, the content is decreased when films with a X value of 0.1 are sintered for 100 h at 850℃. When the pressure treatment is carried out for the calcined films, this decrease can be suppressed. It is assumed that the decrease behavior of the High-Tc phase originates from the evaporation of the Pb component from the film and this Pb evaporationcan be suppressed by pressing the film. It is also assumed that the heating atmosphere formed by the Pb evaporation is reductive, so that the melting temperature of the film is 855℃, lower than that reported in the $Bi_1-Sr_1Ca_1Cu_2O_y$ system by Endo et al.[2]

Production of the High-Tc phase is not improved by pressing the films. The High-Tc phase production may be restrained by production of stable $CaCu_2O_3$ and CuO.[7]

The current density of the films depends upon the contents of the High-Tc phase as shown in Fig 4. This dependency is more important for the pressed films than the non-pressed films. It is assumed that contacts between the High-Tc grains are improved and the current density is increased by the pressure treatment. However, condensations of the impurity phases in the films, which were composed of Sr-Ca-Cu-O phases, could not be suppressed by addition of the pressure treatment. The pressed film whose High-Tc phase content is about 50 % has a current density (Jc) of 330 A/cm².

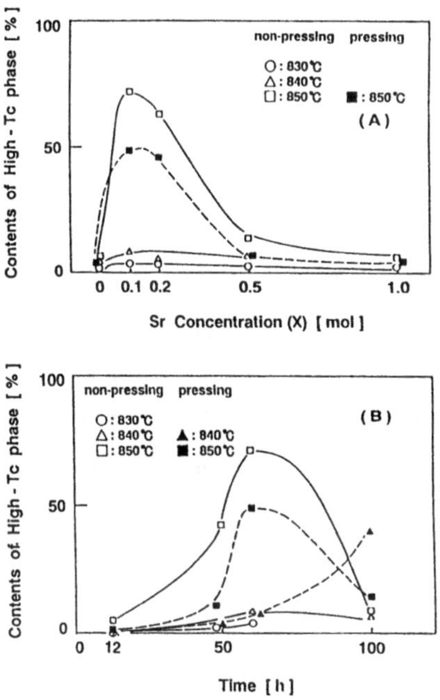

Fig. 3 Dependence of the production of High-Tc phase in the film upon the Sr concentrations (A) in the $PbCa_{1-X}Sr_XCuO_3$ sols and sintering time (B).

$$\text{High-Tc phase contents} = \left(\frac{002_H}{002_H + 002_L} + \frac{115_H}{115_H + 115_L}\right) \div 2 \times 100$$

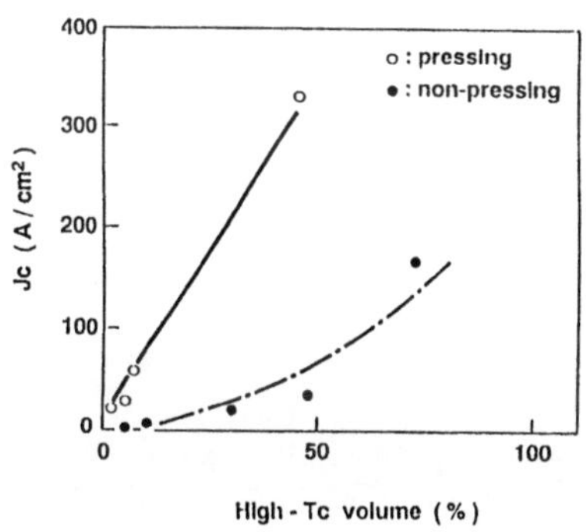

Fig. 4 Dependence of the Jc properties of the films at 77K upon the High-Tc phase contents in the films.

CONCLUSIONS

Using the mixtures of Low-Tc powder and $PbCa_{1-x}Sr_xCuO_3$ sols whose X values were 0.1 and 0.2, High-Tc phase was produced in large amounts in thick films on Ag substrates. From X value dependency of the High-Tc production, it has been found that if Ca_2PbO_4 and CuO were produced more than $CaCu_2O_3$ and if the 2201 phase remained to a small extent in the calcined films, and also if the calcined films were sintered at 850℃, High-Tc phase in the coating films could be increased. It is assumed that a reductive firing atmosphere was formed and the melting temperatures were lowered by evaporation the Pb component from the films. The current density of the pressed films depended more heavily upon the High-Tc phase content than that of the non-pressed films. The current density of the pressed film was 330 A/cm².

REFERENCES

1) M.Takano,J.Takano,K.Oda,H.Kitaguch,Y.Miura,Y.Ikeda,Y.Tomii and H.Mazaki, Jpn.J.App.Phys., 27, L1041 (1988).
2) U.Endo,S.Koyama and T.Kawai, Jpn.J.App.Phys., 27, L1476 (1988).
3) H.Nobumasa,K.Shimizu,Y.Kitanoand T.Kawai, Jpn.J.App.Phys., 27, L846 (1988).
4) Y.Masuda,T.Tateishi,N.Hara,R.Ogawa and Y.Kawate, accepted by Mater.Res. Soc.
5) Y.Masuda, Y.Hukumoto, Y.Kawate and N.Hara, Nippon seramikkusu- kyoukai-nenkaiyokosyu, p.60 (1991).
6) K.Fukushima, Jpn.J.App.Phys., 29, L2195 (1990)
7) H.M.Jang and J.H.Moon, J.Mater.Res., 6, 916 (1991).

Development of Silver Sheathed Bi-Based Superconducting Coil

TASUKU KITAMURA[1], TAKAYO HASEGAWA[1], FUMIO TAKESHITA[1], K. YAMAMOTO[2], S. MURASE[2], and H. OGIWARA[2]

[1] Showa Electric Wire & Cable Co., Ltd., Kawasaki, Kanagawa, Japan
[2] Toshiba Corporation, Kawasaki, Kanagawa, Japan

ABSTRACT

We have successfully fabricated a high magnetic field coil generating 1.15T at 4.2K and 0.19T at 77.3K using Bi-based(2223) powder-in-tube method silver sheathed tape. This was accomplished by the fundamental study in the winding process and the homogenization of the critical current for a long tape sample.

KEY WORDS: Bi-based superconductor, powder-in-tube method, silver sheathed wire, magnetic field, insert coil

INTRODUCTION

Since the high-Tc superconductors were discovered [1] the application has been so far pessimistic become of the low Jc values. However, recently, high Jc values on the order of $10^4 A/cm^2$ at 77.3K,0T have been reported [2],[3] for the powder-in-tube method. Moreover this method makes it possible to prepare long wires, realizing the application for the magnet.

Bi-based 2223 phase ($Bi_2Sr_2Ca_2Ca_3Oy$) is known to be appropriate to the material for powder-in-tube method because the 2223 crystals alligen well during hard cold working such as rolling and pressing.In addition, the critical current density of 2223 phase is high than that in the conventional metallic superconductor in higher fields at 4.2K [4].Therefore, Bi-based superconducting wires will open new applications such as liquid nitrogen temperature coils and high magnetic field magnets at 4.2K.

In this paper, we investigated the fundamental techniques for the preparation for the wire and coil. Thus, we have successfully fabricated the coil producing fields of 1.15T at 4.2K and 0.19T at 77.3K

EXPERIMENTAL

The superconducting powders were prepared by conventional ceramic technique using high quality regents. The calcined powder was packed into silver tube with 7mm in outer diameter and 5mm in inner diameter and was drawn into a wire. The wire was additionally rolled into 0.15mm in thickness, 3.5 mm in width and 11m in length. Subsequently the wire was heat-treated at a given temperature. The tapes were cut at a given length, and were wound with two techniques. One was the "wind-and-react(W&R)": the tape was wound around a silver pipe and then heat-treated. The other was "react-and-wind" (R&W) technique: the tape was heat treated and then wound.

The critical current(Ic) was measured at 77.3K and 4.2K in an external magnet field of 0 to 15T by DC four-probe method. Ic was determined by a criterion of specific resistivity of $10^{-13}\Omega m$. The magnetic field generated by the oxide coil was measured by a hall device at the center of the coil.

RESULTS AND DISCUSSION

1. Homogeneity of long wire

One of the important aim in this work is the enhancement of critical current over the whole length of wire. Figure1 shows the Ic value of 3m long wire at 77K,0T. The wire was cut into a length of 3m from 10m length of as rolled tape. This was wound around an alumina pipe with the outer diameter of 70mm in outer-diameter as a solenoid. In this case, terminals for measurement of voltage were connected to every turn to evaluate the distribution of Ic at each turn. Figure1 shows that the maximum and the minimum Ic were 17.0A and 16.0A, respectively. The distribution was as low as 6%. The Ic of entire 3m long tape was decided by the minimum value. The Ic=16.0A corresponds to the Jc value of $1.6x10^4 A/cm^2$.

2. Bending strain characteristics

Figure2 shows the dependence of critical current density at 77.3K, 0T on bending strain in W&R and R&W techniques. The silver-sheathed tapes were wound around alumina pipes of 15 to 70mm in outer diameter. According to SEM analysis, micro cracks decreasing Jc were observed in the core of each sample. Especially R&W technique, Jc decreased remarkably. The Jc value was decreased to 10% at a bending strain of 1.0%. On the contrary, for W&R technique, high Jc of $1.1x10^4 A/cm^2$ was obtained corresponding to about 70% of the value without strain even at a bending strain of 1.0%. This superior strain tolerance is due to the heat treatment after the winding recovering the micro-cracks of the superconductor.

3. Jc-B characteristics

Figure3 shows the Jc-B characteristic of this wire at 4.2K in the external magnetic field parallel to the tape surface. The Jc value were $8.0x10^4 A/cm^2$ at 0T and $3.5x10^4 A/cm^2$ at 10T. The sample showed less decrease in Jc at high magnetic fields. It appeared that Jc values kept constant above 7T.

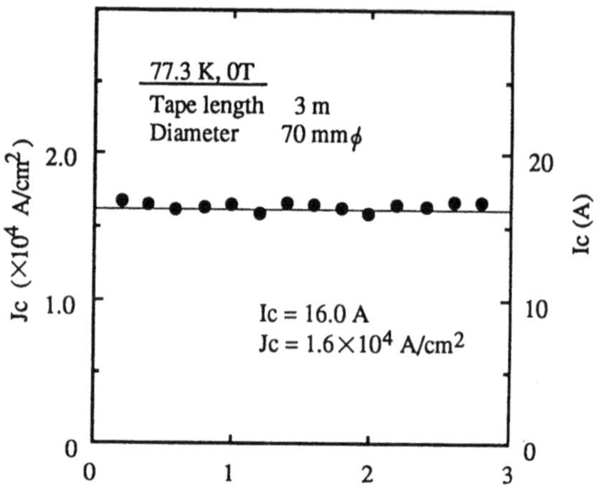

Fig.1 Homogeneity of long wire.

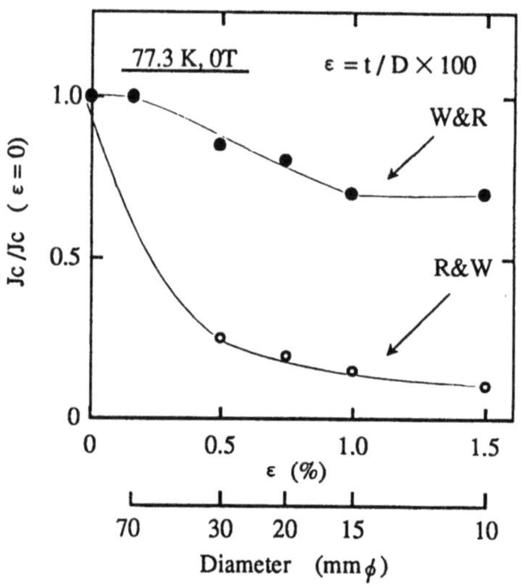

Fig.2 Bending strain characteristics of silver-sheathed wire.

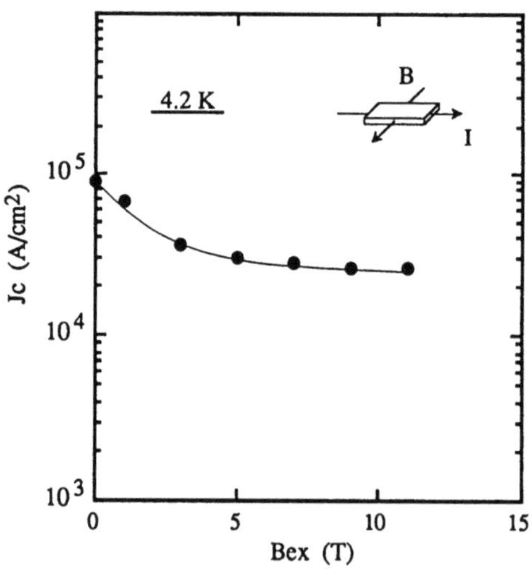

Fig.3 Jc-B characteristics of silver- sheathed wire.

4. Characteristics of stacked pancake coil

The oxide coil in Figure4 was designed and fabricated using the above preliminary results. The specifications are shown in Table1. The oxide coil consists of cylindrical 8 pancakes.The each single pancake coil was electrically connected in series. The generating magnetic field was calculated from the shape and operation current, and expected 1T at 4.2K .

Fig.4 Appearance of stacked pancake coil

Experimental results of the coil are shown in Table2. At 77.3K, the operation current was 25A and Bmax 1900G without external magnetic field. At 4.2K, the operation current was 155A and Bmax 11500G (1.15T). High magnetic field exceeding 1T were successfully generated. Although external magnetic field lowered the current carrying capacity, the coil generated 0.53T with an operating current of 72A even at 7T. This indicates that silver sheathed oxide superconducting coil can be used as an insert coil in a high magnetic fields.

Table 1 Specificasions of stacked pancake coil.

< Tape >

Thickness (mm)	0.15
Width (mm)	3.50
Length (m)	3.0
Parallel number	4

< Coil >

Inner diameter (mmϕ)	15
Outer diameter (mmϕ)	54
Height (mm)	34
Number of coil turns	28
Number of single coils stacked	8
Total number of turns	224
Total length of wires	96

Table 2 Characteristics of stacked pancake coil.

T (K)	4.2					77.3
Bex (T)	0	1	3	5	7	0
Ic (A)	155	140	110	100	72	25
Jc (A/cm^2)	38,800	35,000	27,500	25,000	18,000	6,300
Bo (G)	11,000	10,000	7,800	7,100	5,100	1,800
Bmax (G)	11,500	10,400	8,100	7,400	5,300	1,900

CONCLUSION

We have successfully fabricated a oxide coil generating 1.15T at 4.2K and 0.19T at 77.3K, using powder-in-tube silver sheathed Bi-2223 tapes. Even at an external field of 7T, this coil generated 0.53T at 4.2K. This development proves that silver sheathed tape can be used for high field magnet.

REFERENCES

1. M.K. Wu, J.R.Ashburn, C.J. Trong, P.H. Hor, R.L. Meng, L.Gao, Z.J.Huang, Y.Q.Wang and C.W.Chu : Phys. Rev. Lett, 58,908 (1987)
2. T.Kitamura, T.hasegawa, H.Kobayashi, Y.Kamisada : 43rd meeting on Japanese Cryogenics and Superconductinity, p216.(1990)
3. T.Kitamura, T.hasegawa, Y.Kamisada, K.Yamamoto, S.Murase : 43rd meeting on Japanese Cryogenics and Superconductinity, p217.(1990)
4. K.Togano : OYO BUTURI, 60,5 (1991)

Trial Fabrication of Silver Sheathed Bi2212 Superconducting Coils

KAZUYUKI SHIBUTANI, TOSHIO EGI, SEIJI HAYASHI, RIKUO OGAWA, and YOSHIO KAWATE

Superconducting and Cryogenic Technology Center, Kobe Steel Ltd., Kobe, 651-22 Japan

ABSTRACT

Fabrication of a silver sheathed Bi-2212 superconducting coil has been tried by means of partial melt growth process. Silver sheathed Bi-2212 tapes have been cooled down slowly from the range of 875-890 °C to 850 °C at the rate of 6 °C/h. The characteristics of Jc-H and AC magnetic susceptibility have been measured, and SEM images of cross section have been observed. The pancake type test coil has been fabricated using optimized heat treating condition. This coil generates 0.70T in the external magnetic field of 0T at 4.2K.

KEY WORDS: silver sheathed wire, Bi-2212 oxide superconductor, oxide superconducting coil, partial melt growth process, AC magnetic susceptibility

INTRODUCTION

Recently, several researchers have reported [1-4] that the Bi-2212 (low-Tc phase) phase can be oriented on the silver surface by means of partial melt growth process and the composites thus prepared show very high critical current (Ic) or critical current density (Jc) in high magnetic field. The partial melt growth process consists of partial melting and slow cooling. On the other hand, fabrication of silver sheathed Bi-2223 wire needs the uniaxial pressing for crystal orientation and long time heat treatment for the growth of the Bi-2223 phase from the Bi-2212 phase. Compared to Bi2223, the simple Bi2212 melt process has a greater advantage. So, it is thought that this method is more practical for the fabrication of silver sheathed Bi-2212 superconducting coil.

The purposes of this work are (1) to clarify the optimum heat treatment condition for silver sheathed tapes to realize high Jc in high magnetic field and (2) to investigate the possibility for the fabrication of superconducting magnet by means of partial melt growth process of the silver sheathed tapes.

EXPERIMENTAL PROCEDURES

$Bi_2Sr_2CaCu_2O_8$ powder was prepared by a solid state reaction. The starting materials, Bi_2O_3, $SrCO_3$, $CaCO_3$ and CuO were mixed in Bi:Sr:Ca:Cu=2:2:1:2, ground and calcined twice at 820 °C for 18 hours. After these treatments, in order to suppress the appearance of gases in the silver sheath, these powders were calcined at 865 °C for 2 hours and cooled down to room temperature in the electric furnace. These Bi-2212 powders were packed into silver tubes (4/6 mm in inner/outer diameter and 1500 mm long). These silver tubes were drawn and pressed into the tape shapes of 100 um thickness with 2.6 mm width (for short sample: 30 mm length) and 150 μm thickness with 3.2 mm width (for the test coil: 100 m length) using the roller dice and the rolling machine.

In order to optimize the heat treatment condition, the silver sheathed Bi-2212 tapes (short samples) were heat treated in air using following 4 schedules (see fig.1.(1)-(4)). It is well known that Bi2212 oxide superconductor has the partial melting point near 880 °C [3,4]. However, this partial melting point is varied by the oxygen partial pressure. In this work, the Bi2212 oxide superconducting powder is packed into a silver tube. So the oxygen partial pressure must be lower than the atmospheric oxygen partial pressure. To obtain the best condition for imparting high Jc in silver sheathed wire, the starting temperature at which the slow cooling process (T1) starts should be optimized. Three short samples were bundled into 1 short sample, and these samples were heated to T1=875 °C (schedule 1), 880 °C (schedule 2), 885 °C (schedule 3) and 890

℃ (schedule 4) and cooled slowly to T2=850 ℃ at a rate of 6 ℃/hour to get well aligned Bi2212 grains. And all samples were cooled to room temperature at a rate of 100-30 ℃/hour.

The pancake type superconducting test coil was fabricated using the optimized heat treating condition (T1=890 °C). Table 1 shows the specification of the pancake type superconducting test coil. This test coil consists of twelve series of single pancake coils. Each single pancake coil consists of five layered silver sheathed Bi-2212 tapes (3.2 mm width and 150 μm thickness) which is wound 56 turns. Each single pancake coil used the five silver sheathed tapes which are 20 m long, and then 1200 m length of the silver sheathed tape wire was used in all the test coil. The bobbin and spacer were made of workable ceramics. The tape of fiber glass was employed for the insulator between the conductors. Each single pancake coil was connected by copper leads out of the coil or silver tube which was built in the center pole. The wound coils were heat treated using the schedule (1) in Fig.1.

Table 1 The specification of the pancake type superconducting test coil.

Wire		Coil	
Thickness of tape	150 μm	Inner diameter	$2A_1$= 14mm
Width of tape	3.2 mm	Outer diameter	$2A_2$= 130mm
Total tape length	1200 m	Coil length	2B = 68mm
		Number of turns	56 turns/ pancake

RESULTS AND DISCUSSION

Figure 2 shows the critical current density (Jc) of the short samples at 4.2K as a function of the external magnetic fields up to 6.0T. The magnetic field direction was parallel to the tapes flat surface and perpendicular to the current direction. The criterion for Jc determination was defined as 0.5 μV/cm. On these short samples, the characteristics of Jc-H is divided into two groups; group 1 (T1=875 and 880 ℃) and group 2 (T1=885 and 890 ℃). Group 1 shows considerable decrease of critical current density in the magnetic field region of less than 0.1T, on the other hand, group 2 shows good Jc-H characteristics. Both samples of group 1 (T1=885 and 890 °C with 2.6mm width) show almost the same level of Jc value (1.3-1.7x10^4 A/cm^2) in H=0 at 4.2K and decreasing rate (Jc(H)/Jc(0)) (for example, both samples show 60 % at H=1T and 40 % at H=5T).

Figure 3(a) and (b) show the SEM photographs of the fractured cross section of the tapes which were cooled slowly from T1=875 ℃ and 890 ℃ to T2=850 ℃ at a rate of 6 ℃/hour. The sample slow-cooled from 875 ℃ (Fig. 3(a)) shows weak trace of melting but no oriented grains. For the sample slow-cooled from T1=890 ℃ (Fig. 3(b)), well-oriented grains are observed. The T1=875 ℃ sample (Fig. 3(a)) has no crevice but has many small pores. On the other hand, the T1=890 °C sample (Fig. 3(b) has a large crevice but its crystal has high density. These results suggest

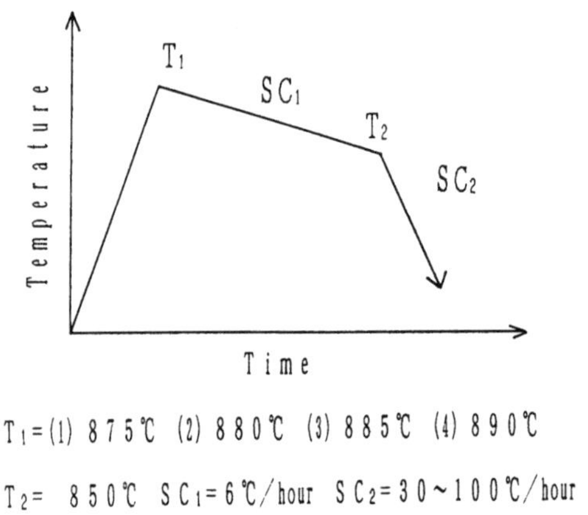

Fig. 1 Schematic explanation of heat treatment schedules for partial melt growth process.

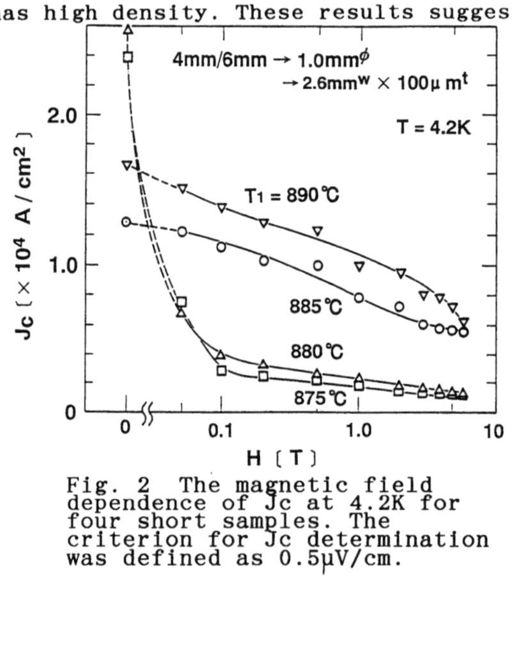

Fig. 2 The magnetic field dependence of Jc at 4.2K for four short samples. The criterion for Jc determination was defined as 0.5μV/cm.

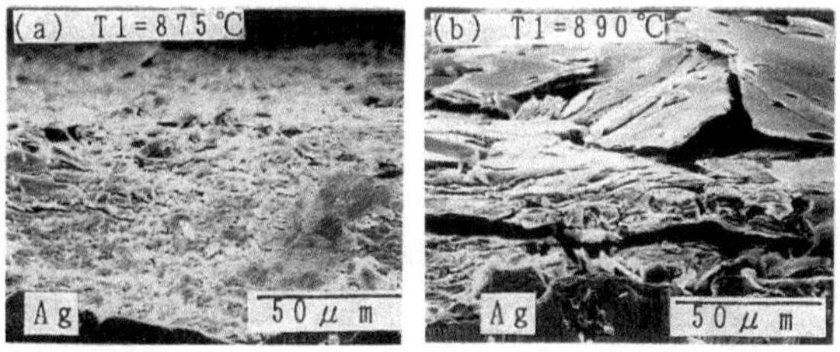

Fig. 3 The SEM photographs of the fractured cross section of the tape which were cooled slowly from (a) T1=875 ℃ and (b) 890 ℃ to T2=850 ℃ at a rate of 6 ℃/hour respectively.

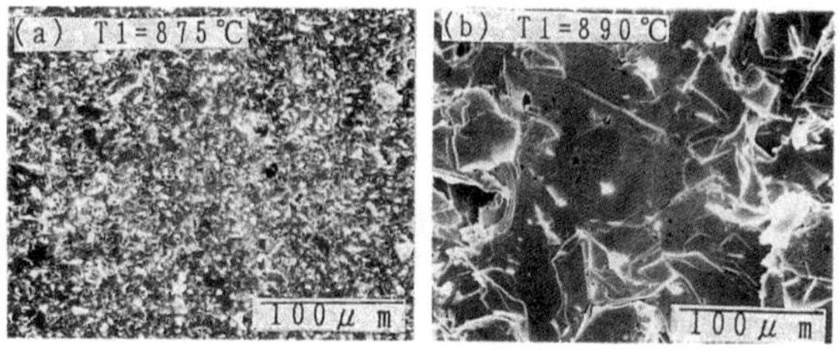

Fig. 4 The SEM photographs of the inside surface of silver sheathed tape which were cooled slowly from (a) T1=875 ℃ and (b) 890 ℃. The observation was carried out after a part of silver sheath was peeled out.

that the effective cross section area of T1=890 ℃ sample is smaller than that of the T1=875 ℃ sample.

Figure 4(a) and (b) show SEM photographs of the inside surface of silver sheathed tapes. The observation was carried out after a part of the silver sheath was peeled out. The sample slowly cooled from T1=875 ℃ (Fig. 4(a)) is covered with the many small grains or flakes larger than a few micron meters in diameter. The sample slowly cooled from T1=890 ℃ (Fig. 4(b)) is covered with the large flat plates of more than 100 um in diameter. These results suggest two features, (1) the T1=890 ℃ sample has well-oriented crystal along the silver surface, and (2) the effect of superconducting weak link junction for the transport of superconducting current in the T1=890 ℃ is considerably less than T1=875 ℃ sample.

From the results of Fig. 2, 3 and 4, the partial melting point of silver sheathed Bi2212 oxide superconductor is in the range of 880 ℃ to 885 ℃. This result is consistent with another report [5].

In order to evaluate the grain boundary weak links, AC susceptibility measurements were carried out. Berg et al.[6] carried out numerical calculations of the imaginary part of the complex susceptibility (χ'') for granular superconductor in which the grains were separated by Josephson junction type weak links. They reported the temperature dependence of χ'' with increasing barrier width of the Josephson junction; namely that the peak shifts to lower temperatures at the same time as it broadens. The temperature dependences of χ' and χ'' for silver sheathed Bi-2212 tapes (T1=875, 880, 885 and 890 ℃) are shown in Fig. 5. The measurements were performed applying the ac field of 0.1mT and 100 Hz perpendicular to the tape plane of the wires. The χ'' peak is related to the coupling loss of the sample, the χ'' peaks of the group 1 (T1=875 and 880 ℃) are broader than those of group 2 (T1=885 and 890 ℃). This means that the grain boundary coupling of group 1 tapes are not good enough compared to those of group 2 and thus coupling loss is observed in a wide temperature range.

Through the observation of SEM images and the measurement of AC susceptibility, well-oriented Bi-2212 system which has few superconducting weak links in the silver sheathed tapes are realized when the T1 is higher than 885 ℃. The reason why the group 1 show higher critical current density only in low magnetic field region is thought to be as follows. The cross section areas of silver sheathed tapes of group 1 and group 2 were measured using the photo microscope images of polished samples. These samples show almost the same value of cross section area; $S=6.23\times10^{-4}$ cm^2. The characteristics of Jc-H (Fig. 2) are plotted using the same value of cross section area. However, the effective cross section area of group 2 must be smaller than that of group 1 because the samples of group 2 were melt grown. This tendency is supported by the cross sectional microscope image of silver sheathed Bi-2212 tapes; the sample of T1=890 ℃ (Fig. 3(b)) has a greater amount of pores or large cracks than that of T1=875 ℃ (Fig. 3(a)).

Figure 6 shows the photo image of the Bi-2212 test coil. In order to measure the generating magnetic fields, the Hall-effect device is mounted in the center pole. After the heat treatment (schedule (1) in Fig. 1, T1=890 ℃), the test coil was immersed in the melting wax and cooled. Inserted figure in the Fig. 6(b) shows the schematic explanation of coil circuit.

The generating magnetic field and critical current were measured in zero magnetic field at 4.2K. The Ic was decided using the electric field criterion of

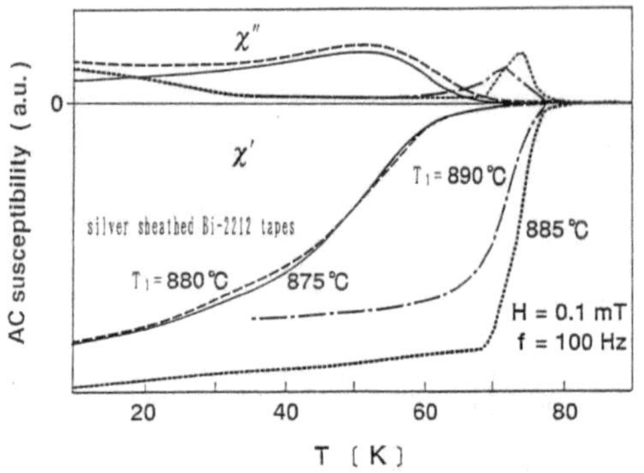

Fig. 5 Temperature dependence of the real part χ' and imaginary part χ'' of complex susceptivility obtained at 100Hz with AC field amplitude of 0.1mT for silver sheathed tapes.

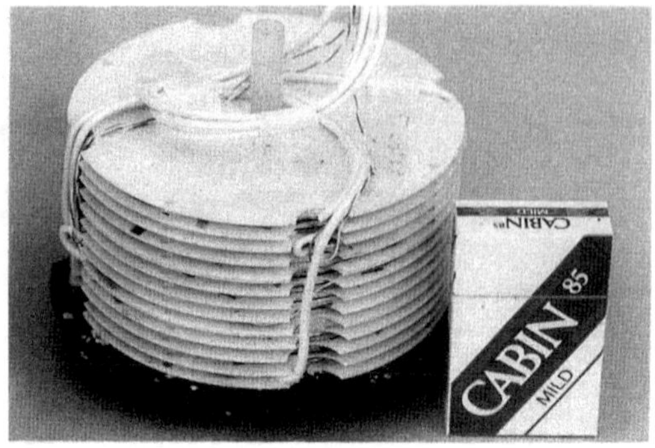

Fig. 6 The photo image of the Bi-2212 test coil. This coil has been connected power and potential lead wires after the heat treatment.

0.5 μV/cm. This corresponds to the voltage of 12 mV over the whole length of 240 m tape (0.5 μV/cm x 240 m=12 mV). Namely the Ic was decided by the same standard in case of short samples. The test coil shows the constant resistance ($8x10^{-4}$ ohm at 4.2K) because each single pancake coil is connected by normal state metals (cooper or silver). The point which showed the deviation of 12 mV from the straight line with constant slope was employed for the determination of critical current. As a result of measurement of the generating magnetic field by means of the Hall-effect device, the amount of generating magnetic field at the center of coil was 0.69T (maximum generating magnetic field was 0.70T). Then the critical current of this coil was 80.0 A in zero external magnetic field at 4.2K.

Generating magnetic field is estimated using the specification of test coil (see Table. 1) as follows. Generating magnetic field at the center of coil; H_0 is given

$$H_0 = A1 \times i(ex.h) \times r \times F(a,b) \times 10^{-6} \ (T) \qquad (1)$$

where, A1 is the inner radius of test coil, i(ex.h) is the over all critical current density in the external magnetic field, r is the occupation factor of silver sheathed tape wire and F(a,b) is Fabry's coefficient;

$$F(a,b) = (4\pi/10) \times b \times \ln((a+\sqrt{a^2+b^2})/(1+\sqrt{1+b^2})) \qquad (2)$$

$$a=A2/A1, \ b=B/A1$$

After calculation, the generating magnetic field is estimated H_0=0.697T for the test coil under zero external field (using values A1=$7x10^{-3}$ m, i(0.7T)=80.0 A/($3.2x10^{-3}$x$1.5x10^{-4}$x5) m^2= $3.33x10^7$ A/m^2, r=3.2x0.15x5x56x12/(68x58)= 0.409 and F(9.29, 4.86)= 7.32). Compared with the estimated value and experimental value, the experimental value is close to the estimated value. This result suggests that the insulation between the silver sheathed tapes was complete.

In summary, it became clear that the simple Bi-2212 partial melt process had a great advantage of simple heat treatment process. It is found that this partial melt growth process is effective for practical use.

REFERENCES

1. Heine K, Tenbrink J and Thoner M, (1989) Appl. Phys. Lett. 55: 2441
2. Asano T, Tanaka Y, Fukutomi M and Maeda H, (1990) Jpn. J. Appl. Phys. 29: L1066
3. Kase J, Togano K, Kumakura H, Dietderich DR, Irisawa N, Morimoto T and Maeda H, (1990) Jpn. J. Appl. Phys. 29: L1096
4. Dietderich DR, Ullmann B, Freyhardt HC, Kase J, Kumakura H, Togano K and Maeda H, (1990) Jpn. J. Appl. Phys. 29: L1100
5. Shibutani K, Hayashi S, Ogawa R and Kawate Y, (1991) Advances in superconductivity 3 (proceedings of ISS'90). Springer Verlag. Tokyo, pp 487
6. Berg P and Kozlowski G, (1989) Mod. Phys. Lett. 3: 1163

Preparation of Bi-2223 Tapes by Partial-Melt Process

D.Y. JEONG, D.W. HA, M.H. SOHN, I.R. HAN, and K.S. RYU

Applied Superconductivity Laboratory, Korea Electrotechnology Research Institute,
Changwon 641-120, Korea

ABSTRACT

The availability of partial-melt process(PMP) in enhancing the densification and the directional grain-alignment was investigated in Ag-sheathed Bi-Pb-Sr-Ca-Cu-O tapes prepared by the powder-in-tube method. The PMP was employed after the Bi-2223 phase was fully formed. The results indicated that the PMP resulted in a dense, layer-like textured micromorphology in a short range order and little grain-texturing in a long range order, resulting in a low Jc. An attempt to extend the locally developed, dense and layer-like structure to a long range order was made by employing the PMP after obtaining a well-textured morphology by using a repetition of the combined process of rolling and sintering. It seemed, however, that the PMP destroyed the previously-made textured morphology.

KEY WORDS : Partial-melt process, Bi-Pb-Sr-Ca-Cu-O superconductor, microstructure, critical current

INTRODUCTION

In order to apply high-Tc superconductors to large-scale high-energy electrical systems, they should be made in a form of wire with Jc of $10^4 \sim 10^6$ A/㎠ at 77 K and under high magnetic fields, as well as with an excellent mechanical flexibility.

Up to now, the highest Jc in the high-Tc superconducting wires was obtained in the Ag-sheathed Bi-Pb-Sr-Ca-Cu-O tapes prepared by repeating a combined process of rolling and sintering, which showed 5.4 x 10^4 A/㎠ at 77 K and 0 T.[1] The process, however, seems to result in a relatively poor densification. Therefore, a new or a modified process is required for further enhancement of Jc, which can result in both the densified and the well grain-aligned morphology. Recently, Jin et al.[2] reported that a partial-melt process employed in the preparation of Bi-Pb-Sb-Sr-Ca-Cu-O ribbon resulted in a dense and layer-like textured microstructure and a Jc of 2.3 x 10^5 A/㎠ at 4.2 K and 8 T, due to a corporative support of partial melting, deformation-induced texturing and substrate-induced texturing.

In the present study, it was attempted to reduce the weak-link which hinders the current-carrying capacity, through an enhanced densification resulting from employing the partial-melt process in the preparation of Ag-sheathed Bi-Pb-Sr-Ca-Cu-O tapes.

EXPERIMENTS

The tapes were prepared with a nominal composition of $Bi_{1.84}Pb_{0.34}Sr_{1.97}Ca_{2.03}Cu_{3.06}O_x$ as follows. The powder mixture of oxides and carbonates with the stoichiometric ratio was calcined at 800 ℃ for 48 hr in air, ground, and filled into a Ag-Tube with 4 mm in in I.D. and 6 mm in O.D. The tube was drawn to a wire with 1.5 mm in diameter. The wire was cut into pieces with 5~6 cm in length and sintered at 834 ℃ for 120 hr in air. Then the wires experienced two different thermomechanical treatments as follows.

(1) The wires were rolled to tapes with 160 μm in thickness. Then the tapes experienced a PMP. The PMP was consisted of heating to 860 ℃, holding for 10 min, slow-cooling to 825 ℃ at a rate of 20 ℃/hr, holding at 825 ℃ for 36 hr, and slow-cooling to room temperature.(T1) Then the tapes were resintered at 825 ℃ for 36 hr.(T2) The heat treatments were conducted in a flowing atmosphere of $Ar:O_2$ = 60:5 ml/min.

(2) The wires were rolled to tapes with 330 μm. The tapes were sintered at 830 ℃ for 72 hr in air, rolled to a thickness of 150 μm, sintered at 834 ℃ for 48 hr (T3), and rolled to a thickness of 100 μm. Then the tapes experienced a PMP. The PMP was consisted of heating to 860 ℃, holding for 10 min, slow-cooling to 825 ℃ at a rate of 20 (T4) or 3.5 ℃/hr (T5), holding at 825 ℃ for 24 hr, and slow-cooling to room temperature. Then the tapes were resintered at 825 ℃ for 24 hr.(T6 and T7) The PMP and the resintering were conducted in a flowing atmorphere of $Ar:O_2$ = 60:5 ml/min.

The measurements of Tc's and Jc's were carried out by four-probe electrical measurements. The test current for Tc was 30 mA. The criterion for Jc's was 1 μV/cm. XRD and SEM analyses were carried out for the phase indentification and the morphology study, respectively.

RESULTS AND DISCUSSION

The wires sintered at 834 ℃ for 120 hr showed broad superconducting(SC) transitions in a range of 109 to 91 K, indicating that they contained relatively large amounts of 2223 phase, in addition to 2212 phase. While the tape (T1) prepared by the PMP in the treatment (1) showed a broad SC transition from 106 to 97 K, the tape (T2) resintered after the PMP showed a relatively sharp transition from 107.5 to 104.5 K and a relatively low Jc of 1,200 A/cm² at 77 K and 0 T, compared to that in the tape prepared by a combined process of rolling and sintering.

Fig.1 shows X-ray diffraction patterns taken in the tapes (T1) and (T2). It shows that the tape (T1) contained larger amount of 2212 phase than 2223 and the tape (T2) contained a little 2212 phase, indicating that the resintering converted large amount of the 2212 phase into 2223. Fig. 2 shows SEM pictures taken in the tapes (T1) and (T2). Both the tapes contained many large-sized pores due to local melting and showed a layer-like microstructure grown-two-dimensionally. While the tapes have a dense and layer-like micromorphology in a short range order, they shows little evidence of grain texturing in a long range order. Indeed it was often observed that the layer-like structures were connected each other with a

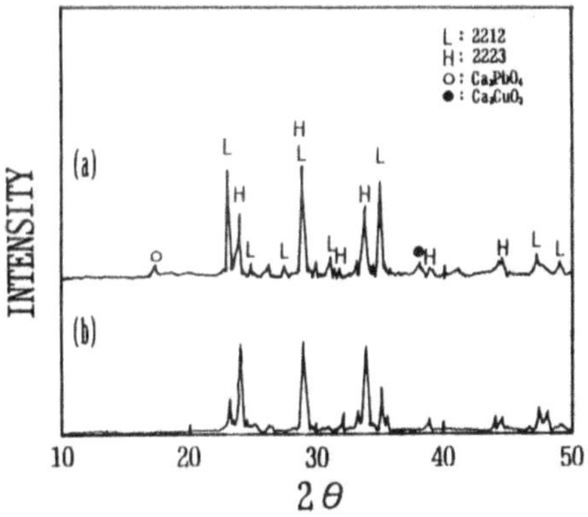

Fig.1. XRD patterns taken in the tapes prepared by (a) PMP (T1) and (b) resintering (T2) in the thermomechanical treatment (1), respectively.

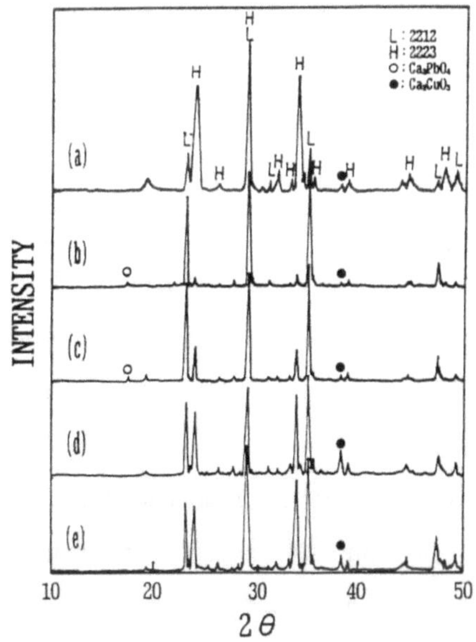

Fig.3. XRD patterns taken in the tapes prepared by PMP with cooling rates of (b) 20 (T4) and (c) 3.5 ℃/hr (T5), and (d, e) resintering (T6, T7) in the thermomechanical treatment (2), respectively. (a) was taken in the tape (T3) sintered at 834 ℃ for 120 hr before the PMP.

(a)

(b)

(c)

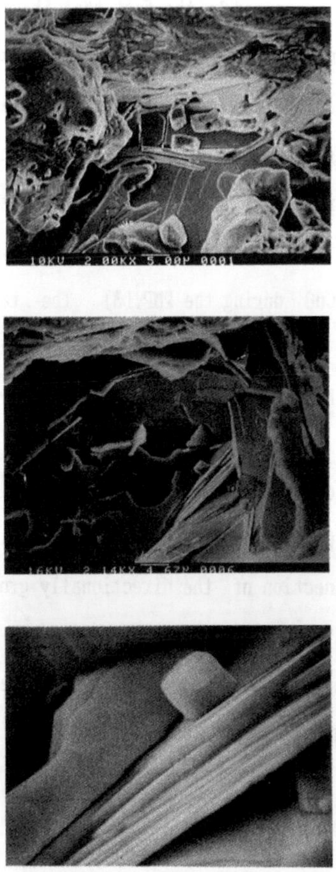

Fig.2. SEM pictures taken in the tapes prepared by (a) PMP (T1) and (b) resintering (T2) in the thermomechanical treatment (1), respectively. (c) was taken near a pore present in the tape (T2).

(a)

(b)

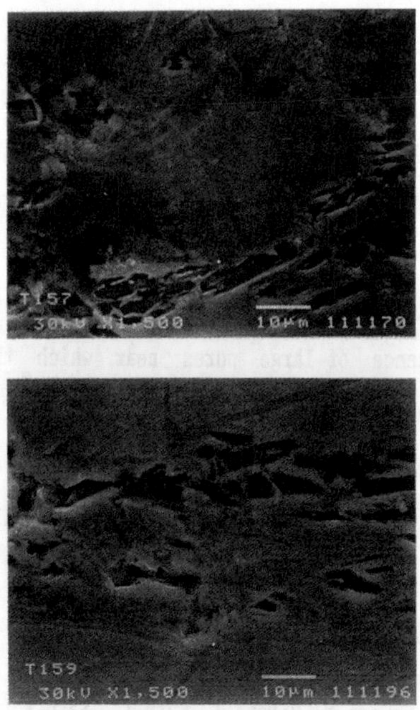

Fig.4. SEM pictures taken in the tapes prepared by resintering after the PMP with cooling rates with (a) 20 (T6) and (b) 3.5 ℃/hr (T7) in the thermomechanical treatment (2), respectively.

high degree of crystallographic misorientation as shown in Fig.2(c). This misorientation seems to be responsible for the observed low Jc. In conclusion, the PMP seemed to hinder the grain texturing in a long range order, contrast to the result of Jin et al.[2]

The thermomechanical treatment (2) was employed as an attempt to extend the locally-developed, dense layer-like micromorphology to a long range order. The purpose for the attempt was to investigate what happens if the PMP is employed after the 2223-phase grains are fully grown and well textured. The tape (T3) prepared by a repetition of the combined process of rolling and sintering showed a SC transition from 107 to 102 K and a Jc of 2,000 A/㎠ at 77 K and 0 T. The tape (T3) was mainly consisted of the 2223 phase as shown in Fig. 3(a). When the tape (T3) was rolled to a thickness of 100 ㎛ and partial-melt processed, the tapes prepared by the PMP with colling rates of 20 (T4) and 3.5 ℃/hr (T5) showed transitions from 104 to 70 K and from 108 to 90 K, respectively. The tape (T5) contained larger amount of 2223 phase than the (T4) did, as shown in Fig. 3. The reason why the tape (T5) showed higher Tc and contained larger amount of the 2223 phase seemed to be caused by the fact that the tape (T5) was heat treated for a longer time in the 2223 phase-stable temperature region of 825 to 840 ℃ during the PMP than the tape (T4) was. When the tapes were resintered, there were no noticeable differences in XRD patterns, Tc and Jc. However, the trace of directional growth of the 2223 phase seems to be a little more pronounced in the tape prepared with a slower cooling rate, as shown in Fig. 4. Another important thing is that the XRD peak for Ca_2CuO_3 appeared after the tapes experienced the PMP. This is due to the decomposition of the 2223 phase into 2212, Ca_2CuO_3 and CuO during the PMP.[3] The tapes (T6) and (T7) resintered for 24 hr showed low Jc's below 1,000 A/㎠ at 77 K and 0 T. This is thought to be due to the resintering time not enough to recover 2223 phase from 2212. Unfortunately, a prolonged resintering was not conducted in this series of tapes.

In conclusion, the attempt to extend the locally developed, dense and layer-like micromorphology to a long range order by obtaining the well-textured morphology prior to the PMP, seems to fail due to the presence of large pores near which the misoriented connection of the directionally grown layer-like structures is prominent. Repetitions of the rolling after the PMP seems to be required to eliminate the pores and to enhance the long-range grain texturing. The studies to use the PMP and the rolling in appropriate stages to get a dense and will-textured morphology are in progress.

REFERENCES

1. Ueyama M, Hikata T, Kato T, Sato K (1991), Jpn. J. Appl. Phys. 30 : L1384
2. Jin S, Graebner J E, Tiefel T H, van Dover R B, white A E, Kammlott G W (1991) Physica C 173 : 75
3. Ullrich M, Heinemann K, Schaper W, Freyhardt H C (To be appeared in Proc. 6th Int'l Workshop on Critical Current, July 8-11, 1991, Cambridge, England)

Observation of Interface Between Bi-Superconducting Oxide and Metals Using HREM and STM

K. SUENAGA, C. IWAMOTO, J. MATSUZAKI, H. ICHINOSE, M. MORI, and Y. ISHIDA

Institute of Industrial Science, The University of Tokyo, Minato-ku, Tokyo, Japan

ABSTRACT

Structure of interfaces between Bi-superconducting oxide(BSCCO) and metals were examined using high resolution electron microscopy(HREM) and scanning tunneling microscopy(STM). The BSCCO-Ag interface was direct with no visible reaction product layer. Epitaxial relationships were noticed both in HREM and STM. Deposited silver atoms tended to align parallel to a axis of BSCCO. The BSCCO-10vol%Ag nanocrystalline composite showed optimum performance in Tc-zero and Jc.
KEYWORDS:interface, nanocrystalline superconducting composites, HREM, STM

INTRODUCTION

When the oxide superconductors are applied to electron devices which contain thin metal interlayers or to superconducting wires which have many interfaces with metals to deform more easily, the structure of interfaces between superconducting oxide and metal is very influential to the electronic and mechanical properties of these materials. High resolution electron microscopy(HREM) is one of the most efficient method to analyze the structure of interfaces, and we can observe cross section of the interfaces by HREM. On the other hand, by scanning tunneling microscope(STM) we can observe not only surface morphology of materials, but process of formation of interfaces. These two methods are used to collect the atomic information to foresee the problems when oxide superconductors are applied practically.

EXPERIMENTS

a)SAMPLE PREPARATION
Nanocrystalline BSCCO was prepared from single crystals by ball-milling and that of silver by gas deposition in Argon at 1 atom. The nanocrystaline composite produced by mixing the two powders by supersonic vibrator, ball-milling the mix, pelletyzing and sintering.[1)]

b)HREM OBSERVATION
Cross sectional HREM of interfaces between BSCCO and Ag was performed after ion-thinning the as sintered and hot pressed nanocrystalline composite with the average train diameter 100nm.

c)STM OBSERVATION
In STM observation we used single crystals of BSCCO prepared by flux method and X-ray diffraction profile confirmed us that as cleaved surface of single crystals was parallel to (001) plane. A small amount of Ag was evaporated on the surface and the morphology was observed by STM, after as cleaved surface was observed. These succesive experiments were performed in UHV chamber without exposure to atmosphere.

RESULTS

We show the dependence of deformability of superconducting nanocomposite on Ag fraction in fig.1. The remarkable increase of deformability are noticed on the samples with more than 10vol%Ag in condition of hot-pressed at 1023K. Therefore we concluded that addition of more than 10vol%Ag to nanosuperconductor is efficient to workability. On the contrary, fig.2 indicated that addition of more than 20vol%Ag causes unfavorable results, that is, temperature-resistivity curves of these samples have tales below transition temperature. This means a decrease in current density as the result. For the reasons mentioned above, we concluded that BSCCO-10vol%Ag is the optimum fraction for both in mechanical and superconducting characteristics.

Fig.3 shows TEM image of nanocomposite which shows the optimum performance. A HREM image of the interface is shown in fig.4. The lattice images of BSCCO(200) and Ag(200) can be seen at the interface, and there is neither visible reaction product layer nor amorphous phase.

We also observed as cleaved surfaces and metal deposited surfaces by STM to investigate the process of formation of the interfaces. Fig.5 shows STM image of BiO-layer of BSCCO which appears on as cleaved (001) surface because of very weak inter-connection of these layers. In this figure, the bright regions are thought to correspond to Bi atoms which locate with 0.54nm intervals, and oxygen atoms are not visible[2]. And we can also see that there are the modulated structures along b axis with periodicity of 2.7nm which corresponds to 5 times as the intervals of Bi atoms[3].

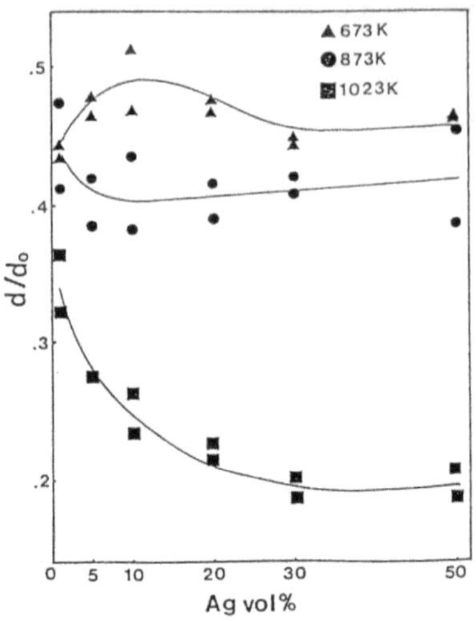

Fig.1 Ag fraction dependence of deformability with three hot press conditions.

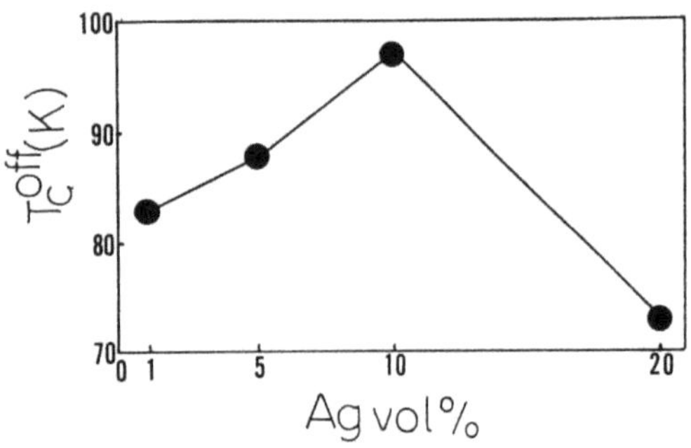

Fig.2 Ag fraction dependence of Tc-offset.

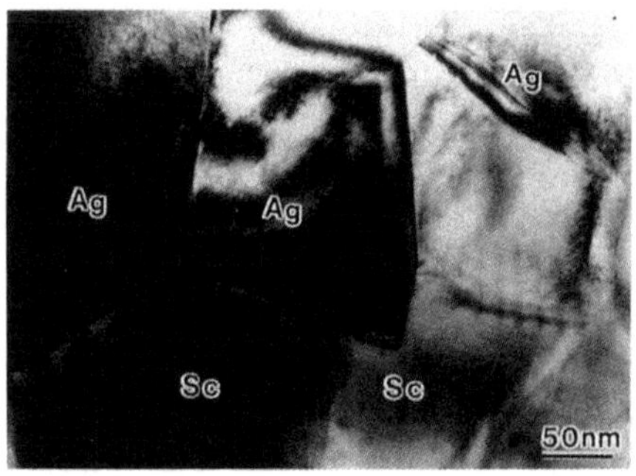

Fig.3 TEM image of nanocrystalline composite.(BSCCO-10vol%Ag)

We tried again to observe the surface of BSCCO after we deposited Ag on the as cleaved surface. Fig.6 shows the morphology of Ag deposited surface observed by STM. In this figure, the bright regions are due to Ag clusters and we found that there is a tendency of them to align parallel to a axis as compared with the as cleaved surface observed before deposition(fig.7).

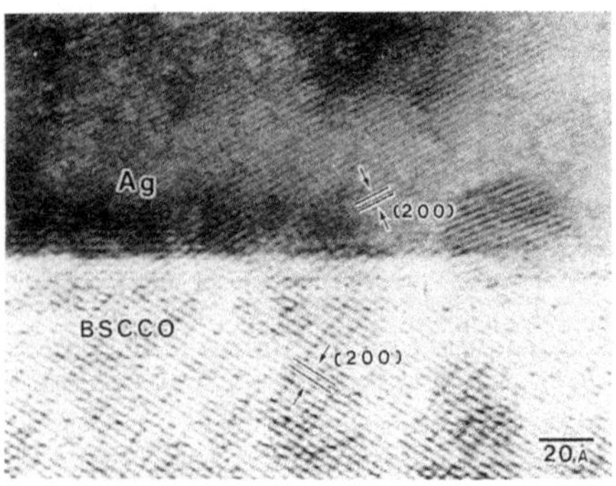

Fig.4 HREM image of the BSCCO-Ag interface.

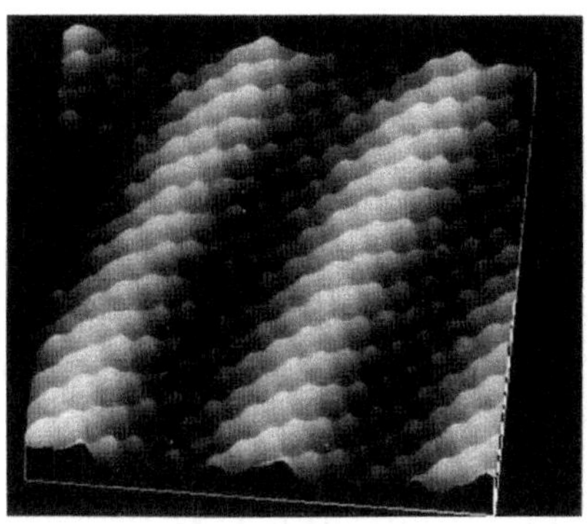

Fig.5 STM image of BiO layer of BSCCO.

Fig.6 STM image of Ag deposited surface.(30nmx30nm)

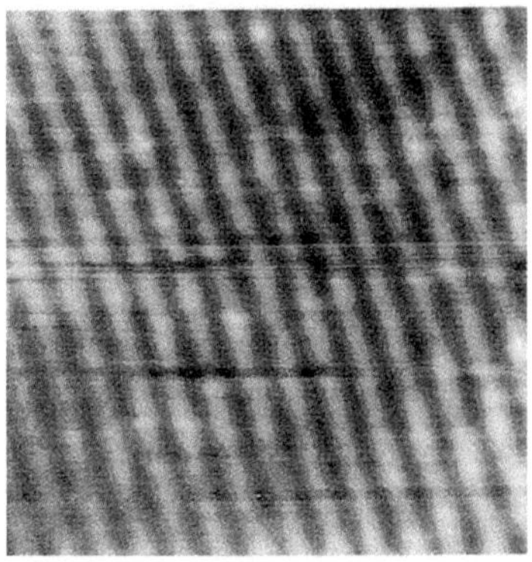

Fig.7 STM image of as cleaved surface of BSCCO. (30nmx30nm)

DISCUSSION

We observed the interfaces of nanocrystalline composites were direct with no reaction products layers or amorphous phase in HREM. In samples before annealing, we occasionally found the amorphous phase, but this phase seemed to disappear during annealing. This is favorable result to resolve many problems which occur when nanocomposites are used as superconducting wires.

In STM observations, it is not clear yet on which site Ag clusters nucleate first; Bi atoms or O atoms, but we found that deposited Ag atoms tend to align parallel to [100] direction of BSCCO, possibly due to influence of the modulated structure. We considered three possibilities which leads to the distinct growth directionality. The influence of (1)the fluctuation of lattice constant of [010] direction accompanied with the modulated structure,(2)the existence of extra oxygen, proposed by Zandbergen et al.[4)](3)the substitution atoms of Bi sites.

CONCLUSION

We made several experiments on interfaces of superconducting oxide and metal, and found that the form of deposited Ag atoms is influenced by the modulated structure of superconducting oxide substrate and that these interfaces are direct without reaction product layer and amorphous phase. These results can be indications to design super-normal-super Josephson junction, and latter result may enhance the possibility to realize superconducting wire made by nanocrystalline composites.

REFERENCES

1)J.Matsuzaki and Y.Ishida:to be published
2)M.D.Kirk, J.Nogami, A.A.Baski, D.B.Mitzi, A.Kapitulnik, T.H.Genelle, C.F.Quate:Science 242,(1988)p1673
3)Y.Matsui, H.Maeda, Y.Tanaka and S.Horiuchi: Jpn.J.Appl.Phys.27(1988)L372
4)H.W.Zandbergen, W.A.Groen, F.C.Mijlhoff, G.van Tendeloo and S.Amelinckx:Physica C 156(1988)p325

5.2 Wires and Tapes — (3) TBCCO

A Superconducting Wire with Strongly Pinned Tl-(1223) Phase in Liquid Nitrogen

KATSUZO AIHARA[1], TOSIYA DOI[1], ATSUKO SOETA[1], SEIJI TAKEUCHI[1], TOYOTAKA YUASA[1], KAZUHIDE TANAKA[1], NAOMI INOUE[1], MASAHIRO SEIDO[2], TOMOICHI KAMO[1], and SINPEI MATSUDA[1]

[1] Hitachi Research Laboratory, Hitachi, Ltd., Hitachi, Ibaraki, 319-12 Japan
[2] Advanced Research Center, Hitachi Cable Ltd., Tsuchiura, Ibaraki, 300 Japan

ABSTRACT

The Tl-Sr-Ca-Cu-O system gives rise to two superconducting phases with a single Tl-O plane, i.e. the (1212) and (1223) phase. Introduction of pinning centers into the Tl-(1223) phase has been performed by modifying the atomic compositions and by optimizing the processing conditions. A strongly pinned (Tl/Pb)-(Sr/Ba)-Ca-Cu-O bulk superconductor has been developed with a magnetic Jc over 43,000 A/cm^2 at 77K and 1T. Au-Pd sheathed tapes with the Tl-(1223) core were fabricated by a drawing-rolling method. The zero-resistance Tc of the tape was 114K in the absence of a magnetic field and 79K at 8T. The transport Jc was 19,000 A/cm^2 in the absence of a magnetic field, and the Jc stayed several 100 A/cm^2 up to 8T at 77K.

KEY WORDS: Tl-system superconductor, introduction of pinning centers, tape shaped wire, magnetic critical current density, transport critical current density

INTRODUCTION

For the application of the high temperature superconductors to magnets it is prerequisite to develop a wire with high current density (Jc) in the presence of significant magnetic fields. In our previous paper[1], the Jc of a tape-shaped wire with the Tl-(2223) phase in the Tl-Ba-Ca-Cu-O (TBCC) system, exceeded 10,000 A/cm^2 at 77K in the absence of magnetic fields, but the Jc decreased very sharply in the presence of low magnetic field below 1 T. The decreases of Jc were thought to be caused by absence of pinning centers. Therefore, it is important to introduce strong pinning centers into superconducting materials.

The Tl-Sr-Ca-Cu-O (TSCC) system gives rise to two superconducting phases, namely the (1212) phase with Tc~75K and (1223) phase with Tc~100K [2]. In contrast to the double Tl-O plane in the TBCC system, the TSCC system has the single Tl-O plane [3]. Although the two superconducting phases of the TSCC system are difficult to synthesize pure, the substitution of Pb into Tl site, Nd into Sr site, Y into Ca site, Ba into Sr site is sometimes effective for improving the purity levels[4].

We have made an extensive study to introduce pinning centers into a derivative of the TSCC system, i.e. (Tl/Pb)-(Sr/Ba)-Ca-Cu-O. We were successful to introduce pinning centers into bulk Tl-(1223) sample [5,6]. Herein, we report mainly on the properties of the strongly pinned Tl-(1223) bulk sample, and preliminary experiments of tape-shaped wire with the Tl-(1223) core fabricated by drawing-rolling method.

EXPERIMENTAL

Bulk samples consisting of (Tl/Pb)-(Sr/Ba)-Ca-Cu-O at typical atomic ratio (.5/.5)-(1.6/0.4)-2-3 were prepared. Appropriate amounts of BaO, SrO, CaO and CuO were mixed and calcined at 870 °C for 20 hours and this resultant Ba-Sr-Ca-Cu-O powder was then further mixed with Tl_2O_3 and PbO, and pressed into pellets. The pellets were sintered at 800-900 °C for 10 hours. The resultant was ground and pressed into pellets. The pellets were heated at 950-1100 °C (partial melt) for 10-120 minutes, then annealed at 850-880 °C for 10-50 hours.
The XRD analysis showed that the major constituents are Tl-(1223), Tl-(1212) and $BaPbO_3$ after partial melt and annealing. In addition the SEM observation revealed the formation of $(Ca,Sr)_2CuO_3$ after the partial melt. The grain size of Tl-(1223) was 50~200 μm.

The electrical resistivity and Jc in magnetic fields applied perpendicular to the sample (1mm x 1mm x 20mm) were measured by the standard four-prove method with indium soldering contacts. The DC magnetization was measured using a vibrating sample magnetometer at a sweep rate of 10 Oe/sec. The measurement of the magnetization was made using a rectangular sample with 5mm length, 4mm width, and 3 ~ 0.12 mm thickness. The thickness was varied successively in the same sample. Magnetic fields were applied in the direction parallel to the thickness. The magnetization was also measured by a SQUID up to 5.5 T. The intragrain Jc was deduced from the hysteresis ΔM in M-H curves using Bean's model.

RESULTS AND DISCUSSION

Figure 1 shows the temperature dependence of electrical resistivities of the bulk sample in the presence of magnetic field up to 8 T. The resistivity dropped sharply and the zero resistance Tc was 117 K in the absence of magnetic field. The partial substitution of Tl by Pb and Sr by Ba increased the Tc to 117 K from ~100 K in the Tl-(1223) phase of the Tl-Sr-Ca-Cu-O system. As the external field is increased, the broadening of the resistive transition curves are observed, zero resistance being 91 K at 8 T. The influence of magnetic fields on the resistive broadening is much smaller than that reported previously on Tl-(2223) and Tl-(1223) samples [7,8].

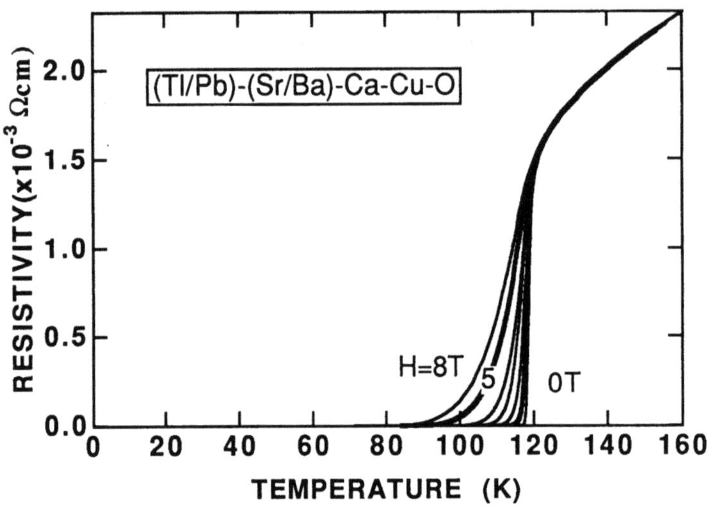

Fig. 1 Temperature dependence of electrical resistivities of the Tl-(1223) bulk sample in the presence of magnetic fields applied perpendicular to the sample. (magnetic field = 8, 5, 2, 1, 0.5, 0.2, 0.1, 0 T)

Figure 2 shows the dependence of the magnetization hysteresis ΔM on sample thickness. The thickness was 3, 1, 0.5, 0.25, and 0.12 mm, the smallest being in order of the superconducting grain size. The ΔM decreased linearly with a slight curvature from 3 to 0.25 mm and a sharp decrease at 0.12 mm. It is supposed that the sample thickness reached the size of the superconducting grain at 0.12 mm. The intragrain Jc was calculated from following relation; Jc=(20xΔM)/d ; where d is the thickness of the sample. Taking ΔM=26 emu/cc and d=0.12 mm, Jc is calculated to be 43,000 A/cm^2 at 77K and 1 T. The Jc value can compare with that obtained in Y-Ba-Cu-O prepared by QMG and MPMG method [9].

Figure 3 shows the temperature dependence of intragrain Jc obtained from magnetization measurements for the sample with 0.12 mm thickness. The intragrain Jc increases to about 600,000 A/cm^2 at 20 K and 1 T.

Using the drawing-rolling method, tape-shaped wires were fabricated with Au-Pd sheath and the Tl-(1223) core. The Tl-(1223) powder was put into a Au-Pd tubing with 6 mm diameter. A composite wire was drawn using a draw bench with outer diameter of 0.9 mm. The wire was cold rolled to tape of 0.3 mm in thickness. After cold working, the tape was heated at 850-960 °C for 1-50 hours.

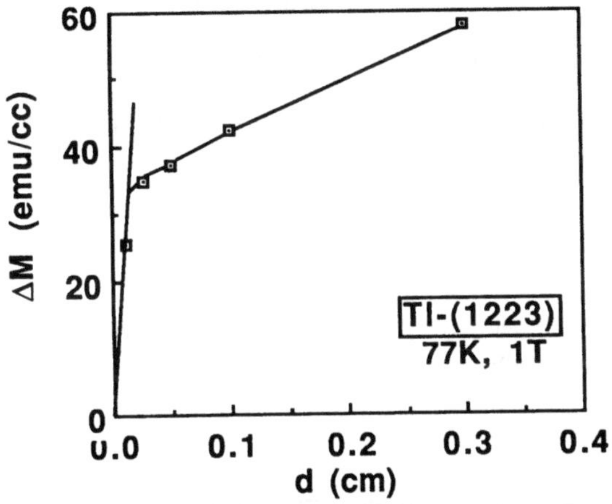

Fig. 2 Dependence of magnetization hysteresis ΔM on sample thickness d at 77K and 1T.

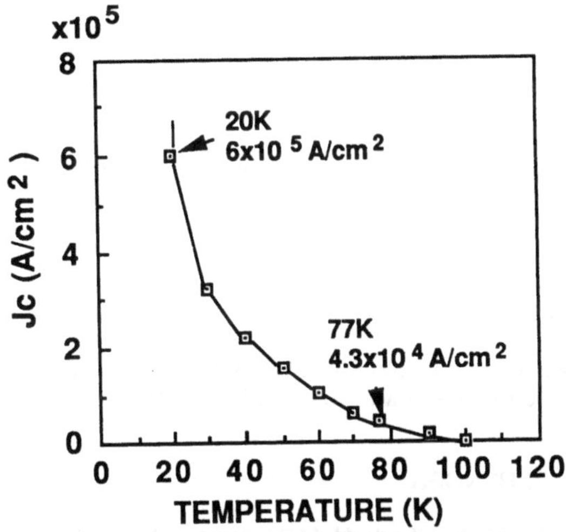

Fig. 3 Temperature dependence of intragrain Jc for the sample with 0.12 mm thickness at a magnetic field of 1 T.

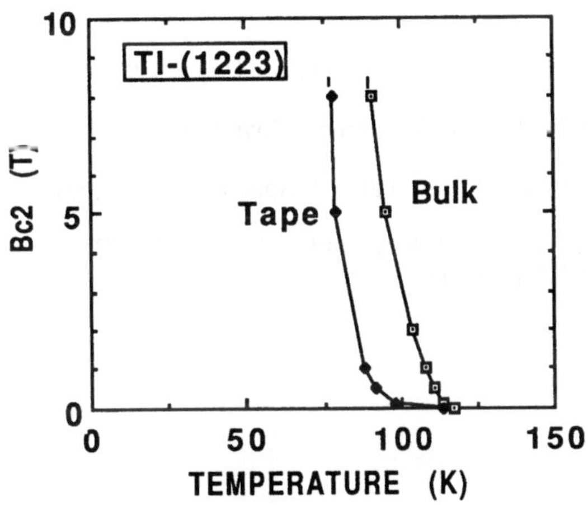

Fig. 4 Critical magnetic field Bc_2 for the Tl-(1223) tape and bulk samples.

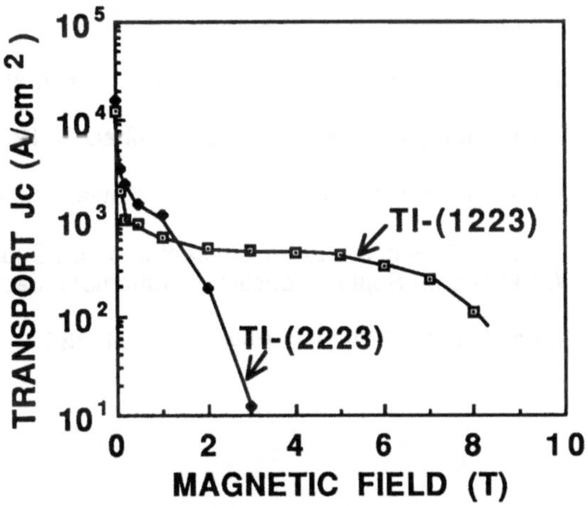

Fig. 5 Transport Jc of Tl-(1223) and Tl-(2223) tapes in magnetic fields at 77 K.

The temperature dependence of resistivity for the Tl-(1223) tape was measured in the presence of magnetic field applied perpendicular to the tape surface. Figure 4 shows the critical field Bc_2 for the Tl-(1223) tape and bulk samples. These data were determined using the zero resistance points of the resistive transitions. The zero resistance Tc of the tape sample was 114 K in the absence of magnetic field and 79 K at 8 T.

Figure 5 compares the transport Jc's between Tl-(1223) and Tl-(2223) tapes at 77K in magnetic field. There was no observable difference in the Jc's of these tapes in the absence of magnetic field. While the Jc of the Tl-(2223) tape becomes zero at ~3 T , the Jc of the Tl-(1223) tape stays relatively high, ~500 A/cm^2 between 1 and 7 T. The sharp decrease in Jc below 0.1T is thought to be caused by weak links between superconducting grains. From the SEM observation there are several non-superconducting phases, e.g., $BaPbO_3$, $CaCuO_2$, and $(Ca,Sr)_2CuO_3$, which could be a cause of weak link. It is also observed that the orientation of Tl-(1223) and Tl-(1212) crystallites are almost random in the present samples. In order to increase the inter-grain Jc we have to reduce drastically the interfering phases between superconducting grains and to improve the orientation.

At present the nature of pinning centers is not clear. It is noted, however, that pinning centers have been successfully introduced into the Tl-(1223) system. It is pointed out that the magnetic flux is not effectively pinned in the Tl-(2223), Tl-(2212), Bi-(2223) and Bi-(2212) systems due to their thick double Tl-O and Bi-O planes. On the other hand, Y-Ba-Cu-O and Tl-(1223) into which pinning centers could be introduced, do not include such insulating double layers. From the SEM observation, there are many non-superconducting particles, but those particle size seem to be too large for pinning centers. Other defect such as vacancies, dislocations, stacking fault, etc. might be pinning centers. Further investigations are necessary.

ACKNOWLEDGEMENT

The authors would like to thank Mr. Takeshi Ozawa and Mr. Hiroyuki Akata for assistance, Dr. Osvaldo F. Schilling and Mr. Tosihide Nabatame for helpful discussions.

REFERENCES

1. Aihara K, Okada M, Matsumoto T, Matsuda S, Hosono F, Seido M (1991) IEEE Trans. Mag. 27: 894

2. Matsuda S, Takeuchi S, Soeta A, Suzuki T, Aihara K, Kamo T (1988) Jpn. J. Appl. Phys. 27: 2062

3. Soeta A, Suzuki T, Takeuchi S, Kamo T, Usami K, Matsuda S (1989) Jpn. J. Appl. Phys. 28: L1186

4. Sheng Z, Herman A (1988) Nature 332: 138

5. Matsuda S, Doi T, Soeta A, Yuasa T, Inoue N, Aihara K, Kamo T (in press) Appl.Phys. Lett.

6. Doi T, Okada M, Soeta A, Yuasa T, Aihara K, Kamo T, Matsuda S (in press) Physica C

7. Togano K, Kumakura H, Mukaida H, Kawaguchi K, Nakao M (1989) Jpn. J. Appl. Phys. 28: L907

8. Liang W "Effect of Carrier Concentration on Superconductivity" Presented in The 3rd ISTEC Workshop on Superconductivity, Kumamoto, Japan (May, 1991)

9. Murakami M (1990) Mod. Phys. Lett. B4: 285

Power Current Leads Using Ag-Sheathed High-T_c Superconducting Tapes

TAKAAKI SASAOKA, TADASHI UMEZAWA, AKIRA NOMOTO, NORIYUKI TATSUMI, and MASAHIRO SEIDO

Advanced Research Center, Hitachi Cable Ltd., Tsuchiura, Ibaraki, 300 Japan

ABSTRACT

To investigate a feasibility of Power current lead using High-Tc superconducting tapes, the assembled conductors stacking Ag-sheathed High-Tc superconducting tapes were fabricated and V-I characteristics were measured. The fabricated conductors had the electrically stable against aging and over current test, whose properties corresponded to that of tape itself considering self-magnetic field resulting from transport current. Calculated results as a function of Current Leads indicate that these conductors can efficiently reduce loss of Liquid He during operation of superconducting facilities.

KEY WORDS: current lead, Tl system superconductor, superconducting tape, contact resistance

INTRODUCTION

Superconducting current leads show promise as a important application for oxide superconductors[1]. Metal-sheathed superconducting tapes are the most practical in utilization as conductors[2]. Because of their high critical current density, smaller size current lead can be obtained by assembling the tapes. The assembled form and length of the conductor can be selected according to user demand. Both electrical and thermal stabilities accommodate joining to other conductors with low-resistance.

The superconducting conductors were fabricated to confirm the desired properties of the metal-sheathed superconducting tapes. The temperature and heat dispersion were calculated with the view to using the power current leads for saving liquid He.

PREPARATION OF ASSEMBLED SUPERCONDUCTING CONDUCTORS

The conductors were fabricated by stacking with the multiple silver-sheathed Tl2223(Tl-Ba/Sr-Ca-Cu-O) superconducting tapes as shown in figure 1 in two different assembly processes. One method involved Construction followed by the heat Reaction(referred to herein as C&R method). The other involved heat Reacting before Construction(the R&C method).

V-I CHARACTERESTICS MEASUREMENTS OF FABRICATED CONDUCTORS.

3.1Superconducting Properties of Stacked Tapes

Investigation into the optimum design was required to establish the relationship between electrical properties of a conductor assembled with multiple superconducting tapes and that of a superconducting tape itself. Assemblies stacked with different cross sectional areas S(cm2) and V(Voltage)-J(A/cm2) measurements were investigated.

Figure 2 shows the experimental relationship between the superconducting critical current Ic(def 0.1μ V/cm) and cross sectional area S of the conductors, where the 77K design line presents the S*Jc(B) values calculated by multiplying S by Jc(B). The Jc(B) is tape critical current density at 77K as determined dependent on the surface self-magnetic field B corresponding to the transport current Ic. The measured Ic of the assembled conductors show almost complete agreement with the design line for both the R&C and the C&R assembly methods

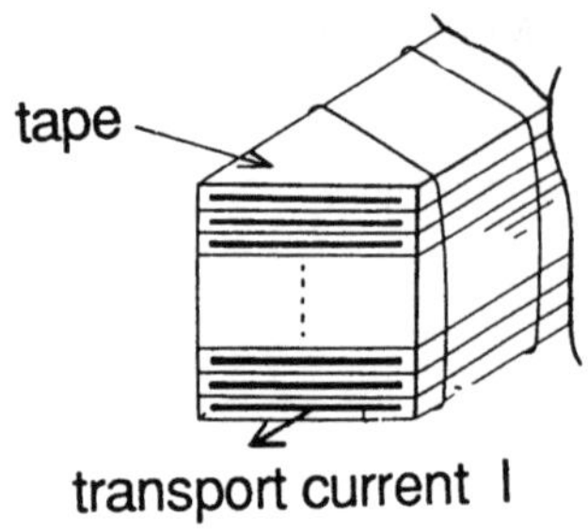

Figure 1 the structure of the assembled superconducting conductor.

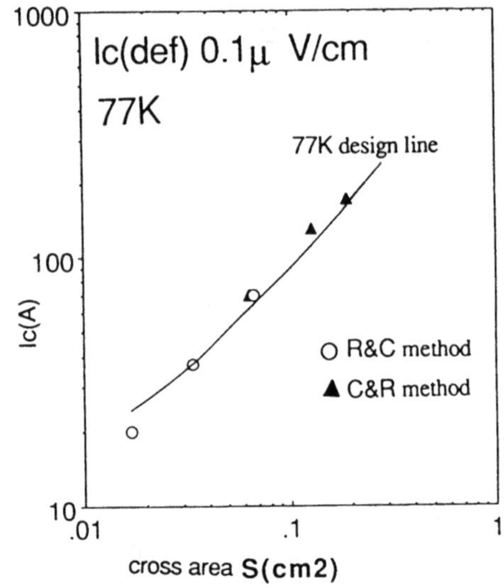

Figure 2 critical current Ic of the fabricated conductors

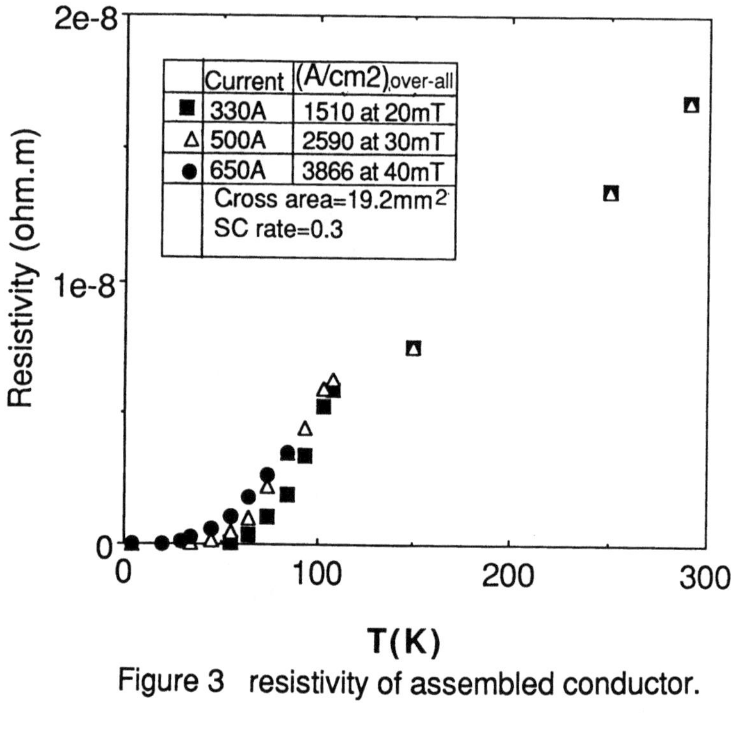

Figure 3 resistivity of assembled conductor.

Figure 3 shows the temperature dependence of the resistivity at transport current I up to 650A for the fabricated assembled conductor stacked with superconducting tapes. In this figure, the data at 77K are measured values for the assembled conductors, while the others data are calculated from the experimental values of V-J-B-T characteristics in the tape itself at each temperature T. Both J>Jc and J<Jc, it was confirmed to be complete agreement measured values of the assembled with estimated values from that of tape. We can thus calculate the V-J-T characteristics of the assembled conductor for given V-J-B-T characteristics as tape itself.

3.2. Current Test of Joint and Assembled Conductors.

A conductor must be assembled to act as a function of a current lead to be joined a metal-lead in a way that each superconducting tape will have a low, almost equal join-resistance. We measured the joint resistance of the assembled conductors connected to Cu-leads using a four-probe method up to 650A at 77K. Figure 4 shows the V-I characteristics in the conductor(S=19mm2), with the joint-resistances estimated to be about 40n ohm per one joint. It was also confirmed that a stable current up to 650A was possible(limitations being caused by the power supply system) through the joint at 77K, though the Ic of the fabricated conductor was 184A at the same temperature.

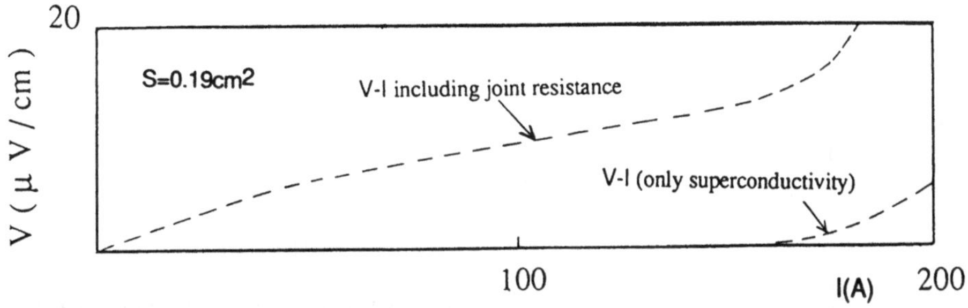

Figure 4 V-I characteristics in the joint resistance of the assembled conductor.

3.3. Heat-Cycle Test of The Conductors

It is important for practical usage of High-Tc superconductors to ensure against any aging the effects resulting from the heat-cycle, atmosphere moisture over the long-term pass as shown in figure 5. Figure 6 shows no aging effects in the superconducting properties, over the test period.

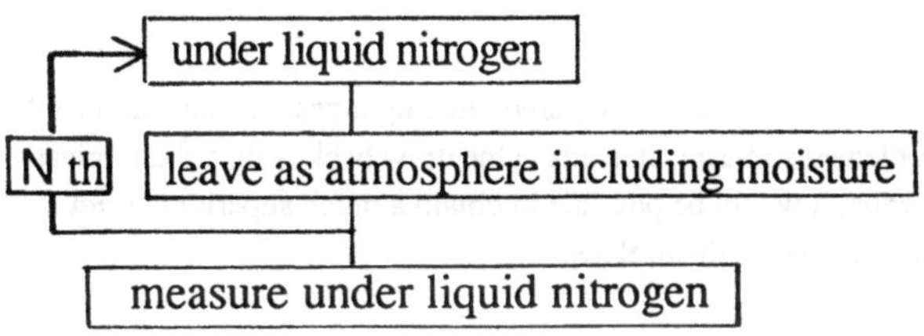

Figure 5 aging cycle test contents against long term pass.

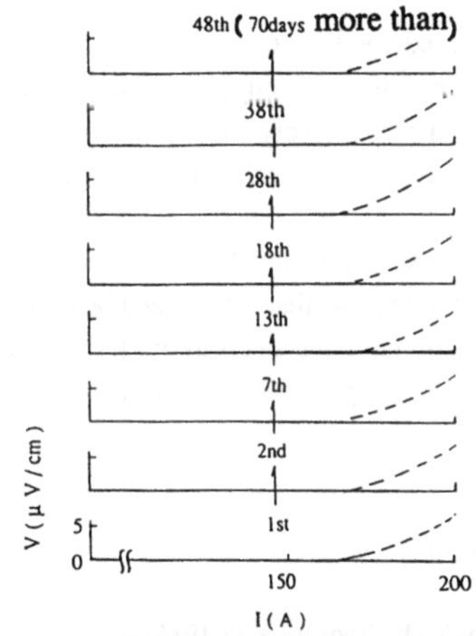

Figure 6 V-I characteristics at 77K under aging test.

CALCULATION OF HEAT DISPERSION UNDER CURRENT OPERATION.

We supposed a power current lead to be utilized at temperatures from 4.2K to 77K as shown in figure 7. Calculations presumed current lead adopting the fabricated superconducting conductors whose electrical characteristics are shown in figure 3. The thermal conductivity of the Ag-sheath was calculated[3] from the residual resistance rate RRR=121(measured at 4.2K) and Debye temperature 180K. It was also considered the heatload caused from joint-resistance 40n ohm. The cooling condition of the conductor was dominated by the thermal conductivity and the He gas caused from only self-heatload[4], having constant temperature 77K for a distance of 1m from the upper surface of the liquid He.

Figure 8 shows the calculated results of temperature distribution of the conductor under current 650A, along with the heat dispersion (Watt/m). As a result of the numerical calculation, it was found below. Temperature distribution of the fabricated conductor was little affected by operating currents of less than 650A, though the current exceed the Ic of the assembled at 77K. Heat intrusion into the low temperature 4.2K was estimated to be 1.2Watt/kA in the case of 650A operation. The heat dispersion from the fabricated joint 40n ohm had no adverse effect in this situation.

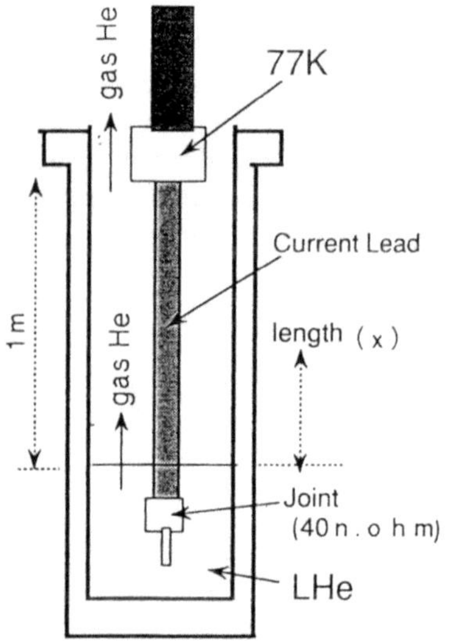

Figure 7 the structure model of the current lead.

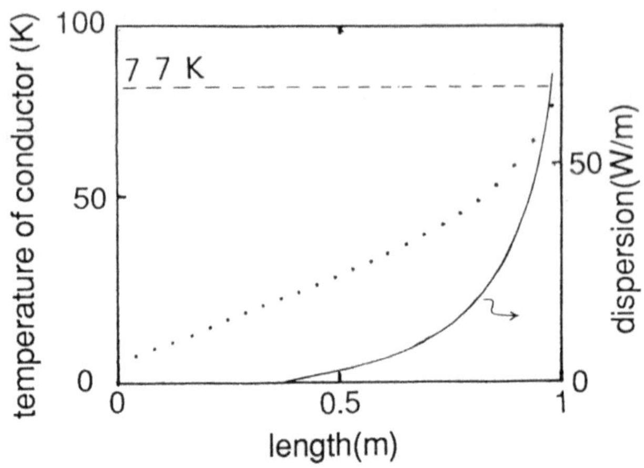

Figure 8 calculated results of temperature and heat dispersion distribution.

CONCLUSION

The fabricated conductors assembled with the multiple silver-sheathed superconducting tapes have the electrically stable properties estimated from the properties of a single tape with consideration self-magnetic field resulting from the transport current. Very low joint-resistance 40n.ohm was obtained under more than 184A at 77K. It was not observed aging effects against long term use.

Calculated results indicate that the conductors assembled with Ag-sheathed superconducting tapes, display almost the same performance as that recognized being the lastest obtained so far,in Cu leads. One draw back is that, the fabricated conductors do not have an optimum form for Current-Leads, It would be possible to obtain a much superior current lead by optimizing the properties and forms of the superconducting tapes themselves.

REFERENCE

1)Buyanov,Yu.L.Cryogenics 25(1985)94.
2)F.Hosono,et al.:Proc.ISS90(Springer,1991)401.
3)Guy K. White: Experimental Techniques in Low-Temperature Physics, pp.284.
4)D.Gusewell and E.-U. Haebel: Proc. IC-EC 3(1970)187.

Properties of Silver-Sheathed Tl-Series Oxide Superconducting Tapes and Coils

MASAHIRO SEIDO[1], AKIRA NOMOTO[1], TADASHI UMEAWA[1], JUNICHI SATO[1], TAKAAKI SASAOKA[1], KATSUZO AIHARA[2], and TOMOICHI KAMO[2]

[1] Advanced Research Center, Hitachi Cable Ltd., Tsuchiura, Ibaraki, 300 Japan
[2] Hitachi Research Laboratory, Hitachi, Ltd., Hitachi, Ibaraki, 319-12 Japan

ABSTRACT

As a preliminary study on the development of Tl-series oxide superconducting coils operated in liquid nitrogen, a twelve-layered pancake coil made of silver-sheathed Tl-2223 tapes was fabricated and the characteristics were measured. This multi-layered coil generated a maximum magnetic field of 510 Gauss at 77K and 1410 Gauss at 4.2K, respectively. To improve characteristics at high magnetic fields over 1T, the properties of the new composition Tl-1223 oxide superconducting bulk and wires were examined. The Tl-1223 bulk has large ΔM in M-H characteristics even over 1T at 77K. Also the Jc value of Tl-1223 wire remains about 500 A/cm^2 in high magnetic fields over 5T at 77K. It was confirmed that Tl-1223 wire has a potential to produce a high Jc in high magnetic fields at 77K.

KEY WORDS: Tl-series superconductor, pancake coil, tape, V-I characteristics

INTRODUCTION

Since high-Tc superconductors with critical temperature (Tc) above the temperature of liquid nitrogen were discovered [1~3], many studies have been carried out to produce superconducting coils and cables for use with liquid nitrogen. Such applications require long and homogeneous wires with high critical current density (Jc). The development of high-Tc superconducting coils, in particular, requires a high Jc(above 10,000 A/cm^2) in magnetic fields and involves various techniques for manufacturing superconducting coils.

In our previous paper [4,5], silver-sheathed Tl-series oxide superconducting long tapes were formed in a continuous pressing method and fabricated into single pancake test coils. The Jc of a tape made from a Tl-2223 (Tl-Ba/Sr-Ca-Cu-O ;TBSCCO) superconducting oxide core exceeded 10,000 A/cm^2 at 77K in the absence of magnetic field, but the Jc deteriorated very sharply on application of magnetic field.

In this paper, long silver-sheathed Tl-2223 tapes were wound into pancake coils and piled in layers of twelve coils, to study the transport properties in long tapes and the performance of the HTS pancake coil. In an attempt to improve the characteristics of Tl-series superconductors, the properties of new composition Tl-1223 (Tl/Pb-Sr/Ba-Ca-Cu-O) oxide superconducting bulk and short tape are reported here in.

FABRICATION OF A TEST PANCAKE COIL WITH TWELVE LAYERS MADE FROM Tl-2223 SUPERCONDUCTING LONG WIRES

Silver-sheathed Tl-2223 (Tl-Ba/Sr-Ca-Cu-O) superconducting tapes were fabricated by the continuous pressing method [4] shown in Fig.1. The superconducting powder was placed in a silver tube. The diameter of the tube was reduced by drawing. The drawn tube was then continuously pressed into tape under high pressure, before sintering in a flowing oxygen atmosphere at a temperature between 1073 and 1173 K for 2 to 10 hours. Microstructural examinations of longitudinal sections of the continuous pressed tape revealed that the superconducting oxide core was uniform and highly densified polycrystalline. The Jc value reached 10^4 A/cm^2 at 77K,0T.

A high-Tc superconducting pancake coil with twelve layers was subsequently fabricated using 210m lengths of silver-sheathed Tl-2223 tape of 10^4 A/cm^2. The specifications and photographs of the pancake coils are shown in Table 1 and Fig.2, respectively.

As the Tl-2223 core is very brittle, the tapes could not be sintered until after the coils had been wound to avoid damage during winding. To make single pancake coils, the tapes were wound on cylindrical bobbins of silver, as indicated in Fig. 2. Thin alumina sheets were used for the insulator between the tapes. The single coils were then heat-treated. Joints to the current supply leads and voltage taps were made by attaching small pieces of Ag foil to the tape before heat treatment of the coils. The sintered single coils were piled twelve layers high, then the current supply leads were joined as shown in Fig.2.

The coil characteristics were measured in liquid nitrogen and liquid helium. Figure 3 shows typical V-I characteristics of the twelve-layered pancake coil (voltage terminal V3). When the coil current is increased, the resistive voltage across the coil rises steeply above 4.1A at 77K or above 11.2A at 4.2K. The critical currents (Ic) of the coil are 4.1A at 77K and 11.2A at 4.2K in the $10^{-13}\Omega$m criterion across the coil (tape length of the each single coil is 17.5m). When the coil current is furthermore increased, the current shifts to the silver-sheath from the superconducting core without coil quench phenomena and the resistive voltage across the coil approaches the value determined by the resistivity of silver-sheath at that temperature. When the coil current decreases, the V-I curves become nearly the same as in the case of the current increasing. Their critical currents are higher than those which are increasing and this difference of Ic's depends on the coil inductance and history effects of the superconducting tapes [6]. These V-I curves reveal that the multi-layered pancake coil was superconducting for the whole length and very stable.

In Fig.4, a comparison of performance is made between the short straight sample and the coil with load lines for maximum and central magnetic fields. The Jc values of the y-axis in Fig.4 were determined using the cross-sectional area of the superconducting core. The Jc of the coil on the load line of maximum field reaches that of the short sample Jc. At this point, the multi-layered coil generates a maximum magnetic field of 510 Gauss and a central field of 485 Gauss in liquid nitrogen. This coil was also tested in liquid helium. The generated magnetic fields are 2.7 times as high as those obtained in liquid nitrogen. The Jc of this coil was restricted by the Jc of the part of the tape where the maximum self-magnetic fields of the coil was induced.

ATTEMPTED CHARACTERISTICS IMPROVEMENT OF Tl-SERIES OXIDE SUPERCONDUCTORS AT 77K

As the generated magnetic fields mainly depend on the Jc-B characteristics of the wire used for the coil, to develop a high magnetic field coil in liquid nitrogen, the top priority is to develop a high Jc wire in high magnetic fields.

To improve the Jc characteristics in high magnetic fields at 77K, we researched and developed Tl-1223 bulk and silver-sheathed tape shaped wires, with properties comparable to Tl-2223 bulk and wire. We succeeded in introducing a strong pinning point similar to that in the MPMG processing of the Y-series [7], by providing a partial fusion heat-treatment at 900 ~ 1100 °C to a bulk material with the composition of $(Tl/Pb)_1(Sr/Ba)_2Ca_2Cu_3O_y$. Using this Tl-1223 bulk powder, we fabricated silver-sheathed short tapes, whose transport Jc-B characteristics were estimated. Figure 5 shows the M-H characteristics measured by a SQUID [8]. Contrary to the almost zero Jc of conventional 2223 phase bulk materials at 77K and 1T, Jc of this 1223 phase bulk material reaches 43,000 A/cm^2, as calculated from M-H characteristics. This material has a ΔM value even at 5T and 77K.

Figure 6 shows the transport Jc-B characteristics of silver-sheathed Tl-1223 and Tl-2223 tapes at 77K. The critical current densities (Jc) at 77K, 0T are 10,000 ~ 20,000 A/cm^2 for both material tapes. Under high magnetic fields over 1T at 77K, the Jc value of Tl-2223 wire goes down to

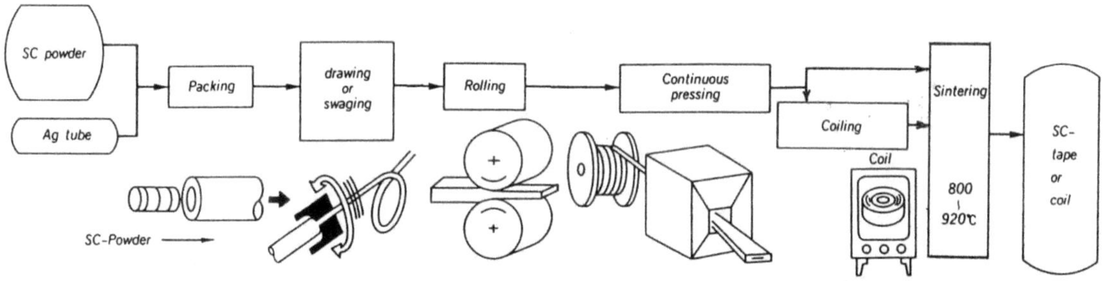

Fig.1 Fabrication process of silver-sheathed oxide superconducting tapes and coils.

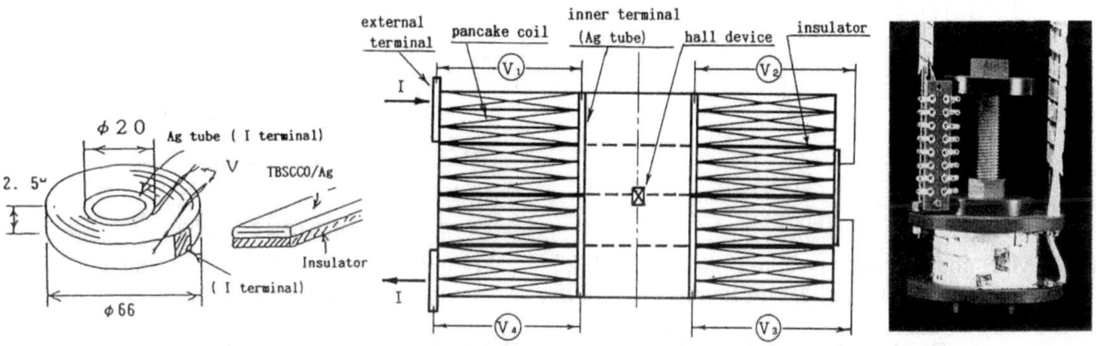

a. single pancake coil b. twelve-layered pancake coil c. twelve-layered pancake coil

Fig.2 Structures and appearance of single and twelve-layered pancake coil.

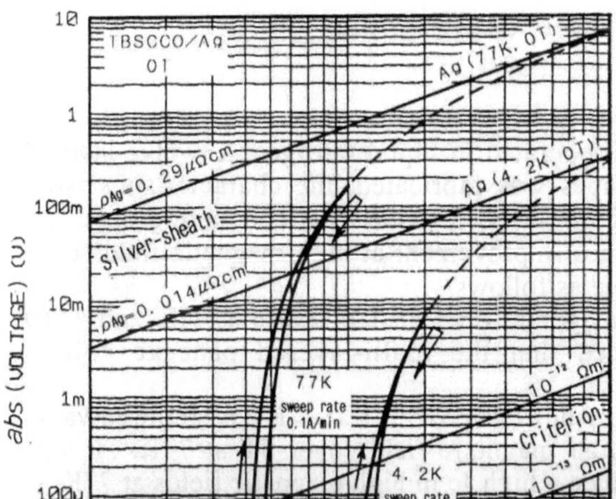

Fig.3 V-I characteristics of twelve-layered pancake coil (super-normal transition characteristics in wide range).

Table 1 Specifications of twelve-layered pancake coil.

Tape wire	Superconducting material	Tl-Ba/Sr-Ca-Cu-O
	Size	0.13mmx2.5mmx17.500mm
	Insulator material	Al_2O_3
	Critical current	8.2 A (at 77K, 0T)
	Critical current density	10^4 A/cm² (at 77K, 0T)
Coil	Inner diameter	φ20 mm
	Outer diameter	φ66 mm
	Height	34 mm
	Winding number	125 turn

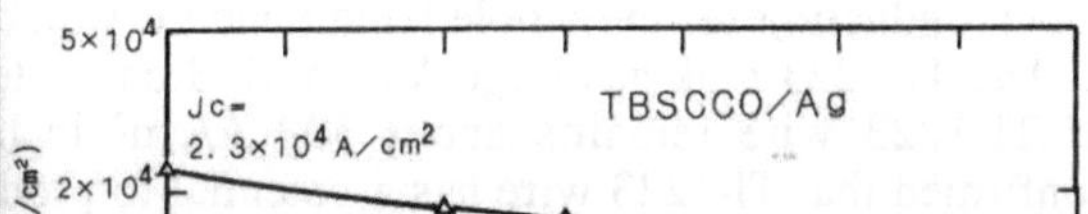

Fig.4 Comparison between performances of fabricated pancake coils and characteristics of short sample tape in magnetic fields.

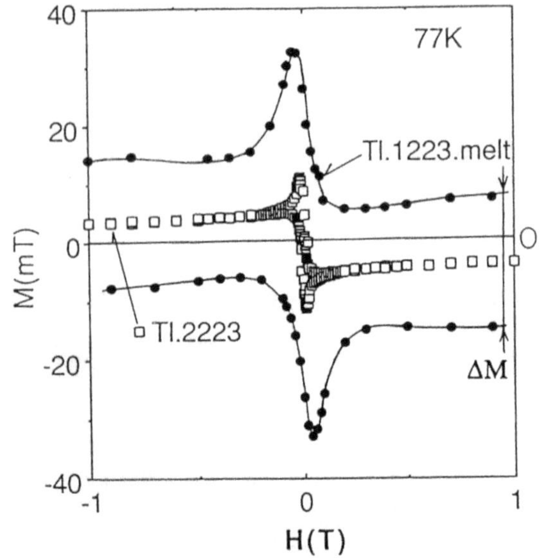

Fig.5 M-H characteristics of Tl-1223 and Tl-2223 superconducting pellets measured by SQUID.

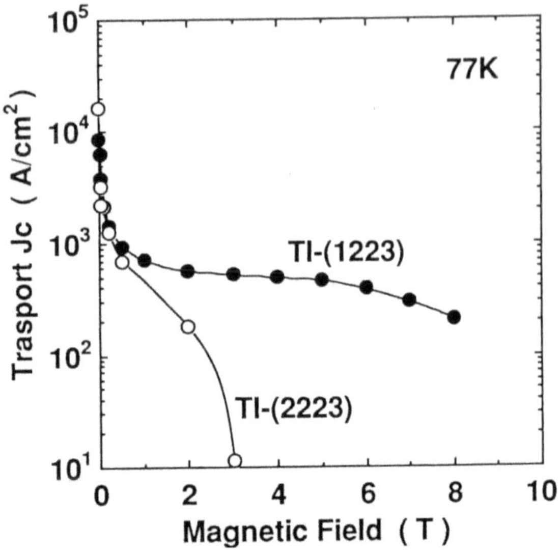

Fig.6 Transport Jc of Tl-1223 and Tl-2223 wire in magnetic field.

nearly zero. On the other hand, the Jc value of Tl-1223 wire remains about 500 A/cm^2 in high magnetic fields over 5T. It is confirmed that there are strong pins in the Tl-1223 material of the tapes.

CONCLUSION

To develop a Tl-series oxide superconducting coil operated in liquid nitrogen, a twelve-layered pancake coil made of silver-sheathed Tl-2223 tapes was fabricated, the characteristics were measured and the properties of the new composition Tl-1223 oxide superconducting bulk and wires were examined as an attempt at characteristics improvement at high magnetic fields over 1T. The conclusions of this study can be summarized as follows.

1. The multi-layered coil generated a maximum magnetic field of 510 Gauss at 77K and 1410 Gauss at 4.2K, respectively. It was confirmed that the multi-layered pancake coil is superconducting for the whole length and very stable.
2. The Tl-1223 bulk has large ΔM in M-H characteristics even over 1T at 77K. Also the Jc value of Tl-1223 wire remains about 500 A/cm^2 in high magnetic fields over 5T at 77K. It was confirmed that Tl-1223 wire has a potential to produce a high Jc in high magnetic fields at 77K.

REFERENCES

1. M. K. Wu, et al.: Z. Phys. B-Condensed Matter 64 (1986)189.
2. H. Maeda, et al.: Jpn. J. Appl. Phys., 27 (1988) 209.
3. Z. Z. Sheng, et al.: Nature, 332 (1988) 138.
4. F. Hosono, et al.: Proc. ISS'90 (Springer, 1991) 401.
5. K. Aihara, et al.: IEEE, Transactions on Magnetics, 27-2 (1991) 894.
6. M. Seido, et al.: Cryogenic Engineering, to be published.
7. H. Fujimoto, et al.: ISS'89 (Springer-Verlag, 1990) 285.
8. T. Sasaoka, et al.: JJAP, to be published.

5.3 Characteristics

ac Losses in $YBa_2Cu_3O_7$ and $Bi_2Sr_2Ca_2Cu_3O_{10}$ Superconductors at Power Frequencies

J. OREHOTSKY[1], K.M. REILLY[1], M. SUENAGA[1], T. HIKATA[2], M. UEYAMA[2], and K. SATO[2]

[1] Department of Applied Science, Brookhaven National Laboratory, Upton, NY 11973, USA
[2] Osaka Research Laboratory, Sumitomo Electric Industries, Ltd., Osaka, Japan

ABSTRACT

ac losses of sintered and melt-textured $YBa_2Cu_3O_7$ bars and powder-in-a-tube processed $Bi_2Sr_2Ca_2Cu_3O_{10}$ tapes were measured at 4.2, 65, 70 and 77 K and at 15-180 Hz. In general, the results are well described by the critical state model for the ac losses. However, the losses for the $Bi_2Sr_2Ca_2Cu_3O_{10}$ tapes at 4.2 K exhibited significant contributions from the eddy currents in the Ag sheath.

INTRODUCTION

After the discovery of the high temperature superconductors, a number of large scale applications of the new superconductors were envisioned. One of these was superconducting power transmission lines at ~77 K. However, it soon became obvious that realization of such applications utilizing the oxide superconductors will not be immediate since the transport critical current densities in these oxides were disappointingly low. The first 60 Hz ac loss measurements of the sintered $YBa_2Cu_3O_7$ bars [1] also confirmed similar difficulties as the losses were substantially higher than the level which are desirable for such applications. These difficulties originate with the inability of the large angle grain boundaries to carry large transport currents. More recently, various oxide fabrication techniques were developed to minimize this effect. One of these for $YBa_2Cu_3O_7$ is the so-called melt-texturing method [2-6] which is essentially a zone-refining method to enhance the directional growth of the oxide. The other is the powder-in-a-Ag sheath for $Bi_2Sr_2Ca_2Cu_3O_{10}$ [7,8]. In this case the crystallographic alignment is accomplished by flattening the composite and growing the oxide in a spatially confined condition in contact with Ag. Since both of these materials show large improvements in the transport J_c, it is of interest to reevaluate ac losses in these textured superconductors. Here, we will report on the losses in $YBa_2Cu_3O_7$ bars [6] which were zone-refined using a quartz-infrared-lamp base furnace and in the powder-in-a-tube $Bi_2Sr_2Ca_2Cu_3O_{10}$ tapes [7]. The results are discussed in terms of the critical state model for ac losses.

EXPERIMENTAL PROCEDURES

The experimental set-up for the ac loss measurements, which was used here is the same as the previous ones for the $YBa_2Cu_3O_7$ bars [1] except for the use of digital integration for calculating the losses [9]. In this arrangement, a specimen with N turns of a pick-up coil was axially placed in a solenoid which was powered by a variable frequency supply. In a such set-up, the ac power losses P (W/cm^2) in the specimen due to the induced surface current density σ (rms A/cm) by the applied field H is given by:

$$P = (1/N\ell\tau)\int_0^\tau H_p(d\phi/dt)dt \qquad (1)$$

where H_p is the peak magnetic field and $H_p = (4\pi\sqrt{2}/10)\sigma$, ℓ is the length of the perimeter of the specimen area which is perpendicular to the applied field direction and τ is the period of the applied field. The pick-up coil for the

$YBa_2Cu_3O_7$ bars was wound directly on the specimen. However, in the case of the Bi(2223)/Ag tapes, four pieces (~10 mm long, ~4 mm wide and ~0.15 mm thick) of the tapes were placed in a collapsible pick-up coil such that each piece of the tapes is placed on the face of a square mandrel. Then, the entire assembly was placed in a solenoid.

In order to study the effect of the variations in critical densities on the ac losses, a number of specimens having a range of critical current densities for both $YBa_2Cu_3O_7$ bars and Bi(2223)/Ag tapes were measured. Furthermore, for investigation of the loss-mechanisms, the dependence of the losses on the frequency of the applied fields was examined for f=15-180 Hz. The losses as a function of temperature were also determined at 4.2, 65, 70 and 77 K. The details of the preparation procedures for the melt-textured $YBa_2Cu_3O_7$ bars [6] and the Bi(2223)/Ag tapes [7] were given elsewhere.

RESULTS AND DISCUSSION

Before presenting the results of the measurements, a brief theoretical description of the ac losses in a Type II superconductor is discussed to facilitate the interpretation of the results. The most commonly used expression for the ac losses of a superconductor is based on the Bean critical state model [10]. According to the model, the hysteretic ac losses in a slab is given by:

$$P = 5\times10^{-7} f H_p^3 / (12\pi^2 J_c) \quad \text{for } H < H^* \text{ and} \quad (2)$$

$$P = 2\times10^{-8} d^2 f H_p J_c \quad \text{for } H > H^* \quad (3)$$

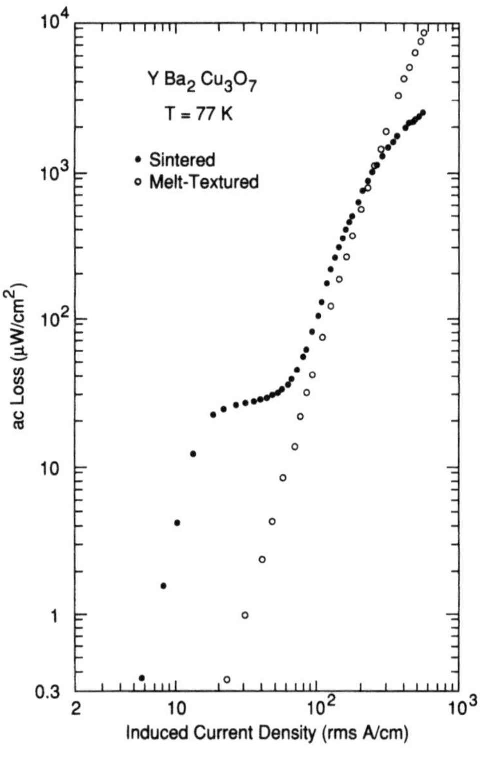

Fig. 1. ac losses as a function of the induced linear current density for a sintered and a melt-textured $YBa_2Cu_3O_7$.

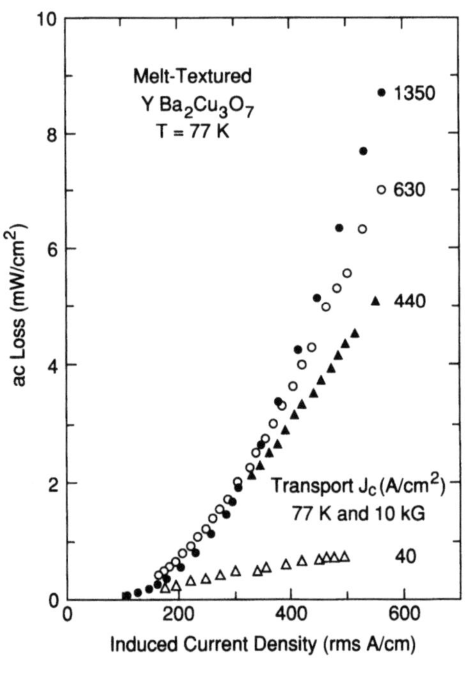

Fig. 2. ac losses as a function of the transport.

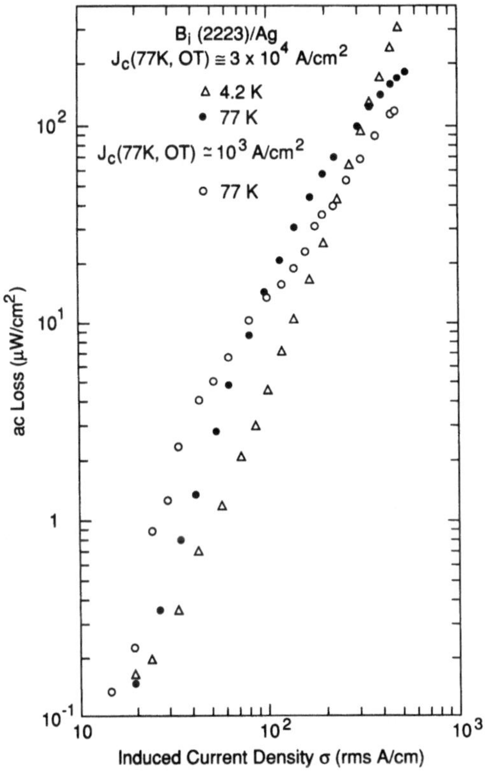

Fig. 3. ac losses in the BiSrCaCuO tapes at 60 Hz.

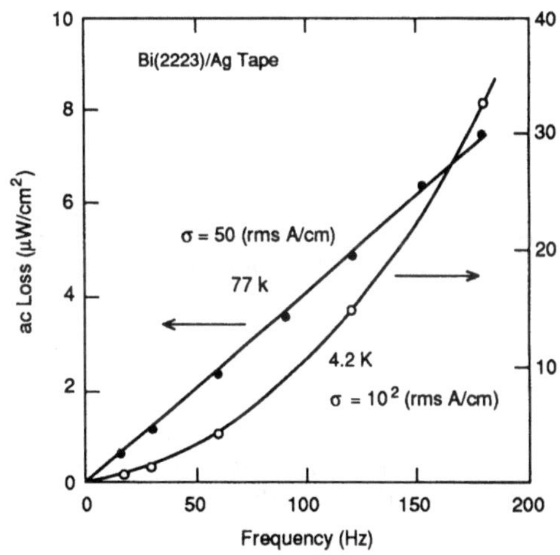

Fig. 4. The frequency dependence of the losses at 4.2 and 77 K.

where f is the frequency in Hz and 2d is the thickness of the slab while H* is the applied field at which the flux fully penetrates the specimen.

In addition, in order to account for the losses in the weakly coupled regime, Müller et al. [11] suggested a critical state model which is based on a field dependent J_c [α $(H_p + H_o)^{-1}$ where H_o is a constant]. Although this $J_c(H_p)$ is the same form as the Kim's model [12], they have arrived at this relationship from an argument based on the Josephson currents for the intergranular currents. Also, such a $J_c(H_p)$ leads to the losses which vary as $H_p{}^4(\alpha\, \sigma^4)$.

The results of the loss measurements for a sintered and a melt-textured bar of $YBa_2Cu_3O_7$ at 77 K and at 60 Hz are shown in Fig. 1. As previously shown [1], the losses in the sintered $YBa_2Cu_3O_7$ raise quickly with σ (P $\alpha\,\sigma^4$) and exhibit a plateau before increasing as P $\alpha\,\sigma^3$ and then P $\alpha\,\sigma$. These trends in the losses with σ follow the prediction by the critical state model which was discussed above [11]. The losses in the melt-textured bar, on the other hand, do not exhibit the plateau in the loss vs. σ plot indicating that there are no obvious weak links in this specimen. This is consistent with the fact that its critical current density was ~1300 A/cm² at 77 K and at 1 T. In order to examine the effect of a variation in J_c on the losses, we have measured a series of textured $YBa_2Cu_3O_7$ bars with varying values of J_c. The results are shown in Fig. 2, and as expected from the Bean model, Eq. (2), the losses increased with increasing critical current densities. Thus, the field and J_c dependences of the losses are well described by the critical state model.

As discussed above, the critical state model works quite well in describing the losses in the $YBa_2Cu_3O_7$. In order to see whether a similar statement can be made for losses in the highly anisotropic Bi(2223) tapes, we have measured the losses for tapes which were prepared by the powder-in-a-tube process [7]. The results for two tapes having critical current densities of ~3 and ~0.9 x 10^4 A/cm^2 are shown in Fig. 3. These do not exhibit the plateau in the loss vs. σ plot indicating that the amount of the intergranularly coupled region is insignificant-if present at all. Again the losses at 77 K follow the power law of σ^3 and σ for the low and high field region, respectively, and the J_c dependences of the losses are shown to be those predicted by Eqs. (2) and (3). However, the losses at 4.2 K deviate from the σ^3 dependence at low fields. In order to seek the cause of this deviation, we have measured the frequency dependence of the losses at 4.2 and 77 K. As shown in Fig. 4, the losses are linear at 77 K but not at 4.2 K with the frequency of the applied magnetic field. Thus, the loss at 77 K is solely due to the hysteresis in the superconductor. However, at 4.2 K, a significant contribution to the loss from the eddy currents in the Ag sheath is clearly observed. This is also confirmed by the measurements of the losses in an annealed Ag tape.

SUMMARY

ac losses in $YBa_2Cu_3O_7$ bars and the powder-in-a-tube processed Bi(2223) tapes are well described by the critical state model. Furthermore, at 4.2 K, the eddy current losses in the Ag sheath can become a substantial part of the total losses of the tapes.

ACKNOWELEDGMENTS

The work performed at Brookhaven National Laboratory was supported by the U.S. Department of Energy, Office of Conservation and Renewable Energy, Office of Utility Technology under Contract No. DE-AC02-76CH00016.

REFERENCES

1. Orehotsky J, Garber M, Xu Y, Wang Y L, and Suenaga M (1990) Appl. Phys. Lett. 67: 1433
2. Jin S, Tiefel T H, Sherwood R C, Davis M E, van Dover R B, Kammlott G W, Fantnacht R A, and Keith H D (1988) Appl. Phys. Lett. 52: 2074
3. Murakami M, Morita M, Doi K, and Miyamoto K (1989) Jpn. J. Appl. Phys. 28: 1189
4. Salama K, Selvamanickam V, Gao Z, and Sun K (1989) Appl. Phys. Lett. 54: 2352
5. McGinn P J, Chen W, Zhu N, Balachandran U, and Lanaga M T (1990) Physica C 165: 480
6. Orehotsky J, Wiesmann H, Moodenbaugh A R, and Suenaga M (1991) IEEE Trans. Magn. MAG-27:
7. Sato K, Hikata T, Mukai H, Ueyama M, Shibuta N, Kato T, Masuda T, Nagata M, Iwata K, and Mitsui T (1991) IEEE Trans. Magn. MAG-27: 1231
8. Flukiger R, Grnf T, Decroux M, Groth C, and Yamada Y (1991) IEEE Trans. Magn. MAG-27: 1258
9. Reilly K M and Morgan G H (submitted) IEEE Trans. Appl. Supercond.
10. Bean C P (1964) Rev. Mod. Phys. 36: 31
11. See for example, Müller K-H (1991) IEEE Trans. Magn. MAG-27: 2174
12. Kim Y B, Hempstead C F, and Strnad A R (1964) Rev. Mod. Phys. 36: 43
13. Orehotsky J, Reilly K M, Suenaga M, Hikata T, Ueyama M, and Sata K (submitted) Appl. Phys. Lett.

ac Transport Properties of High-T_c Superconductive Wires

H. Ishii[1], H. Hoshino[2], H. Hara[1], T. Yamamoto[1], S. Hirano[2], and K. Maekawa[1]

[1] Engineering Research Center, Tokyo Electric Power Company, Chofu, Tokyo, 182 Japan
[2] Power and Substation Department, Kandenko, Minato-ku, Tokyo, 108 Japan

ABSTRACT

The ac transport properties of YBCO films formed by laser ablation and silver-sheathed Bi-2212 wire were measured by two methods, using a locked-in amplifier and a digital oscilloscope. The generated voltage generally consisted of three components, and a resistive voltage component was present in addition to the contribution of the voltage observed in dc transport. This component was also generated in the domain $I<I_c$ (dc), and increased with increasing current and frequency. These results have an important bearing upon applications, and the phenomena concerned require quantitative clarification.

KEY WORDS: ac transport properties, I-V characteristics, resistive voltage

1. INTRODUCTION

The application of superconductivity technology to electrical devices, if possible, would provide advantages such as permitting far greater miniaturization, greater capacities, and higher efficiencies as compared with conventional electrical technology. In particular, if high-capacity ac power transmission cables could be fabricated using high-temperature superconductors, then increased future power demands could be economically satisfied [1]. As the first stage in the application of high-temperature superconductivity to electrical utilities, active research is in progress on topics such as improvement of critical current densities J_c and the associated magnetic field characteristics using various types of process technology. The potentialities of high-T_c superconductivity have been realized in the form of materials permitting J_c values exceeding 10^6A/cm^2 at the temperature of liquid nitrogen [2], and current densities above 2000A/cm^2 have been attained in wires 1m in length [3]. However, all these J_c values have been measured for dc transmission, and at the present time almost no research reports on superconductivity have appeared in connection with the practically important subject of ac transport characteristics. Accordingly, the present authors have measured the ac transport characteristics of YBCO ($YBa_2Cu_3O_y$) films and silver-sheathed $Bi_2Sr_2CaCu_2O_x$ wires at 77.3K under a zero magnetic field, performed detailed analysis of the data, and compared the results obtained with the corresponding direct current characteristics. These studies will be briefly described below.

2. EXPERIMENTS

Measurements of ac electrical characteristics were performed by the conventional 4-terminal method. Since the generated voltages include the contribution of the inductance of the wire, the resistive component must be isolated for purposes of comparison with dc characteristics. Moreover, specimens as long as possible are desirable in order to achieve precise detection. The specimens used in the present study were of two types, i. e., YBCO film wires 5cm in length prepared by applying laser ablation to traveling YSZ substrates [4], and silver-sheathed Bi-2212 wires 10cm in length [5]. The resistive component was measured by two methods; one consisted in independently measuring the resistive and the inductive voltage components at the same time by means of a two-phase lock-in amplifier, while the other consisted in inputting the current and voltage waveforms after averaging over 64 cycles by means of a digital oscilloscope and isolating the resistive voltage component by geometrical analysis. Since microvoltages were being measured, in order to mitigate the effect of noise, appropriately time-averaged data was input at the fixed current in all the measurements, which were performed at frequencies in the range 50 - 500Hz in liquid nitrogen under a zero magnetic field.

3. RESULTS AND DISCUSSION

Figure 1 shows the dc voltage-current characteristic of the specimens used in the present measurements, normalized by the value of I_c determined in accordance with the criterion $\rho_c = 10^{-10}\Omega$cm. The values of I_c, and J_c as well as the dimensions of the respective specimens are shown in Table 1. Figure 2 shows the changes in the 200Hz ac voltage waveforms associated with increasing values of the current in the YBCO film. When the peak current is lower than the critical current I_c (dc), the voltage waveform is nearly sinusoidal.

Table 1. Summary of dimensions, Ic, Jc and so forth of the samples used in the experiments.

item	$YBa_2Cu_3O_y$-film	$Bi_2Sr_2CaCu_2O_x$-tape
fabrication method	laser ablation	Ag-sheathed
length	5 cm	10 cm
width	0.78 cm	0.3 cm †
thickness	1.0 μm	150 μm †
cross-section	7.8×10^{-5} cm^2	1.0×10^{-3} cm^2
distance of voltage electrodes	3.0 cm	6.7 cm
Ic (dc)	0.612 A *	1.87 A *
Jc (dc)	7850 A/cm^2 *	1870 A/cm^2 *

† including silver
* determined using criterion of 10^{-10} Ωcm

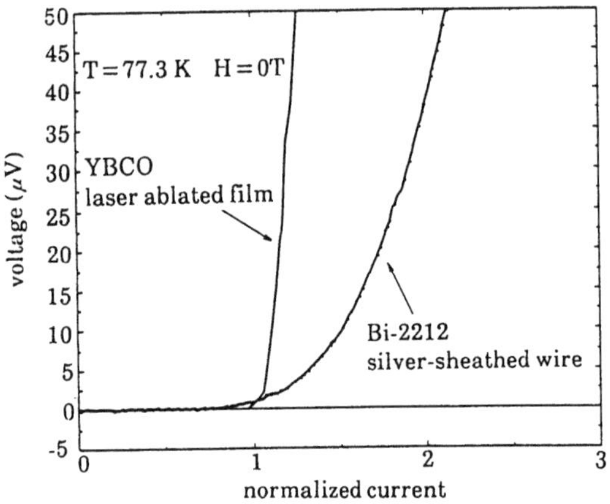

Fig. 1 The dc I-V curves plotted against normalized current. The values of Ic and Jc of each sample are listed on table 1.

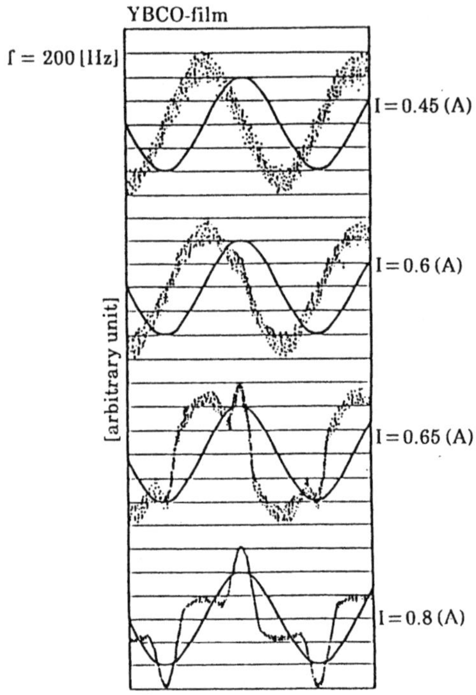

Fig.2 Voltage and current curves of laser ablated YBCO-film for four different currents. Peak values of current curves are indicated.

However, for currents exceeding I_c, owing to the contribution of the resistive component in phase with the current, the overall voltage waveform deviates considerably from sinusoidal form. In this case, the voltage waveform may be regarded as the resultant of the sinusoidal component and the nonlinear resistive component which is observed in dc transmission; the current corresponding to the rise of the resistive component coincides with the value of I_c (dc). That is, during a single cycle of the current waveform, the voltage generated for each current (this voltage component will hereafter be denoted by $\tilde{V}_{dc}$) corresponds to that which is determined by the dc voltage-current characteristic, and thus oscillates temporally in accordance with the said voltage-current characteristic.

On the other hand, the voltage waveform observed in the region $I < I_c$ (dc) is sinusoidal, however, the phase difference between this waveform and that of the current is somewhat less than 90 degree. That is, although the inductive component of the generated voltage is predominant, a resistive component in phase with the current is also present. As will be described below, the results of measurements with a lock-in amplifier also indicated the presence of a resistive component in the domain $I < I_c$.

Summarizing the above-described results, analysis of the voltage waveform $\tilde{V}_{tot}$ observed with the digital oscilloscope revealed that this waveform consisted of three components, i. e.,

$$\tilde{V}tot = \tilde{V}_L + \tilde{V}x + \tilde{V}dc \qquad (1)$$

Here, $\tilde{V}_L$ denotes the inductive component, $\tilde{V}_X$ the resistive component which is also generated for $I < I_c$ (dc), and $\tilde{V}_{dc}$ the resistive component determined by the dc voltage-current characteristic. By setting the current phase to zero with a two-phase lock-in amplifier, the following two components can be measured independently.

$$\tilde{V}_L = \frac{(\text{peak of } \tilde{V}_L)}{\sqrt{2}} \qquad (2)$$

$$\tilde{V}_R = \frac{\sqrt{2}}{T}\int_0^T [\tilde{V}_X + \tilde{V}_{dc}] \cos\left(\frac{2\pi}{T} t\right) dt \qquad (3)$$

The results of these measurements are shown in Figs. 3 (a) and 3 (b), respectively. The value of the inductance as determined from $\tilde{V}_L$ is of the order of 25nH, almost independent of the current and frequency.

On the other hand, as regards the data recorded with the digital oscilloscope, as mentioned above, three voltage components can be isolated by geometrical analysis. The principle of this analysis is illustrated by Fig. 4. The peak value of $\tilde{V}_L + \tilde{V}_X$, the peak value of $\tilde{V}_{dc}$ and the phase angle $\Delta\phi$ between the current and $\tilde{V}_L + \tilde{V}_X$ are determined from the voltage waveform, and the desired components are calculated by the formulae

$$\tilde{V}_X = (\tilde{V}_L + \tilde{V}_X)\cdot\cos\Delta\phi, \quad (4)$$

$$\tilde{V}_L = (\tilde{V}_L + \tilde{V}_X)\cdot\sin\Delta\phi. \quad (5)$$

The results of this evaluation of $\tilde{V}_X$ are shown in Fig. 5. For $I < I_c$ (dc), as anticipated from the fact that $\tilde{V}_{dc} = 0$, the data agree quantitatively with those observed in the region $I < I_c$ (dc) when measurements were performed with the lock-in amplifier (Figure 3 (b)). Thus, the fact that resistive voltage is generated in the ac case even for $I < I_c$ (dc) was verified by two methods of measurement. Furthermore, $\tilde{V}_X$ increases monotonically with current and frequency.

In the case of the silver-sheathed Bi-2212 wire, as one observes from the dc voltage-current characteristic shown in Fig. 1, the generation of resistance is very broad as compared with the case of the YBCO film, and therefore the isolation of the three types of voltage components from the waveform recorded with the digital oscilloscope is difficult; nevertheless, in the domain $I < I_c$ (dc), the values of $\tilde{V}_L$ and $\tilde{V}_X$ can be estimated. The values of $\tilde{V}_X$ as well as the results of measurements with the lock-in amplifier are shown in Figure 6 in the form $\sqrt{2}\,^{V}R$. As in the case of the YBCO film, the value of $\tilde{V}_X$ increases with current and frequency.

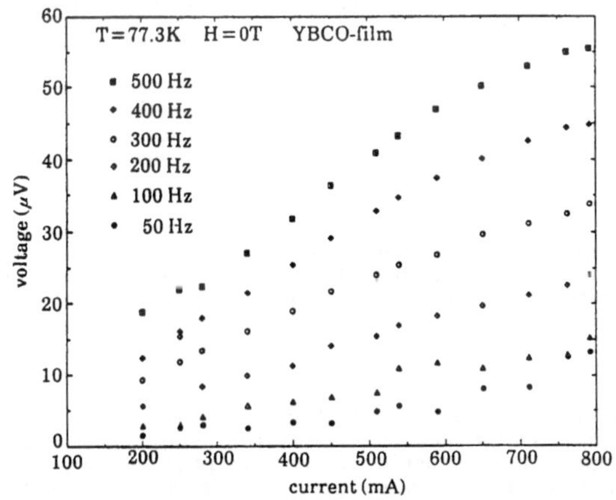

Fig. 3 (a) Inductive component measured by lock-in amplifier.

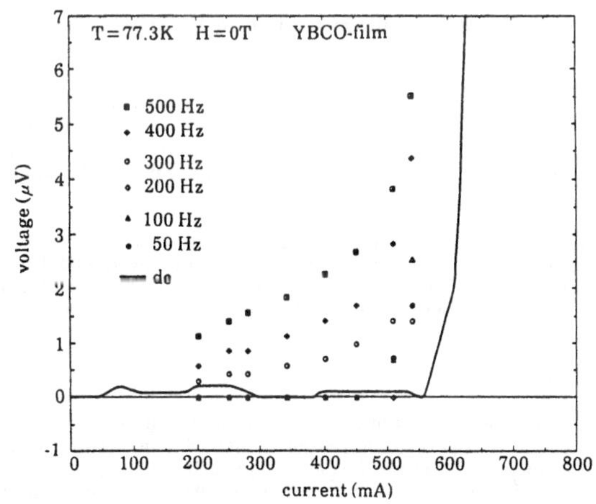

Fig. 3 (b) Resistive component measured by lock-in amplifier

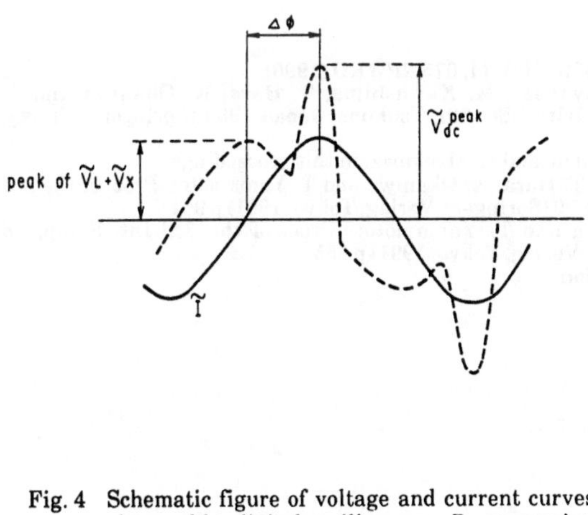

Fig. 4 Schematic figure of voltage and current curves observed by digital oscilloscope. By measuring $\Delta\phi$ and peak value of ($\tilde{V}_L + \tilde{V}_X$), $\tilde{V}_L$ and $\tilde{V}_X$ were estimated (see text.)

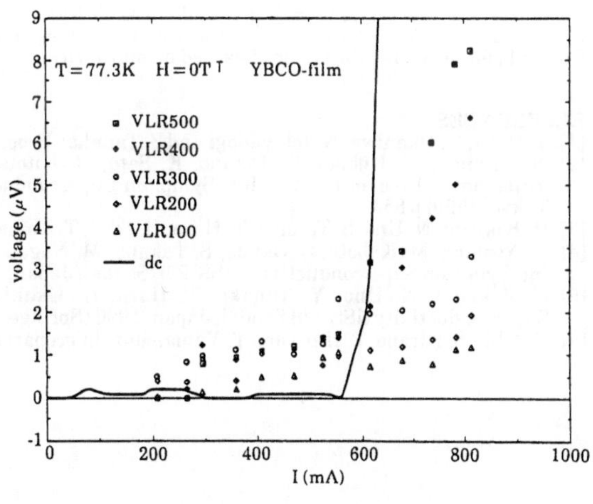

Fig. 5 Plot of Vx-peak values estimated from voltage curves which were measured by digital oscilloscope.

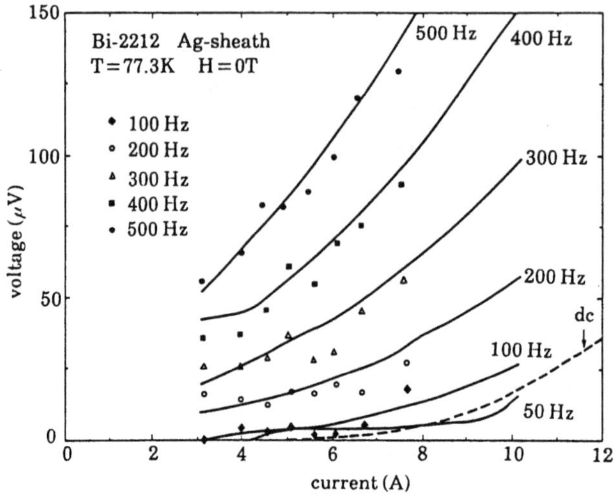

Fig. 6 Resistive voltage induced in Bi2212 Ag-sheathed wire. Lines denote the data obtained by lock-in amplifier whereas symbols indicate that estimated by means of digital oscilloscope.

The mechanism of contribution to $\tilde{V}_X$ is regarded as hysteresis due to magnetic flux pinning. The changes in the self-induced magnetic field created by the current changes the magnetic flux within the superconductor and thereby creates an induced electric field. This electric field can be shown to be opposite in direction to the supercurrent flowing within the superconductor, and of magnitude roughly proportional to both the current and the frequency, or of the order of $1\mu V/cm$ at 1A and 50Hz [6].

These measurement results signify that, in the ac case, owing to the contribution of $\tilde{V}_X$, the resistive voltage component appears at a relatively low current as compared with the dc case, and in general the ac critical current density is lower than the corresponding dc value. Therefore, this characteristic is of great importance with respect to potential applications, hence, the elucidation and quantitative analysis of the mechanisms involved is essential.

4. SUMMARY

Using YBCO films of length 5cm as well as silver-sheathed Bi-2212 wires of length 10cm, the present authors have measured superconductive ac transport characteristics and analyzed the generated voltages in detail. The results obtained showed that the generated voltages consist of three components, and that in addition to the resistive voltage also generated in dc measurements, a resistive component occurring only in the ac case is present. This voltage component increases monotonically with current and frequency. This phenomenon is extremely important with respect to applications of high-temperature superconductivity, and requires quantitative elucidation.

ACKNOWLEDGMENTS

The authors hereby express their gratitude to Mr S. Takano and Mr N. Yoshida of Sumitomo Electrical Industries, Ltd., for providing the YBCO films used in the present studies, and to Mr N. Uno and Mr H. Kikuchi of the Furukawa Electric Co., Ltd., for providing the silver-sheathed bismuth wire.

REFERENCES

[1] T. Hara, K. Okaniwa, N. Ichiyanagi and S. Tanaka: Proc. of IEEE WM, 075-2PWRD (1990)
[2] N. Yoshida, M. Kubota, S. Takano, K. Sato, H. Hitotsuyanagi, M. Kawashima, T. Hara, K. Okaniwa and T. Yamamoto: Proc. of the 2nd Int. Symp. on Superconductivity (ISS' 89), Tsukuba, Japan 1989 (Springer - Verlag, Tokyo, 1990) p 853
[3] H. Kikuchi, N. Uno, S. Tanaka, T. Hara, H. Ishii, T. Yamamoto and K. Maekawa: in this proceedings.
[4] N. Yoshida, M. Kubota, S. Okuda, S. Takano, M. Nagata, T. Hara, K. Okaniwa and T. Yamamoto: Proc. of the 3rd Int. Symp. on Superconductivity (ISS' 90), Sendai, Japan, 1990 (Springer - Verlag, Tokyo, 1991) p 901
[5] H. Kikuchi, N. Uno, Y. Tanaka, T. Hara, K. Okaniwa and T. Yamamoto: Proc. of the 3rd Int. Symp. on Superconductivity (ISS' 90) Sendai, Japan, 1990 (Springer - Verlag, Tokyo, 1991) p 655
[6] H. Ishii, S. Hirano, T. Hara and T. Yamamoto: in preparation

Critical Current Density and Microstructure of Hi-T_c Bi-Pb-Sr-Ca-Cu-O Superconductors

K. HAMADA[1], S. KAMADA[1], K. SUGAWARA[1], A. NAGATA[1], C. TAKAHASHI[2], S. OGOTA[2], M. KOMATSU[2], M. NAGANO[2], and S. HANADA[3]

[1] Department of Materials Engineering and Applied Chemistry, Mining College, Akita University, Akita, 010 Japan
[2] Applied Technology R&D Center, Tohoku Electric Power Co., Inc., Sendai, 981 Japan
[3] Institute for Materials Research, Tohoku University, Sendai, 980 Japan

ABSTRACT

We report measurements of applied magnetic field dependence of J_c and microstructure of Ag-sheathed Bi-Pb-Sr-Ca-Cu-O superconducting tapes which were prepared by a conventional solid state reaction, rolling and pressing. The measurements of J_c at 4.2K and 77K were carried out by a four-probe method and J_c was determined by using a criterion of 1μV/cm. J_c of (5~6)x10^3A/cm^2 at 77K in zero magnetic field was obtained. The hysteresis was observable at 4.2K with increasing and decreasing magnetic field but not at 77K. The anisotropy of J_c, namely, J_c(tape$\perp$H)/J_c(tape$\|$H) firstly increased with increasing magnetic field at both temperatures and then decreased at 4.2K with a further increase of magnetic field above 1T. On the contrary, it increased gradually with increasing magnetic field above 0.01T at 77K.

KEY WORDS: Bi-Pb-Sr-Ca-Cu-O superconductor, Ag-sheathed tape, critical current density, anisotropy, texturing

INTRODUCTION

Recent development of critical current density (J_c) was remarkable for Bi-based high-T_c oxide superconductors, both Bi-Pb-Sr-Ca-Cu-O and Bi-Sr-Ca-Cu-O systems[1~6]. The tapes of J_c above 10^4A/cm^2 at 77K and 10^5A/cm^2 at 4.2K in zero magnetic field were produced by a powder-in-tube method and by a doctor blade method. It has be established that the higher J_c is realized when the tape and superconducting layer thickness is thinner[7]. The thin tapes are, however, easy to deform and to degrade their superconductiong properties and therefore much thicker and higher I_c tapes are desired for a practical use. In the present study, the relationship between the tape thickness and J_c as a function of number of pressing between sintering was investigated for relatively thick tapes. J_c at 4.2K in the magnetic field up to 10T and at 77K up to 0.1T was measured as a function of magnetic field for tapes pressed various times. Anisotropy in J_c as a function of magnetic field will be discussed from a microstructural point of view.

EXPERIMENTALS

Samples with a nominal composition of $Bi_{2.05}$-$Pb_{0.41}$-$Sr_{2.12}$-$Ca_{3.24}$-$Cu_{4.18}$-O_x were prepared from Bi_2O_3, PbO, $SrCO_3$, $CaCO_3$, and CuO powders. These powders were mixed ,calcined and ground twice, and pressed into the rectangular shape. The pressed samples were wrapped with Ag sheet and then rolled to tapes from 0.4 to 0.05mm in thickness. Tapes were sintered for 96h at 840°C in air and then pressing at pressure of 1GPa and sintering for 12h at 840°C in air were repeated for improving J_c.

J_c measurements were carried out at 77K in the magnetic field up to 0.1T and at 4.2K up to 10T as a function of magnetic field orientation. J_c was determined by using a criterion of 1μV/cm. SEM observations were done on the fractured cross-sectional and longitudinal area and oxide surface peeled away Ag sheath. XRD investigations were performed on the oxide superconductor surface.

RESULTS AND DISCUSSIONS

At first, the optimum sintering temperature was investigated for tapes without intermediate pressing and determined to be 840°C for relatively high J_c. In the present study, therefore, this temperature was used for sintering. T_c of these tapes was about 105K. Next, the effect of pressure of intermediate pressing on J_c was studied at 77K without magnetic fields. It was found that J_c increases with increasing pressure of intermediate pressing and therefore 1 GPa pressure for pressing was used in this study.

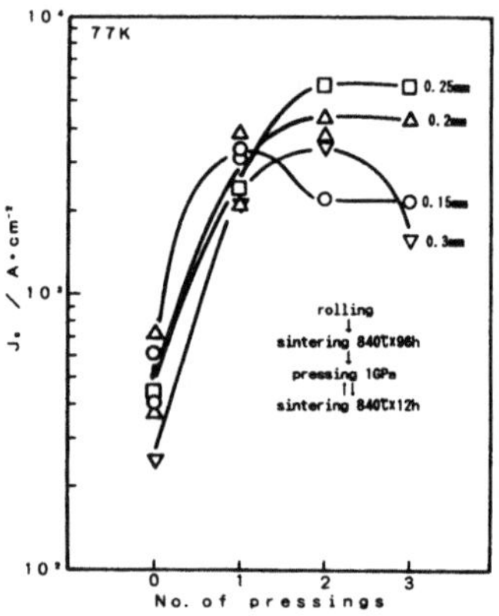

Fig. 1 J_c at 77K without magnetic field as a function of number of pressing

Fig. 2 SEM micrograph for the tape pressed twice (longitudinal, side)

Tapes were rolled to various thickness in order to study the influence of number of intermediate pressing on J_c. The results are shown in Fig. 1. The tape thickness shown in Fig. 1 is that just after rolling and therefore thickness after pressing is much thinner. J_c at 77K was saturated by pressing twice and the pressing is more effective for thicker tapes except 0.3mm thickness in this study. Tapes of 0.20mm and 0.25mm in thickness were used the experiments mentioned hereinafter. Microstructural investigations were carries out by SEM on the fractured cross-sectional and longitudinal surfaces and flat supercondudting surfaces peeled away Ag sheath. Figure 2 shows SEM microgragh of the fractured longitudinal surfaces for the tape 0.2mm in thickness after pressing twice. Since in the microstructure of the tape without pressing the grain alignment was poor, the grain alignment along the rolling direction shown in Fig. 2 is thought to be improved by intermediate pressing. SEM observations on the other surfaces revealed the laminated grain structure. The grain alignment is much improved near the oxide-Ag sheath interface and the much wavy, laminated grain structure is appeared in the center as seen in Fig. 2. XRD patterns taken from the superconducting layer surfacepeeled away Ag sheath showed that the tendency of the c-axis orientation perpendicular to the tape surface became strong by pressing. XRD patterns showedalso the presence of a small amounts of other phases.

J_c measurements at 77K in magnetic field up to 0.1T and 4.2K up to 10T were carried out on the tapes of 0.25mm in thickness as a function of number of intermediate pressing. The magnetic field was applied parallel to the tape surface. Results are shown in Fig. 3. At 77K, the curves of J_c vs magetic field do not show any hysteresis during increasing and decreasing magnetic field as seen in Fig. 3(a). On the contrary, hysteresis is observable in $J_c(H)$ curves at 4.2K as shown in Fig. 3(b). $J_c(H)$ characteristics are improved by pressing twice at both temperatures but the feature of them are not much changed except J_c level. As reported previously [1], the decrease in J_c at 77K with increasing magnetic field is small for the tape of higher J_c. This feature is not clearly observable at 4.2K. J_c decreases sharply up to about 2T and then gradually with increasing field as shown in Fig. 3(b).

Anisotropy of J_c as a function of magnetic field orientation was also studied on the tapes of 0.25mm in thickness produced by pressing twice. Detailed investigations were conducted at 77K in magnetic field up to 0.1T. Figure 4 shows the J_c vs H curves when the magnetic field is applied parallel and perpendicular to the tape surface, respectively. The decrease in J_c(tape⊥H) is steep up to 0.01T and then gradual with increasing field up to 0.1T, while that

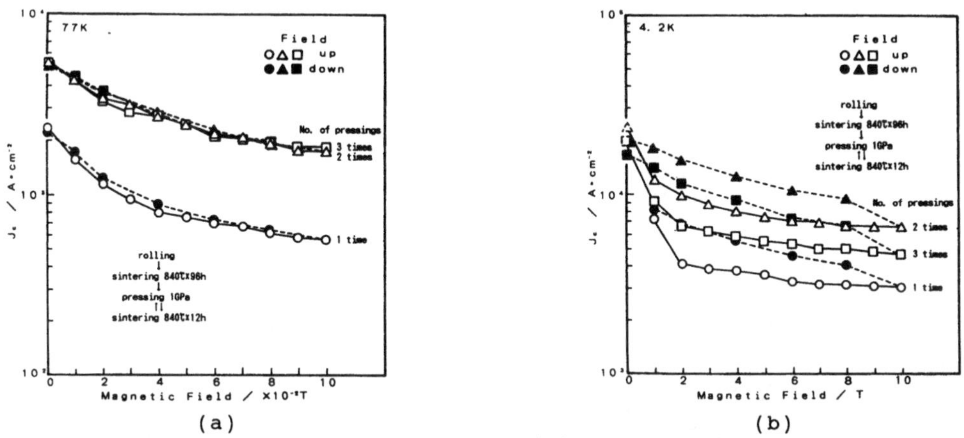

Fig. 3 J_c vs magnetic field curves at 77K (a) and 4.2K (b) for tapes pressed various times

in J_c(tape|H) is not so steep as seen in Fig. 4. Figure 5 shows the anisotropy factor, [J_c(tape⊥H)/J_c(tape|H)] as a function of magnetic field for the tapes with variousnumber of pressing. The anisotroypy drops abruptly with increasing field up to 0.01T and then decreases gradually up to 0.1T, which is almost similar evenwhen J_c levels are different due to different number of pressing. The anisotropy of J_c is thought to increase with a further increase in magnetic field as reported previously [8]. Figure 6 shows the angular dependence of J_c for the tape of 0.25mm in thickness pressed twice. The angle, θ means that between the field direction and the tape surface and J_{cmax} is J_c(tape|H) in various magnetic fields. J_c/J_{cmax} decreases steeply till about 60° and then is almost constant as seen in Fig. 6. The angular dependence of J_c is weak when the applied magnetic field is small. In the higher magnetic field, J_c is sensitive to the magnetic field direction, especially near the direction parallel to the tape surface. With a further increase in magnetic field, the anisotropy of J_c may increase due to the critical field.

For the study of the anisotropy of J_c at 4.2K, J_c(tape|H) and J_c(tape⊥H) were measured in magnetic field up to 10T for the tapes with different number of pressing. Figure 7 shows the anisotropy of J_c which is independendt of the number of pressing. The anisotropy increases steeply till about 0.1T and then decreases with increasing magnetic field up to 10T in the present study. The similar behaviour was reported by Kumakura et al.[9] on (2212) phase superconducting tapes produced by a doctor blade method. Their results indicate that the anisotropy of J_c increases again with a further increase in magnetic field above around 10T. This behaviour is not well understood yet, but in the magnetic field range in this study, J_c(tape⊥H) decreased abruptly up to 2T and

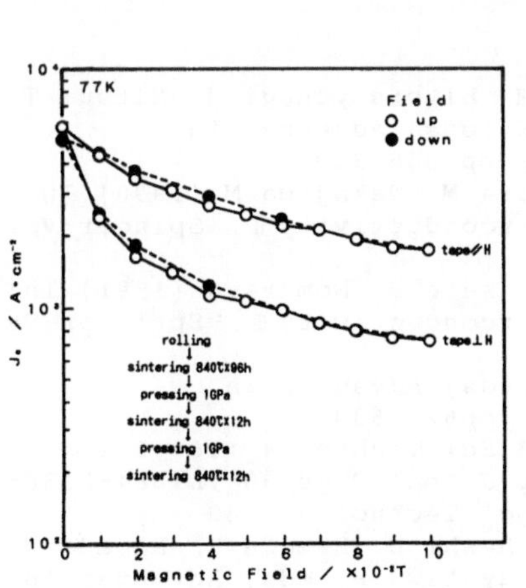

Fig. 4 J_c(77K) vs magnetic field curves as a function of field direction

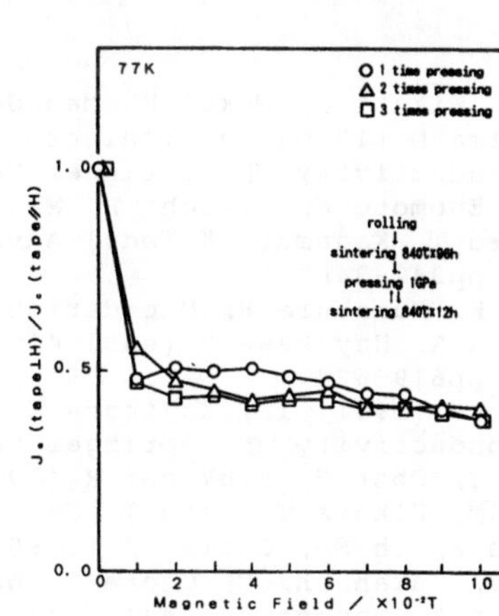

Fig. 5 Anisotropy of J_c(77K) as a function of magnetic field

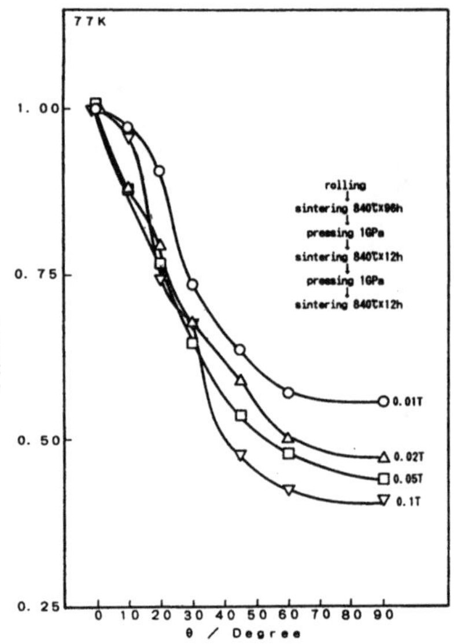

Fig. 6 Angular dependence of J_c(77K) at various magnetic field

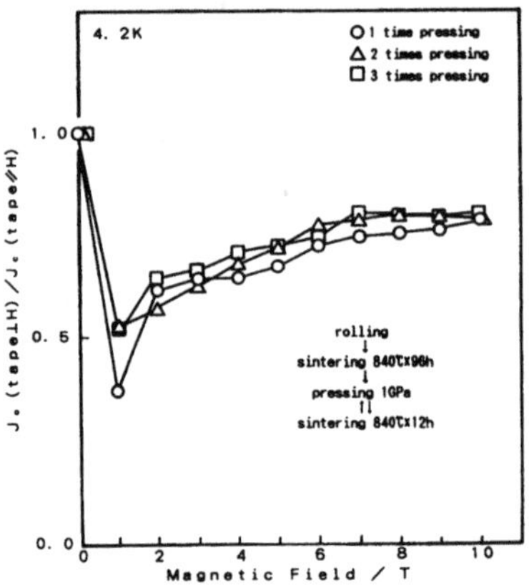

Fig. 7 Anisotropy of J_c(4.2K) as a function of magnetic field

then was almost constant up to 10T. On the contrary, J_c(tape|H) decreased gradually with increasing magnetic field.

In summary, we measured the magnetic field dependence of J_c and the angular dependence of J_c on Ag-sheathed Bi-Pb-Sr-Ca-Cu-O superconducting tapes pressed different times at 77K and 4.2K. The higher pressure up to 1GPa of intermediate pressing improved J_c and the optimum number of pressing was found to be two in the present study. The anisotropy of J_c increased with increasing magnetic field at 77K up to 0.1T. While the anisotropy of J_c at 4.2K increased steeply till 1T and then reduced gradually up to 10T. These behaviours were independent of the number of pressing.

Finally, this work is partially supported by a Grant-in-Aid for Scientific Research on Priority Area (No. 03210102) from the Ministry of Education, Science and Culture, Japan. A part of this work was performed under the inter-university cooperative research program of the Institute for Materials Research, Tohoku University and J_c measurements at 4.2K were carried out by using a high field superconducting magnet of High Field Laboratory for Superconducting Materials, Institute for Materials Research, Tohoku University.

REFERECES

1. Sato K, Hikata T, Mukai H, Masuda T, Ueyama M, Hitotsuyanagi H, Mitsui T, Kawashima M (1990) In: Ishiguro T, Kajimura K (eds) Advances in Superconductivity Ⅱ. Springer-Verlag, Tokyo, pp 335-340
2. Uno N, Enomoto N, Kikuchi H, Matsumoto K, Miura M, Nakajima M (1990) In: Ishiguro T, Kajamura K (eds) Advances in Superconductivity Ⅱ. Spinger-Verlag, Tokyo, pp342-346
3. Togano K, Kumakura H, Dietderich DR, Maeda H, Kase J, Nomura K (1991) In: Kajimura K, Hayakawa H (eds) Advances in Superconductivity Ⅲ. Springer-Verlag Tokyo, pp619-624
4. Enomoto N (1991) In: Kajimura K, Hayakawa H (eds) Advances in Superconductivity Ⅲ. Springer-Verlag, Tokyo, pp625-630
5. Yamada Y, Obst B, Flukiger R (1991) Supercond Sci Technol 4 :165-171
6. Ueyama M, Hikata T, Kato T, Sato K (1991) Jap J Appl Phys 30 :L1384-L1386
7. Osamura K, Oh SS, Ochiai S (1990) Supercond Sci Technol 3 :143
8. Isawa K, Takahashi C, Ogota S, Nagano M, Takahashi H, Hamada K, Kamada S, Sugawara K, Nagata A (1991) In: Kajimura K, Hayakawa H (eds) Advances in Superconductivity Ⅲ. Springer-Verlag, Tokyo, pp839-842
9. Kumakura H, Togano K, Maeda H (1991) Appl Phys Lett 58 :2830-2832

Current Flow in Large Size Bulk Oxide Superconductor

Koichi Numata[1], Akira Hane[1], Kazutomo Hoshino[1], Kenji Tasaki[2], Daisuke Ito[2], and Hiroyasu Ogiwara[2]

[1] Corporate R&D Center, Mitsui Kinzoku, Ageo, Saitama, 362 Japan
[2] Toshiba R&D Center, Kawasaki, Kanagawa, 210 Japan

ABSTRACTS

Dependence of critical current density on sample geometry has been investigated for large size bulk specimens of Bi system oxide superconductor. When the thickness or the cross sectional area of the specimens was reduced, the critical current density increased. Dissipation fields estimated for each specimen agreed with the results obtained from the measurement of the dependence of the critical current density on the magnetic field. In the specimens with high critical density, self-magnetic field exceeded the dissipation field and current along the specimen was supposed to distribute inhomogeneously.

KEY WORDS: Bi system superconductor, critical current density, Self-magnetic field Dissipation field

INTRODUCTION

There is an interest in applying the oxide superconductors to current leads with critical current (I_c) of larger than 1000 A (cross section area of 1 cm^2, at 77K) which will be used in the cryogenic apparatuses, such as superconducting magnet and superconducting current limiter.

While investigating the improvement of critical current density (J_c) of Bi system oxide superconductor, we fabricated large specimens (10 x 40 x 40 mm^3). However it was found that J_c of large specimens was less than that of smaller ones.

There is some literature about the same phenomenon. According to the literature, the phenomenon is explained by the Joule effect due to the measurement current [1] or the self-magnetic field [2]. In Ref. 2, assuming a uniform current distribution along the sample, the dependence of the critical current density on the thickness of the sample was well explained by the self-magnetic field and the dissipation field of 4.47 Oe was obtained which also agreed with the experimental results. In this case, however, bulk specimen and also the critical current were small. The maximum self-magnetic field did not exceed the dissipation field.

In this report, we have measured the dependence of the critical current density with the thickness or the cross section area of samples of Bi system oxide superconductor. The dissipation field was estimated from the equations derived by D'Ovidio [2], and the current distribution along the sample is discussed.

EXPERIMENTAL

In this report, four types of samples were prepared to investigate the dependence of the critical current density on the sample geometry. The fabrication processes of the samples are listed in Table 1.
These samples ((0.5~10) x 10 x 30 mm^3) were fabricated using a uniaxial press followed by a cold isostatic press of powders of the Bi system high T_c phase (nominal composition; Bi:Pb:Sr:Ca:Cu = 0.8:0.2:0.8:1.0:1.4, Tc ~ 103 K). Preparation of the powders from co-precipitated powders is described in another paper [3]. After

sintering the samples, silver foils (20μm in thickness) as contact pads were attached to each end of the samples and pressed isostatically [4]. Then the samples were sintered again. As the contact resistance in this experiment was estimated to be less than $1\mu\Omega\cdot cm^2$, the Joule effect at the contact was negligible.

The critical current of these samples was measured by the four-probe technique. The voltage generation was detected by an X-Y recorder (YOKOGAWA 3036 X-Y RECORDER; 5μV/cm). Magnetic field (0 ~ 500 Oe) was generated by a solenoid coil.

Table 1 Fabrication processes of the samples

Samples	Fabrication processes of the samples
Sample 1	A specimen (10 x 10 x 30 mm^3) was fabricated from a coarse powder (less than 100 μm) and then the thickness was reduced by polishing (see Fig. 1).
Sample 2	Specimens with different thickness ((0.5 ~ 10) x 10 x 30 mm^3) were fabricated from a fine powder (average size was about 10μm).
Sample 3	A specimen (10 x 10 x 30 mm^3) was fabricated from the fine powder and then the cross section area of the specimen was reduced by polishing (see Fig. 2).
Sample 4	Same process as Sample 2.

RESULTS AND DISCUSSION

Figure 1 (a) shows the dependence of the critical current density on the thickness of sample 1 and 2. The critical current density apparently increased as the thickness of the specimens was reduced. The same phenomenon was also observed for Sample 3 when the cross sectional area was reduced (see Fig. 1 (b)).

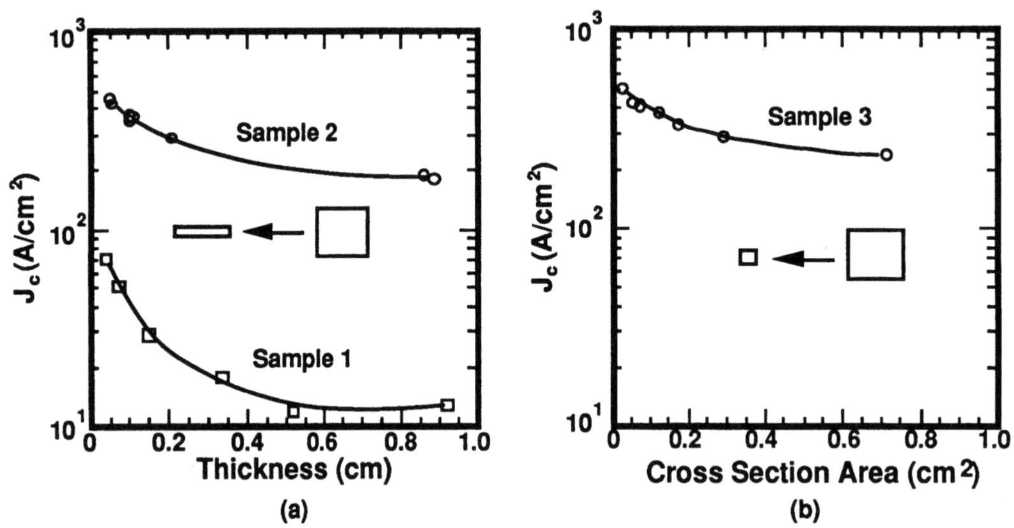

Fig.1 J_c dependence on (a) thickness of Sample 1 and Sample 2 and (b) cross section area of Sample 3.

D'Ovidio et. al. have reported the same phenomenon for the Y-Ba-Cu-O superconductor [2]. They proposed a model based on junction decoupling due to the self-magnetic field, which well agreed with their results assuming a uniform current distribution along the sample. They calculated the dissipation field H_d to be 4.47 Oe. We tried to estimate our experimental data using their model. Table 2 shows the measured and calculated results for Sample 1 and 3 where H_{corner} indicates a self-magnetic field at a corner of a rectangular specimen. Figure 2 shows the plots of J_c v.s. J_c/H_{corner} for Sample 1, 2 and 3. According to their model, the plots of J_c v.s. J_c/H_{corner} should be linear and the values of H_d can be taken from the slope of the plots. The values of H_d calculated from Fig. 2 are 2.4, 13 and 19 Oe for Sample 1, 2 and 3, respectively (the slope is calculated using the least square method).

Table 2. Measured and calculated data

Samples	Cross sectional area (cm^2)	J_c (A/cm^2)	H_{corner} (Oe)	J_c/H_{corner} ($A/cm^2/Oe$)	H_d (Oe)
Sample 1	0.75	13	3.6	3.6	2.4
	0.42	12	2.5	4.8	
	0.27	18	2.8	6.4	
	0.12	29	2.7	11	
	0.059	51	2.8	18	
	0.034	71	2.5	28	
Sample 3	0.71	240	64	3.7	19
	0.30	290	50	5.8	
	0.17	330	45	7.4	
	0.073	410	36	12	
	0.030	510	28	18	

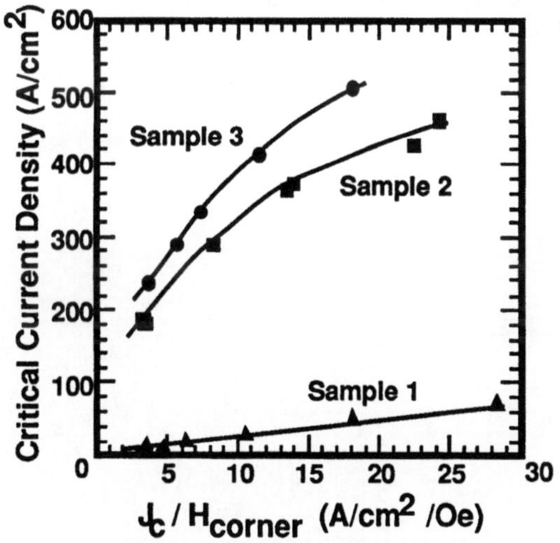

Fig.2 Measured critical current density as a function of Jc/Hcorner.

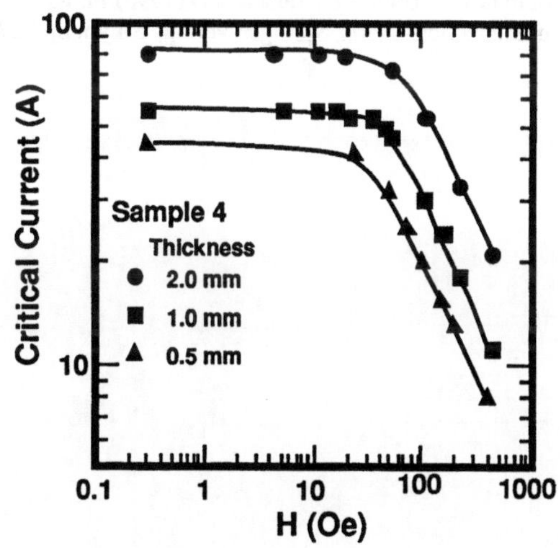

Fig.3 Dependence of the critical current on magnetic field.

It can be seen from Fig. 2 that J_c dependence on the J_c/H_{corner} is linear for Sample 1, and non-linear for Sample 2 and 3. The dissipation field H_d of Sample 1 is nearly equal to the magnetic field at the corner of the specimen, and the linearity of J_c v.s. J_c/H_{corner} was good. On the other hand, in case of Sample 2 and 3, the self-magnetic field H_{corner} is much larger than the dissipation field H_d and the linearity is not good.

Figure 3 shows the dependence of the critical current on magnetic field. The sample was fabricated by the same process used for Sample 2. The critical current degraded rapidly from the magnetic field of 20 ~ 30 Oe.

From these results, the assumption of a uniform current distribution along the sample is supposed to be correct for the low Jc sample in which the maximum self-magnetic field is less than or equal to the dissipation field. In case of a high Jc sample, however, the current distribution is not supposed to be uniform along the sample. Because the dissipation field of 20 ~ 30 Oe exactly exists as shown in Figure 3 and the self-magnetic field H_{corner} exceeds the dissipation field H_d.

In summary, we have measured the dependence of the critical current density on sample geometry for large size bulk specimens of Bi system oxide superconductor. When the thickness or the cross section area of the specimens was reduced, the critical current density increased. Dissipation fields for each specimen were estimated from equations derived by D'Ovidio et. al. The dissipation fields were in agreement with the results obtained from the measurement of the dependence of the critical current density on the magnetic field. In the specimens with high critical current density, self-magnetic field exceeds the dissipation field and the current flow along the specimen was supposed to distribute inhomogeneously.

REFERENCES

1. Imanaka N, Imai H, Adachi G (1989) Jpn. J. Appl. Phys. 28: L2158
2. D'Ovidio CA, Fiscina JE, Esparza DA (1991) J. Appl. Phys. 69: 8265
3. Hoshino K, Ohta H, Sudoh E, Katoh K, Yamazaki S, Takahara H, Aono M (1990) Jpn. J. Appl.Phys. 29: L1435
4. Numata K, Hoshino K, Takahara H (1990) In: Kajimura K, Hayakawa H (eds) Advances in Superconductivity III, Springer-Verlag, Tokyo, pp 827-830

Evaluation of Brittleness of HTSC

S. UCHIDA, T. KATAGIRI, F. KANAI, N. NINOMIYA, M. SHIMABARA, H. SHINYA, Y. MOGI, K. KANNAKA, S. KAWADA, K. GENMA, M. KAWAKAMI, and Y. NISHI

Department of Materials Science, Tokai University, Hiratsuka, Kanagawa, 259-12 Japan

Abstract

We have suggested a new method to evaluate brittleness K_{IC} by use of a standerd hardness tester. We found the K_{IC} value is controlled by the process. This method is useful to evaluate brittleness of small high Tc superconductor specimens.

Introduction

Most high Tc superconductors are brittle. From an engineering point of view, it is important to know mechanical properties. Stress intensity factor (K_{IC}) is used as a standard to evaluate brittleness. Since it is difficult to prepare ceramic specimens, a new method has been suggested to measuer K_{IC} by the use of a standard hardness tester[1]. It is relatively easy to measure on small samples. When cracks form around the Vickers impression under a certain load, it is suggested that a critical deformation energy (E_f) can be defined. Based on a relationship between E_f and K_{IC} [2], K_{IC} can be calculated.

Experimental Procedure

1) Sample Preparation

Sample with nominal composition $YBa_2Cu_3O_{7-y}$ were prepared by high-purity powders of CuO (99.9%), $BaCO_3$ (99.9%), Y_2O_3 (99.99%)[3]. The powders were mixed and sintered in air at 1200 K for 2 h and then air-cooled. After crushing, the pelletized tablet, 2 mm thick and 13 mm diam, was sintered in air at 1200 K for 8 h and air-cooled. The cooling rate was 0.07 K/s at 1000 K. The structure of the samples was an orthorh-ombic structure, which corresponds to oxygen-deficient perovskite.

Samples with nominal composition $Bi_{0.7}Pb_{0.3}SrCaCu_{1.8}O_x$ were prepared from high-purity powders of Bi_2O_3 (99.99%), PbO (99.99%), $SrCO_3$ (99.9%), $CaCO_3$ (99.9%), and CuO(99.9%)[1,4]. The powders were mixed and sintered in air at 1123 K for 6 h and then air cooled. The pellet was additionally sintered in air at 1123 K for 12 h and then furnance cooled. Liquid-quenched glassy foils were prepared from the sintered tablets using a twintype pistone-anvil apparatus[5]. The apparatus was constracted to quench the molten sample. The amount of the sintered tablet melted by an infrared furnance in one run was about 0.2 g. The speed of the piston was about 0.12 m/s. The cooling rate was varied by controlling the sample mass. The molten sample was quenched rapidly.

2) Calculation of K_{IC}

A critical deformation energy of crack was suggested here as a new parameter to evaluate brittleness[2]. The deformation energy (E) could be expressed by the following equation:

$$E=Ph_0$$

Here, h_0 (h_0=3V/S) was the mean depth the Vickers impression, where V and S were the volume and basal plane square of the pyramidal hole. The load P (P=mg) was the force required to deform the hole, where m was the mass, and g was the acceleration of gravity. The critical deformation energies (E_f) could be estimated using the critical value (d_f and P_f) as

$$E_f=Cmgd/3$$

Here, d was the diagonal of the basal plane, C was a constant (C=h/d=0.143) determined by the angle of the pyramid, where h was the depth of the impression.

The values of K_{IC} could be obtained by an impression fracture method. The values of K_{IC} could be determined through the following equations [2]:

$$\log_{10} K_{IC} = 0.44 \log_{10} E_f - 0.63$$

Both nucleation and growth mechanisms of a given crack determined the value of E_f, whereas K_{IC} was controlled by the growth mechanism of the crack. Therefore, the measurement of E_f was an excellent tool to be used in the evaluation of brittleness.

Results and Discussion

Figure 1 shows schematic relations for different hardened specimens. Since the values of the d_f and P_f show opposite resultes for the different hardened specimens, we cannot evaluate the brittleness precisely. Therefore, we suggest a deformation energy of fracture as a new parameter to evaluate the brittleness. Figure 2 shows schematic relations of logarithmic load (P) and logarithmic diagonal (d) for different hardened materials. If the crack occars, the spread of the data can be cound.

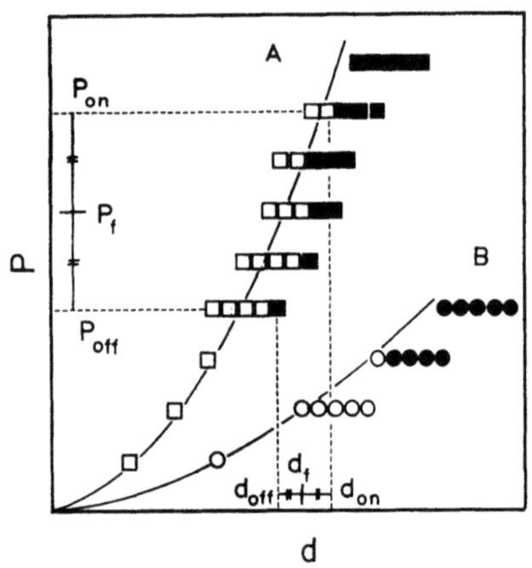

Fig.1 Schematic relations of load (P) and diagonal (d) for different hardened materials.

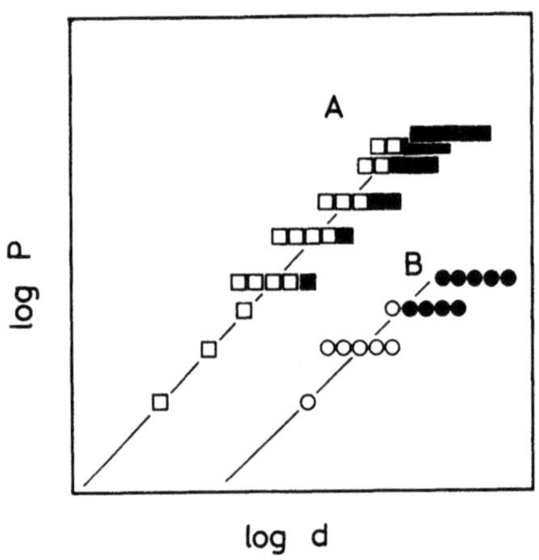

Fig.2 Schematic relations of logarithmic load (P) and logarithmic diagonal (d) for different hardened materials.

Figure 3 shows change in a stress intensity factor (K_{IC}) against sample thickness (D) of liquid-quenched $Bi_{0.7}Pb_{0.3}SrCaCu_{1.8}O_x$ glass. The cooling rate (R) is dominated by the D value just after the liquid-quenching. The higher the cooling rate (the smaller the D value), the lower the K_{IC} value becomes.

K_{IC} of the crystallized specimens before and after annealing are summarized in Fig.4. The crystallization enhances the ductility of the $Bi_{0.7}Pb_{0.3}SrCaCu_{1.8}O_x$ glassy sample. The annealing dereases ductility after the crystallization. The recrystallization (see Fig.4) increases K_{IC}. K_{IC} of the crystallized sample annealed for 12h at 1120K (Tcoff=111.7K;) is about two times larger than that of the glassy sample before the annealing.

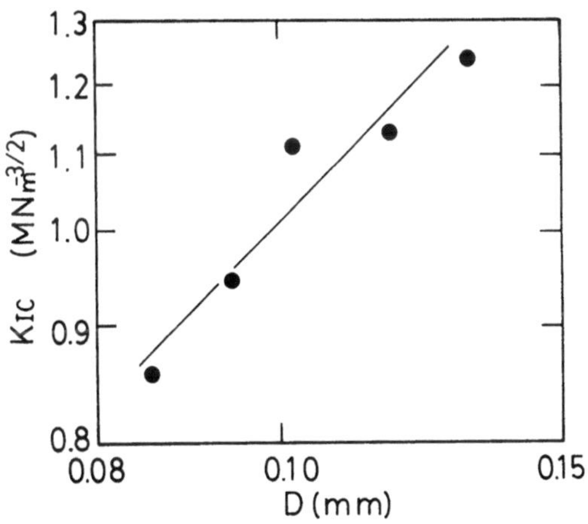

Fig.3 Change in a stress intensity factor (K_{IC}) against sample thickness (D) of liquid-quenched $Bi_{0.7}Pb_{0.3}SrCaCu_{1.8}O_x$.

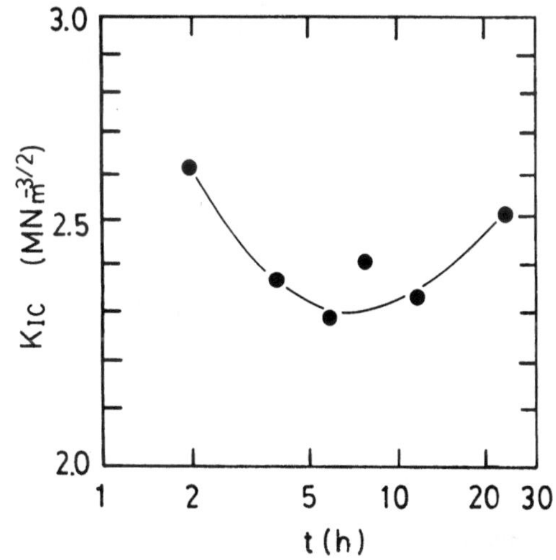

Fig.4 Change in a stress intensity factor (K_{IC}) against annealing time (t) of crystallized $Bi_{0.7}Pb_{0.3}SrCaCu_{1.8}O_x$.

Figure 5 shows relations among K_{IC}, d_f and P_f of the $Bi_{0.7}Pb_{0.3}SrCaCu_{1.8}O_x$ glassy and the crystallized samples together with other materials. The values of the crystallized sample are above that of the Al_2O_3 plate and below those of the sintered superconductive oxides. On the other hand, the value of the glassy sample is near that of the cover glass. Therefore, we conclude that the crystallization of the $Bi_{0.7}Pb_{0.3}SrCaCu_{1.8}O_x$ glass induces ductility. Furthermore, this figure shows that the powder metallurgy is a good tool to induce ductility because of the high K_{IC} value of the sintered oxides.

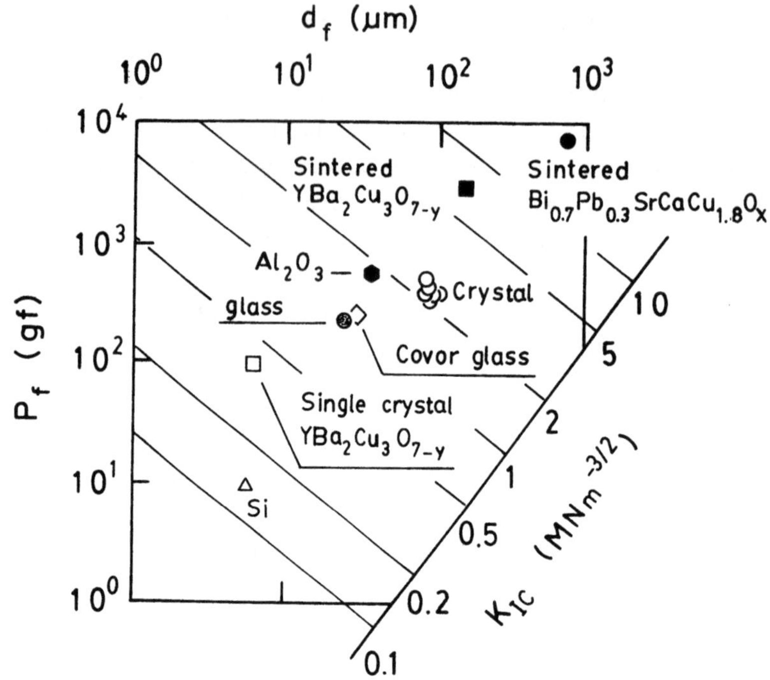

Fig.5 Relations of P_f, d_f and K_{IC}.

Conclusion

In summary, stress intensity factor (K_{IC}) is suggested to be a convenient measurement to determine the ductiliity (brittlness) of new materials by means of new standerd Vickers indentation method. We found that K_{IC} is controlled by the processes.

References

1. Nishi Y, Ninomiya N, Kanai F, Uchida S and Moriya S (1989) J.Appl.Phys. 66:2069.
2. Nishi Y, Katagiri T, Yamano T, Kanai F, Ninomiya N, Uchida S, Oguri K, Morishita T, Endo T and Kawakami M (1991) Appl.Phys.Lett. 58:2084.
3. Nishi Y, Ninomiya N, Oguri K, and Tokunaga S (1989) J.Appl.Phys. 65:3972.
4. Nishi Y, Oguri K, Ohinata H, Tanioka K, Kita Y and Ninomiya N (1990) Phys.Rev.B 41:6520.
5. Nishi Y, Moriya S and Manabe T (1989) J.Appl.Phys. 65:2389.

Mechanical and Magnetic Properties of Zr-YBCO Superconducting Composites

TETSUO OKA[1], YOSHITAKA ITOH[1], YOUSUKE YANAGI[1], HIDEHIKO TANAKA[2], SHINICHI TAKASHIMA[2], and UICHIRO MIZUTANI[2]

[1] Aisin Seiki Co., Ltd., Kariya, Aichi, 448 Japan
[2] Department of Crystalline Materials Science, Nagoya University, Nagoya, 464-01 Japan

ABSTRACT

The mechanism of the enhanced mechanical strength for the Zr-bearing YBCO sintered composites is clarified. The 123-phase in contact with CuO, which is produced by decomposition of YBCO due to addition of Zr, partially melts at 970 ℃. This partial melting contributes to reduce voids, whereas fine $BaZrO_3$ precipitates prevent the sample from collapsing and contribute also to enhance the mechanical strength. The magnetization curves for the Zr-YBCO samples showed that doping Zr up to x=0.1 in Zr_xYBCO increased the values of ΔM_{max} derived from the M-H loops. It is suggested that Zr-precipitates most likely act as pinning centers when its concentration is low.

KEYWORDS: YBCO, sintered composites, flexural strength, magnetization curve

INTRODUCTION

We have reported that the poor mechanical properties inherent to the ceramic high-Tc superconductors can be improved in greater extent by adding Ag and Zr to the $YBa_2Cu_3O_{7-y}$ (Hereafter abbreviated as YBCO or 123-phase)[1-3]. For instance, the flexural strength of about 40MPa for YBCO can be enhanced up to 280MPa by dispersing uniformly Ag particles of the size of about 100 μm in the Zr-bearing YBCO sintered matrix without disrupting superconducting channels. The X-ray diffraction and scanning electron microscope analysis revealed that the Zr-bearing matrix consists of the 123-phase as a major component plus additional minor phases Y_2BaCuO_5(211-phase), CuO and $BaZrO_3$. Ag particles precipitate independently in the sintered body and remain intact with YBCO grains, bestowing the plasticity on composites when strain is applied. It was concluded that the $BaZrO_3$, being finely precipitated along the periphery of the 123-phase and an increase in the sintering density due to the partial melting around the CuO phase are responsible for the increased mechanical strength.

In this paper, making full use of DTA and XRD analyses, we elucidated the details of the chemical reaction occurring in the Zr-YBCO during the sintering process and reinforced our previous proposal on the mechanism for the enhancement in the mechanical strength. Furthermore, the magnetization curves were measured at 77K in order to extract information about the critical current density for our mechanically strengthened composites.

EXPERIMENTAL

The Zr-bearing YBCO composites were synthsized through the sintering process from the starting powders prepared by mixing appropriate amounts of Y_2O_3, $BaCO_3$, CuO and ZrO_2 powders so as to adjust the composition to the formula $Zr_xYBa_2Cu_3O_{7-y}$ with x=0-0.8(abbreviated as Zr_xYBCO). The sintering was carried out in O_2 atmosphere, after calcining in air at 900℃ for several times, with subsequent uniaxial compression into a bar-shape sample, with the size of 1×7 ×20mm. The samples were slowly cooled with the rate of 100℃/h after dwelling time of 12h at the temperatures 900-980·℃. DTA analysis was performed with the heating rate of 20℃/min in O_2 atmosphere up to 1100℃ for Zr-YBCO. XRD analysis was performed for the samples which were quenched to room temperature after holding at various temperatures for 15min. The flexural strength was measured through the 3-point bending test method. The field dependence of magnetization or M-H curve was measured at 77K with the Vibrating Sample Magnetometer(VSM).

The value of ΔM, being defined as the difference in the magnetizations at a given field under its increasing and decreasing modes, was read off from the M-H hysteresis curve.

RESULTS AND DISCUSSION

DTA Analysis

We already revealed[1-3]that the present Zr-YBCO samples consist of the 123-, 211-, CuO and $BaZrO_3$ phases, and that these phases were formed through the following reaction during the sintering process.

$$1/2Y_2O_3+2BaCO_3+3CuO+xZrO_2+(1/4-x/6)O_2 \rightarrow (1-2x/3)YBa_2Cu_3O_{7-y} + x/3Y_2BaCuO_5+xBaZrO_3+5x/3CuO+2CO_2 \quad (1)$$

The DTA thermograms were measured for these Zr-YBCO samples. The results are shown in figure 1. The YBCO sample exhibits a single endothermic peak at 1020℃ due to the partial melting of the 123- into the 211- and liquid phases. When the Zr content exceeds about 0.3, a new extra endothermic peak appears at about 970℃. With increasing Zr content, the original peak at higher temperatures displaces its position to lower temperatures while the newly grown peak remains unchanged. Two peaks apparently merge into one at x=0.8.

The same DTA thermograms as those shown in figure 1 can be obtained when the 123-phase is mixed with different amounts of CuO powders. Furthermore, we found that the Zr concentration dependence of the respective peak temperatures shown by open circles in Figure 2 coincides well with the corresponding CuO dependence for the 123/CuO samples(solid triangles), provided that the scale for the Zr content is 5/3 times expanded relative to the scale for the CuO content. This is, indeed, inferred from equation(1). This demonstrates that the presence of CuO in contact with the 123-phase is responsible for the endothermic peak at 970℃.

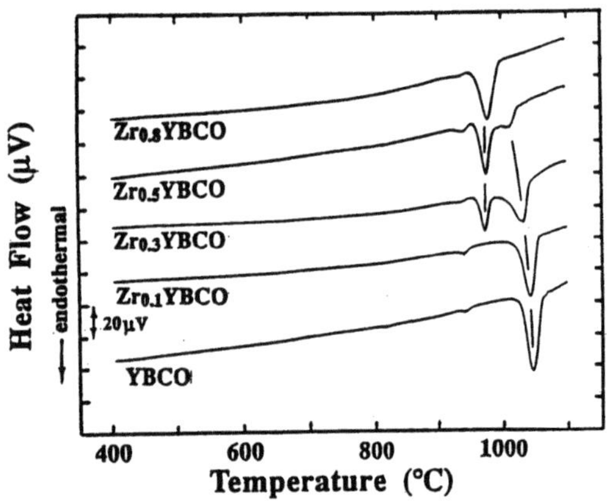

Figure 1 DTA thermograms for Zr-YBCO

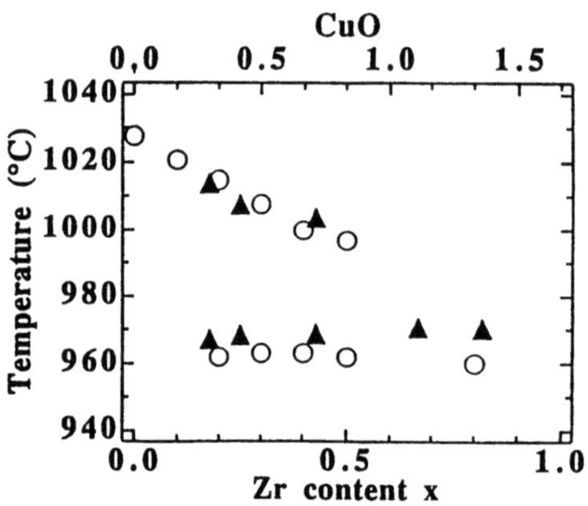

Figure 2. Endothermic temperatures for Zr-YBCO and CuO/YBCO

XRD analysis

The XRD peak heights of the 123- and 211-phases in Zr_xYBCO samples with x=0,0.3 and 0.8 are shown in figure 3 as a function of the heat-treatment temperature. Note here that the samples after heat-treatment for 15min in O_2 atmosphere were rapidly quenched to 300K to retain the phases at high temperatures. The 123-phase decomposes at 1020℃ and yields the 211-phase in the pure YBCO, whereas the same reaction occurs at 970℃ for $Zr_{0.8}YBCO$. The amount of the 211-phase for $Zr_{0.3}YBCO$ increases slightly at 970℃ and more sharply at about 1000℃ at the expense of the 123-phase. These phase transformation occurs in accordance with endothermic peaks observed in DTA runs shown in fig.1. Therefore, it is concluded that the 123-phase in contact with CuO begins to decompose into liquid and 211-phase at 970 ℃. In addition, we revealed that the liquid solidifies into CuO and $BaCuO_2$ without forming the 123- by rapid cooling. This is quite naturally understood in terms of the following reaction:

$$2YBa_2Cu_3O_{7-y} + xCuO \rightarrow Y_2BaCuO_5 + L \quad (970\ ^\circ C) \quad \downarrow \text{(when rapidly cooled)} \quad (2+x)CuO + 3BaCuO_2 \qquad (2)$$

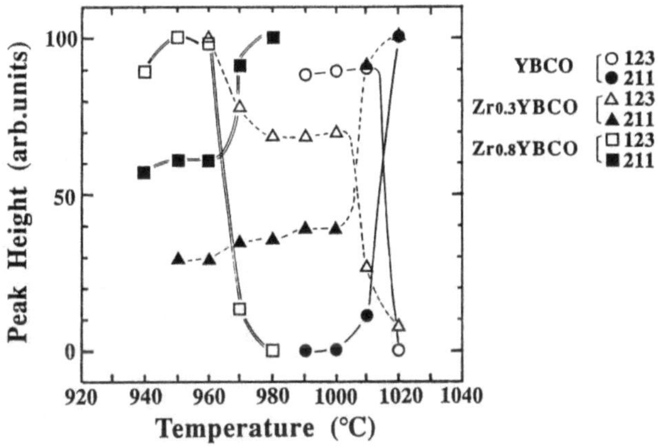

Figure 3. The normalized X-ray peak heights of the 123- and 211-phases for Zr_x YBCO samples with x=0, 0.3 and 0.8

Microstructure

The microstructure on polished surfaces of the $Zr_{0.8}$YBCO composites was studied with the SEM as shown in figure 4. The sample sintered at 960℃ for 12h, which corresponds to the temperature below the partial melting point of 970℃, is found to contain a number of voids due to inappropriate annealing temperature, whereas the sample sintered at 970℃ appears to have a much higher packing density as a result of the partial melting(b).

(a) 960 ℃ (b) 970 ℃

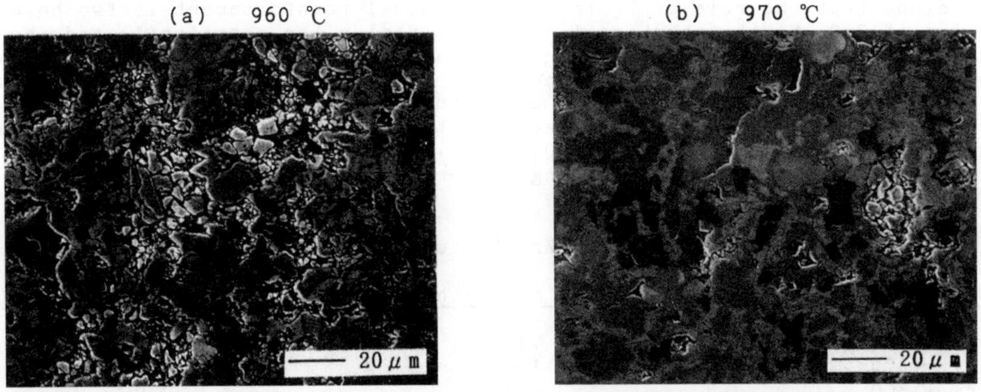

Figure 4. SEM images for $Zr_{0.8}$YBCO after sintering 12h in O_2 atmosphere

Mechanical property

The flexural strength σ_{max} was measured for the $Zr_{0.8}$YBCO composite using the 3-point bending test method and the results are shown in figure 5. The values of σ_{max} was substantially improved when the sintering temperature is raised above 960℃. One can attribute an enhancement in the flexural strength to densification in the microstructure due to the partial melting.

Magnetic property

Figure 6 shows the magnetization curves measured by VSM for pure YBCO (a) and $Zr_{0.04}YBCO$(b), respectively. The M-H curve for the sintered YBCO sample is typical of a granular high-Tc superconductor[4,5]. Value of ΔM_{max} derived from the hysteresis curve is plotted in its inset as a function of Zr content. It is clear that the value of ΔM_{max} increases relative to the pure YBCO in the range of x=0.02-0.08. As is clear from the hysteresis curves, a gain in ΔM_{max} mainly resulted from the different M-H behaviors in the region of less than 0.2T in the decreasing field mode. We believe that the doping Zr, possibly in the form of $BaZrO_3$, partially precipitates in the 123-grains and acts as pinning centers under the weak magnetic field. Further study is in progress to elucidate the role of $BaZrO_3$ in the Zr-bearing YBCO samples from the viewpiont of the capability of conveying superconducting current.
Numerical data on mechanical properties of Zr_xYBCO composites are listed in table 1. It can be seen that the enhancement in the mechanical strength for the YBCO/CuO sample is only one half that for the Zr-YBCO sample. As opposed to the Zr-YBCO, the YBCO/CuO sample collapses as soon as the partial melting starts.

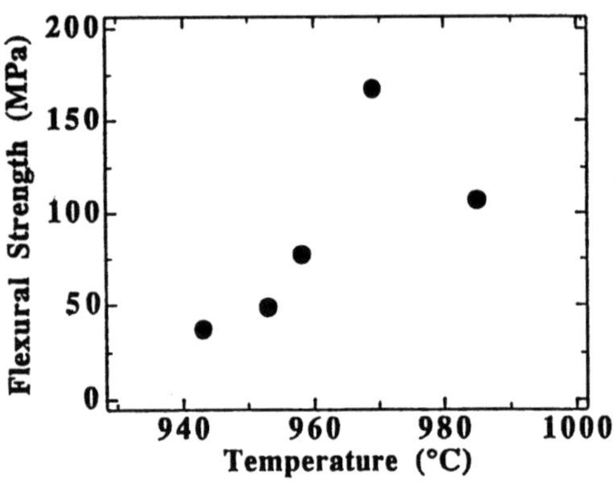

Figure 5. flexural strength of $Zr_{0.8}YBCO$

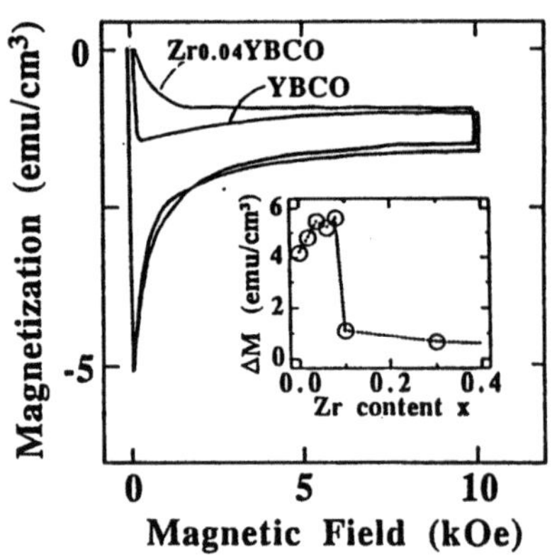

Figure 6. M-H curves and ΔM_{max} measured for YBCO and Zr_xYBCO

CONCLUSIONS

We conclude that the role of Zr in the mechanical reinforcement is two-hold: the addition of ZrO_2 to the starting ingredients yields the reaction(1) and the resulting CuO lowers partial melting temperature to 970 ℃. This is essential to reduce voids. Secondly, the $BaZrO_3$ not only prevents the sample from collapsing during the partial melting but also contributes to enhance the mechanical strength. The value of ΔM_{max} derived from the magnetization curves was enhanced with increasing Zr content up to x=0.1. An increase in ΔM_{max} might be indicative of the presence of pinning centers introduced in the 123-phase.

Table 1. Mechanical properties of Zr_xYBCO composites

samples	flexural strength (MPa)	sintering temperature (℃)
YBCO	40	980
YBCO/CuO(1.75:5)	73	960
$Zr_{0.3}YBCO$	153	970
$Zr_{0.8}YBCO$	168	970
$YBCO/CuO/BaZrO_3$(1.75:5:3)	127	970

REFERENCES

[1] Nishio T,Itoh Y,Ogasawara F,Suganuma M,Yamada Y,Mizutani U(1989)J Mat Sci 24 :3228.
[2] Oka T,Ogasawara F,Itoh Y,Suganuma M,Mizutani U (1990)Jpn J Appl Phys29:1924
[3] Oka T,Itoh Y,Yanagi Y,Tanaka H,Takashima S,Mizutani U (1991)to be publised in Physica C, Prc. of M²S-HTSCⅢ
[4] Kumakura H,Uehara M.Togano K (1987) J Appl Phys Lett 51:1557
[5] Murakami M,Morita M.Koyama N (1988) Jpn J Appl Phys 28:L1125

6 Preparation and Properties of Thin Films

6.1 YBCO and Related Films

Role of Ozone in the Oxidation of $YBa_2Cu_3O_{7-x}$ Thin Films

AKIHITO SAWA[1], SHIN KOSAKA[1], HARUHIKO OBARA[1], KAZUYUKI AOKI[1,*] and JOHN S. EDMENDS[2]

[1] Electrotechnical Laboratory, Tsukuba, Ibaraki, 305 Japan
[2] I.R.C. in Superconductivity, University of Cambridge, Cambridge CB3 OHE, UK

ABSTRACT

The effect of ozone on the oxidation of c-axis oriented $YBa_2Cu_3O_{7-x}$ thin films has been investigated. In-situ resistivity measurements were performed in post-deposition annealing to monitor the diffusion of oxygen in the thin film. It was found that the diffusion coefficient 4.2×10^{-15} cm^2/s at 400 °C of $YBa_2Cu_3O_{7-x}$ thin films in ozone was approximately equal to that in molecular oxygen. In oxidation using ozone, an ozone flux about 10^{16} $cm^{-2}\cdot s^{-1}$ at background pressure 1 mTorr and 400 °C was sufficient to obtain the same steady-state resistivity as in molecular oxygen at 760 Torr and 400 °C.

KEY WORDS: $YBa_2Cu_3O_{7-x}$ thin film, in-situ resistivity measurements, oxygen concentration, ozone, post-deposition annealing

INTRODUCTION

In a $YBa_2Cu_3O_{7-x}$ superconductor, the normal-state resistivity and superconductivity depend strongly on oxygen concentration [1]. In the in-situ film formation, it is important to control oxygen content during deposition and post-deposition oxygenation. This means that a highly activated oxidizing agent is required to achieve in-situ growth of $YBa_2Cu_3O_{7-x}$ films, especially for deposited at low oxygen pressure, such as in a molecular beam epitaxy (MBE) system. Ozone [2] and an electron cyclotron resonance (ECR) plasma sources [3, 4] have been used for this purpose. It is important to understand the oxidation process of thin films in these highly activated oxidizing agent in order to get high quality films.

In $YBa_2Cu_3O_{7-x}$ ceramics and single crystals, oxygen content and diffusion have been measured by using secondary ion mass spectroscopy (SIMS) profiles of tracer [5, 6], thermogravimetric [7] and in-situ resistivity measurements [8]. A phase diagram of oxygen content versus temperature of $YBa_2Cu_3O_{7-x}$ has been reported for molecular oxygen atmosphere [7,9]. Also, oxygen diffusion of YBa2Cu3O7-x thin films during oxidation in molecular oxygen and in ECR plasma has been studied by using in-situ resistivity measurements [10, 11]. In the presence of atomic oxygen supplied by ECR plasma, the phase diagram of oxygen content versus temperatures changed. Ozone is as effective for oxidation as ECR plasma, however effect of ozone for oxidation process has not been studied yet.

In this report, we have studied the effect of ozone for the oxidation of c-axis oriented $YBa_2Cu_3O_{7-x}$ thin films by using in-situ resistivity measurements in post-deposition annealing of films.

EXPERIMENTAL

The c-axis oriented $YBa_2Cu_3O_{7-x}$ thin films used for measurements were sputter deposited [12] or active coevaporated [13] on MgO(100) substrates, and have a superconducting transition temperature Tc(zero) of about 80 K. The thickness of the films were 500 Å ~ 1700 Å.

In-situ resistivity measurements in post-deposition annealing were performed under several condition by using four-point DC transport. Contacts were attached with sliver paste, using gold wire on evaporated gold, and then annealed at 400 °C. A schematic of the general experimental arrangement is shown in Fig. 1.

High purity ozone was liquefied from an oxygen-ozone mixture gas, which was generated by an ozonizer, by using a precisely temperature-controlled still [14]. Ozone flux was controlled by changing the evaporation temperature of the liquefied ozone. Ozone flux at the film surface was estimated by using oxidation rate measurements of silver films evaporated on a quartz crystal rate monitor [15]; a minimum value was used as the estimated value, since ozone which did not react to silver existed.

* on leave from Chiba institute of Technology

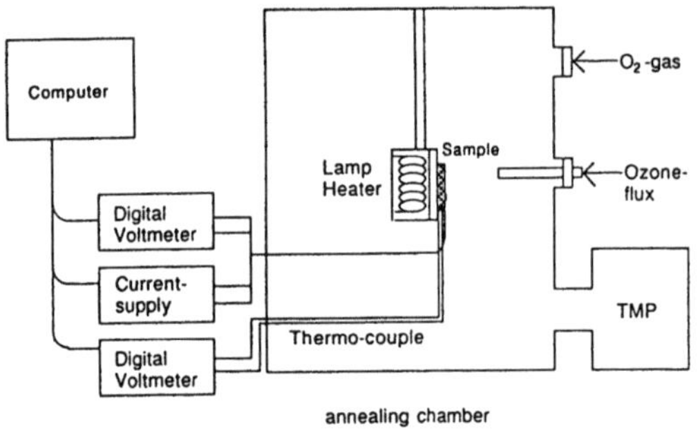

Fig 1. A schematic of the general experimental arrangement.

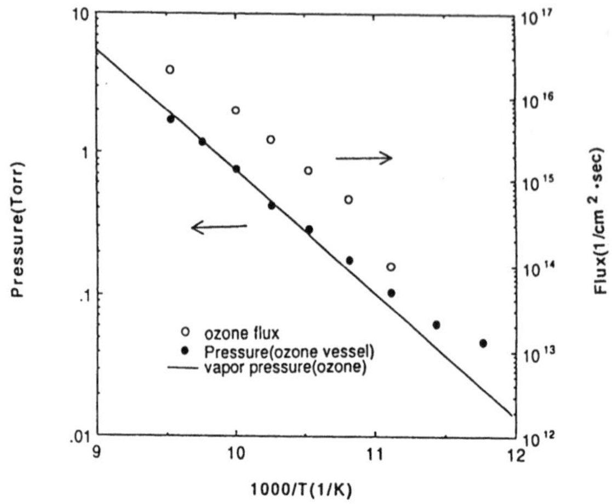

Fig. 2. The ozone flux at the film surface and the pressure in the ozone vessel as a function of the evaporating temperature of the liquefied ozone

RESULTS AND DISSCUSION

The ozone flux at the film surface and the pressure in the ozone vessel was plotted as a function of the evaporating temperature of the liquefied ozone in Fig. 2. Ozone flux could be varied from 10^{14} to 10^{16} $cm^{-2} \cdot s^{-1}$ by adjusting the evaporating temperature of the liquefied ozone from 90 K to 105 K.
Initially, the film was annealed at 400 °C and 5 mTorr in a 1 atm flow of molecular oxygen, and the resistivity of film was allowed to reach equilibrium. Then the oxygen content of the film was set at a certain value determined by the temperature and pressure in molecular oxygen; for a phase diagram of YBa2Cu3O7-x [6,8], oxygen deficiency x was about 0.3 at 400 °C and 5 mTorr. In this way, initial oxygen content approximately reached the same value in every measurement. After the resistivity of film was saturated, and the temperature was rapidly changed to a fixed temperature, the oxygen pressure in molecular oxygen was changed to 10 mTorr at time t=0. We observed the change of resistivity attributed to the change of oxygen content in the film. Figure 3 shows the time dependence of the film resistivity at various film temperatures in molecular oxygen.
In ozone anneal, after equilibrium oxygen content was reached under the same conditions as before, and temperature was changed to a fixed temperature, the molecular oxygen was stopped at time t=0, and an ozone flux of about 3×10^{15} $cm^{-2} \cdot s^{-1}$ was introduced into the chamber at time t=0. Figure 4 shows the time dependence of the film resistivity at various film temperatures in ozone.
The simplest model for oxygen diffusion of thin film is one-dimensional diffusion governed by Fick's law. If we can assume the chemical diffusion coefficient D is approximately constant for a change of oxygen content in the film, the diffusion equation can be solved in the form

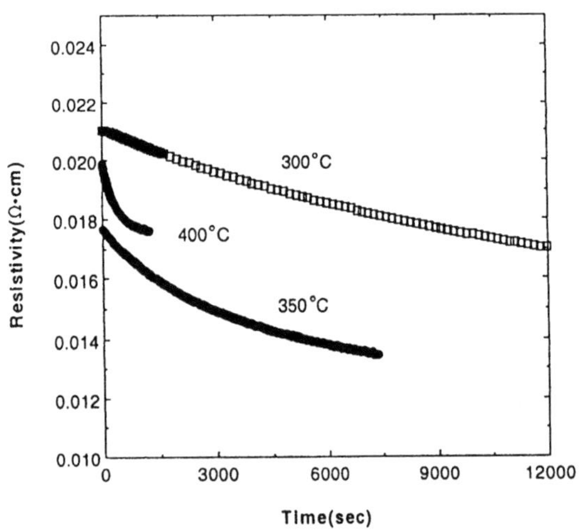

Fig. 3. The time dependence of the film resistivity at various film temperatures in molecular oxygen.

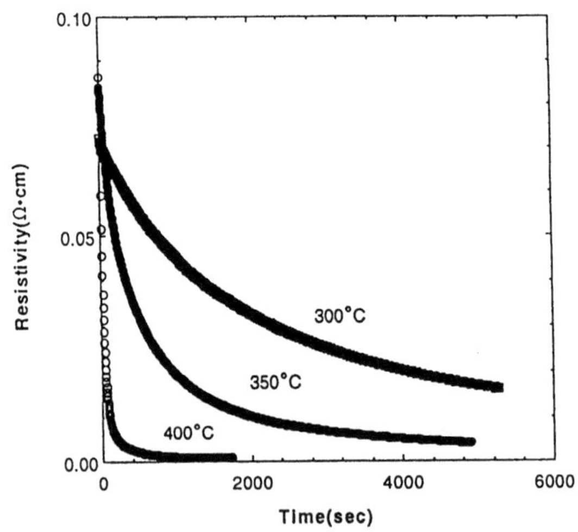

Fig. 4. The time dependence of the film resistivity at various film temperatures in ozone.

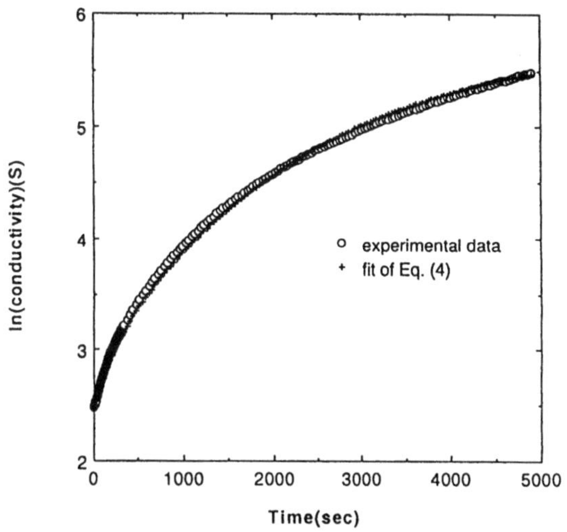

Fig. 5. A fit of Eq. (4) to experimental data for a 500 Å thickness film at 350 °C annealed using ozone.

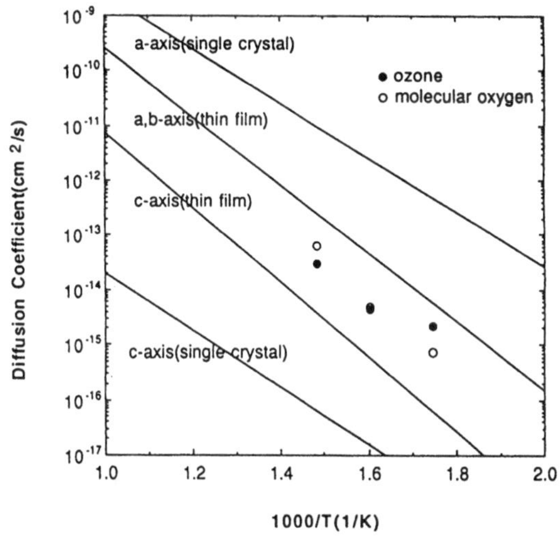

Fig. 6. The Arrhenius plot of the diffusion coefficient versus temperature. The lines represent the diffusion coefficient in the c-axis and a,b-axis of single crystals and thin films in molecular oxygen (see Ref. 6, 11).

$$\frac{C(z,t)-C_0}{C_1-C_0} = 1-\frac{4}{\pi}\sum_{n=0}^{\infty}\frac{(-1)^n}{2n+1}\exp\left(\frac{-D(2n+1)^2 p^2 t}{4l^2}\right)\cos\frac{(2n+1)\pi z}{2l}, \quad (1)$$

where t and z are time and depth, respectively, from film surface, l is the film thickness, C(z,t) is the oxygen concentration, C_0 is an initial oxygen concentration of the film before diffusion anneal (t<0), and C_1 is a oxygen concentration at the film surface during diffusion anneal (t≥0). According to other authors [10], conductivity σ and oxygen deficiency x have a relationship of the form

$$\sigma=\sigma_0\exp(-x/x_0), \quad (2)$$

where $x_0 \approx 0.2$. Conductivity of a sample changes with the change of the local distribution of the oxygen content in the thin film. For large changes in oxygen content, a better approximation is

$$\Delta\ln\sigma \sim \Delta C, \quad (3)$$

where $\Delta\ln\sigma$ is the value of the change of conductivity and ΔC is the value of the change of the oxygen content. For small changes in oxygen content, conductivity changes linearly with the change of oxygen content, so then time dependence of the conductivity of the thin film can be written in the form

$$\sigma(t) = \sigma_0 - \Delta\sigma\frac{8}{\pi^2}\sum_{n=0}^{\infty}\frac{1}{(2n+1)^2}\exp\left(\frac{-D(2n+1)^2\pi^2 t}{4l^2}\right), \quad (4)$$

In this equation, σ_0 is the saturated conductance at t=∞ and $\Delta\sigma$ is the difference between the conductance at t=0 and the saturated value.

The diffusion coefficient D of the thin film was determined by the curve fitting method on the basis of Eq. (4). Figure 5 illustrates a fit of Eq. (4) to experimental data for a film of 500 Å thickness at 350 °C annealed using ozone. In this case, D was 4.2×10^{-15} cm^2/s. Figure 6 shows the Arrhenius plot of the diffusion coefficient versus temperature. In this case, D in ozone was approximately equal to that in molecular oxygen. However, it has been reported that D in atomic oxygen generated by ECR plasma is approximately a factor of 20 higher than that in molecular oxygen [10]. Oxygen diffusion in $YBa_2Cu_3O_{7-x}$ thin films is caused by motion of an oxygen atom from one oxygen site to a vacant oxygen site. If oxygen diffusion in $YBa_2Cu_3O_{7-x}$ is determined by a diffusion limited process, the present result may be expected.

As shown in Fig. 3 and 4, resistivity of the film reached a steady-state value which depended on the presence of the oxygen or ozone for a given temperature. Figure 7 shows the pressure dependence of steady-state resistivity in ozone and molecular oxygen at 400 °C. In case of ozone, pressure was plotted as the background pressure generated from decomposed ozone measured in the annealing chamber. In oxidation using ozone, about 10^{16} cm$^{-2}\cdot$s^{-1} of ozone flux at background pressure 1 mTorr was sufficient to obtain the same steady-state resistivity as in molecular oxygen at 760 Torr and 400 °C. The oxygen content at steady-state resistivity was determined by the amount of atomic oxygen at the surface of the film. This means that the amount of atomic oxygen supplied to the surface of the film was about $10^5 \sim 10^6$ larger in the case of ozone than in the case of molecular oxygen. Further study of the phase diagram of oxygen content in the presence of ozone flux is very important for the understanding of in-situ film growth.

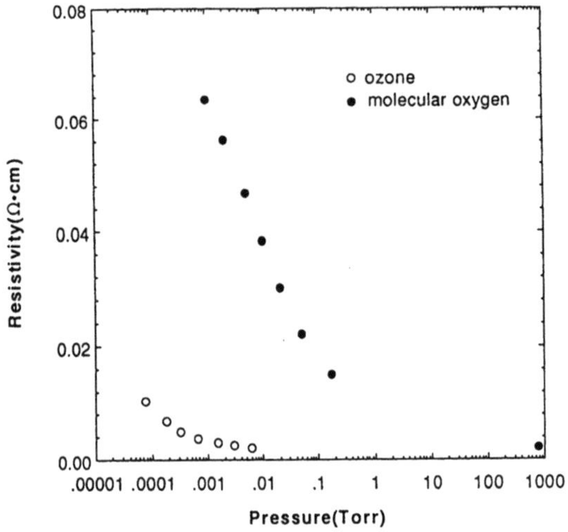

Fig. 7. The pressure dependence of steady-state resistivity in ozone and molecular oxygen at 400 °C.

CONCLUSION

We have studied oxidation of $YBa_2Cu_3O_{7-x}$ thin films by in-situ resistivity measurements. Diffusion coefficient in ozone was approximately equal to that in molecular oxygen. Oxygen diffusion in $YBa_2Cu_3O_{7-x}$ thin films is caused by motion of oxygen atom from one oxygen site to a vacant oxygen site. If oxygen diffusion is determined by a diffusion limited process, this result is expected. It was found that in oxidation using ozone about 10^{16} $cm^{-2} \cdot s^{-1}$ of ozone flux at background pressure 1 mTorr and 400 °C was sufficient to obtain the same steady-state resistivity as in molecular oxygen at 760 Torr and 400 °C. This suggests that ozone oxidizes $YBa_2Cu_3O_{7-x}$ thin films more than molecular oxygen, and it is expected that the phase diagram of oxygen content versus temperature at various oxygen pressure is drastically changed by using ozone. Further experiments by expanding the region of the film temperature and the ozone flux are required to clarify the phase diagram of oxygen content in the presence of ozone.

ACKNOWLEDGMENT

The authors would like to thank M. Matsuda, M. Shoji, H. Sato, and H. Akoh for preparing the thin films, and S. Ichimura for his support in fabricating an ozone generation instrument. We would also like to acknowledge the encouragement from Y. Kimura.

REFERENCES

1. Jorgensen JD, Beno MA, Hinks DG, Soderholm L, Volin KJ, Hitterman RL, Grace JD, and Schuller IK (1987) Phys Rev B36 : 3608-3616
2. Aida T, Tsukamoto A, Imagawa K, Fukazawa T, Saito S, Shinoda K, Takagi K, and Miyauch K (1989) Jpn J Appl Phys 28 : L635-L638
3. Berkley DD, Johnson BB, Anand N, Beauchamp KM, Conroy LE, Goldman AM, Maps J, Mauersberger K, Mecartney ML, Morton J, Tuominen M, and Zahng Y-J (1988) Appl Phys Lett 53 : 1973-1975
4. Nakayama Y, Ochimizu H, Maeda A, Kawazu A, Uchinokura K, and Tanaka S (1989) Jpn J Appl Phys 28 : L1217-L1219
5. Rothman SJ, Routbort JL, Baker JE (1989) Phys Rev B40 : 8852-8860
6. Tsukui S, Yamamoto T, Adachi M, Shono Y, Kawabata K, Fukuoka N, Yanase A, and Yoshioka Y (1991) Proc M^2S-HTSC in Kanazawa, in press
7. Kishio K, Suzuki K, Hasegawa T, Yamamoto T, Kitazawa K, Fueki K (1989) J Solid State Chem 82 : 192-202
8. Tu KN, Yeh NC, Park SI, and Tsuei CC (1989) Phys Rev B39 : 304-314
9. Lindermer TB, Hunley JF, Gates JE, Sutton AL Jr, Bynestad J, and Hubbard CR (1989) J Am Ceram Soc 72 : 1775-1788
10. Kowk HS, and Ying QY (1991) Physica C 177 : 122-128
11. Yamamoto K, Lairson BM, Bravman JC, Geballe TH (1991) J Appl Phys 69 : 7189-7201
12. Masuda M, Shoji A, and Koyangi M : This issue is submitted to this symposium.
13. Matumoto M, Akoh H, and Takada S (1989) J Appl Phys 66 : 3907-3909
14. Hosokawa S, and Ichimura S (1991) Rev Sci Instrum 62 : 1614-1619
15. Kucera JT, Perkins JD, Uwai K, Graybeal JM, and Orlando TP (1990) Rev Sci Instrum 62 : 1630-1632

Growth of Yb-Ba-Cu-O Thin Films by Low-Energy O^+ Ion Beams

RYUSUKE KITA, TAKASHI HASE, TAKESHI KOGA, MASATO SASAKI, TADATAKA MORISHITA, and SHOJI TANAKA

Superconductivity Research Laboratory, ISTEC, Koto-ku, Tokyo, 135 Japan

ABSTRACT

The growth of CuO and Yb-Ba-Cu-O thin films have been investigated using low energy O^+ ion beams as an oxidant. The O^+ ions produced in the Freeman-type ion source was mass-separated and simultaneously irradiated with vapor of metallic components at an acceleration energy of several tens eV. CuO grows epitaxially on MgO(100) at 510°C in a background pressure of $3x10^{-9}$ Torr with 50 eV and 47 $\mu A/cm^2$ O^+ beams. In Yb-Ba-Cu-O system Ba-Cu-O is easy to grow on $SrTiO_3(100)$ at 550°C due to the re-evaporation of Yb.

KEY WORDS : Yb-Ba-Cu-O thin film, low-enrgy, O^+ ion beam

INTRODUCTION

Molecular beam epitaxy (MBE) growth of superconducting oxides involves significant challenges compared to the conventional use of MBE in the growth of semiconductors, since superconducting oxides need to be supplied enough oxygen to compose their crystalline structure and have tendency to decompose in ultrahigh vacuum(UHV). Therefore a powerful oxidant is indispensable to realize in situ superconducting thin films. Several oxidants have been tried: atomic oxygen[1], O_3[2], NO_2[3], ECR oxygen plasma[4]. From thermodynamic calculations[5] O^+ ions are expected to have much greater ability for oxidation among various chemical states of oxygen: O, O_3, O^-, O^{2-}, O^{2+}. One of the advantages to use O^+ ion is that we can control it quantitatively because of the electrical charge, and also use the kinetic energy for migration of adatoms so that the growth of oxides in UHV and low temperatures is expected. The thermodynamic equilibrium line of the tetragonal 123 phase coincides very well with the equilibrium line of CuO[8]. At high temperatures and low oxygen pressures, CuO decomposes and Cu_2O is formed. This coincidence indicates that the growing single-phase CuO is an ideal test to evaluate the performance of O^+ beams so that we tried to synthesize CuO at first and then Yb-Ba-Cu-O films using O^+ beams. In this paper we report the generator of the low-energy and high-density mass-separated O^+ beams and its application to growth of CuO and Yb-Ba-Cu-O films.

EXPERIMENTAL

Figure 1 shows a schematic diagram of a custom-designed low energy O^+ ion beam deposition system, which is composed of the UHV growth chamber and the low-energy O^+ beam generation system. The growth chamber evacuated by two pumps has standard effusion cells, a reflection high energy electron diffraction (RHEED) apparatus and a movable Faraday cup. The O^+ beam generation system consists of an ion source, extraction system, mass-analyzing magnet and deceleration system. The beam line is differentially pumped by three pumps to eliminate the contamination in the growth chamber from the ion source. The base pressure of the growth chamber is typically 5×10^{-11} Torr and the working pressure rises up to 3×10^{-9} Torr during deposition. O^+ ions produced in a Freeman-type hot-cathode-discharge chamber using CO_2 gas were extracted with -30 keV external potential. The mass spectrum of positive ions extracted from the ion source is shown in fig. 2. An ion current of 180 μA are obtained. After mass separation with the resolution (M/ΔM) of over 140 through a 90° analyzing magnet, the O^+ beams were focused with electric quadrupole lenses, deflected by the electrodes, focused again by a einzel lens and finally decelerated just in front of the substrate to low energies. In synthesizing CuO films copper was simultaneously evaporated at 0.06 Å/s from the effusion cell. in synthesizing Yb-Ba-Cu-O films ytterbium, barium and copper were simultaneously evaporated at 0.020, 0.070, 0.022Å/s from the effusion cells, respectively.

RESULTS AND DISCUSSION

The current density of O^+ beams measured by the Faraday cup as a function of the energy of O^+ ion is shown in fig. 3. A current density of 50 μA/cm^2 is obtained independent of O^+ ion energy above 100 eV. Below 100 eV, the ion current density decreases with a decreasing ion energy because of the Coulomb repulsion among O^+ ions and being 35 μA/cm^2 for 30 eV and 47 μA/cm^2 for 50 eV.

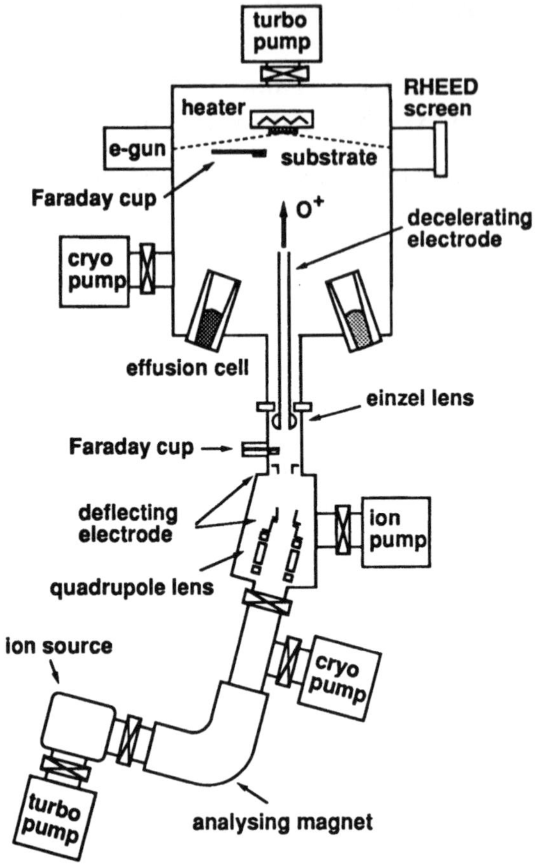

Figure 1 Schematic structure of the low-energy O^+ ion beam deposition system.

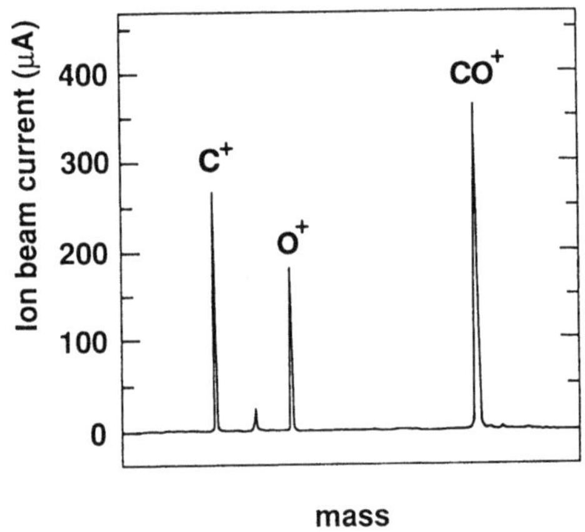

Figure 2 Mass spectrum of extracted positive ions.

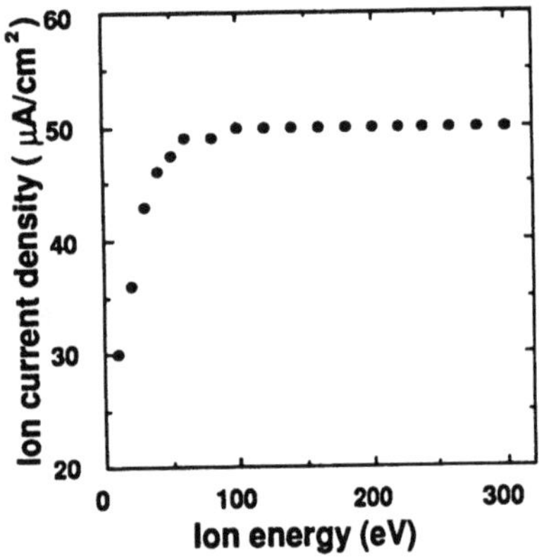

Figure 3 The current density of O^+ beam as a function of the ion energy.

A. Synthesis of CuO films

The O^+ current density of above 10 $\mu A/cm^2$ is required to form CuO on the assumption of unity sticking coefficient of O^+. The O^+ flux density above 10eV is enough to form CuO since the density is several times higher than that of copper. The thickness of CuO is about 150Å after deposition for 1 hr. X-ray photoelectron spectroscopy (XPS) was used to examine the oxidation state of copper atoms. Figure 4 shows XPS spectra of Cu2p core-level for the films deposited on MgO(100) at room temperature and 510 °C, with 50 eV and 47 $\mu A/cm^2$ O^+ beams. This current density corresponds to the oxygen pressure of about $5x10^{-7}$ Torr so that CuO should decompose to Cu_2O at 510°C in the viewpoint of thermodynamics. The existence of the satellite peaks centered at 942 eV indicates the divalent state of copper suggesting the growth of CuO in both substrate temperatures, Ts, even in the non-equilibrium condition. Figure 5 shows the RHEED patterns of the corresponding CuO films in fig. 4. CuO films grow epitaxially on MgO in both Ts. The film deposited at 510°C has different symmetry from the films deposited at room temperature. A lattice spacing of 5.9 Å calculated from the RHEED pattern shown in fig. 5 corresponds to that of CuO films in situ oxidized in a pressure of 10^{-2} Torr using O_2 gas [9].These results indicate that the O^+ ion beams has greater potential for oxidation than O_2 gas in the view point of the necessary flux density to form CuO.

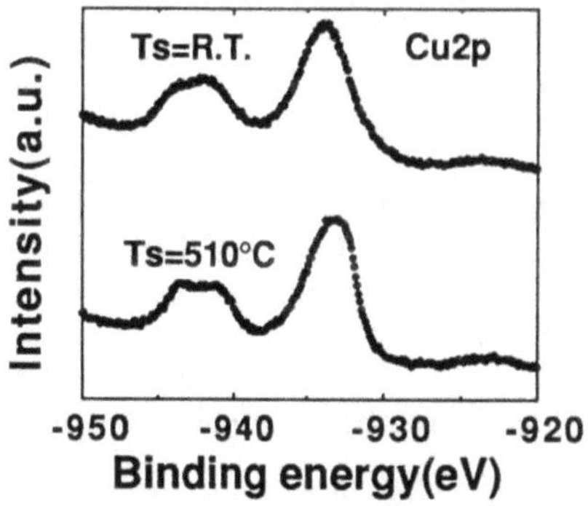

Figure 4 XPS spectra of Cu 2p core-level for the CuO films deposited on MgO(100) using 50 eV O^+ beam at room temperature and 510°C.

(a)

(b)

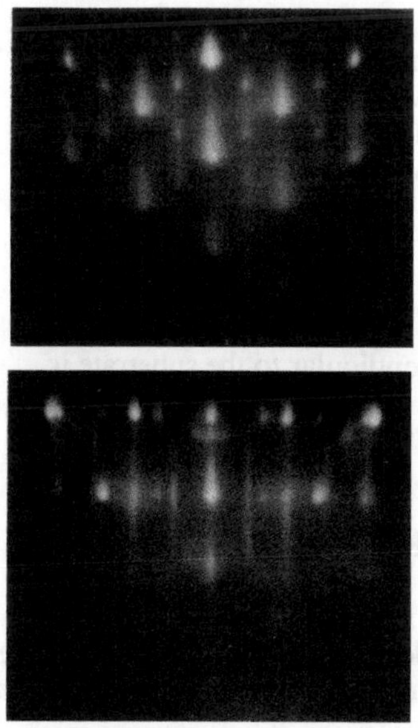

Figure 5 RHEED pattens for the CuO films deposited on MgO(100) using 50 eV O^+ beam (a) at room temperature and (b) 510°C.

B. Synthesis of Yb-Ba-Cu-O

XPS spectra for the $Yb_{1.0}Ba_{2.1}Cu_{3.8}O_y$ film deposited on $SrTiO_3(100)$ at 350°C using 50 eV O^+ beams are shown in fig.6. The Cu 2p spectrum shows Cu^{2+} state of copper atoms in the film and the 529eV peak of the O1s spectrum shows the characteristic of 123 structure. The film, however shows no x-ray diffraction peaks. At higher Ts, we have not yet obtained the films of 123 stoichiometry due to the re-evaporation of Yb from the film.

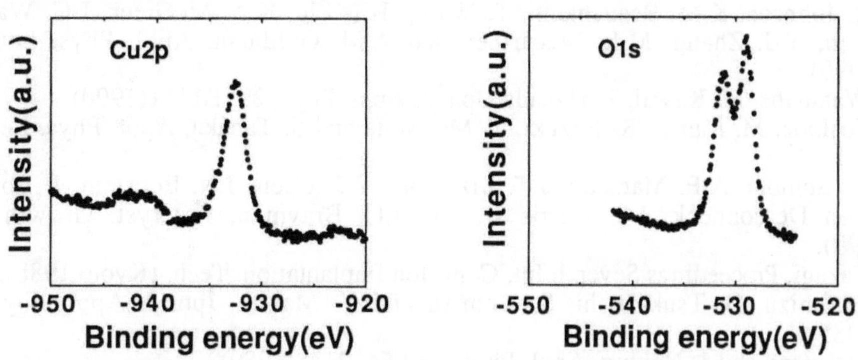

Figure 6 XPS spectra of Cu 2p and O1s core-levels for the Yb-Ba-Cu-O films deposited on $SrTiO_3(100)$ at 350 °C using 50 eV O^+ beams.

Figure 7 shows the RHEED pattern for the film deposited on $SrTiO_3$ at 550°C using 30eV O^+ beams. The composition of the film is Yb:Ba:Cu=~0:2.2:3.0. The distance between the nearest streaks corresponds to the lattice spacing, d, of 6.0Å for the plane perpendicular to the substrate. Figure 8 shows the x-ray diffraction pattern of the corresponding film in fig. 7. The film shows the highly-oriented structure with d=16.0Å normal to the substrate. The crystal structure does not correspond with those of the Ba-Cu-O phases which have been reported.

Figure 7 RHEED pattens for the Yb-poor Ba-Cu-O film deposited on $SrTiO_3(100)$ at 550 °C using 30 eV O^+ beams.

CONCLUSIONS

In conclusion, a new technique for growing oxides in UHV has been developed using low energy and high density O^+ beams. It is applied to the growth of CuO and Yb-Ba-Cu-O films. CuO films epitaxially grows on MgO(100) at 510 °C using 50eV O^+ beams in an ambient pressure of $3x10^{-9}$ Torr. An Ba-Cu-O film which has highly-oriented structure perpendicular to the substrate is easy to grow on $SrTiO_3(100)$ at 550°C.

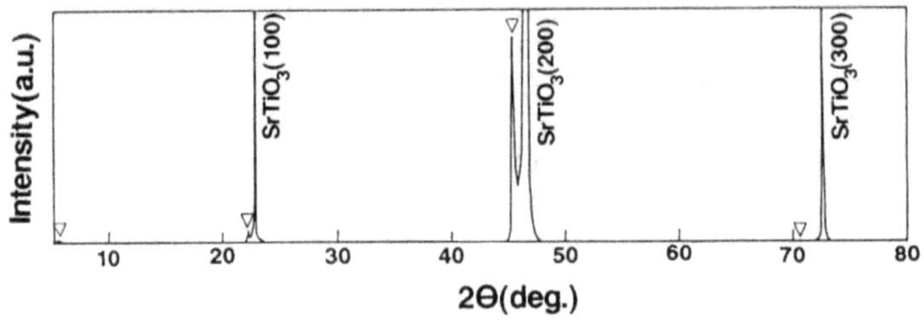

Figure 8 X-ray θ−2θ scan using CuKα radiation for the film shown in fig. 7.
Open triangles indicates the peaks for the film.

ACKNOWLEDGEMENT

The authors like to thank Dr R. Itti for the measurement of XPS and helpful discussion. This work was supported by the New Energy and Industrial Technology Development Organization under the management of the R&D of Basic Technology for Future Industries.

REFERENCES

[1] J.A. Edwads, N.G. Chew, S.W. Goodyear, J.S. Satchell, S.E. Blenkinsop and R.G. Humphreys, J. Less-Commom Metals **164&165**, 414(1990).
[2] B.R. Johnson, K.M. Beauchamp, T. Wang, J-X. Liu, K.A. McGreer, J-C. Wan, M. Tuominen, Y-J. Zhang, M.L. Mecartney and A.M. Guldman, Appl. Phys. Lett. **56**, 1991(1990).
[3] S. Watanabe, M. Kawai, T. Hanada, Jpn. J. Appl. Phys. **29**, L1111(1990).
[4] I. Yoshida, M. Kamei, K. Suzuki, T. Morishita and S. Tanaka, Appl. Phys. Lett. **58**, 654(1991).
[5] D.G. Schlom, A.F. Marshall, J.T. Sizemore, Z.J. Chen, J.N. Eckstein, I. Bozovic, K.E. Von Dessonneck, J.S. Harris Jr. and J.C. Bravman, J. Cryst. Growth **102**, 361(1990).
[6] T. Takagi, Proceedings Seventh Int. Conf. Ion Implantation. Tech. (Kyoto,1988), p1.
[7] S. Shimizu, O. Tsukakoshi, S. Komiya and Y. Makita, Jpn. J. Appl. Phys. **24**, 1130(1985).
[8] R. Bormann and J. Nölting, Appl. Phys. Lett.**54**, 2148(1989)
[9] R. Kita, T. Hase, M. Sasaki, T. Morishita and S. Tanaka, to be published in J. Cryst. Growth.

Low Temperature Synthesis of $YBa_2Cu_3O_{7-x}$ Thin Films at Post Deposition Annealing

TAKASHI HASE, RYUSUKE KITA, TAKESHI KOGA, MASATO SASAKI, TADATAKA MORISHITA, and SHOJI TANAKA

Superconductivity Research Laboratory, ISTEC, Koto-ku, Tokyo, 135 Japan

ABSTRACT

$YBa_2Cu_3O_{7-x}$ thin films with zero resistance temperatures above 80 K have been synthesized at 750℃ from Y-Ba-Cu metallic precursors. The synthesis temperature is about 100℃ lower than those of previous works for synthesis from amorphous oxide precursors without fluoride. Y-Ba-Cu precursor films have been deposited under ultra-high vacuum conditions with Y, Ba and Cu metallic sources then followed by oxidation and annealing. This process is promising for fabrication of $YBa_2Cu_3O_{7-x}$ thin films with expectable homogeneous superconducting properties over a large area due to few geometric limitations for crystallization, and with probably higher purity compared to films of previous process using BaF_2 source.

KEY WORDS : precursor, synthesis temperature, ultra-high vacuum, post-annealing

INTRODUCTION

Most of $YBa_2Cu_3O_{7-x}$ (YBCO) superconducting thin films have been prepared in 'in-situ' process, which means the process for crystallizing the YBCO phase while the constituent materials are deposited on substrates. The YBCO films fabricated by the 'in-situ' process have some advantages as relatively low crystallization temperatures and higher quality of the YBCO crystal than those synthesized from amorphous precursors at a post-annealing. However, the post-annealing process using a furnace for crystallization of YBCO probably enables to prepare samples with more homogeneous superconducting properties over a large area because of few geometric limitations than the 'in-situ' process which needs fairly complicated apparatus.
Recently, Feenstra et al. [1] and Mogro-Campero et al. [2] have reported that amorphous precursors fabricated by coevaporation of Y, BaF_2 and Cu have been transformed into YBCO phase at post-annealing of 750℃ in low oxygen partial pressures such as a few tens of mTorr. The temperature is about 100℃ lower than those for synthesis of YBCO from amorphous Y-Ba-Cu-O precursors requiring decomposition of $BaCO_3$ remained in films at a high temperature [3, 4]. However, the method using BaF_2 requires water-vapor supplying process at which a corrosive HF gas may be produced [5], and moreover fluorine and $Ba(OH)_2$

which might be formed during water-vapor supply possibly remain.

In this paper, we propose a new post-annealing process for fabricating YBCO thin films using Y-Ba-Cu metallic precursors prepared under ultra-high vacuum condition to omit water-vapor supplying process, and to prevent contamination of $BaCO_3$ in the films.

EXPERIMENTAL PROCEDURE

Y-Ba-Cu metallic precursors were fabricated by coevaporation of Y, Ba and Cu up to 400Å in thickness on (100) $SrTiO_3$ single crystalline substrates at 150℃ in the chamber evacuated to ultra-high vacuum less than 5×10^{-10} Torr before each deposition. After deposition, samples were transferred to the lordlock chamber which was evacuated to less than 1×10^{-7} Torr. Then the films were oxidized by introducing an oxygen gas into the lordlock chamber up to 760 Torr and heating up to 200℃. Thus, we have successfully performed formation and oxidation of Y-Ba-Cu precursors under contamination-free conditions.

The YBCO phase was grown from the oxide precursors in a gold furnace. The oxygen partial pressure of the furnace was adjusted to 0.02 Torr at room temperature and then total pressure was increased to 760 Torr by introducing an argon gas. The temperature of the samples was raised to 750℃ and kept at the same value for 120 min followed by treatment at 450℃ for 30 min in an oxygen atmosphere of 760 Torr to transform tetragonal YBCO phase into orthorhombic phase. Thickness of resulting films were approximately increased to 1000Å from 400Å of metallic precursor state.

RESULTS AND DISCUSSION

Figure 1 shows a ternary phase diagram obtained from results of X-ray diffraction measurements for resulting Y-Ba-Cu-O thin films. Metallic composition ratios of the films are determined by an inductively coupled plasma analysis before annealing in the furnace. We have confirmed by several samples that composition ratios are almost the same before and after annealing. It is shown in Fig. 1 that the YBCO phase without an unidentified non-superconducting phase is grown in films with metallic composition which is Cu rich and Ba poor compared with stoichiometric one. We have not clarified why unidentified phase has grown in the film of nearly stoichiometric composition. However, we guess Ba metal which may be phase-separated in Y-Ba-Cu alloy precursors can be easily oxidized and transformed into barium oxide due to its high reactivity. If so, the unidentified phase may be grown from the barium oxide at lower temperature than 750℃ preventing crystallization of YBCO phase.

X-ray diffraction patterns of two films, denoted by① and② in Fig. 1, without the unidentified phase are shown in Fig.s 2(a) and 2(b), respectively. These figures represent the crystallographic orientation of c-axis normal to the film surface. Figure 3 shows resistivity-temperature curves, measured by dc four-probe method, of the two films which exhibit high zero resistance temperature (Tc_0) over 80 K. Furthermore, note that the film of $Y_1Ba_{2.3}Cu_{3.8}O_x$ which is a composition closer to the stoichiometry compared with $Y_1Ba_{1.3}Cu_{2.5}O_x$ film has higher Tc_0 and lower resistivity at normal state. This result reflects $Y_1Ba_{2.3}Cu_{3.8}O_x$ film has a better crystallinity than $Y_1Ba_{1.3}Cu_{2.5}O_x$ film because of the composition closer to the stoichiometry. Figure 4 shows surface morphology of $Y_1Ba_{2.3}Cu_{3.8}O_x$ film observed by the scanning electron microscope. Precipitates which are probably copper oxide are seen, however, matrix region has a smooth surface with roughness less than 100Å. It is expected that we can produce YBCO thin films with a better crystallinity by controlling composition of them more carefully.

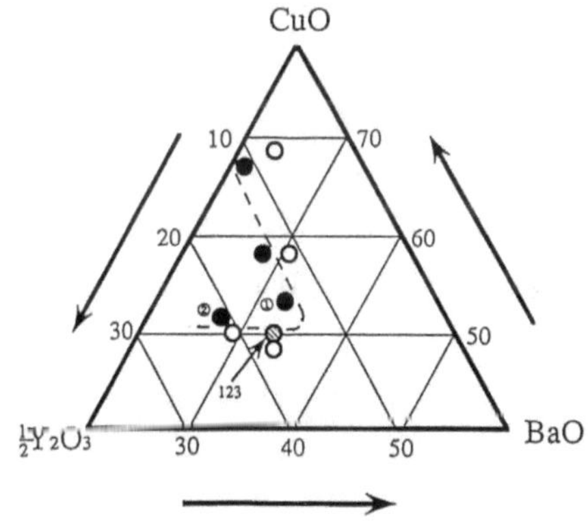

Fig. 1. Ternary phase diagram of Y-Ba-Cu-O thin films. ○ and ● denote YBCO crystallized films with an unidentified non-superconducting phase and without it, respectively. ① ($Y_1Ba_{2.3}Cu_{3.8}O_x$) and ② ($Y_1Ba_{1.3}Cu_{2.5}O_x$) indicate such composition ratios.

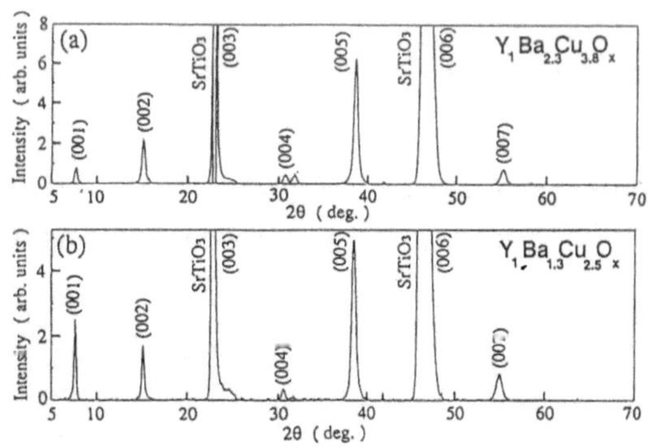

Fig. 2. X-ray diffraction patterns of films with composition ratios of (a) $Y_1Ba_{2.3}Cu_{3.8}O_x$ and (b) $Y_1Ba_{1.3}Cu_{2.5}O_x$.

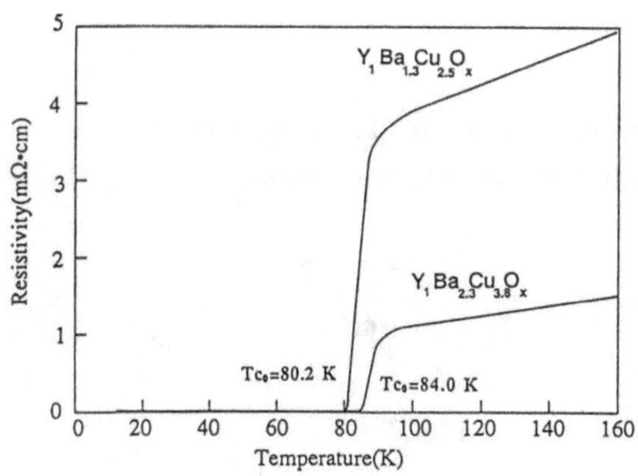

Fig. 3. Resistivity-temperature curves of films with composition ratios of $Y_1Ba_{2.3}Cu_{3.8}O_x$ and $Y_1Ba_{1.3}Cu_{2.5}O_x$.

Fig. 4. Surface morphology of $Y_1Ba_{2.3}Cu_{3.8}O_x$ film.

CONCLUSIONS

We have successfully reduced synthesis temperature of YBCO down to 750℃ by a new process of oxidation and annealing of Y-Ba-Cu metallic precursor, omitting water-vapor supplying process. The temperature is above 100℃ lower than previous works for crystallization from amorphous oxide precursor without fluoride. Resulting films have exhibited higher zero resistance temperatures than 80 K and crystallographic orientation of c-axis normal to the film surface. The YBCO phase has been crystallized during annealing process at a low oxygen partial pressure of 0.02 Torr which is about an order of magnitdue lower at 750 ℃ than thermodynamic critical condition for tetragonal YBCO phase reported by Bormann and Nölting.

ACKNOWLEDGMENT

This work was supported by New Energy and Industrial Technology Development Organization (NEDO) under the management of the R&D of Basic Technology for Future Industries.

REFERENCES

[1] R. Feenstra, T. B. Lindemer, J. D. Budai, and M. D. Galloway, J. Appl. Phys. 69, 6569 (1991).

[2] A. Mogro-Campero and L. G. Turner, Appl. Phys. Lett. 58, 417 (1991).

[3] T. Hase, H. Kajikawa, H. Kannan, M. Okuda, and Y. Kawate, in Extended Abstract of 1989 International Superconductivity Electronics Conference (ISEC'89) (Japan Society of Applied Physics, Tokyo, 1989), 70.

[4] J. Kwo, T. C. Hsieh, R. M. Fleming, M. Hong, S. H. Liou, B. A. Davidson, and L. C. Feldman, Phys. Rev. B36, 4039 (1987).

[5] A. F. J. Levi, J. M. Vandenberg, C. E. Rice, A. P. Ramirez, K. W. Baldwin, M. Aazlowar, A. E. White, and K. Short, J. Crystal Growth 91, 386 (1988).

Superconducting Properties of YBCO Thin Films Deposited on Grainboundary Free Y_2O_3 Substrates

AKIRA OISHI, KATSUMI OHATA, JUN SUGIYAMA, KATSUMI SUZUKI, SHUICHI FUJINO, TSUYOSHI TAKENAKA, SHINJI KAWAMOTO, TADATAKA MORISHITA, and SHOJI TANAKA

Superconductivity Research Laboratory, ISTEC, Koto-ku, Tokyo, 135 Japan

ABSTRACT

Y_2O_3 single-crystal was grown by Verneuil method. The crystallinity ($FWHM^{004}$ = 61") is comparable or superior to the other commercial substrates for superconducting films. The crystalline structure is cubic, and there is no phase transition in the range of 77 - 2,570K. The dielectric constant of Y_2O_3 substrate is smaller than 9, which is comparable to MgO. Not only c-axis, but also a-axis oriented YBCO film have been grown on Y_2O_3 (001) substrates by control of substrate-temperature like MgO, and $SrTiO_3$. The width of the interface between YBCO and Y_2O_3 is less than 100Å, which indicates the interface is very sharp and stable.

KEYWORDS: Y_2O_3 substrate, phase transition, dielectric constant, film orientation, width of interface

INTRODUCTION

Since the discovery of high T_C superconducting materials, there have been many efforts to prepare high-quality thin films. The growth properties of thin films are well known to be sensitively affected by the substrate material [1-3]. The criteria which the substrate materials must meet for the purpose of high-quality epitaxial growth are said to be small mismatch and little or no chemical reactivity with the superconducting materials. Moreover, a low dielectric constant is needed for the sake of electric applications [4].

The crystal structure of Y_2O_3 is cubic. The lattice constant is 10.60Å, and the square unit cell consisting Y in the (001) is excepted to have good lattice matching to the a-b or a-c plane of YBCO (the Y-Y spacing is calculated to be $10.60/2\sqrt{2}$ = 3.75Å). Since Y_2O_3 consists of only Y and O, which are also YBCO components, it is superior to other substrates for making a stable film-substrate interface. Elsewhere, the dielectric constant of Y_2O_3 films is reported to be 9 - 17 [5-7].

In this work, we have estimated the applicability of Y_2O_3 single-crystal substrates for YBCO films from the viewpoints of the crystallinity, phase transition, and the dielectric constant of Y_2O_3 substrate; control of the film orientation, chemical stability of the interface between the film and the substrate, and superconducting properties of the films.

EXPERIMENTAL

The Y_2O_3 substrates were cut from the boules grown by the Verneuil method in the (001) plane. The crystallinity of Y_2O_3 substrate was measured by double crystal method, the existence of phase transition of Y_2O_3 by powder X-ray method at 77K - RT, the dielectric constant of substrates by capacitance method (100Hz - 1MHz, 20K - RT) and cavity method (15GHz, RT). Then, the crystallinity and the dielectric constant of Y_2O_3 were compared with the other commercial substrates for superconducting films. The YBCO thin films were deposited by pulsed laser deposition method onto Y_2O_3 substrates under the conditions: 100 mTorr oxygen partial pressure, and 920 - 970K substrate temperature. The thickness of films is 400 - 500Å. Films were characterized by an X-ray diffractometer with a Cu$K\alpha$ source. The chemical stability of the interface between the film and the substrate was examined by depth plofiling with Auger electron microscopy. As for superconducting properties, we determined the transition temperature and the transition width from the resistivity versus temperature curves.

RESULTS AND DISCUSSION

The rocking curves of MgO (002), $SrTiO_3$ (002), and $LaAlO_3$ (002), and Y_2O_3 (004) peaks by means of double crystal method in Fig.1 shows the crystallinity of each substrates near the surface layer. In the case of commercial substrates, such as MgO, $SrTiO_3$, and $LaAlO_3$, their rocking curves are consisted of two or more peaks, which indicates the existence of sub-grainboundaries. On the other hand, that of Y_2O_3 is consisted of a single peak, indicating grainboundary free. The full width at half maximum (FWHM) of Y_2O_3 (004) peak was 0.017° (61"). Therefore, the crystallinity of Y_2O_3 is comparable or superior to the other commercial substrates for superconducting films.

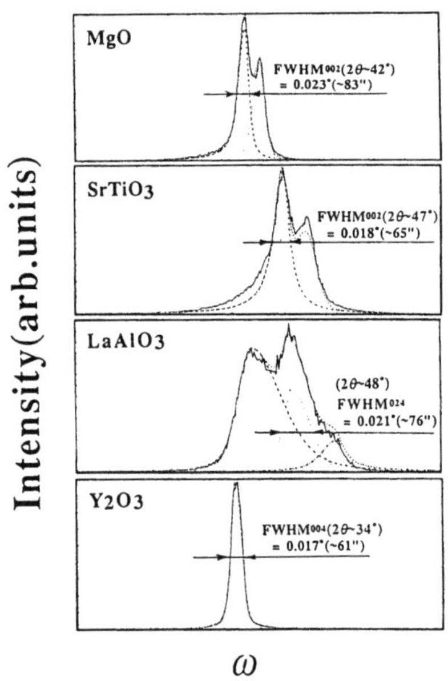

Fig.1 Rocking curves of substrates by means of double crystal method.

The powder X-ray diffraction patterns in Fig.2 show that the crystalline structure of Y_2O_3 is cubic in the range of 77K - RT. Since no phase transition have been reported in the range of RT - 2,570K, the crystalline structure of Y_2O_3 is cubic from the temperature of film-deposited to that of device-used.

Table 1 Dielectric constant of substrates.

substrate	ε_r at RT this study		ref.[8]	ref.[9]	ref.[10]	ref.[5~7]
MgO(100) cubic	7 (15GHz)	11 (~1MHz)	10 (9.5GHz)	10 (15GHz)	9 (~1MHz)	-
$LaAlO_3$(100) pseudo-cubic	24 (15GHz)	30 (~1MHz)	24 (6.1GHz)	24 (15GHz)	25 (~1MHz)	-
Y_2O_3(100) cubic	7 (15GHz)	9 (~1MHz)	-	-	-	film 9~17 (~1MHz)

As shown in Table 1, the dielectric constant of Y_2O_3 substrate is smaller than 9. This is comparable to minimum of the reports for Y_2O_3 film, and MgO substrate. According to the measurement by capacitance method, the dielectric constant of Y_2O_3 is independent of not only frequency but also temperature.

Figure 3 shows X-ray diffraction pattern for YBCO films on Y_2O_3 (001). A 2θ-θ scan shows only the (00ℓ) peaks (ℓ=1 - 9) of the film deposited at 970K, which indicates that the film is oriented with the c-axis perpendicular to the substrate. On the other hand, the (h00) peaks (h=1 - 3) of the film deposited at 920K indicates that the film is oriented with the a-axis perpendicular to the substrate. Thus, it is possible to control the orientation of film by the temperature of substrate like MgO, $SrTiO_3$, and so on.

The Auger depth profile shown in Fig.4 reveals the transition from YBCO to Y_2O_3 occurs over a distance less than 100Å, which includes substrate surface roughness; this indicates the interface is very

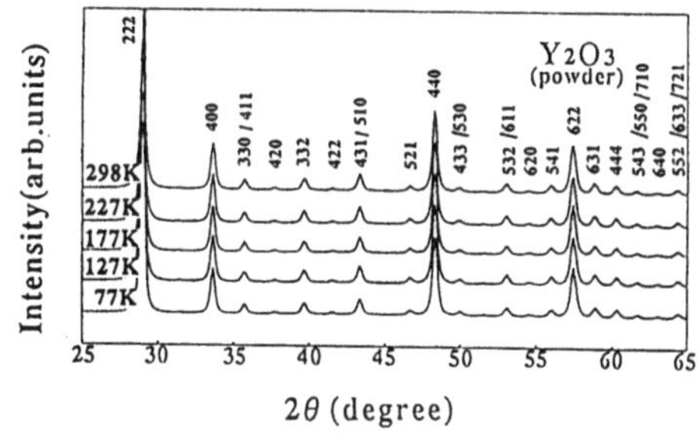

Fig.2 X-ray diffraction patterns of Y_2O_3 powder.

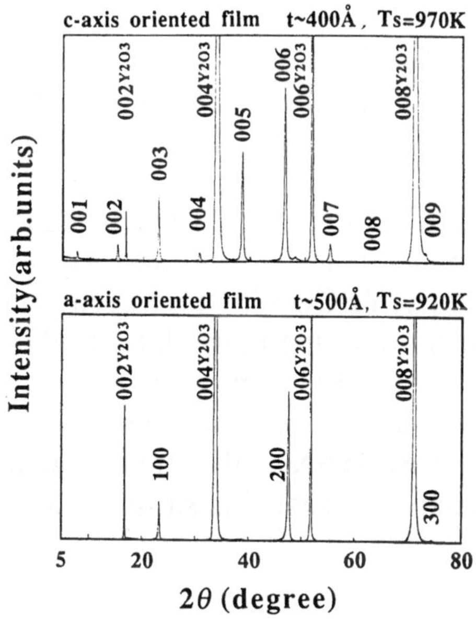

Fig.3 X-ray diffraction patterns for YBCO films on Y_2O_3 (001).

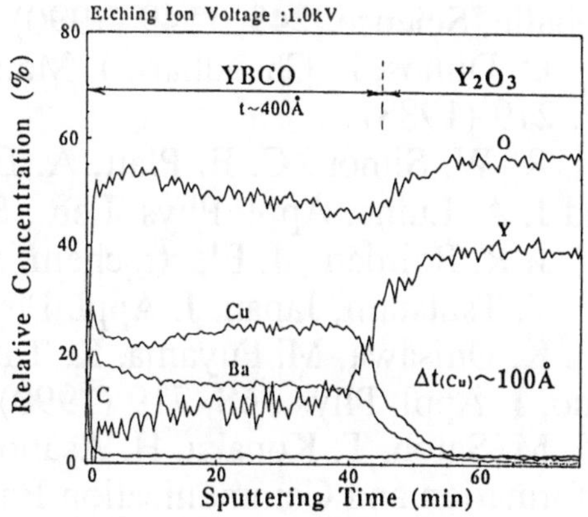

Fig.4 Auger depth profile of YBCO-Y_2O_3 interface.

sharp and stable. Moreover, there was no plateau corresponding to $Y_2Ba_1Cu_1O_x$ phase.

Figure 5 shows the resistivity versus temperature plot for a c-axis oriented YBCO film deposited on the Y_2O_3 (001). The critical temperature (zero resistance) T_C^0 is 83.9K with a transition width ΔT of 1.7K. The extrapolation of the curve intersects near by the origin.

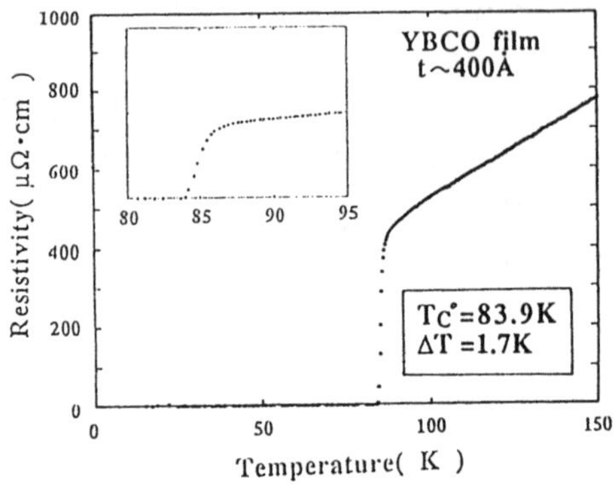

Fig.5 The resistivity versus temperature plot for a c-axis oriented YBCO film deposited on the Y_2O_3 (001).

SUMMARY

The crystallinity of Y_2O_3 grown by Verneuil method was excellent ($FWHM^{004}$ = 61"). The crystalline structure of Y_2O_3 is cubic, and there is no phase transition in the range of 77 - 2,570K. The dielectric constant of single-crystal substrate was smaller than 9, which is comparable to minimum of the reports for Y_2O_3 film, and MgO substrate. Not only c-axis, but also a-axis oriented YBCO film have been grown on Y_2O_3 (001) substrates by controle of substrate-temperature like MgO, and $SrTiO_3$. The width of the interface between YBCO and Y_2O_3 was less than 100Å, which indicates that the interface is very sharp and stable. The critical temperature (zero resistance) T_C^0 was 83.9K with a transition width ΔT of 1.7K. Therefore, Y_2O_3 single-crystal is a promising substrate for high-quality YBCO films.

REFERENCES

[1] B. H. Moeckly, S. E. Russek, D. K. Lathrop, R. A. Buhrman, Jian Li, and J. W. Mayer, Appl. Phys. Lett., **57**, 1687 (1990)

[2]. C. B. Emo, A. F. Marshall, S. S. Laderman, R. D. Jacowitz, and T. H. Geballe, Science, **249**, 1549 (1990)

[3]. D. Dimos, P. Chaudhari, J. Mannhart, and F. K. Legoues, Phys. Rev. Lett., **61**, 219 (1988)

[4]. R. W. Simon, C. E. Platt, A. E. Lee, G. S. Lee, K. P. Daly, M. S. Wire, and J. A. Luine, Appl. Phys. Lett., **53**, 2677 (1988)

[5]. J. R. Rairden , J. Electrochem. Soc., **114**, 75 (1967)

[6]. T. Tsutsumi, Japan. J. Appl. Phys., **9**, 735 (1970)

[7]. K. Onisawa, M. Fuyama, K. Tamura, K. Taguchi, T. Nakayama, and Y. A. Ono, J. Appl. Phys., **68**, 719 (1990)

[8]. M. Satoh, T. Konaka, H. Asano, and S. Kubo, The Institute of Electronics, Information and Communication Engineers (autumn meeting C-318)

[9]. D. Reagor, and F. Garzon, Appl. Phys. Lett., **58**, 2741 (1991)

[10]. G. A. Samara, J. Appl. Phys., **68**, 4214 (1990)

Effects of Oxygen Pressure on Sputtered Y-Ba-Cu-O Films

MITSUMASA SUZUKI[1], HIROSHI MORITA[1], CHOEI TAKAHASHI[2], SEIYA OGOTA[2], MOTOHARU KOMATSU[2], and MITSUGU NAGANO[2]

[1] Department of Electrical Engineering, Tohoku University, Sendai, 980 Japan
[2] Applied Technology R&D Center, Tohoku Electric Power Co., Inc., Sendai, 981 Japan

ABSTRACT

Films of Y-Ba-Cu-O have been prepared in an argon-oxygen atmosphere by dc magnetron sputtering using a single sintered target with a Ba-poorer composition than stoichiometry. The total gas pressure is 300 mtorr and the oxygen pressure is adjusted in the range of 0.1-10 mtorr. As the oxygen pressure is increased, Ba content in films is highly reduced, although Cu content shows little dependence. The zero resistance transition temperature $T_{c,end}$ of the best film reaches 89 K. Effects of oxygen pressure on T_c and composition of films are discussed.

KEY WORDS: Y system superconductor, thin film, sputtering

INTRODUCTION

Thin films of oxide superconductors have been prepared by different techniques such as electron-beam evaporation, sputtering, laser evaporation,and recently chemical vapor deposition[1-4]. First of all, extensive studies on preparation conditions for high quality films have been performed by a sputtering technique, because of the simplicity in constructing deposition apparatus. Particularly, most of Y-Ba-Cu-O films have been made from a single sintered target. However, in the case of sputter-deposition from a sintered target, the formation of negative ions has been reported[5]. The bombardment of negative ions to substrate causes a reduction in the net deposition rate and alters the film composition. In order to reduce such negative ion bombardment, techniques of increasing the sputtering gas pressure and locating the substrate in off-axis positions are used[6,7]. Meanwhile, the deposition rate is strongly reduced by a small addition of oxygen, even in the case of high argon gas pressure, and the film composition which is closely related to T_c might be highly affected.

Up to now, we have directed our efforts to search better sputtering techniques capable of preparing high quality Y-Ba-Cu-O films. In a previous study, we have reported the preparation conditions of high T_c films by dc magnetron sputtering from a single sintered target with a Ba poorer composition than stoichiometry[8]. In this paper, the effects of oxygen pressure on sputtered Y-Ba-Cu-O films are investigated from the viewpoints of superconducting transition temperature T_c and film composition.

EXPERIMENTAL

The preparation of Y-Ba-Cu-O films was carried out using a laboratory-made dc sputtering apparatus. Used targets are disks of 50 mm in diameter and 4 mm in thickness and have been made by a conventional sintering technique. The target compositions are Y:Ba:Cu=1:1.5:3, 1:1.6:3, 1:1.7:3, 1:2:3 and 1:2:3.6. The sputter-deposition was carried out on the polished (100) plane of MgO which was kept at a substrate temperature T_d between 720 and 820 °C by a nichrome sheet heater. The sputtering voltage is approximately 100 V and the current ranges from 400 to 500 mA. The target-substrate(T-S) spacing L is varied between 20 and 50 mm. Argon and oxygen are used as sputtering gases. Each gas has been adjusted independently with Needle valves. The oxygen partial pressure P_{O2} investigated in the present experiment ranges from 0.1 to 150 mtorr and the total gas pressure P_{Ar+O2} is in the range of 150-600 mtorr.

After presputtering for 1.5 h, films 0.1-0.3 μ m thick have been obtained by sputtering for 2-5 h. During cooling after deposition, films have been heat-treated at temperatures $T_{s.h}$ of 550 °C for 0.5 h in an oxygen gas pressure of 160 torr. The optimums of target composition, gas pressures, T-S spacing and heat treatments during cooling have been checked for as-deposited superconducting films.

The T_c of films was measured by a standard four-probe resistance technique using a Pt-resistance thermometer. Metal composition and crystal phases were analyzed by EDX(Energy dispersive X-ray analysis) and XRD (X-ray diffract meter), respectively.

RESULTS AND DISCUSSION

The successful synthesis of high-T_c Y-Ba-Cu-O films has been already reported by several workers using a stoichiometry sintered target at high gas pressure[9-11]. Their optimum gas pressures are considerably high. Namely, the P_{Ar+O2} is in the range of 10^2 to 10^3 mtorr and the P_{O2} seems to exceed 10 mtorr. However, in the range of P_{O2} lower than 10 mtorr, the deposition rate rapidly increases with decreasing P_{O2}. In the present experiment, many deposition runs have been done at $P_{O2}<10$ mtorr and $P_{Ar+O2}=300$ mtorr. Some targets with Ba poorer compositions than stoichiometry have been tested to get high T_c films. Generally, the perovskite phase is seen in films prepared at $T_d>720$ °C and their c-axis parameter ranges from 11.7 to 11.9 Å. Most of films have shown preferred orientation of c-axis perpendicular to film plane.

Figure 1 shows resistive transition curves obtained in several films. Each film indicates the highest T_c value in films which have been deposited from each target under different sputtering conditions. $T_{c,end}$ values higher than 80 K are seen in every film, but there is a tendency of high T_c character in films deposited from Ba poorer targets. The best film shows $T_{c,end}$ exceeding 89 K and has been obtained at $P_{O2}=7.5$ mtorr, $P_{Ar+O2}=300$ mtorr and L=41 mm using a target with the 1:1.5:3 ratio of Y:Ba:Cu.

The $T_{c,end}$ of films deposited at different P_{O2} values are given as a function of P_{O2} in Fig.2. Here, many films have been deposited at $P_{Ar+O2}=300$ mtorr. As the P_{O2} is increased from 0.1 to 10 mtorr, the $T_{c,end}$ of films increases from 40 to 85 K and the maximum $T_{c,end}$ takes place around $P_{O2}=8$ mtorr. Any further increase in P_{O2} causes an appreciable depression of $T_{c,end}$. The target composition seems to have little effect on the relation of $T_{c,end}$ with P_{O2}.

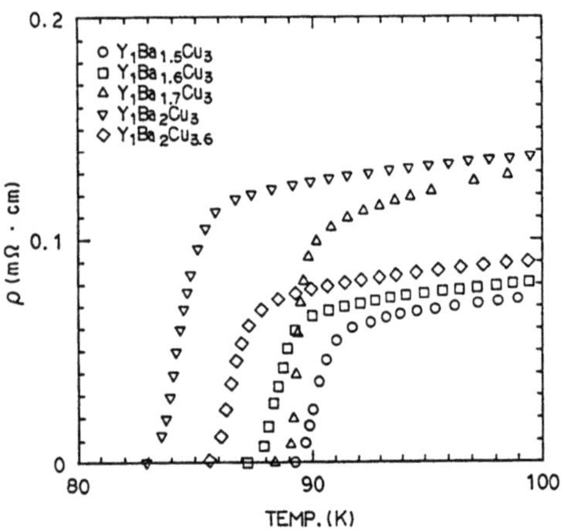

Fig.1 Resistive transitions for films deposited from targets with different compositions.

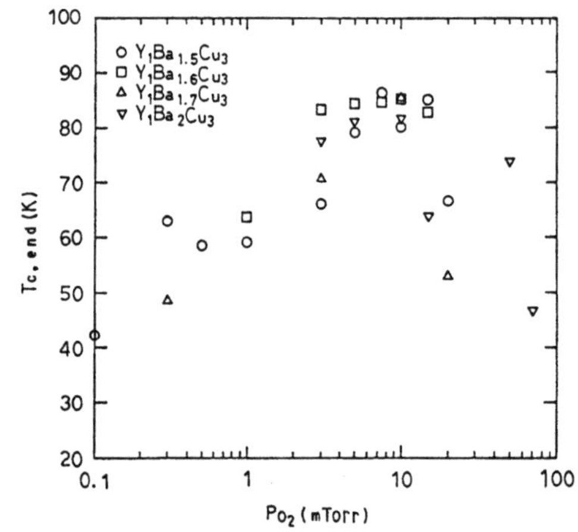

Fig.2 Variation of $T_{c,end}$ with partial oxygen pressure.

Figure 3 shows the results of EDX analysis carried out to investigate the

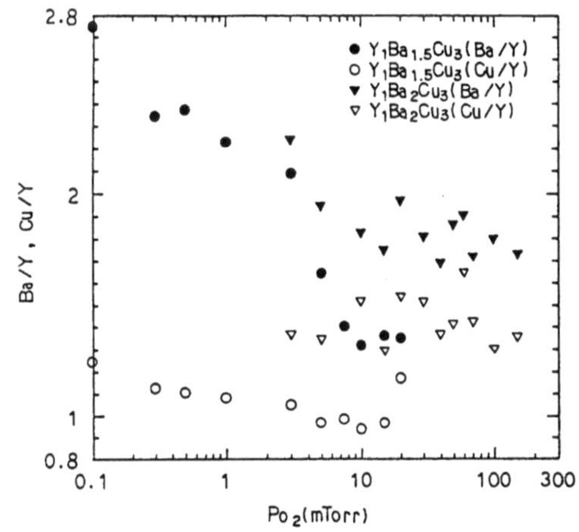

Fig.3 Variation of X-ray intensity ratio Cu/Y and Ba/Y with partial oxygen pressure for some of films.

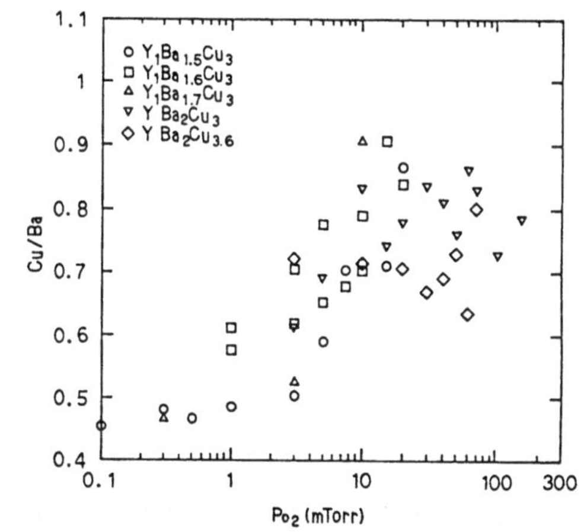

Fig.4 Variation of x-ray intensity ratio Cu/Ba with partial oxygen pressure for some of films.

metal composition of films. X-ray intensity ratio of Ba and Cu to Y is illustrated in the figure. As the P_{O2} is increased, the ratio Cu/Y shows a slight change, while the ratio Ba/Y appreciably decreases. The variation of ratio Cu/Ba with P_{O2} is given in Fig.4. In the range of P_{O2} lower than 10 mtorr, there is an marked increase of ratio Cu/Ba with P_{O2}. On the other hand, any further increase in P_{O2} from 10 mtorr seems to have little influence on ratio Cu/Ba. It should be noted that high T_c films are formed around P_{O2}=10 mtorr where the saturation of ratio Cu/Ba begins. Moreover, the above mentioned variation of ratio Cu/Ba is considered to be mainly caused by an decrease of Ba content in films. Namely, the amount of Ba rather than Cu which should be incorporated in films is highly affected by the magnitude of P_{O2} in the case of sputter-deposition at low P_{O2}.

Figure 5 shows the relation of $T_{c,end}$ to ratio Cu/Ba. The relation of c-axis lattice parameter to ratio Cu/Ba is given in Fig.6. As can be seen in these figures, Both of $T_{c,end}$ and c-axis lattice parameter seems to be closely associated with ratio Cu/Ba. As the ratio Cu/Ba in films is increased up to 0.6, the $T_{c,end}$ monotonically increases up to 80 K or above and the c-axis lattice parameter linearly decreases to 11.7 Å. With further increase in ratio Cu/Ba, both of the $T_{c,end}$ and the c-axis lattice parameter tend to

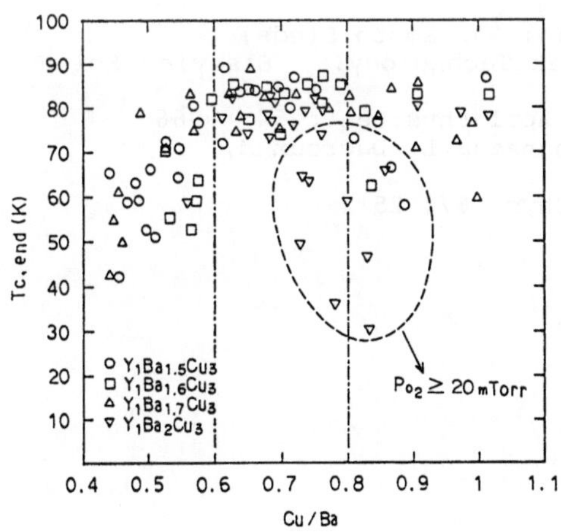

Fig.5 Relation between $T_{c,end}$ and X-ray intensity ratio Cu/Ba.

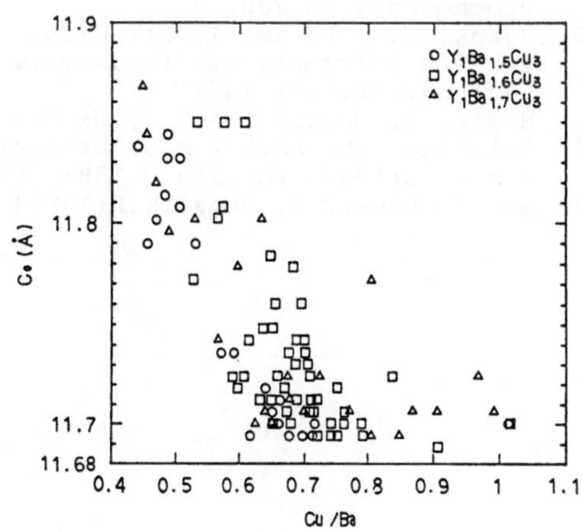

Fig.6 Relation between c-axis lattice parameter and X-ray intensity ratio Cu/Ba

saturate. The ratio Cu/Ba where high T_c films have been reproducibly prepared ranges from 0.6 to 0.8. Some films with ratio Cu/Ba of 0.7-0.8, which have been deposited on the conditions of P_{O2} higher than 20 mtorr, show an appreciable depression of $T_{c,end}$.

Meanwhile, our films with $T_{c,end}$ higher than 85 K have been normally obtained using Ba-poorer targets, while a stoichiometry target has been used by other workers. As compared to their preparation conditions, especially, with regard to P_{O2}, P_{O2} values mainly investigated in the present experiment seems to be lower, and this fact implies that the amount of Ba incorporated in films is relatively high. Accordingly, in order to produce high T_c films using a stoichiometry target, the P_{O2} range which should be investigated must be shifted to an upper side.

SUMMARY

Y-Ba-Cu-O films have been prepared by dc magnetron sputtering using single sintered targets and effects of P_{O2} on $T_{c,end}$ and film composition have been investigated. The amount of Ba rather than Cu which should be incorporated in films is highly affected by a small change in P_{O2} under deposition conditions of low P_{O2}, even in high argon gas pressure. Our best film shows $T_{c,end}$=89 K and have been prepared from a target with the Y:Ba:Cu=1:1.5:3 ratio under conditions of P_{Ar+O2}=300 mtorr and P_{O2}=7.5 mtorr. If a stoichiometry target is adopted, the optimum value of P_{O2} will be shifted to an upper side to reduce the Ba content in films. The experiment on deposition conditions for a stoichiometry target is in progress.

We would like to thank Ms. K. Takahashi of Japan Fine Ceramics Co., Ltd. for preparing sintered targets.

REFERENCES

1. Oh B, Naito M, Aranson S, Rosenthal P, Barton R, Beasley M.R, Geballe T.H, Hammond R.H, Kapitulnik A(1987) Appl.Phys.Lett. 51: 852
2. Enomoto Y, Murakami T, Suzuki M, Moriwaki K(1987) Jpn.J.Appl.Phys. 26: L1248
3. Wu X.D, Dijkkamp D, Ogale S.B, Inam A, Chase E.W, Miceli P.F, Chang C.C, Tarascon J.M, Venkatesan T(1987) Appl.Phys.Lett. 51: 861
4. Berry A.D, Gaskill D.K, Holm R.T, Cukauskas E.J, Kaplan R, Henry R.L(1988) Appl.Phys.Lett. 52: 1743
5. Rossnagel S.M, Cuomo J.J(1988) AIP Conference Proceedings No.165: 106
6. Xi X.X, Linker G, Meyer O, Nold E, Obst B, Ratzel F, Smithey R, Strechalm R, Weschenfelder F, Geerk J(1989) Z.Phys.B 74: 13
7. Yamamoto K, Eom C.B, Sun J.Z, Keith D, Geballe T.H(1990) AIP Conference Proceedings No.200: 87
8. Suzuki M, Uchiyama T, Takahashi K(1990) In: Shimamoto S(eds) in 11th International Conference on Magnet Technology. Elsevier Science, England, pp.1431-1436
9. Hohler A, GuggiD, Neeb H, Heiden C(1989) Appl.Phys.Lett. 54: 1066
10. Triscone J.M, Karkut M.G, Brunner O, Antognazza L, Decroux M, Fisher o(1989) Physica C 158: 293
11. Gao J, Hauser B, Rogalla H(1990) J.Appl.Phys. 67: 2512

Biaxially Aligned YSZ Buffer Layer on Polycrystalline Substrates

Y. Iijima, N. Tanabe, and O. Kohno

Materials Research Laboratory, Fujikura Ltd., Koto-ku, Tokyo, 135 Japan

ABSTRACT

Biaxially aligned YSZ thin films were deposited on polycrystalline metallic substrates by Ion-Beam-Assisted Deposition (IBAD), with inclined ion beam bombardment. In-plane ordered YSZ layers could be formed independent on substrate crystalinity. These films should be effective buffer layers for large-scale application of thin films of oxide superconductors for tape-shaped cables, magnetic shields, microwave apparatuses, and other applications, using polycrystalline substrates.

KEY WORDS: Ion-Beam-Assisted Deposition, yttorium stabirized zirconia (YSZ), biaxial alignment, polycrystalline substrate

INTRODUCTION

High quality $YBa_2Cu_3O_{7-x}$ thin films were routinely obtained by various preparation procedures on suitable single crystalline substrates, such as SrTiO3(100), with prominent superconducting properties.[1-3,9] If the similar structure as YBCO films on these single crystalline substrates is realized on polycrystalline substrates, large-scale superconducting thin films with high transporting current at 77K will be produced as tape-shaped conductors, magnetic shields and so on. There were many works to form c-axis oriented $YBa_2Cu_3O_{7-x}$ films on polycrystalline substrates.[4-9] However, it was difficult to form in-plane aligned YBCO films on them. Sufficient transporting currents can be hardly obtained without in-plane a- and b-axes alignment.[10-12]

For the purpose to deposit in-plane aligned $YBa_2Cu_3O_{7-x}$ films on polycrystalline substrates, biaxially oriented YSZ films were produced on substrates of Hastelloy C276 by Ion-Beam-Assisted Deposition; a method to deposit thin films with concurrent energetic particle bombardment.[13] The influence of the incident angle of the assisting ion beam on the characteristics of films was investigated. Through simultaneous ion bombardments to substrates with the inclined assisting ion beam, biaxially aligned, (100) axis normal to the substrate and (010), (001) axes in-plane aligned YSZ films were produced.

$YBa_2Cu_3O_{7-x}$ films with Jc of the order $10^5 A/cm^2$ have attained on Hastelloy C276 using these biaxially aligned YSZ buffer layers.[14-16]

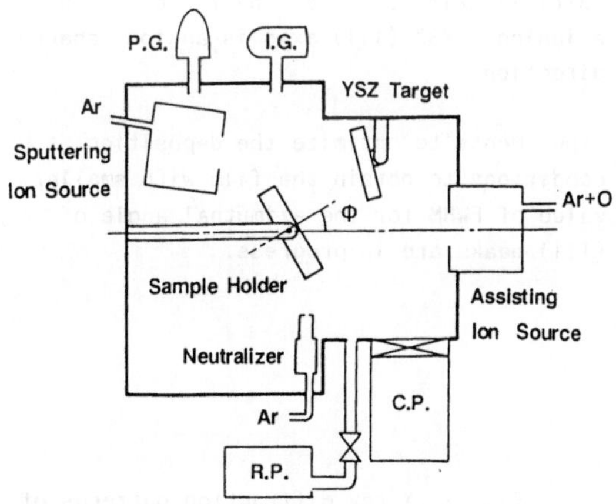

Fig. 1 The schematic drawing of the dual ion beam sputtering system.

EXPERIMENTAL

Films were deposited on substrates of Ni-based alloy, Hastelloy C276. Figure 1 shows a schematic drawing of the dual ion beam sputtering system used in this study. Used ion guns were 5 cm dia. ion sources by Ion Tech Co. An Ar ion beam sputtered a YSZ target (3% Y_2O_3) under $3.0x10^{-4}$ Torr. During the deposition, an assisting Ar+O ion beam bombarded onto the film simultaneously. The energy and the current density of the assisting ion beam set to 300 eV, and 220 microampares/cm^2 respectively. The substrate temperature reached ~100°C during the deposition by the ion beam bombardment. In order to investigate the influence of the incident angle of the assisting ion beam on the characteristics of films, substrates were mounted on holders inclined towards the ion beam direction. The incident angle of the assisting ion beam was defined as the angle between the beam axis and the substrate normal. The deposition rate was 0.05 nm/sec. The film thickness was 0.3 micrometers.

RESULTS AND DISCUSSION

XRD patterns of deposited films are shown in figure 2. Without assisting ion beam bombardment, the resulting YSZ film was randomly oriented. With concurrent ion beam bombardment, (100)-oriented films, identified by the (200) and (400) diffraction peaks, were deposited. (100)-oriented films were also formed with the value of a full width at half maximum (FWHM) of (200) peaks were from 5 to 7 degrees, unless the ion beam incident angle; exceeded approximately 60 degrees.

Pole figure measurements were conducted on (111) peaks to know the distribution of the in-plane (011) axis. When the assisting ion beam bombarded the substrates perpendicular to the surface, there were no alignment of in-plane (011) axis, as shown in figure 3 (a). However, the (111) peak of films deposited with inclined assisting ion beam were found to have in-plane ordering shown in figure 3 (b); = 30 degrees and (c); = 45 degrees. When the assisting ion beam was inclined 65 degrees, the (111) pole figure indicates random polycrystalline structure with a little in-plane textured (100)-oriented grains, which is consistent with the theta - 2 theta result.

In-plane direction of a (111) peak of the every biaxially aligned YSZ film was corresponded to the assisting ion beam. It indicates that YSZ(111) axis is one of the factor for biaxial aligning. YSZ (111) axis is an ion channeling direction.

Experiments to optimize the deposition conditions to obtain the film with smaller value of FWHM for the azimuthal angle of (111) peaks are in progress.

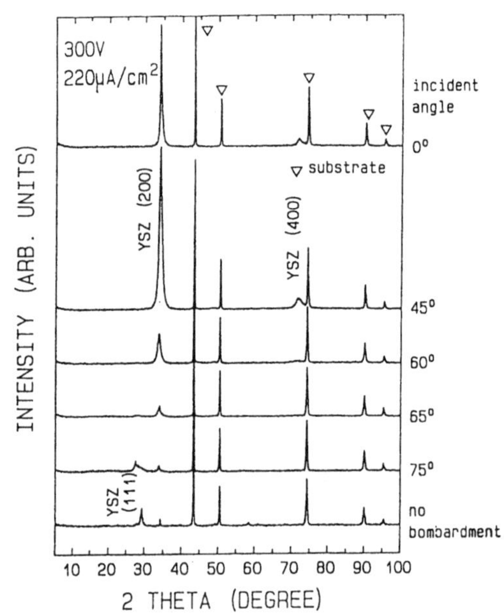

Fig. 2 X-ray diffraction patterns of YSZ films deposited on Hastelloy c276 by IBAD.

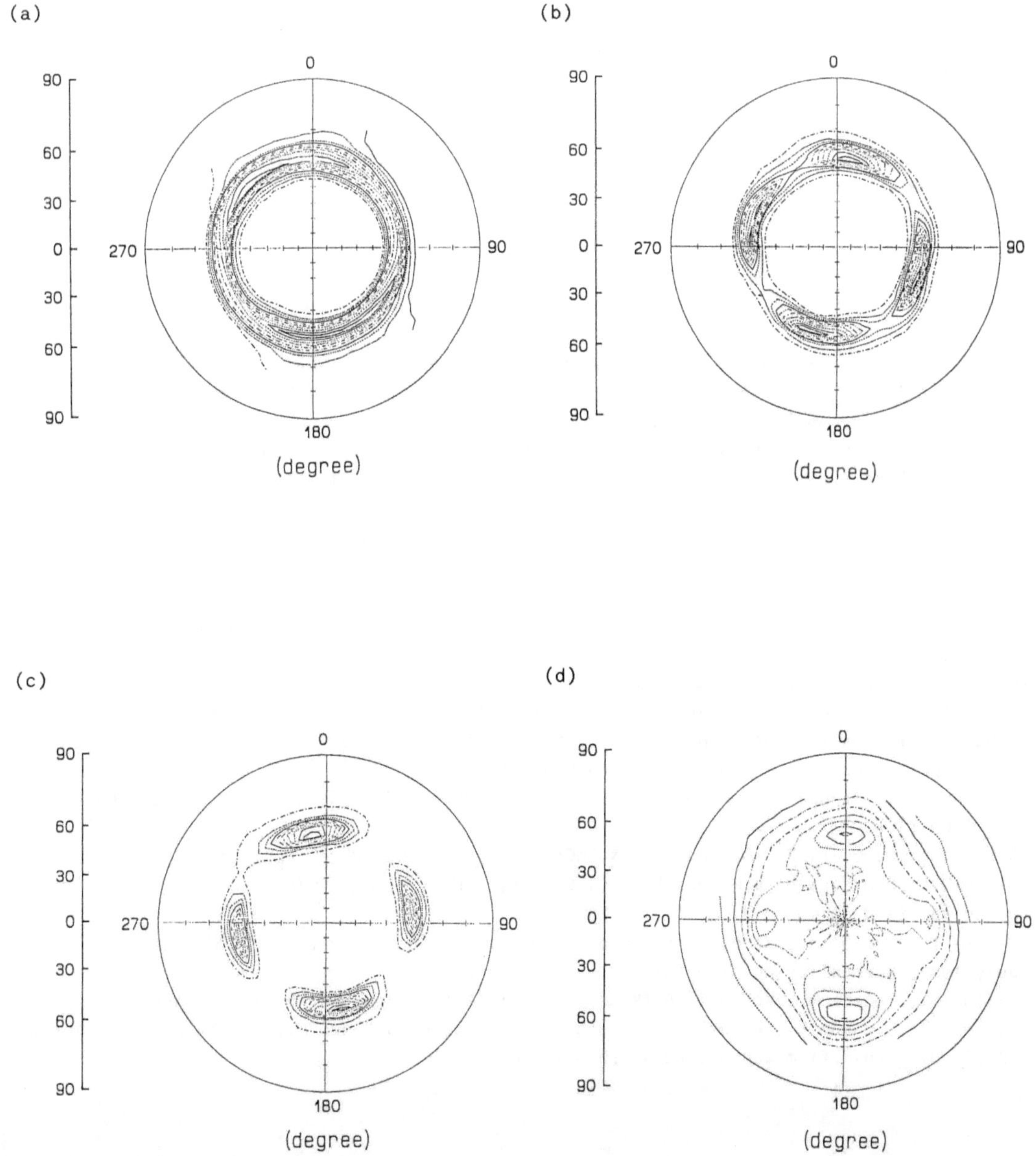

Fig. 3 X-ray pole figures of YSZ (111) peaks deposited on Hastelloy c276 by IBAD, with the incident angle of the assisting ion beam was, (a); ϕ = 0 degree, (b); ϕ = 30 degrees, (c); ϕ = 45 degrees, and (d); ϕ = 65 degrees.

CONCLUSION

Biaxially aligned YSZ films were deposited on polycrystalline metallic substrates by Ion-Beam-Assisted Deposition with the assisting ion beam inclined no more than 60 degrees. In-plane direction of a (111) peak of the every biaxially aligned YSZ film was corresponded to the assisting ion beam. It indicates that YSZ(111) axis is one of the factor for biaxial aligning. YSZ (111) axis is an ion channeling direction.

REFERENCES

1. Roas B., Schulz L. & Endres G. (1988) Appl. Phys. Lett., 53 :1557
2. Venkatesan T., Wu X. D., Dutta B., Inam A., Hedge M. S., Hwang D. M., Chang D. M., Nazar L. Wilkens B. (1989) Appl. Phys. Lett., 54 :581
3. Zheng J. P., Dong S. Y. & Kwok H. S. (1991) Appl. Phys. Lett., 58 :540
4. Narumi E., Song L. W., Yang F., Patel S., Kao Y. H. & Shaw D. T. (1991) Appl. Phys. Lett., 58 :1202
5. Saitoh J., Fukutomi M., Komari K., Tanaka Y., Asano T., Maeda H. & Takahara H. (1991) Jpn. J. Appl. Phys., 30 :L898
6. Witanachchi S., Patel S., Zhu Y. Z., Kwok H. S. & Shaw D. T. (1990)J. Mater. Res. 5 :717
7. Iijima Y., Hayakawa H., Sadakata N. & Kohno O. (1989) Proc. 11th Int'l Conf. Magnet Tech. (MT-11), Tsukuba, Japan, Aug. 1989 :1142
8. Yoshida N., Kubota M., Okuda S., Takano S., Nagata M., Hara T., Okaniwa K. & Yamamoto T. (1990) Proc. 3rd Int'l Symp. Superconductivity (ISS'90), Sendai, Japan :901
9. Onabe K., Iijima Y., Futaki N., Sadakata N., Kohno O. & Ikeno Y. (1990) Proc. 3rd Int'l Symp. Superconductivity (ISS'90), Sendai, Japan :905
10. Chaudhari P., Mannhart J., Dimos D., Tsei C., Oprysko C. & Scheuermann M. (1988)Phys. Rev. Lett., 60 :1653
11. Dimos D., Chaudhari P., Mannhart J. & LeGoues F. K. (1988) Phys. Rev. Lett., 61 :219
12. Dimos D., Chaudhari P. & Mannhart J. (1990) Phys. Rev. B, 41 :4038
13. Rossnagel S. M. & Cuomo J. J. (1987) MRS. Bul.,12 (2) :40
14. Iijima Y., Tanabe N., Kohno O. & Ikeno Y., Proc. 3rd Int'l Conf. M^2S-HTSC, Kanazawa, Japan, July 1991, Special Issue of Physica C, in printing
15. Iijima Y., Tanabe N., Kohno O. & Ikeno Y., submitted to Applied Physics Letters
16. Iijima Y., Onabe K., Futaki N., Tanabe N. Sadakata N. & Kohno O., "A-b plane aligned $YBa_2Cu_3O_{7-x}$ thin film tapes", WB-4, this volume.

High-T_c Superconducting Thin Films on Si with Buffer Layers

HIROAKI MYOREN, TOSHIAKI OKAZAKI, HIDEKI TAMURA, and YUKIO OSAKA

Department of Electrical Engineering, Faculty of Engineering, Hiroshima University, Higashi-Hiroshima, Hiroshima, 724 Japan

ABSTRACT

$Ba_2YCu_3O_x$ thin films have been epitaxially grown on Si(100) substrate using various epitaxially grown buffer layers. (100)-oriented Gd_2O_3, Sm_2O_3 and CeO_2 thin films were epitaxially grown on Si(100) with epitaxially grown Y_2O_3/YSZ thin buffer layers by electron beam evaporation. $Ba_2YCu_3O_x$ thin films sputtered on Si(100) with these buffer layers were epitaxially grown with c-axis oriented normal to the substrate surface. (110)-oriented $SrTiO_3$ thin films were epitaxially grown on Si(100) substrate with epitaxially grown thin buffer layers of Y_2O_3/YSZ, Gd_2O_3/Y_2O_3/YSZ and CeO_2/Y_2O_3/YSZ at 650°C~800°C by rf magnetron sputtering. $Ba_2YCu_3O_x$ thin films sputtered on Si(100) with (110)-oriented $SrTiO_3$ buffer layers were epitaxially grown with (110) or (103)-axis oriented normal to the substrate surface.

KEY WORDS: $Ba_2YCu_3O_x$ thin films, Si substrate, buffer layers

INTRODUCTION

For device application, it is indispensable to develop the technique to fabricate thin films of high-T_c superconducting materials on Si substrate. We have already reported on the successful preparation of as-grown $Ba_2YCu_3O_x$ thin films on Si with buffer layers such as epitaxially grown YSZ, and Y_2O_3/YSZ[1,2]. Especially, as-grown $Ba_2YCu_3O_x$ thin films on Y_2O_3/YSZ/Si(100) have a $T_{c(end)}$ of about 85K and a critical current density of $10^6A/cm^2$ at 77K[2].

The half the lattice constants of Gd_2O_3 ($a/2$=5.407Å) and Sm_2O_3 ($a/2$=5.466Å) and the lattice constant of CeO_2 (a=5.411Å) are larger than the half the lattice constant of Y_2O_3 ($a/2$=5.302Å) and almost nearly equal to the diagonal length of 5.415Å in the a–b plane of the unit cell of $Ba_2YCu_3O_x$. From these facts, it is expected that superconducting properties of $Ba_2YCu_3O_x$ on Si(100) substrates with these buffer layers will be superior to those on Y_2O_3/YSZ/Si(100).

The lattice constant of $SrTiO_3$ (a=3.905Å) is also nearly equal to the lattice constant of $Ba_2YCu_3O_x$ (a=3.823Å, b=3.886Å, $c/3$=3.894Å). $SrTiO_3$ thin films on Si(100) with epitaxially grown thin buffer layers were grown with (110)-axis normal to the substrate surface. This suggests that (103) or (110)-oriented $Ba_2YCu_3O_x$ thin films are grown on Si(100) with the (110)-oriented $SrTiO_3$ buffer layers.

In this work, we report on the epitaxial growth and crystalline properties of Gd_2O_3, Sm_2O_3, CeO_2 and $SrTiO_3$ buffer layers and also report on epitaxial growth and crystalline and superconducting properties of $Ba_2YCu_3O_x$ thin films on Si(100) with these buffer layers.

EXPERIMENTAL

Gd_2O_3, Sm_2O_3 and CeO_2 thin films were deposited on Si(100) with epitaxially grown Y_2O_3(10~20nm)/YSZ(10~20nm) thin buffer layers [3] by electron beam evaporation (EBE) at substrate temperature of 800°C. Detailed deposition conditions are listed in Table I. $SrTiO_3$ thin films were deposited on Si(100) with epitaxially grown thin buffer layers of Y_2O_3/YSZ, Gd_2O_3/Y_2O_3/YSZ and CeO_2/Y_2O_3/YSZ by rf magnetron sputtering at substrate temperature between 600~900°C. Detailed sputtering conditions of $SrTiO_3$ thin films are listed in Table II.

The crystallographic properties of buffer layers on Si(100) were investigated by X-ray Diffraction (XRD) and reflection high energy electron diffraction (RHEED) observation. The crystalline qualities of buffer layers on Si(100) were examined by ion channeling and RBS. Ion channeling and RBS measurements were made on a three-axis goniometer using $^4He^+$ ions at an energy of 2.0MeV with 1-mm-diameter beam size.

The $Ba_2YCu_3O_x$ thin films were deposited on Si substrate with the buffer layers at 700°C by rf magnetron

Table I. Deposition conditions of buffer layers

target	Gd_2O_3, Sm_2O_3, CeO_2 pellet
substrate	Y_2O_3/YSZ/Si(100)
substrate temp.	800 °C
base pressure	$\sim 5 \times 10^{-6}$ Torr
pressure	$3 \sim 6 \times 10^{-5}$ Torr
deposition rate	$\sim$ 10 nm/min
total thickness	50 $\sim$ 100 nm

Table II. Sputtering conditions of $SrTiO_3$

target	$SrTiO_3$ disk
target diameter	4 inches
substrate	Y_2O_3/YSZ/Si(100) Gd_2O_3/Y_2O_3/YSZ/Si(100) CeO_2/Y_2O_3/YSZ/Si(100)
substrate temp.	600~900°C
sputtering gas	Ar+O_2 Ar:O_2=9:1
gas pressure	1.3Pa
rf input power	100W
growth rate	3nm/min

Table III. Random and aligned yield ratio χ_{min} of buffer layers

Material	Substrate	χ_{min}(%)
YSZ(9mol.%)	Si(100)	4.6
Y_2O_3	YSZ/Si(100)	4.1
Gd_2O_3	Y_2O_3/YSZ/Si(100)	5.2
Sm_2O_3	Y_2O_3/YSZ/Si(100)	5.1
CeO_2	Y_2O_3/YSZ/Si(100)	6.7

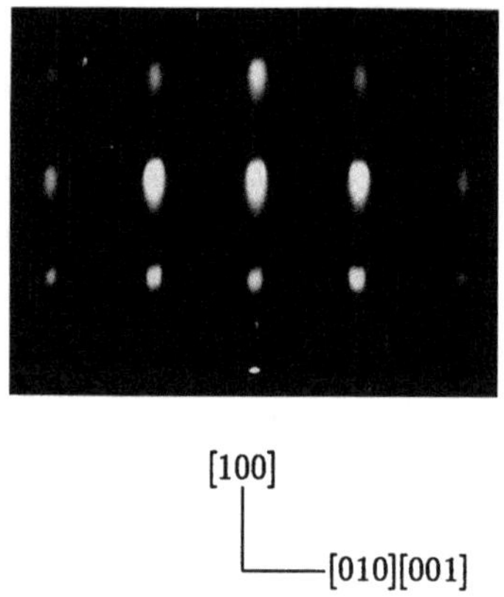

Fig.1. Typical RHEED patterns of CeO_2 thin films on Y_2O_3/YSZ/Si(100) along Si[001] azimuth.

sputtering[1]. The target used was stoichiometric $Ba_2YCu_3O_x$ powder. In order to achieve stoichiometry of the films sputtered on the substrate, we put the substrate at off-centered position for the stoichiometric target[4,5]. The substrate electrode was usually positive biased to provide a suitable amount of oxygen into the deposited films and protect the films from the bombardment of the charged particles[1]. Typical thicknesses of $Ba_2YCu_3O_x$ thin films were 50~100nm.

The superconducting transitions of the films were measured by conventional four probe method with dc current. The critical current densities were measured on about 50μm wide constrictions with a field criterion of 10μV/mm.

RESULTS AND DISCUSSION

Gd_2O_3, Sm_2O_3 and CeO_2 thin films on Y_2O_3/YSZ/Si(100)

XRD patterns of Gd_2O_3, Sm_2O_3 and CeO_2 thin films on Y_2O_3/YSZ/Si(100) showed that these thin films were grown with (100)-axis normal to the substrate surface. Figure 1 shows typical RHEED pattern of CeO_2 thin films on Y_2O_3/YSZ/Si(100) along Si[001] azimuth. As shown in Fig.1, RHEED patterns of these thin films on Y_2O_3/YSZ/Si(100) showed streaky spots. These results reveal that these films are grown epitaxially and have smooth surface.

Ion channeling measurements were carried out for 100~200nm-thick Gd_2O_3, Sm_2O_3 and CeO_2 films on Y_2O_3/YSZ/Si. These measurements showed that the minimum random and aligned yield ratio χ_{min} values of Gd_2O_3, Sm_2O_3 and CeO_2 films were rather small and comparable to those for the perfect single crystals. The measured χ_{min} values for these buffer layers are listed in Table III.

The *c*-axis-oriented $Ba_2YCu_3O_x$ thin films were epitaxially grown on Si(100) with these buffer layers. RHEED patterns of $Ba_2YCu_3O_x$ thin films on these buffer layers were streaky and ion channeling measurements of these $Ba_2YCu_3O_x$ thin films showed that the χ_{min} values were 5~10%. Figure 2 shows temperature dependence of resistance of $Ba_2YCu_3O_x$ thin films on CeO_2/Y_2O_3/YSZ/Si(100). *R–T* curves of $Ba_2YCu_3O_x$ on Si(100) substrate with these buffer layers showed metallic-like behavior as well as Fig.2. Observed $T_{c(end)}$s were 82~85K and critical current densities at 77K were of order of 10^5 ~10^6A/cm^2 for $Ba_2YCu_3O_x$ on Si(100)

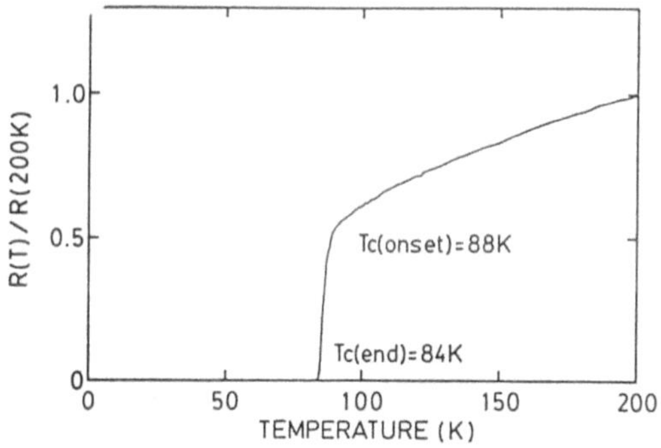

Fig.2. Temperature dependence of resistance of $Ba_2YCu_3O_x$ thin Films on CeO_2 thin films on Y_2O_3/YSZ/Si(100).

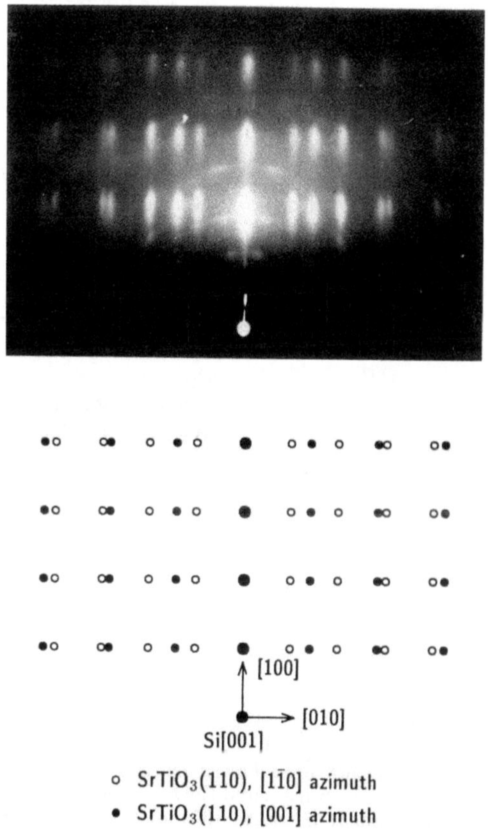

Fig.3. RHEED pattern of $SrTiO_3$ thin films grown on CeO_2/Y_2O_3/YSZ/Si(100) at substrate temperature of 700°C.

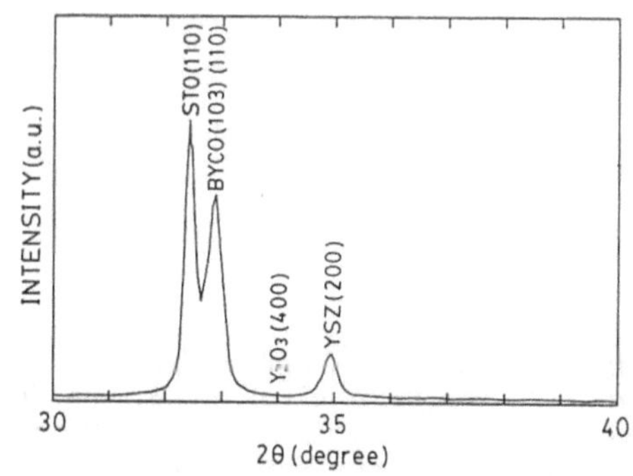

Fig.4. XRD pattern of $Ba_2YCu_3O_x$ thin film on (110)-oriented $SrTiO_3$ thin films grown on Y_2O_3/YSZ/Si(100).

with these buffer layers. Superconducting properties of $Ba_2YCu_3O_x$ on Si(100) substrate with these buffer layers were comparable to those on Si(100) substrate with Y_2O_3/YSZ buffer layers.

<u>$SrTiO_3$ thin films on Y_2O_3/YSZ/Si(100), Gd_2O_3/Y_2O_3/YSZ/(100) and CeO_2/Y_2O_3/YSZ/Si(100)</u>

XRD and RHEED observations of $SrTiO_3$ thin film grown on CeO_2/Y_2O_3/YSZ/Si(100) at ~900°C showed that $SrTiO_3$ thin film were epitaxially grown with (100)-axis normal to the substrate surface. On the other hand, XRD patterns of $SrTiO_3$ thin films grown on Si(100) with epitaxially grown Y_2O_3/YSZ, Gd_2O_3/Y_2O_3/YSZ and CeO_2/Y_2O_3/YSZ at 650°C~800°C showed that (110)-axis of $SrTiO_3$ film preferentially oriented perpendicular to the substrates. Figure 3 shows typical RHEED patterns of (110)-oriented $SrTiO_3$ thin films grown on CeO_2/Y_2O_3/YSZ/Si(100) at 700°C along Si[001] azimuth. This Figure shows that (110)-oriented $SrTiO_3$ films grow into two domains of $SrTiO_3$[001]//CeO_2[001]//Si[001], and of $SrTiO_3$[1$\bar{1}$0]//CeO_2[001]//Si[001].

Figure 4 shows XRD pattern of $Ba_2YCu_3O_x$ thin Film on (110)-oriented $SrTiO_3$ thin buffer layer on Si(100) with epitaxially grown Y_2O_3/YSZ thin film. This figure shows that (110) or (103)-oriented $Ba_2YCu_3O_x$ film grown on Si(100) substrate. RHEED patterns of $Ba_2YCu_3O_x$ thin Film on (110)-oriented $SrTiO_3$ thin buffer layer on Y_2O_3/YSZ/Si(100) showed that (110) or (103)-oriented $Ba_2YCu_3O_x$ thin films were epitaxially grown on Si(100).

Figure 5 shows temperature dependence of resistance of $Ba_2YCu_3O_x$ thin films on (110)-oriented $SrTiO_3$ buffer layer on Y_2O_3/YSZ/Si(100). Observed $T_{c(end)}$ of this film are about 36K. This lower $T_{c(end)}$ than that obtained

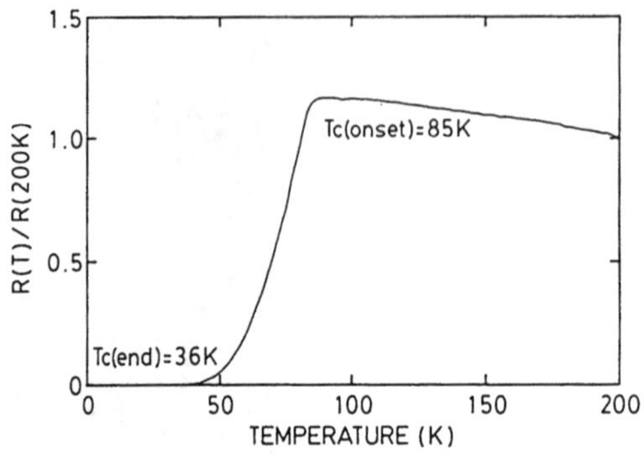

Fig.5. Temperature dependence of resistance of $Ba_2YCu_3O_x$ thin films on (110)-oriented $SrTiO_3$ buffer layer on Y_2O_3/YSZ/Si(100).

for $Ba_2YCu_3O_x$ thin films on Y_2O_3/YSZ/Si(100) indicated that $Ba_2YCu_3O_x$ thin film consists of several domains and the *a–b* planes become discontinuous at the domain boundaries.

SUMMARY

$Ba_2YCu_3O_x$ thin films have been epitaxially grown on Si(100) substrate using various buffer layers. (100)-oriented Gd_2O_3, Sm_2O_3 and CeO_2 thin films were epitaxially grown on Y_2O_3/YSZ/Si(100) by EBE. $Ba_2YCu_3O_x$ thin films sputtered on Si(100) with these buffer layers were epitaxially grown with *c*-axis oriented normal to the substrate surface. Superconducting properties of $Ba_2YCu_3O_x$ on Si(100) substrate with these buffer layers were comparable to those on Si(100) substrate with Y_2O_3/YSZ buffer layers. (110)-oriented $SrTiO_3$ thin films were epitaxially grown on Si(100) substrate with epitaxially grown thin buffer layers of Y_2O_3/YSZ, Gd_2O_3/Y_2O_3/YSZ and CeO_2/Y_2O_3/YSZ at 650°C∼800°C by rf magnetron sputtering. $Ba_2YCu_3O_x$ thin films sputtered on Si(100) with (110)-oriented $SrTiO_3$ buffer layers were epitaxially grown with (110) or (103)-axis oriented normal to the substrate surface.

ACKNOWLEDGMENT

The authors wish to express their sincere gratitude to F. Nishiyama for his skillful aid in the ion–beam channeling experiment.

REFERENCES

1. Myoren H, Nishiyama Y, Fukumoto H, Nasu H, Osaka Y (1989) Jpn. J. Appl. Phys. **28** : 351
2. Myoren H, Nishiyama Y, Miyamoto N, Kai Y, Yamanaka Y, Osaka Y, Nishiyama F (1990) Jpn. J. Appl. Phys. **29** : L955
3. Fukumoto H, Imura T, Osaka Y (1989) Appl. Phys. Lett. **55** : 360
4. Shah SI, Carcia PF (1987) Appl. Phys. Lett. **51** : 2146
5. Terada N, Ihara H, Jo M, Hirabayashi M, Kimura Y, Matsutani K, Hirata K, Ohno E, Sugise R, Kawashima F (1988) Jpn. J. Appl. Phys. **27** : L639

Preparation of YBCO Thin Films Using an Off-Axis Sputtering Technique

KOUKI SATOU, MASAKAZU MATSUI, MISAO SAKANO, TAKASHI SHIMANO, and HIRONORI MATSUBA

Yokohama R&D Laboratories, Furukawa Electric Co., Ltd., Yokohama, 220 Japan

ABSTRACT

$YBa_2Cu_3O_{7-x}$ thin films were deposited on MgO(100) substrates using an off-axis sputtering technique. We investigated the compositional dependence of the substrate position. When the distance between substrate and target center, *x*, was smaller than 50mm, a considerable deviation from stoichiometry was recognized. As *x* increased in the region of *x*>50mm, ratios of Cu/Y and Ba/Y increased slightly, while a ratio of Cu/Ba remained constant. We obtained Tc_0=80K and *Jc*(77K,0T)=$1x10^5A/cm^2$. Surface resistance of the film was 700mΩ at 77K and 9.6GHz.

KEY WORDS: off-axis sputtering, $YBa_2Cu_3O_{7-x}$, thin film, surface resistance

INTRODUCTION

A sputtering technique is one of the conventional and reproductive methods in fabricating superconducting thin films. However, on preparing thin films of high-Tc oxide superconductors using the sputtering technique, a severe bombardment by high-energy particles(O^-, high-energy O) emitted from an oxide target causes compositional changes and considerable damages in the films. To avoid these problems, researches on preparing thin films with appropriate improvements were successfully performed, *e.g.* growing at high gas pressure[1], using a transverse magnetic field to eliminate the bombardment of the electrons[2], introducing an additional electrode and applying a bias voltage to trap charged particles[3,4] and placing the substrates outside the region of head-on negative ion flux, in other words, the substrates were positioned perpendicularly to vis-a-vis the target, so-called an off-axis sputtering method[5-9]. The off-axis sputtering technique is seemed to be a promising method reproducibly to achieve low surface resistance thin films for microwave applications[7,8] as a result of reduction of the bombardment damage by high-energy particles. Therefore, we prepared $YBa_2Cu_3O_{7-x}$(YBCO) thin films *in situ* using the off-axis sputtering technique and discussed the growth conditions of films with low surface resistance.

EXPERIMENTAL

The schematic view of the off-axis sputtering apparatus was illustrated in Fig. 1. A sputtering target was a stoichiometric sintered $YBa_2Cu_3O_x$ with 4 inch in diameter. As indicated in Fig. 1, we defined the *x*-axis as the movable direction of the block heater on which the substrate was clamped, *z*-axis as the direction perpendicular to the target surface and *y*-axis as the direction parallel to the target surface. The center of the target was defined as the origin. In this paper, the films were prepared only by varying *x*, the distance between the substrate and the target center, from 0 to 100mm. The substrate was held at the center of the block heater for thermal uniformity corresponding to

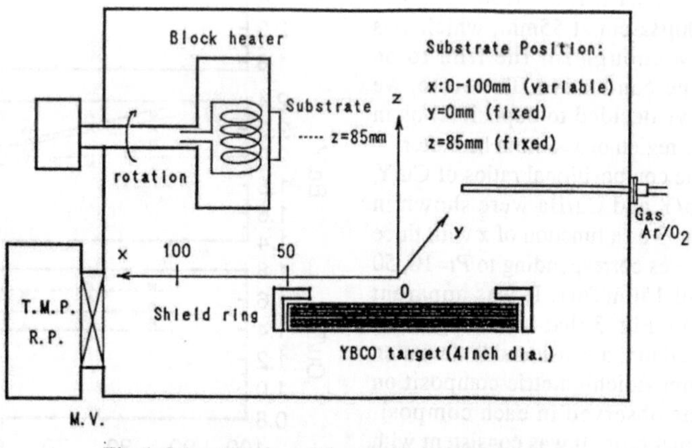

Fig. 1 Schematic view of the off-axis sputtering apparatus. The spatial substrate position is indicated as (*x,y,z*), where *x* is movable from 0 to 100mm, *y* and *z* are fixed at 0mm and 85mm, respectively.

y=0mm and z=85mm. After evacuating the chamber below 1x10^{-5}Torr, the sputtering gas(4:1 mixture of Ar and O_2), whose total flow rate was adjusted using a mass flow controller to be fixed at 10sccm, was introduced into the chamber. The total gas pressure during the film growth(Pt) was controlled by the main valve with the range between 10 and 200mTorr. Then the block heater was raised to 670°C and the YBCO films were grown on MgO(100) substrates. An RF power of 150W gave a deposition rate of about 0.5-1.4nm/min depending on Pt. After deposition, the chamber was immediately vented with 300Torr of O_2 and the films were cooled down to 200°C for 30minutes. The atomic compositions of the films were analyzed by inductively coupled plasma(ICP) emission spectroscopy. The crystallinity, the texture and lattice constant of the films were determined with an X-ray diffractometer(XRD) using CuKα radiation. The critical temperature, Tc_0, and the critical current density, Jc, were measured using a dc four-probe method.

RESULTS AND DISCUSSION

Film Composition

First of all, we explored an appropriate substrate position for obtaining films with low surface resistance by measuring room-temperature resistivity and analyzing compositional ratios of YBCO films deposited at Pt=10mTorr. In Fig. 2, the room-temperature resistivity of the films was plotted as a function of x. It was appeared that the resistivity of the film was so high that the film was not superconductive in the region of x<50mm. Since the radius of the target was 2 inch as shown in Fig. 1, the substrate placed in the x<50mm region was directly exposed to the plasma which included high density of the high-energy particles which caused the bombardment damage to the films. However, it was revealed that in the region of x>50mm, as x increased, the resistivty of the film fell down abruptly and saturated with the value of about 400μΩ·cm at 55mm, which was low enough for the film to be superconductive. Therefore, we have decided to deposit films in the region of x>50mm hereafter.

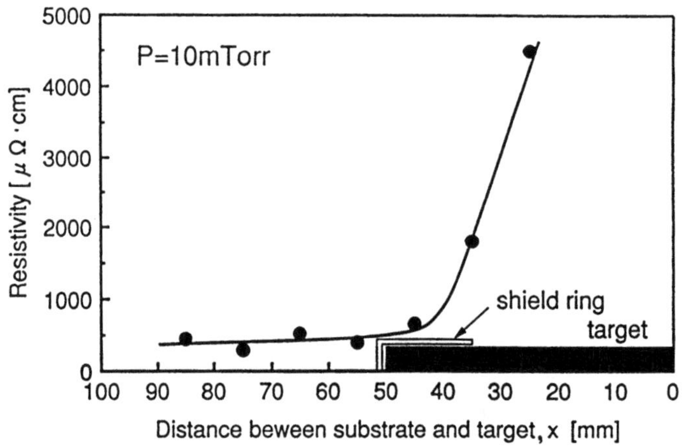

Fig. 2 Room-temperature resistivity of the YBCO thin films grown at Pt=10mTorr as a function of x.

The compositional ratios of Cu/Y, Ba/Y and Cu/Ba were shown in Fig. 3 as a function of x with three curves corresponding to Pt=10, 50 and 150mTorr. It was apparent from Fig. 3 that in the region of x<50mm, a considerable deviation from stoichiometric composition was observed in each compositional ratio. It was consistent with the results of Fig. 2. The high resistivity of the film grown in the region of x<50mm was attributed to this deviation from stoichiome-

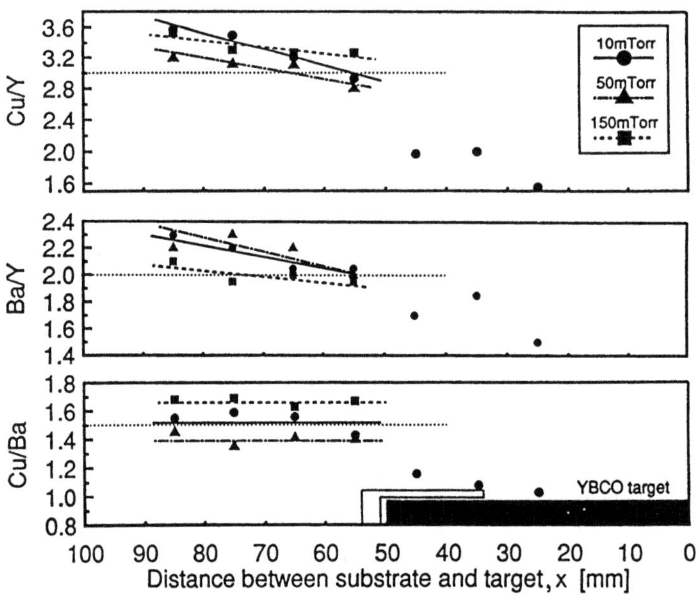

Fig. 3 Compositional ratios of Cu/Y, Ba/Y and Cu/Ba as a function of x.

try. In the region of x>50mm, the ratios of Cu/Y and Ba/Y increased slightly, as x increased, while the ratio of Cu/Ba remained constant with the value of 1.5 independent of x. This behavior agreed well with the results of Matsutani *et. al.*[9]. From these measurements, we selected the region of 60<x<100mm as the substrate position independent of Pt to obtain the YBCO thin films with stoichiometric composition and low resistivity.

Lattice Constant c_0

A c-axis lattice constant c_0, which was calculated from the YBCO(006) diffraction of the XRD pattern, was shown in Fig. 4 as a function of x. It is well known that the c_0 was shortened with increasing Tc[10]. It was revealed that the c_0 became shorter as x increased and became more insensitive to x as Pt was higher. This tendency could be explained by the following discussion. When Pt was low, such as 10mTorr, a collision frequency of the high-energy particles with sputtering gas emitted from the target was so small that a kinetic energy of the high-energy particles was not reduced and could reach the region of x>50mm. And the density of high-energy particles was considered to be decreased as x increased. On the other hand, when Pt was high, such as 200mTorr, the collision frequency was so large that the kinetic energy of the high-energy particles was reduced. These particles were confined to the space above the target and hardly reached the region of x>50mm. As a result, the c_0 grown at low Pt was longer at any x and was more sensitive to x than that grown at high Pt. From the results of Figs. 3 and 4, we could obtain the YBCO thin film grown at Pt=10mTorr and x=95mm with Tc_0 of 80K and Jc at 77K and 0T of 1×10^5A/cm^2.

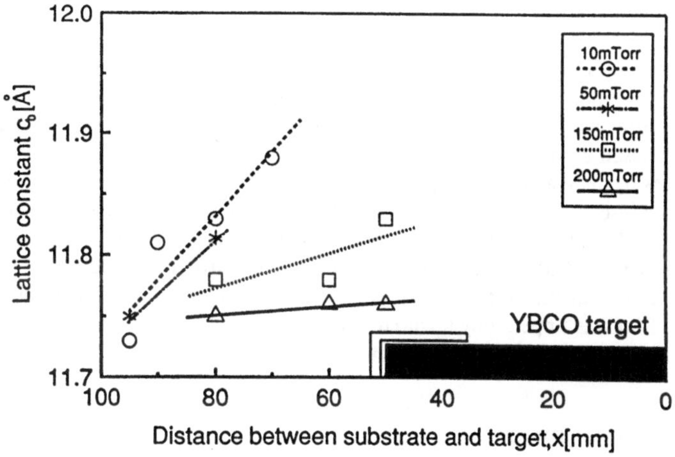

Fig. 4 Lattice constant c_0 as a function of x, calculated from YBCO(006) diffraction.

Surface Resistance

Surface resistance(Rs) was measured at 9.6GHz and the temperature range between 4.2 and 150K with the help of TESTING SERVICE of *SUPERCONDUCTORS TECHNOLOGIES*. The details of Rs measurement were described elsewhere[11]. Tc of the sample was 70K and Jc at 60K and 0T was 5×10^4A/cm^2. The Rs value was 700mΩ at 77K and was constant to be 400mΩ below 60K as shown in Fig. 5. It was three orders of magnitude at 77K and four orders of magnitude at 4.2K higher than the reported values[7,8]. The reason of high Rs is probably interpreted as follows: 1)an imperfect texture of the film; the XRD pattern of the sample showed a slight a-axis orientation with main c-axis oriented texture. 0.45° of full width at half maximum(FWHM) of YBCO(006)

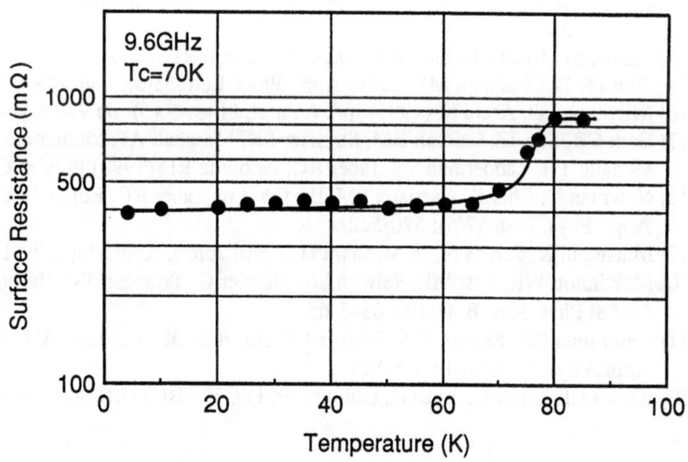

Fig. 5 The surface resistance of the YBCO thin film with Tc=70K and Jc(60K,0T)=5×10^4A/cm^2 as a function of temperature.

diffraction of the sample was larger than that of low *Rs* film (0.21-0.26° using YBCO(005) diffraction) in ref. 8, 2)low *Jc*; weak link between grains was one of reasons for loss of microwave[12]. It was pointed out that the *Rs* of low *Jc* film was 3-4 orders of magnitude higher than that of high *Jc* film[12] and this factor was dominant particularly in the region of about T/*Tc* was 0.2-0.9[12], 3)low *Tc*; an increase in *Rs* around *Tc* observed in the *Rs*-temperature curve should be shifted to higher temperatures if *Tc* became higher, consequently the *Rs* of high *Tc* films at 77K was lowered, 4)impurity and imperfection of surface structure; these were considered to be the main reasons of residual *Rs* at low temperatures, such as T/*Tc* was smaller than about 0.2, as pointed out in ref. 12.

CONCLUSION

$YBa_2Cu_3O_{7-x}$(YBCO) thin films were deposited on MgO(100) substrates by the off-axis sputtering technique. It was revealed that when the substrate was located in the range of x<50mm, a considerable deviation from stoichiometry was observed and the films were not superconductive. In the region of x>50mm, ratios of Cu/Y and Ba/Y slightly increased with x, while a ratio of Cu/Ba remained constant. Therefore, we could obtain the correct stoichiometric films by choosing an appropriate substrate position. The lattice constant c_0 became shorter as x increased. The c_0 was more insensitive to x when the film was deposited at high *Pt* than that when deposited at low *Pt*. As a result, we obtained Tc_0=80K and *Jc*(77K,0T)=1×10^5A/cm^2 grown at *Pt*=10mTorr and x=95mm. The measurement of *Rs* was performed at 9.6GHz with the temperature range between 4.2 and 150K. The *Rs* was 3-4 orders of magnitude higher than those of literatures. The reasons of high *Rs* were estimated that the *Tc* and *Jc* of the film were too low and that the film was imperfectly textured. We are now in progress of fabricating the YBCO thin films with low *Rs* and also constructing a system for measuring *Rs*.

ACKNOWLEDGMENT

The authors would thank *SUPERCONDUCTORS TECHNOLOGIES* for the measurement of the surface resistance of the film as a TESTING SERVICE.

REFERENCES

[1] Michikami O, Asahi M, Asano H (1989) Jpn. J. Appl. Phys. 28(3): pp L448-L451
[2] Terada N, Ihara H, Jo M, Hirabayashi M, Kimura Y, Matsutani K, Hirata K, Ohno E, Sugise R, Kawashima F (1989) Jpn. J. Appl. Phys. 27(4): pp L639-L642
[3] Takechi K, Shiota T, Hatou T, Inoue M, Takai Y, Hayakawa H, Ohbayashi K (1989) Jpn. J. Appl. Phys. 28(3): pp L434-L436
[4] Kisu T, Ariyoshi Y, Enpuku K, Yoshida K, Takeo M, Yamafuji K (1989) Jpn. J. Appl. Phys. 28(8): pp L1385-L1388
[5] Sandstorm RL, Gallagher WJ, Dinger TR, Koch RH, Laibowitz RB, Kleinsasser AW, Gambino RJ, Bumble B, Chisholm MF (1988) Appl. Phys. Lett. 55(5): 444-446
[6] Michikami O, Asahi M (1991) Jpn. J. Appl. Phys. 30(5): pp 939-944
[7] Eom CB, Sun JZ, Lairson BM, Streiffer SK, Marshall AF, Yamamoto K, Angle SM, Bravman JC, Geballe TH, Ladderman SS, Taber RC, Jacowitz RD (1990) Phisica C 171: 354-382
[8] Newman N, Char K, Garrison SM, Barton RW, Taber RC, Eom CB, Gebelle TH, Wilkens B (1990) Appl. Phys. Lett. 57(5): 520-522
[9] Matsutani K, Terada N, Jo M, Ihara H (1990) Jpn. J. Appl. Phys. 29(1): pp 79-80
[10] McKinnon WR, Post ML, Selwyn LS, Pleizier G, Tarascon JM, Barboux P, Greene LH, Hull GW (1988) Phys. Rev. B 38(10): 6543-6551
[11] Hammond RB, Negrete GV, Bourne LC, Strother DD, Cardona AH, Eddy MM (1990) Appl. Phys. Lett. 57(8): 825-827
[12] Konaka T, Sato M, Asano H, Ishii O (1991) OYO BUTURI 60(5): 482-485

Growth and Characterisation of Gd-Ba-Cu-O Films on MgO (100) by Face Target Sputtering (FTS)

W-K LEE, Y. OHUCHI, T. SUGIHARA, and T. TAKESHITA

Central Research Institute, Mitsubishi Materials Corporation, Omiya, Saitama, 330 Japan

ABSTRACT

Gd-Ba-Cu-O films were grown on MgO(100). The film orientation depends on the substrate temperature: a-axis film from 580 to 720 °C ; mix orientation (a+c) film between 720 and 780°C ; c-axis film above 780 °C. The films have high crystalline quality from channeling studies and no film-substrate reaction was detected. The film surface morphologies depended not only on the substrate temperature but also the composition ($Ar:O_2$) of the sputtering gas.

KEY WORDS: 1-2-3 superconductor, channeling, microstructure

INTRODUCTION

For the realisation of utilising high-Tc materials in electronic devices (see, for example, ref. [1]), the growth of high quality thin films is necessary. Besides epitaxy and smoothness, the films should have Tc comparable to that of the bulk material and a high critical current density. Different ways of depositing films have been employed (see, for example, ref. [2]). By far, the most widely used and the simplest way is by sputtering deposition. In this work, films were deposited by Face Target Sputtering (FTS) process [3].

EXPERIMENTAL

Sputtering deposition of Gd-Ba-Cu-O films was carried out in a FTS system as reported before [4]. MgO(100) was used as substrate and it was ultrasonically cleaned in aceton before deposition. After deposition, the films are cooled naturally in 10 Torr of oxygen. The deposition conditions are summarised in Table 1. To study their microstructures and transport properties, the films were characteristised by XRD,SEM, Rutherford Back Scattering (RBS) and 4-point dc measurements.

Table 1. Deposition conditions for the preparation of films.

Substrate:	MgO(100) 15×15×0.5 mm^3
Substrate temperature	580 - 800 °C
Total pressure	1 mTorr
$Ar:O_2$ ratio	1:1, 1:9, 9:1
Deposition time	2 hours
Target composition	1:2:3 + 1:2.1:3.64

RESULTS AND DISCUSSION

All films are black or dark brown in color with highly reflective surfaceses. The average film thickness is about 0.15μm. RBS, with 2.3 MeV He^+ ions, carried out by Tory Research Center, Inc. shows that the average composition of the films is 1:2.1:3.1. No significant variation of the film composition as a function of the deposition temperature was detected. Also, channeling studies give a minimum channeling yield of about 12% (as compared to the theoretical value of 3 to 5% in single crystals [5]) which indicates a considerable degree of epitaxy. The channeling spectrum of a typical film is shown in fig.1. Furthermore, no film-substrate reaction was found.

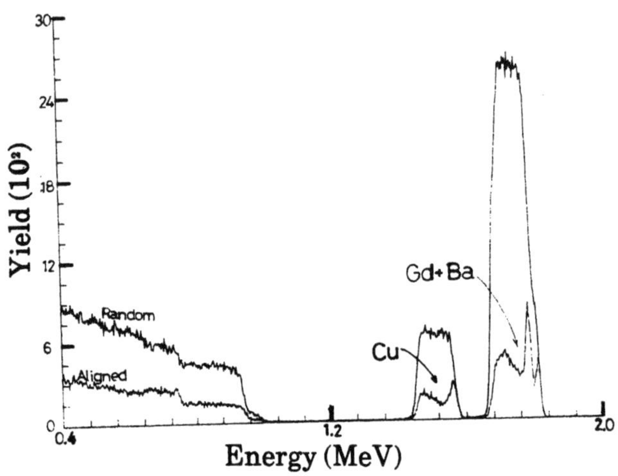

Fig.1 Channeling and random RBS spectra of a film grown at 785 °C.

XRD studies show that the film orientation depends very much on the substrate temperature during growth. As is shown in fig.2, between 580 and 720 °C, a-axis oriented films are grown. From 720 to 770 °C films of mix orientation (a+c) result. Above 780 °C, c-axis film are obtained. No dependence of film orientation on the ratio of $Ar:O_2$ is found at a constant substrate temperature.

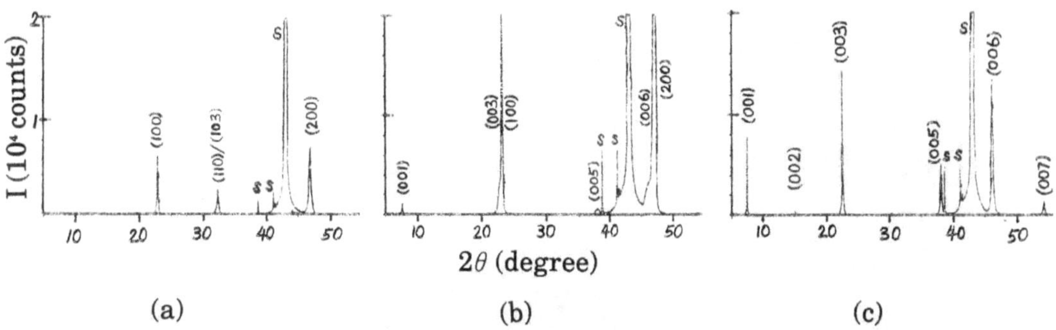

Fig.2 XRD spectra for films grown at (a) 660 °C; (b) 750 °C; and (c) 785 °C. The letter s indicates the substrate peaks.

For films deposited between 580 to 720 °C with $Ar:O_2$ = 1:1, 1:9 and 9:1 respectively, tightly packed circular grains are observed as is shown in fig.3. From 720 to 770 °C, when films of mix orientations were grown, films deposited with $Ar:O_2$ = 1:1 have a weave pattern with elongated grains (fig. 4a). Whereas with $Ar:O_2$ = 1:9 and 9:1, only circular grains are observed (fig.4b,c). Above 780 °C, when only c-axis films are obtained, weave pattern like that in fig.4a was found irrespective of the ratio of $Ar:O_2$, though the density of the long grains is much higher.

Figure 5 is a plot of the superconducting transition temperature Tc as a function of the substrate temperature Ts as determined from 4-point dc measurement. Superconducting films are obtained between 720 and 800 °C. The transition region, between the onset and 10% residual resistivity, are wide (>15K). The highest Tc(zero) is only about 60K. With Ts below 720 °C, only semiconducting films are obtained.

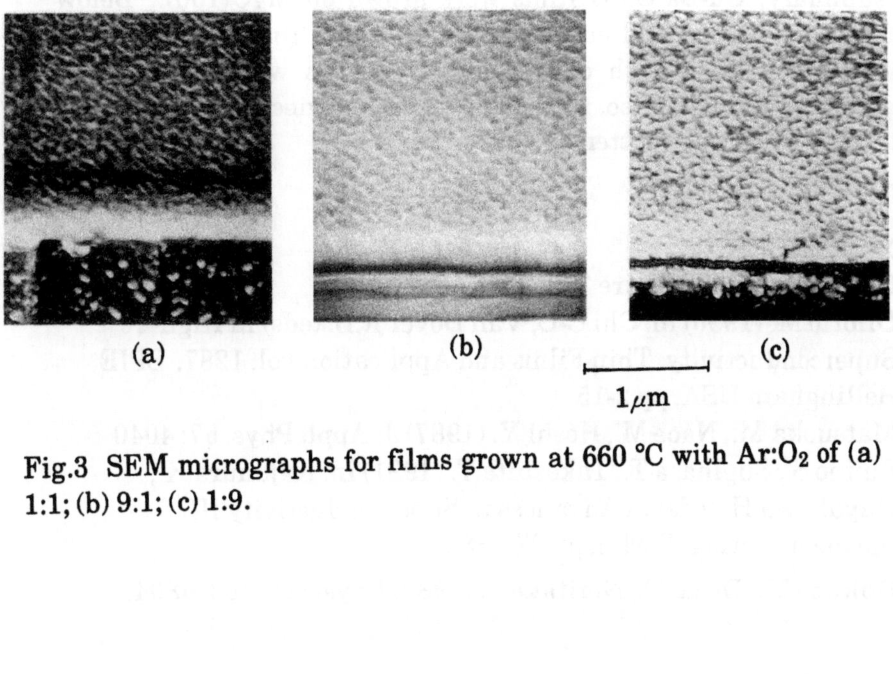

Fig.3 SEM micrographs for films grown at 660 °C with $Ar:O_2$ of (a) 1:1; (b) 9:1; (c) 1:9.

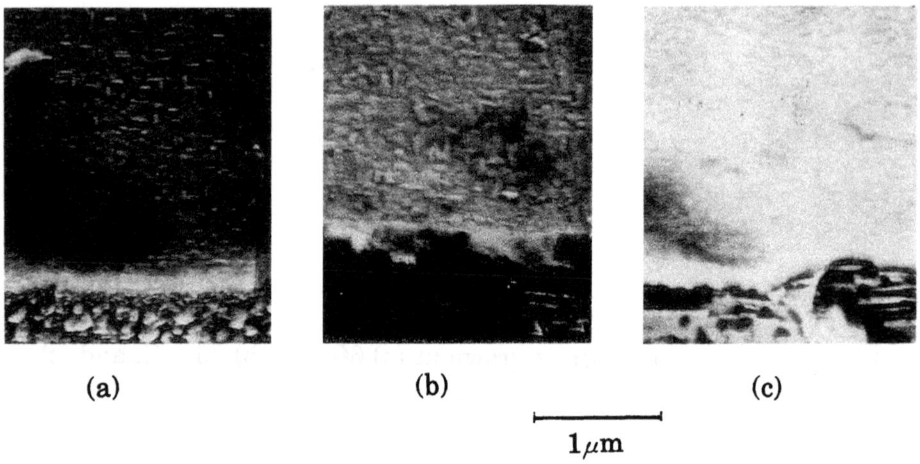

Fig.4 SEM micrographs for films grown at 750 °C with $Ar:O_2$ of (a) 1:1; (b) 9:1; (c) 1:9.

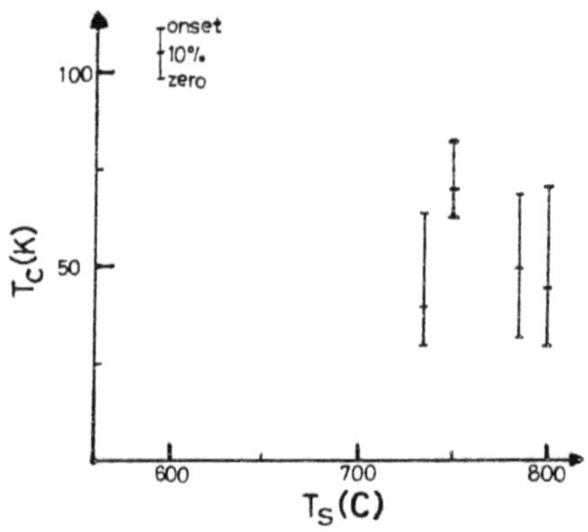

Fig.5 Plot of superconducting transition region Tc against substrate temperature Ts.

In summary, Gd-Ba-Cu-O films were grown on MgO(100). Below 720°C, a-axis films resulted. Above 780°C, c-axis films were obtained. The films have a high degree of epitaxy as was revealed by channeling studies. Also, the surface morphologies depended on the ratio of $Ar:O_2$ of the sputtering gas.

REFERENCES

1. Clarke J. (1988) Nature 333: 29
2. Dilorio M. (1990) in: Chi C-C, Van Dover R.B. (eds) in High-Tc Superconductivity: Thin Films and Application vol. 1287. SPIE, Bellingham USA, pp 2-15
3. Matsuoka M., Naoe M., Hoshi Y. (1987) J. Appl. Phys. 57: 4040
4. Fujino S., Sugihara T., Takeshita T. (1991) in: Kajimura K., Hayakawa H. (eds) in Advances in Superconductivity III. Springer-Verlag, Berlin, pp 877-880
5. Ohkubo M., Dachi T., Noritake T. (1989) Physica C 160: 4804.

A Low Kinetic Energy PVD Process for Low Temperature Synthesis of Y-Ba-Cu-O Films

KIYOSHI ISHII, KAZUSHIGE YASHIRO, and TAKAMICHI NAGAMOTO

Faculty of Engineering, Utsunomiya University, Utsunomiya, 321 Japan

ABSTRACT

A gas-flow-sputtering method in which atoms having very low kinetic energies (<0.1 eV) are deposited on the substrate has been studied to fabricate $Y_1Ba_2Cu_3O_{7-x}$ films. Epitaxial superconducting films are deposited at a substrate temperature of about 580°C. It is observed that polycrystalline films grow at a low substrate temperature of ∿250°C. The result is thought to demonstrate that this sputtering method is very useful for the research on the conditions for in-situ growth of superconducting YBaCuO films at low temperatures.

KEY WORDS: $Y_1Ba_2Cu_3O_{7-x}$ superconducting film, Sputtering, Gas-flow-sputtering

INTRODUCTION

Sputtering has been demonstrated to be an useful method in producing high-Tc oxide superconducting films [1]. However, it is pointed out that the bombardment of the growing film with energetic particles such as negative ions and reflected neutrals has a serious trouble in the formation of stoichiometric and high quality superconducting films[2-5]. To solve this problem, the off-axis sputtering and high-pressure sputtering have been studied [2-6]. The energy range investigated closely is about 10 to 100 eV. The authors consider that even the particles with lower energies (<1 eV) may have a strong influence on the crystal-growth of films because the energy of 1 eV corresponds, in average, to the thermal energy of gas atoms at 11600 K. So, we have developed a high-pressure sputtering method utilizing Ar gas flow, i.e., gas-flow-sputtering (GFS) method[7], to create more sophisticated low kinetic energy process suited to the investigation on YBaCuO film growth.

The fundamental of GFS is shown in Fig.1. The target is in the shape of cylindrical tube. The glow discharge occurs inside the target as hollow cathode discharge and its inner surface is sputter-eroded. Sputtered atoms are carried to the substrate by a viscous Ar flow maintained at a high speed. This mechanism allows the effective deposition of sputtered atoms even at a high pressure of 1 Torr range, contrary to conventional high-pressure sputtering methods. Under such high pressure conditions, energetic particles will lose their initial energies by collisions with sputtering gas atoms and become thermalized to the temperature of the gas in a very short distance. Several

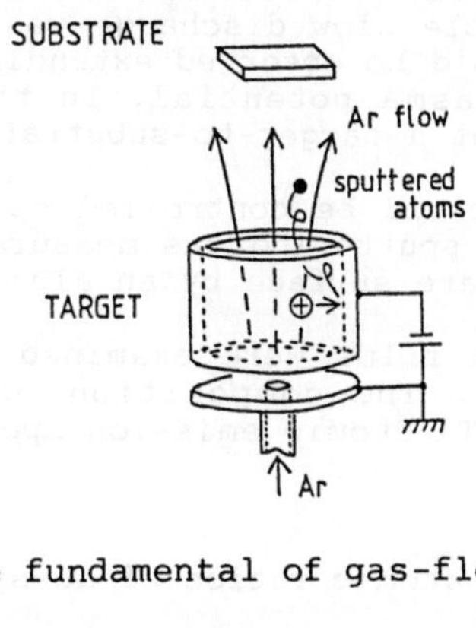

Fig.1 The fundamental of gas-flow-sputtering.

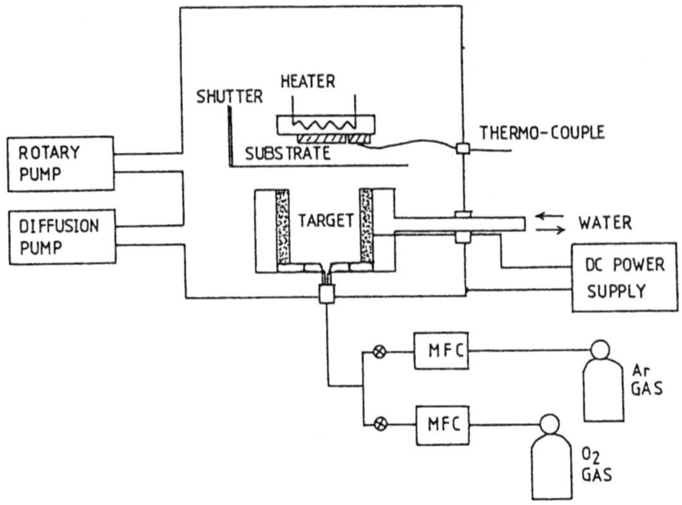

Fig.2 Schematic diagram of gas-flow-sputtering system constructed in this work.

workers have treated this problem of the thermalization of energetic atoms during sputtering process. R.E. Somekh[8] calculated the thermalization of sputtered Al, Cr, W atoms of 20-30 eV and reflected Ar of $\sim$100 eV in Ar as a function of PD, the product of the sputtering gas pressure P and the target-substrate distance D, up to PD=3.7 Torrmm and PD=6.0 Torrmm, respectively. The result shows that their energies decrease to about 0.1 eV at each maximum PD investigated and, moreover, predicts that their thermalization must be complete at PD=10 Torrmm which is satisfied in a typical GFS condition. In GFS process, the sputtering gas flows from target to substrate during sputtering. However, since the flow velocity (a few tens meter per second) is much lower than thermal velocity, the thermalization will not be significantly affected by the flowing of sputtering gas.

A sintered $Y_1Ba_2Cu_3O_y$ tube was used as a target and thin films of 2700 Å thickness were deposited by GFS using Ar-O_2 gas mixture. The structures and superconducting transition temperatures Tc of as-grown films were investigated.

EXPERIMENTAL

Figure 2 shows the GFS system used in this work. The target, placed in the holder cooled by water, was a sintered $Y_1Ba_2Cu_3O_y$ tube of 18 mm inner diameter and 25 mm length. After the chamber was pumped down to $2x10^{-6}$ Torr by the oil diffusion pump, Ar gas (99.9999 %) and O_2 gas (99.9 %) were introduced into it through each mass flow controller. During sputtering the sputtering gas was exhausted by the oil rotary pump. The flow rates of Ar and O_2 were 200 and 50 cm^3/min(at 1 atm, 20 °C), respectively. The total pressure measured at the chamber was 1 Torr. The velocity of sputtering gas streaming out from the target was estimated to be about 25 m/sec. The discharge voltage was 150 V and the current was 0.5 A. The stable glow discharge was well confined within the target, though a weak glow could be observed extending from the target to the substrate, which was at a plasma potential. In these conditions the film deposition rate was 45 Å/min at a target-to-substrate separation of 2.5 cm.

The substrate temperature Ts could be controlled up to $\sim$650 °C by a resistive heater. The temperature during sputtering was measured with a CA thermo-couple which was pasted on the substrate surface by an alumina base cement.

The structures of as-sputtered films were examined by reflection high-energy electron diffraction (RHEED). The composition of films was measured by inductively coupled plasma (ICP) atomic emission spectroscopy.

RESULTS AND DISCUSSION

Figure 3 shows Cu/Y and Ba/Y atomic ratios of as-sputtered YBaCuO films as a

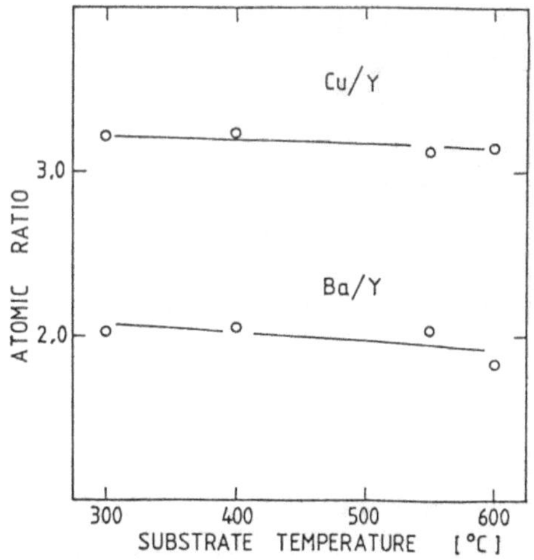

Fig.3 Substrate temperature dependence of composition for as-sputtered YBaCuO films.

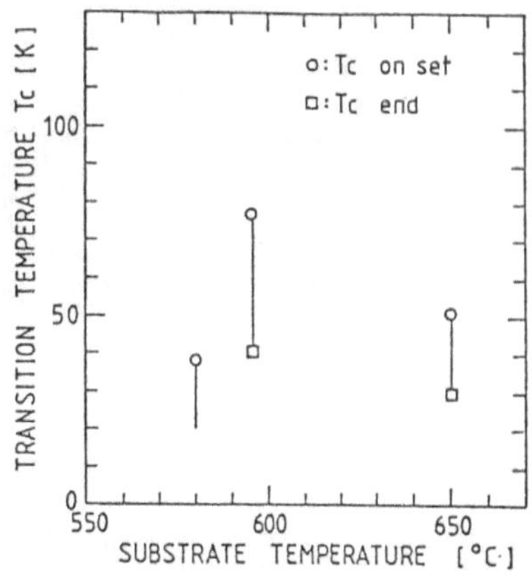

Fig.5 The variation of Tc, transition onset temperature Tc on set and zero resistance temperature Tc end, for as-sputtered YBaCuO films.

function of substrate temperature. This figure indicates that the films deposited by GFS have a composition closer to the target composition than those by conventional sputtering methods. Little deviation in composition between the film and the target is thought to originate from reduced bombardment of energetic particles due to the complete thermalization of ions and atoms.

Figures 4(a), 4(b) and 4(c) show the RHEED patterns of 2700 Å-thick YBaCuO films deposited on (110)$SrTiO_3$ substrates at Ts ∿560°C, ∿580°C and ∿650°C, respectively. The pattern in Fig.4(b) shows tri-layered perovskite structures, and so it is known that this film grew epitaxially and (110) oriented parallel to the substrate plane. As this pattern is somewhat spotty, the film surface is assumed not to be so smooth. Below about 560°C, polycrystalline films were deposited, of which RHEED patterns were like Fig.4(a). The patterns in Fig.4(a) and 4(c) showing polycrystalline structures suggest that epitaxial YBaCuO films on (110)$SrTiO_3$ are obtained in a narrow range of Ts.

The temperature dependence of the resistivity of as-grown films was measured from 300 K to 20 K. The films deposited at Ts higher than ∿570°C showed a superconducting transition. Their transition temperatures Tc(on set) and Tc(end) are plotted in Fig.5.

There have been too few samples examined to discuss the characteristics of the films deposited by GFS. However, in-situ crystallization of the film was observed to occur at considerably low Ts. The RHEED patterns of YBaCuO films deposited at Ts ~ 250 °C and Ts ~ 350 °C are shown in Fig.6. They show

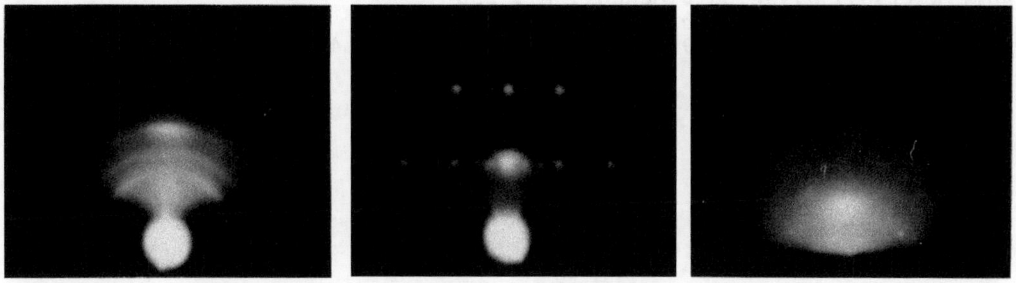

(a) (b) (c)

Fig.4 RHEED patterns of as-grown YBaCuO films sputtered at Ts ∿ 560°C (a), Ts∿ 580°C (b) and Ts∿ 650°C (c).

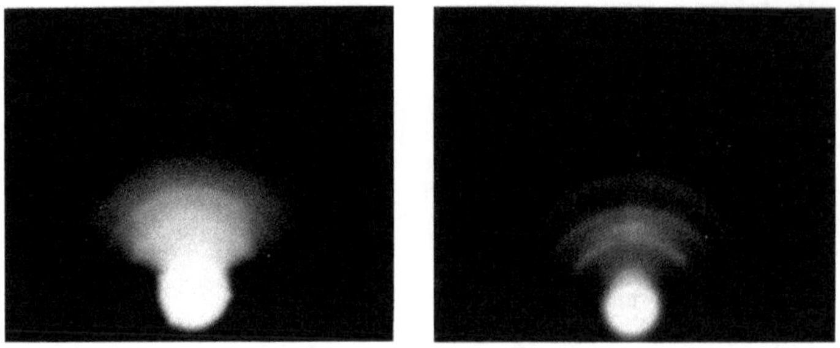

(a) (b)

Fig.6 RHEED patterns of as-grown YBaCuO films sputtered at low Ts, ~250 °C (a) and ~350 °C (b).

polycrystalline structures, though the structures can not be exactly identified. In the deposition of YBaCuO films by conventional sputtering methods, amorphous films are apt to grow at these temperatures [9]. The observed low temperature crystallization may be attributable to the low kinetic energy process of GFS.

SUMMARY

A peculiar high-pressure sputtering method, i.e., gas-flow-sputtering, has been applied to the deposition of YBaCuO films. The characteristics of GFS has never been well known; however, this process must be fruitful in the study of reseaching in-situ growth of YBaCuO films at low temperatures because there are very few particles having kinetic energies much higher than the thermal energy.

ACKNOWLEDGMENTS

We would like to thank Professor G. Oya for helpful discussions on this work, and we would also like thank Mr. N. Aoyama for the cooperation in the measuring a superconducting transition temperature.

REFERENCES

1.Leskela M, Truman JK, Mueller CH, Holloway PH (1989) J.Vac.Sci.Technol. A7: 3147
2.Terada N, Ihara H, Jo M, Hirabayashi M, Kimura Y, Matsutani K, Hirata K, Ohno E, Sugise R, Kawashima F (1988) Jpn.J.Appl.Phys. 27: L639
3.Xiong GC, Wang SZ (1989) Appl.Phys.Lett. 55: 902
4.Eom CB, Sun JZ, Yamamoto K, Marshall AF,Luther KE, Geballe CH, Laderman SS (1989) Appl.Phys.Lett. 55: 595
5.Tanabe K, Lathrop DK, Russek SE,Buhrman RA (1989) J.Appl.Phys. 66: 3148
6.Sakuta K, Iyori M, Katayama Y, Kobayashi T (1989) Jpn.J.Appl.Phys. 29: L611
7.Ishii K (1989) J.Vac.Sci.Technol. A7: 256
8.Somekh RE (1984) J.Vac.Sci.Technol. A2: 1285
9.Tanaka S, Itozaki H (1989) Jpn.J.Appl.Phys. 28: L441

Optical Emission Spectroscopic Studies of O_2+Ar Plasma During the Deposition of YBCO Films by Facing Targets Sputtering Method

WATARU ITO, MUNEYUKI IMAFUKU, HARUO SHIMADA, YASUSHI MIYAMOTO, and MISAO HASHIMOTO

Advanced Materials & Technology Research Laboratories, Nippon Steel Corporation, Kawasaki, Kanagawa, 211 Japan

ABSTRACT

In the preparation of YBCO films by facing targets sputtering method, optical emission spectroscopic analysis was carried out. The phenomenon that the oxidation state proceeded more under the condition of a lower relative concentration of oxygen radical was newly observed by changing the configuration of the magnets. Through this study, we could obtain an as-grown film with high-Tc of 87K.

KEY WORDS: optical emission spectroscopy, oxygen radical, facing targets sputtering, YBCO, magnet

INTRODUCTION

Among the superconductors having Tc above 77K, $YBa_2Cu_3O_x$(YBCO) has several advantages compared to other high-Tc superconductors. Firstly, the single phase can be easily obtained. Secondly, YBCO can be epitaxially grown on the adequately selected substrates. Thus, YBCO thin films have been prepared using various techniques by many researchers. Especially, sputtering method has been widely used from the early stage because of its simplicity and the reliability in the preparation. However, the compositional deviation of the films from the stoichiometric components of the target material is observed in general under the conventional sputtering conditions because of the preferential resputtering of the films by the high-energy negative ions emitted from the target. In order to solve this problem, the method using non-stoichiometric target, whose composition should be carefully adjusted to obtain stoichiometric films, was devised [1]. On the other hand, other methods such as deposition under the high pressure condition (about 600mTorr) were developed so as to reduce the resputtering effect [2]. Lately, off-axis sputtering technique, where the substrate is positioned outside of the plasma, is becoming a standard technique [3]. Facing targets sputtering method (FTS) which belongs to the category of off-axis sputtering technique is excellent in the preparation of the stoichiometric and large-area films from the stoichiometric target [4]. In this technique, the deposition is carried out apart from the plasma confined into a limited region between the facing targets by the magnetic field.

We have successfully obtained the as-grown films formed on MgO substrate with Tc of 82K using FTS (85K for the films on $SrTiO_3$ substrate). However, the optimum oxygen partial pressure during the deposition is about 0.2mTorr for FTS, while several tens mTorr for the conventional sputtering. In this report, we will describe the results of the optical emission spectroscopic studies of O_2+Ar plasma and discuss the relationships between the results and the properties of the obtained films for the purpose of understanding the unique feature of FTS, especially the oxidation mechanisms.

EXPERIMENTAL

The fundamental structure of the equipment used in this study was described previously by Tagaki et al [5]. Two targets, 6-inch diameter x 5mm thick cylindrical pellets made of sintered YBCO powder, were confronted at a distance of 150mm. After evacuated below $3x10^{-6}$ Torr by the turbo molecular pump, argon gas and oxygen gas were introduced into the chamber at the flow rates of 30CCM and 15CCM respectively. Sputtering pressure was controlled ranging from 1 to 10mTorr by adjusting the conductance valve, and RF discharge power was fixed at 250W. As a standard procedure, heating of the substrate was stopped and nitrogen gas was introduced into the chamber up to 1 atm as soon as the formation of the film was finished. After the sample was cooled fully, it was taken out from the chamber and the electrical property of the film was measured with the usual four-probe method.

Optical emission from the plasma was detected through a viewing port made of borosilicate glass with 70mm in diameter and focused to the end of the optical fiber by a quartz lens. The spectra were obtained using a monochromator with 250mm focal length and a 1200 groove/mm grating. This system had a spectral resolution of 1nm. Spatially resolved measurements were made by moving the focussing lens horizontally within a distance of 20mm from the center of the targets and by tilting the lens for covering the range between 30mm and 100mm. We mainly focused on the relative intensity of the spectrum. Therefore, it is not necessary to calibrate the sensitivity of our optical system.

RESULTS AND DISCUSSION

Schematic drawing of the configuration of the magnets sited on the rear of the targets and the photograph of the O_2+Ar plasma generated by RF discharge at a power of 250W are shown in Fig.1. From the spectrum at the center of the targets covering the wave length of 400-830nm, optical emission from O radicals, O_2^+ ions and Ar atoms in the excited states were detected, but not from the other components including the metal species sputtered from the targets. This is presumably because the week optical emission from these species were hidden by the strong emission of Ar. We employed the intensity at the wave length of 772.4nm and that of 777.2nm as a quantitative measure of excited argon and excited oxygen radical respectively. The emission at 772.4nm corresponds to the transitions from $2p_7$ to $1s_5$ and from $2p_2$ to $1s_3$ (according to the formula of Pachen) of argon. And the emission at 777.2nm corresponds to the transition from ^5P to $^5S^0$ of oxygen. The ratio of the intensity at 777.2nm to that at 772.4nm can be considered as a characteristics of oxygen radical concentration. Therefore, the pressure dependence and the spatial distribution of the ratio should give us quite important informations.

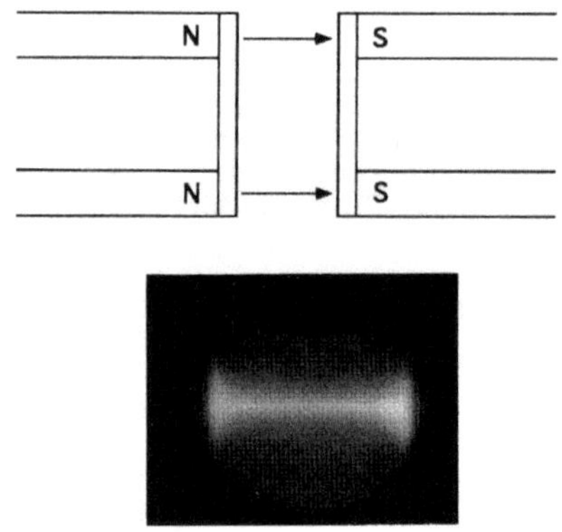

Fig.1 Schematic drawing of the configuration of the magnets and the photograph of the plasma

Figure 2 shows the pressure dependence of the relative intensity of oxygen radical at the distance of 100mm from the center of the targets, where the substrate is located. It can be seen from Fig.2 that the higher pressure leads to the higher relative concentration of oxygen radical. This tendency is almost same for other positions although it is not included in this figure. This pressure dependence can be explained by considering the change of the energy distribution function of electron, ne(E). High energy components of

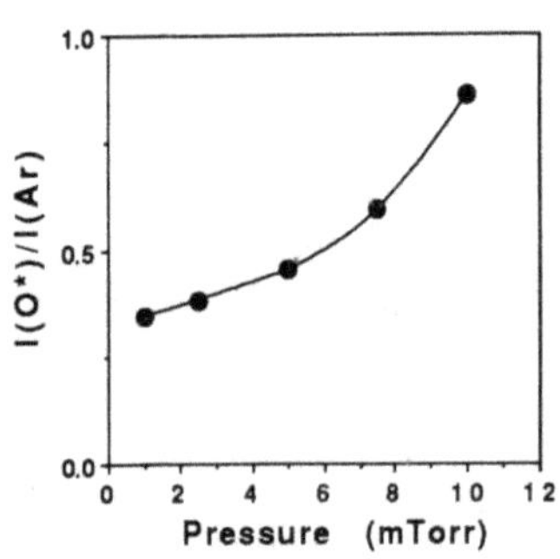

Fig.2 Pressure dependence of the relative intensity of oxygen radical (distance=100mm)

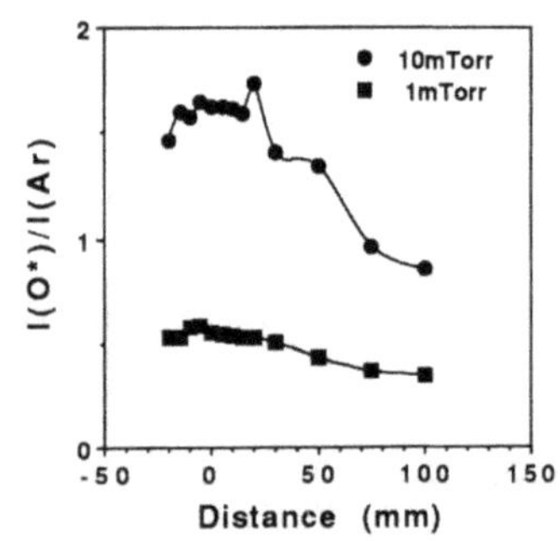

Fig.3 Position dependence of the relative intensity of oxygen radical

ne(E) decreases with the increase of gas pressure. On the other hand, the excitation energy needed for the generation of the excited argon is about 13.3eV, whereas that of oxygen radical is 10.74eV. Therefore, the generation rate of the excited argon should be more heavily affected than that of oxygen radical by reducing the amount of high energy electrons. This explanation was also successfully made for the O_2+He plasma by D.V.Tsu et al [6].

Figure 3 shows the results of the spatially resolved measurements, in which the relative intensity of oxygen radical is plotted against the distance from the center of the targets. Absolute intensities from the excited argon and oxygen radical near the center were extremely strong as speculated from a view of the plasma (see the photograph in Fig.1). The distribution of the ratio decreased as the distance from the center as shown in Fig.3. This tendency can be partly explained by the difference in the life time between excited argon and excited oxygen radical. Among the excited species generated mainly at the center, argon having a long life time of 175ns can move more steadily, thus, distributed more widely than oxygen radical which has a relatively short life time of 29ns [7,8].

It was found that the Tc's of the films formed at various positions did not show the clear dependency on the position. However, we cannot investigate to date the relationships between the Tc and the depositing position over the whole range, especially near the center where more oxygen radical exists, because the position of the substrate is limited only at 75mm or outer due to the geometrical restriction. On the other hand, it is possible to investigate the relationship between the oxygen radical concentration and the Tc by changing the sputtering pressure. Through this course of the study, it was found that the higher Tc was observed for the films prepared under a lower pressure. This result suggests that the properties of the obtained films are not simply determined only by the amount of oxygen radical, but also affected by other deposition condition like total pressure, which reduces the kinetic energy of the metal components arriving at the surface and obstructs the surface reaction. Therefore, it is necessary to carry out the experiment, where only the concentration of the oxygen radical should be changed keeping the other deposition condition fixed in order to clarify the relationships between the Tc and the amount of oxygen radical.

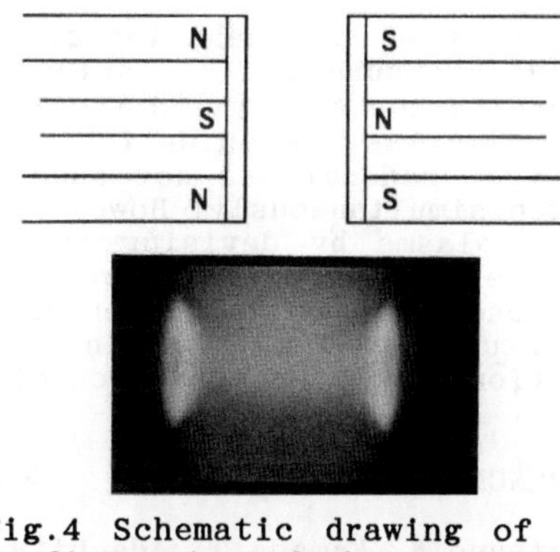

Fig.4 Schematic drawing of the configuration of the magnets and the photograph of the plasma (FTS-type III)

For the purpose of solving this problem, we planned two experiments that rearrange the position of the magnets sited on the rear of the targets. One is the experiment using a type that all magnets are removed, which is called here as "FTS-type II" (consequently, for convenience' sake, former FTS is called as "FTS-type I" hereafter). The other is a type that one more magnet is added to the center of each target as shown in Fig.4, which is called as "FTS-type III". Here, RF power was lowered to 150W for the deposition by type III so as to make the same deposition rate as that by type I. Pressure dependences of the relative intensities of oxygen radical for both type II and III are shown in Fig.5. A great difference in the pressure dependence for the three types can be explained as followings. In type II, oxygen and argon can be excited over a large space because the plasma is dispersed widely. Consequently, the relative intensity is mainly affected by the pressure. On the other hand, in other two types (I and III) which confine the plasma into a limited region, we can imagine that the effect of reduction of high-energy electron which is induced by the higher pressure is comparatively small at the position where the density of the plasma is extremely high. Consequently, the effect of the pressure on the relative intensity of oxygen radical is not so important compared with the case of type II. It can be also speculated that the efficiency of the plasma confinement is higher in type III than type I, if the efficiency can influence the pressure dependence. However, this is not reflected in the position dependence for the type III because the shape of the plasma confinement is different in both types and the data were taken only along the one way, parallel to the target surfaces.

We can estimate the effect of the relative concentration of the oxygen radical on the Tc by comparing type I and type III. Pressure dependences of the Tc's for the films prepared by both techniques are shown in Fig.6. Under the deposition conditions; total pressure of 1mTorr and distance of 100mm, the relative intensities of oxygen radical in the cases of type I and III were 0.35 and 0.32 respectively. While the relative intensity in type I is about 10% higher than that in type III , the Tc of the film prepared by type III is higher as much as 5K. This result does not support the general idea that the oxygen radical may play an important role in the oxidation of the films. As the reasons for this, following explanations could be made; (1) because of the effective confinement of the plasma, the fluxes of the depositing particles or the energies of them at around the substrate surface are suited for the construction of the crystal structure (2) oxygen ions including positive ions or negative ions are the key species for the oxidation process (3) absolute value of the amount of the oxygen radical is related with the oxidation efficiency. The investigation confirming the assumptions (2) and (3) are now in progress; nevertheless, clear explanation for our results cannot be successfully made to date.

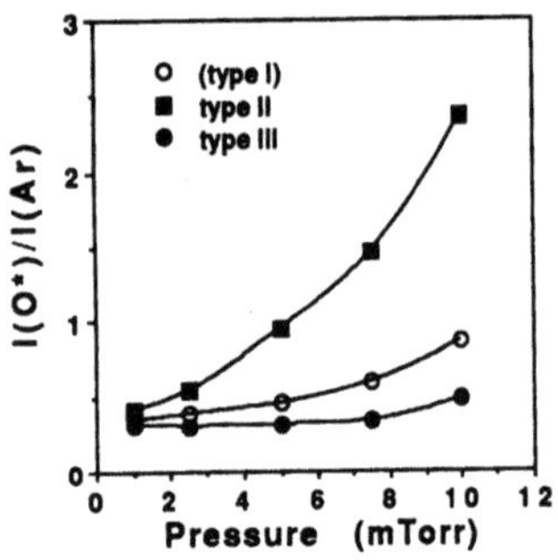

Fig.5 Pressure dependence of the relative intensity of oxygen radical (FTS-type I,II,III, distance=100mm)

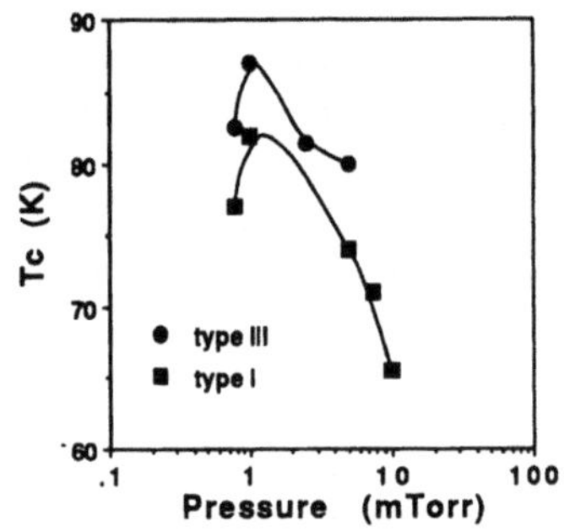

Fig.6 The relationship between Tc and sputtering pressure (FTS-type I,III, distance=100mm)

In conclusion, the behavior of the relative intensity of oxygen radical under various pressure and at various positions could be explained by considering the excitation energy and the life time of oxygen radical and argon. The trial for estimating the oxidation efficiency by monitoring the relative intensity of oxygen radical was not succeeded because other deposition conditions were changed simultaneously. However, from the investigation which change the shape of the plasma by devising the configuration of the magnets, the phenomenon that the oxidation state proceeded more under the condition of a lower relative concentration of oxygen radical was newly observed. Furthermore, through this study, we could obtain the new method and find the optimum deposition condition for the preparation of as-grown films with high-Tc of 87K.

REFERENCES

1. Setsune K, Kamada T, Adachi H, Wasa K: J. Appl. Phys. **64**(1988)1318
   Kawasaki M, Nagata S, Sato Y, Funabashi M, Hasegawa T, Kishio K, Kitazawa K, Fueki K, Koinuma H: Jpn. J. Appl. Phys. **26**(1987)L738
2. Li HC, Linker G, Ratzel F, Smithey R, Geerk J: Appl. Phys. Lett. **52**(1988) 1098
3. Eom CB, Sun JZ, Lairson BM, Streiffer SK, Marshall AF, Yamamoto K, Anlage SM, Bravmann JC, Geballe TH: Physca C **171**(1990)354
4. Imafuku M (1990) In: Proc. of 3rd Int. Symp. on Superconductivity. Springer-Verlag, Tokyo, pp1033-1036
5. Takagi Y, Yamada H, Koyama N (1989) In: Proc. of 2nd Int. Symp. on Superconductivity. Springer-Verlag, Tokyo, pp777-780
6. Tsu DV, Parsons GN, Lucovsky G: J. Vac. Sci. Technol. A **6**(1988)1849
7. Wiese WL, Smith MW, Glennon BM (1966) In: National Standard Reference Data Series, NBS 4, pp77-82
8. Wiese WL, Smith MW, Miles BM (1969) In: National Standard Reference Data Series, NBS 22, pp187-200

YBCO Thin Films with [110] Axis Orientations on $LaSrGaO_4$(110) Substrates

T. Suzuki, T. Ishii, T. Matsui, H. Kimura, K. Tsuda, M. Nagano, and K. Mukae

Fuji Electric Corporation R&D Ltd., Yokosuka, Kanagawa, 240-01 Japan

ABSTRACT

YBCO thin films were grown on $LaSrGaO_4$(LSGO) (100) and (110) substrates.We have obtained YBCO thin films with perfect [110] axis orientation using LSGO (110) substrates. Ordinary X-ray diffraction analysis showed that the films on LSGO (100) substrates had c-axis orientation, and showed the presence of [110] and/or [103] axis orientations in the films deposited on LSGO (110) substrates. This analysis, however, could not distinguish [110] orientation from [103]. X-ray pole-figure method revealed that the axis orientation of the film deposited at 650°C was perfect [110], and [103] axis orientation increased with decreasing substrate temperature.The reason of this epitaxial growth were proposed to be attributed to the misfit of coordination number of oxygen.

KEYWORD: $LaSrGaO_4$, YBCO thin film, pole-figure, axis orientation.

INTRODUCTION

The anisotropic crystal structure of $YBa_2Cu_3O_{7-y}$ (YBCO) leads to a strong unisotropy in its electrical properties[1]. Particularly, the superconductive coherence length in the a,b-axis directions (ξ_{ab}) is an order of magnitude larger than that along the c-axis direction (ξ_c)[2]. When we apply YBCO thin films to Josephson junctions, the non c-axis orientated thin films have an advantage over the c-axis orientated thin films. Numerous researchers have observed non c-axis orientations such as [103][3] and [100] axis orientation[4]. These thin films, however, had a twinning structure which was two directions of c-axis perpendicular to each other. Hence, non c-axis orientated films without the twinning structure were earnestly desired.

In this paper, we report that we successfully fabricated YBCO films with uniform [110] axis orientation, and then discuss mechanism of the epitaxial growth.

EXPERIMENT

The YBCO thin films were prepared by rf magnetron sputtering at a low gas pressure (0.5Pa) as reported in the previous paper [5]. The films were deposited on $LaSrGaO_4$ (100) and (110) substrates at a temperature range of 550-650°C. The LSGO has a tetragonal crystal structure of a=0.3843 and c=1.268nm, where a and c are lattice constant. The oriented substrates were prepared by cutting and mechanical polishing. No special cleaning procedures were used before deposition. X-ray diffraction (XRD) spectra were obtained by Rigaku Gaigerflex RAD-C system using Cu K_α radiation. YBCO axis orientations, which could not be determined by ordinary X-ray diffraction analysis, were investigated by X-ray pole-figure method[4,6]. Resistivity - temperature (R-T) characteristics were measured by a conventional dc four-probe method.

RESULT AND DISCUSSION

Figure 1 shows R-T characteristics of the thin film grown on LSGO(110) substrate at 650°C which has anisotropic behavior. Resistivity at the onset temperature of the film with [110] direction was about thirty times larger than that of [001].

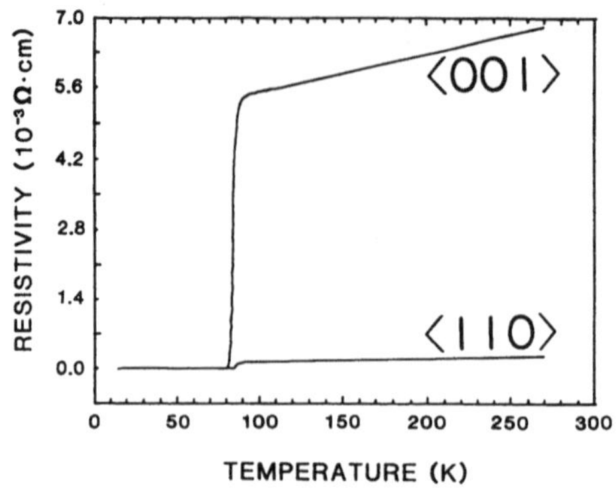

Fig.1 R-T characteristic of the YBCO film grown on LSGO(110) substrate at 650°C.

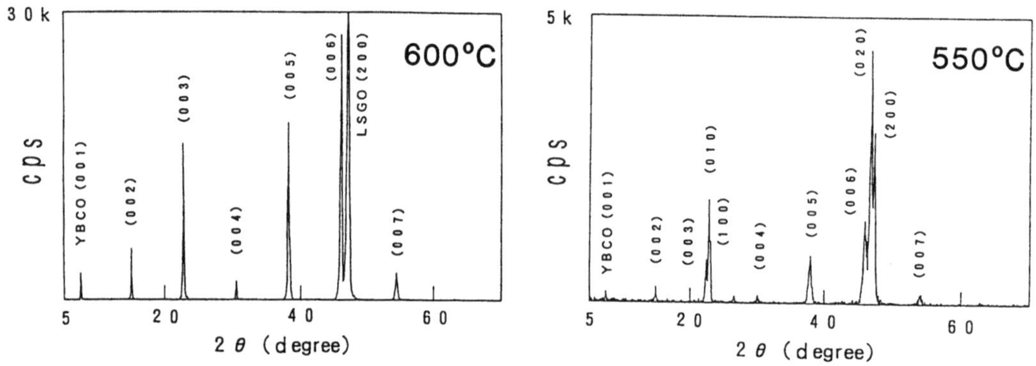

Fig.2 XRD patterns of the YBCO films grown on LSGO(100) substrates.

Figure 2 shows XRD patterns of the films grown on LSGO (100) substrates. A film grown at 650°C had strong c-axis orientation, and the orientation of a film deposited at 550°C was a mixture of a-, b- and c-axis. This tendency was as same as YBCO growth on MgO(100) and $SrTiO_3$ (100). Such deposition of controllable a,b-axis growth by changing the substrates temperature had been reported in previous papers[7-10].
Figure 3 indicates XRD patterns of the films grown on LSGO (110) substrates. This analysis show the presence of [110] and/or [103] axis orientation, and there is no c-axis orientation at the substrate temperature between 550°C and 650°C. This XRD method, however, could not distinguish [110] orientation from [103] because of the same Bragg angles[11]. Therefore we investigated these films by X-ray pole-figure method.

The pole-figure was measured by detecting the reflection from YBCO (012) and (102) planes. Figure 4(a) is theoretical stereographic projection of a YBCO film which has [110] and/or [103] axis orientation. Projections indicated by open circles show the presence of YBCO[110] axis orientation, and the c-axis of this YBCO crystal is parallel to the y-axis of this figure. Projections indicated by closed circles are reflections from YBCO planes except for (102) and (012). This reflections appear when the film has strong orientation of YBCO [110]. Projection indicated by open triangles show the presence of YBCO [103] axis orientation, and the crystal have two growth directions which have angle of 45° and 135° between c-axis of YBCO crystal and x-axis of this figure. Figure 4(b),(c) and (d) show measured stereographic projection of YBCO films grown on the LSGO(110) substrates at each temperature. In these measurement, c-axis of LSGO substrates were parallel to y-axis of these figures. These projections reveal that the axis orientation of the film grown at 650°C is perfect [110], and [103] axis orientation increase with decreasing substrate temperature.

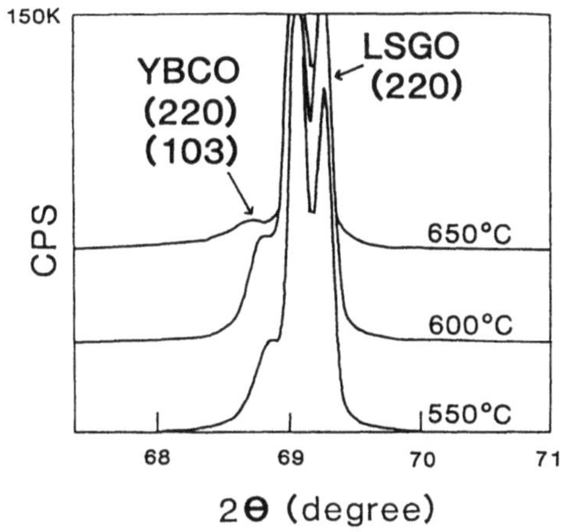

Fig.3 XRD patterns of the YBCO films grown on LSGO(110) substrates.

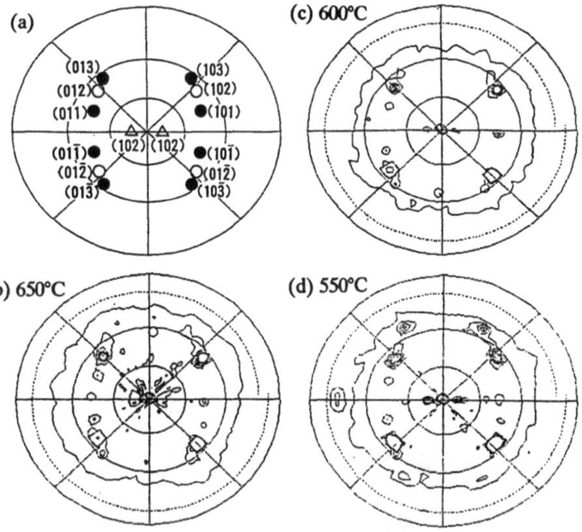

Fig.4 (a) The theoretical stereographic projection of a YBCO film which has [110] and/or [103] axis orientation. (b),(c) and (d) The measured stereographic projection of the YBCO films grown on the LSGO(110) substrates.

From the results as described above, we had discussed the reason why the YBCO doesn't have a [100] orientation on the LSGO (100), but a [110] orientation on the LSGO(110). Several authors explained the mechanism of epitaxial growth from the misfit of axis length[3,4,12,13]. Firstly, we tried to resolve the problem from the same point of view. Table 1 shows the misfit of axis length at 650°C. These calculated values show that the stablest orientation is YBCO [100] on a LSGO (100) and YBCO [110] on a LSGO (110). Since the misfit of axis length didn't agree with the observed results, this view point cannot explain the mechanism of crystal growth in this work.

Table 1 The misfit of the axis length

On the LSGO(100)		On the LSGO(110)	
orientation	misfit(%)	orientation	misfit(%)
YBCO[100]	6.9	YBCO[110]	7.6
[010]	8.5	[103]	8.0
[001]	9.3	[013]	8.8

Then we considered atomic configuration of the substrate and YBCO as follows. Oxygen may combine the cations of substrate with these of YBCO. If this combination is possible, YBCO may show strong epitaxial growth. On the other hand, if it is impossible, YBCO may show natural c-axis orientation.

Figure 5 indicates crystal structures of LSGO and YBCO. The LSGO crystal has K_2NiF_4 structure which is similar to the YBCO. Configuration of oxygen and cation in these both crystal are the same except for coordination number of oxygen around the body centered metal as shown in Fig.5(b) and (c). It was therefore considered that the combination misfit of oxygen and cation was occurred in those parts. Thus we tried fitting between both parts of different structure.

Figure 6 (a) shows an example which is fitted YBCO [100] on LSGO [100]. Oxygen marked with a slash are moved to combine Ga with Y. If a oxygen drawn with a dotted line exists, the coordination number of Ga doesn't change but the number of Y change from 8 to 9. In case of the oxygen defect, the coordination number of Y doesn't change but that of Ga change from 6 to 5. In this way, YBCO [100] orientation on the LSGO (100) substrate causes change of the coordination number. Since the misfit of coordination number could be attributed to unstable YBCO crystal, YBCO didn't grow epitaxialy.

On the other hand, if each four oxygen in YBCO and LSGO marked with a slash is moved as shown in Fig.6 (b), YBCO may grow epitaxialy on the LSGO (110) substrate with sharing these oxygen. In this case, coordination number of oxygen did not change, then the YBCO epitaxial growth might be very stable.

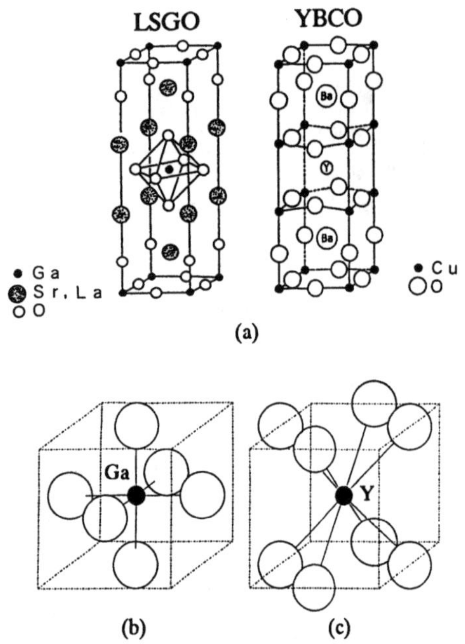

Fig.5 (a) The crystal structures of LSGO and YBCO.
(b) A different part of LSGO crystal structure from YBCO. The body centered cation is Ga, and open circles are Oxygen.
(c) A different part of YBCO crystal structure from LSGO. The body centered cation is Y.

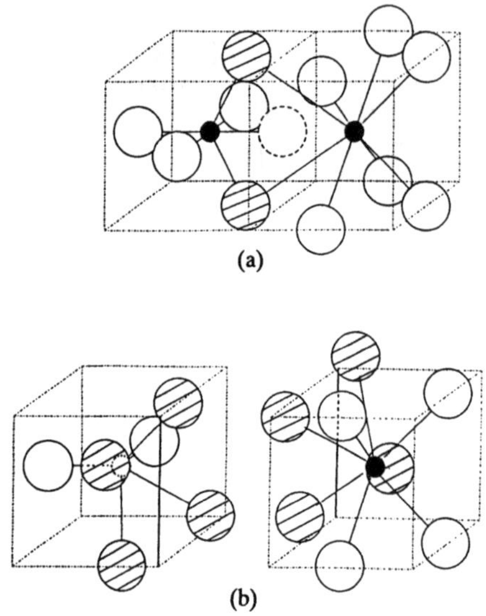

Fig.6 (a) An example of fitting YBCO[100] on LSGO[100].
(b) An example of fitting YBCO[110] to LSGO[110]. Each four oxygen in YBCO and LSGO marked with a slash are moved to fit YBCO[110] on LSGO[110].

CONCLUSION

YBCO thin films grown on the LSGO (100) substrate at 650°C showed c-axis orientation. On the other hand, the films grown on LSGO (110) substrates at 650°C showed the uniform [110], and [103] axis orientation increased with decreasing substrate temperature. The reason of this epitaxial growth were proposed to be the misfit of coordination number of oxygen.

ACKNOWLEDGMENTS

Authors would like to thank Dr. A. Saiki in Tokyo Institute of Technology for his help.

REFERENCES

1. S. W. Tozer, A. W. Kleinsasser,T.Penney, D.Kaiser, and F.Holtzberg (1987) Phys.Rev.Lett. 59 :1768
2. T.K.Worthington, W.J.Gallagher, and T.R.Diger (1987) Phys.Rev.Lett. 59 :1160
3. T.Terashima, Y.Bando, K.Iijima, K.Yamamoto and K.Hirata (1988) Appl.Phys.Lett. 53 :2232
4. Y.-L.He, G.-C.Wang, A.J.Drehman and H.-S.Jin (1990) J.Appl.Phys. 67 :7460
5. M.Muroi, Y.Okamura, T.Suzuki, K.Tsuda, M.Nagano and K.Mukae (1990) Jpn.J.Appl.Phys. 29 :69
6. L.G.Schulz (1949) J.Appl.Phys. 20 :1030
7. Y.Enomoto, T.Murakami, M.Suzuki and K.Moriwaki (1987) Jpn.J.Appl.Phys. 26 :L1248
8. M.Mtsumoto, H.Akoh and S.Takada (1989) J.Appl.Phys. 66 :3907
9. G.Linker, X.X.Xi, O.Meyer, Q.Li and J.Geerk (1989) Solid State Commun. 69 :249
10. A.Gupta, G.Koren, R.J.Baseman, A.Segmuller and W.Holber(1989) Physica C 127 :162
11. T.Terashima, Y.Bando, K.Iijima, K.Yamamoto and K.hirata (1988) Jpn.J.Appl.Phys. 27 :L91
12. X.X.Xi, Q.Li, D.Wu and T.Venkatesan (1990) Appl.Phys.Lett. 57 :1064
13. A.Inam, C.T.Rogers, R.Ramesh, K.Remschnig, D.hart, T.Venkatesan and B.Wilkens(1990) Appl.Phys.Lett. 57 :2484

Effect of an Oxygen Pressure on Crystal Growth of a-Axis Oriented $YBa_2Cu_3O_{7-x}$ Films Prepared by Pulsed Laser Deposition

HIROMI TAKAHASHI, TAKASHI HASE, HIROHIKO IZUMI, KATSUMI OHATA, TADATAKA MORISHITA, and SHOJI TANAKA

Superconductivity Research Laboratory, ISTEC, Koto-ku, Tokyo, 135 Japan

ABSTRACT

The effect of an oxygen partial pressure on growth of a-axis oriented $YBa_2Cu_3O_{7-x}$ films have been investigated by cross-section transmission electron microscopy and scanning electron microscopy. The samples were deposited on $SrTiO_3$(100) substrates at 700°C and oxygen partial pressures of 20, 50 and 100mTorr. Under 20mTorr, a-axis and c-axis oriented domains equally grow over the whole substrate, resulting in the alternate structure of both with a separation of about 150Å in TEM images. The alternate structure continues from the substrate to the film surface. At 50 and 100mTorr, the a-axis oriented domain becomes dominant above a thickness of 40Å and 100Å, respectively. A disordered layer is observed between the a-axis oriented domain and substrate for 50mTorr and the c-axis oriented domain grows on the substrate for 100mTorr. The surface morphology is improved with decreasing the c-axis oriented domain.

KEY WORDS: $YBa_2Cu_3O_{7-x}$, laser deposition, TEM

INTRODUCTION

Since the discovery of superconductive Y-Ba-Cu-O and Bi-Sr-Ca-Cu-O materials, many deposition techniques have been investigated to apply these materials for device applications. The quality of thin films depends on the grain structure and crystalline orientation. Some device applications require a surface smoothness on the atomic scale, which critically depends on the early stage of crystal growth at the substrate surface. The high resolution transmission electron microscopy (HRTEM) is a powerful tool to study a microstructure in thin films. Recently, epitaxy of c-axis oriented Y-Ba-Cu-O films deposited on single-crystal MgO and $SrTiO_3$ substrates has been examined by HRTEM [1-5], especially the growth of a-axis oriented grains in the c-axis oriented domain has attracted much interest.[6-11] However, the dependence of the a-axis oriented growth at the substrate surface on various deposition conditions is scarcely reported.

We have previously reported effects of the oxygen partial pressure on crystalline orientation of films deposited in the range of oxygen partial pressure from 10^{-5} to 5 Torr at a substrate temperature of 700°C. The X-ray measurements showed that a-axis oriented films were preferentially deposited at oxygen partial pressures of 50 and 100mTorr. [12,13]

In this paper, we examine effects of oxygen partial pressure on nucleation and growth of a-axis oriented grains on $SrTiO_3$(100) substrates.

EXPERIMENTAL

The preparation of $YBa_2Cu_3O_{7-x}$ thin films has been previously reported elsewhere [12,13], and is summarized here. ArF excimer laser pulses of 193nm wave length and 10ns duration were used for ablation with a repetition rate of 5Hz. The laser beam was focused on the target down to $1x3mm^2$ to produce a fluence about $1J/cm^2$. Mirror polished $SrTiO_3(100)$ substrates were held at 700°C. After deposition films were cooled in an oxygen pressure of 300Torr. In this experiment, the deposition rates were constant at 1Å/sec (0.2Å/shot) for each condition. The cross-section TEM and scanning electron microscopy (SEM) are used to investigate an interface structure and surface morphology, respectively. Samples suitable for cross-section TEM studies were prepared by a standard procedure of mechanical thinning followed by Ar ion milling at liquid nitrogen temperature. TEM observations were made in a JEM-4000EX microscope capable of achieving a 1.7Å point-to-point resolution. And zero resistance temperatures were determined by a DC four-probe method.

RESULTS AND DISCUSSION

The zero resistance temperatures of films deposited at oxygen partial pressures of 20, 50 and 100mTorr are 76.0, 81.8 and 86.0K, respectively. Figures 1(a), (b), and (c) show cross-section TEM images near the substrate surface for films which were prepared at oxygen partial pressures of 20, 50 and 100mTorr, respectively, and corresponding SEM photographs of the film surface are shown in figs. 2(a), (b) and (c).

In the case of 20mTorr, a-axis and c-axis oriented domains are periodically seen with a separation of about 150Å in fig. 1(a). Judging from this pattern, the volume fraction of a-axis oriented domain is almost constant even near the film surface. There are many defects, in particular, distorted Cu-O chains in the c-axis oriented domains compared with those in the a-axis oriented domains. Furthermore, the texture, as is seen in fig. 2(a), is clearly observed. The pattern of this texture is very similar to what is usually observed in a-axis oriented film [11]. This texture should be formed by coexistence of a-axis and c-axis oriented domains. At this oxygen pressure, the growth rate of a-axis oriented domain would be so small as not to overcome the growth of c-axis oriented domains.

In the case of 50mTorr, the growth of a-axis oriented domains is dominant, restricting the disordered structure domain within 40Å above the substrate surface. The a-axis oriented domains are clearly observed to epitaxially grow from the substrate surface, with a nucleus of about 30Å in diameter as indicated by the arrowhead A in fig.1(b). In this condition, there is not enough oxygen to complete c-axis oriented grains in the early stage of the film growth. We therefore observe disordered structure above the substrate, typically suggested by arrowheads B and C, where the a-axis oriented domain nucleates. It must be noticed that the film surface is very smooth and no grains are observed, as shown in fig. 2(b).

The TEM image of the film deposited at 100mTorr is shown in fig. 1(c). Both the a-axis and c-axis oriented grains grow from the substrate surface like fig.1(a). The volume fraction of c-axis oriented domain however decreases with an increasing thickness. The growth of a-axis oriented domains becomes dominant at a distance of about 100Å from the substrate. In this case, the nucleation and growth process of the c-axis oriented domain would be accelerated by enough ambient oxygen compared with the case of 50mTorr.

Therefore, the c-axis oriented grains are seen below the a-axis oriented domain. The film surface is still smooth, though large particulates of about 2000Å in size are observed as in fig. 2(c).

CONCLUSION

The oxygen partial pressure dependence of growth of a-axis oriented $YBa_2Cu_3O_{7-x}$ films has been investigated by cross-section transmission electron microscopy and scanning electron microscopy. The growth rate of a-axis oriented grain critically depends on the oxygen partial pressure. Under 20mTorr, the a-axis oriented grains grow with a growth rate as well as the c-axis oriented grains do. The a-axis and c-axis oriented domains alternately grow from the substrate surface with a separation of about 150Å. This mixed structure continues from the substrate to the surface of film. For cases of 50 and 100mTorr, the a-axis oriented domain overcomes other kinds of domains above a thickness of 40Å and 100Å, respectively. The disordered structure and the c-axis oriented domain are observed to exist between the substrate and the a-axis oriented domain, respectively. The surface morphology was much improved when the c-axis oriented domain disappeared in films.

REFERENCES

[1] R. Ramesh, D.Hwang, T. S. Ravi, A. Inam, J. B. Barner, L. Nazar, S. W. Chan, C. Y. Chen, B. Dutta, T. Venkatesan, and X. D. Wu, Appl. Phys. Lett., **56** (1990) 2243.

[2] D. M. Hwang, T. S. Ravi, R. Ramesh, Siu-Wai Chan, C. Y. Chen, and L. Nazar, Appl. Phys. Lett., **57** (1990) 1690.

[3] A. Lisa and C. B. Carter, D. K. Lathrop, S. E. Russek, R. A. Buhrman, and J. R. Michael, J. Mater. Res., **4** (1989) 1072.

[4] O. Eibl, H.E. Hoenig, J. M. Triscone, Ø. Fischer, L. Antognazza and O. Brunner, Physica C, **172** (1990) 365.

[5] M. Tomita, T. Hayashi, H. Takaoka, Y. Isshii. Y. Enomoto and T. Murakami, Jpn. J. Appl. Phys., **27**, 636 (1988).

[6] O. Eibl, H.E. Hoenig, J. M. Triscone, Ø. Fischer, L. Antognazza and O. Brunner, Physica C, **172** (1990) 373.

[7] R. Ramesh, C.C.Chang, T. S. Ravi, D. M. Hwang, and A. Inam, X.X.Xi, Q. LI, X. D. Wu, and T. Venkatesan, Appl. Phys. Lett., **57** (1990) 1064.

[8] D. M. Hwang, T. Venkatesan, C. C. Chang, L. Nazar, X. D. Wu, A. Inam, and M. S. Hegde, Appl. Phys. Lett., **54** (1989) 1702.

[9] C. W. Nieh and L. Anthony, Appl. Phys. Lett., **56** (1990) 2138.

[10] Y. F. Hsieh, M. P. Siegal, R. Hull, and J. M. Phillips, Appl. Phys. Lett., **57** (1990) 2268.

[11] H. Takahashi, Y. Aoki, T. Usui, R. Fromknecht, T. Morishita and S. Tanaka, Physica C, **175** (1991) 381.

[12] H. Izumi, K. Ohata, T. Hase, K. Suzuki, and T. Morishita, Proceedings of the 2nd International Symposium on Superconductivity(ISS'90) (1990) 861.

[13] T. Hase, H. Izumi, K. Ohata, K. Suzuki, and T. Morishita and S. Tanaka, J. Appl. Phys.,**68** (1990) 374.

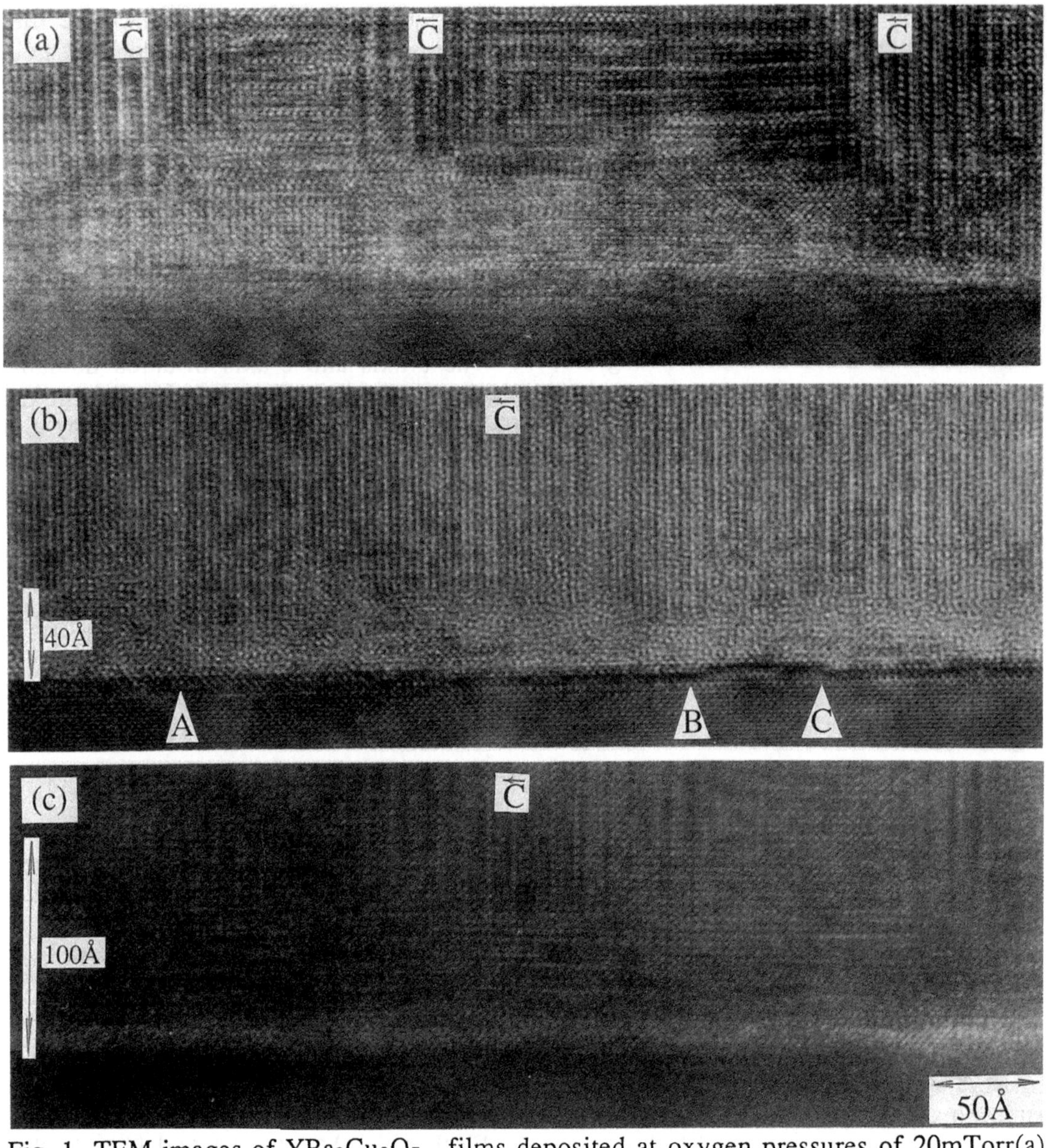

Fig. 1 TEM images of $YBa_2Cu_3O_{7-x}$ films deposited at oxygen pressures of 20mTorr(a), 50mTorr(b) and 100mTorr(c), respectively. These images are taken along the $SrTiO_3$<010> direction. The a-axis oriented domain is clearly observed to epitaxially grow from the substrate surface, with a nucleus of about 30Å in diameter as indicated by the arrowhead A. The distorted structure exists between the arrowhead B and C in fig.1(b).

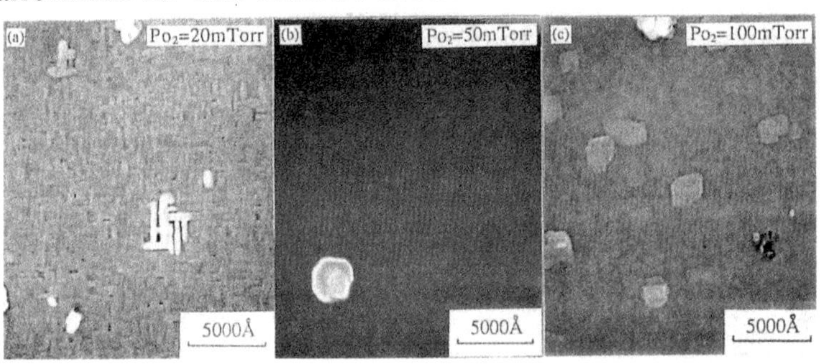

Fig. 2 SEM images of $YBa_2Cu_3O_{7-x}$ films deposited at the oxygen pressures of 20(a), 50(b) and 100mTorr(c), respectively.

Effect of Ozone as Active Oxdizing Gas on YBCO Film Deposition by Laser Ablation

SHUNSUKE HOSOKAWA*, CEHUI LI, EIKI NARUMI, SUSHIL PATEL, and DAVID T. SHAW

New York State Institute on Superconductivity, State University of New York at Buffalo, Buffalo, NY 14260, USA

ABSTRACT

The effect of ozone on YBCO film deposition by laser ablation is tested by introducing ozone containing oxygen gas (ozone:5 mol%). The other conditions were kept same as the ones producing high quality c-axis oriented films by pure oxygen with Tc > 85 K, and Jc > 10^6 A/cm^2. The surface morphology of the films by ozone changed to be smoother. However the films by ozone contained a-axis oriented grains and Jc decreased to 10^5 A/cm^2. It seemed the film growth mechanism was affected by the oxidizing process enhanced by ozone.

KEY WORD: Laser ablation, Ozone, Growth mechanism, Surface morphology

INTRODUCTION

$YBa_2Cu_3O_{7-x}$ (YBCO) films with high superconducting properties of Tc > 85 K, and Jc > 10^6 A/cm^2 are produced by many groups using the laser ablation technique (1). However many applications of YBCO film such as multi-layered and integrated circuit devices require smooth surface. The active oxidizing gases such as N_2O were used to lower the substrate temperature in the laser ablation method (2). Yet it has not reported the enhancement of film quality by active oxidizing gases nor their role on growth mechanism. Ozone has the most powerful oxidizing agent among such gases as N_2O, NO_2, O_2 and atomic oxygen used for the deposition of surperconducting oxides film (3), and pure ozone gas was successfully used in the preparation of superconducting oxide films by the molecular beam epitaxy method (4,5). In the laser ablation method, the deposition chamber pressure is in the range of several tens to several hundreds m·Torr, and the ozone containing oxygen gas produced simply by using the ozonizer can provide enough ozone molecular flux to the substrate. The ozonizer was installed in the gas feed line of the deposition chamber, and the effect of ozone on the film quality was tested, especially aiming to produce films with improved surface morphology, and also examine the change of the film growth mechanism by the oxidizing process.

EXPERIMENTAL

The standard deposition chamber of laser ablation was used to deposit YBCO film on (100) MgO substrates, as shown in Fig. 1. In addition, the ozonizer was installed in the gas feed line, and the quartz tube was used at the end of the gas feed line near the heater block so that the decomposition of ozone on the tube wall was avoided. The ozonizer (Masuda Research Inc., OZC-20S-05-5SW) produced about 5 mol% of ozone from oxygen gas. The effect of ozone could be tested by simply switching the ozonizer on, which is referred as the "ozone mode" in this paper. When the ozonizer was off, the ordinary deposition using pure oxygen could be performed, referred as the "oxygen mode". The MgO substrates used in this experiment were simply mechanically polished and then chemically cleaned by acetone. The deposition conditions are shown in Table 1, which were adjusted for the "oxygen mode". to obtain high superconducting properties (Tc > 85 K, Jc > 10^6 A/cm^2). The laser pulse frequency was changed from 2 Hz up to 60 Hz with keeping the same pulse energy per a shot to alter the deposition rate. As the result, the growth mechanism was tested from the view points of the deposition rate and the oxidizing effect.

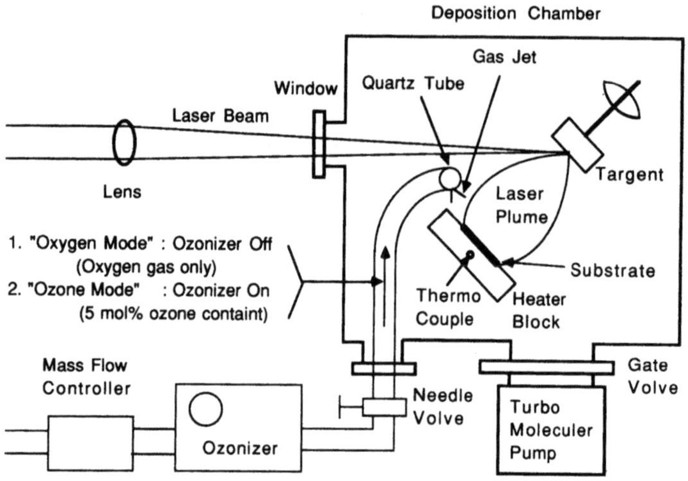

Fig. 1 Schematic diagram of the deposition chamber with an ozonizer

Table 1
Deposition Conditions

Laser	Ar-F (193 nm)
Power	220 mJ
Energy Density	1.5 J / cm^2
Substrate	(100) MgO
Temperature	≈ 600 °C
Target-Substrate Distance	80 mm
Mass Flow Rate of Oxygen Gas	90 sccm
Chamber Pressure	40 mTorr

RESULTS AND DISCUSSION

The physical properties of the films obtained in these series of experiments are summarized in Table 2. All the films prepared by the "oxygen mode" showed reasonably high Tc and Jc as the films deposited on the simply mechanically polished MgO substrates (6). In this case the deposition rate was changed from 1.0 to 26.4 Å/sec by changing the laser pulse repetition frequency. As shown in Fig.2, there is no linear relationship between the deposition rate and the laser pulse repetition frequency. In this figure, also shown the cases of the films deposited by the "ozone mode". It seems that the deposition rate is not only affected by the materal supply rate from the laser plume but also the film growth mechanism such as nucleation and surface migration of molecules in addition to oxidization.

In the case of films deposited by the "ozone mode", Jc decreased in one order magnitude as compared with those by the "oxygen mode" (Table 2). Figure 3 shows the results of the XRD measurements of the films by the "oxygen mode" and the "ozone mode". It is apparent that the film deposited by the "ozone mode" contained a-axis oriented grains, while the one by the "oxygen mode" was perfectly c-axis oriented. A the result, the supur-

Table 2 Summary of physical properties of films

Sample	Deposition Conditions "Mode"	Laser Frequency (Hz)	Deposition Time (min)	Tc (K)	Jc (A/cm^2)	Thickness (nm)	Deposition Rate (Å/s)	SEM Picture
1	Oxygen	2	30	85.6	1.2×10^6	180	1.0	Fig. 4(a)
2	Oxygen	10	10	86.6	5.6×10^5	235	3.9	Fig. 4(b)
3	Oxygen	30	5	84.6	1.9×10^6	290	9.7	Fig. 4(c)
4	Oxygen	60	3	85.2	7.4×10^5	475	26.4	Fig. 4(d)
5	Ozone	30	5	82.0	2.3×10^5	405	13.5	Fig. 5(a)
6	Ozone	60	3	84.1	7.9×10^4	410	22.8	Fig. 5(b)

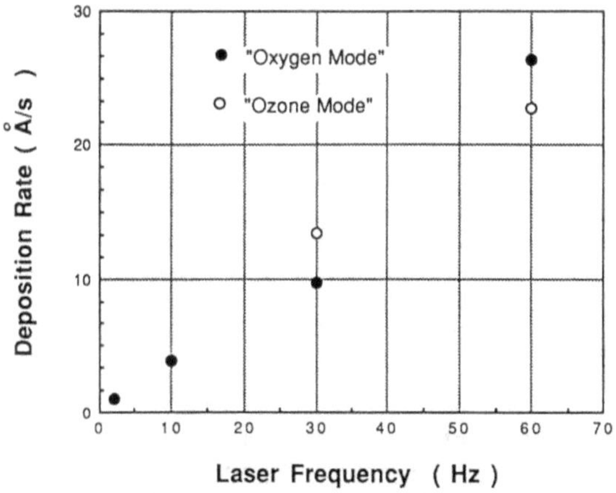

Fig. 2 Deposition rate as a function of laser frequency

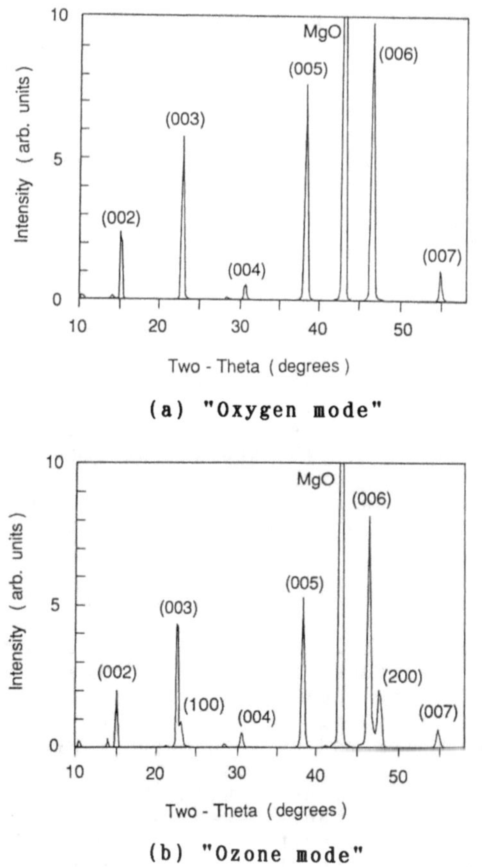

Fig. 3 XRD measurements of films

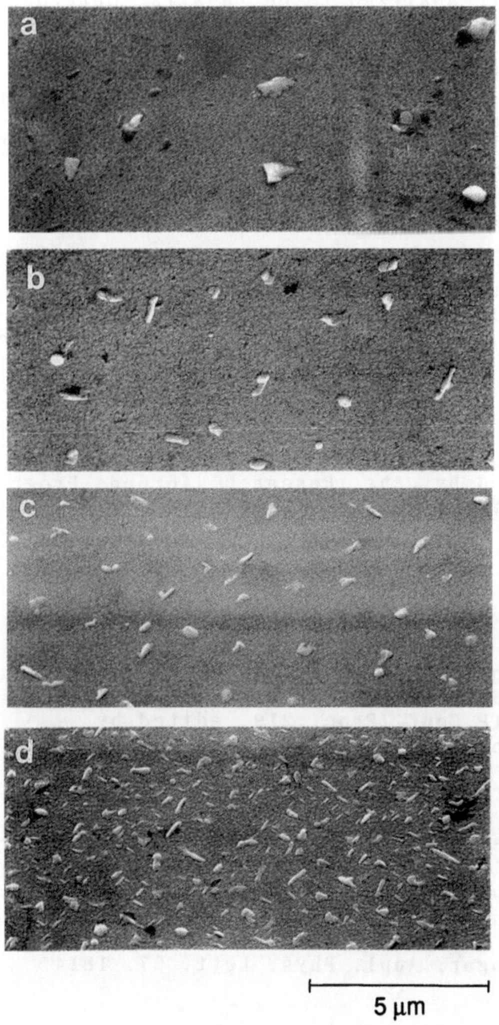

Fig. 4 SEM pictures of films ("Oxygen mode"), (a) 2 Hz, (b) 10 Hz, (c) 30 Hz, (c) 60 Hz

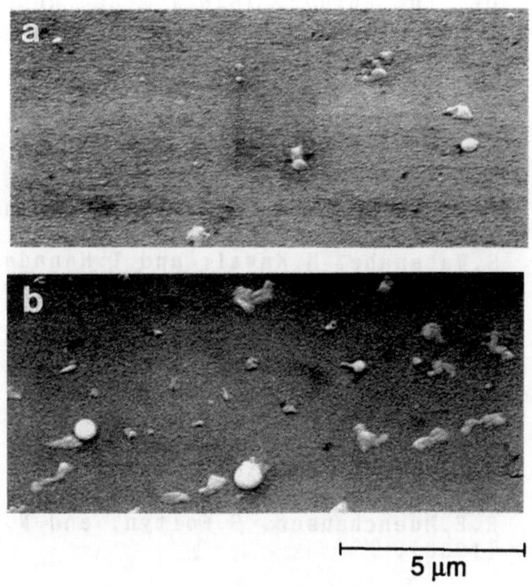

Fig. 5 SEM pictures of films ("Ozone mode") (a) 30 Hz, (b) 60Hz

conducting properties, especially Jc decreased. As mentioned earlier, the deposit conditions were optimized for the "oxygen mode" and were not changed for the "ozone mode". We are not yet optimized the deposition condition for the "ozone mode". However, it is interesting that only 5 mol% of ozone content in the deposition gas affected the growth mechanism. The difference is assumed to be oxidizing power between oxygen and ozone, because the pressure of the deposition chamber were same in both cases of the "oxygen mode" and the "ozone mode", and as a consequence, the laser plume did not change. Even though the 5 mol% ozone contained gas is expected to supply ozone molecules to the substrate at the rate of more than 1 x 10^{17} molecules/$cm^2 \cdot s$. At the surface of the substrate heated to 600 °C, ozone is easily decomposed to oxygen and atomic oxygen which oxidize the metal elements. This oxidizing condition seemed to change the growth mechanism especially the crystal orientation.

Figure 4 shows the SEM pictures of the films deposited by the "oxygen mode". As the deposition rate was increased, the surface became rough. The irregular shaped higher contrast particles in these pictures increased in number with the deposition rate. These particles were not from the target, but "outgrowth" defined by C.C.Chang, et al (7). Especially, at the highest deposition rate, the needle-shaped outgrowth with 90 degrees angle orientation became significant which were analyzed to be a-axis "outgrowth" (7).

Figure 5 shows the SEM pictures of the films by "ozone mode". As compared with those of Figs. 4 (c) and (d), the number of the particles were significantly decreased. It is not clear yet at this stage that the reason of decrease of the particles is originated from the difference of the oxidizing power between oxygen and ozone. However, it can be concluded that the oxidizing process should be taken in account in discussing the growth mechanism and surface morphology of the superconducting oxide film.

In summery, even with the 5 mol% ozone containing oxygen gas changed the growth mechanism such as crystal orientation and surface morphology. The optimum deposition conditions using this ozone containing gas are under developing to clear the effect of ozone, especially aiming to get the good surface.

ACKNOWLEDGEMENT

One of the author (S.H.) was supported by the Research Abroad Program conducted Science and Technology Agency of the Japanese Government.

REFERENCES

(1) See, for example, numerous papers in Appl. Phys. Lett. 53-57.
(2) G.Koren, A.Gupta, and R.J.Baseman, Appl. Phys. Lett. 54, 1920 (1989).
(3) H.Koinuma, M.Yoshimoto, H.Nagata, T.Hashimoto, T.Tsukahara, S.Gonda, S.Watanabe, M.Kawai, and T.Hanada, AIP Conf. Proc. 219, edited by Y.Kawo, P.Corpens, and H.Kwok, American Institute of Physics, New York, 326 (1990).
(4) D.D.Berkley, B.B.Johnson, N.Anand, K.M.Beauchamp, L.E.Conroy, A.M.Goldman, J.Maps, K.Mauersberger, M.L.Mecartney, J.Morton, M.Tuominen, and Y-J.Zahng, Appl. Phys. lett. 53, 1973 (1988).
(5) S.Hosokawa and S.Ichimura, Rev. Sci. Instrum. 62, 1614 (1991).
(6) B.H.Moeckly, S.E.Russek, B.K.Lathrop, R.A.Buhrman, J.Li, and J.W.Mayer, Appl. Phys. Lett. 57, 1687 (1990).
(7) C.C.Chang, X.D.Wu, R.Ramesh, X.X.Xi, T.S.Ravi, T.Venkatesan, D.M.Hwang, R.E.Muenchausen, S.Foltyn, and N.S.Nogar, Appl. Phys. Lett. 57, 1814 (1990).

* On Leave from Electrotechnical Laboratory, 1-1-4, Umezono, Tsukuba, Ibaraki 305 Japan

DC Bias Effect Using Mesh Ring Electrode on Properties of YBCO Superconducting Thin Film by Laser Evaporation

TATSUOKI NAGAISHI and HIDEO ITOZAKI

Itami Research Laboratories, Sumitomo Electric Industries, Ltd., Itami, Hyogo, 664 Japan

ABSTRACT

We examined a dc bias effect of excimer laser pulse evaporation with a mesh electrode. Deposition was performed on MgO (100) substrates at substrate temperature of 600°C. With bias voltage of +75V and +100V on the electrode, significant changes of orientation and improvement of the surface morphology were observed, and the superconducting transition temperature was also increased from 20K up to 75K.

KEY WORDS: YBCO thin film, laser evaporation, dc bias

INTRODUCTION

One of the key issues for the film formation is low deposition temperature. This is because of the problems of diffusion of elements from the substrate into the film or the difference between the expansion coefficient of the substrate and the film. Singh et al. [1] reported that reduction of the substrate temperature (Ts) by about 50°C was achieved by using dc bias voltage of +300V on an interposing ring. Witanachchi et al. [2] also reported Ts reduction by using the same technique.
We report on the dc bias effect at a Ts of 600°C on a (100)MgO substrate by using a mesh ring as an electrode between the target and the substrate. This 600°C is relatively low temperature for the preparation of oxide superconducting thin films. The bias was applied between the mesh ring electrode and the target. Change of crystallinity, surface morphology and also critical temperature of the films were examined.

EXPERIMENTAL

A 193nm ArF excimer laser (Lambda Physik LPX315i) was used for film preparation. The laser pulse duration was 20 nsec and the rate of pulse shot was 1 Hz. The laser beam was focused to about 1.5mm x 5mm on the target, which was a 1.5cm diameter sintered pellet of $YBa_2Cu_3O_{7-x}$. The laser power density was about 0.7J/cm^2 at the surface of the target. A 10mm x 10mm x 1mm thick (100)MgO was used as the substrate. 50 Pa was maintained with oxygen gas during the deposition. Since the gas inlet was sufficiently far from the target and the substrate, it is assumed that the gas pressure was almost the same at both the target and the substrate. The Ts was 600°C measured by a thermocouple clipped on the surface of the substrate. The validity of the Ts in each experiment was doubly checked by the current through the resistive heater. The distance between the target and the substrate was 40mm. A mesh ring electrode made of platinum was located between the target and the substrate. The shape and the size of the mesh ring is shown in Fig. 1. Both the mesh electrode and the target were electrically isolated from the chamber and the ground. The substrate holder was also electrically isolated from the chamber. Two positions for the mesh electrode were chosen for the deposition. One was 10mm from the target and the other was 27mm from the target as shown in Fig. 2. The bias voltage for the mesh ring was varied from -300V to +300V. In this configuration, the plume extended from the target to the substrate through the mesh ring electrode. As Sigh et al. mentioned [1], no extra plasma from the discharge of the gas between the mesh electrode and the target was observed in the above gas pressure region and the geometry of the dc bias system. The deposition was performed for 20 minutes. The film thickness was measured at 200nm to 300nm. After the deposition, Ts decreased to 200°C within 4 minutes with the gas pressure in the chamber reaching 1 atm. At higher Ts of more than 675°C, we can easily obtain a highly c-axis oriented film with Tc above 90K on the (100)MgO substrate. In this experiment, the crystallinity of the film was examined by an X-ray diffractometer. A field emission scanning electron microscope (FE-SEM) was also used for observation of the surface morphology.

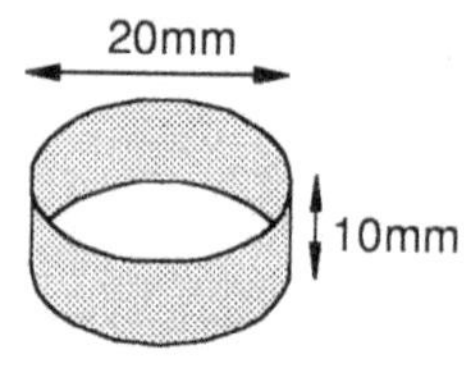

Fig.1. The shape and the size of the mesh ring

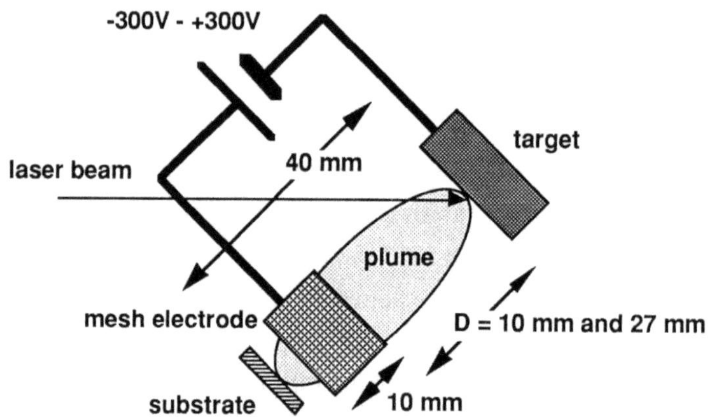

Fig.2. Schematic diagram of dc bias system

RESULTS

The X-ray diffraction spectrum in Fig. 3 shows the crystallinity change for applying the dc voltage. Without dc bias voltage, the film has a mixture of (100), (001) and (103), (013) or (110) orientations. When the bias voltage was increased to +75V, crystal orientation of the specimen became mainly (001). Furthermore, when the bias voltage was increased up to +150V, the orientation of the specimen changed to (100) orientation with small (103), (013) or (110) peaks.

Surface morphology of the specimens with and without a bias voltage of +75V was compared by SEM. The surface morphology of the film deposited without the bias is fairly rough as shown in Fig. 4(a). It indicates that the (100) orientation fragments grow across each other and that the (013), (103) or (110) orientation fragments are triangular. In contrast, the surface morphology deposited with a bias voltage of +75V is rather smooth as shown in Fig. 4(b). The density of the misoriented fragments seems to be

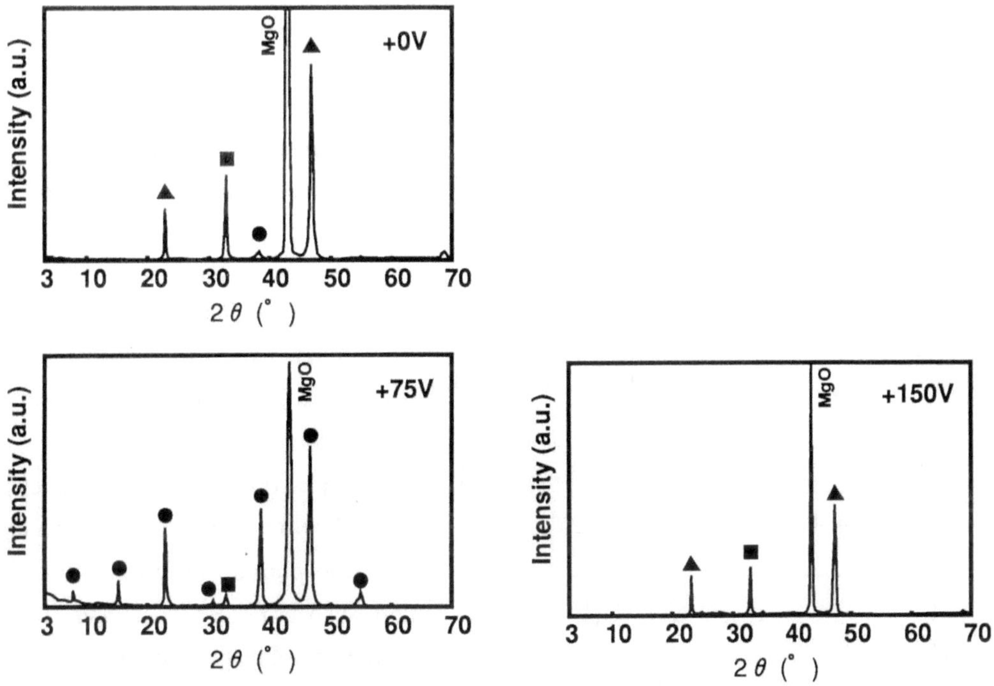

Fig.3. Crystallinity change by various dc bias voltage

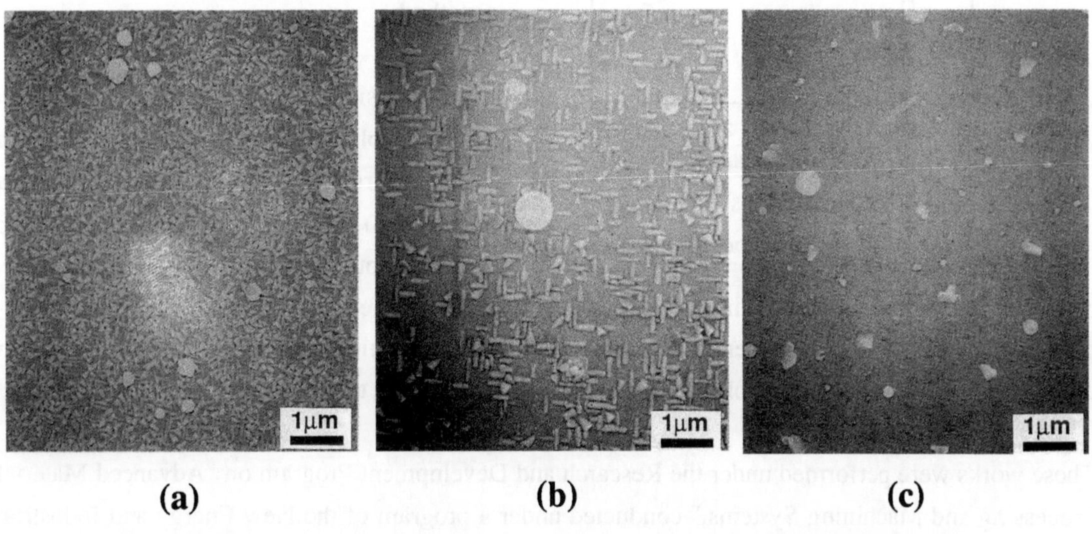

Fig.4. Surface mophology of the specimens: (a) without bias, (b) with a bias voltage of +75V and (c) with a bias voltage of +100V.

reduced compared with the specimen deposited without bias voltage. The specimen with a dc bias of +100V shows a fairly smooth surface as seen in Fig. 4(c).

We also examined the electrical property of the films. Figure 5 shows the temperature dependence of normalized resistance with a bias voltage parameter from +0V to +150V. The mesh electrode is located at 27mm from the target. At the bias voltages of +75V and +100V, Tc improved up to 75K. Tc

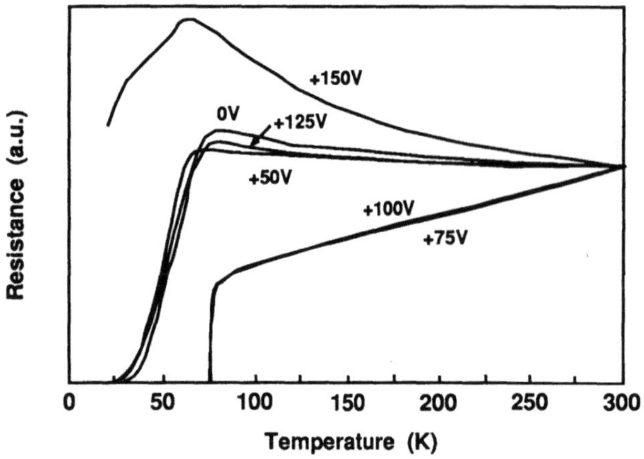

Fig.5. Temperature dependence of normalized resistance with bias voltage parameter

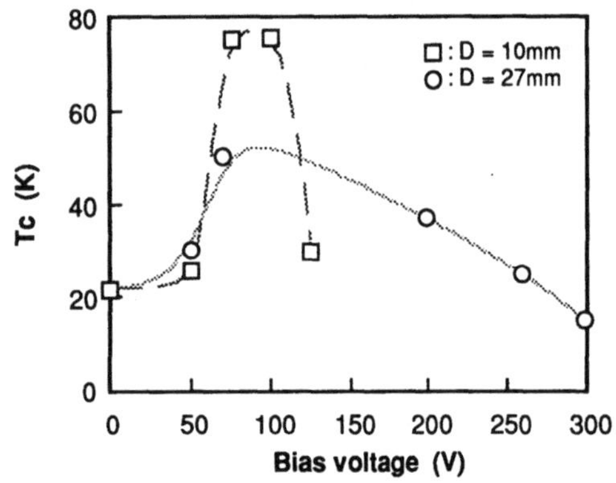

Fig.6 Tc versus bias voltage at two mesh electrode positions

versus bias voltage is plotted in Fig.6 (square) in the above mesh electrode position. Circles in Fig.6 show the data with the mesh electrode 10mm from the target. This result also shows a raise in Tc near the bias voltage of +100V. When the negative bias was applied on the mesh electrode, the specimens showed complete semiconducting behavior.

Since optical emission studies [3, 4] show the existence of the cations in the plume, the elimination of the cations such as Y+, Ba+, Cu+ and YO+ from the target might result in improvement of the film quality at low Ts of 600°C. Further study is needed to elucidate the effect of the dc bias method.

In conclusion, we showed the dc bias effect of laser ablation using a mesh electrode. Tc was improved from 20K to 75K on a (100)MgO substrate at a Ts of 600°C. The Tc was optimized in two different positions of the mesh electrode, both around a bias voltage of +100V. The crystallinity of the film was improved from the mixture of several orientations to c-axis orientation with a small amount of other orientations. The surface mophology was also improved by using the dc bias.

These works were performed under the Research and Development Program on "Advanced Material Processing and Machining Systems," conducted under a program of the New Energy and Industrial Technology Development Organization.

REFERENCES

1. R. K. Singh, L. Ganapathi, P. Tiwari, and J. Narayan (1989) Appl. Phys. Lett. 55: 2351

2. S. Witanachchi, H. S. Kwok, X. W. Wang, and D. T. Shaw (1988) Appl. Phys. Lett. 53: 234

3. X. D. Wu, B. Dutta, M. S. Hegde, A. Inam, T. Venkatesan, E. W. Chase, C. C. Chang, and R. Howard (1989) Appl. Phys. Lett. 54: 179

4. C. Girault, D. Damiani, J. Aubreton, and A. Catherinot (1989) Appl. Phys. Lett. 55: 182

Preparation of a-Axis Oriented YBCO Films by Template Procedure

MIZUSHI MATSUDA, AKIRA SHOJI, HIDEKAZU MATSUHASHI*, and MASAO KOYANAGI

Electrotechnical Laboratory, Tsukuba, Ibaraki, 305 Japan

ABSTRACT

A-axis oriented Y-Ba-Cu oxide (YBCO) films with good crystalline quality and high T_c's comparable to those of c-axis oriented films were successfully prepared on (100) MgO and $SrTiO_3$ substrates by off-axis rf-magnetron sputtering. To grow the YBCO films in the a-axis direction, we adopted a template procedure in which films were grown at a low substrate temperature (T_s) in an initial deposition stage and at a high T_s in the residuary deposition stage. By this procedure, we could obtain a-axis films with critical temperatures (T_c) of 75K and 77K on MgO and $SrTiO_3$ substrates, respectively.

KEY WORDs : a-axis oriented film, template layer, (100) substrate

INTRODUCTION

There has been much interest in the development of multilayered junctions (SNS and SIS junctions) with superconducting YBCO films as electrodes. For such application of YBCO films, it is more attractive to use [100] a-axis oriented or [110] oriented films as the electrode material than [001] c-axis oriented films since the larger coherence length in the ab-planes will prevent a severe degradation of the order parameter near the interface[1]. Up to the present, two approaches have been widely examined to obtain such ab-plane oriented YBCO films. One is to grow films on (110) substrates. In this method, however, films grow on the substrates in both [110] and [103] directions[2]. The existence of an angle of about 45° between the ab-plane and the surface of [103] oriented films will be a disadvantage for junction applications[3]. The other is to grow films on (100) substrates at low substrate temperature (T_s)[4-6]. However, especially on (100) MgO substrates, there have been few successful results which demonstrate production of a-axis oriented films having the same quality as c-axis oriented films. It is likely that problems arise from a defect structure contained in a-axis oriented films grown at lower T_s.
On the other hand, it is reported that once a thin a-axis oriented Pr-Ba-Cu oxide (PBCO) layer is grown on a (100) $SrTiO_3$ substrate, an a-axis oriented YBCO film can be grown on it even at a high T_s where YBCO films usually grow in the c-axis direction[7]. The a-axis oriented PBCO layer is called as a "template" which forces YBCO films to grow in the a-axis direction.
In the present study, we adopted thin layers of YBCO itself as the template to prepare a-axis oriented YBCO films with improved superconducting properties. We also adopted not only (100) $SrTiO_3$ but also (100) MgO as the substrate material because the latter is more attractive for practical applications.

EXPERIMENTAL

YBCO thin films were deposited by rf magnetron sputtering with a sintered target with a stoichiometric composition ($YBa_2Cu_3O_x$). Substrates were mounted on the holder placed off-axis (90°) to the target as sketched in Fig. 1. During the deposition of films, the holder was rotated to

On leave from Chiba Inst. of Tech., 2-17-1 Tsudanuma, Narashino, Chiba, 275 Japan

obtain uniform compositional distribution over the prepared films. The sputtering gas was a mixture of argon and 20% oxygen, and total gas pressure was about 4 Pa. The T_s and rf power densities were varied from 560 °C to 730 °C and from 1.25 w/cm^2 to 2 w/cm^2, respectively.

In the template procedure, the deposition was started at the optimum T_s (590 °C) for a-axis growth without template layers. After the formation of 20-35 nm-thick YBCO films, T_s was increased up to 710 °C and 120-200 nm-thick films were deposited continuously on the initial layers. The deposition rate was 1-2 nm/min for the both steps. After the deposition of films, oxygen gas was let into the chamber up to 1 atm and the films were cooled down to room temperature in 2 hours.

The composition of the films was determined by inductively coupled plasma emission spectroscopy (ICP). The crystallinity of the films was analyzed by X-ray diffraction (XRD) and reflection high energy electron diffraction (RHEED). Surface morphology was observed by scanning electron microscopy (SEM). The electrical resistance of the films was measured by a conventional four-point probe method with silver paste electrodes.

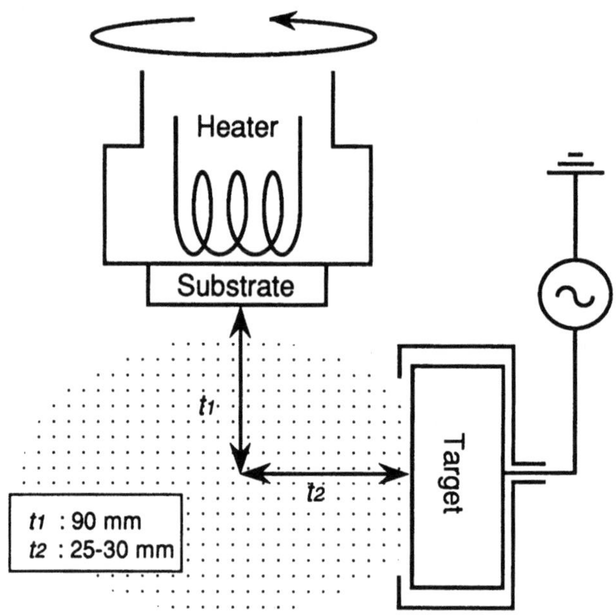

Fig. 1. Schematic drawing of the off-axis sputtering system.

RESULTS AND DISCUSSIONS

First, we investigated the variation of crystal growth orientation with T_s. The relationship between the T_s and the preferential orientation of films on (100) MgO substrates is illustrated in the upper part of Fig. 2. The crystallization temperature is around 570 °C. With increasing the substrate temperature, the orientation of films grown on (100) MgO substrates changes from the a-axis to the c-axis through the [110] orientation including a/c mixture at the temperature range of 610-630 °C. This temperature range is significantly lower than that(660-680 °C) on (100) $SrTiO_3$ substrates. We believe this difference is due to the good lattice matching of $SrTiO_3$ with YBCO. In the case of $SrTiO_3$ substrates, even at relatively high growth temperature, structural coherency at the interfaces of films and substrates may prevent the species from moving to the sites for c-axis orientation and leads to the formation of a-axis oriented films. Figure 2 also shows the dependence of Ba/Y and Cu/Y atomic ratio on T_s. While the Ba/Y ratio remains at the same value, the Cu/Y ratio changes at around 630 °C above which c-axis

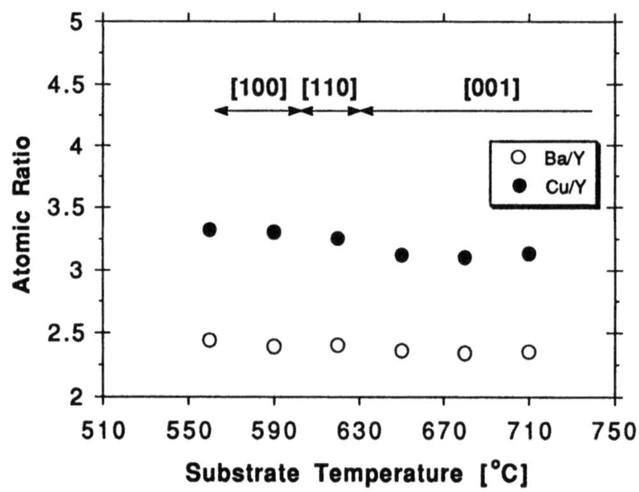

Fig. 2. Variation of composition ratios of Ba/Y and Cu/Y, and the preferential orientation for films deposited on (100) MgO substrates with T_s (substrate temperature).

oriented films were obtained. This change can interpreted as the results of reevaporation during c-axis growth[4].

Figure 3 and Fig. 4 show the XRD patterns and the resistance-temperature dependences for YBCO films deposited on (100) MgO substrates at three different deposition conditions. The film denoted by (a) was deposited at a T_s of 590 °C without using the template procedure. Although the film was grown in the a-axis direction, it has a poor crystalline quality and a low T_c (25K). The film denoted by (b) was deposited by the template procedure. As shown in Fig. 3, the film was grown with a strongly a-axis oriented structure. This shows that the initial deposition layer of the YBCO film acted as a template and made it possible to grow the film at a high T_s with a good crystalline property. The lattice constant determined from the diffraction angle for the [200] plane was 0.382 nm consistent to the lattice constant of tho a dircction for a bulk. The film showed a metallic resistance-temperature dependence like that for a c-axis oriented film (denoted by (c)) which had been deposited on the substrate at a T_s of 710 °C without using the template procedure.

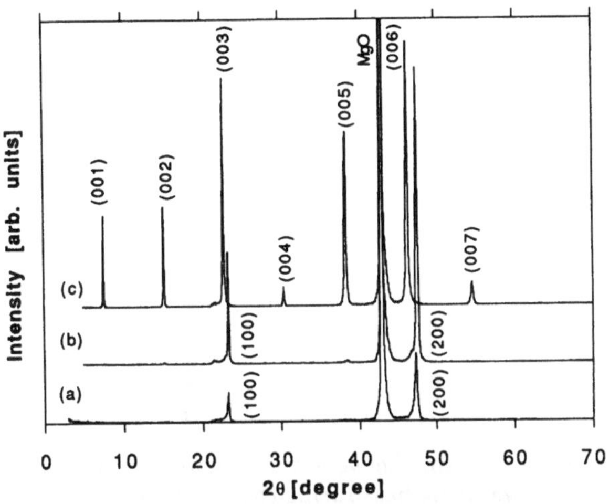

Fig. 3. X-ray diffraction patterns of YBCO films deposited on (100) MgO substrates (a) at 590 °C without using the template procedure, (b) at 710 °C using the template procedure, and (c) at 710 °C without using the template procedure.

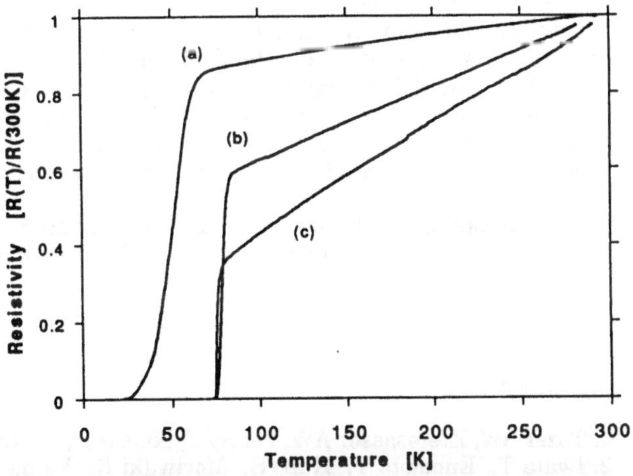

Fig. 4. Temperature dependences of the resistance for YBCO films deposited on (100) MgO substrates (a) at 590 °C without using the template procedure, (b) at 710 °C using the template procedure, and (c) at 710 °C without using the template procedure.

Figure 5 shows the relationship between the T_c values and the rf power densities during the sputter deposition for a-axis films prepared on (100) MgO and $SrTiO_3$ substrates by the template procedure. Critical temperature of the films on MgO substrates increases with increasing rf power in contrast with that on $SrTiO_3$ having no dependence. While the reason why there is a marked difference in T_c values between MgO and $SrTiO_3$ below the rf power of 2 W/cm^2 has not been clear, we infer that the crystalline quality of the template layer would be improved with increasing rf power densities. A-axis oriented films with T_c of 75K and 77K were obtained at the rf power of 2 W/cm^2 on MgO and $SrTiO_3$ substrates, respectively. These T_c values are comparable to those of c-axis oriented films prepared at 710 °C without using the template procedure.

The surface of a-axis oriented films was not smooth especially for films deposited on the MgO substrates. According to the scanning tunneling microscopy measurement in a separate work, a maximum feature height was about 20 nm[8]. The RHEED pattern of an a-axis oriented film is shown in Fig. 6. It looks spotty, indicating that the surface is not smooth. When being rotated by 90° around normal axis, the film showed the same pattern. This indicated that the c-axis of the film is lying on the a-plane with two preferred directions rotated by 90° with respect to each other.

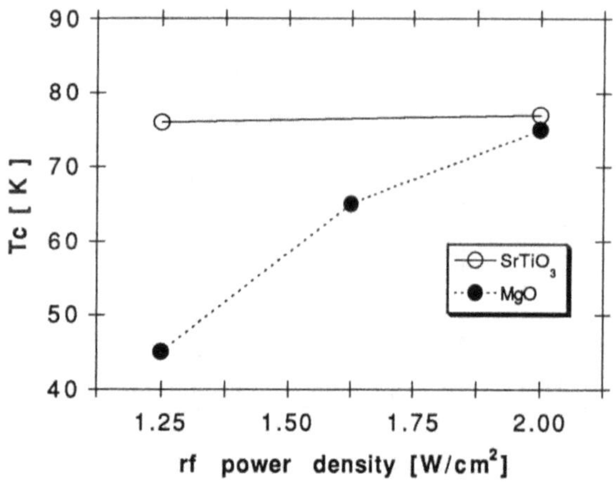

Fig. 5. Relationship between the T_c values and the rf power densities during the sputter deposition for a-axis oriented films prepared on (100) MgO and $SrTiO_3$ substrates by the template procedure.

Fig. 6. Typical RHEED pattern of an a-axis oriented film prepared by the template procedure.

In summary, we found that a-axis oriented YBCO films with good crystalline properties and T_c's (75-77K) as high as those of c-axis oriented films can be grown by a template procedure not only on (100) $SrTiO_3$ substrates but also on (100) MgO substrates. We also found that the sputtering power of the target during the deposition of films is an important parameter for obtaining high T_c's especially for films deposited on the MgO substrates. We believe that the T_c of a-axis oriented films can be more improved by adjusting the composition of films. In fact, we have recently obtained c-axis oriented YBCO films with the maximum T_c of 88K and extremely smooth surfaces by using another target having a less amount of Ba composition. A trial for preparing a-axis oriented films using the new target is currently under way.

We would like to thank Messrs. S. Kashiwaya and F. Hirayama for their effort in the evaluation of film surface by STM measurements. We also thank to Prof. K. Hara of Chiba Institute of Technology for his encouragement.

REFERENCES

1. Tozer SW, Kleinsasser AW, Penny T, Kaiser D, Holtzberg F (1987) Phys. Rev. Lett. **59**: 1768
2. Iwata T, Enomoto Y, Kubo S, Moriwaki K, Yamaji A, in *Proceedings of the second International Symposium on Superconductivity, Tsukuba, 1989* (Springer-Verlag, Tokyo, 1990), p773
3. Lee M, Lew D, Eom CB, Geballe TH, Beasley MR (1990) Appl. Phys. Lett. **57**: 1152
4. Arakawa T, Itozaki H, Harada K, Higaki K, Tanaka S, Yazu S (1990) Jpn. J. Appl. Phys. **29**: L2199
5. Eom CB, Marshall AF, Laderman SS, Jacowitz RD, Geballe TH (1990) Science 249: 1549
6. Asano H, Yonezawa H, Asahi M, Michikami O (1991) IEEE Trans. on Magn. **MAG-27**: 844
7. Inam A, Rogers CT, Ramesh R, Remschnig K, Farrow L, Hart D, Venkatesan T (1990) Appl. Phys. Lett. **57**: 2484
8. Kashiwaya S: *private communication*

123-Structure with Elongated c-Lattice Length in $HoBa_2Cu_3O_x$ Thin Films Prepared at Low Temperatures

AKIRA TSUKAMOTO, MASAHIKO HIRATANI, and SHOICHI AKAMATSU

Central Research Laboratory, Hitachi Ltd., Kokubunji, Tokyo, 185 Japan

ABSTRACT

Structual defects in HBCO films grown at low temperature (500-620 ℃) were investigated by quenching the as-deposited films from the growth temperatures. It is revealed that the quenched films have an extreamly long *c*-lattice length of 1.2 nm. The *c*-lattice length is shortened by incorporating oxygen into the crystal lattice; however, the originally elongated *c*-lattice length cannot be reproduced by quenching the film again from the growth condition. Oxygen ions other than those on the one-dimensional Cu-O chain seem to be partially deficient in the as-deposited and quenched films.

KEY WORDS: $HoBa_2Cu_3O_x$, thin film, *c*-lattice length, oxygen defects

INTRODUCTION

Since the discovery of high-temperature superconducting oxides, there has been much research on growing high-quality thin films for electronic applications. Recently, various thin-film growth techniques such as reactive evaporation, laser ablation, sputtering, and chemical vapor deposition, have been developed for *in situ* growth of superconducting thin films. In general, thin films of $LnBa_2Cu_3O_x$ (Ln=Y, Ho, etc.) superconducting oxides grow in a tetragonal structure with oxygen vacancies on the one-dimensional Cu-O chain, and then the oxygen ions are incorporated during the cooling to room temperature. In low-temperature growth, the structual defects of cations like stacking faults may be introduced as well as the oxygen defects. Moreover, there may be oxygen vacancies other than those on the Cu-O chain, as the oxygen pressure during the growth is lowered. This study investigates the defects of oxygen ions in $HoBa_2Cu_3O_x$(HBCO) films grown at low temperature (500-620 ℃), by quenching the as-deposited films from the growth temperature. As a result, it was revealed that the quenched films have an extremely long *c*-lattice length of 1.2 nm. The *c*-lattice length was shortened by incorporating oxygen into the crystal lattice. However, the originally elongated *c*-lattice length was never reproduced by quenching the film again from the growth condition. Oxygen ions with a higher binding energy, other than reversible ones on the one-dimensional Cu-O chain, seem to be partially deficient in the as-deposited and quenched films.

EXPERIMENTAL

The HBCO thin films were deposited by ECR oxygen plasma-assisted reactive evaporation[1]. The deposition was performed with MgO(100) substrates, a deposition rate of 1.2 nm/min., a film thickness of 50 nm, an oxygen pressure of 10^{-2} Pa and a microwave power of 120 W. The growth temperatures were 500-620 ℃. The substrate was attached to an Au plate (15 × 15 × 0.1 mm) with Ag paste, and the Au plate was attached to a substrate heater with Ag paste. After deposition, the thin films were kept at the growth temperature for an hour in the same plasma as during deposition. Then, the thin films with the Au plate were quenched by pulling them out of

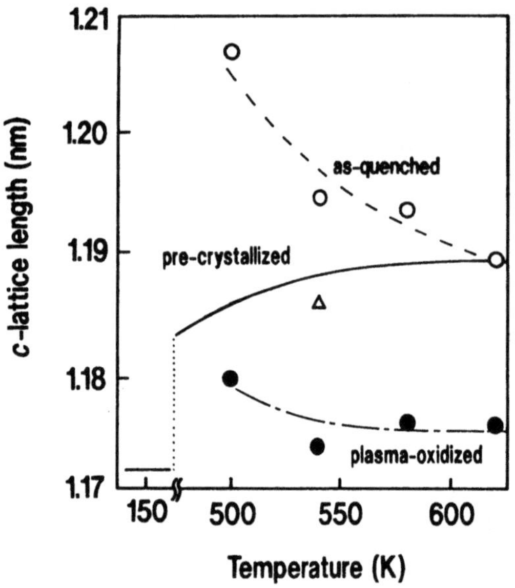

Fig. 1. The relationship between the growth temperature and the c-lattice length. The soild curve shows the dependence on the quenching temperature of the pre-crystallized HBCO film.

the substrate heater. Every quenched film was cut into two pieces, and one was further plasma-oxidized by annealing at 450 ℃ for 30 minutes in oxygen plasma and cooling to 150 ℃ at 10 ℃ /min. This oxidation in plasma has the same effect on the oxidation of the HBCO films as annealing in 1 atm oxygen.

RESULTS AND DISCUSSION

The relationship between the growth temperature and the c-lattice length is shown in Fig. 1. The c-lattice length was determined from the 00l peaks of $\theta - 2\theta$ x-ray diffraction measurement using least-squares fitting. The open and closed circles correspond to the quenched and plasma-oxidized films, respectively. The solid curve shows the dependence on quenching temperature for the pre-crystallized HBCO films. This curve was determined using an HBCO film which was prepared at 580 ℃, annealed at 620 ℃ for 3 hours, then quenched at each temperature [2]. The c-lattice length of the pre-crystallized HBCO film decreases with decreasing quenching temperature, since the oxygen content, in equilibrium with the oxygen plasma, increases with decreasing temperature. Therefore, if the as-deposited and quenched films are in thermal equilibrium, the c-lattice lengths would be shortened as the growth temperature is lowered. However, the c-lattice length of the quenched films increases with decreasing growth temperature, and the film quenched at 500 ℃ has an extremely c-lattice length of 1.207 nm, while the c-lattice length of the quenched film grown at 620 ℃ is close to that of the pre-crystallized film. These results suggest that the films grown at low temperatures do not attain thermal equilibrium. The difference in the c-lattice length between the as-quenched films and the pre-crystallized film reflects the degree of non-equilibrium growth, which increases with decreasing growth temperature.

The c-lattice length of every as-quenched film was shortened by the plasma oxidation, suggesting that the oxygen vacancies were mainly responsible for the extremely long c-lattice length of the as-quenched films. However, even after the as-quenched film was plasma-oxidized, the c-lattice lengths are still longer than those of the pre-crystallized and plasma-oxidized HBCO film (1.172 nm) shown by the solid curve at 150 ℃ in Fig. 1. This residual increase in the c-lattice length by about 0.03-0.08 nm, which is not eliminated by the plasma oxidation, may be caused by structual defects such as cation disordering that is not annealed out at 450 ℃. Moreover, the c-lattice length of the plasma-oxidized film was elongated at a low growth temperature.

Figures 2 and 3 show the dependences on growth temperature of the FWHM (full width at half maximam) of the 003 diffraction peak for θ-2θ scanning, and of the FWHM of the 003 rocking curve by x-ray diffraction measurement, respectively. The circles and triangles correspond to the quenched and the plasma-oxidized films. Both the FWHMs are hardly changed by the plasma oxidation. Therefore, the FWHMs do not reflect the oxygen vacancies but the cation disordering. However, both FWHMs increase with decreasing growth temperature, suggesting that the cation disordering increases with decreasing growth temperature. The elongation of the c-lattice length at low growth themperature can be correlated to the broadening of both FWHMs.

The plasma-oxidized film grown at 540℃ was quenched again from the growth condition, in order to investigate whether the oxygen-defect structure bringing about the elongated c-lattice length of the quenched film is reversible. However, the originally elongated c-lattice length was not reproduced, although it was elongated to 1.185 nm as shown by the triangle in Fig. 1. That is, the redox reaction with the oxygen in the as-deposited and quenched film with elongated c-lattice length is irreversible.

Usually, when the oxygen is removed from the 123 structure, the oxygen ions on the one-dimensional Cu-O chain are preferentially removed because these ions have a smaller binding energy than other oxygen ions [3]. However, with low-oxygen-pressure growth, oxygen ions with higher binding energy such as those on the two-dimensional Cu-O plane are thought to be also partially removed. Once the oxygen ions are incorporated into such oxygen vacancies, oxygen ions with higher binding energy are never removed under the same quenching conditions, and the extremely elongated c-lattice length cannot be reproduced. The structure with oxygen vacancies, other than those on the

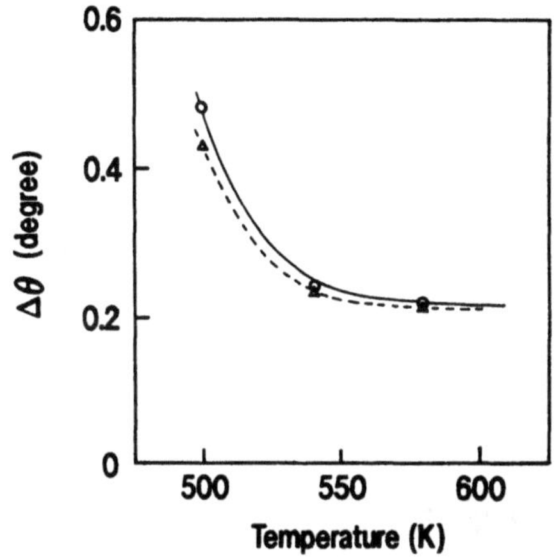

Fig. 2. The dependence on growth temperature of the FWHM of the 003 diffraction peak of θ -2 θ scanning.

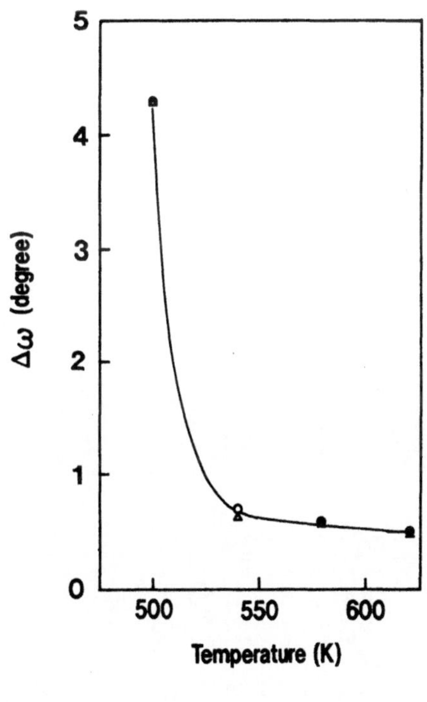

Fig. 3. The dependence on growth temperature of the FWHM of the 003 diffraction peak of rocking curve.

one-dimensional Cu-O chain, seems to be caused by the structual defect of cations.

CONCLUSION

It is revealed that the as-deposited and quenched HBCO films have an extremely elongated *c*-lattice length of about 1.2 nm. Although the *c*-lattice length could be shortened by the plasma oxidation, the originally elongated *c*-lattice length was never reproduced by quenching the film again from the growth condition. These results suggest that under low-oxygen-pressure growth, oxygen ions with higher binding energy, such as those on the two-dimensional Cu-O plane, are partially removed. The structure with excess oxygen vacancies seems to be caused by the structual defect of cations.

REFERENCES

1. T. Aida, A. Tsukamoto, K. Imagawa, T. Fukazawa, S. Saito, K. Shindo, K. Takagi, and K. Miyauchi, Jpn. J. Appl. Phys. 28, L635(1989).
2. A. Tsukamoto, M. Hiratani, and S. Akamatsu, Physica C (to be published).
3. J. D. Jorgensen, B.W. Veal, A.P. Paulikas, L.J. Nowicki, G.W. Crabtee, H. Claus, and W.K. Kwok, Phys. Rev. B 41, 1863(1990).

Texture Development of YBaCuO Film Prepared by Chemical Vapor Deposition

HEE-GYOUN LEE[1], SOON-DONG PARK[1], SUK-WOO YANG[2], HYUNG-SHIK SHIN[2], GYE-WON HONG[1], and DONG-YEON WON[1]

[1] Korea Atomic Energy Research Institute, Daejun 302-353, Korea
[2] Department of Chemical Engineering, Chonbuk National University, Chonju-si, Korea

ABSTRACT

Texture development of YBaCuO films prepared at 650°C and 900°C by chemical vapor deposition method was investigated. Both of the films which were prepared at 650°C and 900°C showed superconductivity above 77K($T_{c,0}$ = 84.5 ~ 88K). The preferred orientation of (001) normal to film surface was developed. The minor texture components such as (100) and (110) and the CuO second phase was included in the films and the amount of them was changed when the composition of vapor was changed. It indicates that the texture and the amount of second phase such as CuO can be easily controlled by changing vapor composition.

KEY WORDS: YBaCuO, Superconductor, CVD, Texture

INTRODUCTION

Since the discovery of high T_c oxide superconductor by Bednorz and Mueller[1], many intensive studies were made to find a new superconductor with higher T_c. Thereby, new high T_c superconductors of YBaCuO[2] (T_c=90K), BiSrCaCuO[3] (T_c=105K) and TlBaCaCuO[4] (T_c=120K) have been found. On the other hand, the preparation techniques of the films of those superconductors have been of great interest to researchers.
Recently, a very promising technique for the preparation of the superconducting thin films has developed in YBaCuO system by Yamane et al.[5] and by Berry et al.[6] and in BiSrCaCuO system by Ihara et al.[7]. More recently, Watanabe et al.[8] succeeded in preparing YBaCuO superconducting thin film with a T_c of 91K and a high J_c of 10^5A/cm^2 (77K) even under high magnetic field of 27 Tesla by low pressure chemical vapor deposition (LPCVD) method. In the previous work[9,10], we reported the effects of reaction pressure and the effects of substrate on the formation of YBaCuO superconducting thin films prepared by LPCVD. In general, it is well known that the vapor composition is one of the important factor in controlling the physical properties of the films prepared by CVD. Therefore, it is very needed to investigated the effects of vapor composition on the formation of YBaCuO film prepared by LPCVD.
Very recently, Yamame et al.[11] reported that the YBaCuO film with a T_c of 84K and a J_c of 10^4A/cm^2 was prepared at the temperature of 650°C on $SrTiO_3$ substrate by CVD using O_2 as an oxidant. We also succeeded in preparing the YBaCuO superconducting film at 650°C by CVD using O_2 as an oxidant[12].

In this work, texture development of the YBaCuO films prepared at 650°C and 900°C by CVD was investigated by observing the microstructure and the crystallographic orientation and by measuring the temperature dependence of electrical resistance.

EXPERIMENTAL

Source materials used to synthesize the YBaCuO superconducting film were beta-diketone chelates of Y(thd), Ba(thd) and Cu(thd) (thd= 2,2,6,6-tetramethyl-3, 5-heptanedionate) (Strem Co., Ltd). These chelates were evaporated at 126-264°C. The source vapors of Y, Ba and Cu were transported into the hot zone by Ar gas and O_2 gas was supplied separately. The CVD reactor was a hot wall type[9].

Deposition of YBaCuO films was carried out on various substrates of MgO(100), $LaAlO_3$(100) and $SrTiO_3$(100) placed normal to a direction of gas stream at 650°C (4 torr), 690°C (4 torr) and 900°C (15 torr). The oxygen partial pressures were 7.5 torr(900°C), 0.1 torr(690°C) and 0.028 torr(650°C). Deposition time of CVD was 30 min. After deposition, the films were cooled to 500°C at a cooling rate of 15°C/min and held for 30 min.
X-ray diffraction (XRD) data were obtained to examine the structure of the films. The microstructures of the films were observed by a scanning electron microscopy(SEM). Electrical resistance was measured by using ac four probe method.
The vapor composition of the reactants was changed by controlling the temperatures of each source materials. The molar ratio of the reactants is summarized in Table I.

Table I. Molar ratio of the evaporated source materials

specimen #	2	3	4	6	7	8	10	L*1	L*2
Ba/Y	1.52	1.63	1.21	1.43	1.16	2.01	1.43	1.04	1.48
Cu/Y	1.28	1.25	1.04	1.03	0.91	1.62	1.60	1.98	1.36

Films were deposited on $SrTiO_3$(100) substrate for specimens 2 - 10.
* L denotes the film deposited on $LaAlO_3$ substrate

RESULTS AND DISCUSSION

The temperature dependence of resistance of the specimens 7, 8, and 10, which were prepared at 900°C on a $SrTiO_3$ substrate, is shown in Fig.1. The zero resistivity($T_{c,0}$ R=0) for the specimens 7, 8, 10 appeared at 86.5K, 88K, 88K respectively and the transition width($T_{c,onset}$ - $T_{c,0}$) was about 1K. $T_{c,0}$ for the other specimens showed the $T_{c,0}$ of 86.5K - 88K with a transition width of 0.5K - 1K. It is very noticeable that the superconducting transition temperature was hardly influenced even though the vapor composition was widely changed as shown in Table I. It is also noticeable that the molar ratio of reactants is greatly deviated from the stoichiometry of Y:Ba:Cu = 1:2:3 for 90K YBaCuO superconductor.
Figure 2 shows the XRD patterns for the specimen 2, 3, and 6. It can be seen that 123 phase and CuO were formed. It can be seen that textures of the specimens are changed with vapor composition. The crystallographic plane of (001) was a major texture component for all the specimens. The peak intensity of other texture components than (001) plane decreased when the Ba concentration increased and the Cu concentration was decreased. XRD peak from CuO second phase, which is very beneficial for the flux pinning in YBaCuO superconductor, was detected at 2θ = 35.6° for the Cu- rich specimen(specimen 2). Texture of the YBaCuO film prepared by CVD method was easily changed by controlling the vapor compostion of source materials.

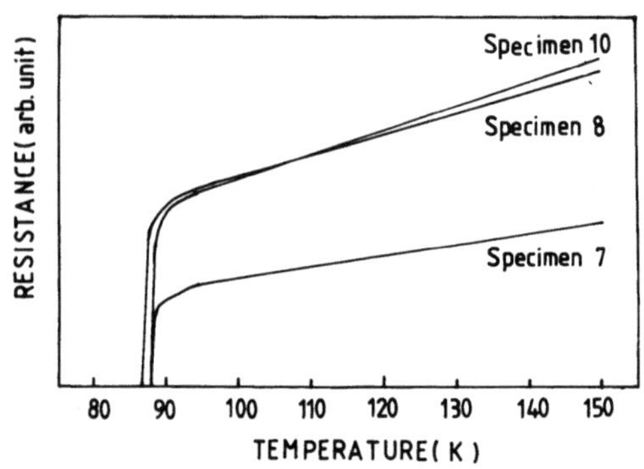

Fig.1 Temperature vs resistance for the specimens 7, 8, and 10.

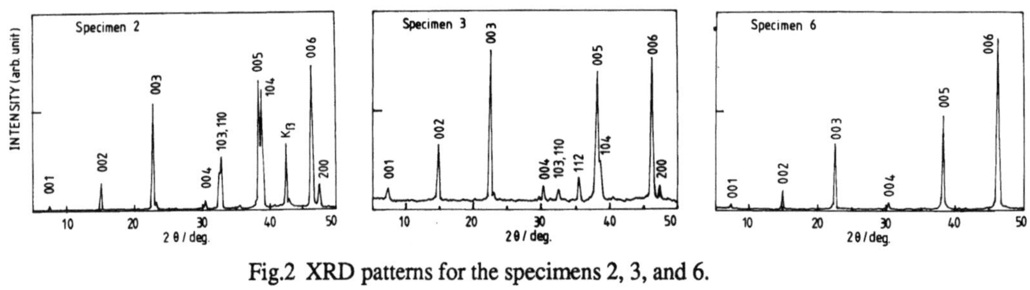

Fig.2 XRD patterns for the specimens 2, 3, and 6.

Figure 3 shows the XRD patterns for the YBaCuO films prepared at 900°C on MgO(100) substrate. Deposition time was changed as 10, 20, and 40 min respectively. It can be seen that the texture of the film was not changed as the deposition time was extended. It means that the YBaCuO crystals formed at the early stage of deposition grew out by maintaining its original orientation during a film growth. It is therefore considered that the texture of the film prepared by CVD method very depends on the nucleation of YBaCuO crystals at the early stage of deposition on the substrate. The XRD peak of (001) plane appeared at higher diffraction angle as the deposition period was extended. It means that the c-axis of the deposited YBaCuO crystal is shortened as the deposition time is prolonged.It was observed that the film thickness linearly increased with the deposition time.

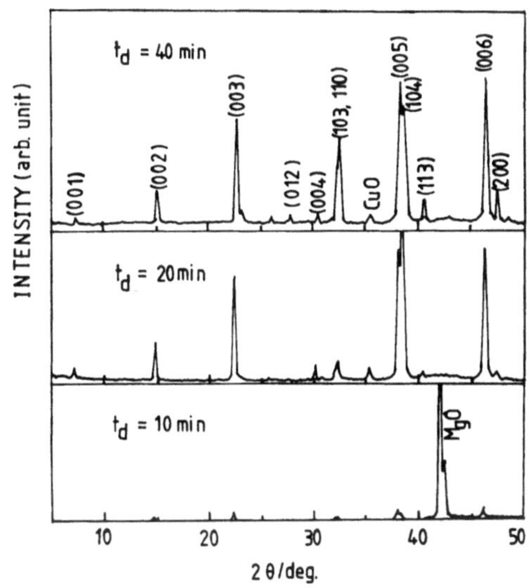

Fig.3 XRD patterns for the YBaCuO films prepared on MgO (100) single crystal. Deposition time(td) was 10, 20, and 40 min, respectively

The temperature dependence of resistance of the s pecimens L1 and L2, which were prepared at 690°C (PO_2=0.1 torr) and 650°C (PO_2=0.028 torr) on a $LaAlO_3$ substrate respectively, is shown in Fig.4. The superconducting transition started as a temperature of 90K. The zero resistivity for the specimens L1 and L2 appeared at 84.5K and 87K respectively.

Figure 5 shows the XRD patterns for the specimens L1 and L2. It can be seen that 123 phase was formed. The crystallographic plane of (001) was a major texture component for the films. The XRD peaks from (110), (113) and (104) planes appeared for the specimen L1 but were not detected for specimen L2. It is noticeable that the XRD peak from CuO second phase was not detected for the specimen L1 in spite of the large value of Cu/Y molar ratio(Cu/Y = 2).

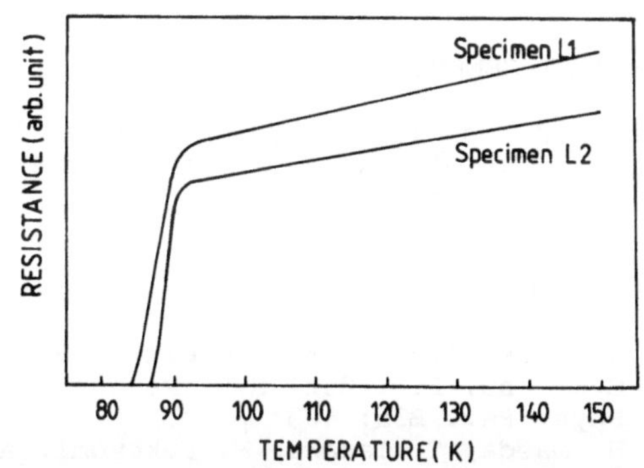

Fig.4 Temperature vs resistance for the specimens L1 and L2.

Fig.6 shows the surface SEM micrographs for the specimen 6 and specimen L2. It can be seen that YBaCuO crystals, which were grown on $SrTiO_3$(100) and $LaAlO_3$(100) substrates are two-dimensionally well aligned. From the observations of XRD patterns and SEM micrograph, it can be recognized that the YBaCuO films deposited on $SrTiO_3$(100) and $LaAlO_3$(100) single crystals were grown out having an epitaxial relationship with the substrates.

Summarizingly, texture development in YBaCuO films prepared at 650°C and 900°C by CVD method was investigated. Both of the films which were prepared at 650°C and 900°C showed superconductivity above 77K($T_{c,0}$ = 84.5 ~ 88K). The preferred orientation of (001) normal to film surface was developed. The minor texture components such as (100) and (110) and the CuO second phase were included in the films and the amount of them was changed when the composition of vapor was changed. It indicates that the texture and the amount of second phase such as CuO were able to be easily controlled by changing vapor composition. It was also observed that the YBaCuO crystals, which were grown on $SrTiO_3$(100) and

$LaAlO_3$(100) substrates are two-dimensionally well aligned.

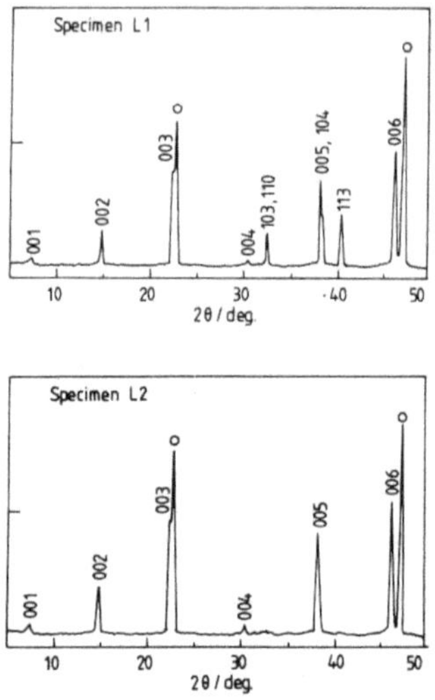

Fig.5 XRD patterns for the specimens L1 and L2.

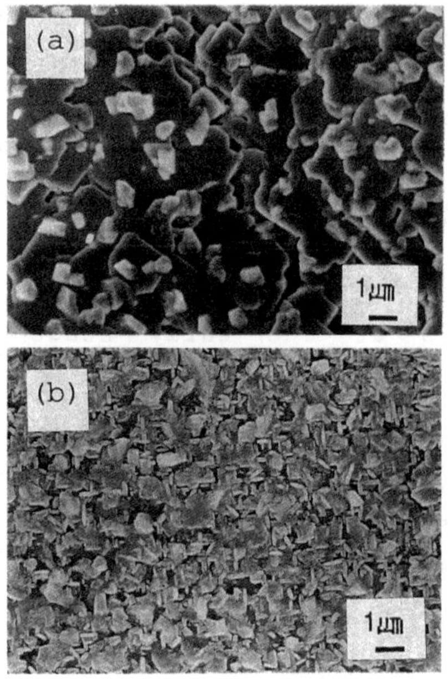

Fig.6 Surface views of SEM micrograph for (a) specimen 6 and (b) specimen L2 which were prepared at 900°C and 650°C respectively.

ACKNOWLEDGEMENT

This work was financially supported by Ministry of Science and Technology. The authors would like to thank Yeon-K. Kim and Man-K. Jeong for XRD and SEM investigations.

REFERENCES

1. J.G. Bednorz and K.A. Mueller(1986) Z. Phys. B 64: 189
2. C.W. Chu, P.H. Hor, R.L. Meng, G. Gao, Z.H. Huang, and Y.Q. Wang(1987) Phys. Rev. B35: 405
3. H. Maeda, T. Tanaka, M. Fukuyomi, and T. Asano(1988) Jpn. J. Appl. Phys. 27: L209
4. Z.Z. Sheng and A.M. Hermann(1988) Nature 332: 138
5. H. Yamane, H. Kurosawa, and T. Hirai(1988) Chem. Lett.: 939
6. A.D. Berry, D.K. Gaskill, R.T. Holm, E.J. Cukauskas, R. Kaplan, and R.L. Henry(1988) Appl. Phys. Lett. 52 : 1743
7. M. Ihara, and T. Kimura(1988) 5th Int'l Workshop on Future Devices, June 2-4, Japan: 137
8. K. Watanabe, H. Yamane, H. Kurosawa, T. Hirai, N. Kobayashi, H. Iwasaki, K. Noto, and Y. Muto(1989) Appl. Phys. Lett. 54: 575
9. H.G. Lee, J.S. Park, H.S. Shin, C.Y. Kim, C.J. Kim, and D.Y. Won (1991) (in press)
10. H.G. Lee, J.S. Park, C.J. Kim, J.C. Lee, G.W. Hong, I.S. Chang, and D.Y. Won(submitted to J. Physics D)
11. H. Yamane, M. Hasei, H. Kurosawa and T. Hirai(1991) Jpn J. Appl. Phys. 30: L1003
12. H.G. Lee, S.D. Park, S.W. Yang, H.S. Shin and D.Y. Won(submitted to Jpn J. Appl. Phys.)

Photo-Assisted Chemical Vapor Deposition of YBaCuO Films

HIDEAKI ZAMA[1], SHUNRI ODA[1], KEN-ICHI SAITO[2], and TAKEO HATTORI[2]

[1] Department of Physical Electronics, Tokyo Institute of Technology, Meguro-ku, Tokyo, 152 Japan
[2] Department of Electrical and Electronic Engineering, Musashi Institute of Technology, Setagaya-ku, Tokyo, 158 Japan

ABSTRACT

Low-temperature growth of YBaCuO films is investigated by photo-assisted chemical vapor deposition with mercury-sensitization technique using β-diketonate complex as starting precursors and N_2O as an oxidizing agent. First, Cu-oxide films are deposited at a substrate temperature T_{sub}=120°C. Highly oriented Cu_2O films are obtained on MgO substrate. Second, YBaCuO films are deposited at T_{sub}=215°C. Film composition analysed by energy dispersive x-ray spectrometer is nearly 1:2:3(=Y:Ba:Cu). The structure of $YBa_2Cu_3O_x$ film is amorphous phase. These facts suggest that UV-light is effective in promoting the decomposition of β-diketonate complex and in enhancing the migration of species on growing surface.

KEY WORDS: photo-assisted chemical vapor deposition, mercury-sensitization, N_2O, Cu_2O film, YBaCuO film

INTRODUCTION

Low-temperature processing is very important for future electron devices with feature size less than submicron scale. Chemical vapor depositon (CVD) is an established method for low temperature growth of semiconductor epitaxial films. CVD of oxide superconductor films is also a very promising method for low temperature growth. Recently, applications of CVD of oxide superconductor films have been reported from a number of research institutions.

However, CVD of oxide superconductor films has several problems including still high deposition temperature and poor surface morphology. To date, several attempts have been made to reduce the deposition temperature without sacrificing characteristics of high T_c $YBa_2Cu_3O_x$ films including the application of plasma[1, 2] or UV-light[3, 4].

In particular, three major effects for low temperature growth are expected for UV-light in the CVD process. These are (i) promotion of decomposition of β-diketonate complex on the basis of strong absorption due to $\pi-\pi^*$ transition, (ii) enhancement of migration of species on the growing surface and (iii) generation of atomic oxygen as an active oxidizing agent. The effects of UV-light irradiation which improved the surface morphology and the crystalline orientation of metal-oxide films were reported by T. Ushida et al. [3] and H. Koinuma et al. [4] The effect of decomposition of β-diketonate complex was reported by our group. [5]

This paper reports the process of photo enhanced decomposition of β-diketonate complex with mercury sensitization using N_2O as an oxidizing agent and the approach of low-temperature photo-assisted CVD of YBaCuO films.

EXPERIMENTAL

Precursors used for CVD of YBaCuO films were β-diketonate complex such as $Y(DPM)_3$, $Ba(DPM)_2$ and $Cu(DPM)_2$. The reactor chamber was a horizontal quartz tube, 3cm in diameter and 20cm in length. Temperatures of precursor vaporizers, gas transfer tubes, a reactor wall and a substrate holder were controlled independently. Five 6W-mercury-lamps (254nm-light) surrounding the reactor chamber and a reflecting cylindrical mirror made from aluminum were equipped.

A mercury-sensitization technique was used to enhance the effect of UV-light. N_2O, as an oxidizing agent, was fed to the reactor chamber through a mercury reservoir whose temperature was varied from room temperature to 70°C. (100) MgO substrates were used in this study.

Thickness of deposited films was estimated by optical transmittance measurement. Crystal structure of deposited films was evaluated by x-ray diffraction (XRD) measurement. Metal composition of deposited films was analysed by energy dispersive x-ray (EDX) spectrometer.

RESULTS AND DISCUSSION

Figure 1 shows UV absorption spectra of 2×10^{-5}mol/l ethanol solution of $Y(DPM)_3$, $Ba(DPM)_2$ and $Cu(DPM)_2$. All of spectra have a peak between 250nm and 300nm in wavelength. Peak values of molar extinction coefficient are as large as 10^4 (l/mol·cm). This strong absorption process is caused by $\pi-\pi^*$ transition related to resonance structure of a metal atom, such as Y, Ba and Cu, and DPM ligands. So, we have chosen to use 254nm-light for the light source of photo-assisted process.

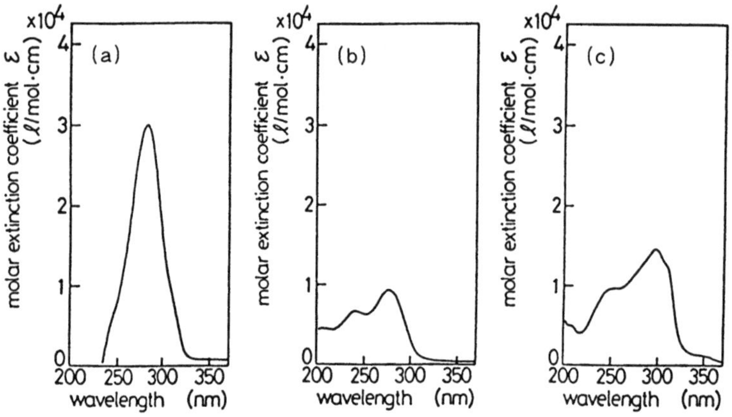

Fig. 1. UV absorption spectra of 2×10^{-5}mol/l ethanol solution. (a) $Y(DPM)_3$, (b) $Ba(DPM)_2$, (c) $Cu(DPM)_2$.

$Cu(DPM)_2$ starts to decompose at the substrate temperature (T_{sub}) of 300°C on the thermal process and at T_{sub}=275°C on the photo-assisted process without mercury-sensitization [5]. We have already reported that a copper film prepared by photo-assisted CVD at T_{sub}=350°C was with highly (111)-orientation and an electrical resistivity of 1.7$\mu\Omega$·cm as low as that of balk crystals. [5]

We have investigated deposition of Cu-oxide films with mercury-sensitized photo-assisted technique using N_2O as an oxidizing agent. As shown in Fig. 2, a deposition rate of Cu-oxide films increased to 0.3A/s at T_{sub}=120°C. While, no deposition took place without mercury sensitization nor UV-irradiation. In this case, the deposition condition is shown in Table I (a). The deposition rate of films decreased with increasing temperature between 120°C and 200°C when the temperature of gas transfer tubes was kept at 150°C and that of a mercury reservoir at 70°C. This negative activation process is associated with photo-assisted decomposition of precursors. The rate may be determined by adsorption/desorption kinetics of precursors or mercury on the growing surface. As shown in Fig. 4 (a), a highly (100)-oriented Cu_2O film on (100) MgO substrate was obtained at T_{sub}=120°C.

When the temperature of gas transfer tubes raised to 235°C, with the other deposition condition shown in Table I (b), the deposition rate of films increased with increasing temperature as shown in Fig. 3. A highly (100)-oriented Cu_2O film as shown in Fig. 4 (b) was obtained at T_{sub}=250°C. Because the split of the peak related to x-ray source of CuKα_1 and CuKα_2 appears at the Cu_2O(200) diffraction peak. Without mercury sensitization, the decomposition activation energy increased as shown in Fig. 3. Without UV-irradiation, little deposition took place as shown in Fig. 3.

Table I. Deposition condition of Cu-oxide films.

	(a)	(b)
Cu(DPM)$_2$ vaporizer temperature:	120°C	110°C
Carrier gas (Ar) flow rate:	20sccm	20sccm
N_2O flow rate:	10sccm	20sccm
Gas-transfer-tube temperature:	150°C	235°C
Reactor pressure:	5Torr	1Torr
Deposition time:	1hour	1hour

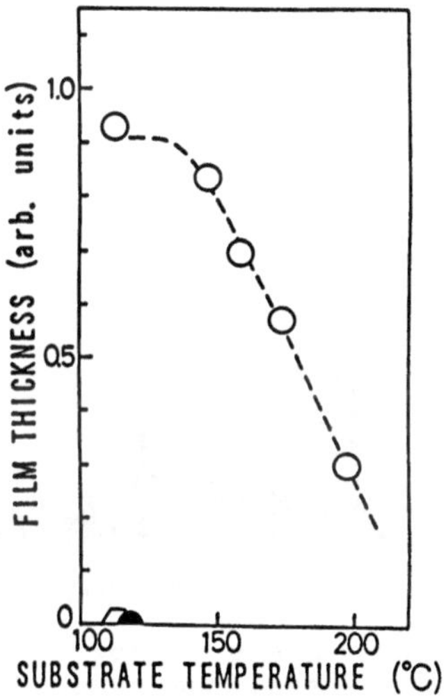

Fig. 2. Film thickness vs substrate temperature for Cu-oxide films with mercury-sensitized UV-light assisted decomposition of Cu(DPM)$_2$.
The deposition condition is shown in Table I(a).
○ : with UV-irradiation and mercury-sensitization (The temperature of a mercury reservoir is 70°C),
○ : without mercury-sensitization,
● : without UV-irradiation.

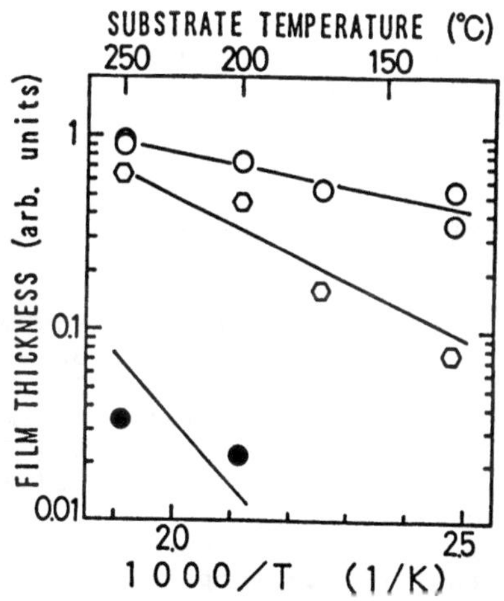

Fig. 3. Film thickness vs substrate temperature for Cu-oxide films with mercury-sensitized UV-light assisted decomposition of Cu(DPM)$_2$.
The deposition condition is shown in Table I(b).
○ : with UV-irradiation and mercury-sensitization (The temperature of a mercury reservoir is 70°C),
○ : without mercury-sensitization,
● : without UV-irradiation.

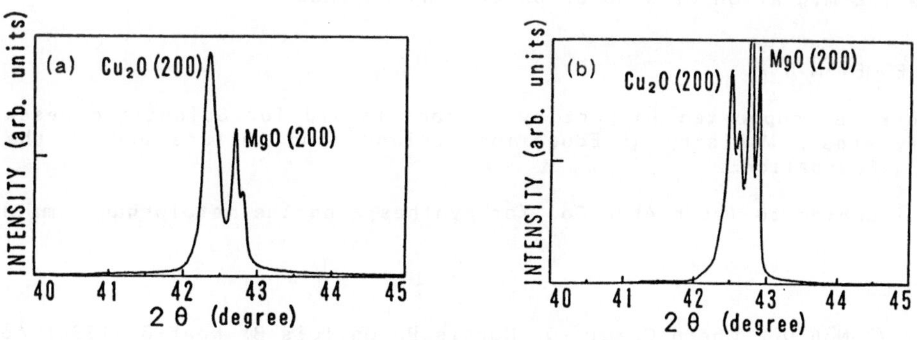

Fig. 4. X-ray diffraction patterns of Cu-oxide films.
(a) T_{sub}=120°C,
(b) T_{sub}=250°C.

$Y(DPM)_3$ and $Ba(DPM)_2$ have similar structure as $Cu(DPM)_2$ and hence have similar decomposition properties. We have deposited the YBaCuO films at T_{sub}=215°C by employing three metal sources simultaneously. We have tried the deposition with various conditions and analysed the chemical composition so as to obtain 123 structure. With increasing the temperature of a mercury reservoir from room temperature to 70°C, the intensity of EDX signals of Y and Cu increased and that of Ba decreased. This is attributable that Ba precursors decomposed before reaching the substrate. The intensity of EDX signals of Ba increased with increasing the carrier gas flow rate of $Ba(DPM)_2$ when the temperature of a mercury reservoir was at 70°C. These results show that $Ba(DPM)_2$ is more sensitive to UV-irradiation than $Y(DPM)_3$ and $Cu(DPM)_2$. $Y(DPM)_3$ and $Cu(DPM)_2$ have similar decomposition properties each other. As shown in Fig.5, a film with metal composition of Y, Ba and Cu of nearly 1:2:3 was obtained by adjusting the deposition condition as shown in Table II. However, it is amorphous phase by XRD measurement in this condition.

Table II. Deposition condition of YBaCuO film.

Vaporizer temperature:	$Y(DPM)_3$	97°C
	$Ba(DPM)_2$	215°C
	$Cu(DPM)_2$	95°C
Carrier gas (Ar) flow rate:	$Y(DPM)_3$	20sccm
	$Ba(DPM)_2$	70sccm
	$Cu(DPM)_2$	20sccm
N_2O flow rate:		140sccm
Mercury reservoir temperature:		70°C
Gas-transfer-tube temperature:		235°C
Substrate temperature:		215°C
Reactor pressure:		3Torr
Deposition time:		4hours

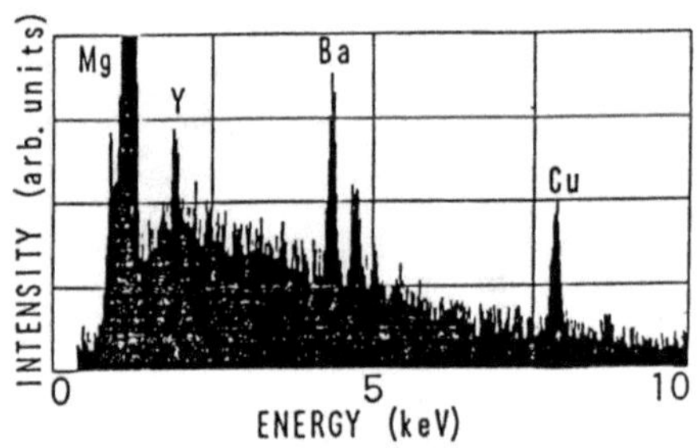

Fig.5. Energy dispersive x-ray spectrum of a YBaCuO film on MgO substrate at T_{sub}=215°C. The deposition condition is shown in Table II.

CONCLUSIONS

We have successfully prepared highly oriented Cu_2O films at a temperature of as low as 120°C on MgO substrate and YBaCuO films at a temperature of as low as 215°C on MgO substrate. However, the structure of YBaCuO films is amorphous phase.

These facts suggest that UV-light of 254nm with mercury-sensitization and N_2O has effects to promote the decomposition of β-diketonate complex and to enhance the migration of species on growing surface.

ACKNOWLEDGEMENTS

This work was supported in part by a Grant-in-Aid for Scientific Research on Priority Areas, Ministry of Education, Science and Culture and by the TEPCO Research Foundation.

We are indebted to Tosoh Akzo Co. for synthesizing the metalorganic materials.

REFERENCES

1. Zhao J, Noh DW, Chern C, Li YQ, Norris P, Gallois B, Kear B (1990) Appl. Phys.Lett. 56: 2342.
2. Sugii N, Saito S, Imagawa K, Saito S, Kanehori K, Takagi K (1991) Jpn.J. Appl.Phys. 30: L1118.
3. Ushida T, Higashiyama K, Hirabayashi I, Tanaka S (1991) Jpn.J.Appl.Phys. 30: L35.
4. Koinuma H, Chaudhary KA, Nakabayashi M, Shiraishi T, Hashimoto T, Kitazawa K, Suemune Y, Yamamoto T (1991) Jpn.J.Appl.Phys. 30: 656.
5. Zama H, Miyake T, Hattori T, Oda S (to be submitted) Jpn.J.Appl.Phys.

Rapid Synthesis of the Y-Ba-Cu-O Tape on Metal Substrate by CVD Technique

SHIN'YA AOKI[1], TAICHI YAMAGUCHI[1], AKIRA KAGAWA[1], OSAMU KOHNO[1], AKIRA SAJI[2], and TOSHIO INOUE[2]

[1] Fujikura Ltd., Koto-ku, Tokyo, 135 Japan
[2] Chubu Electric Power Co., Nagoya, 459 Japan

ABSTRACT

Advanced chemical vapor deposition (CVD) system has been developed for rapid formation of the superconducting oxide tape on long metal substrates. In this system, mixed fine powder of source materials were slowly fed to a vaporizer with carrier gas by a fine powder feeder, and the source gas from the vaporizer were continuously introduced to the reactor with oxygen gas.
And this system has a long substrate conveying equipment. By this system using Hastelloy C-276 tapes with YSZ buffer as substrates, the 1 meter long superconducting Y-Ba-Cu-O (YBCO) tapes could be obtained. These tapes had a critical temperature (T_c) with zero resistivity above 77.3K and the best part of the tape had a T_c at 84K and a critical current density (J_c) of 330A/cm^2 (77.3K, 0T).

KEY WORDS: Y-Ba-Cu-O superconductor, superconducting tape, chemical vapor deposition, metal substrate, thin film

INTRODUCTION

Since the first reports on synthesis of the Y-Ba-Cu-O (YBCO) thin film by CVD in 1988[1,2], high T_c and high J_c YBCO films can be obtained easily by CVD. The result on J_c of 6.5x10^4 A/cm^2 (77.3K, 27T)[3] indicated the possibility of practical use of this material. But in order to practice this oxide superconducting material as long wires on electric power application, it is necessary to get the higher speed of formation and to fabricate on the long poly-crystalline substrates for a long time.
Until now, many techniques[4-6] have been reported to fabricate the high T_c superconducting thin films. Thin film fabricating methods are attractive not only on device application[7,8] but also on electric power application. Because highly oriented or even orientation controlled films can be obtained and therefore the films that have higher J_c can be obtained. In particular, CVD process is promising as main process for application because its deposition rate is superior[9-12] to the other thin film processes.

A new CVD system has been developed for rapid formation and long-time fabrication of the superconducting oxide tape on a long and flexible metal substrate. The CVD source materials of superconducting oxide are solid state under conditions of room temperature and standard pressure. In this system, blended fine powder of source materials were slowly fed to a vaporizer by a fine powder feeder, so the rapid and long-time synthesis of superconducting oxide could be realized. And as this system has a long substrate conveying equipment, the long superconducting tape could be obtained.

In this paper, we report on a new rapid formation technique and properties of long superconducting YBCO tapes on flexible metal substrates by CVD using a fine powder feeder, for supplying blended β-diketonate chelates of Y, Ba and Cu to a vaporizer, and a long substrate conveying equipment.

EXPERIMENTAL

The source materials used in this experiment were β-diketonate chelates of Y, Ba and Cu (2,2,6,6-tetramethyl-3,5-heptanedionate;that is to say thd). But $Ba(thd)_2$ cannot be available in air for a long time because it is unstable under CO_2 and H_2O atmosphere. So we used Bis(2,2,6,6-tetramethyl-3,5-heptanedione)bis(1,10-phenanthroline)Barium [$Ba(thd)_2(phen)_2$] as a Ba source. Figure 1 shows the structure of $Ba(thd)_2(phen)_2$. $Ba(thd)_2(phen)_2$ is much more stable than $Ba(thd)_2$ by the effect of phenanthroline adducts

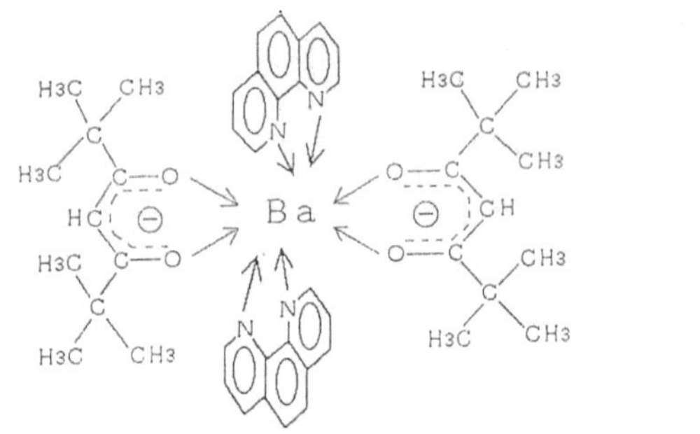

Fig.1 Structure of Bis(2,2,6,6-tetramethyl-3,5-heptanedione)bis(1,10-phenanthroline)Ba[$Ba(thd)_2(phen)_2$].

Figure 2 shows a schematic diagram of the CVD apparatus used in this experiment. Mixed source materials of Y:Ba:Cu = 1.0:3.2:2.4 (at%) were set into the fine powder feeder with anthracene (precursor:anthracene=1:1) and slowly fed to the vaporizer at the rate of 0.1g/min with argon gas used as a powder carrier gas and vaporized at 260°C. A tetrahydrofuran (THF) bubbler was added to the vaporizer. Flow rates of argon gases for the powder carrier and THF bubbling were 150 ml/min and 30 ml/min, respectively. This THF bubbling method could realize stable vaporization of barium source material [13]. Oxygen gases were separately introduced into the mixing chamber and the reactor chamber at the same rate of 150ml/min, respectively.

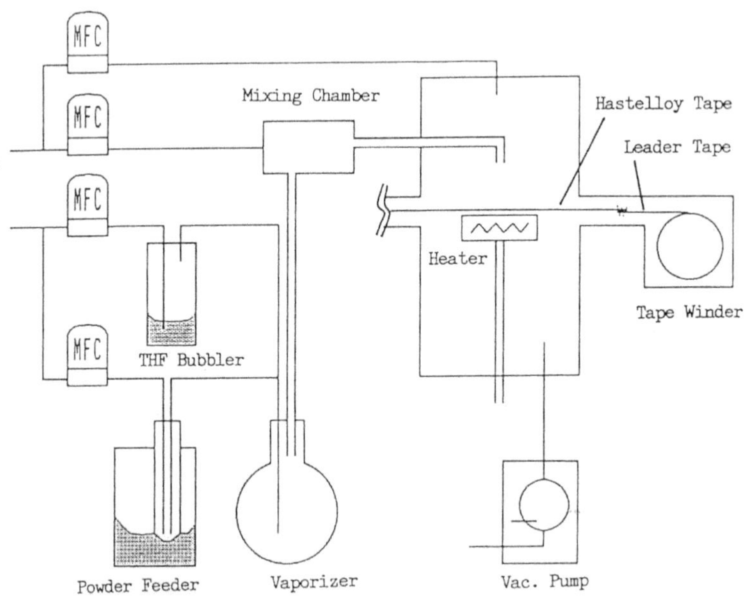

Fig.2 Schematic diagram of a new CVD apparatus.

The substrate temperature was set at 800°C, and the diameter of substrate heater is 140mm. The total gas pressure was kept at less than 0.67kPa during the deposition.
Hastelloy C-276 tapes with yttrium stabilized zirconia (YSZ) buffer were used for long flexible substrates. The dimension of substrates was 0.1mm in thickness, 5mm in width and 1m in length. The YSZ buffer was deposited by the RF magnetron sputtering method and its thickness was 0.5 μm. The traveling speed of substrates was 0.4cm/min. After the deposition, the films were annealed at 500°C for 2 hours under 101.3kPa oxygen.

RESULTS AND DISCUSSION

The 1 meter long superconducting YBCO tapes on metal substrates could be obtained by a new CVD system using a fine powder feeder and a long substrate conveyer. At the traveling speed of 0.8cm /min, the superconducting tape had a Tc at less than 50K and thickness was about 2μm. In case that feeding rate of source powder is fixed, the higher traveling speed is, the lower Tc is. Because the thickness of YBCO tapes becomes thin, so it seems that the diffusion between YBCO and substrate cause. Therefore we decreased the traveling speed of substrate to 0.4cm/min.

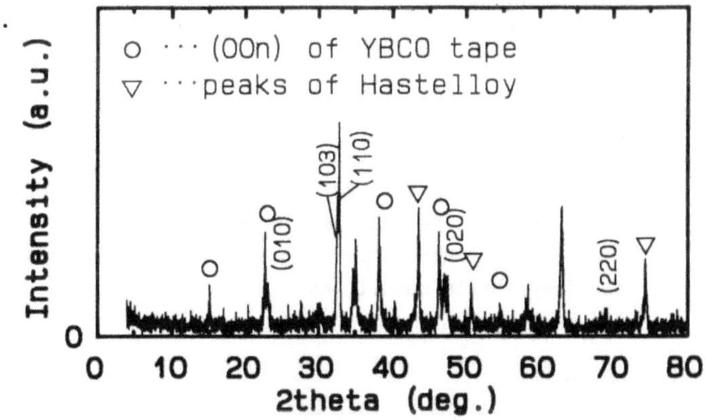

Fig.3 XRD pattern of a YBCO tape on Hastelloy C-276 with YSZ buffer by new CVD system.

Figure 3 shows the X-ray diffraction (XRD) pattern of a long YBCO tape on Hastelloy C-276 substrate with YSZ buffer by the θ-2θ method using Cu-Kα radiation. In this figure, the peaks of YBCO superconducting phase are mainly observed and YBCO(00n) peaks are recognized with YBCO (110), (010) and Hastelloy's peaks. Now we cannot obtain the c-axis oriented YBCO tape. Because the surface of Hastelloy C-276 tape is not smooth. This problem must be resolved in the near future.

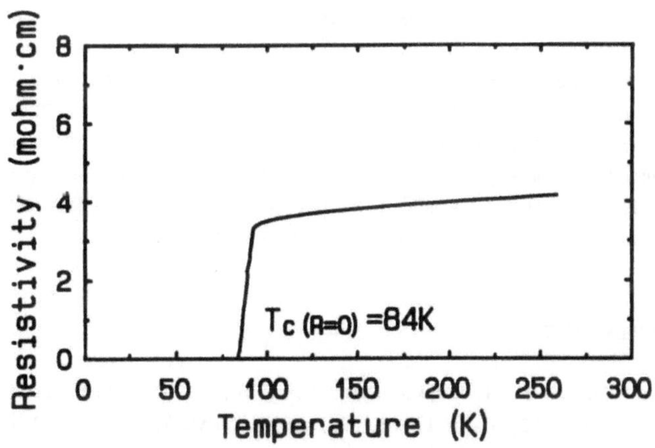

Fig.4 Temperature dependence of resistivity of YBCO tape on Hastelloy C-276 with YSZ buffer by new CVD system.

The temperature dependence of resistivity of this YBCO tape was measured by a direct current four terminal method with indium electrodes. The 1 cm long short samples were picked up from the head and tail part of the long YBCO tape on Hastelloy C-276 substrate, and were etched to small patterns. The dimension of this pattern was 1mm in width and 8mm in length and the thickness of the YBCO tapes was 4μm.

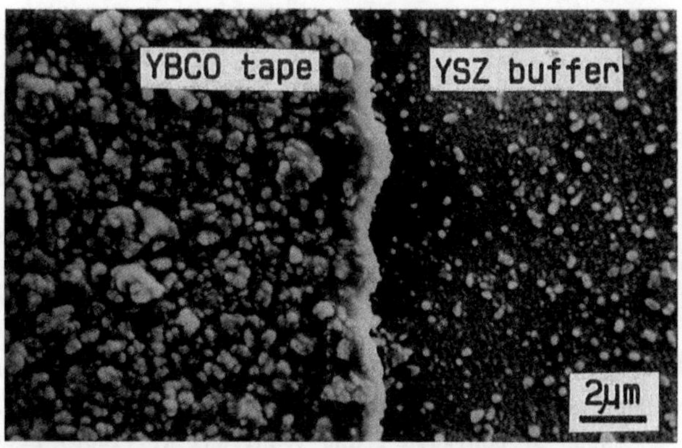

Fig.5 SEM photographs of the YBCO tape by CVD.

Figure 4 shows the temperature dependence of resistivity of this short YBCO sample picked up from the head part of the long YBCO tape. This YBCO sample tape showed an onset superconducting transition at about 90K, and had a Tc with zero resistivity at 84K. Another short sample picked up from the tail part of the long YBCO tape and its Tc was

measured. The tail part sample also had an onset superconducting transition at about 90K, and had a T_c with zero resistivity at 81K. Within the range of 1 meter, homogeneous superconducting YBCO tape that had a T_c above liquefied N_2 temperature could be obtained. Critical current density (J_c) was measured. The head part of long YBCO tape had a J_c of 330A/cm^2 (77.3K, 0T). As compared with the film on single-crystalline substrate, this J_c value was three or four orders of magnitude lower.
Figure 5. shows a microstructure of the YBCO tape on Hastelloy C-276 with YSZ buffer observed by scanning electron microscope (SEM). In this figure, the surface of YSZ buffer on Hastelloy C-276 tape is also showed. The YBCO tape consisted of poly-crystalline pebble-like grains on the Hastelloy substrate. The surface of YBCO tapes was uneven and these tapes had course texture. In order to raise J_c value of these tapes, it is necessary to obtain the more dense texture and more smooth surface of YBCO tapes by more delicate control of synthesis conditions.

CONCLUSION

A new chemical vapor deposition (CVD) system suitable for rapid synthesis of superconducting oxide tape on long flexible substrate has been developed. The new CVD system has a fine powder feeder and a long substrate conveyer. By this method, rapid formation and long-time fabrication of superconducting oxide tape has been realized. The 1meter long superconducting YBCO tapes were prepared on Hastelloy C-276 substrates with YSZ buffer by this CVD using β-diketonate chelates powder as source materials. These tapes had a critical temperature (T_c) with zero resistivity above liquefied N_2 temperature. Within the range of 1 meter, homogeneous superconducting YBCO tapes could be obtained and the best tape had a T_c above 81K. The 1 cm long short sample picked up from the best tape had a T_c at 84K and critical current density (J_c) of 330A/cm^2 (77.3K, 0T).

REFERENCES

1. Yamane H, Kurosawa H and Hirai T, (1988) Chem. Lett.: 939
2. Berry AD, Gaskill DK, Holm RT, Cukauskas EJ, Kaplan R and Henry RL, (1988) Appl. Phys. Lett., 52: 1743
3. Watanabe K, Yamane H, Kurosawa H, Hirai T, Kobayashi N, Iwasaki H, Noto K and Muto Y, (1989) Appl. Phys. Lett., 54: 575
4. Hong M, Liou SH, Kwo J and Davidson B, (1987) Appl. Phys. Lett. 51: 694
5. Yamanishi K, Yasunaga S, Imada K, Sato K and Hashimoto Y, (1988) Mat. Res. Soc. Symp. Proc. 99: 343
6. Wu XD, Dijkkamp D, Ogale SB, Inam A, Chase EW, Miccli PE, Chang CC, Tarascon JM and Venkatesan T, (1987) Appl. Phys. Lett. 51: 861
7. Noge S, Yamashita T, Wang Z, Matsui T, Kurosawa H, Yamane H and Hirai T, (1989) Jpn. J. Appl. Phys. 28: L1581
8. Irie A, Era M, Yamashita T, Kurosawa H, Yamane H and Hirai T, (1989) Jpn. J. Appl. Phys. 28: L1816
9. Yamaguchi T, Aoki S, Sadakata N, Kohno O and Osanai H, (1989) Appl. Phys. Lett. 55 (15): 1581
10. Aoki S, Yamaguchi T, Sadakata N, Kagawa A and Kohno O, (1989) Proc. 2nd Int. Symp. on Superconductivity (ISS'89), Tsukuba, Springer-Verlag Tokyo:p439
11. Matsuno S, Uchikawa F and Yoshizaki K, (1989) Proc. 2nd Int. Symp. on Superconductivity (ISS'89), Tsukuba, Springer-Verlag Tokyo:p877
12. Lackey WJ, Carter WB, Hanigofsky JA, Hill DN, Barefield EK, Neumeier G, O'Brien DF, Shapiro MJ, Thompson JR, Green AJ, Moss,III TS, Jake RA and Efferson KR, (1990) Appl. Phys. Lett. 56: 1175
13. Matsuno S, Uchikawa F and Yoshizaki K, (1990) Jpn. J. Appl. Phys., 29, L947

Growth Mechanism of a-Axis Oriented $YBa_2Cu_3O_x$ Films on MgO(100) by Laser MOCVD

KAZUTOSHI HIGASHIYAMA[1], TAKAHISA USHIDA[1], HIROYUKI HIGA[2], and IZUMI HIRABAYASHI[1]

[1] Superconductivity Research Laboratory, Nagoya Division, ISTEC, Nagoya, 456 Japan
[2] Department of Electrical and Computer Engineering, Nagoya Institute of Technology, Nagoya, 466 Japan

ABSTRACT

UV laser light (λ=193nm) irradiation was effective to convert c-axis to a-axis oriented growth in the deposition of the $YBa_2Cu_3O_x$ films on MgO(100) by MOCVD. This effect was observed also on the substrates of polycrystalline MgO, $SrTiO_3$(100), and YSZ(100). A-axis oriented growth does not proceed from a melt process, because the a-axis orientation disappeared at the high laser power enough to occur surface melting. We expect that the nucleation sites for a-axis oriented growth are formed during deposition on the $YBa_2Cu_3O_x$ films by the laser irradiation.

KEY WORDS : MOCVD, $YBa_2Cu_3O_x$, thin film, UV laser, a-axis orientation

INTRODUCTION

Crystal orientation of $YBa_2Cu_3O_x$ film changes with the growth conditions [1-2]. Detailed knowledge of the microstructure and how to control it are key issues for successful application as well. In MOCVD the $YBa_2Cu_3O_x$ films grown on MgO(100) above 750°C show the c-axis orientation, while the films deposited below 750°C show the mixture of the a-, c- and other orientation. This tendency to form mixed epitaxial grains seriously affects the application of $YBa_2Cu_3O_x$ films because of the anisotropic properties of $YBa_2Cu_3O_x$. Recently we found that the enhancement of a-axis oriented growth was observed in $YBa_2Cu_3O_x$ films deposited on the MgO(100) substrates at 600-750°C using MOCVD with UV laser light (λ=193nm) irradiation [3]. Especially, strongly a-axis oriented films can be grown even on MgO(100) at 650-700°C. It is known that the a-axis oriented films are obtained on $SrTiO_3$(100) at lower temperature [4]. A disadvantage of the a-axis oriented films, however, is lower $T_{c,0}$ of only 70K. Our sample grown at 700°C showed higher $T_{c,0}$ of 82K. In this paper, we discuss the growth mechanism of a-axis oriented films with UV laser irradiation.

EXPRIMENTAL

A schematic diagram of the experimental apparatus is shown in Figure 1. The $YBa_2Cu_3O_x$ films were prepared using 2,2,6,6-tetramethyl-3,5-heptanedione (thd) chelates of Y, Ba, and Cu as source materials and oxygen. The CVD reactor is a cold-wall type and measures 15cm in diameter and 17cm in height. The mixture of the metal chelate vapors with nitrogen carrier gas was introduced into the reactor through the pipe inlet. The oxygen gas was separately introduced into the reactor. The substrates were heated by a ceramic heater. MgO(100) was mainly used as substrates. Polycrystalline MgO, $SrTiO_3$(100), and YSZ(100) were also used to study the effect of the substrates on the growth of $YBa_2Cu_3O_x$ films with irradiation of the UV light. An excimer laser (Lambda physik, LPX110icc) operating at 193 nm (ArF) was used as a light source. The laser beam with a repetition rate of 10 Hz and average power of 100mJ/pulse was introduced at an angle of 45° to the substrates and irradiated a part of the samples directly during the deposition. After each deposition with laser irradiation for one hour, the films were cooled down to about 100°C at a rate of 100°C/min under the oxygen atmosphere of 1 atm.

The films were divided into the laser irradiated and the unirradiated parts to analyze independently. The microstructure of the film surface was observed by scanning electron microscope (SEM) with energy dispersive X-ray spectroscopy (EDX). X-ray diffraction (XRD) patterns of the films were obtained by the standard θ-2θ method.

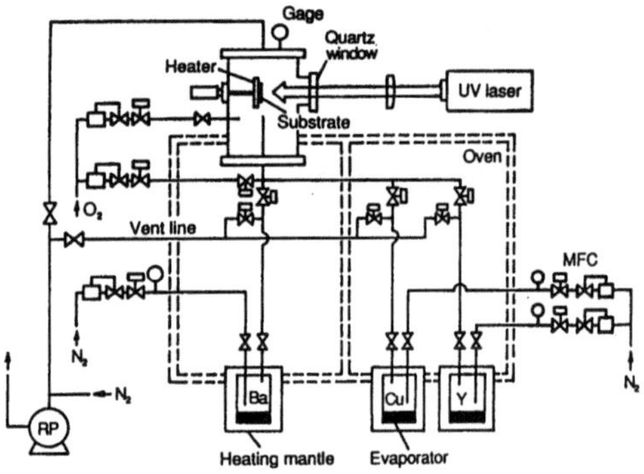

Fig.1 Schematic diagram of CVD apparatus.

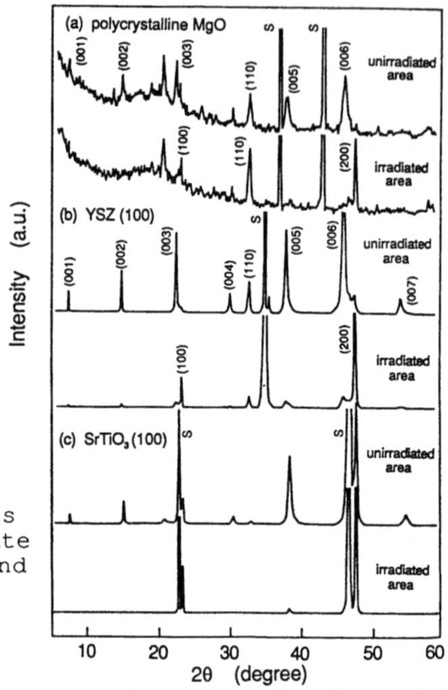

Fig.2 XRD patterns of $YBa_2Cu_3O_x$ films prepared on different substrates with laser irradiation. Substrate temperatures are (a)(b) 700°C and (c) 730°C.

RESULTS AND DISCUSSION

The XRD patterns of $YBa_2Cu_3O_x$ films prepared on the substrates of polycrystalline MgO, YSZ(100), and $SrTiO_3$(100) are shown in Figure 2. In this experiments, the CVD apparatus with a horizontal reactor was used [3]. A laser power density was 70mJ/cm^2 and a repetition rate was 20Hz. In the laser irradiated part of the film on each substrate, the increasing intensity peaks of (100) and (200) showing the a-axis oriented growth are observed. This means that the a-axis oriented growth with laser irradiation is not related to the kind of substrate materials and their orientation.

In order to understand the effects of the UV laser on the growth of a-axis oriented $YBa_2Cu_3O_x$ films, the various studies have been carried out. Details of the studies have been reported elsewhere [3,5]. The important information to discuss the mechanism is summarized in the following;
(1) When the laser light irradiated on a part of the substrate, the film in the only irradiated part showed the a-axis orientation.
(2) According to the cross-section image by TEM, the laser irradiated part on the MgO(100) showed the mixture of the a-, c-, and other oriented grains just above the substrate and a-axis oriented grains become dominant as distant from the substrate surface.
(3) The film deposited at 700°C without laser irradiation onto the MgO(100) which was irradiated for 30min in advance did not show the a-axis orientation.
(4) The a-axis oriented film was grown with laser irradiation also on the c-axis oriented film which covered on the MgO(100).
(5) The irradiation effect was degraded with the decrease of Ba content from stoichiometric composition.
(6) In the deposition of metal oxides, the films including Ba (Y-Ba-O and Ba-Cu-O) showed the marked change of morphology with laser irradiation.

At first, we expected that the a-axis orientation may be a result of the change of surface condition of the substrates, such as steps, etch pits, micro-cracks, by the laser irradiation. Because numerous researchers have observed that a step on the substrate surface is the starting point of a-axis oriented grain [6,7]. The experimental result(3), however, did not support this model. Furthermore the result shown in Figure 2 and (4) indicate that one possible nucleation mechanism of the a-axis oriented growth is not related to the substrate, but to the surface condition of growing $YBa_2Cu_3O_x$ film.

First, we consider the surface melting by the laser irradiation. Nagaya et al.[8] observed that the a,b-axis of $YBa_2Cu_3O_x$ was grown along the moving direction of the melt-solid interface in the rapid melt and solidification of $YBa_2Cu_3O_x$ bulk using IR laser (Nd-YAG laser). If the surface layer of the film is rapidly heated up to melting temperature also by pulsed UV laser beam, a-axis oriented film may be formed because the melt-solid interface moves normal to the surface with the high velocity. The temporal temperature profiles of $YBa_2Cu_3O_x$ thin films flashed by a pulsed KrF excimer laser (λ=248nm) have been studied numerically by Nakamiya et al.[9]. The calculated results showed that the temperature at the film surface increased by 400, 1000, and 1900°C for only 50ns at the laser power density of 40, 100 and 200mJ/cm^2, respectively. Figure 3 shows SEM photographs of the films grown at 700°C with the irradiation at high power density enough to occur a surface melting. In this experiment, the laser beam was focused with a convex lens on a substrate area of 9.2 x 2.2 mm^2. The power of the laser light on the substrate, as measured by a power meter, was 20 mJ/pulse. An average energy density on the irradiated area amounts to about 100 mJ/cm^2. The film at the center of laser irradiated part(e,f) clearly melts and shows the decrease of Ba and Cu concentration as shown in Figure 4. The marked change in morphology is observed at the periphery of melting area. This morphological change probably attribute to the power distribution of the laser beam. The spatial beam profile of the excimer laser output is not flat. The actual energy density is estimated at more than 100 mJ/cm^2 at the center and decreases as being distant from the center. Figure 5 shows the XRD pattern of each area. The laser irradiated part without melting(b-d) shows the strong peaks of (100) and (200), but melting area(e,f) shows no peaks. At the high laser energy density enough to occur a surface melting, the a-axis oriented films are not grown. This result suggests that the a-axis oriented growth does not proceed from the melt process.

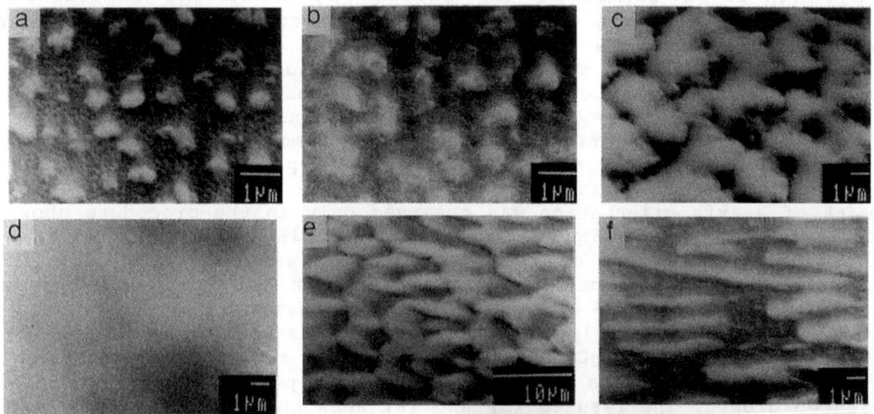

Fig 3 SEM photographs of $YBa_2Cu_3O_x$ film prepared on MgO(100) at 700°C with laser irradiation. Average energy density is 100 mJ/cm^2.

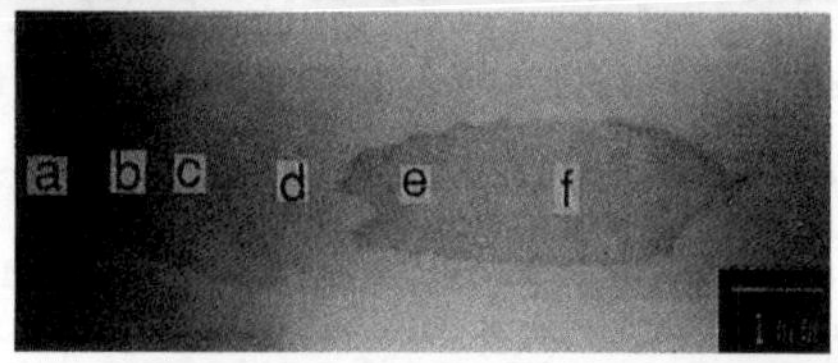

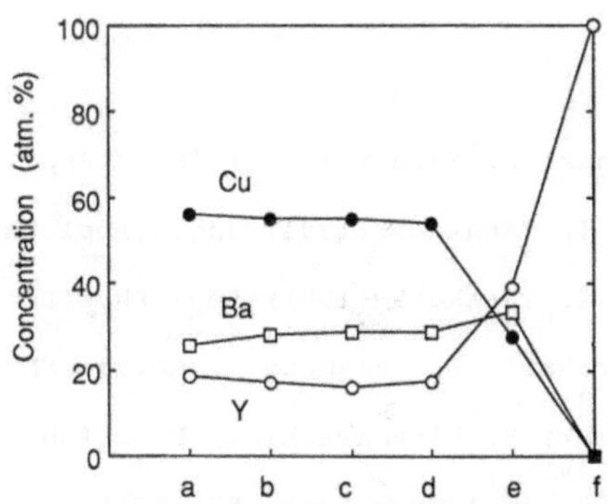

Fig.4 Film composition of each area shown in Fig.3.

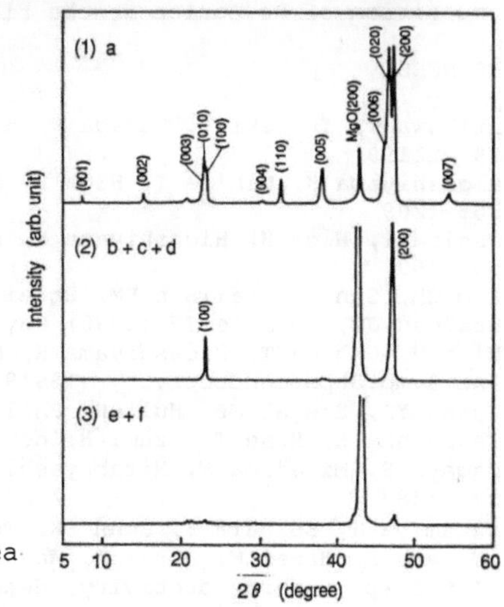

Fig.5 XRD pattern of each area shown in Fig.3.

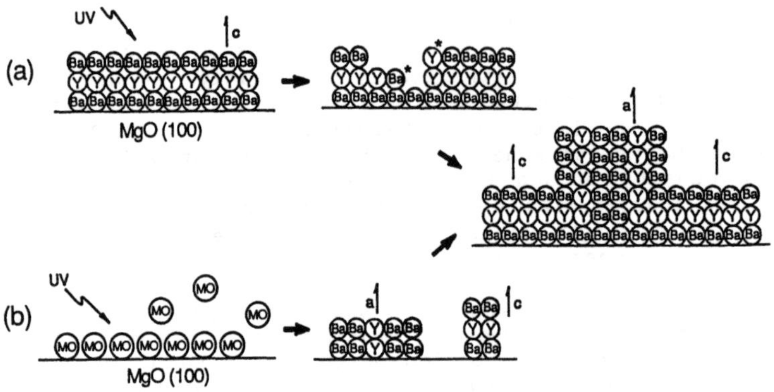

Fig.6 Schematic drawing of nucleation model.

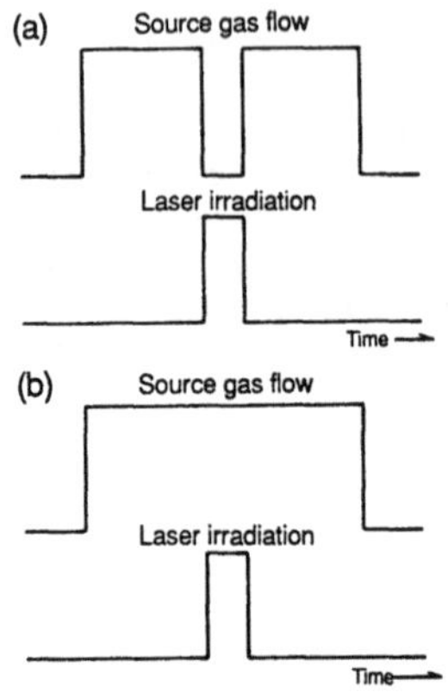

Fig.7 Deposition procedures

Next we consider the possibility providing nuclei for a-axis oriented growth on the films by the laser irradiation. Figure 6 shows the schematic drawing of this model. In case of Figure 6(a), the structural defects are introduced on the surface of $YBa_2Cu_3O_x$ thin layer formed at the initial stage of the deposition with the orientation depending on the temperature. The defects shown in Figure 6(a) provide the nucleation sites for a-axis oriented growth. The result by TEM(2) can be explained by this model. In case of Figure 6(b), nuclei for a-axis oriented growth are formed by the surface reaction of MO gases. In both cases, once the a-axis normal nuclei appear on the surface, a-axis oriented grains propagate because of a faster growing nucleus. If a-axis oriented nuclei are formed according to the former model, the UV laser irradiation have to be effective to prepare the a-axis oriented films in the deposition procedure showed in Figure 7(a) and even in PVD methods. An enhancement of a-axis orientation was observed only in the procedure shown in Figure 7(b). In addition, we attempted to prepare a-axis oriented film also in laser ablation method, the effect of laser irradiation was not observed in the various conditions of laser power density, total pressure, oxygen partial pressure, and substrate temperature. Yamane et al.[10] reported that the a-axis oriented films were selectively prepared on Ba-O layer deposited on the substrate in advance by MOCVD. We observed frequently that $YBa_2Cu_3O_x$ films containing an excess amount of Ba showed a-axis orientation. These facts mean that Ba element play an important role in the a-axis oriented growth. The results (5) and (6) show that Ba is an important parameter for the a-axis growth also in MOCVD with UV laser irradiation. These results appear to support the latter model rather than the former model.

In summary, we focused on the growth mechanism of the a-axis oriented $YBa_2Cu_3O_x$ films in MOCVD with UV laser irradiation. The a-axis oriented growth does not proceed from the change of the substrate surface and the surface melting of the $YBa_2Cu_3O_x$ films. We expect that the nucleation sites for a-axis oriented growth are formed during deposition on the $YBa_2Cu_3O_x$ films by the interplay between the laser irradiation and decomposition of Ba source at the film surface.

REFERENCES

1. Arikawa T, Itozaki H, Harada K, Higaki K, Tanaka S, Yazu S (1990) Jpn.J.Appl.Phys. 29: L2199
2. Higashiyama K, Ushida T, Higa H, Hirabayashi I, Tanaka S (1991) Jpn.J.Appl.Phys. 30: 1209
3. Ushida T, Higa H, Higashiyama K, Hirabayashi I, Tanaka S (1991) Appl.Phys.Lett. 59: 860
4. Eom CB, Sun JZ, Lairson BM, Streiffer SK, Marshall AF, Yamamoto K, Anlage SM, Bravman JC, Geballe TH (1990) Physica C 171: 354
5. Higa H, Ushida T, Higashiyama K, Maeda M, Suzuki S, Hirabayashi I, Proc.4th Int.Symp.Superconductivity (ISS'91), Tokyo, 1991
6. Hsieh YF, Siegal MP, Hull R, Phillips JM (1990) Appl.Phys.Lett. 57: 2268
7. Takahashi H, Hase T, Izumi H,Ohota K,Morishita T,Tanaka S (1991) Physica C 175:381
8. Nagaya S, Miyajima M, Hirabayashi I, Shiohara Y, Tanaka S (1991) IEEE Trans.Magn. 27: 1487
9. Nakamiya T, Ebihara K, John PK, Tong BY Proc.MRS Symp.
10. Yamane H, Hasei M, Hirai T, Watanabe K, Kobayashi N, Kurosawa H, Muto Y Proc,3rd Int.Symp. Superconductivity, Sendai, 1990 (Springer-Verlag, p973 Tokyo 1991)

Growth of Y-Ba-Cu-O Thin Films on Al_2O_3/Si by MOCVD

WON-NAM KANG, SANG-SEOP YOM*, and SANG-SAM CHOI

Korea Institute of Science and Technology, Superconductor Laboratory, Seoul, Korea

ABSTRACT

Superconducting thin films of $YBa_2Cu_3O_{7-x}$ (YBCO) were deposited on Si substrate with Al_2O_3 buffer layer and $LiNbO_3$ single crystal substrates by *in-situ* metalorganic chemical vapor deposition (MOCVD) technique. A resistive heated vertical cold-wall reaction chamber using N_2O gas as an oxidizer was successful to deposit thin films of YBCO and epitaxial Al_2O_3 buffer layer without post annealing. A buffer layer of MOCVD grown Al_2O_3 film on (100) Si is found to be γ phase Al_2O_3 hetero-epitaxial film from the x-ray diffraction analysis. As-deposited YBCO film on Al_2O_3/Si substrate at a substrate temperature of 800 ~ 850° C showed superconducting behavior at $T_{c,zero}$ = 58 K. YBCO films on Z-cut $LiNbO_3$ single crystal substrate by *in-situ* MOCVD growth also showed $T_{c,zero}$ = 85 K. Auger electron spectroscopy (AES) for depth profile analysis is discussed.

KEY WORDS : MOCVD, Si substrate, Al_2O_3 buffer, $LiNbO_3$

INTRODUCTION

The discovery of high temperature superconductors is expected to bring advanced electronic devices. Among wide varieties of electronic applications using high temperature superconductors the most advantageous application would be hybrid electronics of semiconductors with superconductors. Deposition of oxide superconducting thin films directly on the silicone substrates is required for device applications. Although some success has been reported[1], the deposited oxide films usually indicated strong reaction at near interface resulting in severe degradation of the superconducting properties. Beside interdiffusion, microcracks in the high temperature superconducting thin films are also observed due to large lattice mismatch and difference of thermal expansion coefficient between YBCO films and silicone substrates[2,3]. Buffer layers usually introduced between semiconductor and superconductor films to obtain high quality YBCO films. Al_2O_3 is particularly attractive buffer layer material for deposition of YBCO thin films on silicon substrates because of its excellent dielectric properties and low loss tangent at microwave frequency. $LiNbO_3$ substrates are also of special interest for acoustic devices and optoelectronics since they have very high electro-acoustic and electro-optic coefficients. There have been a few reports for successful deposition of YBCO films on Y-cut $LiNbO_3$ by laser ablation[4] and sputtering[5].

In this proceedings, we report successful deposition of YBCO thin films on silicone (100) substrate with MOCVD-grown Al_2O_3 buffer layer and Z-cut $LiNbO_3$ single crystal substrate by in-situ MOCVD technique.

EXPERIMENTAL

The films were grown in vertical cold-wall reaction chamber. β-diketonate complexes of $Y(dpm)_3$, $Ba(dpm)_2$, $Cu(dpm)_2$ (dpm = dipivaloylmethanate), and aluminum isopropoxide [$Al(OC_3H_7)_3$] were used as Y, Ba, Cu, and Al precursors, respectively. Source tanks were immersed in constant temperature controlled oil bathes at 126, 114, 250, and 100° C for Y, Ba, Cu, and Al precursors, respectively. Argon carrier gas which was controlled by a flow rate of mass flow controller was transported the source into the growth chamber. The deposition were carried out at a system pressure of 10 Torr and a substrate temperature of 800~850° C. Substrates were loaded onto resistively heated disk susceptor. The N_2O gas was separately introduced into the chamber as an oxidizer. Pure N_2O gas has been shown to decrease the deposition temperature because of production of highly reactive species by thermal heating or other physical means of excitations[6]. Typical argon flow rate were 10 sccm, 35 sccm, 15 sccm, 30 sccm, 100~500 sccm for Y, Ba, Cu, Al precursors and N_2O gas, respectively. Deposition was usually carried out for 180 minutes under these conditions. After YBCO deposition, films were annealed at

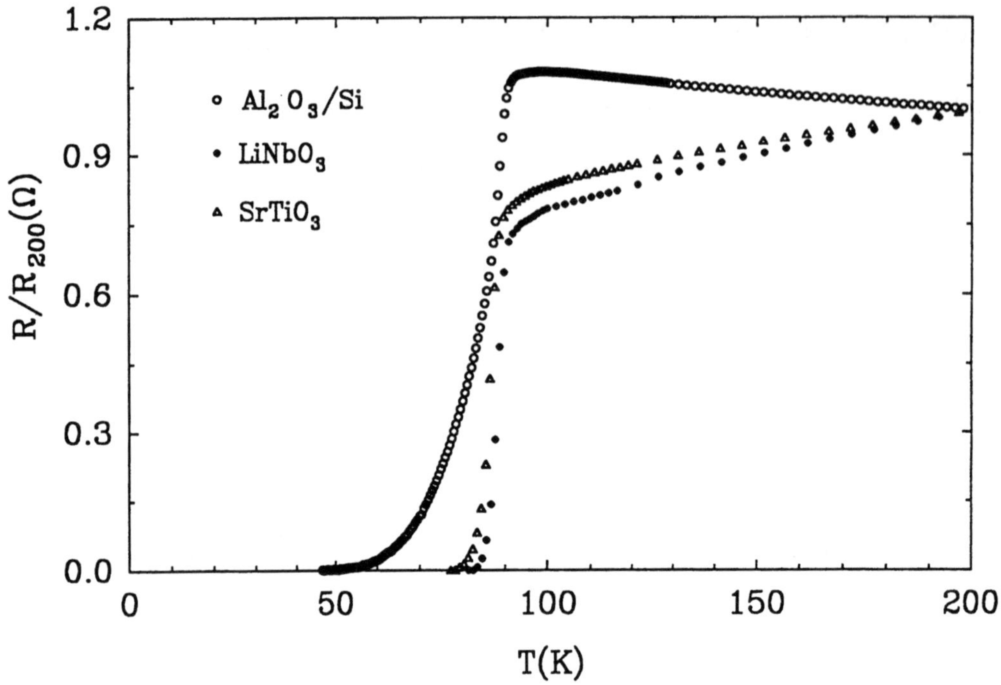

Fig. 1. Temperature dependence of resistance for YBCO thin films on Al_2O_3/Si(open circle), $LiNbO_3$ (closed circle), and $SrTiO_3$(triangle) substrates.

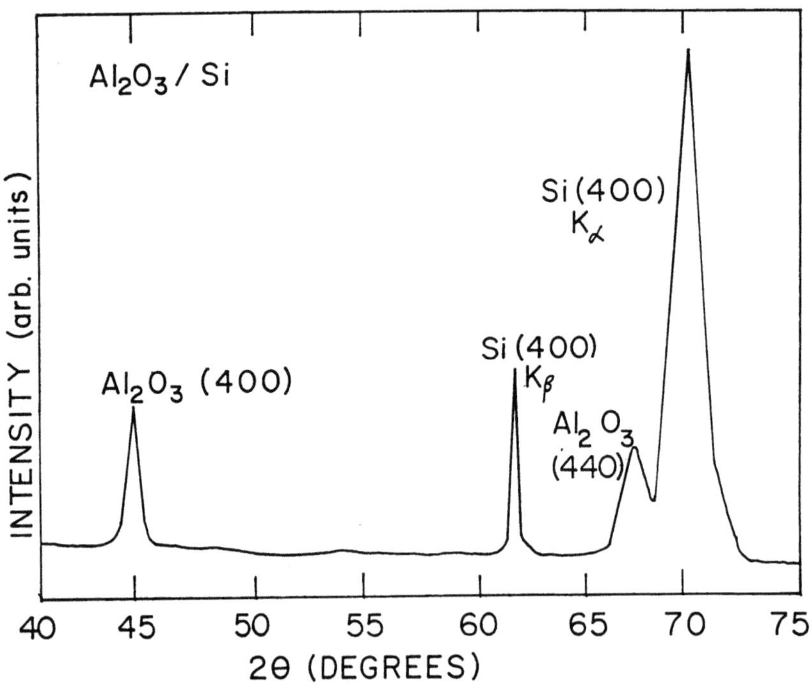

Fig. 2. X-ray rocking curve of γ-Al_2O_3 hetero-epitaxial layers grown on Si (100) substrates.

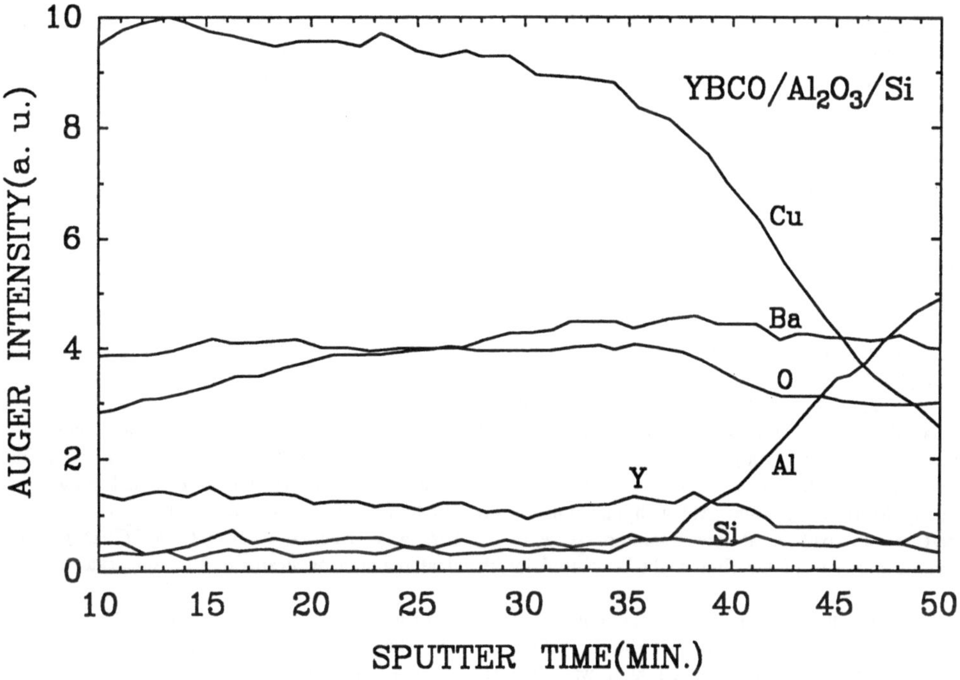

Fig. 3. AES depth profile of $YBCO/Al_2O_3/Si$.

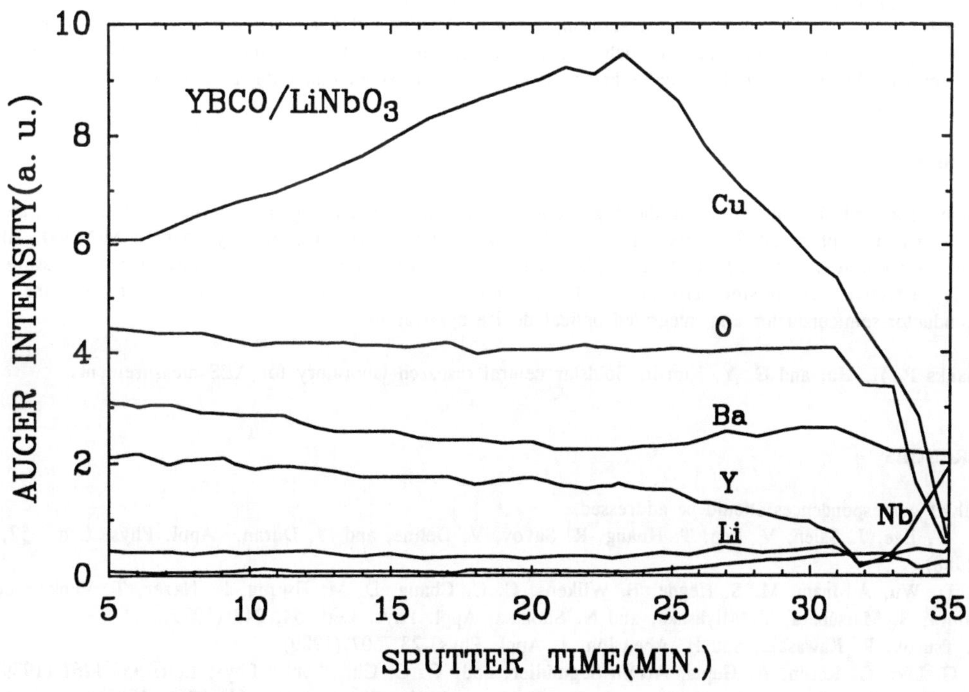

Fig. 4. AES depth profile of $YBCO/LiNbO_3$.

500° C for two hours in one atmospheric pressure of oxygen followed by cooling to room temperature under oxygen. Composition analysis was performed by inductively coupled plasma analysis (ICP), wavelength dispersive spectroscopy (WDS), and Auger electron spectroscopy (AES). X-ray diffraction (XRD) and scanning electron microscopy (SEM) were also used to study structural properties.

RESULTS AND DISCUSSION

YBCO films grown at substrate temperature of 800° C using N_2O gas showed superconducting behavior without post annealing. A buffer layer of Al_2O_3 film on silicone(100) was grown by MOCVD using aluminum isopropoxide and N_2O at substrate temperature of 740° C. In figure 1, the temperature dependence of resistance (R-T) curves of YBCO thin films on Al_2O_3/Si, $LiNbO_3$, and $SrTiO_3$ substrate show onsets at 91 K for all of the films. The transition to zero resistance occur at different temperatures; 58 K, 85 K, and 80 K, respectively. The R-T curve of YBCO/Al_2O_3/Si has a broad transition indicating semiconducting behavior above Tc. XRD analysis (see Fig. 2) indicated that buffer layer of Al_2O_3 film on silicone (100) is found to be γ phase hetero-epitaxial film which is subject to separate publication[7]. Possible lattice matching relationships are (100) γ-Al_2O_3 // Si (100) and (100) γ-Al_2O_3 // Si (110). XRD spectra of all of the films indicated no obvious (001) peaks (not shown). Randomly oriented polycrystalline behavior with no preferentially oriented growth was also confirmed by SEM micrograph. Although no films of preferred orientation films were obtained, we believe that use of N_2O as an oxidizer at low system pressure of 10 Torr was effective to *in-situ* MOCVD growth. The polycrystalline property of the film might be caused by chemical reaction or strains at film-substrate interface.

AES was used to study the interfacial reaction indicating no residual carbon participations. The result of AES depth profile of YBCO/Al_2O_3/Si film is shown in Fig. 3. The film interface between YBCO and Al_2O_3/Si is not clear due to chemical reaction. Aluminum was detected in high sensitivity Auger spectra of the YBCO film. The interfacial reaction severely contributes to diffusion of large amount of aluminum into the film. Figure 4 shows AES depth profile of YBCO/$LiNbO_3$. A Ba-rich interface layer and out-diffusion of Li and Nb from the substrate was observed. These results of interface reaction are consistent with result by Venkatesan et al.[6]. Formation of the Ba-rich layer at the film-substrate interface can be explained by migration of barium toward the interface, presumably to relieve interfacial stress during the film deposition at higher temperature of above 800° C. Copper signal in Fig. 4 increases toward the interface while a sharp decrease of oxygen at near interface is observed. The deposition relationship between YBCO and $LiNbO_3$ substrate is likely to be complex and to change with deposition condition such as substrate temperature and deposition rate. Further work is under way to improve the quality of the samples and to characterize the superconducting properties.

SUMMARY

We have deposited the thin films of the high temperature superconducting YBCO on Si (100) substrates with MOCVD grown γ-phase Al_2O_3 buffer layer and $LiNbO_3$ single crystal substrates by *in-situ* MOCVD. Although interface reaction was observed by AES, replacement of N_2O gas as an oxidizer at low system pressure of 10 Torr is effective to in-situ growth. These thin films are to be useful to future the hybrid superconductor-semiconductor and integrated optical device application.

We thanks P. H. Hur and C. Y. Kim in Goldstar central research laboratory for AES measurements.

REFERENCES

*) All of correspondences should be addressed.
[1] W. Y. Lee, J. Salen, V. Lee, T. Huang, R. Savoy, V. Deline, and J. Duran, Appl. Phys. Lett. 52, 2263 (1988).
[2] X. D. Wu, A. Inam, M. S. Hegde, B. Wilkens, C. C. Chang, D. M. Hwang, L. Nazer, T. Venkatesan, S. Miura, S. Matsubara, Y. Miyasaka, and N. Shohata, Appl. Phys. Lett. 54, 243 (1989).
[3] T. Tsurok, R. Kawasaki, and H. Abe, Jpn. J. Appl. Phys. 28, 607 (1989).
[4] S. G. Lee, G. Koren, A. Gupta, Armin Segmüller, and C. C. Chi, Appl. Phys. Lett. 55, 1261 (1989).
[5] A. Höhler, D. Guggi, H. Neeb, and C. Heiden, Appl. Phys. Lett. 54, 1066 (1989).
[6] T. Venkatesan, C. C. Chang, D. Dijkkamp, S. B. Ogale, E. W. Chase, L. A. Farrow, D. M. Hwang, P. F. Miceli, S. A. Schwarz, and J. M. Tarascon, J. Appl. Phys. 63, 4591 (1988).
[7] S. S. Yom, W. N. Kang, Y. S. Yoon, J. I. Lee, D. J. Choi, T. W. Kim, K. Y. Seo, P. H. Hur, and C. Y. Kim, Thin Solid Films (submitted).

Fabrication of Thick Films of $YBa_2Cu_3O_x$ Superconductors Using Fine Particles Prepared by Mist Pyrolysis Method

HIROAKI NISHIKAWA, JUNYA NISHINO, NOBUHIKO KUBOTA, and MUNEO AYABE

Research Institute, Ishikawajima-Harima Heavy Industries Co., Ltd., Koto-ku, Tokyo, 135 Japan

ABSTRACT

Fine powders of Y-Ba-Cu oxide superconductor were prepared by mist pyrolysis process. Thick films were fabricated by deposition of the powders on MgO(100) substrate. Using Ag as a buffer layer could give the texture with c-axis of 123 phase perpendicular to the surface and sharp superconducting transition(Tc=80K, $Jc=1.8x10^4 A/cm^2$ at 65K) after sintering in Ar flow at 1323K followed by O_2 annealing under 1073K.

It should also be possible to apply the process for fabrication of films using plasma evaporation of fine powders under the condition of ambient pressure and in low power for evaporation.

KEY WORDS: fine powder, Y-Ba-Cu oxide superconductor, mist pyrolysis process, thick film, gas jet atomizer

INTRODUCTION

Mist pyrolysis process[1-5] is one of the simple techniques of chemical solution routes to prepare superconducting powders which are fine and uniform in size. These powders will be able to make ceramics more sinterable and more densely sintered. The process to prepare powders in succession will be applied for fabrication of thin and thick films, tapes and fibers which are oriented towards mass production.

In our previous work[1], fine powders(0.025μm) were prepared by the process using aqueous nitrates as a starting solution. Fabrication of thick films(5μm) was tried by direct deposition of powders on MgO(100) substrate. Sintering in O_2 flow at 1243K gave superconductivity(Tc=84K) in spite of the texture with random orientation.

In this work, we studied the effects of atmosphere in sintering and of Ag as a buffer layer on texture and superconducting properties of films fabricated by direct deposition of fine powders on MgO(100) substrate. Plasma evaporation of fine powders in ambient pressure was tried to fabricate superconducting films.

EXPERIMENTAL METHOD

Two kinds of aqueous solution(0.033mol/l) were used for mist pyrolysis; one is nitrates of Y, Ba and Cu dissolved in water with molar ratio of 1:2:3 and another is acetates. Results of thermal analysis of each starting materials were shown in Fig. 1 and Fig. 2. The experimental apparatus was previously described in detail[1]. The solution was lead into gas jet atomizer to prepare small droplet in mist, the size of which was controlled less than 0.5μm in diameter. The droplet was curried by O_2 gas into electric furnace at 1253K to form fine superconducting powders(0.025μm), TEM and XRD results of which were shown in Fig. 3 and Fig. 4. Superconducting powders were directly deposited on MgO(100) substrate put into the furnace, followed by sintering in Ar or O_2 flow.

Evaporation of fine powders prepared by mist pyrolysis to deposit on MgO(100) substrate was tried using RF plasma apparatus, schematic diagram of which was shown in Fig. 5.

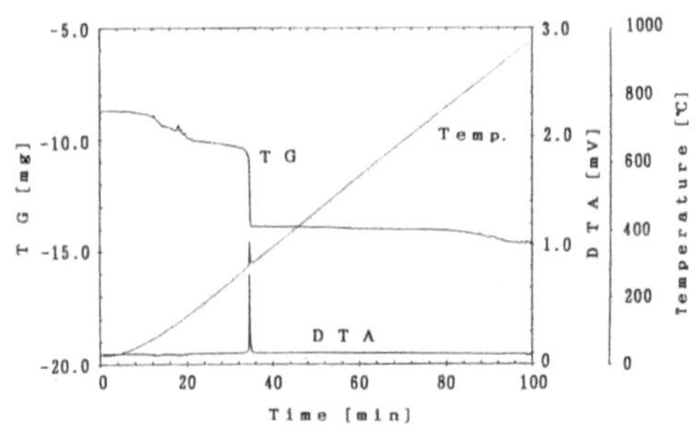

Fig. 1 TG-DTA curve of Y-Ba-Cu nitrates.

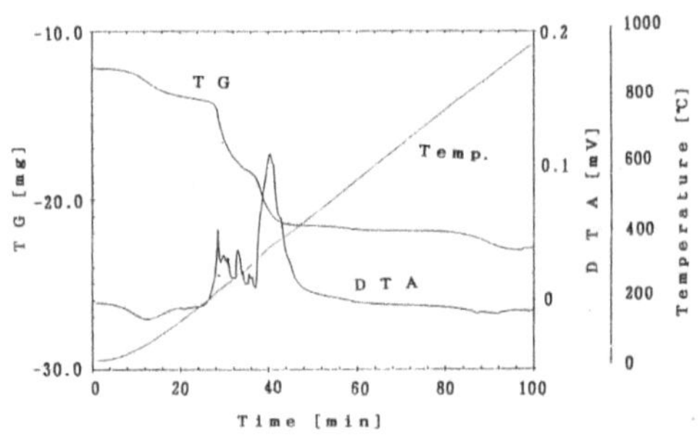

Fig. 2 TG-DTA curve of Y-Ba-Cu acetates.

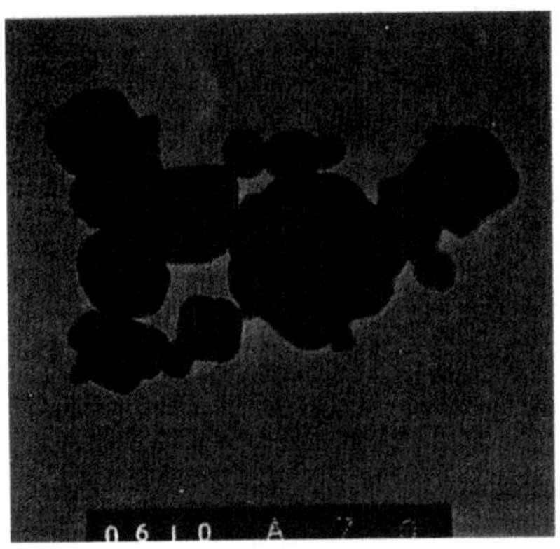

Fig. 3 TEM photograph of powders prepared by mist pyrolysis process(1253K).

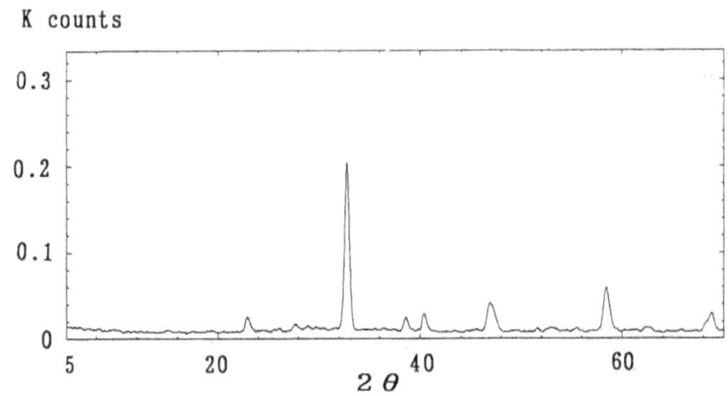

Fig. 4 X-ray diffraction pattern of powders prepared by mist pyrolysis process(1253K).

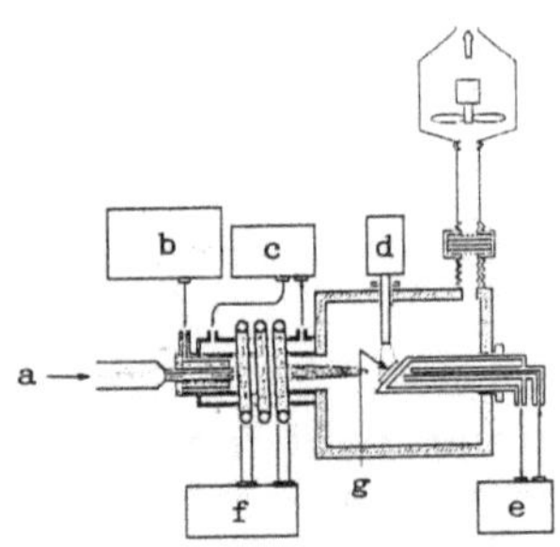

a powder inlet
b plasma gas(Ar)
c water
d image heater
e water
f RF coil
g substrate

Fig. 5 Schematic diagram of RF plasma deposition system.

The compositions and structures of samples were examined by inductively coupled plasma emission spectroscopy(ICP) and X-ray diffraction(XRD). The surface morphology were characterized by scanning electron microscopy(SEM) and transmission electron microscopy(TEM). The temperature dependence of resistivity was measured by a conventional four probe method.

RESULTS AND DISCUSSION

Low temperature dependence of resistivity was shown in Fig. 6. The sharpest superconducting transition and highest Tc(80K) were given by using Ag as a buffer and sintering in Ar flow at 1323K for 10 minutes followed by O_2 annealing under 1073K. Jc was achieved $1.8x10^4 A/cm^2$ at 65K. In the result of XRD(Fig. 7), the structure with c-axis of 123 phase perpendicular to the surface was found, while small amount of 211 phase was included. Deposition on substrate directly without Ag buffer or sintering in O_2 flow instead of Ar flow, with the same temperature cycle gave broader superconducting transition.

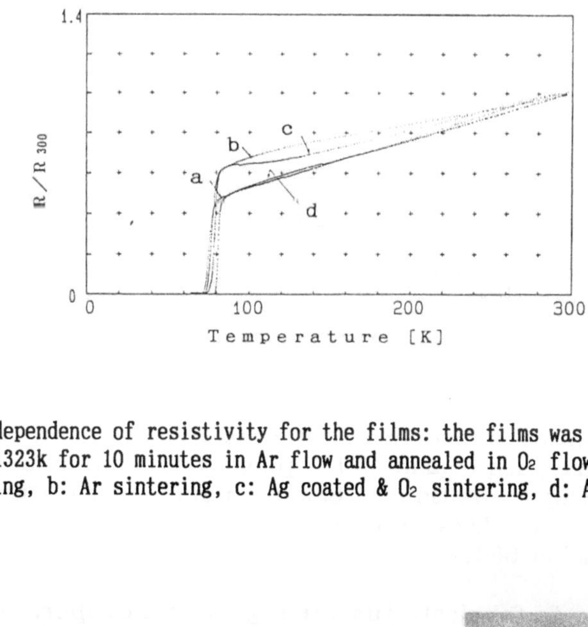

Fig.6 Temperature dependence of resistivity for the films: the films was produced at 1253K, sintered at 1323k for 10 minutes in Ar flow and annealed in O_2 flow under 1073K.
a: O_2 sintering, b: Ar sintering, c: Ag coated & O_2 sintering, d: Ag coated & Ar sintering

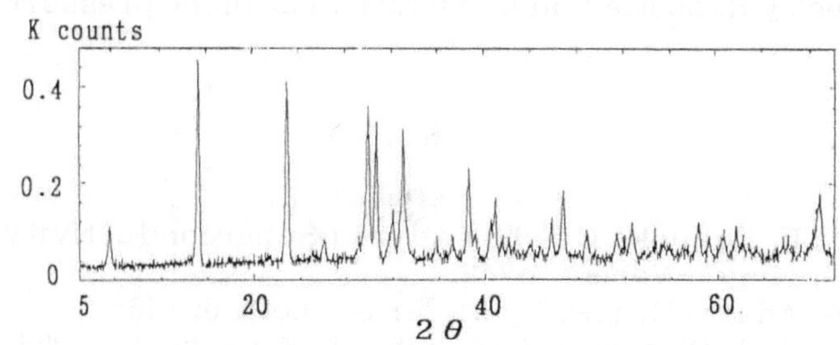

Fig. 7 X-ray diffraction patterns of the film sintered at 1323K in Ar flow followed by O_2 flow under 1023K.

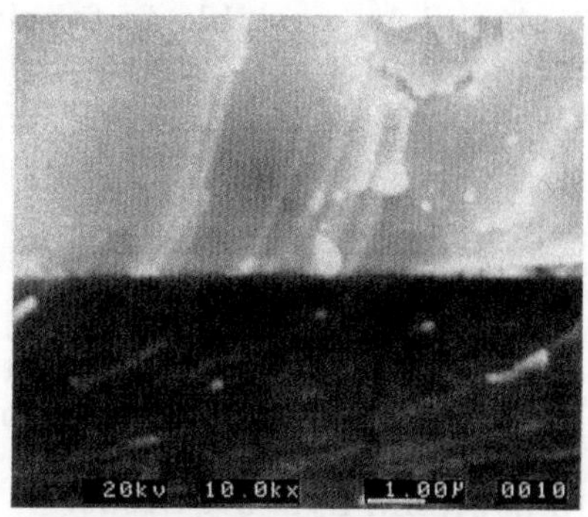

Fig. 8 SEM photograph of cross-sectional morphorogy of the film sintered at. 1323K in Ar flow followed by O_2 flow under 1073K.

Scanning electron micrograph of the film is shown in Fig. 8. Sintering in Ar flow at 1323K followed by O_2 flow under 1072K and using Ag as a buffer gave a dense and granular surface which was the melted texture. The sharp superconducting transition was given in spite of sintering at high temperature to melt. It is probably supposed that both treatments affect with little formation of weak links in grain boundaries, such as non-superconducting materials formed by the reaction of film materials with substrate as D. Dubreuil et. al. reported[6].

Evaporation of fine powders prepared by mist pyrolysis to deposit on MgO substrate was tried to fabricate films using RF plasma in ambient pressure and in low power(1.4KW) for evaporation. The result of XRD showed in Fig. 9. Though the deposition had only given the structure with random orientation and low crystallization, the surface appeared flat and smooth. It is found that the deposition which gives the texture with high orientation and with smooth surface will be easily achieved using such a simple apparatus.

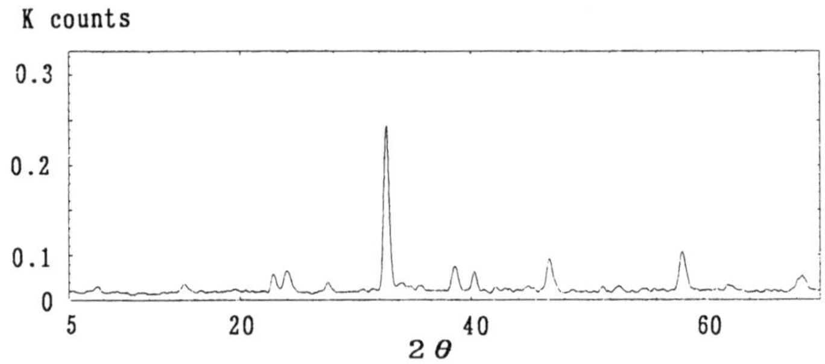

Fig. 9 X-ray diffraction pattern of the film deposited using RF plasma annealed at 973K in O_2 flow.

CONCLUSION

Deposition of fine powders prepared by mist pyrolysis on MgO(100) substrate could give the texture with c-axis of 123 phase perpendicular to the surface and sharp superconducting transition of 80K by using Ag as a buffer layer and by sintering in Ar flow at 1327K followed by O_2 annealing under 1073K. Jc was achieved $1.8x10^4 A/cm^2$ at 65K.

It should be possible to fabricate films using plasma evaporation of fine powders prepared by mist pyrolysis to directly deposited on substrate in ambient pressure and in low power for evaporation.

REFERENCES

[1] Ayabe M, Tatsugae R, Nishikawa H, Fukuoka C (1991) Adv. in Superconductivity -III ed. by Ishiguro T, Kitajima K, Springer-Verlag, p749.
[2] Okuyama K, Kousaka Y, Tohge N, Adachi M, (1987) J.Mater.Sci.Lett. 6: 1466.
[3] Tohge N, Tastumisago M, Minami T, Okuyama K, Adachi M, Kousaka Y (1988) Jpn.J.Appl.Phys. 27: L1086.
[4] Ogawa T (1989) Bull.Chem.Soc.Jpn. 62: 1844.
[5] Takao Y, Awano M, Takagi H (1991) Adv. in Superconductivity -III ed. by Ishiguro T, Kitajima K, Springer-Verlag, p745.
[6] Dubreuil D, Garry G, Lemaitre Y, Rogier L, Dieumegard D (1989) J.Less-Common Metals 151: 303.

Novel Chemical Analysis for Thin Films: Scanning Electron Microscopy & Total-Reflection-Angle X-ray Spectroscopy (SEM-TRAXS)

TOSHIO USUI, MASAYUKI KAMEI, YUJI AOKI, TADATAKA MORISHITA, and SHOJI TANAKA

Superconductivity Research Laboratory, ISTEC, Koto-ku, Tokyo, 135 Japan

ABSTRACT

We have developed a novel chemical analysis method for thin films: scanning electron microscopy & total-reflection-angle X-ray spectroscopy (SEM-TRAXS). In this method, the emission efficiency of characteristic X-rays, excited by the electron beam of a SEM, from elements composing a thin film is drastically enhanced, as compared to that in the conventional SEM-EDX method. As a result, the energy dispersive X-ray spectrum obtained from an 800Å-thick $YBa_2Cu_3O_x$ (YBCO) film on MgO substrate was quite clear and almost the same as that of a thick YBCO film or a bulk YBCO crystal because the intensity of the characteristic X-rays emitted from the MgO substrate was reduced below a detectable intensity.

KEYWORDS: YBCO thin film, SEM, TRAXS, SEM-TRAXS, X-ray chemical analysis

INTRODUCTION

Energy dispersive X-ray spectroscopy in scanning electron microscopy (SEM-EDX) is widely used for nondestructive chemical analysis of thin films. However, this X-ray spectroscopy is generally considered to be not suitable for chemical analysis of substrate-supported thin films less than 1000Å thick because the intensity of characteristic X-rays, excited by the electron beam of a SEM, emitted from the thin film is very weak, and thus insufficient for analysis, compared to that from the substrate.

On the other hand, Sewell *et al.* [1-5] showed that it is possible to identify adsorbates by measuring the characteristic X-rays, excited by the primary electron beam of reflection high energy electron diffraction (RHEED) during RHEED observations. Recently, Ino *et al.* [6-9] obtained much higher surface sensitivity using a Si(Li) solid-state X-ray detector (SSD); the intensity of the AgLα line emitted from a monolayer of Ag on Si(111) was almost equal to that of the SiKα line from the Si substrate. Regarding the surface sensitivity of this X-ray spectroscopy, the intensity of the characteristic X-rays emitted from atoms near a flat surface has a maximum when the X-ray take-off angle (θt) with respect to the sample surface is set at a critical angle (θc) for total external reflection of the expected X-rays [10]. Therefore, this spectroscopy was called "Total-Reflection-Angle X-ray Spectroscopy in RHEED experiments (RHEED-TRAXS)" [11].

In the present study, we have applied this TRAXS technique to X-ray chemical analysis of an 800Å-thick $YBa_2Cu_3O_x$ (YBCO) thin film on MgO(100) in SEM experiments and will discuss the grazing incidence effect of an electron beam of a SEM on the resulting X-ray spectra from the YBCO thin film.

EXPERIMENTAL

Figure 1 shows a schematic system diagram of the scanning electron microscopy & total-reflection-angle X-ray spectroscopy (SEM-TRAXS). The 800Å-thick YBCO film on MgO(100) used was prepared at 630°C by coevaporation of Y, Ba and Cu metals in an oxygen plasma [12]. The chemical composition of the YBCO film was determined as $YBa_{1.9}Cu_{2.8}O_x$ by inductively coupled plasma emission spectroscopy (ICP). A part of the edge of the film was connected to an aluminum holder for SEM observations with carbon tape so that the film did not become electrically charged during SEM-TRAXS experiments. A commercial Si(Li) SSD (Kevex delta) was used as an energy dispersive X-ray detector for the characteristic X-rays emitted from the sample. The organic window 1.8mm in diameter of the SSD corresponded to a 3°-width of θ_t with respect to the sample surface because no slit was used in front of the window. The resolution of the SSD was 0.15keV. In the SEM-TRAXS experiments, the glancing angles (θ_g) of the incident electron beam (20keV) of the SEM (Hitachi S2100) were 90°, 11° and 4° with respect to the film surface, and the θ_t of the SSD was set in the range of 0°-3°, in the vicinity of θ_c for the characteristic X-rays of the elements Y, Ba, Cu and O in the YBCO thin film [10].

RESULTS AND DISCUSSION

Figure 2 shows the change of the resulting energy dispersive X-ray spectra emitted from the 800Å-thick YBCO film on MgO as a function of θ_g of the electron beam of the SEM. CKα(0.28keV) emitted from the carbon tape and AlKα(1.49keV) from the Al holder are seen in Figs.2(b) and 2(c). These characteristic X-rays were excited by the reflected electron beam, a part of the incident beam impinged onto the sample surface, but they are negligible in the following discussion. In the case of θ_g=90°(a), that is, an usual SEM-EDX usage, the intensity of the characteristic X-ray lines YLα(1.92keV) for Y, BaLα(4.47keV), BaLβ1(4.83keV), BaLβ2(5.19keV) and BaLIIIab(5.25keV) for Ba, CuKα(8.04keV), CuKβ(8.80keV) and CuLα(0.93keV) for Cu, and OKα(0.53keV) for O from the 800Å-thick YBCO film was very weak and thus insufficient for chemical analysis, as compared to that of the MgKα(1.25keV) line from the MgO substrate as mentioned earlier. On the contrary, when θ_g=11°(b), the film sensitivity was much improved and the total intensity of the characteristic X-rays emitted from the YBCO film was more than twice as large as that from the substrate. Moreover, when θ_g=4°(c), that is, a typical SEM-TRAXS usage, the resulting X-ray spectrum changed drastically; it is almost the same spectrum as that of a thick YBCO film or a bulk YBCO sample. The characteristic X-rays of the elements Y, Ba, Cu and O composing the YBCO film were clearly observed and their intensity were sufficient for chemical analysis because no characteristic X-rays from the MgO substrate were observed.

This grazing incidence effect of an electron beam as discussed above can be easily explained as follows [6]. When an electron beam of a SEM with an acceleration voltage of 20kV impinges vertically onto the film surface (θ_g=90°), corresponding to Fig.2(a), the electron beam diffuses into a large volume of the substrate and forms a droplet (diameter >1μm) as schematically shown in Fig.3(a). Then the characteristic X-rays are excited and emitted from this entire droplet region. However, the region occupied by the thin film in the droplet is quite small, as compared to that of the substrate. Therefore, this is the reason why most of the characteristic X-rays in energy dispersive X-ray spectra obtained from a substrate-supported thin film are emitted from the substrate.

On the other hand, when the electron beam is incident on a flat and smooth surface at a small glancing angle such as $\theta g=4°$ in Fig.2(c), the electrons are scattered in the surface region, and then towards the outside (reflection) and towards the interior (penetration) of the sample, but the beam does not penetrate into the sample as deeply as shown in Fig.4(b). As a result, the emission efficiency of the characteristic X-rays from a thin film is significantly enhanced as seen from Fig.2(c).

In addition to the grazing incidence of an electron beam, it is also very important to set the X-ray take-off angle at the total external reflection angle of the expected X-rays in order to obtain high surface sensitivity in SEM-TRAXS [10]. This is related to the anomalous surface X-ray emission phenomenon [7-9, 13-20], in which the intensity of characteristic X-rays emitted from the surface region is maximal at θc with respect to the sample surface, reported for the first time by Becker *et al.* for X-ray excitation [13] and by Hasegawa *et al.* for electron excitation [7]. This phenonenon is of great interest from the viewpoint of higher surface sensitivity in SEM-TRAXS. Further SEM-TRAXS experimental data on this phenomenon will be reported elsewhere.

In conclusion, we have developed a novel chemical analysis method of SEM-TRAXS for thin films and have applied this method for X-ray chemical analysis of the 800Å-thick YBCO film on MgO(100); this SEM-TRAXS method was found to be a significantly effective tool for X-ray chemical analysis of very thin films. The SEM-TRAXS system used in the present study was a prototype so that we confirmed the usefulness of the SEM-TRAXS method for chemical analysis of very thin films. Thus, the X-ray sensitivity in SEM-TRAXS can be much improved by optimizing the SEM-TRAXS conditions. In addition, an electron gun of a SEM can be also used as a RHEED gun if a RHEED screen is set on the opposite side of the gun. Therefore, this SEM-TRAXS method will be advanced to be a SEM-RHEED-TRAXS method; surface topographic and morphologic observations by SEM, identification of surface crystalline structure by RHEED, and chemical analysis by TRAXS are possible simultaneously at selected points on a sample.

References

[1] Sewell and M. Cohen, *Appl. Phys. Lett.* **11** (1967) 298.
[2] Sewell and D.F. Michell, *J. Appl. Phys.* **42** (1971) 5879.
[3] Sewell and D.F. Michell, *Surf. Sci.* **55** (1976) 367.
[4] Michell, P.B. Sewell and M. Cohen, *Surf. Sci.* **61** (1976) 355.
[5] Michell, P.B. Sewell and M. Cohen, *Surf. Sci.* **69** (1977) 310.
[6] Ino, T. Ichikawa and S. Okada, *Jpn. J. Appl. Phys.* **19** (1980) 1451.
[7] Hasegawa, S. Ino, Y. Yamamoto and H. Daimon, *Jpn. J. Appl. Phys.* **24** (1985) L.387
[8] Hasegawa, H. Daimon and S. Ino, *Surf. Sci.* **186** (1987) 138.
[9] Ino, Reflection High Energy Electron Diffraction and Reflection Electron Imazing of Surfaces, Eds. P.K. Larsen and P.J. Dobson, p.3 (Plenum Pub. Co., New York, 1988).
[10] In the SEM-TRAXS method, it is very important to estimate the critical angle θc for the total external reflection of the expected X-ray. The theoretical value of θc is approximately given by $\theta c=1.14\sqrt{\rho}/E$ [21], where θc is expressed in degrees, E is the energy of the reflected X-rays in keV, and ρ is the bulk density in g/cm^3. For example, from this equation using $\rho=6.3$ g/cm^3 for YBCO, the values of θc(CuLα-YBCO) and θc(CuKα-YBCO) are calculated as 3.1° and 0.35°, respectively. Note that the value of θc strongly depends on the energy of the expected X-ray.
[11] See ref.[7-9] with regard to TRAXS by electron excitation (RHEED-TRAXS) and also ref.[13-20] with regard to TRAXS by X-ray excitation.
[12] Takahashi, Y. Aoki, T. Usui, R. Fromknecht, T. Morishita and S. Tanaka, *Physica* **C175** (1991) 381.
[13] Becker, J.A. Golovcherko and J.R. Patel, *Phys. Rev. Lett.* **50** (1983) 153.

[14] Patel and J.A. Golovcherko, *Phys. Rev. Lett.* **50** (1983) 1858.
[15] Suzuki, *Phys. Rev.* **B39** (1989) 3393.
[16] Feidenhans'l, *Surf. Sci. Rep.* **10** (1989) 105.
[17] Sasaki and K. Hirokawa, *Appl. Phys.* **A50** (1990) 397.
[18] Sasaki and K. Hirokawa, *Appl. Phys.* **A52** (1991) 28.
[19] Sasaki, *Jpn. J. Appl. Phys.* **30** (1991) L761.
[20] Sasaki and K. Hirokawa, *Appl. Phys. Lett.* **58** (1991) 1384.
[21] Kaelble, *Handbook of X-rays*, chap.48 (Mc-Graw Hill, New York,1969).

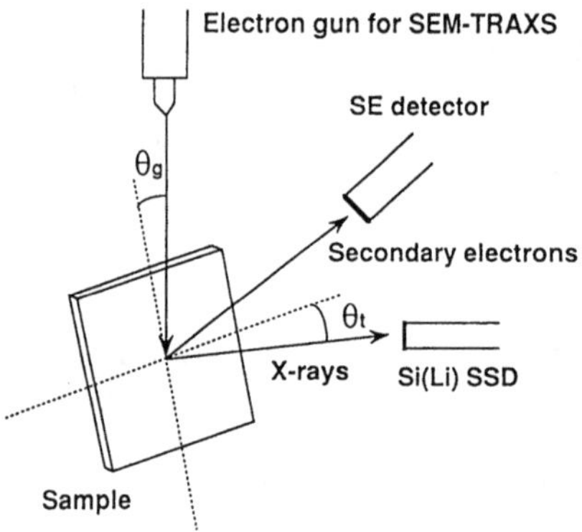

Fig. 1. Schematic system diagram of scanning electron microscopy & total-reflection-angle X-ray spectroscopy (SEM-TRAXS).

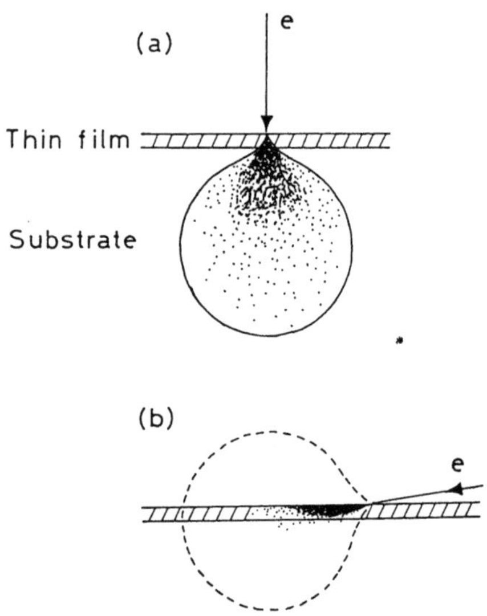

Fig. 3. Schematic electron diffusion range in a substrate-supported thin film of the incident electron beam of a SEM: (a) vertical incidence (a conventional SEM-EDX usage) and (b) grazing incidence (a typical SEM-TRAXS usage).

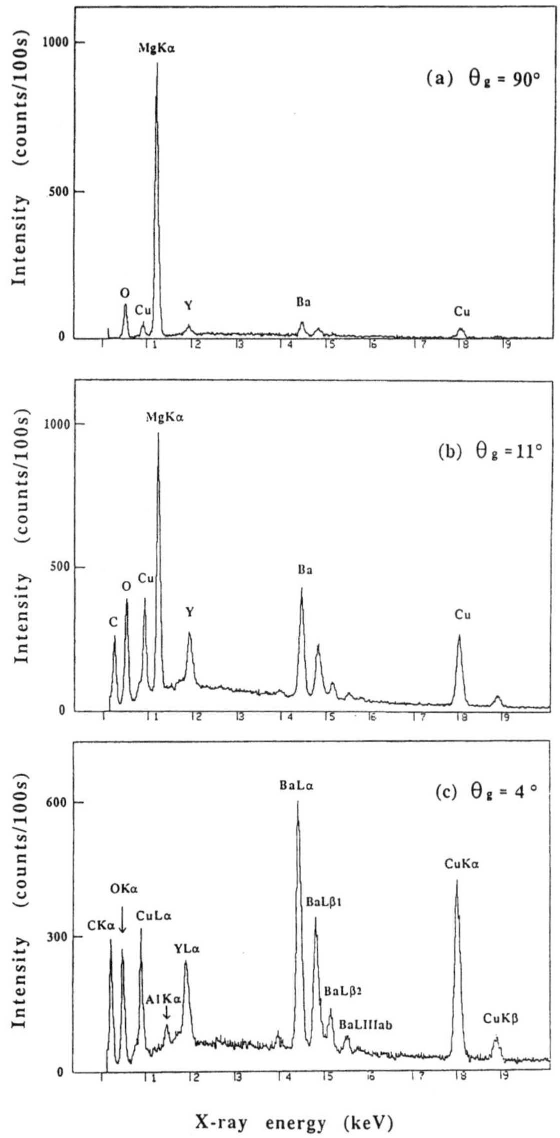

Fig. 2. Change of the resulting energy dispersive X-ray spectra from the 800Å-thick YBCO film on MgO as a function of the glancing angle of the incident electron beam of the SEM: (a)θ_g=90°, (b)θ_g=11° and (c)θ_g=4°.

Magnetic Penetration Depth of $YBa_2Cu_3O_{7-\delta}$ Thin Films Determined by the Power Transmission Method

VERNON O. HEINEN, FÉLIX A. MIRANDA, and KUL B. BHASIN

National Aeronautics and Space Administration, Lewis Research Center, Cleveland, OH 44135, USA

ABSTRACT

A power transmission measurement technique has been used to determine the magnetic penetration depth (λ) of $YBa_2Cu_3O_{7-\delta}$ superconducting thin films on $LaAlO_3$ within the 26.5 to 40.0 GHz frequency range, and at temperatures from 20 to 300 K. Values of λ ranging from 1100 to 2500 Å were obtained at low temperatures. The anisotropy of λ was determined from measurements of c-axis and a-axis oriented films. An estimate of the intrinsic value of λ of 90±30 nm was obtained from the dependence of λ on film thickness. The advantage of this technique is that it allows λ to be determined non-destructively.

KEY WORDS: Magnetic penetration depth, $YBa_2Cu_3O_{7-\delta}$ thin films, Power transmission measurements.

INTRODUCTION

Since the discovery of the high-transition-temperature superconductors (HTS) by Bednorz and Muller [1], and Wu et al. [2] numerous studies have been done to determine the potential of these materials for practical applications and to learn more about the nature of their superconducting state. One of the parameters of most relevance in the evaluation of the suitability of the new HTS thin films for microwave applications is the magnetic penetration depth (λ). This parameter, which is a measure of the distance a magnetic field penetrates into the superconducting material, is important not only because it can provide information about the fundamental mechanisms for superconductivity in these materials but also because of its sensitivity to the quality of the superconductor near its surface. In particular, for thin films of thickness t in which $\lambda \sim t$, λ serves as an excellent figure of merit to determine the quality of the entire film, and provides necessary information for the design of superconductor-based microwave integrated circuits [3]. To date, values of λ for $YBa_2Cu_3O_{7-\delta}$ single crystal and ceramics have been reported using various experimental techniques, such as muon-spin-rotation and low field dc magnetization measurements [4-6]. Others have determined this parameter in HTS thin films from a variety of microwave measurements [7-10]. However, some of the most favored microwave techniques used for the determination of λ usually involve the alteration of the film by way of patterning it into microstrip resonators and other transmission line structures [9,10].

In this paper we present results on the determination of λ for $YBa_2Cu_3O_{7-\delta}$ HTS thin films using a power transmission method. One of the merits of this technique is that it does not alter the film in any way so that the same film can be used for further characterizations or for the fabrication of a microwave device. Values of λ over a wide range of temperatures (T) and film thickness (d) are discussed.

EXPERIMENTAL

The $YBa_2Cu_3O_{7-\delta}$ superconducting thin films considered in this study were deposited by pulsed laser ablation on $LaAlO_3$ substrates. The substrates were approximately 1.0 cm^2, with thickness of 10 to 20 mils and had a (100) crystallographic orientation. The thickness of the films ranged from approximately 800 to 5000 Å, as determined by using a surface profilometer over steps etched at the corner of each film with undiluted phosphoric acid (H_3PO_4). Measurements of dc-resistance (R_{dc}) versus temperature were performed for all the films analyzed in this study using a standard four-point probe method. The temperature at which the measured resistance fell below the resolution of the measuring instrument ($\sim 10^{-8}$ Ω-cm) was defined as the dc-transition temperature (T_c^{dc}). The transition temperature for all c-axis oriented films was between 84 and 91 K. The crystallographic orientation of the films was determined by x-ray diffraction analysis.

The main components of the experimental apparatus are an HP-8510B network analyzer and a closed-cycle helium gas refrigerator, which are both controlled by an HP 9000-216 computer. The network analyzer is coupled to the refrigerator by Ka band (26.5 to 40.0 GHz) rectangular waveguides. The measurement technique compares the transmitted signal against the incident microwave signal to determine the transmission coefficients. The data measured in this way are then stored by the computer for subsequent analysis.

All the measurements were made under vacuum ($<10^{-3}$ torr) in a custom-made aluminum vacuum chamber designed to fit on top of the external shield of the refrigerator and to give access to the waveguides connecting the network analyzer with the refrigerator. Mica windows were used as vacuum windows in the waveguide. The sample was oriented perpendicular to the microwave source by clamping it between two waveguide flanges thermally connected to the cold head of the refrigerator through a Au-plated Cu plate. The film side of the sample was directed toward the incident microwave signal as shown in Fig. 1. The system was calibrated before the beginning of each measurement cycle to account for the impedance of the waveguide network. The noise level for our measurements was determined to be below - 60 dB. Background attenuation and phase corrections were made by subtracting the transmitted power as a function of temperature in the absence of the sample from the data obtained with the sample in place. Measurements of bare $LaAlO_3$ substrates were used to determine any temperature dependence of their dielectric constant. The results of these measurements have been previously reported [11]. The temperature of the film was measured with two silicon diodes placed in a 1/8 inch diameter hole on top of each of the sample's supporting flanges. The accuracy of these diodes is ± 1.0 K from 1.4 to 100 K, and 1% of the actual temperature in the range from 100 to 325 K.

ANALYSIS

The surface reactance (X_s) of a superconductor is defined in terms of λ. Therefore knowledge of either one of these parameters allows the determination of the other. The surface reactance can be expressed in terms of the microwave power transmitted through them. Sridhar [12] was the first to use the microwave power transmitted through Sn thin films, placed at the end walls of a cylindrical copper resonant cavity, to determine the effective surface reactance X_{eff}, as a function of temperature. Therefore, since we were able to measure the microwave transmitted power directly, we have followed Sridhar's method for obtaining X_{eff}. After satisfying the boundary conditions of the problem, the expression for X_{eff} in terms of the transmitted power is

$$X_{eff} = R_N^{MW} \cosh(d/\lambda) \left| P^S_{xmitt}/P^N_{xmitt} \right|^{1/2} \tag{1}$$

where R_N^{MW} is the normal state surface resistance at $T = T_c^{MW}$, which is the transition temperature determined from the power transmission measurements, d is the film thickness, and P^N_{xmitt} and P^S_{xmitt} are the power transmitted in the normal and the superconducting states, respectively. For thin films, X_{eff} is given by

$$X_{eff} = X_s \coth(d/\lambda) \tag{2}$$

where $X_s = \omega\mu_0\lambda$, $\omega/2\pi$ = f is the frequency and μ_o is the permeability of free space. Thus, to determine λ we solved equation (1) for X_{eff} by assuming an initial value of λ = 100 Å. Using the resulting value of X_{eff} in (2), we calculated a new value of λ. This value was then used in equation (1) and the process was repeated until the value of λ changed by less than 0.1% between iterations. This value was taken to be the value of λ at the temperature under consideration.

RESULTS AND DISCUSSION

Figure 2 shows a plot of the power transmission coefficient P versus temperature for one of the films analyzed in this study. For a given film the microwave power transmission coefficient was measured with a repeatability of ± 5% over the entire temperature range. This error was estimated by measuring several films repeatedly with the film being removed and replaced between each set of measurements. The metallic character of the films is evidenced by the positive slope of P for $T > T_c^{MW}$. At T_c^{MW}, the transmitted power decreased abruptly and continued decreasing with decreasing temperature, leveling off at low 1temperatures. We observed that the abruptness of this transition and the temperature at which the transmitted power started to level off was closely related to the film quality, as determined by measurements of the surface resistance (R_s) and the complex conductivity ($\sigma^* = \sigma_1 - j\sigma_2$). Different power levels (~0.1, 1.0, and 16 mW) were used in probing the samples. Note that although for low power levels and low temperatures the data are noisier, the transmitted-to-incident power ratio is the same for all incident powers. This indicates that the superconducting state is not altered by the incident power, at least for the different power levels and frequency range considered in this study. The absence of any appreciable shift of T_c^{MW} for most of the films, within a resolution of 0.5 K supports this argument. This behavior was typical of all the films considered independent of thickness.

As discussed in the last section we have used eqs. (1) and (2) to determine λ for the HTS thin films. Figure 3 shows a plot of λ versus film thickness (d) at 30 K. The anisotropy of λ is indicated in this plot by the large value of λ for the film represented by the ♦ symbol which is a predominantly a-axis oriented film. Taking the ratio between λ for this film and c-axis oriented films of comparable thickness, yields $\lambda_c/\lambda_{ab} \sim 3.8$ and $\lambda_c/\lambda_{ab} \sim 5.7$, respectively. These ratios are consistent with the value of $\lambda_c/\lambda_{ab} \sim 5$ reported by others for single crystals and high purity samples of the $YBa_2Cu_3O_{7-\delta}$ superconductor [4,14-15]. Observations by others [7] have shown that as the film thickness increases it is more difficult to preserve their c-axis orientation, i.e., the number of a-axis oriented grains increases with increasing d. Since the shielding currents along the c-axis oriented grains are smaller than those along the a-b plane, one expects λ to increase with film thickness as seen in Fig. 3.

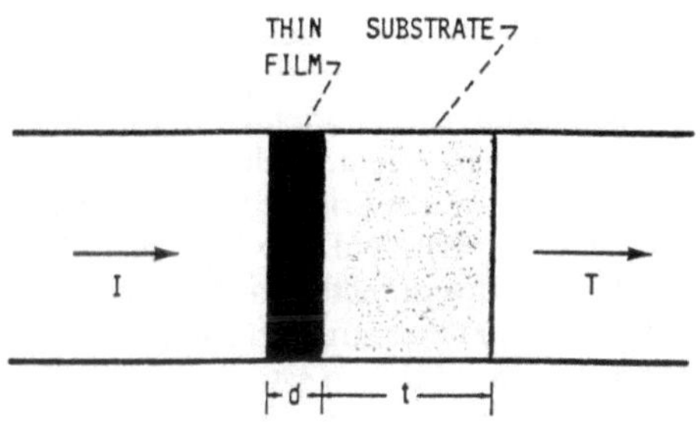

Fig. 1 Schematic of rectangular waveguide propagating the TE_{01} mode with its entire cross section covered by a high-T_c superconducting thin film of thickness d deposited on a dielectric substrate of thickness t and index of refraction n.

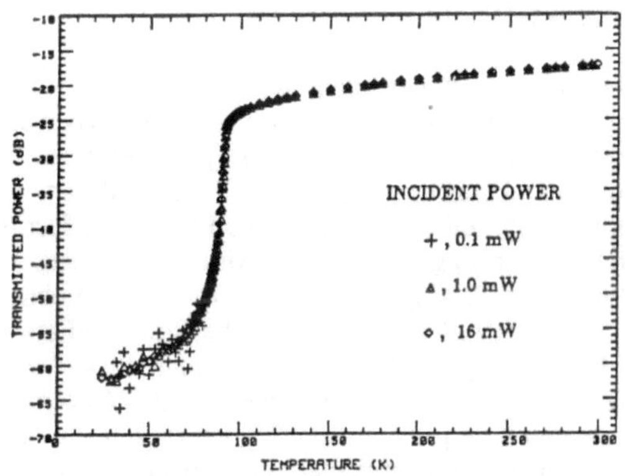

Fig. 2 Power transmission coefficient versus temperature for a laser ablated $YBa_2Cu_3O_{7-\delta}$ thin film (2400 Å) on $LaAlO_3$ (20 mils).

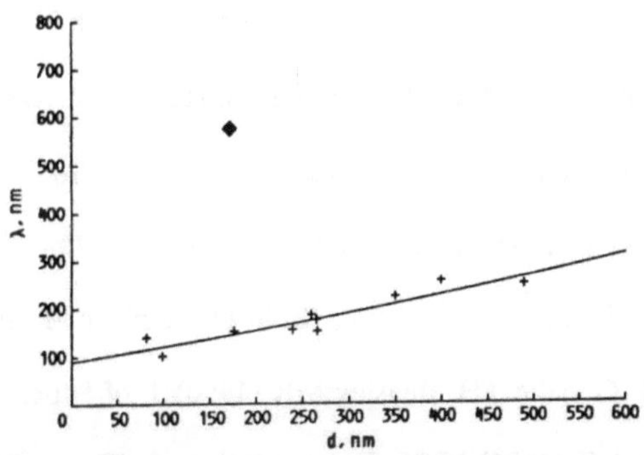

Fig. 3 Magnetic penetration depth (λ) versus film thickness (d) for $YBa_2Cu_3O_{7-\delta}$ thin films on $LaAlO_3$ (20 mils). The ♦ symbol represents an a-axis oriented film and the solid line represents a second degree polynomial fit.

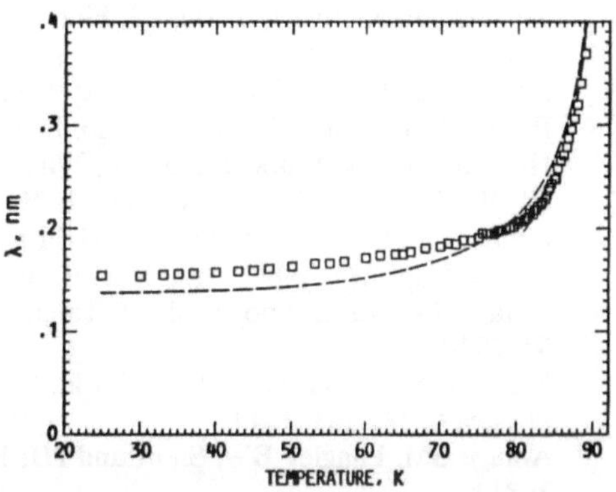

Fig. 4 Magnetic penetration depth (λ) versus temperature for a $YBa_2Cu_3O_{7-\delta}$ thin film on $LaAlO_3$ (20 mils). The dashed (solid) line represents a fit to the data using the temperature dependence for λ according to the two-fluid model (weak coupling BCS theory in the clean limit).

The solid line in Fig. 3 represents a fit to the data using a second degree polynomial regression analysis. This gives a value for the penetration depth corresponding to zero film thickness of $\lambda(d=0) = 90 \pm 30$ nm. This value of λ represents the smallest possible value that can be obtained for a c-axis oriented film free from grain boundary effects and other non-superconducting inhomogeneities; i.e., the intrinsic value of λ at low temperatures for the $YBa_2Cu_3O_{7-\delta}$ superconductor. To our knowledge, this is the first time that this value has been estimated experimentally for any of the HTS materials.

The temperature dependence of λ for a film (d=2400 Å) is shown in Fig. 4. Using the expressions for the temperature dependence of λ in the two-fluid model (TFM) approximation,

$$\lambda(T) = \lambda(0)[1 - (T/T_c)^4]^{-1/2} \tag{3}$$

and according to the weak-coupling BCS theory in the clean limit [14,16],

$$\lambda(T) = \lambda(0)[2(1 - T/T_c)]^{-1/2}, \quad T \rightarrow T_c \tag{4}$$

we have estimated the zero-temperature penetration depth ($\lambda(0)$), by fitting the measured λ to equations (3) and (4) using a standard least-squares fit. For the film in Fig. 4 we obtained $\lambda_{BCS}(0) = 96 \pm 9$ nm and $\lambda_{TFM}(0) = 138 \pm 13$ nm. For another film of d=1769 Å film we found $\lambda_{TFM}(0) = 145 \pm 12$ nm and $\lambda_{BCS}(0) = 113 \pm 8$ nm. These values are consistent with those reported by others for c-axis oriented thin films and single crystals of $YBa_2Cu_3O_{7-\delta}$ [4,17,18].

CONCLUSIONS

The magnetic penetration depth of $YBa_2Cu_3O_{7-\delta}$ HTS thin films on $LaAlO_3$ was determined using a power transmission method. We were able to experimentally verify the intrinsic anisotropy of λ for this HTS superconductor by measuring this parameter in c-axis and predominantly a-axis oriented films. As expected from the anisotropy of these HTS materials, the value of λ for the a-axis oriented film was larger than the values obtained for c-axis oriented films. In fact, the ratio of λ for an a-axis oriented film to that of c-axis oriented films agrees very well with that reported by others for single crystals and very high quality thin films of the $YBa_2Cu_3O_{7-\delta}$ superconductor. We also observed that λ increased with increasing film thickness which is consistent with the increase of the number of a-axis oriented grains and other structural and material defects with increasing film thickness. From the thickness dependence of λ we were able to estimate, for the first time, the intrinsic value of λ for the $YBa_2Cu_3O_{7-\delta}$ superconductor. Finally, the power transmission technique presented here provides a non-destructive technique for the determination of λ for HTS thin films, in contrast to microstrip resonator methods in which the film must be patterned.

REFERENCES

1. Bednorz JG, Mueller KA (1986) Z. Physics B. **64**: 189
2. Wu MK, Ashburn JR, Torng CJ, Hor PH, Meng RL, Gao L, Huang ZJ, Wang YQ, Chu CW (1987) Phys. Rev. Lett. **58**: 908
3. Bhasin KB, Warner JD, Romanofsky RR, Heinen VO, Chorey CM, Kong KS, Lee HY, Itoh T (1990) IEEE MTT-S Int. Microwave Symp. Dig. **1**: 269
4. Harshmann DR, Schneemeyer LF, Waszczak JV, Aeppli G, Cava RJ, Batlogg B, Rupp LW, Ansaldo EJ, Williams DL (1989) Phys. Rev. B **39**: 851
5. Drabeck L, Carini JP, Gruner G, Hylton T, Char K, Beasley MR (1989) Phys. Rev. B **39**: 785
6. Chi H, Nagi AD (1989) Phys. Rev. B **40**: 7361
7. Anlage SM, Sze H, Snortland HJ, Tahara S, Langley B, Eom CB, Beasly MR (1989) Appl. Phys. Lett. **54**: 2710
8. Klein N, Muller G, Orbach S, Piel H, Chaloupka H, Roas B, Schultz L, Klein U, Peiniger M (1989) Physica C **162-164**: 1549
9. Anlage SM, Langley BW, Snortland HJ, Eom CB, Geballe TH, Beasley MR (1990) J. of Supercond. **3**: 311
10. Chorey CM, Kong KS, Bhasin KB, Warner JD, Itoh T (1991) IEEE Trans. Microwave Theory Tech. **39**: 1480
11. Miranda FA, Gordon WL, Bhasin KB, Ebihara BT, Heinen VO, Chorey CM (1990) Microwave and Opt. Tech. Lett. **3**: 11
12. Sridhar S (1988) J. Appl. Phys. **63**: 159
13. Miranda FA, Gordon WL, Bhasin KB, Heinen VO, Warner JD (1991) J. Appl. Phys. **70**: to be published
14. Schilling A, Hulliger F, Ott HR (1990) Physica C **168**: 272
15. Piel H, Muller G (1991) IEEE Trans. on Magn. **27**: 854
16. Muhlschlegel B (1959) Z. Phys. **155**: 313
17. Krusin-Elbaum L, Greene RL, Holtzberg F, Malozemoff AP, Yeshurun Y (1989) Phys. Rev. Lett. **62**: 217
18. Klein N, Chaloupka H, Muller G, Orbach S, Piel H, Roas B, Schultz L, Klein U, Peiniger M (1990) J. Appl. Phys. **67**: 6940

Zero-Field-Cooled, Field-Cooled and Remanent Magnetization of Epitaxial $YBa_2Cu_3O_y$ Films

H. OBARA, S. KOSAKA, Y. YOKOYAMA, M. UMEDA, and Y. KIMURA

Electrotechnical Laboratory, Tsukuba, Ibaraki, 305 Japan

ABSTRACT

Systematic measurements of the temperature, magnetic field and precise sample-size dependence of the magnetization in epitaxial $YBa_2Cu_3O_y$ films have been performed. $YBa_2Cu_3O_y$ films have been prepared by molecular beam epitaxy (MBE). The dc magnetization of films was measured by means of a SQUID magnetometer, and the sample size was precisely controlled by means of a micropatterning technique. We have measured zero-field-cooled, field-cooled and remanent magnetization in various temperatures and fields, and tried a quantitative description of the magnetization of films. Moreover, the flux-density distribution was directly observed through the use of a magneto-optical Faraday effect. The behavior of dc magnetization is discussed in three parts: the almost perfectly diamagnetic region, the transition region and the field-cooled region. Using a uniformly magnetized ellipsoid model and the Bean critical state model, a quantitative description of the magnetization of films was possible and the direct observation of the flux-density distribution confirmed our model. From these experimental results, a quantitative discussion of the magnetization in the oxide superconductor films over almost the entire magnetization curve has become possible.

KEY WORDS: oxide superconductor, $YBa_2Cu_3O_y$ thin film, magnetization

INTRODUCTION

Magnetization measurement is a powerful method for basic experiments of superconductors and important as are conductivity measurements. However, good quality samples are needed to observe the intrinsic properties of superconductors. Thin-film processing technique of oxide superconductors have progressed rapidly, and many good-quality films suitable for the measurements of intrinsic properties have been reported.

On the other hand, the magnetization measurement of thin films is difficult because of the large demagnetization effect and the small volume of thin films.

In the present study, we have measured the temperature, magnetic field and precise sample-size dependence of the magnetization in epitaxial films prepared by MBE. In the high-quality films prepared by MBE, the uniform flux-density distribution was estimated from the magnetization measurements and was confirmed by a direct observation of the flux-density distribution using a magneto-optical Faraday effect. Moreover the H_{c1} of thin films could be determined from the temperature dependencies of the field-cooled (FC) magnetization of films. From these experimental results, a quantitative discussion of the oxide superconductor films has become possible in various temperatures and fields.

EXPERIMENTAL

In the present work, samples were prepared by coevaporation in an ultra-high vacuum (UHV) system for MBE. The background pressure was 1×10^{-10} Torr, and the pressure during deposition was less than 1×10^{-4} Torr. The evaporation sources were three effusion cells (the maximum temperature is 1800°C) and Y, Ba and Cu metal elements were individually evaporated. The deposition rate was typically ~ 0.5 Å/s and the substrate temperature was about 650°C. For the oxidization of the samples, we used an electron cyclotron resonance (ECR) ion source and oxygen gas was introduced to the ECR ion source during the deposition of films. Details of the sample preparation have been described elsewhere [1].

dc magnetization measurements were performed by means of a Quantum Design SQUID magnetometer. We canceled the terrestrial magnetism in the zero-field-cooled (ZFC) process and achieved a very low field ($\sim$ 0.1 Oe), which was measured by means of a Pb shot. For a systematic study of the sample-size effect, films were patterned in a disk shape by using a photoresist mask and ethylenediaminestetraacetic acid (EDTA) for chemical etching [2]. The shaped samples were set in a quartz tube, and a magnetic field was applied perpendicular to the films.

Furthermore, the flux-density distribution of the films was directly observed by means of the magneto-optical Faraday effect of the iron garnet film, and details on experimental technique have been given elsewhere [3]. The observed flux-density distribution was compared with the magnetization data.

RESULTS AND DISCUSSION

Figure 1 shows the sample size dependence of the ZFC magnetization at 10 K of our epitaxial $YBa_2Cu_3O_y$ film. Films were patterned in a group of disks having radii of 1 mm, 48 μm, 23 μm and 10.5 μm. For a sample showing perfect diamagnetism, i.e., $\chi = -1/4\pi$, using a uniformly magnetized ellipsoid model, M reduces to

$$M = -\frac{1}{4\pi}\frac{2R}{\pi d}H_0 \, .$$

This equation shows that the magnetization is proportional to R, and this relation is clearly observed in Fig. 1, except for the smallest sample, in which the penetration depth effect or the magnetization of the substrate and sample holder is not negligible. These ZFC magnetization measurements indicate that the film showed the perfect diamagnetism and the uniformly magnetized ellipsoid model of thin films is available at low temperature and in low magnetic field.

In higher field and temperature, the flux penetrates from the edge of films. The simple Bean critical-state model [4] predicts that the magnetization is proportional to the critical current density, J_c, and the sample radius, R, after the flux meets at the center of sample. This magnetization can be easily measured using a remanent magnetization measurement, i.e., when the sample is field-cooled and field is turn off at low temperature. The sample-size dependence of the remanent magnetization was shown in Fig. 2. The radii of the samples were same as that of Fig. 1. The linear sample-size dependence was clearly observed from 10 μm to 1 mm. This indicate that the shielding current flows throughout the sample and the Bean critical-state model is fully established.

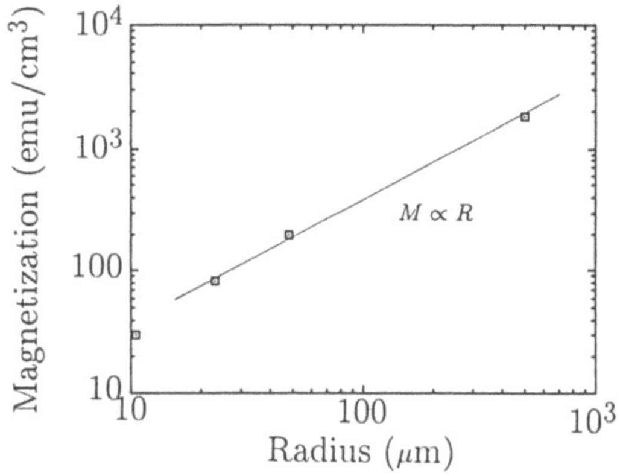

Fig.1 Sample-size dependence of the ZFC magnetization at 10 K.

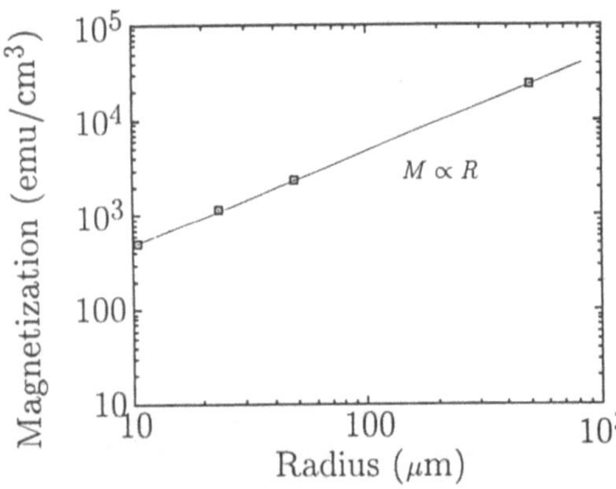

Fig.2 Sample-size dependence of the remanent magnetization.

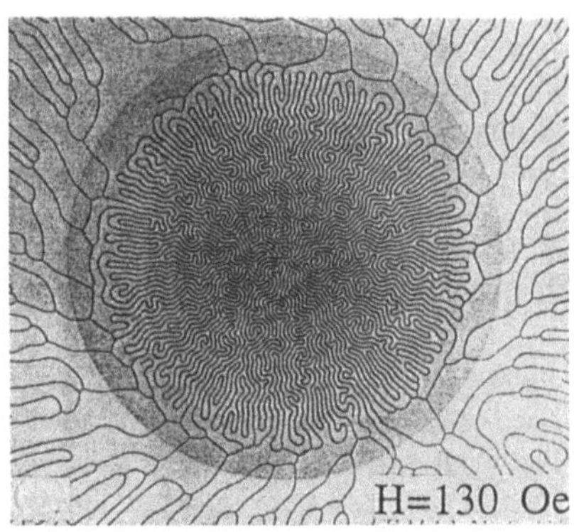

Fig.3 The domain patterns in a iron garnet film on the $YBa_2Cu_3O_y$ film.

The uniform flux penetration was also confirmed using a magneto-optical Faraday effect of the iron garnet film. Figure 3 shows the domain patterns in a iron garnet film on the $YBa_2Cu_3O_y$ film. The film was cooled in a ZFC process and field was applied at 10 K. With increasing the applied magnetic field, the flux penetrated uniformly into the film from the edge. From the sample-size dependence of the remanent magnetization and the direct observation of the flux-density distribution, we concluded that there is no granularity in our epitaxial $YBa_2Cu_3O_y$ films and the J_c of the films can be estimated from the magnetization measurements.

Finally, we have tried to measure the FC magnetization of thin films. The FC magnetization is much smaller than the ZFC magnetization and the measurements of the FC magnetization of films is difficult. Using a SQUID magnetometer, however, the observation of the FC magnetization in thin films was possible. Also in the FC process, we have measured the sample-size dependence of the saturated FC magnetization at low temperature, which is shown in Fig. 4. Compared with the ZFC and the remanent magnetization, the sample-size dependence of the FC magnetization was quite different and no linear sample-size dependence was observed. This results indicates that the flux-density distribution is not simple in the FC process.

Figure 5 shows the magnetic-field dependence of the apparent susceptibility, M/H, where M is the saturated FC magnetization and H is the applied magnetic field. The radii of measured samples A, B and C are 48, 23 and 10.5 μm, respectively. If the sample magnetizes uniformly, the magnetization M expected to be proportional to the magnetic field H, and M/H remains constant irrespective of the field. On the other hand, if there is some flux-density distribution, such as the critical state in a mixed state, M/H is expected to show some field dependence. In the full penetration in the critical state, i.e., when the flux meets at the center of a sample, M is expected to be constant if the field dependence of J_c is negligible. The M/H of sample A, which has the largest radius, was almost constant. In the smaller samples, the reduction of M/H was observed, and M/H of sample C was seen to coincide with the solid line, which shows constant M.

The complex nature of the FC magnetization in thin films can be explained by the large shape effect of thin films and is discussed as follows. In the FC process, the flux is excluded from the superconductor by the Meissner effect. The rapid change of the magnetic induction, B, between the outside and inside of a sample is impossible and is restricted by J_c. The region near the edge of the sample goes into the critical state to maintain the difference of B. In the model of Matsushita et al. [5], the flux-density distribution is considered to be frozen below T_{c1}, when the H_{c1} becomes equal to the applied magnetic field, and the FC magnetization is considered to be due to the frozen flux-density distribution. We estimate the flux-penetration length, which is the region in the critical state that maintains the difference in B between the inside and outside of the sample to be about 8 μm from the edge at 5 Oe. The flux-penetration length is much shorter than the radius of sample A.

On the other hand, the flux penetration is not negligible in sample C (the radius is 10.5 μm), and the sample is entirely in the critical state above 10 Oe. Therefore, if we neglect the field dependence of J_c, the magnetization remains constant irrespective of the field. A constant M was observed for sample C and is clearly shown in Fig. 5. In sample B, the flux penetration is only effective in relatively high fields because the flux penetration length becomes longer in such a high field.

Moreover, we have measured the temperature dependence of the FC magnetization. The onset of the FC magnetization is correspond to the T_{c2}, when the H_{c2} becomes equal to the applied magnetic field. Moreover, the offset of the FC magnetization is correspond to the T_{c1}, because the flux-density distribution freezes below T_{c1}. We plot the T vs dM/dT in Fig. 6. The onset and the offset of the FC magnetization were clearly shown in this plot.

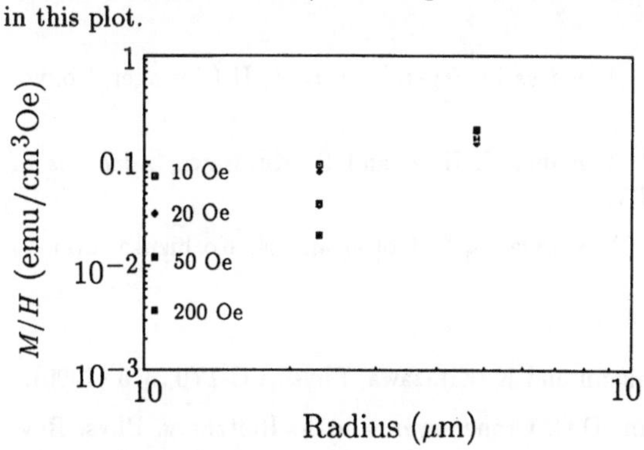

Fig.4 Sample-size dependence of M/H (M is the FC magnetization and H is the applied magnetic field) in several magnetic fields.

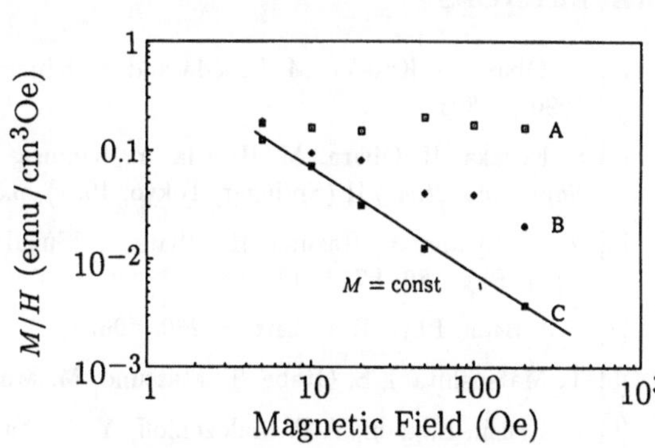

Fig.5 Magnetic field dependence of M/H for different sample sizes. Radii of samples A, B and C are 48, 23 and 10.5 μm, respectively. The solid line shows the constant M.

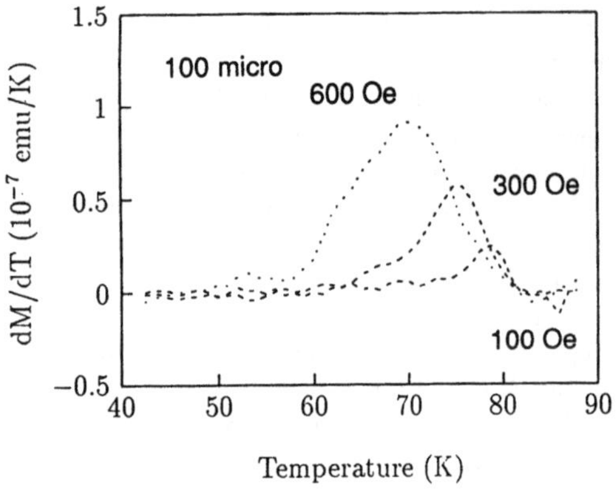

Fig.6 Temperature dependence of dM/dT (M is the FC magnetization and T is the temperature) in several magnetic fields.

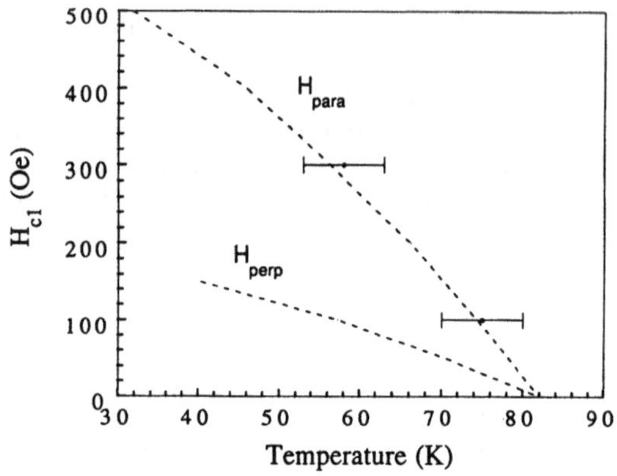

Fig.7 The offset temperature of the FC magnetization. The dashed line shows the parallel and perpendicular H_{c1} to the c axis of a single crystal in Ref. [6].

The onset of the magnetization did not change because the H_{c2} of $YBa_2Cu_3O_y$ is very high compared with the applied magnetic field. On the other hand, the offset of magnetization shows the large field dependence and the field dependence of the offset temperature was plotted in Fig. 7. The temperature dependence of the H_{c1} of a single-crystal $YBa_2Cu_3O_y$ [6] is also shown in Fig. 7, and was in good agreement with the offset temperature of the FC magnetization of our thin film. Therefore, we believe that the determination of the H_{c1} of superconductors is possible from more precise measurements of the FC magnetization of thin films.

CONCLUSION

The magnetization measurement of epitaxial films prepared using MBE have been performed. Samples showed almost perfect diamagnetism, and the magnetization had a sample-size dependence of both the ZFC and the remanent magnetization. The flux-density distribution was directly observed by the magneto-optical Faraday effect and was in good agreement with the magnetization data. Moreover, the FC magnetization indicated that the frozen flux-density distribution below T_{c1} is proper in the epitaxial thin films and that H_{c1} can be estimated from the FC measurements. From these experimental results, we can conclude that a quantitative description of the magnetization in oxide superconductor films over almost the entire magnetization curve is now possible, and the magnetization of thin films is a powerful method for basic experiments and characterization of thin films.

REFERENCES

[1] H. Obara, S. Kosaka, M. Umeda and Y. Kimura, *Advances in Superconductivity* II (Springer, Tokyo, 1990), p.801.

[2] S. Kosaka, H. Obara, M. Umeda, Y. Kimura, H. Yamane, T. Hirai and H. Kurosawa, *Advances in Superconductivity* II (Springer, Tokyo, 1991), p.577.

[3] Y. Yokoyama, Y. Hasumi, H. Obara, Y. Suzuki, T. Katayama, S. Gotoh and N. Koshizuka, Jpn. J. Appl. Phys. **30**, L714 (1991).

[4] C.P. Bean, Phys. Rev. Lett. **8**, 250 (1962).

[5] T. Matsushita, E.S. Otabe, T. Matsuno, M. Murakami and K. Kitazawa, Physica C **170**, 375 (1990).

[6] L. Krusin-Elbaum, A.P. Malozemoff, Y. Yeshurun, D.C. Cronemeyer, and F. Holtzberg, Phys. Rev. **B39**, 2936 (1989).

Optical Response of $YBa_2Cu_3O_{7-x}$ Epitaxial Thin Films

Y. HARADA[1], K. NARUSHIMA[1], M. SANO[1], M. SEKINE[1], Y. HIGASHINO[2], and M. OHTSU[1]

[1] The Graduate School at Nagatsuta, Tokyo Institute of Technology, Yokohama, 227 Japan
[2] Yokogawa Electronic Corporation, Musashino, Tokyo, 180 Japan

ABSTRACT

We have studied the optical response of epitaxial thin films of $DyBa_2Cu_3O_{7-x}$ by irradiation of semiconductor laser at the wavelength of 830 nm. $DyBa_2Cu_3O_{7-x}$ superconducting thin films are prepared on MgO (100) by molecular beam epitaxy (MBE). These are confirmed c-axis oriented epitaxial thin films using X-ray analysis and *in-situ* RHEED observation. We have investigated the output signals (i.e. optical response) under the irradiation of laser light, whose frequency can be modulated up to 100 MHz. We have obtained the evidence of nonbolometric response under irradiation of laser at the modulation frequency of 100 MHz.

KEY WORDS: optical detector, $DyBa_2Cu_3O_{7-x}$ thin film, nonbolometric response

INTRODUCTION

Since the discovery of high-T_C superconductors, the feasibility of optical and infrared detection using these materials has been much investigated as one of the promising electronic applications [1-5]. As they are used as optical detectors, they can be classified into two types; bolometric or nonbolometric devices due to the operation principle. Microbolometers are well known as the typical bolometric devices. Their theoretical and experimental investigations are advanced by P. L. Richards *et al.* [6,7]. They have shown that a high-T_C superconducting bolometer can be potentially two orders of magnitude more sensitive than any other infrared detector operating in the liquid nitrogen temperature at its wavelengths longer than 20 μm. This means that a high-T_C superconducting bolometric detector operate as the broadband detector. However its response time is generally late compared with other nonbolometric superconducting detectors. As regards a nonbolometric high-T_C superconducting detector, a lot of works were reported [3,4,5,8,9]. In general, it is considered that the response time of mostly nonbolometric superconducting devices is faster than that of bolometric ones. This is because their response time is decided by quantum mechanics. However its operating mechanism is still not clear, and it is doubtful whether nonbolometric response of high-T_C superconducting thin films exists.

In this paper, we have investigated optical response of epitaxial $DyBa_2Cu_3O_{7-x}$ superconducting thin films under the irradiation of laser light at the wavelength of 830 nm. $DyBa_2Cu_3O_{7-x}$ superconducting epitaxial thin films are grown by molecular beam epitaxy (MBE). We have investigated the optical response under the irradiation of laser light, whose frequency can be modulated up to 100 MHz. We have obtained the evidence of nonbolometric response.

EXPERIMENTAL

$DyBa_2Cu_3O_{7-x}$ superconducting thin films are deposited on MgO (100) substrate by MBE in an ultrahigh vacuum system [10]. We set the temperature of the effusion cells for Dy, Ba, and Cu to about 950 °C, 600 °C, and 1000 °C, respectively. The substrate is heated to 650 °C with a graphite carbon heater. Activated oxygen species by rf-discharge plasma are introduced to the substrate. The growth rate is 0.06 nm/sec. The critical temperature of these films with a thickness of 30 nm is 88 K measured by standard four-point probe method. From the X-ray diffraction pattern, these films are a high degree of c-axis orientation with a c-axis parameter of 11.71 Å. By means of *in-situ* RHEED observation, fine streak patterns appeared from the initial crystal growth, so it has been proved that an atomically smooth surface is epitaxially grown.

Fabrication process of samples is the following way. $DyBa_2Cu_3O_{7-x}$ superconducting thin films on MgO (100) with the thickness of 100 nm is prepared by MBE as mentioned above. Patterning of sample's configuration is used standard photolithograpy and Ar ion milling. Fabricated samples have constrictions with a size of 20 μm width and 400 μm length. The critical temperature of samples is 65 K. As samples are damaged during the fabrication process, its critical temperature is low compared with that of thin films.

The experimental arrangement for the study of the optical response is shown in Fig. 1. GaAlAs laser diode whose wavelength is 830 nm is used as the light source and its average power is 3.5 mW. A sample is mounted on the stage in the cryostat which can cool down to 20 K. A sample is biased by constant dc current from the current source. Laser light is introduced perpendicularly to a sample through the window of the cryostat. Laser diode can be modulated by a dc-biased sinusoidal signal current and its modulation frequency can be varied from 1 kHz to 100 MHz. We have measured the output electrical signals through the lock-in amplifier under the irradiation of laser light. In this experimental arrangement, the output signals can't be detected even up to the laser modulation frequency of 200 kHz. The output signals at the modulation frequency from 200 kHz to 100 MHz is measured using the another experimental arrangement as shown in Fig.2. In this arrangement, a double balanced mixer (DBM) is used to beet down the output signals to 10 kHz.

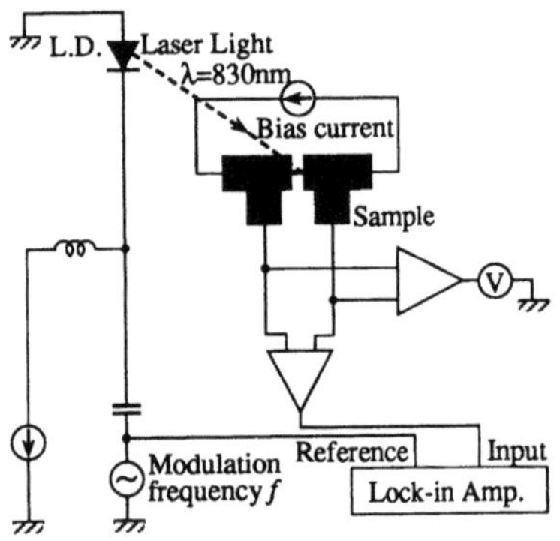

Fig. 1 Arrangement of optical response in $DyBa_2Cu_3O_{7-x}$ superconducting thin films by irradiation of laser light whose frequency can be modulated up to 200 kHz.

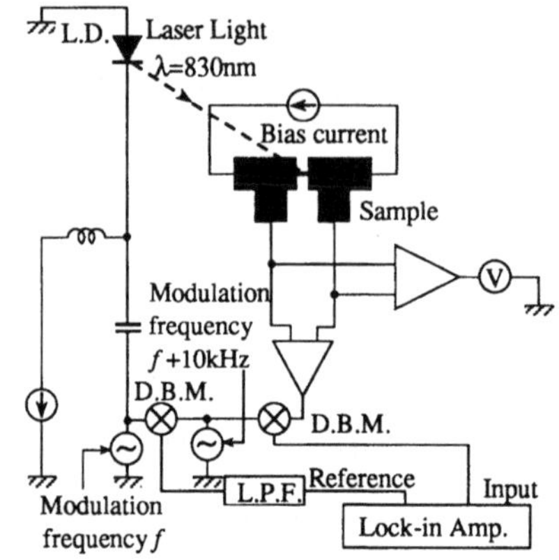

Fig. 2 Arrangement of optical response in $DyBa_2Cu_3O_{7-x}$ superconducting thin films by irradiation of laser light whose frequency can be modulated from 200 kHz to 100 MHz.

RESULTS AND DISCUSSIONS

The upper part of Figs. 3 shows the temperature dependence of resistivity of a sample film (solid line) and output voltage (i.e. photo response) under the irradiation of laser light (break line). The modulation frequency of the laser is 1 MHz, and the biased current of the sample is 0.1 mA. The critical temperature of this sample is found 65 K from this figure. The maximum value of the optical response is obtained at the temperature of 70 K. This optical response includes both the bolometric and nonbolometric ones as mentioned in introduction. This relation is expressed as the following way,

$$\delta V = I_B \frac{\partial R}{\partial T} \Delta T + R \frac{\partial I}{\partial T} \Delta T \quad , \qquad (1)$$

where δV represents the output voltage produced from the laser irradiation, I_B and R are a bias current and the resistance of a sample, respectively. ΔT means the rising temperature by heating effect due to the laser irradiation. The first term of right hand in this equation indicates bolometric response which result from an increase of the sample's temperature by the laser irradiation. On the other hand, the second term of right hand in this equation means nonbolometric response. Recently some theories are suggested as the explanation of the mechanism on nonbolometric response [11-13].

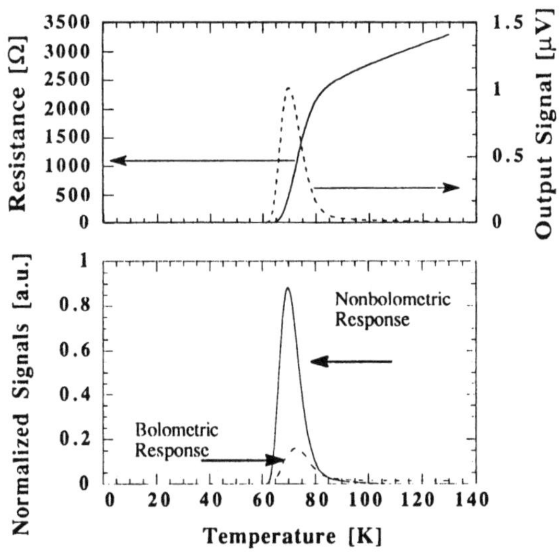

Figs. 3 The optical response of a sample under the irradiation of laser at the wavelength of 830 nm with the modulation frequency of 1 MHz.

The lower part of Fig. 3 shows the magnitude of both the bolometric and the nonbolometric signal under laser irradiation. Both signals are normalized by some factor. These values are calculated according to the expression (1). It is assumed that at the region of high temperature beyond T_C, the total signal equals the bolometric signal. From this figure, it is found that when the modulation frequency of the laser diode is 1

MHz, the nonbolometric response is about 5.5 larger than the bolometric one. It is considered that the bolometric signal is so late that it can't follow the modulation of the laser source.

Figures 4 also show the optical response of the same sample under the irradiation of laser like Figs. 3. But the modulation frequency of laser is different from that of Figs.3, and its frequency is 100 MHz. Clearly, the ratio of a bolometric signal, S_B, to nonbolometric one, S_N, is smaller than that of Figs. 3. In other words, it is considered that the bolometric signal decreases at a higher modulation frequency of the laser.

To confirm this assumption, we have investigated the frequency dependence of the ratio, S_B / S_{TOTAL} ($S_{TOTAL}=S_B+S_N$), as shown in Fig. 5. At the region of low modulation frequency, this ratio is constant. This is because the bolometric signal dominates over the total signals and the nonbolometric signal doesn't contribute because of its rapid response. But it is found that the bolometric signal gradually decrease from the modulation frequency of about 100 kHz and in stead of bolometric signal, nonbolometric one dominates at the higher frequency. According to the theory of nonequilibrium superconductivity, which is based on the BCS theory, it can be calculated that nonbolometric response time is about several ten picosecond, which corresponds to several hundred GHz [14].

As additional facts, it is found that the peak temperature of the nonbolometric response is shifted toward a lower temperature compared with that of the bolometric response. This reason is not evident, but it may be related to the mechanism of the high-Tc superconductivity. We have also studied the bias current dependence of the optical response. Up to now, the evident relations between the bias current and optical response can't be observed.

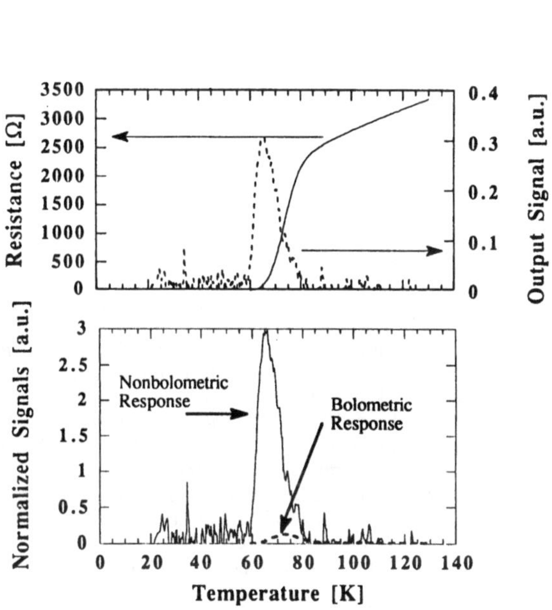

Figs. 4 The optical response of a sample under the irradiation of laser at the wavelength of 830 nm with the modulation frequency of 100 MHz. Bias current of this sample is 0.5 mA.

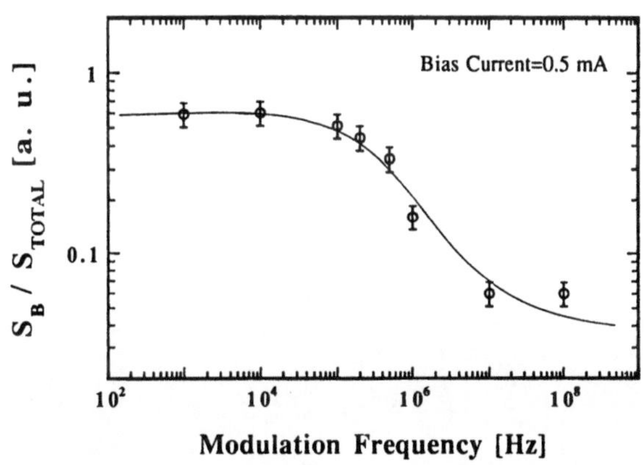

Fig. 5 Modulation frequency dependence of ratio of a bolometric signal to the total one, S_B / S_{TOTAL}.

CONCLUSIONS

We have investigated the optical response of $DyBa_2Cu_3O_{7-x}$ thin films under the irradiation of laser light (λ=830 nm). whose modulation frequency can be varied from 1 kHz to 100 MHz. $DyBa_2Cu_3O_{7-x}$ thin films are prepared by MBE and they are highly c-axis oriented epitaxial thin films. We have observed both bolometric and nonbolometric signal from $DyBa_2Cu_3O_{7-x}$ thin films under the laser irradiation. At the lower modulation frequency, the bolometric signal dominates the total optical response by the irradiation of laser light, but as the modulation frequency goes up, the bolometric signal decreases. At the modulation frequency of 100 MHz, the large part of total optical response is the nonbolometric signal. To confirm this fact, we have also investigated the frequency dependence of the ratio S_B / S_{TOTAL}. It is found that this ratio becomes small as the modulation frequency becomes higher. As the result, we have obtained that the optical response includes the nonbolometric signal in evidence. However origin of the nonbolometric signals is not obvious. From our experiments, it can be stated that the nonbolometric response time is shorter than 10 ns and the bolometric response time of our sample size (constriction with the size of 20 μm × 400 μm) is within several nanoseconds.

REFERENCES

1. Forrester MG, Gottlieb M, Gavaler JR, Braginski AI (1988) Appl. Phys. Lett. 53: 1332-1334
2. Brocklesby WS, Monroe D, Levi AFJ, Hong M, Liou SH, Kwo J, Rice CE (1989) Appl. Phys. Lett. 54: 1175-1177
3. Frenkel A, Saifi MA, Venkatesan T, Lin C, Wu XD, Inam A (1989) Appl. Phys. Lett. 54: 1594-1596
4. Tanabe K, Enomoto Y, Suzuki M, Iwata T, Yamaji A (1990) Jpn. J. Appl. Phys. 29: L466-469
5. Han SG, Vardeny ZV, Wong KS, Symko OG, Koren G (1990) Phys. Rev. Lett. 65: 2708-2711
6. Richards PL, Clarke J, Leoni R, Lerch P, Verghese S, Beasley MR, Geballe TH, Hammond RH, Rosenthal P, Spielman SR (1989) Appl. Phys. Lett. 54: 283-285
7. Hu Q, Richards PL (1989) Appl. Phys. Lett. 55: 2444-2446
8. Chibane F, Bonnel de Longchamp N, Martin R, Perrière J, Hauchecorne G, Kerhervé F, Maneval JP (1989) Solid State Commun. 69: 907-909
9 Zeldov E, Amer NM, Koren G, Gupta A (1989) Phys. Rev. B39: 9712-9714
10 Higashino Y, Umezawa T, Mizobuchi K (1991) Proceedings of the third International Superconducting Electronics Conference (ISEC '91), Glasgow
11. Zeldov E, Amer NM, Gupta A, Gambino RJ, McElfresh MW (1989) Phys. Rev. Lett. 62: 3093-3098
12. Tanabe K, Enomoto Y, Suzuki M, Kubo S, Asano H, Iwata T, Tajima Y, Yamaji A (1990) Proceedings of the third International Symposium on Superconductivity (ISS '90), Tokyo
13. Kadin AM, Leung M, Smith AD, Murduck JM (1990) Appl. Phys. Lett. 57: 2847-2849
14. Kaplan SB, Chi CC, Langenberg DN, Chang JJ, Jafarey S, Scalapino DJ (1976) Phys Rev. B14: 4854-4873

High T_c Superconducting Bandpass Filter

KENJIRO HIGAKI, HIDEO ITOZAKI, and AKIHIRO MOTO

Itami Research Laboratories, Sumitomo Electric Industries, Ltd., Itami, Hyogo, 664 Japan

ABSTRACT

We have measured the surface resistance Rs of a YBCO thin film and fabricated superconducting planar bandpass filters. Planar resonator technique was used for the Rs estimation of the YBCO thin film deposited on a MgO substrate. Considerably low Rs of 0.39mΩ of the YBCO thin film was obtained at 4.4GHz and 77K, which was one thirtieth of that of a gold thin film at 77K. Using such low Rs YBCO thin films, we have also obtained a considerably low insertion loss of about 1dB at 77K with the 10GHz bandpass filters which have a various bandwidth from 0.5% to 10%.

KEY WORDS: Superconducting bandpass filter, High Q resonator, Surface resistance, Low insertion loss

INTRODUCTION

High Tc superconducting materials have a great deal of potential in the field of planar microwave components because of their extremely low loss and no dispersion. It is very important to demonstrate excellent properties of the superconducting microwave components as well as to evaluate the Rs of superconducting thin films for their microwave applications. From this point of view, a lot of Rs measurements and fabrication of microwave components have been performed so far using high Tc superconducting thin films deposited on low dielectric loss LaAlO3 substrates[1~5]. But the LaAlO3 substrate is not so preferable for superconducting microwave circuits because of its high dielectric constant of about 25 and low phase transition temperature of below 500°C. In this paper we present results on planar resonators for the Rs measurements and bandpass filters composed of YBCO thin films sputtered on MgO dielectric substrates, which have been appreciated as low loss substrates comparable to LaAlO3[6].

YBa2Cu3O7 THIN FILM PREPARATION

The YBCO thin films were deposited on MgO(100) single crystals by single target off-axis sputtering. The substrate was heated up to 650°C in 100mTorr O2 and 400mTorr Ar during the deposition. Thickness of the YBCO thin films was 4000Å. The YBCO films prepared this way were superconducting as deposited and required no further heat treatment. Typical Tc and Jc at 77K measured by a four probe method were 85K and $1.0 \times 10^6 A/cm^2$ respectively. Analysis by X-ray diffraction and RHEED indicated that those films had grown epitaxially with their c-axis perpendicular to the substrate.

MEASUREMENT SYSTEM

The measurement system of S-parameters for the superconducting resonators and bandpass filters is shown in Fig.1. These components were cooled down from 300K to 30K in a cryostat and their S-parameters were measured by a network analyzer (HP8510B). In the case of a resonator, loaded Q and insertion loss of the resonant peak as a function of temperature were

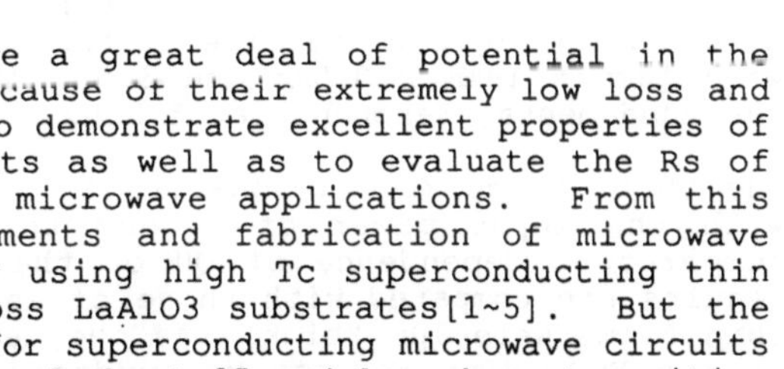

Fig.1 Systematic diagram of the s-parameter measurements for the superconducting microwave components.

measured automatically at each temperature by this computer controlled system. Unloaded Q (Q0) of the resonator and the Rs of the YBCO thin film were calculated from these values and a geometric coefficient of the resonator.

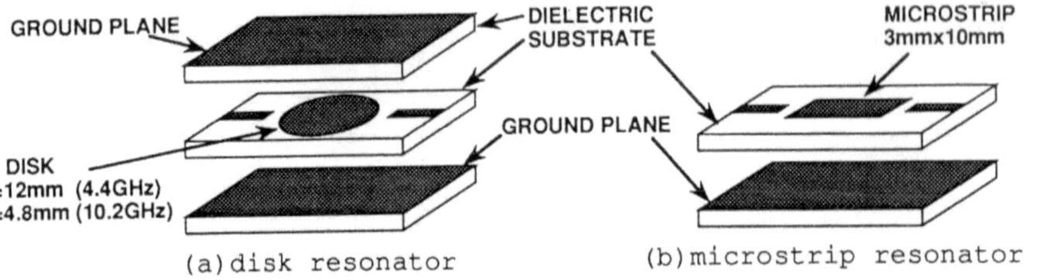

Fig.2 Schematic diagram of (a)disk resonator and (b)microstrip resonator.

SURFACE RESISTANCE MEASUREMENT

The surface resistance of the YBCO thin film was measured with the disk resonator and the microstrip resonator. The disk resonator was composed of a thin film conductor disk, upper and lower ground planes and dielectric layers between them, as shown in Fig.2(a). The superconducting disk resonators were composed of three YBCO thin films deposited on MgO(100) dielectric substrates. The size of the dielectric substrates was 18mm x 18mm x 1mm and the diameters of the disks were 12mm and 4.8mm, which brought about the fundamental resonances (TM110 mode) at 4.4GHz and 10.2GHz, respectively. Also the superconducting microstrip resonator was composed of the YBCO strip and ground plane deposited on MgO(100) dielectric substrates, as shown in Fig.2(b).The size of the dielectric substrates and the YBCO strip were 18mm x 18mm x 0.5mm and 3mm x 10mm, respectively, which brought about the fundamental resonance at 5.5GHz.

The Q0 of these YBCO resonators as a function of temperature were measured with the input power of -15dBm to the resonator. Figure 3 shows the temperature dependence of Q0 on the YBCO disk resonator at 4.4GHz. The results are compared with those of an equivalent Au disk resonator whose Au thin films were 1μm thick. As the temperature of measurement decreased, Q0 of the YBCO disc resonators suddenly increased at their Tc and became the considerably high value of 45000 at 77K. The Rs of the YBCO thin film and the Au thin film were estimated from these measured Q0[7], as shown in Fig.4. The surface resistance of the YBCO thin film was found to be 0.39mΩ at 77K and 4.4GHz, which was 100 times and 20 times lower than that of the Au thin film at room temperature and 77K, respectively. Figure 5 shows the frequency dependence of their surface resistances measured with the disk resonators and microstrip resonator. The surface resistance of the Au thin film is proportional to the square root of frequency due to the skin effect, and that of the YBCO thin film is proportional to the square of frequency, which follows to the two-fluid model[8].

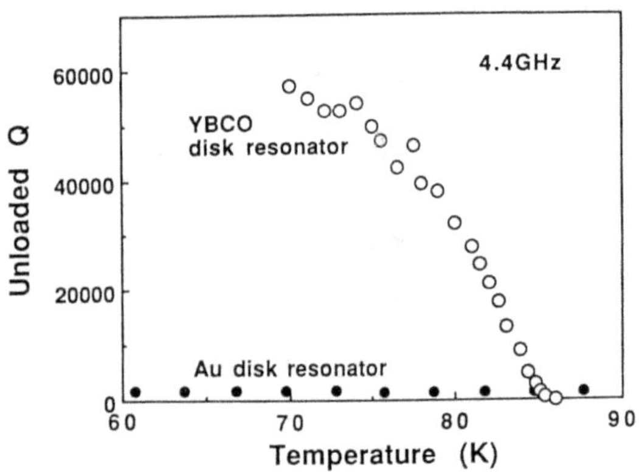

Fig.3 Unloaded Q vs. temperature of the 4.4GHz YBCO and Au disk resonators.

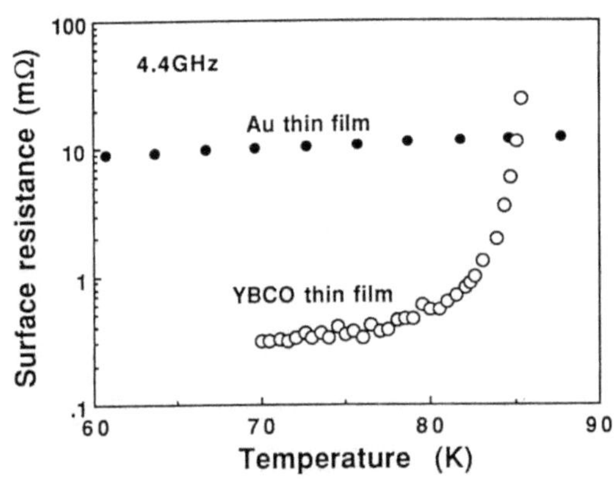

Fig.4 Surface resistance vs.temperature of the YBCO and Au thin films.

We have also investigated input power dependence of the Q0 and the Rs at 77K and 5.5GHz. These values were measured with the YBCO microstrip resonator varying input power from -30dBm to 20dBm, as shown in Fig.6. No significant degradation of the Q0 and the Rs was observed below the input power of 0dBm. As the input power further increased, the Rs of the YBCO thin film increased and became twice larger at 20dBm than that of below 0dBm, which was however still eight times lower than that of Au thin film at 77K and 5.5GHz.

CHEBYSCHEV BANDPASS FILTER

We have also fabricated 10GHz parallel-coupled Chebyschev bandpass filters using the MgO(100) dielectric substrate. Several types of bandpass filters were designed varying bandwidth from 10% to 0.5%. The design for the filter with the bandwidth of 0.5% is shown in Fig.7. Both the coupled transmission lines and ground planes of these filters were composed of YBCO thin films deposited on MgO substrates. Equivalent Au bandpass filters composed of Au thin films (1μm thick) on MgO substrates were also prepared for comparison.

Measurement results of the YBCO bandpass filter with 1% bandwidth at 77K are shown in Fig.8. Considerably low insertion loss of about 1dB and high return loss of about 15dB were obtained. The results are compared with those of the equivalent Au filter at room temperature and 77K. Figure 9 shows the bandwidth dependence of the insertion loss of those bandpass filters. The insertion loss of the Au bandpass filter with narrow bandwidth of 0.5% was about 9dB at room temperature and about 4dB even at 77K. In the case of the YBCO bandpass filter, considerably lower insertion loss of about 1dB was obtained. For the wide bandpass filter with the bandwidth of 10%, the insertion loss of even the Au filter was found to be so small (about 1dB) that a large improvement by using superconducting thin films was not obtained. Therefore, the results actually showed the advantage of the superconducting thin film with a planar bandpass filter having narrow bandwidth. Input power dependence of the insertion loss of the YBCO bandpass filter with 1% bandwidth was also investigated at 77K, as shown in Fig.10. The insertion loss increased according to the increase of the surface resistance of the YBCO thin film. However, it was sustained less than 1.5dB even with the input power of 20dBm, which is due to much lower surface resistance of the YBCO thin film than that of Au thin film even at 20dBm.

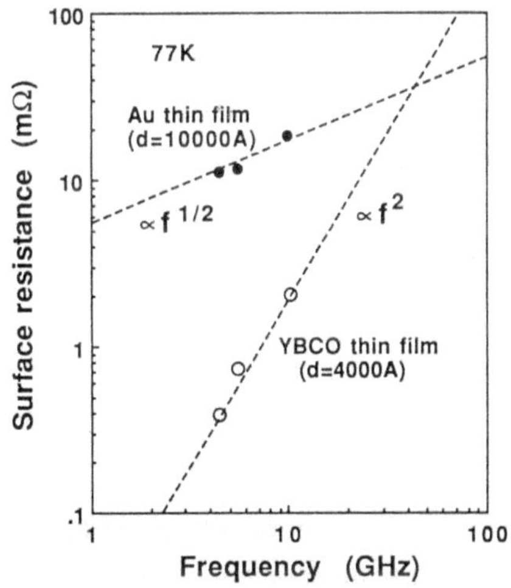

Fig.5 Surface resistance vs. frequency of the YBCO and Au thin films. Results are fitted by the two-fluid model.

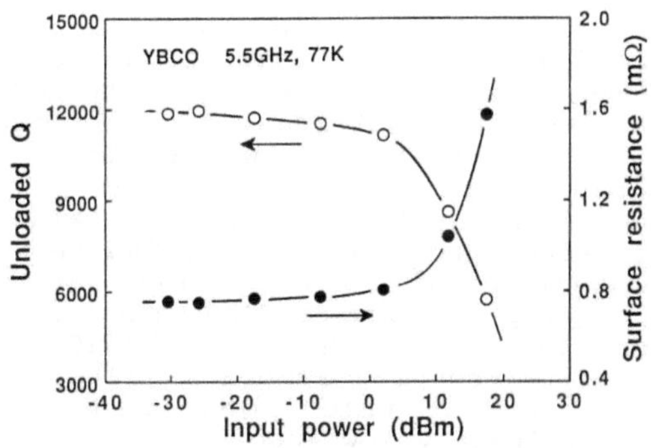

Fig.6 Input power dependence of unloaded Q and surface resistance measured with the 5.5GHz YBCO microstrip resonator at 77K.

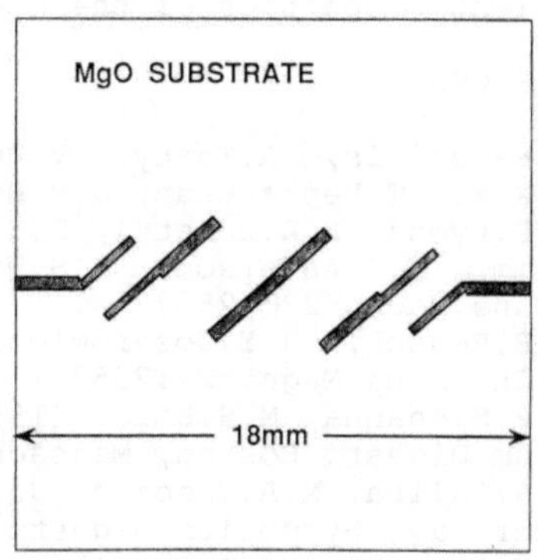

Fig.7 Mask design for the 10GHz Chebyschev bandpass filter with the narrow bandwidth of 0.5%.

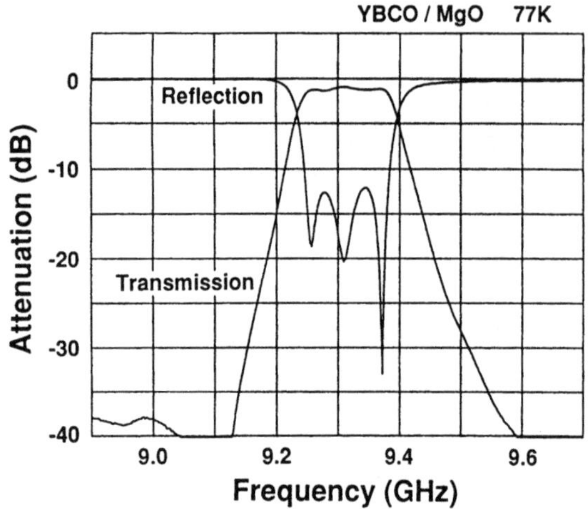

Fig.8 Transmitted power(S21) and reflected power(S11) vs. frequency of the YBCO bandpass filter with 1% bandwidth.

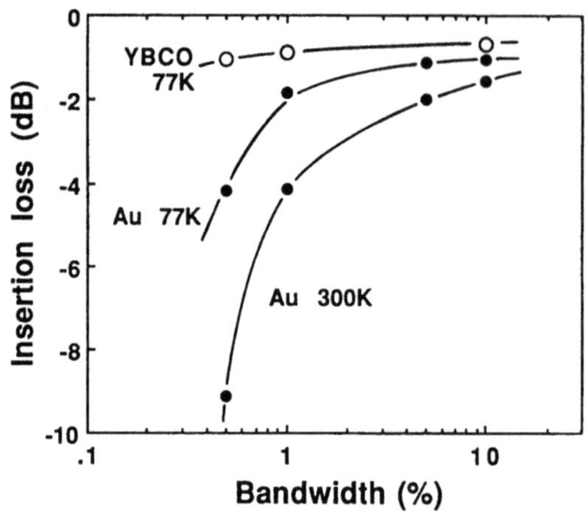

Fig.9 Insertion loss vs. bandwidth of the YBCO and Au bandpass filters.

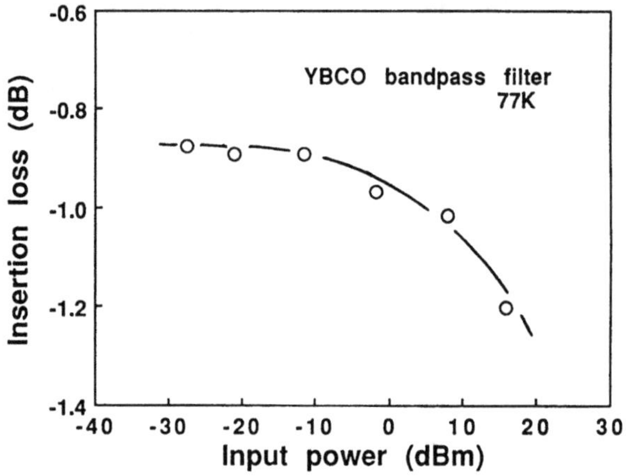

Fig.10 Insertion loss vs. input power for the 10GHz-1% YBCO bandpass filter at 77K.

SUMMARY

We have measured the surface resistance of a YBCO thin film and fabricated the low loss planar bandpass filters composed of the YBCO thin films deposited on the MgO dielectric substrates. The Rs of the YBCO thin film was estimated using the planar resonators. Considerably high unloaded Q value of 45000 and low Rs of 0.39mΩ were obtained at 4.4GHz and 77K. Applying such low Rs YBCO thin films on the MgO substrates to the 10GHz parallel-coupled bandpass filters, we have also obtained a significantly low insertion loss of about 1dB at 77K even with the narrow bandwidth of 0.5%. The insertion loss of the 1% YBCO bandpass filter was less than 1.5dB even with the input power of 20dBm, which is due to much lower surface resistance of the YBCO thin film than that of Au thin film even at 20dBm. The results show extremely low loss in the superconducting thin film and excellent properties of the superconducting microwave components.

REFERENCES

[1]D.Kalokitis, A.Fathy, V.Pendrick, E.Belohoubek, A.Findikoglu, A.Inam, X.X.Xi, T.Venkatesan, J.B.Barner (1991) Appl.Phys.Lett. 58: 537

[2]W.G.Lyons, R.R.Bonetti, A.E.Williams, P.M.Mankiewich, M.L.O'Malley, J.M. Hamm, A.C.Anderson, R.S.Withers, A.Meulenberg, R.E.Howard (1991) IEEE Trans.Magn. 27: 2537

[3]D.B.Rensch, J.Y.Josefowicz, C.W.Nieh, W.Hoefer, F.Krajenbrink (1991) IEEE Trans.Magn. 27:2553

[4]A.P.S.Khanna, M.Schmidt (1991) IEEE MTT-S International Microwave Symposium Digest. Boston, Massachusetts, pp 1239

[5]S.H.Talisa, M.A.Janocko, J.Talvacchio (1991) IEEE MTT-S International Microwave Symposium Digest. Boston, Massachusetts, pp 1325

[6]J.Talvacchio, J.R.Gavaler, J.Greggi, M.G.Forrester, A.I.Braginski (1989) IEEE Trans.Magn. 25: 2538

[7]Y.Kobayashi (1975) Inst.Electron.Commun.Eng. MW75-76: 27

[8]T.Van Duzer, C.W.Turner (1981) Principles of Superconductive Devices and Circuits. Elsevier North Holland, Inc. New York, pp 128

High-Frequency Surface Resistance of Sputtered $EuBa_2Cu_3O_{7-x}$ Thin Films

H. ASANO, M. SATOH, and T. KONAKA

NTT Laboratories, Tokai, Ibaraki, 319-11 Japan

ABSTRACT

We examine the high-frequency surface resistance of sputtered $EuBa_2Cu_3O_{7-x}$ films on MgO, $YAlO_3$, and Sapphire substrates having favorable dielectric properties. Surface resistance measurement is performed as a function of temperature by end-wall replacement in a 50 GHz gold-plated Invar host cavity. For films with good in-plane epitaxial alignment, low surface resistance values of less than 10 mΩ at 77 K are obtained, which is in the same range as that for a classical superconductor of Nb at the same reduced temperature (T/T_c) of 7.7 K. Microstructural study with a transmission electron micrography (TEM) indicates that a number of defects related with in-plane misorientation are observed for films having poor microwave properties.

KEY WORDS: $EuBa_2Cu_3O_{7-x}$, surface resistance, MgO , $YAlO_3$

INTRODUCTION

The microwave properties of high-T_c oxide superconductors are of great interest for both basic research and technological applications. Surface resistance measurements of epitaxial films are suited to clarify the intrinsic properties and ultimate performance of the oxide superconductors at microwave frequencies. Microwave surface resistance is the key parameter which determines the application potential of materials in passive microwave components. Several groups have reported on surface resistance measurements of epitaxial films of the 1-2-3 superconductors on several different substrate materials [1-5]. To date, the best performance (8 mΩ at 77 K and 86.7 GHz) has been obtained with laser-deposited films on such lattice-matched substrates as $SrTiO_3$ and $LaAlO_3$. However, the large dielectric constants of perovskite-type substrate materials (ε=300 for $SrTiO_3$ and ε=26 for $LaAlO_3$) limit their application to microwave devices. Other substrates such as MgO and Sapphire would be excellent materials in terms of low dielectric constant (ε=10) and low loss tangent ($\tan\delta<1\times10^{-5}$) [6]. Higher microwave losses were observed for films grown on these lattice-mismatched substrates, which were mainly due to the incomplete epitaxial nature of the films. Laderman reported that the higher loss could be accounted for by the presence of high-angle tilt boundaries [7]. It is of practical importance to produce films with good microwave performance using these useful substrate materials. Moreover, it is essential to clarify the factors responsible for the microwave losses in high-T_c superconductors.

In this paper, we report results for high frequency surface resistance of sputtered $EuBa_2Cu_3O_{7-x}$ films on the potentially useful substrates of MgO, $YAlO_3$, and Sapphire. It is found that significantly reduced surface resistance can be obtained for films with good in-plane epitaxial alignment. Correlations between dc properties, microstructure and high frequency properties are summarized.

EXPERIMENTAL

The single-target dc magnetron sputtering method was employed for $EuBa_2Cu_3O_{7-x}$ thin film deposition. The apparatus and deposition procedures were described previously [8,9]. Our sputtering process has routinely produced in-situ epitaxially grown, high-Tc films with good uniformity (at least 1 inch square). The substrate materials were (001) MgO and $YAlO_3$, and R-plane Sapphire of 1 inch square. As a substrate surface treatment, a short-time Ar ion beam irradiation or high-temperature annealing were performed. Typical growth conditions were a substrate temperature of 600-650°C, and a total gas pressure of 76 mTorr-Ar+(1-5%)O_2. A number of films approximately 500 nm thick were deposited under different growth conditions.

Surface resistance measurements were performed at 50 GHz using a cylindrical, TE_{013} mode, gold-plated Invar host cavity [10]. When an end wall of the cavity was replaced with a superconducting film sample of ~15 mm x 15 mm, the Q factor of the cavity was measured. Under these conditions, there are no microwave losses induced at the junction between the end wall and the cylindrical body of the cavity. The surface resistance was inferred from the measured Q values for the sample and the standard material, polished oxygen-free high-conductivity (OFHC) copper in this study. The sensitivity range of this system was ~2 mΩ. In the measurements, the rf magnetic field level at the sample surface was below 1 Oe.

RESULTS AND DISCUSSIONS

For the microwave study, several films were deposited on 1 inch square substrates of MgO, $YAlO_3$, and Sapphire. Each film was cut and divided into individual samples for Rs measurement, dc four point resistive measurement, and structural and microstructural analysis. The dc superconducting transition and normal state resistive properties were measured on film patterns of typical dimensions of 10-μm width and 100-μm length. Data, displayed in Fig.1(a), show a sharp transition with a zero resistivity temperature of 93 K. The normal state resistivity shows a linear temperature dependence and an almost zero intercept at 0 K. The films on MgO and $YAlO_3$ exhibited very similar properties in dc resistive measurements; zero resistivity temperatures of 90-94.5 K and normal resistivity (at 300 K) of 200-300 μΩcm. Their critical current densities at 77 K ranged from 5×10^5 to 4×10^6 A/cm^2. In Fig. 1(b), the high-frequency surface resistance of the sample from the same film (Fig. 1(a)) is plotted as a function of temperature at 50 GHz. As shown in the figure, the microwave response curve resemble those of dc resistivity.

In Fig. 2, we present surface resistance (plotted in logarithmic scale) versus temperature curves for three samples on different substrates. The films on MgO and $YAlO_3$ exhibit a sharp drop in R_s at around 90 K, and their values decrease to less than 10 mΩ at 77 K. In contrast, the film on Sapphire had poor microwave properties: broad microwave transition, and higher Rs values of ~400 mΩ at 77 K. Degraded properties of films on Sapphire were also found in dc resistive measurement; zero resistivity temperatures of 87-89 K and critical current densities of as low as 2-6 x 10^4 A/cm^2 at 77 K.

The results of X-ray structural analysis are briefly mentioned in order to understand their correlation with microwave properties. A conventional 2θ/θ scan showed that these films exhibited a complete c-axis normal orientation with respect to the substrate surface with no detectable additional orientations and phases. Measurements of rocking full widths at half-maximum (FWHM) were performed for the (005) reflection of the films, with an experimental resolution of 0.15°. Films on Sapphire had very large FWHM values of 1.4-2.0°, indicating that the films even possessed a polycrystalline nature. The degraded dc and microwave properties observed for films on Sapphire are due to the polycrystalline nature of the films. For films on MgO and $YAlO_3$ substrates, FWHM values of as small as 0.2-0.4° were obtained, indicating that the crystallites were well oriented along normal to the substrate plane. In-plane epitaxial alignment was examined by the ϕ-axis scan of

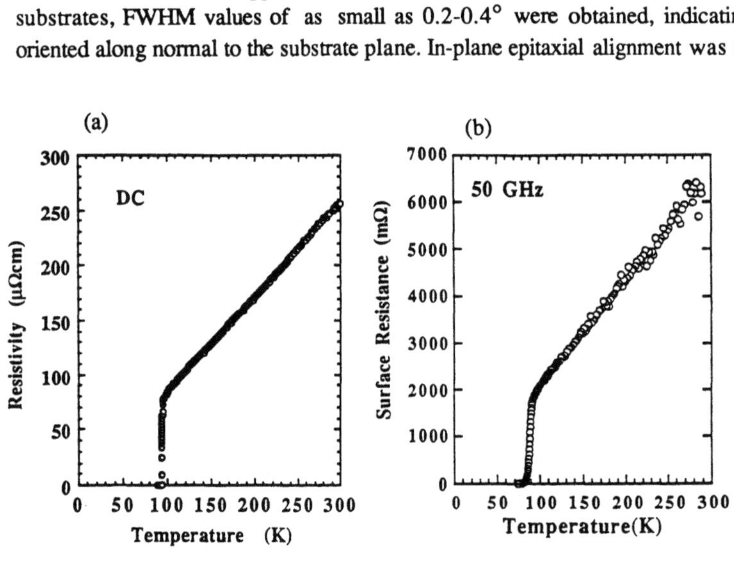

Fig. 1 Representative data for sputtered $EuBa_2Cu_3O_{7-x}$ films. (a) Dc resistivity versus temperature curve, (b) Surface resistance (50 GHz) versus temperature curve.

the (102) reflection with a four-circle goniometer system. For films on MgO and $YAlO_3$ substrates the crystal axes in the film plane were aligned with the crystal axes of substrates. The amount of in-plane misoriented c-axis materialswas within the detection limit of our X-ray analysis (~0.1 %).

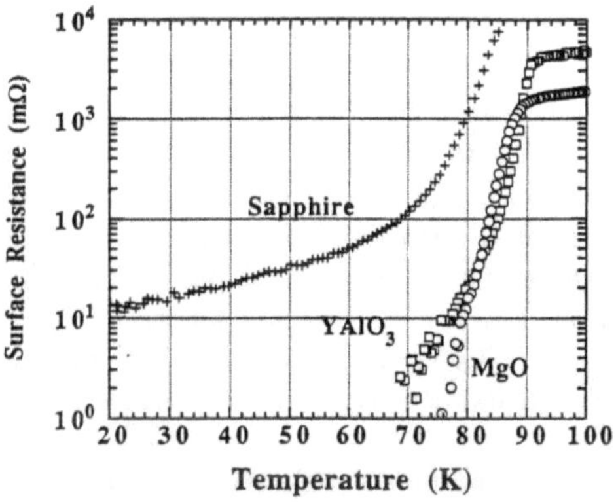

Fig. 2 Temperature dependence of surface resistance (50 GHz) for films on different substrates.

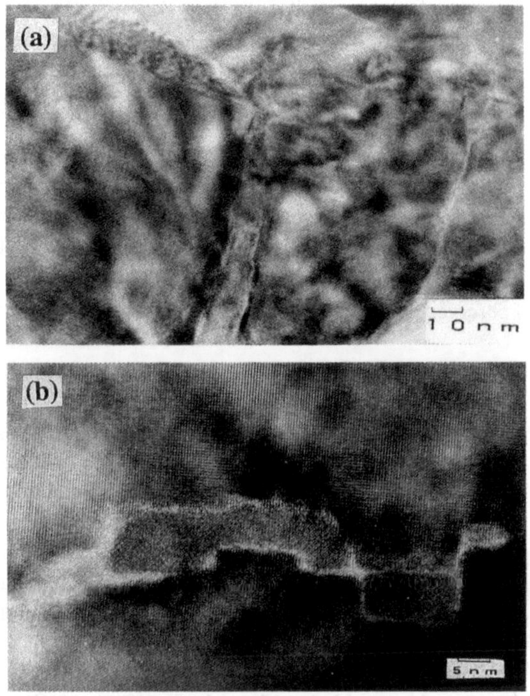

Fig. 4 TEM planar views of films with poor microwave properties. (a) View showing small angle tilt boundaries, (b) View showing 45° in-plane misoriented grains.

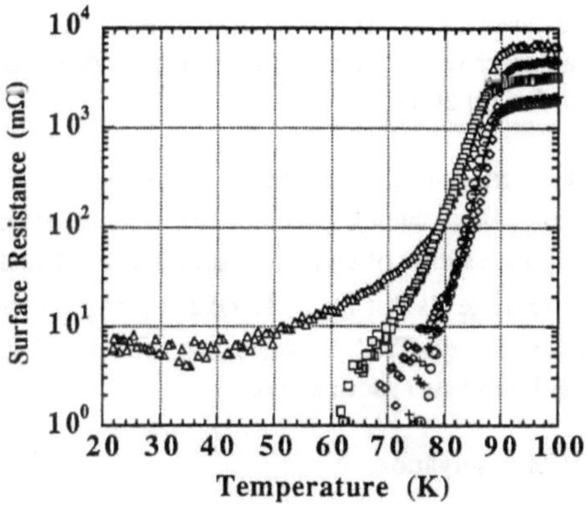

Fig. 3 Temperature dependence of surface resistance (50 GHz) for a number of films .

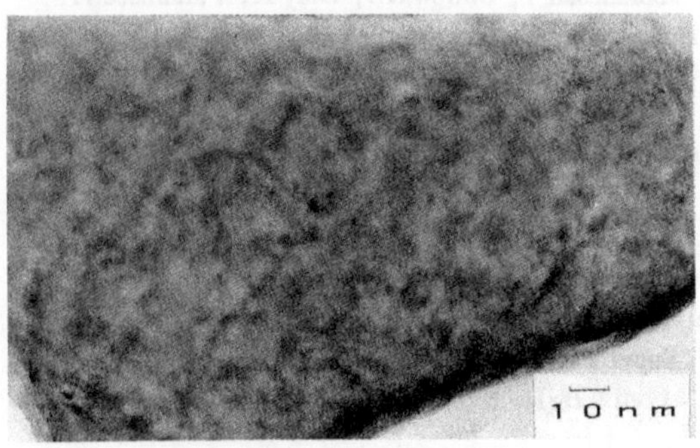

Fig. 5 TEM planar views of films with excellent microwave properties.

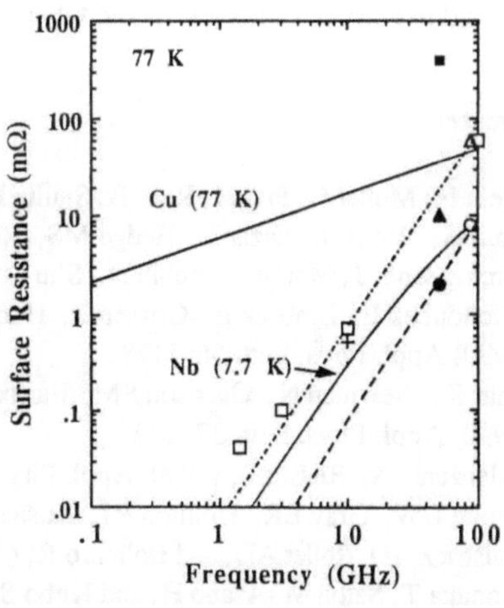

Fig. 6 Surface resistance (77 K) versus frequency plot for the present data (closed circle (on MgO), closed triangle (on $YAlO_3$), and closed square (on Sapphire)) together with previously reported data. See in the text.

Even when films were grown on MgO and $YAlO_3$ substrates, we occasionally observed slightly deteriorated microwave properties. Although these films showed superior dc resistive properties, the microwave properties, that is, transition behavior in the surface resistance and the Rs values at 77 K and at lower temperature, were quite different. In Fig. 3, microwave data are shown for a number of samples. As shown in the figure, these samples exhibit different transition behavior and their Rs values (at 77 K) ranged from 2 mΩ to 80 mΩ. These results mean that the microwave response actually characterizes the homogeneity of the thin film much more stringently than dc resistivity.

To further examine the film structure dependence on the microwave properties, transmission electron micrograph (TEM) analysis was performed on typical samples with different microwave properties. Figure 4 shows TEM planar views of films with an Rs (77 K) of 80 mΩ. In Fig. 4(a), we can clearly observe a lot of tilt boundaries. In TEM images of the large Rs films, we can frequently detect grain boundaries with small tilt angles of 0-5°, and not so frequently detect grain boundaries with a tilt angle of 45° (as shown in Fig. 4(b)). In contrast, in the TEM image (shown in Fig. 5) of film having an Rs (77 K) of 2 mΩ, no apparent grain boundaries can be detected. The TEM analysis and comparison of Rs values lead to conclusion that the observed defects related with in-plane misorientation dominate the microwave properties in these samples.

Finally, in Fig. 6, present Rs data (closed circle (on MgO), closed triangle (on $YAlO_3$), and closed square (on Sapphire)) are plotted against frequency together with data previously reported by Wuppertal (open circle, Ref. 1), Bellcore (open squares, Ref. 2), Conductus (crosses, Ref. 3), and Siemens (open triangle, Ref. 4). For comparison, R_s values for Cu (77 K) and Nb (7.7 K) are also shown. The reported data for films on various substrates [2-4] coincide approximately with the dotted f^2 line. Our best value in this study (closed circle in the figure) is about one order of magnitude lower than the dotted line and lies on the same dashed line as that of Klein's data for $YBa_2Cu_3O_{7-x}$ film on $SrTiO_3$ assuming a quadratic frequency dependence. The surface resistance values of our best high-T_c films are found to be an order of magnitude lower than that of Cu at 50 GHz. Moreover, these values are comparable to that for Nb (7.7 K) at 50 GHz.

In summary, we have examined the high-frequency surface resistance of sputtered $EuBa_2Cu_3O_{7-x}$ films measured in a gold-plated Invar host cavity at 50 GHz. Epitaxial films having good in-plane orientation could be obtained on MgO and $YAlO_3$ and exhibited low surface resistance values of less than 10 mΩ at 77 K and 50 GHz. The best data (~2 mΩ at 77 K and 50 GHz) is in the same range as that for a classical superconductor of Nb at the same reduced temperature. The microstructural analysis with TEM indicated that films with poor microwave properties contained a number of defects related with in-plane misorientation.

We would like to thank Takaoka H, and Satoh E. for their TEM observation.

References

1. Klein N, Muller G, Pial H, Roas B, Shultz L, Klein U, and Peiniger M, (1989) Appl. Phys. Lett. 54: 757.
2. Inam A, Wu XD, Nazar L, Hedge MS, Rogers CT, Venkatesan T, Simon RD, Daly K, Padamsee H, Kirchgessner J, Moffat D, Rubin D, Shu QS, Kalokitis D, Fathy A, Pendrick V, Brown R, Brycki B, Belohoubek E, Drabeck L, Gruner G, Hammond R, Gamble F, Lairsson BM, and Bravman JC, (1990) Appl. Phys. Lett. 56: 1178.
3. Char K, Newman N, Garrison SM, Barton RW, Taber RC, Laderman SS, and Jacowitz RD. (1990) Appl. Phys. Lett. 57: 409.
4. Valenzuela A, Russer P, (1990) Appl. Phys. Lett. 55: 1029.
5. Cooke DW, Gray ER, Houlton RJ, Rusnak B, Meyer EA, Beery JG, Brown DR, Garzon FH, Raistrick ID, Rollet AD, and Bolmaro R, (1989) Appl. Phys. Lett. 55: 914.
6. Konaka T, Satoh M, Asano H, and Kubo S, (1991) J. Supercoductivity (to be published)
7. Laderman SS, Taber RC, Jacowitz RD, and Moll JL, Eom CB, Hylton TL, Marshall AF, Geballe TH, Beasley MR, (1991) Phys. Rev. B 43: 2922.
8. Asano H, Asahi M, and Michikami O, (1989) Jpn. J. Appl. Phys. 28 : L981.
9. Asano H, Yonezawa H, Asahi M, and Michikami O, (1991) IEEE Trans. Mag. MAG-27 : 844.
10. Asano H, Satoh M, and Konaka T, (1991) Appl. Phys. Lett. 58: 2981.

6.2 BSCCO and Related Films

Superconducting Bi-Sr-Ca-Cu-O Ultrathin Films Fabricated by MOCVD

T. SUGIMOTO, H. HAYASHI, H. SAKAI, N. KUBOTA, K. SUGAWARA, R. ITTI, Y. SHIOHARA, N. KOSHIZUKA, and S. TANAKA

Superconductivity Research Laboratory, ISTEC, Koto-ku, Tokyo, 135 Japan

ABSTRACT

The metalorganic chemical vapor deposition of the Bi-Sr-Ca-Cu-O superconductor was found to progress in a two-dimensional mode at the initial stages of deposition, from the measurement of the intensity of photoelectrons from the deposited film and SEM observations. These data suggest that the height of the two-dimensional unit is the half-unit cell of the $Bi_2Sr_2CaCu_2O_x$ phase along the c-axis. From low-field nonresonant microwave absorption and the temperature dependence of the resistivity of the 15-Å thick film, superconductivity may occur in a half-unit-cell thick layer of $Bi_2Sr_2CaCu_2O_x$ phase.

KEY WORDS: Metalorganic chemical vapor deposition, Bi-Sr-Ca-Cu-O superconductor, Ultrathin films, Superconductivity, Growth mechanism

INTRODUCTION

The understanding of the growth mechanism of oxide superconducting thin films, with regard to 2- or 3-dimensional growth, is of great importance in achieving optimal superconducting properties. The growth mechanism has been mainly investigated in physical vapor depositions, such as reactive evaporation [1], magnetron sputtering [2], and laser ablation [3,4]. Terashima et al [1] reported that a superconducting Y-Ba-Cu-O thin film by reactive evaporation grew with the height of the minimum unit which satisfied the chemical composition and electrical neutrality of the oxide, from the RHEED oscillation during the epitaxial growth. Recently, Mukaida et al. [4] reported that two-dimensional lateral growth of superconducting Y-Ba-Cu-O thin film by pulsed ArF laser ablation was clearly observed on a vicinal surface of a lattice-matched NdGaO3 substrate. In a Bi-Sr-Ca-Cu-O thin film, atomic layer-by-atomic layer depositions were achieved by laser ablation [5], shuttered MBE [6], and magnetron sputtering [7]. Matsumoto et al. [8] reported that layer-by-layer epitaxial growth of a $Bi_2Sr_2CuO_x$ thin film on a $Bi_2Sr_2CaCu_2O_x$ single-crystal substrate using a computer-controlled laser MBE technique with a single target. They demonstrated that, from the oscillations of the RHEED intensities, the layered growth occurred with the half-unit-cell thickness having the modulated structure.

The remarkable superconducting properties of high-Tc oxide superconductors are widely considered to be derived from the presence of Cu-O planes. Yet the coherence length in the c-axis direction is shorter than the distance between these planes. It is also an interesting problem whether an isolated single unit-cell layer of an oxide superconductor could exhibit high-Tc superconductivity, and whether the interlayer coupling between Cu-O

Table 1. Deposition conditions of the Bi-Sr-Ca-Cu-O ultrathin films.

MO source	Bi(ph)3	Sr(DPM)2	Ca(DPM)2	Cu(DPM)2
Source Temperature (°C)	83	184	175	95
Flow rate of Ar gas (sccm)	50	50	50	50
Substrate temperature (°C)		800		
Gas pressure (Torr)		10		
Partial pressure of O2 (Torr)		8.8		
Growth rate (Å /min)		3		

planes plays an important role in determining Tc. Recently, Terashima et al. [9] reported that superconductivity can occur in a one-unit-cell thick layer of the Y-Ba-Cu-O superconductor covered with $PrBa_2Cu_3O_7$.

In this paper we report the growth mechanism and superconductivity of ultrathin Bi-Sr-Ca-Cu-O films grown by metalorganic chemical vapor deposition (MOCVD).

EXPERIMENTAL

The MOCVD apparatus consists of a horizontal cold wall quartz reactor, MO feed lines, mass flow controllers, MO source evaporators, and a vacuum system. The MO source evaporators were placed in accurately temperature-controlled ovens and all MO feed lines were also placed in another oven maintained at 200 °C in order to improve the uniformity of line-heating and to prevent the MO vapors from condensing and/or decomposing. The substrate temperature was monitored by a thermocouple placed into the susceptor. Volatile β-dikentonate complexes of Cu, Sr, and Ca with the dipivaloylmethanate ligand (DPM: $(CH_3)_3CCOCHCOC(CH_3)_3$) [that is, $Cu(DPM)_2$, $Sr(DPM)_2$, and $Ca(DPM)_2$] were adopted for MO sources. Triphenylbismuth ($Bi(ph)_3$) was employed as the bismuth source because of its superior volatility and chemical stability. The source temperatures of $Bi(ph)_3$, $Sr(DPM)_2$, $Ca(DPM)_2$, and $Cu(DPM)_2$ were maintained at 83, 184, 175 and 95 °C, respectively. Argon carrier flow rate of 50 sccm was used to transport the MO sources. The argon carrier gas including the MO vapors was mixed with oxygen gas (1200 sccm) prior to entering the reactor. A mechanically polished (100) MgO single crystal substrate (10x10x0.5 mm^3) was degreased with acetone and then placed on the inconel alloy susceptor. The ultrathin films were grown at a substrate temperature of 800 °C under a total pressure of 10 Torr and an oxygen partial pressure of 8.8 Torr. After the growth process, all MO sources were switched to the vent lines and the sample was cooled to 300 °C under the same atmospheric condition during growth. The detailed conditions are shown in Table 1. The chemical composition was analyzed by Inductively Coupled Plasma emission spectrometry (ICP) . The average thickness of the films was estimated from the weight of the deposits on the substrate, using the weight value from the measurements by ICP. The theoretical density of the Bi-Sr-Ca-Cu-O superconductor, 6.55 g/cm^3, was assumed for the films. The deposition rate was maintained at approximately 3 Å/min by controlling the evaporation temperatures of the MO sources. The ultrathin Bi-Sr-Ca-Cu-O films were characterized by x-ray photoelectron spectroscopy (XPS), scanning electron microscopy (SEM), conventional four-probe resistance measurement, and low-field nonresonant microwave absorption (NRMA).

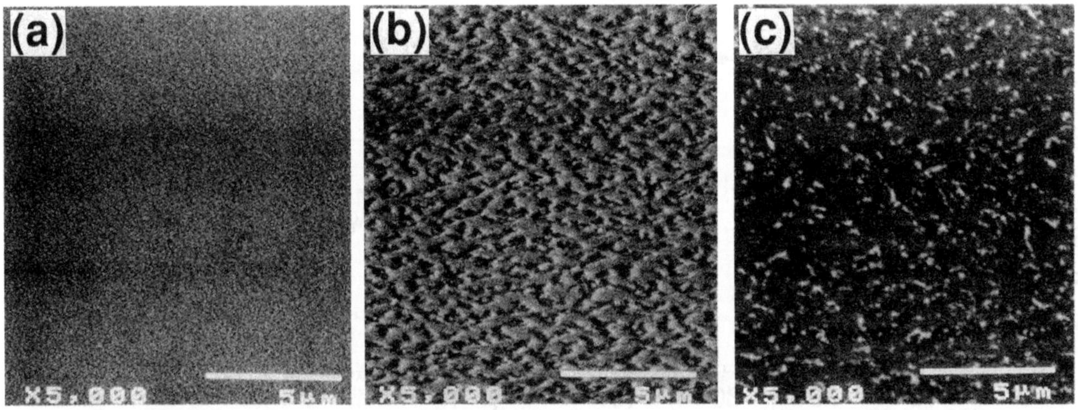

Fig. 1 SEM photographs of the surface with average thicknesses of (a) 5 Å, (b) 10 Å, and (c) 15 Å at the initial stages of deposition. It should be noted that the surface was not coated with a conductive overlayer to prevent the surface from being charged by the electron beam.

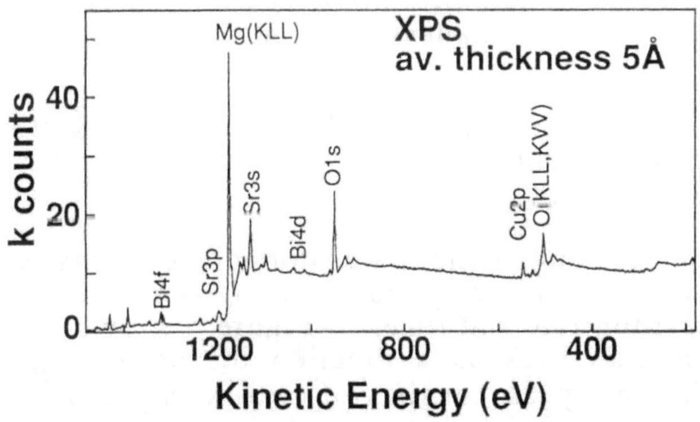

Fig. 2 XPS spectrum of the surface of the ultrathin film with an average thickness of 5 Å.

XPS measurements were carried out in an ultrahigh vacuum (UHV) system equipped with an Al x-ray source ($h\nu$=1486.6 eV). The area probed by the photoelectron energy analyser was approximately 1x1 mm^2. In the samples for four-probe resistance measurements, a Bi oxide film was deposited in situ at 500 °C on the top of the ultrathin Bi-Sr-Ca-Cu-O films by supplying the MO vapor of Bi to the reactor.

RESULTS AND DISCUSSION

Figure 1 shows the surface morphologies by SEM at the initial stage of the deposition. It should be noted that the surface of the samples was not coated with a conductive overlayer to prevent the surface from being charged by the electron beam. For the sample with an average thickness of 5 Å, a featureless morphology was observed. As shown in the XPS spectrum of the sample with an average thickness of 5Å (Fig. 2), however, Bi, Sr, and Cu, as well as Mg derived from the bare MgO substrate, are present at the surface. For the samples with average thicknesses of 10 and 15 Å (Fig. 1(b) and (c)), a characteristic morphology, probably due to the charge in the uncovered region of the surface from the electron beam, were observed. As shown in Fig. 1(b) and (c), the surface was more than half covered with conductive deposits for the sample with a thickness of 10 Å and

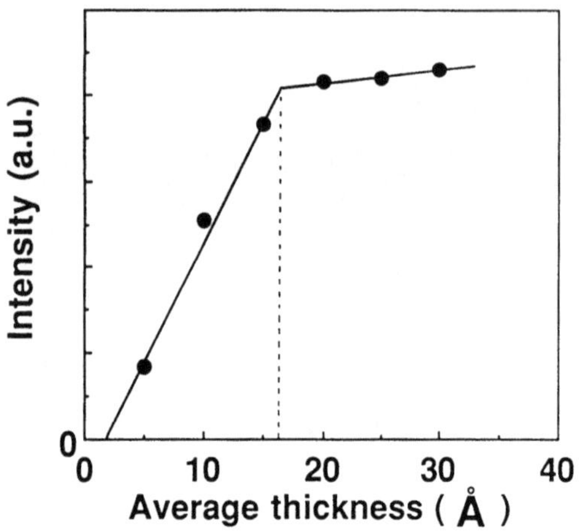

Fig. 3 XPS intensities for Cu $2p_{2/3}$ of the ultrathin films as a function of the average thickness.

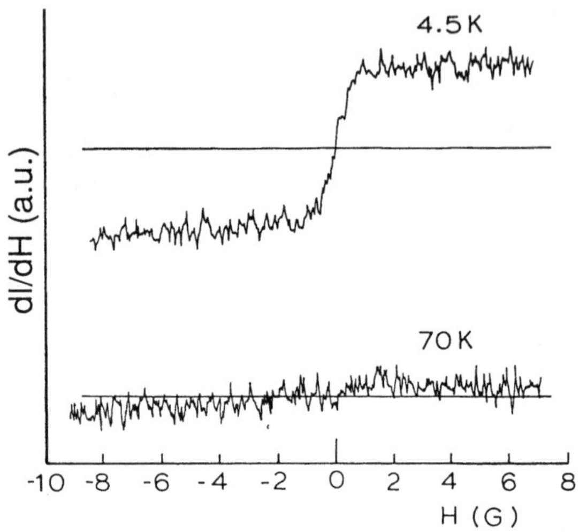

Fig.4 Low-field nonresonant microwave absorption spectra for the ultrathin film with an average thickness of 15 Å.

mostly covered for the sample with a thickness of 15 Å. These results suggest that two-dimensional nucleation and subsequent coalescence occurs at the initial stages of the deposition.

In order to elucidate the two-dimensional nucleation, we plotted the intensity of the photoelectron peaks of the deposits, as a function of the average thickness estimated from the deposition rate and time. Figure 3 shows the intensity of the Cu $2p_{2/3}$ photoelectron peak as a function of the average thickness. The intensity curve of this peak consists of two straight lines with a clear change of slope at an average thickness near 15 Å. A similar tendency was observed in the photoelectron peaks of Bi 4f and Sr $3d_{2/3}$. From these data, we found that the intensity of the photoelectrons from the deposited elements was a function of the surface coverage of the deposited film and that the deposited film completes the first layer at an average thickness of about 15 Å. We reported in an earlier paper [10] that the layer consisting of a $Bi_2Sr_2CaCu_2O_x$ (2212) phase was formed in the early stage of deposition, while the $Bi_2Sr_2Ca_2Cu_3O_x$ (2223) phase was formed via a $Bi_2Sr_2CaCu_2O_x$ (2212) phase, probably due to an annealing effect during the deposition. Hence, these results suggest that two-dimensional growth with a two-dimensional unit of the height of 15 Å corresponding to a half unit cell of the $Bi_2Sr_2CaCu_2O_x$ (2212) phase along the c-axis direction occurs at the initial stage of deposition, and that an ultrathin film with an average thickness of 15Å achieves essentially full coverage. This suggestion is supported by the direct SEM observation of the surface at the initial stages of deposition, as shown in Fig.1. It is reasonable that the straight line of the intensity dose not go through the origin, since the attenuation of the photoelectrons by surface contamination may have occurred (The intensity of the C 1s photoelectron derived from the surface contamination was almost the same for all the samples.).

There remains the problem of whether these ultrathin films consisting of a half-unit cell of the $Bi_2Sr_2CaCu_2O_x$ (2212) phase are superconducting. To understand this problem, we measured the low-field NRMA spectrum of the

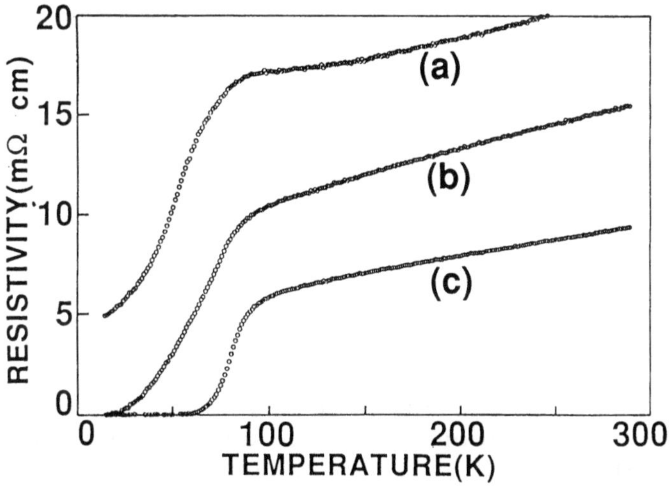

Fig. 5 Temperature dependence of resistivity of the ultrathin films with average thicknesses of (a) 15 Å, (b) 20 Å, and (c) 30 Å. A Bi oxide overlayer was deposited in situ at 500 °C on the top of ultrathin films to prevent surface degradation.

ultrathin film with an average thickness of 15 Å. The ultrathin film was sealed in the sample tube with wax as soon as the sample was taken out of the reactor, in order to prevent surface degradation. Figure 4 shows the low-field NRMA spectra after zero-field cooling. The nonresonant absorption, caused by the reduction of shielding by supercurrents and intergrain Josephson currents, was clearly observed at 4.5 K in increasing the applied field, and very weak absorption was observed even at 70 K. This indicates that the ultrathin film with an average thickness of 15 Å, consisting of a half-unit cell of the $Bi_2Sr_2CaCu_2O_x$ (2212) phase, becomes superconducting below 70 K.

To verify the superconductivity of the ultrathin film with an average thickness of 15 Å, the temperature dependence of the resistivity was measured. A Bi oxide overlayer was deposited in situ at 500 °C on the top of the ultrathin film to prevent surface degradation. Figure 5 shows the temperature dependence of the resistivity of films with average thicknesses of 15, 20, and 30 Å, respectively. As shown in Fig.5, the onset temperatures were the same regardless of the thickness. We showed, from the SEM observation and the XPS measurements, that the ultrathin film with an average thickness of 15 Å consists of a half-unit cell of the $Bi_2Sr_2CaCu_2O_x$ (2212) phase along the c-axis direction from the SEM observation and the XPS measurement. This indicates that the isolated half-unit-cell thick film can have the same onset temperature as a bulk superconductor of the $Bi_2Sr_2CaCu_2O_x$ (2212) phase without intercoupling between the Cu-O double layers. However, the zero-resistivity temperature was not achieved at higher than 15 K in the 15-Å-thick film, although it increased with increasing average film thickness. For the 30-Å-thick film, the zero-resistivity was comparable with that of thicker films [10]. The reduction of the $T_{c,0}$ of these ultrathin films may be mainly due to the crystalline imperfection derived from the surface roughness of the substrates.

At present, we cannot specify the role of the Bi oxide overlayer on the top of the ultrathin films. It is clear that the Bi oxide overlayer prevents the surface degradation of the ultrathin films, since the resistance of the ultrathin films, especially the 15-Å-thick film, increased more and more during setting the four-probe measurement. It has been demonstrated by high-resolution TEM [11] that the cleavage surface of a 2212 single crystal consists of a single Bi-O layer. It is considered that the chemical bonding between adjacent Bi-O double layers is very weak and the surface of a single Bi-O layer is stable. However, it is natural that the surface of a single Bi-O layer is crystallographically imperfect. Therefore the Bi oxide overlayer on the top of the ultrathin films may play a role in completing the structure of the Bi-O surface layer as a hole donor, as well as protecting the structure of the ultrathin films.

CONCLUSION

The metalorganic chemical vapor deposition of the Bi-Sr-Ca-Cu-O 2212 superconductor was found to progress in the two-dimensional mode at the initial stage, from the measurement of the intensity of the XPS photoelectron for the deposited film and the SEM observations. These data suggest that the height of the two-dimensional unit was a half-unit cell of a $Bi_2Sr_2CaCu_2O_x$ along the c-axis. From the low-field nonresonant microwave absorption and temperature dependence of the resistivity of the 15-Å thick film, superconductivity may occur in a half-unit-cell thick layer of $Bi_2Sr_2CaCu_2O_x$.

ACKNOWLEGEMENT

We would like to thank Dr. Yoshida of Div.IV, Superconductivity Research Laboratory for his helpful discussions, and Dr. J.O.Willis of Div.III for helpful discussions and a critical reading of this manuscript.

This work was supported by the New Energy and Industrial Technology Development Organization (NEDO) under the management of the R&D of Basic Technology for Future Industries.

REFERENCES

1. Terashima T, Bando Y, Iijima K, Yamamoto K, Hirata K, Hayashi K, Kamigaki K, Terauchi H (1990) Phys.Rev.Lett. 65: 2684
2. Streiffer SK, Lairson BM, Eom CB, Clemens BM, Bravman JC, Geballe TH (1991) Phys.Rev.B 43: 13007
3. Norton MG, Lisa AT, Summerfelt SR, Carter CB (1989) Appl.Phys.Lett. 55: 2348
4. Mukaida M, Miyazawa S, Sasaura M (1991) Jpn.J.Appl.Phys. 30: L1474
5. Kanai M, Kawai T, Kawai S (1990) Appl.Phys.Lett. 57: 198
6. Schlom DG, Marshall AF, Sizemore JT, Chen ZJ, Eckstein JN, Bozovic I, Von Dessonneck KE, Harris, Jr. JS, Brauman JC (1990) J.Cryst.Growth 102: 361
7. Matsushima T, Ichikawa Y, Adachi H, Setsune K, Wasa K (1990) Solid State Commun. 76: 1201
8. Matsumoto T, Kawai T, Kitahama K, Kawai S, Shigaki I, Kawate Y (1991) Appl.Pyhs.Lett. 58: 2039
9. Terashima T, Shimura K, Bando Y, Matsuda Y, Fujiyama A, Komiyama S (1991) Phys.Rev.Lett. 67: 1362
10. Sugimoto T, Yoshida M, Yuhya S, Baar DJ, Shiohara Y, Tanaka S (1991) J.Appl.Phys. 70: 1600
11. Matsui Y, Maeda H, Tanaka Y, Horiuchi S (1989) Jpn.J.Appl.Phys. 28: L946

MOCVD Growth of High-T_c and High-J_c Superconducting BSCCO Thin Films

KAZUHIRO ENDO, HIROFUMI YAMASAKI, YOSHITAKA MORIYASU*, SHUNJI MISAWA, SADAFUMI YOSHIDA, and KOJI KAJIMURA

Electrotechnical Laboratory, Tsukuba, Ibaraki, 305 Japan

ABSTRACT

Bi-Sr-Ca-Cu-O superconducting films consisting of 2223-high-T_c single phase have been prepared on MgO(100) and $LaAlO_3$(100) substrates without post-annealing by MOCVD. As-deposited film on $LaAlO_3$ exhibited the highest T_c(zero) of 97 K and the highest J_c of 3.8×10^5 A/cm^2 at 77 K under zero magnetic field as reported so far. It also showed little degradation of J_c under high magnetic field up to 8 T at 77 K. The magnetic field dependence of J_c and surface morphology observation strongly suggest that those as-deposited films have no weak links. It was speculated through the measurement of the angular dependence of J_c that the origin of such high J_c is primarily attributed to the intrinsic pinning mechanism.

KEY WORDS: MOCVD, Bi-Sr-Ca-Cu-O superconducting film, 2223-high-Tc phase, zero-resistance critical temperature, critical current density.

INTRODUCTION

The Bi-Sr-Ca-Cu-O (BSCCO) system attracts much attention for its high superconducting critical temperature T_c and superior material stability, compared with the Y-Ba-Cu-O system. The BSCCO system has many superconducting phases showing various T_c values, *e.g.*, the high-T_c phase $Bi_2Sr_2Ca_2Cu_3O_x$ with T_c around 110 K and the low-T_c phase $Bi_2Sr_2CaCu_2O_y$ with T_c around 80 K. However, it has been recognized to be difficult to obtain the 2223-high-T_c single phase films. Much efforts have been devoted to preparing as-deposited films with high T_c and high critical current density J_c, especially for microelectronics. However, most of the films reported so far consisted of the 2212-low-T_c phase in an as-deposited state[1-3], and no films with the 2223-high-T_c phase have been prepared without post-annealing, using any growth techniques.

In our previous study[4], we succeeded, for the first time, in preparing 2223-high-T_c single phase films on MgO substrates without post-annealing by a metalorganic chemical vapor deposition (MOCVD). However, the zero-resistance critical temperature T_c(zero) was as low as 74 K. We carried out *in-situ* Pb-doping during growth by MOCVD using bis (-2, 2, 6, 6-tetramethyl-3, 5-heptanediono)-lead, $Pb(DPM)_2$[5]. However, there were little effects of Pb-doping on the rise of T_c(zero), whereas the high-T_c single phase films doped with Pb were formed.

Our present review is focused on the successful preparation method and properties of 2223-high-T_c single phase films having the highest T_c(zero) and the highest J_c at 77K under magnetic field (1-8 T) without post-annealing by MOCVD as reported so far.

EXPERIMENTAL

Figure 1 is a schematic drawing of the cold-wall type MOCVD apparatus used in this study. The source materials were triphenyl bismuth $Bi(C_6H_5)_3$, bis (-2, 2, 6, 6-tetramethyl-3, 5-heptanediono)-strontium $Sr(DPM)_2$, -calcium $Ca(DPM)_2$ and -copper $Cu(DPM)_2$. They were charged in individual vaporizers, heated to appropriate temperatures depending on the source materials and carried into a quartz reactor using Ar gas. Total flow rate of Ar gas was 770 cc/min. Gas lines were heated to temperatures higher than those of the vaporizers to prevent condensation of source materials. Oxygen gas was separately introduced into the reactor at a flow rate of 675 cc/min. MgO(100) and $LaAlO_3$(100) single crystals, $5\times10\times0.5$ mm^3 in size, were used as substrates. They were placed on an inconel susceptor and then inductively heated to 800 °C. The total pressure in the reactor was kept at 50 Torr during the film growth, while oxygen partial pressure was 23 Torr. The growth rate was 10–500 Å, and the film thickness was 200–5200 Å. The grown films had mirror-like shiny surfaces.

The compositions and the crystal structures of the films were confirmed using inductively coupled plasma spectroscopy (ICP) and X-ray diffraction (XRD), respectively. Surface morphology observed by scanning electron microscopy (SEM) showed no apparent grain boundaries. Resistance and J_c measurements were carried out by a standard four-probe DC method using evaporated Au film pads as electrodes. Critical currents were defined by a criterion of 2 μV/cm electric field for the voltage tap distance of 3 mm. The critical currents were also measured with a pulse current, which showed negligible heating effect at the electrodes.

RESULTS AND DISCUSSION

We have grown BSCCO films on MgO(100) substrates with thicknesses from 200 to 5200 Å at a growth rate around 100 Å/h[6]. Figure 2 shows the effect of the film thickness on T_c(zero). It should be noted that T_c(zero) increases steeply with the increase of film thickness, and then appears to saturate in the large thickness range. The XRD patterns of these films are classified into two cases. One is the 2223-high-T_c single phase pattern: all the XRD peaks were indexed as (00*n*) of the 2223-high-T_c phase. The other is the mixture of the 2223-high-T_c and 2212-low-T_c phases pattern: the XRD peaks were indexed as

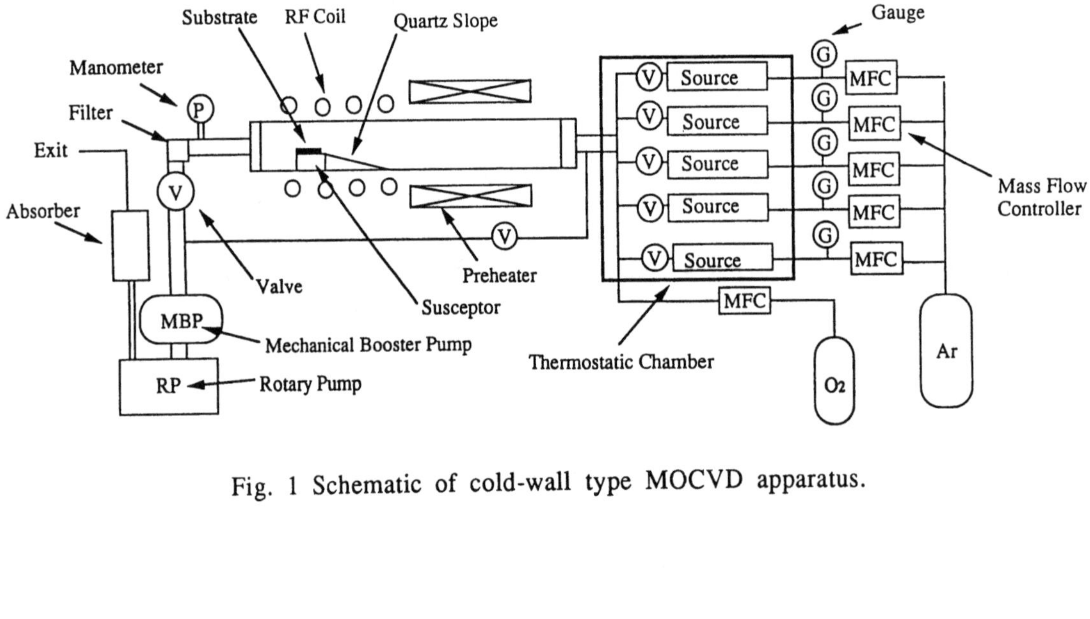

Fig. 1 Schematic of cold-wall type MOCVD apparatus.

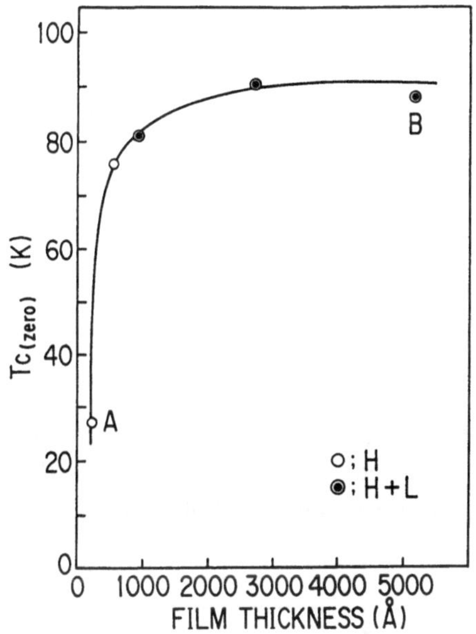

Fig. 2 Effect of film thickness on T_c(zero) for BSCCO films deposited on MgO(100) without post-annealing by MOCVD. Growth rate is 100 Å/h.
○ : 2223-high-T_c single phase;
◉ : the mixture of 2223-high-T_c and 2212-low-T_c phases

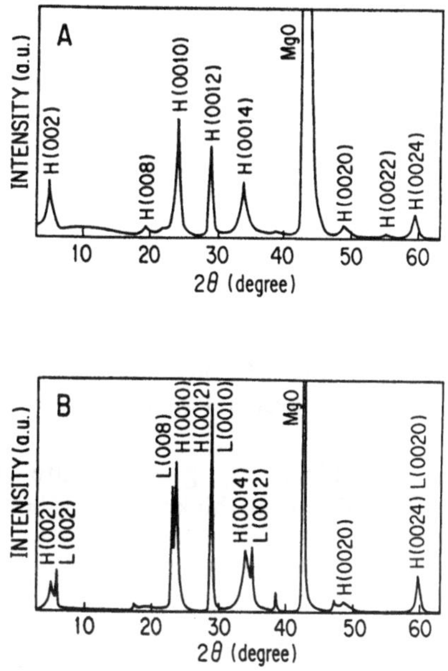

Fig. 3 XRD patterns of samples A and B corresponding to those in Fig. 2. H(00*n*) shows 2223-high-T_c phase peaks and L(00*n*) shows 2212-low-T_c phase peaks.

(00*n*) of the 2223-high-T_c and 2212-low-T_c phases. The thinner films consist of the 2223-high-T_c single phase, and the thicker ones consist of the mixture of the 2223-high-T_c and 2212-low-T_c phases. Figure 3 shows the XRD patterns of samples A(200-Å thick) and B(5200-Å thick), representatives of the thinner and thicker films, respectively. All the films show the crystal orientation with the *c*-axis perpendicular to the substrate surface.

We have also grown the 2223-high-T_c phase films at different growth rates between 10 and 500 Å/h on MgO(100) substrates[6]. Figure 4 shows the effect of the growth rate on T_c(zero) for the films with the thickness of 1800 Å. It is to be noted that T_c(zero) crealy increases with decreasing growth rate.

On the basis of these results, we have grown the films on $LaAlO_3$(100) substrates having very small lattice mismatches with BSCCO at a growth rate around 12 Å/h[7]. Figure 5 shows the XRD pattern of an as-deposited film with the thickness of 960 Å. All the peaks can be indexed as (00*n*) of the 2223-high-T_c phase, *i.e.*, the film is apparently single phase $Bi_2Sr_2Ca_2Cu_3O_x$ with the *c*-axis oriented perpendicular to the substrate surface.

Figure 6 shows the temperature dependence of the resistivity for the as-deposited film. The zero-resistance critical temperature T_c(zero) is 97 K, which is the highest value reported so far for as-deposited BSCCO films.

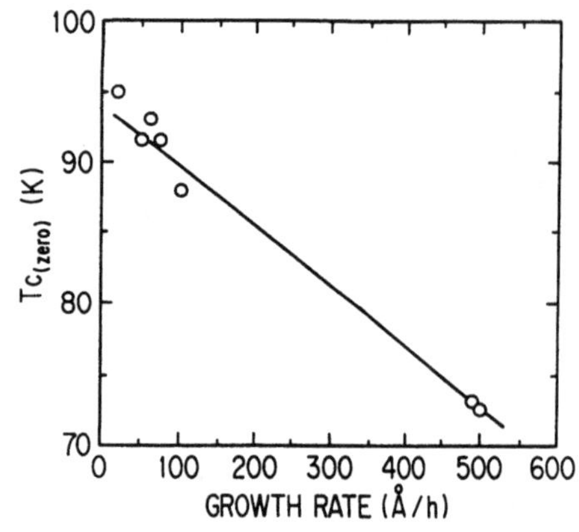

Fig. 4 Effect of growth rate on T_c(zero). The film thickness is 1800 Å.

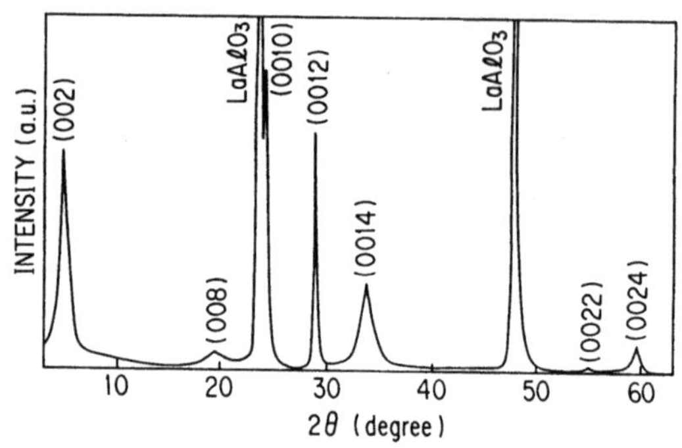

Fig. 5 XRD pattern of a BSCCO film deposited on $LaAlO_3(100)$ without post-annealing by MOCVD.

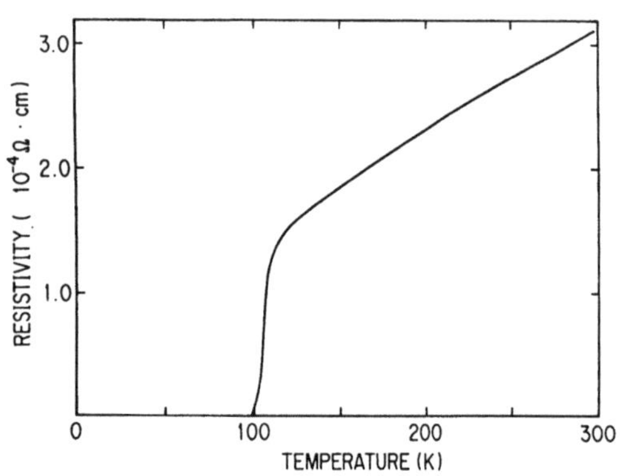

Fig. 6 Temperature dependence of the resistivity for the as-deposited BSCCO film. T_c (zero) is 97 K.

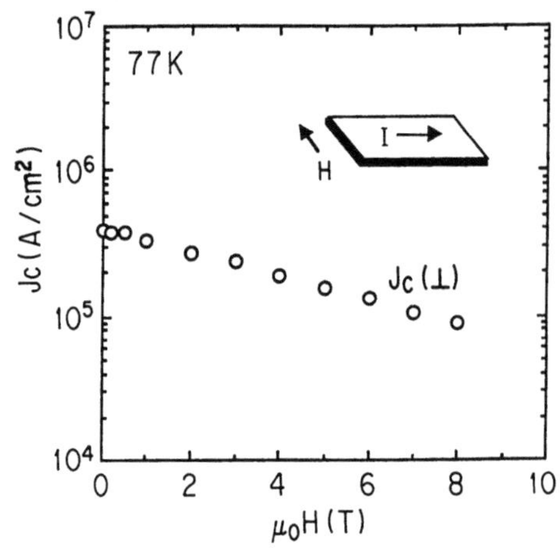

Fig. 7 Magnetic field dependence of the critical current density J_c at 77 K for the as-deposited BSCCO film.

Figure 7 shows the magnetic field dependence of the J_c at 77K. The magnetic field is applied parallel to the film surface and normal to the direction of the current. J_c at 77 K is 3.8×10^5 A/cm^2 in zero field and 3.3×10^5 A/cm^2 in the field of 1 T. These are the highest J_c values as reported so far for BSCCO films. It should be noted that J_c-H curve is very flat, *i.e.*, the degradation of J_c at 77 K is so small that J_c is sustained at 9×10^4 A/cm^2 even under the high magnetic field of 8 T. Such high J_c values are reported for BSCCO films in high magnetic fields for the first time.

There are no apparent grain boundaries observed in the SEM photographs. In addition, J_c decreases very slightly in low fields (0-0.5 T) and no hysteresis is observed in J_c-H curve. These results strongly suggest that there is no sign of weak links in these films.

Such weak-link-free J_c behavior is confirmed by the J_c measurements under the magnetic field applied parallel to the current, $J_c(//)$, which corresponds to the Lorentz-force-free configuration. Figure 8 shows the comparison of $J_c(//)$ with the usual J_c value under magnetic field applied perpendicular to the current, $J_c(\perp)$. The enhancement of J_c, $J_c(//)/J_c(\perp) = 1-2$ was observed under the magnetic field up to 14 T. This feature strongly suggests that J_c was determined by the balance between the flux pinning strength and the Lorentz force and that the system does not consist of weak-link networks. It was reported that $J_c(//)$ is essentially equal to $J_c(\perp)$ at 4.2 K under high magnetic field for textured $Bi_2Sr_2CaCu_2O_y$ tapes[8]. Their interpretation is as follows: the supercurrent flows along zig-zag paths because of the random orientation of the grains, and therefore the volume pinning force is independent of the field direction.

The present results evidently show that weak-link-free BSCCO films have very high J_c with strong flux pinning. In order to obtain more insights into the flux pinning mechanism, effect of the angle of the magnetic field against the *ab*-plane of these films on J_c has been also investigated. As shown in Fig. 9, J_c decreases abruptly from ~10^5 A/cm^2 to zero as the direction of the applied field slightly deviates from the *ab*-plane, implying that the pinning strength is very strong for the magnetic fluxes parallel to the *ab*-plane, and very weak for those parallel to the *c*-axis. Details are given in a separate paper[9]. This result supports the idea of the intrinsic pinning mechanism proposed by Tachiki and Takahashi[10], in which the insulating block layers imbedded among superconducting CuO_2 layers play as strong pinning centers against flux motion.

CONCLUSION

We have grown BSCCO superconducting films consisting of 2223-high-T_c single phase on MgO(100) and $LaAlO_3$(100) substrates without post-annealing by MOCVD. It was found that T_c(zero) increases with increasing the film thickness and decreasing the growth rate. The film grown *in-situ* on $LaAlO_3$ had the highest T_c(zero) of 97 K and the highest J_c of

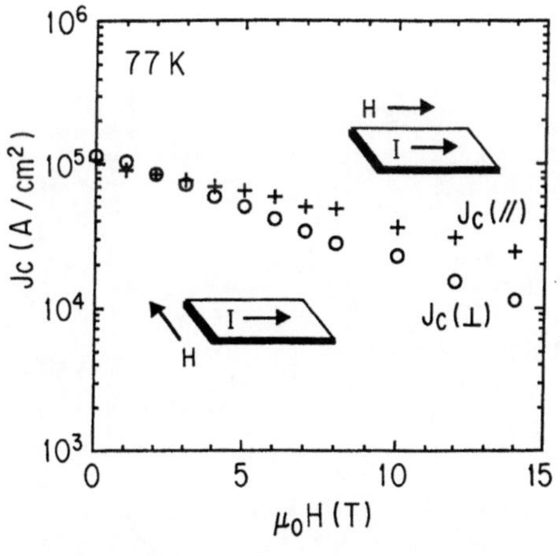

Fig. 8 Effect of the current direction relative to the applied magnetic field on J_c.

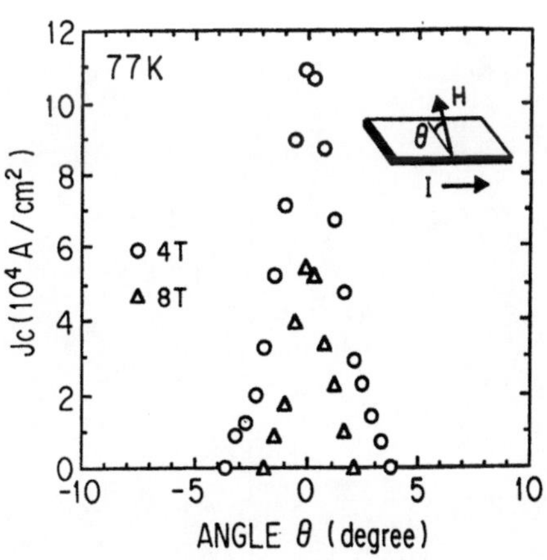

Fig. 9 Effect of the angle of the applied magnetic field against the film surface on J_c.

3.8×10^5 A/cm2 at 77 K ever reported for as-deposited BSCCO films. It also showed little degradation of J_c under high magnetic field up to 8 T at 77 K. The magnetic field dependence of J_c strongly suggests that those as-deposited films have no weak-links, being consistent with the SEM observation. The measurement of the effect of the applied direction of the magnetic field on J_c revealed that J_c dropped abruptly as its direction slightly deviates from the *ab*-plane, which supports the idea of the intrinsic pinning mechanism.

REFERENCES

1. Eckstein JN, Schlom DG, Hellman ES, vonDessonneck KE, Chen ZJ, Webb C, Turner F, Harris Jr JS, Beasley MR, Geballe TH (1989) J Vac Sci Technol B70: 319-323
2. Schmitt P, Schultz L, Saemann-Ischenko G (1990) Physica C168: 475-478
3. Matsushima T, Ichikawa Y, Adachi H, Setsune K, Wasa K (1990) Solid State Commun 76: 1201-1204
4. Endo K, Hayashida S, Ishiai J, Matsuki Y, Ikedo Y, Misawa S, Yoshida S (1990) Jpn J Appl Phys 29: 294-297
5. Endo K, Hayashida S, Nakatsuka K, Ishiai J, Ikedo Y, Misawa S, Yoshida S (1990) In: Ishiguro T, Kajimura K (eds) Advances in Superconductivity II. Springer, Tokyo, pp 912 -915
6. Endo K, Nakatsuka K, Misawa S, Yoshida S (1991) In: Kajimura K, Hayakawa H (eds) Advances in Superconductivity III. Springer, Tokyo, pp1017-1020
7. Endo K, Yamasaki H, Misawa S, Yoshida S, Kajimura K (in press) Nature
8. Kumakura H, Togano K, Maeda H, Kase J, Morimoto T (1991) Appl Phys Lett 58: 2830-2832
9. Yamasaki H, Endo K, Nakagawa Y, Umeda M, Kosaka S, Misawa S, Yoshida S (1991) these proceedings
10. Tachiki M, Takahashi S (1989) Solid State Commun 70: 291-295

*On leave of absence from Asahi Chemical Industry Co., LTD., Fuji, Shizuoka, 416 Japan

RHEED Intensity Monitored Growth of Bi-Sr-Ca-Cu-O Superconductors

PETER BODIN, SHIGEKI SAKAI, YUJI KASAI, and HISAO TANOUE

Electrotechnical Laboratory, Tsukuba, Ibaraki, 305 Japan

ABSTRACT

In order to make Bi-Sr-Ca-Cu-O superconducting thin films, we employ the molecular beam epitaxy (MBE) technique with in situ monitoring by reflection high energy electron diffraction (RHEED). In the MBE system, either co-evaporation or the shuttered growth technique can be used. By co-evaporation we observe RHEED intensity oscillation with a period corresponding to the growth of half a unit cell of $Bi_2Sr_2CaCu_2O_8$ in the c-axis direction. By shuttered growth we observe repetitive RHEED intensity variations throughout the growth of 25 half unit cells.

KEY WORDS: Bi system superconductor, MBE, RHEED, oscillation, thin film

INTRODUCTION

MBE with in situ monitoring by RHEED is a fascinating technology for making oxide superconducting thin films because controlled layered growth on the atomic scale can be made. The Bi-Sr-Ca-Cu-O system consists of 5 elements, and has several different phases related to the number, n, of Ca and CuO_2 planes, expressed by the approximate formula $Bi_2Sr_2Ca_nCu_{n+1}O_{2n+6}$. For this reason shuttered growth where one atomic layer is deposited at a time is a convenient technique for growing well-defined phases[1]. Another method is co-evaporated growth where all elements are irradiated to the substrate simultaneously. Previously co-evaporated growth has been done on the 4 elements YBCO with the c axis length of 11.7 Å[2], and on the 4 element $Bi_2Sr_2CuO_6$ with the c/2 axis length of 12.2 Å[3]. In both cases RHEED intensity oscillations were observed. It is therefore of interest to clarify if this also can be done on the 5 element $Bi_2Sr_2CaCu_2O_8$ (referred to as the 2212 phase) with the longer c/2 axis of 15.4 Å.

In this paper we present our experimental setup, and the results from the two growth methods, coevaporation and shuttering. RHEED data are described in detail, and we show that the RHEED intensity changes are related to the formation of the half unit cell of the 2212 phase with c/2 = 15.4Å. The future perspective of this research could be that RHEED will provide a tool for actively controlling the growth of the superconductor on the atomic layer scale.

EXPERIMENTAL SETUP

Pure ozone was used as an active oxygen source[4]. During growth the background pressure of the MBE chamber was kept at the range of 10^{-8} mbar to $2x10^{-6}$ mbar. This low background pressure was obtained because the ozone flux was as low as 10^{15} to 10^{16} $cm^{-2}s^{-1}$, and the ozone was conducted via a pipe directly to the surface of the substrate. The substrate temperature in the examples to be shown here was 710° C. We used $LaAlO_3$ substrates (produced by Mitsubishi Kasei Corp.) for the growth of the Bi-Sr-Ca-Cu-O 2212 phase because the lattice mismatch is only 0.7%. Pure metal sources were used for the evaporation of Bi, Sr, Ca, and Cu. Every metal source has a shutter whereby the deposition time of each element can be chosen independently. In our RHEED system 30 keV electrons were focused on the

a b c

Fig. 1. RHEED pictures during co-evaporated growth of the 2212 phase. The growth rate was 0.6 Å/s. (a) 20 s after growth start observed along the <110> azimuth of the 2212 phase . (b) At the growth end, 150 s after growth start, observed along the <110> azimuth. (c) After growth and, observed along the <100> azimuth.

surface of the 2212 phase thin film. The RHEED incident angle was typically between 1° and 4° , and fixed for each experiment. To record and analyse the changing RHEED patterns during growth, a video-camera with a subsequent video-processor connected to a computer was used. This enabled us to replay the observed RHEED pattern for detailed analysis.

CO-EVAPORATION

In the case of co-evaporated growth all metal shutters are simultaneously opened whereby all 4 metals arrive at the same time at the substrate. Ozone is supplied continuously. In the example given here the growth rate was estimated to be 0.6 Å/s. The rate and composition ratio were monitored during growth by a quartz crystal oscillator and subsequently decided by Rutherford backscattering (RBS) and inductively coupled plasma (ICP) spectroscopy.

Figure 1 shows the RHEED pictures: observed along the <110> direction of the 2212 phase, (a) 20 s after growth start and (b) at the growth end that was 150 s after growth start; and (c) observed along the <100> direction after growth. The streak spacing for the grown 2212 phase remained identical to substrate streak spacing, indicating that the lattice spacing of the $LaAlO_3$ substrate along the <100> direction and the 2212 phase along the <110> is almost the same. After 20 s of growth, bright spots besides the streaks appeared, showing some additional 3 dimensional growth. Near the time of growth end, the spots weakened and the RHEED pattern became more streaky. By looking carefully at Fig. 1b, a splitting and bending of the streaks can be seen. This is due to a 90° rotation twinning of the a-b plane of the 2212 crystal whose structure is modulated along the b axis direction[5]. Figure 1c shows the RHEED pattern observed along the <100> azimuth of the 2212 film after growth. In this case the streaks are broadened but not bent or split. The broadening can also be attributed to the long range modulation in the b-axis direction, previously observed by Eckstein et al[6]. The 90° rotation twinning with the b-axis long range modulation has been confirmed independently by TEM observation on a sample made in the same way[7].

For the sample discussed in Fig. 1, the intensity change on the specular spot in the RHEED pattern was plotted in Fig. 2. In the figure we see an oscillation with a period of 26 s. Comparing the 26 s period with the growth rate of 0.6 Å/s, we find that the oscillation corresponds nicely to the growth of half a unit cell of 15.4 Å. The total thickness of the film can be directly read from the number of oscillations in Fig. 2 to be 90 Å. During the first 26 s of growth a short period oscillation which might be related to coverage of the atomic layer scale was observed. After 26 s corresponding to growth of half a unit cell, the regular oscillation starts. That the oscillation amplitude decreases with time may be due to increasing surface roughness. After growth end a small recovery of the intensity occurs, most probably due to an improvement of the epitaxy. It is a known phenomenon from

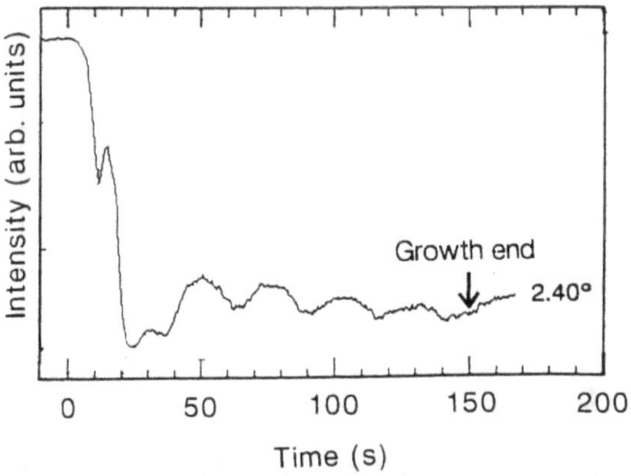

Fig. 2. The RHEED intensity of the specular spot during co-evaporated growth of the 2212 phase. The electron beam incident angle was 2.4° . Intensity oscillation with a 26 s period is found.

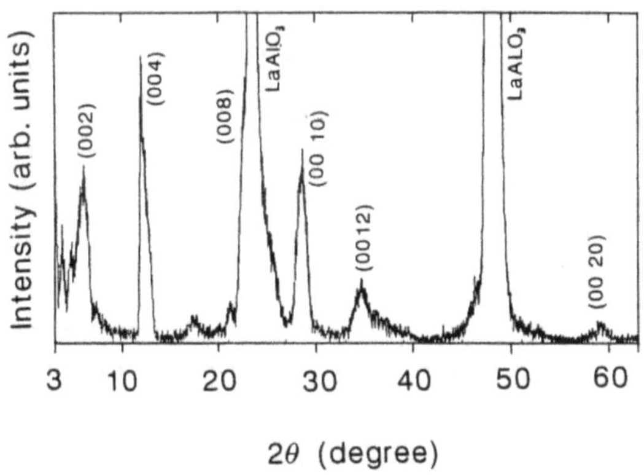

Fig. 3. X-ray diffraction pattern for the 2212 phase film grown by co-evaporation.

growth of Si and GaAs[8] and recently also YBCO[2] , and it usually indicates that the crystal growth conditions were good. The X-ray diffraction pattern in Fig.3 shows that the 90 Å film is a good c-axis orientated 2212 phase film. From ac susceptibility measurement the superconducting transition temperature was 36 K.

SEQUENTIAL SHUTTERING

Another way of growing the Bi-oxide superconductor is to grow it in a layer by layer way - with only one metal source shutter open at a time. By choosing the shutter sequence with the order of Sr, Cu, Ca, Cu, Sr, and Bi according to the sequence of elements of the crystal structure, the crystal can be grown in a controlled way. Ozone is supplied continuously. The growth rate was approximately 0.1 Å/s on the average. Figure 4 shows the RHEED screen pictures from the first half unit cell evaporation sequence. Figure 4a shows the bare $LaAlO_3$ substrate before growth. Figure 4b shows the RHEED pattern after the last Sr and before evaporation of Bi. Figure 4c is the pattern observed 30 s after the start of the first Bi evaporation that closes the first cycle.
Again the splitting of the streaks due to the 90° rotation twinning of the a-b plane of the 2212 phase and b axis modulation is found. At this point we actually see the formation of the minimum thickness (half a unit cell) of the 2212 phase.

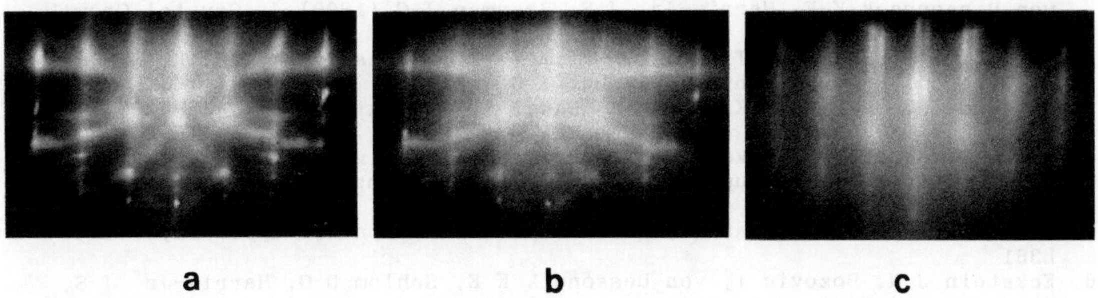

Fig. 4. RHEED pictures from shuttered growth of the first half unit cell of the 2212 phase, observed along the $LaAlO_3$ <100> azimuth: (a) The $LaAlO_3$ substrate before growth. (b) Before the deposition of Bi. (c) During deposition of Bi that ends the first cycle. The growth rate was 0.1 Å/s.

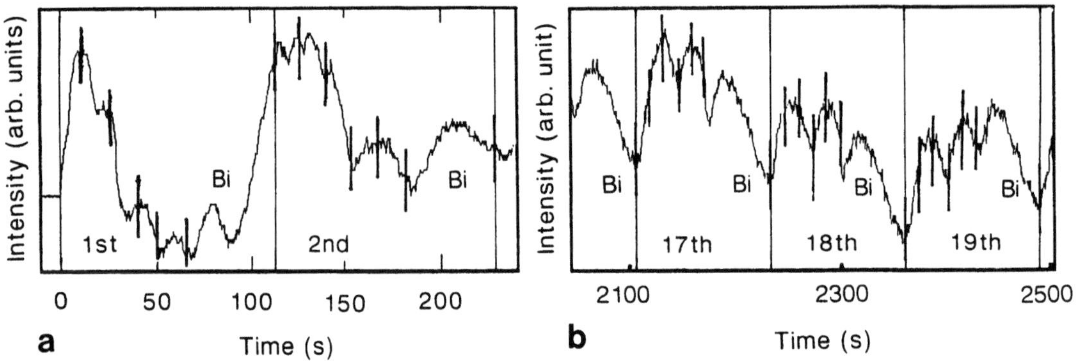

Fig. 5. RHEED intensity at the central streak during shuttered growth of the 2212 phase, (a) from the first and second cycle of evaporation, and (b) from the 17th, 18th, and 19th cycle of deposition. The shutter sequence was Sr-Cu-Ca-Cu-Sr-Bi. The vertical bars on the curves show the boundaries of the shutter changes. Only the Bi deposition sequence is designated in the figures.

Observing the change in the RHEED intensity during the evaporation of each element we obtain Fig. 5. The first evaporation cycle in Fig. 5a is different from the second cycle because the growth is heteroepitaxial during the first cycle, and the growth is homoepitaxial during the second cycle. Evaporation was continued throughout totally 25 half unit cells - and the corresponding film thickness was 390 Å. At cycle number 17, 18, and 19 the RHEED intensity variation with the deposition of each metal was repetitive as shown in Fig. 5b. The superconducting critical temperature was 48 K.

SUMMARY

During co-evaporated growth of $Bi_2Sr_2CaCu_2O_8$ we showed RHEED intensity oscillation with a period of 26 s. By quartz crystal monitoring and the subsequent RBS and ICP analyses, the rates of the individual metals were adjusted and the total rate was estimated to be 0.6 Å/s. By comparing the oscillation period with the growth rate, we conclude that the oscillation period corresponds to the growth of the 2212 phase half unit cell. During shuttered evaporation a repetitive RHEED intensity variation throughout growth of 25 half unit cells was observed. We observed a splitting of the RHEED pattern streaks due to the b-axis modulation and the 90° rotation twinning of the a-b plane of the 2212 phase for both growth methods.

REFERENCES

1. Schlom D G, Marshall A F, Sizemore J T, Chen Z J, Eckstein J N, Bozovic I, von Dessonneck K E, Harris-Jr. J S, Bravman J C (1990) J. Crystal Growth 102: 361
2. Terashima T, Bando Y, Iijima K, Yamamoto K, Hirata K, Hayashi K, Kamigaki K, Terauchi H (1990) Phys. Rev. Lett. 65: 2684
3. Matsumoto T, Kawai T, Kitahara K, Kawai S, Shigaki I, Kawate Y (1991) Appl. Phys. Lett. 58: 2039
4. Sakai S, Kasai Y, Hosokawa S, Ichimura S (1991) In: Kajimura K, Hayakawa H (eds) in Advances in Superconductivity Ⅲ. Springer-Verlag, Tokyo, pp1085-1088
5. Matsui Y, Maeda H, Tanaka Y, Horiuchi S (1988) Jpn. J. Appl. Phys. 27: L361
6. Eckstein J N, Bozovic I, von Dessoneck K E, Schlom D G, Harris-Jr. J S, Bauman S M (1990) Appl. Phys. Lett. 57: 931
7. Matsuhata H, unpublished
8. Neave J H, Joyce B A, Dobson P J, Norton N (1983) Appl. Phys. A31: 1

BiSrCaCuO Thin Films Grown in One Direction by Using Off-Oriented $SrTiO_3(110)$ Substrates

O. WADA[1], K. KURODA[2], J. TANIMURA[1], M. KATAOKA[1], K. KOJIMA[2], T. TAKAMI[2], K. HAMANAKA[2], and T. OGAMA[1]

[1] Materials & Electronic Devices Laboratory, [2] Central Research Laboratory, Mitsubishi Electric Corporation, Amagasaki, Hyogo, 661 Japan

ABSTRACT

BiSrCaCuO thin films consisting of 2212 phase were formed by RF magnetron sputtering on off-oriented $SrTiO_3(110)$ substrates tilted toward the [100] in order to orient *c*-axes only in one direction. Grains with the *c*-axis almost parallel to the [100] of substrate decreased, disappeared and appeared again as the degree of the off-angles increased. These results show that an appropriate off-angle exists in the vicinity of 5° to orient BiSrCaCuO thin films in a single direction almost parallel to the [010] of the substrate. It is considered from the comparison between a normally and a rapidly deposited films that steps on the substrate surface influence the film growth mechanism.

KEY WORDS: BiSrCaCuO superconductor, thin films, sputtering, off-oriented substrate

INTRODUCTION

It is important to obtain long coherence length perpendicular to the film surfaces in order to make Josephson tunneling junctions using high T_c superconductors. It has been reported that *c*-axes of superconducting BiSrCaCuO thin films can be tilted against substrate surfaces by several methods[1-5]. However, these films have many small grains with different crystal orientation. Therefore, we made the *c*-axes aligned almost in one direction by forming a (11n)-oriented BiSrCaCuO thin film on a vicinal surface of $SrTiO_3(110)$ substrate which was cut 5° toward the [100][6].

In order to form a film oriented in a single direction with good crystallinity, it is very important to clarify crystal growth mechanism of the (11n)-oriented BiSrCaCuO films. Thus we investigated the dependence of crystal orientation and microstructure on off-angles of vicinal surfaces of $SrTiO_3(110)$ substrates.

BiSrCaCuO thin films were formed on a just-oriented $SrTiO_3(110)$ substrate and off-oriented $SrTiO_3(110)$ substrates tilted 2°, 5° and 10° toward [100] direction. The crystal orientation and microstructure were examined by reflection high energy electron diffraction (RHEED) and transmission electron microscopy (TEM). It was found from these results that an appropriate off-angle for one directional crystal growth exists in the vicinity of 5°.

Comparison of TEM observation between a rapidly and a normally deposited films suggests that the film oriented in a single direction grows from steps on the off-oriented substrate surface.

EXPERIMENTAL

BiSrCaCuO thin films consisting of 2212-phase were formed on just- and off-oriented $SrTiO_3(110)$ substrates at 660°C in O_2 atmosphere of 300 mTorr by RF magnetron sputtering[1]. The $SrTiO_3(110)$ substrates were cut off by 2°, 5° and 10° toward [100] direction. Figure 1 shows relationship between the substrate surface and crystal orientation. Thickness of the films was about 140 nm when the films were normally deposited. And also a film was rapidly deposited on the 5°-off substrate, where the deposition rate was 3.5 times faster than those of the films normally deposited.

Crystal orientation was measured by RHEED. Microstructure of the films was investigated by cross-sectional TEM. Samples for cross-sectional TEM were prepared by Ar ion thinning method with a sample cooling system.

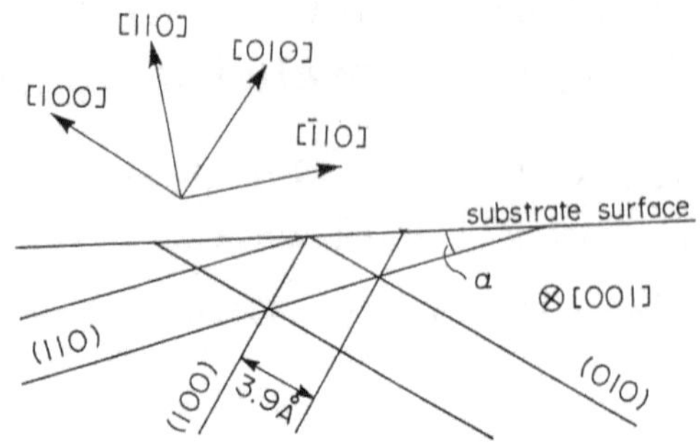

Fig.1 Schematic illustration of crystal orientation of an off-oriented $SrTiO_3(110)$ surface cut α toward [100].

RESULTS AND DISCUSSION

Figure 2 shows RHEED patterns of the BiSrCaCuO thin films formed on the just-oriented $SrTiO_3(110)$ substrate and the off-oriented $SrTiO_3(110)$ substrates tilted 2°, 5° and 10° toward [100] direction, respectively. An incident electron beam was along the [001] direction of $SrTiO_3$. Two kinds of lines of diffracted spots which have the same intensity were observed in the RHEED pattern of the BiSrCaCuO films grown both on the just-oriented substrate as shown in Fig.2(a). Directions of the lines of the spots show that of *c*-axis of the BiSrCaCuO. It was found that the *c*-axes were almost parallel to the [010] and [100] directions of $SrTiO_3$ substrate, which are abbreviated as the $[010]_s$- and $[100]_s$-oriented *c*-axis, respectively hereafter. On the other hand, it was observed that the RHEED patterns of the $[010]_s$-oriented *c*-axes were strong and the $[100]_s$-oriented *c*-axes were weak both in the films formed on the 2°- and 10°-off-oriented substrates as shown in Fig.2(b) and (d). And only the $[010]_s$-oriented *c*-axis was detected in the RHEED pattern of the film formed on the 5°-off-oriented substrate as shown in Fig.2(c). It is considered from these results that the BiSrCaCuO film were formed on the 5°-off-oriented substrate with the *c*-axis in a single direction.

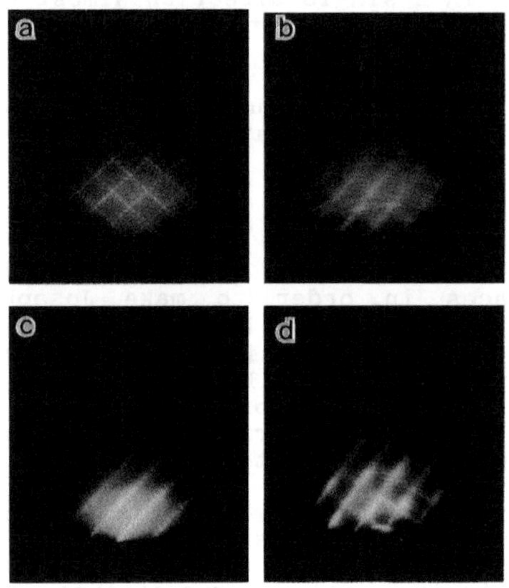

Fig.2 RHEED patterns of BiSrCaCuO films grown on just-oriented (a), 2° (b), 5° (c) and 10° (d) off-oriented $SrTiO_3(110)$.

Microstructure of these films were examined by cross-sectional TEM method. Figure 3 shows cross-sectional TEM images of BiSrCaCuO films formed on the 5°- and 10°-off-oriented substrates. Lattice images corresponding to the (001) lattice planes were observed in both images. It is found from Fig.3(b) that the films formed on the 10°-off-oriented substrates have large grains with the $[010]_s$- and $[100]_s$-oriented *c*-axes. Such two kinds of large grains were also observed in the films grown on the on-oriented and the 2° off-oriented substrates as previously reported [3,7]. These two kinds of grains reach the film surface as shown in Fig.3(c). On the other hand, the grains with the $[100]_s$-oriented *c*-axis are small and do not reach the film surface in the film formed on the 5°-off-oriented substrate as shown in Fig.3(a). These results explain why the RHEED pattern of the film on the 5°-off-oriented substrate does not show the $[100]_s$-oriented *c*-axis, for the RHEED measurement detects only a region near the film surface. Thus these results show that the region near the film surface has a single crystal orientation when the film was formed on the 5°-off-oriented $SrTiO_3(110)$ substrate, and suggest that there is an appropriate off-angle in the vicinity of 5° in order to form a film with only $[010]_s$-oriented *c*-axis.

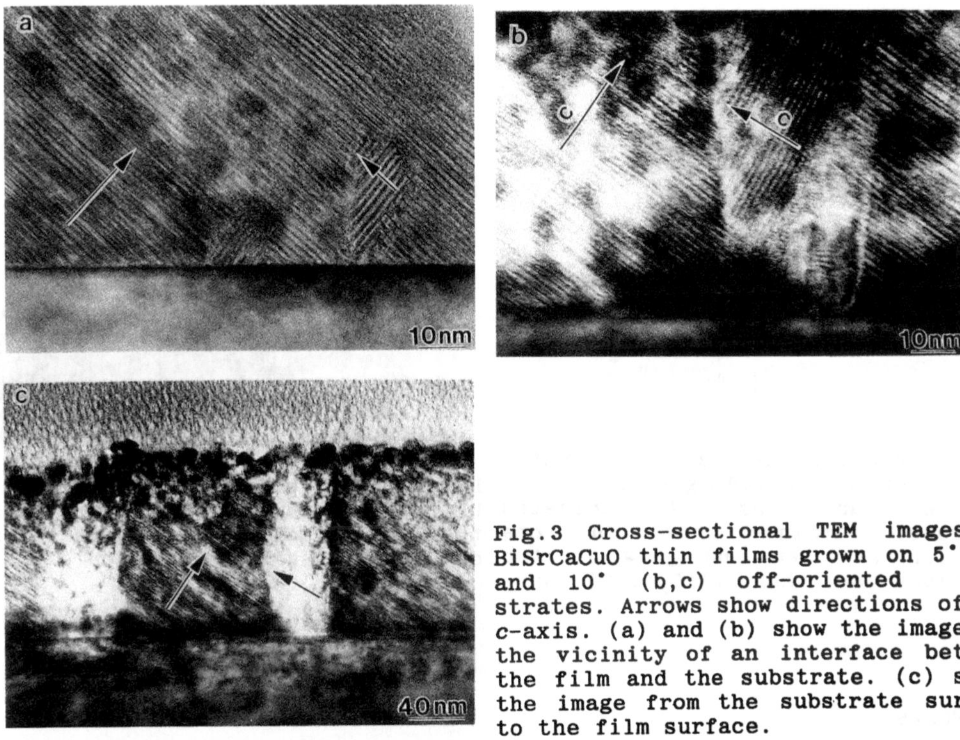

Fig.3 Cross-sectional TEM images of BiSrCaCuO thin films grown on 5° (a) and 10° (b,c) off-oriented substrates. Arrows show directions of the *c*-axis. (a) and (b) show the images in the vicinity of an interface between the film and the substrate. (c) shows the image from the substrate surface to the film surface.

As mentioned above, when the $SrTiO_3$(110) was cut toward the [100], the films has the tendency to grow with the $[010]_s$-oriented *c*-axis. It is most probable that this growth mechanism is due to the existence of steps on the substrate surface. It is considered that atoms on the substrate surface move to the step and the films starts to grow at the step. The interplanar spacing of the (100) planes of $SrTiO_3$ (3.9Å) agrees with that of the (110) plane of BiSrCaCuO (3.8Å). In this case, the film growth in the [100] direction is suppressed as can be seen from Fig.4. Therefore, it is considered that only crystal grains grow with the $[110]_{BSCCO}$ direction almost parallel to the $[100]_{STO}$ direction, that is, the *c*-axis almost parallel to the [010] of $SrTiO_3$. On the other hand, growth nuclei can be formed on the terrace of $SrTiO_3$(110) plane between the steps. In this case the grains may grow with two kinds of orientation in the same way as the film formed on the just-oriented $SrTiO_3$(110) substrate. Fig.4 shows the growth of grains in the two ways. Therefore, when the number of steps increase by increasing the off-angle, the width of the terrace of $SrTiO_3$(110) surface between the steps decreases and the growth nuclei on the terrace decrease. As a result, grains with the $[010]_s$-oriented *c*-axis tend to grow on the 5°-off-oriented substrate. Fig.5 shows a cross-sectional TEM image of the film formed at 3.5 times faster deposition rate. The $[100]_s$-oriented grains which reached the film surface were observed. This result shows that grains with the $[100]_s$-oriented *c*-axis can grow because the number of growth nucleus on the terrace increases by rapid deposition. This supports the growth mechanism above mentioned. However, it is not clear why the $[100]_s$-oriented grain appeared again in the film grown on the 10° off-oriented substrate.

CONCLUSION

Crystal orientation of BiSrCaCuO thin films consisting of the 2212-phase depends on off-angles of $SrTiO_3$(110) substrate. Only the film formed on the 5°-off-oriented substrate has almost a single specific preferred orientation, but the films on 2°- and 10°-off-oriented substrates have two kinds of preferred orientation. These results suggested that there is an appropriate off-angle to make a film with a single orientation. The cross-sectional TEM observation suggests that the steps affect the crystal orientation of the BiSrCaCuO film.

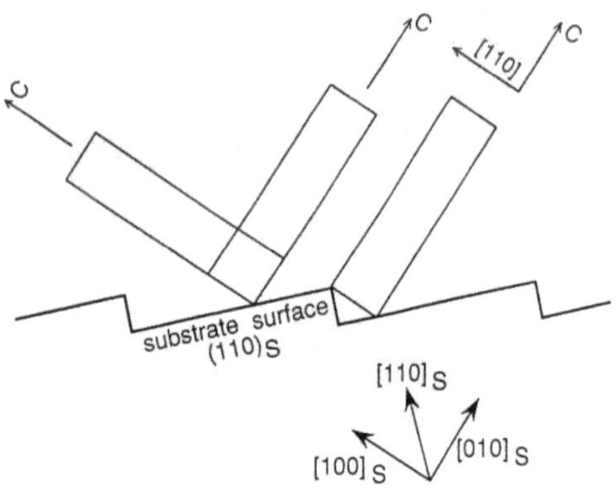

Fig.4 Schematic diagram of crystal orientation of the film growth on off-oriented substrates. S indicates substrate. Grains grow in one direction at step and grow in two directions on the $SrTiO_3$(110) surfaces.

Fig.5 Cross-sectional TEM image of BiSrCaCuO thin film 3.5 times faster than the normal deposition rate formed on 5°-off-oriented substrate.

REFERENCES

1. Kuroda K, Kojima K, Wada O, Tanioku M, Yokoyama K and Hamanaka K (1990) Jpn.J.Appl.Phys. 29:L1816
2. Kuroda K, Wada O, Kojima K, Tanimura J, Kataoka M, Takami T, Yokoyama K, Ogama T and Hamanaka K (1991) Jpn.J.Appl.Phys. 30: L475
3. Wada O, Tanimura J. Kuroda K, Kataoka M, Kojima K, Hamanaka K and Ogama T (1991) Jpn.J.Appl.Phys. 30: L722
4. Ishizuka Y, Terashima Y and Miura T (1991) Jpn.J.Appl.Phys. 30: L1006
5. Kataoka M, Kuroda K, Tanimura J, Wada O, Kojima K, Takami T, Hamanaka K and Ogama T (1991) Jpn.J.Appl.Phys. 30: to be published
6. Kuroda K, Wada O, Tanimura J, Kojima K, Takami T, Kataoka M, Ogama T and Hamanaka K (1991) Jpn.J.Appl.Phys. 30: L582
7. Wada O, Kuroda K, Tanimura J, Kataoka M, Kojima K, Takami T, Hamanaka K and Ogama T (1991) Jpn.J.Appl.Phys. 30: to be published

BiSrCaCuO/BiSrCuO Superlattices: A Prove of Cu-O Plane Superconductivity

MASAO NAKAO, HIROAKI FURUKAWA, HIROMI MUKAIDA, SHUJI FUJIWARA, and RYOHKAN YUASA

Sanyo Tsukuba Research Center, Tsukuba, Ibaraki, 305 Japan

ABSTRACT

The parallel and perpendicular superconducting transport properties of epitaxial superlattices that consist of alternating $Bi_2Sr_2CaCu_2O_8$ (BSCCO) and $Bi_2Sr_2CuO_6$ (BSCO) layers are reported. The interlayer coupling of Cu-O planes within a cell and in adjacent cells is studied by varying the superlattice sequence of different *c*-axis half-unit cells. All of our superlattices show superconductivity in the plane of the layer, including superlattices containing isolated Cu-O double planes of BSCCO in a semiconducting BSCO matrix. The overall shapes of superconducting transitions are found to be quite similar independent of the separation of the BSCCO layer. In the perpendicular direction, the transition width increases with the number of BSCO half-unit cells, but a residual resistivity appears for five half-unit cells. We conclude that superconductivity is essentially independent of the BSCCO and BSCO half-unit-cell sequence, and proximity-effect coupling along the *c*-axis direction cannot occur through the BSCO layer more than two unit cells (49 Å) thick.

KEY WORDS: superconducting superlattice, atomic-layer-by-layor MBE, interlayer coupling, proximity effect, two-dimensional superconductor

INTRODUCTION

In conventional superconductors, superlattices [1] as well as layered film structures containing tunnel junctions or weak links, have been extensively studied for both fundamental research and device applications. In the high-temperature superconductors (HTSCs), material problems have made the realization of such superlattice structures more difficult than in metallic systems. However, all HTSCs have naturally layered crystal structures containing Cu-O planes in common. A consequence of this commonality is fairly good lattice matching in the basal (*a-b*) plane between two different HTSCs. This offers the possibility to grow heteroepitaxial superlattices of two different HTSCs layer by layer using thin-film techniques.

In this paper, we report on the superconducting transport properties of artificially produced superlattices with different *c*-axis half-unit-cell sequences. We have picked $Bi_2Sr_2CaCu_2O_8$ (BSCCO) and $Bi_2Sr_2CuO_6$ (BSCO) as an epitaxial growth pair, because their carrier densities are quite similar but they differ in conductivity. Using the atomic-layer-by-layer molecular beam epitaxy (MBE) technique [2,3] we have been able to tailor the coupling in adjacent BSCCO cells by interposing semiconducting BSCO layers. We find distinctly different superconducting properties depending on the directions of measuring current, i.e., whether in the plane of the layer or in the perpendicular direction.

SAMPLE PREPARATION

Superlattices consisting of alternating BSCCO and BSCO layers have been synthesized *in situ* by sequentially opening and closing the shutters that control the fluxes from evaporation sources in a versatile MBE system that was described earlier [2]. Briefly, oxidation is provided by a flux of atomic oxygen radicals (O*) generated from an rf-excited discharge in a specially designed radical beam source. During growth, the shutters of Bi_2O_3, Sr, Ca and Cu fluxes are operated by a computer to achieve the *c*-axis layering of the desired superlattice structure, usually starting with the

Bi-O bilayer and ending also with the Bi-O bilayer of BSCCO. For each element, the shutter opening time is calibrated in order to deposit exactly one monolayer of the corresponding oxide. All the results presented in this paper have been obtained on (100) $SrTiO_3$. Typical substrate temperature is 700 ℃. The overall growth rate is 0.5 Å/s with a total film thickness of about 1000 Å. After the growth, the films are cooled to room temperature under O^* flux. No further post-deposition annealing is necessary to achieve superconductivity.

The transport resistivity was measured in a standard four-point geometry in Fig. 1(a) using pressed indium contacts. In order to make resistive measurements in the perpendicular direction, we formed a groove across the surface of the superlattices, as shown in Fig. 1(b). The focused ion beam (FIB) apparatus [4] was used for sputter-etching of the grooves. The films are etched by repetitive raster scanning of ions of 40 keV Au^+ emitted from a AuSiBe liquid metal ion source. The probe current of Au^+ FIB is 700 pA. The dimension of the groove is 50 μm wide and 300 μm long. Prior to the etching, we measured the Au^+ dose dependence of the etched depth. Figure 2 shows FIB etching characteristics for a BSCCO single film and a BSCCO/BSCO superlattice. The saturation of the etching rate indicates that the etched depth has reached the substrate. Note that the etching rates for the BSCCO film and the BSCCO/BSCO superlattice are quite similar. To rule out the possibility that the superconducting property of the superlattice is degraded in this etching process, we have fabricated the same structure using the BSCCO single-phase film and checked on its superconducting property.

Figure 3 shows a $\theta - 2\theta$ x-ray diffraction pattern for a BSCCO/BSCO superlattice. The presence of the well defined satellite peaks on the sides of the main peaks demonstrates that superlattice periodicity was achieved in the structure. The overall intensity profile for the superlattice agrees with the caluculated profile and the superlattice periods calculated from the separation of two consecutive satellite peaks also agreed with an ideal value expected from the growth sequence to within a few percent. These confirm that the superlattice is extremely well ordered on a half-unit-cell by half-unit-cell basis.

TRANSPORT PROPERTIES

We describe our superlattices using the notation $(BSCCO)_m/(BSCO)_n$, where m and n are the numbers of BSCCO and BSCO c-axis half unit cells per superlattice period, respectively. A series of superlattices, $(BSCCO)_4/(BSCO)_n$ (n = 2-5), where we kept the BSCCO thickness constant and varied the number of BSCO half-unit cells, has been investigated. In our previous paper [5], we have reported that all of our superlattices showed superconductivity, including superlattices containing isolated Cu-O double planes of BSCCO in a semiconducting BSCO matrix and, moreover, that superconductivity was essentially independent of both the BSCCO and BSCO layer thicknesses. Figure 4 shows the resistivities, calculated using the total thickness of the film, as a function of temperature for $(BSCCO)_4/(BSCO)_3$ and $(BSCCO)_4/(BSCO)_5$ as well as a typical BSCCO single-phase film. Indeed, comparing the overall shapes of superconducting transition for $(BSCCO)_4/(BSCO)_3$ and

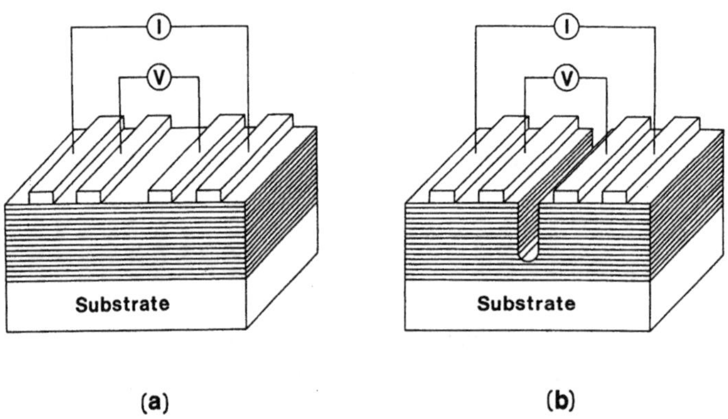

Fig. 1 Schematic diagrams of sample geometry for resistive measurements in parallel and perpendicular directions to the film surface.

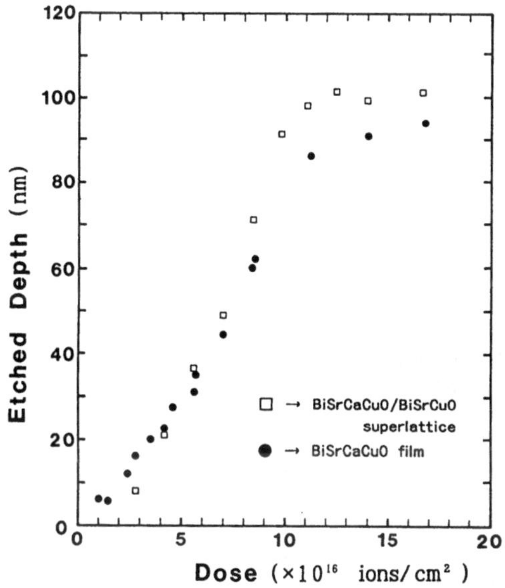

Fig. 2 Eching characteristics of a BiSrCaCuO/BiSrCuO superlattice and a BiSrCuO single film.

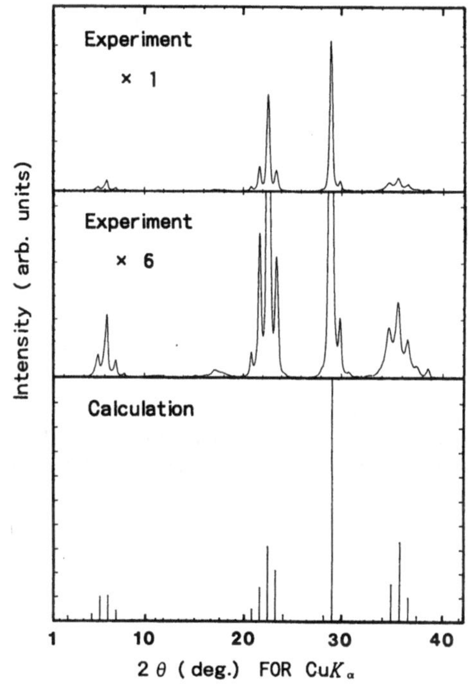

Fig. 3 θ-2θ x-ray diffraction patterns of experimental and caluculated profiles for a $(BiSrCaCuO)_4/(BiSrCuO)_4$ superlattice.

$(BSCCO)_4/(BSCO)_5$ with that for the BSCCO film, we note that there are no essential differences between them. Furthermore, we find that they exhibit similar metallic behavior in the normal state and the resistivity consistently increases with the number of BSCO half-unit cells. This implies that we are actually observing conduction in the plane of the superlattice structures consisting of the metallic BSCCO and semiconducting BSCO layers that are electrically in parallel.

Figure 5 displays the normalized resistivities $\rho(T)/\rho(200\ K)$ measured using the configulation in Fig. 1(b) for $(BSCCO)_4/(BSCO)_2$, $(BSCCO)_4/(BSCO)_4$ and $(BSCCO)_4/(BSCO)_5$. The perpendicular resistive transitions of the superlattices clearly depend on the number of BSCO half-unit cells, in contrast to the parallel transitions. The change from two BSCO to four BSCO separation of BSCCO lead to a shift in T_c and to a broadening of the transition width. Altough we cannot rule out the possibility of mechanical shorting, this may result from proximity-effect coupling along the c direction. The reduction in T_c is associated with a suppression of the superconducting pair wave function amplitude, and can be explained in terms of the proximity effect. However, since the coherence lengths of both BSCCO and BSCO in the c direction are only several angstroms, it is likely that lateral transport paths due to interface roughness are naturally formed. In general, the tansition width represents a lack of long-range phase coherence of the superconducting order parameter. The presence of an array of weakly in-plane-coupled Josephson junctions may accoun for a part of the broadening. When the BSCO is five half-unit cells thick, a residual resistivity appears after a sharp drop of the transition. This strongly suggests that the proximity effect cannot contribute to the enhancement of superconductivity for the BSCO barrier layer of more than two unit cells.

CONCLUSIONS

We have studied the parallel and perpendicular superconductivity in BSCCO/BSCO superlattices consisting of alternating BSCCO and BSCO half-unit cells. In the plane of the layer, our superlattices shows similar superconducting transitions essentially equal to that of a typical BSCCO

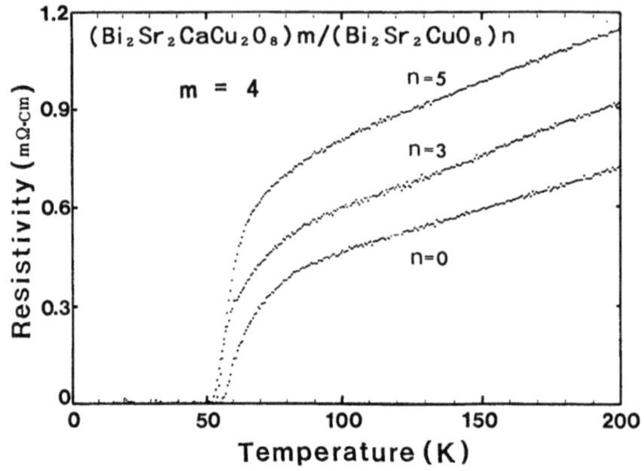

Fig. 4 Resistivity vs temperature for $(Bi_2Sr_2CaCu_2O_8)_4/(Bi_2Sr_2CuO_6)_n$ superlattices. The samples were measured using the configulation shown in Fig. 1(a).

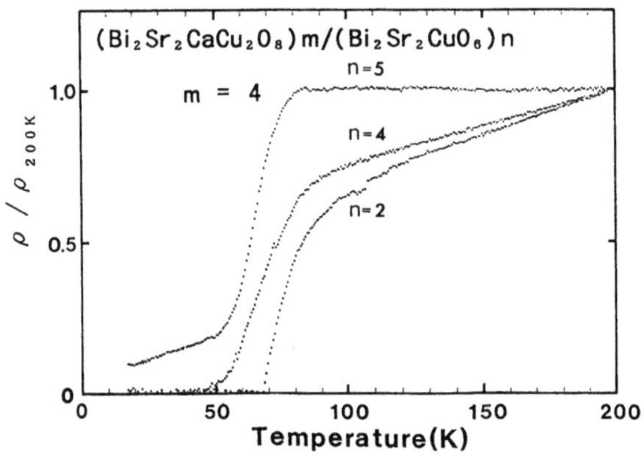

Fig. 5 Normalized resistivity vs temperature for $(Bi_2Sr_2CaCu_2O_8)_4/(Bi_2Sr_2CuO_6)_n$ superlattices. The samples were measured using the configulation shown in Fig. 1(b).

single-phase film itself, independent of the separation of the BSCCO layers. It seems that the coupling of the Cu-O double planes in adjacent cells is not essential for the enhancement of T_c. On the other hand, the transitions measured in the perpendicular direction are affected by the BSCO barrier thickness. A residual resistivity is clearly observed for $(BSCCO)_4/(BSCO)_5$, whereas supercurrents flow along the *c* direction for $(BSCCO)_4/(BSCO)_2$ and $(BSCCO)_4/(BSCO)_4$. The available evidence suggests that Josephson coupling of the BSCCO layers cannot occur through the BSCO barrier layers of more than two unit cells (49 Å). These results are consistent with the conventional proximity-effect formalism.

REFERENCES

1. Jin BY, Ketterson JB (1989) Adv. Phys. 38: 189-366
2. Furukawa H, Tokunaga S, Nakao M (1991) In: Kajimura K, Hayakawa H (eds) Advances in Superconductivity III. Springer-Verlag, Tokyo, pp 1069-1072
3. Furukawa H, Tokunaga S, Nakao M (1991) Ext. Abs. 9th Int. Workshop Future Electron Devices. Kumamoto, pp 72-73
4. Fujiwara S, Yuasa R, Shikichi K, Matsuta Y, Nakao M, Suzuki S (1990) Jpn. J. Appl. Phys. 29: L2307-2311
5. Nakao M, Furukawa H, Yuasa R, Fujiwara S (1991) Jpn. J. Appl. Phys. (in press)

Preparation of (Tl,Bi)$Sr_2CaCu_2O_x$ Thin Films

KATUYA HASEGAWA, MAKOTO ITO, TOSIHIRO KOTANI, HIROMI TAKEI, and KOUJI TADA

Basic High Technology Laboratories, Sumitomo Electric Ind., Ltd., Osaka, 554 Japan

ABSTRACT

The thin film preparation of a high-Tc (Tl,Bi)$Sr_2CaCu_2O_x$ phase with a single Tl-O layered structure has been investigated using a two-step fabrication process which involves a deposition of Bi-Sr-Ca-Cu-O precursor films followed by thallination. The precursor films were deposited onto (100)MgO substrate at room temperature using a high-vacuum evaporator. The as-deposited amorphous films were crystallized to form a (Tl,Bi)$Sr_2CaCu_2O_x$ phase by annealing under thallium oxide vapor. The (Tl,Bi)$Sr_2CaCu_2O_x$ thin film with a smooth surface showed high c-axis orientation and zero resistance at 86K.

KEY WORDS: (Tl,Bi)$Sr_2CaCu_2O_x$ thin film, single Tl-O layered structure

INTRODUCTION

Since the discovery of superconductivity in the Tl-Ba-Ca-Cu-O system, a considerable number of superconducting phases have been synthesized in the Tl-based system. Single and double Tl-O layered structures with the general formula $Tl_m(Ba/Sr)_2Ca_{n-1}Cu_nO_{2n+m+2}$ (m=1,2,n=1-6) have been obtained[1,2]. The Tl-based superconductors are very attractive for thin film investigation, because of their excellent potential for electronic device application. Much of the research to obtain high quality thin film in the Tl-based system has focused on two phases with double Tl-O layered structures of $Tl_2Ba_2CaCu_2O_8$ and $Tl_2Ba_2Ca_2Cu_3O_{10}$[3,4,5], while investigation of thin film preparation for the phases with a single Tl-O layered structure has almost never been carried out.

We have reported in our previous papers that a substitution of Bi for Tl stabilizes the superconducting phases with single Tl-O layered structures of (Tl,Bi)$Sr_2CaCu_2O_x$ and (Tl,Bi)$Sr_2Ca_2Cu_3O_x$[6,7]. In this paper, we will report on the preparation of superconducting thin film of (Tl,Bi)$Sr_2CaCu_2O_x$, whose crystal structure is shown in Fig.1, with a critical temperature of 86K.

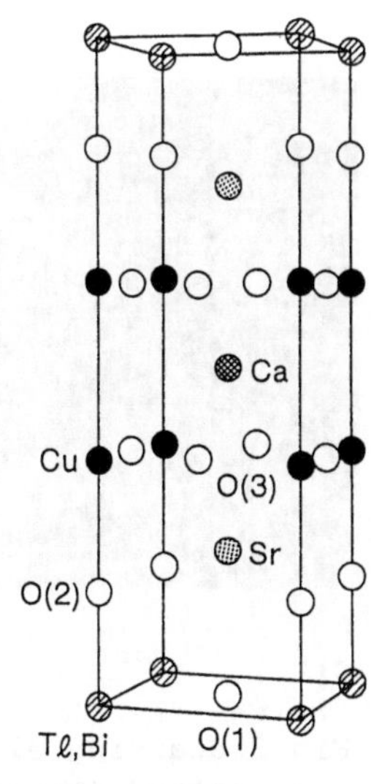

Fig.1 Crystal structure of (Tl,Bi)$Sr_2CaCu_2O_x$

EXPERIMENTAL

(Tl,Bi)$Sr_2CaCu_2O_x$ thin films were prepared by a two-step process which involves the deposition of Bi-Sr-Ca-Cu-O precursor film followed by an ex-situ annealing under thallium oxide vapor. Precursor films were deposited by a high-vacuum evaporation system equipped with three electron beam guns for Bi_2O_3, Ca, and Cu, and a Knudsen cell for Sr. The evaporation rates were monitored by quartz-crystal thickness sensors. During the evaporation, oxygen gas was introduced near the substrates at a back pressure of 5x10^{-6} Torr. MgO(100) single crystals were used as substrates and the substrate temperature was kept at room temperature. The typical deposition rate was about 1Å/s and precursor film thickness was about 0.2um. Precursor films were sealed with (Tl,Bi)$Sr_2CaCu_2O_x$ pellets inside a gold tube and annealed in the temperature range between 750°C and 900°C in flowing oxygen gas.

The compositon of the precursor films was determined by inductivity coupled plasma spectroscopy (ICP). The morphology and the composition of the films were studied by a scanning electron microscope equipped with an energy dispersive X-ray spectrometer (SEM-EDX). The crystal structure of the films was investigated by X-ray diffraction (XRD). Electrical resistance was measured by a standard four probe method.

RESULTS AND DISCUSSION

As-deposited precursor film was amorphous phase with a composition of $Bi_{0.4}Sr_2Ca_1Cu_2O_x$. During the annealing process, Tl volatilizing from bulk (Tl,Bi)$Sr_2CaCu_2O_x$ pellts was transported through the vapor phase to the Bi-Sr-Ca-Cu-O film, and the films were made crystalline to form a superconducting (Tl,Bi)$Sr_2CaCu_2O_x$ phase.

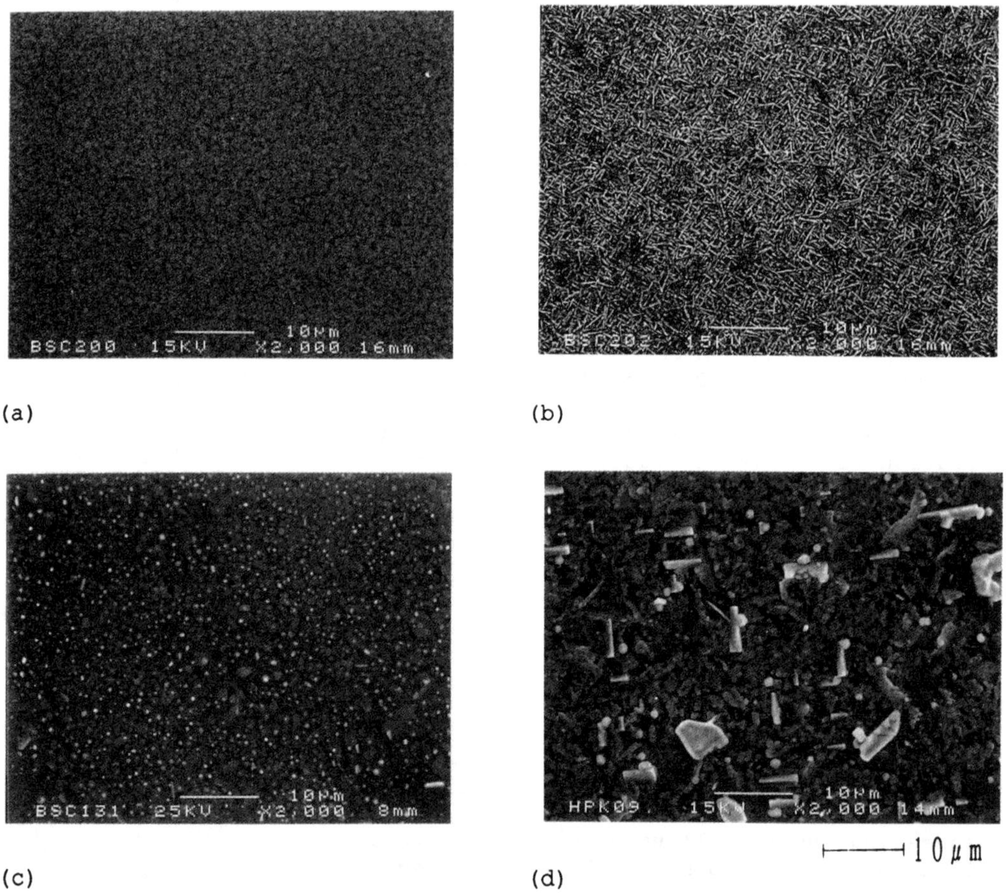

Fig.2 Scanning electron micrographs of thin films on (100)MgO substrate after annealing at (a)750°C, (b)810°C, (c)840°C, and (d)870°C.

Figure 2 shows scanning electron micrographs of the films annealed at various temperatures. The morphology of the films strongly depends on the heat treatment conditions of temperature, time, and thallium vapor supply. The films annealed at 750°C contained Tl, because Tl was transported to the precursor films at temperatures as low as 750°C, while the films did not crystallize at this temperature. When the films were annealed at 810°C, crystallization of the $(Tl,Bi)Sr_2CaCu_2O_x$ phase occurred together with the small needle-like Sr-Ca-Cu-O phase. The films annealed at 840°C showed a smooth surface and consisted almost entirely of the $(Tl,Bi)Sr_2CaCu_2O_x$ phase. When the annealing temperature was raised above 870°C, the superconducting phase began to decompose. In addition to the $(Tl,Bi)Sr_2CaCu_2O_x$ phase, a rod-like Tl-rich phase was formed, and the film surface became rough.

Figure 3 shows an XRD pattern of the film annealed at 840°C. Almost all the peaks can be indexed for (001) peaks of $(Tl,Bi)Sr_2CaCu_2O_x$. The film was almost entirely single phase of highly c-axis-oriented $(Tl,Bi)Sr_2CaCu_2O_x$. The c-axis lattice constant was calculated as c=12.0A .

Figure 4 shows the resistivity versus temperature measurement of a $(Tl,Bi)Sr_2CaCu_2O_x$ thin film. The film showed an on-set temperature of 96K and zero resistance at 86K, and the superconducting transition was sharp. This Tc value coincides well with the homogeneous bulk sample of $(Tl,Bi)Sr_2CaCu_2O_x$ phase made through the sintering process.

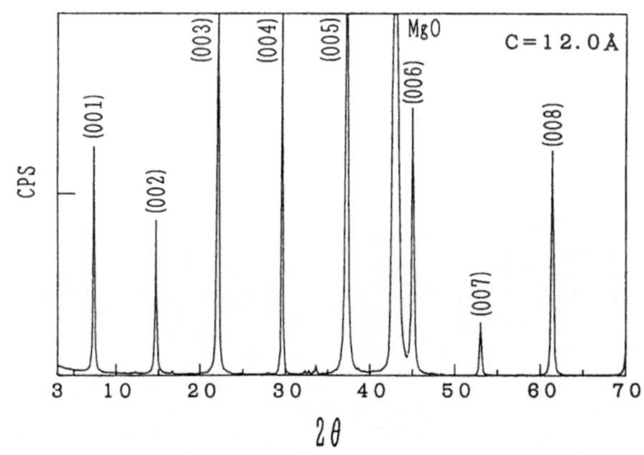

Fig.3 X-ray diffraction pattern of $(Tl,Bi)Sr_2CaCu_2O_x$ thin film annealed at 840°C

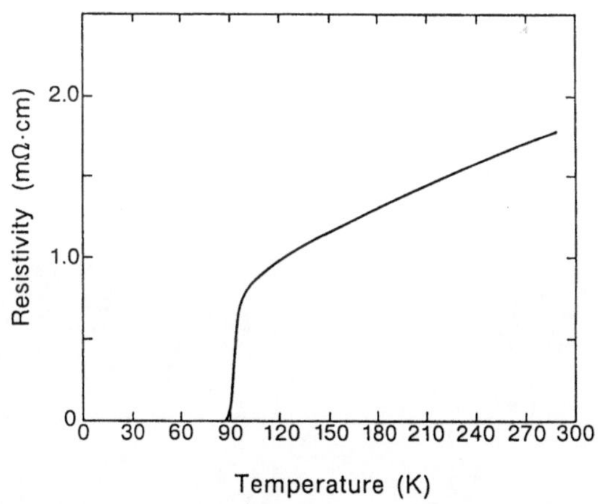

Fig.4 Resistivity vs temperature of $(Tl,Bi)Sr_2CaCu_2O_x$ thin film annealed at 840°C showing zero resistance at 86 K

In summary, a thin film of high-Tc (Tl,Bi)$Sr_2CaCu_2O_x$ phase with a single Tl-O layered structure was prepared by annealing Bi-Sr-Ca-Cu-O precursor film in thallium oxide vapor. The (Tl,Bi)$Sr_2CaCu_2O_x$ thin film has a smooth surface and shows high c-axis orientation. A critical temperature showing zero resistance as high as 86K has been achieved.

REFERENCES

1. Z. Z. Sheng and A. M. Harmann (1988) Nature 332:138.
2. H. Kusuhara, T. Kotani, H. Takei, and K. Tada (1989) Jpn. J. Appl. Phys. 28:L1772
3. W. Y. Lee, V. Y. Lee, J. Salem, T. C. Hung, R. Savoy, D. C. Bullock, and S. S. P. Parkin (1988) Appl. Phys. Lett. 53:329.
4. D. S. Ginley, J. F. Kwak, R. P. Hellmer, R. J. Baughman, E. L. Venturini, M. A. Mitchell, and B. Morosoin (1988) Physica C 156:592
5. S. I. Shah, N. Herron, C. H. Fincher, and W. L. Holstein (1990) Appl. Phys. Lett. 56:782.
6. Y. Torii, T. Kotani, H. Takei, and K. Tada (1989) Jpn. J. Appl. Phys. 28:L1190
7. Y. Torii, H. Takei, and K. Tada (1989) Jpn. J. Appl. Phys. 28:L2192

Film Growth Mechanism for Non-(001) Orientations in Bi-Sr-Ca-Cu-O

YOSHIKI ISHIZUKA, YOSHIAKI TERASHIMA, SUMIO IKEGAWA, and TADAO MIURA

Research and Development Center, Toshiba Corporation, Kawasaki, Kanagawa, 210 Japan

ABSTRACT

Thin films of $Bi_2Sr_2CaCu_2O_y$ were deposited on $SrTiO_3$(110) substrates by coevaporation. It was confirmed that the chemical composition has a crucial effect on the growth orientation. The preferred orientation shifts from (119) to (117) as the Bi composition is increased. The difference in film growth kinetics is discussed from the viewpoint of an epitaxial relationship and crystal structural stability.

KEY WORDS: Bi-Sr-Ca-Cu-O, growth orientation, $SrTiO_3$(110), epitaxy, film growth mechanism

INTRODUCTION

The formation of non-(001) orientations in Bi-Sr-Ca-Cu-O thin films has already been reported [1-3]. This is an important issue not only for fabricating electronic devices, such as Josephson junctions, but also for an understanding of the film growth mechanism.

The (001) orientation, in other words the c-axis orientation, is known to be formed on several kinds of substrates, even if there is no epitaxial relationship between the film and substrate, as in the case of MgO(110). The anisotropic crystal growth kinetics inherent in this type of compound has been thought to influence the growth orientation. However, control of the chemical composition has made it possible to form non-(001) orientations on $SrTiO_3$(110) substrates [4]. Furthermore, non-(001) orientations can be formed by codeposition with a preceding deposition of Sr, Ca, and Cu, when the (001) orientation is normally formed without the preceding deposition [5]. This indicates that the initial growth state is a crucial factor determining the growth orientation. In the case of the $Bi_2Sr_2CaCu_2O_y$ (2212) phase, two non-(001) orientations, (119) and (117), have been reported [1]. However, the difference between them has not been discussed in detail.
In this paper, we discuss the film growth mechanism in Bi-Sr-Ca-Cu-O from the viewpoint of the difference in growth orientation among non-(001) orientations.

EXPERIMENTAL RESULTS

Thin films of the 2212 phase were deposited on $SrTiO_3$(110) substrates by coevaporation with rf oxygen plasma. The experimental configuration and the film preparation process were the same as described previously [6]. The oxygen plasma conditions were kept constant at 50 W rf power with a 0.5 sccm oxygen flow rate.
We investigated the dependence of the growth orientation on chemical composition. Only the Bi composition was varied in this experiment. The chemical compositions of the films were in the range of Bi:Sr:Ca:Cu=(1.2-2.0):2.0:(1.5-1.6):(1.7-1.8), which were analyzed by inductively coupled plasma emission spectroscopy. We confirmed that all of the films showed the superconducting transitions.

Figures 1(a), (b), (c), and (d) show the X-ray diffraction (XRD) patterns of films with different Bi compositions. The (119) peak is observed in Fig. 1(a). The (119) orientation corresponds to an angle of about 48 degrees between the c-axis and the substrate surface. In Fig. 1(c), the (117) peak is observed; this orientation corresponds to about 41 degrees. The (117) and the (00n) peaks are observed in Fig. 1(d). There are no peaks other than the $SrTiO_3$ (110) peak in Fig. 1(b). The reflection high-energy electron diffraction (RHEED) revealed the preferred orientation other than (119) or (117), as shown in Fig. 2. The angle is estimated to be about 45 degrees.

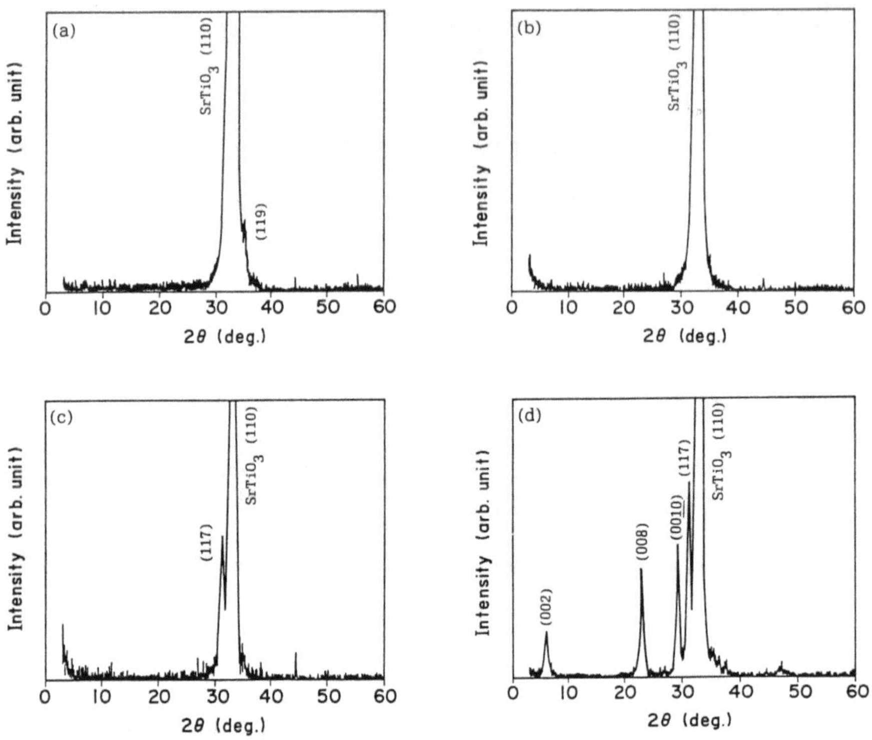

Fig. 1. XRD patterns for non-(001) oriented films with different chemical compositions:
(a) Bi:Sr:Ca:Cu=1.2:2.0:1.6:1.8; (b) Bi:Sr:Ca:Cu=1.3:2.0:1.5:1.7;
(c) Bi:Sr:Ca:Cu=1.8:2.0:1.5:1.8; (d) Bi:Sr:Ca:Cu=2.0:2.0:1.6:1.8.

Fig. 2. RHEED pattern for a film corresponding to Fig. 1(b). The incident electron beam was parallel to the [001] direction of the substrate.

An evaluation of orientation from the viewpoint of superconductivity has also been carried out. We have measured the dependence of resistance (R) on the angle (θ) between the substrate surface and the applied magnetic field [7]. The angle between CuO_2 planes and the substrate surface can be obtained on the basis of the anisotropic superconducting properties. The results are shown in Fig. 3. Figures 3(a), (b), and (c) correspond to the films shown in Figs. 1(a), (b), and (c), respectively. The angle (θo), at which the resistance (R) is a minimun, is identical to the angle between the CuO_2 planes and the substrate surface for each curve. The values of these angles almost correspond to the orientations derived from XRD and RHEED.

These results confirm that the preferred orientation shifts from (119) through (117) to (001) as the Bi composition is increased.

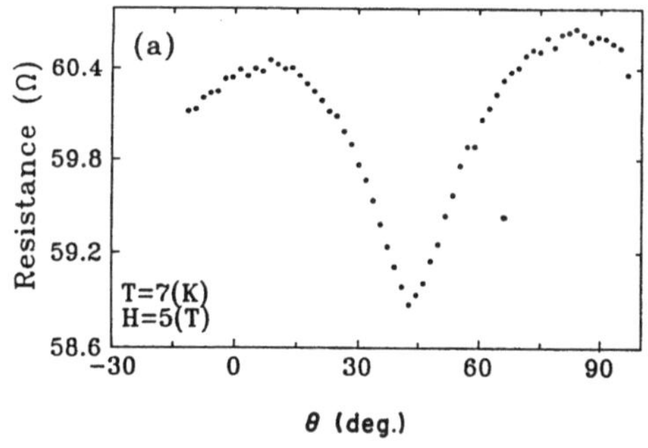

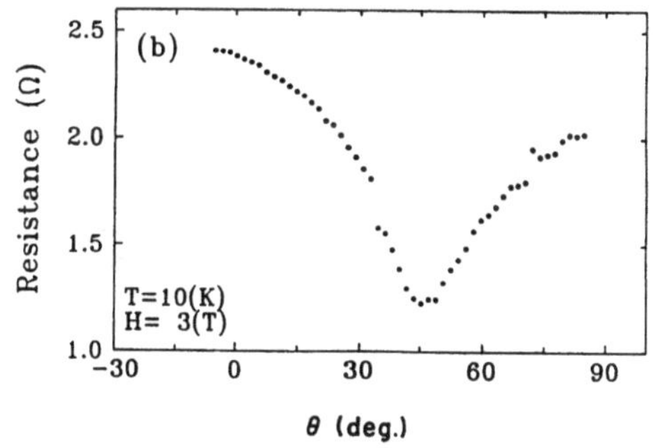

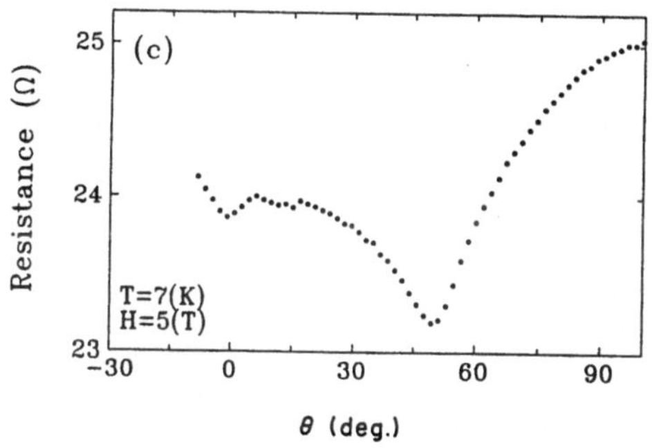

Fig. 3. Angular dependence of magnetoresistance. (a) corresponds to the film shown in Fig. 1(a). (b) corresponds to the film shown in Fig. 1(b). (c) corresponds to the film shown in Fig. 1(c). The values of θo are:
(a) about 42 degrees;
(b) about 45 degrees;
(c) about 49 degrees.
Magnetic field strength (H) and measuring temperature (T) are indicated in each figure.

DISCUSSION

Non-(001) orientations have so far been formed only on $SrTiO_3$(110) substrates in spite of anisotropic crystal growth kinetics. This fact suggests a particular relationship between films with non-(001) orientations and the (110) plane of $SrTiO_3$. To explain this, we have proposed that an epitaxial relationship betweeen perovskite structures in the 2212 phase and the (110) plane of $SrTiO_3$ is a crucial factor in the formation of non-(001) orientations [4]. However, the difference in growth mechanisms among non-(001) orientations has not been discussed in detail.

The (119) orientation is formed under conditions of Bi scarcity, as shown above. Interactions between elements other than Bi - that is Sr, Ca, and Cu - and the substrate are important in this case. The crystal structure of $SrTiO_3$ is perovskite. Thus, perovskite structures are easily formed at the interface through epitaxy. These epitaxially formed perovskite structures are considered to act as nuclei for the formation of the 2212 phase. In other words, the (119) orientation is thought to originate from the growth of the perovskite structures. The (119) planes, shown by solid lines in Fig. 4, support this idea, since they consist of atomic arrangements of perovskite structures. Therefore, the idea of an epitaxial relationship between the perovskite structures and the substrate is appropriate for the growth mechanism of the (119) orientation.
This discussion is based on structural properties. The crystal structure can be divided into two parts, namely the Bi-O sheet and the perovskite structure. Therefore, the preferential formation of the perovskite structures is thought to be possible.
On the other hand, as the Bi composition is increased, stability of a structure containing Bi atoms with the perovskite structure is enhanced. The stability is supported by electrical properties, since the electrical neutrality is important for ionic crystals. Considered from this point of view, a half unit cell is the minimum stable unit, since the perovskite structure has a charge of (-2) and that of the Bi-O sheet is (+1). As a result, a plane which reflects the features of an atomic arrangement of both the half unit cell and the perovskite structure becomes stable. We can confirm that the (117) planes, shown by broken lines in Fig. 4, are contented with the condition. This is the reason for the (117) orientation.

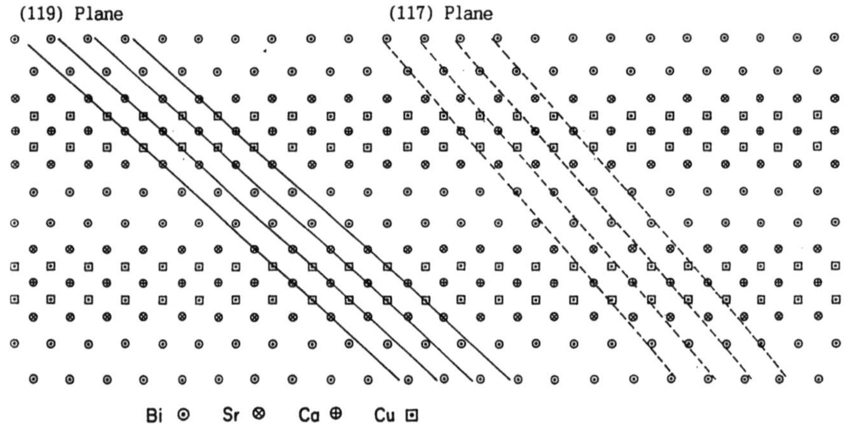

Fig. 4. Atomic arrangement for the 2212 phase [8]. Solid lines correspond to the (119) planes and broken lines correspond to the (117) planes.

The preferential orientation is thought to be determined by the balance between these two effects of the epitaxial relationship and the stability as the half unit cell. As the Bi-O sheet is formed completely, the stability as the half unit cell is enhanced. As a result, the preferential orientation shifts from (119) to (117) through intermediate orientations.
This interpretation is supported by the case of the $Bi_2Sr_2CuO_y$ (2201) phase. The (115) orientation is formed, but (117) is not formed in this phase [4]. The (115) and (117) orientations are similar to the (117) and (119) orientations of the 2212 phase. The 2201 phase has smaller perovskite structures, so the effect of the epitaxial relationship between the perovskite structures and the substrate is thought to be small. On the other hand, it is more stable as a half unit cell, so the effect of the stability is thought to be great. These may be the reasons for the formation of the (115) orientation.
In addition to those mentioned above, the (001) orientation is formed when the effect of the stability as a half unit cell is more enhanced.

SUMMARY

We have confirmed that the preferred orientation shifts from (119) to (117) as the Bi composition is increased. This can be explained by considering the two features of the compound. One is the epitaxial relationship between the perovskite structures and the substrate. This is based on the structural properties. The other is the stability as a half unit cell, which is based on the electrical properties. The growth orientation is thought to be determined by the balance between these two effects.

REFERENCES

1. Ishizuka Y, Terashima Y, Miura T (1991) Jpn.J.Appl.Phys.30: L1006-L1008
2. Kuroda K, Kojima K, Wada O, Tanioku M, Yokoyama K, Hamanaka K (1990) Jpn.J.Appl.Phys.29: L1816-L1818
3. Kuroda K, Wada O, Kojima K, Tanimura J, Kataoka M, Takami T, Yokoyama K, Ogawa T, Hamanaka K (1991) Jpn.J.Appl.Phys.30: L475-L477
4. Ishizuka Y, Terashima Y, Miura T in "Proceedings of the 1991 MRS Spring Meeting", Anaheim, 1991
5. Miura T, Ishizuka Y, Terashima Y, Ikegawa S in "Proceedings of the 3rd International Conference on Materials and Mechanisms of Superconductivity High-Temperature Superconductors", Kanazawa, 1991
6. Ishizuka Y, Terashima Y, Miura T (1990) Jpn.J.Appl.Phys.29: L2045-L2048
7. Ikegawa S, Ishizuka Y, Terashima Y, Miura T in "Proceedings of the 3rd International Conference on Materials and Mechanisms of Superconductivity High-Temperature Superconductors", Kanazawa, 1991
8. Yamamoto A, Onoda M, Takayama-Muromachi E, Izumi F, Ishigai T, Asano H (1990) Phys.Rev.B 432: 4228-4239

Angular Dependence of Magnetoresistance for (11n)-Oriented Bi-Sr-Ca-Cu-O Thin Films Below T_c

SUMIO IKEGAWA, YOSHIKI ISHIZUKA, YOSHIAKI TERASHIMA, and TADAO MIURA

Toshiba Research and Development Center, Kawasaki, Kanagawa, 210 Japan

ABSTRACT

For $Bi_2Sr_2Ca_1Cu_2O_y$ coevaporated films with (11n) orientations, we have measured angular dependence of magnetoresistance (R-vs-θ curve) in a field-induced resistive state. It was confirmed, through measurements for various angles (ϕ) between the sample and the rotating axis, that superconducting CuO_2 planes lie at an angle of ±(40-50)° to the film surface. The temperature dependence of R-vs-θ curves has been measured and is found to be useful to investigate the relation between superconducting transition and growth orientation within a film.

key words: Bi-Sr-Ca-Cu-O film, preferred orientation, magnetoresistance, anisotropy, coherence length.

INTRODUCTION

In high-Tc superconducting films, various preferred orientations are possible. Control of the preferred orientation, as well as evaluation of anisotropy in the superconducting state of these films is important for application in electronic devices. Bi-Sr-Ca-Cu-O (BSCCO) films with (11n) orientations have been synthesized by coevaporation [1] and RF sputtering [2]. These films are considered to have the advantages of greater coherence length perpendicular to the film surface than the (001)-oriented films. The formation of (11n) orientations had been confirmed by crystallographic analyses, i. e. X-ray diffraction (XRD), reflection high-energy electron diffraction, and transmission electron microscopy [1,2]. Recently, we firstly investigated the anisotropic superconducting properties of (11n)-oriented BSCCO films, using resistance (R) measurements as a function of the angle between the film surface and the applied magnetic field (θ) [3]. The angular dependence of magnetoresistance has been extensively studied for (001)-oriented films [4,5], but none had been reported for non-(001) oriented films of superconducting cuprates. This paper reports three-dimensional measurements for the (119)-oriented film, which are R-vs-θ measurements for various settings of the angle (ϕ) between the sample and the rotating axis. The temperature dependence of the R-vs-θ curve for the (11n)-oriented films is also investigated.

EXPERIMENTS

BSCCO thin films were grown by coevaporation, using an MBE machine with an rf oxygen plasma [6]. Films with (11n) orientations were grown on the $SrTiO_3$(110) substrate. Film thickness was typically 80 nm. Details of the preparation were reported previously [1]. The phases present and preferred orientations were determined by XRD.

Resistance measurements were carried out by a dc four-probe method under an applied magnetic field (H) up to 5 T. The sample holder was rotated around the axis perpendicular to the field direction. Resistance (R) was measured as a function of the angle between the film surface and the magnetic field (θ). The sample was attached on the sample holder so that the angle between the [001] axis of $SrTiO_3$ and the rotating axis was ϕ, as shown in Fig. 1. During one sequence of R-vs-θ curve measurements, the temperature was kept constant within ±0.01 K. Typical current density was 20 A/cm^2. For comparison, measurements for (001)-oriented BSCCO films were simultaneously made. To correct instrumental errors in θ, the minimum-resistance angle for the (001)-oriented film was defined as θ=0°.

RESULTS AND DISCUSSION

Figure 2 shows θ dependence of resistance for various ϕ settings, where the resistances of each curve are normalized by the resistance at θ=90°. Current direction for resistance measurements was parallel to the [1,-1,0] axis of $SrTiO_3$, as shown in Fig. 1. Note that the R-vs-θ curve did not depend on the current direction, but on the field direction, as was reported for the (001)-oriented film [4]. In the XRD pattern for the sample shown in Fig. 2, only one peak, probably due to the (119) reflection of the $Bi_2Sr_2Ca_1Cu_2O_y$ phase, was observed besides the peaks due to the substrate. Thus, (119) planes are considered to be aligned parallel to the film surface. In the case of ϕ=0° R-vs-θ curves have minima at $\theta=\pm 42.8°$, which is roughly the same as expected for the "(119)-orientation". This contrasts with the (001)-oriented film case, where a minimum appears at θ=0° [3,4]. In the case of ϕ=90°, R-vs-θ curves have a minimum at θ=0° which is similar to the case of (001)-oriented film. For $0<\phi<90°$, the shape of resistance minimum changes with increasing ϕ. At 70 K (Fig. 2(b)), the minima are broader than those observed at 30 K (Fig. 2(a)). The broadening of the minima will be discussed later.

Superconducting properties of BSCCO have a highly anisotropic nature. Raffy et al. [5] reported for a (001)-oriented $Bi_2Sr_2Ca_1Cu_2O_y$ film that magnetoresistance is only related to the field component perpendicular to the CuO_2 plane ($H\sin\theta$). The angular dependence of magnetoresistance may be caused by both the anisotropy in the flux motion and the anisotropy in H_{c2}, which reflect the anisotropy in coherence length and the layered structure. We also observed similar scaling properties in the R-vs-θ curves for the (001)-oriented $Bi_2Sr_2Ca_2Cu_3O_y$ film. Using the scaling properties and a simple model, R-vs-θ curves for the (119)-oriented film were estimated numerically to understand the ϕ-dependence of the R-vs-θ curves in Fig. 2(a). The explicit procedure is as follows.

The relation between the resistance of the (001)-oriented $Bi_2Sr_2Ca_2Cu_3O_y$ film and $H\sin\theta$ was roughly linear, although the R-vs-$H\sin\theta$ curve had a negative curvature. Therefore,

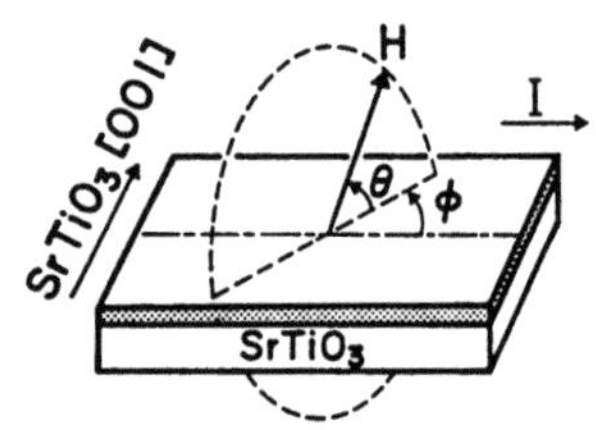

Fig. 1. Schematic drawing of magnetic field direction relative to the sample.

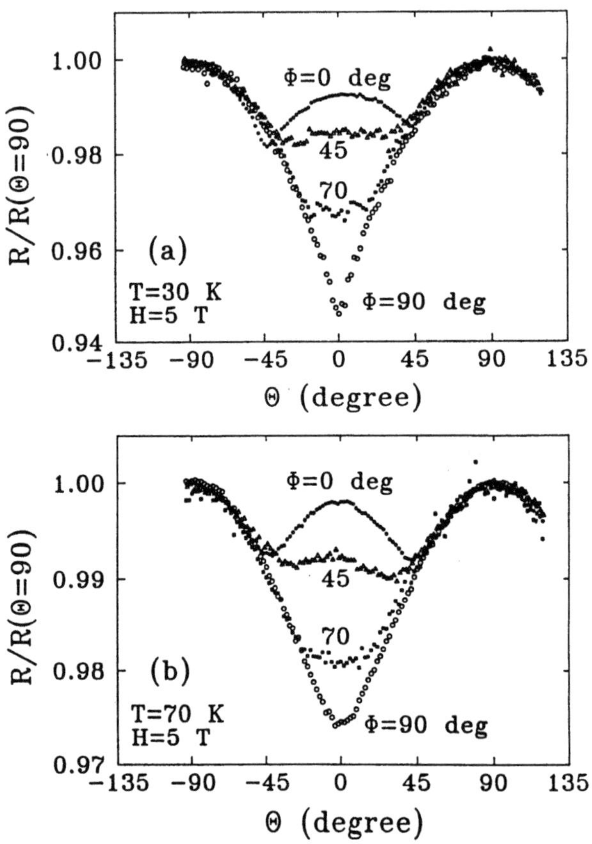

Fig. 2. Angular dependence of magnetoresistance for various ϕ settings. (a) At 30 K. (b) At 70 K. The sample is (119)-oriented BSCCO film.

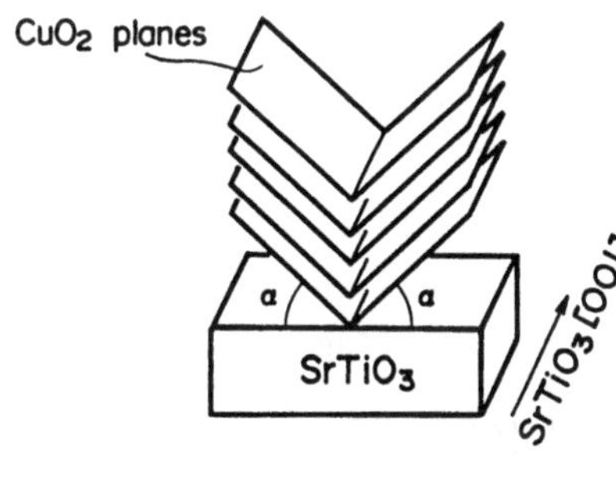

Fig. 3. Schematic drawing of the CuO_2 planes in (11n)-oriented BSCCO film.

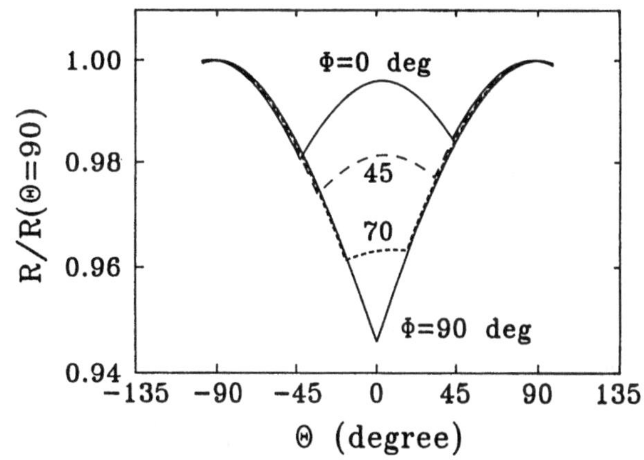

Fig. 4. R-vs-θ curves, derived from the numerical estimation for various ϕ settings.

we assume that resistance of the single-crystal grain, R, is given by $R=AH_p+C$, where H_p is the field component perpendicular to CuO_2 planes in the grain, A is a proportional constant, and C is field-independent resistance. From the crystallographic analyses [1,2], the (119)-oriented film is thought to consist of two sets of grains. The CuO_2 planes in the first set lie at an angle of α to the film surface and those in the second set lie at an angle of $-\alpha$, as shown in Fig. 3. Resistances for the two sets, i. e. $R_1(=A_1H_{p+}+C_1)$ and $R_2(=A_2H_{p-}+C_2)$ where $H_{p\pm}=H|\cos\theta\cos\phi\sin\pm\alpha-\sin\theta\cos\pm\alpha|$, contribute to the measured resistance. The numerical estimations were carried out for both series connection and parallel connection between R_1 and R_2. The series connected model matched experimental results better than the parallel connected model. Figure 4 shows the numerical estimation for the series connected model. It was found that the simple model, which assumes a linear relation between R and H_p and series connection of the two sets of grains, can explain the ϕ-dependence of the R-vs-θ curves. Thus, it was confirmed that the direction of the greatest coherence length is in two planes, which lie at an angle of $\alpha\sim\pm43°$, as shown in Fig. 3. It should be noted that, in Fig. 4, the A_1 and A_2 values were estimated from the data in Fig. 2(a), to account for the observed asymmetry. The small discrepancy between Fig. 2(a) and Fig. 4 may have arisen from an error in setting of ϕ (within $\pm3°$) and the nonlinearity in R-vs-H_p relation.

In Fig. 2, the resistance minima at 70 K were broader than those at 30 K, which was clearly seen in the cases where $\phi=0°$ and 90°. Therefore, we investigated temperature dependence of the R-vs-θ curve for another (119)-oriented film, using the settings $\phi=0°$ and I perpendicular to H. Figure 5 shows temperature dependence of resistance under no applied magnetic field. Two-step superconducting transitions (at around 80 K and 30 K) were observed. Similar two-step transitions have been commonly observed for (11n)-oriented BSCCO films [1,2], contrary to the (001)-oriented film case. The R-vs-θ measurements were carried out at temperatures indicated by arrows in Fig. 5. Figure 6 shows the R-vs-θ curve at 20 K (Fig. 6(a)) and at 60 K (Fig. 6(b)). Each of the R-vs-Θ curves in the temperature range below 70 K has minima at $\theta=\pm42°$. The minima observed in the high-temperature range (T>40 K) were broader than those observed in the low-temperature range (T<40 K).

The following two causes are possible for the broadening of the minimum. First is the effect of broad superconducting transition. In general, anisotropy is weak near the onset of superconducting transition. The second cause is fluctuation in orientation. In the case of the (001)-oriented film, even when the superconducting transition was broad, the resistance minimum was sharp in the temperature range where the resistance fell below 80% of normal-state resistance. For the sample shown in Fig. 6, resistance in the magnetic field is less than 80% of normal-state resistance at temperatures below ~65 K. Therefore, the first cause is deemed not likely. The broadening of the minimum may be caused by the existence of grains having fluctuated orientations around the (119) orientation. The broadening of the minima is conspicuous at temperatures above 40 K. This may be related to the two-step transitions at around 80 K and 30 K. The (11n)-oriented film may consist of two kinds of grains. The first kind has higher T_c (~80 K) and slightly disordered orientation, and the second kind has lower T_c (~30 K) and well-defined orientation. This suggests that T_c is closely related to the formation mechanisms of (11n) orientation.

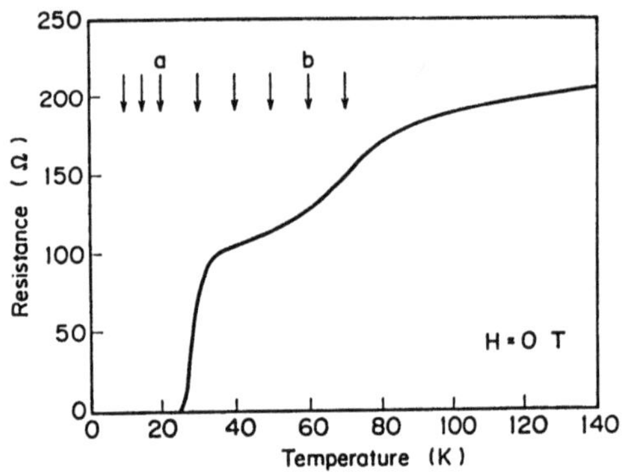

Fig. 5. Temperature dependence of resistance for (119)-oriented BSCCO film.

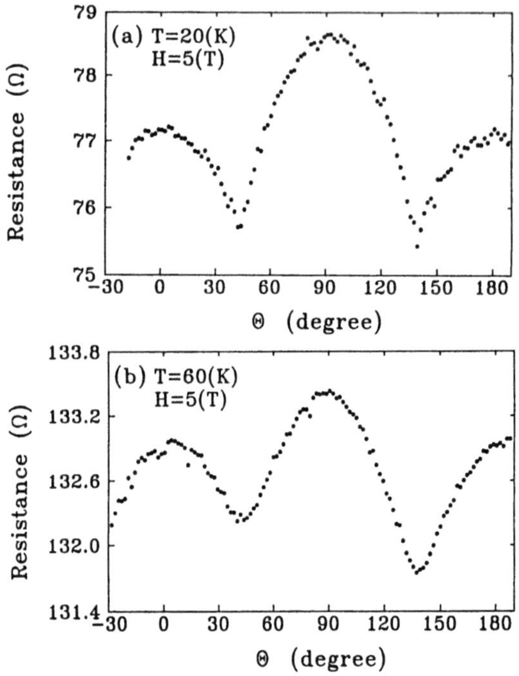

Fig. 6. Angular dependence of magnetoresistance for $\phi=0°$. (a) At 20 K. (b) At 60 K. The sample is the same as in Fig. 5.

The temperature dependence of the R-vs-θ curve was investigated for another (11n)-oriented sample, whose XRD pattern contained a small trace of the (001)-oriented $Bi_2Sr_2Ca_1Cu_2O_y$ phase. Minima at $\theta=\pm46°$ were observed in the temperature range of 10-70 K. At temperatures less than 50 K, a minimum at $\theta=0°$ was also observed, which is due to (001)-oriented grains contained in the sample. This result indicates that the superconducting transition temperatures for different grains with different orientations are detected separately [3]. Such a separate detection is impossible by a standard technique for T_c, e. g. temperature dependence of resistance or magnetic susceptibility.

SUMMARY

Magnetoresistance for the (119)-oriented BSCCO film on the $SrTiO_3(110)$ substrate was measured as functions of the two angles (θ and ϕ). It was demonstrated that the direction of the greatest coherence length was in two planes, which lay at an angle of $\sim\pm43°$ to the film surface and were perpendicular to the (001) plane of the $SrTiO_3$ substrate. The temperature dependence of R-vs-Θ curves was found to give details for the superconducting transition. This measurement is useful to investigate the relation between growth orientation and superconducting properties within a film.

REFERENCES

1. Ishizuka Y, Terashima Y, Miura T (1991) Jpn. J. Appl. Phys. **30**: L1006-L1008; In Proceedings of the 1991 MRS Spring Meeting, Anaheim, 1991.
2. Kuroda K, Kojima K, Wada O, Tanioku M, Yokoyama K, Hamanaka K (1990) Jpn. J. Appl. Phys. **29**: L1816-L1818.
3. Ikegawa S, Ishizuka Y, Terashima Y, Miura T (1991) In: Tachiki M, Muto Y, Shono Y (eds) Proceedings of the Third International Conference on Materials and Mechanisms of Superconductivity High-Temperature Superconductors. Kanazawa, Japan, (Physica C, in press).
4. Iye Y, Nakamura S, Tamegai T, Terashima T, Yamamoto K, Bando Y (1990) Physica C **166**: 62-70.
5. Raffy H, Labdi S, Laborde O, Monceau P (1991) Phys. Rev. Lett. **66**: 2515-2518.
6. Ishizuka Y, Terashima Y, Miura T (1990) Jpn. J. Appl. Phys. **29**: L2045-L2048.

XPS Studies of As-Deposited Superconducting Bi-Sr-Ca-Cu-O Thin Films with Critical Temperature of 98K

KAZUSHIGE OHBAYASHI[1], KENJI YOSHIDA[2], MASATOSHI ANMA[2], YOSHIAKI TAKAI[2], and HISAO HAYAKAWA[2]

[1] Research and Development Center, NGK Spark Plug Co., Ltd., Nagoya, 467 Japan
[2] Department of Electrical Engineering, Nagoya University, Nagoya, 464-01 Japan

ABSTRACT

As-deposited highly oriented BSCCO high-Tc phase superconducting films with various critical temperatures that were prepared by changing the Bi content have been studied by X-ray photoemission spectroscopy (XPS). The Bi 4f core level electron spectrum shows that the line width of the Bi 4f main peak gradually increases, and the binding energy shifts to a lower binding energy side with an increase of $Tc_{(zero)}$. The examination of the Cu 2p peak resulted in the following facts that the main doublet peaks were observed with their satellites, and noticeable shoulders of Cu^{3+} were observed in the spectra of the films with $Tc_{(zero)}$ of 86K and 98K.

KEY WORDS:BSCCO,thin film,high-Tc phase,Bi content,XPS

INTRODUCTION

We have reported[1] that BSCCO highly oriented as-grown superconducting high-Tc phase films with zero resistivity of 95K can be deposited by preparing the BSCO buffer layer on the MgO(100) substrate and regulating the Bi content in the film. We described in the paper, the critical temperature of high-Tc phase is quite sensitive to the Bi content of the film. By the way, XPS has been extensively used to study high-Tc superconductors and has yielded interesting results[2-4]. In this paper, BSCCO system thin films with various critical temperatures were prepared, and XPS core level spectra have been measured to clarify the effect of the slight changes of the Bi content (1.97-2.01) on the critical temperature of high-Tc phase.

EXPERIMENTAL

The films were grown on MgO(100) substrates using RF magnetron sputtering, at the substrate temperature of 765°C. The compositions of the targets are Bi_2O_3, $Bi_{2.0}Sr_{2.0}Cu_{1.0}O_x$ and $Ca_{1.0}Cu_{1.0}O_x$. In the first step, BSCO buffer layer was deposited on the MgO substrate about 70Å, then BSCCO high-Tc phase film

Table I. Sputtering conditions of Bi-Sr-Ca-Cu-O film

	Target	Sample (a)	(b)	(c)	(d)	(e)*
Input power (W)	Bi_2O_3	23.5	23.0	22.5	21.5	23.5
	$Bi_2Sr_2Cu_1O_x$	50.0	50.0	50.0	50.0	50.0
	$Ca_1Cu_1O_x$	48.0	48.0	48.0	48.0	48.0
Shutter opening time (sec.)	Bi_2O_3	50x2	50x2	50x2	50x2	50x2
	$Bi_2Sr_2Cu_1O_x$	50x2	50x2	50x2	50x2	50x2
	$Ca_1Cu_1O_x$	85	85	85	85	85
$Tc_{(zero)}$ (K)		98.3	86.0	76.4	66.4	<20
Bi content (estimated values)**		2.01	2.00	1.99	1.97	2.01

* :without buffer layer
** :assuming that the Sr content was 2.00

(700Å) was deposited. Details of sputtering conditions are listed in Table I. After deposition, air was introduced to the chamber for immediate cooling of the films. These samples were transferred to the XPS cell from sputtering chamber after cooling instantly, and XPS measurements were carried out without surface treatments. It was reported that Ar ion sputtering damaged the superconducting properties[5]. The XPS data were obtained with Mg-Kα radiation and peaks of binding energies were referenced to the carbon 1s peak at 285.0eV. These samples showed superconducting transitions at 98K, 86K, 76K and 66K, but the film without the buffer layer did not show $Tc_{(zero)}$ above 20K (sample e). The XRD analysis identified these films as of the high-Tc phase and showed a c-axis orientation except sample e, but the compositions of these samples including e were determined approximately $Bi_{2.0}Sr_{2.0}Ca_{2.0}Cu_{3.0}O_x$ by inductively coupled plasma (ICP) analysis. The c-axis unit cell lengths of these films except e were determined to be 37.2Å.

RESULTS AND DISCUSSION

In Fig.1, XPS spectra of the Bi 4f core levels are shown with a spline fitting background subtraction, and with leastsquares fits assuming Gaussians adjustable width and intensity. The Bi core level electron spectrum shows that the line width of the Bi 4f main peak gradually increases and the binding energy shifts to a lower binding energy side, with an increase of Tc. The spectra are decomposed into two components, one is the component indicated by H at 158.7eV which is actually identical with the spectra of pure Bi_2O_3 (158.7eV), and the other is the lower binding energy component indicated by L which is observed at an energy lower than the component H by about 1eV. Dharmadhikari et al. measured BiO and Bi_2O_3 $4f_{7/2}$ binding energies and the values were 159.1 and 160.1eV, respectively[6]. The correlation between Tc and the intensity ratio L/(L+H) of the Bi 4f is shown in Fig.2. It is evident from the figure that Tc increases with an increase of the ratio of L/(L+H). We can estimate from these results that the component L has an intimate relation with the superconducting properties, and it seems reasonable that the peak L is assigned to BiO. These results indicate a possibility of the excess oxygen of BSCCO system are incorporated in Bi-O sheets. We could not find these phenomena at the other constituent metals as like the release of the excess oxygen from the Bi-O sheets with an increase of Tc. On the other hand, the presence of Bi^{5+} ions is reported at a higher binding energy nearby 159.6eV in 2212 phase[7]. In the same manner, we measured the XPS spectra of Bi_2O_5. However, as in Fig.1, any evidences showing the presence of Bi^{5+} ions have not been obtained in a higher binding energy side of the H component. From these considerations it is difficult to consider Bi^{5+} ions are contained in the BSCCO superconducting films and that they are related to the critical temperature of the high-Tc phase.

The Cu 2p electron spectra are shown in Fig.3. These Cu 2p spectra were decomposed into three components; monovalent, divalent and trivalent of Cu oxides, which were observed in the Cu $2p_{3/2}$ spectrum of Cu_2O, CuO and $NaCuO_2$, respectively[8]. Each curve was composed of two Gaussians distribution curves. The Cu $2p_{3/2}$ peak positions of Cu^+, Cu^{2+} and Cu^{3+} states were found to be 932.8, 933.8 and 935.4eV, respectively. In the spectra of the films with Tc's of 86K and 98K, noticeable shoulders of Cu^{3+} were observed, and the amounts of Cu^{3+} species were approximately 10% and 15% of the total Cu, respectively. Nevertheless the width of the Cu $2p_{3/2}$ main lines of the film are still large compared to Cu^{2+} compounds. This is explained from the fact that about 8% of the Cu ions are

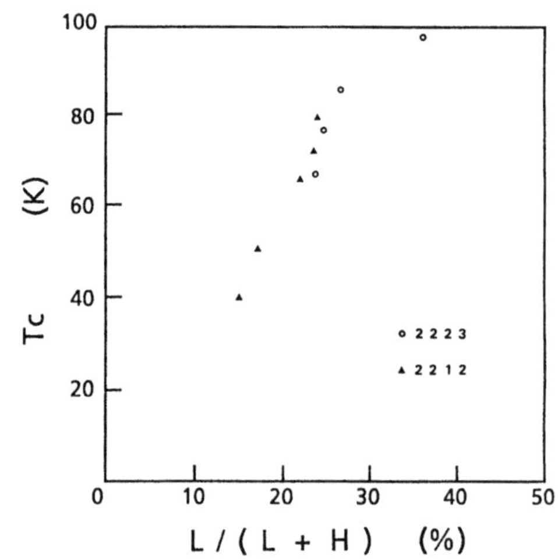

Fig.2. The correlations between Tc and L/(L+H) of Bi 4f.

present as Cu^+ or Cu metal in the film with Tc of 98K. The spectrum of Cu_2O and Cu metal are quite narrow and have positions overlapping with the spectra of CuO[8]. The binding energy of Cu^+ is close to that of Cu metal practically, however there is a difference between the kinetic energies of Cu^+ and Cu metal in auger spectra[9]. Judging from the auger spectra, it seems reasonable to assume that the peak at 932.8eV is caused by Cu^+ compounds. So the mean valence for Cu corresponds in the film with Tc of 98K is 2.07. Furthermore, in other samples, sample c with Tc of 86K and sample e which behaves as like a semiconductor in resistive properties, Cu mean valence corresponds to 2.03 and 2.00 respectively.

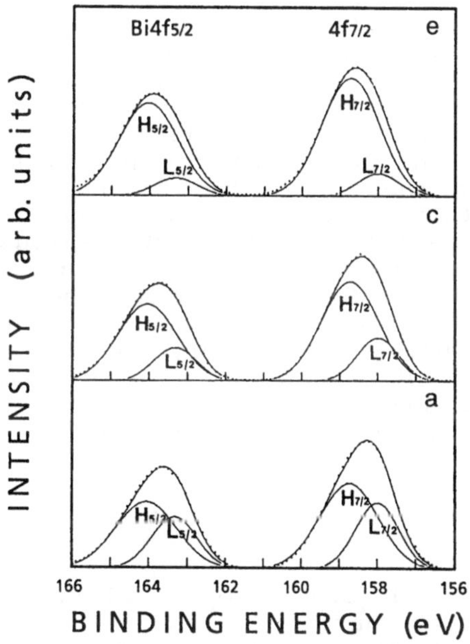

Fig.1. X-ray photoemission spectra of Bi 4f core level of sample a, c and e.

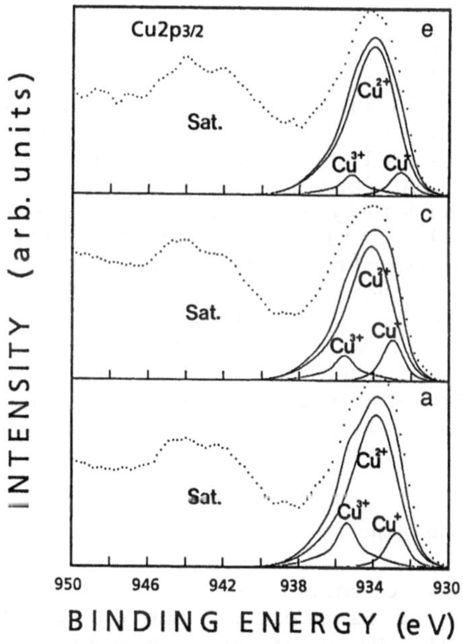

Fig.3. X-ray photoemission spectra of Cu $2p_{3/2}$ core levels of sample a, c and e.

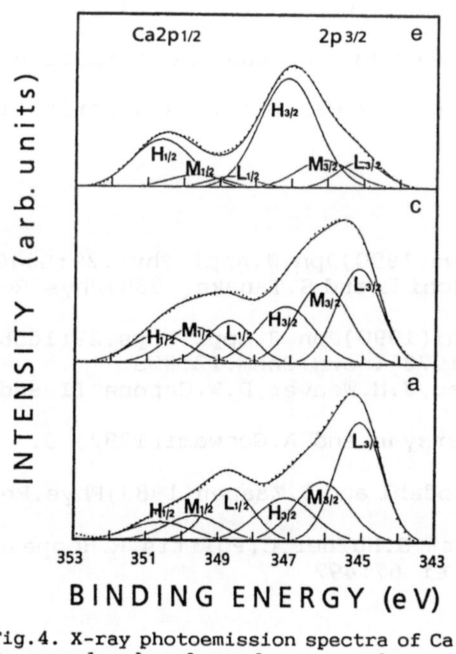

Fig.4. X-ray photoemission spectra of Ca 2p core levels of sample a, c and e.

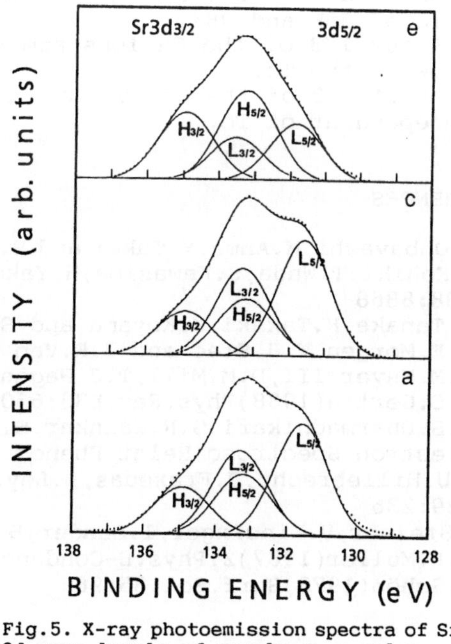

Fig.5. X-ray photoemission spectra of Sr 3d core levels of sample a, c and e.

The Ca 2p spectra can be decomposed into three components at 345.1, 346.1 and 347.1eV respectively as shown in Fig.4 referring to the $2p_{3/2}$ lines. The lowest binding energy component L can be associated with the superconducting properties, and the intensities of the peak L are to 49, 37 and 13% of total Ca intensity, respectively for the films with Tc of 98K, 86K and under 20K. Whereas Tc decreases with an increase of the intensity of the highest binding energy peak which is indicated by H in Fig.4. The middle peak M is not related with Tc, and the ratio of M/(L+M+H)=1/3 is constant for all specimens except sample e. In this manner, the main peak of the Ca gradually sifts to a high binding energy side with a decrease of Tc. Consequently the peak L of the Ca 2p spectrum can readily be assigned to the Ca atom in the site which sandwiched between the CuO sheets. On the other hand, it is considered that the peak H of the Ca 2p spectrum can be assigned to CaO, which is detected in the XRD pattern of sample e, as a strong peak at 37.5°. The Sr 3d spectra can be decomposed into two components, these peaks are identified at 131.6 (indicated by L) and 133.2eV (indicated by H) respectively referring to the $3d_{5/2}$ lines. The Sr 3d spectra are shown in Fig.5. A feature of spectra of Sr 3d is that the ratio of H/(L+H) is approximately 1/3, irrespectively of Tc. This value is in a good agreement with the ratio value of M in the Ca 2p spectra as described previously. These results show that the Ca 2p spectrum M should be related to substituted Ca atoms occupying Sr sites, and the Sr 3d spectrum H should be related to substituted Sr atoms occupying Ca sites. It is reported that Sr and Ca substitutes Ca and Sr sites, respectively[2]. Moreover, these results show that the substitution of Sr and Ca has not effects on Tc. From a point of view described above, the peak L of Sr spectrum corresponds to the Sr site between the BiO and CuO sheets. While in the spectrum of sample e the ratio H/(L+H) is significantly high, 2/3.

CONCLUSIONS

We have obtained XPS core level spectra of $Bi_2Sr_2Ca_2Cu_3O_x$ to clarify the effect of Bi content, and our study indicates that:
(1) The Bi core level electron spectrum shows that Tc increases with an increase of the ratio of BiO, and with a decrease of the ratio of Bi_2O_3. These results indicate a possibility of the excess oxygen of Bi system are incorporated in Bi-O sheets.
(2) Noticeable shoulders of Cu^{3+} were observed in the spectra of the films with Tc of 86K and 98K.
(3) About 1/3 of the Ca ions are substituted by Sr, and the formation of CaO lowers the Tc.
(4) About 1/3 of the Sr ions are substituted by Ca, but this substitution is independent of Tc.

REFERENCES

1. K.Ohbayashi,M.Anma,Y.Takai and H.Hayakawa(1990)Jpn.J.Appl.Phys.29:L2049
2. S.Kohiki,T.Wada,S.Kawasima,H.Yakagi,S.Uchida and S.Tanaka(1988)Phys.Rev. B38:8868
3. K.Tanaka,H.Takaki,K.Koyama and S.Noguchi(1990)Jpn.J.Appl.Phys.29:1658
4. W.E.Morgen,W.J.Stec and J.R.Van Wazer(1973)Inorg.Chem.12:953
5. H.M.Mayer III,D.M.Mill,T.J.Wagener,Y.Gao,J.H.Weaver,D.W.Capone II and K.C.Gertta(1988)Phys.Rev.B38:6500
6. V.S.Dharmadhikari,S.R.Sainkar,S.Badrinarayan and A.Goswami(1982) J. Electron Spectrosc.Relat.Phenom.25:181
7. F.U.Hillebrecht,J.Fraxedas,L.Ley.H.J.Trodahl and J.Zaanen(1989)Phys.Rev. B39:236
8. P.Steiner,V.Kinsinger,I.Sander,B.Siegwart,S.Hufner,C.Politis,R.Hoppe and H.P.Muller(1987)Z.Phys.B-Condenced Matter 67:497
9. G.Schön(1973)Surf.Sci.35:96

I-V Characteristics and Transport Critical Current of Bi(Pb)-Sr-Ca-Cu-O Thin Films on $PbTiO_3$

HIDEAKI FUKUDA and MASAHIRO AKINAGA

Department of Physics, Fukuoka University of Education, Munakata, Fukuoka, 811-41 Japan

ABSTRACT

The I-V characteristics of superconducting Bi(Pb)-Sr-Ca-Cu-O thin films prepared on piezoelectric $PbTiO_3$ film/MgO substrate were measured and the temperature dependence of critical current were investigated. In I-V curves at low temperature, a first detectable voltage drop appeared at a small current (I_{c1}). With still more increasing the current, voltage showed drastic change to an ohmic behavior at a relatively large current (I_{c2}). We suppose that I_{c1} is due to the intergrain weak links and I_{c2} corresponds to the intrinsic critical current inside the grain. The results of SEM, XRD and temperature dependence of I_{c1} analysis support the situation.

KEY WORDS: $PbTiO_3$ thin film, Bi system superconductor, I-V characteristics, critical current

INTRODUCTION

The fabrication of superconducting film on a functional substrate, for example on piezoelectric thin fim, is important for application or acoustic investigation of High-T_c superconductivity. Since the elastic energy of the surface acoustic wave (SAW) concentrates within a layer with the width of its wave length from the surface, SAW measurement is quite suitable for acoustic measurement in a superconducting film with a few μm thickness. Thus we have successfully fabricated superconducting Bi(Pb)-Sr-Ca-Cu-O thin films on c-axis oriented piezoelectric $PbTiO_3$ film and investigated their electric properties as a step toward SAW measurement in high-T_c superconducting thin film.

On the other hand, a very important transport property of a superconductor is its critical current density (J_c). This property depends to a great extend on the quality of the material. Although the critical temperature and critical magnetic field of high-T_c oxide superconductor are superior to those of conventional metal superconductor, J_c is too low for practical use such as wires or tapes. It seems that the low J_c is caused by not-artificial weak links at grain boundaries. In order to study the correlation between the critical current and the presence of grain boubdaries, we have investigated the I-V characteristics and the temperature dependence of J_c in the sputtered Bi-compound films on $PbTiO_3$ film consisting of many c-axis oriented grains, on single crystal MgO and MOCVD-grown films without grain boundaries.

EXPERIMENTAL

The perovskite type $PbTiO_3$ films were grown on single crystalline MgO (100) substrate by RF-sputtering. The details of the preparation method have been described in ref. [1]. Figure 1 shows a typical X-ray powder diffraction (XRD) pattern of c-axis oriented $PbTiO_3$ film. The figure shows the c-axis orientation rate α = 86 %, which is calculated from the intensity of peaks (001) and (100) [2]. SEM analysis revealed that the surface of a $PbTiO_3$ film consists of many regular mosaic cell with a size of the order of 0.1 μm.

The superconducting Bi(Pb)-Sr-Ca-Cu-O thin films on $PbTiO_3$ film/MgO (100) and MgO (100) substrates were prepared by RF-sputtering method. The target with composition of $Bi_{0.8}Pb_{0.2}Sr_{0.8}Ca_{1.0}Cu_{1.4}O_y$ was prepared by mixing powders of Bi_2O_3, $SrCO_3$, CaO and PbO, then heating for 72 hours at 740 °C in air and pressed into a shallow copper dish. The sputtering conditions of substrate temperature 350 °C; Ar/O_2 (90/10) gass pressure 4.4 Pa; sputtering time 5 hours; plate voltage 1.2 kV were used. After deposition as-deposited films were annealed in air at 825 °C for 1.5 hours on a pellet ceramic whose composition was similar to that of the target. The film surface was directly in contact with the pellet to prevent the composition of Pb and Bi elements in the film from reducing [3]. Growing superconducting Bi-compound film on $PbTiO_3$/MgO substrate was difficult. However, the fabrication of the c-axis oriented $PbTiO_3$ films with more highly α-value made it possible to grow the superconducting films.

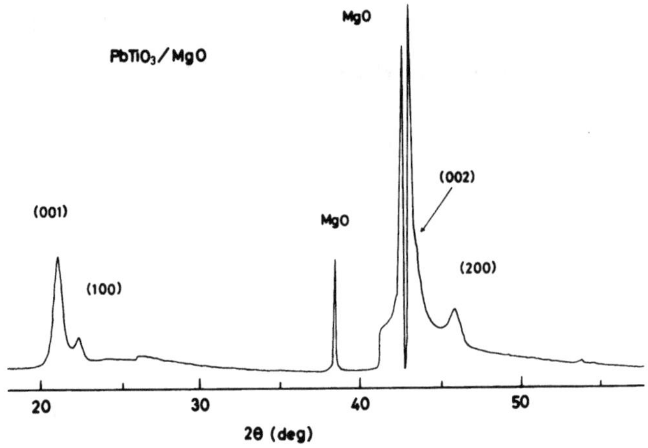

Fig.1 A typical X-ray powder diffraction pattern of the $PbTiO_3$ film on MgO (100) substrate.

The experimental I-V plots were acquired by a four-probe configuration point by point in the stationary state for both current directions and then calculated as the average in both directions. Critical current I_c (I_{c1}) was experimentally defined as the current at which the voltage drop between the voltage probes of the sample rises to above 2 μV/cm.

For comparison, I-V curves and the temperature dependences of I_c were measured for the Bi(Pb)-Sr-Ca-Cu-O thin films prepared on MgO (100), $Bi(Pb)_2Sr_2Ca_2Cu_3O_y$ bulk samples with bridge-junction structure [4] and $Bi_2Sr_2CaCu_2O_y$ thin films prepared by MOCVD [5] in the same way.

RESULTS AND DISSCUTION

Figure 2 shows the typical I-V plots for the Bi(Pb)-Sr-Ca-Cu-O thin film (labeled as PT34B) on $PbTiO_3$ film/MgO substrate at various temperatures. A dc resistance measurement indicated that this film consists of 2212 phase with T_{c-on} = 80 K and T_{c-end} = 58 K. The I-V characteristics for high voltage regime are presented in Fig. 2. Figre 3 shows log V versus log I data at low voltage regime near I_c (I_{c1}) for the same sample. The I-V curves at relatively low temperature in Figs. 2 and 3, there are two breaks in I-V behavior. In Fig. 2 the first detectable voltage drop appeared at a small current value I_{c1}, when a transport current was

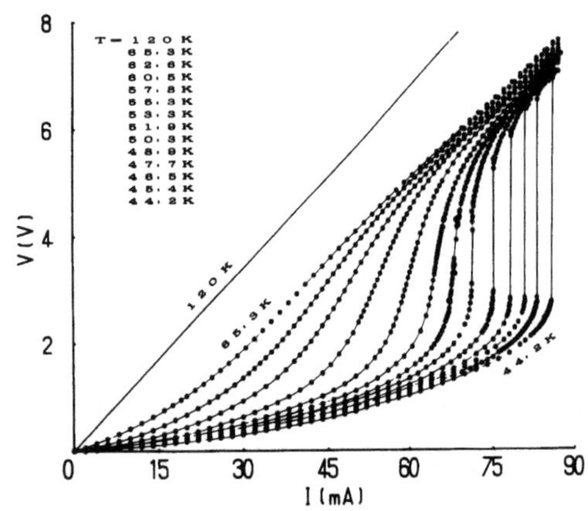

Fig.2 I-V curves at various temperatures in high voltage regime for the Bi(Pb)-Sr-Ca-Cu-O film on $PbTiO_3$/MgO. The solid straight line is an ohmic line at 120 K.

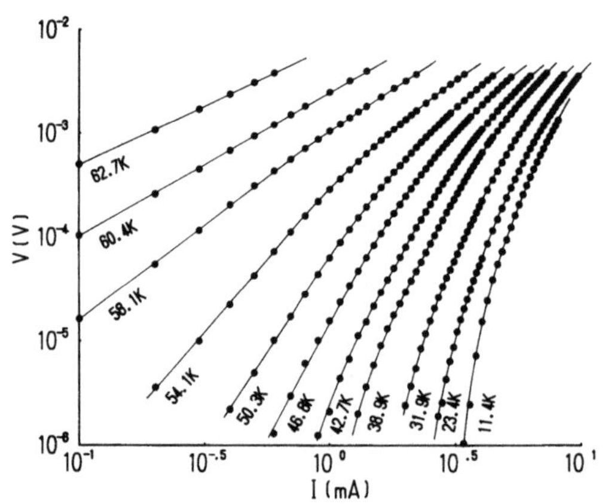

Fig.3 I-V curves at various temperatures near critical current I_{c1} for the same sample shown in Fig.2.

exceeding just the ordinaly critical current. Still more increasing the current, at first, the voltage gradually increases, next rapidly (I_{c2}) and finally approaches an ohmic behavior at a large current value. In Fig. 2 at low temperature, between 44.2 K and 55.3 K, it seems that the slope of the I-V curves from I_{c1} to I_{c2} keeps approximately constant. In addition, the value of I_{c2} is higher than that of I_{c1} at 44.2 K by two orders of magnitude. Therefore we considered that the voltage appearance near I_{c1} is caused by presence of weak links in the film. Furthermore, the straight line from the normal state in the I-V curves does not pass through the origin and then the existence of the excess current that appears in the traditional SNS junction is suggestive.

In order to check that the I_{c1} may be associated with weak link behavior in the film, we have analyzed temperature dependence of I_{c1} using some proposed models. In Fig.4 we show the normalied critical current I_{c1} versus $(1-t)^2$ plots near T_c of the same sample (PT34B), where $t = T/T_c$. For comparison, the data of the film sputtered on MgO (100) (labeled as BT2B), a $Bi_2Sr_2CaCu_2O_y$ thin film prepared by MOCVD (labeled as CV31A) and a $Bi(Pb)_2Sr_2Ca_2Cu_3O_y$ bulk sample with bridge-junction structure (labeled as BP44B) are also shown in the figure. The model of an array of SNS-type junctions [6], that predicts $I_c \propto (1-t)^2$ near T_c, is in reasonable agreement with the results shown in Fig. 4 except the bulk sample (BP44B). For the sample BP44B, which consists of many grains, temperature dependence of I_c is similar to the that of the SIS-type junction array model where the formula $I_c \propto (1-t)$ near T_c [7]. If we support these model, the different temperature dependence of I_c may be resulted in the different nature of the grain boundaries which is caused by different sample preparation method.

Though the sample PT34B was prepared on a house-made not-single crystalline $PbTiO_3$ film, the temperature dependence of I_c shows almost same behavior as the sample on single crystalline MgO substrate (BT2B). These data for the sputtered films showed a intermediate type indicating that probably a kind of SNS- and SIS-type junctions are present in the film. By XRD and SEM analysis it was found that the sputtered samples consist of many layered grains with almost c-axis orientation as shown in Fig. 5. Here, we can easily imagine that the superconducting grains are connected by weak links made of normal metal or insulator, which act as the SNS-or SIS-junction network.

The sample CV31A showed the best fitting to the SNS-type array model for $d/\xi_N(T_c) << 1$, where d is the thickness of the normal metal barrier and $\xi_N(T_c)$ the penetration distance of the electron pair into the barrier. Moreover there were no visible grain boundaries by SEM observation. Then it is reasonable that superconducting regions have very good contacts. Extremely speaking, the SNS-type behavior is possible at the intragrain.

In I-V curves of the MOCVD-grown films at low temperatures the slope from I_{c1} to I_{c2} was gentle as compared with those of the sample PT34B shown in Fig. 2 or another sputtered samples. In addition, any long slopes or the I_{c1} described above were not existent in the I-V curves for the $YBa_2Cu_3O_7$ single crystal carried out by Yeh et al. [8] and only one break uniquely defined

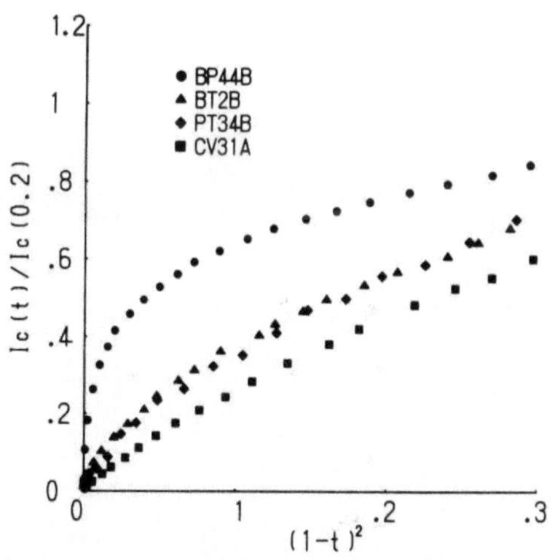

Fig.4 Normalized critical current I_{c1} versus $(1-t)^2$ plots for $Bi(Pb)_2Sr_2Ca_2Cu_3O_y$ bulk sample (BP44B), the sputtered film on MgO (BT2B) and on $PbTiO_3$/MgO (PT34B), and MOCVD-grown film (CV31A).

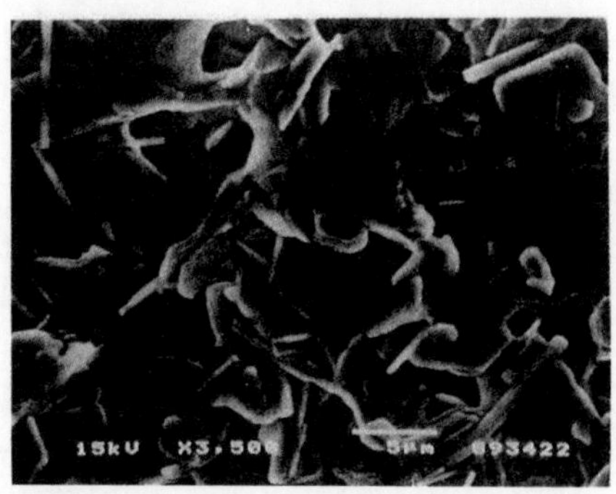

Fig.5 A typical SEM photograph of the Bi(Pb)-Sr-Ca-Cu-O film on the $PbTiO_3$/MgO substrate.

as critical current I_c was existent. Then it can be thought that the long slopes from I_{c1} to I_{c2} are implicitly related to the presence of weak links in the samples. We gess that I_{c1} is due to the intergrains with weak links and I_{c2} may correspond to the intrinsic critical current in the intragrains.

ACKNOWLEDGEMENT

The authors would like to express thanks to Profs. T. Fukami, T. Aomine, S. Suzuki and H. Sakashita for their valuable discussions and encouragement. First trial for preparation of $PbTiO_3$ films was performed by Prof. T. Fukami using sputtering apparatus in The Center of Advanced Instrumental Analysis, Kyushu University. One of the authors (H. F) would like to thank Prof. I. Kimura for his continuous interest in this work and his encouragement. This work was partially supported by Grant-in-Aid for Scientific Research from the Ministry of Education, Science and Culture, Japan.

REFERENCES

1. Akinaga M, Fukuda H (in press) J. Crystal Growth
2. Matsubara S (1988) Kagakukogyo (Chemical Technology) 52: 655-658 (in Japanese)
3. Yotsuya T, Suzuki Y, Ogawa S, Imokawa H, Yoshikawa M, Naito M, Takayama R, Otani K (1989) Japn. J. Appl. Phys. 28: 972-974
4. Akinaga M, Yasumoto K, Rinderer L (1991) Proc. 9th Int. Workshop on Future Electron Devices, Kumamoto, Japan (R & D Association for FED, Tokyo, 1991): 112-113
5. Hayashida S, Nakai T, Kobayashi I, Sugimori Y, Tsuyama T, Koshiba S (1990) Proc. 3rd Int. Symp. Superconductivity (ISS'90) TFP-58
6. De Vries JWC, Stollman GM, Gijs MAM (1989) Physica C 157: 406-414
7. Clem JR, Bumble B, Raider SI, Gallagher WJ, Shih YC (1987) Phys. Rev. B 35: 6637-6642
8. Yeh NC, Tsuei CC (1989) Phys. Rev. B 39: 9708-9711

Selective Growth of BiSrCaCuO Thin Films with (11*n*) and (001) Orientations

KEN'ICHI KURODA[1], TETSUYA TAKAMI[1], OSAMU WADA[2], JUNJI TANIMURA[2], KAZUYOSHI KOJIMA[1], MASAYUKI KATAOKA[2], TETSUO OGAMA[2], and KOICHI HAMANAKA[1]

[1] Central Research Laboratory, [2] Materials and Electronic Devices Laboratory, Mitsubishi Electric Corporation, Amagasaki, Hyogo, 661 Japan

ABSTRACT

A BiSrCaCuO thin film was sputtered on an off angle vicinal surface of $SrTiO_3$(110) substrate partially covered with a MgO buffer layer. From cross-sectional trnsmission electron microscopy(TEM) observations, it was confirmed that the film region directly grown on the substrate surface is oriented in the [11*n*] direction and that grown on the MgO layer is oriented with the *c*-axis. In addition, the boundary line between the (11*n*) and (001) oriented regions was clearly observed by cross-sectional TEM.

KEY WORDS: BiSrCaCuO, superconductor, thin film, sputtering, film growth

INTRODUCTION

As recently reported [1-5], BiSrCaCuO thin films grown on $SrTiO_3$(110) substrates can be oriented in the [11*n*] direction, whereas films grown on MgO(100) or $SrTiO_3$(100) substrates are oriented with the *c*-axis perpendicular to the substrate surface [6-10]. However, using a buffer layer, such as MgO, a BiSrCaCuO film grown on a $SrTiO_3$(110) substrate is expected to have a *c*-axis orientation. Thus, to form artificial grain boundaries, we tried to selectively grow a BiSrCaCuO thin film both in the [11*n*] and [001] directions at the same time on a substrate by using a $SrTiO_3$(110) substrate partially covered by a MgO layer.

In this paper, we report on the selective film growth of BiSrCaCuO in the [11*n*] and [001] directions, and the high-resolution cross-sectional TEM obsevations of the film.

EXPERIMENTAL

First, a MgO buffer layer with a thickness of 15nm was deposited on an off angle vicinal surface of $SrTiO_3$(110) substrate at 350-450°C in a neat O_2 atomosphere of 300mTorr by RF magnetron sputtering. To limit the directions of the film growth, a substrate oriented off the [110] azimuth by approximately 5° was used in this study[3]. Then, the MgO layer was masked with conventional photoresist and was partially removed by chemical wet etching with dilute phosphoric acid to form 5μm wide lines. Lastly, a 120-150nm thick BiSrCaCuO thin film was grown to form (11*n*) and (001) oriented regions; the film was formed at a substrate temperature of 660°C in a neat O_2 atomosphere of 300mTorr by sputtering a single $Bi_2Sr_{2.1}Ca_{2.2}Cu_{3.4}O_x$ target.

The crystalline orientation of the film was analyzed by X-ray diffraction(XRD) measurements and reflection high-energy electron diffraction(RHEED) observations. Further, to comfirm the film orientation, high-resolution cross-sectional TEM observations were performed. The surface morphology was observed by scanning electron microscopy (SEM).

RESULTS AND DISCUSSION

A MgO buffer layer and a BiSrCaCuO layer sputtered on an off angle-vicinal surface of a $SrTiO_3$(110) substrate were observed by cross-sectional TEM after the BiSrCaCuO layer was deposited; a MgO layer grown on a $SrTiO_3$(110) substrate had been examined beforehand by XRD measurements and been found to have a (110) orientation. In Fig.1, the interface between the substrate surface and the MgO buffer layer is clearly observed. The (110) plane of the MgO layer is tilted against the substrate surface and is nearly parallel to the (110) plane of the substrate. Thus, the MgO layer is confirmed to be epitaxially grown on the substrate.

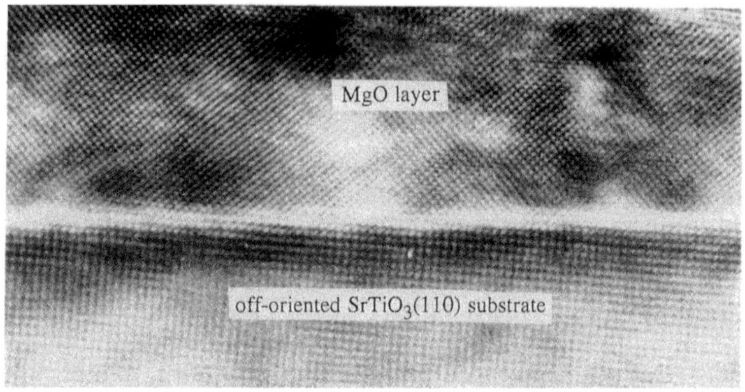

Fig.1. A cross-sectional TEM image of the MgO layer grown on the off-oriented substrate.

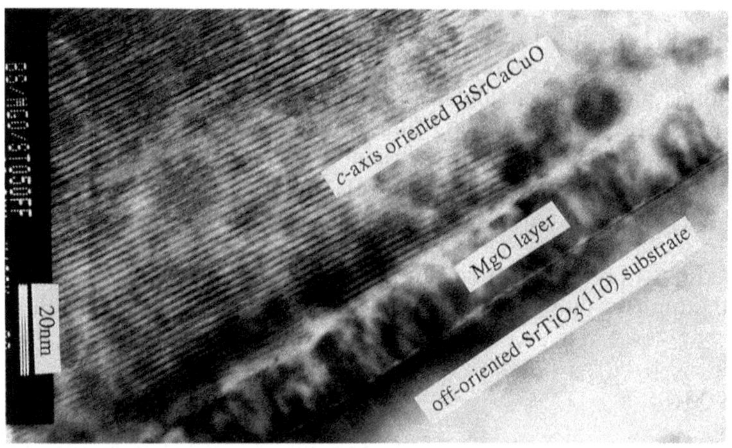

Fig.2. A cross-sectional TEM image of the BiSrCaCuO layer grown on the MgO layer.

A cross-sectional TEM picture of a BiSrCaCuO layer on the buffer MgO layer is shown in Fig.2. As shown in the picture, the BiSrCaCuO layer is oriented with the *c*-axis, but it is noticeable that the *c*-axis tilted against the normal direction of the substrate surface by about 5°. RHEED observation also showed the *c*-axis of the BiSrCaCuO layer tilted by 5°. Thus, the *c* plane of the BiSrCaCuO layer is also parallel to the (110) plane of the substrate. On the other hand, the BiSrCaCuO film directly grown on the substrate was not oriented with the *c*-axis, as shown Fig.3. In the figure, no regions with *c*-axis orientations were observed. In adition, the *c*-axis was uniformly tilted against the substrate surface in only one direction. The angle between the *c*-axis and the substrate surface was measured to be about 50°. Therefore, the *c*-axis was estimated to meet the (110) plane of the substrate at an angle of almost 45°.

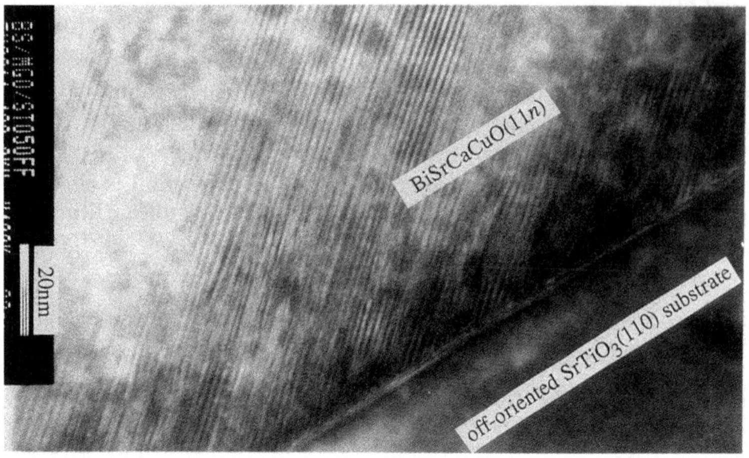

Fig.3 A cross-sectional TEM image of the BiSrCaCuO layer directly grown on the off-oriented $SrTiO_3$(110) substrate.

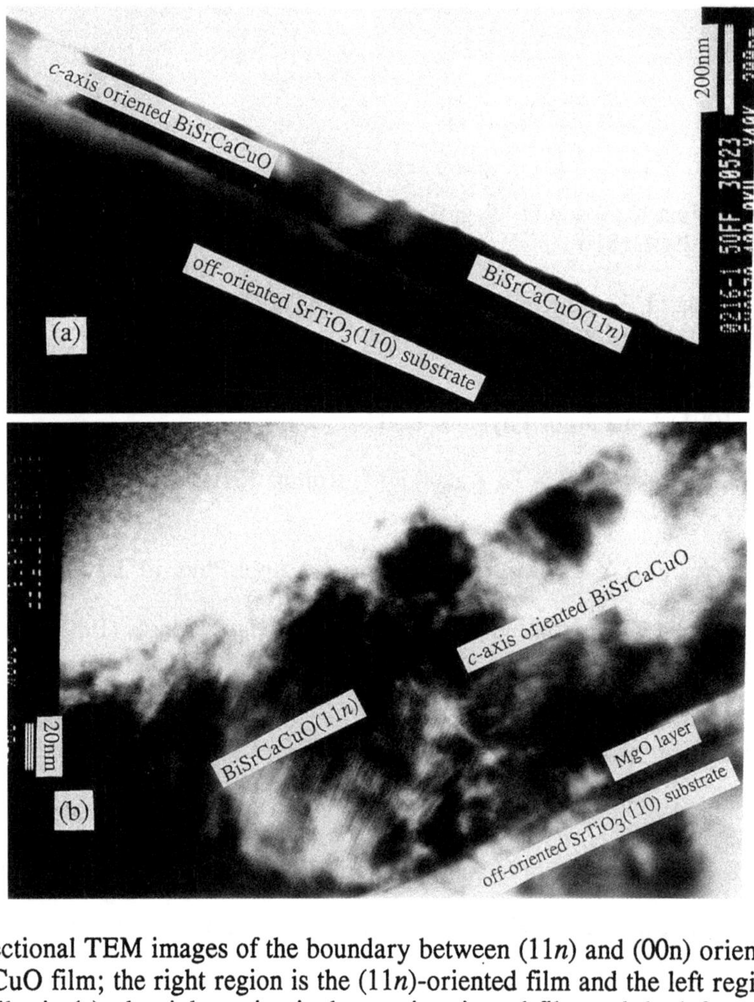

Fig.4 Cross-sectional TEM images of the boundary between (11*n*) and (00n) oriented regions of the BiSrCaCuO film; the right region is the (11*n*)-oriented film and the left region is the *c*-axis oriented film in (a), the right region is the *c*-axis oriented film and the left region is the (11*n*)-oriented film in (b).

Figures 6(a) and 6(b) show cross-sectional TEM pictures of the boundary between (11*n*) and (001) oriented regions of the BiSrCaCuO film. The boundary lines are inclined toward the film region directly grown on the substrate in both figures; however, the slope of the boundary line in Fig.6(a) is gentler than that in Fig.6(b). These experimental results indicate that the horizontal growth rate of the *c*-axis oriented film was higher than that of the (11*n*)-oriented film. In addition, the vertical growth rate of the *c*-axis oriented film is considered to be lower than that of the (11*n*)-oriented film, because the *c*-axis oriented film is about 10% thinner than the (11*n*)-oriented film in Fig.6(b). Figure 8 is an SEM picure of the film surface. In the figure, the boundary is also clearly observed.

CONCLUSIONS

The as-grown BiSrCaCuO thin film was grown at 660°C on the off angle vicinal surface of the $SrTiO_3$(110) substrate partially covered by the MgO buffer layer using RF magnetron single-target sputtering. Cross-sectional TEM observations exhibited that the BiSrCaCuO layer directly grown on the substrate is uniformly oriented in the [11*n*] direction, and that grown on the MgO buffer layer is oriented with the *c*-axis. Further, the boundary lines of the BiSrCaCuO film with the (11*n*) and (001) orientations were also observed by cross-sectional TEM.

REFERENCES

1. Kuroda K, Kojima K, Wada O, Tanioku M, Yokoyama K and Hamanaka K (1990) Jpn.J.Appl.Phys. 29: L1816

2. Kuroda K, Wada O, Kojima K, Tanimura J, Kataoka M, Takami T, Yokayama K, Ogama T and Hamanaka K (1991) Jpn.J.Appl.Phys.30: L475

3. Kuroda K, Wada O, Tanimura J, Kojima K, Takami T, Kataoka M, Ogama T and Hamanaka K (1991) Jpn.J.Appl.Phys.30: L582

4. Wada O, Tanimura J, Kurada K, Kataoka M, Kojima K, Hamanaka K and Ogama T (1991) Jpn.J.Appl.Phys.30: L722

5. Isizuka Y, Terashima Y and Miura T (1991) Jpn.J.Appl.Phys.30: L1006

6. Kuroda K, Kojima K, Tanioku M, Yokoyama K and Hamanaka K (1989) Jpn.J.Appl.Phys.28: 1586

7. Schlom D.M, Marshall A.F, Sizemore J.T, Chen Z.J, Eckstein J.N, Bozovic I, von Dessonneck K.E, Harris Jr J.S. and Bravmann J.C (1990) J.Cryst.Growth 102: 361

8. Fujita J, Yoshitake T, Igarashi H and Satoh K (1990) Appl.Phys.Lett.56: L295

9. Ohbayashi K, Anma M, Takai Y and Hayakawa H (1990) Jpn.J.Appl.Phys.29: L2049

10. Sugimoto T, Yoshida M, Yamaguchi K. Sugawara K, Shiohara Y and Tanaka S (1990) Appl.Phys.Lett.57: 928

Preparation of BiSrCaCuO Superconducting Films on Ag Tapes by MOCVD Method

H. TANAKA, T. SUGIHARA, and T. TAKESHITA

Central Research Institute, Mitsubishi Materials Corporation, Omiya, Saitama, 330 Japan

ABSTRACT

BiSrCaCuO superconducting films were grown on Ag tapes by metal organic vapor chemical deposition (MOCVD) using $Bi(ph)_3$, $Cu(DPM)_2$, $Ca(DPM)_2$ and $Sr(DPM)_2$ as source materials. The as-grown film on Ag tape showed mainly Bi-2212 phase with c-axis orientation. The critical temperature Tc(zero) and the critical current density Jc of this film were 67 K and 80,000 A/cm^2 at 4.2 K, respectively.

KEY WORD : Bi-Sr-Ca-Cu-O, MOCVD, superconducting films, Ag tape, critical current density

INTRODUCTION

Since the discovery of BiSrCaCuO superconductor [1], many studies have been performed in order to fabricate superconducting films and wires which are important for the application of this new material in electronic devices and electric power systems. Many techniques have been tried for obtaining superconducting films and wires, e.g., sputtering [2, 3], electron beam evaporation [4, 5], laser ablation [6, 7] and MOCVD [8, 9]. MOCVD offers advantages of high deposition rate, wide range of oxygen pressure, large scale processing and complicated shape substrate, in comparison with physical vapor deposition (PVD) such as sputtering and electron beam evaporation.

In this paper, we report preparation of BiSrCaCuO superconducting films on Ag tapes by MOCVD method and effects of cooling rate to electrical properties of the films on Ag tapes. Possible merits of Ag existance in the superconducting wire fabrication process are also discussed.

EXPERIMENTAL

Figure 1 is a schematic diagram of the MOCVD apparatus used in this study. The source materials were triphenl bismuth $Bi(ph)_3$, bis-(dipivaloylmethanate) -copper $Cu(DPM)_2$, -calcium $Ca(DPM)_2$ and -strontium $Sr(DPM)_2$. $Bi(ph)_3$, $Cu(DPM)_2$, $Ca(DPM)_2$ and $Sr(DPM)_2$ were sublimed at 115 °C, 123 °C, 190 °C and 230 °C, respectively. Each source was transported by Ar carrier gas at a rate of 50 sccm. Oxygen gas was separately introduced into a reactor at the rate of 300 sccm. Total gas pressure was kept at 15 Torr during the film growth and the substrate temperature was 800 °C. Ag tapes and single crystal MgO (100) were used as substrates. After the deposition, the BiSrCaCuO films were cooled in 1 atm of oxygen. The films were characterized by using X-ray diffraction (XRD) with Cu-Kα radiation, scanning electron microscopy (SEM) and electron probe microanalysis (EPMA). The conventional DC four probe technique was used for determining the zero resistance critical temperature and the critical current density of the as-grown films.

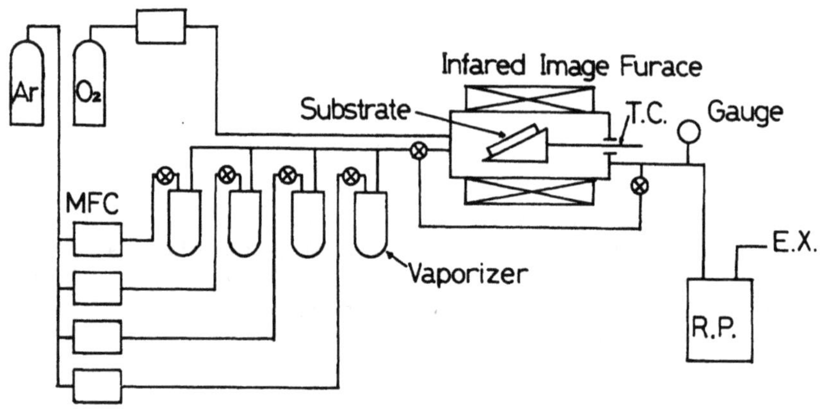

Fig. 1 A schematic diagram of the MOCVD apparatus.
MFC : mass flow controller
T.C. : themocouple

Table 1 Deposition conditions

	$Bi(ph)_3$	$Sr(DPM)_2$	$Ca(DPM)_2$	$Cu(DPM)_2$
Temperature of source vaporizer(°C)	115	230	190	123
Ar gas flow rate	50 SCCM			
O_2 gas flow rate	300 SCCM			
Total pressure	15 Torr			
Temperature of substrate	800 °C			
Substrate	Ag, MgO (100)			
Deposition time	60 min			

RESULTS AND DISCUSSION

Figure 2 shows the XRD patterns of the films on different substrates cooled at a rate of 10 °C/min. In the XRD pattern of sample A which was grown on MgO substrate, the peaks are indexed as the (0 0 n) of $Bi_2Sr_2CaCu_2O_x$ (2212) phase. This indicates preferred orientation of the c-axis in direction perpendicular to the substrate surface. On the other hand, sample B, which was grown on Ag tape, shows the (0 0 n) peaks of 2212 phase together with weak peaks of $Bi_2Sr_2CuO_x$ (2201) and $Bi_2Sr_2Ca_2Cu_3O_x$ (2223) phase. Figure 3 shows the XRD patterns of the films cooled at 2 °C/min. As was mentioned before, the XRD pattern of sample C which was grown on MgO substrate shows only the peaks of 2212 phase. Sample D, which was grown on Ag tape, shows both the peaks of 2201 phase and 2223 phase.

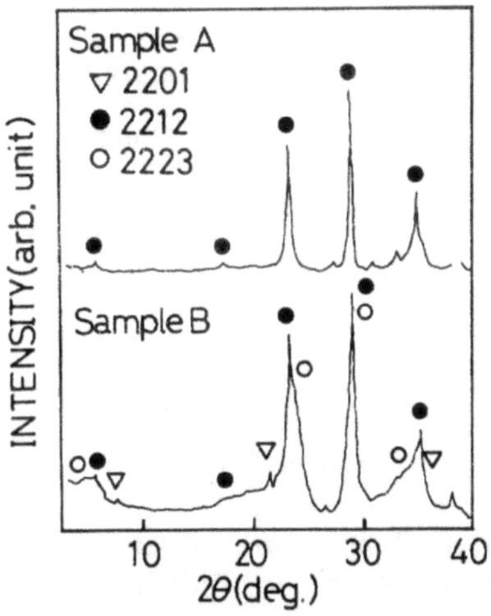

Fig. 2 X-ray diffraction patterns of BiSrCaCuO films.
Cooling rate : 10 °C/min

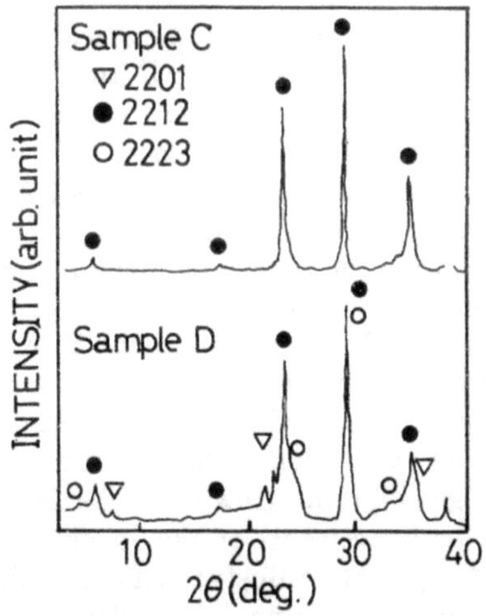

Fig. 3 X-ray diffraction patterns of BiSrCaCuO films.
Cooling rate : 2 °C/min

Figuer 4 and figure 5 show the SEM photographs of the surface of each sample. Sample A and C, which were deposited on MgO substrate, consist of many particle-like grains which are stacking on the substrate. However, sample B on Ag tape consisted of poly-crystalline rectangular grains. For sample D, the surface of the film is smooth, dense and continuous. It seems that slow cooling rate of films on Ag tape is effective in improving their flatness.

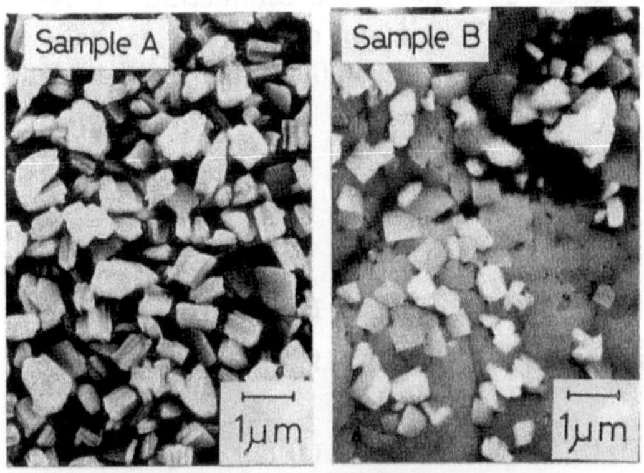

Fig. 4 The SEM photographs of the surface of BiSrCaCuO films.
Cooling rate : 10 °C/min

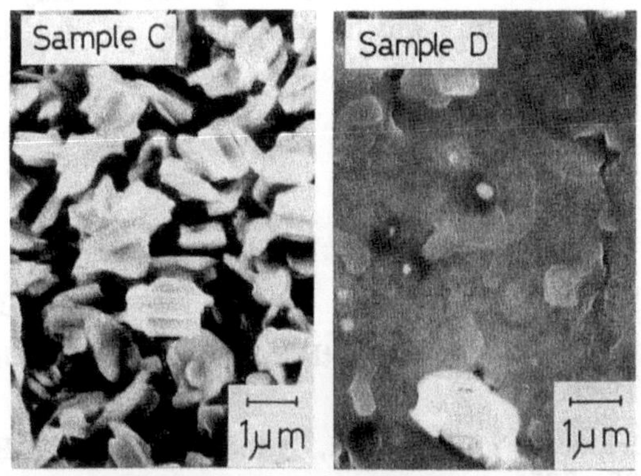

Fig. 5 The SEM photographs of the surface of BiSrCaCuO films.
Cooling rate : 2 °C/min

Figure 6 shows temperature dependence of resistance for each sample. The critical temperature are listed in Table 2. All the samples are metallic in the normal state and have a resistance drop at around 80 K. For example, the critical temperature of sample D which was grown on Ag tape and cooled at 2 °C/min of cooling rate is 67 K.

The critical current density was measured

Table 2 Characterization of samples

Sample No.	$T_{c(zero)}$ (K)	J_c (A/cm^2)*
A	65	1,520
B	55	7,800
C	72	1,630
D	67	80,000

* at 4.2 K

at 4.2 K under zero applied field, and the results are summarized in Table 2. The critical current densities of films on Ag tapes are higher than those of sample A (1520 A/cm²) and sample C (1630 A/cm²) which were deposited on MgO substrate. In particular, the critical current density of sample D which was cooled at a slow rate of 2 °C/min is fairly high, ie., 80,000 A/cm². This result is thought to be due to a smooth, dense and continuous structure of the film, which seems to a result of good grain growth. It is believed that the enhancement of the grain growth is caused by interdiffusion between Ag and oxide elements during the slow cooling. Therefore, we expected that superconducting wires of more high critical current density would be obtained by using Ag tapes.

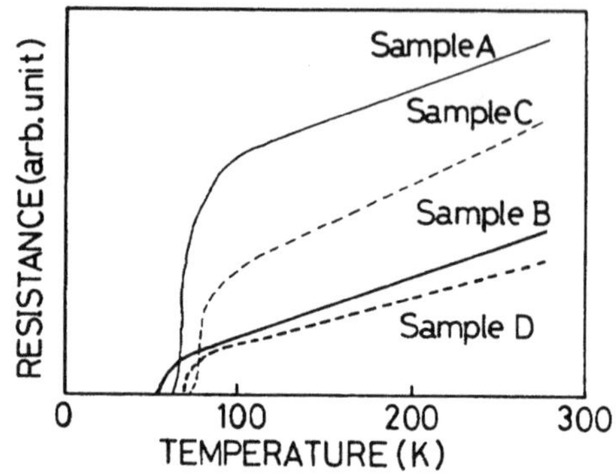

Fig. 6 Temperature dependence of the resistance of BiSrCaCuO films

CONCLUSION

BiSrCaCuO superconducting films were grown on Ag tapes by MOCVD method. We have found that the electrical properties of the films on Ag tapes largely depend upon the cooling rate. It was found that the slow cooling is effective in preparing smooth and dense films with high critical current density. Thus, we think that the films on Ag tapes may be used as the superconducting wires. The best film which was grown on Ag tape with 2 °C/min of cooling rate, has a critical temperature at 67 K and a critical current density of 80,000 A/cm² at 4.2 K.

REFERENCES

1. Maeda H, Tanaka Y, Fukutomi M and Asano T (1988) Jpn. J. Appl. Phys. 27 : L209
2. Kang J. H, Kampwrith R. T, Grey K. E, Marsh S and Huff E. A (1988) Phys. Lett. A 128 : 102
3. Ichkawa Y, Adachi H, Hirochi K, Setsune K, Hatta S and Wasa K (1988) Phys. Rev. B 38 : 765
4. Kuroda K, Mukaida M, Yamamoto M and Miyazawa S (1988) Jpn. J. Appl. Phys. 27 : L625
5. Steinbeck J, Tsaur B-Y, Anderson A. C and Strauss A. J (1989) Appl. Phys. Lett. 54 : 466
6. Kanai M, Kawai T, Kawai M and Kawai S (1988) Jpn. J. Appl. Phys. 27 : L1293
7. Deshmukh S, Rothe E. W, Reck G. P, Kushida T and Xu Z. G (1988) Appl. Phys. Lett. 53 : 2698
8. Yamane H, Kurosawa H, Hirai T, Iwasaki H, Kobayashi N and Muto Y (1988) Jpn. J. Appl. Phys. 27 : 1495
9. Zhang J, Zhao J, Marcy H. O, Tonge L. M, Wessels B. W, Marks T. J and Kannewurf C. R (1989) Appl. Phys. Lett. 54 : 1166

Thermal Stability of Source Materials Used for MOCVD of Bi-Sr-Ca-Cu-O Thin Film Superconductors

JUNICHI ISHIAI, SINYA YAMADA, and SHUJI YOSHIZAWA

Central Research Laboratory, Dowa Mining Co., Ltd., Hachiouji, Tokyo, 192 Japan

ABSTRACT

The volatility and thermal stability of source materials were very important factor for preparation of thin film superconductors using MOCVD method. In general, the source materials of Bi system were triphenyl bismuth $Bi(C_6H_5)_3$, bis(-2,2,6,6,-tetramethyl-3,5-heptandion) -strontium $Sr(DPM)_2$, -calcium $Ca(DPM)_2$ and -copper $Cu(DPM)_2$. In this study, we have fabricated single thin films of these components. The deposition rate of these films was estimated by using inductively coupled plasma spectroscopy (ICP). In the temperature dependence of the deposition rate of $Sr(DPM)_2$, the abrupt increase was recognized from 210 to 230°C. The deposition rate of $Sr(DPM)_2$, therefore, can be controlled by changing the source material temperature. The growth rate was dependent on the source material temperature in the temperature range up to 230°C, which shows the stability of source material.

KEY WORDS: thin film, superconductor, MOCVD method, Bi-system

INTRODUCTION

Since the discovery of the high Tc superconductors, many studies have been made to obtain thin films especially for microelectronic devices and microwave devices. In the fabrication process of oxide thin film, oxidation of elements is very important factor. Characteristics of metalorganic chemical vapor deposition (MOCVD) method [1, 2] compared with sputtering [3] and MBE [4] are possibility of large scale processing and wide range of oxygen pressure. We have fabricated Bi-system superconducting thin films by MOCVD method. It was reported that in the film the zero resistance increased to 86K and the critical current was over $10^4 A/cm^2$, and that the preparation of $Bi_2Sr_2Ca_{n-1}Cu_nO_{2n+4}$ (n=1, 2, 3 and 4) films was controlled by changing only the temperature of the Ca and Cu source vessels [5, 6]. In MOCVD method, the reproducebility and the stability of superconducting property in films depends largely on the stability of source materials. In this paper, we reports on the thermal stability of source materials for Bi-system superconducting thin film by MOCVD method.

EXPERIMENTAL

An MOCVD apparatus (DOWA MOCVD-0110) used in this study is classfied as a hot-wall type MOCVD, where the substrates were heated by an electric furnace. The source materials were triphenylbismuth $Bi(C_6H_5)_3$, bis(-2,2,2,6,6,-tetramethyl -3,5-heptandione)-strontium $Sr(DPM)_2$, -calcium $Ca(DPM)_2$, and -copper $Cu(DPM)_2$. Argon gas was used as a carrier. The flow rates of argon gas for $Bi(C_6H_5)_3$, $Sr(DPM)_2$, $Ca(DPM)_2$ and $Cu(DPM)_2$ were 50, 200, 200 and 25 cc/min, respectively. Oxygen gas was separately introduced into the quartz reactor at the rate of 200cc/min. During the deposition, the total pressure in the quartz reactor was maintained to be 50 Torr and the substrate temperature was 800°C for $Sr(DPM)_2$, $Ca(DPM)_2$ and $Cu(DPM)_2$. In the case of $Bi(C_6H_5)_3$ the substrate temperature was 700°C, because at 800°C no film deposition was observed. Single crystal MgO (100) was used as a substrate. ICP analysis revealed that the single crystal MgO (100) contained Ca element. Thus single crystal $SrTiO_3$ (100) was used as a substrate for $Ca(DPM)_2$. The films were characterized by using ICP and X-ray diffraction (XRD).

RESULTS AND DISCUSSION

Figure 1 illustrates the XRD patterns of the films. Each of the XRD patterns was assigned to be Bi_2O_3, SrO, CaO and CuO, respectively. The preferred orientation of the a-axis normal to the substrate surface was observed in SrO and CaO while Bi_2O_3 film was polycrystal structure. CuO film oriented to (111).

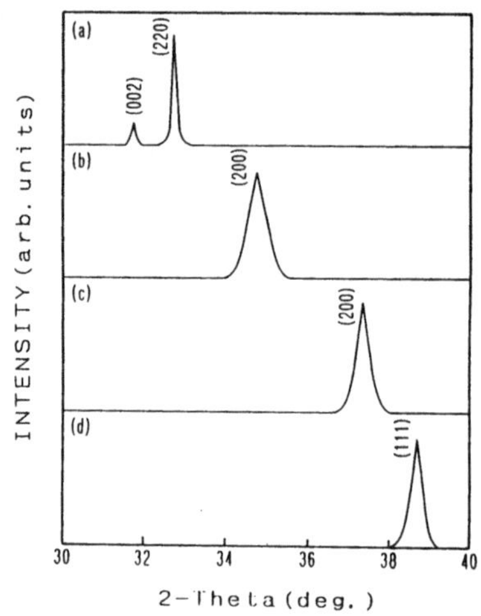

Fig.1. X-ray diffraction patterns of (a) Bi_2O_3 film, (b) SrO film, (c) CaO film and (d) CuO film.

The relationship between the passage time in the initial alteration of material sublimation and deposition rate is shown in Fig.2. The temperature of source vessels for $Bi(C_6H_5)_3$, $Sr(DPM)_2$, $Ca(DPM)_2$ and $Cu(DPM)_2$ materials were 75, 220, 170 and 110 °C, respectively. The total passage time of preparing films was sammed up to 30 hours. The deposition rates of Bi, Ca and Cu in oxide films was decreased with the passage time at 75, 170 and 110°C of source temperature, respectively, and then, saturated around 20 hour for Bi_2O_3 film, around 15 hour for CaO film and around 16 hour for CuO film. This suggests that the initial alteration in the deposition rate was partly caused by the impurity and/or the hydrate molecules. In the case of SrO the deposition rate was constant from the beginning.

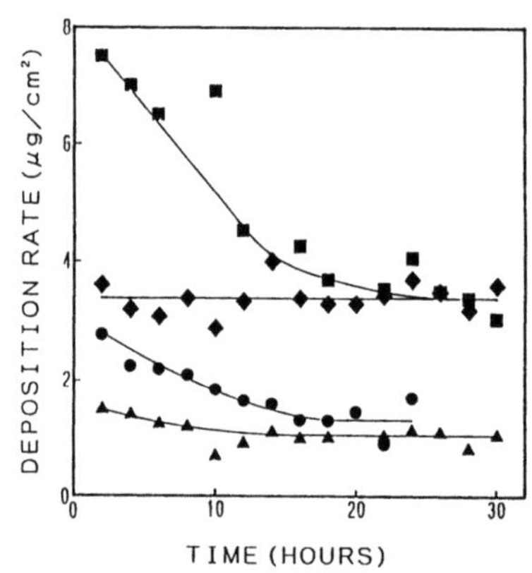

Fig.2. Deposition rate of Bi, Sr, Ca and Cu in oxide films vs passage time in the initial alteration of material sublimation. ■: Bi, ◆: Sr, ▲:Ca, ●: Cu

Figure 3 shows the source temperature dependences of deposition rate. The temperature of source vessels for $Bi(C_6H_5)_3$, $Sr(DPM)_2$, $Ca(DPM)_2$ and $Cu(DPM)_2$ materials were 60 to 80, 190 to 230, 140 to 180 and 90 to 130°C, respectively. The deposition rate of element in Bi_2O_3 film showed the constant with the source

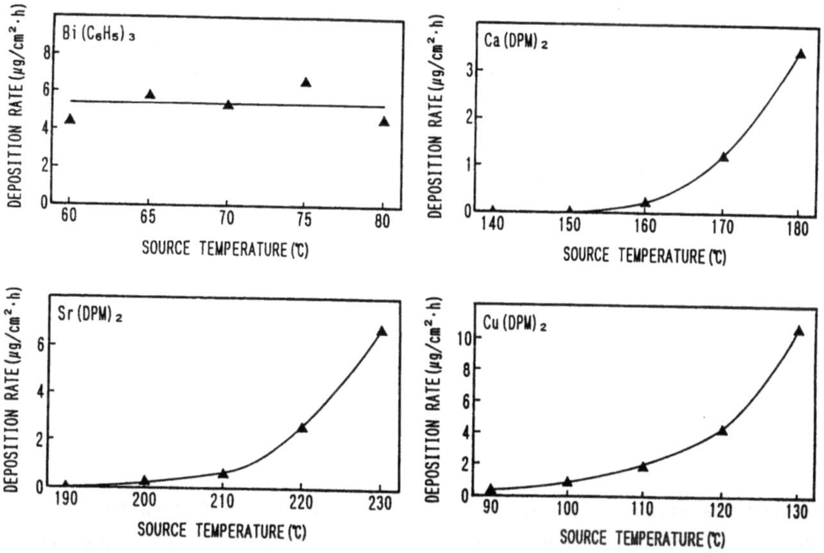

Fig.3. Deposition rate of Bi, Sr, Ca and Cu in oxide films vs source temperature.

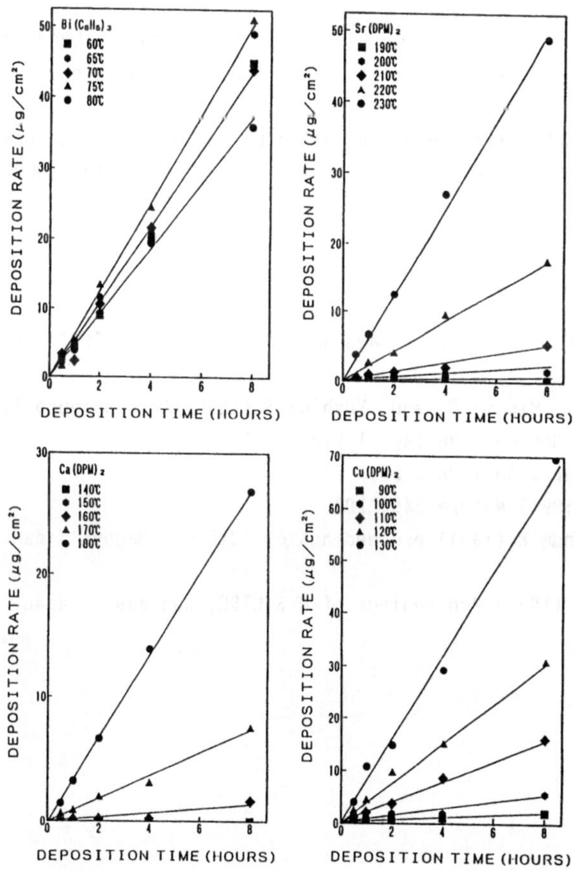

Fig.4. Deposition rate of Bi, Sr, Ca and Cu in oxide films vs deposition time.

temperatures. It was thought that the sublimation was acconpanied by the decomposition of $Bi(C_6H_5)_3$. This indicates that the preparation of film can not be changing the source temperature. In the temperature dependence of the deposition rates of each element in SrO, CaO and CuO, the abrupt increase was recognized from 210 to 230℃ for SrO, from 160 to 180℃ for CaO and from 110 to 130℃ for CuO. The deposition rate of each element in SrO, CaO and CuO can be controlled within the range of 0.6 to 6.8, 0.2 to 3.5 and 1.9 to 10.7 $\mu g/cm^2 \cdot h$ by adjusting the source temperatures, respectively.

Figure 4 shows the deposition time dependences of deposition rate. The deposition times were 0.5, 1, 2, 4 and 8 hours. The deposition rates of all films were propotional to the deposition time. This reveals that the volatitility of source materials are constant under these experimental conditions. The film thickness can be controlled by the deposition times.

The deposition rate of each element in $Bi_2Sr_2Ca_2Cu_3O_y$ film and each element in Bi_2O_3, SrO, CaO and CuO films is compared in Table 1. The deposition rates of each element in $Bi_2Sr_2Ca_2Cu_3O_y$ film was not equal to the single films of source materials. It should be mentioned that the amount of deposited element depends on the fablication order of $Bi_2Sr_2Ca_2Cu_3O_y$ film, although Bi, Sr, Ca and Cu source material were supplied simultaneously in the case of $Bi_2Sr_2Ca_2Cu_3O_y$film

Table 1. Comparison of deposition rate of each element in $Bi_2Sr_2Ca_2Cu_3O_y$ film with each element in Bi_2O_3, SrO, CaO and CuO films.

	Bi	Sr	Ca	Cu
Source Temp. (℃)	76	220	169	110
Deposition Rate of BSCCO ($\mu g/cm^2 \cdot h$)	3.8	1.5	0.8	1.8
Deposition Rate of each elements	6.6※	2.2	1.0	1.9

※ Substrate temperature : 700℃

CONCLUSIONS

We have fabricated the films of Bi_2O_3, SrO, CaO and CuO by MOCVD method using $Bi(C_6H_5)_3$, $Sr(DPM)_2$, $Ca(DPM)_2$ and $Cu(DPM)_2$ as source materials. The results can be summarized as follows.

1) The preferred orientation of the a-axis normal to the substrate surface was observed in SrO and CaO while Bi_2O_3 film was polycrystal structure. CuO film oriented to (111).

2) The deposition rate of Bi_2O_3 film was uncontrollable by source temperature. Under a constant deposition condition, however, the constant deposition rate can be obtained.

3) The deposition rate of SrO, CaO and CuO films were controlled by changing the source materials temperatures. $Sr(DPM)_2$, $Ca(DPM)_2$ and $Cu(DPM)_2$ materials were suitable for MOCVD source material.

REFERENCES

1. Endo K, Hayashida S, Ishiai J, Matsuki Y, Ikedo Y, Misawa S and Yoshida S (1990) Jap. J.Appl.Phys.29: L294.
2. Endo K, Hayashida S, Ishiai J, Ikedo Y, Misawa S and Yoshida S (1990) In: Ishiguro T, Kajimura K (eds) in Superconductivity II. Springer-Verlag, Tokyo, p 911
3. Nakamura K, Sato J and Ogawa K (1990) Jap.J.Appl.Phys.29: 77
4. Kawai T, Egami Y, Tabata H and Kawai S (1991) Nature.349: 200
5. Ishiai J, Yamada S, Yoshizawa S and Endo K (1991) proceedings of ICVGE-7, Nagoya Japan. J.Cryst.Growth (to be pablished)
6. Ishiai J, Yamada S, Yoshizawa S and Endo K (1991) Proceedings of M^2S-HTSC, Kanazawa Japan. Phisica C (to be pablished)

Resistive Transition of $YBa_2Cu_3O_{7-\delta}$ Ultrathin Films Down to One Unit Cell Thickness in Magnetic Fields

YUJI MATSUDA[1], AKIRA FUJIYAMA[1], SUSUMU KOMIYAMA[1], TAKAHITO TERASHIMA[2], KEN-ICHI SHIMURA[2], and YOSHICHIKA BANDO[2]

[1] Department of Pure and Applied Sciences, The University of Tokyo, Meguro-ku, Tokyo, 153 Japan
[2] Institute for Chemical Research, Kyoto University, Uji, Kyoto, 611 Japan

Abstract

Resistive transition of high-T_c superconductors have been studied by changing the thickness of $YBa_2Cu_3O_{7-\delta}$ down to one-unit-cell thickness. We show that superconducting transition occurs at finite temperature even in one-unit-cell thick YBCO sample. Interlayer coupling is suggested to be unimportant for achieving a T_c of 90 K. Discussed on the basic of the data in magnetic fields are collective behavior of two-dimensional vortex motion in *ab*-plane, correlation length of flux line along *c*-axis, flux flow nature near T_c and striking enhancement of H_{c1}.

Key Words: Ultrathin Film, Transport, Vortex Lattice Melting

It is interesting to clarify the role of interlayer coupling in determining T_c of high-T_c superconductors. The interlayer coupling is also important in the properties of vortex motion in the mixed state. A convenient system to study interlayer coupling is ultrathin films and artificially layered materials, in which interlayer coupling strength can be controlled.

We studied two types of samples, both produced by activated reactive evaporation method. In our samples, the crystal growth is monitored and rigorously controlled on one unit cell scale by RHEED oscillations [1]. The first type of samples is *n*-unit-cells thickness (*n*-UCT) ultrathin films. 6-UCT $PrBa_2Cu_3O_{7-\delta}$ (PrBCO) are (001) epitaxially grown on (100) surface of $SrTiO_3$. Then well defined *n*-UCT $YBa_2Cu_3O_{7-\delta}$ (YBCO) are grown on top of it. The surface of YBCO is finally covered by 6-UCT PrBCO. PrBCO is semiconducting and its resistivity is more than 500 times larger than that of YBCO below 100K. The second type of samples is a YBCO/PrBCO superlattice, where a multiple layer of 1-UCT YBCO and 1-UCT PrBCO are repeated for 20 times. The sheet resistance $R_\square$ at the onset temperature is 2600, 500, 200, 110 and 150Ω for *n*-UCT films with *n*=1,3,5,10 and superlattice(*n*=1), respectively. When these values of $R_\square$ are normalized by the sample thickness, the resistivity of the present samples is only by a factor of 3~6 times larger than the best values of bulk single crystals. This indicates surprisingly excellent quality of the sample.

Fig.1 shows thickness dependence of the resistive transition in the absence of magnetic field [2]. Three interesting features are noted: Firstly, superconducting transition occurs at finite temperature even in the 1-UCT sample. Although a 1-UCT sample contains two CuO_2 planes separated by ~3.3Å, it can be regarded as two dimensional. In purely two dimensional system, it is expected that the phase part of the order parameter $\psi=|\psi|exp(i\theta)$ diverges and superconducting transition does not occur at finite temperature. Therefore, the present two dimensional superconductivity should be described by the Kosterlitz-Thouless (KT) theory of phase transition.

Secondly, T_c^{onset} decreases as *n* decreases. This T_c^{onset} reduction has been reported by many authors through the studies of superlattices, which have similar structure to our superlattice sample. They concluded that T_c^{onset}-reduction is due to the decreases of interlayer coupling strength between conducting CuO_2 planes [3]. However, we derive a different conclusion as follows. Fig.2 shows temperature dependence of Hall number $n_H(=1/eR_H)$ for *n*-UCT samples with *n*=1,5 and 100 and superlattice (*n*=1). In *n*=5 sample, where T_c^{onset} is 85 K, Hall number is almost the same as that of the *n*=100 sample, for which T_c=90 K. However, n_H decreases in the *n*=1 superlattice with T_{conset}=78K. In the *n*=1 sample with T_c^{onset}=68 K, n_H is about an half of bulk samples with T_c=90 K. Therefore, we suppose that T_c reduction in 1-UCT samples are mainly caused by decrease of carrier number; namely, interlayer coupling is not important for T_c of 90 K.

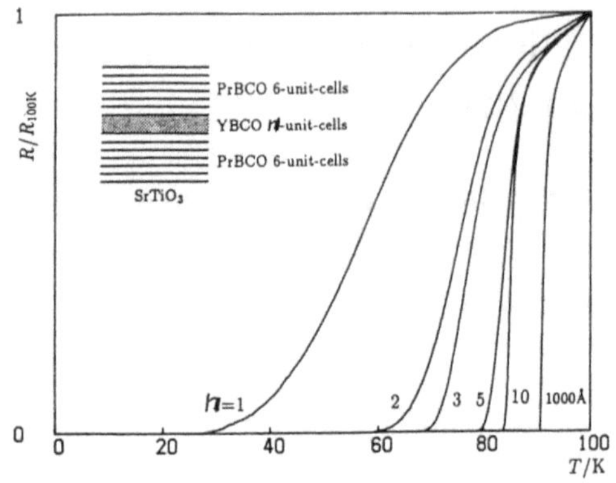

Fig.1 Thickness dependence of the resistive transition in the absence of magnetic field. n is the number of YBCO layers.

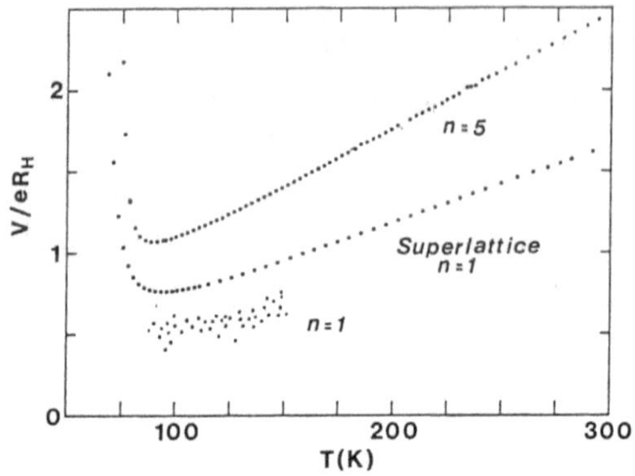

Fig.2 Temperature dependence of Hall number for n–UCT samples with n=1,5,and n=1 superlattice.

Thirdly, resistive transition becomes broad, as n decreases. There are several possible reasons for the broadening. It is possible that KT transition temperature is located far below mean field transition temperature T_c. Also, very slight inhomogeneous distribution of oxygen deficiency and/or of other local randomness would lead to a significant broadening of T_c–distribution when the system thickness is extremely small. It is not clear which mechanism is dominant in the present system.

Figs. 3a)–c) show the Arrhenius plot of the sheet resistance in magnetic field perpendicular to ab–plane for n–UCT samples of n=10,3 and 1 [4]. Instead of the n=1 sample, we show the data of the n=1 superlattice. Activation type behavior of the resistive transition is evident in all the samples in a lower temperature range. Activation energy decreases as n decreases. To obtain the activation energy from these data, we assume the relation

$$R_{\square}(T,H) = R_{\square}^{0}\exp\left(-\frac{U(T,H)}{k_B T}\right) \quad (1)$$

with T– and H–dependent activation energy $U(T,H)$, where $U(T,H)=U_0(H)(1-T/T_c(H))^{1.5}$ near T_c. Fig.4 shows the thickness dependence of the U_0 for the present samples. The activation energy U_0 increases linearly with n until n=10 is reached, beyond which U_0 is almost saturated. This indicates that the correlation length of the flux line along c–axis is approximately equal or slightly larger than 10 unit cell length in the present system; namely, the vo
rtex in YBCO is not pan-cake like below 10 tesla. These results indicate that the vortex motion in YBCO film with a thickness less than 10–UCT is purely two dimensional like.

U_0 in the n=1 superlattice is about 300 K under 10 tesla. This value is close to that in BSCCO, where anisotropy is much larger than YBCO. Therefore, the vortex motion in isolated YBCO one unit cell layer is suggested to be similar to that in each layer of BSCCO [5].

Layered superconductors are suggested to become two–dimensional when H exceeds $H_{2D}=\phi_0/\Gamma d^2$, where $\Gamma(=\sigma_{ab}/\sigma_c)$ and d are anisotropic factor and interlayer distance,respectively. We obtain $\Gamma>400$ in our superlattice, which is compared to ~50 and ~3000 in YBCO and BSCCO, respectively.

Recently, melting transition of the vortex state from a solid phase to a liquid has been discussed by many authors. We suppose that the present T–range where the activation type resistivities are observed are well above the melting transition temperature T_m because of the following reasons. Firstly, we observed Ohmic I–V characteristics in T–range where the activation type resistivities are observed. Secondly, the melting of the pure two–dimensional lattice is known to be caused by KT transition. T_m estimated from the KT melting condition is T_m~30 K for n=3 sample. Thirdly, T_m/T_c is less than 0.6 according to the mechanical oscillation experiments by Gammel et al. [7] on InO_2/In superconducting thin films with the sheet resistance of ~500 Ω, which corresponds to our n–UCT sample.

The activation type behavior in a *liquid* phase indicates that large rigid volume of two-dimensional flux bundles are coherently pinned even in the vortex liquid state. This favors models invoking large viscous liquid state with finite order or amorphus solid phase [6].

Recently, a phase separation line within the *liquid* phase has been proposed, which distinguishes between the flux-flow- and activated-regimes. Fig.5 shows the Hall resistivity ρ_{xy} normalized by H for 5-UCT sample. ρ_{xy} reverses its sign near the separation line. This effect is well known in bulk high-T_c materials. We stress the similar effect is present in the present ultrathin system [8]. The origin is not clear at present.

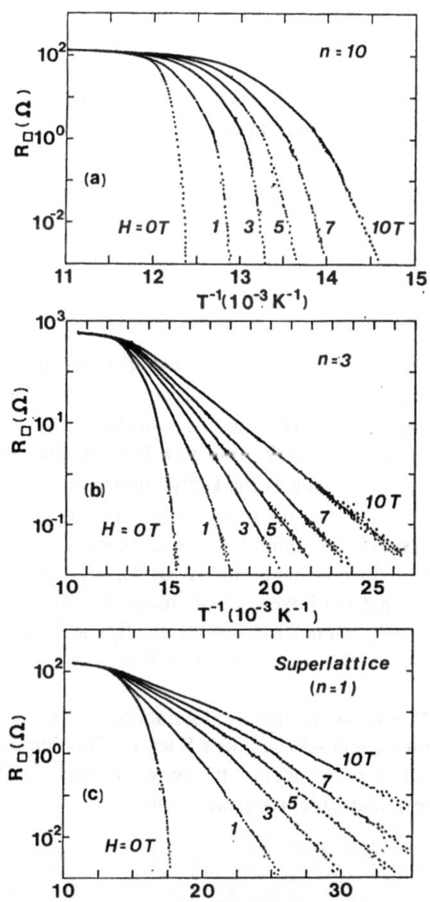

Fig.3a)–c) Arrhenius plot of the sheet resistance in magnetic fields perpendicular to *ab*-plane for *n*=10,3 and *n*=1 superlattice.

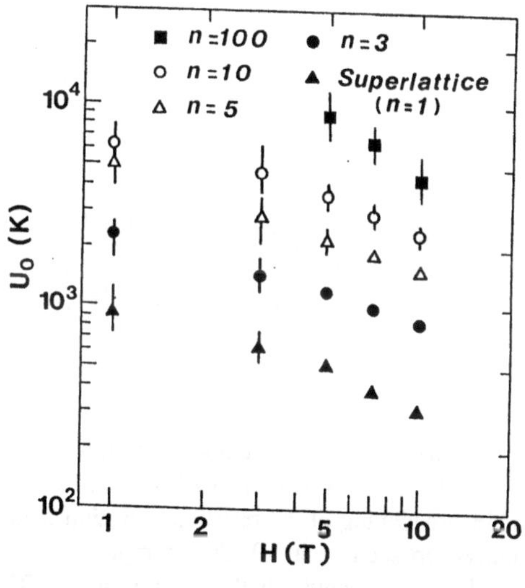

Fig.4 Thickness dependence of activation energy.

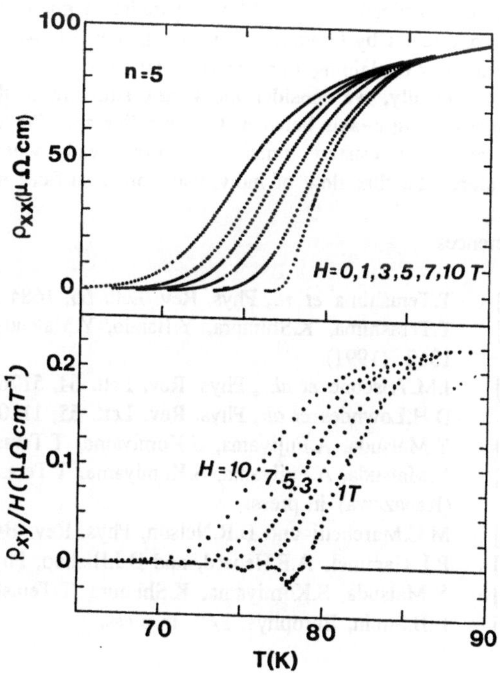

Fig.5 Temperature dependence of Hall resistivity ρ_{xy} for *n*=5 sample.

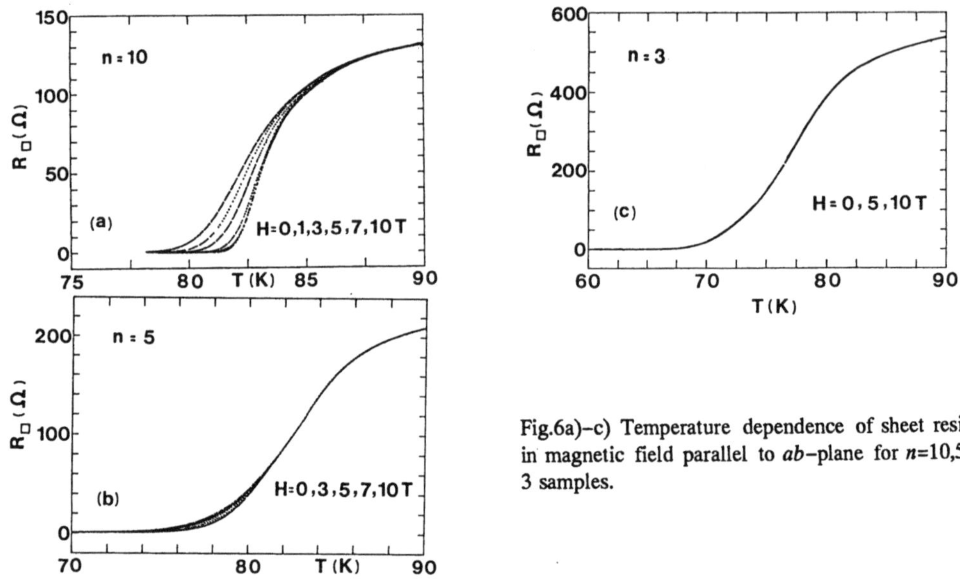

Fig.6a)–c) Temperature dependence of sheet resistance in magnetic field parallel to *ab*–plane for *n*=10,5 and 3 samples.

Figs.6a)–c) show the temperature dependence of the sheet resistance in magnetic fields parallel to *ab*–plane for *n*–UCT films with *n*=10,5 and 3. Currents are applied perpendicular to magnetic fields. The remarkable broadening of the resistive transition is known in the case of bulk YBCO in this configuration. The field induced broadening in 10–UCT sample is qualitatively similar to that in bulk sample with the magnitude of broadening being about a half of bulk samples. The field induced broadening is markedly suppressed in the *n*=5 sample. In the 3–UCT sample, even the highest magnetic field of 10 tesla does not cause any appreciable effect. These results indicates that vortices are not formed unless *n* exceed 5. In thin film superconductors where the thickness is much shorter than the magnetic penetration depth, field penetrates almost uniformly into the sample already before the vortex formation occurs. This indicates that the lower critical magnetic field H_{c1} increases drastically in thin film in parallel magnetic fields. Recently, Sonin has calculated H_{c1} in layered superconductors by Lowrence–Doniach model [9]. According to his calculation, H_{c1} is about 8 tesla in 3–UCT sample, thus explaining our experimental data.

Finally, we consider the vortex structure in the 10–UCT sample in this configuration. The mean distances along *c*–axis between two flux line are ~70 Å under 10 tesla and ~50 Å under 5 tesla. This implies that the broad resistive transition occurs already when the system accommodate one vortex along *c*–axis. Therefore, the flux flow of individual vortex suffices to cause the remarkable resistance near T_c.

References

[1] T.Terashima *et al.*, Phys. Rev. Lett. 65, 1684 (1990).

[2] T.Terashima, K.Shimura, Y.Bando, Y.Matsuda, A.Fujiyama, and S.Komiyama, Phys. Rev. Lett. 67, 1362 (1991).

[3] J.M.Triscone *et al.* , Phys. Rev. Lett. 64, 5176 (1990), Q.Li *et al.*, Phys. Rev. Lett. 64, 3086 (1990), D.H.Lowndes *et al.*, Phys. Rev. Lett. 65, 1160 (1990).

[4] Y.Matsuda, A.Fujiyama, S.Komiyama, T.Terashima, K.Shimura, and Y.Bando, a preprint.

[5] Y.Matsuda, A.Fujiyama, S.Komiyama, T.Terashima, K.Shimura, and Y.Bando, Proc. M^2S conference (Kanazawa) in press.

[6] M.C.Marchetti and D.R.Nelson, Phys. Rev. B42, 9938 (1990).

[7] P.L.Gammel, A.F.Hebard, and D.J.Bishop, Phys. Rev. Lett. 60, 144 (1988).

[8] Y.Matsuda, S.Komiyama, K.Shimura, T.Terashima, and Y.Bando, in preparation.

[9] E.B.Sonin, Europhys. Lett. in press.

Microwave Properties in Bi-Based Superconducting Thin Films

ATSUSHI TANAKA and NOBUO KAMEHARA

Fujitsu Laboratories Ltd., Atsugi, Kanagawa, 243-01 Japan

ABSTRACT

Microwave transmission loss and phase velocity of Bi-Sr-Ca-Cu-O(BSCCO) microstrip lines with a gold ground plane were measured as a function of temperature and frequency. Microstrip lines were prepared using as-deposited high-Tc phase films and patterned to a meander line 78 mm long and 200 μm wide on a 0.25-mm-thick MgO substrate. We studied their transmission properties at frequencies up to 18 GHz using a vector network analyzer. Through-reflect-line(TRL) calibration was done for accurate measurement. Loss in the BSCCO microstrip line was 1.5 dB at 10 GHz and less than that in geometrically equivalent gold lines throughout the frequency range measured. The phase velocity was slightly slower than that of gold lines at 1.18×10^{10} cm/s at 54 K and 1 GHz and decreased as temperature increased and kinetic inductance of microstrip line increased.

KEYWORDS: Bi-based superconductor, high-Tc phase, microstrip line, transmission loss, phase velocity

INTRODUCTION

A thorough characterization of the microwave response of oxide superconductors has far-reaching implications both as a probe of the superconducting mechanism and also as a means of assessing the capability of these materials to high-frequency applications. To date, most work on the microwave properties of oxide superconductors has used cavity perturbation techniques, focusing mainly on Y-Ba-Cu-O (YBCO) [1,2,3,4]. Results have also been reported for Tl-Ba-Ca-Cu-0 [5,6]. Although Bi-Sr-Ca-Cu-O (BSCCO) is generally more difficult to synthesize [7], its high-Tc phase has several potential advantages over YBCO. In addition to being somewhat more durable than YBCO, its higher Tc implies less loss due to thermally excited quasiparticles.

We have reported microwave transmission through polycrystalline BSCCO double-sided microstripline [8,9]. In this paper, we report microwave measurements on a meander microstrip line using as-deposited high-Tc phase BSCCO films. We produced samples on a thin MgO substrate and measured transmission loss and phase delay versus temperature for frequencies up to 18 GHz. We compared our results with results for geometrically identical gold microstrips.

SAMPLE PREPARATION AND CHARACTERIZATION

Since high-frequency properties of superconductors are extremely sensitive to the presence of intergrowth and impurity phases, fabrication conditions must be carefully monitored and accurately controlled to inhibit the formation of such secondary phases.
Films were prepared on 15 by 15 mm MgO substrates by RF magnetron sputtering from a composite BSCCO target (Bi:Sr:Ca:Cu = 3:1.9:2:3.5). Substrates were secured to a stainless block heated to 760°C during deposition. Deposition was 0.5 to 1 nm/min and films were about 200 nm thick. Post-deposition annealing at 850°C was to improve superconducting properties. The X-ray diffraction patterns showed the films were nearly

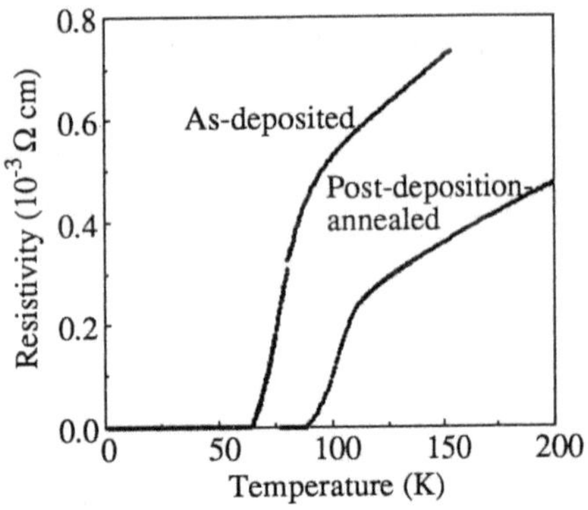

Fig. 1. Resistivity vs. temperature for as-deposited film and annealed film. Onset temperatures are 97 K and 119 K and Tc (R=0) are 64 K and 89 K.

Fig. 2. BSCCO microstrip line mounted on a measurement stage. Midsection was made of invar alloys. BSCCO meander line is connected to gold microstrip line of launcher blocks by gold ribbons.

single-phase high-Tc materials. Figure. 1 shows the dependence of the film resistivity on temperature. Tc (R=0) was 64 K before annealing and improved to 89 K after annealing. Annealed films were patterned to meander lines 78 mm long and 200 μm wide by wet etching. The temperature dependence of meander line resistance was measured by a simple two-point probe. The onset temperature was 94 K and Tc (R=0) was about 60 K.

MICROWAVE MEASUREMENTS

High-frequency transmission measurements were made for both loss and phase delay using a Hewlett-Packard HP-8510 vector network analyzer operating from 50 MHz to 18 GHz. Calibration was done using a through-reflect-line (TRL) technique at room temperature. TRL calibration has the advantage that reference planes are exactly at the ends of the meander line. This allows accurate transmission loss and phase measurements, the magnitude is unity and the phase is zero at the sample meander line input end at every frequency. The microstrip lines were mounted on a gold-plated stage used for TRL measurements and consisted of three components, a midsection and two launcher blocks (Fig. 2). The meander line was connected to the launcher block stripline using silver paint and gold ribbons. The midsection was made of invar alloys to avoid thermal expansion differences between the stage and substrate. Full two-port calibration was used for time domain measurements. The sample stage was put on a cold stage in a vacuum chamber and cooled from room temperature to 40 K by a refrigerator. Transmission properties were compared with those of geometrically equivalent 360-nm-thick gold lines on MgO substrates. The resistivity of the gold line was 3.0 μΩ cm at 300 K and 1.0 μΩ cm at 95 K.

RESULTS AND DISCUSSION

Frequency-dependent transmission loss (Fig. 3) rapidly decreased below the onset temperature. The losses at 10 GHz were 3.1 dB, 1.9 dB, and 1.5 dB at 63.8 K, 59.2 K, and 54.6 K. The dips around 6.5 GHz and 13 GHz were attributed to the resonance of the meander line because it has high impedance parts at its curvatures. The transmission loss in the meander line has very weak frequency dependence below 6 GHz. Below 59.2 K, the transmission loss did not significantly decrease.
The transmission loss in BSCCO was compared with that of geometrically equivalent gold lines. The observed BSCCO loss is less than that of gold

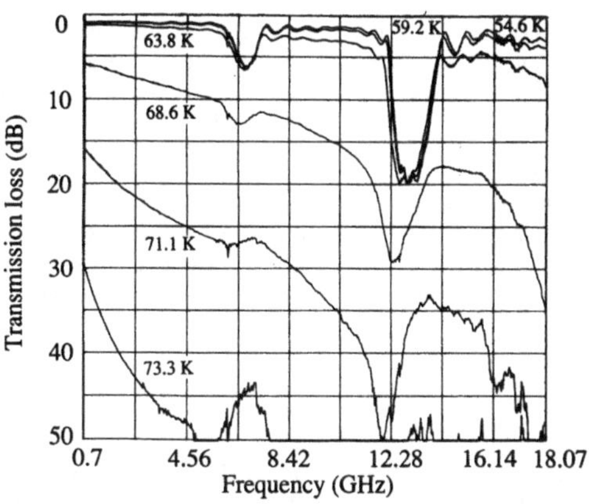

Fig. 3. Frequency dependence of transmission loss of BSCCO meander line below onset temperature. Drops around 6.5 GHz and 13 GHz are due to the line's geometrical resonance.

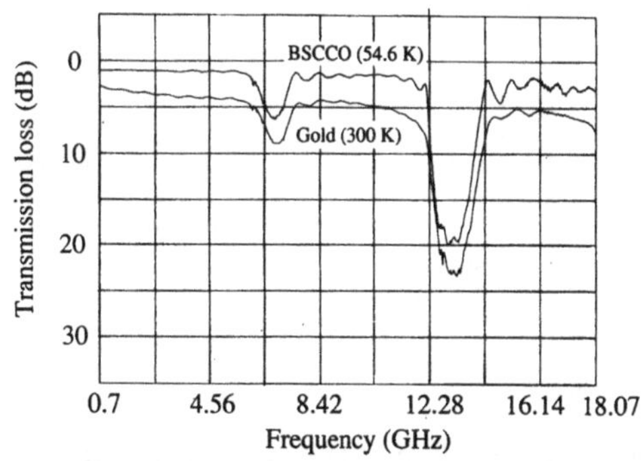

Fig. 4. Transmission loss in BSCCO meander microstrip line at 54.6 K and gold line at 300 K.

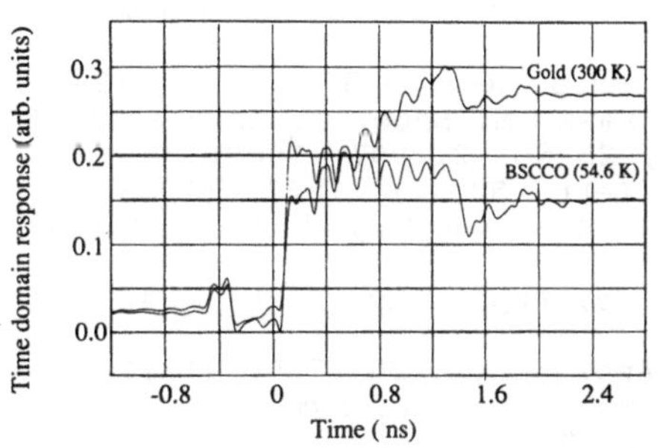

Fig. 5. Time domain response to a step function through gold and BSCCO meander lines. Several reflection peaks were observed corresponding to meander line curvatures.

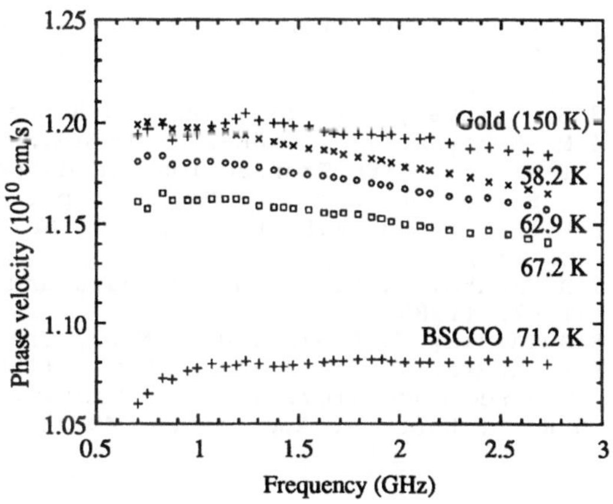

Fig. 6. Temperature dependence of phase velocity for BSCCO microstrip line. Also shown is the phase velocity for a gold line.

lines up to 18 GHz (Fig. 4).

Figure 5 shows the time domain response. Plots exhibited similar jump due to an impedance mismatch at the ends of meander lines. During signal propagation through a meander line, several reflection peaks were observed corresponding to meander line curvatures. The gold line has a DC resistance of 4.2 Ω/cm at 300 K and its time domain response has a positive gradient slope, not present in the superconducting meander line response. The averaged propagation velocities were estimated to be 1.17×10^{10} cm/s for BSCCO and 1.18×10^{10} cm/s for gold.

Phase velocity were estimated from the measured phase delay at several frequencies below 3 GHz. In contrast to the phase velocity in superconducting material, the phase velocity in normal metal is frequency-dependent because the skin depth depends on frequency. This phase velocity in superconducting material is a function of penetration depth as predicted by the two-fluid model [10]; it is constant for frequencies well below the frequency corresponding to the superconducting gap frequency.

Figure 6 shows the temperature dependence of BSCCO phase velocity compared with that of gold. Below the onset temperature, the phase velocity increased rapidly, but is slightly lower than that of gold even at 58.2 K. This is partially explained by the temperature dependence of the kinetic inductance which depends on the penetration depth in the superconducting line. Phase velocity versus frequency in the superconducting line at 58.2 K was more linear than in gold, but decreased slightly as frequency increased. As mentioned above, the measured meander line exhibited geometrical resonance at about 6.5 GHz, an effect not eliminated even at this frequency. Further study is needed using a different BSCCO meander line pattern.

SUMMARY

We fabricated nearly single-phase 110 K Bi-Sr-Ca-Cu-O meander microstrip lines on MgO substrates by RF magnetron sputtering. Their transmission loss and phase velocity were measured up to 18 GHz. BSCCO showed a lower loss than gold. The phase velocity in BSCCO is highly temperature-dependent and little frequency-dependent. This can be explained by changes of kinetic inductance, or penetration depth, in the BSCCO line. Further study is needed using a sample free from the geometrical resonance.

REFERENCES

1 J. C. Gallop, Supercond. Sci. Technom **3**, 20 (1990).
2 A.M. Awasthi, J.P Carini, B. Alavi and G. Grüner, Sol. S. Commun., **67**, 373 (1988).
3 S. Sridhar, C.A. Shiffman and H. Hamdeh, Phys. Rev. **B 36**, 2301, (1987).
4 M. Kobayashi, K. Okabe, S. Tanuma, I. Sankawa, M. Sato, T. Konaka and K. Ishihara, IEEE Transactions on Magnetics, **25** No. 2, 997 (1989).
5 J.S. Martens and J.B. Beyer, Appl. Phys. Lett. **52** (21) 1822, (1988).
6 F.A. Miranda, K.B. Bhasin, V.O. Heinen, R. Kwor and T.S. Kalkur, Physica C, 168, 91, (1990).
7 A.Tanaka, T. Machi, N. Kamehara and K. Niwa, Appl. Phys. Lett. **54**, (14) 1362, (1989).
8 A Tanaka, J. Crain and K. Niwa, Jan. J. Appl. Phys. **30** L700, (1991).
9 A Tanaka, J. Crain and K. Niwa, (1991), Proc. 3rd Int. Symp. Superconductivity. Springer-Verlag, Tokyo, 1991, pp. 1179
10 R. L. Kautz J. Appl. Phys. **49** (1), 308(1978).

Anomalous Behavior of Low Field Microwave Absorption in Ultra-Thin Bi-Sr-Ca-Cu-O Films Fabricated on MgO by MOCVD

KAZUSHI SUGAWARA, TSUNEMI SUGIMOTO, YUH SHIOHARA, and SHOJI TANAKA

Superconductivity Research Laboratory, ISTEC, Koto-ku, Tokyo, 135 Japan

ABSTRACT

Non-resonant microwave absorption (NRMA) has been studied on a 100Å thick $Bi_2Sr_2Ca_2Cu_3O_y$ film fabricated on an MgO(100) substrate by MOCVD. Temperature dependence of the absorption shape, linewidth and angular dependence between the applied magnetic field and the film surface have been studied.Three different kinds of signals (S_1, S_2, S_3) have been observed below T_c. These signals become maximum at around 55K, 40K and 30K, respectively. The absorption linewidths of S1 and S2 signals increase at temperatures close to T_c ($\simeq$65K) and 50K, respectively, a possible indication of vortex interaction.

KEY WORDS: non-resonant microwave absorption, Bi-system superconductor, MOCVD thin film

INTRODUCTION

Non-resonant microwave absorption (NRMA) studies have been done recently on Bi-Sr-Ca-Cu-O film (350Å thick) fabricated on MgO(100) by MOCVD. The intensity and linewidth of the NRMA depend on temperature and orientation of the sample with respect to the applied magnetic field[1]. The effects of NRMA are usually treated in terms of weakly coupled superconducting domains, and can be used for the characterization of superconducting properties such as the micro-Josephson loop size[2]. In this study, we have done NRMA studies on an ultra-thin(100Å thick) film fabricated on a single-crystal substrate of MgO(100) , and found novel phenomena in the NRMA behavior.

EXPERIMENTAL

The film was fabricated by the MOCVD method. The details of film formation are described elsewhere[3]. Triphenyl bismuth $Bi(C_6H_5)_3$ and dipivaloylmethanates (DPM) of $Sr(DPM)_2$, $Ca(DPM)_2$,and $Cu(DPM)_2$ were used as MO gas sources. The Bi-Sr-Ca-Cu-O film was grown on an MgO substrate with size 10x10x0.5mm^3 at 850°C in flowing mixed gases (MO gases and oxygen). The growth rate of the film was about 3Å/min. The thickness of the film was measured by SEM and ICP analysis. The latter was also used for the analysis of chemical composition of the film. The thickness is about 100Å and composition is $Bi_2Sr_2Ca_2Cu_3O_y$.

X-ray diffraction (using CuK_α radiation) confirmed that the film contains mostly high-T_c phase with a small amount of low-T_c phase. The resistivity studies revealed that the superconducting onset,T_{on}, and zero-resistance temperatures were 110K and 65K, respectively. For the NRMA study, a sample with size 3x4x0.5 mm^3 was prepared by cutting the original as-fabricated substrate. The sample was inserted into a standard ESR quartz sample tube with ID=3mm and length=30 cm. A small amount of DPPH (2,1-diphenyl-2-picryl-hydrazyl) was put at the bottom of the sample tube. A conventional ESR spectrometer, JEOL RE3, operating at about 9 GHz was employed. A cylindrical cavity with TE_{011} mode was used. The modulation amplitude and microwave power were 0.2 Gauss and 1 mW, respectively.

RESULTS AND DISCUSSION

Typical NRMA signals (dP/dH) at several temperatures are depicted in Figs.1 and 2, in which magnetic fields were applied perpendicularly to the sample surface. Hereafter, the angle between applied magnetic field and film

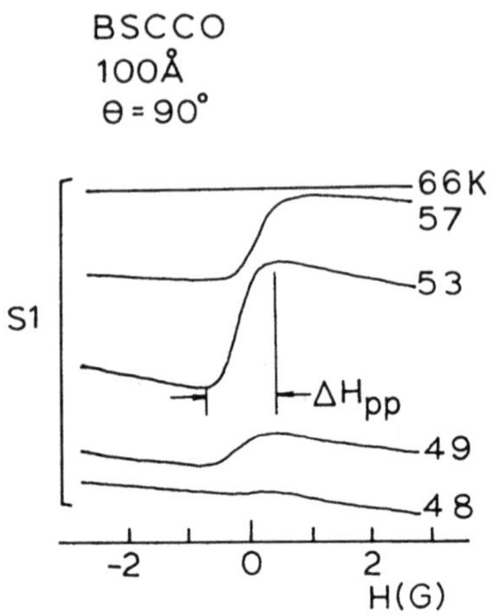

Fig.1. NRMA signals observed at several temperatures.

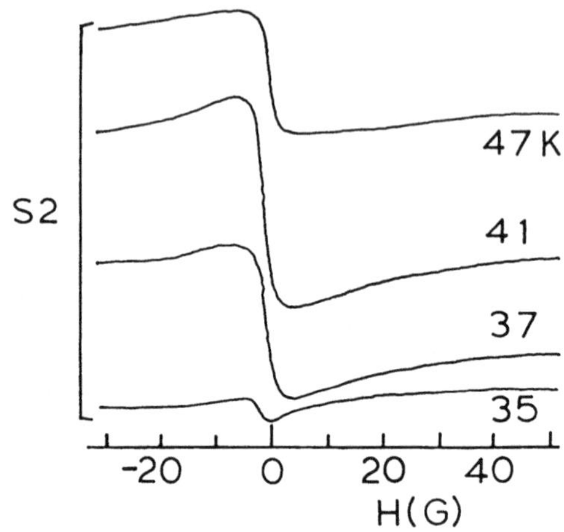

Fig.2. NRMA signals observed at several temperatures.

surface is defined as θ. In Figs.1 and 2, it should be noted that the phase of dP/dH of the NRMA signal changes at around 48K. The NRMA signals shown in Figs.1 and 2 are denoted by S1 and S2, respectively. At temperatures below about 34.5 K, NRMA signals (S3) with the same phase as S1 were observed. However, these S3 signals depend strongly on the sweeping direction of the applied magnetic field. On the contrary, the S1 and S2 signals exhibit very weak hysteresis. Thus, in the following, only S1 and S2 signals are discussed.

The NRMA intensity,I, of the S1 and S2 signals are depicted in Fig.3 as a function of temperature. The S1 signal gradually disappears with decreasing temperature and the S2 signal appears. At temperatures near 49-51K, both S1 and S2 signals exist. The peak-to-peak linewidth, ΔH_{pp} is defined in Fig.1 for the S1 signal. ΔH_{pp} is similarly defined for S2 signals. $\Delta H_{pp}/2$ gives a

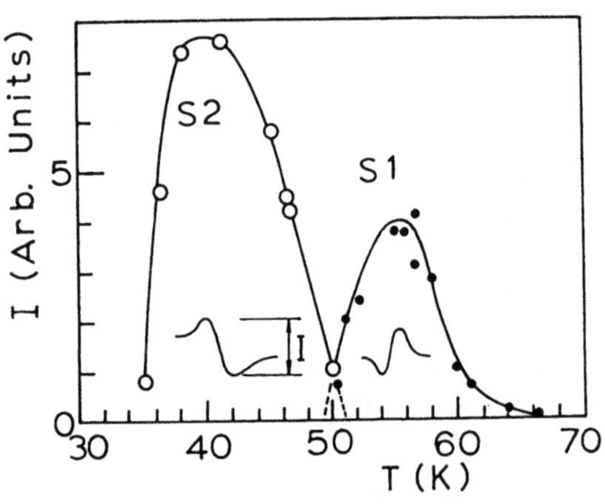

Fig.3. Temperature dependence of NRMA intensity of S1 and S2 signals.

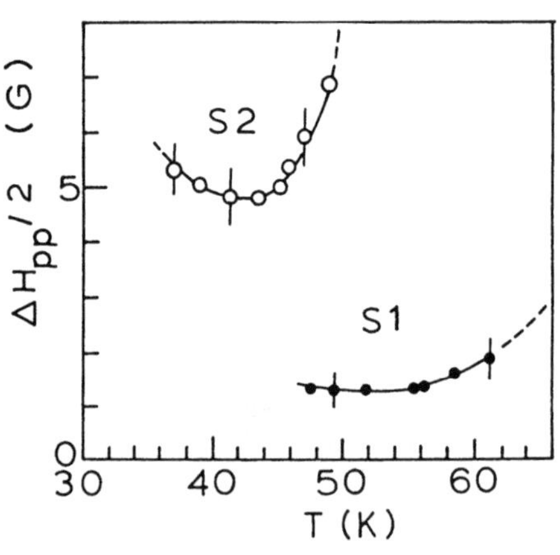

Fig.4. Temperature dependence of $\Delta H_{pp}/2$ of S1 and S2 signals.

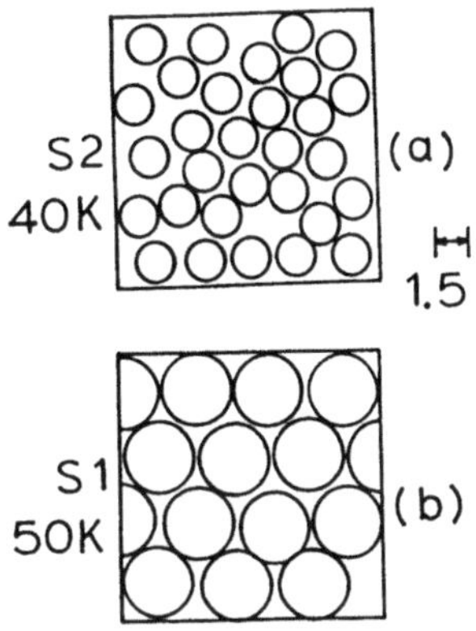

Fig.5. Qualitative distribution of vortices of S2 and S1 signals at 40K and 50K, respectively.

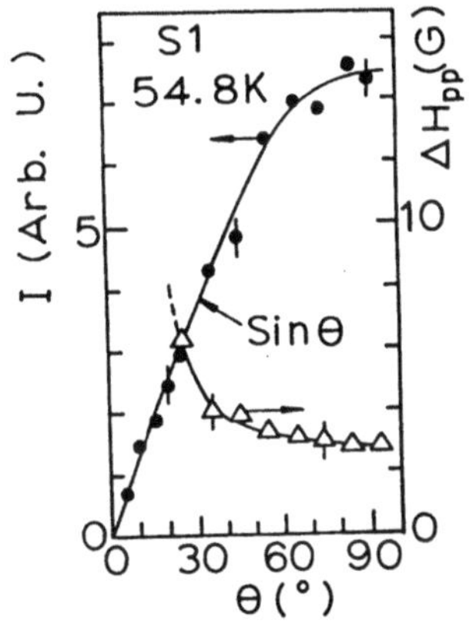

Fig.6. Angular (θ) dependence of NRMA intensity I and ΔH_{pp}.

measure of the lower critical field H_{c1}^* of a type II superconductor. The average size of a micro-Josephson loop can be estimated from H_{c1}^* (or from $\Delta H_{pp}/2$). In Fig.4, $\Delta H_{pp}/2$ is plotted against temperature. At temperatures near the mixed region ($\simeq$50K), a substantial broadening in $\Delta H_{pp}/2$ is found for the S2 signals. For example, $\Delta H_{pp}/2$ is greater than about 20 Gauss at 50.1K. In Fig.5, the vortices corresponding to S1 and S2 signals are qualitatively represented.

The intensity,I, of NRMA is dependent on the direction of applied magnetic field. A typical example is shown in Fig.6 for the S1 signal at 54.8K. As can be seen in Fig.6, experimental data (solid circles) follows Sinθ. Similar θ dependence was also observed for the S2 signals. The linewidth, ΔH_{pp} is also θ-dependent as shown in Fig.6 (see the triangles).

We have also done NRMA studies on several MOCVD Bi-Sr-Ca-Cu-O films with different thicknesses. From these studies we have found that NRMA signals like S2 were often observed for thinner films less than about 350Å. On the contrary, for thicker films, no clear signal like S2 was observed. As mentioned before, the phase of S2 is opposite to that of the S1 signals. The S2 signals may be due to anti-vortices (or anti-fluxons). The physical origin of the co-existence of these different vortices is not well understood presently.

Finally the micro-Josephson loop sizes of the S1 and S2 signals are estimated. The average area,S, of the micro-Josephson loops is given by $S= \phi_0/2(\Delta H_{pp}/2)$ (cm^2), where ϕ_0 is an elementary flux quantum of $2x10^{-7}$ Gauss-cm^2. The approximate radii determined from the S1 and S2 signals at about 49K are about 1.5 microns and 0.7 microns, respectively.

ACKNOWLEDGEMENTS

This work is supported by the New Energy and Industrial Development Organization under the management of the R&D of Basic Technology for Future Industries.

REFERENCES

1. K. Sugawara, T. Sugimoto, K. Yasuike, D.J. Baar, Y. Shiohara, and S. Tanaka, (1991) Modern Physics Letters B5:667.
2. K.W. Blazey, K.A. Müller, J.G. Bednorz, W. Berlinger, G. Amoretti, E. Buluggiu, A. Vera,and F.C. Matacotta, 81987) Phys. Rev. B36:7241.
3. T. Sugimoto, M. Yoshida, K. Yamaguchi, K. Sugawara, Y. Shiohara and S. Tanaka, (1990) Appl. Phys. Lett. 57:928.

Enhanced Anisotropy of Resistive Transition and Critical Current Density in Bi-Sr-Ca-Cu-O Thin Films

SHIGENORI YUHYA, KOICHI NAKAO, DAVID J. BAAR*, TSUNEMI SUGIMOTO, and YUH SHIOHARA

Superconductivity Research Laboratory, ISTEC, Koto-ku, Tokyo, 135 Japan

ABSTRACT

Resistivity and critical current density of Bi-Sr-Ca-Cu-O films of thickness 5 to 300 nm prepared by MOCVD have been investigated. All films displayed strong anisotropy for magnetic fields applied parallel and perpendicular to the film surface. This strong anisotropy is thought to result from the effects of the thinness, as well as from the intrinsic anisotropy of the Bi-based superconductors. By calculation of the free energy in the limit of high κ, the effective lower critical field for the case of a magnetic field parallel to the film was shown to increase above that for a bulk sample substantially with decreasing the sample thickness. In addition, it was shown that it was difficult for a flux line to move through the film because of the surface barrier.

KEY WORDS: Bi system, thin film, anisotropy, H_{c1}, surface barrier

INTRODUCTION

High-T_c superconductors are well known to show a pronounced broadening of resistive transition and a strong anisotropy in resistivity, critical current, and critical field in a magnetic field.[1] Recently, it had been reported that superlattices and isolated ultra thin films of YBCO with thickness of from one to a few unit cells in thickness had been prepared by physical vapor deposition and the broadening of resistive transition in the magnetic field applied parallel to the film surface had been small.[2]

We have succeeded in preparation of a Bi-based ultra-thin film of unit cell thickness with high transition temperature by MOCVD.[3] In this report, we present the superconducting properties for a film of near-unit cell thickness, and also those of thicker films and discuss the enhanced anisotropy by the film thinness.

EXPERIMENTAL

Bi-based superconducting films were grown on (100) MgO substrates using a susceptor temperature of 800 °C (detailed growth procedures have been described previously[3]). ICP analysis showed that the relative proportions of Bi:Sr:Ca:Cu were almost 2:2:2:3 for every sample thickness we studied. X-ray diffraction patterns showed that all samples were mixed phase containing both 2212 and 2223, and were highly oriented with the c-axis normal to the substrate.

We have performed electrical transport measurement on the samples using a conventional DC four terminal method with a semi-rotating sample holder in a superconducting magnet. Magnetic fields up to 6 Tesla (T) were applied parallel (B//f) and perpendicular (B⊥f), to the film surface. For ρ-T measurement, the measuring current density was about 500 A/cm^2 and the criterion for zero resistance temperature (T_{c0}) was a resistivity of 10^{-8} Ω·cm. For J_c measurement, the critical current was taken at an electric field of a 10 μV/cm. The current was always perpendicular to the magnetic field.

*presently, University of British Columbia, Canada

RESULTS AND DISCUSSION

Figure 1 shows the ρ-T curves of 10 nm sample for a range of magnetic fields. In the case of B⊥f, a large broadening of the transition was observed. However in the case of B//f, the ρ-T curves scarcely changed in the applied field, and the degradation of T_{c0} was less than 1 K in fields up to 6 T. These results were almost the same for the other samples. The degradation of T_{c0} due to a magnetic field of 6 T applied parallel to the film surface for each sample were 0.8, 0.7, 2.4 and 2.8 K for films with thicknesses of 5, 10, 100 and 300 nm, respectively.

Figure 2 shows the J_c-B properties of 10 nm sample and Fig. 3 shows the J_c-B properties at 50 K for all samples. J_c was nearly independent of magnetic fields up to 6 T in the parallel field while it was decreased significantly in the perpendicular field.

The anisotropy observed here is much more extreme than that for Bi-based single crystals.[4] In particular, the insensitiveness of T_c and J_c on magnetic fields applied perpendicular to the c-axis (parallel to the film surface) is different from the observations for single crystals.

We suspect that this difference originates from a unique character of thin films[5] in addition to the intrinsic anisotropy of Bi-based superconductors. For the type-II superconductors, the relation between the internal magnetic field (B) and the lattice spacing of the flux lattice (a_0) is $B \propto a_0^{-2}$, and a_0 is about 50 and 20 nm at the internal magnetic fields of 1 and 5 T, respectively, if the fluxons form an Abrikosov lattice. In the case of magnetic fields of 5 T applied parallel to the film surface, it is expected that a flux line cannot exist in the ultra-thin films of thickness less than 20 nm. In other words, it is estimated that the effective lower critical field parallel to the film ($H_{c1}^{//}$) increases above that for bulk sample with decreasing the film thickness.

A.A.Abrikosov had calculated the Gibbs free energy for an isolated flux line at the center of a thin film taking into account of the existence of surfaces by introducing a infinite array of image vortices.[6] The change of the Gibbs free energy (ΔG) due to the entry of a flux line is expressed by,

$$\Delta G = \frac{\Phi_0}{8\pi}\left[2H_{ex}\left\{\frac{1}{\cosh(d/2\lambda)}-1\right\}+\frac{\Phi_0}{2\pi\lambda^2}\left\{\sum_{n=-\infty}^{\infty} K_0\left(\frac{|2d\cdot n|}{\lambda}\right)-\sum_{m=-\infty}^{\infty} K_0\left(\frac{|2d\cdot m+d|}{\lambda}\right)\right\}\right],$$

$$\because n=0 \Rightarrow K_0\left(\frac{|2d\cdot n|}{\lambda}\right)=\ln\kappa$$

where Φ_0 is a flux quantum, λ is the magnetic penetration depth, d is the film thickness, K_0 is the 0-th order modified Bessel function and H_{ex} is the external magnetic field applied parallel to the superconducting area. $H_{c1}^{//}$ is determined by ΔG=0. Figure 4 shows the calculated $H_{c1}^{//}$ normalized by the value of bulk case (H_{c1}) as a function of d/λ. $H_{c1}^{//}$ increased above H_{c1} substantially when d/λ is less than 10, and is nearly proportional to d^{-2} when the d/λ is less than 3 and κ is larger than 100.

The calculation for ΔG is easily extended to the case where a flux line is located at a general position. Figure 5 shows the dependences of ΔG on the position of the flux line (x/d) for the case of $d=2\lambda$, κ=100 and $H_{ex}=n\cdot H_{c1}$ (n=1 to 5). In this figure, ΔG is equal 0 for n=2.7 and x=0. The open circle in Fig.4 corresponds to this case. A large potential barrier can be seen near the film surface when ΔG is negative for a flux line located at x=0. Accordingly, it is expected that a flux line needs a large driving force to go through the film surface and to induce a voltage drop.

For the case of the existence of an transport current, the form of ΔG in fig.5 inclines because of the induced self-field. A revised form of ΔG is calculated and expressed by,

$$\Delta G = \frac{\Phi 0}{8\pi}\left[2\left(H_{ex} + \frac{H_i}{2}\right)\left\{\frac{\cosh(x/\lambda)}{\cosh(d/2\lambda)} - 1\right\} + H_i\left\{1 - \frac{\sinh(x/\lambda)}{\sinh(d/2\lambda)}\right\}\right.$$
$$\left. + \frac{\Phi 0}{2\pi\lambda^2}\left\{\sum_{n=-\infty}^{\infty} K0\left(\frac{|2d\cdot n - x|}{\lambda}\right) - \sum_{m=-\infty}^{\infty} K0\left(\frac{|2d\cdot m + d - x|}{\lambda}\right)\right\}\right],$$

where $H_i=(4\pi/c)\cdot I$, and I is the total transport current in the CGS gauss unit.[7] It can be shown that for a film with λ=1500Å and $d=2\lambda$=3000Å, for example, a transport current as large as 10^7 A/cm^2 shifts the stable position of flux line only by 10% of the film thickness from the center under a external field of $H_{ex}=4\cdot H_{c1}$. This current density is much larger than that of the observed J_c. In our samples, it has been shown that J_c is determined by the weak link behavior in the result for the temperature dependence of J_c[8]. Accordingly, it is considered that the insensitiveness of Jc on magnetic fields applied parallel to the film surface is due to the large effective $H_{c1}{}^{/\!/}$ and the surface barrier although the absolute value of Jc is limited to the weak link behavior.

These considerations are qualitatively consistent with our experimental results. For the ultra-thin films of 5 and 10 nm thickness, the observed degradation of T_{c0} in magnetic fields applied parallel to the film surface are probably due to misalignment between the magnetic field and film surface and/or the intrinsic misorientation in the films due to misfit dislocations at the substrate.[9]

CONCLUSION

In summary, we observed strong anisotropy of the resistive transition and critical current density in superconducting BSCCO thin films as compared with that of single crystals. It was considered to originate from a unique character for the thin films. Our experimental results were qualitatively consistent with the calculated estimation that the effective lower critical field in a parallel magnetic field to the film surface increase above that for the bulk case and the potential barrier exist near the film surface (surface barrier).

ACKNOWLEDGMENT

This work was supported by the New Energy and Industrial Technology Development Organization (NEDO) under the management of the R&D of basic Technology for Future Industries.

REFERENCES

[1] for example: A.P.Malozemoff, T.K.Worthington, Y.Yeshrun, F.Holtzberg, and P.H.Kes, Phys.Rev. B38, 7203 (1988)
[2] for example: J.-M.Triscone, Ø.Fischer, O.Brunner, L.Antognazza, A.D.Kent, and M.G.Karkut, Phys.Rev.Lett. 64, 804 (1990)
[3] T.Sugimoto, M.Yoshida, Y.Shiohara, and S.Tanaka, Appl.Phys.Lett. 58, 1103 (1991)
[4] T.T.M.Palstra, B.Batlogg, R.B.Van Dover, L.F.Schneemeyer, and J.V.Waszczak, Phys.Rev. B41, 6621 (1990)
[5] to be published in Appl.Phys.Lett., S.Yuhya, K.Nakao, T.Sugimoto, D.J.Baar, Y.Shiohara and S.Tanaka
[6] A.A.Abrikosov,"Fundamentals of the Theory of Metal", 18 (North-Holland, 1988)
[7] K.Nakao, unpublished
[8] submitted in J.A.P., D.J.Baar, S.Yuhya, T.Sugimoto, K.Sugawara, Y.Shiohara, S.Fujita,

and S.Tanaka
[9] T.Sugimoto, M.Yoshida, K.Yamaguchi, Y.Yamada, K.Sugawara, Y.Shiohara, and S.Tanaka, J.Crystal Growth 107, 692 (1991)

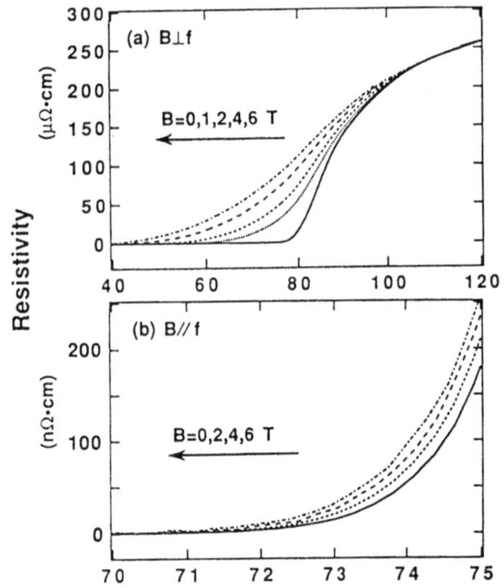

Fig.1 ρ-T curves of 10 nm thickness sample: a) magnetic field perpendicular to the film surface and b) field parallel to the film (expanded near T_{c0})

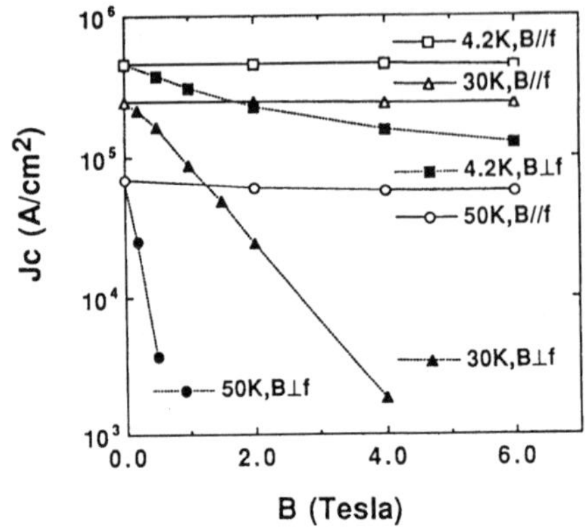

Fig.2 J_c-B properties of 10 nm thickness sample: solid lines show the case of magnetic fields applied parallel to the film surface and dotted lines show the case applied perpendicular to the film

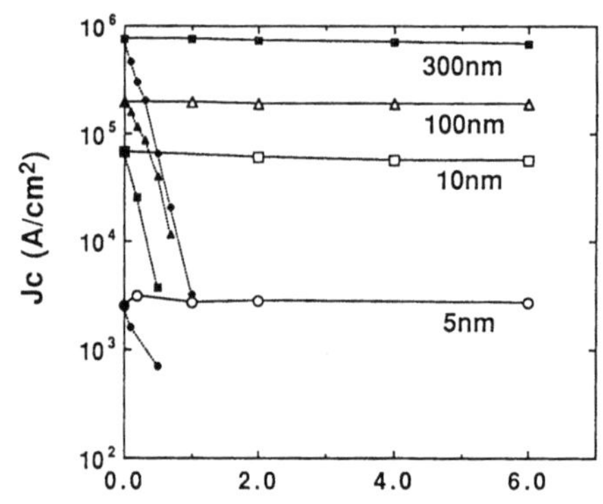

Fig.3 J_c-B properties at 50 K for all samples: solid lines show the case of magnetic fields applied parallel to the film surface and dotted lines show the case applied perpendicular to the film

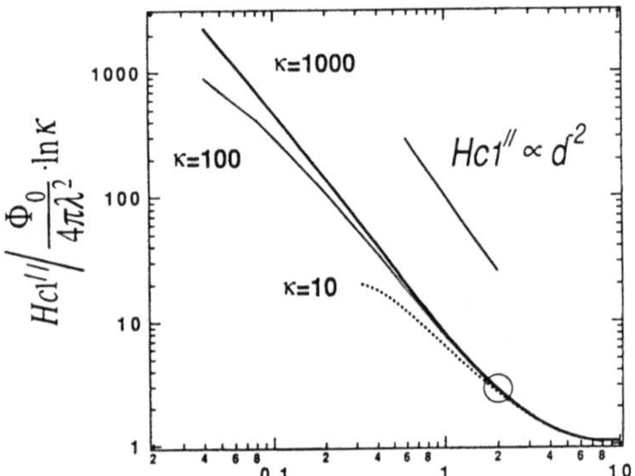

Fig.4 Calculated H_{c1}'' as a function of d/λ normalized to the bulk value

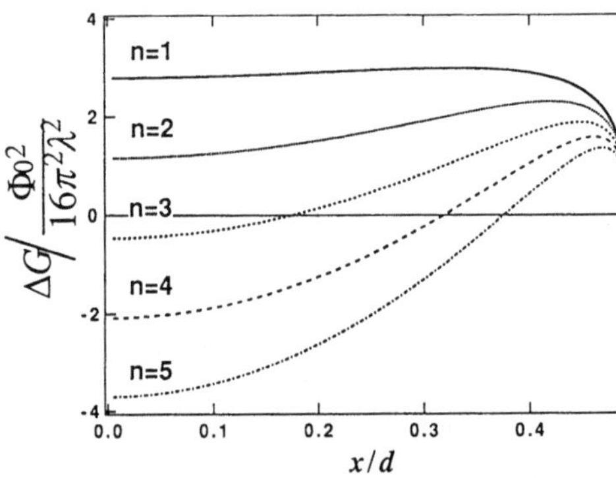

Fig.5 Calculated ΔG as a function of x/d in the case of $d=2\lambda$, $\kappa=100$ and H_{ex} of n times as much as the H_{c1} of the bulk case

LT-SEM Studies of Superconducting Bi-Sr-Ca-Cu-O Film

CHIKAKO YOSHIDA, TOMOJI NAKAMURA, SHIGEO KODAMA, HIDEKI YAMAWAKI, and MASARU IHARA

Fujitsu Laboratories Ltd., Atsugi, Kanagawa, 243-01, Japan

ABSTRACT

Electron beam induced voltage images, which directly indicate the distribution of the critical current density, are observed in the superconducting strips of $Y_1Ba_2Cu_3O_{7-\delta}$ and Bi-Sr-Ca-Cu-O films by Low Temperature Scanning Electron Microscopy (LT-SEM). Band patterns appear in several places within the strip of $Y_1B_2C_3O_{7-\delta}$ whereas the vague homogeneous voltage contrast extends throughout the strip of Bi-Sr-Ca-Cu-O. Voltage images of 90° misoriented grain-boundaries yield dark contrast which is considered to arise from semiconductor-like behavior of the grain boundary resistivity.

KEY WORDS: $Y_1B_2C_3O_{7-\delta}$, Bi-Sr-Ca-Cu-O, LT-SEM, grain-boundary

INTRODUCTION

Recently, Low Temperature Scanning Electron Microscopy (LT-SEM) has been used to observe the critical current density for $Y_1Ba_2Cu_3O_{7-\delta}$ films[1,2] and $Y_1Ba_2Cu_3O_{7-\delta}$ bicrystals[3]. In the LT-SEM measurement, a superconducting film is mounted on a cold stage of a SEM and is kept at a certain constant temperature below T_c and biased with a current close to the critical current. During the scanning process, the beam irradiation causes a small local thermal perturbation at the coordinate point (x,y) where the beam is focussed[1,2], and an elctron beam induced voltage signal $\Delta V(x,y)$ is generated at this point. Since the voltage signal $\Delta V(x,y)$ is related to the local critical current density $J_c(x,y)$ and local critical temperature $T_c(x,y)$, the two dimensional voltage images directly display information on the local homogeneity of the critical current density and temperature of the sample.

In previous work, we have characterized the size and orientation of grain-boundaries for Bi-Sr-Ca-Cu-O films and have investigated the influence of grain-boundaries on the temperature dependence of the electrical resistivity[4]. It is the purpose of the present paper to investigate the influence of grain-boundaries on superconducting properties of both $Y_1Ba_2Cu_3O_{7-\delta}$ and Bi-Sr-Ca-Cu-O films.

EXPERIMENTAL PROCEDURE

The c-axis oriented $Y_1Ba_2Cu_3O_{7-\delta}$ films with thickness of 2000 Å were deposited by off-axis sputtering[5] on (001)$LaAlO_3$. The c-axis oriented Bi-Sr-Ca-Cu-O films (by X-ray diffraction these films contain a the mixture of high-T_c(2223) and low-T_c(2212) phases) with thickness of 2500 Å were grown on (001)MgO substrates using halide open-tube CVD[6]. For LT-SEM measurements, the films are fabricated into narrow strips with dimensions of 10 μm × 200 μm using standard photoresist lithography and dry etching.

In order to detect small voltage signals, ΔV~1 nV, the electron beam is pulsed at a frequency of 110 kHz by a beam-blanking unit and beam-induced ac voltage signal is measured using a lock-in amplifier. In our system, a Germanium thermometer which is fixed at the sample holder is used to monitor the temperature from ~20 K to 100 K with an accuracy of about 0.1 K.

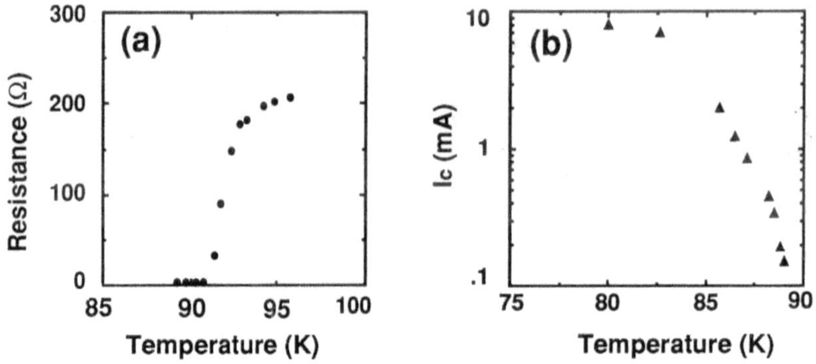

Fig. 1 (a) Temperature dependence of electrical resistance and (b) critical currentof the line of YBa2Cu3O7-δ film.

RESULT AND DISCUSSION

The temperature dependence of resistance for the $Y_1Ba_2Cu_3O_{7-\delta}$ strip is shown in Fig.1(a): a sharp transition occurs just above 90 K. The critical current, which is determined from the *I-V* curves, is shown Fig.1(b). The backscattered electron image for this specimen is shown in Fig.2. From the surface morphology, a high density of projections can be observed throughout the strip and any existing grain-boundaries cannot be recognized. The ECP measurements indicate that the a- and b-axis directions of this film is are <100> in the film plain.

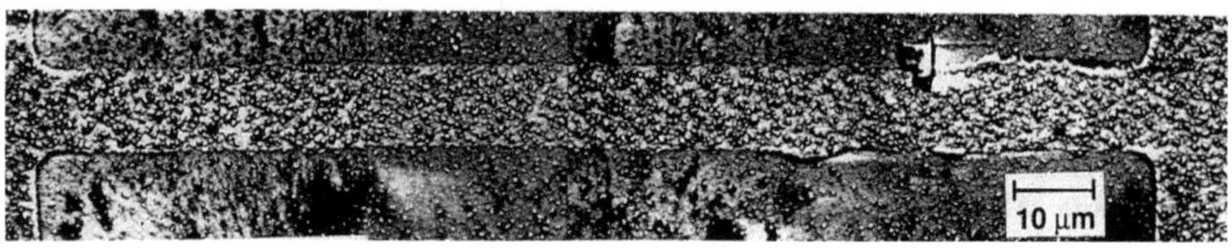

Fig.2 The back-scattered electron image for the $Y_1Ba_2Cu_3O_{7-\delta}$ strip.

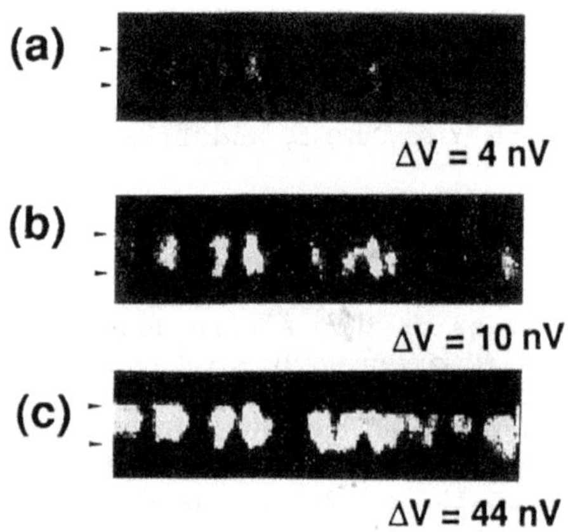

Fig.3 Voltage images of the $Y_1Ba_2Cu_3O_{7-\delta}$ strip at *T*=87.5 K at the following bias currents: (a)0.75,(b)1.12, and (c)1.51 mA. Electron beam parameters are 20 keV and 3 nA.

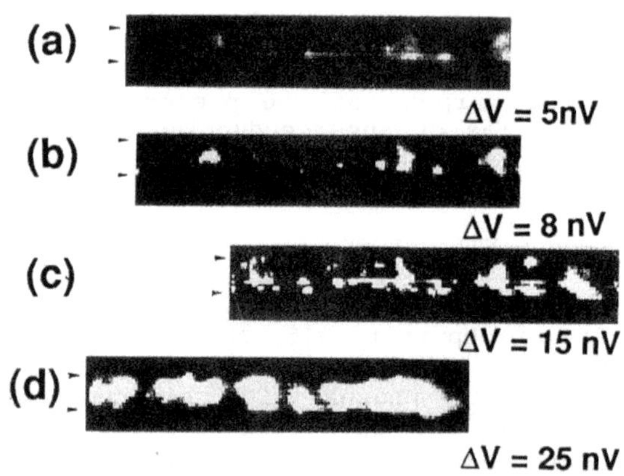

Fig.4 Voltage images of the $Y_1Ba_2Cu_3O_{7-\delta}$ strip at *T*=89.5 K at the following bias currents: (a)0.034,(b)0.067, (c)0.10, and (c)0.13 mA. Electron beam parameters are 20 keV and 3 nA.

Figures.3(a),(b),(c) show two-dimensional electron beam induced voltage images at different bias-currents I_B=0.75, 1.12, 1.51 mA at T=87.5 K. Here the bright regions correspond to a large voltage signal. The arrows in these figures mark the film boundaries. According to Fig.1(b), the critical current of this strip at this temperature is about 0.75 mA. When $I_B \approx I_c$(Fig.3(a)), the voltage contrast appears as small bright spots corresponding to the weak areas where superconductivity is easily destroyed. As the bias current increases, the bright spots change into the bands and extend across the whole strip. The maximum value of voltage signal is also increases (ΔV=4, 10, 44 nV) with increasing I_B. As the temperature is increased, voltage images can be observed at smaller bias-currents. As shown in Fig.4(a-d), images similar to those in Fig.3(a-c) can be observed at T=89.5 K at I_B=0.034, 0.067, 0.10, 0.13 mA, respectively. According to Fig.1(b), $I_c \lesssim 0.1$ mA at this temperature. These images directly indicate the inhomogeneous nature of critical current density along the strip. The origin of this inhomogeneity, however, could not be determined, despite careful analysis of the SEM images.

To investigate the influence of grain-boundaries, a Bi-Sr-Ca-Cu-O film which consists of large grains with dimensions of about 50 µm, was examined. The secondary electron image and its schematic illustration are shown in Fig.5. The arrows in the illustration indicate the a-axis direction of the grain and the gray curves show 90° misoriented grain-boundaries. By using a polarizing optical microscope together with ECP and TEM techniques, we can determine the orientation of the Bi-Sr-Ca-Cu-O film[7]. The temperature dependence of electrical resistance is shown in Fig.6 : a wide transition occurs between 100 K and 75 K.

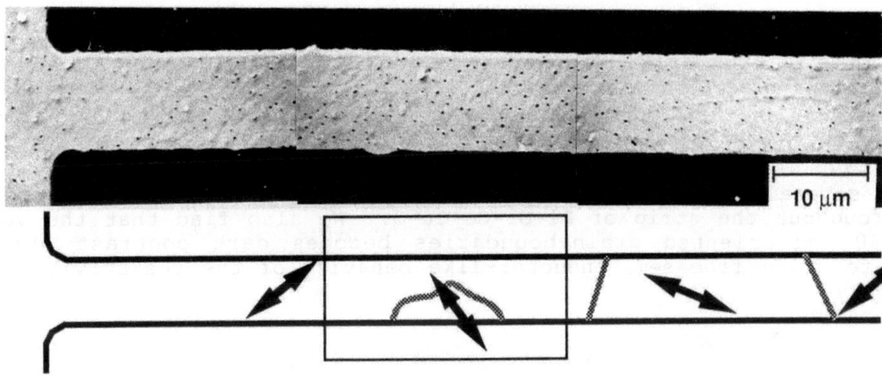

Fig.5 The secondary electron image for Bi-Sr-Ca-Cu-O strip and its schematic illustration. The square in this figure corresponds the area which is observed by LT-SEM.

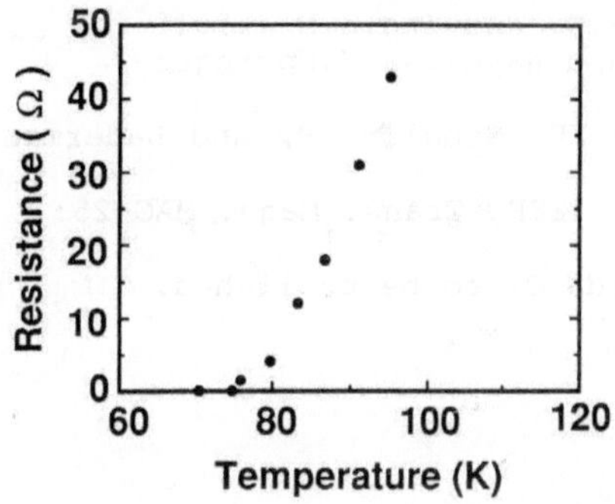

Fig.6 The temperature dependence of electrical resistance of Bi-Sr-Ca-Cu-O film.

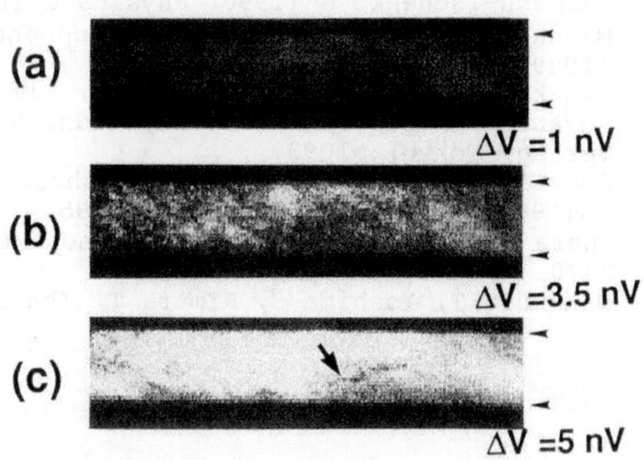

Fig.7 Voltage images of the Bi-Sr-Ca-Cu-O strip at T=76.3 K at the following bias currents: (a)0.12,(b)0.19, and (c)0.26 mA. Electron beam parameters are 20 keV and 580 pA.

As shown in Fig.7(a),(b),(c), the voltage images for specimen are taken at T=76.3 K and I_B=0.12, 0.19, 0.26 mA. In contrast to the $Y_1Ba_2Cu_3O_{7-\delta}$ case, the voltage contrast appears vague throughout the strip. The voltage signal increases with increasing I_B. Let us concentrate on Fig.7(c), at the region of dark contrast indicated by the arrow. Comparing with Fig.5, this region is seen to correspond to a 90° misoriented grain-boundary.

To explain this contrast, we propose that the temperature dependence of the electrical resistance of the 90° misoriented grain boundary is semiconductor-like. As shown in Fig.8, the resistivity for such a grain boundary increases with increasing temperature while that of the superconducting area drops sharply. The electron beam induced thermal heating ΔT causes the change of resistance, and hence, a change of voltage signal $-\Delta V'$ for the grain boundary and ΔV for the superconducting area. Consequently, the voltage image of the superconducting area becomes brighter and that of grain boundary becomes darker.

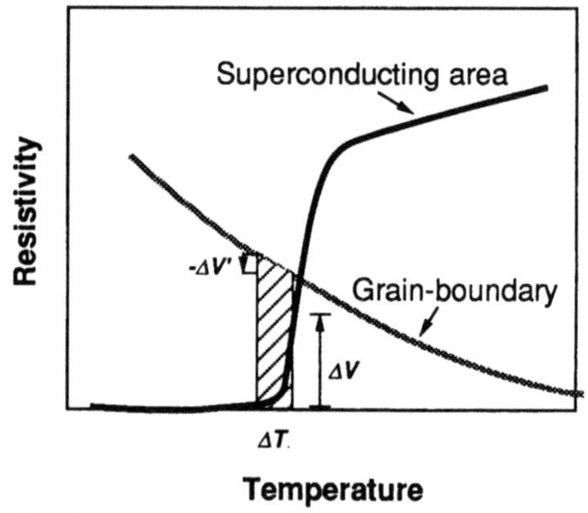

Fig.8 The temperature dependence of resistivities.

In summary, electron beam induced voltage images, which directly indicate the distribution of the critical current density, have been observed in superconducting strips of $Y_1Ba_2Cu_3O_{7-\delta}$ and Bi-Sr-Ca-Cu-O films by LT-SEM. We show that the bright contrast (the voltage signal) increases within the strips as bias current increases or temperature approaches T_C. We find a major difference between voltage images for the two films: in $Y_1Ba_2Cu_3O_{7-\delta}$ the voltage contrast appears as several bright bands, whereas vague homogeneous voltage contrast extends throughout the strip of Bi-Sr-Ca-Cu-O. We also find that the voltage image of 90° misoriented grain-boundaries becomes dark contrast which is considered to arise from semiconductor-like behavior of the resistivity of the grain boundary.

REFERENCES

1. Gross R, Hipler K, Mannhart J, Huebener RP, Chaudhari P, Dimos D, Tsuei CC, Schubert J, and Poppe U(1989) Appl. Phys. Lett. 55: 2132.
2 Kolle D, Kober F, Hartmann M, Gross R, Huebener RP, Roas B, Schultz L, and Saemann-Ischenko G (1990) Physica C 167: 79.
3. Mannhart J, Gross R, Hipler K, Huebener RP, Tsuei CC, Dimos D, and Chudhari P (1989) Science 245: 839.
4. Tanaka C, Nakamura T, Yamawaki H, Kimura T, Ueda O, and Ihara M (1991) Advances in Superconductivity, eds. Kajimura K and Hayakawa H(Springer-Verlag, Tokyo) p1093.
5 Eom CB, Sun JZ, Yamamoto K, Marshall AF, Luther KE, Geballe TH, and Laderman SS(1989) Appl. Phys. Lett. 55: 595.
6. Ihara M, Kimura T, Yamawaki H, and Ikeda K(1989) IEEE. Trans. Magn. MAG-25: 2470.
7 Nakamura T, Yoshida C, Kimura T, Ihara M, and Ueda O, to be published.

6.3 Other Superconducting Films

Molecular Beam Epitaxy of $Ba_{1-x}K_xBiO_3$ Films and Heterostructures

E.S. HELLMAN[1], E.H. HARTFORD, JR.[1], S. MARTIN[1], D.J. WERDER[1], G.M. ROESLER, JR.[2], and P.M. TEDROW[2]

[1] AT&T Bell Laboratories, Murray Hill, NJ 07974, USA
[2] Francis Bitter National Magnet Laboratory, Massachusetts Institute of Technology, Cambridge, MA 02139, USA

ABSTRACT

$Ba_{1-x}K_xBiO_3$ with x≈0.4 is a very attractive material for superconducting tunnel junctions. We briefly describe the molecular beam epitaxy growth technique as applied to $Ba_{1-x}K_xBiO_3$. After a high temperature nucleation step, (1 0 0) oriented $Ba_{1-x}K_xBiO_3$ films with good superconducting properties can be grown on MgO at quite low substrate temperatures (200-300°C). The films are oxygen deficient during growth, and the films must be annealed in oxygen to optimize superconducting properties. Oxidation induced shrinkage can result in cracking. We have used a new insulating phase, $BaBi_2O_y$, along with $Ba_{1-x}K_xBiO_3$ to build superlattices and sandwich junction heterostructures. We briefly discuss the electrical properties of these heterostructures.

KEYWORDS: BKBO; MBE; thin films; tunneling

INTRODUCTION

The cubic bismuthates, $Ba_{1-x}K_xBiO_3$(BKBO) and $Ba_{1-x}Rb_xBiO_3$(BRBO), are the highest temperature superconductors (30K) apart from the layered cuprates.[1, 2] The long (≈50Å), isotropic coherence length of these simple perovskites compared to the layered cuprates makes them potentially of technological interest for superconducting electronics. The recent observation of conductance with the Bardeen-Cooper-Schrieffer form in $Ba_{1-x}K_xBiO_3$/gold tunnel junctions[3-5] and near-ideal Josephson-junction behavior in $Ba_{1-x}K_xBiO_3$/$Ba_{1-x}K_xBiO_3$ point contacts[6] has made clear the desirability of all thin-film Josephson structures in this material system. Realization of such a structure requires a technology for producing consistently high quality superconducting films and which can in addition produce a good insulating overlayer capable of supporting epitaxial overgrowth of superconducting BKBO.

In this paper, we summarize the molecular beam epitaxy (MBE) growth technique as applied to BKBO by focusing on the techniques used to grow heterostructures and then briefly discuss the electrical properties of the heterostructures which have been made.

MBE GROWTH

MBE of $Ba_{1-x}K_xBiO_3$ and $Ba_{1-x}Rb_xBiO_3$ is relatively straightforward. The potassium source is K_2O, which liberates potassium vapor upon heating. The K_2O was synthesized by heating potassium superoxide to 300°C in a sealed quartz ampule with excess potassium metal. The resulting brown powder is put in a magnesia crucible fitted to a platinum shielded effusion cell. We find that the choice of alkali metal (K vs. Rb) does not appear to have a significant effect on either the film growth or the film properties. Substrates, mostly MgO but also $SrTiO_3$, are soldered onto a 3" block with indium. Substrate rotation during growth provides good compositional uniformity. Improved MgO substrate surfaces have been obtained by furnace annealing the substrates in oxygen at 900°C overnight and by exposing them to an oxygen plasma prior to growth.

The relationship between substrate temperature and oxidizing ambient for optimum growth of BKBO is similar to that found for growth of $YBa_2Cu_3O_{7-x}$ epitaxial films.[7] We use an oxygen plasma source [8] using powers between 15 and 18 W and a source pressure of 80 mTorr. The oxygen pressure at the substrate is about $1x10^{-4}$ Torr during growth. The choice of substrate temperature is crucial to obtaining superconducting films. At temperatures which are too high the potassium fraction of the film tends to be too low, even for excess potassium flux during growth. A range of growth temperatures and pressures have been reported in the literature for growth of $Ba_{1-x}(K,Rb)_xBiO_3$ films by various methods.[9-14] Figure 1 shows a "Hammond-Borrman" plot of the reported conditions for film growth showing

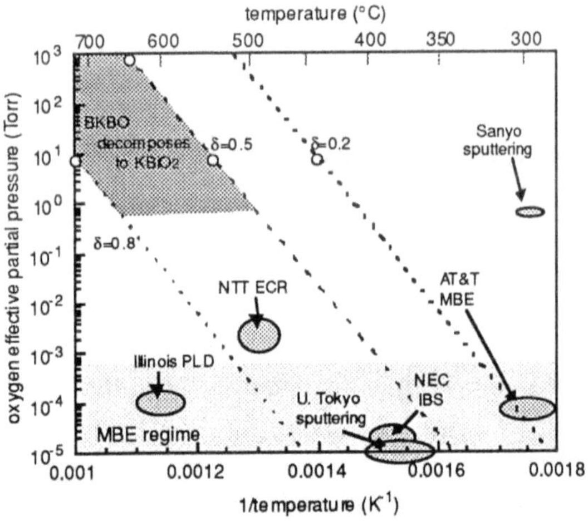

Figure 1. Effective oxygen partial pressure versus temperature diagram showing extrapolated oxygen deficiencies δ for (x=0.5) and growth conditions for optimal *in situ* growth in the $Ba_{1-x}K_xBiO_{3-\delta}$ system. The composition lines here are extrapolated from the 4 data points shown, which are from Pei et. al.[15]. Growth conditions for B(K,R)BO from the literature are plotted.

extrapolated oxygen deficiencies δ. Most of the reported growth conditions should result in oxygen deficient films during growth, based on the oxygen deficiencies extrapolated from bulk studies.[15] Note that the film synthesis conditions have been four orders of magnitude lower in oxygen pressure than any of the bulk synthesis conditions.

Successful fabrication of a heterostructure, particularly a tunneling structure with ultra-thin barrier layers, demands single orientation epitaxial growth. Although the lattice mismatch between MgO and superconducting $Ba_{1-x}K_xBiO_3$ or $Ba_{1-x}Rb_xBiO_3$ is only 1.4% for the (1 0 0) orientation, it is the (1 1 0) orientation which grows on MgO for substrate temperatures below 350°C.[16] Two different in-plane orientations occur, and superconducting films with this microstructure do not have high critical currents. We have found that we can induce single orientation growth at lower temperature by first nucleating the film without alkali metal at high temperature, 450-550°C. This nucleation layer is single orientation and exhibits very streaky RHEED patterns with Kikuchi lines. It is relatively insensitive to variations in the Ba and Bi fluxes, and is thus very reproducible. RBS and x-ray diffraction measurements of films grown at this high temperature indicate that these nucleation layers consist of a new phase[17] with chemical formula $Ba_{1-x'}M_xBi_2O_y$ (where M= K or Rb, and x<0.2). A determination of the unit cell for one oriented $Ba_{1-x'}Rb_xBi_2O_y$ thin film with x=0.19 by x-ray diffraction suggested a tetragonal cell with a=17.32Å and c=22.30Å. This was verified by TEM. A full determination of the structure awaits the bulk synthesis of this phase. Figure 2 shows a cross-section transmission electron micrograph of a $Ba_{1-x}K_xBiO_3$ epitaxial film grown using a high temperature nucleation layer. The nucleation/buffer layer is about 450Å thick. After a nucleation layer is grown, the substrate temperature can be lowered to as low as 200°C while still obtaining single orientation epitaxial growth. This low temperature synthesis may prove useful for integration with semiconductors; indeed, we have demonstrated growth of non-epitaxial superconducting BRBO films on both GaAs and Si.[18]

Figure 2 also shows one of the structural problems of epitaxial $Ba_{1-x}K_xBiO_3$ films, cracking. A crack is seen to extend from the buffer layer through the $Ba_{1-x}K_xBiO_3$ up to a third layer. We have observed cracking in both $Ba_{1-x}K_xBiO_3$ and $Ba_{1-x}Rb_xBiO_3$ films of good crystalline quality with doping fraction x greater than 0.3. The cracking is due to oxygen deficiency of the as-grown films; when they take up oxygen, either on cooling or during a post-growth oxygen anneal, they shrink. The lattice parameters of $Ba_{1-x}K_xBiO_{3-\delta}$ have been observed to shrink by 0.34%. for a change in δ from 0.25 to 0.18.[19] This should be compared to the thermal expansion strain of less than 0.008% for films on MgO cooled from 350°C to room temperature. In general, the films with the best crystalline quality have cracked most severely. To reduce the cracking, we have followed three strategies: lower temperature growth (to reduce oxygen deficiency during growth), thinner films (to reduce the total strain) and buffer layers. We have found that once nucleated, $Ba_{1-x}K_xBiO_3$ growth can be sustained at as low as 200°C. The buffer layer approach was suggested by the observation that cracks often occurred only in the lower layer of sandwich junction heterostructures, as seen in Figure 2.

HETEROSTRUCTURES

Once crack-free single-orientation epitaxial $Ba_{1-x}K_xBiO_3$ films can be grown reproducibly, it would be desirable to be able to fashion them into useful devices. The Josephson junction is one possible device, but is one which puts stringent demands on the growth process. To obtain a supercurrent in one of these devices requires that regions of superconducting material must be separated by an insulating material within a length comparable to the superconducting coherence length. For a BKBO based heterostructure junction, this requirement means that good BKBO growth must be established within a few monolayers of the insulating material. At the same time, the barrier material must be sufficiently insulating and pinhole free that leakage currents do not degrade the junction characteristics. Since $BaBi_2O_y$ is insulating and proved to be a good nucleation layer, it was an obvious candidate for a barrier material. The a-b plane of $BaBi_2O_y$ is a good lattice match to BKBO (a/4=4.33Å vs. a=4.28 for x=0.4). Since the barium/bismuth ratio in $BaBi_2O_y$ is quite close to that in superconducting BKBO, superconductor/insulator heterostructures can be conveniently made by opening and closing the potassium shutter. We demonstrated this technique by fabricating superlattices with 80Å periods.[17] Clear third order superlattice diffraction peaks were observed both in the as-grown sample and the same sample after a 375°C anneal in oxygen.

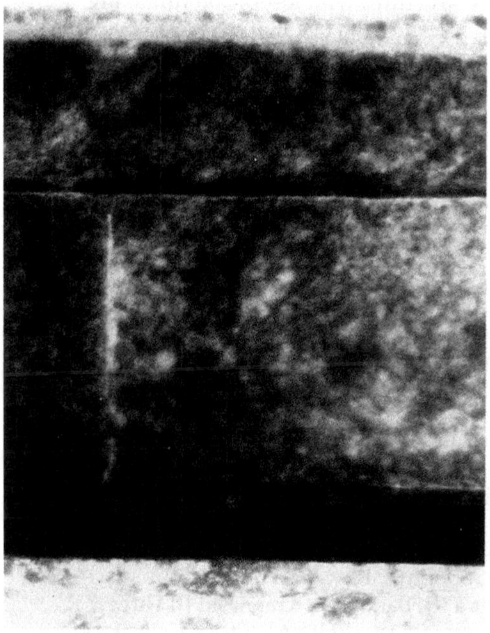

Fig. 2. Cross-section transmission electron micrograph of a $Ba_{0.5}K_{0.5}BiO_3/BaBi_2O_y$ epitaxial sandwich-junction heterostructure grown on MgO using a high temperature nucleation layer. The nucleation/buffer $BaBi_2O_y$ layer is about 450Å thick, the bottom $Ba_{0.5}K_{0.5}BiO_3$ layer is 1700Å and the $BaBi_2O_y$ barrier layer is about 75Å. The top $Ba_{0.5}K_{0.5}BiO_3$ layer has been partly eroded by the sample preparation. A crack is visible on the left.

SANDWICH JUNCTIONS

Using the techniques described above, superconductor–insulator–superconductor structures with excellent crystallinity can be grown. Studies of their electrical properties are the first step on the road to heterostructure Josephson junctions. Figure 3 shows I-V curves from one such structure with a ~40Å barrier. The junctions are fabricated using conventional photolithographic techniques. The sandwich structure is Ar ion milled into stripes of BKBO. Mesas are then milled on the stripes to define the junction areas. Silver contacts to both top and bottom layers are then evaporated and patterned by lift-off, and the chip is annealed at 375°C for two hours in oxygen. Areas of photoresist are baked on at 200°C to prevent shorting on junction edges, and a second layer of silver is used to connect the top layer contacts to wire bonding pads. Junctions with sizes between 20x20μm and 80x80μm were tested in a four point measurement to eliminate the effects of the contact resistance. The data shown in Fig. 3 is from an 80x80μm junction. A 20x20μm junction on the same chip had junction resistance roughly 80 times higher, with similar characteristics. Since the I-V curves remain non-linear above the T_c of the film (20K) it is clear that this is **not** superconducting pair or quasiparticle tunneling. (There may have been ephemeral indications of quasiparticle tunneling which disappeared early in the measurement.) The curves can all be fit well by the expression $I = I_o \sinh([V/V_o]^{\alpha})$ where α is 0.3±0.1. Such an expression is appropriate for conduction by variable-range-hopping, which is characteristic of non-metallic $Ba_{1-x}K_xBiO_3$[20]. Thus, it is possible that the conduction in fig. 3 is limited by a layer of non-superconducting $Ba_{1-x}K_xBiO_3$ adjacent to the $BaBi_2O_y$ barrier layer. Improved growth procedures may be effective in eliminating such a dead layer.

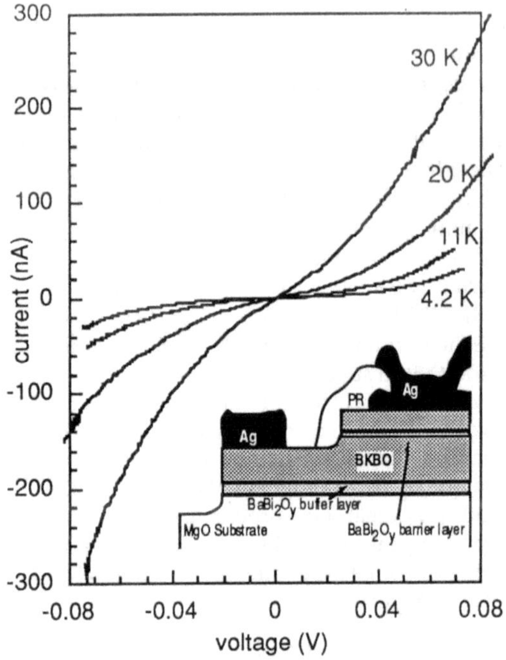

Fig. 3. Current-voltage characteristics for a BKBO/$BaBi_2O_y$/BKBO sandwich junction heterostructure The junction area is 80x80μm. Note that the curves remain non-linear above the T_c of the BKBO films. The inset shows a schematic of the device structure.

CONCLUSIONS

In this paper we have briefly discussed the progress and problems in the area of epitaxial $Ba_{1-x}K_xBiO_3$ films. While we have not yet made heterostructures with Josephson tunneling characteristics, the fact that we have been able to make 40Å barrier layers with high resistances for junctions as large as 80x80μm is encouraging in itself. We remain confident that Josephson junction structures in this system have great potential as a technology for superconducting electronics operating at 20K.

ACKNOWLEDGEMENTS

The authors would like to acknowledge T. A. Fulton for making the masks and F. Sharifi for helpful advice. Part of this work was supported by the Consortium for Superconducting Electronics.

REFERENCES

1. Mattheiss LF, Gyorgy EM, Johnson DW Jr. (1988) Phys. Rev. B 37: 3745-6
2. Cava RJ, Batlogg B, Krajewski JJ, Farrow R, Rupp Jr. LW, White AE, Short K, Peck WF, Kometani T (1988) Nature 332: 814-6
3. Sharifi F, Pargellis AN, Dynes RC, Miller B, Hellman ES, Rosamilia JM, Hartford EH (1991) unpublished.
4. Sato H, Takagi H, Uchida S (1990) Physica C 169: 391
5. Huang Q, Zasadzinski JF, Gray KE, Richards DR, Hinks DG (1990) Appl. Phys. Lett. 57: 2356-2358
6. Pargellis AN, Sharifi F, Dynes RC, Miller B, Hellman ES, Rosamilia JM, Hartford EH (1991) Appl. Phys. Lett. 58: 95-96
7. Hammond RH, Bormann R (1989) Physica C 162-164: 703-704
8. Spah RJ, Hess HF, Störmer HL, White AE, Short KT (1988) Appl. Phys. Lett 53: 441-3
9. Sato H, Tajima S, Takagi H, Uchida S (1989) Nature 338: 241-3
10. Enomoto Y, Murakami T, Moriwaki K (1989) Jpn. J. Appl. Phys. 28: L1355-7
11. Hellman ES, Hartford EH (1990) J. Vac. Sci. Technol. 8: 332-335
12. Moon BM, Platt CE, Schweinfurth RA, VanHarlingen DJ (1991) Appl. Phys. Lett. 59: 1905-1907
13. Iyori M, Kamino M, Takahashi K, Yoshisato Y, Nakano S to be published in Physica C
14. Fujita J, Yoshitake T, Satoh T, Igarashi H, Takeuchi I, Miura S, Tsai JS, Tsuge H to be published in Physica C
15. Pei SY, Jorgensen JD, Hinks DG, Zheng Y, Richards DR, Dabrowski B, Mitchell AW submitted to J. Sol. St. Chem.
16. Hellman ES, Hartford EH, Fleming RM (1989) Appl. Phys. Lett. 55: 2120-2122
17. Hellman ES, Hartford EH Jr., Fleming RM, Marsh P, Werder DJ to be published in Physica C
18. Hellman ES, Hartford EH, Brandle CD, Berkstresser GW, Jenssen HP, Cassanho A, Gabbe D to be published in MRS Proc.
19. Ueki K, Tokiwa A, Kikuchi M, Suzuki T, Nagoshi M, Suzuki R, Kobayashi N, Syono Y (1989) In: Ishiguro T, Kajimura K (eds) Advances in Superconductivity II: Proceedings of the 2nd International Symposium on Superconductivity (ISS'89), November 14-17, 1989, Tsukuba. Springer-Verlag, Tokyo, pp 489-492
20. Hellman ES, Hartford EH, Miller B, Baldwin KW, Rosamilia JM to be published in Phys. Rev. B

Formation of $(Ba,Rb)BiO_3$ Superconducting Thin Films by Molecular Beam Epitaxy

M. OGIHARA, F. TODA, R. KAWASAKI, T. YAMADA, and H. ABE

Research and Development Group, Oki Electric Industry Co., Ltd., Hachioji, Tokyo, 193 Japan

ABSTRACT

(Ba,Rb)BiO3 (BRBO) superconducting thin films were prepared on MgO(100) substrates by molecular beam epitaxy using distilled ozone at the substrate temperature of~370°C. The BRBO superconducting thin film of ~1000Å with good crystalline is mainly c-axis oriented and superconducting transition was obtained at 21K. In order to examine interdiffusion of the elements between the semiconductor substrate and the oxide superconducting thin film, the BRBO thin film was preliminarily grown on the Si (100) substrate with Ag-overlayer. It is concluded that interdiffusion of Si into the BRBO thin film is small.

KEY WORDS: (Ba,Rb)BiO3 superconductor

1 INTRODUCTION

Fabrication of good interface between the oxide-superconductor and the semiconductor is important technique for application of the oxide-superconductor. In order to form the abrupt interface between the oxide-superconductor and the semiconductor, the growth temperature of the superconducting thin film should be as low as possible. It was reported that the (Ba,Rb)BiO3 (BRBO) superconducting thin film was prepared at the substrate temperature lower than ~400°C by Hellman et al. [1]. Therefore the BRBO is one of the promissing materials to prevent interdiffusion of the elements at the interface between the semiconductor and the oxide-superconductor. In this paper, we report preparation of the BRBO superconducting thin film on the MgO(100) substrate by molecular beam epitaxy (MBE). It is also reported that the BRBO thin film was formed preliminarily on Si(100) with Ag-overlayer to examine interdiffusion of Si into the BRBO thin film.

2 EXPERIMENT

2.1(Ba,Rb)BiO3 thin film

It is important for formation of good BRBO superconducting thin film to determine the optimum growth conditions for the BiBaO3 (BBO) thin films[2]. In this respect, good BBO thin film was prepared on the MgO(100) substrate at first and next the BRBO superconducting thin films were formed by MBE using distilled ozone. Bi, Ba and Rb2O sources were used. Especially we paied special attention to control Rb2O source stably. The BBO thin films were synthesized at various substrate temperature (Ts) and ozone pressure (PO3) in order to determine the optimum conditions for good crystalline. The BRBO superconducting thin film was prepared by adjusting the Rb2O beam flux intensity. The BRBO superconducting thin film was fabricated at Ts~370°C. The substrate temperature was calibrated at the melting point of Bi (~271°C) and In (~157°C). The conditions for formation of the BRBO superconducting thin films on the MgO(100) substrates was listed in table 1. The X-ray diffraction pattern (XRD) of the BBO thin films and the BRBO thin films were measured to check the quality of the crystalline and its orientation. The composition ratio of the thin films were measured by energy dispersive x-ray analysis (EDX) and Rutherford back scattering (RBS) to adjust the beam flux intensities. The resistance-temperature curves were measured to evaluate superconducting transition temperature. Surface morphology of the thin films were also observed by scanning electron microscope (SEM).

2.2(Ba,Rb)BiO3 thin film on Si with Ag-overlayer

In order to examine preliminarily interdiffusion of the elements at the interface between the BRBO layer and the semiconductor substrate, the BRBO thin film was formed on the Si substrate although lattice mismatch between Si (d=5.431Å) and BRBO (d~4.30Å) is very large. Cleaning of the n-type Si (100) was carried out following the standard cleaning method [3]. After cleaning the Si(100) substrate, no contamination on the substrate surface was observed by auger electron spectroscopy (AES)[4] . About 2 monolayers of Ag was deposited on the cleaned Si substrate. On the Si(100) substrate with Ag-overlayer (Ag/Si) [5] , the BRBO thin film was formed at Ts~370°C under the same growth conditions for the BRBO superconducting thin film on MgO(100) substrate. The film thickness is 1000~1500 Å. The XRD pattern of the BRBO thin film on Ag/Si (BRBO/Ag/Si) was measured and surface morphology of BRBO/Ag/Si was also observed. In order to investigate interdiffusion between BRBO and Si, the depth profile of the composition ratio of BRBO/Ag/Si was measured by AES with Ar sputtering.

Table 1

Sources
Bi 500°C
Ba 551°C
Rb2O 550°C
Substrate
MgO(100)
Ts~370°C
Deposition rate ~20Å/min
Film Thickness ~500Å

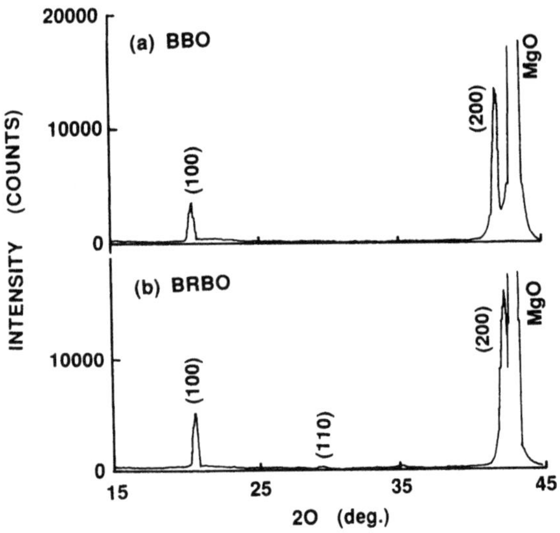

Fig.1 XRD pattern of the BRBO thin film of ~500Å ((a)) and the BRBO superconducting thin film of ~ 1000Å ((b)) .

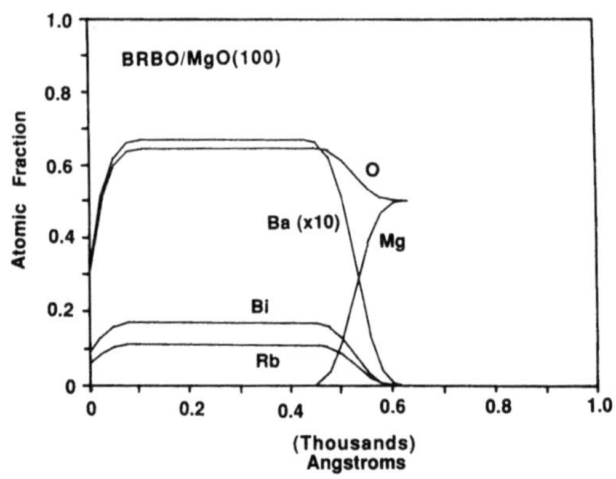

Fig. 2 Depth profile of composition ratio of the BRBO thin film grown on MgO(100) measured by RBS.

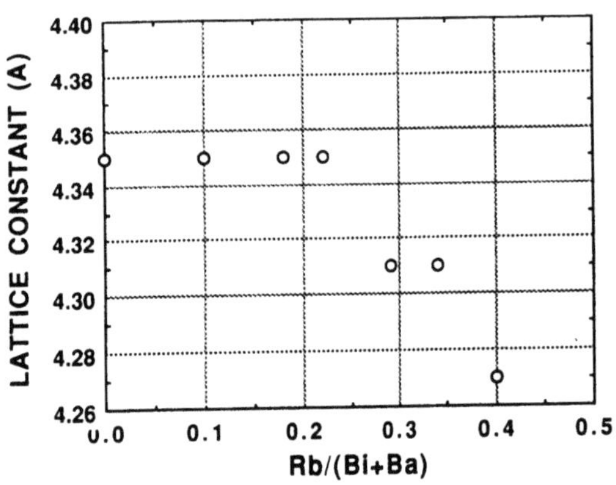

Fig. 3 Lattice constant of the BRBO thin film deduced from the XRD pattern as a function of Rb-composition ratio.

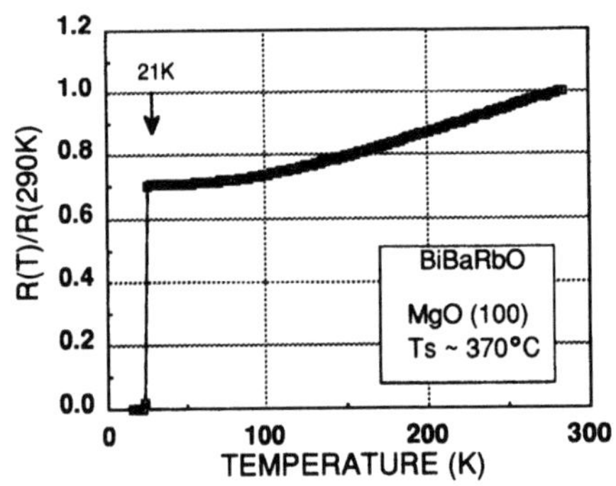

Fig. 4 R(T) curve of the BRBO superconducting thin film formed on MgO (100) at Ts~370°C. Superconducting transition is obtained at Tc(0)=21K.

3 RESULT AND DISCUSSION

3.1 BRBO superconducting thin film

Figure 1 shows the XRD patterns of the BBO thin film formed at Ts~550°C ((a)) and the BRBO superconducting thin film formed at Ts~370°C ((b)). As shown in Fig. 1, the BBO thin film and the BRBO thin film were c-axis oriented. The small peak for (110) plane also appears in the XRD pattern of the

BRBO thin film. The XRD pattern of the BBO thin film formed at Ts~350 °C was much worse than that of the BBO thin film formed at Ts~550°C. The XRD pattern of the BRBO thin film (Fig. 1(b)) indicates that good crystalline of the BRBO thin film formed at Ts~370°C is obtained.

Figure 2 shows the depth profile of the composition of the BRBO thin film measured by RBS. As shown in Fig. 2, the composition ratio of the BRBO thin film is roughly constant although the atomic fraction of O, Bi, Ba, Rb at the surface decreases. Decrease of the atomic fraction of the elements at film surface is under examination.

Figure 3 shows the lattice constant as a function of the Rb-composition ratio. As shown in Fig. 3, lattice constant decreases with increasing Rb-composition ratio. Peak intensity of the (100) plane in the XRD pattern of the BRBO thin film also depends on the Rb-composition ratio. The ratio of the peak intensity of the (100) plane to that of the (110) plane increases with increasing Rb-composition ratio in the range of Rb/(Bi+Ba)<0.5. Resistance of the BRBO thin film depends on the Rb-conposition ratio. Resistance of the BRBO thin film decreases with increasing Rb-composition ratio in the range of Rb/(Ba+Bi)<0.5. By adjusting Rb-composition ratio, the BRBO superconducting thin film was prepared. Figure 4 shows the resistance-temperature curve (R(T)-curve) of the BRBO superconducting thin film formed at Ts~370°C. Superconducting transition was obtained at Tc(0)=21K. In order to increase the superconducting transition temperature, further adjustment of growth conditions such as beam flux intensity, Ts, PO3 are required.

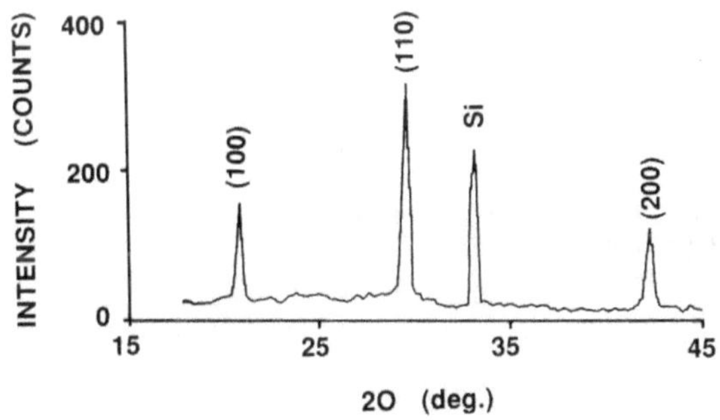

Fig.5 XRD pattern of the BRBO thin film formed on Si (100). Thickness of the thin film is 1000~1500Å.

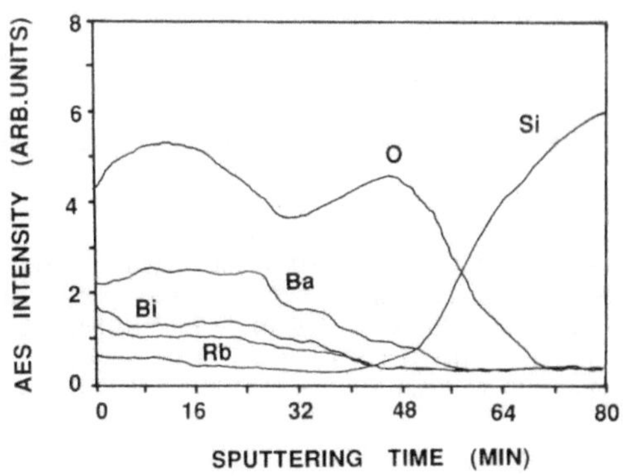

Fig.6 Depth profile of AES intensity for BRBO/Ag/Si(100).

Fig.7 Photograph of surface of BRBO/Ag/Si obserbed by SEM.

3.2 BRBO/Ag/Si

Figure 5 shows the XRD pattern of BRBO/Ag/Si. Figure 5 indicates that the peak intensity of the (110) plane is larger than that of the (100) plane in contrast to the BRBO superconducting thin film on

MgO(100). The crystalline of BRBO/Ag/Si is also worse than that of the BRBO superconducting thin film formed on MgO(100). The peak intensity is much smaller. In order to qualify the crystalline of BRBO/Ag/Si, further optimization of the growth conditions for BRBO/Ag/Si is desired. Large mismatch of the lattice constant between Si and BRBO may be one of the reasons for bad crystalline.

Figure 6 shows the depth profile of the BRBO/Ag/Si measured by AES. Depth profile of the elements other than oxygen is roughly independent of depth of the BRBO thin film. The reasons of large fluctuation of O signal is under examination. As seen in Fig. 6, interdiffusion of Si into the BRBO thin film is small. In Fig. 6, small signal of Si was observed at the surface of BRBO/Ag/Si. But small signal of Si might not be due to interdiffusion of Si into the BRBO thin film. Figure 7 shows the photograph of the film surface measured by SEM. As seen in Fig. 7, many cracks among the plate-like crystals of ~ 1 micrometer were observed. The size of the electron beam of AES is ~ 0.5 mm. Therefore it is possible that auger electrons from the Si substrate surface passing through many cracks contribute to the spectrum measured at the BRBO thin film surface. Further analysis is in progress.

4. CONCLUSIONS

The BRBO superconducting thin film with good crystalline was prepared on MgO(100) substrate by molecular beam epitaxy at Ts~370°C. Superconducting transition temperature of Tc(0)=21K was obtained. The lattice constant, orientation of the crystal and resistance of the BRBO thin film depend on Rb-composition ratio. Rb-composition ratio of the thin film also has a large influence on stability of the BRBO thin film.

The BRBO thin film was preliminarily formed on the Si substrate with Ag-overlayer. AES analysis of BRBO/Ag/Si indicates that interdiffusion of Si into the BRBO thin film is small.

ACKNOWLEDGEMENT

This work was supported by NEDO under the management of FED.

REFERENCES

[1] Hellman ES, Hartford EH (1990) J.Vac.Sci.Technol. B8 :332 and references therein
[2] Hellman ES, Hartford EH, Gyorgy EM (1991) Appl. Phys. Lett. 58:1335
[3] Ishizaka A, Shiraki Y (1986) J. Electrochem. Soc. 133:666
[4] Yamada T, Abe H (1990) Jpn. J. Appl. Phys. 29 :L1192
[5] Yamada T, Ogihara M, Abe H (1991) 3rd FED WORKSHOP :62

Laser MBE of Tetragonal $BaCuO_{2-x}$ Films and Their Characterization

J.P. GONG, M. YOSHIMOTO, H. NAGATA, S. GONDA, H. OHKUBO, and H. KOINUMA

Research Laboratory of Engineering Materials, Tokyo Institute of Technology, Yokohama, 227 Japan

ABSTRACT

$BaCuO_{2-x}$ films were epitaxially grown on $SrTiO_3$ (001) substrates by pulsed laser MBE. The formation of tetragonal $BaCuO_{2-x}$ films was confirmed by RHEED and XRD under such deposition conditions as the substrate temperatures ranging from 550°C to 600°C and the ambient NO_2 gas pressure ranging from 10^{-7} Torr to 10^{-5} Torr. The as-grown tetragonal $BaCuO_{2-x}$ film showed conductivity as high as 10^3 S/cm at room temperature.

KEY WORDS: Laser MBE, (AE)CuO_2, Epitaxial growth, RHEED

INTRODUCTION

The infinite stacking of (CuO_2) and alkaline earth metal planes gives the "parent structure" of hitherto discovered high Tc cupurates[1]. Several attempts have been made to synthesize and characterize the tetragonal alkaline earth (AE) cupurates of (AE)CuO_2 by both bulk sintering and thin film processes. Although bulk $CaCuO_2$ has not been synthesized in a single phase by the conventional sintering process, $(Ca_{1-y}Sr_y)CuO_2$ bulks with y=0.09-1 were sintered successfully at 1000°C and higher under high oxygen pressures of 0.1-6GPa[1-3]. Pure tetragonal $CaCuO_{2-z}$ was prepared in the form of highly oriented film by the sputtering method[4]. In a prior work we succeeded in the epitaxial layer-by-layer growth of tetragonal $(Ca_{1-y}Sr_y)CuO_{2-z}$ films (y=0.14, 0.5, 1) on $SrTiO_3$(001) substrates by the Laser MBE method, i.e. pulsed laser deposition in ultra high vacuum (UHV) [5]. Although partial Ba substitution in bulk $CaCuO_2$ and $SrCuO_2$ was achieved by Takano's group, a complete Ba substitution has not been successful yet. Cubic phase(a=1.83nm) is known to be thermodynamically stable for $BaCuO_2$. We can imagine that a tetragonal $BaCuO_2$ should have a c-axis length larger than that of $CaCuO_2$ and $SrCuO_2$ because the ionic radius of Ba^{2+} is larger than those of Ca^{2+} and Sr^{2+}. By substituting Ba for Ca or Sr, it should be possible to change the distance between the $(CuO_2)^{2-}$ planes. In relation to the elucidation of the superconductivity mechanism of high Tc cupurates as well as to the exploration of new superconductors, we have been examining laser MBE technology as a suitable means for the epitaxial growth of tetragonal (AE)CuO_{2-x} (AE=Ca,Sr,Ba). Here, we report on the success in synthesizing tetragonal phase of $BaCuO_{2-x}$ by epitaxial growth on $SrTiO_3$(001) substrate with our laser MBE method.

EXPERIMENTAL

A sintered disk target of cubic-type $BaCuO_2$ was ablated by ArF pulsed excimer laser beam (20ns, 2Hz, 1J/cm^2) to deposit films on $SrTiO_3$(001) substrates heated at temperatures between 500°C and 700°C measured by an optical pyrometer. In order to facilitate the oxidation, NO_2 gas was flushed onto the substrate under pressures of 10^{-7} Torr-5x10^{-5} Torr during the deposition. The surface structure of the growing films was in situ monitored by reflection high energy electron diffraction(RHEED). The grown films were characterized by in situ X-ray photoelectron spectroscopy(XPS) and ex situ X-ray diffraction(XRD) using a monochromatic Cu Kα probe beam. The rate of film deposition was about 0.3nm/min. The electrical resistivity was measured by the standard four-probe method.

RESULTS AND DISCUSSION

The film structure and properties were very sensitive to the growth conditions; tetragonal $BaCuO_{2-x}$ films were obtained by optimizing the deposition temperature and NO_2 gas pressure into relatively narrow ranges of 550°C-600°C and 10^{-7} Torr-10^{-5} Torr, respectively. Figure 1 shows in situ RHEED patterns for the $SrTiO_3$(001) substrate at [010] azimuth of $SrTiO_3${001} (a), the film grown at 550°C and $1x10^{-6}$ Torr NO_2 (b), and the film grown at 650°C and $1x10^{-6}$ Torr NO_2 (c), respectively. As soon as the deposition started, the Kikuchi lines disappeared and the streaks became broader, but the distance between the streaks, i.e., the reciprocal lattice spacing along the a,b-axis was unchanged during the film growth as shown in Fig. 1(b). Additional streaks of (0 1/2) reciprocal lattice rods appeared about 4 min after the deposition had started. When the incident beam was at [110] azimuth of $SrTiO_3${001}, the RHEED pattern was similar to Fig. 1(b), but the spacing between the reciprocal rods was $\sqrt{2}$ times larger than that in Fig.1(b). These results indicate that the film was epitaxially grown on $SrTiO_3$ substrate with its a- and b-lattice parameters same as the a-lattice parameter (=0.39nm) of cubic $SrTiO_3$ substrate. In Fig. 1(c), new streak pattern with narrower intervals is observable. The spacing between the reciprocal rods was about 3.5 times smaller than that of $SrTiO_3$ substrate. This fact suggests that a cubic phase was formed with its axis 45° rotated within the growth plane.

Figure 2 shows the X-ray diffraction data of $BaCuO_{2-x}$ films having the RHEED patterns shown in Fig. 1(b) and Fig.1(c). XRD pattern of bulk $BaCuO_2$ in the cubic phase is filed, but no structural information on tetragonal $BaCuO_2$ has been reported in JCPDS file. The XRD pattern in Fig. 2b is assigned well to the c-axis oriented cubic single crystal $BaCuO_2$[6]. The lattice parameter is 1.81nm which is very close to that of bulk (a=1.83nm). From the in situ XPS analysis, the valence state of Cu in the as-grown film of Fig.1(b) and Fig.2(a) was evaluated to be 2+. From the analogy to the XRD and RHEED data for the (Ca,Sr)CuO_{2-x} epitaxial films reported in our previous paper[5], we assign Fig. 2(a) to the XRD pattern of tetragonal $BaCuO_{2-x}$ film [5]. The c-axis length of the $BaCuO_{2-x}$ film was calculated from the 002 XRD peak to be 0.406nm. This result is consistent with the estimated value 0.412nm for a tetragonal structure which will be discussed in detail later.

Ba-Cu-O films with thicknesses of about 20nm were grown under various conditions by the laser MBE to give the phase diagram shown in Fig.3. From the RHEED, XPS and XRD analyses, it is evident that the tetragonal $BaCuO_{2-x}$ films were epitaxially grown on $SrTiO_3$(001) substrates at the substrate temperatures of 550°C-600°C and NO_2 pressures of 10^{-7} Torr-10^{-5} Torr. When the substrate temperature was higher than 600°C, cubic $BaCuO_{2-x}$ film with the

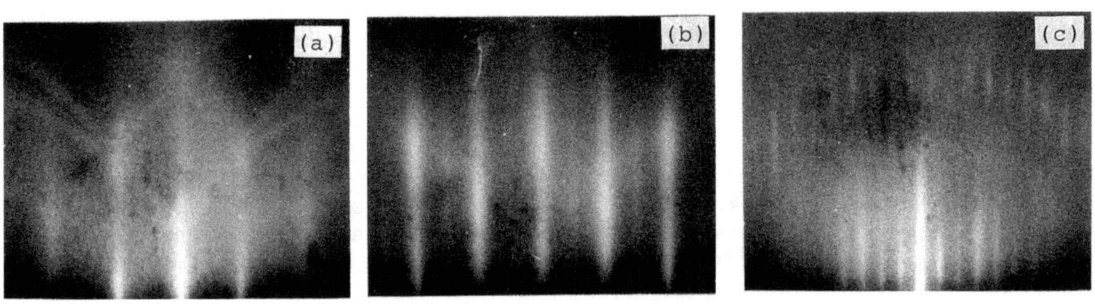

Fig.1 In situ RHEED pattern of (a) $SrTiO_3$ (001) substrate, (b) $BaCuO_{2-x}$ film grown at 550°C, $1x10^{-6}$ Torr NO_2, 60min deposition, and (c) $BaCuO_{2-x}$ film grown at 650°C, $1x10^{-6}$ Torr NO_2,60min deposition, from [010] azimuth of $SrTiO_3${001}.

orientation of $BaCuO_{2-x}(001)//SrTiO_3(001)$ and $BaCuO_{2-x}[110]//SrTiO_3[010]$ was formed. The film prepared at 550°C and 10^{-5} Torr NO_2 showed spotty RHEED pattern which was almost overlapped by the streaks of $SrTiO_3$ substrate. From the XRD analysis, this film was assigned to $Ba(NO_3)_2$ (cubic phase a=0.811nm). At the substrate temperatures below 550°C and the NO_2 pressures higher than 10^{-7} Torr, $Ba(NO_3)_2$ mixed with BaO (cubic phase a=0.552nm) was formed judging from the XRD pattern.

Figure 4 shows the variation of RHEED intensity at the (0 1/2) and near (0 2/7) reciprocal lattice rods with the progress of the $BaCuO_{2-x}$ film deposition at 650°C. The RHEED pattern of this film with a thickness of 20nm was shown in Fig. 1(c). As shown in Fig. 4(a) and 4(b), the RHEED intensity of (0 1/2) streaks originated from tetragonal phase reached maximum after 7 min deposition, then decreased with further progress of deposition. Instead, the intensity of streaks originated from cubic phase started increasing at 7 min deposition and saturated in 35 min. This phenomenon indicated that the surface structure of the growing film changed from tetragonal to cubic when the film thickness became larger than a critical value of about 2nm. The existence of critical film thickness for tetragonal film growth seems to depend largely on the substrate temperature. When the NO_2 gas pressure was fixed at $1x10^{-6}$ Torr, and the substrate temperature decreased to 550°C, the tetragonal phase prevailed throughout the deposition for 120min and no phase transition was observed.

Figure 5 shows the relationship between the a and c-axis lengths of tetragonal $(AE)CuO_{2-x}$ (AE=Ca, Sr, Ba) and the alkaline earth ionic radius. Data for the bulk compounds were taken from the reports by Takano et al[3]. The longest a-axis is 0.393nm and the alkaline earth site is not substituted by an ion with its ionic radius larger than 0.1313nm. Apparently, the Cu-O bond length, which is half of the a-axis length, can not be longer than 0.1965nm. Ba^{2+} has a radius (0.142nm) so significantly larger than Sr^{2+} (0.126nm) and Ca^{2+} (0.112nm) that the complete substitution of Ba^{2+} for Sr^{2+} or Ca^{2+} would make the formation of adjacent CuO_2 lattice intolerable in the thermodynamically equilibrium process. The transformation to cubic phase is considered to occur for the stress relaxation in the film. In the non-equilibrium thin film process by laser MBE, however, the tetragonal $BaCuO_{2-x}$ could be heteroepitaxially grown on $SrTiO_3$ substrate.

In Fig. 5, the c-axis length of tetragonal $(AE)CuO_{2-x}$ thin film is shown to be compared with the bulk values[3] as well as with the values calculated by the following equation;

$$c^2=(2d)^2-a$$

where d is the AE-O bond length evaluated as the sum of the ionic radii of alkaline earth in eight coordination state and oxygen. The a-axis was assumed to be constant at 0.39nm, since the films were epitaxially grown on the $SrTiO_3(001)$ substrate without a notable change in the RHEED inter-streak

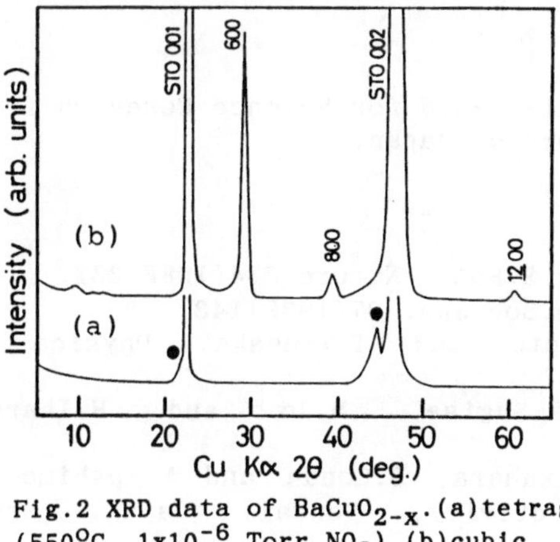

Fig.2 XRD data of $BaCuO_{2-x}$.(a)tetragonal (550°C, $1x10^{-6}$ Torr NO_2),(b)cubic (650°C; $1x10^{-6}$ Torr NO_2).

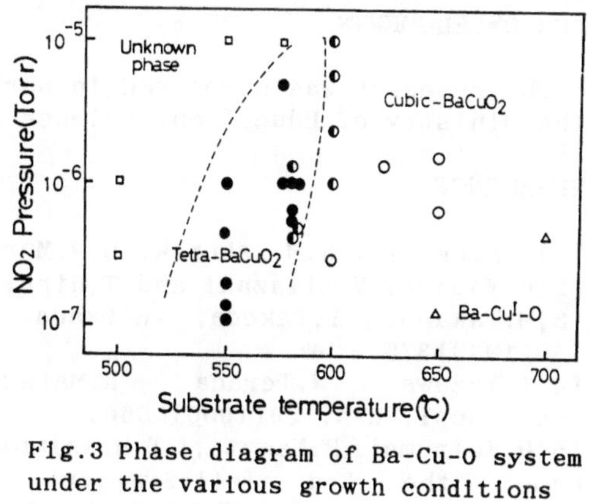

Fig.3 Phase diagram of Ba-Cu-O system under the various growth conditions by the Laser MBE.

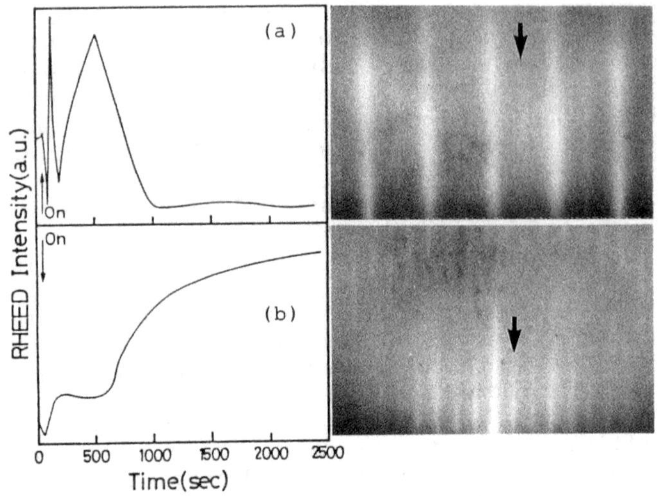

Fig.4 Variation of RHEED intensity with time at (0 1/2)(a) and near (0 2/7)(b) reciprocal lattice rods, for the film prepared at 650°C, $1x10^{-6}$ Torr NO_2 flushing.

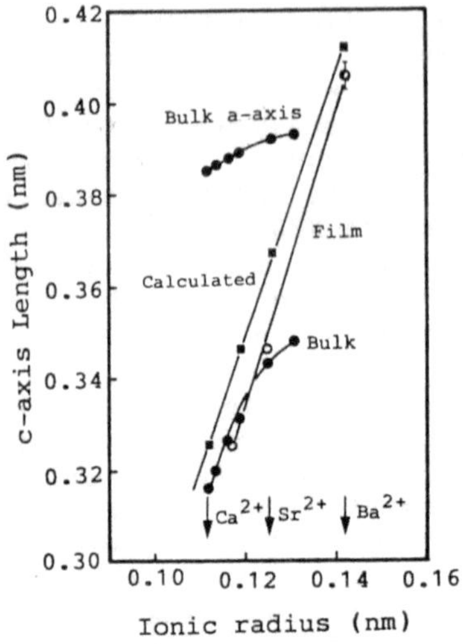

Fig.5 Relations between the alkaline earth ionic radius and the a-axis/c-axis length of tetragonal $BaCuO_{2-x}$.

distance. The calculated c-lattice parameters agree well with the values of bulk samples in the ionic radius range from 0.112nm (Ca^{2+}) to 0.126nm (Sr^{2+}). However, in the ionic radius range exceeding 0.126nm, the value for $(Ba,Sr)CuO_x$ bulk tends to deviate from the calculated value. On the other hand, the value determined experimentally from our $(AE)CuO_{2-x}$ epitaxial films[5] are very close to the calculated values in the whole ionic radius range from 0.112nm (Ca^{2+}) to 0.142nm (Ba^{2+}). This coincidence should reinforce our assignment of tetragonal phase to the epitaxial $BaCuO_{2-x}$ film deposited by the laser MBE.

The tetragonal film has blackish color, different from brownish cubic film. The as-grown tetragonal films were so unstable in air that they changed color from blackish to transparent in a few minutes and became insulative. To protect the film from the deterioration in air, an insulative amorphous CeO_2 film of 30nm thick was in situ deposited at the room temperature immediately after the deposition of the $BaCuO_{2-x}$ film. The CeO_2 coated tetragonal $BaCuO_{2-x}$ film had a conductivity as high as 10^3S/cm at room temperature and showed a metallic conduction behavior at the temperatures ranging from 50K to 150K. It appears to be the general tendency that tetragonal $(AE)CuO_{2-x}$ is metallic and has high conductivity even at room temperature.

ACKNOWLEDGMENTS

The research was supported in part by Grant-in-Aid for Science Research from the Ministry of Education, Science and Culture of Japan.

REFERENCE

[1]T.Siegrist, S.M.Zahurak, D.W.Murphy and R.S.Roth, Nature 334(1988)231.
[2]H.Yamane, Y.Miyazaki and T.Hirai, J.Ceram.Soc.Jpn. 97(1989)143.
[3]M.Takano, Y.Takeda, H.Okada, M.Miyamoto and T.Tsusaka, Physica C 159(1989)375.
[4]I.Yazawa, N.Terada, K.Matsutani, R.Sugise, M.Jo and H.Ihara, Jpn.J.Appl.Phys. 29(1990)L566.
[5]H.Koinuma, H.Nagata, T.Hashimoto, T.Tsukahara, S.Gonda and M.Yoshimoto, Proc. the 3rd Intl.Symp.on Superconductivity, Sendai, Japan, Sep. 1990(Springer-Verlag,Tokyo).
[6]E.F.Paulus, G.Miehe, H.Fuess, I.Yehia and U.Lochner, J. Solid State Chem.,90(1991)17.

7 Device Application

7.1 Multilayer Structures and Junctions

Bi-Epitaxial Josephson Junctions in $YBa_2Cu_3O_7$ with Controlled 45-Degree Grain Boundaries

K. CHAR, M.S. COLCLOUGH, L.P. LEE, and G. ZAHARCHUK

Conductus, Inc., Sunnyvale, CA 94086, USA

ABSTRACT

We will report structural and electrical properties of bi-epitaxial grain boundary Josephson junctions in $YBa_2Cu_3O_7$, created at predetermined locations by controlling the in-plane epitaxy of $YBa_2Cu_3O_7$ with a seed layer and a buffer layer on a base layer. Reproducibility and uniformity of the junctions is described. SQUIDs made from bi-epitaxial junctions show voltage modulation at 77 K with reasonably low noise. Complete magnetometers coupled with multi-turn flux transformers have magnetic field sensitivity at 77 K sufficient to be useful for many applications. Recent success in fabricating integrated SQUIDs by combining the SQUID and multi-turn flux transformer on a single substrate is described.

KEY WORDS: grain boundary, bi-epitaxial junctions, seed layer, buffer layer, SQUID

INTRODUCTION

The discovery of high T_c oxide superconductors brought a great deal of interest in the possible applications of these unusual materials. It ranges from large scale application based on superconducting wires to small scale application based on superconducting thin films. Along with their high intrinsic anisotropy, the grain boundaries of these materials have been the major obstacle to development of commercially viable technology. This problem was partially overcome by aligning grains in the case of wire and by developing epitaxial structures in the case of thin films. However, the properties of high angle grain boundaries have not been completely understood and still remains to be the area of large interest and investigation.

In this paper we will report our development of multilayer structure in order to create controlled 45 ° grain boundaries by means of standard photolithography. Properties of these *bi-epitaxial* grain boundary junctions will be reported. As an example of devices made with bi-epitaxial junctions, we will report dc SQUIDs (Superconducting QUantum Interference Devices) made using two bi-epitaxial junctions. Our recent demonstration of the first integrated SQUID magnetometers operating at 77 K by combining flux transformers with SQUIDs on a single wafer shows the possibility of future integrated circuits made with high T_c superconducting oxides.

BI-EPITAXIAL PROCESS

The elegant work by IBM [1] that studied the weak link behavior of high angle grain boundaries on $SrTiO_3$ bi-crystals made us realize the importance of in-plane alignment of $YBa_2Cu_3O_7$ (YBCO) thin films. Later it was observed that the superconducting properties such as critical current density and microwave surface resistance were affected by the presence of high angle grain boundaries in YBCO thin films on substrates like Al_2O_3 [2], MgO [3] and Yttria stabilized Zirconia (YSZ) [4]. All these substrates do not have a good lattice match with YBCO, and that is the reason why the high angle grain boundaries are present. The randomly occurring 45 ° grain boundaries on MgO substrates were well studied by Cornell group [5]. Their behavior was well described by a resistively shunted junction (RSJ) model. However, their technological usefulness was hampered by their random presence in the film.

Ways to control grain boundaries were not found until the observation of two distinct in-plane epitaxial orientations of YBCO on YSZ substrate [4] at Conductus. We thought that, if one can grow YBCO in two different epitaxial direction on a single crystal substrate, it should be possible to create 45 ° grain boundaries at predetermined locations on single crystal substrates. We also learned that the epitaxial relations between YBCO and substrate materials are affected by deposition parameters and surface condition of the substrates. Instead of relying on deposition parameters and substrate condition that are often irreproducible, we decided to use very thin *seed* and *buffer* layers.

Figure 1 depicts the general concept of the *bi-epitaxial* process. There are two different epitaxial relations between the *base* layer and the *buffer* layer. One of the epitaxial relations is created by the

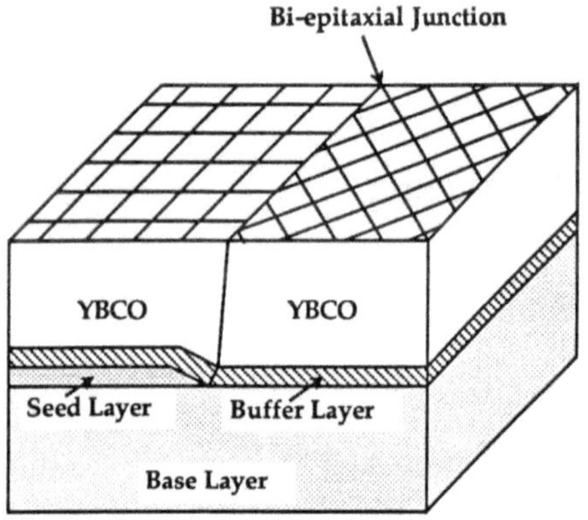

Fig. 1 Generic bi-epitaxial process involving *base*, *seed* and *buffer* layers.

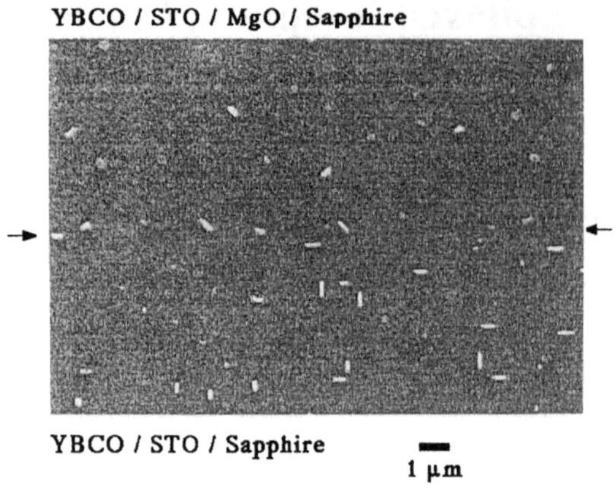

Fig. 2 SEM image of 45 ° grain boundary.

seed layer. It should be noted that the buffer layer (or layers) should have only single epitaxial relations with YBCO. The YBCO on top of the buffer layer will have 45 ° grain boundaries wherever the seed layer ends. And the YBCO, except at the predetermined locations, should not have any other high angle grain boundaries.

Our first example of the bi-epitaxial process [6] used r-plane sapphire as a base layer, MgO as a seed layer and $SrTiO_3$ as a buffer layer. The epitaxial relations, in the growth plane, are $SrTiO_3$ [110] // Al_2O_3 [11$\underline{2}$0] and $SrTiO_3$ [100] // MgO [100] // Al_2O_3 [11$\underline{2}$0]. Figure 2 is a scanning electron microscope (SEM) image of such a bi-epitaxial junction. The line connecting the arrows is the 45 ° grain boundary junctions. The a-axis oriented grains on each side of the boundary are aligned at 45 ° to each other.

Recently, by using $SrTiO_3$ as a base layer, MgO as a seed layer and CeO_2 as a buffer layer, we extended our bi-epitaxial junction process. [7] Since the $SrTiO_3$ base layer can be grown on a variety of substrates such as $LaAlO_3$, MgO and YSZ, the bi-epitaxial junction process no longer has to rely on r-plane sapphire substrates. In addition, as illustrated in Fig. 3, the bi-epitaxial junctions can be fabricated on top of a YBCO ground plane without creating grain boundaries in the YBCO ground plane. The in-plane epitaxial relations between $SrTiO_3$, MgO and CeO_2 are $SrTiO_3$ [100] // CeO_2 [110] and $SrTiO_3$ [100] // MgO [100] // CeO_2 [110]. The epitaxial relation between the CeO_2 buffer layer and the YBCO is always YBCO [100] // CeO_2 [110]. The ability to create bi-epitaxial junctions at any level of multi-level structures will offer great flexibility in fabrication and design of devices.

FABRICATION

All the oxide layers were deposited by laser ablation in a chamber where six different targets can be used in a sequence without breaking the vacuum. Substrates were glued on a heater block with silver paste and placed about 5 cm in front of the target. Optical photolithography was used to pattern all the oxide layers. Ar ion beam milling was used for etching. Care was taken to keep the substrates cool during milling. Deposition temperature was 400 - 800 C and oxygen pressures of 2 - 500 mTorr were used. The fluence used was 2 - 4 J/cm^2.

TRANSPORT PROPERTIES OF BI-EPITAXIAL JUNCTIONS

The current-voltage (I-V) characteristics of bi-epitaxial grain boundary Josephson junctions are qualitatively similar to other types of grain boundary junctions, such as IBM's grain boundary junctions on $SrTiO_3$ bi-crystals [1] and Cornell's random grain boundary junctions on MgO substrates [5]. At low temperatures the I-V curves are well described by a resistively shunted junction model. As the temperature approaches the transition temperature, the I-V curves are significantly rounded due to thermal noise. These thermal roundings are well described by thermally activated phase slippage across the junction [8, 9]. Close to the transition temperature, the junction does not have zero resistance current, even though the I-V curve is non-linear.

The critical current densities of the bi-epitaxial junctions are in the range of 10^3 -10^4 A/cm^2 at 4.2 K and 10^2 -10^3 A/cm^2 at 77 K. We found that critical current densities, J_c, and the specific junction resistance, $1/\sigma_n$, are affected by the types of substrate material the junctions are fabricated on. Figure 4 illustrate

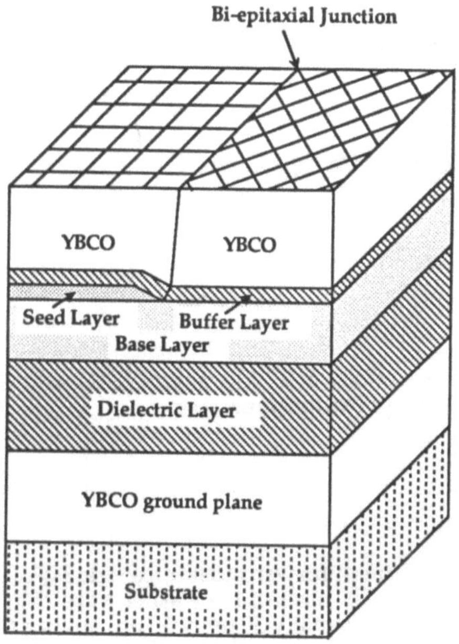

Fig. 3 Extended bi-epitaxial process on YBCO ground plane.

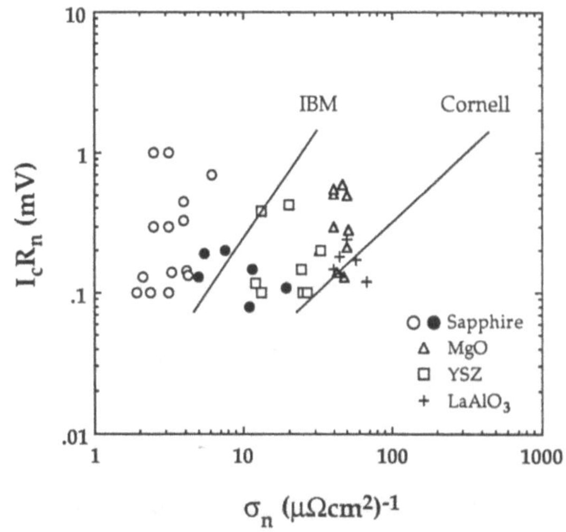

Fig. 4 I_cR_n versus σ_n of bi-epitaxial junctions compared to other types of grain boundary junctions.

the effect of substrate on these junction parameters. The circular data points are from the structure made with sapphrie substrate as a base layer, MgO seed layer, and $SrTiO_3$ buffer layer. When the MgO was made by ablating an Mg metal target, the data points are represented by empty circles. On the other hand, solid circles represent the case when an MgO single crystal target was used; this yielded smoother YBCO films, as shown by the scanning electron microscope. The remaining data points represent junctions made using our new process: $SrTiO_3$, MgO and CeO_2. Junctions on MgO substrates are shown by triangular points, junctions on YSZ substrates are shown by square points and those on $LaAlO_3$ are shown by crosses. The lines represent data from IBM's $SrTiO_3$ bi-crystal grain boundaries [10] and Cornell's randomly occurring 45 ° grain boundaries on MgO [5]. One can immediately notice that the range of I_cR_n for all three types of junctions is between about 100 µV and 1 mV. However, there is a large range of conductance, depending on the type of junction. One possible explanation for this is that the active regions of the junctions are the same kind of grain boundary junction in each case, and the differences lie in the fraction of the junction width that is actually connected, depending on the surface morphology and stress in the film caused by the differing substrates.

The critical currents are approximately a linearly decreasing function of temperature, approaching zero near T_c, and the resistances of the junctions are almost independent of temperature. The critical currents and the resistances scale with the width of the junctions with a 30 % variation on a single chip after taking account of thickness variation across the chip, even though the junctions are wider than the Josephson penetration depth. This suggests that, on the length scale of tens of µm, the bi-epitaxial grain boundary junctions are parallel arrays of Josephson junctions.

The variations of critical current density and conductance of the junctions on separately processed chips are usually larger than those of the junctions on a single chip, suggesting that the thickness of the seed layer, the step height of the seed layer, the angle of the step, and surface condition of the step may have an effect on the junction properties. Since these parameters are not easily accessible, detailed correlation between these parameters and the junction properties will require more effort. By optimizing the processing steps, such as deposition of each layer and etching of the seed layer, we believe that the variations in the bi-epitaxial junction parameters can be reduced further.

Uniformity of the Josephson current have been studied by measuring critical current dependence on magnetic field. Because of its planar geometry the effective junction area is modified by flux focusing effect due to the presence of large superconducting banks next to the junctions. The effective junction area was calculated [11] to be $0.543 \cdot w^2$ instead of usual $w \cdot (2\lambda + d)$, where w is the width of the junction, λ is the penetration depth of the superconductor, and d is the thickness of the barrier layer. Figure 5 shows the magnetic field dependence of a 10 µm wide bi-epitaxial junction when the junction is current biased. The minimum critical current is about 10 % of maximum critical current.

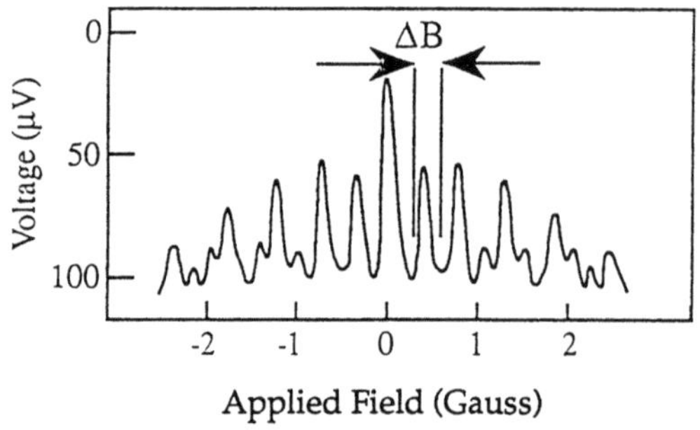

Fig. 5 Magnetic field diffraction pattern of a 10 μm wide bi-epitaxial junction.

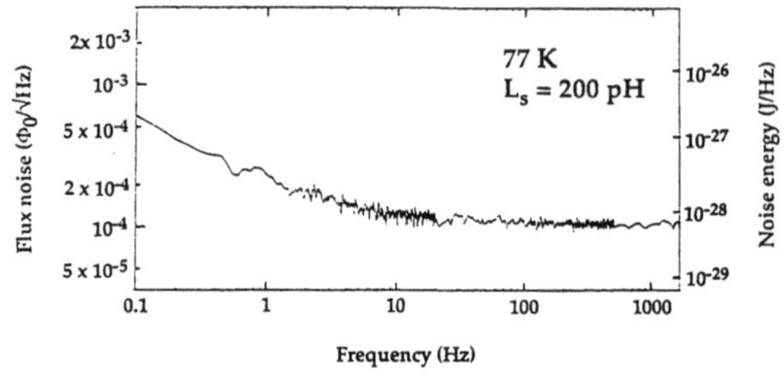

Fig. 6 Flux noise at 77 K of a dc SQUID with about 200 pH inductance.

SQUIDS USING BI-EPITAXIAL JUNCTIONS

Since the bi-epitaxial junctions are planar type junctions, it is straightforward to fabricate dc SQUIDs. We have been using the Jaycox-Ketchen washer geometry [12]. The sizes of the washer were 250 μm x 250 μm, 500 μm x 500 μm, or 1 mm x 1 mm and the inner hole was usually 30 μm x 30 μm, giving rise to a SQUID inductance of about 100 - 200 pH including the inductance from the slit. The typical critical current of SQUIDs at 77 K were 10 - 20 μA, yielding $\beta_L = 2 I_c L/\Phi_0 = 1.5 - 3$. The usual modulation depth is in the range of 1 - 5 μV at 77 K.

The figure of merit for SQUIDs is their noise data. Figure 6 shows the flux noise, $S_\Phi(f)$, of a dc SQUID on a sapphire substrate as a function of frequency at 77 K. The size of the washer was 1 mm x 1 mm, leading to the SQUID inductance of about 200 pH. It has the white noise level of $1 \times 10^{-4}\ \Phi_0/\sqrt{\mathrm{Hz}}$ and significant 1/f noise below 100 Hz. The noise spectra are similar to what has been reported on bi-crystal dc SQUID [13], both qualitatively and quantitatively. At low temperatures, the noise is dominated by 1/f noise in the frequency range up to 5 kHz. As the temperature increases, the white noise level begins to show up and it is approximately at the theoretical Johnson noise level of $S_\Phi(f) = 16k_BTL^2/R$. In many cases, we observed that the 1/f noise has a minimum near the transition temperature. This has been observed by other groups in their grain boundary SQUIDs. [13, 14] Recently the origin of this 1/f noise have been investigated on bi-crystal grain boundary junctions. [15] The 1/f noise when the bias current was near the critical current was attributed to critical current fluctuations. As we understand more about 1/f noise, it is very likely that some reduction in 1/f noise will be made in the future.

The dc SQUIDs with above noise specta can be made into very sensitive SQUID magnetometers when they are coupled with multi-turn flux transformers. This was achieved at U. C. Berkeley (UCB) by combining a Conductus bi-epitaxial SQUID and a UCB flux tranformer. [16] This UCB/Conductus magnetometer had sensitivites at 77 K of 1 pT/√Hz at 1 Hz and 0.2 pT/√Hz at 100 Hz. Recently this "flip chip" magnetometer was successfully used to measure heartbeats of humans. [17]

We would like to note that the sensitivity of the SQUID magnetometers can be further improved by changing geometries. These are a larger pick-up loop, more turns in the input coil and tighter coupling between the input coil and the SQUID body. Of course the ideal way to increase the coupling is to build a flux transformer and SQUID on a singe wafer with a thin dielectric layer in the middle, the integrated SQUID magnetometer.

INTEGRATED SQUID MAGNETOMETERS

The integrated SQUID magnetometer itself is going to be an important device, because at 77 K it will be sensitive enough to measure even the smallest magnetic signal, except perhaps the one from the human brain. More importantly, the processing steps needed to fabricate the integrated SQUID magnetometer are identical to those of most high-T_c integrated circuits one can imagine. They involve creating Josephson junctions, opening a window in the dielectric layer, and making superconducting contacts between two YBCO layers.

Figure 7 shows the geometry of the 10 turn input coil with its crossover. The crossover was designed to cover the slit in the SQUID layer in order to reduce the parasitic inductance from the slit. The pick-up loop was 6 mm x 9 mm and the width of the input coil was 10 μm. The SQUID washer was 500 μm x 500 μm, and the width of the junction was 10 μm.

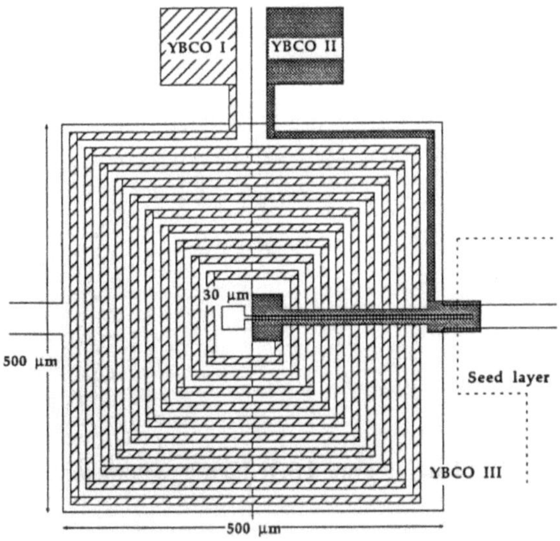

Fig. 7 Plan view of a 3 superconducting layer integrated SQUID magnetometer.

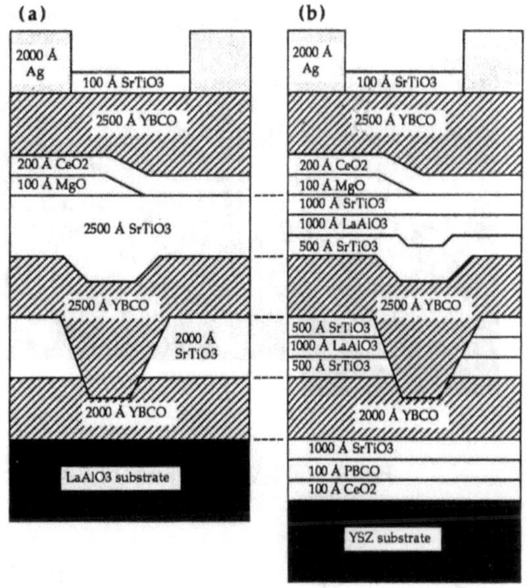

Fig. 8 Cross sectional view of a 3 superconducting layer integrated SQUID magnetometer (a) on $LaAlO_3$ substrates and (b) on YSZ substrates.

Figure 8 (a) depicts the cross sectional view of three YBCO layer SQUID magnetometers on $LaAlO_3$ substrates we successfully built recently [18]. The first YBCO layer was patterned into the pick-up loop and multi-turn input coil. A dielectric layer was deposited and windows in the dielectric layer were opened by lithography and Ar ion milling. The second YBCO layer was patterned to work as a crossover layer for the input coil. The next step was to put down another insulating layer (base layer) and seed layer. The seed layer was patterned to create bi-epitaxial junctions. Finally the third YBCO was deposited and patterned into a SQUID. The entire process requires five separate high temperature depositions and ex-situ patterning.

On YSZ substrates, as illustrated in Fig. 8 (b), CeO_2 and PBCO layers were used to grow good epitaxial $SrTiO_3$ layers. Instead of single $SrTiO_3$ dielectric layers, $SrTiO_3/LaAlO_3/SrTiO_3$ sandwiched layers were used in order to reduce the capacitance of the structure. After the last YBCO deposition a thin $SrTiO_3$ layer was added to passivate the surface of the device.

After each deposition step, the quality of each layer was non-destructively checked with a four circle x-ray diffractometer and/or mutual inductance probe. Since all the layers have to be heteroepitaxial or bi-epitaxial, the x-ray scan after each deposition confirmed the presence of the 45 ° grain boundary as well as the epitaxial qualities of each layer. The mutual inductance probe was able to measure the T_c and J_c of the unpatterned film, T_c and J_c of the flux transformer, and even the J_c of the bi-epitaxial junctions before they are patterned [19].

The electrical performance of an integrated SQUID magnetometer in liquid nitrogen is presented in Fig. 9. When an integrated SQUID magnetometer works, it is so sensitive that it requires multiple mu metal shields. The modulation depth at 77 K was about 1 µV, and the single flux quantum was created by a 1 nT magnetic field, yielding an effective SQUID magnetomer area of about 1.9 mm^2. This effective area corresponds to a gain of about 125, which is about the estimated value from the geometry.

SUMMARY

We have described our efforts to build superconducting integrated circuits using the high T_c oxide superconductors. We invented a bi-epitaxial process, a way of creating 45 ° grain boundary in a controlled fashion. We also built multi-layer structures of YBCO and epitaxial dielectric layers such as $SrTiO_3$, $LaAlO_3$, MgO and CeO_2. Combining these two technologies by optical photolithography and Ar ion beam milling, we were able to fabricate three YBCO layer integrated SQUID magnetometers.

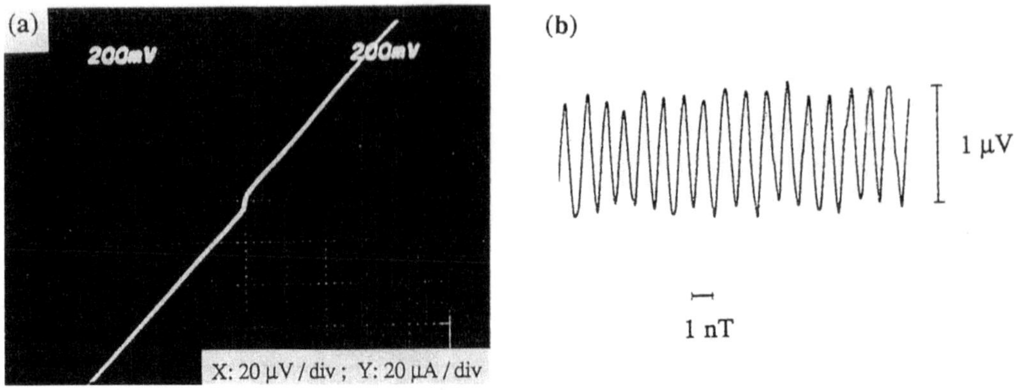

Fig. 9 (a) I-V curve and (b) V-Φ curve at 77 K of an integrated SQUID magnetometer.

During the last couple of years, we learned how important the epitaxial relations between all the oxide layers are, how each layer behaves mechanically and chemically when processed for patterning, and, most importantly, how these structural and chemical effects affect the electrical behavior of the superconducting YBCO layers. We believe during the next two years we will witness significant progress towards building really practical devices based on high T_c thin films.

REFERENCE

1. Dimos D, Chaudhari P, Mannhart J, LeGoues FK (1988) Phys. Rev. Lett. 60: 1653
2. Char K, Newman N, Garrison SM, Barton RW, Taber RC, Laderman SS, Jacowitz RD (1990) Appl. Phys. Lett. 57: 409
3. Laderman SS, Taber RC, Jacowitz RD, Moll JL, Eom CB, Hylton TL, Marshall AF, Geballe TH, Beasley MR (1991) Phys. Rev. B 43: 2922
4. Garrison SM, Newman N, Cole BF, Char K, Barton RW (1991) Appl. Phys. Lett. 58: 2168
5. Russek SE, Lathrop DK, Moeckly BH, Buhrman RA, Shin DH, Silcox J (1990) Appl. Phys. Lett. 57: 1155; Lathrop DK, Moeckly BH, Russek SE, Buhrman RA (1991) Appl. Phys. Lett. 58: 1095
6. Char K, Colclough MS, Garrison SM, Newman N, Zaharchuk G (1991) Appl. Phys. Lett. 59: 733
7. Char K, Colclough MS, Lee LP, Zaharchuk G (in press) Appl. Phys. Lett.
8. Ambegaokar V, Halperin BI (1969) Phys. Rev. Lett. 22: 1364
9. Gross R, Chaudhari P, Dimos D, Gupta A, Koren G (1990) Phys. Rev. Lett. 64: 228
10. Gross R, Chaudhari P, Kawasaki M, Gupta A (1990) Phys. Rev. B 42: 10735
11. Rosenthal PA, Beasley MR, Char K, Colclough MS, Zaharchuk G (in press) Appl. Phys. Lett.
12. Ketchen MB, Jaycox JM (1982) Appl. Phys. Lett. 40: 736
13. Gross R, Chaudhari P, Kawasaki M, Ketchen M, Gupta A (1990) Physica C 170: 315
14. Foglietti V, Koch RH, Gallagher WJ, Oh B. Bumble B, Lee WY (1988) Appl. Phys. Lett. 54, 2259
15. Kawasaki M, Chaudhari P, Gupta A (submitted) Phys. Rev. Lett.
16. Miklich AH, Kingston JJ, Wellstood FC, Clarke J, Colclough MS, Char K, Zaharchuk G (1991) Appl. Phys. Lett. 59: 988
17. Miklich AH, Kingston JJ, Wellstood FC, Clarke J, Colclough MS, Char K, Zaharchuk G (1991) Nature 352: 482
18. Lee LP, Char K, Colclough MS, Zaharchuk G (in press) Appl. Phys. Lett.
19. Char K (unpublished)

Electric Properties of YBCO/PBCO/YBCO Junctions

TATSUNORI HASHIMOTO, MASAYUKI SAGOI, YOSHIHISA MIZUTANI, JIRO YOSHIDA, and KOICHI MIZUSHIMA

Advanced Research Laboratory, Toshiba R&D Center, Kawasaki, Kanagawa, 210 Japan

ABSTRACT

A-axis oriented YBCO/PBCO/YBCO junctions have been fabricated and tested to investigate the possibilities of an all-oxide Josephson junction experimentally. Almost all the fabricated junctions exhibited not only an RSJ-like *I-V* characteristic but also a Fraunhofer-like magnetic field dependence of the critical current. The obtained maximum values of the critical current density and the I_cR_n-value were 110 A/cm^2 and 80 μV, respectively, for 50 nm-thick PBCO interlayer at 3 K.

KEY WORDS: YBCO/PBCO/YBCO junction, Josephson junction, Fraunhofer pattern

INTRODUCTION

All-oxide Josephson junctions are desired not only for a high operation temperature but also for a large I_cR_n-value. Recently, many experimental studies on all-oxide junctions including the grain boundary junction, planar-type junction, sandwich-type junction, etc. have been fabricated and tested for high-Tc device application[1-6]. Among these various types of junctions, the sandwich-type junction, which is expected to be easy to realize small parameter-deviation of junctions on the same chip, is considered to be most suitable for integration.

A sandwich-type $YBa_2Cu_3O_{7-y}/PrBa_2Cu_3O_{7-z}/YBa_2Cu_3O_{7-y}$ (YBCO/PBCO/YBCO) junction was first demonstrated by the Bellcore group[1]. They have reported the electric properties of c- and a- axis orientation junctions[1,4]. However, a Fraunhofer-like magnetic field dependence characterizing the Josephson junction has not been reported, to our knowledge.

We have developed a-axis oriented YBCO/PBCO/YBCO Josephson junctions showing not only an RSJ-like *I-V* characteristic but also a Fraunhofer-like magnetic field dependence to investigate the potential and possibilities of an all-oxide Josephson junction. The electric properties of the developed Josephson junctions are reported in this paper.

YBCO/PBCO/YBCO TRILAYER PREPARATION

A-axis oriented YBCO/PBCO/YBCO trilayers were prepared on (100) $SrTiO_3$ substrates by a multi-target sputtering method[7] using Y, Pr, Ba_2CuO_3, and Cu targets. The individual layers of the trilayer were consecutively deposited in the same chamber by controlling shutters over the Y and Pr targets, where the substrate temperature was 680 °C, the deposition rate was 2.5 nm/min, and the sputtering gas pressure was 1.1 Pa of 50%-O_2/Ar. The thicknesses of the upper YBCO, PBCO interlayer, and lower YBCO were 300, 50-100, and 450 nm, respectively. After the deposition, the substrate was cooled down to room temperature in O_2 at atmospheric pressure. Then, a 300 nm-thick Au layer was deposited by a sputtering method using 100%-Ar gas at room temperature so as not to degrade the trilayer surface in the following processes.

JUNCTION FABRICATION

Figure 1 shows a schematic structure of the fabricated YBCO/PBCO/YBCO junction. YBCO/PBCO/YBCO junctions whose dimension ranged from 100x100 μm to

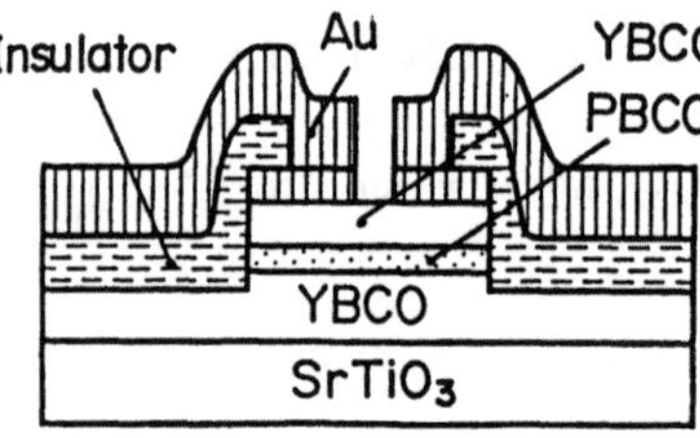

Fig.1 Schematic figure of a YBCO/PBCO/YBCO junction

30x30 µm were fabricated using a photolithography process as follows. First, the trilayer covered with a Au layer was etched by an Ar-ion milling technique until the lower YBCO layer was exposed. A 1 µm-thick interlayer-insulation layer including contact holes for wiring was made of a negative photoresist. After an Ar-plasma cleaning procedure of the Au layer surface to completely eliminate any resist residue, a 1.2 µm-thick Au wiring-layer was deposited by a thermal evaporation method using a metal mask. Finally, the Au wiring-layer was separated by the Ar-ion milling technique in order to obtain potential- and current- leads for four-point measurement.

EXPERIMENTAL RESULTS AND DISCUSSION

Figure 2 shows typical dc *I-V* curves of the fabricated junctions at 3 K. RSJ-like *I-V* curves were observed, where the critical current density j_c and the normal resistance normalized by the junction area were 4 A/cm^2 and 4.7 µΩcm^2, respectively, for 100 nm-thick PBCO interlayer, and 110 A/cm^2 and 0.7 µΩcm^2, respectively, for 50 nm-thick PBCO interlayer. The obtained maximum value of I_cR_n was 80 µV in the latter case.

Junction-area dependences of the critical current I_c and the normal resistance R_n were checked to confirm the spatial uniformity of the current flow through the PBCO interlayer. As shown in Fig. 3, both I_c and R_n were scaled well against the junction area *A*. This result implies that both the super- and normal- current flow were uniform. In other words, both I_c and R_n ascribe not to the junction edge, which may be degraded in the fabrication process, but to the whole junction-area.

If the current flow through the PBCO layer is completely uniform over the junction, the PBCO resistivity can be calculated as 0.1-0.5 Ωcm using the R_n-value. This value is much smaller than the resistivity of bulk PBCO ($\approx 10^7$ Ωcm). Such a small PBCO resistivity may ascribe to the film-deposition condition[8].

Figure 4 shows the magnetic field dependences of I_c for junctions with the same 100 nm-thick PBCO interlayers but different dimensions. As shown in

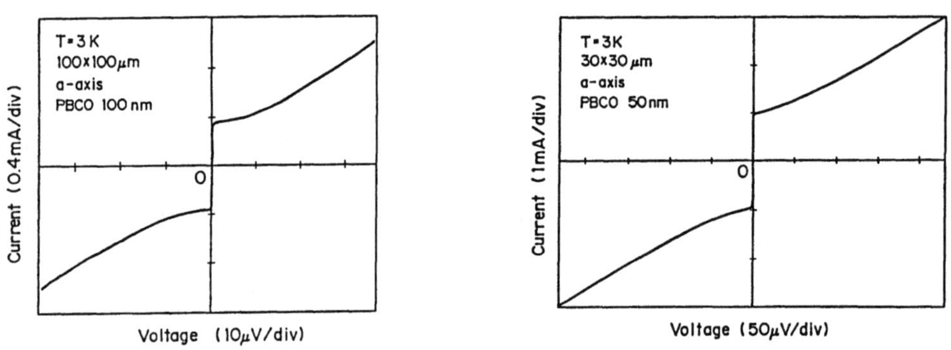

Fig.2 DC I-V curves of YBCO/PBCO/YBCO junctions

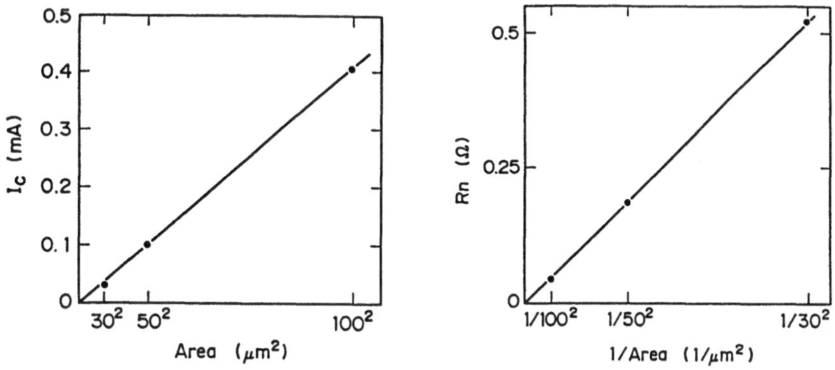

Fig.3 Junction-area dependences of I_C and R_n for junctions with a 100 nm-thick PBCO-interlayer

this figure, Fraunhofer-like dependences were obtained, where the interval of the minima was proportional to the side length of the junction, as theoretically expected. If we assume that the a-axis magnetic penetration depth is nearly equal to or larger than one-half the thicknesses of the YBCO layers, a total penetration depth of d=475 nm is obtained using the Josephson junction theory. On the other hand, the experimental value calculated using the intervals was about 430 nm. Thus, the theoretical value was in reasonable agreement with the experimental one.

Here, the Josephson penetration depth λ_J of this junction will be roughly evaluated to check the junction-length dependence of the magnetic field dependence. As shown in Fig.4(b), the self-field effect was observed, which implies that this junction was large compared to twice the Josephson penetration depth. On the other hand, this junction was not larger than 4 times the Josephson penetration depth because the periodicity of the pattern remained. Using these results, the Josephson penetration depth λ_J was roughly obtained to be 30 µm in the 100 nm-thick PBCO interlayer case.

Figure 5 shows the magnetic field dependence for a thinner PBCO interlayer(50 nm thick). Using j_C=110 A/cm^2, the λ_J-value was obtained to be 5 µm for this junction. The junction length 30 µm is large compared to 4 times the λ_J-value. This result is consistent with the obtained pattern of a large Josephson junction. Consequently, the electric properties concerning the magnetic field coincided well with the Josephson junction theory.

Figure 6 shows the temperature dependence of I_C and R_n for the 100x100 µm junction with a 100 nm-thick PBCO interlayer. The obtained I_CR_n-value was in

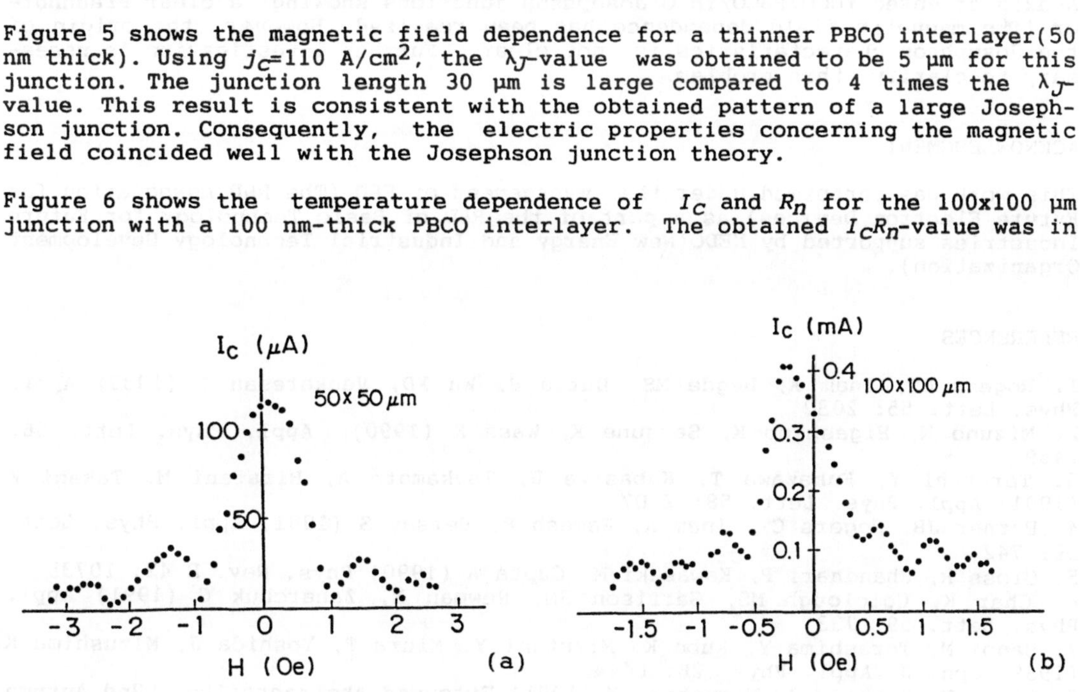

Fig.4 Magnetic field dependences of I_C for junctions with 100 nm-thick PBCO interlayers and various dimensions

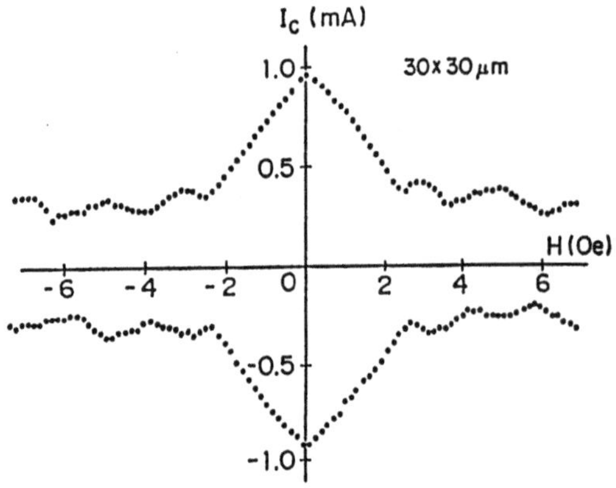

Fig.5 Magnetic field dependence of I_c in the case of a 50 nm-thick PBCO interlayer

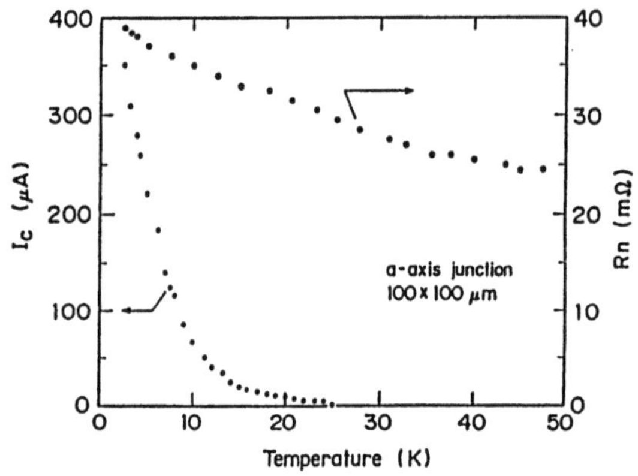

Fig.6 Temperature dependences of I_c and R_n for a 100x100 μm junction with a 100 nm-thick PBCO interlayer

proportion to $\exp(aT^{-1/2})$, where a and T are a constant and temperature, respectively. Therefore, the fabricated Josephson junction can be regarded as an SNS-type one. If we calculate the normal coherence length of the PBCO layer ξ_N by using a, ξ_N=30 nm is obtained at 4.2 K. On the other hand, the ξ_N-value was calculated as 1 nm at 4.2 K using the weak-link theory[9] and the resistivity obtained from the value of R_n. The two values of normal coherence length did not agree. No evidence accounting for the origin of the observed Josephson characteristics has been obtained.

CONCLUSION

A-axis oriented YBCO/PBCO/YBCO Josephson junctions showing a clear Fraunhofer-like magnetic field dependence has been realized. However, the origin of the Josephson characteristics is not clear. Further investigation is necessary to clarify this problem.

ACKNOWLEDGMENT

This work was performed under the management of FED (The R&D Association for Future Electron Devices) as a part of the R&D of Basic Technology for Future Industries supported by NEDO(New Energy and Industrial Technology Development Organization).

REFERENCES

1. Rogers CT, Inam A, Hegde MS, Dutta B, Wu XD, Venkatesan T (1989) Appl. Phys. Lett. 55: 2032
2. Mizuno K, Higashino K, Setsune K, Wasa K (1990) Appl. Phys. Lett. 56: 1469
3. Tarutani Y, Fukazawa T, Kabasawa U, Tsukamoto A, Hiratani M, Takagi K (1991) Appl. Phys. Lett. 58: 2707
4. Barner JB, Rogers CT, Inam A, Ramesh R, Bersey S (1991) Appl. Phys. Lett. 59: 742
5. Gross R, Chaudhari P, Kawasaki M, Gupta A (1990) Phys. Rev. B 42: 10735
6. Char K, Colclough MS, Garrison SM, Newman N, Zaharchuk G (1991) Appl. Phys. Lett. 59: 733
7. Sagoi M, Terashima Y, Kubo K, Mizutani Y, Miura T, Yoshida J, Mizushima K (1989) Jpn. J. Appl. Phys. 28: L444
8. Sagoi M, Yoshida J, Mizushima K (1991) Extended Abstracts(The 52nd Autumn Meeting, 1991). The Japan Society of Applied Physics, p.125 (in Japanease).
9. Likarev KK (1979) Rev. Mod. Phys. 51: 101

Electrical Characteristics of All-Oxide S-N-S Junctions

YOSHINOBU TARUTANI, TOKUUMI FUKAZAWA, UKI KABASAWA, AKIRA TSUKAMOTO, MASAHIKO HIRATANI, MITSUO SUGA, and KAZUMASA TAKAGI

Central Research Laboratory, Hitachi Ltd., Kokubunji, Tokyo, 185 Japan

ABSTRACT

All-oxide S-N-S (superconductor-normal layer-superconductor) junctions having a planar structure are fabricated and their electrical characteristics in the normal state are investigated. The superconducting electrodes are composed of $HoBa_2Cu_3O_{7-x}$, and the normal layer is $La_{1.5}Ba_{1.5}Cu_3O_{7-y}$ or $PrBa_2Cu_3O_{7-z}$. Conductance characteristics are measured for the $La_{1.5}Ba_{1.5}Cu_3O_{7-y}$-coupled junction and explained by the geometrical resonance of quasiparticles. The resistances of $PrBa_2Cu_3O_{7-z}$-coupled junctions are also measured systematically and compared to that of a single $PrBa_2Cu_3O_{7-z}$ film.

KEY WORDS: S-N-S junction, $HoBa_2Cu_3O_{7-x}$, $La_{1.5}Ba_{1.5}Cu_3O_{7-y}$, $PrBa_2Cu_3O_{7-z}$, geometrical resonance,

INTRODUCTION

All-oxide S-N-S junctions in which both superconducting and normal layers are made of cuprate films, are very useful as Josephson junctions and expected as an element for three-terminal devices. High junction performance can be expected since the superconducting wave function may not be as much attenuated at an S-N interface as that of metal-coupled junctions. In addition, the all-oxide junctions help to make clear the electrical properties of the S-N interface including the superconducting proximity effect. The superconducting behavior of a cuprate itself may also be understood by researching the S-N coupling characteristics, since superconductivity can be controlled using the proximity effect.

Several all-oxide S-N-S junctions have been fabricated and their superconducting characteristics investigated. Oxide materials used for the coupling layer are $PrBa_2Cu_3O_{7-z}$[1], $Bi_2Sr_2CuO_v$[2], $La_{1.5}Ba_{1.5}Cu_3O_{7-y}$[3-8], and $La_{1-w}Ca_wMnO_3$[9]. Supercurrent has been detected flowing through these normal layers for distances of several tens or hundreds of nanometers, which are two orders longer than expected[10-12]. The elongation of decay length may be particular to cuprate materials.

It is necessary to investigate superconducting characteristics much more deeply and analyze the influence of S-N coupling on quasiparticle properties for understanding the nature of proximity in cuprate materials. For example, carriers in the $PrBa_2Cu_3O_{7-z}$ are known to be localized at low temperatures, and its resistivity is close to that of compensated semiconductors at 4.2K[13,14]. Consequently, superconducting characteristics cannot be expected in $PrBa_2Cu_3O_{7-z}$-coupled junctions from these normal conductance behaviors[15]. Quasiparticles in the $PrBa_2Cu_3O_{7-z}$ layer have to be speculated to change their properties by the S-N coupling.

In this report, S-N-S junctions comprising superconducting electrodes of $HoBa_2Cu_3O_{7-x}$ and normal layers of $La_{1.5}Ba_{1.5}Cu_3O_{7-y}$ or $PrBa_2Cu_3O_{7-z}$ are fabricated and their electrical properties in the normal states are investigated. Conductance characteristics are measured for the $La_{1.5}Ba_{1.5}Cu_3O_{7-y}$-coupled junction and the fine structures are explained by the geometrical resonance of quasiparticles. The normal resistances of $PrBa_2Cu_3O_{7-z}$-coupled junctions are also measured systematically and compared with that of a single $PrBa_2Cu_3O_{7-z}$ film. The elongation of the superconducting decay lengths is discussed with reference to these normal state characteristics.

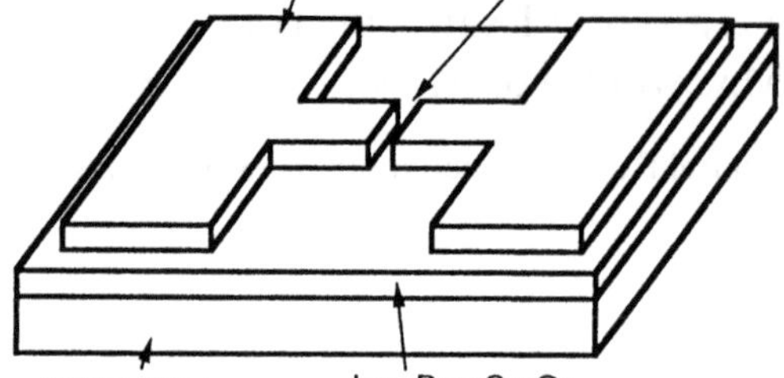

Fig. 1 Planar-type junction structure.

EXPERIMENTAL METHODS

The S-N-S junctions had a planar structure in which both superconducting electrodes are put on a normal layer and with a spacing of less than one micron as shown in Fig. 1. The junctions were fabricated by the following steps. First, the $La_{1.5}Ba_{1.5}Cu_3O_{7-y}$ or $PrBa_2Cu_3O_{7-z}$ normal layers were deposited by an RF magnetron sputtering method. The substrate material used was $SrTiO_3$(110). The film thickness was 300 nm. Second, the $HoBa_2Cu_3O_{7-x}$ superconducting layer was deposited by a reactive coevaporation method. The substrate was heated at 590 ℃ and placed in an atmosphere of oxygen plasma at 0.01 Pa excited by a 2.45 GHz microwave during deposition[16].
Third, an electrode pattern was formed for the $HoBa_2Cu_3O_{7-x}$ layer. The resist pattern including the submicron groove between the electrodes was formed by electron beam lithography. The resist pattern was transferred to the $HoBa_2Cu_3O_{7-x}$ layer by an ion beam etching method with a gas mixture of Ar+50%O_2. Fourth, the $HoBa_2Cu_3O_{7-x}$ layer was exposed to oxygen plasma under 0.01 Pa at 450 ℃ to recover the critical temperature deteriorated during the pattern formation step.

RESULTS AND DISCUSSION

$HoBa_2Cu_3O_{7-x}$-$La_{1.5}Ba_{1.5}Cu_3O_{7-y}$-$HoBa_2Cu_3O_{7-x}$ Junctions

The superconducting characteristics of the $La_{1.5}Ba_{1.5}Cu_3O_{7-y}$-coupled junctions wider than 5 μm have been measured. The superconducting decay length for the $La_{1.5}Ba_{1.5}Cu_3O_{7-y}$ has been derived from the dependence of critical current on electrode spacing to be 65 nm at 4.2 K. This value is about 50 times that calculated by the conventional dirty-limit formula. The normal-state resistance is two or three times that calculated from the value of a single $La_{1.5}Ba_{1.5}Cu_3O_{7-y}$ film. The resistivity difference between a junction and a single film may not be due to some barrier formed by the interdiffusion at the S-N interface but due to the current distribution in the $La_{1.5}Ba_{1.5}Cu_3O_{7-y}$ cross section.
A junction with a width of 0.5 μm and an electrode spacing of 0.46 μm was fabricated for measuring the normal-state characteristics. The supercurrent for this junction was estimated to be 10^{-7} A from the junction dimension and the dependence of critical current on electrode spacing. The current of this level was unmeasurable and may not have overlaped the normal current at voltage state and deformed conductance spectrum. The supercurrent was not really observable for this junction.
The differential conductance dI/dV and its derivative with respect to voltage d^2I/dV^2 are shown in Fig. 2. The measurement temperature was 4.2 K. Periodic peaks were detected in the conductance curve in addition to the zero-voltage peak. The peak positions were determined as the voltages at which the d^2I/dV^2 curve crosses the zero from positive to negative as voltage increases, and are shown by arrows. The peak positions were almost symmetrical with respect to the origin.
One possible origin for the periodic peaks in the conductance curve may be the geometrical resonance of quasiparticles in the $La_{1.5}Ba_{1.5}Cu_3O_{7-y}$ as schematically shown in Fig. 3. The pair potential is assumed to extend from the $HoBa_2Cu_3O_{7-x}$ electrodes into the $La_{1.5}Ba_{1.5}Cu_3O_{7-y}$ layer by the proximity effect. Quasiparticles emitted from one pair potential edge pass through the $La_{1.5}Ba_{1.5}Cu_3O_{7-y}$ layer and reach another pair potential edge[17-19].

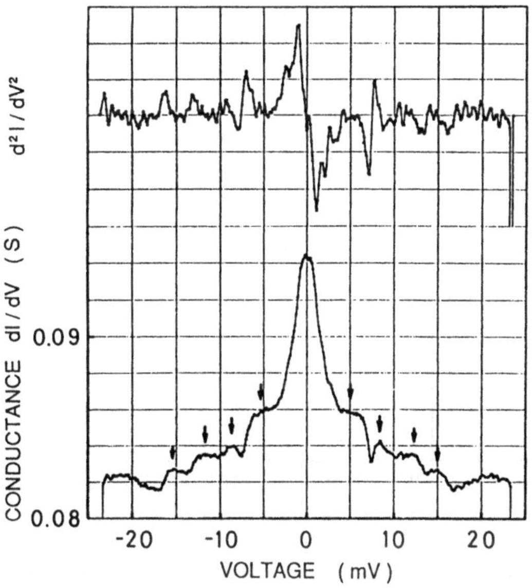

Fig. 2 Conductance dI/dV, and d^2I/dV^2 vs. applied voltage for the $HoBa_2Cu_3O_{7-x}$-$La_{1.5}Ba_{1.5}Cu_3O_{7-y}$-$HoBa_2Cu_3O_{7-x}$ Junction.

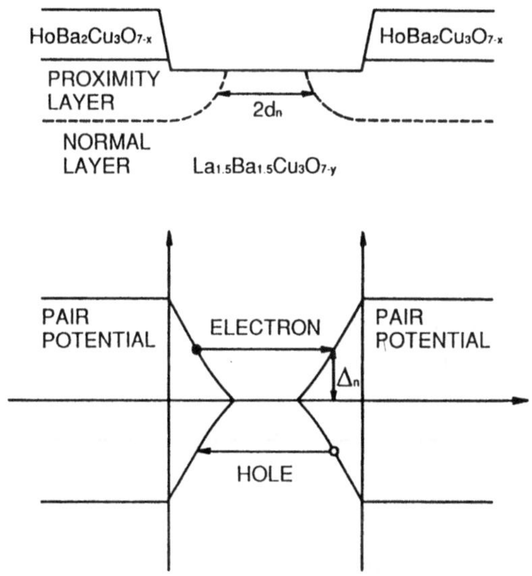

Fig. 3 Geometrical resonance model in a proximity-S-N-S junction.

A superconducting pair is formed by coupling with a carrier having the characteristic state of opposite wave vector and spin and enter into the $HoBa_2Cu_3O_{7-x}$ electrode. A hole is left behind in the same state when the quasiparticle is an electron and an electron is left behind after a hole is removed. Such exchanged quasiparticles come back along the same path as the incident quasiparticles[20-22].

Such basic processes happen repeatedly at both pair potential edges in the $La_{1.5}Ba_{1.5}Cu_3O_{7-y}$ layer when an integer times the quasiparticle wavelength is equal to the distance between the pair potential edges. The wavelength can be derived from the quasiparticle energy, which is defined as the mean square of pair potential and kinetic energy. The quasiparticle energy at which geometrical resonance occurs is equivalent to the applied voltage at the conductance peaks in Fig. 3.

The quasiparticle peak energy E_n is shown in Fig. 4 with respect to the structure number n. The E_n values are almost equally spaced, but not exactly since both the pair potential and edge position change with the quasiparticle energy level. The geometrical resonance corresponds to the Rowell oscillation since the E_n values are smaller than the typical energy gap of the $HoBa_2Cu_3O_{7-x}$ electrode.

The pair potential Δ_n and the effective interval d_n between the pair potential edges were calculated from E_n and n values. The pair potential distribution in the $La_{1.5}Ba_{1.5}Cu_3O_{7-y}$ layer can be derived from these Δ_n and d_n values, and is depicted in Fig. 5. The Fermi velocity for the $La_{1.5}Ba_{1.5}Cu_3O_{7-y}$ was assumed to be 3.1×10^5 m/s. The energy gap of $HoBa_2Cu_3O_{7-x}$ Δ_s has been measured to be near 20 meV from tunneling data.

The Δ_n value near the $HoBa_2Cu_3O_{7-x}$ electrodes could not be determined since the detectable quasiparticle energy was lower than 15 meV. The pair potential distribution is rather coincident with the exponentially decaying curve with a decay length of 65 nm as derived from the dependence of critical current on electrode spacing. These two pair potential distributions coincide even if the pair potential is discontinuous at the S-N interfaces. This may explain the anomalously long superconducting decay length for the $La_{1.5}Ba_{1.5}Cu_3O_{7-y}$.

The geometrical resonance phenomenon only occcurs for quasiparticles which do not suffer from inelastic scattering in the normal layer. The existence of such quasiparticle components may also explain the long decay length. However, the calculated mean free path for the $La_{1.5}Ba_{1.5}Cu_3O_{7-y}$ is only 1 nm, which contradicts the

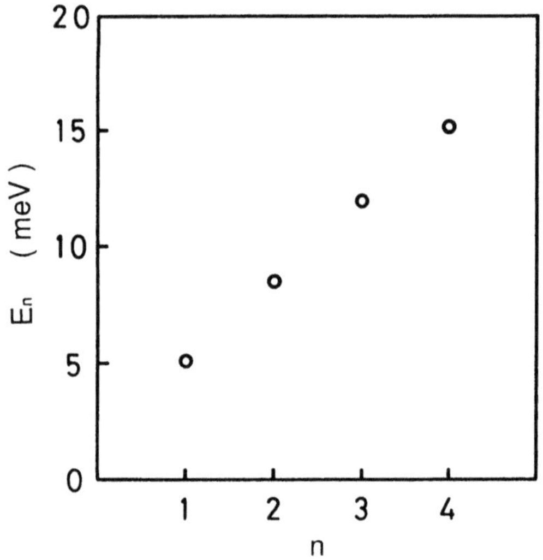

Fig. 4 Quasiparticle enrgy E_n vs. structure index n.

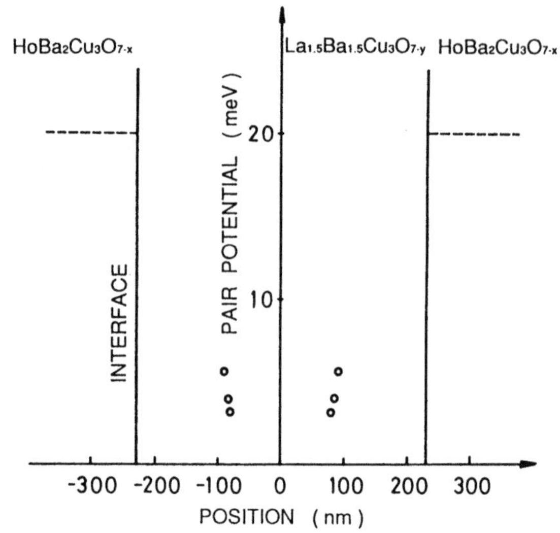

Fig. 5 Estimated pair potential distribution in the junction region.

resonance phenomenon. Carriers in the $La_{1.5}Ba_{1.5}Cu_3O_{7-y}$ are slightly localized since the resistivity has a negative dependence on temperature. The existence of localized states is a possible cause of the short mean free path. An additional conduction mechanism, by which carriers are not trapped in the localized states and extend to the submicron range without inelastic scattering, may occur at low temperatures at the S-N interface. Such a conduction mechanism may be connected with a kind of tunneling which couples with wave functions in the superconducting electrodes.

$HoBa_2Cu_3O_{7-x}$-$PrBa_2Cu_3O_{7-z}$-$HoBa_2Cu_3O_{7-x}$ Junctions

Conductance characteristics of the $HoBa_2Cu_3O_{7-x}$-$PrBa_2Cu_3O_{7-z}$-$HoBa_2Cu_3O_{7-x}$ junctions were investigated and their dependences on temperature and electrode spacing were measured. The dependences of junction resistance on temperature T are shown in Fig. 6. The electrode spacings for these junctions are 0.1, 0.2 and 2 μm. The horizontal scale is the cube root of temperature $T^{-1/3}$. The resistivity of a single $PrBa_2Cu_3O_{7-z}$ film below room temperature obeys the $T^{-1/3}$ dependence which indicates variable-range-hopping conduction[23]. The junction resistances also follow the $T^{-1/3}$ law in the higher temperature ranges and the slope is almost the same as that of the single $PrBa_2Cu_3O_{7-z}$ film. These temperature dependences indicate that carriers in the $PrBa_2Cu_3O_{7-z}$ layer are in the localized states.
The resistance of the junction having an electrode spacing of 2 μm followed the $T^{-1/3}$ law down to 4.2 K. However, the resistances of the junctions with spacings of 0.2 and 0.1 μm deviated from the $T^{-1/3}$ law and saturated at lower temperatures. The temperatures of the resistance deviation were 20 and 39 K for spacings of 0.2 and 0.1 μm, respectively. The junctions with a spacing longer than 0.2 μm did not show superconducting behavior at all. The junction with the spacing of 0.1 μm showed a current near to that for a zero voltage, with a small resistance component due to thermal fluctuation. The current at 4.2 K was 2 μA which was small enough to be suppressed by the thermal noise. A junction with a spacing smaller than 0.1 μm may show superconducting characteristics.
An example of the junction conductance characteristic is shown in Fig. 7. The electrode spacing of was 0.3 μm and the measurement temperature was 4.2 K. The minimum conductance at zero voltage was not due to the superconducting energy gap. The nonlinearity of the conductance was more pronounced as the temperature was lowered and

the electrode spacing decreased. However, the conductance was independent of the applied voltage in the temperature range where the junction resistance obeyed the $T^{-1/3}$ law. The conductance component whose temperature dependence deviated from the $T^{-1/3}$ law may obey some other conduction mechanism such as tunneling between energy levels in the $PrBa_2Cu_3O_{7-z}$ layer. This additional conductance component did not replace the variable-range-hopping one, but was dominant at lower temperatures and may be free from trapping at localized states.
Superconducting characteristics have been detected for the $PrBa_2Cu_3O_{7-z}$-coupled junctions having a layered structure[1]. Two important conditions must be fulfilled for supercurrent to exist. One is that the junction conductance must be increased to detect the critical current level. The second is that the carrier components must not suffer from inelastic scattering, as in the $La_{1.5}Ba_{1.5}Cu_3O_{7-y}$-coupled junction. These two properties may originate from the same conduction mechanism, although this demands further studies.

In conclusion, all-oxide S-N-S junctions were fabricated and their electrical characteristics in the normal state were investigated. The conductance of the $La_{1.5}Ba_{1.5}Cu_3O_{7-y}$-coupled junction showed geometrical resonance of quasiparticles. The pair potential distribution in the $La_{1.5}Ba_{1.5}Cu_3O_{7-y}$ was derived from the resonance structure. The resistivity of the $PrBa_2Cu_3O_{7-z}$-coupled junctions differed from that of a single $PrBa_2Cu_3O_{7-z}$ film and saturated at low temperatures. These conductance characteristics were influenced by the S-N coupling and had a close correlation with the proximity effect in cuprate materials.

This work was performed under the management of FED (the R & D Association for Future Electron Devices) as a part of the R & D of Basic Technology for Future Industries supported by NEDO (New Energy and Industrial Development Organization). The authors are indebted to Dr. U. Kawabe and Dr. K. Miyauchi for their encouragement through this work, and Dr. T. Nishino and Dr. M. Hatano for their valuable discussion. They also wish to thank Mr. S. Akamatsu for preparing oxide films.

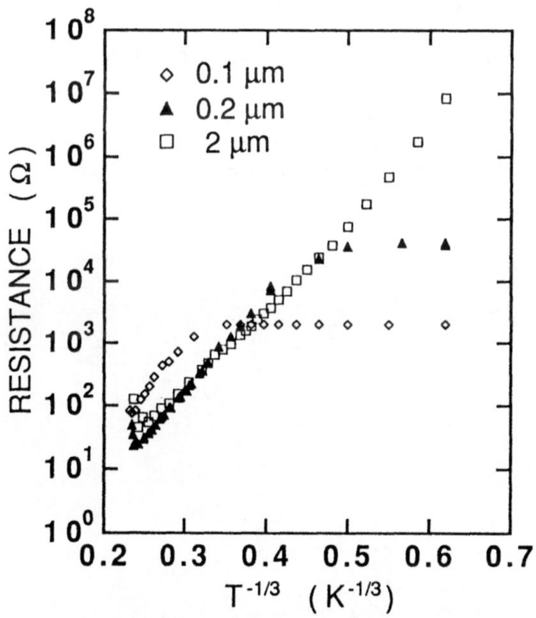

Fig. 6 Temperature dependence of the resistance for $HoBa_2Cu_3O_{7-x}$-$PrBa_2Cu_3O_{7-z}$-$HoBa_2Cu_3O_{7-x}$ junctions.

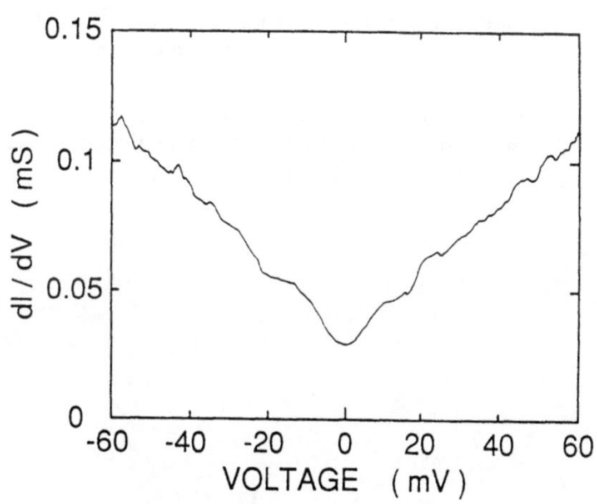

Fig. 7 Conductance vs. applied voltage for the junction with electrode spacing of 0.3 μm.

REFERENCES

1. Rogers CT, Inam A, Hedge MS, Dutta B, Wu XD, and Venkatesan (1989) Appl Phys Lett 55: 2032
2. Mizuno K, Higashino H, Setsune K, and Wasa K (1990) Appl Phys Lett 56: 1469
3. Saitoh S, Fukazawa T, Tarutani Y, and Takagi K (1991) Advances in Superconductivity III Springer Verlag, p 1187
4. Tarutani Y, Fukazawa T, Kabasawa U, Tsukamoto A, Hiratani M, and Takagi K (1991) Appl Phys Lett 58: 2707
5. Tarutani Y, Fukazawa T, Kabasawa U, Tsukamoto A, Hiratani M, Suga M, and Takagi K (1991) Int Conf Materials and Mechanisms of Superconductivity, High-Temperature Superconductors III, Kanazawa
6. Suga M, Nishino T, Tarutani Y, and Takagi K (1991) Int Conf Materials and Mechanisms of Superconductivity, High-Temperature Superconductors III, Kanazawa
7. Kabasawa U, Tarutani Y, Fukazawa, Tsukamoto A, Hiratani M, and Takagi K (1991) Jpn J Appl Phys 30: 1670
8. Suga M, Nishino T, Tarutani Y, and Takagi K, to be published in Physica C
9. Kasai M, Ohno T, Kanke Y, Kozono Y, Hanazono M, and Sugita Y (1990) Jpn J Appl Phys 29: L2219
10. Seto J and Van Duzer T (1974) Low Temperature Physics LT-13. Plenum, New York, p.328
11. Likharev KK (1979) Rev Mod Phys 51: 101
12. de Gennes PG (1964) Rev Mod Phys 36: 225
13. Peng JL, Klavins P, Shelton RN, Radousky HB, Hahn PA, and Bernardez L (1989) Phys Rev B 40: 4517
14. Kebede A, Jee CS, Schwegler J, Crow JE, Mihalisin T, Myer GH, Salomon RE, Schlottmann P, Kuric MV, Bloom SH, and Guertin RP (1989) Phys Rev B 40: 4453
15. Norton DP, Lowndes DH, Sales BC, Budai JD, Chakoumakos BC, and Kerchner HR (1991) Phys Rev Lett 66: 1537
16. Aida T, Tsukamoto A, Imagawa K, Fukazawa T, Saito S, Shindo K, Takagi K, and Miyauchi K (1989) Jpn J Appl Phys 28: L635
17. Hollcomb M, Collman JP, Little WA (1991) Int Conf Materials and Mechanisms of Superconductivity, High-Temperature Superconductors III, Kanazawa
18. Rowell JM and McMillan WL (1966) Phys Rev Lett 16: 453
19. Nishino T, Hatano M, Hasegawa H, Kure T, and Murai F (1990) Phys Rev B 41: 7274
20. Andreev AF (1964) Sov Phys JETP 19: 1228
21. Blonder GE, Tinkham M, and Klapwijk TM (1982) Phys Rev B 25: 4515
22. van Son PC, van Kempen H, and Wyder P (1987) Phys Rev Lett 59: 2226
23. Mott NF (1974) Metal-Insulator Transitions. Taylor & Francis, London

Long-Range Josephson Coupling in BiSrCaCuO-Based Coplanar Junctions

MASAO NAKAO, RYOHKAN YUASA, MASAAKI NEMOTO, HIROAKI FURUKAWA, and SHUJI FUJIWARA

SANYO Tsukuba Research Center, Tsukuba, Ibaraki, 305 Japan

ABSTRACT

We present measurements of the characteristics of superconducting weak links prepared in epitaxial bilayer structures of $Bi_2Sr_2CuO_6$ (BSCO)/$Bi_2Sr_2CaCu_2O_8$ (BSCCO) by focused ion beam lithography. The bilayer films with each layer thickness of 150 nm have been grown *in situ* by atomic-layer-by-layer molecular beam epitaxy and then patterned by implanting Si^+ ions to destroy the superconductivity of the BSCCO top layer. We find S-N-S-like *I-V* characteristics for junctions in which the separation between the two superconducting BSCCO banks is about 300-400 nm. Shapiro steps in the nonhysteretic *I-V* curves appear in response to microwave radiation. We further find that the critical currents I_c for all the weak links fit well with the form given by the Ginzburg-Landau theory, $(1 - T/T_c')^{3/2}$, where T_c' is estimated to be 26-33 K. We conclude that our junctions are not proximity-effect-type but microbridge-type S-S'-S weak links. This characteristics can be interpreted as the formation of filamentary superconducting paths in BSCO.

KEY WORDS: weak link, Josephson effect, oxide barrier material, S-S'-S junction

INTRODUCTION

Various efforts have been concentrated on the fabrication of reliable and reproducible Josephson weak links using high-temperature superconductors (HTSCs) instead of tunnel junctions. Several groups recently have reported on anormalous long-range Josephson coupling in both trilayer and coplanar structures with a normal conductor of a perovskite-related oxide compound, including $YBa_2Cu_3O_{7-x}$/$PrBa_2Cu_3O_{7-y}$ [1], $Bi_2Sr_2CaCu_2O_x$/$Bi_2Sr_2CuO_y$ [2], $Bi_2Sr_2CaCu_2O_x$/$Bi_2Sr_2NdCu_2O_y$[3], $HoBa_2Cu_3O_{7-x}$/$La_{1.5}Ba_{1.5}Cu_3O_{7-y}$ [4], $YBa_2Cu_3O_{7-x}$/$La_{1-y}Ca_yMnO_z$ [5] and $YBa_2Cu_3O_{7-x}$/$SrTi_{1-x}Nb_xO_3$ [6] systems. The oxide layers are believed to become superconducting by proximity-effect coupling, even though the separation between superconducting electrodes is substantially longer than their superconducting coherence lengths $\xi_N(T)$. For a superconductor-normal-metal-superconductor (S-N-S) junction, the order parameter in the normal region decays exponentially over a characteristic length $K_N^{-1} = \xi_N(T)$. This novel phenomenon has some analogy with previously reported "long-range proximity effect" in aluminum thin films with spatially modulated-T_c superlattice structures [7].

In this paper, we describe the properties of superconducting weak links prepared in epitaxial bilayer structures of $Bi_2Sr_2CuO_6$ (BSCO)/$Bi_2Sr_2CaCu_2O_8$ (BSCCO) which demonstrate that the coplanar junctions obtained are microbridge-type S-S'-S weak links. We point out, moreover, that replacing the BSCO barrier layer by $PrBa_2Cu_3O_{7-\delta}$ (PBCO), we failed in observing this anormalous long-range Josephson effect. In contrast to BSCO/BSCCO and PBCO/YBCO, PBCO/BSCCO is free from the problem that superconductivity might be enhanced by local interdiffusion or alloying in the normal regions. The origin of the S-S'-S weak links is discussed in connection with *a-b* twinned structures of the orthorhombic films and a realistic model to explain this phenomenon is presented.

SAMPLE PREPARATION

The epitaxial bilayer films of BSCO (150 nm thick)/BSCCO (150 nm thick) were prepared *in situ* by atomic-layer-by-layer molecular beam epitaxy (MBE) using a reactive oxygen source which has been described in earlier publications [8]. Briefly, the elements are evaporated successively (layer-by-layer), and a beam of atomic oxygen radicals is injected towards the substrate during the entire process. During growth, the substrates are held at elevated temperatures, typically around 700 ℃, to provide adequate surface mobility for the deposited atoms. We mainly use $SrTiO_3$ or MgO substrates. The films are *c*-axis oriented as layered and currently reach zero resistance around 60 K. The focused ion beam (FIB) machine [9] was then used to define the S-N-S geometry. The BSCO/BSCCO bilayer films were patterned by implanting 100-KeV Si^+ ions emitted from a AuSiBe liquid metal ion source at a typical line dose of 3×10^9 ions/cm. A beam diameter is 100 nm at 75 pA beam current. Implantation with Si^+ ions destroys the superconductivity in the BSCCO top layer and forms a narrow insulating region. Thus, after implantation, two BSCCO superconducting banks are separated by the narrow region, so that current should flow through the underlying BSCO layers. The dimension of the implanted region along the current direction will be referred to as the nominal length *L*. In this experiment, *L* ranges from 300 nm to 400 nm and the width *W* ranges from 5 μm to 100 μm.

As for the bilayer structures of PBCO/BSCCO, the PBCO layers were first deposited by rf magnetron sputtering before being loaded into the MBE apparatus. The PBCO films are *c*-axis oriented and shows semiconducting behavior. This was followed by the atomic-layer-by-layer MBE growth of BSCCO directly on the PBCO layers. Prior to the MBE growth, no special surface treatment is performed. Figure 1 shows a θ-2θ x-ray diffraction pattern for the PBCO/BSCCO bilayer film, indicating that both PBCO and BSCCO layers are highly *c*-axis oriented. The PBCO/BSCCO bilayer films were then patterned using the same FIB procedures as described above.

MATERIAL PROPERTIES

The transport resistivity was measured in a standard four-point geometry using pressed indium contacts. Figure 2 shows typical resistivities as a function of temperature for the BSCO thin film, the as-layered BSCO/BSCCO film and the coplanar structure obtained after the Si^+ implantation. It is well known that BSCO can be modified from a superconductor (~ 15 K) to a semiconductor, depending on its oxygen content. In this experiment, however, the MBE growth was carried out under such conditions that the BSCO layer is semiconducting, whereas the BSCCO layer is superconducting in order to utilize BSCO as a barrier material. The epitaxial bilayer film shows zero resistance around 60 K without being degraded by layering. Surprisingly, our coplanar junction also exhibits a superconducting transition quite similar to the as-layered film, so that supercurrent flowing through the BSCO barrier layer is clearly observed up to around 60 K. To rule out the possibility that the superconductivity of the BSCCO layer survives despite the insulation by implanting Si^+ ions, we have patterned a BSCCO single-phase film using the same FIB procedures. Consequently, the resistivity of the insulating region was estimated to be over 30 Ω cm.

On the other hand, typical resistivities for the PBCO thin film, the as-layered PBCO/BSCCO film and the coplanar structure obtained after the Si^+ implantation are plotted as a function of temperature in Fig. 3. Since PBCO is less conductive than BSCO, the coplanar junction demonstrates a relatively high resistivity. We see a clear drop in resistivity around 60 K for the junction, but, in this case, the resistivity does not go to zero.

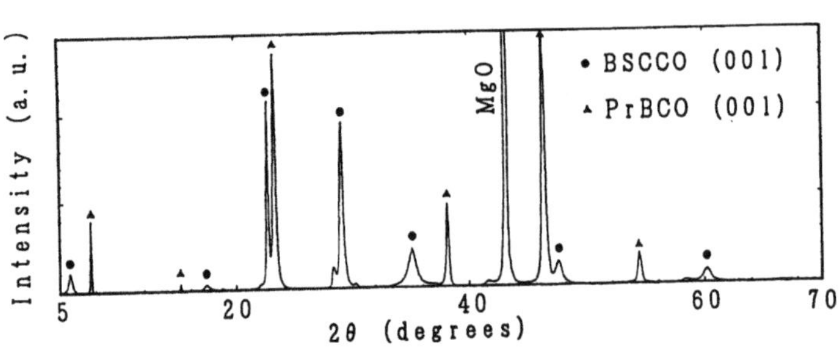

Fig. 1. θ-2θ x-ray diffraction pattern for a PBCO/BSCCO bilayer film.

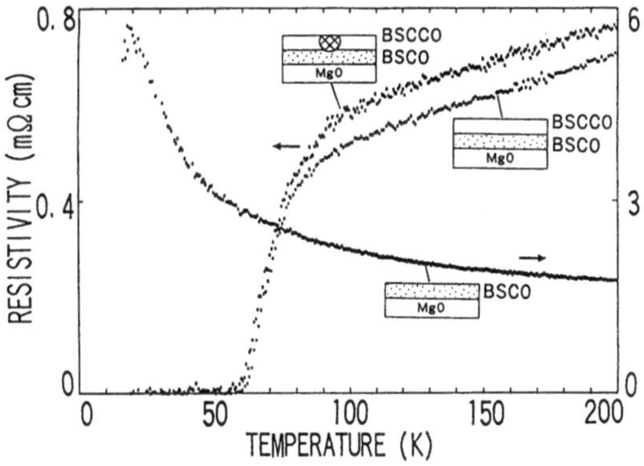

Fig. 2. Resistivity vs temperature for a BSCO thin film, an as-layered BSCO/BSCCO film and a coplanar structure obtained after Si^+ implantation.

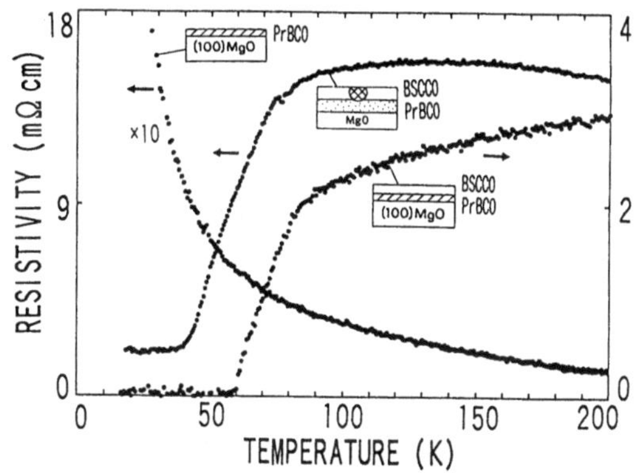

Fig. 3. Resistivity vs temperature for a PBCO thin film, an as-layered PBCO/BSCCO film and a coplanar structure obtained after Si^+ implantation.

WEAK-LINK CHARACTERISTICS

Next, we turn to observation of the Josephson effect. Figure 4 shows typical current-voltage (I-V) characteristics displaying S-N-S-like behavior for the BSCO/BSCCO junction. We note that a zero-voltage supercurrent is clearly observed in the I-V curve. The critical current I_c of the junction is 8.1 mA at 16 K and 4.6 mA at 30 K. The typical low-temperature $I_c \cdot R_N$ product is estimated to be about 1.5 mV at 16 K, where R_N is the normal-state resistance of the junction. Furthermore, when an rf signal at 10.8 GHz was applied, steps appeared in the I-V curves at voltages $V = nfh/2e$, where n is an integer [10]. We could not see any subharmonic steps.

To better understand the origin of these weak links, we have measured the temperature dependence of I_c. The data for junctions with W of 20 μm and 100 μm are plotted in Fig. 5, indicating that I_c follows the Ginzburg-Landau form, $I_c/I_0 = (1 - T/T_c')^{3/2}$ at low temperatures, but with a tail extending to temperatures well above T_c'. Here, T_c' is the temperature obtained by extrapolating the curve to $I_c = 0$ and estimated to be 26 K and 33 K for the junctions with W of 20 μm and 100 μm, respectively. The values of I_0 for the 20-μm and 100-μm junctions are 0.2 mA and 8.6 mA, respectively. Thus, we find that these features are well described by the behavior of microbridge-type S-S'-S weak links.

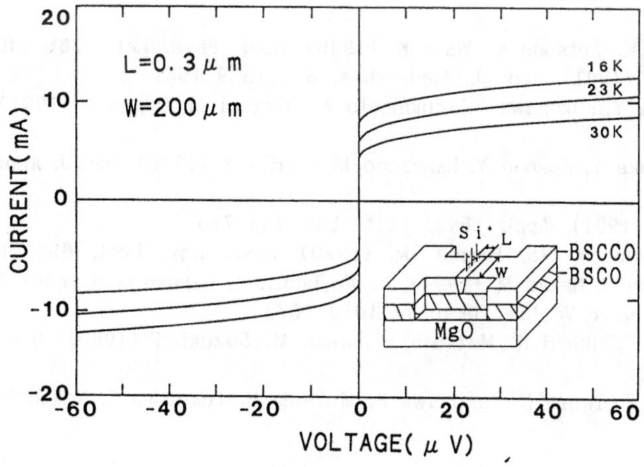

Fig. 4. Typical I-V characteristics for a BSCO/BSCCO junction.

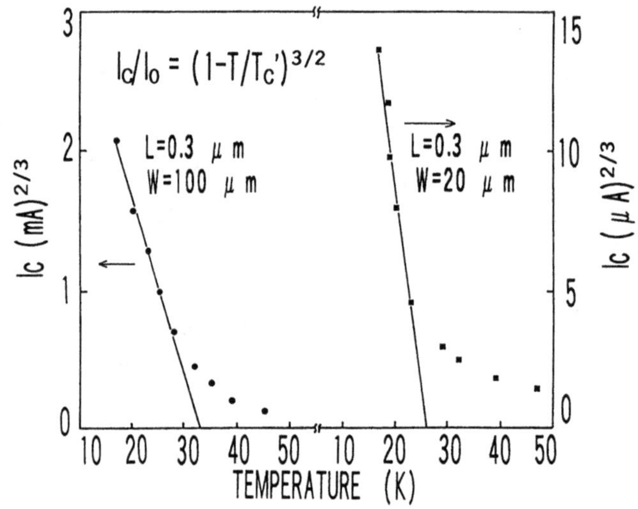

Fig. 5. Temperature dependence of critical current (I_c) for two junctions. I_c follows the Ginzburg-Landau form, $I_c/I_0 = (1 - T/T_c')^{3/2}$ at low temperatures, but with a tail extending to temperatures well above T_c'.

Instead of assuming long-range proximity-effect models, we propose that this weak-link behavior indicates the importance of film-growth mechanisms in perovskite-related oxides from a materials point of view. While microscopic defects introduced by the multilayer growth are minimized, mechanical shorting is still possible. The BSCCO and BSCO compounds are well known to have the orthorhombic crystal structure with incommensurate modulation. Thus, all BSCCO and BSCO thin films grown on cubic or tetragonal substrates have in-plane 90° misorientation defects, i.e., *a*-*b* twinning. Supposing that Ca in the BSCCO layer readily diffused into the BSCO layer along the twin boundaries, filamentary superconducting paths with enhanced T_c' would be formed in the BSCO layers. This network of filamentary superconducting paths in BSCO account for the S-S'-S weak-link behavior for our coplanar structures as well as trilayer structures. Also, there are different possibilities to explain the enhancement of local superconductivity such as misfit strain or microcracks. Additional experiments, however, are needed in order to understand this phenomenon.

REFERENCES

1. Rogers CT, Inam A, Hegde MS, Dutta B, Wu XD, Venkatesan T (1989) Appl. Phys. Lett. 55: 2032-2034
2. Mizuno K, Higashino K, Setsune K, Wasa K (1990) Appl. Phys. Lett. 56: 1469-1471
3. Mizuno K, Setune K (1991) Jpn. J. Appl. Phys. 30: 1559-1561
4. Kabasawa U, Tarutani Y, Fukazawa T, Tsukamoto A, Hiratani M, Takagi K (1991) Jpn. J. Appl. Phys. 30: L1670-1675
5. Kasai M, Ohno T, Kanke Y, Kozono Y, Hanazono M, Sugita Y (1990) Jpn. J. Appl. Phys. 29: L2219-2222
6. Chin DK, Duzer TV (1991) Appl. Phys. Lett. 58: 753-755
7. Kwong YK, Lin K, Isaacson MS, Parpia JM (1990) Phys. Rev. Lett. 65: 2905-2908
8. Furukawa H, Tokunaga S, Nakao M (1991) In: Kajimura K, Hayakawa H (eds) Advances in Superconductivity III. Springer-Verlag, Tokyo, pp 1069-1072
9. Fujiwara S, Yuasa R, Shikichi K, Matsuta Y, Nakao M, Suzuki S (1990) Jpn. J. Appl. Phys. 29: L2307-2311
10. Yuasa R, Nemoto M, Fujiwara S, Furukawa H, Mukaida H, Tokunaga S, Nakao M (1990) Physica C (in press)

Tunneling Spectroscopy of $Bi_2Sr_2CaCu_2O_x$ Single Crystals and Analysis Using a Superlattice Model

K. SHIMAOKA, Y. YAMANO, K. TAKAHASHI, T. USUKI, Y. YOSHISATO, and S. NAKANO

Functional Materials Research Center, Sanyo Electric Co., Ltd., Hirakata, Osaka, 573 Japan

ABSTRACT

The results of tunneling spectroscopy of Bi2Sr2CaCu2Ox (BSCCO) superconductor are analyzed using a model developed by Tachiki et al. [1]. We typically obtained two types of tunnel conductance in the experiments. We believed the difference between these tunneling conductances to be cased by differences of the crystal plane in which quasi-particles were injected. Based on this supposition, the behavior of the tunneling conductance is qualitatively explained, although the value of $2\Delta/k_BT_c=10.2$ obtained from the experiment is much larger than the 3.2 obtained from the calculation under BCS approximation.

KEY WORDS: tunneling spectroscopy, Bi2Sr2CaCu2Ox, superlattice model, single crystal

INTRODUCTION

We studied the electronic structure of BSCCO superconductors for the purpose of application to a quasi-particle injection-type superconducting base transistor. In such a device, a switching effect is caused by the existence of an energy gap in a superconductive base layer [2]. For this reason the electronic structure of the superconducting base material is very important. However despite considerable research in this area [3]~[5] the real electronic structure for BSCCO yet has to be discovered. To investigate this electronic structure as thoroughly as possible we have conducted tunneling spectroscopy experiments using Nb/MgO/BSCCO tunnel junctions fabricated on cleavage planes of highquality BSCCO single crystals. Details of these experiments has been previously reported [6]. In these experiments we typically obtained two types of tunnel conductance curves. One is similar in shape to BCS-type tunnel conductance and the other has several small conductance peaks on a linear background. As a results of SEM (scanning electron microscope) observation of the cleavage planes of BSCCO single crystals, we found regions existed which include many step-like defects

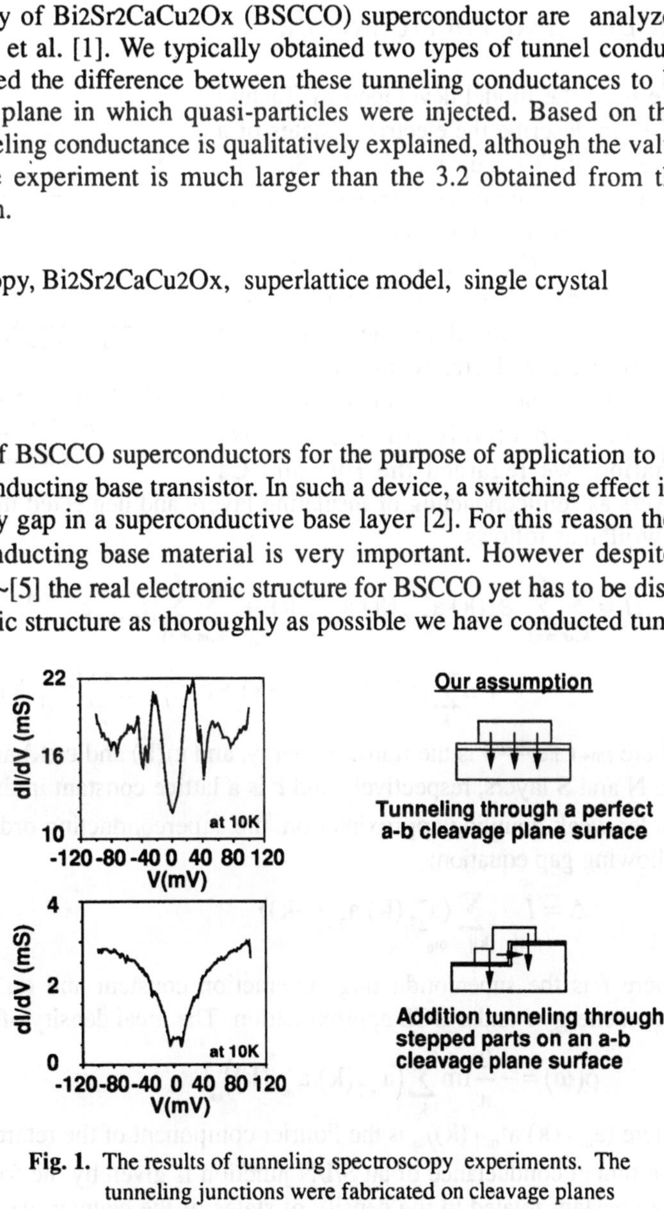

Fig. 1. The results of tunneling spectroscopy experiments. The tunneling junctions were fabricated on cleavage planes of BSCCO single crystals (5 mm X 5 mm, Tc = 91 K). See Ref. 6

(typically several-hundred nm in height) on the cleavage planes of the BSCCO single crystals. It is well known that BSCCO superconductors have a very anisotropic layered crystal structure. It is therefore expected that their electronic structure should have strong anisotropy, too. Thus, we assumed that these differences in tunnel conductance curves were dominated by the direction of injection of the quasi particles, i.e. in one sample, injection of quasi-particles into the a-b plane of BSCCO crystal occurs, in the another sample the injection into the a-c or b-c plane occurs.

To confirm this theoretically, we calculated the tunnel conductance of a layered superconductor using a formula developed by Tachiki et al. [1] and compared the experimental results with the calculated ones. In this paper we report the results of this comparison.

MODEL AND FORMULATION.

We used the model developed by Tachiki et al. to describe the electronic states of a layered BSCCO superconductor (See Fig. 2). BSCCO superconductors have a strong layered structure constructed of a stack of BiO, SrO, CuO_2 and Ca layers. We simplified this structure as an N/S/S/N periodic structure along the z axis with equal spacing. Here, N and S represent normal-conducting and superconducting layers, respectively. In this approximation, we regarded the BiO and Ca layers as semiconducting or insulating layers and neglected them. The Hamiltonian of this system is written as follows:

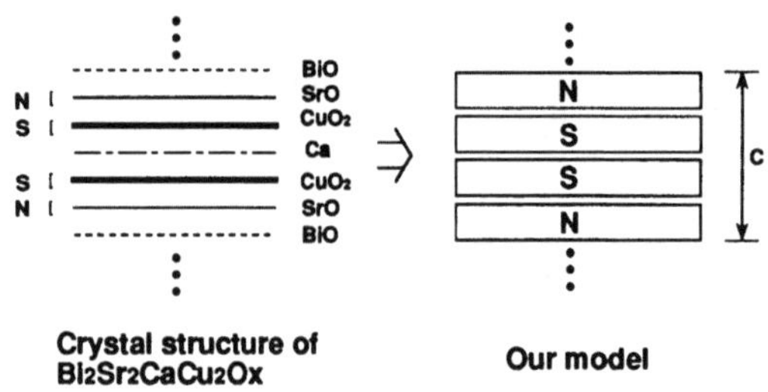

Fig. 2. A N/S/S/N periodic superlattice model of BSCCO crystals. N is normal-conducting and S is the superconducting layer.

$$H = \sum_{\mathbf{k},\sigma}\sum_{m=1}^{4} \varepsilon_m(\mathbf{k})\, a^{+}_{m,\sigma}(\mathbf{k})\, a_{m,\sigma}(\mathbf{k}) + \sum_{\mathbf{k},m}\sum_{m=1}^{4}\left(t_{m+1,m} e^{-ikz/4} a^{+}_{m+1,\sigma}(\mathbf{k})\, a_{m,\sigma}(\mathbf{k}) + \mathrm{c.c.}\right)$$
$$+ \Delta\sum_{\mathbf{k}}\left(a^{+}_{2,\uparrow}(\mathbf{k})\, a_{2,\downarrow}(-\mathbf{k}) + \mathrm{c.c.}\right) + \Delta\sum_{\mathbf{k}}\left(a^{+}_{3,\uparrow}(\mathbf{k})\, a_{3,\downarrow}(-\mathbf{k}) + \mathrm{c.c.}\right) + \frac{2\Delta}{I^2} \quad (2.1)$$

where $t_{m+1,m}e^{-ikzc/4}$ is the transfer energy, and $\varepsilon_1(\mathbf{k})$ and $\varepsilon_2(\mathbf{k})$ are two-dimensional energy bands for the N and S layers, respectively and c is a lattice constant in the direction of the z axis (See Fig.2). Under weak-coupling approximation, the superconducting order parameter Δ is determined by the following gap equation:

$$\Delta = I \sum_{|\varepsilon_1(\mathbf{k})| \le \omega_0} \langle a^{+}_{2\uparrow}(\mathbf{k})\, a_{2\downarrow}(-\mathbf{k})\rangle \quad (2.2)$$

where I is the superconducting interaction constant and ω_0 is a cut-off parameter of the interaction energy I under BCS approximation. The local density of states at the m th layer is given by

$$\rho(\omega) = -\frac{2}{\pi}\mathrm{Im}\sum_{\mathbf{k}}\left(a_{m\uparrow}(\mathbf{k}), a^{+}_{m\uparrow}(\mathbf{k})\right)_{\omega} \quad (2.3)$$

where $(a_{m\uparrow}(\mathbf{k})\, a^{+}_{m\uparrow}(\mathbf{k}))_\omega$ is the Fourier component of the retarded Green function with energy ω. The tunnel conductance of an S/I/N junction is given by the formula (2.4), where the coefficient A is a constant related to the density of states of the counter electrode (normal metal) and the nature of the junction.

$$G_m(V) = A\frac{e}{4k_BT}\int_{-\infty}^{\infty} d\omega\ \rho(\omega)\mathrm{sech}^2\left(\frac{\omega + eV}{2k_BT}\right) \tag{2.4}$$

RESULTS OF NUMERICAL CALCULATIONS AND COMPARISON WITH EXPERIMENTAL RESULTS.

We supposed the following parabolic energy bands for $\varepsilon_1(\mathbf{k})$ and $\varepsilon_2(\mathbf{k})$.

$$\varepsilon_1(\mathbf{k}) = \frac{\hbar^2}{2m_1}(k^2 - k_1^2)\ ,\ \varepsilon_2(\mathbf{k}) = \frac{\hbar^2}{2m_2}(k^2 - k_2^2)\ ,\ \varepsilon_0 \equiv \frac{\hbar^2 k_1^2}{2m_1}\ ,\ k = \sqrt{k_x^2 + k_y^2} \tag{3.1}$$

In numerical calculations, we used the following values for the parameters.

$$\left|t_{1,2}\right| / \varepsilon_0 = 0.07\ ,\ \left|t_{2,3}\right| / \varepsilon_0 = 0.07\ ,\ \left|t_{1,4}\right| / \varepsilon_0 = 0.03\ ,\ \Delta / \varepsilon_0 = 0.1$$

$$\frac{m_1}{m_2} = 1.22\ ,\ \left(\frac{k_2}{k_1}\right)^2 = 2.57 \tag{3.2}$$

Essentially the value of the superconducting order parameter Δ should be determined by solving the gap equation (2.2) self-consistently. In fact, we solved the gap equation and got the temperature dependence of the superconducting-order parameter $\Delta(T)$. As a result , we obtained a value of 12.5 meV for the order parameter at 0 K. This value is, however, considerably smaller than the experimentally obtained value, as shown later. For this reason, we employed a value of the superconducting order parameter about three times as large as the calculated one in the subsequent calculations of the tunnel conductance curve.

Using the formulas (2.3) and (2.4), we can obtain the following tunnel conductance curves for an S/I/N junction fabricated on this system:

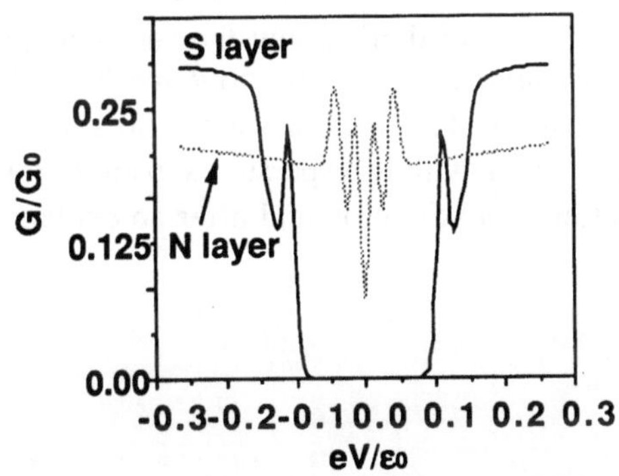

Fig. 3. The calculated tunneling conductances of normal-conducting and superconducting layer. (ε_0 = 400 meV, T = 10 K)

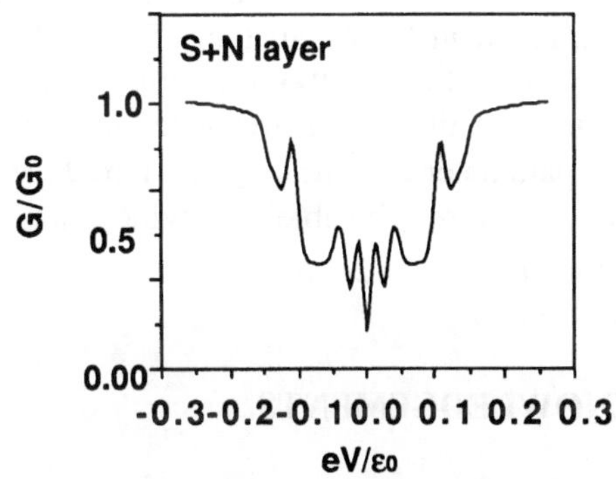

Fig. 4. The calculated tunneling conductance of simultaneous tunneling into normal-conducting and superconducting layer.

As shown in Fig. 3 the conductance of the normal-conducting layer behaves as if the layer had a narrow energy gap compared with the supposed value Δ/ε_0 = 0.1. If the quasi particles are injected into these superconducting and normal-conducting layers at once, the conductance will change into a kind of mixture of the previous ones. Conductance in this situation is shown in Fig. 4. At first sight, this conductance seems to have several peaks on a linear background.

Considering these results, we compared the calculations with the experimental findings. In Fig. 5, the experimental conductance of the superconducting layer C is computed by subtracting about half the conductance B from the conductance of A. This corresponds to a subtraction of the normal

-conducting component from the mixed conductance. Here the value the of super-conducting order parameter is supposed to 40 meV. From this graph, we can see that the result of these calculations qualitatively explains the experimental results. The bias voltage at the main peak of calculated conductance C' is almost equal to that of the experimental conductance C. The bias voltage at the main peak of calculated conductance C' is almost equal to the supposed value of the superconducting order parameter divided by an electron charge, 40 mV. Thus, we can estimate the real value of the superconducting order parameter to be about 40 meV.

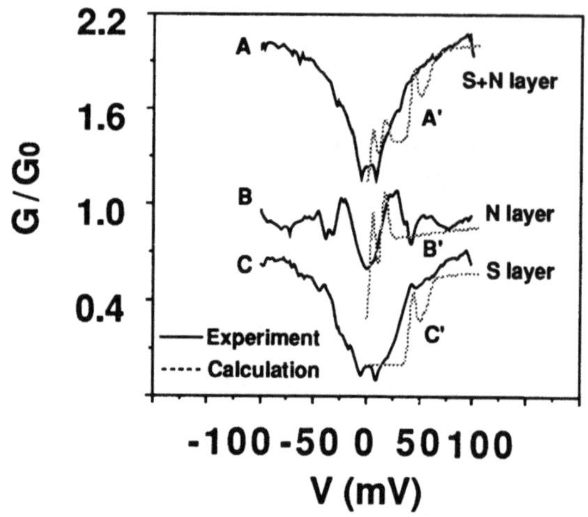

Fig. 5. Comparison between the calculations and the experimental results. The conductance A and A' are shifted up by a normalized conductance of 1.0 for easy viewing.

DISCUSSION AND CONCLUSION

Using a model developed by Tachiki et al. we calculated the tunneling conductance of a layered superconductor and compared it with the results of tunneling spectroscopy experiments. As a result, we found this model qualitatively explains the pattern of tunnel conductance. However the value of a superconducting order parameter Δ obtained from the comparison (40 meV, $2\Delta/k_BTc$ = 10.2) is considerably larger than the calculated value (12.5 mV, $2\Delta/k_BTc$=3.2). We suspect this is because BSCCO is a strong-coupling superconductor, and we cannot employ the BCS approximation in the calculation of Δ in this system. Since we don't know the real interaction which causes the superconducting state of BSCCO superconductors, now we have little information on the real superconducting coupling parameter, I. However, if we can obtain a value for this parameter in some way and introduce a new factor, $I_{\mathbf{k}}$ with **k** dispersion substitute for the constant I, we may be able to provide a somewhat better approximation and offer an explanation for the large value of $2\Delta/k_BTc$.

ACKNOWLEDGEMENTS

This work was performed under the management of FED (the R & D Association for Future Electron Devices) as a part of R & D of Basic Technology for Future Industries sponsored by NEDO (New Energy and Industrial Technology Development Organization).

REFERENCES

1. M.Tachiki, S.Takahashi, F.Steglich and H. Adrian: Z.Phys.B-Condensed Matter, **80**,161(1990)
2. D.J.Frank, M.J.Brady and A. Davidson: IEEE Trans. Magn. Mag-21, 721 (1985).
3. Z.Shiping, T.Hongjie, C.Yinfei, Y.Yifen and Y.Qiansheng: Solid State Comm., 67, 1179(1989)
4. T.Ekino and J.Akimitsu: Phys. Rev., **B40**, 6902 (1989).
5. M.Lee, D.B.Mitzi, A.Kapitulink and M.R.Beasley: Rhys.Rev., **B39**, 801 (1989)
6. K. Takahashi, K, Shimaoka, K. Yamano, T. Usuki, Y. Yorinobu and S. Nakano: Ext. Abs. of 3rd FED Workshop on High-Temperture Electron Devices 182 (1991) .

Properties of Junctions with Layered Structure

TOSHIYUKI MATSUI, TAKESHI SUZUKI, HIROSHI KIMURA, KOICHI TSUDA, MEGUMI NAGANO, and KAZUO MUKAE

Fuji Electric Corporate Research and Development, Ltd., Yokosuka, Kanagawa, 240-01 Japan

ABSTRACT

The relation between surface roughness of the YBCO films and uniformity of a barrier layer was studied using AFM and tunnel junctions with Au-LSGO-YBCO layered structure. AFM images showed that the surface roughness of both (001) and (013) oriented films had strong dependence on their thickness, and the roughness decreased with reducing the thickness. When the LSGO barrier was formed on these YBCO films, the barrier height increased from 24 mV to 1.3 V with reducing the thickness of the YBCO from 400 nm to 70 nm. It suggests that the uniformity of barrier layer was improved by reducing the roughness of the YBCO films.

KEY WORDS: AFM, surface roughness, tunnel junction, uniformity of barrier layer

INTRODUCTION

Tunnel type Josephson junctions using high-critical temperature (Tc) superconductors have a potential not only for the high operation temperature and high speed devices, but also for the novel devices with new concepts. In this point of view, many types of tunnel junctions have been reported with artificial barriers [1-5] or native barriers [6,7], however they did not show the same characteristics on each other. This is probably due to not only electric problems of surface of the high-Tc superconductors, but also problems of barrier layers. Although the former was studied with several techniques such as ESCA [8] and SNS type junctions [9,10], the latter was little reported.

Uniformity of the barrier layer is mainly determined by two factors; one is a difference of surface energies between the base electrode and the barrier material, and the other is a surface roughness of base electrode, that is to say step-coverage. Since the high-Tc superconductors trend to form large crystal grains, the latter seems important factor at present. Moreover, recently, it was reported that screw dislocations exist even on epitaxial films [11]. Hence it is necessary to study a relation between the surface morphology of the base YBCO electrode and the uniformity of the barrier layer. In this paper, we will report the results about the observation of surface roughness of the YBCO films using atomic force microscope (AFM) and will discuss the relation between the surface roughness of the YBCO films and the uniformity of the barrier layer using tunnel junctions with Au-$LaSrGaO_4$ (LSGO)-YBCO layered structure.

EXPERIMENT

YBCO films were prepared by an rf magnetron sputtering system. A sintered $YBa_2Cu_4O_x$ disc was used as the target. A substrate was located 28 mm above the center of the target to avoid a bombardment of high energy particles [12]. The sputtering gas was a mixture of 50%Ar and 50%O_2 and the total gas pressure was 0.5 Pa. The substrate temperature was kept at 650 °C during the deposition process of YBCO. After the process, oxygen of 1 atm was introduced into the chamber and cooled below 50 °C. MgO and $SrTiO_3$ single wafers were used as the substrates. X-ray diffraction (XRD) analysis revealed that highly (001) and (013) oriented films were formed on MgO(001) and $SrTiO_3$ (110)

substrates respectively. The surface roughness of these films was investigated with AFM in the air at the room temperature.

In order to estimate the uniformity of the barrier layer formed on the YBCO films, tunnel junctions were fabricated using LSGO as the barrier layer. The LSGO layer was formed on the YBCO film with an in-situ process. The substrate temperature was 300 °C for LSGO and the total gas pressure was 1 Pa. After that Au was formed on it as a top electrode in another chamber with a metal mask. The junction area was 2x1.8 mm^2. Au was selected as the top electrode to prevent a formation of native barrier. The thickness of both YBCO and LSGO was estimated from the deposition rate and the deposition time.

RESULTS AND DISCUSSIONS

Although all YBCO films had mirror-like surfaces, AFM images showed they had large crystal grains. Figure 1 shows the AFM images of the YBCO films with the (001) oriented structure for the various thickness. Using these data, magnitude of the mean roughness was plotted as a function of the YBCO thickness as shown in Fig.2. The mean surface roughness was about 8 nm (magnitude of peak to valley was 59 nm) for the 400 nm thickness and it decreased with reducing the thickness of the YBCO films. This is probably due to a formation of the crystal grains and it results in the increase of surface roughness. This trend was also obtained about the (013) oriented films. Although the (013) oriented films had flatter surfaces than the (001) oriented one, partially they had sharp grains as shown in Fig. 3. Then magnitude of the mean roughness was almost the same as that of the (001) oriented films. XRD analysis and scanning auger microprobe (SAM) revealed that the grains were not impurities but YBCO grains with the (013) structure. In contrast to the improvement of the surface roughness, the Tc drastically dropped below the film thickness of 100 nm and superconductivity was no longer observed below 10 nm thickness. Finally we could reduce the surface roughness to about 1 nm (magnitude of peak to valley was about 10 nm) on the YBCO films of 70 nm thickness. This value was equal to that of Nb films prepared by an electron beam evaporation system.

Tunnel junctions of Au-LSGO-YBCO layered structure were fabricated to estimate the uniformity of the LSGO barrier layer. The YBCO films of 400 nm and 70 nm thickness were used as the base electrodes. Junctions(a) were fabricated on the YBCO films of 400 nm and junctions(b) were on that of 70 nm thickness. The Tc's of the films were 70 K and 32K respectively.

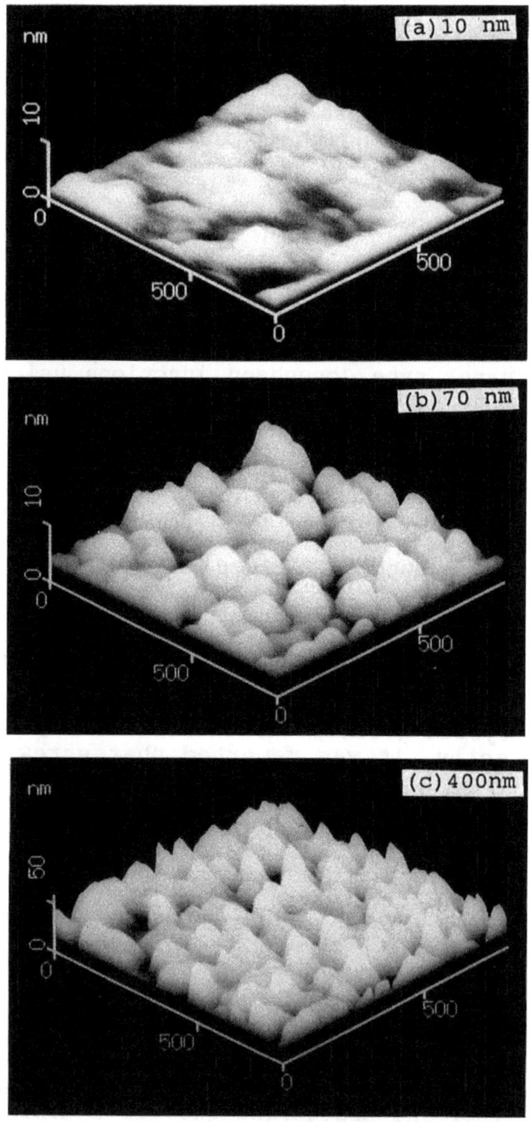

Fig.1 AFM images of YBCO films with (001) structure for the various thickness.

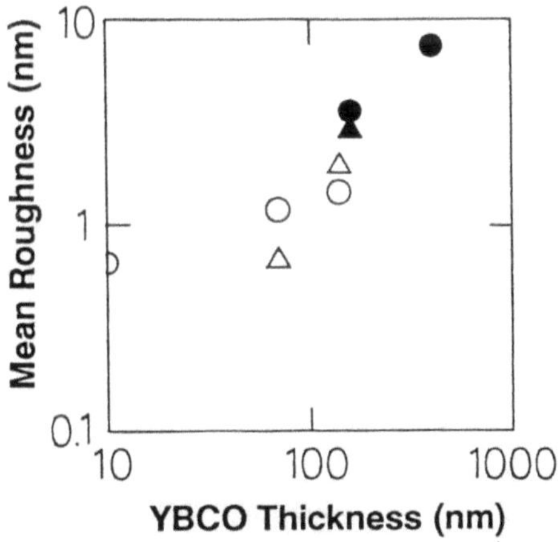

Fig.2 Mean roughness of the YBCO surfaces as a function of the YBCO thickness. Open and closed circles show the surface roughness about the (001) structure formed with the deposition rate of 4.7 nm/min and 13.3 nm/min respectively, and triangles show about the (013) structure.

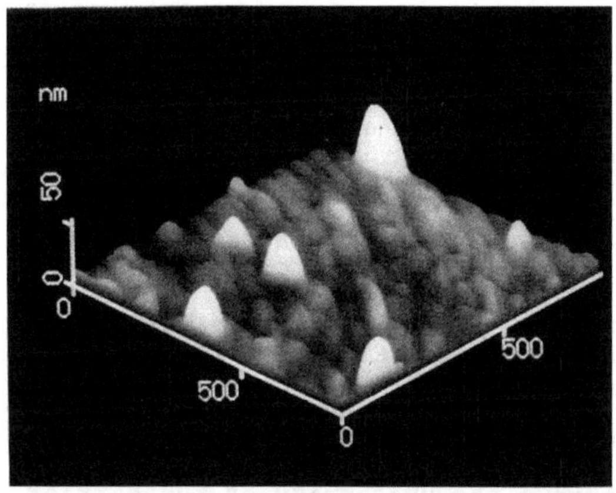

Fig.3 AFM image of the YBCO film with the (013) structure. The thickness was 140 nm.

According to the tunneling theory at very low voltage region, tunnel resistance has exponential dependence on the product of the barrier thickness and the square root of the barrier height. We can, therefore, estimate the barrier uniformity from the magnitude of the barrier height. Figure 4 shows a variation of junction resistances as a function of the barrier thickness. The junction resistance was defined as the resistance at 40 mV since dV/dI-V characteristics of the junctions shows parabolic curve even above the Tc. The junction resistance exponentially increased with the increase of the barrier thickness on both junctions, as shown in Fig. 4. It revealed that the current flowed across the junction with the tunnel process. However the barrier height, which is estimated from the slope of the curve, was only 24 mV for the junctions(a). This value is almost the same as a previous report [1]. It suggests the presence of the regions where the barrier is very thin or has pinholes since the optical gap of LSGO is 3-4 V. On the other hand, the junctions(b) had larger junction resistance and slope than those of junctions(a). The barrier height of junctions(b) was estimated to be about 1.3 V. Figure 5 shows the logarithmic conductivity of a junc-

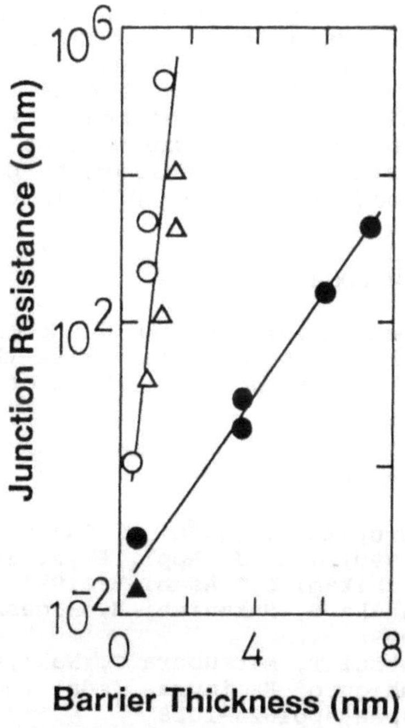

Fig.4 Junction resistance of YBCO-LSGO-Au tunnel junctions as a function of the barrier thickness. Open and closed characters show the YBCO thickness of 70 and 400 nm, and circles and triangles show YBCO structures with (001) and (013) orientation, respectively. These measurements were performed at 10 K.

tion(b) with the barrier thickness of 3.6 nm as a function of bias voltage. Using this technique, since the position of maximum logarithmic conductivity allows the barrier height, we can determined it independently of the other parameters on the theory [13]. The maximum point was clearly observed at around 1.6 V, which corresponds to the barrier height between YBCO and the LSGO barrier. This value agrees well with the barrier height estimated from Fig. 4.

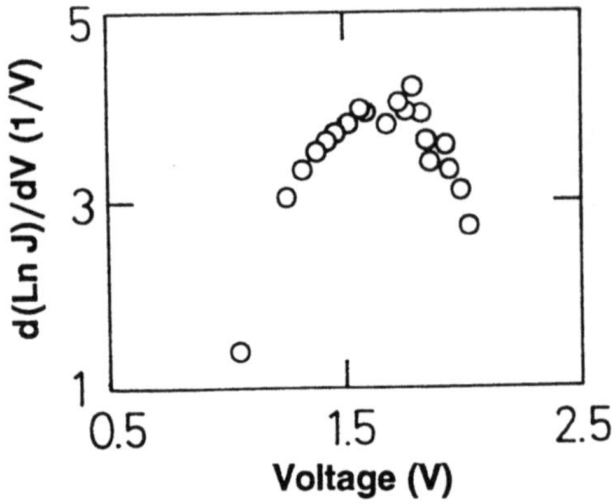

Fig.5 Logarithmic conductivity of junction(b) with the barrier thickness of 3.6nm as a function of the voltage. This measurement was performed at 10 K.

Hence reducing the surface roughness of the base YBCO electrode, we could improve the uniformity of the barrier layers.

CONCLUSION

The relation between surface roughness of the YBCO films and uniformity of the barrier layer was studied using AFM and tunnel junctions with Au-LSGO-YBCO layered structure. AFM images showed that the surface roughness of YBCO films decreased with reducing the film thickness. When the LSGO barrier layer was formed on the YBCO films, the barrier height increased with reducing the YBCO thickness. It suggests that the uniformity of the barrier layer was improved by reducing the roughness of the YBCO films.

ACKNOWLEDGMENTS

Authors would like to thank Prof. Y. Okabe in the Univ. of Tokyo for his useful discussions. They also thank Mr. K. Suzuki and M. Kimura in the Univ. of Tokyo for the preparation of Nb films.

REFERENCES

1. Furuyama M, Iguchi I, Shirai K, Kusumori T, Ohtake H, Tomura S, Nasu M (1990) Jpn. J. Appl. Phys. 29: L459
2. Michikami O, Asahi M (1991) Jpn. J. Appl. Phys. 30: 466
3. Tanaka S, Nakanishi H, Higashi K, Itozaki H (1990) Jpn. J. Appl. Phys. 29: 1059
4. Matsui T, Matsubara G, Nakayama A, Satoh N, Mukae K, Okabe Y (1989) In: Ishiguro Kajimura (eds) 2nd Int. Sympo. on Superconductivity. Tsukuba, Japan, pp1025-1028
5. Matsui T, Miyazaki M, Kurosaka T, Matsubara G, Nakayama A, Mukae K, Okabe Y (1990) In: Kajimura Hayakawa (eds) 3rd Int. Sympo. on Superconductivity. Sendai, Japan, pp1195-1198
6. Geerk J, Xi XX, Linker G (1990) Z. Phys. Lett. 56: 1487
7. Gurvitch M, Valles JM, Jr., Cucolo AM, Dynes RC, Garno JP, Schneemeyer LF, Waszczak JV (1989) Phys. Rev. Lett. 63: 1008
8. Parmigian F, Samoggia G, Calandra C, Manghi F (1989) J. Appl. Phys. 66: 5958
9. Mizushima K, Sagoi M, Yoshida J (1991) Extended Abstract of 9th Int. Workshop on FED, Kumamoto, Japan, pp178.
10. Fujimaki A, Takai Y, Hayakawa H (1991) IEEE Trans. Magn. 27: 1353
11. Hawley M, Raistrick ID, Beery JG, Houlton RJ (1991) Science 251: 1587
12. Muroi M, Okamura Y, Suzuki T, Tsuda K, Nagano M, Mukae K (1990) Jpn. J. Appl. Phys. 29: 69
13. Gundlach KH, Holzl J (1971) Surface Science 27: 125

Artificial Grain Boundary in BiSrCaCuO Films

TETSUYA TAKAMI[1], KEN'ICHI KURODA[1], KAZUYOSHI KOJIMA[1], OSAMU WADA[2], JUNJI TANIMURA[2], MASAYUKI KATAOKA[2], TETSUO OGAMA[2], and KOICHI HAMANAKA[1]

[1] Central Research Laboratory, [2] Materials and Electronic Devices Laboratory, Mitsubishi Electric Corporation, Amagasaki, Hyogo, 661 Japan

ABSTRACT

Artificial grain boundary junctions in sputtered BiSrCaCuO films on $SrTiO_3$ (110) substrates have been fabricated by controlling the orientation of the films using a MgO buffer layer. A grain boundary with large twist misorientation is formed along the edge of the MgO buffer layer that is placed perpendicular to the [001] direction of the $SrTiO_3$ substrate. The grain boundary junctions thus obtained exhibited clear Shapiro steps in response to microwave irradiation (15.7 GHz). This indicates that the artificial grain boundaries obtained by controlling the orientation of the BiSrCaCuO films behave as a weak link.

KEY WORDS: superconductor, grain boundary, Josephson effect, BiSrCaCuO, MgO,

INTRODUCTION

There have been many reports concerning various methods of fabricating artificial grain boundary junctions using Y-based superconductor [1-6]. The critical temperature of bulk for a Bi-based superconductor, however, is about 107K, which is higher than that of a Y-based superconductor (about 90K). Recently there have been a few reports regarding the succesful preparation of BiSrCaCuO thin films with Tc above 90K by RF magnetron sputtering and the CVD methods[7,8]. Thus, it is important to fabricate the artificial grain boundary junctions using a Bi-based superconductor. It is well known that BiSrCaCuO films are grown with an (11*n*) orientation on SrTiO3 (110) substrates by RF sputtering[9,10]. On the other hand, BiSrCaCuO films are grown with a (001) orientation on thin MgO buffer layers grown on SrTiO3 (110) substrates.

In this paper, we describe the fabrication and characteristics of artificial grain boundary junctions in BiSrCaCuO films formed on a $SrTiO_3$ (110) substrates where the orientation of the deposited films are controled and when using MgO buffer layers.

EXPERIMENTAL

MgO buffer layers were prepared by the RF magnetron sputtering method using a sintered MgO target. The films were deposited on the $SrTiO_3$(110) at a substrate temperature of 350-450 °C. The deposition was carried out in a neat oxygen ambient of 300 mTorr. The deposition rate and thickness were about 0.5 nm/min and 15 nm, respectively. MgO buffer layers were patterned using a photolithographic process and wet etched with phosphoric acid to open a strip line like window with a width of 10-100μm and exposing the surface of the $SrTiO_3$ (110) substrate. The BiSrCaCuO high-Tc superconducting thin films were prepared using the RF magnetron sputtering method. Detailed conditions are described elsewhere[9,10]. The thickness of the BSCCO films was typically 150 nm. The composition of the films was close to that of the high-*Tc* phase (2223).

In order to form films with only a single preferred orientation, the $SrTiO_3$ substrates were oriented off the (110) azimuth toward the [100] direction by approximately 5°[11]. On this substrate, BiSrCaCuO thin films with only the *c*-axis oriented toward [010] were observed by X-ray diffraction (XRD) and reflection high-energy electron diffraction (RHEED). In these experiments, 5° off substrates were mainly used.

Three different grain boundary geometries are expected between the film with the *c*-axis and (11*n*) orientation, in accordance with the direction of the edge of the MgO buffer layer. We arranged the edge of the MgO buffer layer in a perpendicular direction to the [001] direction of the $SrTiO_3$ substrate. In this case, current flows in the *a-b* plane of the (11*n*) oriented BiSrCaCuO thin films as well as the *c*-axis oriented films. Figure 1 shows the schematic diagram of the artificial grain boundary (twist boundary). Microbridges, 10 and 20 μm wide and 20 μm long, were fabricated and placed across the edge of the MgO buffer layer.

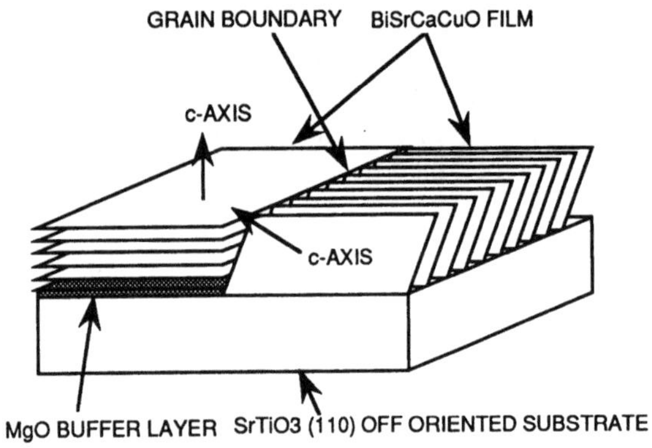

Fig. 1. Schematic diagram of an artificial grain boundary.

RESULTS AND DISCUSSION

MgO buffer layers were found by XRD to be (110) oriented on a $SrTiO_3$ (110) substrates under our sputtering conditions. Figure 2(a) shows the XRD patterns of the BiSrCaCuO films sputtered on normally oriented $SrTiO_3$ (110) substrate and Fig. 2(b) the MgO buffer layer formed on the same substrate. The film grown on $SrTiO_3$ substrate has the main peak diffracted from (11*n*) plane and weak (00*n*) peaks. On the other hand, although the XRD pattern of the film grown on MgO buffer layer shows (00*n*) peaks of BiSrCaCuO and a (220) peak of MgO, the (11*n*) peak is not observed. From this results, it is found that the crystalline orientation of the BiSrCaCuO films grown on $SrTiO_3$ (110) substrate can be controlled from (11*n*) to (00*n*) by using a MgO buffer layer which a thickness of only 15 nm. Next, to investigate the orientation of the films grown on the off-oriented $SrTiO_3$ (110) substrates, the films were analyzed by RHEED observation. Figure 3(a) shows the RHEED patterns of BiSrCaCuO films grown on the off-oriented $SrTiO_3$ (110) substrate and Fig. 3(b) the films grown on the MgO buffer layer formed on the same substrate. RHEED pattern of the film grown on the $SrTiO_3$ substrate shows slanting streaks in one direction, which are almost parallel to the [010] direction of the $SrTiO_3$ substrate. On the other hand the RHEED pattern of film grown on the MgO buffer formed on the $SrTiO_3$ substrate shows streaks almost perpendicular to the substrate surface. The angle between the streaks and the normal to the substrate is about 5°. This result indicates that the BiSrCaCuO (001) and MgO (110) surface are grown on $SrTiO_3$ (110) surface epitaxialy in spite of relatively large lattice mismatch. These facts were confirmed by TEM observation. The twist angle of the grain boundary was found to be about 45° by XRD, RHEED and TEM observations.

The artificial grain boundaries showed a Tc of 25-30 K. The critical current of the artificial grain boundaries were 10^3-10^4 A/cm^2 at 10 K. The I-V characteristics of the grain boundary junctions often exhibited asymmetric critical current when the current flows in the opposite direction, and Ic value was easily changed by inducing extremly large current. Although the

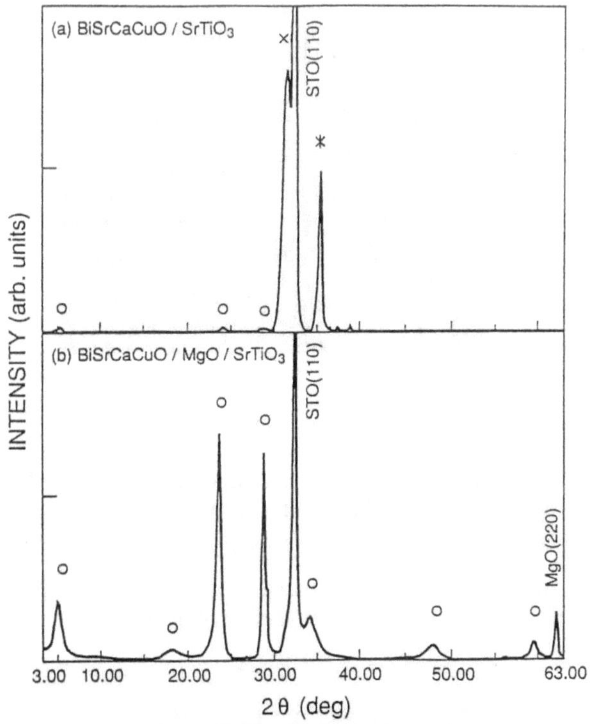

Fig. 2. XRD patterns of the BiSrCaCuO films grown on the SrTiO3 (110) off oriented by 5° (a) and on the MgO buffer layer (b). Two peaks in Fig.2 (a) , marked by a cross and asterisk, can be assigned to the (11*n*) peaks and the peaks marked by open circles can be assigned to the (00*n*) peaks. Each peak in Fig. 2(b) can be assigned to the (00*n*) peaks.

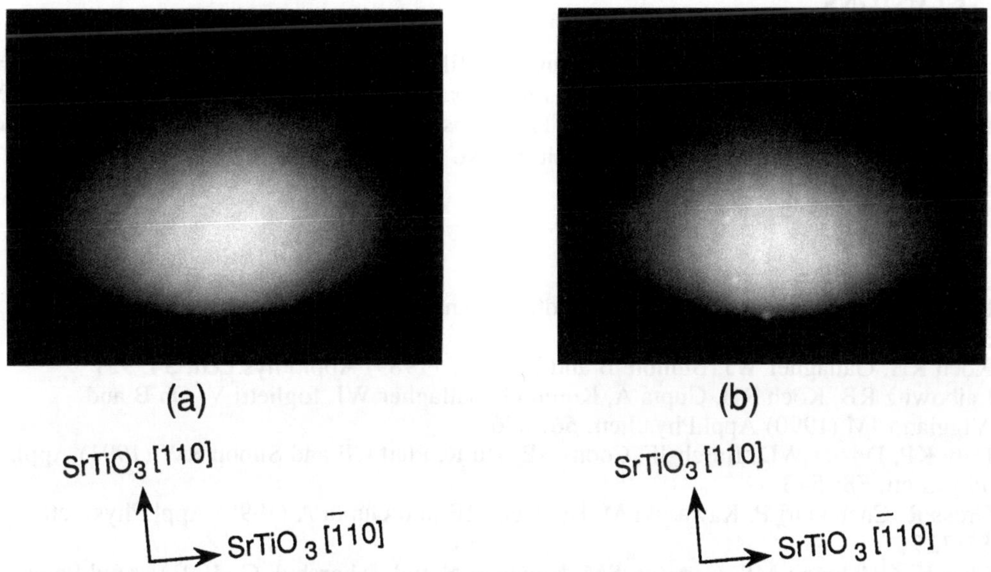

Fig. 3. RHEED patterns of the BiSrCaCuO films grown on the SrTiO3 (110) off oriented by 5° (a) and on the MgO buffer layer (b).

higher voltage region of the I-V characteristics had a slightly positive curvature; predicted from the resistive shunted junction (RSJ) model, the region just above the critical current of the I-V characteristics had a complex structure and instability. These behaviors are probably due to the presence of a trapped flux as mentioned by D.Dimos *et.al.*[12]. Figure 4 (a) shows a typical I-V characteristic for a 10 μm wide, 20 μm long microbridge that spans an artificial grain boundary at 5 K. This characteristic was obtained in the presence of a trapped flux, and its Ic was half of the untrapped characteristic. The ac Josephson effect was also observed on the artificial grain boundary junctions. A typical I-V characteristic of the junction, with 15.7 GHz microwave

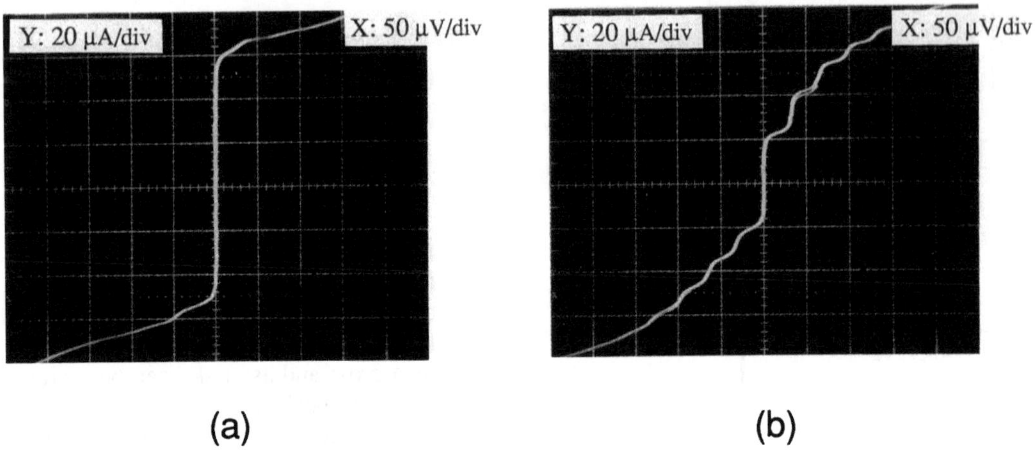

Fig. 4. Typical Current-voltage characteristics of an artificial grain boundary junction at 5 K. (a) Without microwave irradiation ; (b) with microwave irradiation at 15.7 GHz.

irradiation applied at 5 K, is shown in Fig. 4 (b). The microwave-induced voltage steps (Shapiro steps) are spread by 32.5 μV, which agrees with the well-known V=(h/2e)f, where f is the frequency of the microwave.

CONCLUSIONS

The artificial grain boundaries in BiSrCaCuO thin films with large twist misorientations were fabricated by controlling the orientation of the films. The grain boundary formed along the MgO buffer layer, which was sputtered on the $SrTiO_3$ substrate, exhibited the ac Josephson effect on the I-V characteristics in response to the microwave irradiation (15.7 GHz).

REFERENCES

1. Koch RH, Umbach CP, Clark GJ, Chaudhari P and Laibowitz RB (1987) Appl.Phys.Lett. **51**: 200
2. Koch RH, Gallagher WJ, Bumble B and Lee WY (1989) Appl.Phys.Lett. **54**: 951
3. Laibowitz RB, Koch RH, Gupta A, Koren G, Gallagher WJ, foglietti V, Oh B and Viggiano JM (1990) Appl.Phys.Lett. **56:** 686
4. Daly KP, Dozier WD, Burch JF, Coons SB, Hu R, Platt CE and Simon RW (1991) Appl. Phys.Lett. **58:** 543
5. Gross R, Chaudhari P, Kawasaki M, Ketchen MB and Gupta A (1990) Appl.Phys.Lett. **57:**727
6. Char K, Colclough MS, Garrison SM, Newman N and Zaharchuk G (1991) Appl.Phys. Lett. **59:** 733
7. Ohbayashi K, Anma M, Takai Y, and Hayakawa H (1990) Jpn.J.Appl.Phys.29: L2049
8. Endo K, Misawa S, Moriyasu Y, Yamasaki H and Yoshida S (1991) Proc. of FED-HiTcSc-ED 3rd Work shop: L74
9. Kuroda K, Kojima K, Wada O, Tanioku M, Yokoyama K and Hamanaka K (1990) Jpn.J. Appl.Phys. **29**: L1816
10. Kuroda K, Wada O, Kojima K, Tanimura J, Kataoka M, Takami T, Yokoyama K, Ogama T and Hamanaka K (1991) Jpn.J.Appl.Phys. **30:** L475
11. Kuroda K, Wada O, Tanimura J, Kojima K, Takami T, Kataoka M, Ogama T and Hamanaka K (1991) Jpn.J.Appl.Phys. **30:** L582
12. Dimos D, Chaudhari P and Mannhart J (1990) Phys.Rev.B **41**: 4038

Fabrication of Au/MgO/$Ba_{1-x}K_xBiO_3$ Tunnel-Type Junctions Using the PMQ-Processed $Ba_{1-x}K_xBiO_3$ Substrate

Maruo Kamino, Masahiro Iyori, Kazuhiko Takahashi,
Yorinobu Yoshisato, and Shoichi Nakano

Functional Materials Research Center, SANYO Electric Co., Ltd., Hirakata, Osaka, 573 Japan

ABSTRACT

Au/MgO/$Ba_{1-x}K_xBiO_3$ tunnel-type junctions have been fabricated for the first time using PMQ-processed $Ba_{1-x}K_xBiO_3$ substrate and MgO artificial barriers. High density, homogeneous $Ba_{1-x}K_xBiO_3$ superconductors with a Tce of 28.5 K were successfully synthesized by the plasma-arc melting and rapid-quenching (PMQ) method. The highest relative density was 97% of the theoretical density. The dI/dV-V characteristics of Au/MgO/$Ba_{1-x}K_xBiO_3$ junctions were qualitatively close to the BCS density of states. A gap energy Δ of 2.7 meV was obtained. The coefficient of tunneling resistance at zero bias to the tunnel resistance at $V>\Delta e$ was 11, which indicates that quasi-particle tunneling occurred with low leakage.

KEY WORDS : $Ba_{1-x}K_xBiO_3$, substrate, artificial barriers, junctions, gap energy

INTRODUCTION

The reproducible fabrication of tunnel junctions for oxide superconductors is very important for superconducting devices. There are many experimental reports on the fabrication of junction structures using Cu-O based superconductors. However, adequate junction characteristics have not been achieved despite tremendous efforts. One difficulty is the very short and anisotropic coherence length of Cu-O based superconductors. The other difficulty is that it is not easy to prepare reproducible junction structures free of pinholes because of non-homogeneity and irregularity in the interfaces.

Recently, some tunnel experiments have been performed using Ba-K-Bi-O superconductors and native barriers [1-4]. $Ba_{1-x}K_xBiO_3$ superconductors have a cubic perovskite structure, thus they have no anisotropic electrical properties. Furthermore, their superconducting coherence length is estimated to be 4~7 nm, which is longer than those of Cu-O based superconductors. Therefore, $Ba_{1-x}K_xBiO_3$ superconductors will likely be effective for fabricating junction structures.
However, it is difficult to synthesize high-quality $Ba_{1-x}K_xBiO_3$ superconductors because of the evaporation of potassium in the preparation process [5-7]. We synthesized high-quality oxide superconductors using a homogeneous solid solution created by plasma-arc melting and rapid quenching (PMQ) method [8),9)].

In this paper, we report on the synthesis process for bulk $Ba_{1-x}K_xBiO_3$ superconductors using the PMQ method. The tunnel characteristics of Au/MgO/$Ba_{1-x}K_xBiO_3$ junctions using PMQ-processed high-quality $Ba_{1-x}K_xBiO_3$ substrate and MgO artificial barriers are also discussed.

EXPERIMENTAL

A mixed powder of BaO, K_2CO_3 and Bi_2O_3 with nominal composition was melted in a plasma-arc under low pressure Ar gas. The molten materials were then quenched on a cooled copper hearth by the PMQ method [8)]. The quenched sample was annealed according to the two-step procedure described by Hinks et al [10)]. First, the quenched sample was heated at 660°C for 1 hour in flowing

N_2, then rapidly cooled to room temperature. Second, the sample was heated without any air contact up to 475°C in flowing O_2, then slowly cooled to room temperature.

Subsequently, tunnel-type junctions with MgO artificial barriers were fabricated using the PMQ-processed $Ba_{1-x}K_xBiO_3$ substrate.

Figure 1 shows the fabrication process for Au/MgO/$Ba_{1-x}K_xBiO_3$ junction structures. First, an Au film was fabricated on the polished $Ba_{1-x}K_xBiO_3$ surface as an electrode for electrical measurement. Then, an MgO barrier layer with a thickness of 4 nm was deposited onto the polished $Ba_{1-x}K_xBiO_3$ substrate by electron beam evaporation. An Au normal-conducting thin film with a thickness of 200 nm was fabricated on the MgO insulating layer. Finally, an Au electrode and MgO buffer layer, 500 nm and 300 nm thick, respectively, were fabricated to measure electrical characteristics through the Au/MgO/$Ba_{1-x}K_xBiO_3$ junction. Typical junction area was 0.5 mm x 0.5 mm.

The current-voltage characteristics were measured by a four-probe method.

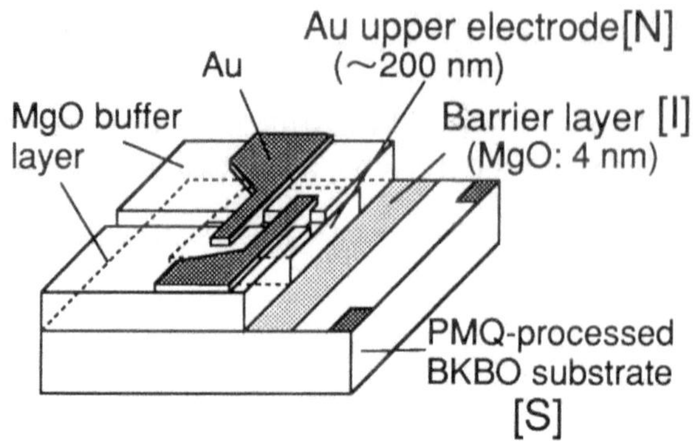

Fig. 1. Typical junction structure of Au/MgO/$Ba_{1-x}K_xBiO_3$.

RESULTS AND DISCUSSION

The temperature dependence of the resistance for the PMQ-processed $Ba_{1-x}K_xBiO_3$ sample is shown in Fig. 2. Here, the resistance was normalized to that of 300 K. A sharp superconducting transition is observed at 28.5 K with a transition width of less than 0.5 K. Figure 3 shows the surface morphology of the polished $Ba_{1-x}K_xBiO_3$ substrate. The surface of the polished $Ba_{1-x}K_xBiO_3$ substrate is very smooth and of high density over a wide range. The highest relative density measured by the Archimedes method was 97% of the theoretical density. Judging from the Tc measurement and optical micrograph results, the PMQ-processed $Ba_{1-x}K_xBiO_3$ samples were of good homogeneity.

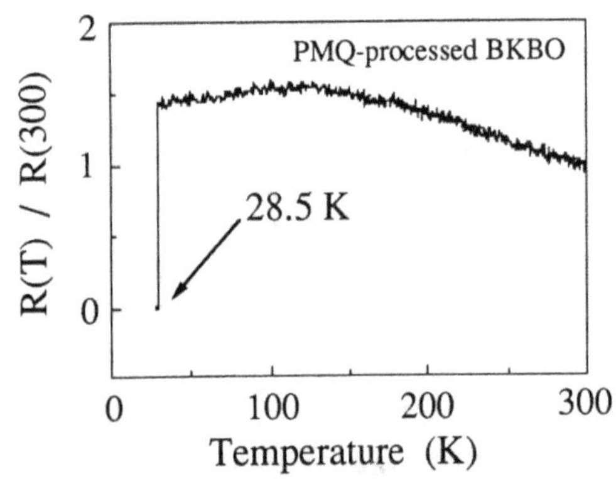

Fig. 2. Temperature dependence of the resistance for the PMQ processed $Ba_{1-x}K_xBiO_3$ sample.

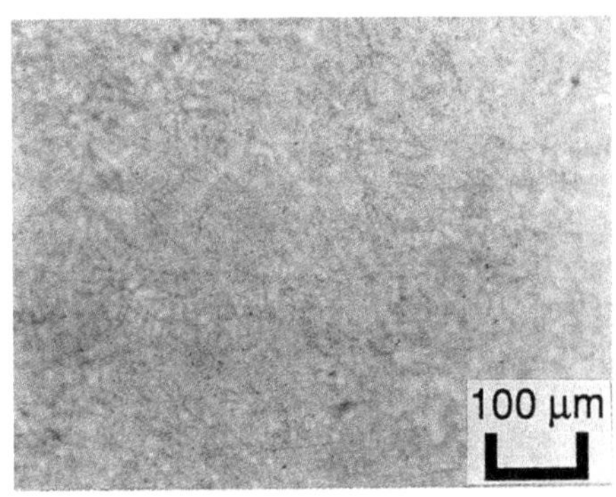

Fig. 3. Optical micrograph of the PMQ processed $Ba_{1-x}K_xBiO_3$ surface.

Subsequently, tunnel-type junctions with MgO artificial barrier were fabricated using the PMQ-processed $Ba_{1-x}K_xBiO_3$ substrates. The junction resistance at room temperature was typically 1-2 Ω.

The temperature dependence of the resistance through the Au/MgO/$Ba_{1-x}K_xBiO_3$ junctions is shown in Fig. 4. Here, the resistance was normalized to that of 300 K. The resistance of the junction was found to decrease sharply at the critical temperature of $Ba_{1-x}K_xBiO_3$. The resistance then increases sharply with decreasing temperature below 19 K. If the $Ba_{1-x}K_xBiO_3$ enters the superconducting state, the gap opening begins at the same time, resulting in an increase of junction

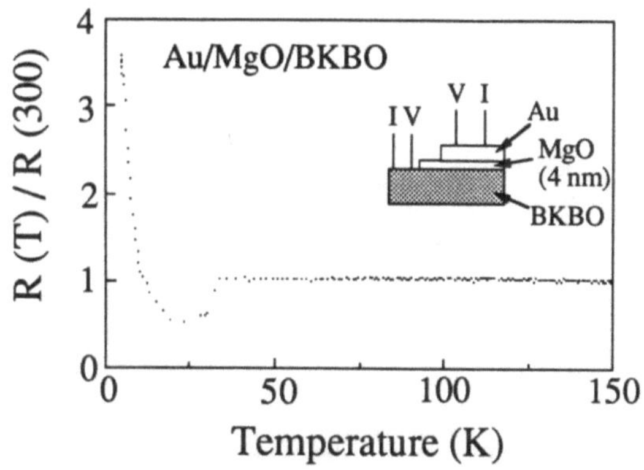

Fig. 4. Temperature dependence of the resistance for Au/MgO/$Ba_{1-x}K_xBiO_3$ junction.

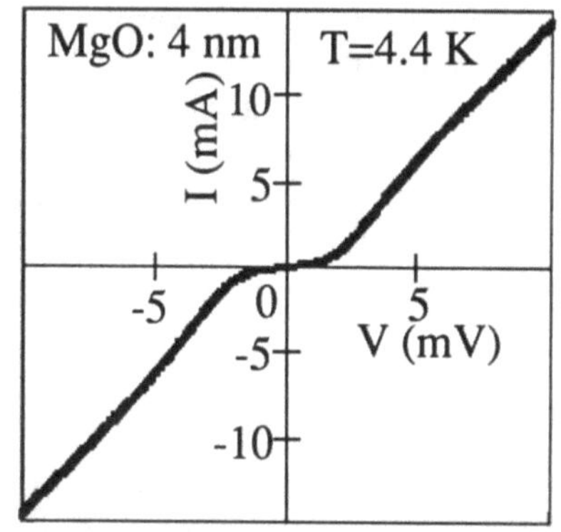

Fig. 5. I-V characteristics of Au/MgO/$Ba_{1-x}K_xBiO_3$ junction at 4.4 K.

resistance. Our sample showed this behavior. The typical I-V characteristics of junctions with an MgO artificial barrier with a thickness of 4 nm at 4.4 K are shown in Fig. 5. Non-linearity was clearly observed at about 3 mV corresponding to the energy gap of $Ba_{1-x}K_xBiO_3$ superconductors. Non-linearlity of the tunnel characteristics is not observed above 19 K. This clear non-linearity suggests that MgO artificial films on the PMQ-processed $Ba_{1-x}K_xBiO_3$ substrate can be good, low-leakage barriers.

Tunnel conductance, dI/dV-V, calculated from the I-V curve is presented in Fig. 6. The result was reproducible among the several different junctions, and resembles that reported for junctions using $Ba_{1-x}K_xBiO_3$ thin films and native barriers [2]. The dI/dV-V characteristics for the junctions were qualitatively close to the BCS density of states, showing a pair of sharp conductance peaks at 2.5 mV. The energy gap Δ estimated from R.C. Dynes' equation using this data is about 2.7 mV, which agrees with the value reported by B. Batlogg et al. [11]. Furthermore, the coefficient of tunneling $\{(dI/dV)_{\text{at } V>\Delta e} / (dI/dV)_{\text{at } V=0}\}$ obtained from this data is 11, which indicates that quasi-particle tunneling occurred with low leakage [12].

Figure 7 shows the temperature dependence of the tunnel conductance. The gap energy decreases with increasing temperature. The gap structure appears below 19 K. The gap energy at various temperatures were estimated by Dynes fitting, as shown in Fig. 7. Figure 8 shows the temperature dependence of the gap energy. The solid line represents the fitting curve according

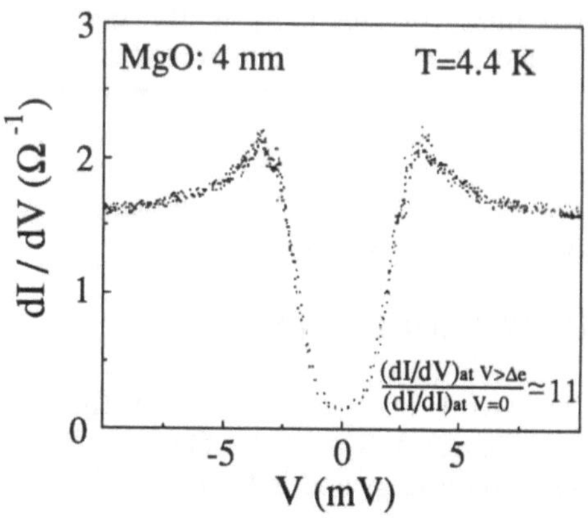

Fig. 6. Tunnel conductance for Au/MgO/$Ba_{1-x}K_xBiO_3$ junction at 4.4 K.

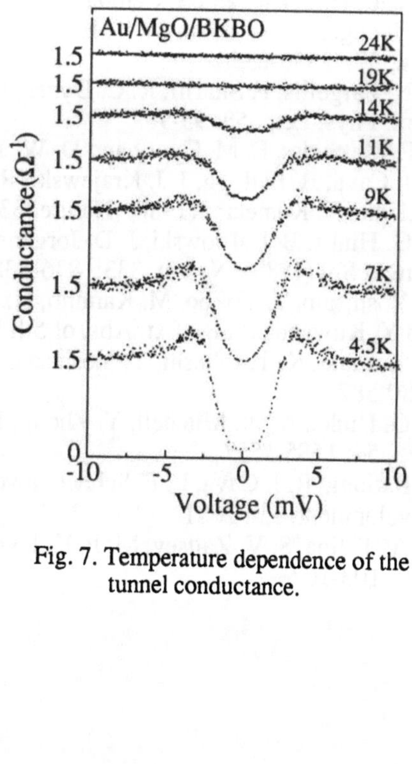

Fig. 7. Temperature dependence of the tunnel conductance.

to the BCS theory. The experimental data were coincided well with the calculated line, when the critical temperature of $Ba_{1-x}K_xBiO_3$ substrate was assumed to be 19 K. From the fit to the BCS theory, $2\Delta/k_BTc=3.4$ was obtained. These results show that $Ba_{1-x}K_xBiO_3$ is a conventional BCS superconductor.

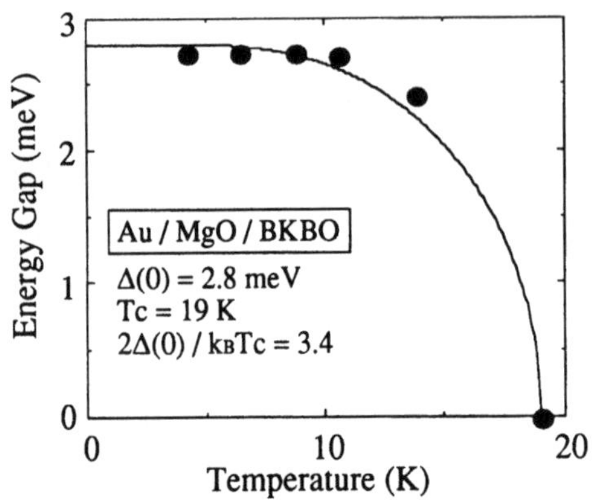

Fig. 8. Temperature dependence of the gap energy for $Au/MgO/Ba_{1-x}K_xBiO_3$ junction.

CONCLUSION

We have fabricated $Au/MgO/Ba_{1-x}K_xBiO_3$ tunnel-type junctions using PMQ-processed $Ba_{1-x}K_xBiO_3$ substrate and MgO artificial barriers for the first time. In I-V characteristics of the junction, non-linearity was clearly observed at about 3 mV, corresponding to the energy gap of $Ba_{1-x}K_xBiO_3$ superconductors. The dI/dV-V characteristics of the junctions were qualitatively close to the BCS density of states. The gap energy Δ obtained was 2.7 meV. From the fit to the BCS theory, $2\Delta/k_BTc=3.4$ was obtained. The coefficient of tunneling resistance at zero bias to the tunnel resistance at $V>\Delta e$ was 11. This shows the quasi-particle injection to be of high efficiency. It is believed that the tunnel junction using the PMQ-processed $Ba_{1-x}K_xBiO_3$ substrate is effective for superconducting devices relying on nonequilibrium quasi-particles.

ACKNOWLEDGMENTS

This work was performed under the management of FED (the R & D Association for Future Electron Devices) as a part of the R & D of Basic Technology for Future Industries sponsored by NEDO (New Energy and Industrial Technology Development Organization).

REFERENCES

1. Y. Enomoto and K. Moriwaki (1990) Jpn. J. Appl. Phys., 29: L1455-1457
2. H. Sato, H. Takagi and S. Uchida (1990) Physica C, 169: 391-395
3. Q. Huang, J. F. Zasadzinski, K. E. Gray, D. R. Richards and D. G. Hinks (1990) Appl. Phys. Lett., 57: 2356-2358
4. A. N. Pargellis, F. Sharifi, R. C. Dynes, B. Miller, E. S. Hellman and J. M. Rosamilia (1991) Appl. Phys. Lett., 58: 95-96
5. L. F. Mattheiss, E. M. Gyory and D. W. Johnson (1988) Phys. Rev., B37: 3745-3746
6. R. J. Cava, B. Batlogg, J. J. Krajewski, R. Farrow, L. W. Rupp Jr, A. E. White, K. Short, W. F. Peck and T. Kometani (1988) Nature, 332: 814-816
7. D. G. Hinks, B. Dabrowski, J. D. Jorgensen, A. W. Mitchell, D. R. Richards, Shiyou Pei and Donglu Shi (1988) Nature, 333: 836-838
8. Y. Yoshisato, T. Yokoo, M. Kamino, H. Suzuki, T. Ikemachi, Y. Takai, T. Ienaga, S. Nakano and Y. Kuwano (1988) Ext. Abs. of 5th Int. Ws. on Future Electron Devices: 209.
9. M. Kamino, K. Takahashi, Y. Yoshisato and S. Nakano (1990) Jpn. J. Appl. Phys., 29: L580-582
10. D. G. Hinks, A. W. Mitchell, Y. Zheng, D. R. Richards and B. Dabrowski (1989) Appl. Phys. Lett., 54: 1585-1587
11. B. Batlogg, R. J. Cava, L. F. Schneemeyer and G. P. Espinosa (1989) IBM J. Research and Development, 33 : 2081
12. N. A. Tulina, S. V. Zaitsev, M. P. Kulakov, R. K. Nikoleav and N. S. Sidorov (1990) Physica C, 171: 103-107

All High-T_c Sandwich-Type Josephson Junctions of Bi-Systems Using a New Barrier Material

KOICHI MIZUNO and KENTARO SETSUNE

Central Research Laboratories, Matsushita Electric Industrial Co., Ltd., Moriguchi, Osaka, 570 Japan

ABSTRACT

Sandwich-type junctions using a Bi-based oxide barrier were successfully fabricated. $Bi_2Sr_2Ca_1Cu_2O_{8+x}$ films were used for superconducting electrodes and $Bi_2Sr_2Nd_1Cu_2O_{8+x}$ films were used for the barrier layers. All materials had the same crystalline structure and nearly the same lattice constant with each other. The conductivity of $Bi_2Sr_2Nd_1Cu_2O_{8+x}$ films were rather low than ordinary metals or $Bi_2Sr_2Cu_1O_{6+x}$ and its temperature dependence can be explained by the variable range hopping (VRH) model. We observed S/N/S-type Josephson effects in the junctions. Those junctions also exhibited negative resistive behaviors on the I-V curves at higher current biased region. These characteristics were evaluated from a local heating mechanism using VRH model.

KEY WORDS:Bi-Sr-Ca-Cu-O, Bi-Sr-Nd-Cu-O, S/N/S junction, negative resistance

INTRODUCTION

Since the discovery of the high-Tc oxide superconductors, many people challenged to observe Josephson effects and tunneling phenomena in these new superconductors. We have also investigated such phenomena and basic characteristics of junction structures especially using the combinations of a high-Tc Bi-system and Bi-system oxides [1-3]. In those investigations, 2212 phase $Bi_2Sr_2Ca_1Cu_2O_{8+x}$ (BSCCO) was used for the superconducting electrodes. We tried to use some Bi-system oxides for the barrier layer, for example 2201 phase $Bi_2Sr_2Cu_1O_{6+x}$ (BSCO) and 2212 phase $Bi_2Sr_2Nd_1Cu_2O_{8+x}$ (BSNCO). The electrical conductivity of BSCO film is suitable for studies of proximity effects or S/N/S junctions with the high-Tc Bi-systems but is not enough low to observe tunneling operations. On the other hand, 2212 phase Bi-systems in which the Ca of the BSCCO is substituted by one of rare-earth elements have lower conductivity than that of BSCO. It is interesting to compare some characteristics of those junctions which barrier have different characteristics especially on the conductivity. Nd was applied as the substitution element for Ca in 2212 phase BSCCO for the junction construction. It is because that the ionic radius of Nd is rather close to that of Ca than other rare-earth elements.

We observed Josephson effects in both BSCCO/BSCO/BSCCO and BSCCO/BSNCO/BSCCO systems. Each junctions exhibited S/N/S type characteristics and clear Shapiro steps on its I-V curves. Those results were reported in previous articles [1,3]. Moreover, we observed negative resistive region on I-V curves of BSCCO/BSNCO/BSCCO samples. In this paper, we report such anomalous phenomena of $Bi_2Sr_2Ca_1Cu_2O_{8+x}$/$Bi_2Sr_2Nd_1Cu_2O_{8+x}$/$Bi_2Sr_2Ca_1Cu_2O_{8+x}$ (BSCCO/BSNCO/BSCCO) junctions.

EXPERIMENTAL

All oxide films, superconducting BSCCO electrodes and rare-earth substituted 2212 phase Bi-system oxides, were deposited on MgO(100) substrates from powder targets of mixed compounds by an rf magnetron sputtering technique. Each films had c-axis oriented crystalline structure. Our sputtering chamber has four targets which can be independently sputtered.

The BSCCO and BSNCO films were sequentially deposited. The multi-layered film structures of our junctions were fabricated by a fully in-situ deposition process. Detailed sputtering conditions and fabrication process for the junction structure are described in ref. 1.

Sintered sample of rare-earth substituted 2212 phase Bi-systems, for example BSNCO, were reported as an insulator [4]. On the other hand, our sputter deposited film samples were electrical conductors. It was mainly caused by nonstoichiometry of the composition. Temperature dependence of resistivity for some Bi-system oxide films are shown in fig. 1. These resistivity values are measured perpendicular to the c-axis. BSNCO and BSECO represent Bi-Sr-Nd-Cu-O and Bi-Sr-Er-Cu-O, respectively. Each indices of 2212 and 2201 express the crystalline structure of 2212 phase and 2201 phase Bi-systems. In fig. 1(a), resistivity increase for decreasing temperature. In general, some of such behaviors can be explained by variable range hopping (VRH) model [5,6]. In these model, the resistivity relates to the temperatures as $r=r_0\exp(T/T_0)^{-1/4}$ where r is resistivity and T is temperature. T_0 and r_0 are constants. We replotted the raw data of resistivity in logarithmic scale for $(T)^{-1/4}$. These results are shown in fig 1(b) and there are good linear dependence. The conduction mechanism in our rare-earth substituted Bi-systems also can be explained by VRH model.

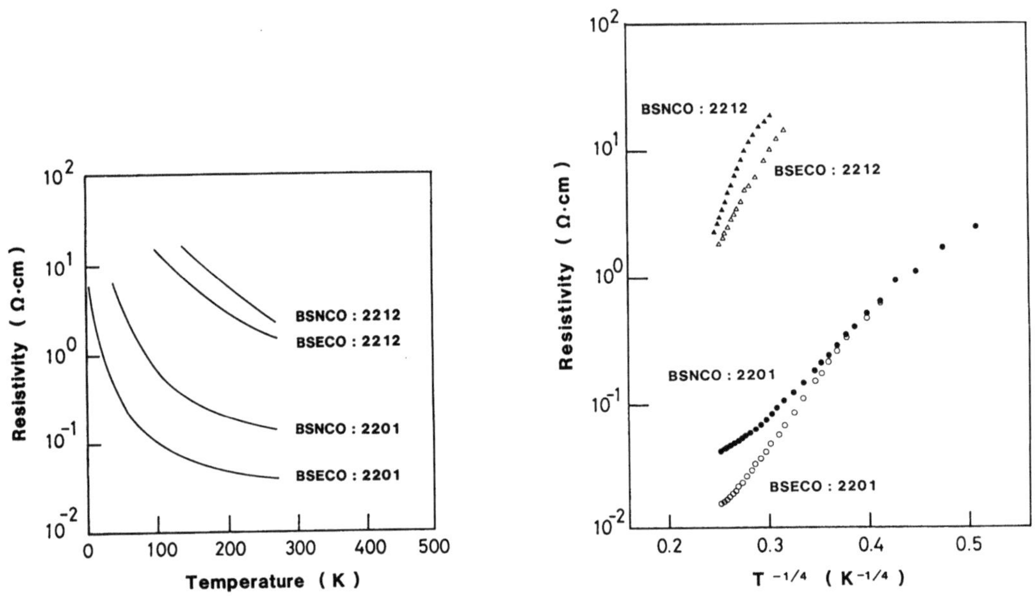

(a) (b)

Fig.1 Temperature dependence of resistivity for some Bi-based oxide films under (a) linear scale, and (b) repplotted as log(r) versus $(T)^{-1/4}$. BSNCO and BSECO represent Bi-Sr-Nd-Cu-O and Bi-Sr-Er-Cu-O respectively. Moreover, each indices of 2212 and 2201 means the crystalline structure of 2212 phase or 2201 phase Bi-systems.

RESULTS AND DISCUSSION

A schematic configuration of the junction is shown in fig. 2. Current-voltage (I-V) characteristics were measured by conventional four probe method. The junctions of 5X10, 10X10, 10X20 and 20X20 μm^2 were fabricated. The barrier thickness were varied from 3 nm to 120 nm. The junctions

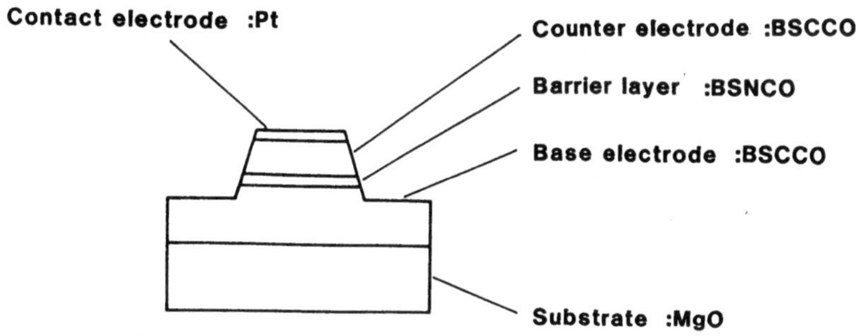

Fig.2 Schematic configuration of a BSCCO/BSNCO/BSCCO Josephson junction.

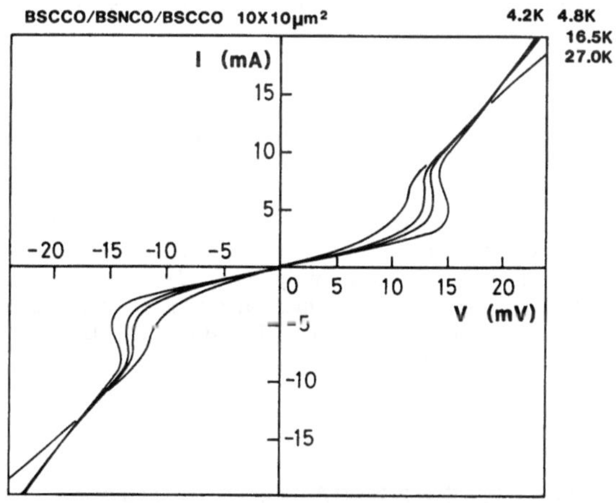

Fig.3 Current-voltage curves for BSCCO/BSNCO(60nm)/BSCCO junction of various temperatures.

which barriers were in the range from 11nm to 100nm thick operated as S/N/S or S/S'/S-type Josephson junction. We observed Shapiro steps on each I-V curves under the irradiation of microwaves. On the same time, we also observed an anomalous behavior in those junctions. Figure 3 shows the I-V curves of a BSCCO/BSNCO/BSCCO junction in which BSNCO thickness and junction area size were 60nm and 10X10 μm^2, respectively. There are negative resistive regions for some current biased range. The values of those negative resistance are increased in lower temperatures. All other junctions of this BSCCO/BSNCO/BSCCO system exhibited the similar behaviors.

A simple model was introduced to make a qualitative explanation for these phenomena. We considered a local heating effect caused by current flow through the barrier layer. There are many complicated factors to estimate the temperature at the barrier layer. They are, for examples, generated power at the junction, heat transfer coefficients between the junction and ambient gas or MgO substrate which relates to structural factor, thermal capacitance of the barrier layer and so on. It is difficult to take all these conditions into the account. We introduced a simple approximation that the dissipated power at the junction (I^2r) heated the barrier layer. The deviation of the temperature (ΔT) from the ambient one (T) was in proportion to the dissipated power in the equilibrium state. The temperature dependence of resistivity of the barrier layer should be dominated by VRH mechanism as already shown in fig.1. It can be expressed by the relation of $r=r_0\exp(T+\Delta T/T_0)^{-1/4}$. These assumption derives the current dependence of resistivity as $r=r_0\exp(T/T_0+C_0(I^2r/dS)/T_0)^{-1/4}$; where I is

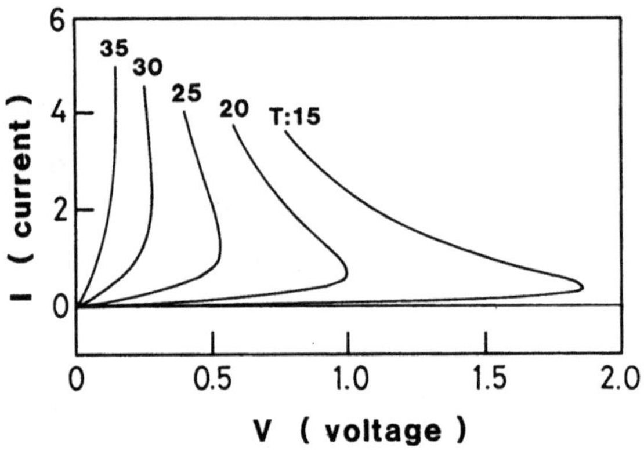

Fig.4 Calculated current-voltage curves from an equation derived using VRH mechanism in the barrier layer. Variables in the figure correspond to the temperature. Negative resistive regions are obvious in lower temperatures.

current, r is resistivity of the barrier layer, d is the thickness of the barrier and S is the area size of the junction. The C_0 is a constant and corresponds to the thermal capacitance of the barrier materials.

Figure 4 shows calculated I-V curves for some values of the parameter "T" which corresponds to the temperature. In these calculations we put appropriate values for the constants, T_0, C_0, d, and S. Negative resistive region as observed in the fig. 3 was appeared and its negative resistance is more large in the case of small T value. It is also agreed with the behavior of the measurements.

CONCLUSION

We have investigated BSCCO/BSNCO/BSCCO junctions. S/N/S-type Josephson effects were confirmed in lower current biased region. We also observed negative resistance in the range of some current values. Such anomalous phenomena were tried to be explained by local heating mechanism. In those consideration, the VRH model was successfully applied to derive the relation between current and resistivity of the barrier material.

ACKNOWLEDGMENT

The authors would like to thank Dr. T. Nitta and Dr. K. Wasa for their continuous encouragements and also thank Prof. S. Kuriki of Hokkaido University for his useful discussions.

REFERENCES

1. K. Mizuno, H. Higashino, K. Setsune and K. Wasa, Appl. Phys. Lett. 56, 1469 (1990)
2. T. Matsushima, Y. Ichikawa, H. Adachi, K. Setsune and K. Wasa, Solid State Commun. 76, 1201 (1990)
3. K. Mizuno and K. Setsune, Jap. J. Appl. Phys. 30, L1559 (1991)
4. A. Groen, R. Steens, H. W. Zandbergen, M. W. Dirken, F. M. Mulder and R. C. Thiel, J. Less-Common Met. 155, 133 (1989)
5. N. F. Mott, Phil. Mag. 19, 835 (1969)
6. N. Apsley and H. P. Hughes, Phil. Mag. 30, 963 (1974)

Electric Characteristics of High T_c Superconductor/Semiconductor Junction

R. KAWASAKI, F. TODA, K. HASHIMOTO, M. OGIHARA, T. YAMADA, and H. ABE

Research and Development Group, Oki Electric Industry Co., Ltd., Hachioji, Tokyo, 193 Japan

ABSTRACT

We formed BiSrCaCuO and YBaCuO / Nb doped $SrTiO_3$ (STO(Nb)) junctions. We measured current - voltage (I-V) characteristics and capacitance - voltage (C-V) characteristics of both junctions by varying temperature from 300K to 25K. Both samples clearly show the diode characteristics. Near room temperature the I-V curves of forward bias voltage show exponential dependence and the $1/C^2(V)$ decrease linearly with voltage, thus following the usual Schottky relation. At lower temperature, however, the characteristics of samples differ from the Schottky relation.

KEY WORDS: Schottky junction, I-V characteristics, C-V characteristics, BiSrCaCuO, YBaCuO, STO (Nb)

INTRODUCTION

The heterojunction of a high Tc superconductor and semiconductor with Schottky barrier is a basis for various superconductor devices - super Schottky diode, superconducting base hot electron transistor[1-3]. We prepared BiSrCaCuO and YBaCuO thin films on STO(Nb) substrates. Here, we report on characteristics of these junctions.

A high Tc superconductors of high quality were prepared on $SrTiO_3$. The metal - STO(Nb) junction is known to exhibit electric properties of a Schottky-like diode. The STO(Nb) is suited for the present purpose. We present I-V and C-V characteristics of the junctions with YBaCuO and BiSrCaCuO / STO(Nb) measured from 300K to 25K.

SAMPLE PREPARATION

BiSrCaCuO thin films on (100) STO(Nb) substrates were formed by molecular beam epitaxy[4]. The critical temperature of BiSrCaCuO thin film was 65K. YBaCuO thin films were formed by organometallic chemical vapor deposition[5]. The critical temperature of YBaCuO thin film was 89K. The XRD patterns of BiSrCaCuO and YBaCuO thin films show (001) orientation. We patterned these films by usual photo lithography and Ar dry ethcing. Fig. 1 is a schematic diagrams of the diodes. We decide the 2mm diameter junction by the C-V measurement accuracy. We deposited Au on superconductor and Al on STO(Nb) as electrodes. Both electrodes were ohmic contacts with low contact resistance. The quality of the superconducting thin films were kept in these processes.

MEASUREMENT

I-V and C-V characteristics were measured by the typical four probe method. Fig. 2 shows a schematic diagram of experimental setup. Samples were set in He gas flow cryostat. The temperature of samples was controlled from 300K to 25K. In I-V measurement, we used a source measure unit: Keithley 236. The applied voltage was

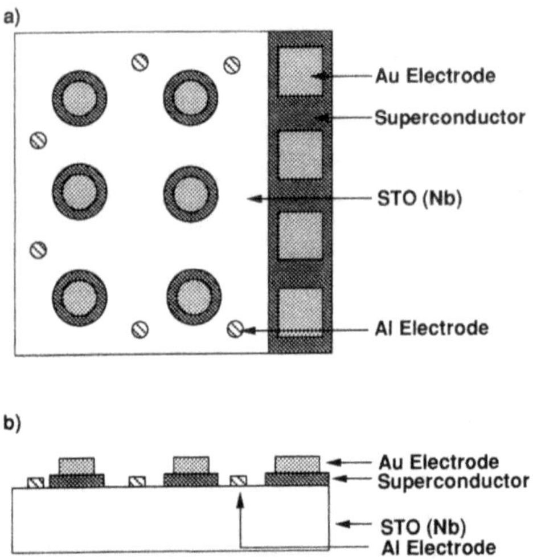

Fig. 1 Fabricated device structure a) top view, b) cross section

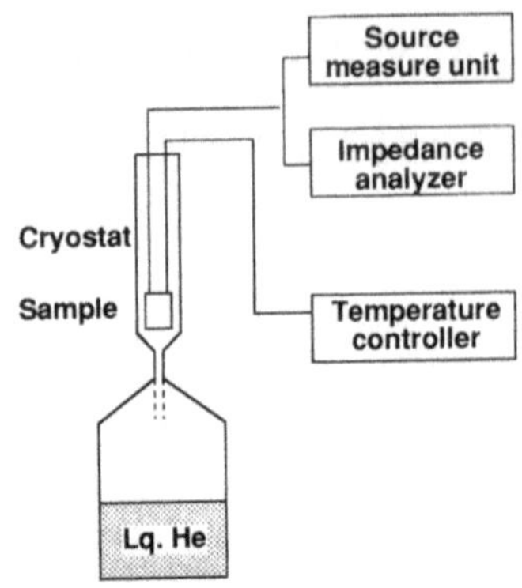

Fig. 2 Schematic diagram of experimental setup.

from -5V to 3V. We used pulse sweeps measurement to avoid the damage of samples. Its pulse duration of voltage was 1 msec. In C-V measurement, we used an impedance analyzer: HP4192A. The applied voltage was from -1V to 0.5V. The measurement frequency was from 100Hz to 100K Hz.

RESULTS

Fig. 3 show I-V characteristics of BSCCO and YBCO / STO(Nb) junctions, which clearly show the diode characteristics. Fig. 4 show C-V characteristics of these samples. Near room temperature the I-V curve show exponential dependence in the subthreshold region of forward current. $1/C^2(V)$ decreases linearly with voltage. These characteristics satisfy the usual Schottky relation[6]. However, the n values of the I-V curves are found to be about 2 - 4, which makes it difficult to estimate the Schottky barrier height ϕ_B.On the other hand, as shown in Fig. 4, the C-V result give ϕ_B unambiguously as the voltage where extrapolated lines of $1/C^2$ intersect with the line of $1/C^2=0$. Estimated ϕ_B are found to be 1.5 - 1.6 V for both junctions. The carrier concentration estimated from C-V data is about $5E10^{18}$ /cm^3. This value is very close to the carrier concentration of STO (0.05% wt Nb) obtained from the Hall measurement[7].

At lower temperature we found characteristics different from those discussed in the preceeding paragraph. The logarithm of subthreshold forward current deviates from a linear dependence on the voltage obtained near room temperature. Moreover, the ϕ_B obtained from C-V curves decreases from the values of 1.5 - 1.6 V by about 0.3 V. A more detailed investigation indicates that these transition occur around 100K. Changes of I-V and C-V characteristics of both samples suggest a structural transition of STO(Nb). The structural transition of $SrTiO_3$ in fact occurs in STO at 105K[8]. Clarifying the relation between the structural transition and the electric properties of the diode is an interesting problem left in future.

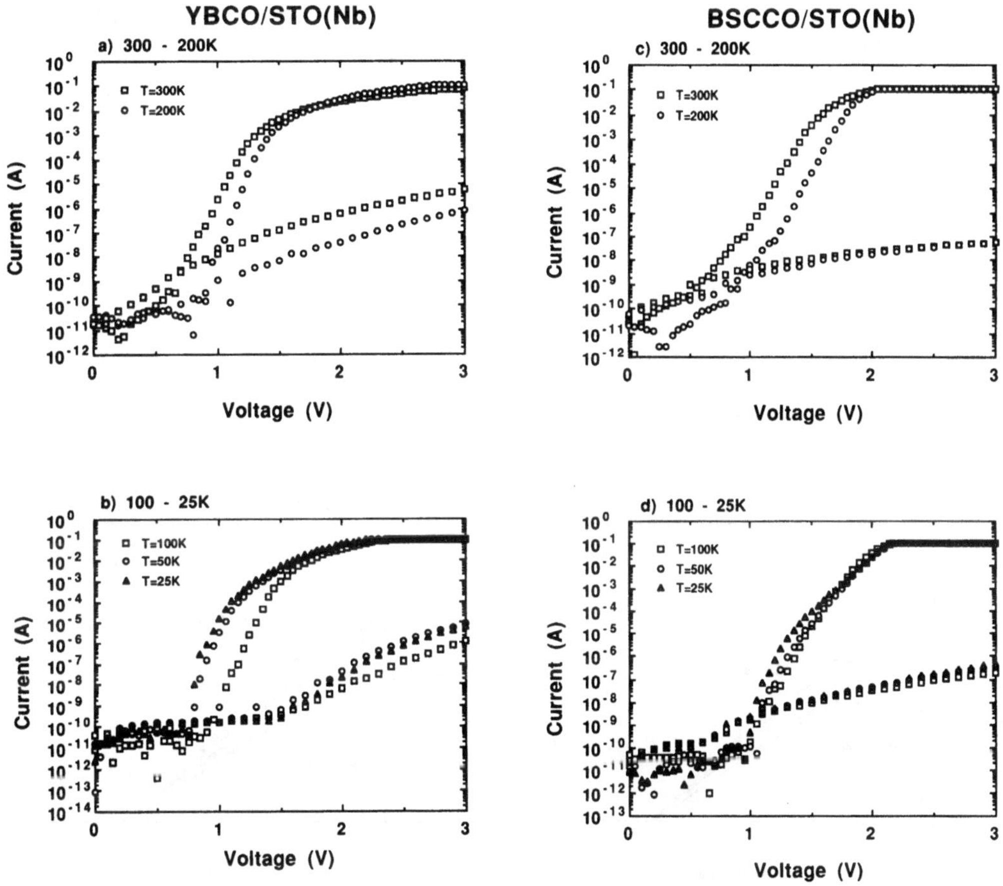

Fig. 3 Forward and reverse currents versus voltage measured at various temperature are shown for YBCO / STO (Nb) junction (a and b), and for BSCCO / STO (Nb) one (c and d). Forward currents are obtained when plus voltage is applied to the oxide superconductors.

The I-V curves of BiSrCaCuO / STO(Nb) have the kink about 1.1V, while these of YBaCuO / STO(Nb) were smooth. This suggest the difference of the density of states between BiSrCaCuO and YBaCuO. The analysis of this left in the future.

SUMMERY

(1) Schottky junctions of oxide superconductors on STO(Nb) with barrier height of about 1.5 eV were obtained.
(2) The electric characteristics of the STO(Nb) heterojunction change bellow about 100K, which may be associated with the structural transition of $SrTiO_3$.
(3) In the I-V characteristics of BiSrCaCuO / STO(Nb) the kink about 1.1V was observed.

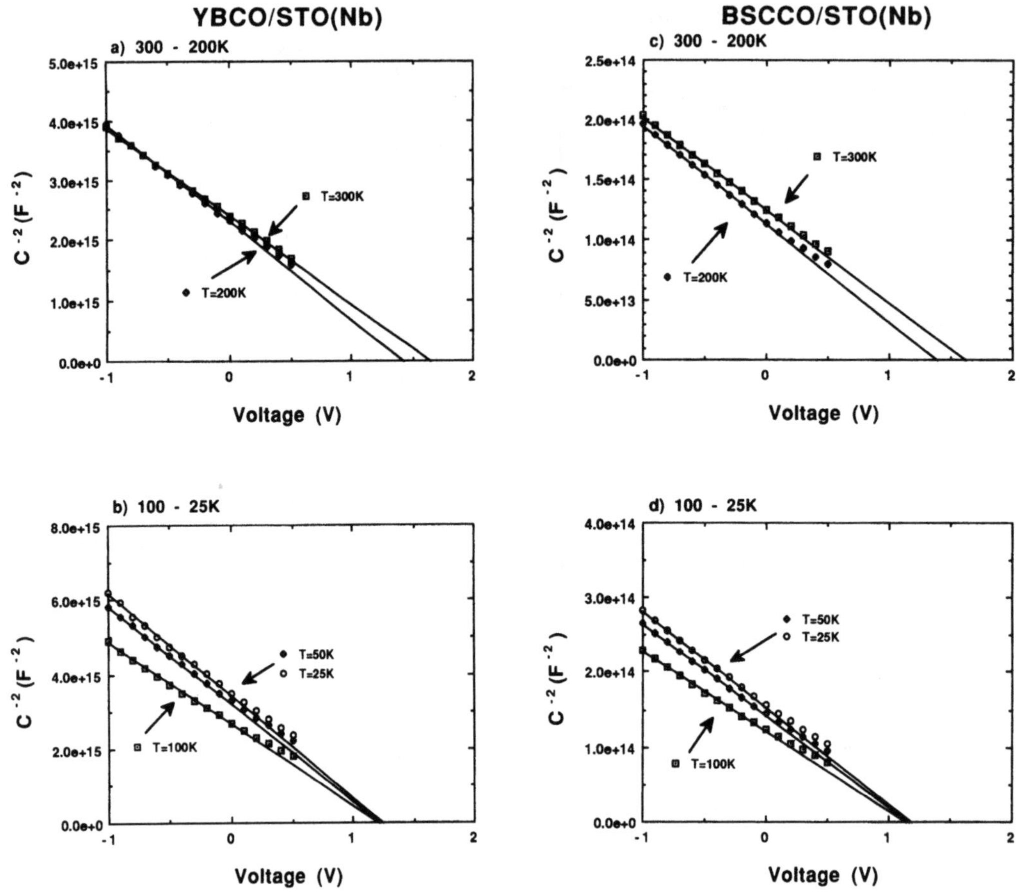

Fig. 4 $1/C^2$ are plotted as a function of voltage at various temperature. a) and b) are the results for YBCO / STO (Nb) and c) and d) are for BSCCO / STO (Nb), respectively. The Schottky barrier heights are estimated as the voltage where extrapolated lines of $1/C^2$ intersect with the line of $1/C^2 = 0$.

ACKNOWLEDGMENT

This work was performed under the management of FED (the R & D Association for Future Electron Devices) as a part of the R & D of Basic Technology for Future Industries supported by NEDO (New Energy and Industrial Technology Development Organization).

REFERENCE

1. Kobayashi T, Sakai H and Tanouchi M (1986) Electron Lett. 22: 659
2. Yoshida A, Tamura H, Takauchi H and Hasuo S (1991) Appl. Phys. Lett. 59: 1242
3. Kawasaki R, Hashimoto H and Abe H (1991) IEICE Trans. 74E: 2004
4. Ogihara M, Yamada T, Tsuruoka T and Abe H (1990) Extended Abstracts of the 22nd (1990 International) Conference on Solid State Device and Materials, Sendai: 581
5. Abe H, Nakamori T, Kanamori T and Shibata S (1988) J. Cryst. Growth 93: 929
6. Sze SM, Physics of Semiconductor Device, John Wiley & Sons
7. Hasegawa H, Fukazawa T and Aida T (1989) Jpn. J. Appl. Phys. 28: L2210
8. Unoki H and Sakudo T (1967) J. Phys. Soc. Jpn. 23: 546

Fabrication of the Superconducting Devices Using the 80K-Phase Bi-Based Single Crystal

MAKOTO CHIHAYA, SATORU KISHIDA, HEIZO TOKUTAKA, KATSUMI NISHIMORI, NAGANORI ISHIHARA, and HIDEKI KINOSHITA

Department of Electrical and Electronic Engineering, Tottori University, Tottori, 680 Japan

ABSTRACT

We prepared the Pb/Y_2O_3/Bi-Sr-Ca-Cu-O (BSCCO) specimens using the 80K-phase BSCCO single crystals and investigated the current-voltage characteristics. From the results, we found that Y_2O_3 was useful as an artificial barrier and that the native barrier was obtained by cleaving the BSCCO single crystals in air. The changes of the dI/dV-voltage characteristics of the Pb/air-cleaved surface(native barrier)/BSCCO specimen as well as the Pb/Y_2O_3(50Å)/BSCCO specimen were observed around ±3mV and ±25mV. The native barrier, which was removed by heating the BSCCO single crystals in vacuum at 300°C, is thought to be formed by the adsorption of gases.

KEY WORDS : Bi-Sr-Ca-Cu-O single crystal, junction device, I-V characteristics

INTRODUCTION

The studies on superconducting devices using the thin films of high-Tc superconducting oxides are carried out. However, the devices do not show reproducible current-voltage (I-V) characteristics. This may be due that the surface of the thin films deteriorates during the fabrication of devices and that the thin film with good surface morphology can not be prepared easily. In order to clarify the effect of the interface between superconductors and insulators on device characteristics, it is necessary to fabricate the devices using the Bi-Sr-Ca-Cu-O (BSCCO) single crystal, because the top-most layer of the cleaved BSCCO single crystal are well defined[1]. In this study, we prepared the S_1 (BSCCO single crystal)-I-S_2 (Pb) and S_1-I-N (Au) type devices using the 80K-phase BSCCO single crystal, and measured the I-V characteristics of the specimens.

EXPERIMENTAL

The BSCCO single crystals prepared by a self-flux method[2], were cleaved in air and heated in vacuum (10^{-4}Pa, 300°C, 30min) and then, Bi_2O_3 and Y_2O_3 with the thickness from 20Å to 100Å were deposited on them at room temperature using an electron-beam deposition method. Next, 1000Å Pb or 1000Å Au film was deposited in situ using a resistive heating method. Finally, we formed electrodes by depositing Au on the surfaces. The junction area of the devices was about 2 X 2 mm^2. Thus, S_1-I-S_2 type and S_1-I-N type devices were prepared. The I-V characteristics of the specimens were measured by a 4-point-probe method at the temperatures from 4.2K to 77K.

RESULTS AND DISCUSSION

Figure 1 shows the I-V characteristics of the Pb/Y_2O_3(20Å)/BSCCO specimen at (a)4.2K, (b)10K and (c)30K. With decreasing temperatures, the I-V characteristics of the specimen became more non-linear. Especially, the characteristics at 4.2K were different from those at 10 or 30K. Figure 2 shows the dI/dV-voltage characteristics of the specimen shown in Figs.1(a) and (b). The abrupt change of dI/dV was observed around ±3mV at 4.2K, as shown by arrow signs (↓) in Fig.2(a). However, the change of dI/dV around ±3mV was not

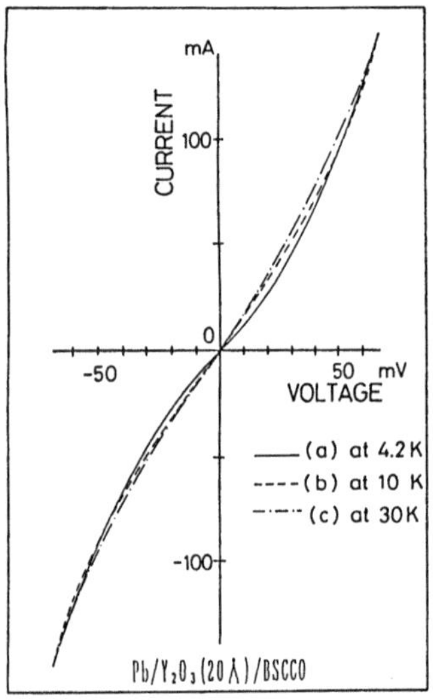

Fig.1 The current-voltage characteristics of the Pb/Y_2O_3(20Å)/BSCCO specimen at (a)4.2K, (b)10K and (c)30K.

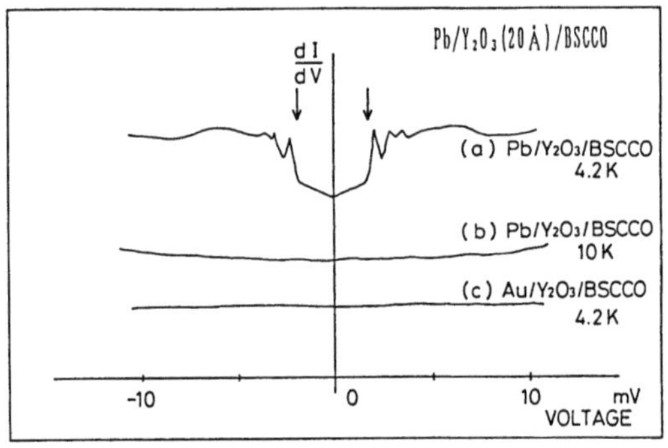

Fig.2 The dI/dV-voltage characteristics of the Pb/Y_2O_3(20Å)/BSCCO specimen at (a)4.2K and (b)10K, and (c) the Au/Y_2O_3(20Å)/BSCCO specimen at 4.2K.

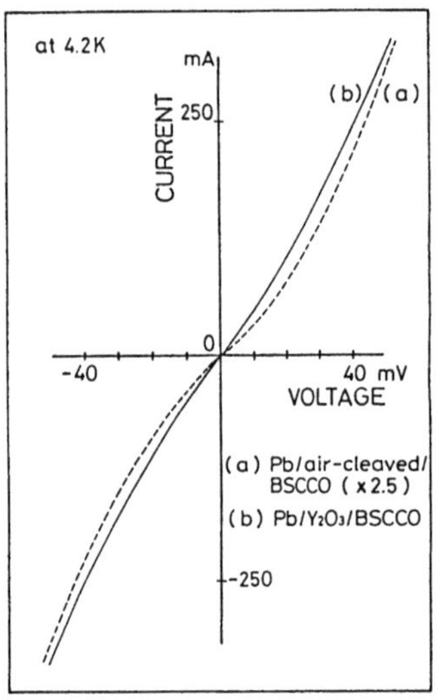

Fig.3 The current-voltage characteristics of (a) the Pb/air-cleaved surface/BSCCO specimen at 4.2K and (b) the Pb/Y_2O_3(50Å)/BSCCO specimen.

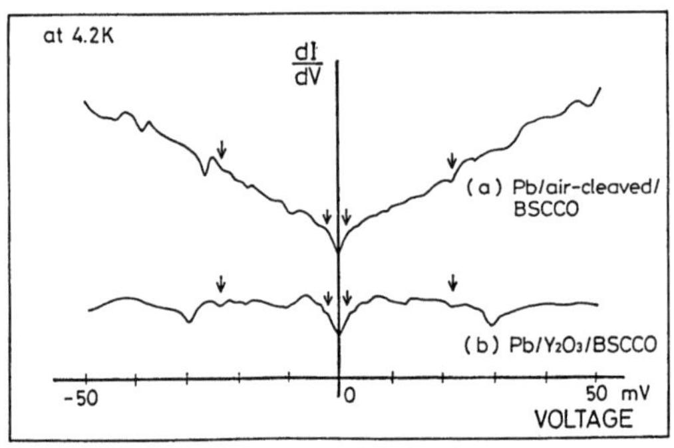

Fig.4 The dI/dV-voltage characteristics of (a) the Pb/air-cleaved surface/BSCCO specimen at 4.2K and (b) the Pb/Y_2O_3(50Å)/BSCCO specimen.

observed at 10K, which is higher than the critical transition temperature Tc of Pb (7.2K). We also prepared the Au/Y_2O_3/BSCCO specimens by depositing Au in place of Pb on the insulating layer. As shown in Fig.2(c), the change of dI/dV around ±3mV was not observed at 4.2K. The results may indicate that the abrupt change of the dI/dV-voltage characteristics around ±3mV in the Pb/Y_2O_3(20Å)/BSCCO specimens corresponds to the superconducting energy gap of Pb. In addition, we have previously reported that Y_2O_3 was homogeneously deposited on flat surface of the air-cleaved BSCCO single crystal[3]. From the results, Y_2O_3 is thought to be useful as an artificial barrier in the superconducting devices with S-I-S structure.

The specimens were prepared by depositing 1000Å Pb on two kinds of the surfaces : the surface obtained by cleaving BSCCO single crystal in air (native barrier : Pb/air-cleaved surface/BSCCO), and the surface after depositing 20Å Y_2O_3 on the clean surface of the BSCCO single crystals, which were cleaved in air and heated in vacuum at 300°C (artificial barrier : Pb/Y_2O_3/BSCCO). Figure 3 shows the I-V characteristics of the specimens at 4.2K. The I-V characteristics of both specimens became non-linear. Figure 4 shows the dI/dV-voltage characteristics of the specimens shown in Fig.3. A number of the abrupt changes of dI/dV were observed in Fig.4. Since they randomly occurred in each specimen, we can not identify their origins satisfactorily. However, two abrupt changes were observed around ±3mV and ±25mV as shown by arrow signs (↓) in Fig.4, and occurred at the same positions in both specimens. The change of dI/dV around ±3mV corresponds to the energy gap of Pb from the result of Fig.2. The change of dI/dV around ±25mV seems to correspond to the energy gap of BSCCO superconductors, as reported in other paper[4]. The results indicated that the air-cleaved surface (native barrier) acted as an insulating layer as well as Y_2O_3 and that the native barrier was formed by cleaving the BSCCO single crystals in air. Although the X-ray photoelectron spectroscopy spectra from the air-cleaved surface were not shown, the results indicated that the air-cleaved surface of BSCCO single crystals was covered with an impurity layer. Therefore, the native barrier is thought to be formed by the adsorption of gases on the air-cleaved surface of the BSCCO single crystal. In addition, the specimen was also prepared by depositing 1000Å Pb on the clean surface, that is, the surface after heating the air-cleaved BSCCO single crystal in vacuum at 300°C. Although the I-V characteristics of the specimen were measured from 0 to 2A, the output voltage was not observed and the specimens remained superconductive. This showed that Pb on the clean surface of the BSCCO single crystal acted as good electrodes. From the results, we found that the native barrier was removed by heating the BSCCO single crystal at 300°C in vacuum. The result is consistent with the results obtained by low-energy electron diffraction and Auger electron spectroscopy[5]. The two specimens, the Pb/air-cleaved/BSCCO specimen and the Pb/Y_2O_3/BSCCO specimen, showed different dI/dV-voltage characteristics as the voltage changed from 0 to ±50mV, except the result that the changes of dI/dV occurred around ±3mV and ±25mV. The reasons can not be clarified because of the large junction area of the prepared specimens.

SUMMARY

We prepared the Pb/Y_2O_3/BSCCO specimens using the 80K-phase BSCCO single crystals and investigated the I-V characteristics. From the results, we found as follows.

(1) Y_2O_3 is useful as an artificial barrier in the superconducting devices with S-I-S structure.
(2) The native barrier was obtained by cleaving the BSCCO single crystals in air.
(3) The changes of the dI/dV-voltage characteristics of the Pb/air-cleaved surface(native barrier)/BSCCO specimen as well as the Pb/Y_2O_3(50A)/BSCCO specimen were observed around +3mV and +25mV.
(4) The native barrier, which was removed by heating the BSCCO single crystals in vacuum at 300°C, is thought to be formed by the adsorption of gases.

ACKNOWLEDGMENTS

The present research was supported by a Grant-in-Acid for Scientific Research on Priority Areas from the Ministry of Education, Science and Culture, Japan.

REFERENCES

1. Nakanishi S, Fukuoka N, Nakahigashi K, Kogachi M, Sasakura H, Minamigawa S, Yanase A (1989) Jpn. J. Appl. Phys. 28 : L71
2. Kishida S, Tokutaka H, Nakanishi S, Fujimoto H, Nishimori K, Ishihara N, Watanabe Y, Futo W (1990) J. Crystal Growth 99 : 937
3. Chihaya M, Kishida S, Tokutaka H, Nishimori K, Ishihara N (1991) J. Physica (in press)
4. Miyakawa N, Shimada D, Kido T, Tsuda N (1990) J. Phys. Soc.Jpn. 59 : 2473
5. Kishida S, Tokutaka H, Nakanishi S, Nishimori K, Ishihara N, Fujimoto H (1989) Jpn. J. Appl. Phys. 28 : L406

Preparation and Characterization of a Superconductor-Semiconductor Junction: $Ba_{1-x}K_xBiO_3$/Nb-Doped $SrTiO_3$

HIROSHI SUZUKI, MASAHIRO IYORI, KAZUHIRO SHIMAOKA, KOUJI YAMANO, TETSUYA YAMAMOTO, MARUO KAMINO, KAZUHIKO TAKAHASHI, TATSURO USUKI, YORINOBU YOSHISATO, and SHOICHI NAKANO

Functional Materials Research Center, SANYO Electric Co., Ltd., Hirakata, Osaka, 573 Japan

ABSTRACT

Electrical properties of Au/native-barrier/BKBO tunnel junction and BKBO/$SrTiO_3$:Nb(0.1wt%) heterojunction were studied. The differential conductance *dI/dV* spectrum of the Au/native-barrier/BKBO tunnel junction had a clear gap structure and was very close to the ideal BCS form with $\Delta(0)$=3.2 meV. BKBO/$SrTiO_3$:Nb heterojunction fabricated at 350°C showed tunneling-like *I-V* characteristics. In the *dI/dV* spectrum of the high-Tc superconductor/semiconductor junction, we observed a superconducting gap structure for the first time.

KEY WORDS: BKBO thin film, SrTiO3:Nb, differential conductance, energy gap

INTRODUCTION

The oxide high-Tc superconductor $Ba_{1-x}K_xBiO_3$ (BKBO) with a cubic perovskite-type structure has attracted much attention because of its interesting superconducting properties and the possibility of widespread use for superconducting devices. BKBO has the highest Tc(≈30 K at x=0.4) among the oxide superconductors not containing Cu ion.[1-3] In order to fabricate the superconducting base transistor,[4,5] a metal/insulator/superconductor (N/I/S) tunnel junction and a superconductor/semiconductor (S/SE) heterojunction are necessary. Some problems have become apparent for those junctions fabricated of high-Tc cuprates, such as YBCO and BSCCO. One was the very short and anisotropic coherent length,[6] and the other was the lack of BCS-like gap structure[7,8] which was possibly due to surface degradation[9] or an intrinsic property of high-Tc oxide superconductors containing Cu ion.

However, BKBO compounds have substantial advantages over the cuprates. Those compounds are isotropic and coherent length is estimated to be ~40-70 Å, which is larger than that of cuprates. In spite of these features, there have been very few studies on the epitaxial growth of BKBO thin films for the N/I/S tunnel junction and S/SE heterojunction.

In this paper, we reported on the fabrication process of the N/I/S tunnel junction and S/SE heterojunction : Au/native-barrier/BKBO and BKBO/Nb-doped $SrTiO_3$. Electrical properties of each junction were measured. The differential conductance *dI/dV* for the N/I/S tunnel junction shows BKBO to be a BCS-type superconductor. In the *dI/dV* spectrum of the S/SE heterojunction, a superconducting gap structure was for the first time observed for high-Tc superconductor/semiconductor systems.

EXPERIMENTAL

BKBO thin films were deposited on niobium-doped $SrTiO_3$(110) (hereafter referred to as $SrTiO_3$:Nb) by rf-magnetron sputtering under the conditions shown in Fig.1. $SrTiO_3$:Nb is used as a substrate. S/SE junction is fabricated once sputtering has been completed, because $SrTiO_3$:Nb was known as an n-type semiconductor.[10,11] The formation of BKBO thin film was performed at a substrate temperature of 300~400 °C, which is appreciably lower than that of YBCO and BSCCO.

Low-temperature deposition is considered to be very important for fabrication of the S/SE heterojunction. To fabricate the N/I/S tunnel junction, BKBO thin films were exposed to dry oxygen for 1 hour. A native barrier is very easily created by exposure to form the insulating tunnel barrier. Then Au was evaporated onto the film.

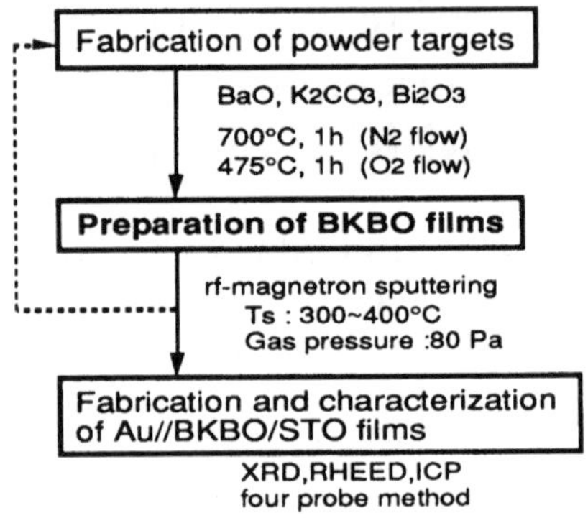

Fig. 1 Sputtering conditions for BKBO thin films.

RESULTS AND DISCUSSION

Quality of the BKBO thin film

Temperature dependence of the resistivity for the as-grown BKBO thin films, fabricated under various substrate temperatures Ts, is shown in Fig. 2. A sharp superconducting transition is observed at 28 K for a Ts=400 °C sample, at 16 K for a Ts=350 °C sample and at 13 K for a Ts=300 °C sample. The value of Tc=28 K is the maximum transition temperature for BKBO thin film as far as we know from the literature. The current-voltage characteristics in the superconducting state had no features associated with a the grain boundary Josephson junction, indicating that there is no grain boundary Josephson junction electrical effect in these thin films. These results indicate that good quality BKBO thin films can be fabricated by our high-pressure reactive rf-magnetron sputtering.

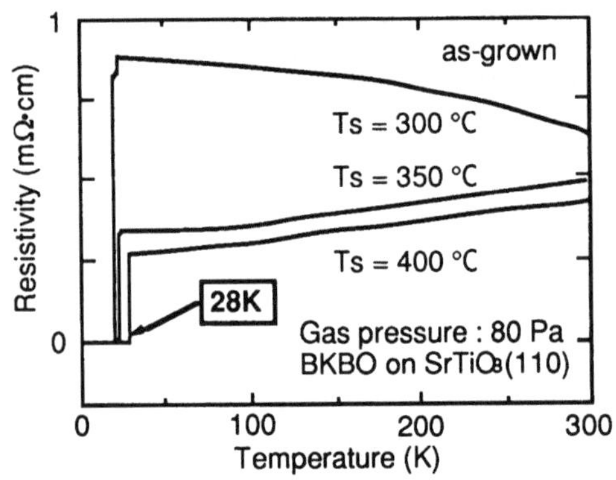

Fig. 2 Temperature dependence of the resistivity of as-grown BKBO thin films. Ts denotes the substrate temperature.

Au/native-barrier/BKBO tunnel junction

Figure 3 shows the *I-V* characteristics of an Au/native-barrier/BKBO junction with sputtering Ts=400 °C (contact Tc=19.5 K) at various temperatures. Nonlinearity is clearly observed around 3 meV corresponding to the energy gap of BKBO. This suggests that high quality tunnel junctions can be manufactured by Au deposition at room temperature. The tunnel barrier fabricated by our method may be either a natural oxide barrier or a region of depleted oxygen from the BKBO.

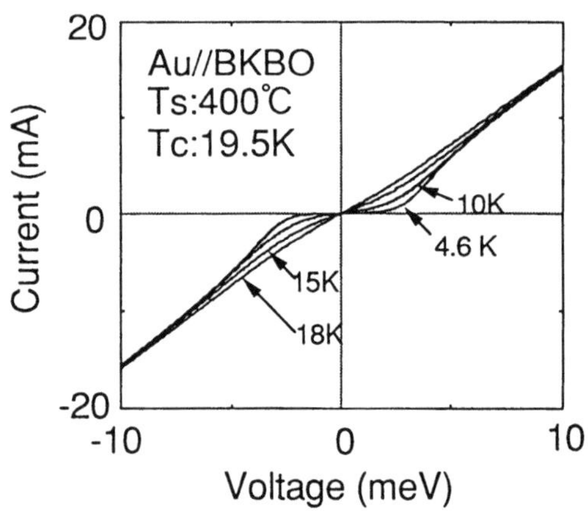

Fig. 3 I-V characteristics of an Au/native-barrier/BKBO junction at the temperatures reported. Tc is the transition temperature of BKBO and Ts is the substrate temperature.

For finite temperatures, the *I-V* characteristics of a N/I/S tunnel junction are given by

$$I(V) = CN_0 \int_{-\infty}^{\infty} N_s(E)[f(E,T) - f(E+V,T)]dE \quad (1)$$

where V is the applied voltage at the junction, N_0 the normal metal density of states, C the constant which contains all the tunneling probabilities, f the Fermi function at temperature T, and $N_s(E)$ the superconducting density of states. In the weak coupling BCS approximation,

$$N_s(E) = \frac{|E|}{\sqrt{E^2 - \Delta^2}} \quad (2)$$

Δ is the superconducting energy gap . The first derivative dI/dV of these *I-V* characteristics yields a thermally smeared measure of $N_s(E)$ results.

The results of the differential conductance dI/dV measurement of an Au/native-barrier/BKBO tunnel junction at various temperatures are shown in Fig. 4. Also shown is a dI/dV determined from Eq. (1) assuming the BCS form for $Ns(E)$ with $\Delta(0)$=3.2 meV. This fit is excellent at any temperature and gives us confidence in the BCS form of density of states. We also obtained $2\Delta(0)/kT$=3.8, which is very near to the BCS value and is in agreement with the result obtained by Sato et al.[12] From these results the material appears to be a BCS superconductor.

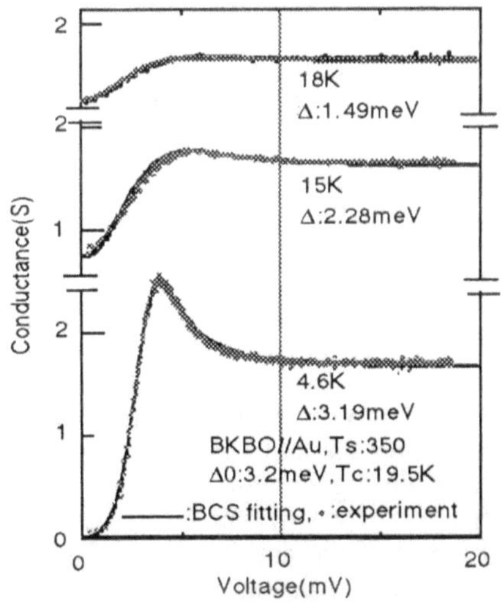

Fig. 4 Differential conductance spectrums for an Au/native-barrier/BKBO tunnel junction at various temperatures. The solid line is the curve calculated by the BCS density of states with Δ(0)=3.2 meV. The closed circles (•) are the experimental data.

BKBO/$SrTiO_3$:Nb heterojunction

The BKBO/$SrTiO_3$:Nb(0.1wt%) junction resistance strongly depends on the sputtering temperature. A sample of Ts=300 °C usually has 2 kΩ junction resistance and that of a Ts=400 °C shows 2MΩ junction resistance. These high resistances are thought to be due to the existence of an unknown layer in the junctions. However, a sample of Ts=350 °C has a resistance of just several tens of ohms. Good quality BKBO/$SrTiO_3$:Nb electrical junctions were obtained at a substrate temperature of 350 °C. It seems that the oxide-superconductor/$SrTiO_3$:Nb contact may be very sensitive to temperature conditions during deposition.
It is generally said that a Schottky barrier is constructed at the interface between superconductor and semiconductor.[13] For complex oxide mixtures such as BKBO, we consider that some other oxide layer exists in the S/SE heterojunction, especially for samples deposited at high temperatures.

Figures 5(a) and (b) show the temperature dependence of the I-V characteristics of the BKBO/SrTiO3:Nb junction fabricated at Ts=300 °C and 350 °C, respectively. In Fig. 6(a), junction resistance is about 1 kΩ. The rectifying characteristic was observed at T=200 K and some asymmetric structure in the I-V curves suggested that a structure like a Schottky barrier was produced at the junction interface. There are no signs of a structure which would imply superconductivity in the I-V curves because of the high-resistance non-superconducting layer in the junction. Since the mean free path of conduction electrons in $SrTiO_3$:Nb increases with decreasing temperature, avalanche breakdown voltage decreases with the temperature as shown in Fig. 5(a). In

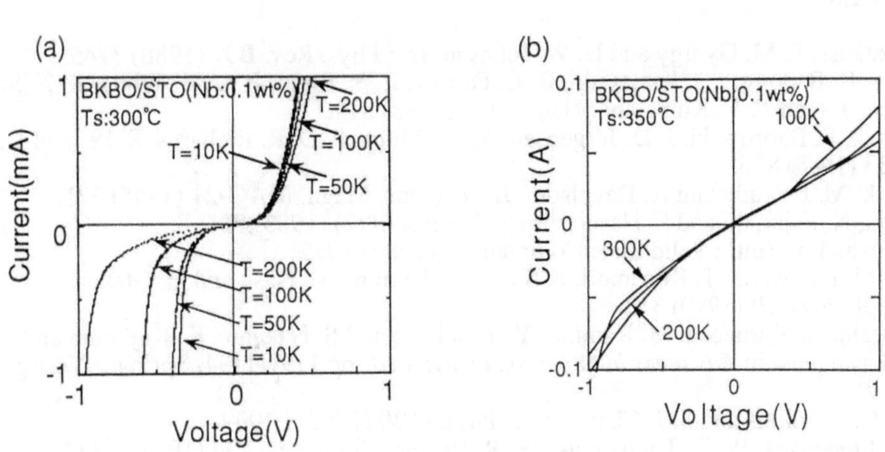

Fig. 5 Temperature dependence of the I-V characteristics of the BKBO/$SrTiO_3$:Nb(0.1 wt%) fabricated at (a) Ts=350 °C and (b) Ts=300 °C.

Fig. 5(b), on the other hand, the ohmic *I-V* characteristics are observed in the low energy regions (-0.2~0.2 V). The *I-V* curves depend less on temperature than in Fig. 5(a). These results suggest this junction resembles an N/I/N tunnel junction, where the first N stands for a $SrTiO_3$:Nb (carrier density N_d=2x10^{19} cm^{-3}), I is a thin Schottky barrier and the last N is a BKBO in the normal state.

The differential conductance *dI/dV* of sample as shown in Fig. 5(b) was also measured. The result can be seen in Fig. 6. Flat and symmetric *I-V* characteristics just like in the case of an N/I/N tunnel junction were observed at T=20 K. With decreasing temperature, the gap structure reflecting the superconductivity of BKBO appeared near zero biasing conductance for the first time. The *dI/dV* characteristics of this S/SE heterojunction fabricated at Ts=350 °C resemble those of an N/I/S tunnel junction under low biasing area. This is caused by the thin Schottky barrier with sharp S/SE heterojunction and large coherent length of BKBO.

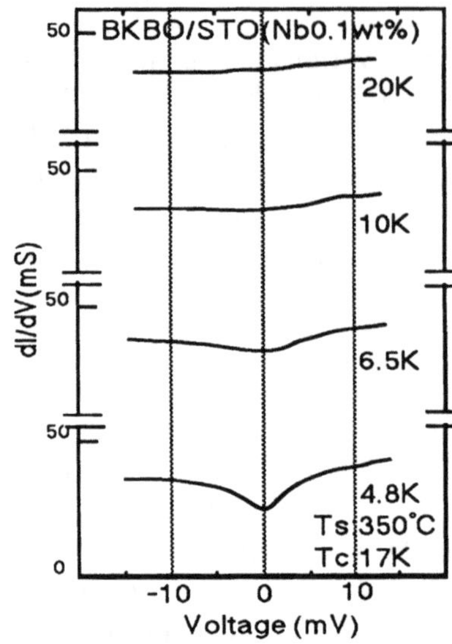

Fig. 6 Differential conductance for BKBO/$SrTiO_3$:Nb heterojunction at various temperatures.

CONCLUSIONS

An Au/native-barrier/BKBO (N/I/S) tunnel junction and a BKBO/$SrTiO_3$:Nb (S/SE) heterojunction were fabricated by using the rf-sputtering. Electrical properties of these junctions were studied. The differential conductance *dI/dV* spectrum of the N/I/S tunnel junction showed a clear gap structure and was very close to the ideal BCS form with $\Delta(0)$=3.2 meV. In the case of the BKBO/$SrTiO_3$:Nb heterojunction, the junction resistance strongly depends on the substrate temperature Ts. We first observed the superconducting gap structure in the *dI/dV* spectrum of the high-Tc superconductor/semiconductor junction. Since BKBO is isotropic and has a clear gap, it is advantageous to study basic superconducting devices.

ACKNOWLEDGEMENT

This work was supported by NEDO under the management of FED.

REFERENCES

1) L. F. Mattheiss, E. M. Gyorgy and D. W. Johnson, Jr. : Phys. Rev. **B37** (1988) 3745.
2) R. J. Cava, B. Batlogg, J. J. Krajewski, R. C. Farrow, L.W. Rupp, Jr., A.E. White, K. T. Short, W.F. Peck, Jr. and T. Y. Kometani : Nature **332** (1988) 814.
3) D. G. Hinks, R. Dabrowski, J. D. Jorgensen, A. W. Mitchell, D. R. Richards, S. Pei and D. Shi : Nature **333** (1988) 836.
4) D. J. Frank, M. J. Brady and A. Davidson : IEEE Trans. Magn. **MAG-21** (1985) 721.
5) H. Tamura, N. Fujimaki and S. Hasuo : J. Appl. Phys. **60(2)** (1986) 711.
6) G. Briceno and A. Zettl : Solid State Commun. **70** (1989) 1055.
7) I. Iguchi, M. Furuyama, T. Kusumori, K.Shirai, S. Tomura, M. Nasu and H. Otake : Jpn. J. Appl. Phys. **29** (1990) 313.
8) K. Takahashi, K. Shimaoka, M. Kamino, Y. Yoshisato and S. Nakano : K. Kajimura and H. Hayakawa (eds) in *Advances in Superconductivity III* pp.1191-1194, Springer-Verlag, Tokyo 1991.
9) A.Edger, C. J. Adkins and S. J. Chandler : J. Phys. C **20** (1987) L1009.
10) H. P. R. Frederikes, W. R. Thurber and W. R. Hosler : Phys. Rev. **134** (1964) A442.
11) O. N. Tufte and P. W. Chapman : Phys. Rev. **155** (1967) 796.
12) H. Sato, H. Takagi and S. Uchida : Physica C **169** (1990) 391.
13) H. Tamura, A. Yoshida, S. Muto and S. Hasuo : Jpn. J. Appl. Phys. **26** (1987) L7.

7.2 Application to Microwave Devices and Antennas

High Temperature Superconductors in High Frequency Fields — Fundamentals and Applications

H. PIEL, H. CHALOUPKA, and G. MÜLLER

Institute of Material Sciences, University of Wuppertal, D-56 Wuppertal, Germany

ABSTRACT

The surface resistance R_s characterizes the quality of high temperature superconductors for their application in passive microwave devices. In this report R_s data of high quality epitaxial $YBa_2Cu_3O_{7-\delta}$ thin films and textured polycrystalline thick films are discussed. The thin films show the best R_s improvement ratio compared to copper and give at 77 K for frequencies below 100 GHz a sound basis for first applications in communication electronics. A condensed reference is given to a variety of passive microwave devices suggested by different research groups world wide. Special attention is paid to a novel electrically small but relatively broadband HTS antenna.

KEY WORDS: Surface resistance, passive microwave components, miniaturized antennas

INTRODUCTION

Microwave experiments with high-T_c superconductors (HTS's) were performed shortly after their discovery and soon demonstrated that the surface resistance R_s of epitaxially deposited $YBa_2Cu_3O_{7-\delta}$ films at 77K and at frequencies below 100 GHz was significantly lower than that of copper at the same temperature [1]. This observation made high-T_c superconductors attractive for microwave applications already quite early during the exciting development of high-T_c superconductivity. Looking at possible applications of superconductors in high frequency fields one finds that microwave cavities fabricated from classical superconductors like niobium play an important role in particle accelerators. In contrast to these large scale applications high quality HTS films can presently be prepared only on single crystal substrates. Polycrystalline thick films show a much higher surface resistance especially at high magnetic surface fields [2]. Miniaturized planar microwave components in communication and sensor technologies are therefore considered a first field for microwave applications and are discussed in this report. Other interesting applications of high-T_c superconductors in high frequency fields like bolometers or devices based on individual Josephson junctions like SIS mixers are covered in other contributions to this conference.

For classical superconductors like niobium the surface resistance and its frequency and temperature dependence can be calculated within the framework of the BCS theory [3]. The applicability of this theory to the HTS's is still discussed controversially. It does not describe the influence of the anisotropy of the cuprate superconductors on the conduction process, it does not include the granularity and it contains no information on the dependence of R_s on magnetic fields. Therefore the considerations of potential applications have to be based on experimentally obtained informations.

THE SURFACE IMPEDANCE.

Among the various oxide superconductors with transition temperatures above 80 K, YBCO is at present the most promising HTS for microwave applications and has been investigated at many laboratories [1]. YBCO films epitaxially grown on high quality single crystalline substrates with low dielectric losses and only small mismatches in their crystallographic parameters (e.g. $LaAlO_3$, $NdGaO_3$ and MgO) were found to show the best microwave performance. We therefore want to discuss the rf properties of such "thin films". All microwave experiments performed with other 1:2:3 compounds where Yttrium is replaced by another rare earth element display a very similar behaviour. Data on highly textured but polycrystalline YBCO thick films on technical substrates are included in this report as they offer the only option for applications where large or curved surfaces are required. Very promising in this respect are also polycrystalline thick films of Tl-Ba-Ca-Cu-O [4] especially because of their expected superior performance at temperatures of 77 K and above. This report, however,will concentrate on the 1:2:3 compounds and their applications.

In YBCO thin films the penetration depth λ depends mainly on the density of the superconducting charge carriers in the CuO_2 planes. Values for $\lambda_{ab}(0)$ between 140 and 170 nm are reproducibly obtained [5]. Considering the slightly increased λ value at 77 K, a film thickness between 400 and 800 nm is sufficient for most microwave applications. In polycrystalline thick films the effective penetration depth is increased [1]. This and the intergranular losses lead in part to the increased surface resistance of polycrystalline HTS materials [6].

The temperature dependence of the experimentally observed surface resistance of selected samples of YBCO films (a thick polycrystalline and two thin films) are shown in comparison to Nb and Nb_3Sn in Fig 1. R_s is plotted as a function of T_c/T in order to facilitate the comparison of materials of different transition temperatures. The surface resistance of a superconductor results from the microwave absorption by the unpaired charge carriers within the field penetration depth. Therefore a sharp drop of R_s just below T_c due to the field expulsion is observed. The data of the laser ablated film in Fig.1. are characteristic for the performance of high quality thin films. It is noteworthy that below $T/T_c = 0.8$ they show a much higher resistance than niobium or Nb3Sn. This "excess resistance" may hide the exponential decrease of R_s due to the expected reduction in the number of the thermally exited quasi particles. The results of very recent experiments with electron-beam coevaporated [7] and with sputtered [8] YBCO films show that R_s can still be significantly reduced at least for temperatures below 50 K. This is attributed to a well controlled micro-stoichiometry and a reduced impurity content of these films.

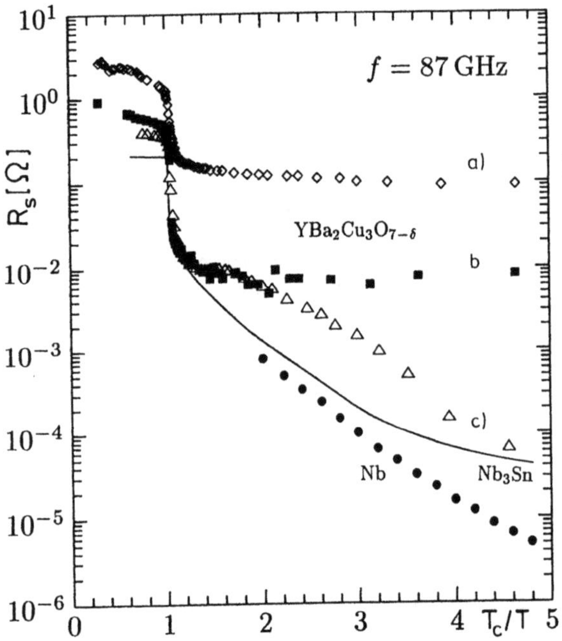

Fig.1 The temperature dependence of the surface resistance of a polycrystalline YBCO sample (a) [1], a standard thin film (b) [1] and recently published thin film data (c) [8] in comparison to Nb and Nb_3Sn.

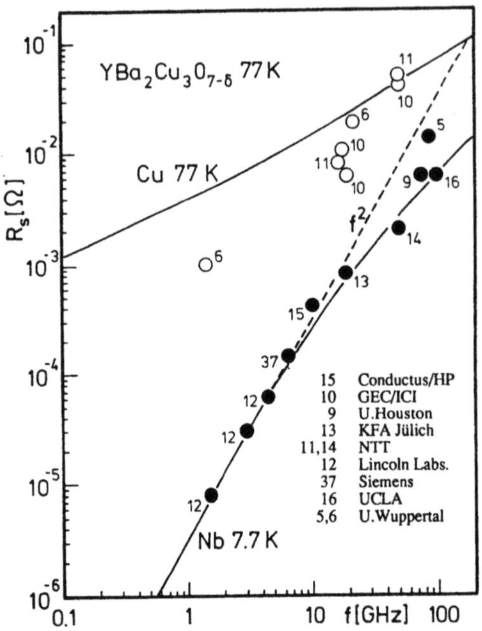

Fig.2 Best published results of R_s of thin films of YBCO on $LaAlO_3$ (full symbols) in comparison to the best results obtained for textured YBCO thick films (open symbols).The numbers refer to the corresponding references.

A summary of the best published data on R_s for high quality YBCO thin and thick films is given in Fig.2. One recognizes that for frequencies below about 100 GHz the surface resistance of YBCO thin films is lower than R_s for pure copper at 77K and close to the results for classical superconductors measured at the same reduced temperature T_c/T. This is encouraging with regard to microwave applications because those are most likely relying on liquid nitrogen temperatures. Below 77 K, however, there remains a big margin for improvements in materials processing. A quadratic frequency dependence as expected from the two fluid model [1] seems to be confirmed between 1 and 100GHz. Recent measurements using stripline techniques [12] have added the lower frequency data. With thick films produced using the elctrophoretic technique [3] or by melt texturing one obtains results superior to those of Cu for frequencies below 10 GHz and the R_s data show a weaker, approximately linear, frequency dependence.

Important for the basic understanding of the residual resistance in YBCO films and for their applicability in high power microwave devices is the study of the dependence of R_s on the magnetic surface fields H_s. Some of the best R_s results for different types of YBCO samples investigated at Wuppertal at high microwave power levels are compared in Fig.3.

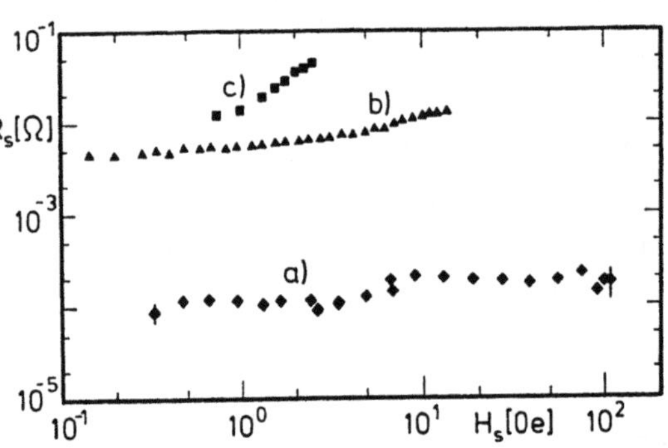

Fig.3. Magnetic surface field (H_s) dependence of the surface resistance of a YBCO thin film (1 inch in diameter) coevaporated on a MgO single crystal (a), a highly textured electrophoretic thick film (b), and an untextured electrophoretic sample (c)-ref.[7] All data are taken at a frequency of 21.5GHz and a temperature of 4.2 K.

The epitaxial film shows a much lower R_S compared to the textured and electrophoretically deposited polycrystalline film and a negligible field dependence. The field dependence of the highly textured electrophoretic film is considerably weaker than the one displayed by the untextured thick film and as found in earlier experiments with untextured polycrystalline YBCO samples [7]. The thin film [7] displays an approximately constant surface resistance up to 120Oe. It is furthermore remarkable that in the experiment with the thin film the achieved field level was limited by a thermal instability of the niobium host cavity and not by the performance of the YBCO sample.
In summary one can conclude that the present state of the art of preparation of thin and thick YBCO films on suitable substrates offers high quality materials for medium power applications of high temperature superconductivity to microwave technology with, of course, important differences between the two film categories. The following paragraphs will be devoted to the present applications of thin films which are considerably more advanced than the applications of thick films (one example of the latter is described in [23]).

GENERAL WORK IN PASSIVE HTS MICROWAVE COMPONENTS.

As soon as the first measurements of the microwave surface impedance of HTS materials showed their comparatively low losses in the microwave frequency regime applications to passive microwave devices were suggested [17,18] and meanwhile a large number of prototype components have been developed. For a large class of linear passive microwave component like bandpass filters and delay lines the realization of "optimum" multiport properties is restricted by dissipative losses in conductors and substrates. By replacing normal conductors with low loss HTS's a lower insertion loss and therefore a higher signal-to noise level and a "sharper" frequency response can be obtained. Furthermore, the size of the components can be considerably reduced without a degradation in the electrical performance,e.g.by replacing bulky normal conducting waveguide-circuits by miniaturized planar HTS circuits. For bandpass filters with a given relative bandwidth and geometric size the minimum achievable attenuation at the centre frequency and the sharpness of the skirt is limited by dissipation losses. In a typical planar microstrip, stripline or coplanar filter the losses in the conductors (gold etc.) dominate the dielectric losses in the substrate. A four - pole microstrip filter at 4.8 GHz with $\Delta f/f_o = 0.013$ may serve as an example. With gold at room temperature and silver at 77 K the insertion loss at centre frequency cannot be reduced below about 10dB and 6dB, respectively. But with YBCO at 77 K a value of about 1dB is achieved [28]. Miniaturized small-bandwidth bandpass filters may find appliations in multiplexing filter banks required in communication systems. Nondispersive and dispersive (e.g. chirp filter) delay lines potentially offer a much higher bandwidth than surface acoustic wave devices and can therefore be used to extend the capability of analog signal processing structures like matched filters and correlators. The utilization of HTS's allows with respect to maximum tolerable attenuation and dispersion the realization of much larger time-delays as in the case of normal conducting structures [18]. High-Q HTS-resonators can be used to reduce the phase-noise in microwave oscillators [60]. The world wide activities are summarized in table 1.They concentrate on components using microstrip and coplanar designs for linear devices like resonators, bandpass filters, delay lines, interconnects, antennas

Laboratory	Resonators	Filters	Delay lines, Interconnects	Antennas	Switches Limiters, Mixers	References
AT&T		X^3				25,28
Argonne N.L.		X^2				42
Bellcore		X^1				41
U. Birmingham	X^3			X^2		24,48
COM DEV		X				39
COMSAT		X^3				25
E.I. du Pont	X		X			35,45
FUBA	X					21
General Electric		X				36
GEC Marconi				X^2		53
Hughes R.L.	X					20,22
HYPRES			X			29
ICI	X^3			X^2		24
INTERATOM				X^1		52
JPL		X,X^2				34,42
Lincoln Lab.	X	X^3	X			25,28
Naval Weapons C.				X^3		50
Northeastern Univ.		X^2				42
T.U. München	X^1					37
NASA Lewis	X^4					22
NTT, Tokai	X	X				44,46
NRL	X^2	X				23,26
Rutgers University		X^1				41
Saitama University			X			30,31
D. Sarnoff Res. C.		X^1				41
Siemens, München	X^1					37
STI	X	X				38
U.St.Petersburg					X	47
Sverdrup Techn.	X^4					22
TRW	X					19,43
UCLA		X				40
Westinghouse		X		X^3		27,55
C. William & Mary	X^2					23
U. of Wisconsin					X	32
U. of Wuppertal	X^2			X^1		23,52

Table1. Overview over the present work in passive microwave components X-marks in the same columns which have the same superscript refer to the same device and indicate collaborations between the corresponding laboratories

and nonlinear components like mixers, limiters and microwave switches. Many of these devices are integrated with a cryogenic refrigerating and data acquisition system to form a space package, which will be launched in the "High Temperature Superconductivity Space Experiment" (HTSSE) in late 1992 [33]. The preparation of this experiment has been very stimulating for the development of miniaturized microwave components.

Table 1 shows that most of the present work is focussed to resonators and filters. This because one expects an early use of such devices for space based communication systems. We have concentrated our work in Wuppertal on the applications of HTS thin films to antennas [52] or antenna systems and our present component development in this field is described in the next chapter.

APPLICATIONS TO ANTENNAS

There are mainly three areas in antenna techniques which may take considerable advantage of HTS material, namely electrically small antennas, superdirective arrays and millimeterwave-arrays [48,49]. With respect to the limited space the present paper is restricted to electrically small antennas with special emphasis to the enhancement of the frequency bandwidth of small HTS antennas.

The power P received at the port of an antenna which is illuminated by a plane wave with power density S_o is given by the effective antenna aperture A_e of the (theoretically) lossless structure and furthermore by the radiation efficiency η and the reflection coefficient Γ at the antenna port which account for the dissipative losses and the mismatch losses, respectively:

$$P = (1 - \Gamma^2)\eta A_e S_o$$

If the external noise does not dominate the internal noise [49,50] or if the external noise itself is the quantity to be determined (radiometry), a reduction of P by a low efficiency η and/or a high mismatch leads to an unwanted reduction of the signal-to-noise ratio.

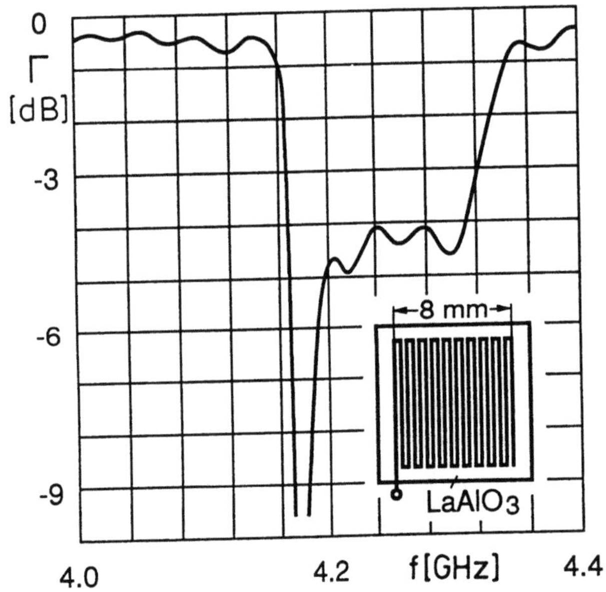

Fig. 4 **Meander Antenna**
Frequency response of the input reflection coefficient Γ with a 3dB-bandwidth of about 4%. Linewidth of YBCO strip conductors 0.1 mm, substrate thickness 1 mm, gold ground plane.

Electrically small antennas are mostly designed to exhibit a "dipol-like" radiation pattern which leads to a size independent value $A_e = 3\lambda_o^2/8\pi$ (half the value for half-space antennas). For $\Gamma \approx 0$ a matching network has to be added to or integrated into the antenna (resonant antenna structure). In the following a possible matching network is assumed to be part of the antenna. It is important to note that from a theoretical point of view a lossless antenna $\eta=1$ could be made "arbitrary" small without reducing the received power. For a real and therefore lossy antenna the miniaturization is governed by a decreasing efficiency η. If the overall size L of the matched antenna (including matching network) is chosen to be smaller than $\lambda_o/2\pi$ ("electrically small") the ratio of the mean energy stored in the near field and matching network to the received energy during one period increases sharply with λ_o/L. Since the dissipation losses P_{diss} are directly related to the stored energy, the radiation efficiency $\eta = P/(P+P_{diss})$ drops down to low values for electrically small antennas. Conductor losses occur in the case of short dipoles mainly in the tuning coil (matching network) [51, 54], but in the case of loop-antennas and electrically short resonant antennas [52] in the antenna itself. If the conductor losses dominate the dielectric losses the radiation efficiency can be considerably enhanced by replacing the normal conductors with HTS's. This was experimentally demonstrated by means of an electrically short wire dipole at 550 MHz made from polycrystalline YBCO [53], by a miniaturized ("Stepped impedance resonator") microstrip patch antenna [52] at 2.4 GHz using an epitaxial YBCO thin film on $LaAlO_3$ and by thin film half-loop antennas [55].

Besides the radiation efficiency η the relative frequency bandwidth $B=\Delta\omega/\omega$ is another important parameter of electrically small antennas. In a number of papers concerned with the "fundamental limits" of small antennas [56,57] B is assumed to be determined by the local curvature of the frequency response of Γ^2 at the centre frequency ω_o. With this assumption the "intrinsic" bandwidth B_i of a (theoretically) lossless antenna structure (including the matching network) can be shown to fall very sharply if its overall size L is reduced to values below $\lambda_o/2\pi$. If losses are present, the so defined bandwidth increases with decreasing efficiency according to $B=B_i/\eta$. Therefore the improvement in efficiency gained by using HTS structures is obtained at the cost of a reduced frequency bandwidth [58,49]. A more detailed analysis shows that the "technical" bandwidth (e.g. the 3db-bandwidth with $\Gamma^2<0.5$) can only in special cases be estimated by the "local" behaviour of Γ^2, namely when the antenna is operated in an "isolated" resonant mode and can be modelled in a "sufficiently large" frequency regime by a single resonant circuit. In the case of multiresonant structures with closely neighboured frequencies only a "global" consideration of the frequency response of Γ^2 leads to an appropriate estimation of B. The utilization of novel electrically small HTS antenna structures which exhibit a multiresonant frequency response and where the radiation resistances of the modes are in the same order of magnitude is therefore a possibility to overcome the problem of obtaining both a small size and a relatively high bandwidth.The "HTS Meander Antenna" [59], shown in Fig. 4 serves as an example for an antenna which offers with respect to its size ($L/\lambda_o \approx 0.1$) and radiation efficiency ($\eta>50\%$) a relatively large 3dB bandwidth ($B\approx4\%$). The open ended microstrip meander structure is build from N=20 line sections In the resonant mode with the centre frequency f_o the section length L is about half a transmission line wavelength and the contribution of all line currents to the radiated field are adding up "in phase". The resonant frequencies of the adjacent modes are about $f_o/2N$ apart from f_o and the radiation from the corresponding current distribution is also dominated by "in phase" summation. The transmission bandwidth of this antenna is therefore extended over at least 3 different modes, resulting in a bandwidth $B\approx 1/N$. Although the bandwidth of about 4% shown in Fig. 4 is large with respect to the electrical size an enhancement to more than 10% can be expected as a result of a further optimization and the utilization of inhomogenous meander structures.

ACKNOWLEDGEMENTS

This work was funded by the German Federal Minister for Research and Technology (BMFT) under the contract number 13 N 5502 A.

REFERENCES

1. H. Piel and G. Müller (1991) IEEE Trans.Magn. MAG -27: 854
2. J.R. Delayen, C.L. Bohn, and C.T. Roch (1990) Journal of Superconductivity, 3: 243
3. D.C.Mattis and J. Bardeen (1958) Phys. Rev. 111: 412
4. D.W. Cooke, P.N. Arendt, E.R. Gray, and A.M. Portis (1991) IEEE Trans. Microwave Theory Tech. 39: 1539
5. N. Klein, H. Chaloupka, G. Müller, S. Orbach, H. Piel, B. Roas, L. Schultz, U. Klein and M. Peiniger (1990) J. Appl. Phys. 67: 6940
6. M. Hein, S. Kraut, E. Mahner, G. Müller, D. Opie, H. Piel, L. Ponto, D. Wehler, M. Becks, U. Klein, and M. Peiniger (1990) Journal of Superconductivity 3: 323
7. M. Hein, S. Hensen, G. Müller, S. Orbach, H. Piel, M. Strupp, N.G. Chew, J.A. Edwards, S.W. Goodyear, J.S. Satchel, R.G. Humphreys, " contr. to the E-MRS Spring Meeting, Strasbourg, May (1991)
8. N. Klein, U. Dähne, U. Poppe, H. Soltner, B. Kabius, K. Urban, J. Schubert, W. Zander, S. Orbach, S. Hensen, G. Müller, H. Piel, B.P. Gorshunov, G.V. Kozlov, S.P. Lebedev, A.A.Volkov, contr. to M^2S.HTS III, Kanazawa, July (1991)
9. M.F. Davisand, J. Wosik, K. Forster , S.C. Deshmukh, H.R. Rampersad, S. Shah, P. Siemsen, J.C. Wolfe, D.J. Economou (1991) J. Appl. Phys. 69: 7182
10. N.McN. Alford, T.W. Button, M.J. Adams, S. Hedges, B. Nicholson, W.A. Phillips (1991) Nature 349: 680
11. O. Ishii, T. Konaka, M. Sato, Y. Koshimoto (1990) Jap. J. of Appl. Phys. 29: 2177
12. D.E. Oates, A.C. Andersen, D.M. Sheen, S.M. Ali (1991) IEEE Trans. Microwave Theory Tech. 39: 1522
13. N. Klein, private communication, Sep 1991
14. H. Asano, M. Satoh, and T. Konaka (1991) Appl. Phys. Lett., 58, 2981 (1991)
15. N. Newman, B.F. Cole, S.M. Garrison, K. Char, R.C. Taber (1991) IEEE Trans. Magn. MAG-27: 1276
16. L. Drabeck, K. Holczer, G. Grüner, Jhy-Jiun Chang, D.J. Scalapino, and T. Venkatesan (1990) Journal of Superconductivity 3: 317
17. H. Chaloupka, H. Piel (1988) Proc. of the Third Workshop on RF Superconductivity, ed. by K.W.Shepard, Argonne ANL-PHY-88-1 (1988)
18. R.S. Withers and R.W. Ralston (1989) Proc. IEEE 77: 1247
19. J.I. Takemoto, C.M. Jackson, R. Hu, J.F. Burch, K.P. Daly (1991) IEEE Trans. Magn. MAG-27: 2549
20. D.B. Rensch, J.Y. Josevowicz, P. Macdonald, C.W. Nieh, W. Hoefer and F. Krajenbrink (1991) ibid ref.19: 2553
21. M. Kuhn, M. Klinger, A. Baranyak, J.H. Hinken (1991) ibid ref.19: 2809
22. C.M. Chorey, K.B. Bhasin, J.D. Warner, J.Y. Josevowicz, D.B. Rensch, C.W. Nieh (1991) ibid ref.19: 2940

23 D. Opie, H. Schone, M. Hein, G. Müller, H. Piel, D. Wehler, V. Folen, S.A. Wolf (1991) ibid ref.19: 2994
24. A. Porch, M.J. Lancaster, T.S.M. Maclean (1991) ibid ref. 19: 2948
25. W.G. Lyons, R.R. Bonetti, A.E. Williams, P.M. Mankiewich, M.L. Malley, J.M.Hamm, A.C.Anderson, R.S.Withers, A. Meulenburg, R.E. Howard (1991) ibid ref 19: 2537
26. H.S.Newman, D.B. Chrisey, J.S. Horwitz (1991) ibid ref. 19: 2540
27. S.H.Talisa, M.A. Janocko, C. Moskowitz, J. Talvacchio, J.F. Billing, R. Brown, D.C. Buck, C.K. Jones, B.R.Mcavoy, G.R. Wagner, and D.H. Watt (1991) IEEE Trans. Microwave Theory Tech. 39: 1448
28. W.G. Lyons, R.S. Withers, J.M. Hamm, and Alfredo C. Anderson, P.M. Mankiewich, M.L. O'Malley, and R.E.Howard (1991) ibid ref. 19: 2932
29. E.K. Track, G.K.G. Hohenwarter, L.(Rao) Madhavrao, R. Patt, R.E. Drake, and M. Radparvar (1991) ibid ref. 19: 2936
30. Mititada Morisue, Joji Asahina, Wenchich Lin, Koki Yo and Norihiro Komine (1991) ibid ref. 19: 2801
31. M. Morisue, S. Furusawa, J. Asahina and A. Kanasugi (1991) ibid ref. 19: 2805
32. J.H. Thompson, J.B. Beyer and J.E. Nordman (1991) ibid ref. 19: 2817
33. J.C. Ritter, M. Nissenoff, G. Price and S.A. Wolf (1991) ibid ref.19: 2533
34. W. Chew, A.L. Riley,D.L. Rascoe, B.D. Hunt, M.C. Foote, T.W.Cooley,and L.J. Bajuk (1991) IEEE Trans Microwave Theory Tech.39: 1455
35. C. Wilker, Zhi-Juan Shen, P. Pang, D.W. Face, W.L. Holstein, A.L. Matthews, and D.B. Laubacher (1991) IEEE Trans Microwave Theory Tech.39: 1462
36. M.R. Namordi, A. Mogro-Campero, L.G. Turner, and D.W. Hogue (1991) IEEE Trans Microwave Theory Tech. 39: 1468
37. A.A. Valenzuela, B. Daalmans, and B. Roas (1991) Electron. Lett. 25: 1435
38. M.S. Schmidt, R.J. Forse, R.B. Hammond, M.M. Eddy, and W.L. Olson (1991) IEEE Trans Microwave Theory Tech. 39: 1475
39. R.R. Mansour, and A. Zybura (1991) IEEE Trans Microwave Theory Tech. 39: 1488
40. D. Zhang, M. Matloubian, Tae Wan Kim, H.R. Fetterman, K. Chou, S. Prakash, C.V. Deshpandey, R.F. Bunshah, and K. Daly (1991) IEEE Trans Microwave Theory Tech. 39: 1493
41. D. Kalokitis, A. Fathy, V. Pendrick, and E. Belohoubek, A. Findikoglu, A. Inam, X.X. Xi, and T. Venkatesan, J.B. Barner (1991) Appl. Phys. Lett. 58: 537
42. C. Zahopoulos and S. Sridhar, J.J. Bautista and G. Ortiz, M. Lanagan (1991) Appl. Phys. Lett. 58: 977
43. J.H. Takemoto, C.M. Jackson, H.M. Manasevit, D.C.St. John, J.F. Burch, K.P. Daly, and R.W. Simon (1991) Appl. Phys. Lett. 58: 1109
44. M. Sato, T. Konaka, H. Asano, S. Kubo, and Y. Nagati (1991) IEICE Transactions. E74: 1980
45. Z.Y. Shen, P.S.W. Pang, W.L. Holstein, C. Wilker, S.Dunn, D.W. Face, D.B. Laubacher (1991) IEEE-MTT-S, Boston, Digest: 1235
46. Y. Kogami, Y. Kobayashi, T. Konaka, and M. Sato (1991) IEEE-MTT-S, Boston, Digest: 1345
47. O.G. Vendik, M.M. Gaidukov, S.G. Kolesov, A.B. Kozyrev (1991) Proc. 21.st. Europ. Microwave Conference, Stuttgart, 72
48. R.C. Hansen (1991) IEEE Trans. Aerospace Electron. Syst. 26: 345
49. R.J. Dinger (1990) J. of Superconductivity 3: 287
50. R.J. Dinger, D.R. Bowling and A.M. Martin (1991) IEEE Trans. Microwave Theory Tech. 39: 1498
51. R.C. Hansen (1991) IEEE Trans. Microwave Theory Tech.39, 1508
52. H. Chaloupka, N. Klein, M. Peiniger, H.Piel, A. Pischke (1991) IEEE Trans. Microwave Theory Tech., 39: 1513
53. S.K. Khams, M.J. Mehler, T.S.M. Maclean, C.E. Gough, N.McN. Alford, and M.A. Harmer (1988) Electron. Lett. 26: 654
54. S.K. Khamas (1990) Electron. Lett. 26: 654
55. R.J. Dinger, D.R. Bowling, A.M. Martin, and J. Talvacchio (1991) IEEE-MTT-S, Boston, Digest: 1243
56. H.A. Wheeler (1947) Proc. IEEE 35: 1479
57. H.A. Wheeler (1975) IEEE Trans. Antennas Propagat. AP-23: 462
58. R.J. Dinger and D.J. White (1990) IEEE Trans. Antennas Propagat. 38: 1313
59. H. Chaloupka, B. Fortyr, M. Peiniger, A. Pischke, "Planar Meander Antennas", European Patent Application 911106920
60. A.P.S. Khanna, M.Schmidt (1991) IEEE-MTT-S, Boston, Digest: 1239

Surface Resistance in High-T_c Superconductors and Their Application to Microwave Devices

TSUNEO KONAKA[1], MAKOTO SATO[1], HIDEFUMI ASANO[2], and YASUHIRO NAGAI[2]

[1] NTT Telecommunication Field System R&D Center, Tokai, Ibaraki, 319-11 Japan
[2] NTT Interdisciplinary Laboratories, Tokai, Ibaraki, 319-11 Japan

ABSTRACT

The surface resistance of high-T_c superconducting c-axis oriented $EuBa_2Cu_3O_x$ thin film sputtered on MgO is greatly lowered by reducing in-plane misorientation. A coplanar delay line with the delay time of 6.1 nsec and a insertion loss of 0.03 dB/nsec at 77 K and 1 GHz are fabricated using the $EuBa_2Cu_3O_x$ thin film. We summarize recently reported data on high-T_c superconducting resonators, bandpass filters and delay lines.

KEY WORDS: High-T_c superconductor, Surface resistance, Microwave device

INTRODUCTION

High-T_c superconductors are expected to have great potential for microwave applications because they have low-loss and low-dispersion properties and a high working temperature. The surface resistance of these superconductors is of particular interest and its qualitative correlations with other properties are widely studied [1,2].The surface resistances of c-axis oriented thin films can be effectively lowered by reducing in-plane misorientation [3,4]. Many passive microwave devices are fabricated with high-T_c superconductors and tested in many laboratories [5].

In this paper, we discuss the correlation between surface resistance and structure in $EuBa_2Cu_3O_x$ thin films sputtered on MgO. A coplanar delay line is fabricated using the $EuBa_2Cu_3O_x$ thin film and its properties are measured. We also summarize recently reported data on high-T_c superconducting resonators, bandpass filters and delay lines, and discuss their performances.

SURFACE RESISTANCE

The surface resistance of high-T_c superconductor depends strongly on surface morphology and orientation. We lowered the surface resistance of c-axis oriented $EuBa_2Cu_3O_x$ thin films by improving surface morphology and reducing in-plane misorientation.

C-axis oriented $EuBa_2Cu_3O_x$ thin films were fabricated on (001) MgO substrates by the single-target dc magnetron sputtering method [6,7]. Typical growth conditions were a substrate temperature of 600-650 °C, and a total gas pressure of 76 mTorr-Ar+(1-5%)O_2. Surface resistance was measured at 50 GHz with a cylindrical, TE_{013} mode, gold-plated Invar host cavity [8]. Surface morphology and orientation were observed by SEM, TEM and X-ray diffraction.

Figure 1 shows SEM photographs of the surfaces of films A, B and C with different surface resistances. The surface resistance of film A with a smooth surface morphology is lower than those of films B and C which have a porous surface morphology or other phases. Film A has a high J_c value of 1.2×10^6 and a high c-axis orientation (a small FWHM value of 0.23° for the (005) reflection). There was no great difference in either the dc-properties or the c-axis orientation of the films with lower surface resistances than several tens of mΩ at 77 K. No in-plane misorientation could be observed within the detection range of our X-ray analysis (0.1%). On the other hand, films such as films B and C which have a porous surface morphology or other phases exhibit surface resistances of more than 200~300 mΩ at 77 K. The J_c and c-axis orientation of these films were also low.

We used TEM to observe the films with low surface resistances of 2 and 80 mΩ at 77 K [9]. The value of 2 mΩ at 50 GHz and 77 K is the same as that of the lowest surface resistance reported [10], if a quadratic dependence of surface resistance is assumed. No apparent grain boundaries could be detected in 2 mΩ film. In contrast, grain boundaries with tilt angles of both 0-5° and 45° were observed in 80 mΩ film. Grain boundaries with a tilt of 0-5° were frequently detected. This result shows that the decrease in the observed in-plane misorientations results in lowering the surface resistance of these films in a similar way to that described in Ref. 1.

We fabricated a coplanar resonator using the film with the surface resistance of 2 mΩ to observe response of surface resistance on a static low magnetic field [11]. An increase (several tens of μΩ) in surface resistance was observed. Though the increase was much smaller than that in the the high-T_c bulk samples (several tens of mΩ) [12], it was thought that the grain boundaries still exist in the film. There is a possibility of lowering the surface resistance of thin film by further decreasing in-plane misorientation.

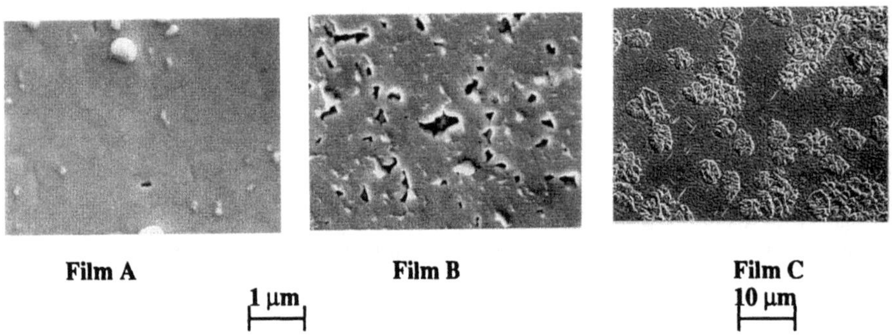

Fig. 1 SEM photographs of $EuBa_2Cu_3O_x$ thin films with surface resistances of 40 mΩ, 250 mΩ and 350 mΩ at 50 GHz and 77 K.

DELAY LINE

It is expected that a compact, low loss high-T_c delay line with a long delay time will be fabricated. However, the transmission line widths of stripline and microstrip delay lines are fixed by the thickness and dielectric constant of the substrate for a given characteristic impedance Z. The distance between the transmission lines is also limited by rf field coupling with the nearest transmission line. Figure 2 shows calculated delay times per square inch and losses per nsec for coplanar, stripline and microstrip delay lines when the line separation is 3 times ground separation and the impedance is 50 Ω. The coplanar line on MgO exhibits the longest delay time and the microstrip line on MgO has the lowest loss. These results are due to the differences in the ground separation and the transmission line width. With a coplanar line, it is possible to realize a longer delay time by adopting a narrower line width while maintaining Z at 50 Ω. The delay time and loss are calculated based on Refs. [13], [14] and [15] and the calculation conditions are shown in Table 1.

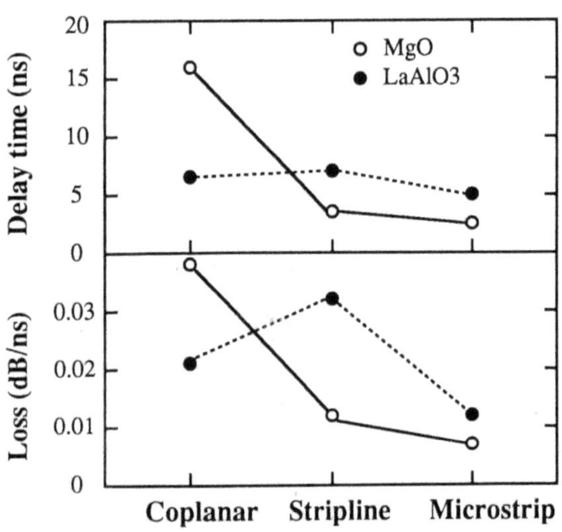

Fig. 2 Delay time per square inch and loss per nsec for coplanar, stripline and microstrip delay lines. Conductor width of coplanar is 100 μm.

Table 1 Calculation conditions for delay time and loss.

Frequency	10 GHz	
Film thickness	0.5 μm	
Substrate (0.5mm thick)	MgO	$LaAlO_3$
Dielectric constant	9.1 [15]	27 [16]
Loss tangent [17]	$6.5x10^{-6}$	$6.6x10^{-6}$
Surface resistance	184 μΩ	
(calculated from 28 μΩ at 3.9 GHz [18] assuming f dependence)		

Figure 3 is a photograph of a 761 mm long coplanar delay line on a 22 mm square MgO substrate. The film used was a 0.4 μm thick $EuBa_2Cu_3O_x$ film fabricated by in-situ dc magnetron sputtering on a MgO substrate 1 inch square and 0.5 mm thick. The film was patterned with photolithography and Ar ion milling. The impedance was designed as 50 Ω with a dielectric constant of 10 which was estimated from coplanar resonator experiments [20]. The line width and the ground separation are 100 μm and 42 μm, respectively. The distance between lines is 400 μm. The properties of this coplanar delay line are compared with reported results in Table 2.

Figure 4 shows the results of Time Domain Reflectometry (TDR) measurement for the coplanar delay line. The impedance is almost the same over the whole length of the delay line and serval ohms higher than 50 Ω. However the spike due to the mismatch between the line and connector is small. The impedance tends to become small as temperature decreases. This temperature dependence is due to the penetration depth change with temperature. A delay time of about 6 nsec is obtained.

The cyclic change in delay time (a cycle of about 3.5 GHz and an amplitude of about 0.6 nsec) was observed against frequency. The cycle corresponds to the resonant frequency of a half-wave coplanar line resonator with the same length as the distance between the corners of the meander-line. This result and the TDR result show that weak impedance mismatch exists at the corners of the meander-line. Figure 5 shows the insertion loss from 0.13 to 10 GHz for the coplanar delay line. The loss becomes small at low temperatures and 0.1 dB at 1 GHz and 19 K. Many loss peaks are observed above about 1.8 GHz and there is a large increase loss. The peaks are thought to be caused by the reflection at the corners where weak impedance mismatch exists. The high frequency properties of the coplanar delay line can be improved by reducing the impedance mismatch at the corners.

Table 2 Properties of high-Tc planar delay lines

Structure	Film size (mm)	Conductor width (mm)	Delay time (ns)	Loss at 77K and 1GHz (dB/ns)	Reference
Coplanar EuBaCuO/MgO	22x22	100	6.1 (761mm)	0.03	This work
Stripline EuBaCuO/ MgO/EuBaCuO-ground	20x20	160	4 (400mm)	0.07	[21]
Micostrip YBaCuO/ MgO/Cu-ground	25.4x25.4	150	5 (600mm)	0.2	[22]
Coplanar TlBaCaCuO/LaAlO3	25.4x25.4	30	11 (---)	0.1	[23]

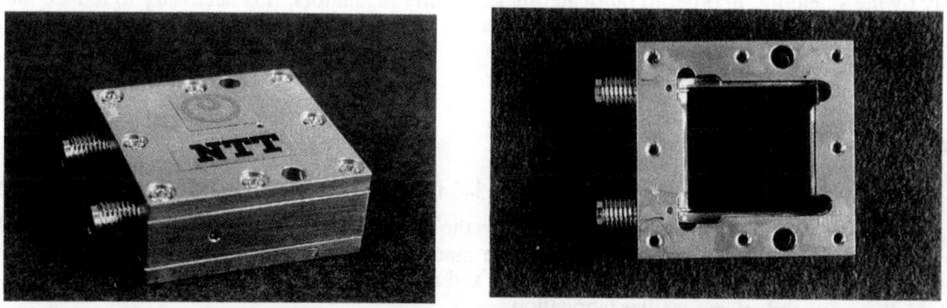

Fig. 3 Housed coplanar delay line on 22 mm square MgO.

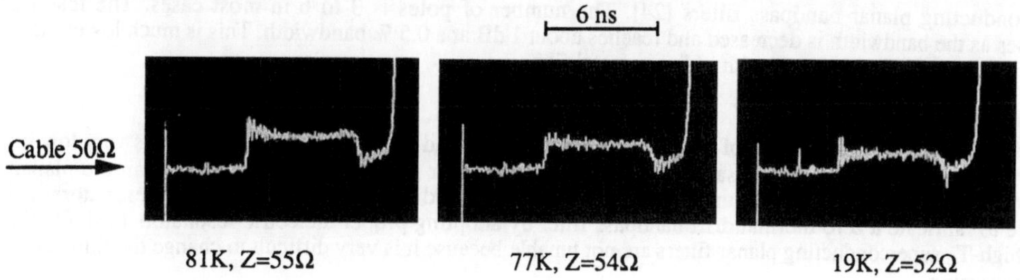

Fig. 4 TDR results for the coplanar delay line.

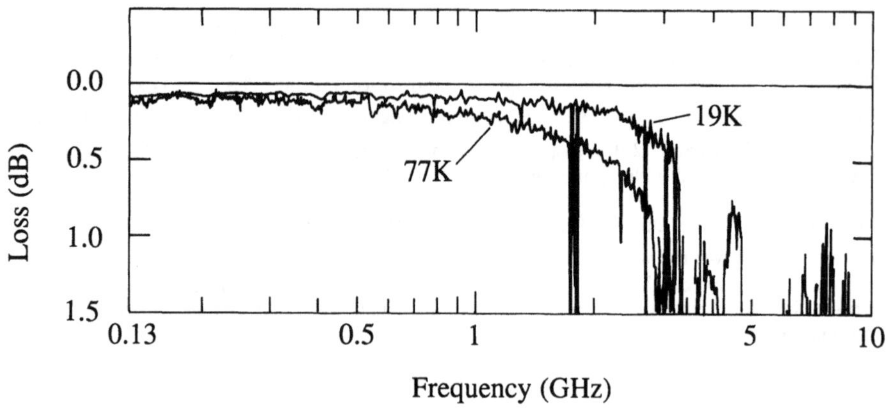

Fig. 5 Frequency dependence of insertion loss for the coplanar delay line.

RESONATORS

Stripline, microstrip and coplanar resonators have been fabricated using high-T_c superconducting films. The types of high-T_c superconducting film employed are TlCaBaCuO, YBaCuO and EuBaCuO thin films. The substrate materials used are $LaAlO_3$ and MgO. Figure 6 shows the capability of currently available high-T_c resonators by plotting their quality factors as a function of frequency [24]. The quality factors are more than one order higher than those of ordinary metal planar resonators. The quality factors increase at lower frequencies and they become higher than 10^4 below 10 GHz.

The unloaded quality factor Q_0 of a planar resonator is given as $1/Q_0=1/Q_c+1/Q_r+1/Q_d$. Q_c, Q_r and Q_d are the quality factors due to conductor loss, radiation loss and dielectric loss in a substrate, respectively. The loss tangents tan δ (~ $1/Q_d$) of $LaAlO_3$ and MgO are lower than 10^{-5} below 77 K as shown in Fig. 7 [18]. Therefore, dielectric loss dose not currently limit the quality of the planar resonators. The radiation loss depends strongly on the resonator structure and the shielding quality of the resonator case. In an open system, the radiation loss of an end-coupled microstrip resonator is particularly large and Q_0 is limited by Q_r. For example, the Q_r of a microstrip resonator on 0.5 mm thick MgO is 567 at a conductor width of 0.52 mm (Z=50 Ω) and at 10 GHz. The radiation loss can be reduced considerably by the shielding. In order to fabricate higher-quality resonators, it is necessary to reduce the radiation loss by improving the housing technique as well as to reduce the conductor loss by improving the fabrication and patterning processes of high-T_c films.

BANDPASS FILTERS

The insertion loss α of a bandpass filter is given as $\alpha=4.343n/(\omega Q_0)$ [33]. Here Q_0 is the unloaded quality factor of the coupled resonators (poles) of the bandpass filter, ω is the fractional bandwidth and n is the numbers of poles. A bandpass filter with a very steep skirt property can be made by adopting large numbers of poles. Very high-Q resonators are needed in order to achieve both low-loss and narrow-band performance. This is the main reason for adopting high-T_c superconductors for filter fabrication.

Figure 8 shows the current capability of high-T_c superconducting filters by plotting the insertion losses of superconducting planar bandpass filters [24]. The number of poles is 3 to 6 in most cases. The insertion loss increases as the bandwidth is decreased and reaches about 1 dB at a 0.5 % bandwidth. This is much lower than that of planar bandpass filters fabricated with ordinary metal films.

A hybrid bandpass filter composed of a high-T_c superconductor and dielectric resonators exhibits very low insertion loss at a narrow bandwidth. The disadvantage of the hybrid filter is that it is larger in size than the planar filter. However, the hybrid filter can be tuned by changing the size and position of the dielectric resonators. It is also possible to fabricate a zero thermal drift bandpass filter by adopting proper dielectric resonators [43]. On the other hand, high-T_c superconducting planar filters are not tunable because it is very difficult to change the film pattern after initial patterning.

CONCLUSION

Low-loss delay line, high-Q resonators and low-loss bandpass filters fabricated with high-T_c superconductors have been demonstrated. However, we have not yet taken the advantage of the intrinsic potential of high-T_c superconductors in their microwave applications. There are many important areas still to be studied. These include the reduction of the surface resistance and patterning damage, and the improvements in housing techniques and designing methods for high-T_c microwave circuits. Moreover, it is important to clarify the low-temperature dielectric properties of substrate materials. The quality of high-T_c superconducting devices will be progressively improved and they will be widely applied when a low cost refrigerator with high cooling efficiency is realized.

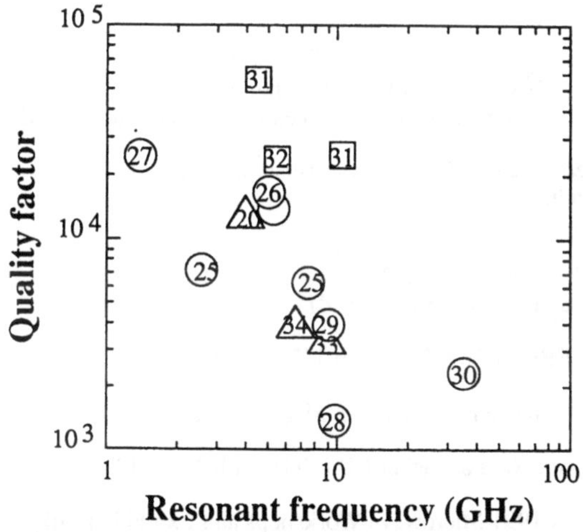

Fig. 6 Quality factors of high-T_c planar resonators. (□);stripline, (○);microstrip, (△);coplanar. Numbers in fig. are reference numbers.

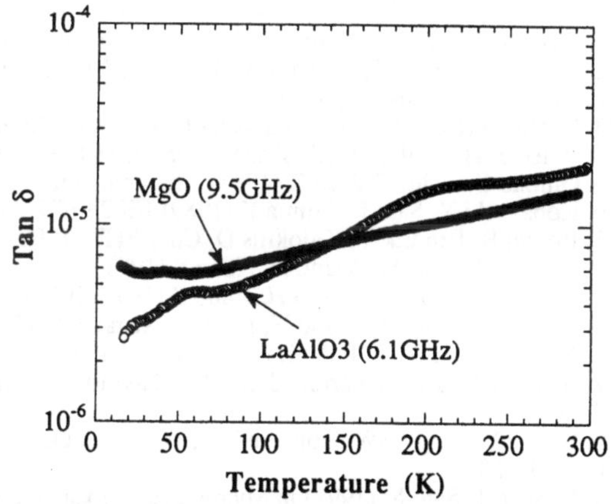

Fig. 7 Temperature dependence of loss tangent for MgO and $LaAlO_3$.

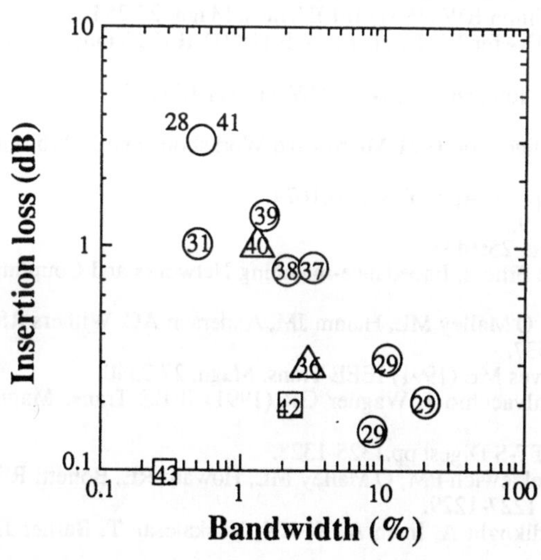

Fig. 8 Insertion loss of high-T_c superconducting bandpass filters. (○);planar type with center frequencies of 9 to 14 GHz, (△);planar type with center frequencies of 4 to 5 GHz, (□);hybrid type composed of high-T_c superconductor and dielectric resonators with center frequencies of 9 to 13 GHz. Numbers in fig. are reference numbers.

References

1. Laderman SS, Taber RC, Jacowitz RD, Moll JL,.Eom CB, Hylton TL, Marshall AF, Geballe TH, Beaseley MR (1991) Phys. Rev. B 43:2922.
2. Alford NMcN, Button TW, Peterson GE, Smith PA, Davis LE, Penn SJ, Lancaster MJ, Wu Z, Gallop JC (1991) IEEE Trans. Magn. 27:1510.
3. Char K, Newman N, Garrison SM, Barton RW, Taber RC, Laderman SS, Jacowitz RD (1990) Appl. Phys. Lett. 57:409.
4. Asano H, Satoh M, Konaka T (1991) Appl. Phy. Lett. 58:2981.
5. for example (1991) IEE Trans. Magn. 27:2533-2954.
6. Asano H, Asahi M, Michikami O (1989) Jpn. J. Appl. Phys. 28;L981.
7. Asano H, Yonezawa H, Asahi Michikami O (1991) IEE Trans. Magn. 27:844.
8. Konaka T, Sato M, Asano H (1990) Jpn. J. Appl. Phys. 29:L561.
9. Asano H, Satoh M, Konaka T (1991) ISS'91.
10. Klein N, Muller G, Piel H, Roas B, Schultz L, Klein U, Peiniger M (1989) Appl. Phys. Lett. 54:757.
11. Sato M, Konaka T, Asano H (1991) in Conference Bookelt M^2S-HTSC III Kanazawa p.416.
12 Sato M, Konaka T, Ishihara K (1990) In: Ishiguro T, Kajimura K (eds.) in Advances in Superconductivity II. Springer-Verlag, Tokyo, pp.509-512.
13. Hoffman RK (1987) in Handbook of Microwave Integrated Circuits. Artec House, Boston, chapters 6, 11, 12.
14. Wheeler HA (1977) IEEE Trans. Microwave Theory & Tech. 25:631.
15. Ghione G, Naldi C, Zich R (1983) Alta Frequenza 52:191.
16. Kobayashi Y, Sato J, Yajima K (1989) IEICE E72:290.
17. Brown R, Pendrik V, Kalokitis D, Chai BHT (1990) Appl. Phys. Lett. 57:1351.
18. Konaka T, Sato M, Asano H, Kubo S (1991) J. Superconductivity (to be published).
19. Sato M, Konaka T, Asano H, Kubo S (1991) IEICE Trans. E74:1980.
20. Konaka T, Sato M, Asano H, Kubo S, Nagai Y (1991) in IEEE MTT-S Digest pp.1377-1340.
21. Nagai Y, Suzuki N, Itoh K, Michikami O (1991) ISS'91.
22. Track EK, Hohenwarter GKG, Madhavrao L, Patt R, Drake RE, Radparvar M (1991) IEE Trans. Magn. 27:2936.
23. Shen ZY, Pang PSW, Holstein WL, Wilker C, Dunn S, Face DW, Laubacher DB (1991) in IEEE MTT-S Digest pp.1235-1238.
24. Konaka T, Sato M (1991) in Microwave Workshop Digest of 1991 Microwave Workshops and Exhibition Tokyo pp.163-168.
25 Hammond RB, Negrete GV, Schmidt MS, Moskowitz MJ, Eddy MM, Strother DD,.Skoglund DL (1991) in IEEE MTT-S Digest, pp.867-870.
26. Young KH,.Negrete GV, Hammond RB, Inam A, Ramesh R, Hart DL, Yonezawa YY (1991) Appl. Phys. Lett.58:1789.
27. Kuhn M, Klinger M, Baranyak A, Hinken JH (1991) IEEE Trans. Magn. 27:2809.
28. Takemoto JH, Jackson CM, Hu R, Burch JF, Daly KP, Simon RW (1991) IEEE Trans. Magn. 27:2549.
29. Rensch DB, Josefowicz JY, Macdonald P, Nieh CW, Hoefer W, Krajenbrink F (1991) IEEE Trans. Magn. 27:2553.
30. Chorey CM, Bhasin KB, Warner JD, Josefowicz JY, Rensch DB, Nieh CW (1991) IEEE Trans. Magn. 27:2940.
31. Itozaki H, Higaki K (1991) in Microwave Workshop Digest of 1991 Microwave Workshops and Exhibition Tokyo pp.157-161.
32. Nagai Y, Sato M, Asano H, Kubo S, Konaka T (1991) Jpn. J. Appl. Phys. 30:1679.
33. Valenzuela AA, Russer P (1989) Appl. Phys. Lett. 55:1029.
34. Valenzuela AA, Daalmans B, Roas B (1989) Electron. Lett. 25:1435.
35. Matthaei G, Young L, Jones EMT (1980) in Microwave Filters, Impedance-matching Networks and Coupling Structure Artech House Dedham pp.149-155.
36. Lyons WG, Bonetti RR, Williams AE, Mankiewich PM, O'Malley ML, Hamm JM, Anderson AC, Withers RS, Meulenberg A, Howard RE (1991) IEEE Trans. Magn. 27;2537.
37. Newman HS, Chrisey DB, Horwitz JS, Weaver BD, Reeves ME (1991) IEEE Trans. Magn. 27;2540.
38. Talisa SH, Janocko MA, Jones CK, McAvoy BR, Talvacchio J, Wagner GR (1991) IEEE Trans. Magn. 27;2544.
39. Talisa SH, Janocko MA, Talvacchio J (1991) in IEEE MTT-S Digest pp.1325-1328.
40. Lyons WG, Withers RS, Hamm JM, Anderson AC, Mankiewich PM, O'Malley ML, Howard RE, Bonetti RR, Williams AE, Newman N (1991) in IEEE MTT-S Digest, pp.1227-1229.
41. Kalokitis D, Fathy A, Pendrick V, Belohoubek E, Findikoglu A, Inam A, Xi XX, Venkatesan T, Barner JB (1991) Appl. Phys. Lett. 58;537.
42. Curtis JA, Fiedziuszko SJ, Holme SC (1991) in IEEE MTT-S Digest pp.447-450.
43. Kogami Y, Kobayashi Y, Konaka T, Sato M (1991) in IEEE MTT-S Digest pp.1345-1348.

Advanced High T_c Superconducting Active Antennas

T. OHNUMA, T. KUROKO, Y. TANAKA, K. YOSHIDA, Y. MATSUURA, and A. MARUMOTO

Department of Electrical Engineering, Tohoku University, Sendai, 980 Japan

ABSTRACT

High Tc superconducting (SC) active antennas, which are new types of antennas with a use of active properties of high Tc superconductors, are investigated. The SC active antennas were made from Y-type thin films, which were prepared with magnetron (sputtering) method. As for detections of electromagnetic waves (f=9.55GHz, f=50.4GHz and 474THz), the sensitivities of the SC active antennas were found to depend on polarizations of radiated electromagnetic waves, applied currents, the annealing temperature, the strength of electromagnetic waves and structures of antennas et al.

KEYWORDS: Superconducting active antennas, Electromagnetic waves, Sputtering plasmas, SC thin films, Terahertz light waves

INTRODUCTION

New types of superconducting active antennas for receiving electromagnetic waves et al were reported by previous report.[1)] In this report, the advanced properties of the superconducting active antennas are described. Properties of sputtering plasmas and the improved measurement system are also investigated.

EXPERIMENTAL

High Tc superconducting thin films (Y-type) for SC active antennas were prepared with a magnetron (sputtering) method. The relations between the sputtering plasmas and the thin films were also investigated. The electromagnetic properties of the superconducting active antennas were obtained with a radiation of 9.55GHz, 50.4GHz and 474THz electromagnetic and light waves.

RESULTS AND DISCUSSIONS

(A) RELATIONS OF MAGNETRON PLASMAS TO SUPERCONDUCTING THIN FILMS

Spatial properties of magnetron plasmas (the electron temperature and the plasma density) for preparing superconducting thin films were investigated. The typical spatial distributions of the electron temperature and the plasma density are shown in Fig.1. The corresponding spatial properties of superconducting thin films are indicated in Fig.2, in which the resistances of superconducting thin films produced by the magnetron plasmas are shown. The thin films ($R \sim 0\Omega$) of the center area used as the following SC active antennas, that is, the plasma parameters in the central regions show the sputtering conditions of the thin films for SC active antennas.

(B) ELECTROMAGNETIC DETECTIONS BY SUPERCONDUCTING ACTIVE ANTENNAS

Electromagnetic waves of 9.55GHz and 50.4GHz were radiated to SC active antennas which consist of SC thin films (prepared with the method mentioned above) through horn antennas with the fixed polarizations (E and H directions). The sensitivities of the SC active antennas were measured as V (the output voltage) or ΔV (the increment of the voltage) with parameters of the applied current (I) and the applied current density (J) in Figs.4-6.

(B-1) CURRENT-CONTROLLED DETECTIVITIES OF SC ACTIVE ANTENNAS

The structure of the typical superconducting active antennas is shown in Fig.3, which consists of structured superconducting thin films on MgO substrate. The thickness of thin films is $\sim 1\mu m$. The typical detectivities of the SC active antennas against the applied current are shown in Fig.4 for 9.55GHz electromagnetic waves. The detectivities are indicated to increase with an increase of the applied currents for small currents and decrease for larger currents.

(B-2) DETECTIVE SENSITIVITIES OF ACTIVE ANTENNAS VS ANNEALING TEMPERATURE

The sensitivities of the superconducting active antennas can be controlled in the process of the fablication of the SC thin films. That is, the sensitivities of the SC active antennas could be controlled by the annealing temperature of the thin films. The typical relation between the sensitivities vs. the annealing temperature is shown in Fig.5 for the case of the detections of 9.55GHz electromagnetic waves. The dependencies on the radiated electromagnetic power (P) are also indicated.

(B-3) REFLECTOR-CONTROLLED DIRECTIVITIES AND SENSITIVITIES OF SC ACTIVE ANTENNAS FOR 50GHz

With a use of a reflector the SC active antennas, the directivity and sensitivity of the SC active antennas are found to be controlled easily. Improved directivities and sensitivities of the SC active antenna with a corner reflector are shown in Fig.6 for the case of the detections of 50.4GHz electromagnetic waves.

CONCLUSIONS

Superconducting active antennas (new types of electromagnetic antennas), which use essential properties of superconductors were improved in the sensitivities and the directivities for detecting 9.55GHz, 50.4GHz and 474THz electromagnetic waves. The spatial profiles of sputtering plasmas for the thin films of SC active antennas are also investigated.

REFERENCES

(1) Ohnuma T. (1991) JJAP, Vol.29, 11(1990), p.2415-2416

(2) Ohnuma T.,T.Kuroko,H.Tomi,M.Shimegi,Y.Tanaka and K.Yoshida (1990) 3rd ISS'90, ADP-1, p.1167-1170

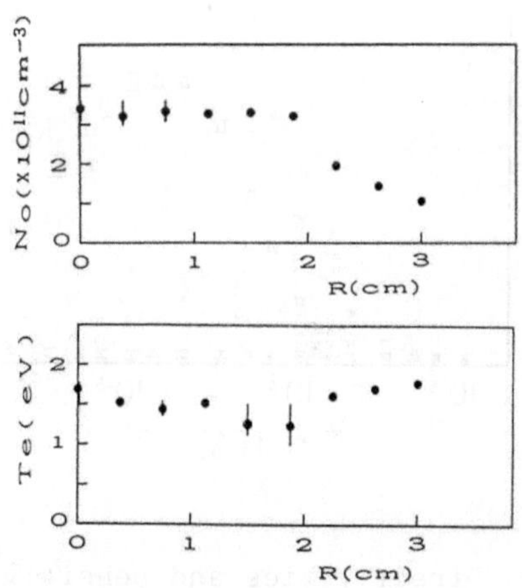

Fig.1 Spatial profiles of sputtering plasmas (the electron temperature and the plasma density) for the fabrications of superconducting thin films

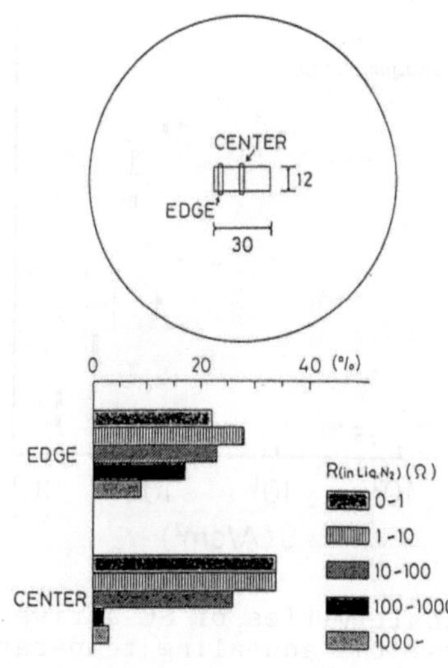

Fig.2 Resistances of superconducting thin films vs. positions of substrates

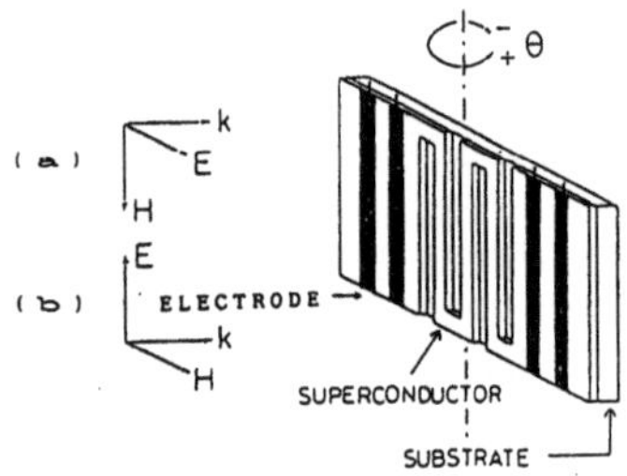

Fig.3 Typical superconducting active antenna

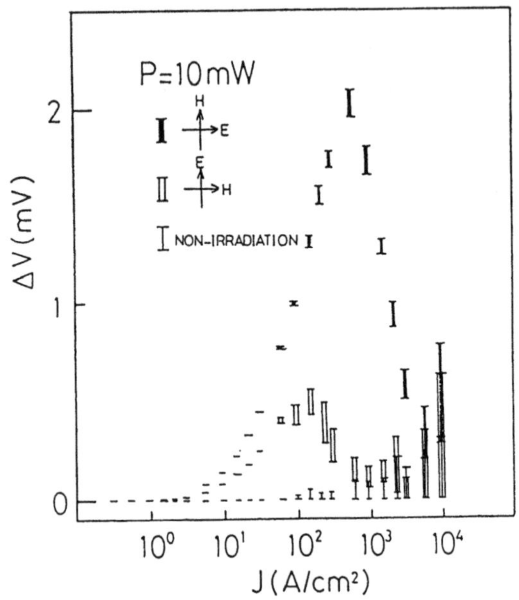

Fig.4 Typical detectivities of the SC active antenna

(A)

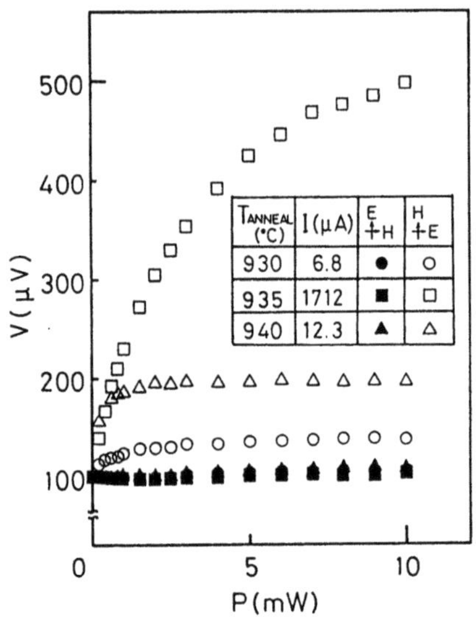

(B)

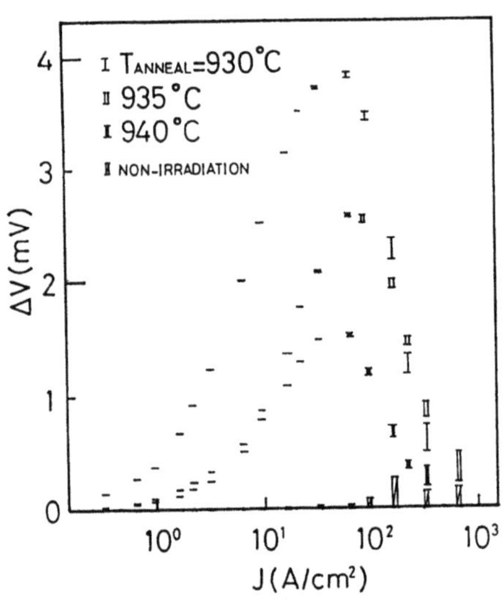

Fig.5 Sensitivities of SC active antenna vs. the annealing temperature

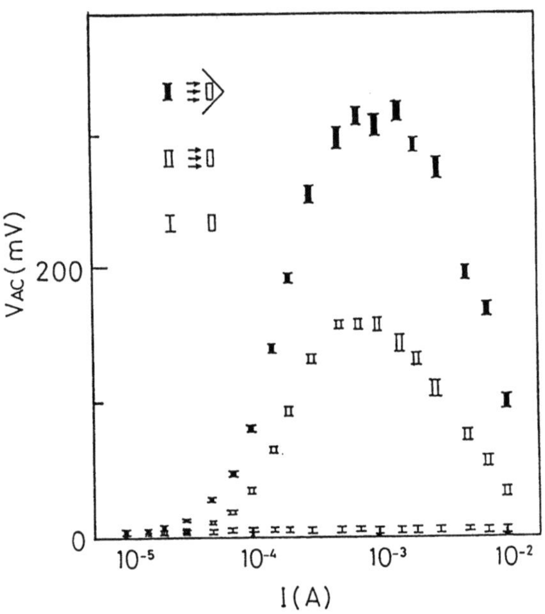

Fig.6 Directivities and sensitivities with a reflector of SC active antenna (f=50.4GHz)

Small Helical Antennas Made of $YBa_2Cu_3O_{7-x}$ Ceramics

KEIICHIRO ITOH[1], OSAMU ISHII[2], YASUHIRO KOSHIMOTO[1], and KEIZO CHO[2]

[1] NTT Interdisciplinary Research Laboratories, Tokai, Ibaraki, 319-11 Japan
[2] NTT Radio Communication Systems Laboratories, Yokosuka, Kanagawa, 238-03 Japan

ABSTRACT

The use of high-Tc superconducting ceramics offers the potential for achieving highly efficient small antenna systems. In this paper, small antennas which employ 1/45 and 1/70-wavelength radiators resonating around 500 MHz and quarter-wave matching circuits were made of sintered $YBa_2Cu_3O_{7-x}$ bulk materials and thick films with zirconia cores, and their surface resistances were estimated by measuring radiation characteristics at 80 K. As a result, it is shown that the surface resistance of the $YBa_2Cu_3O_{7-x}$ is $0.05 \sim 0.21$ lower than that of copper at room temperature.

KEY WORDS: $YBa_2Cu_3O_{7-x}$ superconductor, small antenna, ohmic loss, helical radiator, quarter-wave matching circuit

INTRODUCTION

The radiation efficiency of a small antenna made of a normal conductor is very low because of its low radiation resistance. Moreover, it is difficult to achieve exact impedance matching between a small radiator and a feed line because of the large difference in their impedances. Therefore, to obtain highly efficient small antenna systems, it is important to realize low-loss radiators and low-loss impedance matching circuits.

The use of superconductors offers the potential for achieving highly efficient small antenna systems[1]–[4]. However, it is difficult to achieve high efficiency when coils are used, and to widen the bandwidth when stub matching circuits are used, because of the losses of the coils[5] and the narrow bandwidths of the stubs. It should be noted that the impedance of a self-resonating radiator can be matched to the feed line impedance without any stubs and coils, and that a quarter-wave matching circuit can match only a pure resistance but has a wider bandwidth than a stub matching circuit[6]. Accordingly, high-Tc superconducting small antennas which employ self-resonating helical radiators and quarter-wave matching circuits[7],[8] are described, and the relations between their antenna gains and their component surface resistances are examined, in this paper.

ANTENNA STRUCTURE AND DESIGN

Figure 1 shows a perspective view of the antenna. The helical radiator, which has only pure resistance at the resonant frequency, works as an electric dipole and is joined to one end of a quarter-wave matching circuit composed of parallel transmission lines. The microstrip feed line is joined to the other end of the quarter-wave matching circuit. Dielectric sheets (thickness of h_m and h_f) are positioned between the parallel lines and between the microstrip line and ground plane.

The helical radiators are designed by the moment method[9]. Figure 2 shows number of turns N, pitch P, and radiation resistance R_r as functions of radiator length L_r, for a helical radiator, whose diameter D and wire diameter d are 10 and 0.5 $mm\phi$, designed to resonate

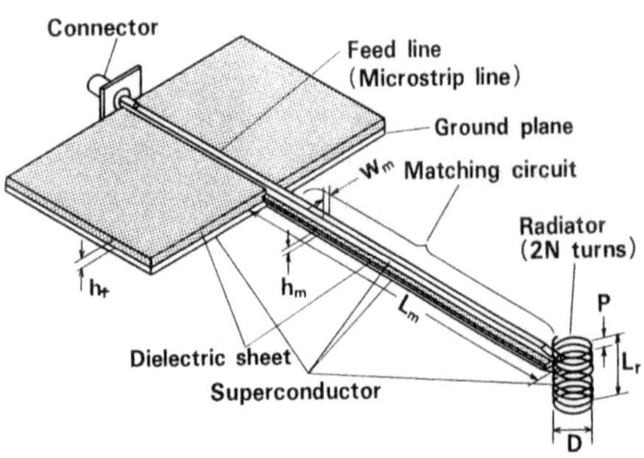

Fig. 1. Perspective view of the antenna.

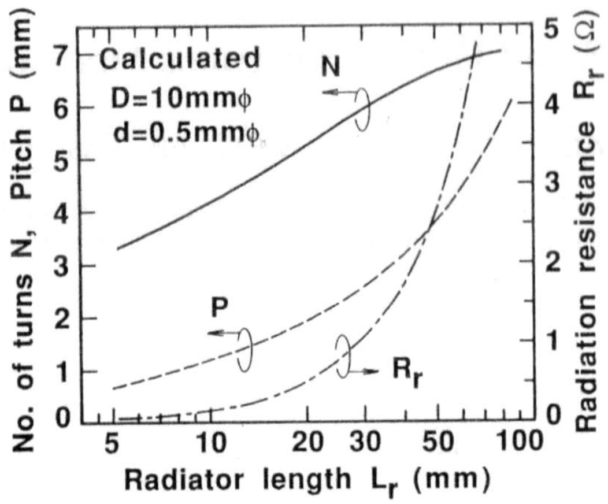

Fig. 2. Dimensions and radiation resistance of a helical radiator designed to resonate at 500 MHz as functions of L_r.

at 500 MHz. If the matching circuit length L_m is chosen as $\lambda_g/4$ (λ_g is the wavelength in the circuit) and the surface resistances of the radiator and circuit are both small, then the characteristic impedance of the circuit Z_c^m is given by the following equation to achieve perfect matching[6].

$$Z_c^m = \sqrt{Z_c^f \cdot R_r}\ ,$$

where Z_c^f is the characteristic impedance of the feed line.

To examine the losses of this type of antenna, the ohmic losses of the radiators and the matching circuits are calculated using the moment method considering surface resistance[10] and the transmission equation[7], respectively. Figure 3 shows the total ohmic losses as functions of radiator length L_r, for the antenna which has a radiator, whose dimensions are shown in Fig. 2, and a matching circuit, whose width w_m and length L_m are 2.0 and 100 *mm*. The losses increase greatly as L_r is reduced, when surface resistance R_s is large.

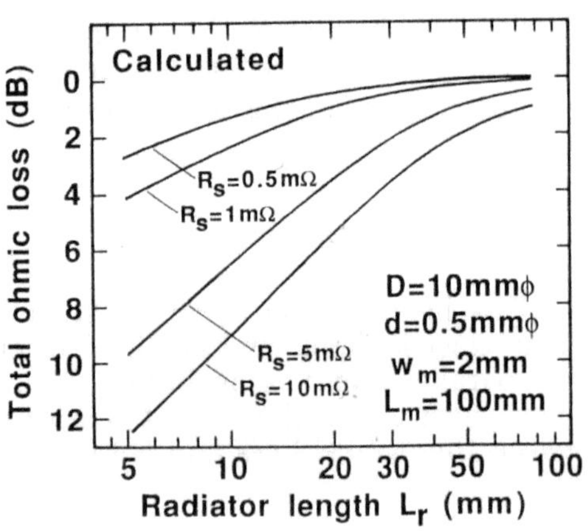

Fig. 3. Total ohmic losses of the antenna as functions of L_r.

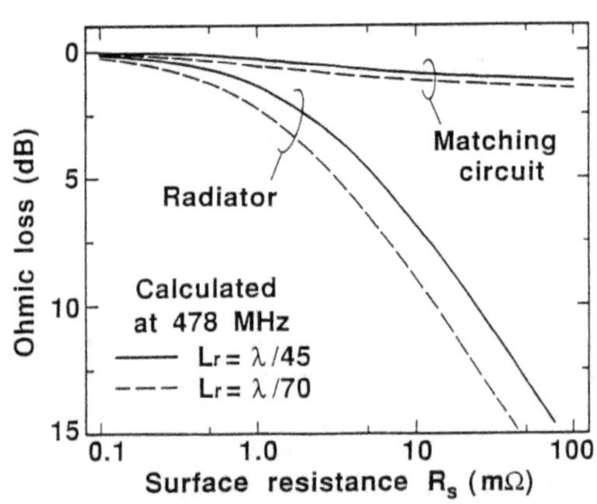

Fig. 4. Ohmic losses of the radiators and matching circuits as functions of R_s.

Antennas were designed to have a resonant frequency of about 500 MHz. The dimensional discrepancies between designed radiators shown in Fig. 2 and fabricated ones were less than 20 %. The radiator lengths L_r were 14.0 and 9.0 *mm* which mean $\lambda/45$ and $\lambda/70$ (λ is free space wavelength) at the actual resonant frequency of 478 MHz, respectively. Teflon sheets (relative dielectric constant ϵ_r of 2.6) were used as the dielectric sheets. The dimensions of the matching circuits w_m, L_m and h_m were 3.8, 97 and 0.15 *mm* for the $\lambda/45$-length radiator, and were 2.0, 100 and 0.05 *mm* for the $\lambda/70$-length radiator, respectively.

Figure 4 shows the calculated ohmic losses of the fabricated radiators and matching circuits as functions of surface resistance R_s. The radiator losses increase greatly, but the circuit losses remain small, as R_s increases. This indicates that the radiation efficiency of this type of antenna is improved by using a low surface resistance radiator rather than a low surface resistance matching circuit.

EXPERIMENTS

Components (the helical radiator, matching circuit and microstrip line) of the antenna with $\lambda/45$-length radiator were made of sintered $YBa_2Cu_3O_{7-x}$ bulk materials (YBCO, transition temperature T_c of about 90 K), and those of the antenna with $\lambda/70$-length radiator were made of YBCO thick films[8]. The thick films were fabricated by wrapping zirconia cores with YBCO about 30 μm thick. The components were joined with indium solder. Copper (Cu) antennas with the same geometries were also prepared for comparison.

The antenna was set inside a glass vessel which was filled with helium gas, and cooled by dipping the vessel into liquid nitrogen. Antenna gains were measured by evaluating the radiated field strength using a standard dipole antenna. The distance between the antennas was about 3λ, and both antennas were surrounded by radio-wave absorbers.

RESULTS AND DISCUSSIONS

The measured actual gains G_{ac} and input reflection coefficients Γ of antennas are summarized in Table 1. These results indicate that the present YBCO antennas have higher gains than Cu antennas. The gain of the YBCO antenna with a $\lambda/70$-length radiator is lower than that of the YBCO antenna with a $\lambda/45$-length radiator. One reason for this lower gain is that the radiation resistance of a $\lambda/70$-length radiator is lower than that of a $\lambda/45$-length radiator.

Table 1. Actual gains G_{ac}, input reflection coefficients Γ, available gains G_{av}, total losses, and average surface resistances R_s of the antennas.

L_r	Material	Temp.	G_{ac}(dBi)	Γ (dB)	G_{av}(dBi)	Loss (dB)	R_s(mΩ)
$\lambda/45$	YBCO	80 K	-1.5	-9.2	-0.9	1.1	0.64
$\lambda/45$	Cu	80 K	-7.4	-7.0	-6.4	6.6	7.2
$\lambda/45$	Cu	RT	-9.0	-12.7	-8.8	9.0	13
$\lambda/70$	YBCO	80 K	-6.5	-7.8	-5.7	5.3	2.5
$\lambda/70$	Cu	80 K	-8.8	-14.1	-8.6	8.2	6.0
$\lambda/70$	Cu	RT	-11.4	-18.0	-11.3	10.9	12

The high gains of the YBCO antennas are attributed to the low losses of their components. The total loss of the antenna is estimated as the difference between the directive gain and the available gain. The directive gain G_d is calculated using the moment method. G_d values for the $\lambda/45$ and $\lambda/70$-length radiators are calculated at 0.2 and −0.4 dBi, respectively. The available gain G_{av} is derived from the actual gain considering the mismatching loss as follows;

$$G_{av} = G_{ac} / \{1 - |\Gamma|^2\} .$$

The loss of the antenna is considered to be the sum of the radiator loss and the matching circuit loss, and the microstrip line loss is ignored because its characteristic impedance (50 Ω) is much higher than its surface resistance. Therefore, the average surface resistances (including joint resistances) of the YBCO and Cu are estimated by fitting the total losses of the antennas to the sum of the radiator loss and the matching circuit loss in Fig. 4.

Table 1 also shows the derived available gains G_{av}, total losses, and estimated average surface resistances R_s of the antennas. The estimated R_s values of the Cu at 80 K and RT for the antenna with a $\lambda/45$-length radiator are almost the same as those for the antenna with a $\lambda/70$-length radiator. The estimated R_s value of the YBCO bulk is almost equal to that measured for YBCO with the cavity method[11]. On the other hand, the R_s of the YBCO thick film is 3.9 times higher than that of the YBCO bulk. This is possibly because the thick film is not uniform over the wide area of the sample, and because indium is unsuitable for joining films which are much thinner than bulk.

However, the R_s values of the YBCO bulk and thick films are estimated to be respectively 0.05 and 0.21 times lower than that of the Cu at RT. These results indicate that the R_s of YBCO is much lower than that of metal, and that YBCO is useful for fabricating highly efficient small antenna systems.

ACKNOWLEDGMENTS

The authors would like to thank Dr. E. Sugita and Dr. K. Kagoshima for their advice about antenna analyses and measurements, and T. Mikazuki for his help in transmission line analyses. The authors would also like to thank Dr. N. McN. Alford of ICI Advanced Materials, H. Kuwajima and S. Yamana of Hitachi Chemical Co., Ltd. for their assistance in preparing the YBCO materials.

REFERENCES

1. Walker GB, Haden CR, (1969) J Appl Phys 40: 2035–2039
2. Adachi S, Kimura S, Ohnuki S, Mohri Y, Onodera Y, (1976) Trans IECE of Japan 59-B: 299–300
3. Adachi S, Mohri Y, Ohnuki S, Ashida K, (1978) Summaries of Papers, 1978 Int. Symp. Antenna and Propagation, Japan : 109–112
4. Khamas SK, Mehler MJ, Maclean TSM, Gough CE, Alford NM, Harmer MA, (1988) Electron Lett 24: 460–461
5. Hansen RC, (1990) IEEE Trans AES-26: 345–355
6. Kraus JD, (1950) Antennas. McGraw-Hill Book
7. Itoh K, Ishii O, Koshimoto Y, Cho K, (1991) Trans IEICE of Japan (in press)
8. Itoh K, Ishii O, Koshimoto Y, Cho K, (1991) J Superconductivity (in press)
9. Harrington RF, (1968) Field computation by moment method. RE Krieger Publishing
10. Khamas S, Cook GG, Kingsley SP, Woods RC, (1990) Electron Lett 26: 654–655
11. Alford NM, Button TW, Peterson GE, Smith PA, Davis LE, Penn SJ, Lancaster MJ, Wu Z, Gallop JC, (1991) IEEE Trans MAG-27: 1510–1518

14GHz Y-Ba-Cu-O Microstripline Resonators

Katsumi Suzuki[1], Shuichi Fujino[1], Tsuyosi Takenaka[1], Kazuo Imai[2], Fumiyasu Suginoshita[2], Keiichi Yamaguchi[1], Katsumi Ohata[1], Kohichi Nakao[1], Norihiko Yazawa[2], Mineo Kobayasi[3], Tadataka Morishita[1], and Shoji Tanaka[1]

[1] Supercoductivity Research Laboratory, ISTEC, Koto-ku, Tokyo, 135 Japan
[2] NHK Science and Technical Research Laboratory, Setagaya-ku, Tokyo, 157 Japan
[3] High-Energy Physics Laboratory, Tsukuba, Ibaraki, 305 Japan

ABSTRACT

High-Tc superconductor(HTS) resonators are reported in this paper. The $YBa_2Cu_3O_{7-\delta}$(YBCO) films in this work were pulsed laser deposited onto 0.5mm thick [100] MgO substrates. We have developed superconducting 13.5GHz microstrip straight-line resonators which use Au as ground planes. The resonator structure is the best candidate for hybrid superconducting-semiconducting microwave circuits and for a characterization tool to precisely measure HTS film quality. The microstip resonator has the highest unloaded Q factor ever reported using the similar structure. We also report magnetic penetration depth of the YBCO films and Ohmic contact metallization process on the films using microwave micostripline and cavity characterization. We found it out that Rs(0K) is inversely proportional to λ(0K).

KEY WORDS: YBCO, superconductor, microwave devices, microstripline,penetration depth

INTRODUCTION

High-temperature oxide superconducting thin films have important potential applications in microwave devices because these films are predicated on achieving low-loss and low-dispersion that are superior to Au at 77K. Two areas of interest for using high-temperature superconducting (HTS) microwave devices, such as coplanar devices of resonators and filters,are space applications and cellular telecommunications systems for mobile radio telephones. From the near future application point of view, it became clear that three terminal semiconductor devices might be suitable for incorporation with the HTS passive microwave devices which operate at 77K[1]. Figure 1 shows an example of the hybrid superconducting-semiconducting receiver circuit. What is the most suitable HTS passive device structure for the hybrid concept?. There are several kinds of coplanar resonator. The planar structures are stripline,microstrip and coplanar. The superconductor patterns, which mainly determine resonant frequencies, are line, ring and disk. And, ground plane materials are normal metals and superconductors. The hybrid device, as shown in Fig.1, is the microstrip line structure with Au on backside plane. Therefore, we are beginning to develop the YBCO microstrip straight-line resonators with the Au backside ground metal taking account of the compatibility with fabrication process of semiconductor microwave devices.

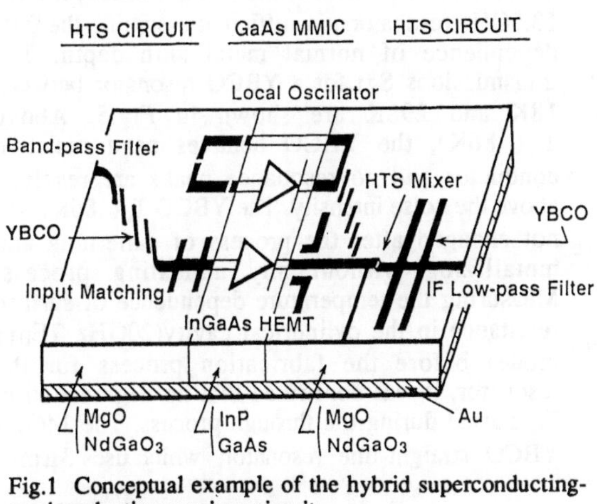

Fig.1 Conceptual example of the hybrid superconducting-semiconducting receiver circuit.

The HTS resonator is also a convenient tool to sensitively measure the the temperature dependence of the magnetic penetration depth, $\lambda(T)$,in superconducting thin films[2]. We can precisely measure small changes in $\lambda(T)$ on the order of ten angstroms by this technique, allowing a characterization of HTS thin film quality. The simple configuration of straight-line and Au backside plane meets to need of characterizing HTS film and dielectric substrate qualities.

EXPERIMENTAL

The $YBa_2Cu_3O_{7-\delta}$(YBCO) films with the C-axis parallel to the growth direction used in this work were ArF pulesd-laser-deposited onto 0.5mm thick MgO substrates at a temperature of 700C and in O_2

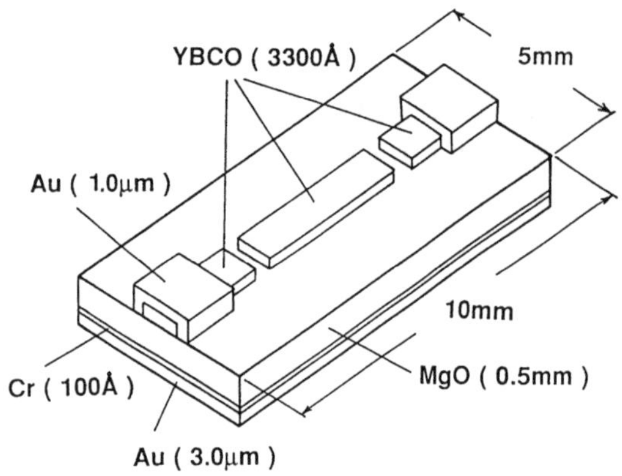

Fig.2 Geometry of the YBCO microstrip straight-line resonator.

gas pressure of 100mTorr[3]. The MgO substrate, which has a low dielectric constant(ε_r=10),was used despite its large lattice mismatches with YBCO. The YBCO critical temperature T_c(~86K) was not changed after patterning and metallizing. From measuring the temperature dependence of surface resistance in a silver plated cylindrical cavity(20GHz TE_{011} mode), we have observed that no degradation in T_c caused during the fabrication process for the resonator. The resonator circuit is formed with a straight-microstripline of 4.0mm length and 0.5mm width. Both ends of the line are open-circuited and coupled with feed lines through series gaps of 0.5mm length, as shown in Fig.2. These lines were patterned using conventional photo-resist processes and Ar ion-beam etching techniques. We were successful in depositing Au contact metal on the YBCO feed lines by Ar sputtering without post-annealing. An adequate thickness of the Au contact metal was determined by measuring surface resistance using the cylindrical cavity. The final step in fabricating HTS devices is the deposition of the ground plane metallization. Cr (100Å) and Au(3μm) were deposited in order onto the backside of the MgO

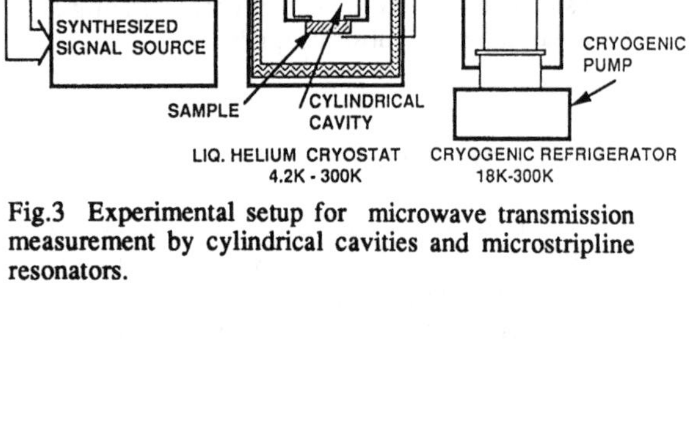

Fig.3 Experimental setup for microwave transmission measurement by cylindrical cavities and microstripline resonators.

substrate. The microstripline resonators of 3μm thick with the same configuration were fabricated to compare them with the YBCO resonators. The experimental setup for microwave transmission measurement by cylindrical cavities and microstripline resonators is shown in Fig.3. These YBCO and Au microstrip samples were mounted in microwave packages with K-connectors and cooled down to 18K with a closed-cycle He refrigerator. Copper semi-rigid coaxial 3.5mm cables are used in the refrigerator. Resonant frequency(f_0),loaded factor Q(Q_L) which is obtained by 3dB bandwidth, and insertion loss(I_L) of the microstrip resonator,as a function of temperatures,were measured by an HP-8510B network analyzer with an HP8360 synthesized signal generator. Unloaded Q(Q_0) is determined by the equation $Q_0=Q_L/(1-10^{I_L/20})$.

RESULTS AND DISCUSSION

After the surface of the substrate was oxygen plasma cleaned in a reactive ion etcher,Au contact metal was deposited by Ar ion sputtering and was patterned by a metal mask. Figures 4(a) and (b) show the thickness dependence (t_{Au}= 0,50,100,210,900Å) of Au contact metal, deposited onto YBCO film, on the surface resistance at temperatures; (a)between 4.2K and 300K, (b)between 82K and 88K. The Au contact metal was sequentially deposited onto YBCO by Ar sputtering. The surface resistances were measured by the silver plated cylindrical cavity(20GHz TE_{011} mode). The parallel displacement of the curves arises from the altered current distribution within the Au and YBCO films. We find that the surface resistance is proportional below 84K and inversely proportional above 85K,to the Au thickness,t_{Au}. The metallization process for Ohmic contacts turns out to be very well because the critical temperature T_c is fixed for each thickness. And we could observe the opposite dependence of t_{Au} on R_s above and below the T_c, due to no degradation of the YBCO surface. Au thickness of 900Å metallization practically shields YBCO. Therefore, one micron thick Au, we adopted for 13.5GHz resonators, is sufficient assuming the $f^{-1/2}$ dependence of normal metal skin depth. The transmissions S_{21} for a YBCO resonator between 18K and 298K are shown in Fig.5. Above T_c(~86K), the YBCO behaves as a resistive conductor and no resonance peaks are resolved above the noise intensity. The YBCO T_c(~86K) was not changed after the process of patterning and metallizing without any annealing process. Measuring the temperature dependence of surface resistance in the cylindrical cavity(20GHz TE_{011} mode) before the fabrication process for the resonator, it was confirmed that no degradation in T_c caused during the through process. The 14GHz YBCO straight-line resonator, which uses 3μm thick Au as its ground plane, has unloaded $Q(Q_0)$ factors of 4400 at 18K and 2800 at 77K, respectively. Figure 6 shows two features of Q_0 measurements made on YBCO and Au films. The values of YBCO Q_0 at 13.5GHz are significantly higher than those of normal metal Au. For example, at 77K, Au film resonator has a Q_0=700 and a slowly-varying temperature dependence of Q_0 in the vicinity of 77K. Figure 7 shows the representative Q_0 obtained in this work and those of previous reports [4],[5],[6],[7] at different resonant frequencies. To permit a comparison of measurements at different frequencies to be made by different laboratories, it was assumed that Q_0 scales with frequency as $(f_0)^{-2}$, where f_0 is the resonant frequency. Since the Q_0 is inversely proportional to the surface resistance, R_s, and the R_s is proportional to f^2, the frequency dependence of the Q_0 of superconducting YBCO varies by a power law, f^{-2}. Therefor, the unloaded figure Q=2800 at 13.35GHz and 77K, which was obtained in this work, is the highest value in the YBCO microstrip line resonator with Au on the backside. The magnetic penetration depth, $\lambda(T)$, in superconducting thin films is a good parameter to characterize HTS quality. The above mentioned HTS end-coupled resonator is a splendid tool to sensitively measure the temperature dependence of the magnetic penetration depth of YBCO films. We can precisely measure small changes in $\lambda(T)$ on the order of ten angstroms by measurering the resonant frequency at each temperature, using the equations, $(f_0(T)-f_0(20K))/f_0(20K)=(F(T)-F(20K))/F(20K)$, where $F(T)=[1+\lambda(T)/d*\coth(t/\lambda(T))]^{-1/2}$ is the phase velocity function, $\lambda(T)=\lambda(0K)[1-(T/T_c)^4]^{-1/2}$ is the Gorter-Casimir two-fluid model function, d is the MgO substrate thickness and t is the YBCO film thickness. Figure 8 shows the experimental and calculated data of the normalized resonant frequency shift, $(f_0(T)-f_0(20K))/f_0.(20K)$, versus the resonator temperature. From the fitting parameters of superconducting penetration depth at 0K, $\lambda(0K)$ =3900Å, and the critical temperature T_c=86K, a resonant frequency shift was calculated being in good agreement to the measurement. The resonant frequency $f_0(T)$ is rapidly-varying temperature dependence in the vicinity of T_c. Taking account of the reproducibility of HTS films'T_c, it is clear that the HTS microstrip line resonator structure with normal metal backside ground plane is the best candidate for precisely-measureing HTS film qualities in comparison with the other

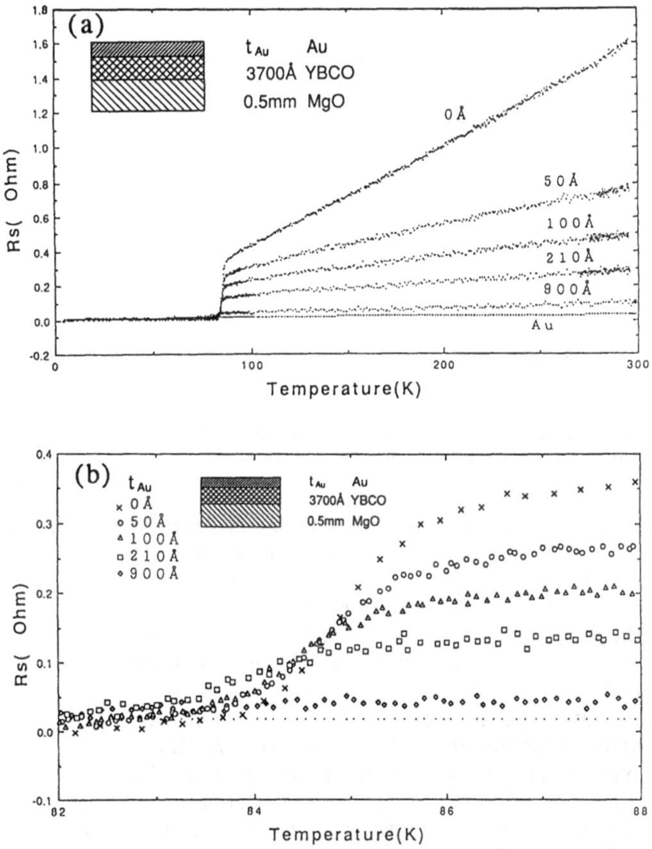

Fig.4 Surface resistance, R_s, of Au/YBCO/MgO structure samples for different values of the Au film thickness, t_{Au}(0,50,100,210,900Å), at 20GHz and temperatures; (a)between 4.2K and 300K, and (b)between 82K and 88K.

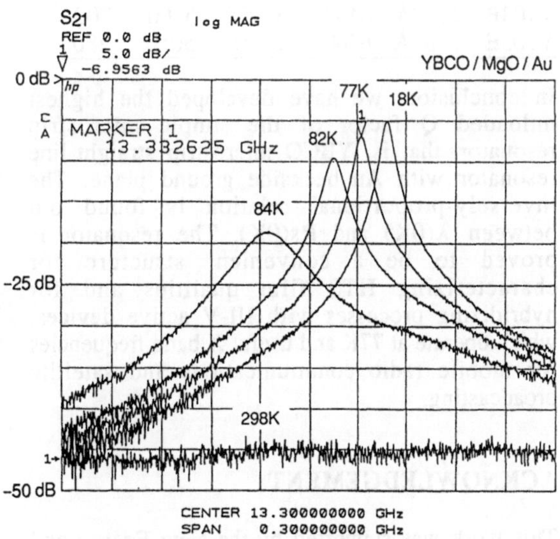

Fig.5 Transmissions S_{21} for a YBCO(T_c=~86K) straight-line resonator with Au on the backside described in text measured between 18K and 298K.

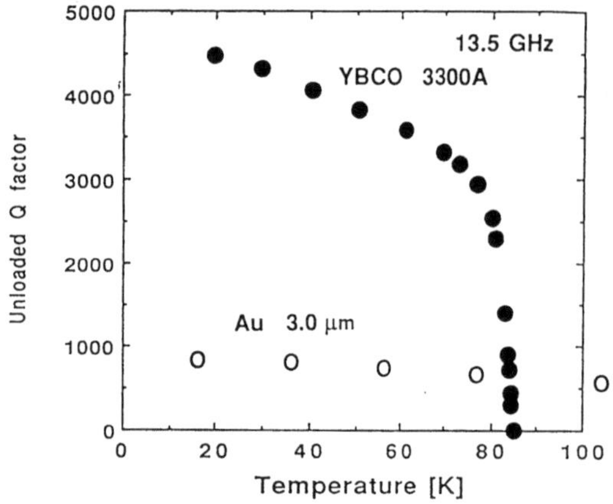

Fig.6 Relative performance of Unloaded Q factors of the YBCO and Au resonators with the same structure.

configurations; the microstrip with an HTS film as backside ground plane and the stripline structure with an HTS film used as the second ground plane. Table 1 shows the data of λ(0K) ,T_c, unloaded Q at 20K and 77K, for our four YBCO resonators. And, the key YBCO characteristic which meets with the need of microwave device operation at 77K is the magnetic penetration depth, λ(0K), in the superconducting thin films. However, λ(0K) is inversely proportional to the unloaded Q at 20K which is corresponding to the temperature independent residual loss from an unidentified mechanism at lower temperatures.

Table 1 Film thickness, λ(0K) ,T_c, unloaded Q at 20K and 77K, of four YBCO resonators in this work.

Sample	thickness	λ(0K)	T_c	20K	77K
Y090A	3300Å	3900Å	86.0K	4400	2800
Y090E	1500Å	3900Å	88.4K	4300	2650
Y103B	2800Å	6400Å	84.5K	5300	700
Y103E	780Å	6400Å	84.5K	5050	280

In conclusion, we have developed the highest unloaded Q factor of the simple microstrip resonator, that is, YBCO microstrip straight-line resonator with Au backside ground plane. The inversely-proportional relation is found out between λ(0K) and Rs(0K). The resonator is proved to be a convenient structure for characterizing HTS film qualities and for hybridizing processes with III-V active devices, which operate at 77K and useful X band frequencies for mobile radio communications and satellite broadcasting.

ACKNOWLEDGEMENT

This work was supported by the New Energy and Industrial Technology Development Organization under the management of the R&D of Basic Technology for Future Industries.

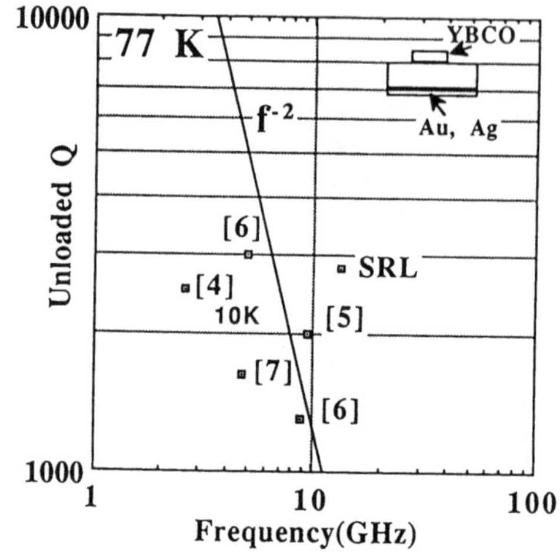

Fig.7 Comparison between representative values of the unloaded Q factor of YBCO microstrip line resonators with the normal metal backside ground plane.

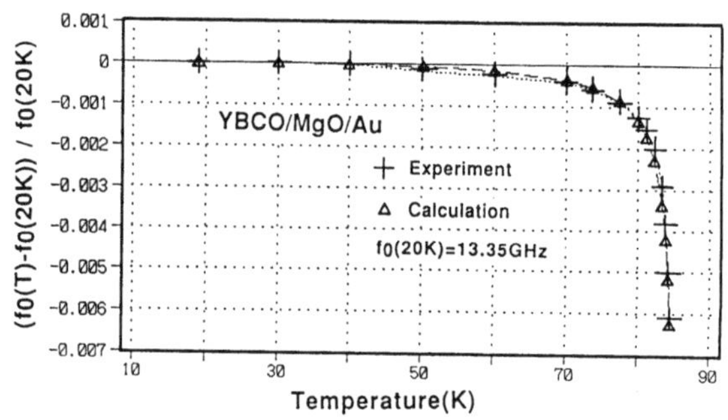

Fig.8 Experimental and calculated data of the normalized resonant frequency shift,($f_0(T)$-$f_0(20K)$)/$f_0(20K)$, versus the resonator temperature.

REFERENCES

[1] R.F.Leonard, D.J.Connolly, K.B.Bhasin, J.D.Warner, and S.A.Alterovitz, SPIE Vol.1394 Progress in High-Temperature Superconducting Transistors and Other Devics(1990).

[2] S.M.Anlage, H.Sze, H.J.Snortland, S.Tahara, B.Langley, C.B.Eom, and M.R.Beasley Appl. Phys.Lett.54(26) June 1989 pp2719-2712.

[3]H.Izumi,K.Ohata,T.Hase,K.Suzuki,T.Morishita, and S.Tanaka, J.Appl.Phys.68(12), 15 December 1990,pp6331-6335.

[4] A.Enokihara,K.Setune and K.Wasa, Proc. 3rd FED Workshop on HTS-ED 1991 pp130-131.

[5] G.Liang,X.Dai,D.David,F.Hebert,T.V.Duzer, N.Newman and B.F.Cole, IEEE Trans.on Applied Superconductivity,Vol.1,No.1,March 1991. pp58-66.

[6] G.Solkner,R.Dill, Proceeding IEEE MTT-S, Dallas 1990.

[7] M.R.Namordi, A.Morgro-Campero, L.G.Turner,and D.W.Hogue IEEE Trans. Microwave Theory and Tech.,Sept.1991

Properties of Superconductive Delay Line Using EuBaCuO Films on MgO (100)

YASUHIRO NAGAI, NAOBUMI SUZUKI, KEIICHIRO ITOH, and OSAMU MICHIKAMI

NTT Interdisciplinary Research Laboratories, Tokai, Ibaraki, 319-11 Japan

ABSTRACT

This paper reports on the properties of superconductive 40 cm delay lines fabricated using EuBaCuO films and several new techniques. The insertion loss is 0.068 dB/nsec at 77 K and 1 GHz, and the delay time is 4 nsec independent of frequency. The reflection coefficient and VSWR are less than -15 dB and 1.5 at frequencies below 3 GHz.

KEY WORDS: Microwave component, Superconductive delay line, Low insertion loss

INTRODUCTION

Superconductive microwave components are likely to be some of the earliest applications of high-*Tc* superconductivity. In particular, passive components with much higher electrical performance and greater compactness can be expected for frequencies below 100 GHz because of low loss and low dispersion [1]. Thus, microwave components such as resonators [2-4], filters [5-7], delay lines [8-10] and antennas [11-13] have been fabricated to investigate their performance.

In order to realize low loss delay lines, low loss joints and large-area patterning are important in addition to low loss superconducting films. To this end, we have developed a technique for preparing uniform film with over a large area, a large-area patterning technique and a low loss joint technique for superconductive delay lines using high-*Tc* superconducting films. This paper reports the properties of superconductive 40 cm delay lines, composed of EuBaCuO and Cu films, fabricated with the above techniques.

DESIGNED PARAMETERS AND EXPERIMENTAL PROCEDURE

The center conductor pattern is shown in Fig. 1. We put a 40 cm long and 160 μm wide meander-line into an area of 2x2 cm^2 on a MgO (100) substrate. A strip-line structure was adopted for the delay line because of its low conductor loss, very low radiation and low modal dispersion. The geometrical specifications are summarized in Table I.

Table I Geometrical specifications of the delay line

Structure	Material	Line pattern
Strip-line, meander	0.5μm-EuBaCuO film (conductor) 4μm-Cu film (ground-plane)	40cm long, 160μm wide line 840μm space between lines

As we adopted a line width of 160 μm and a MgO substrate 0.5 mm thick with a dielectric constant of 9.4, the characteristic impedance was 50 Ω. The capacitance and inductance of this structure were 2.03×10^{-10} F/m and 5.17×10^{-7} H/m, respectively. Thus, the delay time was expected to be 4 nsec in this 40 cm delay line. Additionally, as shown in Fig. 1, some parts of the meander line formed parallel coupled lines, of which the coupling and the center frequency were 28.4 dB and 2.7 GHz.

High-quality, 0.5 μm-thick EuBaCuO superconducting films, with a T_c and J_c of 90 K and 2×10^6 A/cm^2 at 77 K, were uniformly prepared on a 1 in. square MgO (100) substrate by *in-situ* magnetron sputtering [14]. Ag film electrodes were coated around the joint, and then a 1 μm-thick amorphous fluoropolymer (*AF*) was spin-coated as a process protection layer [15]. The protective layer made it easy to fabricate long and large-area patterns, which were free from any degradation in their superconducting properties which might have been caused by the patterning processes. A 40 cm-long and 160 μm-wide meander line was fabricated by conventional photolithography and the argon dry etching of *AF*/Ag/EuBaCuO films. The line and two Cu-film ground-planes were clamped together as a tri-layer delay line, and mounted in a metal case with *SMA* connectors. A Cu delay line with the same geometrical parameters was also fabricated to compare the properties.

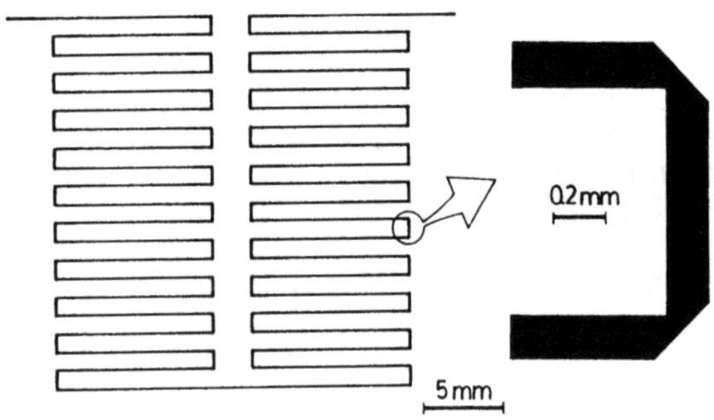

Fig. 1 Center conductor line and bend patterns of the delay line.

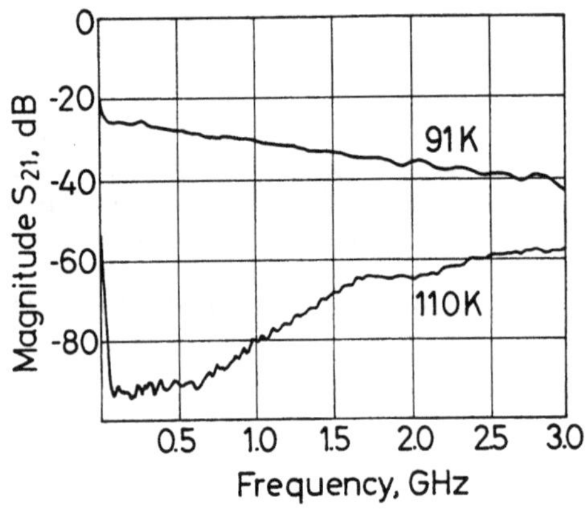

Fig. 2 Frequency dependence of transmission (S_{21}) properties at temperatures higher than the critical temperature.

RESULTS AND DISCUSSION

The resistance of the EuBaCuO line was approximately 20 kΩ at 300 K. This is identical to that estimated from a resistivity of 0.4 mΩcm and the geometrical parameters. Thus, it was confirmed that there was no degradation in resistivity after patterning.

The transmission properties (S_{21}) of the EuBaCuO delay line are shown in Fig. 2. Above the critical temperature, signals cannot pass through the delay line because of the large insertion loss. In contrast, the insertion loss becomes very low below the

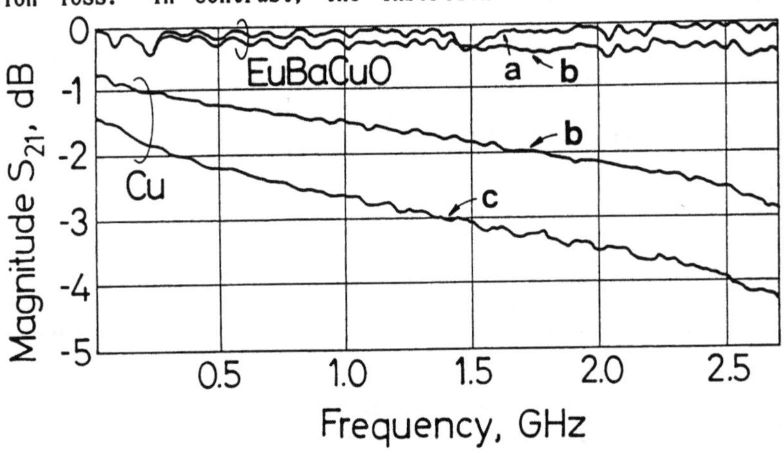

Fig. 3 Frequency dependence of transmission (S_{21}) properties. *a*, *b* and *c* are measured at 30 K, 77 K and 300 K, respectively. Input power was -13 dBm.

critical temperature, as shown in Fig. 3. Although the insertion loss depends on both temperature and frequency, the insertion loss in the EuBaCuO delay line is much lower than that in the Cu delay line. For example, the insertion loss in the EuBaCuO delay line is 0.27 dB at 77 K and 1 GHz, and this value is, respectively, 1/5 and 1/10 those at 77 K and 300 K in the Cu delay line. The insertion losses at each temperature are summarized in Table II. This result means that a EuBaCuO delay line exhibits sufficiently low conductor and joint losses even though Cu films are used as groundplanes.

The delay time is shown in Fig. 4 as a function of frequency. The delay time in the EuBaCuO delay line is approximately 4 nsec independent of frequency, although the delay time in the Cu delay line greatly increases as frequency decreases below 3 MHz. This result means that group and phase velocities show less dependency in the EuBaCuO superconducting line than in the Cu delay line. The normalized insertion loss is therefore 0.068 dB/nsec at 77 K and 1 GHz. However, a slight frequency dependence of delay time was observed in the frequency range below several-10 GHz. The variation in delay time was almost the same in both the EuBaCuO and Cu delay lines, and the periodicity was confirmed in the delay time variation. If the variation was caused by changes in the dielectric constant, the dielectric constant would increase by 20 % in the frequency range below 3 GHz. This value is too large, and the periodicity cannot be explained by the mechanism. Based on the fact that the maximum delay time was first observed at approximately 2.6 GHz and the maximum seemed to appear at intervals of 5.6 GHz, the weakly dispersive properties [16] are thought to be induced by the parallel coupled lines in the delay line.

Table II Insertion loss at each temperature (300kHz~3GHz)

Material	Temperature	Insertion loss
EuBaCuO	30 K	< 0.3 dB
EuBaCuO	77 K	< 0.5 dB
Cu	77 K	0.8~2.8 dB
Cu	300 K	1.5~4.3 dB

Table III VSWR at each temperature in delay lines (300kHz~3GHz)

Temperature	EuBaCuO	Cu
300 K	——	< 1.2
80 K	< 1.5	< 1.2
30 K	< 1.4	——

The reflection properties (S_{11}) are shown in Fig. 5 as a function of frequency. The reflection coefficients of S_{11} and S_{22} were identical. The reflections are less than

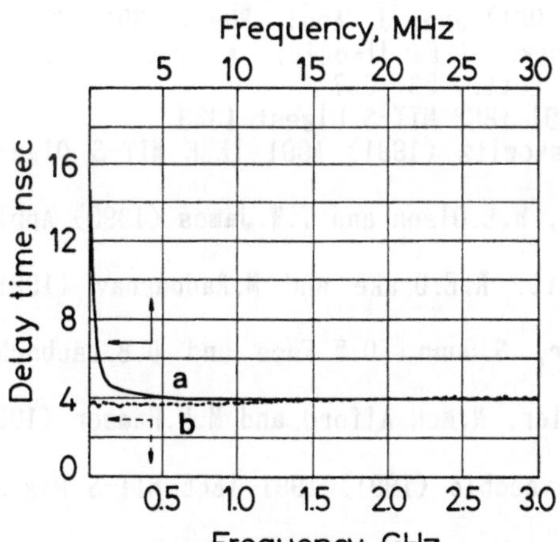

Fig. 4 Delay time as a function of frequency. *a* and *b* correspond to the Cu delay line at 300 K and the EuBaCuO delay line at 77 K, respectively.

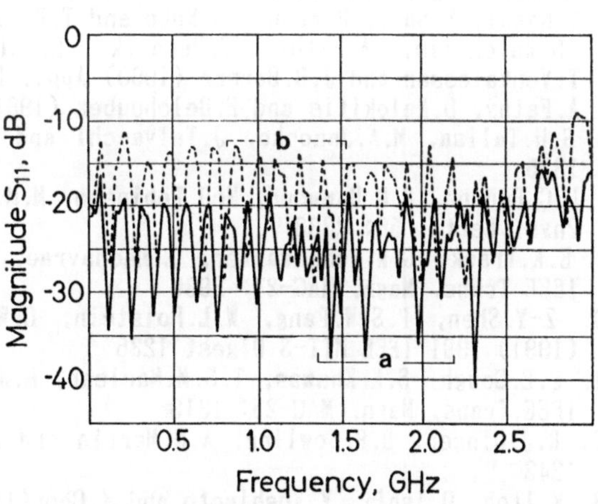

Fig. 5 Frequency dependence of reflection (S_{11}) properties at 80 K. *a* and *b* are Cu and EuBaCuO delay lines, respectively. Input power was -13 dBm.

-20 dB and -15 dB for the Cu and EuBaCuO delay lines, respectively. However, these reflection patterns imply very weak resonance. The interval of the absorbed peaks is approximately 120 MHz, and this interval corresponds well to the resonant frequency of a 40 cm long λ/2 linear resonator. Thus, the very weak resonance seems to be caused by slight mismatching around joints. The voltage standing wave ratio (VSWR) at each temperature is shown in Table III. Additionally, although the difference of 5 dB between the EuBaCuO and Cu delay line reflections can probably be attributed to the difference in practical contact resistance, it might be caused merely by a slight difference in each joint when the Ag and/or Cu electrodes and *SMA* connectors are joined with silver paste. This is because the practical contact resistances in both delay lines were almost of the same order, and the contact resistance in the EuBaCuO delay line had a low value of less than 6 mΩ.

CONCLUSION

Superconductive 40 cm delay lines were fabricated using high-quality EuBaCuO films, and new patterning and joint techniques. From the transmission properties, it was found that the insertion loss was 0.27 dB at 77 K and 1 GHz, and that the delay time was approximately 4 nsec independent of frequency. However, the weakly dispersive properties were confirmed in a wide frequency range. The reflection coefficient and VSWR were, respectively, less than -15 dB and 1.5 below 3 GHz. Also, the very weak resonance seemed to be caused by slight mismatching around the joints.

ACKNOWLEDGEMENTS

The authors wish to thank Dr. Y. Koshomoto and Dr. E. Sugita for their continuous encouragement. Also, the authors thank Dr. M. Sato and Dr. T. Konaka for valuable discussions.

REFERENCES

1. R.S.Withers and R.W.Ralston (1989) Proc. IEEE 77: 1247
2. D.E.Oates and A.C.Anderson (1991) IEEE Trans. Magn. MAG-27: 867
3. M.Sato, T.Konaka, H.Asano, S.Kubo and Y.Nagai (1991) Trans. Inst. Electron. Inf. & Commun. Eng., E74: 1980
4. Y.Nagai, M.Sato, H.Asano, S.Kubo and T.Konaka (1991) Jpn. J. Appl. Phys., 30: 1679
5. D.Kalokitis, A.Fathy, V.Pendrik, E.Belohoubek, A.Findikoglu, A.Inam, X.X.Xi, T.Venkatessan and J.B.Barner (1990) Appl. Phys. Lett., 58: 537
6. A.Fathy, D.Kalokitis and E.Belohoubek (1991) 1991 IEEE MTT-S Digest 1329
7. S.H.Talisa, M.A.Janocko, J.Talvacchi and C.Moskowitz (1991) 1991 IEEE MTT-S Digest 1325
8. L.C.Bourne, R.B.Hammond, McD Robinson, M.M.Eddy, W.L.Olson and T.W.James (1990) Appl. Phys. Lett., 56: 2333
9. E.K.Track, G.K.Hohenwarter, L.Madhavrao, R.Patt, R.E.Drake and M.Radparnav (1991) IEEE Trans. Magn. MAG-27: 2936
10. Z-Y.Shen, P.S.W.Pang, W.L.Holstein, C.Wilker, S.Dunn, D.W.Face and D.B.Laubacher (1991) 1991 IEEE MTT-S Digest 1235
11. C.E.Gough, S.K.Khamas, T.S.M.Maclean, M.J.Mehler, N.McN Alford and M.M.Hammer (1989) IEEE Trans. Magn. MAG-25: 1313
12. R.J.Dinger, D.R.Bowling, A.M.Martin and J.Talvacchio (1991) 1991 IEEE MTT-S Digest 1243
13. K.Itoh, O.Ishii, Y.Koshimoto and K.Cho (1991) to be published in J. Supercond.
14. O.Michikami and M.Asahi (1991) Jpn. J. Appl. Phys., 30: 939
15. Y.Nagai, N.Suzuki, M.Sato and T.Konaka (1991) to be published in Jpn. J. Appl. Phys.
16. D.A.Gandolfo, A.Boornard and L.C.Morris (1968) J. Appl. Phys., 39: 2657

Hybrid HTS and III-V Microwave Oscillators

D.B. LAUBACHER, Z.Y. SHEN, P.S. PANG, C. WILKER, and W.L. HOLSTEIN

Central Research and Development Department, Du Pont Co., Wilmington, DE 19880-0304, USA

ABSTRACT

Low-noise oscillators operating as pure microwave frequency sources have many critical real world uses in modern radar, communications, and information processing applications. One such application would be as a local oscillator in an advanced microwave front-end receiver. Hybrid oscillators consisting of III-V FETs and high temperature superconducting (HTS) passive microwave devices have been fabricated and tested at cryogenic temperatures at Du Pont. Initial devices operating in the C-band at 4.7 GHz have utilized commercially available High Electron Mobility Transistor (HEMT) unpackaged chips. Wide band amplifiers with bandwidth from 2 to 8 GHz were fabricated and tested at 77 K show 17 dB gain. This is 3 dB higher than amplifiers operating at room temperature. Hybrid oscillators were fabricated using these wide-band amplifiers with feed back provided by a $Tl_2Ba_2CaCu_2O_8$ microstrip resonators. Resonators were fabricated from two-side coated HTS films on $LaAlO_3$ substrates and had loaded Qs of 5000 with insertion loss of -6 dB. Oscillators show phase noise of less than -106 dBc (1 Hz) at 10 KHz offset from operating frequency of 4.7 GHz.

INTRODUCTION

Microwave frequency sources are a possible area for the first practical real-world application of high temperature supercondcuting (HTS) materials and devices. Low phase noise oscillators find extensive use in modern radar, communications, and information processing applications. Critical characteristics of resonators utilized in microwave oscillators include the quality and operational frequency product (Q-f product) and sensitivity to environmental conditions such as temperature and vibration [1]. Dielectric resonator oscillators (DROs) are extensively used in low phase noise applications. These resonators have high Q-f products (more than 10^{13}) and tood temperature stability. However, they can be sensitive to vibration. HTS microstrip resonators can be produced with very similar Q-f products and possibly reduced sensitivity to vibration. Additional gain can be obtained if the amplifier utilized in the oscillator is cooled to cryogenic temperatures. The purpose of this paper is to report the demonstration of a low phase noise microwave oscillator operating in the C-band at 4.7 GHz utilizing a cooled III-V FET chip amplifier and an HTS resonator providing feedback.

RESONATOR FABRICATION

$Tl_2Ba_2CaCu_2O_8$ Film Fabrication and Surface Resistance

The primary technique used for thallium microwave films at Du Pont is a post anneal process to produce $Tl_2Ba_2CaCu_2O_8$ [2]. A precursor film is fabricated sequentially on two sides of the substrate using off-axis RF magnetron sputtering of $Ba_2CaCu_2O_x$ in a dedicated vacuum system. The purcursor film stoichiometry has been shown to be within 2% of the desired final film stoichiometry over a three inch wafer. The precursor film in then annealed in an alumina crucible at between 800°C and 900°c in a partial pressure of thallium oxide. Depending on the composition of the inital precursor film and the thallium oxide partial pressure, a number of phases of mixture of phases of superconducting films can be formed. The $Tl_2Ba_2CaCu_2O_8$ phase is easily formed over a wide process window and is very stable.

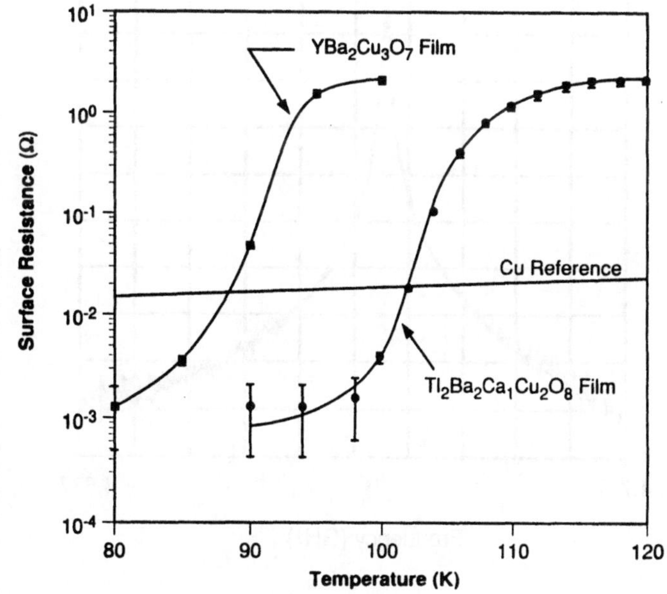

Fig. 1 Microwave surface resistance versus temperature of $YBa_2Cu_3O_{7-x}$ and $Tl_2Ba_2CaCu_2O_8$ films deposited on (100) $LaAlO_3$. R_s was measured by end-wall replacement in a 20 GHz TE_{011} copper cavity.

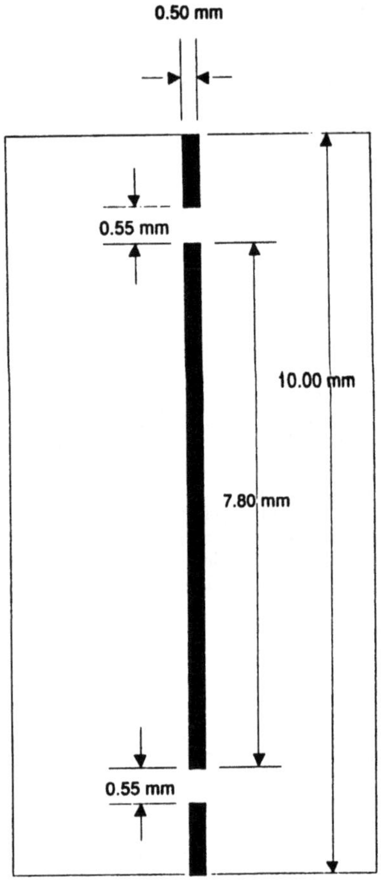

Fig. 2 Design of a 4.7 GHz microstrip resonator on (100) $LaAlO_3$.

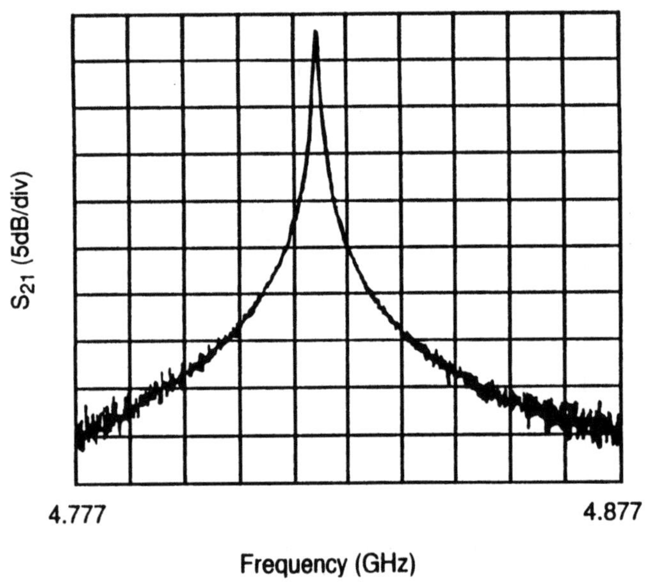

Fig. 3 Performance of a $Tl_2Ba_2CaCu_2O_8$ microstrip resonator at 80 K.

Since the T_c of such a film is approximately 15°C higher than that of a $YBa_2Cu_3O_{7-x}$ film, for devices operating near 77 K, the advantages are apparent. The microwave surface resistance comparison of Du Pont $Tl_2Ba_2CaCu_2O_8$ and $YBa_2Cu_3O_{7-x}$ films deposited on lanthanum aluminate ($LaAlO_3$) is shown in Figure 1. R_s was measured using a TE_{011} end-wall replacement cavity calibrated using a polished oxygen-free-high-conductivity (OFHC) copper standard.

Patterned Resonators

Figure 2 shows the design of the fabricated microstrip resonators. A completely dry process has been developed for fabrication of the resonator chip. Photolithographic patterning of the chip is accomplished by applying a PMMA barrier layer to the HTS surface followed by standard positive photoresist patterning. The HTS film is etch by argon ion milling. Final photoresist removal is accomplished by oxygen reactive ion etching (RIE). Silver contacts were applied to the resonator chips using metal shadow masks followed by an anneal in oxygen at approximately 500°C. DC contact resistances of $<10^{-8}\Omega/cm^2$ are standardly produced using this process.

Resonator Performance

The performance of the resonator utilized in the hybrid oscillator is shown in Figure 3. The device had a loaded Q of nearly 5000 at 77 K with -6 dB insertion loss when tested in a standard microwave test fixture. Further information regarding the performance of similar resonators is found in Reference [3].

HYBRID OSCILLATOR

Low Noise Amplifier

One critical component necessary for the hybrid oscillator is a low noise high gain amplifier. The design for this amplifier using a high electron mobility transistor (HEMT) is shown in Figure 4. This design was fabricated on alumina using standard electroplated gold for the bias and microwave circuitry. The FET chips were die attached using 80Au/20Sn solder. Since the carrier mobility of a HEMT at 77 K is approximately 10 times higher than at room temperature, the resulting gain is >3 dB higher. The model and actural responses for this device are show in Figure 5. The gain of the amplifier at 4.7 GHz is approximately 17 dB.

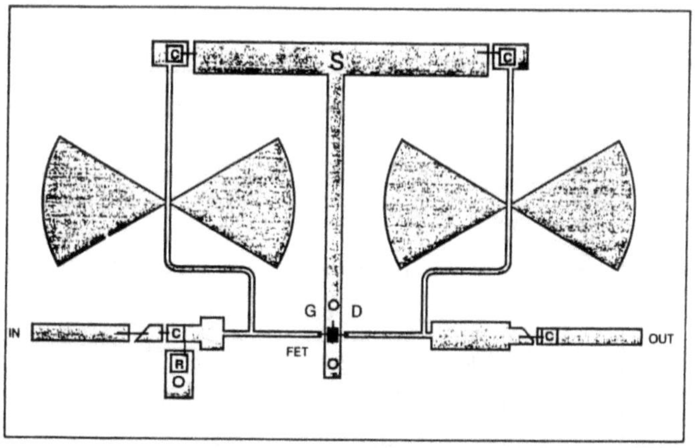

Fig. 4 Design of a wide band HEMT amplifier. The bias and microwave input and output are fabricated in gold denoted by lighter shading.

Oscillator

Layout for a hybrid III-V / HTS oscillator is shown in Figure 6. Again, the bias and microwave circuitry were fabricated on alumina using electroplated gold. The HTS resonator chip was secured and ground contact made to the package case using Delrin® clamps placed as far away as possible from the resonant structure. Contacts to the resonator were made using 0.075 mm ribbon bonds. Both the oscillator and the resonator were cooled to cryogenic temperatures and total noise measurements were made on an microwave spectrum analyser. The noise floor on the spectrum analyser is approximately -115 dBc (1Hz) at 10 KHz from center frequency. Output from the spectrum analysers is shown in Figure 7. At 70 K, the hybrid oscillator phase noise was less than -106 dBc (1Hz) at 10 KHz from the center frequency. This phase noise is over 6 dB lower than previously measured and probably results from also cooling the FET amplifier.

CONCLUSION

Hybrid low-noise oscillators have been fabricated using cooled III-V FETs and HTS passive microwave devices. Such devices have been packaged in a single 25 mm square package and have phase noise of less than -106 dBc (1Hz) at 10 KHz offset from the operational frequency of 4.7 GHz. Further work is underway to look at near-carrier spectra and evaluation of the resonator sensitivity to vibration. Additionally, improved resonators are being fabricated to improve the loaded Qs at C-band and X-band frequencies.

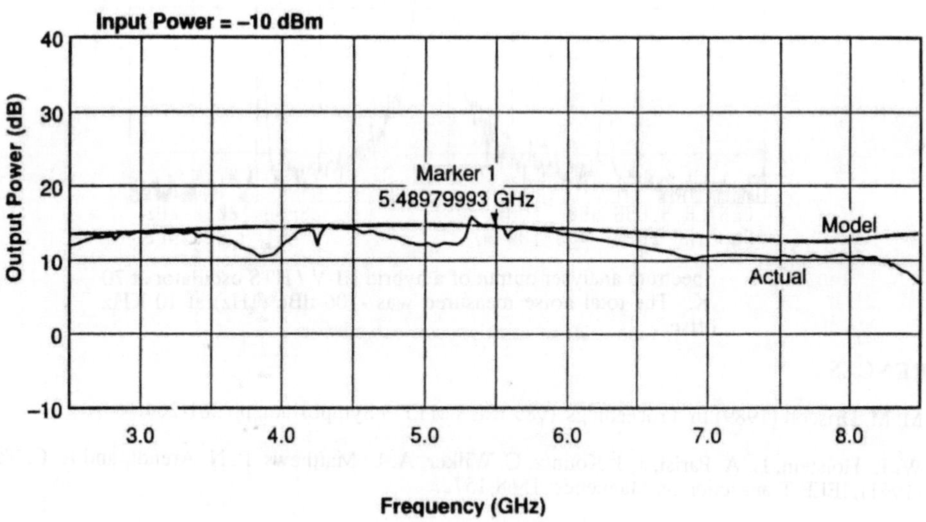

Fig. 5 Response of the wide band HEMT amplifier at 70 K.

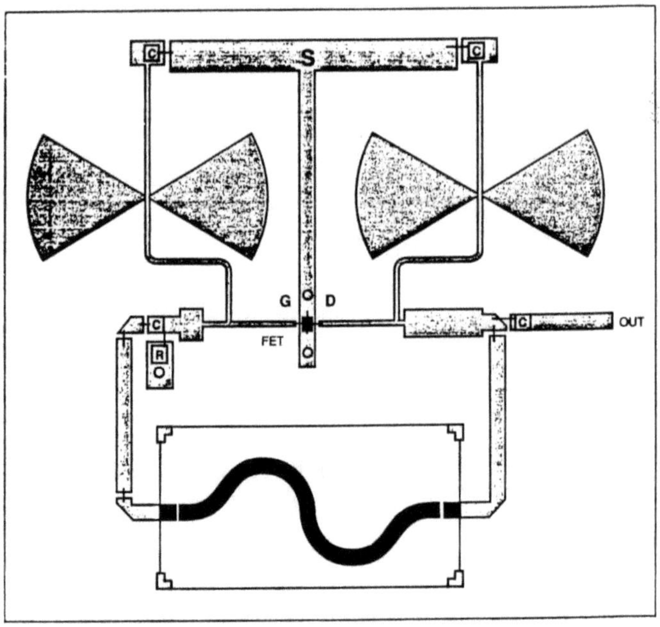

Fig. 6 Design of a hybrid III-V / HTS oscillator. The HTS oscillator is denoted by the darker shading.

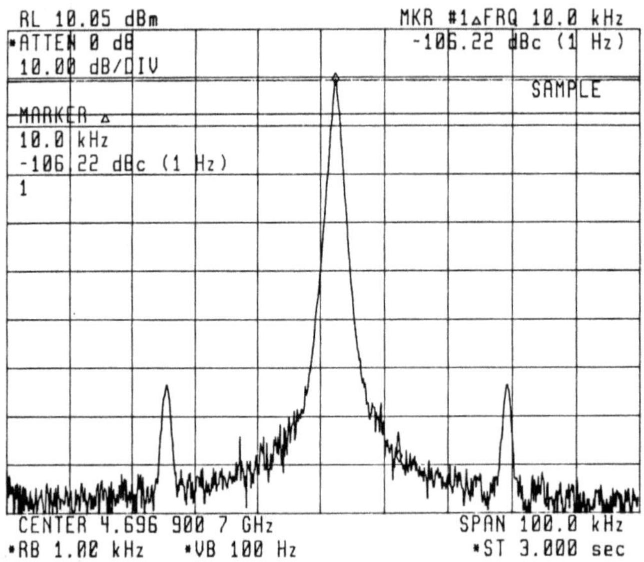

Fig. 7 Spectrum analyser output of a hybrid III-V / HTS oscillator at 70 K. The total noise measured was -106 dBc (1Hz) at 10 KHz offset.

REFERENCES

[1] M. M. Driscoll (1989) In: Proceedings 1989 IEEE MTT-3 Symposium, pp 261-264.

[2] W. L. Holstein, L. A. Parisi, J. J. Kountz, C. Wilker, A. L. Matthews, P. N. Arendt, and R. C. Taber (1991), IEEE Transaction on Magnetics: 1568-1572.

[3] C. Wilker, Z. Y. Shen, P. Pang, D. W. Face, W. L. Holstein, A. L. Matthews, and D. B. Laubacher (1991), IEEE Transactions on Microwave Theory and Techniques: 1462-1467.

[4] A. P. S. Khanna, M. Schmidt, and R. B. Hammond (June, 1991), Microwave Journal: 127-130.

Application of Superconducting Thin Films to RF Cavities

HIROSHI NAKAI[1], JOJI SHINOHARA[1], MITSURU OGOSHI[2], HISAHARU SAKAE[2], and YASUYUKI MIYAUCHI[2]

[1] Research Institute, [2] Energy & Chemical Plants, Ishikawajima-Harima Heavy Industries Co., Ltd., Koto-ku, Tokyo, 135 Japan

ABSTRACT

We have studied RF accelerating cavities using Niobium low-Tc superconducting thin films and YBCO high-Tc superconducting thin films. The size of cylindrical cavity was chosen so that the TM_{010} mode oscillated at 2800 MHz. This gave the dimensions 80 mm diameter and 20 mm length. These thin films were coated on the whole inner surfaces of the cavities using RF magnetron sputtering method. We measured RF properties of these cavities, as compared with no coated cavity, using the networkanalyzer from room temperature to liquid He temperature.

KEY WORDS: YBCO thin films, Nb thin films, RF properties, Metallic substrate

INTRODUCTION

RF accelerating cavities should be supplied high electric power in order to achieve high accelerative electric field. Consequently innersurface of cavity is heated by the resistance. So accelerative electric field is limited by the residual RF losses, though RF accelerating cavities are made of a good conductor for example copper.

Recently several studies are reported about application of the superconducting materials to RF accelerating cavities(1),(2). The aim of this attempt is reduction of RF losses based on reduction of resistance at critical temperature. Especially high temperature superconductors are expected to have high property with low cost because its critical temperature is higher than the temperature of liquid nitrogen.

We have studied RF accelerating cavities as application of superconducting thin films. Besides we succeed that YBCO high temperature superconducting thin films were coated on metallic substrates with ceramics buffer layers(3). In this paper we report RF properties of three kinds of cavities coated low-Tc and high-Tc superconducting thin films.

EXPERIMENTAL

Figure of Cavities

We made two kinds of cavities in shape. One is cylindrical and the other is more practical shape. Cylindrical cavities consist of two disks and one cylinder. Two different types of cylindrical cavities in materials are made.

One is made of copper to coat niobium thin films. And the other is made out of stainless steel to coat YBCO thin films. Both of the two cavities made by lathing. These outlines and main dimensions are given in Fig.1.

The other type of cavity is more practical figure than the former. We call it model cavity. This cavity's outline and main dimensions are given in Fig.2. This is made out of copper using electrical casting method except for the both sides flanges to fix the antenna.

Measurement

The all sizes of these three cavities are chosen so that the TM_{010} mode oscillated at 2800 MHz. RF properties are measured by networkanalyzer as the cavities are refrigerated in cryostat from room temperature to liquid He temperature. Composition of equipment are given in Fig.3. We measured loaded Q-value(Q_L) and coupling coefficient (β), then calculated unloaded Q-value (Q_0) from them according to the following formula:$Q_0=Q_L(1+2\beta)$.

Coating of Thin Films

The cylindrical cavities are coated Nb and YBCO thin films using RF magnetron sputtering method. Niobium is sputtered under a pressure of 0.67 Pa at a power density of 5.7 W/cm^2. The sputter gas is Argon. Two disks and one cylinder which are made of copper are set on the substrate holder and coated at one time. Then the other side of cylinder are coated again to make film uniformly.

The stainless steel cylindrical cavity is set the way of copper one. YBCO thin films are coated on the two disks and both side of cylinder under a pressure of 0.67 Pa at a power density of 1.1W/cm^2 in Argon, after MgO buffer layer coating. Target compound is $Y_1Ba_{2.5}Cu_{3.5}O_X$, moreover few pellets of $BaTiO_3$ compound is put on the YBCO target in order that Ti is doped in the films to improve the quality of superconducting.(3)

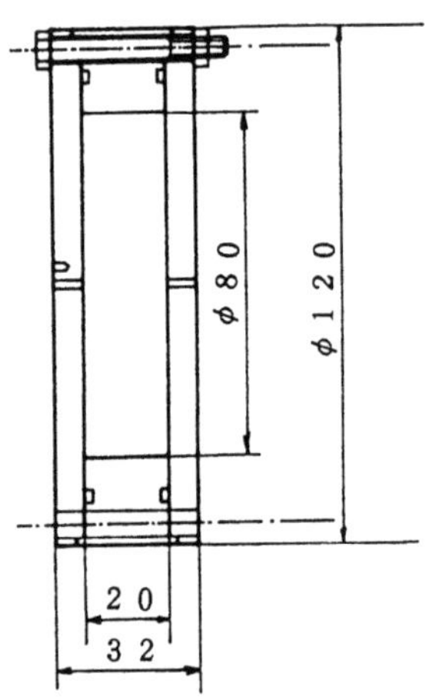

Fig.1 Schematic diagram of the cylindrical cavity

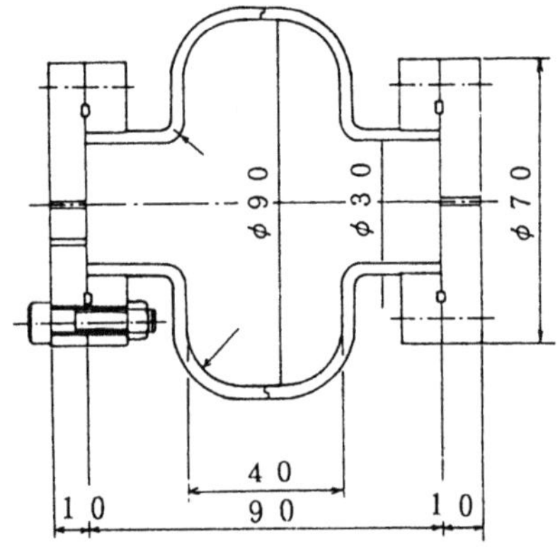

Fig.2 Schematic diagram of the model cavity

The model cavity is coated with niobium thin films after coating of nickel-chrome intermediate layer. These targets of nickel-chrome and niobium are sticks in shape. The dimensions of these are 10 mm in diameter both of them. The target is inserted into the cavity just as the outline which is given in Fig.4 when the cavity are coated. Nickel-chrome are sputtered under a pressure 0.67 Pa at a power density of 5.7 W/cm^2 in Argon. The coating conditions of niobium thin films are the same as those of nickel-chrome except that RF power is 1.4 W/cm^2.

RESULTS AND DISCUSSION

Fig.5 shows the temperature dependence of unloaded Q-value which are the copper cylindrical cavity coated niobium film and the no-coated cavity. Q_0 of niobium coated cavity shows a half value of reference at the room temperature. As the cavity are cooled, it increases gently and jumps up steeply at about 9.22 K which is superconducting transitional temperature of niobium. This results mostly fit in with theory but the final Q_0 at 4.2 K is lower than calculative value. Calculative Q_0 of niobium cavity at 4.2 K is about 1×10^8 but experimental data is a hundredth of that. This is caused by contact the resistance which occuers at the face between the disks and the cylinder.

Fig.6 shows the temperature dependence of Q_0 which are the model cavity coated niobium film and the no-coated model cavity. The surface morphology of the niobium film looks like uniformly and Q_0 value of that are fine from room temperature to liquid He temperature. However the superconducting transitional behavior is not observed clearly. The film thickness of niobium is estimated about 1 μm according to the monitor. This thickness is enough compared with the skin-depth of niobium at 4.2 K. So we guess that the no-coated or very thin place exist in vertical plane against the axis of target.

Q_0 of YBCO coated cavity are shown in Fig.7. These values does not change as the temperature decreases and the superconducting tansitional behavior is not observed. After annealing of YBCO film, a few points in the surface of film comes off. According to this matter,all the YBCO film does not become super-

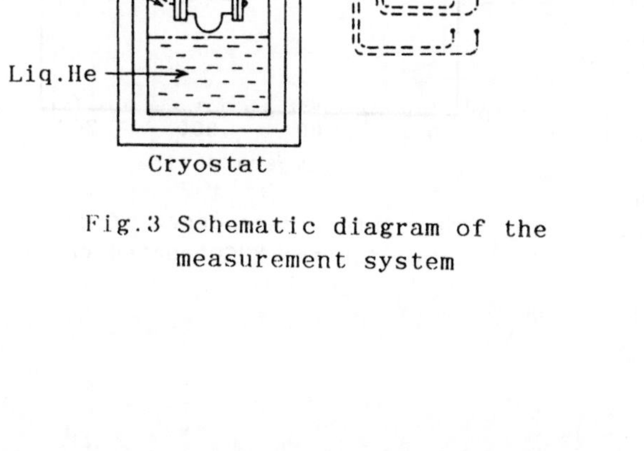

Fig.3 Schematic diagram of the measurement system

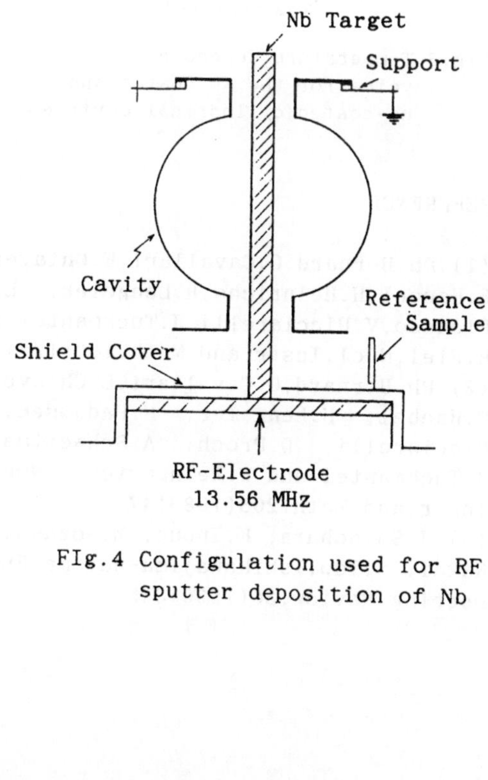

FIg.4 Configulation used for RF sputter deposition of Nb

conductor because of diffusion into MgO layer or disorder of the elemental ratio of Y:Ba:Cu. Q_0 of the cavity are influenced by the ratio of the superconducting area to normal conducting area. The superconducting transitional behavior of the cavity is not observed because of existing normal conducting part.

CONCLUSION

1. Q-value of the niobium coated cylindrical cavity is about 1×10^6 at 4.2 K. It is necessary to improve how to join cylinder to disk in order to realize higher Q-value.
2. The model cavity does not have superconducting transition clearly though Q-value is fine above the niobium critical temperature. It is subjects to measure the real distribution of Nb film thickness and observed micro morphlogy of the surface.
3. The superconducting transitional behavior of YBCO coated cavity is not observed. It is necessary to make further examinations to deposit superconductive film uniformly on large area.

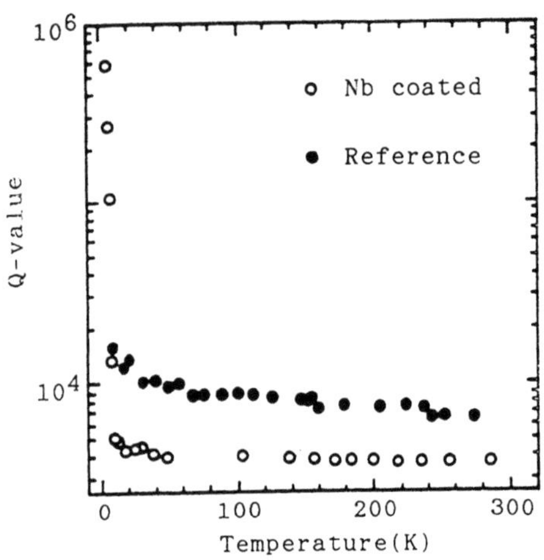

Fig.5 Temperature dependence of Q-value for the Nb coated and n^-coated cylindrical cavities

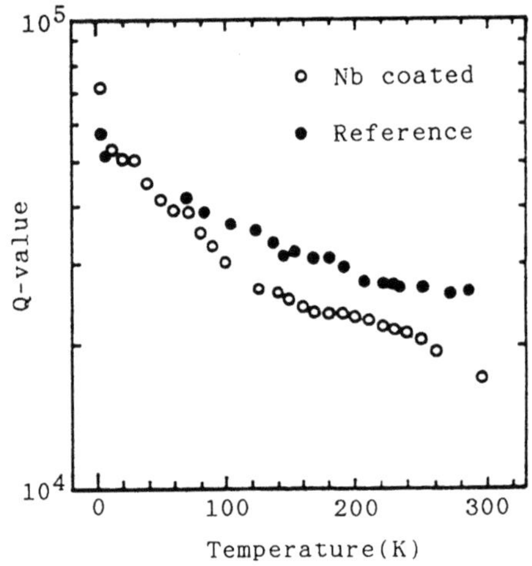

Fig.6 Temperature dependence of Q-value for the model cavities

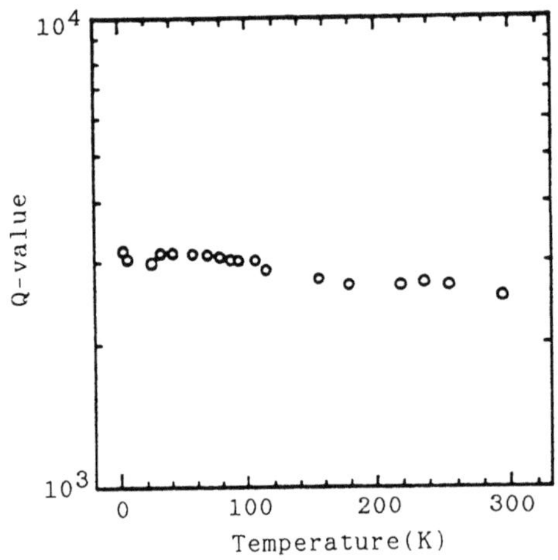

Fig.7 Temperature dependence of Q-value for the YBCO coated cavity

REFERENCE

(1) Ph.Bernard,G.Cavallari,E.Chiaveri, E.Haebel,H.Heinrichs,H.Lengeler, E. Picasso,V.Picciarelli,J.Tuckmantel and H.Piel.Nucl.Instr.and Meth.190(1981)257

(2) Ph.Bernard,G.Cavallari,E.Chiaveri, E.Haebel, H.Lengeler, H.Padamsee, V. Picciarelli, D.Proch, A,Schwettman, J.Tuckmantel and W.Weingarten. Nucl. Instr.and Meth.206(1983)47

(3) J.Shinohara, K.Inoue, M.Nozawa, H. Nakai, S.Ido.Ishikawajima-Harima Engineering Review.30(1990)345

Low Surface Resistance of Ag-Added $YBa_2Cu_3O_{7-x}$ Ceramics

OSAMU ISHII[1], YOSHIHIRO KIMACHI[2], KEIICHIRO ITOH[2], TSUNEO KONAKA[3], MAKOTO SATO[3], and UICHIRO MIZUTANI[4]

[1] NTT Radio Communication Systems Laboratories, Yokosuka, Kanagawa, 238-03 Japan
[2] NTT Interdisciplinary Research Laboratories, Tokai, Ibaraki, 319-11 Japan
[3] NTT Telecommunication Field System R&D Center, Tokai, Ibaraki, 319-11 Japan
[4] Department of Crystalline Materials Science, Nagoya University, Nagoya, 464-01 Japan

ABSTRACT

The surface resistance Rs of sintered $YBa_2Cu_3O_{7-x}$ (YBCO) samples is investigated using cavities of 17 and 50 GHz. Marked decreases in Rs are realized by adding Ag. The optimum Ag-fraction is determined to be about 75 vol%. where the lowest values at 77K (10 mΩ at 17 GHz and 18 mΩ at 50 GHz) are achieved. This improvement in Rs is attributed to the grain growth and ab-plane orientation of the YBCO grains and also to the increase in the density of the samples. The 50 vol%Ag-YBCO sample also exhibits much lower contact resistance with In solder than the YBCO sample, showing that Ag-addition inhibits semiconductor phase formation.

KEY WORDS: oxide superconductor, surface resistance

INTRODUCTION

Among the high Tc oxide superconductors, $YBa_2Cu_3O_{7-x}$ (YBCO) was the first material found to have a Tc higher than liquid nitrogen (77K) indicating a significant potential for device application[1]. From the many possible applications of YBCO, microwave devices (i.e. antennas, resonators, filters and delay lines) seems to be the most promising candidates[2-5]. To this end, it is very important to reduce the surface resistance (Rs) of YBCO. We have already reported that a small addition of Ag greatly reduced the Rs of sintered YBCO bulks [6].

It has also been reported that the onset of the superconductivity occurred for the Ag-YBCO bulk systems containing more than 80 vol% and that the mechnical strength increases with increasing Ag fraction [7]. YBCO bulks containing large amounts of Ag are therefore considered to be worthy of investigation. In addition to low surface resistance, soldered joints having low contact resistance (Rc) are also required when fabricating actual devices with complicated shapes such as antennas. This paper studies in detail the Rs, the Rc, crystal grain morphology, the ab-plane orientation and the density of the Ag-YBCO bulk systems over the wide volume fractions of 0 to 100 %Ag.

EXPERIMENTAL

Ag-YBCO pellets (26.5 mm in diameter and 1-2 mm thick) were fabricated by the sintering method. Ag was added to them by mixing YBCO powders and Ag powders in appropriate proportions. The powders were pressed into pellets at 1 GPa and were sintered under an oxygen flow at 890°C for 12 hrs. The fabrication processes were the same as those in ref.7.

The Rs of the samples was measured by the cavity method from 30 to 300 K. Both TE_{011} and TE_{013} resonating cavities of 17 and 50 GHz were used to measure the loaded Q values. Other measuring conditions and the Rs calculation method from the loaded Q values have been reported in ref. 8. To evaluate Rc, the resistance of Ag-YBCO rods (1x2x20 mm), the center of which were joined with In solder, was measured by the fourprobe method. The accuracy of the Rc measurement was 0.2 $\mu\Omega/cm^2$.

A scanning electron microscope (SEM) was used to observe the crystal grain morphology. Auger electron spectroscope (AES) observation was carried out to analyze the Ag distribution. X-ray diffractometry was also used to evaluate the orientation of the YBCO grains. The orientation factor (P) was calculated from (022) and (013)(103)(110) diffarction intensities[9].

The density of the samples was defined as the ratio of the actual measured density to the ideal one, which is given by the sum of appropriate proportion of the crystallographic densities of pure Ag (10.5 g/cc) and YBCO (6.3 g/cc).

RESULTS AND DISCUSSION

Figure 1 shows the temperature dependences of Rs at 50 GHz for samples A (0 vol%Ag), B (25 vol%Ag), C (50 vol%Ag), D (75 vol%Ag) and E (100 vol%Ag). Resonating curves for sample A could be measured only below 80 K because of its high Rs. These high Rs values should be attributed to the low sintering temperature (890°C), at which the YBCO grains could not grow sufficiently. The Rs values in both normal and superconducting states are decreased as the Ag fraction increases. Abrupt decreases in Rs at the transition temperatures (Tc=90-91 K) also become large with increasing Ag fraction. The Rs of sample D was almost half that of sample E at liquid nitrogen temperature (77K).

Figure 2 shows the Rs values at 77K as a function of Ag fraction. Filled and open circles show data at 50 and 17 GHz, respectively. As the Ag fraction increases, both Rs values decrease and reach a minima at 75 vol%Ag. The Rs values at 75 vol% are almost half of those of sintered Ag. If the Rs values of Ag and YBCO were constant for all Ag-YBCO systems, the Rs should change monotonously as a function of the Ag fraction. The fact that the Rs values show a minima at 75 vol%Ag indicates that the Rs of YBCO grains in the 75 vol%Ag-YBCO system is much lower than that of sintered Ag.

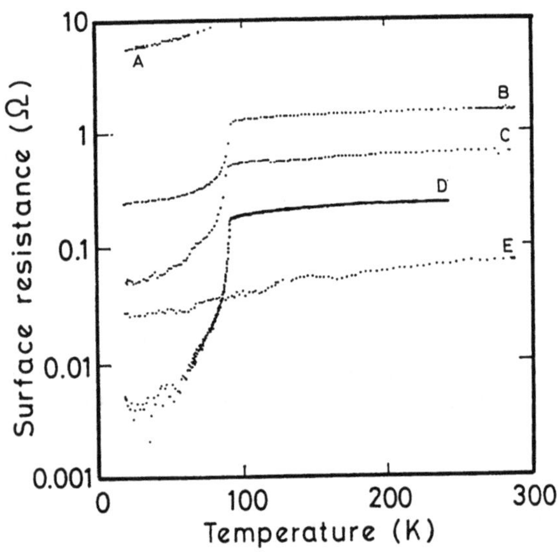

Fig. 1 The temperature dependencies of Rs at 50 GHz for samples A (0 vol%Ag), B (25 vol%Ag), C (50 vol%Ag), D (75 vol%Ag) and E (100 vol%Ag).

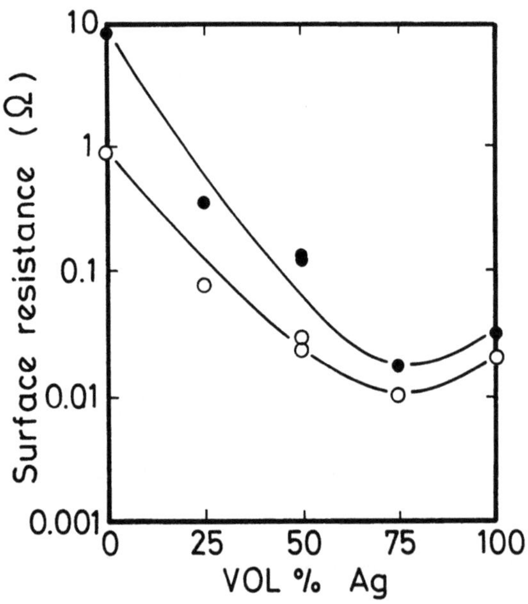

Fig. 2 The Rs values at 77K as a function of Ag fraction.

Figure 3 shows (a) SEM-image and (b) Ag- (c) O- (d) Y- (e) Ba- and (f) Cu-AES-mappings for sample C (50 vol%Ag-YBCO). The Ag did not contain any constituents of YBCO which means that the Ag was not oxidized. Even after sintering at 890°C in the oxygen atmosphere, the Ag remained dispersed among the YBCO crystals. As a result of Rs measurements and AES observation, it is speculated that Ag improved the YBCO structure which leads to lower Rs values.

Figure 4 shows SEM photographs of samples A,B,D and E. As Ag-fraction increases, the grain size increases. It can be also seen that the porosity of the samples decreases with increasing Ag fraction. The density of the Ag-YBCO system is shown in Fig.5 together with reported data[7]. The density tends to increase with the Ag fraction, which is consistent with the SEM observation. Hence it is speculated that YBCO grain boundaries closely contact each other and the influence of weak links on Rs is reduced when Ag is added to YBCO.

Figure 6 shows the Ag fraction dependence of ab-plane orientation factor (P). While the pure YBCO sample exhibits a low P, Ag-added YBCO samples show high P values that are almost double. Consequently, it is confirmed that the addition of Ag increases grain size, ab-plane orientation of YBCO grains and the density of the samples. All of these crystallographic improvements are considered to improve the surface resistances of YBCO samples.

Figure 7 shows the temperature dependencies of Rc for (a) pure YBCO- and (b) 50 vol%Ag-YBCO-samples. The Rc of the pure YBCO sample increased with decreasing temperature, showing that semiconductor phase was formed at the interface during the soldering. On the other hand, 50 vol%Ag-YBCO sample exhibited extremely low resistance (Rc<0.2 $\mu\Omega/cm^2$) and superconductivity at the boundaries of YBCO and In solder remains without any trace of deterioration.

Fig. 3 (a) SEM-image and (b) Ag- (c) O- (d) Y- (e) Ba- and (f) Cu-AES-mappings for sample C (50 vol%Ag-YBCO).

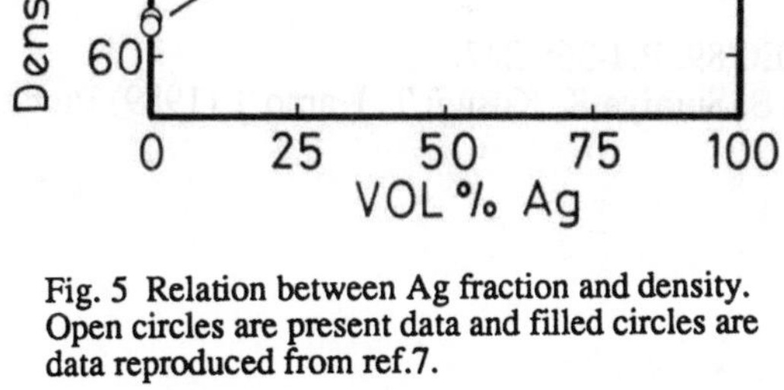

Fig. 5 Relation between Ag fraction and density. Open circles are present data and filled circles are data reproduced from ref.7.

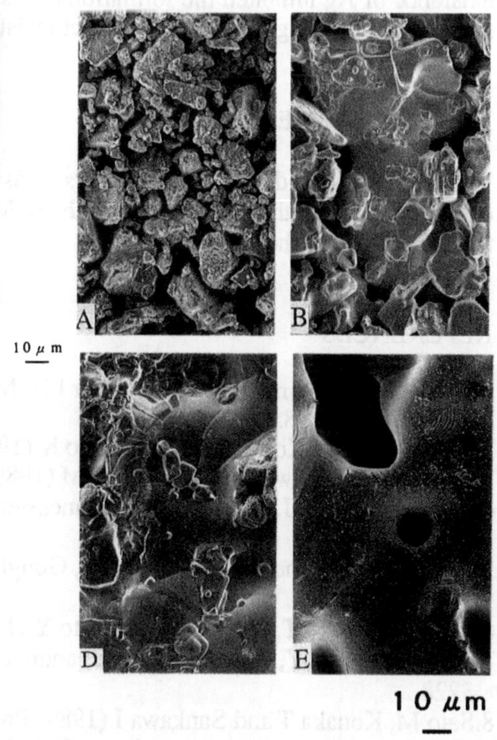

Fig. 4 SEM photographs of samples A, B, D and E.

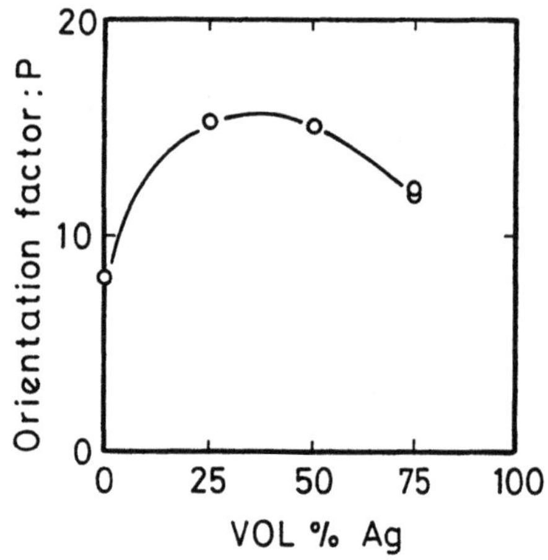

Fig. 6 The Ag fraction dependence of ab-plane orientation factor (P).

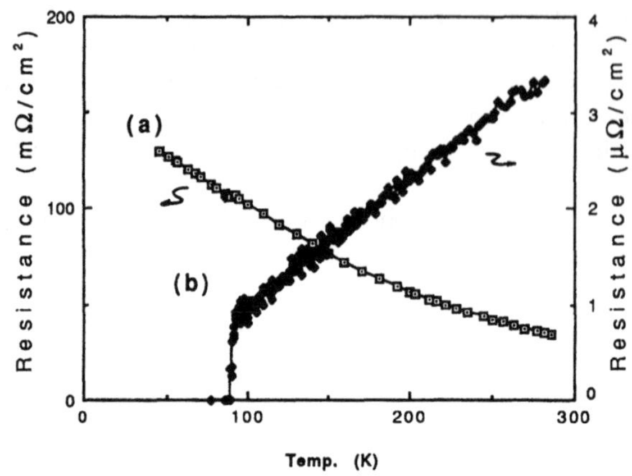

Fig. 7 The temperature dependencies of Rc for (a) pure YBCO- and (b) 50 vol%Ag-YBCO-samples.

In summary, the frequency dependencies of the Rs of sintered Ag-YBCO systems reveal that the systems act as an mixture of metals and superconductors. The 75vol%Ag-YBCO system, moreover, shows extremely low Rs values at 77K (10 mΩ at 17 GHz and 18 mΩ at 50 GHz); almost half those of pure Ag. Such a marked improvement in the Rs is attributed to the improved grain growth and ab-plane orientation of YBCO grains and an increase in sample density. It is also revealed that the existence of Ag inhibited the formation of a semiconductor phase at the boundaries of YBCO and In solder thus resulting in very low contact resistances.

ACNOWLEDGMENTS

The authors would like to thank Drs. Yasuhiro Koshimoto and Etsuji Sugita for their helpful discussion. The authors would like to thank Mr. Tetsuo Oka at Aishin Seiki Co.,Ltd. for assistance in the fabrication of the YBCO samples.

REFERENCES

1. Wu MK, Ashburn JR, Torng CJ, Hor PH, Meng RL, Lao L, Huang ZJ, Wang YQ, Chu CW(1978) Phys Rev Lett 58: 908.
2.Itoh K, Ishii O, Koshimoto Y and Cho K (1991) to be published on J. Supercon.
3.Peterson GE, Stawicki RP, Alford NM (1989) Appl Phys Lett 55:1798.
4.Wu Z, Mehler MJ, Maclean TSM, Lancaster MJ, Gough CE, Alford NM (1989) Physica C162-164: 385.
5.Khamas SK, Meher MJ, Maclean TS, Gough GE, Alford NM, Harmer MA (1988) Elect Lett 24: 460.
6.Ishii O, Konaka T, Sato M, Koshimoto Y (1990) Jpn J Appl Phys 29: L1075.
7.Nishino T, Itoh Y, Ogasawara F, Suganuma M, Yamada Y, Mizutani U (1989) J Mater Sci 24: 3228.
8.Sato M, Konaka T and Sankawa I (1989) Proc ISEC'89, PH-36: 247.
9.Yamana S, Kuwashima H, Ashizawa T, Shimoda S, Sumiya K, Kosugi T, Kamo T (1989) Proc ISS'89: 305.

Surface Resistance of the BSCCO Microstrip Lines

SHIGETOSHI OHSHIMA[1], KATSURO OKUYAMA[1], KUNIO SAWAYA[2], and KEISUKE NOGUCHI[2]

[1] Department of Electronic and Information Engineering, Yamagata University, Yonezawa, Yamagata, 992 Japan
[2] Department of Electrical Engineering, Tohoku University, Sendai, 980 Japan

1. ABSTRACT

The surface resistance, Rs, of BSCCO (2212 Phase) microstrip lines prepared by the doctor blade method was measured by the resonance method. $Tc_{R=0}$ of these microstrip lines measured by the usual dc current method was about 78 K. The temperature dependence of Rs is very similar to the superconducting transition curves measured by dc method. The Rs values were about 12~50 mΩ at 65 K in the frequency range of 0.6~2.5 GHz. It was clarified from SEM observation and X-ray analysis that Rs values of the microstlip lines consisting of the textured material were smaller than those of the polycrystalline material.

The Rs values which are expected to be proportional to f^2 in the two-fluid model were proportional to $f^{1.0}$~$f^{1.1}$. The difference between experiment and theory can be explained by the existence of weak junctions which can easily transform from superconducting to normal state under the high frequency electric field.

KEY WORDS: surface resistance, doctor blade method, Bi system, two-fluid model

2. INTRODUCTION

From the viepoint of applications of the high Tc oxide superconductors for high frequency devices, e.g, superconducting antenna, mixers, detectors of ultraviolet light etc, it is important to obtain films with small surface resistance values at high frequency. In addition, for such applications it is necessary to form fairly large-scare samples.Screen printing can easily offer large-scare samples but their Rs values are rather large[1~3]. Also, large-scare $Bi_2Sr_2CaCu_2O_{8+\delta}$ (BSCCO) thick films with good superconducting properties can be prepared by the doctor blade method [4].
However, there are only few reprts on Rs measurements of BSCCO thick films prepared by this method.

In this paper, we quote Rs values of BSCCO (2212 phase) thick films prepared by the doctor blade method. For high frequency devices, it is very important to use substrates with small dielectric losses. Alumina has a small dielectric loss compared with YSZ, STO and LAO, however, alumina easily reacts with BSCCO. So, thereforce we used Ag films as a buffer layer. We measured the temperature and frequency dependence of Rs and discuss the values by the two-fluid model.

3. EXPERIMENTAL

The strip lines of BSCCO were prepared using the green sheets made by the doctor blade method. The green sheets were set on alumina and on Ag coated alumina substrates and then annealed at various temperatures in air. The Ag film was formed by vacuum deposition with a thickness of about 1 μm. The annealing conditions were similar to those reported by Enomoto et al.[5]. In order to obtain c-axis oriented films, the microstrip lines were annealed close to the melting temperature. Grains orientation and surface morphology were measured by X-ray diffraction and scanning electron microscopy (SEM). The Rs measurements were carried out in the same way as reported previously[3]. The quality factor Q of the microstrip lines which were set in the cavity made by copper was measured by network analyser. We measured Q values as a function of temperature and frequency. The Rs values are calculated from the Q facrors.

4. RESULTS AND DISCUSSION

4-1 Grain orientation and surface morphology

It is well known that the critical current density, Jc, in the c plane is larger than Jc in the a-b plane for BSCCO. So, therefore it is necessarry to obtain c axis oriented thick BSCCO films for small Rs values, and it is also important to have flat surfaces for high frequency devices. For obtaining smooth surfaces we varied the annealing conditions, temperature and time. Fig.1 shows X-ray diffraction patterns and SEM photographs of BSCCO thick films annealed at 860℃ and 870℃ for 10 minutes. The substrates used were Ag coated alumina. The figure renotes that the film annealed at 860 ℃ has no strong peaks of the BSCCO (2212 phase) but very strong peaks from the Ag buffer layer. In addition, the grains show a plate-like configuration and are separated from each other. On the other hand, the film annealed at 870 ℃ has very strong peaks of BSCCO (2212), especially (00L) lines indicating c axis texture. The surface quality is also improved and the grains are quite large. The film annealed in the temperature range of 870~900℃ are very similar to those of films annealed at 870 ℃, but films annealed above 920℃ decomposed.

Fig.2 shows X-ray diffraction patterns of BSCCO films on pure alumina and Ag coated alumina substrates annealed at 880℃ for one hour. It is seen that the film on Ag coated alumina substrate has oriented grains, while the film on pure alumina is non-oriented and similar to the films annealed at 860 ℃ as desplayed in Fig.1 (A). Fig.3 shows SEM photographs of the samples shown in Fig.2. It can be seen that the film on Ag coated alumina substrate is partially melted and the surface is rather smooth, on the other hand, films on pure alumina is not melted. It is also observed that the films on pure alumina substrate annealed at 900 ℃ for one hour were melted and the surface morphology was similar to those of the films on Ag coated alumina annealed at 880 ℃. From this appearance, we conclude that the Ag buffer layer decreases the melting temperature of BSCCO (2212) and help to form a grain growth.

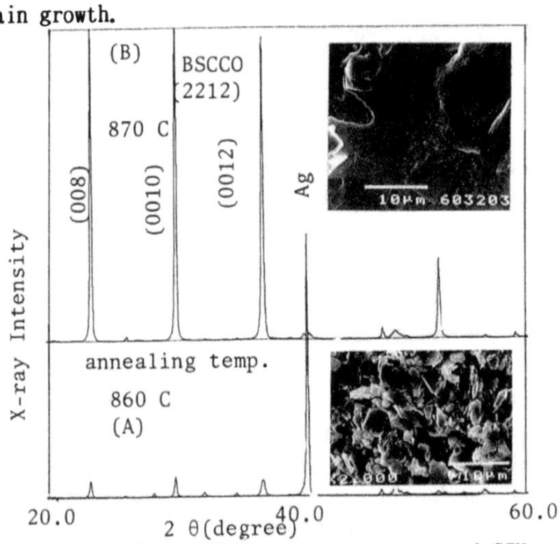

Fig.1. X-ray diffraction patterns and SEM photographs of BSCCO thick films.

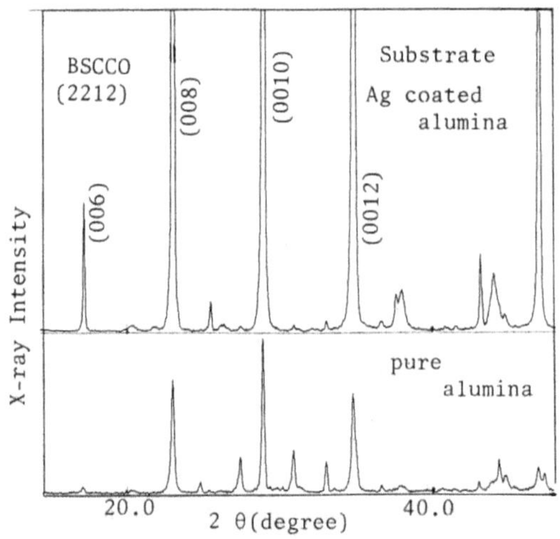

Fig.2. X-ray diffraction patterns of BSCCO thick films on pure and Ag coated alumina.

4-2 Superconducting properties and surface resistance

Fig.4 shows the superconducting transition curves of the BSCCO thick films annealed at 860, 880 and 900 ℃ measured by the standard four-probe method. The $Tc_{R=0}$ of the film annealed at 900 ℃ is about 78 K, which is close to the value of the bulk BSCCO (2212). On the other hand, the film annealed at 860 ℃ has a lower $Tc_{R=0}$, a broader transition width and a larger normal resistace compared with the film annealed at 900 °C.

Fig.5 shows the temperature dependence of Rs for BSCCO microstrip lines annealed at 900 ℃ measured at frequencies of 0.63, 1.30, 2.01 GHz. The Rs curves are similar to the superconducting transition curves shown in Fig.4. The Rs values in the normal and superconducting state increases with increasing frequency. No drastic change of Rs was observed in the film annealed at 860 ℃ between 60 K and 120 K.

Fig.3. SEM photographs of BSCCO.

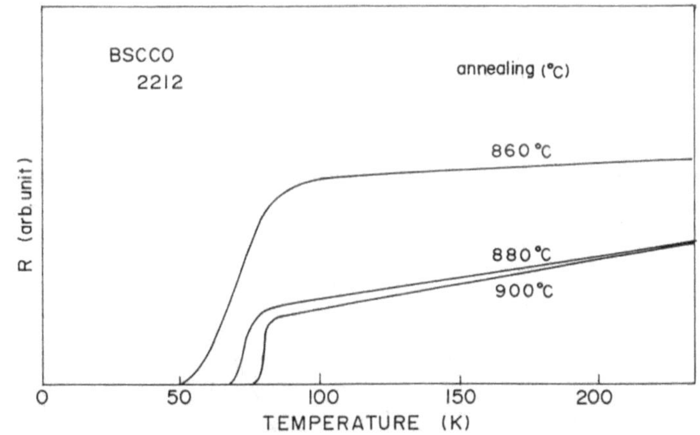

Fig.4. Superconducting transition curves of the BSCCO thick films.

The frequency dependence of Rs measured at 70 K for BSCCO strip lines is shown in Fig. 6. It was found that Rs is almost proportional to $f^{1.0}\sim f^{1.1}$ From the two-fluid model the Rs value of superconducting is given by

$$Rs_{super} = \omega^2\mu^2\sigma^2\lambda^2/2 \qquad \text{for } \omega\mu\sigma_n\lambda^2 \ll 1 \qquad (1)$$

where ω, μ, σ_n and λ are the angular frequency, magnetic susceptibility, electrical conductivity in the normal state and penetration depth, respectively. So, Rs in the superconducting state can be expected to be proportional to f^2.

The difference between our experimental observation and the prediction of the two-fluid model may be caused by weak junctions which can easily transform from the superconducting to the normal state under the application of a high frequency electrical field. We calculated the Rs value of a model film consisting of a series circuit of normal conductors and superconductors. The surface resistance of the normal conductor can be written by,

$$Rs_{normal} = (\mu\omega/2\sigma_n)^{1/2} \qquad (2)$$

We propose for the total surface resistance, the following expression

$$Rs_{tortal} = (1-x)\ Rn + x\ Rs \qquad (3)$$

where x is the volume parcentage of superconducting material in the films.

Figs.7 and 8 show the frequency dependence of Rs calculated using eqs (1)~(3). In Fig.7, we show the frequency dependence of Rs as a funetion of x. If the films contain only 1 % normal conductor the

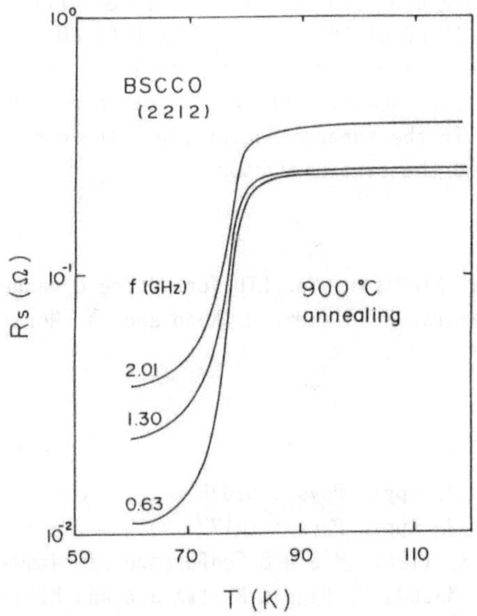

Fig.5. Temperature dependence of Rs for BSCCO microstrip lines annealed at 900 °C.

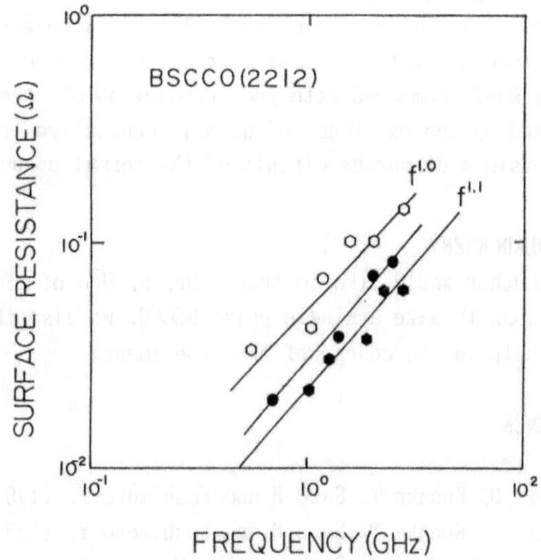

Fig.6. Frequency dependence of Rs measured at 70 K for BSCCO striplines.

Rs value is proportional to $f^{0.5} \sim f^{1.2}$ in the frequency range between 0.5 and 5 GHz. In fig.8 we show the frequency dependence of Rs as a function of relative temperature, T/Tc. The figure shows that the frequency dependence of Rs changes with measuring temperature, ie, if films have some phases with different Tc values, the frequency dependence of Rs may be changed. Recently, Y. Kobayashi and T. Imai [6] proposed a three-fluid model to explain the conduction mechanism of highTc oxide superconductors They assumed residual normal electrons which are independent of temperature. That is, the normal electrons move in the superconductors even at low temperature. They could explain the temperature dependence of Rs in the low temperature region. Their model is very similar to our idea, to treat films as series circuit of the normal conductor and the superconductor.

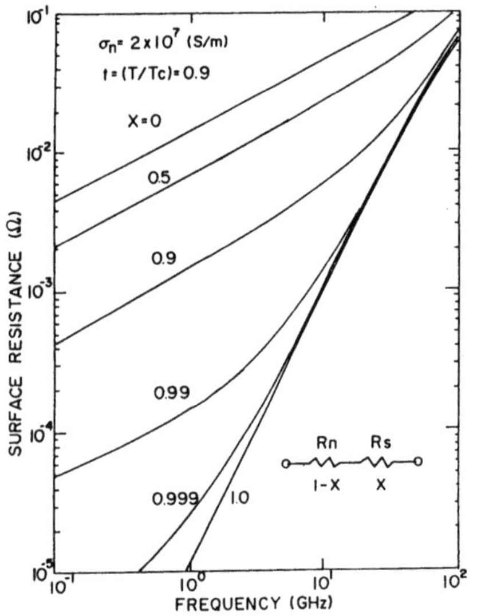

Fig.7. Frequency dependence of Rs as a function of X calculated by the two-fluid model.

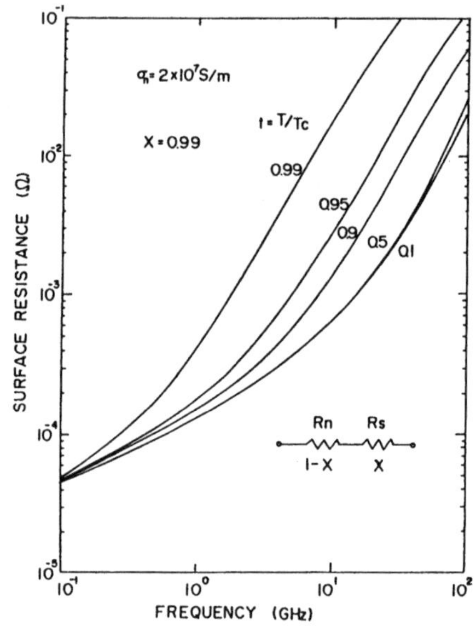

Fig.8. Frequency dependence of Rs as a function of relative temperature.

5. CONCLUSION

BSCCO (2212 phase) microstrip lines were prepared by the doctor blade method. The grain orientation of the films was very sensitive to the annealing temperature and c axis oriented films were obtained by annealing above 870 ℃. The Rs value was about 15 mΩ at 65 K measured at 0.63 GHz and the value increased with increasing frequency. The Rs value was almost proportional to $f^{1.0}$. The exponent is small compared with the prediction of the two-fluid model, while the difference can be explained by the existence of normal conductive regions in the superconductors, ie, the microstrip line consists of series circuit of the normal conductor and the superconductor .

ACKNOWLEDGEMENT

The author would like to thank Dr. N. Uno of Furukawa Electric Co., LTD for advice of annealing conditions to make a single phase BSCCO. We also thank Messrs. K. Aizawa, H. Kono and K. Morita for their help in the course of the experiments.

REFERENCES

1. Ishii O, Konaka T, Sato M and Koshimoto Y. (1990) Jpn. J. Appl. Phys.: L1075
2. Ishii O, Konaka T, Sato M and Koshimoto Y. (1990) Jpn. J. Appl. Phys.: L2177
3. Ohshima S, Okuyama S, Sawaya K, Maejima H and Noguchi K. (1991) M^{2}S-HTS Conference , Kanazawa
4. Sato K, Hikata T, Mukai H, Ueyama M, Shibata N, Kato T, Masuda T, Nagata M, Iwata K and Mitsui T (1990) Applied Superconductivity Conference, Colorado
5. Enomoto N, Uno N and Doi K. (1990) Applied Superconductivity Conference, Colorado
6. Kobayasi Y and Imai T (1991) IEICE Trans. 1986

Responsivity and Noise Characteristics for a Microwave Detector Using High-T_c Superconductors

MINORU TAKAI, KAZUYA NIKI, SHUICHI YOSHIKAWA, YOSHIAKI NAKATANI, YORINOBU YOSHISATO, and SHOICHI NAKANO

Functional Materials Research Center, SANYO Electric Co., Ltd., Hirakata, Osaka, 573 Japan

ABSTRACT

Practical responsivity and noise equivalent power (NEP) were mesured using a granular-type Josephson microwave detector with YBCO-BiO composite superconductors prepared by Bi_2O_3 doping into a $YBa_2Cu_3O_x$ superconductor for the first time. It was found that the normal resistance (Rn) of the detector increased with increased Bi_2O_3 doping and the measured Rv increase exceeded the square of the Rn value for the detector. The measured responsivity R_Mv of 180 V/W and NEP of $1.4x10^{-14}$ $W/\sqrt{Hz}$ were obtained at 53 K using the detector with the Rn of 13 ohms.

KEY WORDS : grain boundary, normal resistance, noise equivalent power

INTRODUCTION

One of the most practical applications for superconductors in electronics is high-frequency electromagnetic radiation detectors, due to their high sensitivity and high frequency response [1]. A Josephson effect detector, which works on the princple of Josephson pair tunneling current, has the potential for use as a detector of microwave or millimeterwaves. However, high-frequncy active devices based on high-Tc superconductor Josephson junction technology proved less successful [2]. The small coherence length of the oxide superconductor causes major difficulties in artificial junction fabrication. Moreover, the coupling of incident microwaves with the detector is insufficient because the junction impedance or normal resistance (Rn) of the usual microbridge-type Josephson junction is quite low in comparison with the impedence of incident microwaves from free space and follwing amplifier.

Only granular superconductors which include naturally formed Jpsephson junction networks at the grain boundary have potential as sensitive microwave detectors. We have developed microwave detectors with an Rn of about 0.1 to 0.5 ohm using YBCO bulk granular superconductors with fine particles. Though, these Rn values are larger than those reported for stacked-type SNS junctions or single-grain boundary Josephson junctions [3][4], it is even small to offer good coupling with microwaves and obtain high responsivity based on the Josephson effect.

Here, we report on new YBCO-BiO composite superconductor prepared by Bi_2O_3 doping into $YBa_2Cu_3O_x$. This new superconductor has a large normal resistivity (ρ) and high superconductiving zero-resistant temperature (Tc). We also describe the experimental results of the practical responsivity and noise characteristics tests for a detector using this material. These results suggest a strong potential for practical application of this granular-type microwave detector.

EXPERIMENTAL

Preparation of YBCO-BiO composite superconductor

$YBa_2Cu_3O_x$ superconducting powder for YBCO-BiO composite ceramics was prepared by a coprecipitation method and sintering process, where the pelletized samples were sintered at a temperature of 945°C for 9 hrs in flowing oxygen gas and were ground into individual grains with

sizes of about 5 μm. Then, Bi_2O_3 fine powders were mixed with $YBa_2Cu_3O_x$ powders. The mixed powders with proportions of 0%, 5%, 10%, 15% and 20% (Bi_2O_3 : $YBa_2Cu_3O_x$ by weight) were pressed into pellets with dimensions of 2x5x20 mm. The resintering was carried out at 940°C for 3 hrs in oxygen followed by a slow cooling.
The temperature dependence of the resistivity of YBCO-BiO composite ceramic samples was measured by a four-probe method. The microstructure of the ceramics were investigated using a scanning electron microscope (SEM) and X-ray diffraction.

<u>Measurement of microwave response of the detector</u>

The YBCO-BiO ceramics were fabricated for a detector with a small microbridge by mechanical procedures. The width, length and thickness of the superconducting line for the detector were 100 μm, 300 μm and 100 μm, respectively. Figure 1 shows the structure of the detector. The details of detector fabrication have been reported previously [5][6].
The microwave response (video-detection) was measured by a four-probe method, where microwaves of about 10 GHz were irradiated from a horn antenna or a coaxial antenna, and the detector was cooled to 53 K by the Gifford McMahan cycle. The microwave power was measured at the output of the generator and calibrated. Noise characteristics were analyzed using a video-detection system and spectrum analyzer.

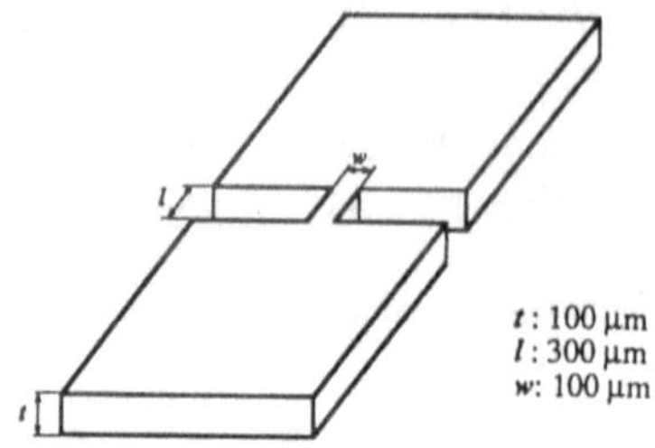

Fig.1 Schematic diagram of microwave detector

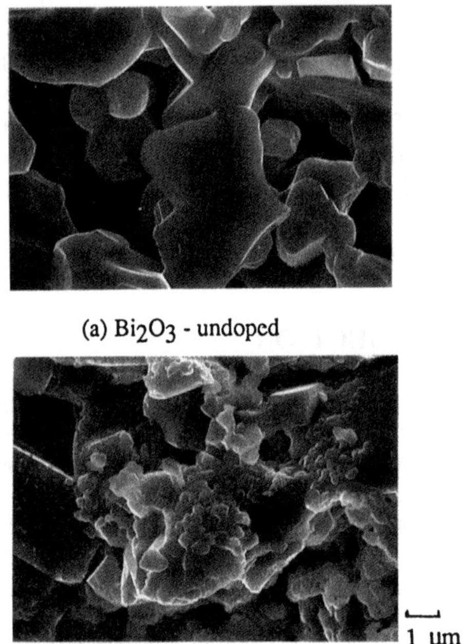

(a) Bi_2O_3 - undoped

(b) Bi_2O_3 10 wt. %

Fig. 2 Scanning electron micrographs showing (a) undoped and (b) 10 wt.% Bi_2O_3 doped $YBa_2Cu_3O_x$ bulk

RESULTS AND DISCUSSION

<u>Characteristics of YBCO-BiO composite ceramics.</u>

Figure 2 shows SEM photographs of the YBCO-BiO composite ceramic samples with a wt.% Bi_2O_3 content of (a) 0 % and (b) 10 %, respectively. It was observed that large $YBa_2Cu_3O_x$ particles (2~5 μm) were surrounded by small particles (0.2~0.3 μm) in sample (b). X-ray measurement showed that sample (a) was $YBa_2Cu_3O_x$ single phase. Sample (b) was found to include some impurities, such as small amounts of Bi-Ba-O and some Bi-oxide families in the $YBa_2Cu_3O_x$ phase. The peaks of impurity increased with increased Bi_2O_3 doping , however, the $YBa_2Cu_3O_x$ phase was not reduced even in the sample with 20% Bi_2O_3 doping. This means that these impurites coated the surface of the

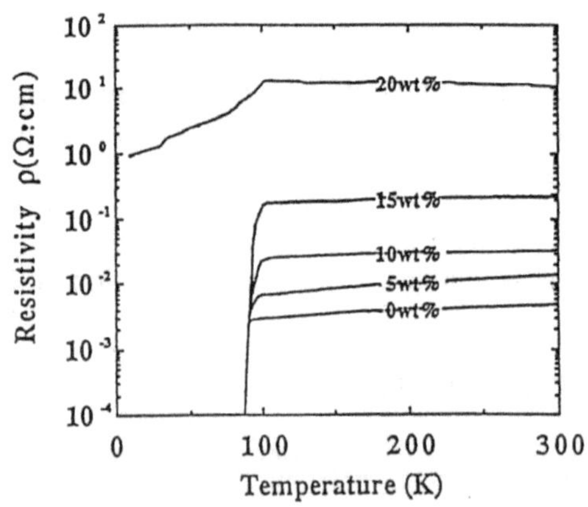

Fig. 3 Temperature-dependent resistivity of YBCO-BiO composite bulks plotted on a semilogarithmic scale, with Bi_2O_3 concentration as a parameter.

$YBa_2Cu_3O_x$ particles, as shown in Fig. 2 (a).
Figure 3 shows the temperature dependence of the resistivity of the samples. The normal-state resistivity (ρ) at 300 K increased logarithmically when the amount of Bi_2O_3 was increased, and the superconducting state was not destroyed even by a doping ratio of 20% Bi_2O_3. Furthermore, the zero-resistivity temperature (Tc = 90 K) was not reduced in samples with Bi_2O_3 doping of less than 10%. This means that the surface coating on $YBa_2Cu_3O_x$ particles by the impurties seems to increase the resistivity of the sample through the enhancement of normal resistance (Rn) of each grain-boundary Josephson junction.
This suggest that the granularity of the YBCO superconductor is enhanced by the addition of Bi_2O_3 and that YBCO-BiO composite ceramics would be suitable for a granular-type microwave detector, where the network of the Josephson junction is the essential rule for microwave response.

Microwave response of the detector

Table 1 shows the Rn value and measured responsivity (R_Mv) of the detectors fabricated using YBCO-BiO composite ceramics. Measurements were taken at a temperature of 53 K and at a microwave frequency of about 10 GHz. In this measurement, we did not use any matching circuit between incident microwaves and the detector. Although the value of R_Mv included a loss factor due to insufficient coupling with the microwaves and was smaller than that of true-responsivity Rv, (defined as the ratio of the output voltage to the absorbed microwave power), it was useful for estimating the practicality of the detector.
It was shown that the Rn values of YBCO-BiO detectors were quite large compared with the those of the usual YBCO granular bridge previously reported. The largest Rn value obtained for a YBCO detector was about 0.5 ohm. This system used fine $YBa_2Cu_3O_x$ particles with a grain size of less than 1 μm and a smaller microbridge (width, length and thickness were 50 μm, 300 μm and 50 μm, respecrively) than YBCO-BiO detectors. These results show the effect of enhanced of resistance through Bi_2O_3 doping.
Figure 4 shows the measured responsivity (R_Mv versus Rn values) for various detectors, including data for conventional YBCO granular detectors. It is clear that the YBCO-BiO granular detectors are more sensitive than YBCO granular type and that R_Mv increases squarely with the increase in Rn value. The maximum responsivity obtained was 180 V/W. This value is more than ten times as

Table 1. Normal resistance Rn and measured responsivity R_Mv of the YBCO-BiO detectors.

Sample	Bi_2O_3 (wt%)	Rn (Ω)	Rv(V/W)
A	10	2.6	10
B	10	1.8	11
C	15	13.0	180

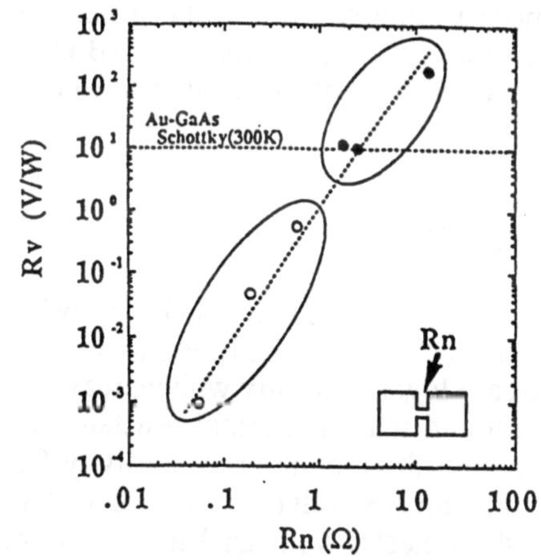

Fig. 4 Normal resistance (Rn) dependence of measured responsivity (R_Mv) the detectors at 53 K. The open circles represents the data for the detectors with YBCO ceramics,the solid circles with YBCO-BiO composite ceramics.

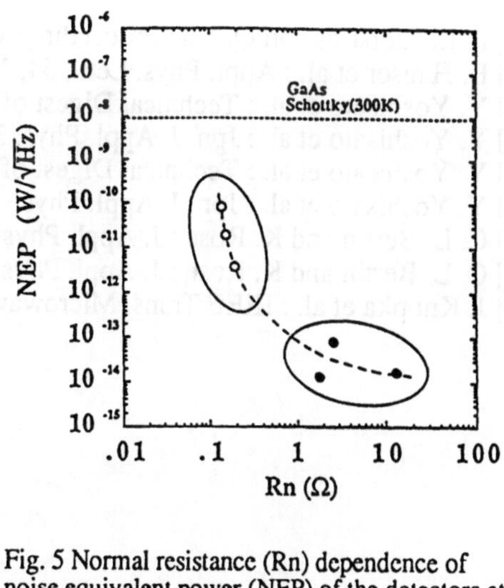

Fig. 5 Normal resistance (Rn) dependence of noise equivalent power (NEP) of the detectors at 53 K. The open circles represents the data for the detectors with YBCO ceramics,the solid circles with YBCO-BiO composite ceramics.

sensitive as the current Au-GaAs Shottky-type semiconductor detector.
Figure 5 shows the normal resistance dependence of noise equivalent power (NEP) for these detectors. The NEP value also clearly decrease with increases in the Rn value. The best NEP value of $1.4x10^{-14}$ W/$\sqrt{Hz}$ was obtained for the detector with an Rn of 13 ohms at 53 K. This value was also quite low compared with that of Au-GaAs detectors and is by no means inferior to that of Nb Josephson junction at liquid Herium temperature.
The physical processes responsible for the observed microwave responsivity of granular-type detectors composed of many SNS-type grain boundary Josephson junction networks are not clearly understood. Nevertheless, the behavior of the detectors is quite similar to the enhanced mode of detection, introduced by Bertin and Rose using granular Sn films [7]. They reported that microwave response is enhanced by taking advantage of the high-resistance granular superconducting films and low noise equivalent power of this type of detector [8][9]. The existence of the normal phase distributed between superconducting grains of $YBa_2Cu_3O_x$ particles by doping of Bi_2O_3 in YBCO-BiO composite ceramics seemed to lead to high responsivity to microwaves due to the enhanced Josephson effect of the grain boundary.
These findings suggest that YBCO-BiO granular type detectors have good potential for application due to their high impedence, attributable to a series of junction arrays, and high temperature operability.

CONCLUSION

We have developed YBCO-BiO composite superconductor for a granular-type Josephson microwave detector for the first time. It was found that the normal resistivity (ρ) of the superconductors at 300 K increase logarithmically with increases in the amount of Bi_2O_3 doping.
It was found that the normal resistance (Rn) of the detector also increased with increased Bi_2O_3 doping, and the measured responsivity (R_Mv) increase exceeded the square of the Rn value of the detector. The best values of R_Mv of 180 V/W and NEP of $1.4x10^{-14}$ W/$\sqrt{Hz}$ were obtained at 53 K using the detector with an Rn of 13 ohms, where Bi_2O_3 of 15% wt. % was used to dope the $YBa_2Cu_3O_x$.
These results suggest a strong potential for the practical application of granular-type microwave detector.

REFERENCES

[1] P. L. Richards and Q. Hu : Proceedings of IEEE 77, 1233 (1989)
[2] B. Hauser et al. : Appl. Phys. Lett., 54, 1368 (1989)
[3] Y. Yoshisato et al. : Technical Digest of 9th Sensor Symposium 59 (1990)
[4] Y. Yoshisato et al. : Jpn. J. Appl. Phys. 30, 661 (1991)
[5] Y. Yoshisato et al. : Technical Digest of 8th Sensor Symposium 243 (1989)
[6] Y. Yoshisato et al. : Jpn. J. Appl. Phys. 29, 1080 (1990)
[7] C. L. Bertin and K. Rose : J. Appl. Phys. 39, 2561 (1968)
[8] C. L. Bertin and K. Rose : J. Appl. Phys. 42, 631 (1971)
[9] J. Knopka et al. : IEEE Trans. Microwave, 38, 160 (1990)

Properties of Josephson Mixer Fabricated by High T_c YBaCuO Film

YOSHITO FUKUMOTO, HIROSHI KAJIKAWA, RIKUO OGAWA, and YOSHIO KAWATE

Superconducting & Cryogenic Technology Center, Kobe Steel Ltd., Kobe, Hyogo, 651-22 Japan

ABSTRACT

Properties of Josephson mixers fabricated by high Tc YBaCuO film were investigated. Microbridge junctions were fabricated by high Tc YBaCuO film deposited on a MgO(100) substrate which had a step-edge on its surface, and were mounted in a copper case to build a mixer. Current-voltage (I-V) characteristics under microwave irradiation showed that the mixer could respond to the wide frequency range from 11GHz microwave to 340GHz submillimeter-wave. IF signals with frequency of DC~2GHz could be observed. The mixer had an RF sensitivity down to -47dBm (20nW) with a small LO signal (90μW). The power linearity between RF input and IF output ranged over 23dB. Using this mixer as a frequency converter, the Broadcasting Satellite (BS) signal was successfully received and a clear picture was obtained on television.

KEY WORDS: YBaCuO, step-edge, Josephson junction, microwave communication

INTRODUCTION

Many concerns have been paid to the applications of high Tc oxide superconductors to microwave and millimeter-wave components. Passive devices such as resonators and band-pass filters were fabricated and showed very high Q-values and small insertion loss at liquid nitrogen temperature, reflecting the low RF surface impedance of epitaxial films [1-3]. On the other hand, because of difficulties in fabricating a Josephson junction using oxide superconductors, studies of active devices such as mixers and oscillators still remain at the basic stage. We have previously reported the fabrication of the Step-Edge microbridge Josephson Junction (SEJJ) fabricated by sputter-deposited YBaCuO films and the basic mixing experiments with K-band microwaves.[4,5]. It was shown that AC and DC characteristics of SEJJs showed a good correlation with the current activated Resistively Shunted Junction (RSJ) model and that heterodyne mixing was successfully achieved by using SEJJs.

In this study, we measured the device properties of SEJJ mixers such as the operating RF/IF frequency ranges, RF input sensitivity and relationship between the RF input power and the IF output power. In addition, in order to examine the practical feasibility of the SEJJ mixer in the microwave communication, it was incorporated to a BS-television system as a frequency converter.

EXPERIMENTAL

The fabrication procedure of the SEJJ was as follows: A step-edge of 200nm high was formed on the surface of MgO(100) substrate employing the common photolithograph and wet-etching techniques [6]. Then, YBaCuO thin film was deposited on the substrate by the off-axis sputtering method. A microbridge of 5μm wide and 30μm long was fabricated so as to cross the step-edge. Finally, four silver electrodes were formed by vacuum evaporation method followed by oxygen annealing at 500℃. Fig.1 shows a SEM photograph of the junction. The SEJJ was then packaged in a silver coated copper case to build up a single-end mixer. Figs.2 and 3 show an outward appearance and the internal circuit in the package. The mixer package includes two SMA connectors for RF/LO input and IF output, two resistors for impedance matching (R_1) and for junction protection from the excessive current (R_2), a inductor (L) to avoid RF signals flowing into the DC circuit and a capacitor (C) for microwave coupling with the SEJJ. Details of fabricating this mixer were described elsewhere [4,5].

I-V characteristics of the SEJJ were measured under irradiation of microwaves of X-band (8.2~12.4GHz) and K-band (18~26.5GHz). Mixing experiments were carried out by using K-band RF and LO signals which were coupled by a directional coupler with nominal coupling of -6dB. The IF output signal taken from the output connector was observed by a spectrum analyzer. The powers of RF, LO inputs and IF output were measured at the RF/LO input connector and the IF output connector, respectively. The bias current was set to get the maximum of IF output.

Fig.4 shows the block diagram of the experiment of receiving BS signal. The BS signal was caught by a parabolic antenna of 0.5m in diameter (gain 35dB) and amplified by two amplifiers with gains of 35dB and 30dB. The signal was then coupled with a LO signal of 10.678GHz by a directional coupler with nominal coupling of -20dB and converted to 1~1.3GHz IF signals by the SEJJ mixer. The IF signal was sent to a BS-tuner, of which the output signal is further connected with a television monitor.

RESULTS & DISCUSSIONS

RF Frequency Range

The I-V curves of the SEJJ under microwave irradiation were shown in Figs.5 and 6. The sample temperature was 10K and the microwave frequencies were 11.00GHz and 23.08GHz, respectively. On each curve, typical constant-voltage steps (Shapiro steps) were visible at $V=nhf/2e$ (h: Plank constant, e: electron charge, n: an integer). In Fig.6, the 15th step could be observed. This result shows that AC Josephson current with the frequency of 340GHz could be detected by the SEJJs. Therefore the mixer is expected to work for high-frequency signals from 11GHz microwave to the 340GHz submillimeter-wave. These microwave-induced steps could be observed up to 69K though they lost sharpness as the temperature increased.

IF Frequency Range

Figure 7 shows the frequency spectra of the IF output. The IF signals corresponding to the difference between the RF and LO frequencies were observed in each spectrum. The IF output was obtained up to 2GHz (data is shown until 1.25GHz in figure). The upper limit of IF frequency range might be determined by the reflection loss or transmission loss inside the package, and might not be due to the junction nature. In the range from 100MHz to 300MHz, the IF output power level varied within $\pm$3dB from the averaged level .

RF Sensitivity and Dynamic Range

In Fig.8, the relationship between the RF input power and the IF output power was shown. The RF and LO frequencies were 23.95 GHz and 23.85 GHz, respectively. The LO power was 90μW and the sample temperature was 10K during the experiments. The IF signal of 100MHz could be detected even when the RF signal was -47dBm (20nW). RF signal exceeding -17dBm was not tried because it was the maximum power of the RF oscillator. The SEJJ mixer showed a good lineality in the relationship between the RF input power and the IF output power at the range of -40dBm to -17dBm. From these results, it was shown that the SEJJ can detect a very small RF signal with the small LO power compared with the most sensitive mixers using semiconductor diodes. However, the IF output was extremely small compared with the RF input. The gross conversion efficiency was about -60dB ($\sim 1/10^6$). This might also included the reflection of the signal and/or the transmission loss inside the package in addition to the native efficiency of the Josephson mixer.

Experiment of Receiving BS-signals

By optimizing the direction of the parabolic antenna, LO frequency and LO power, a clear picture was successfully obtained on television. The power density emitted from the satellite (BS-2) is -117dBm/m^2 [7]. The RF power which is introduced to the SEJJ mixer was calculated to be about -17dBm. This level lies in the dynamic range of SEJJ (Fig.9) and correlates with the gross sensitivity measured by using K-band microwave. As the conversion efficiency is being improved, the BS signal will be received using a preamplifier with a smaller gain.

REFERENCES

1) Young K, Negrete GV, Sun JZ: Jpn. J. Appl. Phys. **30** (1991) L1355-1358.
2) Nagai Y, Sato M, Asano H, Kubo S, Konaka T: Jpn. J. Appl. Phys.**30** (1991) L1679-1684.
3) Itozaki H, Higaki K: 1991 Microwave Workshop and Exhibition pp157-161.
4) Fukumoto Y, Kajikawa H, Ogawa R, Kawate Y: 1991 International Conference on Solid State Devices and Materials pp444-446
5) Fukumoto Y, Kajikawa H, Ogawa R, Kawate Y: to be appeared in Jpn. J. Appl. Phys.
6) Kajikawa H, Fukumoto Y, Ogawa R, Kawate Y: 1990 International Symposium on Superconductivity pp893-896.
7) "BS Receiving Technology": Dempa publications, Inc. (1987) (in Japanese)

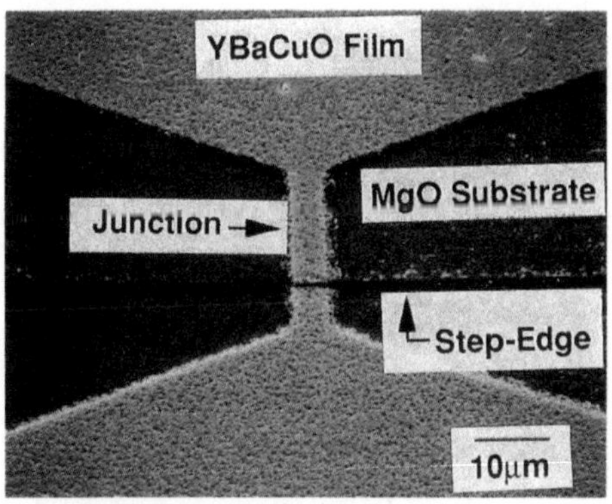

Fig.1 SEM photograph of the step-edge junction.

Fig.2 Outward appearance of packaged SEJJ mixer head.

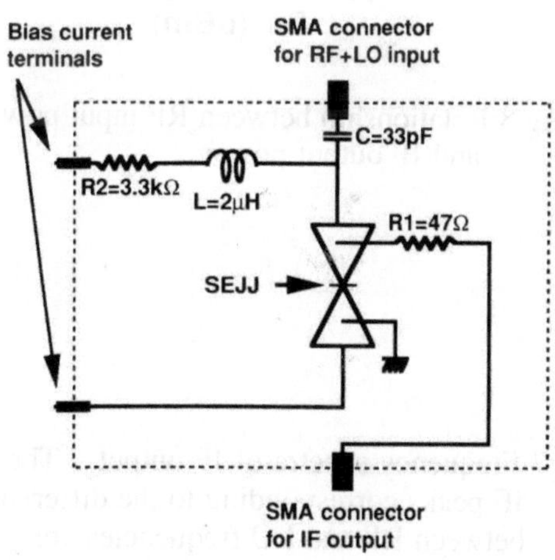

Fig.3 Internal circuit included in the mixer package.

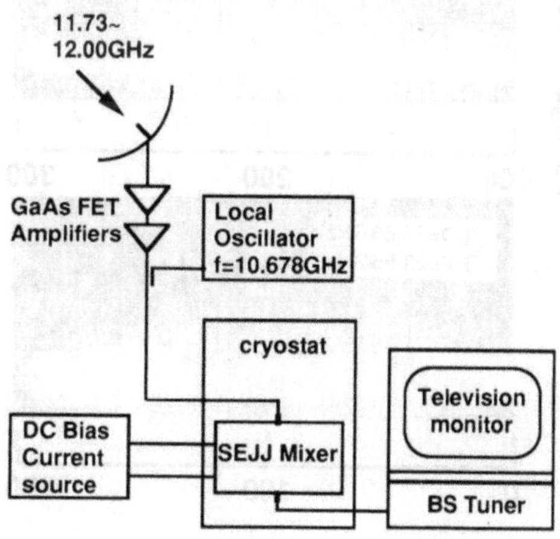

Fig.4 Block diagram of the experiment of BS-signal receiving.

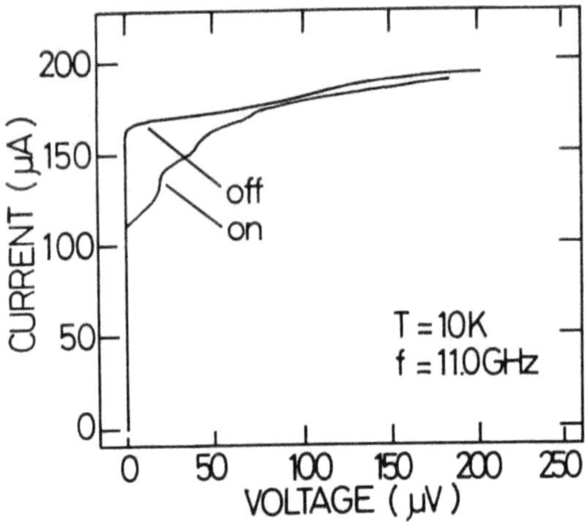

Fig.5 I-V characteristics of the SEJJ under irradiation of 11.00GHz microwave.

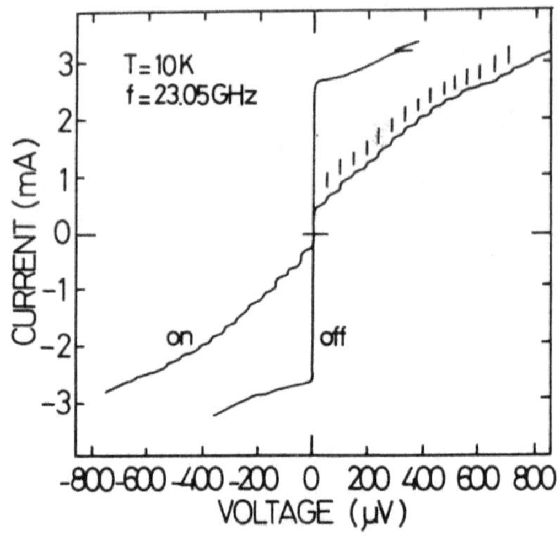

Fig.6 I-V characteristics of the SEJJ under irradiation of 23.08GHz microwave.

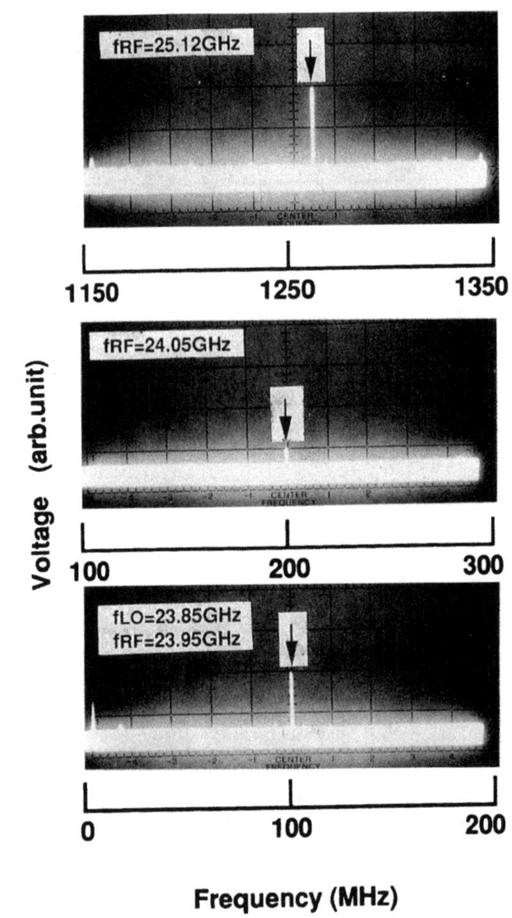

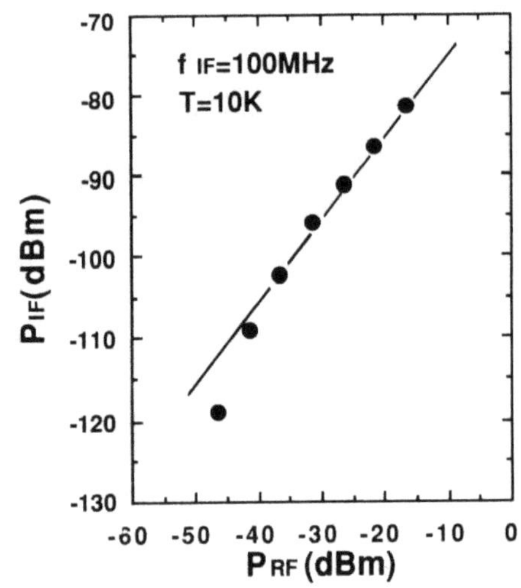

Fig.8 Relationship between RF input power and IF output power.

Fig7 Frequency spectra of IF output. The IF peaks corresponding to the difference between RF and LO frequencies are clearly seen.

7.3 Other Superconducting Devices and Basic Properties

The Basic Study of HTS/Semiconductor Hybrid Device

QIXIN HUANG, NOBUYUKI YOSHIKAWA, and MASANORI SUGAHARA

Faculty of Engineering, Yokohama National University, Yokohama, 240 Japan

ABSTRACT

Experimental study is represented concerning electrical property of BSCCO/Te interface and device characteristics of SUBSIT structure. It is found that the BSCCO/Te interface reveals almost no Schottky barrier property . The SUBSIT structure made of BSCCO, BSCO, Te and Au layers is fabricated and transistor characteristics are measured.

KEY WORDS: high-Tc superconductor, superconductor/semiconductor contact, SUBSIT, superconducting three-terminal devices.

INTRODUCTION

There have been various studies on low-temperature superconductive three-terminal devices of injection type such as SUBSIT and super-HET, where most of them suppose the use of the semiconductor/superconductor interface [1-3]. One of the difficulties of the hybrid devices is the appearance of the Schottky Barrier on the semiconductor/superconductor interface which disturbs the carrier transport across the interface.

In previous experiment we studied the BSCCO/p-Si interface showing barrier lower than that of BSCCO/n-Si [4]. At present, however, it is very difficult to obtain a good high-Tc superconductor film with desirable interface property when it is deposited on semiconductor substrate such as Si and GaAs. In order to compose a superconductor/semiconductor hybrid device with SUBSIT structure, we select BSCCO and Te as the materials of the junction. BSCCO is supposed to have something better surface property than YBCO. Te is known to be a degenerate p-type semiconductor. Especially in previous experiments, we found that the deterioration of superconductivity of BSCCO material was fairly small in case of partial doping of Te in BSCCO so long as the doping level was moderate[5]. Therefore it is considered to cause less harmful effect to BSCCO by interdiffusion than other semiconductors such as Si. In this report, we represent our experimental results of the transistor properties of the SUBSIT structure with BSCCO/BSCO/BSCCO/Te.

CONDUCTION CHARACTERISTICS OF THE BSCCO/Te INTERFACE

The semiconductor Te layer is evaporated on the as-sputtered surface of the BSCCO film by vacuum-deposition, forming a junction of BSCCO/Te. The detail of the fabrication process can be seen in reference[6]. Figure 1 shows current-voltage characteristics of the junction whose BSCCO layer makes superconducting transition. It is found with temperature decrease, that(i) the junction resistance increases mainly due to the rising of resistance of semiconductor Te itself, and that (ii) the asymmetry and the nonlineality feature of the current-voltage characteristics increase. In comparison with a Pb/Te/Au junction where the Schottky barrier plays the dominant role in the junction characteristics [7], nonlinear characteristics in superconducting state of the BSCCO/Te/Au junction reveal completely different feature with little Schottky barrier property. The lack of the property may be explained by the fact that both of BSCCO and Te are materials characterized by degenerated hole carriers. From the current- voltage curve, the barrier height at 10K is estimated to be about 15.7 meV. This fact may show the adaptability of this HTS/semiconductor interface to the applications of the semiconductor/superconductor hybrid devices.

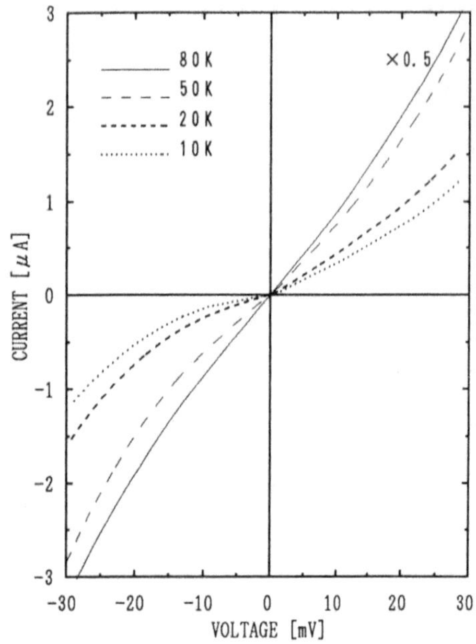

Fig .1.Temperature dependence of I-V characteristics for BSCCO/Te interface.

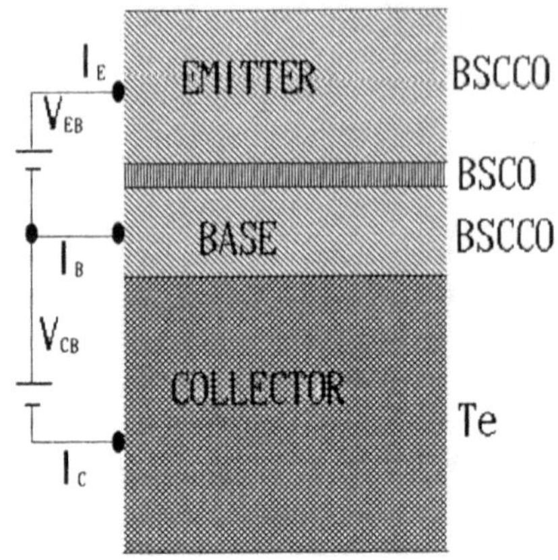

Fig.2. Schematic structure of the SUBSIT composed of BSCCO,BSCO and Te.

SUBSIT STRUCTURE AND FABRICATION

Figure 2 shows a schematic diagram of the SUBSIT structure made of the high-Tc superconductor BSCCO and the semiconductor Te, where SIS emitter junction is composed of BSCCO superconductors and BSCO insulator, collector junction is composed of Te semiconductor and BSCCO superconductor. The sample was fabricated in the following way. First, BSCCO/BSCO/BSCCO junction were formed by the RF sputtering on the MgO(100) substrate under the conditions shown in Table.1, the thicknesses of layered films are 200nm,10nm and 150nm respectively. After sputtering, the heating treatment was made at 800°C for 10 minutes. Then, semiconductor Te layer of thickness 600nm and electrode Au were formed on the surface of upper BSCCO layer by evaporation at room temperature. The metal masks were used in all the pattern formation. The areas of emitter and collector are $2.5\times 10^{-4} cm^2$, $1 \times 10^{-3} cm^2$ respectively.The electrical characteristics were measured by four-probe method.

Table 1. Sputtering conditions of BSCCO/BSCO/BSCCO layers.

	$Bi_2Sr_2CaCu_2O_x$	$Bi_2Sr_2Cu_1O_x$
Target	$Bi_{1.6}Sr_{0.92}Ca_{0.57}Cu_{1.05}O_x$	$Bi_{1.6}Sr_{0.92}Cu_{0.52}O_x$
Substrate temperature (°C)	740 ~ 780	740
Sputtering gas (Ar/O_2)	8/1 ~ 6/1	6/1
Sputtering pressure (Pa)	6 ~ 8	6
RF power (W)	80 ~ 100	90

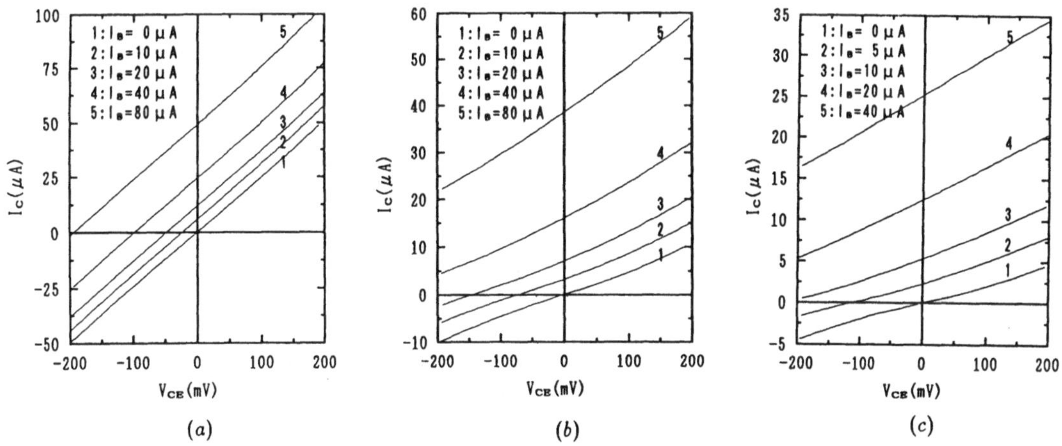

Fig.3. Current and voltage characteristics of SUBSIT structure in common emitter configuration at different temperature at (a) 300K, (b) 77K and (c) 4.2K.

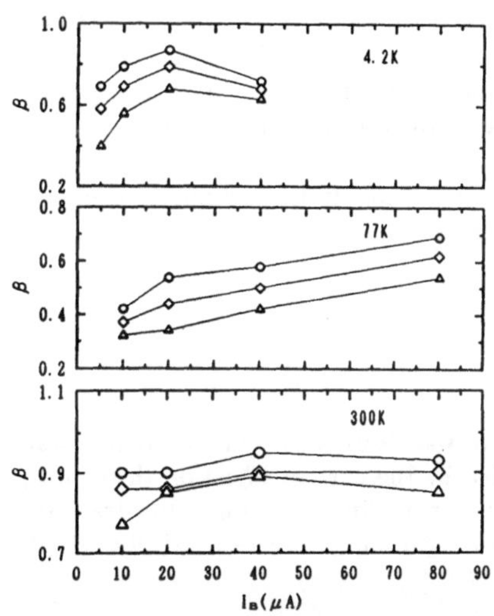

Fig.4. Relation of current-gain coefficient in common-emitterconfiguration versus base current at different collector voltage(○ : $V_{CE} = 180mV$, ◇ : $V_{CE} = 60mV$, △ : $V_{CE} = -60mV$).

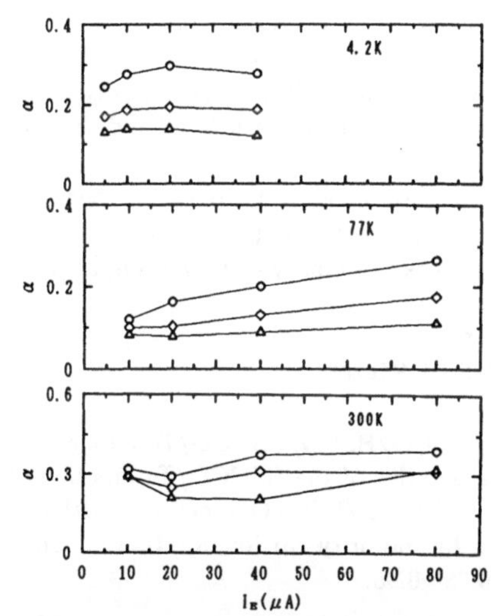

Fig.6. Relation of current-gain coefficient in common-base configuration versus emitter current at different collector voltage (○ : $V_{CB} = 180mV$, ◇ : $V_{CB} = 60mV$, △ : $V_{CB} = -60mV$).

RESULTS AND DISCUSSION

Figure 3 shows the collector current-voltage characteristics of the sample with the SUBSIT structure in common-emitter configuration(CEC). Figure 4 is the base current dependence of the current gain coefficient in CEC with the collector voltage as a parameter. From Fig.3 and Fig.4, it is found that the collector current can be modulated by base current in this structure and the modulation of collector current increases

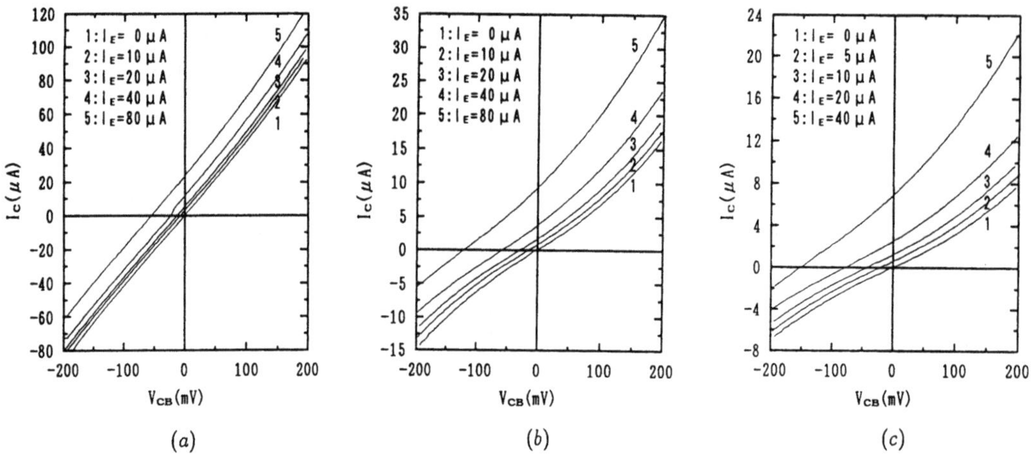

Fig.5. Current and voltage characteristics of SUBSIT structure in common base configuration at different temperature at (a) 300K, (b) 77K and (c) 4.2K.

with temperature decrease. At given temperature , the modulation is considerably larger when the Te layer is negatively biased relative to the emitter layer than when it is positively biased. This means that the injected holes are collected by the p-type Te layer more easily at a negative bias voltage than at a positive bias voltage. Current gain $\beta < 1$ indicates that the device still lacks current amplification ability.

Figure 5 shows collector current-voltage characteristics in common base configuration. Figure 6 is emitter current dependence of gain coefficient α with the collector voltage as a parameter. Since the $\alpha < 0.5$, the device lacks the ability of power amplification.

CONCLUSION

The BSCCO/BSCO/BSCCO/Te triode with the SUBSIT structure was fabricated. The current-voltage characteristics of the BSCCO/Te junction of the device are modified by the current injection through the BSCCO/BSCO/BSCCO junction. Although the experiment is still in preliminary stage, the transistor action by minority carrier injection is confirmed. The gain coefficients α and β are respectively 0.1~0.4 and 0.35~0.95.

REFERENCES

1. Frank D J, Brady M J and Davidson A, (1985) IEEE Trans. Magn.,MAG-21, L1985.
2. Tamura H, Hasuo S and Yamaoka T, (1985) Jpn. J. App. Phys., 24 L709.
3. Tonouchi M, Sakai H, Kobayashi T and Fujisawa K,(1987) IEEE Trans. Magn., MAG-23, L1647.
4. Huang Q X, Akeyoshi T, Yoshikawa N and Sugahara M (1990) Abstracts of 3th international Symposium on Superconductivity (ISS'90) L164.
5. Otsuka K, Akeyoshi, Kwasaki and Sugahara M,(1989) ISEC'89 Tokyo Japan: L119.
6. Huang Q X, Yoshikawa N and Sugahara M, (1989) Proc. of FED- HiTcSc-ED 3th Workshop, Kumamoto, Japan, L102
7. Seto J, Van Duzer(1971) Appl. Phys. Lett. 19,L488.

Experimental Study of HTS FET Device Based on Superconducting Phase Transition

XIAOYI HAN, JIANFEI JIANG, and MASANORI SUGAHARA

Faculty of Engineering, Yokohama National University, Yokohama, 240 Japan

ABSTRACT

An experimental study is made on HTS FET device based on superconducting phase transition. The sample of the three-terminal device is made of BSCCO thin film formed on (100) MgO substrate with $SrTiO_3$ gate-insulating layer. Appreciable conduction modulation is caused by the field effect with the modulation of superconducting phase transition at helium temperature.

KEY WORDS: BSCCO superconductor, FET effect, phase transition,three-terminal device

INTRODUCTION

There have been many proposals and trials concerning superconducting FET with low Tc superconductor(LTS) channel.[1-3] Since the discovery of high Tc superconductors(HTS), this material with lower carrier density has been expected to show higher field effect than that of LTS. However, the field effect of HTS devices observed in experiment is much smaller than the expected level.[4-5] Some field effect was observed in MISFET structure using YBCO channel.[6-7] In this report we represent an experiment on MISFET structure with a channel of BSCCO where appreciable field effect is observed. The superconductivity of HTS is known to have fairly strong carrier-density dependence. We made a calculation of the feasibility of FET device based on the dependence.[8] The observed effect is discussed with use of the theory.

DEVICE STRUCTURE AND PREPARATION

The sample structure and the measurement circuit are shown in Fig. 1. BSCCO film is deposited on well polished (100) MgO single crystal substrate at 780°C by using rf magnetron sputtering method under 8 Pa of Ar atmosphare. After that in situ annealing is made under 2 Torr of oxygen atmosphare for 10 minutes. $SrTiO_3$ film and Al film are evaporated on the BSCCO film as insulator layer and gate electrode respectively. The thickness of BSCCO film is 1500 Å and the $SrTiO_3$ film is 3500Å. The gate area is 1mm × 1mm. Four-probe method is used in measurement to prevent the influence of the contact resistance. $SrTiO_3$ is chosen as the gate insulator because of its good insulation property with high dielectric constant, especially at low temperature.

EXPERIMENT

The leak characteristic between the gate and channel at room temperature is shown in Fig. 2. It is seen that even at the gate voltage V_{GS}=20 V the leak current I_{GS} is no more than 0.3 nA. The insulation becames better at 4.2 K, when no leak curren was observed even at V_{GS}=50 V. This means that the leak current could be negligible. Channel current-voltage $I_{DS} - V_{DS}$ characteristics at room temperature and at 4.2 K are shown in Fig. 3. Calculated from $I_{DS} - V_{DS}$ characteristic, the channel resistances are 880 Ω and about 60 Ω at room temperature and at 4.2 K respectively. The influence of leak curren is so small even at room temperature that V_{GS}= 20 V causes channel current change corresponding to V_{DS}=0.3nV. Figure 4 shows the R-T characteristic of channel. It is clear that the channel does not show zero resistivity down to 4.2 K.

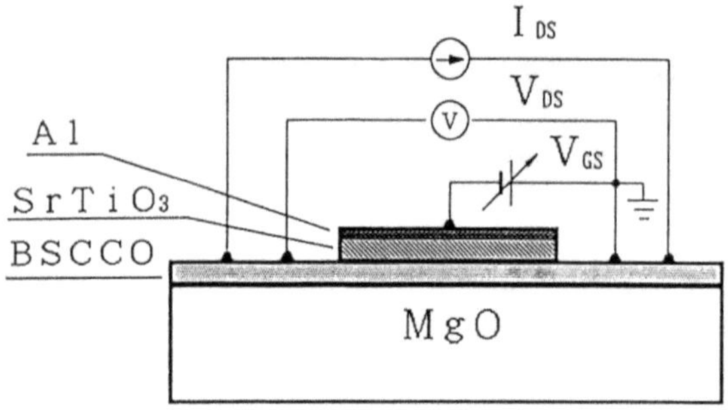

Fig. 1 MISFET structure (cross section) and measurement circuit

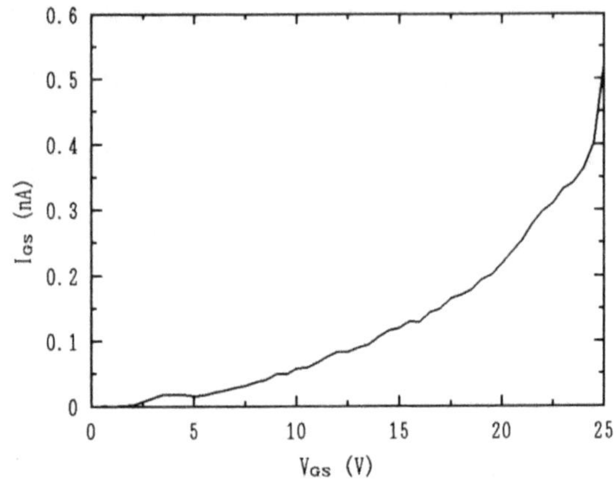

Fig. 2 Gate-channel $V_{GS} - I_{GS}$ leak characteristic

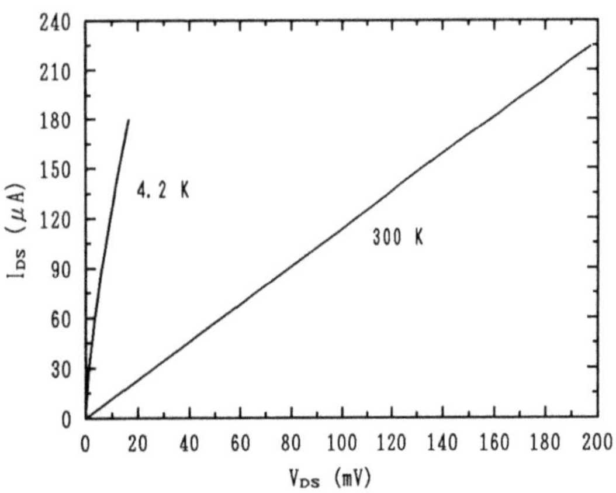

Fig. 3 $I_{DS} - V_{DS}$ characteristic at 300 K and 4.2 K

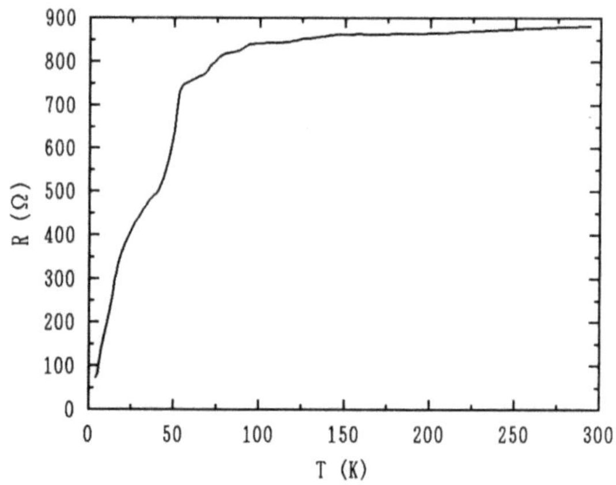

Fig. 4 Channel R-T characteristic

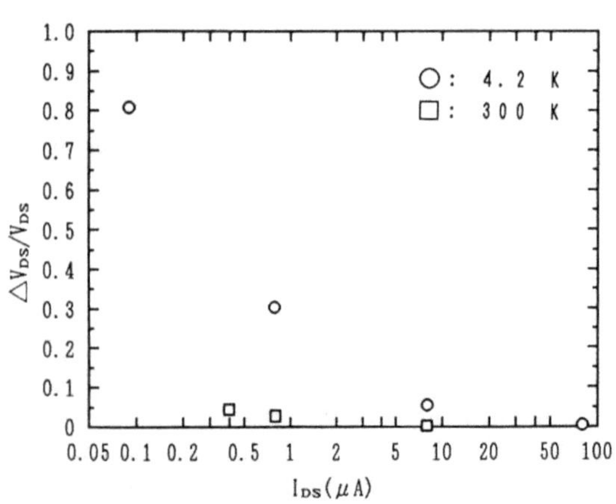

Fig. 5 I_{DS} dependence of $\Delta V_{DS}/V_{DS}(V_{GS}=0)$ where $\Delta V_{DS} = V_{DS}(V_{GS}=-20V) - V_{DS}(V_{GS}=0V)$

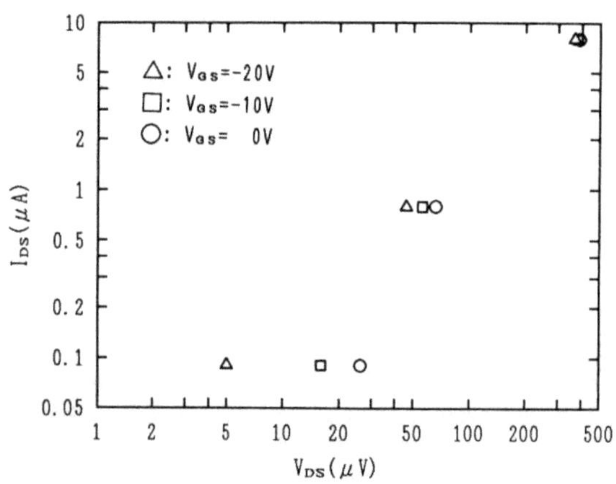

Fig. 6 The relation of $I_{DS} - V_{DS}$ at 4.2 K with V_{GS} as parameter

The carrier in BSCCO is considered to be hole. When a positive V_{GS} is applied, the hole density on channel will decrease. Contrary, when a negative V_{GS} is applied, the density will increase. In the experiment, when a constant current I_{DS} is applied to the channel, the change of carrier density should cause the change of V_{DS}. Figure 5 shows the change of V_{DS} caused by an applied V_{GS} at room temperature and at 4.2 k. These data of V_{DS} are normalized to $V_{DS}(V_{GS}=0)$. Figure 6 is displayed the $I_{DS} - V_{DS}$ relation for the various V_{GS}. On the other hand the V_{GS} dependence of σ_s-I_{DS} relations at 4.2 K is also obtained where σ_s is channel surface conductivity(fig. 7). Obviously, appreciable field effect is observed.

As reported before[8], if the field induced carrier density reaches the level large enough to cause superconductivity, the channel of carrier deficient HTS will change from nonsuperconducting phase into superconducting phase. The carrier density increase in the channel occurs at negative gate voltage. Such a transition should make vanishing V_{DS} at finite I_{DS}. We observed very sharp resistance drop at low temperature and at 4.2 K with the vanishing V_{DS}. Figure 8 shows the $V_{DS} - V_{GS}$ characteristic at 4.2 K at I_{DS} of 0.09μA. The vanishing V_{DS} at V_{GS} < -23 V means that the superconducting transition is occurring or at least about to occurring by the field effect..

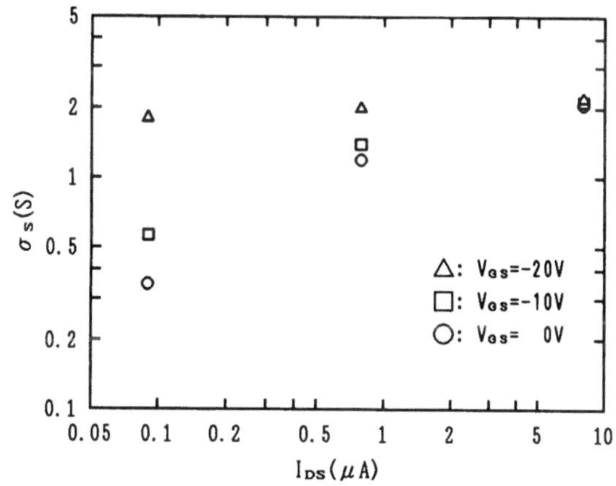

Fig. 7 The dependence of channel surface conductivity σ_S on I_{DS} at 4.2 K

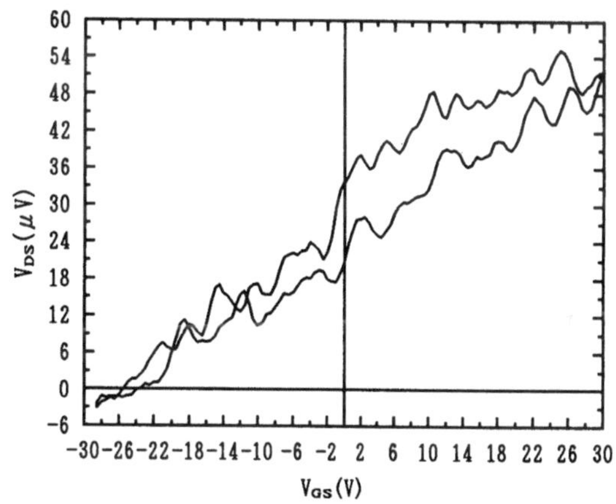

Fig. 8 $V_{DS} - V_{GS}$ characteristic when I_{DS}=0.09 μA at 4.2 K

DISCUSSION AND CONCLUSION

We made an experimental study on the HTS FET superconducting phase transition, where the change of the channel conductance is directly controlled by the variation of the field induced carrier density of the BSCCO channel.

Our experiment results show the followings:
1. Appreciable field effect modulation of the channel carrier density is observed at 4.2 K and at room temperature.
2. The field induced superconducting phase transition is observed at low temperature.
3. The high dielectric constant of the $SrTiO_3$ of the gate insulating layer is considered to be effective for channel-carrier increase enough to make the superconducting transition.

REFERENCES

1. T.D.Clark, R.J.Prance and A.D.C. Grassie (1980) J. Appl. Phys.51: 2736
2. T.Nishio, M.Miyake, Y.Harada and U.kawabe (1985) IEEE Trans. Elect. Device Lett. EDL-6: 297
3. H.Takayanagi and T.Kawakami(1985) Tech. Digest IEDM 85, Washington D.C., p.98
4. U.Kabasawa, K.Asano, T Kobayashi (1990) Japan J. Appl. Phys. 29: L86
5. A.T.Fiory, A.F.Hebard, R.H.Eick, P.M. Mankievich, R.E. Howard and M.L. O'Malley (1990) Phys. Rev. Lett. 65: 3441

6. J.Mannhart, J.G.Bednozr, K.A. Muller and D.G. Schlom (1991) Z. Phys. B- Cond. Matt. 83: 307
7. T.Kobayashi, K.Sakuta, K.Asano, S.Arisawa and Y.Sakaguchi (1991) Extended Abstract of ISEC'91, Glasgow 1991, p.436
8. J.F.Jiang, N.Yoshikawa, X.Y.Han and M.Sugahara (1991) Extended Absteact of ISEC'91, Glasgow 1991, p.420

AC Impedance of Thin-Film Bi-Sr-Ca-Cu-O

P.E. PHELAN[1], T. OHMORI[2], K. HIJIKATA[1], and M. IHARA[3]

[1] Department of Mechanical Engineering Science, Tokyo Institute of Technology, Meguro-ku, Tokyo, 152 Japan
[2] Ishikawajima-Harima Heavy Industries Co., Ltd., Yokohama, 235 Japan
[3] Fujitsu Laboratories Ltd., Atsugi, Kanagawa, 243-01 Japan

ABSTRACT

In high-T_c superconducting electronics, the superconducting thin film can be subjected to a large alternating current (AC) rather than to a direct current. This paper reports the results of experiments in which the AC impedance (z) of a thin-film Bi-Sr-Ca-Cu-O bridge is measured using a four-point technique. A function generator supplies the input voltage for frequencies up to 1 MHz; and the voltages are measured with an oscilloscope. In this way, we are able to determine both the amplitude and phase of z for current densities as high as 5×10^5 A cm^{-2}. The experimental results indicate z becomes nonzero for frequencies greater than or equal to about 500 kHz. The nonlinear dependence of z on the current density can distort the shape of the voltage drop profile. Finally, the contact impedance between the Au contact pads and the Bi-Sr-Ca-Cu-O film is found to be mostly resistive at these frequencies.

KEY WORDS: Bi-Sr-Ca-Cu-O, impedance, alternating current, high-temperature superconductors

INTRODUCTION

In many electronic devices based on high-temperature superconductors (HTS), the thin-film HTS are invariably subjected to an alternating (AC) or pulsed current, making it necessary to investigate the AC specific impedance (z), not only to establish the operating parameters and characteristics of the devices, but also to determine the AC loss power for thermal design of the electronic package.

In previous investigations of the AC impedance of Y-Ba-Cu-O, the authors measured the normal and superconducting resistivity (ρ) [1], and the normal-state conductivity (σ_n) [2], but only for relatively low effective (rms) current densities (J). Behrooz and Zettl did not give J [2], but Miguelez et al. specified that $J \leq 25$ A cm^{-2} [1] . These studies concluded that there is no appreciable frequency (ν) dependence of σ_n for $\nu \leq 1$ GHz [2], nor of ρ for $\nu \leq 120$ kHz [1]. In other words, σ_n and ρ were found to be equal to their direct-current (DC) values. In a similar investigation of Bi-Sr-Ca-Cu-O (BSCCO) thin films, however, Ho et al. found a significant difference between the superconducting DC and AC complex conductivity for frequencies near 60 GHz [3]. As before, though, J was relatively small.

In contrast to the earlier studies, in this paper we report some qualitative results for *large* J of limited experimental data of z and its phase φ, of thin-film BSCCO ($Bi_1Sr_{.8}Ca_1Cu_{3.7}O_x$). An additional result of our investigation is an estimation of the specific contact impedance magnitude ($|z_c|$) and its phase (φ_c). The rms current density is varied from 1.1×10^5 to 4.7×10^5 A cm^{-2}, the temperature (T) range is 4 K $\leq T \leq 86$ K, and the frequency range is $0 \leq \nu \leq 1$ MHz. The experiments are performed using a straightforward 4-point method. It is found that in the superconducting state, z becomes nonzero for $\nu \geq 500$ kHz.

EXPERIMENT

The experimental sample is an epitaxial film of BSCCO deposited on an MgO substrate by chemical vapor deposition, and was supplied through the courtesy of Fujitsu Laboratories. Details of the film deposition process are given elsewhere [4]. The critical temperature is about 75 K, the film thickness is 160 nm, and the bridge dimensions are 50 μm x 50 μm. Electrical contacts to the film were made through circular vacuum-deposited Au electrodes of 150 μm diameter.

A schematic diagram of the experimental circuit is given in Fig. 1. A function generator supplied the AC voltage, and the output voltages V^+, V^-, and V_i, where V^+ is the positive-end voltage, V^- is the negative-end voltage, and V_i is the voltage across a shunt resistor used to determine the sample current, were measured consecutively with a two-channel oscilloscope.

The voltage drop along the bridge could not be directly measured because of the 'leakage' current from the cables to their grounded shields. In order to extract the actual impedance of the BSCCO bridge from the measured voltages, the experimental circuit of Fig. 1 was modelled as the equivalent circuit shown in Fig. 2. The currents I^+ and I^- are the leakage currents between the positive and negative voltage cables and their respective shields. The capacitances C^+, C^-, C_s, and C_i are the capacitances between the shields and the various cables. The resistance R_s is the known shunt resistance. The voltages V_H and V_L are the actual voltages just upstream and downstream of the bridge, while Z_i and Z_v represent the impedances (in Ω) of the current and voltage taps, and are given by the following expressions:

$$Z_i = aZ + Z_c \quad , \quad Z_v = bZ + Z_c \tag{1a,b}$$

where a and b are constants that are determined by the geometry of the current and voltage taps, Z is the bridge impedance (in Ω), and Z_c is the contact impedance (in Ω) between the Au pads and the BSCCO film. Here, the values of a and b are $a = 0.5$ and $b = 15$.

Two crucial assumptions are inherent in Eqs. (1a,b). The first is that Z_c is assumed to be the same at all four taps, which is reasonable given that the room temperature DC contact impedances are roughly equivalent. The second is that the dependence of Z on the current density is neglected in Eq. (1b), since the voltage taps consist of long, narrow (20 μm) strips that in general will experience a current density somewhat smaller than that of the bridge itself. This assumption, however, does not alter our qualitative conclusions, since taking the extreme case where $b = 0$ tends to increase Z and Z_c.

The remaining equations governing the circuit of Fig. 2, and whose solution leads to Z, φ, Z_c, and φ_c, are given below:

$$V_H - V_L - (I - I^+)Z = 0 \tag{2}$$
$$V_H - j\omega C^+V^+Z_v = V^+ \tag{3}$$
$$V_L - j\omega C^-V^-Z_v = V^- \tag{4}$$
$$V_L - Z_i(I - I^+ - I^-) = 0 \tag{5}$$

where the currents are given by

$$I = \frac{V_i(1 + j\omega C_s R_s)}{R_s} \quad , \quad I^+ = j\omega C^+V^+ \quad , \quad I^- = j\omega C^-V^- \tag{6a,b,c}$$

and the angular frequency $\omega = 2\pi\nu$ and $j = \sqrt{-1}$. Using the known values of R_s and of the capacitances, which range between 850 and 1200 pF, the desired impedances can be directly determined from the measured V^+, V^-, and V_i by solving Eqs. (1) - (5) simultaneously.

The experimental uncertainty is relatively high. That, coupled with the scarcity of our data, allows only qualitative conclusions regarding the dependence of $|z|$ and φ on ν, J, and T. The estimated maximum uncertainties in the measured voltage amplitude and phase of ±0.01 V and ±5°, respectively, lead in some cases to estimated maximum uncertainties in $|z|$ and φ as high as ±80% and ±45°,

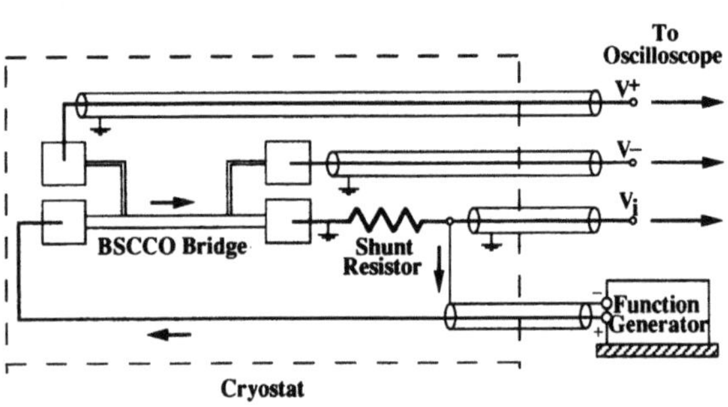

Fig. 1 Schematic diagram of the experimental circuit.

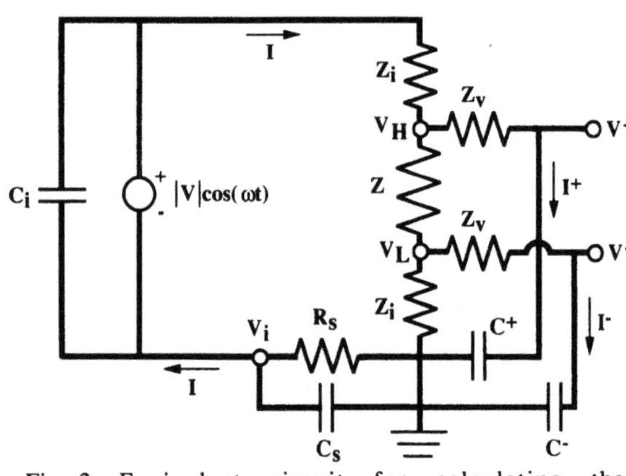

Fig. 2 Equivalent circuit for calculating the film and contact impedances.

respectively. For $|J|$ and $|z_c|$, however, the measurements are within $\pm1\%$ and $\pm20\%$, respectively. These values, in fact, do not take into account any uncertainty inherent in the measurement of the cables' capacitances. Still, although assuming a worst-case scenario where the capacitance of each cable is varied between half and twice its measured value leads to substantial quantitative changes, our qualitative conclusions remain the same. Finally, the uncertainty in the temperature measurement is estimated to be less than ±1 K.

EXPERIMENTAL RESULTS AND DISCUSSION

Table 1 contains all the experimental results. One normal-state point, at 86 K and 100 kHz, is included and is marked with an asterisk. At both 34 K and 4 K, the DC impedance for $J \le 6.25 \times 10^5$ A cm^{-2} is zero. The DC critical current density (J_c) at 4 K and 34 K is approximately 5×10^6 A cm^{-2} and 3.3×10^6 A cm^{-2}, respectively [5].

The most striking result concerning the film impedance $|z|$ is that below 500 kHz, $|z| = 0$ within our measurement sensitivity and for the given range of T and J. Furthermore, the single normal-state datum for $|z|$ (2.28 Ω µm) is nearly an order of magnitude larger than those in the superconducting state. The phase angle, φ, for this point is also much smaller than for the others, indicating a greater amount of resistive dissipation as compared to the superconducting state.

In every case where a finite $|z|$ is determined, φ is found to be positive, which indicates the presence of both resistive and inductive components. If the superconducting bridge is modelled as a resistor connected in parallel with an inductor, the equivalent resistance ranges from 5 to 43 Ω, and the equivalent inductance ranges from 0.4 to 1.2 µH.

A plot of the impedance magnitude $|z|$ as a function of rms current density $|J|$, normalized by $J_c(T)$, is given in Fig. 3. Only the nonzero, superconducting-state data points are shown, and the uncertainties of the measurements are included as error bars. Figure 3 reveals two qualitative trends in the data: $|z|$ increases with decreasing $|J|/J_c$, and $|z|$ increases with increasing ν.

The basis of the finite values for $|z|$ is assumed to be the movement of vortices resulting from the Lorentz force F_l from the transport current density and its self-field. An estimate of $|z|$ based on an approximate form of the Mattis-Bardeen theory of the frequency-dependent conductivity [6], normally applied for far-infrared frequencies, yields a value for $|z|$ some six orders of magnitude less than the experimental results in Table 1. Thus, $|z| > 0$ must derive from some mechanism other than the inertia of the charge carriers or pair breaking.

It seems reasonable to suppose that the film, for this range of J, is in the mixed state. One can not, however, be certain of this, since the lower critical field H_{c1} has not been measured for this particular film. A very rough estimate for $B_{c1} = \mu_o H_{c1}$, where μ_o is the magnetic permeability of free space, is 0.001 T [7,8]. The self-fields of the current densities in Table 1 are the same magnitude or slightly less, so the mixed state is a plausible assumption. Taking this to be true, there should be no measurable

Table 1 Values of the bridge (z,φ) and contact (z_c,$φ_c$) impedance magnitudes and angles.

ν (Hz)	T (K)	$\|J\|$ (x 10^5 A cm^{-2})	$\|z\|$ (Ω µm)	φ (deg)	$\|z_c\|$ (x 10^{-3} Ω cm^2)	$φ_c$ (deg)
1 k	4	1.12	0	—	4.1	0
1 k	34	1.26	0	—	3.0	0
10 k	4	1.12	0	—	4.0	-1
10 k	34	1.26	0	—	2.8	-1
100 k	4	1.14	0	—	6.6	3
100 k	34	1.27	0	—	3.0	2
100 k	86*	1.33	2.28	5	10.2	8
500 k	4	1.51	0.59	74	3.9	-6
500 k	34	1.47	0.43	81	2.7	2
1 M	4	1.07	0.95	82	2.6	-10
1 M	4	2.34	0.61	81	2.4	-4
1 M	34	1.97	0.37	63	2.6	6
1 M	35	4.68	0.45	75	2.0	-14

* normal state

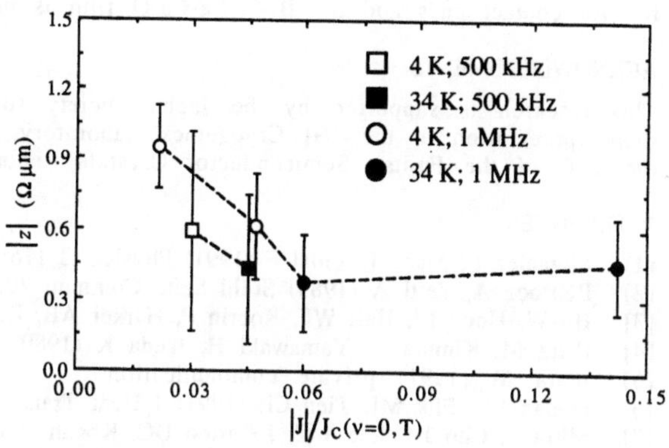

Fig. 3 Impedance vs. current density. The curves are drawn to aid the eye.

NORMAL
T = 86 K
$|J| = 1.33 \times 10^5$ A cm^{-2}

SUPERCONDUCTING
T = 34 K
$|J| = 1.27 \times 10^5$ A cm^{-2}

Fig. 4 Comparison of the voltage drop $V^+ - V^-$ profiles for the normal (left) and superconducting (right) states. The frequency is 100 kHz.

impedance until F_l exceeds the pinning force. After the vortices have begun moving, z is expected to be proportional to the vortex speed and hence to the frequency ν [9], as our limited experimental data indicate in Fig. 3. Thus, the 'critical frequency' of 500 kHz described in the discussion of Table 1 perhaps represents the frequency at which there is sufficient vortex movement to generate a measurable impedance. Without more data, it is difficult to elaborate further, especially with regard to the dependence of $|z|$ on $|J|/J_c$, where presumably $|z|$ would tend to near zero (for these frequencies) as J is decreased to the point where no vortices are present in the superconductor.

An indication of the nonlinear nature of z in the superconducting state is provided in Fig. 4, in which the voltage drop $V^+ - V^-$ (Figs. 1 and 2) is compared for normal and superconducting temperatures. The normal-state curve is a smooth sinusoid like the imput voltage, but the superconducting-state curve shows significant distortion. The appearance of the two profiles suggests that nonlinearities are present in the superconducting state, but not in the normal state.

The last result to be discussed is the specific contact impedance z_c between the Au pads and the BSCCO film (Table 1). Apparently, z_c is primarily resistive. The uncertainty in φ_c is about ±10°, and so we are unable to determine accurately if z_c is slightly inductive, or slightly capacitive. As for $|z_c|$, the values shown in Table 1 are at least two orders of magnitude higher than expected [10]. The cause for this abnormally high contact resistance is not known. Finally, the data of $|z_c|$ do not display any significant trend other than that the normal-state impedance is higher than the rest.

CONCLUSIONS

Limited experimental data of the AC impedance of a superconducting Bi-Sr-Ca-Cu-O bridge indicate that for current densities as high as one tenth the DC critical current density, the impedance becomes nonzero for frequencies of 500 kHz and above. The impedance magnitude appears to increase with both decreasing normalized current density, and with increasing frequency. The phase angle of the impedance is positive, indicating the superconducting bridge has both resistive and inductive components. Nonlinearities present in the superconducting state, but not in the normal state, distort the shape of the voltage drop trace from a pure sinusoid. Finally, the AC contact impedance between the Au contact pads and the Bi-Sr-Ca-Cu-O film is primarily resistive.

ACKNOWLEDGMENTS

This research is supported by the Japan Society for the Promotion of Science. The authors express their appreciation to the IHI Cryogenics Laboratory, especially Y. Hasegawa and A. Yamanishi, and to the staff of the Fujitsu Semiconductor Crystals Laboratory.

REFERENCES

[1] Miguelez F, Maza J, Vidal F (1991) Physica C 176: 49-56
[2] Behrooz A, Zettl A (1989) Solid State Commun 70: 1059-1063
[3] Ho W, Hood PJ, Hall WF, Kobrin P, Harker AB, DeWames RE (1988) Phys Rev B 38: 7029-7032
[4] Ihara M, Kimura T, Yamawaki H, Ikeda K (1989) IEEE Trans Magn 25: 2470-2473
[5] Ihara M (1991) private communication
[6] Phelan PE, Flik MI, Tien CL (1991) J Heat Trans 113: 487-493
[7] Mitra S, Cho JH, Lee WC, Johnston DC, Kogan VG (1989) Phys Rev B 40: 2674-2677
[8] Worthington TK, Gallagher WJ, Dinger TR, (1987) Phys Rev Lett 59: 1160-1163
[9] Matsushita T (1990) Cryogenics 30: 314-323
[10] Goodrich LF, Bray SL (1990) Cryogenics 30: 667-677

Dependence of the Inductance Characteristics of Bi System on the External Magnetic Field

KATSUYOSHI HOTTA, YOSHIRO SUGIYAMA, HIDEKAZU MAGOME, TOSHIMASA SUZUKI, and HARUO HIROSE

Nippon Institute of Technology, Miyashiro, Minami-Saitama, 345 Japan

ABSTRACT

We made the experiment on the samples, each of which is with two coils wound tightly on a bar type core of the superconductor, Bi system, and examined them on self-inductance, mutual-inductance, coupling coefficient and frequency dependence at the liquid nitrogen temperature under external magnetic field. As the result, the inductance L increases non-linearly with increasing external magnetic field H. The relation between L and log H is expressed by a linear line for a certain range. A similar result is obtained for the mutual-inductance with changing coupling coefficient.

KEY WORD: Self-inductance, Mutual-inductance, Coupling coefficient, External magnetic field, Fluxoid

INTRODUCTION

Since the high-Tc superconductor was found, many researches have been done[1-3], and the application to various fields is expected. It is well known that the second type superconductor begins to introduce the magnetic flux as the form of fluxoid into the body when put into higher magnetic field than the critical magnetic field H_{c1}, thereafter, the superconductor is in the so called mixed state until the higher critical magnetic field H_{c2} that defines that the whole body is out of superconductive state[4]. We report the results of the relevant experiment and discussion on the various types of samples. One is with two solenoidal coils wound tightly on a bar (cylinder) type core superconductor of Bi system and the second one is made of coils with the core in the center part of the coil plane, and the third one is with coil wound tightly on a hollow cylinder core of Bi system. They were examined on self-inductance [5], mutual-inductance[6], coupling coefficient and frequency dependence at the liquid nitrogen temperature under external dc magnetic field.

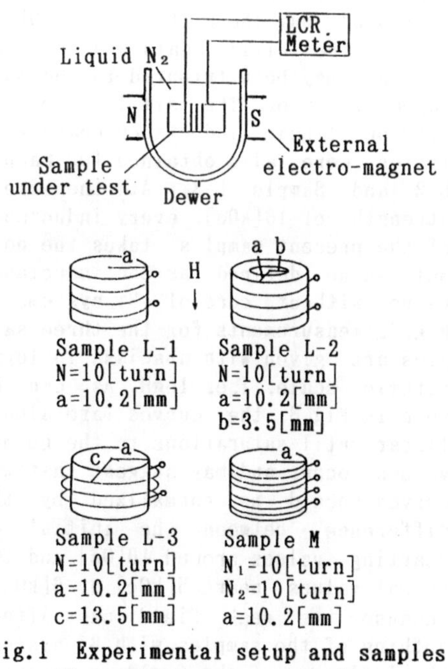

Fig.1 Experimental setup and samples for self-inductance and mutual-inductance measurement.

EXPERIMENTAL

For the samples in this experiment, the starting materials are high purity (more than 99.9[%]) powders of Bi_2O_3, PbO, $SrCO_3$, $CaCO_3$, CuO. These powders were weighed to the composition of $Bi_{0.8}Pb_{0.2}SrCaCu_{1.8}O_y$ and mixed for 3 hours in a mortar for solid phase reaction method. The mixed powder was pressed into a form of bar a (24[mm] in diameter, 4[mm] thick) at the pressure 49[MPa]. The result was semi-sintered in the atmosphere at the temperature 800[℃] for 12 hours, then pulverized and filtered through 325 mesh. This is again formed into bar (10.2[mm] in diameter, 4.8[mm] thick) at the pressing pressure 196[MPa].The bar samples were sintered at 843[℃] for 50 hours. The hollow cylinder core is then shaped by making a hole in the center of the core. These core samples were measured for resistance-temperature characteristics by dc four terminal method. The samples with Tc=94[K] are used for inductor core material. Figure 1 illustrates the measurement method for self-inductance and the various types of samples where the used wires for coil are the covered wire of diameter 0.2[mm]. External dc magnetic field is applied to the sample by an electro-magnet. Inductance is measured by an LCR meter. Frequency for the measurement is 10[kHz]. Mutual-inductance was obtained as $M=(L_a-L_b)/4$, where L_a is the inductance of two coils in summing connection and L_b is that of two coils in differential connection, so that coupling coefficient is obtained by $K=M/(L_1 \cdot L_2)^{1/2}$. The magnetic field was measured by a gauss meter.

RESULTS AND DISCUSSION

Figure 2 shows typical result of self-inductance vs. external magnetic field measurement at the liquid nitrogen temperature for Sample L-1. It was noted that the self-inductance L increases with the increasing magnetic field, and for the range from 0[Oe] to 20[Oe] the change is very rapid however it is slow for the relatively larger field. The measurements were so obtained that first (initial magnetization), the magnetic field strength was raised from 0[Oe] to 16[kOe], then reduced to 0[Oe], second, the field was again raised up to 16[kOe] to be reduced to 0[Oe]. This kind of measurement was to clarify the hysteresis behavior of the core materials of Bi system[3]. But the difference >100[Oe] between the initial data and the second ones may be attributed to the measurement error. The similar results with non-linear inductance characteristics were also obtained for Sample L-2 and Sample L-3. At the field strength of 16[kOe], every inductance of the present samples takes the normal value defined as the inductance value without core of Bi system. In Fig.3, measurements for the three samples are given with abscissa in logarithmic scale, i.e. logH. As can be seen in Fig.3, the curves are almost linear until saturations to the normal values occur and may suggest that the curves should be normalized by the difference between the initial or starting values around 10[Oe] and the normal values near 5[kOe] or 6[kOe]. Because the weak field must affect nothing of the samples with Bi system, the influence of the field is considered to begin at the starting points and end at the normal value points in

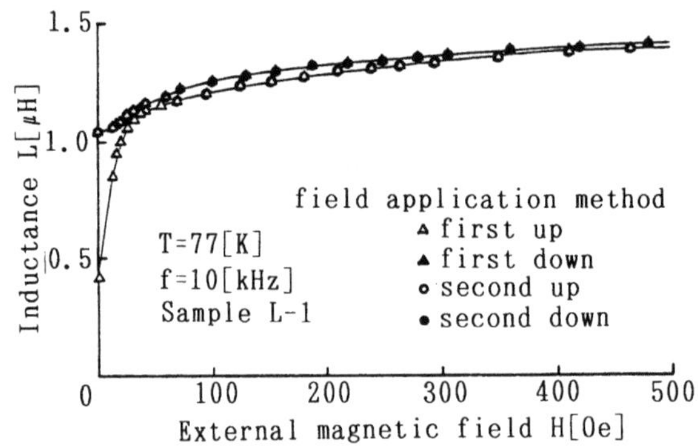

Fig.2 Self-inductance vs. external dc magnetic field characteristics for Sample L-1.

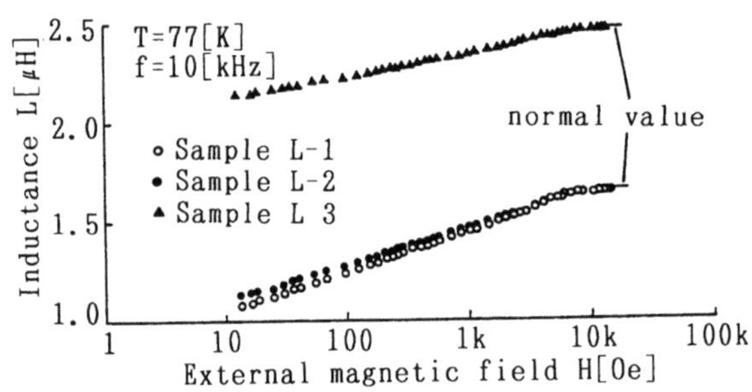

Fig.3 Self-inductance vs. external dc magnetic field characteristics with logarithmic scale.

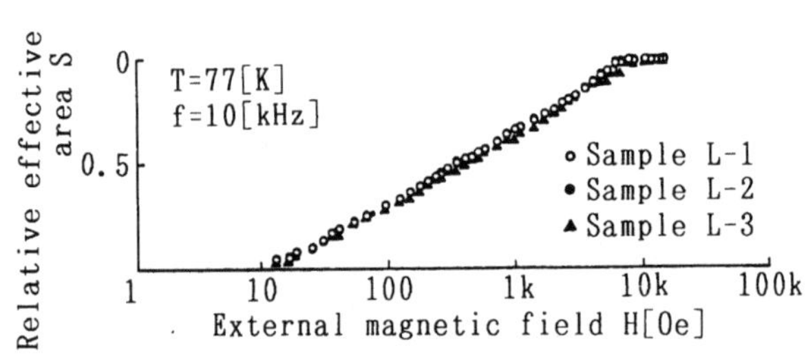

Fig.4 Relative effective area S vs. external dc magnetic field characteristics.

the figure. The starting point value of inductance is taken as the leakage inductance that the core of superconductor can not affect while the final value or the normal value means that the influence of the core is no more effective at the field strength. Figure 4 illustrates the same data as given in Fig.2 in a redrawn form for percentage expression. As is expected the curves seem to converge into a line and may be expressed by only one line for a certain range of the external dc magnetic field strength. It is noted here that the superconductor used for the core material of the inductors is non-magnetic at room temperature or at the normal state. So that the external magnetic field could transform parts of the core into $\mu=1$ regions by invading into the core body to change the state. Thus the result given in Fig.4 is reasonable. Figure 5 shows mutual-inductance and derived coupling coefficient of Sample M accompanied by the original measurements of inductance in a summing connection of two coils L_a and in a differential connection L_b. Each of the curves is in a good agreement with Fig.4 except for the coupling coefficient K. K exceeds the normal value over an narrow range of field strength. This means that some magnetic corridors guiding the flux by the inductor's first coil to the second coil might exist in the core body. However the measuring method using the induced voltage across the second coil could not confirm the existence of the corridors. So the derivation of K is possibly accompanied by the error because of the calculation among the original measurements with low accuracy. It is very difficult to discuss the result by certain calculations because the situation is non-linear and so complicated although simple inductors are treated. But the experimental formula for the dependence of inductance on the external dc magnetic field will be obtainable by modeling the percentage expression. Our interpretation for the behavior of inductance toward external magnetic field is that the effective cross section of inductor coil is reduced by the superconductor core cross section at the superconductive state so that the inductance will be reduced accordingly. The well-known be-

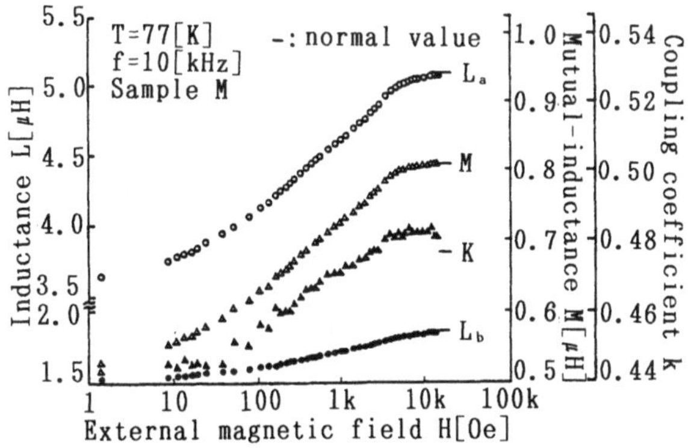

Fig.5 Mutual-inductance M and coupling coefficient K vs. external dc magnetic field for Sample L-1.

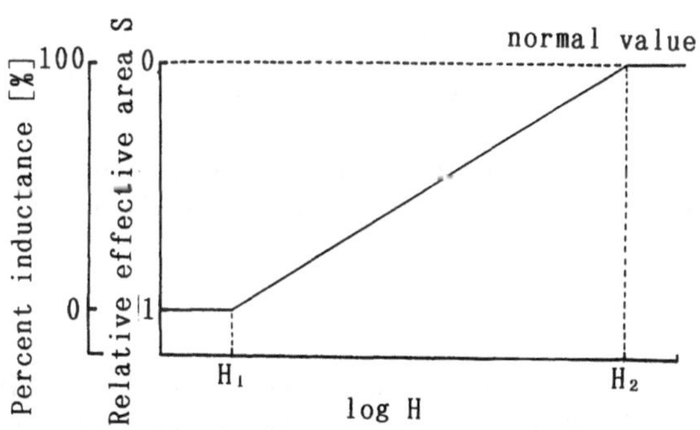

Fig.6 Modeling experimental data.

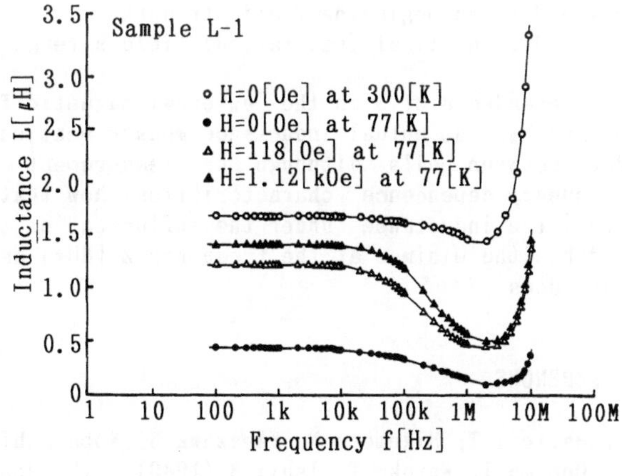

Fig.7 Frequency dependence of self-inductance.

havior of flux density of the second type superconductor is given as the susceptibility change toward magnetic field strength. In reality, the pinning sites will play a important role and their potential distribution may result in the logarithmic dependence of inductance or effective cross section on the external magnetic field. Therefore it is assumed here that the external magnetic field with certain strength generates the non-superconductive regions in the core which become flux passes giving the same susceptibility as the vacuum while the pinning sites that have persistent potential to the external force will limit the widening of the normal state regions. On the assumption that this situation can be interpreted solely by the cross sectional change of inductor coil (for instance, Sample L-1), the relative area S is given as,

$$S=(\log H/H_2)/(\log H_1/H_2)$$

where parameters are given in Fig.6 that represents the model.

The field strength H_2 may characterize the pinning potential while H_1 will be related to the beginning of fluxoid invasion. It should be noted that the reversibility is obtained for the certain range of external magnetic field. These characteristic fields for the inductors will give us indices for designing variable inductor using superconductor. For the mutual-inductance measurement, if it could exceed the normal value although the accuracy does not deserve to confirm, some magnetic corridors can exist. And the existence of the corridors may be very probable. Here our measurement can not stand for the existence but it will explain how the non-superconductive regions develop in the body of the superconductor. However, more accurate measurements and ingenious approaches are necessary to discuss more in details. Figure 7 shows frequency dependence of the self-inductance for Sample L-1. Every case, the inductance is not dependent on frequencies from 100[Hz] to 5[kHz]. But at 2[MHz], the minimum inductances appear, followed by rapid increases caused by parasitic resonances. For these behaviors, no detailed discussions have been done so far because only the role of measuring frequency is concerned.

CONCLUSION

We made the relevant experiments on the samples with solenoidal coil(s) wound tightly or loosely on the various cores of the superconductor of Bi system and examined them on self-inductance, mutual-inductance and coupling coefficient at liquid nitrogen temperature and at room temperature under external dc magnetic fields. By assuming that the effective coil cross sectional area S in percent is varied by the external field H, the result were represented by the experimental formula as,

$$S=100(\log H/H_2)/(\log H_1/H_2)\ [\%]$$

where H_1: the beginning field strength
H_2: the final (saturation) field strength.

The dependence of S on the external magnetic field is reversible for the range between H_1 and H_2. The mutual-inductance measurement suggests that there can be magnetic corridors between coils although our measurement accuracy can not deserve to confirm. The frequency dependence characteristics show that the frequency range from 100[kHz] to 5 [kHz] the inductance under the influence of the external magnetic field keeps constant and has the minimum at the frequency 2 [MHz] as that without the field at room temperature does.

REFERENCES

1. Hattori T, Higemoto N, Kanazawa S, Kobayashi M (1988) Jpn.J.Appl.Phys.27:L1120-L1122
2. Ohnuma T, Kuroko T, Ishii M (1989) J.IEE.Jpn. MAG 89: 59-67
3. Uchiyama T, Mohri K, Ozeki A, Shibata T (1990) J.Mag.Soc.Jpn. 14: 509-514
4. F.C.Moon (1990) Appl.Electromagnetics in Materials 1: 29-35
5. Hotta K, Sugiyama Y, Suzuki T, Hirose H (1990) J.Mag.Soc.Jpn. 14: 337
6. Hotta K, Sugiyama Y, Suzuki T, Hirose H (1990) J.IEE.Jpn. 3: 175

Characteristics of Josephson Junctions Using the Cracks in YBCO Thin Films

HIROSHI KIMURA[1], MASAYUKI MIYAZAKI[1], KOICHI TSUDA[2], and YOICHI OKABE[1]

[1] RCAST., The University of Tokyo, Meguro-ku, Tokyo, 153 Japan
[2] Fuji Electric Corp. R&D Ltd., Yokosuka, Kanagawa, 240-01 Japan

ABSTRACT

We studied (013) oriented YBCO thin films on $SrTiO_3$ (110) substrates. Although the surfaces of the films were smooth enough, there were parallel and straight cracks in the film with thickness of more than 150 nm because of the difference of the thermal expansion between YBCO and substrate. We utilized these cracks to fabricate Josephson junction devices. In order to control properties of these cracks, the cracks were produced by changing thickness of the films. The observed voltage of the first Shapiro step was about three times as large as the theoretical value, because there were three junctions in series in the device.

KEY WORDS: cracks, YBCO thin film, crack junction, thermal expansion

INTRODUCTION

In order to fabricate Josephson device using high critical temperature superconductors, the YBCO thin films should satisfy the following requirements. First, the surface of the YBCO films must be excellently smooth, because very thin insulator layer should be deposited on the films. Second, non-c-axis orientation is preferable, because it is difficult to fabricate Josephson junction using c-axis oriented film owing to the shorter coherence length of these surfaces.

We have reported that (013) oriented film with excellently smooth surface was obtained using $SrTiO_3$ (110) substrates by the sputtering method at a low working gas pressure.[1] In this paper, we will give further information about the (013) oriented YBCO thin films on $SrTiO_3$ (110) substrates and report that new type Josephson devices were fabricated using these films.

EXPERIMENTAL

The YBCO thin films were deposited on $SrTiO_3$ (110) substrates by rf magnetron sputtering method under the conditions as shown in Table 1.[2] After the deposition, the films were cooled from 650°C to room temperature in an oxygen gas pressure of 1 atm. Structural information of the films was obtained by X-ray diffraction (XRD). Surface morphology of the films, including their crack morphology, was observed by scanning electron microscope (SEM). Resistance-temperature (R-T) characteristics and current-voltage (I-V) characteristics were measured using the standard four-probe technique.

Table 1. Sputtering conditions.

Substrates	$SrTiO_3$(110) 15×15×0.5 mm
Substrate Temperature	650℃
Total Gas Pressure	0.5Pa($Ar:O_2$=1:1)
Target	Y:Ba:Cu=1:2:4
Target-Substrate Distance	28mm
Applied Power	300W
Deposition rate	10nm/min
Deposition time	7-35min

RESULTS AND DISCUSSION

Crack formation in (013) oriented YBCO thin film

XRD showed that the film had (013) orientation. Figure 1 shows a scanning electron micrograph of the film. Although the surface of the film was excellently smooth, the parallel and straight cracks which propagated along the (001) planes of YBCO were observed in the film. E. Olsson et al.[3] reported that similar cracks were formed in the film, however, their crack spacing was different from ours. These cracks were formed during cooling, since the difference in the coefficient of the thermal expansion between the film and the substrate induces a tension along the c axis. On the other hand, no crack was observed in the film where the surface was rough or the thickness was less than 150 nm, because the thermal stress of these films were not enough to form the cracks.

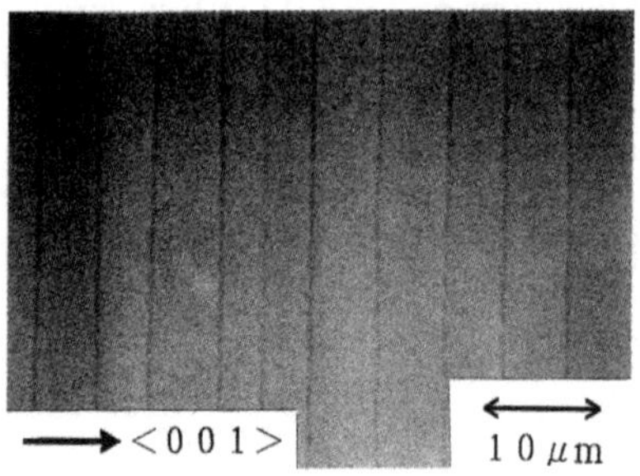

Fig. 1. SEM photograph of the film on $SrTiO_3$ (110) substrate. The arrow indicates the direction of c-axis of YBCO.

Figure 2 shows the dependence of resistance on the thickness of the film at 250K. Above 150 nm, the resistance of the film increased along with the thickness. This result suggests that the cracks ran into the film and the number of cracks increased with the film thickness, because the amount of thermal stress increases. Critical temperature (Tc) decreased and ΔTc increased with thickness of the film. However, the value of Tc onset did not change as shown in Fig. 3. These results show that Tc of the film did not depend on the thickness and that the part under the crack in the film formed a weak-link junction.

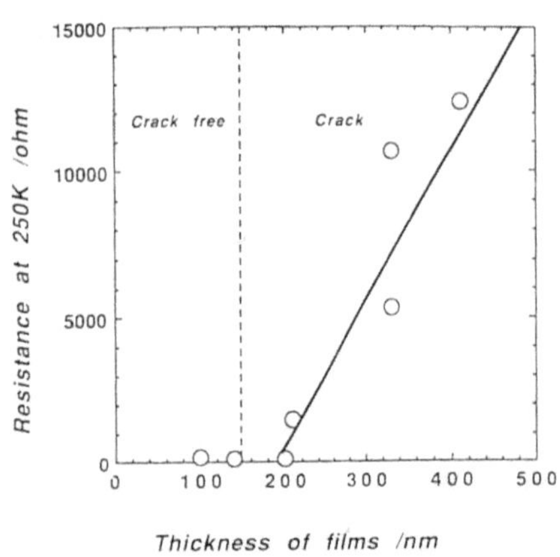

Fig. 2. Dependence of the resistance on the thickness of the film at temperature 250K. No crack was observed in the films with thickness of less than 150 nm by SEM.

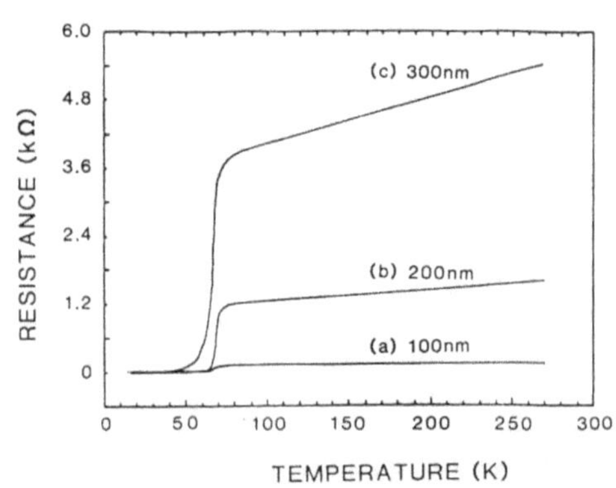

Fig. 3. R-T characteristics of the films with thickness of (a)100nm, (b)200nm and (c)300nm.

Josephson junction device using the cracks

We utilized these cracks to fabricate Josephson junction devices. In order to control properties of these cracks, the cracks were produced by changing thickness of the film. Therefore the following films were used, i.e., (a) 100nm thick YBCO film where no crack was observed by SEM, (b) 200nm thick film and (c) 300 nm thick film. Figure 4 shows a schematic illustration of the crack Josephson junction device fabricated in our experiments. Current path was perpendicular to the crack direction which is parallel to the (001) plane of YBCO.

I-V characteristics under microwave irradiation of these devices were measured at 4.2K. The critical current (Ic) of the device(a) was very large compared with two other devices and no microwave response was observed. When the 8.35GHz microwave radiation was applied to the device(b), the critical current of it was changed from 2.0mA to 1.4mA. However, no Shapiro step was observed. Ic value of the device(c) was 240 μA and this value was smaller than that of device(b). The device(c) exhibited particularly strong microwave response and Shapiro steps were observed under 8.35 GHz microwave irradiation. The observed voltage of the first step was about 15 μV. This value considerably corresponded to the theoretical value (17 μV) calculated using the following equation for a single Josephson junction.

$$V = hf/2e$$

Because Shapiro steps were not clear and the only first step was observed at 4.2K, we have tried to improve these characteristics by the following method. The amorphous YBCO film (100nm) was deposited in situ on the cracked YBCO film (300nm) to prevent the surface of the crack from deterioration by the air as shown in Fig. 5.

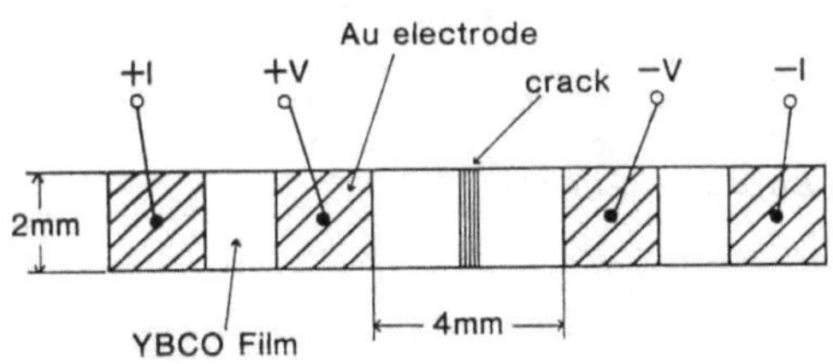

Fig. 4. Schematic illustration of the crack Josephson junction device. Current path was perpendicular to the crack direction.

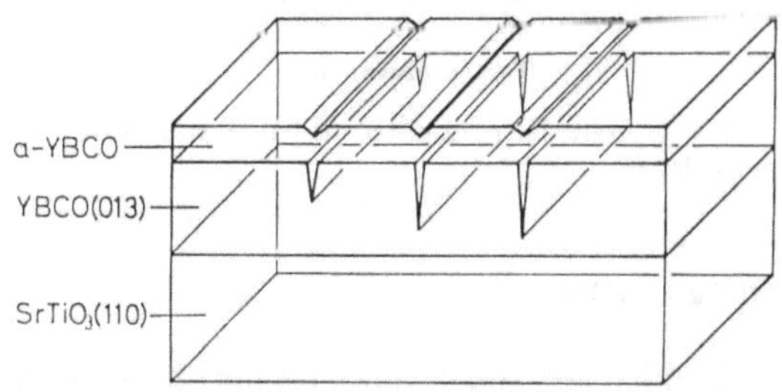

Fig. 5. Schematic illustration of the double layer films on SrTiO3substrate. Amorphous YBCO film prevents YBCO film and the surface of the cracks from deterioration by the air.

As a result, Shapiro steps became clear and the second step was observed at 4.2K. The first Shapiro step was, however, 50 μV at 22.6K under 8.26GHz microwave irradiation as shown in Fig. 6. This value is about three times as large as the theoretical value of 17 μV. Therefore this device has three series Josephson junctions whose characteristics are almost the same. Similar phenomenon can be observed for metal superconductor Josephson devices.

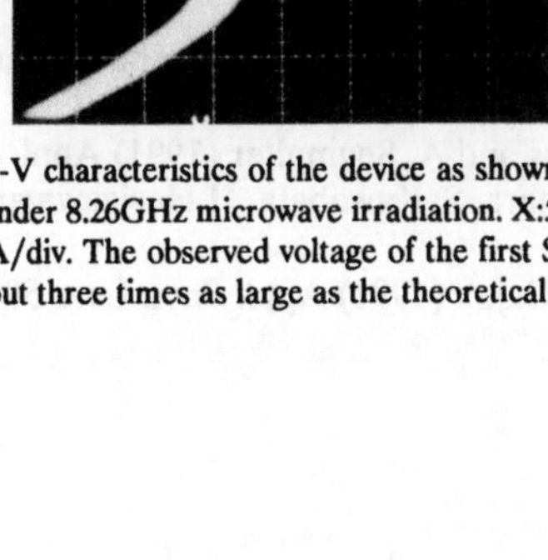

Fig. 6. I-V characteristics of the device as shown in Fig.5 at 22.6K under 8.26GHz microwave irradiation. X:50 μV/div. Y:40 μA/div. The observed voltage of the first Shapiro step was about three times as large as the theoretical value.

Although there were several junctions in the grain boundary type device[4] or the step-edge type device[5] using YBCO films, these junctions behaved just like as "one" junction. The difference of these devices from ours was the junction spacing, since there are three junctions for a space of 4 mm in our device as shown in Fig. 4. On the other hand, for example, the distance between two junctions in the grain boundary type device is a few μm which depends on the grain size. The characteristics of the series junctions in our device is probably caused by the long distance between junctions compared with grain boundary devices.

The temperature dependence of the critical current for the device is shown in Fig. 7. A linear relationship appeared in a considerably wide temperature range of about 10-40K. The Ic value deviated from a linear relationship near Tc. Figure 8 shows the temperature dependence of the critical current for this device near Tc. The critical current Ic value of the device near Tc changed in proportion to $(1\text{-}T/Tc)^{1.5}$. The relationship agrees with Ginzburg-Landau theory for microbridges. Hence, the device fabricated using the cracks shows typical microbridge junction characteristics.

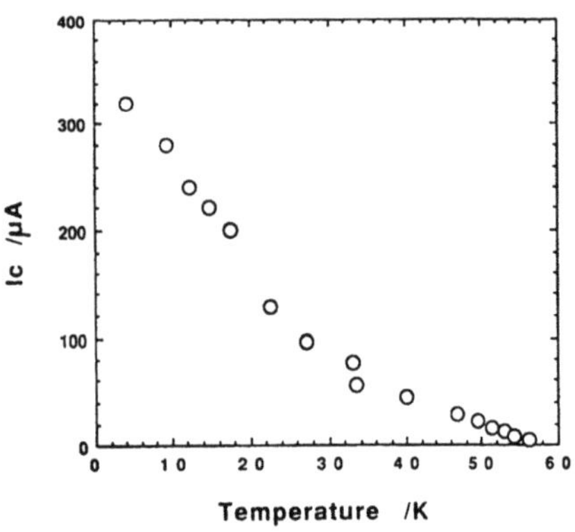

Fig. 7. Temperature dependence of the critical current (Ic) for the device.

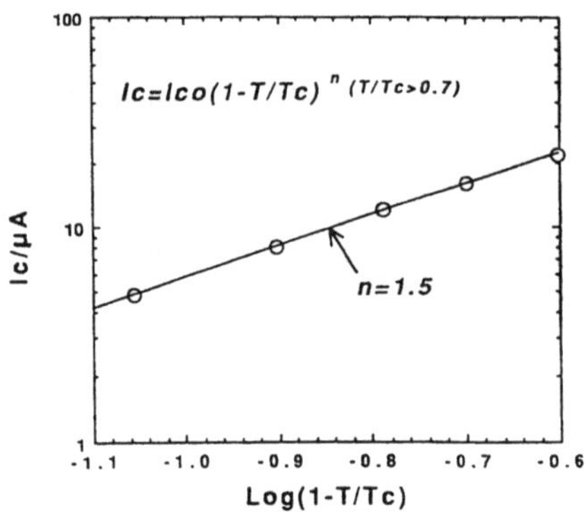

Fig. 8. Temperature dependence of the critical current (Ic) for the device near Tc. Ic's value of this device near Tc changed in proportion to $(1\text{-}T/Tc)^{1.5}$.

CONCLUSION

In conclusion, we have obtained the films with excellently smooth surface on the SrTiO3 substrate, however, there were the cracks in the film with thickness of more than 150 nm because of the difference in the coefficient of the thermal expansion between YBCO and substrate. Josephson junction devices were fabricated using these cracks. In order to control properties of these cracks, the cracks were produced by changing thickness of the film. These devices can be considered as typical microbridge junctions. The observed voltage of the first Shapiro step was about three times as large as the theoretical value. It suggests that there were three junctions in series in these devices.

ACKNOWLEDGEMENTS

Authors would like to thank to Mr. K. Mukae, T. Ishii, T. Matsui and T. Suzuki in Fuji Electric Corp. R&D Ltd. for their useful advice.

REFERENCES

1. K. Tsuda, Y. Okamura, T. Suzuki, T. Matsui, M. Nagano and K. Mukae (1990) Proceedings of the 3rd ISS'90, (Sendai) : 881
2. M. Muroi, Y. Okamura, T. Suzuki, K. Tsuda, M. Nagano and K. Mukae (1990) Jpn. J. Appl. Phys. 29 : 69
3. E.Olsson, A.Gupta, M. D. Thouless and A. Segmuller (1991) Appl. Phys. Lett. 58 :1682
4. T. Shinohara, H. Enami, N. Kawahara, S. Kawabata, H.Hosizawa and T. Imura (1991) Jpn. J. Appl. Phys. 30 : 929
5. B. D. Hunt, M. C. Foote and L. J. Bajuk (1991) Appl. Phys. Lett. 59 :982

Single Layer HTS-DC-SQUIDs with Large Flux-To-Voltage Conversion at 77K

G. FRIEDL[1,2], M. VILDIC[1,3], M. RÖMHELD[1], B. ROAS[1], D. UHL[1], F. BÖMMEL[1], G.M. DAALMANS[1], B. HILLENBRAND[1], and H.E. HOENIG[1]

[1] SIEMENS Research Laboratory, Erlangen, Germany
[2] Also University Augsburg
[3] Also University Erlangen

ABSTRACT

Single layer YBCO-DC-SQUIDs have been prepared on $SrTiO_3$- and $LaAlO_3$-substrates. The Josephson junctions are of grain boundary type grown on a step in the substrate. The YBaCuO-layers, deposited by laser ablation, were patterned by wet etching as well as by an inhibit structuring process. The junctions are about 5 µm wide and the SQUID-hole is a square of about 5 µm. The SQUID-inductance is estimated to be about 20 pH. Very regular flux-to-voltage conversion has been obtained over many flux quanta with an amplitude in the order of 10 µV pp. The equivalent flux noise of good devices as measured in the flux locked loop mode is well below $1*10^{-4}$ Fluxquanta/$Hz^{1/2}$ at 1 Hz.

KEY WORDS : YBCO superconductor, DC-SQUID, single layer, 77 K operation, low 1/f noise

INTRODUCTION

Shortly after the discovery of high temperature superconductivity by Bednorz and Müller the phase coherence in SQUID-structures made of bulk material and polycrystalline thin films was demonstrated [1], using the grain boundaries between crystallites as weak links. Noise measurements on these SQUIDs revealed huge amounts of noise, especially at the lower frequencies. Ferrari et al. [2] showed that rings made of polycristalline material produce noise due to flux motion in the superconducting structures. It was expected therefore to obtain better noise performance for SQUID-structures made of quasi-epitaxial thin films. However, in order to make use of such films, artificial weak links had to be

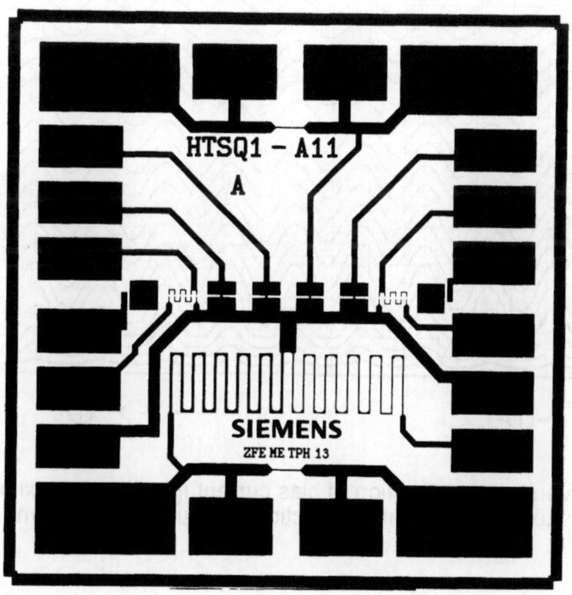

Fig.1 Design of the SQUID-chip containing DC-SQUID structures, weak links and jc-bars

prepared. The introduction of well defined grain boundaries in YBCO thin films, deposited on bicrystalline substrates [3], allowed to prepare weak links in such quasi-epitaxial thin films. SQUIDs made by this preparation technique showed excellent performance in the white noise regime [4]. For the lower frequencies, however, a high level of 1/f noise still existed.

In this paper it is demonstrated that the 1/f noise level can be reduced substantially by introducing artificial grain boundary weak links grown over steps in the substrate.

SQUID- AND CHIP-DESIGN

The SQUID-chip is designed for substrates of 10 by 10 mm and contains DC-SQUID structures, weak links structures and jc-bars. Figure 1 shows the arrangement of the different structures. The 500 μm long and 20 μm wide jc-bars, positioned at both sides of the step in the substrate, allow to determine the superconducting properties of the superconducting film. The 5 μm wide weak link structures are used to determine the critical current of the weak links, with and without magnetic field applied. The SQUID-ring is composed of two 5 μm wide weak links in between two massive superconducting banks, enclosing a SQUID-hole of about 5 μm by 5 μm. The superconducting banks of 500 μm by 250 μm serve as flux focussing areas.

PREPARATION TECHNIQUES

A step, nominally 0.3-0.4 μm deep, is etched into the substrate of $SrTiO_3$ or $LaAlO_3$ by ion milling. The masking material is photo resist (AZ 5218). The step angle is prepared larger than 60 deg in order to provide optimum conditions for the growth of a double GB at the step [5]. After removing the photoresist by a solvent and an oxygen plasma process, either YBCO is deposited directly onto the substrate or a photoresist film is layed down and patterned for preparing inhibit- or reaction-patterning. In the case, the YBCO has been deposited first, the patterning is performed by photolithography in combination with wet etching in dilute H_3PO_4 or a saturated solution of EDTA in water. In the case of reaction patterning, a 100 nm thick SiO_x-film is used as masking material. The YBCO is deposited by a laser ablation technique [6]. The thickness of the YBCO-films is adapted to the step height and is typically one half of the step height. The contact pads to the devices are covered by a 300 nm thick gold layer, deposited by e-beam evaporation through a photoresist mask. An ion beam etching step in between assures a good electrical contact. The devices are contacted with aluminum wire by ultrasonic wedge bonding.

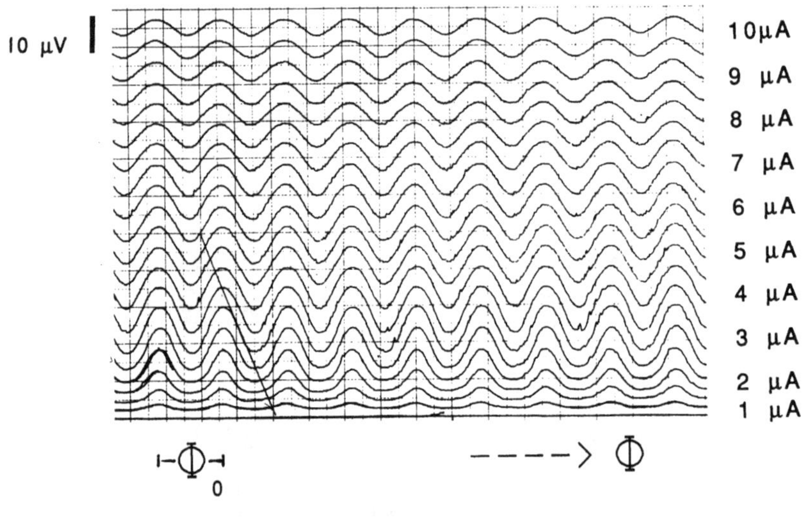

Fig.2 Flux-to-voltage conversion as a function of bias current I_B at 77 K. Maximum voltage amplitude is 15 μV, corresponding to a flux to voltage transferfunction of 50 μV/Fluxquantum.

MEASUREMENTS

The measurements are restricted to films with critical current densities well above $10^6 A/cm^2$ at 77 K and transition temperatures of about 90 K with a transition width of about 1 K. The critical current density of the weak links at 77 K ranges typically between 10^3 A/cm^2 and 10^4 A/cm^2, resulting in critical currents of a few μA to a few hundred μA. The Josephson penetration depth lies in between 3 μm and 10 μm for the current densities mentioned above. The 5 μm wide weak links should be interpreted as being small if scaled on the Josephson penetration depth. The I-V curves are RSJ-like up to voltages of at least 100 μV with substantial noise rounding at the critical current especially for lower values of the critical current. Cooling down to a temperature of 4 K results in much higher critical currents and flux flow type I-V characteristics with excess supercurrents.

The I-V curves of the SQUIDs are of RSJ-type at 77 K for current densities in between 10^3 A/cm^2 and 10^4 A/cm^2 as expected from the results for the single junctions. The I_cR-product ranges between 10 μV and 50 μV at 77 K. The amplitude of the voltage modulation due to the flux penetration into the SQUID-ring is typically 1/2 of the I_cR-product for bias currents of about the critical current. The best value obtained so far for the voltage modulation is about 15 μV and is shown in Figure 2. It corresponds to a value of the flux to voltage transfer function dV/dΦ of 50 μV/Flux quantum. For most of the SQUIDs a very regular flux-to-voltage characteristic has been obtained for several hundreds of flux quanta.

Noise measurements are performed with the cryostat placed in a Faraday cage. Metal cryostats as well as fiberglass cryostats are used. The cryostat is inserted into a μ-metal can at room temperature, then it is filled with liquid nitrogen and the SQUID is immersed. The SQUID is surrounded by a niobium screen, non superconducting at 77 K. It is operated in the well known flux-locked-loop circuit. The impedance transformation from the low ohmic SQUID to the high ohmic preamplifier is realized by means of a cooled LC-resonance circuit at 145 kHz. The voltage output of the loop is analyzed with a fast Fourier spectrum analyzer. In the white noise regime usually an equivalent flux noise level below $5*10^{-5}$ Fluxquanta/Hz$^{1/2}$ has been obtained corresponding to a noise temperature smaller than 10 times the bath temperature. The voltage noise in the white noise regime usually is in between 0.1 nV/√Hz and 1 nV/√Hz requiring an impedance transformation to the amplifier input. The rise of the noise level at lower frequencies starts below 10 Hz for good devices measured under appropriate conditions.

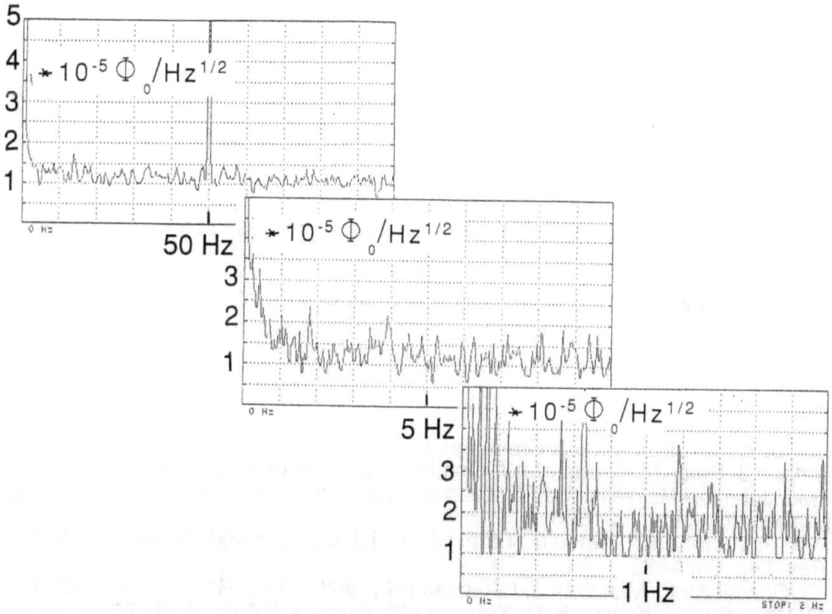

Fig.3 Equivalent flux noise at 77 K of the SQUID as characterized in Fig.2 as a function of frequency in different resolutions. The SQUID is operated in the flux-locked-loop mode.

The noise spectrum of the best device measured so far is shown in Figure 3. The bias current is near the critical current where dV/dΦ is at maximum. The flux to voltage characteristics for this device are given in Figure 2. The device shows an equivalent flux noise of $1*10^{-5}$ $\Phi_0/\sqrt{Hz}$ down to about 5 Hz, corresponding to an energy resolution of $1*10^{-29}$J/Hz. The 1/f-noise component is very low resulting in an energy resolution of $2*10^{-29}$J/Hz at 1 Hz.

DISCUSSION

Compared to other DC-SQUID noise data published in the literature, we obtained a very low 1/f-noise level. However RF SQUIDs prepared with similar weak links [7] show about the same 1/f-noise level. We don't know why devices with step edge junctions behave different. At least two sources of noise, the flux motion of depinned flux and the noise in the weak links, play an important role in existing devices. For instance, we could easily increase the 1/f-noise level of our devices by cooling down in higher magnetic fields. From these experiments we learned that it is rather easy to obtain devices that are noise limited in a frequency range up to 500 Hz by the motion of depinned flux. The white noise level of our best devices is not far away from the thermal limit. The scaling factor between the noise temperature and the bath temperature is comparable to what has been obtained for excellent niobium SQUIDs at 4 K [8] with a minimum value of about 4. Finally we like to mention that our best SQUID has already a better performance at 77 K than a commercial RF-SQUID at 4 K (Figure 4).

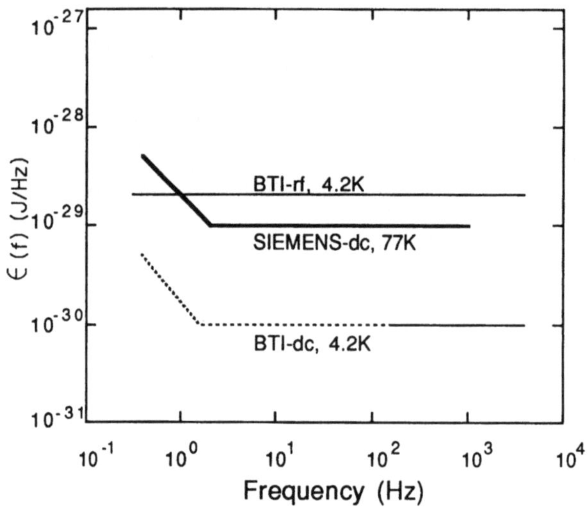

Fig.4 Energy sensitivity of the best 77 K-SQUID compared to commercial 4 K-SQUIDs.

The question to be answered in the future will be, whether the SQUID-performance at 77 K will be comparable to the performance of a commercial DC-SQUID at 4 K or not.

REFERENCES

[1] Koch RH, Umbach CP, Clark GJ, Chaudhari P, Laibowitz RB (1987), Appl. Phys. Lett. 51: 200
Zimmermann J, Beall JA, Cromar MW, Ono RH (1987), Appl. Phys. Lett. 51: 617
[2] Ferrari MJ, Johnson M, Wellstood FC, Clarke J, Rosenthal PA, Hammond RH, Beasley MR (1988), Appl. Phys. Lett. 53: 695
[3] Chaudhari P, Mannhart J, Dimos D, Tsuei DD, Chi J, Oprysko MM, Scheuermann M (1988) Phys. Rev. Lett. 60: 1653
Dimos D, Chaudhari P, Mannhart J, LeGoues FK (1988), Phys. Rev. Lett. 61: 1653
[4] Gross R, Chaudhari P, Kawasaki M, Ketchen MB, Gupta A (1991), IEEE Trans. on Magnetics, MAG-27: 2565
[5] Friedl G, Roas B, Schultz L, Hillenbrand B (June 1991), Proceedings ISEC 3, 127, Strathclyde
Herrmann K, Zhang Y, Cui GJ, Schubert J, Zander W, Braginski AI (June 1991) Proceedings ISEC 3: 357 (Strathclyde)
[6] Roas B, Schultz L, Endres G (1989) J. of Less Com. Metals 151: 413
[7] Zhang Y, Mück HM, Herrmann K, Braginski A, Heiden C, submitted to Appl. Phys. Lett.
[8] Daalmans GM, Bär L, Bömmel FR, Kreß R, Uhl D (1991), IEEE Trans. on Magn. MAG-27: 2997

Work supported by the German BMFT.

8 System Application

8.1 Application for Electric Power System

Present State of Alternating Current Superconductors and Its Applications to Power System Apparatus in France

JEAN-LOUIS SABRIÉ

GEC ALSTHOM, F 90018 Belfort, France

ABSTRACT

The purpose of the today French effort in AC superconductivity is to check whether commercial frequency SC equipment could be competitive. The method to that aim consists of 4 steps:

- to develop ACSC wires and cables
- to demonstrate the feasibility of equipments in the laboratory
- to test demonstration machines in operation, and finally
- to test prototypes in service.

Present state in France is as follows:

- ACSC wires are industrially produced but wire assemblies are not yet satisfactory.
- 20 kVA fully SC synchronous generator continues tests,
- 200 kVA single phase SC transformer has been tested in Marcoussis,
- 25 kV 200/11000 A fault current limiter has been tested in Lyon.

Evaluations of all these ACSC equipments and small SMES are under way in order to determine aims, costs and duration of the demonstration phases.

Finally, it is likely that demonstrations will take place by the year 2000 especially for fault current limiters, small SMES then synchronous machine.

Commercial equipments could come at the beginning of next century.

INTRODUCTION

Today, the only existing market for superconducting equipments is in direct current, the main segments of this being:

- High Energy Physics (particle accelerator and detectors)
- Thermo-nuclear Fusion (Tokamaks, Heliotrons)
- Magnetic Resonance Imaging for medical application.

In 1983 appeared in France the first Alternating Current SuperConducting wire (AC SC wire). This could result in the penetration of superconductivity in the market of AC industrial equipment, much larger than the DC machines market. From this time, studies and experiments dealing with ACSC wires and applications started in parallel with continuing works on DC wires and applications.

At present time only NbTi wires and cables can be considered for applications to the power system apparatus (Table I) so, all the equipments described in this paper use NbTi.

TABLE I

Main Superconducting Materials

CRITERIA (cf text)	MATERIALS			
	NbTi	A15 Nb_3Sn	Chevrel Phases	HTS
Tc (K)	9	18	15	126
Bc (T)	14	25	55	> 100
Jc(A/mm²)	3 000(5T)	1 500(10T)	200(20T)	10(0T)
Ac	+	-	?	?
Q	+	-	?	-
W	+	-	--	---

AC = possibility to operate on commercial frequency
Q = possibility to operate in Quench condition
W = workability at factory.

The purpose of the French works in ACSC is to determine whether commercial frequency SC apparatus could be competitive. The way chosen consists of four steps:

- to develop ACSC wires and cables
- to demonstrate the feasibility of the machines in the laboratory
- to test demonstration machines in operation
- to test prototypes designed for specific service.

This paper gives the status of this activity for France.

STATUS OF ACSC wires and cables

The status of French works in this field can be expressed under three titles:
- wires - cables - Quench protection.

- ACSC wires

Since 1983 the development of new ACSC strands has been guided by the following considerations. According to analysis of AC loss power generation (1), losses may be split into two contributions: the first one is due to the hysteresis of the superconducting material magnetization, the second contribution is due to eddy currents in the matrix.

The hysteretic contribution is lowered by reducing the filament diameter but while reducing the filament diameter one has to keep spacing between filaments larger than a few times the coherence length in the matrix metal.

When proximity effects are avoided, by a proper choice of the filament spacing, hysteretic losses of ultrafine filaments in low magnetic induction are actually lower than expected from the simple proportionality to diameter because of the reversible flux motion.

In order to lower the eddy current coupling losses, a 30% Ni content coppernickel matrix is used. The twist pitch is set to five times the strand diameter.

The present French conclusion, with respect to overall critical current density, proximity effects and reversible flux motion is that the filament diameter should be around 0.15 μm, with about the same spacing (2), whatever the strand diameter and number of filaments are.

Electromagnetic stability studies (3,4) give the upper limit of strand diameter along with the number of filaments. It mainly depends on the magnetic induction experienced by the superconductor in its application. As a consequence, for industrial scale manufacture, three different strands (5), referred to as "L", "T" and "S", have been developed (Table II).

TABLE II

S, T, L wires

	Diameter of strand (mm)	Number of filaments	Diameter of filaments (n.m.)	Diameter of enamelled strand polyesterimide type (mm)	Weight Kg/Km	Twisting pitch (mm)	% Cu	% CuNi	% SC	Field of Application (approx)
S	0.3	920 304	136	0.349	0.59	1.5	1.3	79.6	19.1	1T
T	0.25	597 102	136	0.297	0.41	1.25	2.3	79.1	18.6	0.5T
L	0.2	377 982	136	0.244	0.26	1.0	0	81.3	18.7	self-field

As far as AC losses are concerned, the calculation depends strongly on the coils geometry. Formulas and values are available in the literature. Measured values are given for well defined test coils.

- ACSC cables

ACSC strand cannot generally hold high currents needed for application and it is necessary to assemble it in various kinds of cables.

Up to now the strand insulation is satisfactory only for 6+1 and $(6+1)^2$ assemblies. The behaviour of such assemblies is studied on test coils such as the following (Table III).

TABLE III

Test coils for ACSC cables

Conductor type	6+1	$(6+1)^2$	$(6+1)^2$
basic strand	"L"	"L"	"S"
Overall diameter	2.16	2.16	3.15
Inner diameter (mm)	26.4	92.9	92.9
Outer diameter (mm)	48.6	127.4	134.4
Winding height (mm)	90.0	90.0	90.0
Number of layers	6	4	4
Turns per layer	124	47	33
Number of turns	744	188	132
Load line (mT/A)	9.7	1.70	1.15
Self-inductance(mH)	5.8	2.5	1.22
50 Hz impedance (Ω)	1.82	0.79	0.38

Performances of these coils are given in (6). Tests continue on various types of coils to precise the behaviour of ACSC cables.

- DC and AC quench currents
- Commercial frequency loss power
- AC loss power versus frequency

- ACSC winding quench protection

Even though the stored energy of ACSC windings is much lower than the one of DC coils, their behaviour during quench is more critical because of their high resistivity matrix and high current density. The French works are oriented towards two directions for the protection of ACSC coils using cables and not monolith conductors (7):

a) . active protection involves an ultra-rapid quench detection. It is based on the measurement of the current passing through the central resistive wire, and/or of unbalanced currents in the different superconducting wires. About 20 milliseconds after detection, a fast circuit-breaker switches off the current.

b) . a complementary passive protection is provided by the resistance developing during normal phase propagation. This behaviour naturally induces a certain amount of current reduction.

Active protection needs a detection time lower than half a period since in AC the current can be cut in a few half periods.

There are many methods of detection, which present different quicknesses and reliabilities. The simplest is probably to use again the central insulated wire in the case of 6+1 and $(6+1)^2$ cables.

Due to quasi-perfect magnetic coupling, the current flows in the different wires of the strand, proportionally to their conductances.

Thus, the current in the cental wire is very near zero in regular conditions, and rapidly rises when a normal phase grows in the superconducting wires. Comparing this current with a low threshold is a relevant method of quench detection. Tests have confirmed this point of view.

In the most usual mode of protection, a large part of the conductor is made of copper (or aluminum), with a very low resistivity at low temperature ($\sim 10^{-10}\,\Omega$.m). Active protection detects the quench, and deviates part of the stored energy out of the cryostatic zone, for example in a resistance or to the network. However, the speed of current decay is limited by the maximum acceptable inductive voltage of the winding.

An essential parameter for protection is the normal zone propagation velocity: high velocities are favourable, insofar as they aid the current decay, and reduce the levels of localized overheating and stress. The classical mechanism of propagation combines Joule heating in the normal zone, and thermal diffusion along the wire, across the propagation front. The velocity in that case is less than the adiabatic limit, which gives values of 30-100 m/s

In the hypothesis when a large contact resistance does exist between the strands it is clear that during a quench, a large part of the current is carried by the central strand into the normal zone and that the central strand current is zero in the superconducting zone: there is a transfer zone, in which occurs heating that increase quench propagation velocity.

Calculations and tests agree on propagation velocities in excess of 1000 m/s.

When the central strand is insulated but connected to other strands from place to place a sort of "mass" transition may occur in a few milliseconds.

Association of passive and active protection, as suggested, has been experimented to an impregnated coil (4 layers, L = 2.3 mH, conductor length = 70 m). The conductor was made of six strands around an insulated CuNi wire, used for quench detection. Each strand was made of 6 superconducting copper free wires (Ø 0.2 mm), around one insulated CU-CuNi wire, used for mass transition initiation.

Numerous quenches have been performed without degradation (mass transition in 7 ms after partial quench initiation and current switch off after 25 ms).

All these results on strands, cables and winding protection looks encouraging and it seems likely that ACSC applications could be developed in the next decade.

STATUS OF ACSC EQUIPMENTS

French works in this field consist of laboratory experiments, factory tests, design studies and evaluations.

Laboratory experiments

Since 1983, date of the first ACSC wire production, researches have been undertaken at laboratory level, to evaluate the characteristics of ACSC machines. The main experiments have been the following:

- fully superconducting 18 kVA SC synchronous machine (Grenoble 1989) (8)
- superconducting transformer (Marcoussis 1989) (9)
- SC fault current limiter (Belfort, Marcoussis and Lyon 1989) (10)

These experiments have demonstrated the physical feasibility of ACSC equipments. The fully SC synchronous machine continues tests in Grenoble. The fault current limiter was installed in a conventional circuit breaker test facility in the Villeurbanne factory: it limited tens of times under 400 A fault currents which should have exceeded 10 kA. A new fault current limiter is under construction in Belfort.

Factory test

The only factory test in France up to date took place in the Belfort factory in the context of the development of SC field winding generators. A 12 tons SC rotor, 8 m long, 1.06 m in diameter was built and tested. Experiments lasted 5000 hours rotating among which 1500 hours at rated temperature of 4.5 K. Cooling down the winding needed 5 days, rated current was 5000 A. At full speed test with current varying linearly from 0 to 5000 A to 0 have shown that the winding did not quench for variation speed under 25.000 A/s.

The program was terminated by the EDF and French Administration in 1983 due to the energy crisis.

Design Studies

Many design studies have been made.

Fully superconducting generators and motors from 1.1 MW up to 300 MW have been considered from the point of view of weight, volume and efficiency.

Transformer and fault current limiters are considered for rated power from 4 to 330 MVA with rated voltages of 6.6, 20, 60 and 400 kV. A fault current limiter of 20 kV, 1 kA limiting the fault currents around 3300 A has been designed for the EDF.

The most thorough design was done for a 300 MJ, 350 MVA SMES in collaboration with three companies specialised in electric equipment, power electronics and cryogenics. This SMES is specified for two modes of operation as follows (11).

- Mode 1: absorb or supply 10 MW during 10 s, then 300 MVars permanently.
- Mode 2: sinusoïdal modulation of active power at frequency between 0.5 and 2 hertz with stored energy limited between 200 and 300 MJ.

Evaluations

The works mentioned above lead to three conclusions:

a) ACSC equipments are technically feasible
b) In comparison with conventional machines, the main advantages of SCAC machines are strong reductions of:- weight - volume - losses - noise and vibrations - response time.
c) Economical feasibility must be demonstrated.

The proof of economical feasibility requests first the test in operation of demonstration machines in order to precise the performances and the costs of construction, installation and operation (including maintenance and reliability). After such a phase, prototypes devoted to specific uses are requested before the first commercial orders can be made.

It is clear that such proofs request from ten to twenty years and are very expensive according to the types of machines (generators, motors, transformers, fault current limiters, SMES ....). The time for development money back is larger than 10 years and the private companies generally cannot afford it. North America and Europa have no policy for the future of electrical engineering. Japan has such strategy with long-term, well funded programs able to lead to a world monopole for superconducting industrial equipments (Super GM, Maglev,...).

On the other hand the development of ACSC equipment should strongly affect the economy of power systems.

Break even points for generators are estimated 800 MW for SC field winding generator, 400 MW for fully SC generator and 250 MW for fully SC generator with integrated transformer and SMES for speed and voltage regulation.

A second effect is that the use of ACSC equipment, including SMES and fault current limiters should increase the stability margin or the capacity of existing systems.

A third effect is that introduction of ACSC equipments should start a new approach of system design due to a natural low limit of fault currents and improve means to control voltage and frequency specially during transients.

CONCLUSION

Finally the position of French manufacturers and users of electrical equipment can be summarized in four points:

1. ACSC equipments should appear in power systems around the year 2000.
2. The first commercial ACSC power devices would likely use low temperature superconductors.
3. It is worthwhile to continue the developments of SC equipments intended for power systems.
4. In close relation with system studies, manufacturers should perform demonstrations in order to determine operational characteristics, constructions and installation costs, operating costs, maintenance and reliability.

New technology development is long and expensive. For alternating current superconducting power equipment it is essential to establish a close cooperation between research centers, electric utilities and manufacturers as it is done in Japan. Finally more and more people think of international cooperation and it is likely very worthwhile to establish such a large cooperation to define future world system equipments.

ACKNOWLEDGMENTS:

This paper resumes works made by GEC ALSTHOM, ALCATEL ALSTHOM RECHERCHE, CEGELEC, L'AIR LIQUIDE, ELECTRICITE DE FRANCE, CENTRE NATIONAL DE LA RECHERCHE SCIENTIFIQUE, UNIVERSITE DE GRENOBLE, COMMISSARIAT A L'ENERGIE ATOMIQUE, MINISTERES DE LA RECHERCHE ET DE L'INDUSTRIE. HYDRO-QUEBEC AND IREQ of CANADA helped for SMES definition. CRIEPI of JAPAN helped for ACSC coils definition.

REFERENCES

(1) A. Fevrier, "Losses in a twisted multifilamentary SC composite submitted to any space and time variations of the electromagnetic surrounding", Cryogenics, 1983, 23, p. 185.
(2) J. R. Cave, et al. "Reduction of AC losses in ultrafine multifilamentary NbTi wires", IEEE Trans. Mag., vol. 25, n° 2, March 1989, p. 1945.
(3) A. Fevrier, et al. "Thermo-electromagnetic stability of ultrafine multifilamentary superconducting wires for 50/60 Hz use". IEEE Trans. Mag., Vol. 25, n° 2, March 1989, p. 1496.
(4) A. Gueraud, et al. "Thermo-electromagnetic stability of ultrafine multifilamentary superconducting cables for industrial frequency use", Proc. MT 11.
(5) A. Lacaze, et al. "Coils performances of superconducting cables for 50/60 Hz applications", IEEE Trans. Mag., vol. 27, n° 2, Mars 1991, p. 2178.
(6) A. Lacaze, et al. "Coil performances of SC cables for AC applications". MT 12 Proc. St Petersbourg, June 1991.
(7) T. Verhaege, et al. "Protection of SCAC windings", MT 12 Proc., St Petersbourg, June 1991.
(8) P. Tixador, et al. "Electrical tests on a fully SC synchronous machine" IEEE, trans-Mag., vol. 27, n° 2, March 1991, p. 2256.
(9) A. Fevrier "Transformer" in High Technology information for specialists - Applied Superconductor technology: Now and Future, CMC Corp. Lim. publisher, TOKYO, March 1988.
(10)E.Thuries, et al. "Towards the SC fault current limiter" IEEE-PER V11 N4 April 91 p.76.
(11)J. L. Sabrié, G. Ruelle "Electric energy storage technology - SC Magnetic Energy Storage" Proc. CSEE/IEEE Int. Conf. POWER SYSTEM TECHNOLOGY 1991, Beijing, China, September 1991.

Stability of Rotating Superconducting Magnet Under Large Centrifugal Acceleration

TOSHITADA ONISHI, KATSUYUKI KAIHO, NOBORU HIGUCHI, and ITARU ISHII

Electrotechnical Laboratory, Tsukuba, Ibaraki, 305 Japan

ABSTRACT

The stabilities of a rotating magnet were studied under centrifugal accelerations up to 2,000 g (3,600 rpm) using a test rotor. Stability energy margins (minimum energy required for quench), propagating characteristics of normal zone and cooling heat flux were measured, and both steady state and transient stabilities were analysed. It was clarified that both minimum film boiling and maximum nucleate boiling heat fluxes were considerably increased with increasing centrifugal acceleration. The former at 3600 rev min^{-1} (2000 g) was about 7 times that at zero speed, while the latter at 3600 rev min^{-1} was about 3 times that at zero speed. It was shown that both steady state and transient stabilities of the rotating magnet was improved much under large centrifugal acceleration, and that the Maddock's equal area criterion was satisfied at more than around 1000 g, even against an overall current density as large as 120 A/mm^2 (1400 A) at 2.1 T.

KEY WORDS: Stability, Boiling heat flux, Magnet, Superconducting generator

INTRODUCTION

Since a rotating magnet is apt to suffer wire motions under large centrifugal acceleration and huge electromagnetic force, the stability is one of the most important problems to be solved in a superconducting generator. It has not been studied yet under high rotating speed, at which a large centrifugal acceleration will be generated and the cooling characteristics of liquid helium are much different from the one under static environment (one acceleration of gravity). In the present paper, the stabilities of a rotating magnet were investigated under large centrifugal acceleration up to 2000 g using a test rotor of which maximum rotating speed was 3600 rev min^{-1}. In the stability experiments, cooling heat fluxes requested for stability analyses were measured at various rotating speeds. Both the amount of thermal disturbances forcing a superconductor to go normal and the characteristics of normal zone propagation were also measured and analysed from viewpoints of steady-state and transient stabilities.

EXPERIMENTAL SETUP AND PROCEDURES

Since the detail of the test rotor used in the experiment is described in an another paper(1), the main parameters will be briefly shown here. The schematic illustration of the inner structure of the rotor and cross sectional view of the rotating magnet are shown in Fig.1 and Fig.2, respectively. The magnet is composed of both N pole and S pole coils, each of which consists of saddle-shape coils and a race track coil. The conductor for the N pole coil has the same electrical characteristics and size as the one for the S pole coil, except the electrical insulation of the basic strands. The former consists of 15 0.7mm-basic strands insulated with 0.025 mm thick Formvar (denoted as an insulated conductor), while the latter consists of 15 0.7mm-non-insulated basic strands (denoted as a bare conductor). The rating current is 2000 A . Those main parameters are listed in Table 1. They are wrapped with semi-cured GFRP tape whose size is 5 mm wide and 0.3 mm thick, at an interval of about 3 mm, and hence the wetted perimeter of the conductor is about 37 %.

A heater is wound non-inductively along the length of 3 mm on a conductor of the race track coil and thermally insulated along a length of 10 mm. A 0.05 mm-thick Kapton tape is interleaved between the conductor and heater for insulation. Its total resistance is about 9 ohms, including a lead wire resistance of about 1 ohm. A pulse current with a decay time constant of about 0.9 ms is supplied to the heater by discharging a 0.1 mF capacitor.

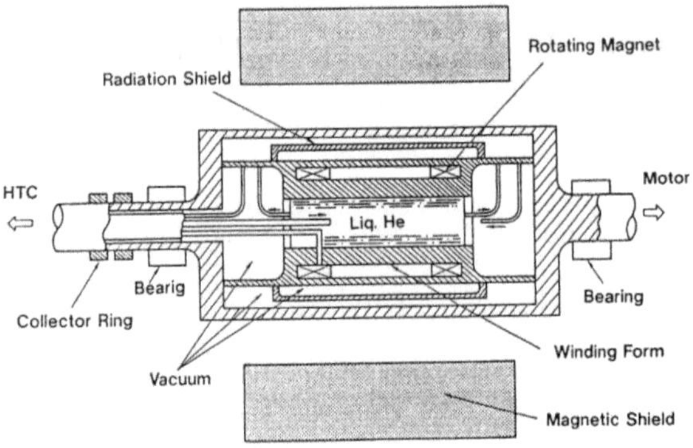

Fig.1 A schematic illustration of a rotor.

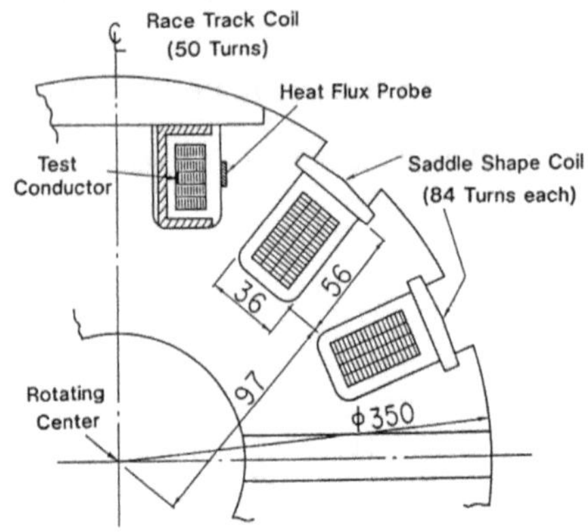

Fig.2 A schematic illustration of the cross section of the windings.

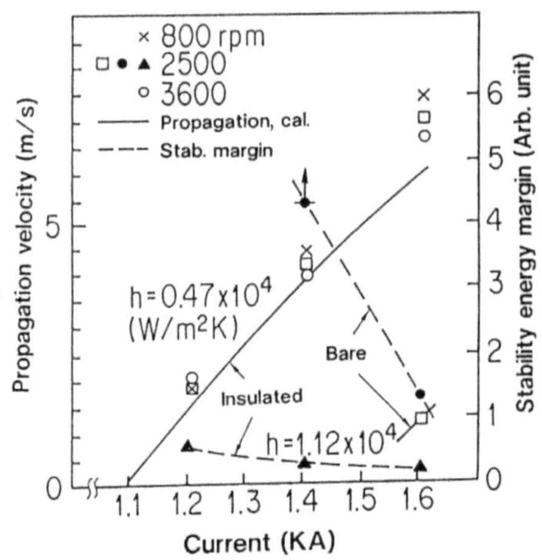

Fig.3 Propagation velocities of normal zone and stability energy margins.

Table 1. Parameters of Cable

	Back-up Coil	Test Coil
Strand		
Dia.	0.72 mm	0.70 mm
SC/Cu/CuNi	~1/1.3/0	~1/2/1
Filament dia.	~15 μm	~6 μm
Twist pitch	9 mm	7 mm
Ic(4T)	>400 A	260 A
cable		
Size	5.4mmx1.3mm	5.3mmx1.3mm
Twist pitch	55 mm	60 mm
Ic(4T)	>6000 A	3900 A

EXPERIMENTAL RESULTS

Heater-induced quench experiments were carried out by applying heat pulse to the conductors at straight section of the race track coil at which the magnetic field is 3 T at 2000 A. The propagation velocities of normal zone and stability energy margins were measured as a function of transport current and rotating speed for both N pole and S pole coils. The result is shown in Fig.3. The solid lines are the values calculated by Dresner's expression[(2)] of the propagation velocity.

DISCUSSIONS

Steady state stabilities

From Fig.3, a minimum propagation current is estimated roughly to be 1500 A. In the heater-induced quench experiments, the bare conductor of which wetted perimeter is about 37 % showed a normal transition in a range of about 30 mm at 2500 rev min^{-1} for the case where I_t = 1400 A ($J_{c,overall}$: about 120 A/mm^2). But the normal zone disappeared in about 0.19 sec after the heat pulse. This means that the Maddock's equal area criterion may be satisfied and hence cryostatic stability be realized at a value of the transport current. It will be discussed below using heat fluxes under large centrifugal acceleration.

The heat flux of liquid helium was measured not in an open bath but in a cooling channel within a slot where the conductors were actually wound[3]. Not only the maximum nucleate boiling but also minimum film boiling heat flux from a clean copper surface was found to be extremely increased under large centrifugal acceleration as compared with the fluxes in a static environment. The heat flux measured at various accelerations is shown in Fig.4(a) and (b). The relation between the heat flux "Q" and acceleration "a" is obtained from the result.

$Q \propto a^{0.27}$, for the maximum nucleate boiling heat flux,

$Q \propto a^{0.31}$, for the minimum film boiling heat flux.

Ogata et al.[4] showed that the maximum heat flux was approximately proportional to $a^{1/3}$ in the acceleration range of less than 100 g. The present result is well consistent with their data.

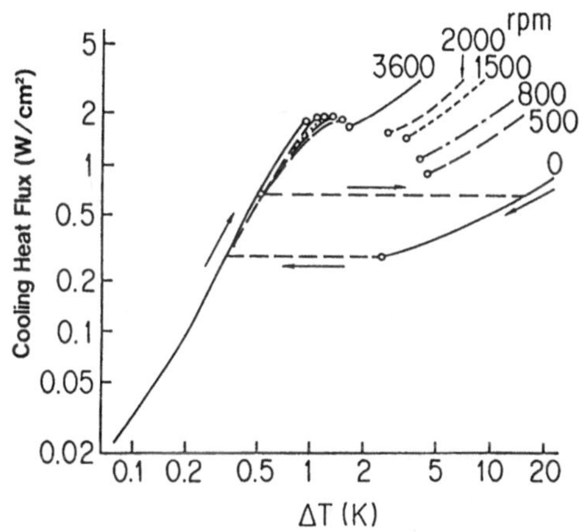

Fig.4(a) Dependence of the cooling heat flux on accelerations.

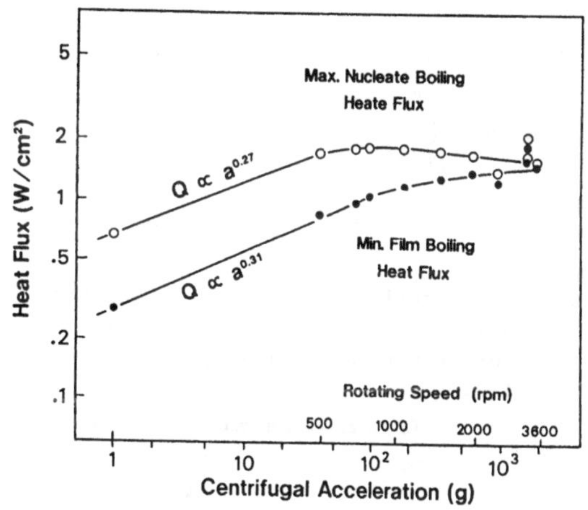

Fig.4(b) Maximum nucleate and minimum film boiling heat fluxes.

The heat generation curves normalized with the wetted area of the present conductor is shown in Fig. 5, together with the heat flux at a few values of acceleration. The heat generations above Tc were calculated using the conductor resistances measured at a temperature a little higher than Tc and magnetic fields corresponding to 1400 A and 1600 A, and they were assumed to remain constant. Both current-sharing and critical temperatures were calculated using a relation between Tc and B, given as $Tc(B) = 9.14 - 0.44B$[5], where Tc and B are respectively critical temperature in Kelvin and magnetic field in Tesla. In a current-sharing temperature range, a critical current density was assumed to be proportional to T. The broken lines of cooling heat flux are not measured values, but an extension of the solid lines connecting the measured values. If the heat flux characteristics shown by the broken line for 1400 A are assumed to be true and the thermal conductivity of the conductor is also assumed not to depend on temperature, since the two shaded areas in Fig. 5 are approximately equal, the Maddock's equal area criterion is expected to be satisfied under an acceleration of over 1000 g, even against an overall current density as large as about 120 A/mm^2 (1400 A). This will be the reason why the heater-induced quench did not occur. In this case, The Maddock's effective heat flux Qe is as large as about 4.8 W/cm^2, although a magnetic field is low and 2.1 T.

Transport currents satisfying the equal-area criterion, that is, the stabilized currents will be increased for higher rotating speeds, because of improvement of heat flux. Such values calculated in the same manner as the above-mentioned one are listed in Table 2, together with the Maddock's effective heat fluxes. They are calculated at a magnetic field of 2.1 T. The Qe depends on the magnetic filed strength. Because Tc and current sharing temperature depends on magnetic field. The dependence of Qe of the present bare conductor on magnetic field is shown in Fig. 6. These results mean that a fairly large Qe will be expected under high centrifugal accelerations.

Transient stabilities

The stability energy margins of the bare conductor are much larger than those of the insulated conductor at the rotating speed of 2500 rev min^{-1}, as shown in Fig.3. such an excellent improvement is supposed to result from large heat flux under cetrifugal acceleration. However, for the case of the insulated conductor, they did not depend much on the speeds.

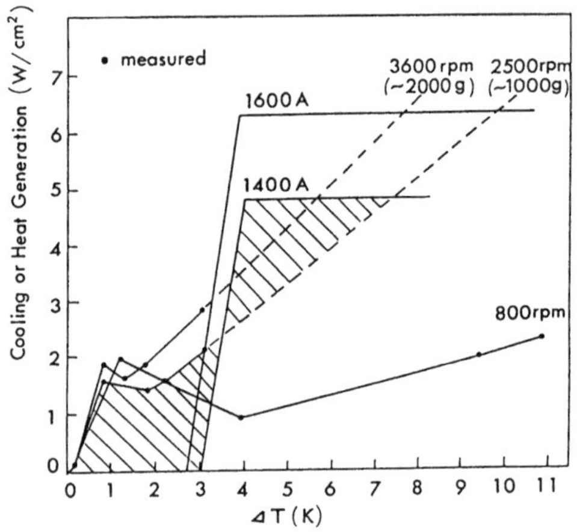

Fig.5 Heat generation per unit wetted surface area and cooling heat flux to liquid helium at various centrifugal acclerations.

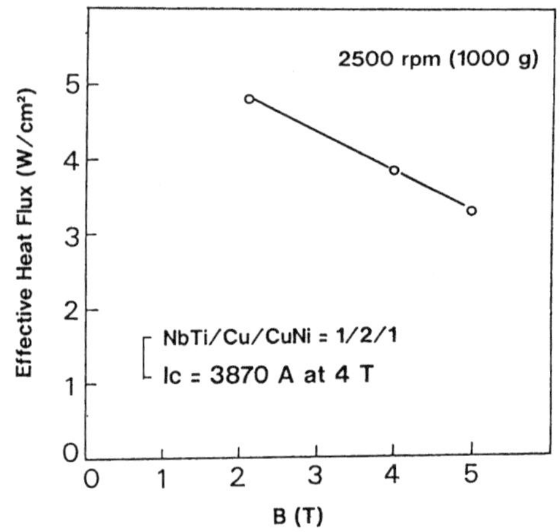

Fig.6 Maddock's effective cooling heat flux versus magnetic field.

Table 2. Dependence of the effective heat flux Qe on acceleration

Accel. (g)	Qe (W/cm²)	Imp (A) Jmp (A/mm²)	J / J_{800rpm}
100 g (800 rpm)	2.4	990 (155)	1.0
1000 g (2500 rpm)	4.8	1400 (220)	1.4
2000 g (3600 rpm)	6.3	1600 (250)	1.6

(B : 2.1 T)

The minimum propagation zone (MPZ) of the bare conductor was calculated for the case where I_t = 1600 A, B= 2.4 T and 2500 rev min^{-1}, usig the following one-dimensional steady state equation of heat conduction.

$$\frac{d}{dx}\left(Ak \frac{dT}{dx} \right) = PH(T) - AG(T)$$

where G(T) is a heat generation, H(T) a cooling heat flux, P a cooling perimeter and A a cross-sectional area of copper stabilizer. k is the thermal conductivity and assumed not to depend on temperature. In the current sharing temperature (Tcs) range, the critical current density is assumed to be proportional to temperature and then G(T) is given as,

$$G(T) = \begin{cases} 0 & \text{for } T < Tcs \\ Gc\,(T - Tcs)\,/\,(Tc - Tcs) & \text{for } Tc > T > Tcs \\ Gc & \text{for } T > Tc \end{cases}$$

The value of GcA/P is calculated to be about 6.3 W/cm^2. Tc and Tcs are 7.1 K and 6.9 K, respectively. The algebraic expression of cooling heat flux is approximated from the data at 2500 rev min^{-1} in Fig.5 and given as,

$$H(T) = \begin{cases} 1.870T - 7.850 & \text{for } T < 5.03 \\ -0.135T + 2.230 & \text{for } 5.03 < T < 6.14 \\ 0.603T - 2.305 & \text{for } 6.14 < T \end{cases}$$

The result is shown in Fig.7. If the MPZ is regarded as the zone in a temperature higher than that where the current sharing starts, it is obtained to be about 13 mm from Fig.7.
Heater-induced super-to-normal transition was observed in the present experiment. It is shown in Fig. 8(a) and (b). A value of 2 mV in the vertical line corresponds to about 10 mm of the conductor which turns into full normal state. In Fig.8(a), a voltage slightly smaller than 2 mV appeared but the normal zone dissappeared in about 35 msec after the pulse. On the other hand, in Fig.8(b) a voltage approximately equal to 2 mV appeared and eventually the zone was built up to quench. This phenomena indicate that the MPZ will be of about 10 mm. Therefore, the calculated value is well consistent with the experimental one. The value of MPZ is fairly larger than a MPZ calculated in an adiabatic condition. This may be the reason why the stability margins of the bare conductor are much larger than those of the insulated conductor.

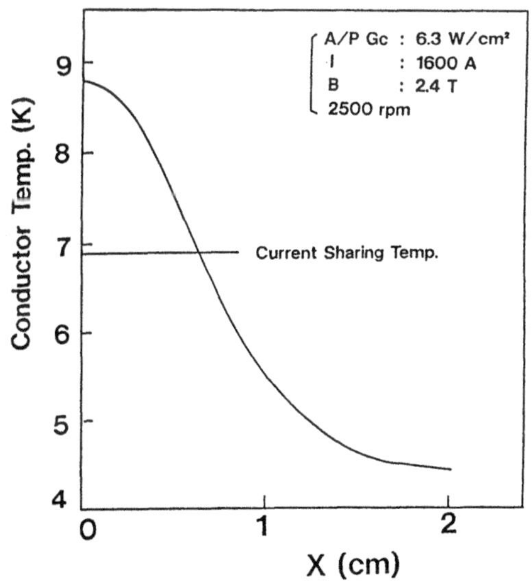

Fig.7 Temperature profile along the conductor Satisfying the Maddock's equal area criterion.

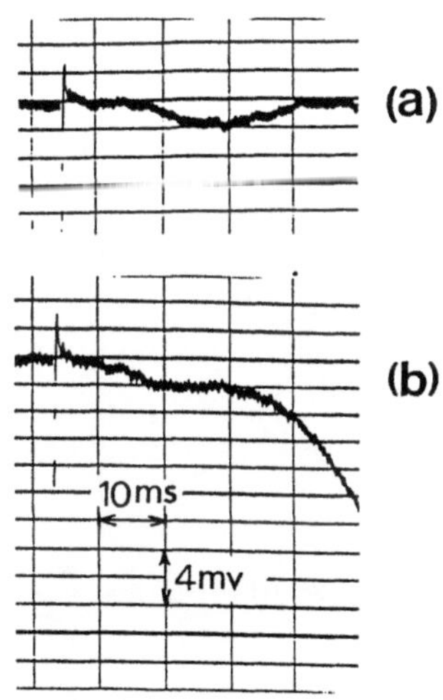

Fig.8 Super-to-normal transition due to heat pulse.

CONCLUSIONS

The stabilities were evaluated by applying heat pulses with a time constant of 0.9 ms to the conductors under large centrifugal acceleration up to 2000 g. It was clarified that the cooling heat flux was considerably increased under large centrifugal acceleration and accordingly contributed to the stabilities of the rotating magnet. It was shown that under large centrifugal acceleration, the Maddock's effective heat flux was extremely increased and that the stability energy margins in the bare conductor were considerably larger than those in the insulated conductor.

ACKNOWLEDGEMENTS

The authors would like to express their gratitude to Mr. H. Satoh, Director for development program of the Moonlight Project of the Agency of Industrial Science and Technology for his support and encouragement and to Toshiba Corp. for cooperation with the rotor fabricaton.

REFERENCES

(1) Onishi, T. et al. : Cryogenics, Vol. 31 (July, 1991) p.664
(2) Dresner, L. : IEEE Trans. on Mag., Vol. MAG-15, No.1 (Jan., 1979)
(3) Ishii, I. et al. : To be published in Journal of Japanese Cryogenic Engineering
(4) Ogata et al. : Cryogenics, Vol.17, No.8 (1977) P.461
(5) Polak, H. et al. : Cryogenics (Dec., 1973) p.702

Basic Study on Current Limiter Using Thin Film High T_c Superconductor

OSAMI TSUKAMOTO[1], WON-KEUN YOO[1], and HIROSHI SATOH[2]

[1] Faculty of Engineering, Yokohama National University, Yokohama, 240 Japan
[2] Sony Corporation Sibaura Technology Center, Minato-ku, Tokyo, Japan

ABSTRACT

We are investigating a super-normal transition type current limiter using superconductor thin film. The thin film has large cooling surface to volume ratio, and, therefore, it is expected that very quick super-normal transition can be obtained. Two types of samples, one without a current by-pass and another with a current by-pass of normal metal thin film were made and the characteristics of the super-normal transition of the samples were experimentally investigated by applying 50 Hz AC current at 77K. The experiments showed that the recovery time of the superconducting state in both types of the samples was short enough that the transition between the resistive and superconducting states was completed during one cycle of the 50 Hz AC, provided that the sample current was not so high.

INTRODUCTION

Superconducting power apparatuses, such as superconducting generators, cables and transformers have inherent possibility of quench, when they are subject to over-current due to a fault in the electrical power grid. Usually, the power apparatuses are operated at high voltage of several tens ~ several hundreds of kV, and the superconducting windings of the apparatuses have heavy electric insulation materials. The electric insulation materials take significantly long time to be cooled, when they are warmed. Therefore, it takes a time for the superconducting apparatuses to recover the superconducting state, once they are quenched. Generally, a superconductive apparatus is of large scale and the suspension of its function for significantly long time gives a seriously bad effect on the power grid. Therefore, development of a current limiting device to prevent a quench caused by the over current is essential to introduce the superconducting apparatuses in the power grid. Usually, a fault causing the over-current is cleared by switching gears in a few cycles and the current limiting device should limit the current until the switching gears function. Several types of the current limiters have been investigated[1 ~ 5]. Those current limiters are divided mainly in two categories; a device with iron cores[1 ~ 3] and one using the super-normal transition[4, 5]. The former can switch to the current limiting state from the normally operating state and vice versa at fast transition speed. However, this device has iron cores which produce additional losses and significantly increases the size and weight of the device. The latter has a simple, compact and light structure . Its transition to the current limiting state is fast enough but generally its recovering time to the superconducting state is long.

We are investigating a super-normal transition type current limiter using superconductor thin film[6]. The merits to use the high Tc superconductor thin film to the current limiter are ; (a) high current density at superconducting state and high resistivity at the resistive state, (b) quick recovery of superconducting state because of large ratio of cooling surface area to volume of superconductor. We have made small samples of the current limiter and investigated current limiting characteristics. It has been shown that the limiter became resistive instantaneously when the sample current I exceeded the threshold current Ic and recovered superconductivity instantaneously also when I decreased below Ic, when an excess current to Ic was not large. The results shows that the current limiter using high Tc superconductor thin film is very promising.

EXPERIMENT

Experiment samples

We made two types of samples, one without a current by-pass and another with a normal metal (gold thin film) current by-pass.

i) Sample without current by-pass

YBCO thin film was formed on a MgO base by the electron beam vacuum deposition. The thickness of the thin film was controlled by controlling the deposition time and 0.3 μ m. YBCO thin film with MgO base was cut into a rectangular shape of 2 mm wide and 12 mm long. The sample was arranged as is shown in Fig. 1. The YBCO sample was placed on a phenolic plate and the electrodes to supply current to the sample and those to measure voltage across the sample were mechanically attached to the surface of the thin film. Indium was inserted in interfaces between the electrodes and the thin film.

ii) Sample with current by-pass

Configuration of the sample with a current by-pass is shown in Fig. 2 a). When the YBCO thin film becomes resistive, a part of current supplied to the sample flows through the current by-pass of gold thin film, and the Joule heating in the YBCO thin film is reduced, which improves the recovering time to the superconducting state. The sample was made as follows : i) YBCO thin film was formed on MgO base by electron-beam vacuum deposition. ii) The thin film was shaped to a desired patten by photo-chemical edging. iii) The gold thin film some areas of which was deposited over the YBCO thin film was formed by vacuum deposition with a mask patten. The electrodes for the current supplying and voltage measurement were arranged as is shown in Fig. 2 b). The contacts between the YBCO film and the electrodes were made mechanically by inserting indium between the thin film and electrodes as was same as the sample without by-pass, and insulator was inserted between the gold thin film and the electrodes for measuring the voltage.

Experiment results

A schematic experimental arrangement is shown in Fig. 3. The sample were immersed in liquid nitrogen. Voltage across the sample was controlled by a slide autotransformer. Sample voltage and current were recorded by a transient recorder.

i) Characteristics of the sample without current by-pass

Voltage and signal traces of the sample for the case that the sample current was 525 m Arms and 50 Hz AC is shown in Fig. 4 a). The YBCO thin film had smooth transition to resistive state from the superconducting state, when the sample current exceeded the threshold current. The reason for this smooth transition is considered due to the local fluctuations in the critical current of the thin film. As is obvious from Fig. 4 b), the threshold current for the transition between resistive and superconducting states was 0.37 A. When the Joule heating rate in the resistive state was small, the superconducting state recovered almost instantaneously when the sample current was below the threshold current, and, as is shown in Fig. 4 a), voltage appeared across the sample only during the period when the sample current exceeded the threshold. This result demonstrates that the superconductor thin film can be used as a power frequency current limiting device which has a response quick enough that the transition between the current limiting and superconducting modes is reversible in one cycle. Figure 5 shows time traces of the current and voltage signals of the sample for larger sample current, I=1.1 Arms and 50 Hz. In this case, the Joule heating was high and the heat was not cooled by liquid nitrogen during one cycle, so that, the temperature of sample increased gradually until the whole thin film became normal.

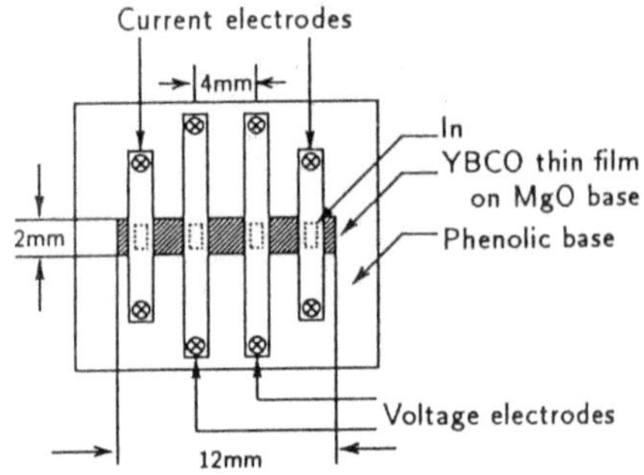

Fig1. Sample without current by-pass

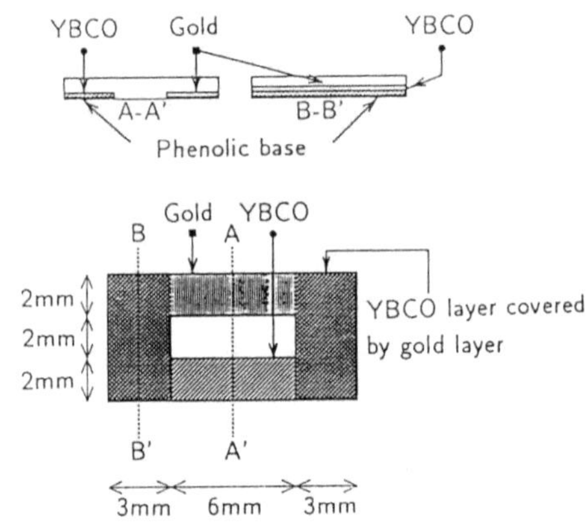

a) Configuration of the sample

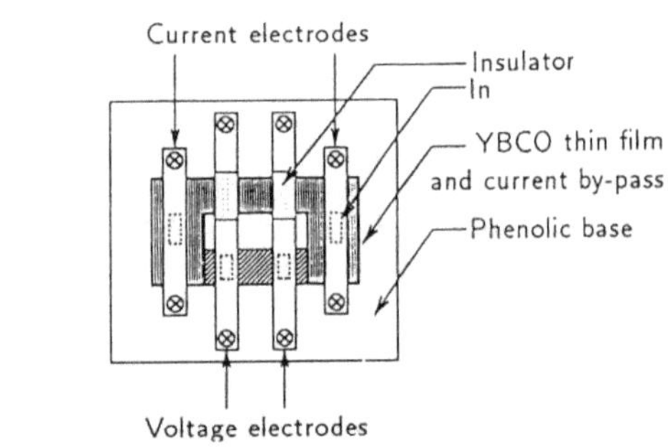

b) Sample arrangement

Fig2. Sample with current by-pass

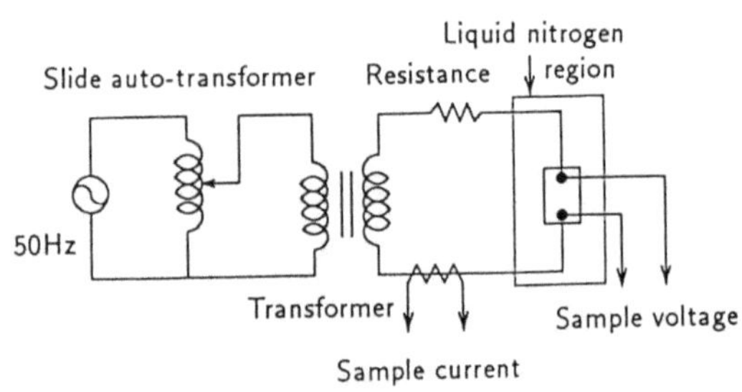

Fig3. Schematic experimental arrangement

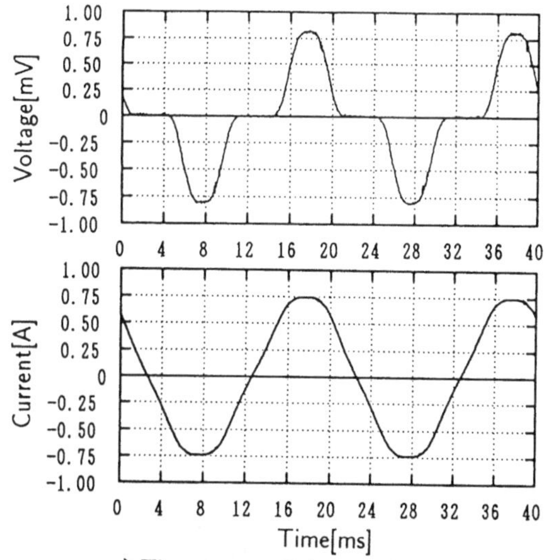

a) Time traces of the sample Voltage and current

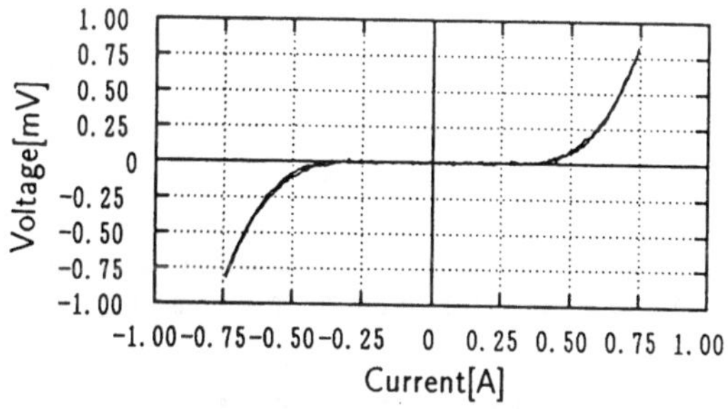

b) Voltage vs. current curve

Fig4. Voltage and current characteristics of the sample current for I=525 mA rms and 50 Hz AC

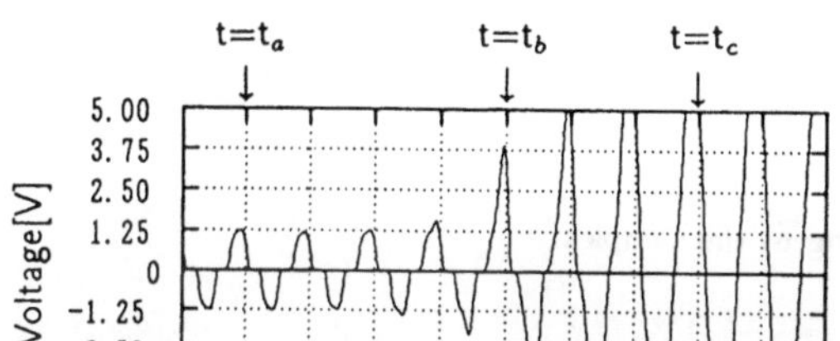

Fig.5 Time traces of sample voltage and current for I=1.1 Arms and 50Hz AC

Voltage vs. current curves at time $t=t_a$, t_b and t_c in Fig. 5 are shown in Fig. 6 a), b) and c). At the early period, $t=t_a$, transition between the resistive and superconducting states occurred during one cycle. However, there appeared hysteresis because it took a time for the thin film to be cooled and recover the superconducting state. At $t=t_b$, the hysteresis became more remarkable because the heat was accumulated in the sample and it took more time for the thin film to recover the superconducting state. Finally at $t=t_c$, the thin film did not recover the superconducting state and become a pure resistance.

ii) Sample with current by-pass

Figure 7 a) as shows voltage and current traces of the sample with the current by-pass for the sample current I=1.15 Arms and 50 Hz, and Fig. 7 b) is the voltage vs. current curve. At this current level, obviously from the figure, the transition without hysteresis between the resistive and superconducting state was obtained at steady state, while the sample without the current by-pass did not recover the superconducting state at the same level of current. This result shows that the current by-pass reduced the current through the thin film and the excessive Joule heating and that the sample had the non-hysteresis transition at higher current than the sample without the current by-pass.

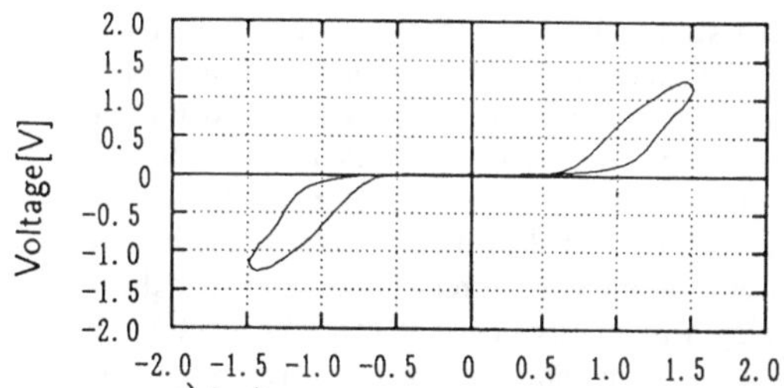

a) $t=t_a$ Current[A]

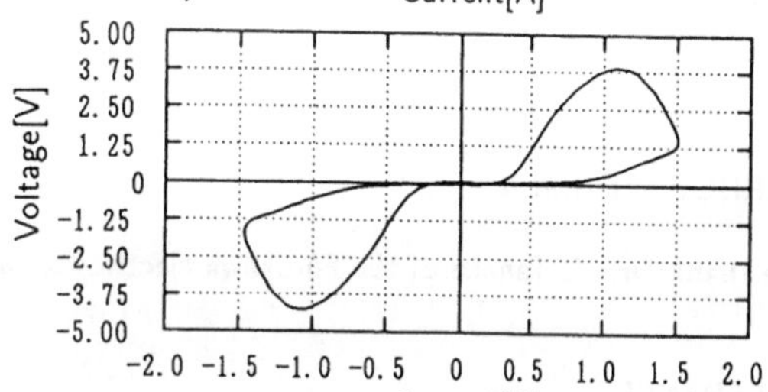

b) $t=t_b$ Current[A]

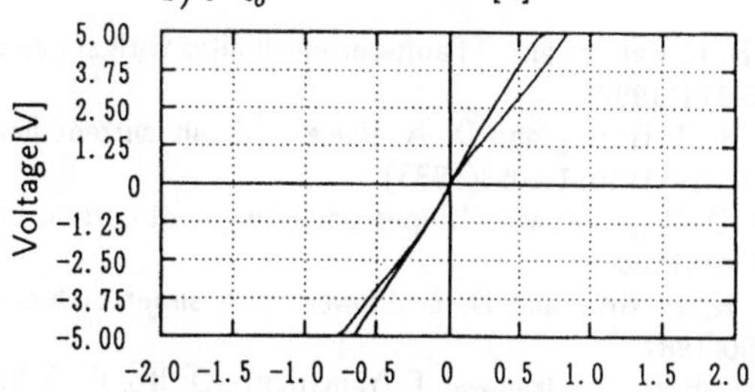

c) $t=t_c$ Current[A]

Fig.6 Voltage vs. current Characteristics of the sample without current by pass for I=1.1 Arms and 50Hz AC

DISCUSSION AND CONCLUSION

To have an idea about of size of a current limiter which may be used in a electrical power distribution line, size of a 66MVA(66kV/1kA) current limiter is discussed. The experiment mentioned above shows that the superconducting state can be recovered in one cycle, provided that the peak heat flux of the thin film in the resistive state is less then about 2×10^4 watt/m^2. Using this data we design a current limiter. To limit the current under 1.3kA in a short-circuit fault and to recover superconducting state in one cycle, after the fault is cleared, the necessary size of thin film is 0.31m wide, 295m long and 0.2μ thick, provided that the critical current density and resistivity of the YBCO thin film are 3×10^{10} A/m^2 and 10^{-6} Ωm. The volume of the current limiter is estimated about 0.5 $\sim$ 1m^3 including cooling channel but excluding the volume for bushings and current leads. The addition of the current by-pass requires additional amount of the thin film to compensate the decrease of the resistivity of the current limiter due to the by-pass. However, it protects superconductor thin film in the current limiting mode and ensures the quick recovery of superconducting state.

In the study, we show that a current limiter using superconductor thin film has quick response and is potentially promising to be introduced to the electrical power grid. However, it should be noted that the development of superconductor thin film of large area, high current density and uniform quality is essential to a practical current limiter and that the development of large current capacity junctions between the thin film and current leads is also very important.

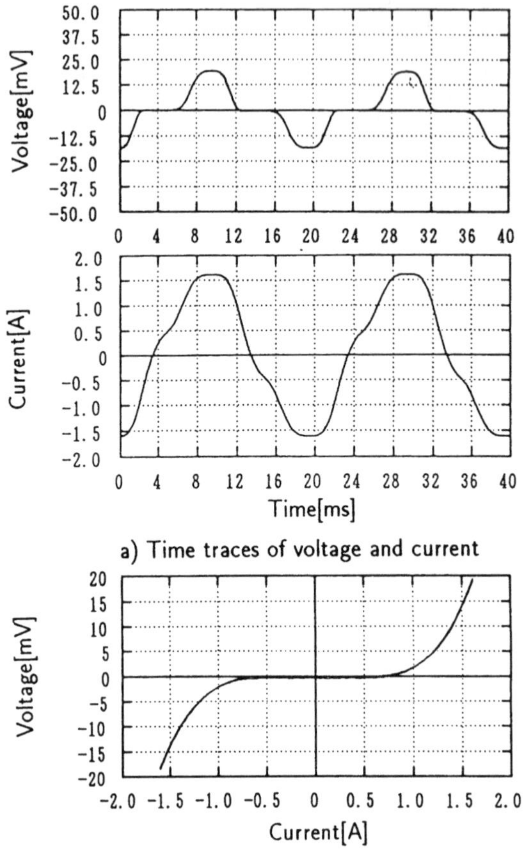

a) Time traces of voltage and current

b) Voltage vs. current curve

Fig.7 Voltage and current characteristics of the sample with current by pass for I=1.15 Arms and 50Hz AC.

ACKNOWLEDGMENT

We thank Dr. Y. Tanaka of the Furukawa electric for supplying us the samples.

REFERENCES

1. B. P. Raju et al. : "Fault-current limiter with superconducting DC bias", IEEE Proc., Vol. 129, Pt.C, No.4, p. 166-171(1982)
2. H. J. Boenig and D. A. Paice : "Fault current limiter using a superconducting coil", IEEE Trans. Magn., MAG-19, P.1051-1053(1983)
3. J. D. Rogers et al. : "Superconducting fault current Limiter and inductor design", IEEE Trans. Magn., MAG-19, p.1504-1508
4. K. E. Gray and D. E. Flower : "A superconducting fault-current limiter", J. Appl. Phys.,49(4), p.2546-2550(1987)
5. T. Hara, K. Okaniwa, T. Yamamoto, D. Ito, E. S. Yoneda, K. Tsurunaga, Y. Sugiyama : "6.6 kV class superconducting fault current limiter", 12th, International conference on magnet technology, June. 1991, Leningrad, USSR
6. H. Satoh, O. Tsukamoto : "Superconducting switching device using high-Tc superconductor thin film", Advances in superconductivity II, Springer-verlag Tokyo, p. 1231-1234(1991)

Development of 6.6kV/1.5kA-Class Superconducting Fault Current Limiter

KENJI TASAKI[1], KAZUYUKI TURUNAGA[1], DAISUKE ITO[1], TAKAMITSU TADA[1], TSUKUSHI HARA[2], TAKESHI OKUMA[2], KAZUO MAEKAWA[2], and TAKAHIKO YAMAMOTO[2]

[1] Toshiba Corporation, Kawasaki, Kanagawa, 210 Japan
[2] Tokyo Electric Power Company, Chofu, Tokyo, 182 Japan

ABSTRACT

The authors have developed and tested a 6.6kV/1.5kA-class superconducting fault current limiter, wound with an AC superconducting wire having ultra-fine NbTi filaments in Cu-10%Ni matrix. In experiments, voltages up to 6.6kV were applied to the limiter. The limiter was able to successfully limit the fault current to 1.8kA from the 55kA short-circuit current that would flow in a circuit without a limiter. Interestingly, the limiting current peak value depends on the applied voltage. This transient phenomenon was analysed based on a model including flux flow resistance contribution. Experimental results were explained with the model.

KEY WORDS: fault current limiter, AC superconducting wire, flux flow state

INTRODUCTION

Accompanying increasing electric power demands, a stable electric power supply becomes a still more important problem. Thus, even if a fault occurs, the fault currents must be suppressed before fault influence spreads. At present, circuit breakers are used for suppressing fault current. However, as fault current becomes larger with increasing the electric power system scale, instantaneous operation of the breaker becomes more difficult, owing to arcing. Therefore, electric power companies have wanted fault current limiters.

Various kinds of current limiters have been developed[1][2][3][4][5]. However, these limiters don't have sufficient current capacity. The authors developed a 6.6kV/1.5kA class limiter and succeeded in suppressing the fault current with the limiter.

This paper reports the performance of and a current limiting process for the 6.6kV/1.5kA class superconducting current limiter.

EXPERIMENTAL

(1) Operation principle

As shown in Fig.1, the superconducting fault current limiter consists of two kinds of coils: a trigger coil and a limiting coil. The trigger coil is wound non-inductively with a 42-strand AC superconducting wire, which has ultra-fine NbTi filaments in CuNi matrix, in order to reduce AC losses. The limiting coil is wound with Cu wire and is connected in parallel with the trigger coil.

Under normal operating conditions, most of the circuit current flows through the trigger coil, because the trigger coil has almost zero inductance. Moreover, the trigger coil windings have no electric resistance in a superconducting state, therefore the voltage drop at the limiter is almost zero, under normal operating conditions.

On the other hand, if a fault occurs and a current, exceeding the critical current of the trigger coil windings, flows through the circuit, transition from a superconducting state to a normal conducting state occurs in the windings and the trigger coil resistance appears instantaneously. The resistance suppresses the fault current. If the normal state resistance for the trigger coil were to be chosen much higher than the limiting coil impedance, the fault

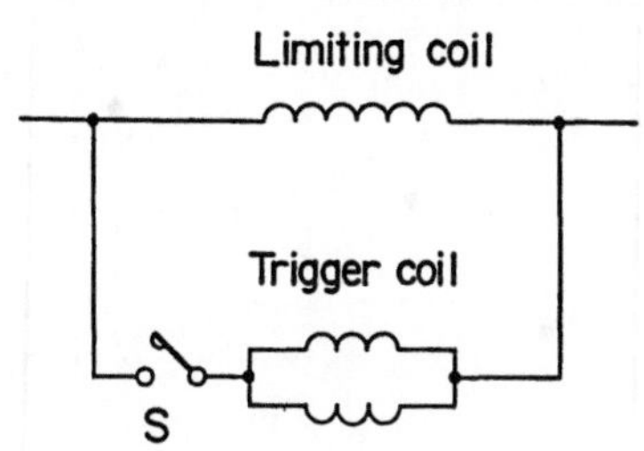

Fig.1 6.6kV/1.5kA class fault current limiter configuration

current is able to be transferred to the limiting coil immediately after the fault occurrence. The fault current is finally limited by the limiting coil inductance. In order to reduce heat generation at the trigger coil in liquid helium, switch S, shown in Fig.1, should be opened as fast as possible after the fault occurrence.

(2) Test and results

The 6.6kV/1.5kA class superconducting fault current limiter was tested in the circuit shown in Fig.2. Voltages up to 6.6kV were applied to the circuit with a 200MVA generator, under conditions in which switch S_1 and S_2 were closed and S_3 was opened. This is a steady-operating state and several hundred ampere current, which was below the critical current for the trigger coil, flows through the circuit.

In order to simulate a fault, loads R_r and L_r were short-circuited by closing S_3. When the applied voltage and short-circuit impedance (consisting of a regulating reactor inductance L and the internal impedance for transformer Tr) were set at 6.6kV and 0.17Ω, respectively, and the peak short-circuit current, which would flow through the circuit without the limiter, was 54.5kA. In the circuit with the limiter, however, the fault current was successfully suppressed to 1.87kA (limiting current), as shown in Fig.3. Moreover, the fault current was decreased to 250A during 200μsec after the fault occurrence and a fast limiting operation performance for the limiter was also demonstrated.

DISCUSSION

Figure 3 shows the limiting current dependence upon the applied voltage. As shown in Fig.3, the limiting current tends to rise with increasing applied voltage. This was an unexpected result for the authors. A detailed analysis of current limiting performance is necessary to explain the result. Therefore, the following computer simulation was carried out.

(1) Computer simulation model

Figure 4 shows an equivalent circuit for the test circuit shown in Fig.2. In Fig.4, Ll represents the short-circuit inductance, L_1 and L_2 are the self-inductances for the trigger coils, M is the mutual inductance between the trigger coils, L_3 is the self-inductance for the limiting coil, and r_1 and r_2 are resistances for the trigger coils. From Fig.4, the circuit equations are given by

$$\begin{pmatrix} Ll-L_1 & Ll+M & Ll \\ Ll-M & Ll+L_2 & Ll \\ Ll & Ll & Ll+L_3 \end{pmatrix} \begin{pmatrix} dI_1/dt \\ dI_2/dt \\ dI_3/dt \end{pmatrix} = \begin{pmatrix} V-V_{r1} \\ V-V_{r2} \\ V \end{pmatrix} \quad (1)$$

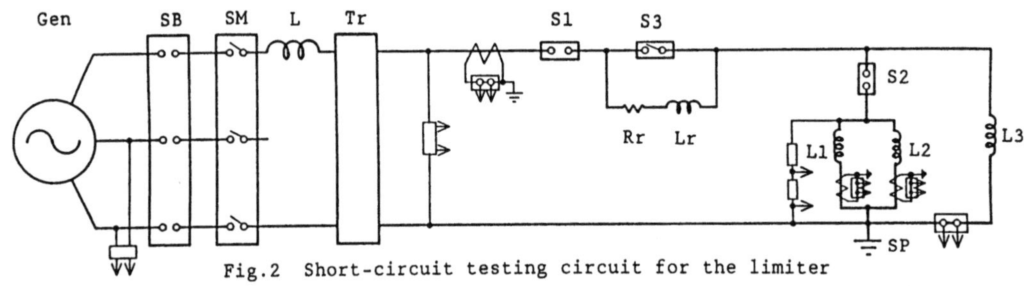

Fig.2 Short-circuit testing circuit for the limiter

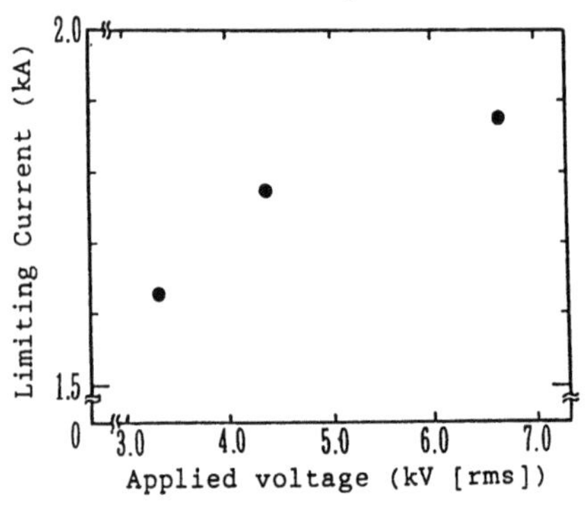

Fig.3 Applied voltage dependence for the limiting current

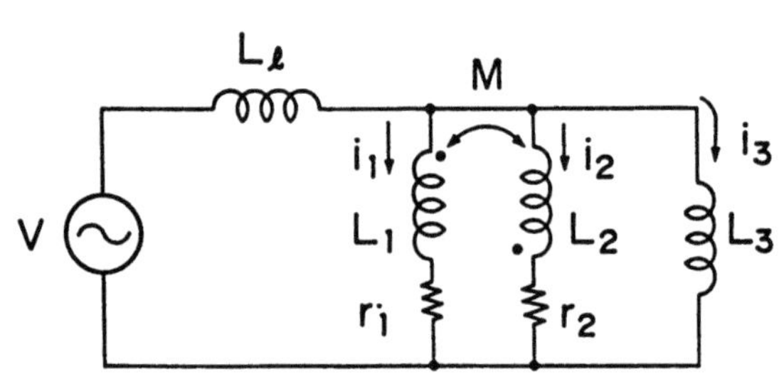

Fig.4 Equivalent circuit for testing

where V_{r1} and V_{r2} are voltage drops due to resistances r_1 and r_2, respectively. These voltage are shown as the following equation:

$$V_{rm} = \int_0^{l_{nm}} \rho_{nm}(T)\ I_m/S\ dx + \int_0^{l_{fm}} \rho_{fm}(T)\ (I_m - I_{cm}(x))/dx \quad (m=1,2) \qquad (2)$$

where l_{nm} and l_{fm} represent the lengths of the normal and the flux flow regions, and I_m is current flowing through the trigger coil. In this equation, it is assumed that the critical current of the windings is not uniform and has a certain distribution, i.e. $I_{cm}(x)$. ρ_{nm} and ρ_{fm} indicate the average longitudinal resistivities for the windings in the normal conducting state and in the flux flow state, and are given by

$$\rho_{nm} = \lambda\rho_{NbTi} + (1-\lambda)\rho_{CuNi} \qquad (3)$$
$$\rho_{fm} = \lambda\rho_{ff}(T) + (1-\lambda)\rho_{CuNi} \qquad (4)$$
$$\rho_{ff}(T) = \rho_{NbTi}(B/B_{c2}(T)) \qquad (5)$$
$$B_{c2}(T) = B_{c2}(0)[1-(T/T_c)^2] \qquad (6)$$

where ρ_{NbTi}, ρ_{ff} and ρ_{CuNi} respectively represent resistivities for NbTi in the normal conducting state, for NbTi in the flux flow state and for CuNi. λ is a volume fraction of superconducting filaments and $B_{c2}(T)$ is a temperature dependence of upper critical magnetic flux density.

Since the resistivity in the flux flow state is a function of temperature, the following heat balance equation should be introduced, in order to calculate the temperature in the trigger coil windings:

$$(C(T)\cdot dT/dt)\cdot V = Q\cdot V - H\cdot S \qquad (7)$$

In Eq.(7), the thermal conduction term is neglected, because current limiting process is extremely short, i.e. on the 100μsec order. C(T) represents overall volumetric specific heat capacity in the trigger coil windings, Q is the heat generation per unit volume for the windings, H is the cooling heat flux per unit area, V is the volume for the windings and S is the effective cooling area. Here, heat generation Q is represented by the following equation:

$$Q = \begin{cases} (I/S)^2 & (T>T_c \quad \text{: Normal conducting state}) \\ (I-I_c)I/S^2 & (T<T_c \text{ and } I>I_c \quad \text{: Flux flow state}) \end{cases} \qquad (8)$$

(2) Calculated results

The circuit equation (1) and the heat balance equation (7) were simultaneously solved by computer simulation. Figures 5,6 and 7, respectively, show time dependences for the total resistance in the trigger coils, the total current flowing through the trigger coils and the voltage between the limiter, under the 4.4kV applied voltage condition. In these figures, the solid lines represent experimental values and the broken lines represent calculated values. The calculated curves are close to the experimental curves. Therefore, the foregoing computer simulation model is shown to be mostly correct.

(3) Current limiting process

On the basis of the calculated results, current limiting process of the limiter is discussed in this section.

As shown in Fig.6, the current rapidly increases immediately after a fault occurred. With the current rising, the superconducting state is gradually disrupted to the flux flow state starting from the lower critical current regions. As a result, the trigger coil resistance increases (① in Fig.6). Flux flow resistivity near 4.2K is much smaller than the normal resistivity. Thus, even if large flux flow regions generate, the resistance does not become sufficient to make the current decrease (② in Fig.6). With further temperature rising in the flux flow regions, the trigger coil resistance rapidly increases. Therefore, the current begins to decay (③ in Fig.6). The resistance continues to increase, even after the current decreases, since the transition from a flux flow state to a normal conducting state occurs in the trigger coils (④ in Fig.6)

(4) Applied voltage dependence for the limiting current

As the computer simulation results, the authors confirm that the cooling effect on the windings is closely related to the applied voltage dependence for the limiting current. The limiting current values are more than 300A above the critical current in Fig.6. This is because the windings don't change to a normal conducting state directly. Instead, for a while, they stay in a transient low resistance state, i.e., flux flow state owing to the cooling effect of liquid helium, so that the current doesn't decay, even after the trigger

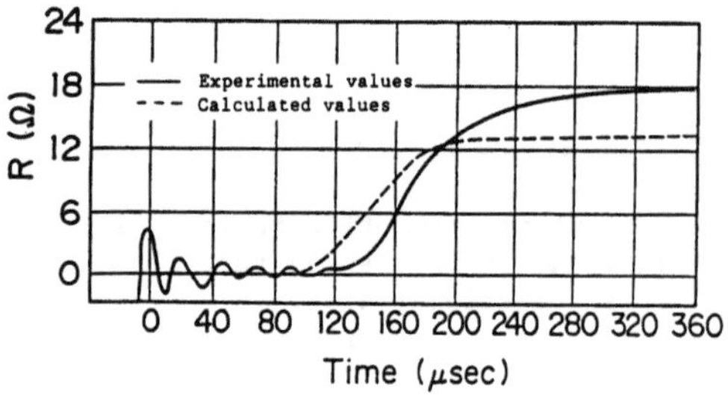

Fig.5 Time dependence for total resistance in trigger coils, under 4.4kV applied voltage

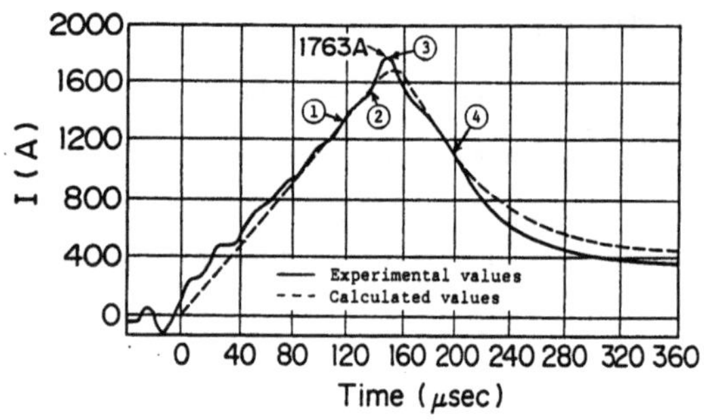

Fig.6 Time dependence for the total current flowing through trigger coils, under 4.4kV applied voltage

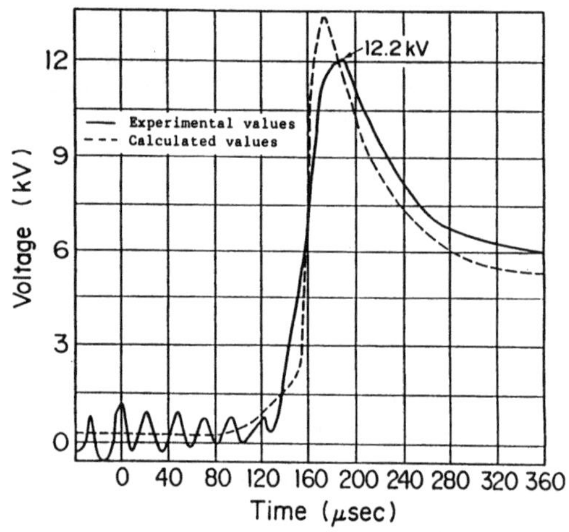

Fig.7 Time dependence for voltage drop between the limiter, under 4.4kV applied voltage

coil current exceeds the critical current.

Here, it is assumed that the cooling capacity is constant and that the temperature scarcely rises before the heat flux time integral becomes equal to the cooling capacity. With the increase in dI/dt value, the current can increase higher before the temperature begins to rise. The dI/dt value is proportional to the applied voltage value. Therefore, the limiting values depend upon the applied voltages.

CONCLUSION

The authors developed and tested a 6.6kV/1.5kA class superconducting fault current limiter. The obtained results confirmed that it operated successfully, as designed. In this test, however, the applied voltage dependence for the limiting current was observed, which was an unexpected result. Thus, the authors carried out a computer simulation, in order to explain the result. As a result, it was found that the calculated results agree closely with the experimental values and the reason of the unexpected result become clear.

REFERENCES

[1]. H. Nakayama, et al., Proc. of IEEE WM, 225-5 (1988)
[2]. K.E.Gray et al., J. Appl. Phys., 49, 4, 2546 (1978)
[3]. A. Ulbricht, IEEE Trans. Mag. MAG-15, 172 (1979)
[4]. D.Ito et al., Adv Cryo. Eng., 35-A, 653 (1990)
[5]. T. Verhaege et al., Proc. 11th Int. Conf. on Magnet Technology (MT-11), 413, Tsukuba, (1989)

A Study on a High-T_c Superconducting Fault Current Limiter

TATSUYA HAYASHI[1], TEIJIRO MORI[1], YUUICHI WADA[1], SHIGERU MATSUNO[2], SHIN-ICHI KINOUCHI[2], and FUSAOKI UCHIKAWA[2]

[1] Central Research Laboratory, Mitsubishi Electric Corp., Amagasaki, Hyogo, 661 Japan
[2] Materials and Electronic Devices Laboratory, Mitsubishi Electric Corp., Sagamihara, Kanagawa, 229 Japan

ABSTRACT

The application of a high-Tc superconductor with a high critical current density (Jc) to a fault current limiter has been investigated. When the superconductor changes to the normal conductor, it has a resistance which can limit fault currents. Experiments were made using a current limiting element of a YBaCuO film 0.25μm thick. The element has a current limiting portion with 1mm wide and 2mm long to limit fault currents. For a half cycle, an alternating voltage (10Vp,50Hz) was applied to the element. A typical current limiting waveform was obtained in the experiment. The fault current, with a crest value of 420A, was limited to the crest value of 8A. After that, the current was rapidly decreased and was less than 1.6A.

KEY WORDS: YBaCuO film, fault current limiter, short-circuit current, peak let-through current

INTRODUCTION

When a short circuit in an electric power system occurs, a large short-circuit fault current flows in the circuit. This fault current is usually interrupted by a circuit breaker connected in series to the circuit. Since a large fault current mechanically and thermally damages electrical equipment in the system, it is desirable that the fault current is limited by functional devices called fault current limiter (FCL). There are many types of FCLs such as current-limiting reactors, permanent power fuses(1)(2), current-limiting wires, and superconducting fault current limiters(3). In recent years the appearance of high-Tc superconductors has led many researchers to investigate their applications to devices in electric power systems(4). We have also carried out research on the application of high-Tc superconductors to the FCL. When the superconductor exceeds a critical point due to a fault current, it changes from a superconducting state to a normal conducting state, causing a resistance which brings the fault current within an acceptable level. The rate of resistance change in the high-Tc superconductor is much larger than that of any other FCLs. Consequently, the size of the

high-Tc superconducting FCL can be made smaller than that of other FCLs. An experiment has been conducted on future S-N (superconductivity-normalconductivity) transition-type FCL, using a high-Tc superconducting film with a high Jc.
In this paper, we will report the experimental results.

EXPERIMENTAL

A YBaCuO thin film superconducting element used in the experiment as a FCL is shown in Fig.1. The film had a critical current density (Jc) of $8.8 \times 10^5 A/cm^2$ at 77K, and a critical temperature (Tc) of 91K. The film's thickness was 0.25μm. The element had a current limiting portion which was 1mm wide and 2mm long, and a critical current of 2.2A. Terminals for current input and voltage measurement were formed by sputtering to create several 1mm×1mm gold films. Current and voltage leads were connected to these terminals by the pressure welding of indium. An experimental circuit is shown in Fig.2. The circuit had a resistance R1 to regulate the current flow. For a half cycle of 50Hz, an alternating voltage of 10Vp was applied to the element. Isolation amplifiers to prevent grounding were used to measure current and voltage. In the experiments the values of R1 were set at 0,0.2,0.5,1.0 and 2.0Ω.

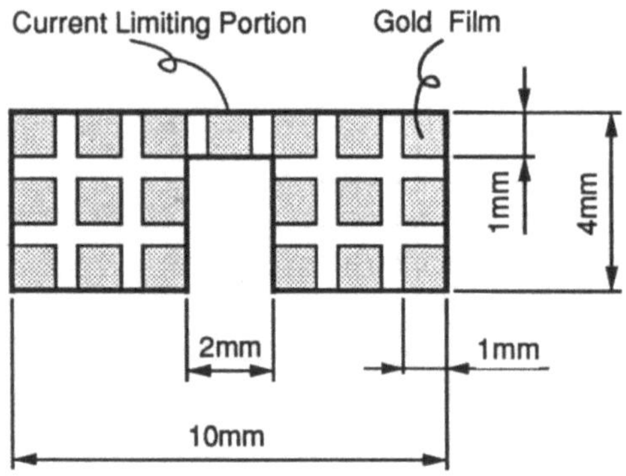

Fig.1 Superconducting FCL Element

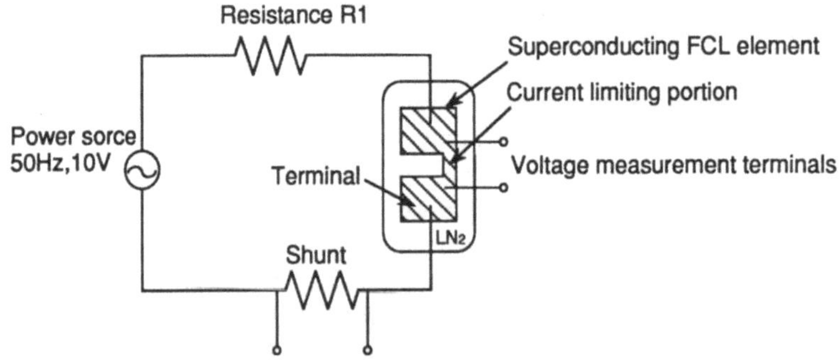

Fig.2 Experimental circuit

RESULTS AND DISCUSSION

A prospective short-circuit current in the experimental circuit when R1=0Ω was about 420A, at an applied voltage of 10Vp. A current limiting waveform when R1=0Ω is shown in Fig.3. A typical current limiting waveform, which had an abrupt change in the current, was obtained. The peak let-through current was 8A, which was about one-fiftieth of the prospective short-circuit current. After that, the current decreased rapidly. The instantaneous value of the current was about 1.5A when the voltage between voltage measurement terminals was maximum during the current limiting operation. The resistance of the current limiting portion of the element was about 6.8Ω at that time. In this experiment the resistance at resistive state of the element was considerably larger than the high-Tc superconducting bulk element used in our previous experiment(5).

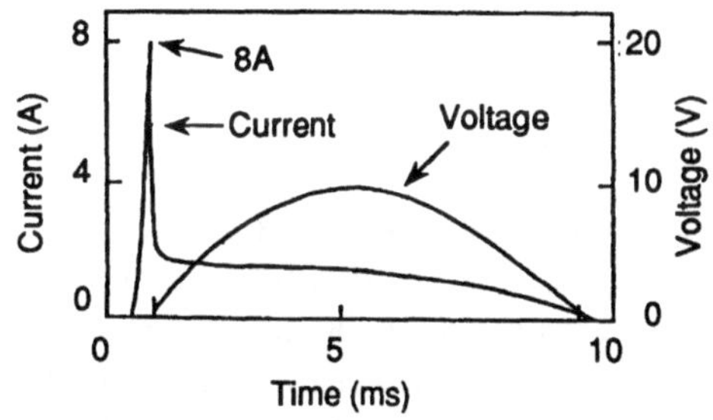

Fig.3 Current limiting waveform

Fig.4 shows the relation between the peak let-through current and the prospective short-circuit current when R1=0,0.2,0.5, 1.0 and 2.0Ω. When the prospective short-circuit current exceeded 10A, values of the peak let-through current remained constant, at nearly 8A. The resistivity changes with respect to prospective short-circuit currents are shown in Fig.5. As the prospective short-circuit current increased, the value of resistivity approached nearly 0.09mΩ · cm.

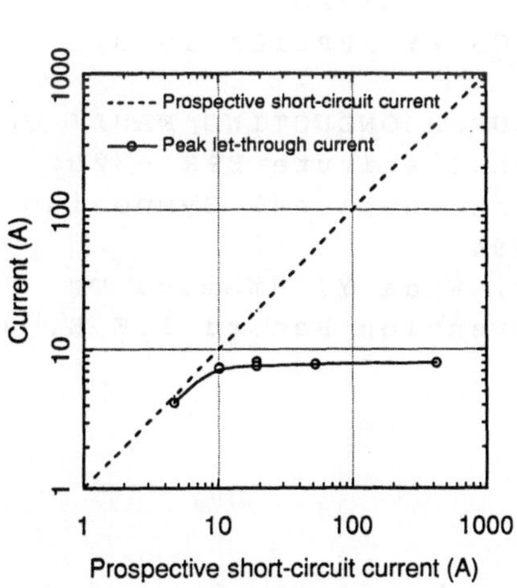

Fig.4 Relation between peak let-throgh current and prospective short-circuit current

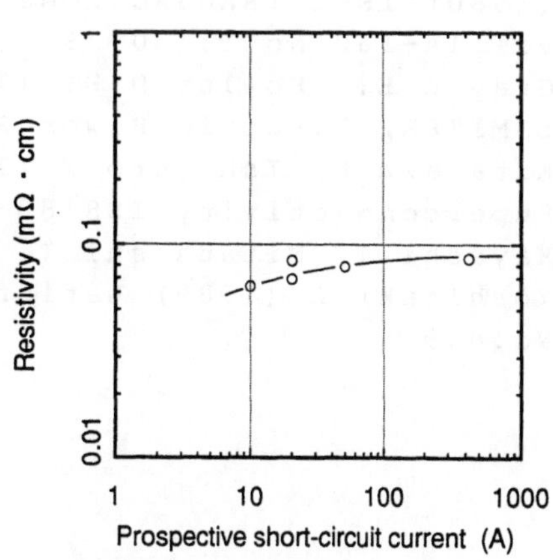

Fig.5 Resistivity change with respect to prospective short-circuit current

We believe the reason why the typical current limiting waveform with an abrupt change in current was obtained is as follows: The high-Tc superconductor used in the experiment changed to the normal conducting state with high resistance when it was quenched. This was because the superconducting film had a high Jc value, and the FCL element was fabricated with a high length to cross-sectional area ratio.

CONCLUSIONS

Experiments on a fault current limiter were conducted using a high-Tc YBaCuO thin film superconducting element. The 0.25μm thick element had a current limiting portion which was 1mm wide and 2mm long, and a critical current of 2.2A. An alternating voltage (10Vp,50Hz) with a prospective short-circuit current of 420A was applied to the element for a half cycle. The results of the experiments were as follows:

1. The peak let-through current was about 8A, which was one-fiftieth of the prospective short-circuit current.
2. The instantaneous value of the current was about 1.5A when the voltage between the voltage measurement terminals was maximum after quenching of the element.
3. The maximum resistivity of the element during the current limiting operation was nearly 0.09mΩ · cm.
4. A typical current limiting waveform which showed an abrupt change in the current was obtained.

REFERENCES

(1) Itoh T, Miyamoto T, Wada Y, Mori T, Sasao H (1973) IEEE TRANSACTIONS ON POWER APPARATUS AND SYSTEMS,Vol.PAS-92, No.4: 1292-1297

(2) Wada Y, Takagi Y, Mori T, Hamano S, Miyamoto T, Itoh T (1980) IEEE TRANSACTIONS ON INDUSTRY APPLICATIONS, Vol.IA-16, No.1: 30-38

(3) Gray K.E., Fowler D.E. (1976) SUPERCONDUCTING FAULT CURRENT LIMITER, Electric Power Research Institute EPRI-329

(4) Kitazawa K, Ishiguro T (1988) International Symposium on Superconductivity ISS'88: 691-694

(5) Hayashi T, Mitsuhashi T, Mori T, Wada Y, Umemura T, Yoshizaki K (1989) National Convention Record I.E.E. JAPAN, No.839

Quenching Characteristics of a 400V-100A Class Superconducting Fault Current Limiter

TAKESHI OKUMA[1], TSUKUSHI HARA[1], TAKAHIKO YAMAMOTO[1], and HARUHIKO HOSHINO[2]

[1] Engineering Research Center, Tokyo Electric Power Co., Chofu, Tokyo, 182 Japan
[2] Power and Substation Department, Kandenko, Tokyo, 108 Japan

ABSTRACT

The suppression of large fault currents in large electric power systems constitutes an important problem for solution. The authors have previously proposed a conceptually new type of superconducting fault current limiter and its application to model distribution substations. In the research described by the present paper, the operating characteristics of a 400V-100A class model superconducting current limiter were evaluated in detail in consideration of the restrictions upon application to the above model system. The characteristics of quench propagation were also assessed. The results demonstrated that the limiting current remained almost constant even when the magnitude and phase of the fault current were varied, and even after repeated operation of the system. The quench propagation velocity in this limiter is estimated as 80m/sec. Based on the above results, no problems arose as regards the applicability of the present type of current limiter to the model system.

KEY WORDS: superconductivity, electric power system, fault current, current limiter, AC superconducting wire

INTRODUCTION

Electric power systems are being expanded and becoming more complex with increasing of power demand, and accordingly the occurrence of short-circuit currents in electric power systems is also increasing. It is therefore a task of critical importance to suppress such fault currents. In view of this situation, we have proposed an approach whereby a new fault current limiter shown in Fig. 1(a), which takes full advantage of the phenomenon known as quenching in the superconductor, is installed on the secondary bus tie at the distribution substation to suppress short-circuit currents(1).

Normally, current flows through the non-inductively wound superconducting trigger coil, which has almost zero inductance. AC niobium titanium (NbTi) wires with filaments of submicron diameter are employed in order to reduce AC losses. If a short-circuit current exceeding the rated current value should flow, then the trigger coil manifests a transition to the normal conductive state, and an appreciable electric resistance develops rapidly, thereby suppressing the current, which is diverted to the limiting coil(2). The diverted current is finally limited by the inductance of the limiting coil. As shown in Fig. 1(b), the coaxial winding of the two coils permits a compact assembly.

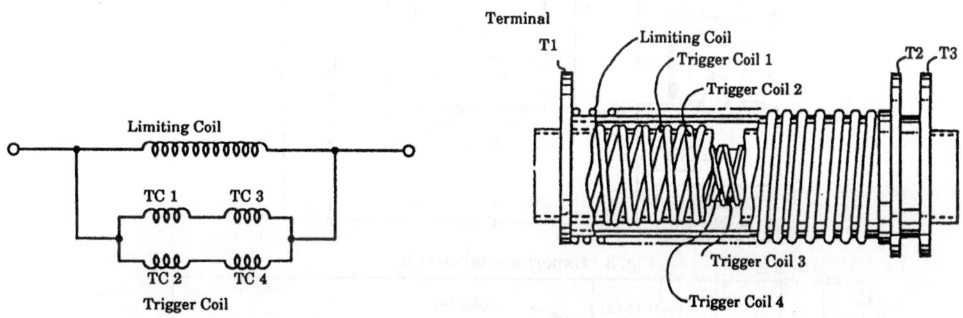

(a) Scheme of operation

(b) Configulation of both trigger coil and limiting coil

Fig.1 Principle and configuration of a new superconducting current limiter

Figure 2(a) shows a model system representing such a substation. The substation is composed of three 66kV/6.6kV transformers. If the individual transformers were linked by means of 6.6kV bus bars in order to maintain a proper balance among the respective loads and ensure efficient operation, then a fault in the 6.6kV system would result in a short-circuit current of 30.2kArms in the 6.6kV circuit breaker. Since this current exceeds the 20kV short-circuit current rating of the devices, the aforesaid linkage is not possible. Hence, this suggests the installation of a superconducting current limiter at the post where the individual transformers are linked.

Let I_0, I_ℓ, and I_s denote the normal current, limiting current, and short-circuit current, respectively, and define $K_1 = I_\ell/I_0$, $K_2 = I_s/I_\ell$. According to previous considerations(3),(4), the values of K_1 and K_2 are clearly confined to the following narrow ranges in the model system of Figure 2.

$$\left.\begin{array}{l} 1.0 \leqq K1 \leqq 6.7 \\ 1.0 \leqq K2 \leqq 10.1 \end{array}\right\} \quad \cdots\cdots\cdots\cdots \quad (1)$$

In the light of the foregoing considerations, a 400V-100A class model superconducting current limiter was fabricated, and current limiting tests were applied to this device over a wide range of varied short-circuit current magnitudes and phases for which verification would be difficult at the voltage level of actual power systems.

In order to reproduce the conditions likely to be encountered in an actual system, the test was conducted within the restricted range indicated by the above inequalities (1). Next, in order to elucidate the quench propagation characteristics which have hitherto remained largely obscure, the voltage generated in each part of the trigger coil was measured by means of a channel waveform detector, after which the quenching state and quench propagation velocity were determined from the measured data.

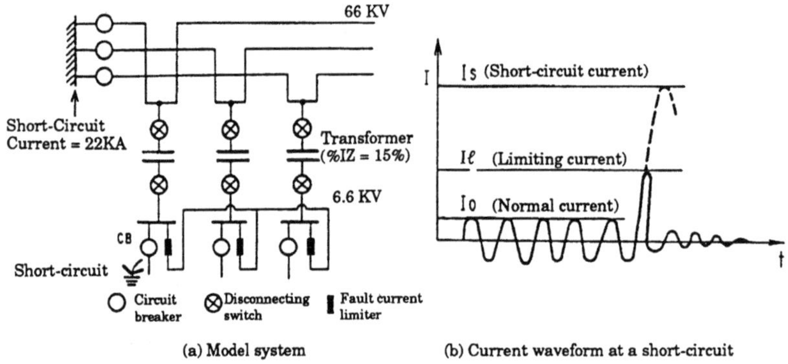

Fig. 2 Model system for application of superconducting current limiters and current waveform following occurrence of short-circuit

EXPERIMENTAL STUDIES

Figure 3 shows the experimental circuit. The measurement sampling time was set at 1.2μsec in order to analyze the fine details of the wave forms. The design value of the critical current was 120A for the said combination of two parallel coil pairs. The resistances Z and R1 were selected so that the foregoing condition (1) could be created, by varying the voltage, whereupon experimental evaluation studies were conducted with respect to the following items.

① Dependence of limiting current I_ℓ and quench current Iq upon crest value of short-circuit current Is
② Changes in I_ℓ due to repeated operation and long-term passage of current
③ Dependence of I_ℓ upon short-circuit phase θ at a short circuit
④ Normal conductivity transition characteristics of trigger coil in terms of resistance created
⑤ Quench propagation velocity in trigger coil

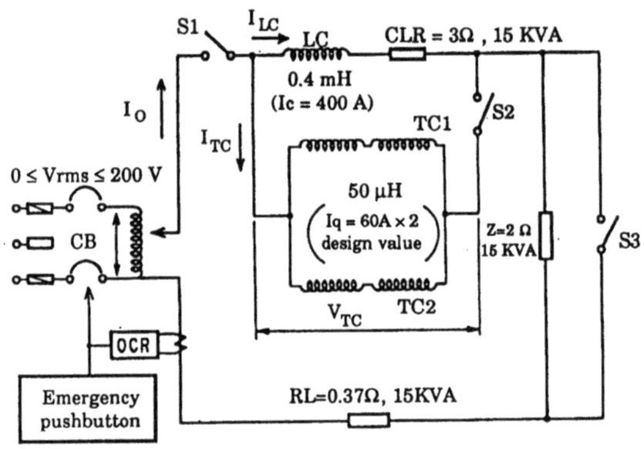

Fig. 3 Experimental circuit

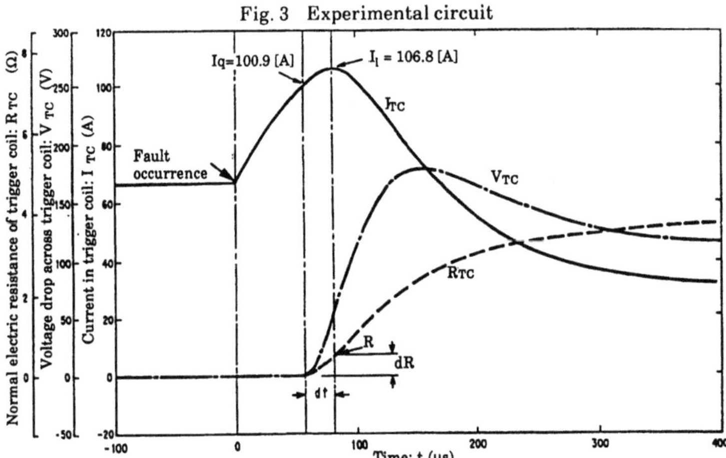

Fig. 4 Measured waveforms in trigger coil

Figure 4 shows the various measured wave forms of the trigger coil for a normal current of Io = 67.0A (applied voltage Vo = 119.3Vrms). When a short circuit occurs, the current in the trigger coil rises abruptly, the limiting current of 106.8A is attained just 82μsec after the occurrence of the short-circuit, and current limiting commences. Since the speed of quench propagation along the trigger coil is finite, even after the current in the trigger coil at the transition to flux-flow or normal conductivity (i.e., the quench current Iq) has been exceeded, the current continues increasing, and the current is finally suppressed at the instant when a limiting current I_ℓ somewhat greater than Iq has been attained.

RESULTS AND ANALYSIS

Concerning evaluation item ①

Experiments were performed in the range $1.1 \leqq K_1 \leqq 2.9$, $2.0 \leqq K_2 \leqq 5.8$, which are within the conditions of actual operation indicated in inequalities(1). The results obtained are shown in Fig. 5.

The number of data corresponding to each indicated value of Is was at least 20, so that the evaluation was conducted with respect to at least 100 data. The settings were chosen so that the short-circuit phase θ would be 0° or 90°, however, the actual phases determined by wave form analysis were in the ranges $0° \leqq \theta \leqq 22°$, $85° \leqq \theta \leqq 112°$, and $155° \leqq \theta \leqq 180°$. As shown by Fig. 5, I_ℓ and Iq displayed almost invariant characteristics despite changes in Is. The mean value of I_ℓ was 115A, with an error less than ±15A, which would constitute no problem in actual practice. As will be explained in the following discussion of evaluation item ③, this error arises from the dispersion in the values of Il due to the short-circuit phase.

Concerning evaluation item ②

As regards the case $K_1 = 1.6$ and $K_2 = 3.9$, when 30 continuous transitions to normal conductivity were created at intervals of 30 to 60sec, the limiting current displayed a maximum value of 116.4A, a minimum value of 109.0A, and a mean value of 116.4A, thus, the experiment demonstrated that I_ℓ was maintained at what could in actual practice be regarded as a nearly constant value, despite repeated actuation of the current limiter.

Furthermore, as regards long-term passage of current, in the case $K_1 = 1.4$, $K_2 = 4.7$, 15 transitions to normal conductivity had occurred after current flow up to a maximum of 3 minutes. I_ℓ remained virtually constant, with an error within the range ± 5A.

Concerning evaluation item ③

Figure 6 shows the relation between the limiting current I_ℓ and the short-circuit phase θ in the ranges $1.1 \leqq K_1 \leqq 2.2$, $2.5 \leqq K_2 \leqq 5.4$. From this figure, one observes that I_ℓ remained at values which could for all practical purposes be regarded as constant, irrespective of the phase θ.

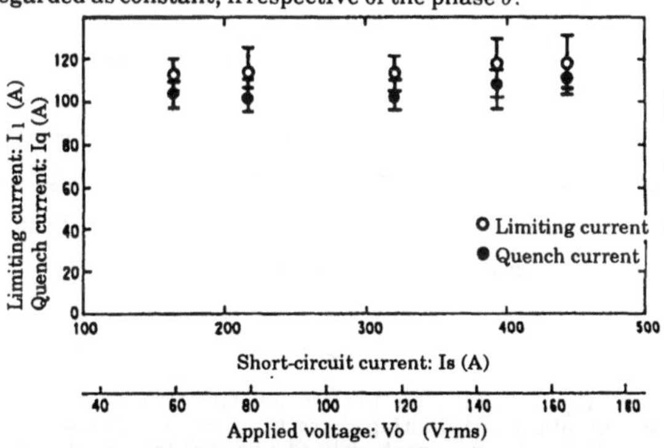

Fig. 5 Short-circuit current dependence of limiting and quench currents

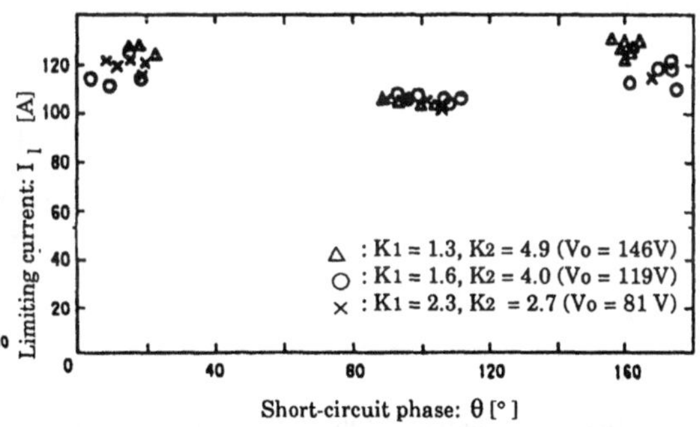

Fig. 6 Short-circuit phase dependence of limiting current

Examination of the details shows that when short-circuits were created in the vicinity of the crest current, i.e., for θ in the vicinity of 90°, the value of I_ℓ tended to be approximately 10% lower than the corresponding values observed for short-circuits at phases θ in the vicinity of 0° or 180°. As indicated by Fig. 4, the current following a short-circuit continues to increase even after the quench current Iq of the superconducting wire is exceeded, and current limiting action commences only when the electric resistance of the trigger coil has attained a value comparable with the impedance of the limiting coil. Consequently, I_ℓ is generally somewhat higher than Iq. As will be explained in the ensuing discussion of evaluation item ④, the diversion of current from the trigger coil to the limiting coil occurs at an even smaller current value, since the resistance created tends to increase rapidly with proximity of the short-circuit phase to 90°.

Concerning evaluation item ④

In order to assess the characteristics of transition to normal conductivity during the initial quenching phase of the trigger coil, the mean rate of increase of electric resistance (dR/dt) during the time interval between the occurrence of the short circuit and the instant of current limiting was determined for various short-circuit currents I_s and short-circuit phases θ. The quantity dR/dt is illustrated among the wave forms in Fig. 4, while the results in the ranges $1.1 \leqq K_1 \leqq 2.9$, $2.0 \leqq K_2 \leqq 5.8$ are shown in Fig. 7. dR/dt displayed relatively higher values for larger values of I_s and values of θ in closer proximity to 90°. This seems to be attributed to the fact that larger values of Is and value of θ closer to 90° are associated with larger rates of change of the trigger coil current (dI_{TC}/dt) between the instant immediately following the occurrence of the short circuit and the limitation of current.

Concerning evaluation item ⑤

Figure 8 shows the longitudinal distribution of voltage and resistance generated in the trigger coil by quenching. Since the resistance generated in the entire coil was 25Ω, the quenched length constituted approximately 17% (25/150Ω) of the total length of the coil. The generated voltage V1 reached approximately 4Ω in 10msec, and subsequently became almost constant owing to the diversion of current to the limiting coil. From this result, considering the fact that the resistance of each superconducting wire is 300Ω, calculation of the quench propagation velocity yielded the value 80m/sec, hence, during a 10msec time interval, the distance of propagation would be at most 0.8m. Since the length between the terminals of the coil is 6.3m, the fact that the quench does not propagate to the other parts of the coil may possibly be attributed to diversion of current to the limiting coil simultaneously with the quenching of the trigger coil, thereby slowing the quench propagation velocity.

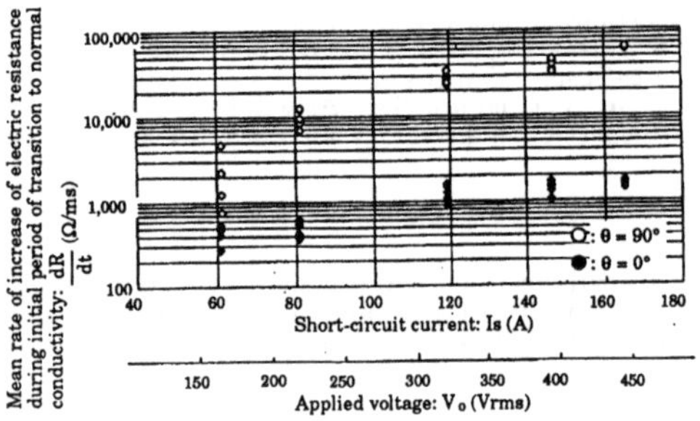

Fig. 7 Mean rate of increase of electrical resistance during initial period of transition to normal conductivity as function of short-circuit current

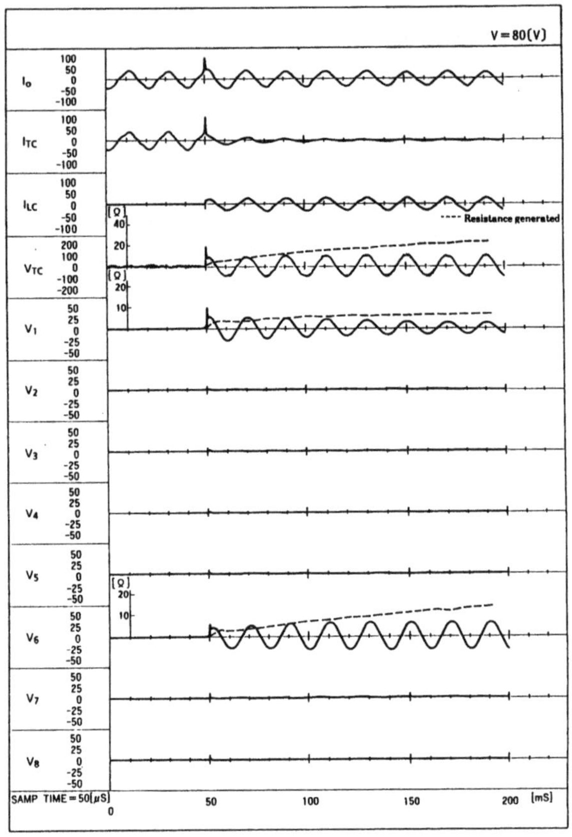

Fig. 8 Examples of observed wave forms

SUMMARY

Using the above-mentioned 400V-100A class model superconducting current limiter, the phenomena which could arise in an actual system were simulated, and experiments for assessment of current limiting characteristics were conducted in detail. These results demonstrated in principle that the value of the limiting current realized when the magnitude of the short-circuit current was varied as well as the reliability after repeated use constituted no particular problems, and that the present superconducting current limiter possessed performance characteristics adequate for use in the model system.

On the other hand, these experiments revealed for the first time that the quenched length of the trigger coil was small, constituting less than 20% of the total length of the coil, while the quench propagation velocity was 80m/sec. Hereafter, the authors intend to conduct more detailed studies of the mechanism of quench propagation in the present superconducting current limiter.

ACKNOWLEDGMENT

The authors hereby express their gratitude to Dr. Ito and other staff members of Toshiba, Ltd., for constructing the 400V-100A class superconductive current limiter used in the present studies.

REFERENCES

(1) T. Yamamoto, T. Hara, K. Okaniwa, Y. Sugiyama, D. Ito, K. Tsurunaga: IEEJ General Convention, No.894 (1990-3)
(2) D. Ito, E. Yoneda, K. Tsurunaga, T. Tada, T. Hara, T Okuma, T. Yamamoto: Proc. of 12th Conf. on Magnet Technology, Leningrad (1991)
(3) K. Okaniwa, T. Hara, T. Yamamoto: IEEJ General Convention, No.893 (1990-3)
(4) k. Okaniwa, T. Hara, T. Yamamoto, H. Hoshino: Proc. of ISS'91, Sendai (1990)

Test Results of the SMES System in the Ariuragawa Hydropower Station

OSAMU KATAHIRA[1], FUMIHIKO FUKUI[1], HIDEHIKO OKADA[2], TADAO EZAKI[2], KOKICHI OGAWA[2], HIROMI KOBA[2], MASAKATU TAKEO[3], SEIKI SATO[3], FUJIO IRIE[4], MASAHIRO TAKAMATSU[5], and TOSHIKAZU SHIMOJO[6]

[1] Kyushu Electric Power Co., Inc., Japan
[2] Oita University, Oita, Japan
[3] Kyushu University, Fukuoka, Japan
[4] Kyushu Engineering Department, Kinki University, Japan
[5] Kobe Steel, Ltd., Hyogo, Japan
[6] West Japan Engineering Consultants, Inc., Japan

ABSTRACT

In light of high energy storage efficiency and very quick response in energy delivery, superconducting magnetic energy storage(SMES) systems are expected to find many opportunities of application for power system stabilization and for load leveling in place of pumping-up power plants. The authors have continued studies on SMES application to the power system since 1982[1-3] and recently carried out a field tests, mainly relating to system stabilization, with a SMES connected to Ariuragawa hydropower station of Kyushu Electric Power Company. This paper describes some results of the test.

KEY WORDS : SMES , power system , system stabilization

OUTLINE OF THE TEST

Date: November 20-22, 1990
Site: Ariuragawa power station (Fujihira, Genkai-cho, Saga Prefecture)
Items: System stabilization test, Momentary voltage drop compensation test, etc

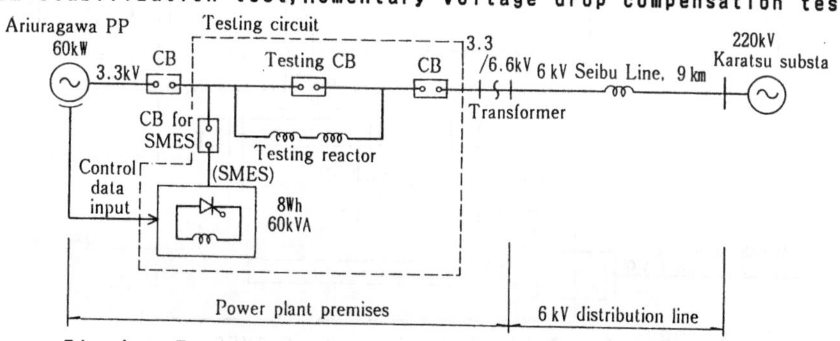

Fig.1 Testing circuit for system stabilization

Table 1 Technical data of major components

SMES	Power System
Superconducting magnet:	Line connected for testing:
Storage capacity - 30kJ(8Wh)	A 6.6kV general distribution line,
Maximum current - 100A	9 km long to Karatsu substation,
Inductance - 6.1H	belonging to 220kV Karatsu Network
Superconducting material - Nb-Ti	Ariuragawa hydropower station:
Type of coil - Solenoid	Rated output - 60kW
Cooling - Helium (pooled)	Available water flow - 0.22m³/s
Converter:	Net head - 38.485m
Capacity - 60 kVA(30kVA×2)	Turbine: Power - 67kW
Switching devices - GTO thyristors	Speed - 1,200RPM
(2500V, 300A)	Generator: Output - 72kVA
Type of control: PWM	Voltage - 3,300V

Testing Circuit and Method

The testing circuit is shown in Fig.1. Ariuragawa power station is a small hydro power plant, belonging to 220kV Karatsu Network. Its generator has a terminal voltage of 3.3kV and it feeds the power to the 6.6kV general distribution line through a step-up transformer in the premises. For this testing, a part of the 3.3kV bus of the power station was temporarily sectioned so that it provided an independent bus for connecting the testing circuit which consisted of a testing reactor, a SMES system and its switchgear and other testing equipment.

System stabilization tests were carried out in this way : with the generator operating at a constant power, the testing circuit breaker was opened in order to cause an oscillation of the generator, and then the SMES' control was applied to the system for suppressing that oscillation. Momentary voltage drop compensation tests went on as this : a three-phase short circuit through the testing reactor was caused at the end of the distribution line for the duration of about 1 sec and the voltage drop appearing at the middle point of the reactor was compensated by means of the SMES which was connected at that point.

Technical data of major components of the testing circuit and the schematic diagram of the SMES system are shown in Table 1 and Fig.2. The superconducting magnet consisted of three sets of magnets designed for system stabilization and connected in series, which had a total inductance of 6.1H and an energy storage capacity of 30kJ. The converter system was made up of two converter units of 30 kVA each. The type of control for this system was a pulse width modulation, in which six current pulses appeared every half cycle.

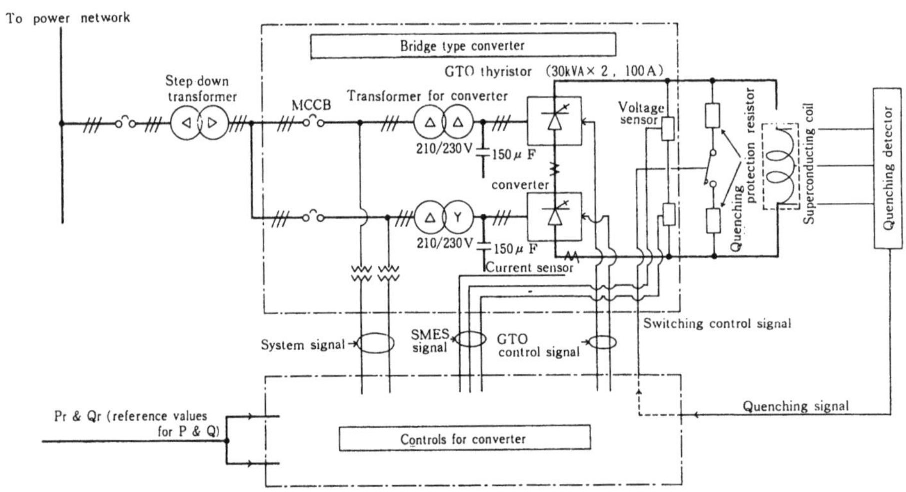

Fig.2 Schematic diagram of SMES system

TEST RESULTS

System Stabilization Test

With the generator operating at a power of 50kW and the SMES system out of use, the oscillation of the generator power as shown in Fig.3. This oscillation was featured by the period of about 1.5 sec and the duration of about 10 sec .

In contrast, shown in Fig.4 is the result of a similar test except that the SMES system was in use so as to damp down the oscillation of the generator, referring to the deviation of its velocity($\Delta\omega$). In this case, the oscillation reduced quickly to get stable within about 1 sec after a disturbance, which was brought about by a feedback gain of 114 kW/Hz and resulted in a maximum power of 23.0kW actually exerted by the SMES that served as a load, representing a ratio of about 46 % of the generator power. The direct current of the SMES increased gradually, implying absorption of energy, from an initial value of 70 A up to 86.6 A, resulting in a total amount of absorbed energy of about 10kJ. The proportion of the energy in two streams coming into the SMES, one from the generator and the other from the distribution line, is determined by the conditions of the system involved. In thie case, about 80 % of the total energy absorbed by the SMES came from the generator and the remainder or 20 % from the distribution line.

Fig.5 shows a result of system stabilization test carried out with the SMES under a restricted output. This test was designed based on a possibility that a relatively small SMES may be installed for system stabilization. In fact, a SMES output of ±4 kW to be used with a generator of 50 kW, representing about 8 % of the generator power, proved satisfactory through this test.

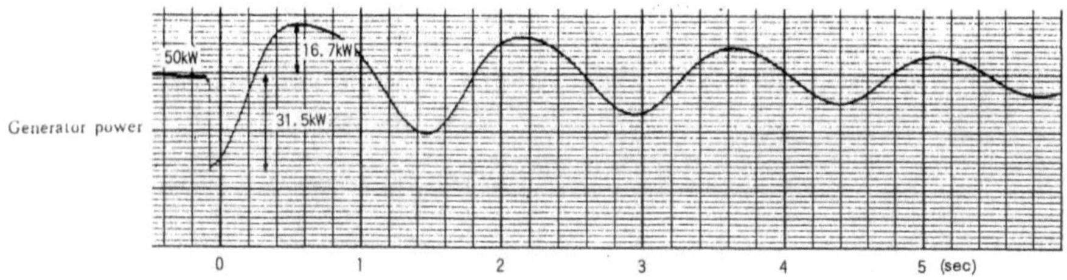

Fig.3 System stabilization test - SMES out of use

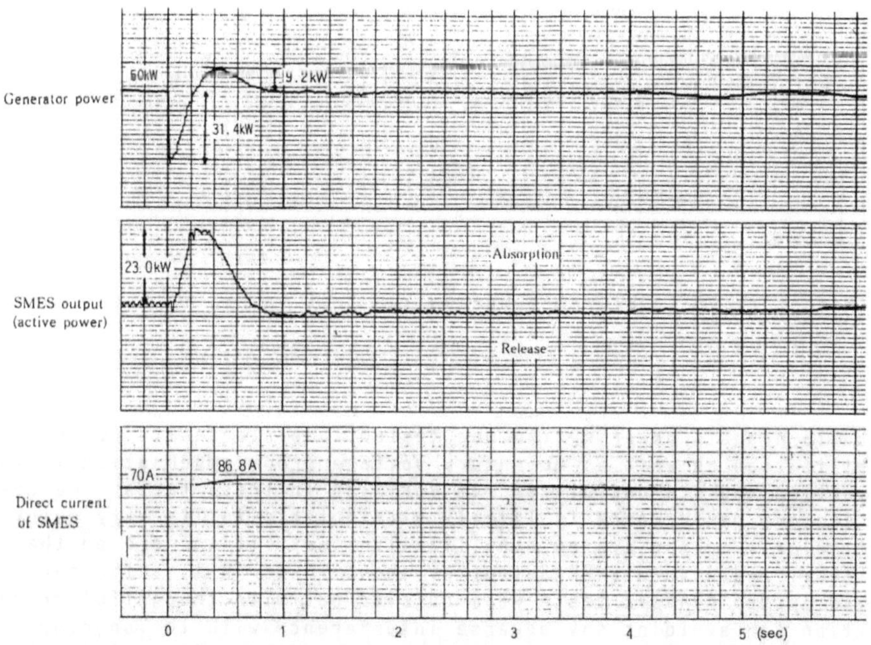

Fig.4 System stabilization test - SMES in use ($\Delta\omega$ detection - P control)

Momentary Voltage Drop Compensation Test

A result of momentary voltage drop compensation test is shown in Fig.6 . The voltage at the middle point of the testing reactor dropped by about 51% when a three-phase short circuit was given at the end of the distribution line. However this drop was compensated by the SMES to attain an ultimate level for compensation in such short time as about 0.1 sec. The value of the voltage compensation was about 240V, which was caused by reactive power of 15 kVar lead controlled by the SMES.

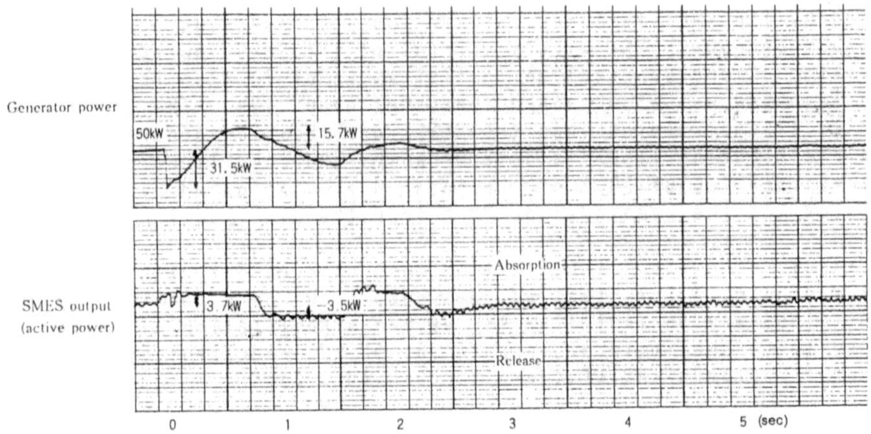

Fig. 5 System stabilization test - SMES in use with a restricted output ($\Delta\omega$ detection - P control)

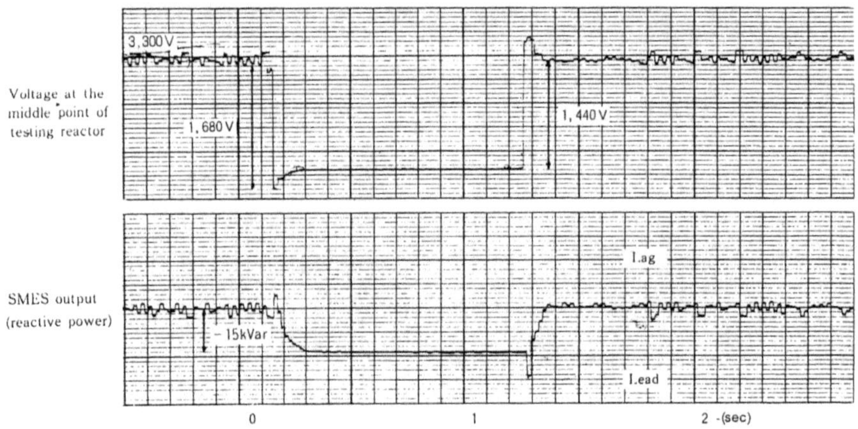

Fig. 6 Momentary voltage drop compensation test - SMES in use

CONCLUSION

A SMES connected to a small power system was tested to evaluate its performance for system stabilization as well as momentary voltage drop compensation in terms of dynamic properties and its usefulness was verified in these tests. We deem that such satisfactory performance is largely attributable to its very quick response. Particularly in the system stabilization test, it was proved that the generator was stabilized with a SMES output of about 10 to 50 % of the genarator capacity . These system dynamic tests were carried out with thoughtful precautionary preparation for avoiding any adverse interference with the ongoing commercial supply , resulting in a satisfactory course of development in this regard.

REFERENCES

1. F. Irie, Workshop on SMES as satellite meeting of ICEC10, 1984.
2. H. Okada, T. Ezaki, K. Ogawa, H. Koba, M. Funaki, S. Sato, F. Irie, J. Chikaba, K. Terazono, M. Takamatsu, N. Kawakami and M. Hirano, "Experimental study of SMES with dc inter tie for power line stabilization,"PESC'88 Record, Kyoto, Japan, 1988, pp. 326-333.
3. H. Okada, T. Ezaki, K. Ogawa, H. Koba, F. Irie, M. Takeo, S. Sato, K. Terazono and M. Takamatsu, "Study of SMES system with dc intertie for power line stabilization," Proc. MT-11, Tukuba, Japan, 1989, pp. 396-401.

Development of a 1 MJ SMES System with Use of a Superconducting Magnet Having Function of "Enthalpy Protection"

YUJI KOBAYASHI[1], MISAO JOUSHO[1], HITOSHI HONMA[1], KATSUHIKO ASANO[2], MASAICHI ENDO[2], and YUKIO ISHIGAKI[2]

[1] Tohoku Electric Power Co., Inc., Sendai, 981 Japan
[2] Hitachi Ltd., Hitachi, Ibaraki, 317 Japan

ABSTRACT

A 1MJ superconducting magnetic energy storage (SMES) system has been developed by a joint reserch of Tohoku Electric Power Co. Inc., and Hitach Ltd.. A magnet applied to the SMES has a function of "Enthalpy Protection". A new concept, that is, QI simultaneous control algorithm has been adopted to the main control system. The aspects mentioned above have verified through a system test carried out. Especially, the first test in the world concerning to the "Enthalpy Protection"has been successfully achieved.

KEY WORDS: power system, stability, enthalpy protection, superconductivity

1. INTRODUCTION

A superconducting technology is a promising one for an electric equipments. It is expected to improve an efficiency of electric power machinery, install the system more compact, and reduce the operating costs. A SMES system has superior features such as high efficiency, high-speed response, and simultaneous control of active and reactive power. Besides, various applications are possible corresponding to the capacity of the system.
As the first step of the development of superconducting equipments to be used in future power systems, the authors manufactured a 1MJ SMES system and have started demonstoration tests. The SMES system is featured by the function of "enthalpy protection", that is, the stored energy is absorbed in the superconducting coil when a quench occures.
PQ/QI simulataneous control is also adopted to control the SMES system for power system stabilization.

2. ENTHALPY PROTECTION SYSTEM

When a quench occurs, the stored energy is discharged to a dumping resistor equipped outside the magnet in a conventional protection methods (a dumping resistor protection system). In the enthalpy protection system, when the coil quenches, the stored energy is absorbed in the coil.
The superconducting magnet with the enthalpy protection function under development will be available to a large-scaled SMES system to prove the reliability. The comparison between these two systems mentioned above is shown in Table 1.

Dumping Resistor Protection System	Enthalpyprotection System
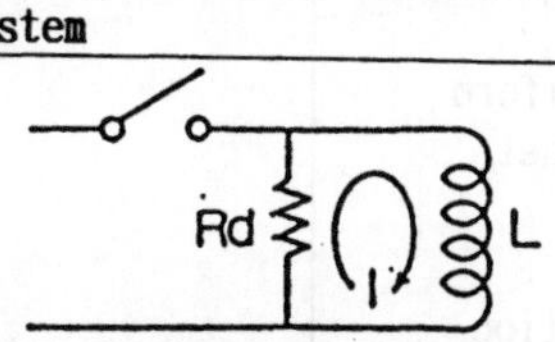	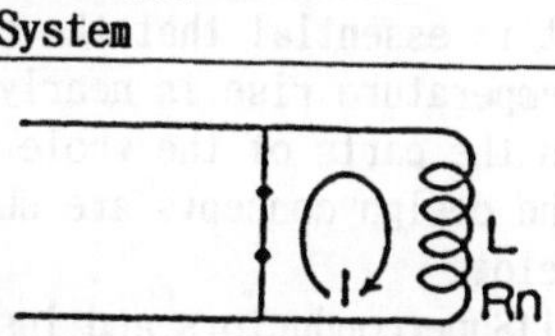
Swich should be open when the coil quenched Stored energy is absorbed in external dumping resistor	Stored energy is absorbed superconducting magnet itself

Table 1 Quench protection Method

The following relations are required to the superconducting magnet design with the enthalpy protective function.

$$E < \int^{M_{max}} \int^{T_{max}} C \, dT \, dM = \int^{M_{max}} H \, dM \qquad (1)$$

E : stored energy
M : mass of superconducting magnet coil
C : specific heat
H : cdT : enthalpy
T : temperature

3. PQ/QI SIMULTANEOUS CONTROL SYSTEM

The SMES system has the capability of stabilizing disturbances generated in the power system by means of a simultaneous control of the active power (P) and the reactive power (Q). However, an operation method in a phase of preservation dose not have been studied yet.
In QI simultaneous control method which is different method from PQ simultaneous control one, Q and the magnet current (I) can be controlled simultaneously. This method is more practical for the operation of the SMES. Therefore, the PQ/QI simultaneous control system is adopted to the SMES system.

4. SYSTEM CONFIGULATION AND FUNCTIONS

The SMES system is consisted of a superconducting magnet, GTO thyristor converter and control system. The system configuration is shown in Figure 1 and the main specifications are also shown in Table 2.

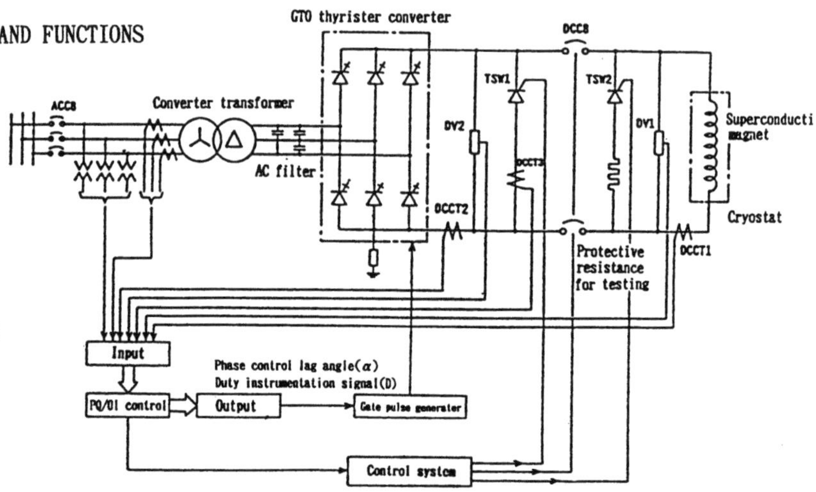

Fig 1 System configuration

The superconducting magnet, the stored energy of which is to 1MJ, can charge and discharge its energy according to pulse operation.
To make the enthalpy protection, it is essential that the coil temperature rise is nearly uniform in the parts of the whole magnet, The design concepts are shown below.
(1)Superconductors and insulation spacers are used for energy absorbing material.
Temperature rise is restricted below 100 K.

Equipment	Item	Specification
Superconducting Magnet	Stored energy	1.0 MJ
	Inductance	2.0 H
	Rated current	1,000 A
	Coil dimention	
	Inner diameter	360 mm
	Outer diameter	814 mm
	Average diameter	587 mm
	Hight	334 mm
	No. of turns	2,400 turns
	Magnetic field (center)	4.6 T
	Magnetic field (maximum)	5.3 T
	Current density	31 A/mm
	Cryostat outer diameter	1.2 m
	Cryostat material	SUS 304
GTO thyristor converter	Capacity	25 kW
	Current	1,000 A
	Voltage	25 V

Table 2 System specification

(2)The superconductor is composed of 14 strands. Each strand contains 6 μm NbTi filament containing copper and copper-nickel alloy to reduce AC losses. As a core material, copper is used to conserve enthalpy.

(3)To increase normal propergation velocity for coil protection, liquid helium dumping method is used.

Table 3 shows the design value of a superconductor, and Figure 2 shows the cross-section of a superconductor.

Item	Design value
Conductor hight	6.4 mm
Conductor width	2.6 mm
Number of strand	14
Strand diameter	1.0 mm
Number of filament	104,000
Filament diameter	6 μm
Cu/CuNi/NbTi	4.4:0.9:1
Residual resistivity ratio	≧80
Critical current	3,130(5T, 4.2K)

Table 3 Design value of superconductor

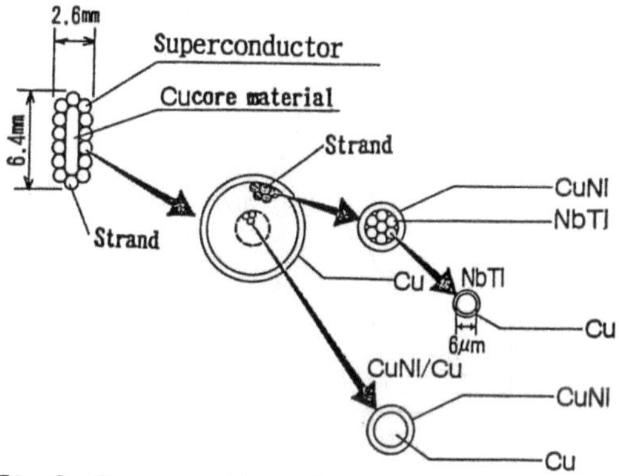

Fig. 2 Cross-section of superconductor

The top and sides of the coil are covered with a coil-cover, and it is kept in the cryostat.
Figure 3 shows the cross-section of the magnet. The coil-cover and the liquid helium vessel are equipped with separating tubes to emit helium gas. When the coil quenches, the helium gas is fed into the coil-cover to remove the liquid helium around the coil. In this way, cooling conmditions are weakened in a short time and the normal conductivity area increases quickly.

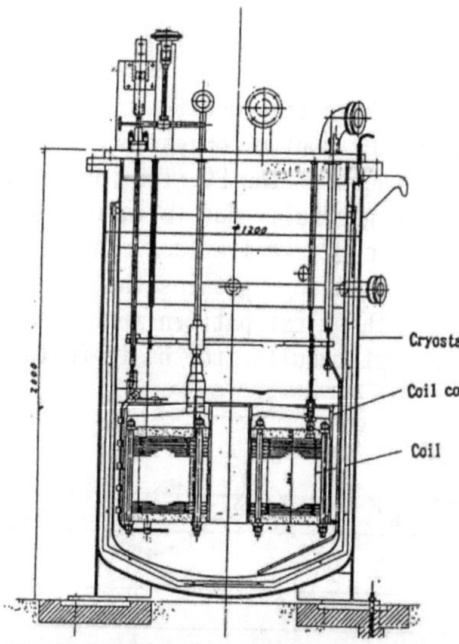

Fig. 3 Vertical Cross-section of Super conducing Magnet

5. TESTS

5.1 VERIFICATION OF THE ENTHALPY PROTECTION SYSTEM

Verification test was carried out under the rated current of 1000A (corresponding to 1 MJ), and the superconducting coil is forced to quench by a heat adhered to the coil. After the quench detected, liquid helium dumping system is begin to operate. Figure 4 shows the temperature rise change of the superconducting coil. The temperature of the intermidiate portion starts to rise bythe heater. The temperature rise of the upper portion is caused by the exposure rate of the coil to the helium gas port .
Temperature rise does not start simultaneously on both portions. However, the final temperature rise reaches almost the same level. Figure 7 shows the propergation velocity of the normal conductivity area.The liquid helium was dissipated in a short time (about 10 seconds), and the normal conductivity area was rapidly extended, which certifies that the liquid helium is effectiveon the extention of normal conductivity area .
As these test results, the enthalpy protective function of the superconducting magnet is verified.

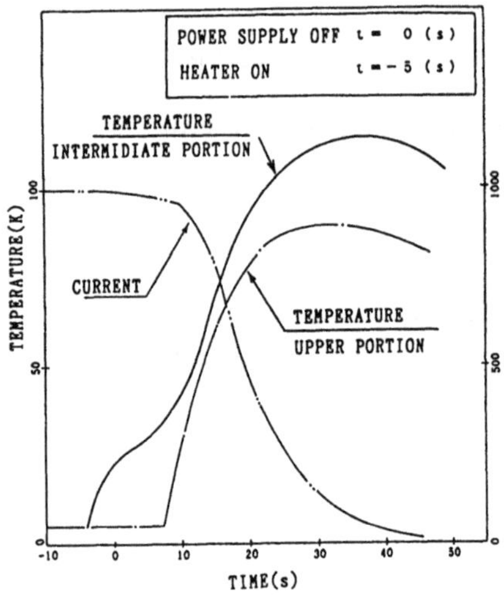

Fig. 4 TEMPERATURE RISE OF THE COIL

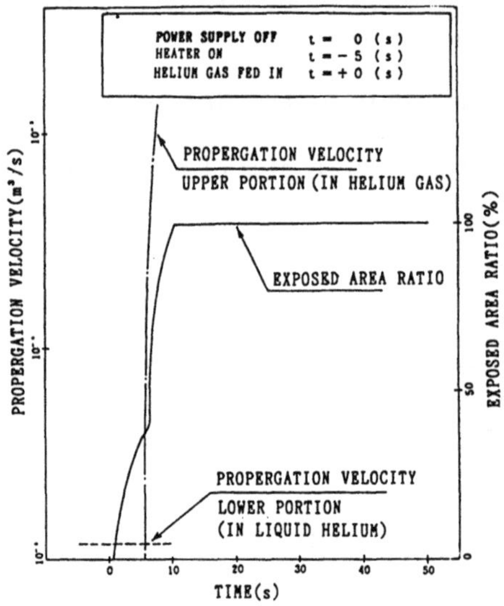

Fig. 5 PROPAGATION VELOCITY

5.2 QI SIMULTANEOUS CONTROL TESTS

Figure 6 shows the results obtained from PQ-QI-PQ simultaneous control tests. In the PQ simultaneous control operation, P and Q values follow the references of Pr and Qr. Namety the effectiveness of the PQ simultaneous control is verified. In this operation, the coil current is dumped gradually due to various losses in the circuit.
Switched to QI simultaneous control, the coil current increases to the reference value Ir. However, Q value maintains the reference value Qr. Switched back again to PQ simultaneous control, the first pattern mentioned above is observed.
As the test results, the basic feature of QI simultaneous control is successfully confirmed.

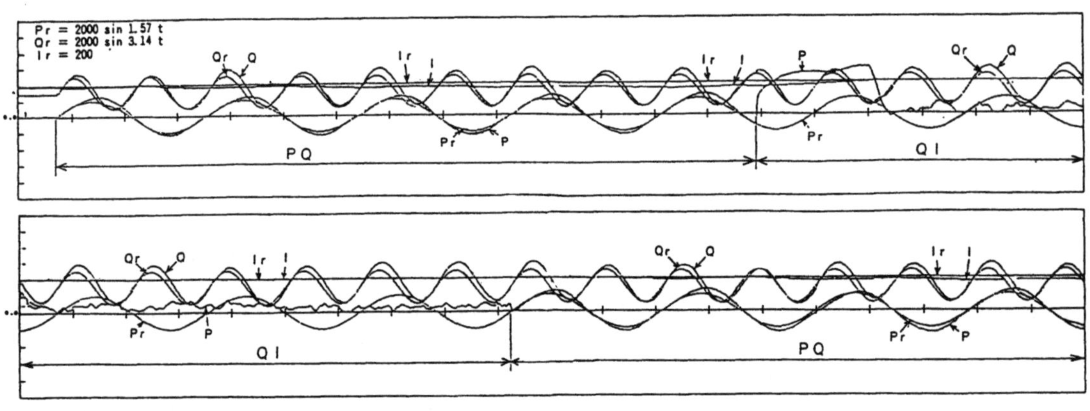

Fig. 6 PQ-QI SIMULTANEOUS CONTROL TEST

6. CONCLUSION

Through these various tests, the enthalpy protection system under coil quench and QI simultaneous control function of the SMES system are respectively verified.
We are planning to carry out demonstration tests, such as transient stabilization test, dynamic stabilization test, voltage fluctuation test, load leveling fluctuation test, short time power failure test, etc. by connecting this SMES system with simulated transmission utility.
It will be sured by these next step tests, that will verify that the SMES system contributes to power system stabilization.

Temperature Rise in a Superconducting Transformer Winding Overheated by Quenching

BONG HWAN OH[1], KAZUAKI ISHIKAWA[1], NAOKI HAYAKAWA[1], HITOSHI OKUBO[1], YUKIO KITO[1], and MASANORI YAMAMOTO[2]

[1] Department of Electrical Engineering, Faculty of Engineering, Nagoya University, Nagoya, 464 Japan
[2] Research & Development Center, Takaoka Electric MFG. Co., Ltd., Nishibiwajima, Nishikasugai, Aichi, 452 Japan

ABSTRACT

The temperature rise in a superconducting winding after its quenching is investigated with a developed 6600/210 V, 100 kVA hybrid-type superconducting transformer. In an overcurrent test, the low-voltage superconducting winding was quenched and ruptured due to partial overheating. The induced temperature rise is estimated from the distortion of Nb-Ti ultra-fine filaments in the cross-section of the superconducting wire in the vicinity of the broken point. The progress of the temperature rise and its propagation along the superconducting wire is revealed through a computer simulation. The derived resistance increase in the wire is supported by the actual experimental data.

KEY WORDS: superconducting transformer, quench, overheat, temperature rise

1. INTRODUCTION

In superconducting transformers, there is a possible quenching (superconducting- to normal-state transition) of windings by an overcurrent due to short-circuit faults or inrushes following energization. The superconducting transformer would fail in its function by the quenching. Thus, the quenching is to be a very important factor on operation of superconducting transformers. Nevertheless, quite few experiments on the quenching characteristics have been reported in actual superconducting transformers.

We have already developed a 6600/210 V, 100 kVA hybrid-type superconducting transformer (H-SCT) [1]. It is composed of a superconducting winding for its low-tension side and a cryoresistive copper winding for its high-tension side. The quenching characteristics of the superconducting winding in the H-SCT are described in this paper. In an overcurrent test, the superconducting winding was partially overheated due to excessive ohmic heating by post-quenching current during 1 sec, until it was ruptured at three points. The sample wires with short length are taken out of the superconducting winding after the test. The cross-sections of these samples are observed by means of a SEM (scanning electron microscope) and an X-ray microanalyzer. From these microscopic inspections, the geometric arrangement of Nb-Ti ultra-fine filaments is found to be more distorted as close to the ruptured point. On the other hand, the critical current of a short sample picked out of the burned portion becomes lower than that of the original superconducting wire. On the basis of these experimental data, it is expected that the partial, steep temperature gradient was developed around the ruptured points and propagated along the wire. The temperature distribution along the superconducting winding is derived with the lapse of time after the quenching by a computer simulation, where the ohmic heating in the copper stabilizer, thermal conduction along the winding and also the thermal dissipation into the liquid helium are taken into account.

2. OVERCURRENT TEST OF HYBRID-TYPE SUPERCONDUCTING TRANSFORMER (H-SCT)

The structure of the developed superconducting transformer is shown in Fig. 1 [1]. It has a liquid-helium-cooled superconducting winding rated as 476 A for 210 V winding and a liquid-nitrogen-cooled cryoresistive copper winding rated as 15.2 A for high-voltage, 6600 V winding. The iron-core is placed outside the cryostat vessel and kept at the room temperature. Thus, this superconducting transformer is referred to as the "H-SCT": hybrid-type superconducting transformer. The low-voltage, high-current winding is a single-layer coil with 46 turns of superconducting wire wound on FRP former. The total length of the superconducting wire is 49 m. The wire consists of 7×7 strands. The cross-section of a single strand is shown in Fig. 2. This photograph shows that the Nb-Ti ultra-fine filaments are arranged in an annular zone in the copper stabilizers. Dimensions of the superconducting wire are summarized in Table 1.

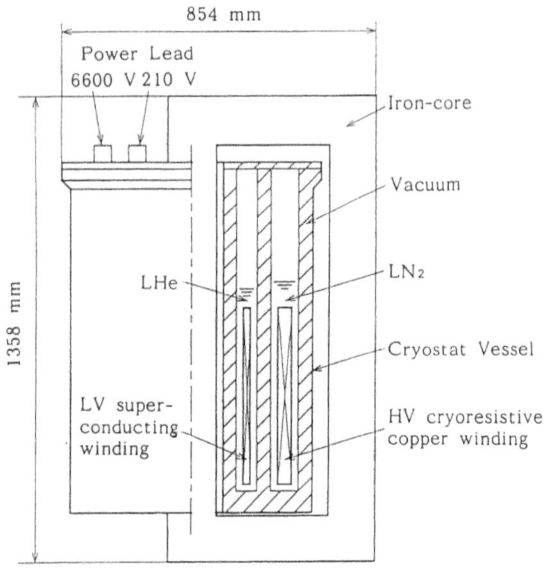

Fig. 1. Structure of the H-SCT. (Ref. [1], © 1991 IEEE)

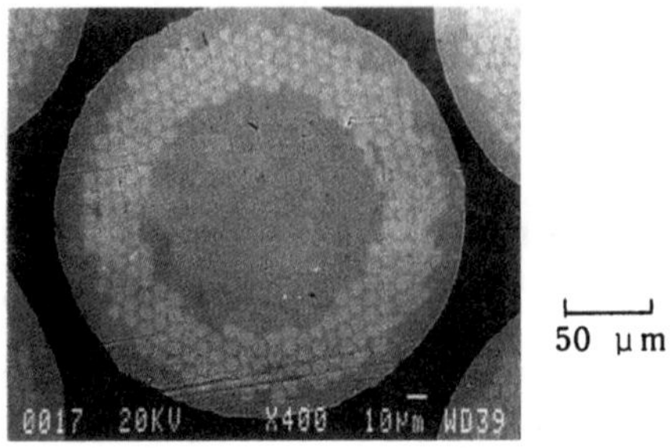

Fig. 2. Cross-sectional view of a single strand of the superconducting wire.

Table 1. AC superconducting wire.

Outside diameter	2.18 mm
Number of strands	7×7
Strand diameter	0.22 mm
Diameter of Nb-Ti ultra-fine filament	0.55 μm
Nb-Ti/Cu/Cu-Ni	1/1.2/1.7

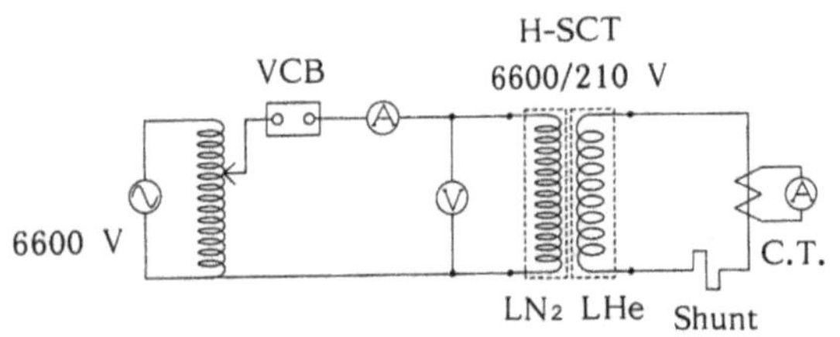

Fig. 3. A circuit diagram of the overcurrent test.

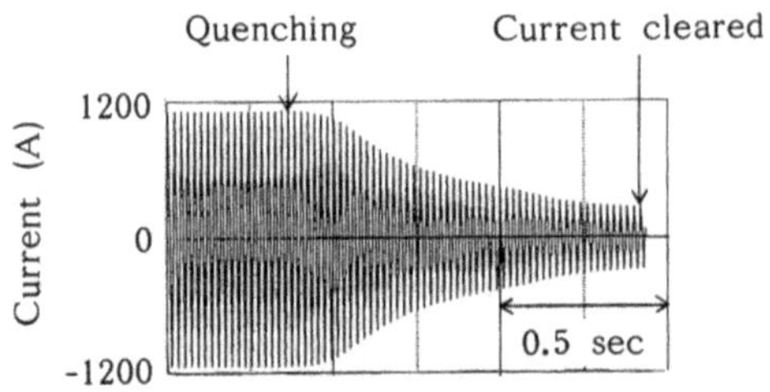

Fig. 4. Current waveform before and after quenching in the overcurrent test.

An overcurrent test of the H-SCT is carried out in a circuit as shown in Fig. 3, where the low-voltage superconducting winding is short-circuited. The H-SCT is energized from the high-voltage side with gradual increase of its applied voltage by an induction regulator. The current waveform before and after the quenching is shown in Fig. 4. The critical current level is 1120 A(peak), i.e. 790 A(rms). The post-quenching current is slowly decayed with the lapse of time and finally interrupted by a backup circuit breaker in 1 sec after the quenching. This period has been taken a little longer in order to measure the decay process of the post-quenching current. Liquid helium evaporated as much as 11 liters during this period. As the result of this experiment, the superconducting wire happened to be broken.

The H-SCT was disassembled and the superconducting winding was taken out of the cryostat. The winding was found to be partially burned into brown. The brown-colored length was 1.4 m in 4th and 5th turns from the uppermost turn. The superconducting wire was also ruptured at three points on them. The other part of the wire was apparently not damaged.

3. THERMAL DISTORTION OF Nb-Ti FILAMENTS IN THE SUPERCONDUCTING WIRE

The geometric arrangement of Nb-Ti filaments is found to be thermally distorted in the cross-section of the burned wire through SEM observations. The metallic composition of the superconducting wire is also inspected by an X-ray microanalyzer. The cross-sectional views of the centered 7-strand are shown in Fig. 5(a) to (c) at the locations of 0.175 m, 0.075 m and 0.01 m from a ruptured point, respectively. It is pointed out that the geometric arrangement of superconducting fine filaments is distorted more irregularly at the locations closer to the ruptured point. As shown in Fig. 5(a), the annular arrangement of Nb-Ti filaments is likely to remain unchanged, but the ultra-fine structure begins to collapse. In Fig. 5(b), some strands are partially fused. In Fig. 5(c), at the distance of 0.01 m off from the ruptured point, the whole strands are to be mutually fused and the dendrite structure of Nb is found through the inspection by an X-ray microanalyzer. Thus, the wire

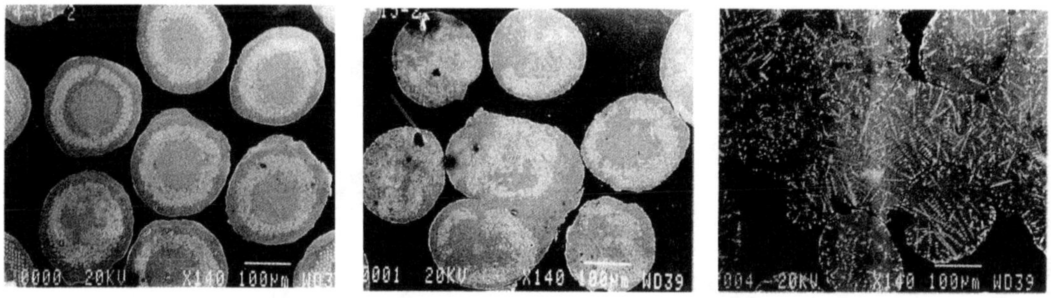

(a) x=0.175 m (b) x=0.075 m (c) x=0.01 m

100 µm

Fig. 5. Cross-sectional views the burned superconducting wires.
(x: Distance from the ruptured point)

temperature at the ruptured point is considered to have reached higher than the melting temperature of copper, 1356 K, at least. No distortion of Nb-Ti filaments is recognized in the cross-section of the wire at the distance more than 0.175 m off from the ruptured point. These results suggest that the steep temperature gradient was developed around the ruptured point and propagated along the wire.

4. REDUCTION OF CRITICAL CURRENT VALUE IN THE BURNED WIRE

The sample wires of 13 pieces with each length of 0.15 m are taken out of the superconducting winding. Each sample shows the superconductivity when it is re-cooled to liquid helium temperature. Then, the critical current value of each sample is measured. An ac current of 2000 A(peak) with a ringing frequency of 63 Hz is supplied into the samples from a capacitor bank of 450 µF through a reactor of 14.2 mH. The critical current values of the sample wires are summarized in Fig. 6 as a function of the distance from the ruptured point. The critical current value of the burned wire at the distance of 0.6 m from the ruptured point is 1220 A(peak). This value is 70% of the critical current value of the original superconducting wire [1]. The critical values, on the other hand, are kept nearly constant as high as 1730 A(peak) at the distance more than 0.9 m from the ruptured point. This value is in good agreement with that of the original wire. The critical current value of a superconducting wire is affected by the heat treatment temperature [2] in its manufacturing process and considered to be reduced when the wire is overheated beyond the temperature. In the case of our superconducting wire, the heat treatment temperature has been about 700 K. Thus, the wire temperature could be raised to higher than 700 K at least within the distance of 0.6 m from the ruptured point.

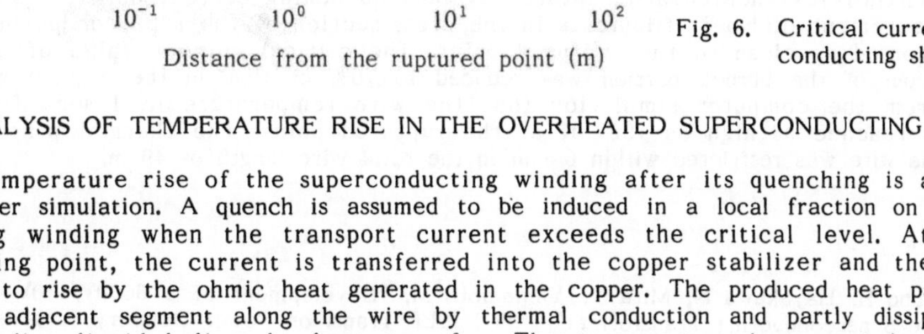

Fig. 6. Critical currents of superconducting short samples.

5. ANALYSIS OF TEMPERATURE RISE IN THE OVERHEATED SUPERCONDUCTING WIRE

The temperature rise of the superconducting winding after its quenching is analyzed by a computer simulation. A quench is assumed to be induced in a local fraction on the superconducting winding when the transport current exceeds the critical level. At this quench initiating point, the current is transferred into the copper stabilizer and the temperature begins to rise by the ohmic heat generated in the copper. The produced heat partly transmits to an adjacent segment along the wire by thermal conduction and partly dissipates into the surrounding liquid helium by heat transfer. The temperature increment is calculated by solving the the heat balance equation [3].

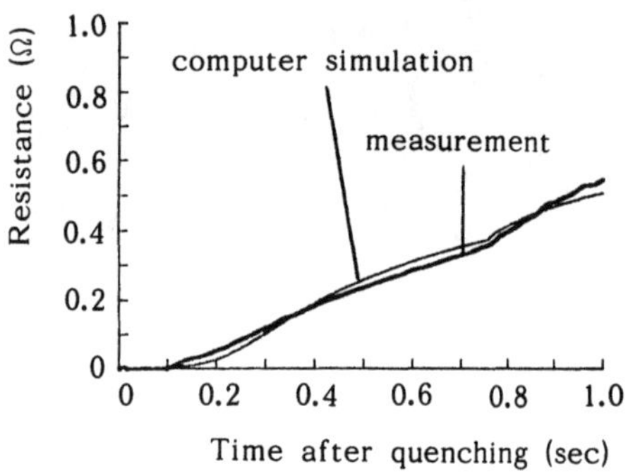

Fig. 7. Resistance increase after quenching.

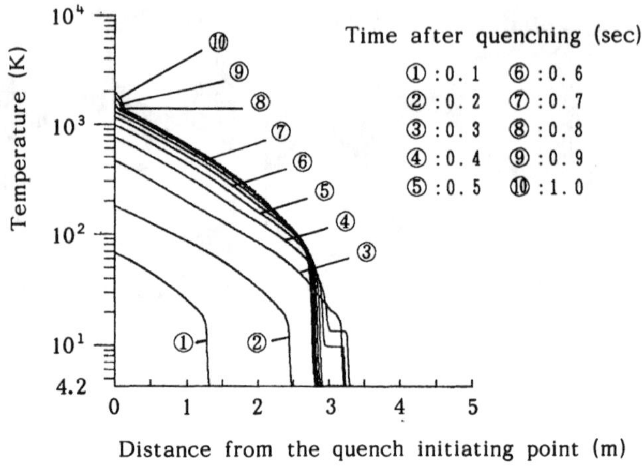

Fig. 8. Progress of temperature distribution after quenching.

The resistance increase of the superconducting winding is calculated as shown in Fig. 7. The simulated curve is drawn by a thin solid line, while the experimental data by a thick solid line. The simulated curve is satisfactorily coincident with the experimental data. Especially, sudden increases of the resistance are recognized on both curves at t=0.75 sec. This is brought about by discontinuous increase of the resistivity of copper at the melting temperature, 1356 K. The progress of the temperature distribution along the superconducting wire is represented in Fig. 8. The temperature distribution is regarded as taking a symmetrical configuration in both sides of the quench initiating point, which is taken as the origin (x=0 m) of abscissa. The temperature at x=0 reaches as high as 1967 K within 1 sec. This temperature is much higher than the melting temperature of copper, 1356 K and nearly equal to that of Nb-Ti, 2173 K. The calculated temperature rise is high enough to result in rupture of the wire due to melting at this location. The temperatures are estimated as 1960 K at x=0.01 m, 1716 K at x=0.075 m and 1326 K at x=0.175 m, respectively. These values justify the distortion of geometric arrangement of superconducting fine filaments shown in Fig. 5(a) to (c). The temperature reaches 930 K at x=0.6 m and 705 K at x=0.9 m. These temperature rises support the reduction of critical current values of the superconducting wire shown in Fig. 6. On the other hand, the steep temperature rise is developed within 3.3 m from the quench initiating point as shown in Fig. 8. Thus, the thermal propagation length along the wire is restricted within 6.6 m in both sides of the quench initiating point in the total wire length of 49 m.

The curve ① of Fig. 8 shows the temperature distribution in 0.1 sec after the quenching. The temperature at the quench initiating point, x=0, is only 67 K at that time. The superconducting winding will not be damaged, if the overcurrent is interrupted within 3-5 cycles of power frequency as adopted in conventional circuit breakers.

6. CONCLUSIONS

Experimental data on the quenching characteristics of the superconducting winding were obtained by using our 100 kVA hybrid-type superconducting transformer. The superconducting winding was partially overheated and ruptured by post-quenching current during 1 sec. The geometric arrangement of Nb-Ti filaments in the cross-section of the superconducting wire was more distorted as close to the ruptured point. The critical current value of a short sample picked out of the burned portion was reduced to 70% of that in the original wire. It is revealed from the computer simulation that the wire temperature in 1 sec after the quenching has reached as high as 1967 K at the ruptured point. The thermal propagation length along the wire was restricted within 6.6 m in the total wire length of 49 m.

References

[1] Kito Y, Okubo H, Hayakawa N, Mita Y, Yamamoto M, "Development of a 6600V/210V 100kVA Hybrid-Type Superconducting Transformer" (1991) IEEE Trans. on Power Delivery 6: 816-823
[2] Neal D.F, Barber A.C, Woolcock A, Gidley J.A.F (1971) Acta Metall. 19: 143-149
[3] Lvovsky Yu.M, Lutset M.O (1982) Cryogenics 22: 639-647

8.2 Application of Electric Railway and Other Industry

Levitation Studies in High T_c Superconductors at Lower Temperatures and High Fields

FRANCIS C. MOON

Mechanical and Aerospace Engineering, Cornell University, Ithaca, NY 14853, USA

ABSTRACT

Dramatic increases in magnetic levitation forces between magnetic field sources and $Y\,Ba_2\,Cu_3\,O_7$ have been achieved by new processing methods such as melt textured and melt-quenched materials. This has raised levitation magnetic pressures to 4-5 N/cm^2 with small magnetic field magnets of 0.4 T at 78° K. In this paper we report on further increases in force capability from experiments conducted on a melt-powder-melt-growth YBCO sample from the ISTEC laboratories conducted at temperatures of 4.2° K and the interaction with a small low T_c solenoid that produces fields up to 2.0 T. Results show that magnetic pressures up to 60 N/cm^2 and higher can be achieved. These results suggest that higher loads for superconducting bearings may be achievable in the near future.

KEY WORDS: magnetic levitation, high-T_c superconducting bearings

INTRODUCTION

The use of high-T_c superconductors has been proposed for the use of passive magnetic bearings [1-3]. Initial studies of levitation forces were conducted in this laboratory, but showed marked hysteresis in the force-height relation as well as low values of magnetic pressure. These tests were conducted with small permanent magnetic dipoles of 3-6 mm in diameter and with a maximum flux density of $B = 0.4$ T on the surface. Classical analysis of magnetic forces shows that the magnetic force between two magnetic bodies can be calculated using the Faraday-Maxwell concept of magnetic pressure $P_m = B^2/2\mu_0$, where B is the tangential component of the field acting on a plane surface between the two surfaces

$$F = \int_\Sigma (B^2/2\mu_0)\, dA$$

We define an effective levitation magnetic pressure $P_L = F/A_m$, where A_m is the projected area of the test magnet. Early tests on sintered YBCO samples showed values of $P_L \sim 0.2$ N/cm^2, which is well below similar bearing pressures of 10 N/cm^2 for gas bearings. Nonetheless, small rotors (< 10 gm) were levitated with superconducting journal bearings and spun at speeds up to 120 RPM or surface speeds of 40 m/s [2]. The potential application for these small devices might be cryocooler turbines and small gyros. However, rapid improvements in processing of superconducting materials using so-called melt growth or melt-quench processes dramatically improved the magnetic levitation pressures to where at 78° K average bearing pressures were around 5 N/cm^2. These results place superconducting bearings on a par with gas bearings. Two laboratories that have consistently produced samples with these values were from ISTEC in Tokyo (Dr. M. Murakami), and the Catholic University of America in Washington, D.C. (Dr. Hamid Hojaji). In the last year several other laboratories were reported to produce similar results.

With such progress, one naturally can ask if further improvements in levitation force can be achieved by either material processing or magnetic design. Two control parameters immediately come to mind, short of new material processes, when one seeks to increase magnetic forces. These are temperature and applied magnetic field. It is well known that J_c and the effective magnetization of high-T_c superconductors increases with decrease in temperature. Thus, it is natural to try and directly measure the total magnetic force between a magnet and superconductor as a function of temperature from 4.2° K - 78° K. Also, one can see from equation (1) that the force is limited by the strength of the applied magnetic field. Typically the test magnets are rare earth based ferromagnetic materials which have an intrinsic magnetization field on the surface of 1-1.2 Tesla. However, in a free air path, with no magnetic circuit, the demagnetization effect drops the maximum value to less than 0.5 Tesla. Another method to increase applied field is to use an air core electromagnetic solenoid. If reliable high-T_c wire were available, such experiments could be run at 78° K. However, long lengths of high T_c wire with 10^5 A/cm^2 in a 1 Tesla field are not available, so we opted to build a small test solenoid from Nb-Ti wire and run the test at 4.2° K in order to significantly raise the applied field up to 2 Tesla.

Another method to increase the magnetic levitation force is to increase the magnetic field gradient. Some results in this direction have been reported informally by Chu at the University of Houston, and also by our Cornell laboratory in a contractor report to the U.S. NASA.

Levitation Force vs. Temperature

These measurements are reported in a paper by P.-Z. Chang and the author in a Cornell University doctoral dissertation [3-4]. Unlike experiments at 78° K, this experiment requires the design of a special cryostat as shown in Fig. 1. In this design the forces are carried from the test magnet in the cryostat at 4.2° K, to a measuring apparatus at the top of the cryostat at room temperature.

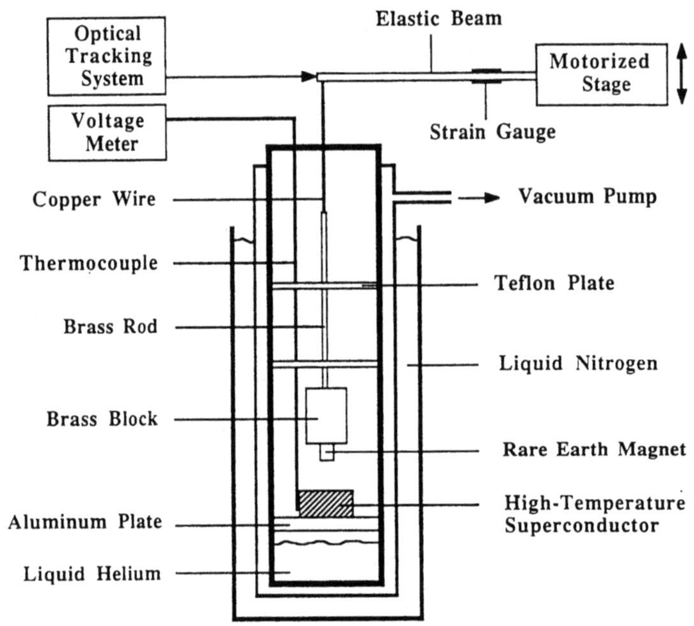

Fig. 1. Sketch of the device used to measure the levitation force between a permanent magnet and a superconductor from 4.2° K to 78° K.

A superconductor was placed in an aluminum sample holder, which was fixed in a liquid-helium dewar. A small rare earth permanent magnet (3.2 mm diam, 0.35 T) was press-fitted into a cylindrical 600 gram brass block. The brass block was connected to an aluminum beam through a Teflon-coated brass rod and a copper wire. A motorized stage was used to move the beam vertically, which in turn moved the magnet through the wire and the rod. Two strain gages glued on the top and bottom surfaces of the beam were used to measure the bending strain, which, after calibration, was used to measure the levitation force between the magnet and the superconductor. The position of the magnet was measured indirectly by an optical tracking system outside the dewar by measuring the position of a reference tag attached to the tip of the beam. The superconductor was cooled by helium gas evaporated from the liquid helium deposited on the lower portion of the dewar. The temperature of the superconductor was monitored by a gold-chromel thermocouple directly attached to the surface of the superconductor.

The results for a sintered specimen (S), a melt-quenched specimen (MQ) from Catholic University, Washington, D.C. and the melt-powder-melt-growth (MPMG) specimen from ISTEC, Tokyo are shown in Fig. 2. All levitation forces increase almost monotonically toward the low-temperature end. For the sintered specimen, the peak levitation force increases linearly from 77° K to 40° K and from 40° K to 4.2° K with a higher rate. For the melt-quenched sample, peak levitation force increases linearly from 77° K to 58 K and, with some exceptions, from 58 K to 20 K with a smaller rate with similar behavior for the MPMG sample. The increase in peak levitation force from 77° K to 4.2° K is a ratio of 12.3 for the sintered specimen compared to a ratio of 7.1 for the melt-quenched one, and 3.6 for the ISTEC MPMG specimen. It is noted that the relative force hysteresis in all specimens decreased at 4.2° K compared with measurements at 77° K.

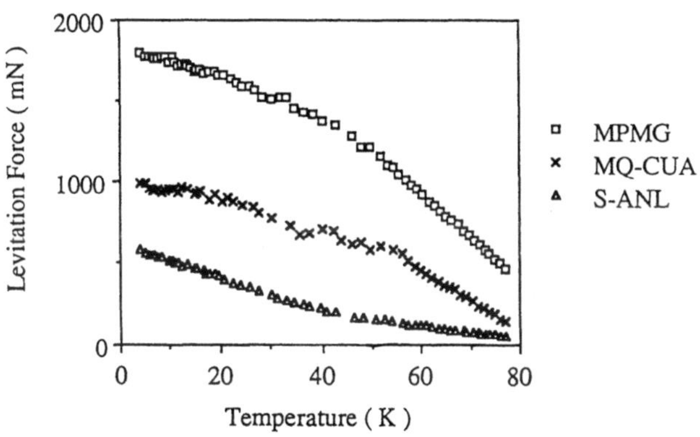

Fig. 2. Peak levitation force versus temperature for two melt-quenched specimens and a sintered specimen.

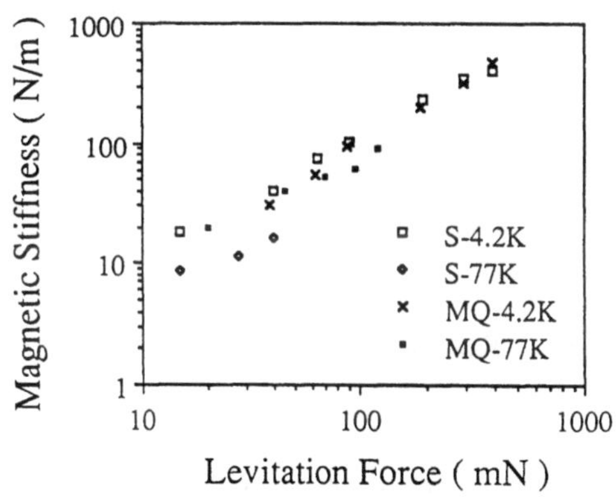

Fig. 3. Magnetic stiffness versus levitation force for the Catholic University melt-quenched specimen and the sintered specimen at 4.2° K and at 78° K.

Magnetic stiffness is usually defined as the slope of the minor loop on the force-versus-distance curve [1]. Due to the friction force in the measurement device, the slope of the minor loop is hard to define in this case. Since the hysteresis was small, we used the slope of the major loop as the measure of magnetic stiffness. Magnetic stiffness versus levitation force for the sintered and melt-quench specimens at 4.2° K and at 77° K are shown in Fig. 3. At 4.2° K, the results for both specimens almost fall on the same straight line. Since the slope is close to unity, the magnetic stiffness is approximately proportional to the levitation force. At 77° K, the magnetic stiffness in the melt-quenched specimen is twice that in the sintered sample.

Conclusion

Levitation forces for sintered and melt-quenched $YBa_2Cu_3O_7$ specimens increase monotonically with temperature decrease from 77° K to 4.2° K. The magnetic pressure on the projected magnet area is of the order of 10-20 N/cm^2 which suggests that $YBa_2Cu_3O_7$ may find early application in magnetic bearings for low temperature space applications.

Levitation Force vs. Magnetic Field

Most experiments to date on magnetic levitation force use small rare earth magnets. Although these magnets have intrinsic magnetic fields of around 1 Tesla in a closed, high permeability magnetic circuit, in a free space magnetic environment, the maximum operating field is less than 0.5 Tesla. In order to test magnetic levitation forces at higher fields, we designed a small low-temperature superconducting flat solenoid to generate fields between 1-2 Tesla at 4.2° K. These results are reported in greater detail in a paper soon to be published by Dr. C. Golkowski and the author [5].

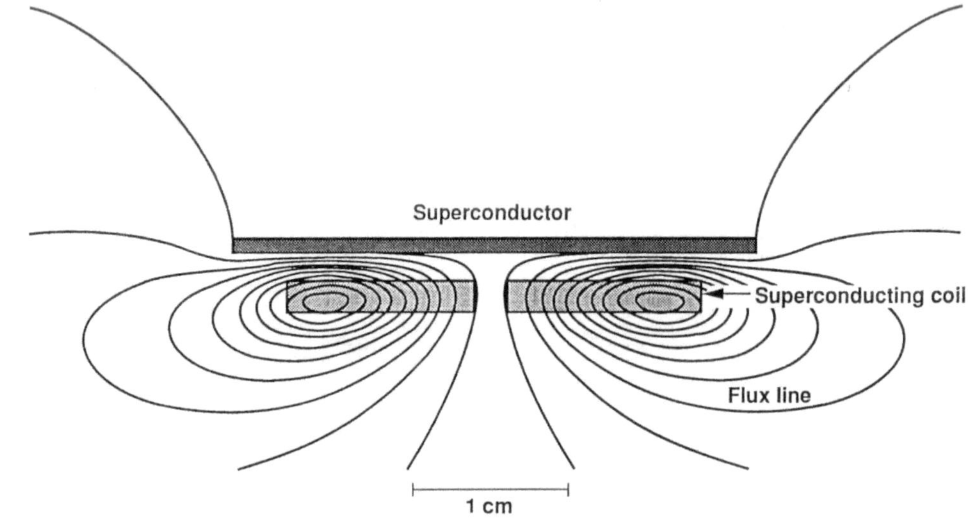

Fig. 4. Resultant magnetic flux lines of the coil and the superconductor.

The schematic of the experimental set-up is shown in Fig. 4. A superconducting coil of Niobium Titanium wire consists of 453 turns and 28 layers, with 3.2 mm ID and 23.5 mm OD, was the source of the magnetic field. The coil was calibrated with a Gauss-meter at liquid nitrogen. The magnetic field at the surface of the coil is related to the current as

$$B[G] = 209.9 \bullet I[A]. \tag{1}$$

Thus a current of 100 A produces a field of over 2 Tesla.

The coil was mounted on a load cell, which measured the force acting on the coil. Four strain gages were mounted to the load cell to eliminate any torque from the measurements. The displacement of the superconducting sample was measured with a thin beam attached to one end of the structure of the experiment; the other end was attached to the movable part. The superconducting sample was placed in a sample holder which was attached to a moveable rod. The other end of the rod was attached to a motorized translation stage. After cooling to the temperature of liquid helium the load cell and the displacement gages were calibrated before the experiment was run.

The superconducting sample used in the experiment was processed by the melt-powder-melt-growth method at ISTEC by Dr. M. Murakami [6] in the shape of a disk, with the diameter of 35 mm and thickness of 12.8 mm. The sample was placed above the coil in such a way that their axes of symmetry were aligned. The sample was cooled in zero magnetic field. After the temperature of liquid helium was reached and the gages were calibrated, the current was turned on through the coil.

For 100 A per turn, a nominal field of 2 T produced a magnetic levitation force in the YBCO of 300 N capable of lifting over 30 kgm of mass. The force acting on the coil does not show any hysteresis (it has the same shape for decreasing and increasing distance), which means that the sample behaves as if it were a type I superconductor. The experimental conditions were also simulated numerically assuming that the sample was a perfect superconductor. The simulation was done using an axi-symmetric superconducting boundary magnetic field solver. First the magnetic flux produced by the coil was calculated and compared with the experimental measurement. The measured and calculated magnetic fields were the same. Then the code calculated the force acting on the superconductor placed above the superconducting coil. The total flux lines produced by the coil in the presence of the superconducting sample are shown in Fig. 4. The force between the coil and the sample versus current is shown in Fig. 5, for both the

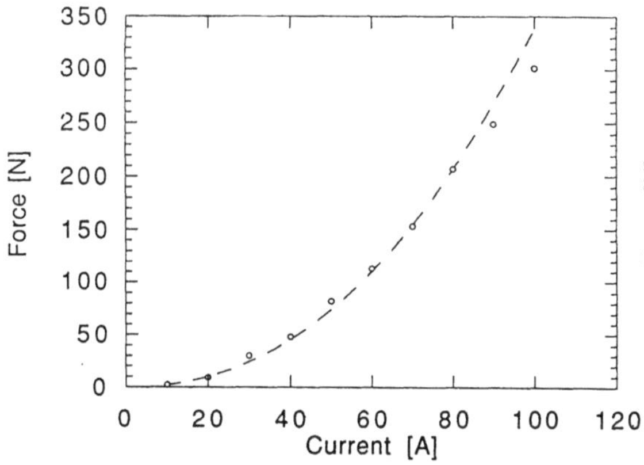

Fig. 5. Force versus current (4.2° K for a distance of 1.5 mm between the coil and the superconductor, the dotted line is the parabola $F = 0.01424\ x^{2.187}$.

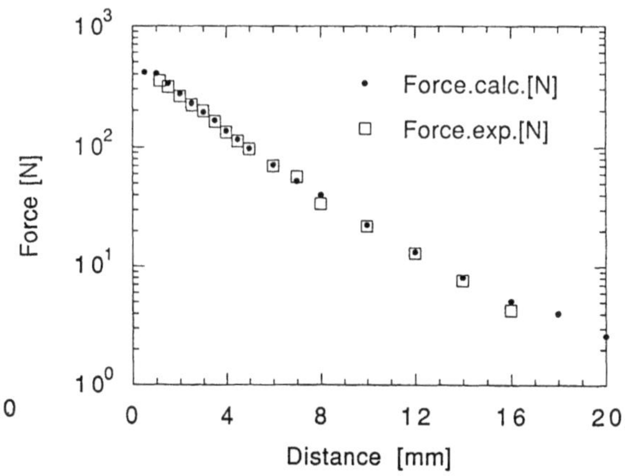

Fig. 6. Force versus distance for 100 A current through the coil; comparison of the experimental and calculated results.

experimental and simulated results. This data shows that the magnetic force is nearly proportional to the square of the current in the coil. The dependence of force versus distance (Fig. 6) can be approximated in the measuring range by

$$F = a \bullet exp(- b \bullet y), \tag{2}$$

where a and b are constants for a given current in the coil. If the force F is expressed in [N] and y in [mm] then for 100 A the experimental results fit the equation (2) with a = 461.4 and b = 0.3014, while for the calculated results for the same conditions have a = 456.6 and b = 0.2988. Thus the coefficients for both results are very close and the agreement between the simulation and the experiment is excellent.

On the basis of these results the stiffness can be calculated either differentiating the equation (2) with given constants or by a graphical differentiation of the experimental results such as in Fig. 2. For the simulated results the magnetic stiffness κ is

$$\kappa = - \frac{\partial F}{\partial x} = a \bullet b \bullet exp(- b \bullet x), \tag{3}$$

which can be written as

$$\kappa = b \bullet F. \tag{4}$$

This means that the stiffness is proportional to the force. For the experimental results the stiffness was calculated by graphical differentiation. The comparison of the stiffness from the experimental and simulated results is shown in Fig. 7. And again the agreement between both of them is very good.

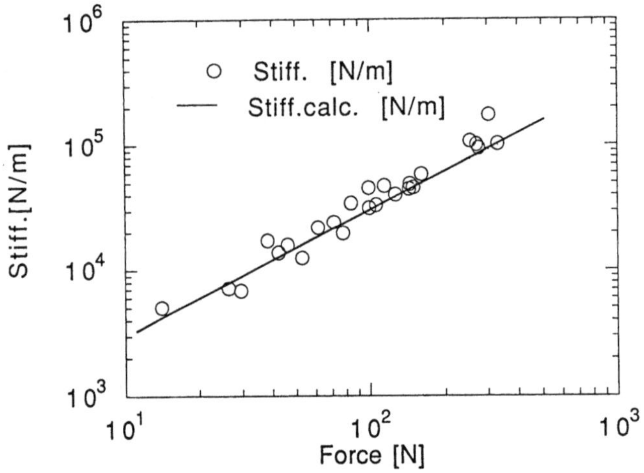

Fig. 7. Magnetic stiffness; comparison of the experimental with the calculated results.

ACKNOWLEDGMENT

This research was sponsored in part by the U.S. NASA Goddard Space Flight Center.

REFERENCES

1. Moon FC (1990) Appl. Electromag. in Mater. 1: 29-35
2. Moon FC, Chang PZ (1990) Appl. Phys. Lett. 56: 397-399
3. Chang PZ (1991) Ph.D. dissertation, Cornell University.
4. Moon FC, Chang PZ (1991) "Low Temperature Measurements of Levitation Forces in $YBa_2Cu_3O_7$; 4.2 K to 77 K," Cornell University Report.
5. Golkowski C, Moon FC, (1991) "Measurement of Levitation Forces in Bulk $YBa_2Cu_3O_7$ in a High Magnetic Field," Cornell University Report.
6. Murakami M, Morita M, Koyama N (1989) Jpn. J. Appl. Phys. 28: L1125.

Magnetic Shielding by HT_c Superconductor

HIRONORI MATSUBA

Furukawa Electric Co., Ltd., Yokohama, 220 Japan

ABSTRACT

As an application of high Tc superconductors, we have developed a design concept to make a superconducting cylinder shielding system. The system gives a very low magnetic field environment which allows to detect biomagnetic signals from the brain by SQUID magnetometers. Using a 3 channel SQUID magnetometer, we demonstrate the magnetic shielding performance with a shielding enclosure as large as 1/10 of the design. We made a big superconducting cylinder of 65cm diameter and 200cm length in collaboration with NGK insulators Ltd.. The cylinder exhibited the designed shielding performance. Besides, we describe a development of a superconducting shielding panel for other shielding applications.

KEY WORDS: HTc superconductors, superconducting shield, Biomagnetic measurements

INTRODUCTION

HTc superconductors have zero resistivity and this propery enable to make a superconducting shield which gives a very low magnetic field space, while being incapable by a ferromagnetic shield. One of the most promising applications is a superconducting shield for biomagnetic measurements. Biomagnetic measurements have been carried out with gradiometers, mostly, in a μ-metal shielded room, because a magnetometer works only in a highly shielded room. The μ-metal shielded room which has been constructed so far [1-2] for the magnetometer measurement had to be composed of more than 3 shells. Main interest of neural activity of the brain is now focused on a low frequency range. And this brings about a heavy and large space occupied shielded room because of its inferiority of shielding property in the low frequency range. Magnetic shielding by a superconducting hollow cylinder is expected to give an ideal low noise environment even in the low frequency range with low cost, light weight and small size. This could allow detection of deeper and/or weaker biomagnetic sources and enable people to study neuroral activities of the brain in the lower frequency close to DC.

We have developed a design concept to give a design of a high temperature superconducting shielded enclosure of 3m length and 1m diameter with keeping the attenuation 10^{-8} for a magnetometer MEG(Magnetoencephalography). In this paper, we demonstrate the magnetic shielding performance of the design with a 1/10 sized model. Then we made a big superconducting cylinder of 65cm diameter and 200cm length (named as "1/2 model" .) in collaboration with NGK insulators Ltd. The cylinder exhibited the designed shielding performance. Besides, we also developed a superconducting shielding panel and a cooling dewar of 1m square size which is expected to shield a magnetic field generated from a magnet, a high current generator or a some other disturbance source.

HTSC Shield for Biomagnetic measurements

Magnetic field in a Superconducting hollow cylinder

The magnetic field in a superconducting cylinder is determined by a ratio of the distance from the open end to its radius [3]. The magnetic field outside the cylinder attenuates along the depth from the open end of the cylinder and the resultant magnetic field along the axis of the cylinder is shown in Fig.1. Fig.1 proves that an external transverse magnetic field remains at a deeper site in the cylinder.

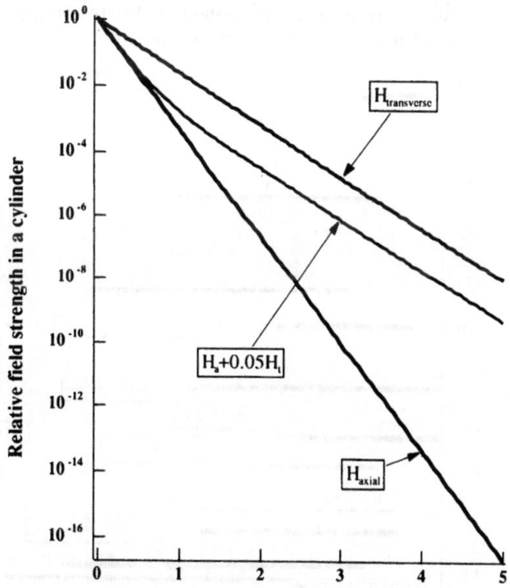

H_{axial} : Field by an external transverse field
$H_{transverse}$: Field by an external axial field
$H_a+0.05H_t$: $H_{axial}+0.05H_{transverse}$

Fig.1 Attenuation of magnetic field in a superconducting cylinder

The magnetic field gradient in a cylinder is so large that a gradiometer set in a superconducting cylinder gives almost the same response as a magnetometer. Fig.2 shows a calculation of the response ratio of a gradiometer to a magnetomater located in a superconducting cylinder.This reveals that it does not make sense to use a gradiometer in a superconducting cylinder for eliminating noise from a remote source.

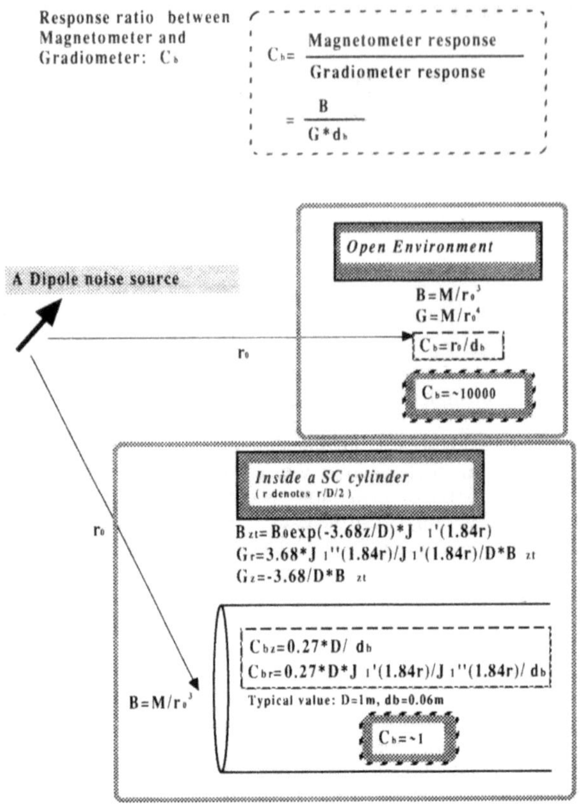

Fig.2 Response ratio of a gradiometer to a magnetomater located in a superconducting cylinder compared with an open environment.

<u>A design of a superconducting shielding enclosure for biomagnetic mesurement</u>

Fig.3 shows magnetic field strengths from neural activities of a human body compared with ambient magnetic fields. This indicates that an attenuation by a superconducting shield for biomagnetism measurement is required to be 10^{-8}. To put a human body inside a superconducting shielding cylinder, the diameter must be larger than 1m. Then the length results in more than 5m to attenuate the external field by 10^{-8}. The length is too long to be an practical equipment. We developed a design concept to shorten the length by coupling ferromagnetic with the superconducting shield.

DESIGN CRITERIA

<u>Criterion 1. Magnetic field between two semi-infinite hollow cylinders</u>

We considered shielding properties in a gap between two semi-infinite coaxial arranged hollow cylinders, which may be superconducting or ferromagnetic as illustrated in Fig.4. We assume that the superconductor exhibits perfect diamagnetism and that the ferromagnetic shows infinite permeability. Then the boundary condition at the surface of the superconductor is $H_r=\partial Ø/\partial r=0$, and the boundary condition at the surface of the ferromagnetic is $H_\theta=\partial Ø/\partial\theta=0$ and $H_z=\partial Ø/\partial z=0$. The general solution which gives magnetic potential Ø between the two semi-infinite cylinders can be written as:

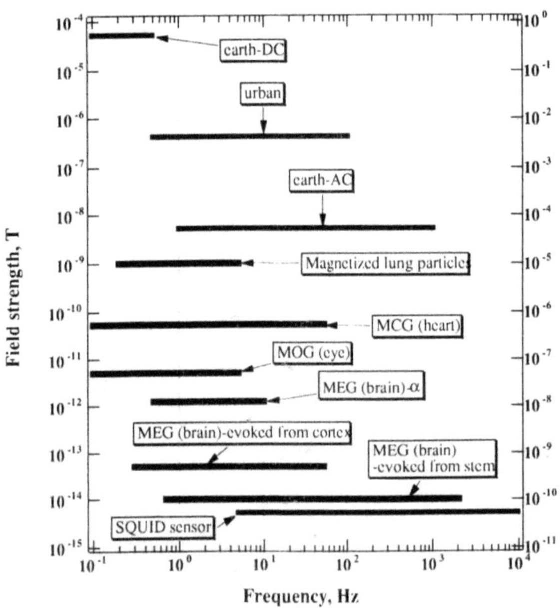

Fig3 Magnetic field strengths from neural activities of a human body compared with ambient magnetic fields.

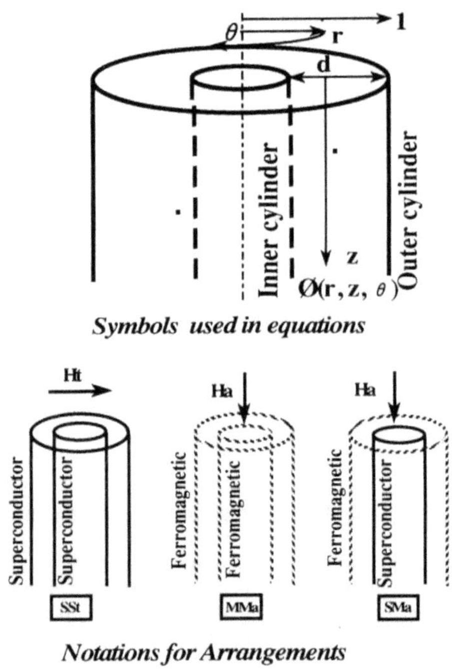

Fig.4 Two semi-infinite coaxial arranged hollow cylinders which may be superconducting or ferromagnetic.

$$\varnothing(r,z,\theta)=\sum_{m=0}^{\infty}\sum_{n=0}^{\infty}\{J_m(K_{mn}r)+C_{mn}Y_m(K_{mn}r)\}\cdot\exp(-K_{mn}z)\cdot\{A_{mn}\sin(m\theta)+B_{mn}\cos(m\theta)\} \qquad \text{------ (1)}$$

where K_{mn}, A_{mn}, B_{mn} and C_{mn} can be determined by the boundary conditions. Considering external applied homogeneous axial or transverse field, we obtained the following useful expression which describes the dominant term of the magnetic field in the gap:

$$H \propto \exp(-K \cdot z/d), \qquad \text{------ (2)}$$

where K is a constant depending on the direction of external field, the gap d between the two cylinders and the materials of cylinders. Fig.5 shows the values of K, where d=1 expresses the internal field of a single hollow cylinder.

Criterion 2. Finite permeability of ferromagnetic cylinders

We took into account the finite permeability of the ferromagnetic in a practical design. The finite permeability limits the shielding property of the two cylinders system. We obtained the limit by calculating the field among finite permeability cylinders of infinite length.

DESIGN

Considering the fact that permeability of a ferromagnetic decreases as the external field decreases, we designed the ferromagnetic cylinders to reduce the field noise from environment level to 10^{-4}. Then the superconducting cylinder reduces it from 10^{-4} to 10^{-8}. We separated the ferromagnetic arrangement into two sets. We placed the one at the open end of the cylinder and the other, inside it. As the field in the cylinder is subjected by the field at the open end, we designed the one set of ferromagnetic cylinders, placed at the open end, to decrease the field at the open end of the superconducting cylinder.
Then we designed the other ferromagnetic set, placed inside the cylinder, to enlarge the decay of transverse field in the superconducting cylinder by shorting out the magnetic path in the cylinder. We determined the diameters of the ferromagnetic cylinders by the criterion 1, while the lengths, by the limit from the criterion 2. Finally we optimized the arrangement by a numerical method. The optimized design of a shielding enclosure for biomagnetism measurements is illustrated in Fig.6. This indicates that a superconducting cylinder of 1m D and 3mL not including the end cap makes able the field attenuation to be 10^{-8}. We also designed the arrangement of the ferromagnetic cylinders so as to give minimum cost and to allow a well observation of inside from the outside.

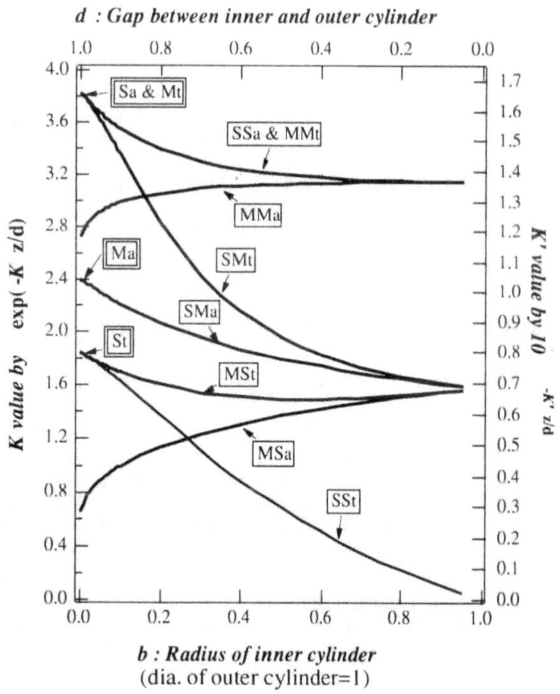

Fig.5 Calculated K values for magnetic field between two semi-infinite coaxial arranged hollow cylinders.

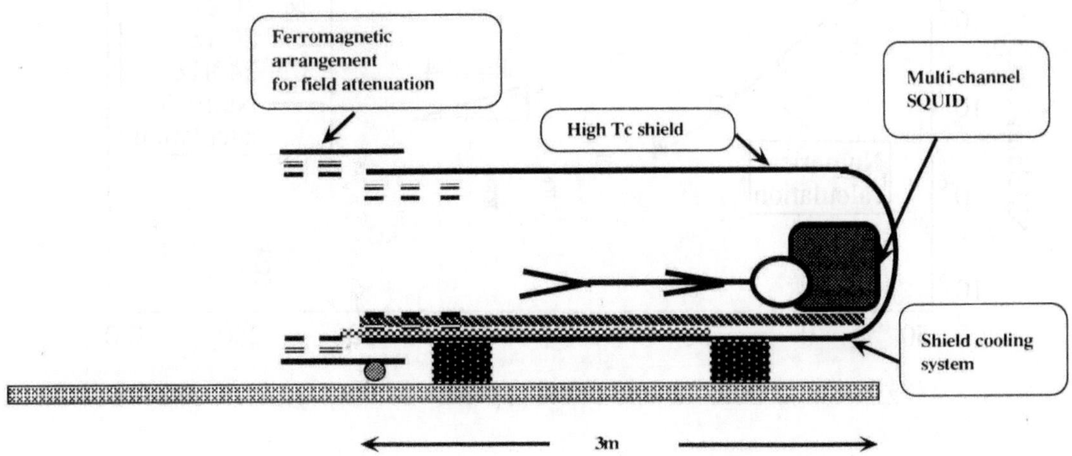

Fig.6 A design of a shielding enclosure for biomagnetism measurement

Table 1 Dimensions of the cylinders of a 1/10 scale model of a superconducting/ferromagnetic shielding enclosure for biomagnetism measurements
Unit: (mm)

	diameter	length	thickness
superconducting cylinder	100	450	0.3
arragement of ferromagnetic cylinders	60	10	1
	80	10	1
	100	10	1
	120	110	1
3-layered ferromagnetic shielding bucket	171	300	1
	200	400	1
	230	500	1

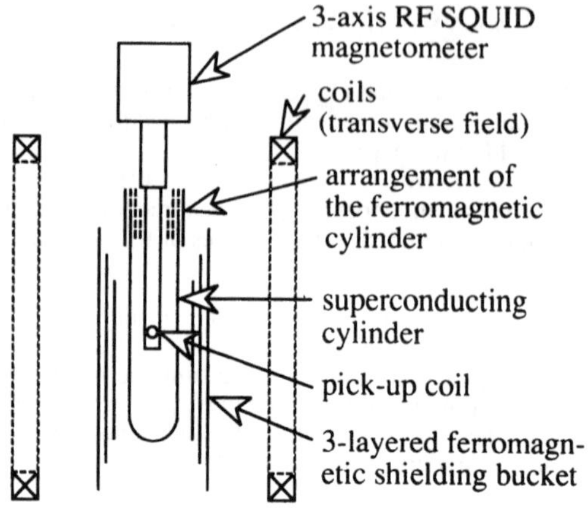

Fig.7 Experimental set up for measuring magnetic field in a 1/10 scale model

SHIELDING PERFORMANCE OF 1/10 SCALE MODEL

Using a 3 channel SQUID magnetometer, we evaluated the magnetic shielding performance with a shielding enclosure as large as 1/10 of the design. The superconducting hollow cylinder was made of BSCCO 2212 phase superconductor coated on a silver substrate. The dimensions of the components are shown in Table 1. We cooled down the shielding enclosure to 77K under "zero" field less than 10^{-8} T DC field. The experimental set up is shown in Fig.7. We measured the attenuation of the field with applying an external uniform transverse sinusoidal field of 1.3×10^{-5} T_{pp}. We calculated shielding effect 1/S as a ratio of the field strength in the shielding encolsure to the external field strength without the enclosure. Fig.8 shows the results. The field in the shielding enclosure decayed rapidly at the ferromagnetic arrangement as the position from the end increased, and then the decay followed the same gradient as the theoretically calculated decay (K=1.84) of the superconducting cylinder. The results agreed with a numerical calculation so far as 1×10^{-5} of its calculation limit. Shielding effect at more than 240mm position was as low as 1.3×10^{-7} for 0.1Hz, as low as 7×10^{-8} for 1Hz and 10Hz, and as low as 4×10^{-8} for 80Hz and 240Hz.

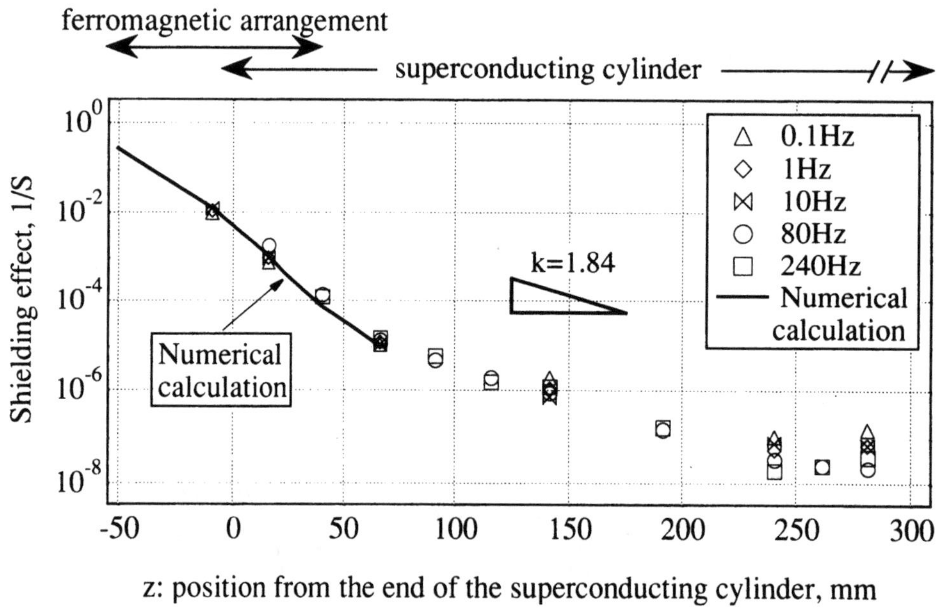

Fig.8 Shielding effect of a 1/10 scale model of a superconducting shield coupled with ferromagnetic. The bold solid line indicates a numerically calculated result of the shielding enclosure.
External uniform field: 1.3×10^{-5} $T_{p\text{-}p}$

1/2 Model

We made a big superconducting cylinder of 65cm diameter and 200cm length in collaboration with NGK insulators Ltd. The shielding performance of a superconducting cylinder depends on a surface magnetic field along the superconductor. And the maximum shielding magnetic field B of a superconductor of thickness d and of critical current density Jc is described by the following equation:

$$B=\mu_0 * Jc * d. \qquad ------(3)$$

The equation (3) suggests that a large superconducting cylinder with wall thickness d has the same maximum shielding magnetic field as a small analogous cylinder with the same wall thickness d does. The other shielding performances are expected to be the same between the both cylinders. Therefore we made the big superconducting cylinder with the same wall thickness as the 1/10 model cylinder. Fig.9 shows the big superconducting cylinder (1/2 model) and a dewar for cooling it in liquid nitrogen.

Evaluation of the 1/2 model

Using a long tail 3 axes SQUID magnetometer , we measured magnetic field distributions in the superconducting cylinder at 77.3K under external magnetic fields generated by big rectangular coils. The experimental arrangement is shown in Fig. 10. The results are described in Fig.11. The experimental results agreed well with a calculated field distribution by equations derived from equations (1) and (2) and with a field distribution obtained by a numerical method. To confirm a shielding performance of the superconducting shield coupled with ferromagnetic by this big superconducting cylinder, we set ferromagnetic arrangements at the both open ends of the cylinder and measured the attenuations of external magnetic fields with the same set up as shown in Fig.10. The results shown in Fig.12 agreed with numerical calculations.

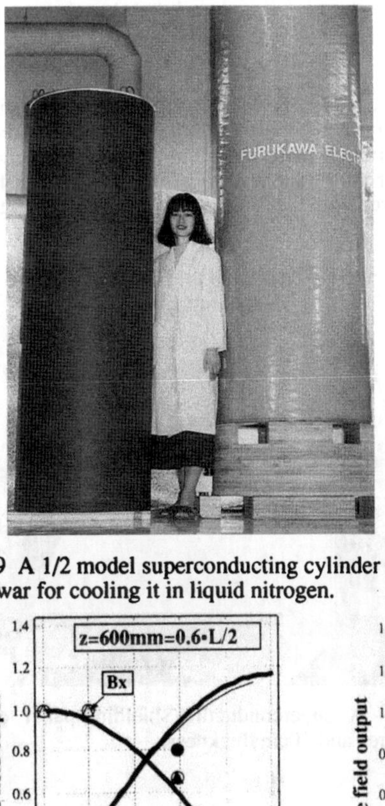

Fig.9 A 1/2 model superconducting cylinder and a dewar for cooling it in liquid nitrogen.

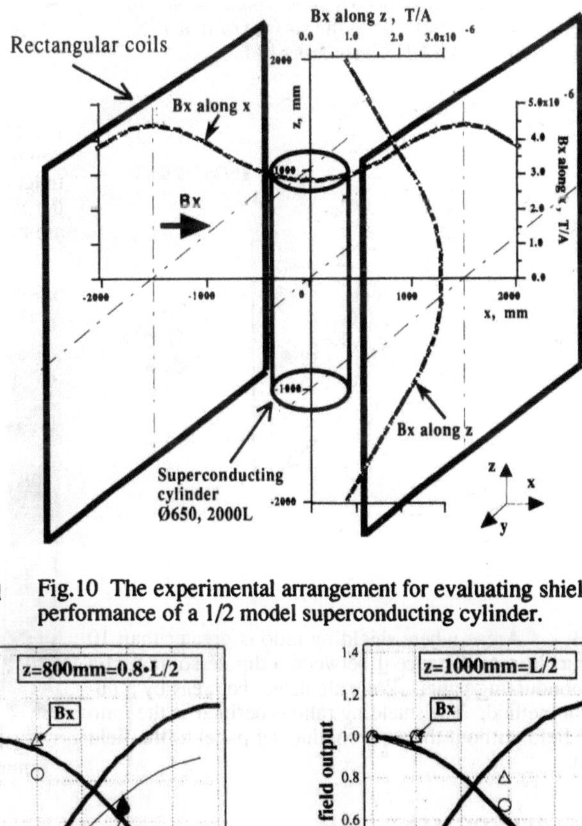

Fig.10 The experimental arrangement for evaluating shielding performance of a 1/2 model superconducting cylinder.

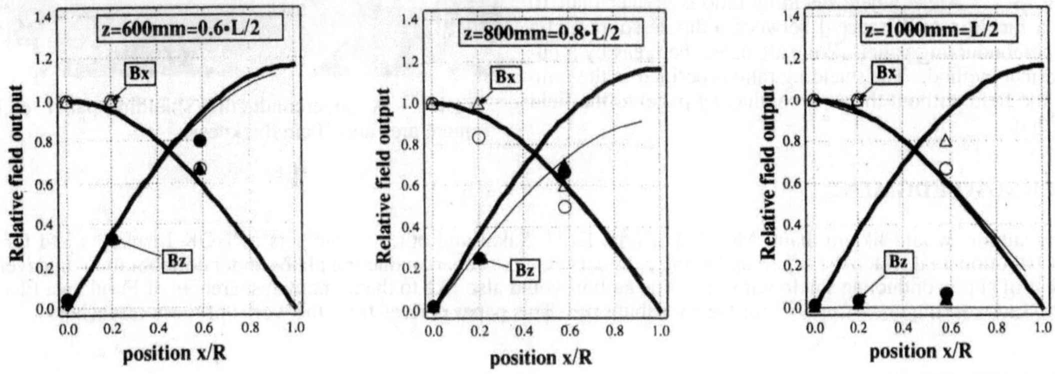

Caluculated distribution by equations — Distribution by a numerical method

		80Hz	240Hz
Field measurements	Bx	△	○
	By	▲	●

Fig.11 Magnetic field distribution in a 1/2 model superconducting cylinder.

SUEPRCONDUCTING SHIELDING PANEL

We developed a superconducting shielding panel and a cooling dewar of 1m square size and 10cm thickness. The dewar contained 30 litre liquid nitrogen and a superconductor in the dewar kept superconductivity during two days without supply of liquid nitrogen. We also developed a liquid nitrogen circulating system. We connected it to the shielding panel and carried out a long time operation test successfully.
The shielding panel is expected to shield a magnetic field generated from a magnet, a high current generator or a some other disturbance source. Fig.13 describes areas where shielding ratio is greater than 10 as a function of distance d between a magnetic dipole soure and a superconducting panel, where the shielding ratio is defined as the ratio of the field without the superconducting panel to the field with it.
Fig.14 shows a photograph of the shielding panel.
The maximum shielding magnetic field depends on a superconducting plate contained in the panel.
Currently we have developed a plate which can shield 50 Gauss magnetic field along the surface of it at 77K, in collaboration with NGK Insulators Ltd..

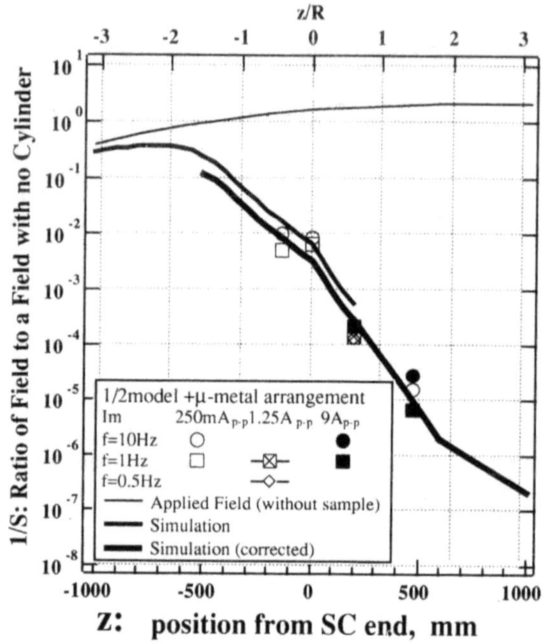

Fig.12 A shielding performance of a 1/2 model superconducting shield coupled with ferromagnetic. Ferromagnetic arrangements are set at the both open ends of the cylinder and the attenuations of external magnetic fields are measured with the same set up as shown in Fig.10.

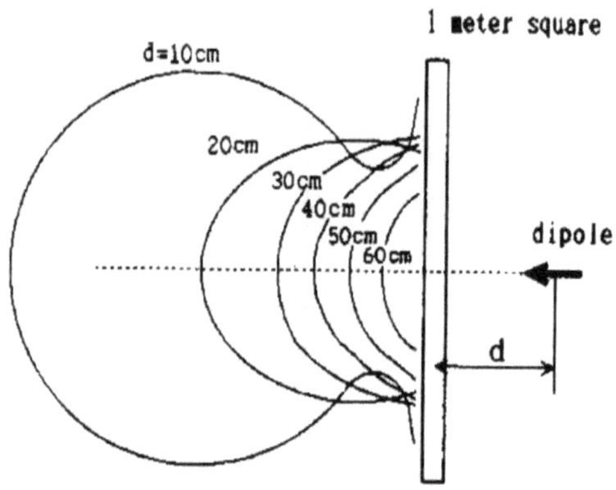

Fig.13 Areas where shielding ratio is greater than 10 as a function of distance d between a dipole soure and a superconducting panel. We calculated the areas by a numerical method. The shielding ratio is defined as the ratio of the field without the superconducting panel to the field with it.

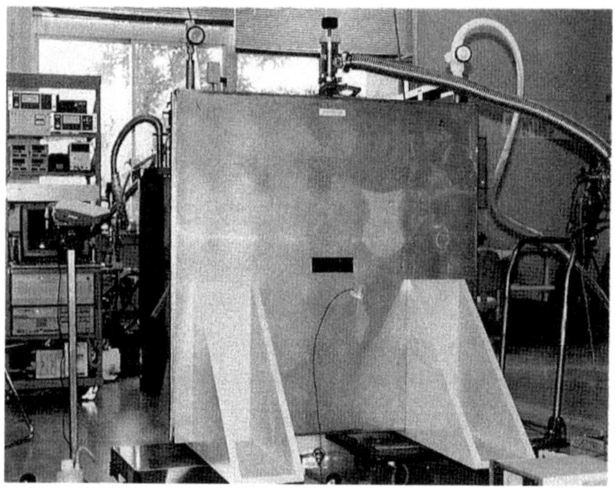

Fig.14 A superconducting shielding panel of 1m square area and 10cm thickness.

ACKNOWLEDGMENT

The author would like to thank Mr. H. Yoshida, Dr.H. Sakai and other researchers of NGK Insulators Ltd for the contribution to develop the superconducting cylinders and the superconducting plates under collaboration on development of superconducting shield with us. The author would also like to thank many researchers of Furukawa Electric Co., Ltd working on this project for the contribution. This paper derives from the works of these reseachers.

REFERENCES

1 Erne S N et al, in: "Biomagnetism'81": 79-87
2 Kelha V O, in: "Biomagnetism'81": 33-50
3 Cabrera B, Doctor thesis of Stanford Univ., Mar. 1975

Static and Utility Frequency Magnetic Shielding by High-T_c Superconductors

SUSUMU TORII, HIROYUKI OHSAKI, and EISUKE MASADA

Department of Electrical Engineering, The University of Tokyo, Bunkyo-ku, Tokyo, 113 Japan

ABSTRACT

Oxide superconductors are discovered, and hoped to be used for many applications. Magnetic shielding is one of their aspects because of the diamagnetism of superconductors. The behavior of the superconductors is studied, on the basis of Finite Element models. Numerical results are compared with the results of the experiment in static magnetic field. We here demonstrated a model including shielding current, with which reasonable results are obtained. We proposed an model also using shielding current to express the behavior in alternating field, and results of calculation are shown with utility frequency and even with the harmonics. The result shows good applicability of superconductors to magnetic shields in higher frequency.

KEY WORDS: magnetic shielding, oxide superconductor

INTRODUCTION

Oxide superconductors are found recently, and hoped to be used in many aspects of application[1]. Some of them are the ones using diamagnetism of superconductor. Magnetic shielding is one of these[2], and tools to express its complicated magnetic behavior are needed for these kind of researches, to evaluate the shielding effect and compare with other methods.

We have studied about the static shielding effect and the uniqueness of oxide superconductors. First, shielding effects were measured experimentally. Using these results, modeling and approximation of the behavior of superconductor has been studied with finite element analysis.

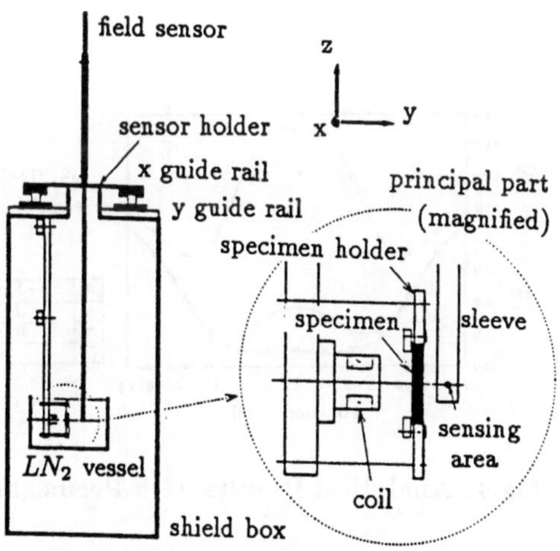

Fig.1. Experimental Equipment for Magnetic Shielding

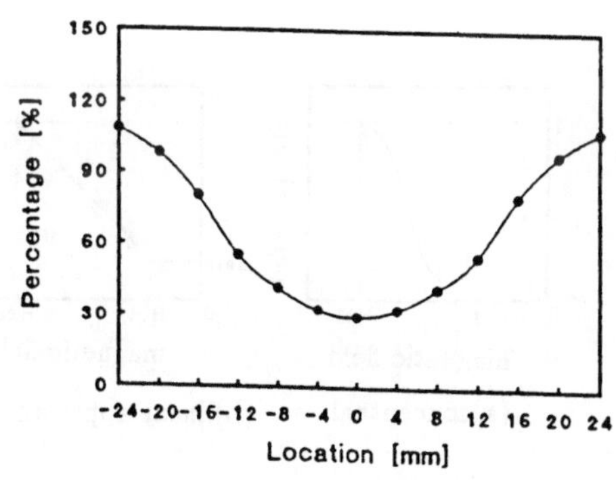

Fig.2. Experimental Results

STATIC MAGNETIC SHIELDING EXPERIMENT

Fig.1 shows the experimental equipments. The outer box is made of iron (t=6mm) to shield the environmental magnetic field. Superconducting specimen is BiSrCaCuO, diameter is 30[mm]. Coil is wound with 0.8ϕ Cu, and constant current was applied. Others such as holders are all non-magnetic, nonconductive. Coil, sensor and superconductor are dipped into liquid nitrogen. Measurements have done with zero-field cooling. When the sample is in superconducting state, backside magnetic field decreases from that with normal conducting. Fig.2 shows the decrease in percentage, and it shows the typical diamagnetic characteristics of the sample. Many measurements was done changing the measuring point and the thickness of specimen, and good resemblance were obtained.

MODELING WITH PERMEABILITY AND ITS PROBLEM

If we assume that we can describe these characteristics with permeability, field equation of static magnetic field is eq.(1).

$$\nabla \times (\nu \nabla \times \mathbf{A}) = \mathbf{J_0} \tag{1}$$

Numerical analysis has been made in 3-Dimension axi-symmetric field with finite element method.
First, we calculated with constant permeability, and obtained some flat-top characteristics in comparison with Fig.2. It means that flux penetrates heavily near the edge, and constant permeability cannot simulate the difference.
Magnetic behavior of superconductors are usually described like Fig.3(a). The flux penetration increases with field strength, and it must be better to use this field-dependent permeability model. We approximate this curve linearly like Fig.3(b), because the relations needed are only narrow width. Iterations are needed to obtain final state. Initial state with perfect diamagnetism shows the good exponential convergence, which proved this iteration to resemble to the time dependent simulation of flux penetration. The best results are shown in Fig.4 with filled square. Some differences are still remaining especially in the edge.
The line with filled triangle in Fig.4 shows that flux penetrates in the center or the specimen like Fig.5(a). However, ring-shaped superconductors are known to inhibit the flux across the contour like Fig.5(b). This contradiction is caused by the concept of permeability itself, which can describe the current loop of atomic size in uniform distribution. Superconductors have static shielding current loop of specimen size, therefore other method is preferred. Method itself also has a problem. This method needs μ_{min}, H_{c1}, H_{c2} as the parameters. We need parameter-fitting with trial-and-error, and it causes inconsistency of analysis.

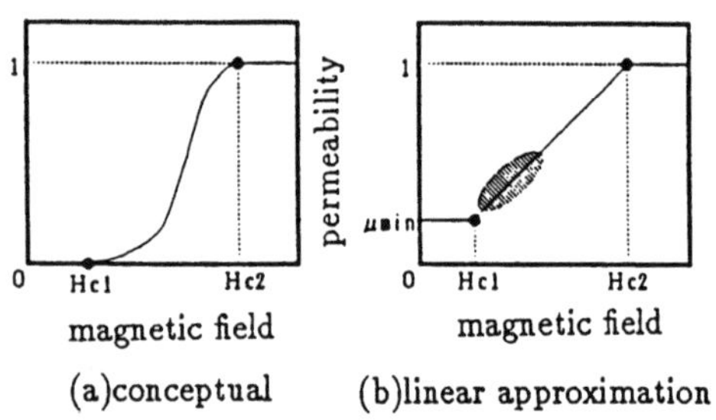

Fig.3. Approximation with Permeability

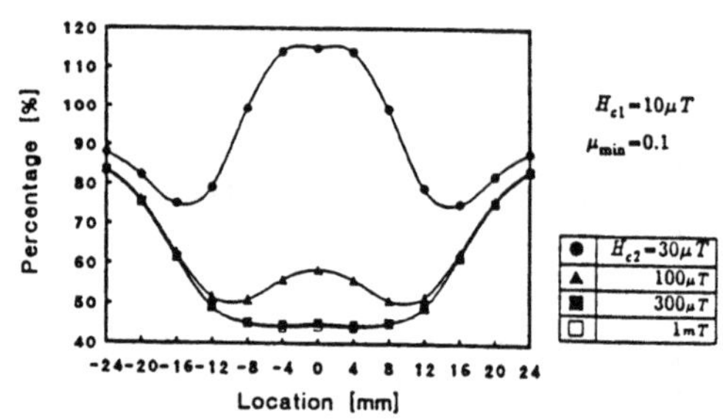

Fig.4. Analytical Results with Permeability

MODELING WITH SHIELDING CURRENT

According to the results above, we proceed to the shielding current. We cannot use the normal time-dependent expression because of the basic characteristics of superconductor $\sigma = \infty$. We assumed the forced shielding current proportional to the radius. Still we used the permeability, to express the spatial effect of the sintered specimen. The currents are iteratively calculated, and if any current density gets higher than J_c, the current is forced down to zero and so as the permeability to 1. Apparently, this process simulates the destruction of superconducting state. Fig.6 shows the results, and it shows good resemblance with Fig.2. Alternatively, We have developed another modeling, using point-fitting of the magnetic field. It perfectly does not need the permeability, but still needs the iteration.

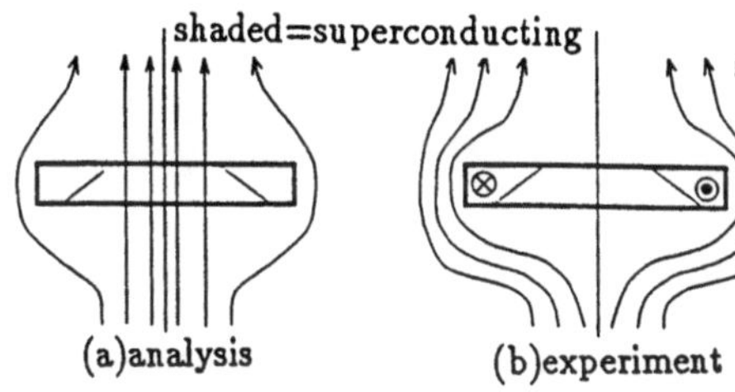

Fig.5. Problem of Modeling with Permeability

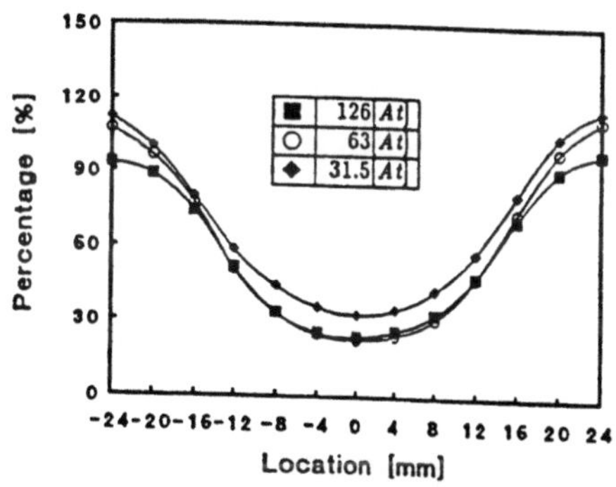

Fig.6. Analytical Results with Shielding Current

CONCLUSION

Magnetic shielding behavior of oxide superconductors were measured, and several numerical modelings were proposed. Conventional modeling with only equivalent permeability was proved insufficient to express its behavior, because of the lack of evaluation about the major-loop current. Modeling with forced shield current was proposed, and sufficient results have obtained. Also suggested was the alternative method using only current loops.

REFERENCES

[1] (1990) "Superconductivity: Dealing in Futures", EPRI Journal, pp.18-27, Jan/Feb

[2] Hayes WF (1987) "agnetic Field Shielding for Electrodynamic MAGLEV Vehicles", Proc. of 9th Int. Conf. on Maglev and Linear Drives, 53-66

Large Vessels of Bi Oxide Superconductor for Magnetic Shield

ATSUSHI KOIKE[1], KAZUTOMO HOSHINO[1], HIROFUMI KOTAKA[1], EICHI SUDOH[1], KAZUHIKO KATOH[1], and HIROSHI OHTA[2]

[1] Corporate R&D Center, Mitsui Mining & Smelting Co., Ltd., Ageo, Saitama, 362 Japan
[2] The Institute of Physical and Chemical Research, Wako, Saitama, 351 Japan

ABSTRACT

We have developed a very large magnetic shielding vessel of high-Tc Bi-Pb-Sr-Ca-Cu-O superconductor and experimentally confirmed that this vessel can reduce a weak magnetic field to -100 dB or $1/10^5$. The size of this vessel is 32 cm in diameter, 64 cm in depth and 2.5 cm in thickness. The shielding effect of the vessel does not go down even at such a low frequency as 0.2 Hz. The high attenuation ratio in the very low frequency range indicates that magnetic shielding of high-Tc superconductor is promising for neuromagnetic measurements using SQUIDs.

KEY WORDS: oxide superconductor, Bi-Pb-Sr-Ca-Cu-O, magnetic shield, SQUID

INTRODUCTION

Magnetoencephalography (MEG) including neuromagnetic measurements usually requires a carefully designed shielding room which consists of five or six-layer shells of Permalloy and aluminum. Magnetic fields from human brains are several tens of femto Tesla at such a low frequency as 1 Hz. Recently developed SQUID magnetometers with second-order gradiometers can have a sensitivity of 20 $fT/Hz^{1/2}$. Since the magnetic field of the earth is 3×10^{-5} T, a heavy magnetic shielding is essential for MEG. However, a conventional magnetic shielding of a Permalloy is quite difficult to shield low-frequency magnetic fields below 10 Hz[1]. The minimum magnetic field inside a Permalloy shield is limited by residual magnetic fields due to mechanical stress in the Permalloy composing the shielding room.

Recently, we have experimentally proved that magnetic shield vessels of Bi-Pb-Sr-Ca-Cu-O superconductors can reduce a weak magnetic shield like the geomagnetic field to -100 dB or 10^{-5} at such a low frequency as 1 Hz, where the size of the vessel was 15 cm in diameter and 32 cm in depth[2,3]. The attenuation ratio of 10^{-5} is promising for MEG.

In the present study, we report the magnetic shielding effect of a very large vessel of high-Tc superconductor. We have successfully fabricated very large magnetic shielding vessel of the Bi-Pb-Sr-Ca-Cu-O system, whose size is 32 cm in diameter, 64 cm in depth and 2.5 cm in thickness. The magnetic field inside the vessel in a superconducting state at 77 K was compared with that at room temperature using an rf SQUID magnetometer.

EXPERIMENTAL

Superconducting vessels were shaped out of fine powder of Bi system high-Tc phase (nominal composition; Bi:Pb:Sr:Ca:Cu=0.8:0.2:0.8:1.0:1.4) using a cold isostatic press, and were sintered. The "large" and "middle" vessels were fabricated in this experiment, whose sizes were 32 and 15 cm in diameter, 64 and 32 cm in depth and 2.5 and 1 cm in thickness, respectively. Figure 1 shows the large vessel after sintering. The critical temperature Tc of 103 K and the critical current density Jc of 140 A/cm^2 were obtained in the pellet samples prepared by the same process as the vessels.

Figure 2 shows a schematic arrangement for measuring the shielding effect with a SQUID magnetometer. A Helmholtz coil of 94 cm in diameter generated homogeneous external magnetic fields (B_{ex}) outside the vessel at liquid nitrogen temperature. The rf SQUID of an all-niobium quasi-planer Josephson junction was used as a magnetometer, where the second-order gradiometers were detached and the single-turn coil of 25 mm diameter was directly coupled with the SQUID. The noise level of the SQUID is less than 20 $fT/Hz^{1/2}$[4].

The SQUID magnetometer was inserted into the vessel and the shielding effect was measured as a function of the ratio of L/D, where L is the distance from the top and D is the diameter of the vessel. The signals obtained by the SQUID were analyzed using by a digital spectrum Analyzer. The shielding effect for DC magnetic field was also measured by the flux-gate magnetometer.

Alternating currents at various frequencies of 0.2 through 130 Hz and DC current were applied to the exciting coil to generate external magnetic fields. The shielding factor S is usually defined as the ratio of the external field H_{ext} to the field inside the shield H_{int}:

$$S = H_{ext} / H_{int}. \qquad (1)$$

In this experiment, we assume that S is the ratio of a magnetic field inside the vessel in a normal state at room temperature around 290 K to that in a superconducting state at 77 K. The external magnetic field was evaluated from calculations based on the electromagnetic theory using the values of the current applied to the exciting coil and the number of turns of

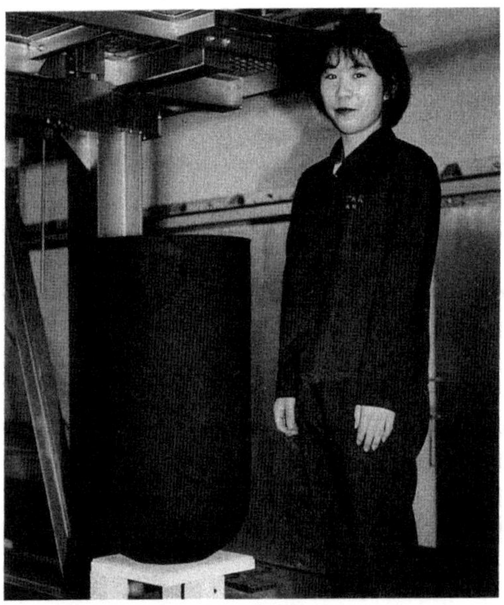

Fig. 1 A large superconducting vessel. (inner diameter: 320mm, depth: 640 mm, thickness: 25mm)

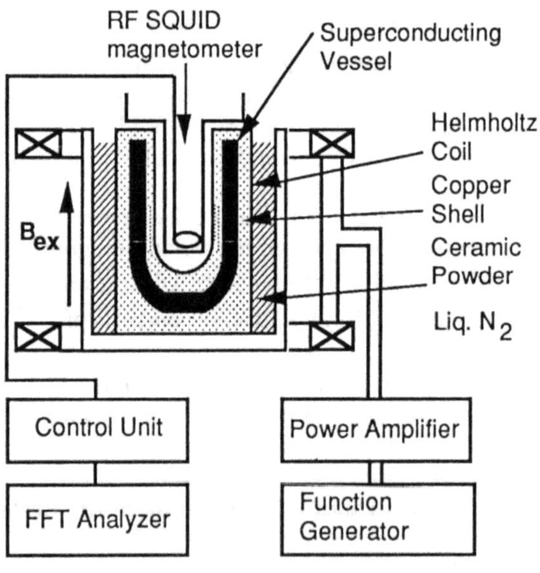

Fig 2 A schematic arrangement of the SQUID magnetometer, coils and a superconducting vessel.

the coil. The magnetic field inside the vessel in a superconducting state was estimated from the output voltage and the field sensitivity of the SQUID. All the experiments were carried out in the electromagnetic shielding room.

RESULTS AND DISCUSSION

The experimental results about our very large shielding vessel indicates that high-Tc oxide superconductors are highly promising for the application to MEG. First of all, we want to emphasize that the SQUID magnetometer can work in our superconducting vessels without a second- order gradiometer.

According to the theoretical calculation for a perfect semi-infinite superconducting tube exposed to a homogeneous external magnetic field, the interior magnetic field of the tube is expressed as

$$H_{//} = H_0 \exp(-7.6\, L/D) \qquad (2)$$

$$H_{\perp} = H_0 \exp(-3.6\, L/D), \qquad (3)$$

where $H_{//}$ and $H_{\perp}$ represent the axial and transverse interior magnetic fields, respectively. H_0 is the external magnetic field[5]. These equations assume that the tube is an ideal superconductor and has an infinite current density. The experimental results were discussed based on the eq. (2) as a function of L/D.

Figure 3 shows the frequency dependence of shielding factors for the large and middle vessels, where the external magnetic field (B_{ex}) was about 3×10^{-5} T. The observed S in these superconducting vessels was independent on the frequency of magnetic fields at very low frequencies (0.2 - 1 Hz). Since the shielding behavior seems to be independent on frequencies, the frequency dependence of S above 10 Hz could be attributed to the shielding effect due to the eddy current on the double inner copper shells, which support the superconducting vessel, as shown in Fig. 2. Moreover, in the case of the measurement of the large vessel, the double-wall

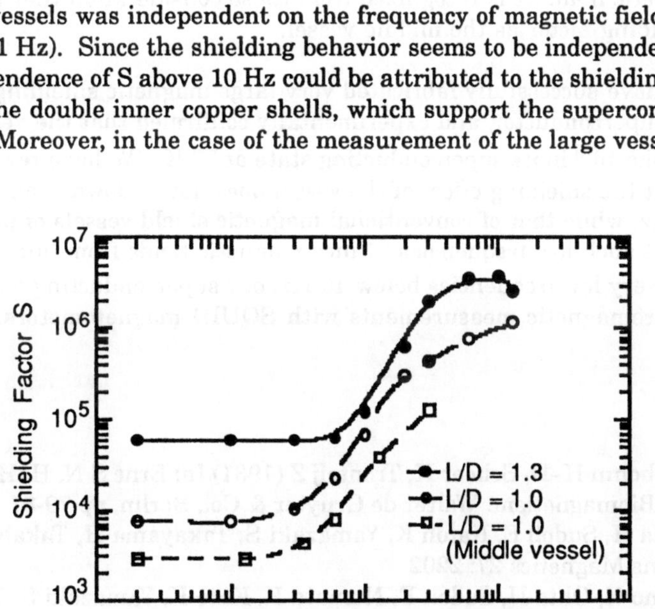

Fig. 3 The frequency dependence of the shielding factor S for the large and middle vessels.

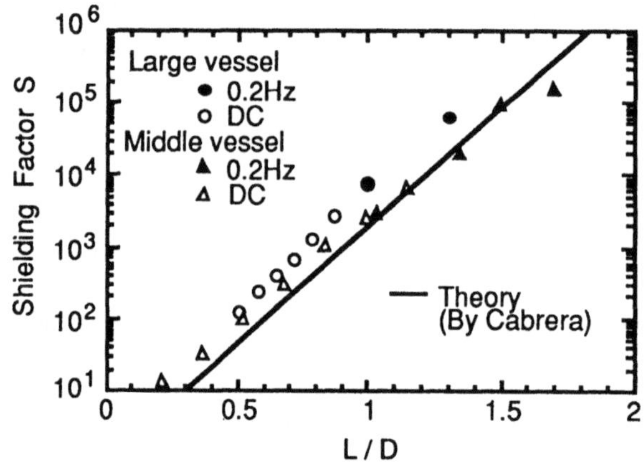

Fig. 4 The relationship between S and L(depth)/D(diameter) for the large and middle vessel.

stainless steel container was used as a liquid nitrogen cryostat.

Figure 4 shows the relationship between S and L/D for those vessels. The theoretical line estimated from eq. (2) is also indicated in this figure. The measured S in the middle vessel almost coincide with the theoretical estimation. On the other hand the measured S in the large vessel is three times higher than the theoretical one. Since the stainless steel was slightly magnetized, the container can work as a magnetic shielding shell even for the external DC magnetic field. It is recognized from these consideration that the large vessel has the same shielding factor as the middle vessel.

In summary, we have successfully fabricated very large magnetic shielding vessels of the high-Tc Bi oxide superconductor, and experimentally confirmed that the vessel has very high attenuation ratio like 10^{-5} in its superconducting state at 77 K. We have revealed from SQUID measurements that the shielding effect of the vessel does not go down even at such a low frequency as 0.2 Hz, while that of conventional magnetic shield vessels of permalloy goes down very much at such low frequencies. Since magnetic fields from human brains are as weak as 10^{-13}T at very low frequencies below 10 Hz, our superconducting vessel is highly promising for neuromagnetic measurements with SQUID magnetometers.

REFERENCE

1. Erne S N, Hahlbohm H -D, Scheer H, Trontelj Z (1981) In: Erne S N, HaHlbom H -D, Lubbig H (eds) Biomagnetism. Walter de Gruyter & Co., Berlin, pp 79-87
2. Hoshino K, Ohta H, Sudoh E, Katoh K, Yamazaki S, Takayama H, Takahara H, Aono M (1991) IEEE Trans Magnetics 27: 2202
3. Koike A, Hoshino K, Ohta H, Sudoh E, Numata K, Kato K, Yamazaki S, Takahara H, Aono M (1991) In: Kajimura K, Hayakawa H (eds) Advances in Superconductivity III. Springer-Verlag,. Tokyo, pp 591-594
4. Ohta H, Takahata M, Takahashi Y, Shinada K, Yamada Y, Hanasaka T, Uchikawa Y, Kotani M, Matsui T, Komiyama M (1989) IEEE Trans Magnetics 25: 1018
5. Cabrera B (1975) Ph.D.Thiesis, Stanford Univ.

High Field Magnetic Shielding Using Superconducting Rings

M. KOJIMA, S. KOHAYASHI, Y. ISHIKAWA, and S. YOSHIZAWA

Central Research Laboratory, Dowa Mining Co., Ltd., Hachioji, Tokyo, 192 Japan

ABSTRACT

It is possible to apply superconducting rings to magnet (in F.C condition) and magnetic shielding (in Z.F.C condition). In this study, we prepared magnetic shielding unit using superconducting rings and discs to be arranged along the center axis. Magnetic flux density in the inside and outside of the unit was measured by using a hall probe. For external magnetic field H_{ex} parallel to the center axis, zero-field space was formed up to $H_{ex}\sim125$(G) in the inside of the rings between the second and the third one. In the shielding unit using higher Jc superconductor, the shape effect of shielding unit influenced by the winding magnetic flux was observed.

KEY WORDS: high-Tc superconductor, Bi-system, high field magnetic shielding, superconducting rings

INTRODUCTION

Since the discovery of high-temperature oxide superconductors, there have been many studies on the fundamental and application field on the magnetic shielding. In the AC low magnetic field shielding[1], uniformity of superconductor and shape relating to the effect of demagnetic field of shielding vessel are important problems. In the DC high magnetic field shielding[2], in addition, high critical current density Jc(Ic) value of sample is required. In this paper, we report the Jc dependence on the shielding characteristics of the shielding unit consist of Bi-system superconducting rings and discs.

EXPERIMENTAL

Bi-Pb-Sr-Ca-Cu-O powder was prepared by the ordinary solid state reaction method. The powder was pressed into rings(48mm outer diameter, 34mm inner diameter, and 5mm thickness), and then, sintered in air at 850 ℃ for 30 hours. In order to change Jc value of samples, we applied an intermediate pressing process. Three kinds of samples were finally prepared which have different Jc values. The Jc was measured by a four-prove method.

A schematic figure of the experiment was illustrated in Fig.1. The magnetic shielding unit using four superconducting rings was arranged along the center axis. External DC magnetic field was generated by an electro-magnet parallel to the center axis of the shielding unit. The magnetic flux density in the inside of the unit was measured with a hall probe which was put into the center of diameter between the second and the third rings.

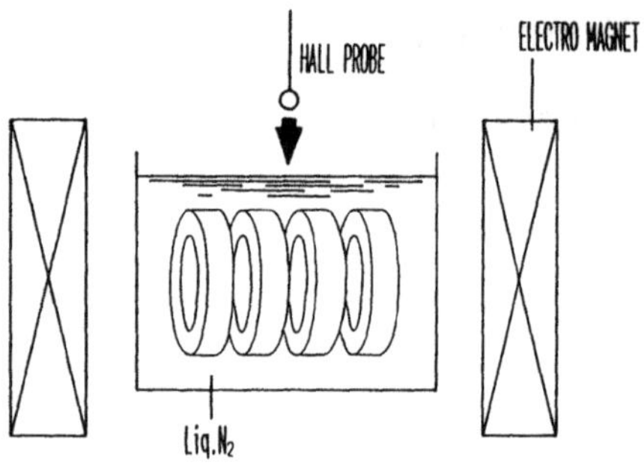

Fig.1 Schematic figure of the experiment. A hall probe is put into the center of diameter between the second and the third rings.

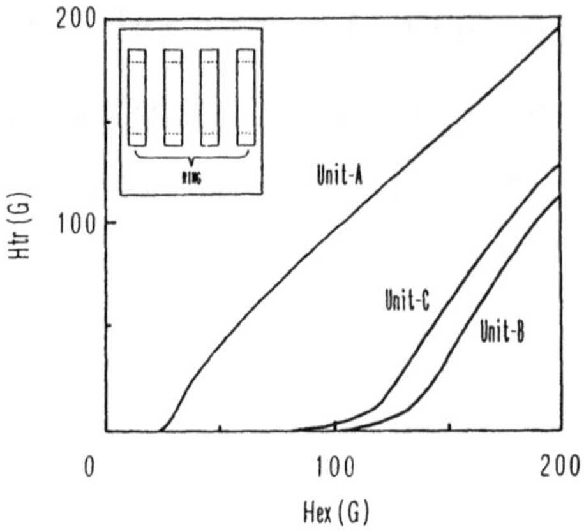

Fig.3 The magnetic shielding characteristics in the shielding units composed of four rings.

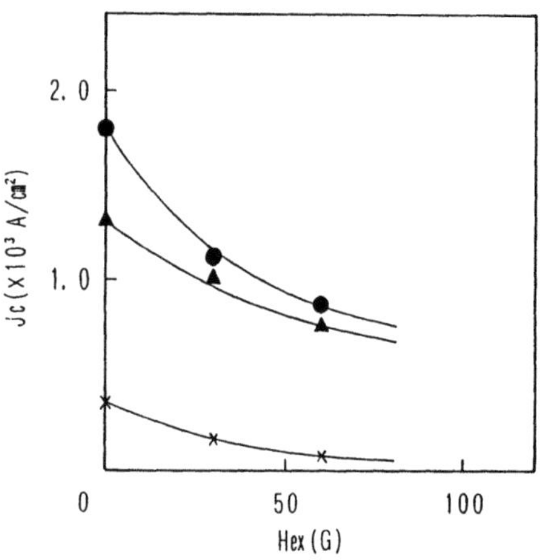

Fig.2 The magnetic field dependence of critical current density Jc for three kinds of rings.

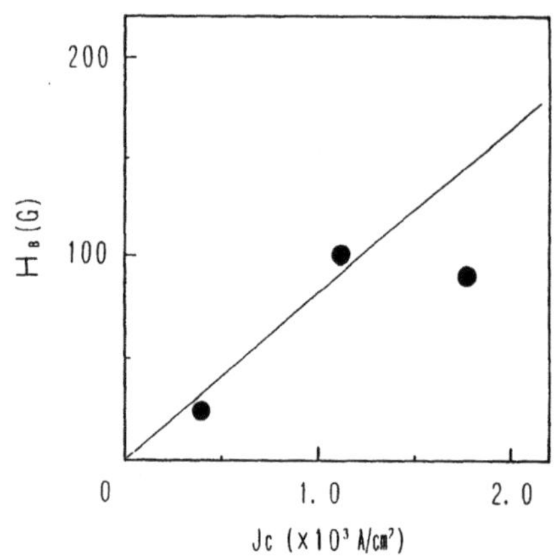

Fig.4 The relation between Jc of rings in zero magnetic field and the external magnetic field H_B at which the zero-field shielding effect was broken in the shielding unit of four rings.

RESULTS AND DISCUSSION

Figure.2 shows the magnetic field dependence of Jc for each ring. The Jc values in zero magnetic field were 390 A/cm^2, 1120 A/cm^2, and 1760 A/cm^2, respectively. The decrease of Jc observed by exposure of external magnetic field, where the Jc decreasing rate of higher Jc ring was larger than that of lower Jc ring. This suggests that the flux pinning force of Bi-system is very weak. The magnetic shielding characteristics of the shielding units is shown in Fig.3. Figure.4 shows the relation between Jc of rings in zero magnetic field and the external magnetic field H_B at which the zero-field shielding effect was broken in Fig.3.

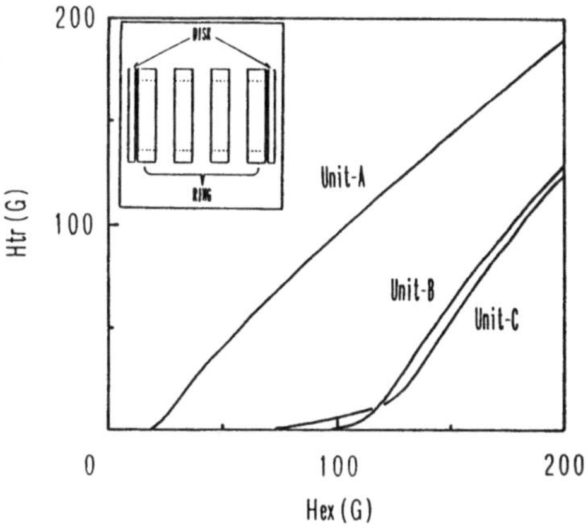

Fig.5 The magnetic shielding characteristics in the shielding units composed of four rings with discs in the ends.

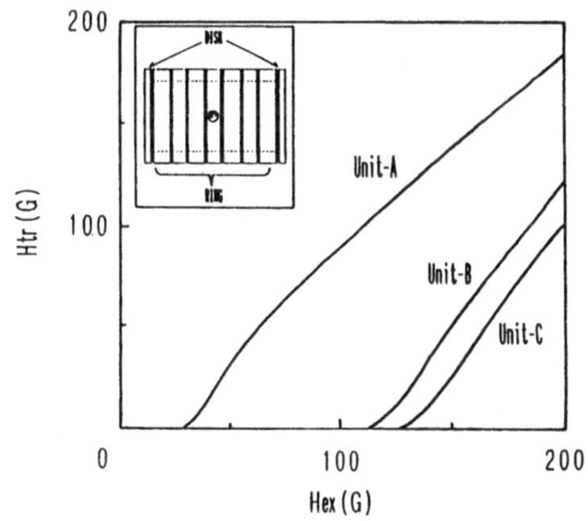

Fig.6 The magnetic shielding characteristics in the shielding units where seven rings and discs in the ends are attached each other.

The shielding effect of Unit-C using higher Jc rings was worse than that of Unit-B using middle-Jc rings. This result is not in accord with expectation based on the Jc-H_{ex} relation in Fig.2. Thus we investigated about leak of winding magnetic flux from each opened end and/or gaps between rings.
First, we set superconducting disks which had the same superconducting properties of the shielding unit at both opened end in order to prevent the winding magnetic flux invading into the shielding unit. As shown in Fig.5, however, the gradual increase of magnetic flux density in the inside of the unit-C was observed under 100(G) region.

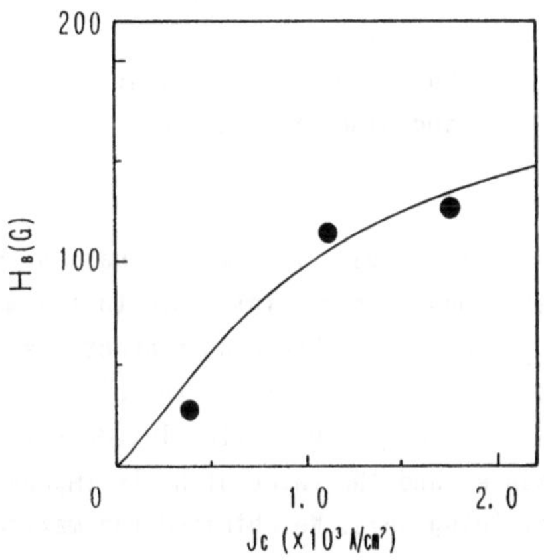

Fig.7 The relation between Jc of rings in zero magnetic field and the external magnetic field H_B at which the zero field shielding effect was broken in Fig.6.

Secondly, we inserted three superconducting rings, which had same superconducting properties and size of the rings, between rings of the shielding unit illustrated in Fig.5 in order to prevent the winding magnetic flux invading into the shielding unit from gaps. Inside space was perfectly covered by superconductors except a small hole so that we can put a hall probe into the shielding unit. Figure.6 shows the magnetic shielding characteristics of the shielding unit and Fig.7 shows relation between Jc in zero magnetic field and the external magnetic field H_B at which the zero-field shielding effect was broken. In this case for Unit-C, gradual increase of magnetic flux density in the inside of the unit around 100(G) region seen in Fig.5 was disappeared. It is found that the magnetic shielding characteristics in the shielding unit preventing the winding magnetic flux invasion depend on the result from Jc-H_{ex} relation. The magnetic shielding characteristics mentioned in Fig.3, 5 and 6 was summerized from the viewpoint of the Jc of materials. As shown in Fig.8, in the case of Unit-A using lower Jc superconductor, there was little change in shielding effect

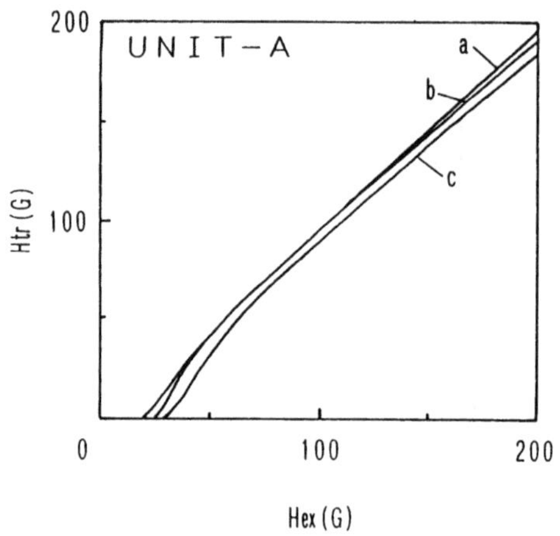

Fig.8
The magnetic shielding characteristics of the Unit-A in the three types of shielding unit.
a: composed by four rings,
b: composed by four rings and discs in the ends,
c: composed by seven rings and discs in the ends.

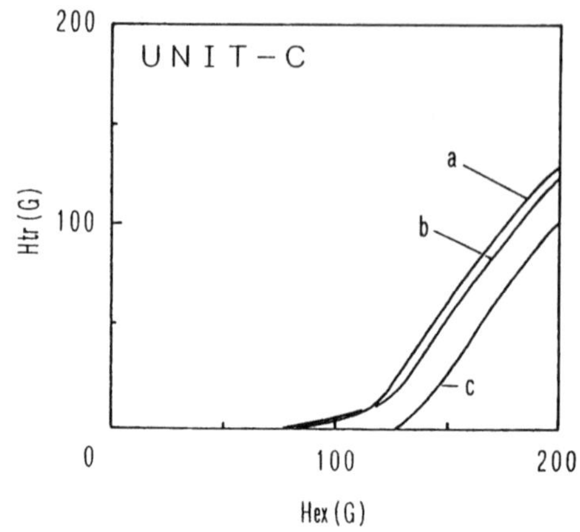

Fig.9
The magnetic shielding characteristics of the Unit-C in the three types of shielding unit.
a: composed by four rings,
b: composed by four rings and discs in the ends,
c: composed by seven rings and discs in the ends.

for the three types. This means that the winding magnetic flux from each opened end and gaps between rings does not influence to the magnetic shielding when the Jc value is small. Figure.9 shows the magnetic shielding characteristics of Unit-C using higher Jc superconductor. In the type of shielding unit where two discs in the ends was set and seven rings was attached, the gradual increase of magnetic flux density in the inside of the unit around 100(G) region is disappeared and the value of H_B is improved preventing the winding magnetic flux invading into the shielding unit. We obtained the maximum shielding value of 125(G) in this case.

CONCLUSION

The effect of Jc of superconductor and the shape of shielding unit comprising of superconducting rings and discs was studied on the magnetic shielding characteristics.
The winding magnetic flux from each opened end and gaps between rings does not influence to the magnetic shielding when Jc is small. In the shielding unit using higher Jc superconductor, the shape effect of shielding unit influenced by the winding magnetic flux was recognized, and the maximum shielding value H_B of 125(G) was obtained.

REFERENCES

1. Hoshino K, Ohta H, Sudoh E, Katoh K, Yamazaki S, Takahara H, Aono M (1990) Jpn.J.Appl.Phys.29: L1434.
2. Seo K, Nishijima S, Okada T, Katagiri K (1991) Cryogenics Vol 31: 524

Magnetic Shielding in High-T_c Bi-Pb-Sr-Ca-Cu-O Superconducting Cylinder

YUUICHI ISHIKAWA[1], SHUJI YOSHIZAWA[1], KENICHI TENYA[2], and HIDEKI MIYAJIMA[2]

[1] Central Research Laboratory, Dowa Mining Co., Ltd., Hachioji, Tokyo, 192 Japan
[2] Department of Physics, Faculty of Science and Technology, Keio University, Yokohama, 223 Japan

ABSTRACT
The superconducting shielding of AC magnetic fields was investigated by measuring the field distribution in a cylinder made from high-T_c Bi-Pb-Sr-Ca-Cu-O superconductors. Two kinds of superconducting cylinders were prepared by different methods; the one was a cylinder made of sintered $Bi_{1.85}Pb_{0.35}Sr_{1.90}Ca_{2.05}Cu_{3.05}O_x$ powder, and the other was a MgO cylinder coated with the superconducting powders and paste. The shielding attenuation rate depends exponentially on the shape factor. The attenuation rate is decreased with the critical current density and volume fraction of the superconducting components, but the rate in the sintered cylinder is independent of the magnitude of the applied field below 10 Oe.

KEY WORDS : Bi-Pb-Sr-Ca-Cu-O superconductor, magnetic shielding, cylindrical shielding, shielding attenuation rate

INTRODUCTION

The magnetic shielding is important when very weak fields produced in human body are determined precisely. At the time, the geomagnetic field or fields resulting from an electric instrument need to be excluded. The superconducting shielding can offer the possibility of obtaining truly zero magnetic fields, since no magnetic flux can penetrate the shield. There is great interest in using high-T_c superconductors in shielding applications for the reasons of the availability and low cost. The ceramic superconductors such as polycrystalline Y-Ba-Cu-O, however, do not always shield the external magnetic flux because of the imperfect flux pinning, flux creep, and weak links. Therefore, it is significant to prepare better impermeable materials and also to determine the shape factor so as to obtain the complete shielding. The purpose of this study is to investigate the relation between the shielding effect and the superconducting properties in Bi-Pb-Sr-Ca-Cu-O systems.

In this study, two types of superconducting cylinders were prepared. The one was a cylinder made up from Bi-Pb-Sr-Ca-Cu-O of different critical current densities and magnetizations, and the other was a MgO cylinder coated with Bi-Pb-Sr-Ca-Cu-O powder and paste. AC magnetic fields were applied along the axis of these cylinders, and the transmitted magnetic field into the cylinder was measured with a pick up coil as functions of the internal position along the axis of the cylinder, the shape factor, the magnitude of the applied field and the frequency of the AC field. The shielding attenuation rate, defined by the ratio of transmitted and applied fields, is discussed in connection with the superconducting characteristics.

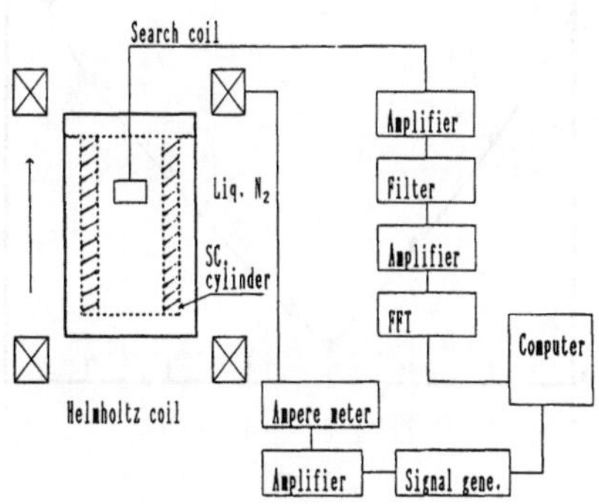

Fig. 1: Measurement of shielding

EXPERIMENTAL PROCEDURE

(1) Preparation of Superconducting Cylinder by Coating Method

The powder of $Bi_{1.8}Pb_{0.4}Sr_2Ca_2Cu_3O_x$ (x was undetermined) was mixed with an organic binder in solvent to prepare the paste, and it was coated on the surface of a MgO cylinder of 47 mm in outer diameter and 150 mm in length. The thickness of the paste was about 100 μm. After that, the cylinder was calcined at 850 °C in air for 10 - 30 min.

(2) Preparation of Superconducting Cylinder by Sintering Method

By using a cold isostatic pressing method, the powder of $Bi_{1.85}Pb_{0.35}Sr_2Ca_2Cu_3O_x$ was pressed under a pressure of 1-3 ton/cm^2 so as to form a cylinder. The inner and outer diameters of the cylinder were 20 mm and 30 mm, respectively, and the length was 110 mm. The cylinder was sintered in air at 830 - 850 °C for 50 h, and then cooled to room temperature at the rate of 1 °C/min. This heat treatment was repeated two or three times in order to improve the superconducting properties.

(3) Measurement of Superconducting Shielding

The volume fraction of the superconducting component in the specimen was evaluated from the magnetization curves. The specimen was cooled down from room temperature to 77 K in the absence of magnetic fields, and then the magnetization was measured by using a vibrating sample magnetometer.
The transport critical current density was measured with a four-probe method.
The magnetic shielding in the superconducting cylinder was measured in the following procedures. The schematic diagram of the measurement is shown in Fig. 1. The cylinder was immersed into liquid nitrogen and was placed at the center of a Helmholtz coil. The coil applies an external magnetic field, H_{ex}, parallel to the axis of the cylinder. A search coil to detect the transmitted magnetic field, H_{tr}, was located along the axis of the cylinder. The position of the coil was varied along the axis. The instruments were installed in a permalloy box so as to prevent the influence of the geomagnetic field. The magnitude of H_{ex} was changed from 3x10^{-3} Oe to 15 Oe in the frequency range between 10 Hz and 1 kHz.
In this paper, the magnetic shielding attenuation rate is defined by the ratio of the transmitted and applied magnetic fields (H_{tr}/H_{ex}).

RESULTS AND DISCUSSION

Figure 2 shows the relation between the position of the coil, z, and the transmitted field, H_{tr}, when an AC magnetic field of 1 Oe (f=826 Hz) is applied parallel to the axis of the cylinder, where the origin of the z-axis is set at the open-end of the cylinder. The value of H_{tr} decreases exponentially as z approaches the center of the cylinder. According to Hamilton [1] and Ogasawara et al [2], the attenuation rate is given by

$$H_{tr}/H_{ex} = \exp(-7.6z/D), \quad (1)$$

where D is the diameter of the cylinder. The value for the exponent constant estimated from the experiment is 5.52, which is smaller than the theoretical value. The relationship between the shielding attenuation rate and the shape factor, L/D, shows also an exponential variation, as shown in Fig. 3.

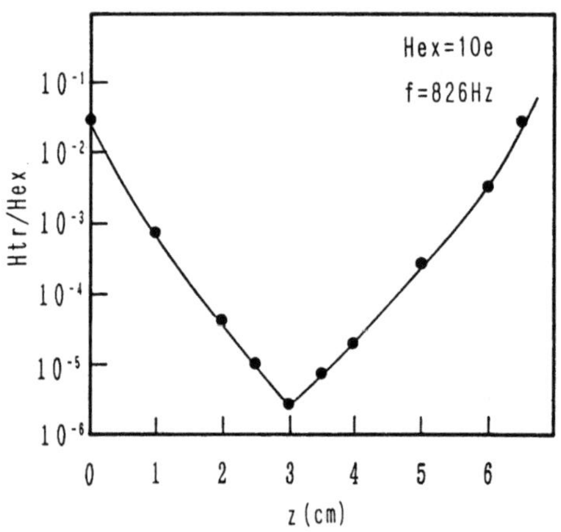

Fig. 2: Position dependence of the attenuation rate

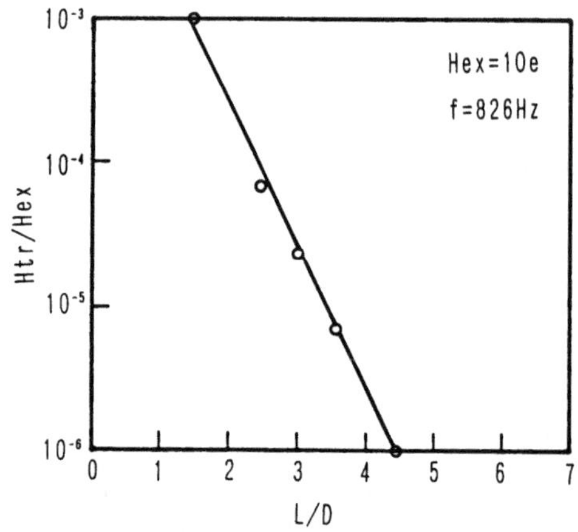

Fig. 3: Shape factor dependence of the attenuation rate

The magnetic field and frequency dependences of the attenuation rate in the sintered and coated cylinders are illustrated in Figs. 4 and 5, respectively. The frequency dependence is not clearly observed in both cylinders. The attenuation rate of the sintered cylinder is almost independent of the magnitude of the field. The rate of the coated cylinder, however, is constant below H_{ex}=1 Oe, while H_{tr} is increased steeply with increasing H_{ex} higher than about 1 Oe. The magnetic field variation of the shielding in the coated cylinder can be classified into two field regions; $H_{ex}<H_{c1}$ and $H_{ex}>H_{c1}$. In $H_{ex}<H_{c1}$, the shielding is mainly determined by the London penetration depth, λ, but for $H_{ex}>H_{c1}$ the flux penetration becomes dominant. By Bean model, the attenuation rate in two regions is expressed by

$$H_{tr}/H_{ex}=\exp(-7.6z/D)+\exp(-d/\lambda) \qquad H_{ex}<H_{c1} \qquad (2)$$

$$=\exp(-7.6z/D)+1-J_c\cdot d/H_{ex}, \qquad H_{ex}>H_{c1} \qquad (3)$$

where d is the thickness of the coating. The field at which H_{tr} begins to increase rapidly is lower than H_{c1} [3,4]. Therefore, the transmitted magnetic flux is the flux penetrating through superconducting and non-superconducting areas.

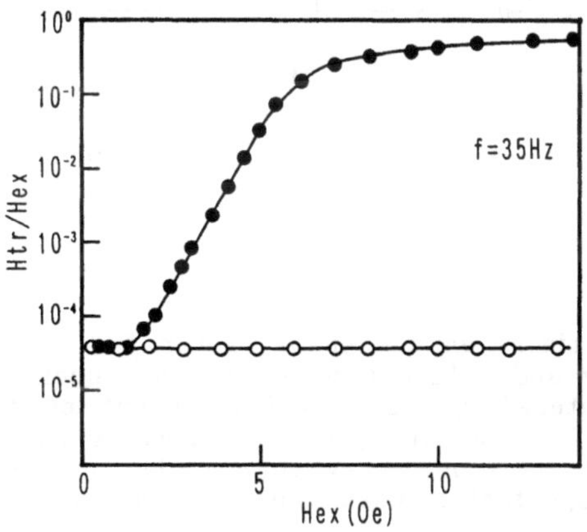

Fig. 4: Magnetic field dependence of the attenuation rate.
open circle: sintered cylinder
solid circle: coated cylinder

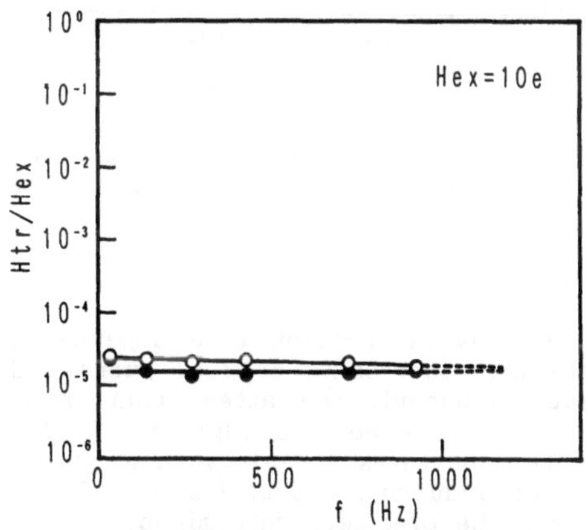

Fig. 5: Frequency dependence of the attenuation rate.
open circle: sintered cylinder
solid circle: coated cylinder

Figure 6 shows the relation between the shielding attenuation in the coated cylinder and magnetization at H_{ex}=1 Oe. As is seen in the figure, the attenuation rate does not depend lineally on the magnetization. As the magnetization may be induced by both intra-grain current and inter-grain current, the magnetization is not directly related to the rate.
The critical current density and initial susceptibility dependences are shown in Figs. 7 and 8, respectively, in the case of f=826 Hz and H_{ex}=1 Oe. The value of the initial susceptibility, $4\pi\chi$, is smaller than -1 because the demagnetization in the cubic sample was neglected. The transmitted field H_{tr} is decreased with transport critical current density and initial susceptibility which is proportional to the volume fraction of the superconducting component. As already mentioned, the attenuation rate is almost independent of the magnetization, while it depends on the transport critical current density. Therefore, the inter-grain current

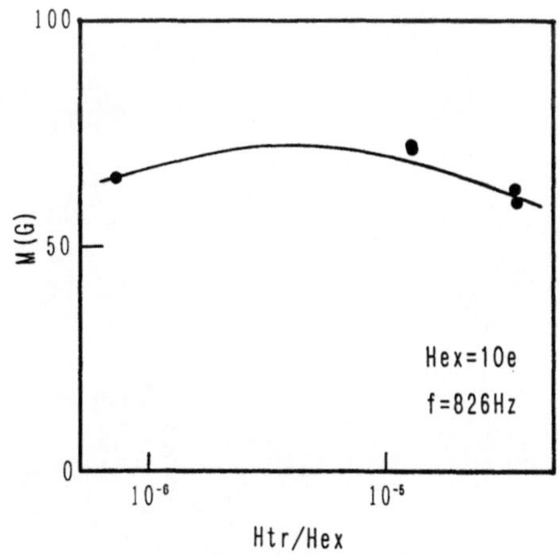

Fig. 6: Relation between the magnetization and the attenuation rate

is dominant in the magnetic shielding, and it needs to reduce the weak coupling between the grains.

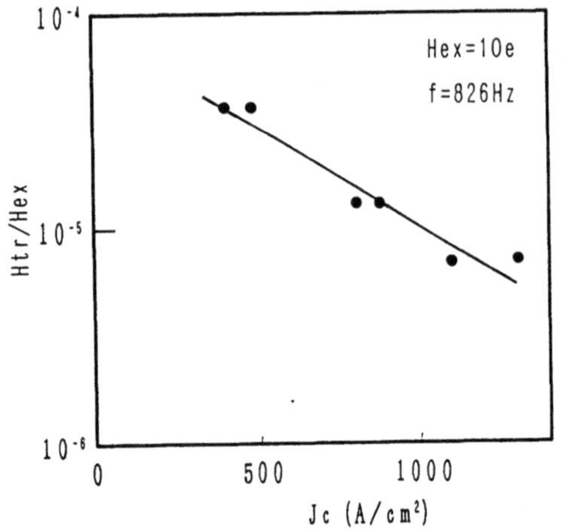

Fig. 7: Transport critical current density dependence of the attenuation rate

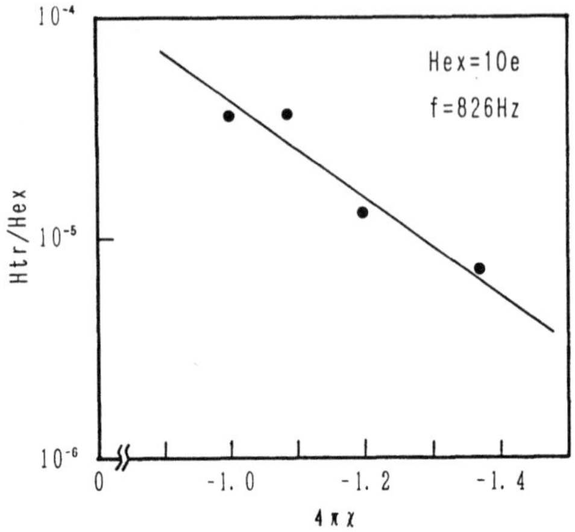

Fig. 8: Initial susceptibility dependence of the attenuation rate

CONCLUSION

The shielding attenuation rate depends exponentially on the shape factor and it decreases with increasing critical current density and volume fraction of the superconducting component. The attenuation in the sintered cylinder is independent of the magnitude of the applied magnetic field, while that in the cylinder coated by superconductor increases steeply with magnetic field. It is necessary to consider not only the superconducting properties but also the shape factor, when the shielding attenuation is to be precisely determined.

REFERENCES

[1] Hamilton W. O. (1970) Rev. Phys. Appl. 5;41-48
[2] Ogasawara T., Boesten L. (1973) J. Phys. Soc. Jpn. 28;1041-1049
[3] Mazaki H. (1988) J. Jpn. Powder and Powder Metall. 35:811-816
[4] Mazaki H. (1988) Jpn. J. Appl. Phys. 27:91-93

Numerical Analysis and Evaluation of the Stability of Superconducting Magnets in Maglev System

SUSUMU TORII, HIROYUKI OHSAKI, and EISUKE MASADA

Department of Electrical Engineering, The University of Tokyo, Bunkyo-ku, Tokyo, 113 Japan

ABSTRACT

Linear Synchronous Motor (LSM) driven transportation system was studied. Superconducting magnets are planned to be used for the field magnet of LSM, and also work as the repulsive levitation magnets. Magnets are exposed in electromagnetic disturbance from the primary and levitation coils on the ground. We have studied the electromagnetic phenomena, and numerically evaluated the stability of the magnets.

Magnetic field at the SC magnet plane are calculated, and rather high spatial harmonics are observed. Using the results above, Flux across the SC magnets are obtained under running condition. Rather small deviation of the currents, but higher dI/dt were shown. Sensitivity analysis of this dI/dt on the horizontal displacement is also studied.

Assuming the eddy current elements (conductor plate of cryostats, levitation coils on the ground) can be described as shorted coils, time-variant equivalent circuit equation system are derived. Solving these equations, unknown current in the whole system was determined, and consequently electromagnetic behavior and ohmic losses of the conductive elements can be calculated.

KEY WORDS: magnetic levitation, superconducting linear synchronous motor

INTRODUCTION

Magnetic levitation (Maglev) is one of the application of superconductivity. Superconducting windings are used as strong magnets, and serve for both the levitation magnets and the secondary of the synchronous motor. Magnets are used in the bogies of the car as shown in fig.1, and exposed in electromagnetic disturbance from the primary and levitation coils on the ground. Magnets should be used in these most severe condition, nevertheless we need higher stability of them for the request of safety. We must separate these problems into two — one about the direct electromagnetic interference into superconductor, the other about the heat caused by the eddy current induced in the vessels of the cryostat.

The first problem is estimated by conventional studies. Magnetic field and flux at the magnet must be calculated. To treat the second one, Electromagnetic analysis is needed to estimate the vibrations of the magnets. Field calculations are not adequate for this purpose, because of the large 3-D analysis region. We have studied about these two problems, and were possible to evaluate the stability of the superconductor.

SPATIAL HARMONICS OF ARMATURE WINDINGS

First, we calculate the spatial harmonics of armature windings, to evaluate the spatial configuration of the LSM without ferromagnetics. There exists only some distributed currents, and we can easily calculate the surrounding magnetic field with Biot-Savart's law.

$$B = \frac{\mu_0}{4\pi} \oint_C \frac{(r_2 - r_1) \cdot i}{r^3} ds \qquad (1)$$

where $r = |r_2 - r_1|$

Fig.2 shows the longitudinal magnetic field distribution at SC magnets plane. Ideally it must be sinusoidal, and this distorted waveform means that it has spatial harmonics. To improve this, we separated the two layer of armature. Fig.3 shows the components, and referring this we can reconstruct the configuration of armature windings.

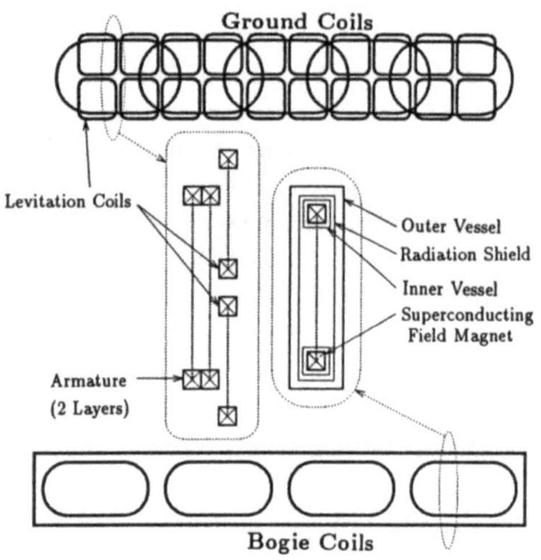

Fig.1. Dimension of the Maglev-LSM Bogie

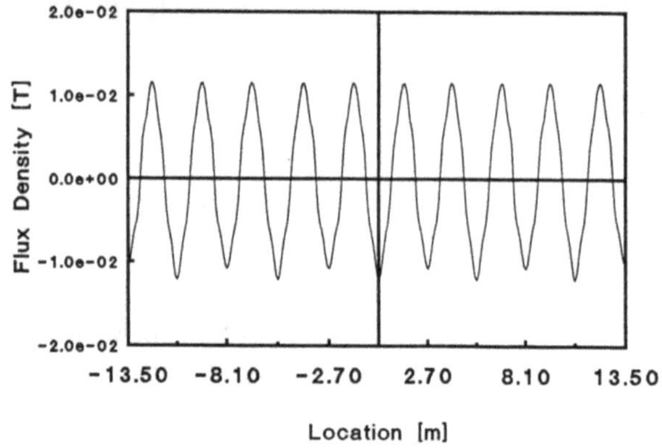

Fig.2. Flux Density at the SCM Plane

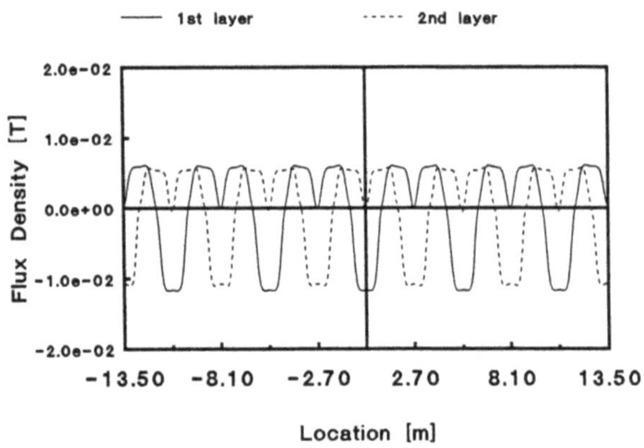

Fig.3. Components of Flux Density

FLUX DEVIATION IN SUPERCONDUCTING MAGNETS

Flux across the SC magnets can be obtained under running condition. Of course we can calculate the flux by numerical integration of the Biot-Savart's law, but more convenient technique is to use Neumann integration.

$$\phi = \frac{\mu_0}{4\pi} \int_{C_1} \int_{C_2} \frac{ds \cdot dr}{r} \tag{2}$$

Using equation(2), total flux crossing the SC magnet closed loop was calculated. the LSM is driven with $phaseangle = \pi$, so ideal total flux is 0. Though for the spatial harmonics, there exists certain deviation. Fig.4 shows the flux deviation. It is calculated with rectangular and racetrack coil configuration. Rather small deviation of the flux, but higher $d\phi/dt$ were shown. Think that superconductor has zero conductivity, $d\phi/dt$ is proportional to dI/dt of the magnet. The dI/dt causes the internal eddy currents in superconductor. and in higher rate of dI/dt, it may quench. Thus, sensitivity analysis of this dI/dt on the horizontal displacement is also studied, and fig.5 shows the results. They are the important index to determine the configuration of the magnet.

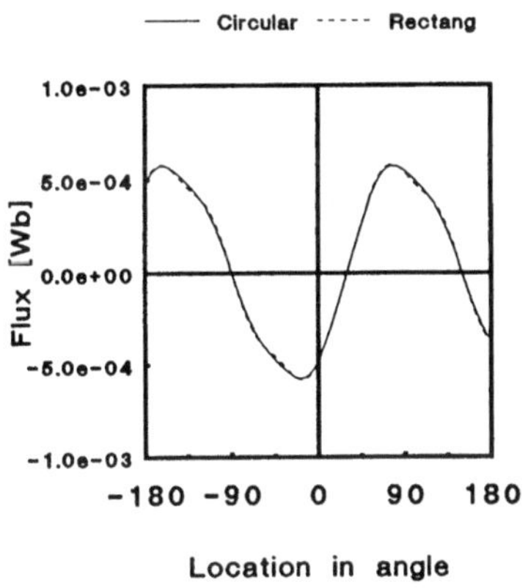

Fig.4. Deviation of Flux through SCM

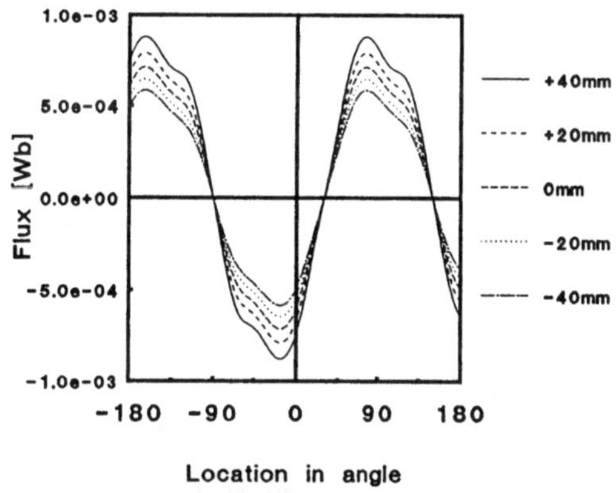

Fig.5. Horisontal Displacement Dependancy of the Deviation

FORCE CALCULATION WITH EDDY CURRENT ELEMENTS [1][2]

Assuming the eddy current elements (conductor plate of cryostats, levitation coils on the ground) can be described as a set of shorted coils, time-variant equivalent circuit equation system are derived.

$$R \cdot I = -\frac{d}{dt}\phi \tag{3}$$

$$\phi(t) = M_e \cdot I_e + M_a \cdot I_a + M_s \cdot I_s \tag{4}$$

The dimension of the system equals to the number of shorted coils, and coefficients are the self or mutual impedance. Subscript e means the eddy current elements, a for armature, and s for superconducting magnets. Solving these equations, unknown current in the whole system was determined, and consequently electromagnetic behavior and ohmic losses of the conductive elements can be calculated. Fig.6 and fig.7 shows the thrust force and the levitation force of the bogie, respectively. The waveform of vibration in these figures are caused by the levitation cils on the ground. Fig.8 shows the ohmic loss on the outer vessel of the cryostat. This is the eddy current loss caused by the spatial harmonics, and have to be very small in the inner vessel. The loss on the outer vessel has the same meaning as the shielding ability, thus fig.8 means the resistivity dependance of the shielding effect. Fig.9 is one of the example of the dependancy upon the analysis parameters. In general, this method of analysis consequently has the danger of resonance, and must be modified in future.

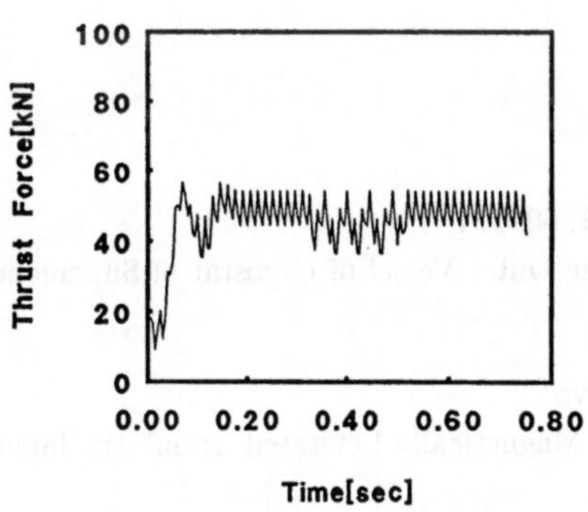

Fig.6. Thrust Force on the Bogie

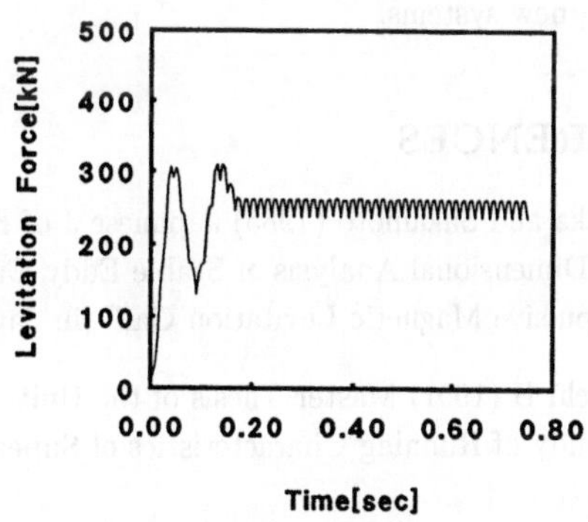

Fig.7. Levitation Force on the Bogie

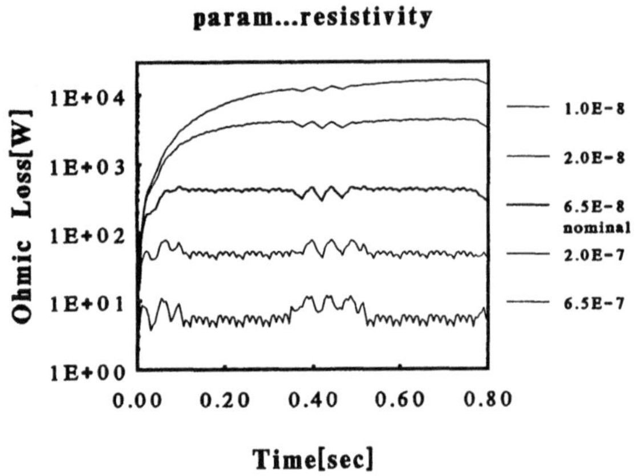

Fig.8. Ohmic Loss of Cryostat

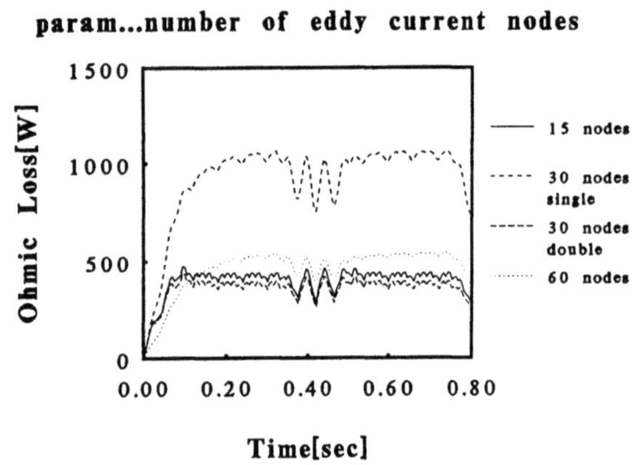

Fig.9. Ohmic Loss in Varied Number of Nodes

CONCLUSION

Mechanical and electromagnetic disturbances are analyzed. Using this method, the stability of superconducting magnets in MAGLEV system are carefully studied. The results obtained from the analysis are useful in designing new systems.

REFERENCES

[1] Nonaka and Sakamoto (1985) Japanese J of Elec. Eng. B, 60-B24
"3-Dimensional Analysis of Stable Eddy Currents in the Outer Vessel of Cryostat of Superconducting Repulsive Magnetic Levitation Car" (in Japanese)

[2] Deguchi H (1991) Master Thesis of the University of Tokyo
"Study of Running Characteristics of Superconducting Magnetically Levitated Train" (in Japanese)

A Conceptual Design of a Superconducting Magnet for a Magnetic Levitated Train Using a High T_c Oxide Superconducting Wire

Kenji Shimohata[1], Toshie Takeuchi[1], Shoichi Yokoyama[1], Hideto Yoshimura[1], Shirou Nakamura[1], Tadatoshi Yamada[1], and Shin Utsunomiya[2]

[1] Central Research Laboratory, Mitsubishi Electric Corporation, Amagasaki, Hyogo, 661 Japan
[2] Materials and Electronic Devices Laboratory, Mitsubishi Electric Corporation, Sagamihara, Kanagawa, 229 Japan

ABSTRACT

A conceptual design of a superconducting magnet for magnetic levitated train (MAGLEV) using a silver sheathed wire of high Tc oxide superconductor is studied. In the conceptual design, the superconducting magnet is operated by persistent current at 20K. The efficiency and the reliability of refrigerator are high at 20K. The critical current density of $6x10^4$ A/cm^2 is required at 5T for the conductor. The coil is stable against disturbance, because the minimum quench energy is large. On the other hand, the quench propagation velocity of the coil is low. The coil needs active quench protection.

KEY WORDS : Superconducting magnet, Oxide superconductor, Critical current density, MAGLEV

INTRODUCTION

The superconducting magnet of MAGLEV is made of the NbTi/Cu wire, and it is operated at liquid helium temperature[1]. The oxide superconductor was discovered in 1986. The critical temperature (Tc) of the high Tc oxide superconductor exceeds liquid nitrogen temperature. The coil using high Tc superconductor does not require liquid helium, and its advantages seem great. The critical current density of high Tc superconductor is improved and is close to practical level. In this paper, operating temperature, cooling system, current density, stability and protection of the superconducting magnet using high Tc oxide superconducting wire for MAGLEV is studied in comparison with the magnet of MLU002.

CONCEPTUAL DESIGN

OPERATING TEMPERATURE : Figure 1 shows one example of temperature dependence of normalized critical current density (Jc) for $(BiPb)_2Sr_2Ca_2Cu_3O_y$ silver sheath tape. The Jc is normalized by the Jc at 4.2K and zero magnetic field ($2.2x10^4$ A/cm^2). The difference of Jc is small between the Jc at 4.2K and that at 20K. The efficiency of the refrigerator at 20K is theoretically five times as high as that at the liquid helium temperature. The refrigerator of Gifford-MacMahon (GM) type has a running record of over 10,000 hours, and it is highly reliable[2]. Therefore in the conceptual design, superconducting magnet for MAGLEV which is operated at 20K with GM refrigerator is studied.

SCHEMATIC STRUCTURE OF MAGNET : Figure 2 shows a schematic structure of a high Tc superconducting magnet for a MAGLEV using the GM refrigerator. The outer shape of the coil is same as that of the MLU002. The conductor is composed of five layers of 0.2 mm x 2 mm Bi silver sheath tape. The specification of the two-stage GM refrigerator is shown in Table 1. It is designed to cool the 80K shield plate at the first stage, and liquid nitrogen for cooling the 80K shield is not necessary.
Table 2 shows the weight and required Jc for average current density of the coil. The magnetomotive force of 700kAT and the ratio of stabilizer and superconductor of 1.0 are supposed. The coil sectional area is setted as same as that of MLU002. The thickness of the insulatin of the conductor is 0.08mm. In this case, the Jc of over $6x10^4$ A/cm^2 is required for the high Tc superconductor. If the Jc is lower than $6x10^4$ A/cm^2, for example Jc is $3x10^4$ A/cm^2, the weight of the coil is more than 250 kg heavier than MLU002.

The MAGLEV coil is operated in the persistent current mode. Since obvious limit of magnetic field attenuation in MAGLEV system is not clear yet, we suppose the maximum current attenuation to be 0.5%/h (10 %/day). From the inductance of 3.04H, allowed resistance is calculated to be under $4.3\times10^{-6}\ \Omega$. Hence the superconductivity criterion of $10^{-13}\ \Omega$-cm is required. Even a slight resistance not reaching the quench of the coil is not permitted. Besides, since the operation temperature is high as 20K, it seems necessary to consider also the magnetic field attenuation due to flux creep.

STABILITY OF COIL : Figure 3 shows the magnetic field distribution of the coil section. In this study, the coil sectional area was supposed to be 45 x 71 mm^2. The maximum magnetic field applied parallel to the tape is 5T, and that applied vertically is 3T. As known from Fig. 1, normalized Jc is 0.2 in both cases. Therefore a superconductive quench may not always occur at the maximum magnetic field position in the coil using high Tc superconductor owing to the anisotropy of the wire. This tendency is more obvious as the operation temperature becomes higher.
Table 3 indicates the minimum propagation zone (MPZ) and the critical energy (Q_{MPZ}) at various operating point. Since the MPZ is sufficiently larger than the conductor sectional area, the temperature in the conductor is kept uniform, and the V_{MPZ} is set at [MPZ]x[conductor sectional area]. For example, if the Jc of the superconductor is $6.7\times10^4\ A/cm^2$, the operating point of the superconducting coil is 0.9, and at this time the Q_{MPZ} is 20 mJ. This value is sufficiently larger than that of the Cu/NbTi coil, and the coil stability using the high Tc superconductor is extremely high. It means that the high Tc superconducting coil may be used nearly at the limit of Ic of the wire.

COIL PROTECTION : In Figure 4, it is shown that the temperature rise profile in the case of quench of the coil. It is assumed that quench propagate only in the longitudinal direction of the wire. Case A is an ordinary case in which only one wire is quenched. It is known from this figure that the wire may be burned out unless protected at the time of quench. The propagation velocity of quench is given in the following formula, in which J is the current density of wire, γ is the specific heat, ρ is the resistivity, κ is the thermal conductivity, Ts is the intermediate temperature of current shearing temperature and superconducting critical temperature, and To is the bath temperature [3].

$$V = J/\gamma \times (\rho\kappa/Ts-To)^{1/2}$$

Hence in the case of high Tc superconducting coil, the quench propagation velocity is estimated at 0.5 m/sec. This value is 1/40 that of NbTi coil, and in the high Tc superconducting coil, a hot spot is likely to be formed at the time of quench. Case B shows the result of forced quench at four positions in the coil surface. In this case, too, the wire is burned out. Case C is the result of forced quench of the entire surface of the coil. The result of quench of the entire coil is shown as case D. It is found from these results that the coil can be protected at the time of quench by forced quenching of the entire coil surface. It is also important to consider the quenching propagation in the coil sectional direction from the viewpoint of protection of coil at the time of quenching.

CONCLUSION

In the high Tc oxide superconducting magnet operating at 20K for MAGLEV, stability of the coil and reliability of the refrigerator are improved extremely. The magnetic field attenuation due to flux creep and coil protection must be investigated precicely.

REFERENCES

[1]H. Nakao, M. Yamaji, K. Kurosawa, Y. Jizo, H. Nakashima and H. Tsuchishima Proceedings of 11Th International Conference on Magnetically Levitated Systems and Linear Drives (MAGLEV'89) July7-11, 1989, Yokohama, Japan, P229
[2]T. Inaguchi, M. Nagao and H. Yoshimura Proceedings of 16Th International Symposium On Space Techinology and Science, Sapporo, Japan, 1988, P921
[3]M. Wilson Superconducting Magnet, Clarendon Press Oxford, P200

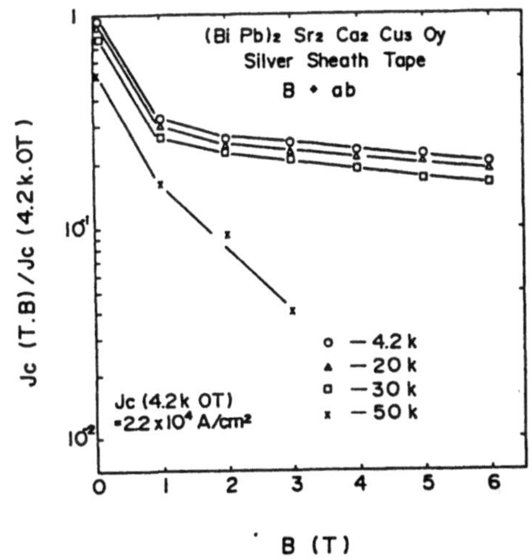

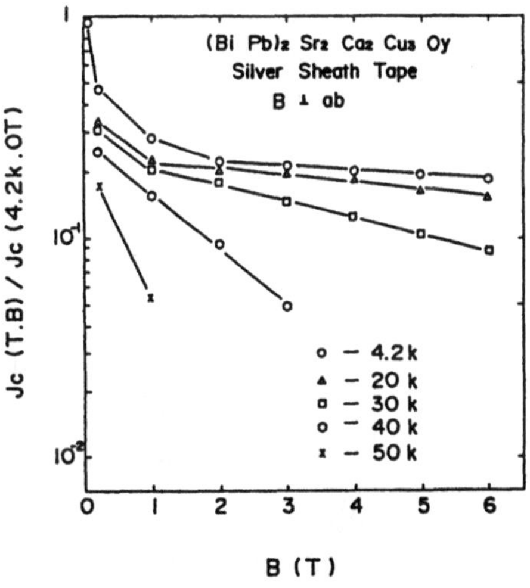

Fig.1. Temperature and field dependence of normalized Jc for $(BiPb)_2Sr_2Ca_2Cu_3O_y$ tape

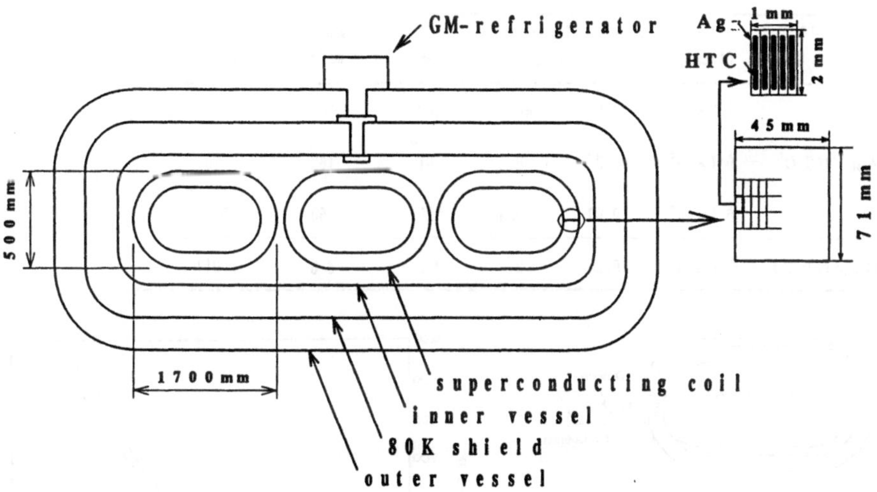

Fig.2. Schematic structure of HTc superconducting magnet for MAGLEV

Table 1. Major parameter of refrigerator

Cycle	2-stage GM
Cycle speed	144 rpm
Compressor	
Discharge pressure	30atm
Suction pressure	15atm
Mass	120kg
Input power	8kW
Expander	
Refrigeration capacity	200W at 80K 5w at 20K
Cylinder diameter	65mm/25mm
Stroke	30mm
Mass	30kg

Table 2. Mass of HTc superconducting coil and required Jc for some current density

CONDUCTOR	Scoil (mm^2)	Jcoil (A/mm^2)	Jc (A/cm^2)	MASS (kg)
Ag / HTC	45x71	219	6.0×10^4	255
	63x99	110	3.0×10^4	510
Cu / NbTi	45x71	219	$\sim2\times10^5$	231

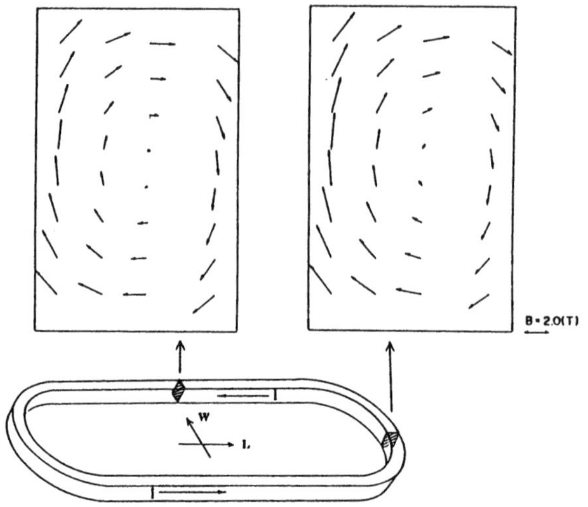

Fig.3. Magnetic field distribution of cross section of the coil

Table 3. Critical energy for Jc of the conductor

CONDUCTOR	Jc (A/cm²)	Jop/Jc	Temperature margin (K)	MPZ (mm)	V_{MPZ} (mm³)	Q_{MPZ} (mJ)
Ag/HTC	1.2×10^5	0.5	40	182	364	1270
	6.7×10^4	0.9	8	81	162	20
	6.1×10^4	0.99	0.8	25	50	0.37
Cu/NbTi	2.0×10^5	0.25	2	4.3	8.6	0.017

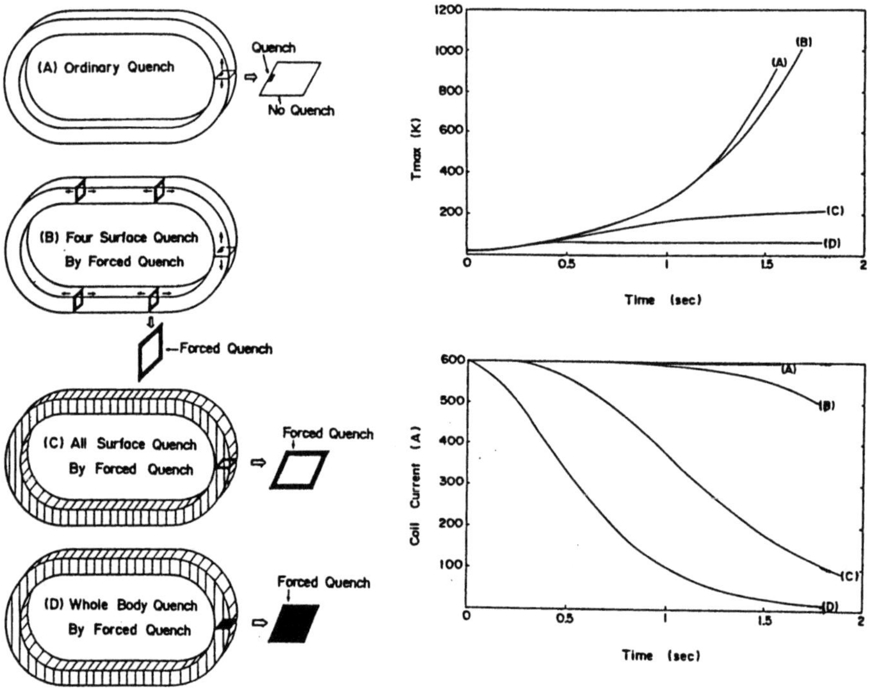

Fig.4 Quench simulatin of HTc superconducting coil

Advanced Design for Load Capable Superconducting Bearings

M. SHIBAYAMA[1], H. HIGASA[1], H. ISHIKAWA[1], T. OYAMA[1], T. TADA[1], T. YAMADA[2], S. NAKAMURA[2], S. YOKOYAMA[2], T. SHIMODA[3], and N. ITO[3]

[1] Shikoku Research Institute Inc., Takamatsu, 761-01 Japan
[2] Mitsubishi Electric Corp., Amagasaki, Hyogo, 661 Japan
[3] Seiko Epson Corp., Suwa, Nagano, 392 Japan

ABSTRACT

Superconducting bearings have some room for improvement in loading capacity and cost, the authors examined the effects of gaps between magnets and superconductors, width and thickness of magnets, thickness of superconductors, and combination of plural ring magnets. Four advanced design rules were obtained.

KEYWORDS: Bearing, Superconductivity, Design, Stiffness, Magnet

INTRODUCTION

The properties of superconducting magnetic bearings (SMB) such as high speed rotation without control, small decay torque and in vacuum, are big advantages for their application to fly-wheel energy storage systems[1,2]. However, the rather low loading capacity and low stiffness seem to be barriers to this application.

In the previous study, the authors proposed two countermeasures; a split-and-coupled fly-wheel and a floating installation method[3]. Figure 1 displays the loading capacity of SMB of 0.01～0.05 kgf/cm², which were made by Ichinosekogyo Co.Ltd., Koyo Seiko Co.Ltd., ISTEC, and Mitsubishi Electric Corp. in those days.

In the present study, a strong effort is contributed to developing the capability of superconducting bearings.

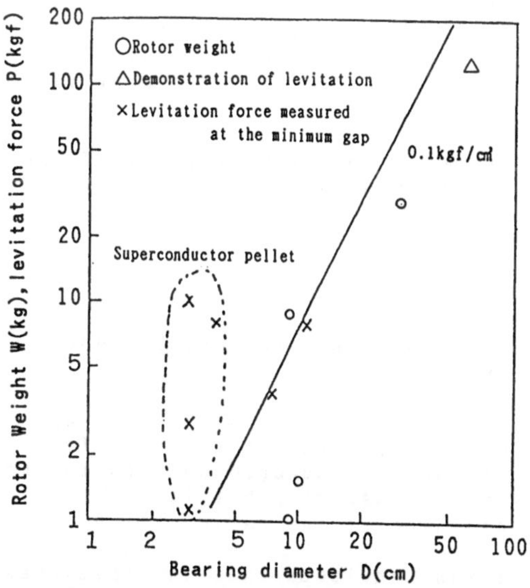

Fig.1 Loading capacity of SMB reported before this summer.

SPECULATION

SMBs are usually constructed in a "field cool" process. Loading capacity of SMB is nearly equal to the value of stiffness which is expressed in kgf/mm unit. The magnetization of superconductors during stiffness-measurement is similar to the C-B part of the M-B curve in Fig.2.

By considering the mentioned conditions to magnetic reaction rule; F = grad (M·H), a formula for the stiffness of SMB can be derived as,

$$K = m\ d(H_z^2)/dz \quad \text{------} \quad (1)$$

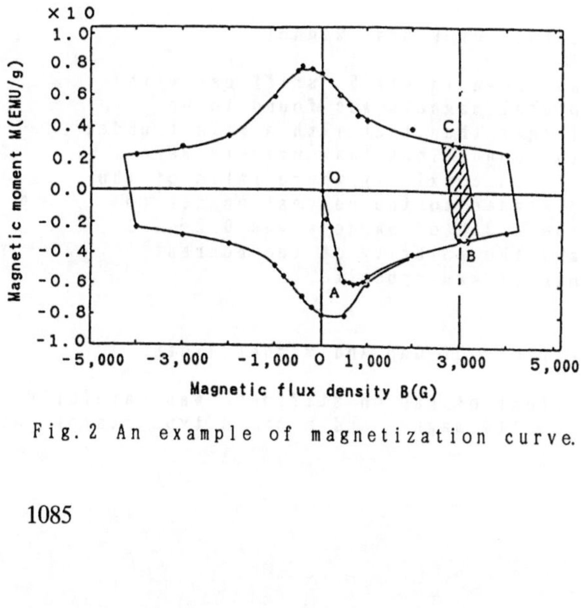

Fig.2 An example of magnetization curve.

where m is the magnetic susceptibility.

Equation (1) suggests that by optimizing the size of magnet, the arrangement of the magnet, and the size of superconductors, the loading capacity of SMB can be increased.

EXPERIMENTAL PROCEDURE

Experiments on stiffness was performed by using dummy rotors containing ring permanent magnets, dummy stators containing superconductors, and a loading machine with a displacement measuring device. Five different sizes of magnets were attached singly or doubly to the rotors. Three types stator were used in the experiments; solid type for the accurate measurement of stiffness in the wide ranging gap, open type for the effects of the thickness of superconductors on stiffness, semi-open type for the effect of plural magnets on stiffness.

Figure 3 shows the overview of the experimental setup.

Fig.3 Setup of stiffness test.

RESULTS

Effect Of The Thickness Of Superconductors

As shown in Fig.4, stiffness was found to increase the ratio of the thickness of superconductors to the width of magnets, but it was saturated at the ratio of 1.0.

It was concluded that the thickness of superconductors should be from 0.5 to 1.0 times of the width of magnets.

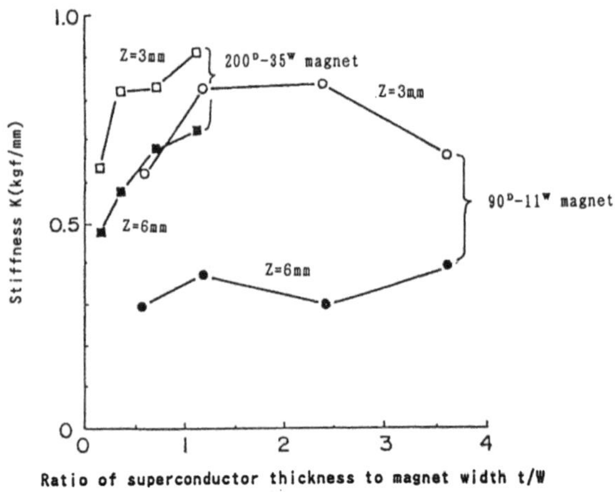

Fig.4 Effect of superconductor thickness on stifness.

Effect Of Plural Magnet

As shown in Fig.5, stiffness with plural magnets was found to be larger than that with a magnet under the conditions that magnets were placed coaxially, where ratio of the distance to the nearest magnet to the width of magnets was 0.2～1.0, and the polarity of the nearest magnet was opposite.

Effects Of Gap And Magnet Width

Effect of gap on stiffness was carefully examined by using a rotor with 90D-11W magnet and a solid type stator. Stiffness of the SMB was expressed as,

$$K = 23\ (1 + z)^{-1.5} \qquad \text{-------- (2)}$$

where K in kgf/mm, and z in mm.

The effect of magnet width on stiffness was examined by using two rotors containing 90D-11W and 200D-35W magnets. Open type stator was used for the test.

Measured stiffness are shown in Fig. 6 by solid marks. As the magnetic flux density and the volume of superconductor in the case of larger rotor was less than those in the case of smaller rotor, data was corrected by using equation (1). The corrected values are plotted with open marks in the figure.

A relation between stiffness K and the elemental dimensions : magnet diameter(D), magnet width(W), surfacial magnetic flux density (B_0), the distance of gap (z), was found as,

$$K = a\,\pi\,D\,B_0^2\,W\,(0.1 + z/W)^{-1.5} \qquad \text{-------- (3)}$$

where $\alpha = 0.1 \sim 0.3$.

As bearings need surfacial area proportional to $2\pi DW$, the stiffness per unit area "effective stiffness k" was defined as, $k = K / 2\pi DW$. The derived equation is expressed as,

$$k = \left[\frac{aB_0^2}{2z^{1-\alpha}}\right]\left[\left(\frac{z}{W}\right)^{1-\alpha}\left(0.1+\frac{z}{W}\right)^{-1.5}\right] \qquad \text{-------- (4)}$$

The term "geometry parameter P_g", of the right hand of the equation, expresses how k depends on W under a given gap condition.
With the help of the Fig. 7, it was concluded that the size of magnets was reasonably selected to be 5~10 times of the distance of gaps.

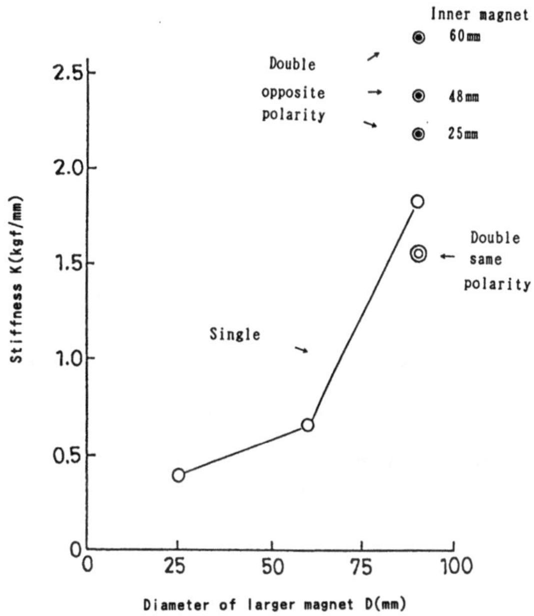

Fig. 5 Effect of plural magnets on stiffness.

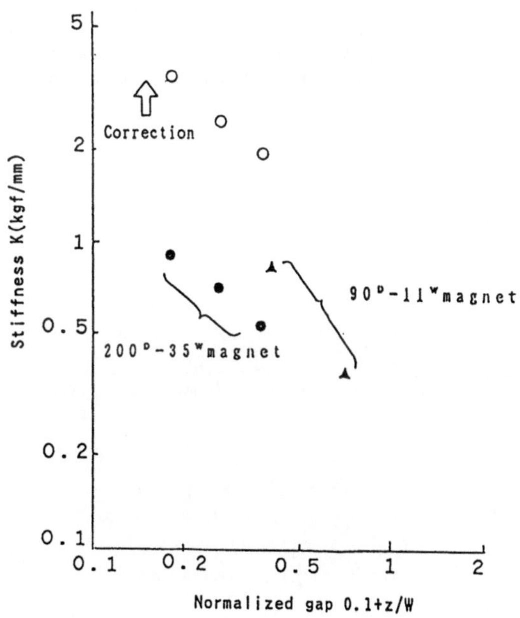

Fig. 6 Chart for formulation of stiffness.

CONCLUSIONS

(1) Use of a ring-shaped magnet of a large width is effective in improving the stiffness, when the structure requires a wide gap. A magnet of a width (thicness, too) 5 to 10 times the gap is proper for a case such as above.

(2) With a magnet optimized as described above, superconductors of thickness 0.5 to 1.0 times the magnet width, work well.

(3) Arranging ring-shaped magnets concentrically on the rotor is effective in increasing the stiffness per unit cross sectional area of the bearing. In this case, better results can be obtained by combining the magnets in opposite

poles with a gap between magnets from 0.2 to 1.0 times the magnet width.

When compared to a superconducting magnetic bearing designed without due considerations to the factors examined in this research, design with the above considerations increase the stiffness about two folds by the first term, 2 to 3 folds by the third term, resulting in 5 to 6 fold increase as a total. Consequently, stiffness per unit cross sectional area is increased from 0.05 kgf/mm·cm² at present to 0.2 kgf/mm·cm².

The authors predict that more developed superconductors and magnets may multiply 10 to the present best values in the near future.

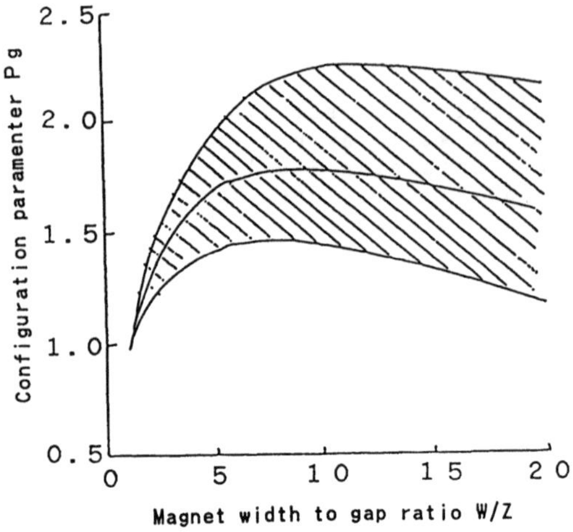

Fig.7 Effect of magnet size on surface stiffness.

ACKNOWLEDGEMENTS

Authors appreciate valuable discussions made by Mr.Takahata of KoyoSeiko Co. Ltd., Mr.Isihara and Mr. Ohtani of Shikoku Electric Power Co. Ltd.

REFERENCES

1. Moon FC, Chang PZ (1990) Appl.Phys.Lett. 56(4): 397
2. Takahata R, Yotsuya T (1991) IEEE Trans. on Magn. 27-2: 2423
3. Shibayama M, Takaichi H, Okada S, Takahata R, Isihara T (1991) Annual reports of Shikoku Electric Power 57,: 52 (June,1991)

Characterization of Superconducting Magnetic Bearings (Vibration Damping and Hysteresis of Magnetic Force in Superconductor)

RYOICHI TAKAHATA and HIROCHIKA UEYAMA

Koyo Seiko Co., Ltd., Kashiwara, Osaka, 582 Japan

ABSTRACT

Superconducting magnetic bearings using flux pinning forces have a stability, permanent magnet bearings have not, without any control. High-Tc superconductor $YBa_2Cu_3O_X$ as a stator and a permanent magnet for a rotor were assembled into a superconducting magnetic bearing. Its stiffness and vibration damping coefficient in axial direction were measured. These properties depended on an axial gap between superconductor and permanent magnet. As a flux pinning force of $YBa_2Cu_3O_X$ increased, bearing stiffness also increased but vibration damping coefficient decreased. The damping effect was related to the area of the hysteresis loop measured in a function of magnetic force versus axial gap.

KEY WORDS : superconductor, bearing, magnetic force, magnetic stiffness, vibration damping

INTRODUCTION

Since a high-Tc superconductor was discovered, many efforts have been made on developing a superconducting magnetic bearing (SMB) [1-5]. Recently a melt process on bulk $YBa_2Cu_3O_X$ has been improved [6,7] and a levitation pressure has also become large [8]. The SMB has been able to support a rotor weighing about 1 kg, rotating at a speed of 5,000rpm [3]. The SMB has been estimated using several factors, levitation pressure, dynamic stiffness [2], damping coefficient [4], and the rotational speed [1,5].

In this paper dynamic properties of SMB were discussed. Bearing stiffness and vibration damping coefficient in axial direction were measured. One of the purpose of this work was to examine the vibration damping and the hysteretic magnetic force of SMB . They are very significant for bearing design.

EXPERIMENT

A superconducting magnetic bearing were assembled by using bulk $YBa_2Cu_3O_X$ and a permanent magnet.

Three pieces of bulk $YBa_2Cu_3O_X$ whose repulsive force acting on permanent magnet were different were used for samples in this test. Physical and levitation parameters of the samples were shown in Table 1. The authors prepared the samples MQ-1,2 by MPMG process [6]. The sample MQ-3 was made by QMG process [7] by Nippon Steel Corporation. The levitation pressure of MQ-3 was much larger than that of MQ-1,2.

The permanent magnet has 9mm in diameter and 12mm in thickness and its surface flux density was 0.45T.

The apparatus shown in Fig.1 was constructed in order to measure the vibration damping and the stiffness of SMB in axial direction. A shaft having permanent magnets on its lower side was located vertically above the bulk $YBa_2Cu_3O_X$. The shaft was able to be supported with no mechanical contact by air bearings in radial direction and by SMB in axial direction. To prevent the damping effect brought about by eddy current, conducting materials were located as far from permanent magnets as possible. The total mass of the shaft was 35g which was light enough to stabilize the axial vibration of the shaft by MQ-1 whose pinning force was smaller than others.

Table 1 Physical and levitation parameters of samples

Sample	Diameter (cm)	Thickness (cm)	Density (g/cm^3)	Levitation Pressure*1 (N/cm^2)	Process
MQ-1	3.8	1.3	5.9	0.9	MPMG*2
MQ-2	3.5	1.2	5.6	1.9	MPMG
MQ-3	4.3	1.2	5.7	3.4	QMG*3

*1; 0.1mm distance between sample and magnet(Nd-Fe-B,Ø30 x 12,0.45T) at zero magnetic field cooling
*2; Melt Powdering Melt Growth
*3; Quench and Melt Growth

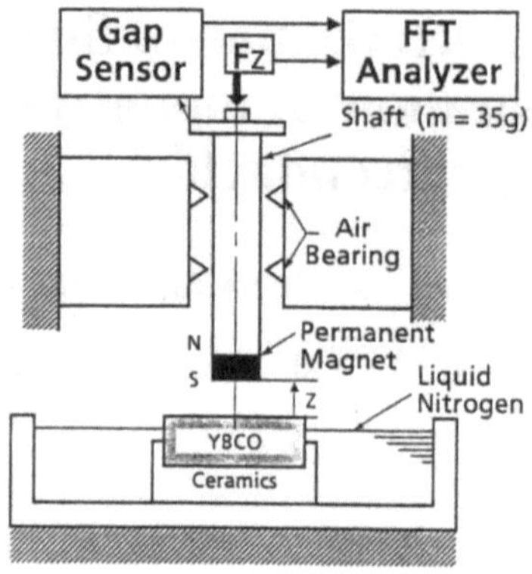

Fig.1 Dynamic properties measuring apparatus

The shaft was supported by air bearings in radial direction and fixed at initial gap Z_0 between the superconductor surface and the bottom surface of the permanent magnet in axial direction, then the bulk $YBa_2Cu_3O_X$ was cooled down to about 77K by liquid nitrogen. After above field cooling, the shaft was released and supported with no mechanical contact. At that time an axial gap of SMB decreased to $Z<Z_0$. That is the equilibrium position. A vibration of the shaft in axial direction was measured by gap sensor when an external impulse force Fz acted on the shaft.

RESULT AND DISCUSSION

An axial vibration of the shaft is shown in Fig.3(a). Fig.3(b) shows a vibration of the shaft supported by a permanent magnet bearing (PMB) by the same procedure. The vibration damping of SMB was larger than that of PMB using repulsive forces of permanent magnets. Therefore a SMB model in one degree of freedom can be shown in Fig.3(b). The motion of this model can also be described by equation (1).

$$m\ddot{z} + c\dot{z} + kz = Fz \qquad (1)$$

In this equation m is the mass of the shaft, k is the spring constant and c is the damping coefficient respectively.

When the shaft was struck by a impuls-hammer in axial direction, the axial vibration of it were measured at several gaps of Z corresponded to the initial gap Z_0 = 1.0, 2.0, 3.0, 4.0, 5.0 mm respectively. Z did not change after the hammering.

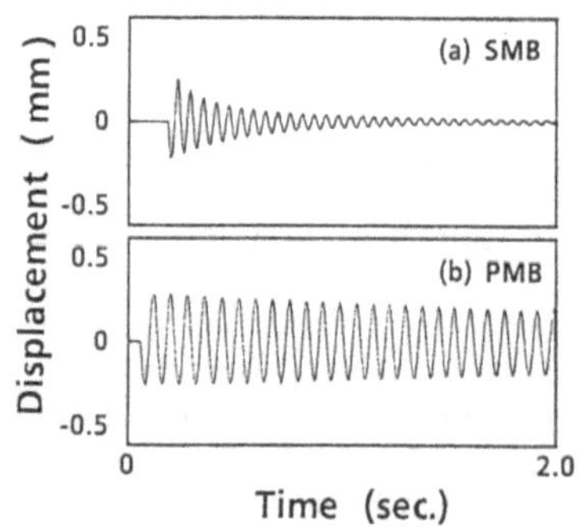

Fig.2 Vibration of shaft in axial direction; (a) in the case of SMB, (b) in the case of PMB

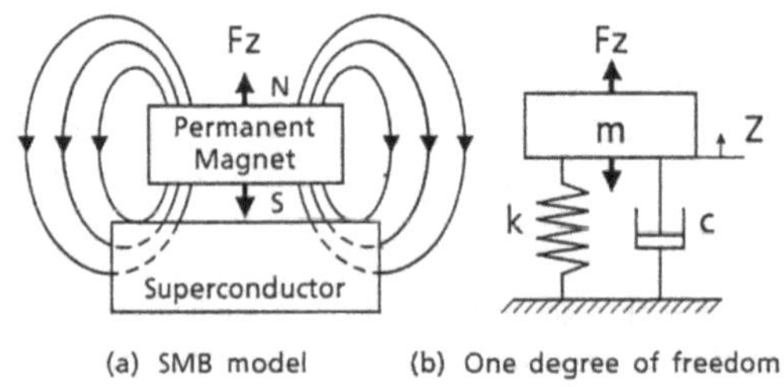

Fig.3 SMB model in 1 degree of freedom

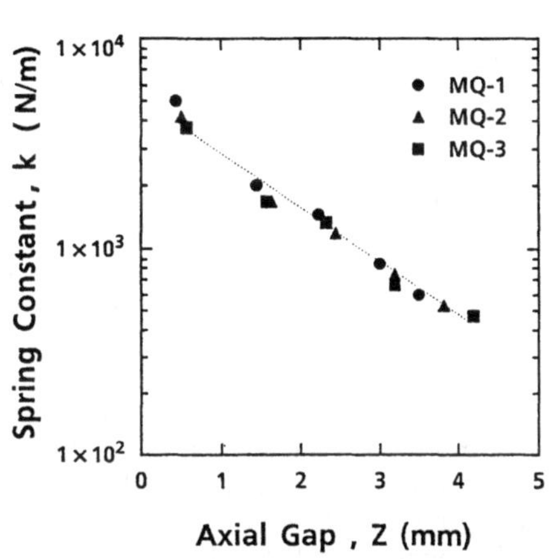

Fig.4 Spring constant vs axial gap

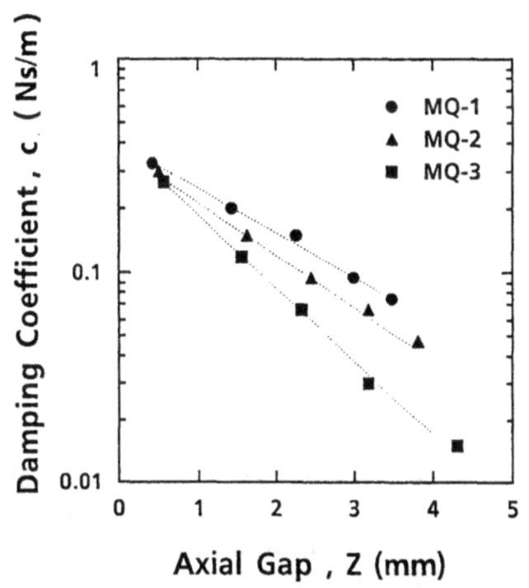

Fig.5 Damping coefficient vs axial gap

The spring constant of axial direction is able to be given by equation (2).

$$k = 4\pi^2 \cdot m \cdot f^2 \tag{2}$$

where f is the vibration frequency.
The value of spring constant k calculated from equation (2) using the measured f is shown in Fig.4. The spring constant increased as the axial gap decreased. The spring constant shows the stiffness of SMB. The value of stiffness of the three samples is approximately equal at same gap Z respectively.

It is also possible to estimate damping coefficient c from the decreasing characteristics of vibration amplitude. Damping coefficient c can be described by equation (3).

$$c = 2m \cdot \beta \tag{3}$$

where β is a natural logarithm of damping ratio per second in vibration amplitude.
Measured damping coefficient c is shown in Fig.5. Damping coefficient c also increased as an axial gap decreased.
But in this test equipment, measured c included the intrinsic damping coefficient in addition to the damping coefficient due to SMB. It caused by air viscous resistance acting on the shaft. It was clear from the fact that the vibration of the shaft supported by permanent magnet bearing, not having a damping factor, had actually a little damping as shown in Fig.2(b). As the intrinsic damping were very small in this experiment, the measured damping coefficient c is considered as the damping due to SMB.

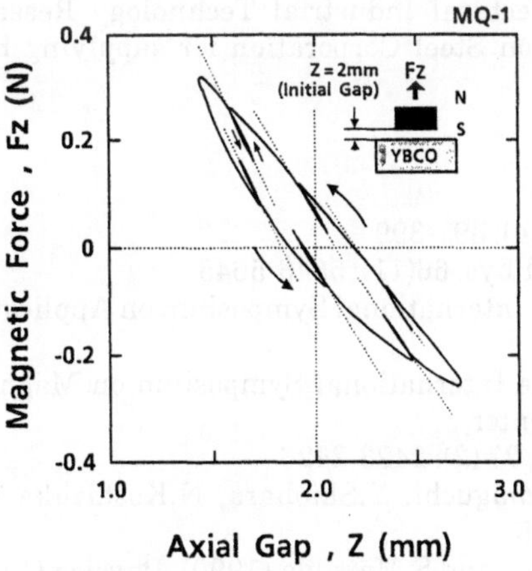

Fig.6 Magnetic force vs axial gap
for static motion of magnet

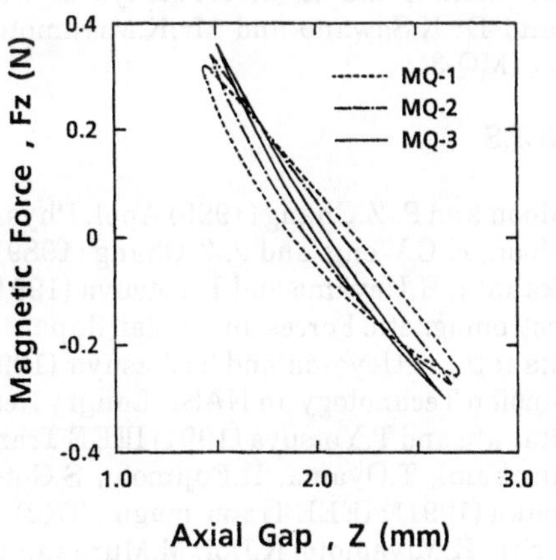

Fig.7 Magnetic force vs axial gap
for three samples

Damping coefficients of the three samples were not the same as shown in Fig.5. It seems to be caused by the difference in the area of hysteretic magnetic force [2] in superconductor. A hysteretic magnetic force in the sample MQ-1 were measured when the magnet were oscillated by 1mm amplitude slowly in axial direction near Z_0=2mm as shown in Fig.6. Some small hysteresis loops were contained in the large hysteresis loop. It caused by small vibrations (about 0.2mm amplitude) added at different points on 1mm oscillation. In the case of the sample MQ-1 at Z_0=2mm, damping measurement as shown in Fig.5 is considered as the small vibration at Z=1.6mm in Fig.6. Therefore It is believed that vibration damping is related to the area in the small hysteresis loop at Z=1.6mm. This area in miner loop depended on that of large loop.
The large hysteresis loops of the three samples at Z_0=2mm are shown in Fig.7. Static stiffness of MQ-3 whose damping coefficient was smaller than others was largest in judging from slop of hysteresis loop. It is obvious that $YBa_2Cu_3O_X$ whose damping coefficient is large has large hysteresis loop and has small static stiffness. Since the vibration frequency of these samples are the same at Z=1.6mm as shown in Fig.6, the damping effect was due to the area in hysteresis loop of magnetic force in superconductor. It is considered that the large hysteretic magnetic force is concerned with weak pinning force of superconductor.

It is necessary for practical SMB to improve both magnetic stiffness and damping effect. But it seems that it is difficult for superconductor to be improved these properties at the same time. Therefore a improvement of optimizing magnetic stiffness and damping factor in $YBa_2Cu_3O_X$ material is expected for one of the ways to develop a practical SMB. If such improvement in material would be impossible, it would be necessary for SMB to be added another damping factor in cryogenic environment.

CONCLUSIONS

Dynamic characteristics of the superconducting magnetic bearing, using high-Tc superconductor and a permanent magnet in axial direction were measured.

(1) Dynamic stiffness and damping coefficient of SMB were dependent on an axial gap.
(2) The vibration damping is related to the area in the hysteresis loop in a function of magnetic force versus axial gap. It is obvious that $YBa_2Cu_3O_X$ whose damping coefficient is large has large hysteresis loop and has small static stiffness.
(3) The optimizing magnetic stiffness and damping factor of superconductor is expected for one of the ways to develop a practical SMB.

ACKNOWLEDGMENTS

The authors wish to thank Mr.T.Yotsuya of Osaka Prefectural Industrial Technology Research Institute, and Dr.K.Sawano and Mr.K.Miyamoto of Nippon Steel Corporation for supplying bulk $YBa_2Cu_3O_X$ (MQ-3).

REFERENCES

1. F.C.Moon and P.-Z.Chang (1990) Appl. Phys. Lett. 56 (4): 397-399
2. F.C.Moon, K-C.Weng, and P.-Z.Chang (1989) J. Appl. Phys. 66(11): 5643-5645
3. R.Takahata, H.Ueyama and T.Yotsuya (1991) In: The International Symposium on Application of Electromagnetic Forces, in Sendai, Japan
4. R.Takahata, H.Ueyama and T.Yotsuya (1991) In: The International Symposium on Magnetic Suspension Technology in NASA Langry Research Center
5. R.Takahata and T.Yotsuya (1991) IEEE Trans. magn., 27 (2): 2423-2426
6. M.Murakami, T.Oyama, H.Fujimoto, S.Gotoh, K.Yamaguchi, Y.Shiohara, N.Koshizuka and S.Tanaka (1991) : IEEE Trans. magn., 27(2): 1479-1486
7. M.Morita, K.Miyamoto, K.Doi, M.Murakami, K.Sawano and S.Matsuda (1990) Physica C 172 : 383-387
8. K.Miyamoto, M.Morita, M.Tanaka, S.Takabayashi, K.Kimura, M.Kimura and K.Sawano (1990) Advance in superconductivity III: 727-731

Superconducting Magnetic Bearing Using MPMG YBaCuO

H. FUKUYAMA[1], K. SEKI[1], T. TAKIZAWA[1], S. AIHARA[1], M. MURAKAMI[2], H. TAKAICHI[2], and S. TANAKA[2]

[1] Tribology Research & Development, Research Center, NSK Ltd., Fujisawa, Kanagawa, 251 Japan
[2] Superconductivity Research Laboratory, ISTEC, Koto-ku, Tokyo, 135 Japan

ABSTRACT

A large superconducting magnetic bearing with MPMG YBaCuO material was built in this study. The bearing was driven with a high frequency motor. A pair of the MPMG YBaCuO rings($\phi 40 \times \phi 65 \times$ t20) were fixed in the cooling spaces which were partitioned with thin seal plate. The spaces were cooled with liquid nitrogen.

The shaft was centerized with a specially designed centerizing device. Without the device the rotational shaft will be fixed at arbitrary position. The weight of a rotor of the bearing was 2.4kg and it could be rotated safely at 30000 rpm.

INTRODUCTION

Composing bearings with superconductors has been a much talked idea. And when it is realized it would be an excellent magnetic bearing without regulating device and with very small friction loss. However, these conventional superconducting magnetic bearings using only Meissner effect have very small load carrying capacity, so they are far from realistic bearings for wide use.[1),2),3),4)] On the other hand, Ag-doped Y-Ba-Cu-O superconducting pellets fabricated by the Melt-Powder-Melt-Growth(MPMG) process developed in ISTEC Superconductivity Research Laboratory have very large flux pinning effect and display large magnetic force compared with Meissner effect.[5)] So, MPMG YBaCuO seems to become actual magnetic bearing material, and several examples utilizing repulsive force of MPMG YBaCuO have been published.[6),7)]

However, in the application of MPMG YBaCuO to superconducting magnetic bearings, if only the repulsive force is used, load capacity would not be enough to carry large load and it would not be realized that the shaft supported by superconducting magnetic bearings can rotate at high speed beyond the critical speed of the bearing system.

In this paper, utilizing not only repulive force but also attractive force caused by sophisticated bearing system deseign, and applying peculiar centerizing device composed of ball bearings, it is reported that the development of superconducting magnetic bearings using MPMG YBaCuO has enabled the shaft weighing 2.4kg to rotate safely at 30000 rpm beyond the critical speed of the bearing system.

DESEIGN

The structure of bearing rig is shown in Fig. 1. A pair of superconducting bearing components were fixed in two places axially separated from each other in the rig. These superconducting bearing components cooperate with magnets on the shaft and the superconductivity appears below the critical temperature(Tc).

Centerizing devices were set at the upper portion and bottom portion of the rig. The shaft was rotated by a high frequency motor. Radial displacements of the shaft were measured both at the upper and the bottom measuring points in X-Y directions as shown in Fig.1.

The differences of this bearing from conventional published examples are as follows :

① Centerizing device was provided for centerizing the shaft resulting in perfect non-contact levitation for high speed rotation beyond the critical speed of the bearing system.
② Superconductive materials were machined into rings by grinding in high precision and acted as both axial and radial bearings .
③ Superconductor rings were set in the spaces partitioned with thin shield plates and were cooled effectively by circulating liquid nitrogen.
④ Thin film of epoxi resin was coated over the superconducter rings so as to prevent deterioration of superconductor due to frost during cooling and heating cycles.

⑤ Using two superconductors, namely upper and bottom rings, and two sets of magnet rings, the rotating shaft was driven by a high frequency motor, and with touch down bearings, this rig was almost an actual magnetic bearing spindle.

The dimension of the superconducting magnetic bearings is shown in TABLE 1.

TABLE 1. DIMENSION OF SUPERCONDUCTING MAGNETIC BEARING

superconductor	$\phi 40 \times \phi 65 \times t20$
magnet	
for axial bearing	$\phi 40 \times \phi 65 \times t\ 8$
for radial bearing	$\phi 36 \times \phi 25 \times t20$
seal plate	
inner diameter	$\phi 38$
thickness	0.6 mm
clearance	
radial of magnetic	2.0 mm
radial of touch-down	0.35mm
axial of magnetic	6.5 mm
axial of touch-down	5.5 mm

A photgraph of the superconducting bearings and the shaft is down in Fig. 2, and the aspect of this bearing apparatus is shown in Fig. 3.

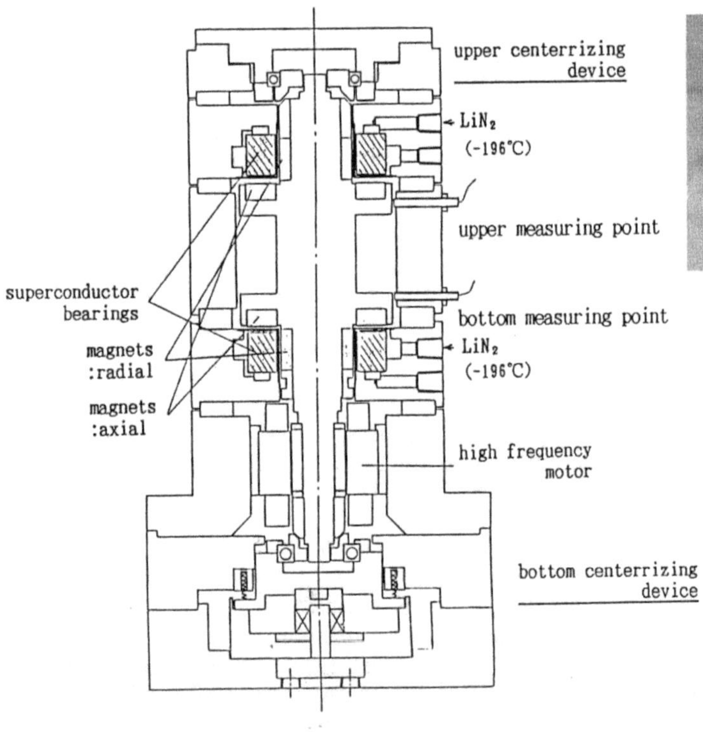

Fig. 1 Superconductor Magnetic Bearing

Fig. 2 Superconducting Bearing Component and Rotor Shaft

Fig. 3 Entire View of the Bearing

CENTERIZING DEVICE

The structure and operation of the centerizing device are described in more detail below.

Under field cooling operation, the centerizing device is needed so that the the rotor shaft should be kept in the right axis. Without such a centerizing device, superconducting bearings would pin flux even with the shaft contact anywhere in the housing when it was cooled. In this case, superconducting magnetic bearings can not operate at perfect noncontact rotation.

The Centerizing device consists of upper centerizing part and bottom centerizing part as shown in Fig.1. The lower centerizing part has air-cylinder function consists of bellows and coil springs, and it could move the shaft up and down within the span of 4 mm. The shaft was pushed up to the upper centerizing part and becomes centerized.

Under centerized condition, the shaft was rotated and supported by ball bearings over the crytical speed of the bearing system, and then the superconducting magnetic bearings were cooled with liquid nitrogen.

In this state, air pressure of lower centerizing part was released. The shaft fell down to the position where the axial load capacity of superconducting magnetic bearings was equal to the shaft weight, and rotated without contact. In this case, the levitation force was mainly due to the attractive force. That is, the suspension of this bearing apparatus was carried by not only the repulsive force of axial surface of bottom bearing but also due to the attractive force of axial and radial surfaces of upper bearing and of radial surface of bottom bearing.

The tapered inner rings of the centerizing device also play a role of touch-down bearings in the case of high speed rotation.

EXPERIMENTAL RESULTS

Tests were carried out up to the rotational speed of more than 30000 rpm. The rsults of run test are shown in Figs. 4 to 6. Fig.4 shows the amplitude of vibration versus rotational speed of the first and second experiment. In the first experiment, the amplitude became nearly constant 120 μm peak to peak at upper measuring point beyond 11000 rpm. In the second test, the amplitude became minimum at 16000~18000 rpm both at upper and bottom measuring point. Beyond 26000 rpm, the amplitude reached nearly constant 120 μm peak to peak at upper measuring point and nearly constant 75 μm peak to peak at bottom measuring point.

Fig.5 and Fig. 6 are the oscillograms of vibration of the shaft in the second test. Fig.5 shows the rotational speeds before and after releasing the function of centerizing device. Fig.5 (a) is when the centerizing device was acting and the shaft was supported by ball bearings. In Fig.5(a), the setting speed of high frequency motor was about 8100 rpm, and actual speed of the shaft was 4317 rpm. Fig.5(b) is when the centerizing device was set free and the shaft was levitated in perfect noncontact state by superconducting magnetic bearings. Rotational speed increased rapidly from 4317 rpm to 8000 rpm. This assures that supporting situation changed from ball bearing support with large friction to superconducting magnetic bearing support with low friction.

Fig.6 shows the displacements of shaft and the pulses of rotation at 17700 rpm (a) and at 30840 rpm (b). The shaft was found rotating in whirling motion with amplitude of 120 μm peak to peak with 1/5 ~1/7 period of shaft rotation. The phase of whirling motion was the same at the upper measuring point and at the bottom measuring point, so the whirling motion was in cylndrical motion. At present, it is not clear whether this long whirling period is due to high frequency motor or due to native characteristics of superconducting magnetic bearings.

CONCLUSIONS

1. The superconducting magnetic bearing with a structure similar to a conventional magnetic bearing, utilizing a large levitation force due to pinning effect between MPMG-YBaCuO superconductors and magnets enabled a 2.4 kg shaft to rotate safely at over 30000 rpm.

The key points of this bearing apparatus are as follows.

① Centerizing device with ball bearings is used to centerize the shaft and the device is set free to result in perfect noncontact situation at high rotational speed supported only by superconducting magnet bearings, beyond the critical speed of bearing system.

② Setting superconductor in the space partitioned with thin seal plate, the superconductor has been cooled effectively by circulating liquid nitrogen.

③ Machining the superconductor into ring(ϕ40×ϕ65×t20) in good precision, both axial and radial bearing function have made progress.

④ Thin film coating of epoxi plastics over the superconductor has evaded the effect of having frost on superconductor during cooling and heating cycles.

2. Compared with conventional superconducting magnetic bearing using the repulsive force alone, this bearing has utilized mainly the attractive force due to flux pinning. By utilizing this attractive force, it has been possible to accomplish effective levitation in a closed limited machinery space and to achieve noncontact high speed rotation .
3. It has been proved that YBaCuO superconductors fabricated by MPMG can be produced in large size and they can be applied to precise machining without spoiling their superconducting characteristics.
4. The present design is not sufficient for radial bearing stiffness and this causes large amplitude of whirling. In order to suppress the amplitude, it is necessary to increase in radial stiffness with more sophisticated bearing design.

REFERENCES

1) G.Binning et al, Appl.Phys.Lett.,40(2),15(1982),178.
2) M.Okano et al, 32nd Meeting on Cryogenics and Superconductivity,b2-13(1984),105
3) F.C.Moon and P.Z.Chang, Appl.Phys.Lett.56(4),22(1990),397
4) R.Takahata and T.Yotsuya, IEEE Trans. Magn. 27,2,Mar.(1991),2423
5) M.Murakami et al, Jap. J. Appl. phys.,29(1990)L1991
6) T.Yotuya, M.Shibayama and R.Takahata: Advances in Cryogenic Engineering, vol.36,1991
7) T.Yokoyama, Paper Presented at Annual Autumn Appl. phys. Meeting(1990)

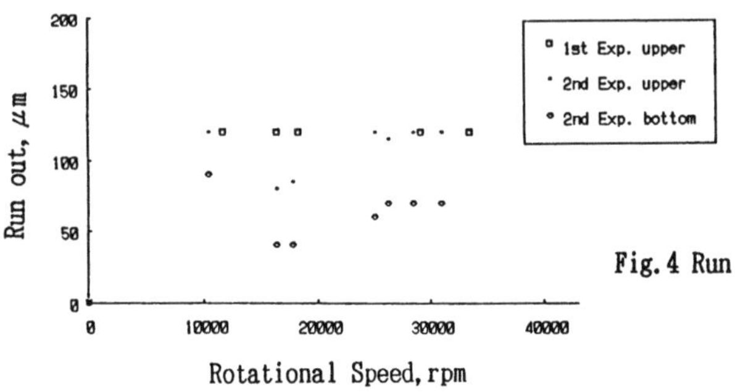

Fig.4 Run out with Rotational Speed

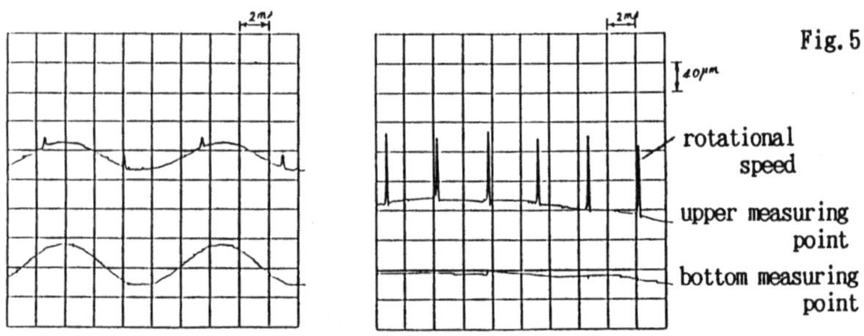

(a)4317rpm (b)7999rpm

Fig.5 Increase of Rotational Speed

(a)with rolling bearing

(b)superconducting magnetic bearing alone

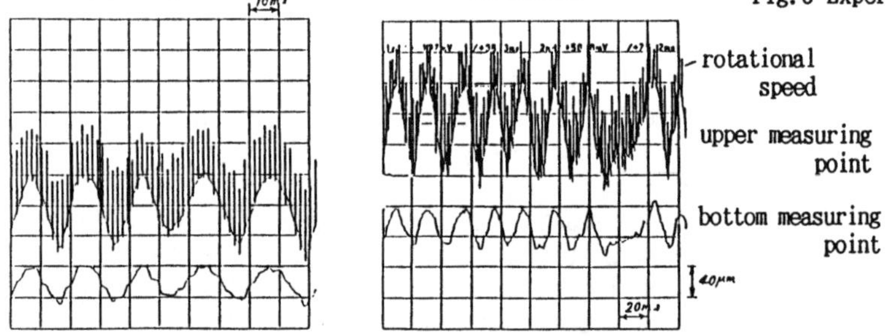

(a)17700rpm (b)30840rpm

Fig.6 Experimental Record at High Speed

Induction Linear Accelerators with a High-T_c Bulk Superconductor Lens (Supertrons)

HIDENORI MATSUZAWA, HARUHISA WADA, SATOSHI MORI, TADASHI YAMAMOTO, and TETSUYA AKITSU

Faculty of Engineering, Yamanashi University, Kofu, 400 Japan

ABSTRACT

High-T_c powder-pressed bulk superconductor lenses (Supertrons) were applied to induction linear accelerators. Conventional induction accelerators have solenoidal coils to focus and guide electron beams. The present accelerator which consisted of only one-stage accelerating gap, contrarily, had no coils but Bi-based superconductor lenses. The accelerator postaccelerated injected relativistic electron beams to ∿400 keV, ∿0.35 kA, and ∿10 ns of duration time. The results indicate that Supertrons can replace solenoidal coils in beam-handling devices.

KEY WORDS: powder-pressed bulk superconductors, lenses, electron beams, induction linear accelerators, Supertrons

INTRODUCTION

Induction linear accelerators [1] have often been applied to free-electron lasers [2]. In the present article, we propose and demonstrate an induction accelerator which has high-T_c bulk suoperconductor lenses (Supertrons) [3-8] instead of solenoidal coils. Conventional induction accelerators need power sources and timing controllers for their coils. The superconductor lenses, on the other hand, need only be cooled below their zero-resistance critical temperatures (T_c's).

The principle of the lenses has been described in previous papers. When intense electron beams are injected into small apertures of superconducting tubes (lenses), the self-magnetic field of the beams is confined in a space between the wall of the tubes and the beams. The enhanced field then focuses the beams. On the basis of this principle, we have already experimentally shown high-T_c bulk superconductor wigglers [9] for free-electron lasers. The induction accelerator in the present article is another example of applications of Supertrons.

EXPERIMENTAL DEVICES

Figure 1 shows the principle of induction linear accelerators which accelerate charged particles with the electric field produced over the secondary terminal gaps. Block diagram of the experimental devices is shown in Fig. 2.

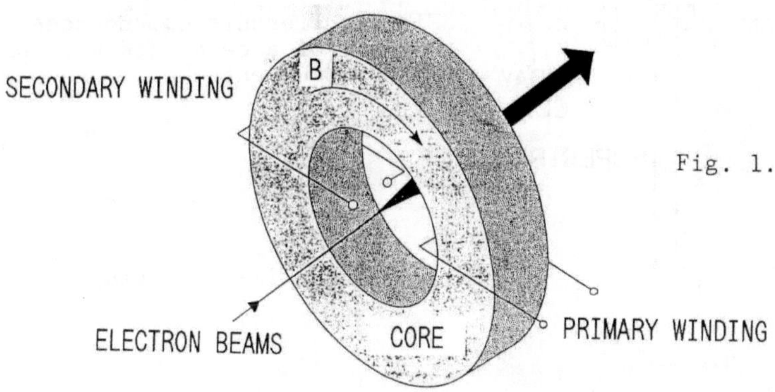

Fig. 1. Principle of induction linear accelerator.

Figure 3 shows detailed structures of the one-stage induction accelerator. Electron beams were injected into the induction accelerator from the electron source (YES-II) [10]. The accelerator was pulse-modulated with a power source (YES-III) [11] of coaxial Marx type. The acceletrator had no solenoidal coils, but a bulk superconductor lens (Supertron). The lens was made of high-T_c Bi-based powder-pressed bulk superconductors (T_c = 105 K). Bi-compounds were chosen because they are stable against water and are nontoxic. The lens was 155-mm long and 2-mm thick, and was glued onto the inner surface of the copper-made heat sink with electrically conducting epoxy resin.

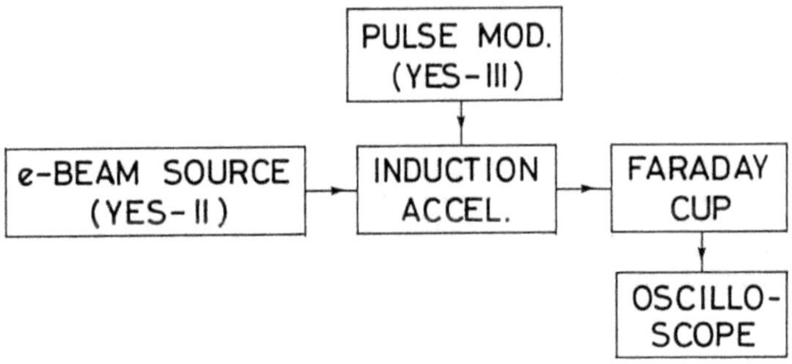

Fig. 2. Block diagram of experimental devices.

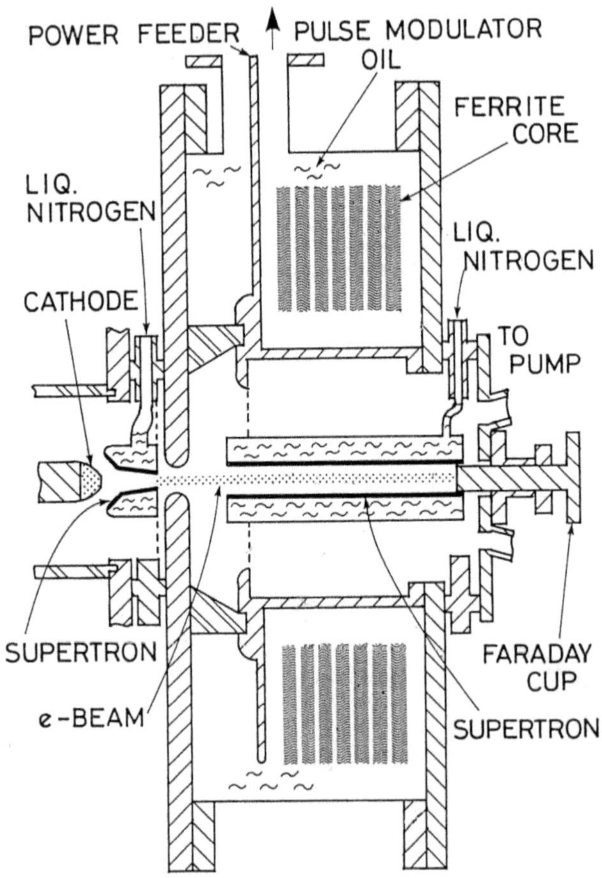

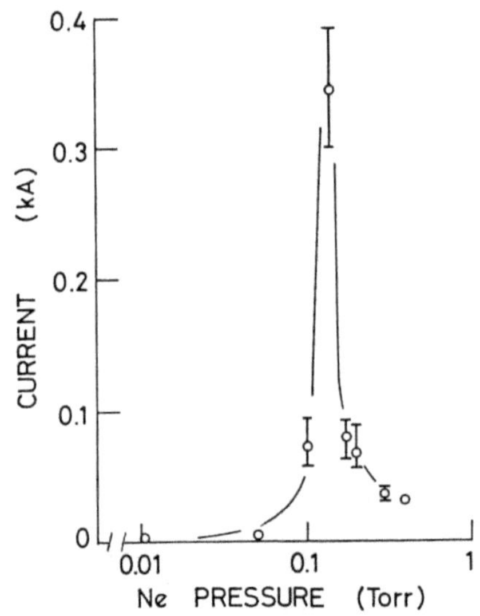

Fig. 4. Pressure dependences of accelerated elctron currents.

Fig. 3. Experimental setup.

The accelerator contained seven pieces of toroidal Ni-Zn-compound ferrite cores (Hitachi Metals, ST-314). The cores were immersed in insulating oil inside the aluminum casing.

Neon gases were introduced into the beam diode and into the accelerating gap of the accelerator to generate high electron currents and to realize space-charge neutralization, respectively. Neon gas was used from among H_2, He, and Ne, which do not liquefy at liquid nitrogen temperature, because of its low thermal conductivity. The electron current was 340 keV, ∿1 kA, and ∿10 ns of duration time at the exit of the diode anode. The superconductor lenses were cooled with liquid nitrogen to 85 ∿ 96 K.

The inlet of the accelerator was placed just downstream of the diode anode. This is because the electron beams tend to diverge after the beam ejects from the diode anode. The accelerating gap was 30 mm in the axial direction to keep its electrical insulation. Postaccelerated electron currents were detected with a compact movable Faraday cup 38 mm upstream from the exit of the tube lens. Detected signals were displayed on a high-speed storage oscilloscope (Tektronix 7913, 500 MHz).

EXPERIMENTAL RESULTS

We observed Ne-pressure dependences of accelerated electron beams (Fig. 4). The optimum neon pressure was 0.14 Torr. Open circles in Fig. 4 indicate averaged values for ten shots and vertical bars show the maximum and the minimum values for each condition. When no pulse-modulator power was applied to the ferrite cores, background levels of elecctron beams were low, as in Fig. 5(a), because the electron beams diverged just downstream of the diode anode. When the accelerator was pulse-modulated, signals such as Figs. 5(b) and 5(c) were obtained, with Fig. 5(b) representing the case where one sheet

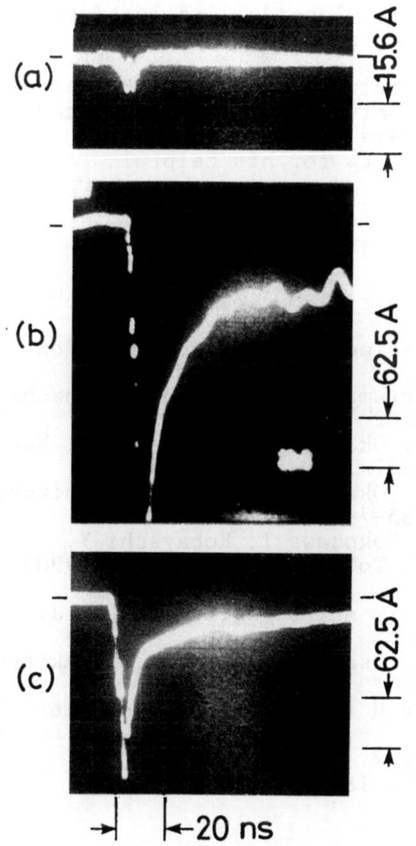

Fig. 5. Electron currents detected with Faraday dup. (a) no modulation power and no aluminum foil, (b) a sheet of 70-μm-thick aluminum foil and (c) three sheets of the foil in front of Faraday cup for power-modulated cases, respectively.

of 70-μm-thick aluminum foil was placed just in front of the Faraday cup, and Fig. 5(c) that with three sheets of aluminum foil.

Waveforms in Figs. 5(b) and 5(c) have long decaying tails because the Faraday cup was not evacuated but was in an atmosphere of neon gases. The cup, accordingly, detected secondary charged particles in addition to accelerated primary electrons.

To estimate kinetic energies of the postaccelerated electron currents, the currents were observed as a function of the thickness of aluminum foils stacked in front of the Faradauy cup. The results mostly fitted a theoretical curve for kinetic energies of 400 keV. Therefore, electron beams were postaccelerated by about 60 keV (= 400 - 340 keV) with the induction accelerator.

We did not monitor voltage waveforms which were applied to the accelerator. The charging voltage of each stage of the eight-stage pulse modulator was about 25 kV. The output voltage of the modulator was, at most, 200 kV. When impedance matching is realized between the modulator and the accelerator, the ideal voltage appearing over the accelerating gap would be 1 KV, which is the desired value.

The discrepancy in kinetic energy between the estimated and the desired value is possibly due to the rapid short output voltages (∿15 ns) of the pulse modulator. The voltages which appeared over the accelerating gap would be less in amplitude then those which were applied to the accelerator, mainly because of the frequency characteristics of the ferrite cores.

CONCLUSIONS

The one-stage induction accelerator postaccelerated and guided relativistic electron beams with a bulk superconductor lens (Supertron). These results indicate that Supertrons can replace conventional solenoidal coils to focus and transport charged-particle beams in beam-handling devices.

The authors would like to thank the members of the workshop of the Research Institute of Electrical Communication of Tohoku University for welding the aluminum casing. Thanks are also due to A. Tohyama and T. Nagakura for adjusting the YES-III and to Professor S. Suganomata for his helpful discussions.

REFERENCES

1. Humphries S Jr (1986) Principles of charged particle acceleration. John Wiley. New York
2. Luchini P, Motz H (1990) Undulators and free-electron lasers. Clarendon. Oxford
3. Matsuzawa H, Ohmori O, Yamazaki H, Ueno J, Furumizu A, Saito A, Takahashi T, Akitsu T (1989) J. Appl. Phys. 65: 2596-2602
4. Matsuzawa H, Osada T, Ishibashi Y, Irikura K, Okamoto K, Mochizuki A, Wada H, Akitsu T (1989) Jpn. J. Appl. Phys. 28: L717-719
5. Matsuzawa H, Ishibashi Y, Osada T, Irikura K, Okamoto K, Wada H, Mochizuki A, Akitsu T (1990) Jpn. J. Appl. Phys. 29: 785-791
6. Ishibashi Y, Osada T, Matsuzawa H, Marumo H, Yokozawa T, Kobayashi Y, Kawamura M, Ohishi M, Morimoto N, Nagakura T, Tohyama A, Akitsu T (1990) Jpn. J. Appl. Phys. 29: L1120-L1123
7. Matsuzawa H, Osada T, Ohya M, Tsuji T, Chino Y, Akitsu T (1991) Jpn. J. Appl. Phys. 30: L841-L843
8. Matsuzawa H, Miyamoto T, Chino Y, Fukutake E, Osada T, Ishibashi Y, Akitsu T (1991) Jpn. J. Appl. Phys. 30: L1390-L1392
9. Matsuzawa H, Ishibashi Y, Ariizumi M, Saegusa H (1991) Appl. Phys. Lett. 59: 141-142
10. Matsuzawa H, Akitsu T (1985) Rev. Sci. Instrum. 56: 2287-2289
11. Matsuzawa H (1983) J. Vac. Sci. Technol. A 1: 1880-1881

Author Index

Key Word Index

MIX
Papier aus verantwortungsvollen Quellen
Paper from responsible sources
FSC® C105338

If you have any concerns about our products,
you can contact us on
ProductSafety@springernature.com

In case Publisher is established outside the EU,
the EU authorized representative is:
Springer Nature Customer Service Center GmbH
Europaplatz 3, 69115 Heidelberg, Germany

Printed by Libri Plureos GmbH
in Hamburg, Germany